The Joinery Compendium

Learning from
Traditional
Woodworking

With English, German,
and Japanese
Terminology

Ruby Press

Sascha Bauer
Daniel Pauli

Preface

Throughout history and until the nineteenth century, the fundamental principles of architectural construction remained largely the same. There was no compelling reason for change, given the limited choice of building materials—primarily wood, natural stone, and bricks. Due to the limitations of hand tools and manual production, a high level of craftsmanship was achieved.

Notably, wood held a distinct position. Unlike stone, which was largely used to build wall-like structures, timber beams were harmonized in intricate spatial frameworks. In the twelfth century, the introduction of half-timbered houses in central Europe brought a revolutionary shift in wood construction. Different from the log houses, pile dwellings, and post construction methods prevalent since the Bronze Age, this innovation vastly improved the structural protection of wood and facilitated easier repair and replacement of load-bearing components. It allowed buildings of up to six floors, prefabricated under sheltered conditions and swiftly erected on-site.

Even if the conditions and production techniques have changed over time, the principle of "prefabrication and quick assembly" still applies today. Then as now, the key lies in the precise joining of structural elements and their ability to transmit forces effectively.

The tradition of wood joinery has existed since the introduction of half-timbered construction. To craft strong wooden connections, one must consider the inherent properties of wood and its ability to distribute forces. While traditional wood joints remain relevant in construction and furniture making, since the mid-twentieth century they have increasingly been replaced by metal fasteners.

In European half-timbered construction, wood joints are primarily subjected to vertical forces—essentially compression—with horizontal forces absorbed by diagonal struts. All the structural load-bearing elements of a story-high facade, from the sill to the studs and diagonal struts to the frame, are linked together in such a way that they can be regarded as a braced wall.

In contrast, the Japanese evolution of timber construction was influenced by Chinese building techniques introduced in the sixth century. The fundamental system involves a post-and-beam framework with a non-load-bearing infill and an expansive cantilevering roof. Its optimal application in a rainy climate, characterized by moist summers and dry winters, is the reason it became part of the Japanese culture and lifestyle as we know it today.

Unlike European timber framing, Japanese wooden architecture must withstand substantial horizontal forces due to frequent earthquakes.

This necessity significantly influenced the properties of wood joints, ultimately shaping the architecture. The Japanese timber house is a post-and-beam structure without braces or diagonal struts. The overall stiffness of the structure relies solely on the wood joints. This has resulted in a variety of highly complex wood joints and node solutions tailored to different structural demands. To withstand extreme stresses, these joints have to be crafted with precision and skill. As a result, they appear object-like, abstract, and artistic, possessing a unique aesthetic.

Wood joinery, esteemed for its aesthetics and craftsmanship, enriches timber constructions and furniture, exemplifying efficient material use and structural integrity while serving as a model for sustainability in various applications.

This publication aims to provide a comprehensive overview of traditional wood joinery gathered from around the world, possibly in a format not seen before. It comes at the right time, because contemporary timber constructions can benefit from it. Today, centuries-old wood joinery from traditional woodworking and timber construction methods can be easily produced with high precision, utilizing cutting-edge modern technology and high-strength timber materials.

It is very likely that we will encounter traditional wood joints in one form or another in the near future.

Prof. Peter Cheret
University of Stuttgart, Germany

Introduction

Editor's Note

Introduction

New technical solutions are changing the way timber constructions are built and manufactured. Traditional artisanship today competes with modern production techniques. Detailed knowledge of the craftsmanship of timber joints and their application can help us understand and challenge these potential changes in manufacturing and construction. Since wood was first used as a construction material, various methods have been developed to join individual parts. Over time, many types of connections have emerged. The mid-nineteenth century saw the development of timber-steel structures with large spans in train stations, market halls, and gymnasiums, as well as a shift in the relative costs of labor and materials, with the former rising and the latter falling. Taken together, that significantly contributed to the addition of mechanical fasteners such as steel connections, nails, and nail plates. The introduction of machinery into manufacturing has further diminished the relevance of timber connections that have been used in everyday practice.

Yet joinery and timber structures are regaining popularity. The growing understanding that wood is an important renewable resource has hastened a shift toward timber construction. This development, though, is threatened as increasingly dry and over-logged forests are destroyed by wildfires and insect infestations. The availability of renewable resources is increasingly restricted, and fast-growing varieties are thinning out old-growth forests. Given the threat of climate change, we are once again starting to appreciate the importance of resources, repairability, and approaches that preserve rather than destroy. For a resource-efficient and reusable approach of buildings and construction elements, we must relearn the repairability of traditional wood joints and adapt them to make them future-proof.

In times when there is a shortage of skilled workers, this in turn requires expertise in craftsmanship shared by many instead of specialized machine knowledge for a few. Craftsmanship can be internalized through experience and doing. A central feature of craftsmanship is that it requires both physical and mental acuity and can be learned through observation.

For example, if an artisan repeatedly encounters similar damage when repairing or deconstructing, this understanding can be incorporated into the work and corrected accordingly.

Therefore, working with traditional timber connections means that the right proportions must be found between the resources of working time, materials, production methods, and energy, as well as the conceptual design decisions. The balance in terms of appropriateness has a corresponding impact on the design.

This experiential knowledge can be passed from generation to generation through rituals or traditions. The formalized, ritualized framework of traditions is intended to restrict the further development of forms and techniques to safeguard knowledge over time. In fact, however, traditions can change as new tools and methods are introduced into production.

At the same time, mistakes, oversights, or experiments, as well as the preferences of the respective master artisans, also play a role in the production of wood joints. So in addition to technical requirements, there are cultural, functional, and social factors that accompany the development of these joints.

Changing environmental conditions, observations, or external influences, for example, from traveling artisans, have led to further adaptations of wood joints. In the broader context of material consumption, handcrafted wood joints can help reform construction by encouraging repair instead of replacement.

The current upswing in timber construction and the growing importance of wood as a building material contribute to this compendium's relevance. Technical possibilities and production techniques can serve not only as a guide for direct reproduction but may also offer a starting point for innovative connections and concepts. Without claiming to be exhaustive, this book aims to present as many traditional wood joints from different cultures as possible and make them accessible and easy to understand using simple drawings and guides to manufacturing processes.

The attempt to create an initial and uniform terminology for wood joints from different regions is intended to facilitate further research into analog and digital media. The resulting publication should contribute to a greater understanding of historical constructions and building methods and their maintenance and preservation. It should foster new, independent, and outstanding solutions across disciplines.

How to Use *The Joinery Compendium*

The Joinery Compendium makes the research and design of joinery projects easy. The joints are divided into chapters according to their geometric orientation and complexity. For easy access, the visual glossary offers a quick overview of the content and makes it simple to locate the appropriate connection.

The 400-plus wood joints are provided with their English, German, and Japanese names to make it easy to communicate with project partners and offer a starting point for further research.

For better understanding, the terminology chapter explains the basic lexicon of joinery. There is also an illustrated, step-by-step guide to creating each joint. The simple two-colored drawings have significant blank space that offers room for notes. *The Joinery Compendium* deepens the understanding of traditional joinery and encourages work on historical studies and building traditions or new connections with contemporary techniques.

Joint Selection

The historical development of wood joints shows various types and applications worldwide, resulting in different dimensions, orientations, and adaptations. Limiting the types of joints and their variations was necessary to keep this book manageable. Connections from half-timbered houses, timber frames, post-beam structures, and furniture making are displayed. The focus is on handcrafted joinery from all over the world made with traditional tools and excluding mechanical fasteners. We didn't include natural wood joints such as tree fork joints or archaic types like lashing, as those usually do not involve processed timber. The principles of log joinery can be reduced to connections such as lap joints and cogged joints. Furthermore, we sought to look at the joint types without cultural decorations or other additions.

The book, therefore, concentrates on the principle of wood joints at the exact point where different parts meet. We can hardly call this book complete. Even with hundreds of examples, one publication cannot cover the myriad combinations of so many basic principles. Nevertheless, we have settled on a standard denomination of the most common connections.

Joint Illustrations

This book aims to provide an easy way to better understand wood joints. The illustrations are deliberately abstracted to square cross sections on the scale of carpentry. This means that most of the joint principles can be considered independent of round, rectangular, natural shapes, or any decoration. Consideration can, first and foremost, be detached from scale, manufacturing process, orientation, loads, raw materials, and application.

The illustration shows principles rather than specific, detailed applications. Dimensional deviations for assembly or wood movements are excluded to simplify the geometry and allow flexibility according to scale. Furthermore, additions such as wedges, keys, and pegs have been abstracted to the scale of carpentry joints. This abstraction is continued to the scale of furniture making, even if machine tools for pins and dowels are used in practice.

To keep the illustrations simple, most joints are illustrated in their typical orientation. However, there are specific splicing and corner joints that can be used horizontally or vertically, depending on the application. The obligatory orientation for some Japanese or classic wood joints had to be abstracted.

In general, the indication of the wood grain shows the usual orientation of end grain and long grain symbolically and will need to be adapted to the available material and its application.

Step-by-Step Manual

The step-by-step manual shows one of many ways to craft the joint. The approach might differ depending on the size of the workpiece, available tools, and the joiner's experience. Indicating only simple hand tools makes a cultural exchange possible. The tool icon represents the category of tools and can be selected or supplemented more precisely according to personal preference. We have limited the step-by-step instructions to fifteen separate actions. Therefore, some steps with lower priority are combined or left out to be able to show the whole production process of the connection. Specific supporting elements were omitted in order to keep the manufacturing process as personalized as possible.

Proportions and Dimensions

Historical dimensions of wood joints are based on the evaluation of existing connections and adapted over time. The dimensions are a cultural and technical development—strongly influenced by the available resources, loads, orientation, application, and tools for precise production. While there are numerous joints shared between carpentry and furniture making, *The Joinery Compendium* shows the connection principle with the dimensions of carpentry to make them comparable. Only the case joints are shown in the scale of furniture making. The illustrations show common proportions of the joints, but they must be reconsidered in terms of scale and application. Various cultural regions may also employ the same joint in different dimensions. As we have excluded dimensional deviations for assembly or wood movement, there are no numerical dimensional ratios in the illustrations—allowing us to cover the broadest possible use. Precise measurements must be calculated among the numerous influencing factors for each unique case. It is impossible to propose an ideal dimension recommendation by combining different regulations and cultural backgrounds. Therefore, the book aims to illustrate the connection principles of joinery.

Prospects for the Future

Historically, a few handcrafted wooden joints have been so neatly crafted that they have remained stable in load-bearing applications for over 1,000 years. Without glue or another auxiliary, they still fit tightly today. These connections were produced with artisanal expertise using appropriate tools and got their character through the selection and availability of wood types. The material that is available in various regions, in turn, led to different traditional and regional processing techniques and tools. However, only a fraction of traditional wood joints demonstrates this precision in practical and everyday use.

Since the introduction of industrial production, the requirement for structural calculations of timber connections and the necessary fatigue testing of materials has led to a critical dispute between engineering and traditional joinery. The tolerances and predictions resulting from manual production, and the joy artisans derive from this natural material, can hardly be controlled in the empirical world. This led to oversized cross sections with corresponding material consumption to compensate for calculated tolerances. Calculations and verifications have become an integral part of our modern world. But there are technical changes that promise to make wood a more viable option for an increasing number of projects. New processing methods adapted to a material's behavior can be accounted for in the planning stage, allowing for cost-effective prefabrication of timber components. Combined with new adhesives, other auxiliaries, and the careful selection of individual timber components, wood can become a standardized and reliable building material. This significantly expands the range of applications and designs for this renewable resource. In a formalized environment, the mostly informal value of hands-on experience has lost relevance, but the principle of craftsmanship through trial and error, reinterpretation, and learning has always remained an integral part of every production process. We've learned that despite varying manufacturing techniques, the connection principles remain similar. Power woodworking doesn't replace craftsmanship; it complements it.

Thanks to automated and precise processing, traditional wood joints are gaining new momentum. New digital and computer-controlled production techniques can minimize tolerances in wood joints, resulting in slim and aesthetic constructions. These new production techniques also lessen the constraints of serial production of complex three-dimensional traditional joints.

Digitally controlled and individualized production techniques offer interesting possibilities and boost the momentum for developing new wood connections. We can look forward to applying these unique solutions and the resulting potential yet to be discovered. Throughout this yearslong project, the opportunity to study wood joints and share our knowledge of the subject has provided endless satisfaction. We hope you enjoy this book.

Wood, Tools, Terminology

Wood Quality

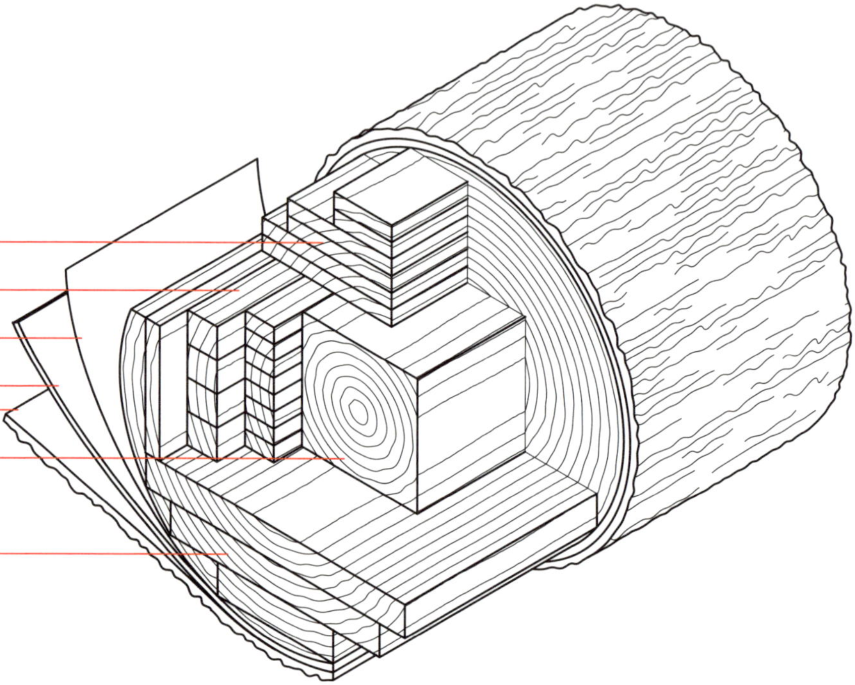

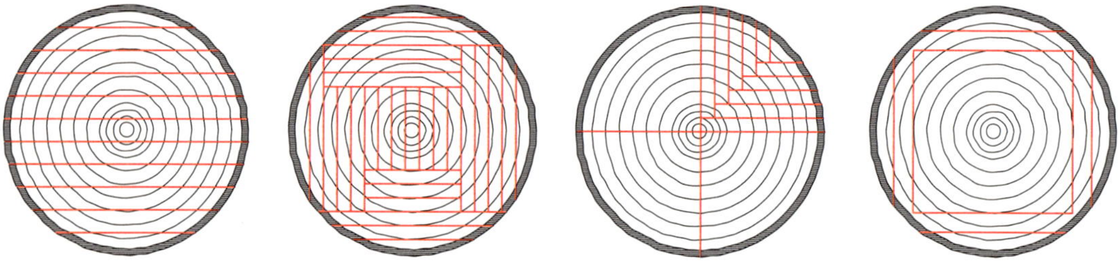

Live Sawn   Plain Sawn   Quarter Sawn   Whole Piece

The structure of the tree—thicker and denser at the base and thinner and lighter at the crown—is naturally designed to resist gravity. But the characteristics of the wood, and therefore its usability, depend on where the tree was grown, as various environments obviously mean different sun, wind, soil, altitude, and moisture conditions. Trees grown at higher elevations and in deserts often produce denser, more resilient wood owing to challenging climates and slower growth rates. However, the most vigorous trees grow above the midpoint of a mountain, where they receive the best sunlight and air circulation. Trunk diameters are thicker since trees there can rapidly branch out and develop the necessary girth to support many leaves or needles. Simultaneously, the wood of more exposed trees is impacted by the wind, resulting in higher density and an altered fiber structure on the compression and tension side.

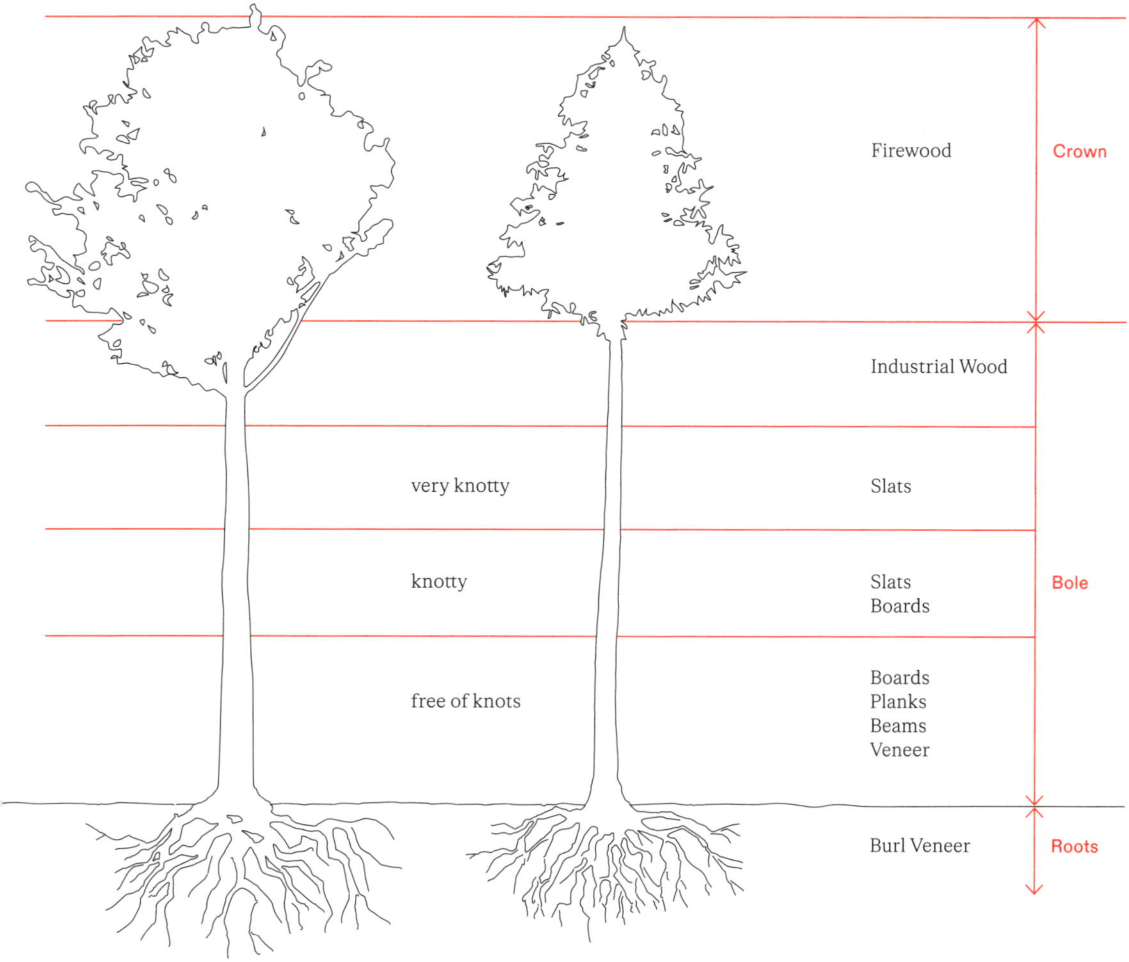

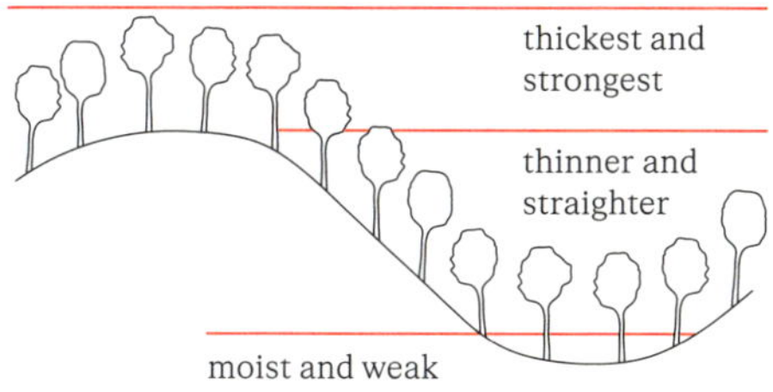

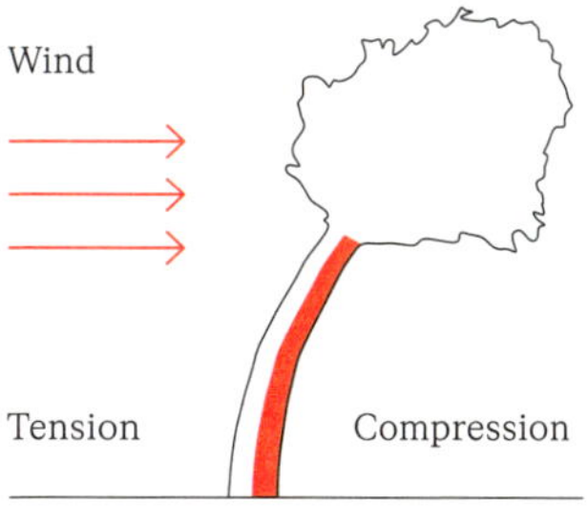

This can affect the wood's strength and workability. On the other hand, trees that grow below the midpoint, where they have to compete for sunlight with their brethren higher on the slope, sprout higher before branching, produce thinner trunks that are straighter and relatively free of knots. Trees that grow in valleys provide lower-quality timber with a high moisture content, but they can nevertheless be used for secondary construction elements.

Accompanying all deviations, the type and quality of soil a tree grows in can affect its overall health and wood quality. Nutrient-rich soil can promote healthy and fast growth, while clayey and sandy soils can leave traces in the wood cut from those trees.
For this reason, the choice of wood and its growing conditions is an essential component to consider when producing wood joints.

Shrinkage and Movement

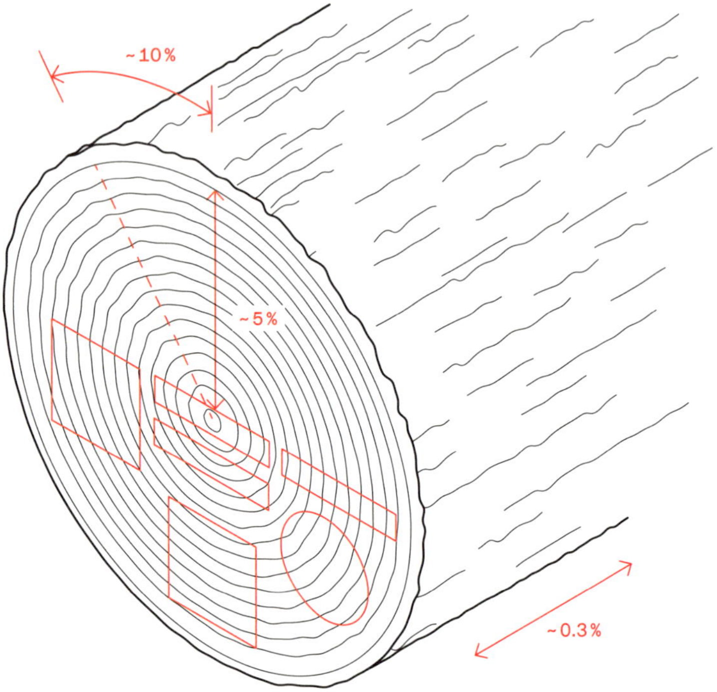

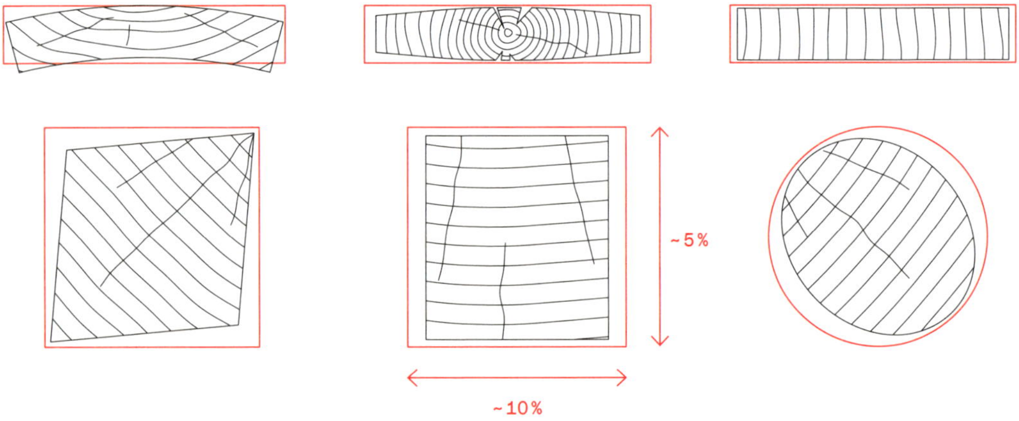

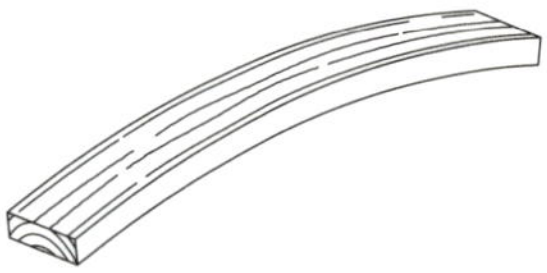

Bow

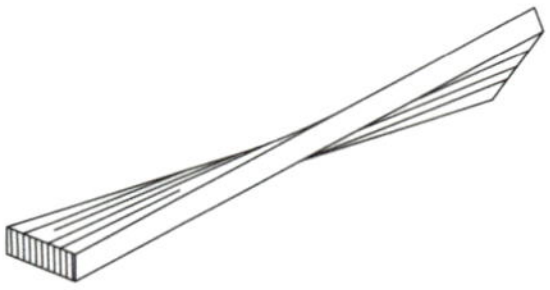

Twist

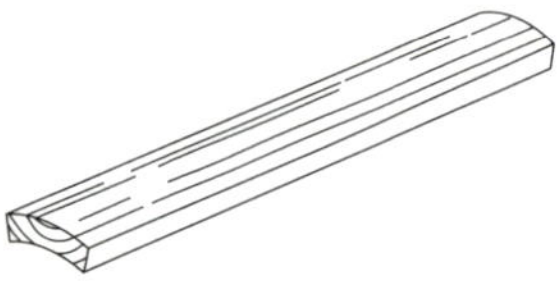

Cup

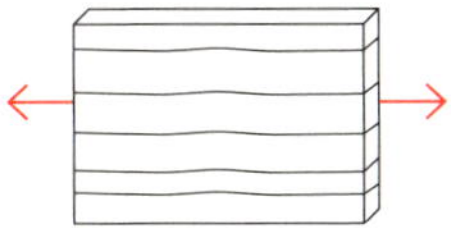

Tension along Grain

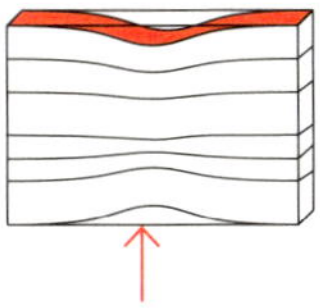

Compression against Grain

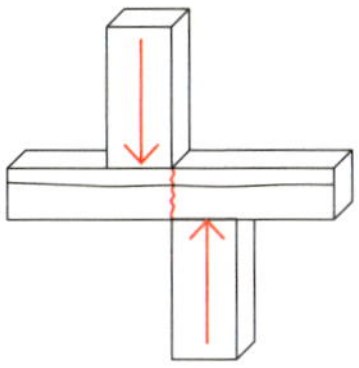

Shear against Grain

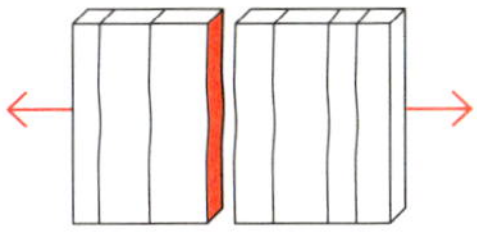

Tension against Grain

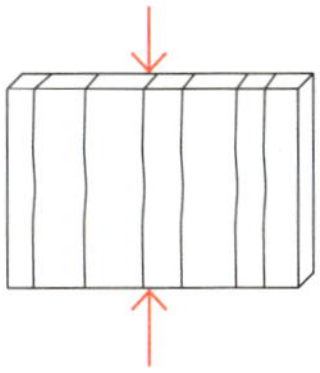

Compression along Grain

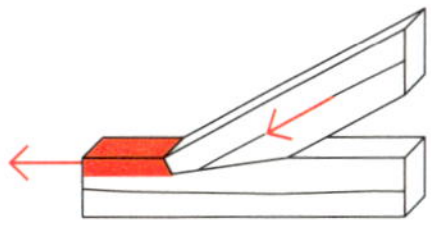

Shear along Grain

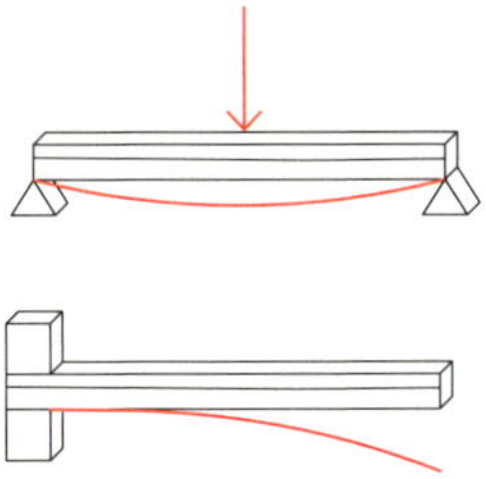

Bending

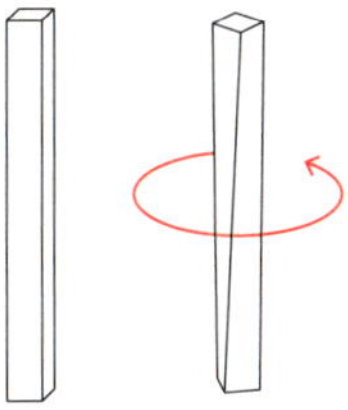

Torsion

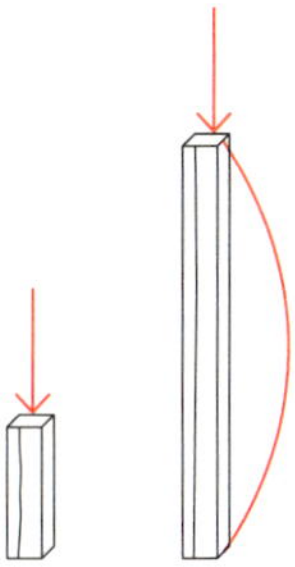

Buckling

Shrinkage and movement are inherent properties of wood that affect its dimensional stability and structural integrity. Wood movement occurs primarily because of changes in humidity and temperature. It varies depending on its fiber orientation and is approximately 10 % tangential, 5 % radial, and 0.3 % longitudinal. This differential shrinkage and swelling often leads to deformations such as buckling, warping, bowing, and twisting, which can affect the performance of wood products and lead to cracks and other damage. The delicate balance between the wood's moisture content and the environment's relative humidity is a criterion for the subsequent location of the wood joint. The choice of wood and its orientation must also correspond to the application to account for tension, pressure, shear, bending, torsion, and buckling in the structure.

Wood is generally more resilient to tensile forces along the grain rather than perpendicular to the grain. When pressure is applied perpendicular to the grain, it is advisable to use hardwood to minimize deformation. Sophisticated joinery techniques for controlling movement, such as breadboard ends, floating panels, or sliding dovetail battens, are just some of the possibilities for dealing with the movement of the wood in a way that suits the material. Understanding the mechanisms of timber movement and stresses is critical to designing and constructing wooden structures and products to mitigate potential problems and ensure their durability and functionality. However, taking account of deteriorating components can also be a conscious approach to sustainable design.

Measuring and Marking

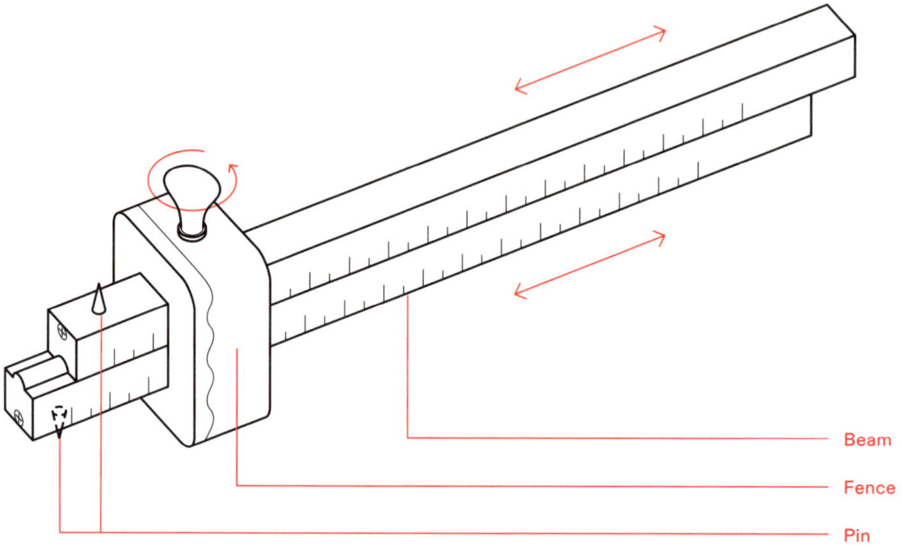

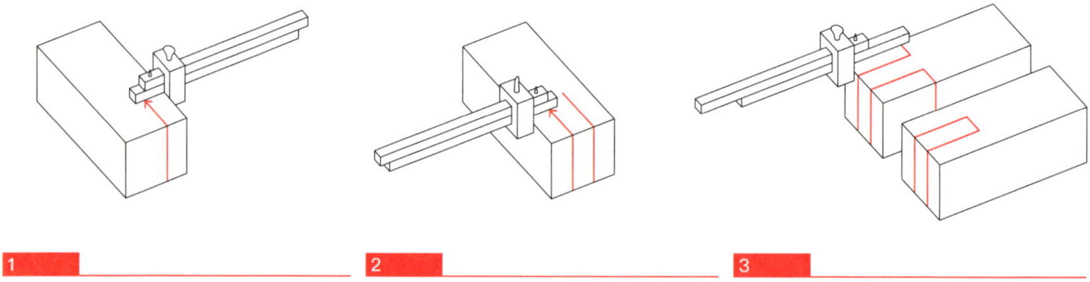

Accurate layout is crucial in crafting wood joints, requiring precision in marking and preparing wood. Various tools, influenced by cultural backgrounds and craft evolution, cater to different materials. Rules, squares, bevels, and gauges, complemented by a marking knife or pencil, form the essential tool kit for marking. But the marking gauge is the optimal tool for repetitive marking with high consistency and precision—allowing you to measure and mark simultaneously with a single device. Consisting of a fence, beam, and marking tip, it secures multiple measurements with a screw. By gently guiding the fence alongside the edge of the workpiece, the marking tip draws a parallel line on the wood's surface. Some marking gauges have marking blades or cutting wheels instead of pins. These are especially designed to ensure tear-free marking of the end and cross grain of the workpiece.

Pencil

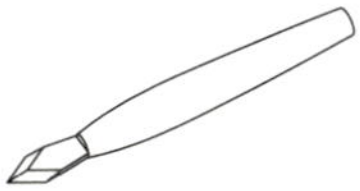

Marking Knife

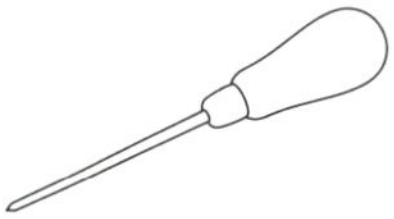

Pricking Awl

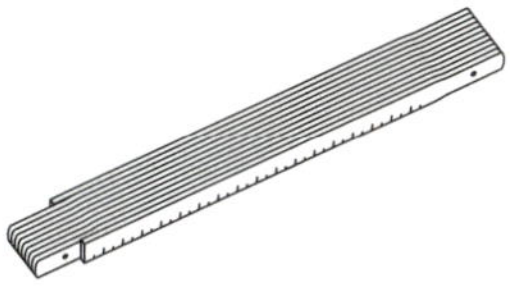

Folding Rule

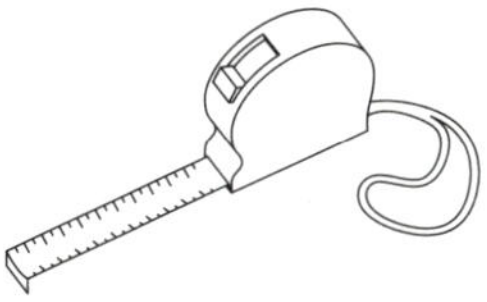

Tape Measure

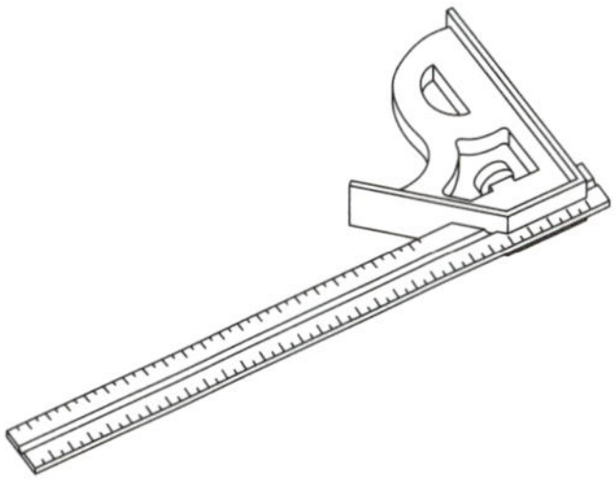

Combination Square

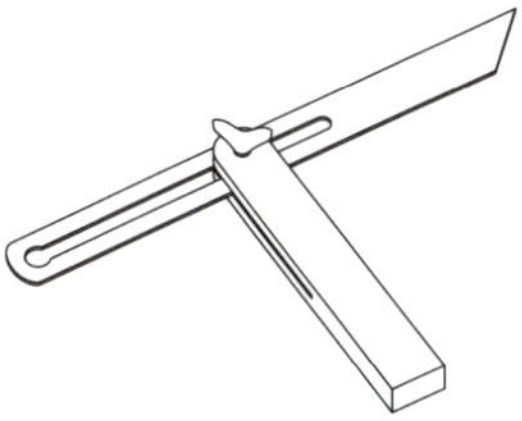

Sliding Bevel

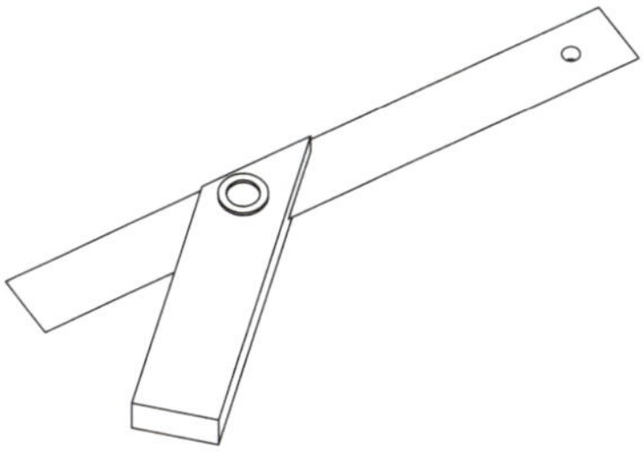

Miter Square

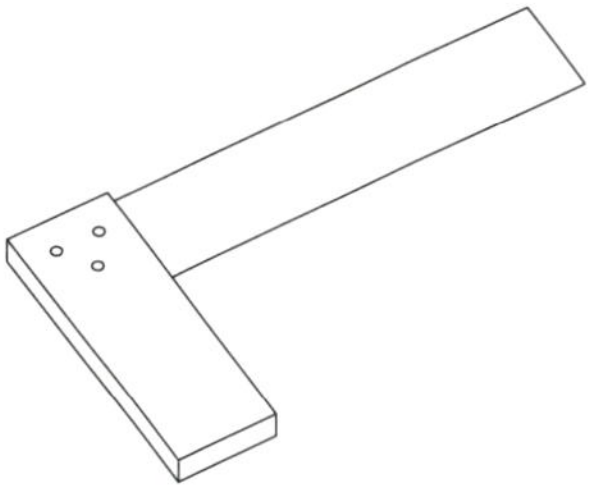

Tri Square

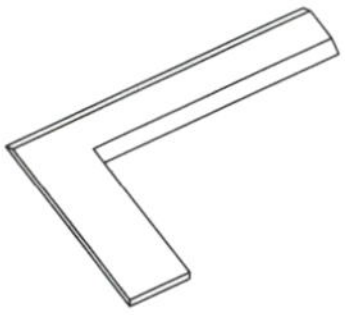

Knife Edge Square

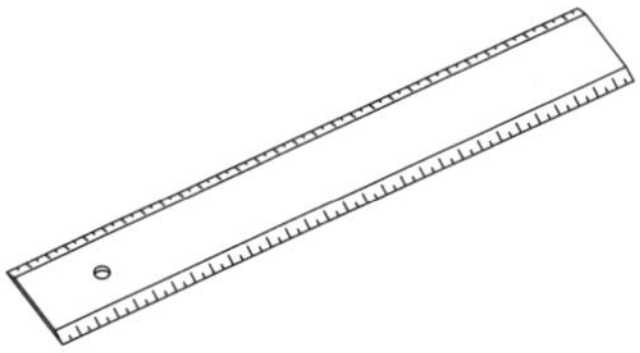

Ruler

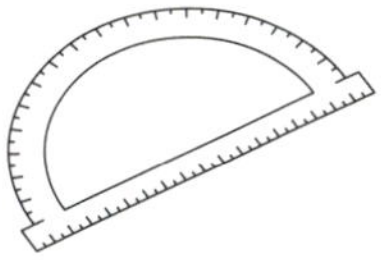

Protractor

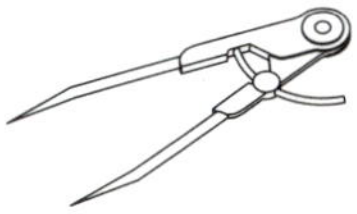

Wing Compass

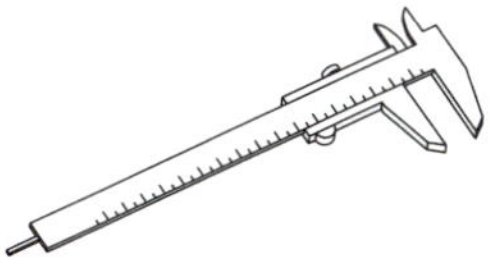

Calliper

Sumitsubo

Chisels

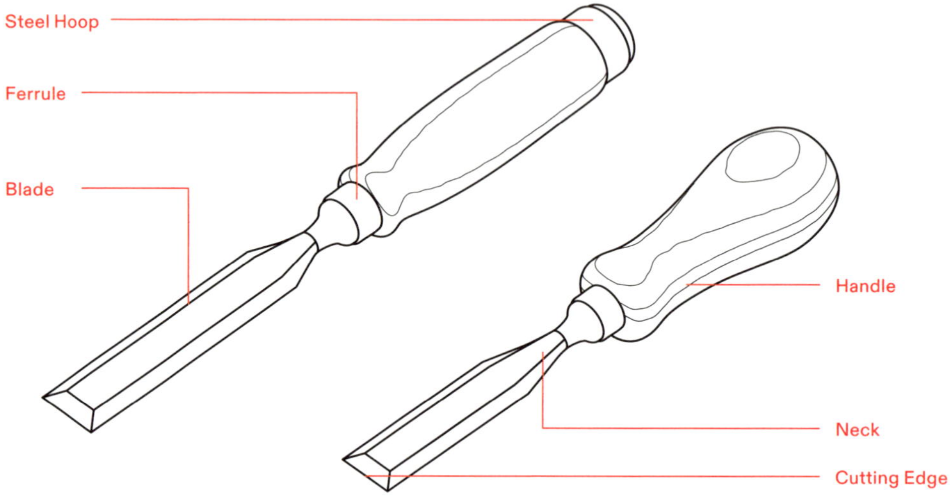

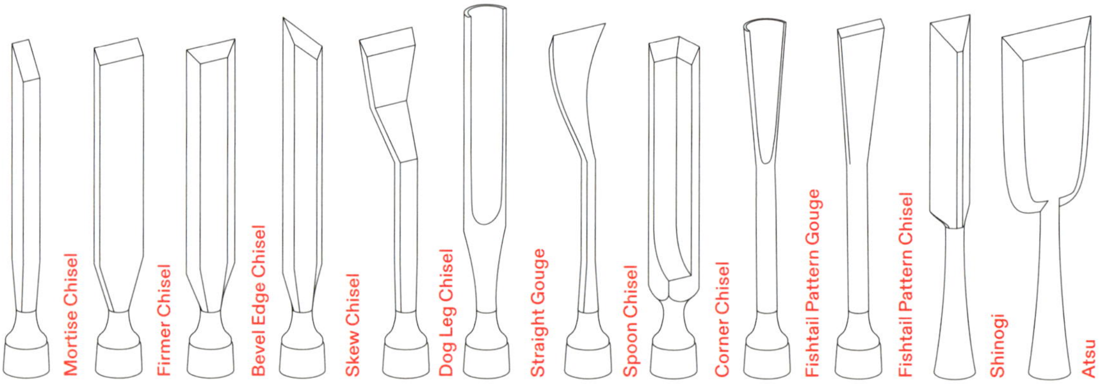

The chisel is a steel tool with a handle and a sharpened cutting edge. There are numerous shapes and sizes for specific applications. The chisel is crucial in creating wood joints and is often used to cut the wood in various directions. Working parallel to the grain and at right angles both require similar forces depending on the wood density. However, working on the end grain of the workpiece requires an especially sharp chisel, a more delicate application of force, and smaller cuts to prevent tearing and crushing of the end grain. The shape and size of the required chisel will depend on the task and the preferences of the craftsperson. Bevel edge chisels are for general woodworking tasks such as paring, splitting, chopping, and shaping. They are ideal for precise cuts and clean joinery. Mortise chisels feature thick, sturdy blades designed to withstand the force of chopping out mortises.

Carving chisels are equipped with thin, delicate blades for the intricate carving and shaping of wood. Chisels are made of various types of steel, depending on the brand and application. Different steels and steel treatments enable the adjustment of the tool to the raw material to be processed. In combination, laminated blades of high-carbon steel can yield an exceptionally sharp edge, but are more brittle and must be handled with care. The typical bevel angle of a chisel is 25–30° but can vary depending on the cultural region, the specific application, and the wood being worked on. Chisels are often sharpened with natural or synthetic whetstones with various grits, but other methods are also used in some situations. The sharpening frequency depends on how often the chisel is used, which types of wood are cut, and the types of cuts that are made.

Chisel Cut

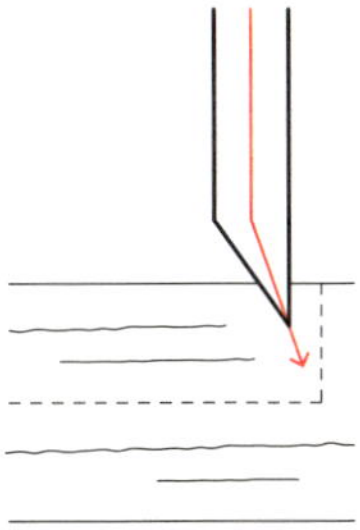
1

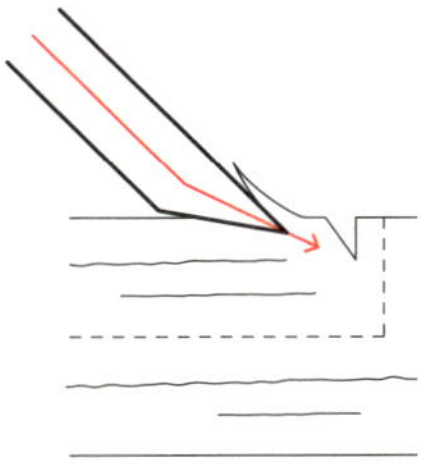
2

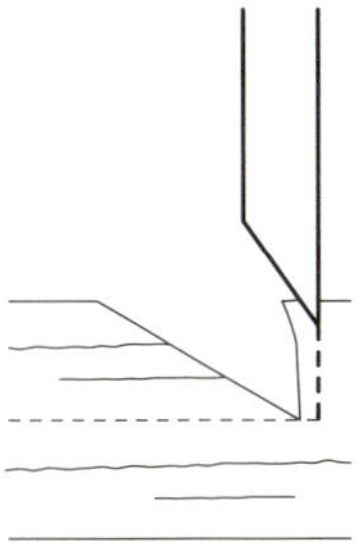
3

For precise processing of the workpieces, the focus is on the final edge to be produced. The chisel cut drifting through the cutting bevel must be given sufficient space. A first cut, deviating from the final edge, can remove a first chip up to the tool thickness. The pressure on the cutting bevel can be relieved by repeating the process. The final edge is now processed along the back side of the chisel with low counter forces from the cutting bevel.

Working Out

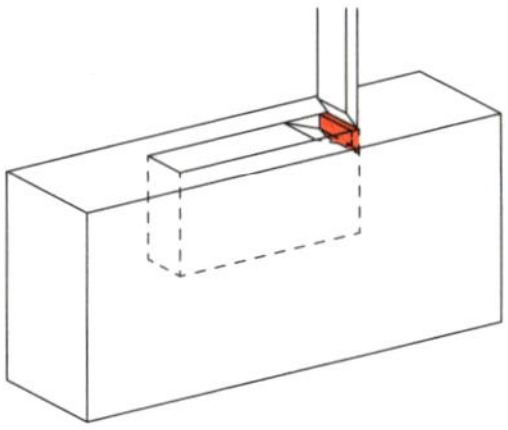
A1

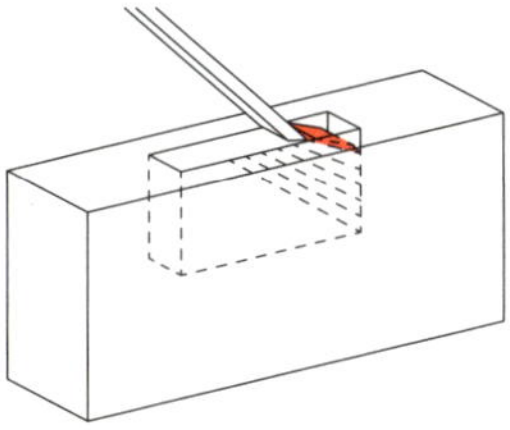
A2

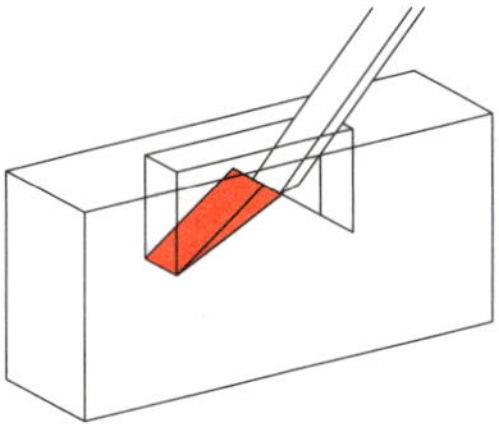
A3

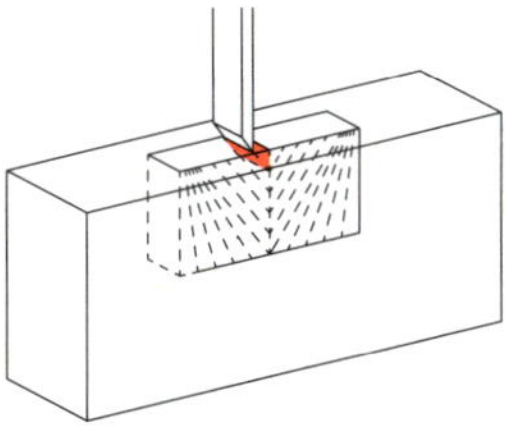
B1

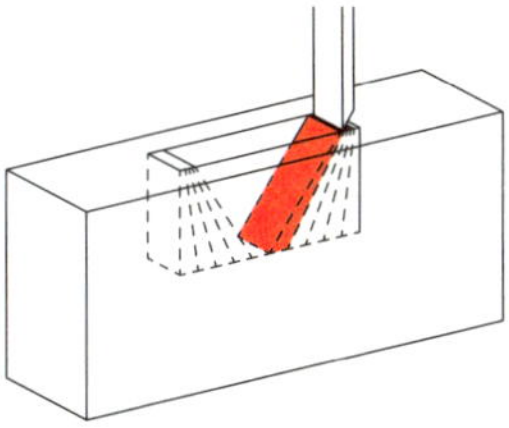
B2

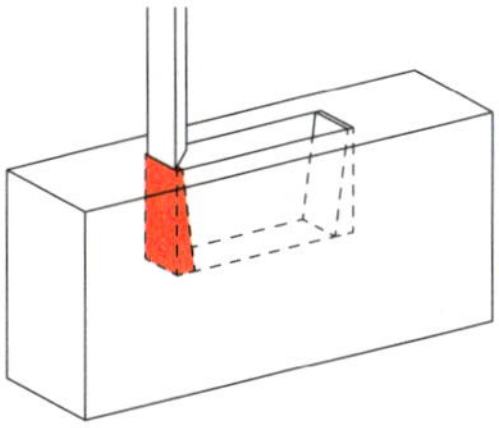
B3

There are two possible methods for removing material in mortises and recesses using traditional tools. When working from the outside in, the wood is removed from the outer edge one chip at a time. Chips are removed step by step from the final outer edges, depending on the tool's thickness. It's essential to ensure that the chisel does not drift owing to excessive pressure on the cutting bevel. The final edge is now cut straight down along the mirror side of the chisel. During this process, attention must be paid to the final outer edge, but the processing can be done in larger steps depending on the type of wood. When removing material from the inside to the outside, the final edges remain protected until the end, but the work is carried out in smaller steps due to the limited space available.

Planes

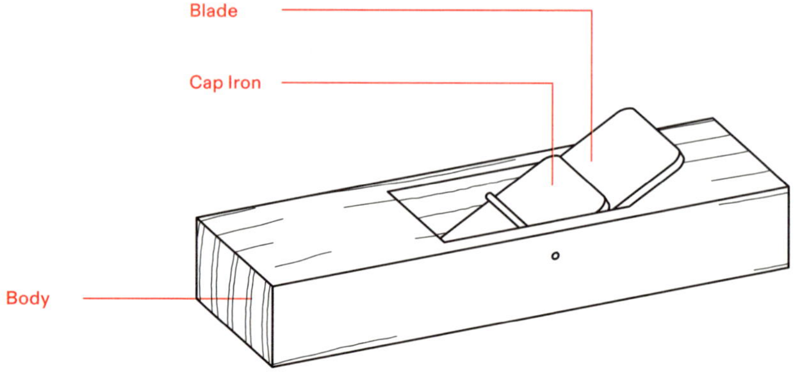

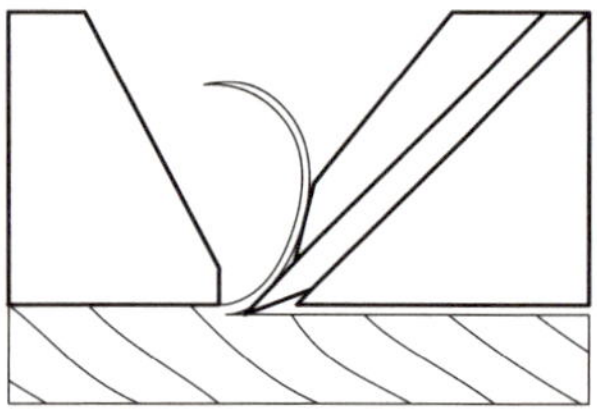

Chip Breaker

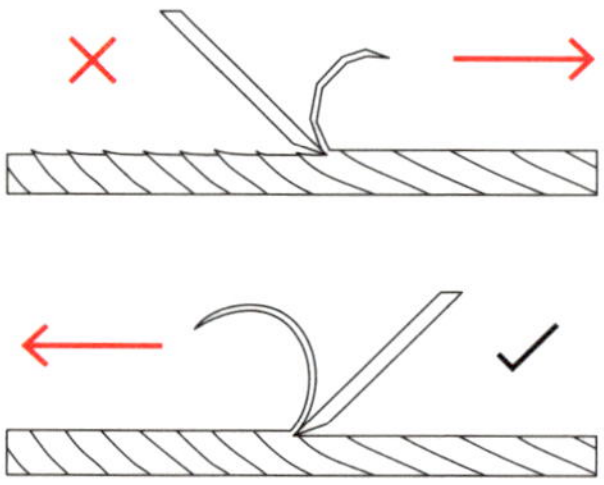

Shaving Direction

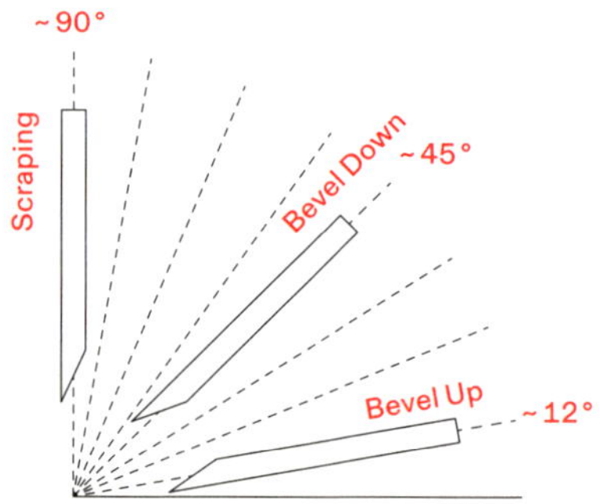

Angles

Planes are standard hand tools in various cultures. The main differences are the construction of the plane, the cutting direction, and the cutting angle of the iron with respect to the orientation of the bevel. These are combined depending on the processing requirements, the property of the wood, and the intended surface quality. The direction of operating the plane can also vary in different cultures. Most Japanese planes cut on the pull stroke, whereas western planes typically cut on the push stroke. The direction of the plane's cut—whether push or pull—won't necessarily affect the quality of the surface in the end. Depending on the plane's design, the plane iron is clamped with a bevel up or down. Though those with the bevel down usually need to be sharpened less often. A flat cutting angle of a block plane allows fast material removal by taking advantage of the pre-cutting of the wood, applying less stress on the blade tip. Low angle planes are excellent for many challenging timbers and produce an excellent finish with a shaving motion.

Steeper angled bench planes or scraping planes are intended for denser or irregular woods and enable finer surfaces by merely scraping rather than cutting. A chip breaker is a valuable addition in some planes. The surface quality is improved by bending and breaking the planed chip at the edge of the chip breaker to reduce pre-splitting. The finer the processing, the sharper the blades should be, and the chip thickness as well as the edge of the chip breaker should be set more accurately. Better developed planes in different cultures enable the subsequent adjustment of the mouth and a chip thickness limitation, resulting in an excellent surface quality. Achieving precise alignment between the workpiece and the plane is crucial, as careful consideration of the grain direction during planing is essential for avoiding tear-out damage, demanding meticulous attention from the craftsperson, mainly when dealing with irregular wood grain patterns.

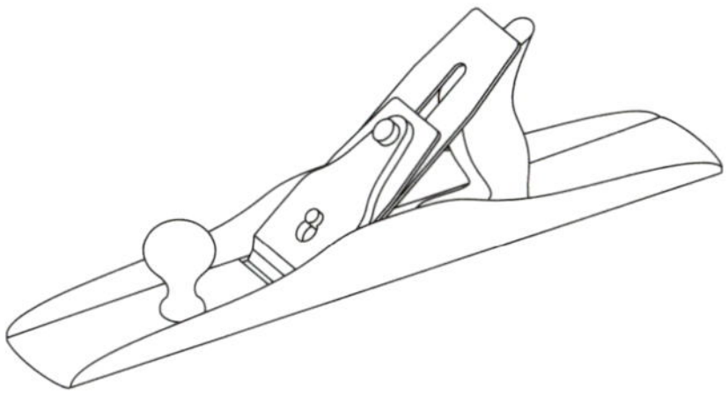

Jointer Plane

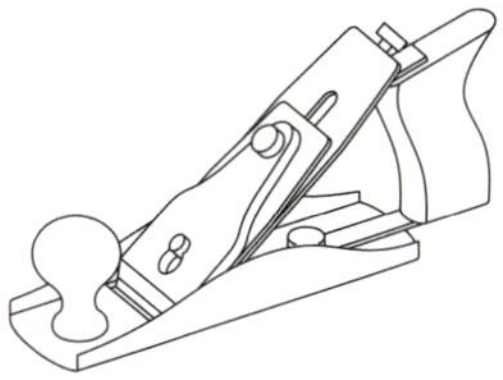

Smooth Plane

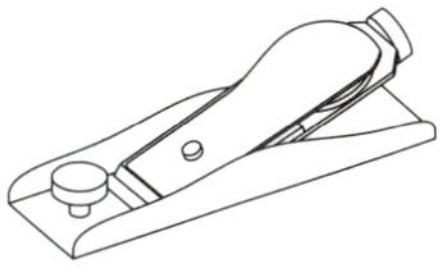

Block Plane

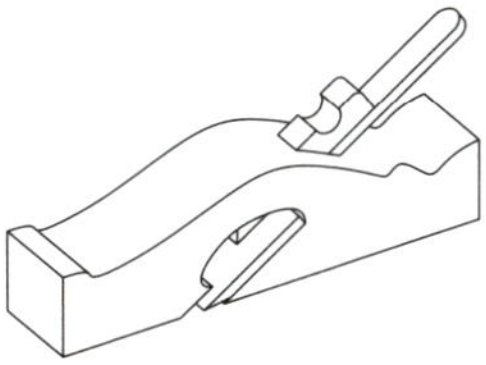

Shoulder Plane

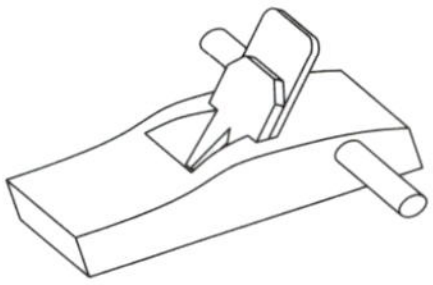

Bench Plane

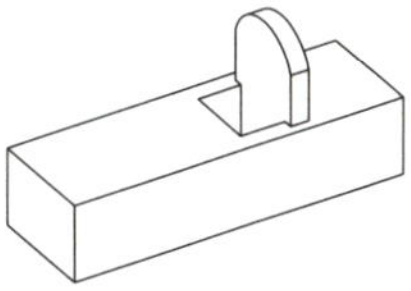

Scraping Plane

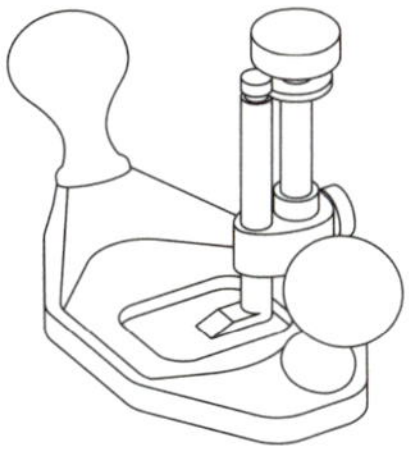

Router Plane

Finger Plane

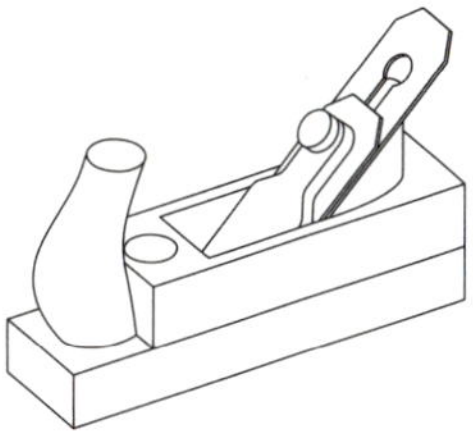

Smoothing Plane

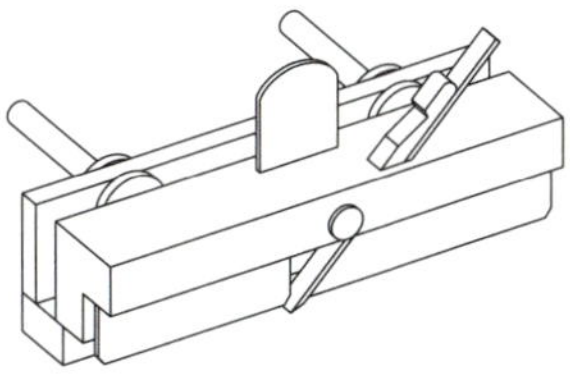

Grooving Plane

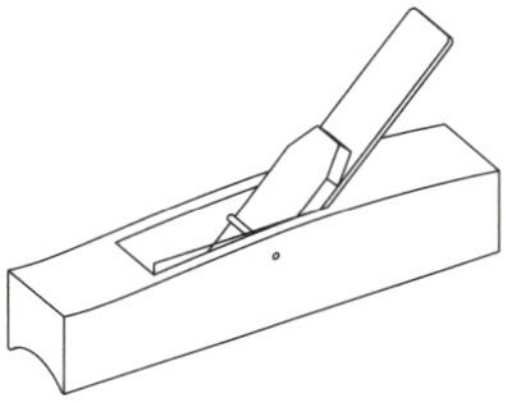

Radius Plane

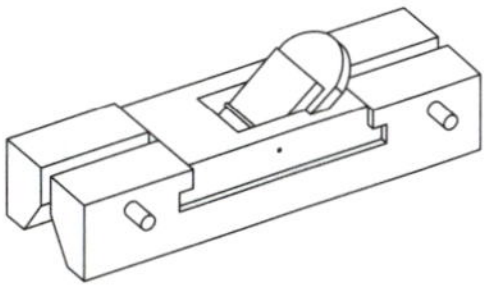

Chamfer Plane

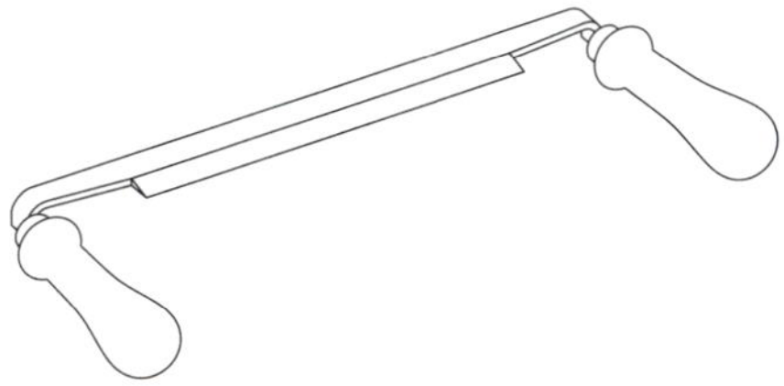

Drawknife

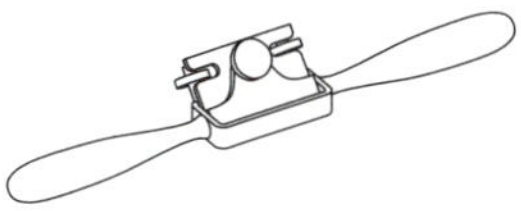

Spokeshave

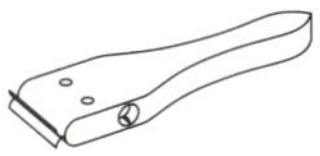

Wood Scraper

Saws

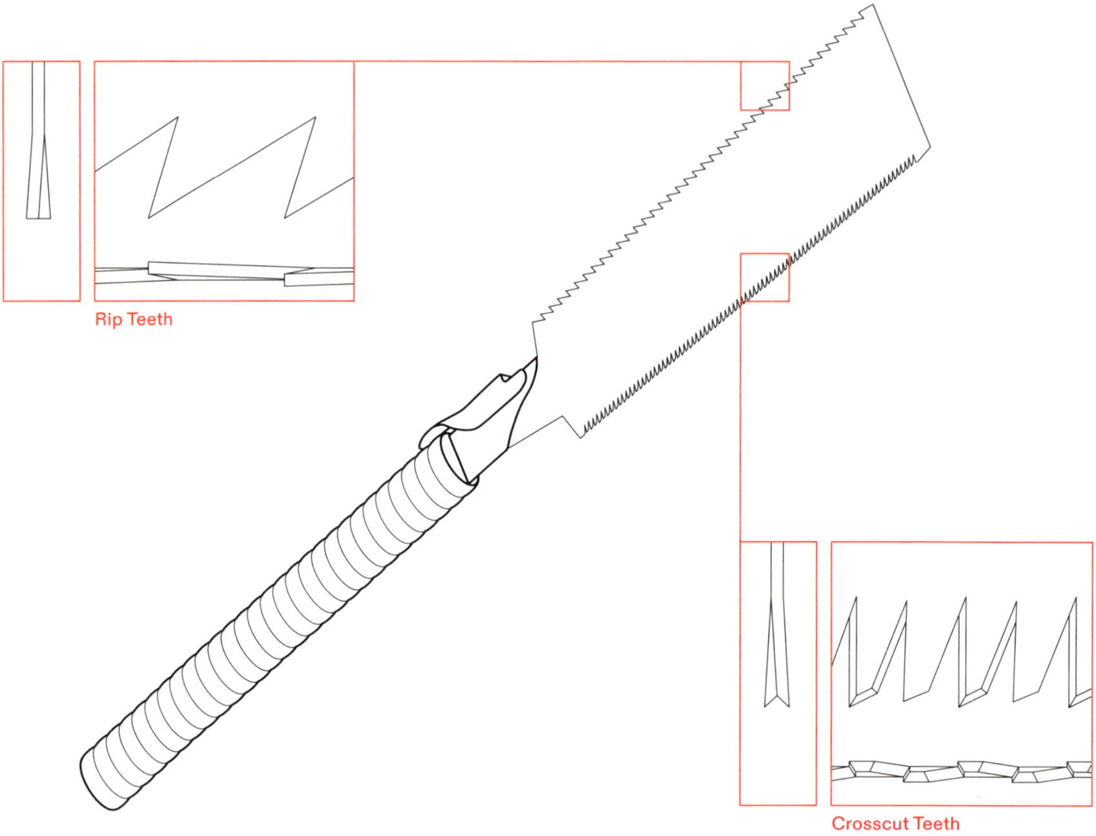

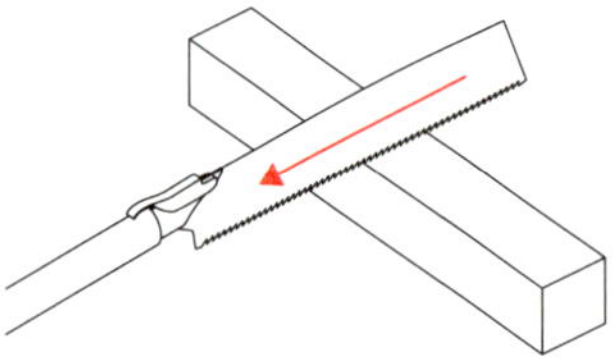
Pull Saw

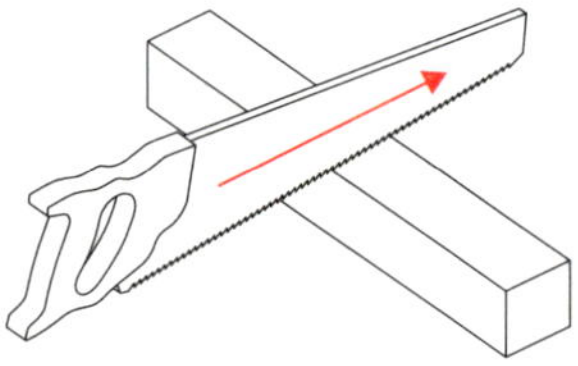
Push Saw

Saws are essential for woodworking, and the choice of saw depends on the specific task, density, fiber content, fiber direction, irregular wood growth, and the individual preferences of the craftsperson. The type of force applied in combination with the shape of the saw teeth influences the cut. Push saws are known for their sturdy construction, engineered to resist blade bending. Those featuring slender blades are fortified with a rigid spine running along the blade's back, ensuring optimal stability. Pull saws have thin blades and offer precise and delicate cutting performance. The pulling motion minimizes the bending of the saw blades for more accurate cuts without great effort. Japanese saws typically cut on the pull stroke rather than the push stroke (unlike most of their Western counterparts). Japanese saws come in various styles, including dozukis for precise joinery, ryobas for versatile crosscut and rip cuts, and katabas for heavy-duty tasks like timber framing. Saw teeth are primarily differentiated by geometry, cutting direction, cutting bevels, and how they interlock. Crosscut saws have more and finer teeth and are designed to cut across the grain and wood fibers. Rip saws are used to cut parallel to the grain. A combination of different saws in a workshop offers a wide range of possible applications, especially for different types of processing on a workbench in a standing or sitting position.

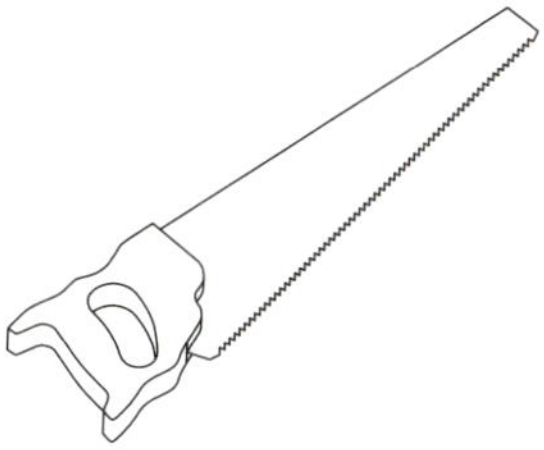

Rip Cut Saw

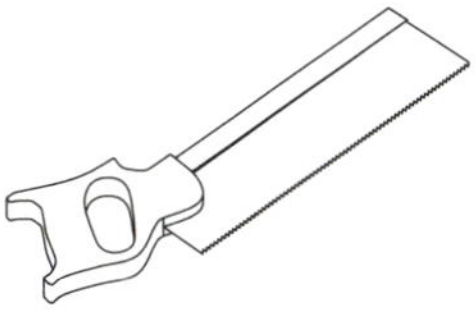

Back Saw

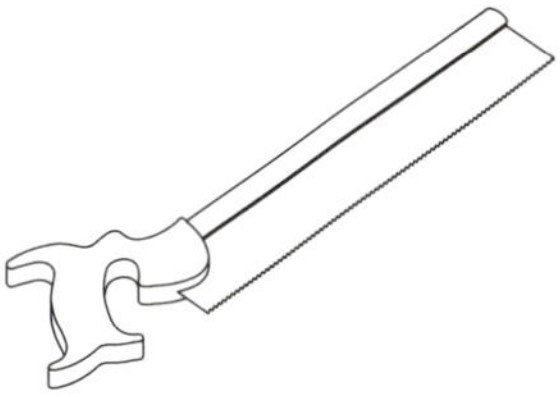

Carcass Saw

Fret Saw

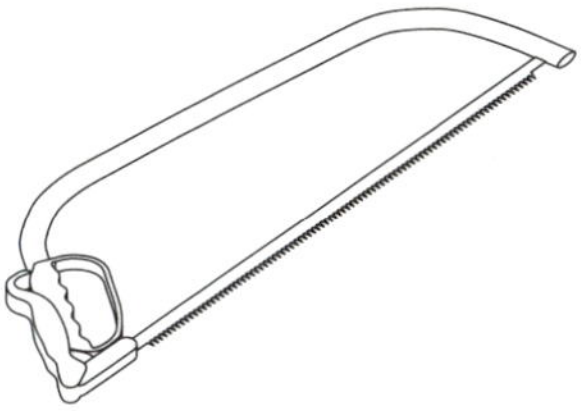

Bow Saw

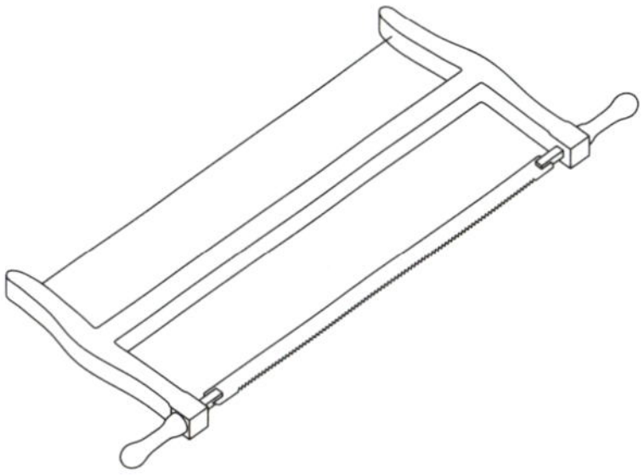

Frame Saw

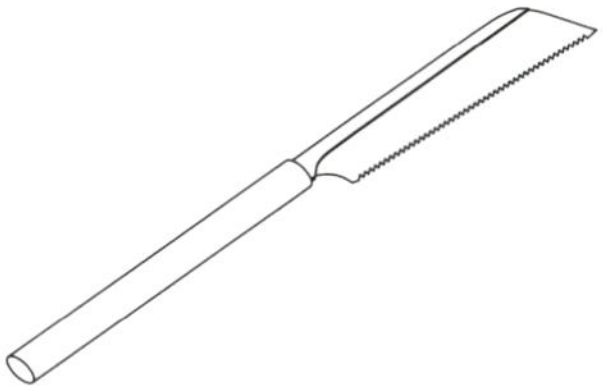

Dozuki

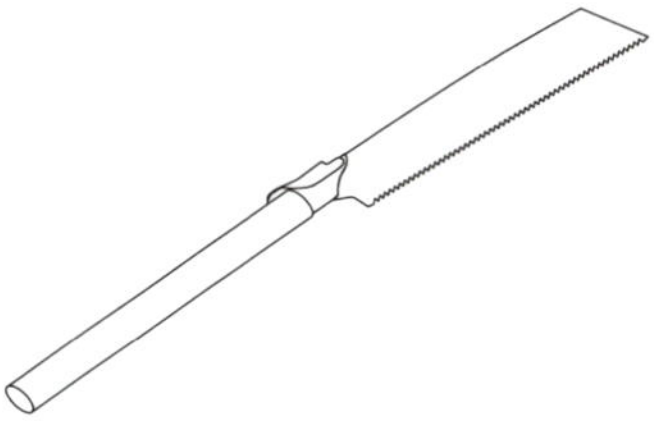

Kataba

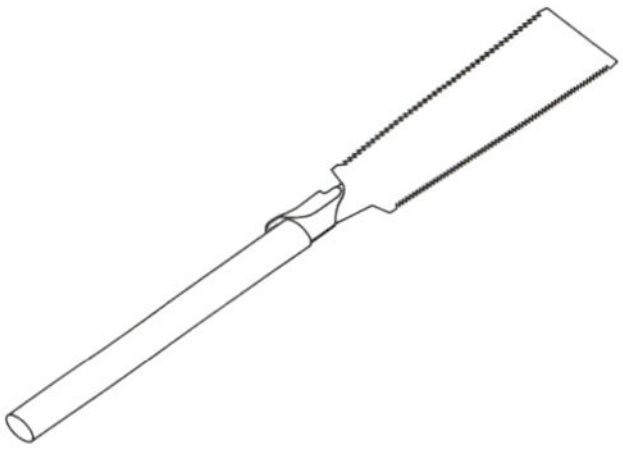

Ryoba

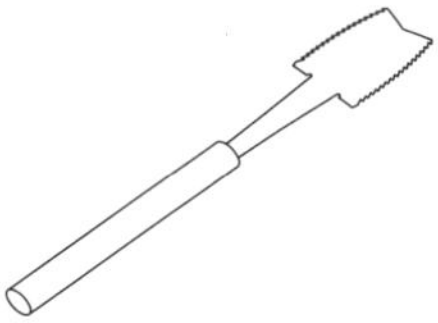

Azebiki

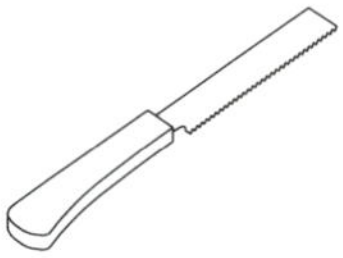

Kugihiki

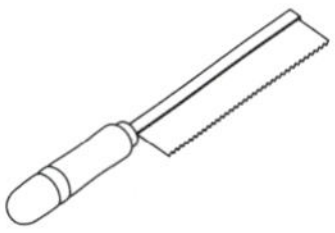

Fine Saw

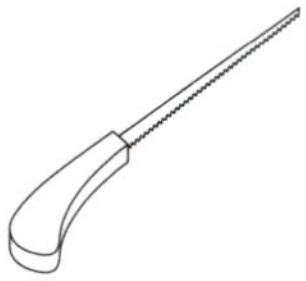

Keyhole Saw

Hammers

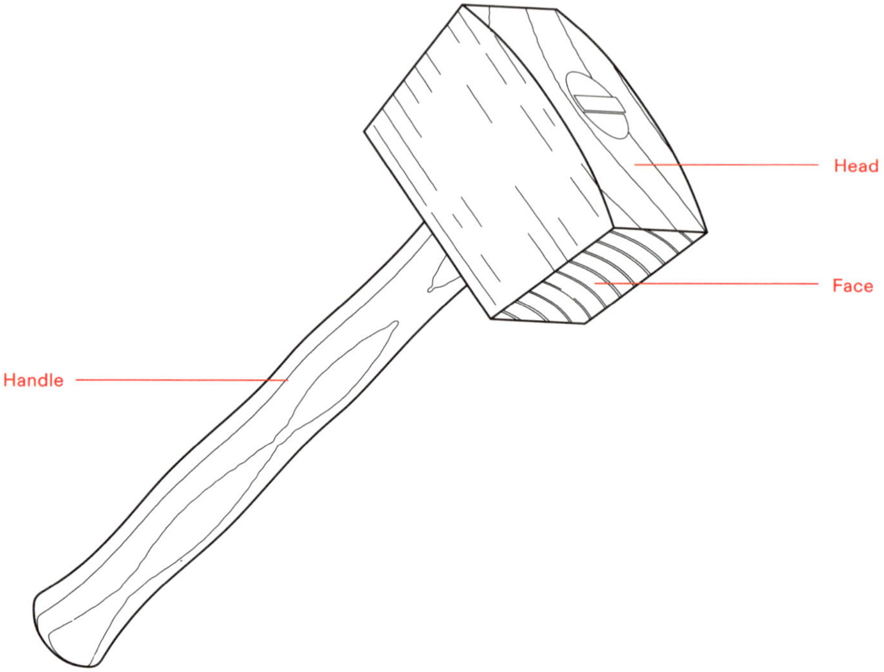

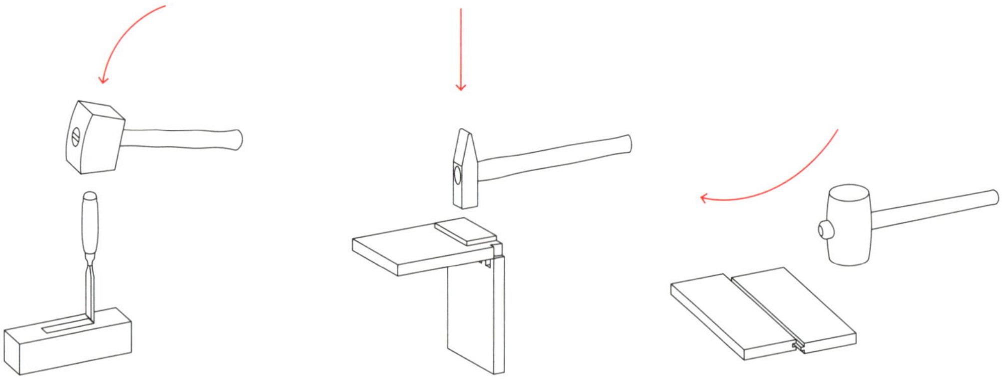

As one of the oldest tools, the hammer has undergone an impressive evolution over time—from simple stone to highly specialized for various applications. Selecting a suitable striking tool plays a crucial role. In woodworking, there are specialized carpenter hammers and mallets tailored for use with a chisel. The carpenter's claw hammer excels at nail removal, while soft face hammers and rubber mallets are ideal for joining workpieces without damaging the wood. If the hammering requires a greater force, a sacrificial piece of wood has to be used to prevent denting.

Different striking techniques enable flexible adaptation to force, precision, and efficiency requirements. Depending on the application, there is a distinction between a straight blow, struck from above for precise control, the overhead strike for maximum force in heavy work, and the sweeping motion for covering a large area when tapping flooring into place. Overall, a comprehensive knowledge of the various types of hammers and their respective techniques contributes significantly to optimizing manual work.

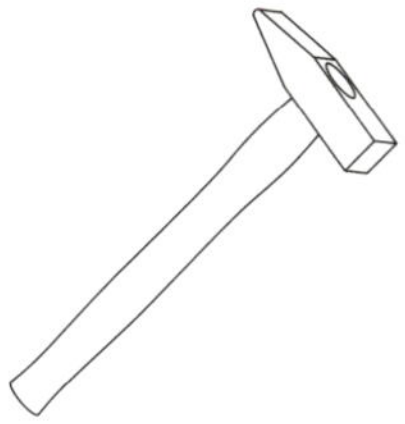

Machinist Hammer

Cross and Straight Pein

Claw Hammer

Carpenter's Mallet

Carving Mallet

Brass Mallet

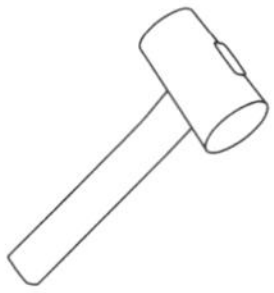

Kozuchi

Hattori Genno

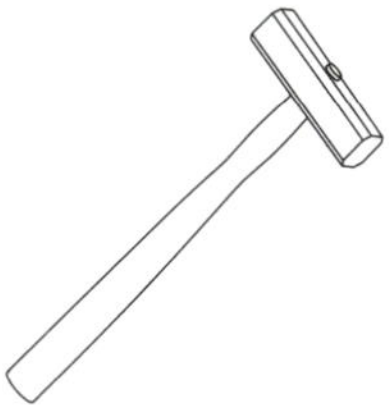

Hakkaku Genno

Soft Face Hammer

Rubber Mallet

Veneer Hammer

Stone Tool

Terminology

About Naming Wood Connections

The naming of the wood joints refers to a set of rules developed collaboratively within numerous discussions, consulting publications, and the active assistance of experts from various cultural backgrounds. Depending on the region, dialect, and publication, individual parts of a joint are named differently. The terminology in this book is intended to use a descriptive approach to overcome regional differences and describe the elementary parts of the connection principle. However, cultural differences in naming were of great importance. For instance, the English dovetail, Japanese *ari* (ant), and German *Schwalbenschwanz* (swallow's tail) all refer to the same joint type. Some designations are figurative but should be understood as a cultural achievement and are therefore reflected in the terminology. We chose standard terms for description throughout the naming process to maintain uniformity across the book. In some cases as a name specific to a certain joint is commonly used, this name has been explicitly adopted for the respective joint. To indicate common usage, some joint names are marked with an asterisk [*]. Not all wood joints usually share the same cultural origin, so some lack corresponding names. In such cases, we rely on the collaboratively developed set of rules. The descriptive names aim to provide designations for these joints for the first time. Given this publication's wide scope, we couldn't cover all aspects, but the naming process serves as a basis for discussion and future discovery.

Comments on the Japanese Transcription

The Latin transcription of the Japanese expressions follows the revised Hepburn romanization and is largely based on "Regeln für die Umschrift des Japanischen in schriftlichen Arbeiten" (Rules for the Transcription of Japanese in Written Works) by Wolfgang Schamoni, edited by Hans Martin Krämer of the Institute of Japanese Studies at Heidelberg University. Apostrophes have been omitted. The consistent hyphenation of the Japanese expressions into their semantic-morphological components is done for reasons of clarification, standardization, pronunciation, reading flow, and any future indexing. Due to the multilingual nature of the book and the assumed multilingualism of the intended readership, no explanations of the pronunciation rules are given here. Readers are kindly requested to research these themselves.

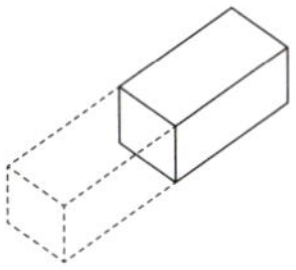	butt joint	stumpfer Stoß	*tsuki-tsuke* 突付	A butt joint is simply two flat wood surfaces meeting without interlocking parts. However, it is not a connection in the proper sense.
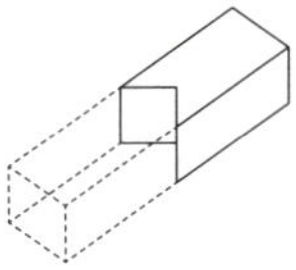	birdsmouth	Gratstoß (Führungsstoß / Gratschnitt)	*yahazu-mechi* 矢筈目違	A birdsmouth is a triangular notch cut into the end of a workpiece.
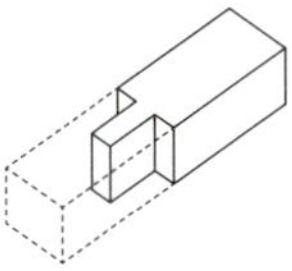	tenon	Zapfen (Zapfenstoß / Scherzapfen)	*hozo* 枘	A tenon is a tongue-shaped piece that projects out from the end of a workpiece, designed to fit into a corresponding mortise (a slot or cut in the piece being joined).
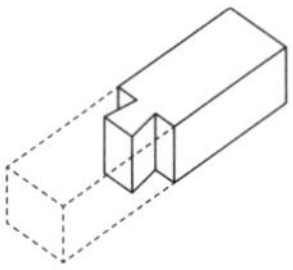	dovetail tenon	Schwalben-schwanzzapfen	*ari-hozo* 蟻枘	A dovetail tenon is a trapezoidal-shaped tenon fitting into a corresponding mortise. It can also transfer low tensile forces.
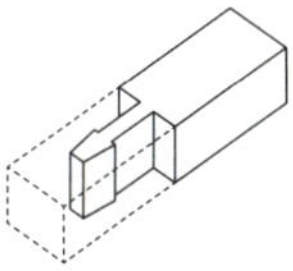	gooseneck tenon	Sichelzapfen	*kama-hozo* 鎌枘	A gooseneck tenon is a ribbed end of a long tenon with a gooseneck shape and a corresponding mating mortise cut into the other member. It can also transfer tensile forces.
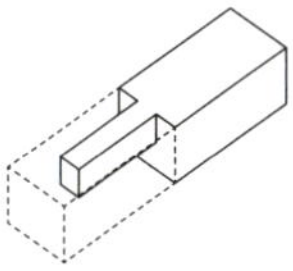	rod tenon (pole tenon)	Stabzapfen (Stangenzapfen)	*sao-hozo* 竿枘	A rod tenon, or pole tenon, is a three-shouldered and elongated form, often employed in Japanese carpentry for creating two-, three-, or four-way post connections.
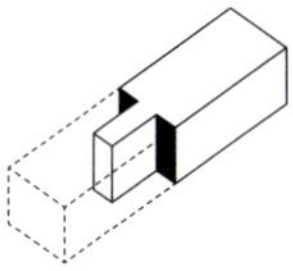	shoulder	Brüstung (Brust)	*dō* 胴	A shoulder is a perpendicular and distinct surface found at the sides of a tenon or joint, transferring loads and providing structural support and alignment.

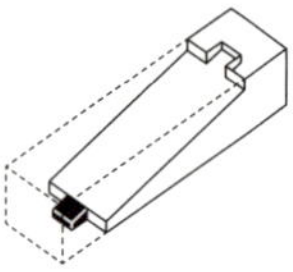	stub tenon (tenoned)	Brüstungszapfen (Stirnzapfen)	*mechi-hozo* 目違枘	A stub tenon is a shortened protrusion on the end of a workpiece designed to fit into a corresponding mortise.
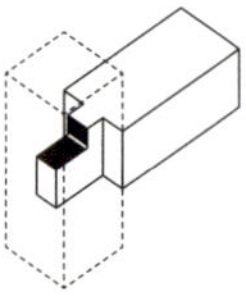	haunched tenon	Stufenzapfen (Brustzapfen / Nutzapfen)	*kone-hozo* 小根枘	A haunched tenon is a tenon with an extra partial section to reinforce the joint and prevent twisting.
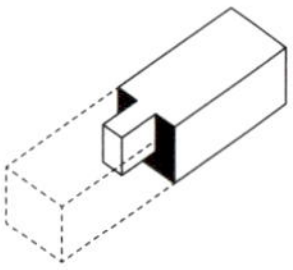	triple-shouldered tenon	abgesetzter Zapfen (abgesteckt / geächselt)	*sanpō-dō-tsuki-hozo* 三方胴付枘	A triple-shouldered tenon is a regular tenon with a third shoulder, mostly used in corner joints.
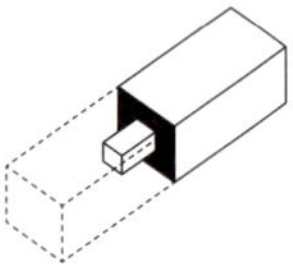	fully shouldered tenon	beidseitig abgesetzter Zapfen (beidseitig / allseitig abgesteckt)	*shihō-dō-tsuki-hozo* 四方胴付枘	A fully shouldered tenon is a tenon with two additional shoulders to increase the area for load transfer.
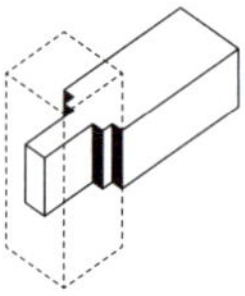	blind-housed tenon (tenon with stepped shoulder)	Zapfen mit gestufter Brüstung (abgesetzter Brüstung / Brust)	*ko-dō-tsuki* 小胴付	"Blind-housed" refers to a shoulder on a joint that features multiple levels, enhancing the structural strength by increasing the bearing surface.
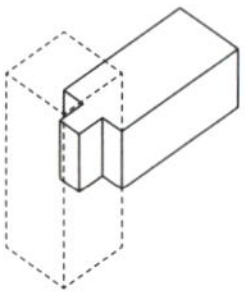	blind tenon	verdeckter Zapfen (eingelassener / gestemmter / einfacher Zapfen)	*tsutsumi-komi-hozo* 包込枘	A blind tenon is a tenon designed to fit into a corresponding mortise without extending through the counterpart.
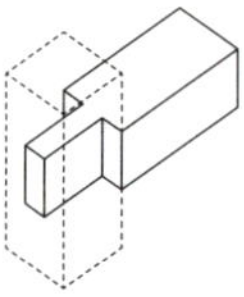	through tenon	durchgehender Zapfen (durchgestemmter Zapfen)	*tōshi-hozo* 通枘	A through tenon extends through the counterpart, showing the end of the tenon.

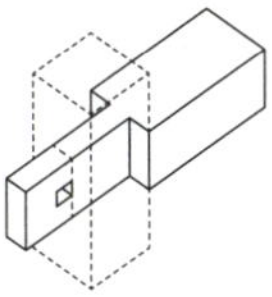	tusk tenon	durchgesteckter Zapfen (durchgestoßener Zapfen)	*uchi-nuki-hozo* 打抜枘	A tusk tenon extends through the counterpart and sticks out to accept a wedge or pin.
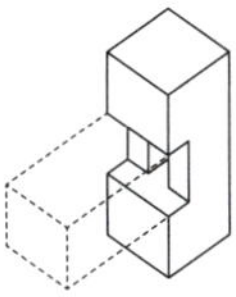	housed mortise and tenon	eingelassener Zapfen	*oire* 大入、追入	The term "housed" refers to a joint where the end of one member is secured into a housing or recess cut into the face of the adjoining piece.
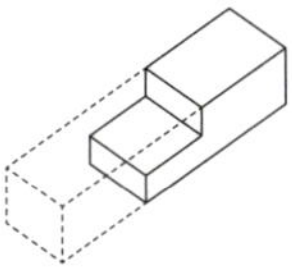	half lap (halved scarf / halving joint / lap joint)	gerades Blatt (Verblattung, Überblattung)	*koshi-kake* 腰掛け	A half lap joint connects two pieces of wood by overlapping and removing half of the material from each, creating a flush connection.
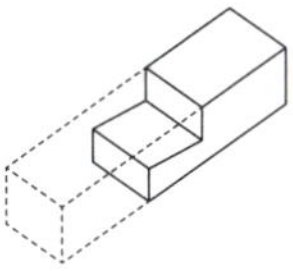	beveled lap (dovetailed scarf joint)	Zugblatt (Druckblatt)	*sogi-tsuki-koshi-kake* 殺付腰掛け	A beveled lap joint is formed by interlocking inclined wood pieces with halved dovetail projections. It requires pressure from above to withstand tensile forces.
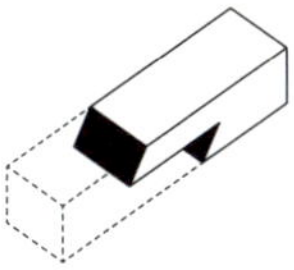	undersquinted (squint-butted, undercut)	hinterschnittene Brüstung (schräger Schnitt / schräger Stoß)	*kōbai-tsuki-koshi-kake* 勾配付腰掛け	Undersquinted in woodworking terminology refers to a joint where the ends or shoulders of the workpiece are cut at an oblique angle.
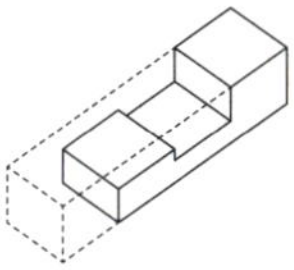	cogged lap joint (tabled lap joint)	gerades Hakenblatt	*dan-tsuki-koshi-kake* 段付腰掛け	A cogged lap joint involves notching and overlapping at the ends of two pieces, creating a hook to secure the joint.
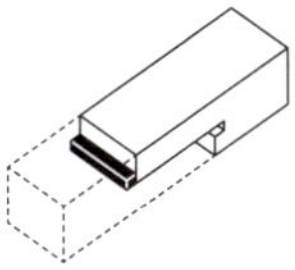	rabbeted (bladed / lipped)	Steckfalz (Falz / gestufte Stirn)	*mechi-tsuki-koshi-kake* 目違付腰掛け	The term "rabbeted" for joints signifies the incorporation of a step along the mating surfaces, enhancing the fit and security of the joint by limiting the directional movement.

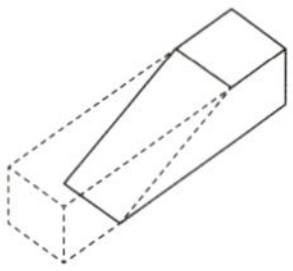	splayed scarf (oblique scarf)	Schäftung (schräger Stoß)	*sogi* 殺ぎ	The splayed scarf joint resembles a butt joint with angled cuts. It increases the contact area for a better gluing surface.
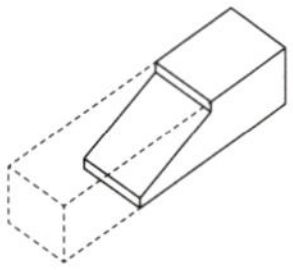	stop-splayed scarf	schräges Blatt (Gerberstoß)	*tome* 留め	A stop-splayed scarf is a joint characterized by an angled cut that stops short of the edge, providing positional stability.
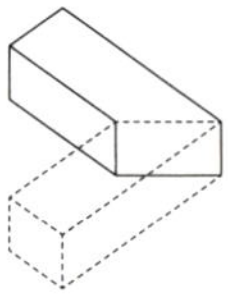	miter	Gehrung	*kane* 矩	A miter is made by dividing the angle of a corner joint where two members intersect, typically creating two 45-degree angles in a right-angle configuration.
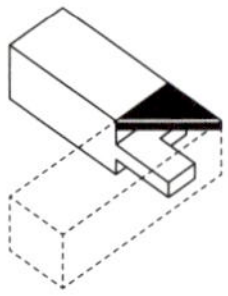	lapped miter (barb)	Gehrungsblatt	*binta* 鬢太	A lapped miter is a thin, beveled overlapping element in a joint, enhancing the aesthetic appeal of visible connections while structural strength is maintained through the primary connection.
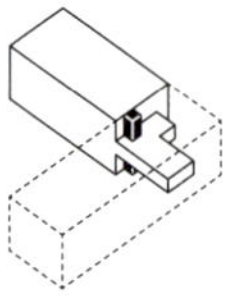	collar	Führungszapfen (Brustzapfen)	*eriwa* 襟輪	A collar in woodworking is an additional stub tenon fitted around the main tenon of the joint.
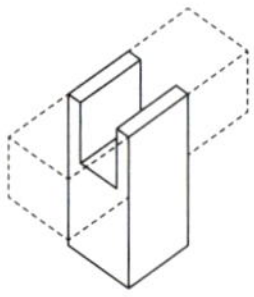	T-bridle joint	Einhälsung (Gabeleinblattung)	*nagiwa* 薙輪	A T-bridle joint is an open slot mortise and a perpendicular notched counterpart. This joint is the evolution of a tree fork.
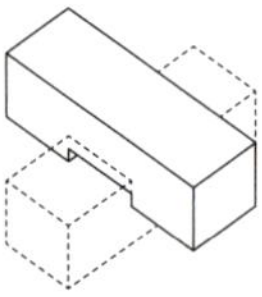	notched joint	Kamm (Verkämmung)	*ai-kaki-gumi* 相欠組み	A notched joint is defined by a slight recess on one or both parts that secures the position when staggered.

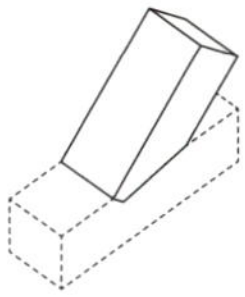	oblique thrust joint	Versatz (Versatzung)	*ochi-gakari* 落掛	An oblique thrust joint is a connection where the thrust or pressure is applied at an angle to the axis of the receiving member.
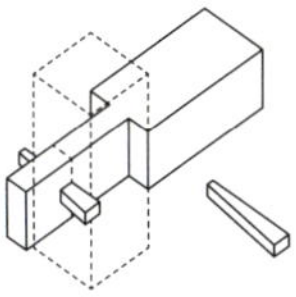	loose wedge	Keilschloss (Spannschloss / Spannkeil)	*komi-sen* 込栓	A loose wedge is a tapered wooden element that tightens a joint securely but retains the flexibility of being removable when needed.
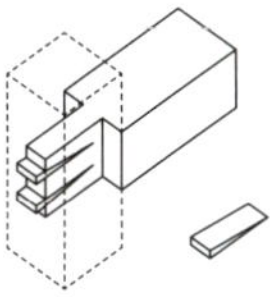	fixed wedge	Keil/verkeilt	*wari-kusabi-sen* 割楔栓	A fixed wedge is a triangular wooden element driven into a kerf in the end grain of a tenon, providing a strong and permanent fastening solution.
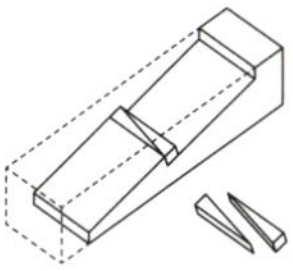	folding wedges	verkeilt	*kai-kusabi* 飼楔	Folding wedges are a pair of opposing tapered elements incorporated to tighten a joint.
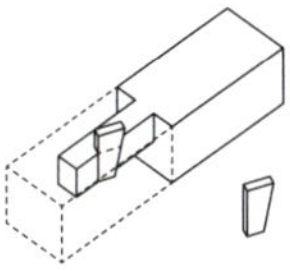	tapered key	Spannkeil	*shachi-sen* 車知栓	A tapered key is a fastener characterized by a gradually narrowing rectangular profile, commonly utilized to tighten Japanese rod joints.
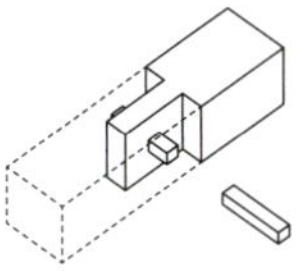	peg	Dübel (Holznagel)	*dabo* 太枘	A peg in woodworking is a square-shaped or cylindrical fastener, typically made of hardwood, used to secure or tighten a joint.
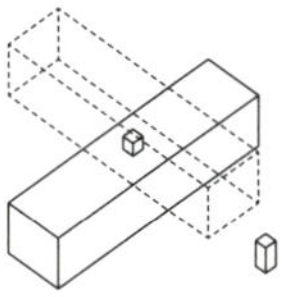	blind peg	Verdollung (Dolle / Dübel)	*dabo-gumi* 太枘組み	A blind peg refers to a short square-shaped or cylindrical wooden fastener inserted into a joint without protruding through the surface.

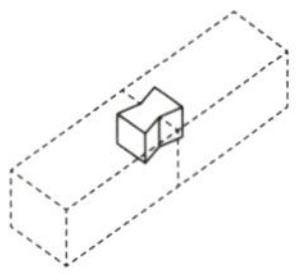	butterfly key (butterfly spline / bowtie spline)	Schwalben-schwanzeinlage (abgesetzter Schwalben-schwanzdübel / falsches Schwalben-schwanzblatt)	*chigiri-sen* 千切栓	A butterfly key is a thin element shaped like back-to-back dovetails, also known as a "bowtie." It elegantly reinforces weak or cracked areas, especially in planks, but can also be used in butt or miter joints.
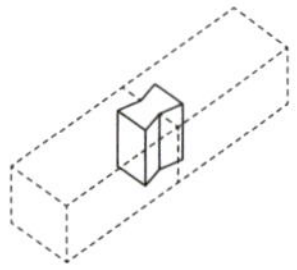	floating butterfly tenon (loose butterfly tenon)	eingesetzter Schwalben-schwanzzapfen (durchgehender Schwalben-schwanzdübel / falscher Schwalben-schwanzzapfen)	*chigiri-ōdome-sen* 千切大留栓	A floating butterfly tenon is a through element shaped like back-to-back dovetails. It can be used to reinforce butt joints.
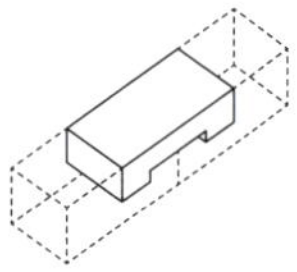	floating piece	eingesetztes Hakenstück (eingesetztes/r Schloßstück/ Stoßstück/ Stück/Haken)	*yatoi* 雇い	A cogged lap joint with a floating piece is a connection with an additional notched element that connects two cogged ends of the joint.
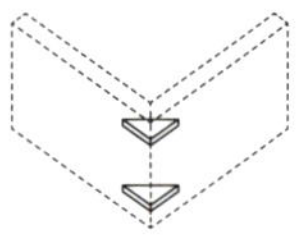	feather spline	eingelassene Feder	*hiki-komi-zane* 挽込実	A feather spline is a thin, flat element made from cross-grain wood. It is inserted into grooves at the corner of miter joints to strengthen the connection.
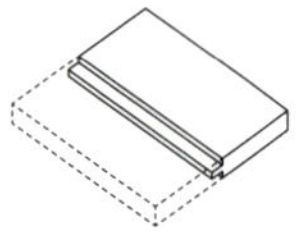	tongue and groove	Nut und Feder (Spundung)	*koana + yatoi-zane* 小穴＋雇実	A groove is the corresponding slot for a long, narrow tenon known as a tongue, typically located at the edge of a board and parallel to the grain.
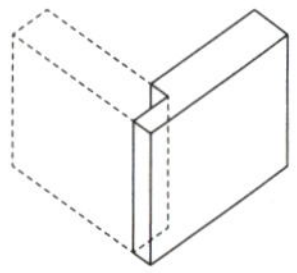	rabbet	Falz	*tsutsumi* 包み	A rabbet is a rectangular stepped recess cut along the edge or end of a board, commonly used to cover the end grain and overlap with another board.
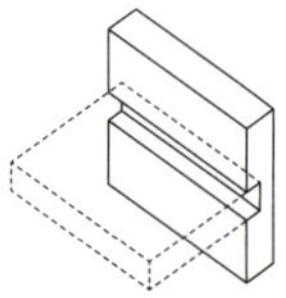	dado	Nutverbindung	*koana-tsugi* 小穴接ぎ	A dado is a rectangular slot cut across the grain, providing a space to accommodate the edge or end of another piece.

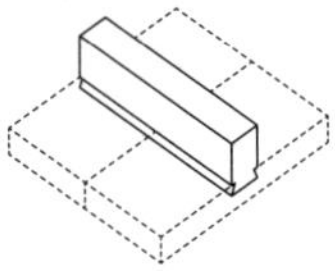	sliding dovetail batten	Gratleiste (Einschubleiste / Querleiste)	*sui-tsuki-ari* 吸付蟻	A sliding dovetail batten joins boards perpendicular to each other, forming a level surface for a dovetail-shaped dado, which can allow for sliding the corresponding batten.
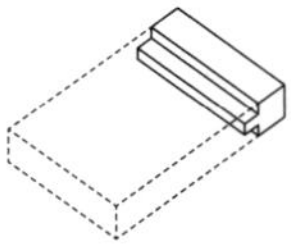	breadboard end	Hirnleiste	*hon-zane* 本実	A breadboard end is attached to the ends of boards to enhance stability and prevent warping. It is commonly used in tables.

Chapter

Timber Joints

Splicing Joints

Half lap joint

koshi-kake-tsugi [*]
腰掛継ぎ

gerades Blatt

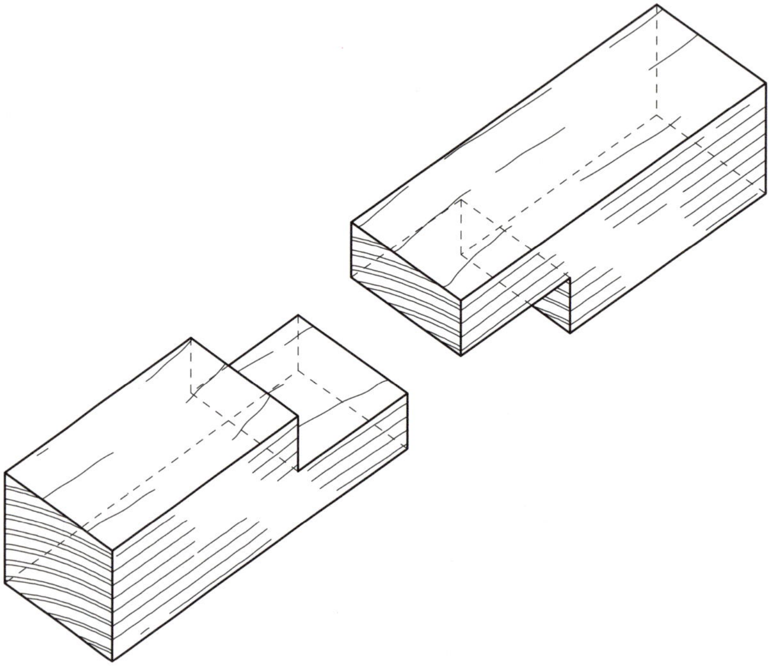

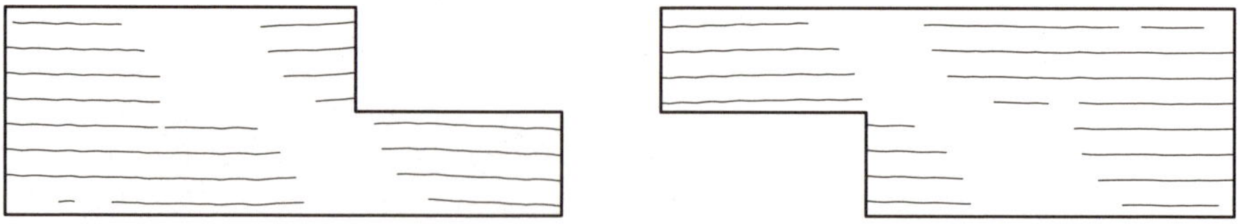

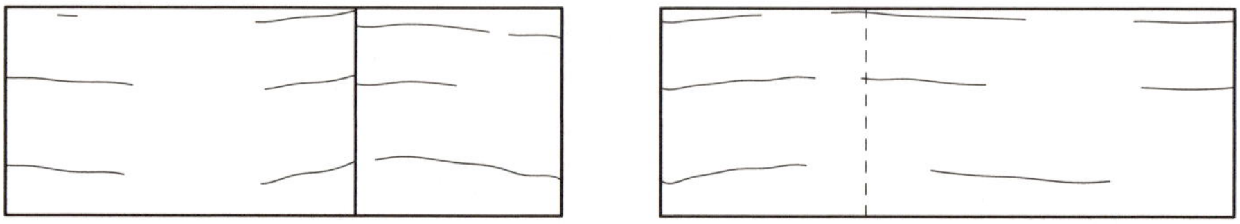

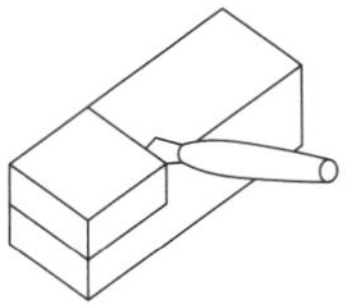

A1

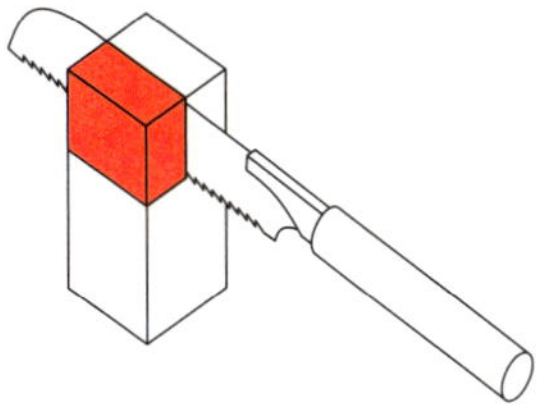

A2

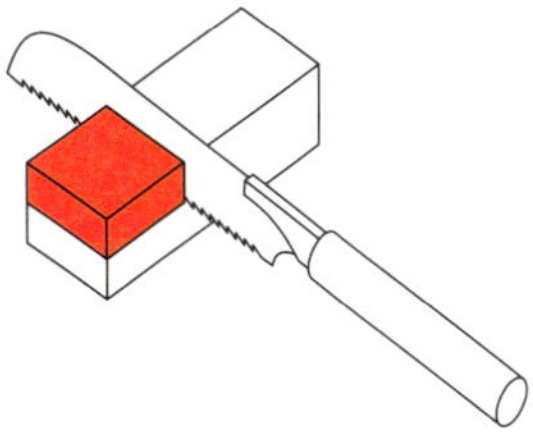

A3

2×

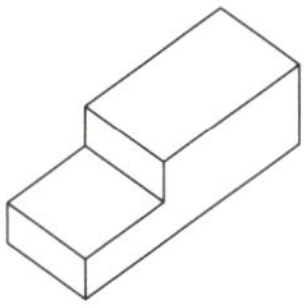

A

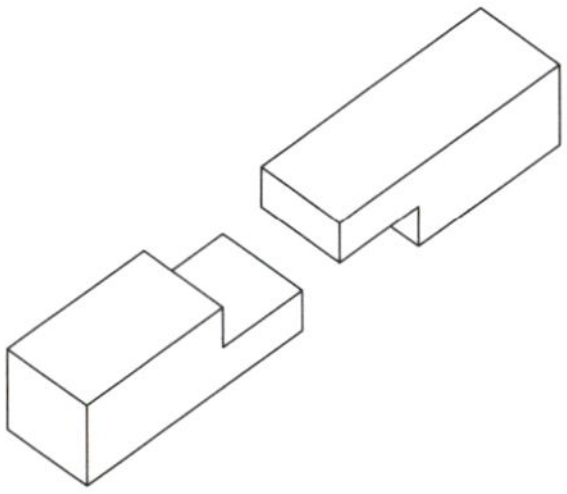

A+A

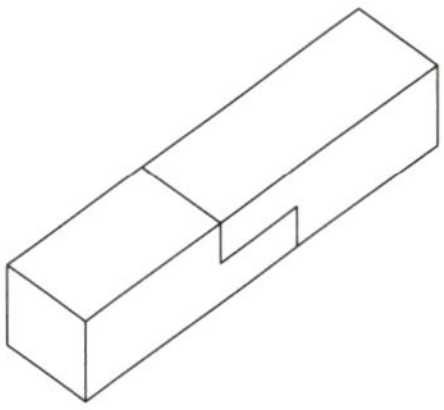

AA

Splicing Joints

Mitered half lap joint

koshi-kake-sogi-tsugi
腰掛殺継ぎ

gerades Blatt mit
einseitig schrägem Stoß

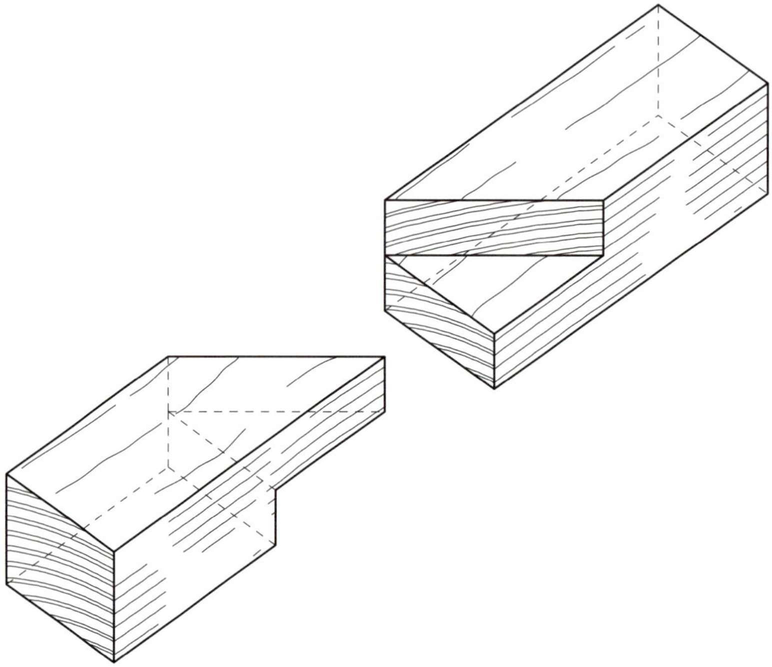

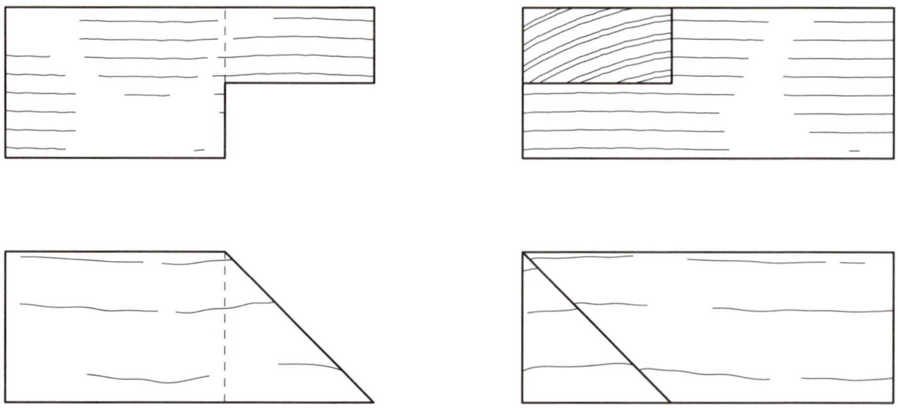

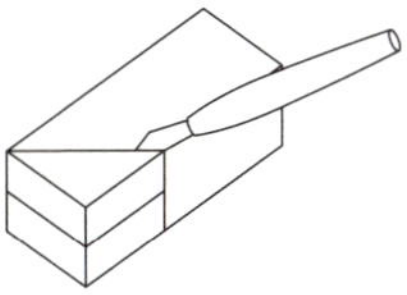

A1

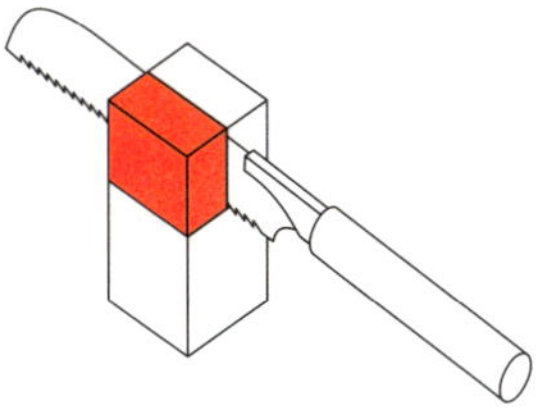

A2

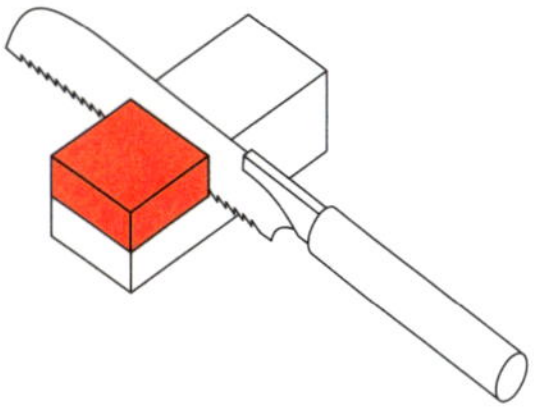

A3

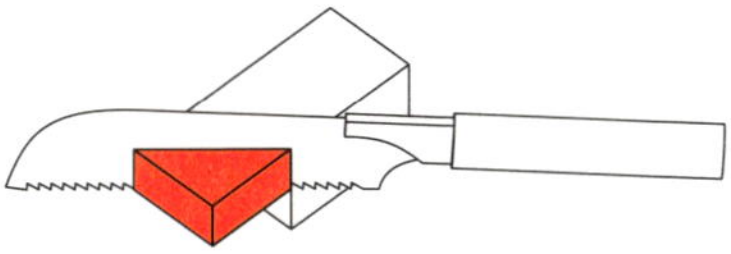

A4

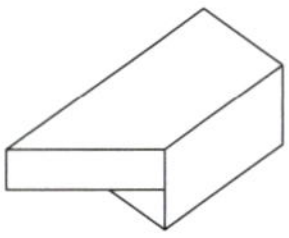

A

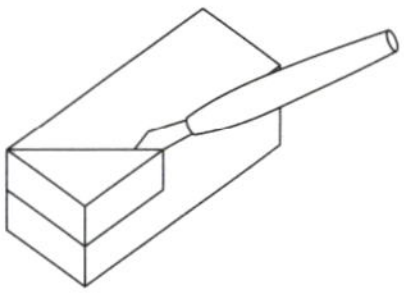

B1

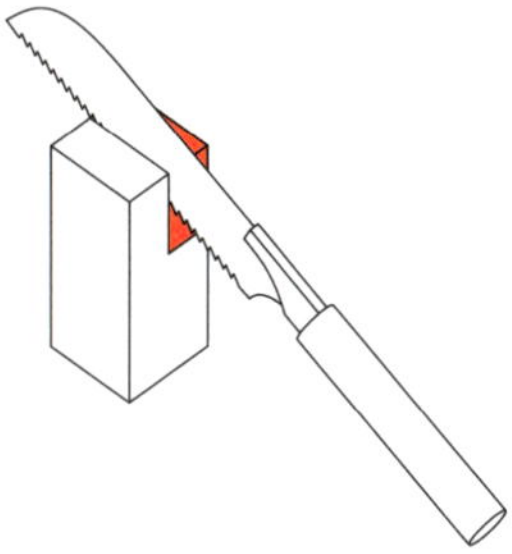

B2

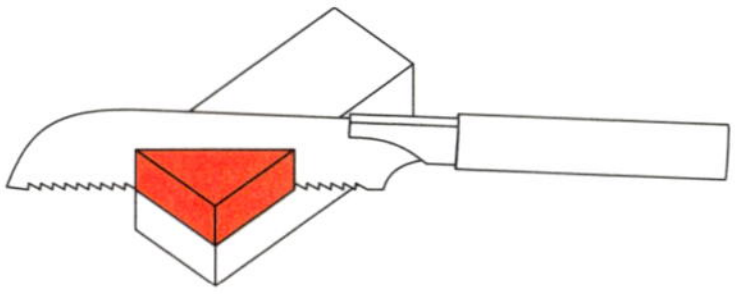

B3

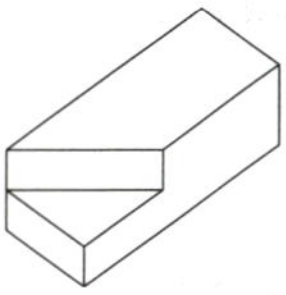

B

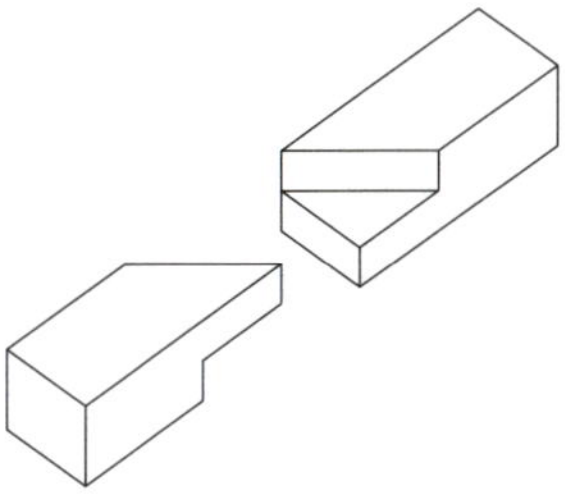

A+B

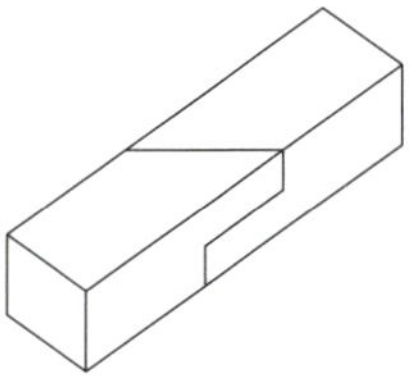

AB

Splicing Joints

Beveled lap joint

koshi-kake-tsugi (kōbai-tsuki)
腰掛継ぎ(勾配付)

Zugblatt

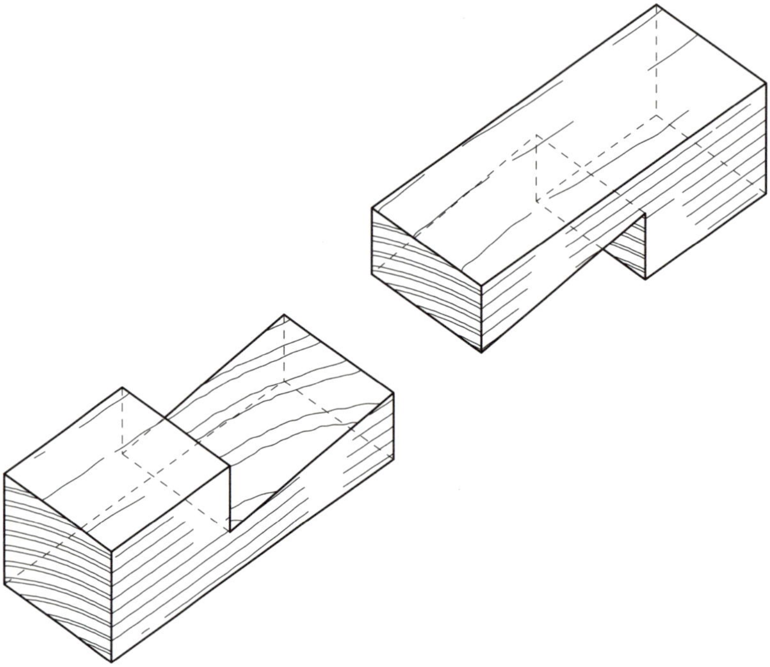

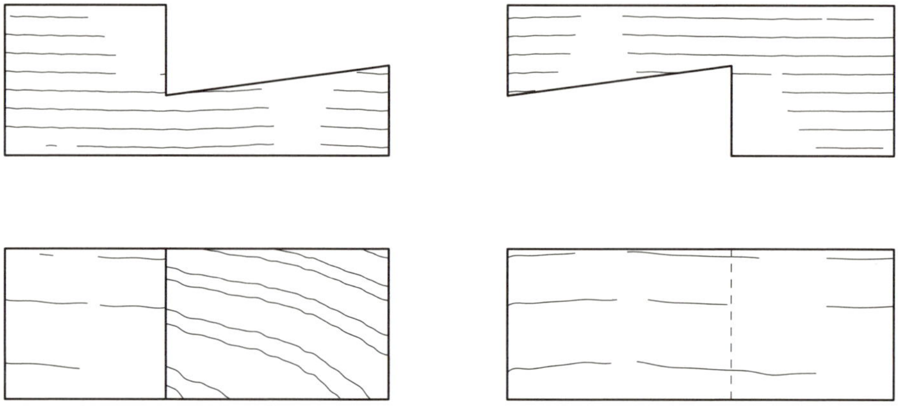

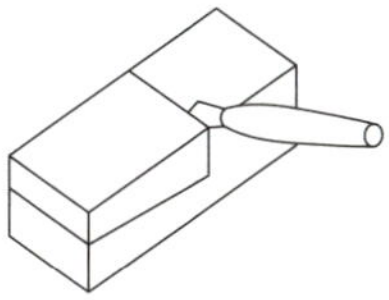

A1

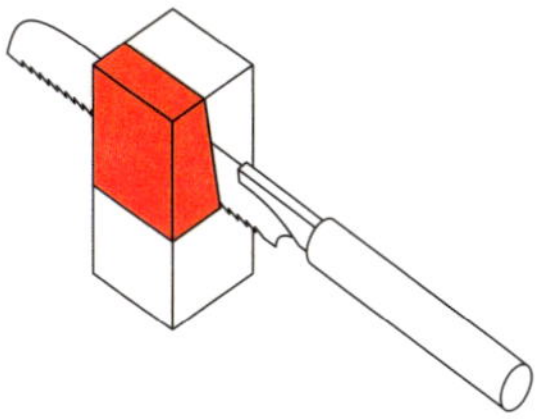

A2

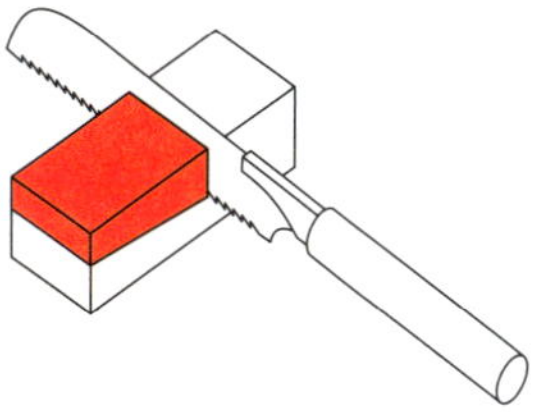

A3

2×

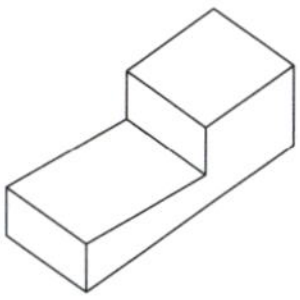

A

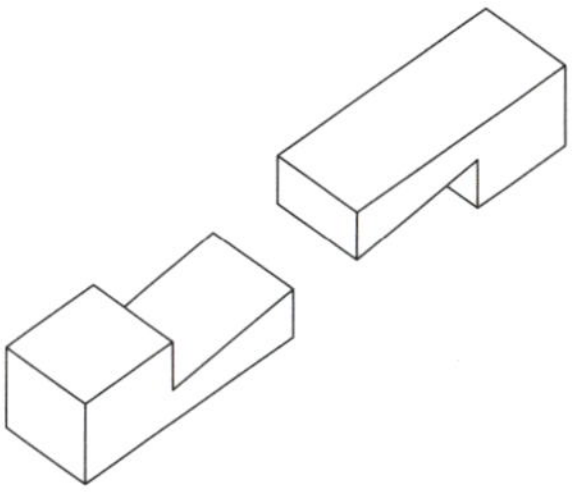

A+A

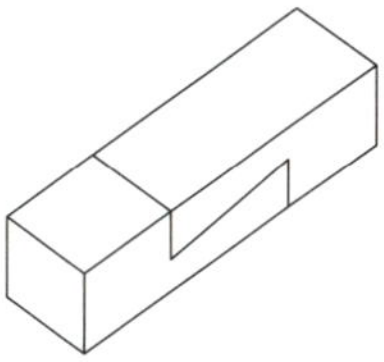

AA

Splicing Joints

Beveled lap joint with stub tenons

naga-koshi-kake-mechi-tsugi (kōbai-tsuki)
長腰掛目違継ぎ（勾配付）

Zugblatt mit Brüstungszapfen

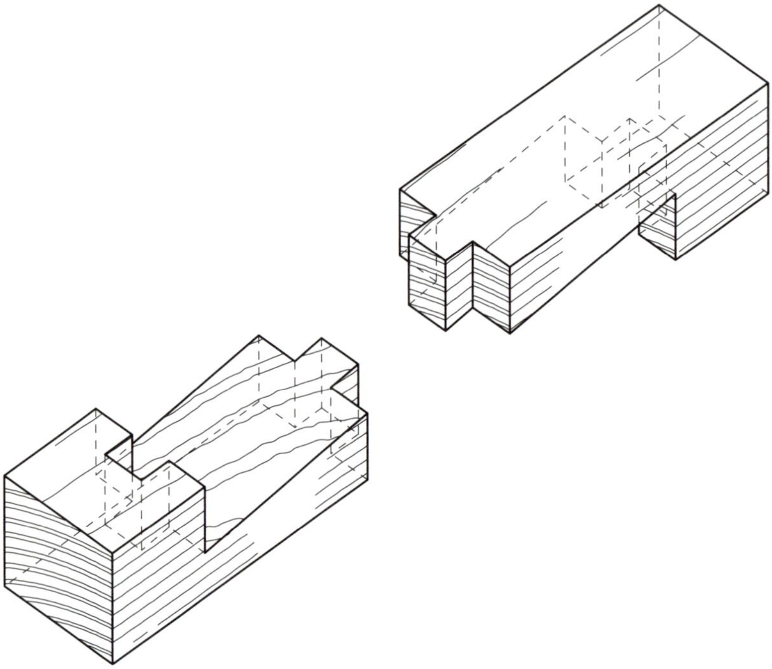

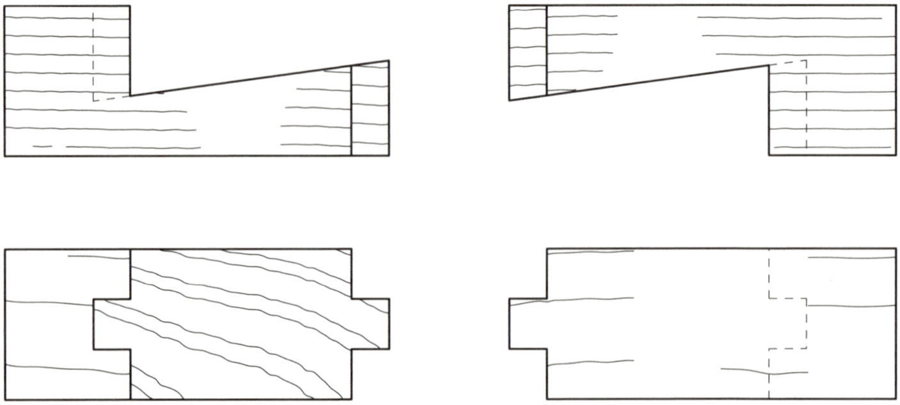

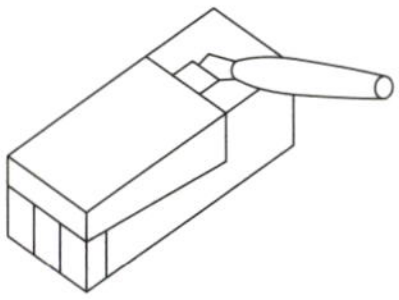

A1

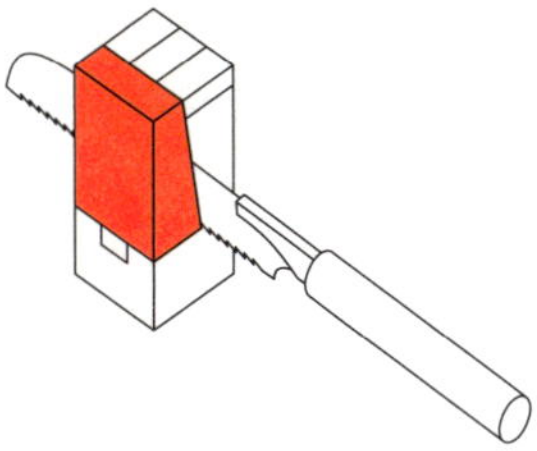

A2

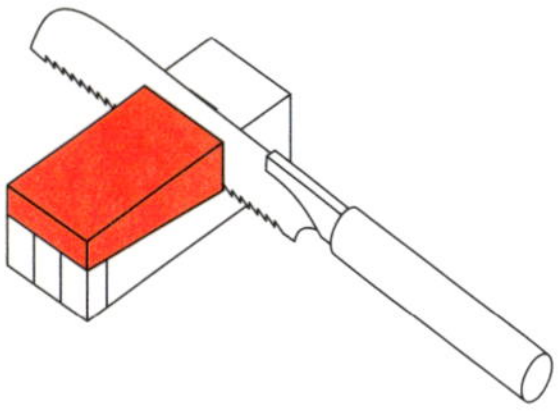

A3

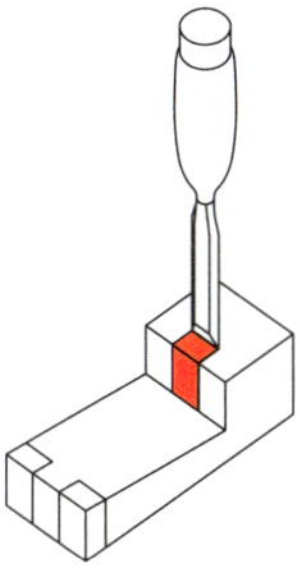

A4

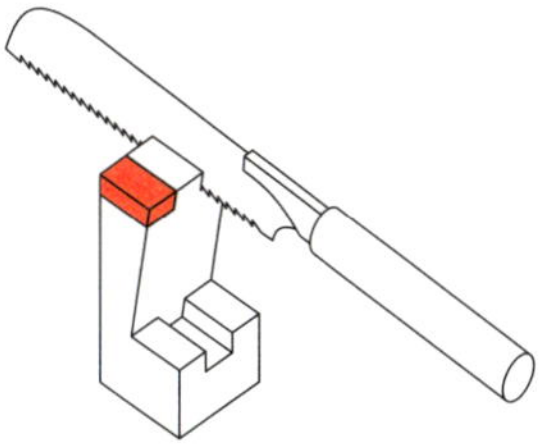

A5

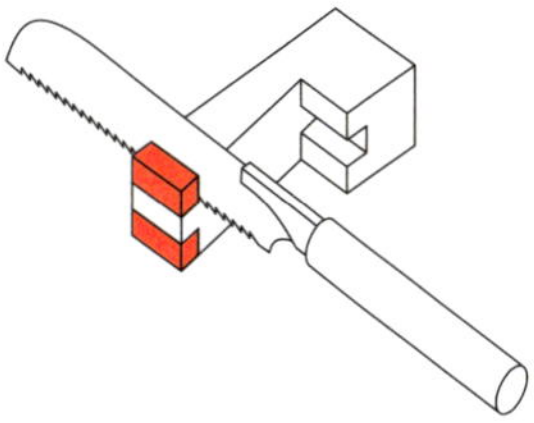

A6

2×

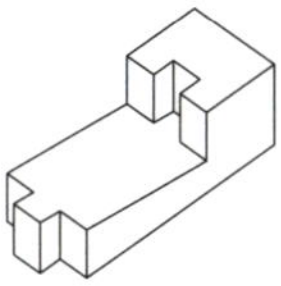

A

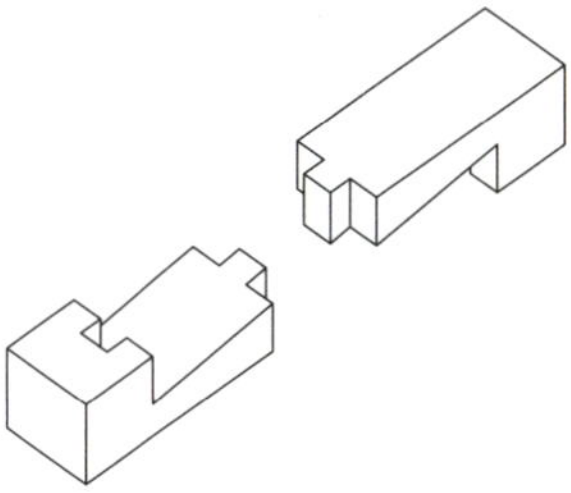

A+A

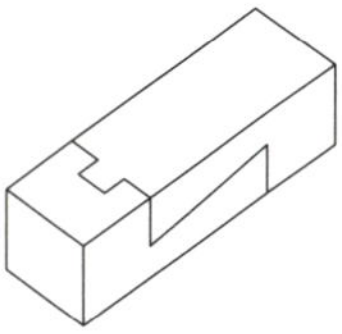

AA

Splicing Joints

Half lap joint with stub tenon

naga-koshi-kake-mechi-tsugi
長腰掛目違継ぎ

gerades Blatt mit einseitigem Zapfen

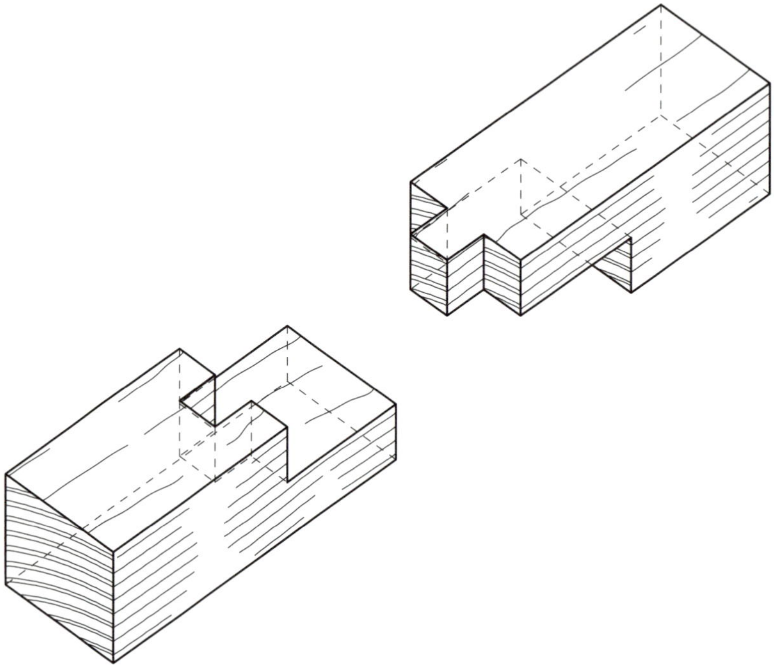

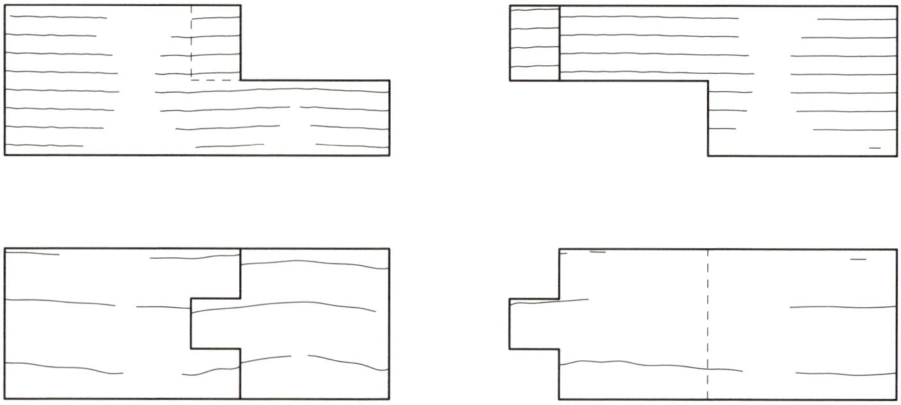

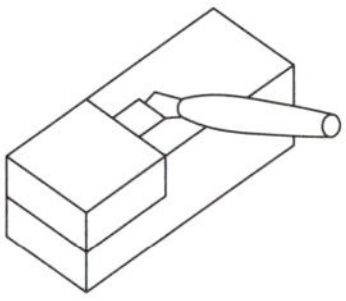

A1

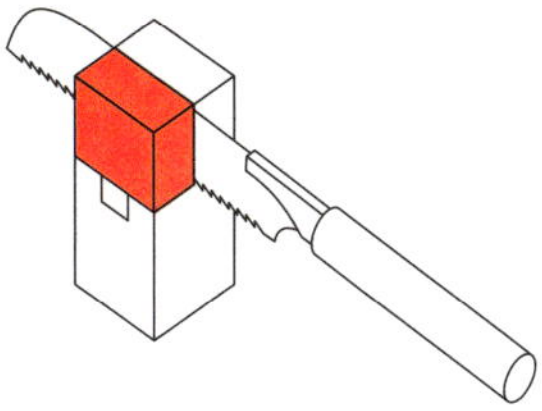

A2

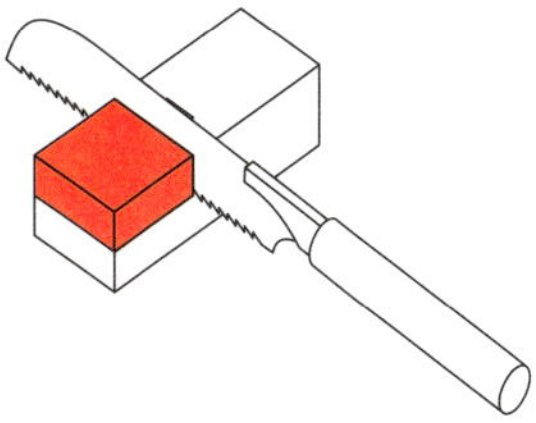

A3

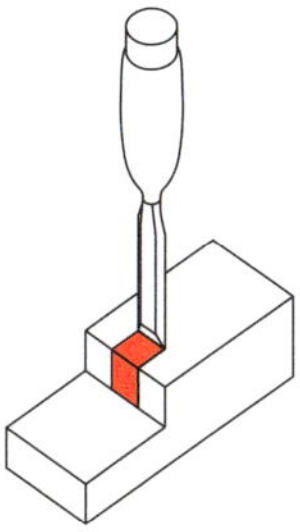

A4

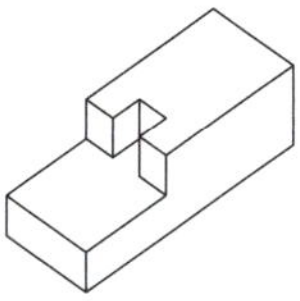

A

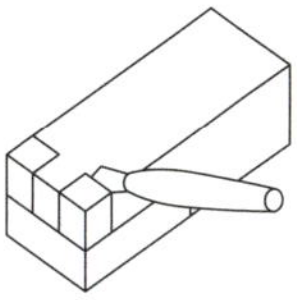

B1

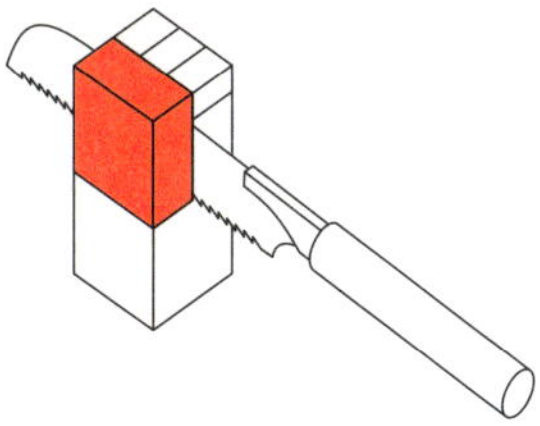

B2

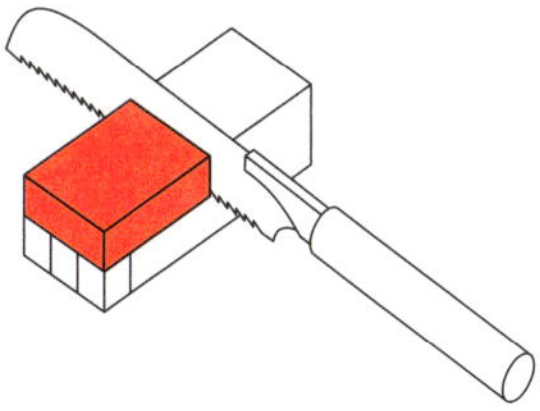

B3

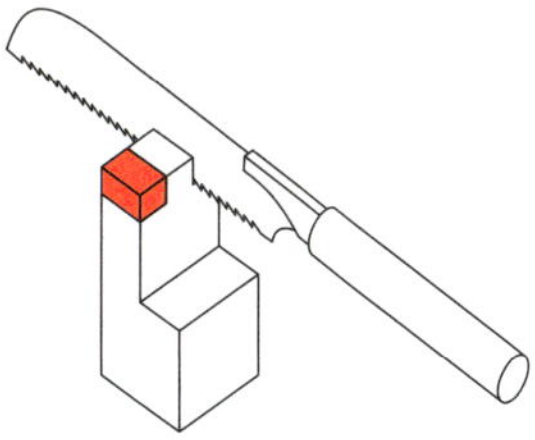

B4

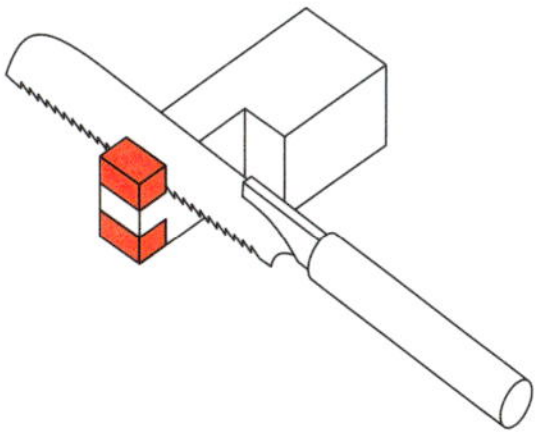

B5

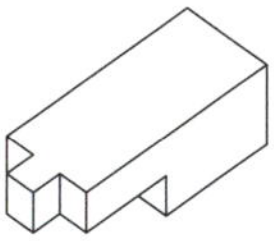

B

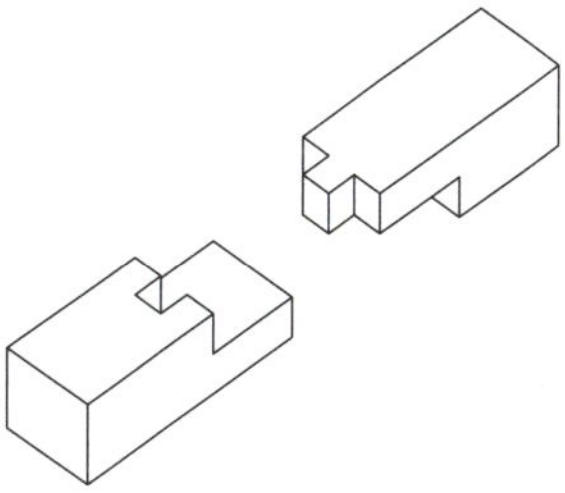

A+B

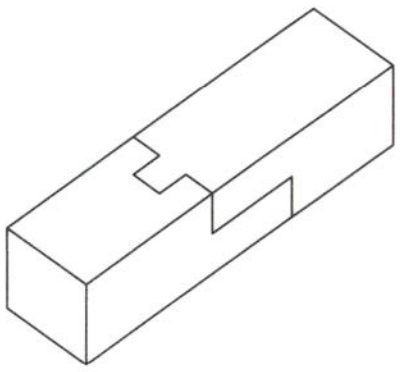

AB

Splicing Joints

Half lap joint with haunched tenon

koshi-ire-mechi-tsugi
腰入目違継ぎ

gerades Blatt mit Stufenzapfen

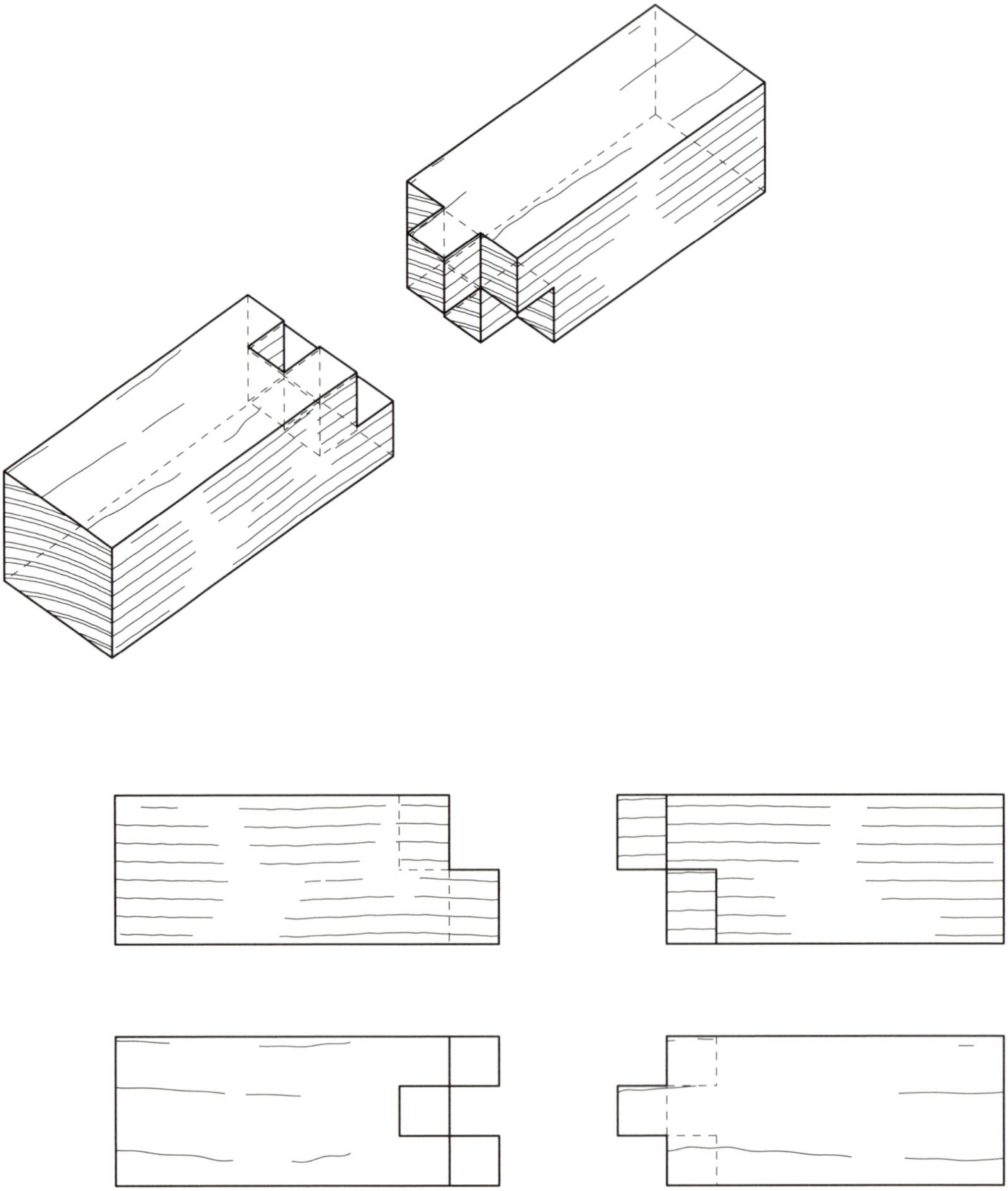

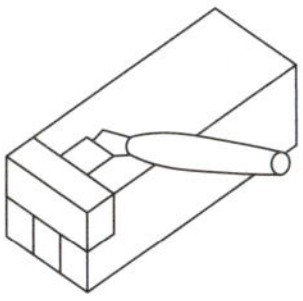

A1

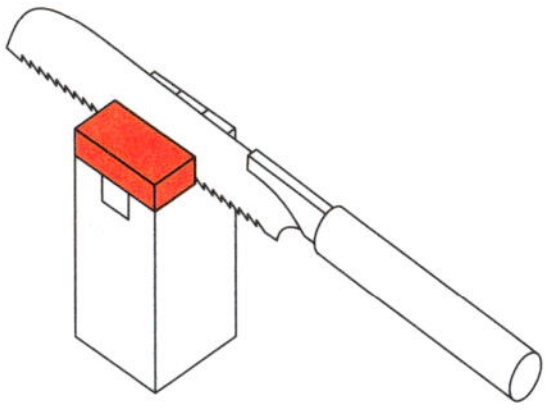

A2

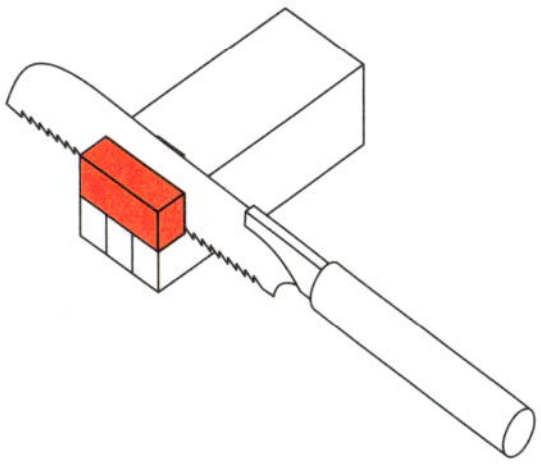

A3

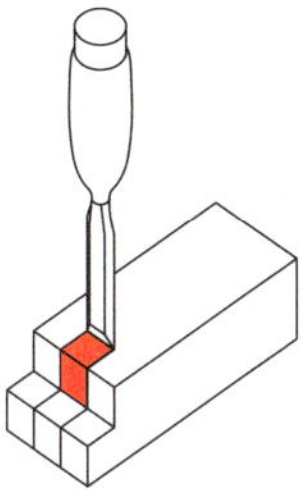

A4

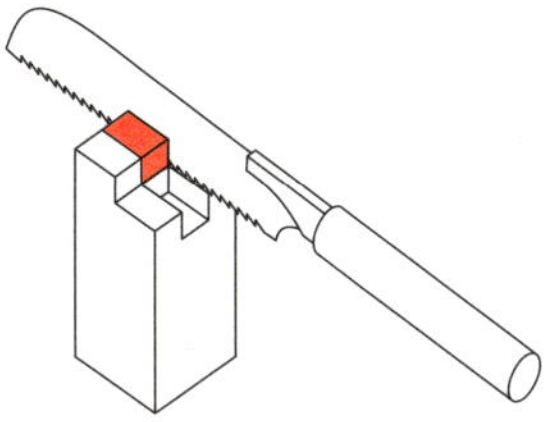

A5

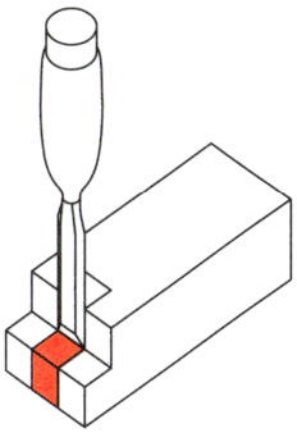

A6

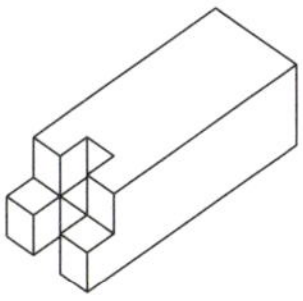

A

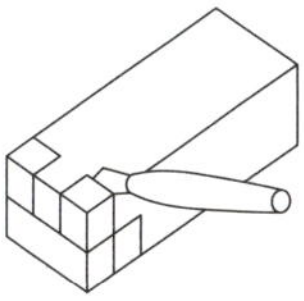

B1

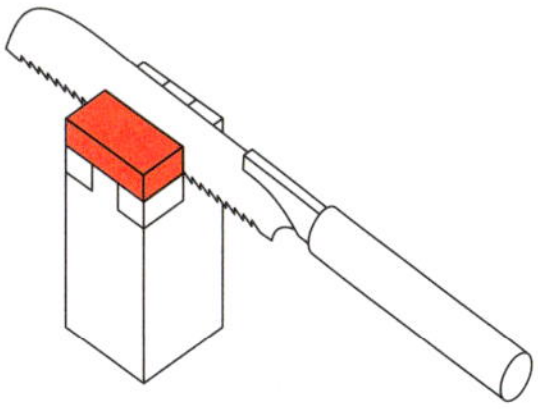

B2

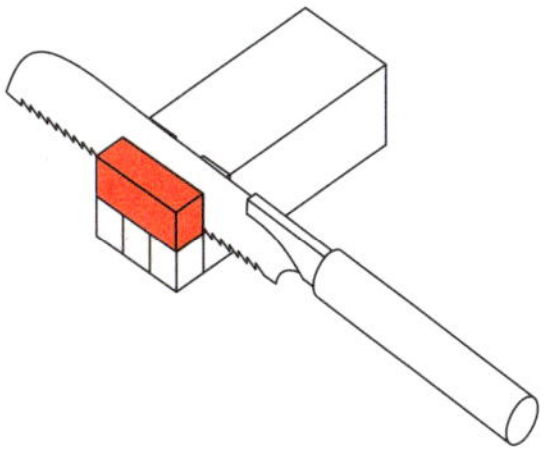

B3

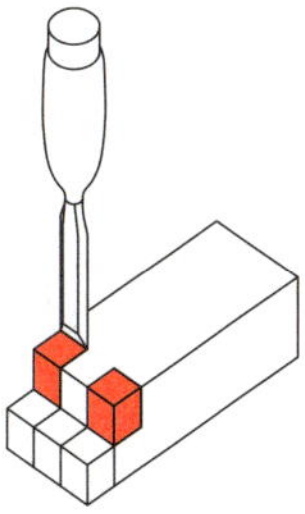

B4

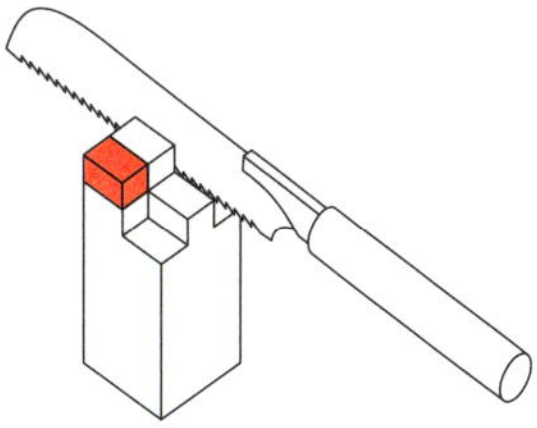

B5

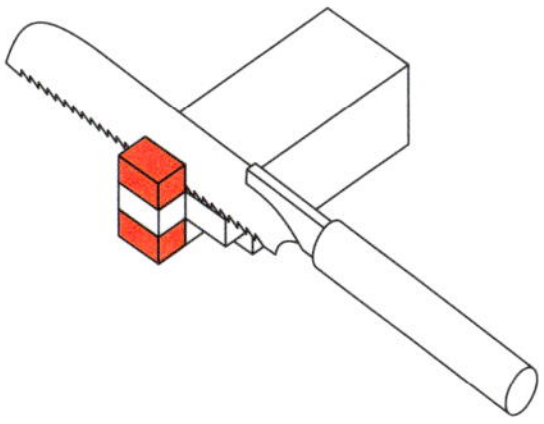

B6

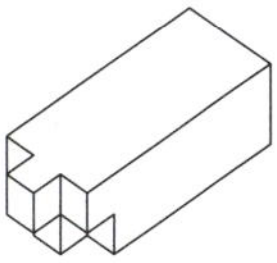

B

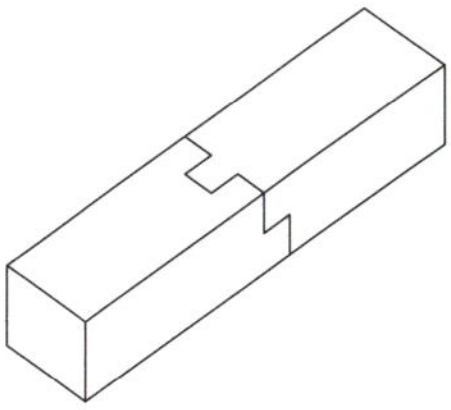

AB

Splicing Joints

Half lap joint with stub tenons

naga-koshi-kake-ai-mechi-tsugi
長腰掛相目違継ぎ

gerades Blatt mit beidseitigem Zapfen

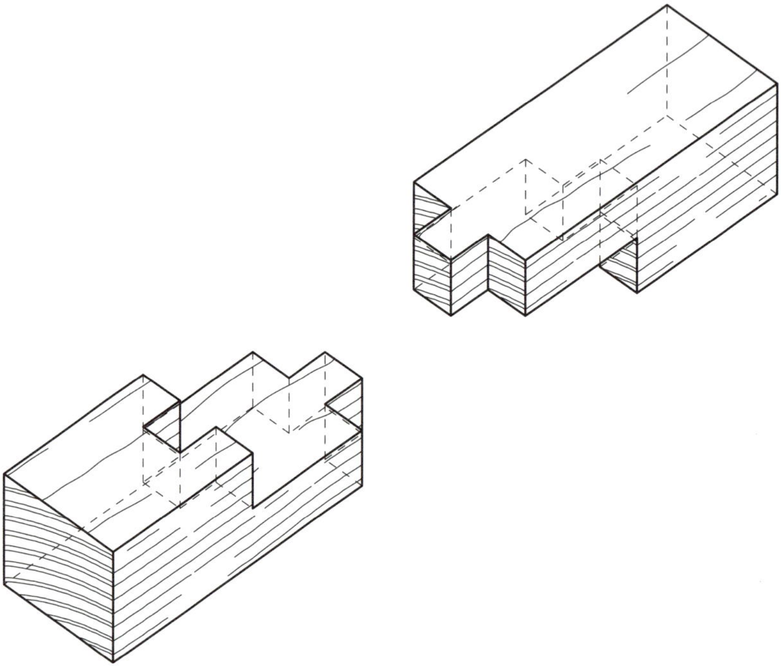

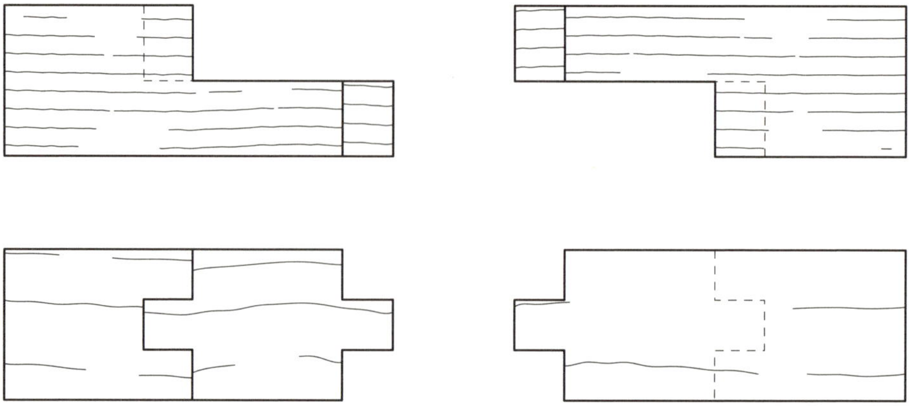

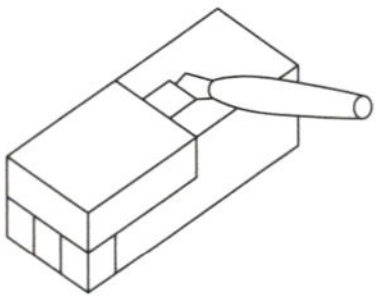

A1

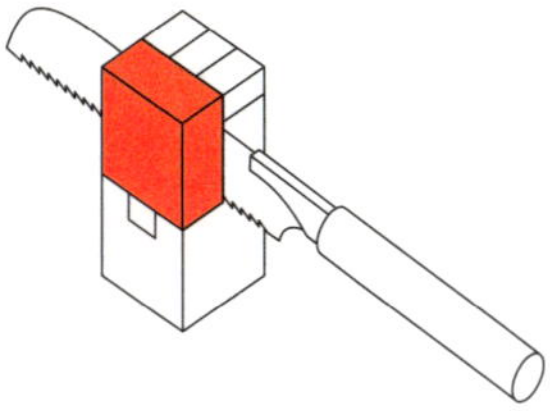

A2

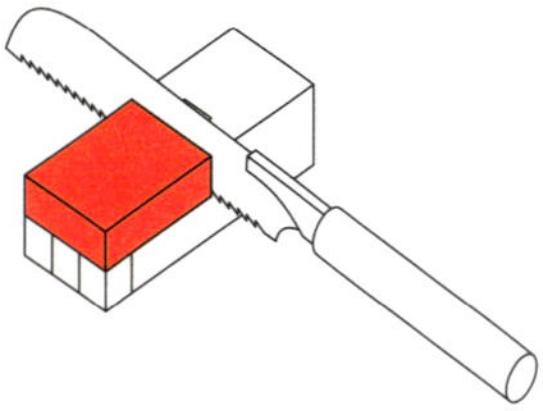

A3

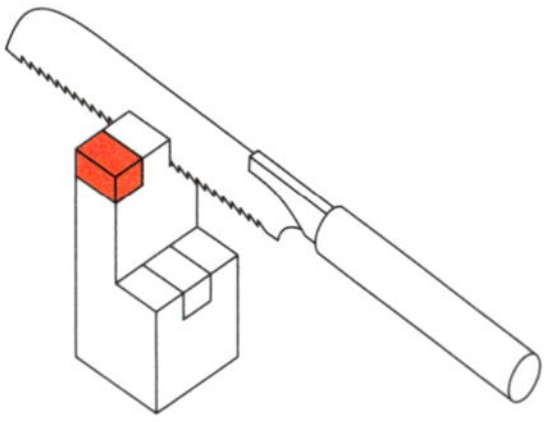

A4

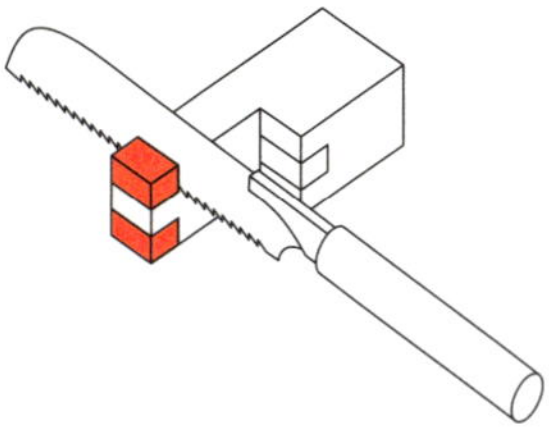

A5

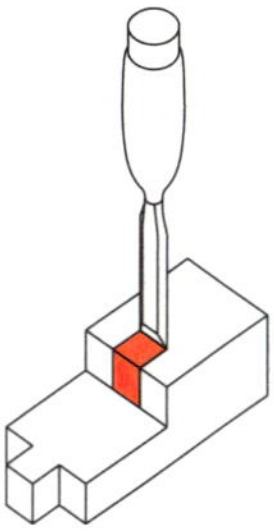

A6

2×

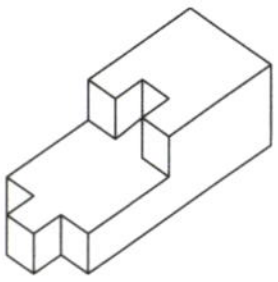

A

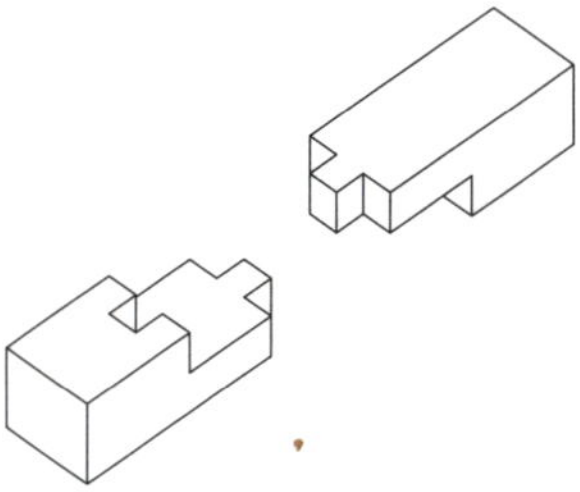

A+A

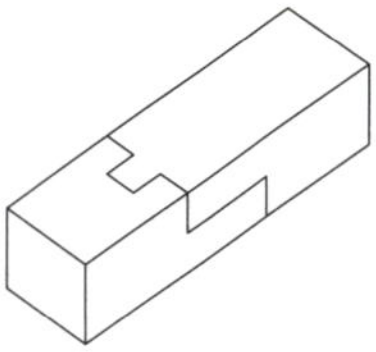

AA

Splicing Joints

Undersquinted half lap joint

naga-koshi-kake-tsugi (sogi-tsuki)
長腰掛継ぎ(殺付)

gerades Blatt mit
hinterschnittener Brüstung

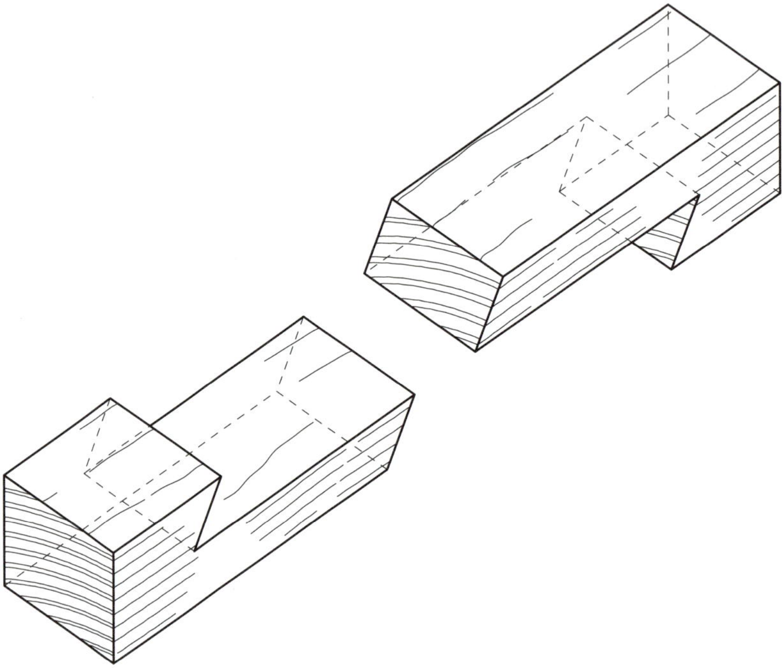

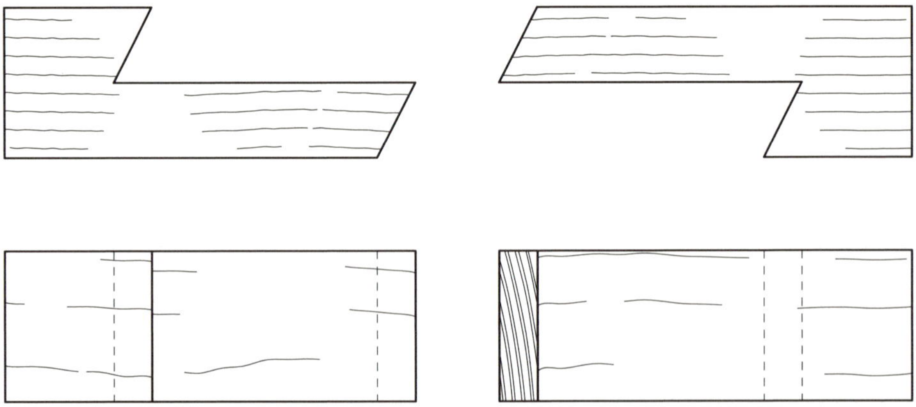

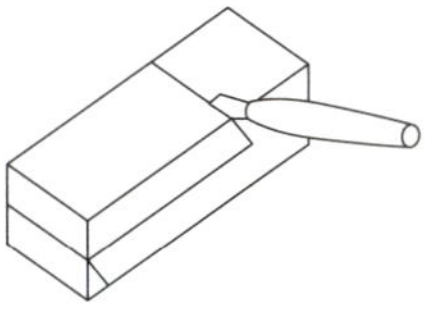

A1

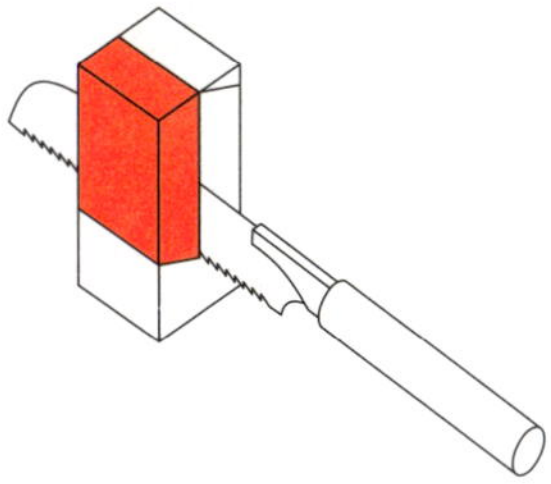

A2

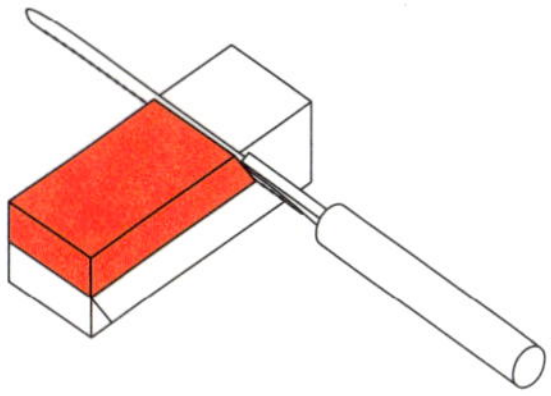

A3

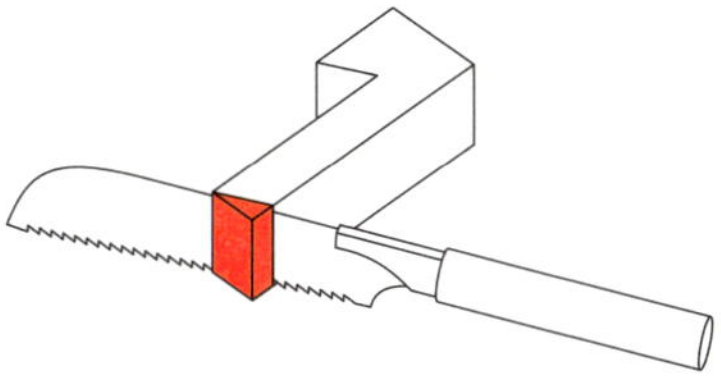

A4

2×

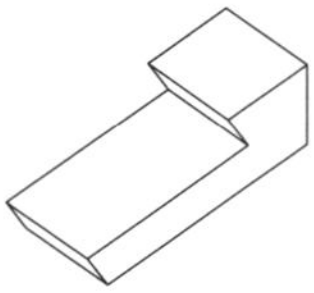

A

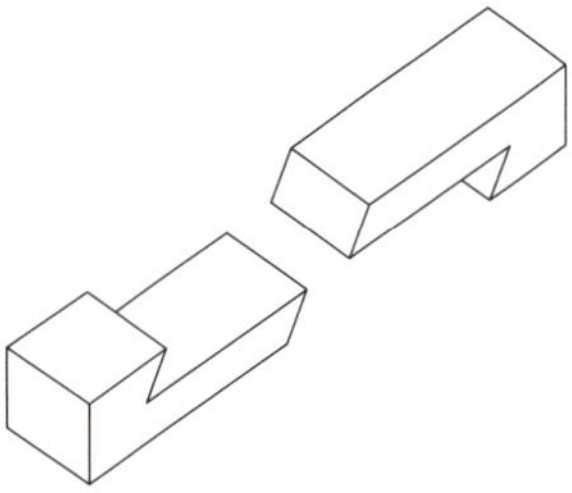

A+A

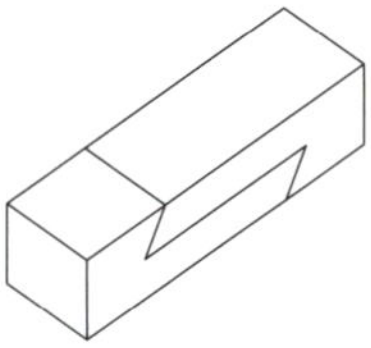

AA

Splicing Joints

Pegged undersquinted half lap joint

naga-koshi-kake-tsugi (sogi-tsuki, kusabi-shime)
長腰掛継ぎ(殺付・楔締)

gerades Blatt mit hinterschnittener Brüstung und Dübelsicherung

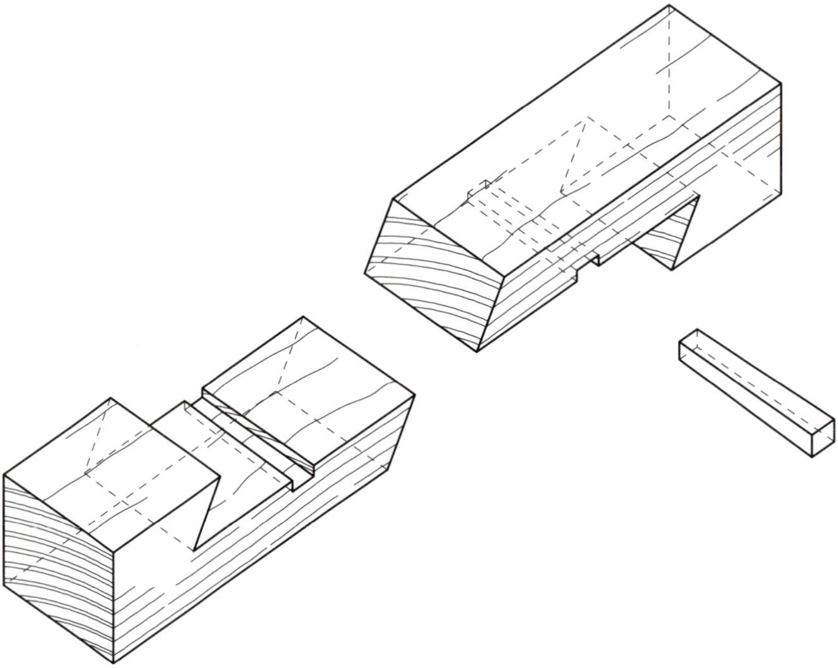

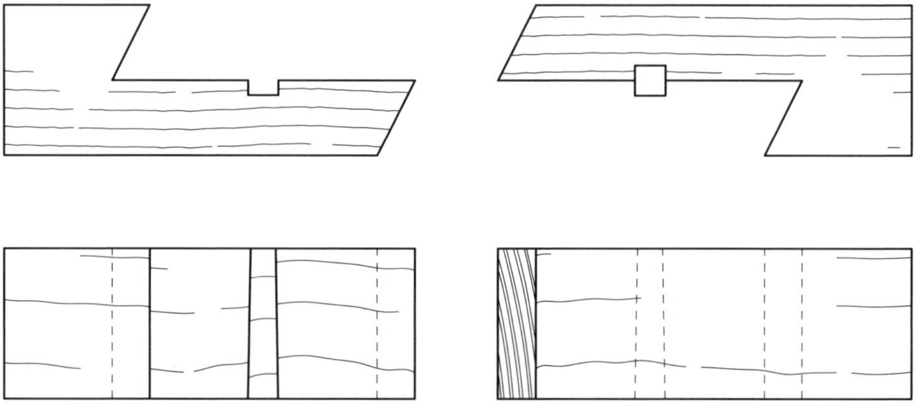

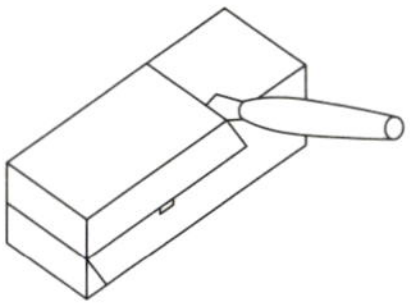

A1

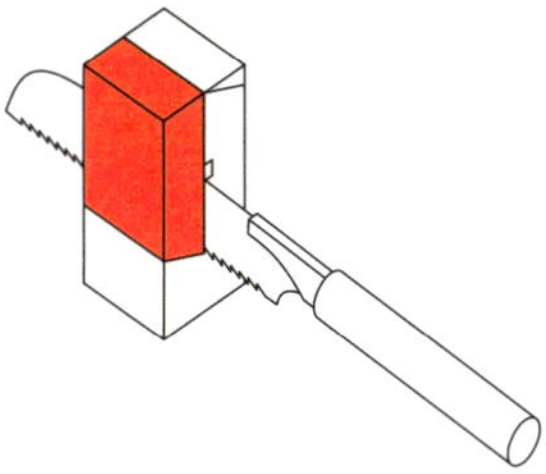

A2

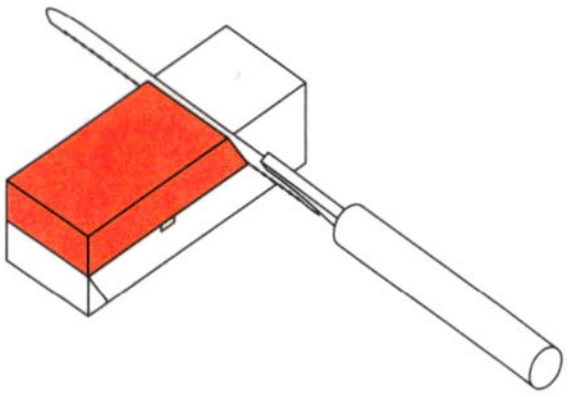

A3

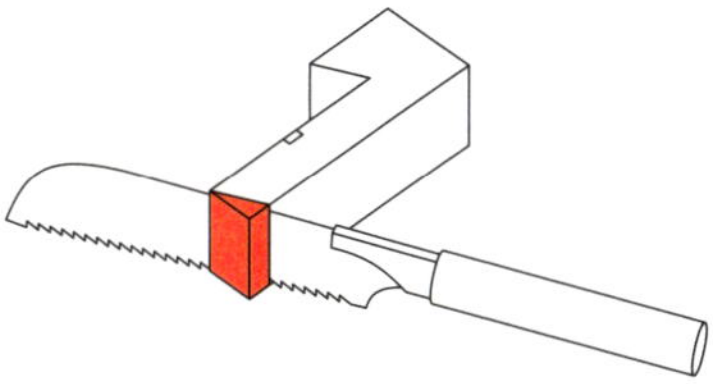

A4

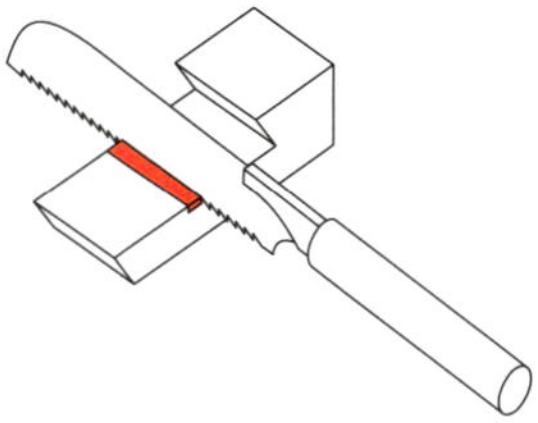

A5

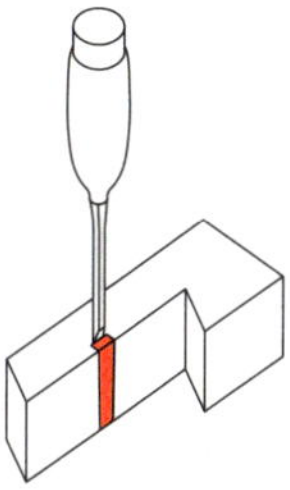

A6

2×

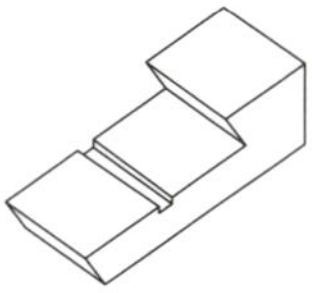

A

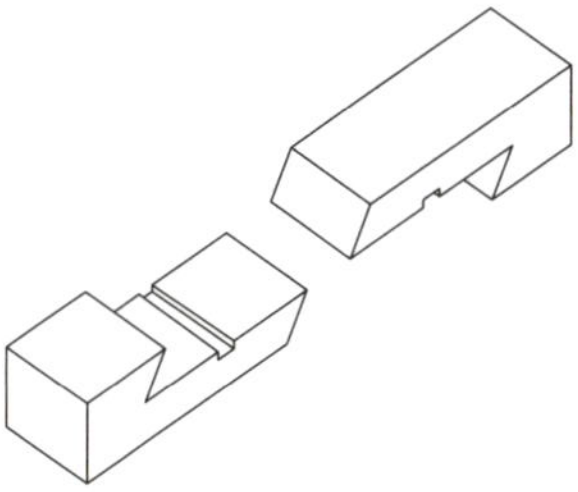

A+A

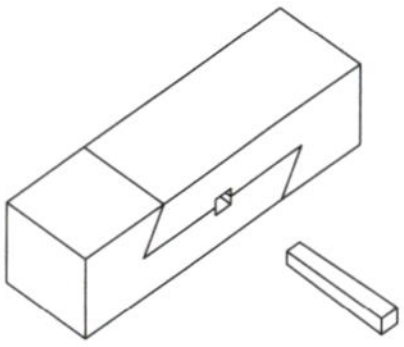

A+A

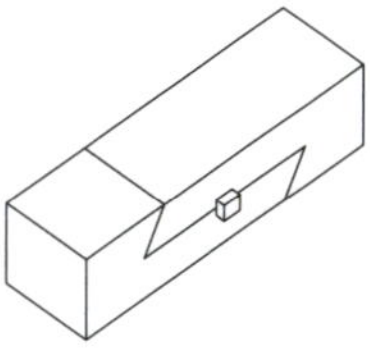

AA

Splicing Joints

Pegged undersquinted half lap joint

naga-koshi-kake-tsugi-dai-sen-uchi (sogi-tsuki)
長腰掛継大栓打ち(殺付)

gerades Blatt mit hinterschnittener Brüstung und zweifacher Dübelsicherung

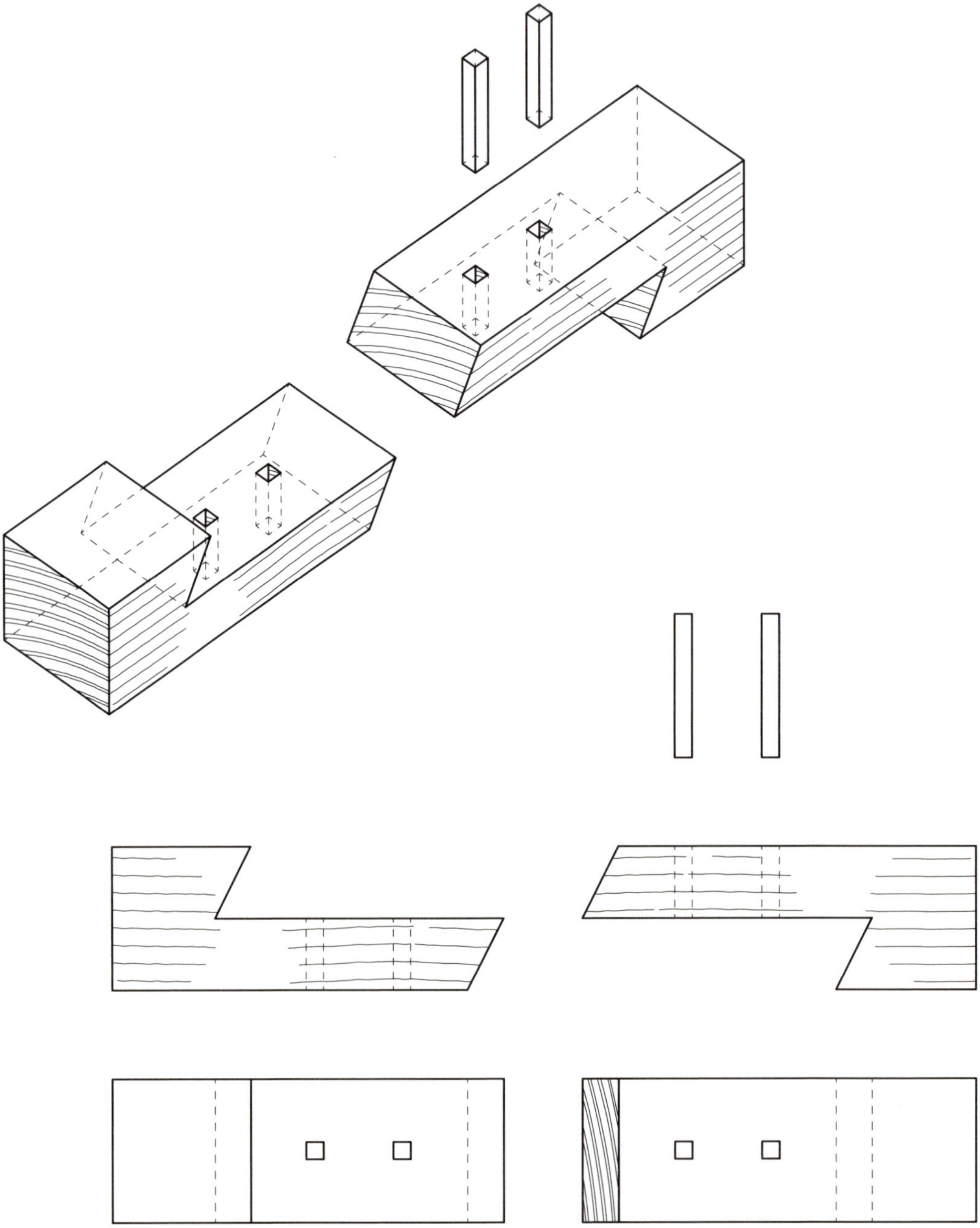

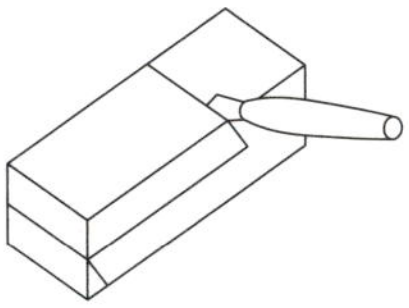

A1

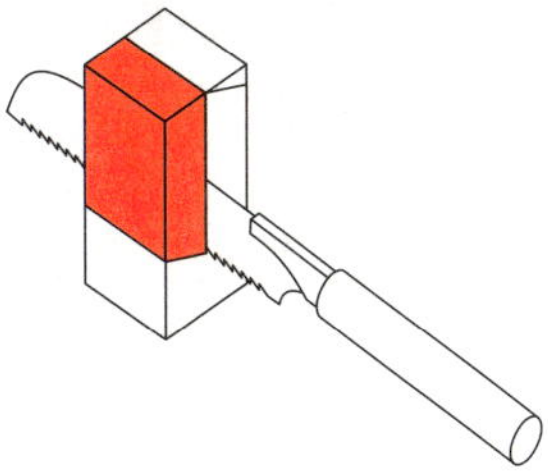

A2

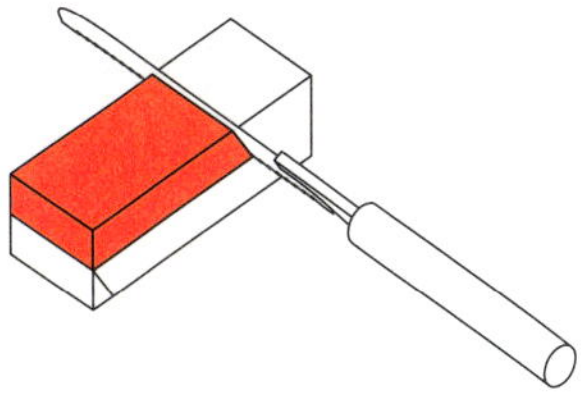

A3

2×

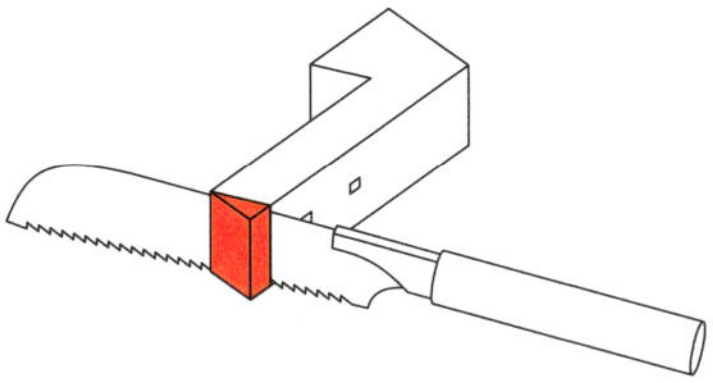

A4

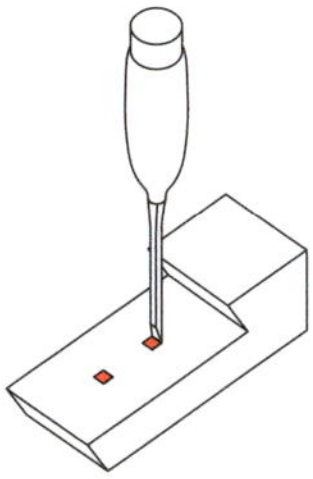

A5

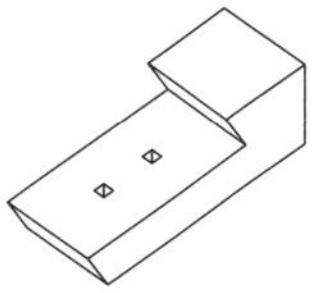

A

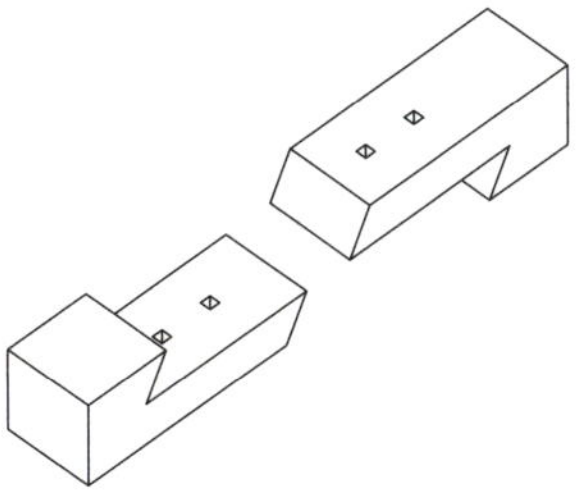

A+A

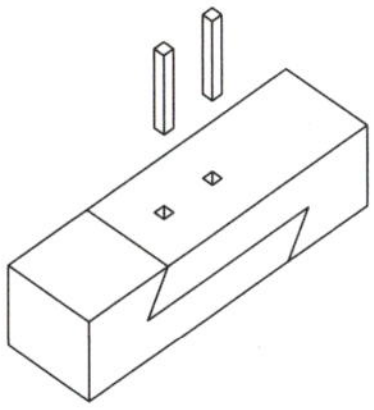

A+A

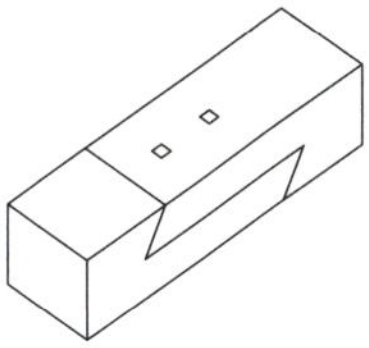

AA

Splicing Joints

Spear point half lap joint

naga-koshi-kake-yahazu-tsugi
長腰掛矢筈継ぎ

gerades Blatt mit Gratschnitt

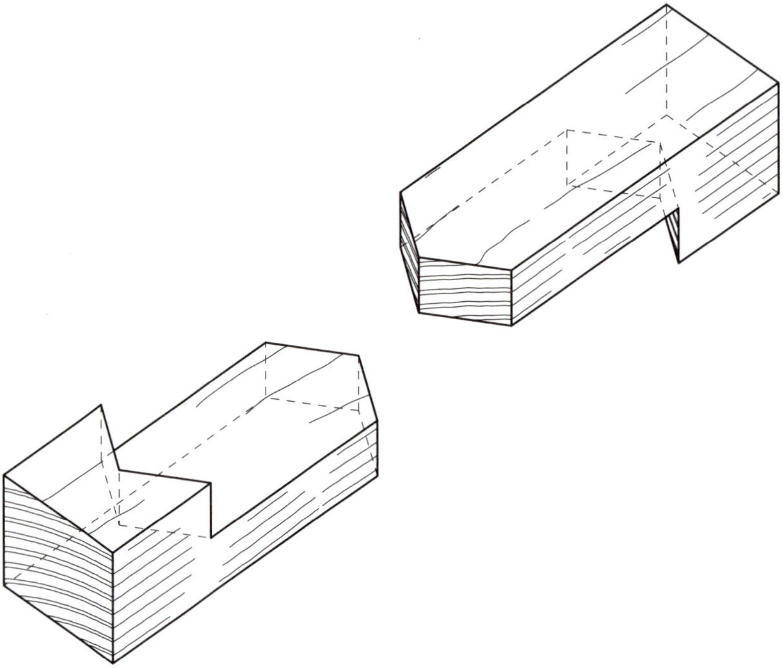

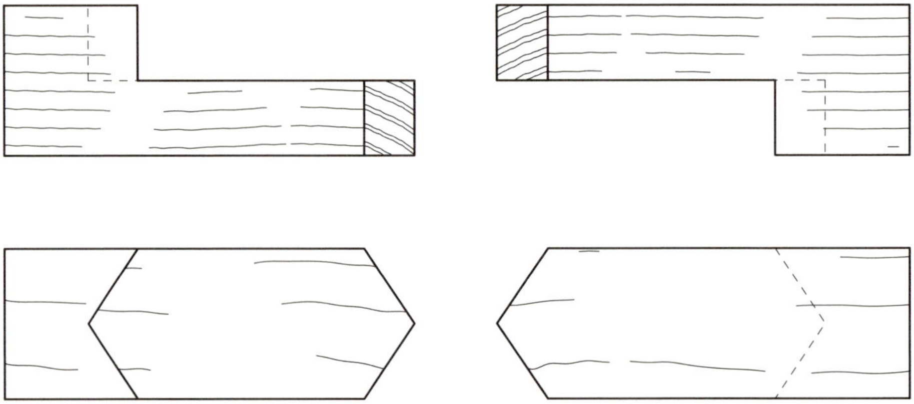

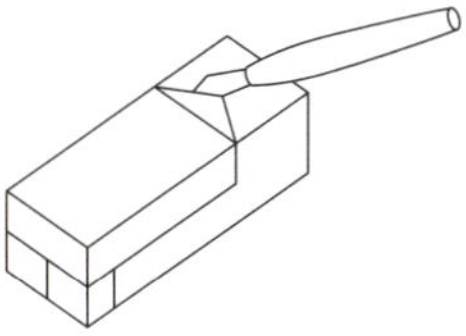

A1

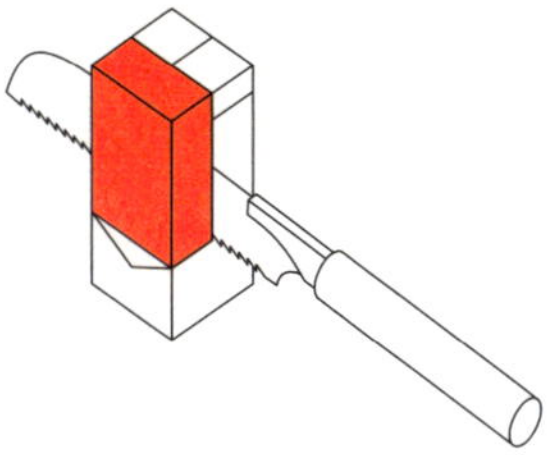

A2

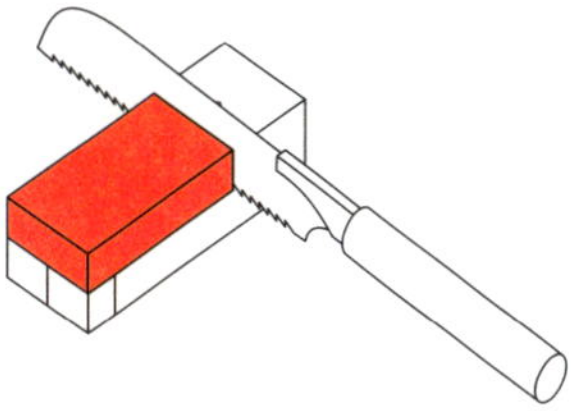

A3

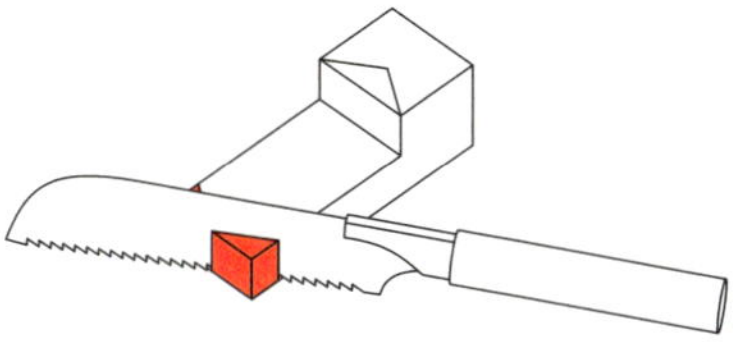

A4

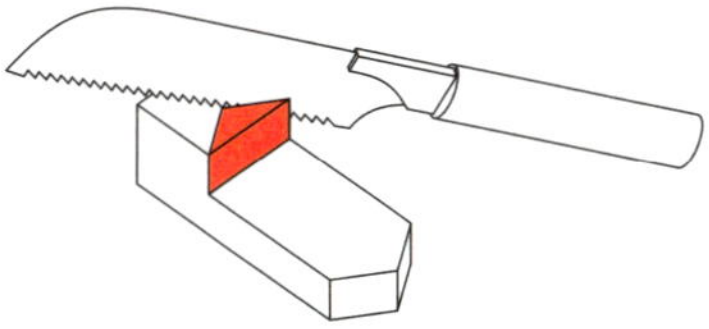

A5

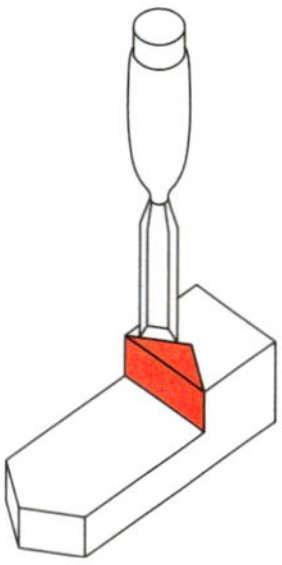

A6

2×

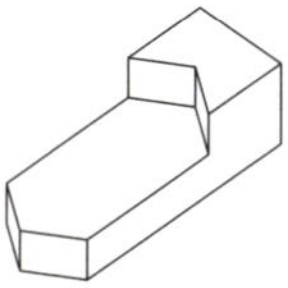

A

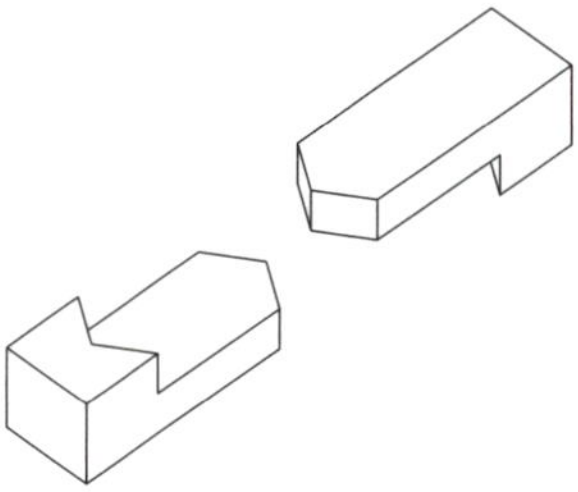

A+A

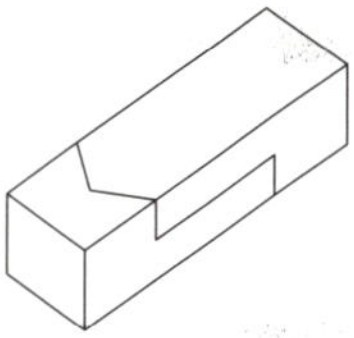

AA

Splicing Joints

Beveled spear point lap joint

naga-koshi-kake-yahazu-tsugi (kōbai-tsuki)
長腰掛矢筈継ぎ（勾配付）

Zugblatt mit Gratschnitt

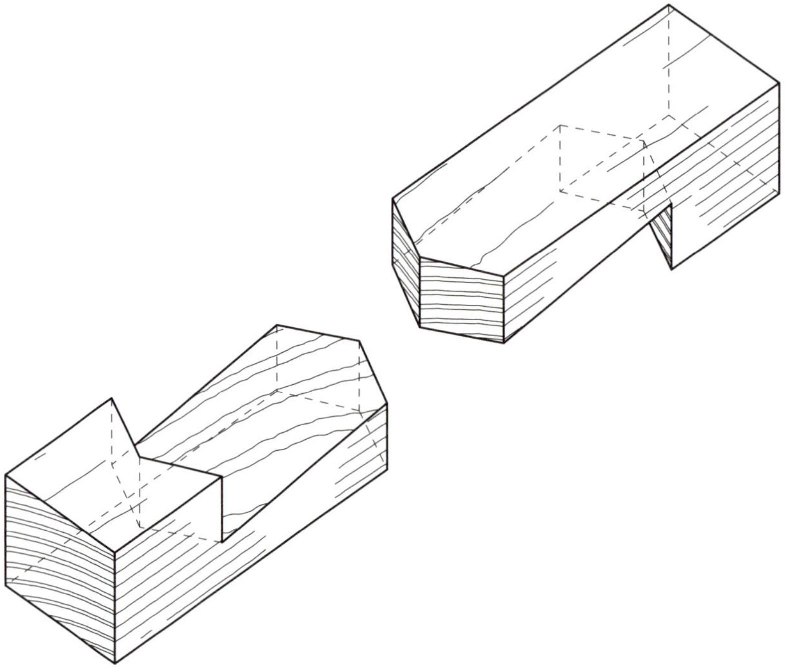

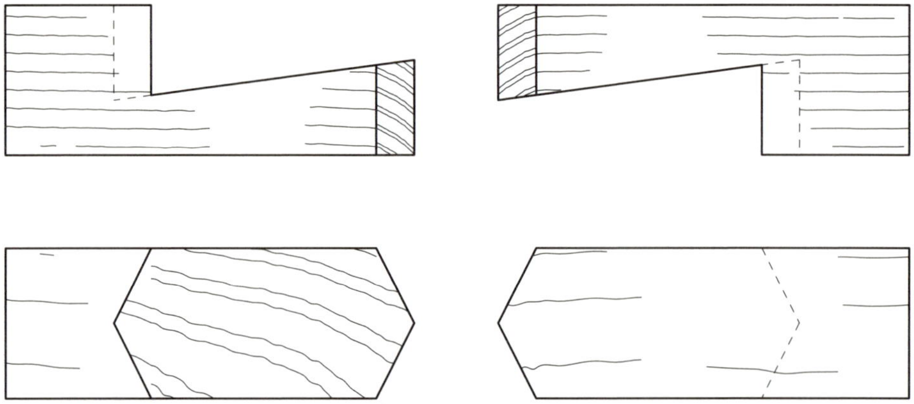

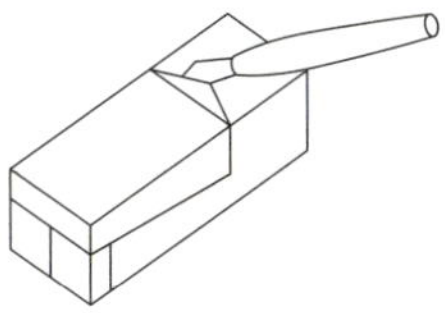

A1

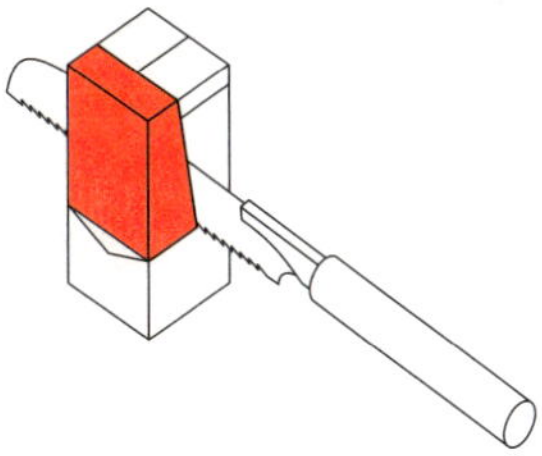

A2

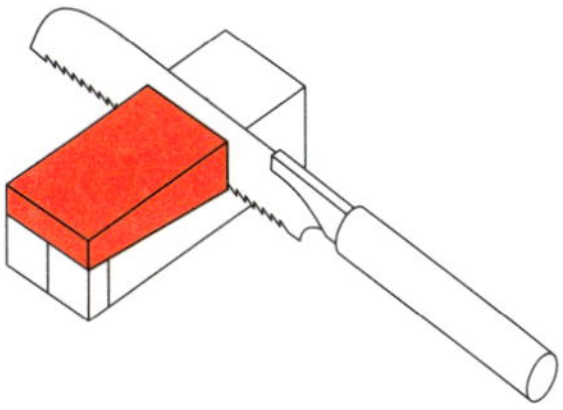

A3

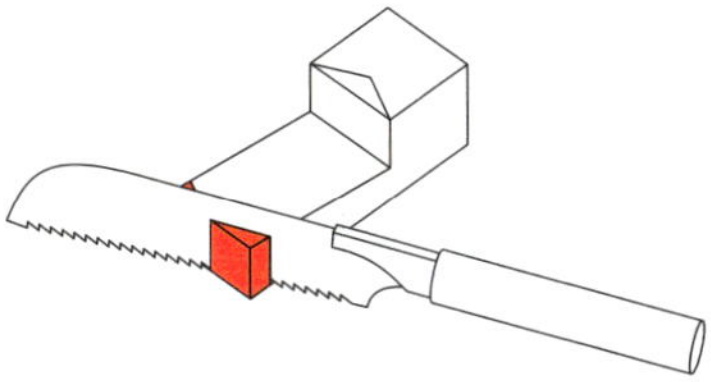

A4

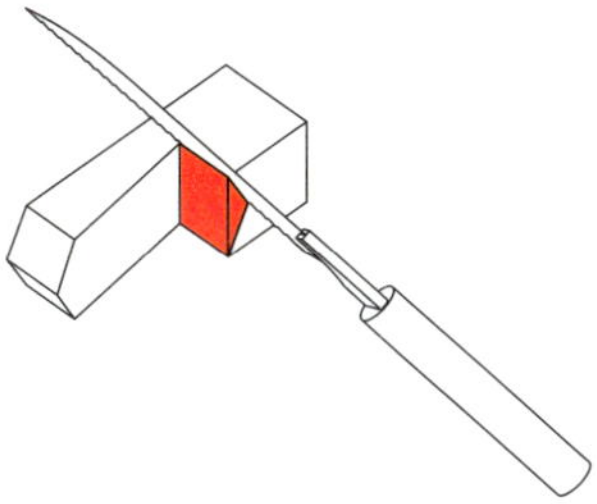

A5

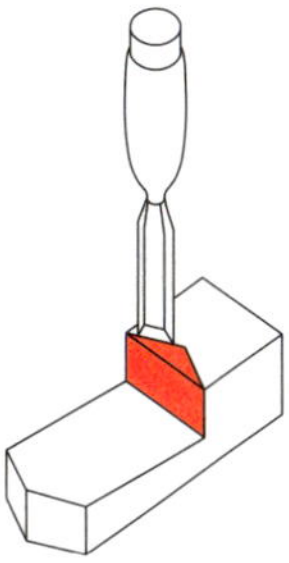

A6

2×

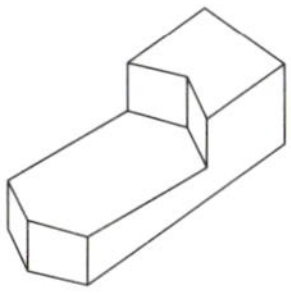

A

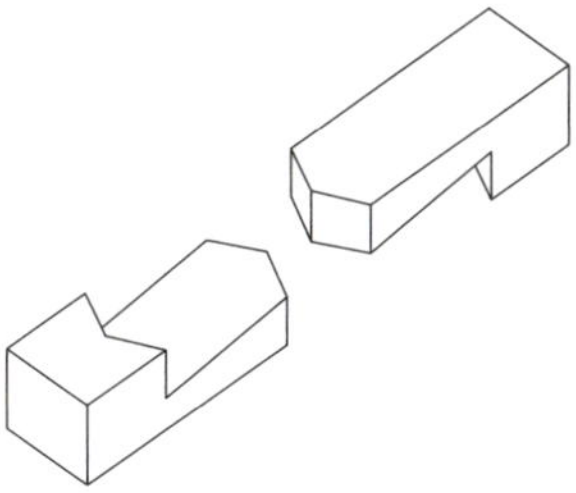

A+A

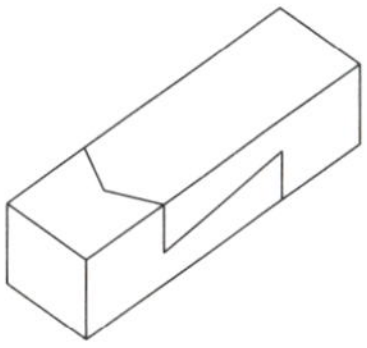

AA

Splicing Joints

Undersquinted spear point half lap joint

naga-koshi-kake-yahazu-tsugi (sogi-tsuki)
長腰掛矢筈継ぎ(殺付)

gerades Blatt mit schrägem Gratschnitt

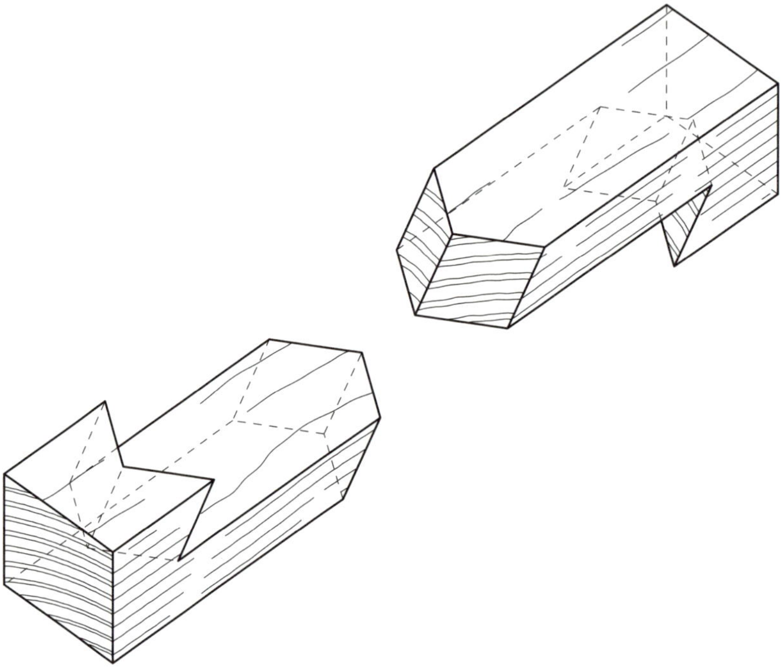

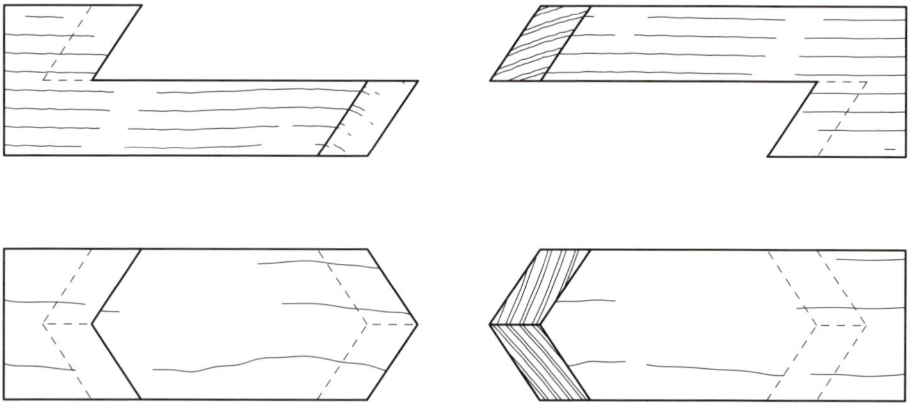

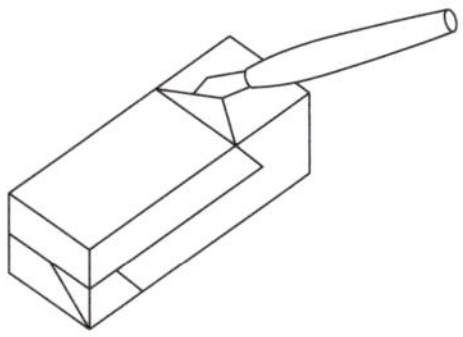

A1

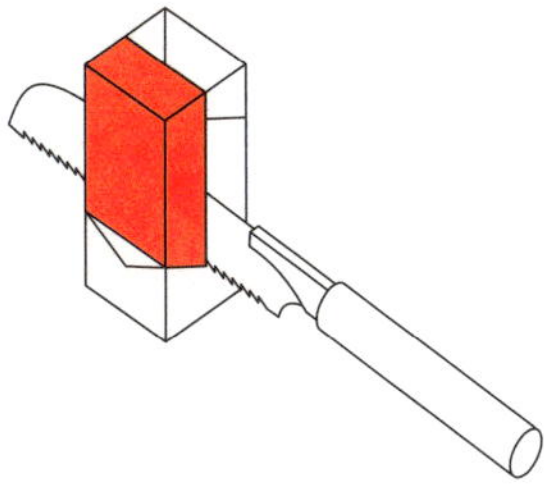

A2

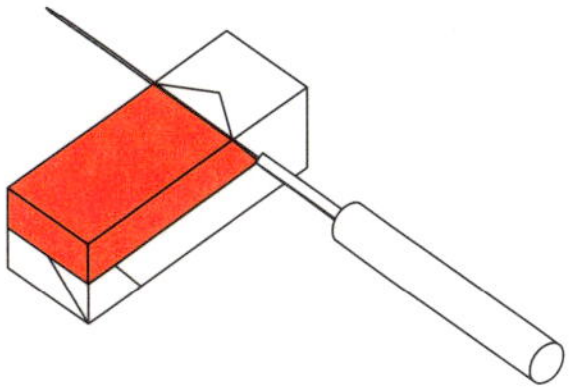

A3

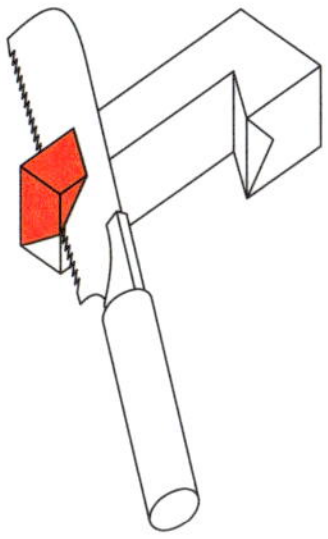

A4

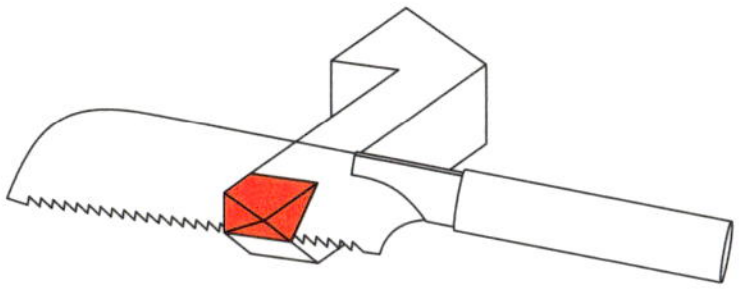

A5

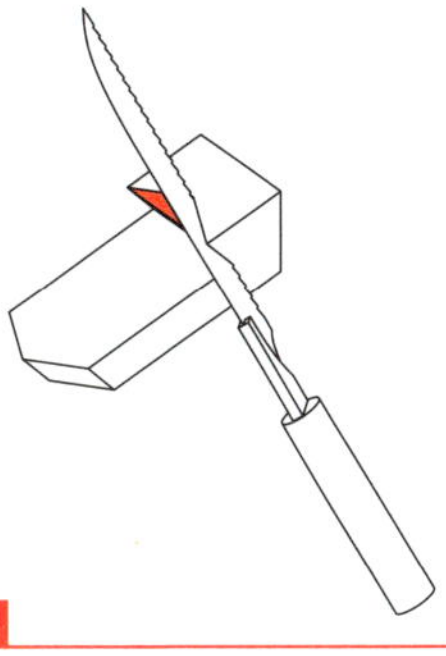

A6

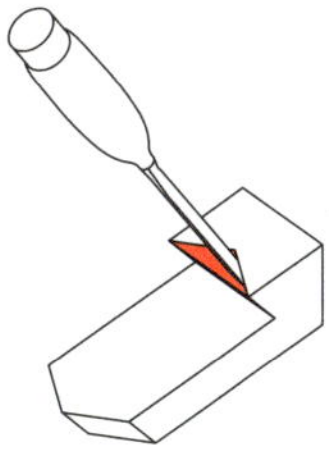

A7

2×

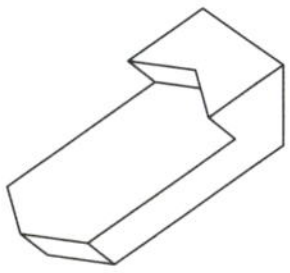

A

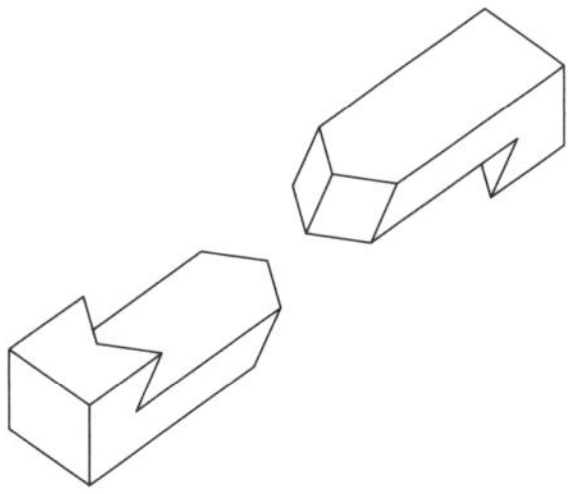

A+A

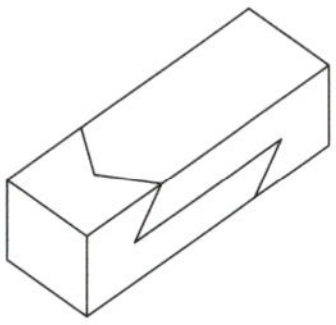

AA

Splicing Joints

Pegged undersquinted spear point half lap joint

naga-koshi-kake-yahazu-tsugi-komi-sen-uchi (sogi-tsuki)
長腰掛矢筈継込栓打ち（殺付）

gerades Blatt mit schrägem Gratschnitt und Dübelsicherung

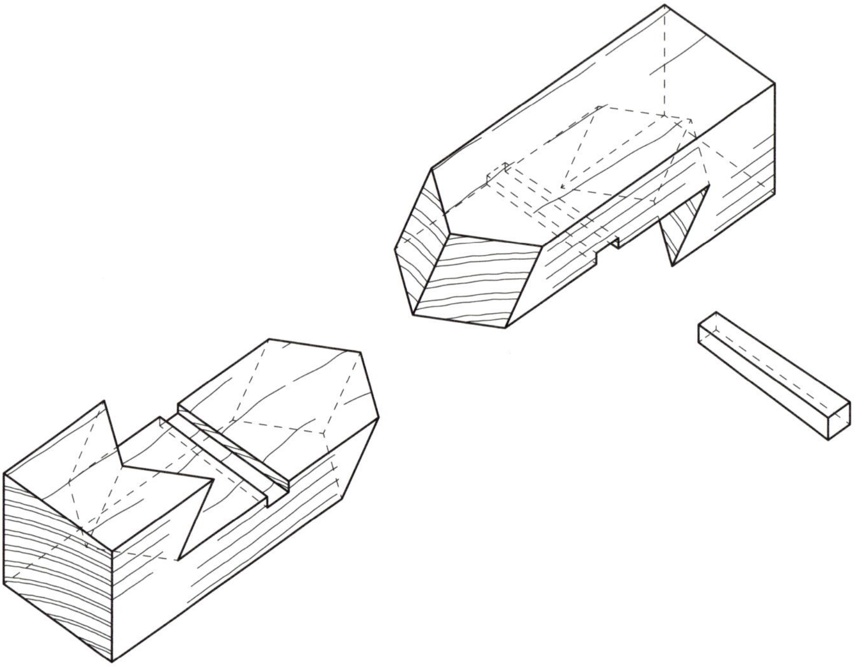

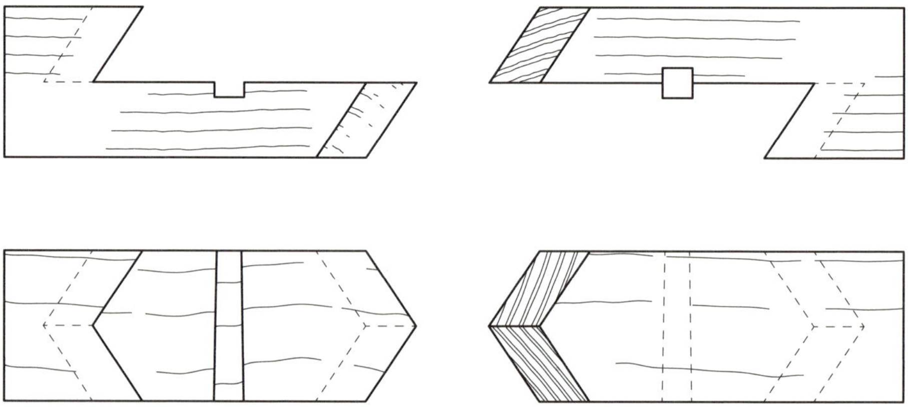

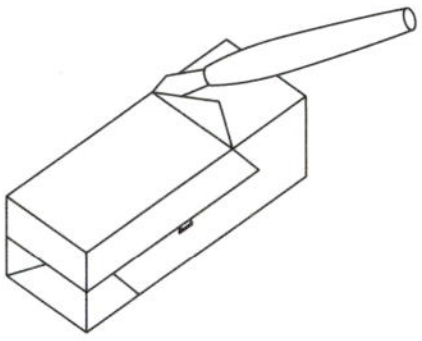

A1

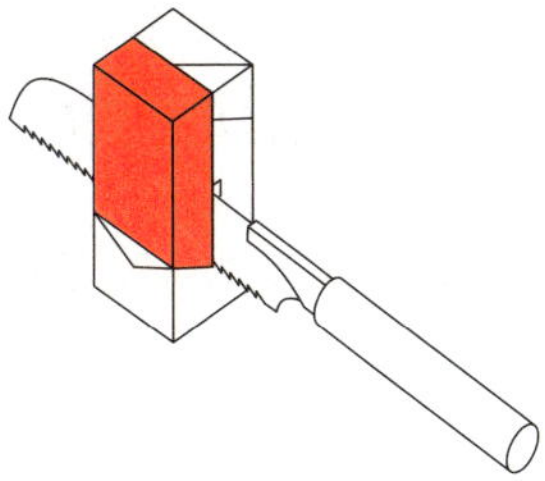

A2

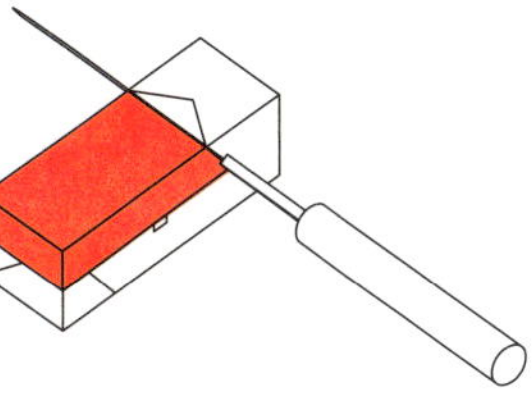

A3

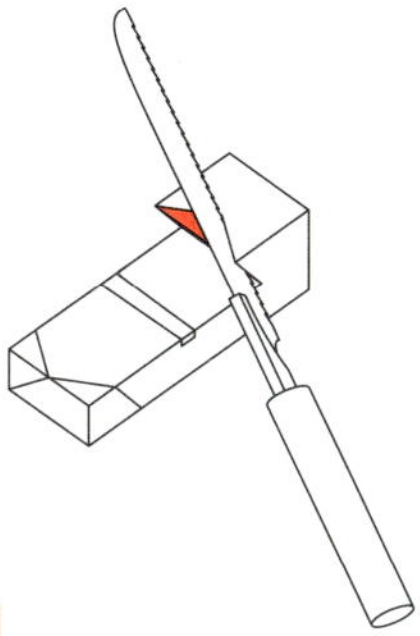

A4

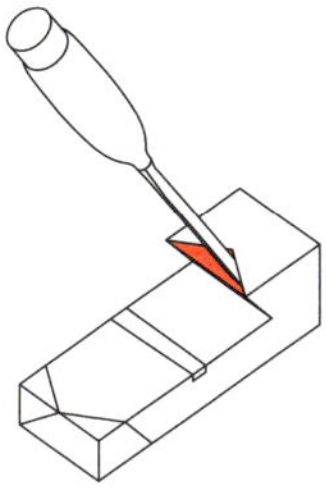

A5

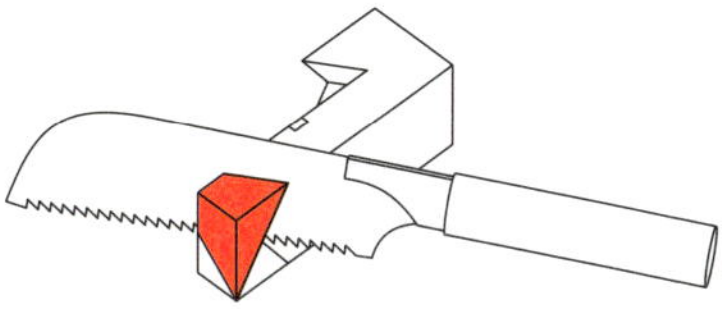

A6

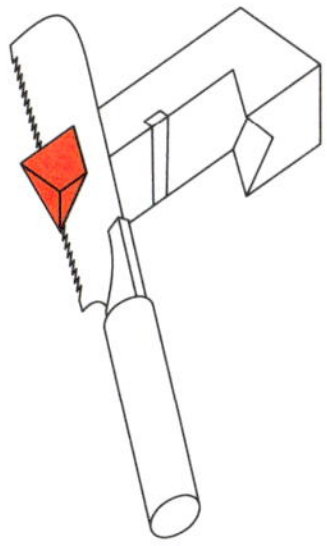

A7

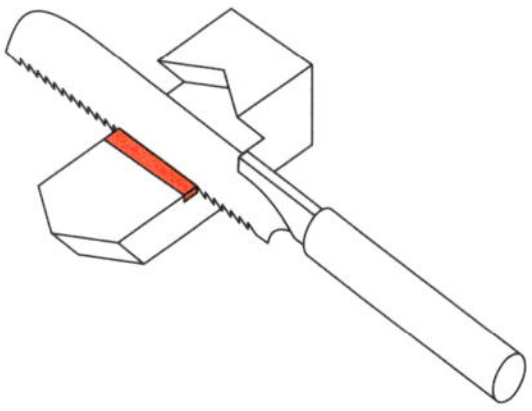

A8

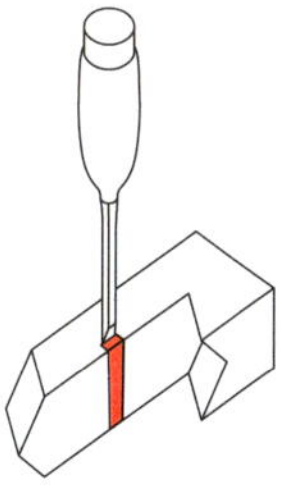

A9

2×

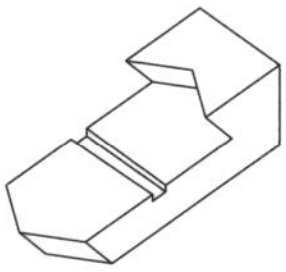

A

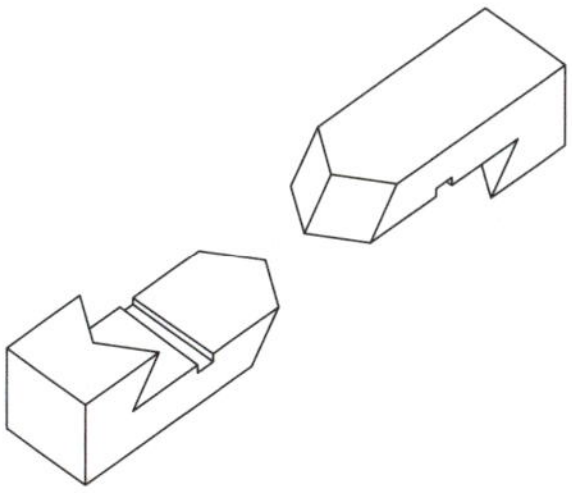

A+A

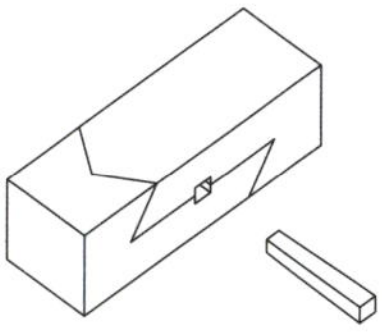

A+A

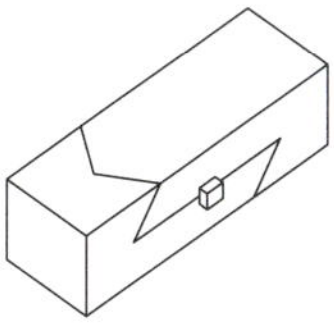

AA

Splicing Joints

Pegged undersquinted spear point half lap joint

naga-koshi-kake-yahazu-tsugi-dai-sen-uchi (sogi-tsuki)
長腰掛矢筈継大栓打ち（殺付）

gerades Blatt mit schrägem Gratschnitt und zweifacher Dübelsicherung

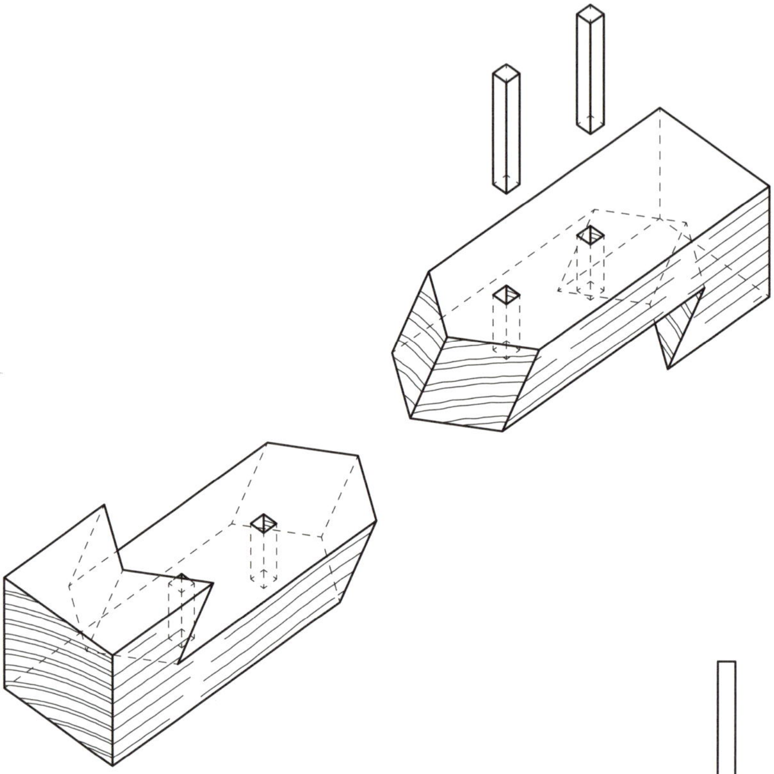

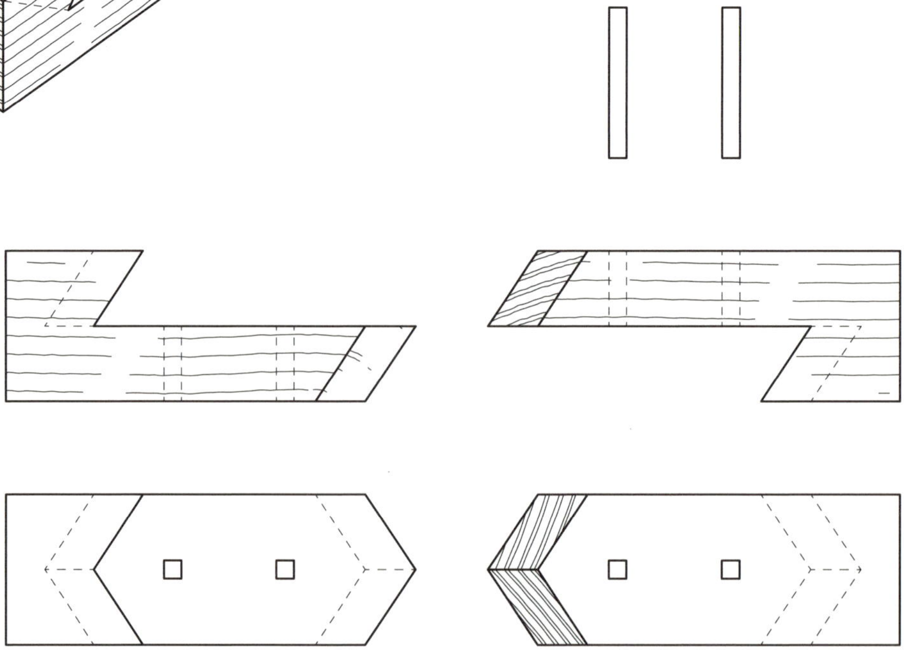

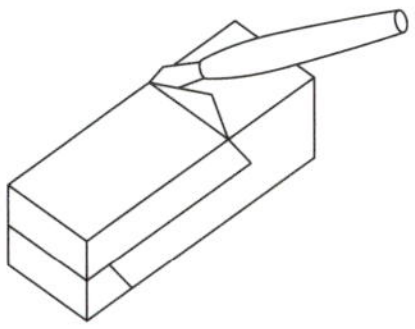

A1

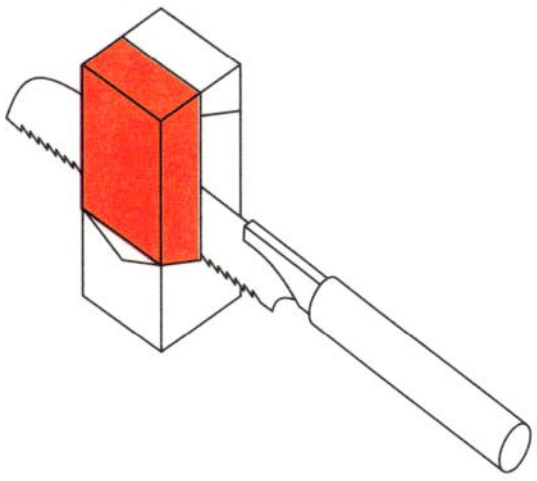

A2

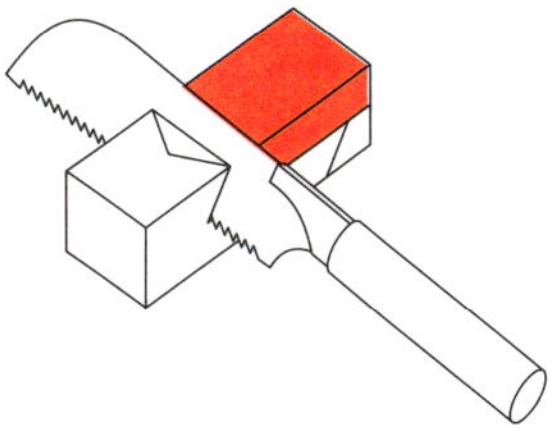

A3

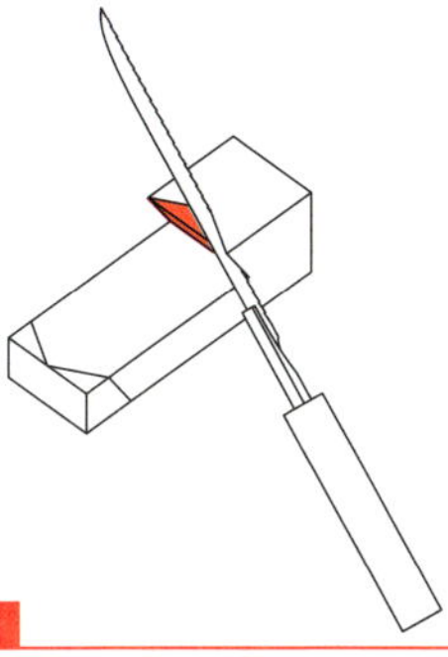

A4

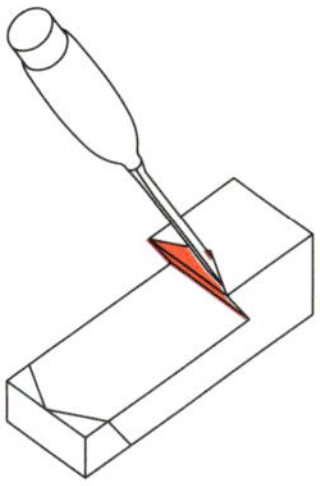

A5

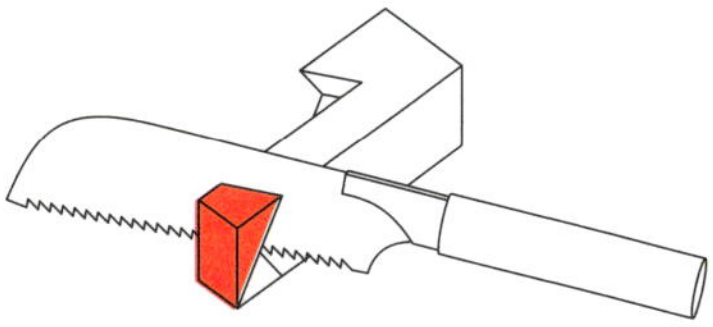

A6

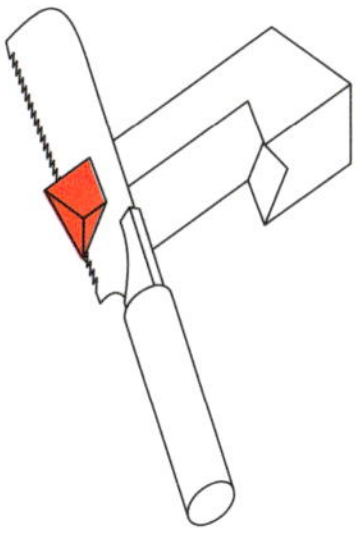

A7

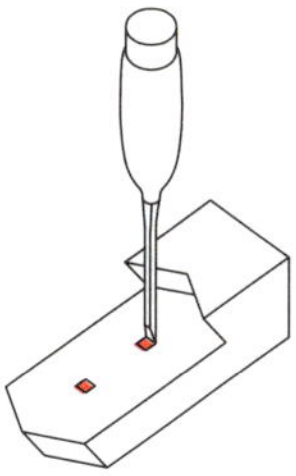

A8

2×

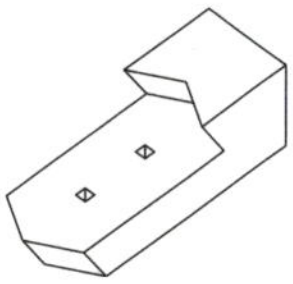

A

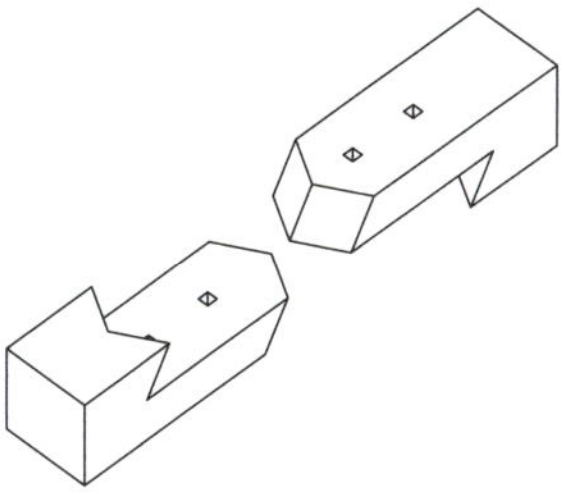

A+A

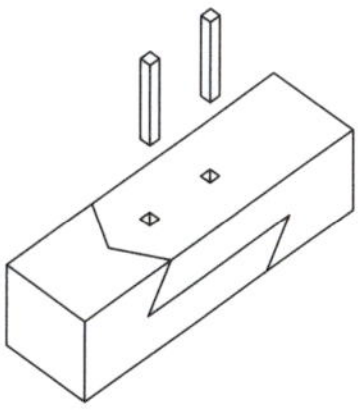

A+A

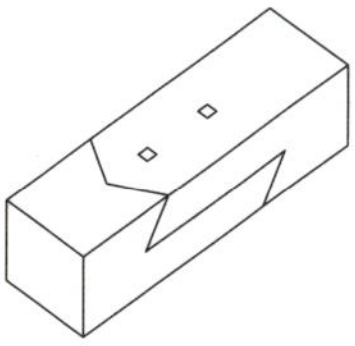

AA

Splicing Joints

Cogged lap joint

dan-koshi-kake-tsugi
段腰掛継ぎ

gerades Hakenblatt

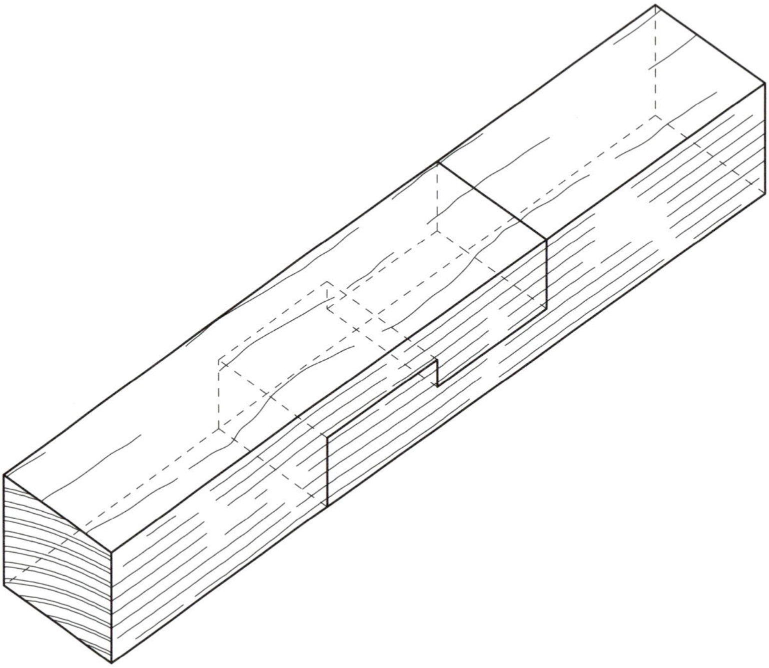

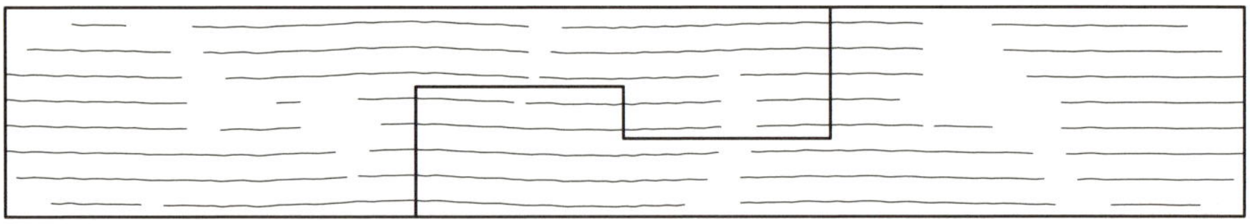

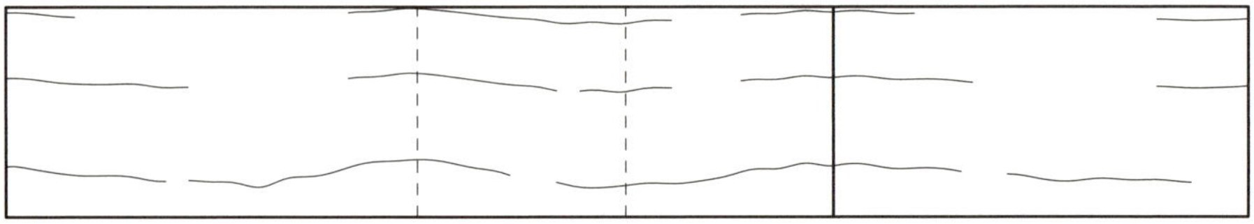

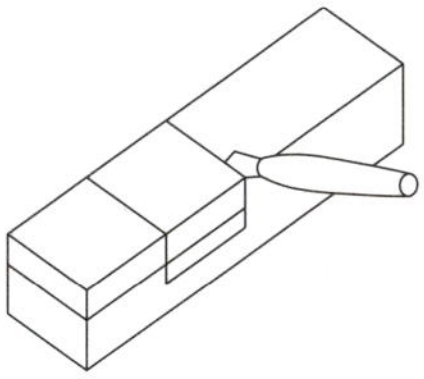

A1

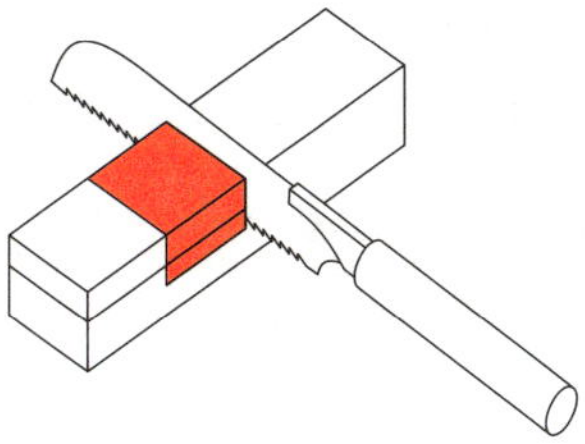

A2

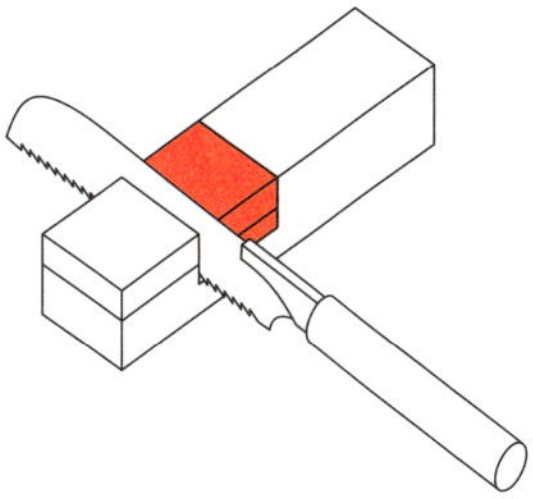

A3

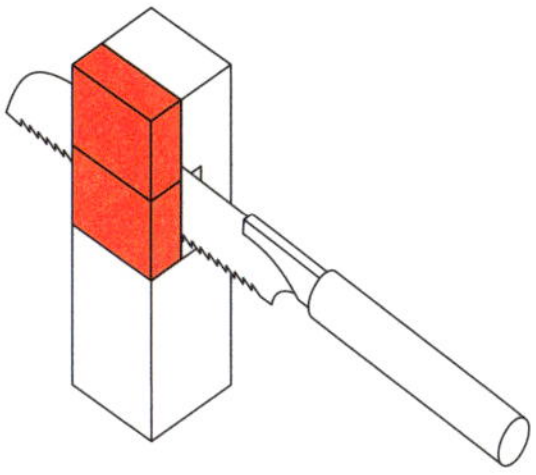

A4

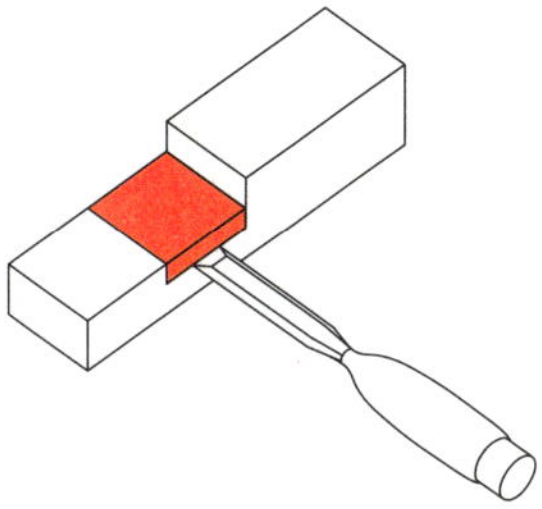

A5

2×

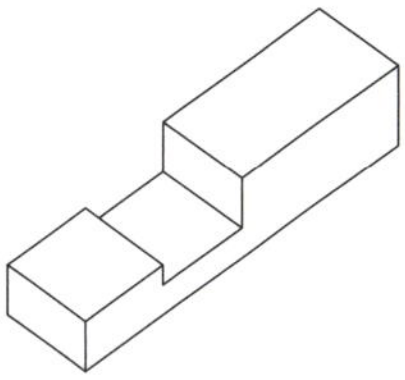

A

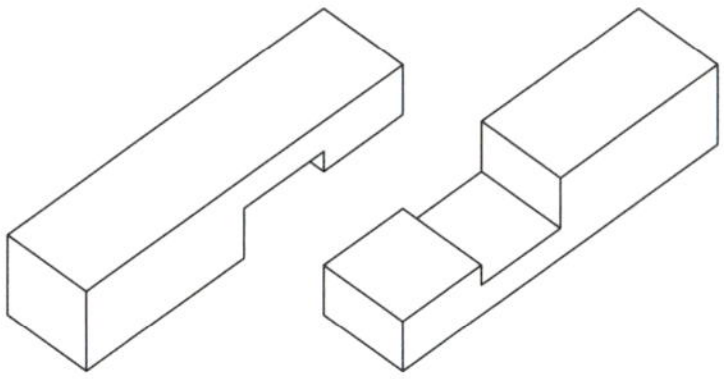

A+A

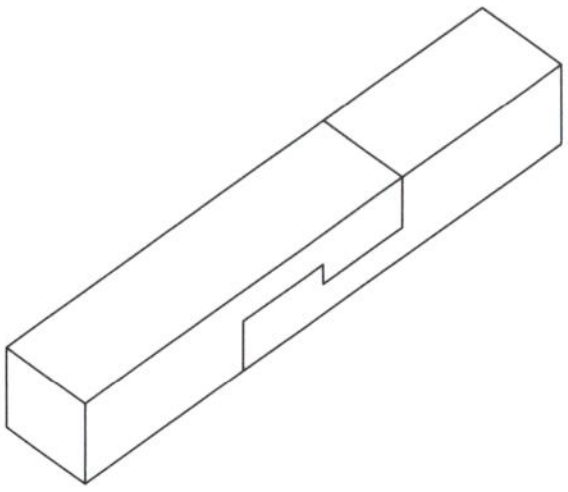

AA

Splicing Joints

Cogged lap joint with folding wedges

dan-koshi-kake-tsugi (kusabi-shime)
段腰掛継ぎ（楔締）

verkeiltes gerades Hakenblatt

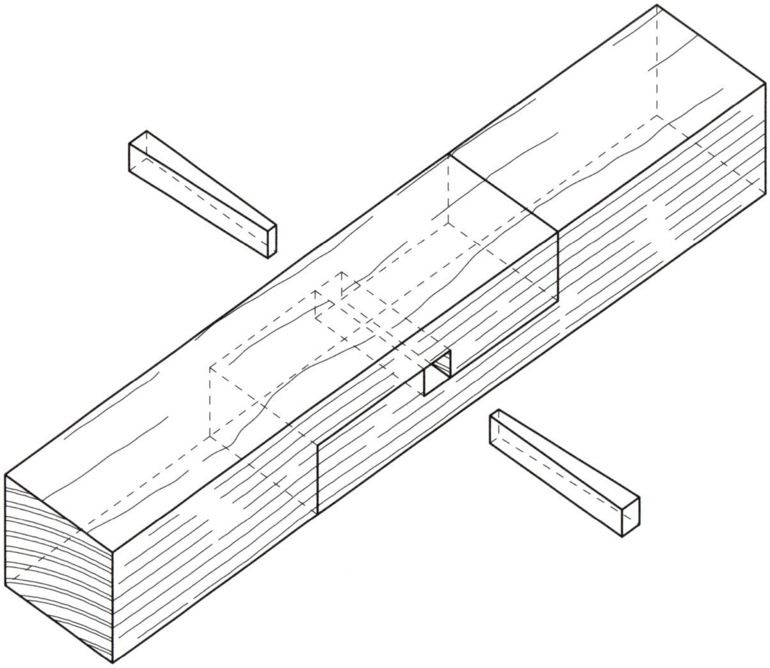

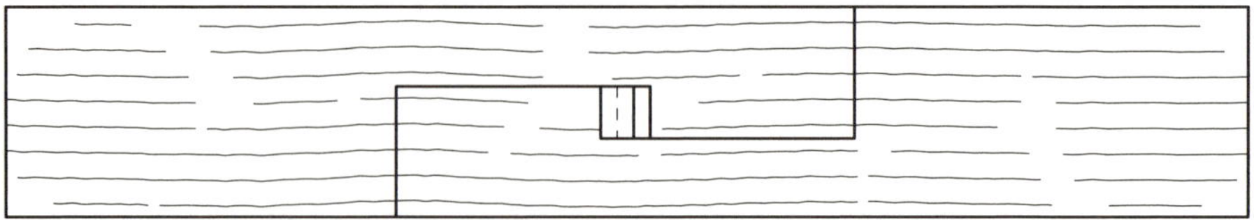

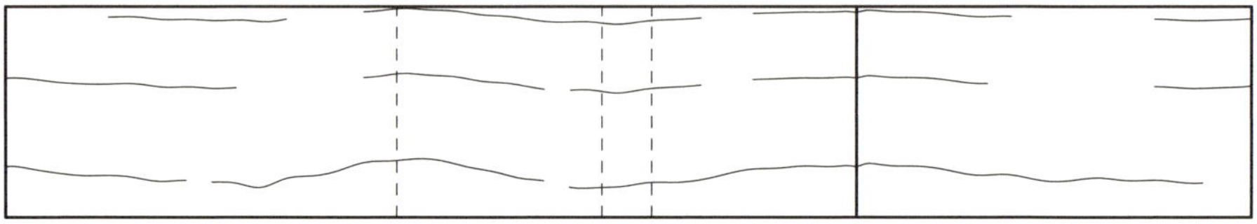

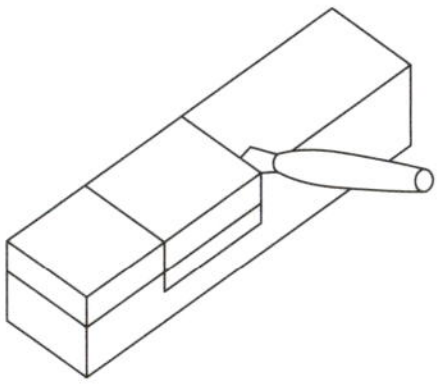

A1

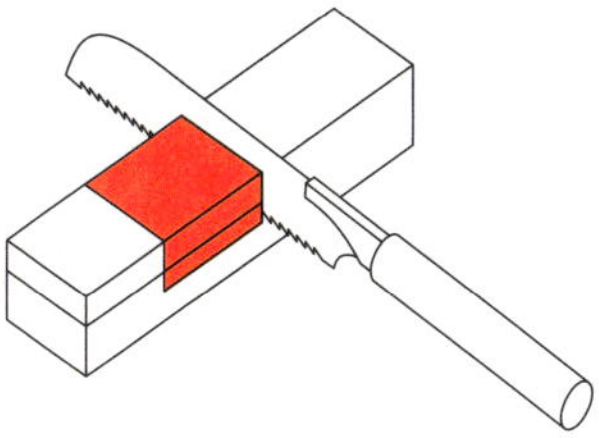

A2

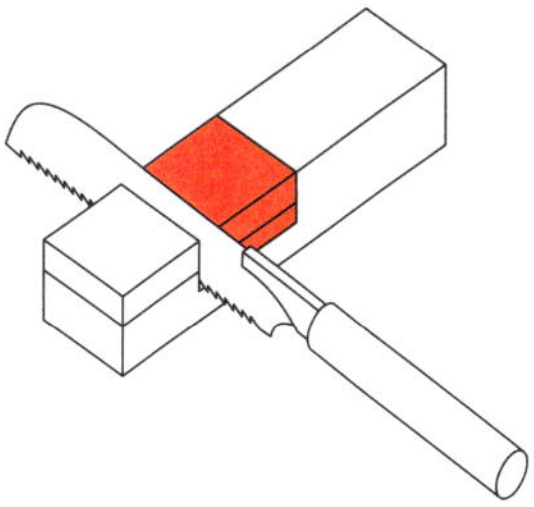

A3

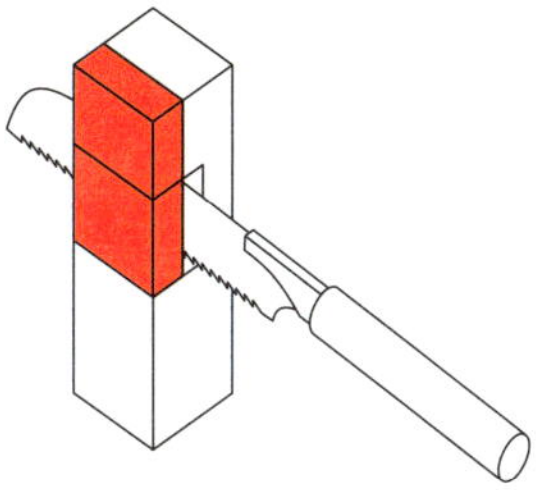

A4

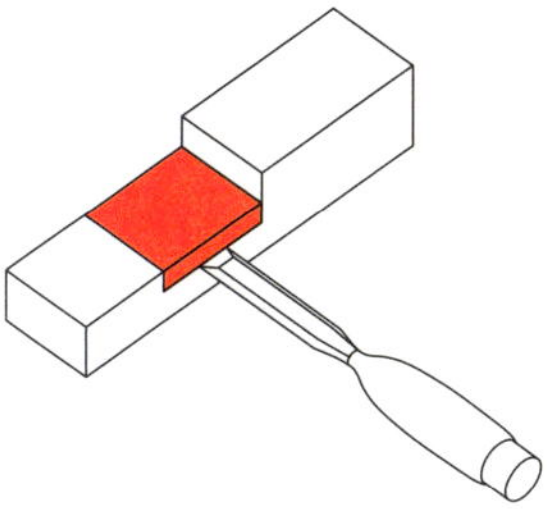

A5

2×

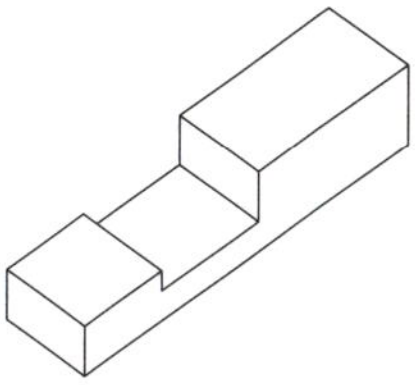

A

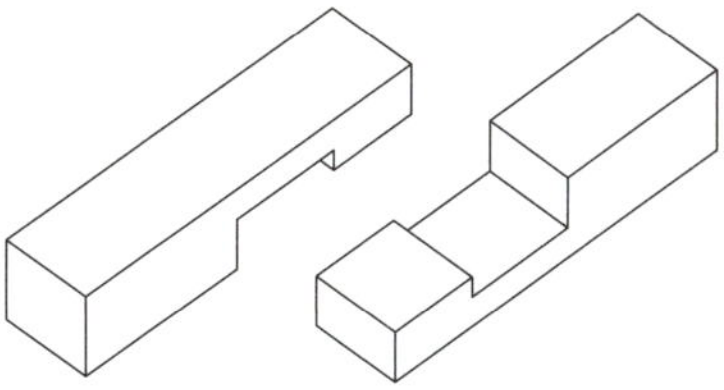

A+A

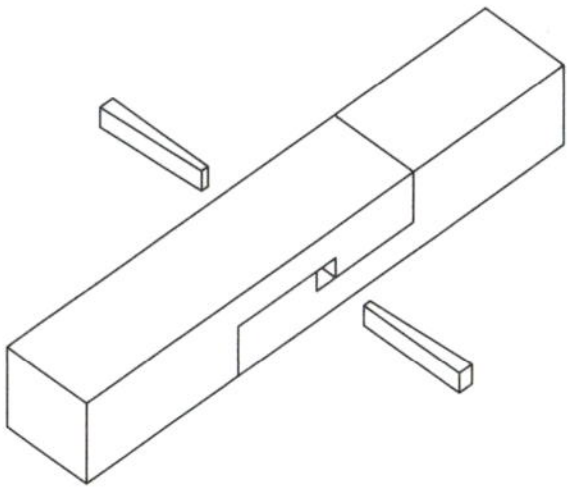

A+A

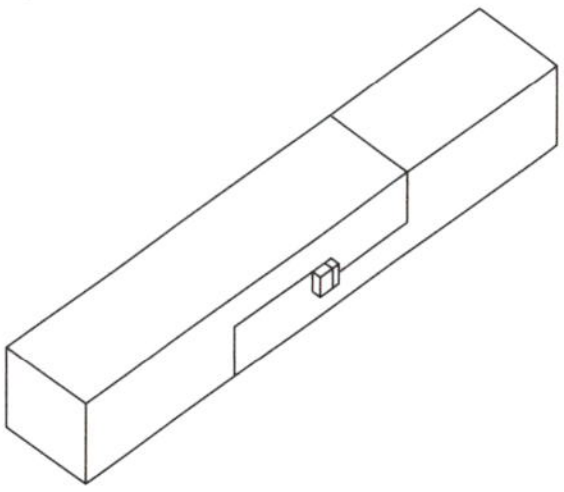

AA

Splicing Joints

Undersquinted cogged lap joint with folding wedges

dan-koshi-kake-tsugi (sogi-tsuki, kusabi-shime)
段腰掛継ぎ(殺付・楔締)

verkeiltes gerades Hakenblatt mit hinterschnittener Brüstung

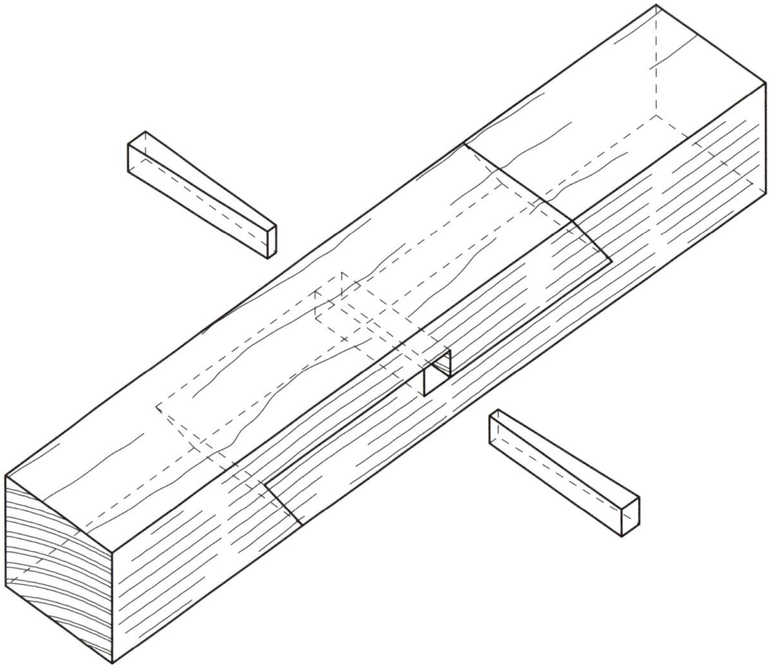

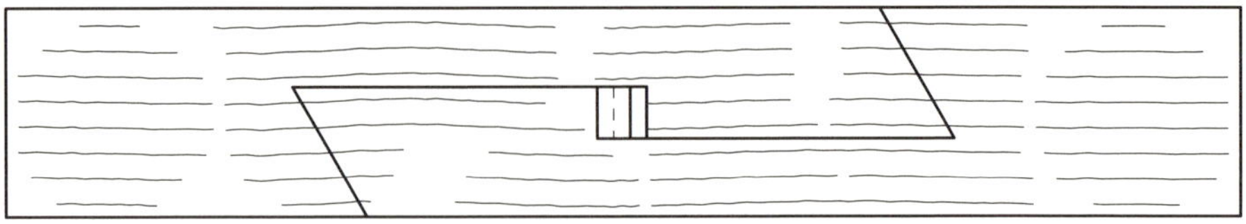

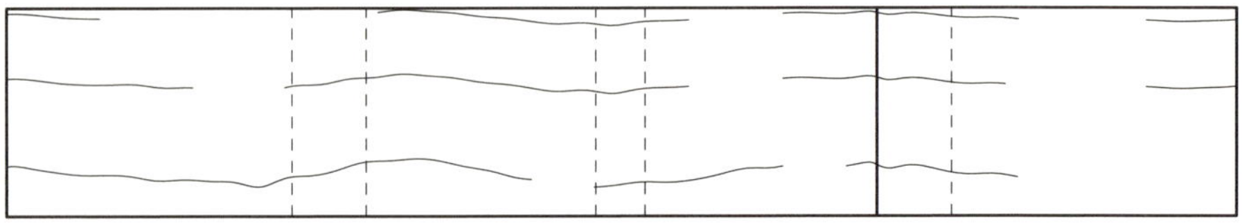

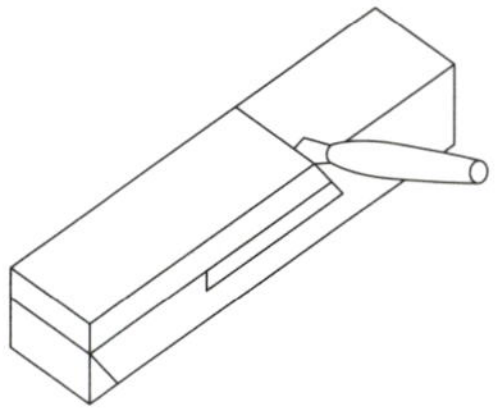

A1

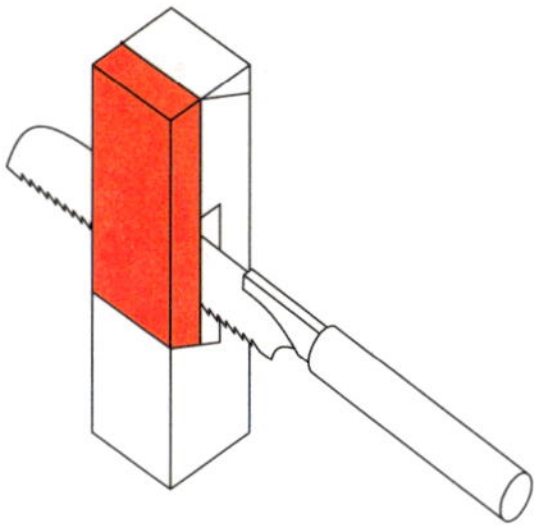

A2

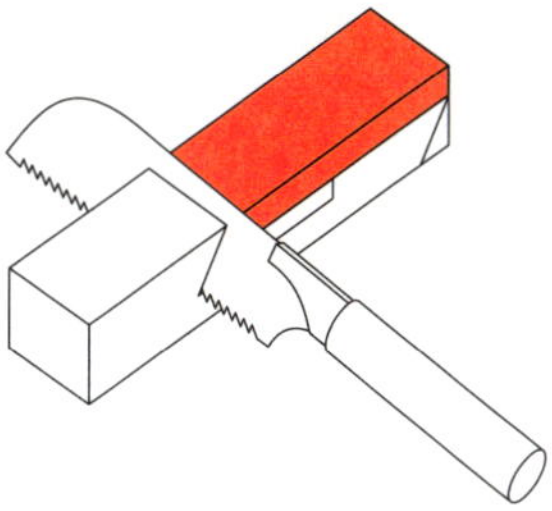

A3

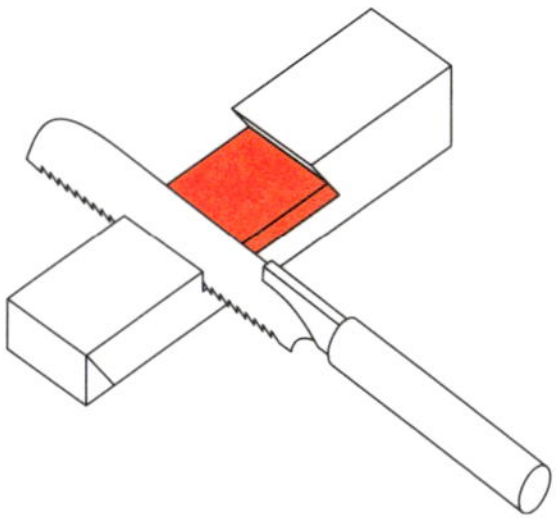

A4

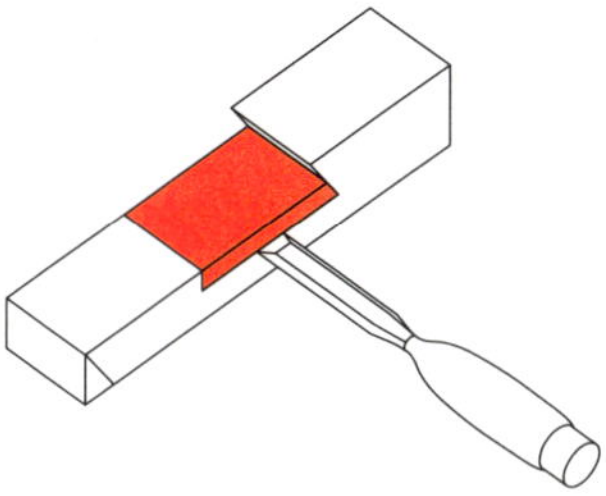

A5

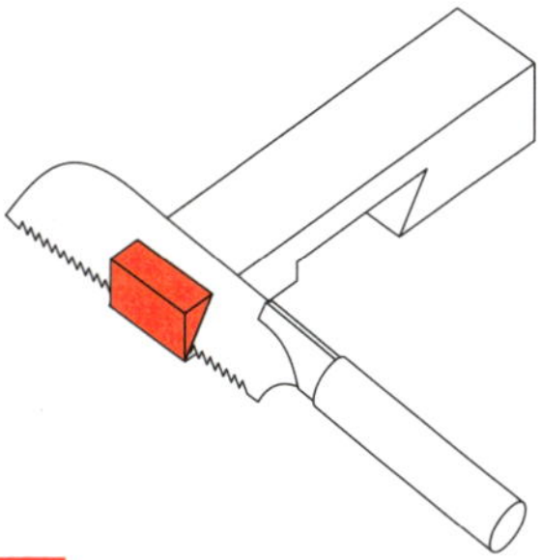

A6

2×

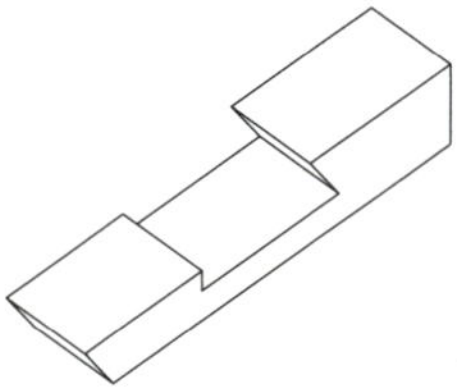

A

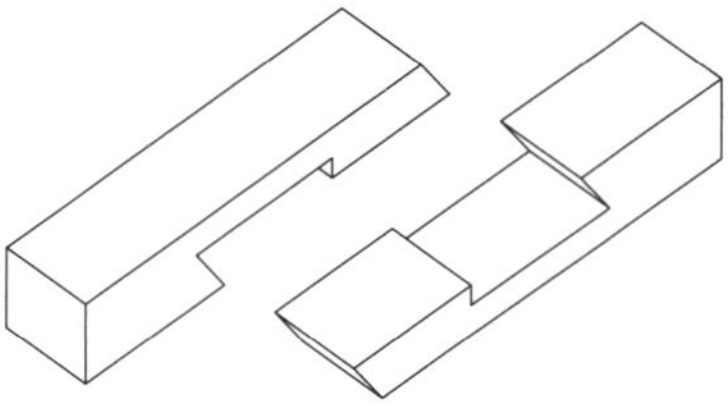

A+A

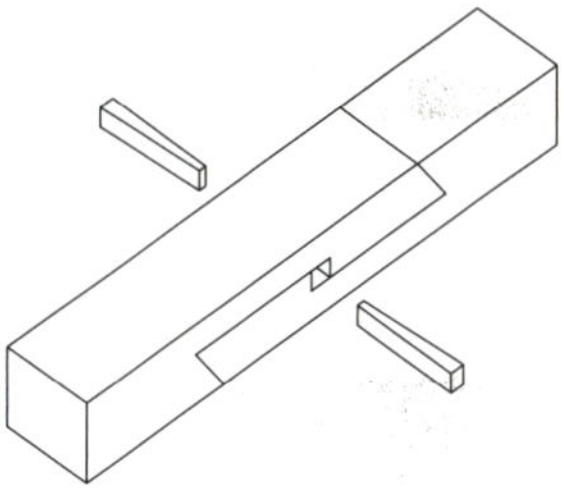

A+A

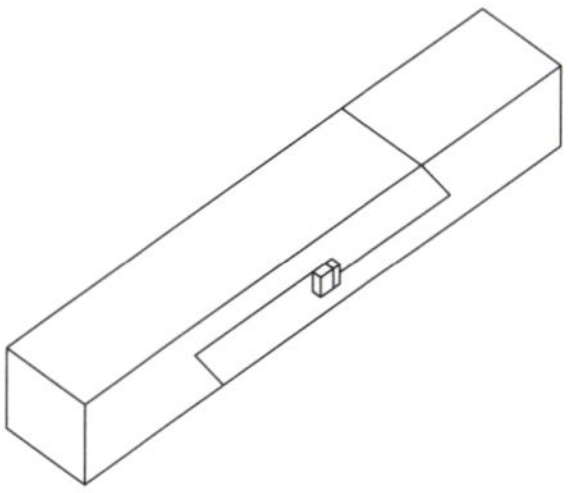

AA

Splicing Joints

Spear point cogged lap joint

dan-koshi-kake-yahazu-tsugi
段腰掛矢筈継ぎ

gerades Hakenblatt mit Gratschnitt

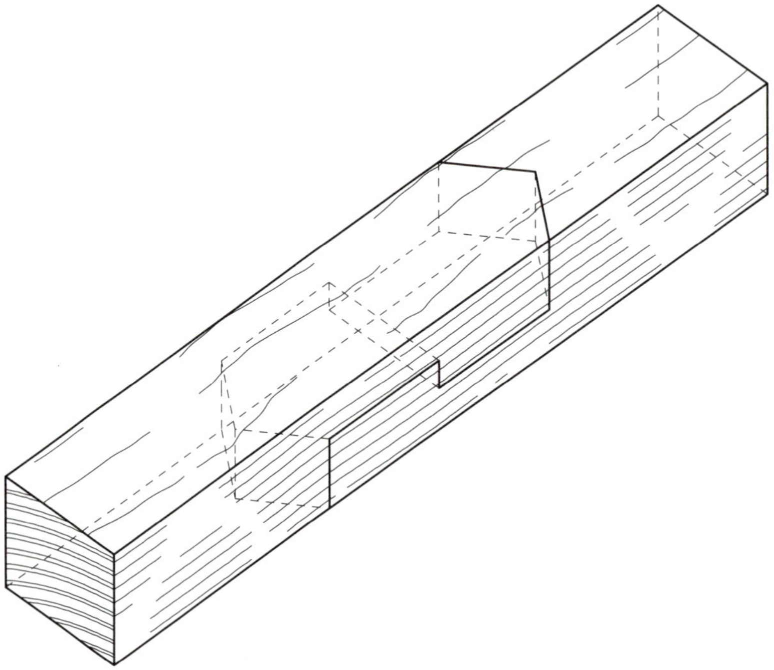

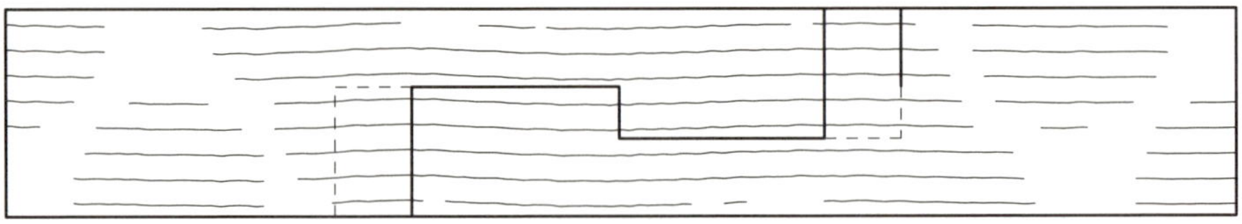

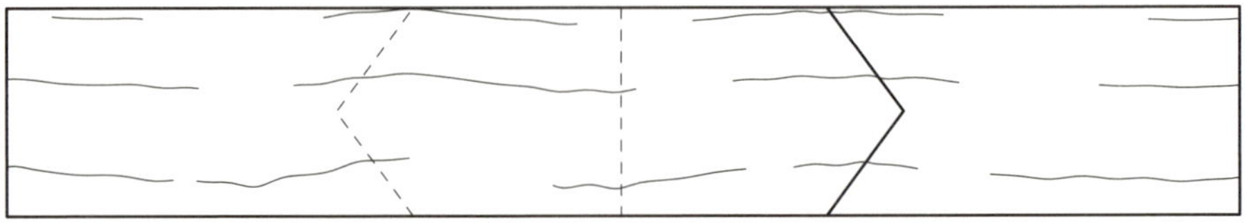

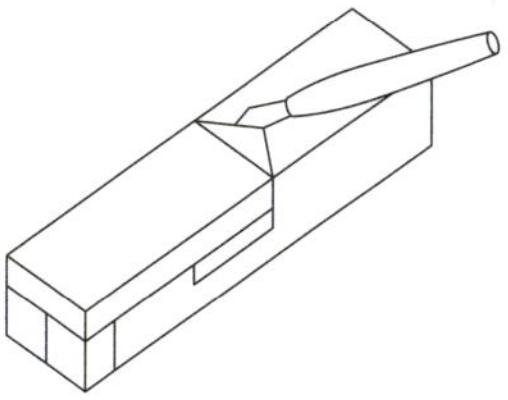

A1

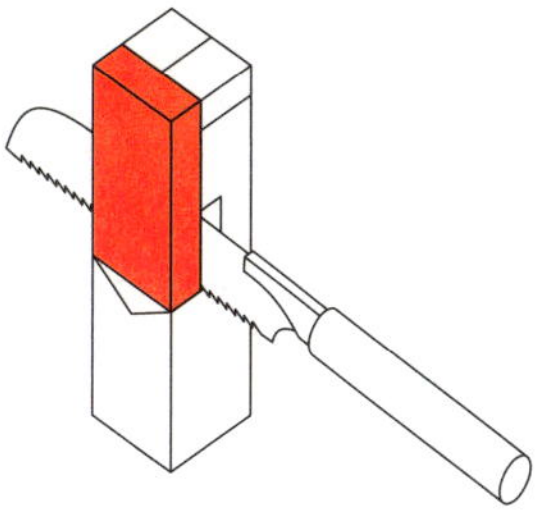

A2

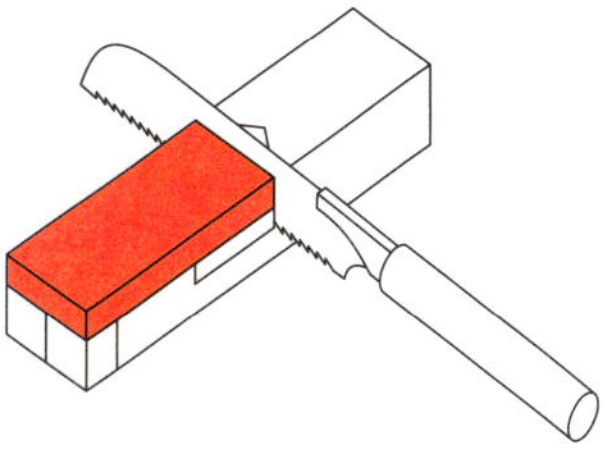

A3

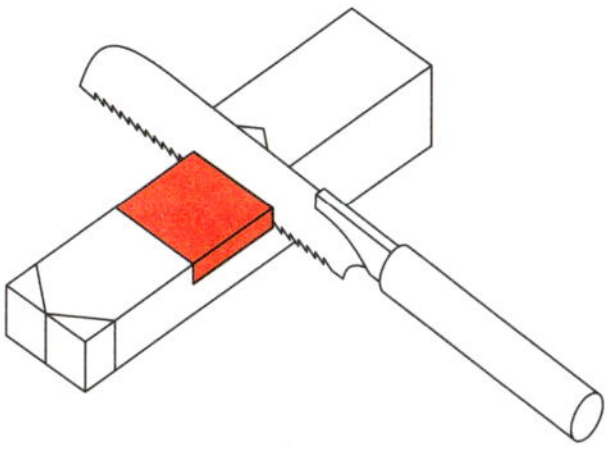

A4

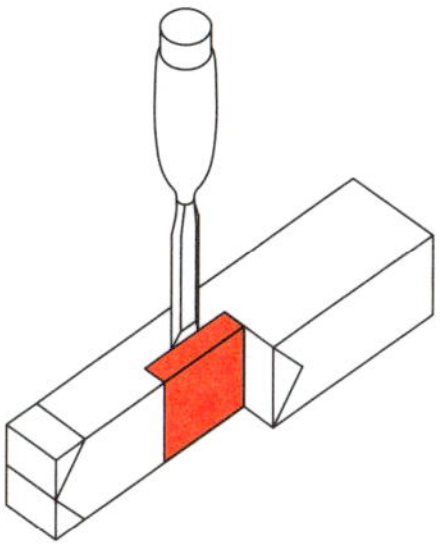

A5

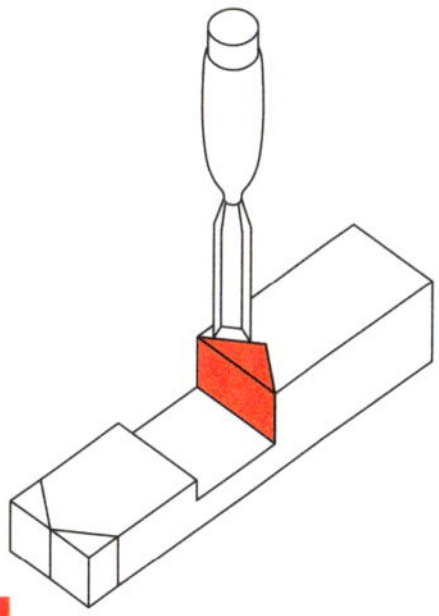

A6

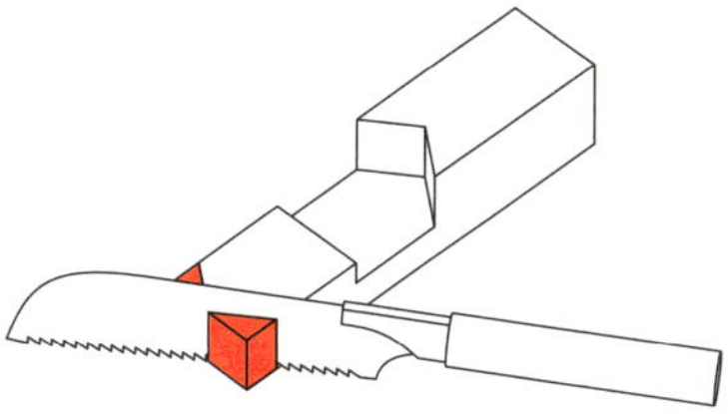

A7

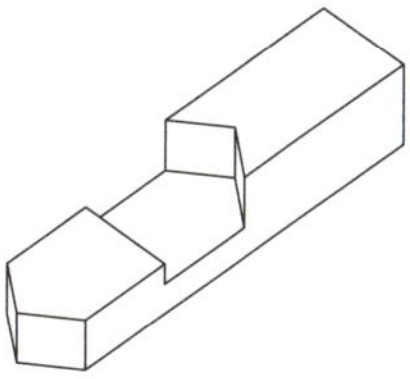

2×

A

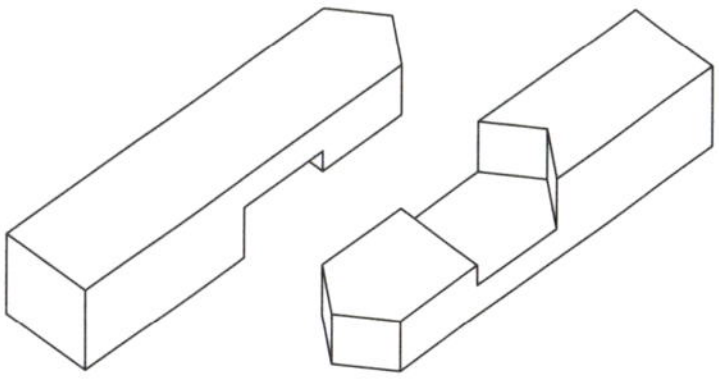

A+A

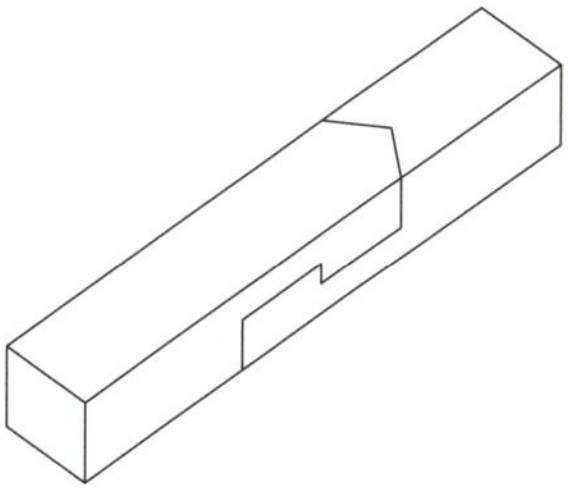

AA

Splicing Joints

Spear point cogged lap joint with folding wedges

dan-koshi-kake-yahazu-tsugi (kusabi-shime)
段腰掛矢筈継ぎ(楔締)

verkeiltes gerades Hakenblatt mit Gratschnitt

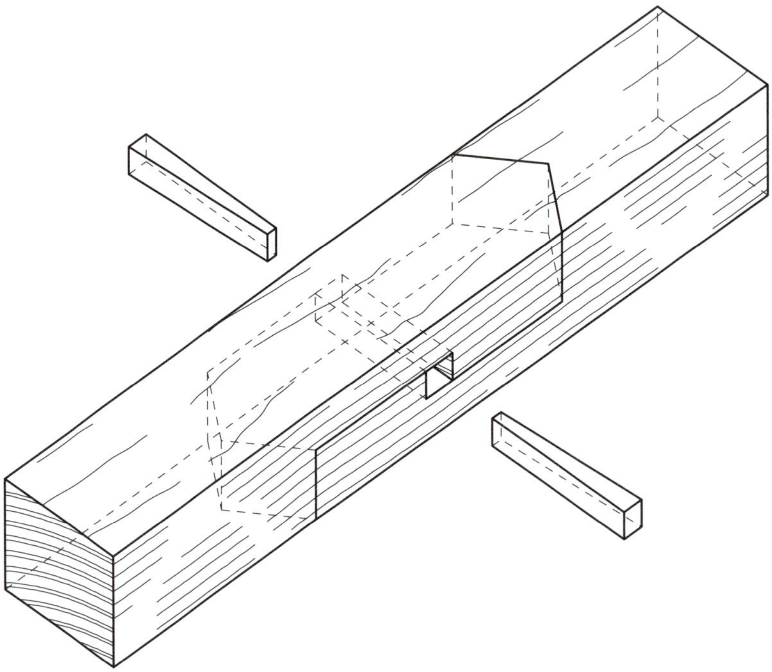

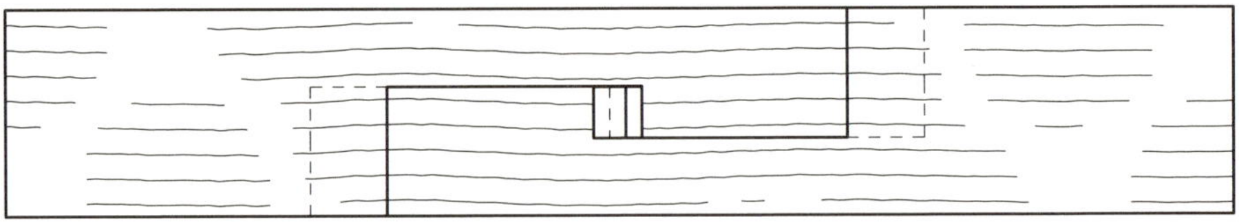

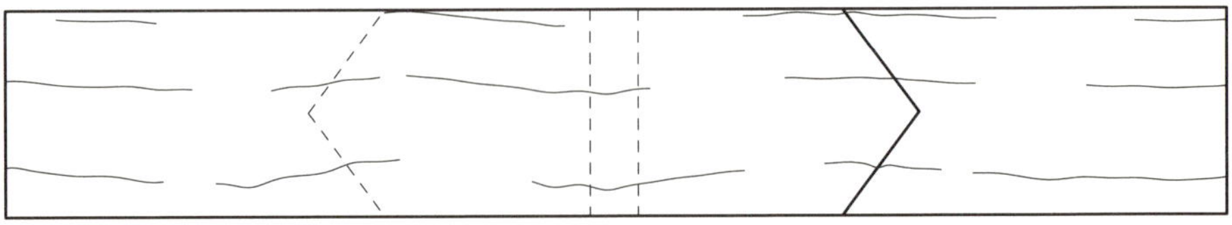

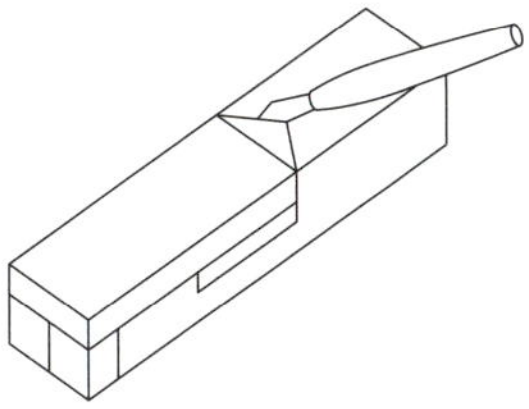

A1

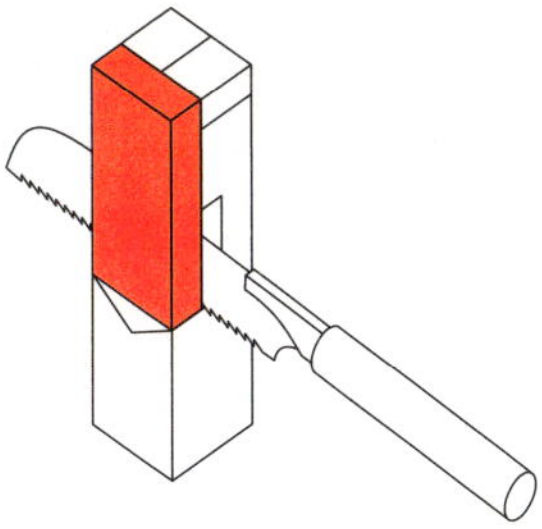

A2

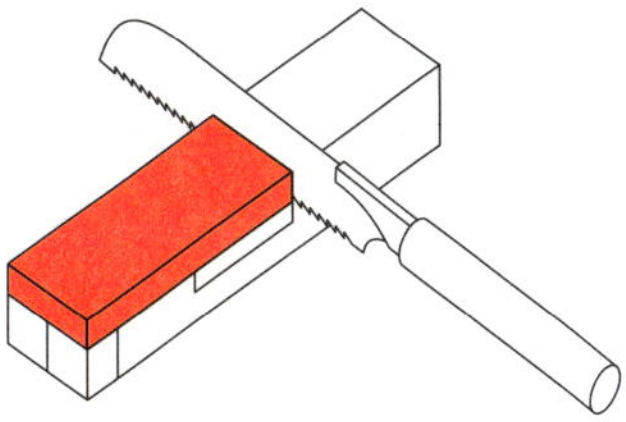

A3

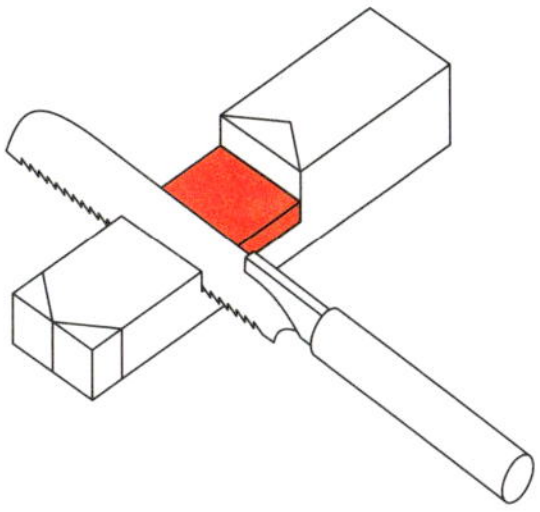

A4

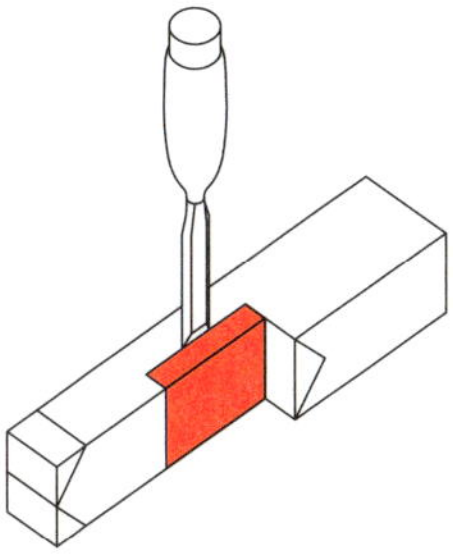

A5

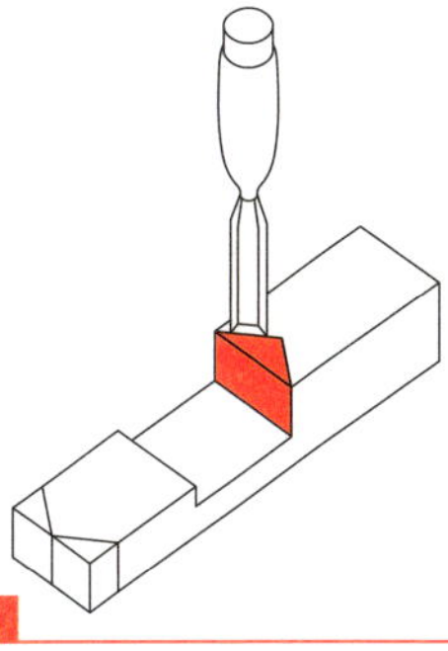

A6

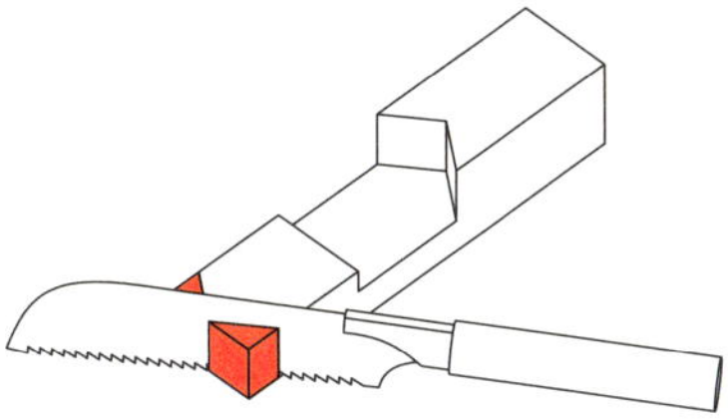

A7

2×

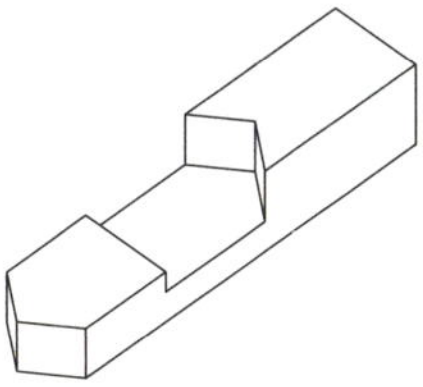

A

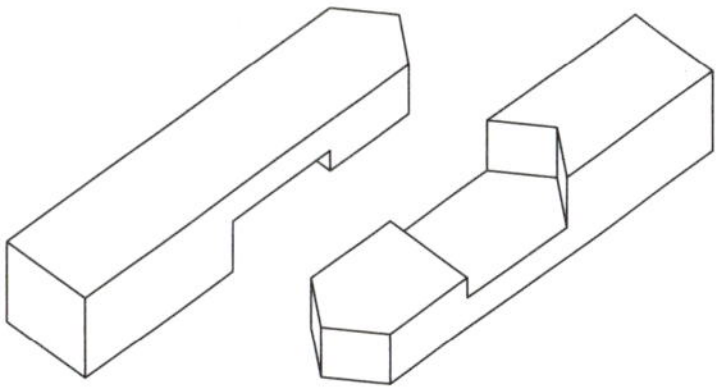

A+A

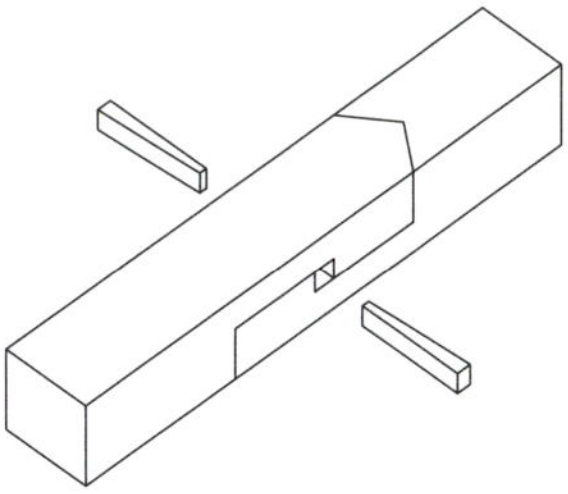

A+A

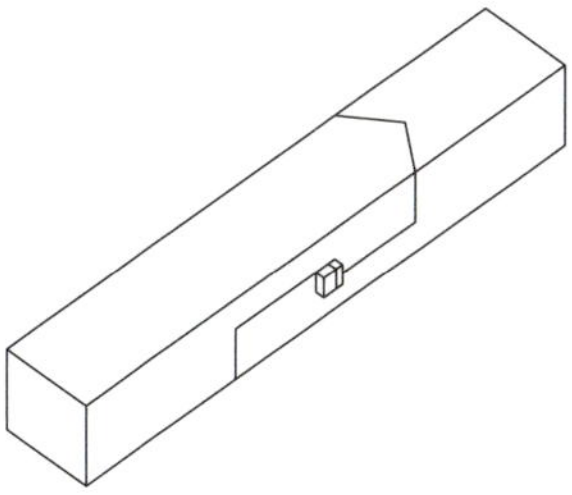

AA

Splicing Joints

Undersquinted spear point cogged lap joint with folding wedges

dan-koshi-kake-yahazu-tsugi (sogi-tsuki, kusabi-shime)
段腰掛矢筈継ぎ（殺付・楔締）

verkeiltes gerades Hakenblatt mit schrägem Gratschnitt

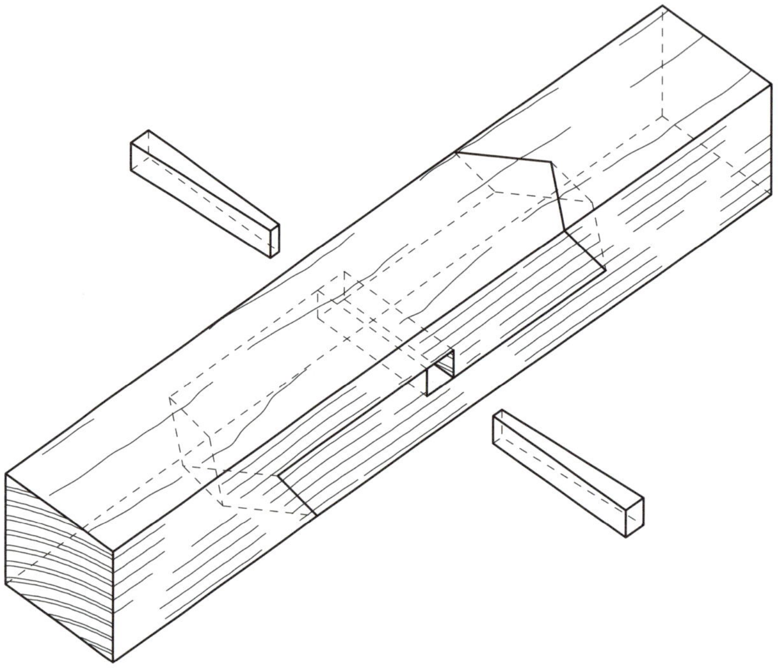

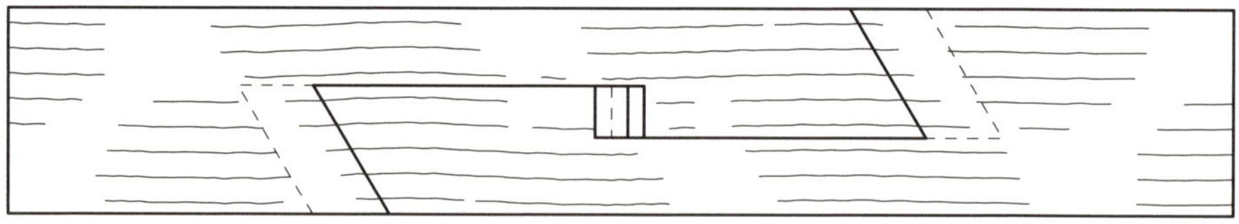

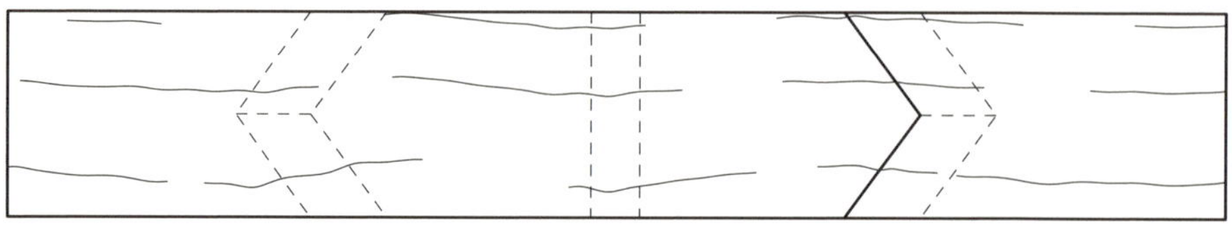

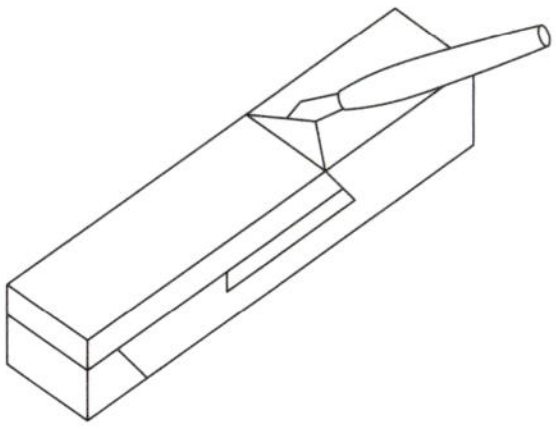

A1

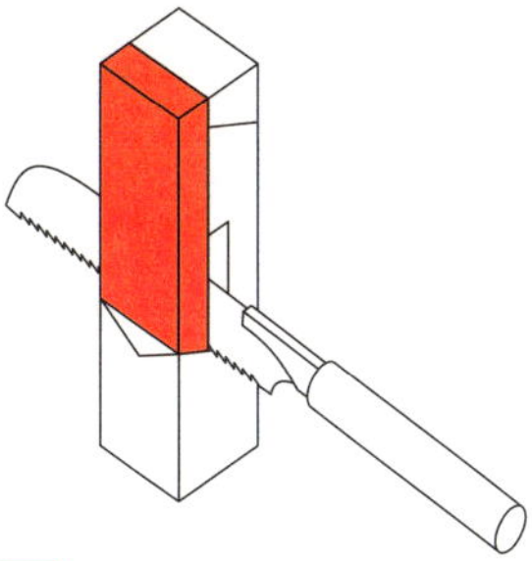

A2

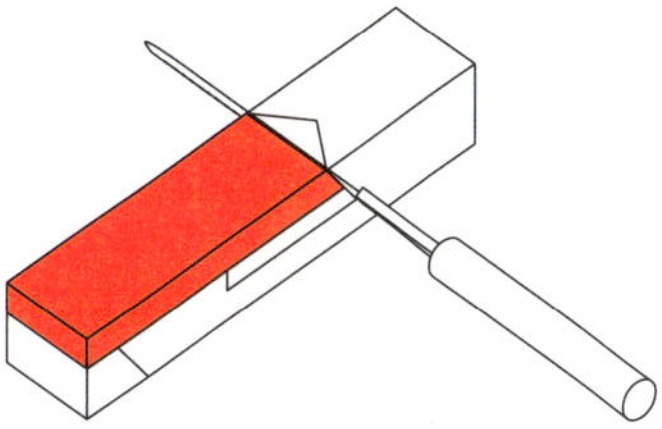

A3

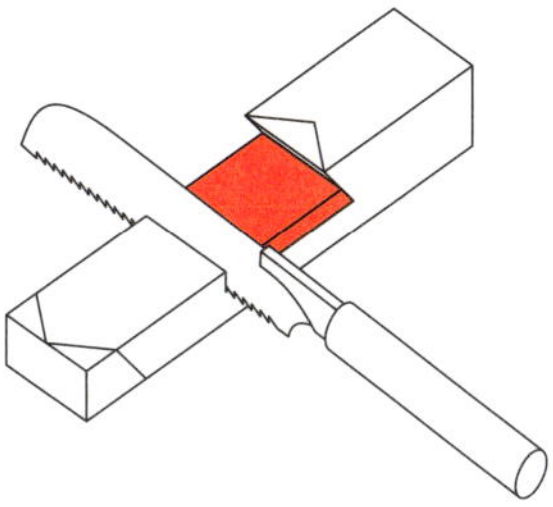

A4

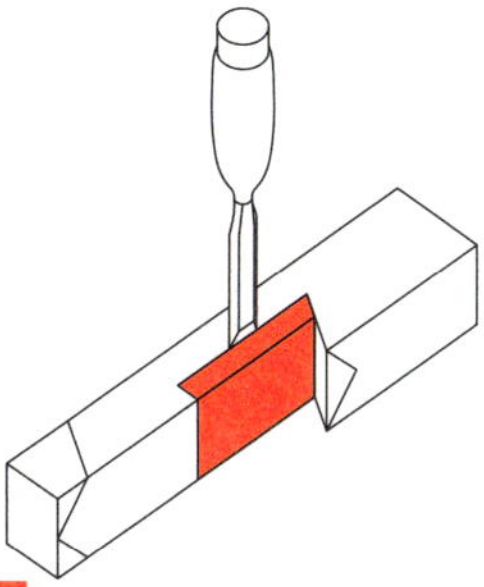

A5

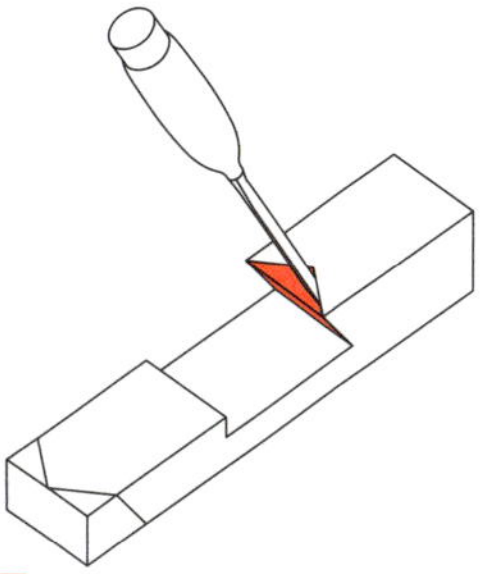

A6

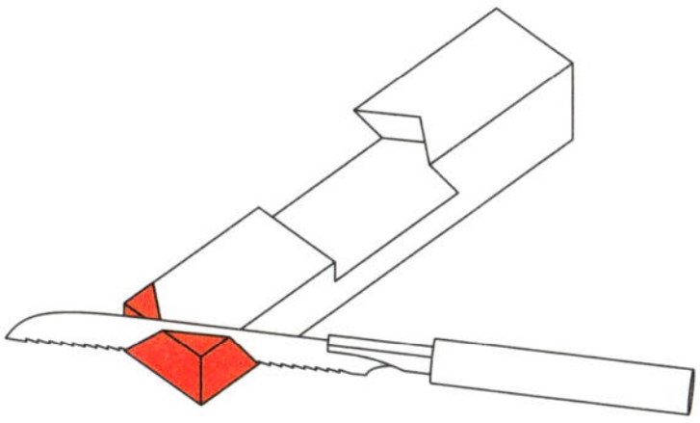

A7

2×

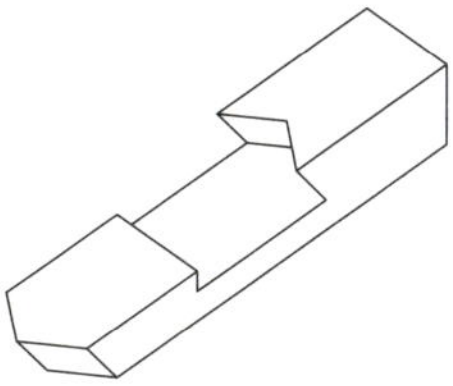

A

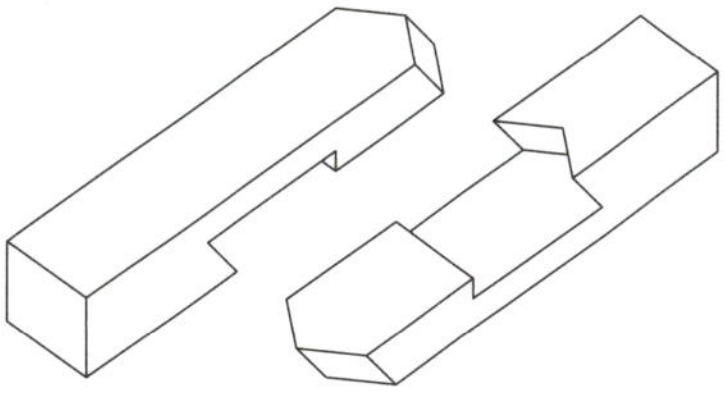

A+A

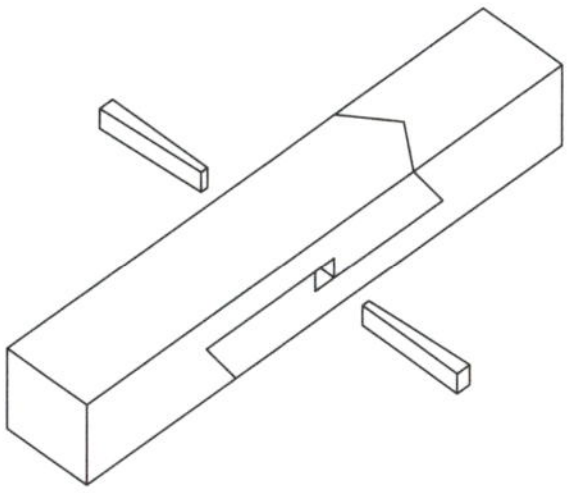

A+A

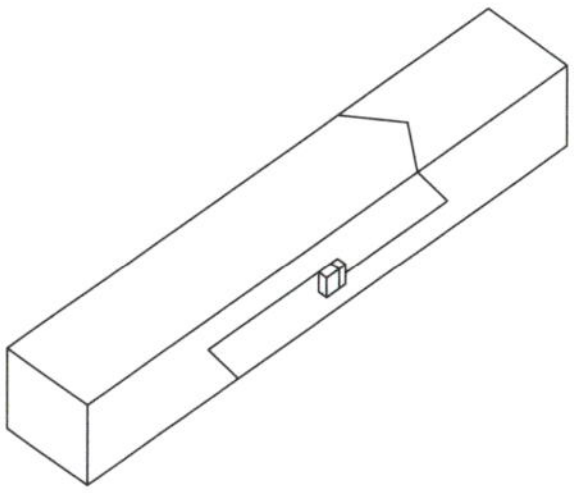

AA

Splicing Joints

Half lap joint with floating piece

yatoi-kabuse-tsuki-tsuke-tsugi
雇被突付継ぎ

gerades Blatt mit eingesetztem Mittelstück

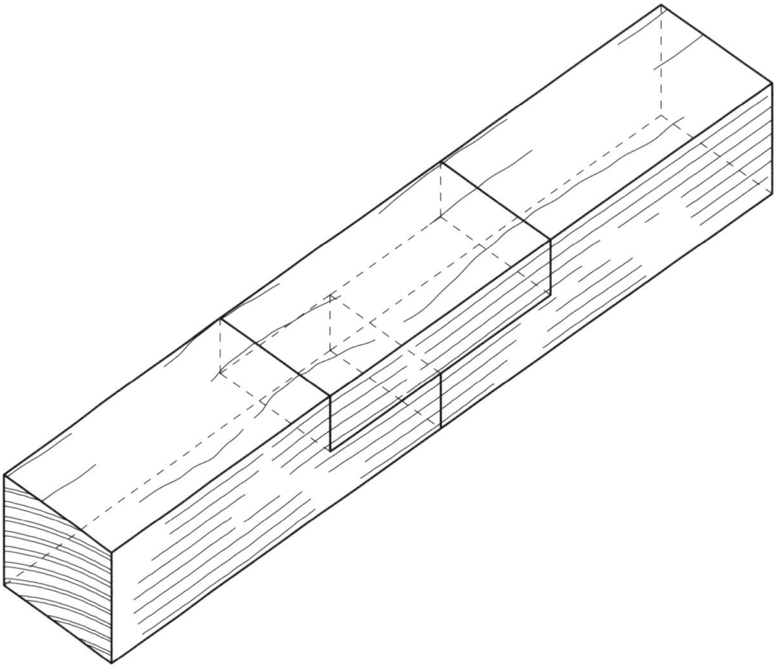

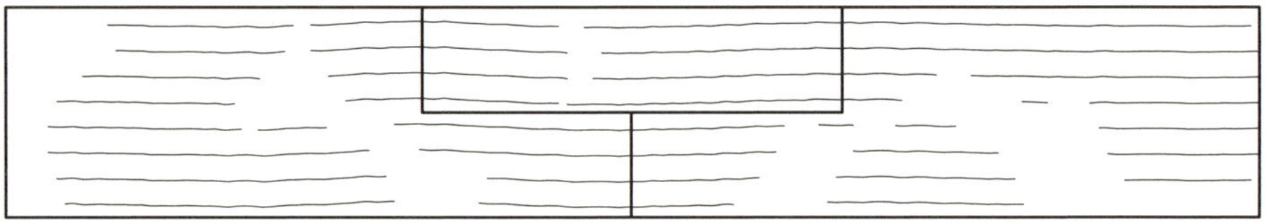

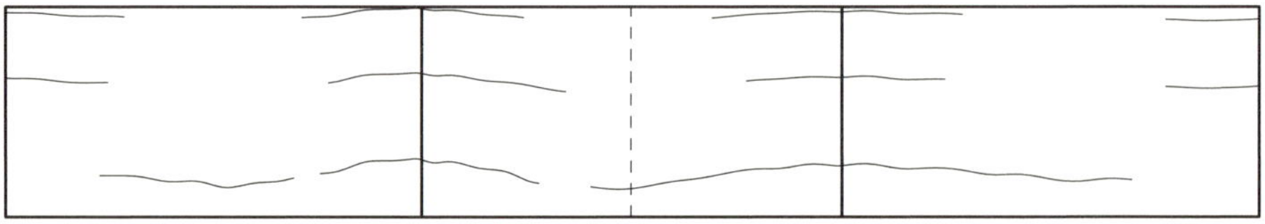

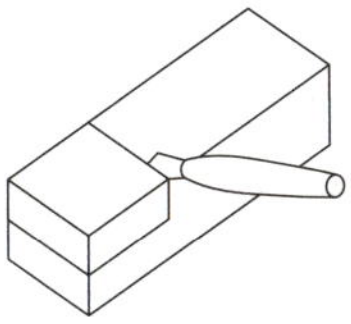

A1

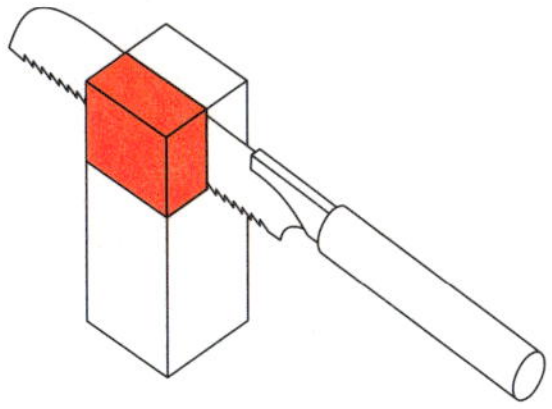

A2

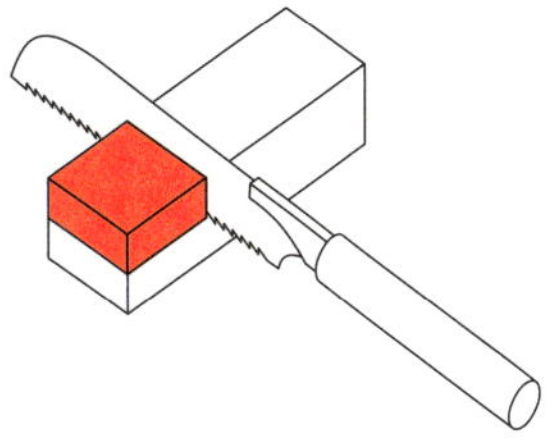

A3

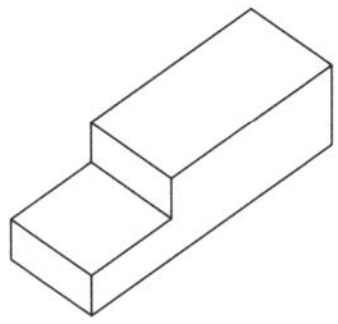

2×

A

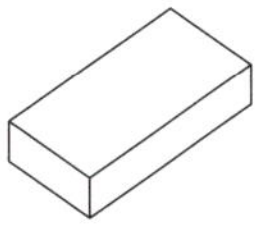

B

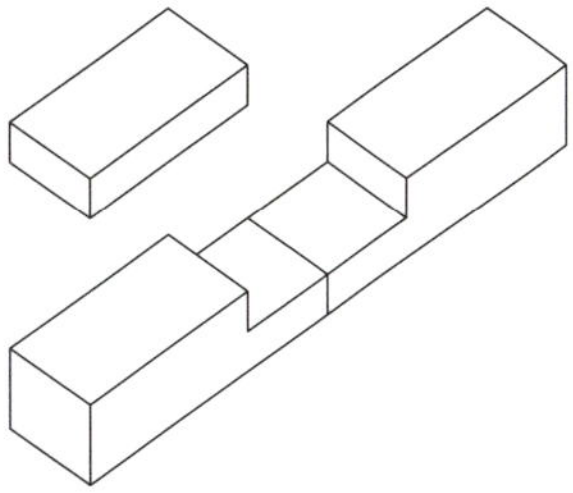

A+B+A

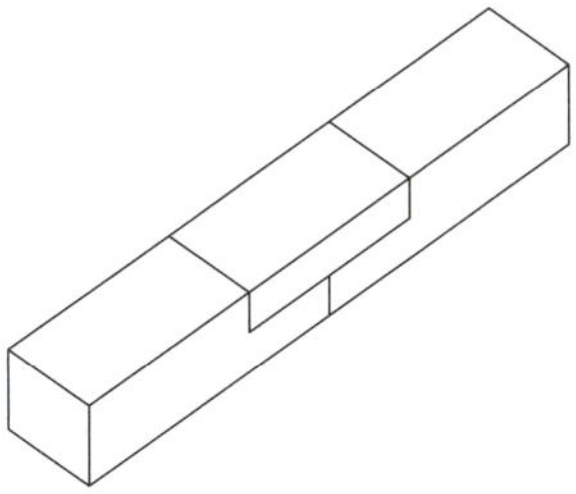

ABA

Splicing Joints

Undersquinted half lap joint with floating piece

yatoi-kabuse-tsuki-tsuke-tsugi (sogi-tsuki)
雇被突付継ぎ(殺付)

gerades Blatt mit hinterschnittener Brüstung und eingesetztem Mittelstück

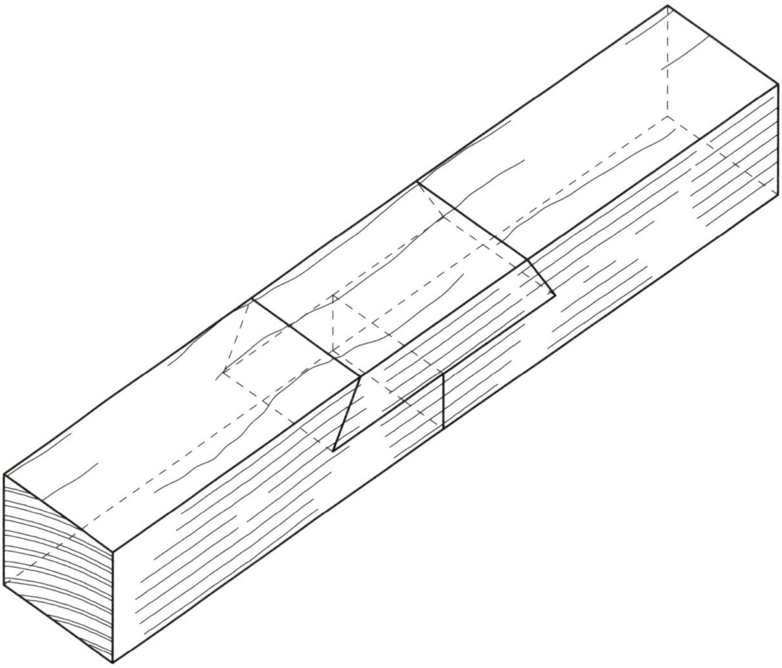

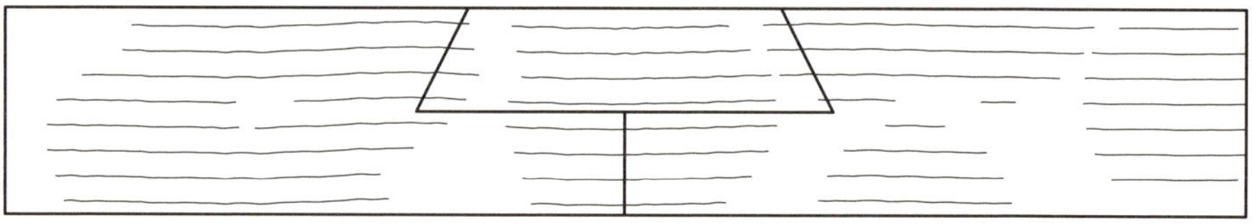

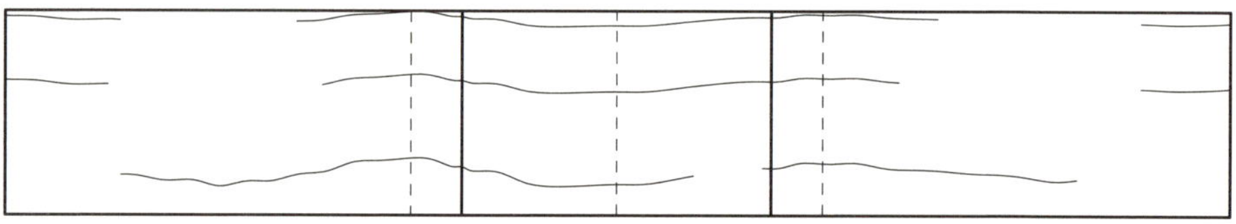

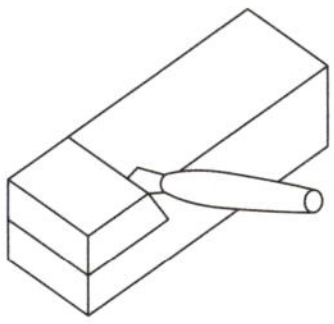

A1

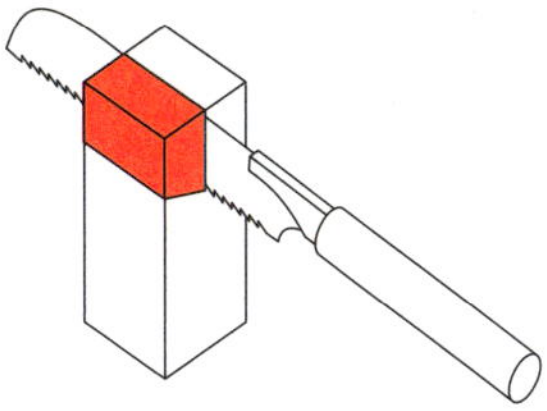

A2

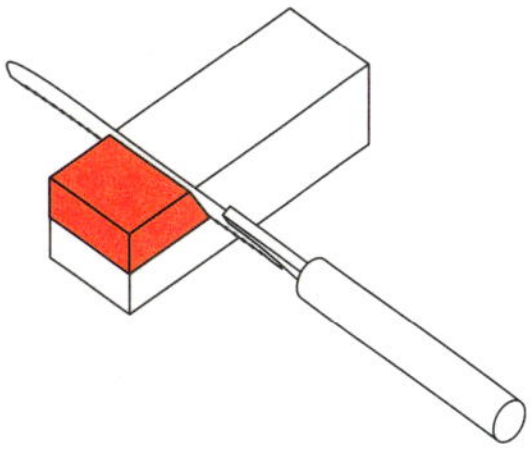

A3

2×

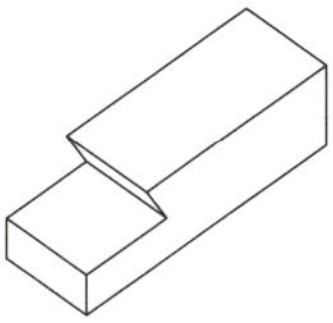

A

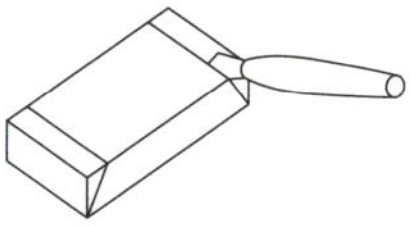

B1

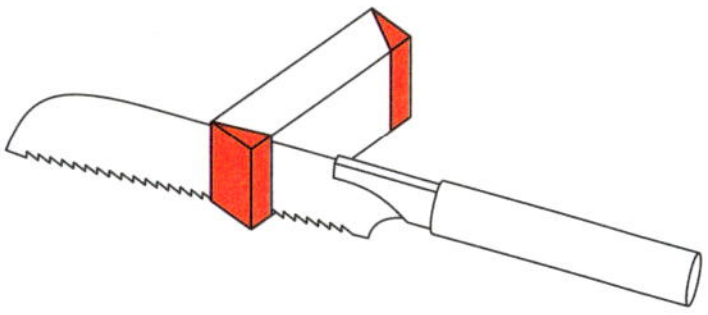

B2

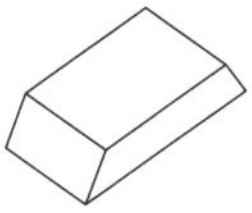

B

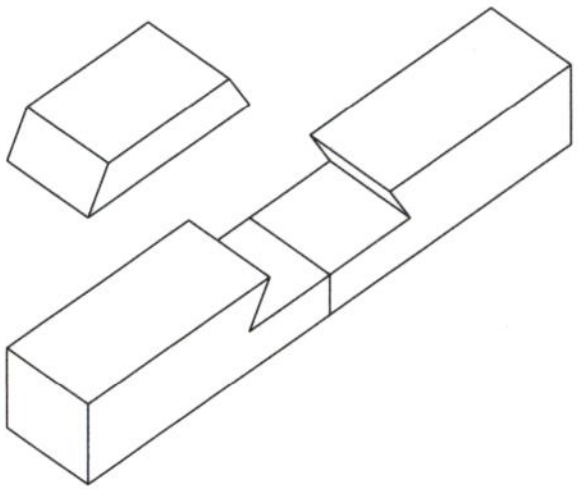

A+B+A

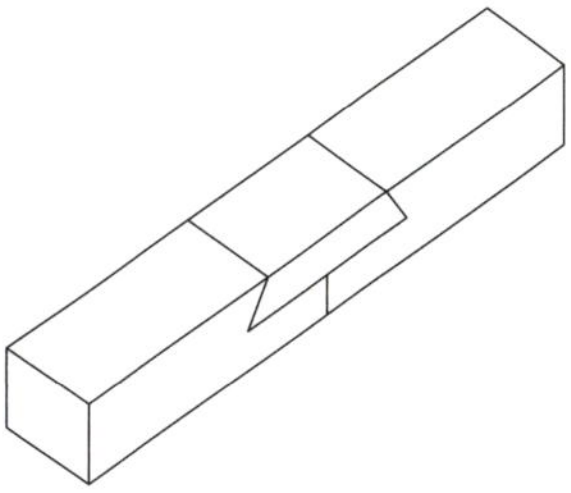

ABA

Splicing Joints

Cogged lap joint with floating tenoned piece

yatoi-kabuse-dan-tsuki-tsuke-mechi-tsugi
雇被段突付目違継ぎ

gerades Hakenblatt mit eingesetztem Hakenstück mit Zapfen

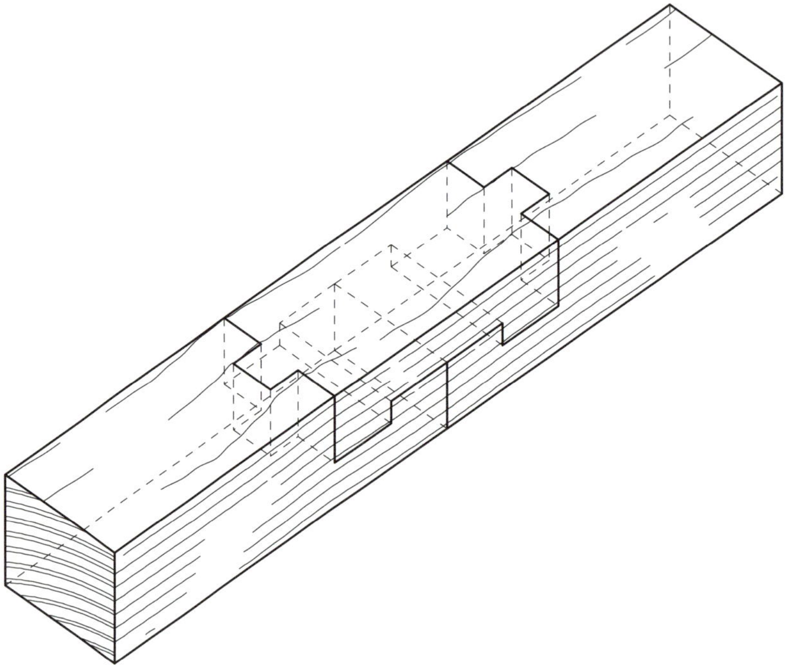

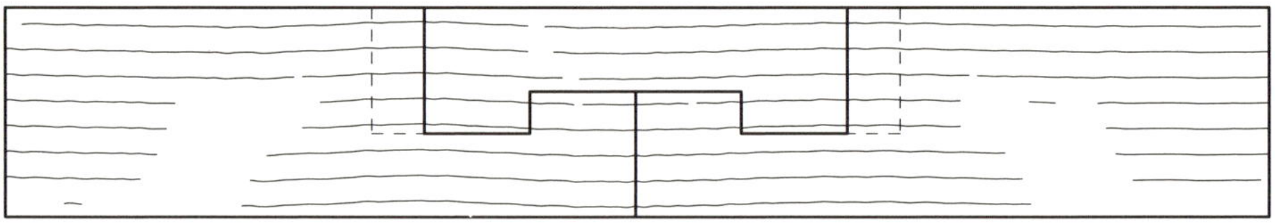

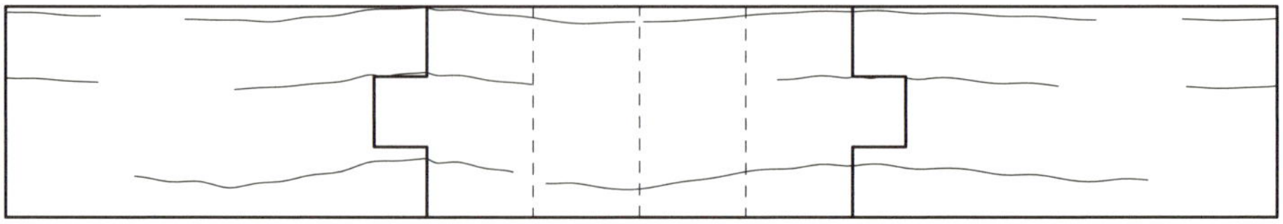

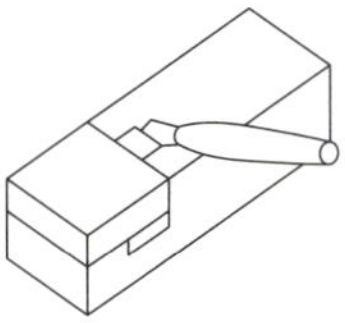

A1

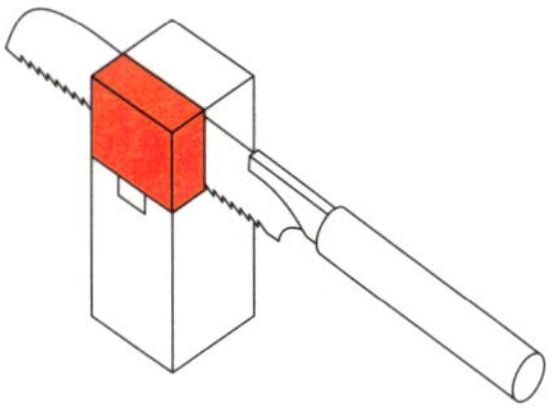

A2

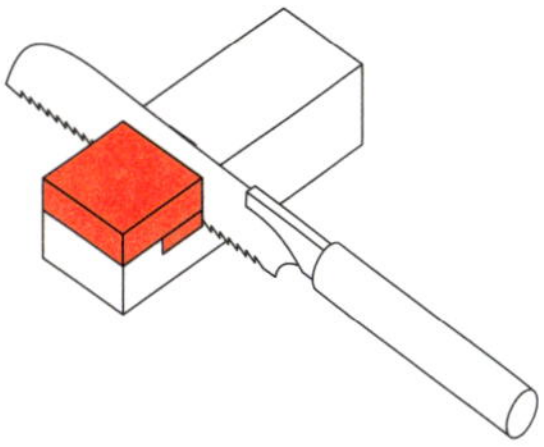

A3

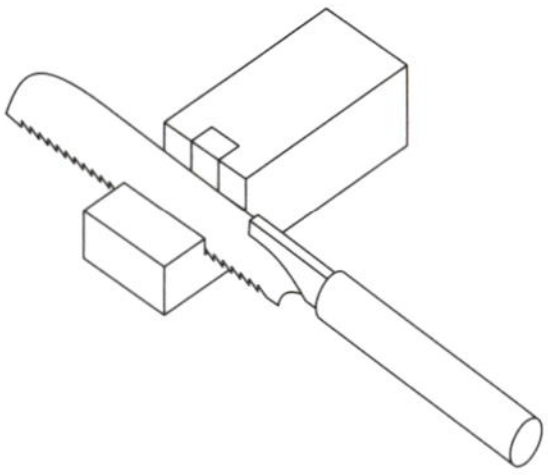

A4

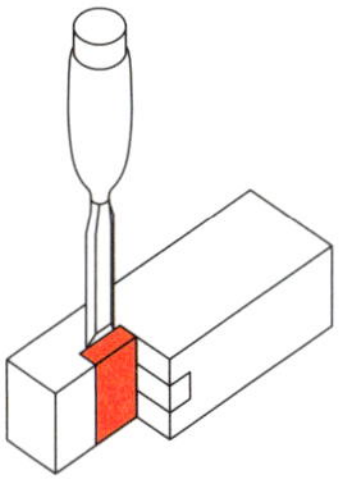

A5

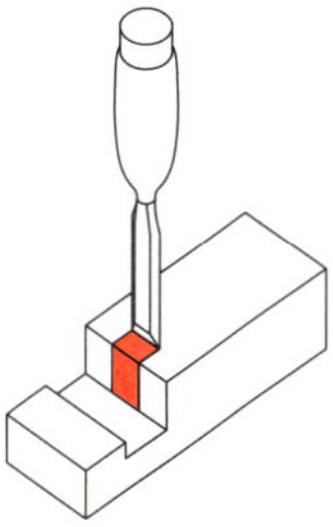

A6

2×

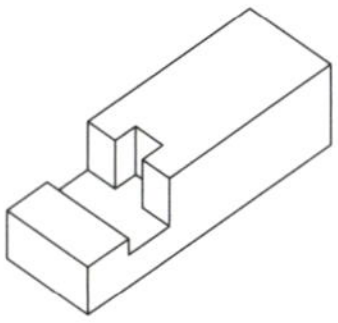

A

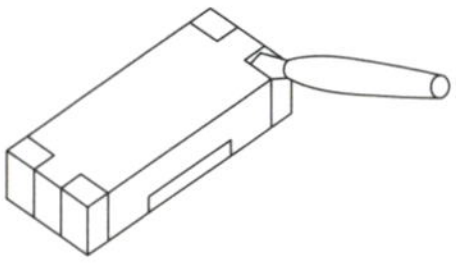

B1

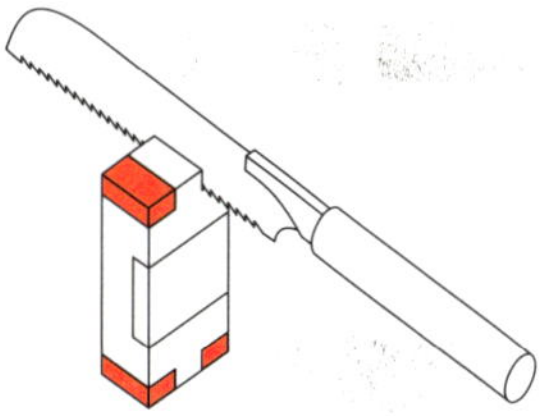

B2

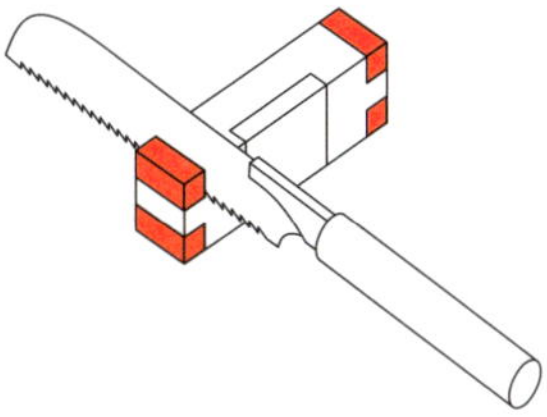

B3

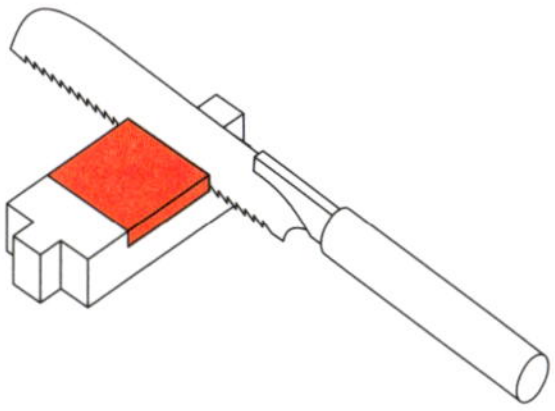

B4

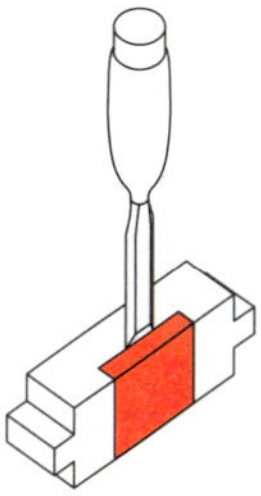

B5

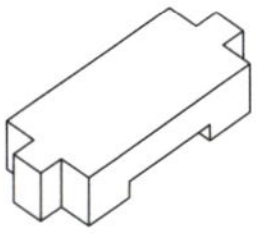

B

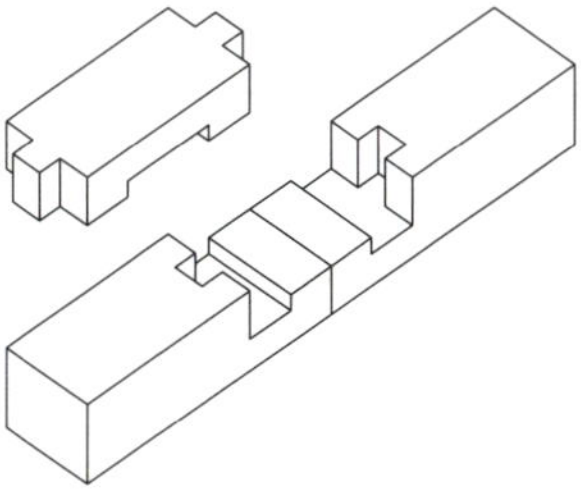

A+B+A

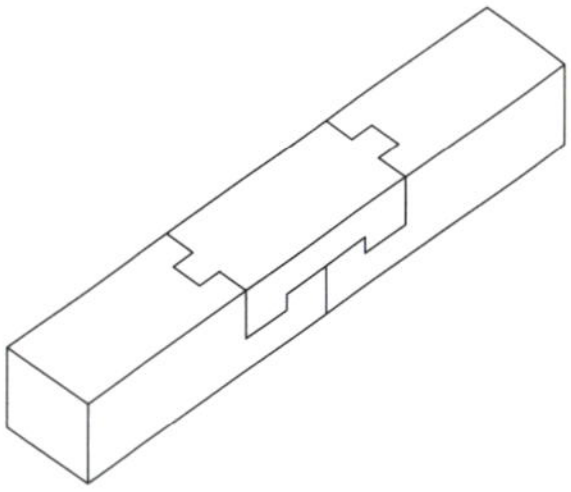

ABA

Splicing Joints

Cogged lap joint with floating piece

yatoi-kabuse-naga-dan-tsuki-tsuke-tsugi
雇被長段突付継ぎ

gerades Hakenblatt mit eingesetztem Hakenstück

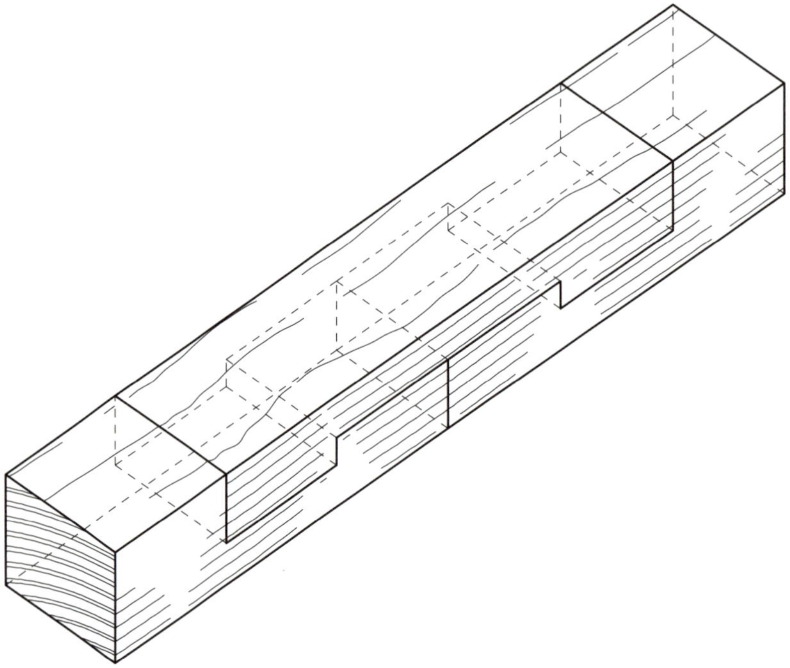

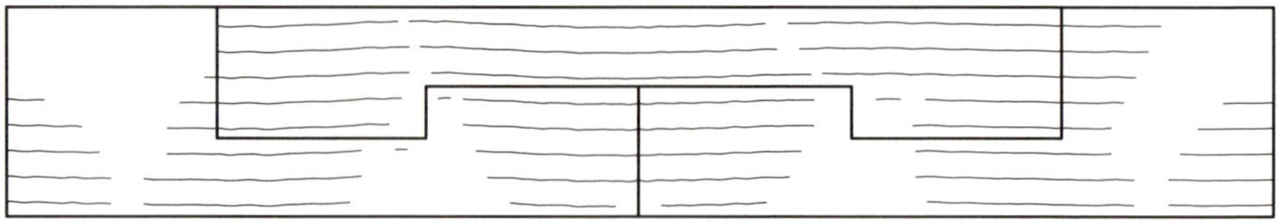

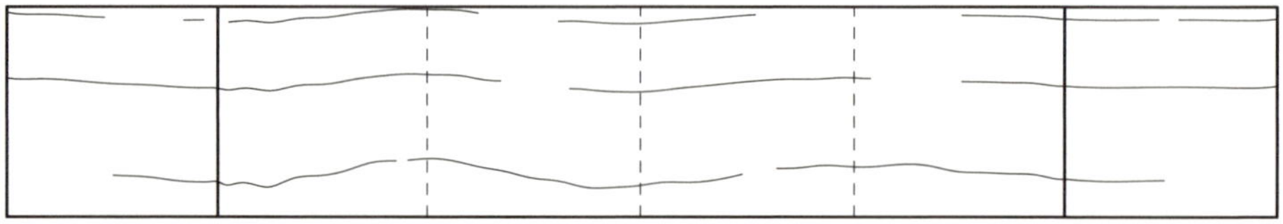

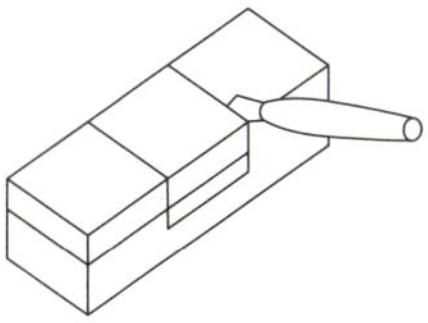

A1

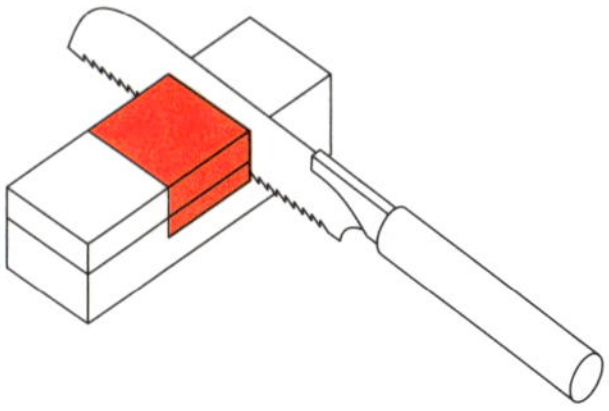

A2

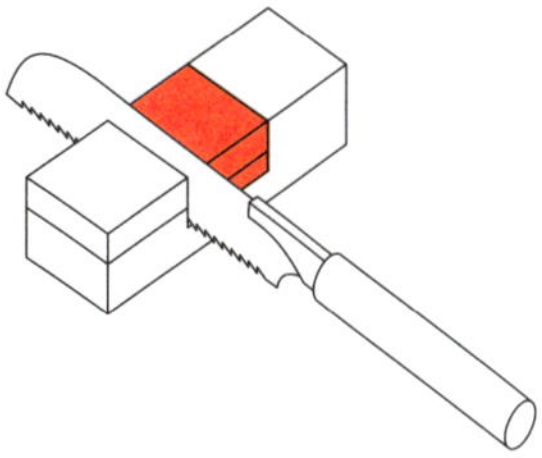

A3

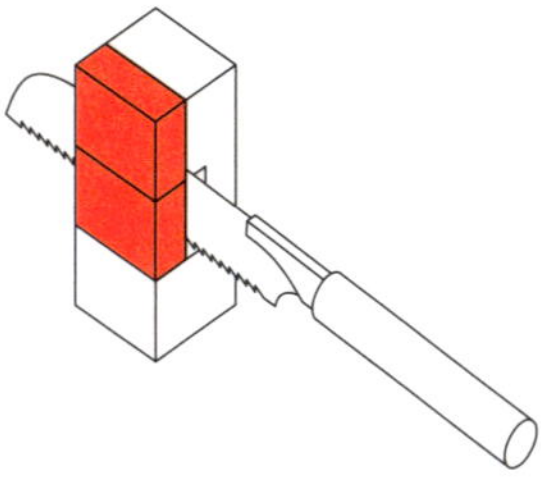

A4

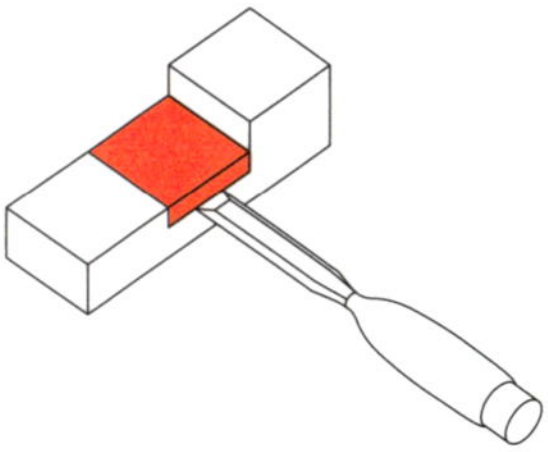

A5

2×

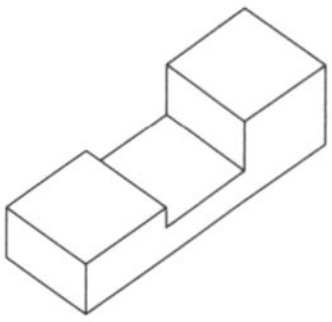

A

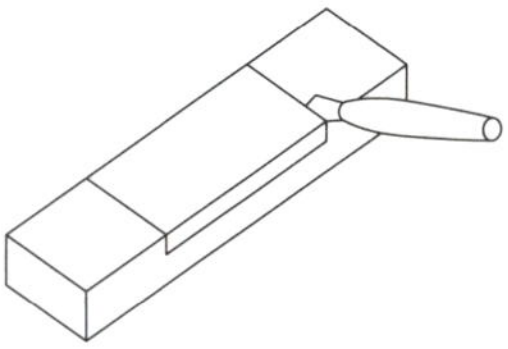

B1

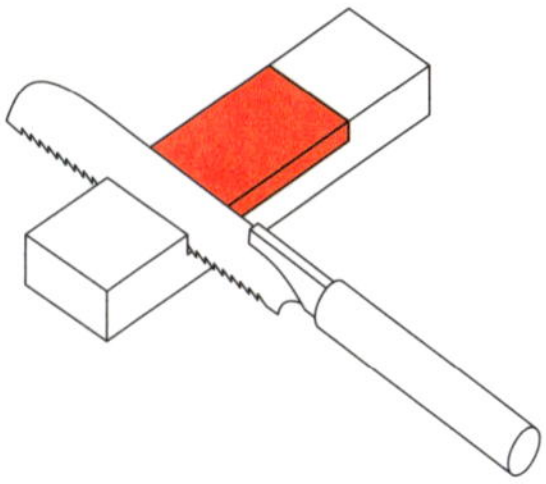

B2

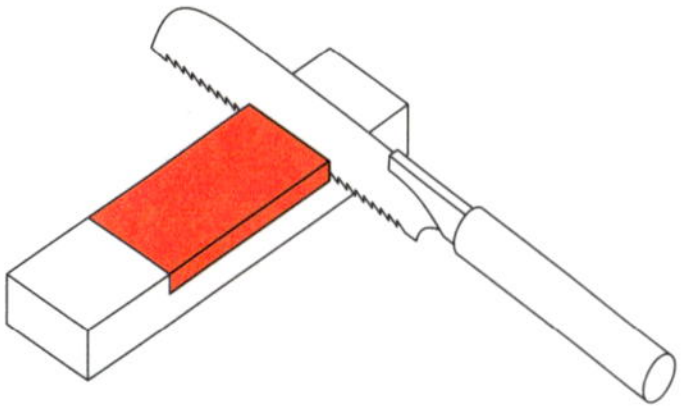

B3

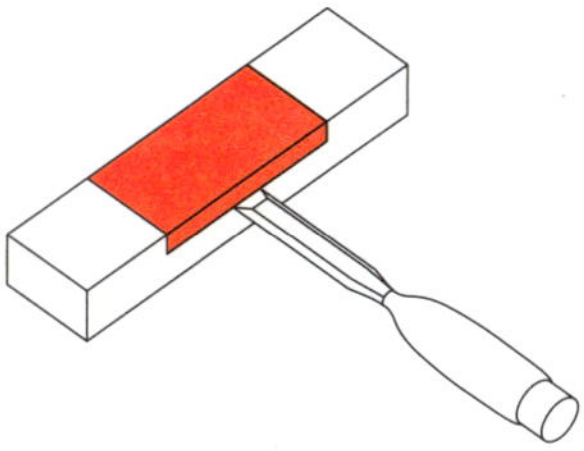

B4

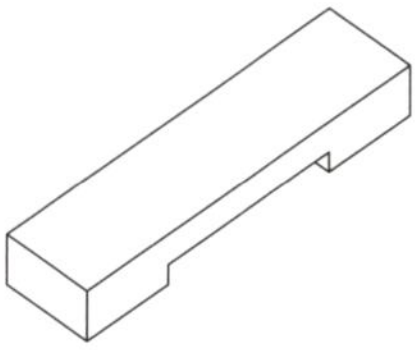

B

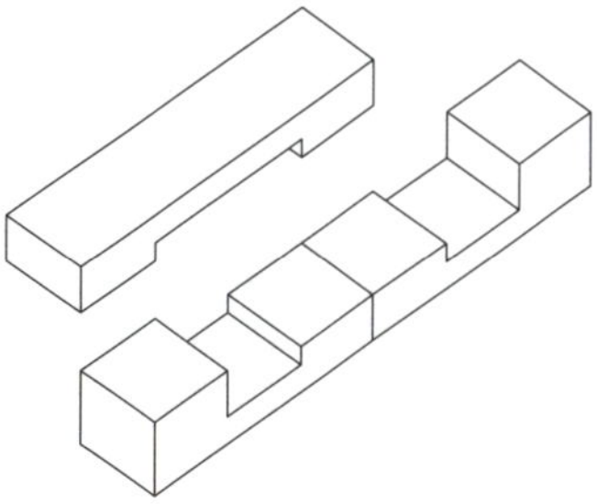

A+B+A

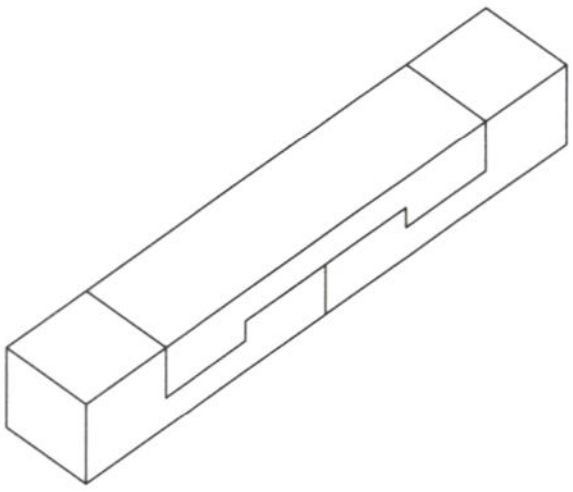

ABA

Splicing Joints

Cogged lap joint with floating piece and folding wedges

yatoi-kabuse-naga-dan-tsuki-tsuke-tsugi (kusabi-shime)
雇被長段突付継ぎ(楔締)

verkeiltes gerades Hakenblatt mit eingesetztem Hakenstück

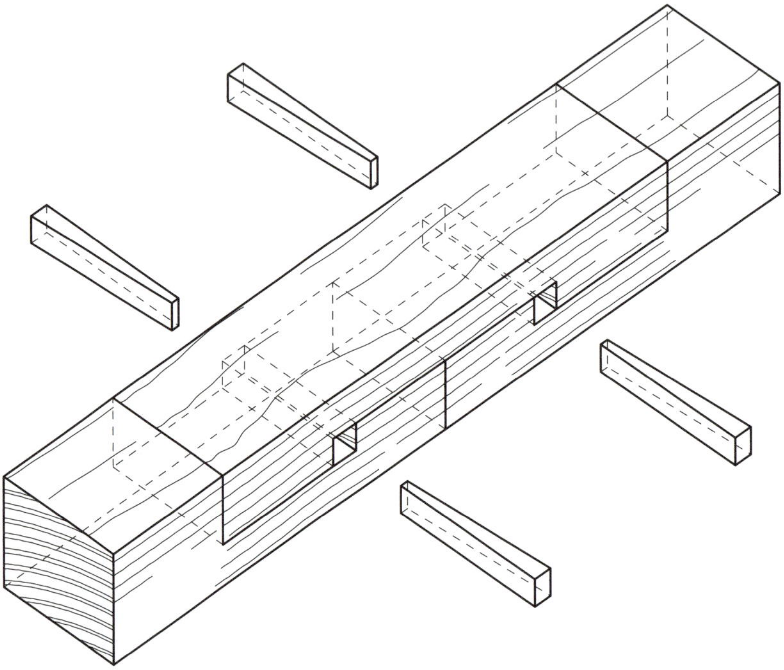

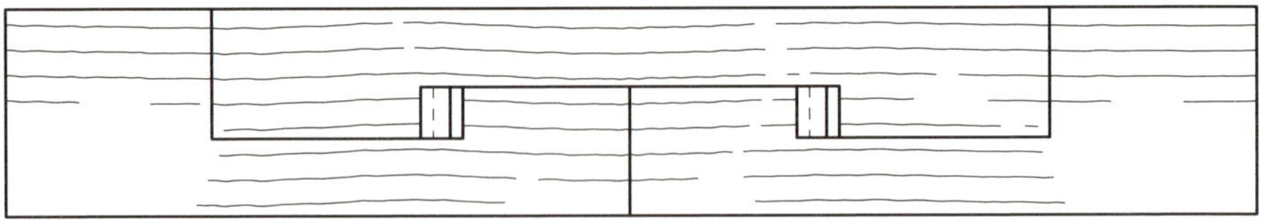

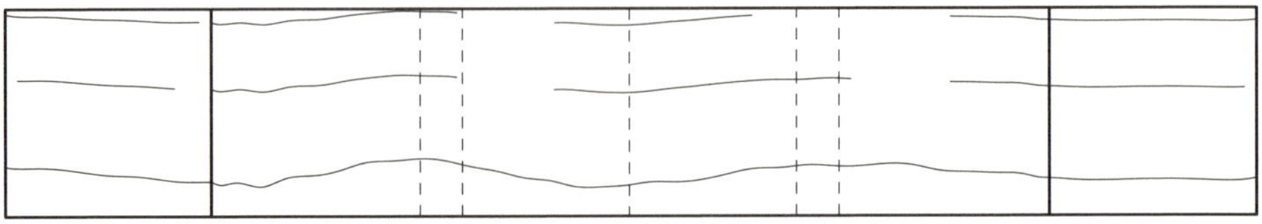

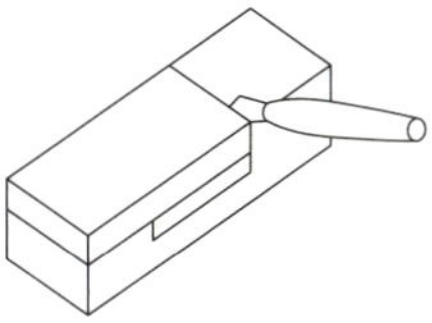

A1

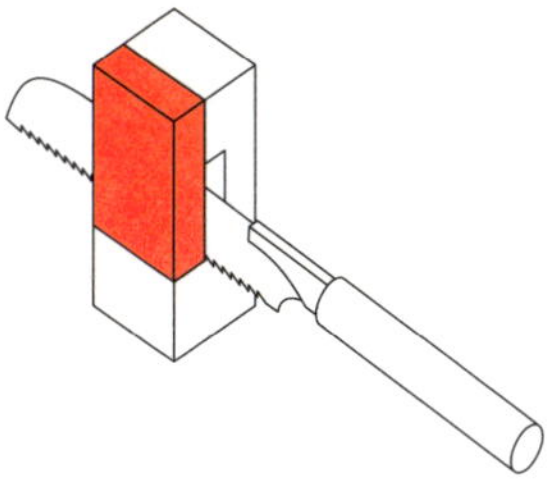

A2

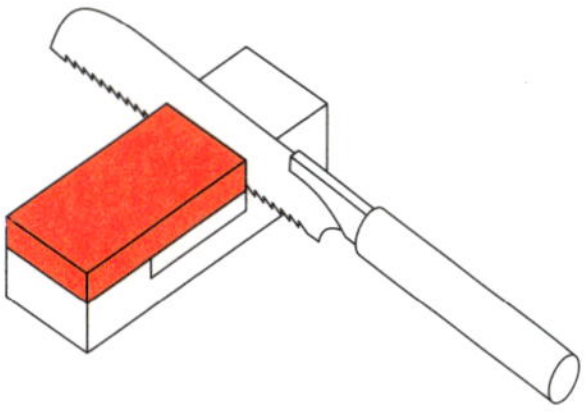

A3

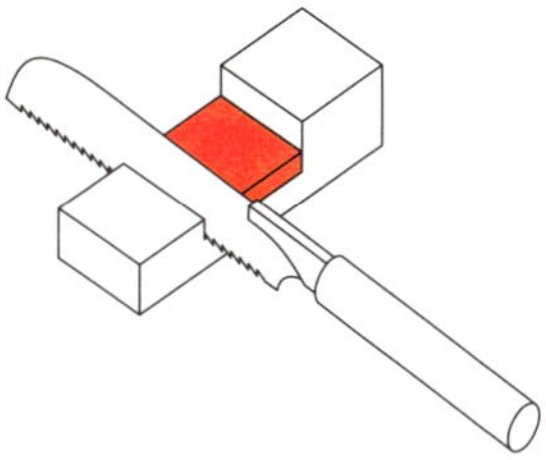

A4

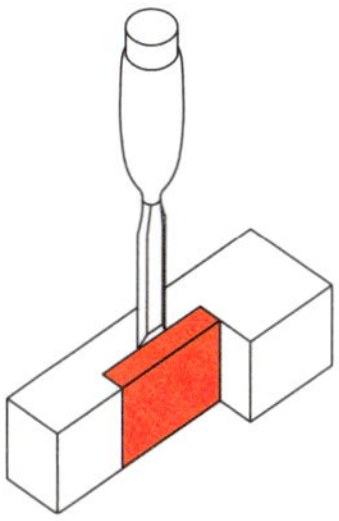

A5

2×

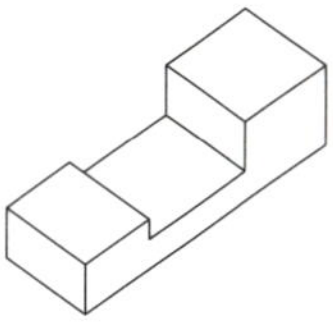

A

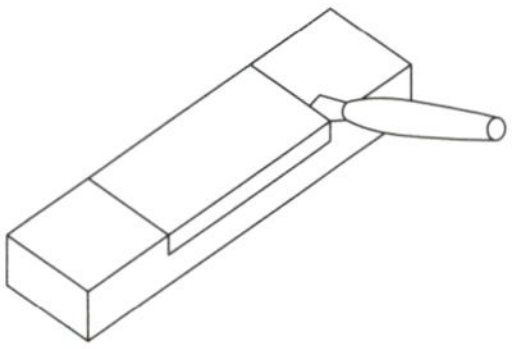

B1

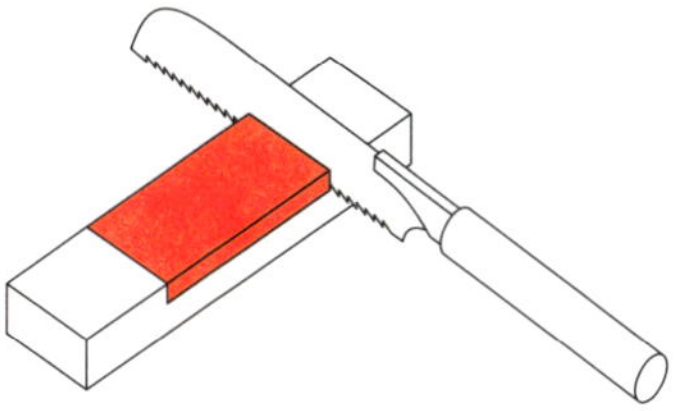

B2

B3

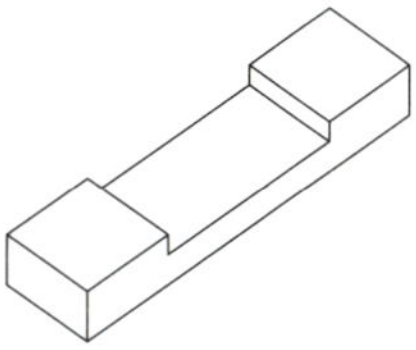

B

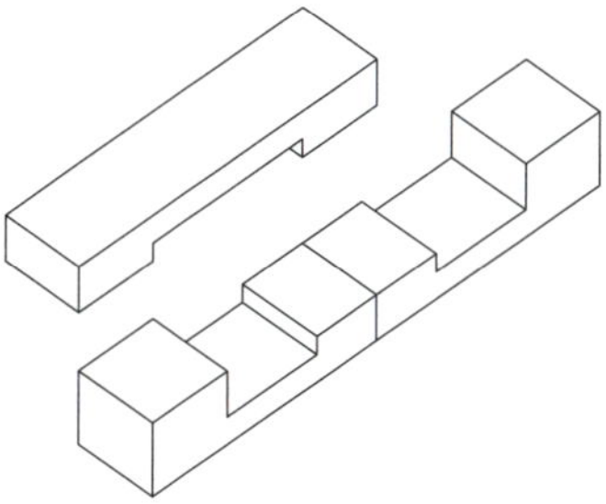

A+B+A

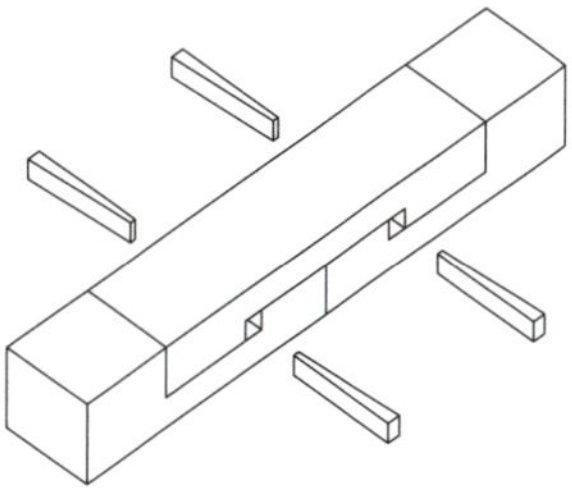

A+B+A

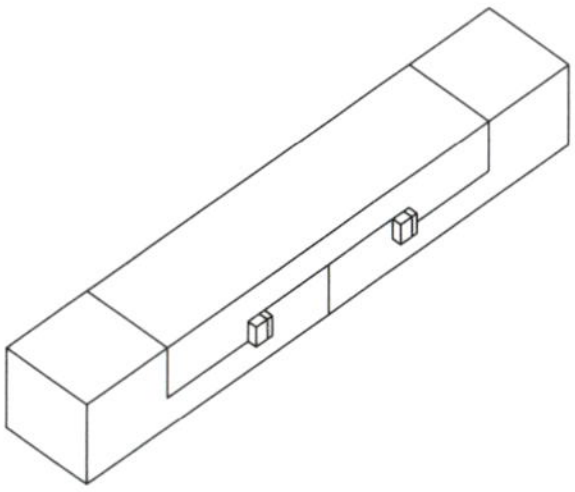

ABA

Splicing Joints

Undersquinted cogged lap joint with floating piece and folding wedges

yatoi-kabuse-naga-dan-tsuki-tsuke-tsugi (sogi-tsuki, kusabi-shime)
雇被長段突付継ぎ（殺付・楔締）

verkeiltes gerades Hakenblatt mit hinterschnittener Brüstung und eingesetztem Hakenstück

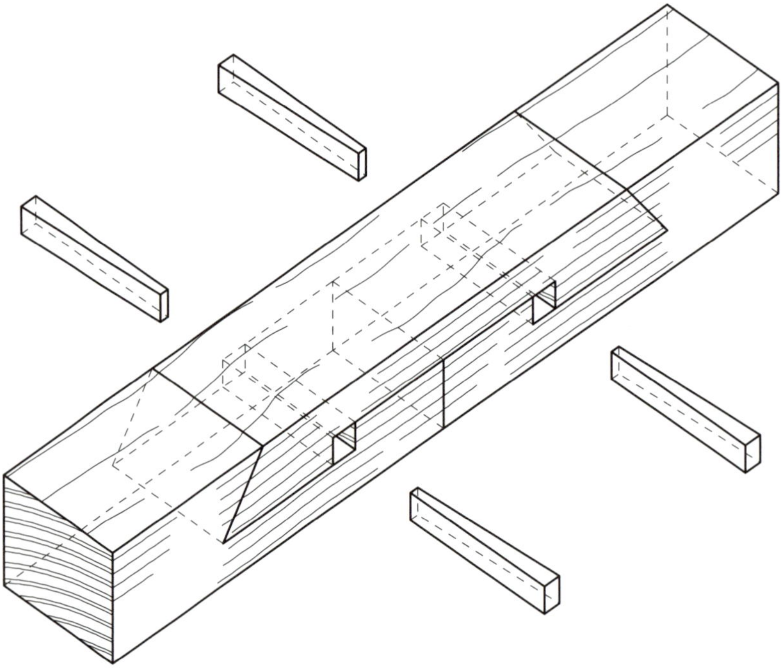

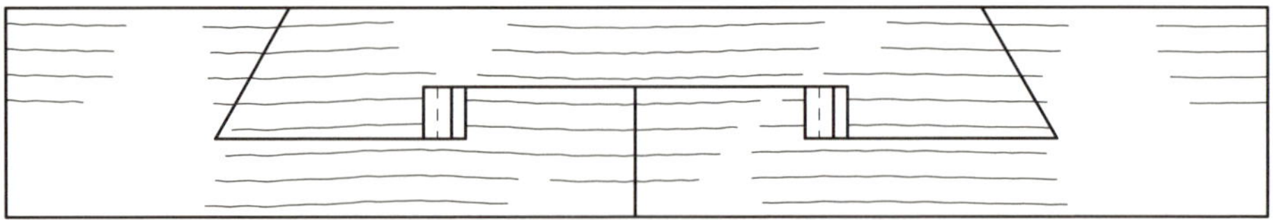

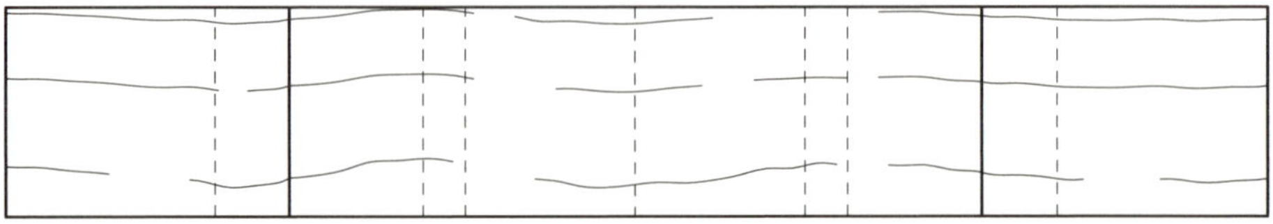

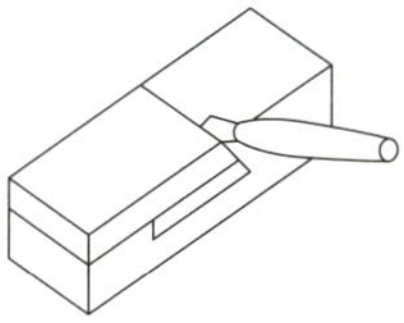

A1

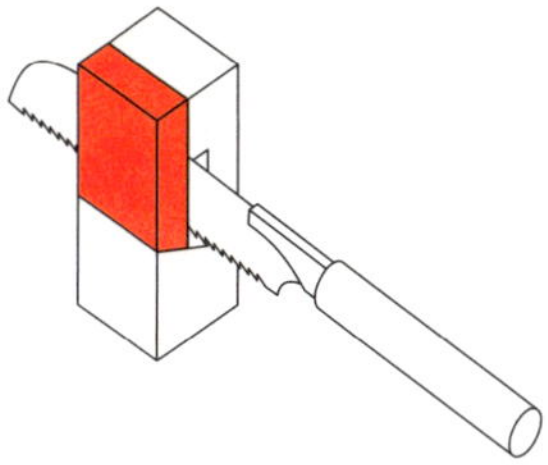

A2

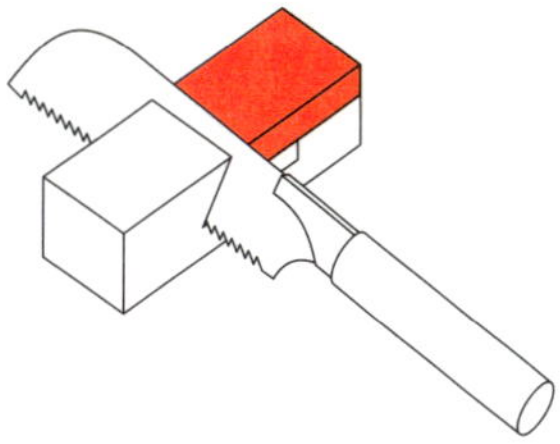

A3

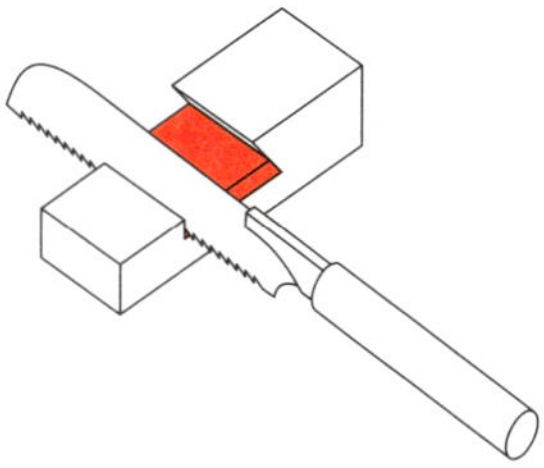

A4

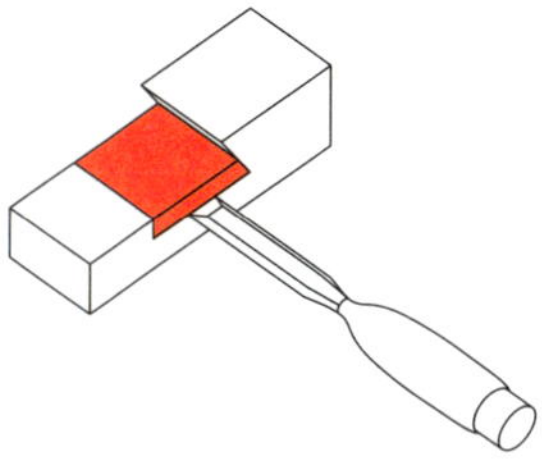

A5

2x

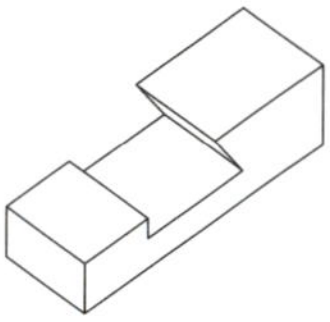

A

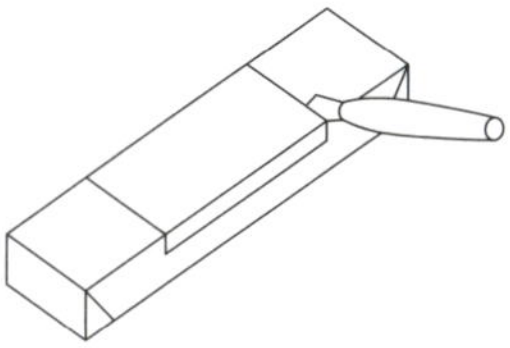

B1

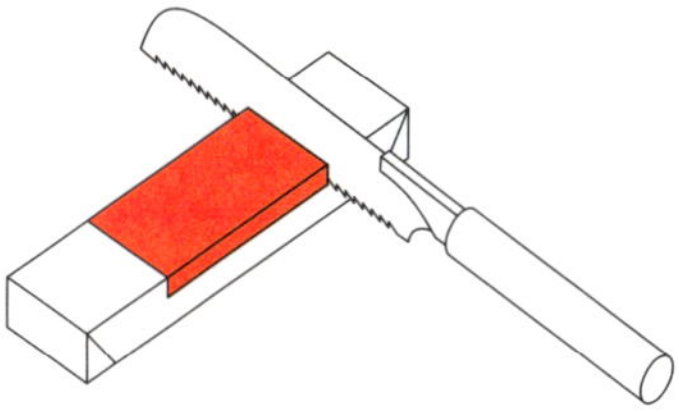

B2

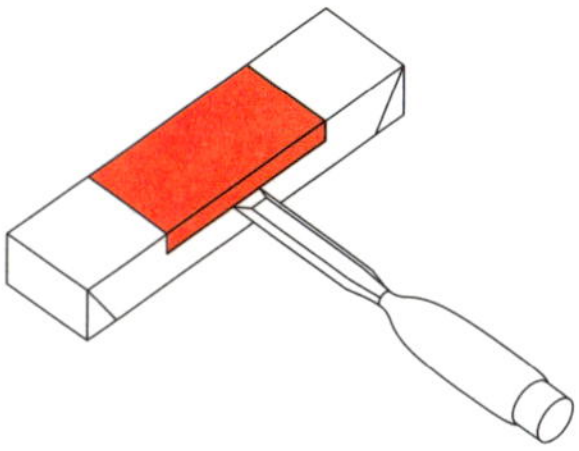

B3

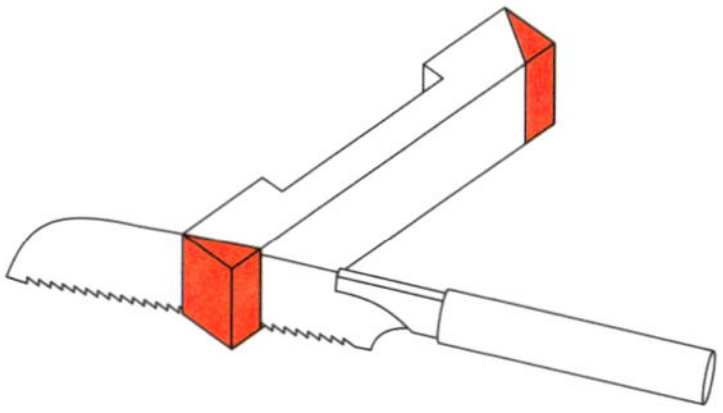

B4

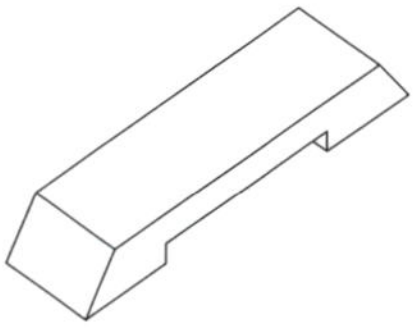

B

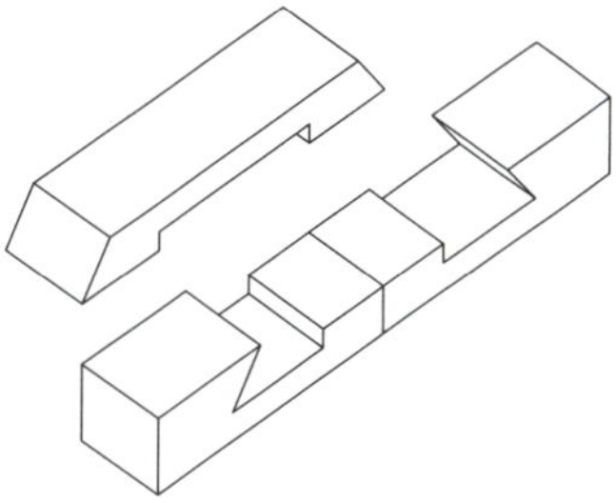

A+B+A

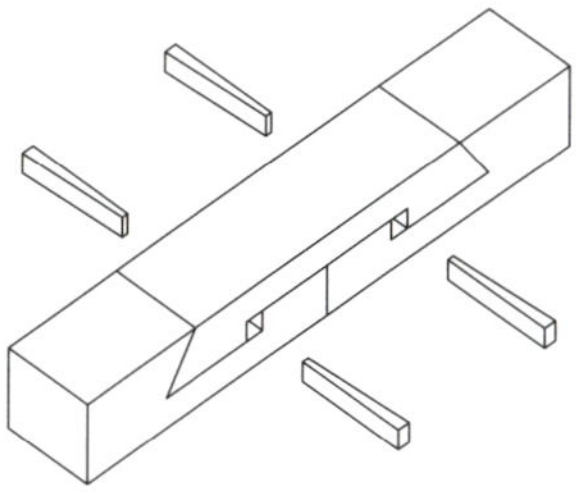

A+B+A

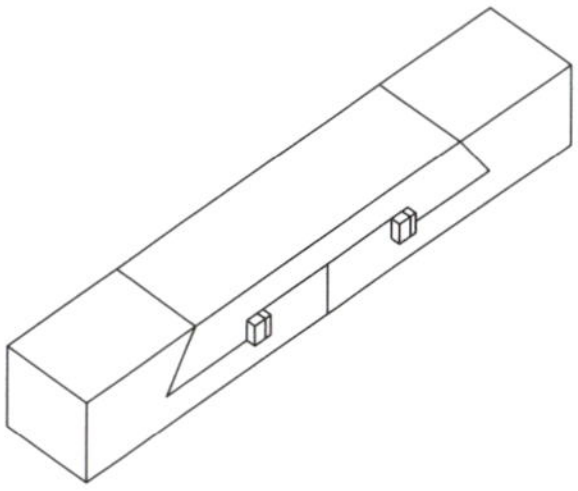

ABA

Splicing Joints

Stepped cogged lap joint with floating piece

yatoi-kabuse-naga-nidan-tsuki-tsuke-tsugi
雇被長二段突付継ぎ

gestuftes Hakenblatt mit eingesetztem Hakenstück

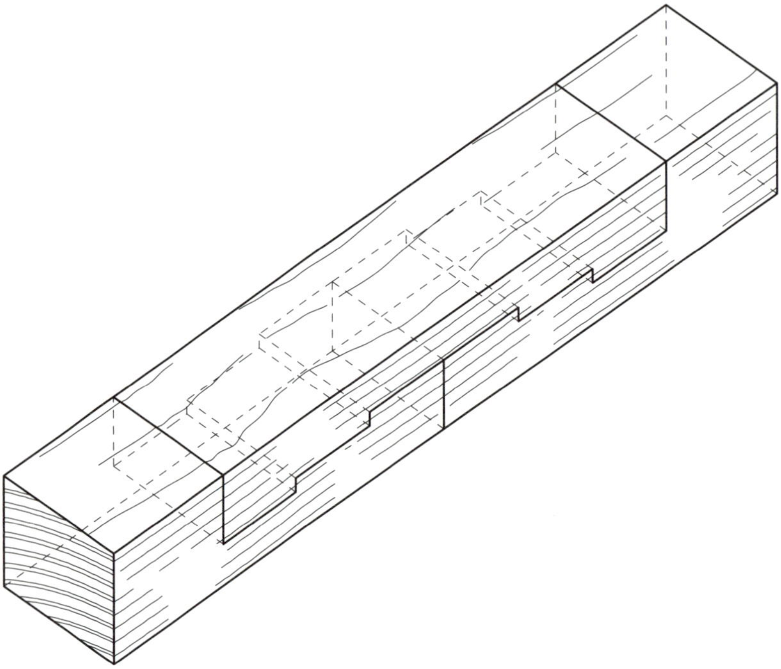

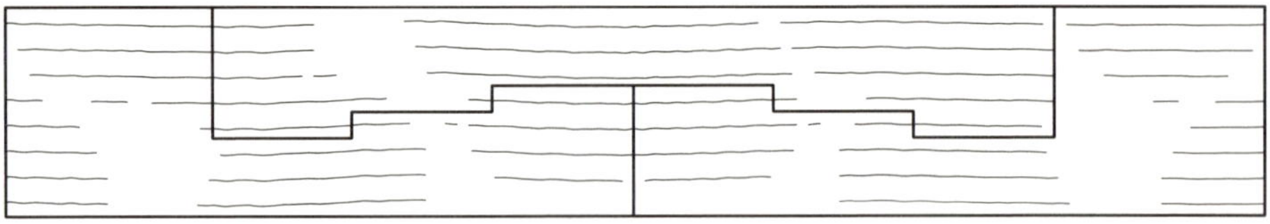

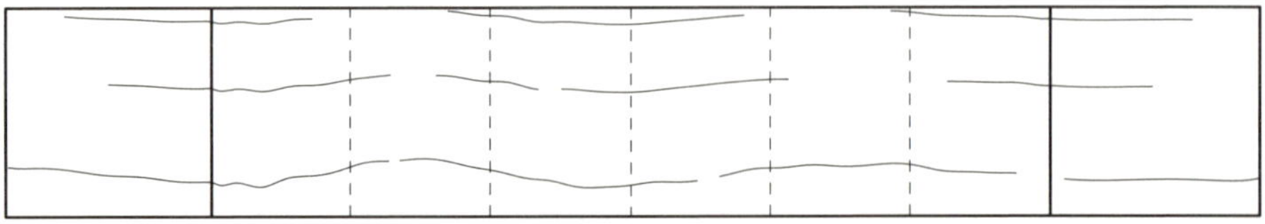

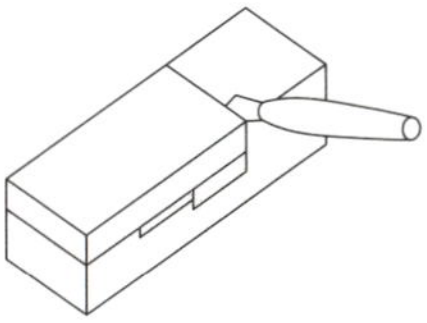

A1

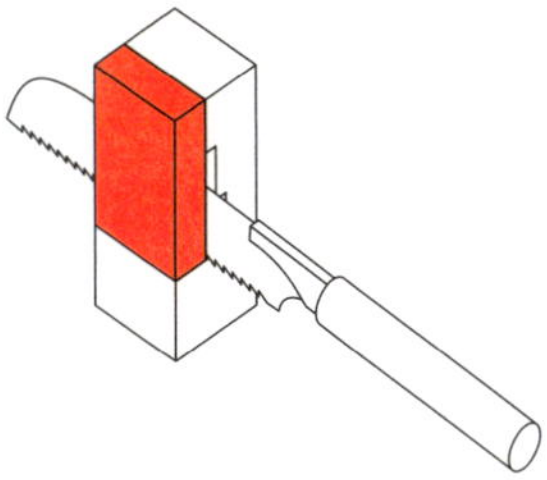

A2

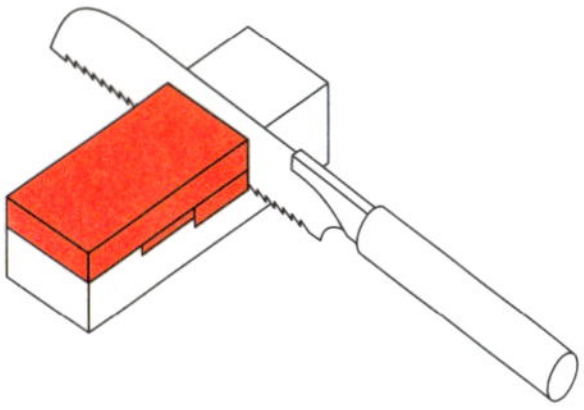

A3

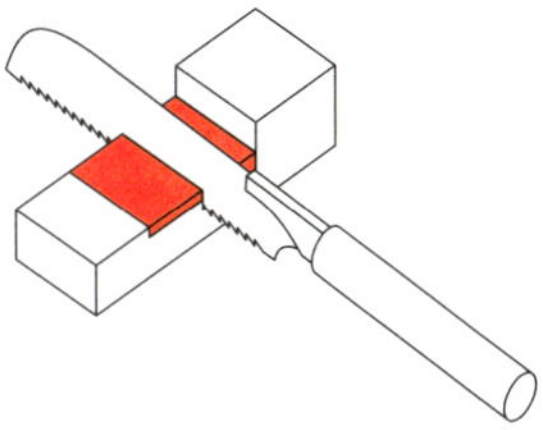

A4

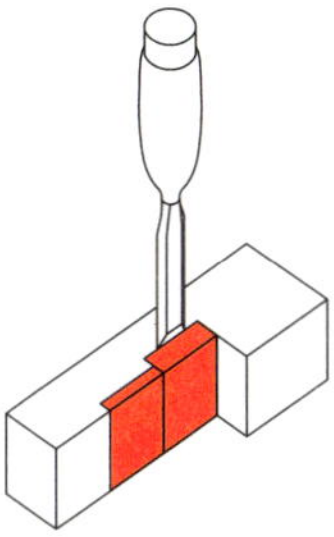

A5

2×

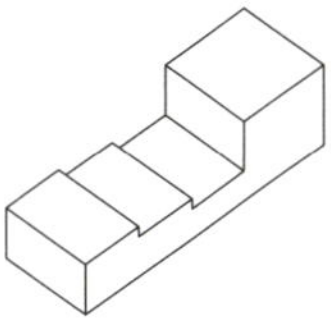

A

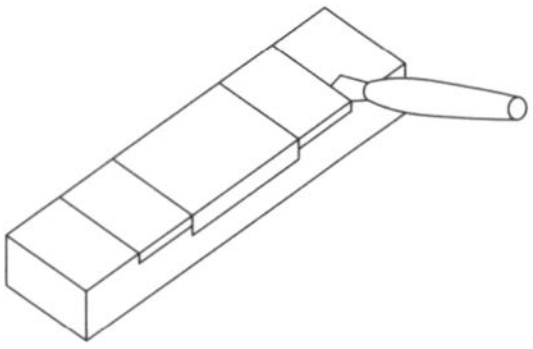

B1

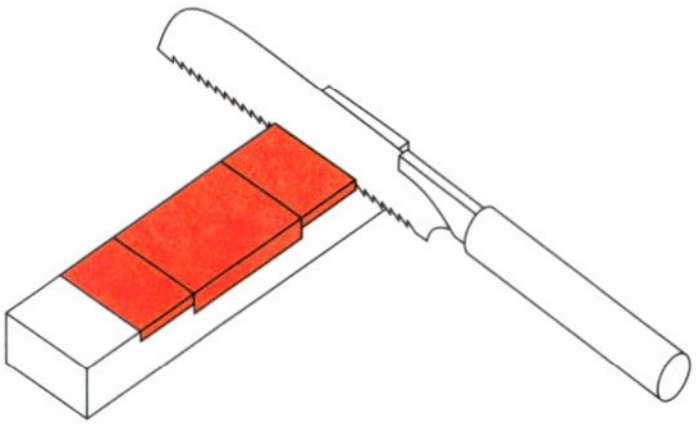

B2

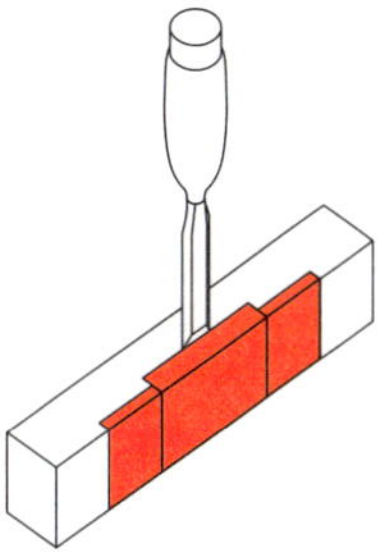

B3

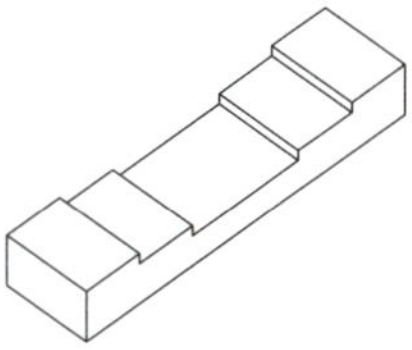

B

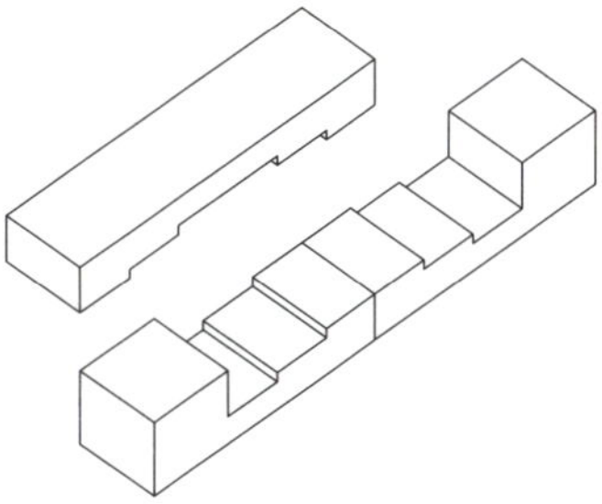

A+B+A

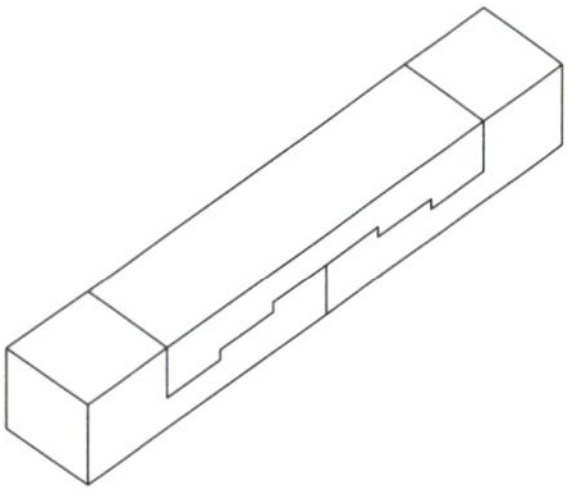

ABA

Splicing Joints

Stepped cogged lap joint with floating piece and folding wedges

yatoi-kabuse-naga-nidan-tsuki-tsuke-tsugi (kusabi-shime)
雇被長二段突付継ぎ（楔締）

verkeiltes gestuftes Hakenblatt mit eingesetztem Hakenstück

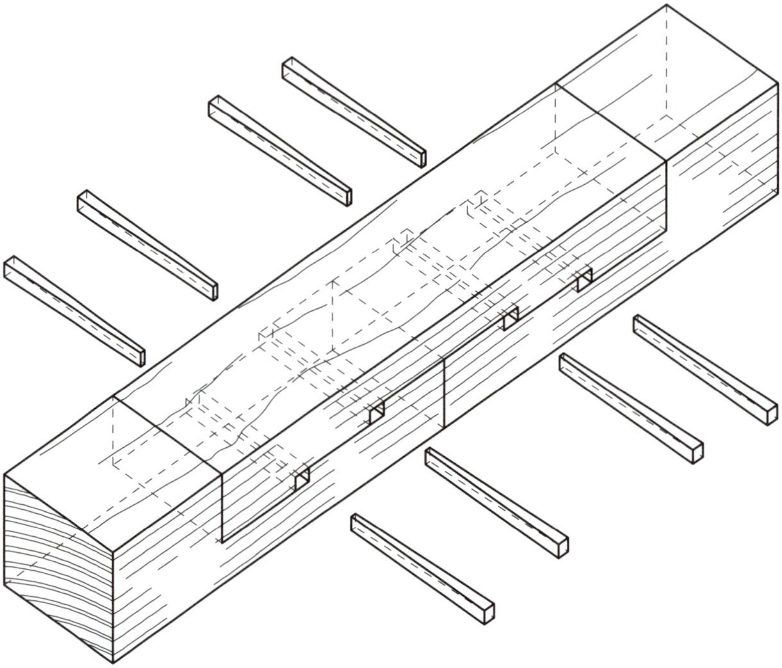

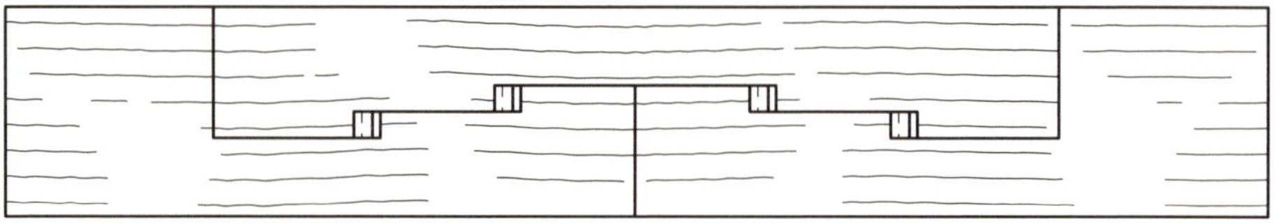

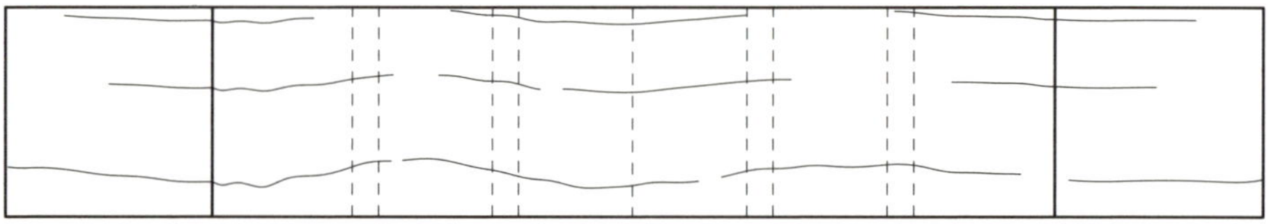

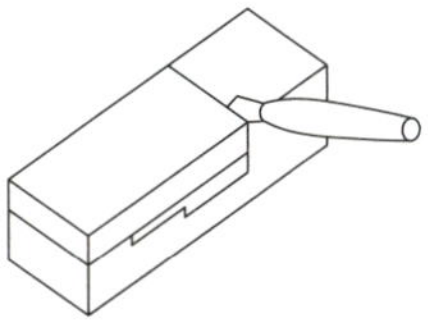

A1

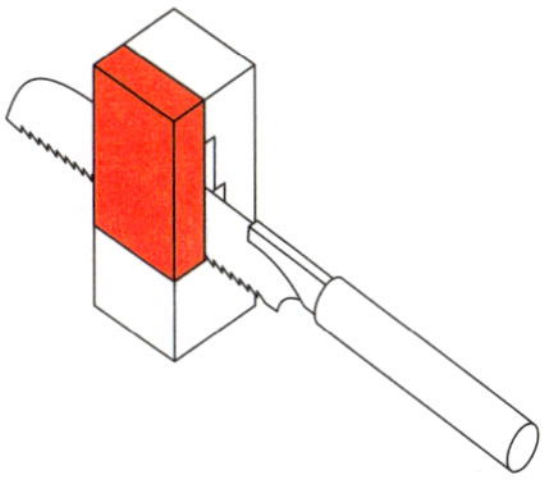

A2

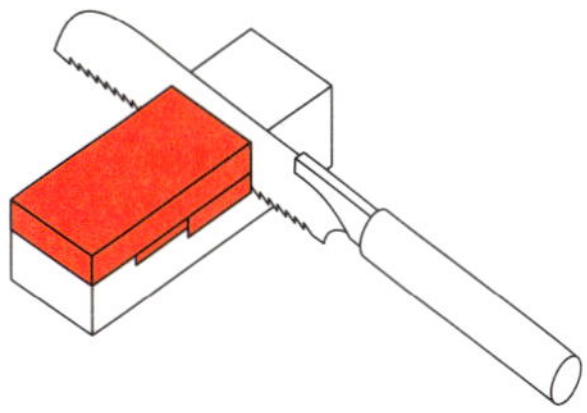

A3

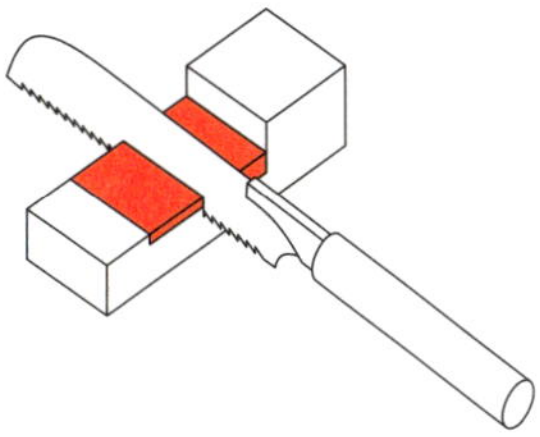

A4

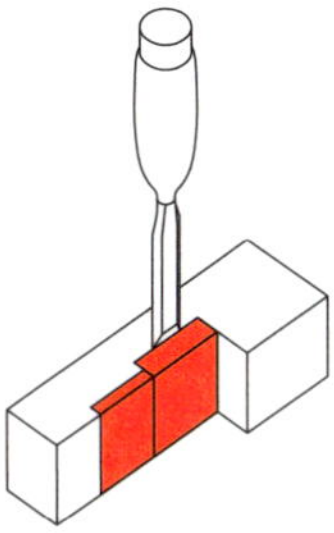

A5

2×

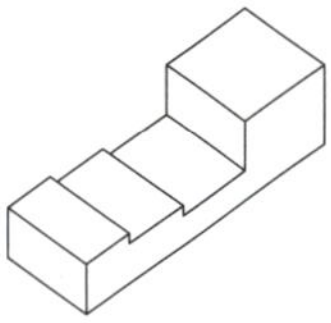

A

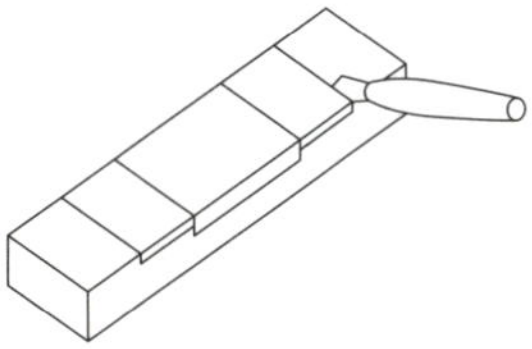

B1

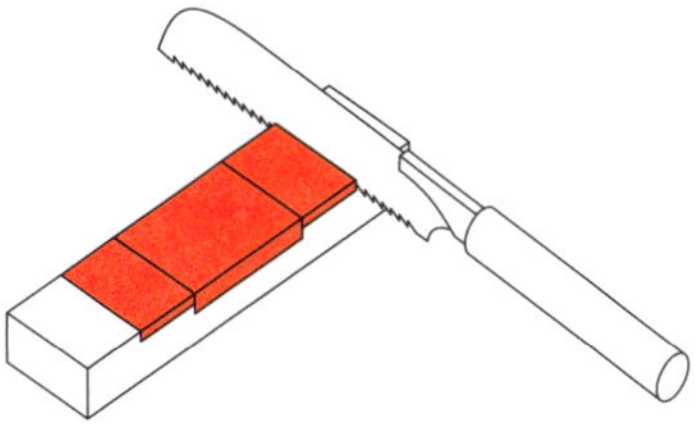

B2

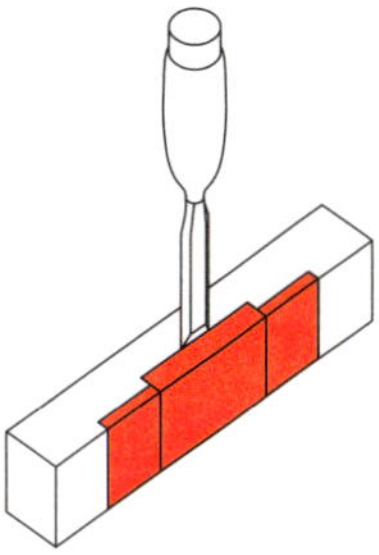

B3

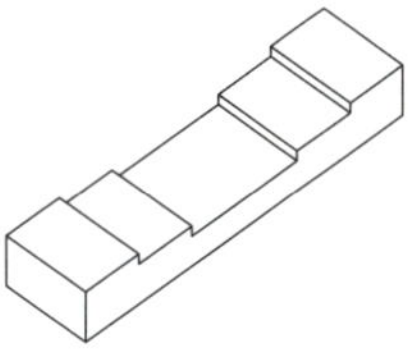

B

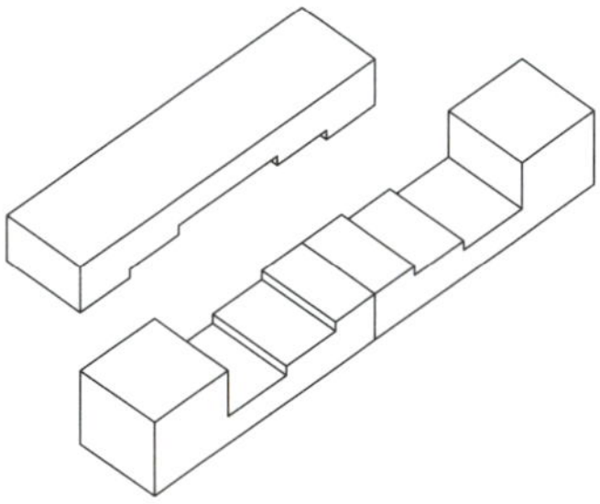

A+B+A

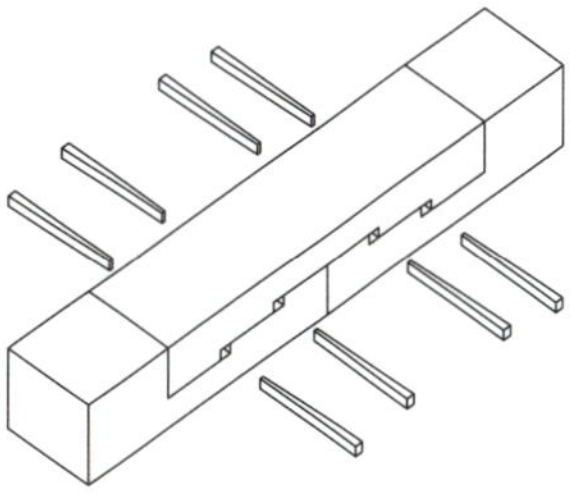

A+B+A

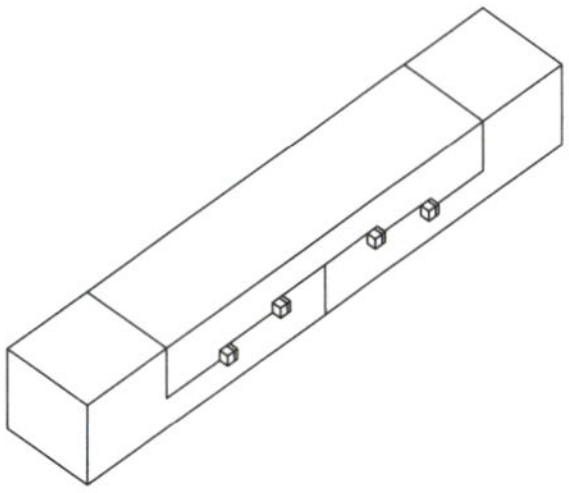

ABA

Splicing Joints

Undersquinted stepped cogged lap joint with floating piece and folding wedges

Japanese

yatoi-kabuse-naga-nidan-tsuki-tsuke-tsugi (sogi-tsuki, kusabi-shime)
雇被長二段突付継ぎ(殺付·楔締)

German

verkeiltes gestuftes Hakenblatt mit hinterschnittener Brüstung und eingesetztem Hakenstück

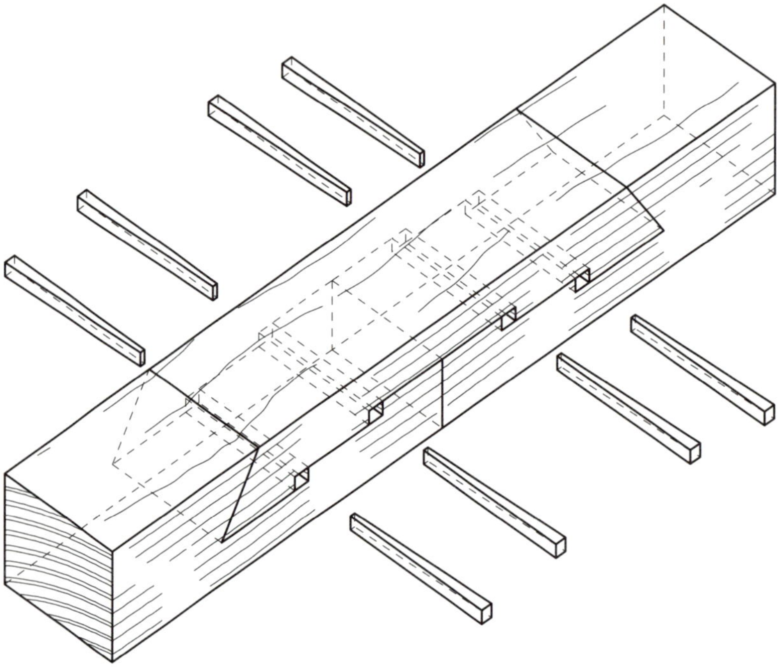

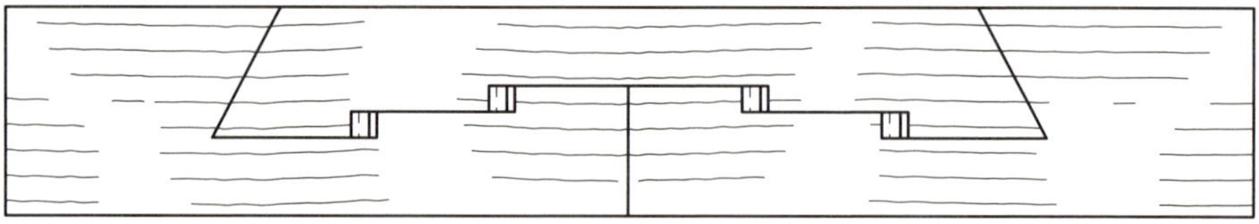

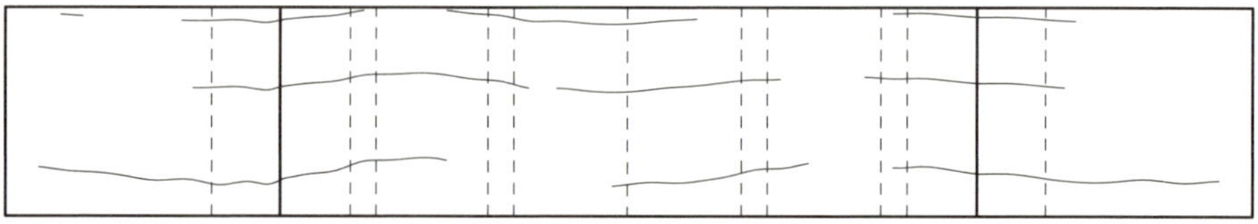

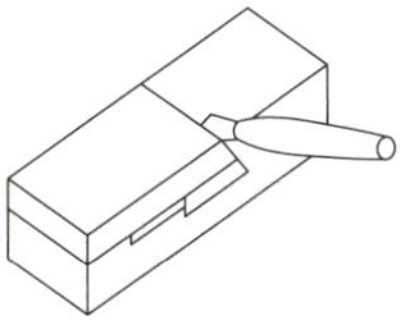
A1

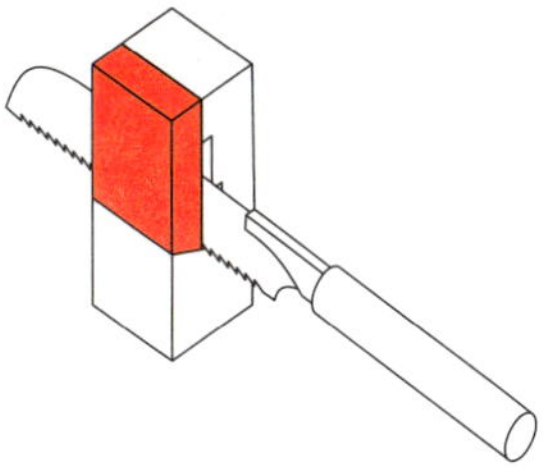
A2

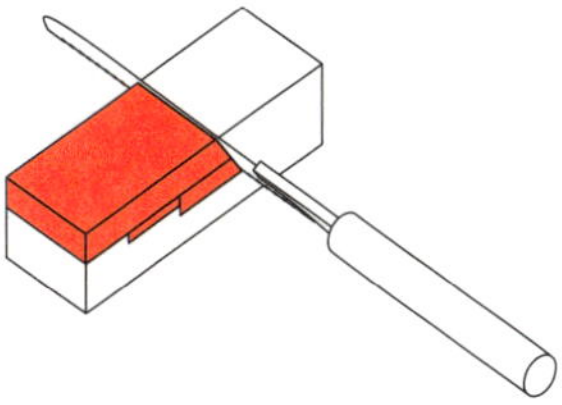
A3

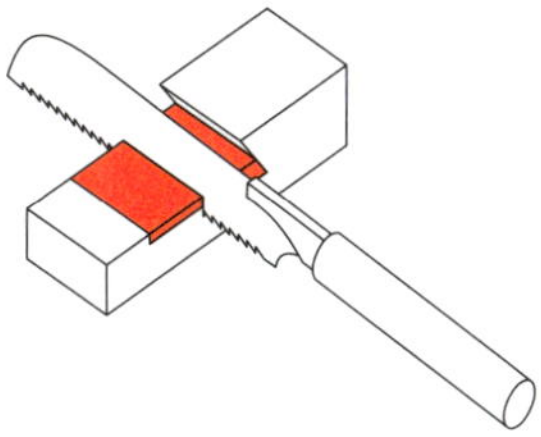
A4

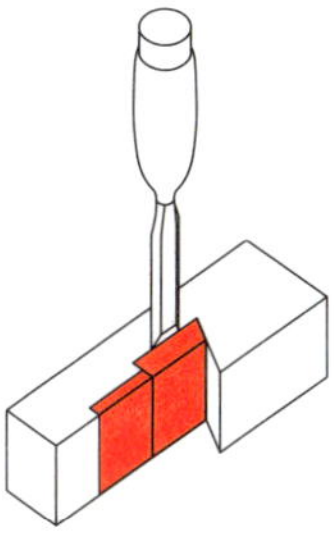
A5

2×

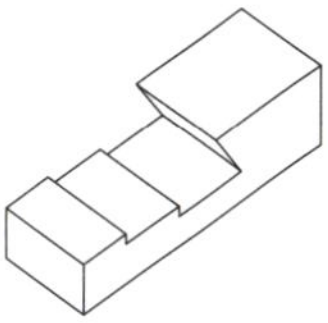
A

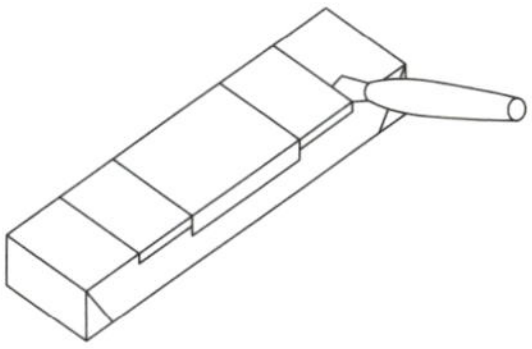
B1

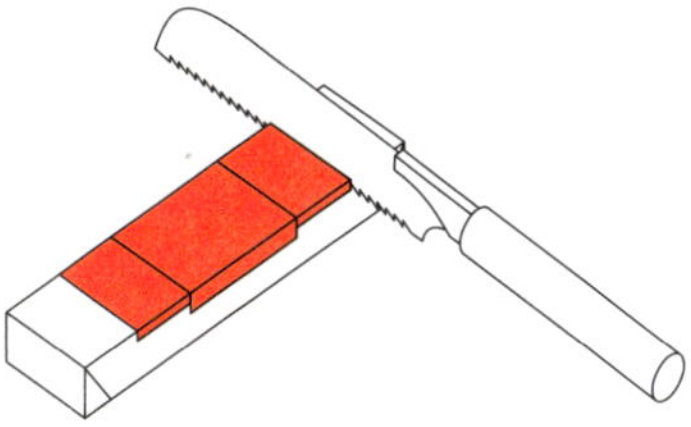
B2

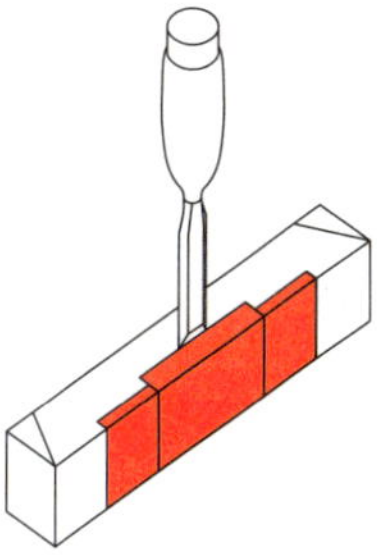
B3

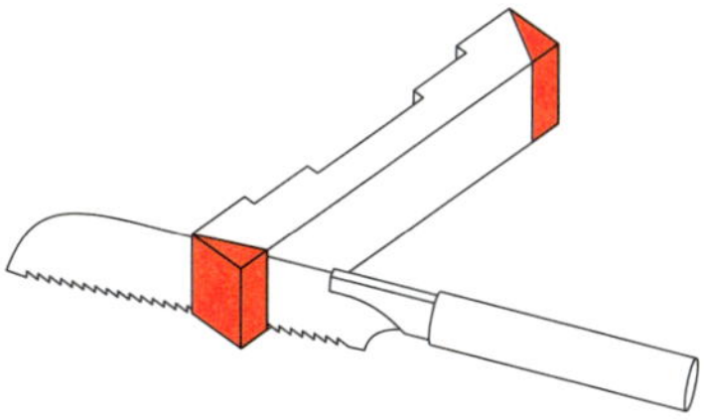
B4

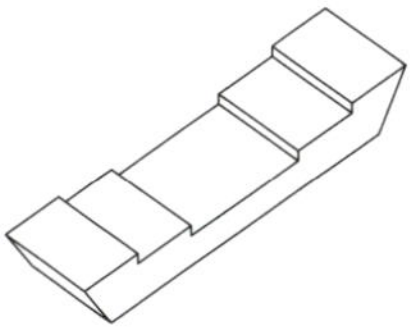
B

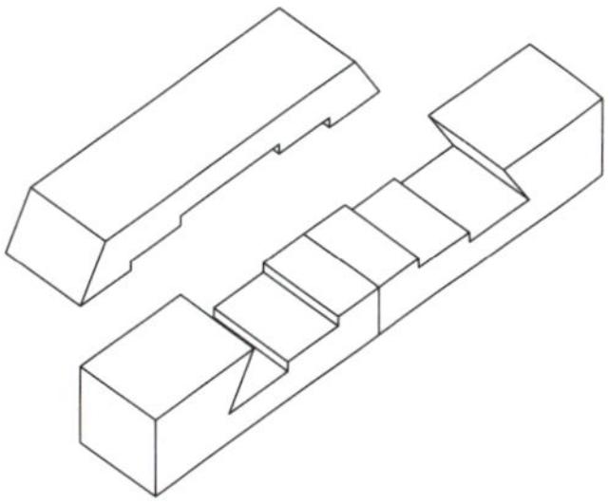
A+B+A

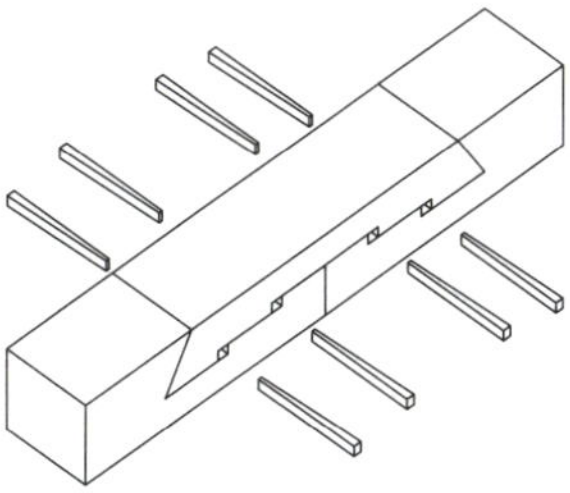
A+B+A

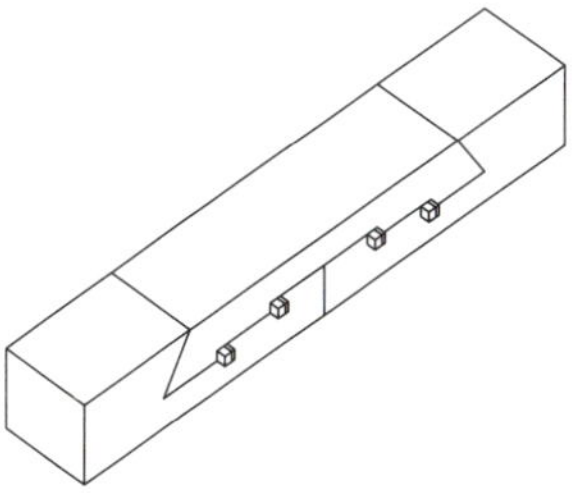
ABA

Splicing Joints

Triple-cogged lap joint

naga-dan-koshi-kake-ai-kaki-tsugi
長段腰掛相欠継ぎ

gerades Hakenblatt
mit dreifachem Haken

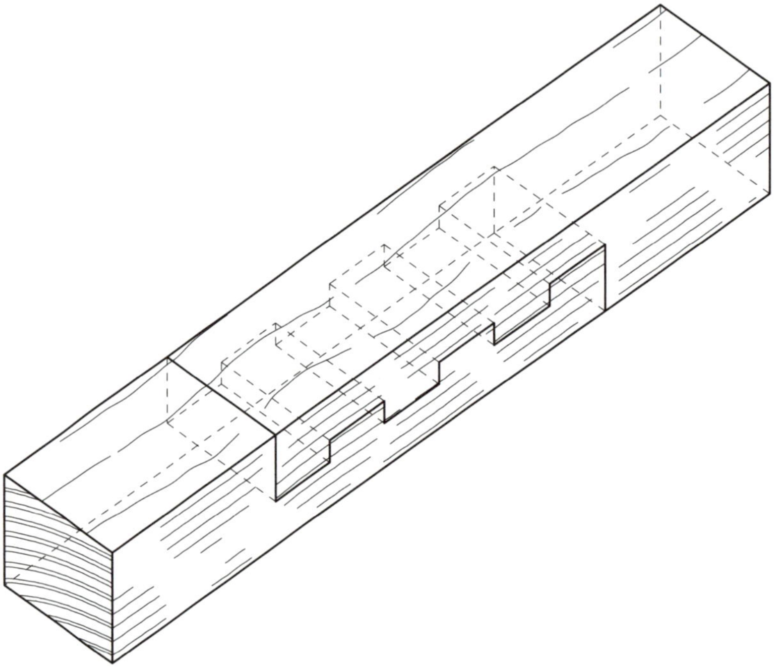

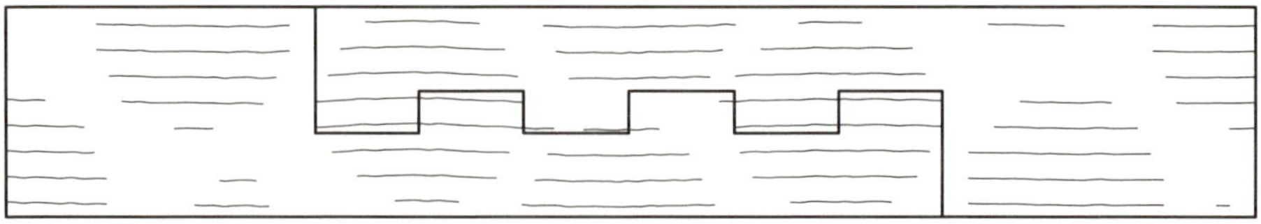

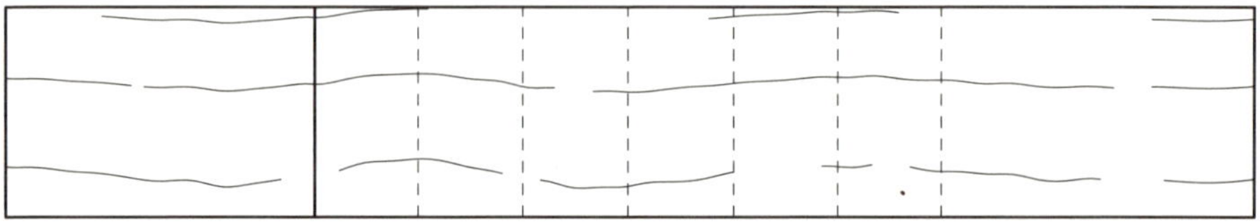

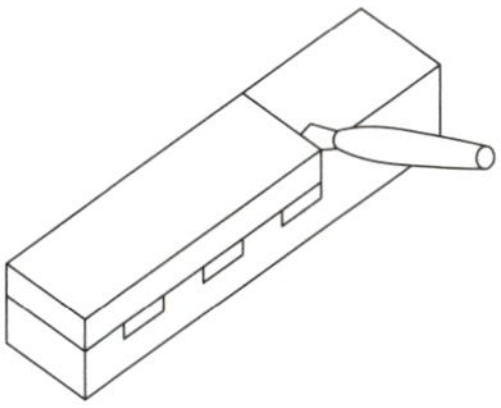

A1

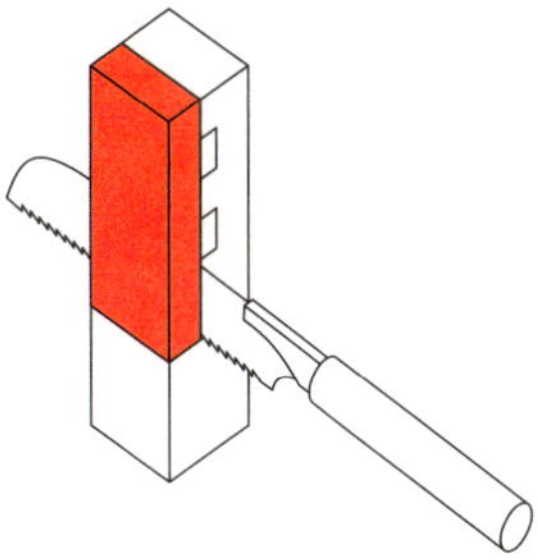

A2

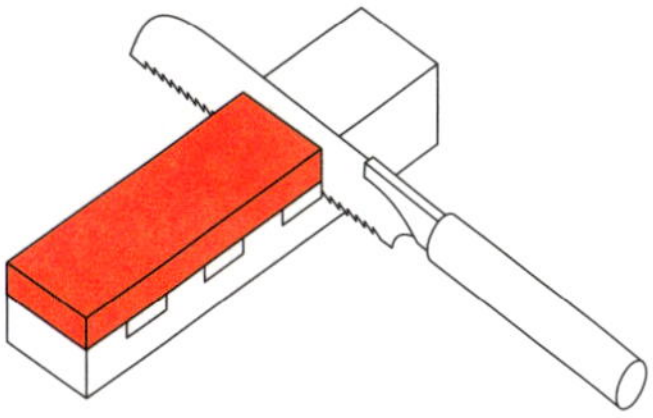

A3

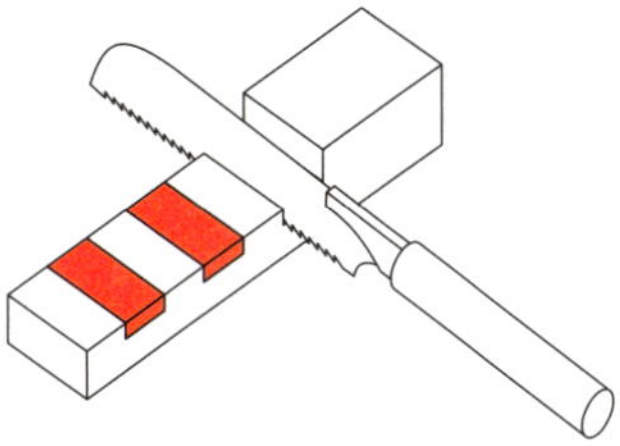

A4

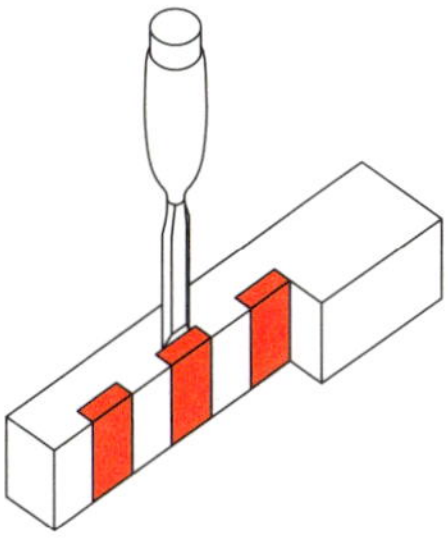

A5

2×

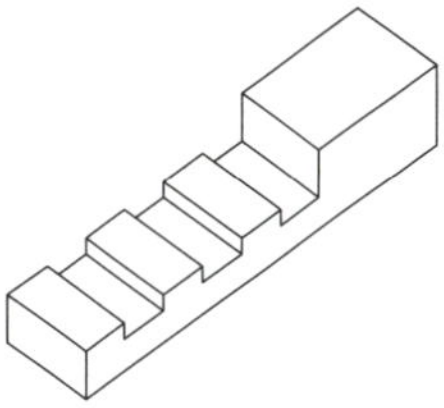

A

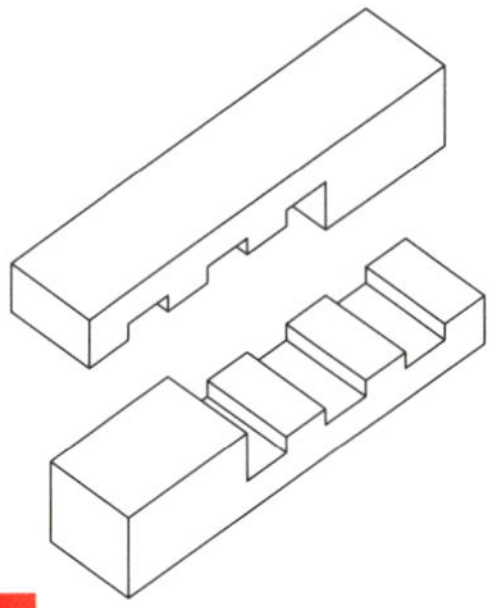

A+A

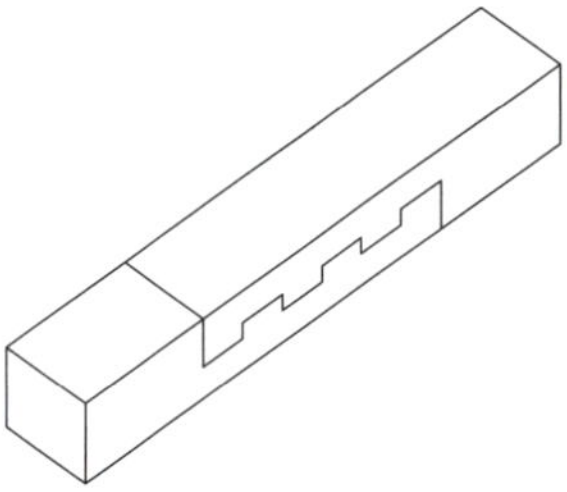

AA

Splicing Joints

Rabbeted half lap joint

naga-koshi-kake-mechi-tsugi
長腰掛目違継ぎ

gerades Blatt mit Steckfalz

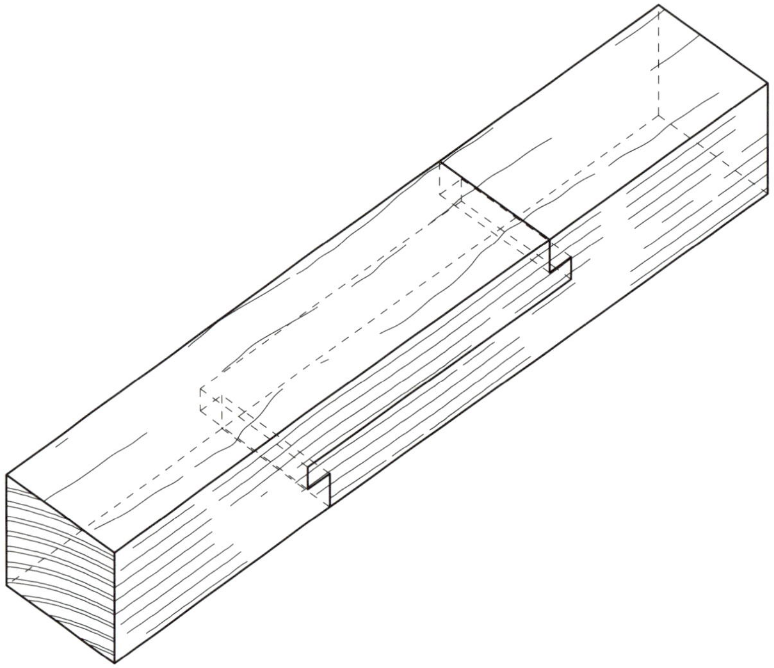

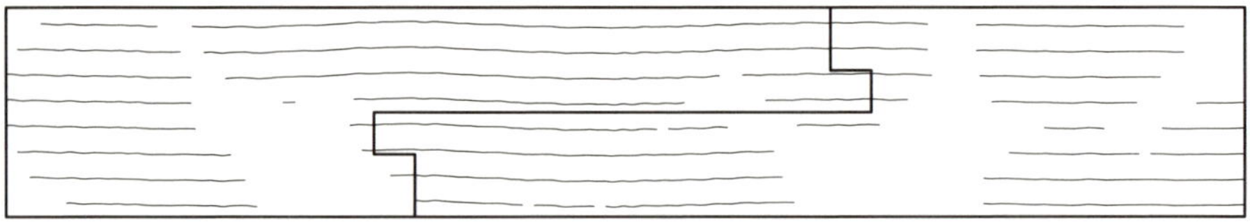

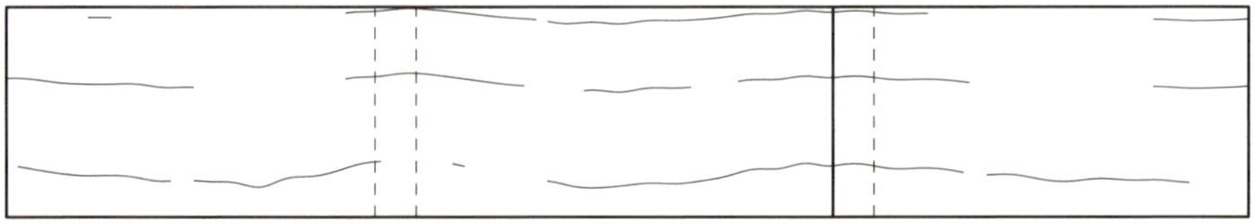

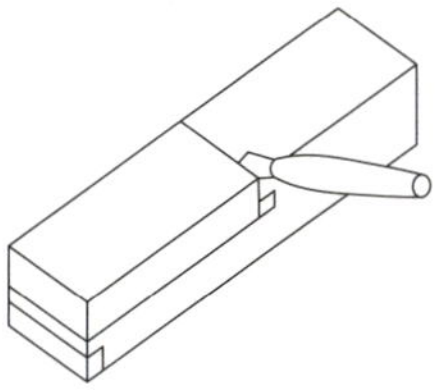
A1

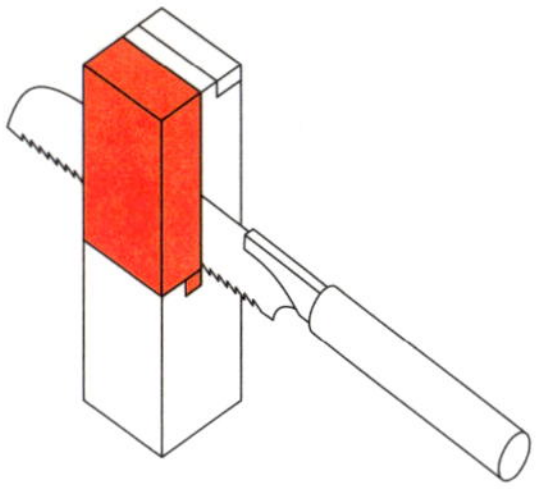
A2

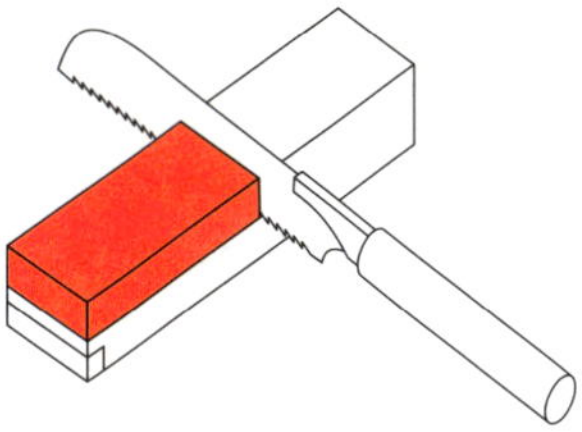
A3

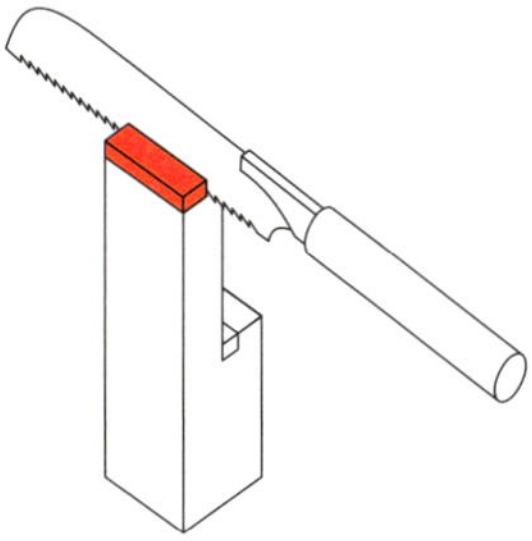
A4

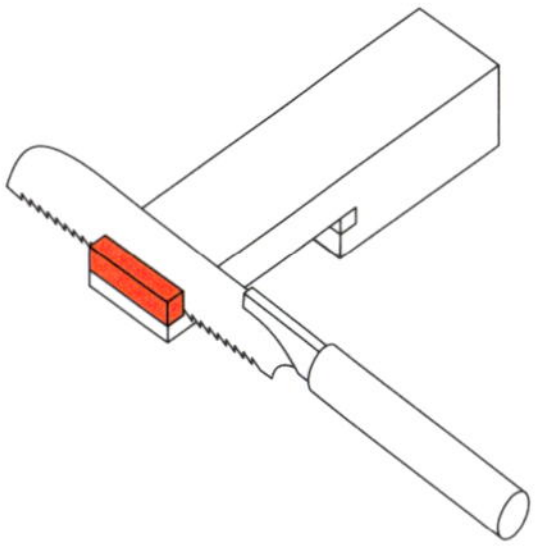
A5

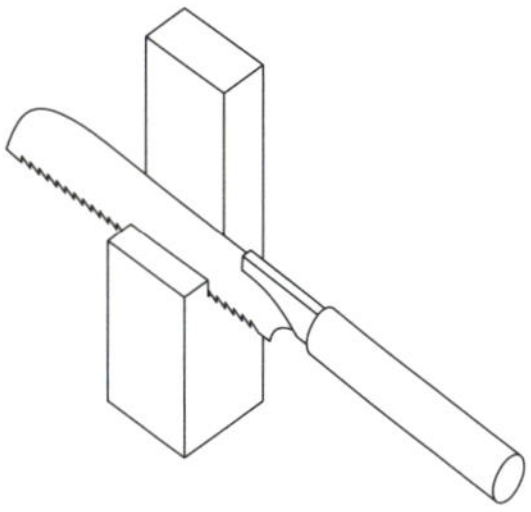
A6

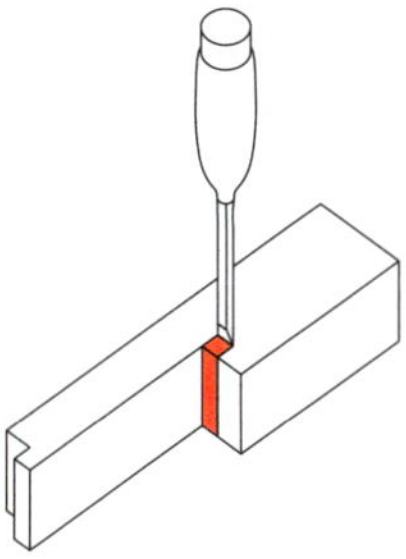
A7

2×

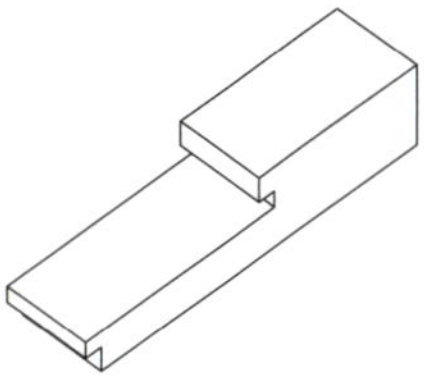
A

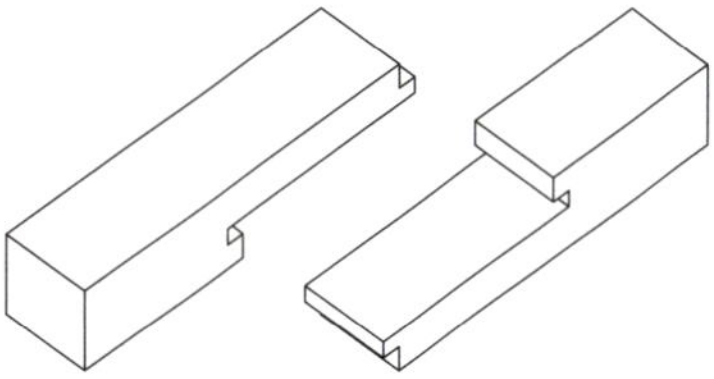
A+A

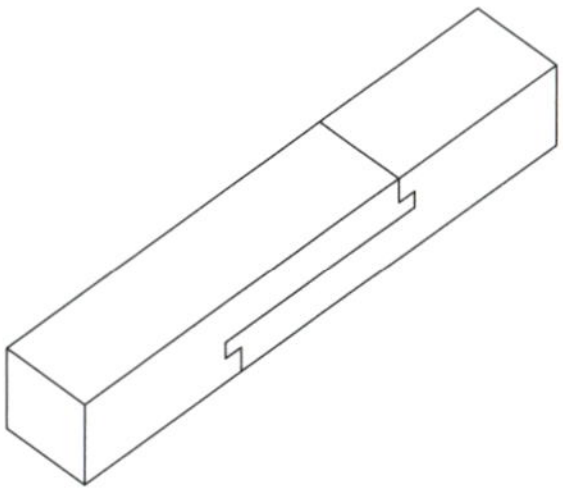
AA

Splicing Joints

Rabbeted cogged lap joint

naga-dan-koshi-kake-mechi-tsugi
長段腰掛目違継ぎ

gerades Hakenblatt mit Steckfalz

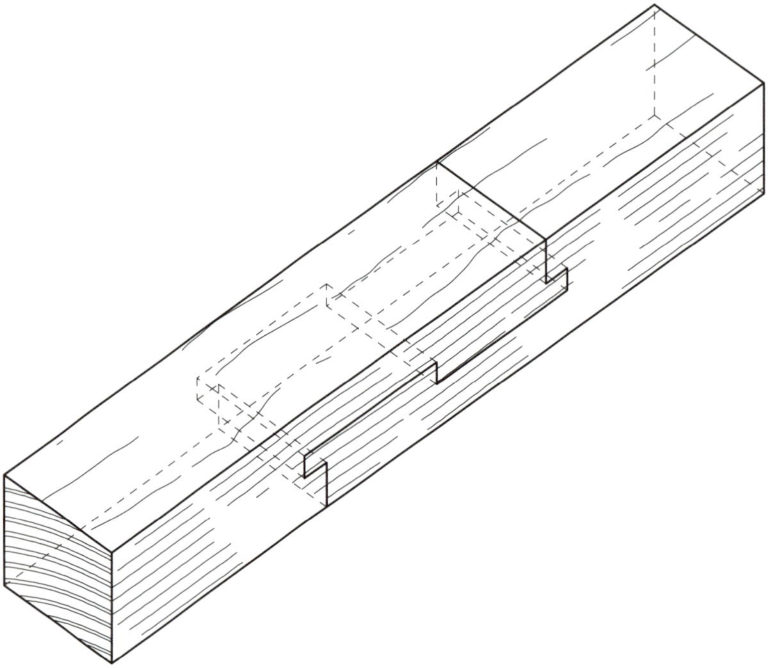

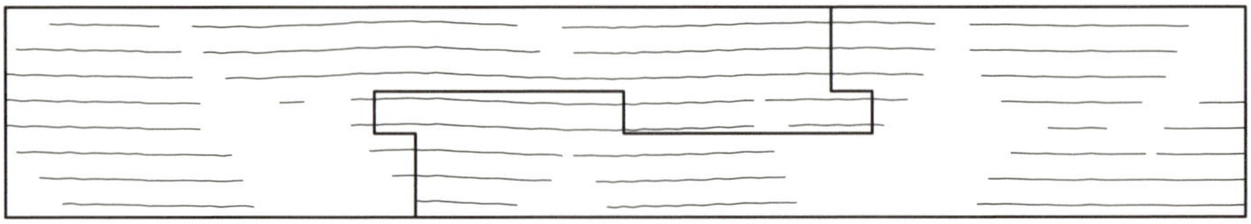

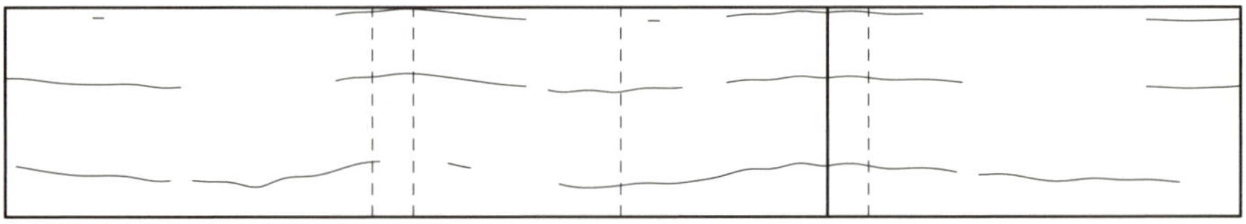

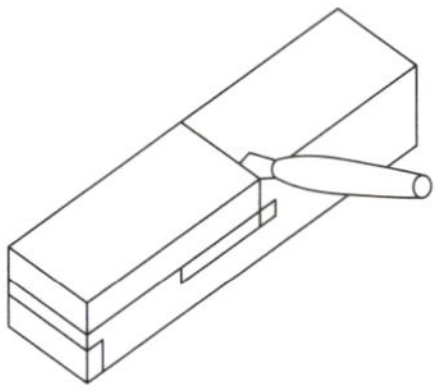

A1

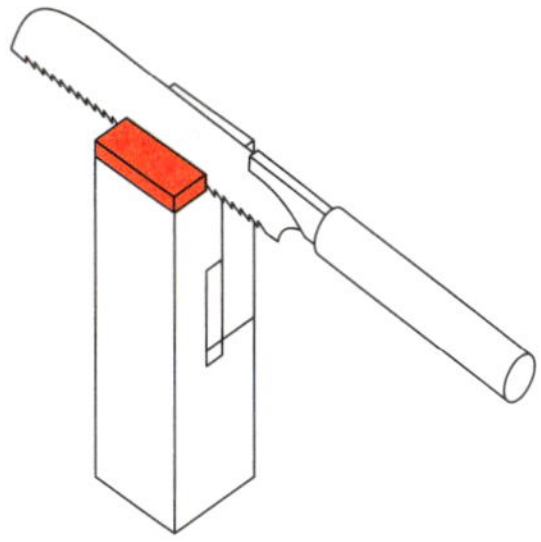

A2

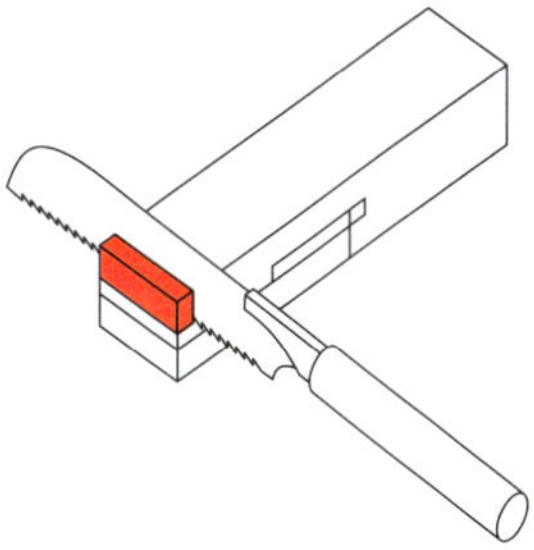

A3

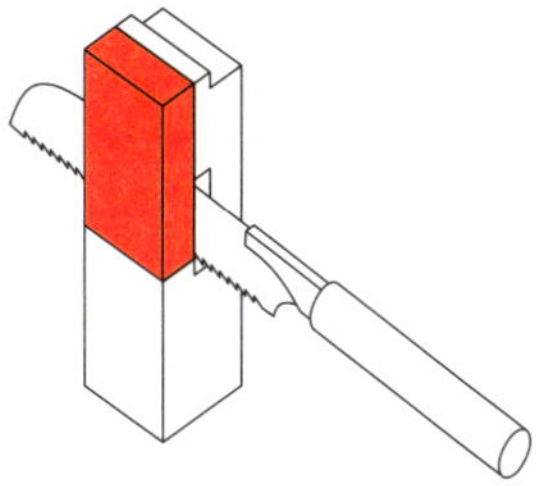

A4

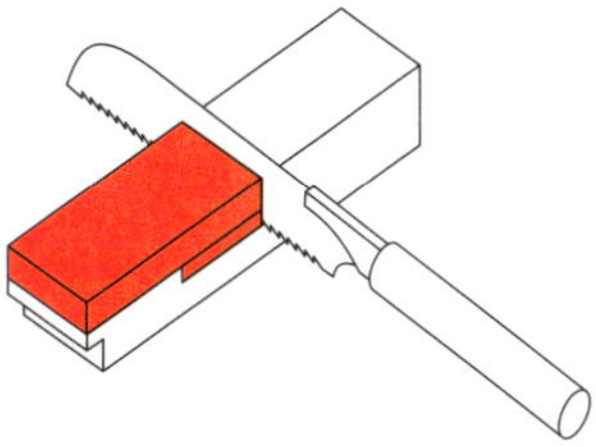

A5

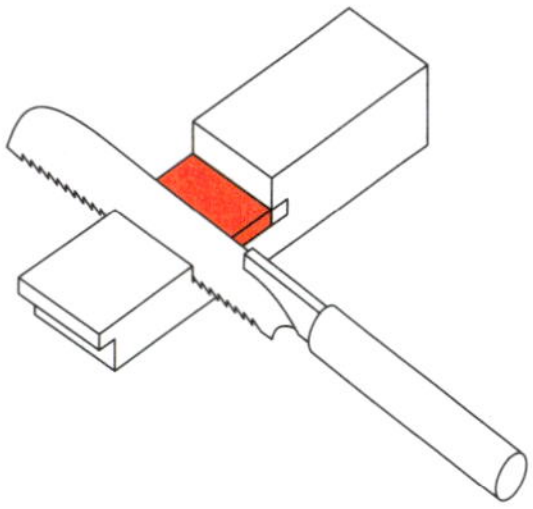

A6

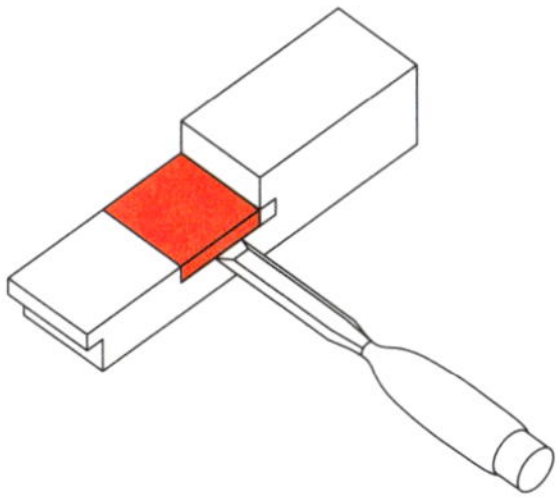

A7

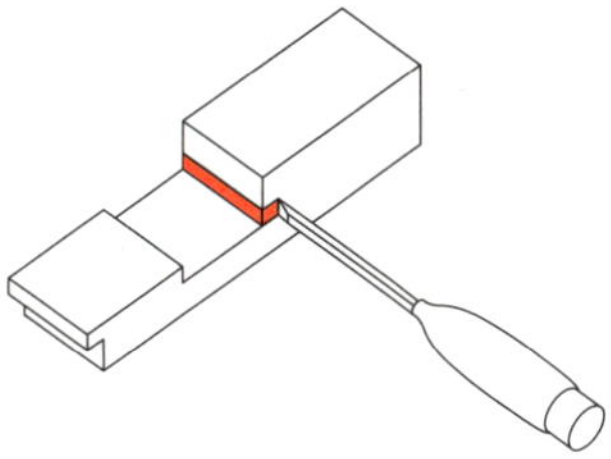

A8

2×

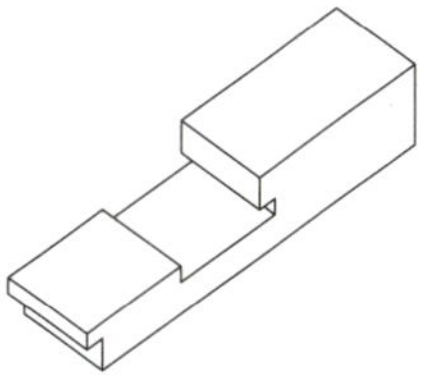

A

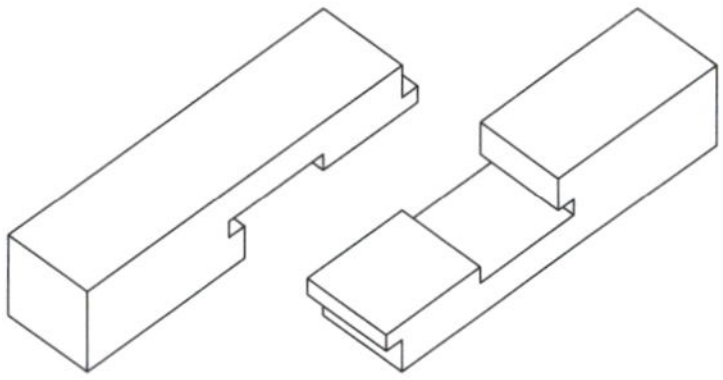

A+A

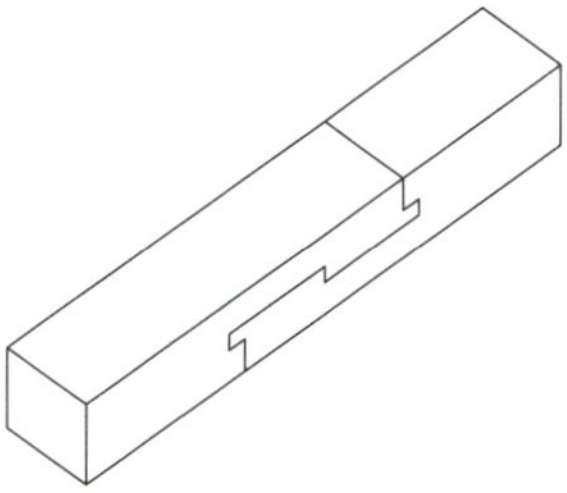

AA

Splicing Joints

Rabbeted cogged lap joint with folding wedges

naga-dan-koshi-kake-mechi-tsugi (kusabi-shime)
長段腰掛目違継ぎ(楔締)

verkeiltes gerades Hakenblatt mit Steckfalz

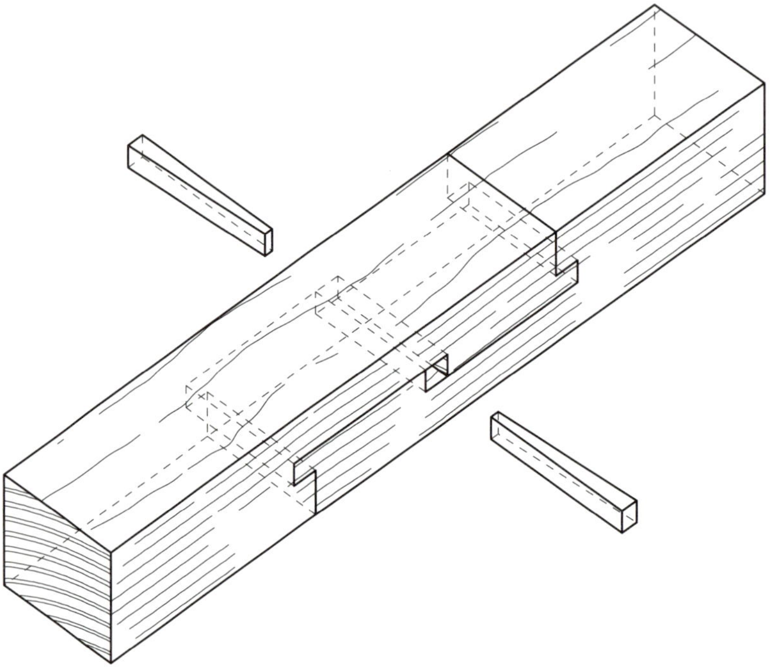

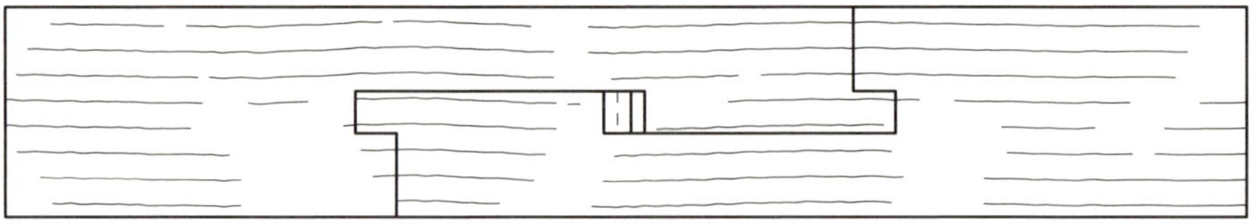

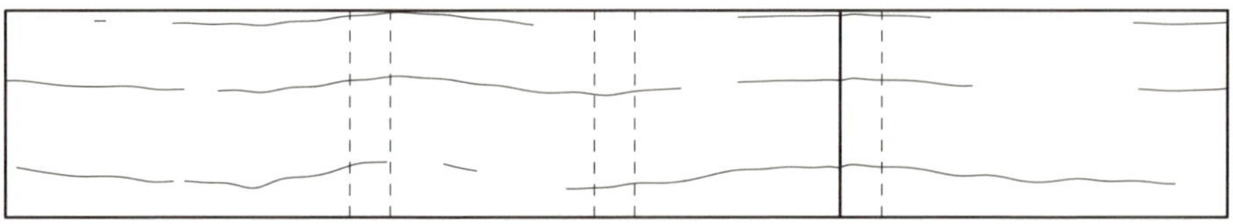

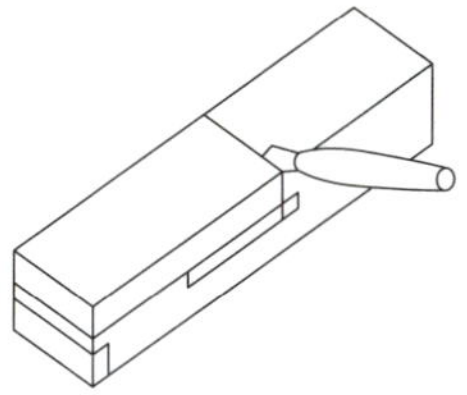

A1

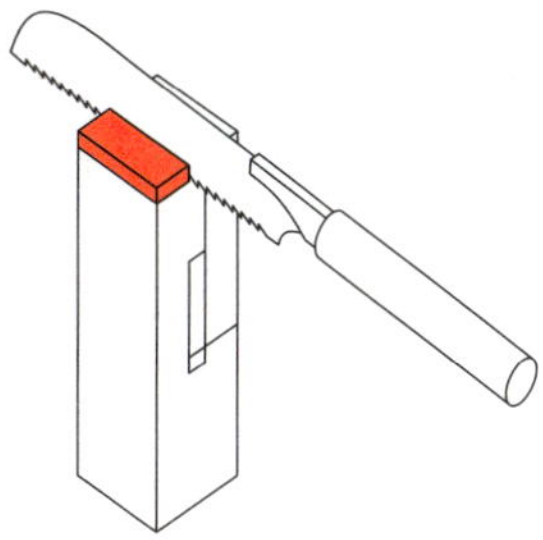

A2

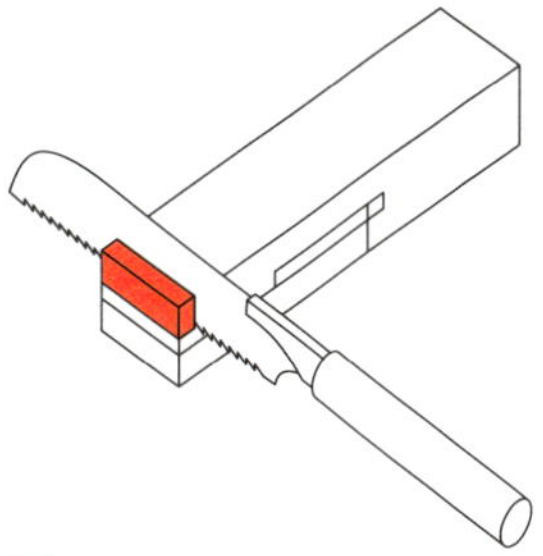

A3

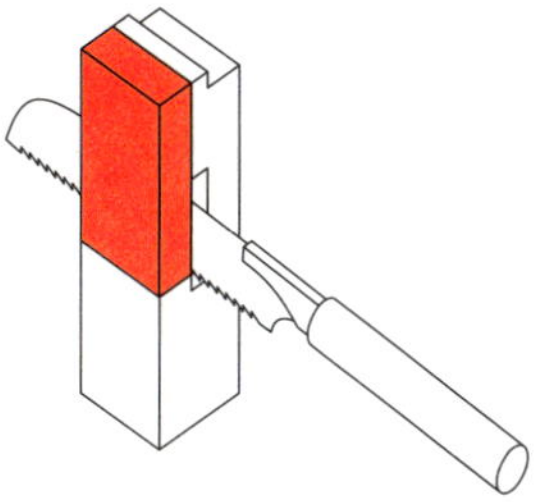

A4

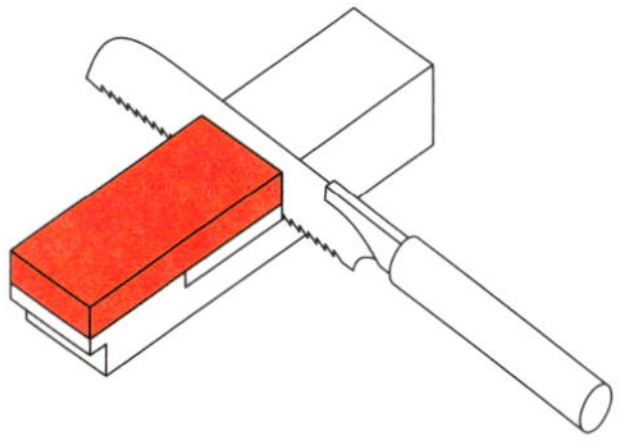

A5

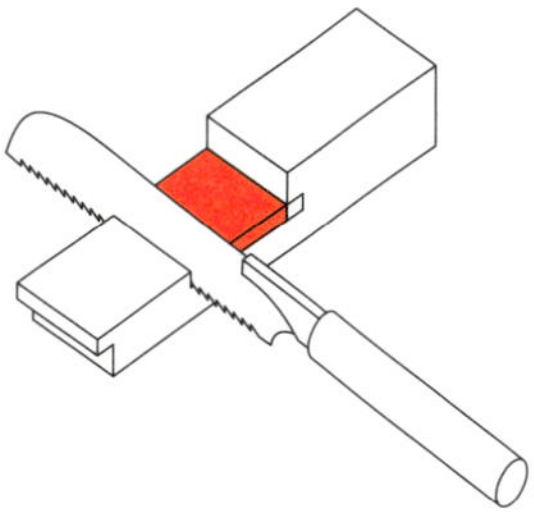

A6

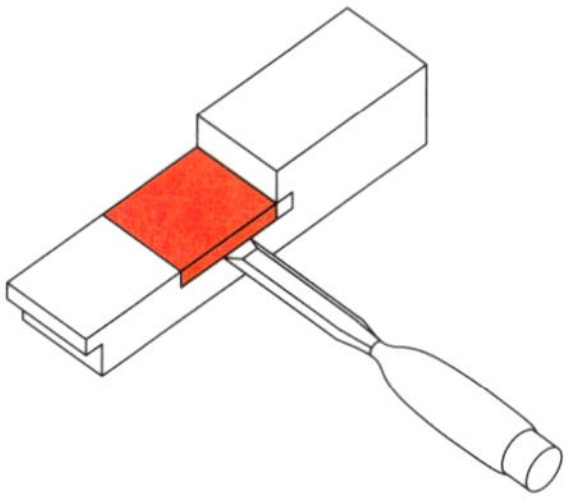

A7

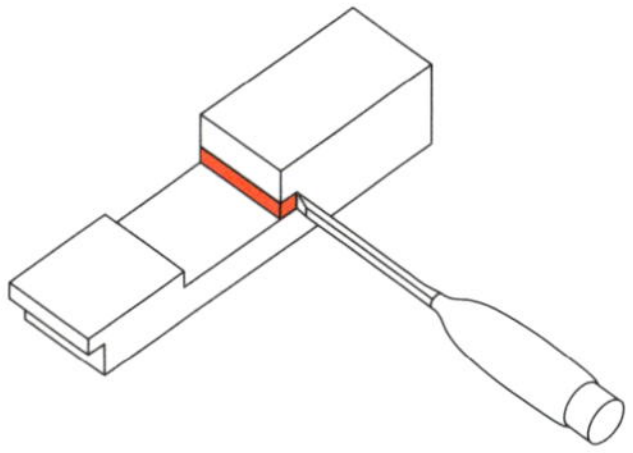

A8

2×

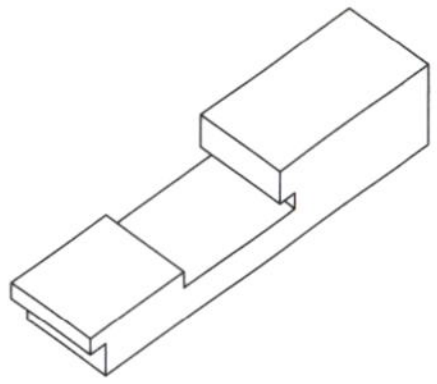

A

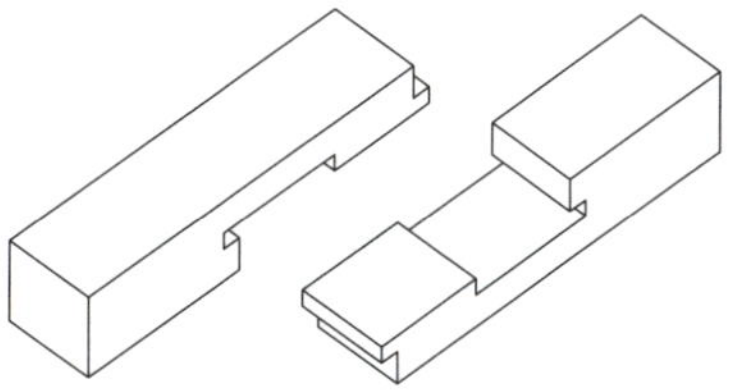

A+A

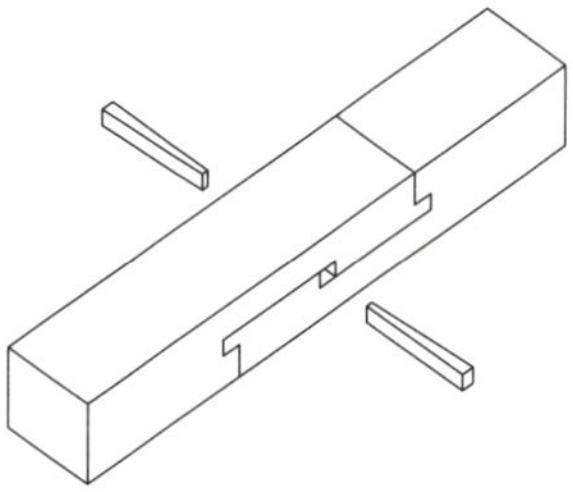

A+A

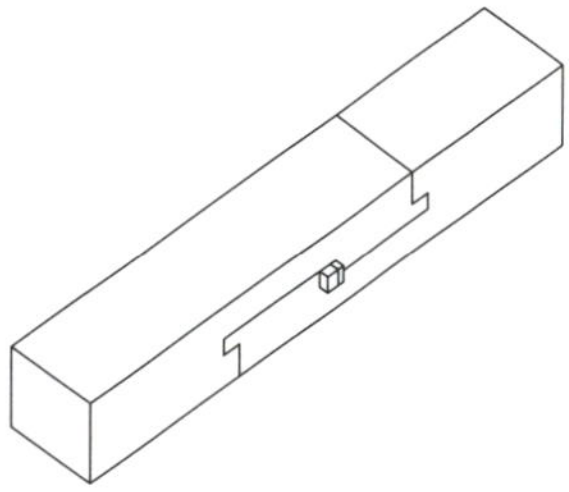

AA

Splicing Joints

Rabbeted cogged lap joint with stub tenons and folding wedges

naga-dan-koshi-kake-teiji-mechi-tsugi (kusabi-shime)
長段腰掛丁字目違継ぎ(楔締)

verkeiltes gerades Hakenblatt mit Steckfalz und Brüstungszapfen

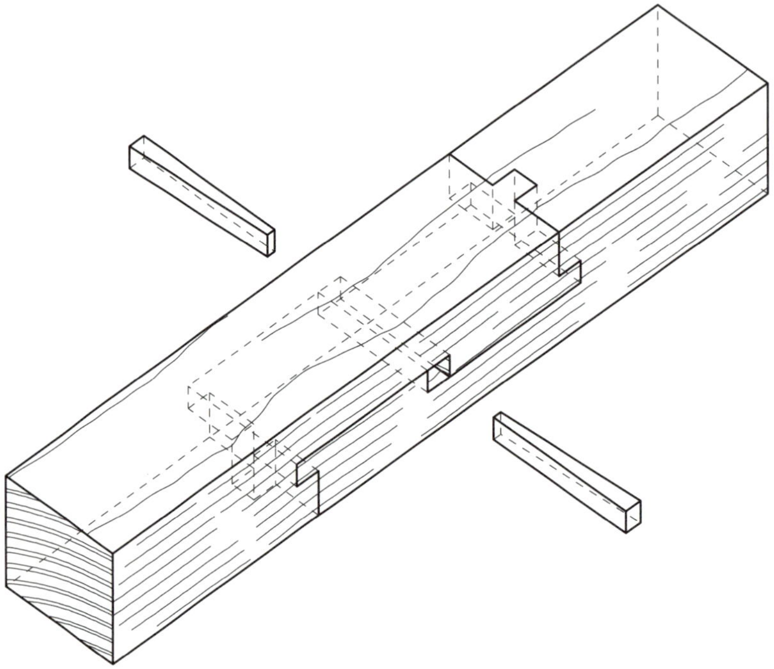

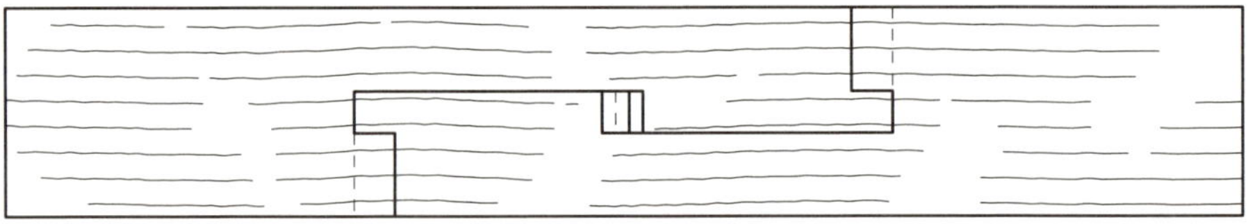

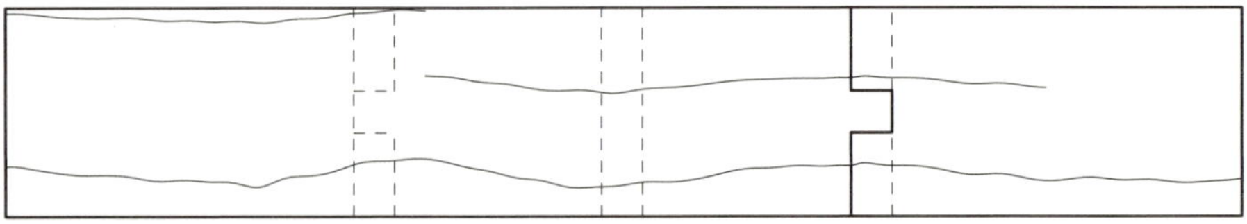

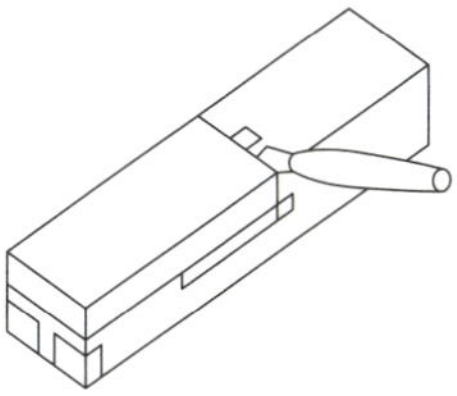

A1

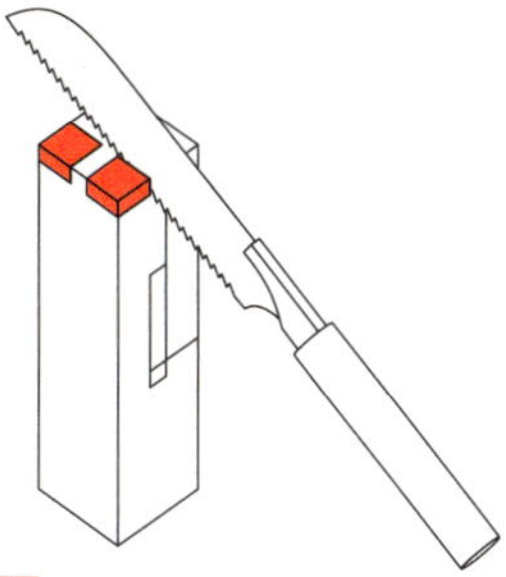

A2

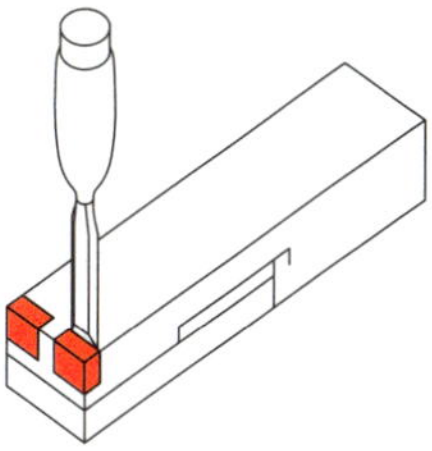

A3

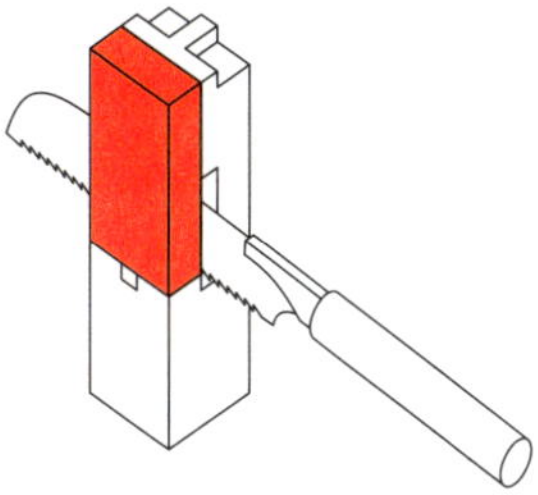

A4

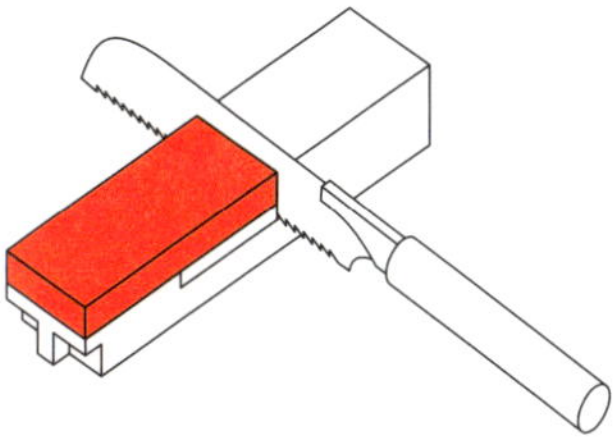

A5

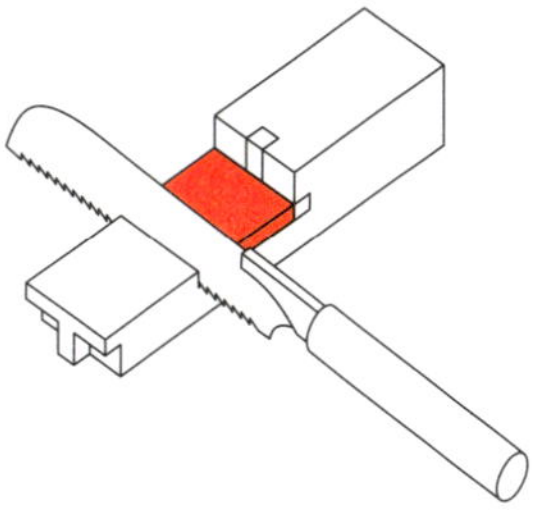

A6

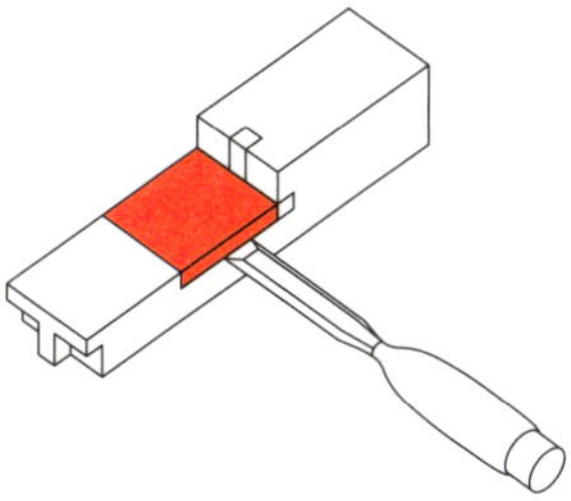

A7

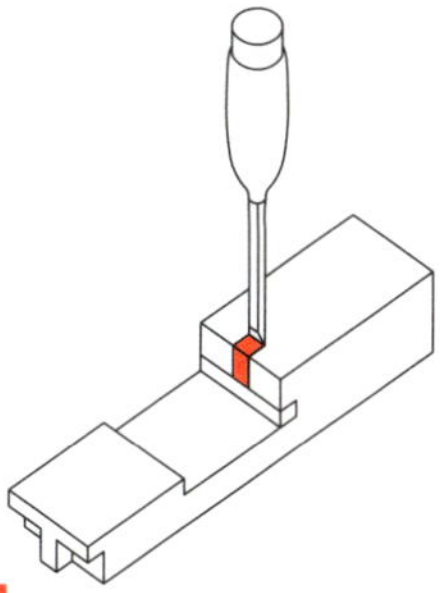

A8

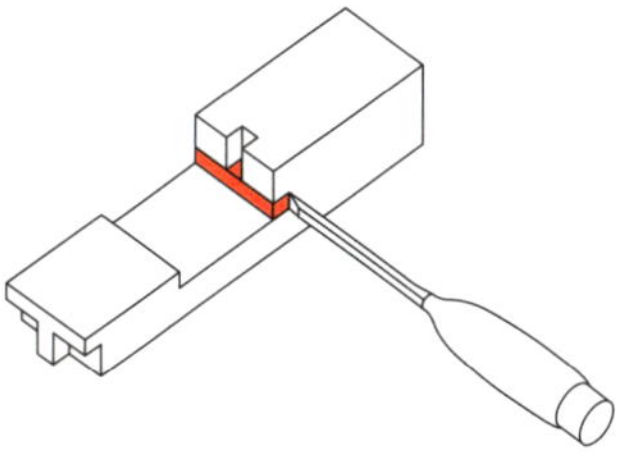

A9

2×

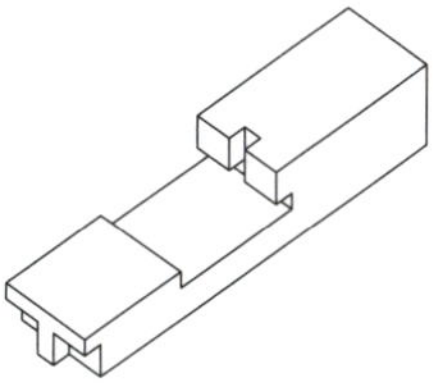

A

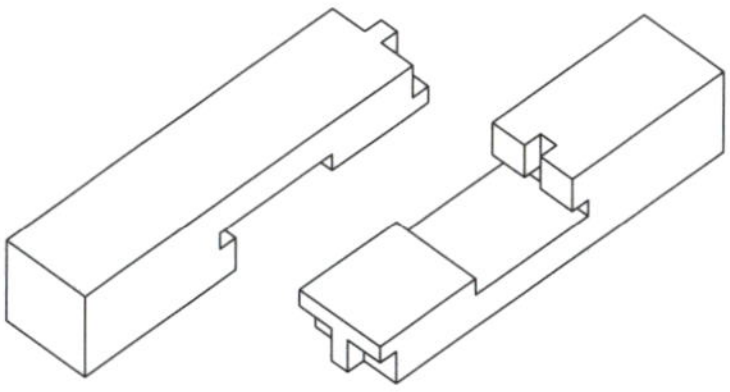

A+A

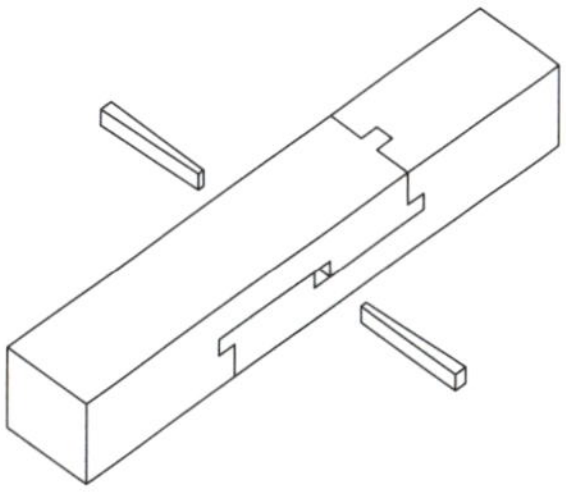

A+A

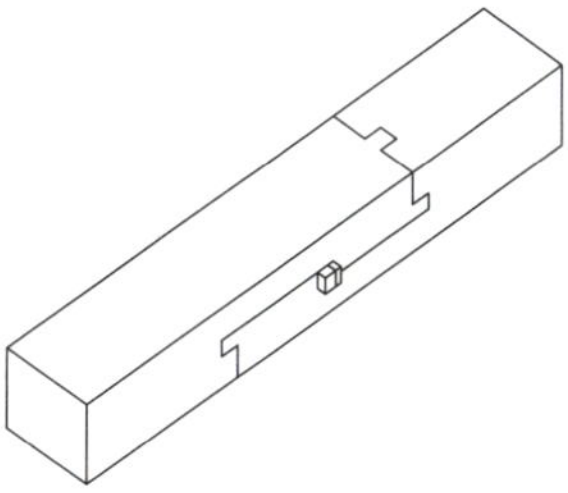

AA

Splicing Joints

Half-blind rabbeted lap joint with tapered keys

hako-shachi-tsugi [*]
箱車知継ぎ

halbverdecktes gerades Blatt
mit Steckfalz und Spannkeilen

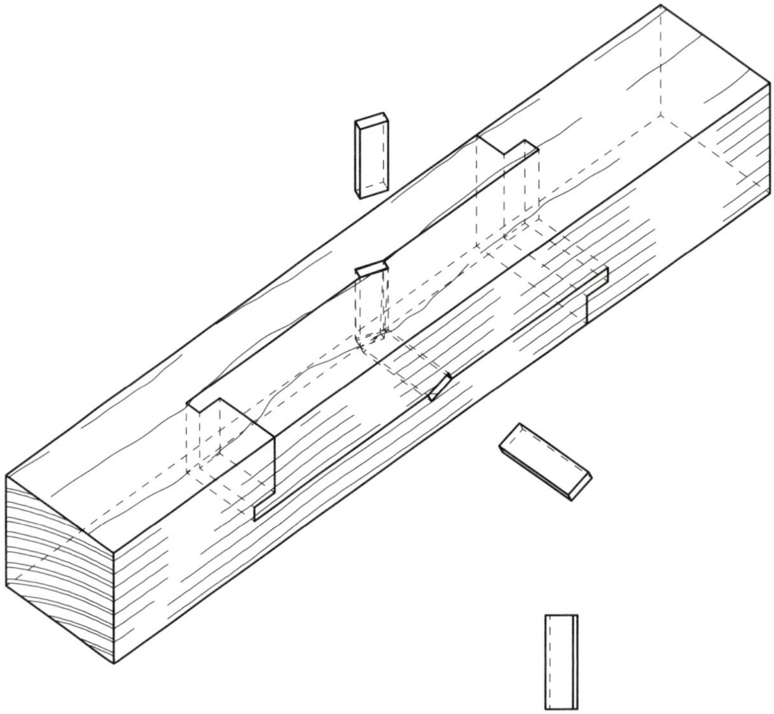

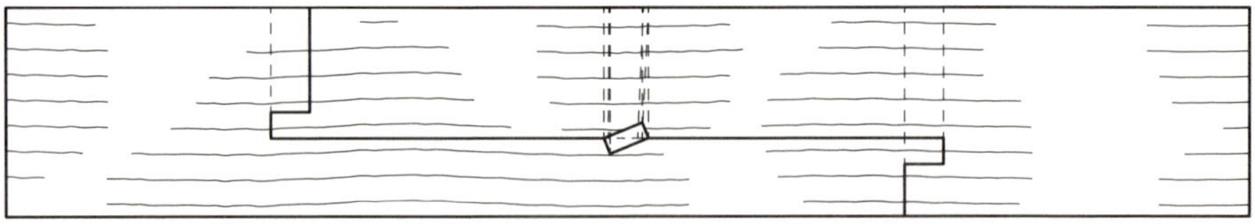

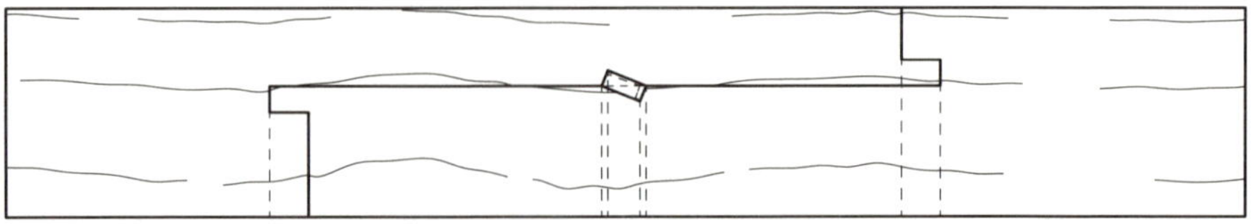

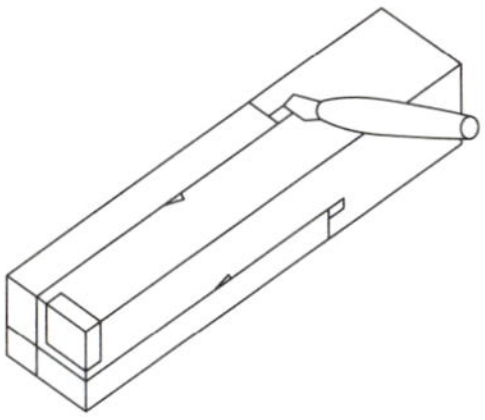
A1

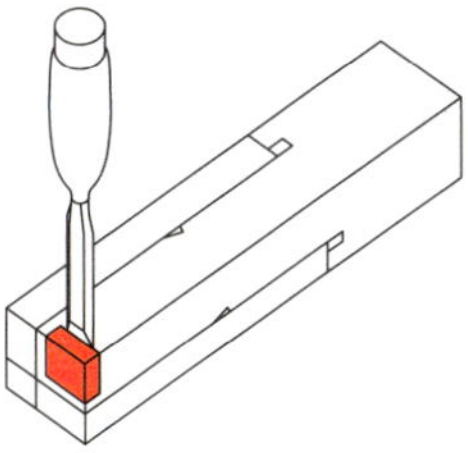
A2

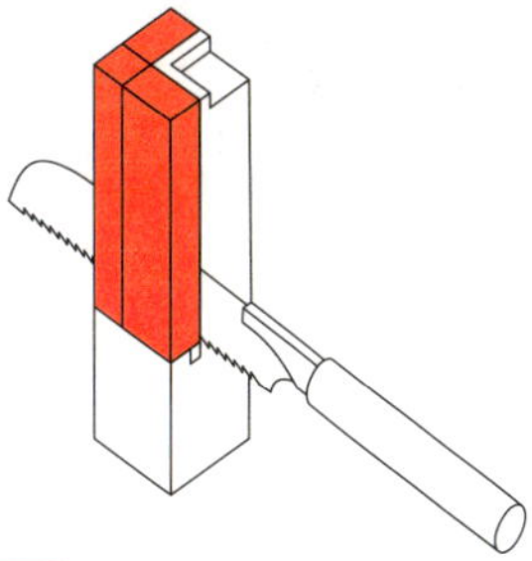
A3

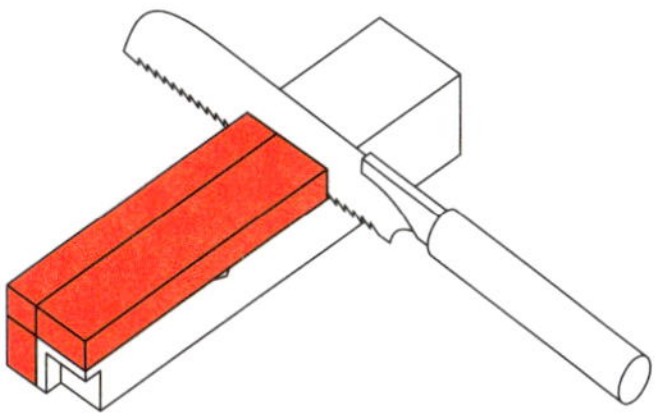
A4

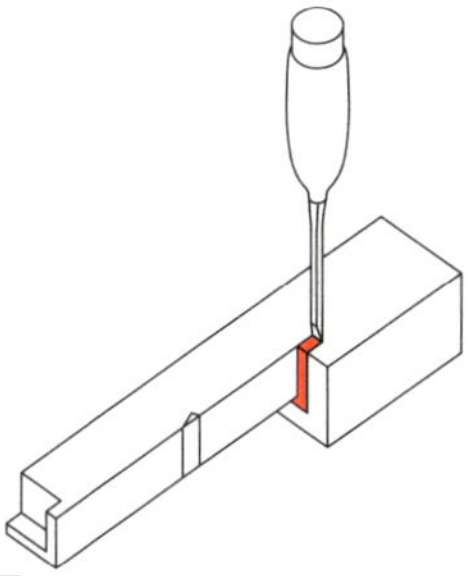
A5

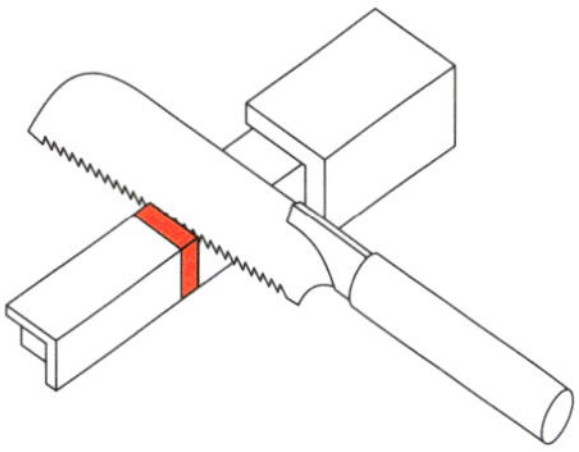
A6

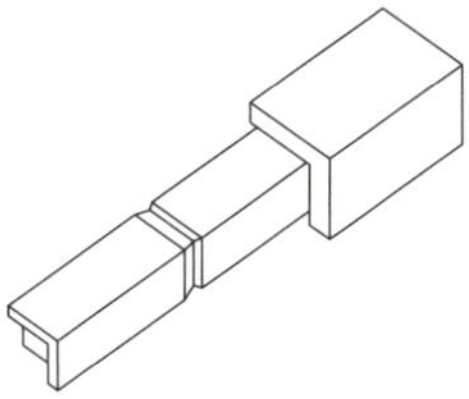
A

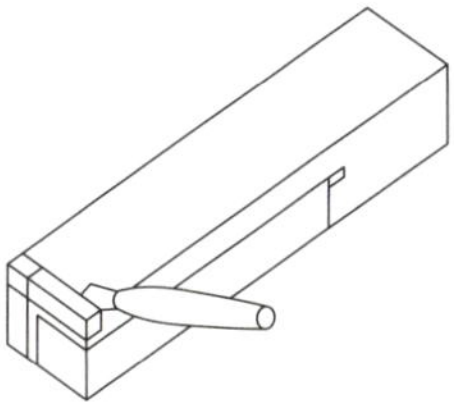
B1

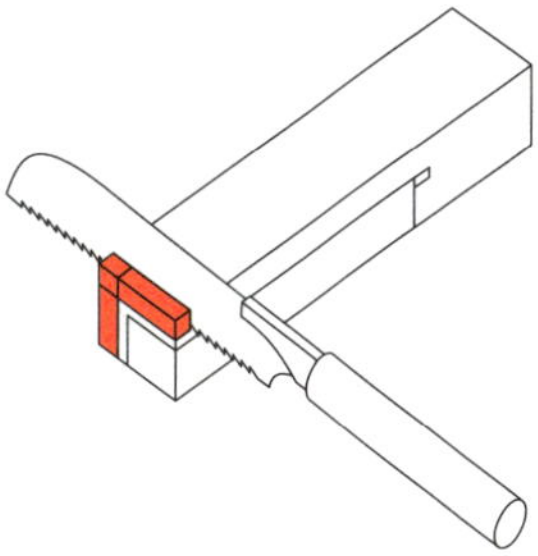
B2

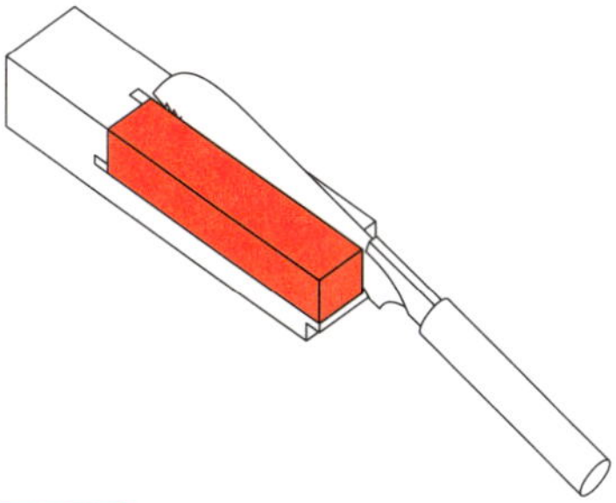
B3

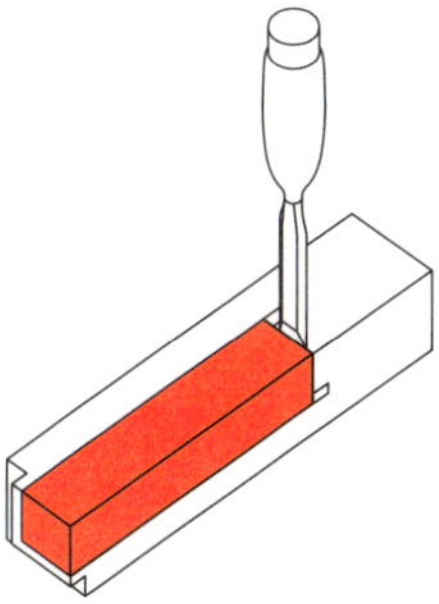
B4

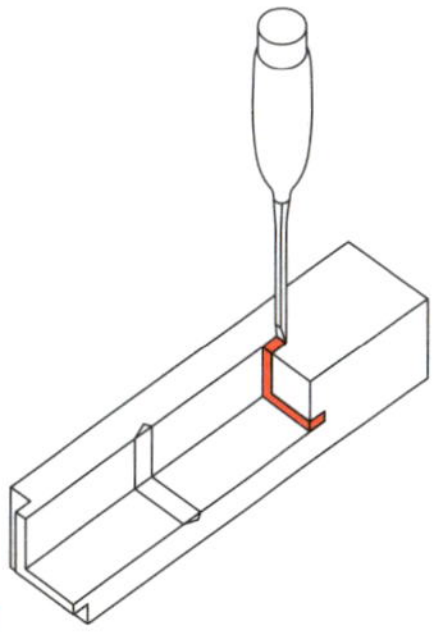
B5

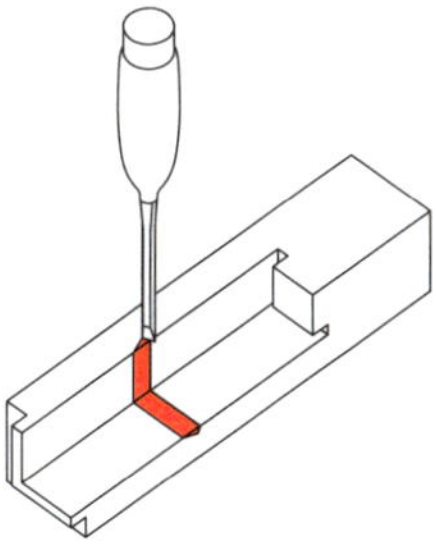
B6

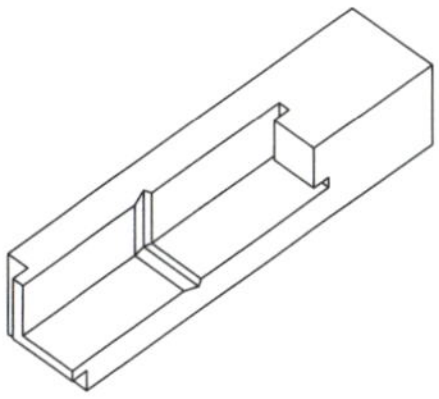
B

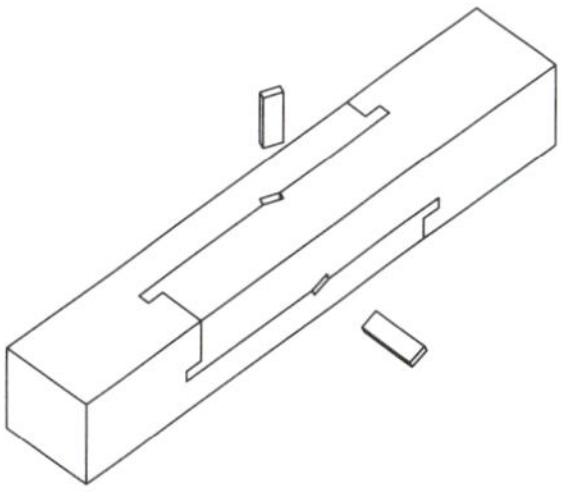
A+B

Splicing Joints

Stop-splayed scarf joint

tsuki-tsuke-sogi-tsugi
突付殺継ぎ

schräges Blatt

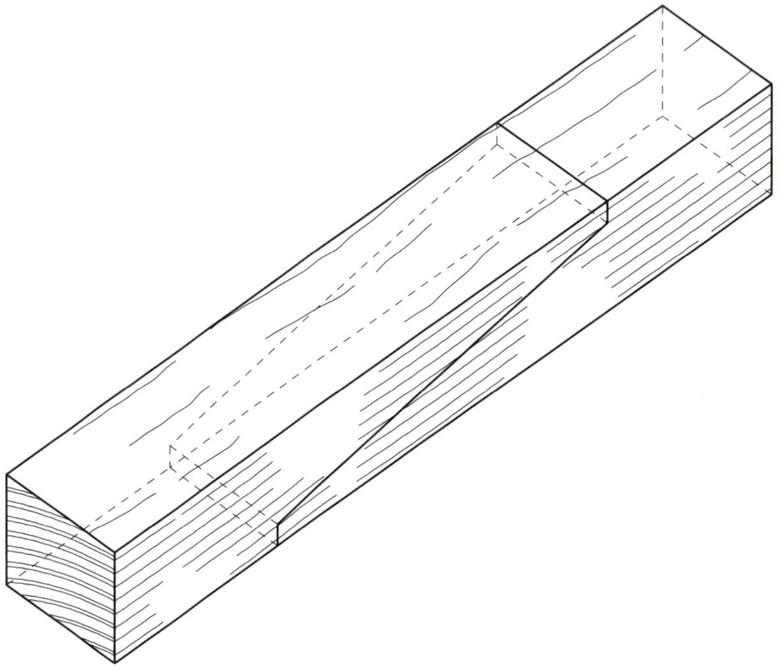

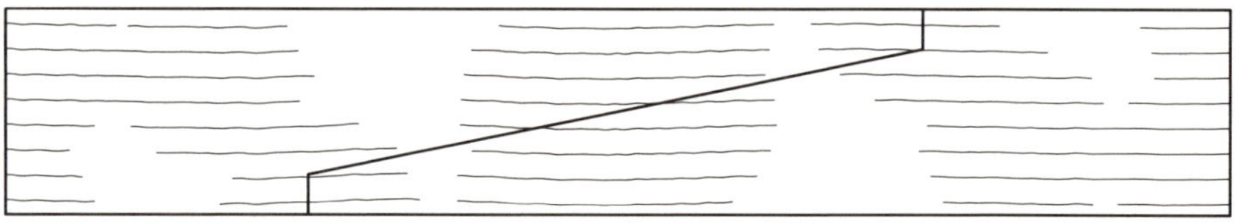

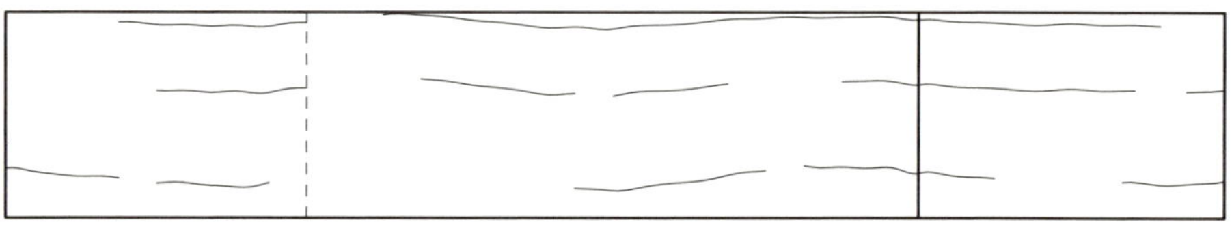

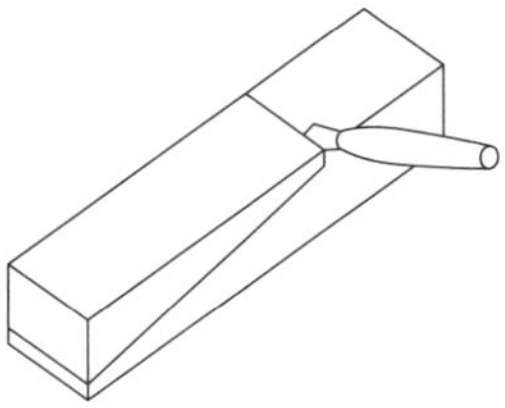

A1

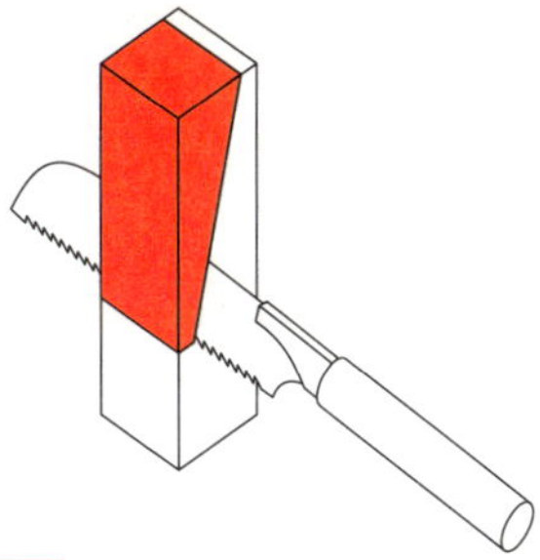

A2

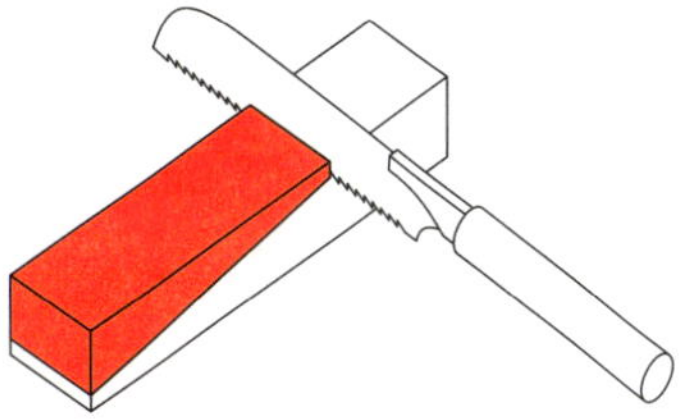

A3

2×

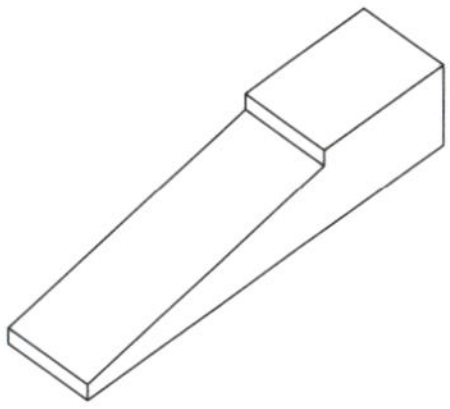

A

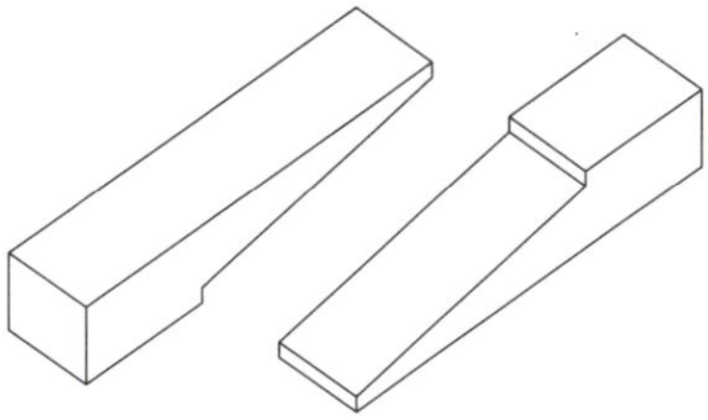

A+A

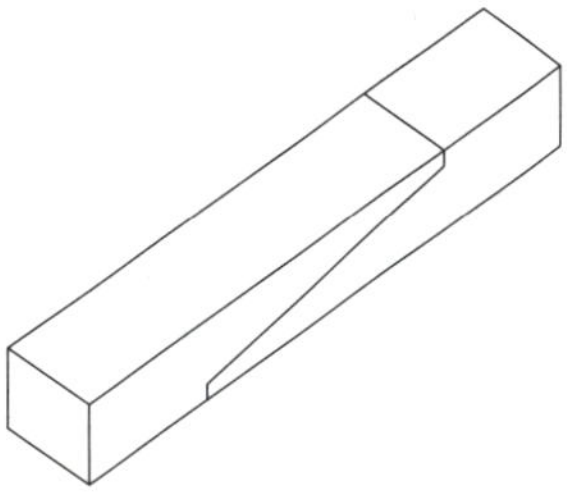

AA

Splicing Joints

Undersquinted stop-splayed scarf joint

tsuki-tsuke-sogi-tsugi (sogi-tsuki)
突付殺継ぎ(殺付)

schräges Blatt mit schräger Brüstung

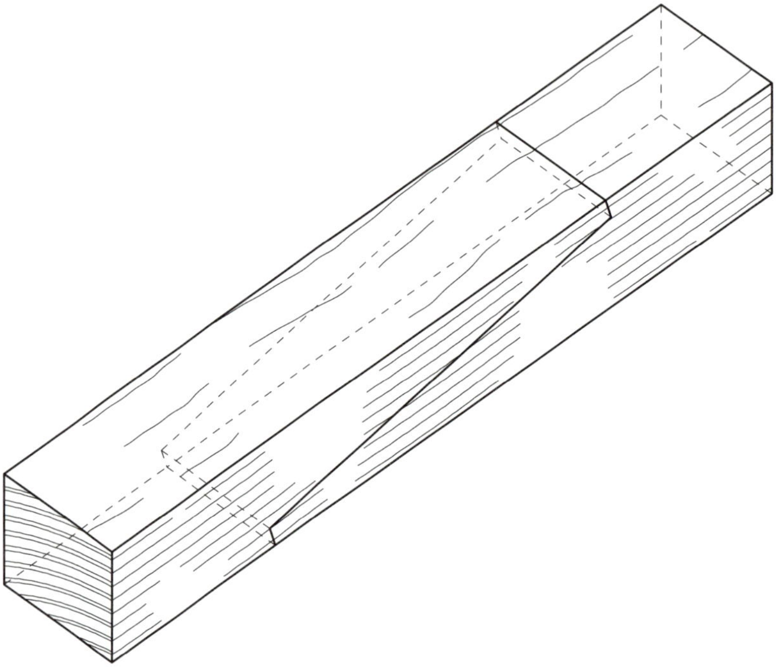

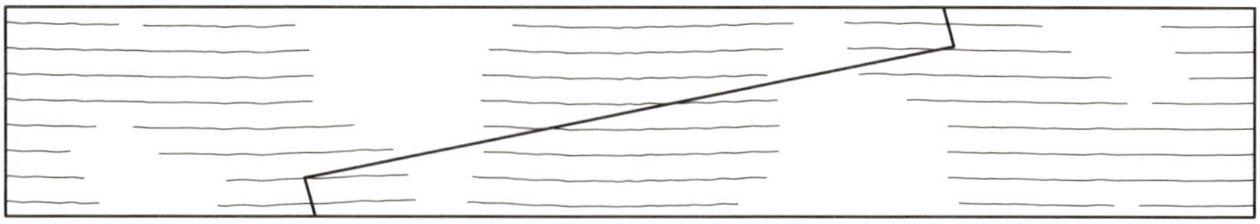

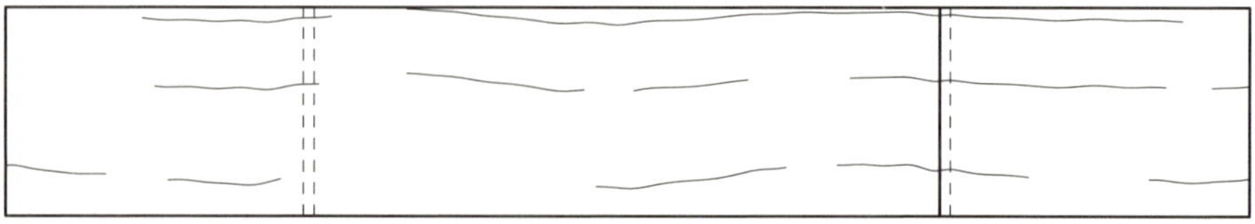

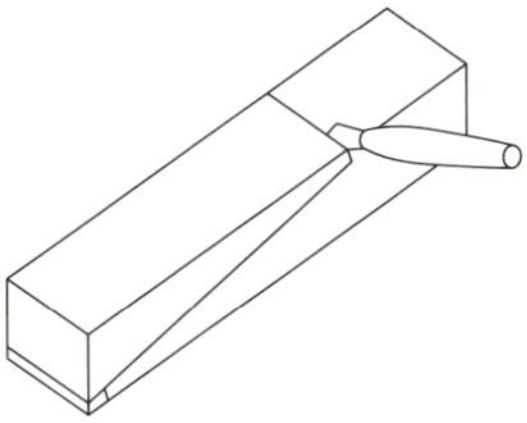

A1

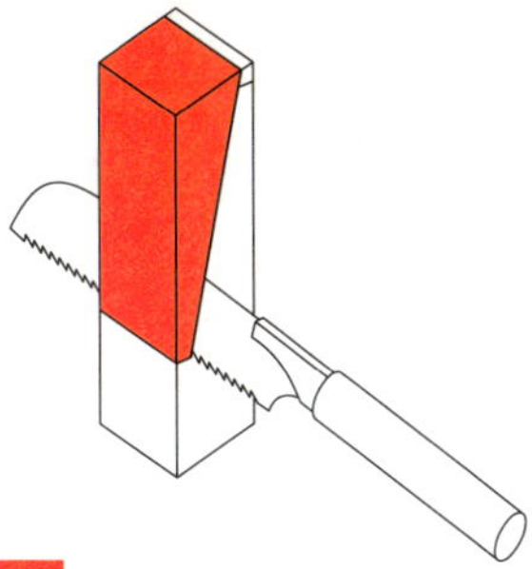

A2

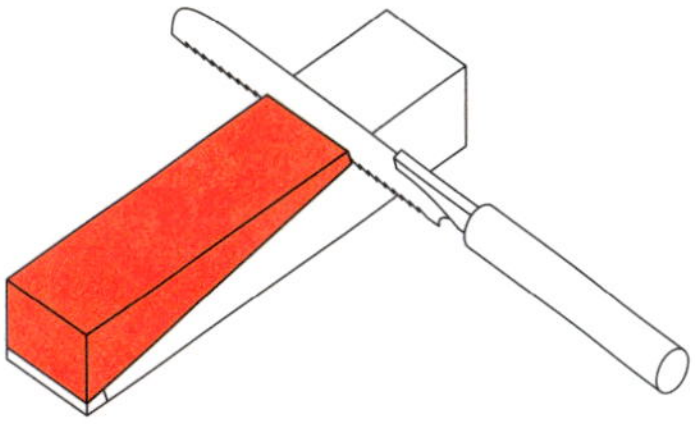

A3

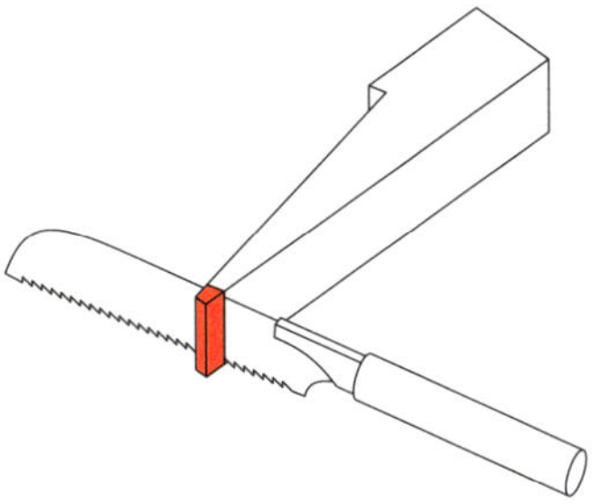

A4

2×

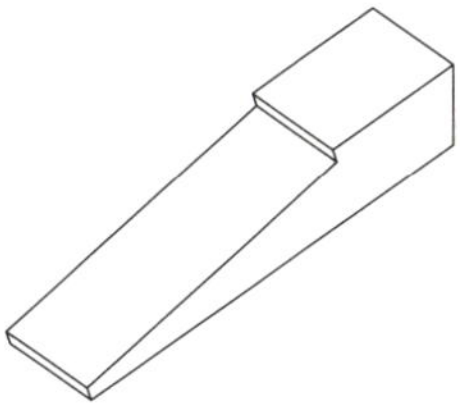

A

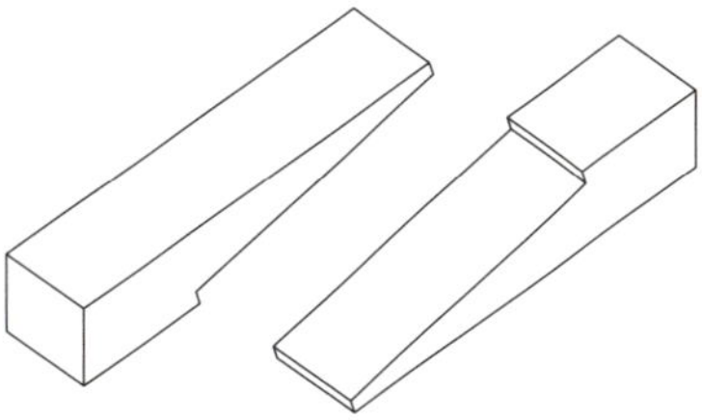

A+A

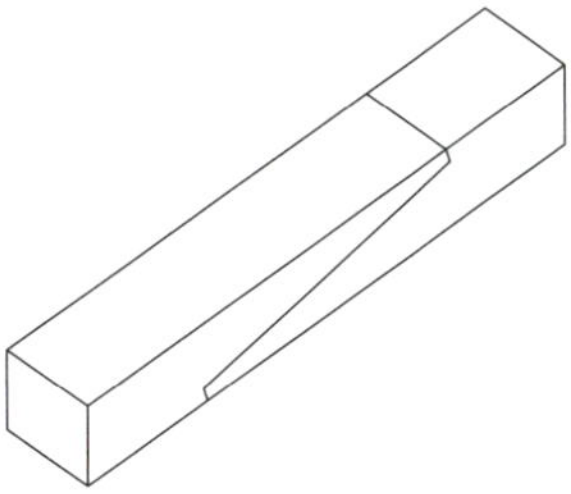

AA

Splicing Joints

Stop-splayed scarf joint with stub tenons

tsuki-tsuke-mechi-sogi-tsugi
突付目違殺継ぎ

schräges Blatt mit Brüstungszapfen

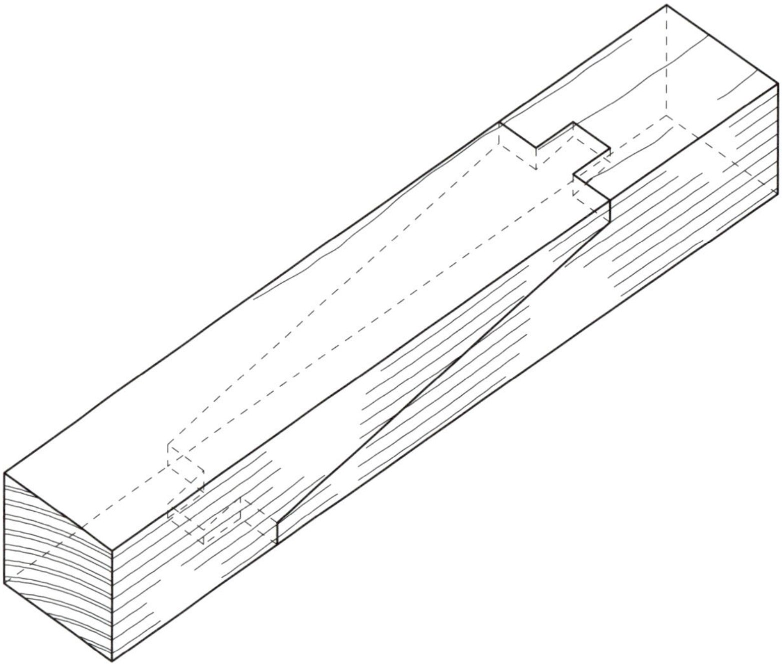

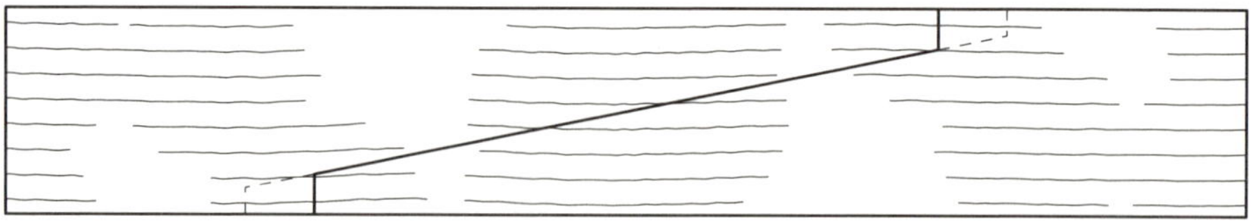

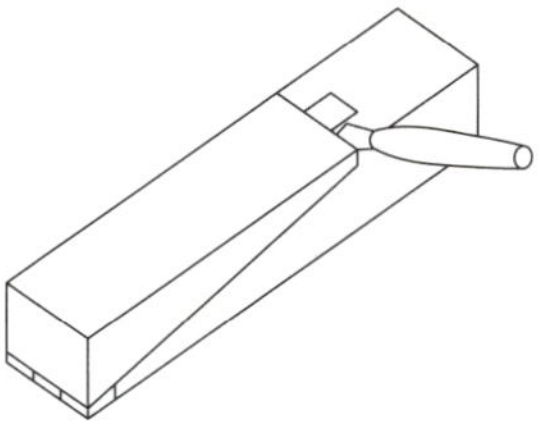

A1

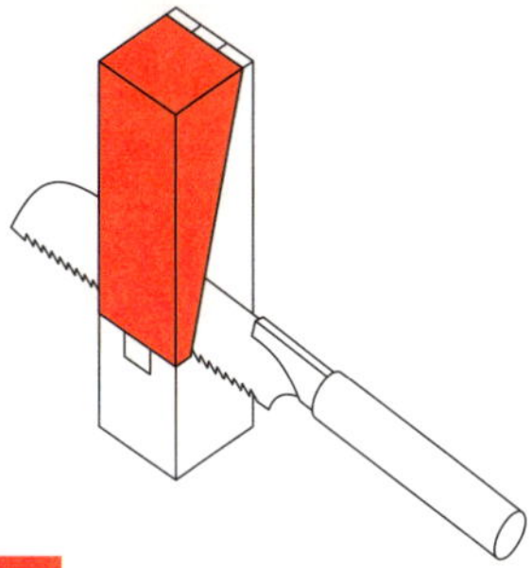

A2

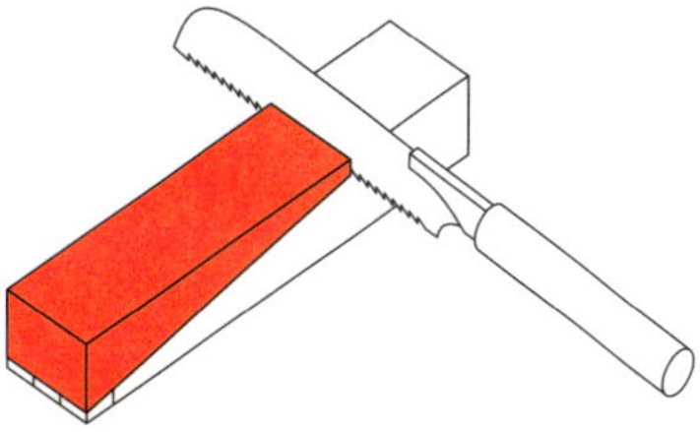

A3

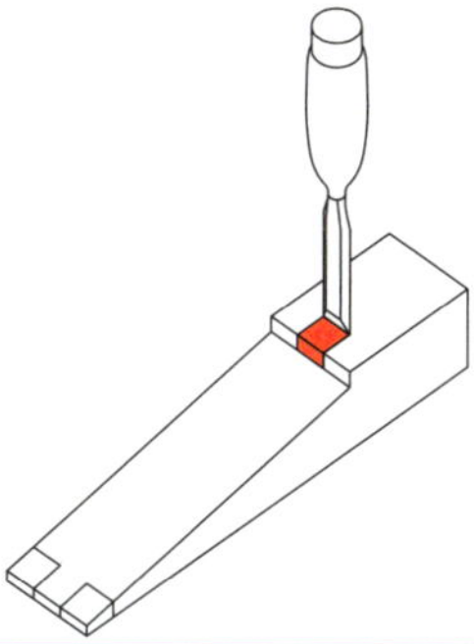

A4

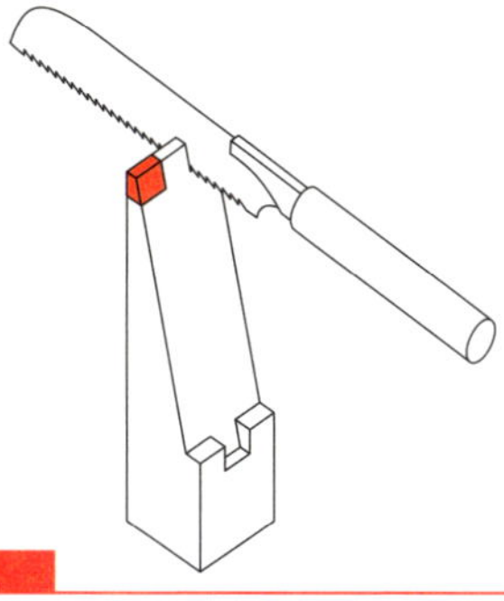

A5

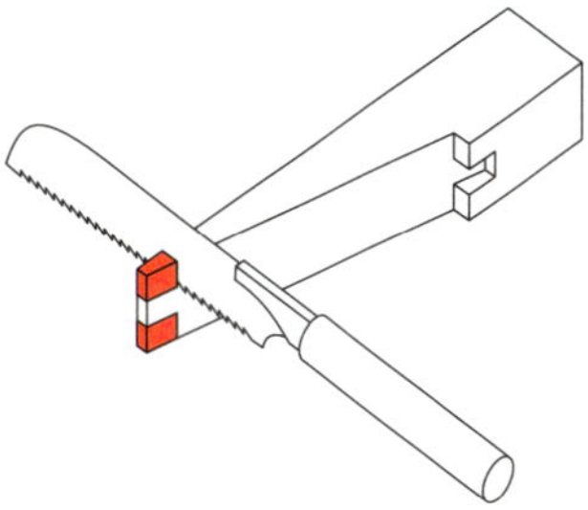

A6

2×

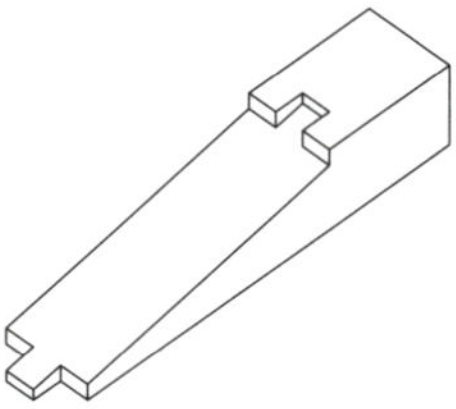

A

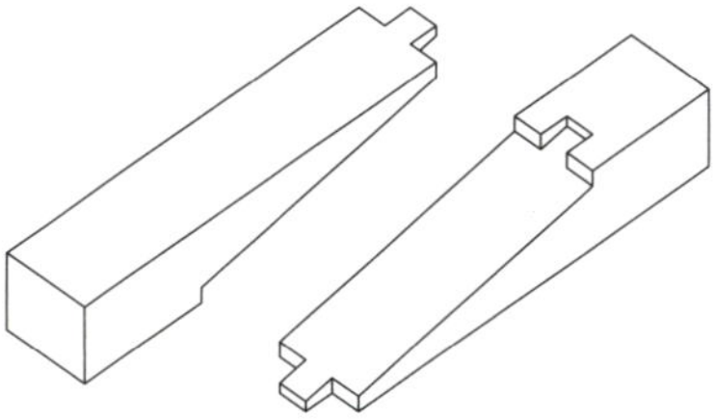

A+A

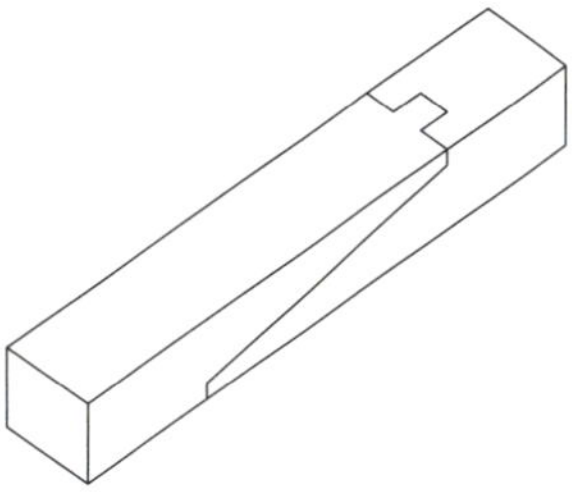

AA

Splicing Joints

Pegged stop-splayed scarf joint

tsuki-tsuke-sogi-tsugi-komi-sen-uchi
突付殺継込栓打ち

schräges Blatt mit Dübelsicherung

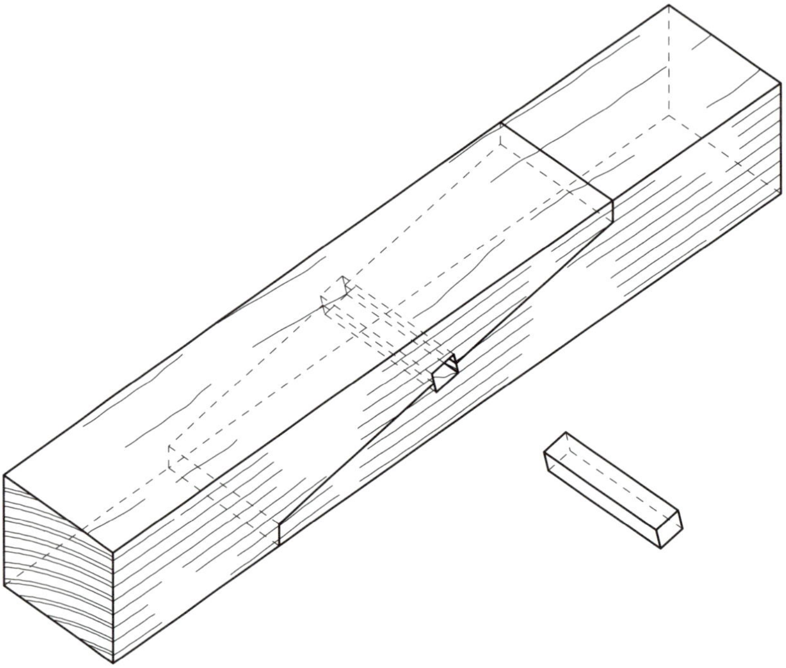

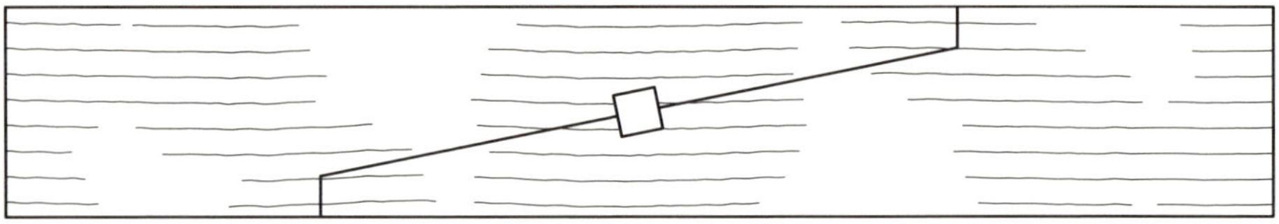

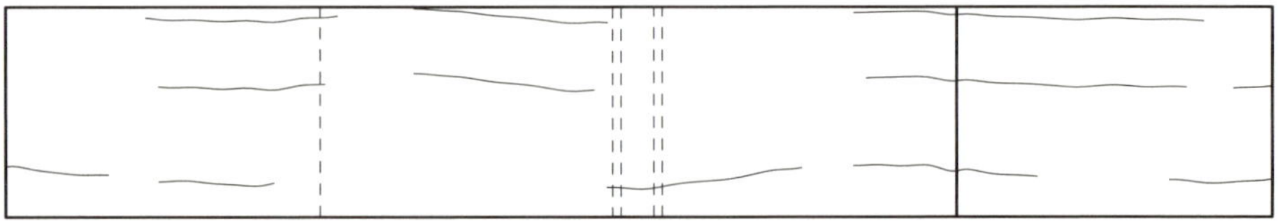

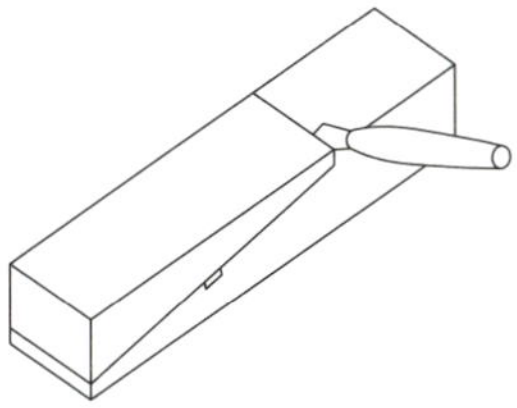

A1

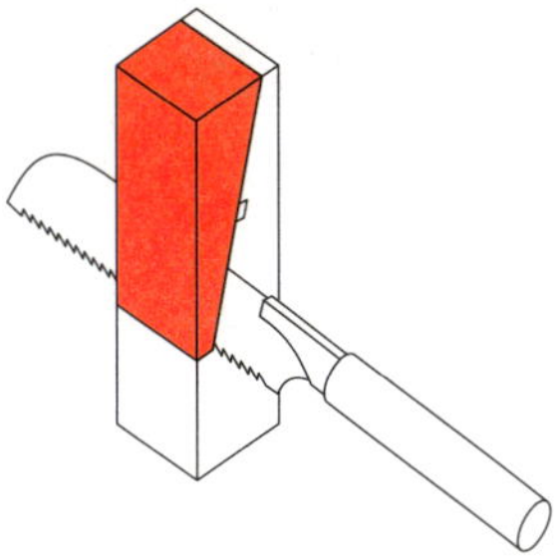

A2

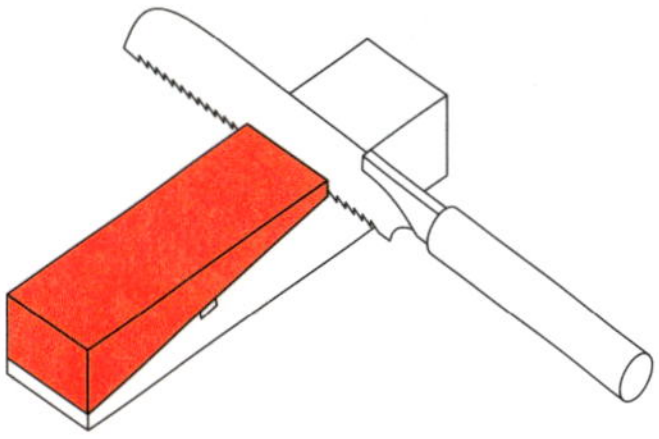

A3

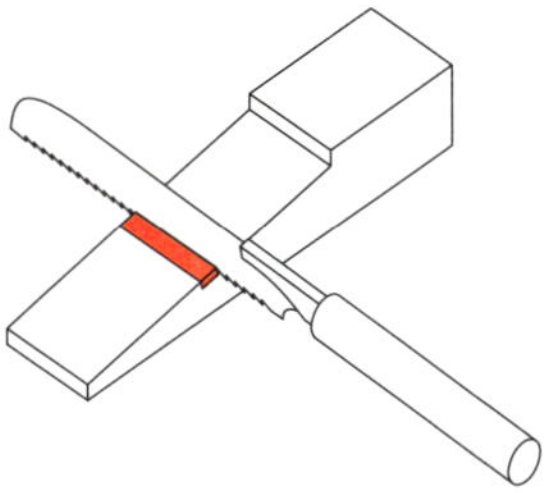

A4

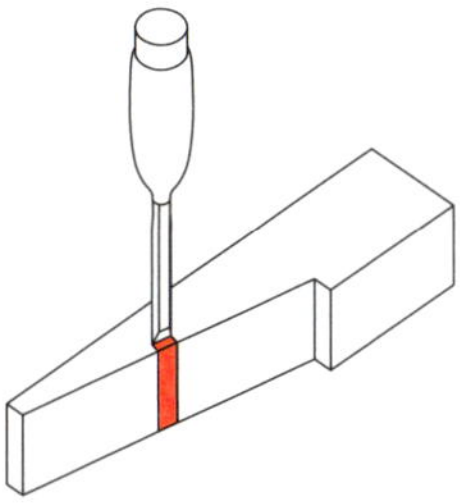

A5

2×

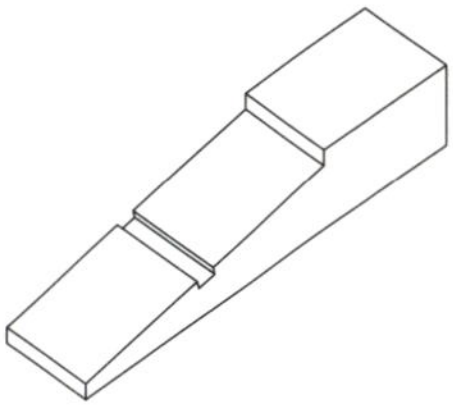

A

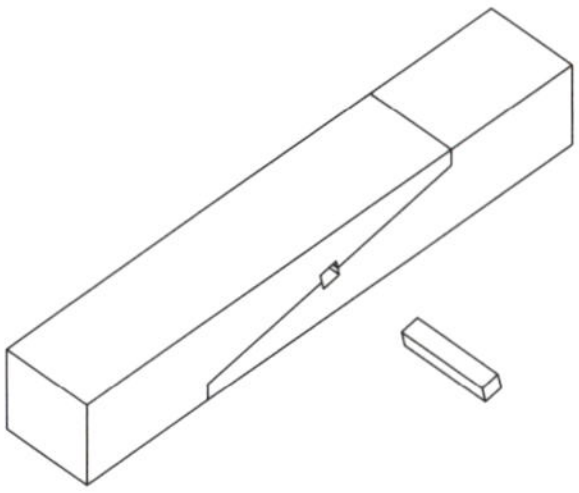

A+A

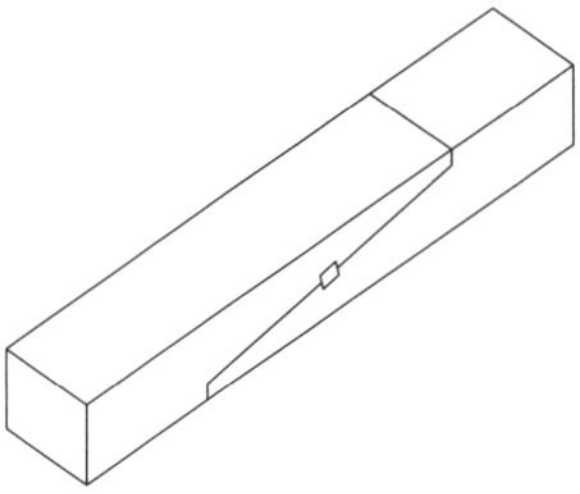

AA

Splicing Joints

Stop-splayed cogged scarf joint

ryaku-kama-tsugi
略鎌継ぎ

schräges Hakenblatt

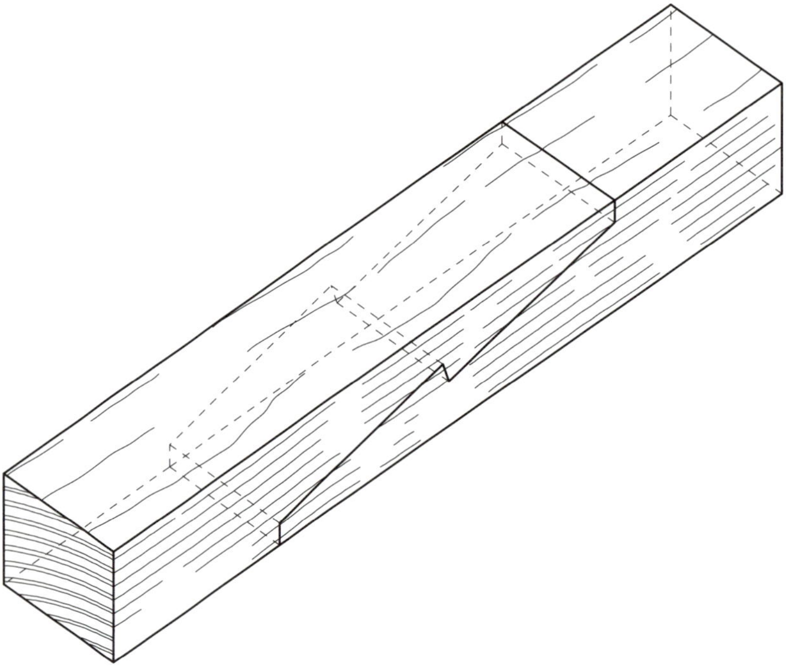

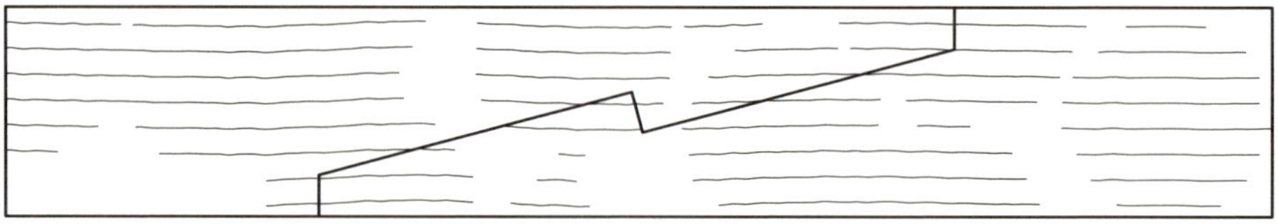

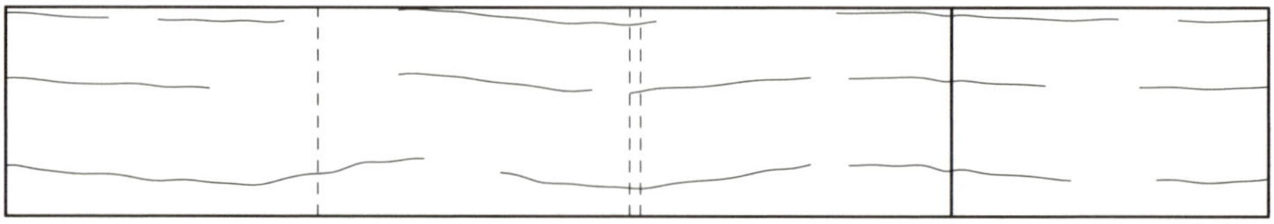

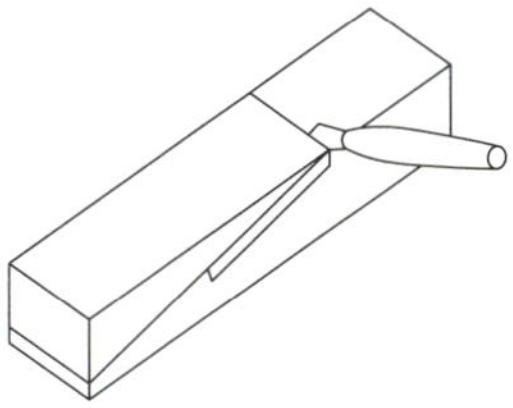

A1

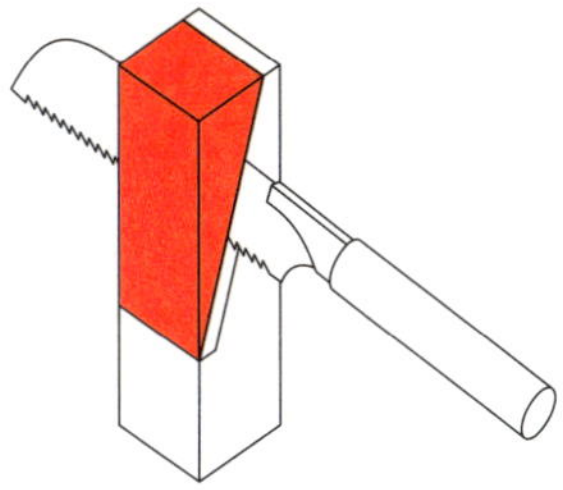

A2

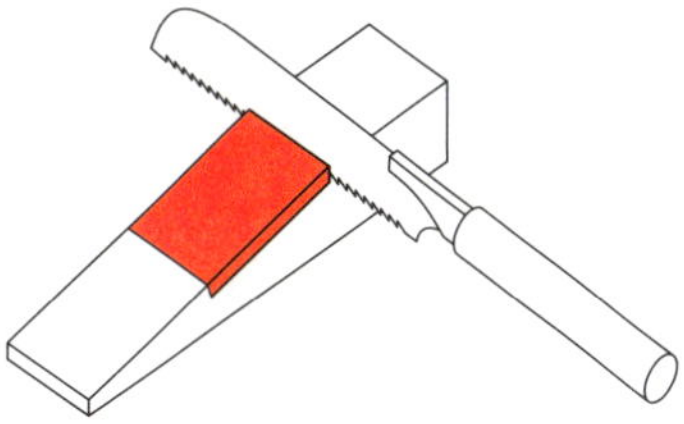

A3

2×

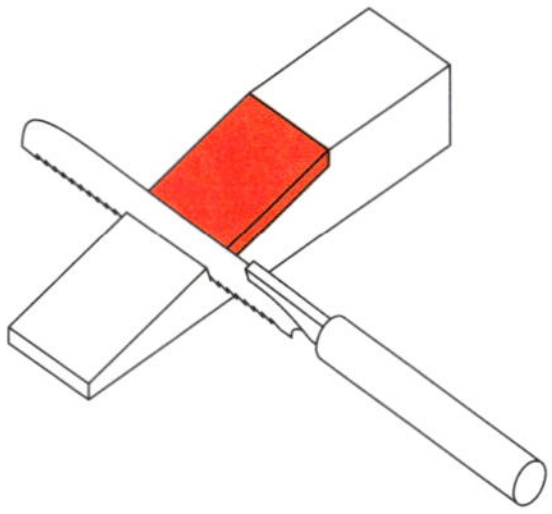

A4

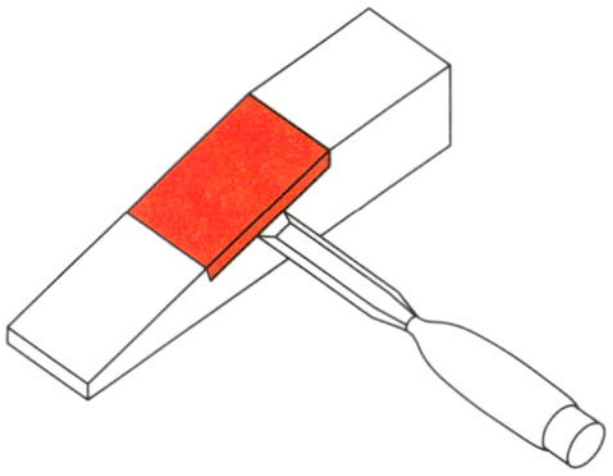

A5

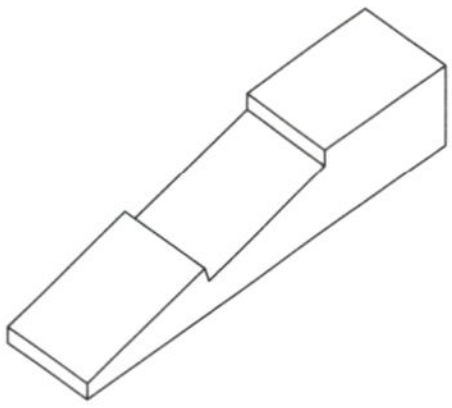

A

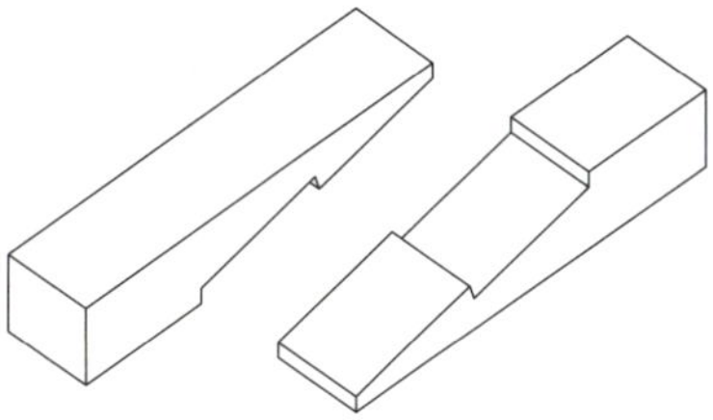

A+A

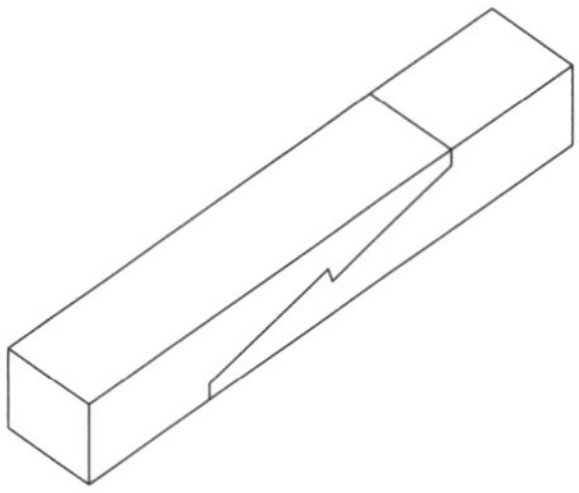

AA

Splicing Joints

Undersquinted stop-splayed cogged scarf joint

ryaku-kama-tsugi (sogi-tsuki)
略鎌継ぎ(殺付)

schräges Hakenblatt
mit schräger Brüstung

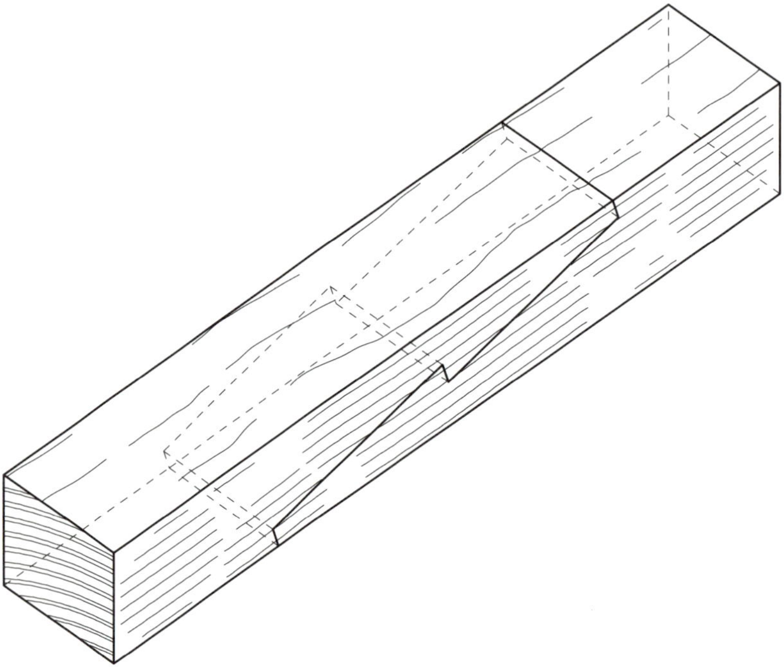

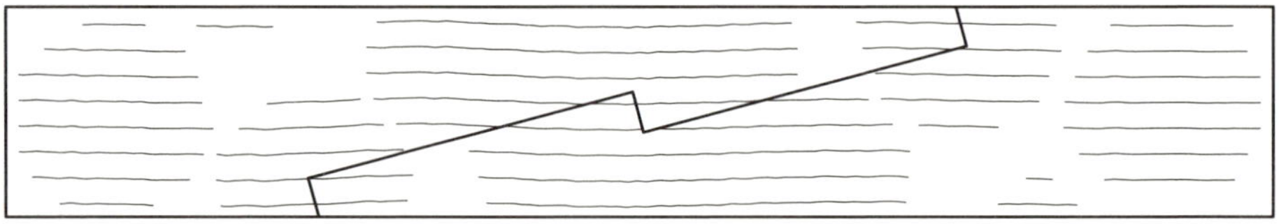

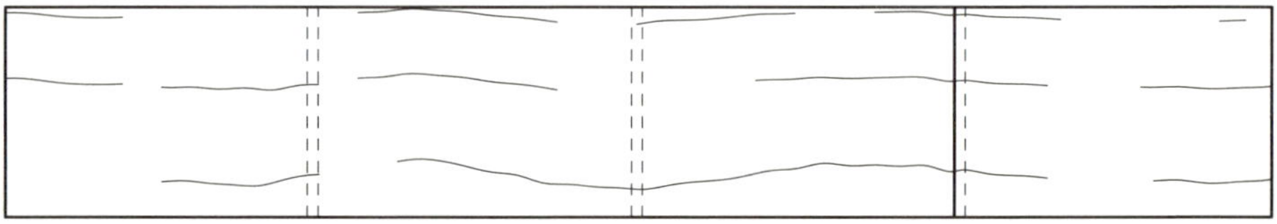

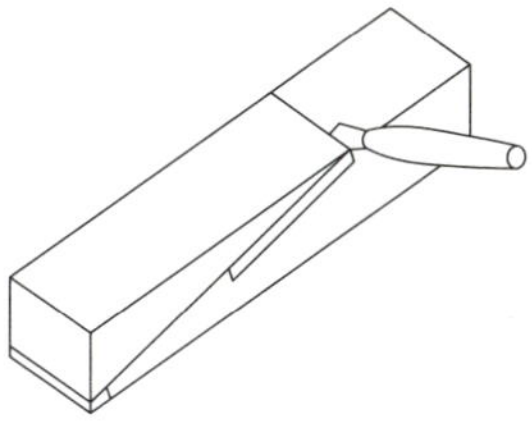

A1

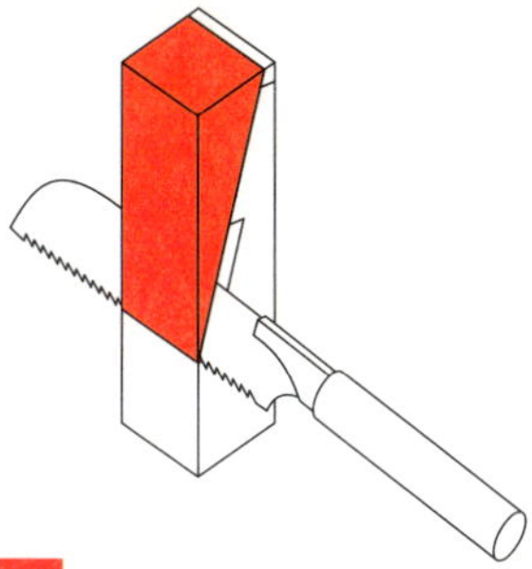

A2

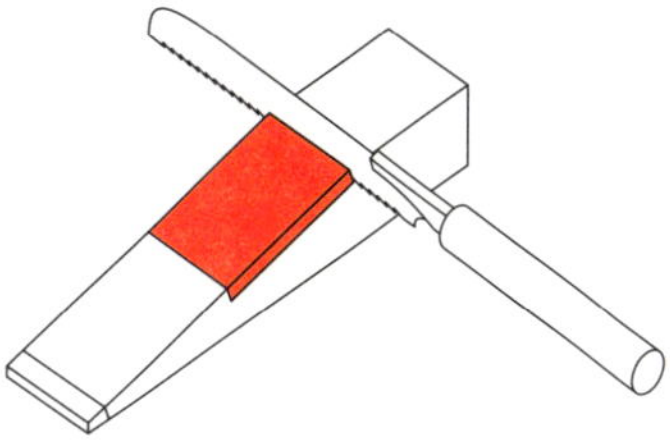

A3

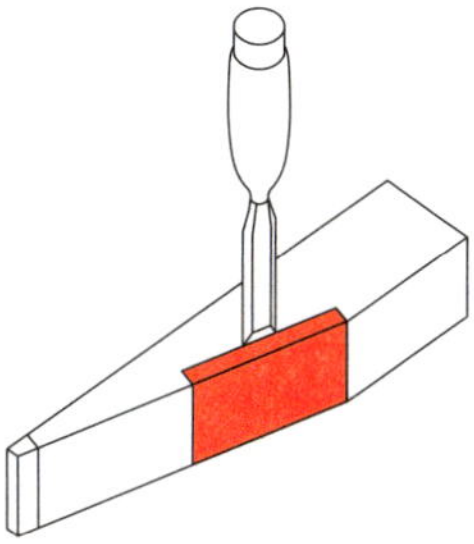

A4

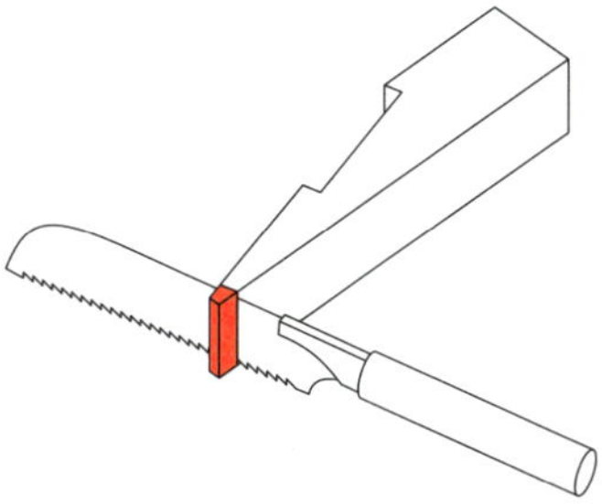

A5

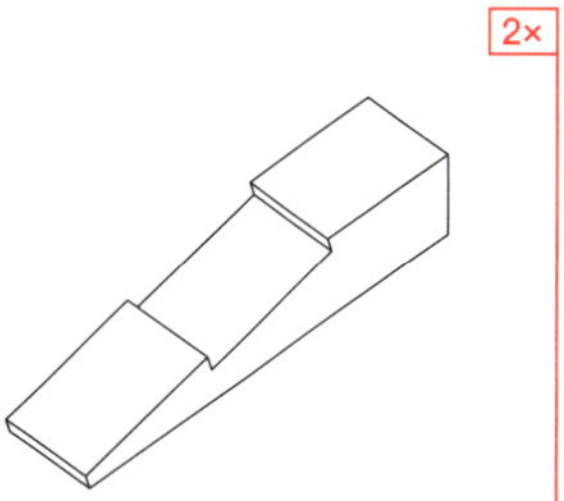

A

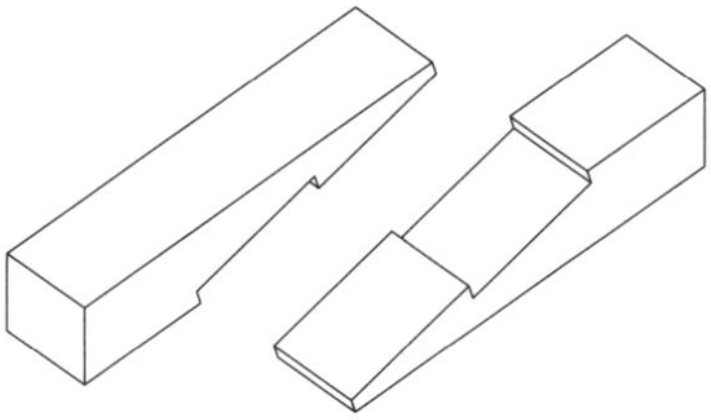

A+A

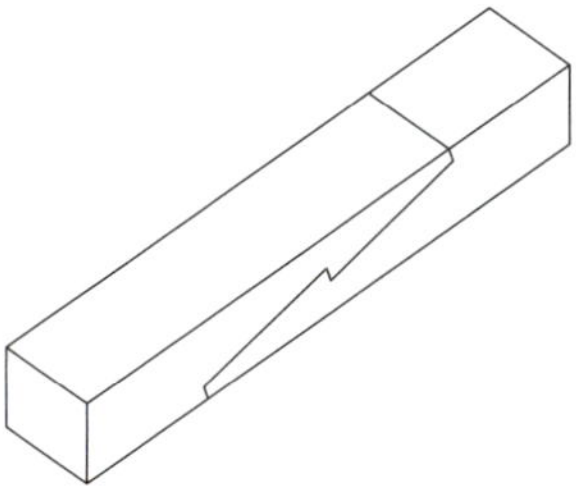

AA

Splicing Joints

Stop-splayed cogged scarf joint with folding wedges

ryaku-kama-tsugi (kusabi-shime)
略鎌継ぎ(楔締)

verkeiltes schräges Hakenblatt

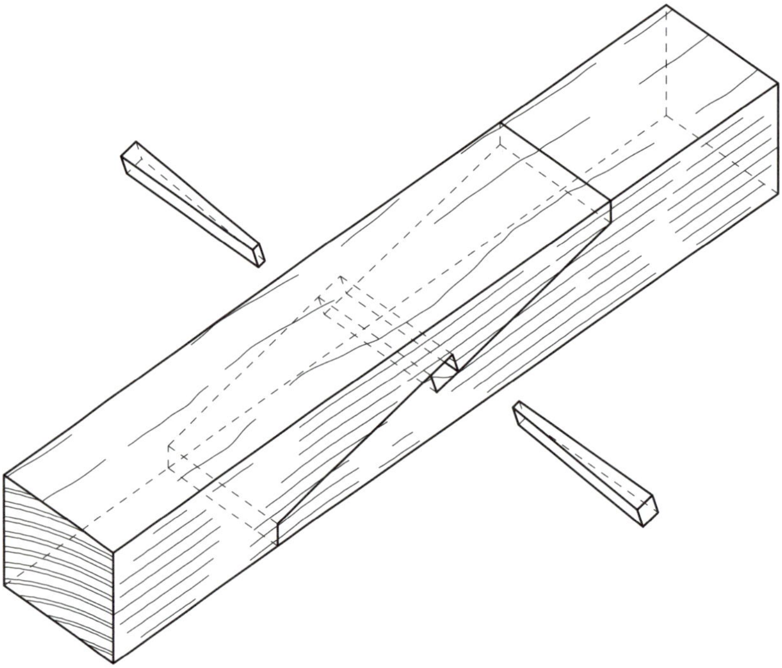

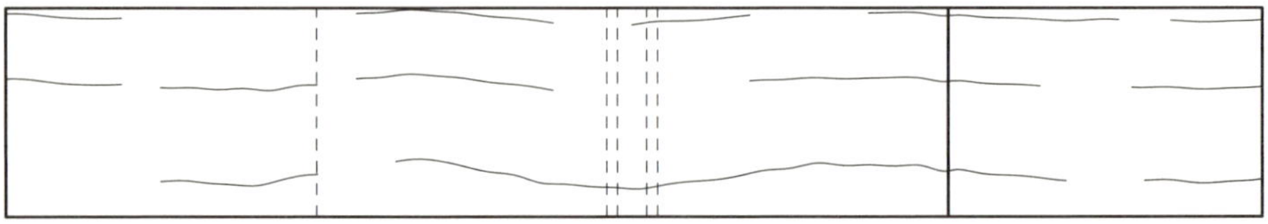

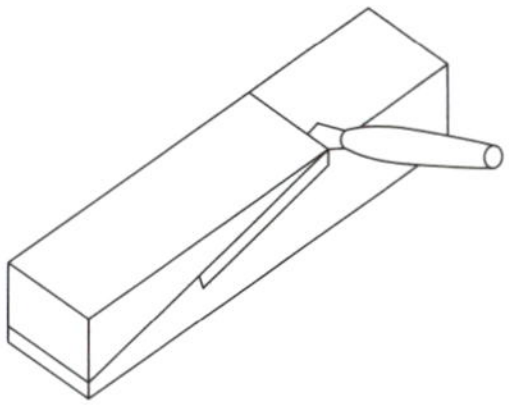
A1

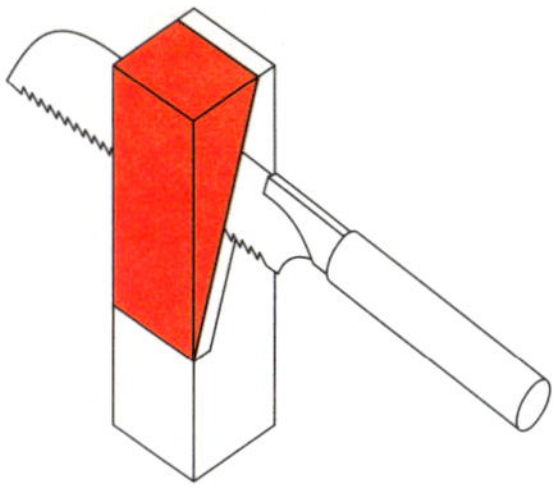
A2

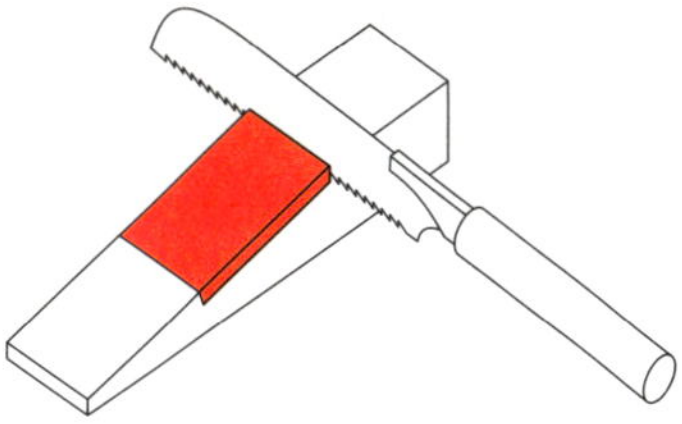
A3

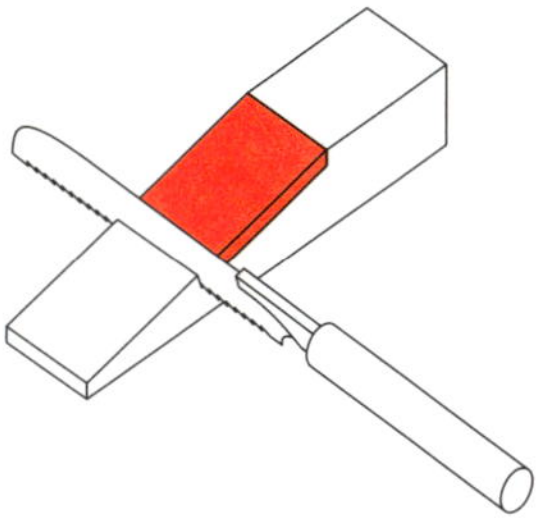
A4

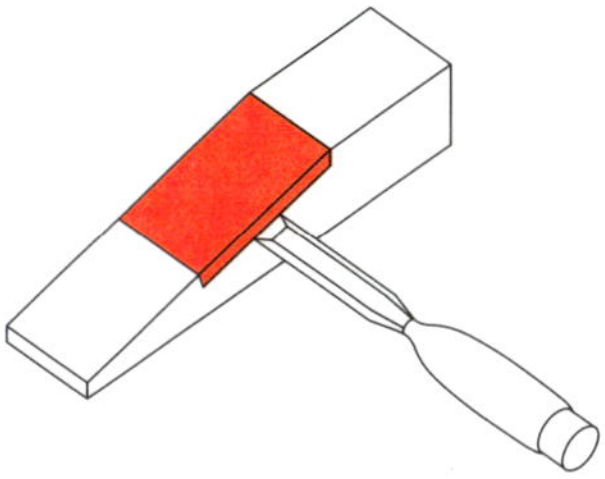
A5

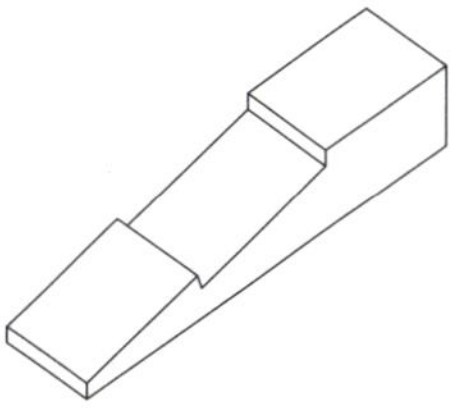
2×

A

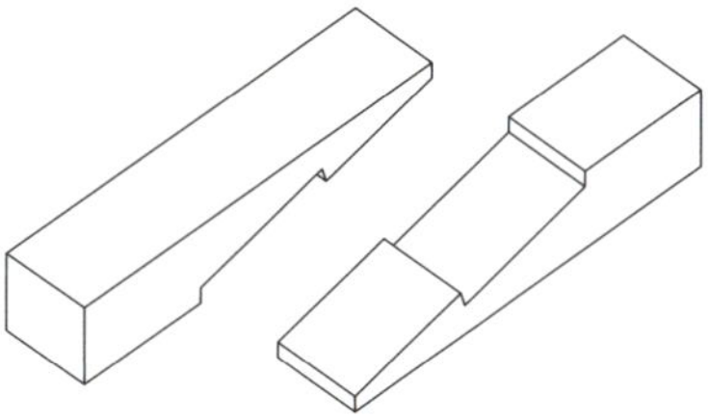
A+A

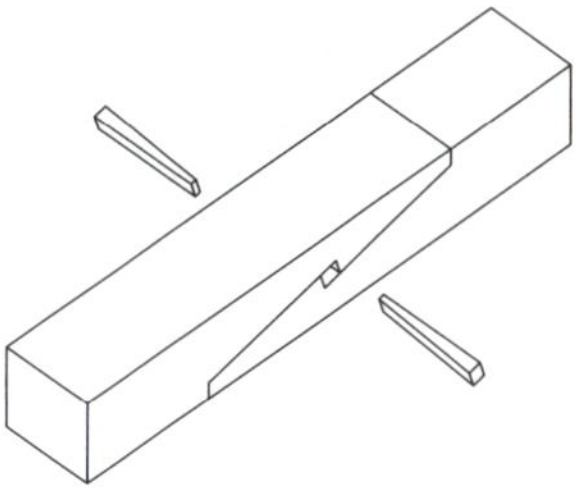
A+A

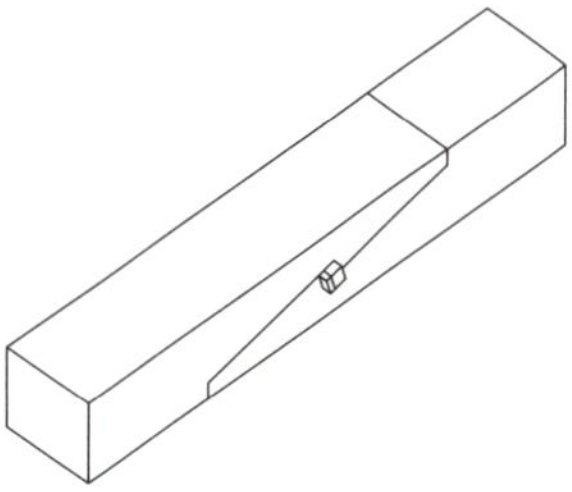
AA

Splicing Joints

Undersquinted stop-splayed cogged scarf joint with folding wedges

ryaku-kama-tsugi
(sogi-tsuki, kusabi-shime)
略鎌継ぎ(殺付・楔締)

verkeiltes schräges Hakenblatt mit schräger Brüstung

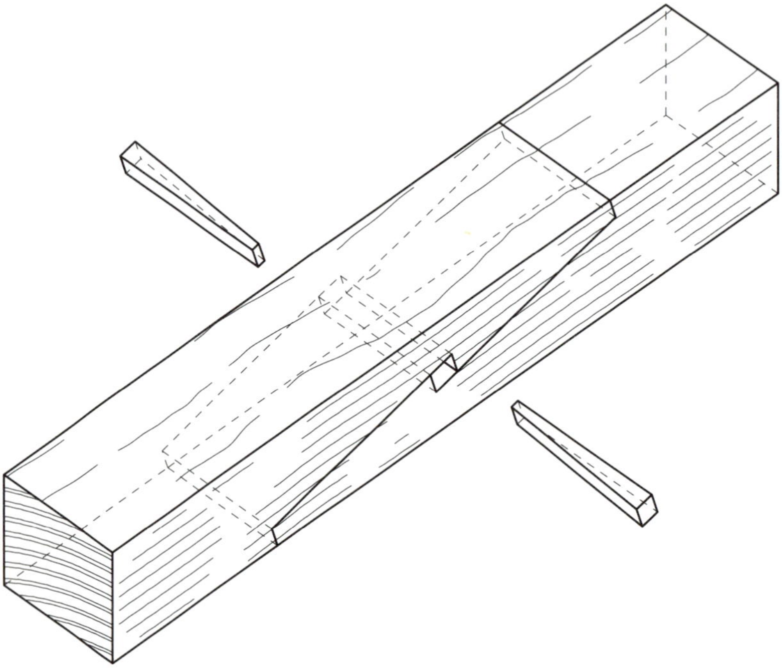

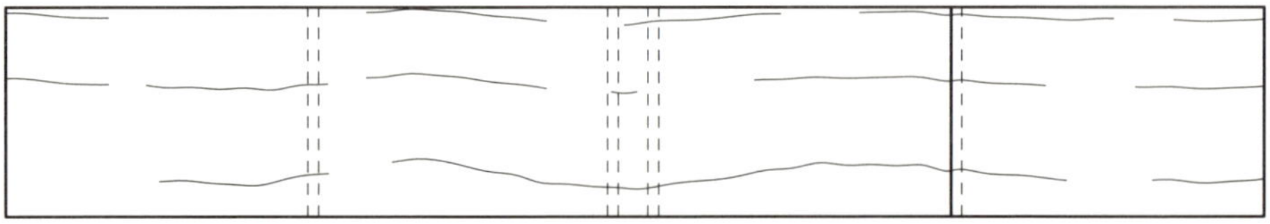

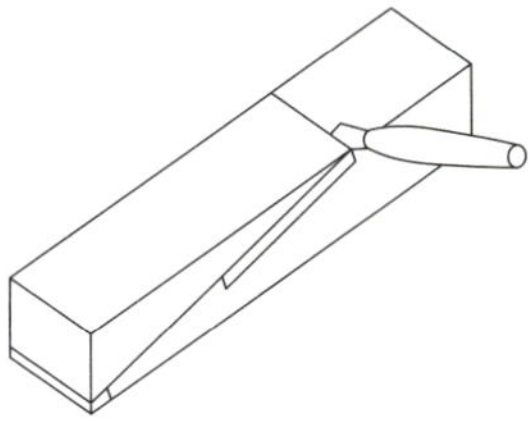

A1

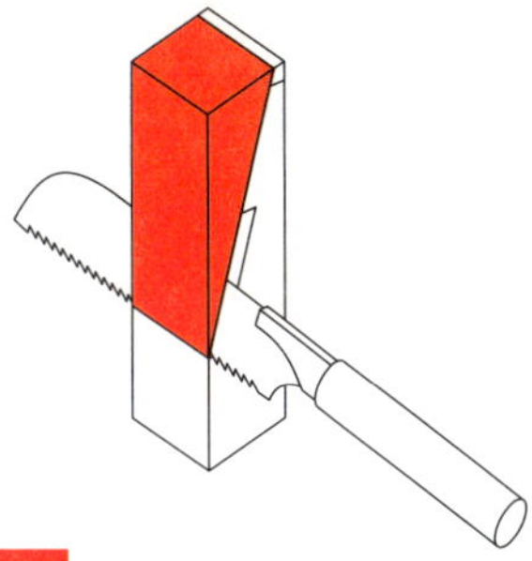

A2

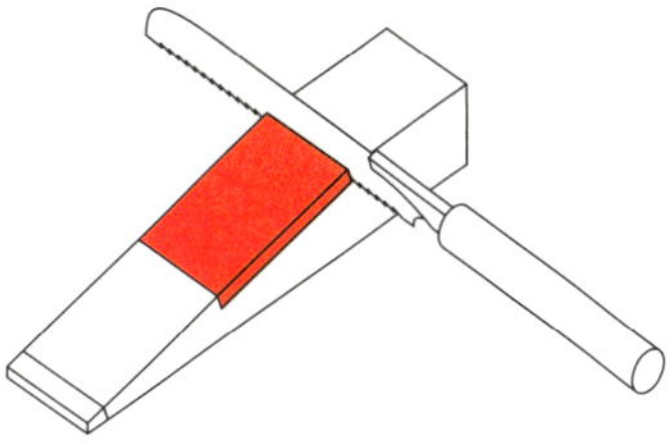

A3

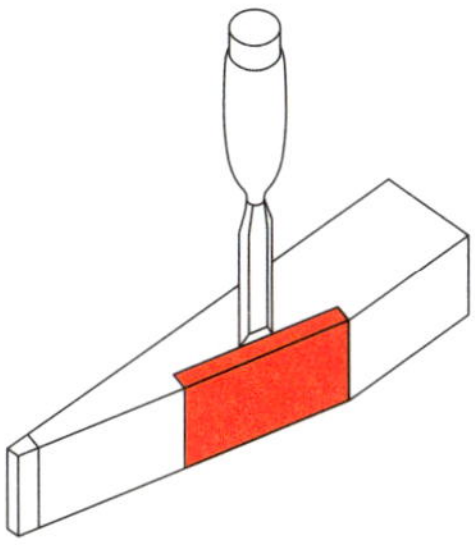

A4

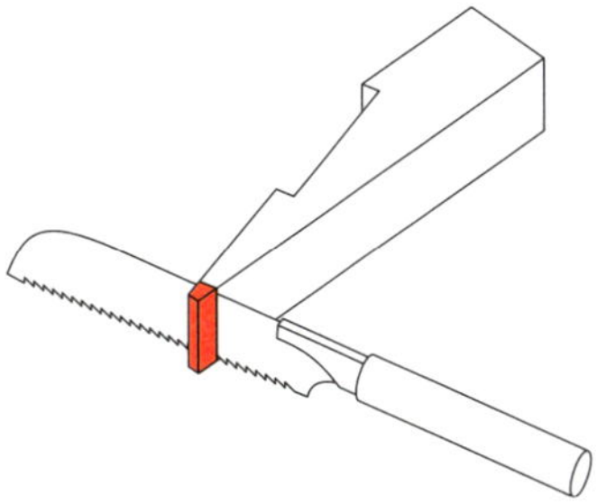

A5

2×

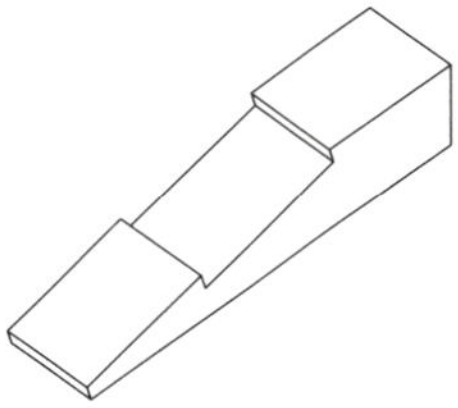

A

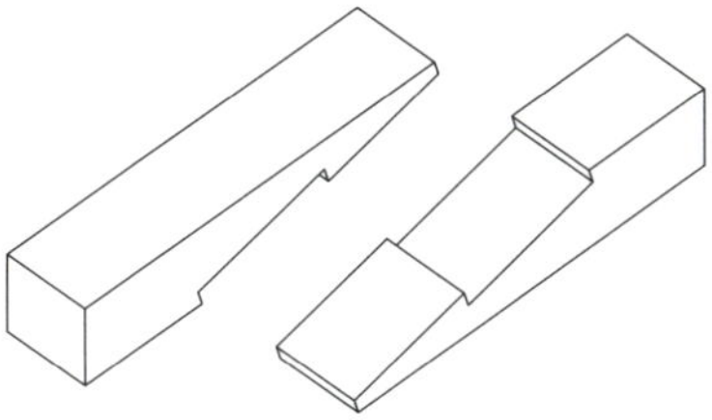

A+A

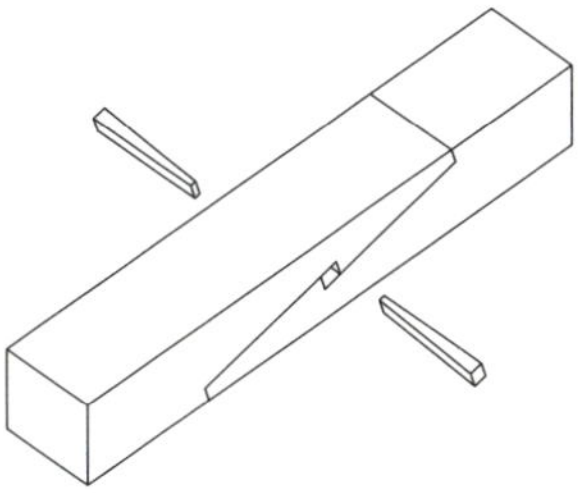

A+A

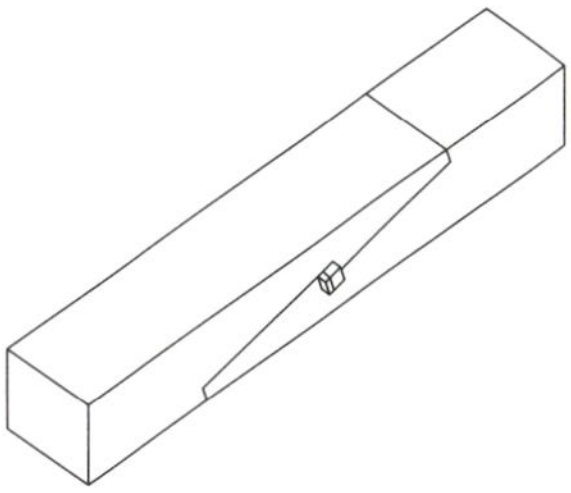

AA

Splicing Joints

Pegged stop-splayed cogged scarf joint

ryaku-kama-tsugi-dai-sen-uchi
略鎌継大栓打ち

schräges Hakenblatt
mit Dübelsicherung

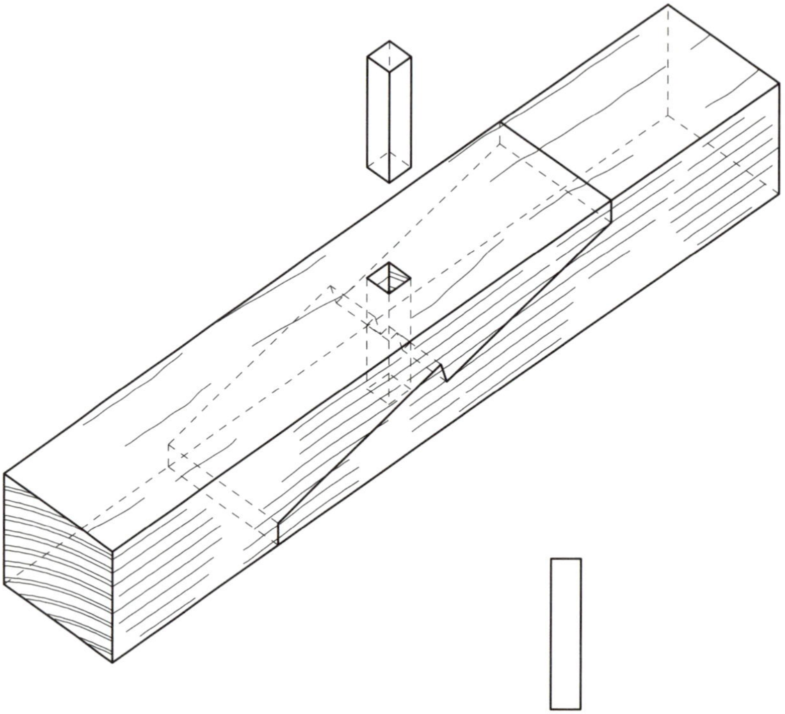

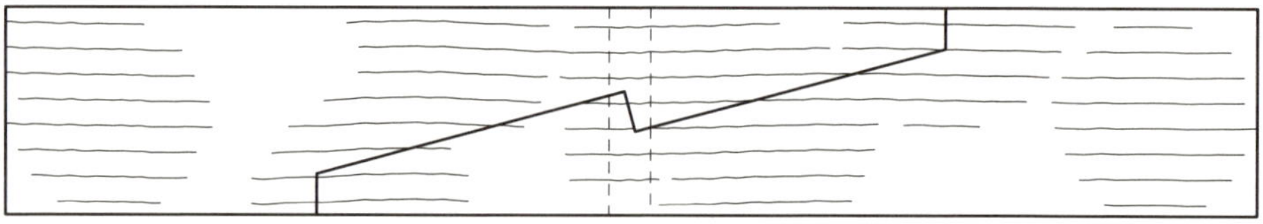

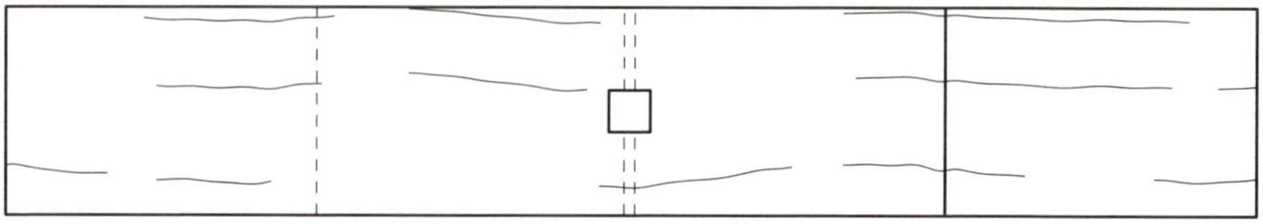

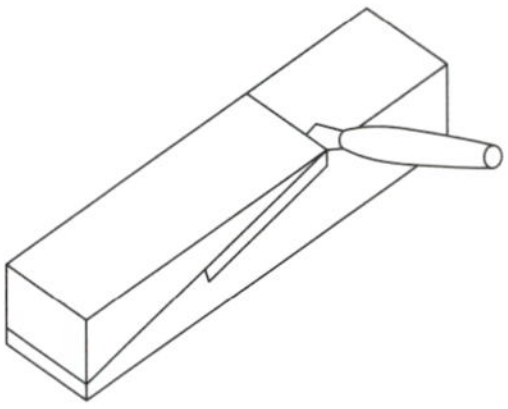

A1

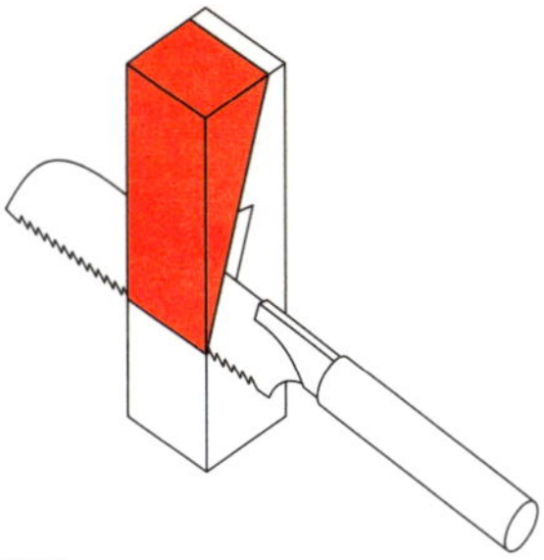

A2

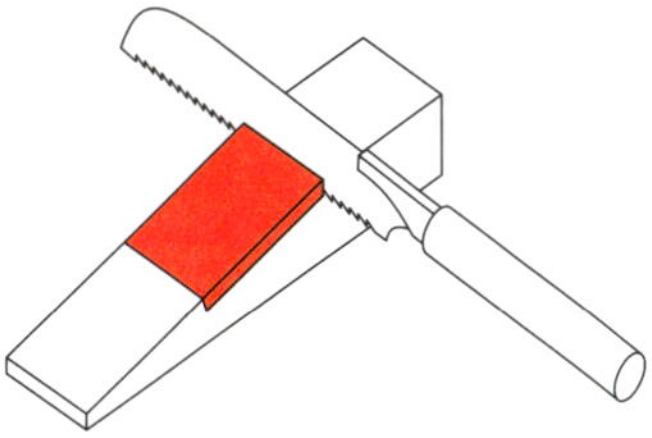

A3

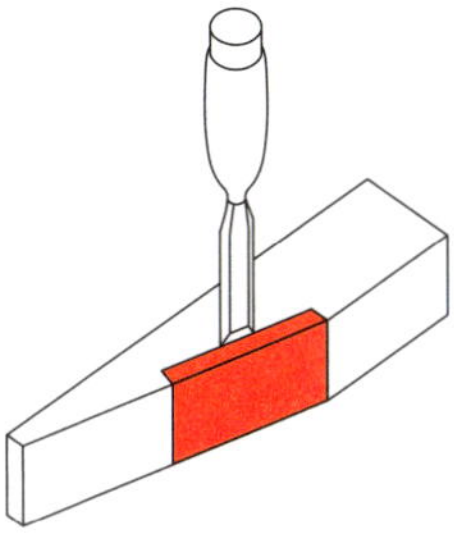

A4

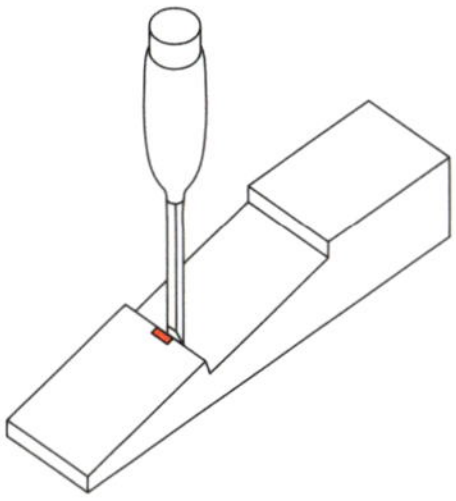

A5

2×

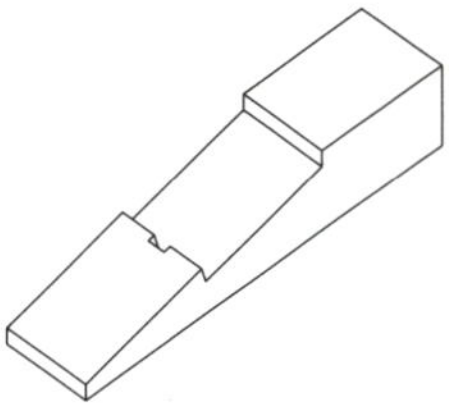

A

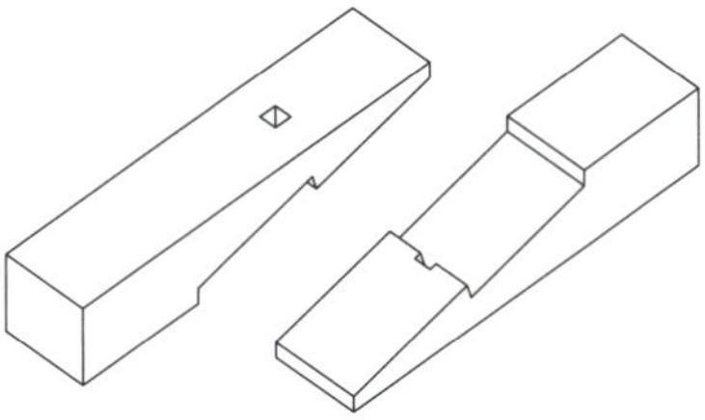

A+A

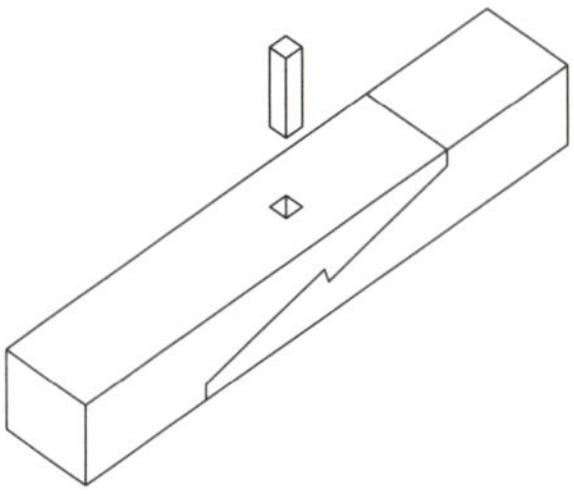

A+A

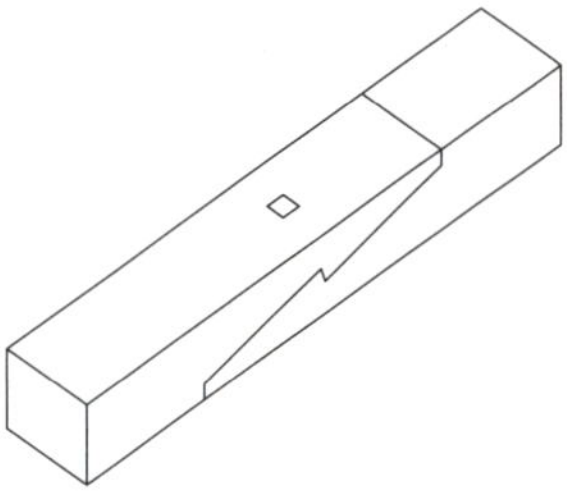

AA

Splicing Joints

Stop-splayed cogged scarf joint with stub tenons and blind pegs

daimochi-tsugi [*]
台持継ぎ

schräges Hakenblatt
mit Brüstungszapfen und Dollen

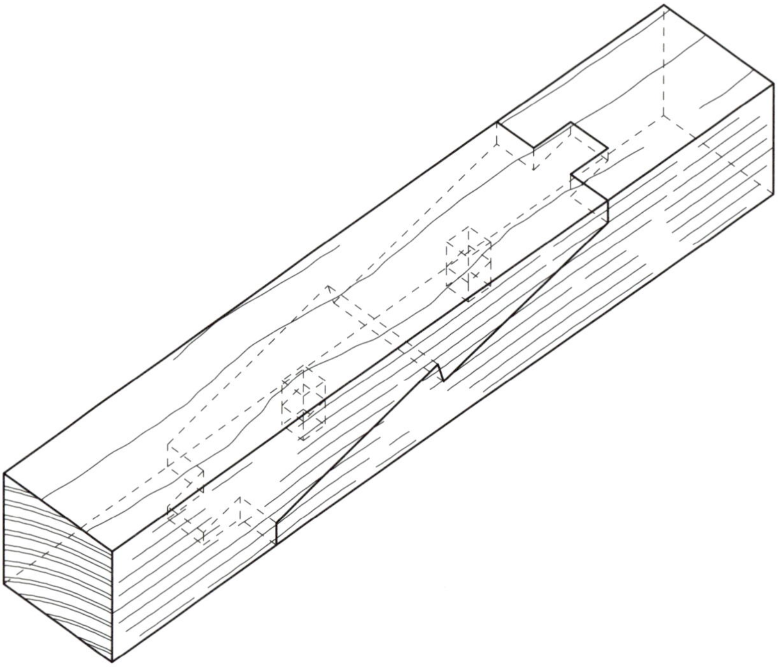

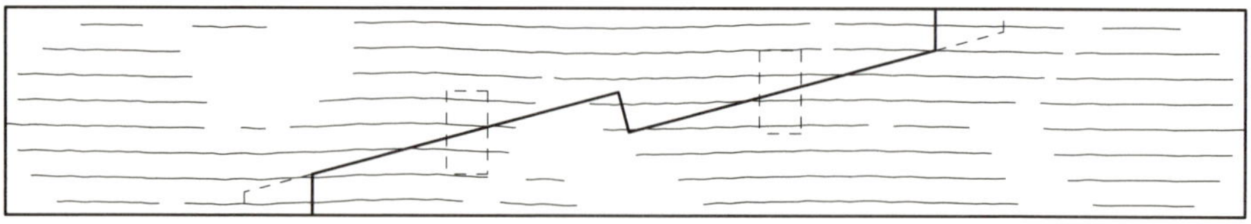

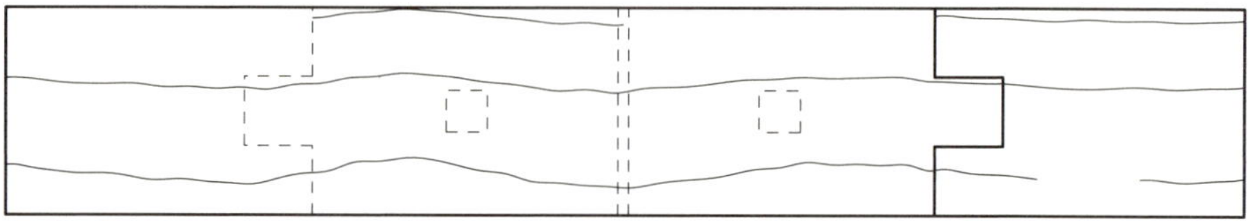

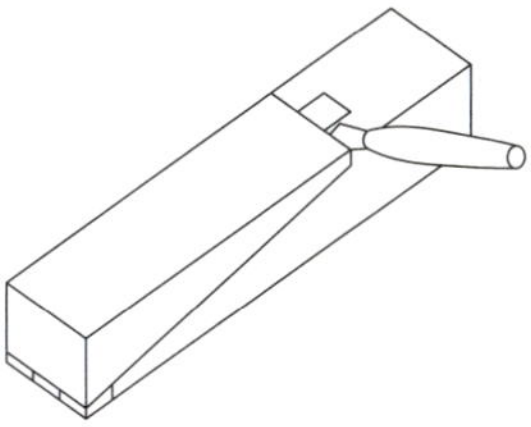

A1

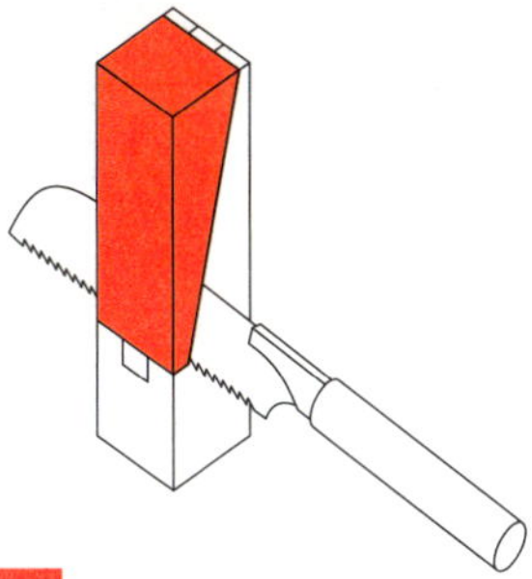

A2

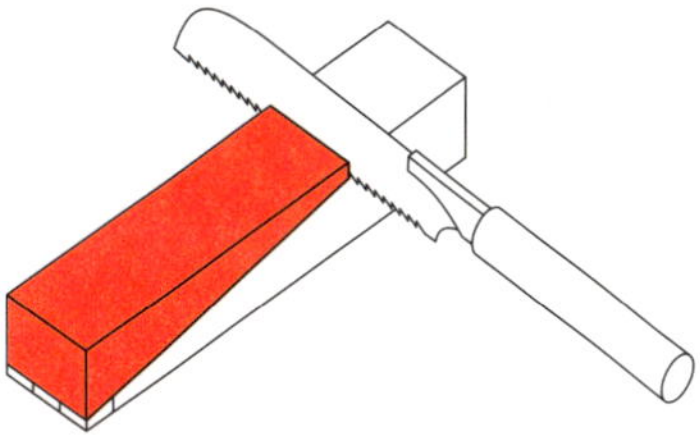

A3

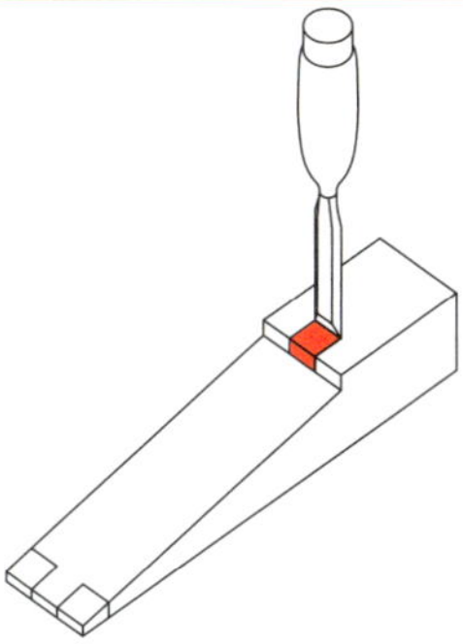

A4

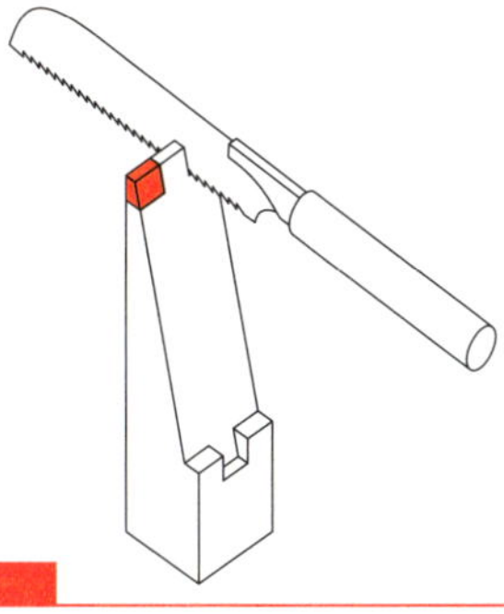

A5

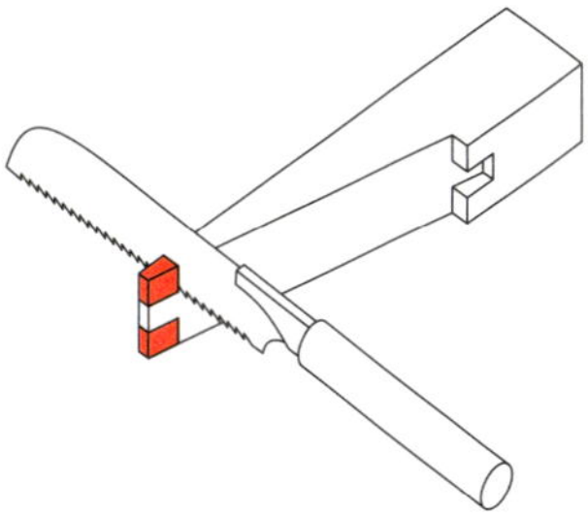

A6

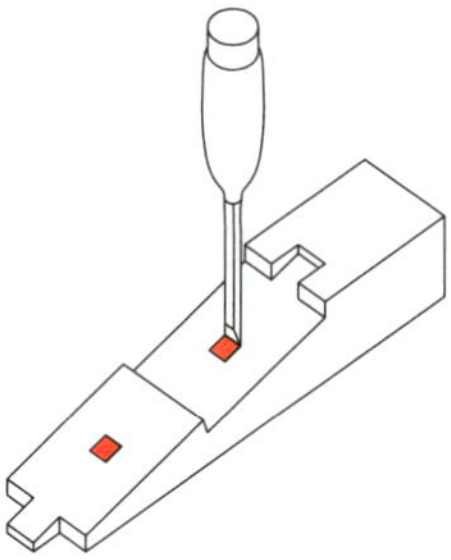

A7

2×

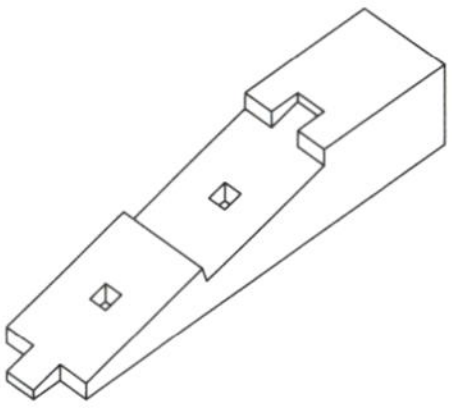

A

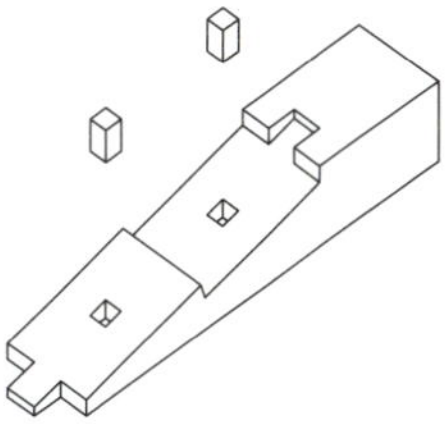

A+A

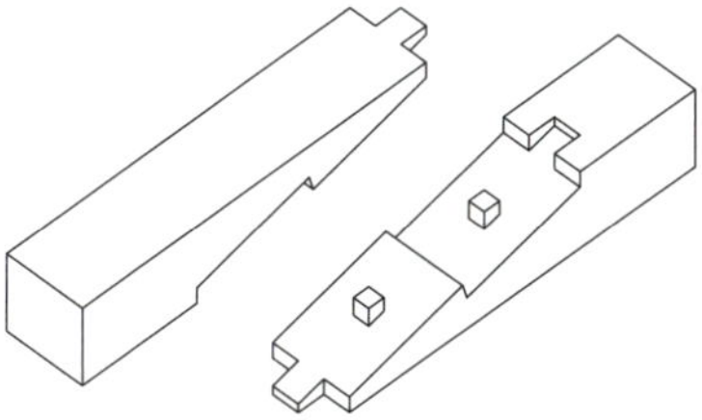

A+A

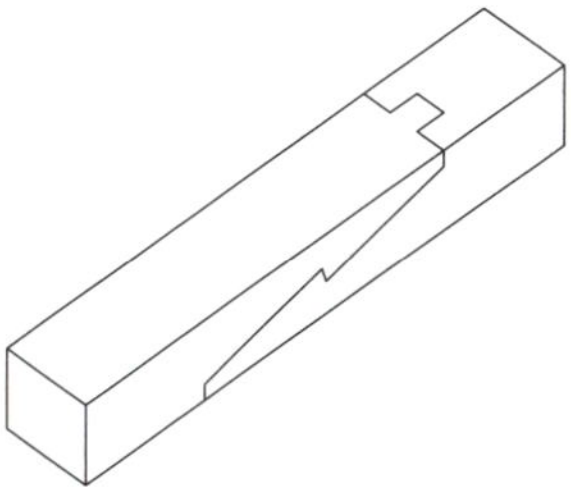

AA

Splicing Joints

Spear point stop-splayed cogged scarf joint with folding wedges

yahazu-ai-kaki-tsugi (kusabi-shime)
矢筈相欠継ぎ(楔締)

verkeiltes schräges Hakenblatt
mit Gratschnitt

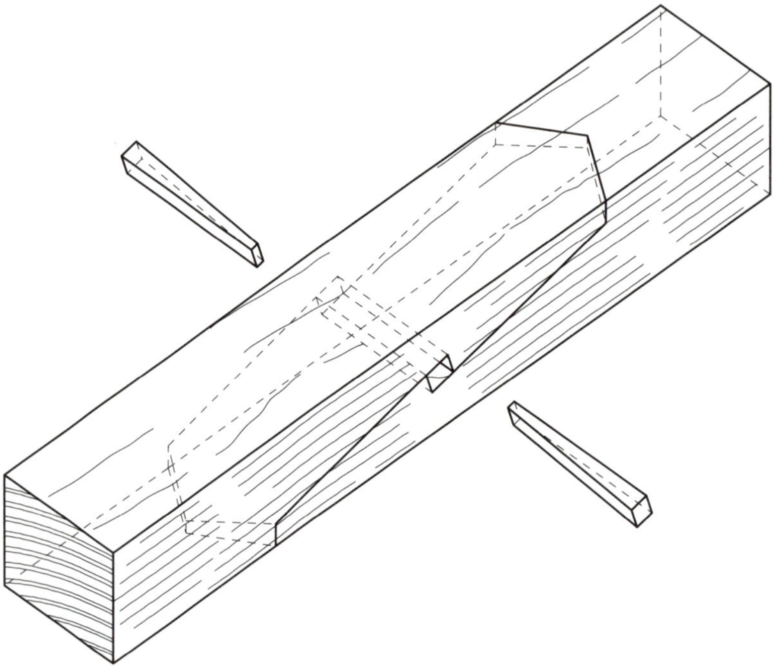

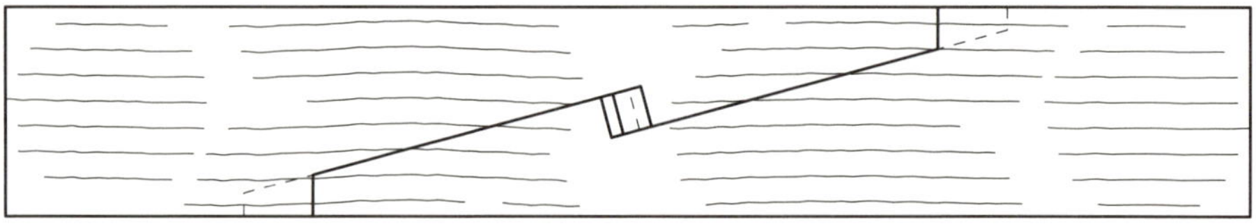

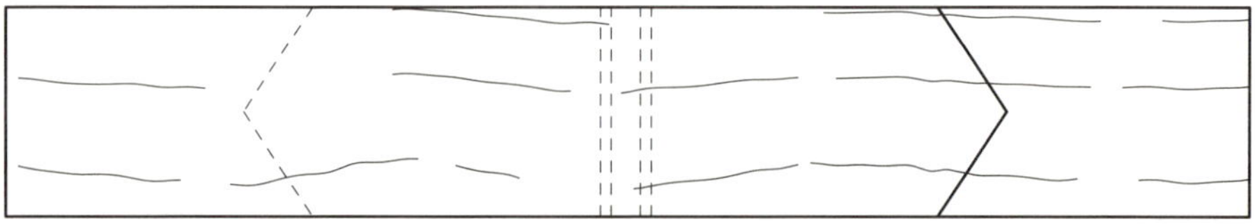

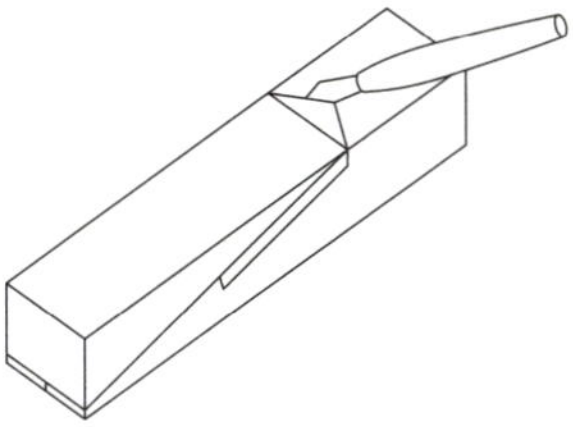

A1

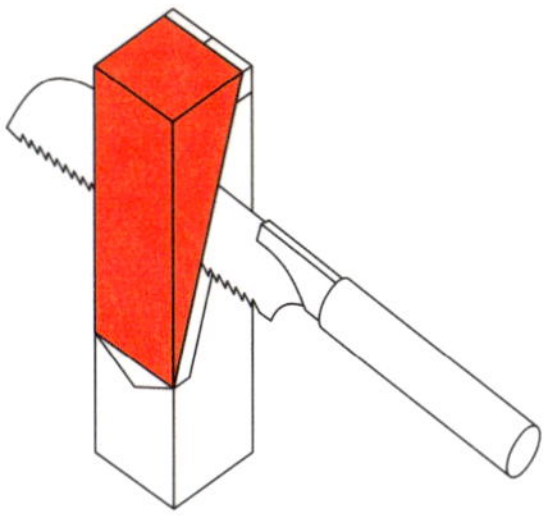

A2

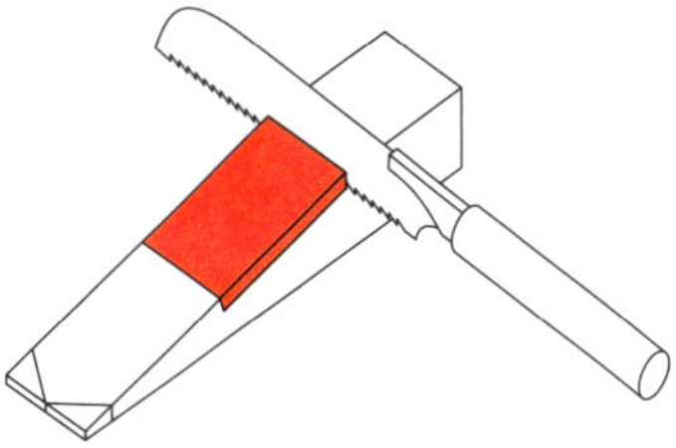

A3

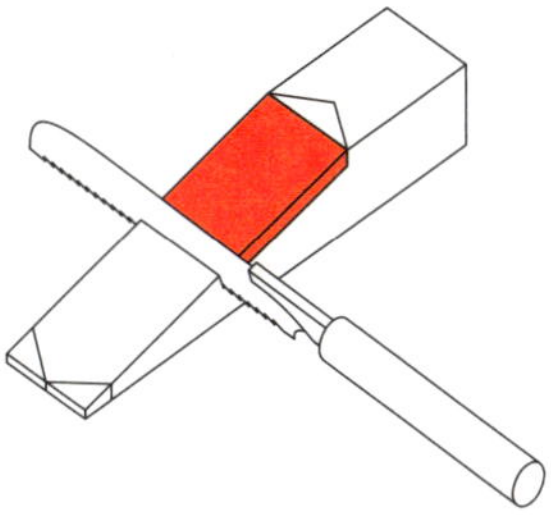

A4

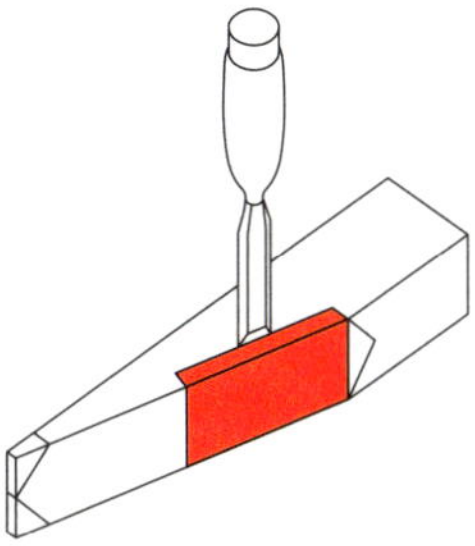

A5

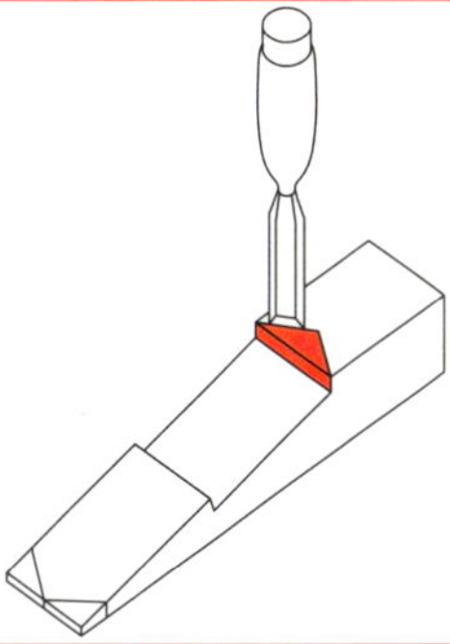

A6

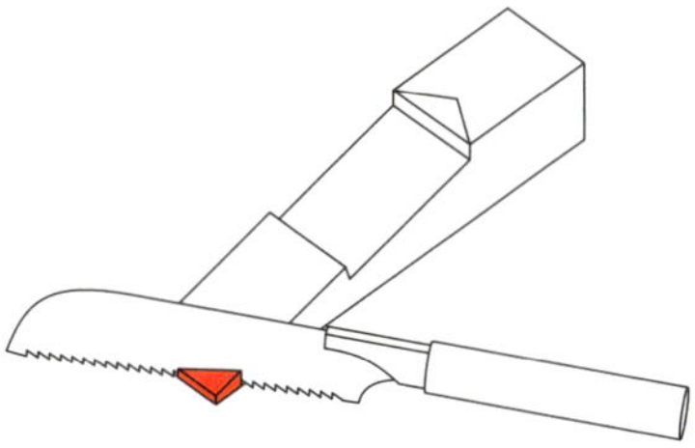

A7

2×

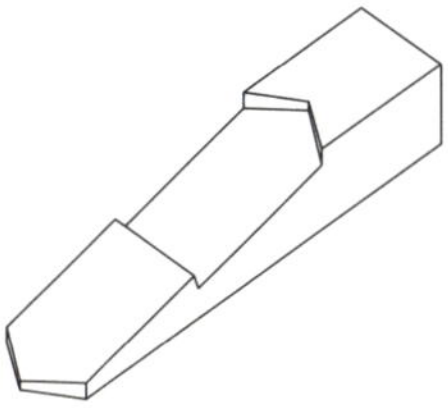

A

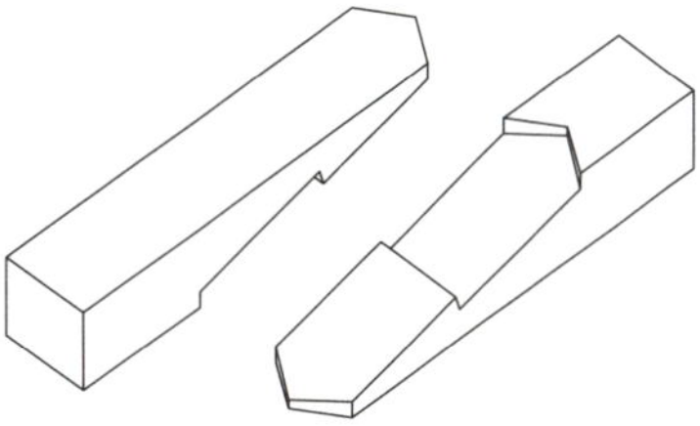

A+A

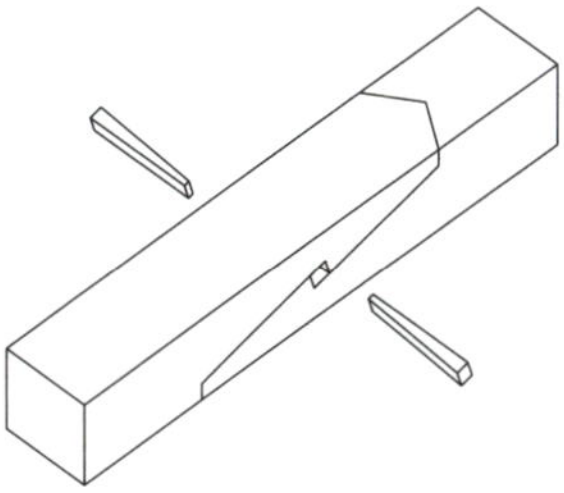

A+A

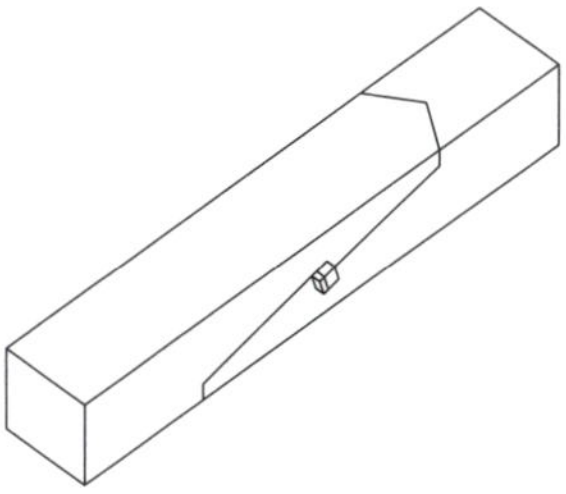

AA

Splicing Joints

Stop-splayed scarf joint with hidden hook

kakushi-yahazu-ai-kaki-sogi-tsugi
隠矢筈相欠殺継ぎ

schräges Blatt mit verdecktem Haken

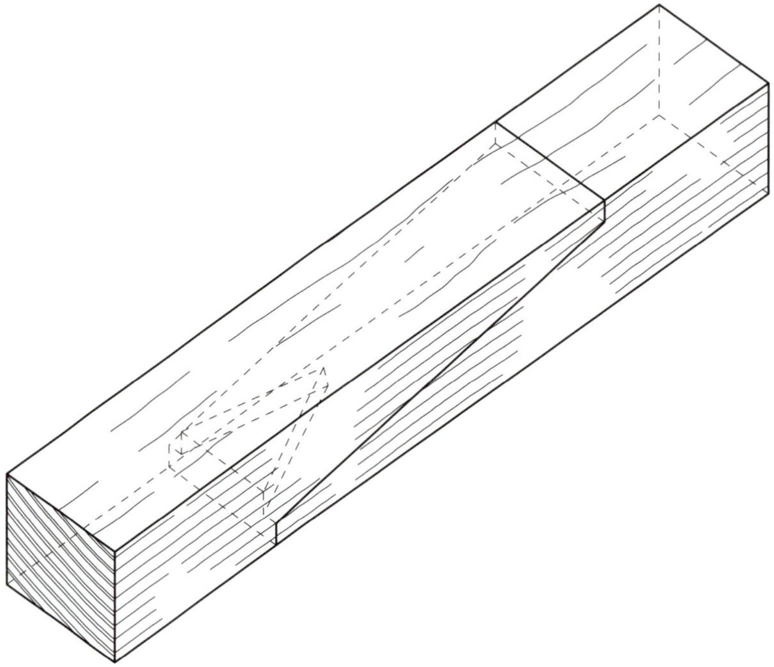

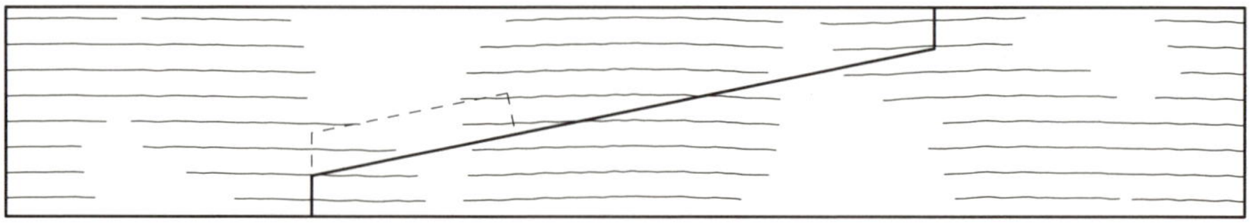

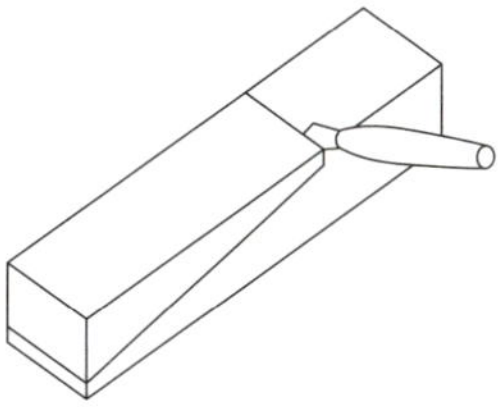

A1

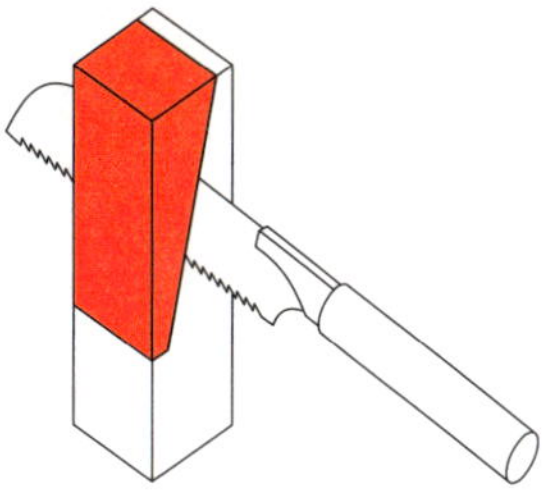

A2

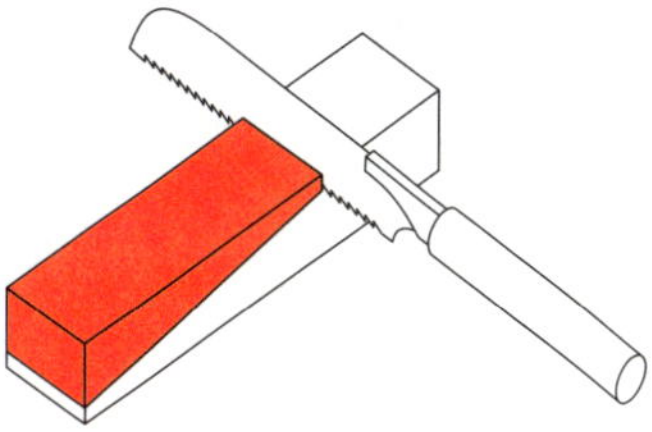

A3

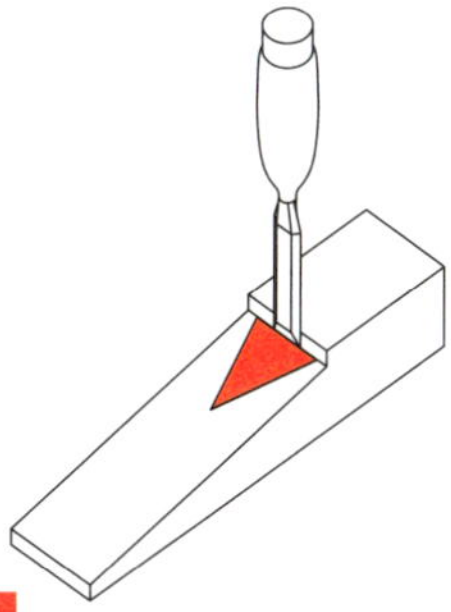

A4

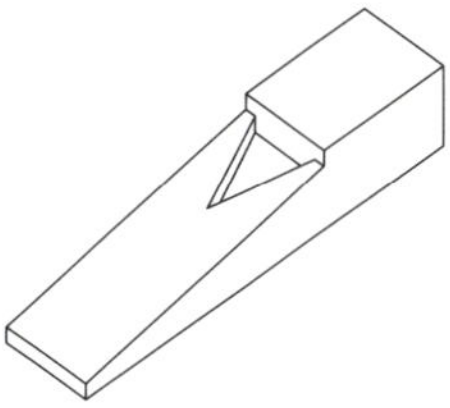

A

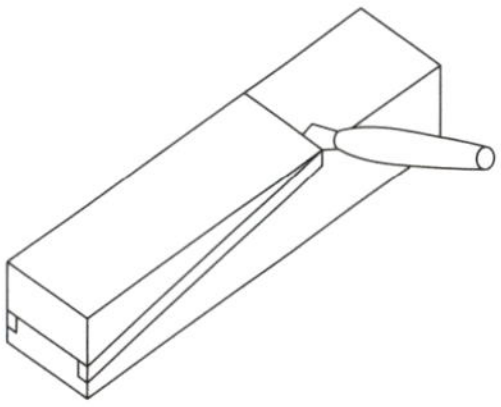

B1

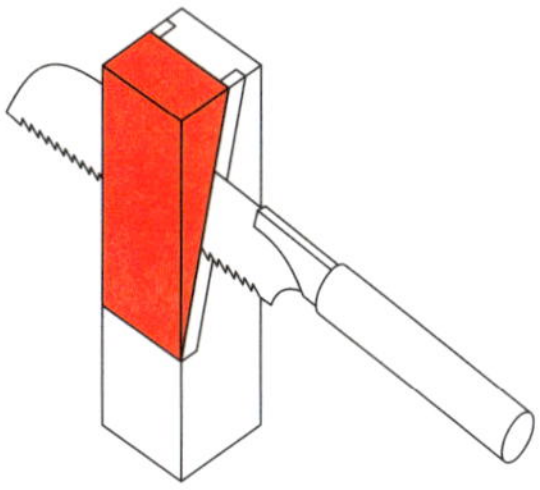

B2

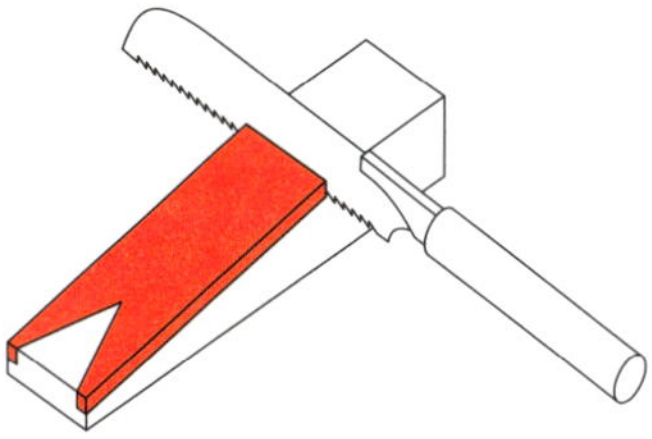

B3

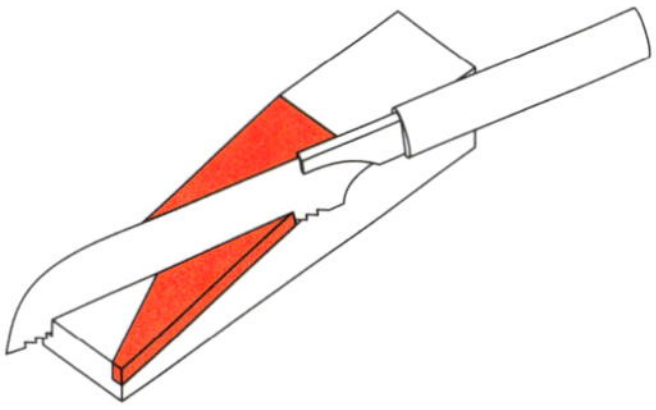

B4

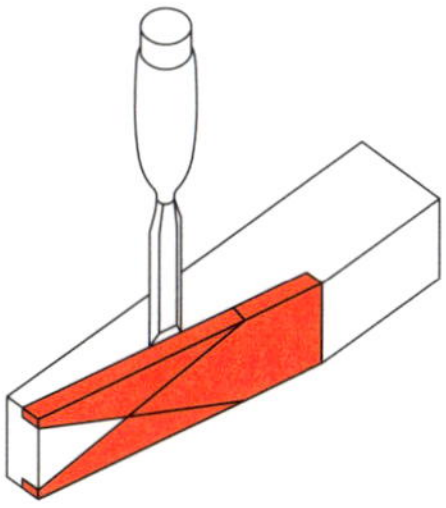

B5

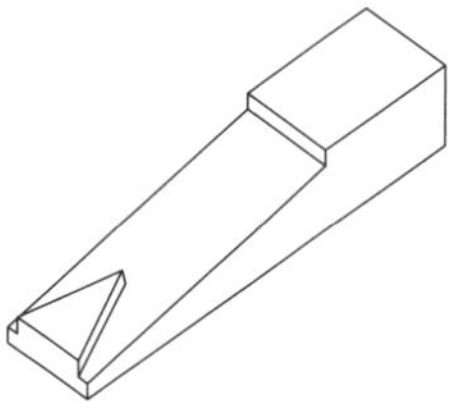

B

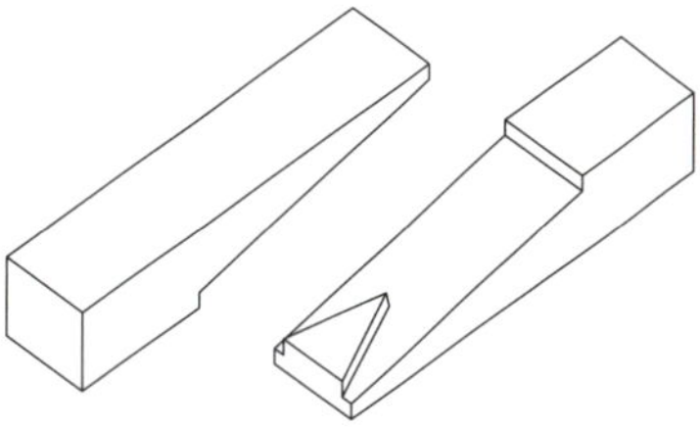

A+B

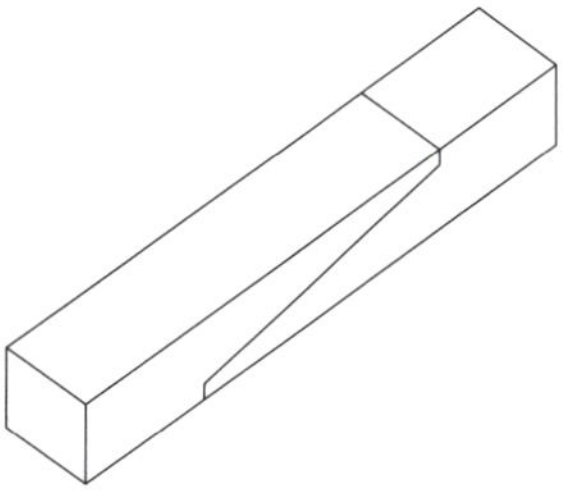

AB

Splicing Joints

Rabbeted stop-splayed cogged scarf joint

mechi-ryaku-kama-tsugi
目違略鎌継ぎ

schräges Hakenblatt mit Steckfalz

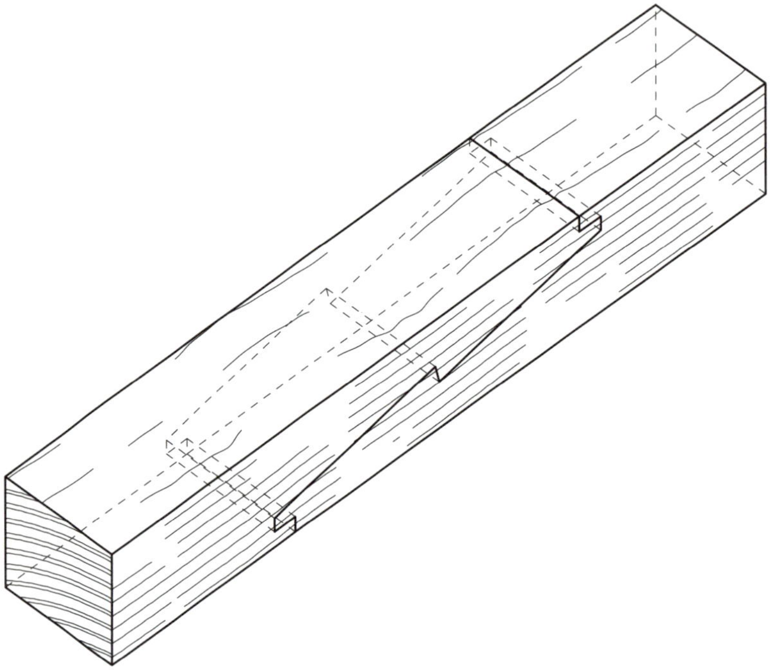

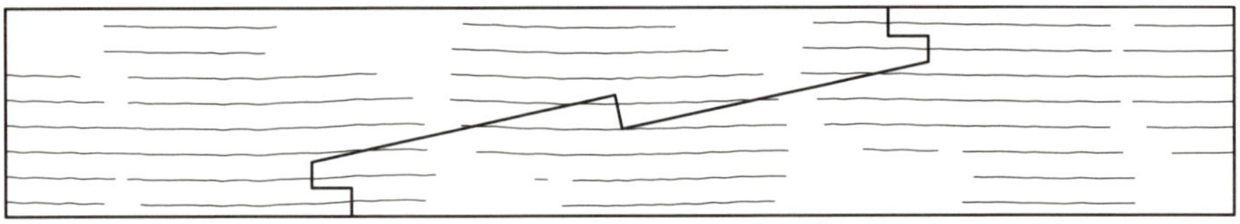

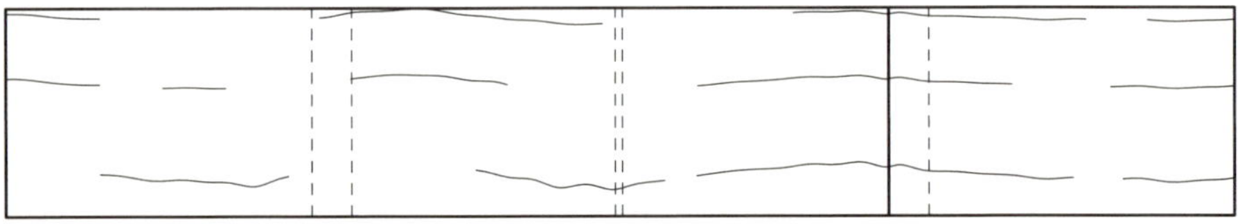

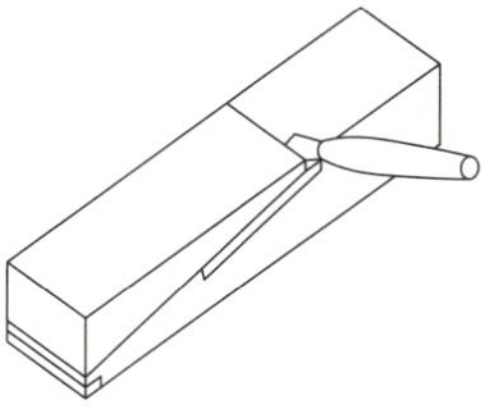

A1

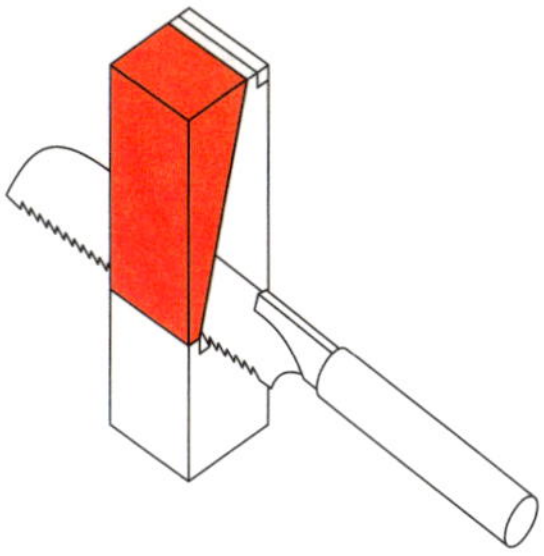

A2

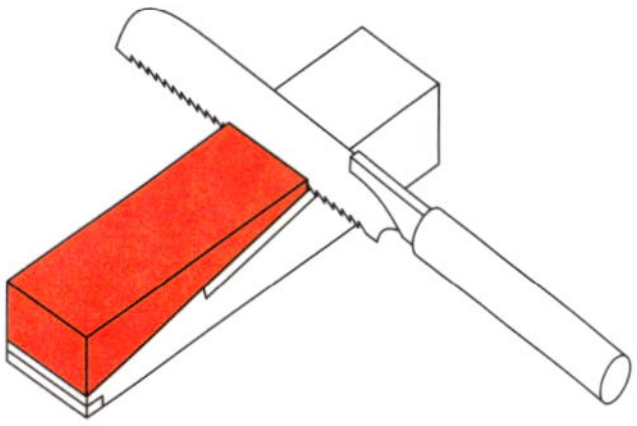

A3

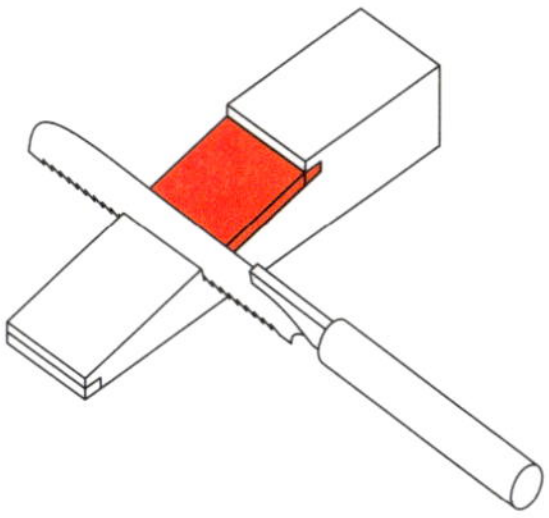

A4

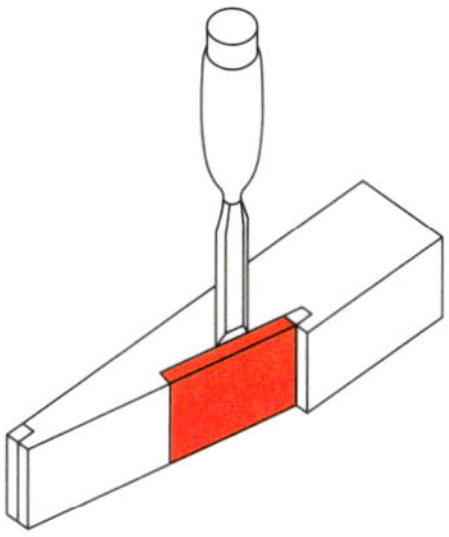

A5

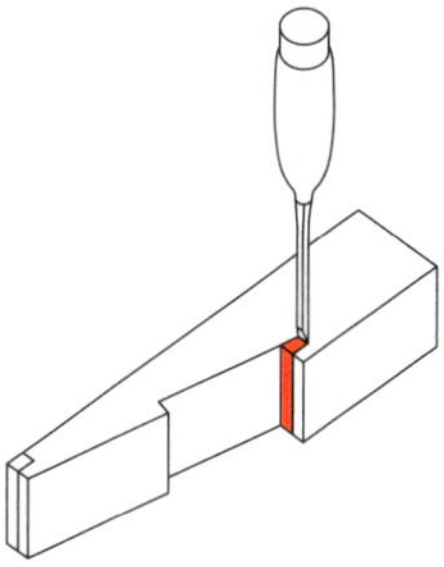

A6

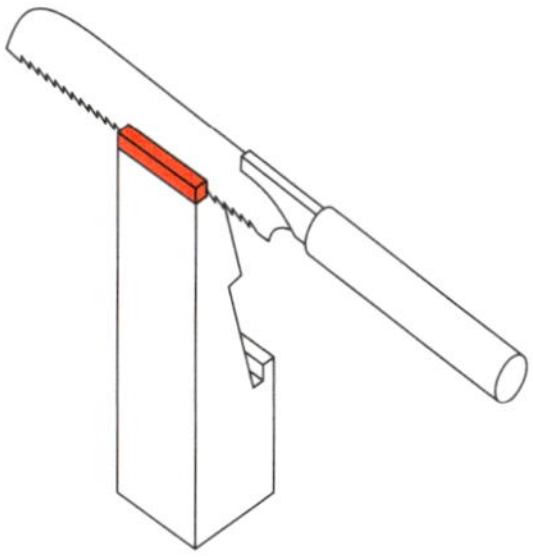

A7

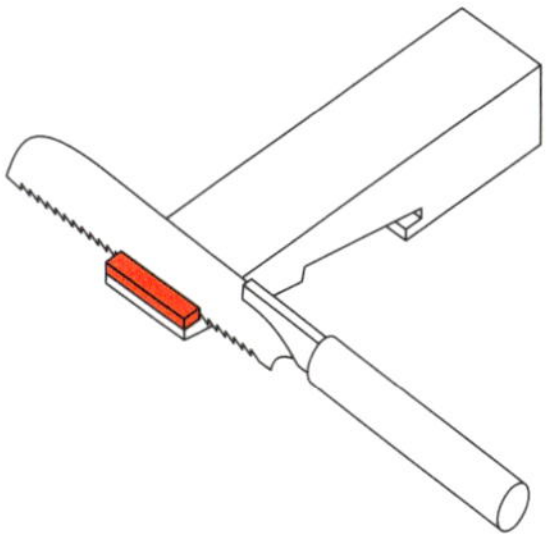

A8

2×

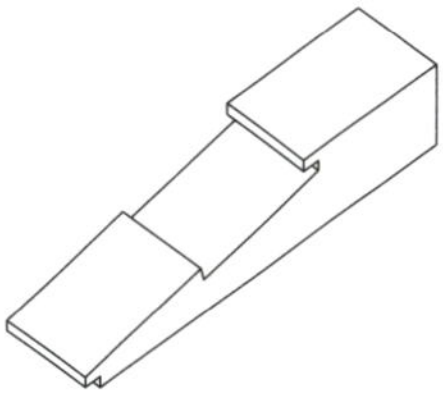

A

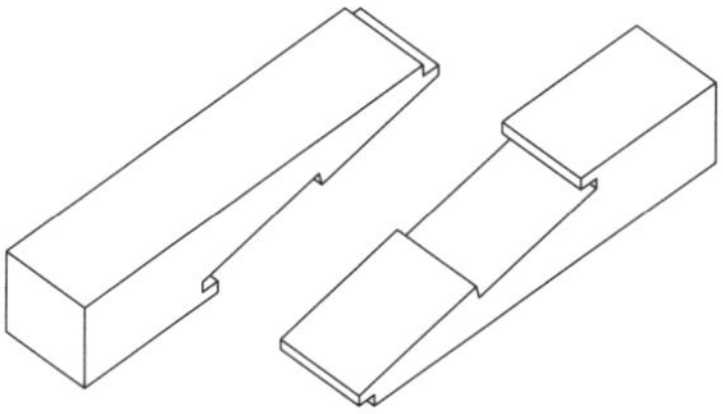

A+A

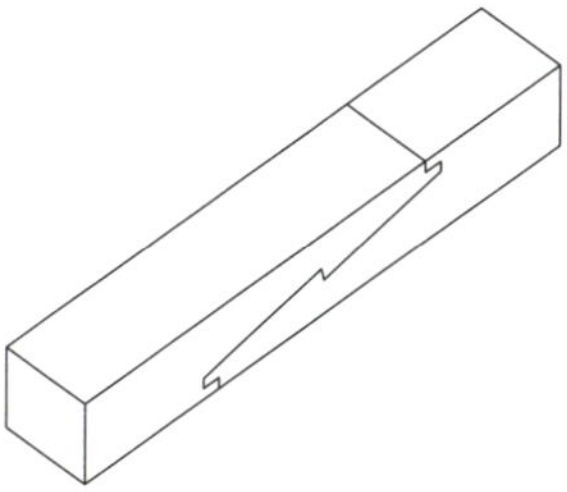

AA

Splicing Joints

Pegged rabbeted stop-splayed cogged scarf joint

mechi-ryaku-kama-tsugi-dai-sen-uchi
目違略鎌継大栓打ち

schräges Hakenblatt mit Steckfalz und zweifacher Dübelsicherung

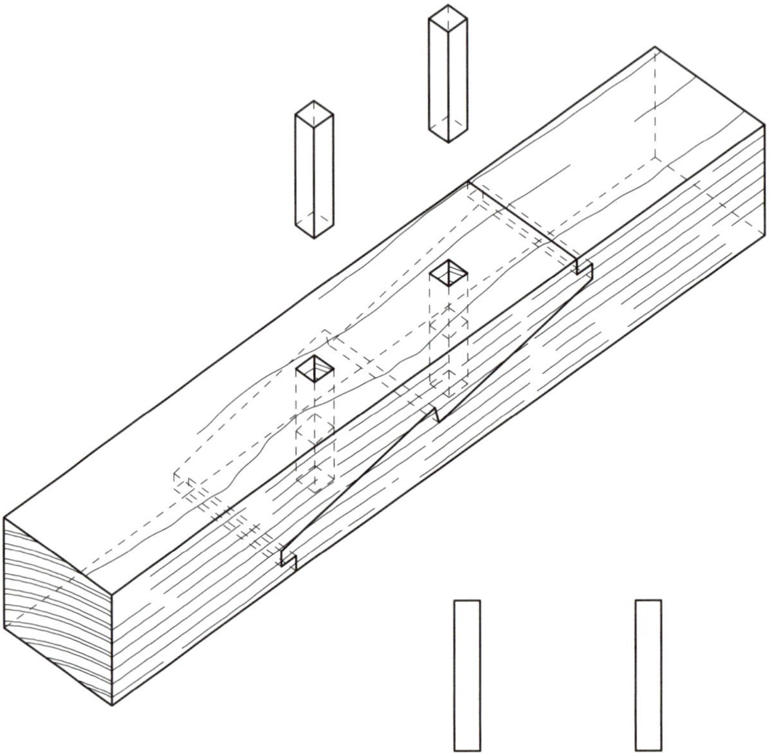

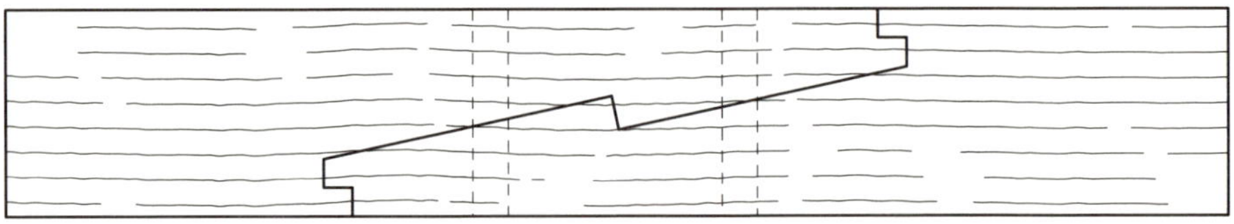

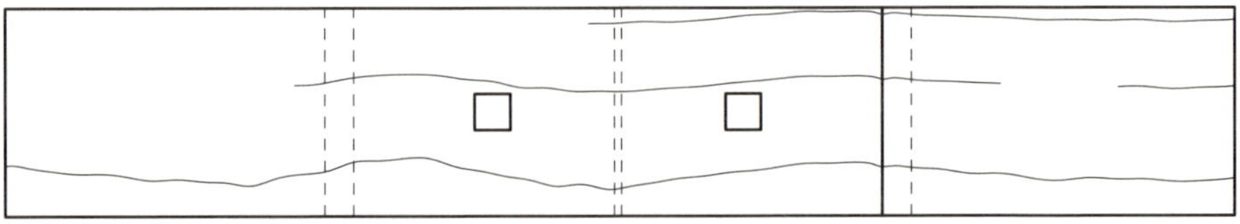

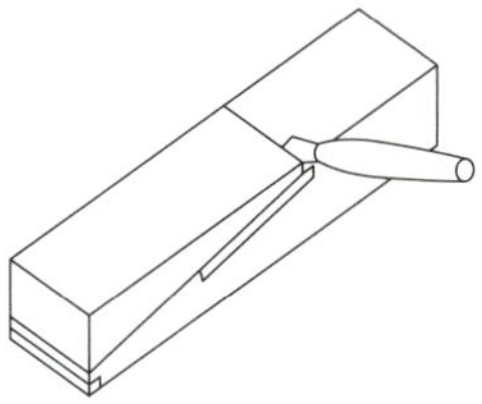
A1

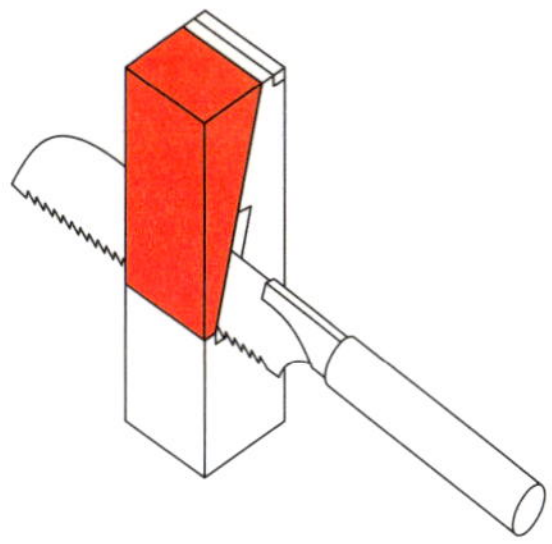
A2

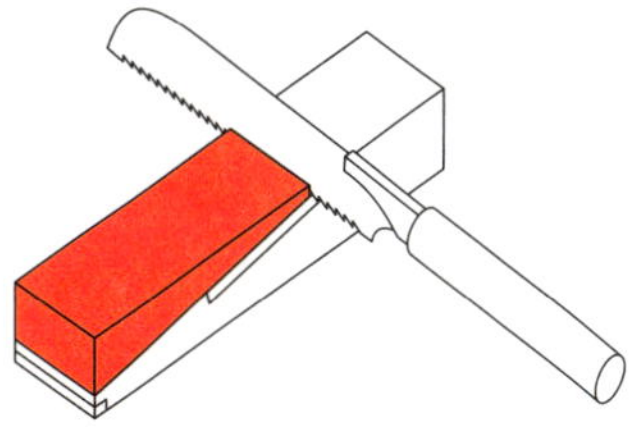
A3

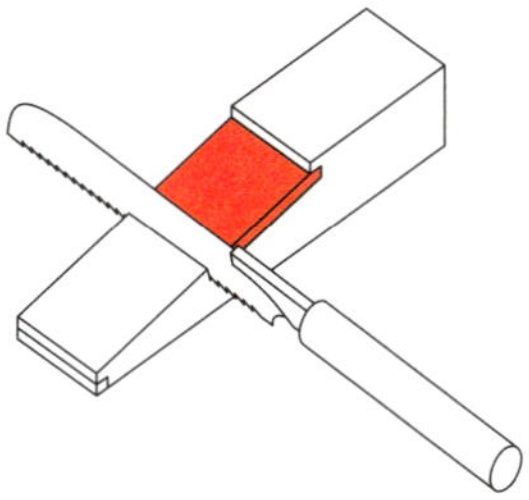
A4

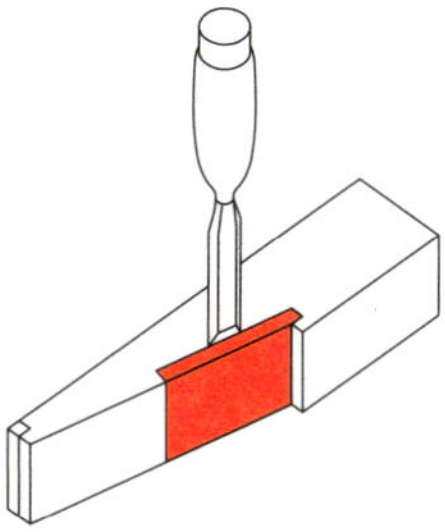
A5

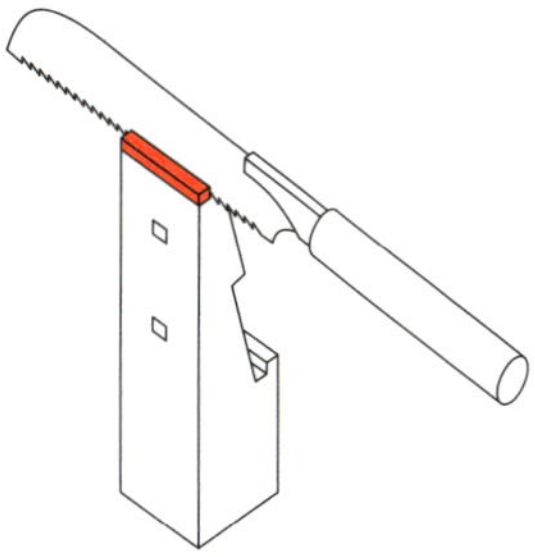
A6

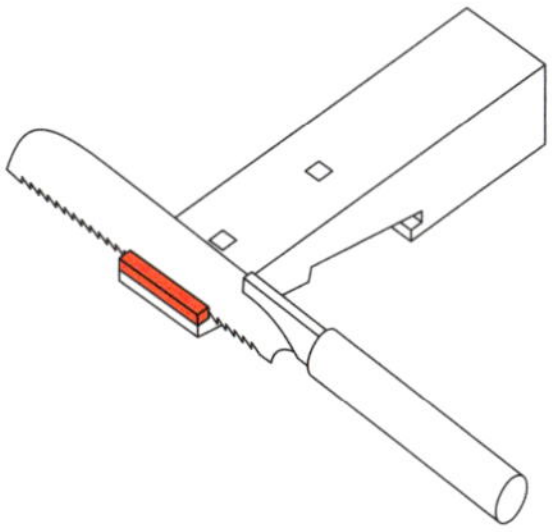
A7

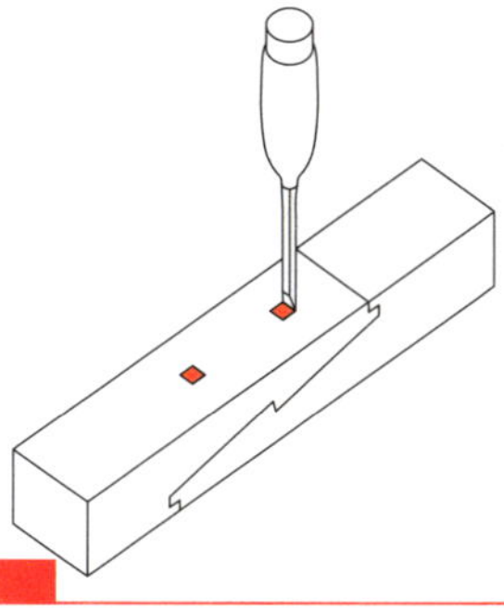
A8

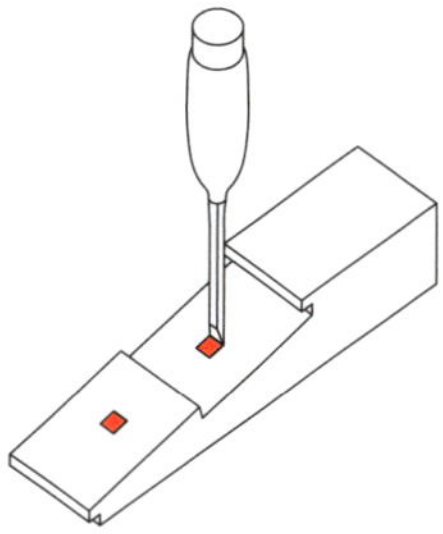
A9

2×

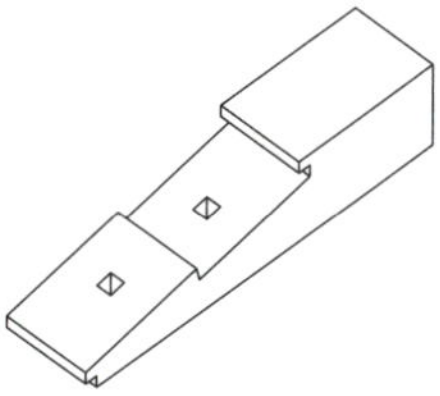
A

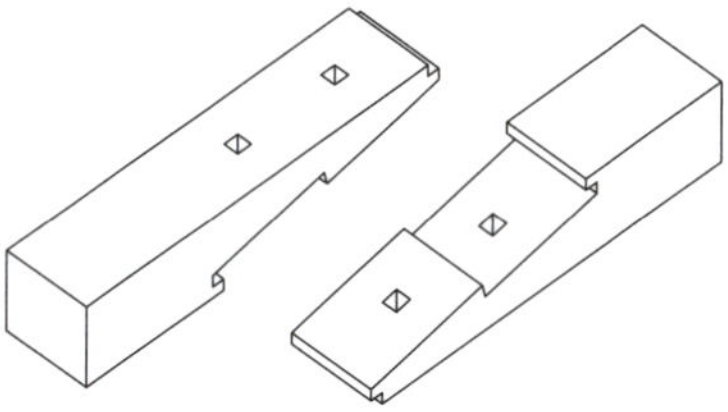
A+A

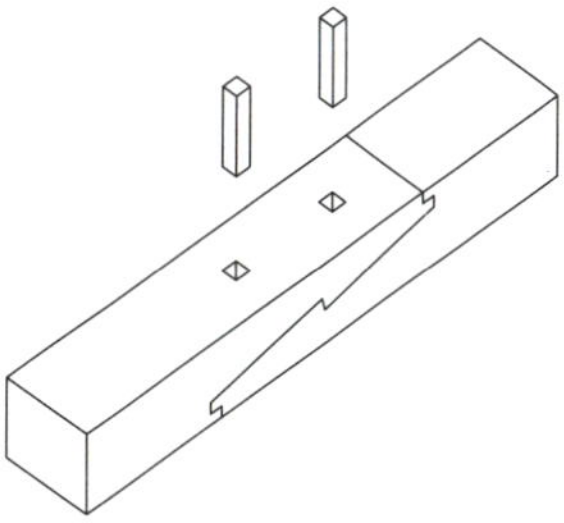
A+A

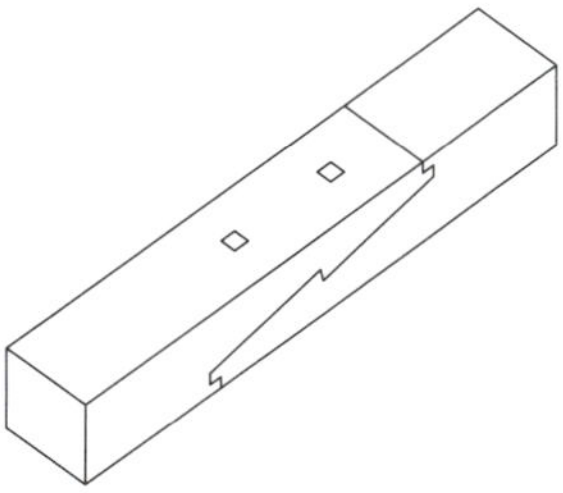
AA

Splicing Joints

Rabbeted stop-splayed cogged scarf joint with folding wedges

mechi-ryaku-kama-tsugi-komi-sen-uchi
目違略鎌継込栓打ち

verkeiltes schräges Hakenblatt
mit Steckfalz

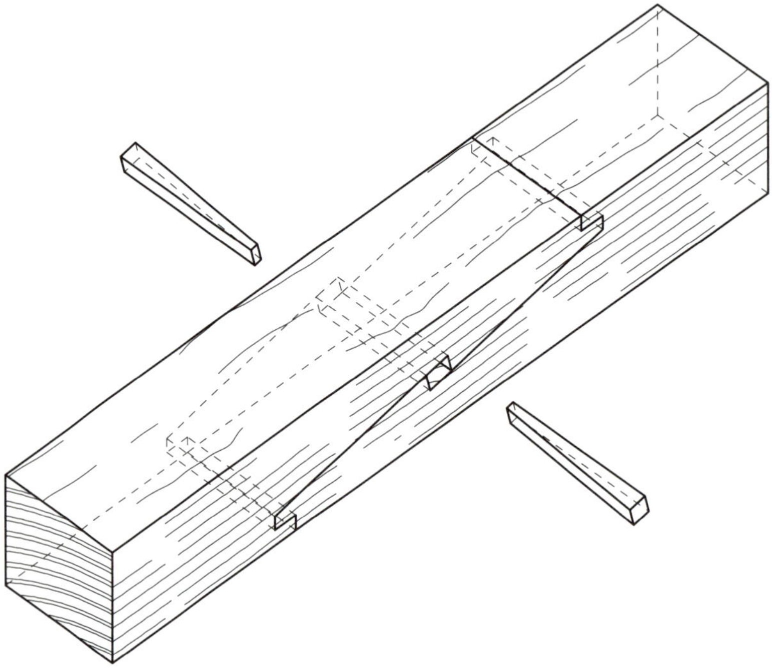

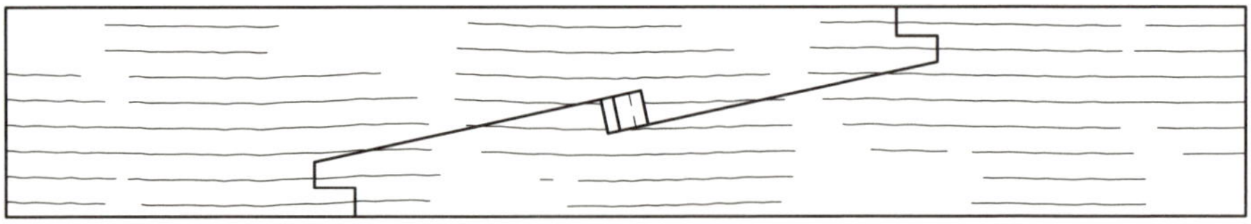

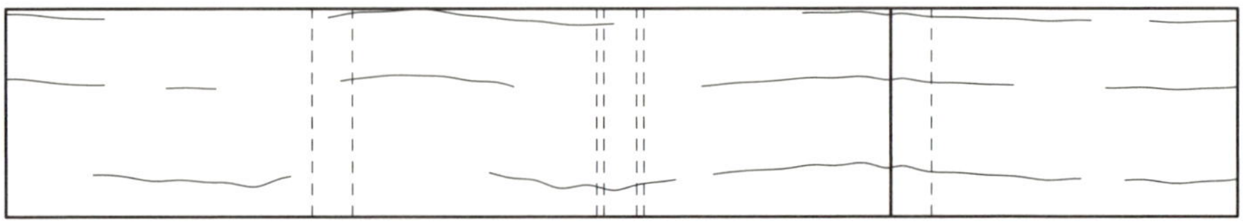

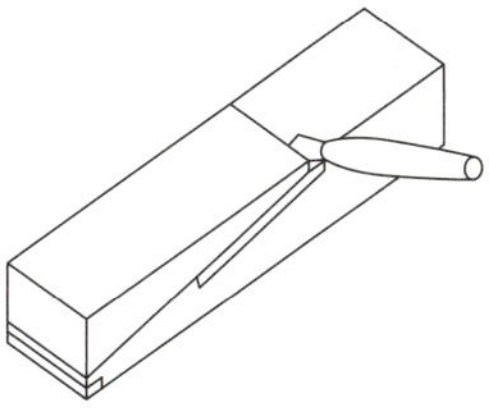

A1

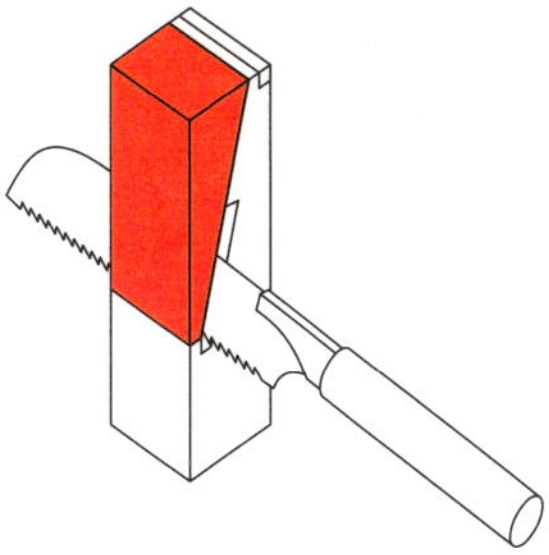

A2

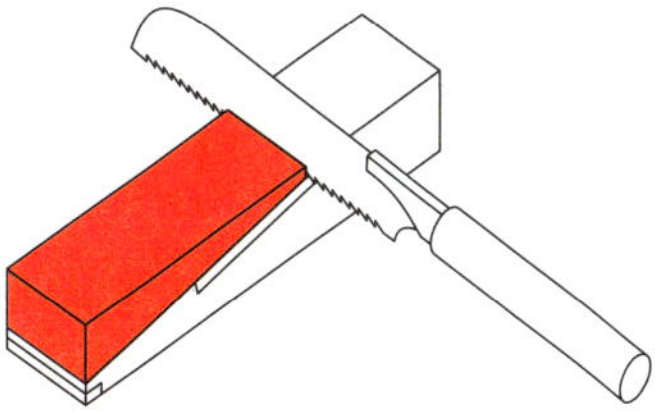

A3

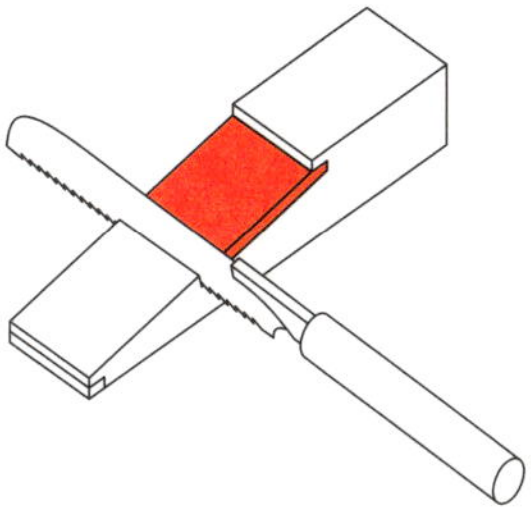

A4

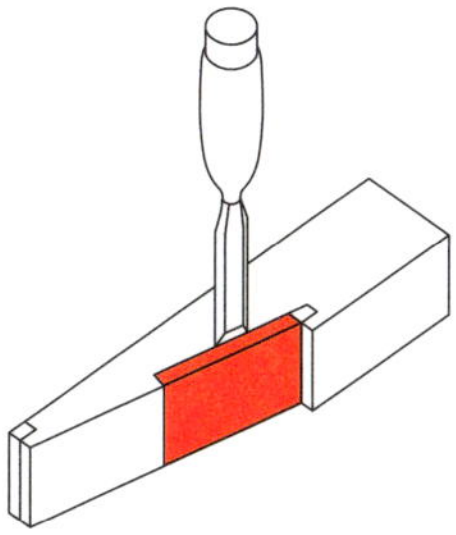

A5

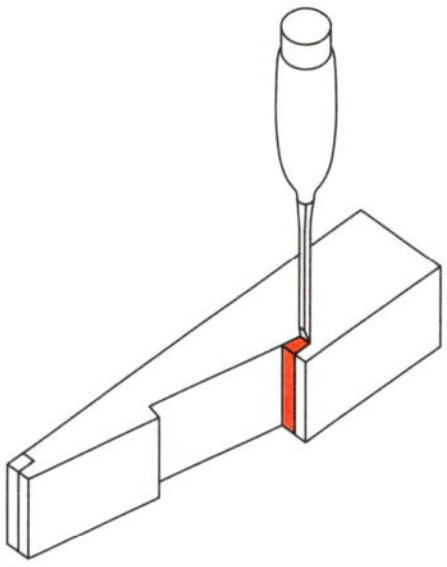

A6

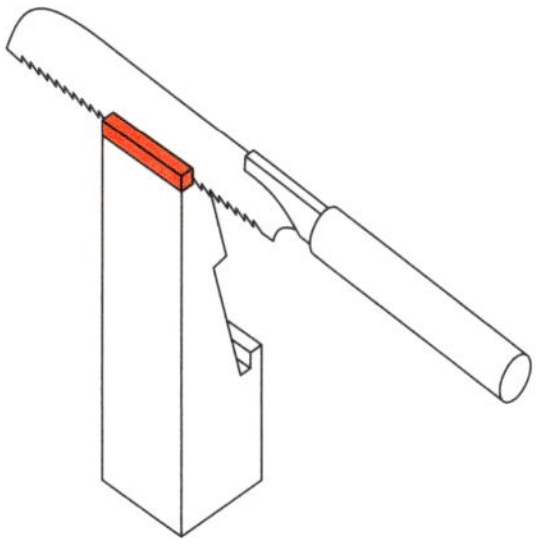

A7

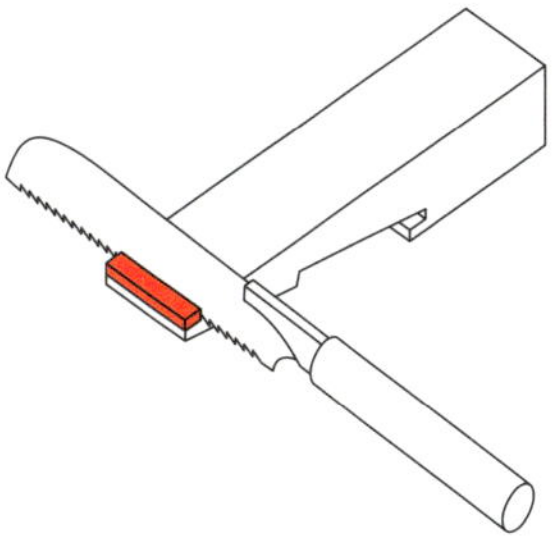

A8

2×

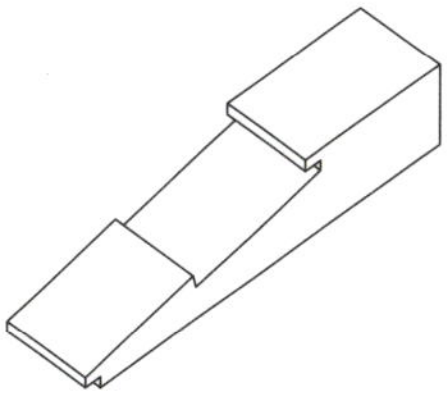

A

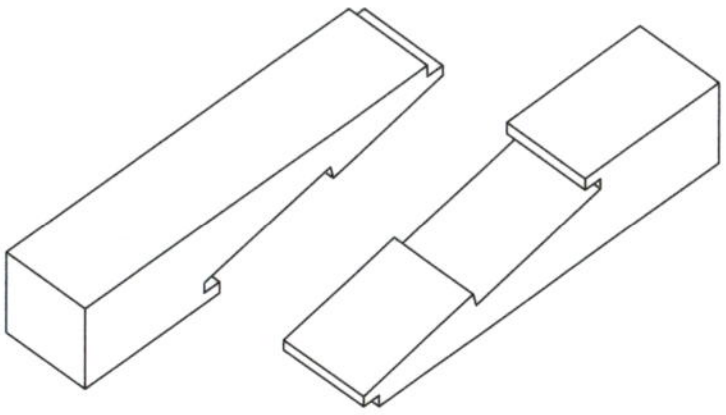

A+A

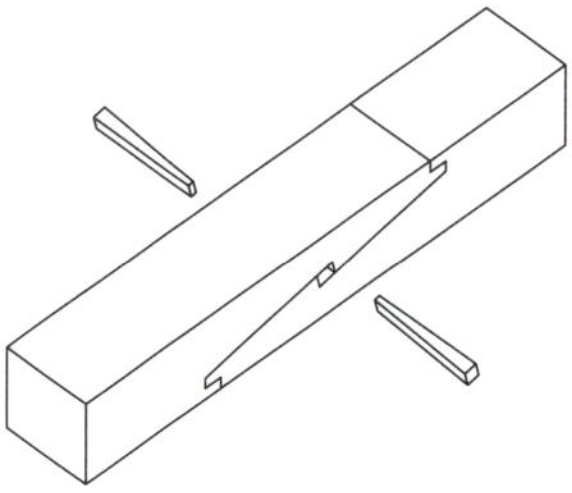

A+A

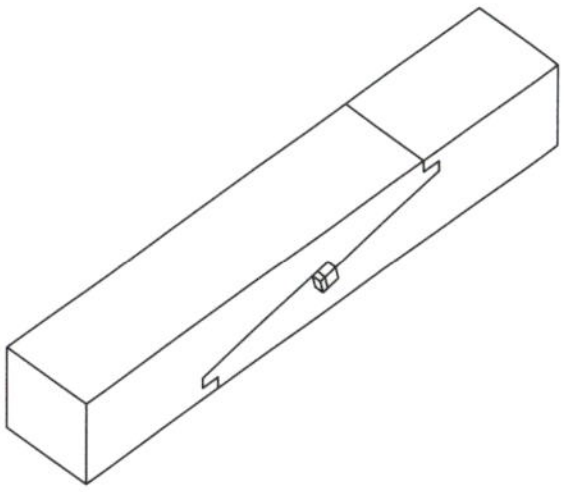

AA

Splicing Joints

Rabbeted stop-splayed cogged scarf joint with stub tenons and folding wedges

teiji-mechi-ryaku-kama-tsugi (kusabi-shime)
丁字目違略鎌継ぎ(楔締)

verkeiltes schräges Hakenblatt mit Steckfalz und Brüstungszapfen

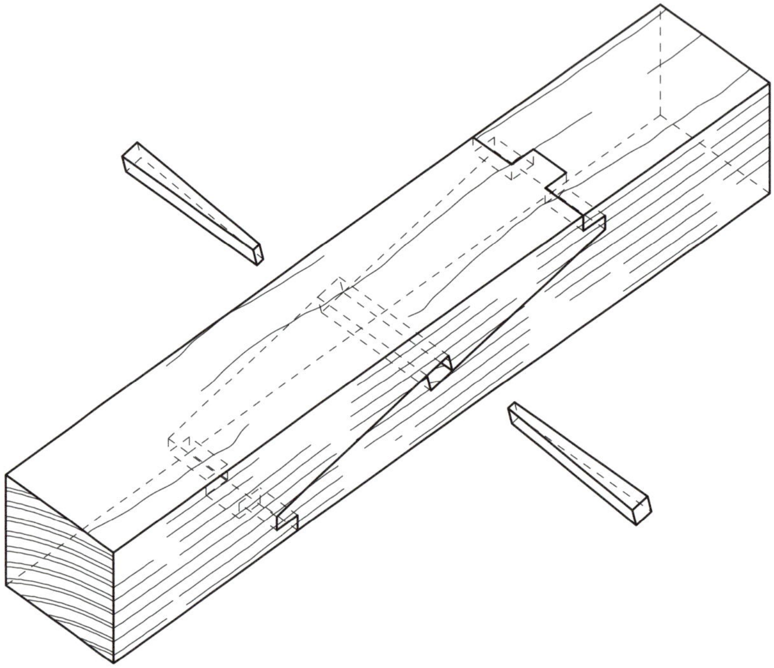

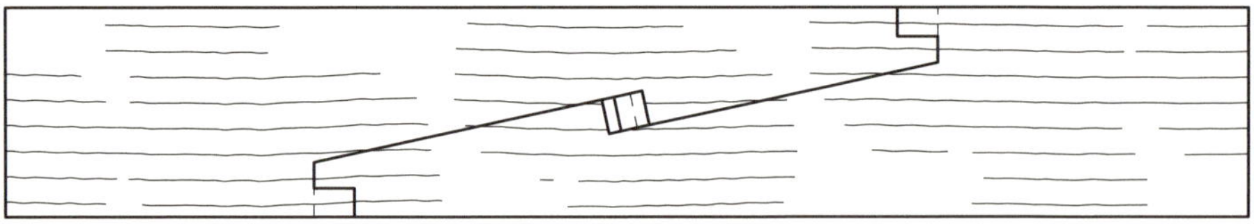

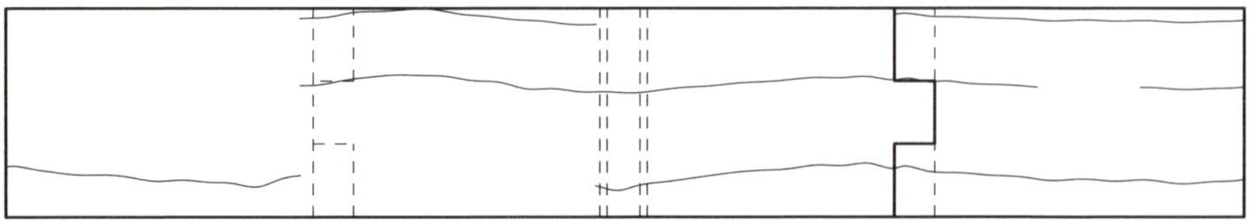

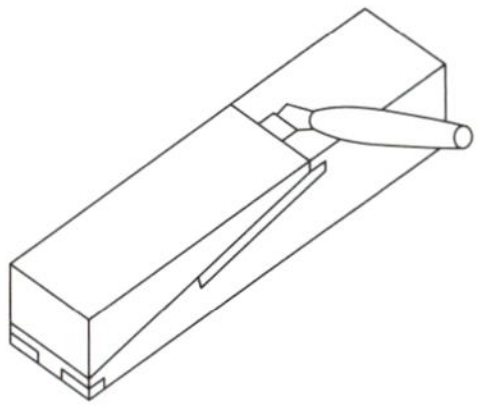
A1

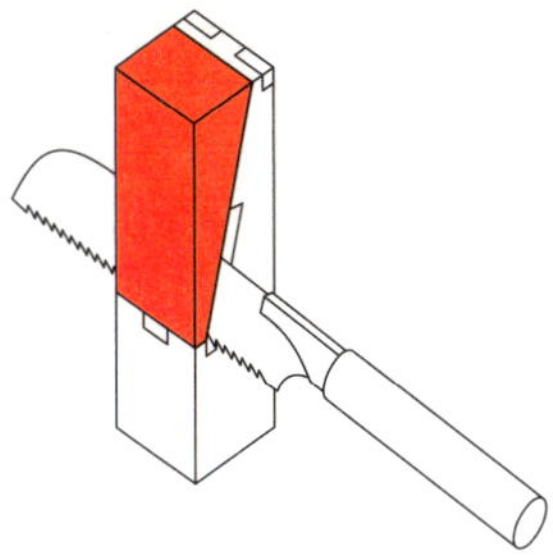
A2

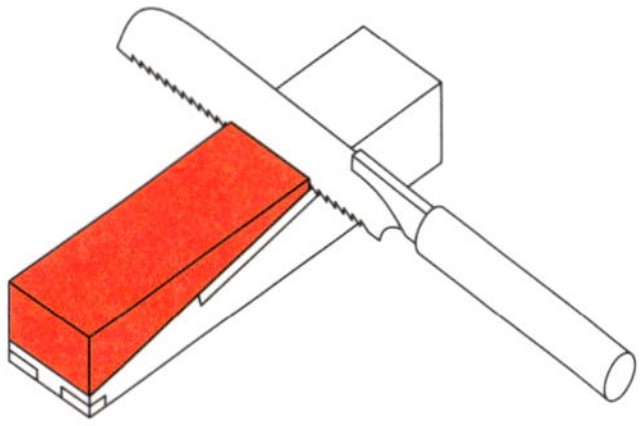
A3

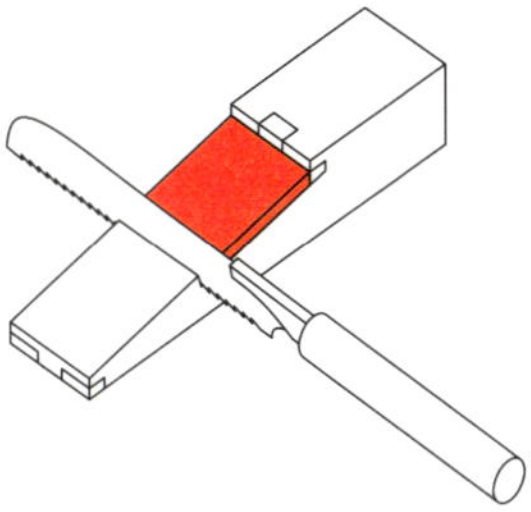
A4

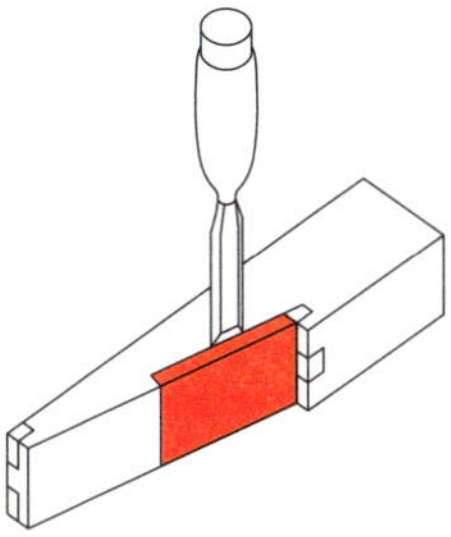
A5

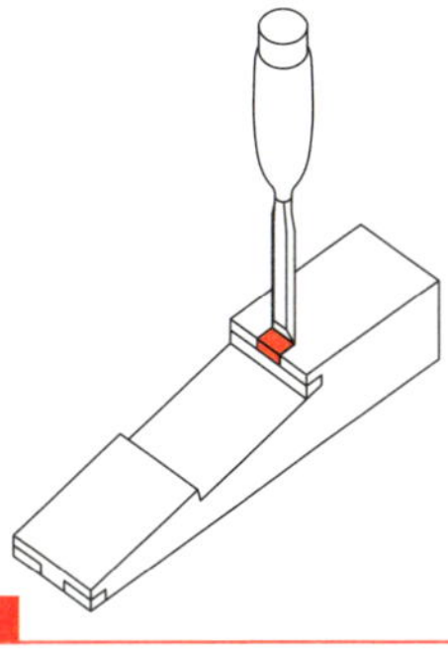
A6

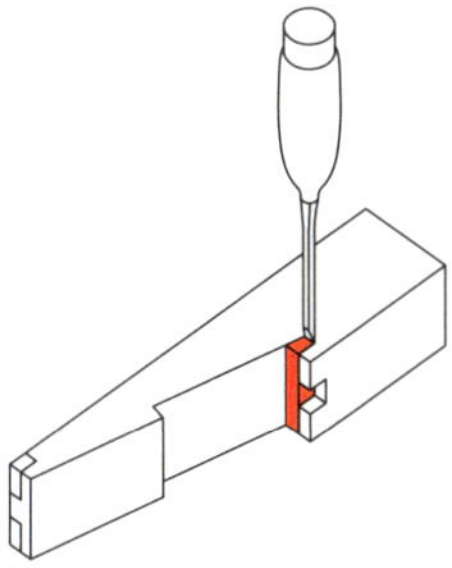
A7

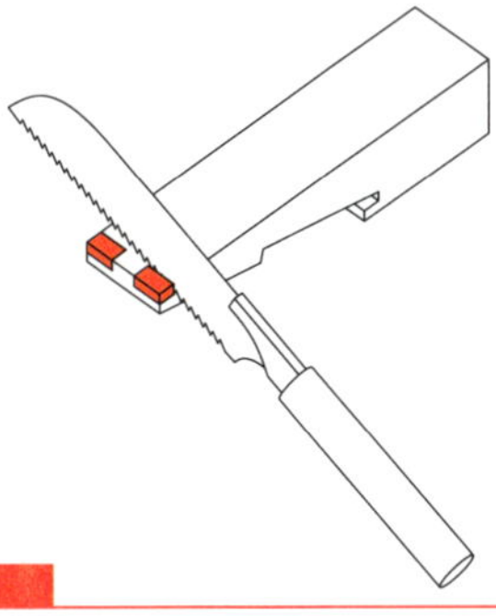
A8

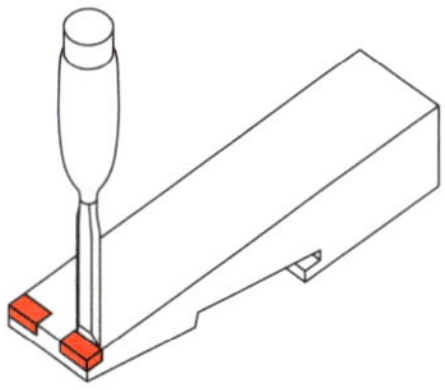
A9

2×

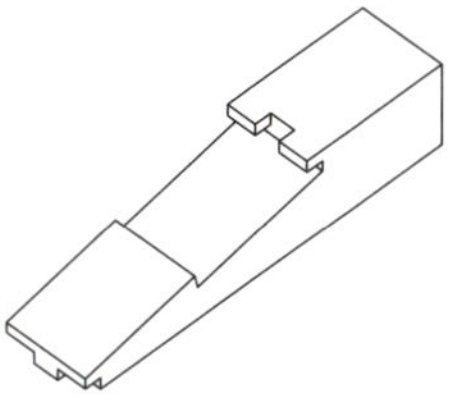
A

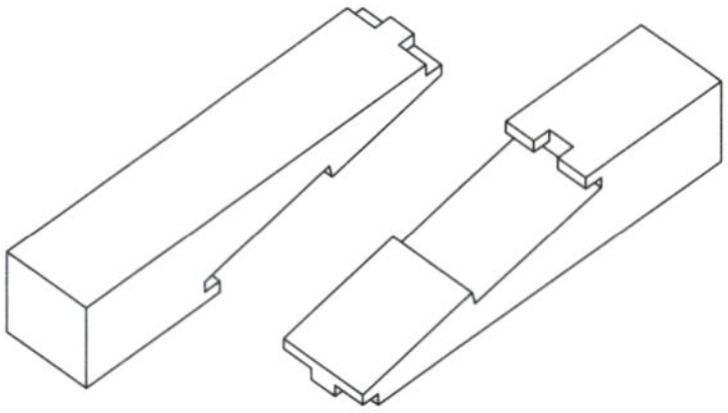
A+A

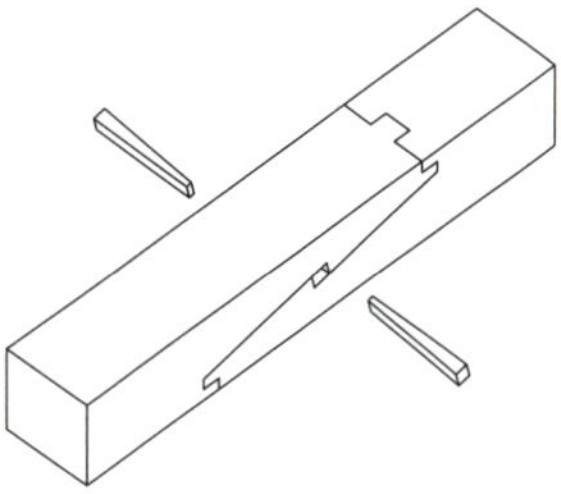
A+A

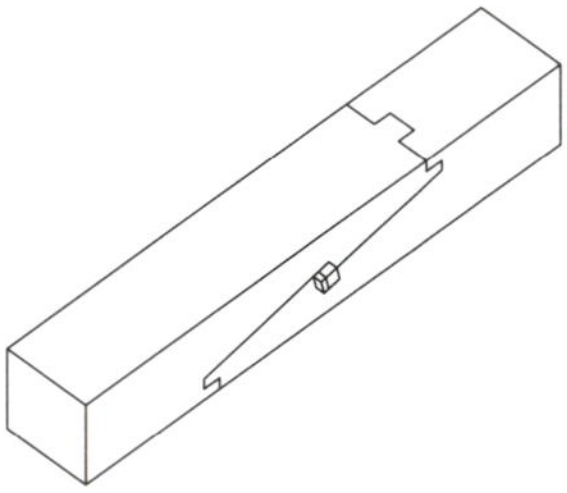
AA

Splicing Joints

Pegged rabbeted stop-splayed sloped cogged scarf joint

okkake-dai-sen-tsugi (*)
追掛大栓継ぎ

stehendes schräges Hakenblatt mit Steckfalz und doppelter Dübelsicherung

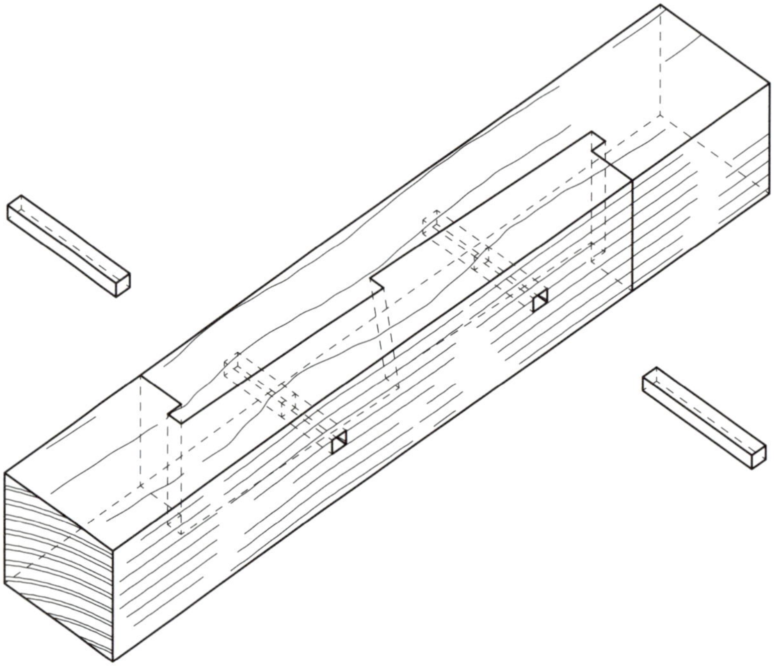

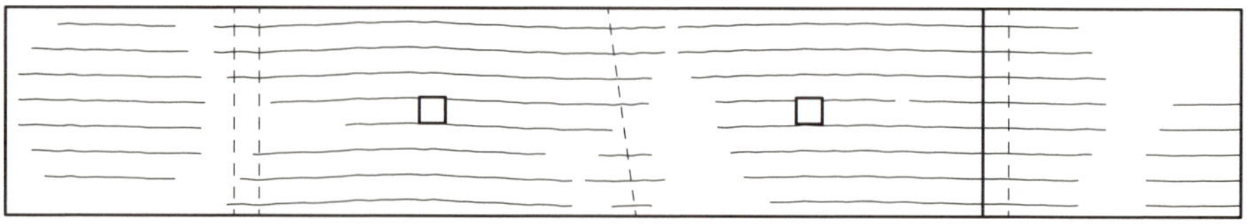

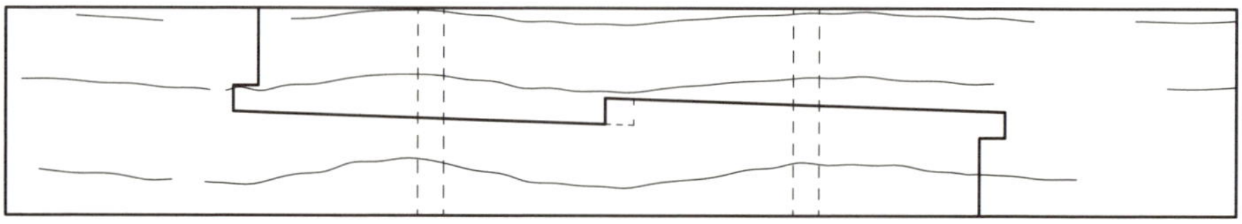

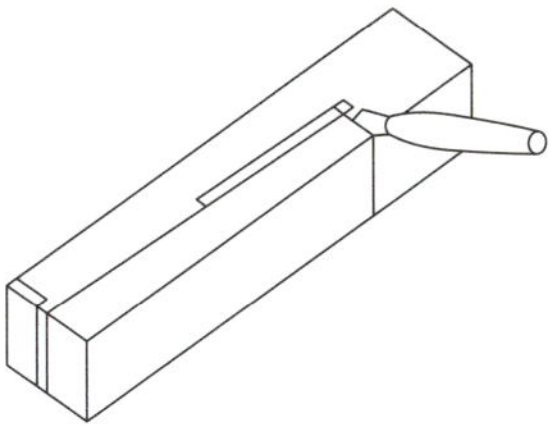

A1

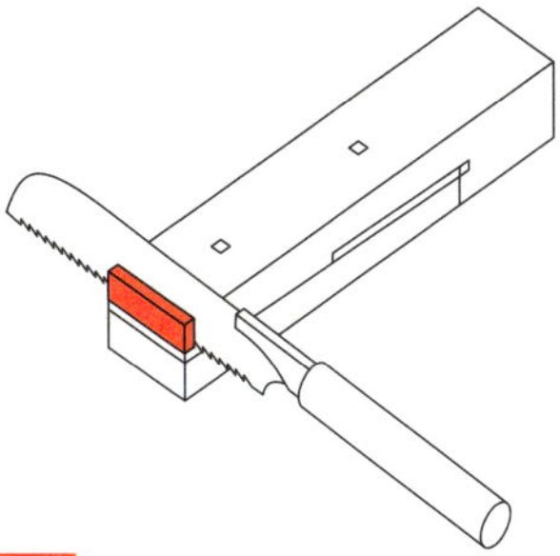

A2

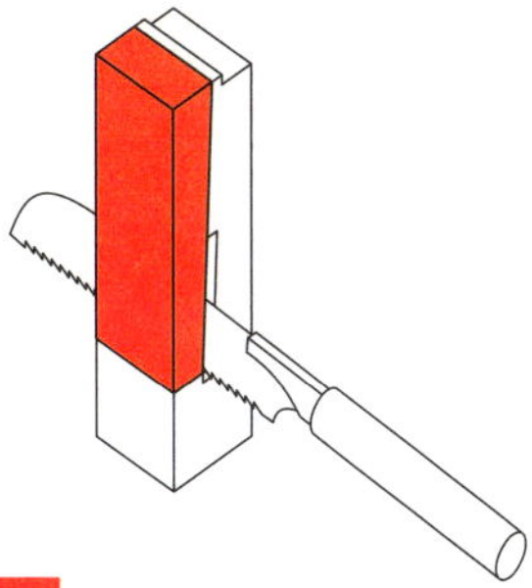

A3

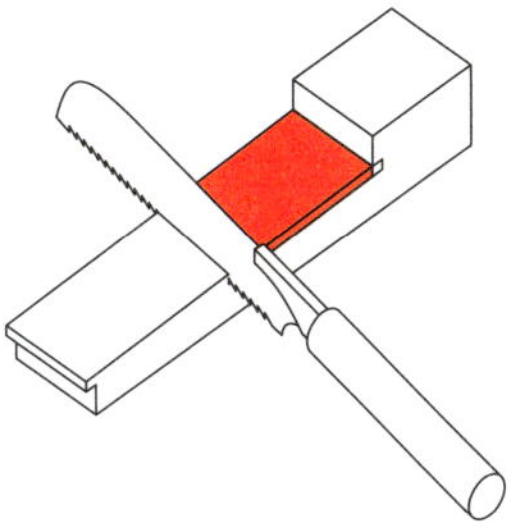

A4

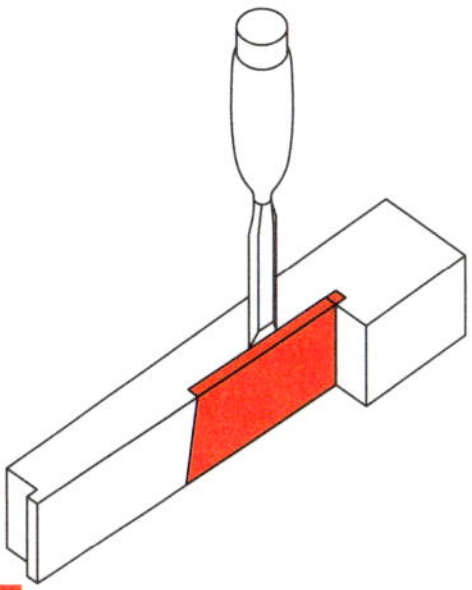

A5

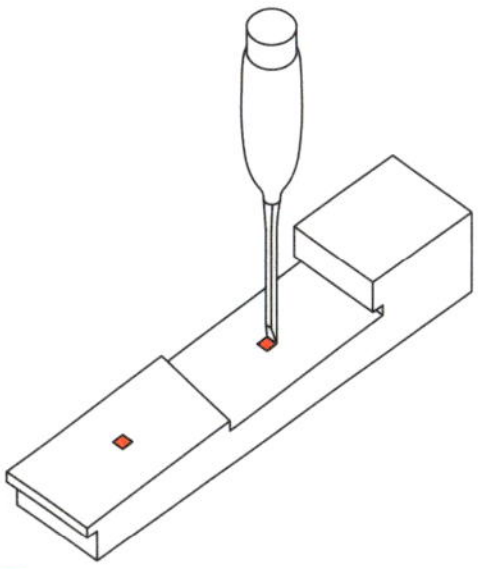

A6

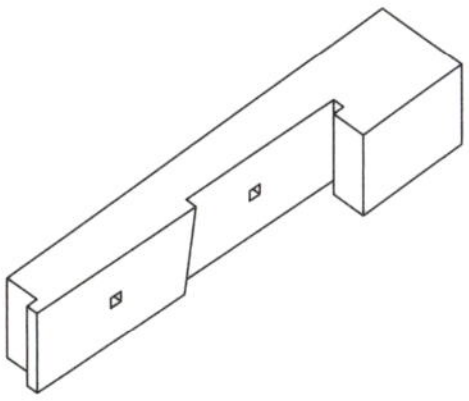

A

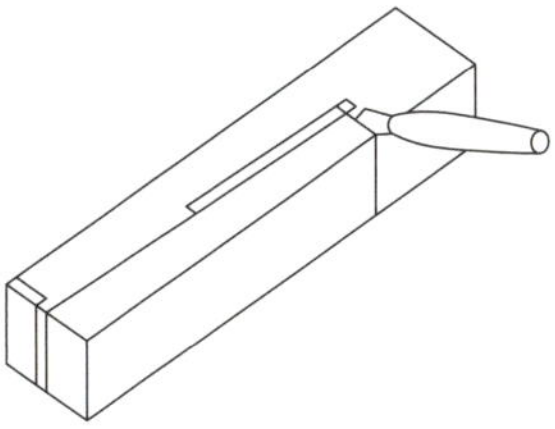

B1

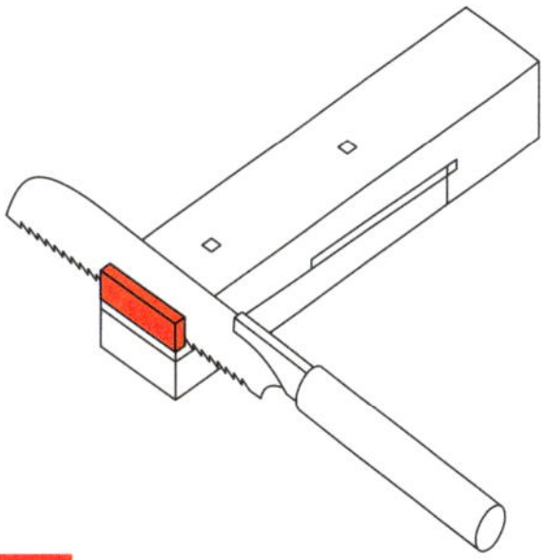

B2

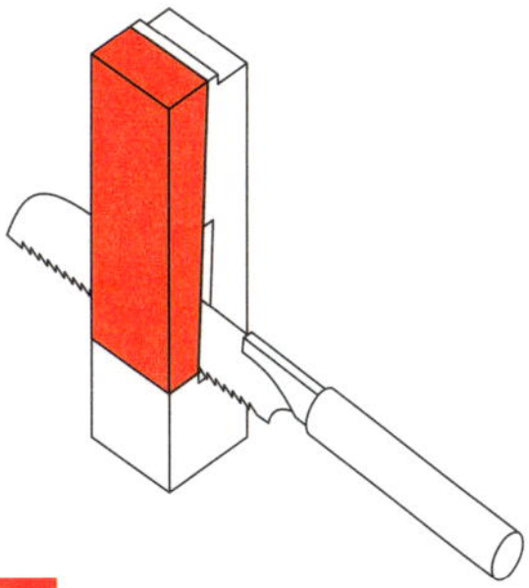

B3

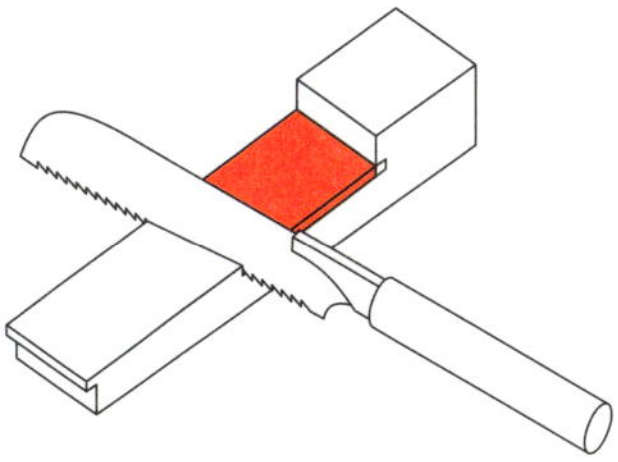

B4

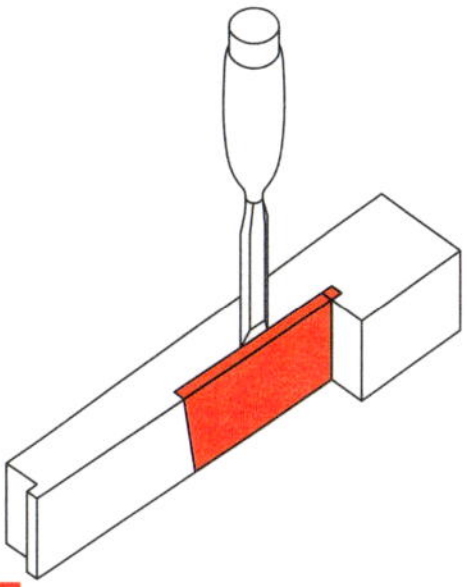

B5

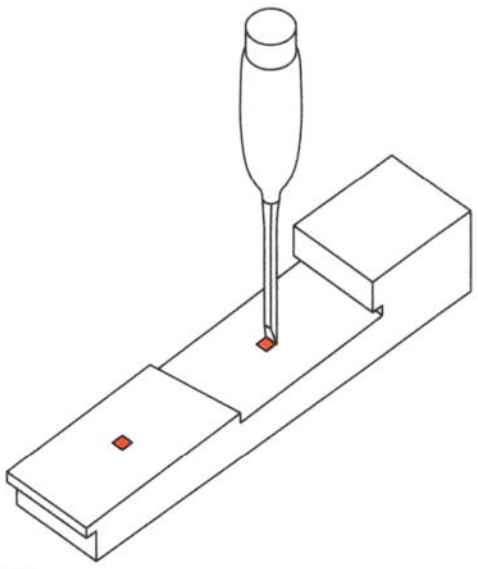

B6

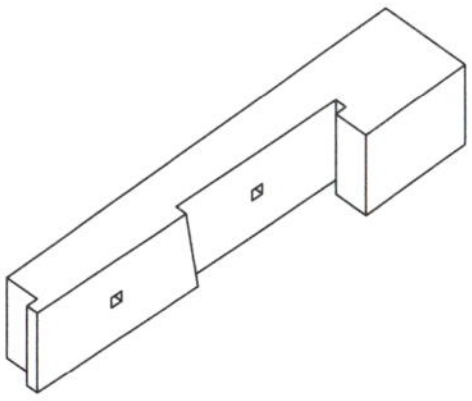

B

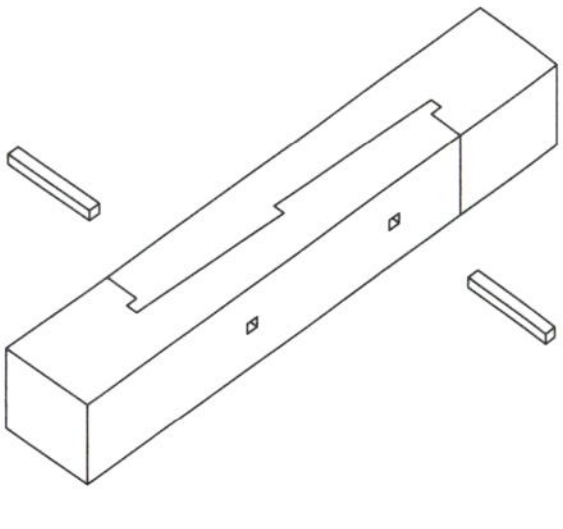

A+B

Splicing Joints

Pegged rabbeted notched stop-splayed cogged scarf joint

shippasami-tsugi [*]
尻挟継ぎ

stehendes schräges Hakenblatt mit unterbrochener Steckfalz und Dübelsicherung

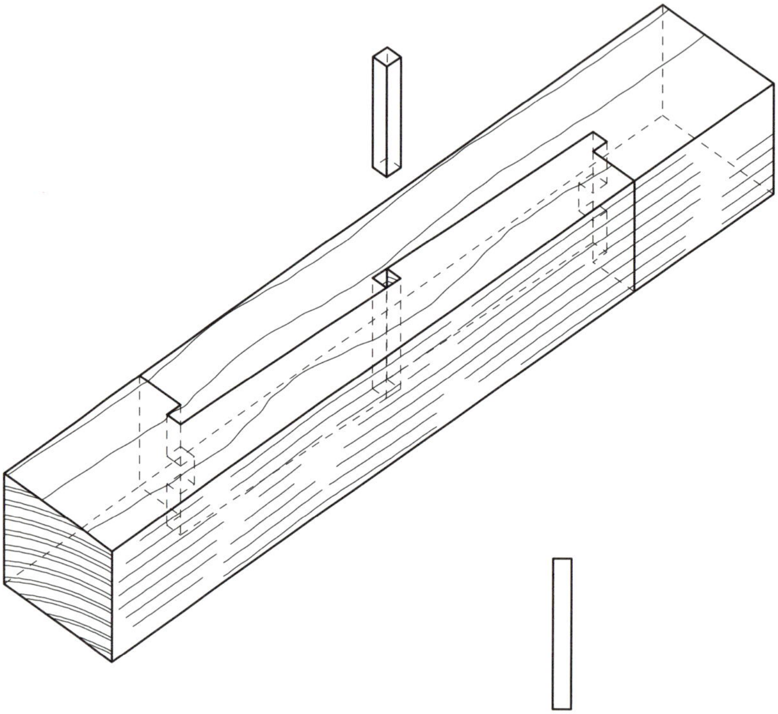

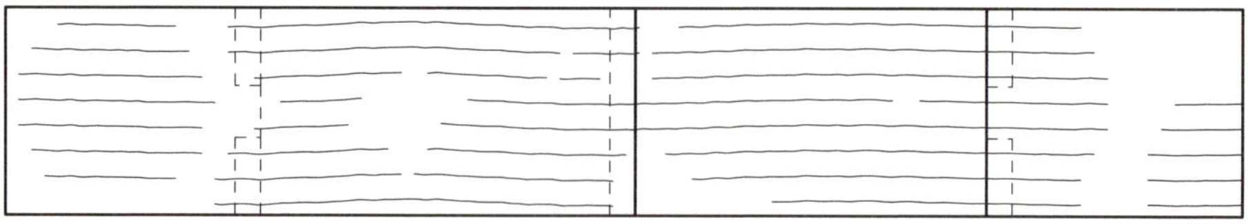

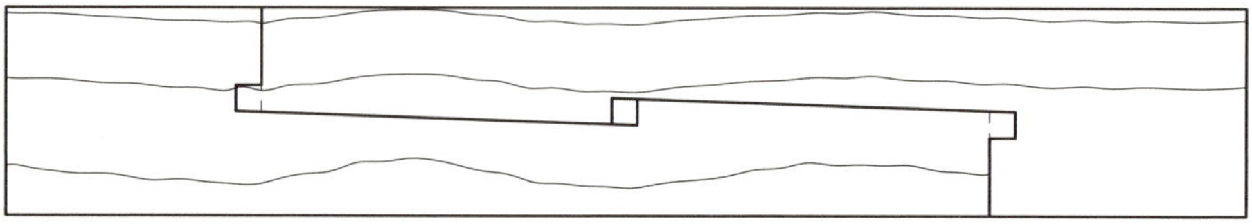

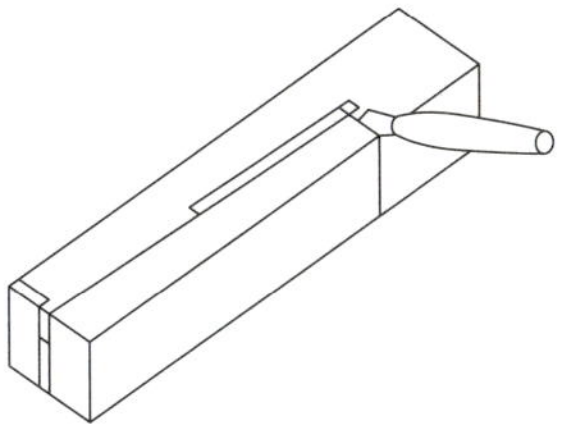

A1

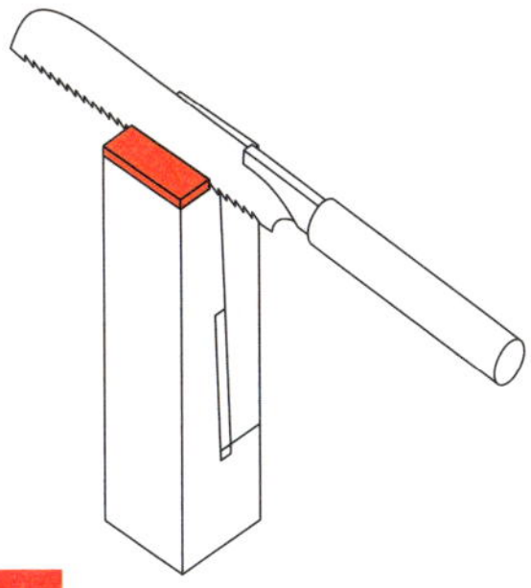

A2

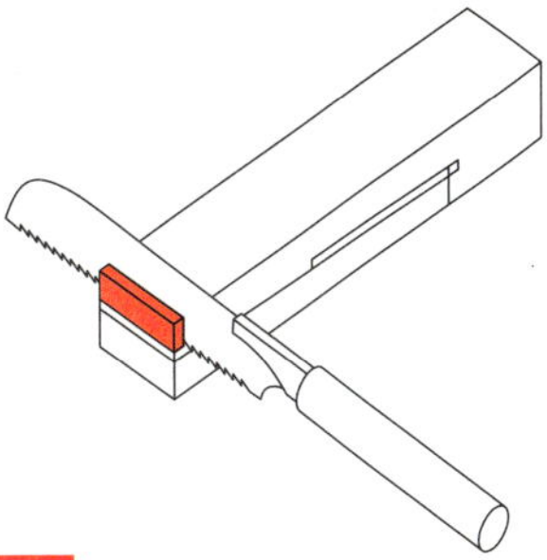

A3

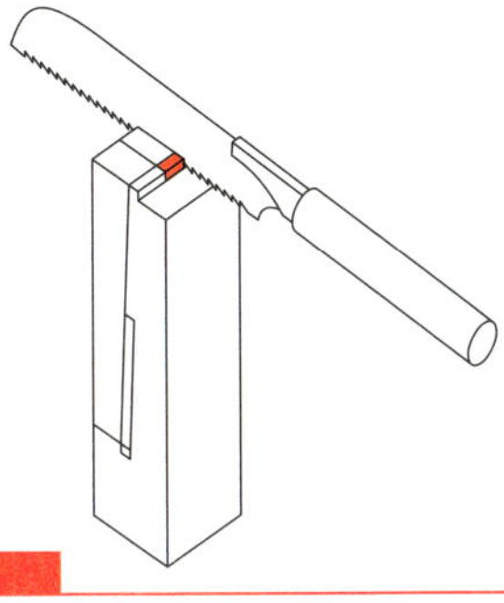

A4

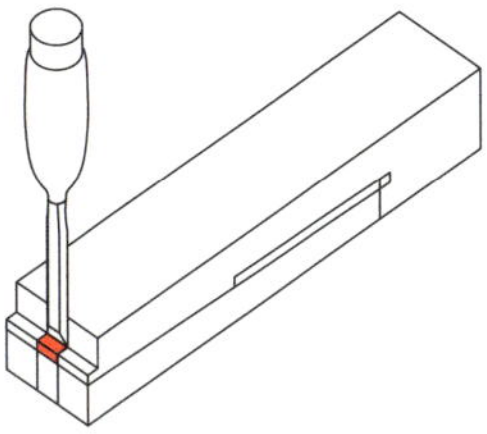

A5

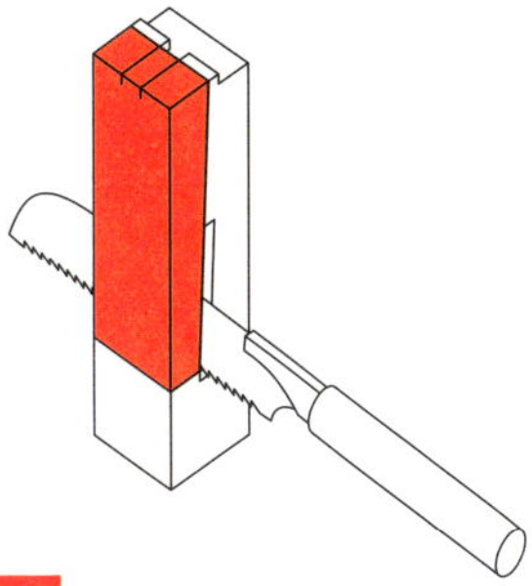

A6

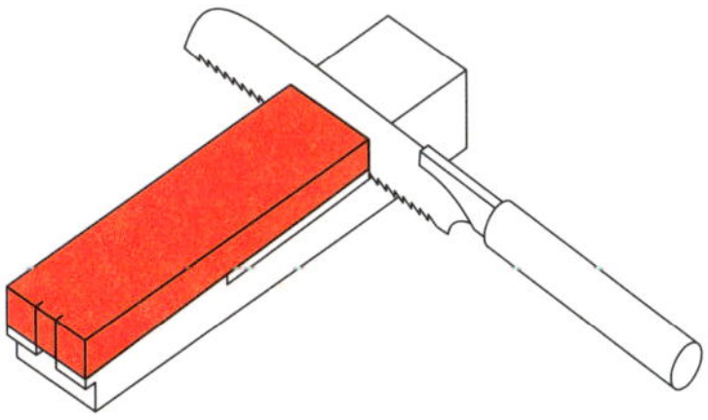

A7

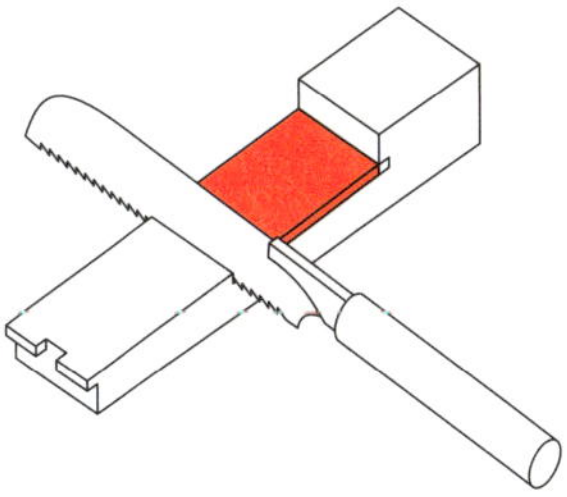

A8

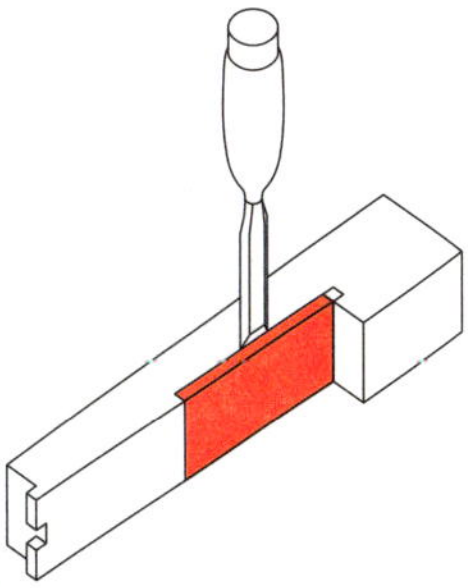

A9

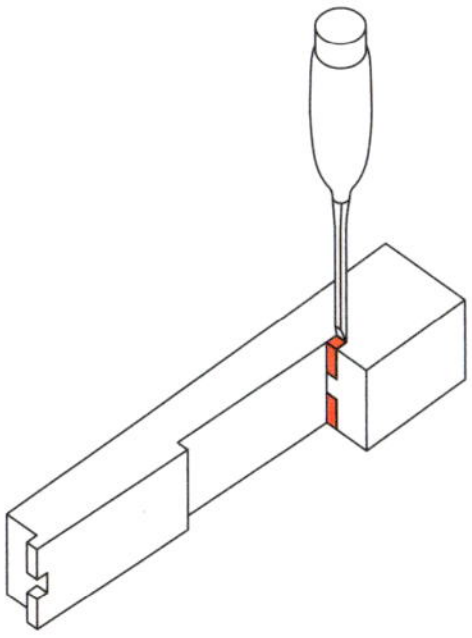

A10

2×

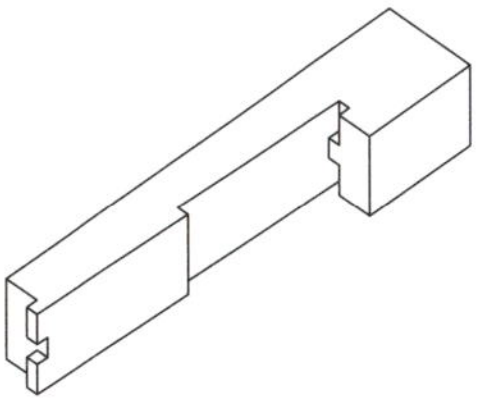

A

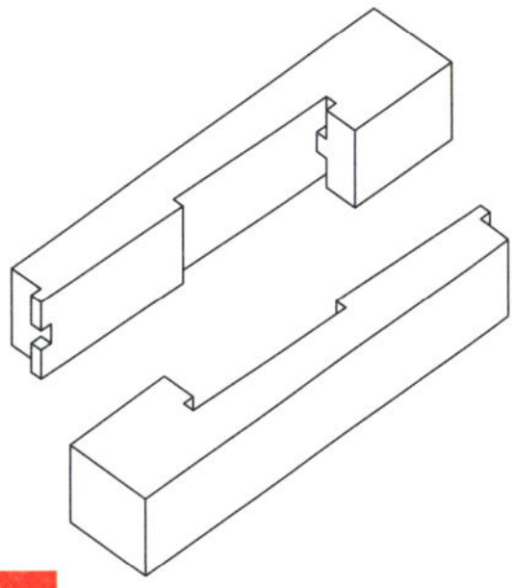

A+A

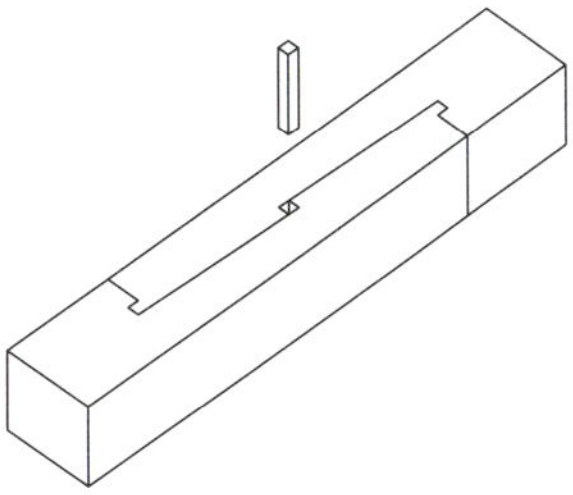

A+A

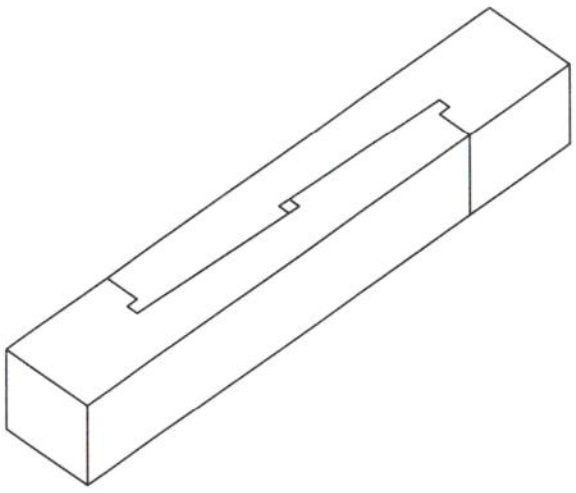

AA

Splicing Joints

Pegged rabbeted stop-splayed cogged scarf joint with stub tenons

kanawa-tsugi [*]
金輪継ぎ

stehendes schräges Hakenblatt mit Steckfalz, Brüstungszapfen und Dübelsicherung

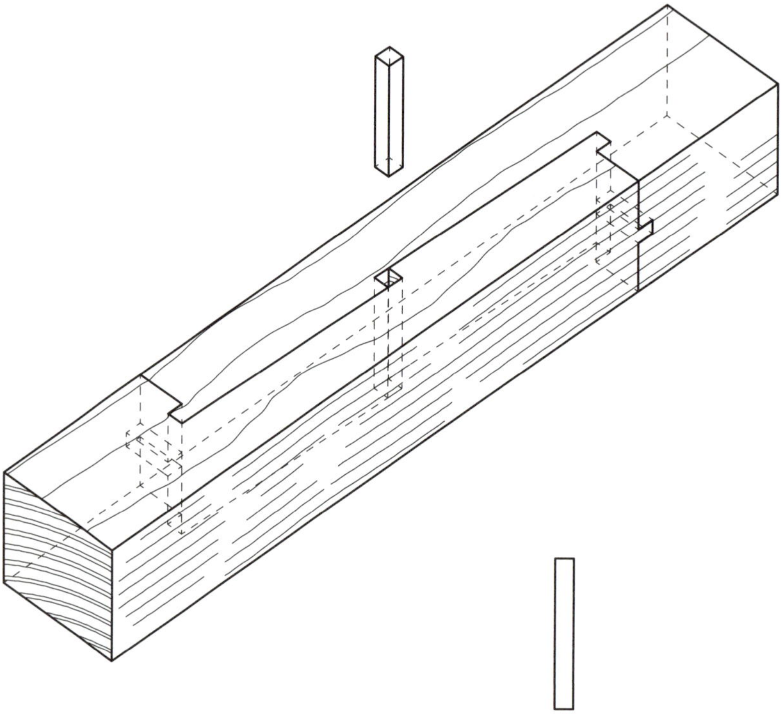

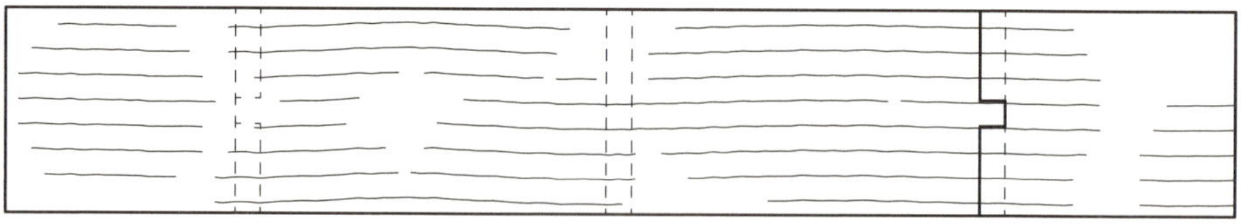

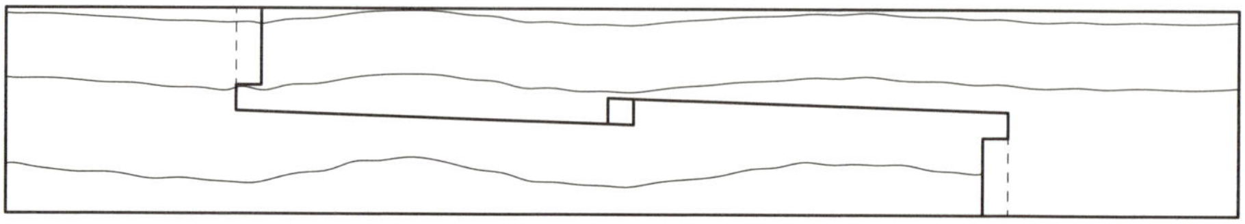

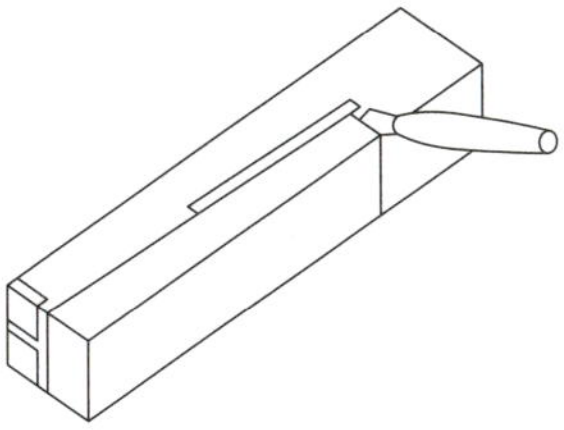
A1

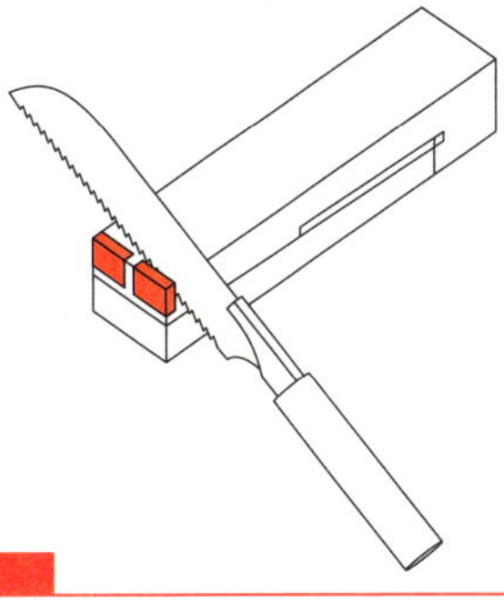
A2

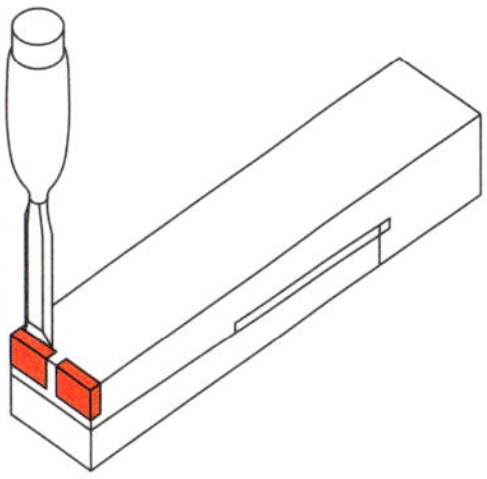
A3

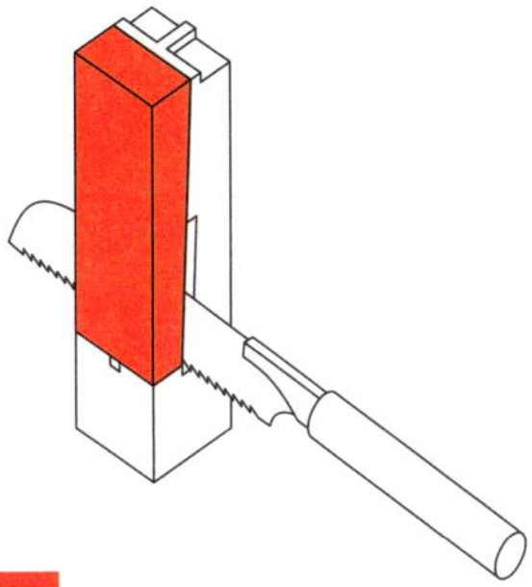
A4

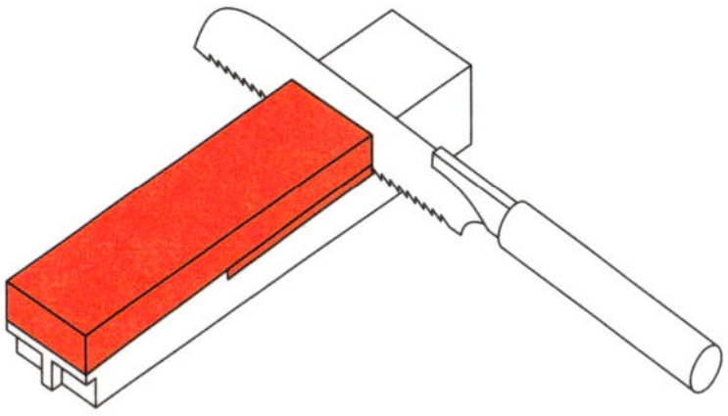
A5

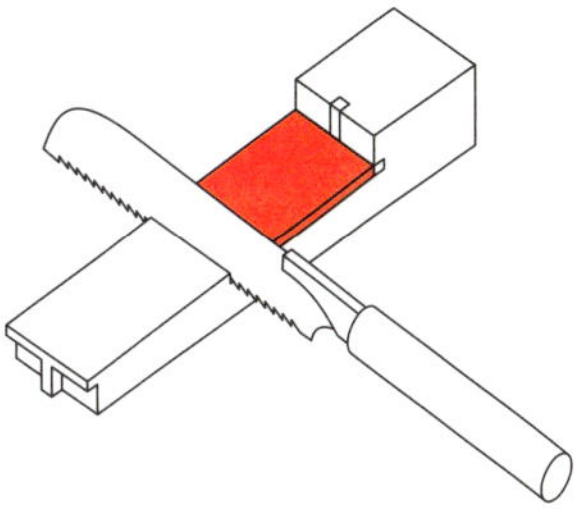
A6

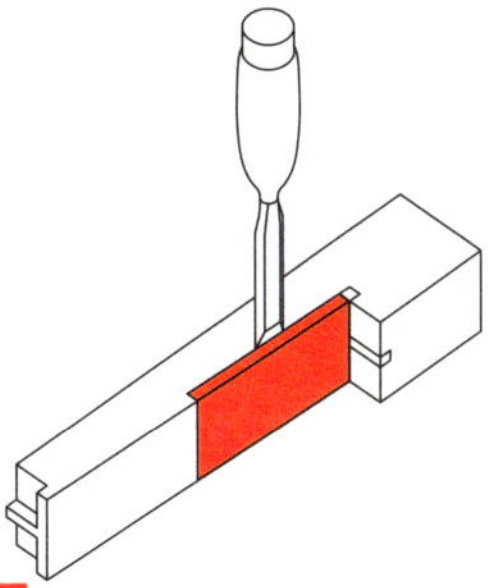
A7

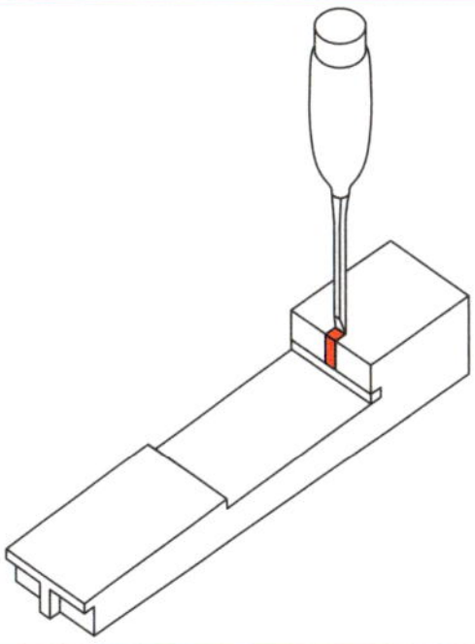
A8

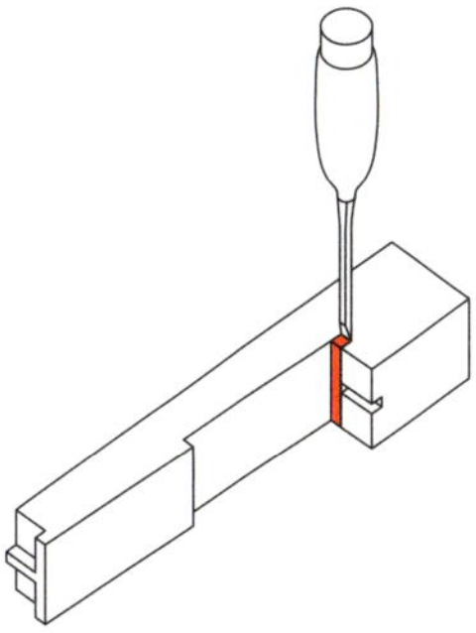
A9

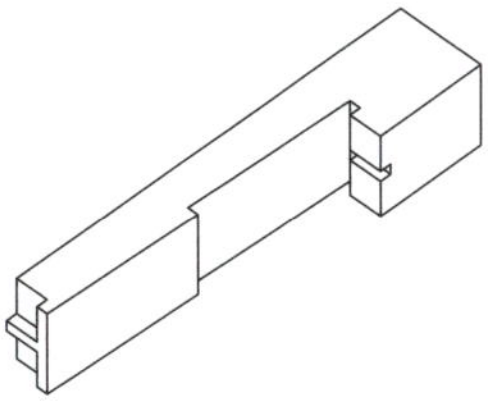
2×
A

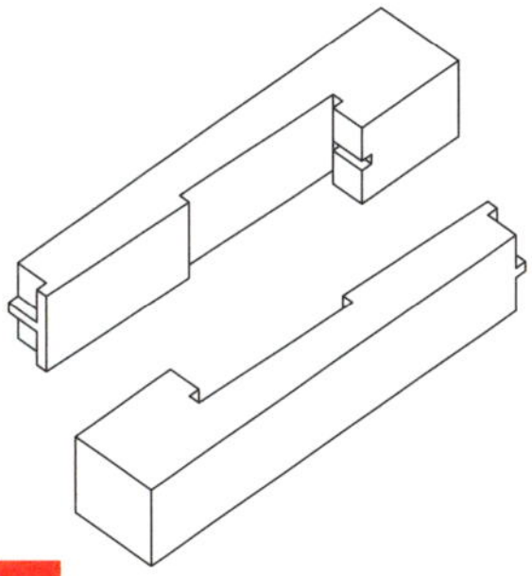
A+A

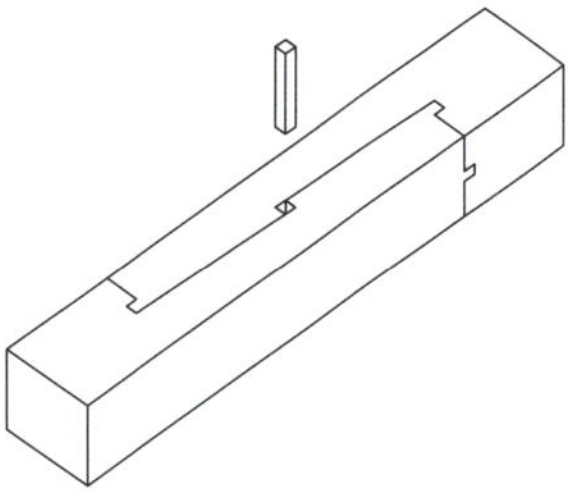
A+A

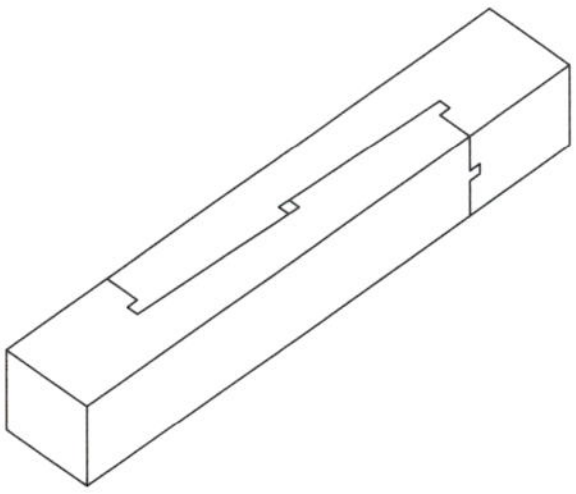
AA

Splicing Joints

Pegged half blind rabbeted stop-splayed cogged scarf joint

hako-sen-tsugi [*]
箱栓継ぎ

halbverdecktes schräges Hakenblatt mit Steckfalz und Dübelsicherung

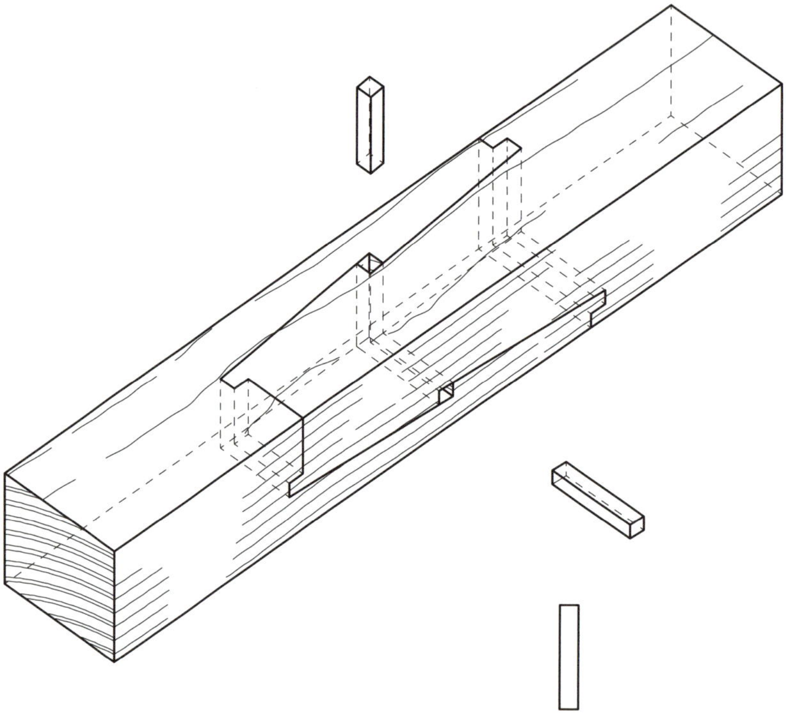

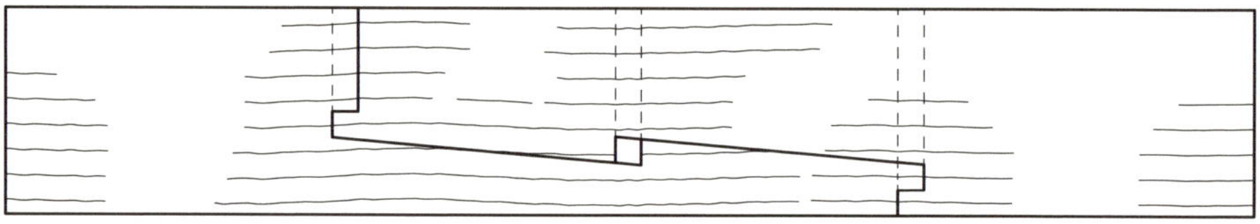

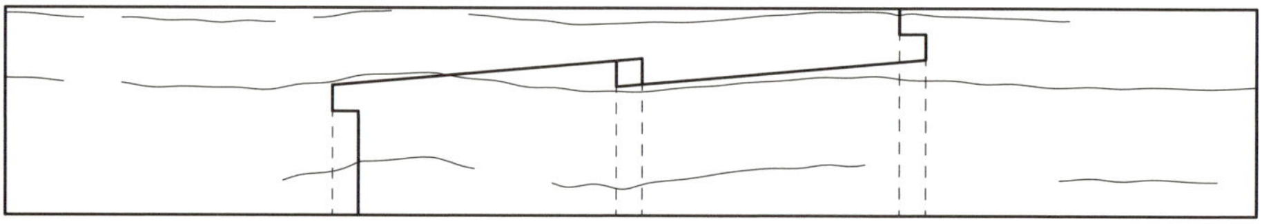

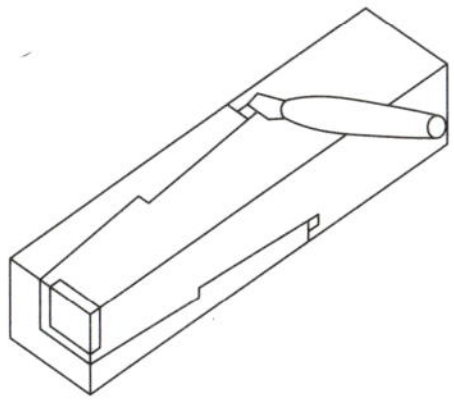

A1

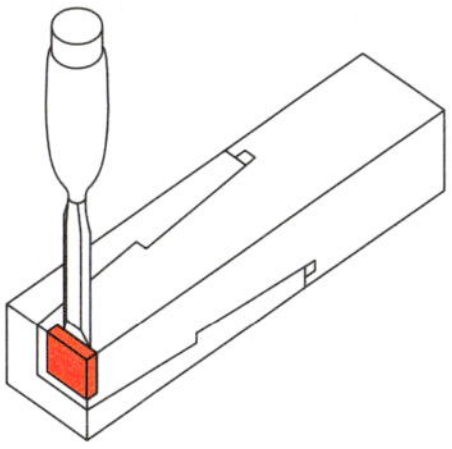

A2

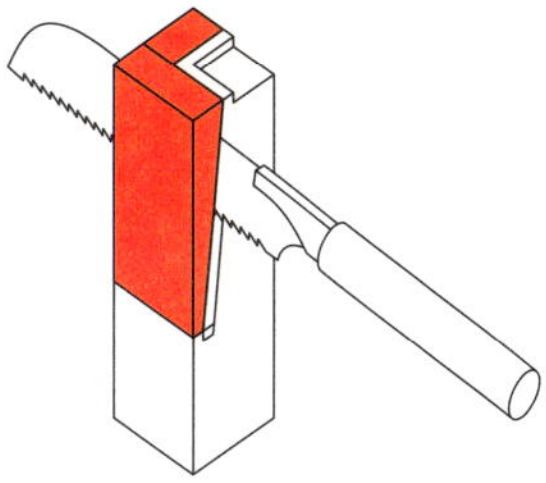

A3

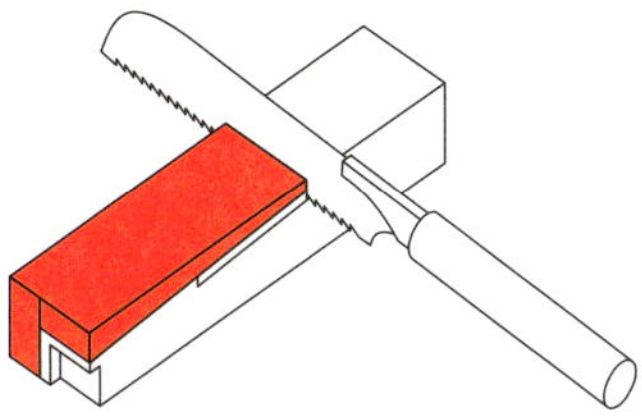

A4

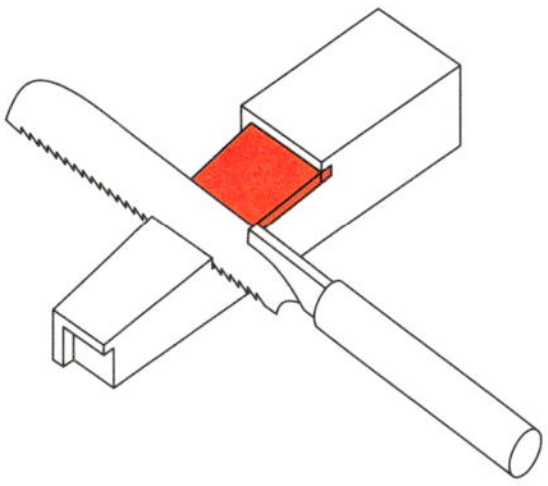

A5

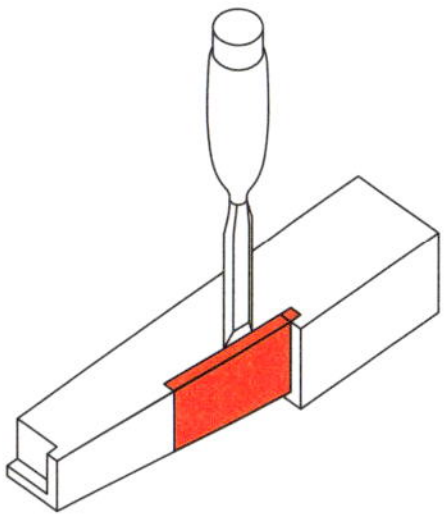

A6

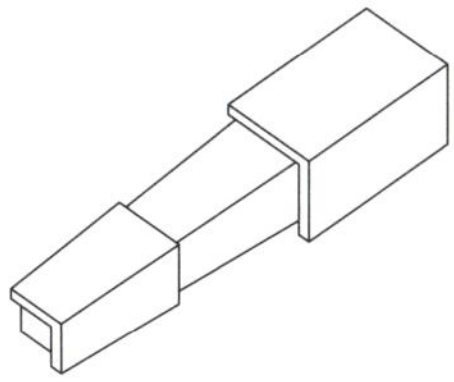

A

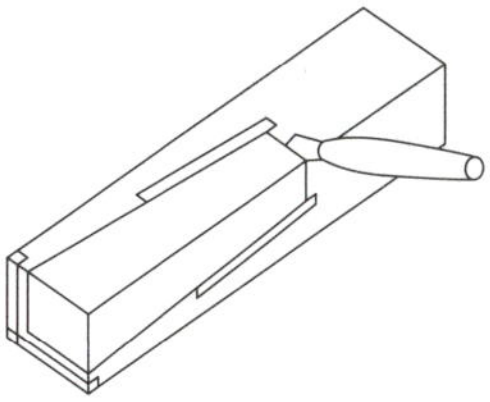

B1

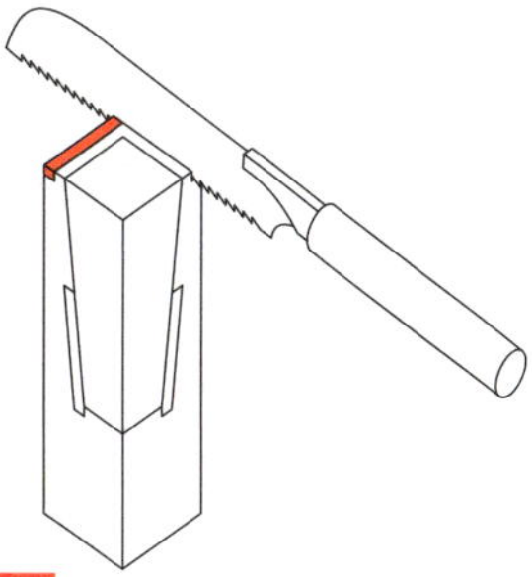

B2

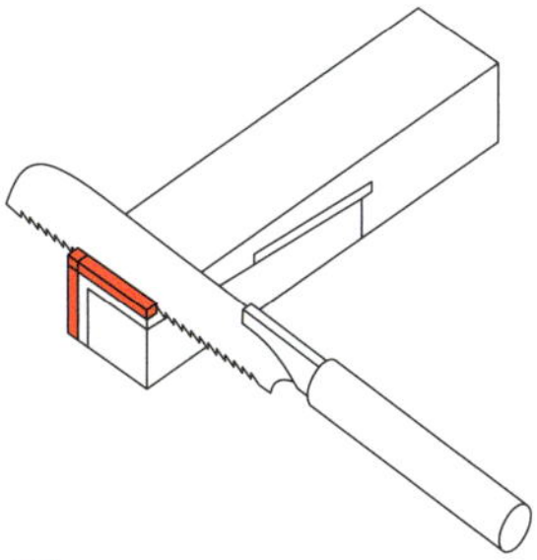

B3

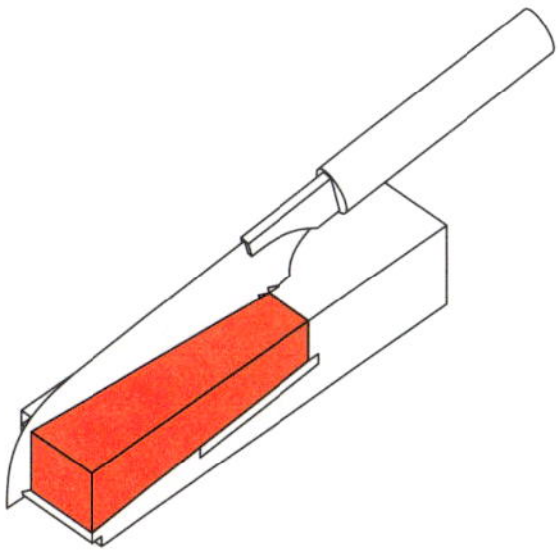

B4

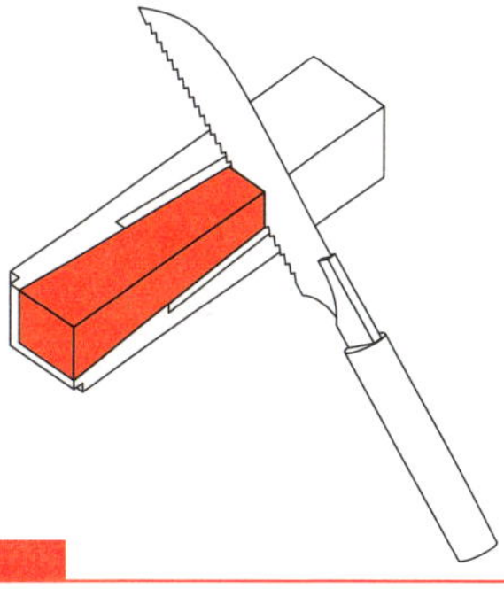

B5

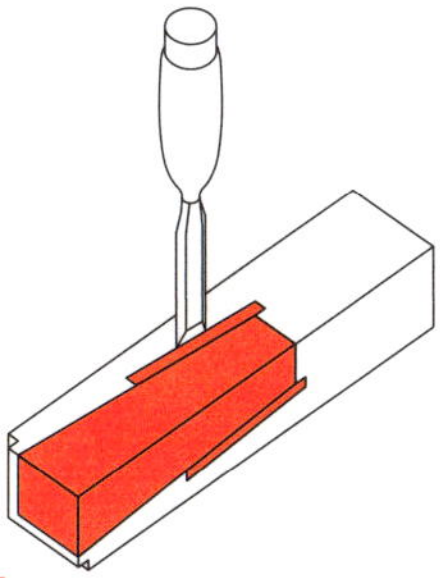

B6

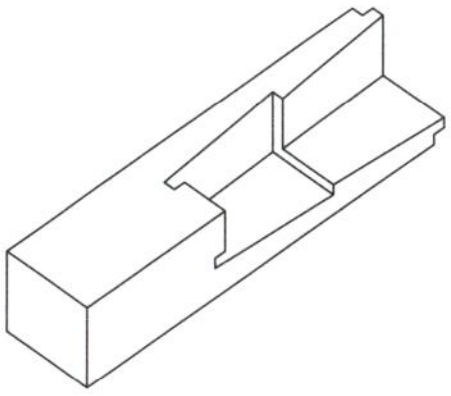

B

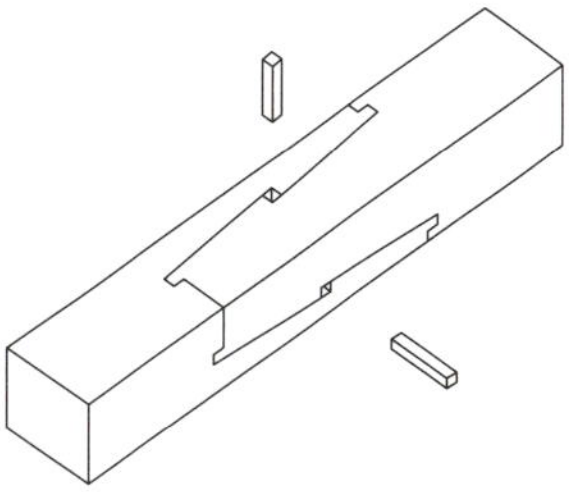

A+B

Splicing Joints

Blind rabbeted diagonally cut half lap joint with a key

hako-tsugi [*]
箱継ぎ

diagonales Blatt mit verdecktem Steckfalz und Spannkeil

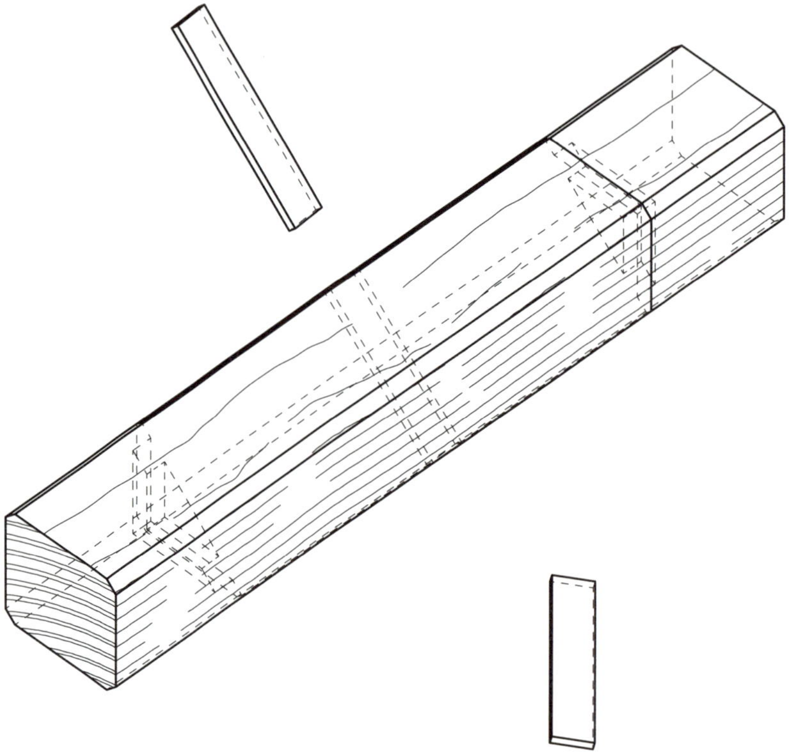

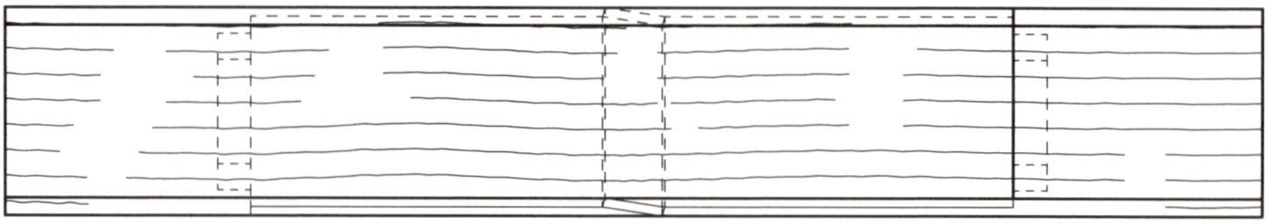

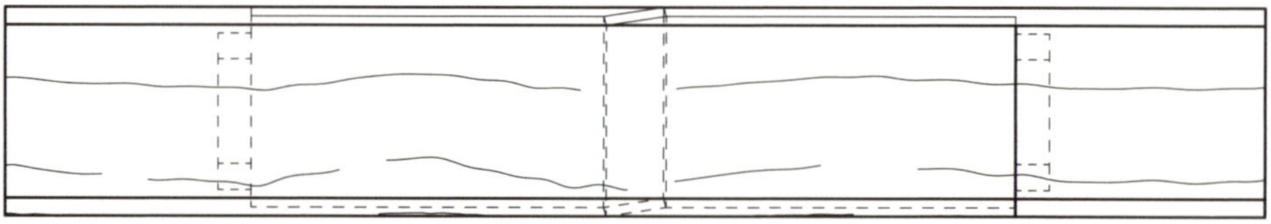

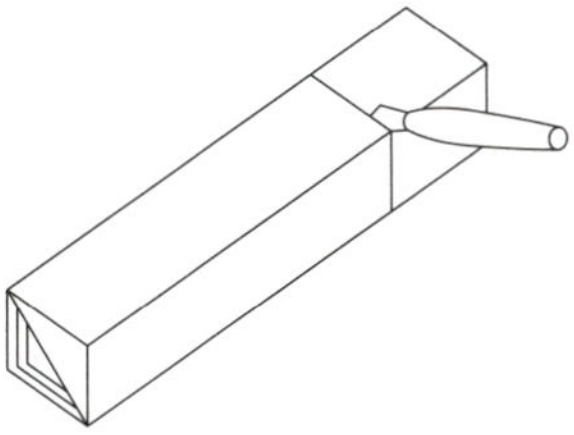

A1

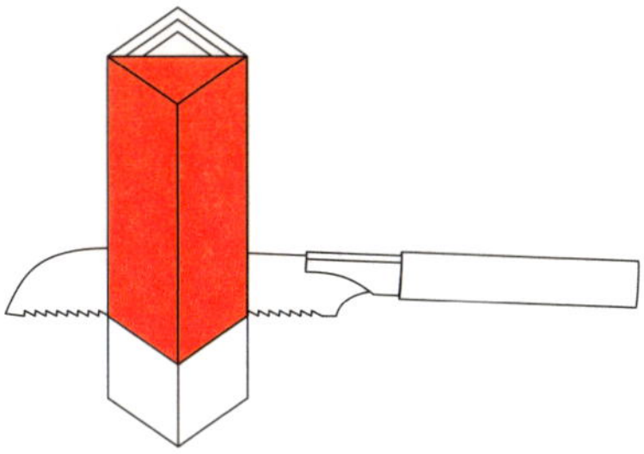

A2

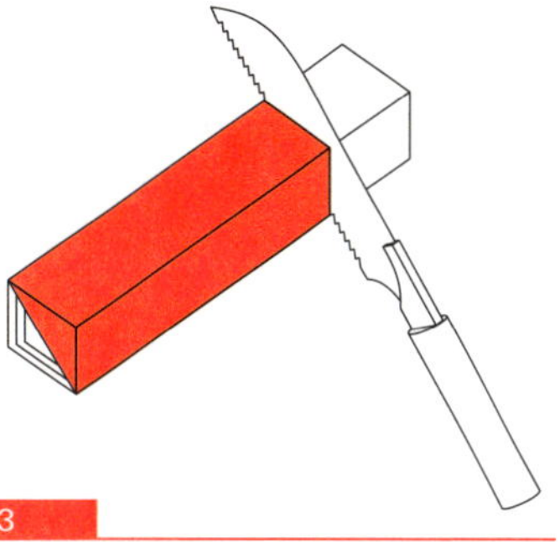

A3

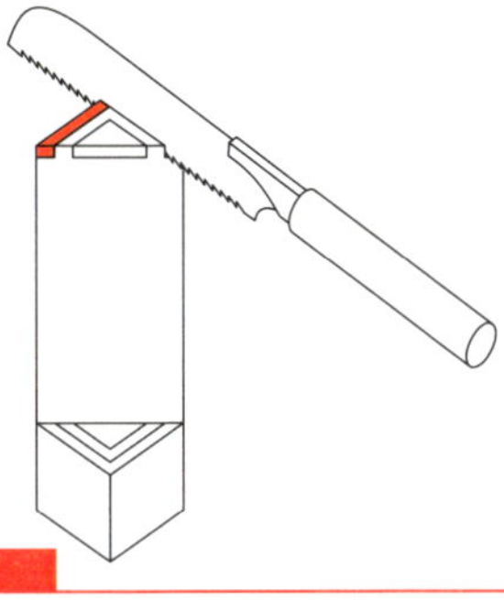

A4

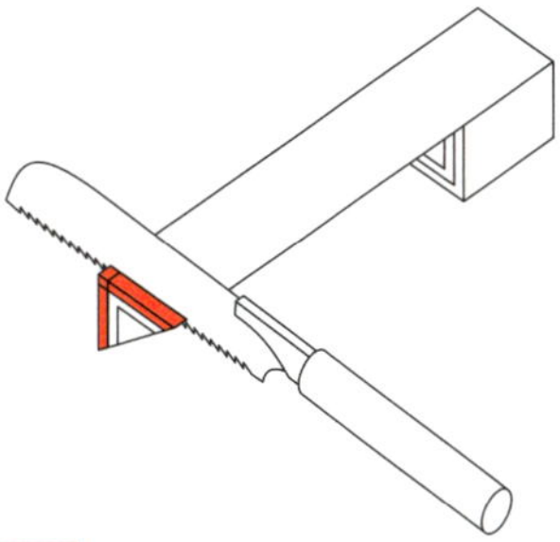

A5

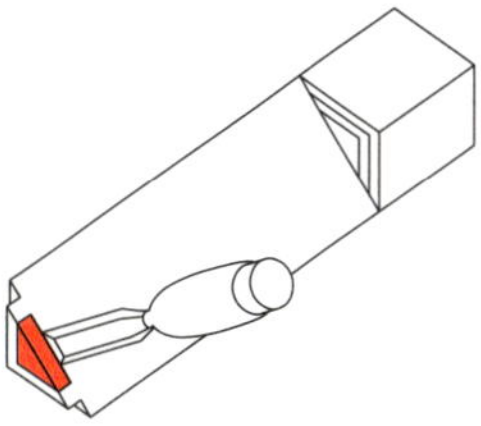

A6

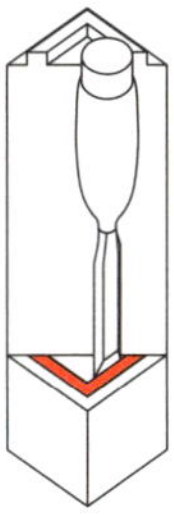

A7

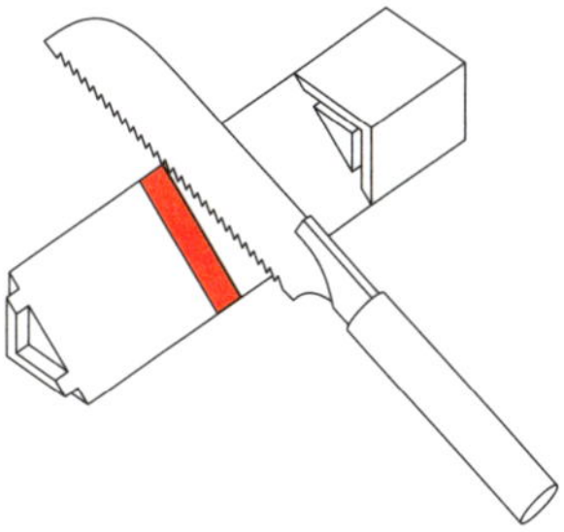

A8

2×

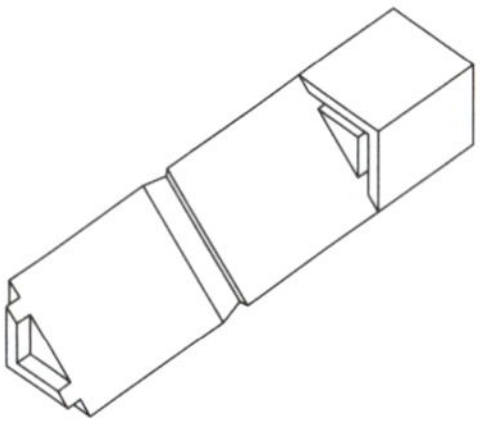

A

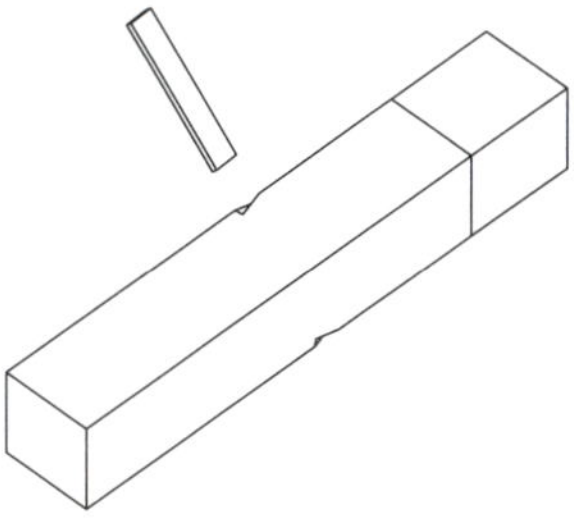

A+A

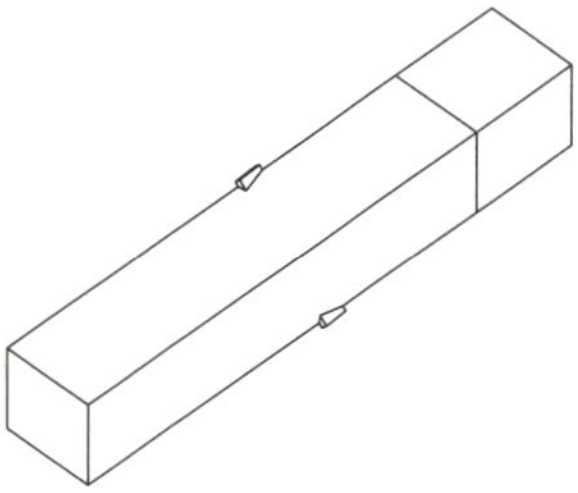

AA

Splicing Joints

Splayed scarf joint

sogi-tsugi [*]
殺継ぎ

Schäftung

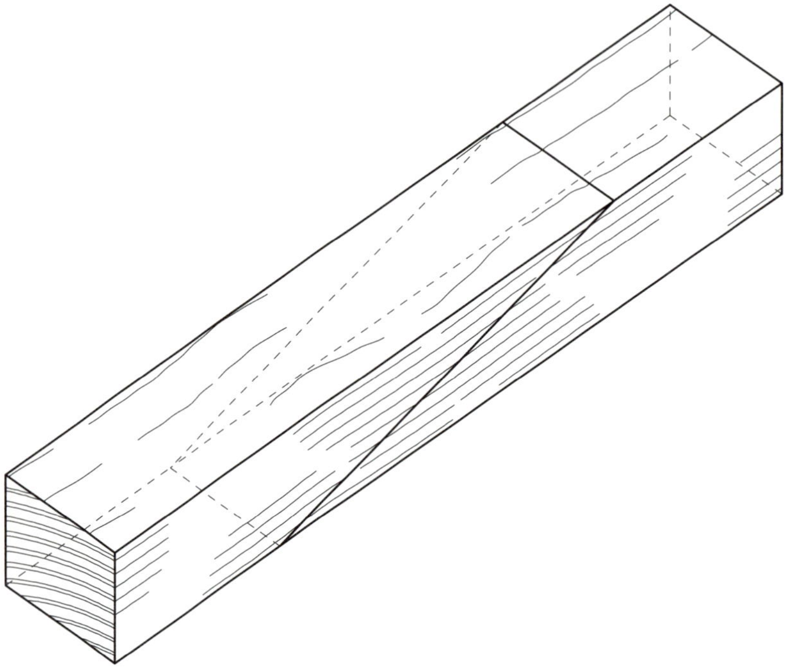

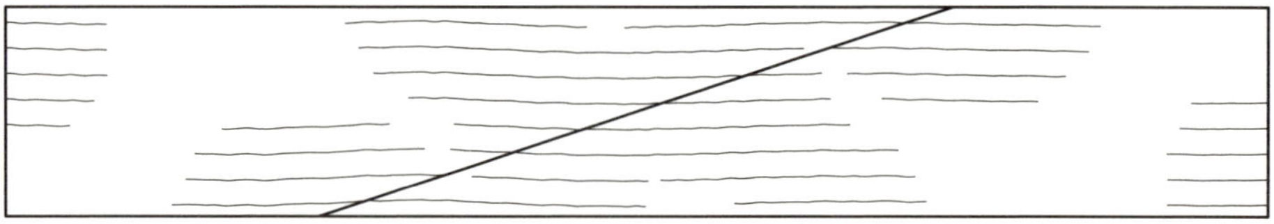

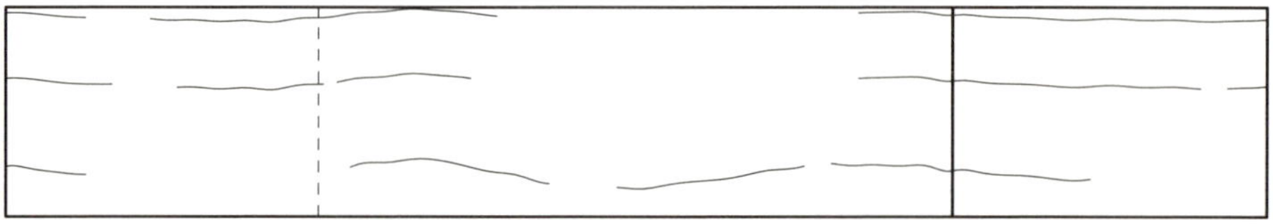

2×

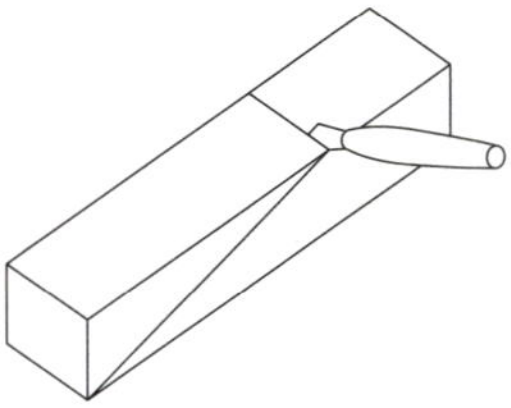

A1

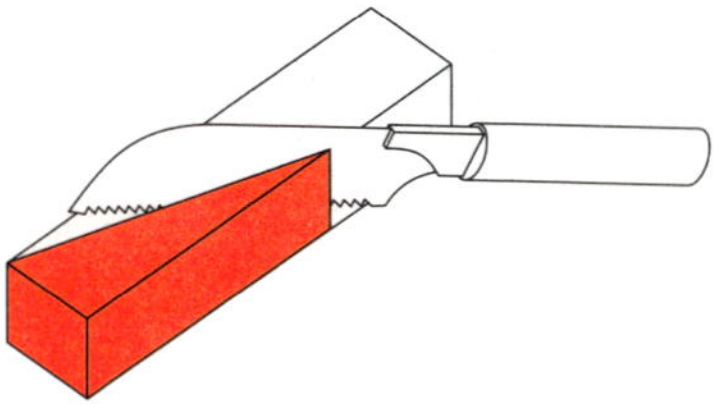

A2

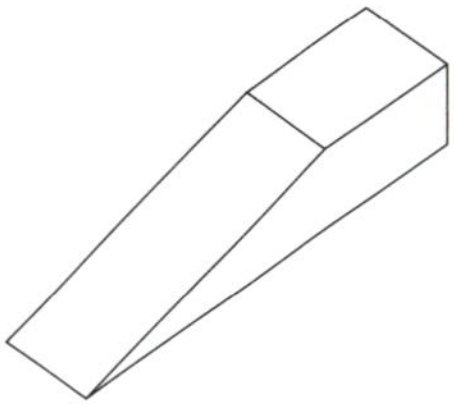

A

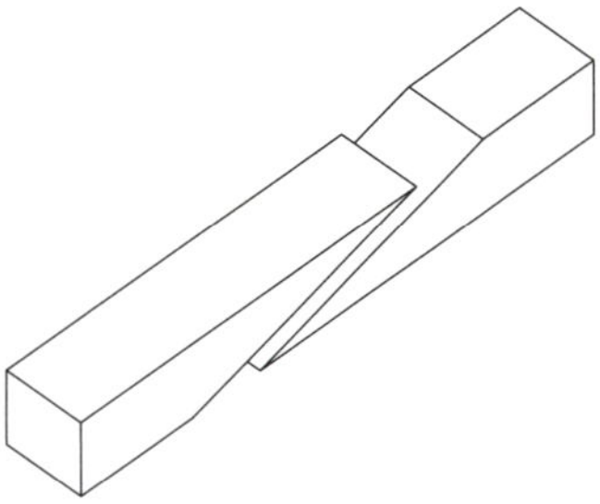

A+A

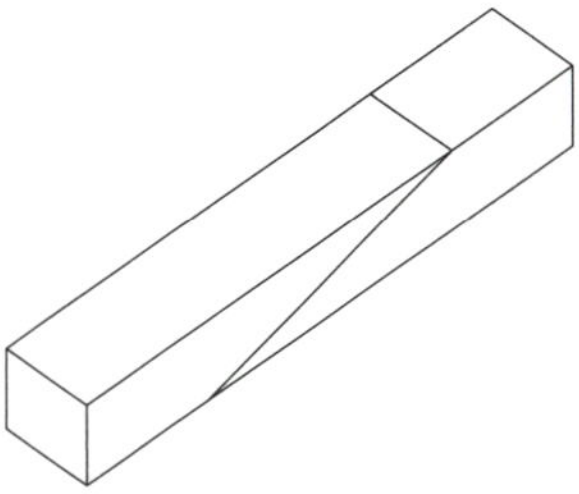

AA

Splicing Joints

Half lap joint with mitered ends

ai-sogi-tsugi
相殺継ぎ

stehendes gerades Blatt
mit beidseitig schrägem Stoß

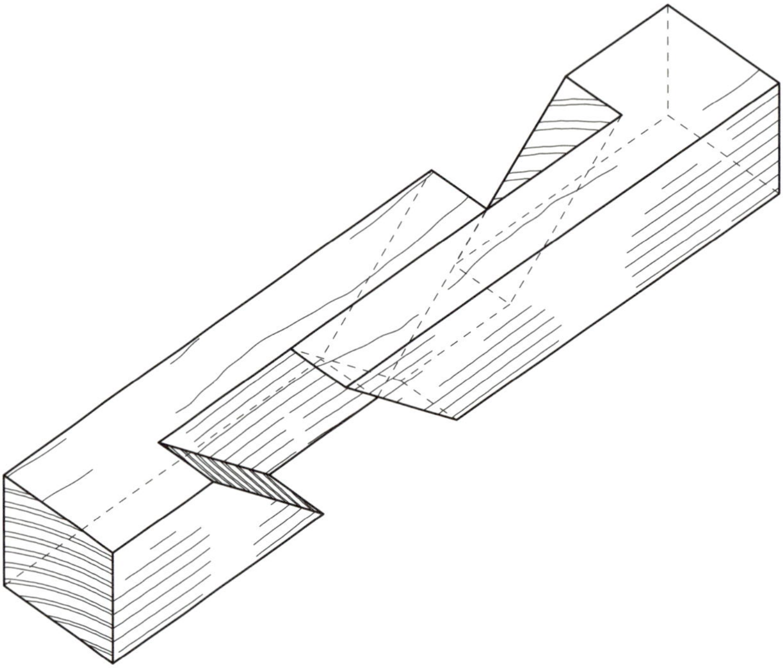

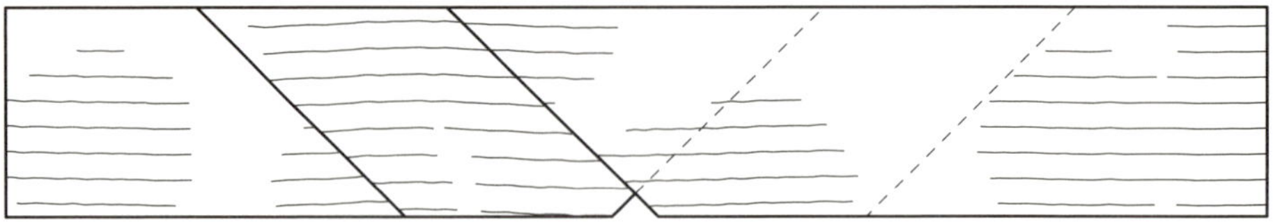

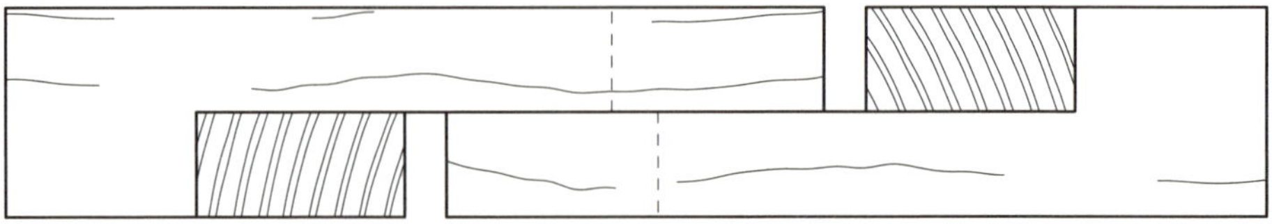

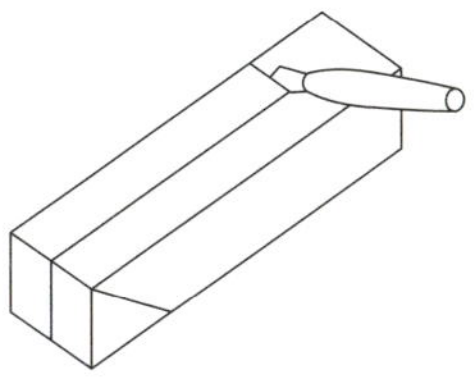

A1

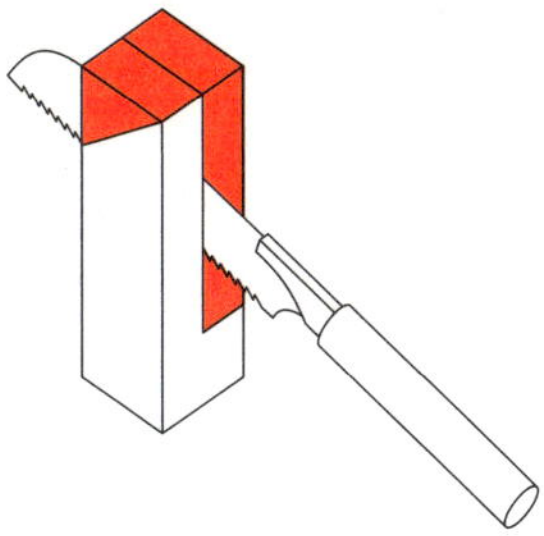

A2

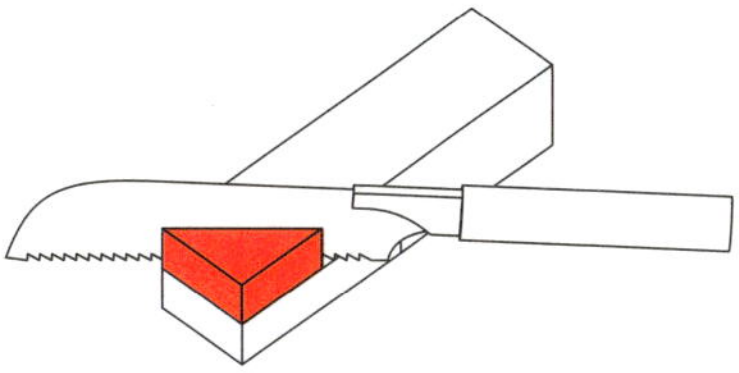

A3

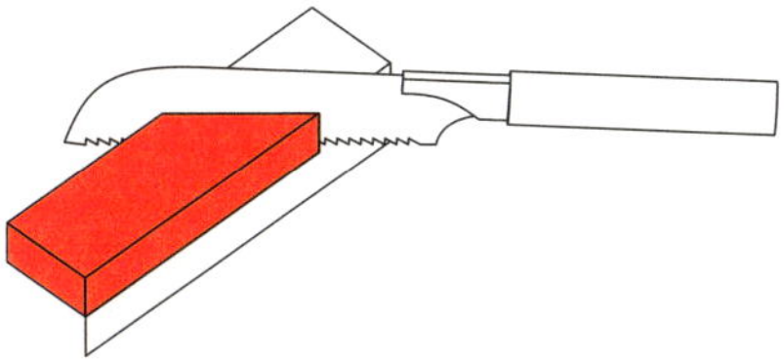

A4

2×

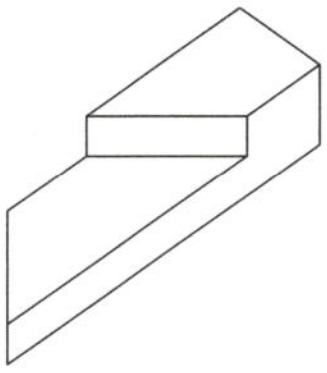

A

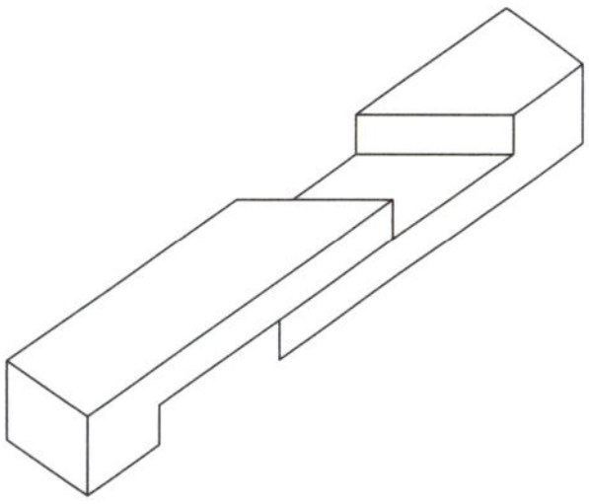

A+A

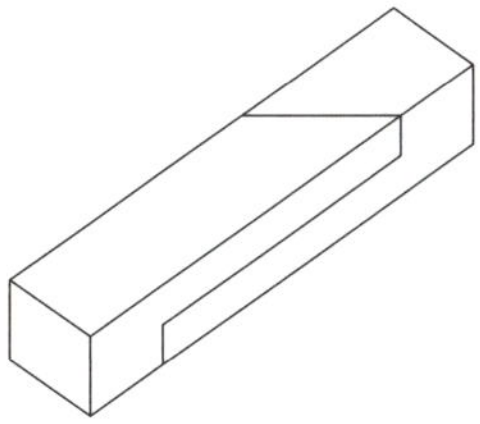

AA

Splicing Joints

Triple-splayed scarf joint

nihō-isuka-tsugi
二方鶍継ぎ

dreifache Schäftung

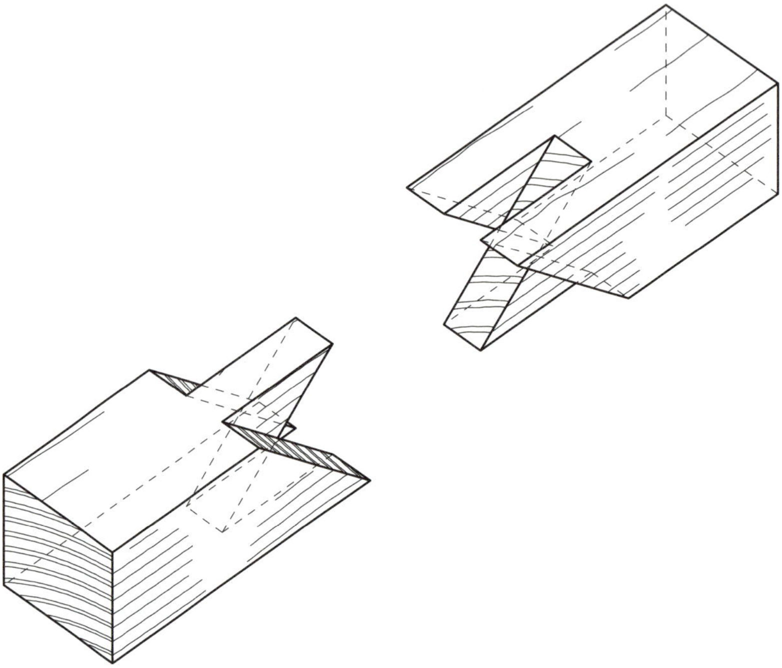

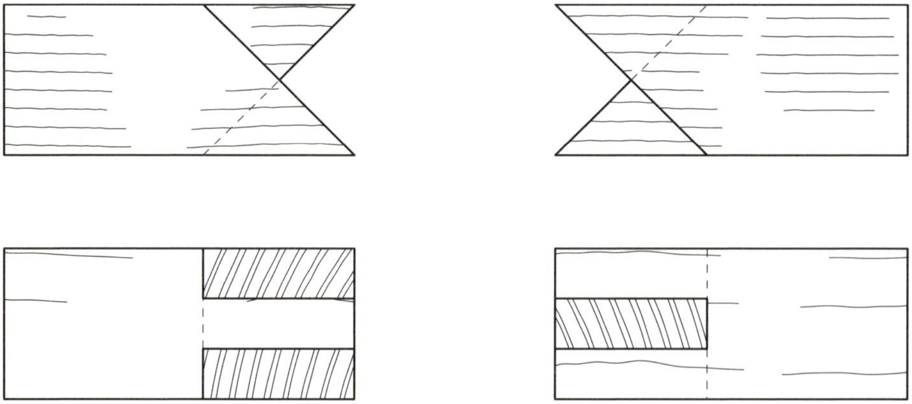

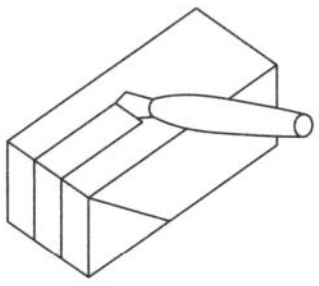

A1

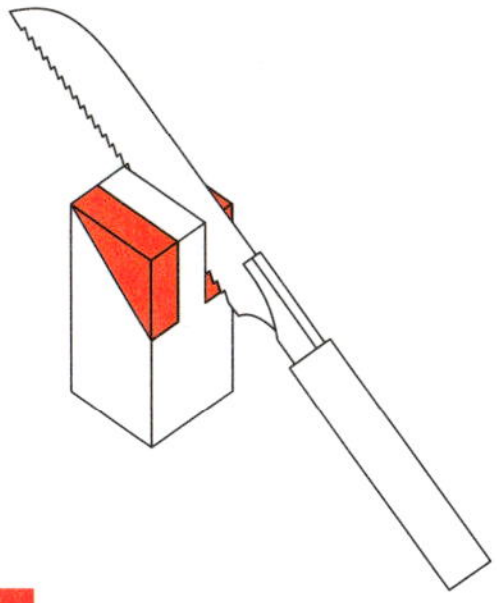

A2

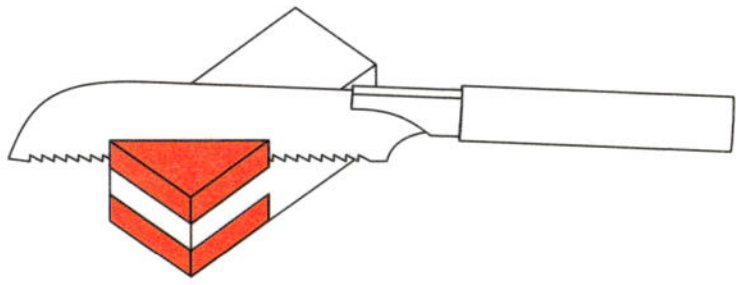

A3

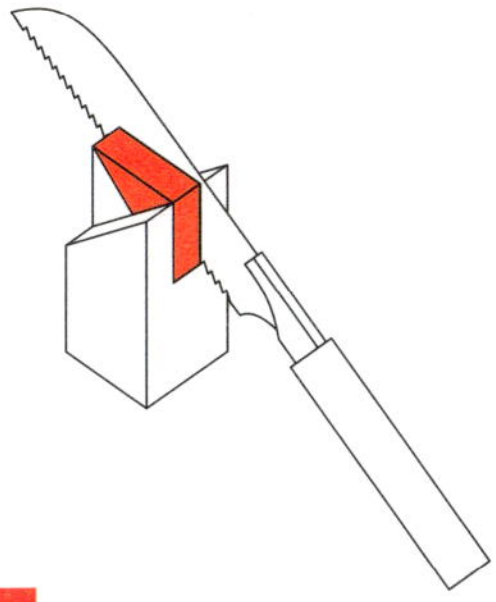

A4

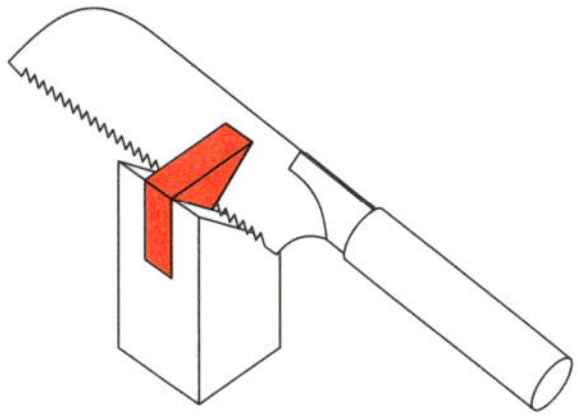

A5

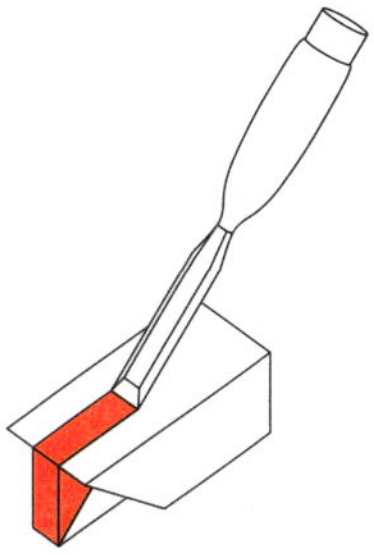

A6

2×

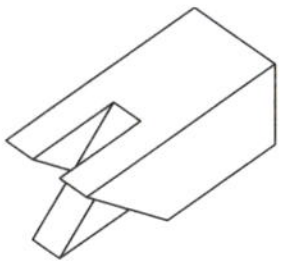

A

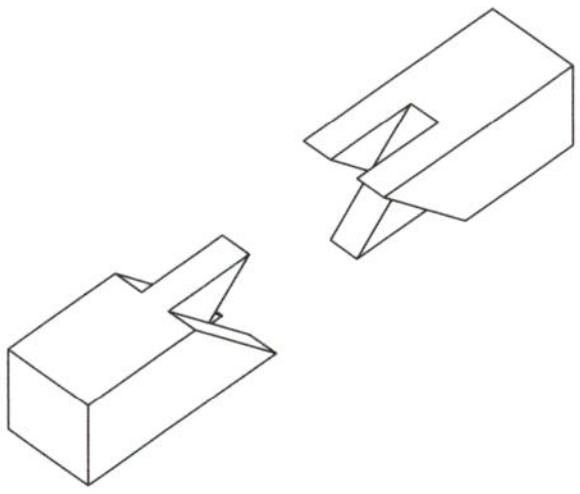

A+A

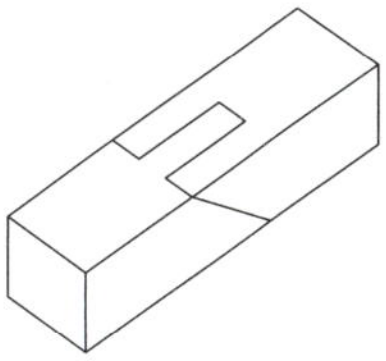

AA

Splicing Joints

Halved splayed scarf joint

isuka-tsugi [*]
鶍継ぎ

doppelte Schäftung

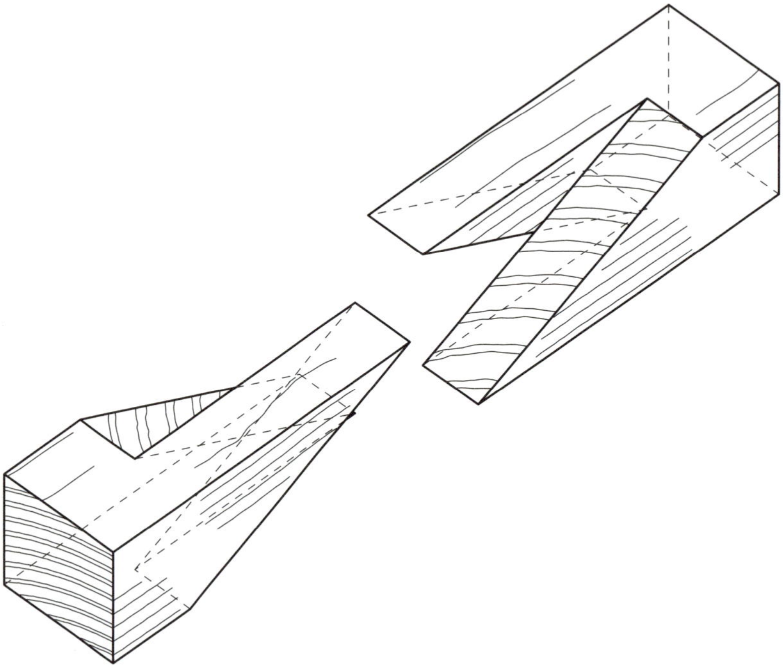

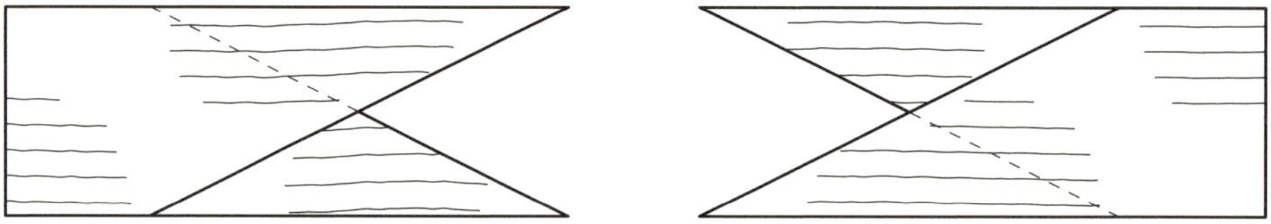

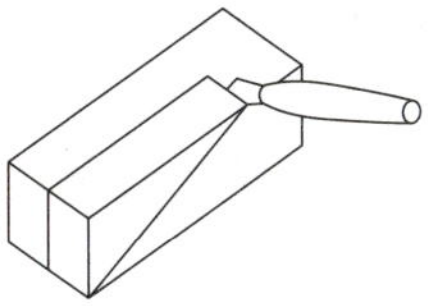

A1

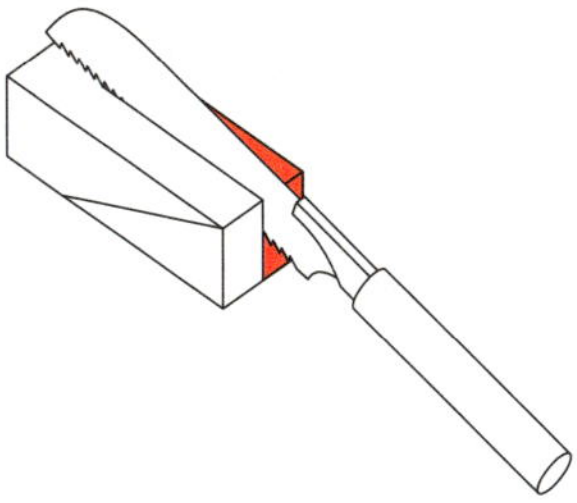

A2

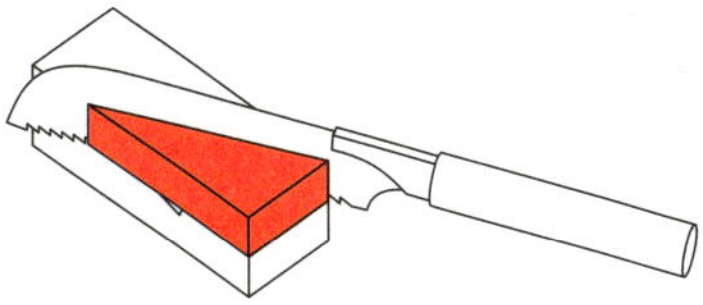

A3

2×

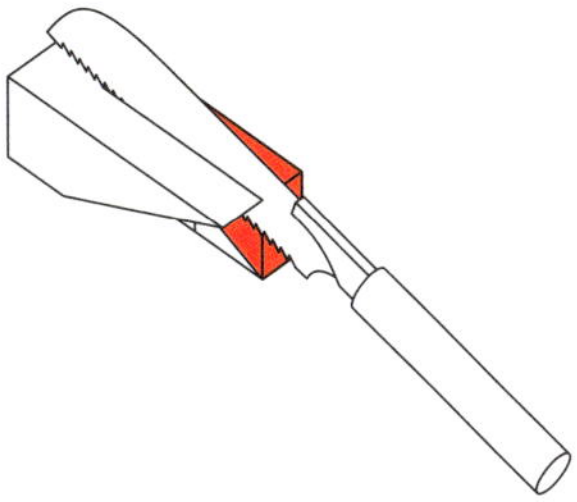

A4

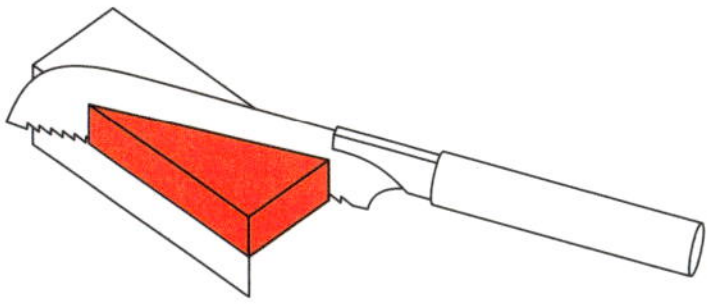

A5

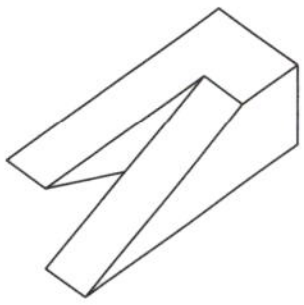

A

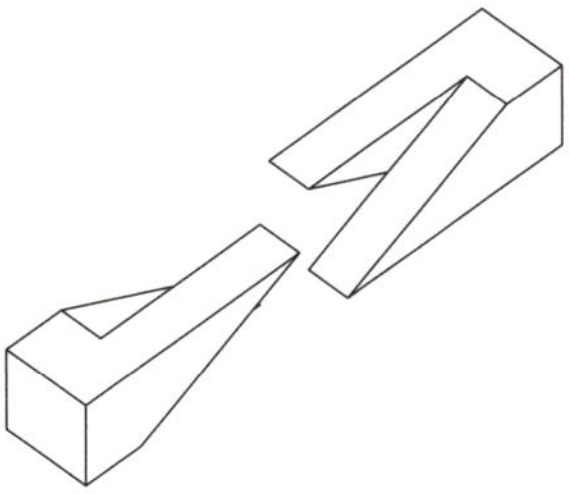

A+A

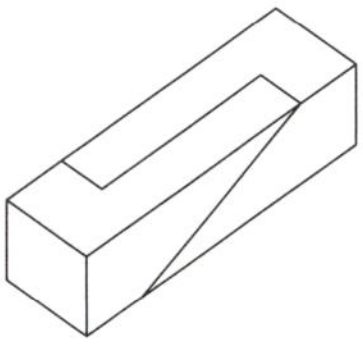

AA

Splicing Joints

Halved splayed rabbeted scarf joint with tapered key

sumi-kiri-isuka-tsugi [*]
隅切鶍継ぎ

doppelte Schäftung
mit Brüstungszapfen und Spannkeil

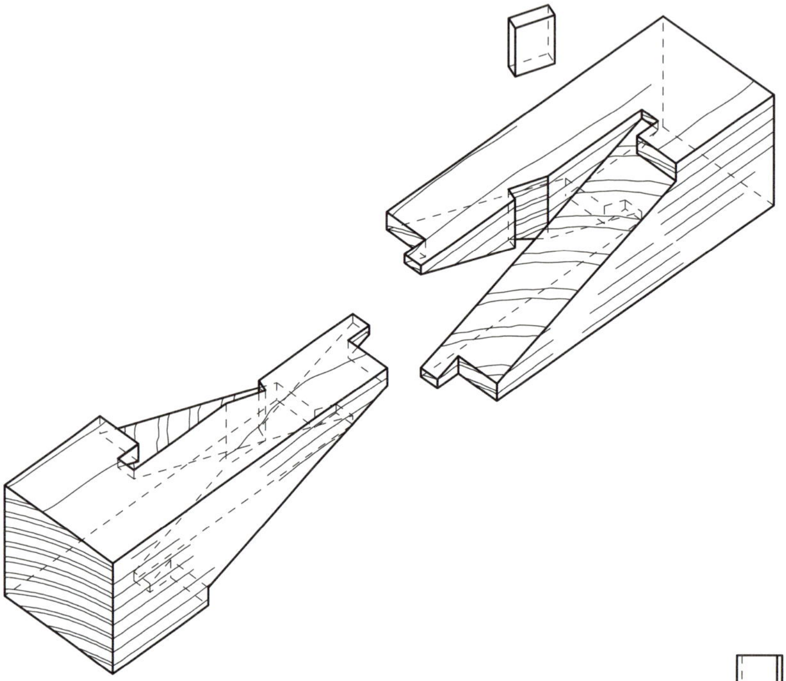

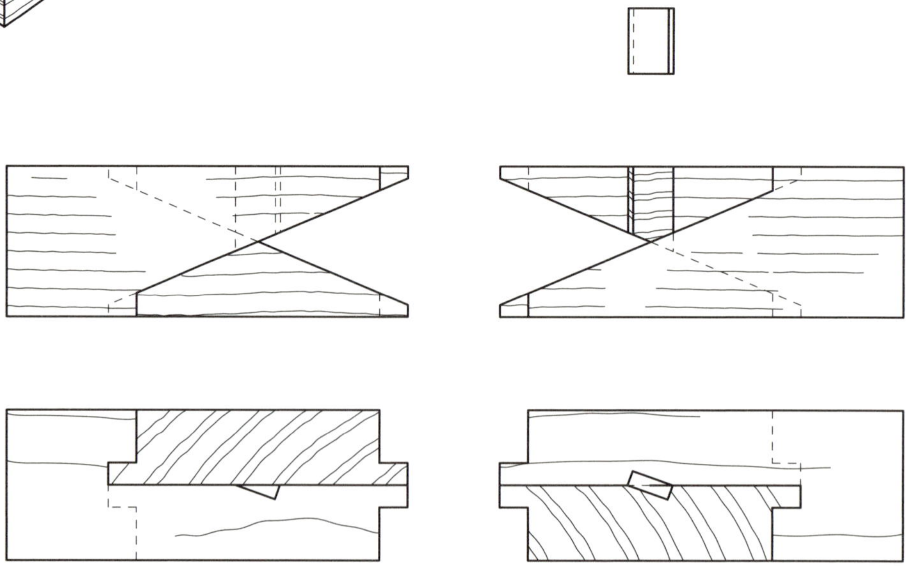

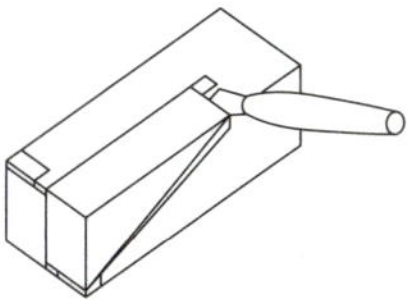

A1

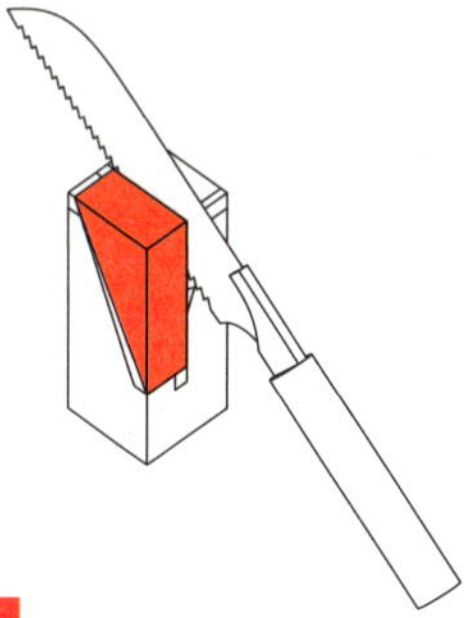

A2

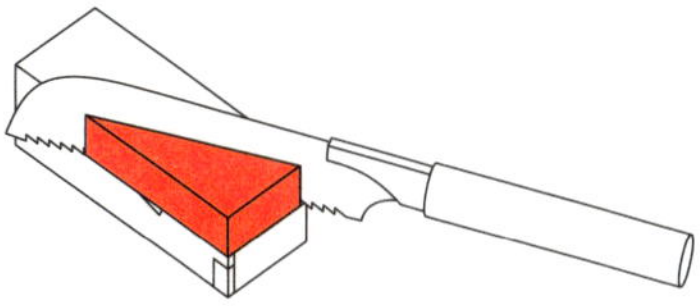

A3

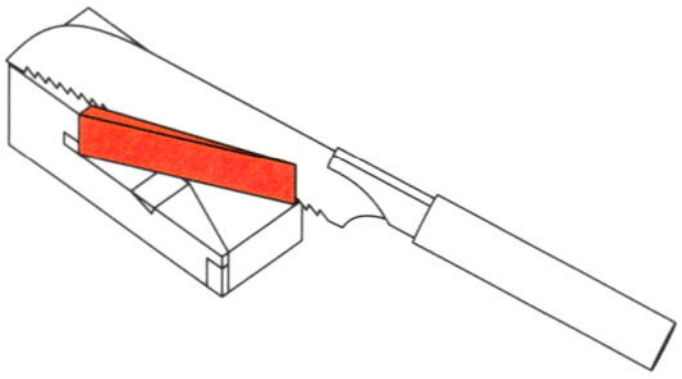

A4

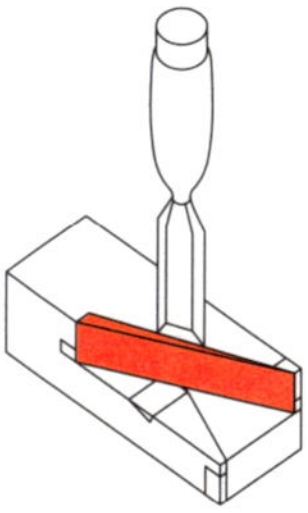

A5

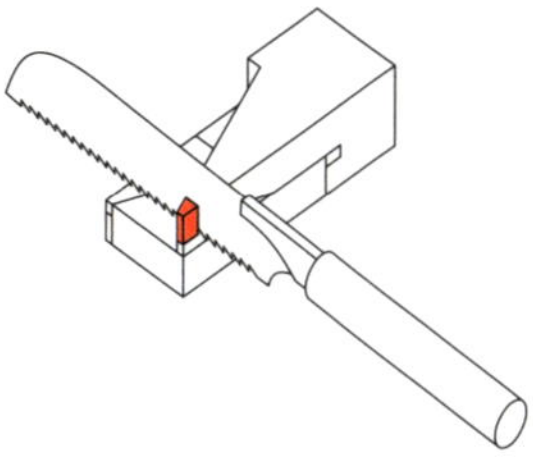

A6

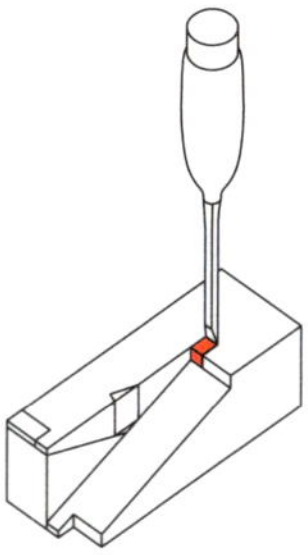

A7

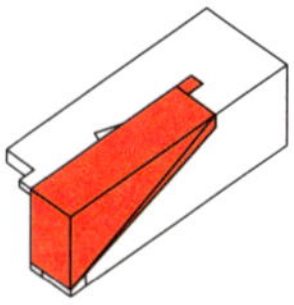

A8

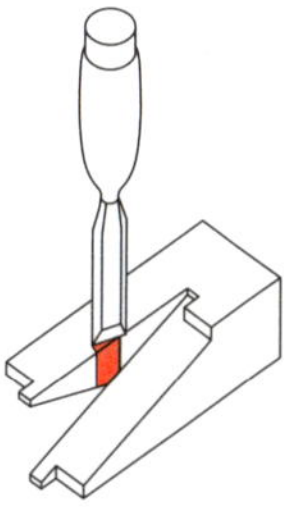

A9

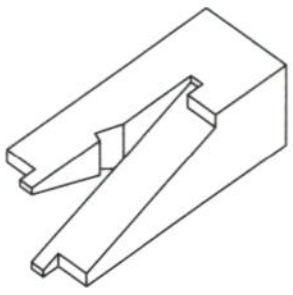

2×

A

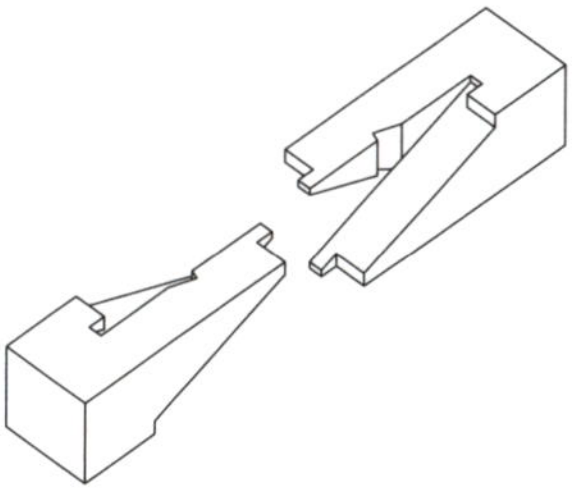

A+A

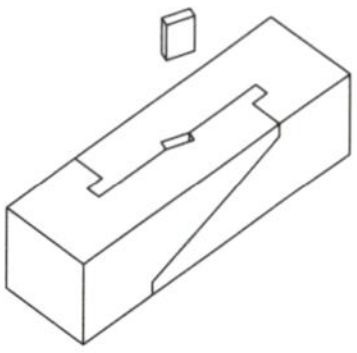

A+A

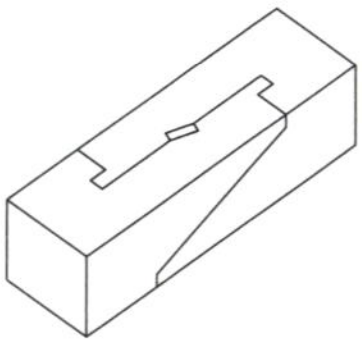

AA

Splicing Joints

Halved splayed scarf joint with oblique mating surfaces and tapered key

Japanese

Miyajima-*tsugi* [*]
宮島継ぎ

German

schräg angeschnittene doppelte Schäftung mit Spannkeil

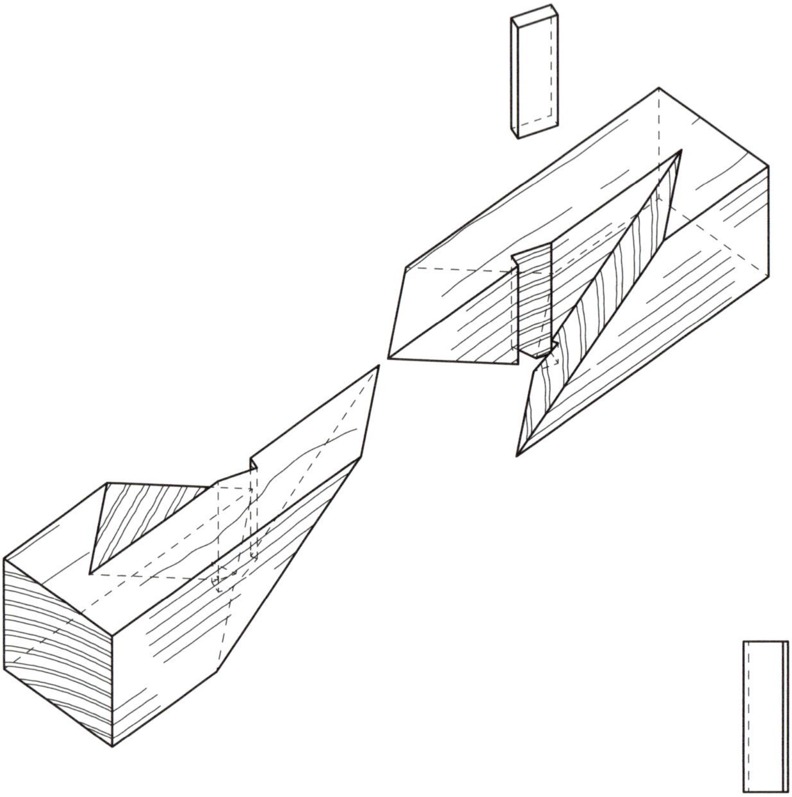

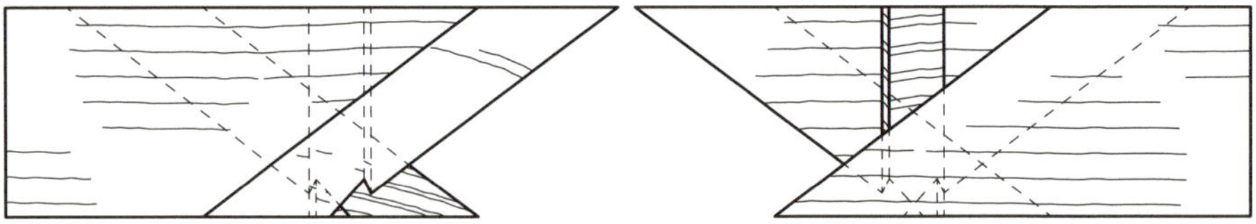

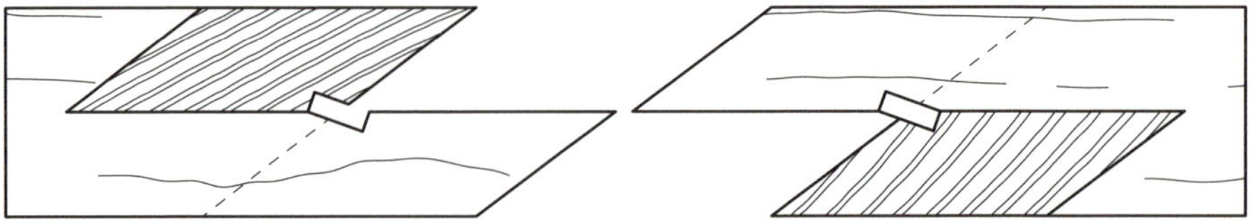

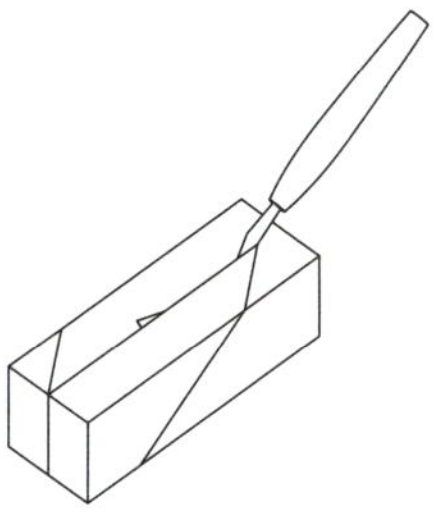

A1

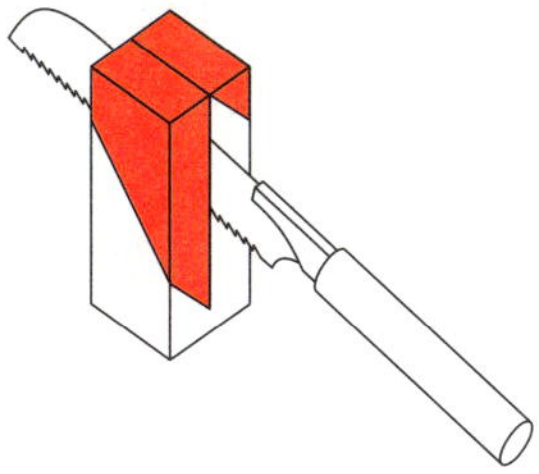

A2

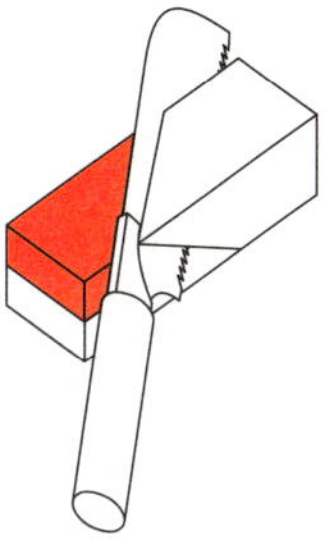

A3

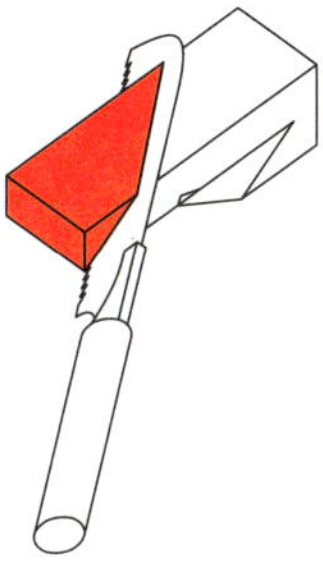

A4

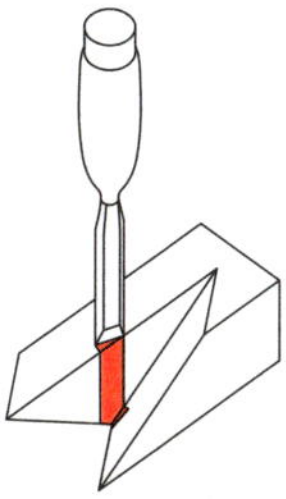

A5

2×

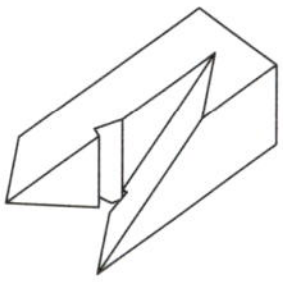

A

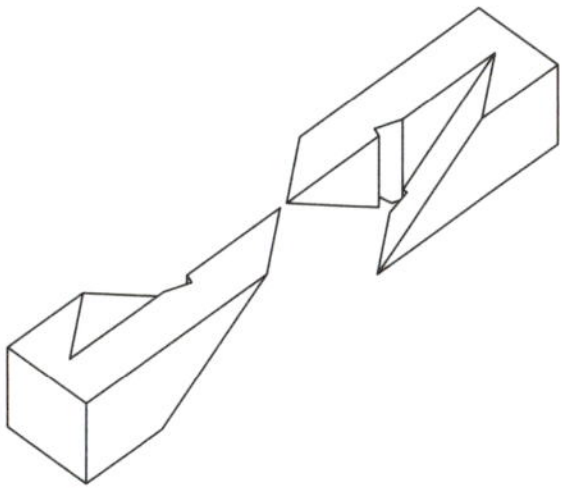

A+A

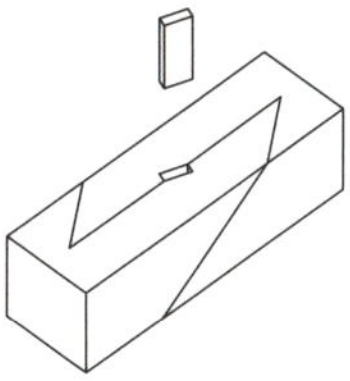

A+A

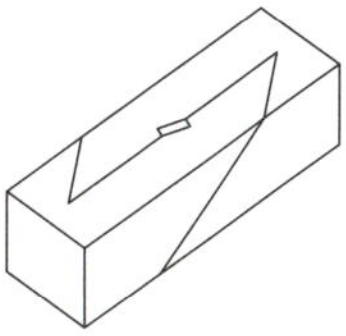

AA

Splicing Joints

Birdsmouth

yahazu-mechi-tsuki-tsuke-tsugi
矢筈目違突付継ぎ

Gratstoß

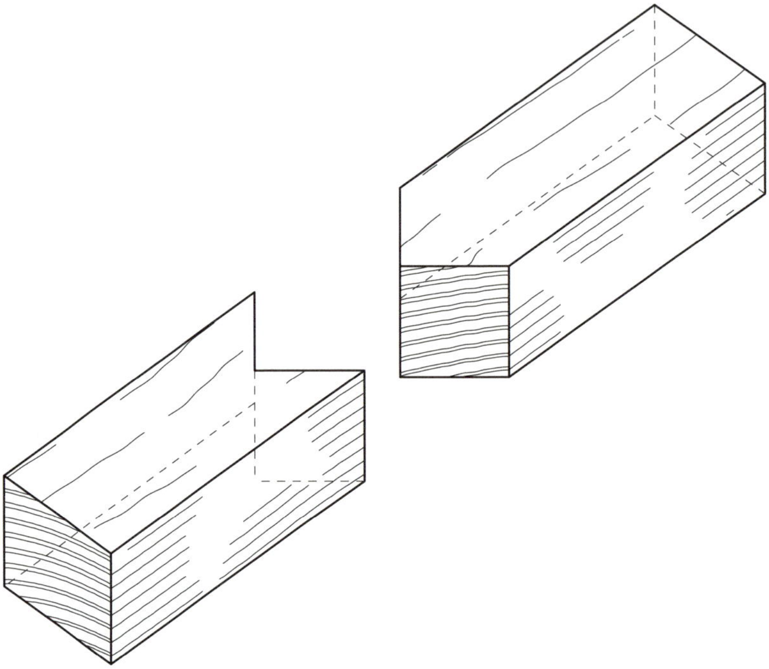

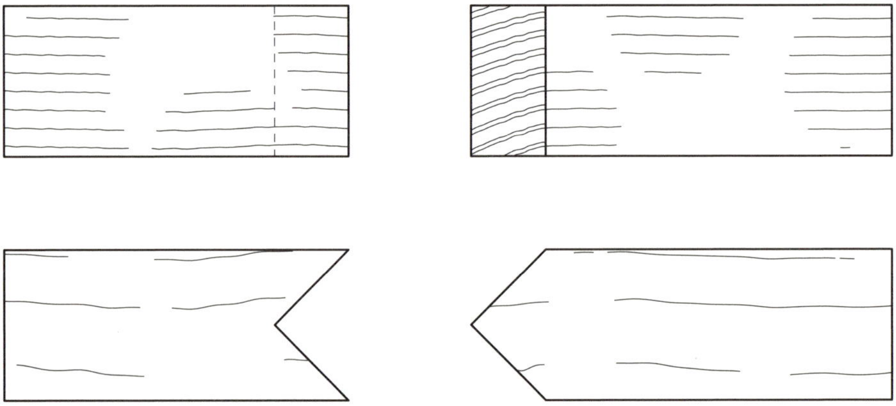

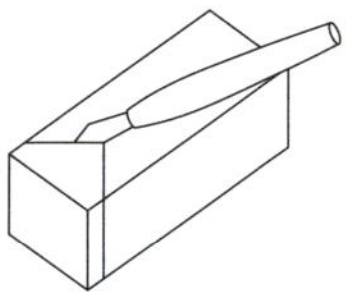

A1

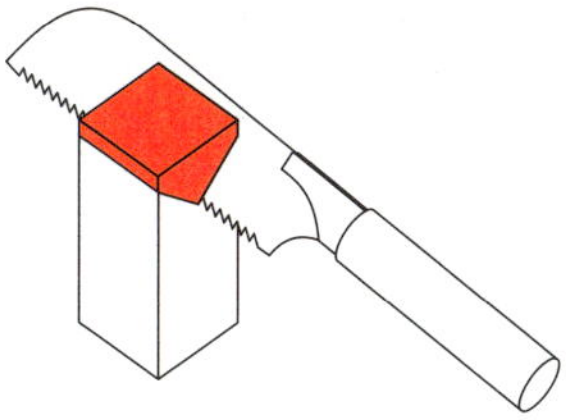

A2

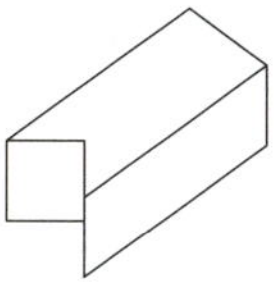

A

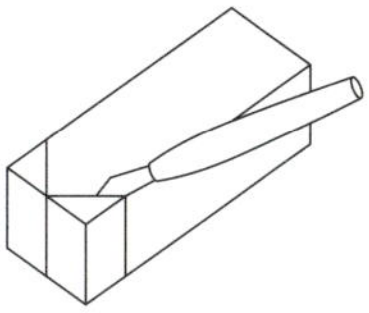

B1

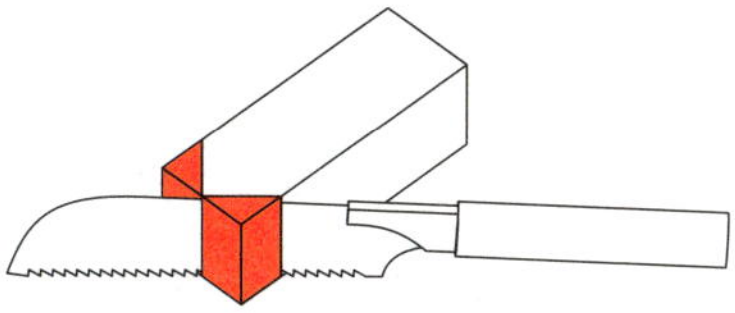

B2

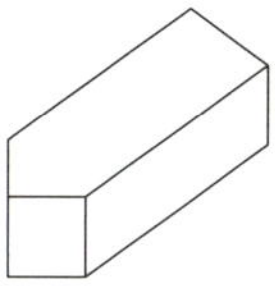

B

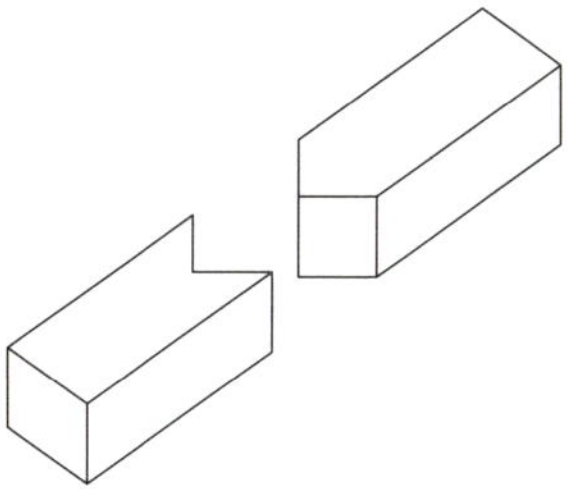

A+B

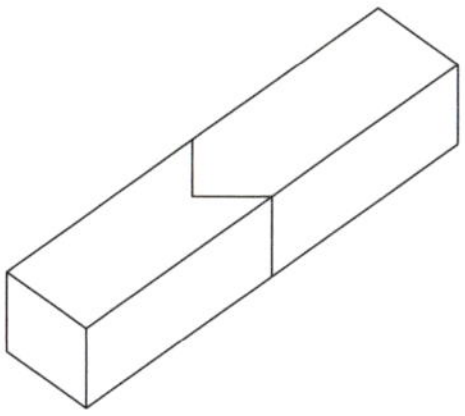

AB

Diagonally cut birdsmouth

ha-gata-shimen-isuka-tsugi
刃形四面鶍継ぎ

diagonale doppelte Schäftung

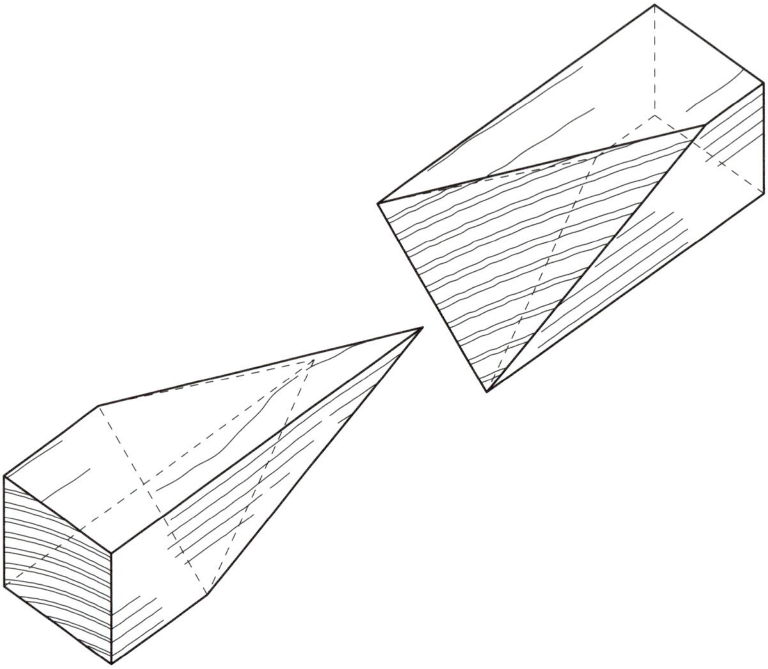

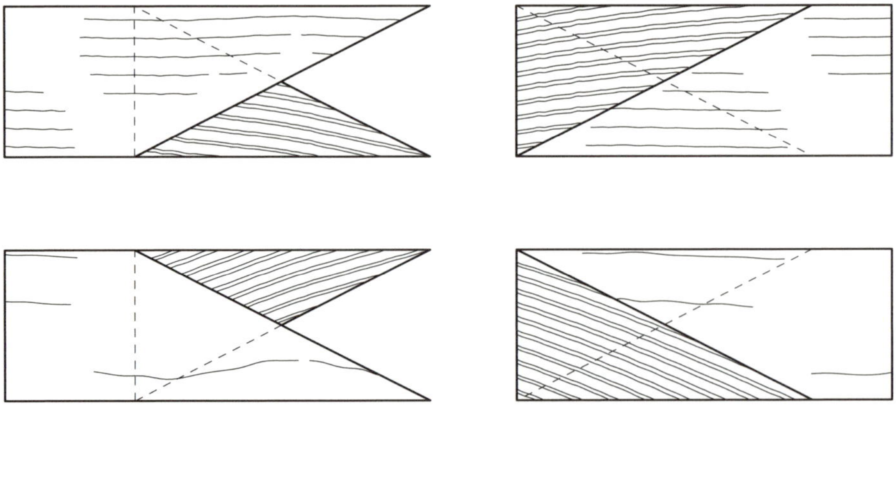

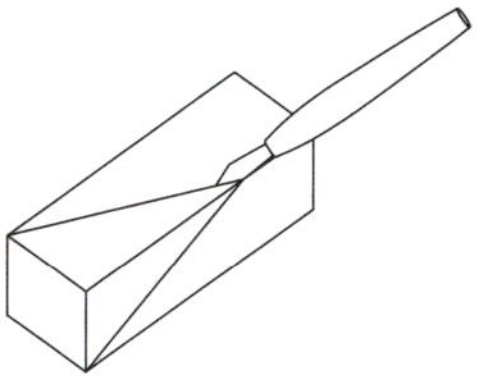

A1

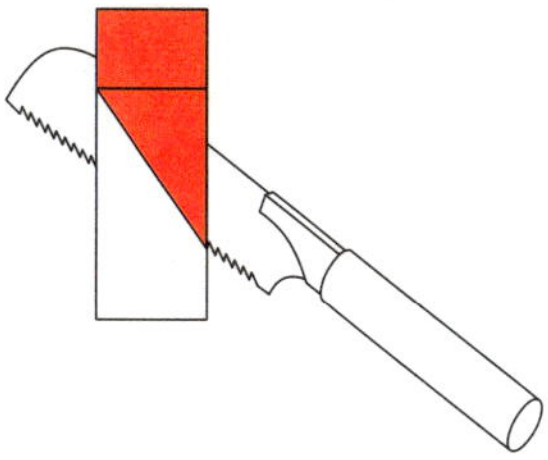

A2

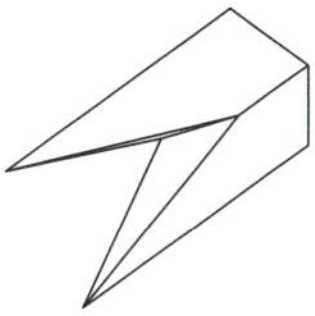

A

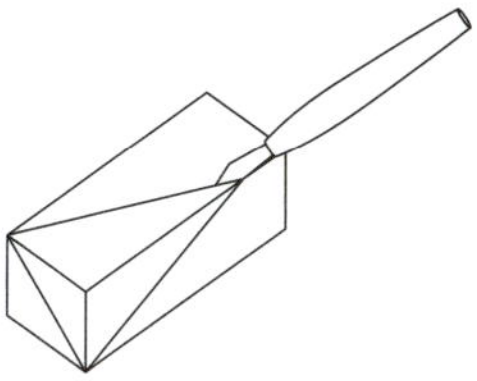

B1

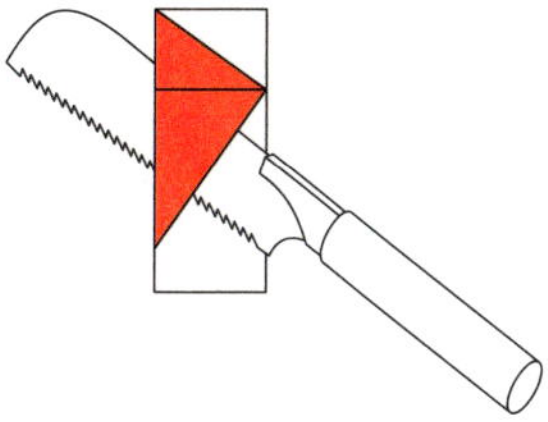

B2

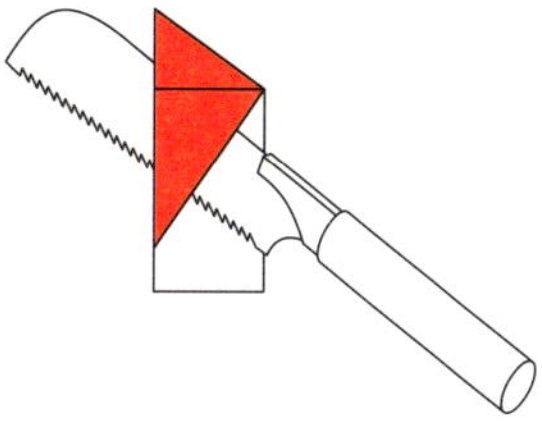

B3

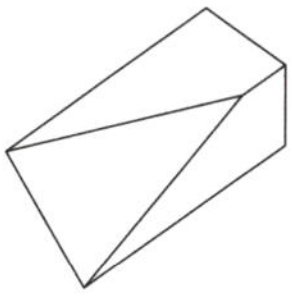

B

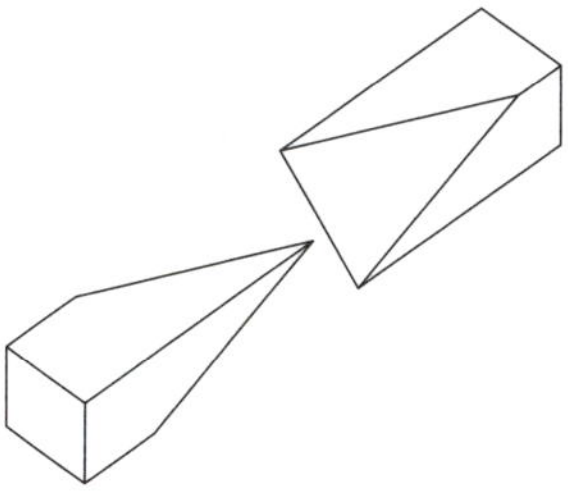

A+B

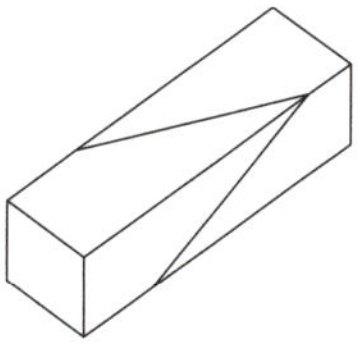

AB

Splicing Joints

Triple-faced splayed scarf joint

sanmen-isuka-tsugi
三面鶍継ぎ

einseitig schräg angeschnittene doppelte Schäftung

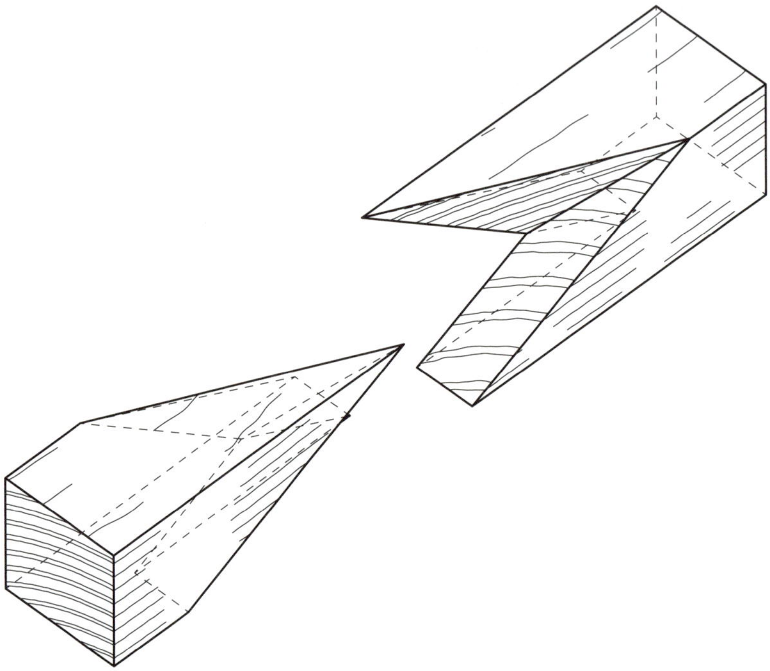

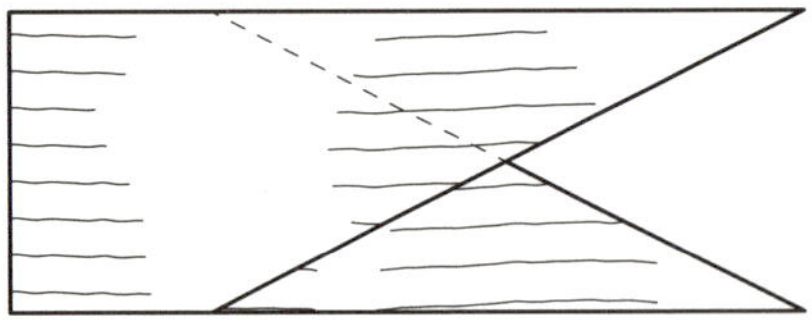

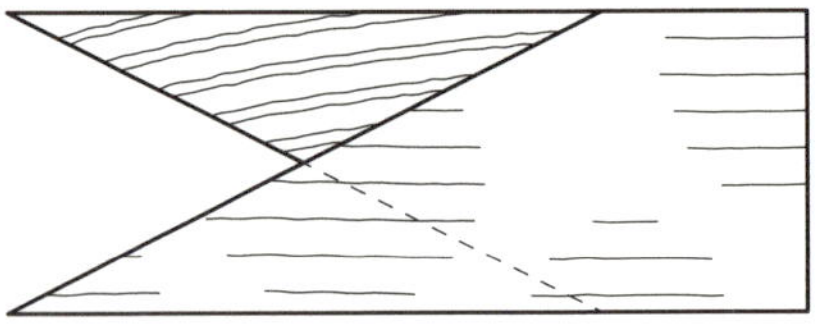

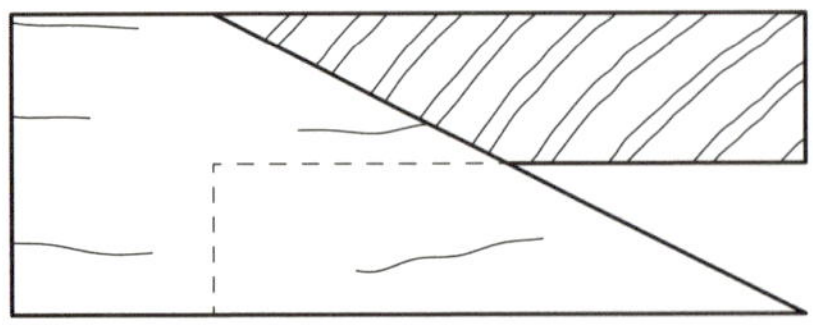

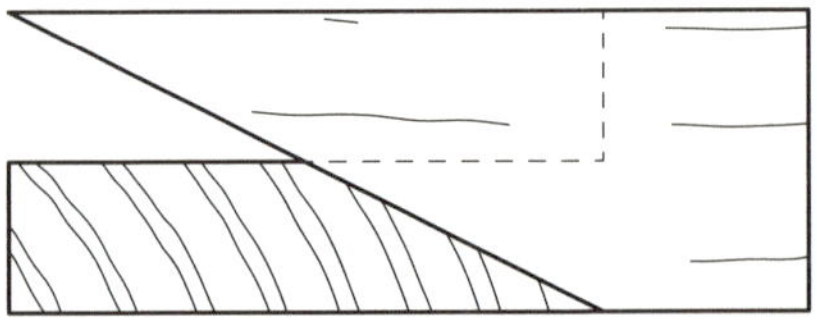

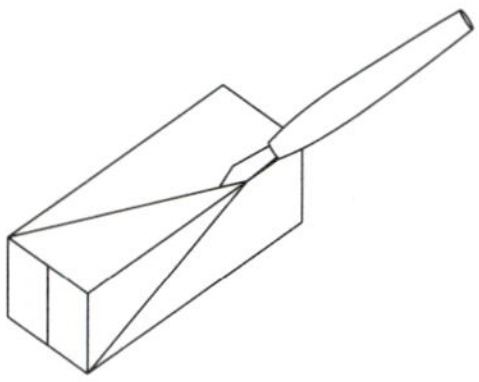

A1

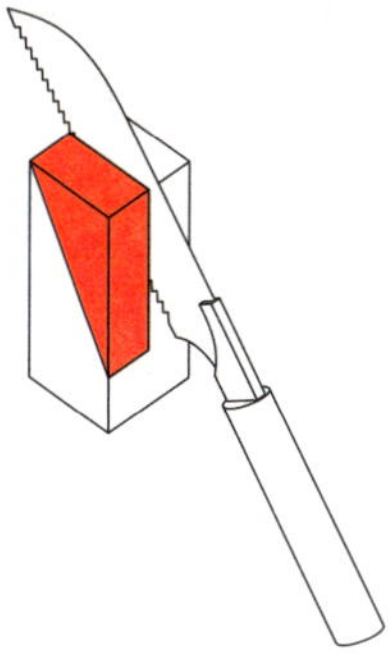

A2

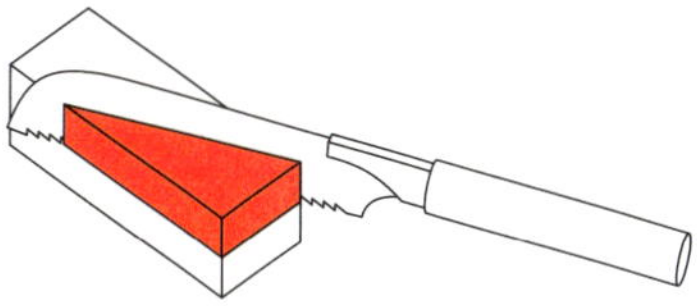

A3

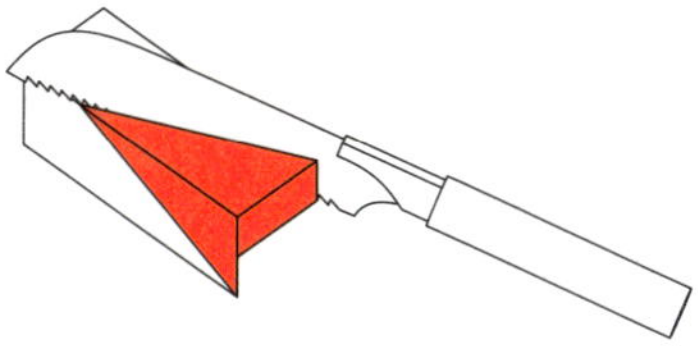

A4

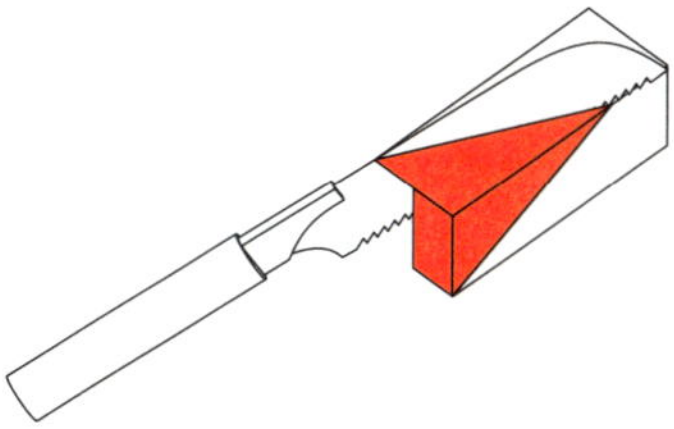

A5

2×

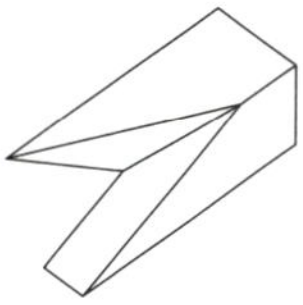

A

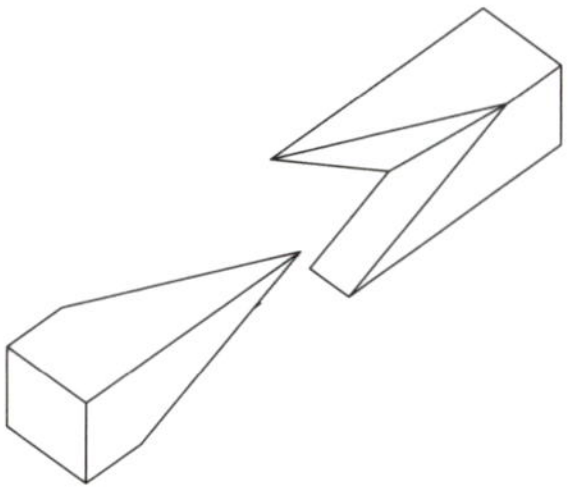

A+A

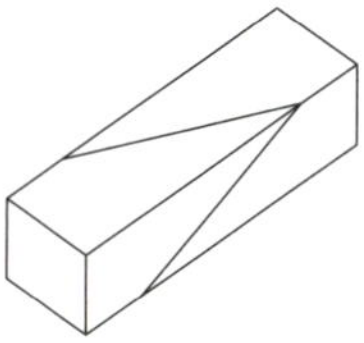

AA

Splicing Joints

Four-faced splayed scarf joint

shimen-isuka-tsugi [*]
四面鶍継ぎ

vierseitige Schäftung

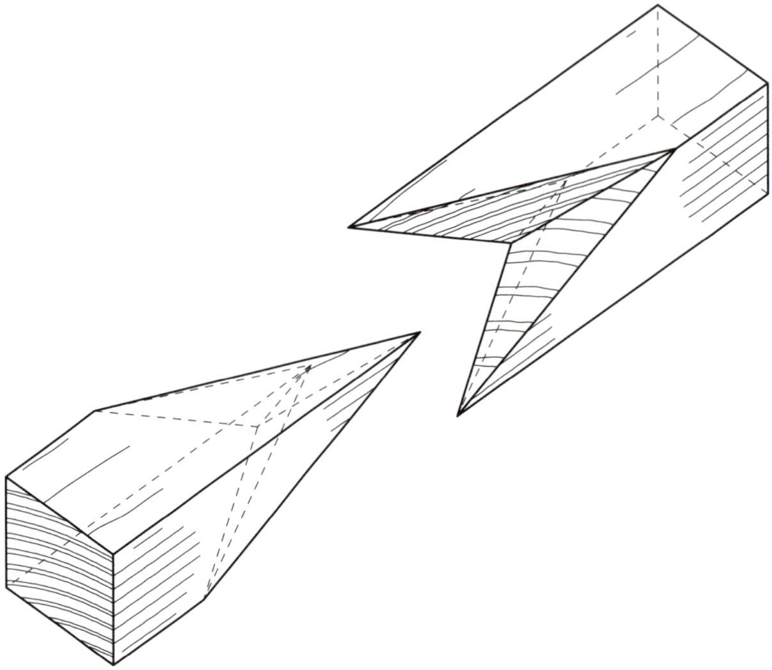

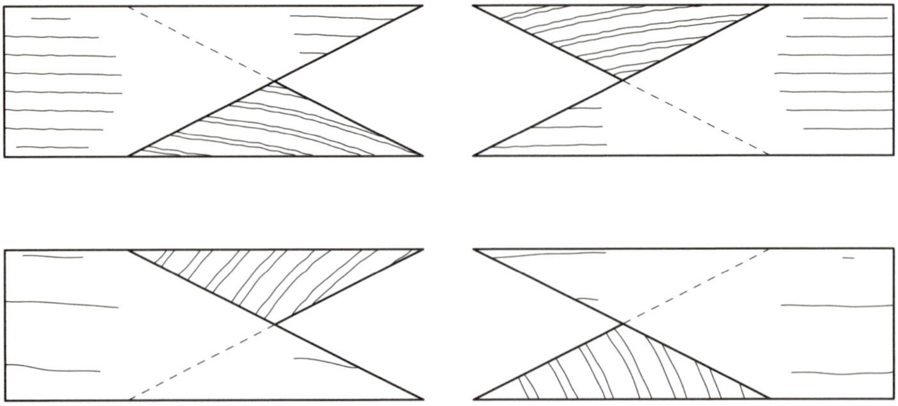

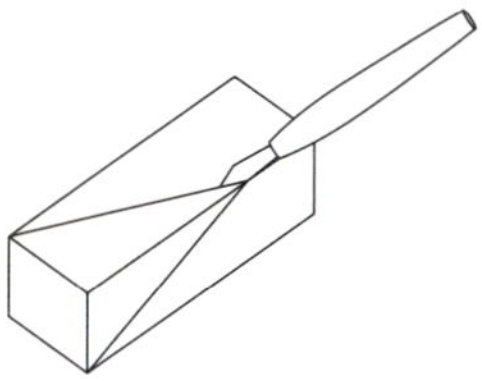

A1

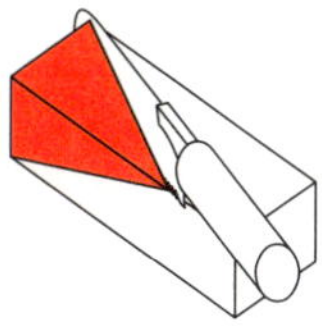

A2

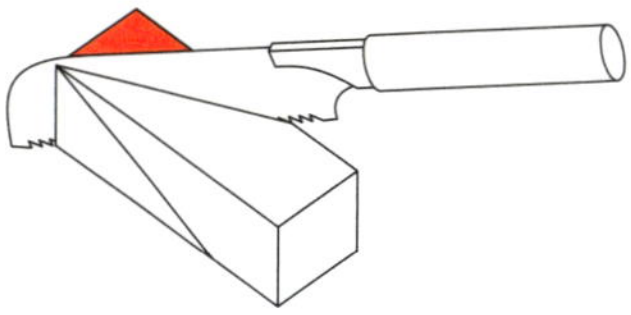

A3

2×

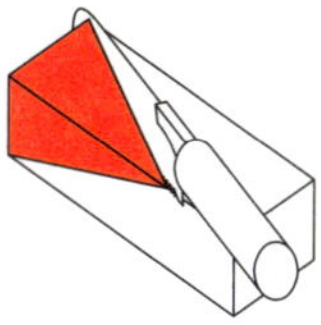

A4

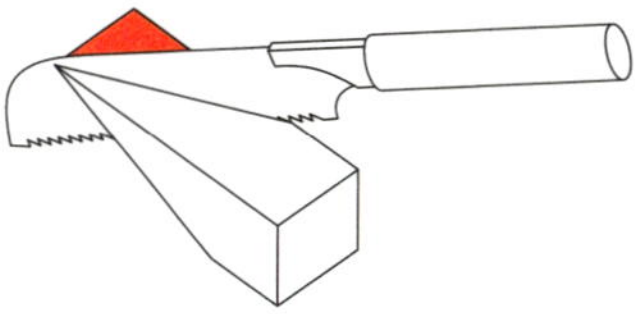

A5

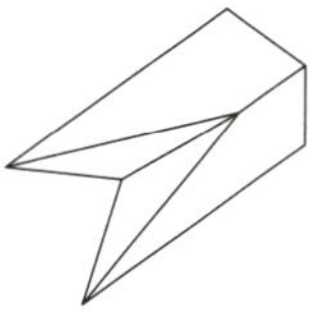

A

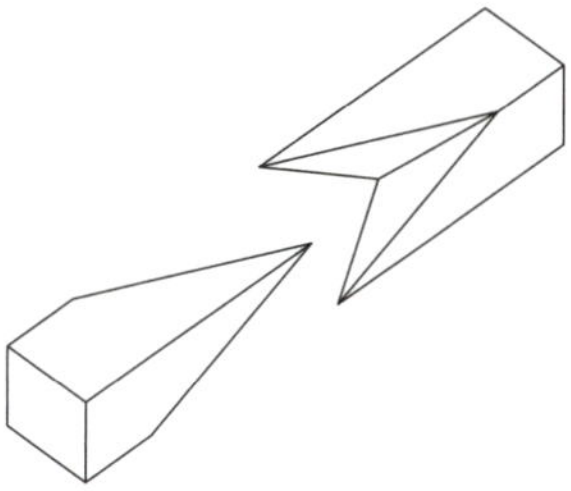

A+A

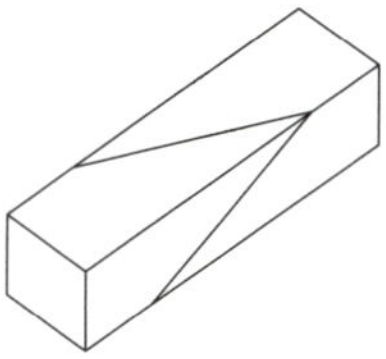

AA

Splicing Joints

Stub tenon joint

inro-tsugi
印籠継ぎ

Zapfen

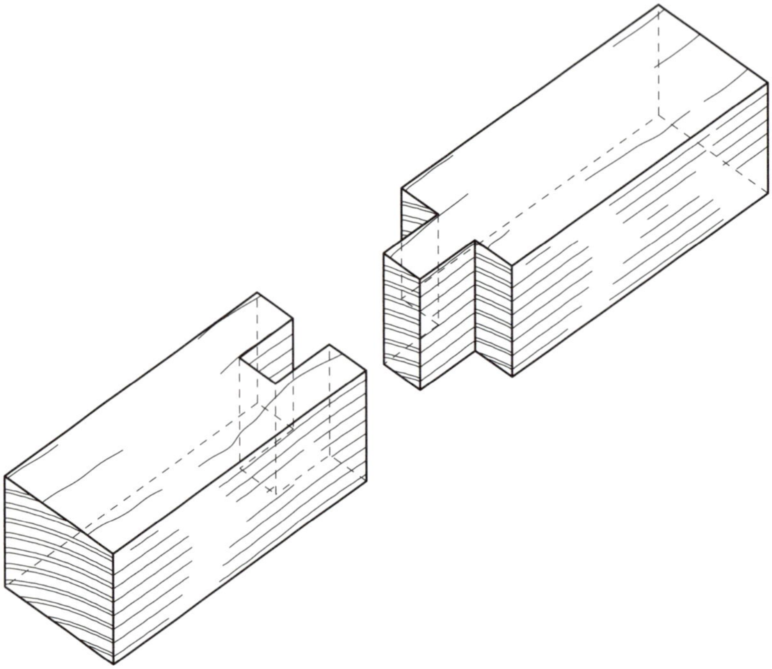

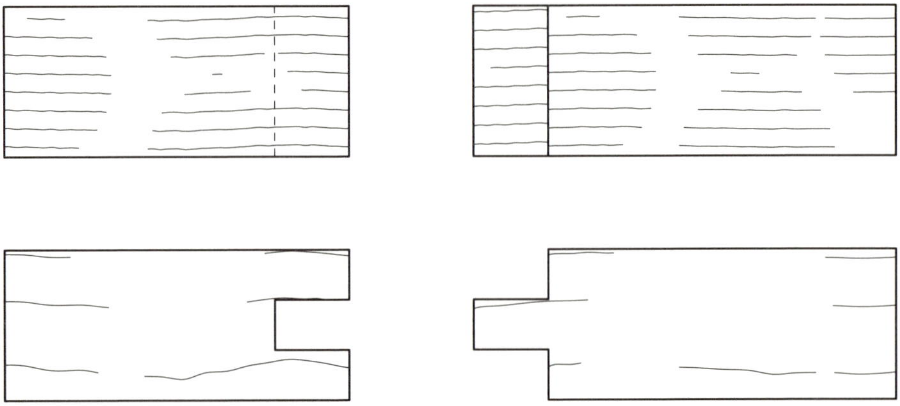

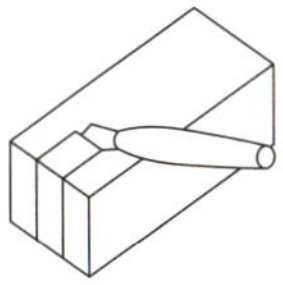

A1

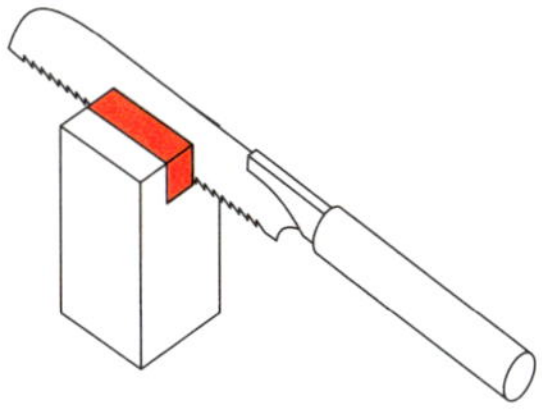

A2

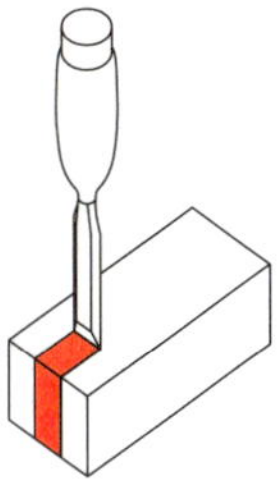

A3

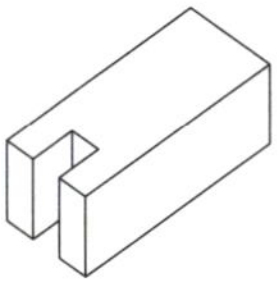

A

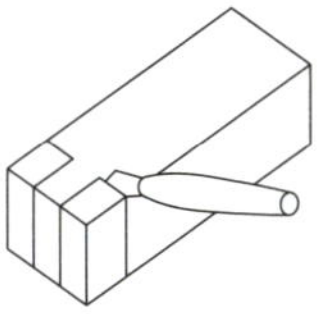

B1

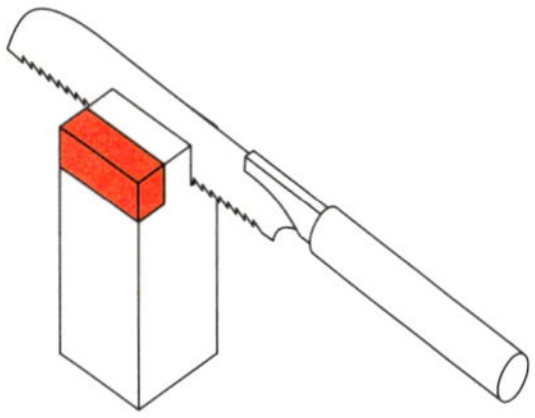

B2

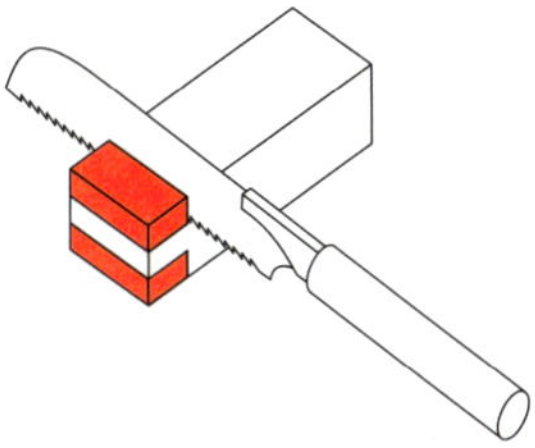

B3

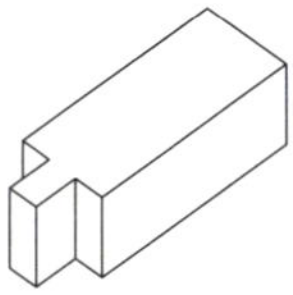

B

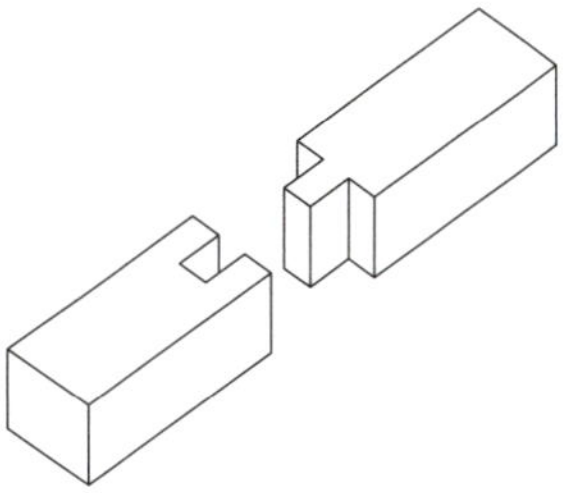

A+B

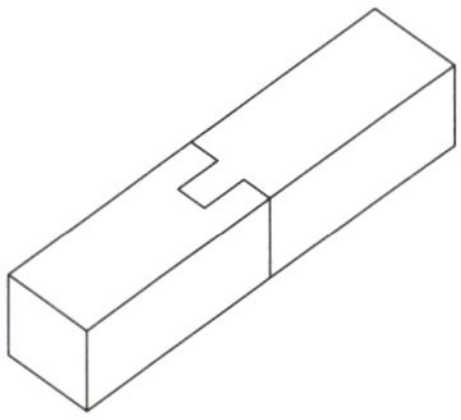

AB

Splicing Joints

Half-blind stub tenon joint

sanpō-dō-tsuki-tan-hozo-tsugi
三方胴付短枘継ぎ

einseitig abgesetzter Zapfen

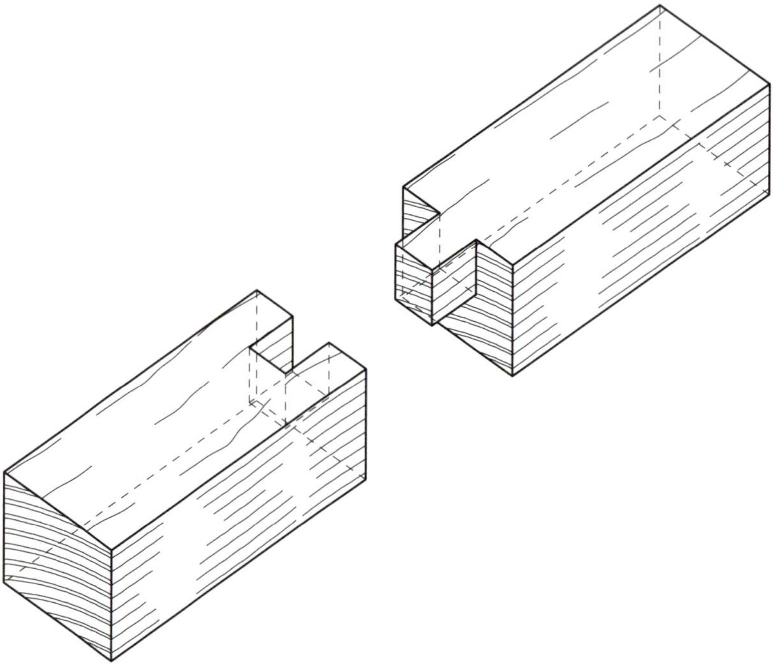

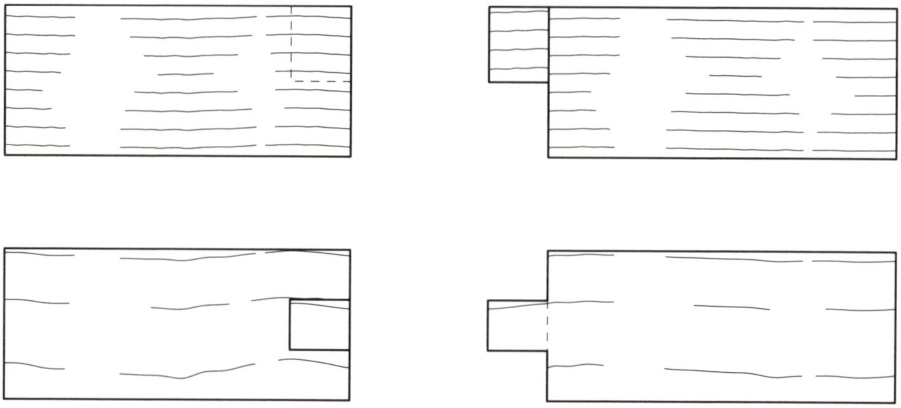

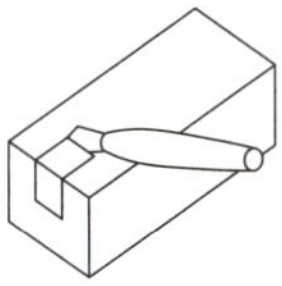
A1

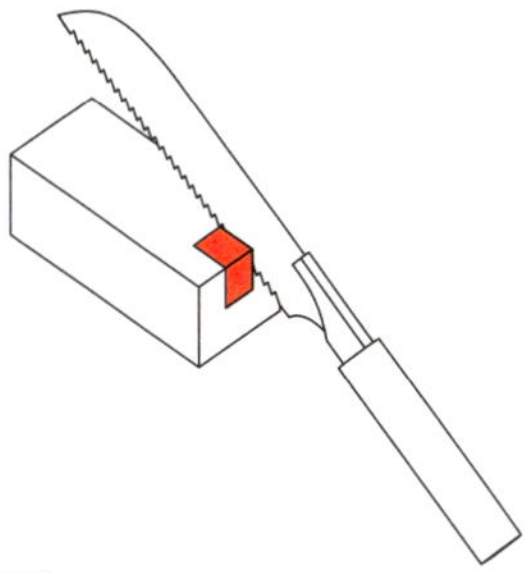
A2

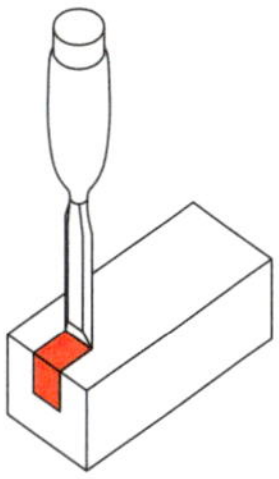
A3

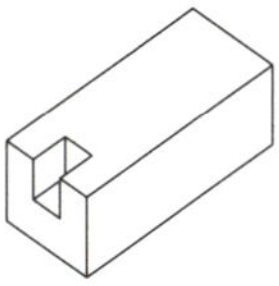
A

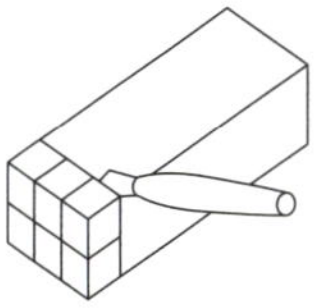
B1

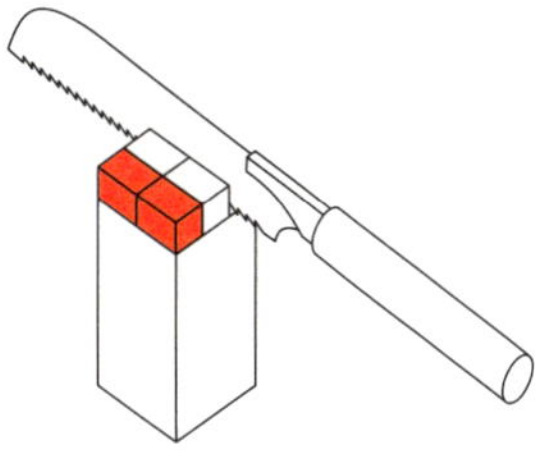
B2

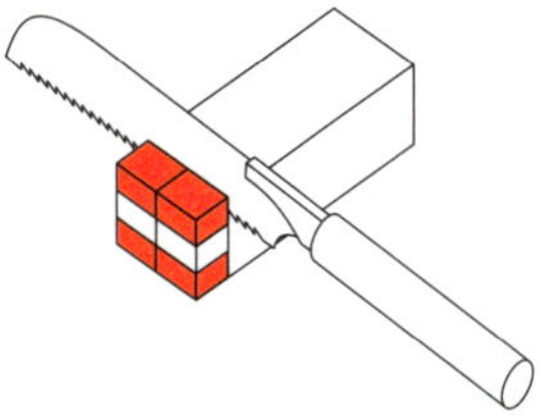
B3

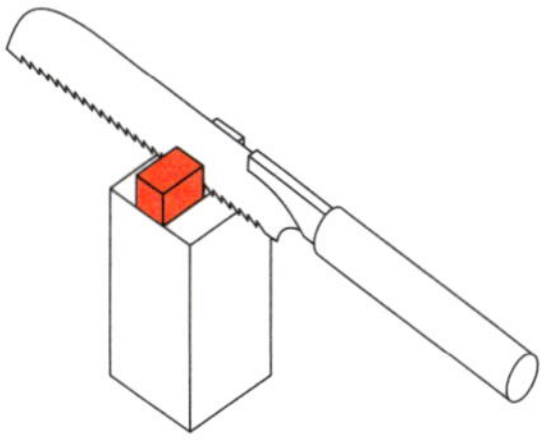
B4

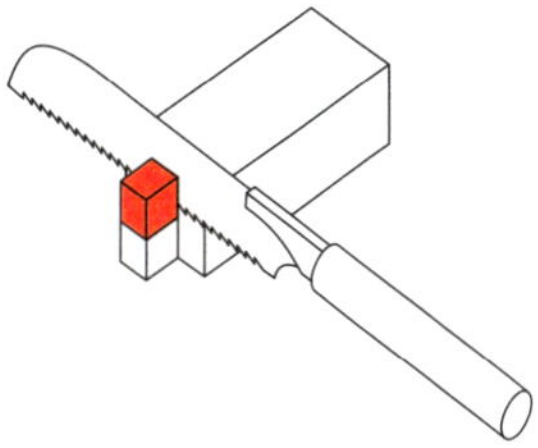
B5

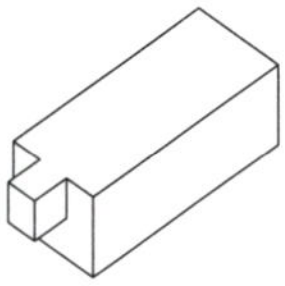
B

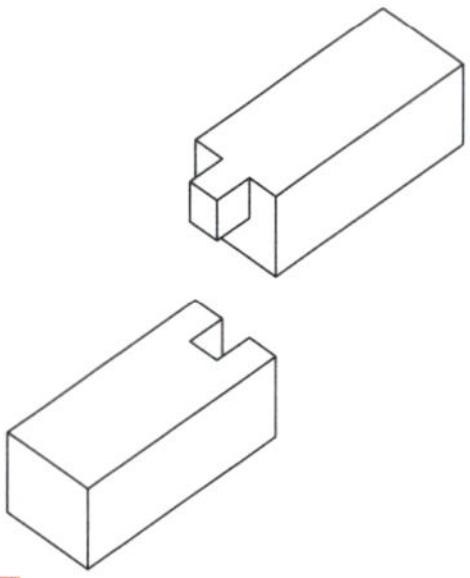
A+B

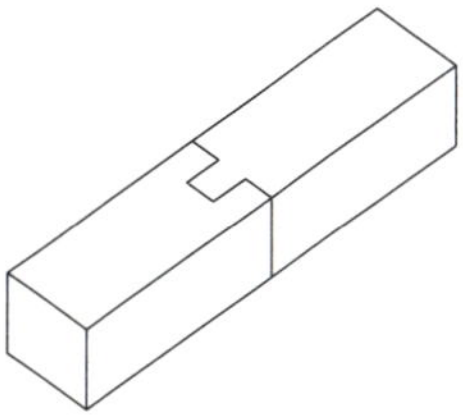
AB

Splicing Joints

Floating tenon joint

yatoi-hozo-imo-tsugi
雇枘芋継ぎ

eingesetzter Zapfen

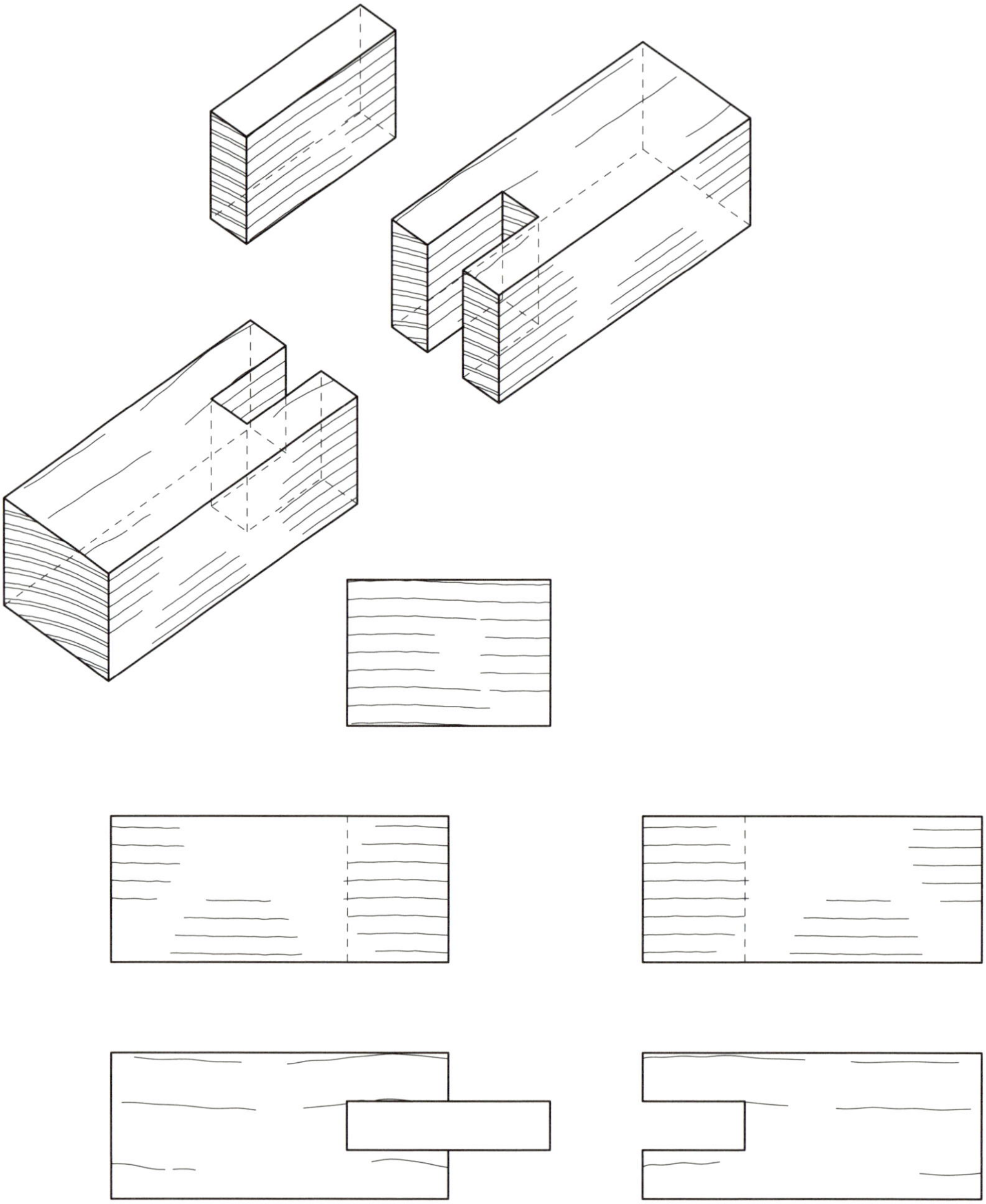

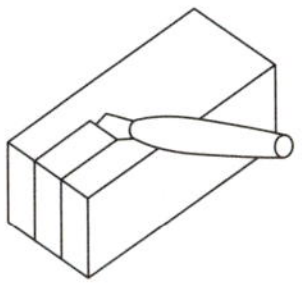

A1

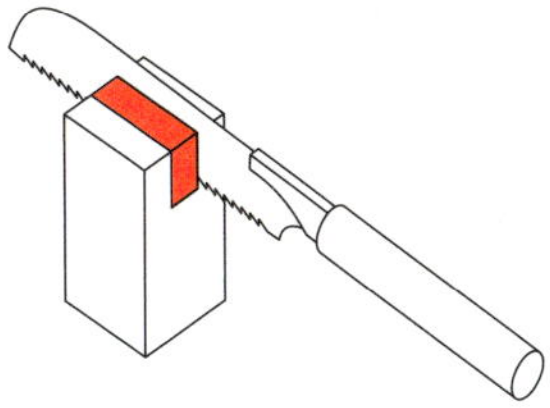

A2

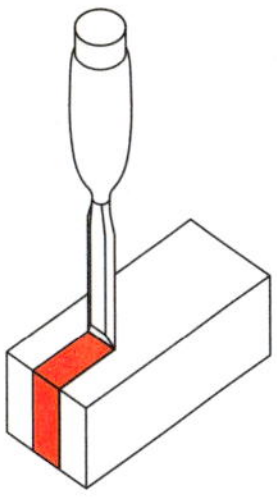

A3

2×

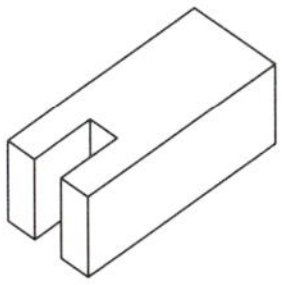

A

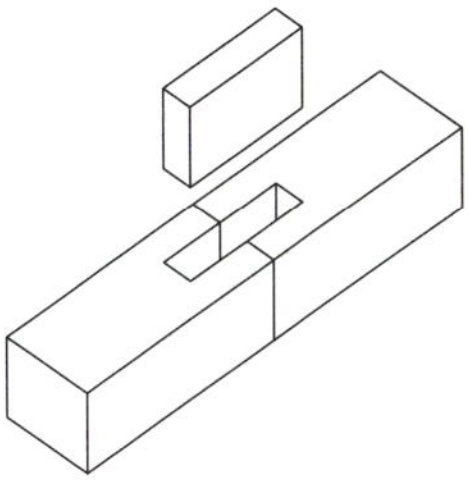

A+B+A

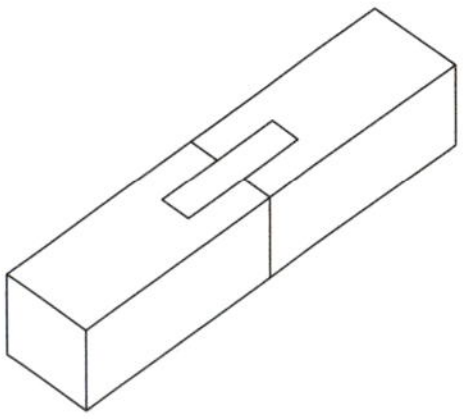

ABA

Splicing Joints

Pegged bridle joint

hira-hozo-mechi-tsugi-komi-sen-uchi
平枘目違継込栓打ち

Zapfen mit Dübelsicherung

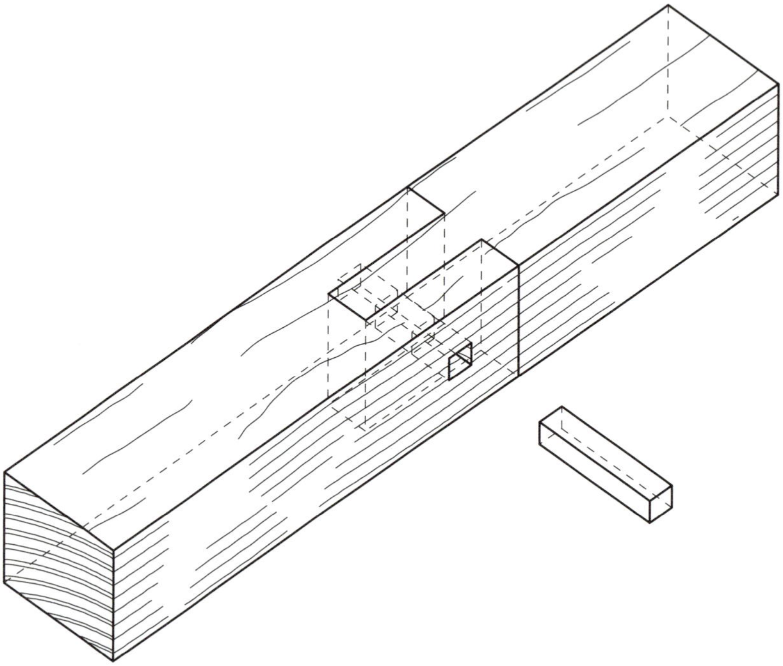

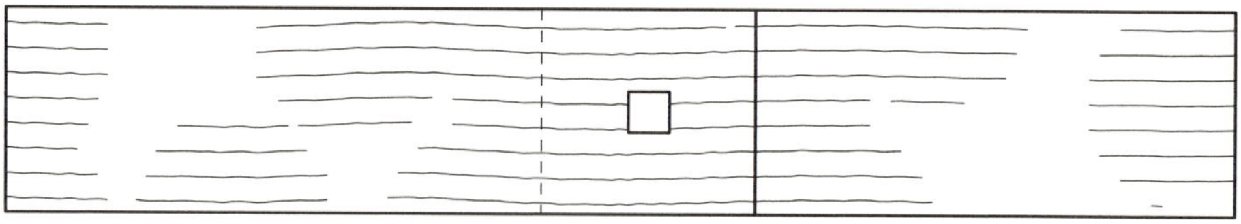

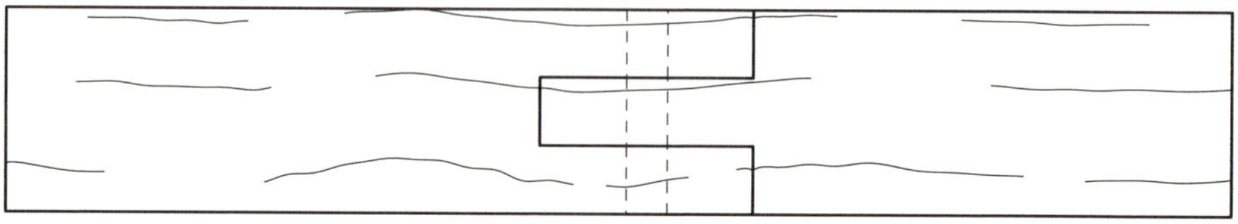

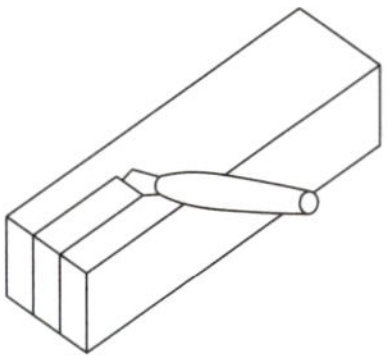

A1

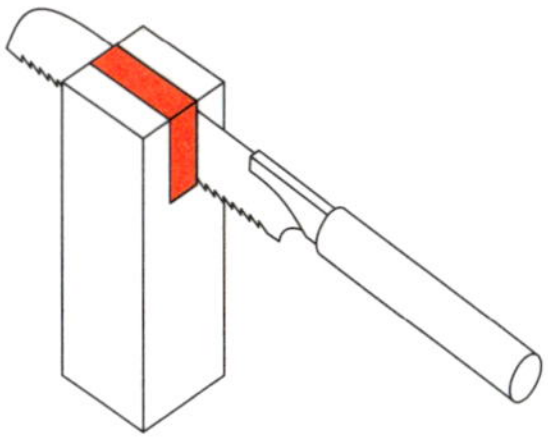

A2

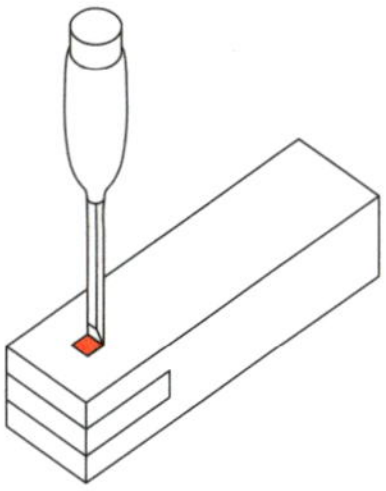

A3

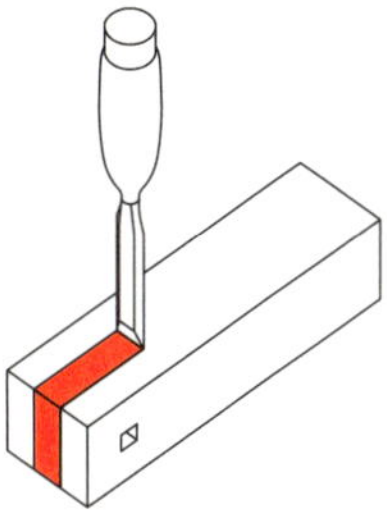

A4

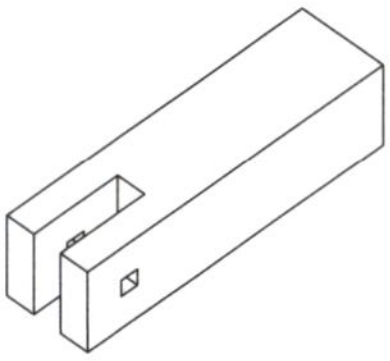

A

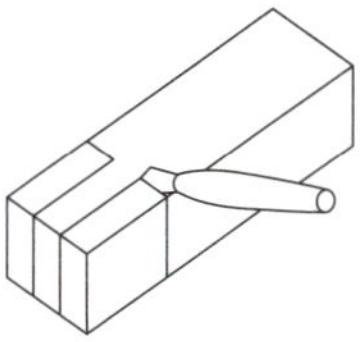

B1

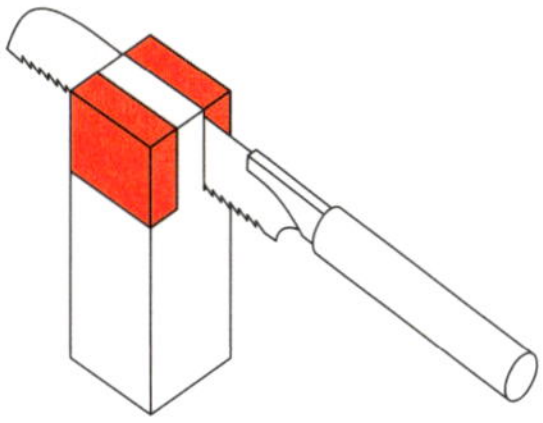

B2

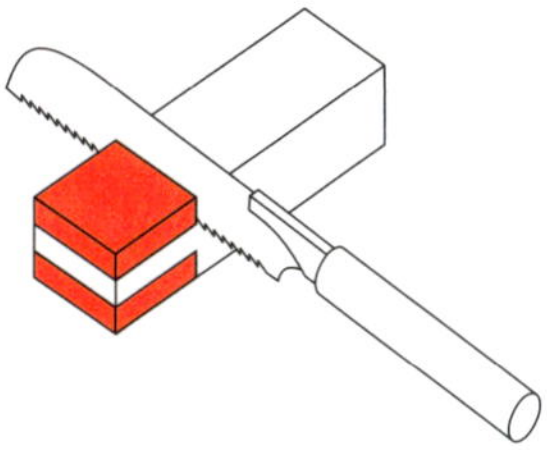

B3

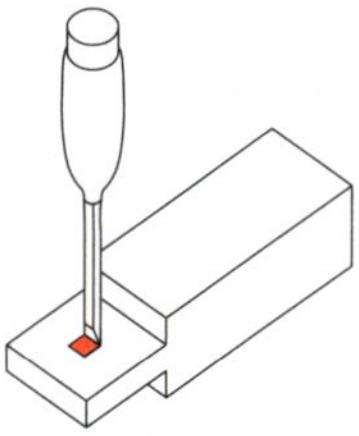

B4

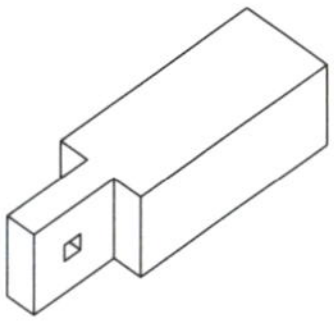

B

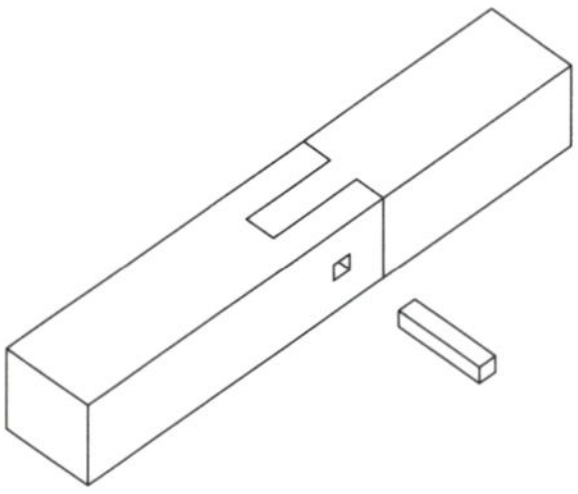

A+B

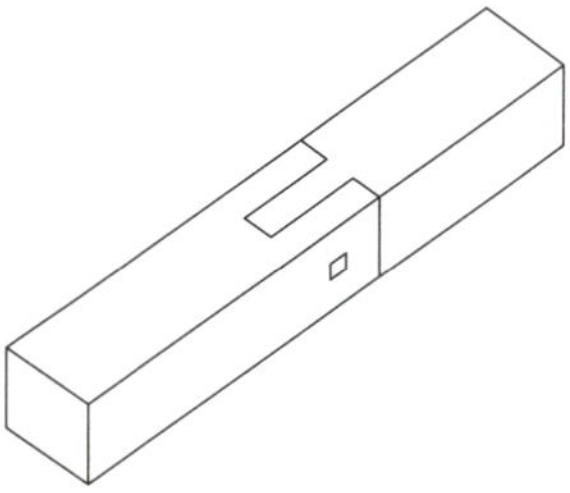

AB

Splicing Joints

Mortise and tenon with V-ends and undercut shoulders

yahazu-mechi-sanmai-kumi-tsugi (sogi-tsuki)
矢筈目違三枚組継ぎ(殺付)

Zapfen mit hinterschnittener Brüstung und v-förmig eingeschnittener Stirn

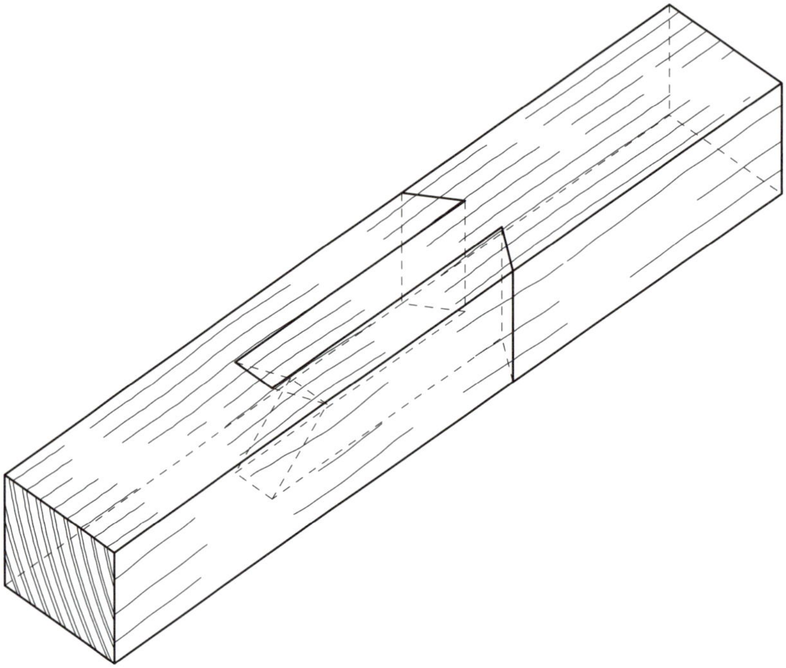

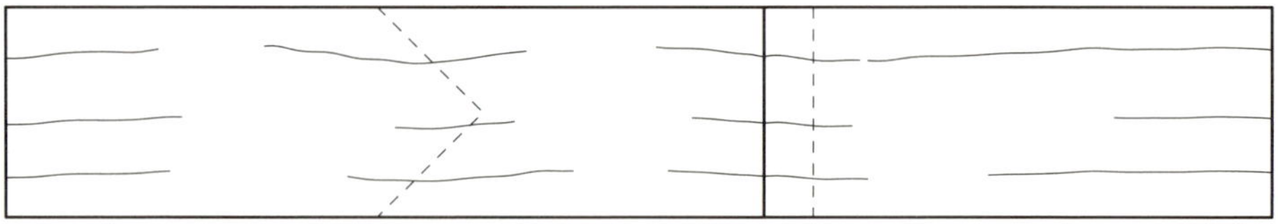

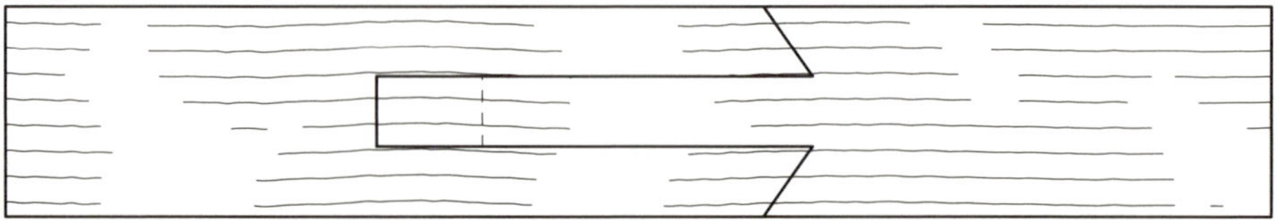

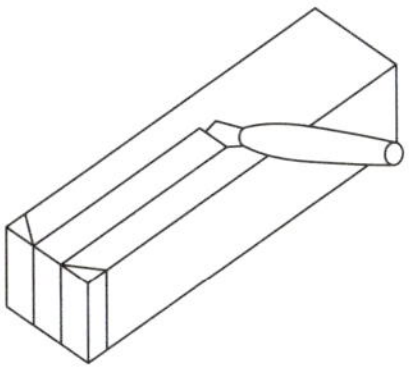

A1

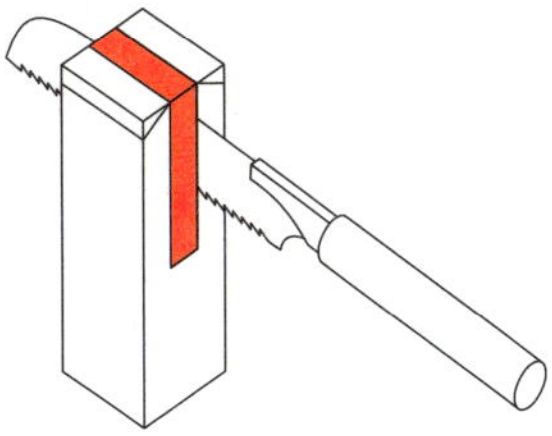

A2

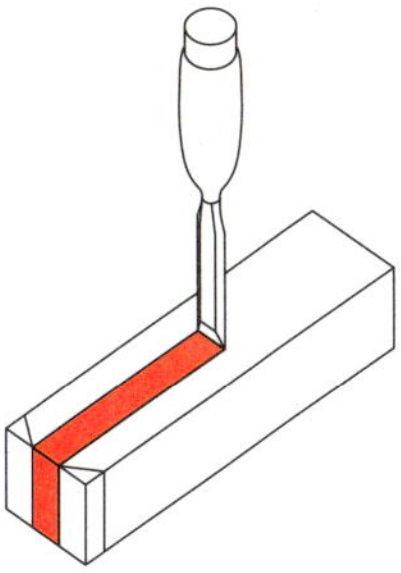

A3

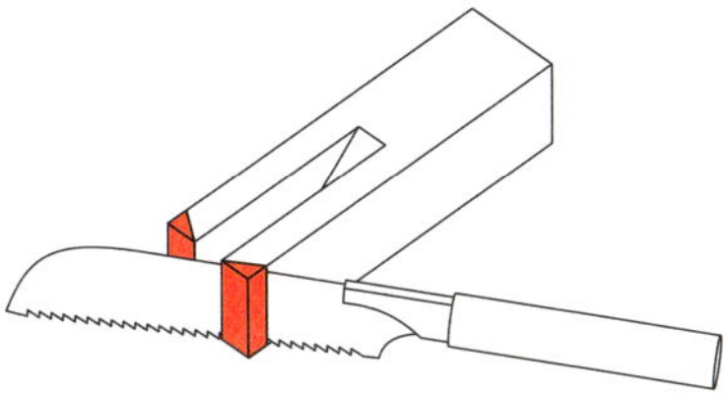

A4

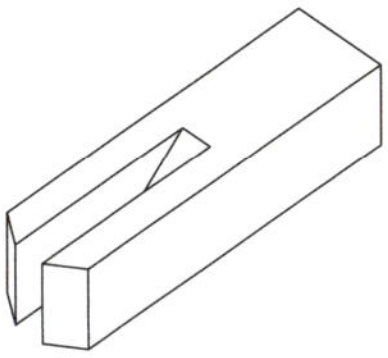

A

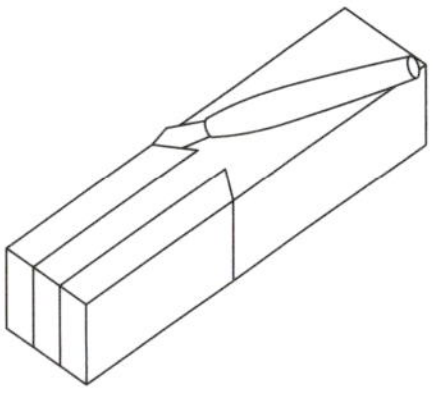

B1

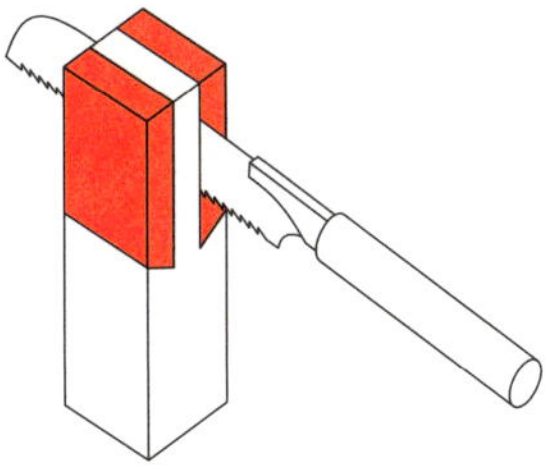

B2

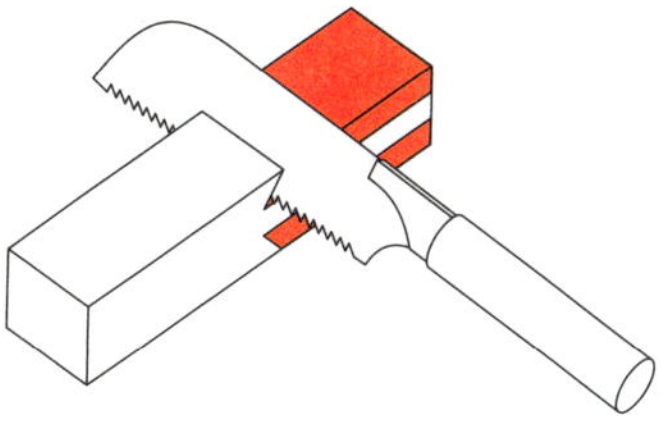

B3

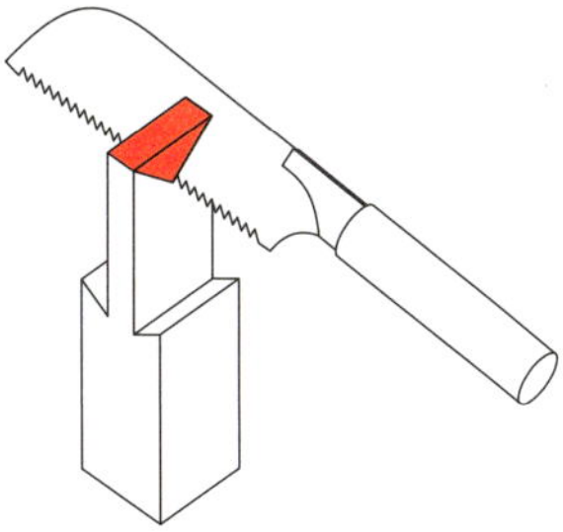

B4

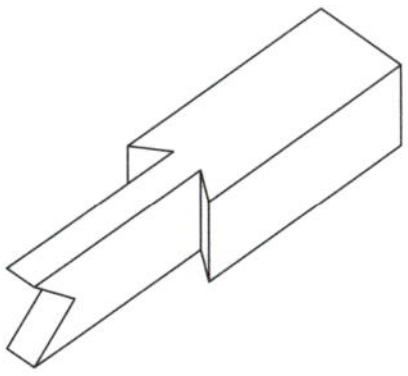

B

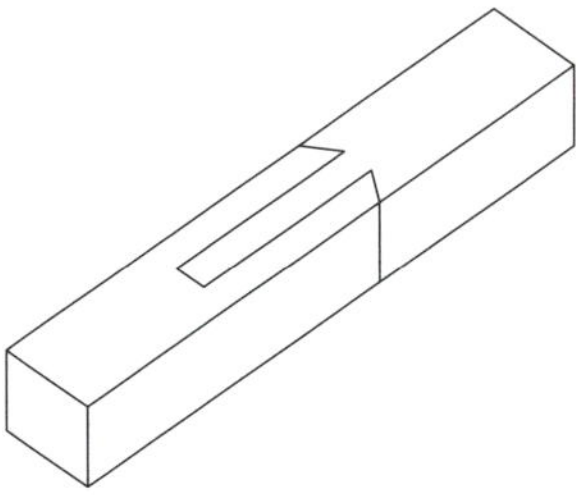

AB

Splicing Joints

Pegged mortise and tenon with undercut shoulders

sanmai-kumi-tsugi-komi-sen-uchi (sogi-tsuki)
三枚組継込栓打ち(殺付)

Zapfen mit hinterschnittener Brüstung und Dübelsicherung

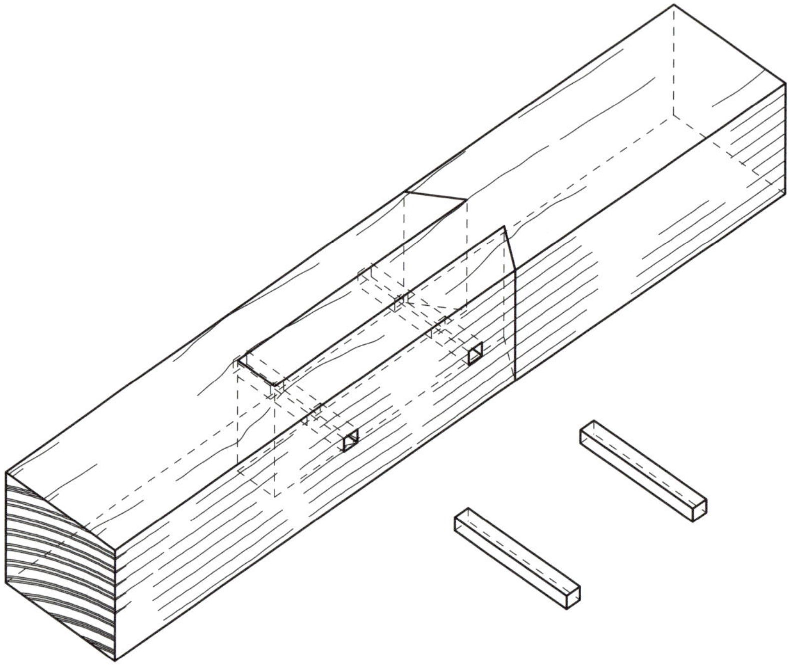

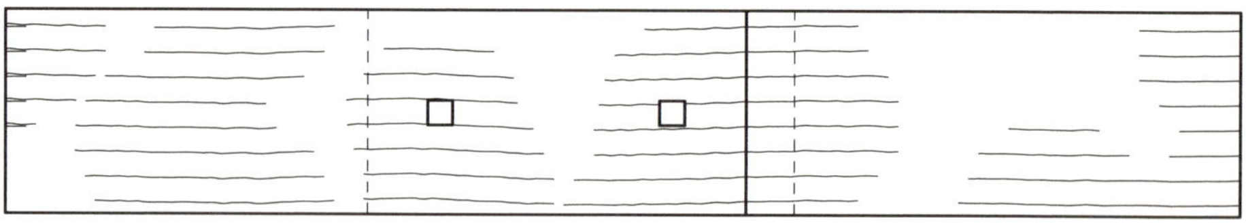

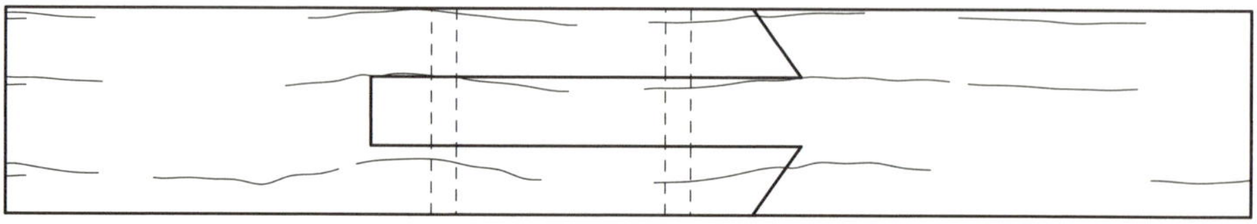

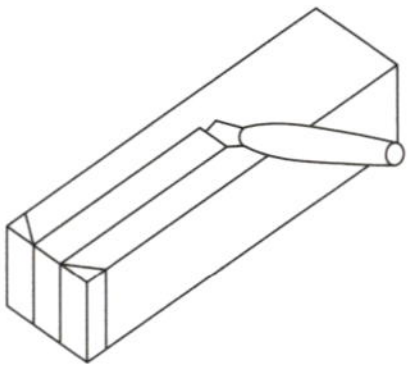

A1

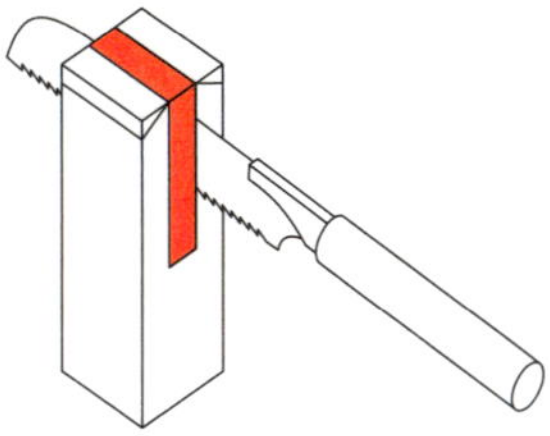

A2

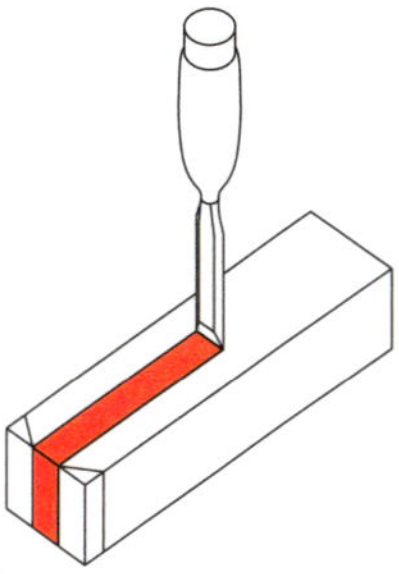

A3

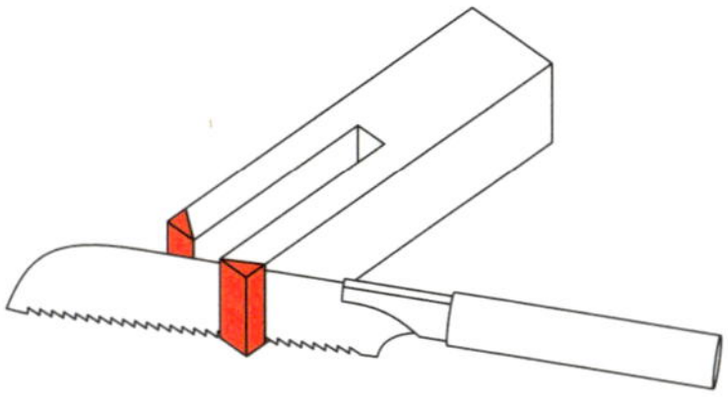

A4

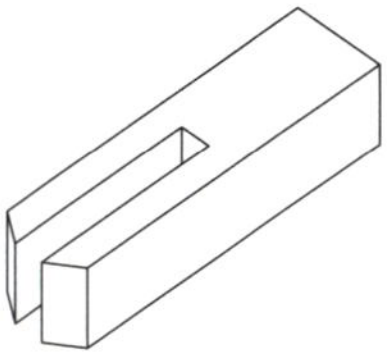

A

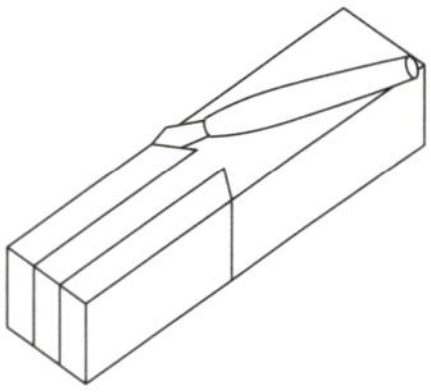

B1

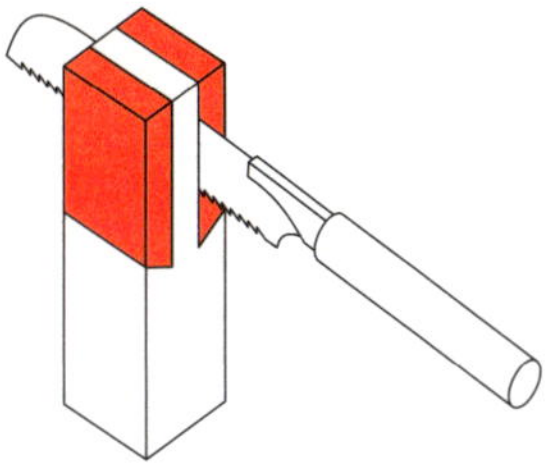

B2

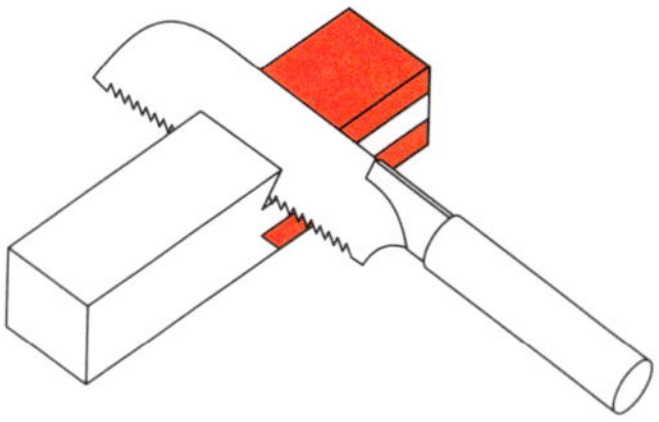

B3

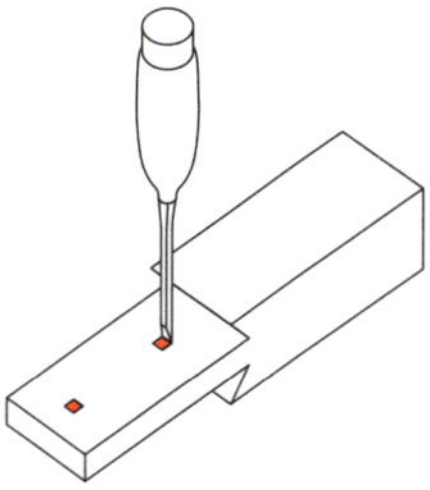

B4

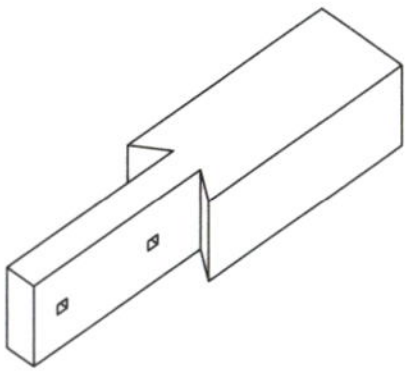

B

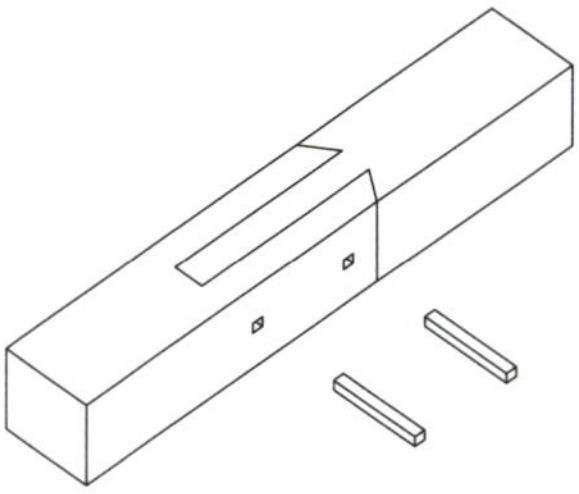

A+B

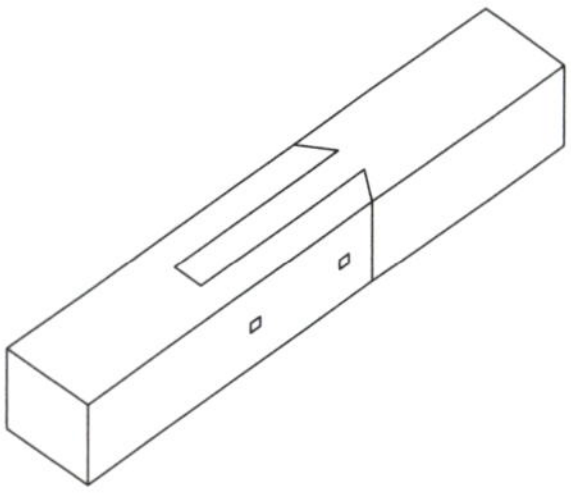

AB

Splicing Joints

Spear point mortise and tenon with beveled shoulders

yahazu-mechi-tsugi (sogi-tsuki)
矢筈目違継ぎ（殺付）

Zapfen mit Gratschnitt und schräger Brüstung

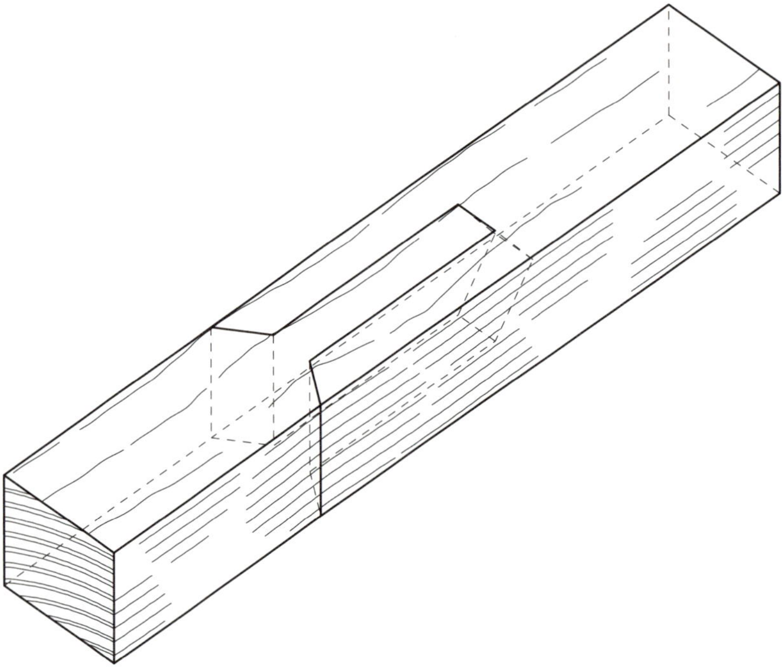

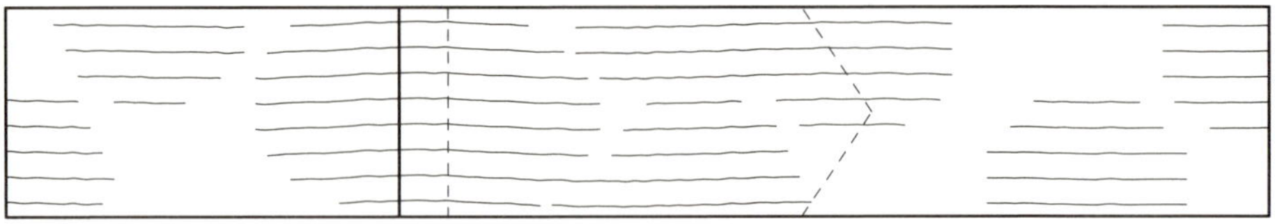

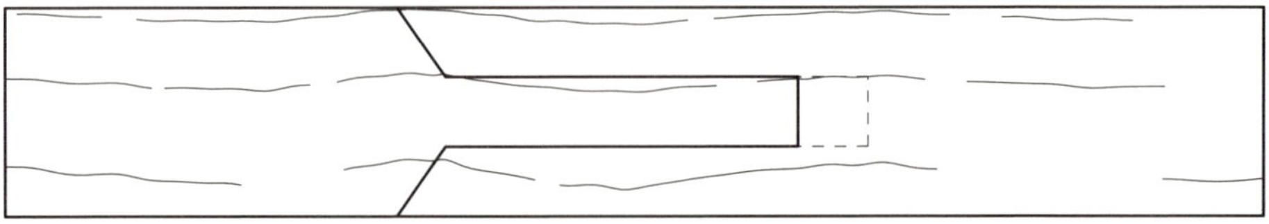

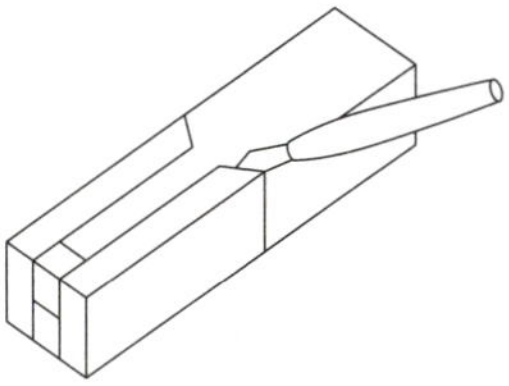

A1

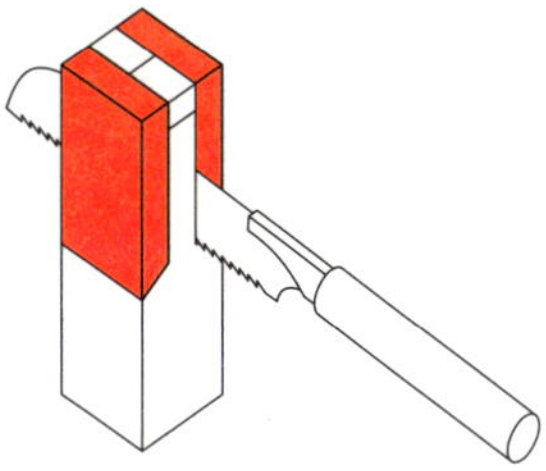

A2

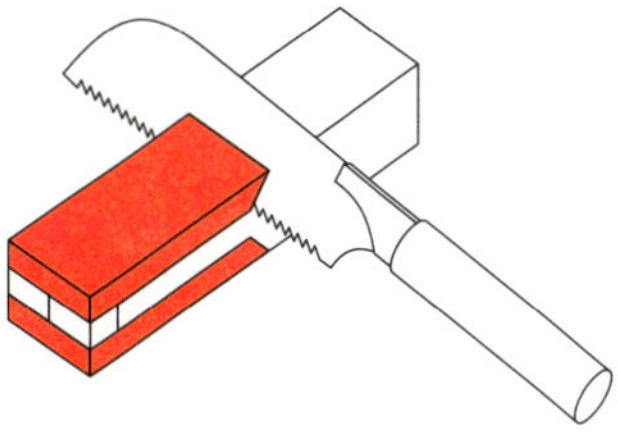

A3

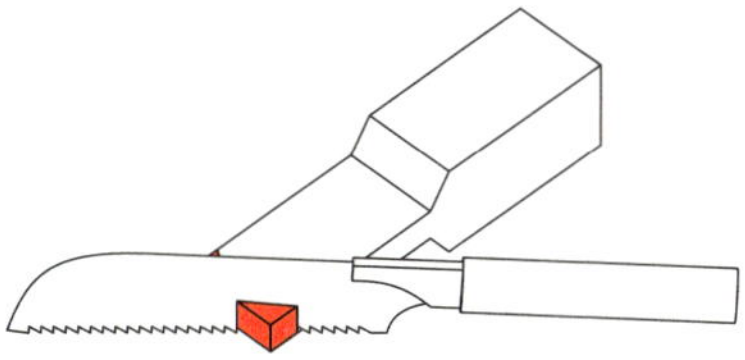

A4

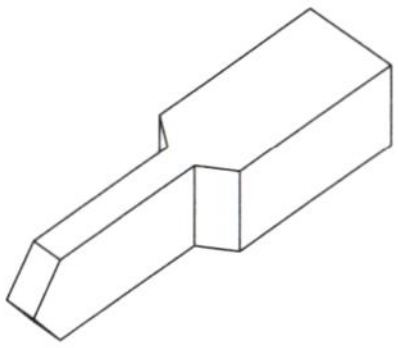

A

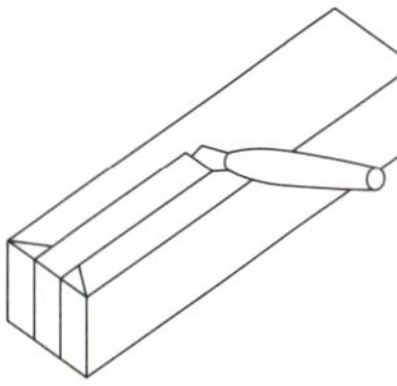

B1

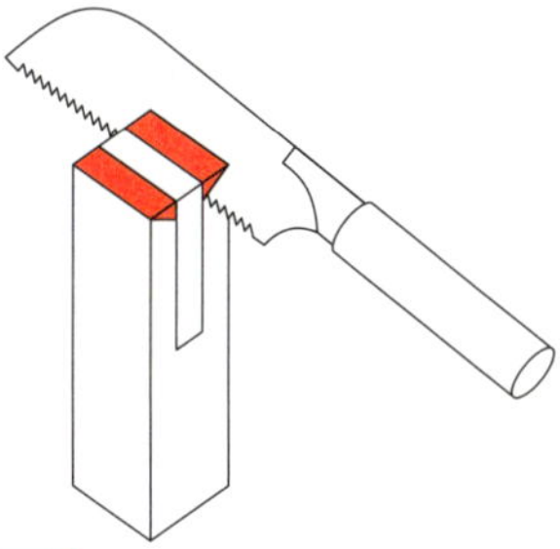

B2

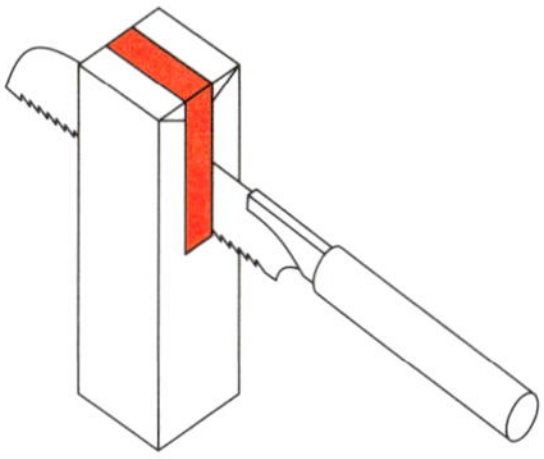

B3

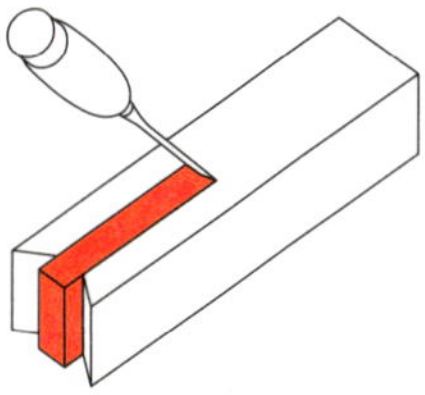

B4

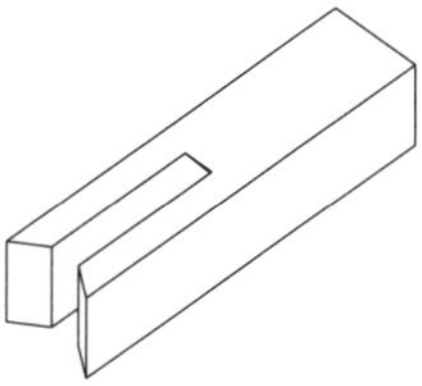

B

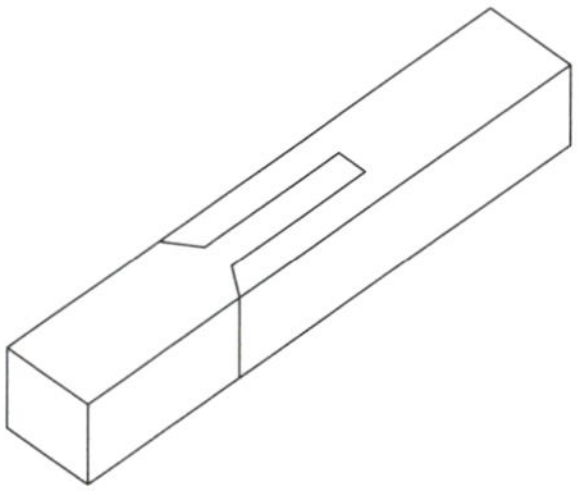

AB

Splicing Joints

Pegged spear point mortise and tenon with V-shaped shoulders

yahazu-mechi-ken-dome-tsugi-komi-sen-uchi
矢筈目違剣留継込栓打ち

Zapfen mit Gratschnitt, v-förmig eingeschnittener Brüstung und Dübelsicherung

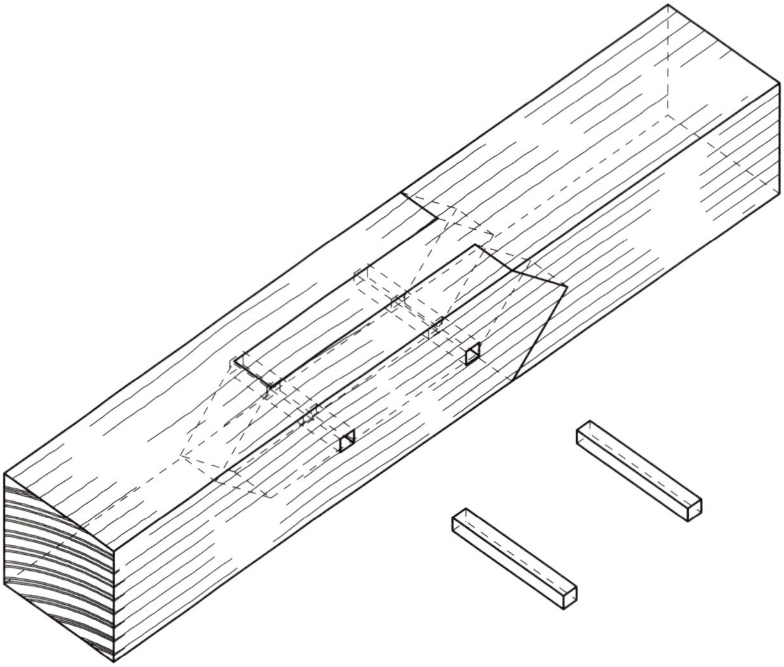

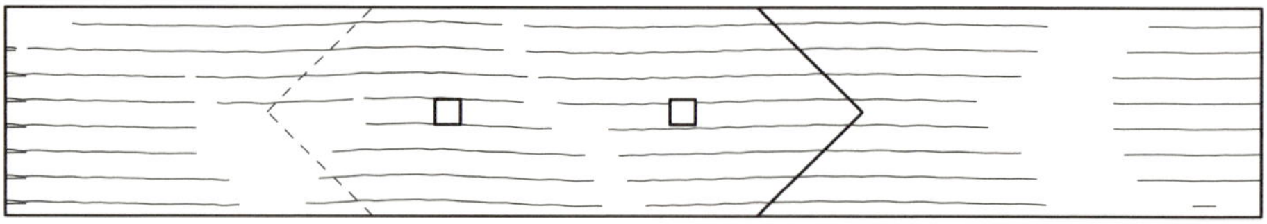

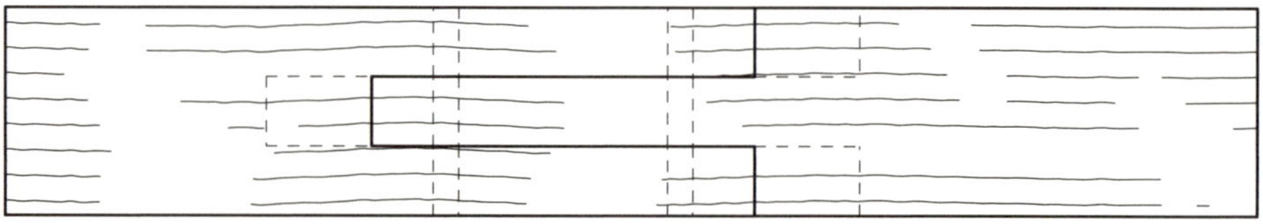

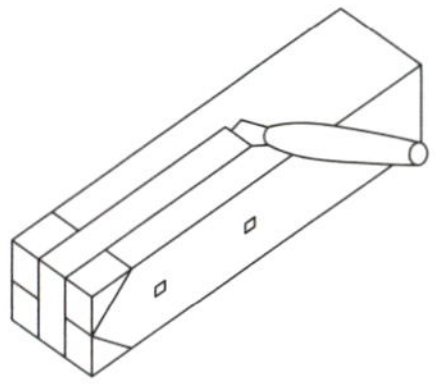

A1

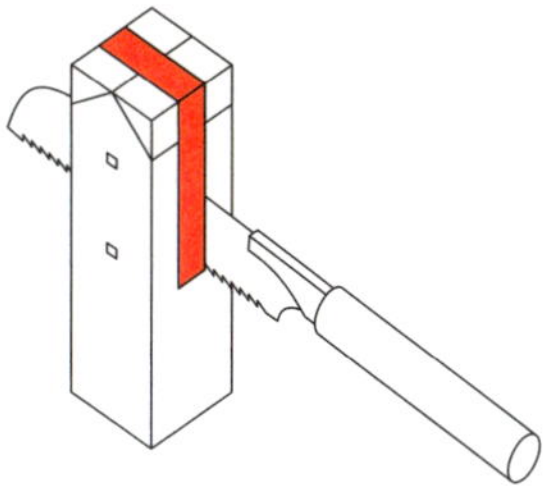

A2

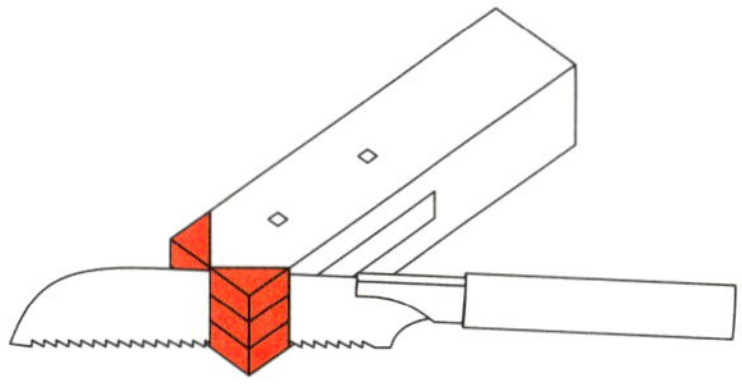

A3

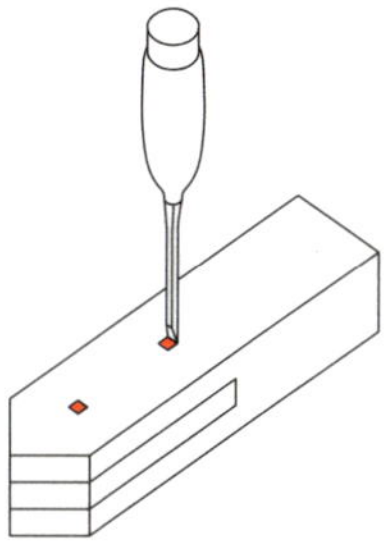

A4

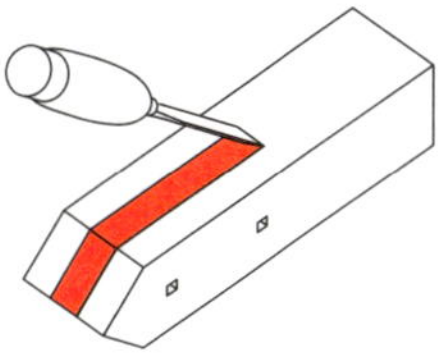

A5

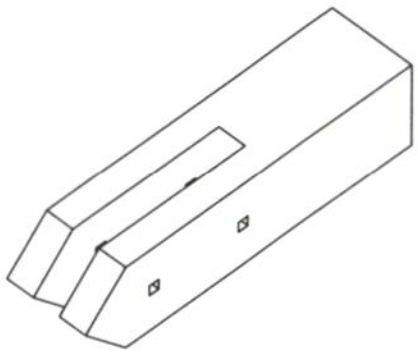

A

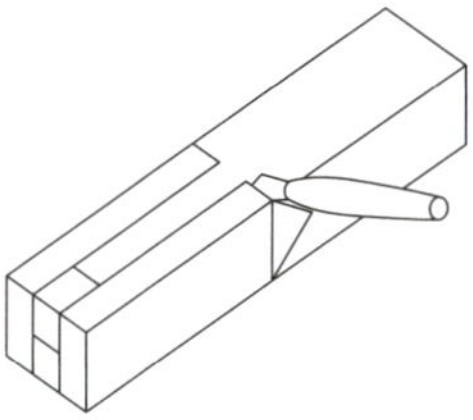

B1

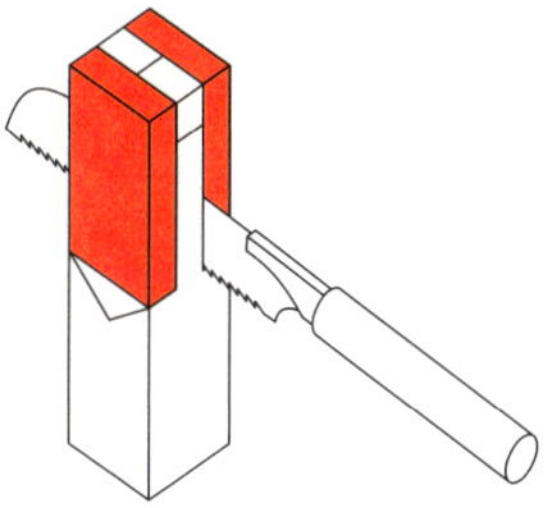

B2

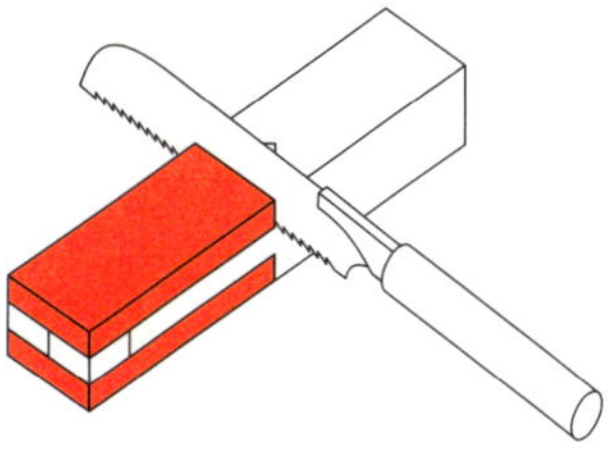

B3

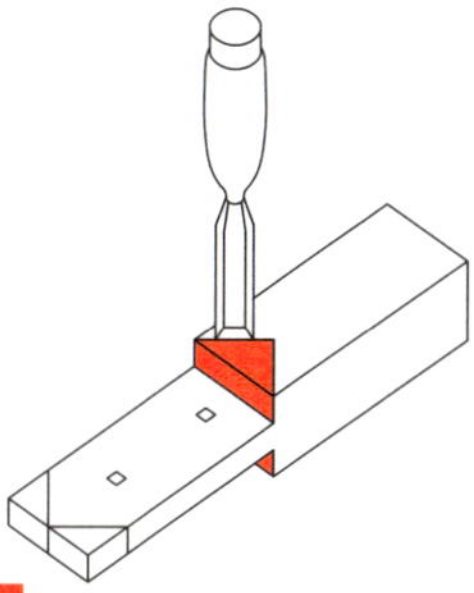

B4

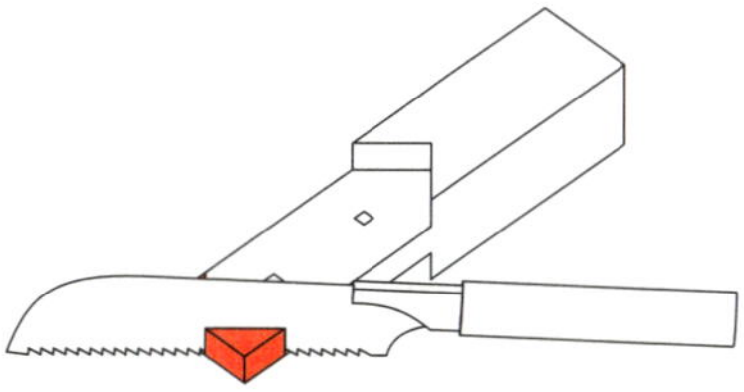

B5

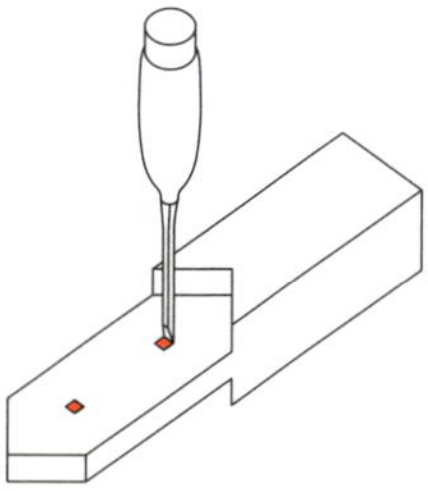

B6

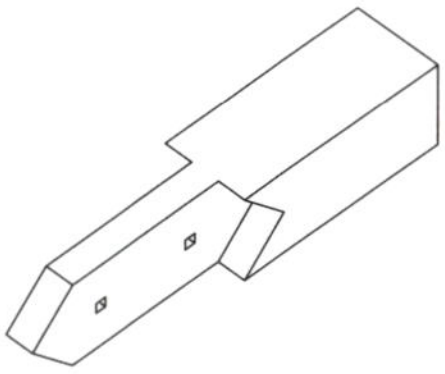

B

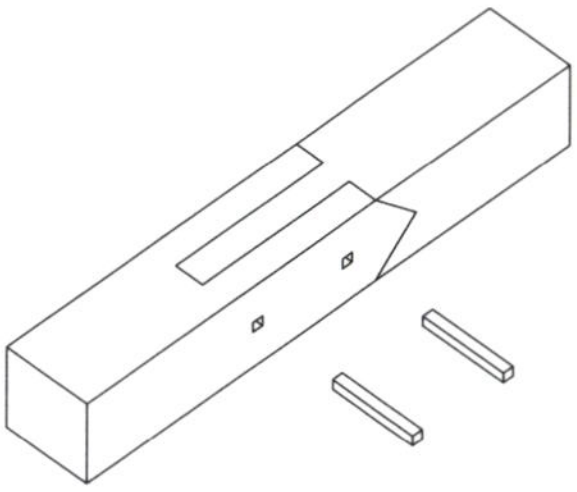

A+B

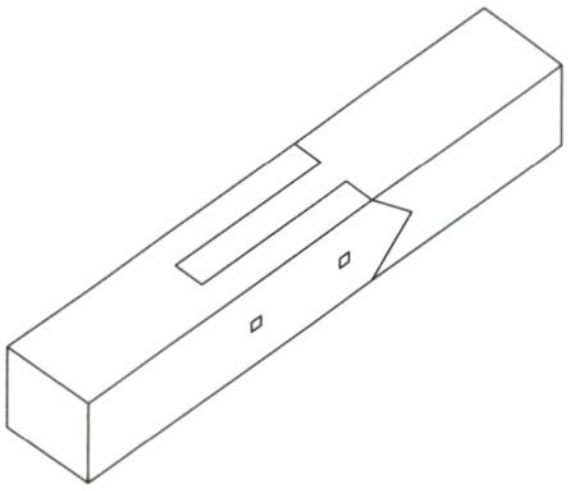

AB

Splicing Joints

Pegged rabbeted mortise and tenon joint

sao-hozo-komi-sen-tsugi
(hako-mechi-tsuki)
竿枘込栓継ぎ（箱目違付）

Zapfen mit Steckfalz
und Dübelsicherung

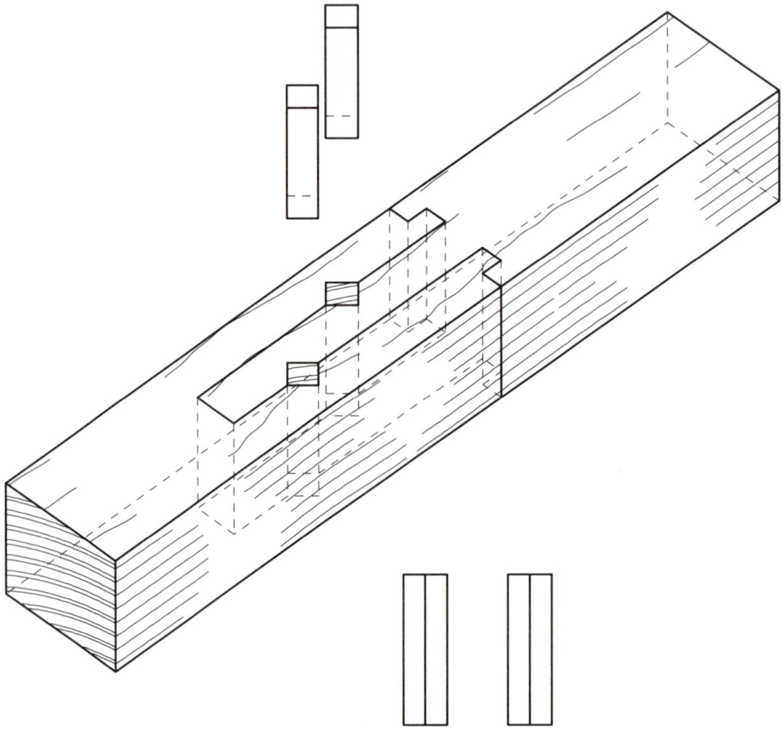

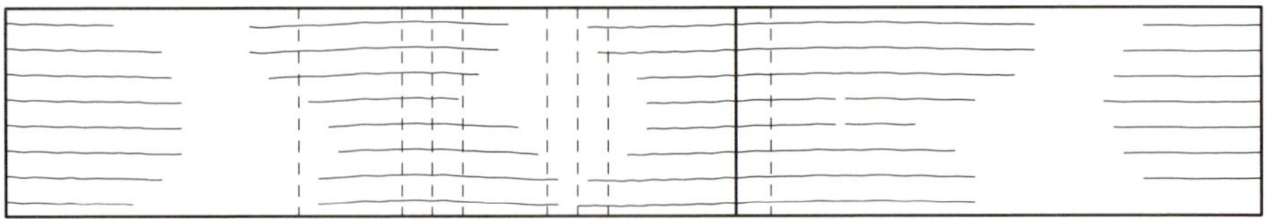

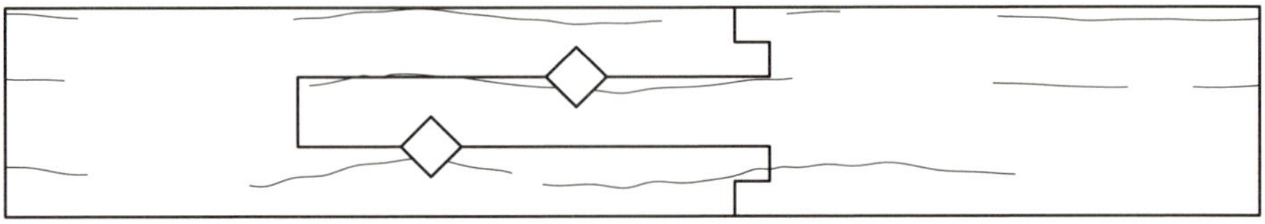

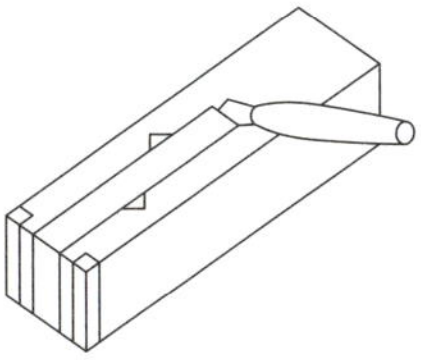
A1

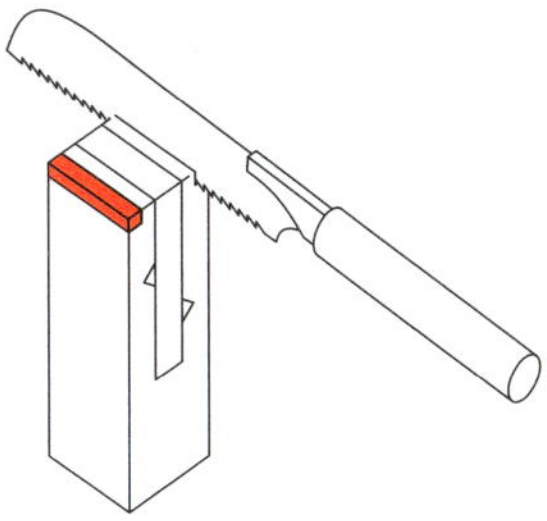
A2

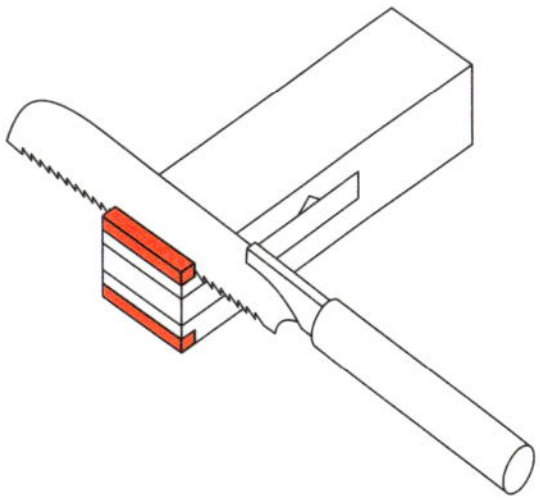
A3

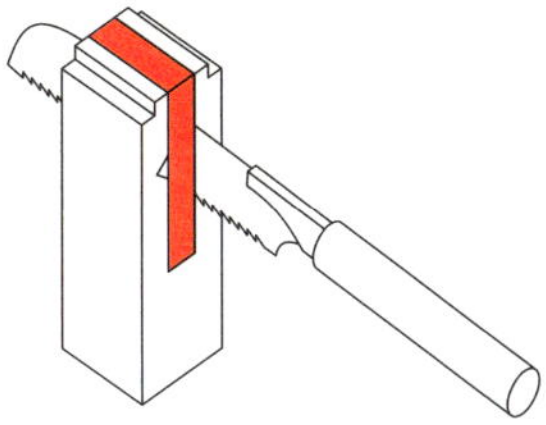
A4

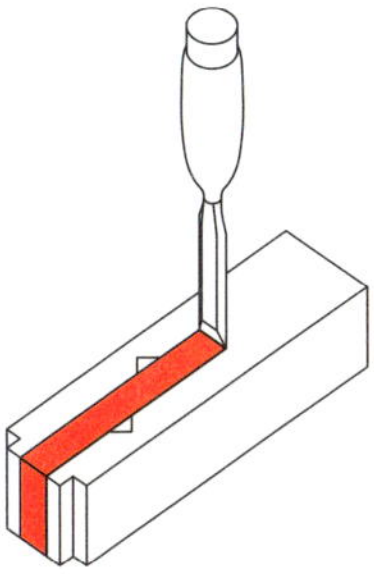
A5

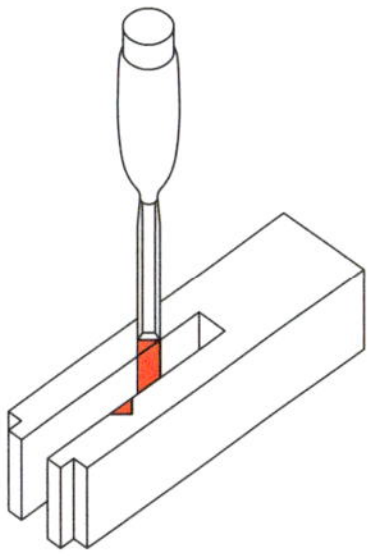
A6

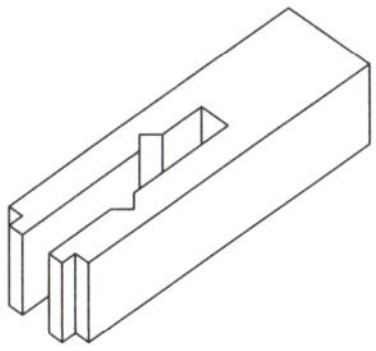
A

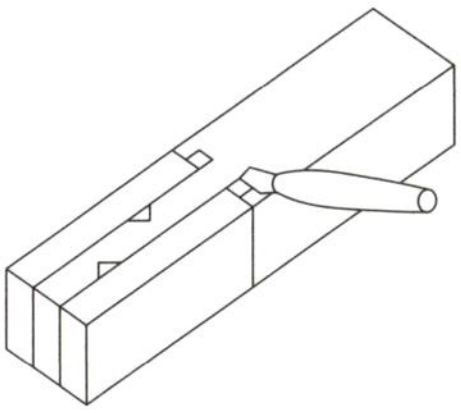
B1

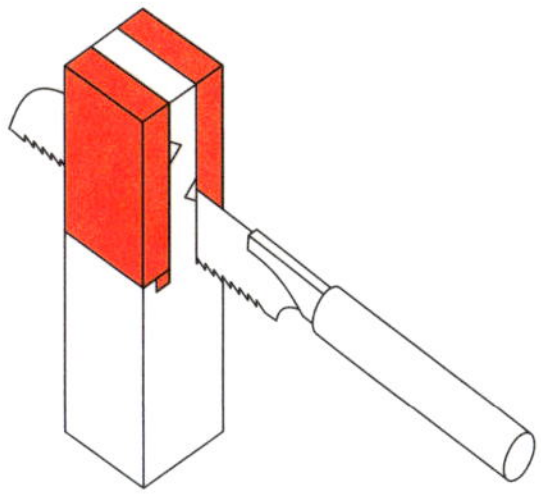
B2

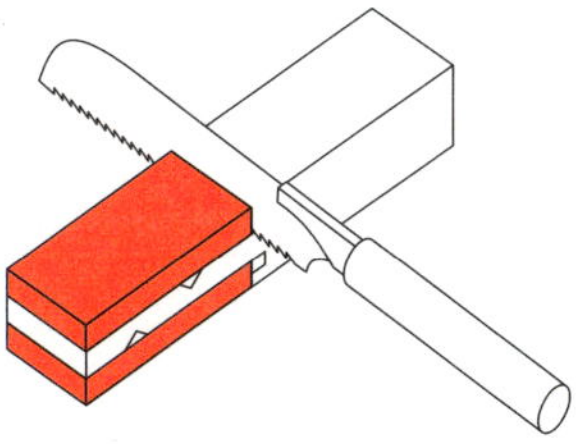
B3

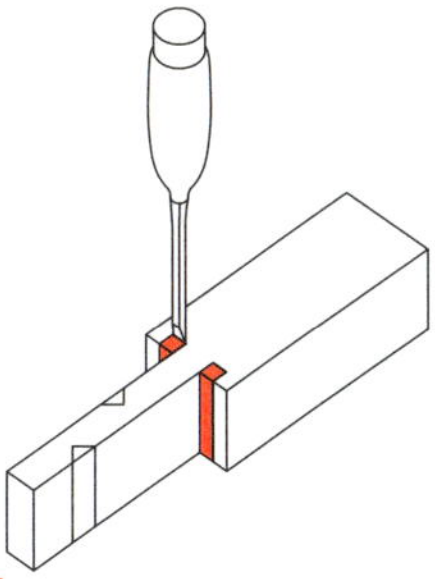
B4

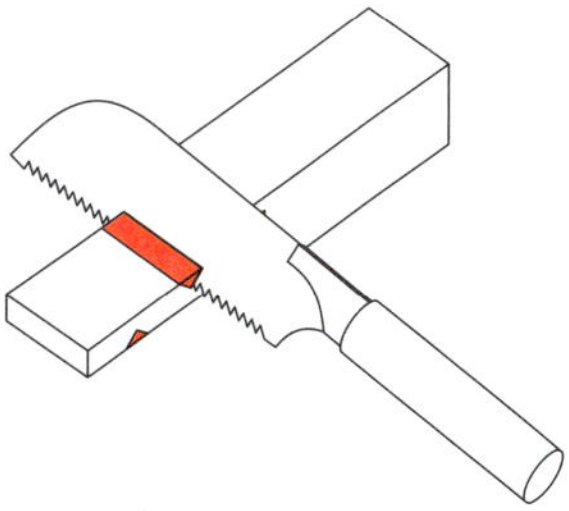
B5

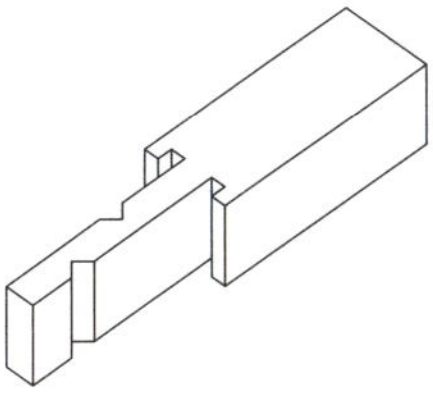
B

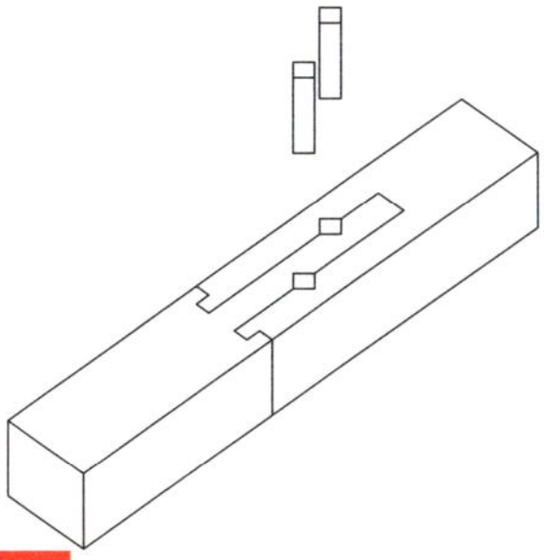
A+B

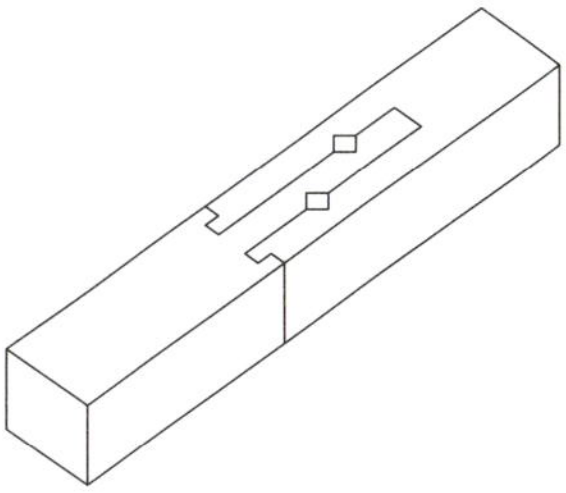
AB

Splicing Joints

Rabbeted rod tenon joint with tapered keys

sao-hozo-shachi-sen-tsugi (hako-mechi-tsuki) [*]
竿枘車知栓継ぎ(箱目違付)

Stabzapfen mit Steckfalz und Spannkeilen

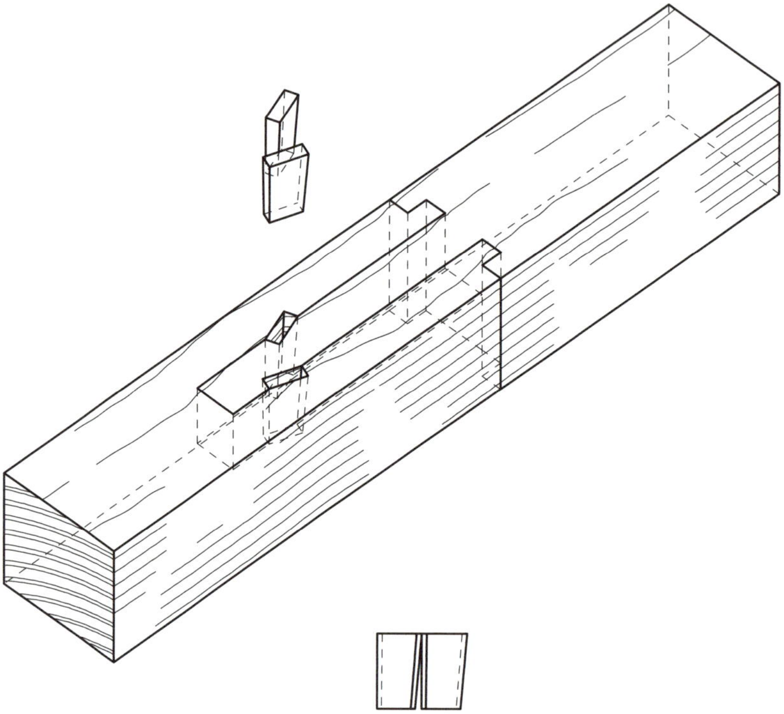

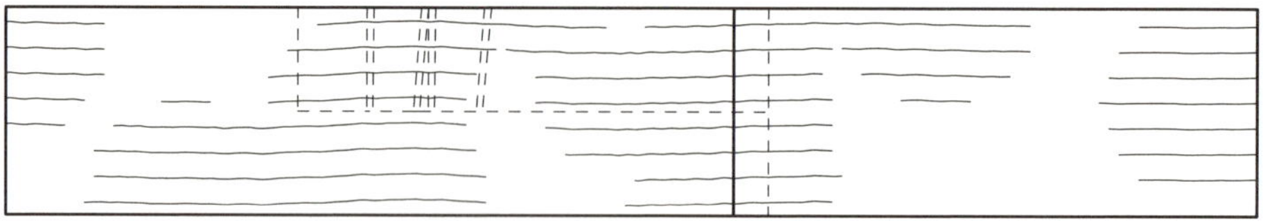

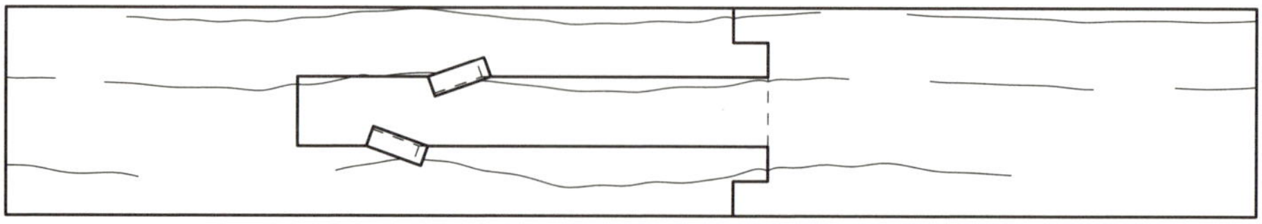

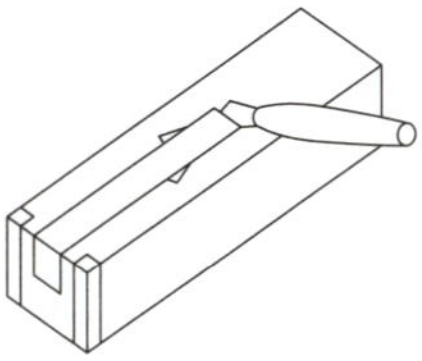
A1

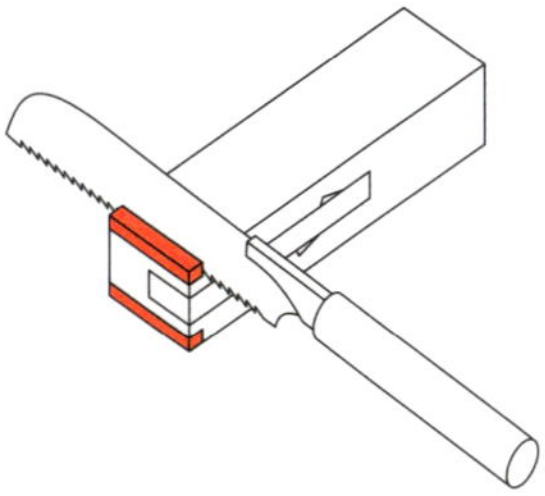
A2

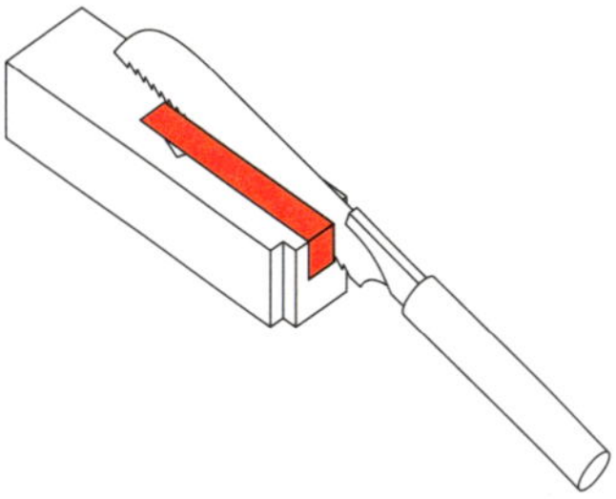
A3

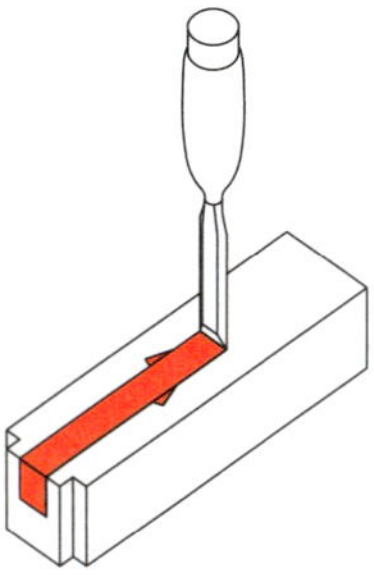
A4

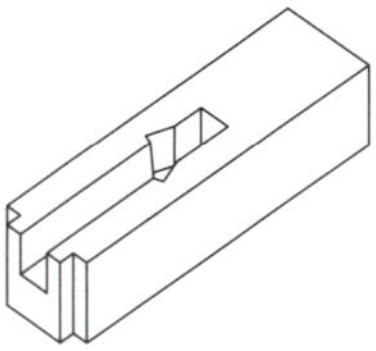
A

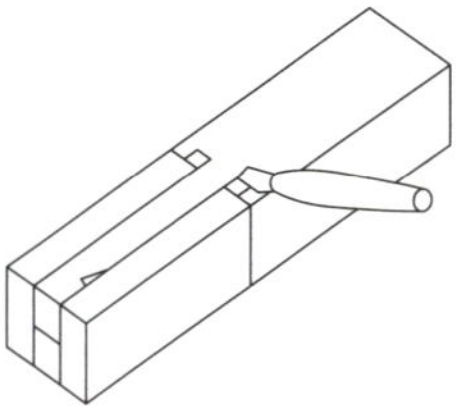
B1

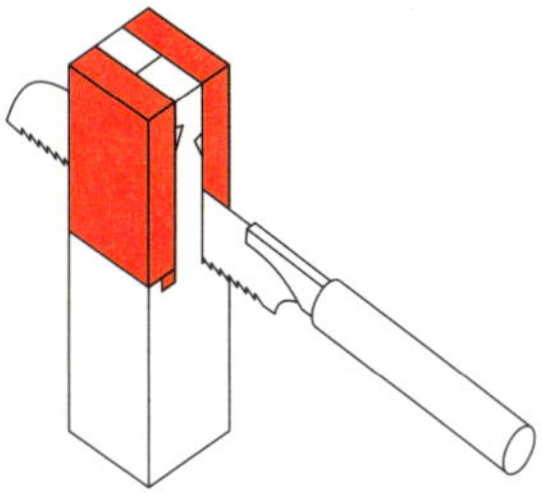
B2

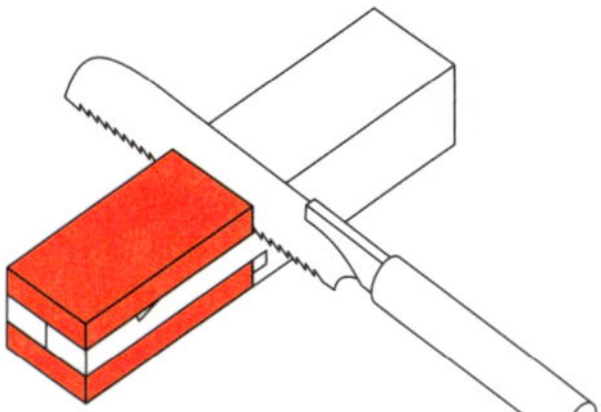
B3

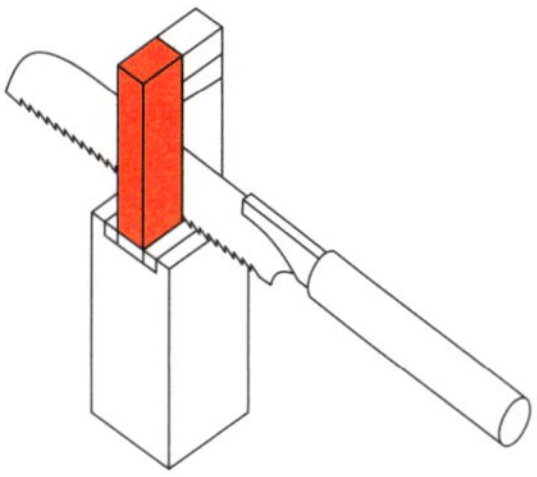
B4

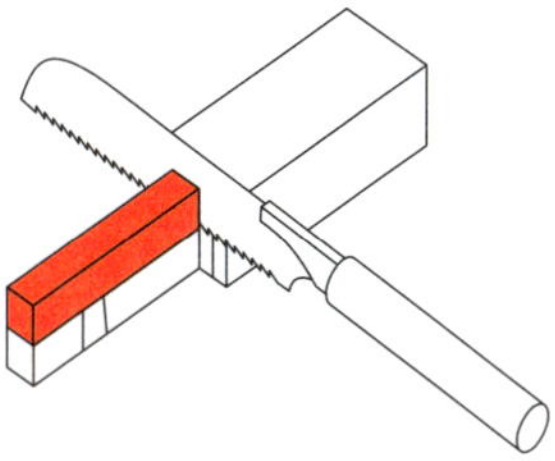
B5

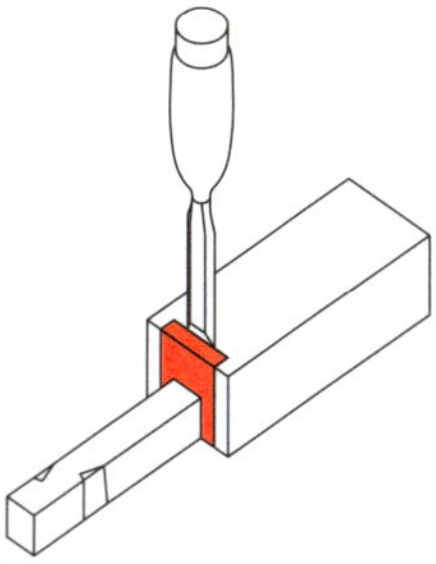
B6

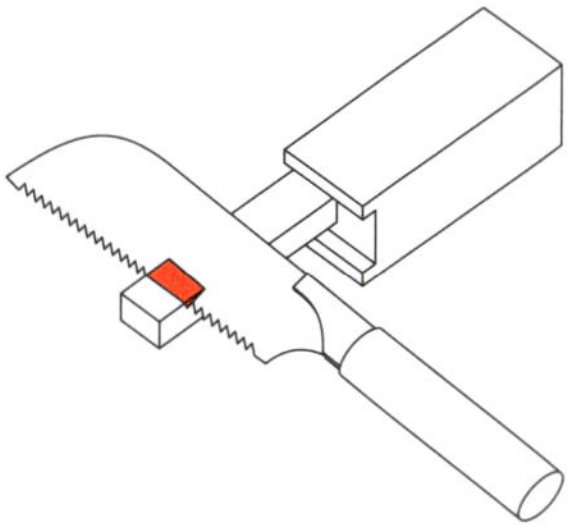
B7

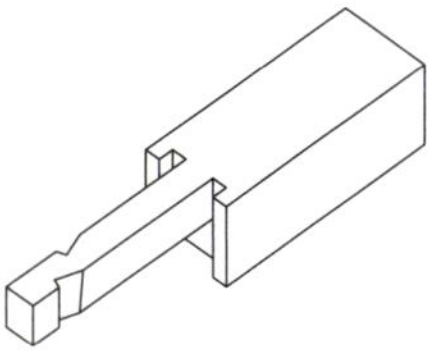
B

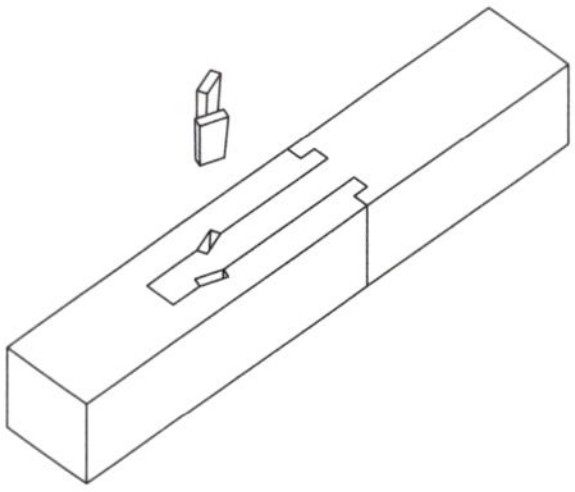
A+B

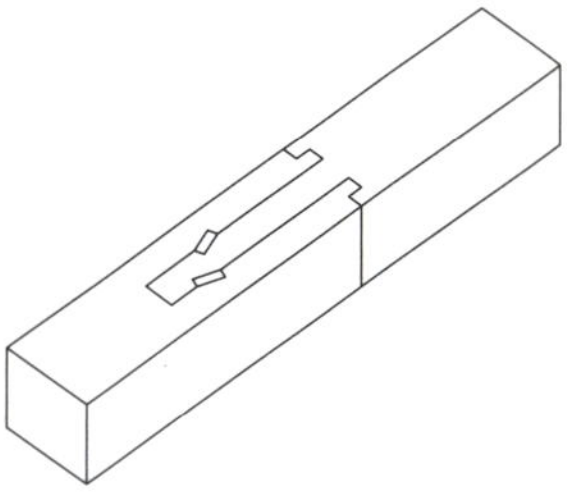
AB

Rod joint with two half rods and tapered keys

ai-sao-hozo-shachi-sen-tsugi
相竿枘車知栓継ぎ

beidseitiger Stabzapfen
mit Spannkeilen

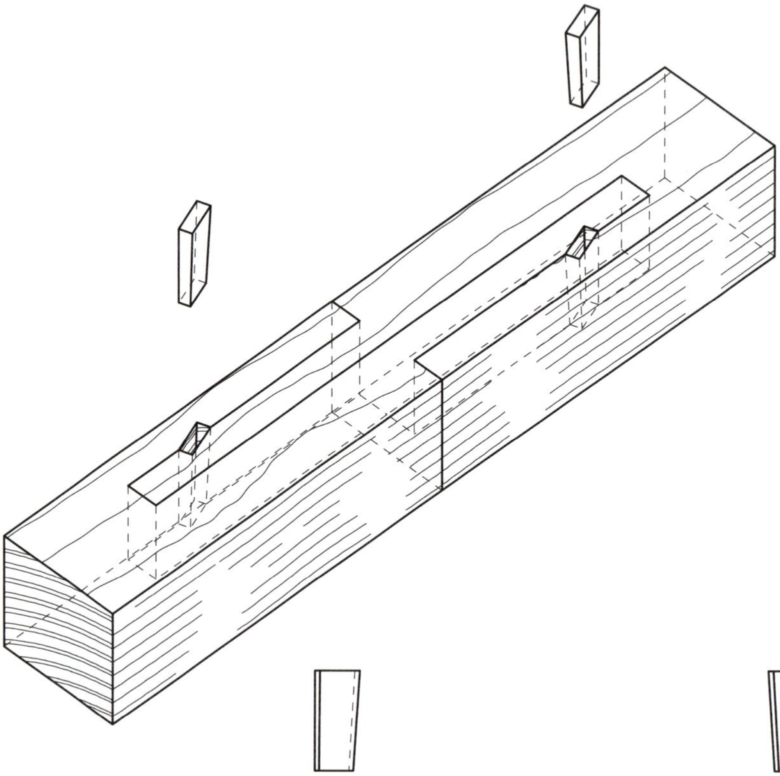

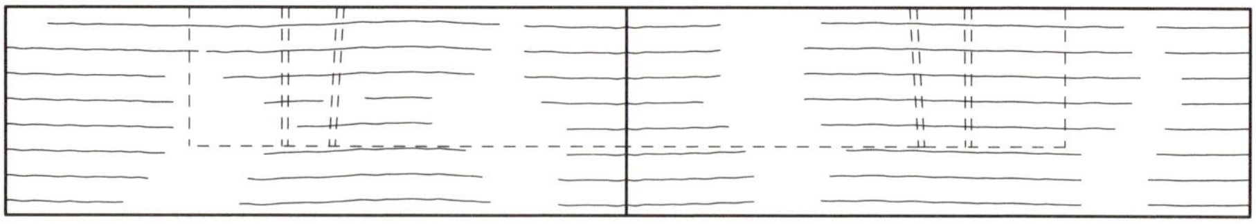

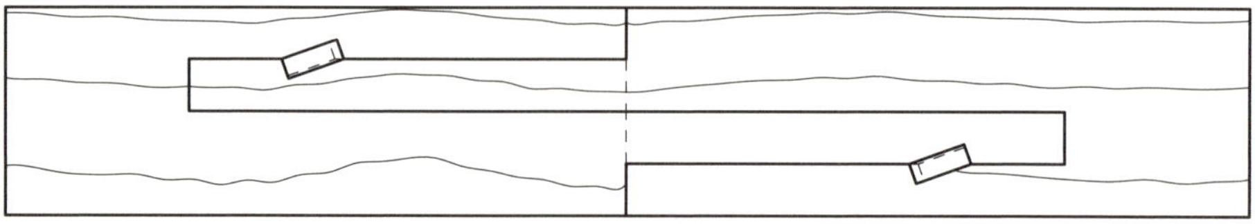

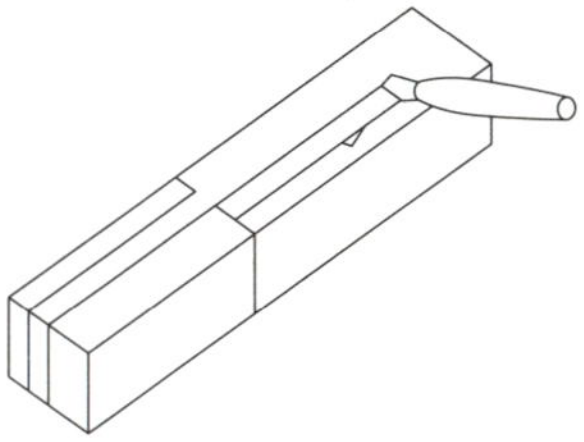
A1

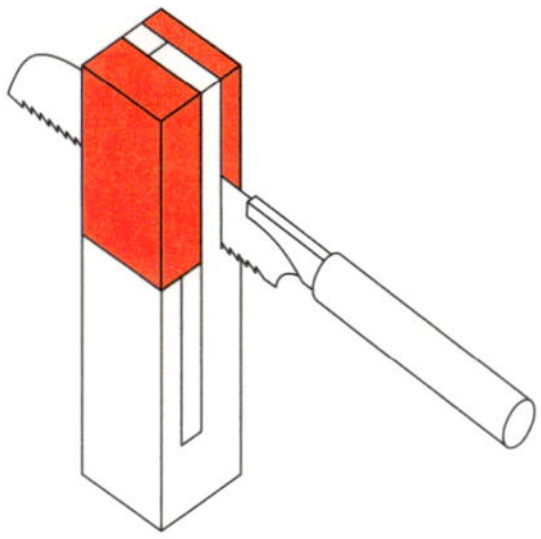
A2

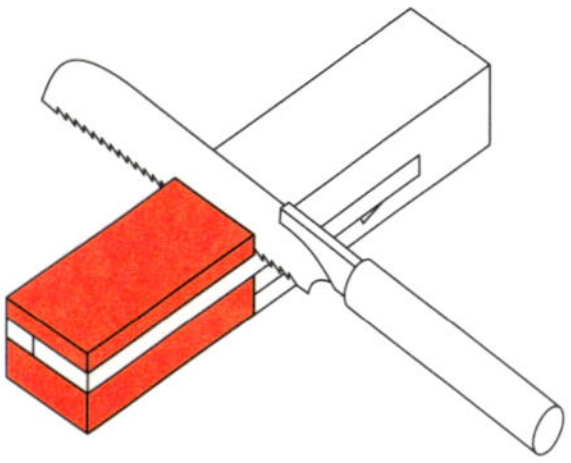
A3

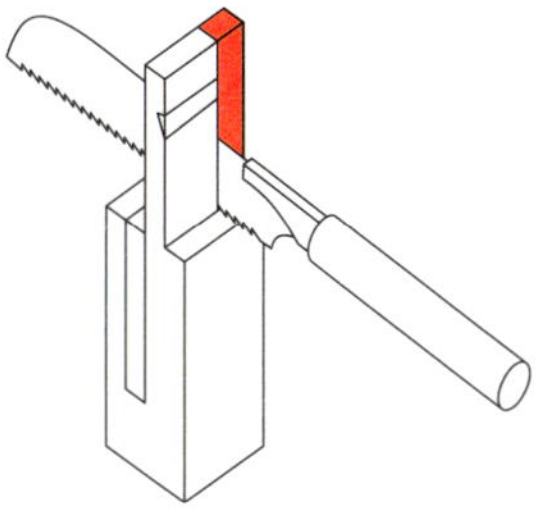
A4

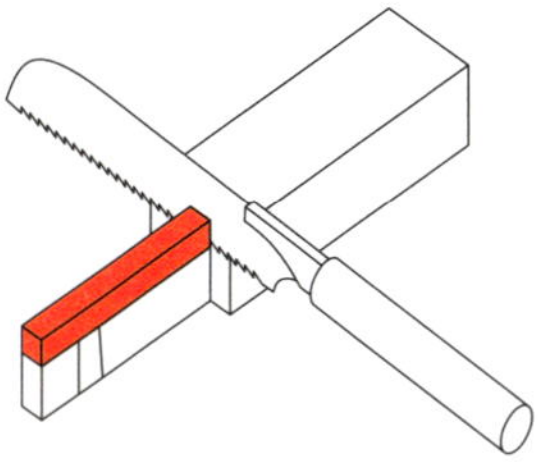
A5

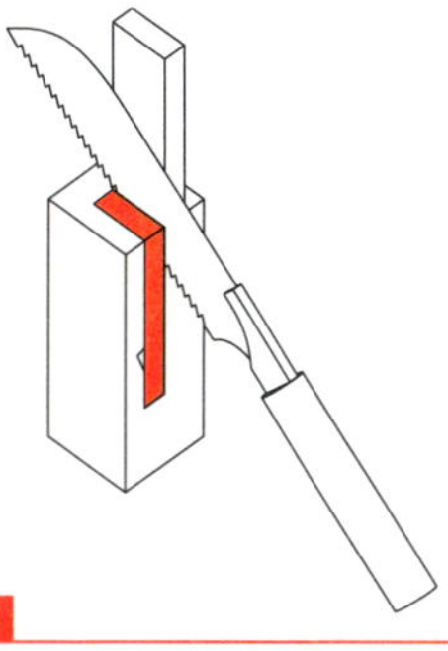
A6

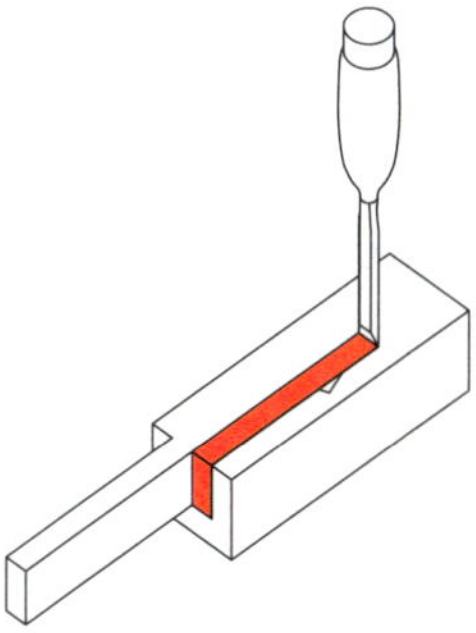
A7

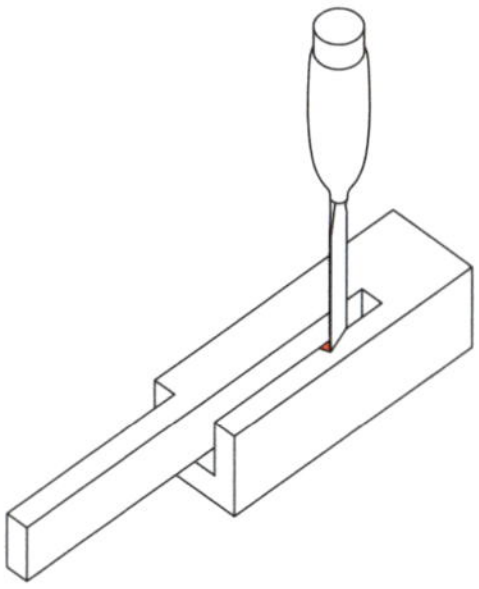
A8

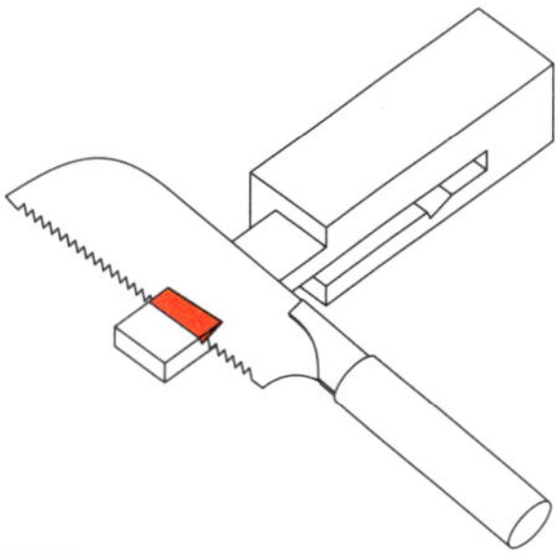
A9

2×

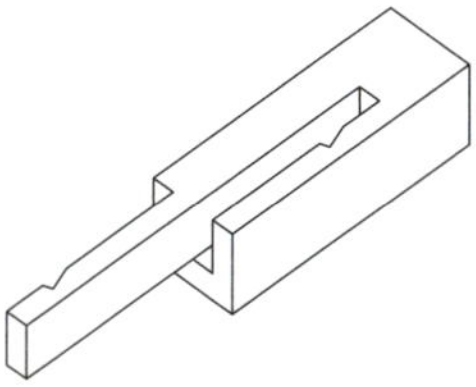
A

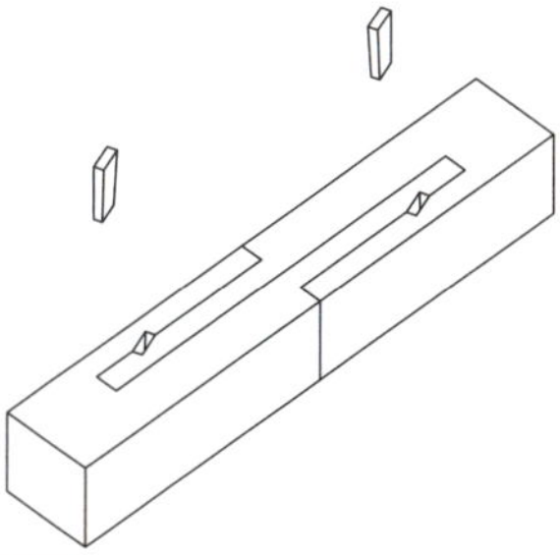
A+A

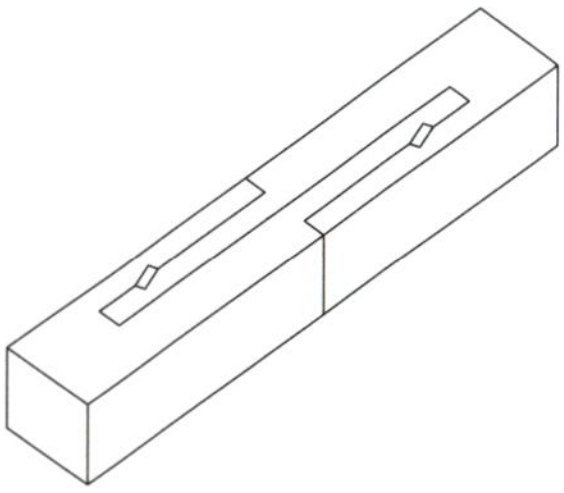
AA

Splicing Joints

Double tenon joint

nimai-hozo-tsugi
二枚枘継ぎ

doppelter Zapfen

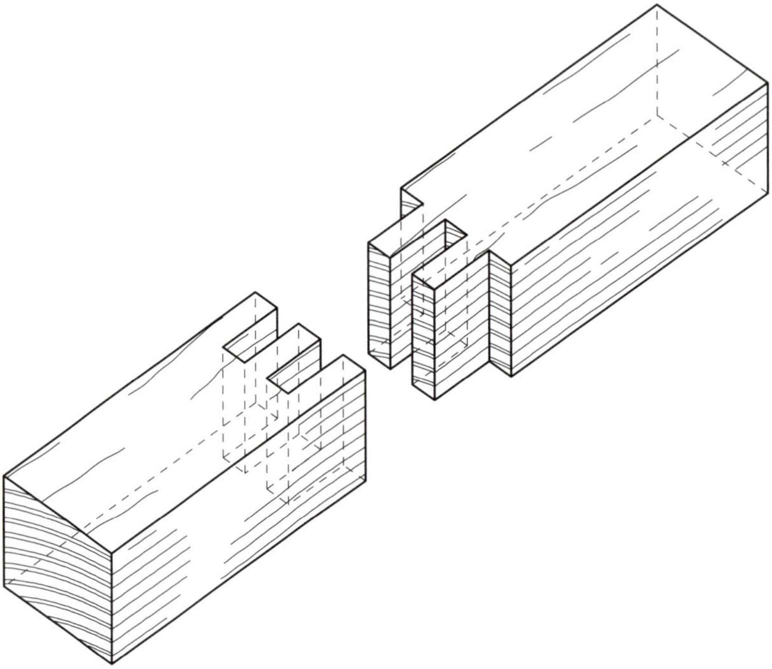

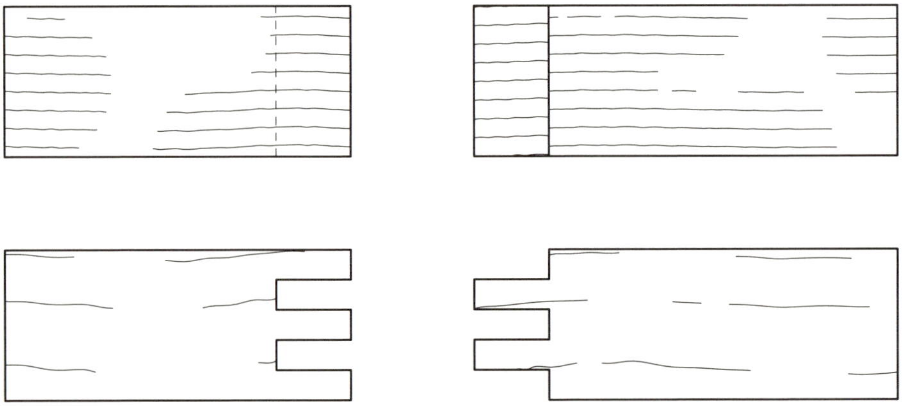

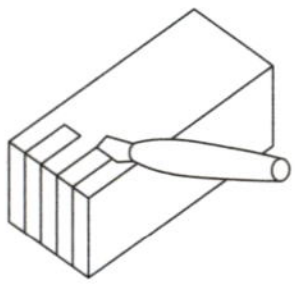

A1

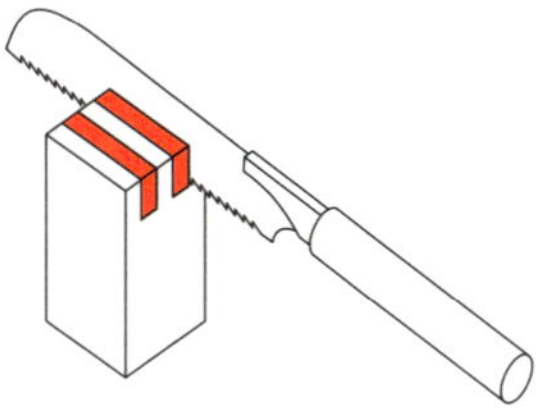

A2

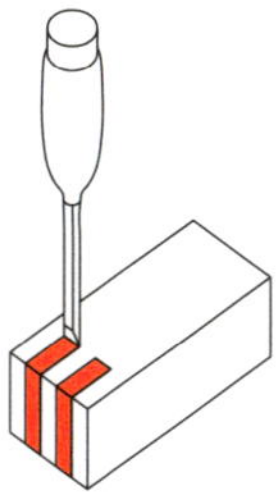

A3

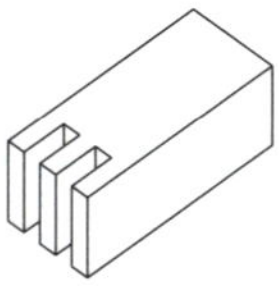

A

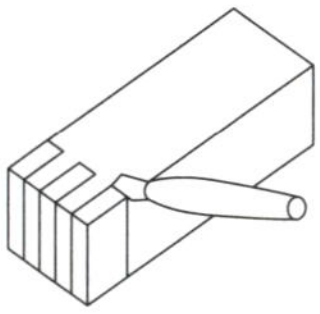

B1

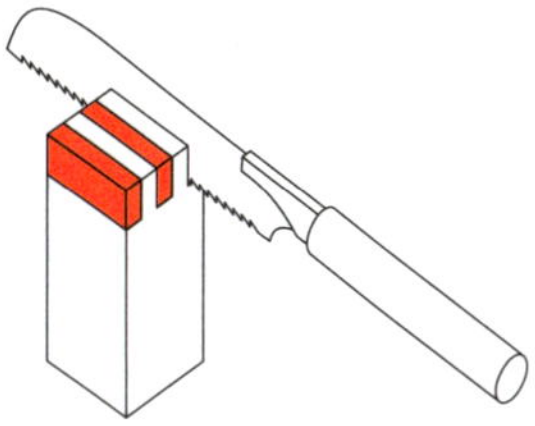

B2

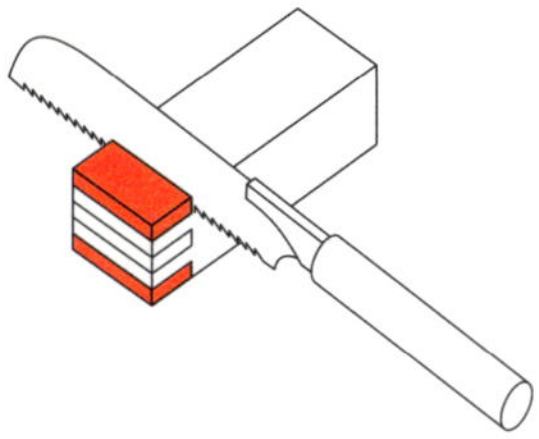

B3

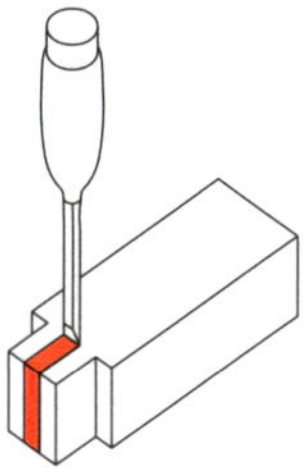

B4

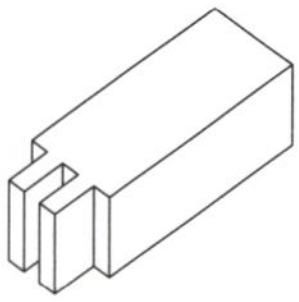

B

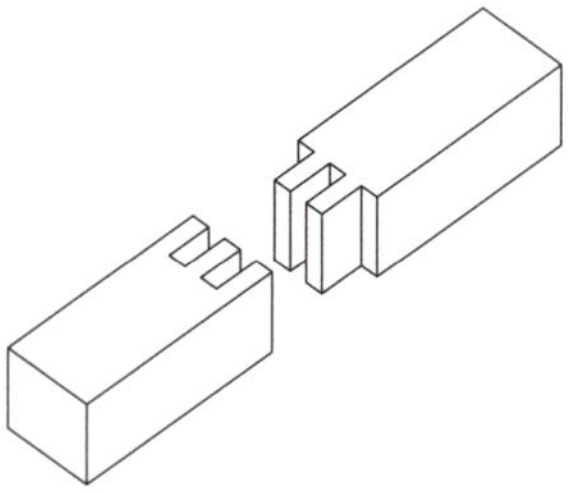

A+B

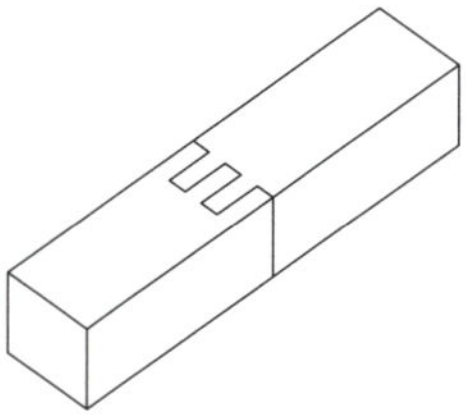

AB

Splicing Joints

Half-blind double tenon joint

kakushi-nimai-hozo-tsugi
隠二枚枘継ぎ

einseitig abgesetzter doppelter Zapfen

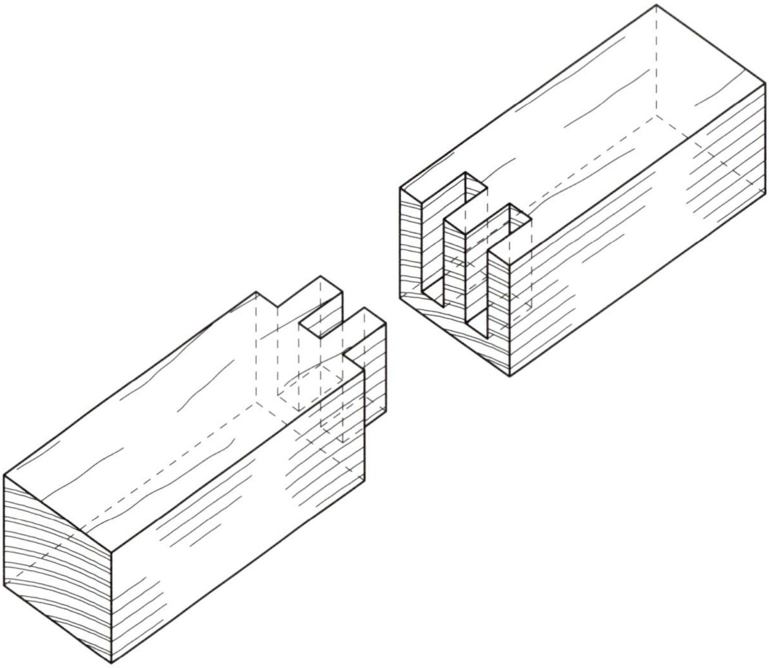

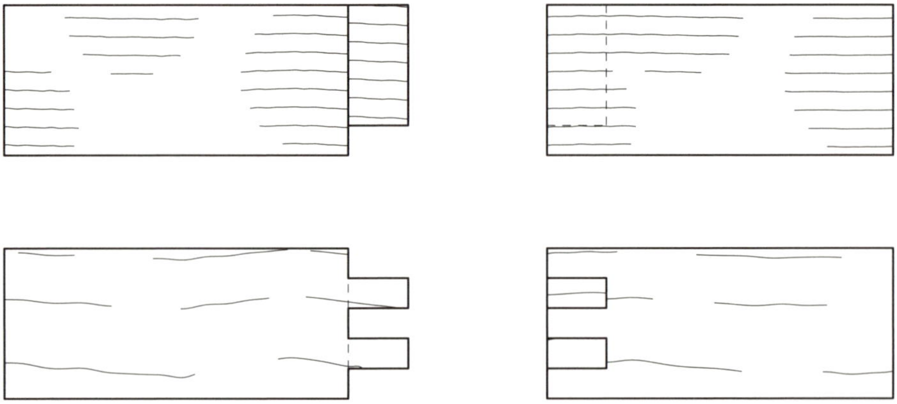

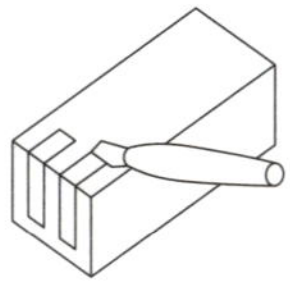

A1

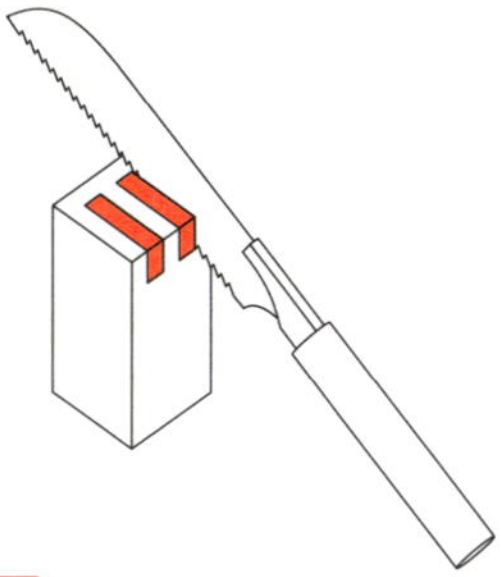

A2

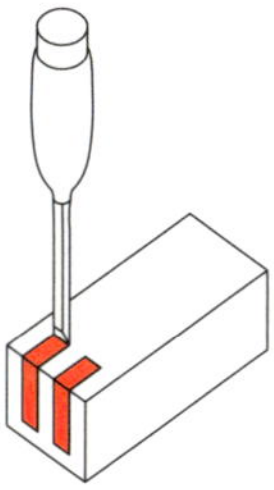

A3

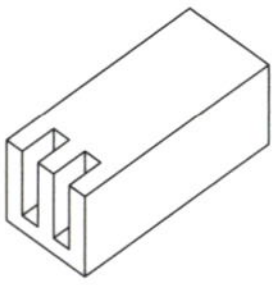

A

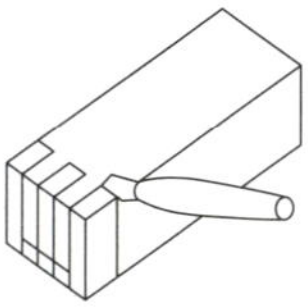

B1

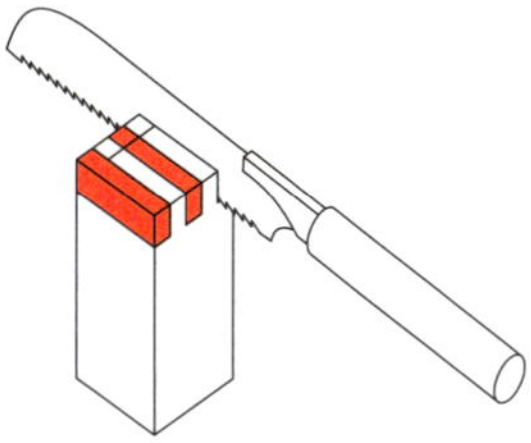

B2

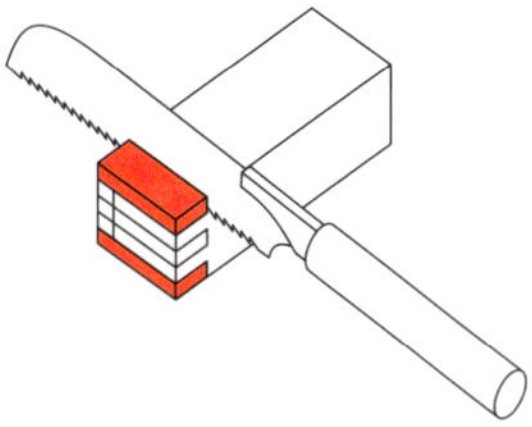

B3

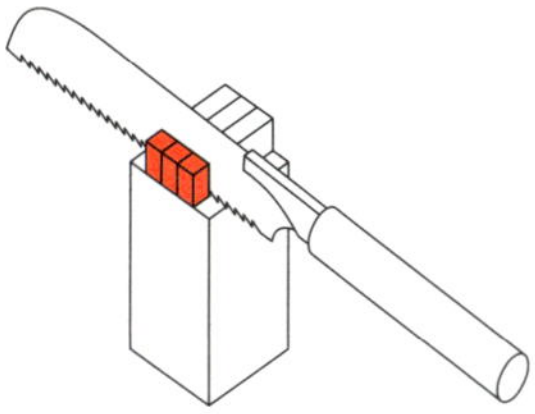

B4

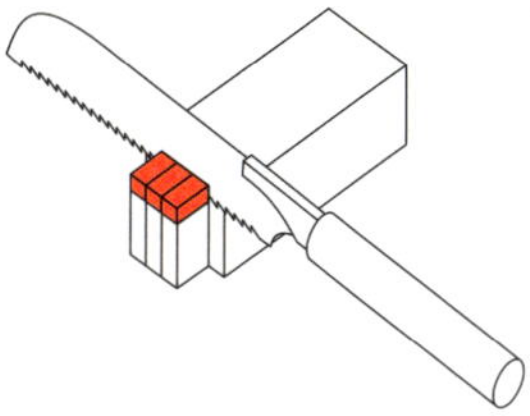

B5

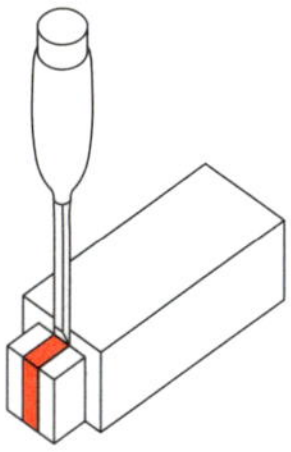

B6

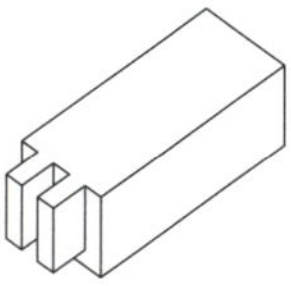

B

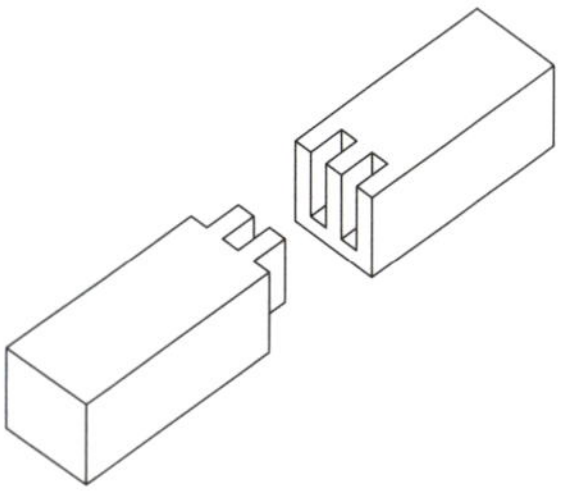

A+B

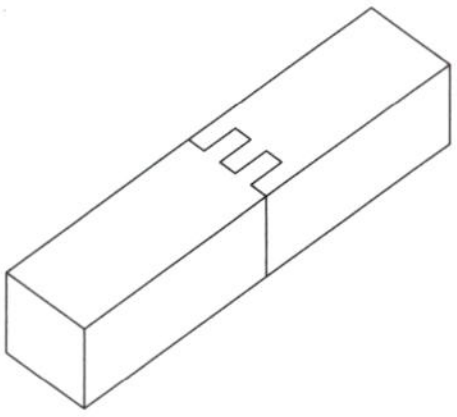

AB

Splicing Joints

Blind square stub tenon joint

shihō-dō-tsuki-dabo-tsugi
四方胴付太枘継ぎ

verdeckter beidseitig abgesetzter Zapfen

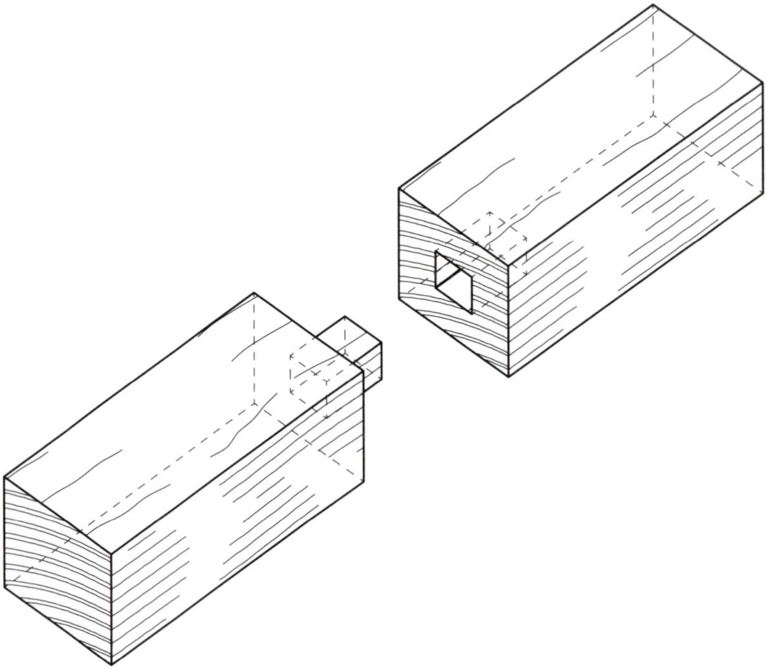

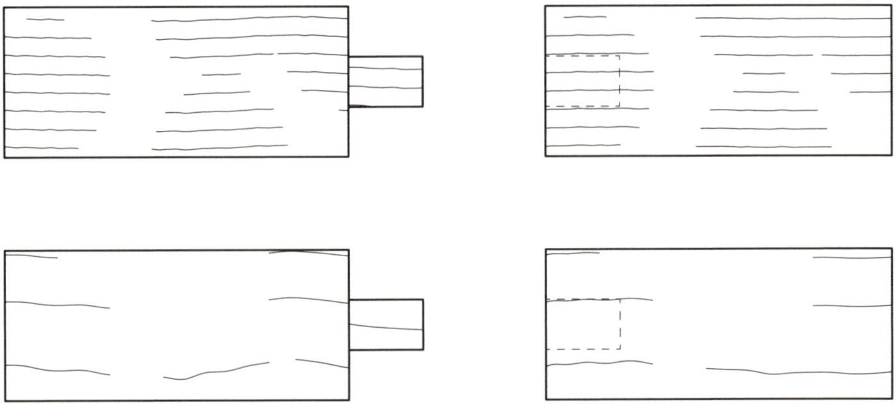

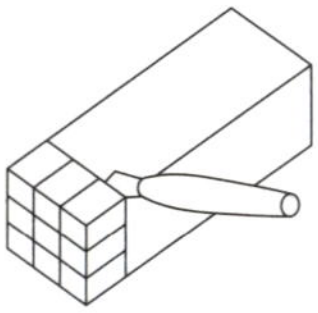

A1

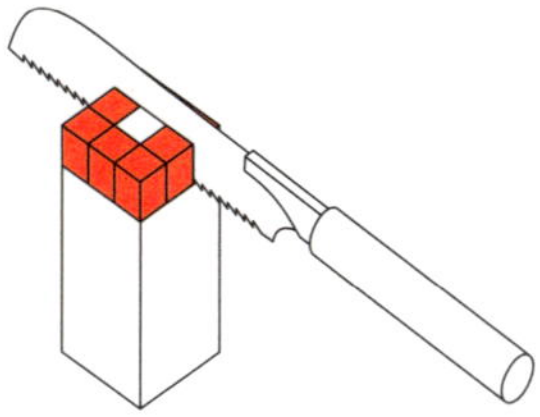

A2

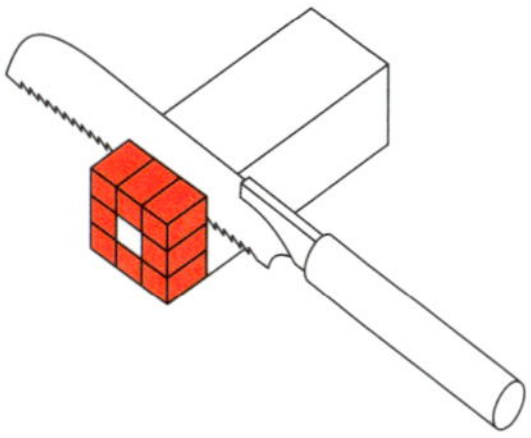

A3

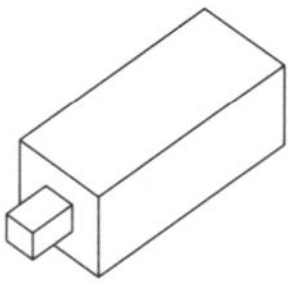

A

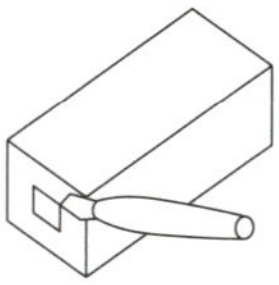

B1

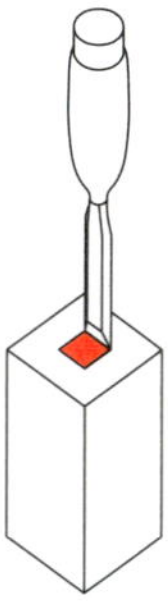

B2

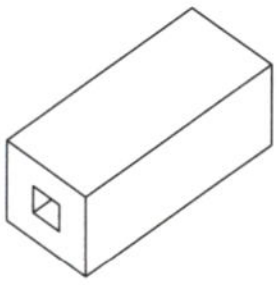

B

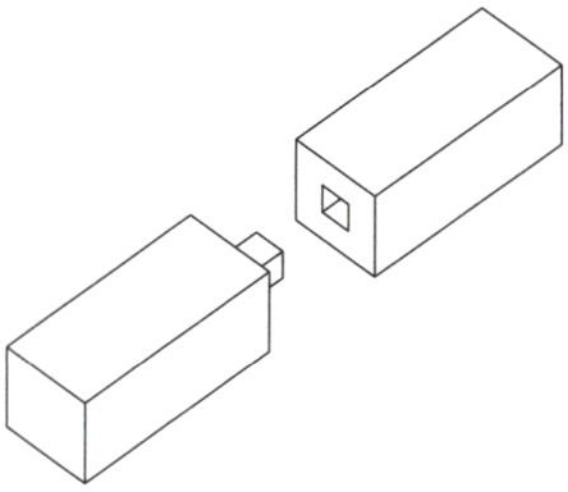

A+B

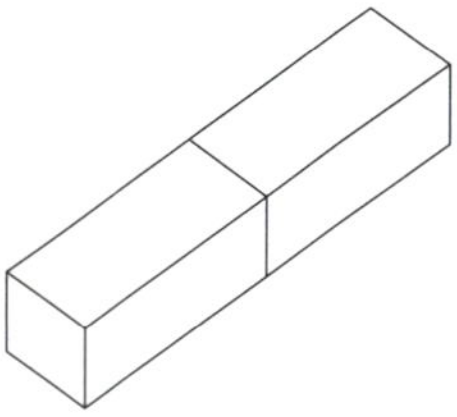

AB

Splicing Joints

Blind fan-shaped stub tenon joint

ōgi-hozo-tsugi
扇枘継ぎ

verdeckter trapezförmiger Zapfen

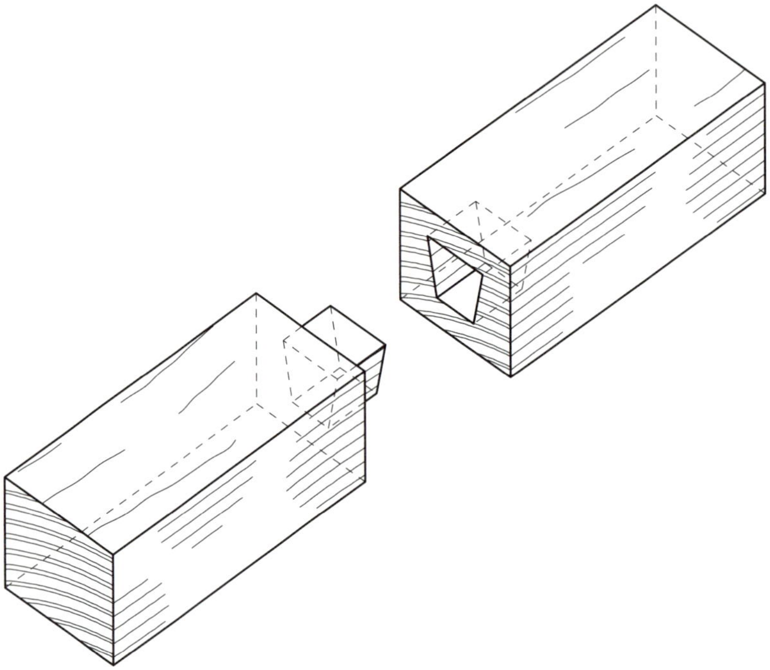

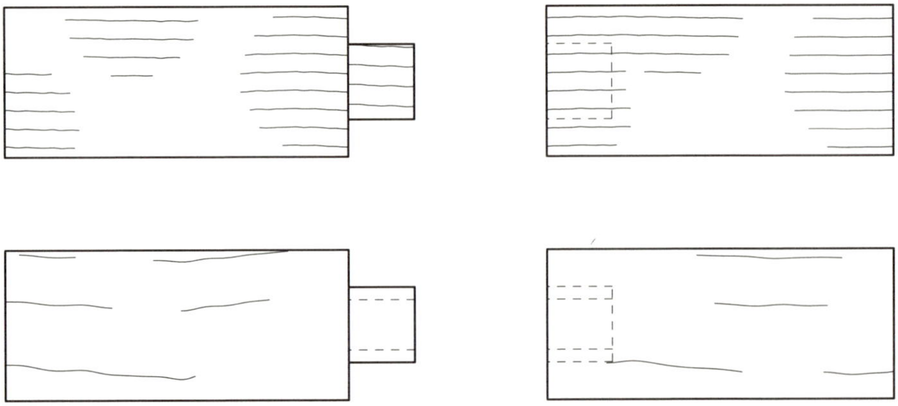

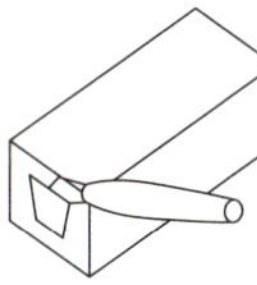

A1

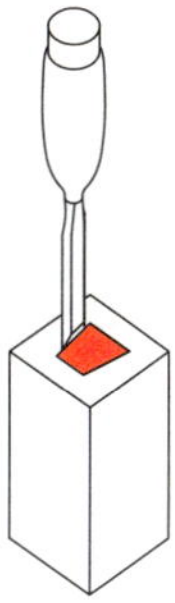

A2

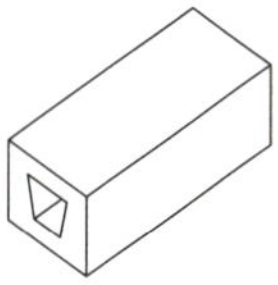

A

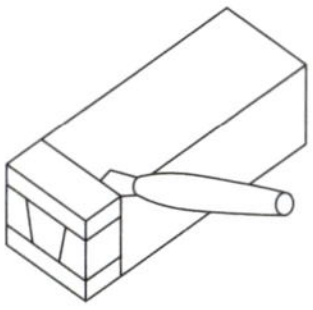

B1

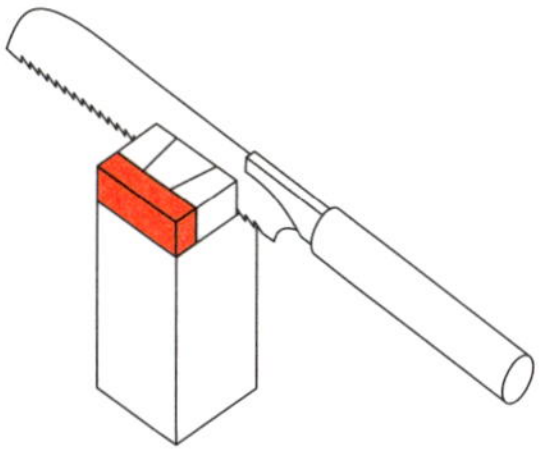

B2

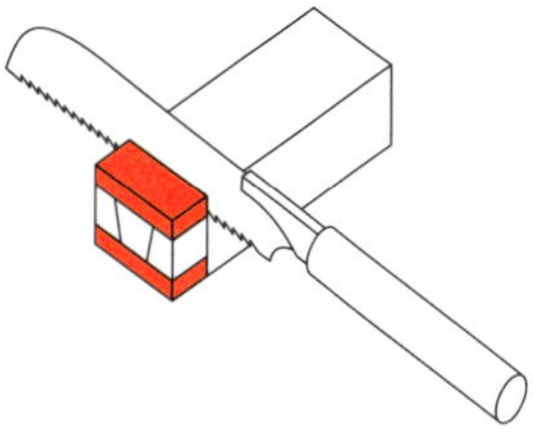

B3

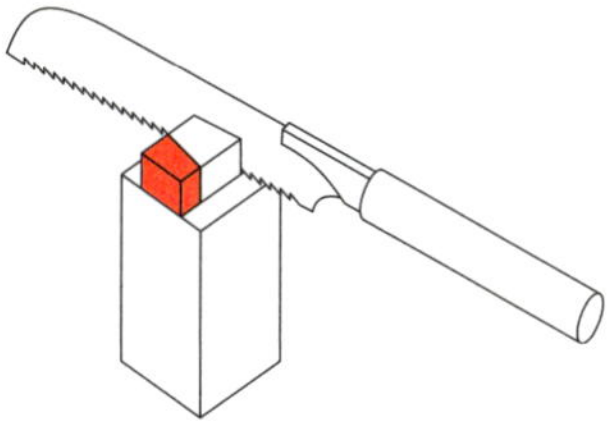

B4

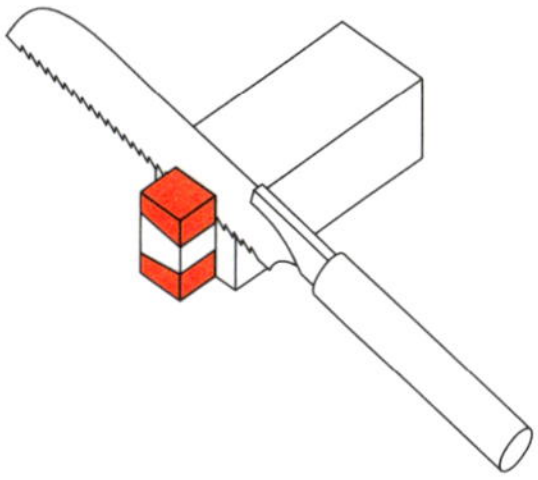

B5

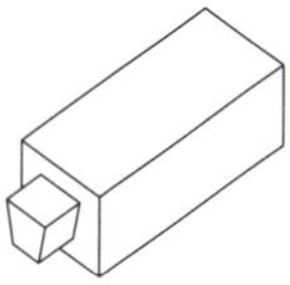

B

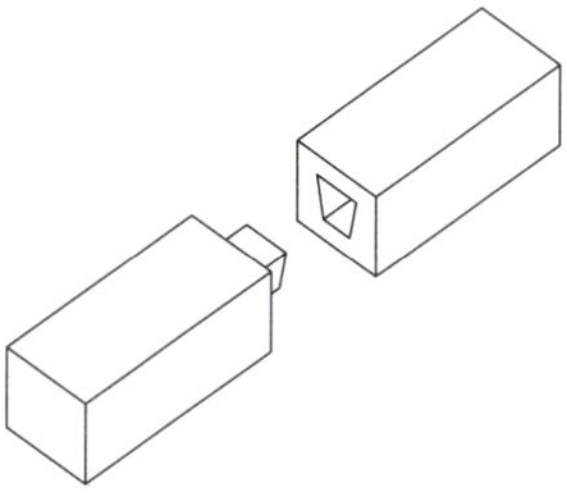

A+B

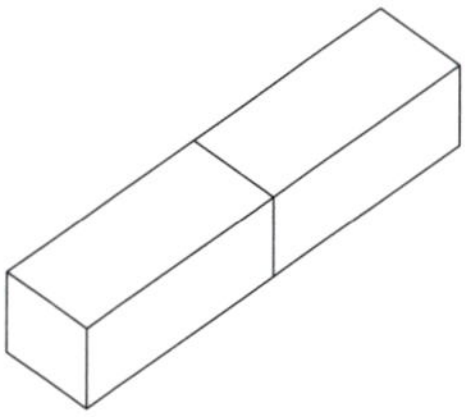

AB

Splicing Joints

Half-blind L-shaped stub tenon joint

kane-ori-mechi-tsugi
矩折目違継ぎ

Winkelzapfen

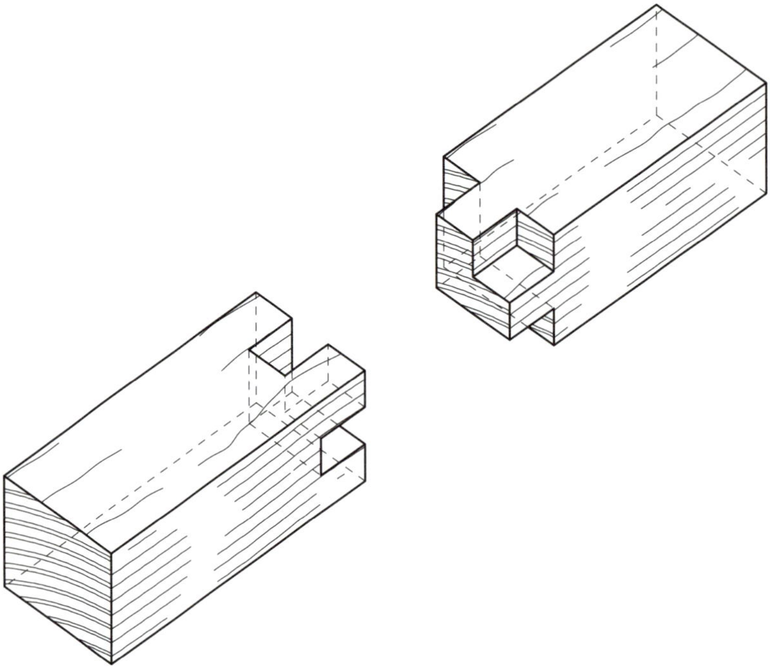

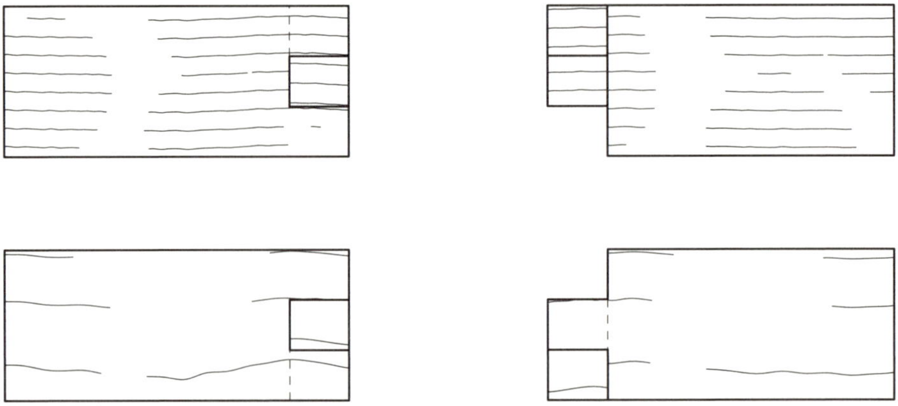

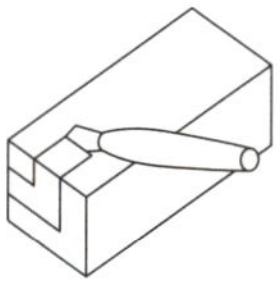

A1

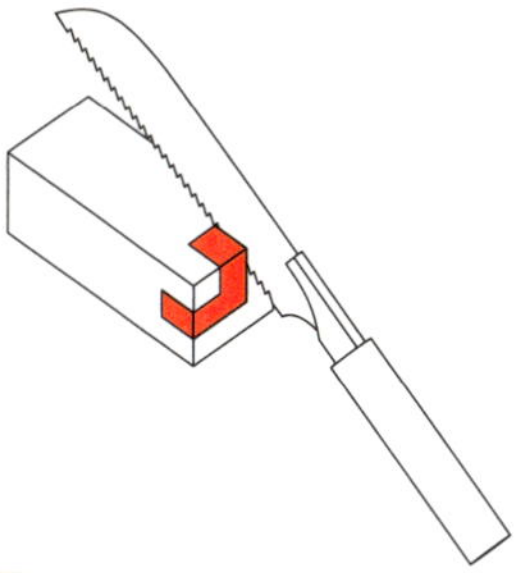

A2

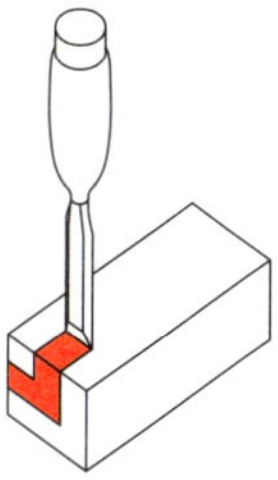

A3

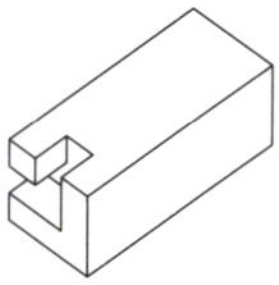

A

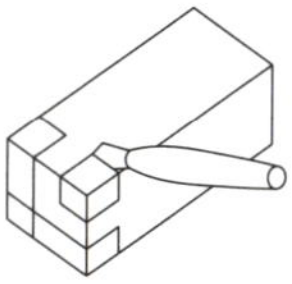

B1

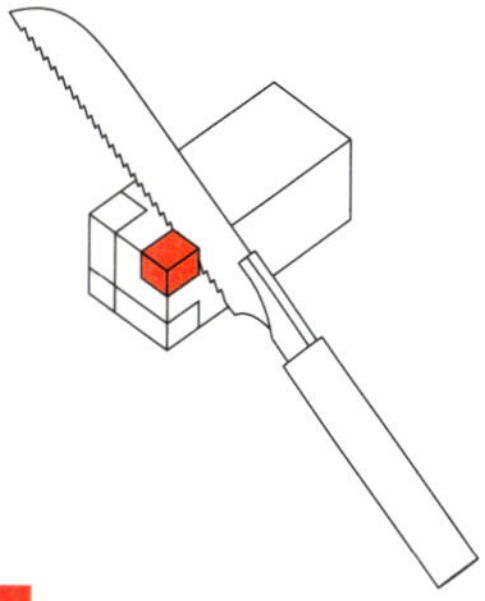

B2

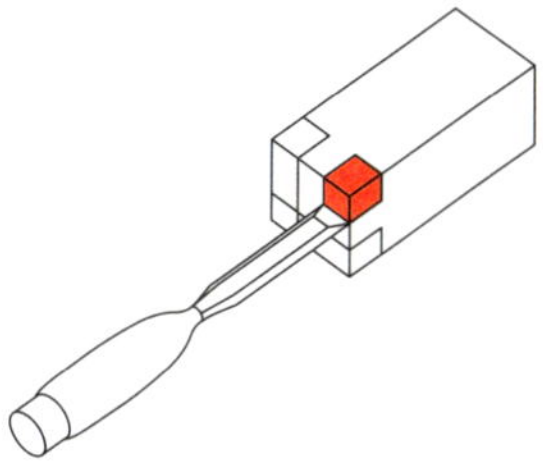

B3

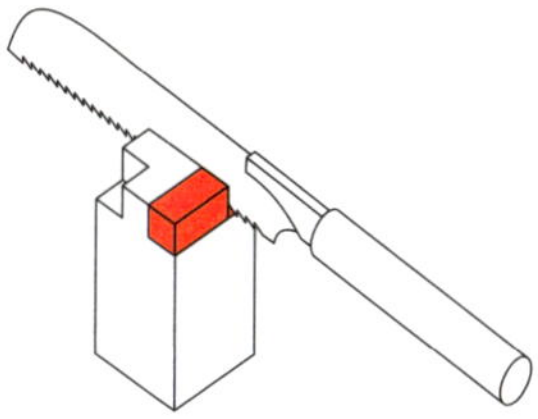

B4

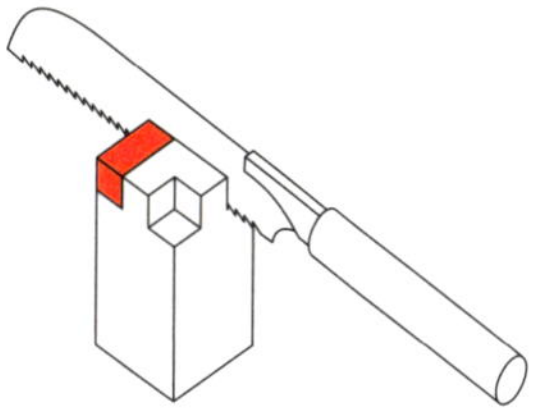

B5

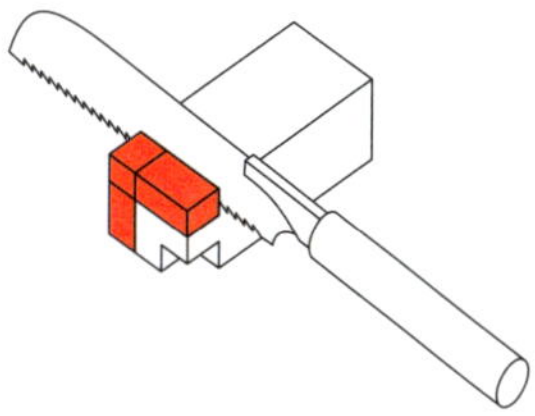

B6

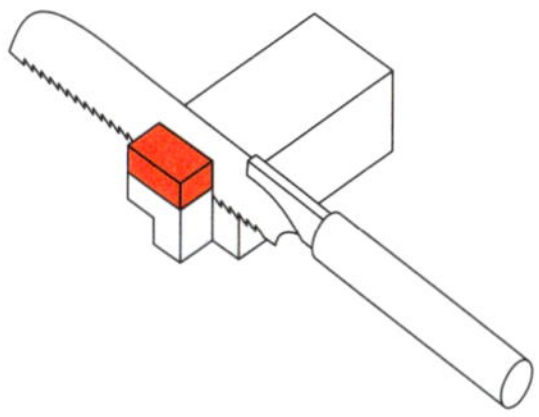

B7

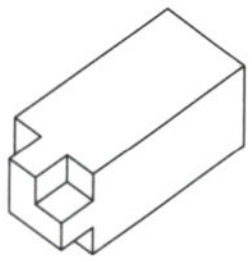

B

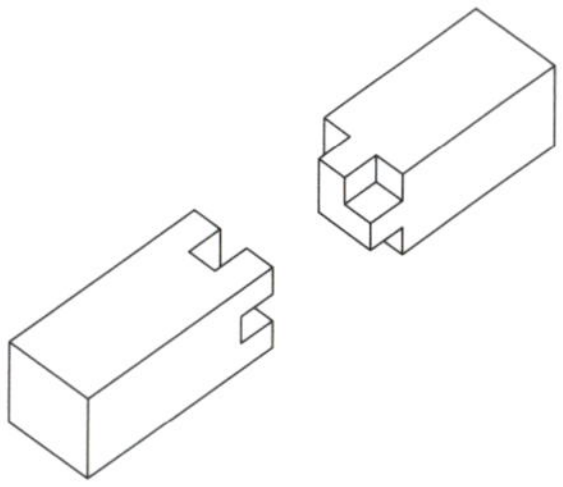

A+B

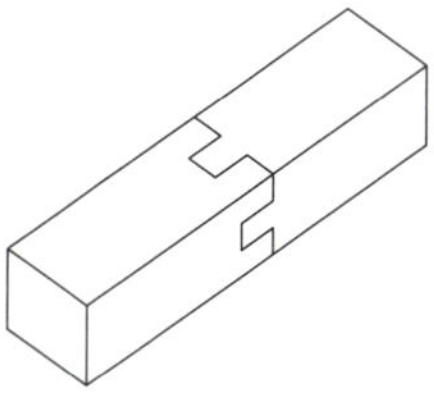

AB

Splicing Joints

Blind U-shaped stub tenon joint

hako-mechi-tsugi
箱目違継ぎ

verdeckter u-förmiger Zapfen

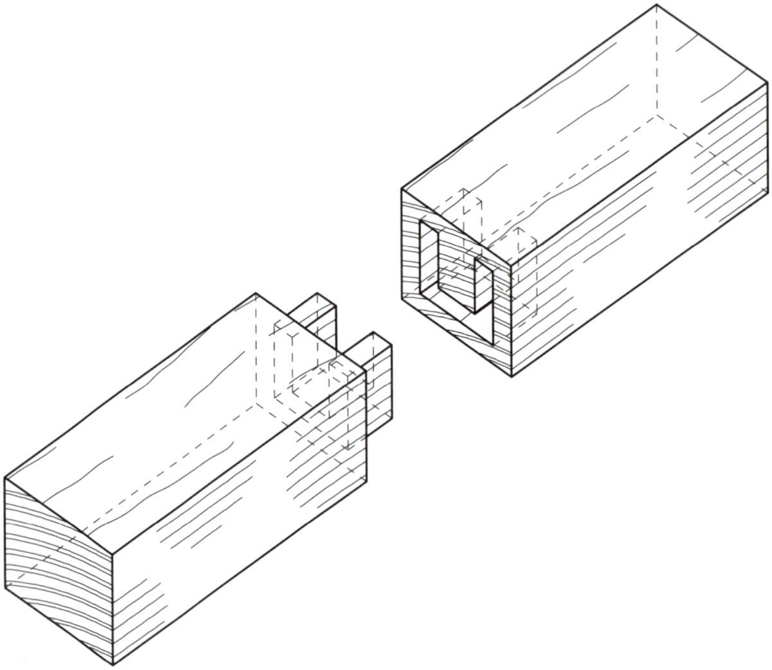

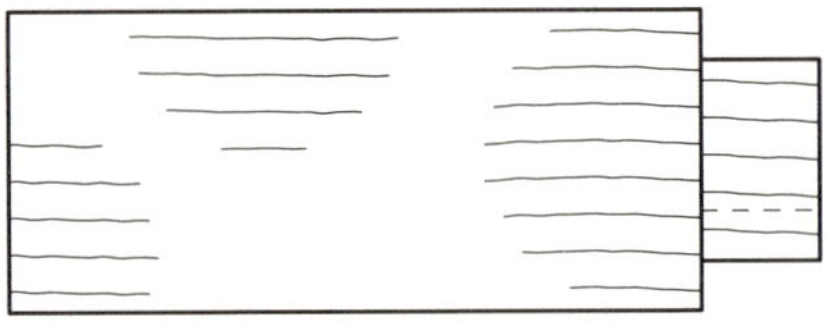

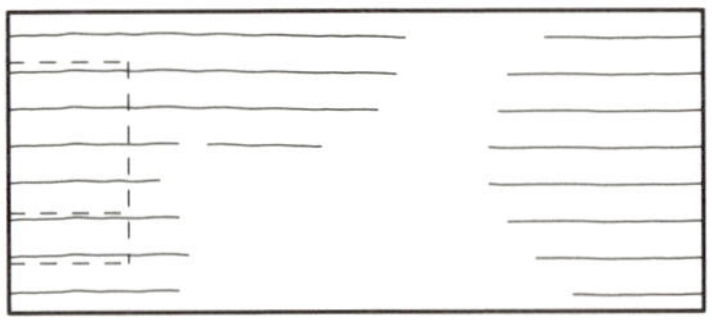

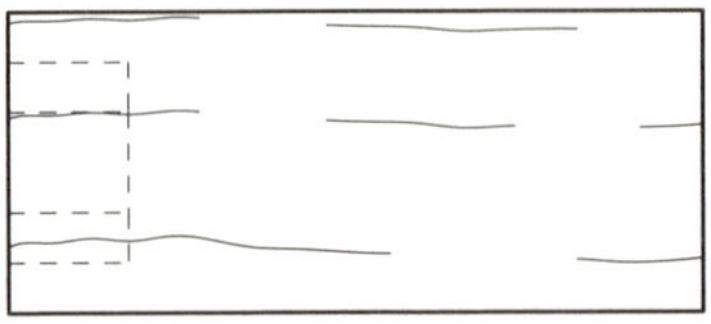

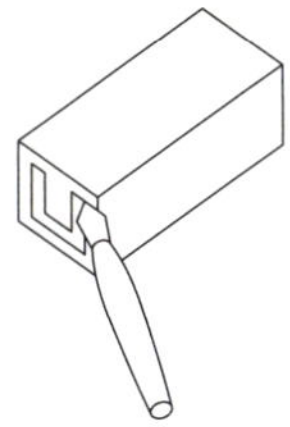

A1

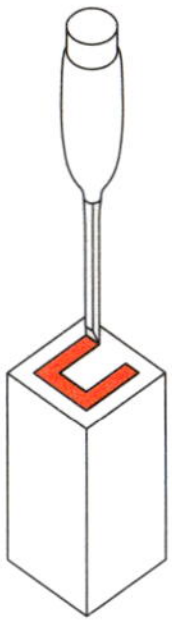

A2

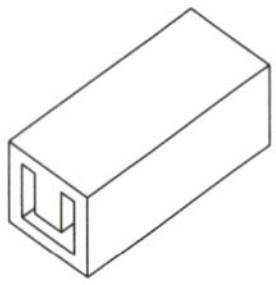

A

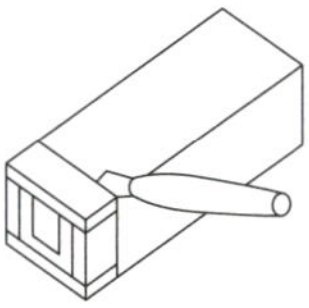

B1

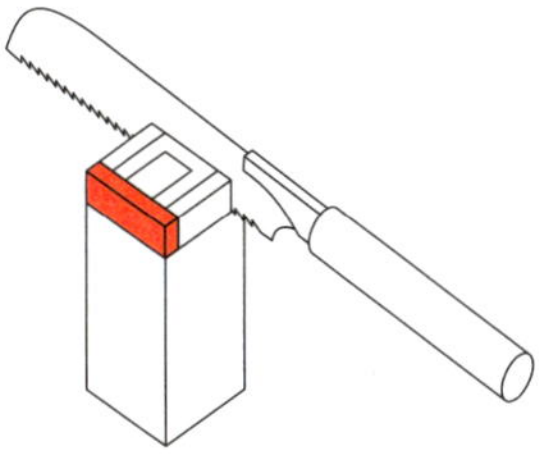

B2

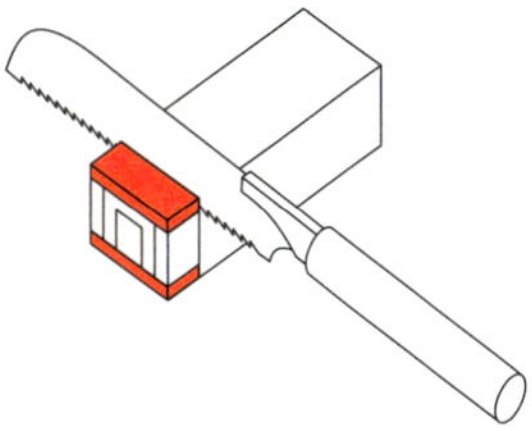

B3

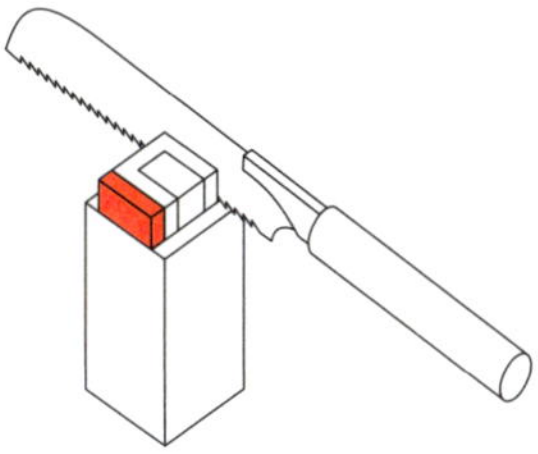

B4

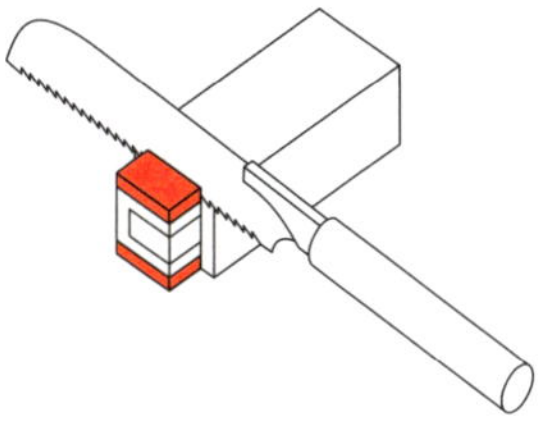

B5

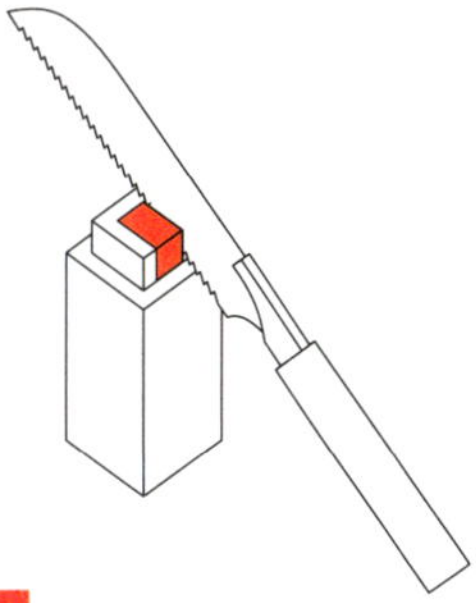

B6

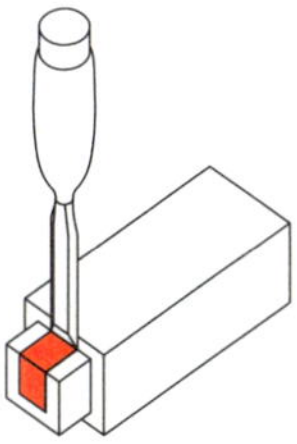

B7

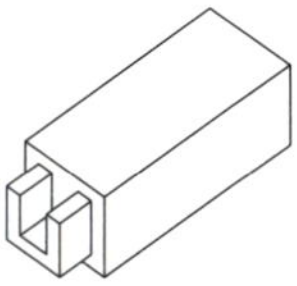

B

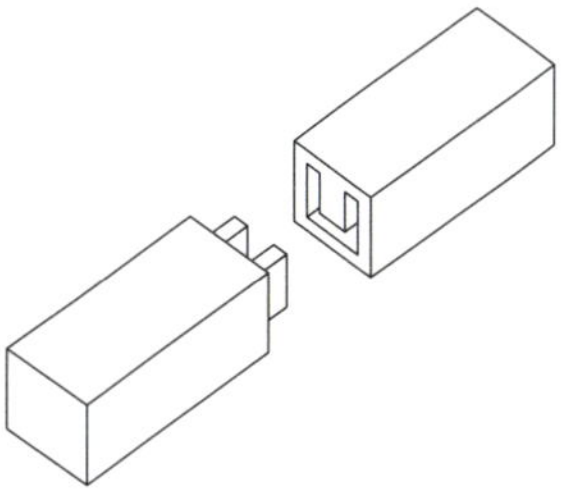

A+B

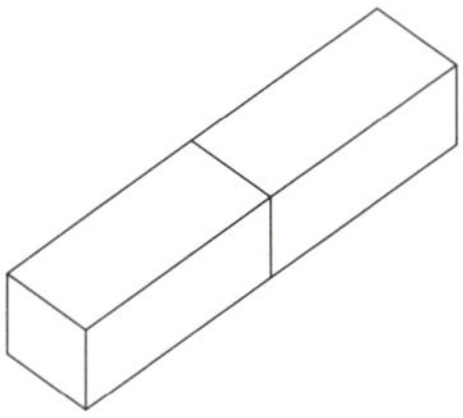

AB

Splicing Joints

Double halved lap joint

kainokuchi-hako-tsugi
貝口箱継ぎ

diagonal versetztes abgesetztes Blatt

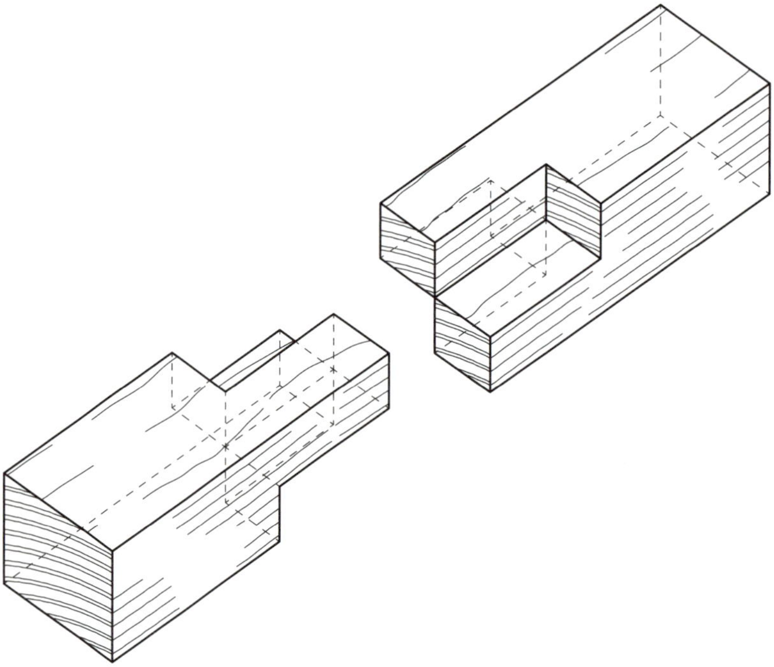

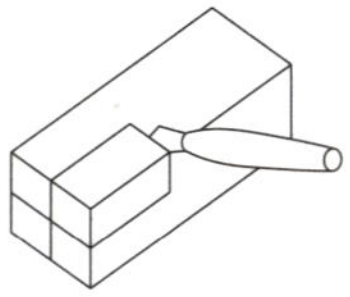

A1

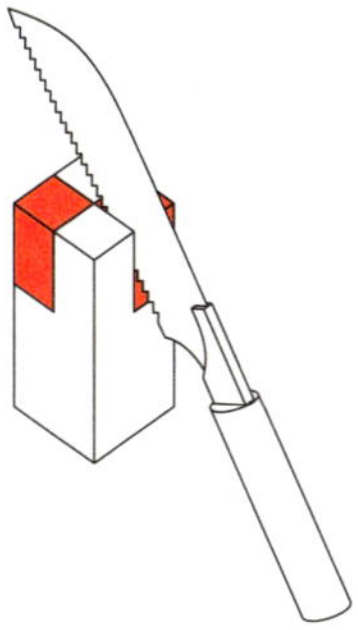

A2

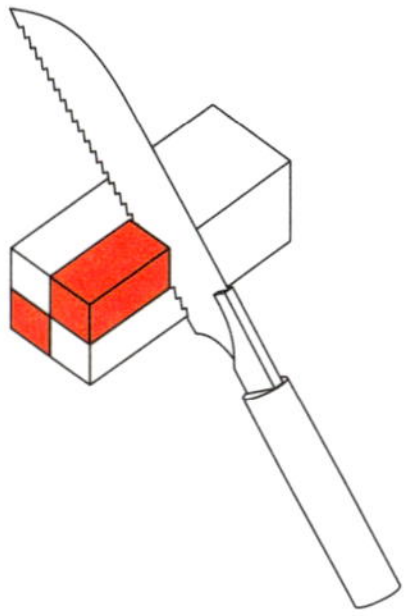

A3

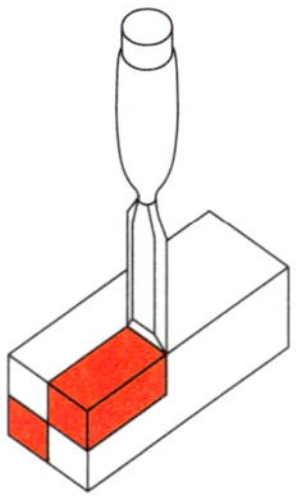

A4

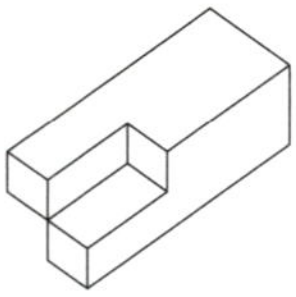

2×

A

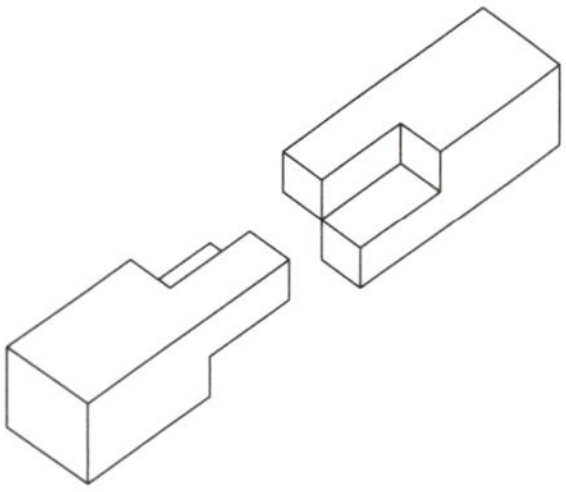

A+A

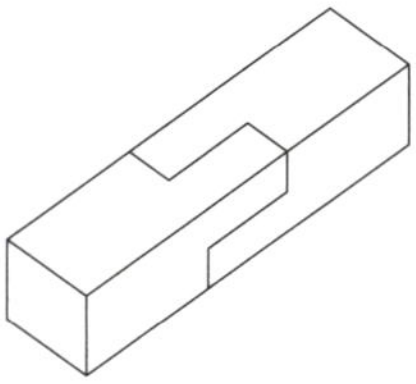

AA

Splicing Joints

X-shaped tenon joint

kainokuchi-tsugi
貝口継ぎ

Kreuzzapfen

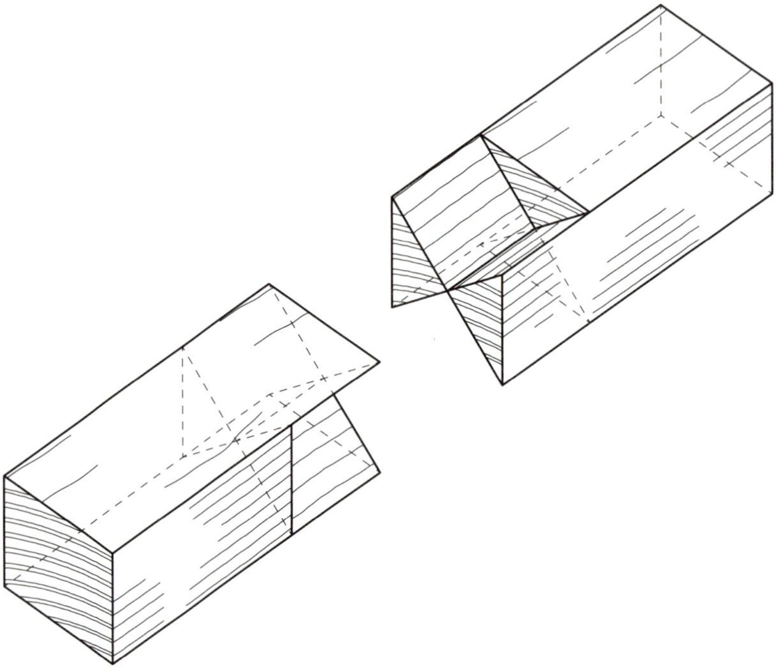

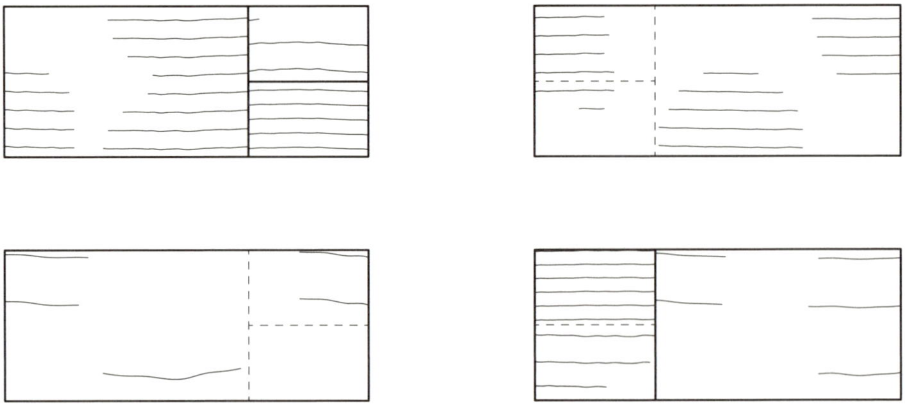

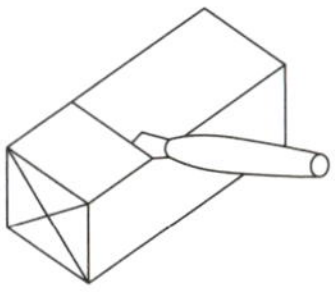

A1

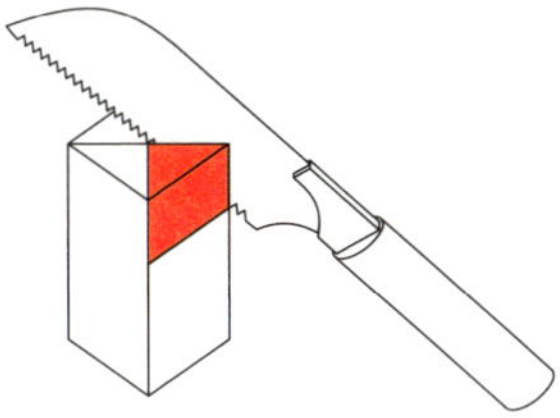

A2

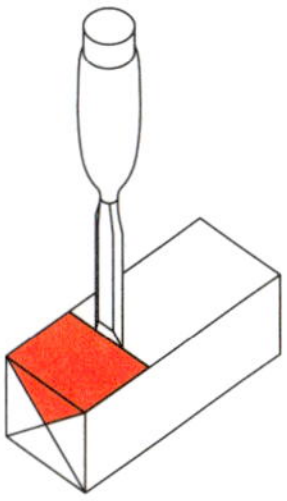

A3

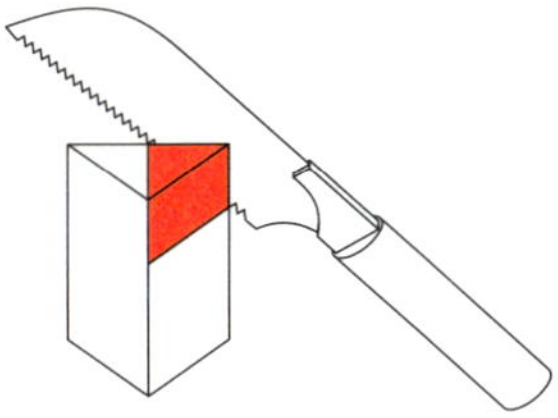

A4

A5

2×

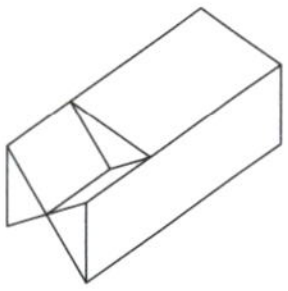

A

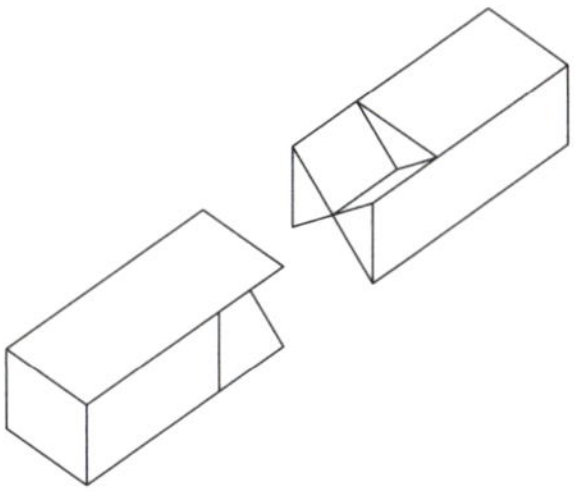

A+A

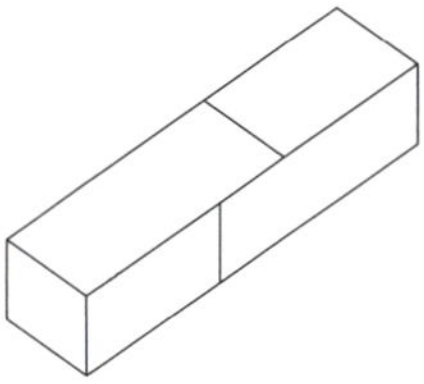

AA

Splicing Joints

Double mortise and stub tenon joint

mata-jūji-mechi-tsugi
股十字目違継ぎ

geschlitzter Zapfen
mit beidseitigem Zapfenloch

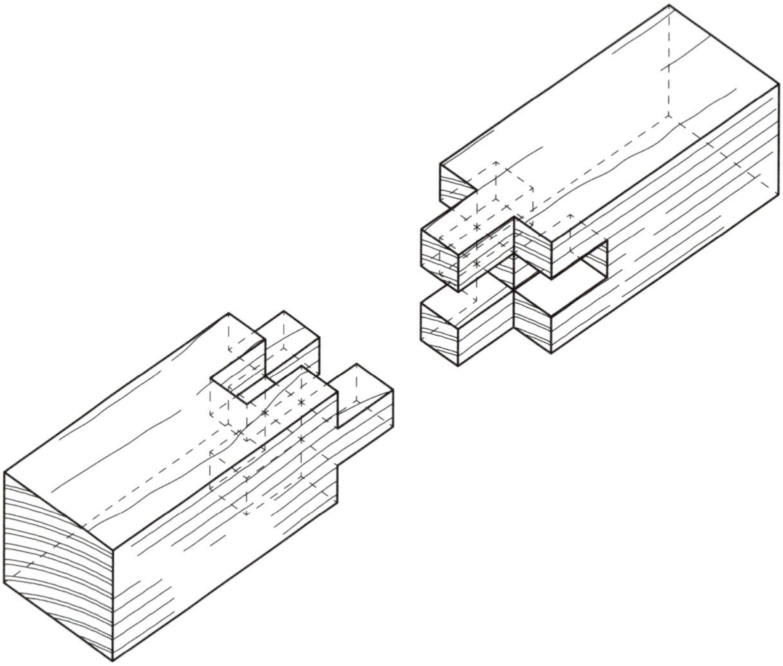

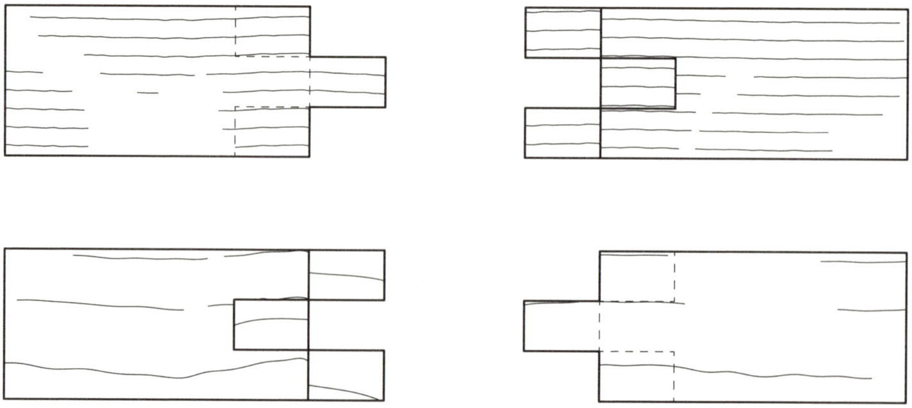

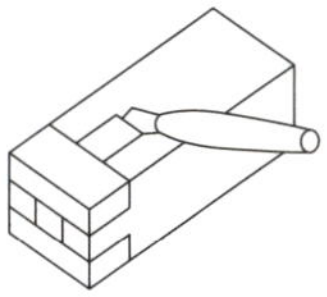

A1

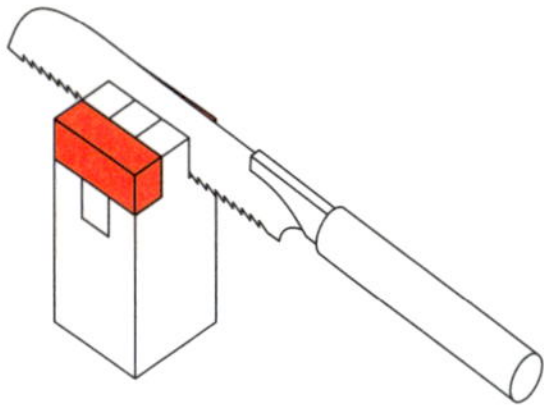

A2

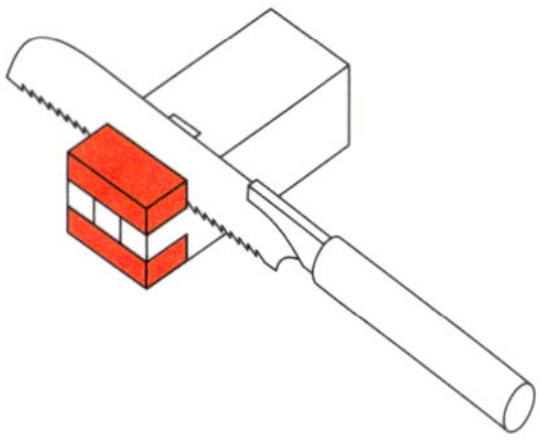

A3

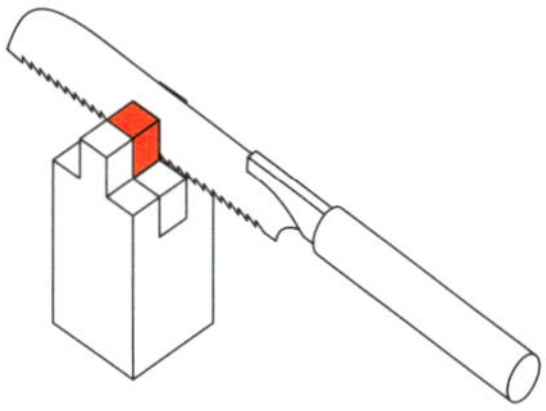

A4

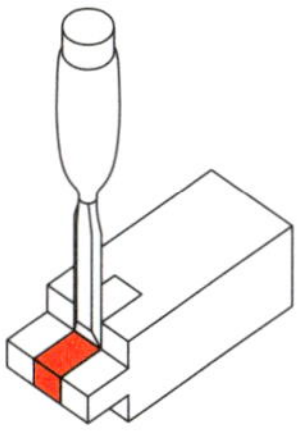

A5

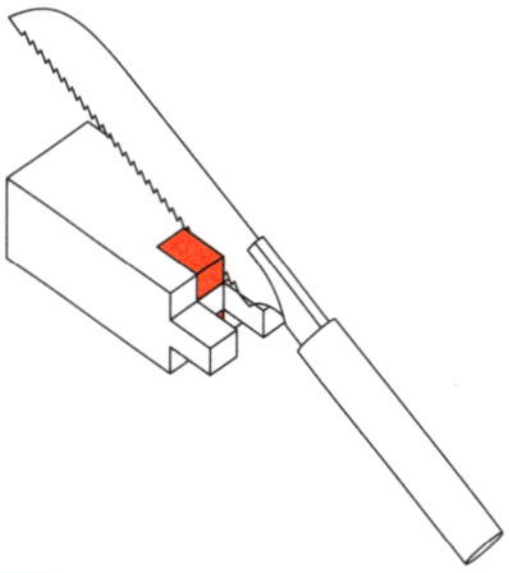

A6

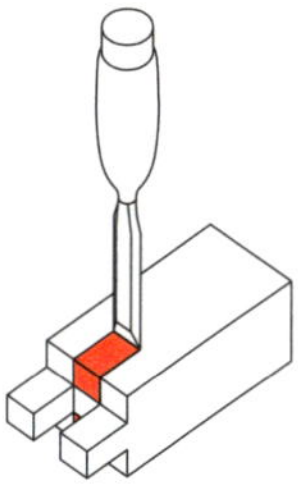

A7

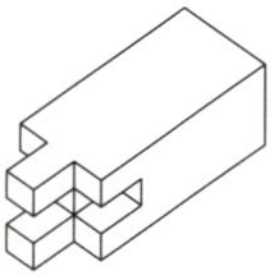

A

2×

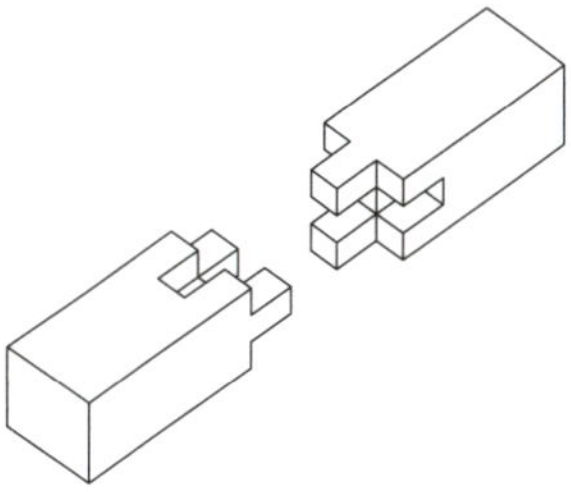

A+A

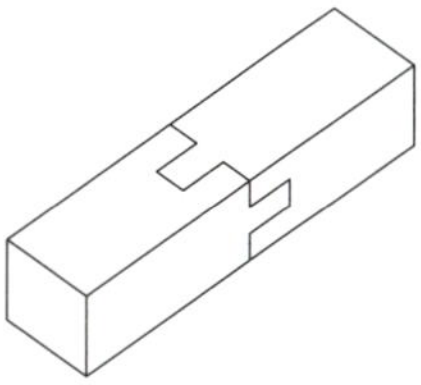

AA

Splicing Joints

Cross-shaped stub tenon joint

jūji-mechi-tsugi
十字目違継ぎ

kreuzförmiger Zapfen

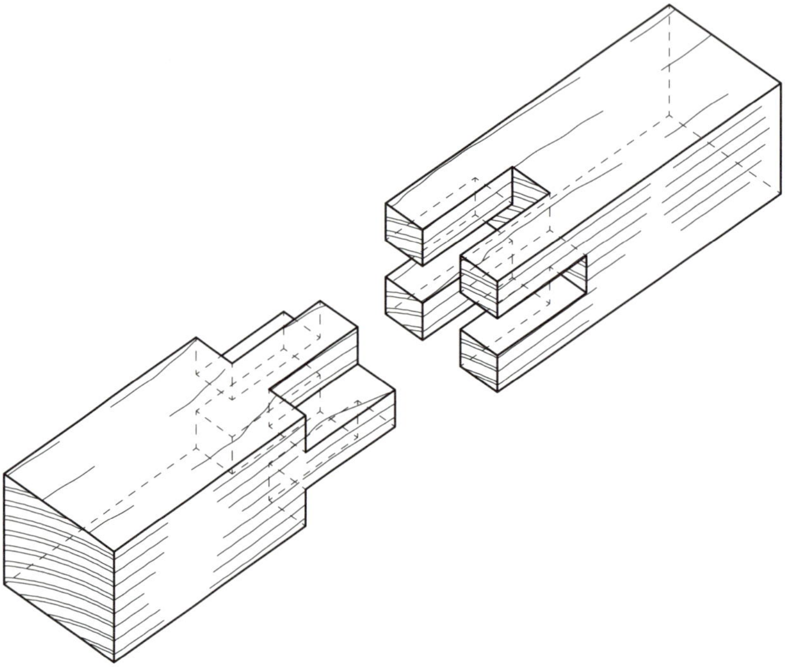

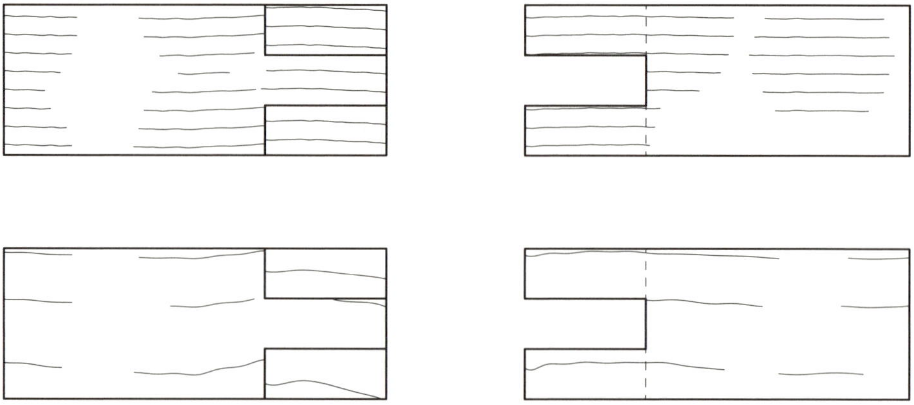

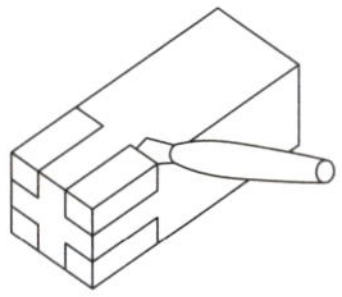

A1

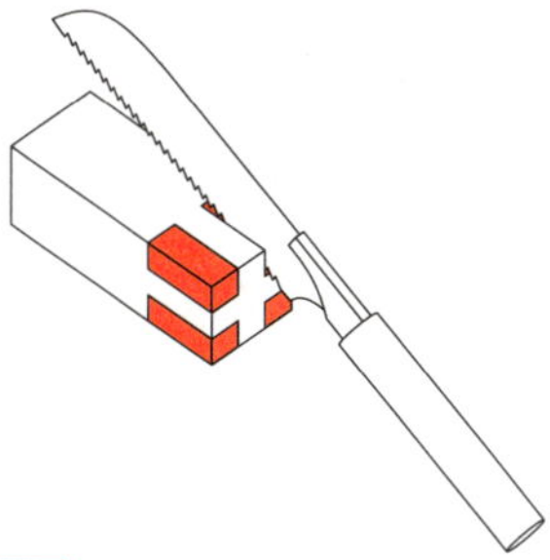

A2

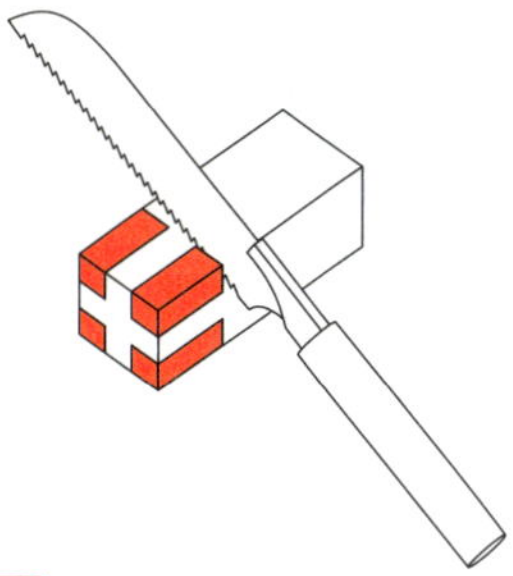

A3

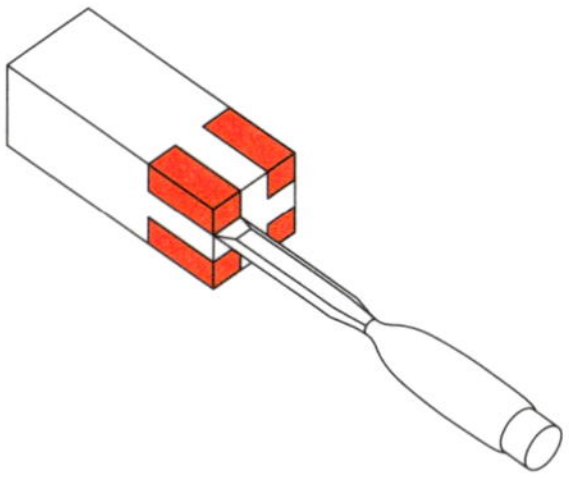

A4

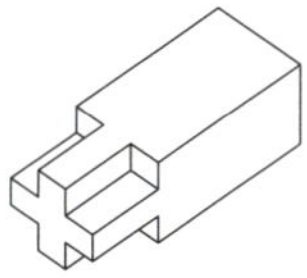

A

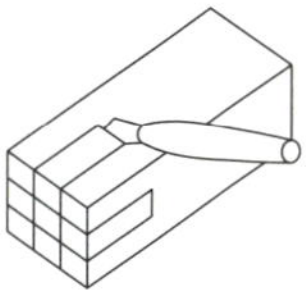

B1

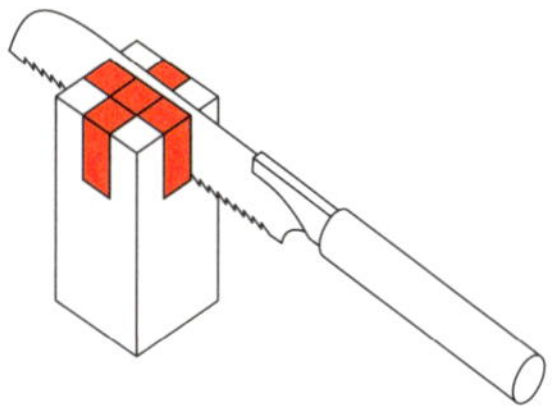

B2

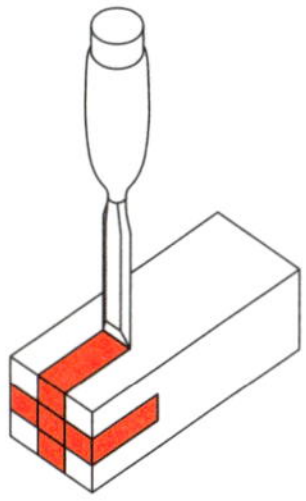

B3

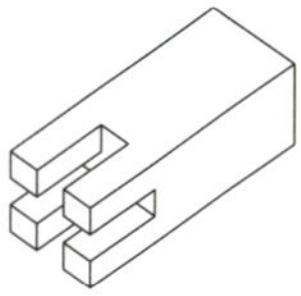

B

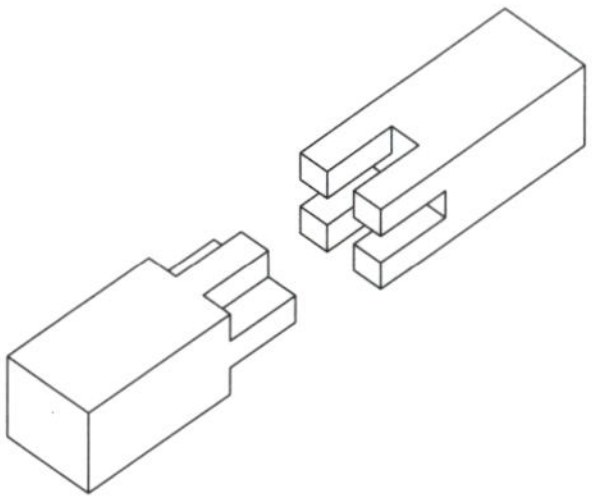

A+B

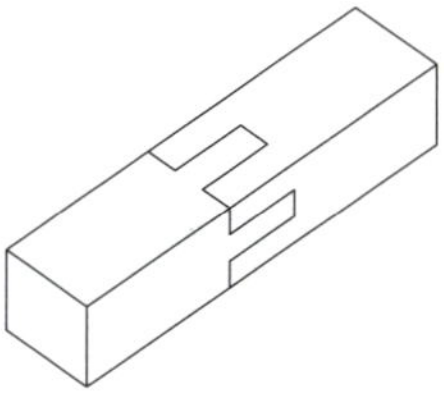

AB

Splicing Joints

Spear point cross-shaped stub tenon joint with undercut shoulders

jūji-yahazu-mechi-tsugi
十字矢筈目違継ぎ

kreuzförmiger Zapfen mit Gratschnitt und hinterschnittener Brüstung

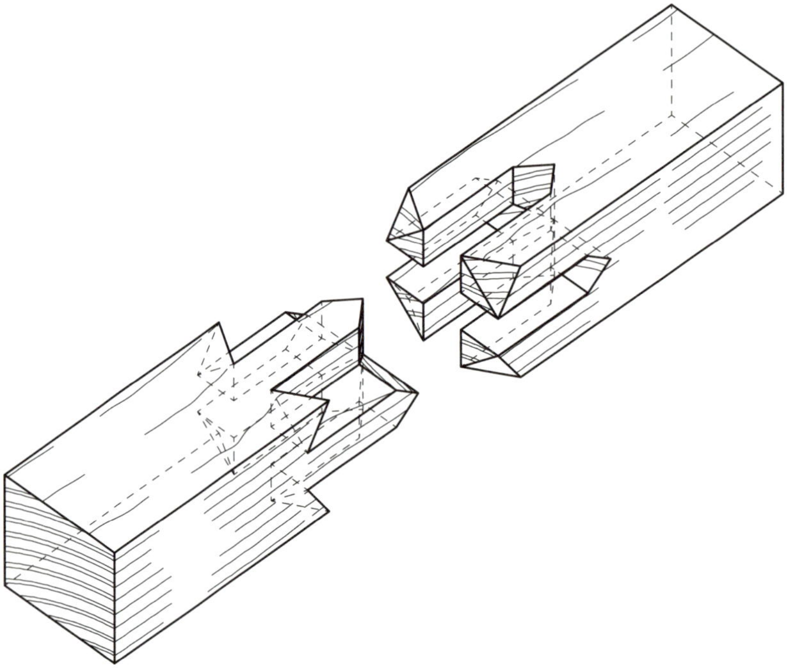

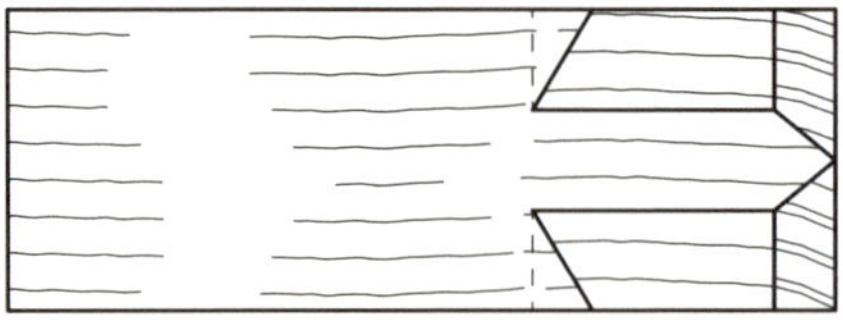

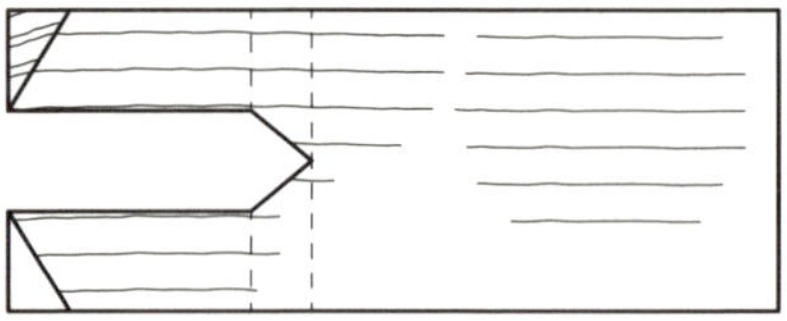

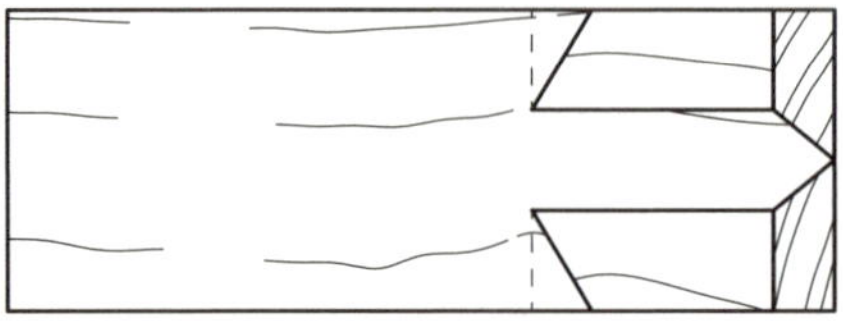

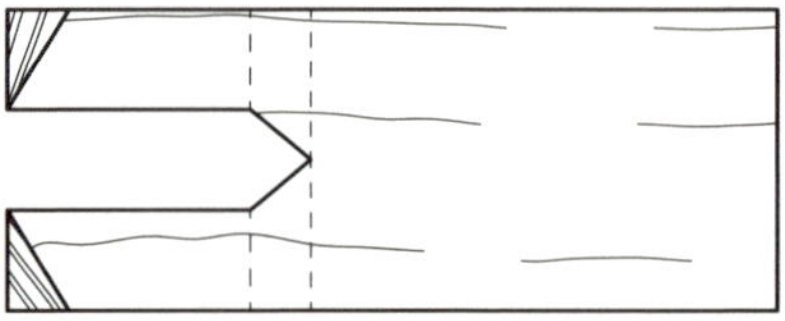

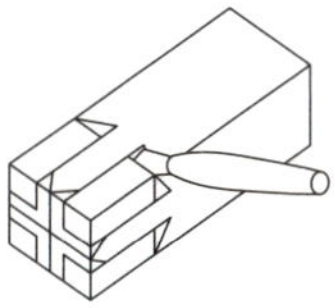

A1

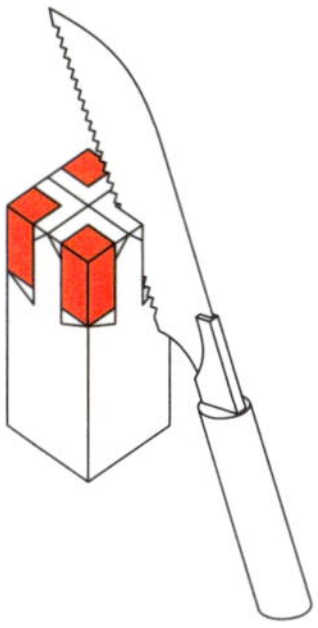

A2

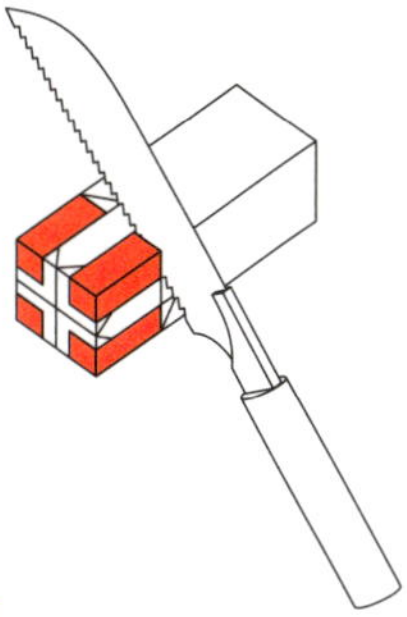

A3

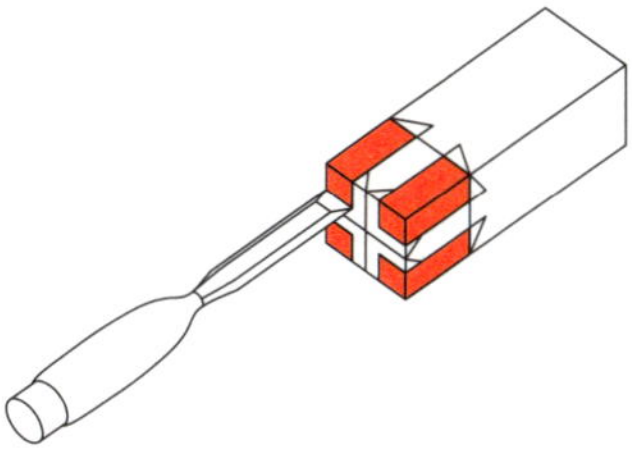

A4

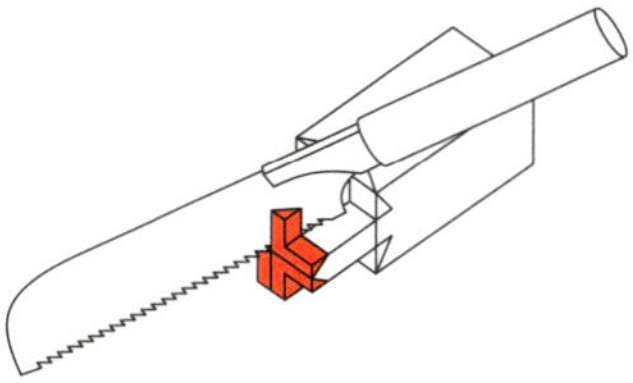

A5

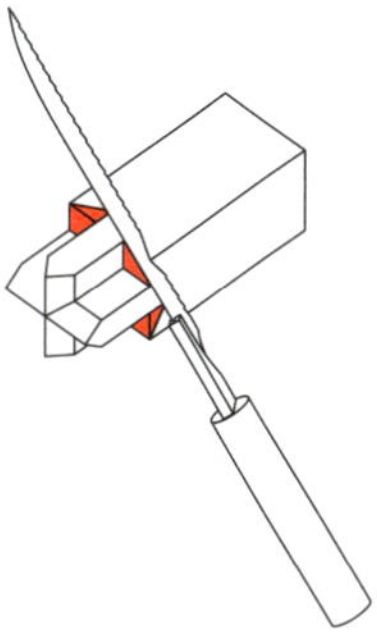

A6

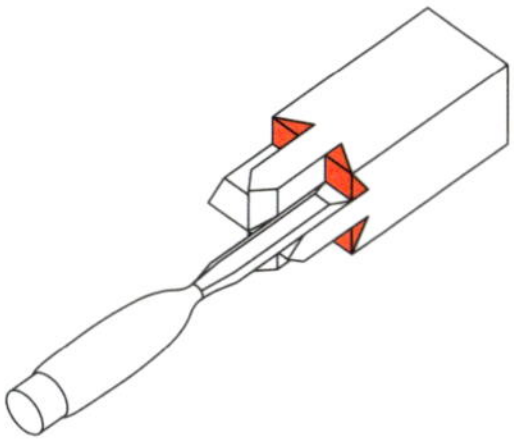

A7

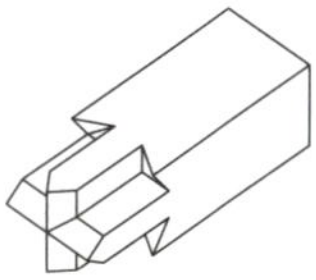

A

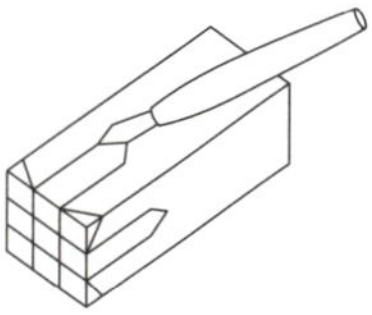

B1

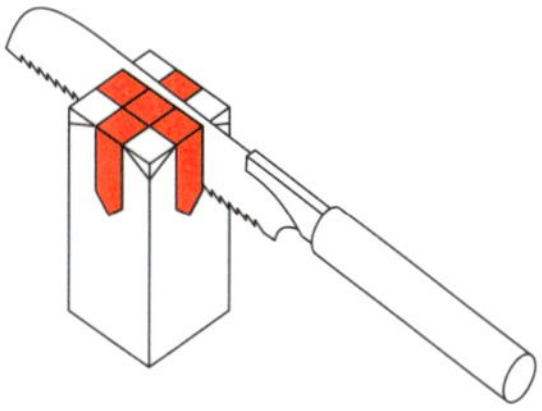

B2

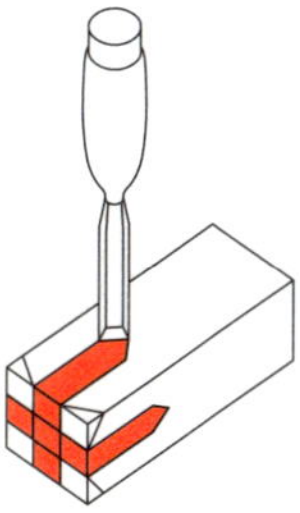

B3

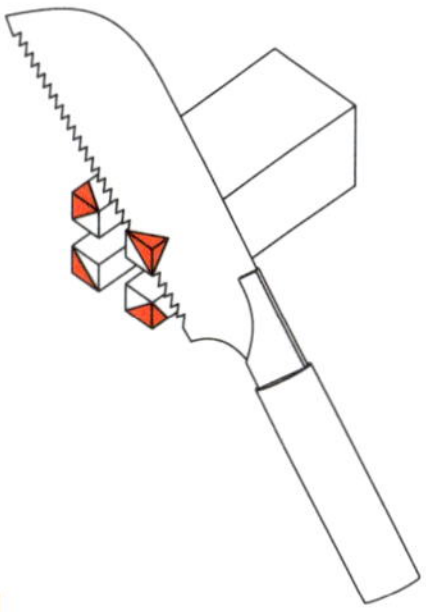

B4

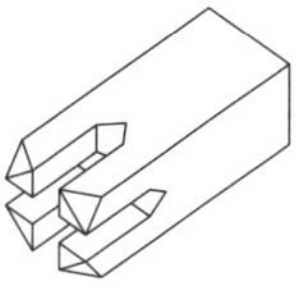

B

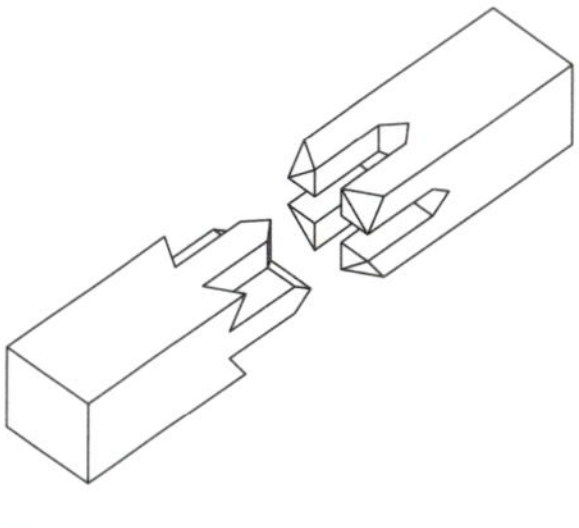

A+B

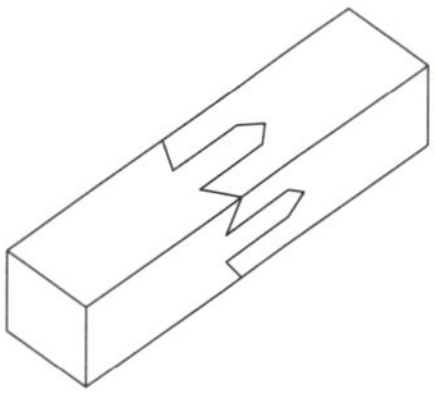

AB

Splicing Joints

Dovetail tenon joint

otoshi-ari-tsugi
落蟻継ぎ

Schwalbenschwanzzapfen

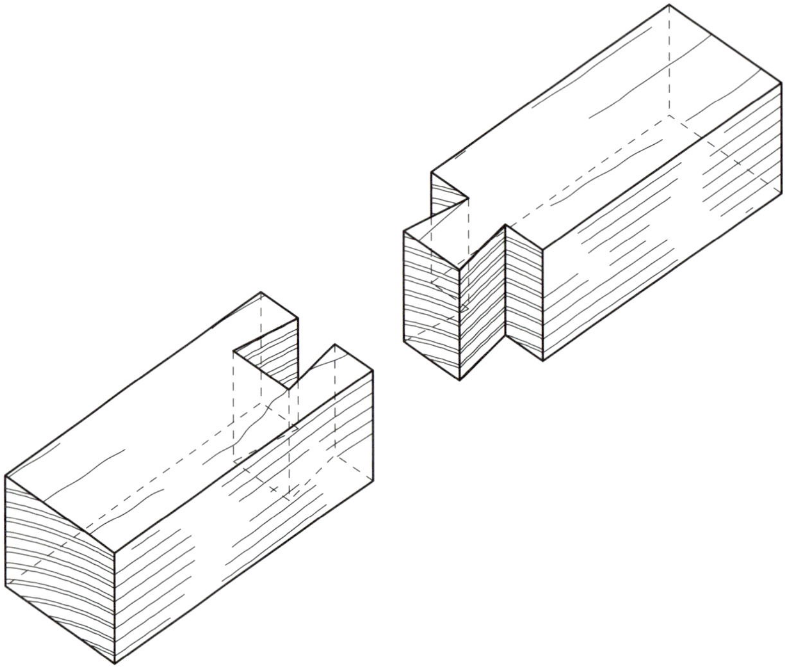

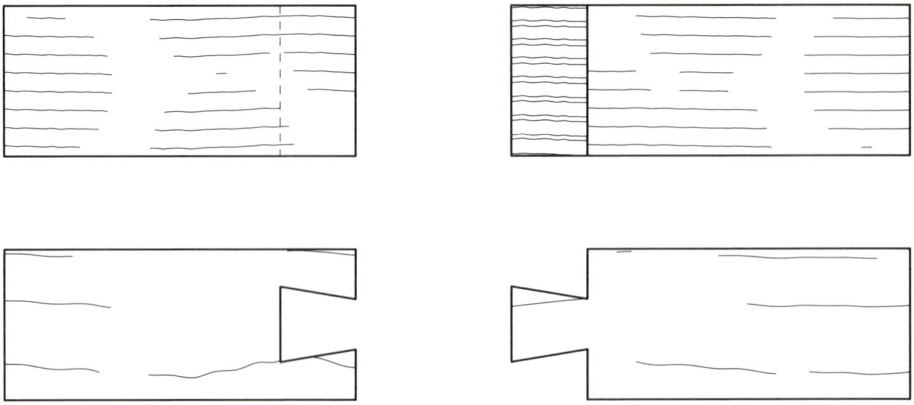

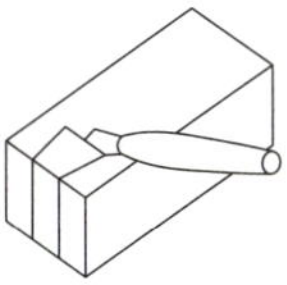

A1

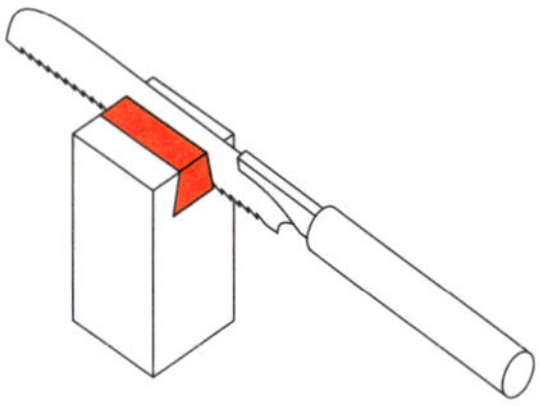

A2

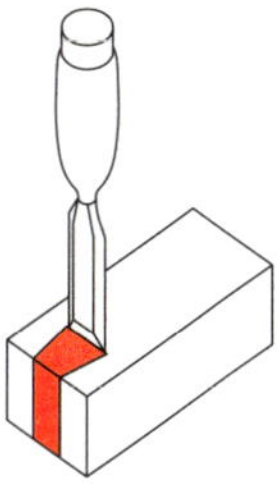

A3

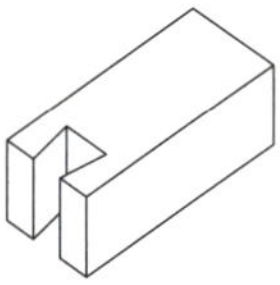

A

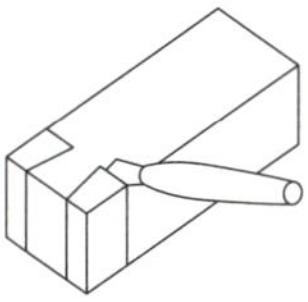

B1

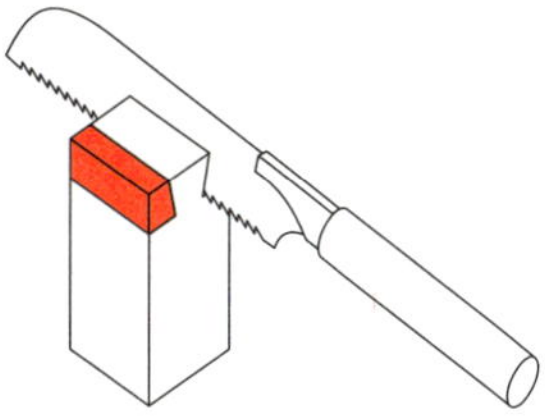

B2

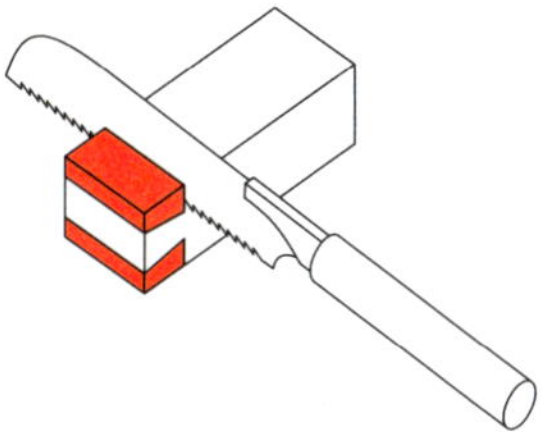

B3

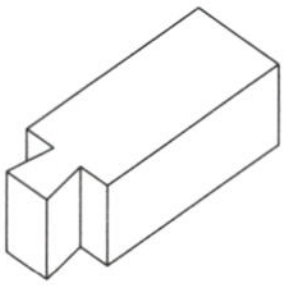

B

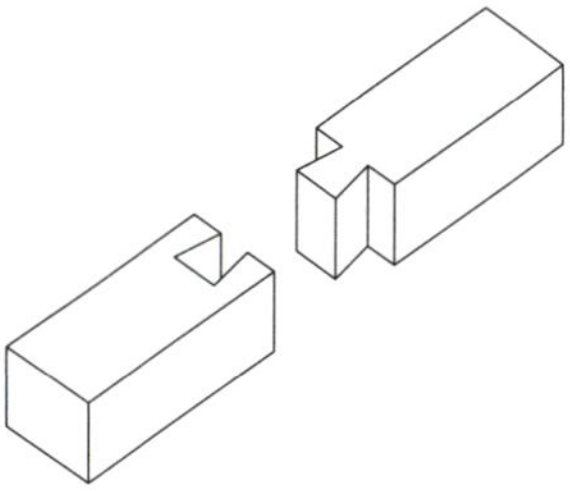

A+B

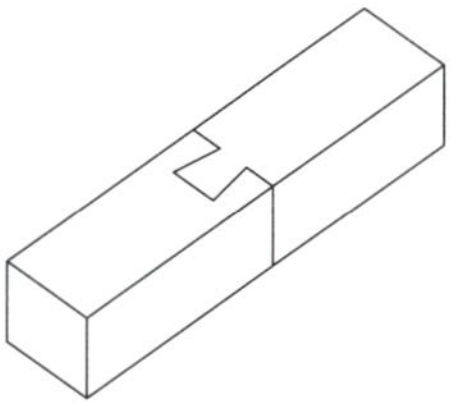

AB

Splicing Joints

Triple-shouldered dovetail tenon joint

ari-tsugi
蟻継ぎ

abgesetzter Schwalbenschwanzzapfen

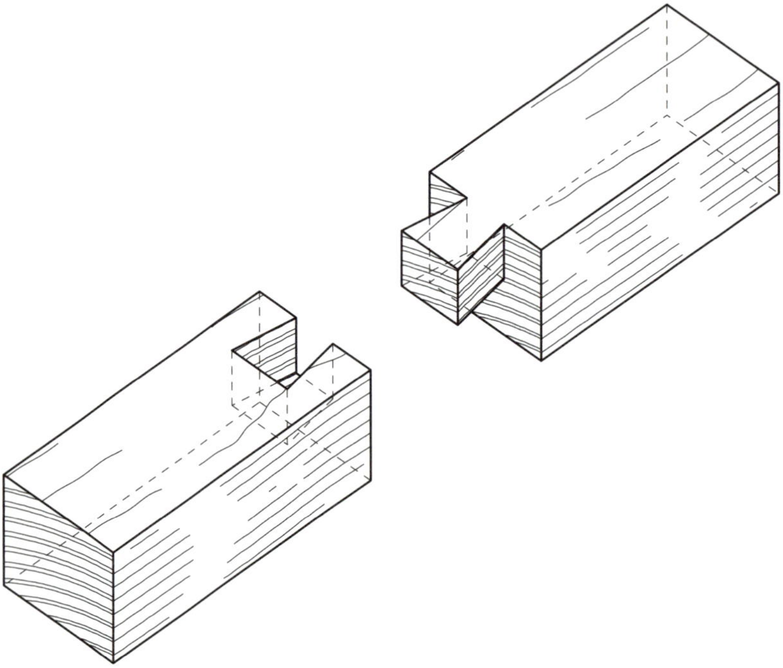

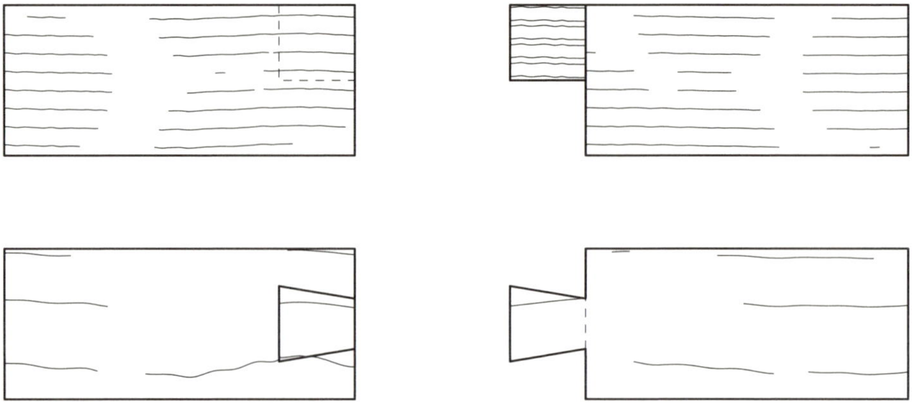

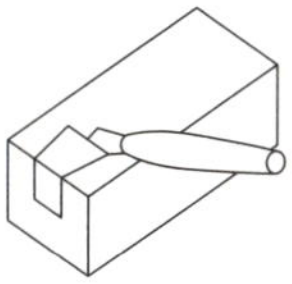

A1

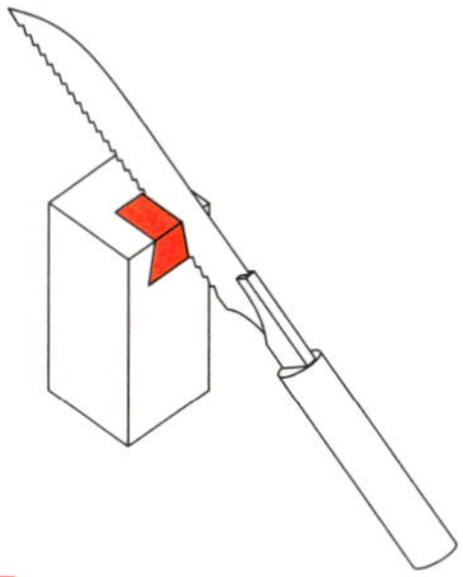

A2

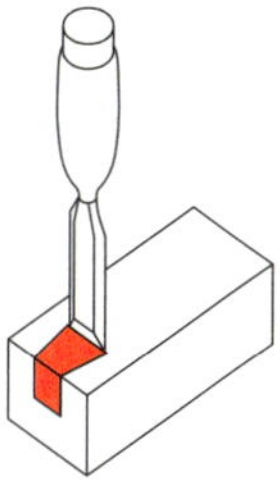

A3

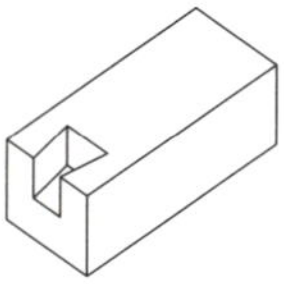

A

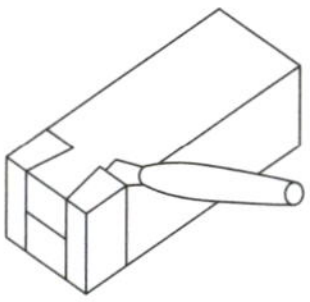

B1

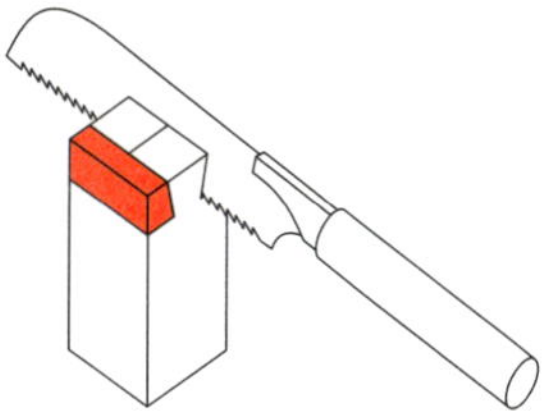

B2

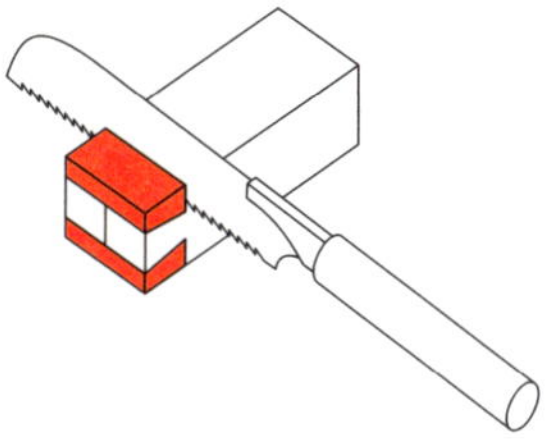

B3

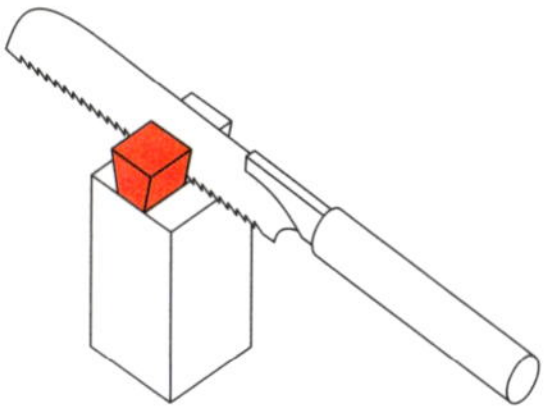

B4

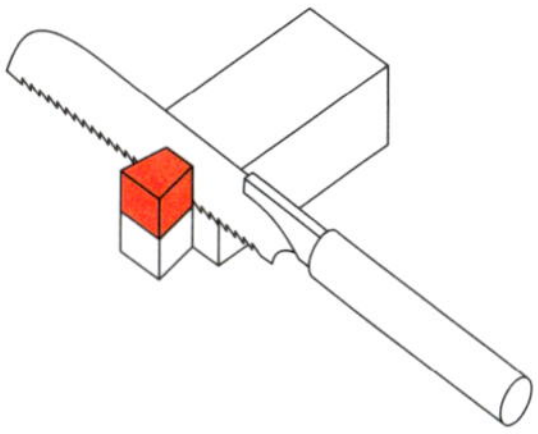

B5

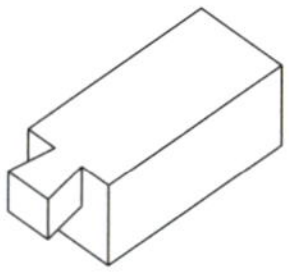

B

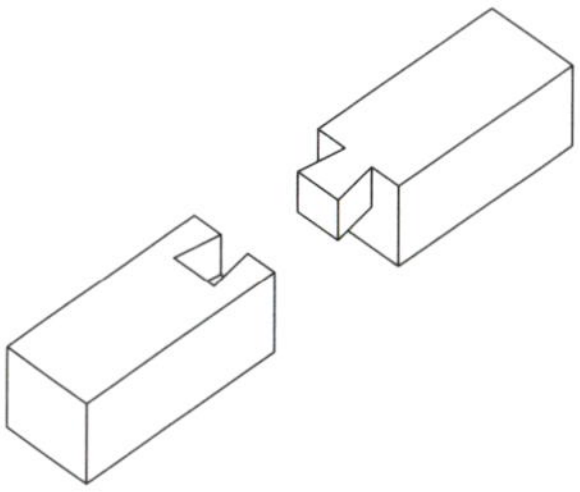

A+B

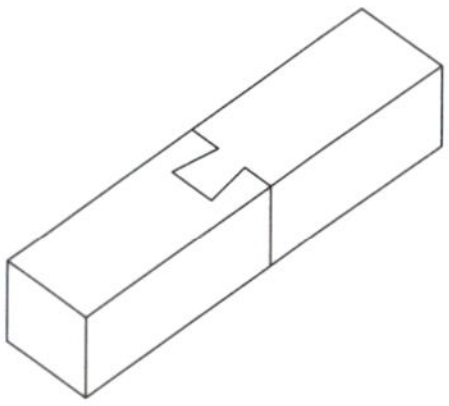

AB

Splicing Joints

Lapped dovetail joint

koshi-kake-ari-tsugi [*]
腰掛蟻継ぎ

Schwalbenschwanzzapfen mit geradem Blatt

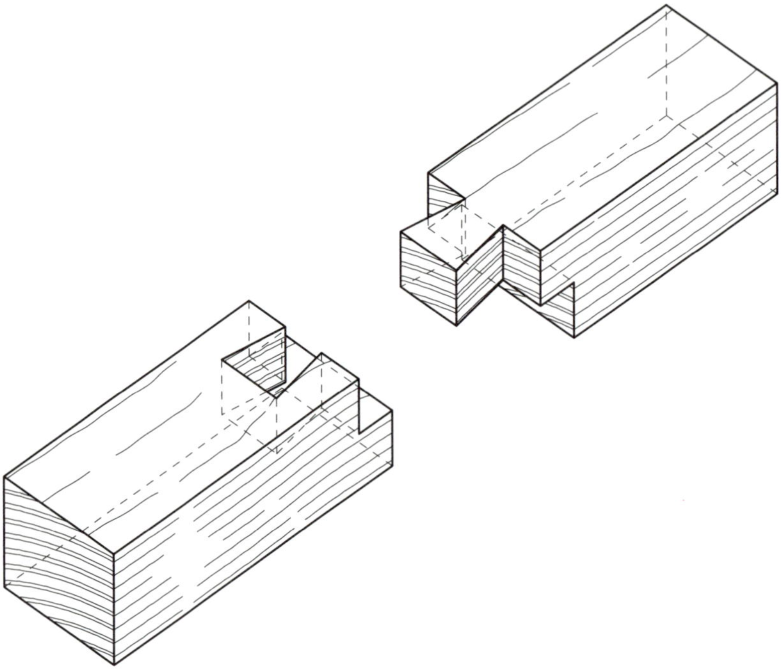

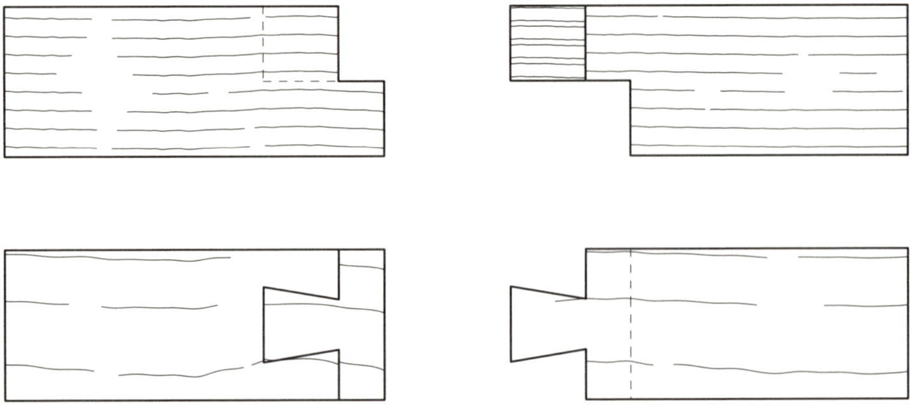

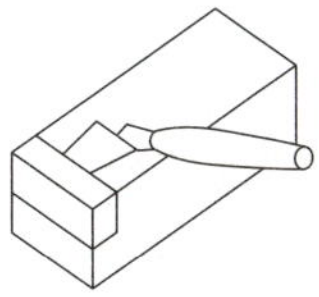

A1

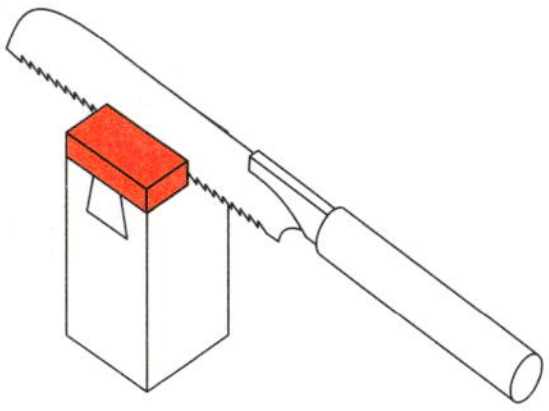

A2

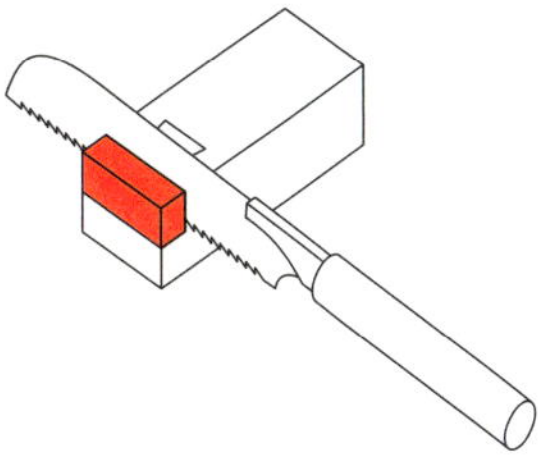

A3

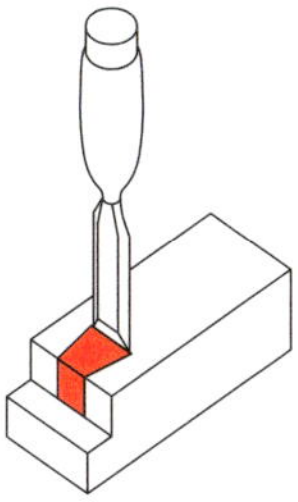

A4

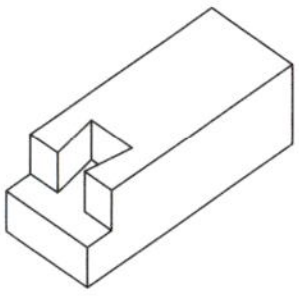

A

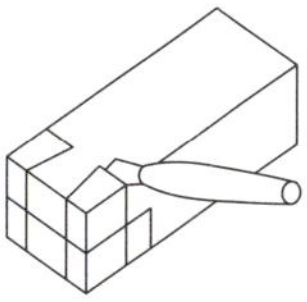

B1

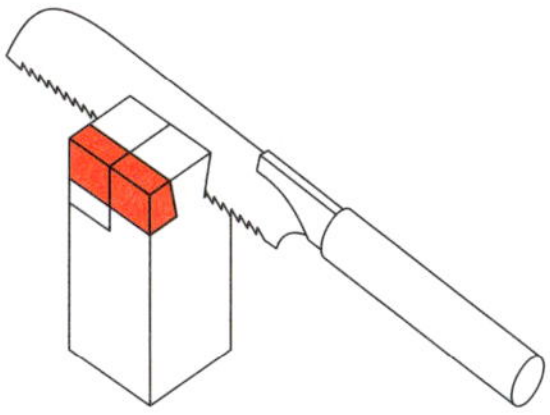

B2

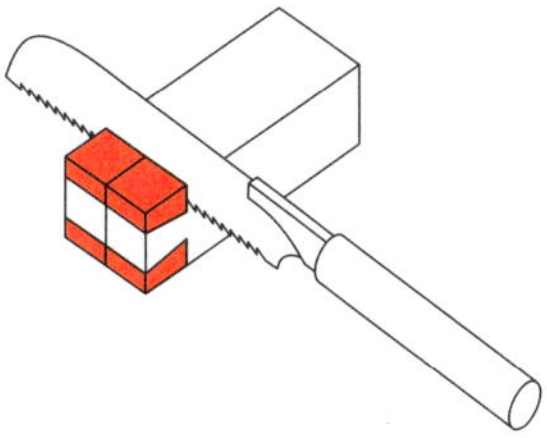

B3

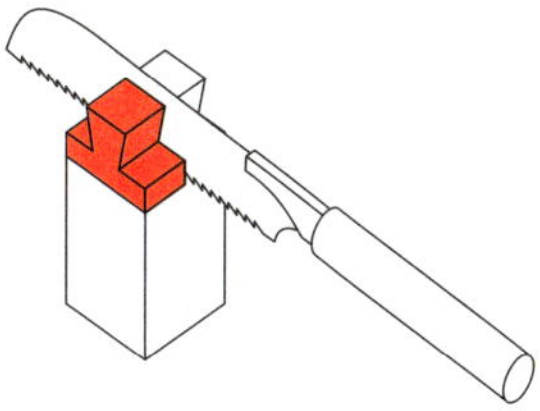

B4

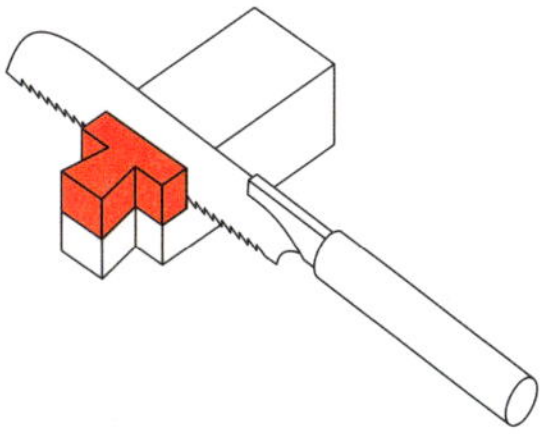

B5

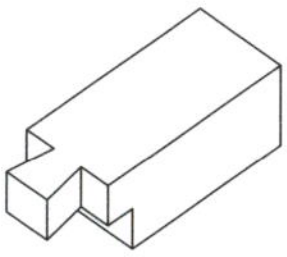

B

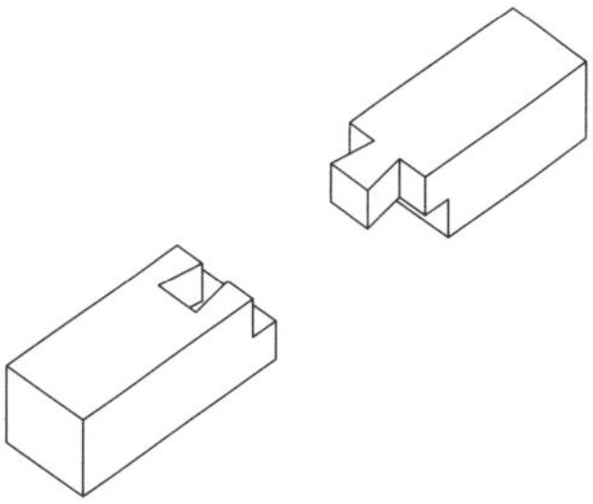

A+B

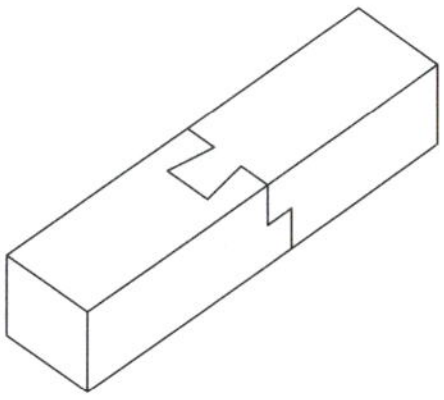

AB

Splicing Joints

Lapped dovetail joint with stub tenon

koshi-ire-mechi-ari-tsugi [*]
腰入目違蟻継ぎ

Schwalbenschwanzzapfen mit geradem Blatt und Brüstungszapfen

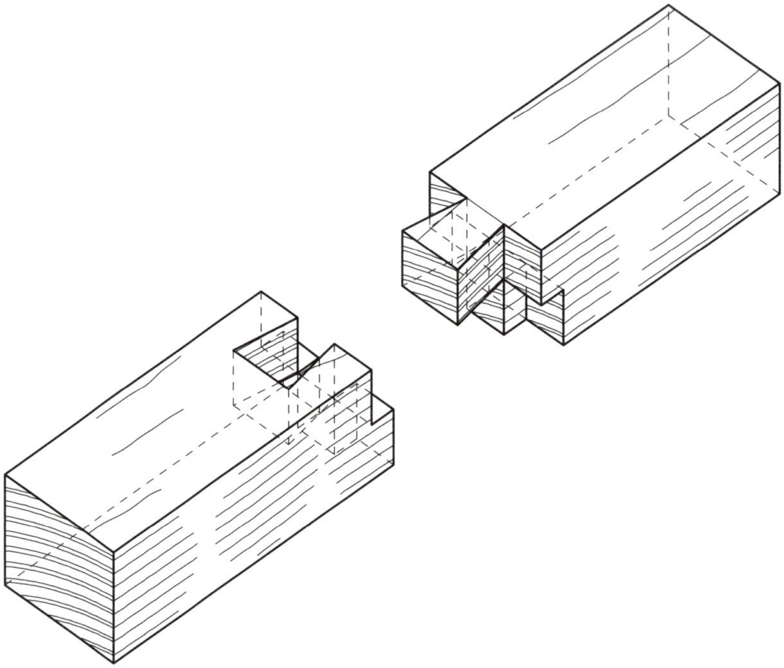

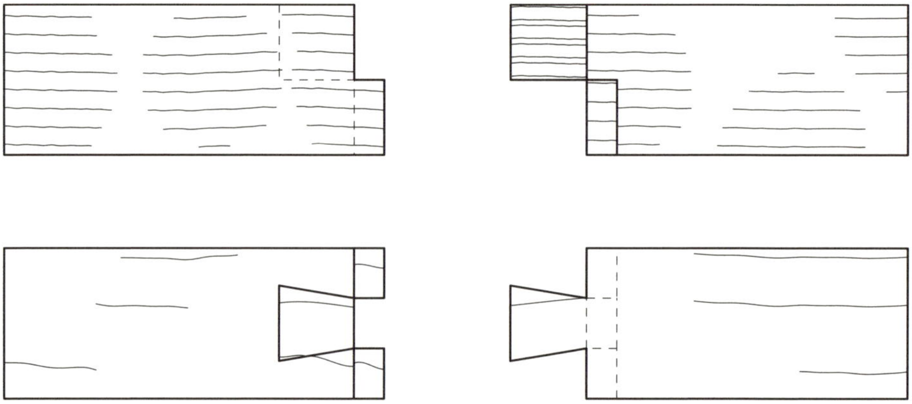

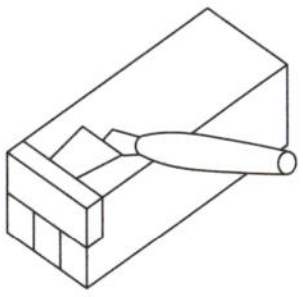

A1

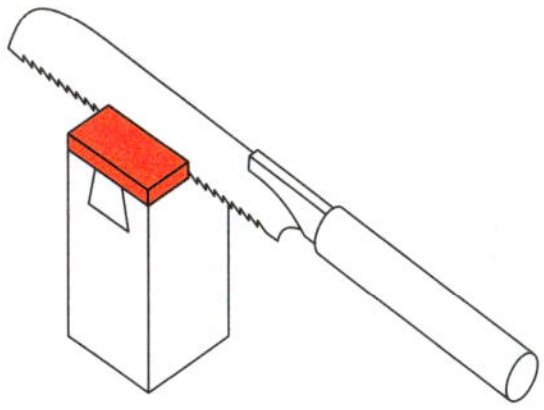

A2

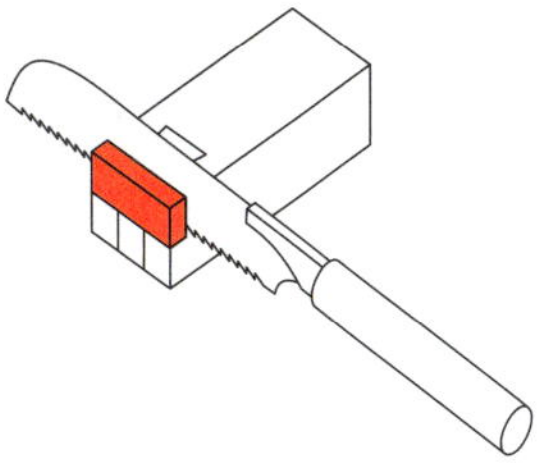

A3

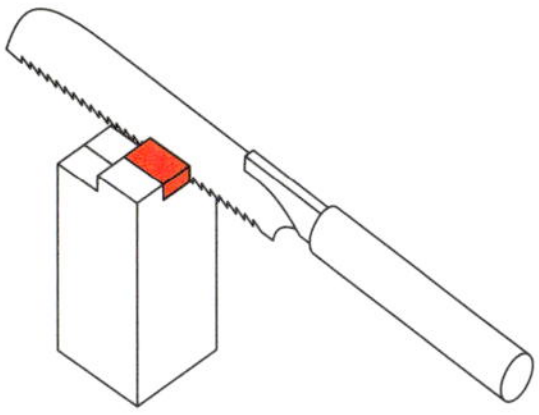

A4

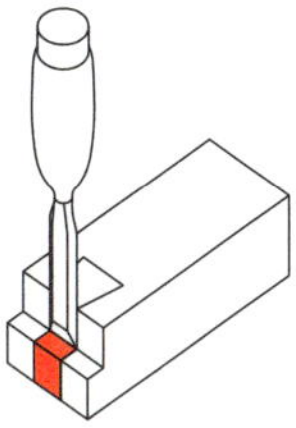

A5

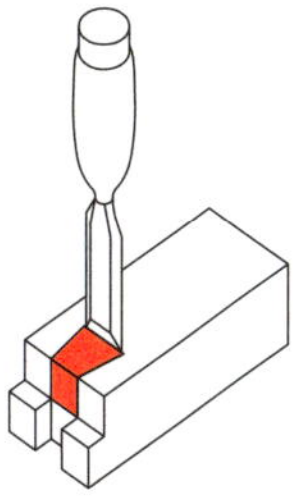

A6

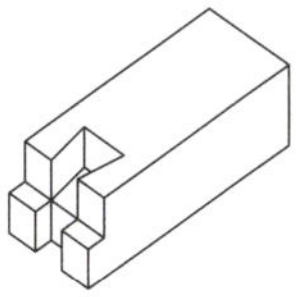

A

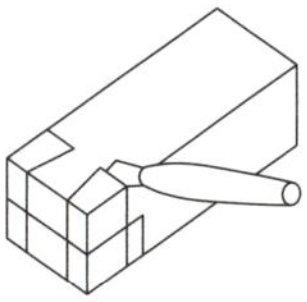

B1

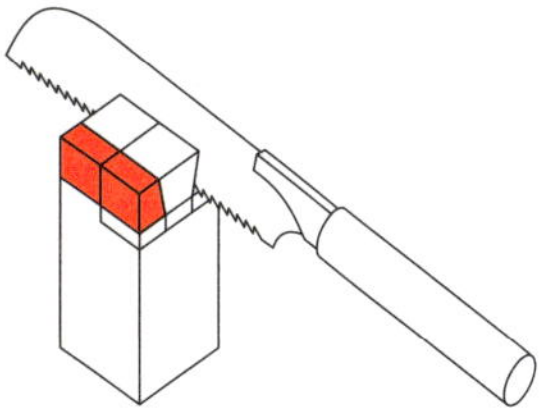

B2

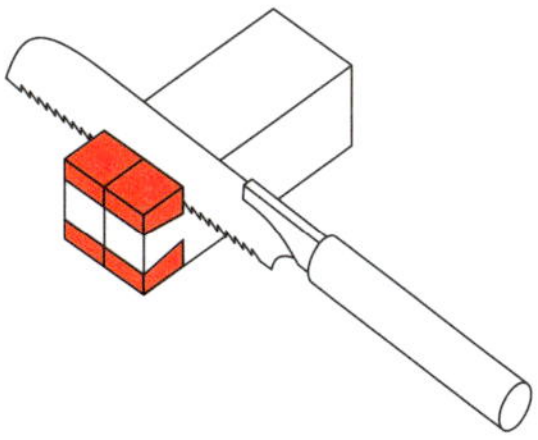

B3

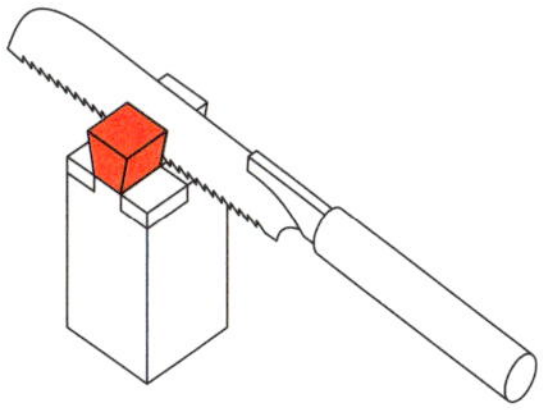

B4

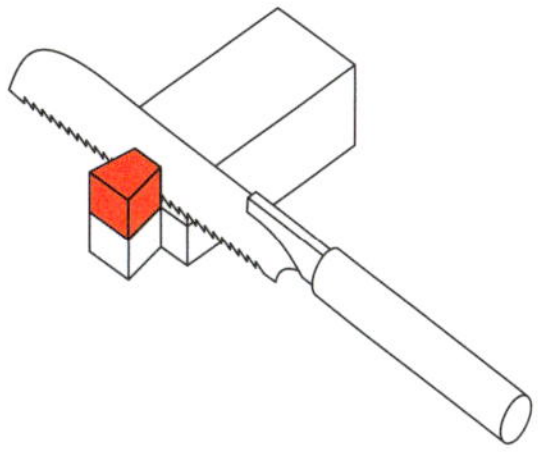

B5

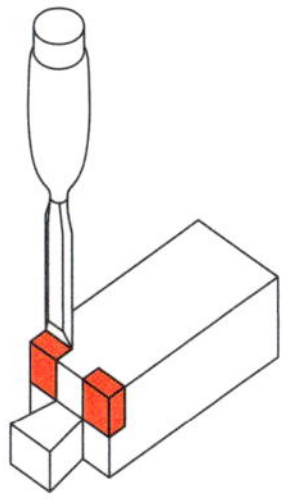

B6

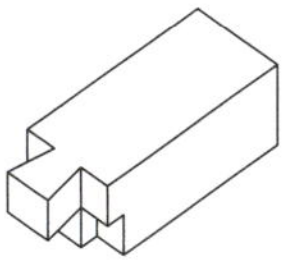

B

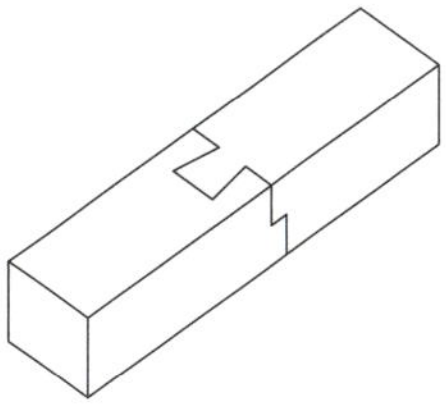

AB

Splicing Joints

Lapped double dovetail joint

ai-ari-tsugi
相蟻継ぎ

beidseitiger Schwalbenschwanzzapfen mit geradem Blatt

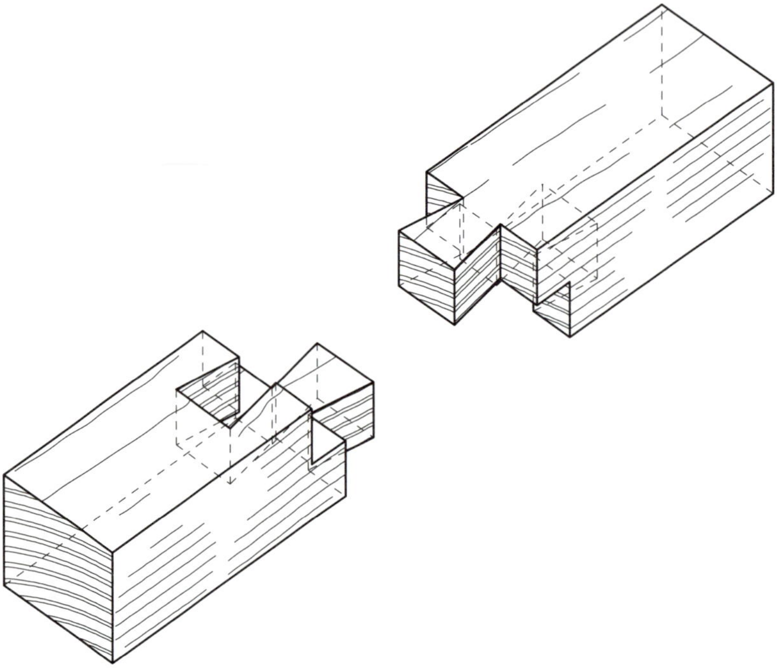

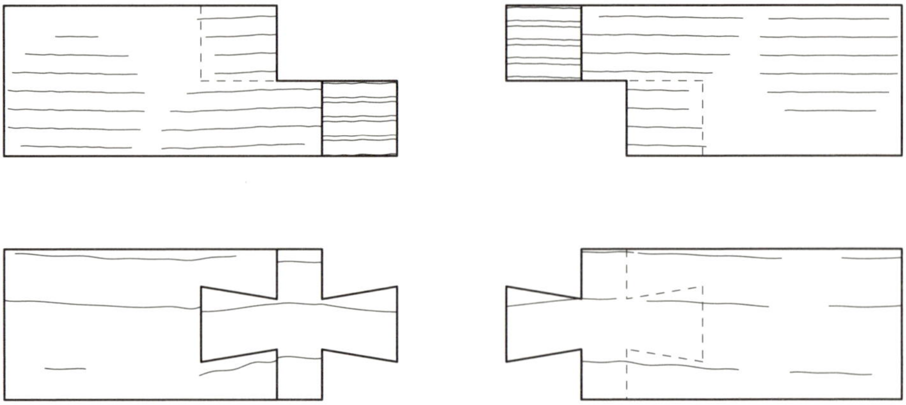

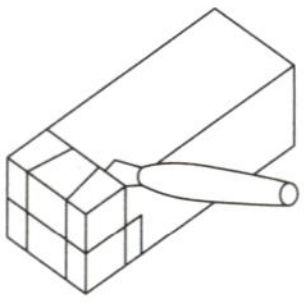

A1

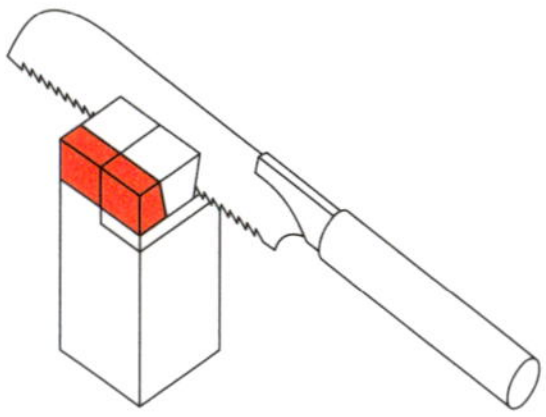

A2

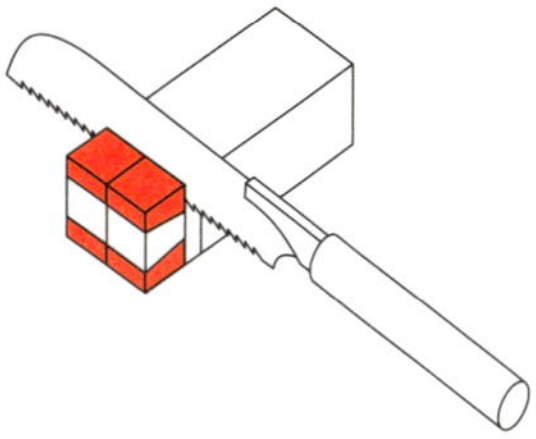

A3

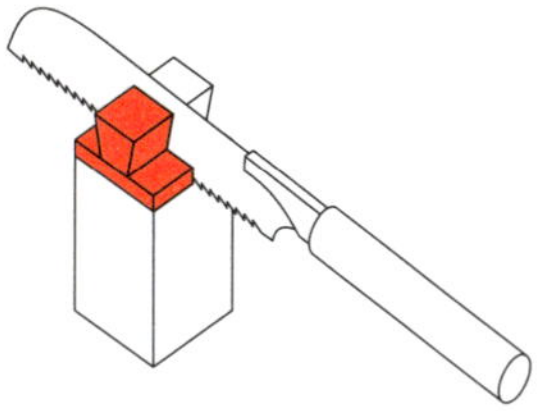

A4

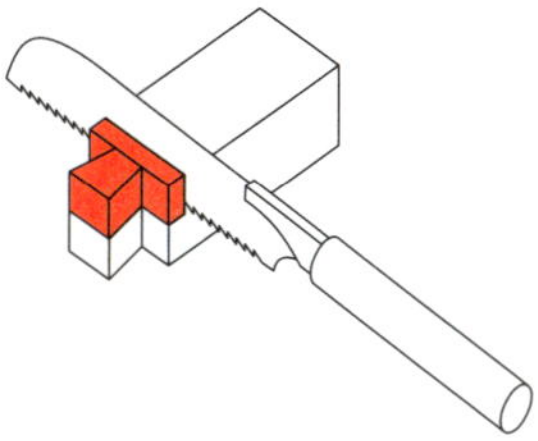

A5

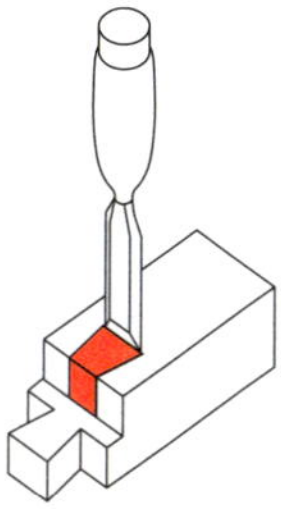

A6

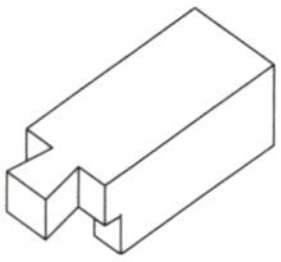

A

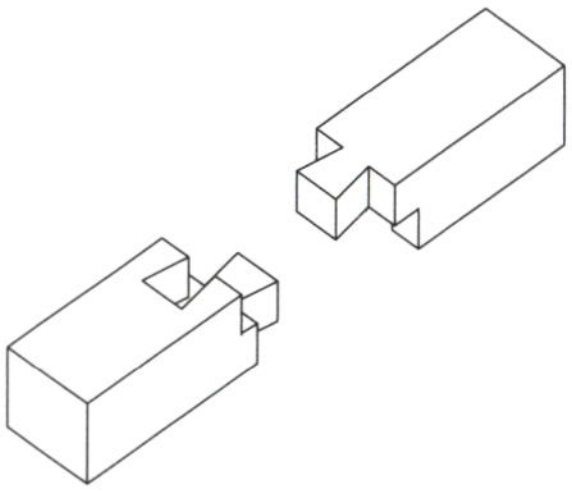

A+A

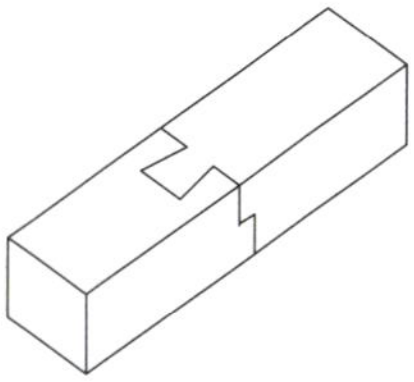

AA

Splicing Joints

Butt joint with floating butterfly tenon

tōshi-chigiri-tsugi
通千切継ぎ

eingesetzter Schwalbenschwanzzapfen

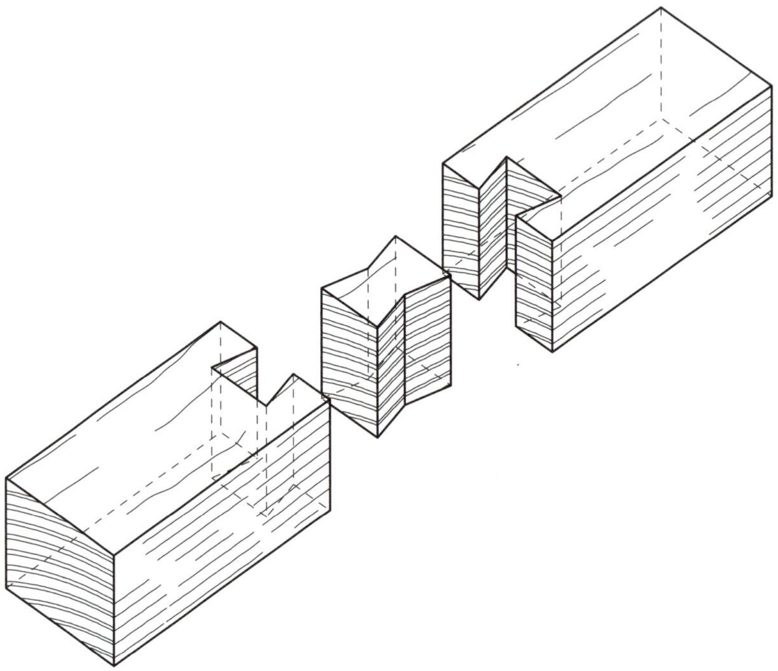

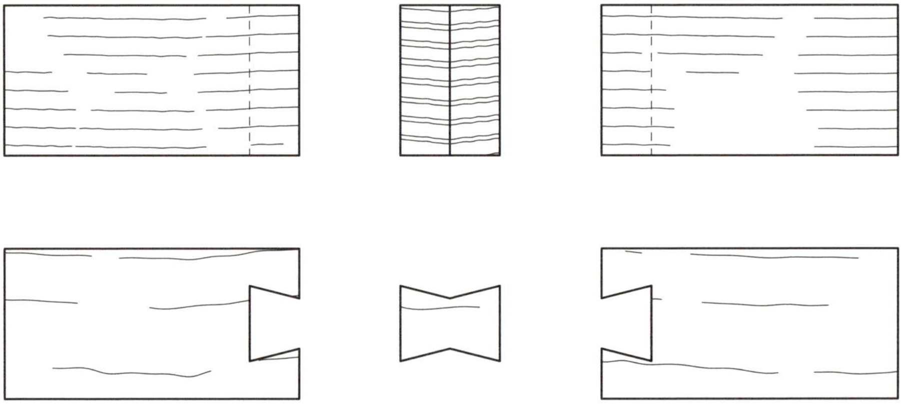

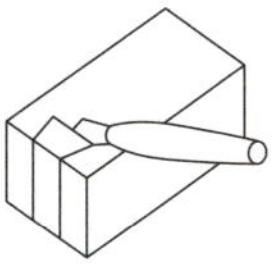

A1

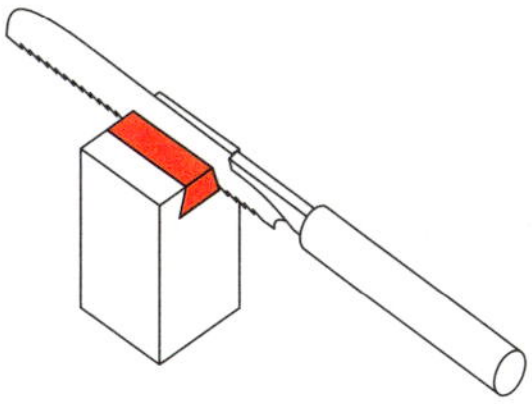

A2

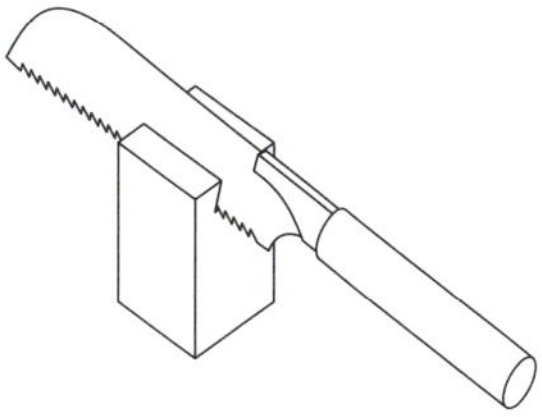

A3

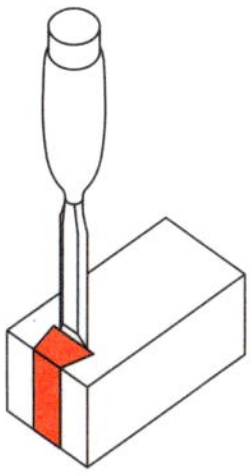

A4

2×

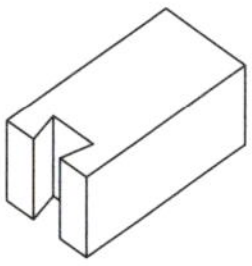

A

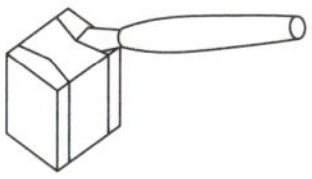

B1

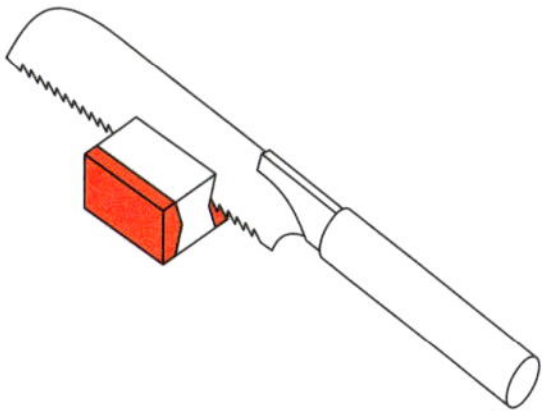

B2

B

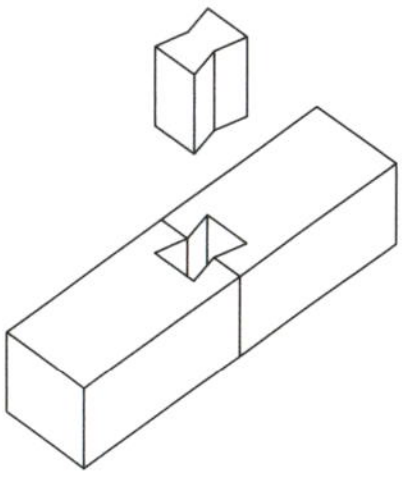

A+B+A

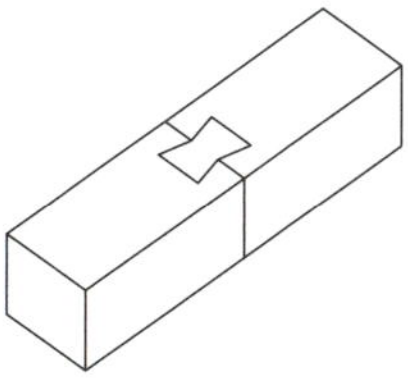

ABA

Splicing Joints

Butt joint with butterfly key

chigiri-tsugi
千切継ぎ

Schwalbenschwanzeinlage

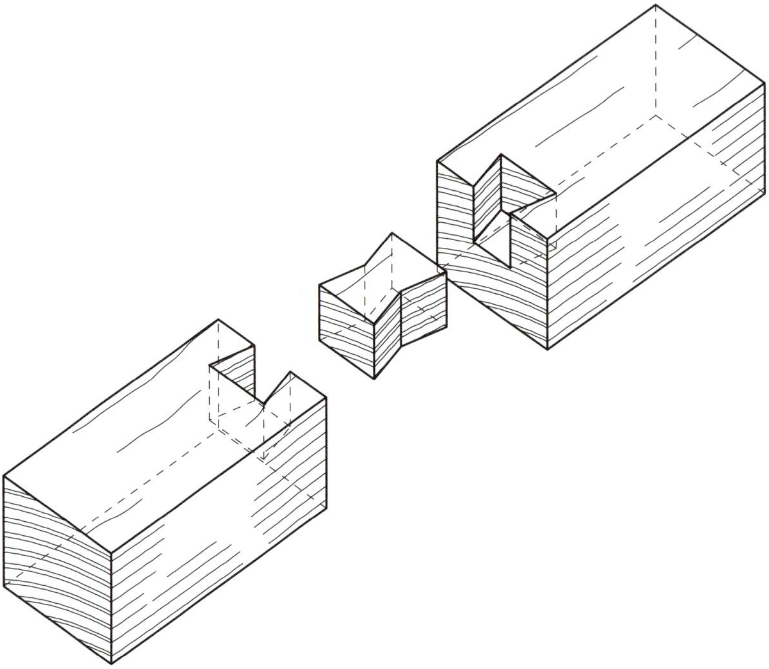

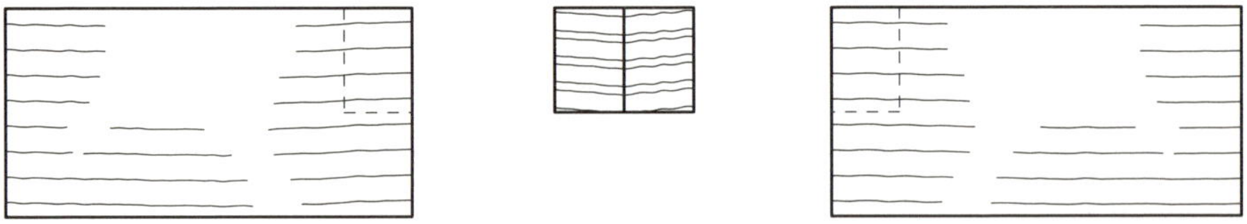

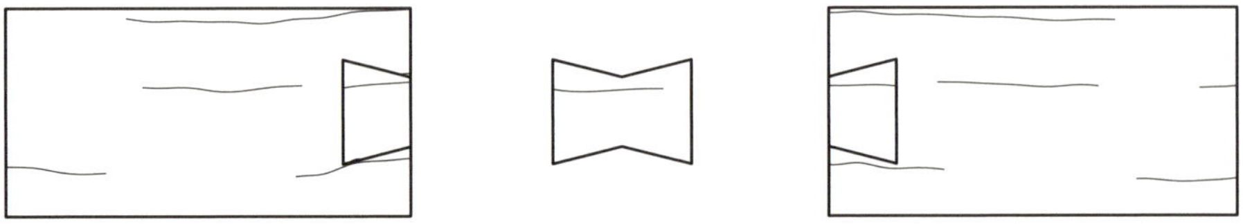

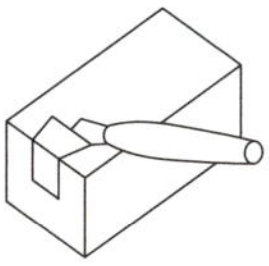

A1

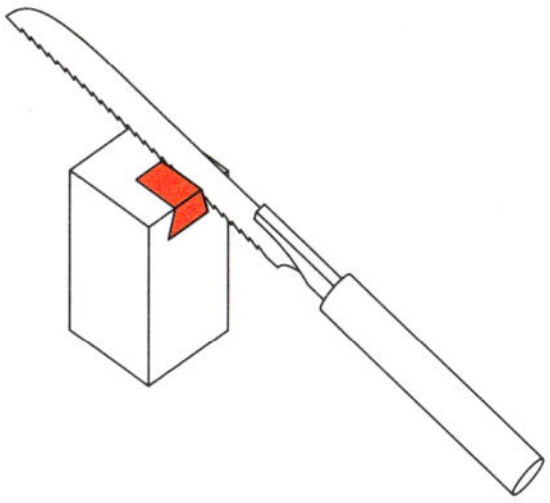

A2

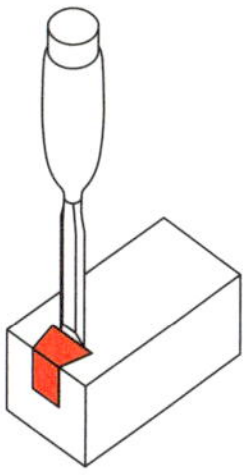

A3

2×

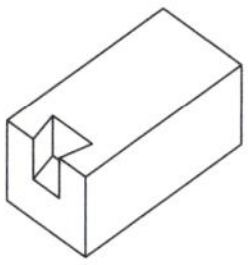

A

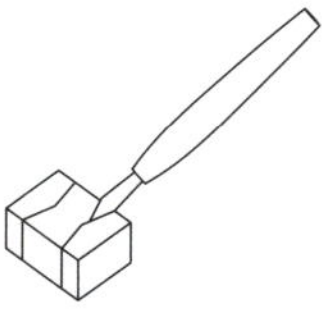

B1

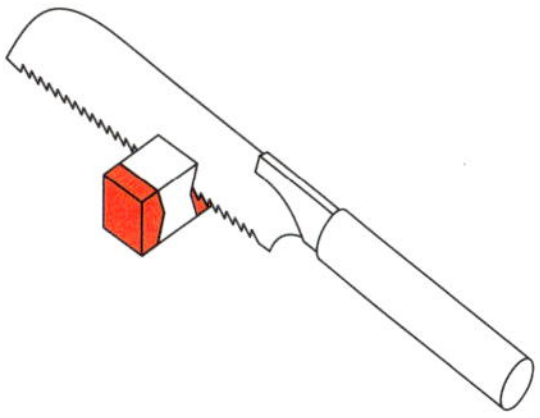

B2

B

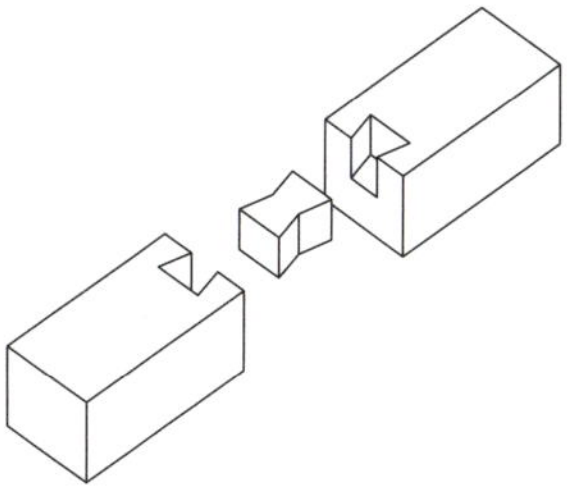

A+B+A

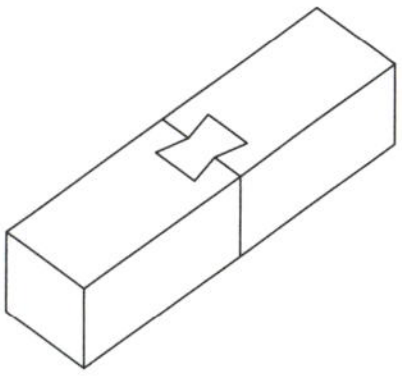

ABA

Splicing Joints

Half lap joint with butterfly key

koshi-kake-chigiri-tsugi
腰掛千切継ぎ

gerades Blatt mit Schwalbenschwanzeinlage

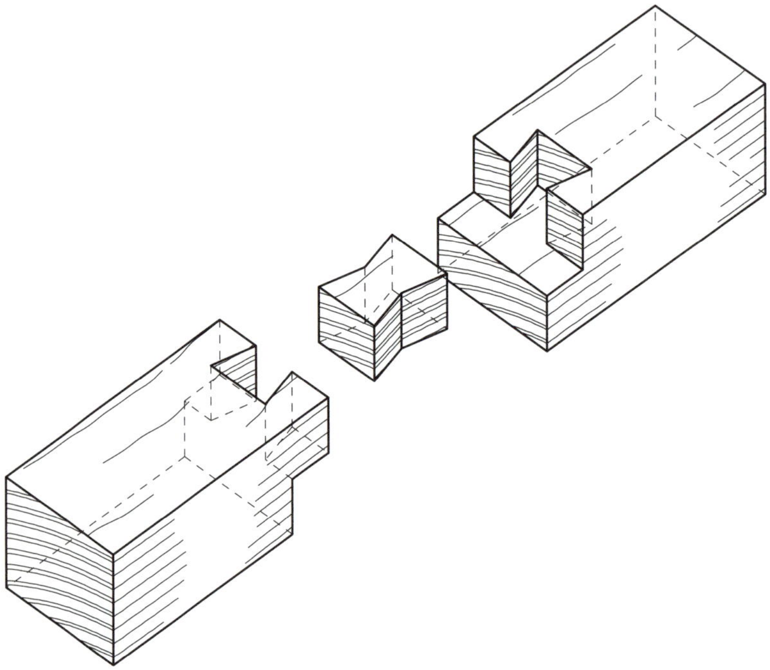

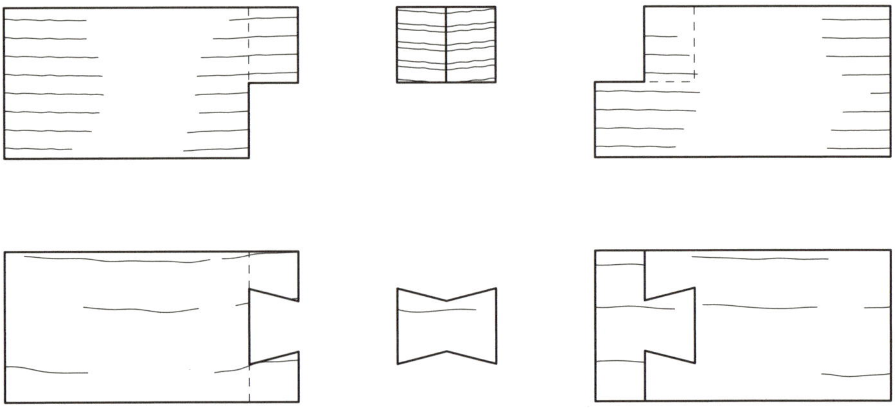

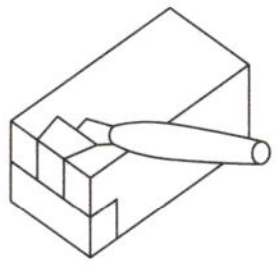
A1

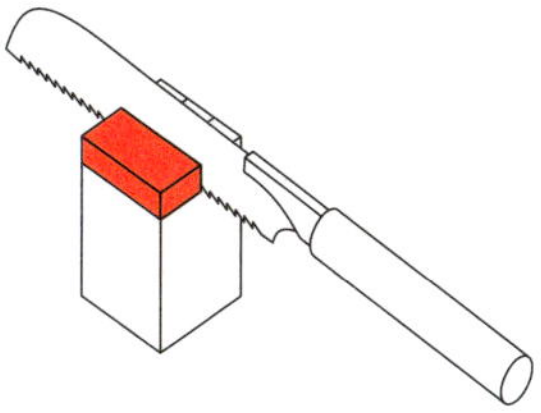
A2

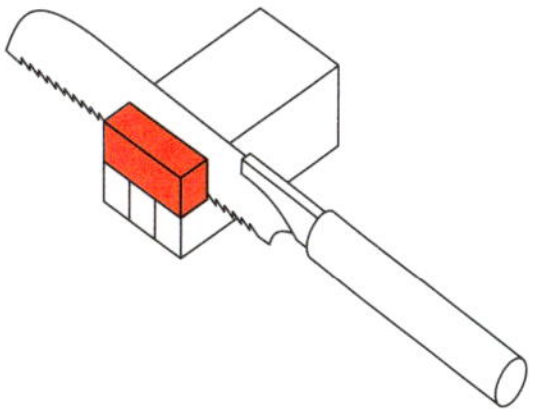
A3

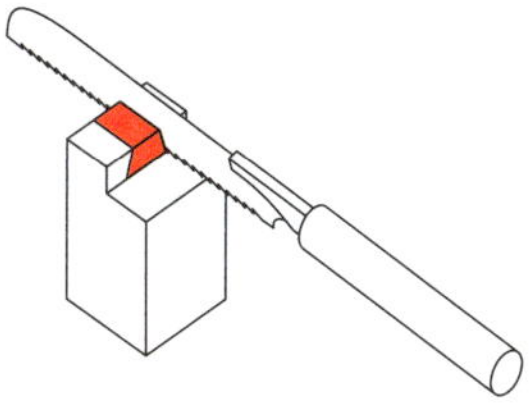
A4

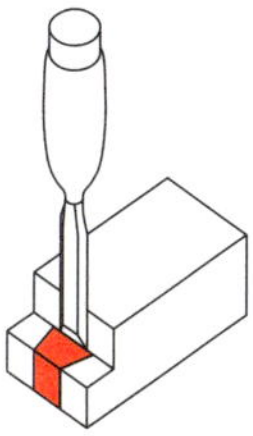
A5

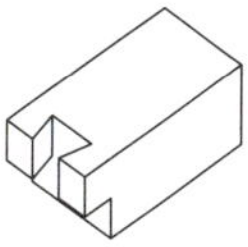
A

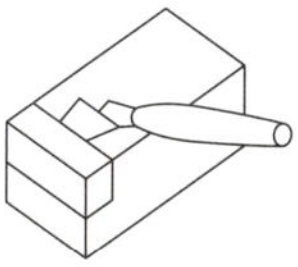
B1

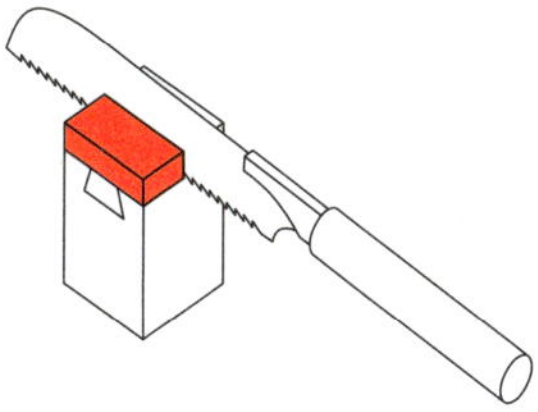
B2

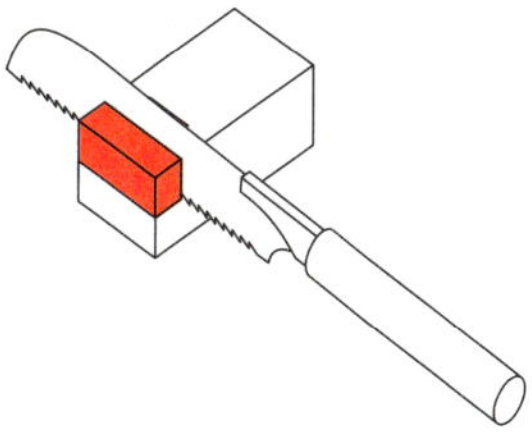
B3

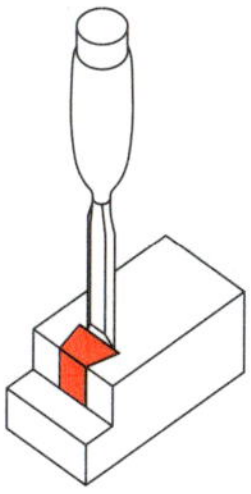
B4

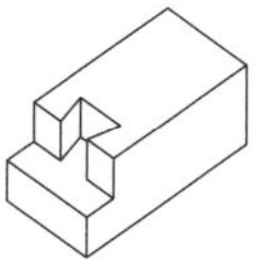
B

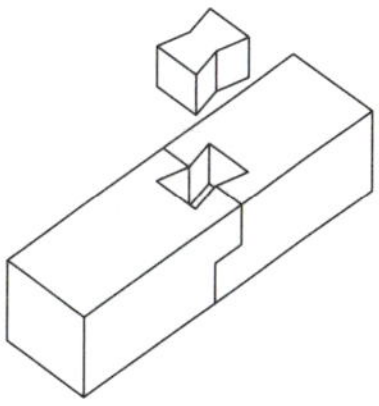
A+B+C

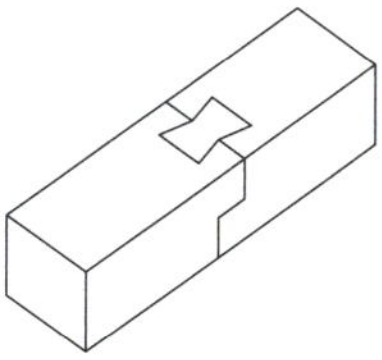
ABC

Splicing Joints

Double dovetail tenon joint

ari-mechi-tsugi
蟻目違継ぎ

doppelter Schwalbenschwanzzapfen

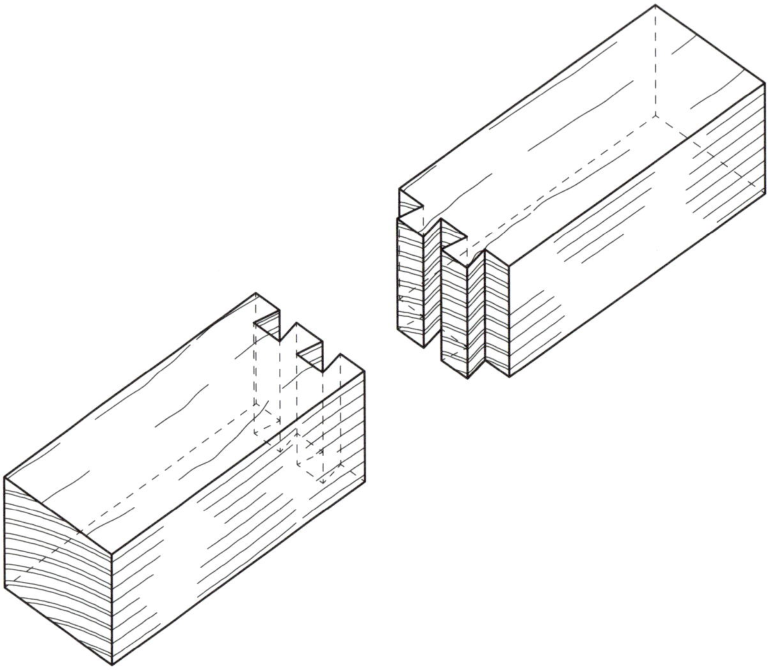

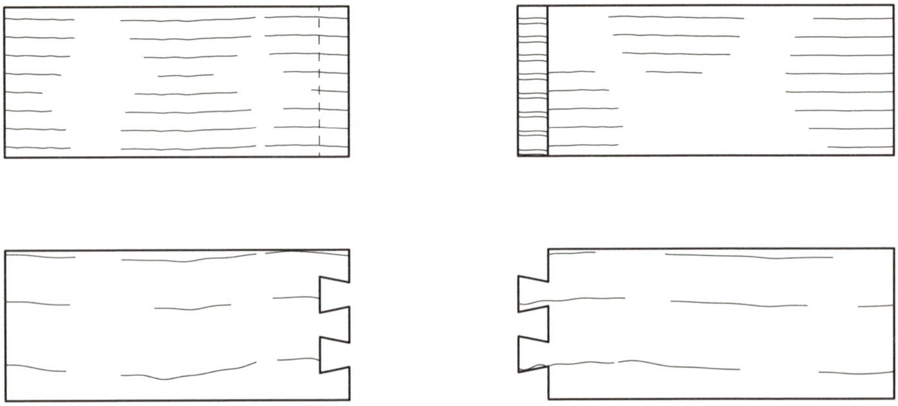

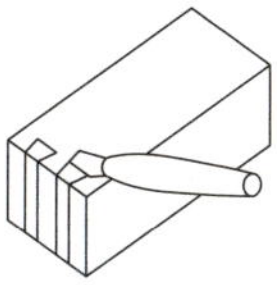
A1

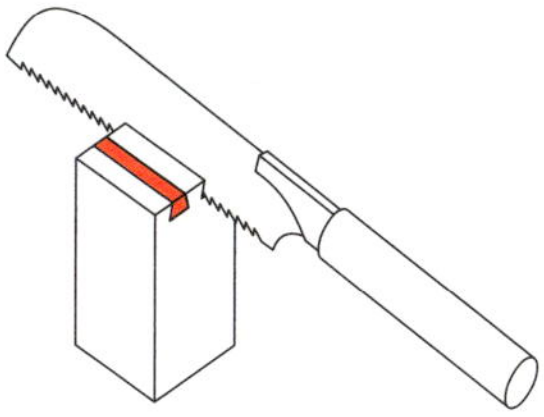
A2

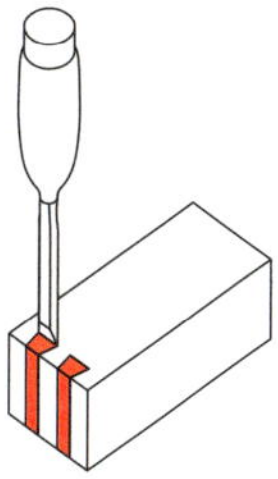
A3

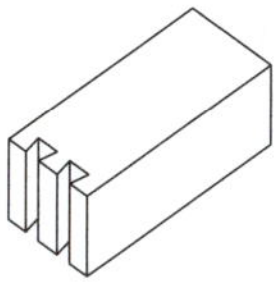
A

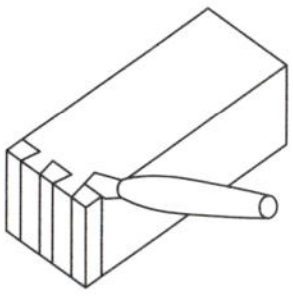
B1

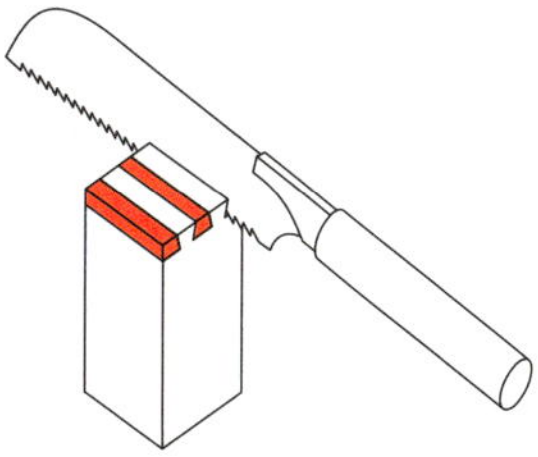
B2

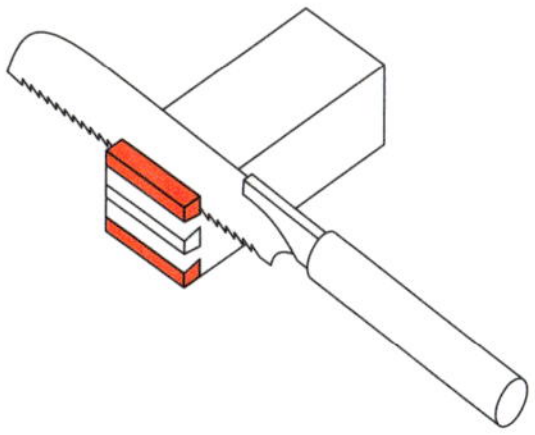
B3

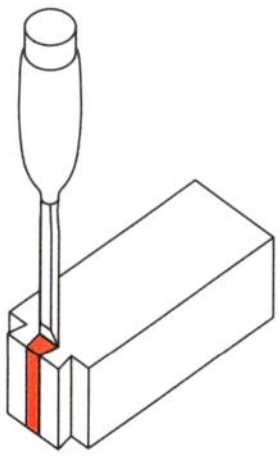
B4

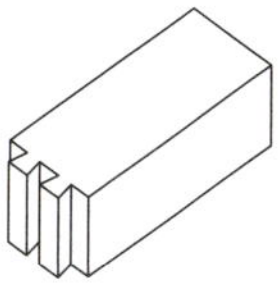
B

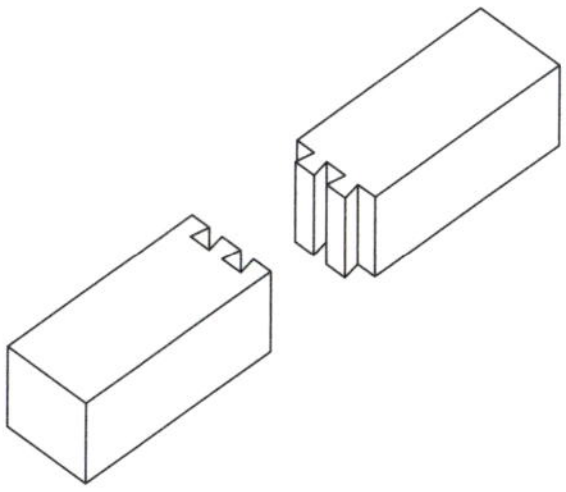
A+B

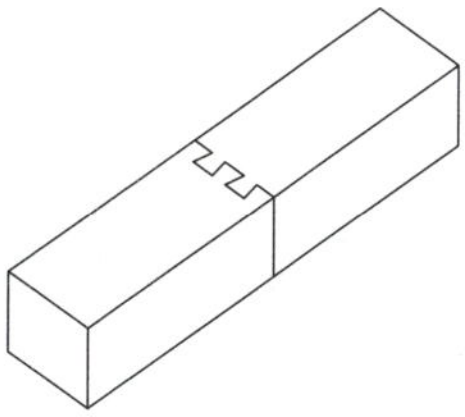
AB

Splicing Joints

Four-faced dovetail joint

shihō-ari-tsugi [*]
四方蟻継ぎ

diagonaler doppelter Schwalbenschwanzzapfen

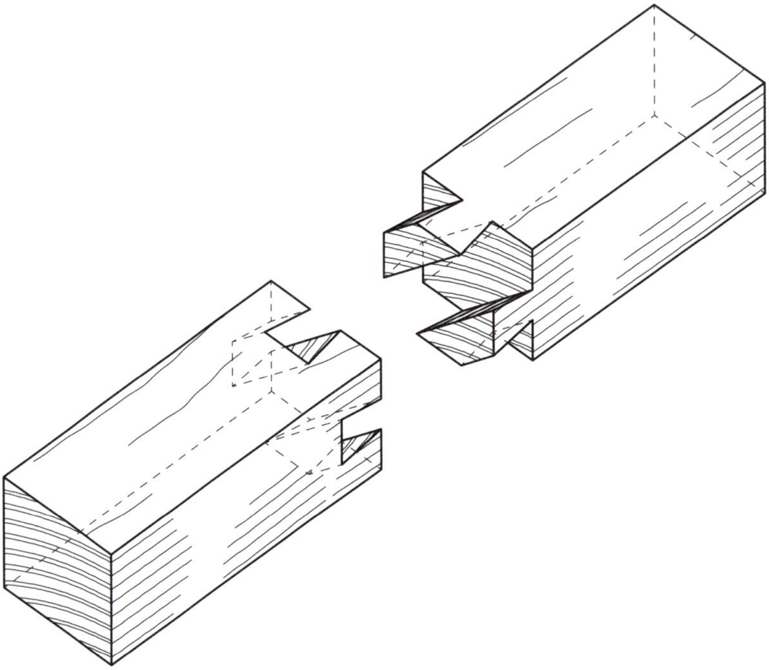

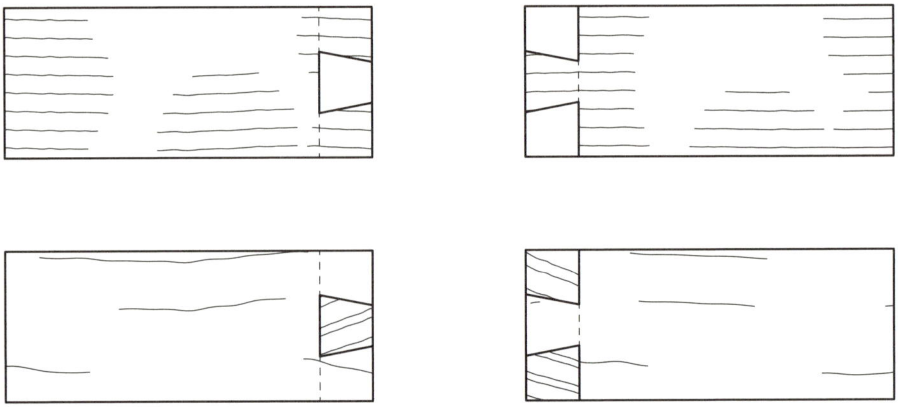

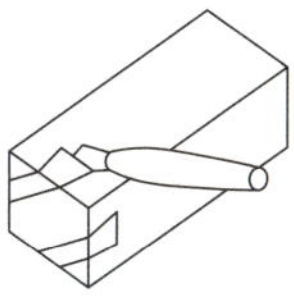

A1

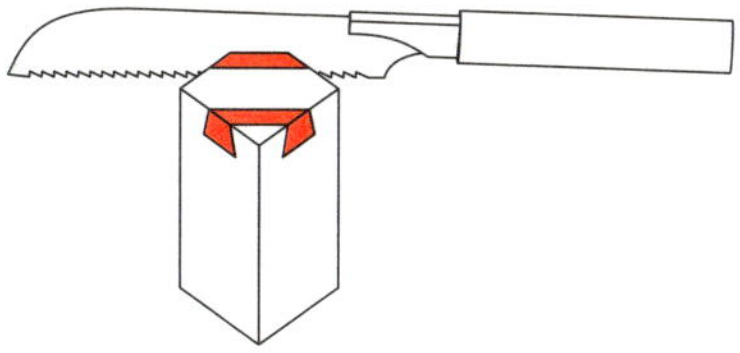

A2

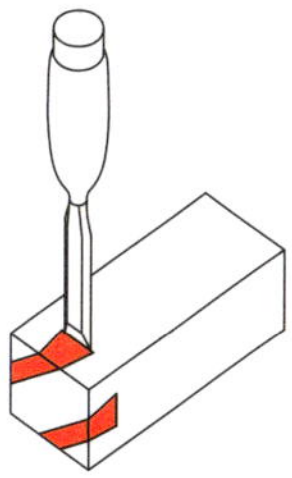

A3

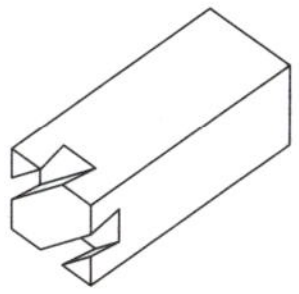

A

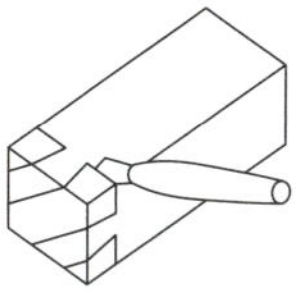

B1

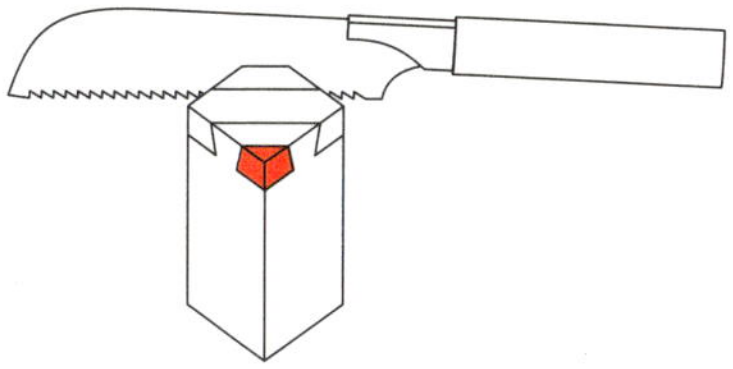

B2

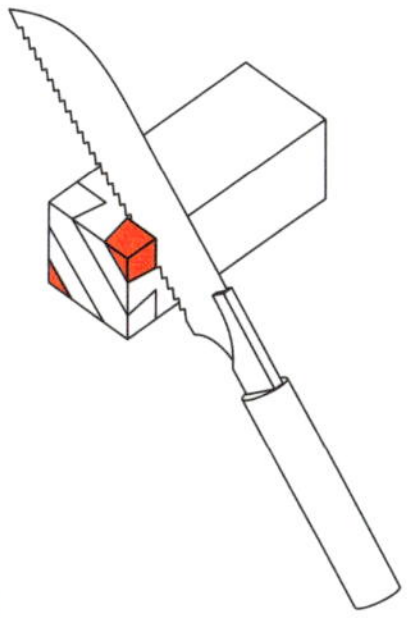

B3

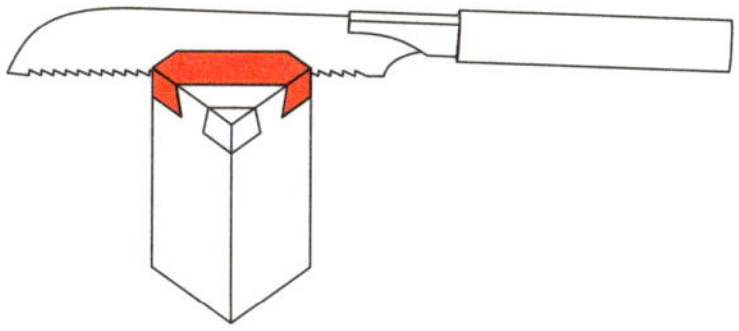

B4

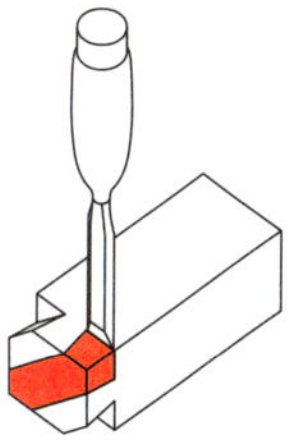

B5

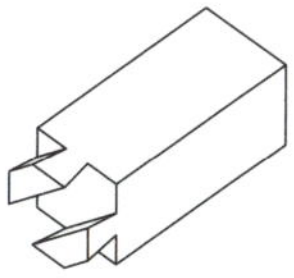

B

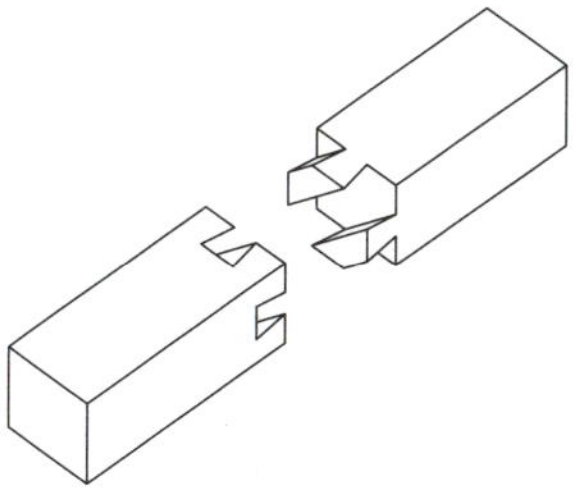

A+B

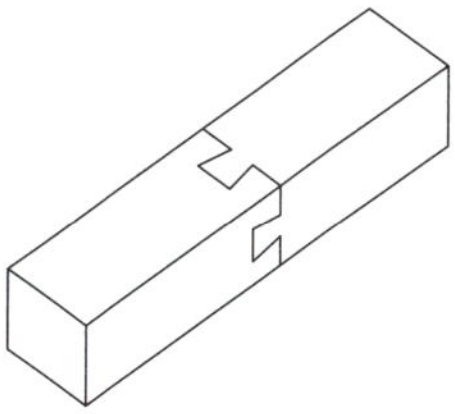

AB

Splicing Joints

Tapered dovetail tenon joint with mitered shoulders

basara-tsugi
婆娑羅継ぎ

Stoß mit Gratschnitt und keilförmigen Schwalbenschwanzzapfen

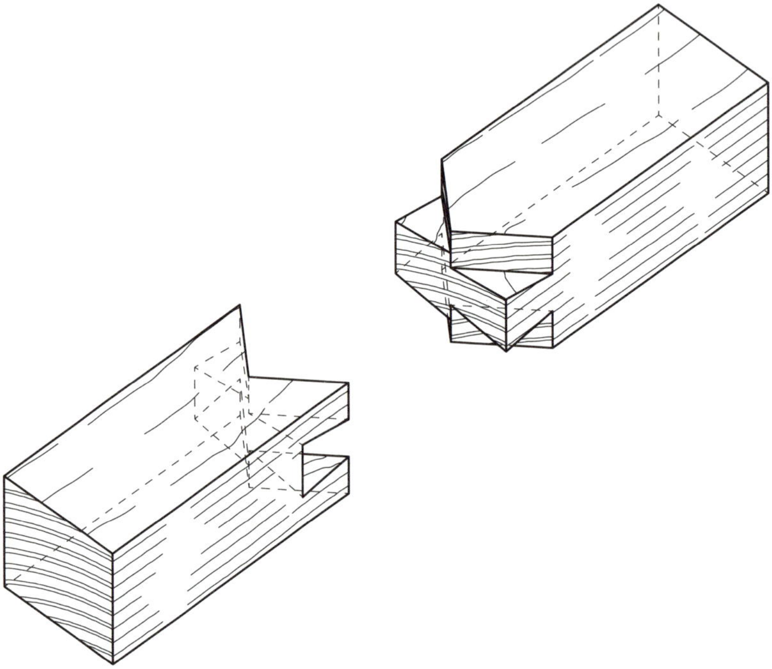

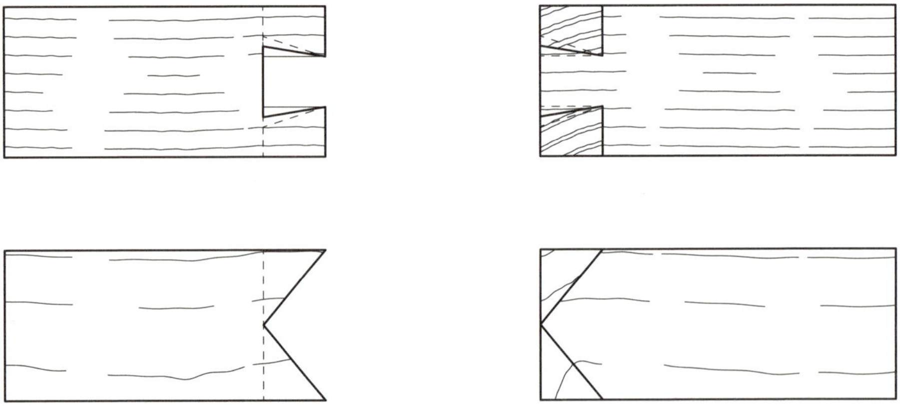

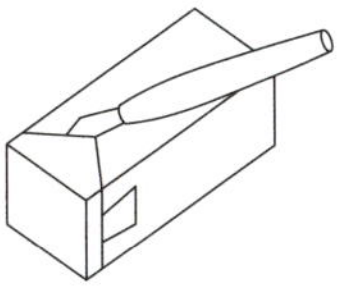
A1

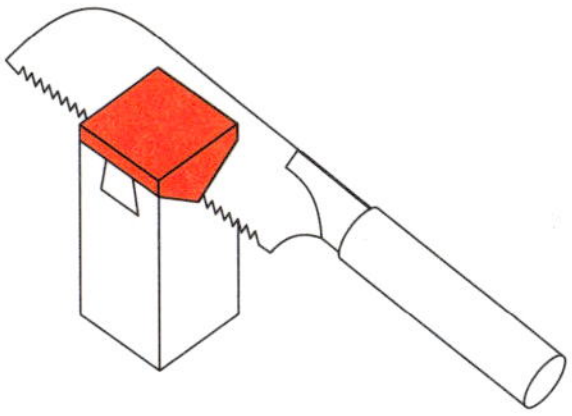
A2

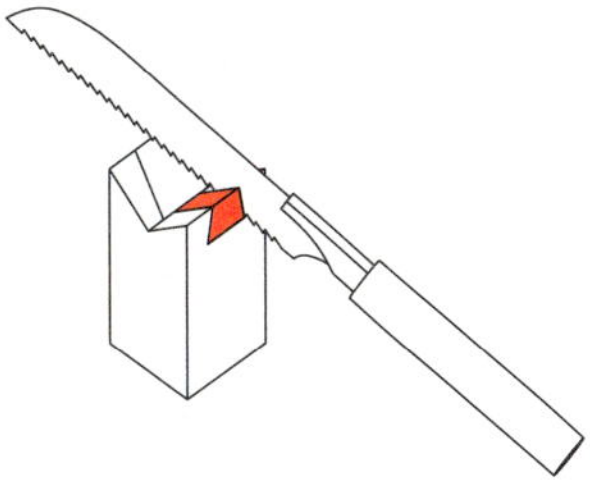
A3

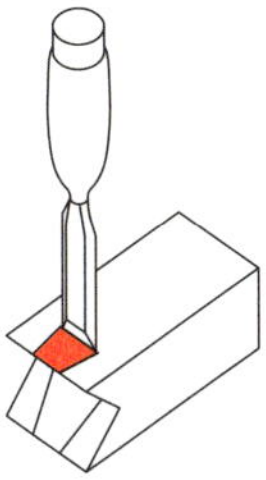
A4

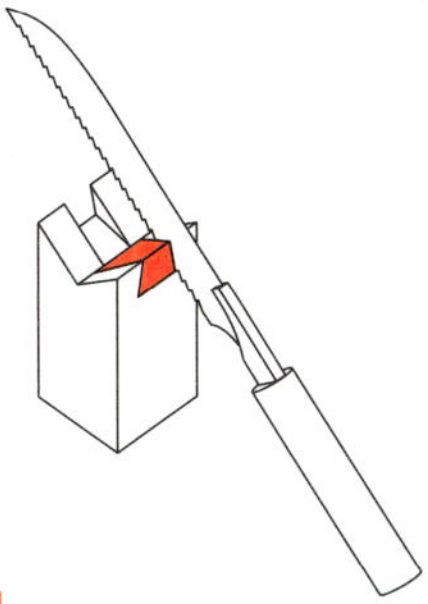
A5

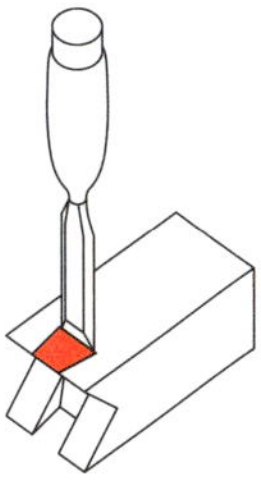
A6

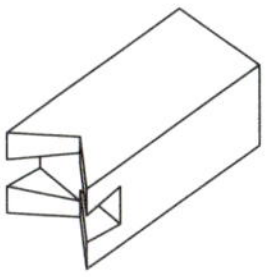
A

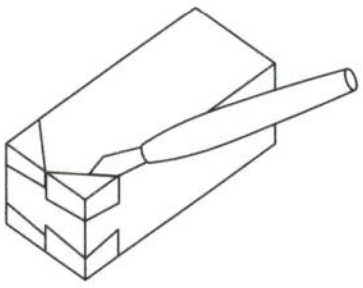
B1

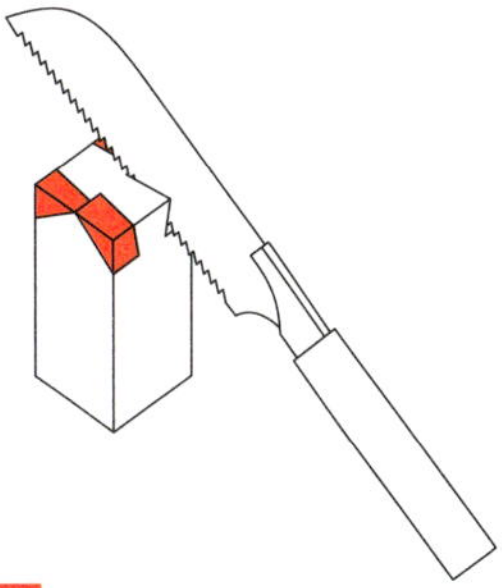
B2

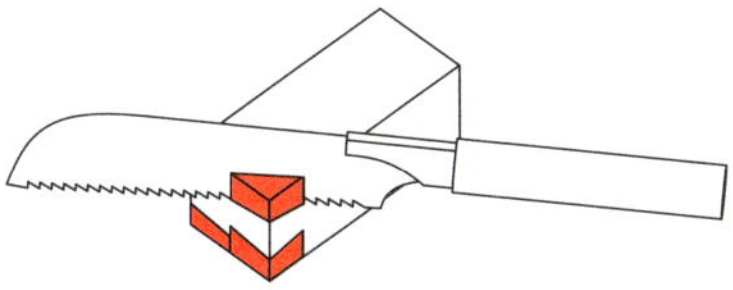
B3

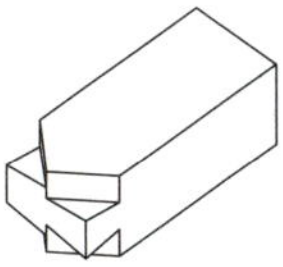
B

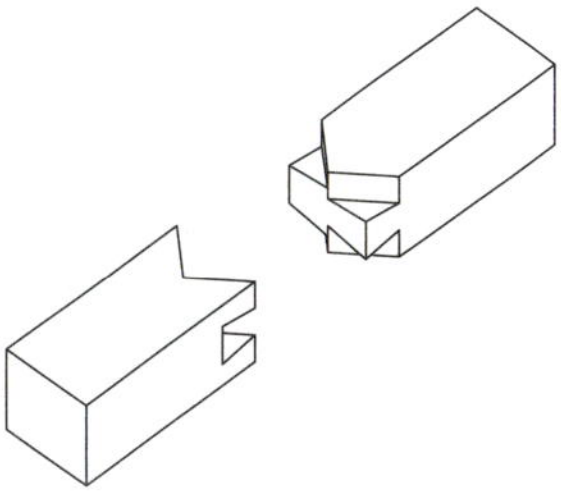
A+B

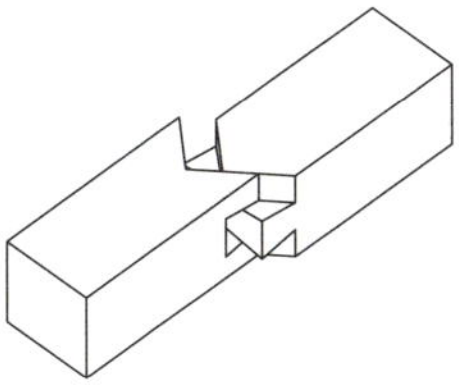
A+B

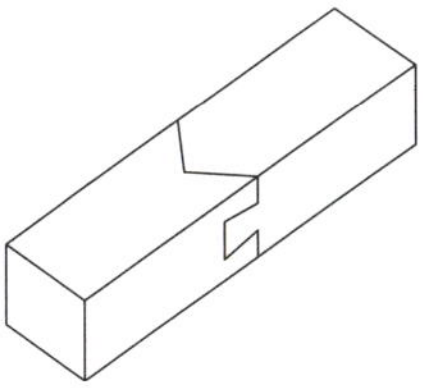
AB

Splicing Joints

Kawai joint

kawai-tsugite
河合継手

diagonal angeschnittener Stoß mit dreifacher polyedrischer Ausstemmung

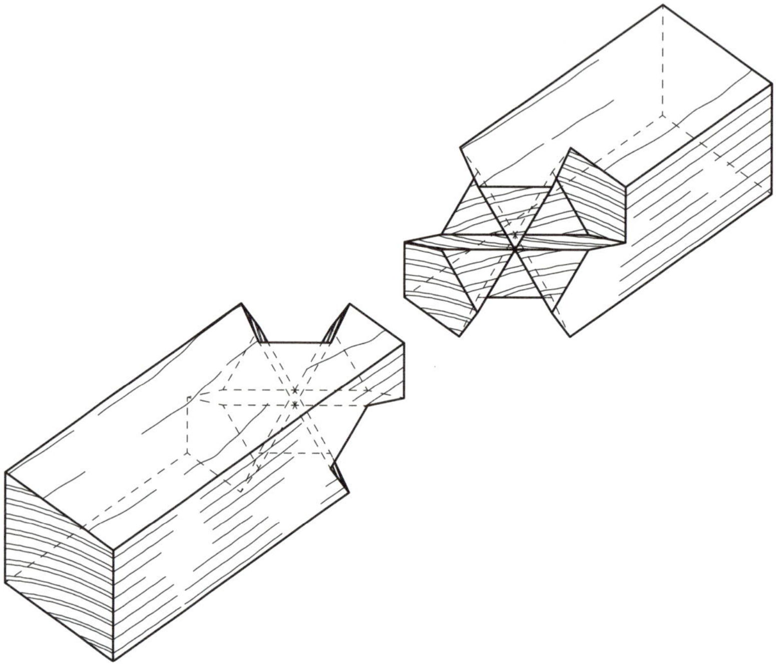

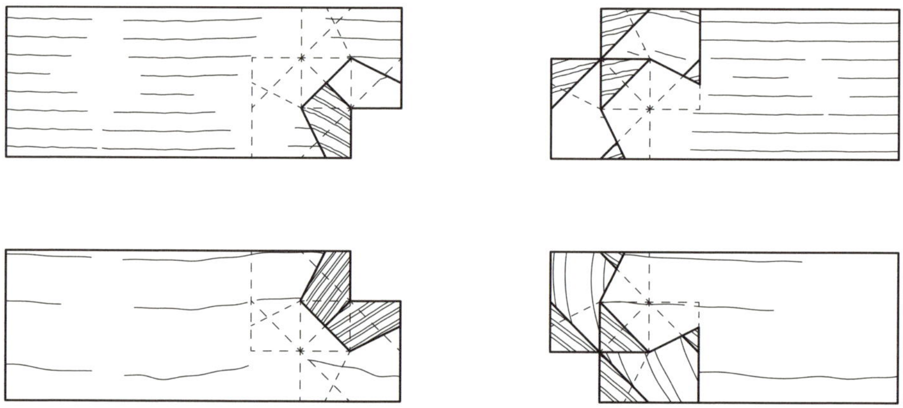

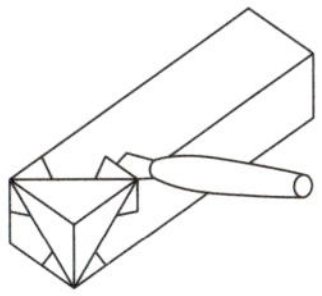

A1

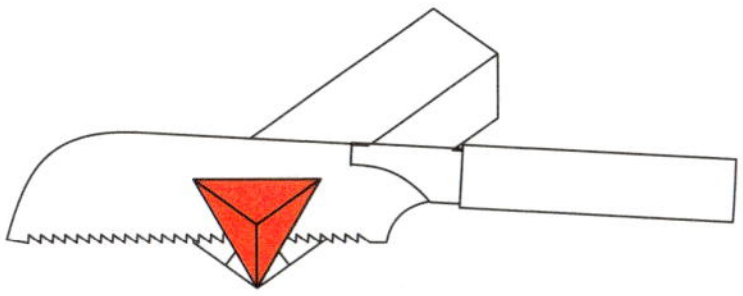

A2

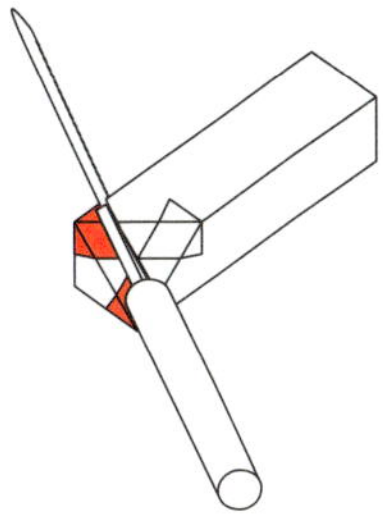

A3

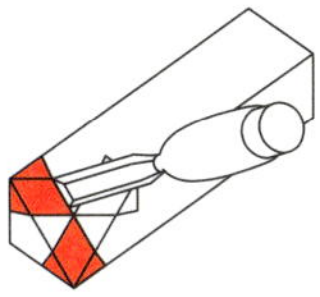

A4

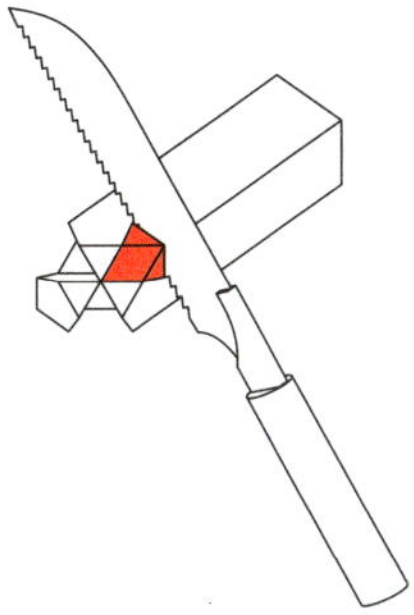

A5

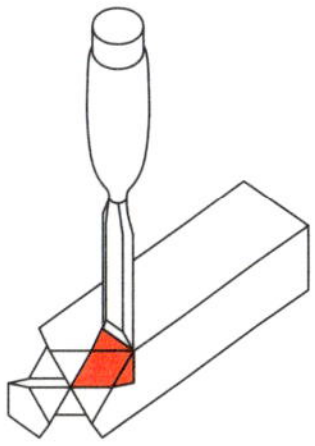

A6

2×

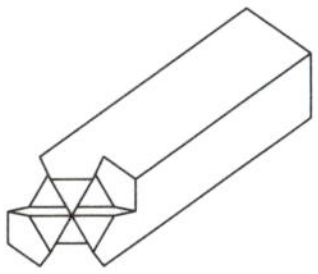

A

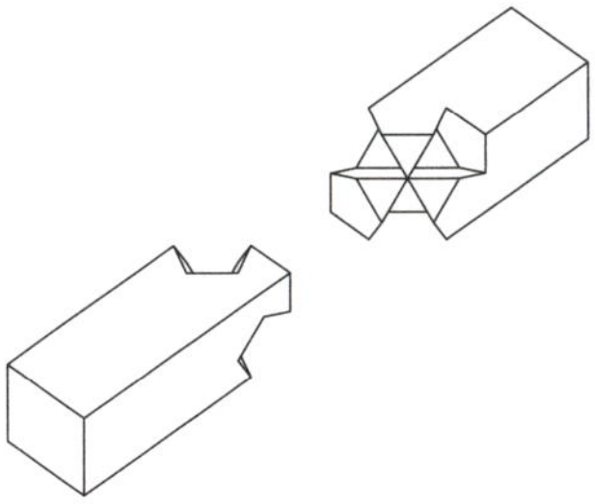

A+A

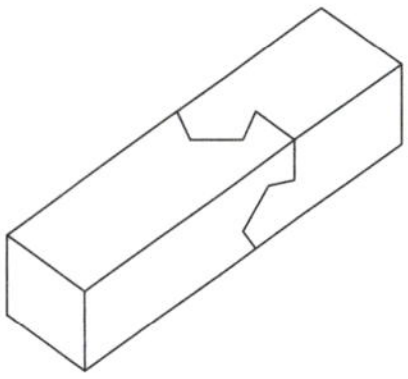

AA

Gooseneck tenon joint

shiki-men-kama-tsugi [*]
敷面鎌継ぎ

Sichelzapfen

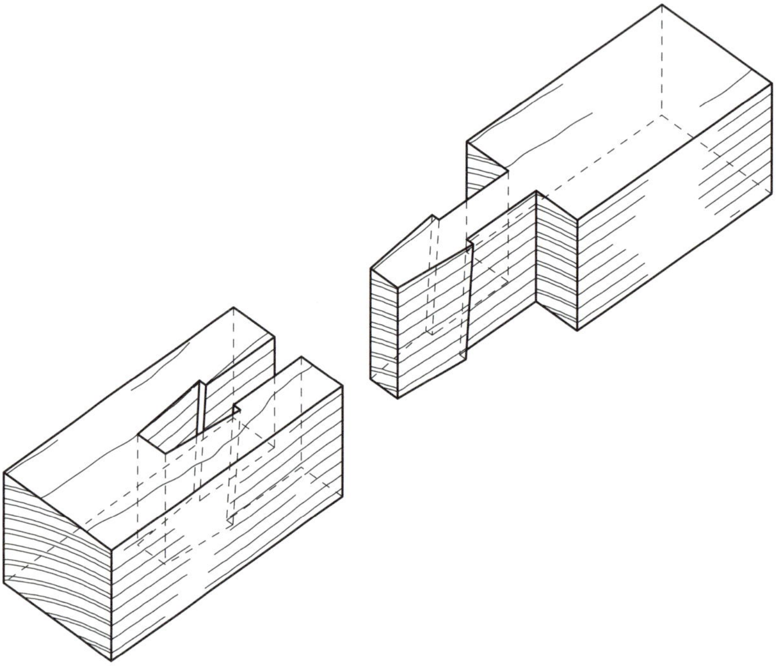

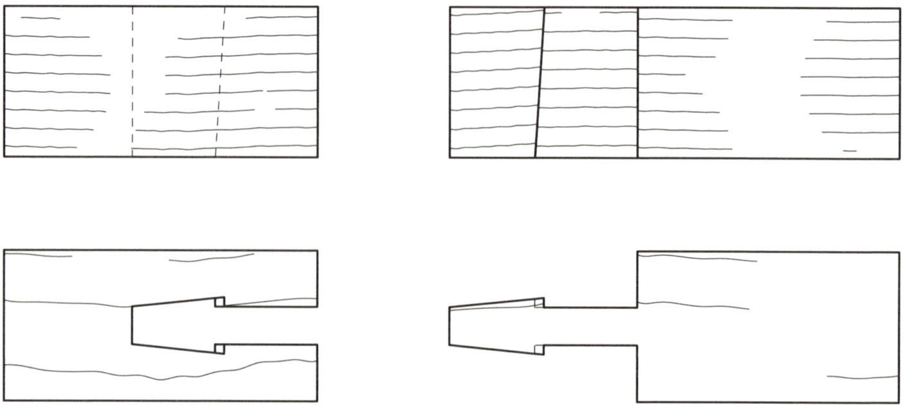

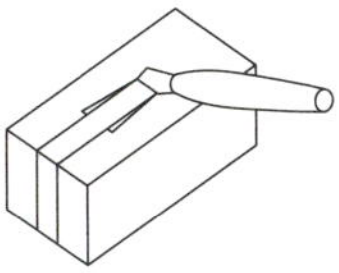

A1

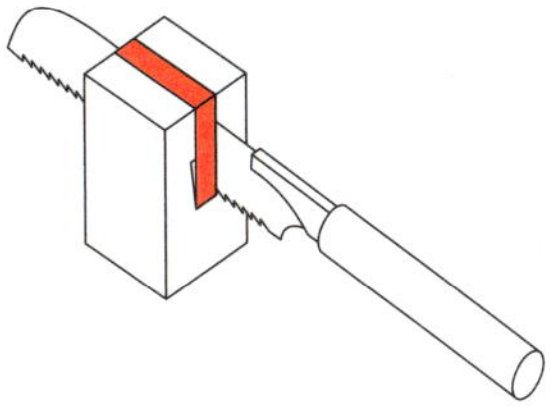

A2

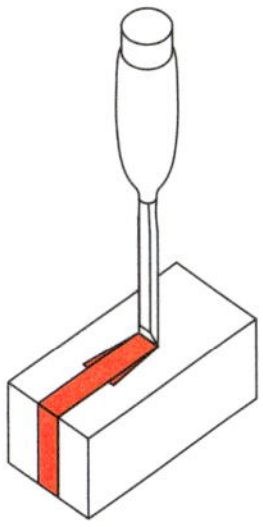

A3

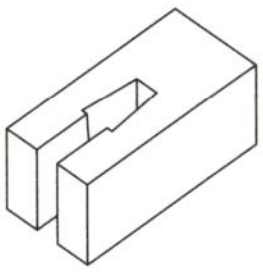

A

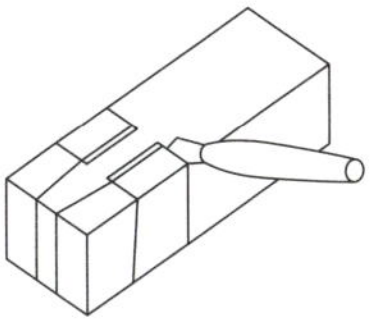

B1

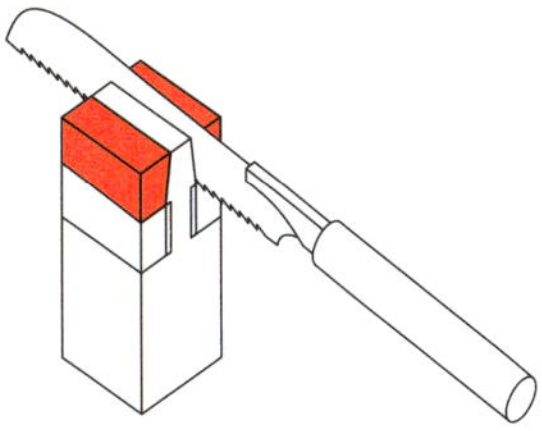

B2

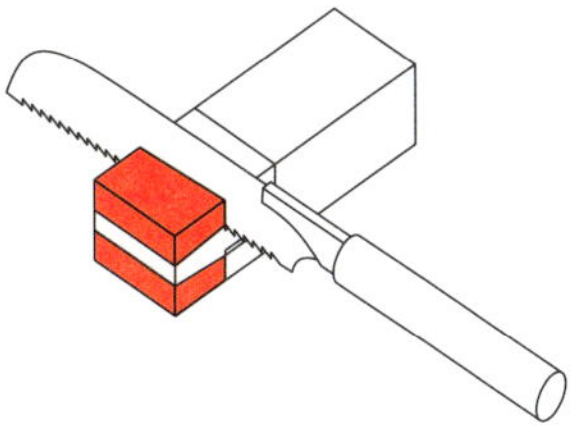

B3

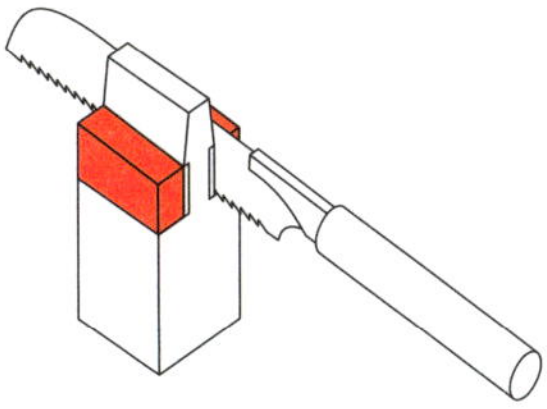

B4

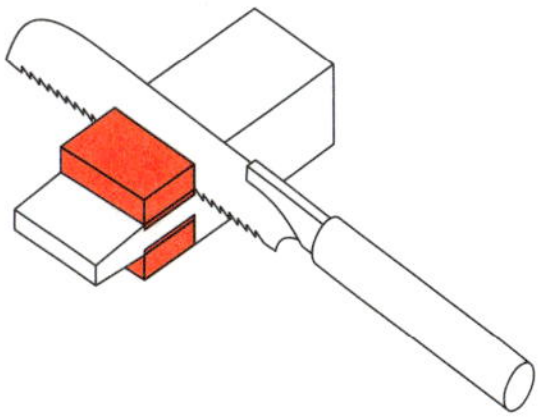

B5

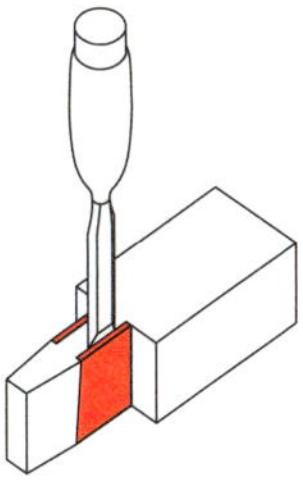

B6

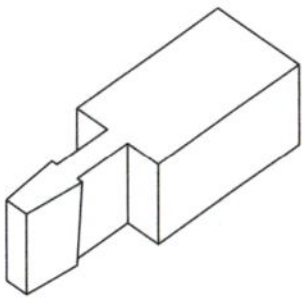

B

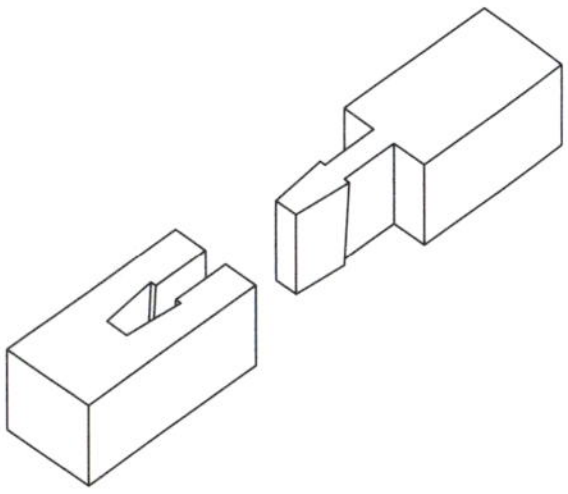

A+B

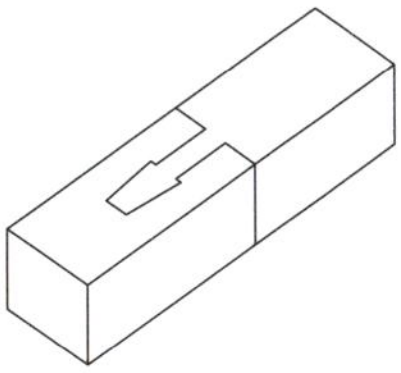

AB

Splicing Joints

Triple-shouldered gooseneck tenon joint

kama-tsugi [*]
鎌継ぎ

abgesetzter Sichelzapfen

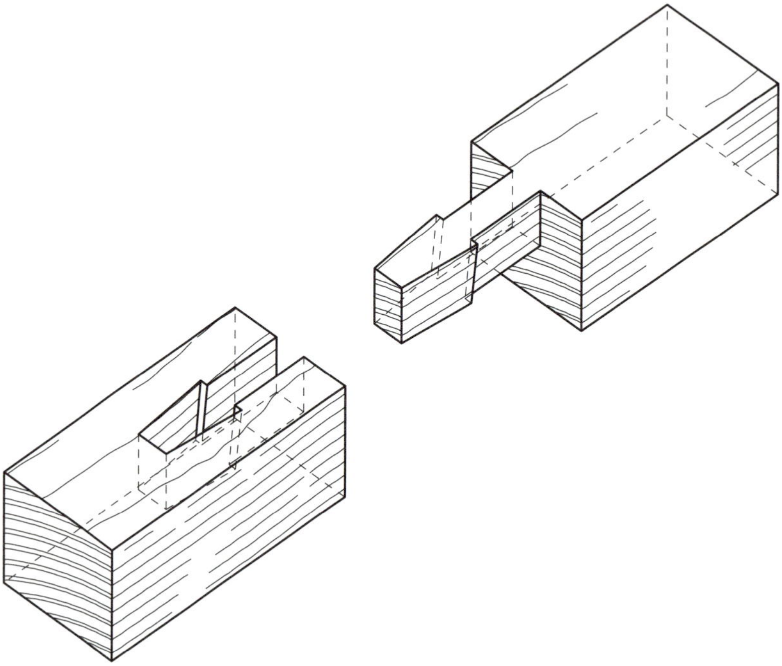

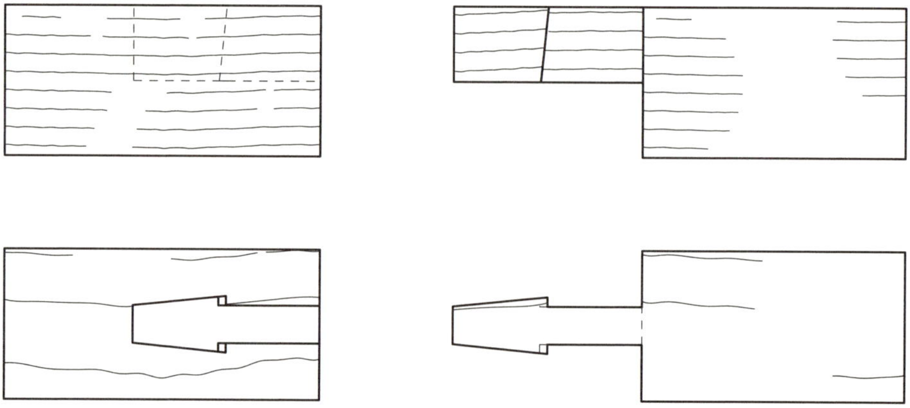

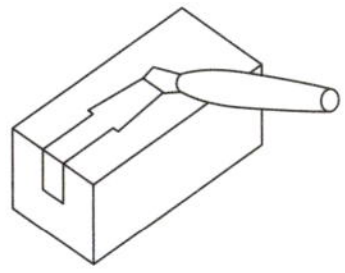
A1

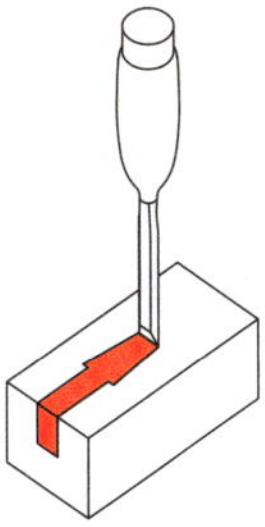
A2

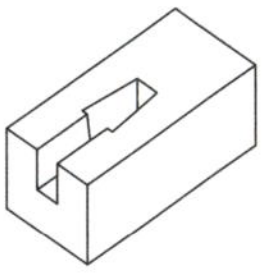
A

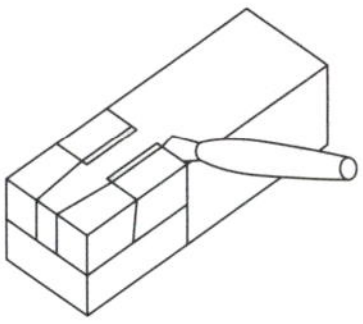
B1

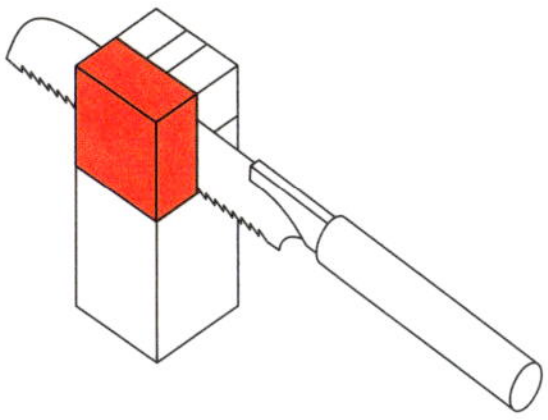
B2

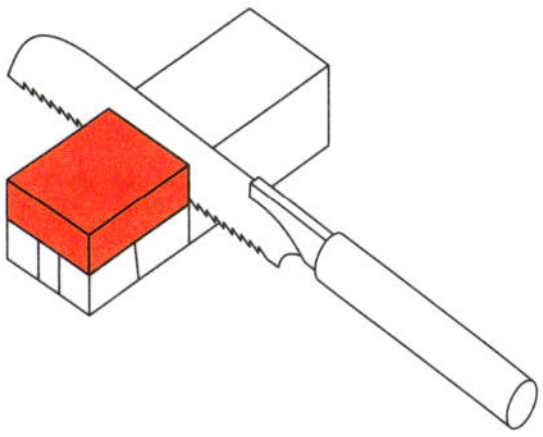
B3

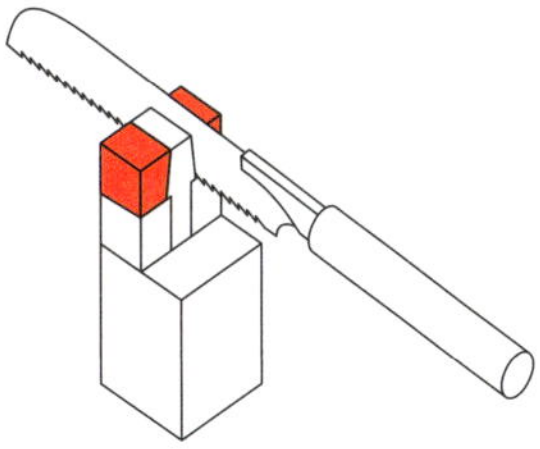
B4

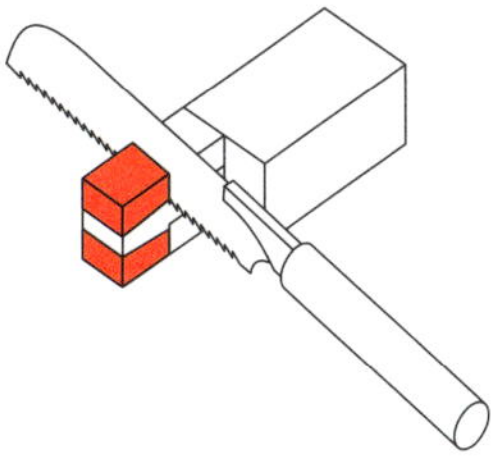
B5

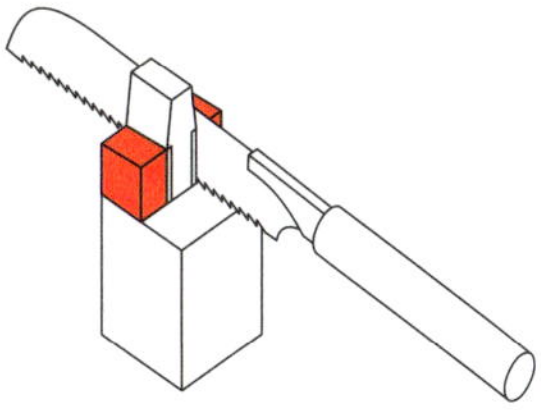
B6

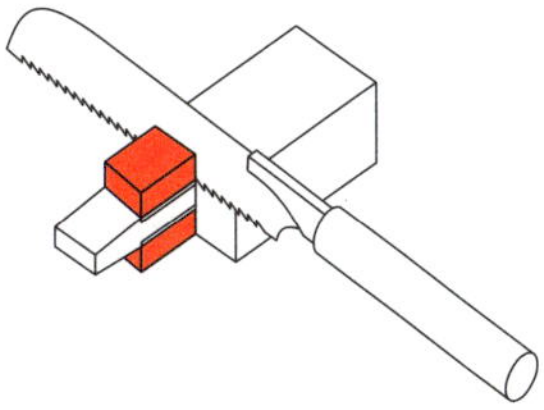
B7

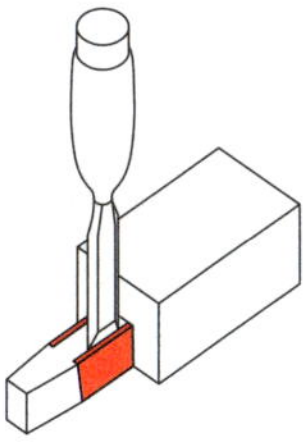
B8

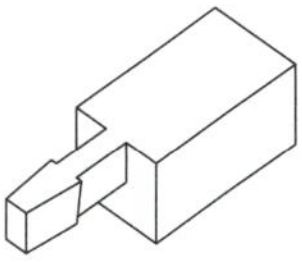
B

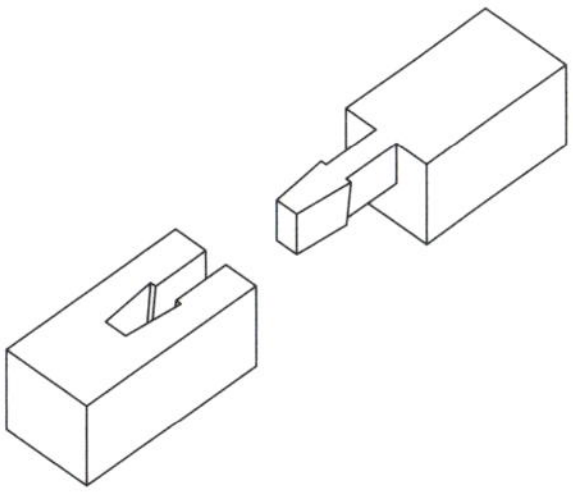
A+B

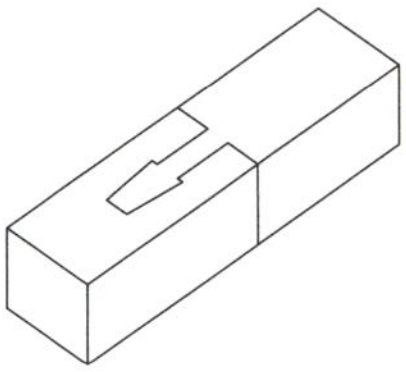
AB

Splicing Joints

Lapped gooseneck joint

koshi-kake-kama-tsugi [*]
腰掛鎌継ぎ

Sichelzapfen mit geradem Blatt

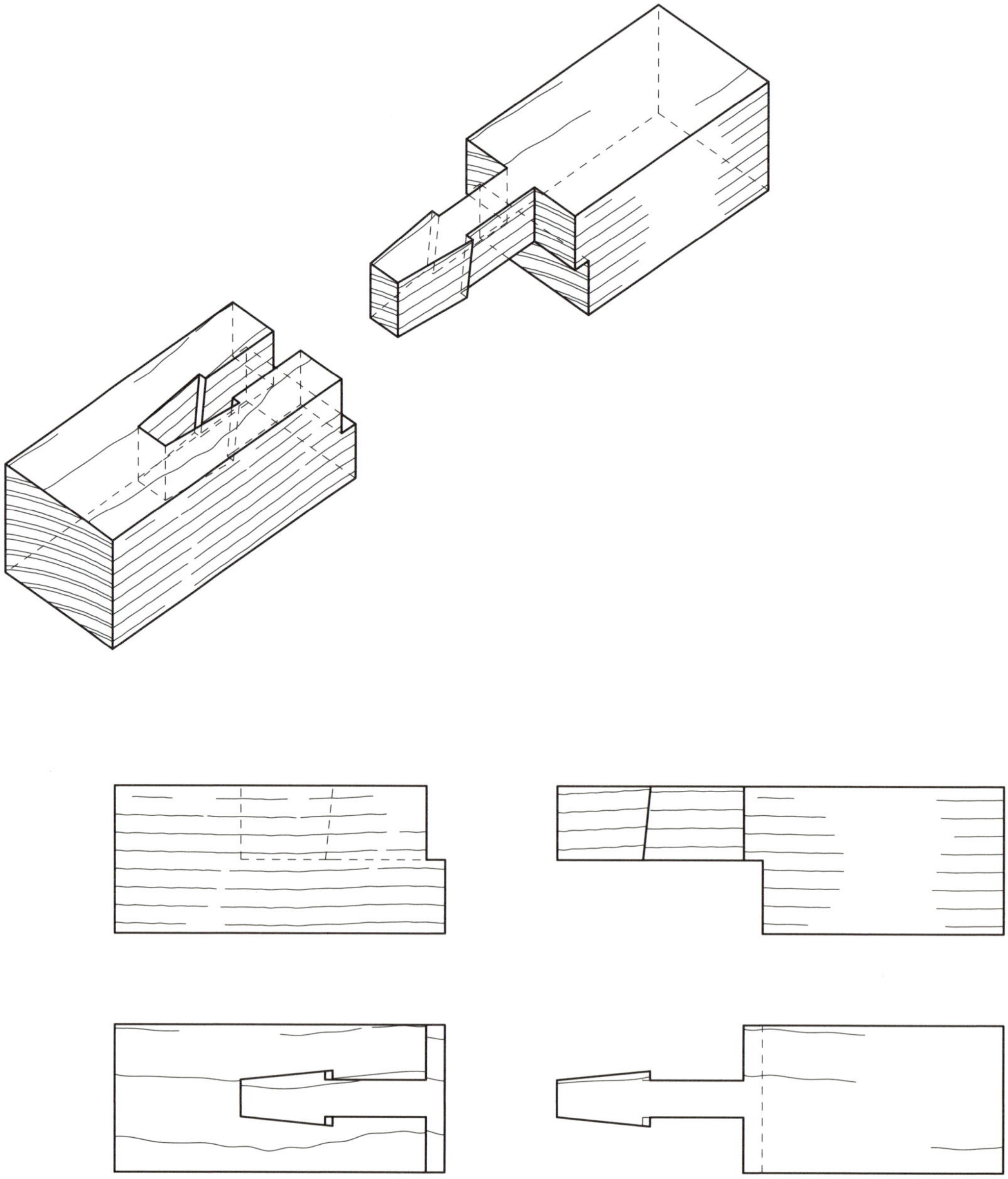

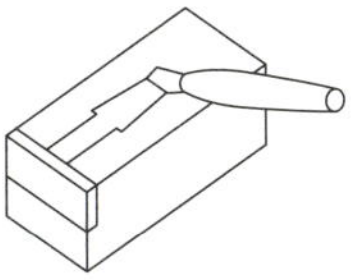

A1

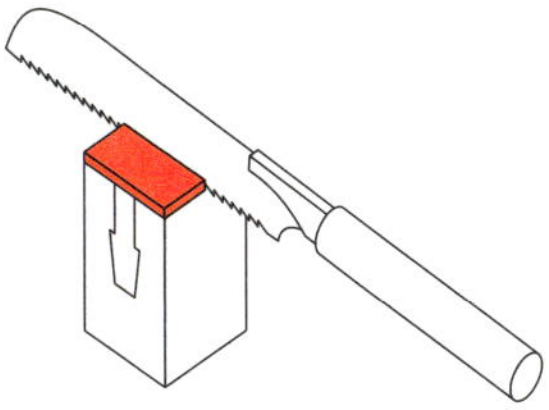

A2

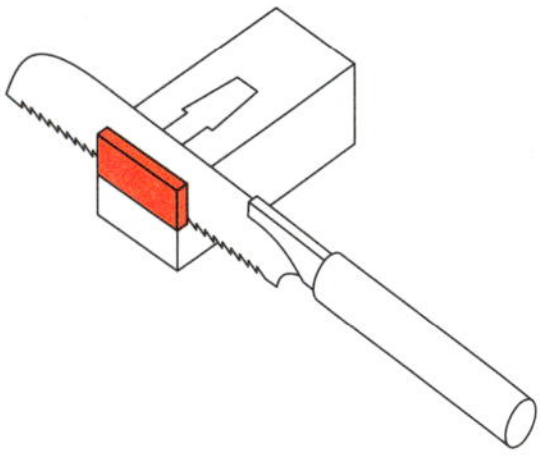

A3

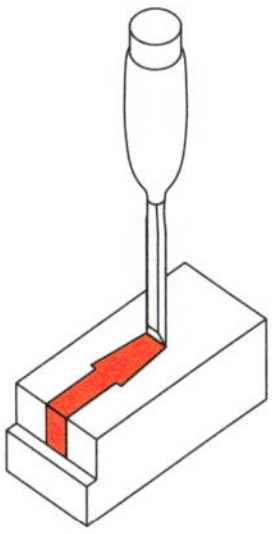

A4

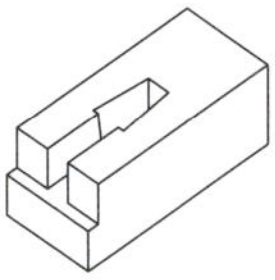

A

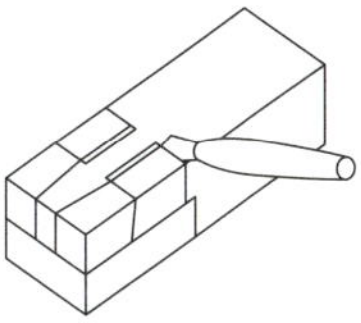

B1

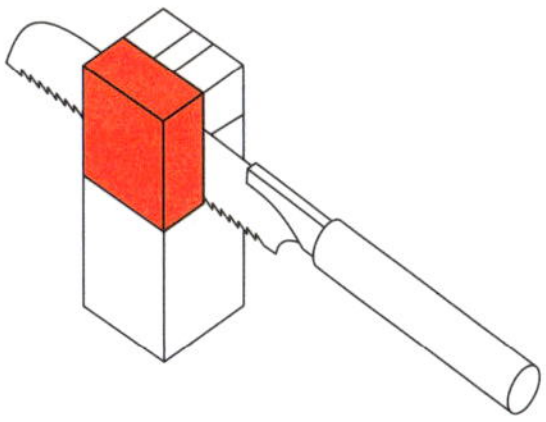

B2

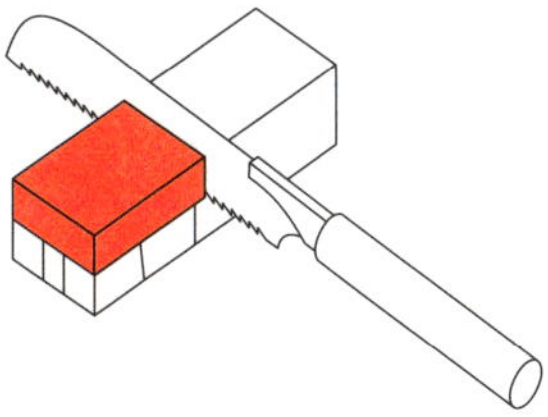

B3

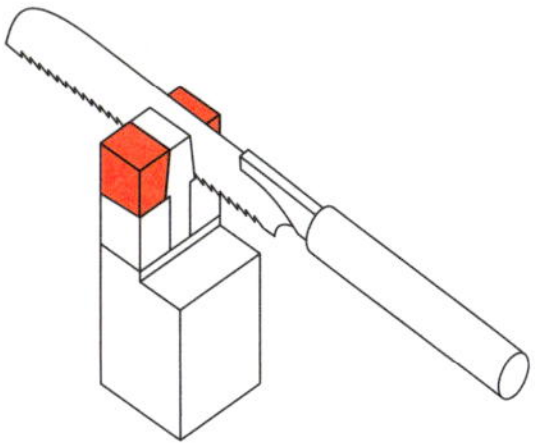

B4

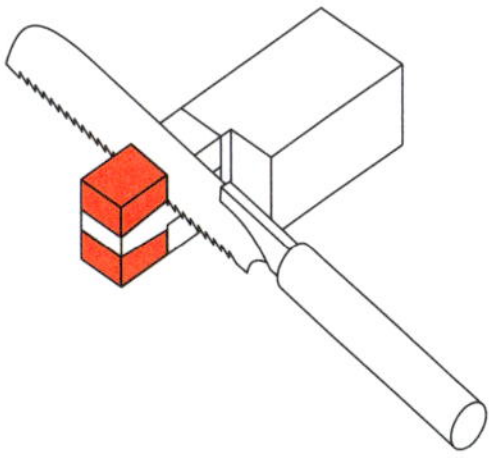

B5

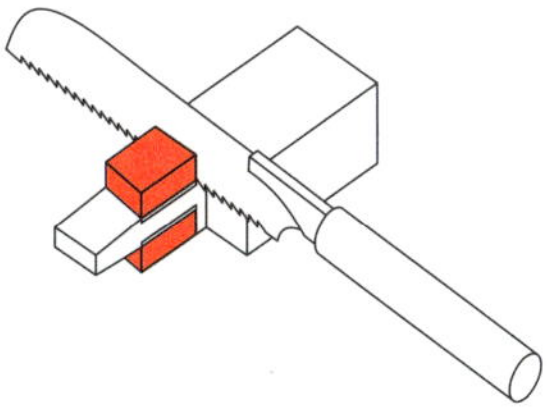

B6

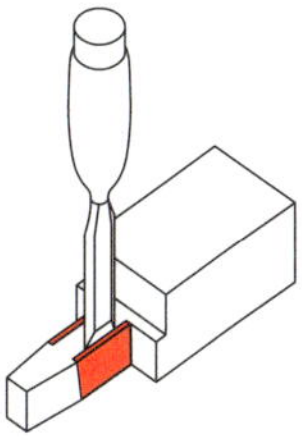

B7

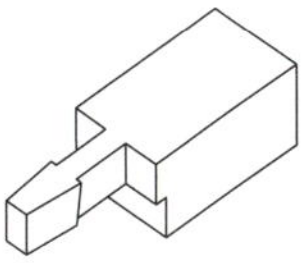

B

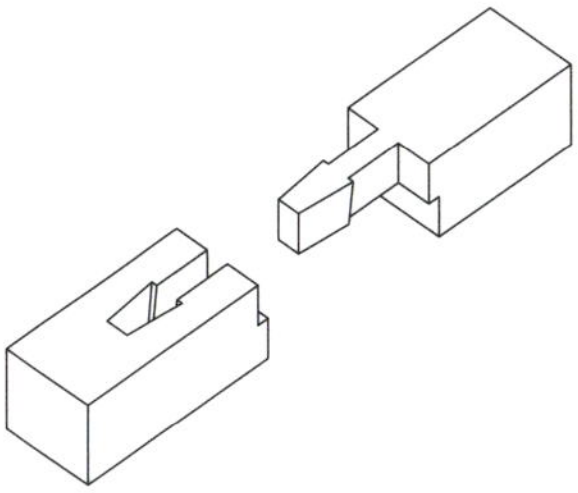

A+B

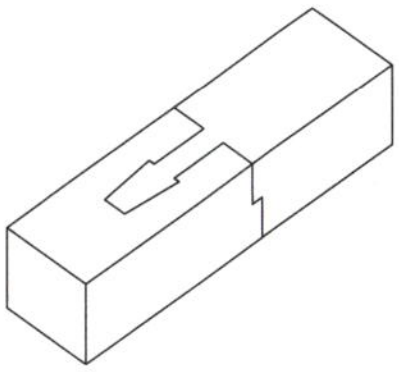

AB

Splicing Joints

Lapped gooseneck joint with stub tenon

koshi-kake-koshi-ire-kama-tsugi
腰掛腰入鎌継ぎ

Sichelzapfen mit geradem Blatt und Brüstungszapfen

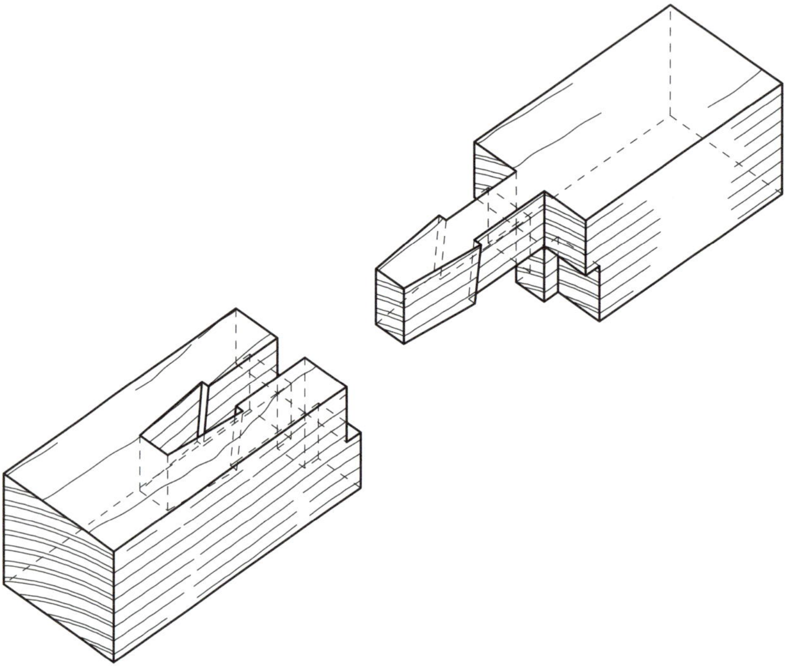

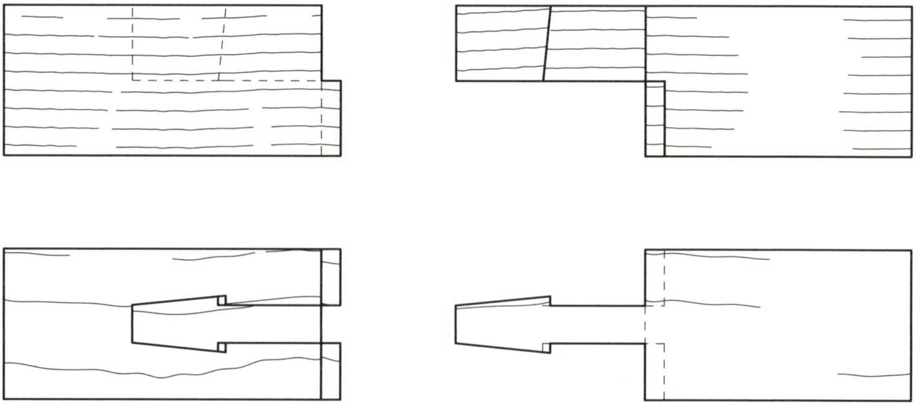

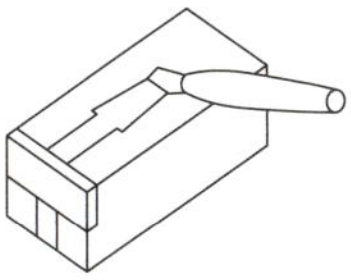

A1

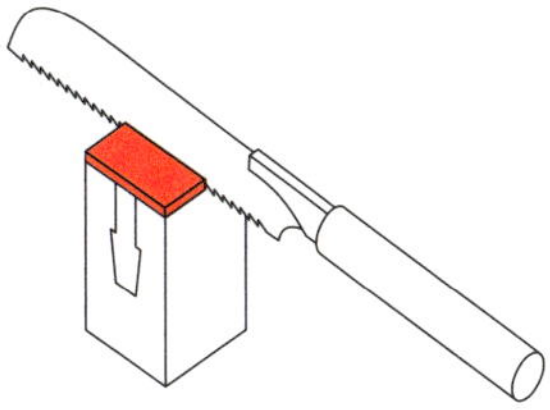

A2

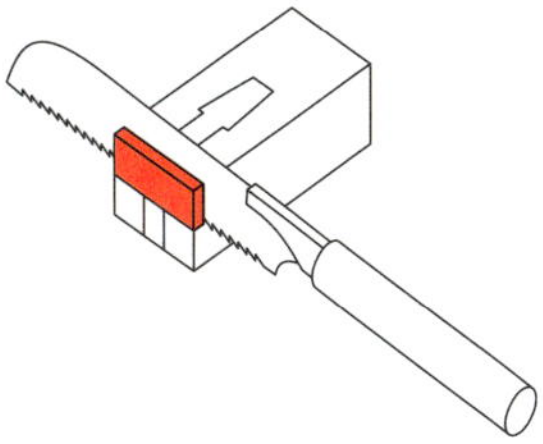

A3

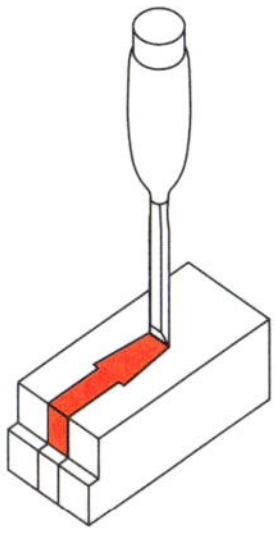

A4

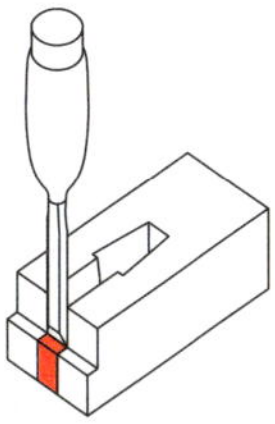

A5

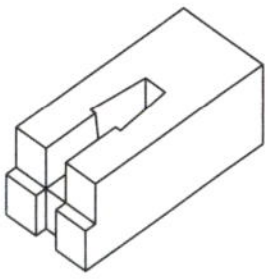

A

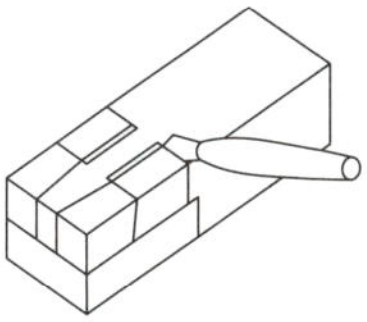

B1

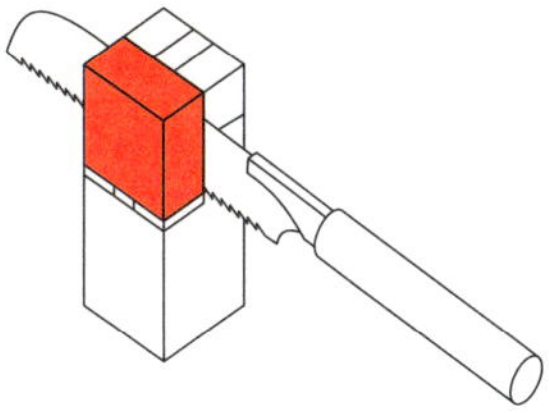

B2

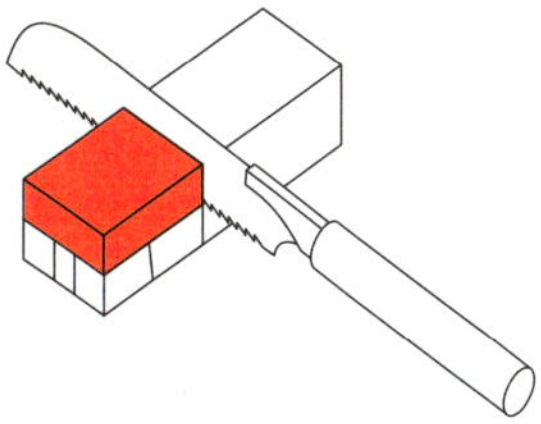

B3

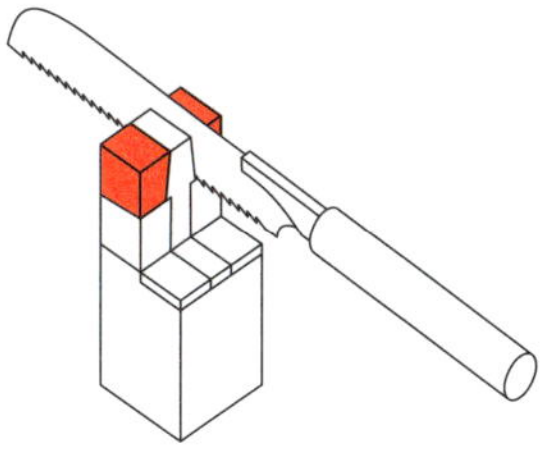

B4

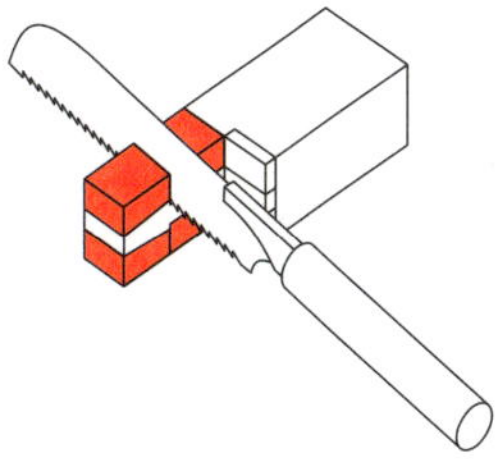

B5

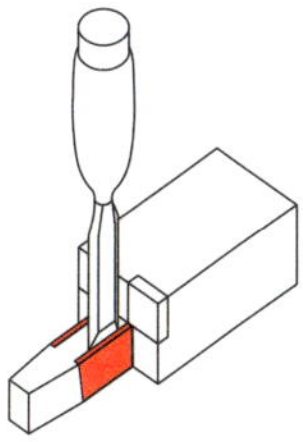

B6

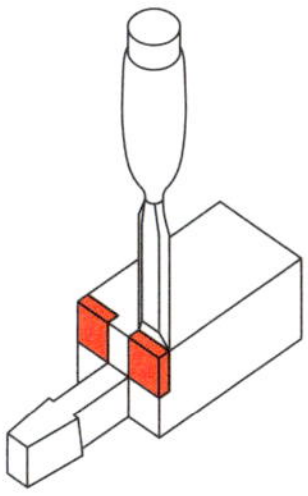

B7

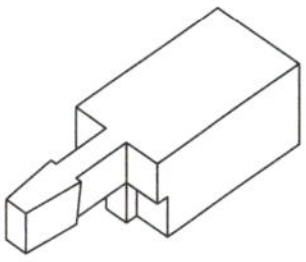

B

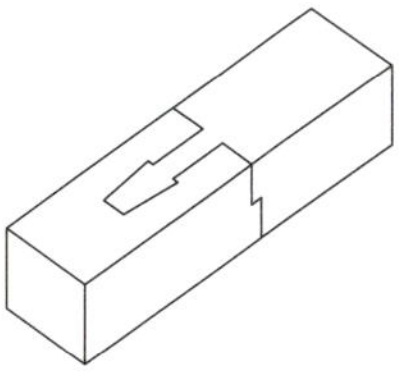

AB

Splicing Joints

Haunched gooseneck tenon with stub tenons

koshi-kake-koshi-ire-kama-tsugi (ryō-mechi-tsuki)
腰掛腰入鎌継ぎ(両目違付)

Sichelstufenzapfen mit doppeltem Brüstungszapfen

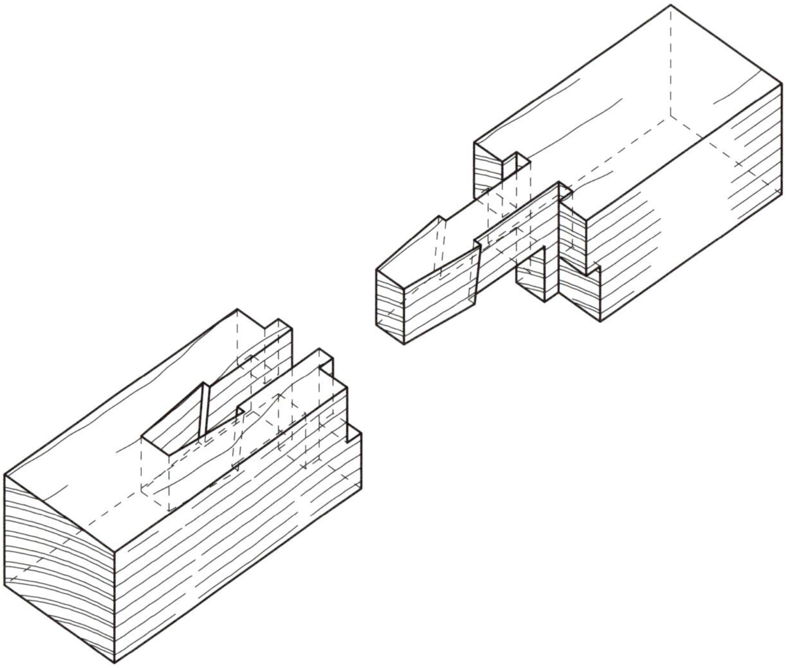

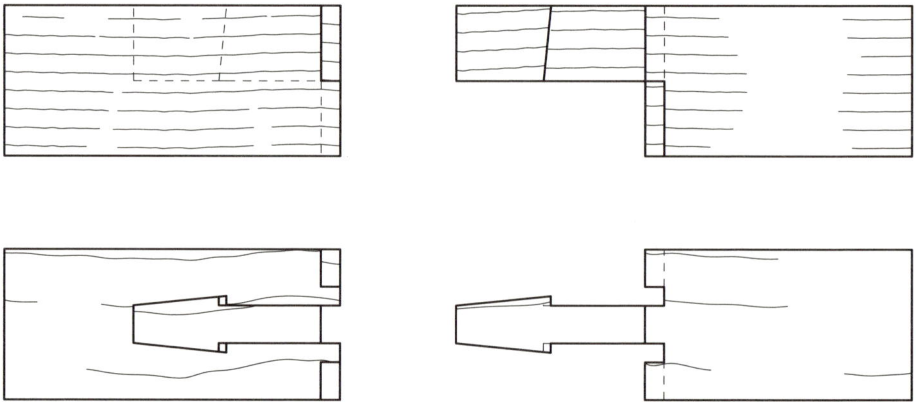

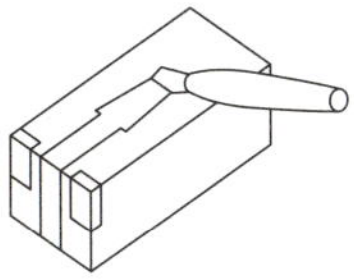
A1

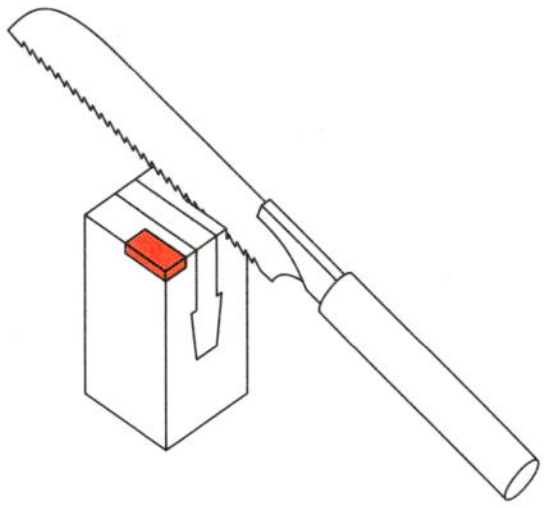
A2

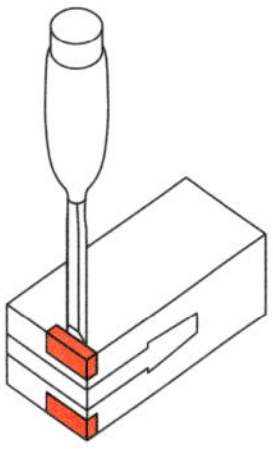
A3

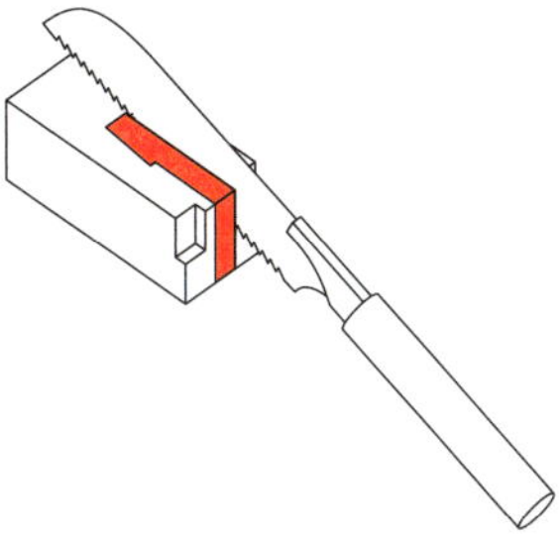
A4

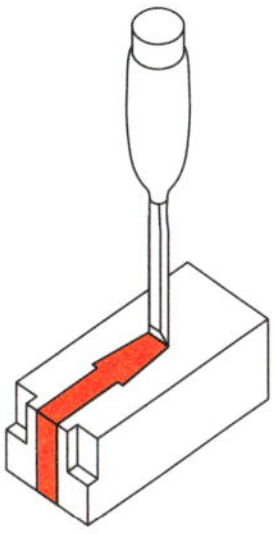
A5

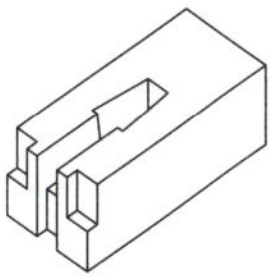
A

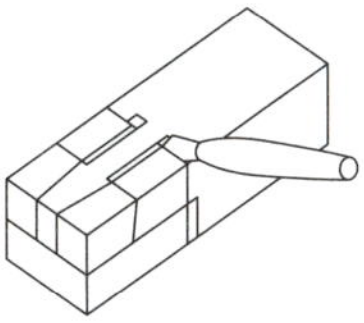
B1

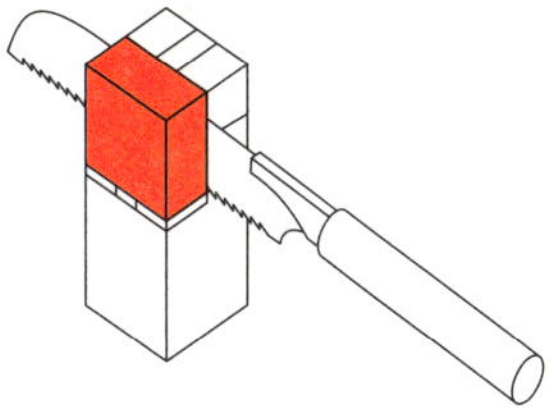
B2

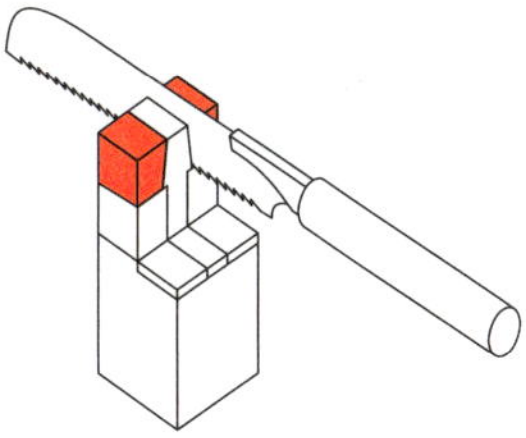
B3

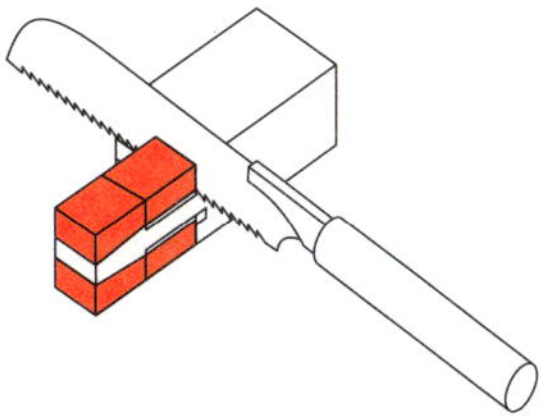
B4

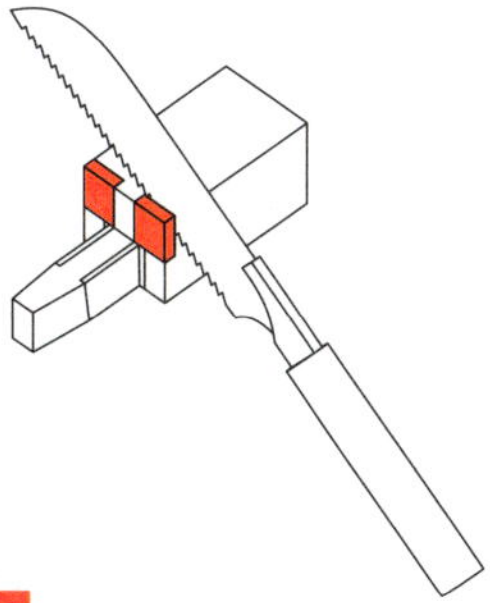
B5

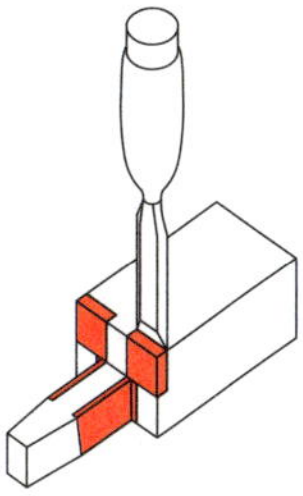
B6

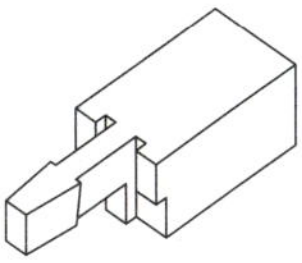
B

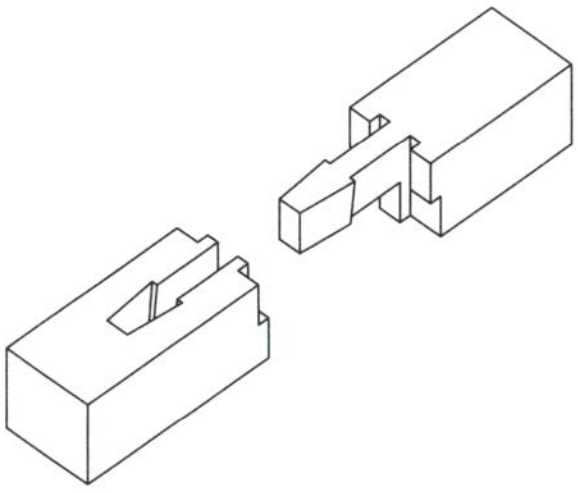
A+B

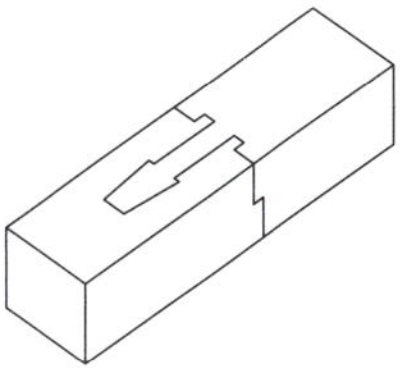
AB

Splicing Joints

Rabbeted gooseneck joint with barb

ryō-mechi-tsuki-kayaoi-kama-tsugi (noge-tsugi)
両目違付茅負鎌継ぎ(芒継)

abgesetzter Sichelzapfen mit Steckfalz und abgesetztem Gehrungsblatt

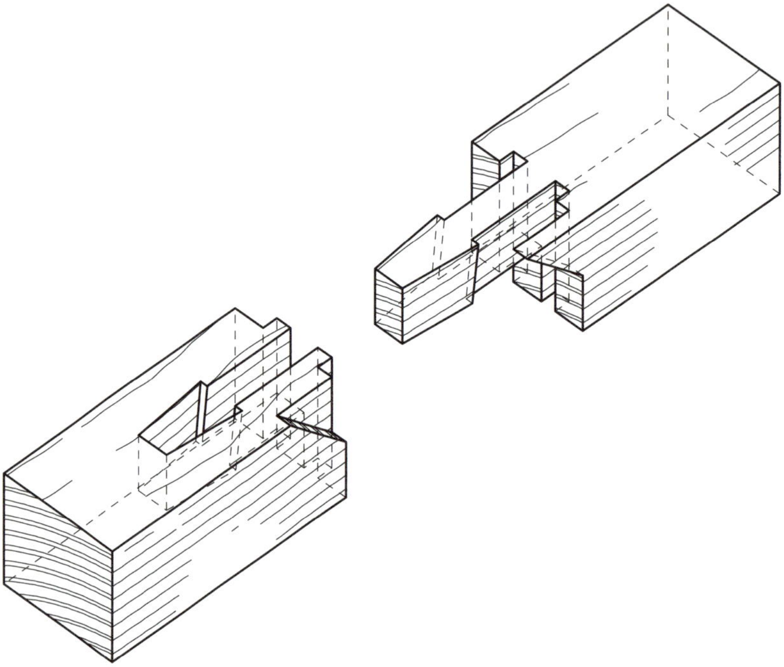

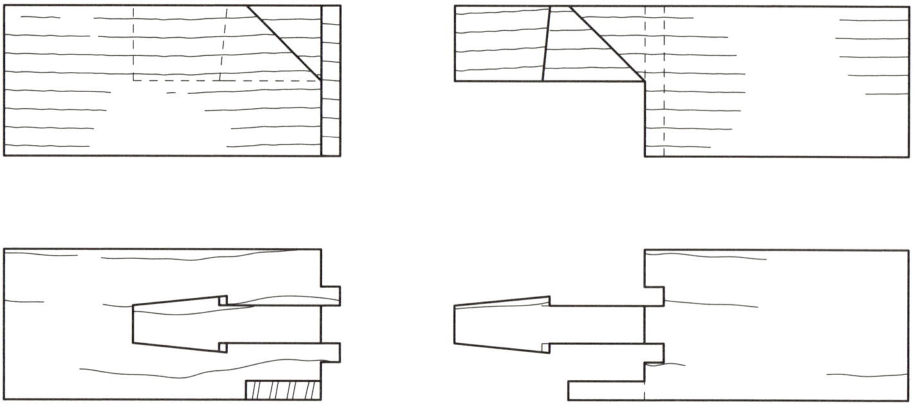

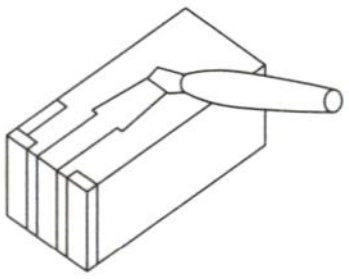

A1

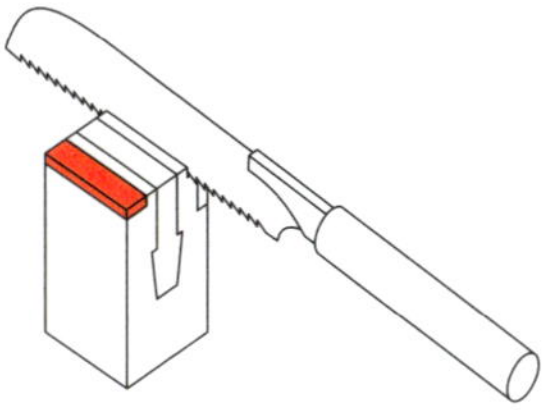

A2

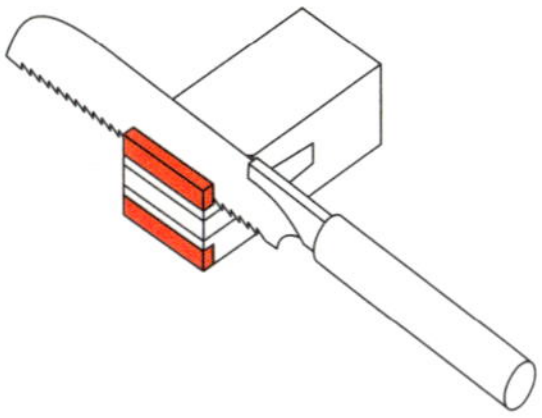

A3

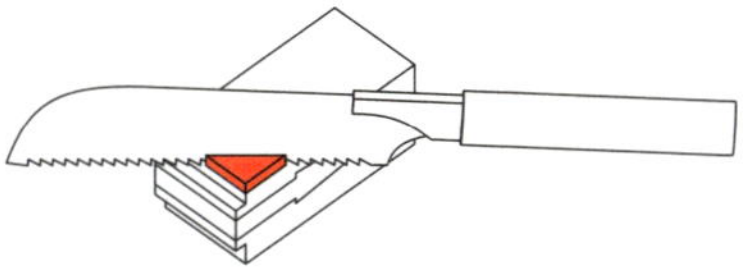

A4

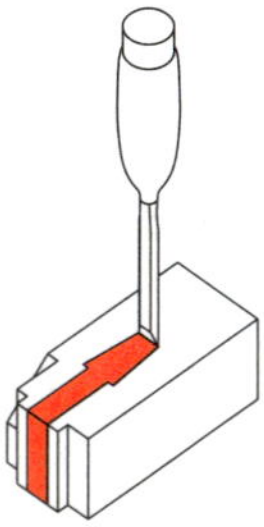

A5

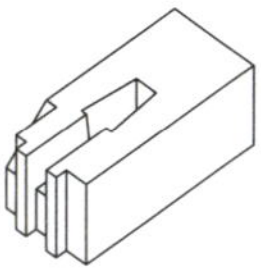

A

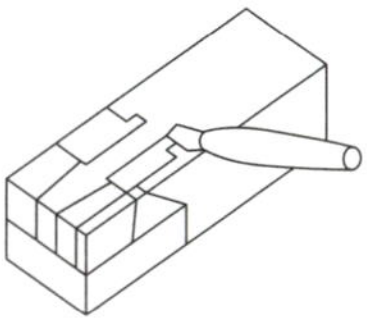

B1

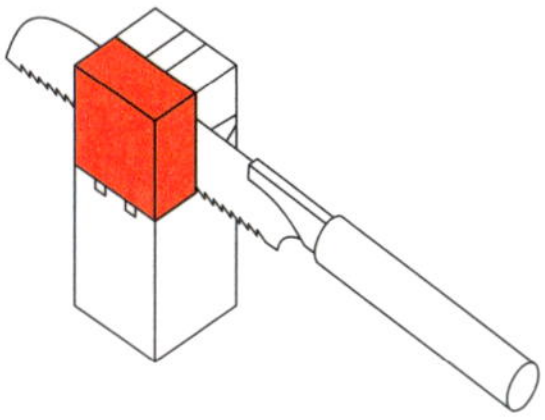

B2

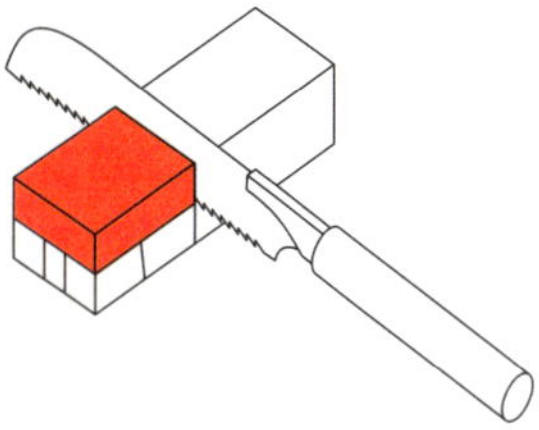

B3

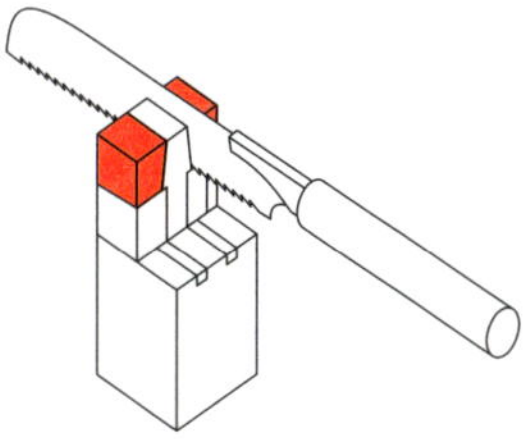

B4

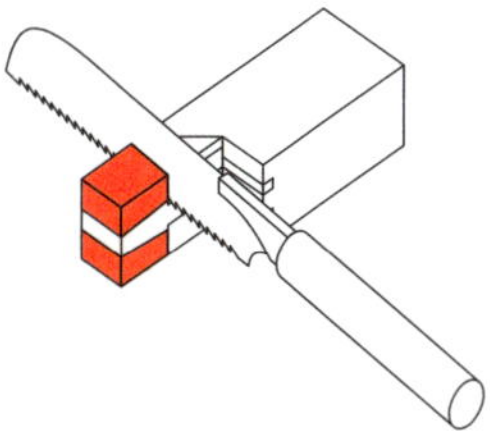

B5

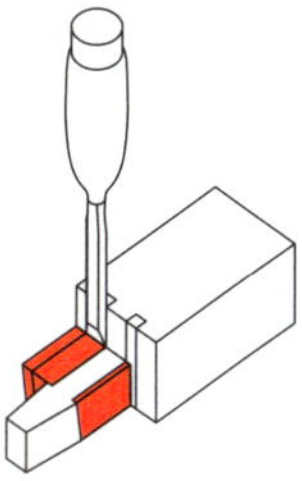

B6

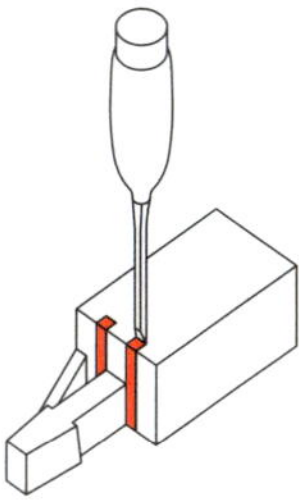

B7

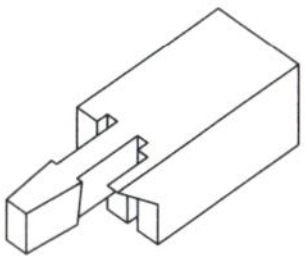

B

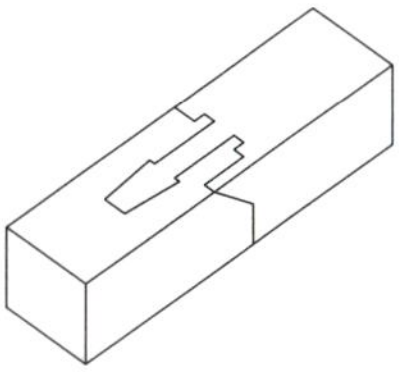

AB

Splicing Joints

Butt joint with floating gooseneck tenon

kine-gata-chigiri-tsugi
杵形千切継ぎ

Stoß mit eingesetztem Sichelzapfen

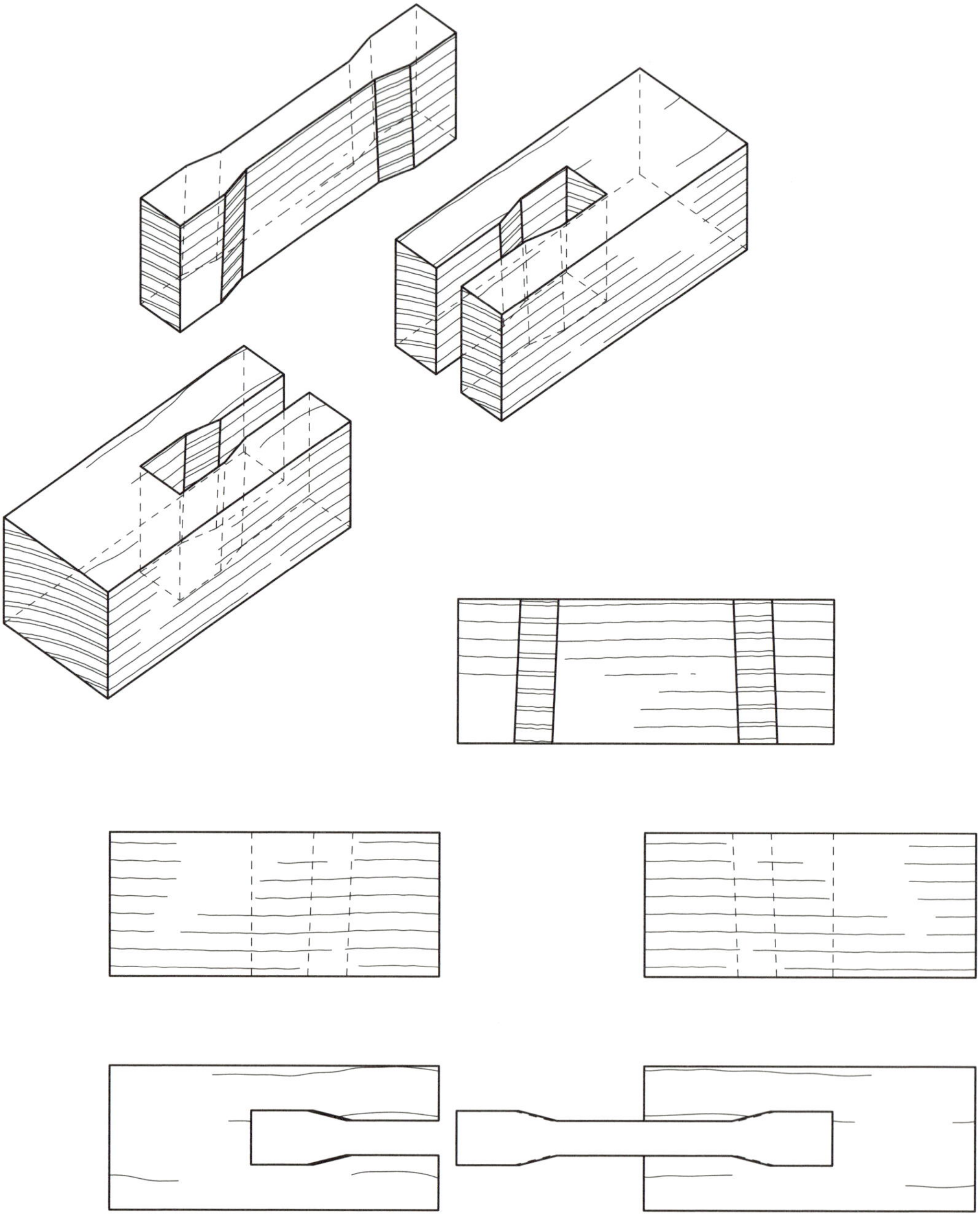

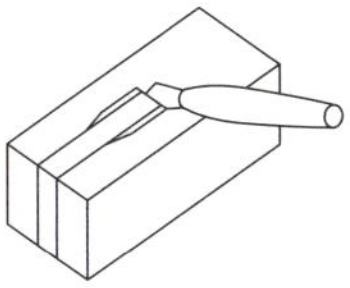

A1

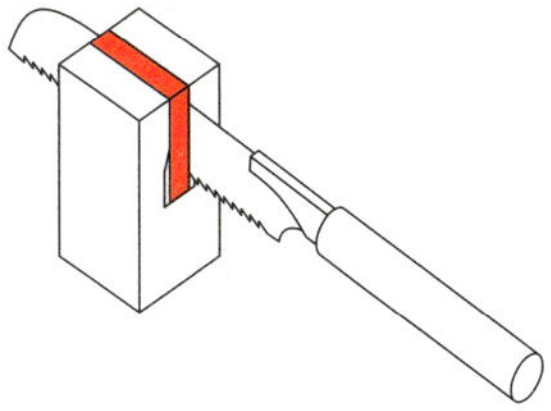

A2

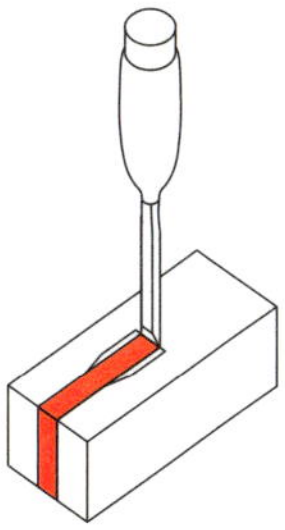

A3

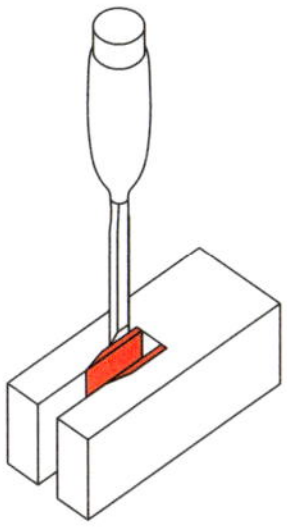

A4

2×

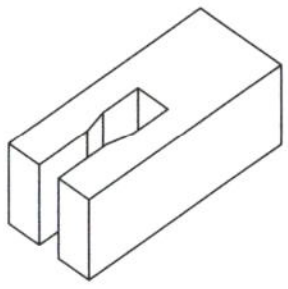

A

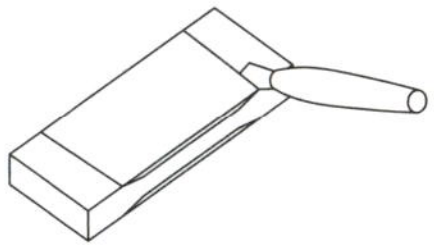

B1

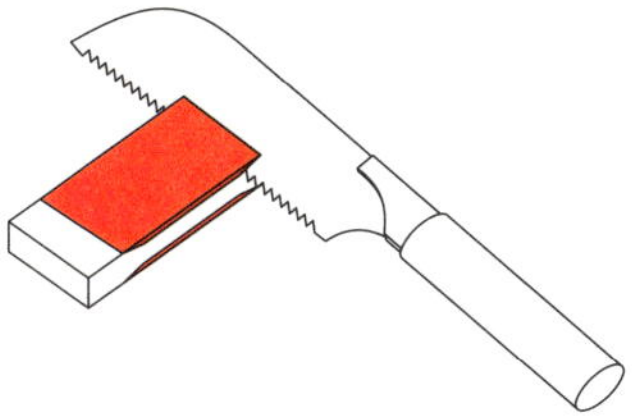

B2

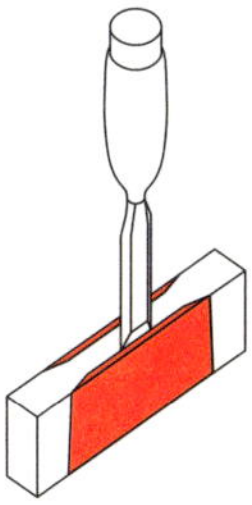

B3

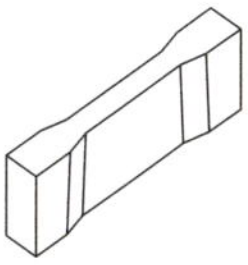

B

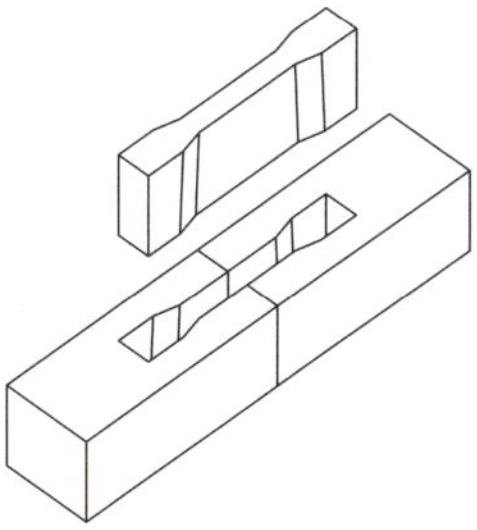

A+B+A

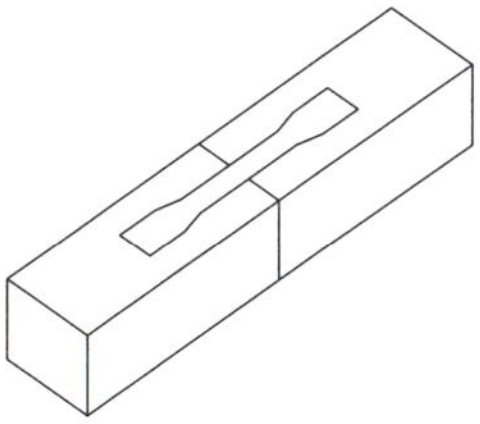

ABA

Splicing Joints

Butt joint with a gooseneck key and stub tenons

kine-gata-chigiri-tsugi (dabo-sashi)
杵形千切継ぎ（太枘差）

Stoß mit Sichelzapfeneinlage
und doppelten Brüstungszapfen

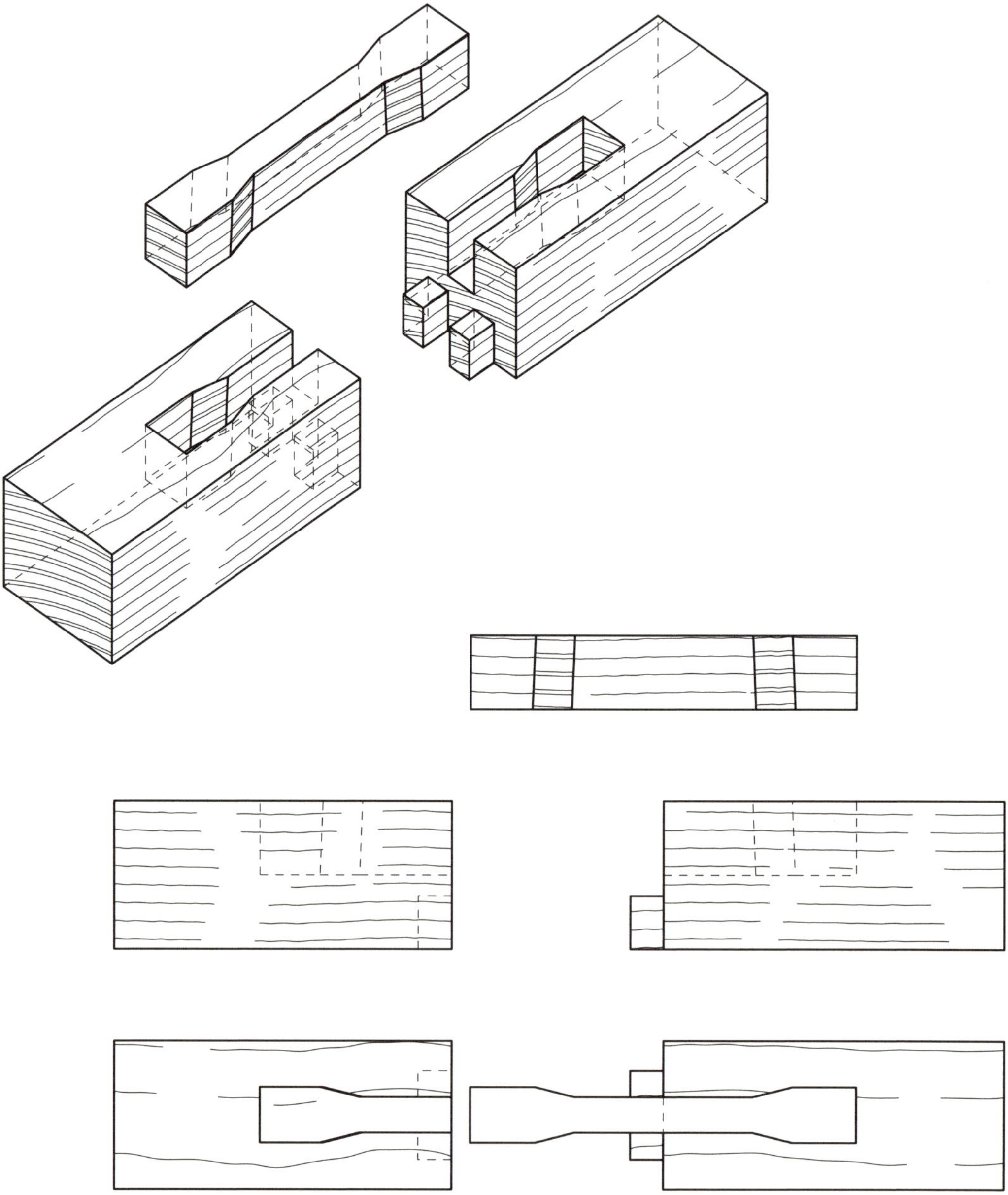

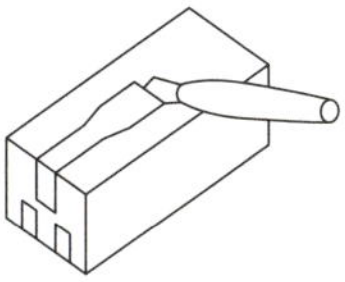

A1

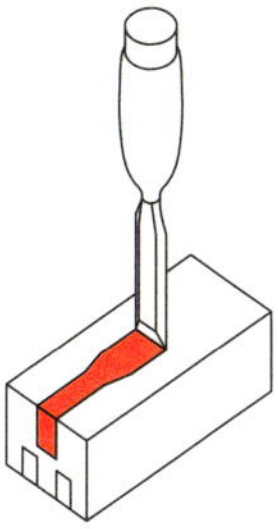

A2

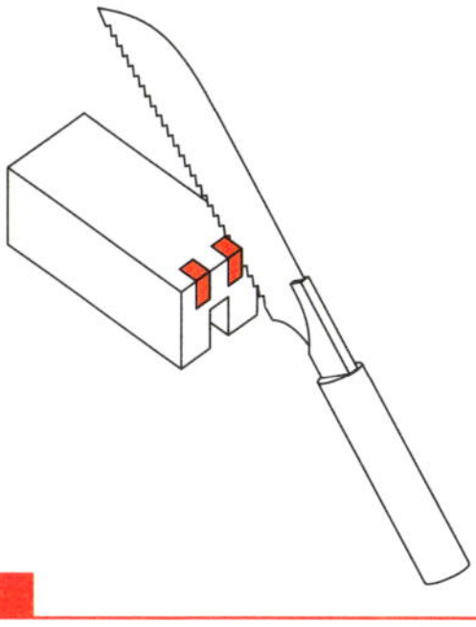

A3

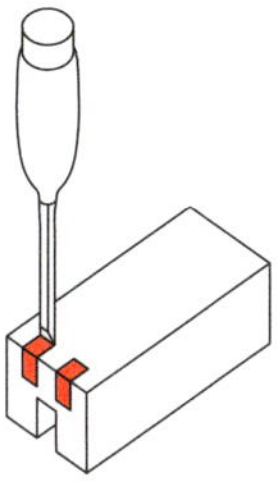

A4

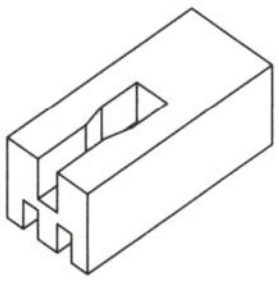

A

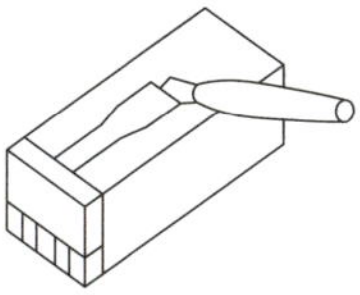

B1

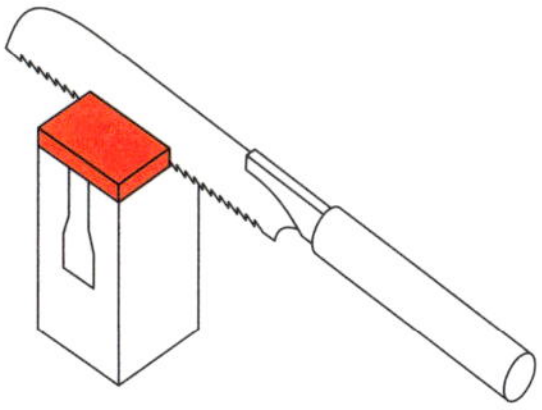

B2

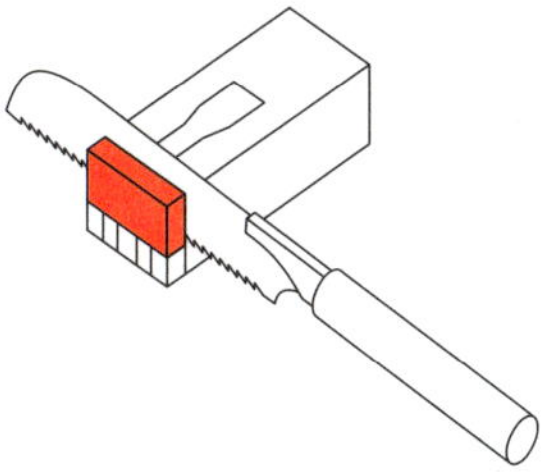

B3

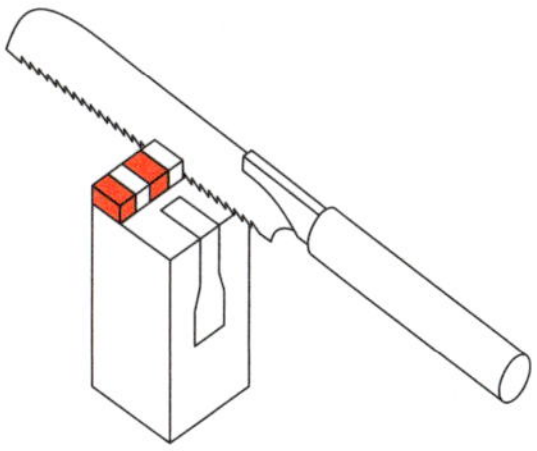

B4

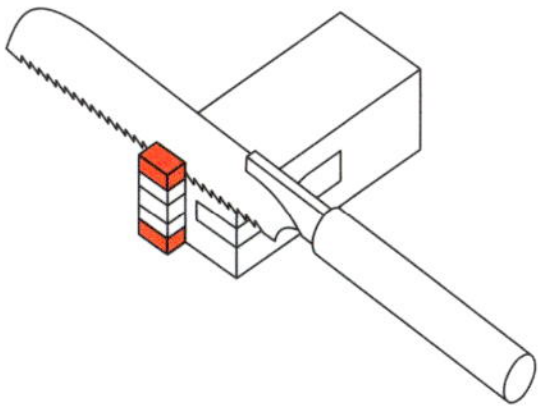

B5

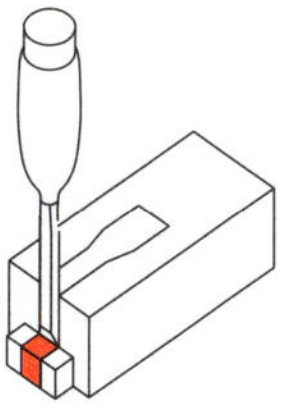

B6

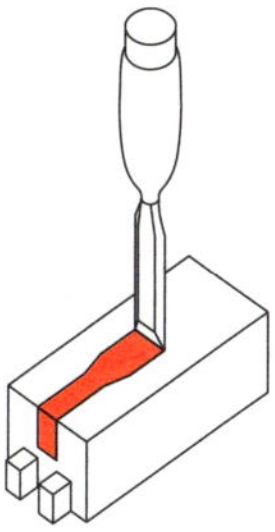

B7

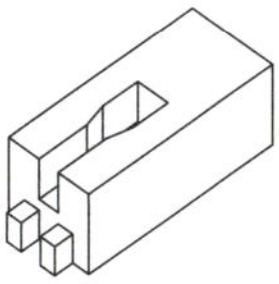

B

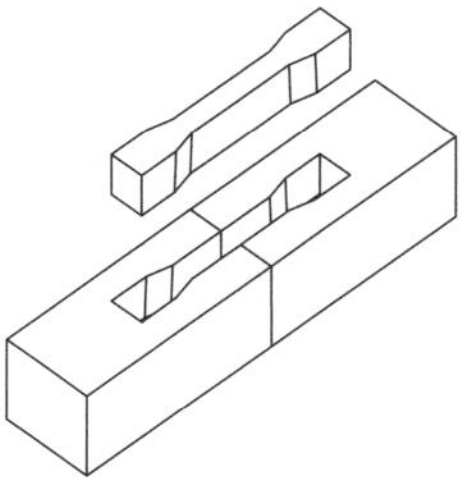

A+B+C

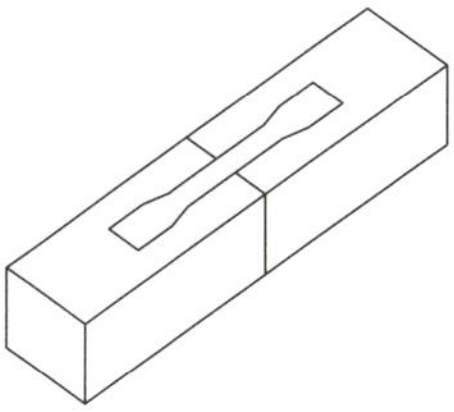

ABC

Splicing Joints

Four-faced gooseneck joint

shihō-kama-tsugi [*]
四方鎌継ぎ

diagonaler doppelter Sichelzapfen mit Steckfalz

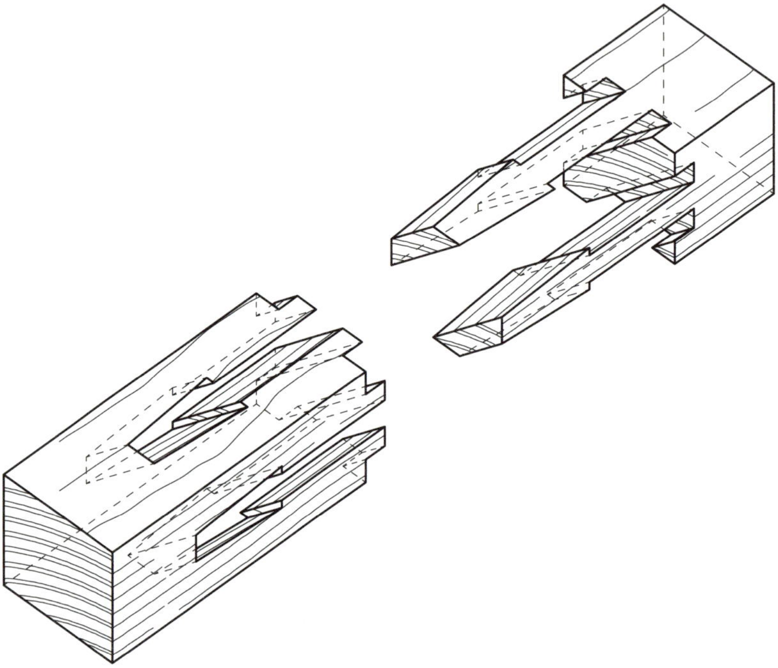

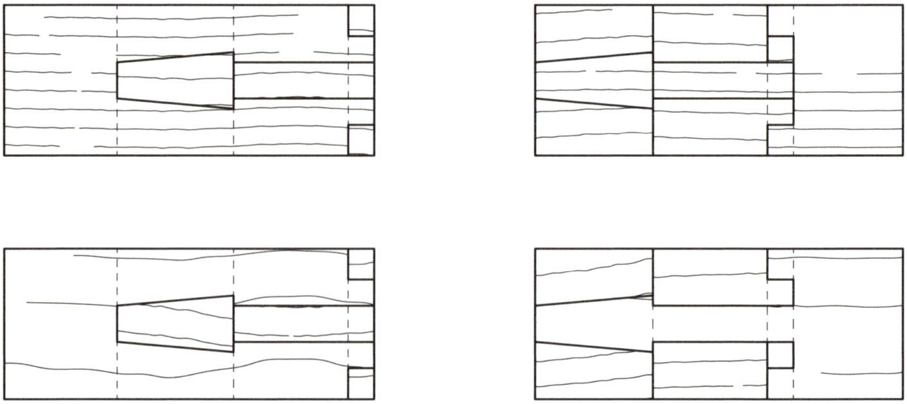

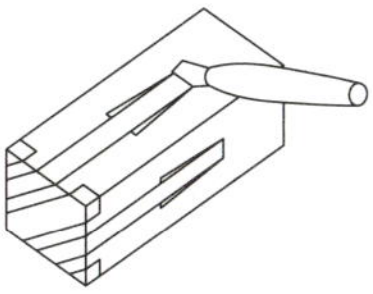
A1

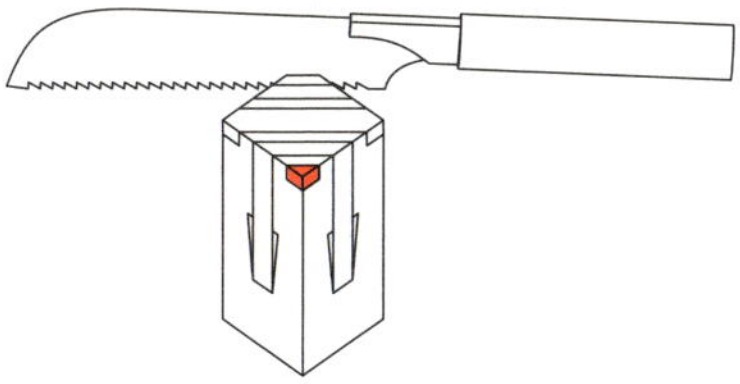
A2

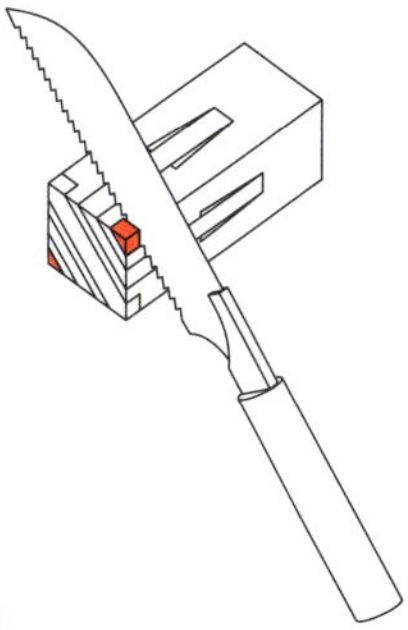
A3

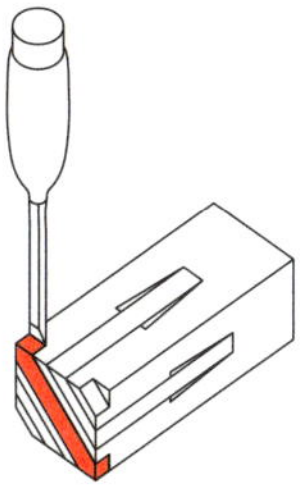
A4

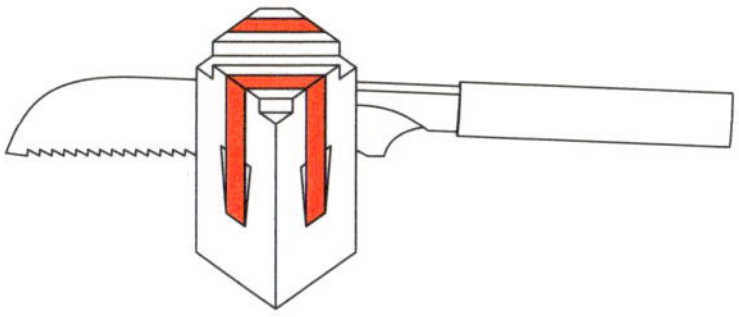
A5

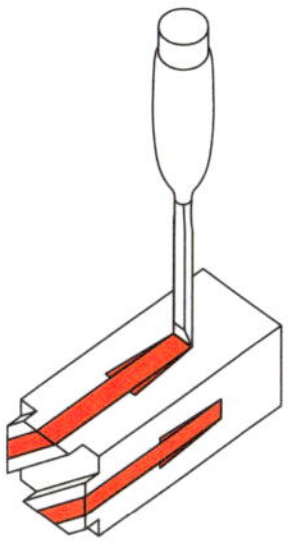
A6

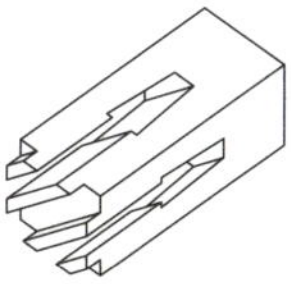
A

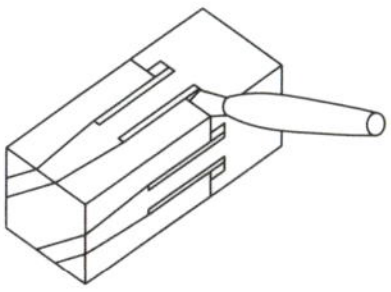
B1

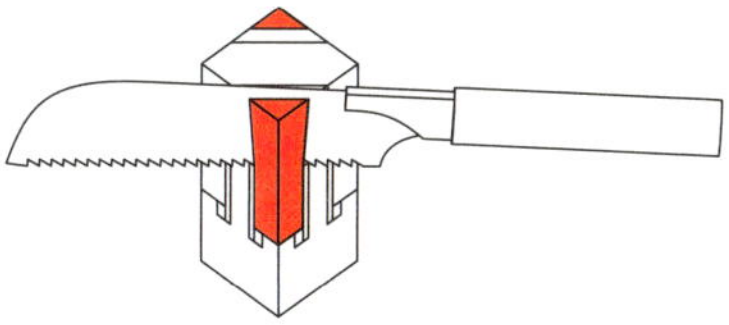
B2

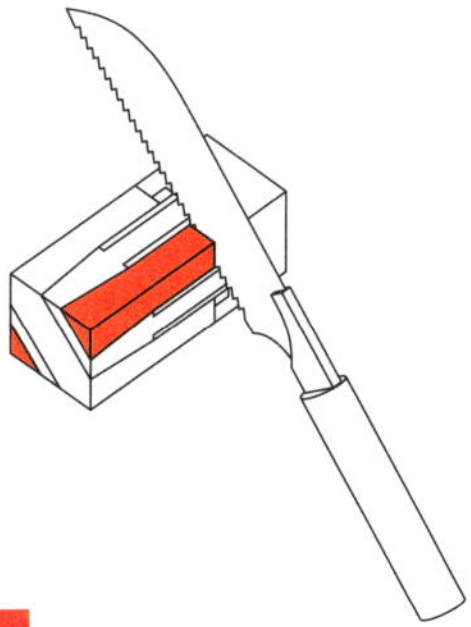
B3

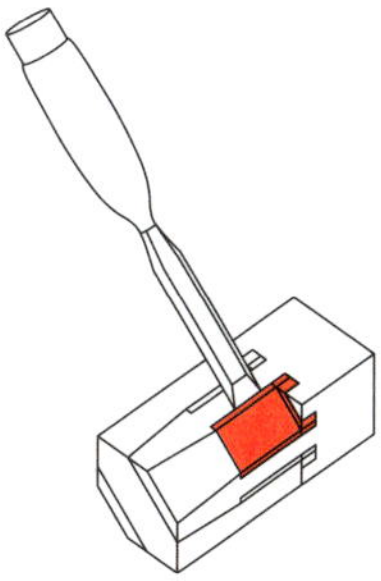
B4

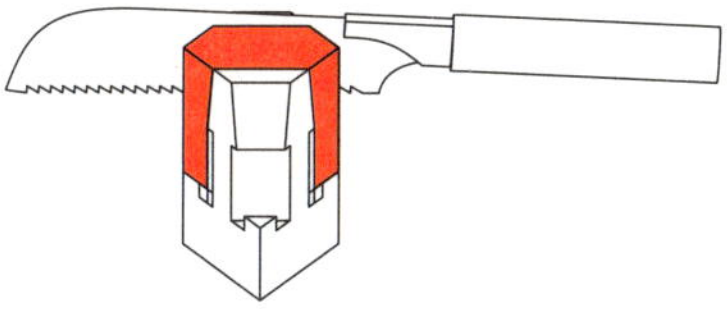
B5

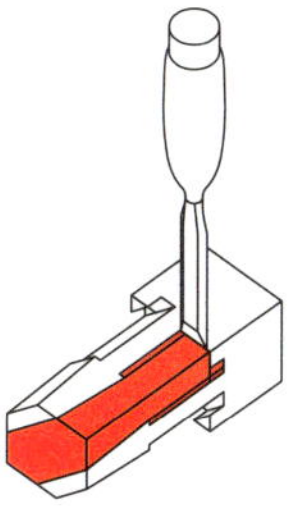
B6

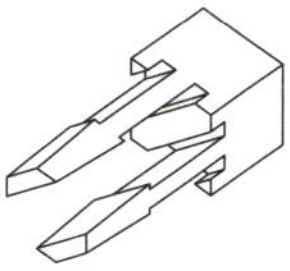
B

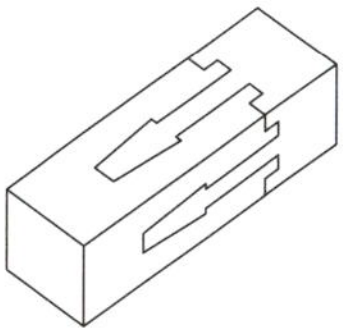
AB

Corner Joints

Corner lap joint

kane-ai-kaki-tsugi
矩相欠接ぎ

Eckblatt

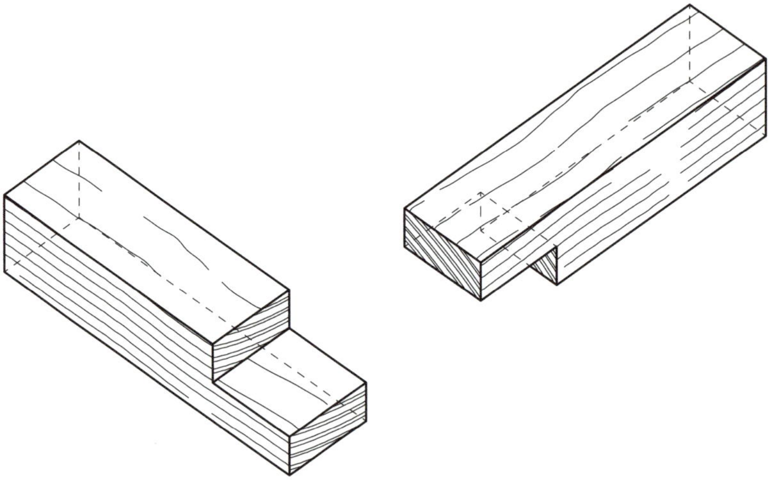

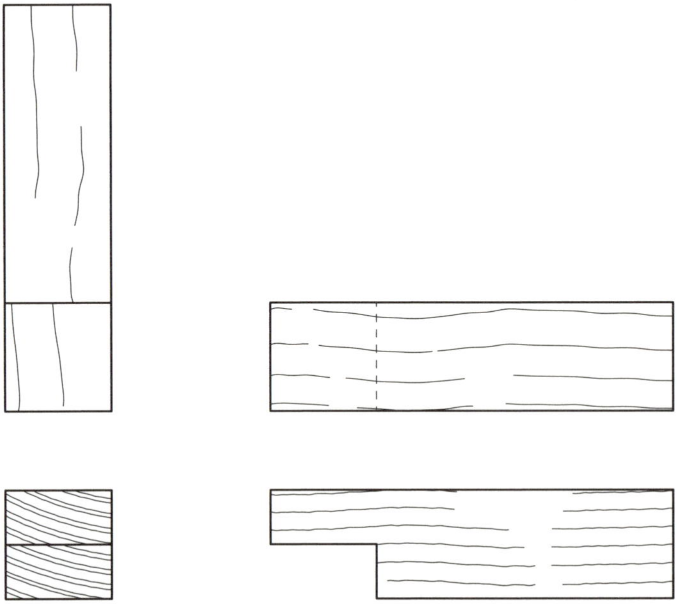

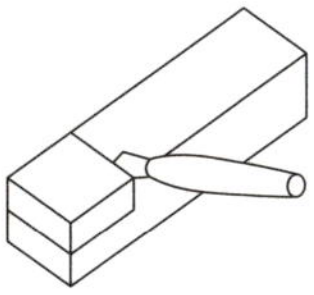

A1

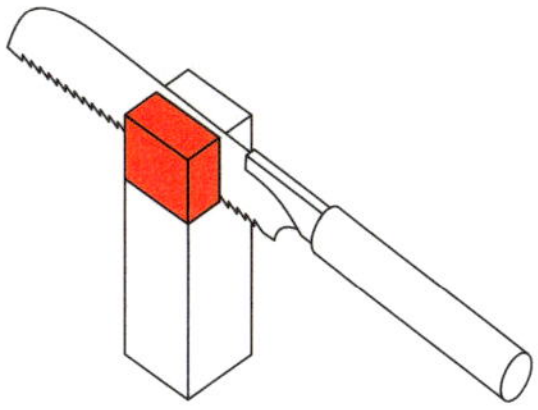

A2

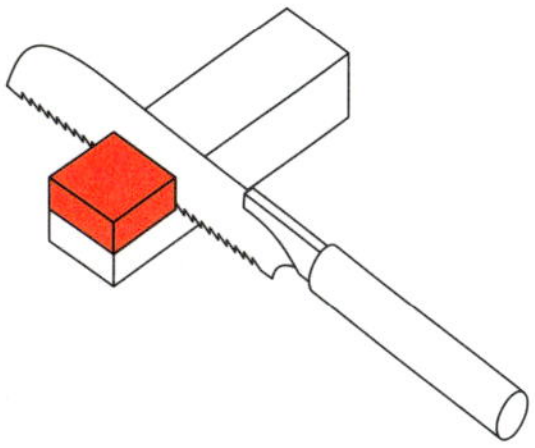

A3

2×

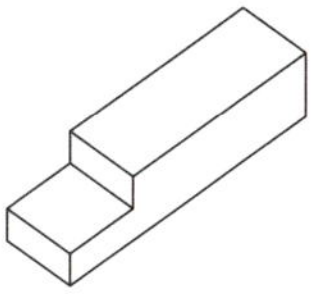

A

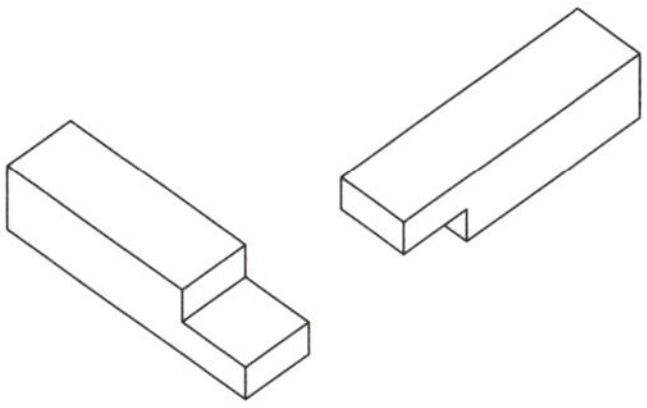

A+A

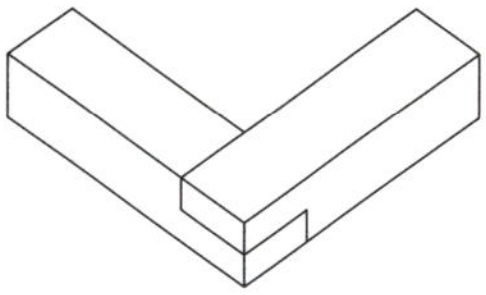

AA

Corner Joints

Blind mitered corner lap joint

kakushi-kane-ai-kaki-tsugi
隠矩相欠接ぎ

verdecktes Eckblatt mit Wange auf Gehrung

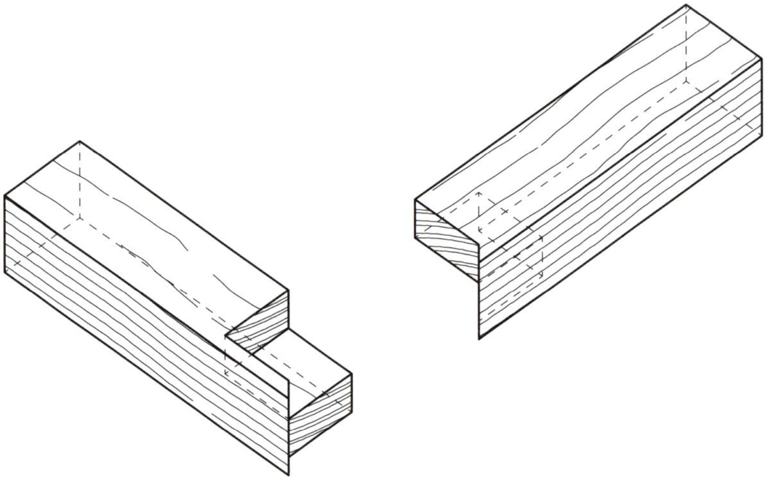

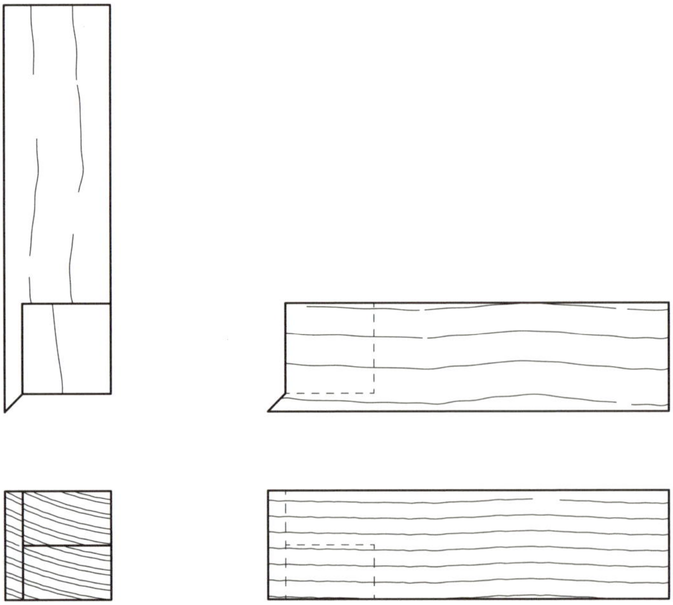

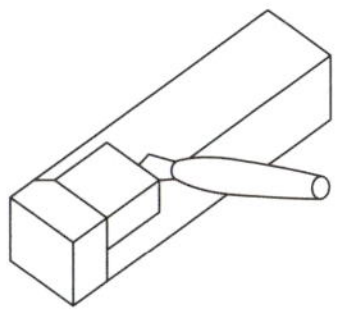

A1

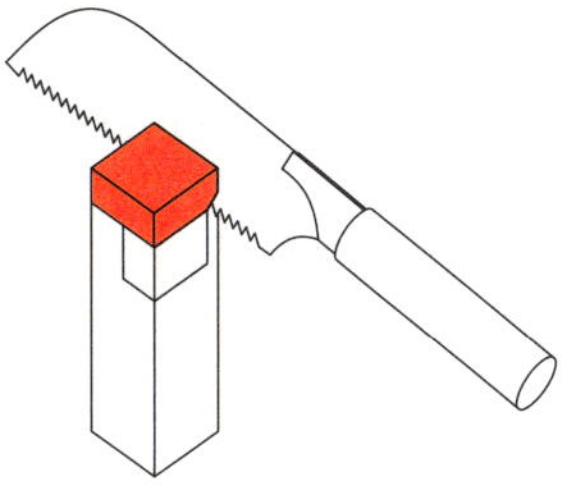

A2

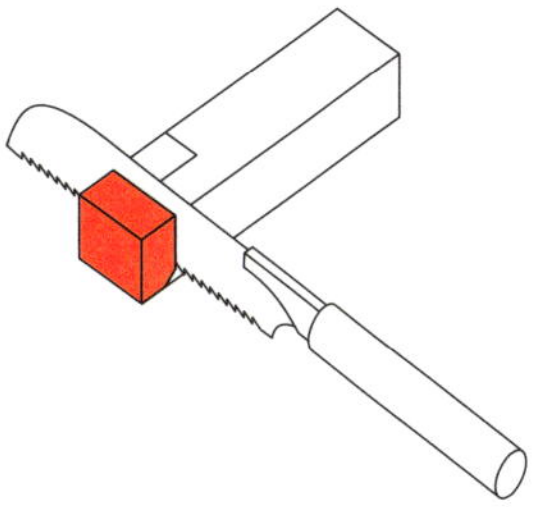

A3

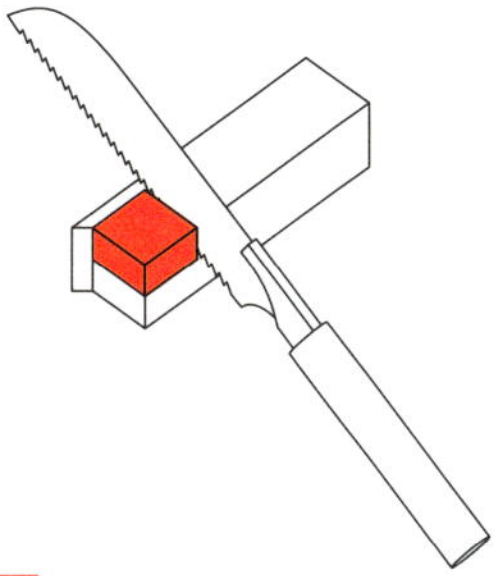

A4

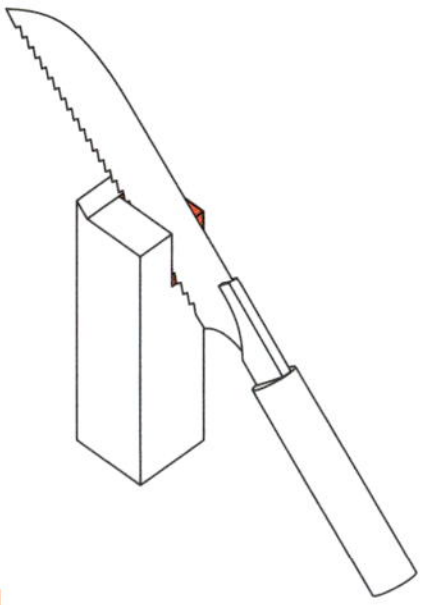

A5

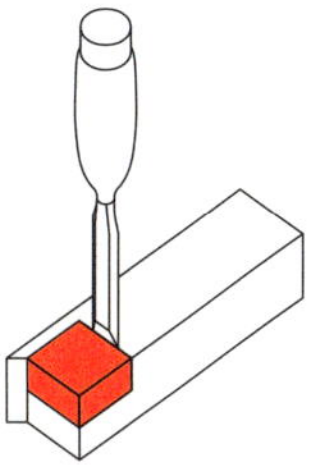

A6

2×

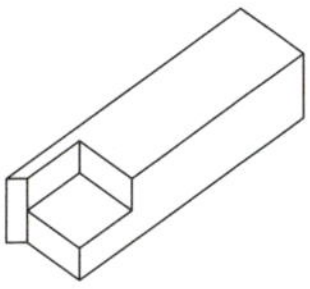

A

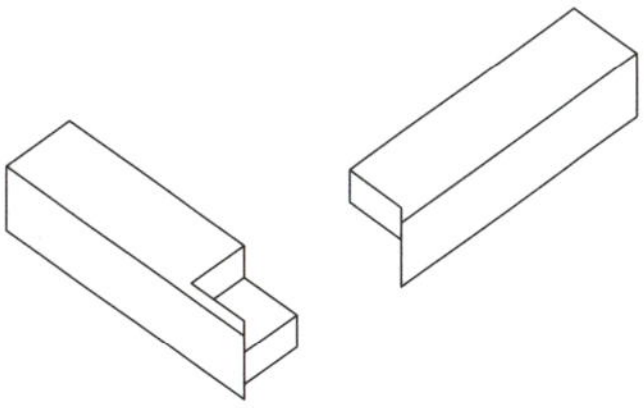

A+A

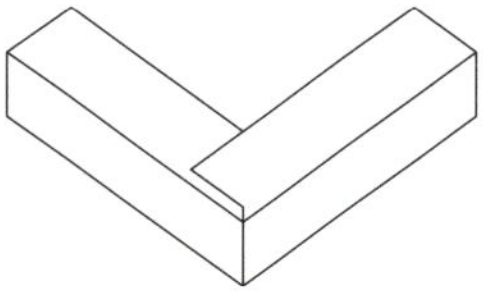

AA

Corner Joints

Dovetailed corner lap joint

ari-kata-ai-kaki-tsugi
蟻形相欠接ぎ

Eckblatt mit schräger Blattfläche

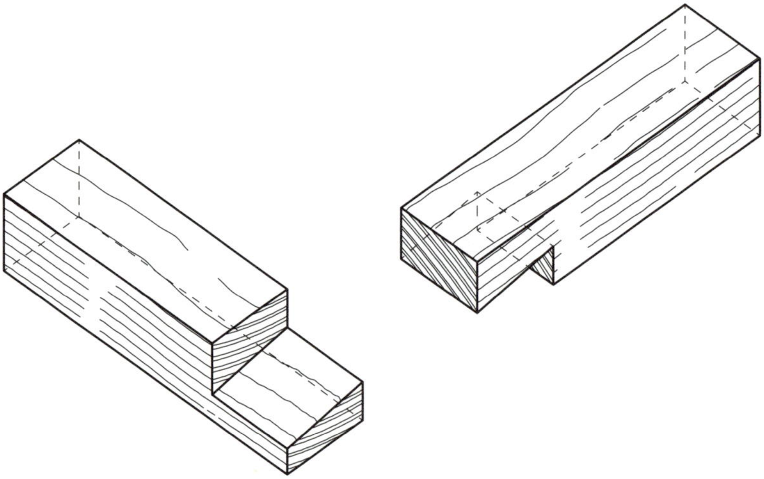

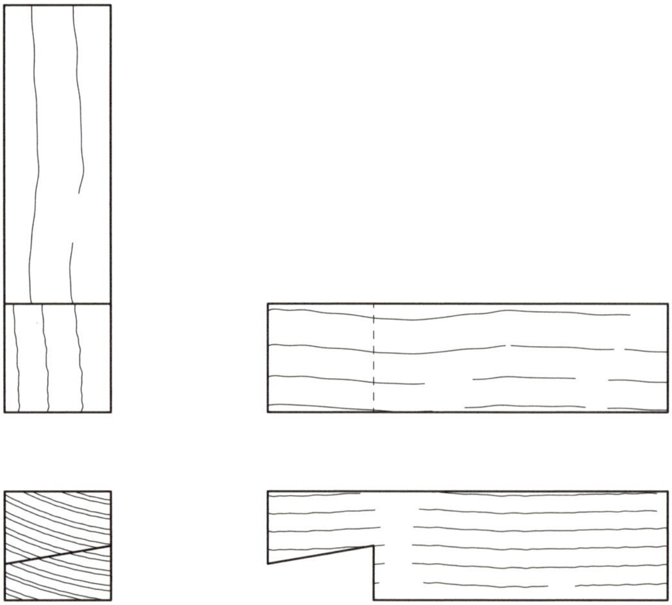

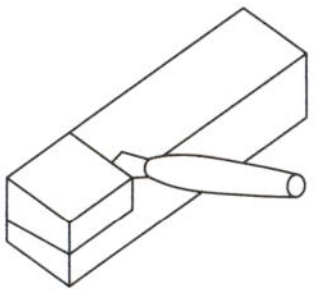

A1

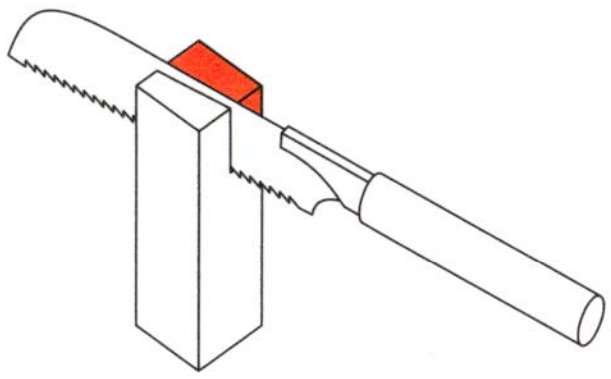

A2

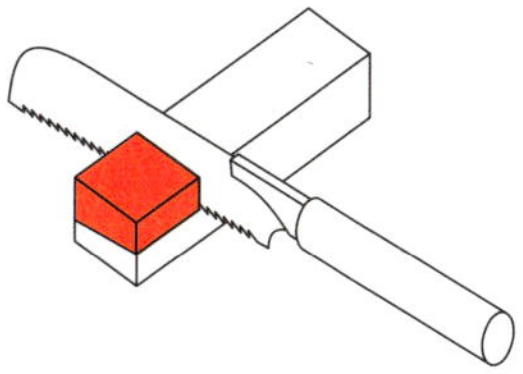

A3

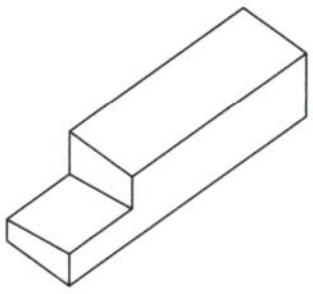

A

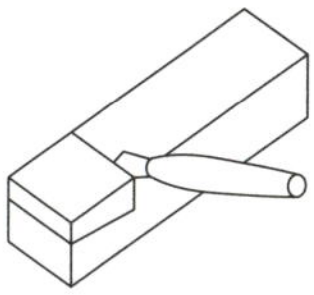

B1

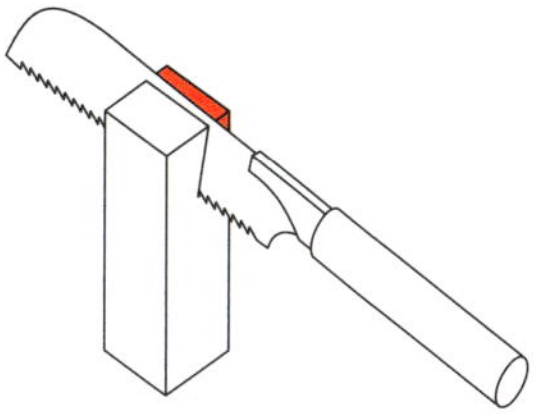

B2

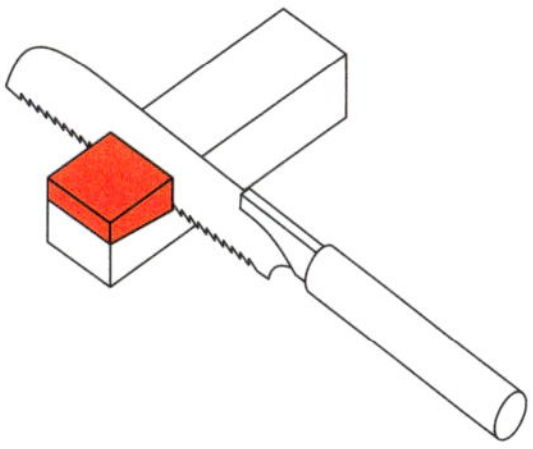

B3

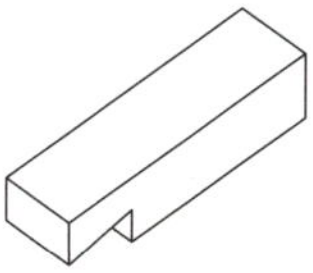

B

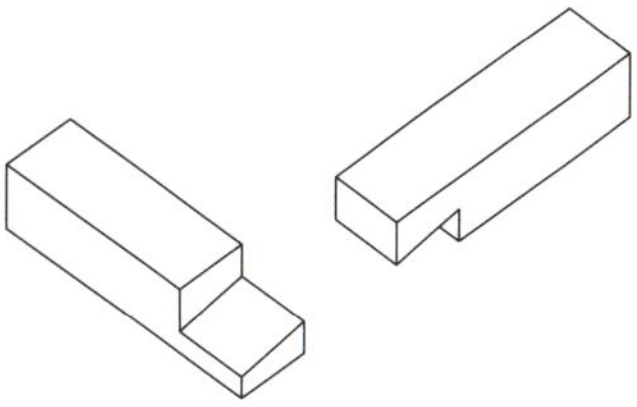

A+B

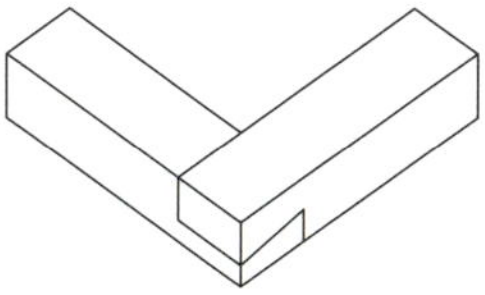

AB

Corner Joints

Beveled corner lap joint

ari-kata-ai-kaki-tsugi (sogi-tsuki)
蟻形相欠接ぎ(殺付)

Eckblatt mit zweifach schräger Blattfläche

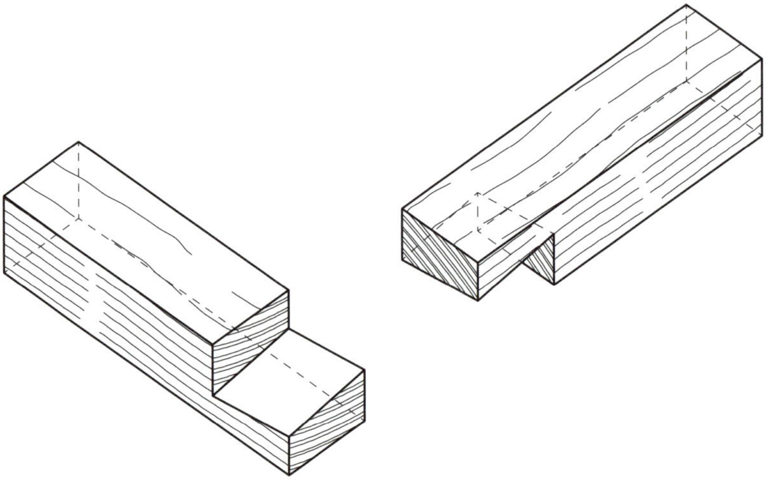

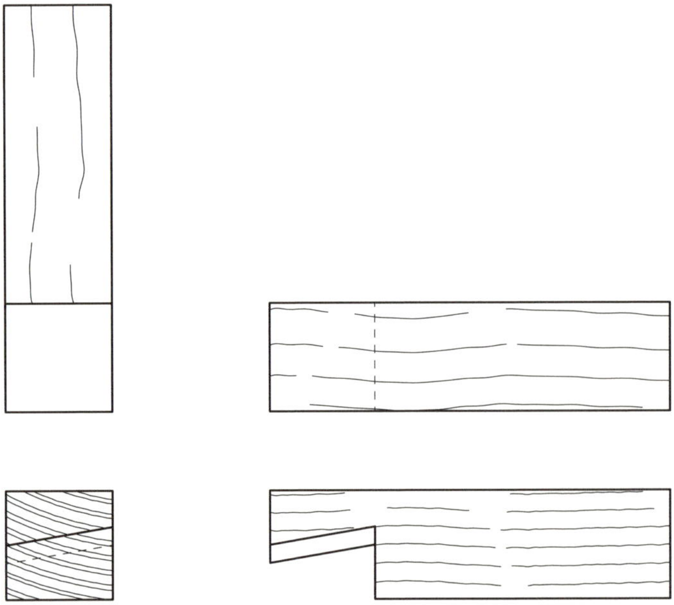

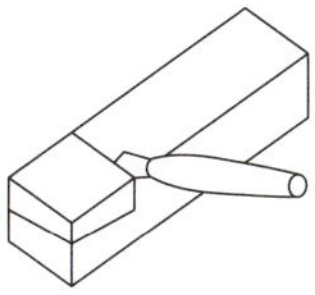

A1

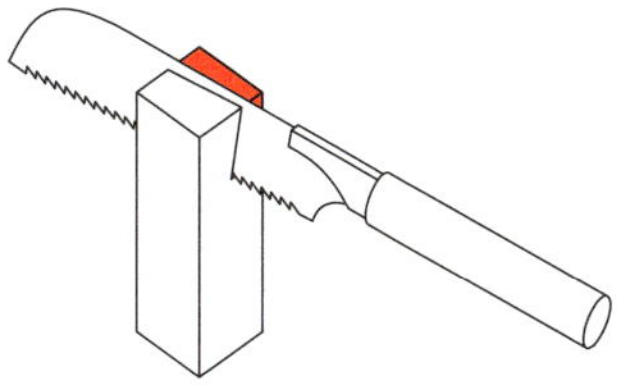

A2

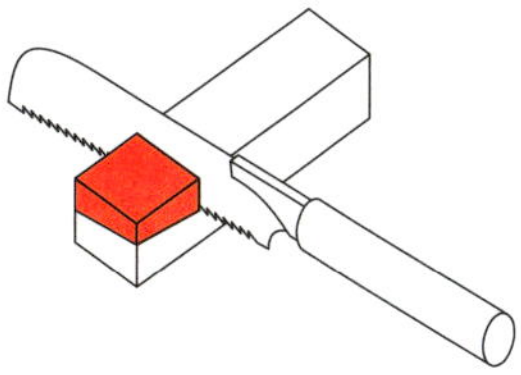

A3

2×

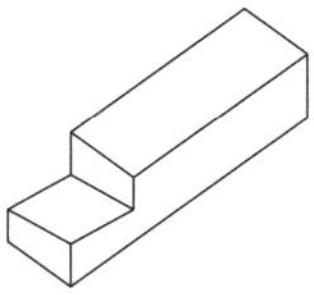

A

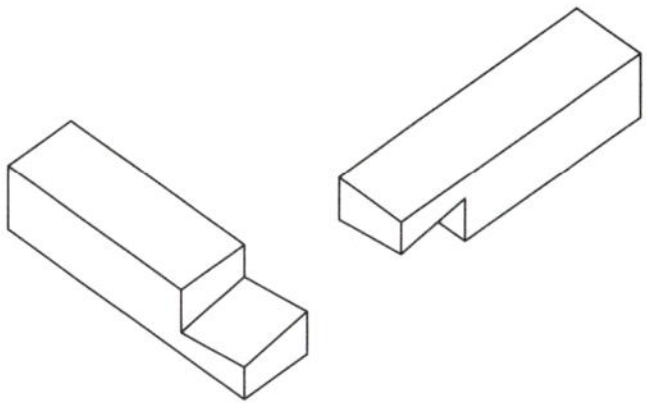

A+A

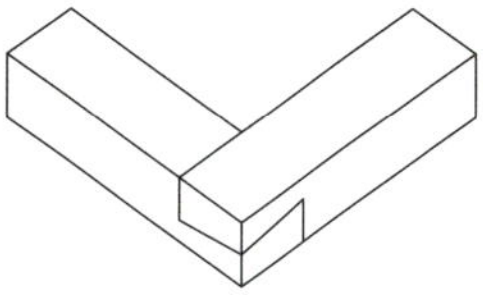

AA

Corner Joints

Blind mitered beveled corner lap joint

kakushi-ari-kata-ai-kaki-tsugi
隠蟻形相欠接ぎ

verdecktes Eckblatt mit zweifach schräger Blattfläche und Wange auf Gehrung

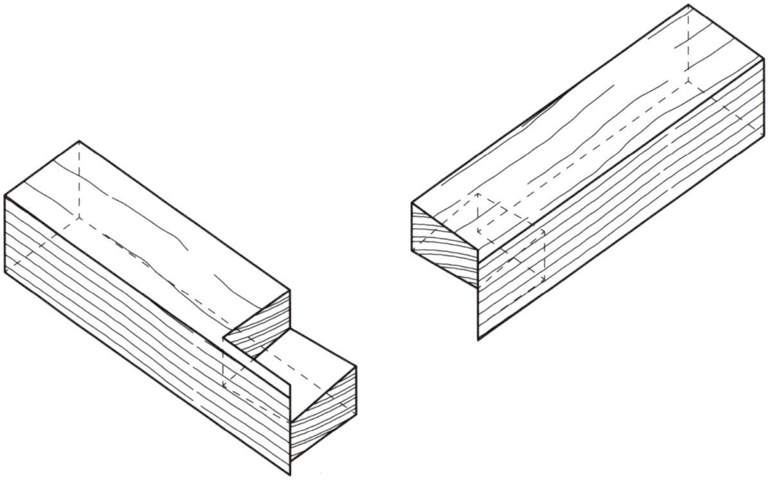

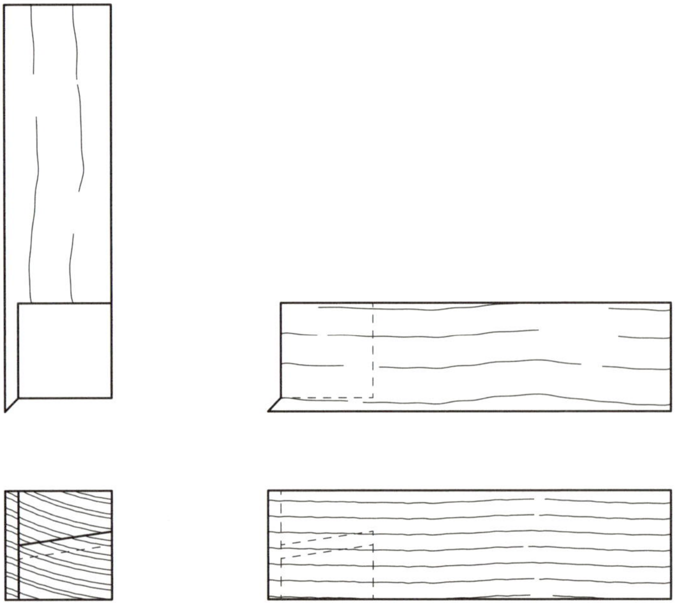

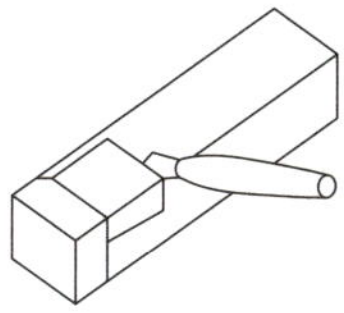
A1

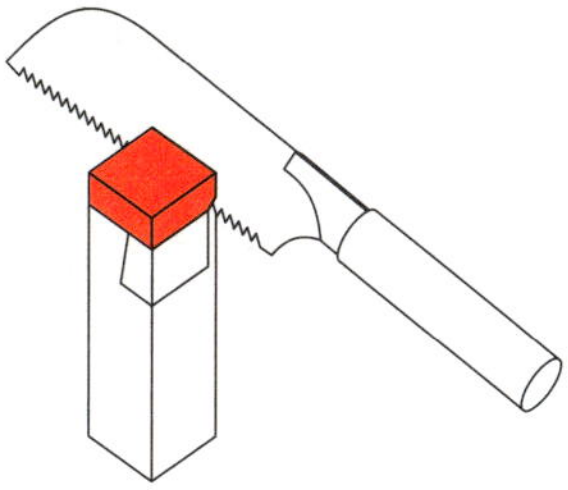
A2

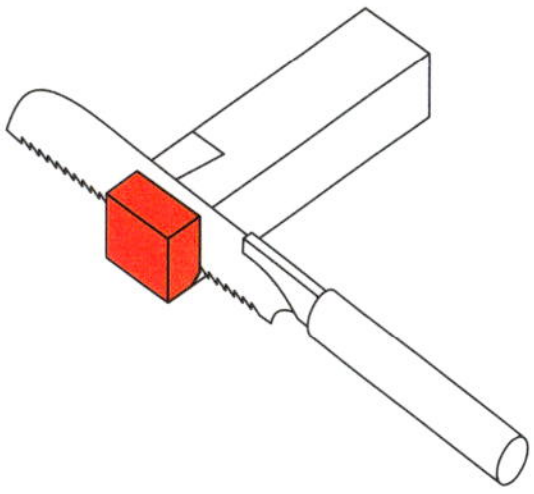
A3

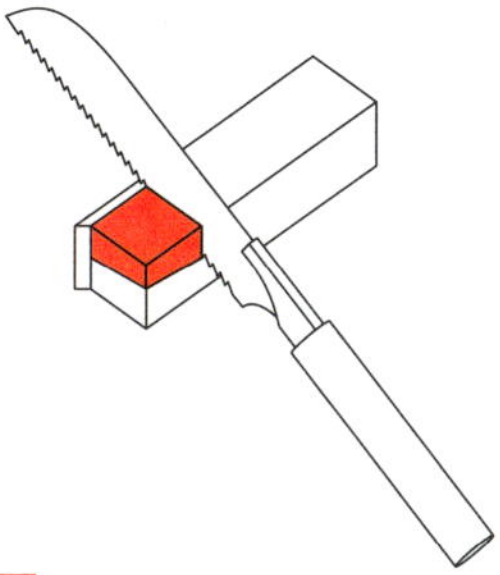
A4

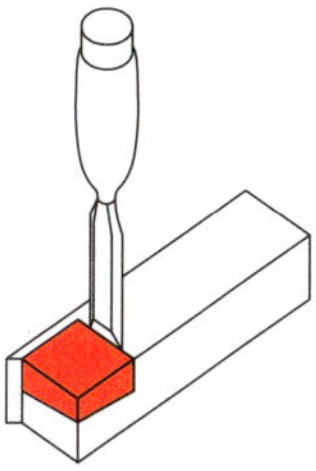
A5

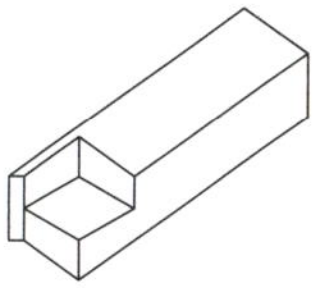
2×

A

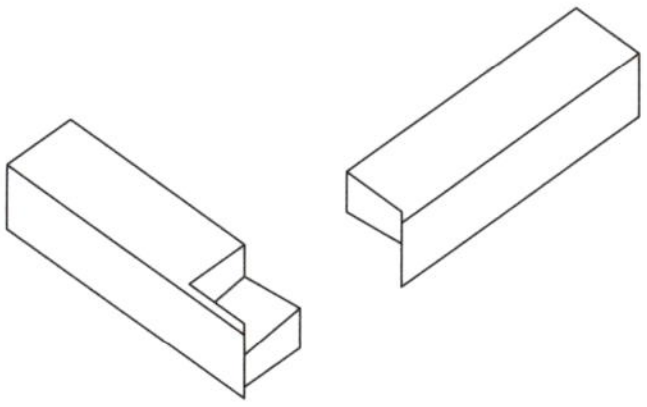
A+A

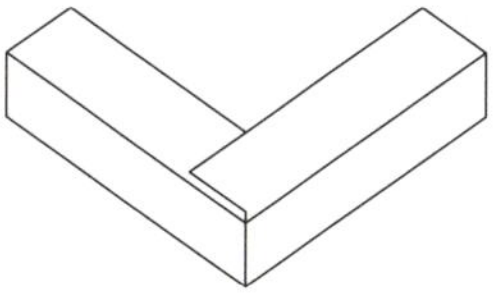
AA

Corner Joints

Cogged corner lap joint

koshi-kake-ai-kaki-tsugi
腰掛相欠接ぎ

Hakeneckblatt

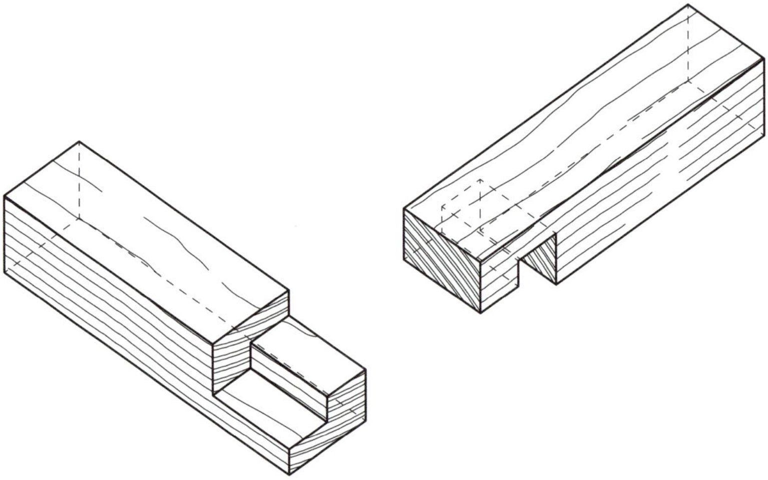

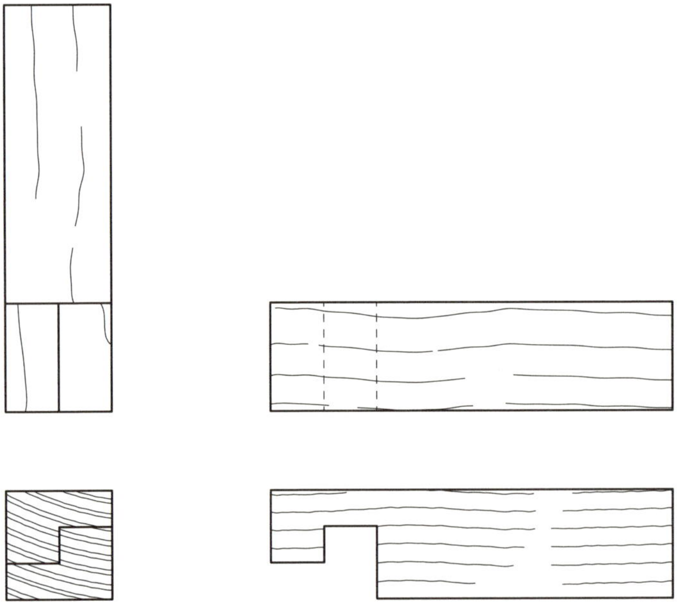

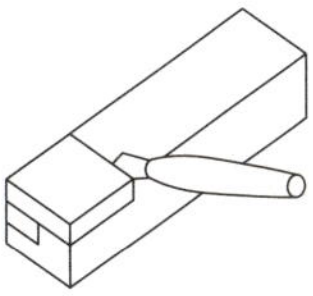

A1

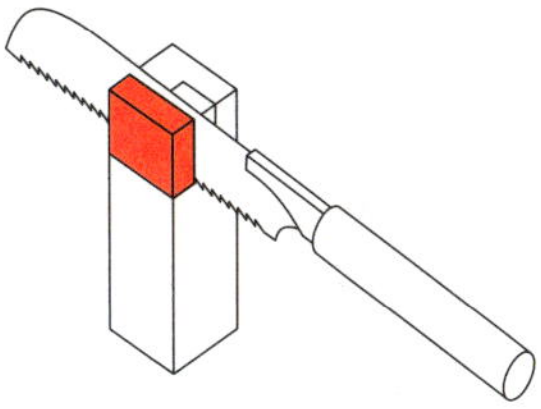

A2

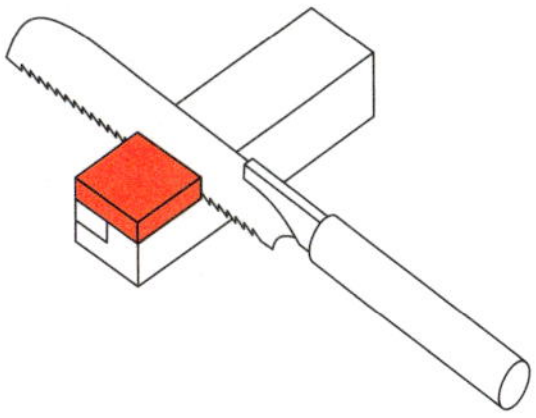

A3

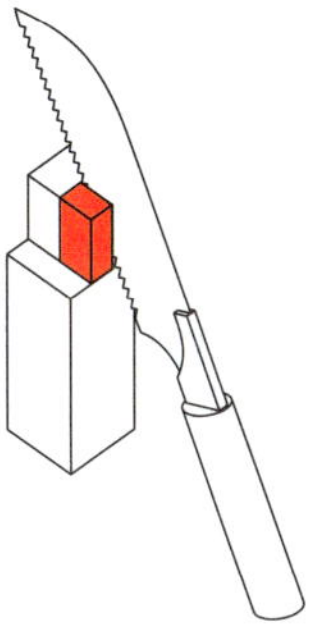

A4

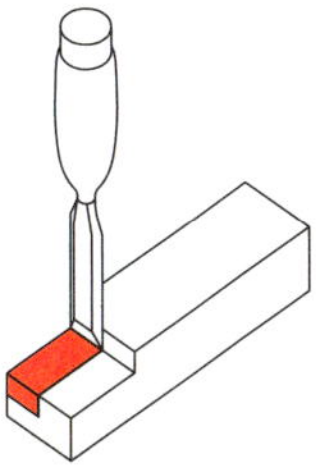

A5

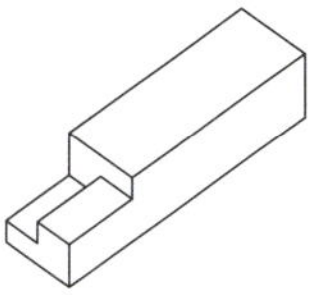

A

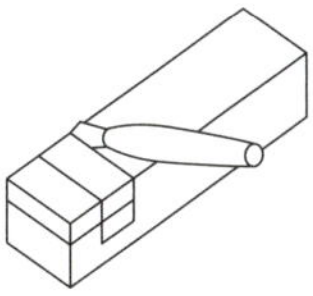

B1

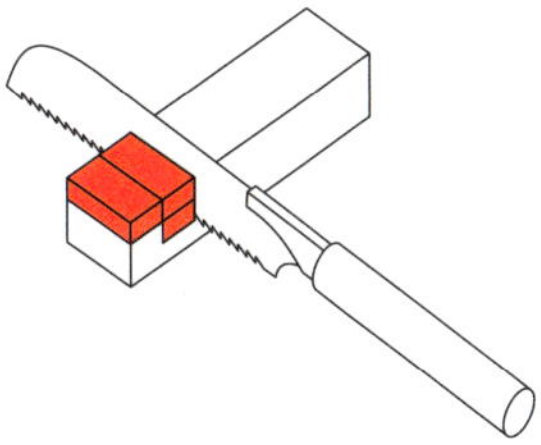

B2

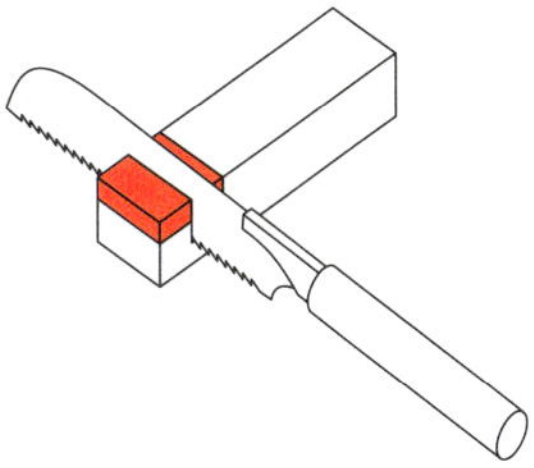

B3

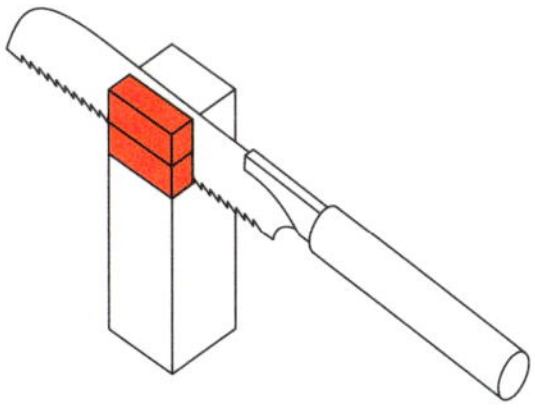

B4

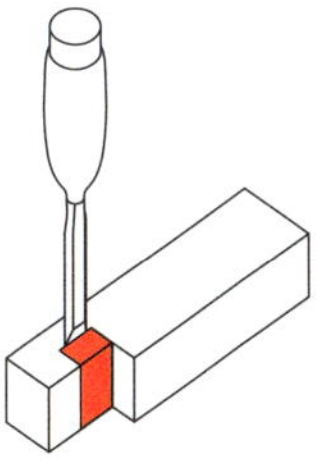

B5

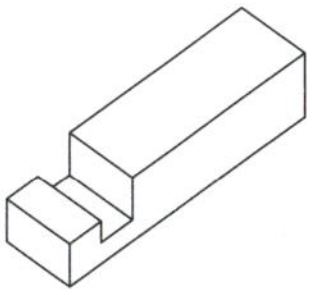

B

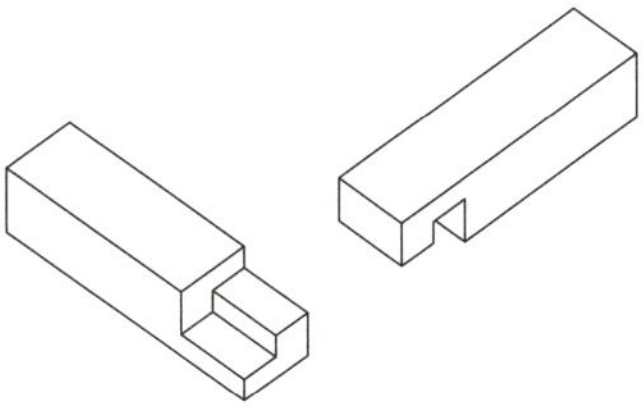

A+B

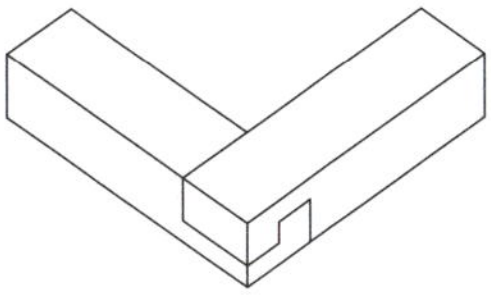

AB

Corner Joints

Sloped cogged corner lap joint

ari-kake-ai-kaki-tsugi
蟻掛相欠接ぎ

Hakeneckblatt mit schrägem Haken

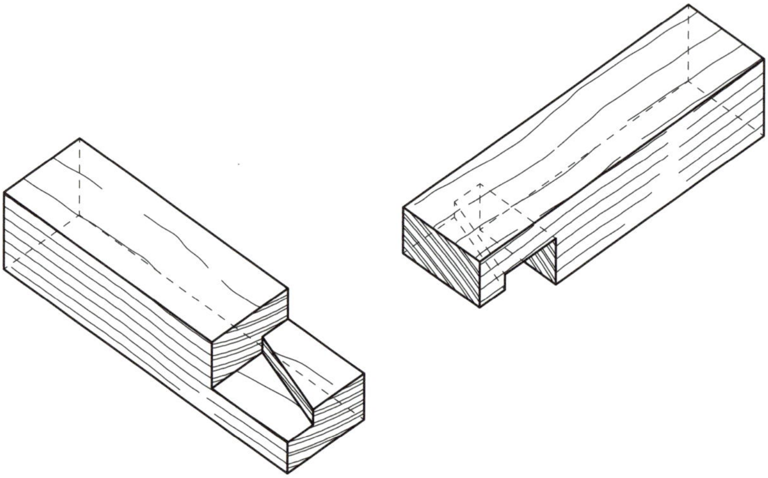

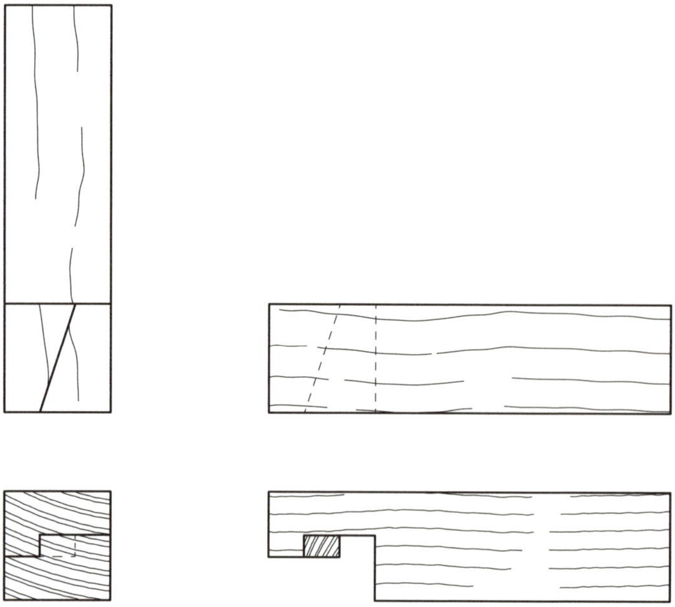

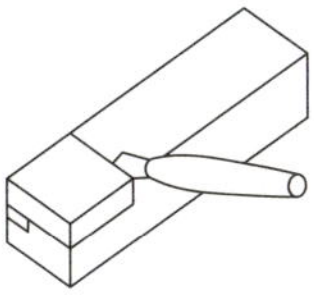

A1

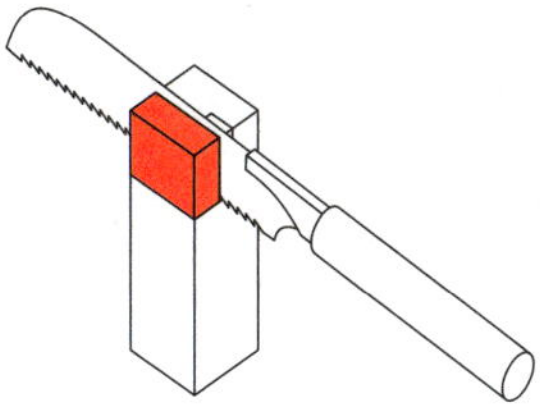

A2

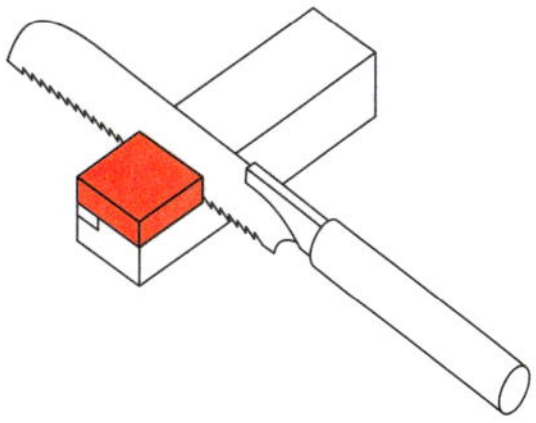

A3

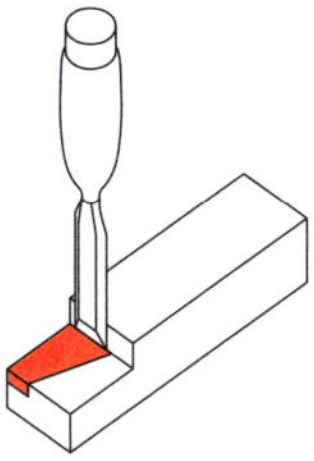

A4

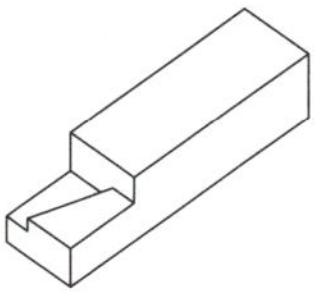

A

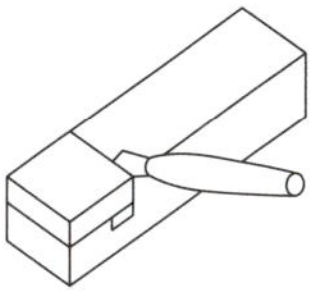

B1

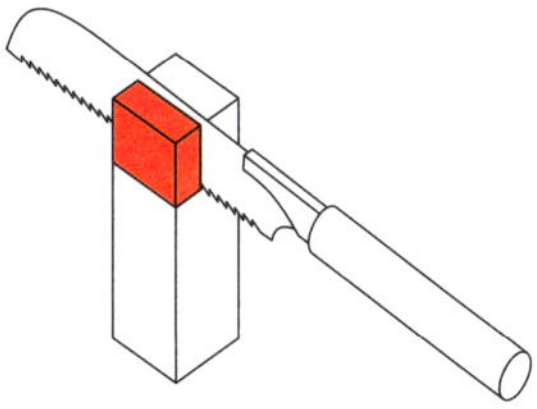

B2

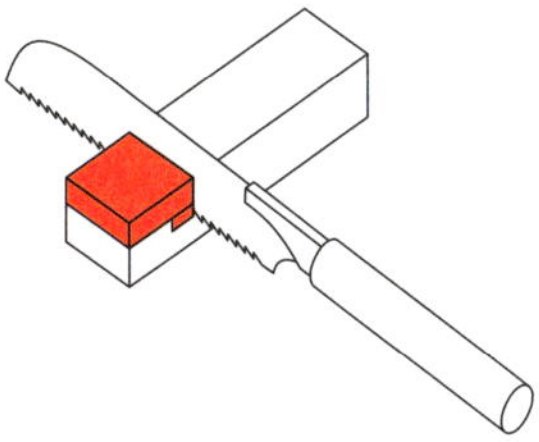

B3

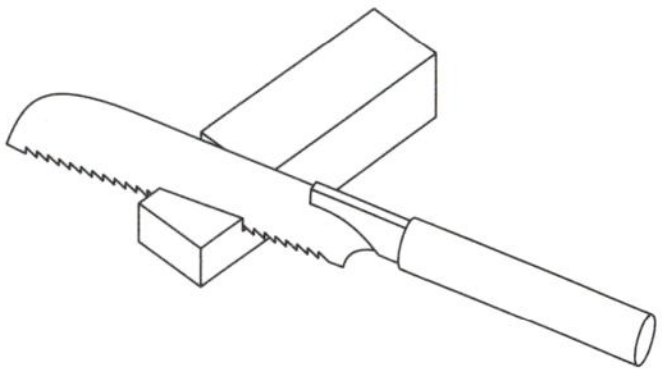

B4

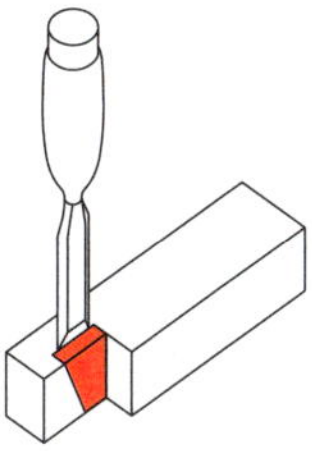

B5

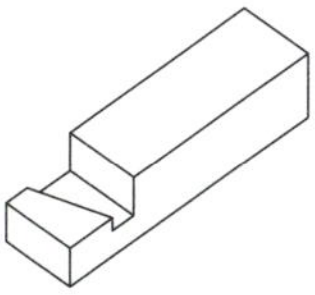

B

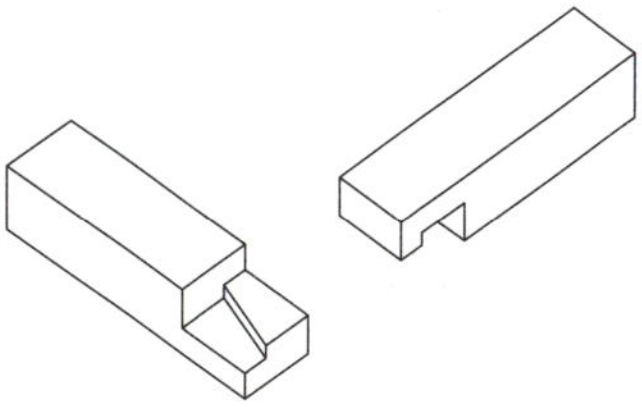

A+B

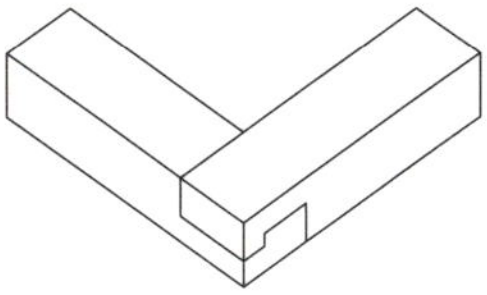

AB

Corner Joints

Double cogged corner lap joint

dō-tsuki-koshi-kake-tsugi (watari-ago)
胴付腰掛接ぎ(渡顎)

Hakeneckblatt mit mittigem Haken

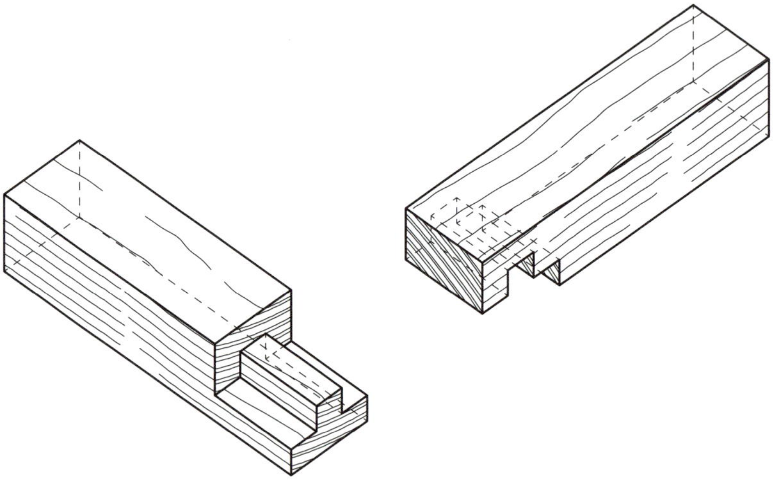

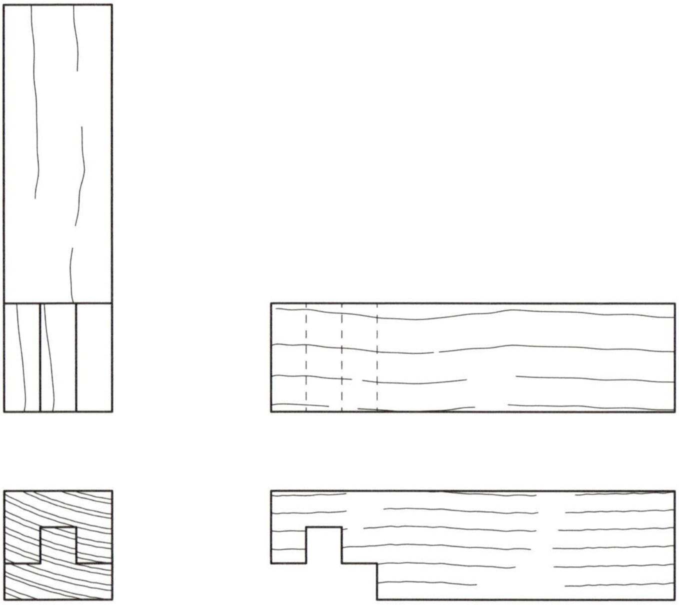

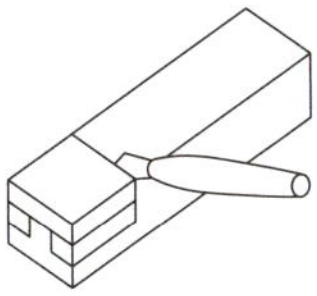

A1

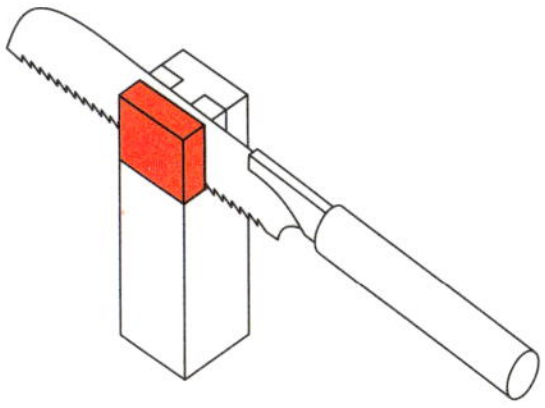

A2

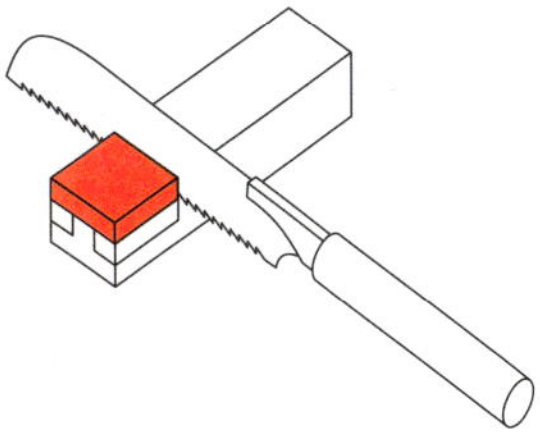

A3

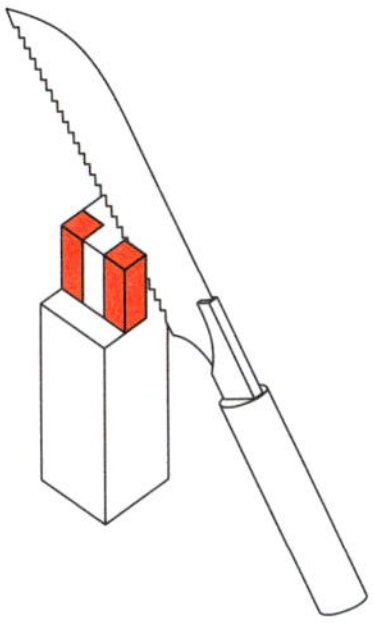

A4

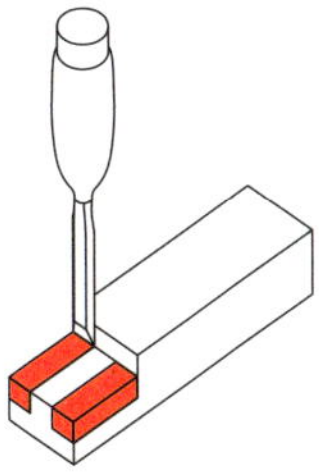

A5

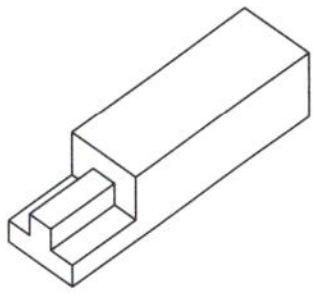

A

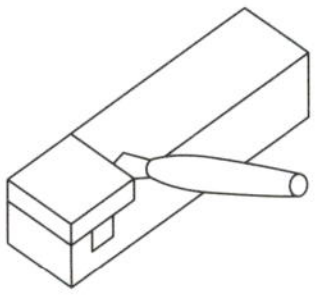

B1

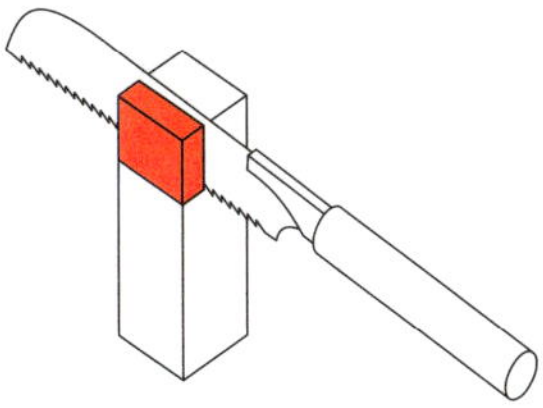

B2

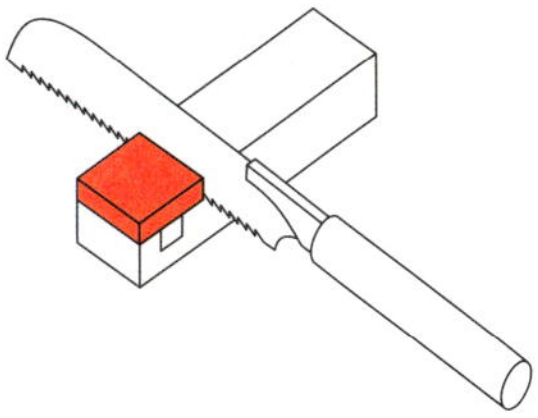

B3

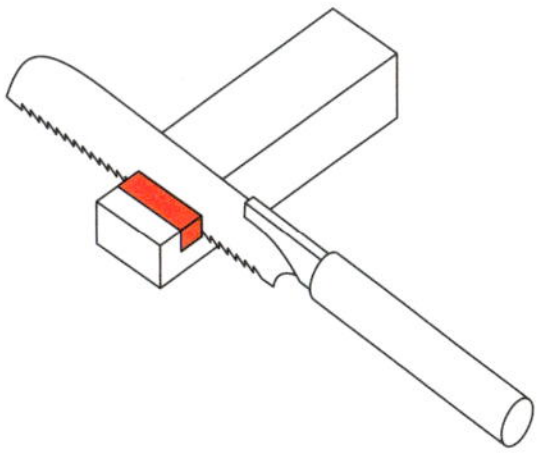

B4

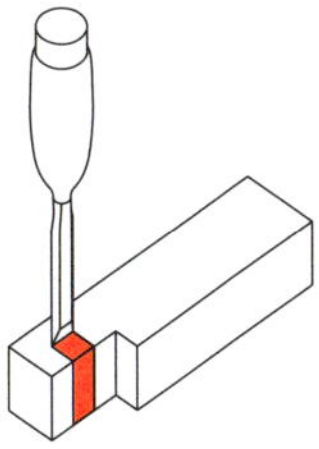

B5

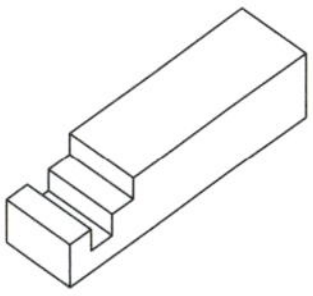

B

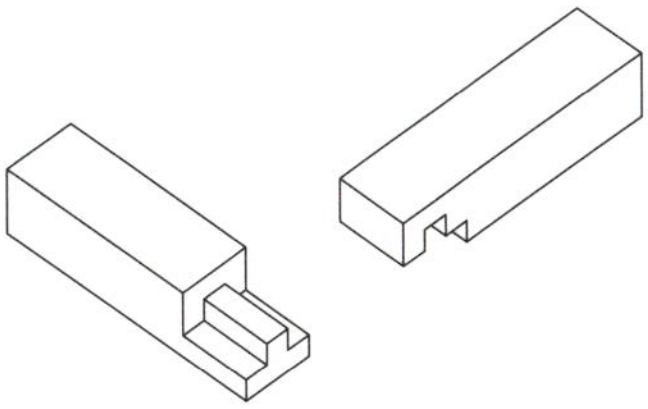

A+B

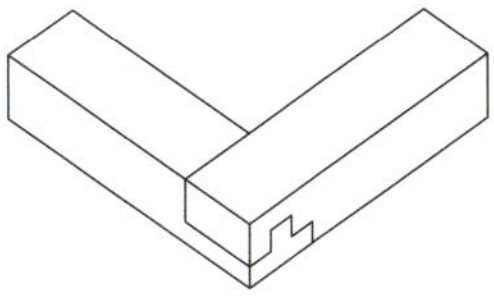

AB

Cogged dovetailed corner lap joint

dō-tsuki-ai-kaki-tsugi (ōari-watari-ago)
胴付相欠接ぎ(大蟻渡顎)

Hakeneckblatt mit schwalbenschwanzförmigem Haken

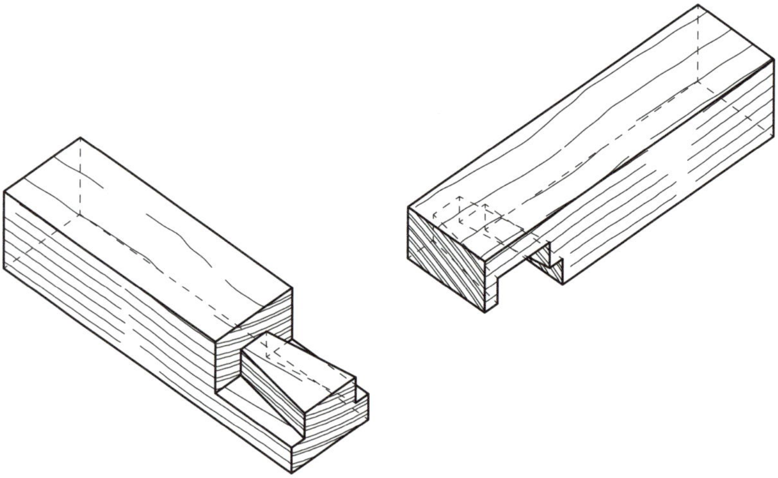

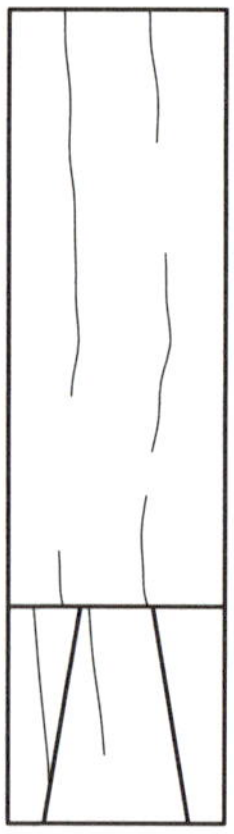

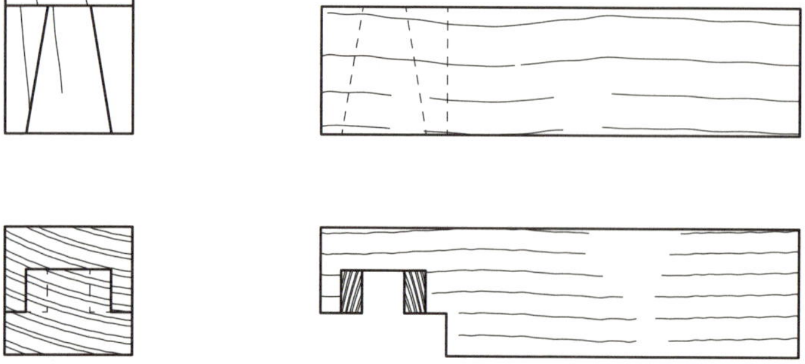

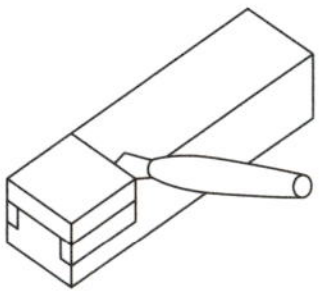
A1

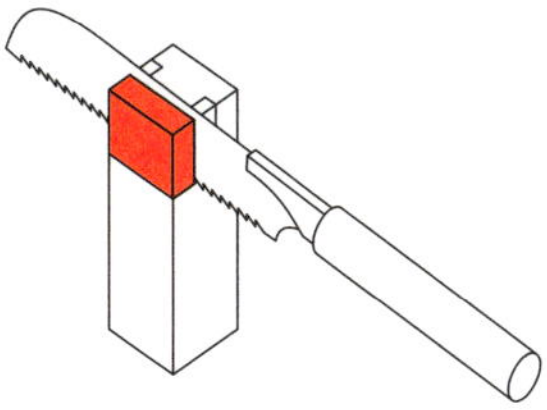
A2

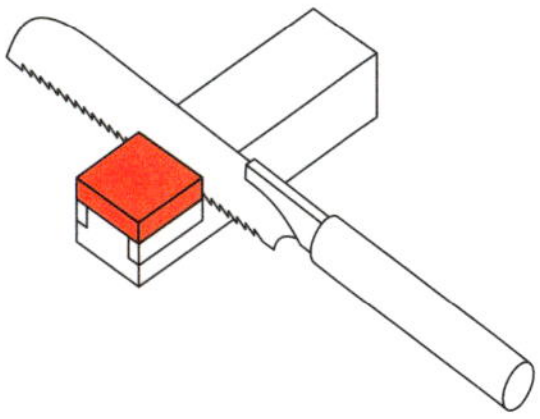
A3

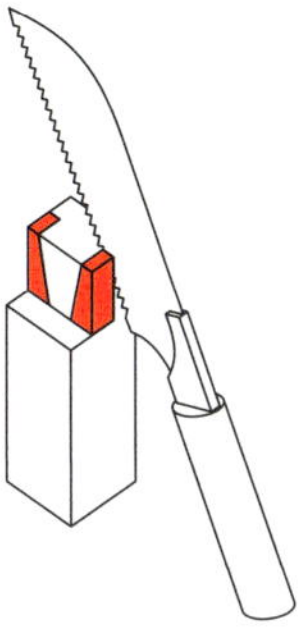
A4

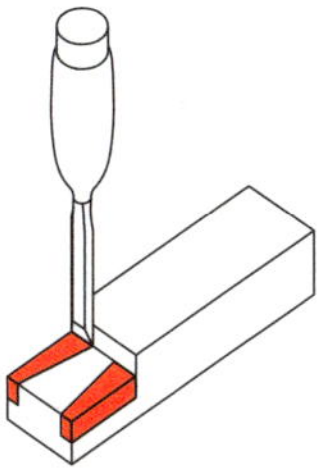
A5

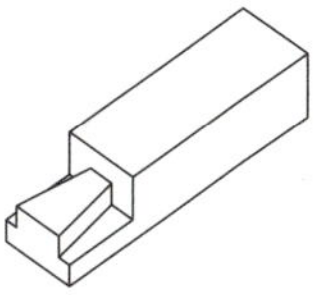
A

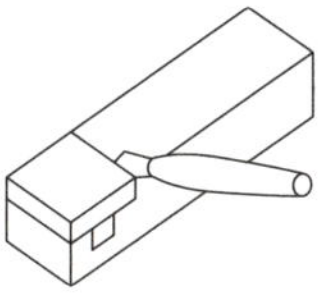
B1

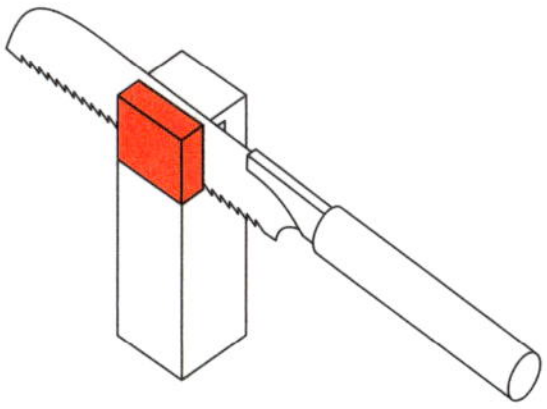
B2

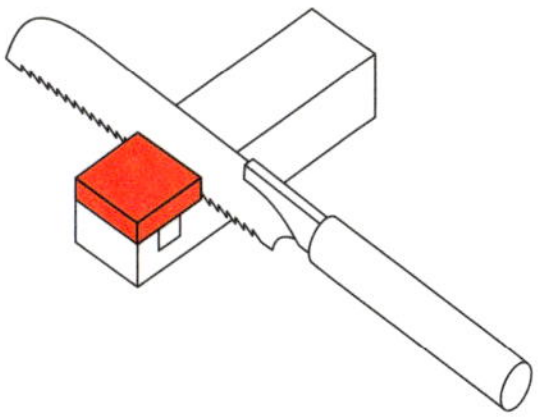
B3

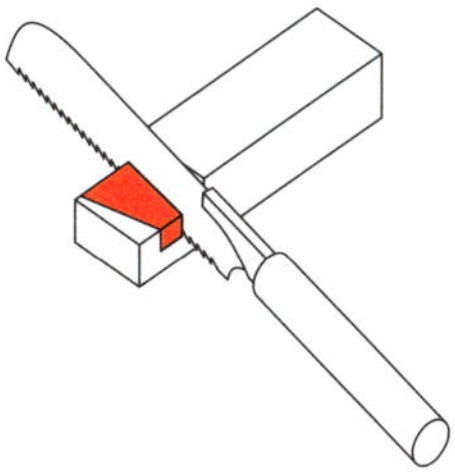
B4

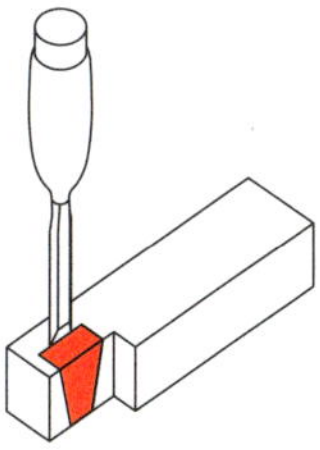
B5

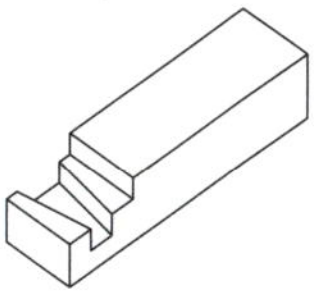
B

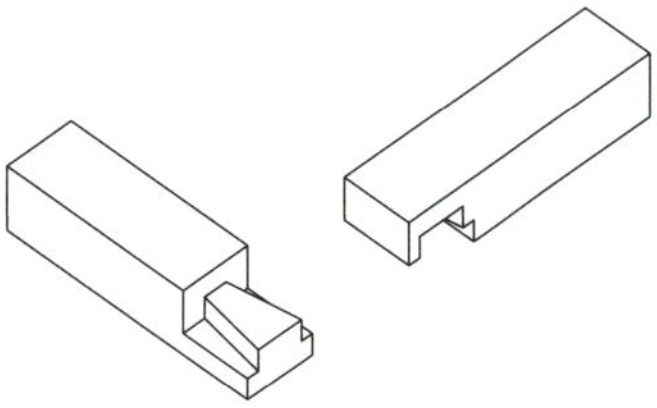
A+B

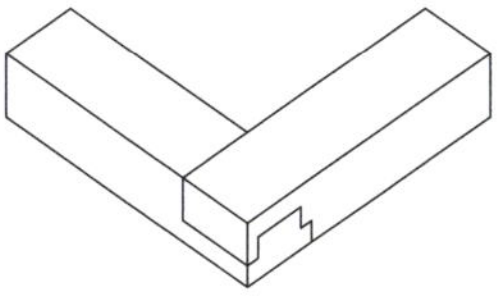
AB

Corner Joints

Half-blind cogged corner lap joint

koshi-kake-ai-kaki-tsugi
腰掛相欠接ぎ

halbverdecktes Hakeneckblatt
mit Wange

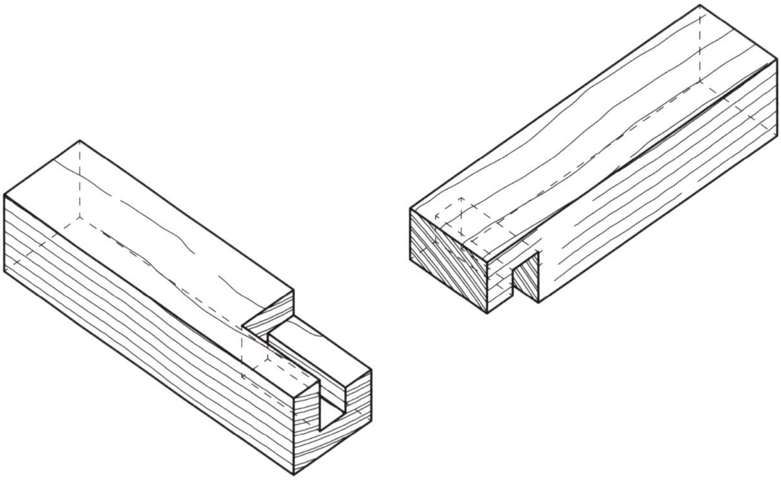

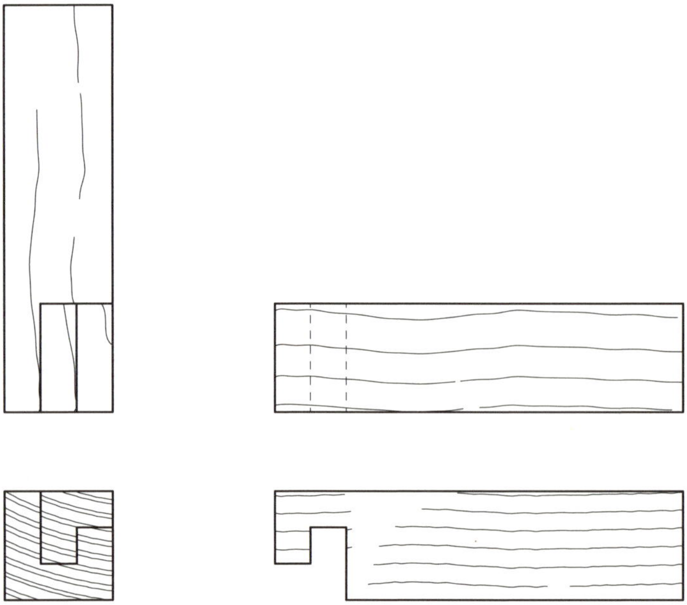

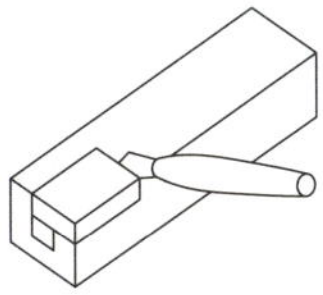

A1

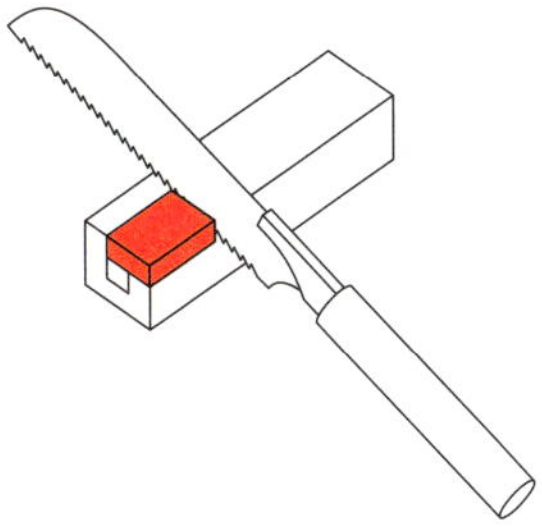

A2

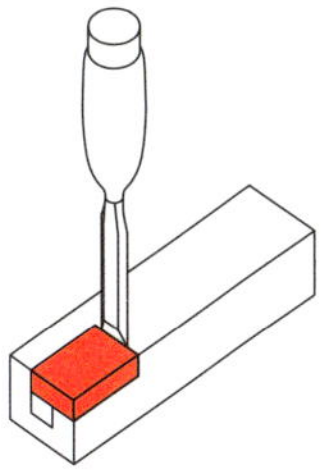

A3

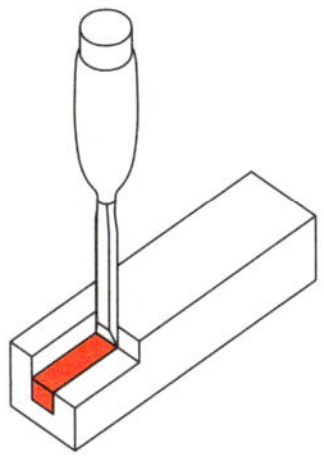

A4

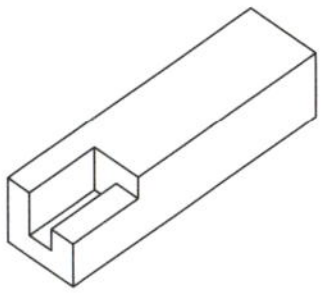

A

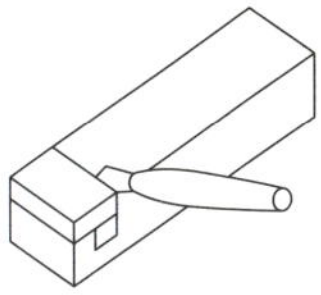

B1

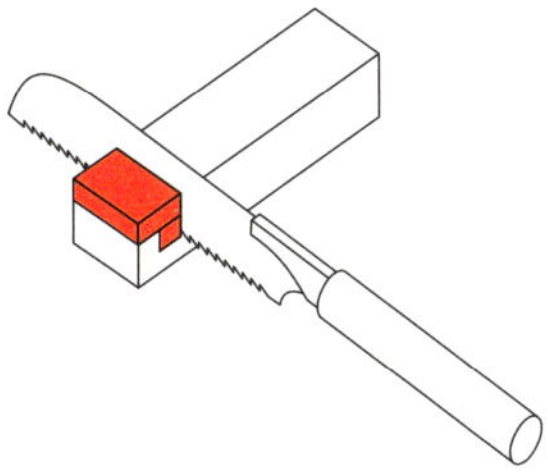

B2

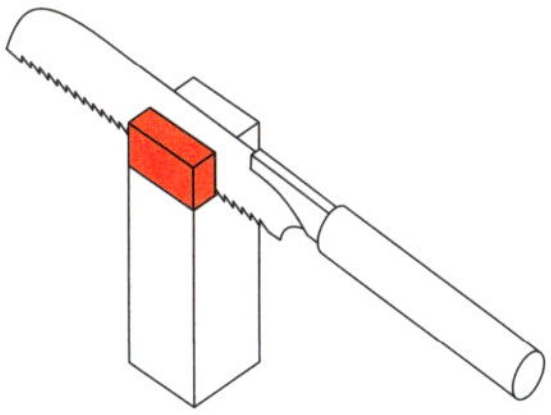

B3

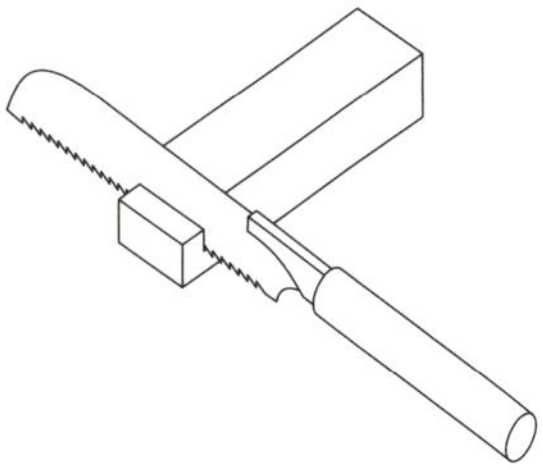

B4

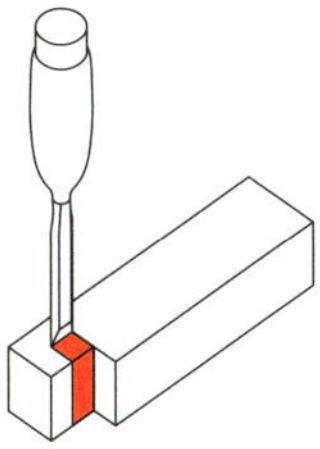

B5

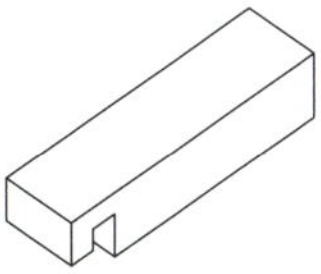

B

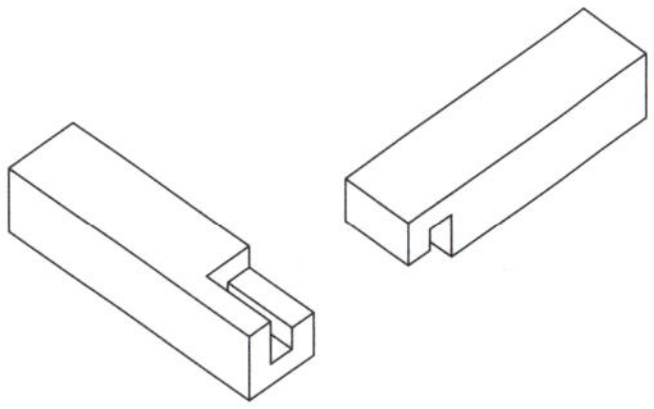

A+B

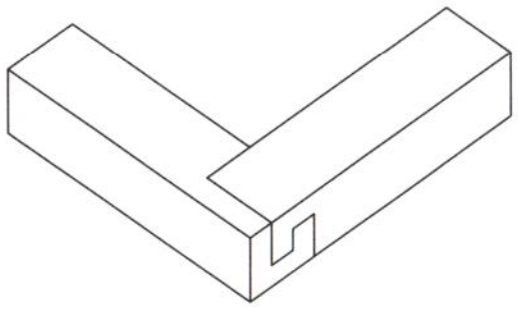

AB

Corner Joints

Half-blind mitered sloped cogged corner lap joint

kakushi-ari-kata-tsugi
隠蟻形接ぎ

halbverdecktes schräges Hakeneckblatt mit Wange auf Gehrung

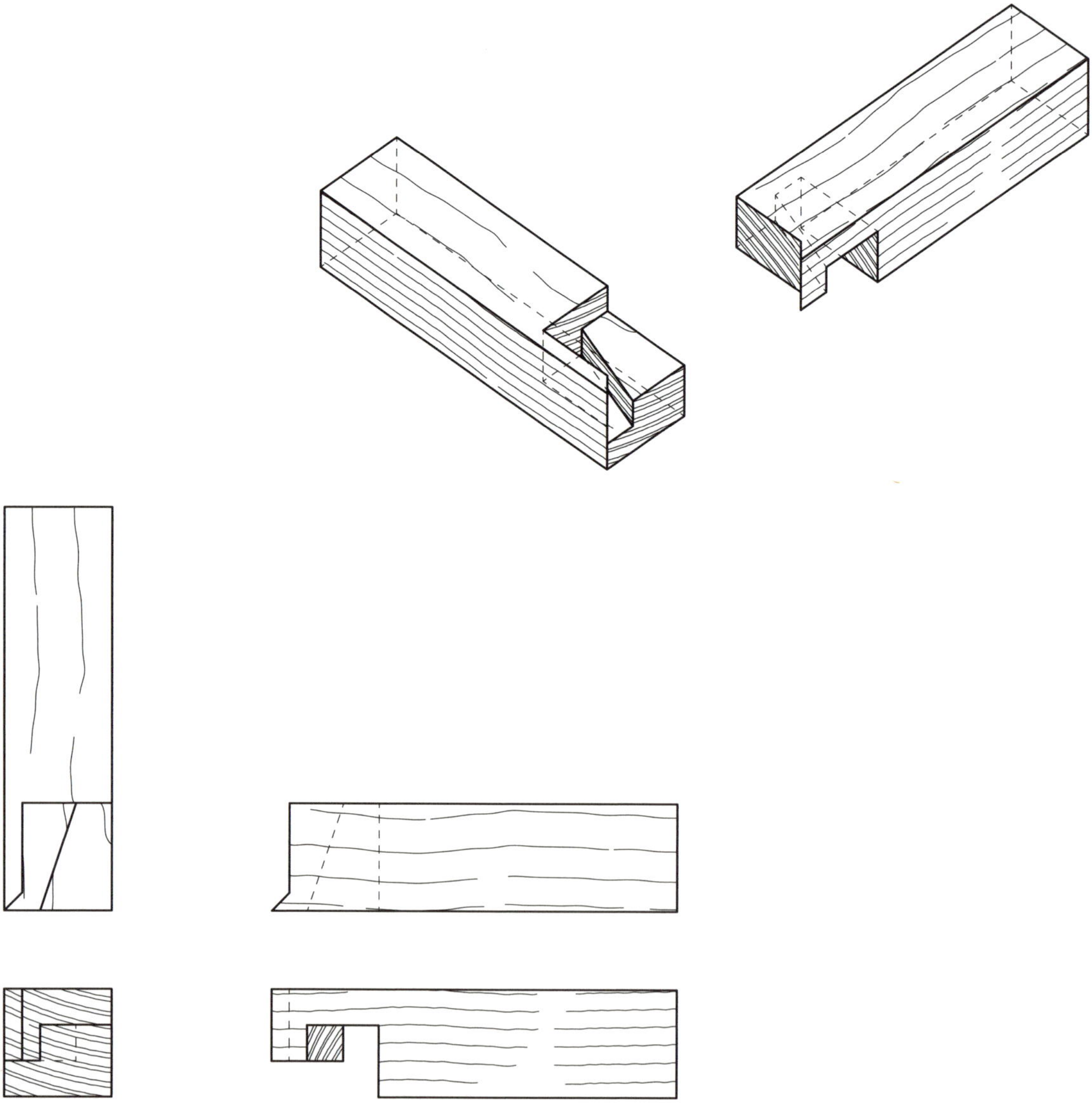

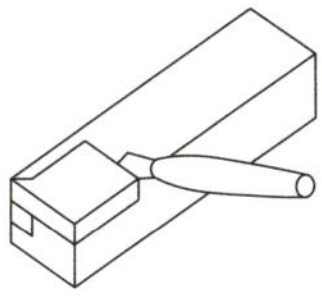

A1

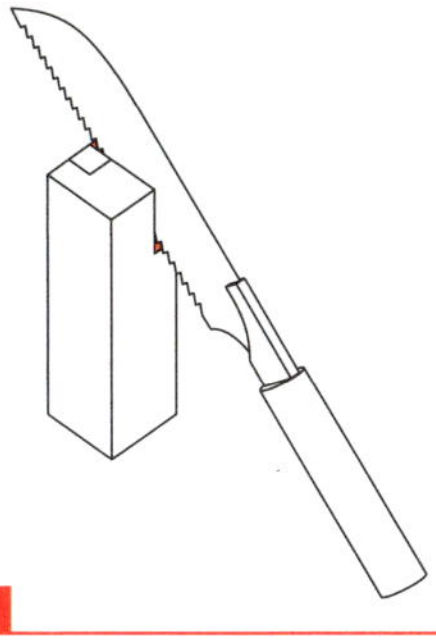

A2

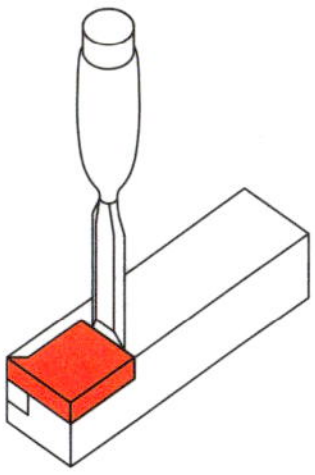

A3

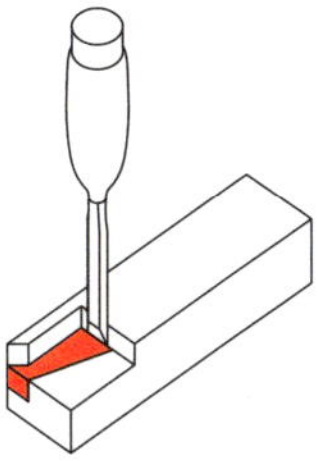

A4

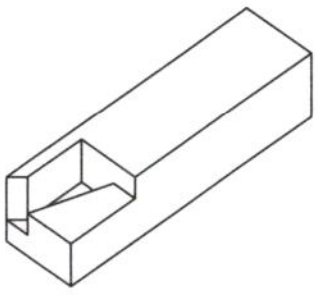

A

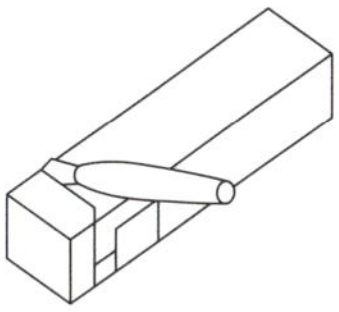

B1

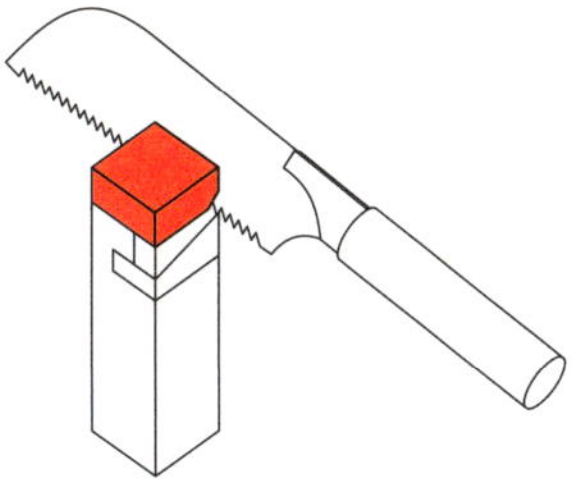

B2

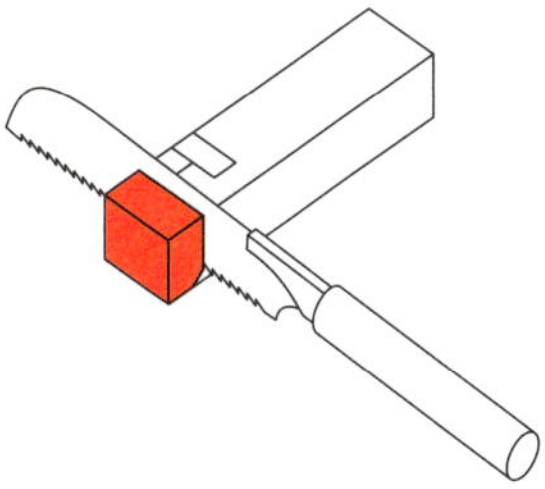

B3

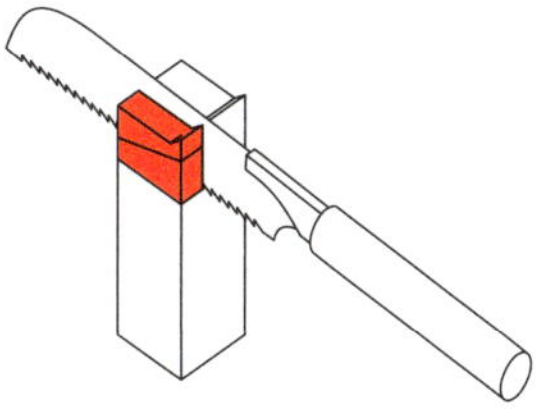

B4

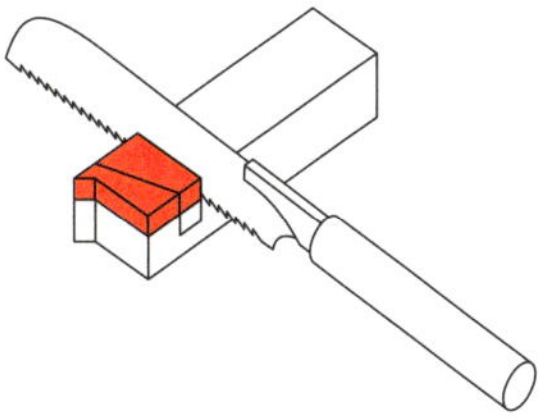

B5

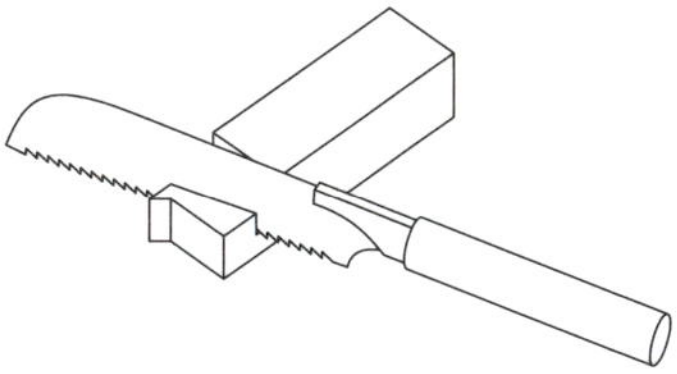

B6

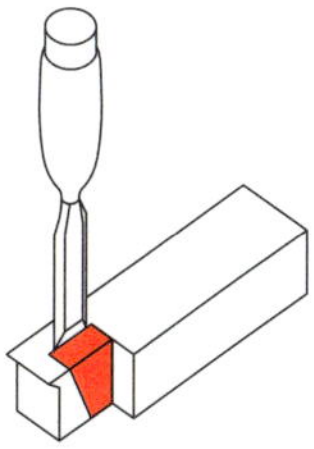

B7

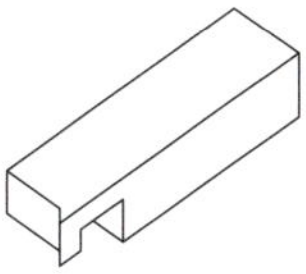

B

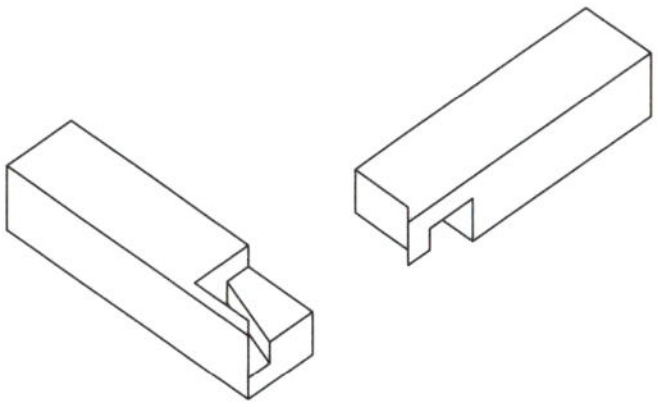

A+B

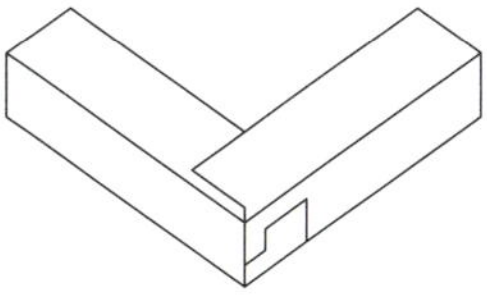

AB

Corner Joints

Blind mitered cogged corner lap joint

kakushi-ai-kaki-tsugi
隠相欠接ぎ

verdecktes Hakeneckblatt mit Wange auf Gehrung

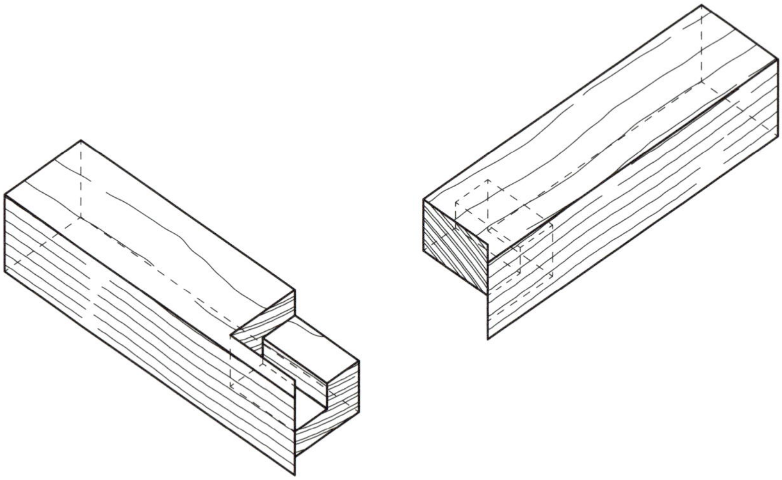

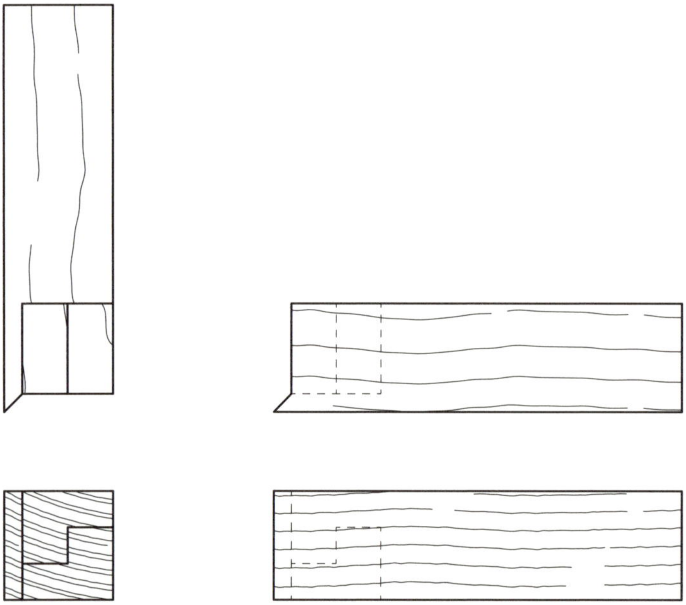

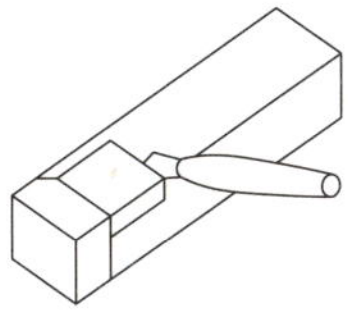
A1

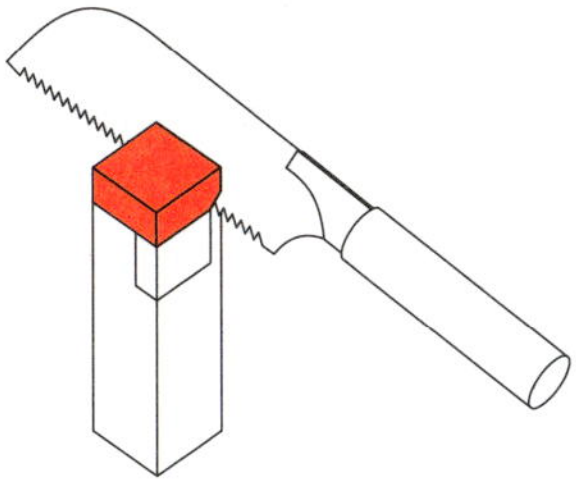
A2

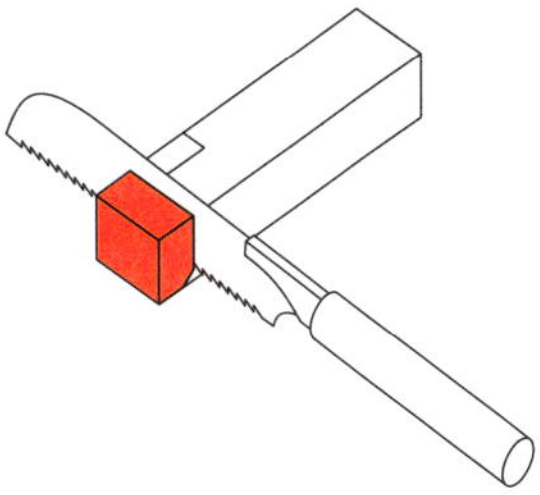
A3

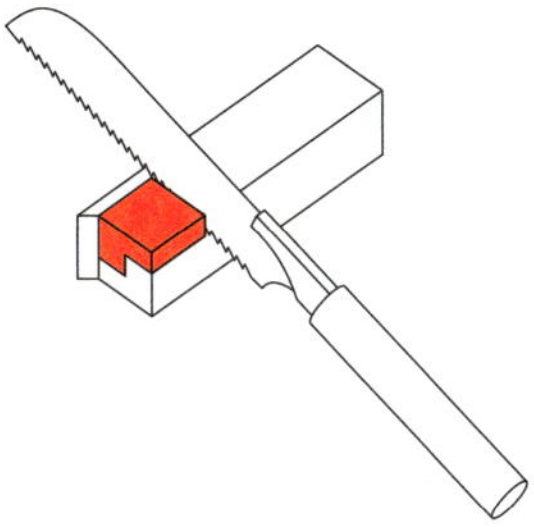
A4

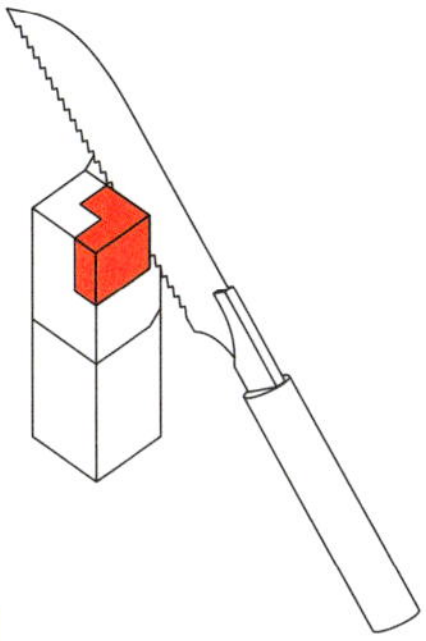
A5

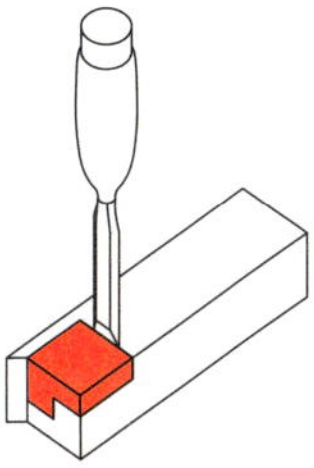
A6

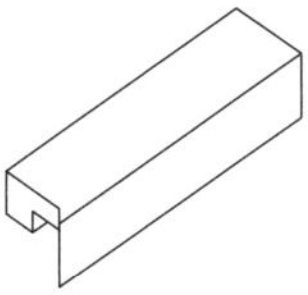
A

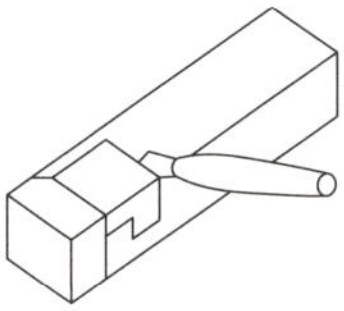
B1

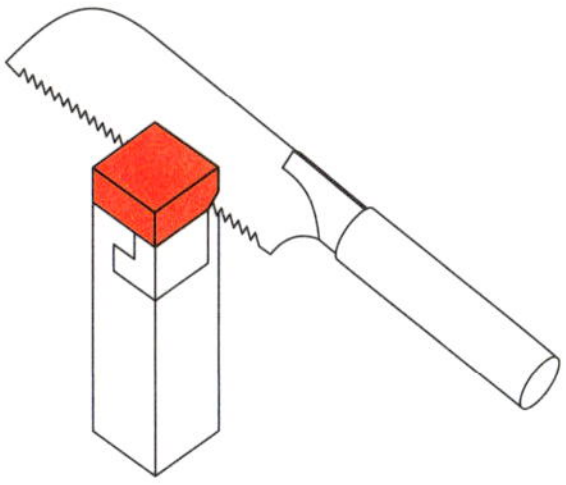
B2

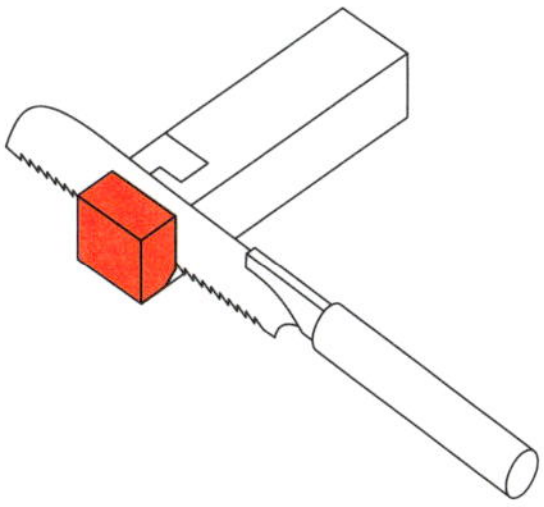
B3

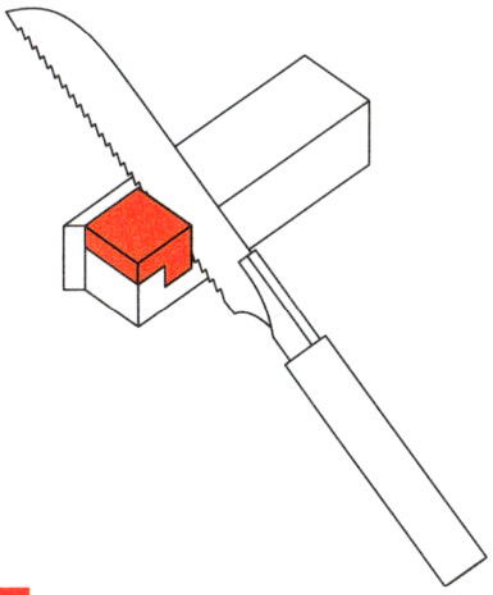
B4

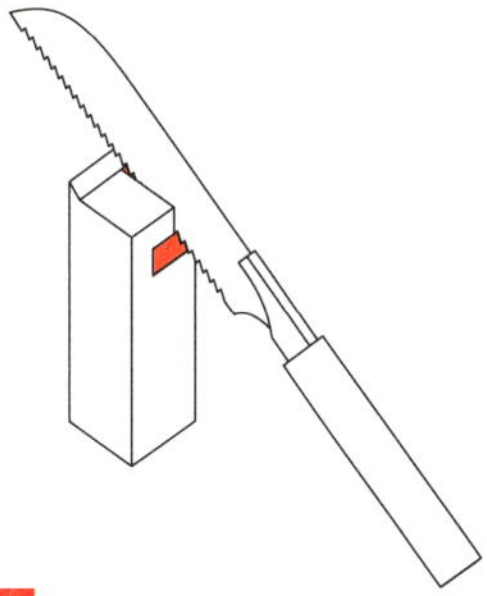
B5

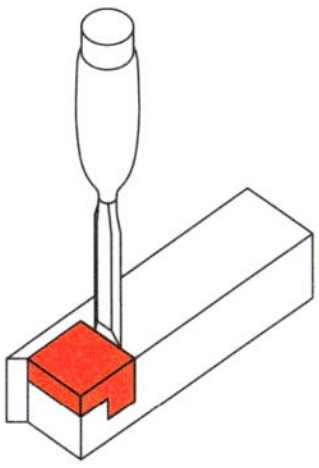
B6

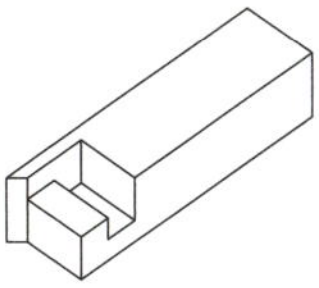
B

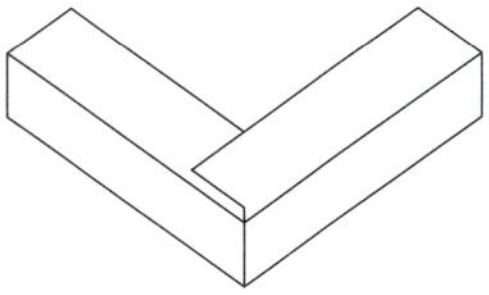
AB

Corner Joints

Half-blind dovetailed corner lap joint

ari-tsugi
蟻接ぎ

halbes Schwalbenschwanzeckblatt

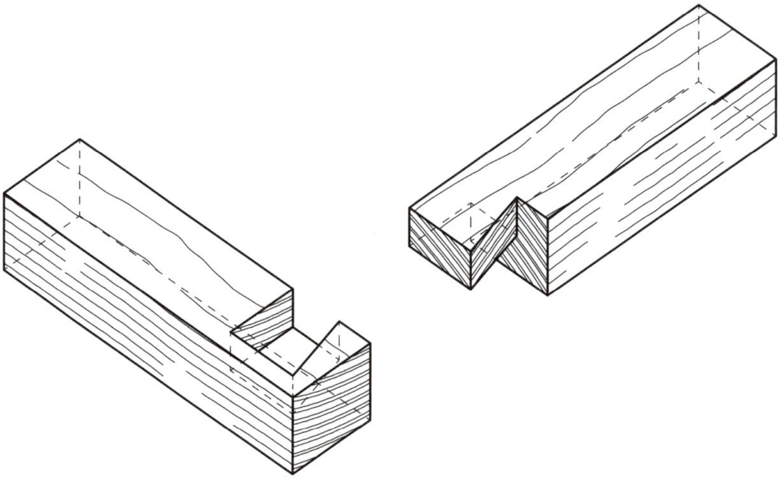

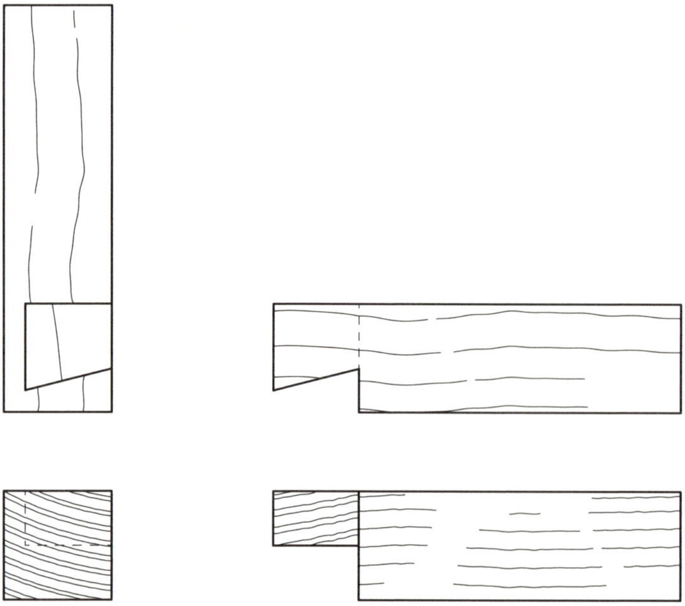

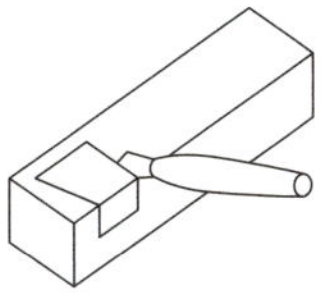

A1

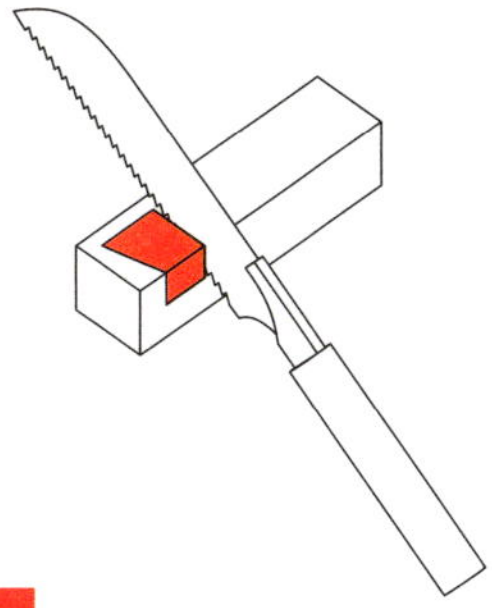

A2

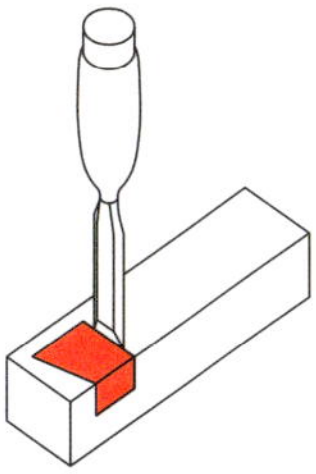

A3

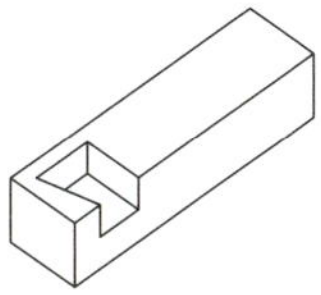

A

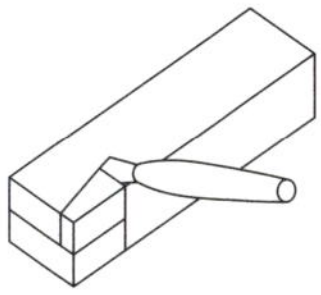

B1

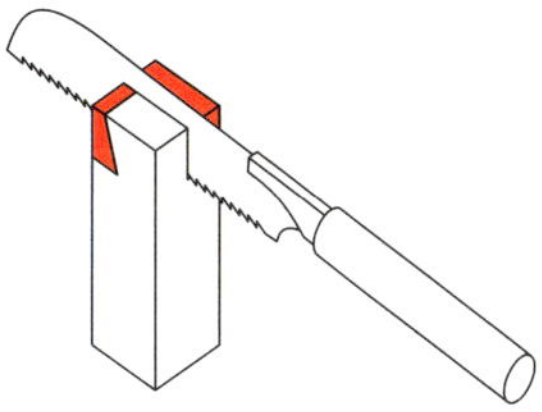

B2

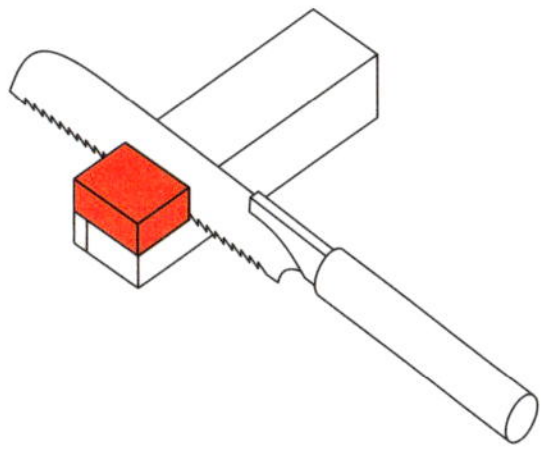

B3

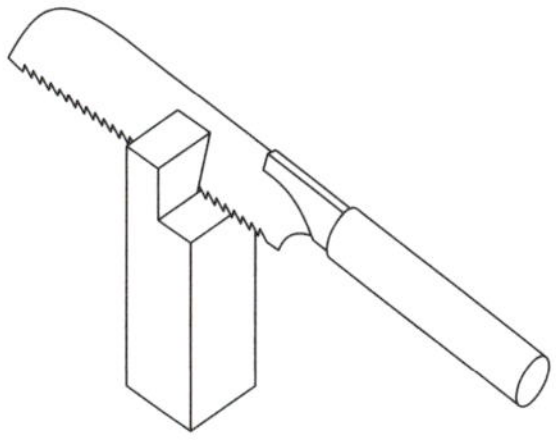

B4

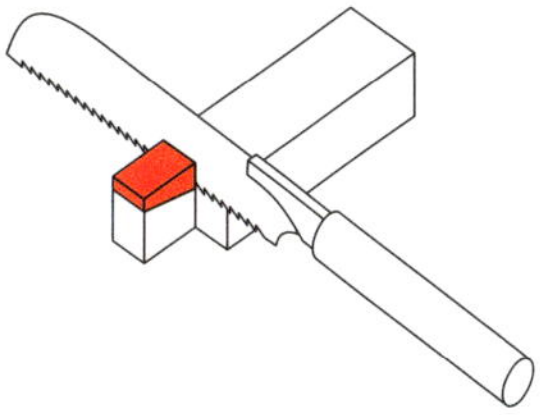

B5

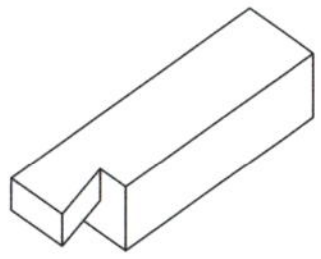

B

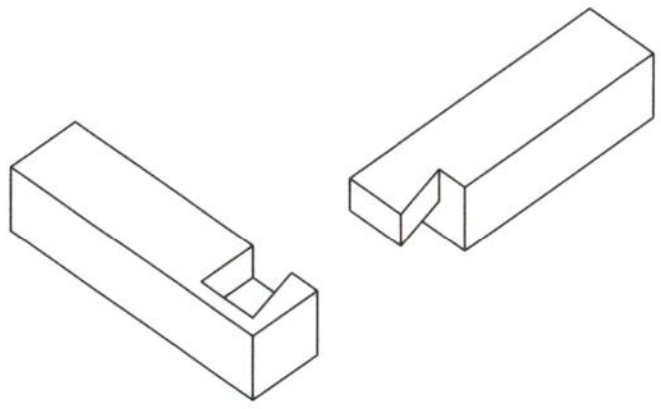

A+B

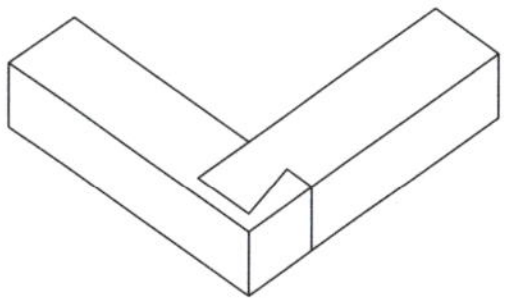

AB

Corner Joints

Rabbeted tongue and dado corner joint

tsutsumi-hame-komi-uchi-tsuke-tsugi
包嵌込打付接ぎ

Eckstoß mit doppeltem Führungszapfen

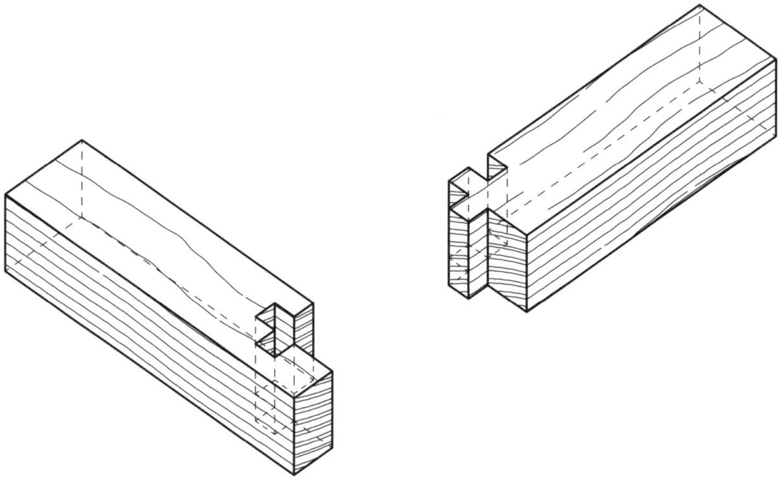

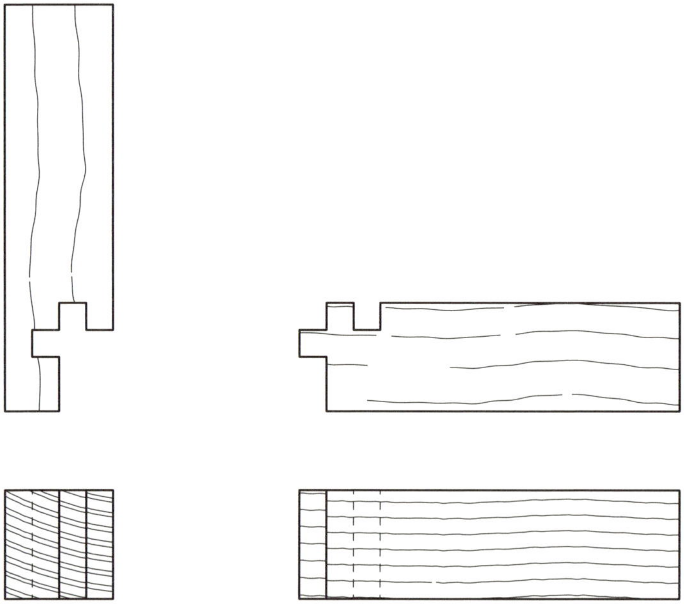

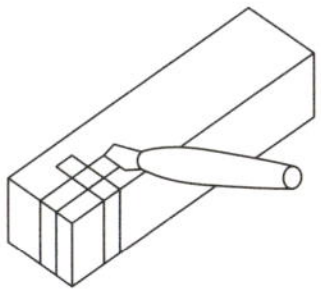
A1

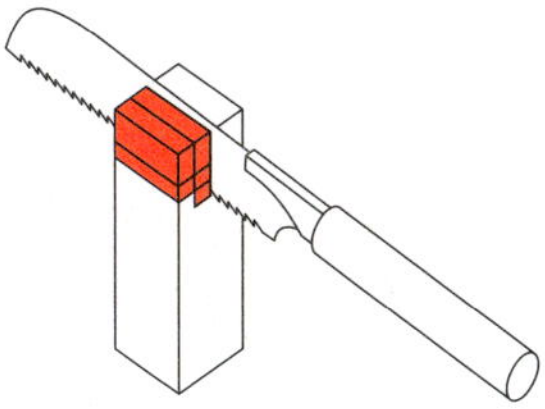
A2

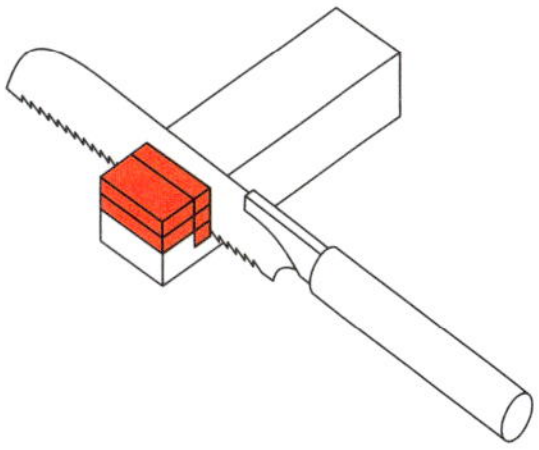
A3

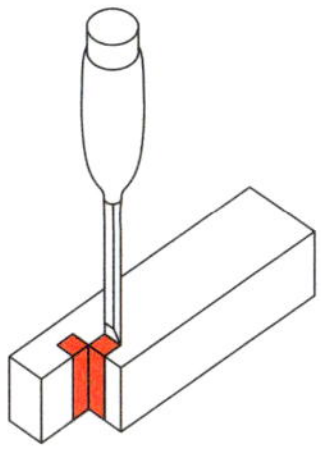
A4

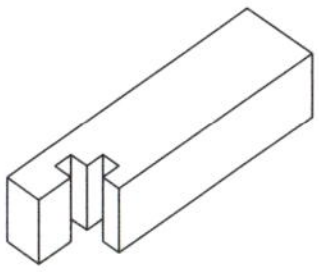
A

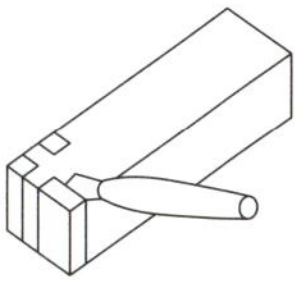
B1

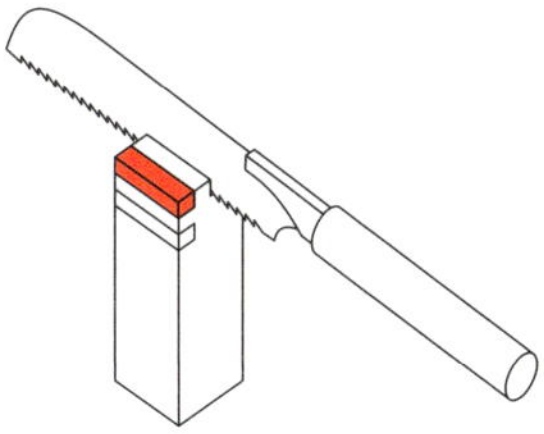
B2

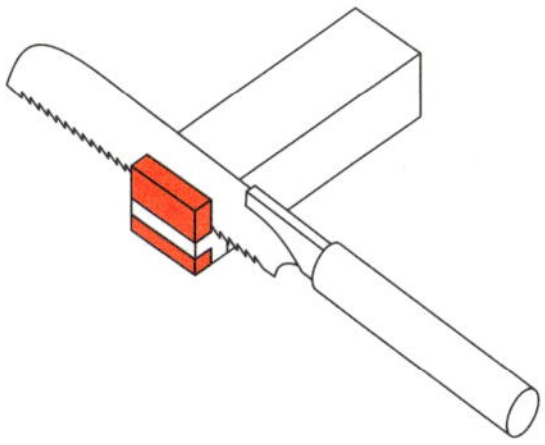
B3

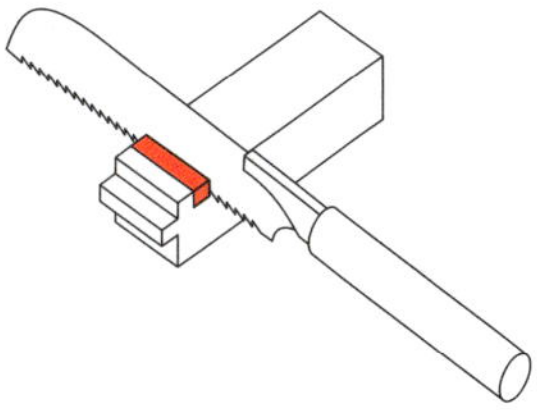
B4

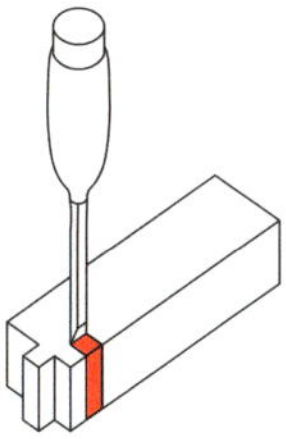
B5

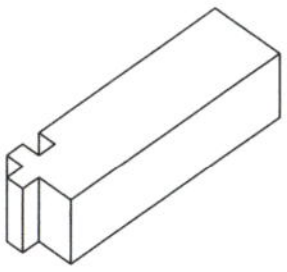
B

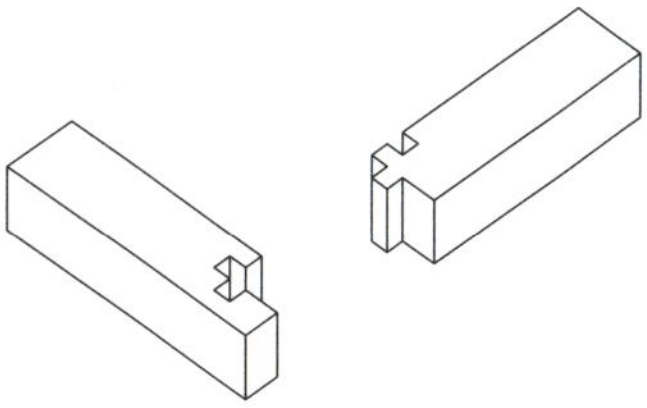
A+B

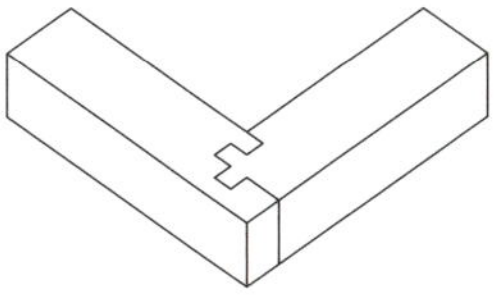
AB

Corner Joints

Rabbeted tenoned corner joint

tsutsumi-hame-komi-uchi-tsuke-tsugi (binta-tsuki)
包嵌込打付接ぎ(鬢太付)

abgesetzter Eckstoß
mit Führungszapfen

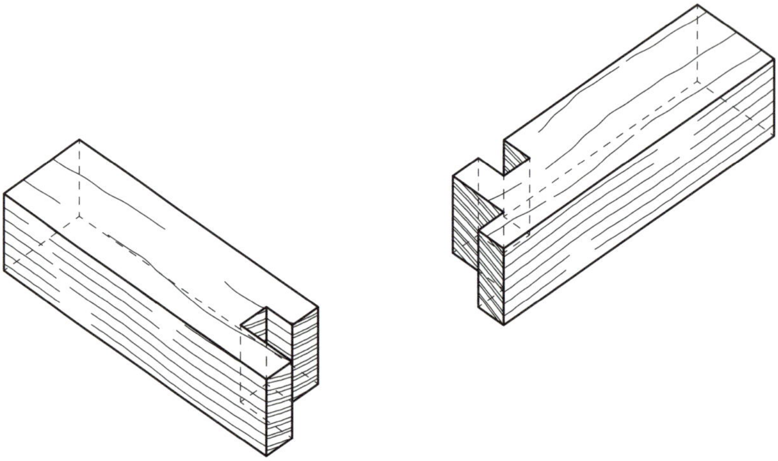

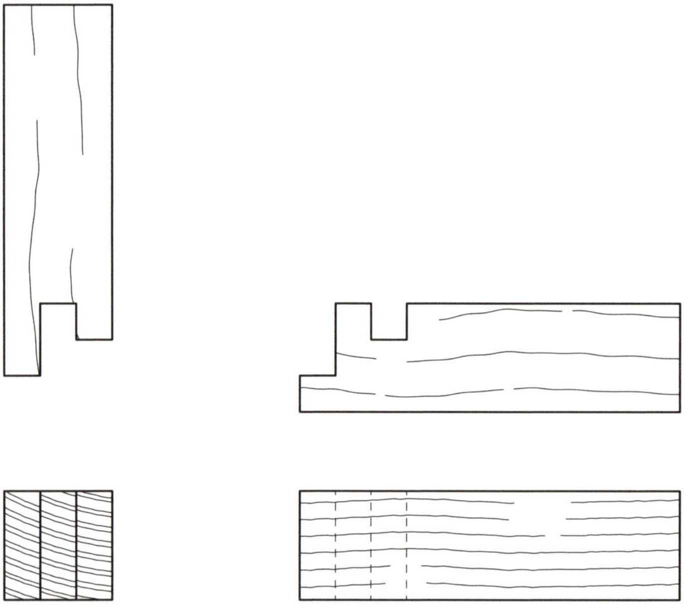

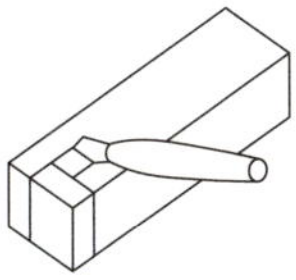

A1

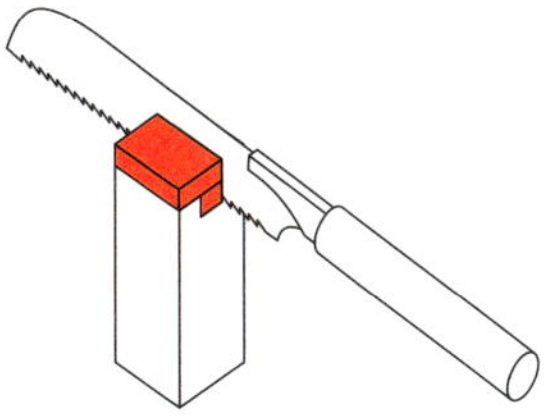

A2

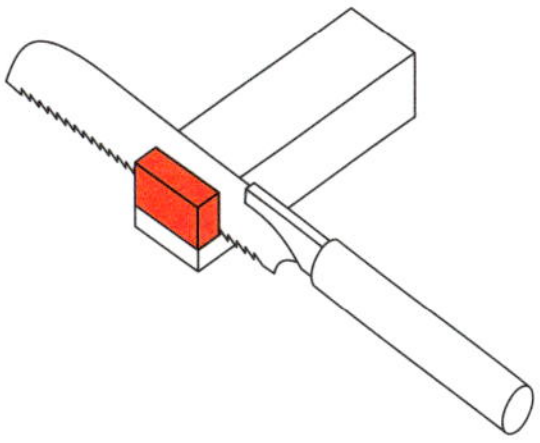

A3

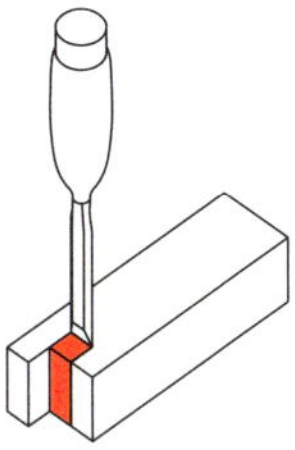

A4

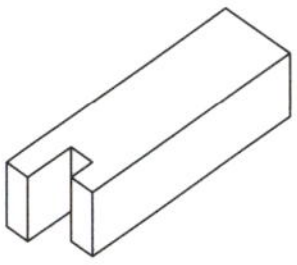

A

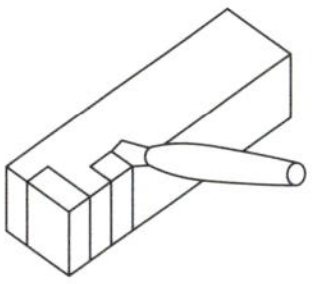

B1

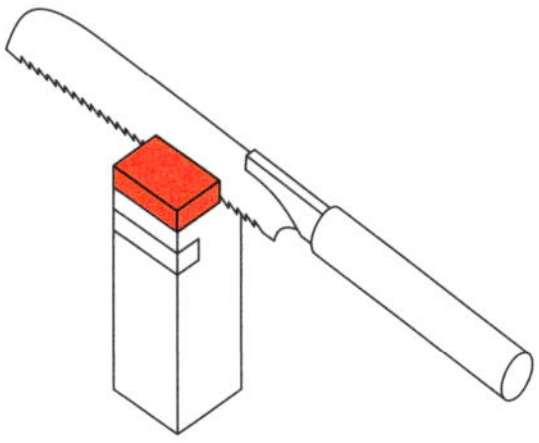

B2

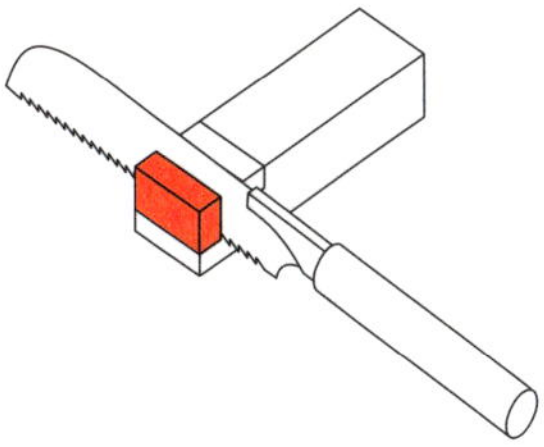

B3

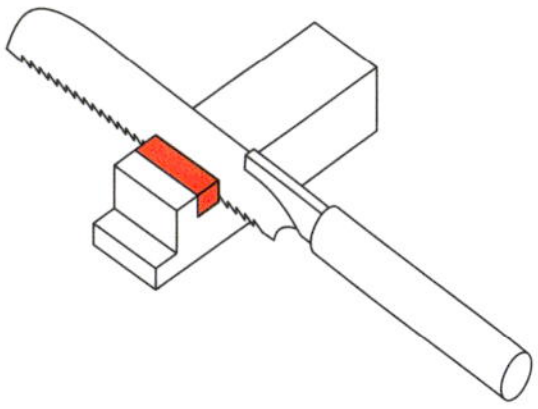

B4

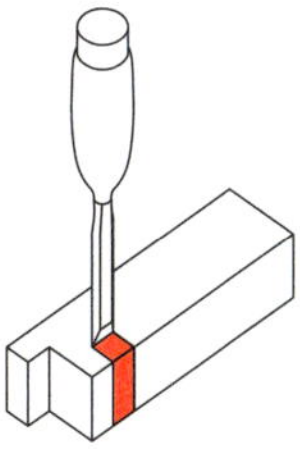

B5

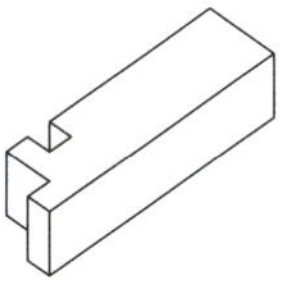

B

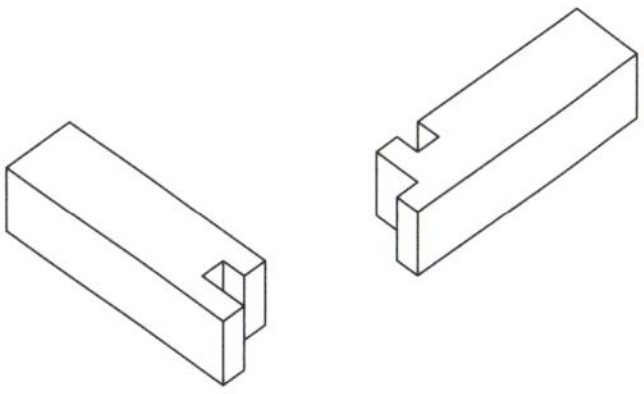

A+B

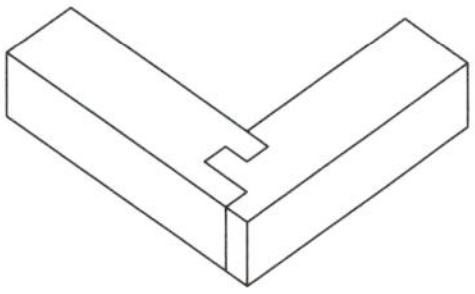

AB

Corner Joints

Rabbeted tenoned miter joint

hako-dome-tsugi
箱留接ぎ

abgesetzter Eckstoß auf Gehrung
mit Führungszapfen

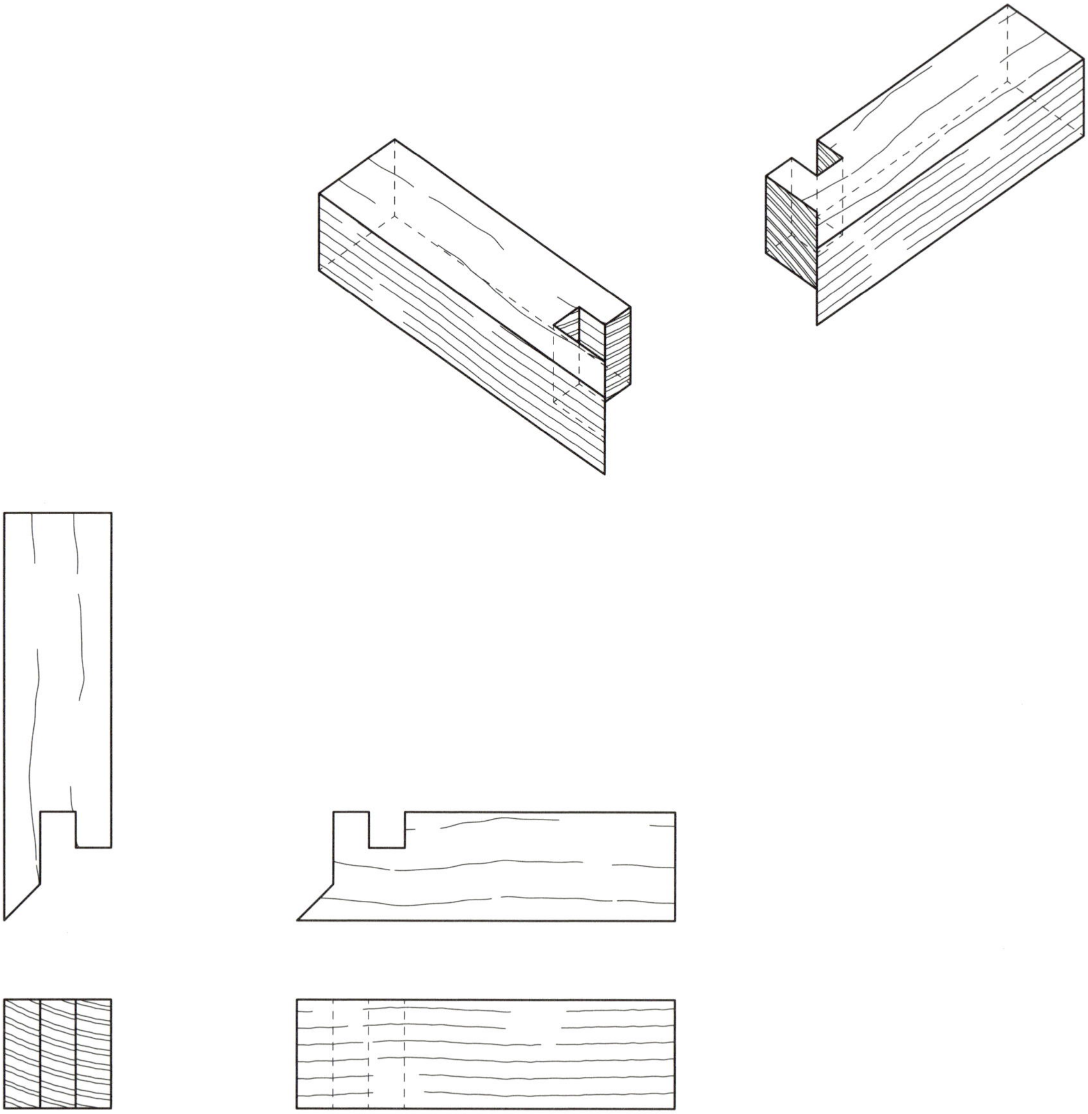

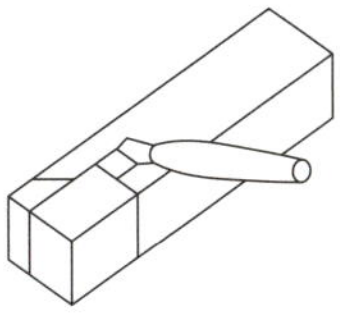
A1

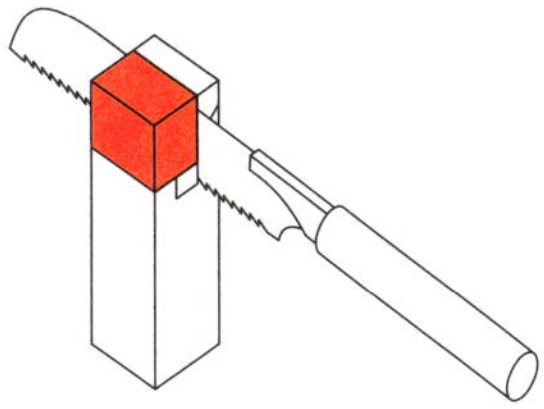
A2

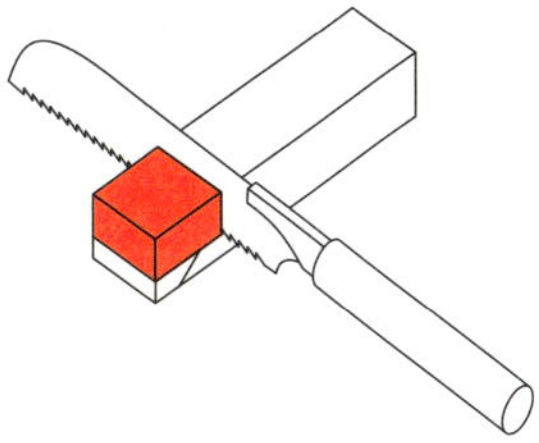
A3

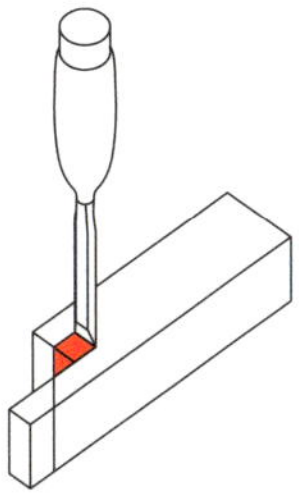
A4

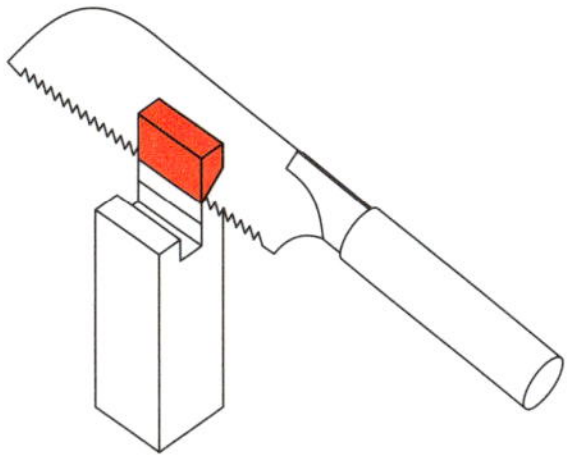
A5

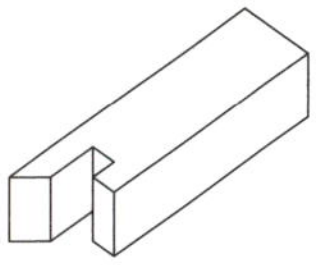
A

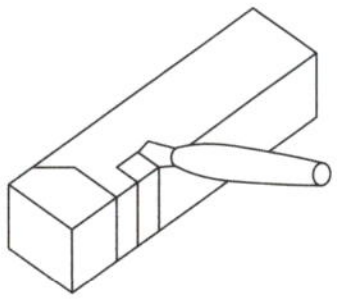
B1

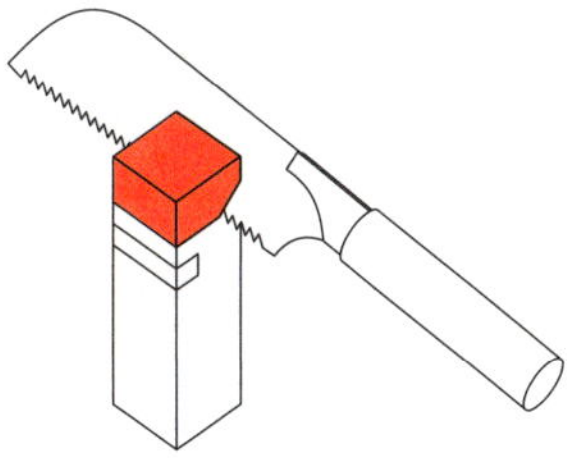
B2

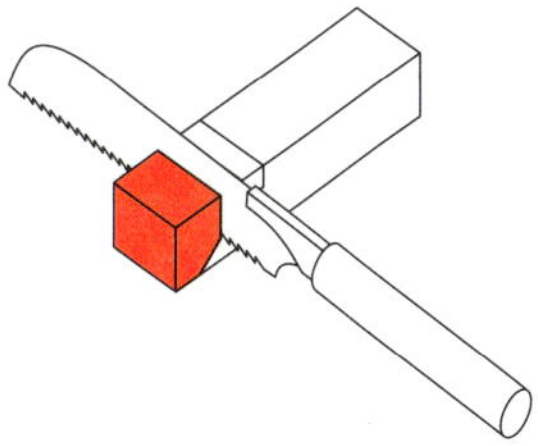
B3

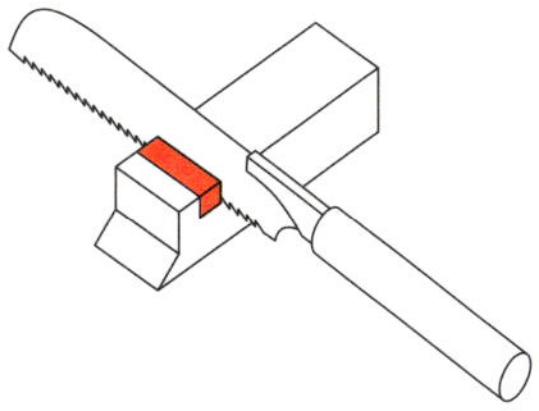
B4

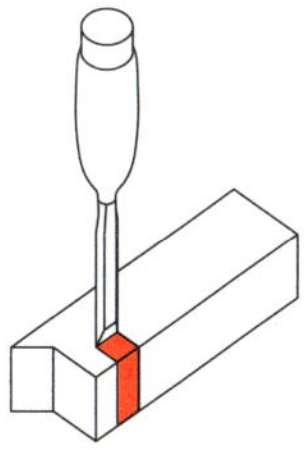
B5

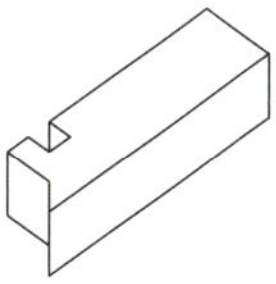
B

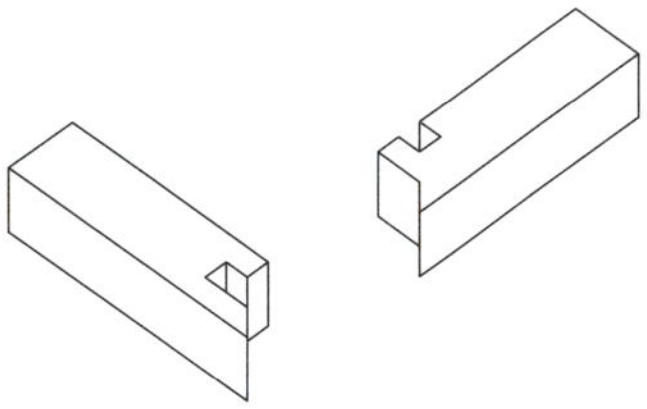
A+B

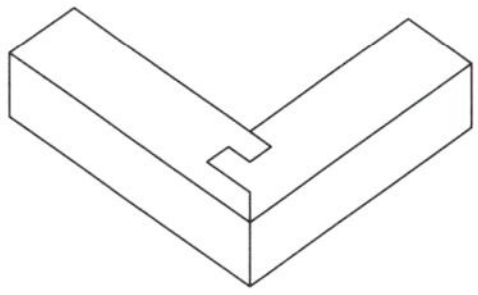
AB

Corner Joints

Bridle joint

wanagi-komi-tsugi
輪薙込接ぎ

Scherzapfen

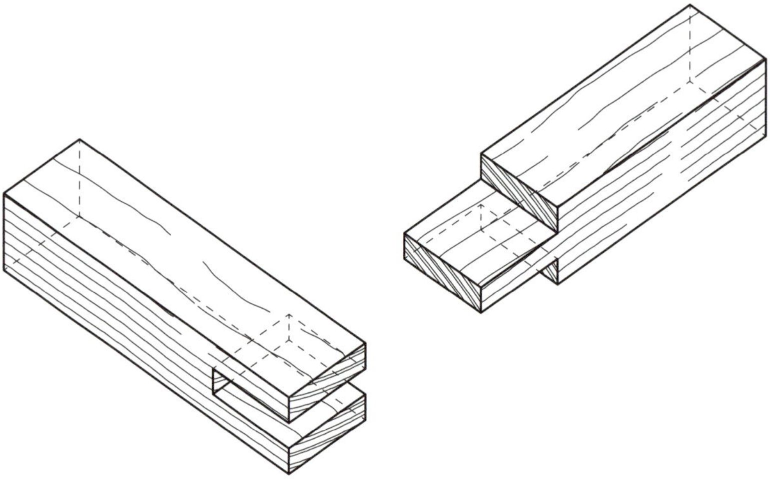

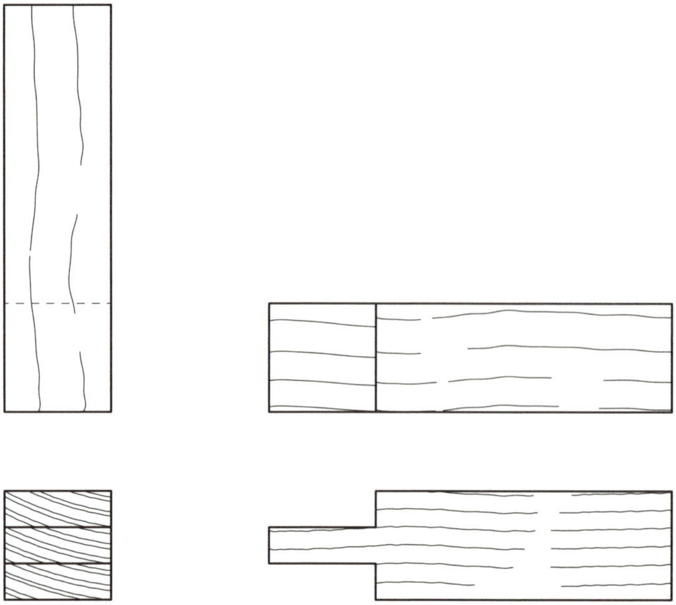

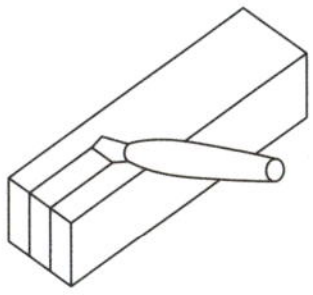

A1

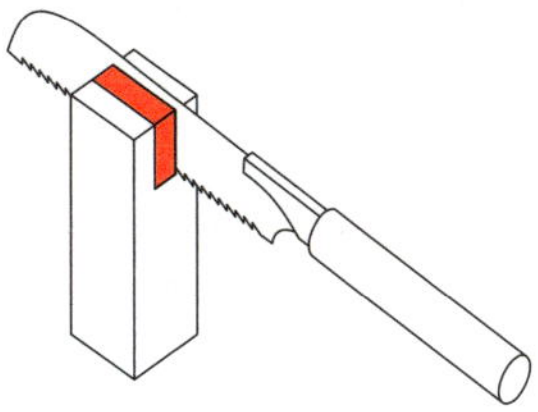

A2

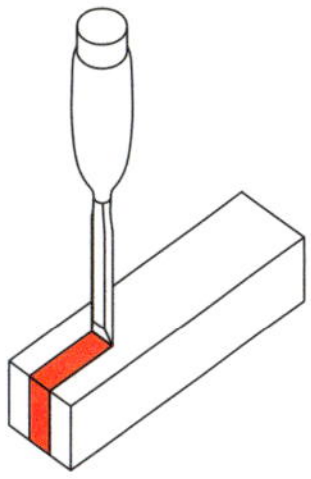

A3

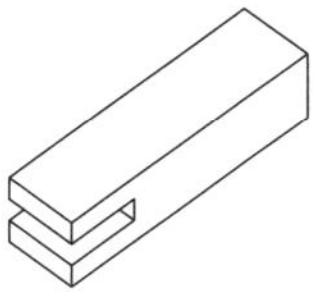

A

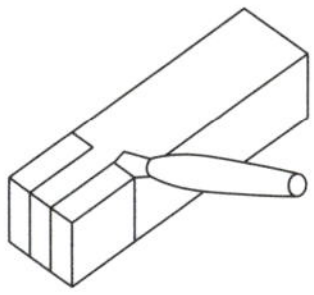

B1

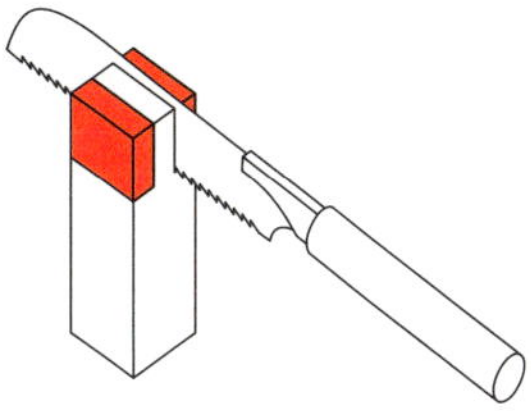

B2

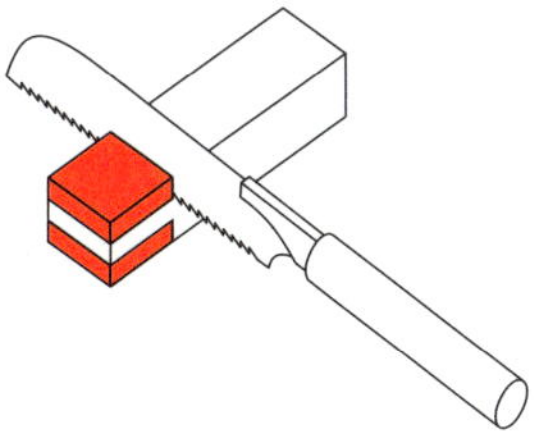

B3

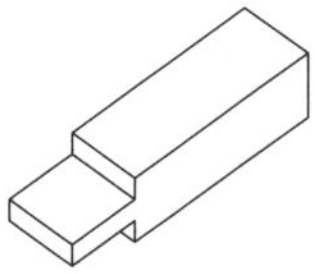

B

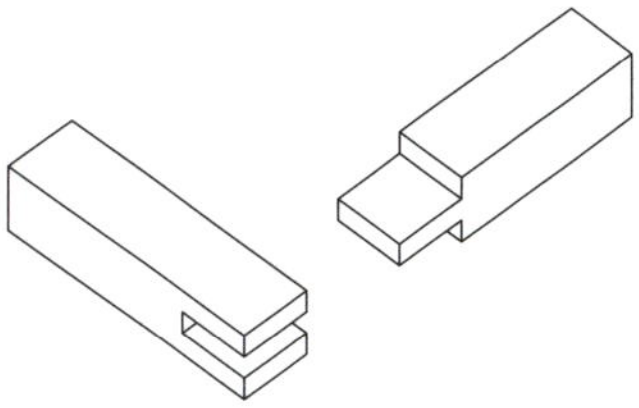

A+B

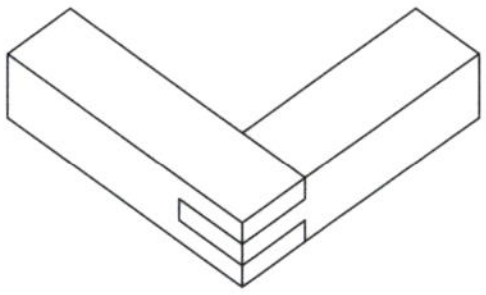

AB

Corner Joints

Twin bridle joint

wanagi-komi-nimai-hozo-tsugi
輪薙込二枚枘接ぎ

doppelter Scherzapfen

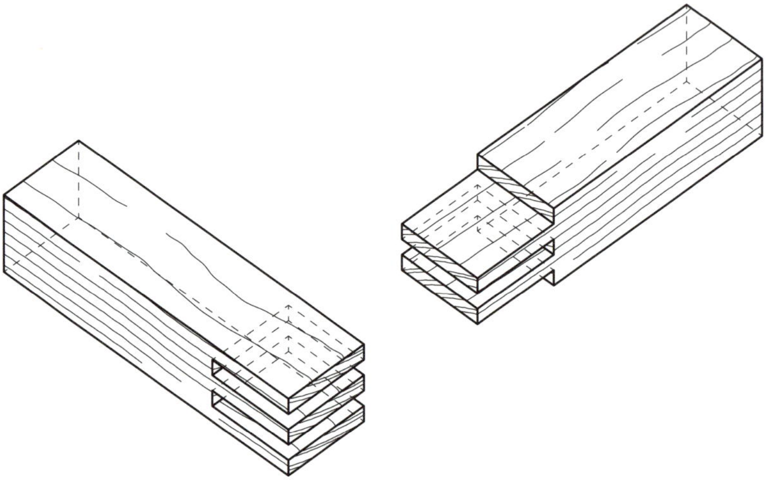

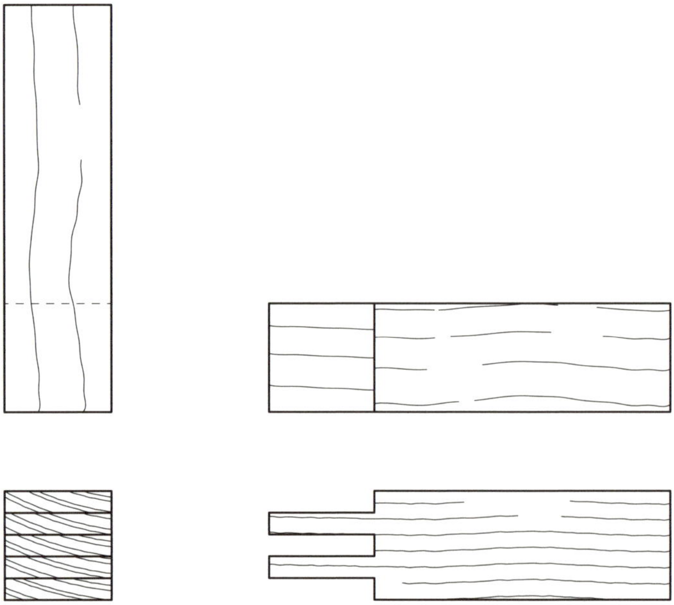

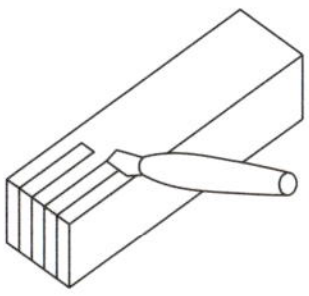

A1

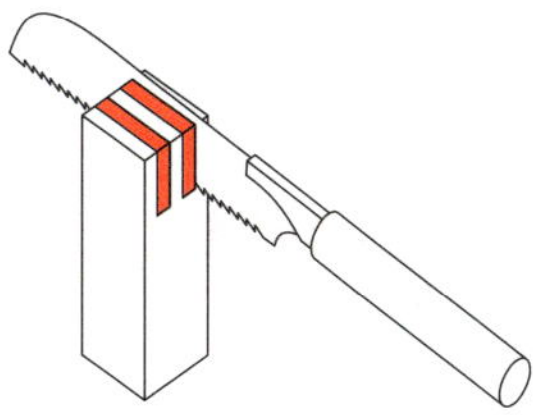

A2

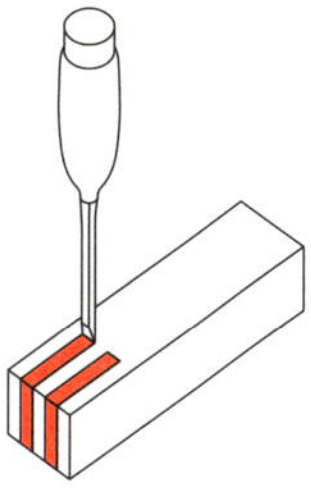

A3

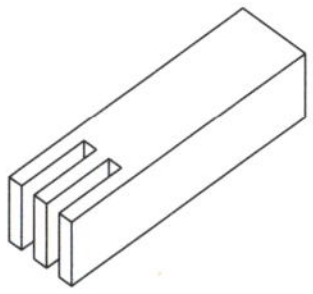

A

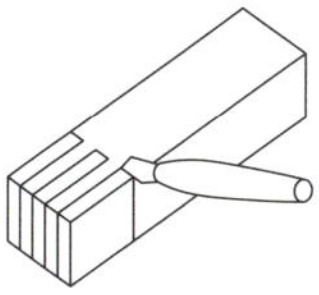

B1

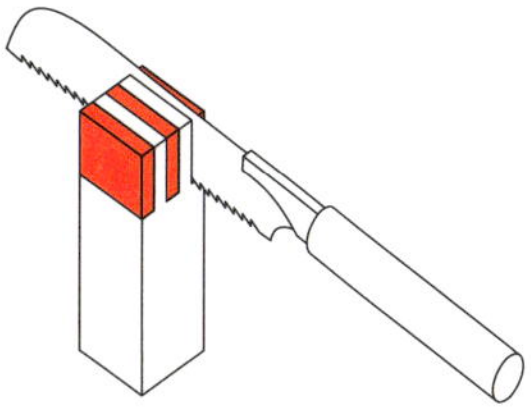

B2

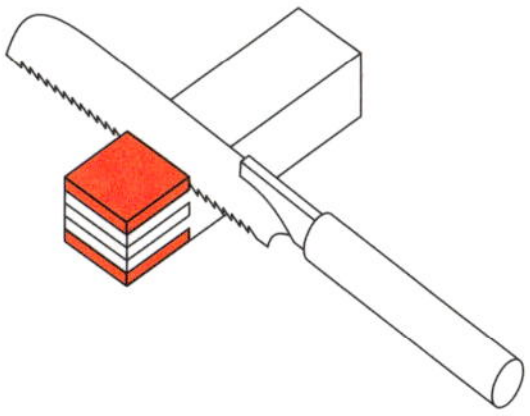

B3

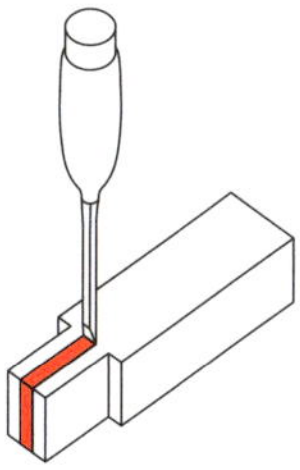

B4

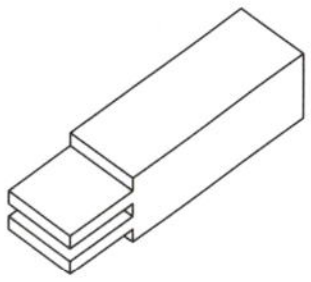

B

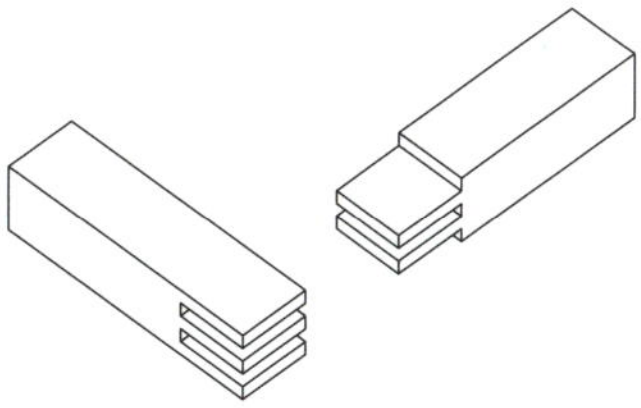

A+B

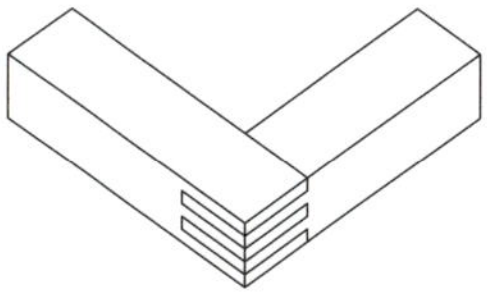

AB

Corner Joints

Through dovetail tenon joint

ōgi-hozo-tsugi
扇枘接ぎ

durchgehender Schwalben-
schwanzeckzapfen

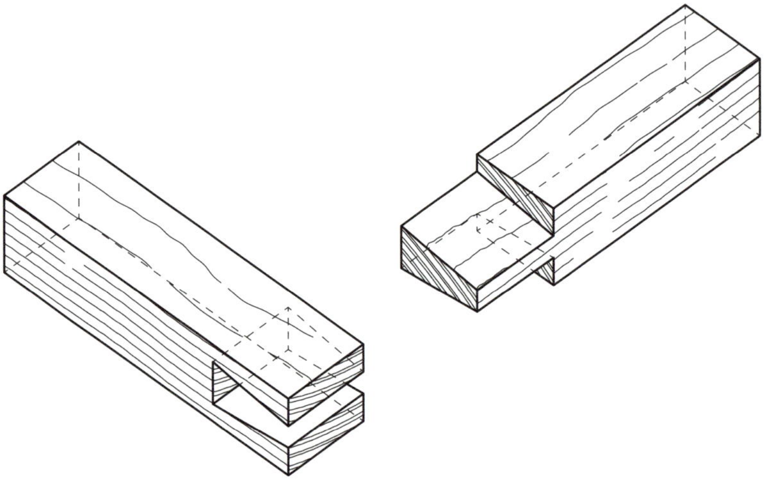

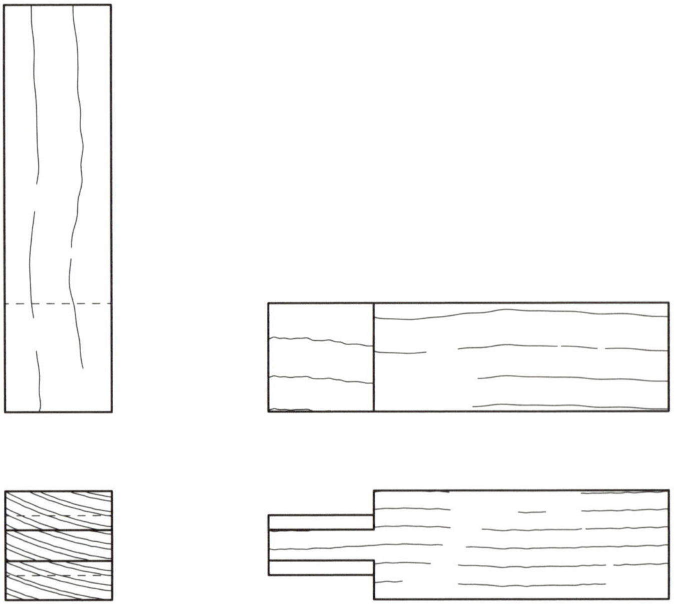

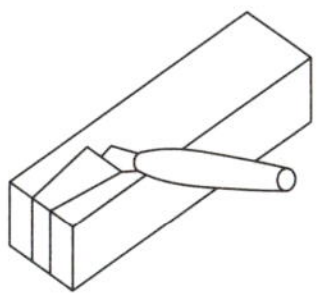

A1

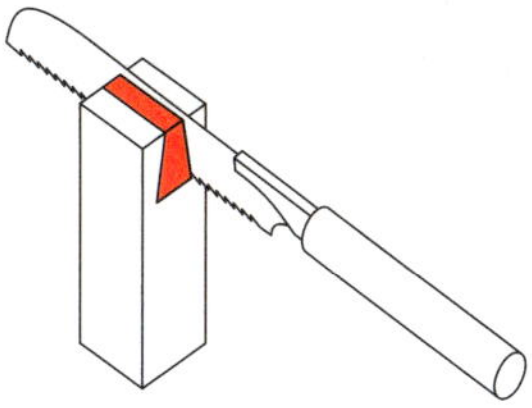

A2

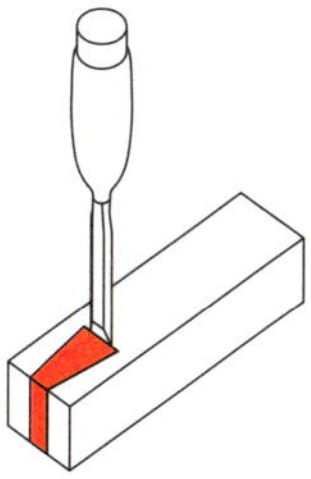

A3

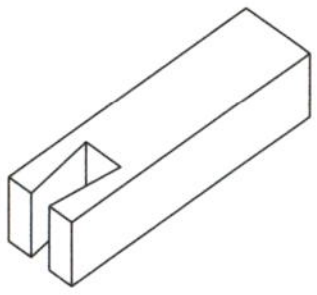

A

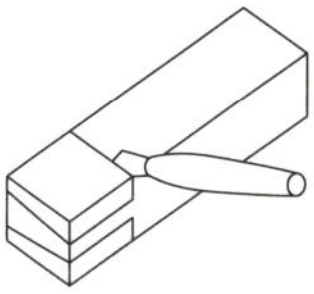

B1

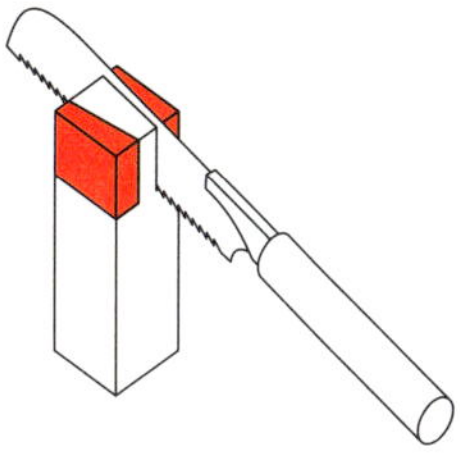

B2

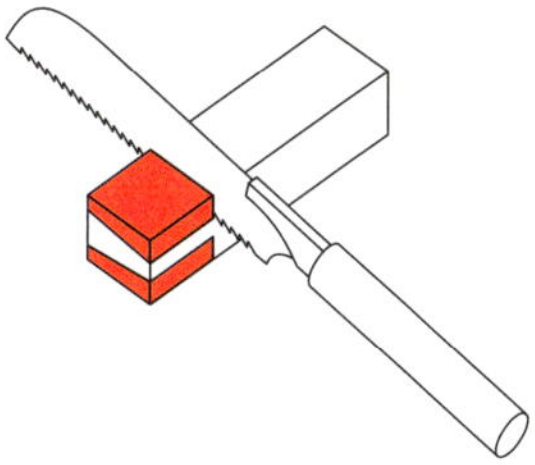

B3

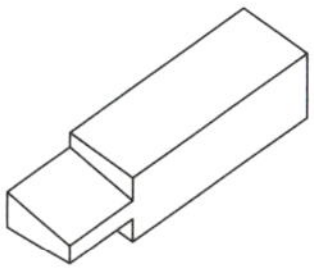

B

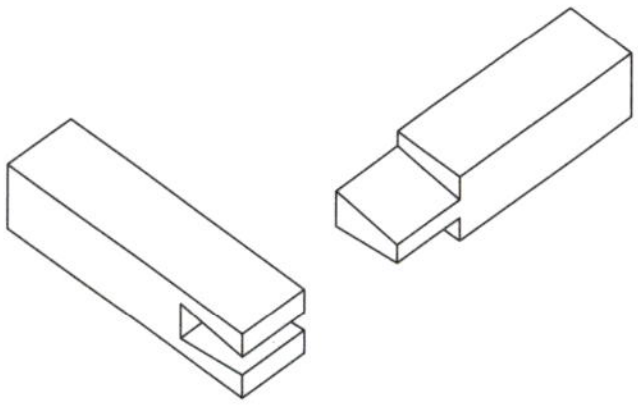

A+B

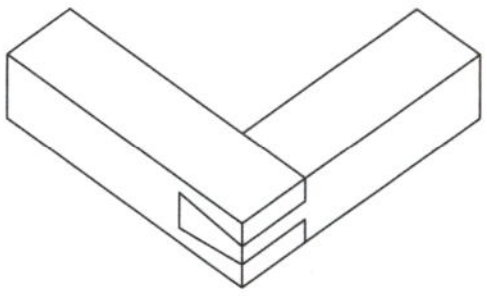

AB

Corner Joints

Pegged through dovetail tenon joint

ōgi-hozo-komi-sen-uchi
扇枘接ぎ（込栓打）

durchgehender Schwalbenschwanz-eckzapfen mit Dübelsicherung

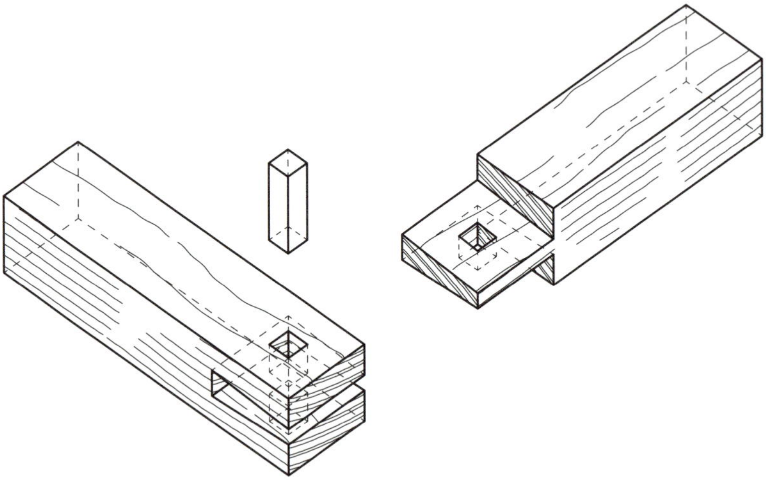

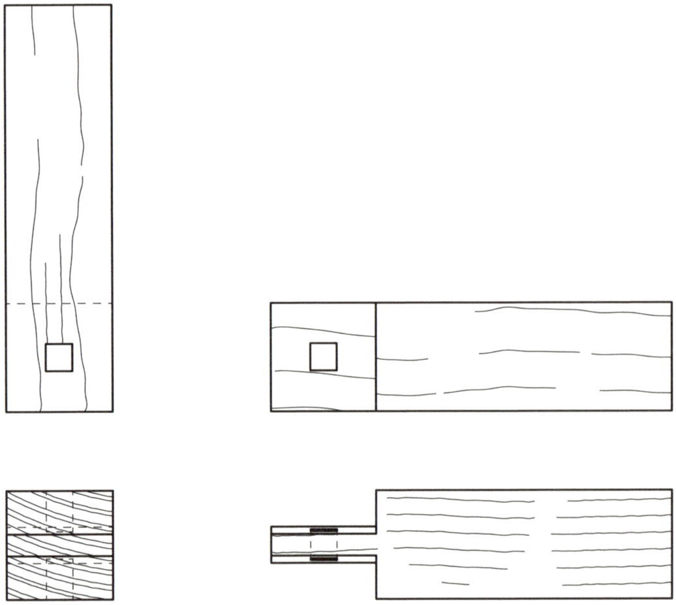

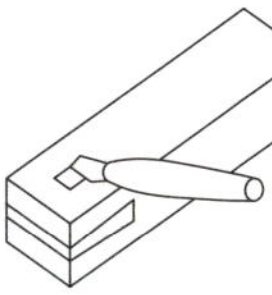

A1

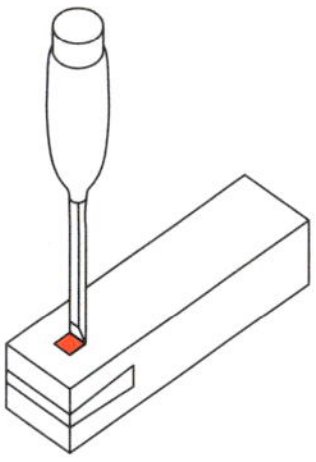

A2

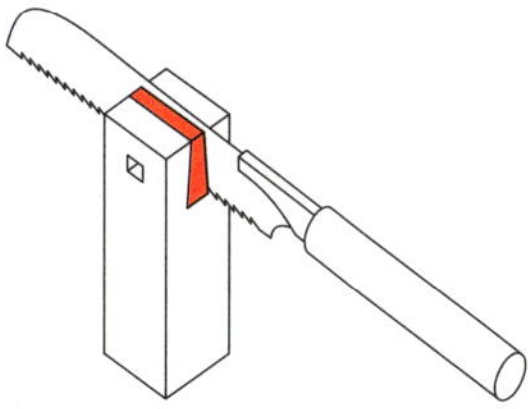

A3

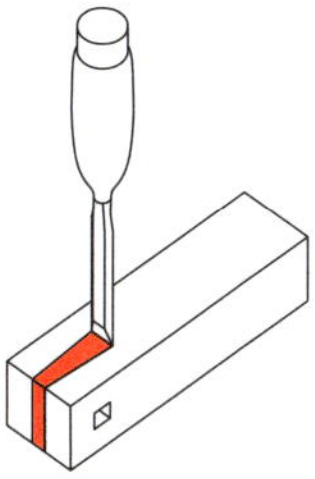

A4

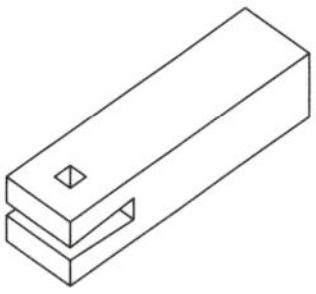

A

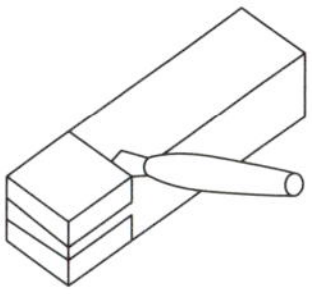

B1

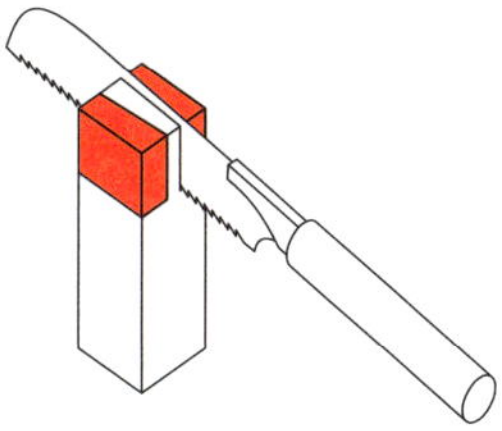

B2

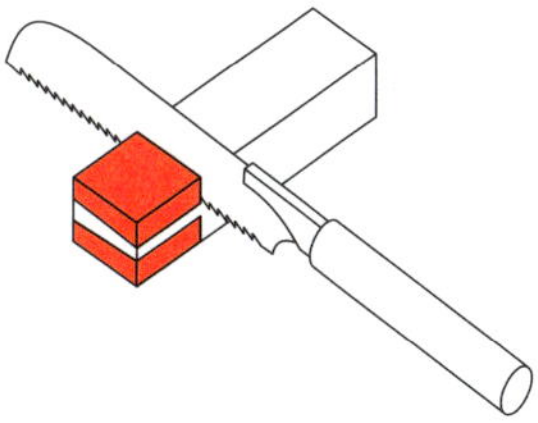

B3

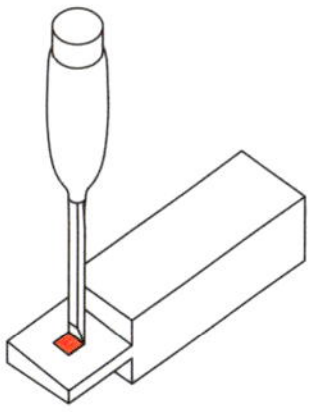

B4

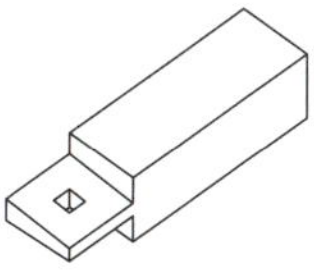

B

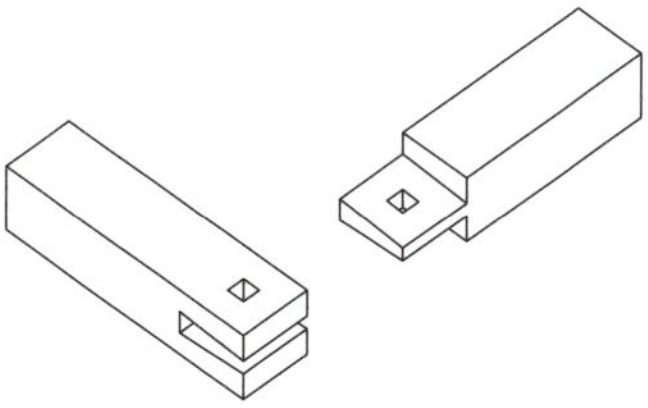

A+B

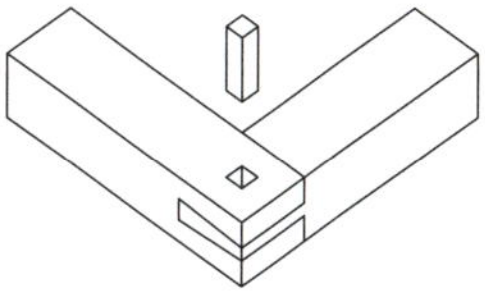

A+B

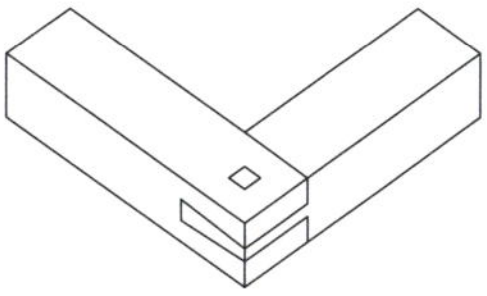

AB

Corner Joints

Half-blind bridle joint

kakushi-hozo-tsugi
隠枘接ぎ

halbverdeckter Scherzapfen

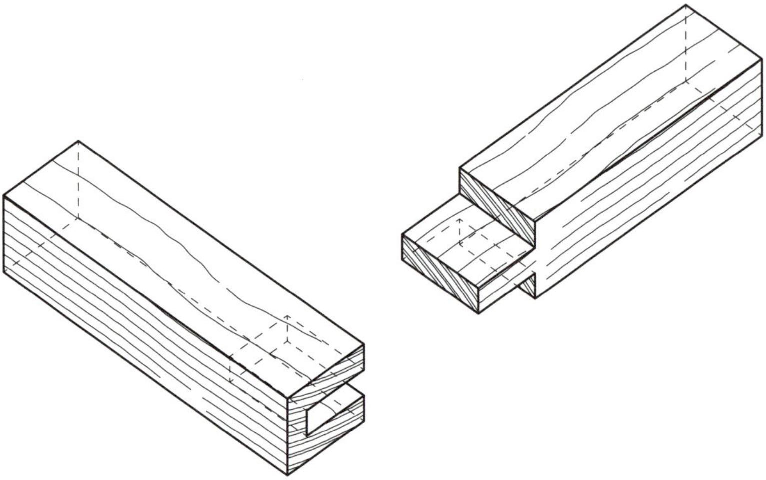

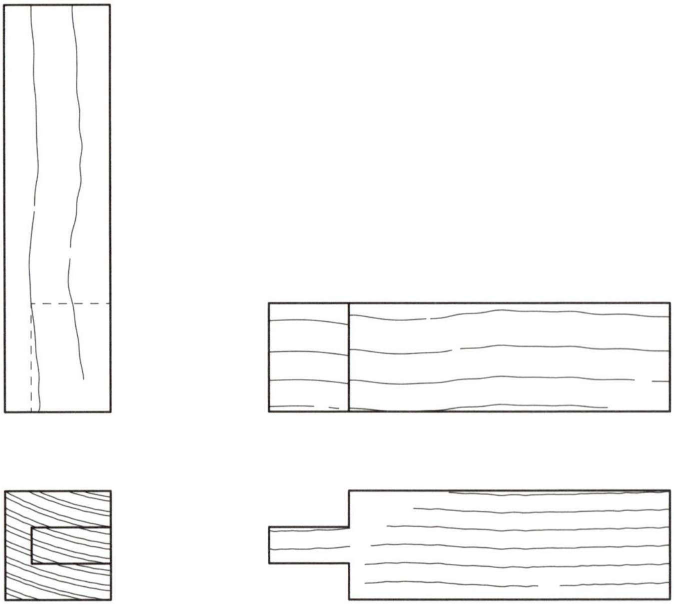

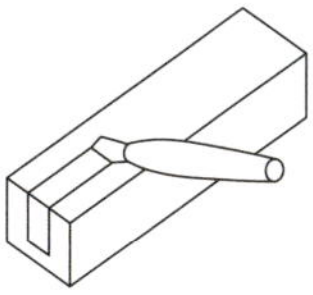

A1

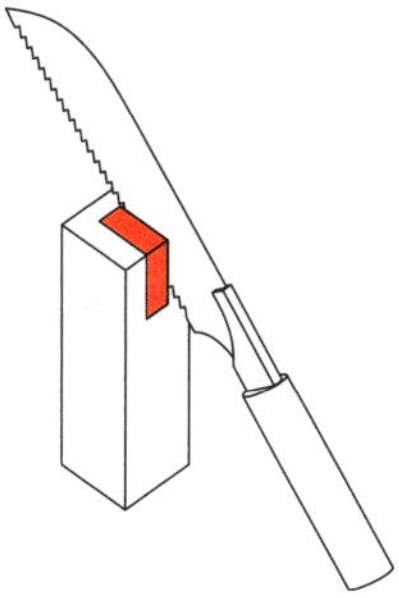

A2

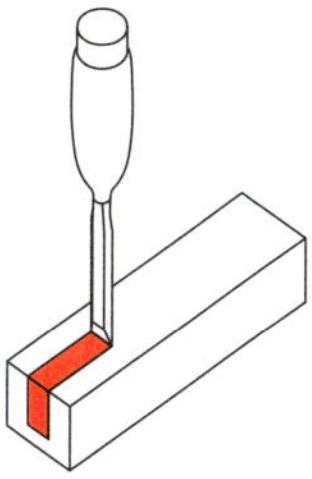

A3

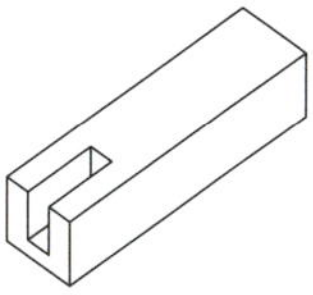

A

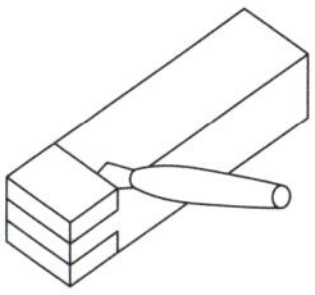

B1

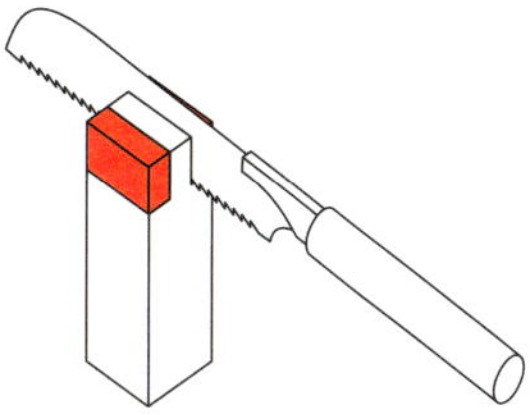

B2

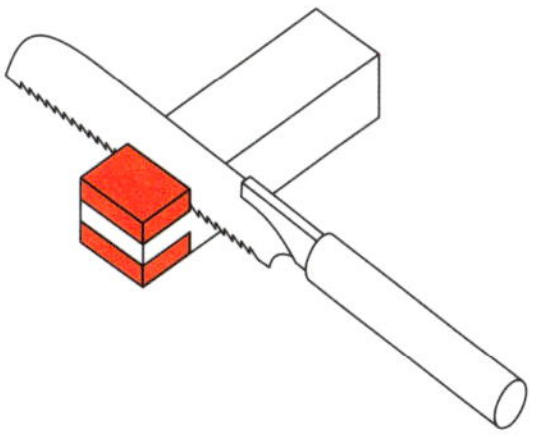

B3

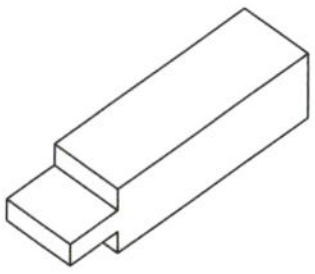

B

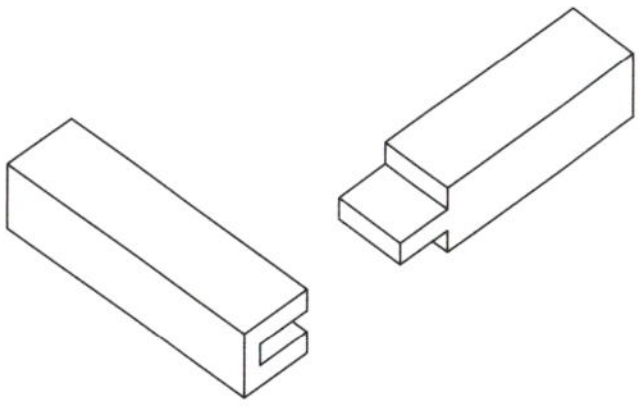

A+B

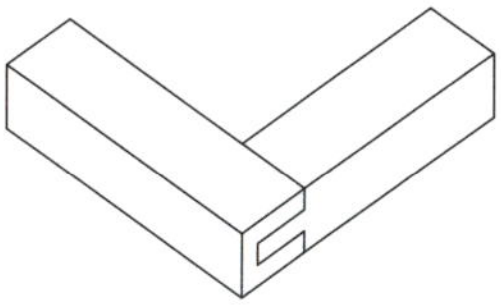

AB

Corner Joints

Half-blind haunched mortise and tenon joint

kiba-hozo-tsugi
牙枘接ぎ

halbverdeckter Stufeneckzapfen

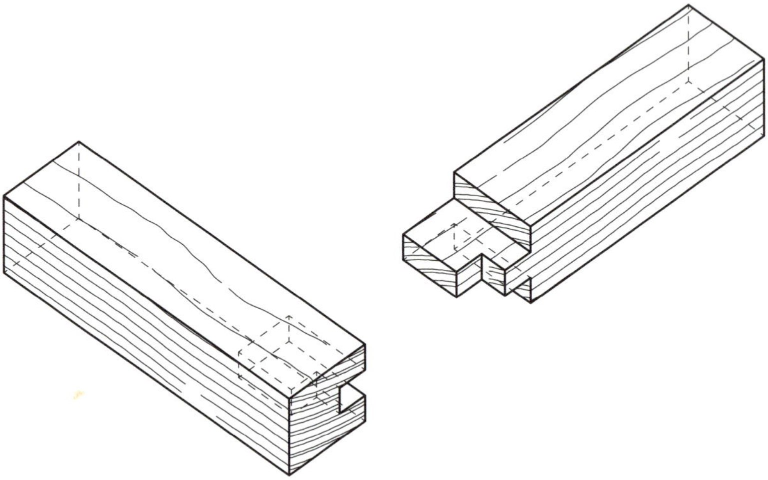

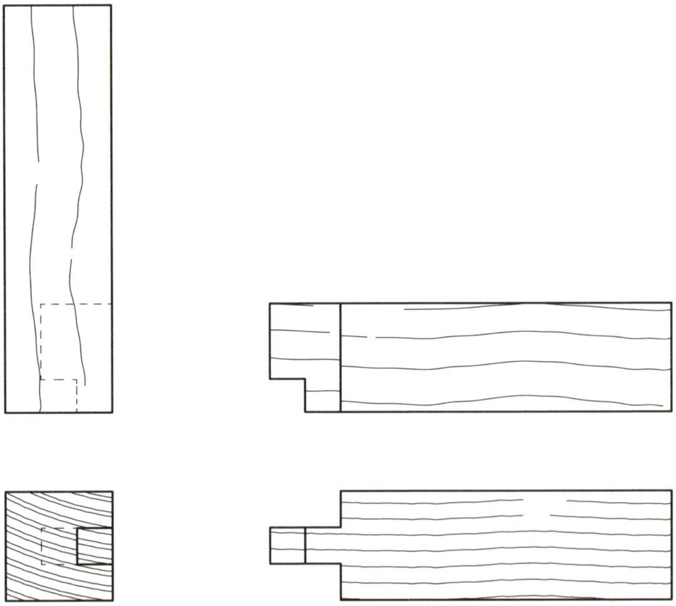

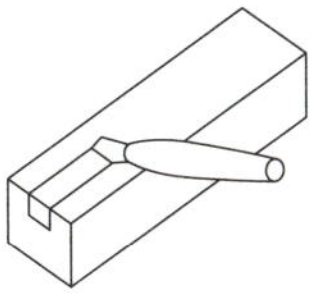

A1

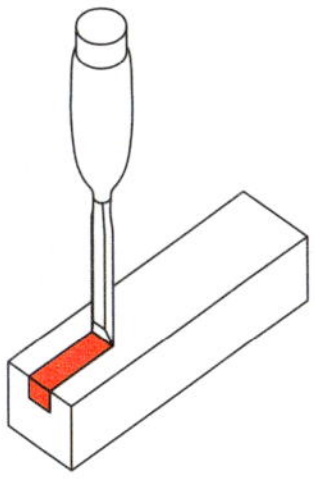

A2

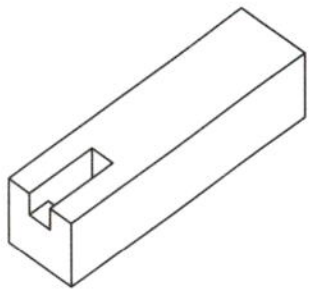

A

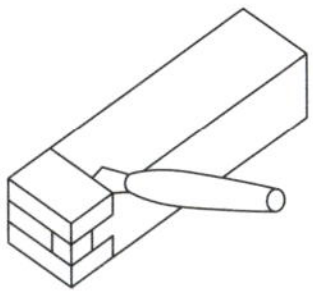

B1

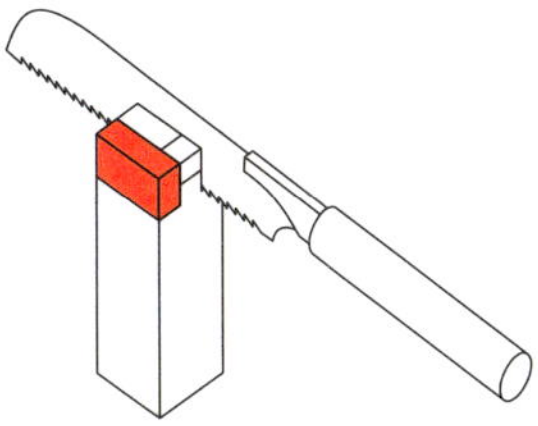

B2

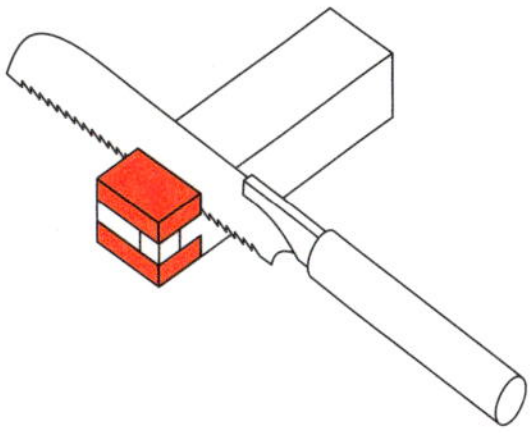

B3

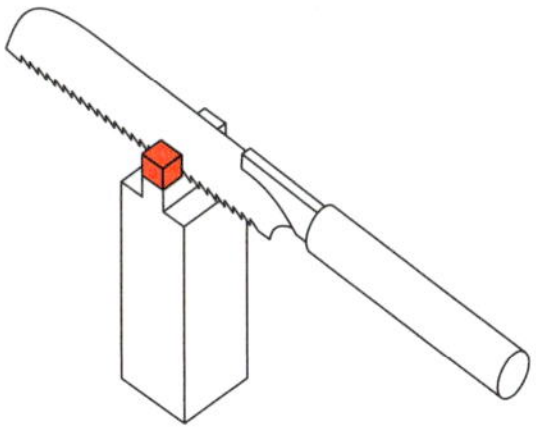

B4

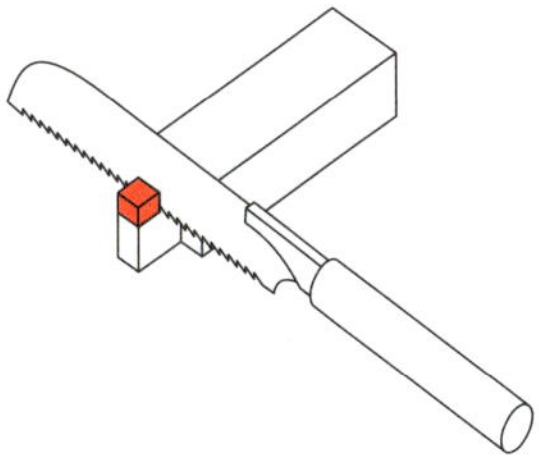

B5

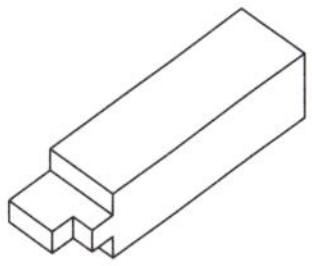

B

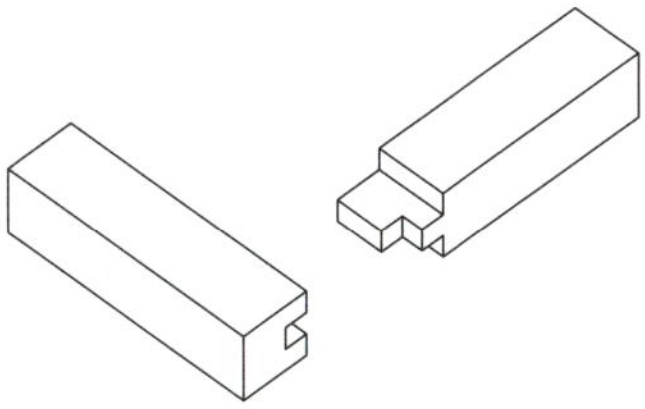

A+B

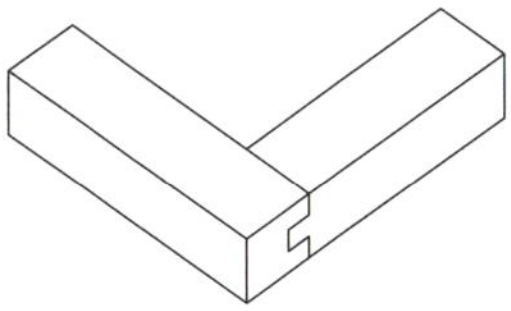

AB

Corner Joints

Wedged haunched through mortise and tenon joint

kiba-hozo-sashi-kumi-tsugi-komi-sen-uchi
牙柄差組接込栓打ち

verkeilter durchgehender Stufeneckzapfen

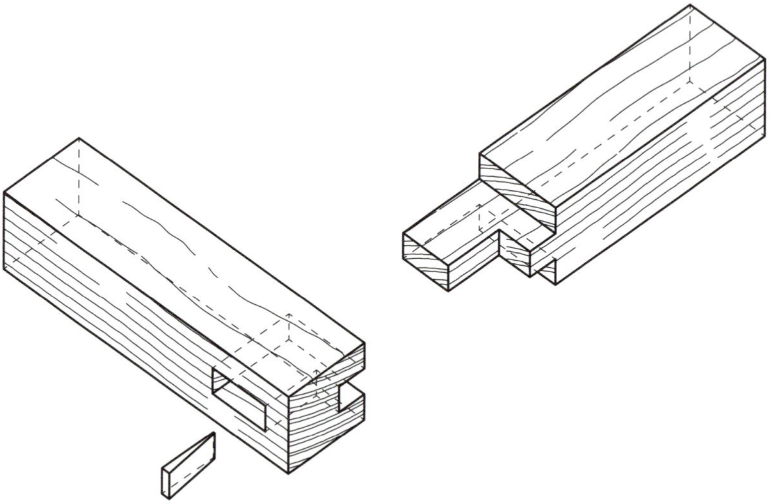

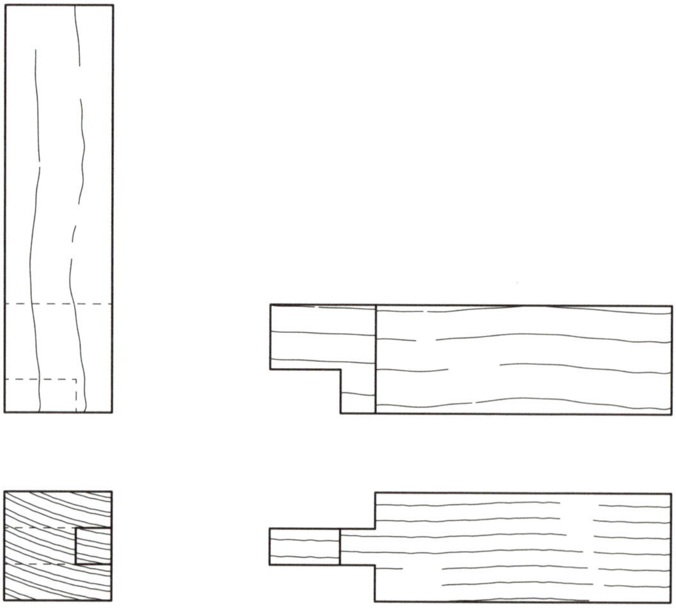

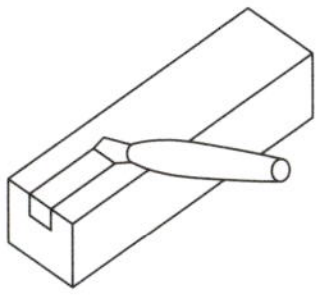

A1

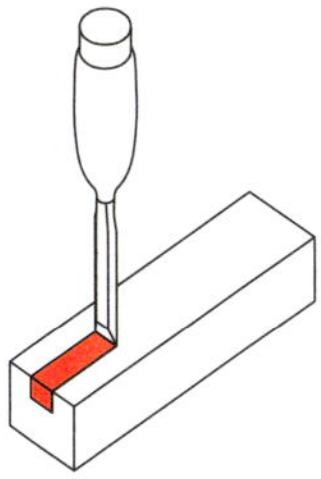

A2

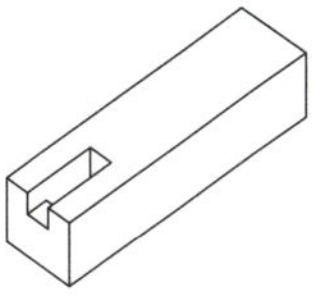

A

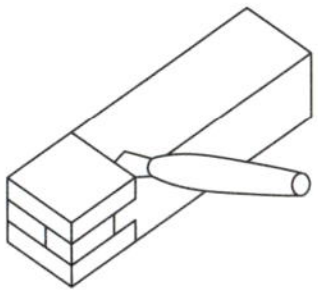

B1

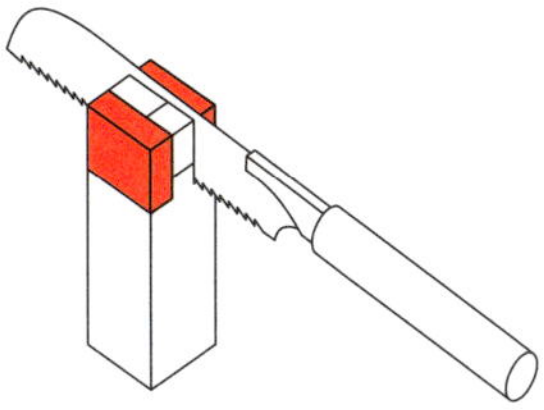

B2

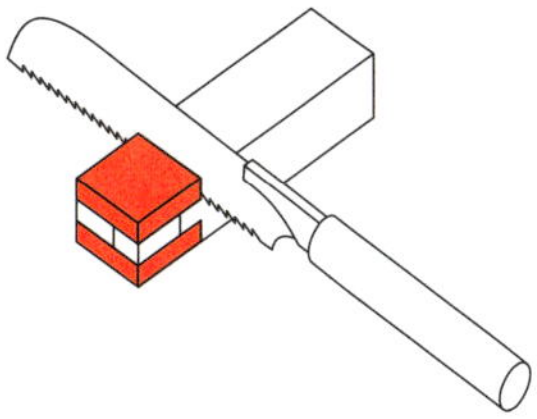

B3

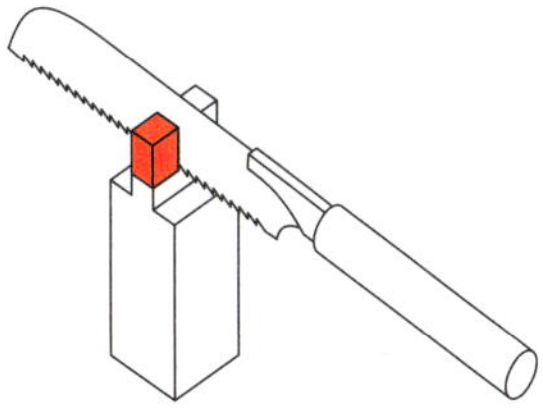

B4

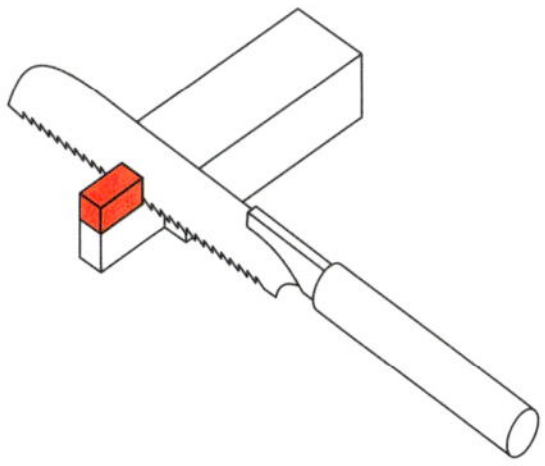

B5

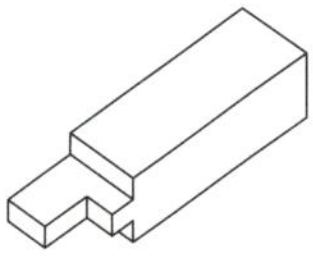

B

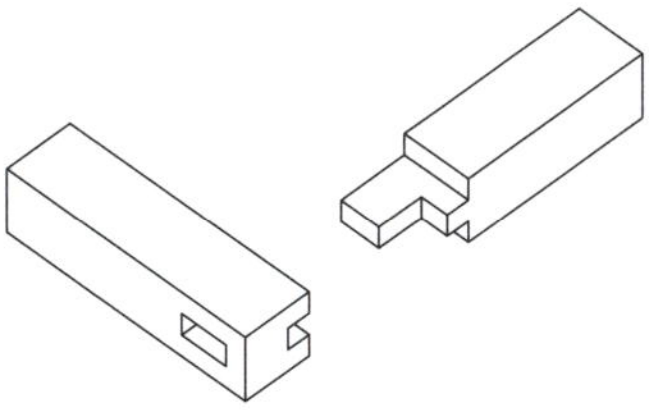

A+B

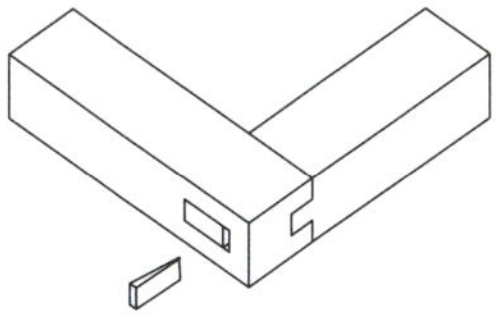

A+B

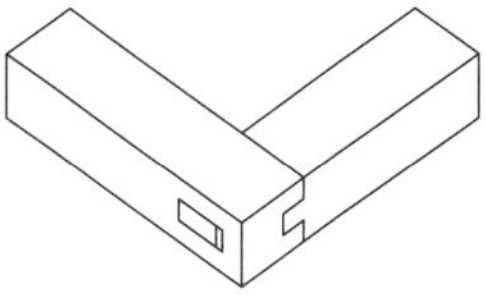

AB

Corner Joints

Triple-shouldered through mortise and tenon joint

uchi-nuki-hozo-sashi-kumi-tsugi
打抜枘差組継ぎ

durchgehender abgesetzter Eckzapfen

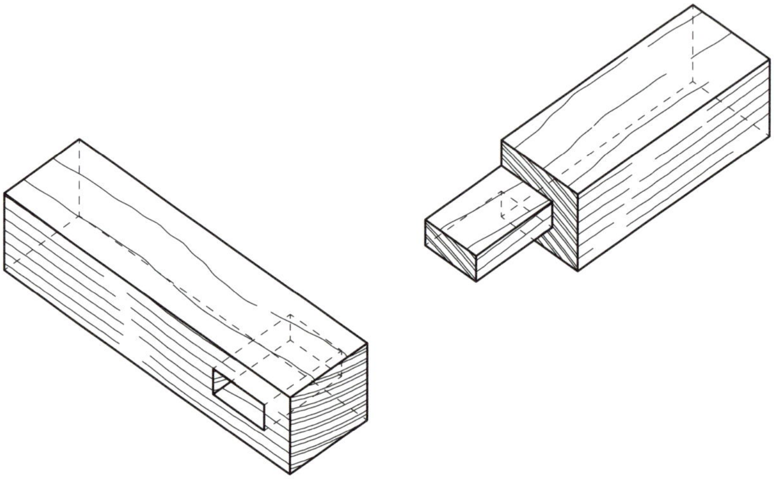

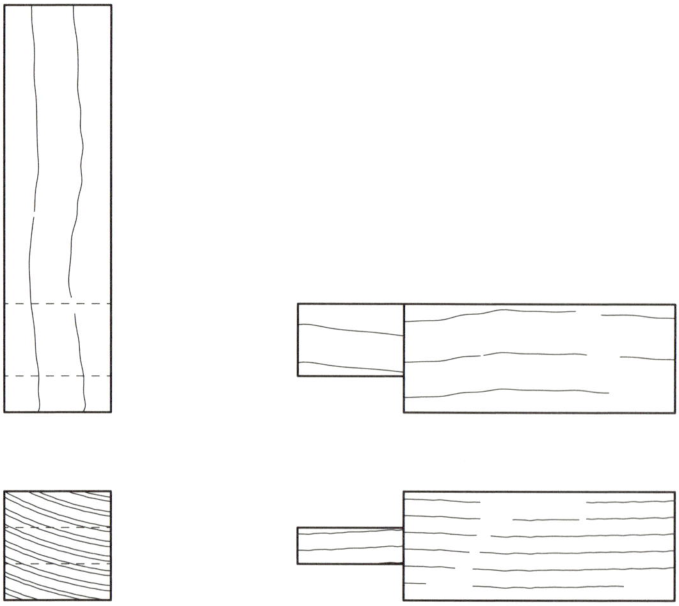

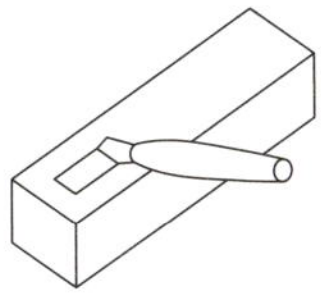

A1

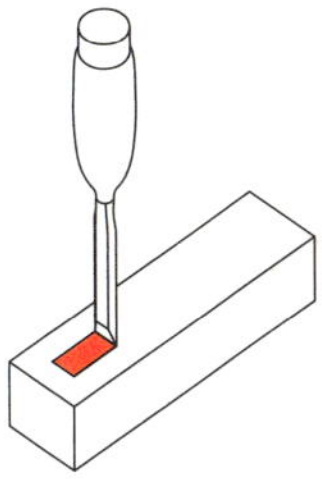

A2

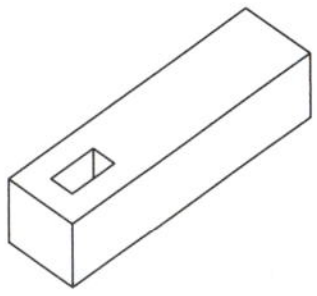

A

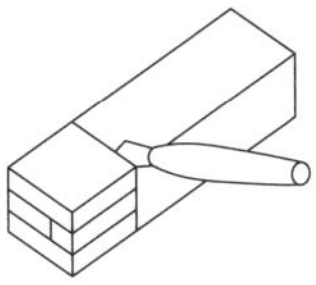

B1

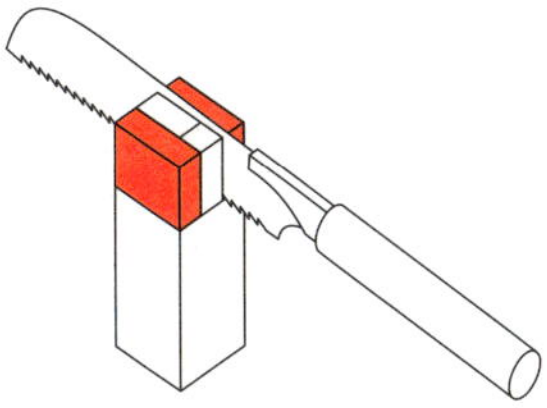

B2

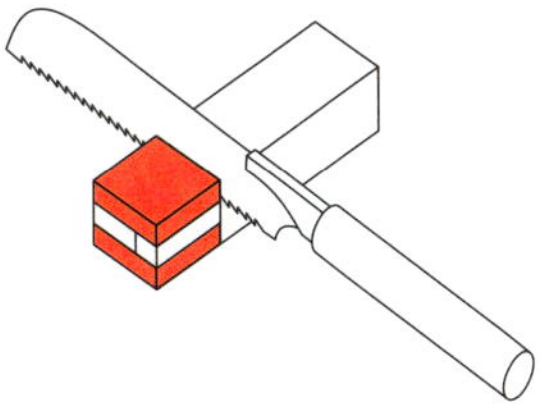

B3

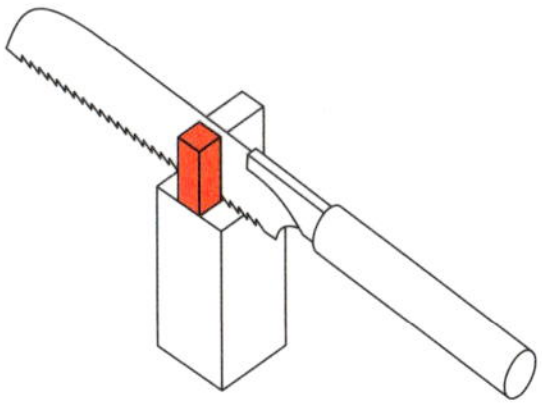

B4

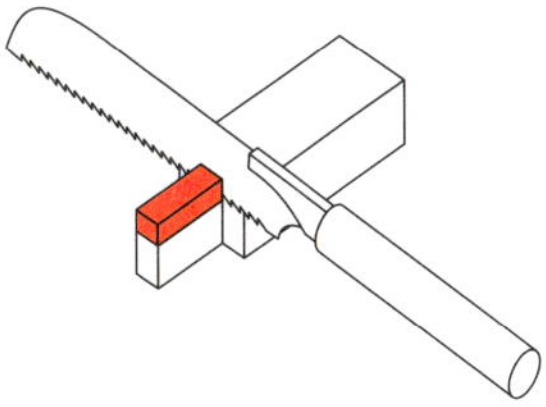

B5

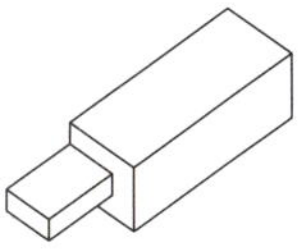

B

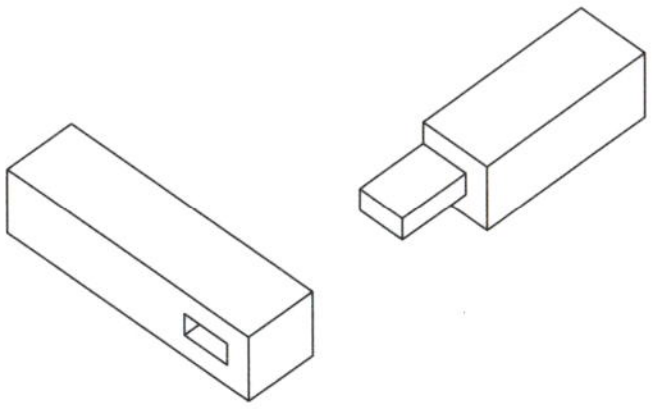

A+B

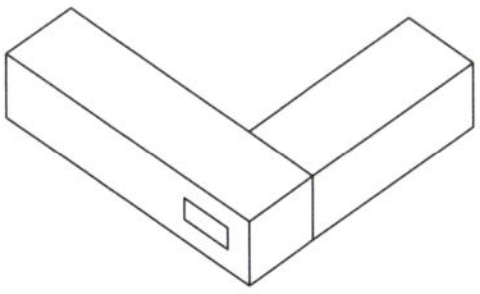

AB

Corner Joints

Wedged triple-shouldered through mortise and tenon joint

uchi-nuki-hozo-sashi (kusabi-shime)
打抜枘差し(楔締)

durchgehender abgesetzter Eckzapfen mit Keilschloss

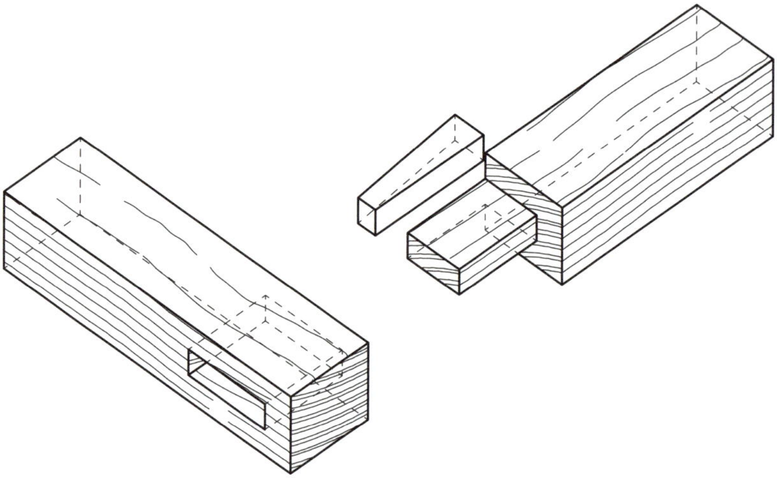

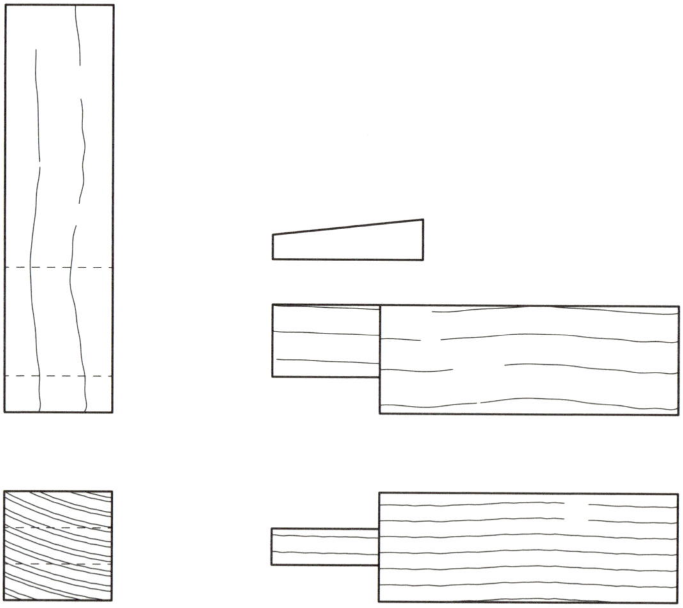

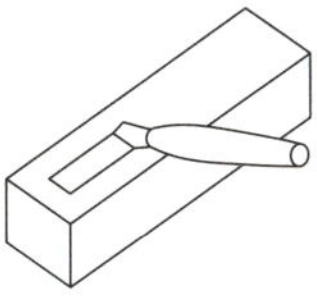

A1

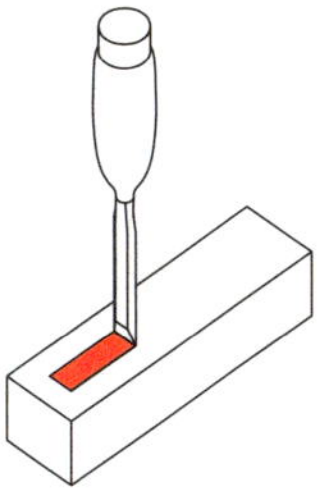

A2

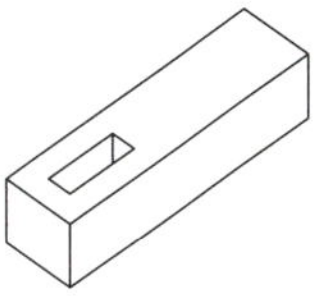

A

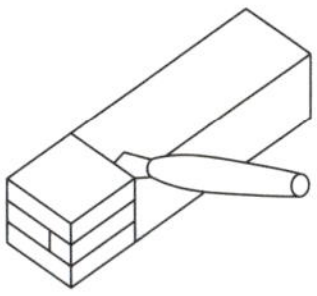

B1

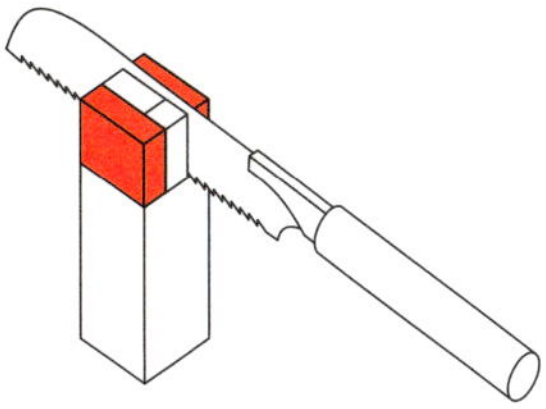

B2

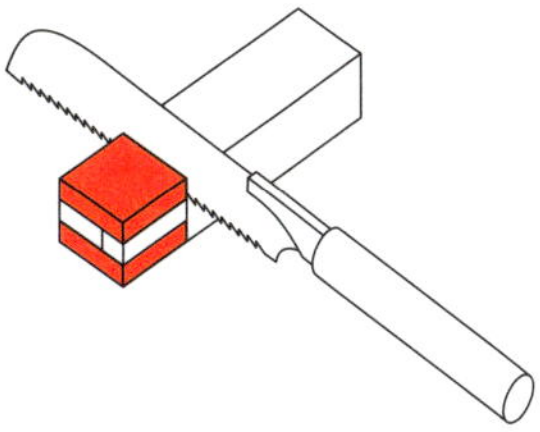

B3

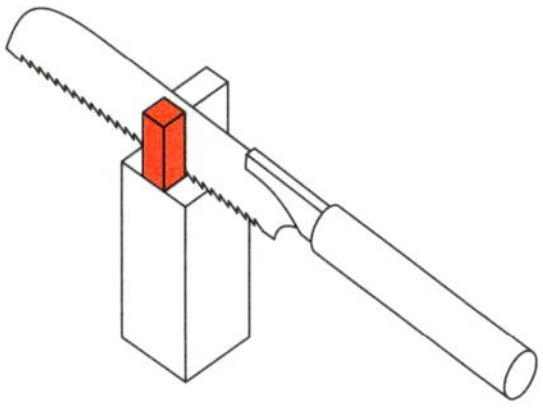

B4

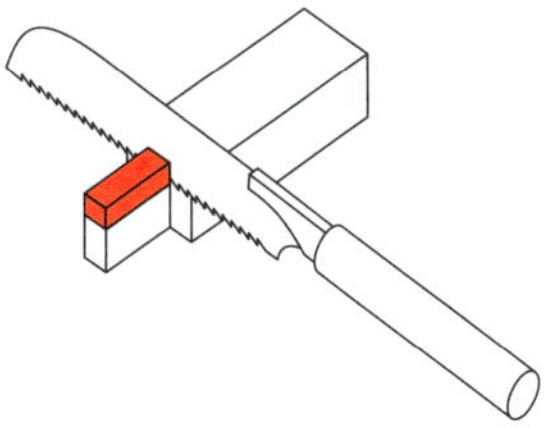

B5

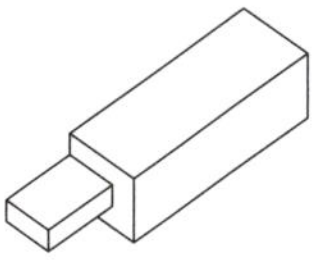

B

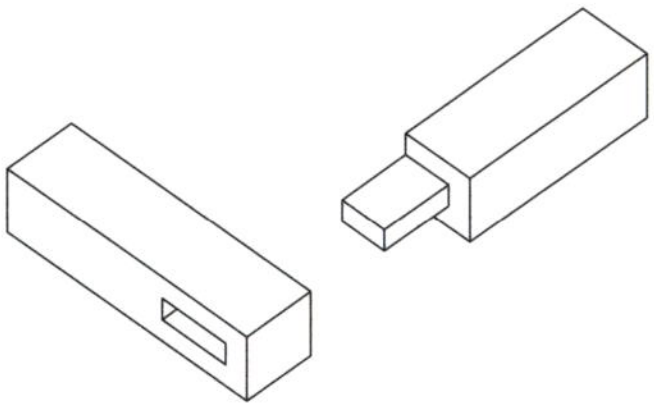

A+B

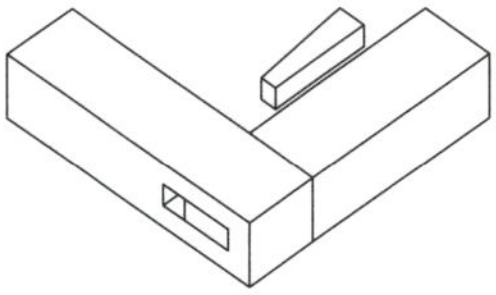

A+B

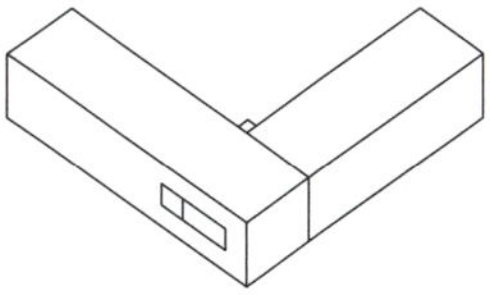

AB

Corner Joints

Through mortise and tenon with sloped haunch

kataki-koshi-tsuke-hira-hozo-tsugi
傾腰付平枘接ぎ

durchgehender abgesetzter Eckzapfen mit schräger Stufe

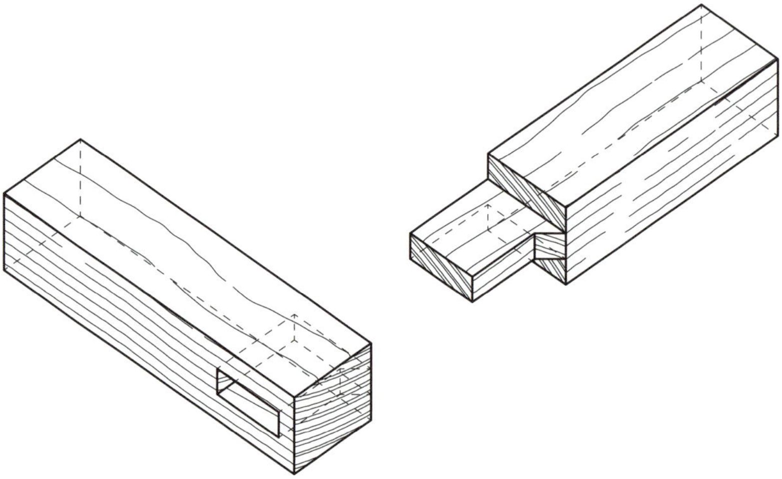

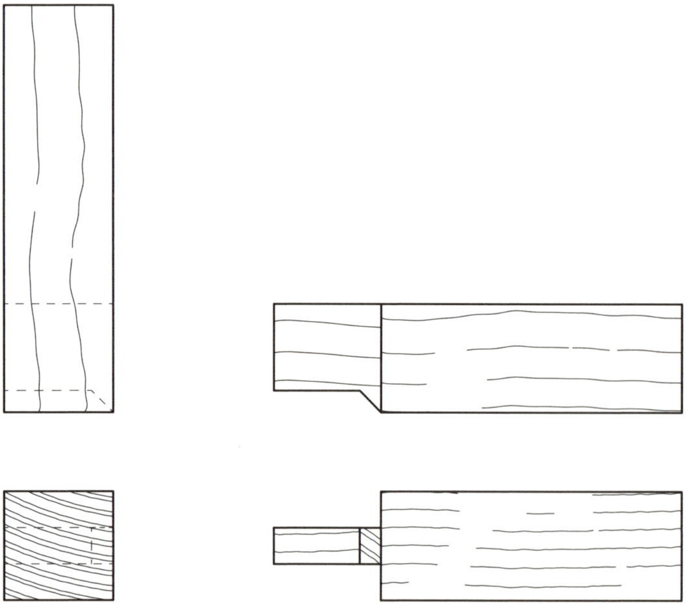

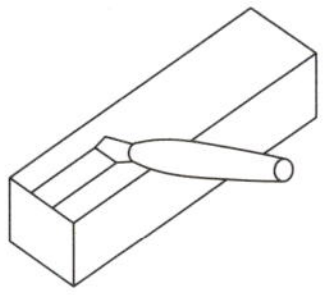

A1

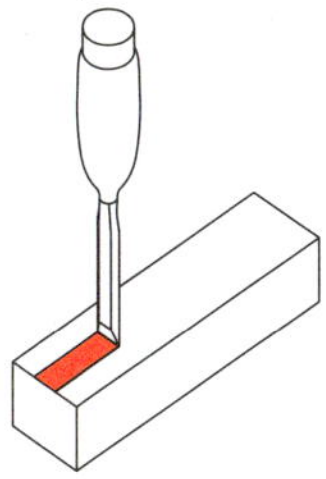

A2

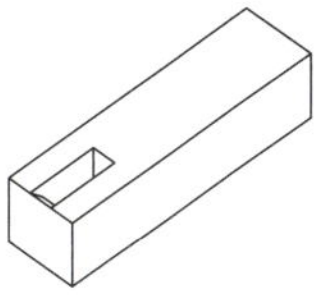

A

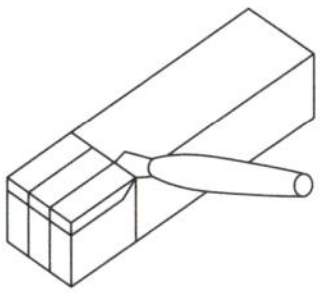

B1

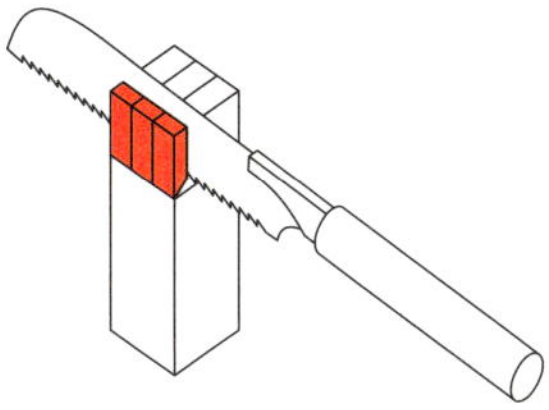

B2

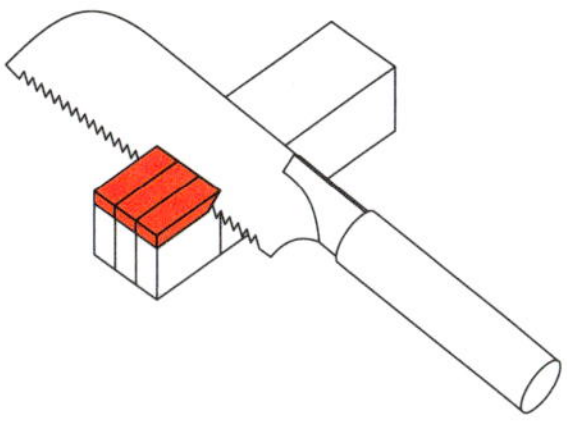

B3

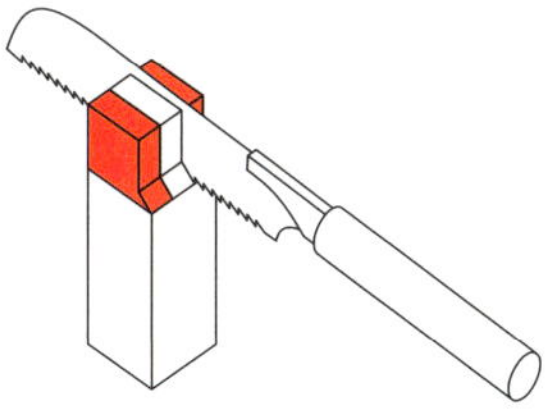

B4

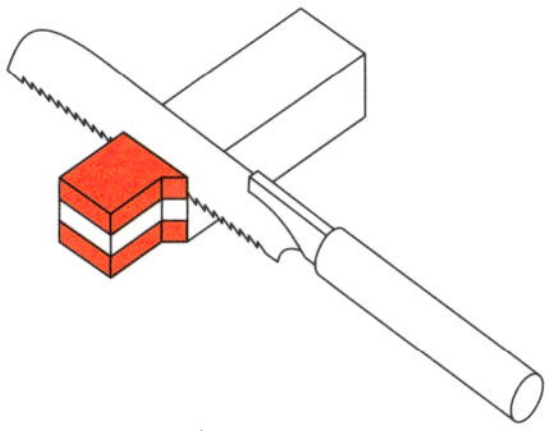

B5

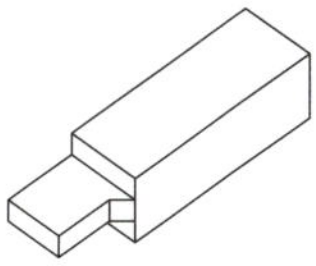

B

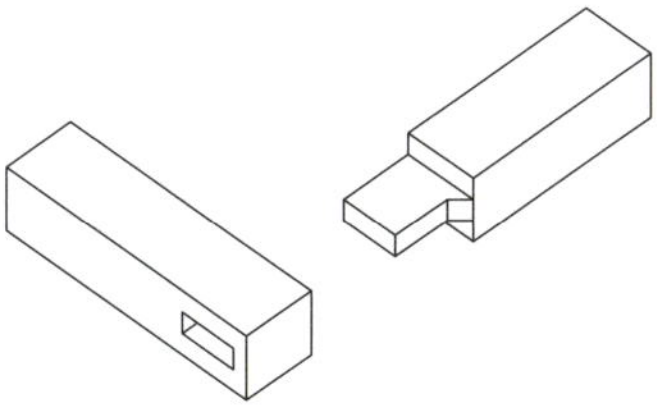

A+B

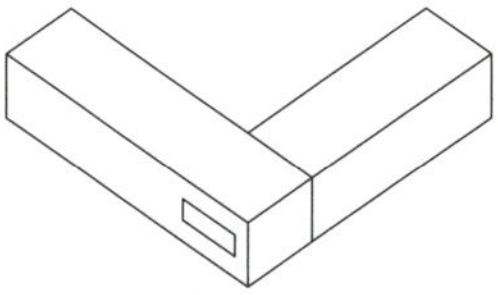

AB

Corner Joints

Beveled triple-shouldered mortise and tenon with housing

katagi-oire-hozo-tsugi (kōbai-tsuki)
傾追入枘接ぎ(勾配付)

verdeckter abgesetzter Eckzapfen mit schräger Brust

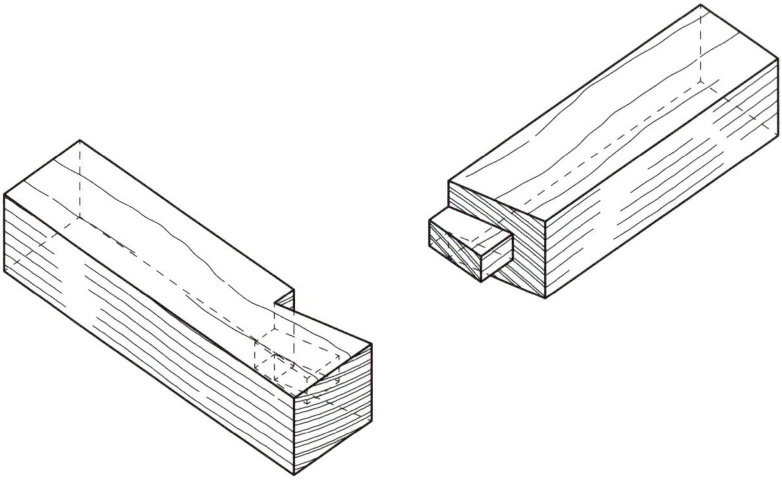

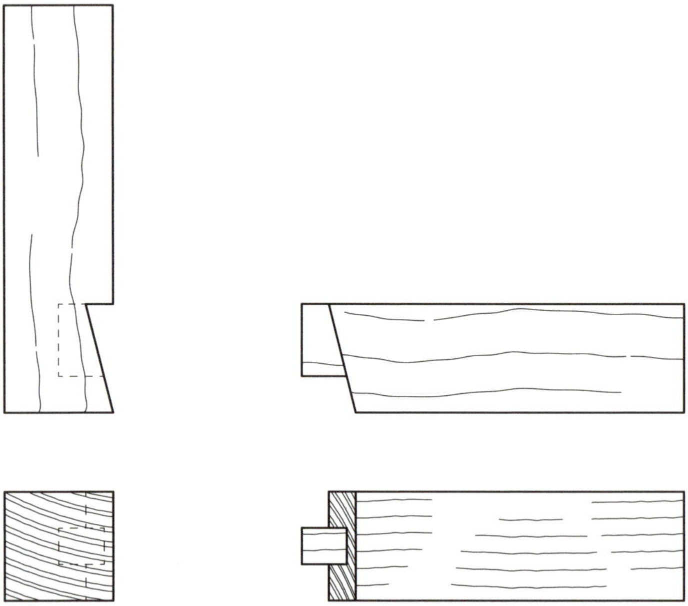

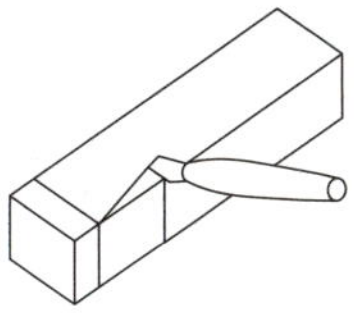
A1

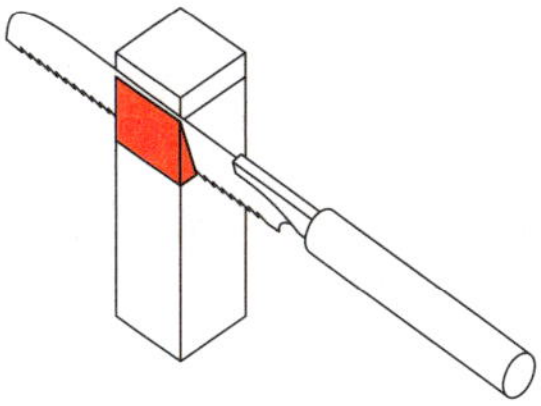
A2

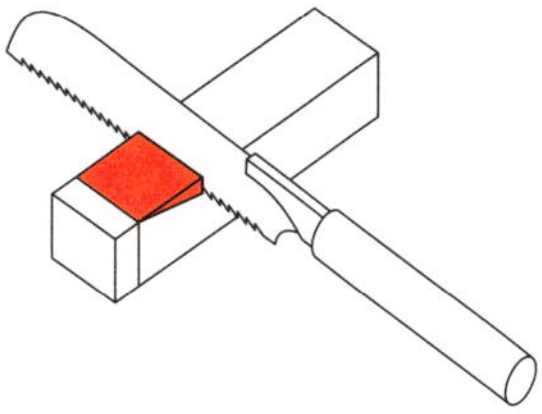
A3

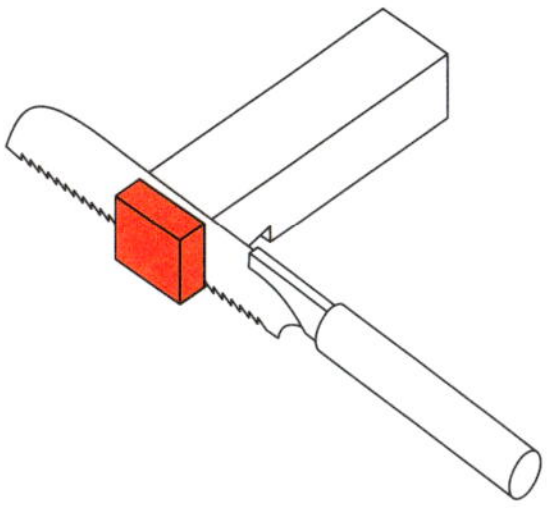
A4

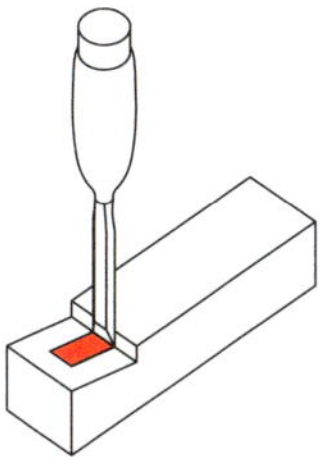
A5

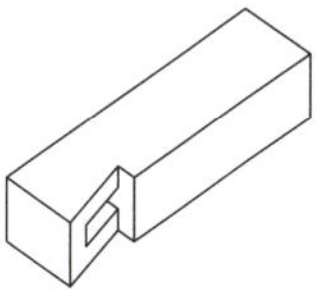
A

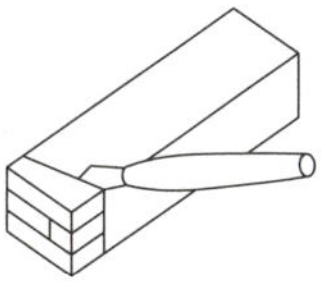
B1

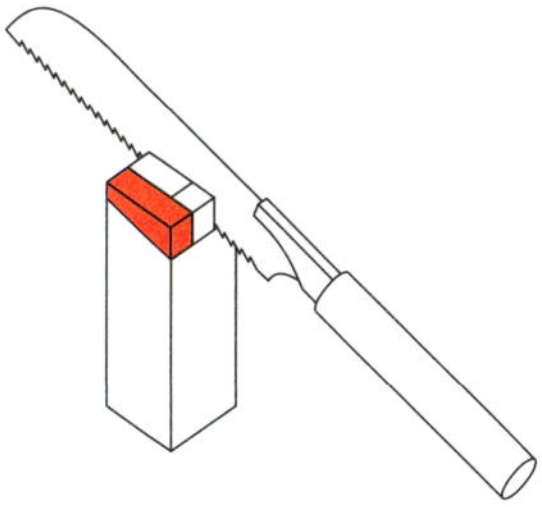
B2

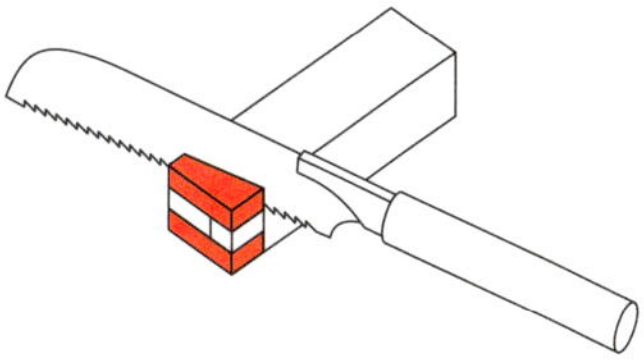
B3

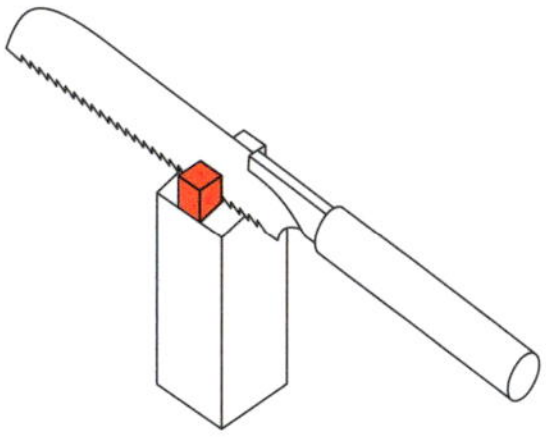
B4

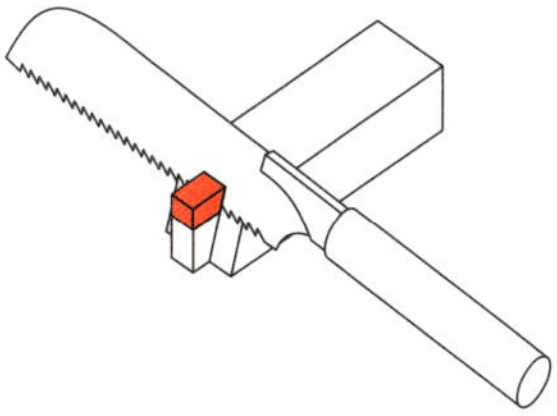
B5

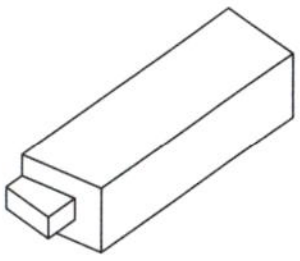
B

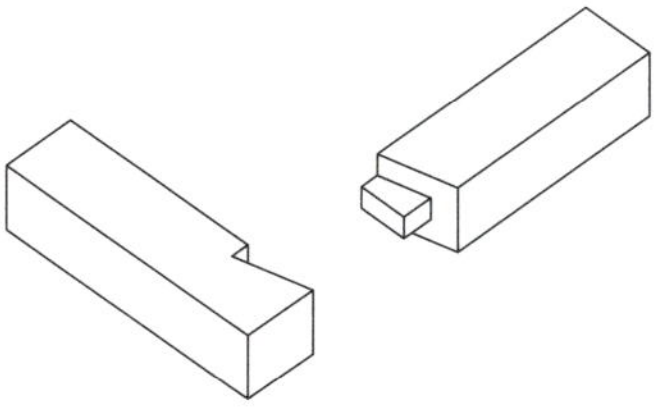
A+B

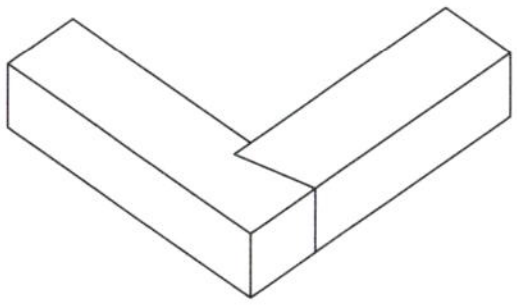
AB

Corner Joints

Triple-shouldered and sloped through mortise and tenon

hiki-komi-hira-hozo-tsugi
挽込平枘接ぎ

abgesetzter durchgehender schräger Eckzapfen

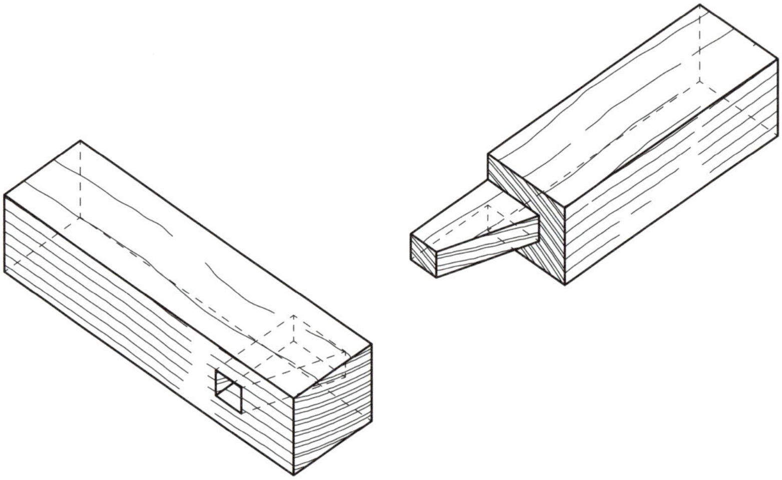

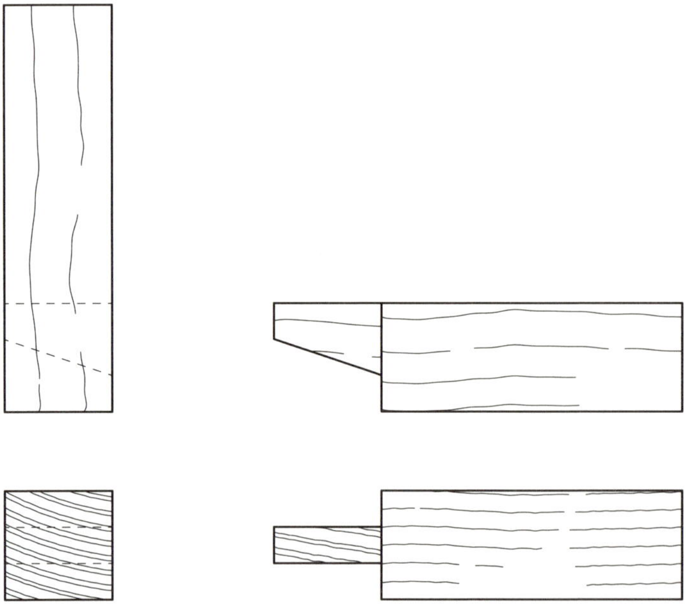

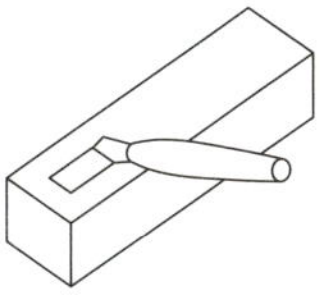

A1

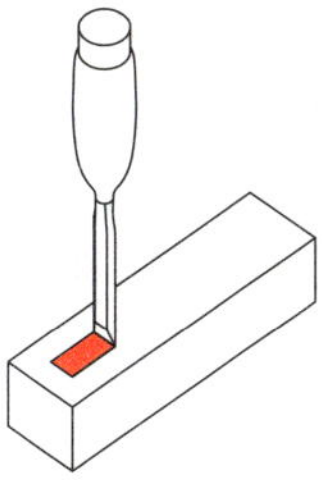

A2

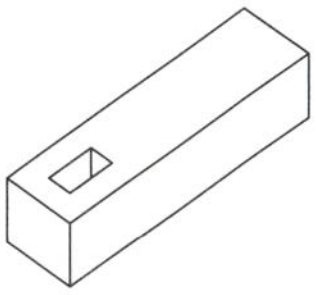

A

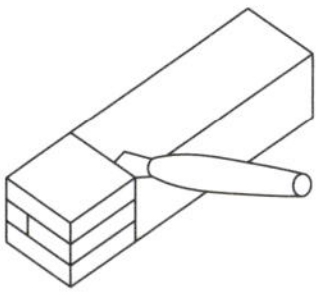

B1

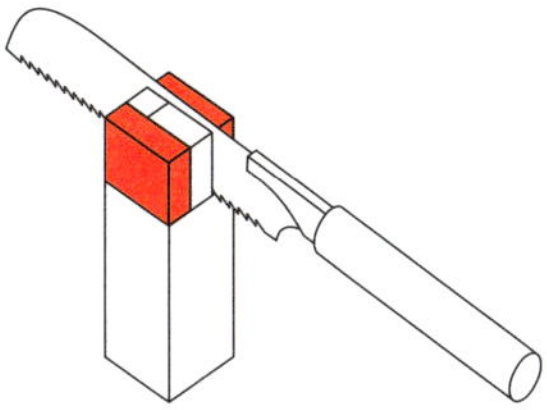

B2

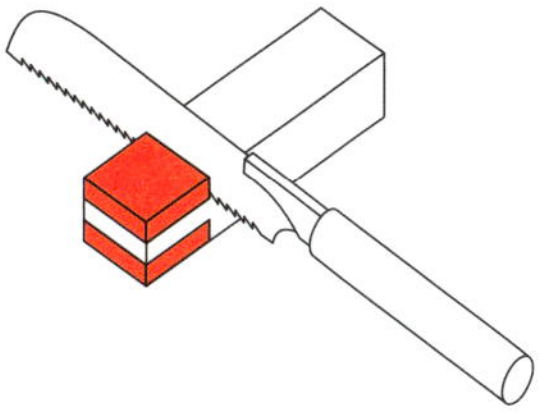

B3

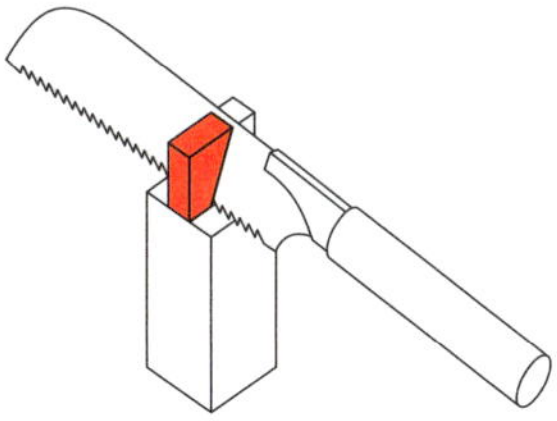

B4

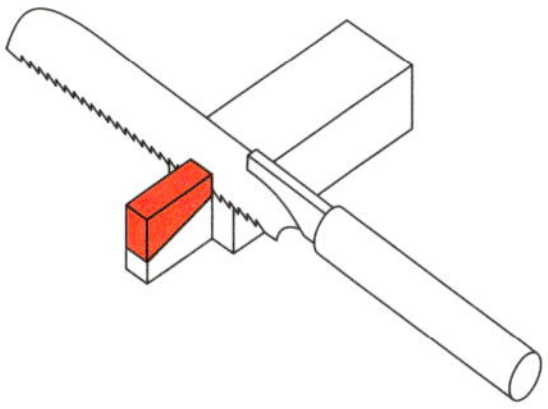

B5

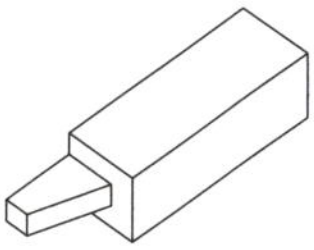

B

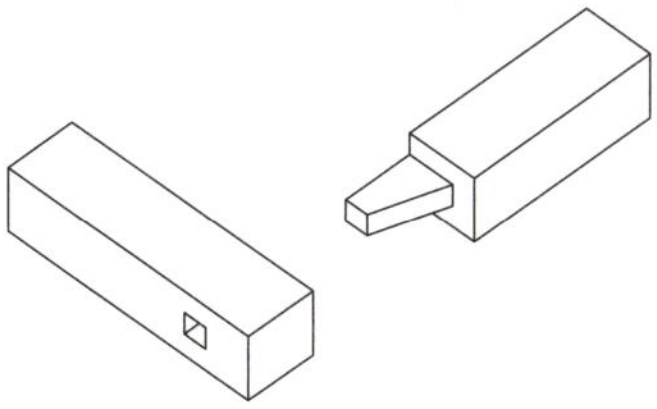

A+B

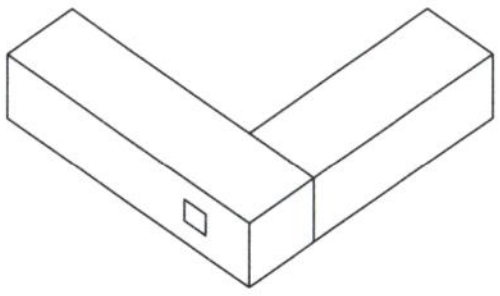

AB

Corner Joints

Fully shouldered through mortise and tenon

tōshi-kone-hozo-tsugi
通小根枘接ぎ

beidseitig abgesetzter
durchgehender Eckzapfen

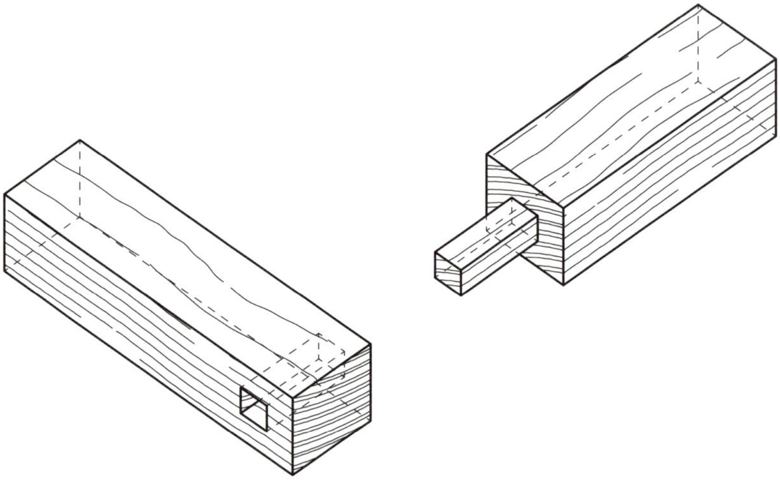

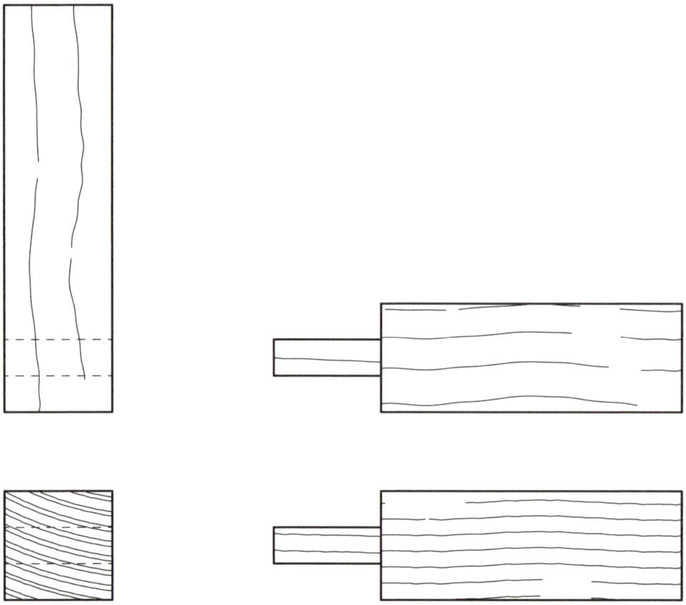

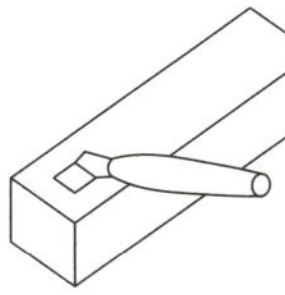

A1

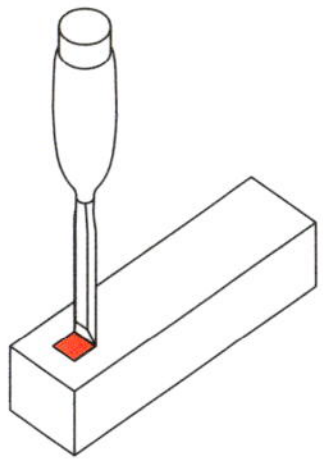

A2

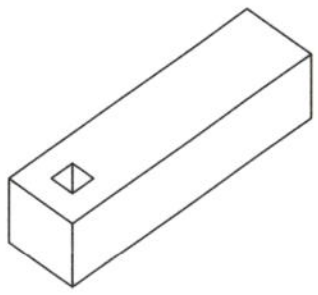

A

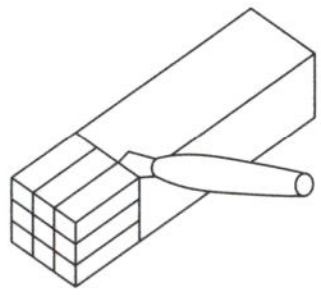

B1

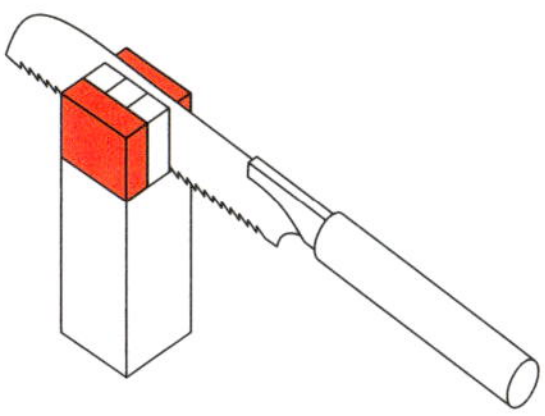

B2

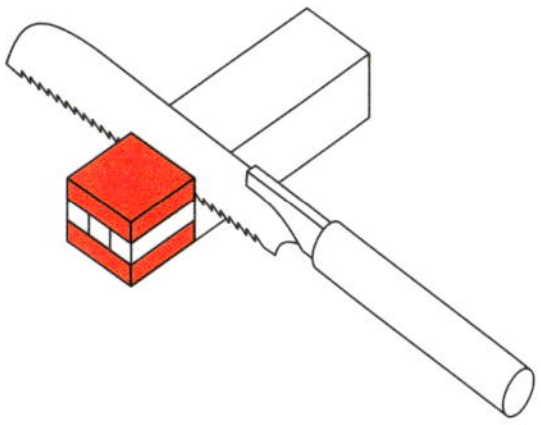

B3

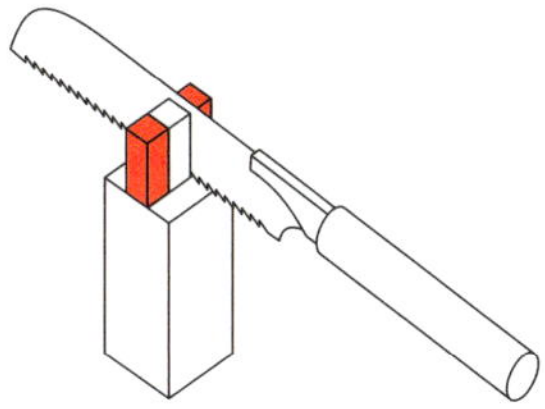

B4

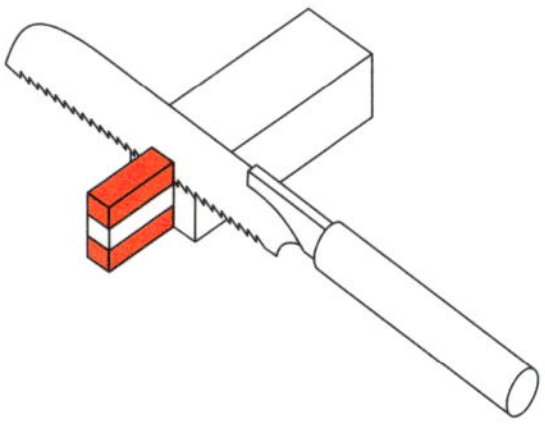

B5

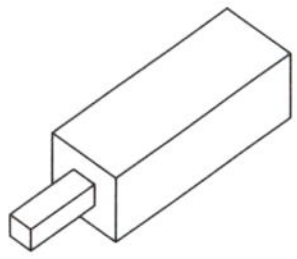

B

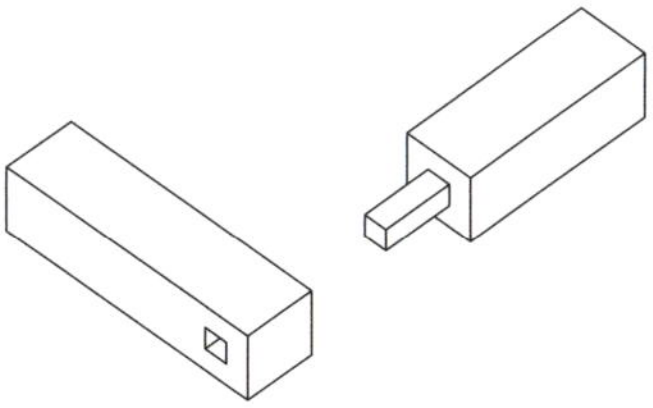

A+B

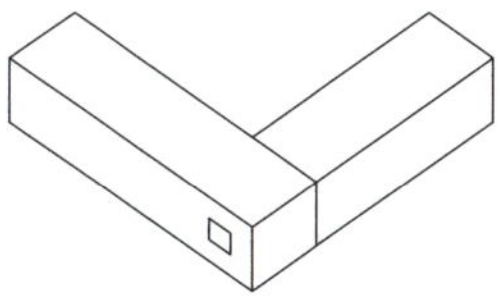

AB

Corner Joints

Blind bridle joint

tsutsumi-komi-sumi-dome-sanmai-hozo-tsugi
包込隅留三枚枘接ぎ

verdeckter Scherzapfen mit Wange auf Gehrung

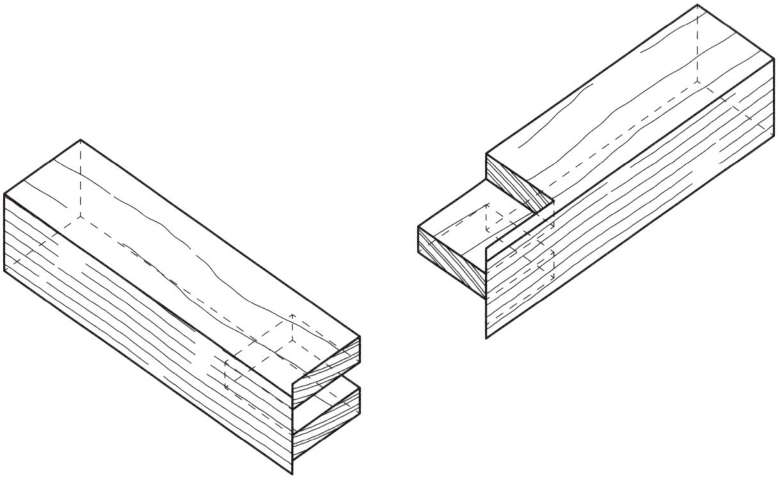

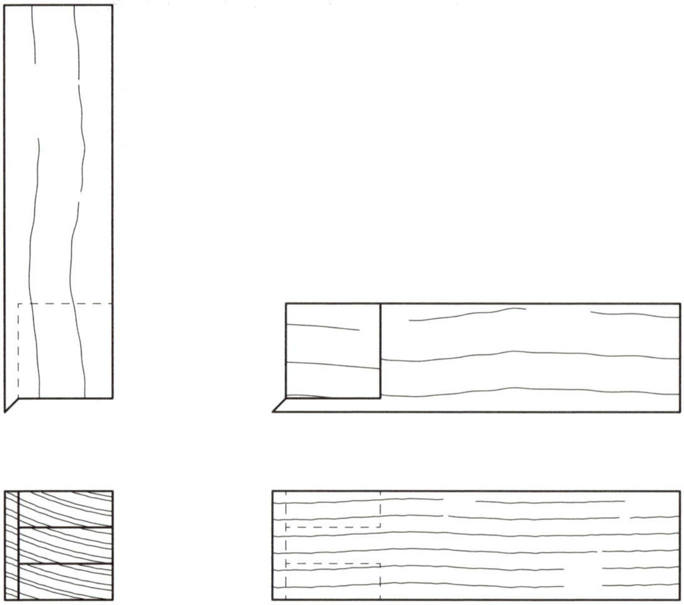

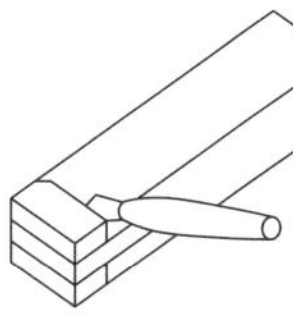

A1

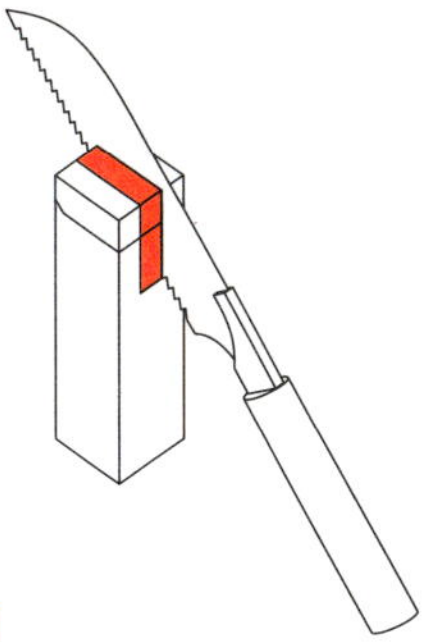

A2

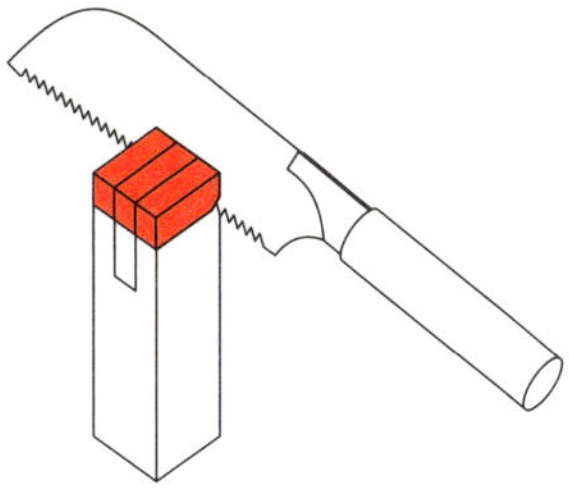

A3

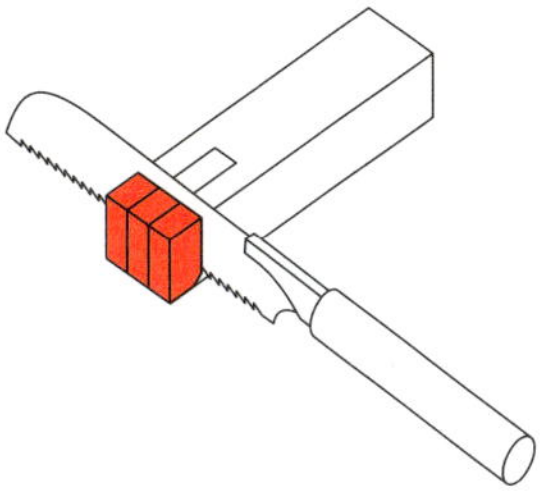

A4

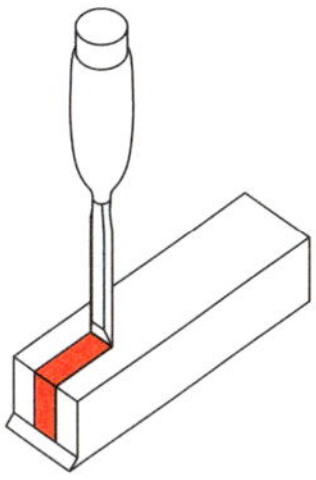

A5

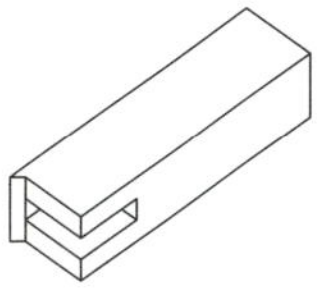

A

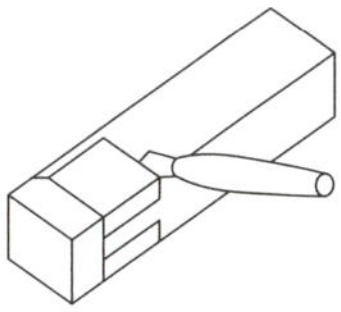

B1

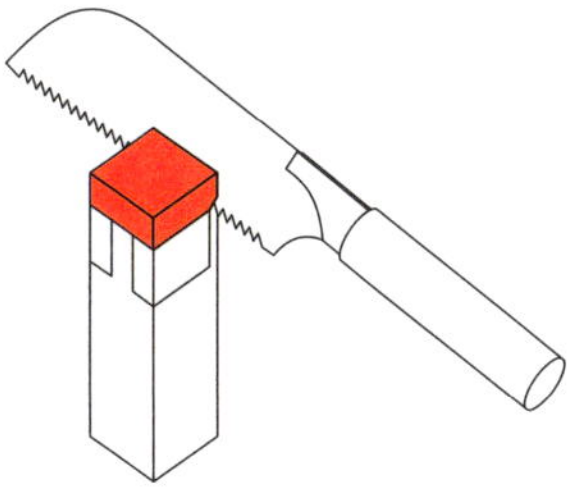

B2

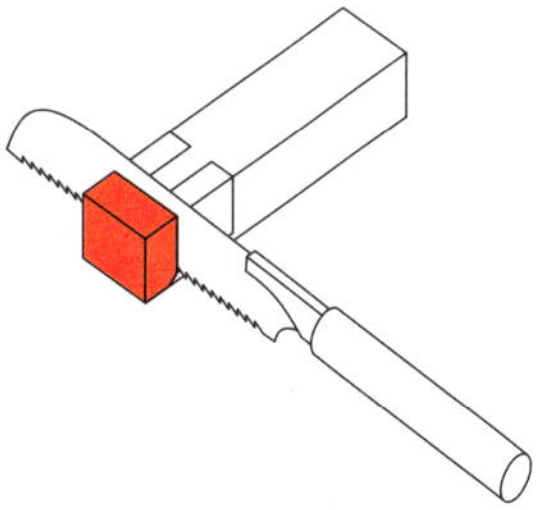

B3

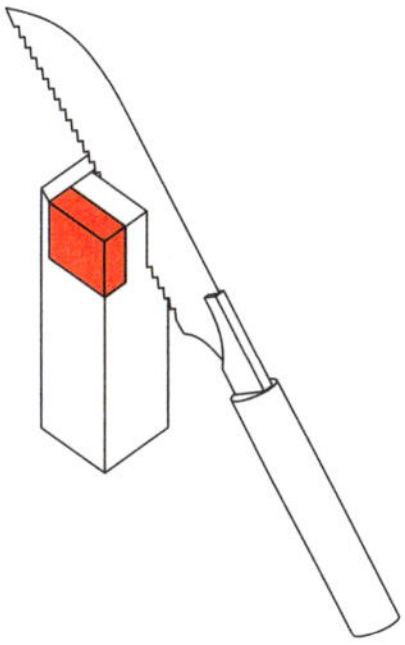

B4

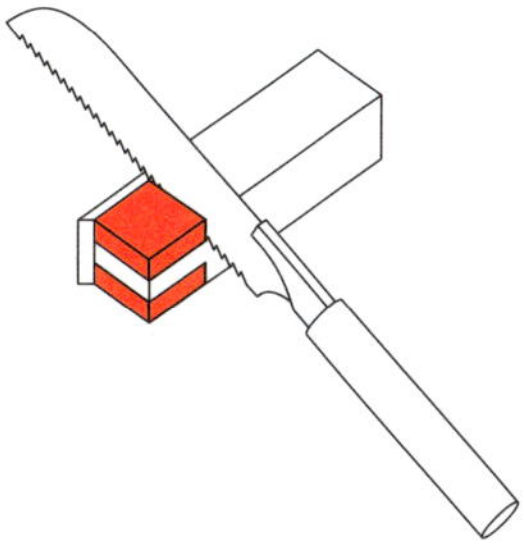

B5

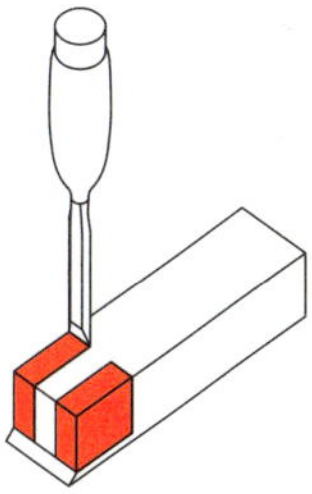

B6

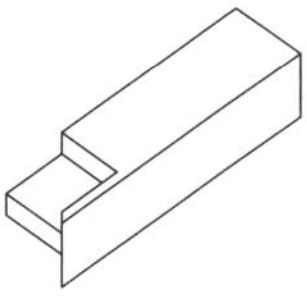

B

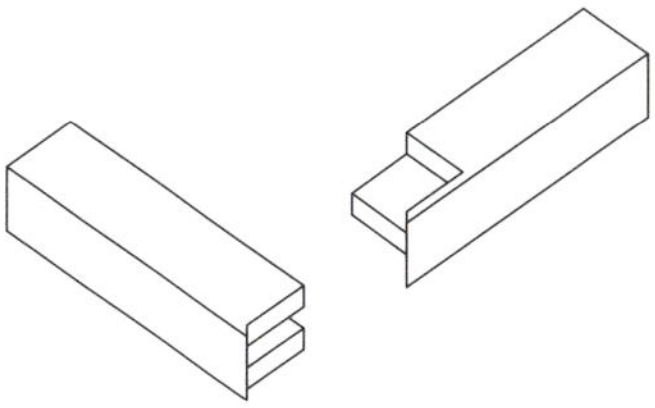

A+B

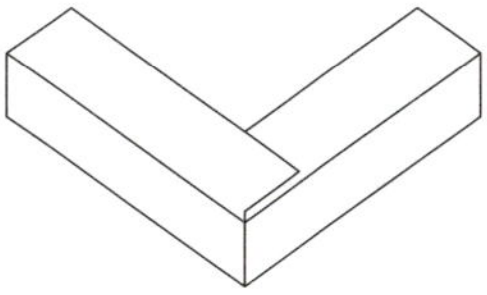

AB

Corner Joints

Blind dovetail tenon

tsutsumi-komi-sumi-dome-ōgi-hozo-tsugi
包込隅留扇枘接ぎ

verdeckter Schwalbenschwanzeckzapfen mit Wange auf Gehrung

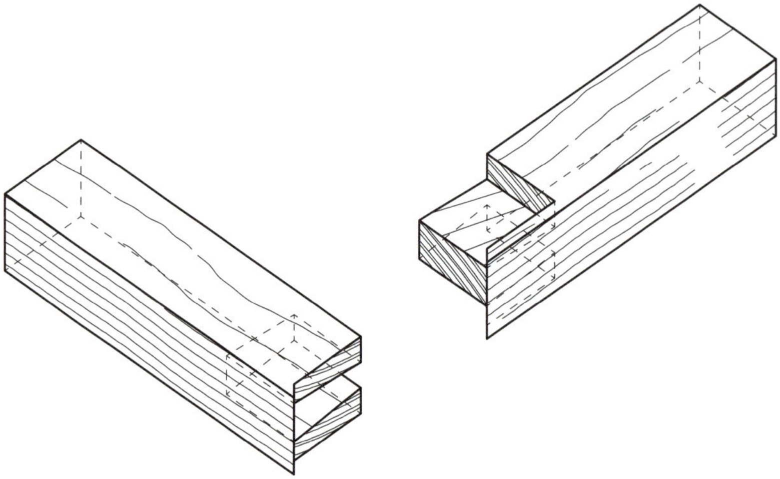

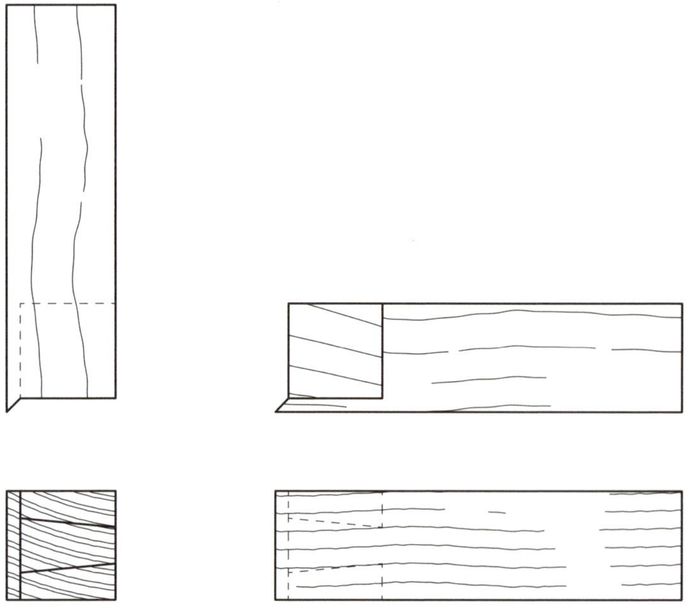

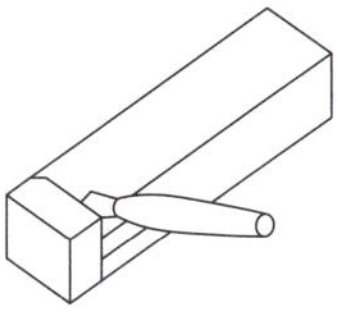

A1

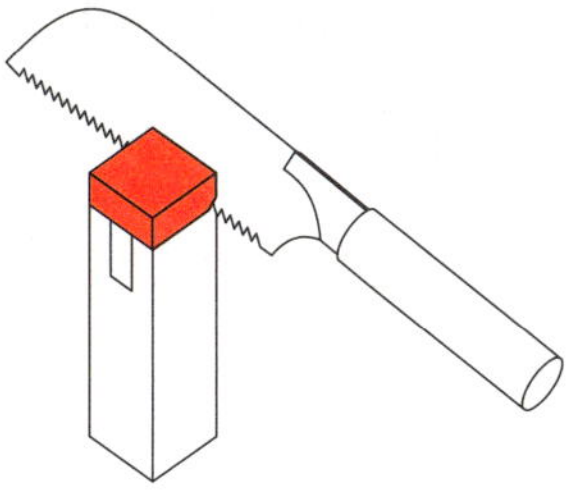

A2

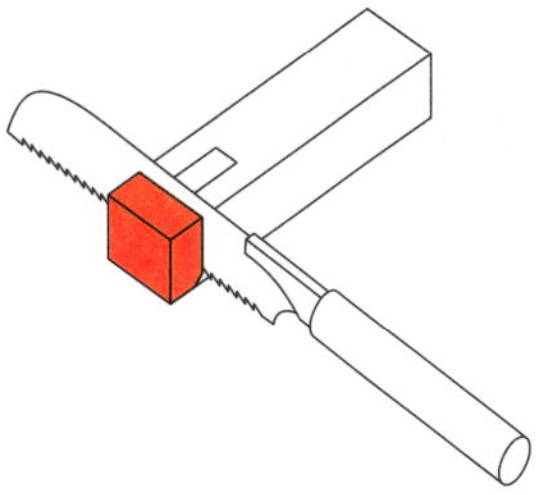

A3

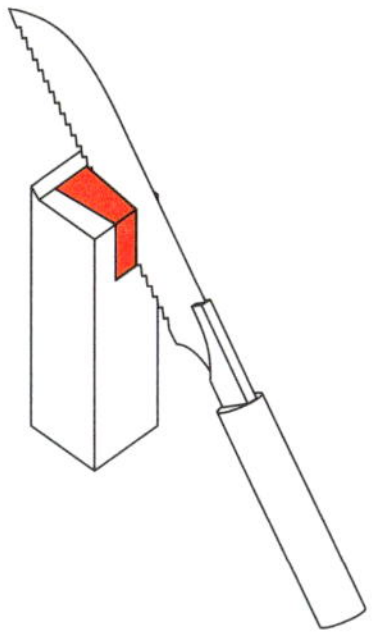

A4

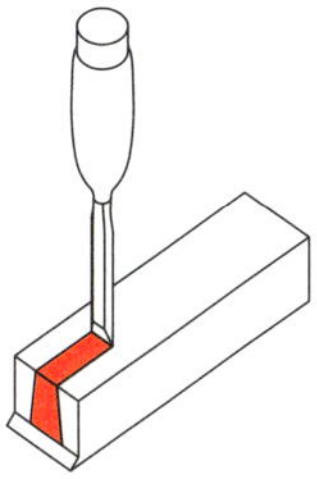

A5

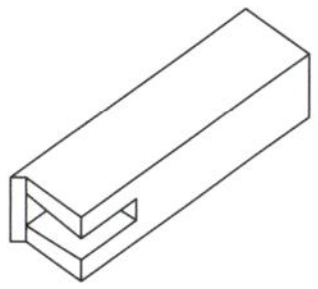

A

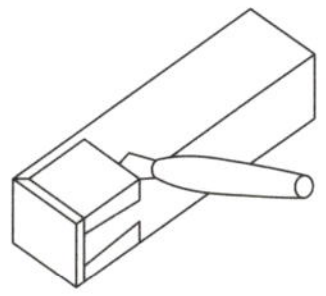

B1

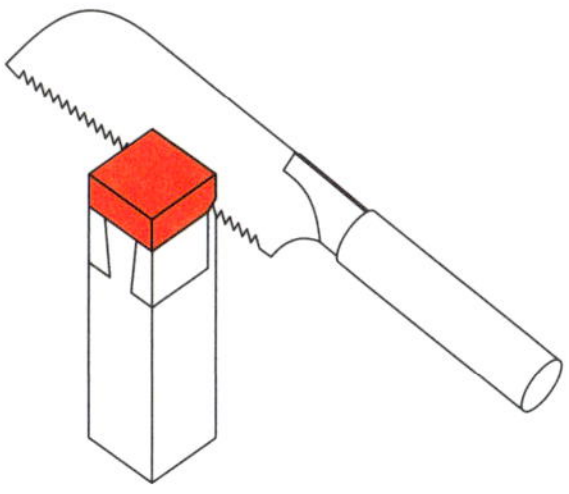

B2

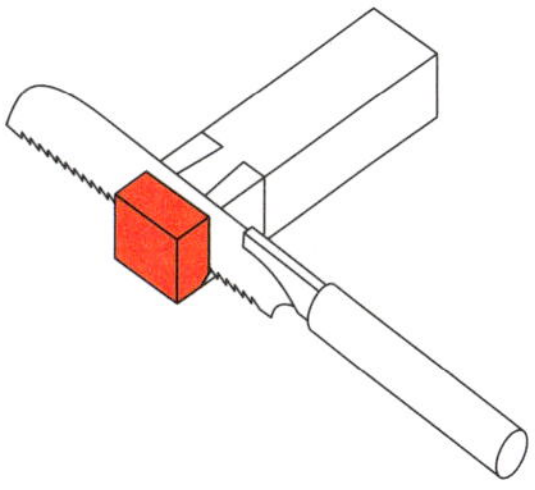

B3

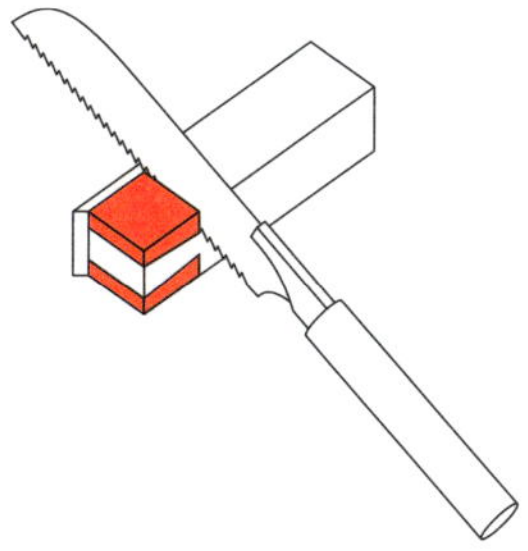

B4

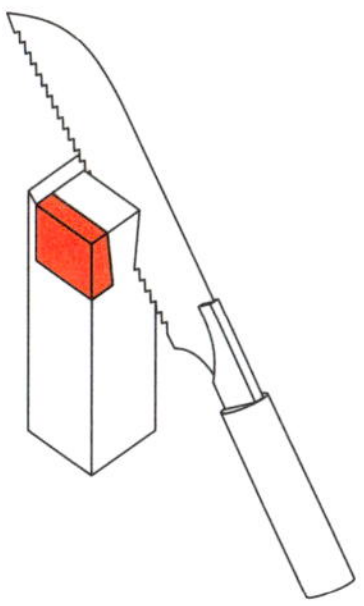

B5

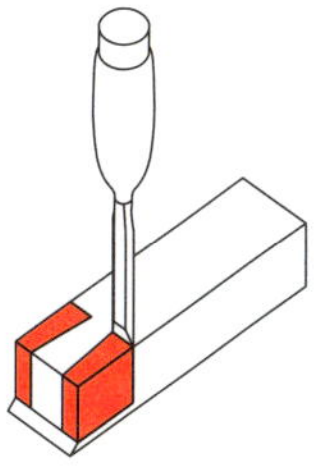

B6

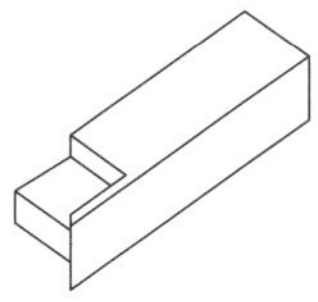

B

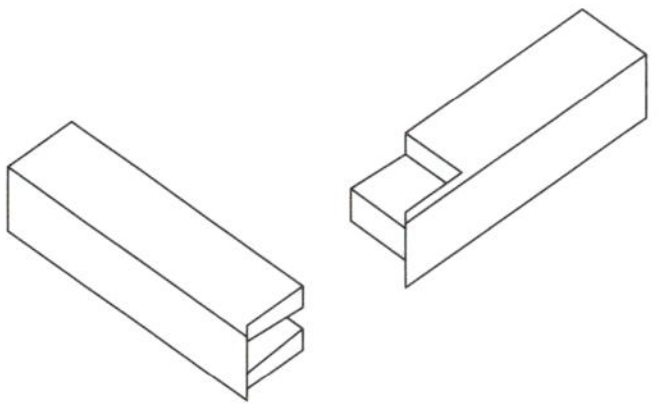

A+B

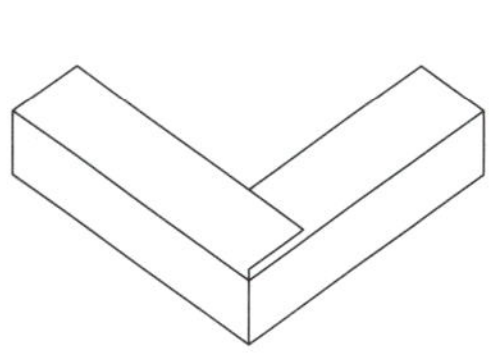

AB

Corner Joints

Blind triple-shouldered through mortise and tenon

uchi-nuki-kone-hozo-sashi-sumi-dome-tsugi
打抜小根枘差隅留接ぎ

durchgehender abgesetzer Eckzapfen mit Wange auf Gehrung

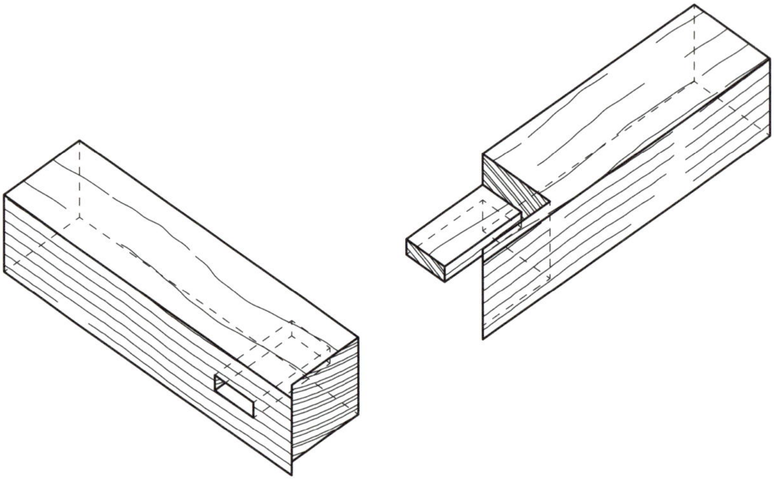

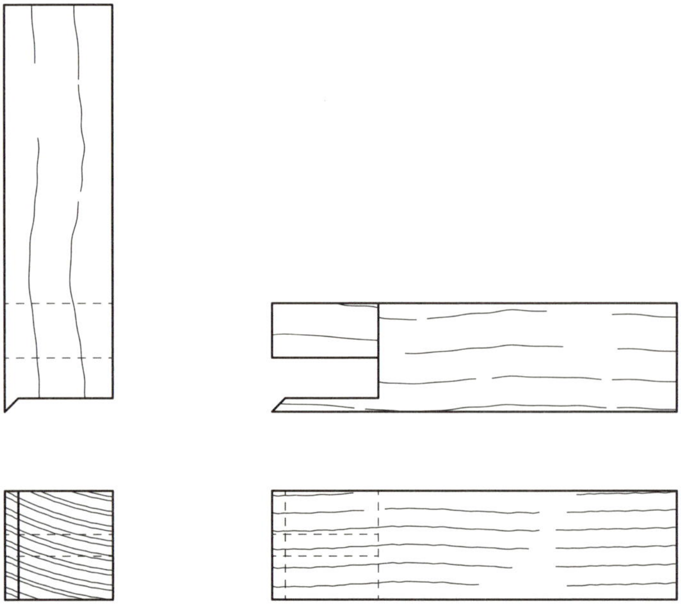

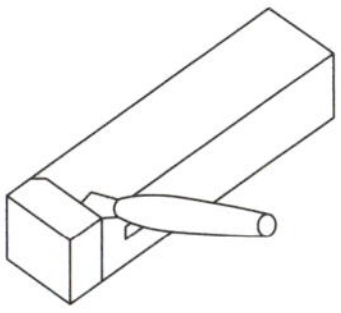

A1

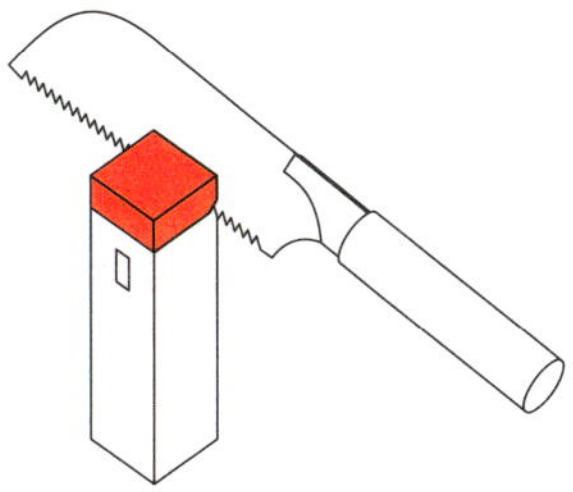

A2

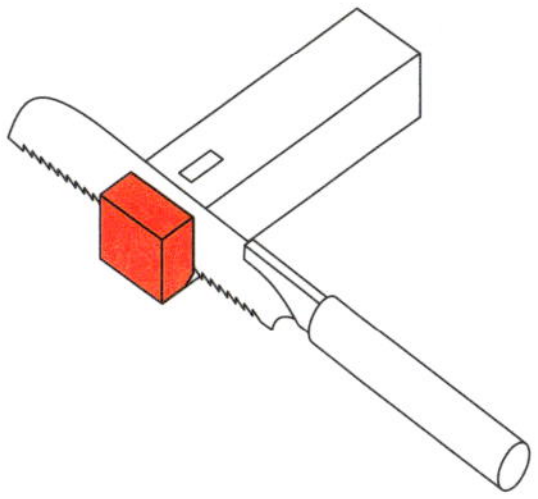

A3

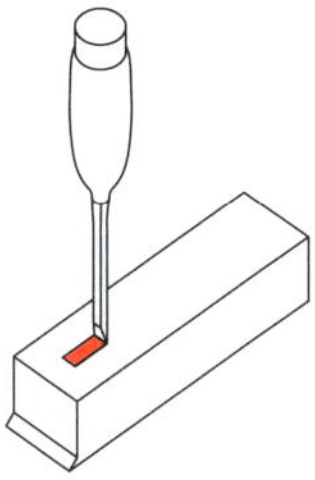

A4

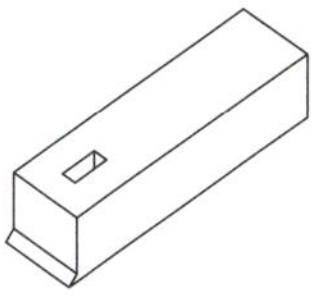

A

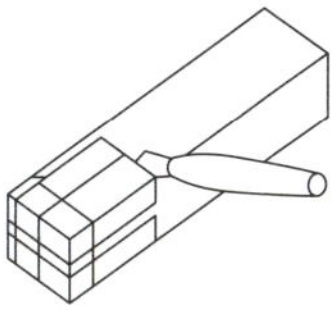

B1

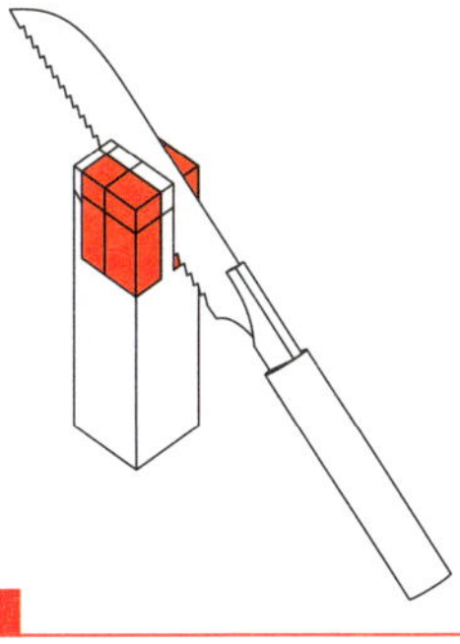

B2

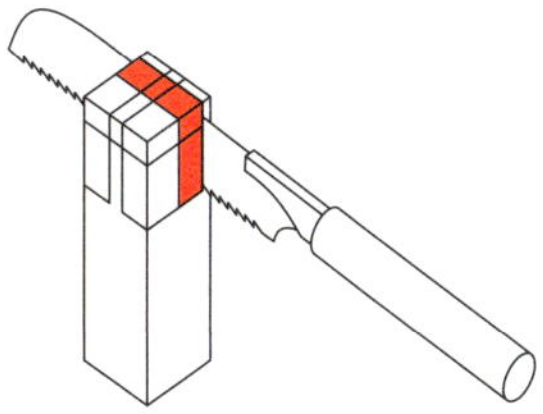

B3

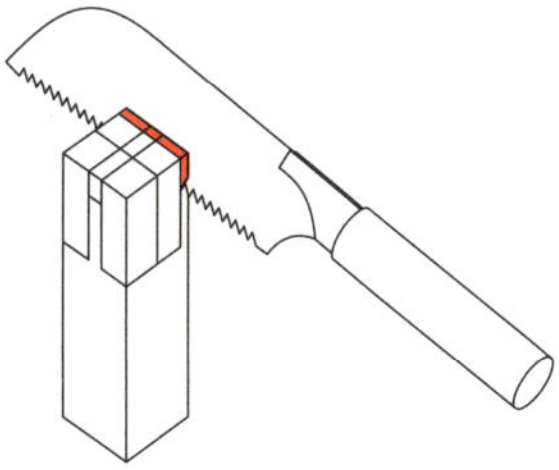

B4

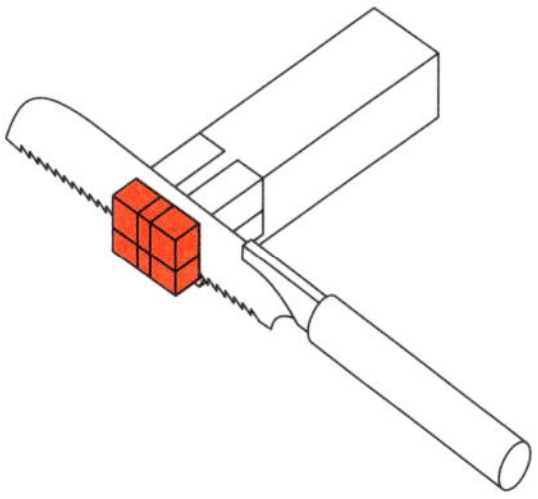

B5

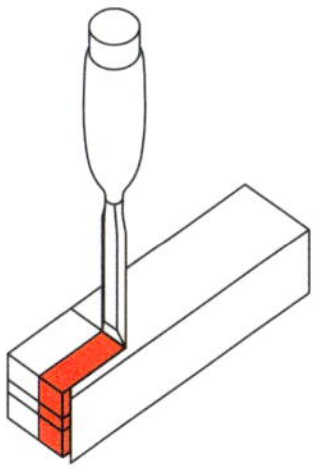

B6

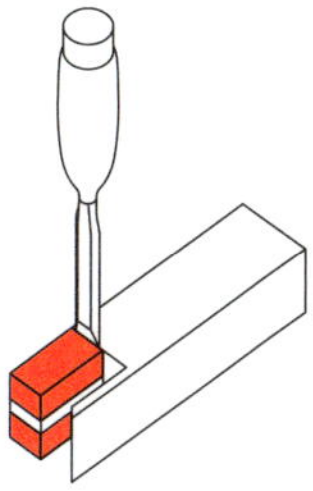

B7

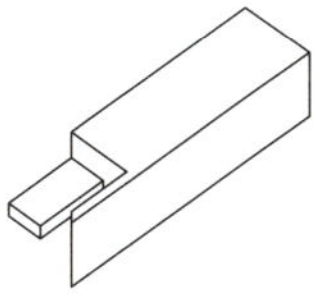

B

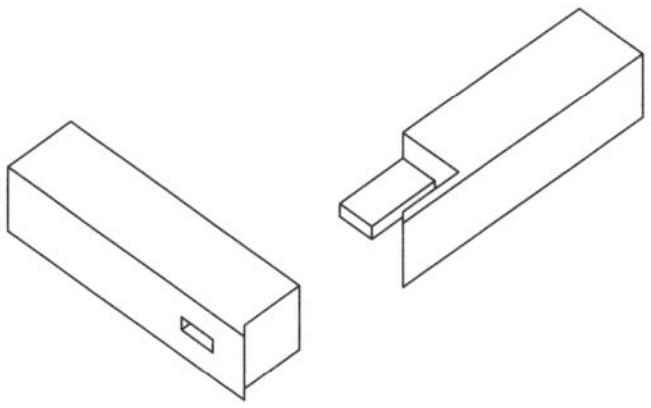

A+B

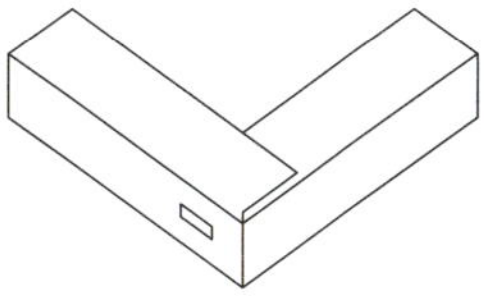

AB

Dado and rabbet joint

eriwa-tsugi
襟輪接ぎ

Eckstoß mit Führungszapfen

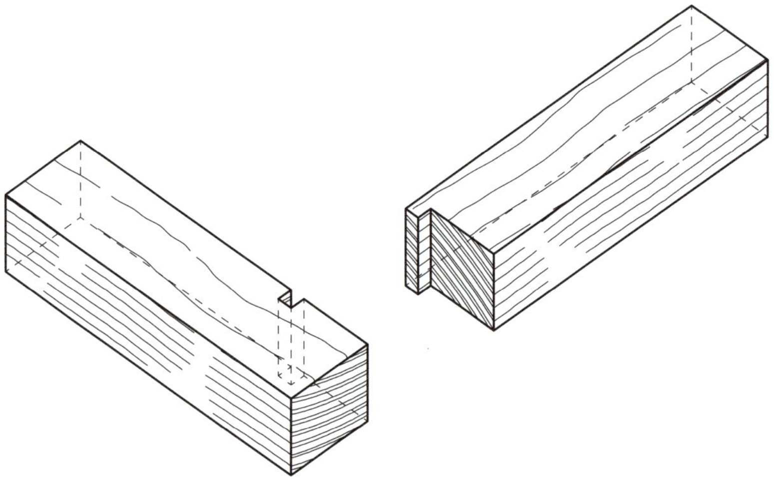

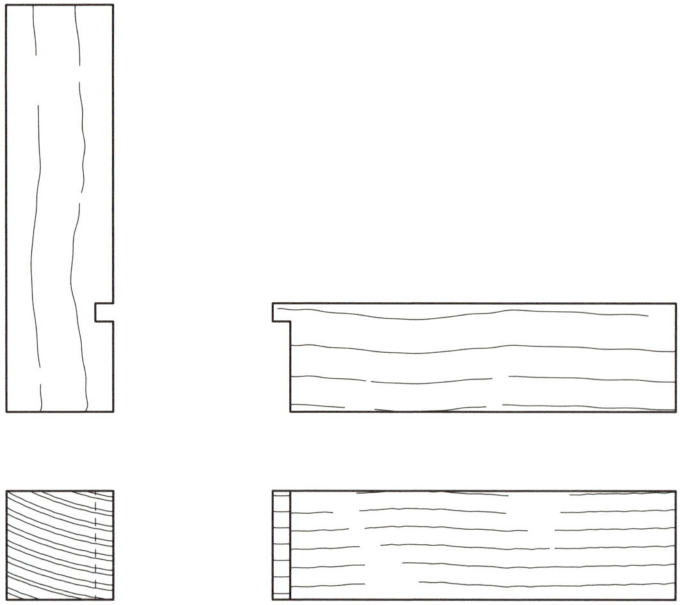

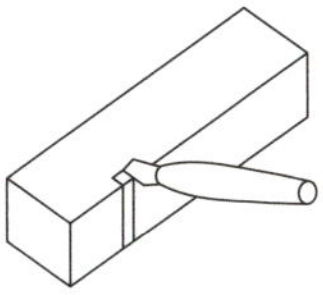

A1

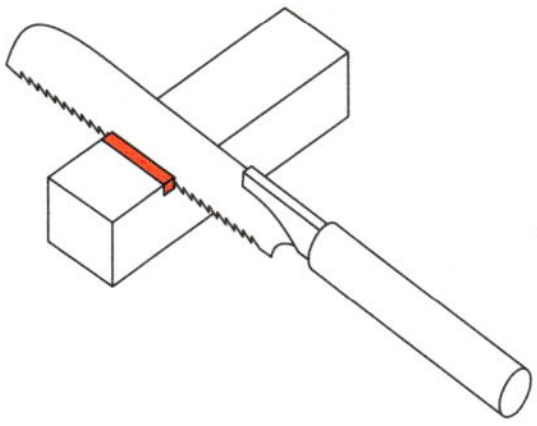

A2

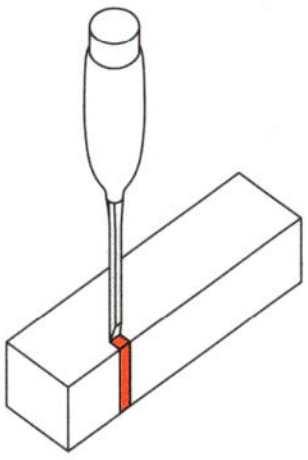

A3

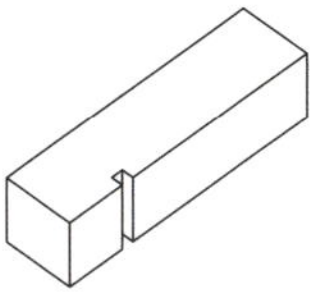

A

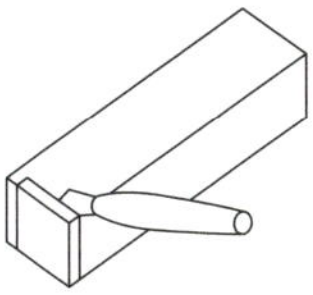

B1

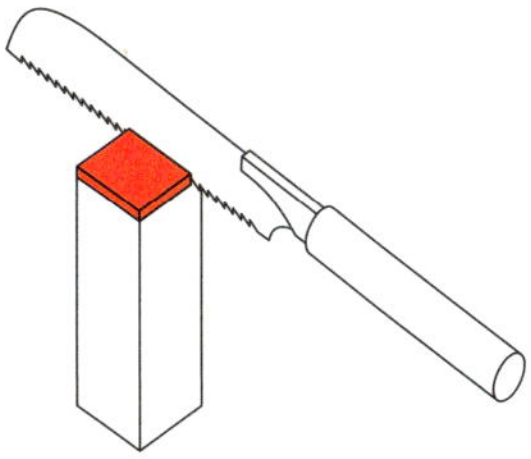

B2

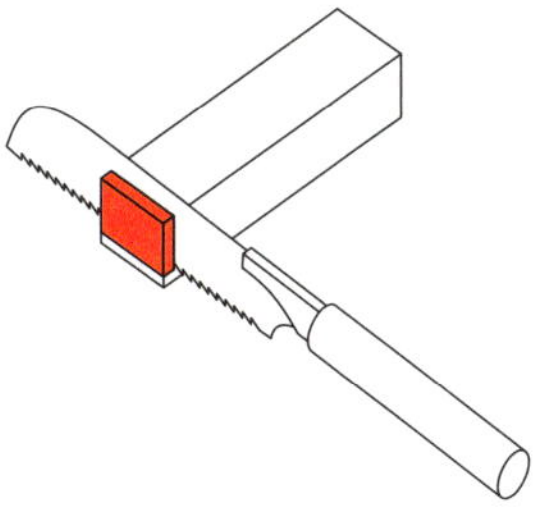

B3

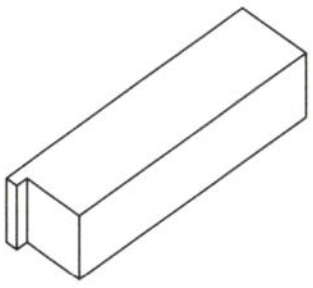

B

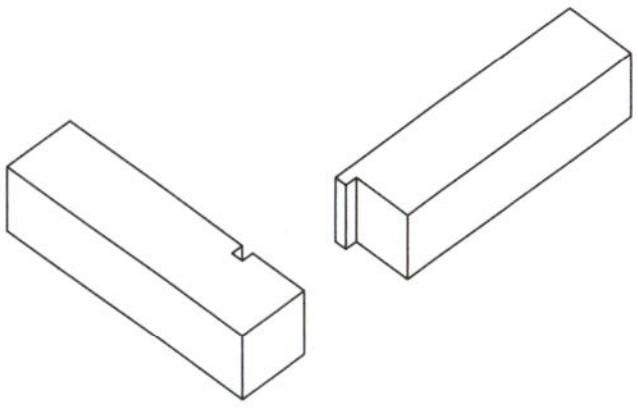

A+B

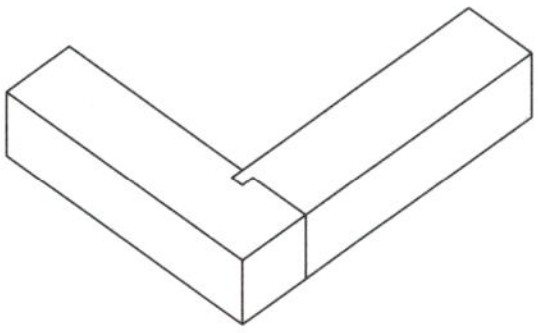

AB

Corner Joints

Blind dado and rabbet joint

han-dome-eriwa-tsugi
半留襟輪接ぎ

Eckstoß mit Führungszapfen
und Wange auf Gehrung

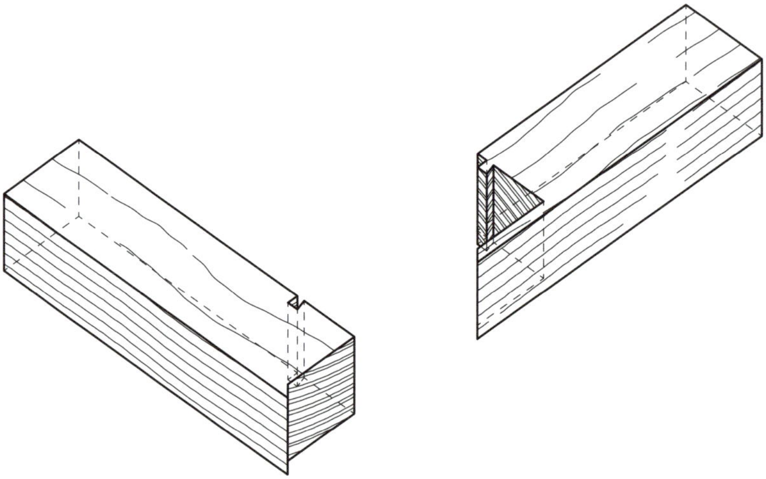

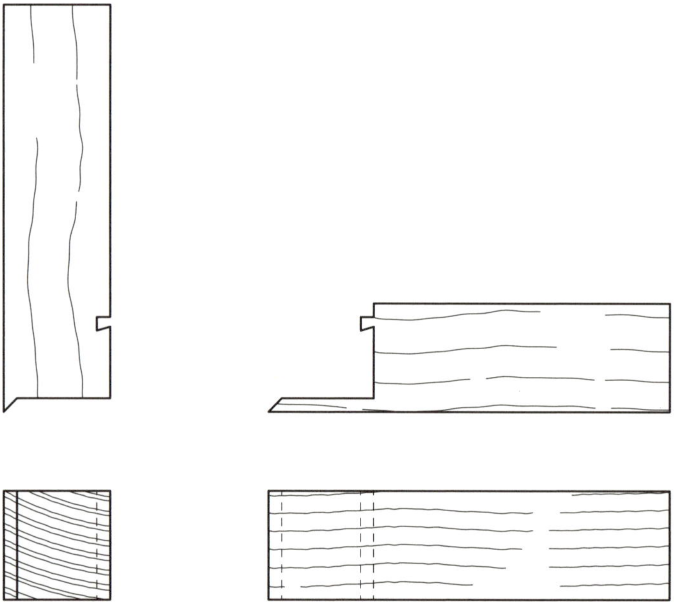

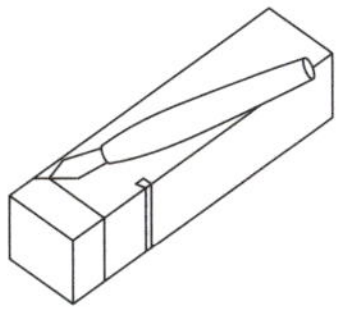
A1

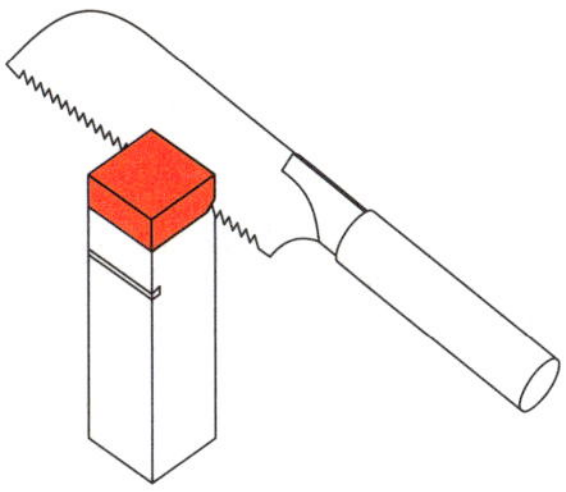
A2

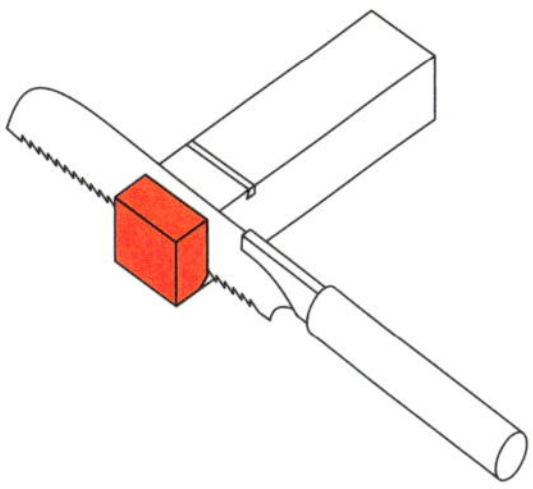
A3

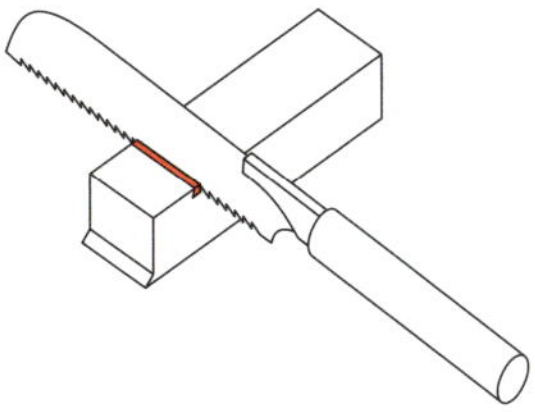
A4

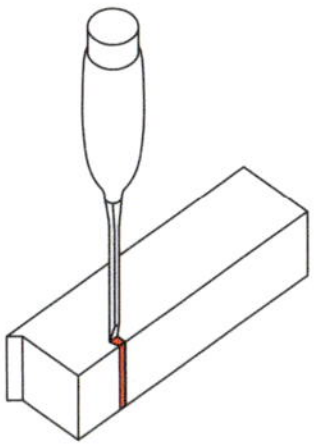
A5

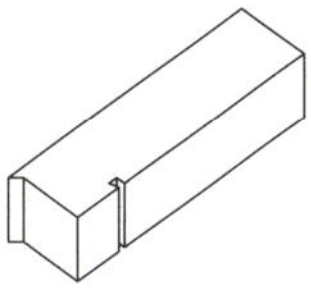
A

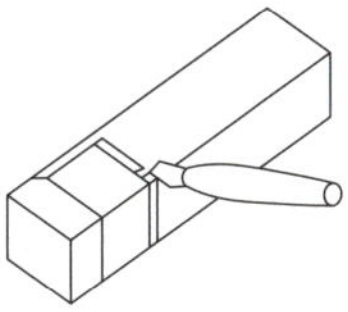
B1

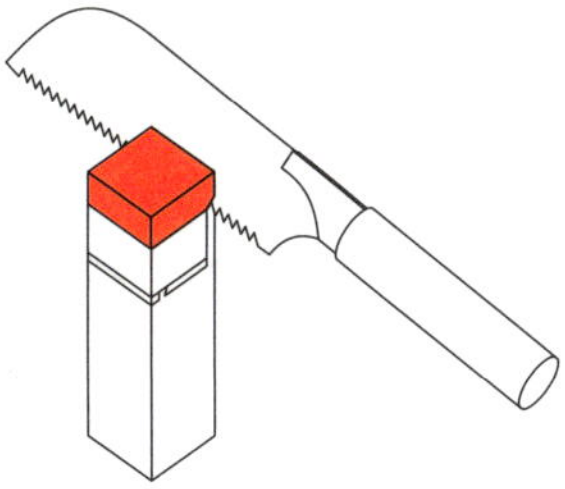
B2

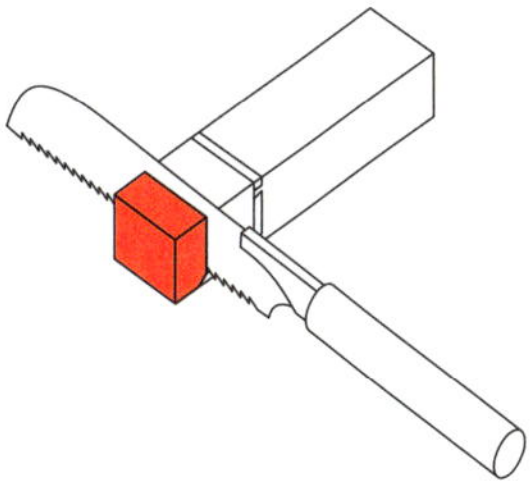
B3

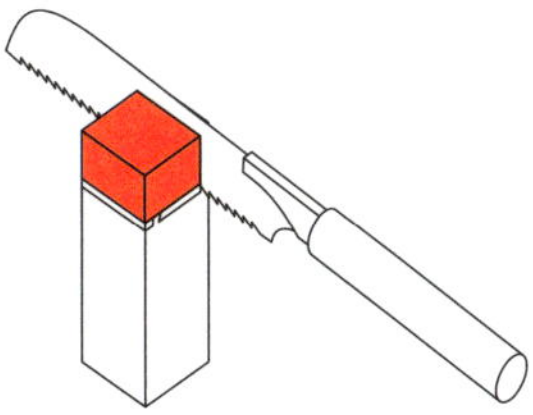
B4

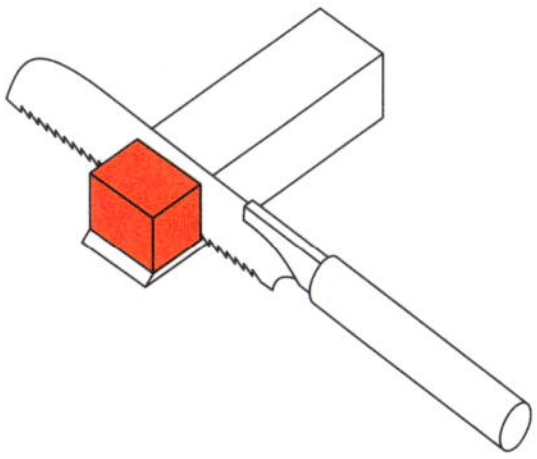
B5

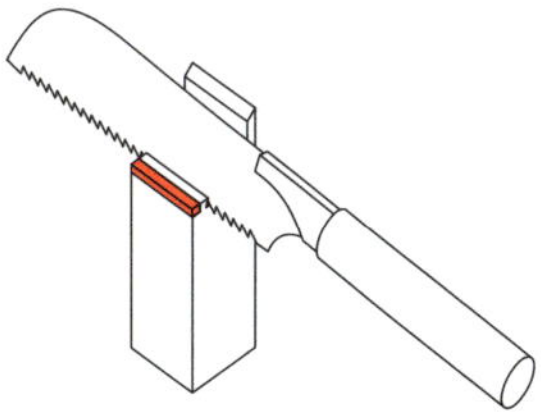
B6

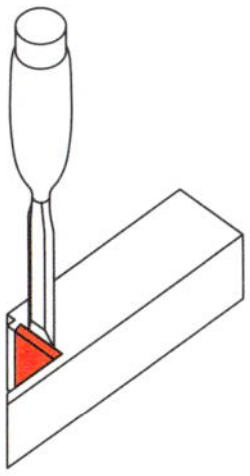
B7

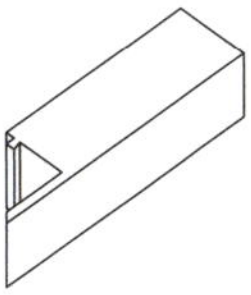
B

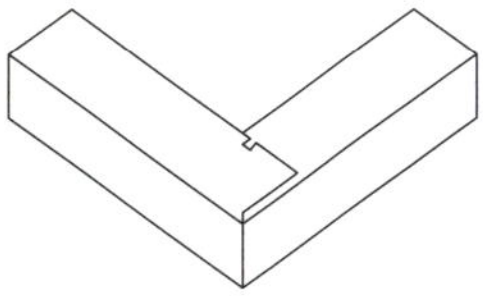
AB

Corner Joints

Collared haunched mortise and tenon joint

eriwa-kone-hozo-sashi-tsugi
襟輪小根枘差接ぎ

durchgehender abgesetzter Stufeneckzapfen mit Führungszapfen

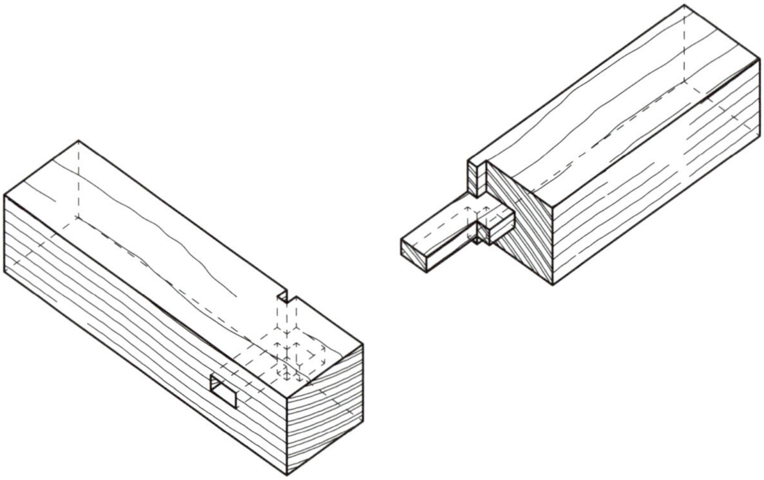

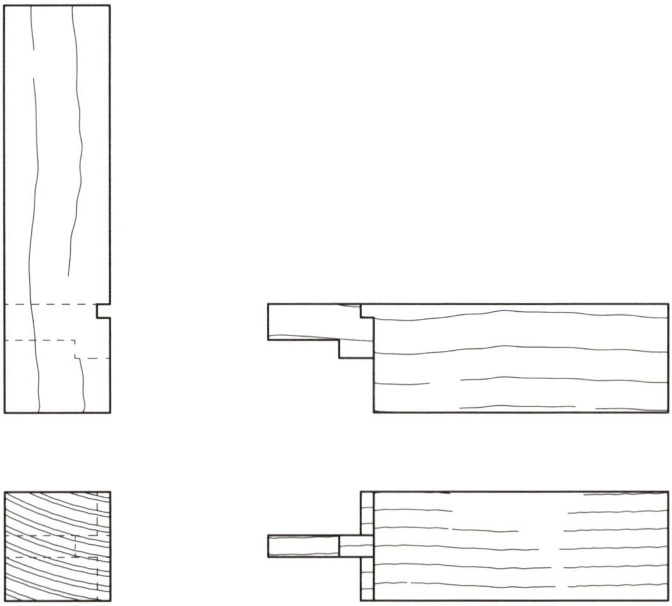

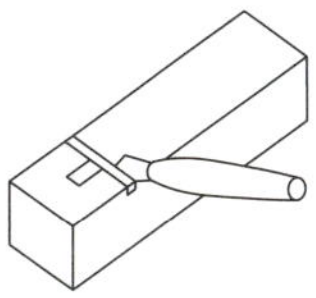
A1

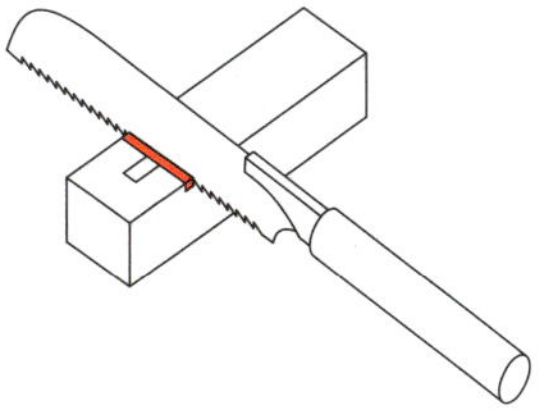
A2

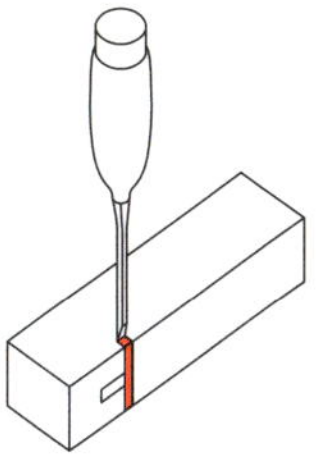
A3

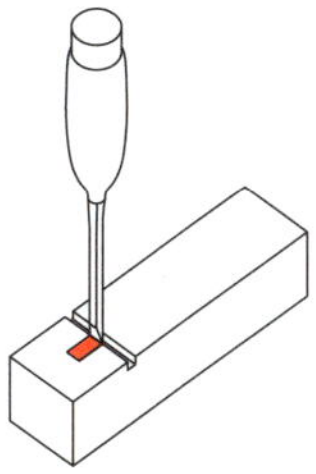
A4

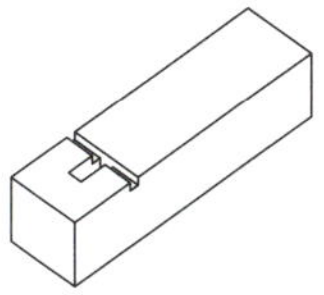
A

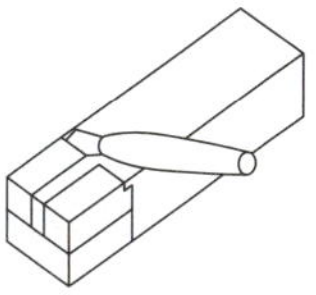
B1

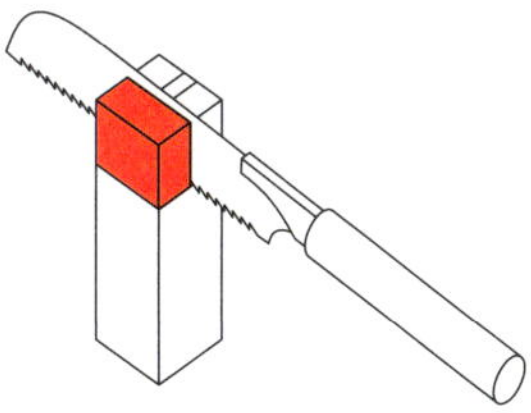
B2

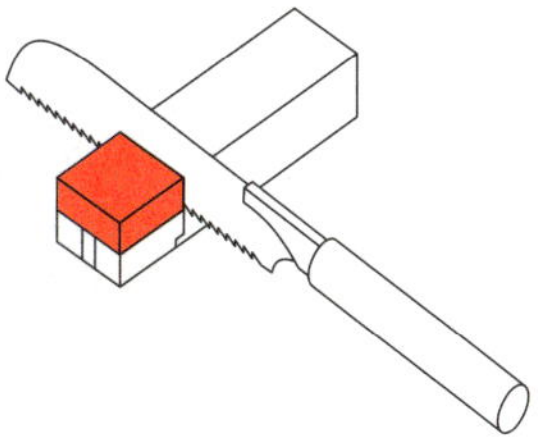
B3

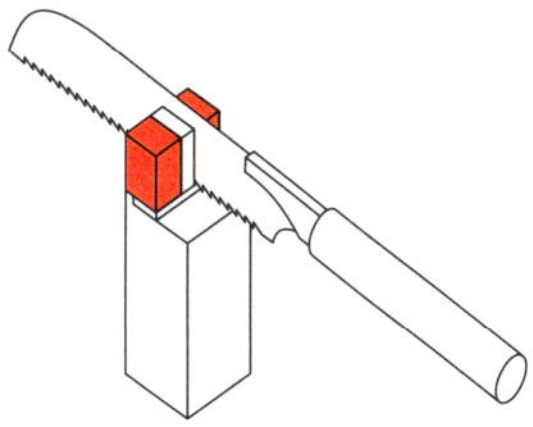
B4

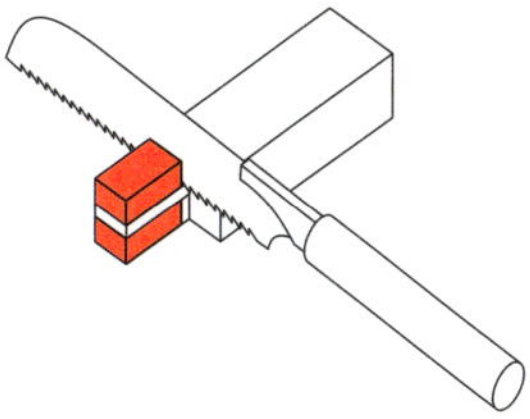
B5

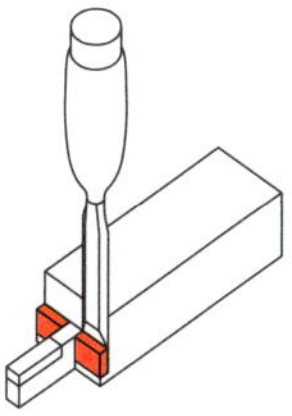
B6

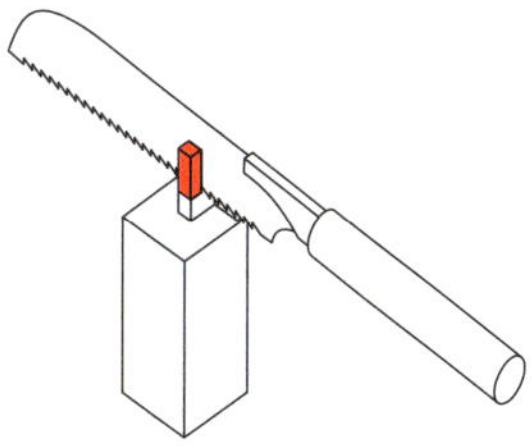
B7

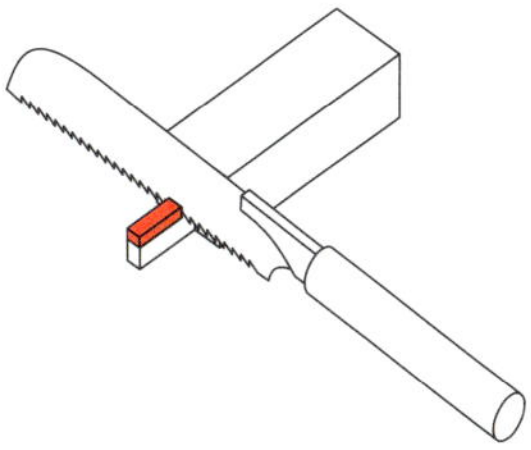
B8

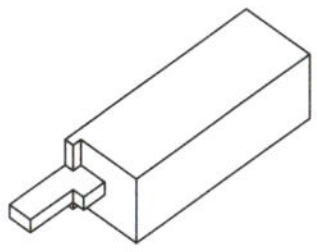
B

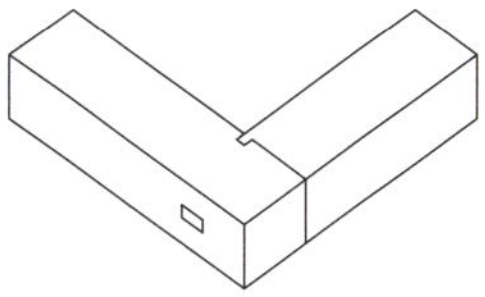
AB

Corner Joints

Wedged collared blind mortise and tenon

sumi-dome-hozo-sashi-tsugi-komi-sen-uchi (mechi-tsuki, binmen-dome) [*]
隅留枘差接込栓打ち(目違付·鬢面留)

verkeilter durchgehender abgesetzter Eckzapfen mit Führungszapfen und Wange auf Gehrung

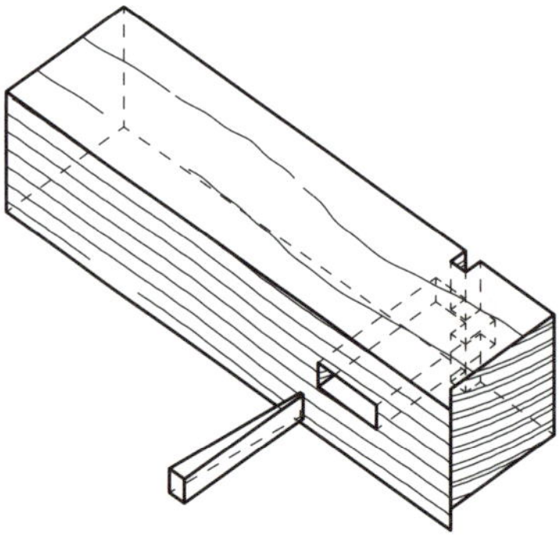

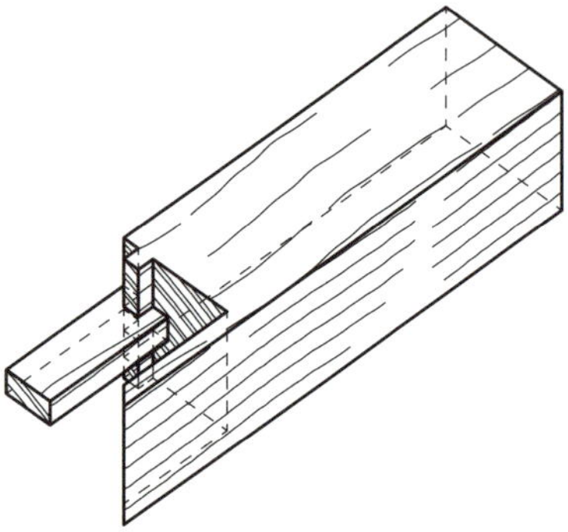

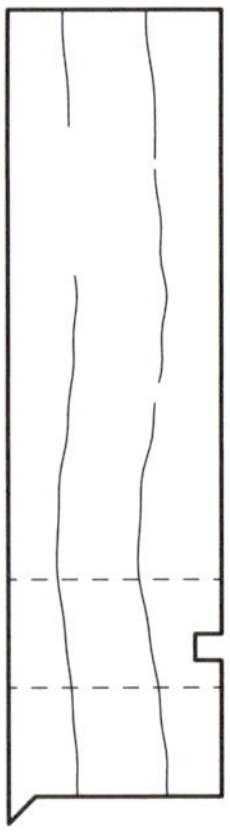

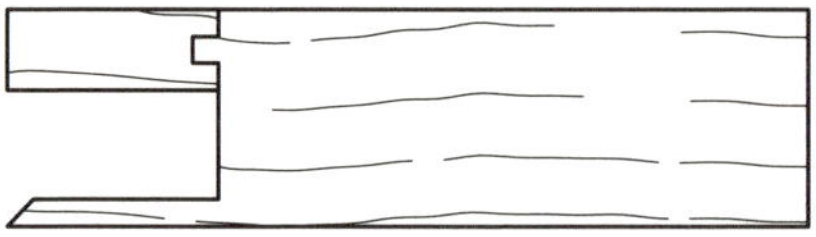

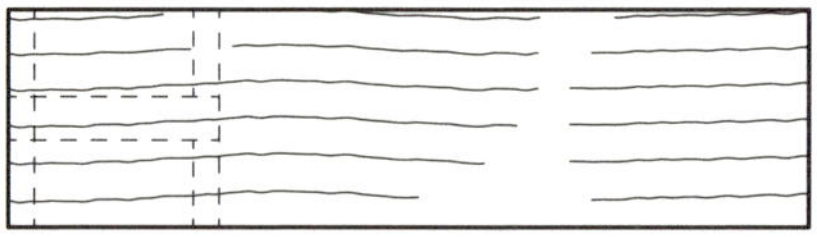

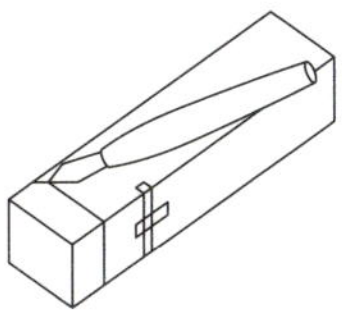

A1

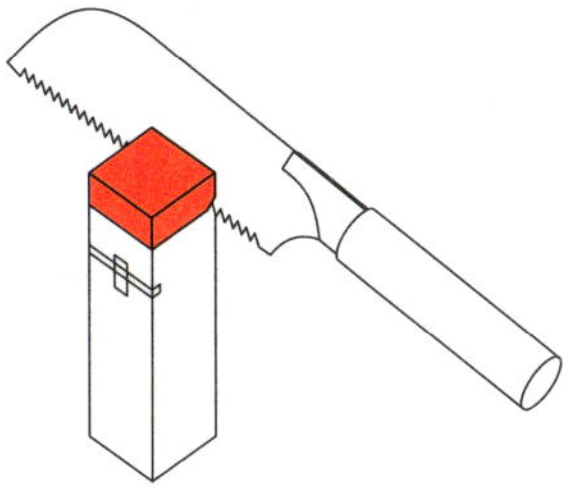

A2

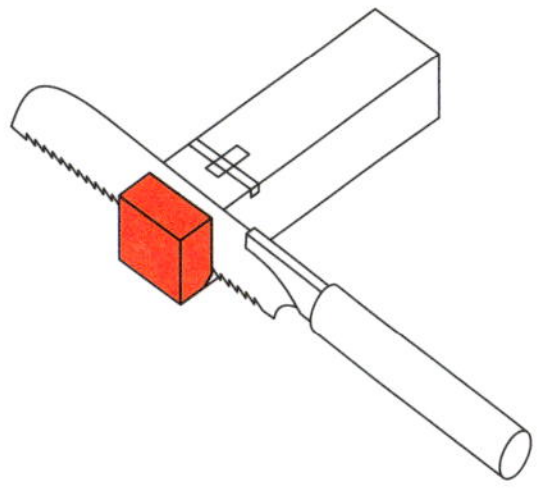

A3

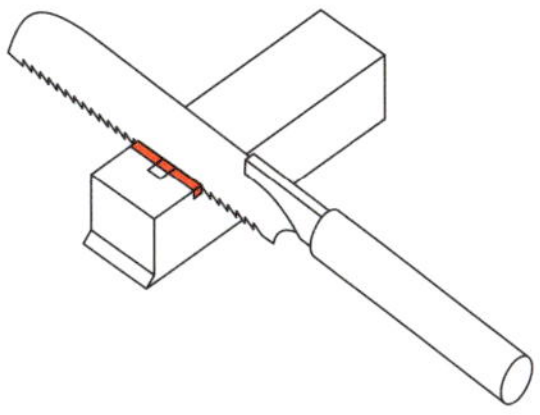

A4

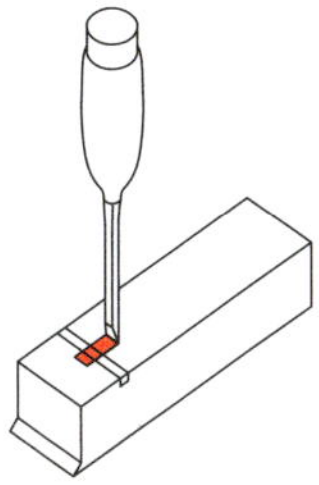

A5

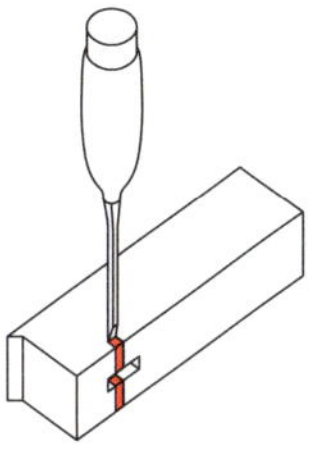

A6

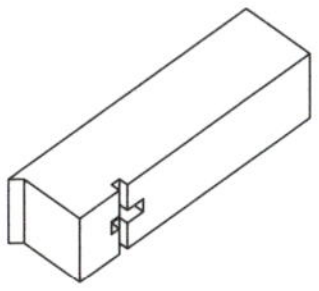

A

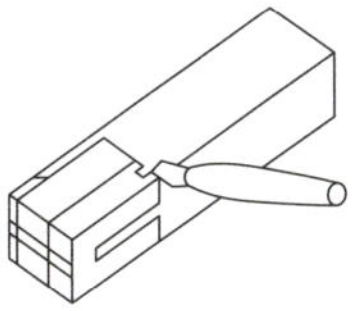

B1

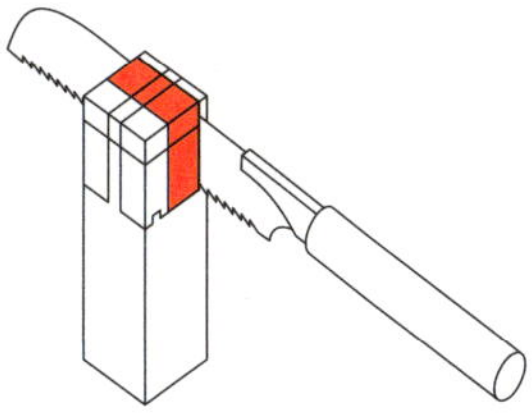

B2

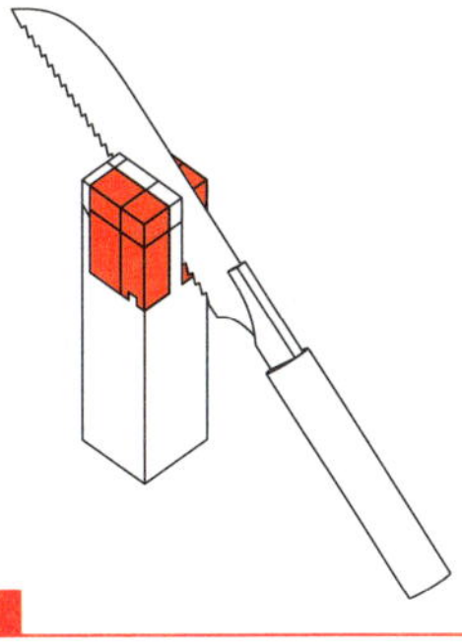

B3

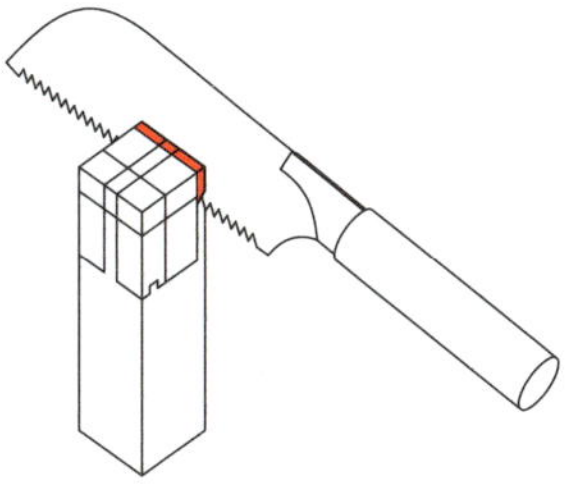

B4

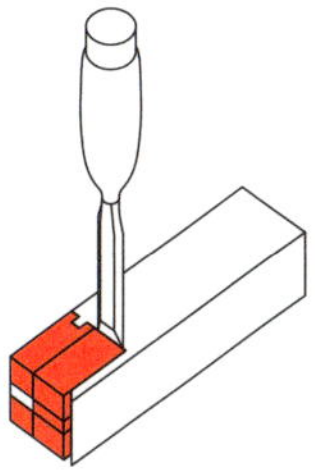

B5

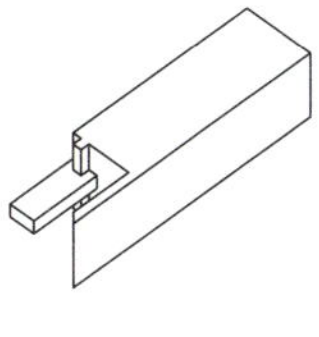

B

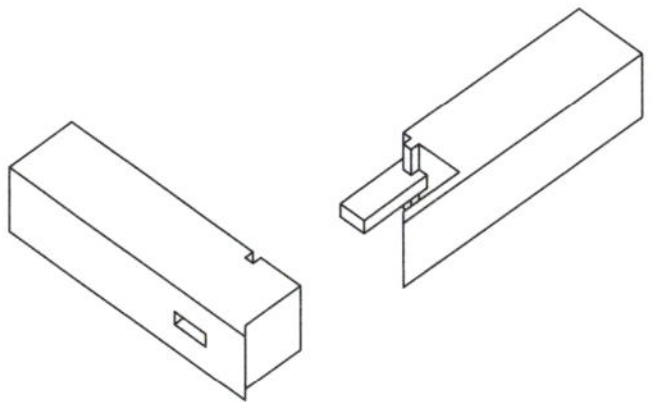

A+B

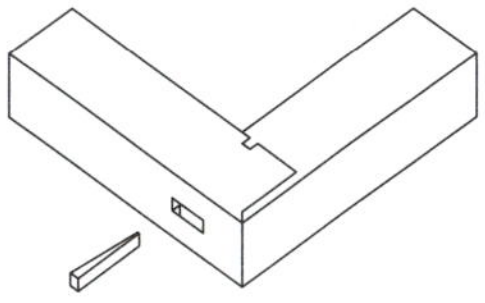

AB

Corner Joints

Plain miter

fure-dome-tsugi
振留接ぎ

stumpfer Gehrungstoß

2×

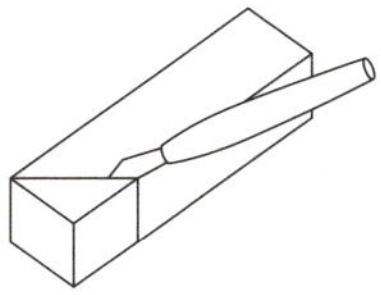

A1

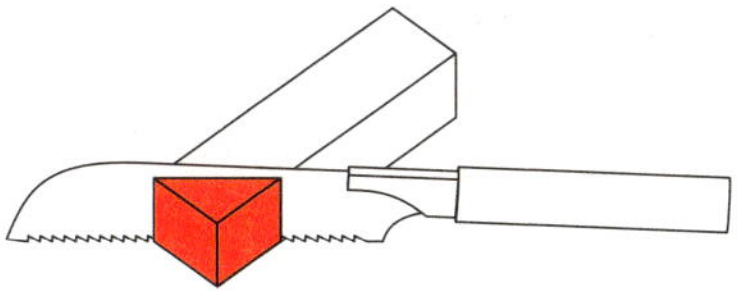

A2

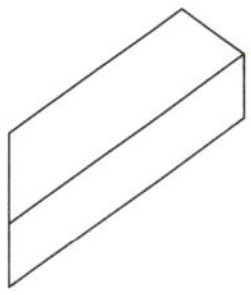

A

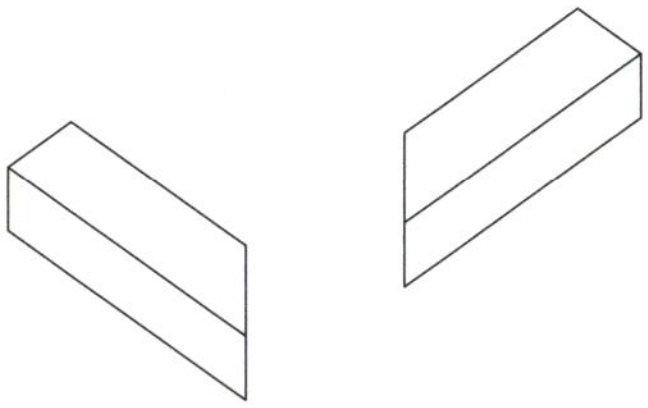

A+A

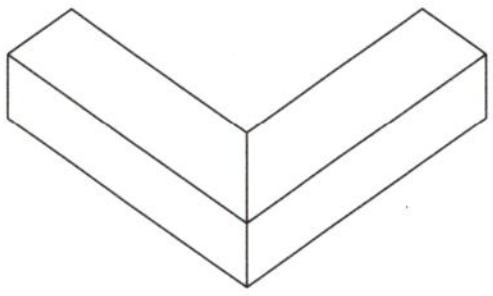

AA

Corner Joints

Plain miter with spline key

hiki-komi-dome-tsugi
挽込留継ぎ

Gehrungsstoß mit Feder

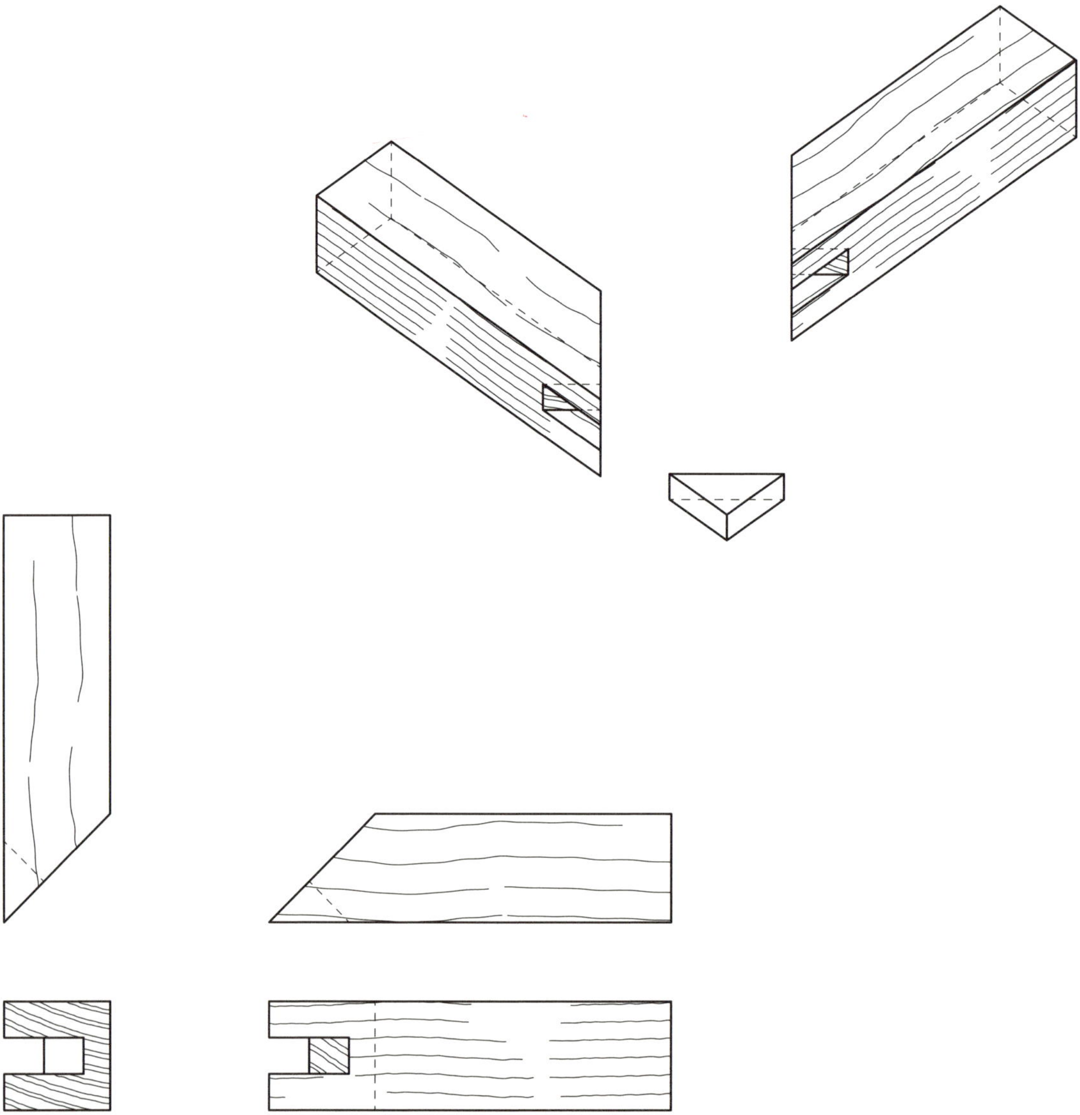

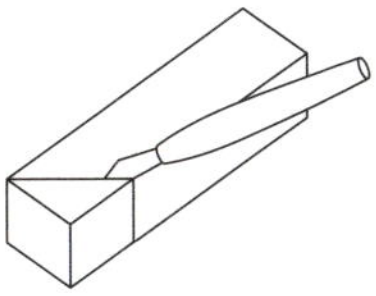

A1

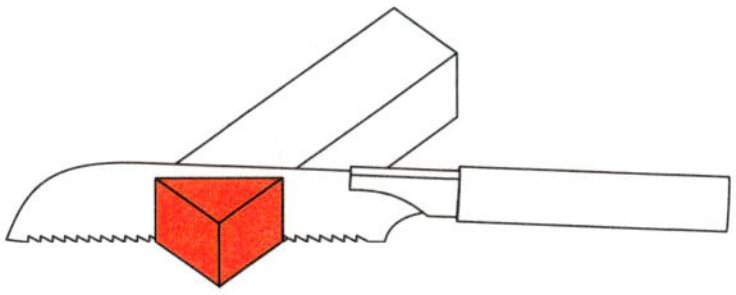

A2

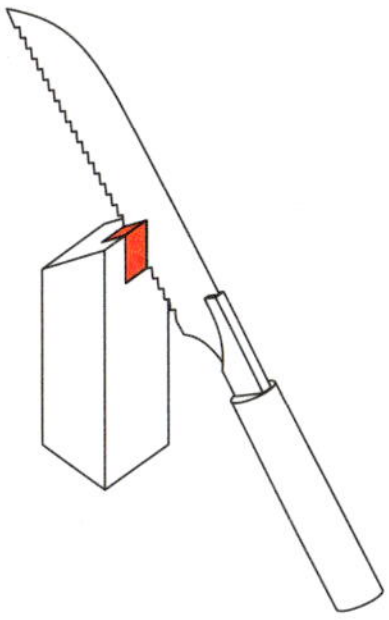

A3

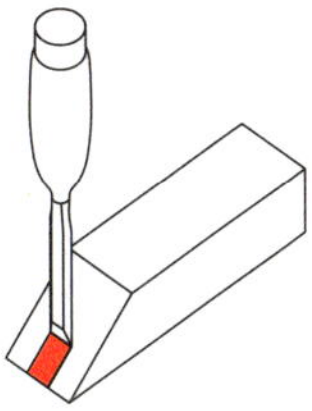

A4

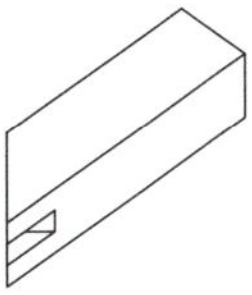

A

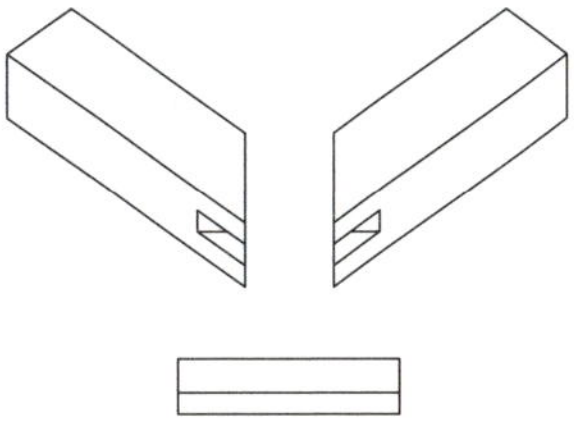

A+B+A

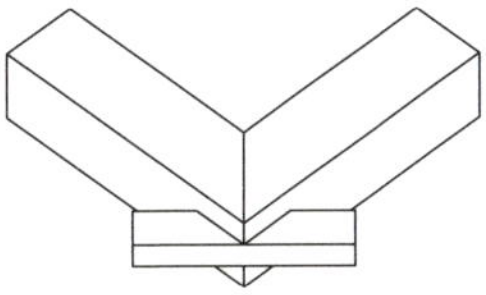

A+B+A

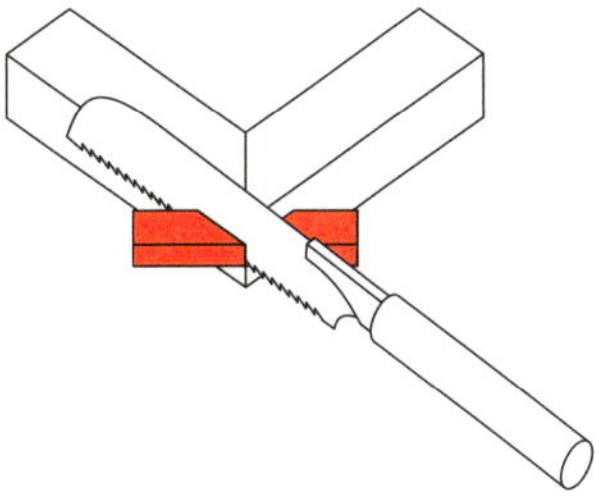

A+B+A

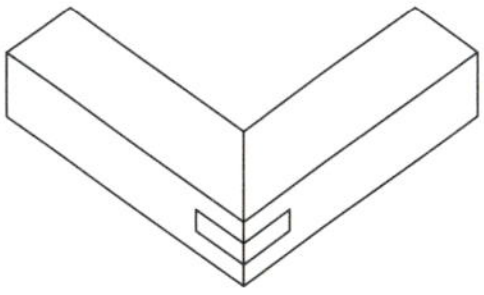

ABA

Corner Joints

Plain miter with spline

yahazu-yatoi-zane-dome-tsugi
矢筈雇実留継ぎ

Gehrungsstoß mit Federstäbchen

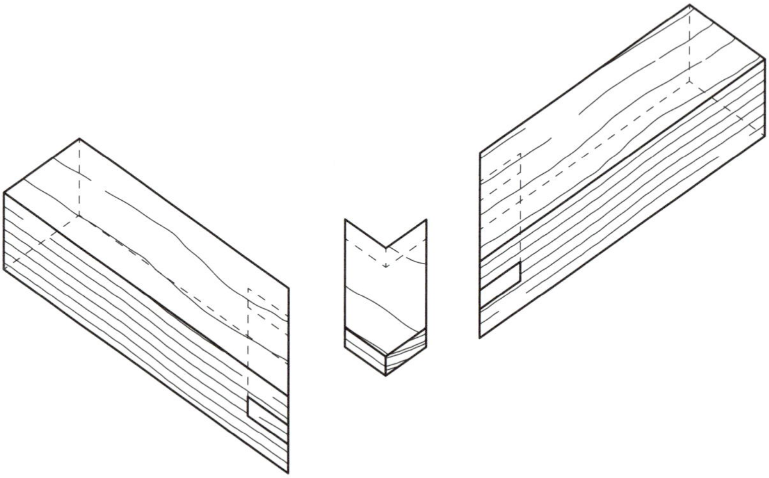

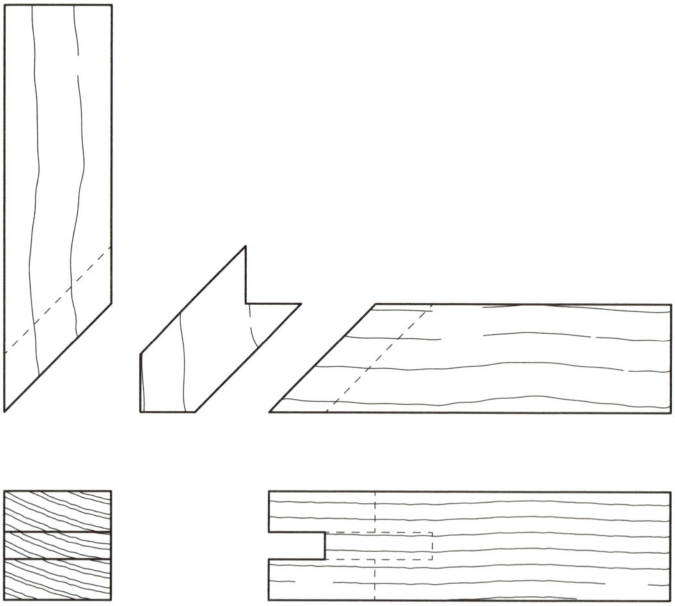

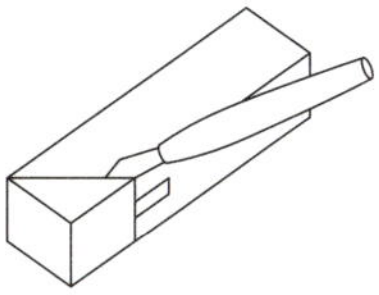

A1

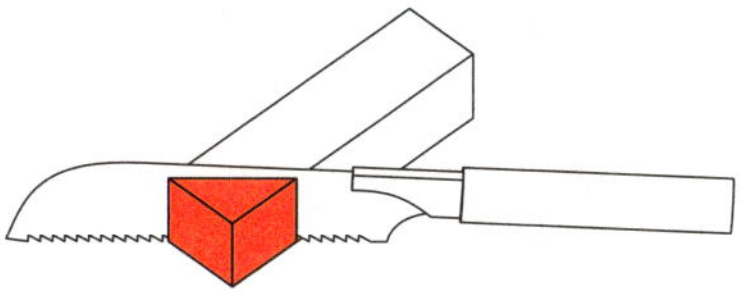

A2

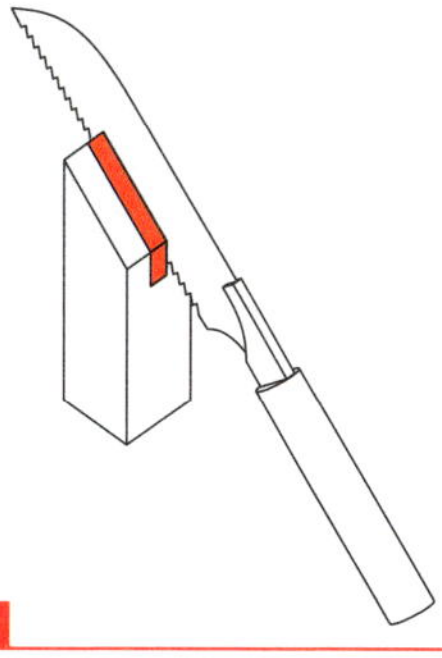

A3

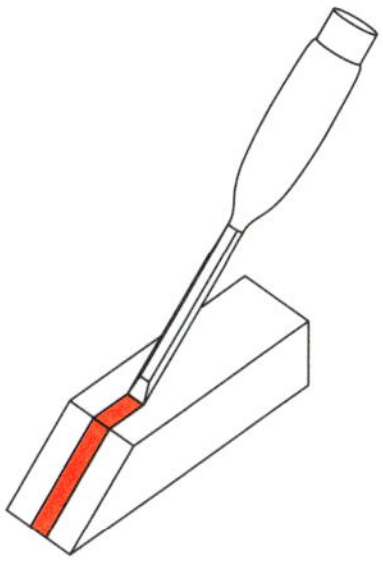

A4

2×

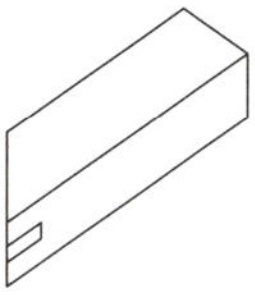

A

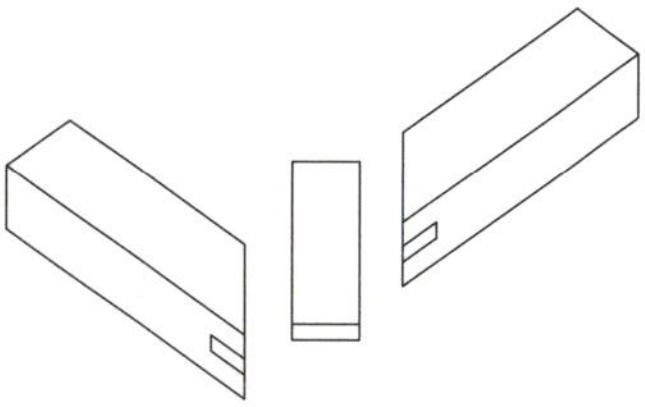

A+B+A

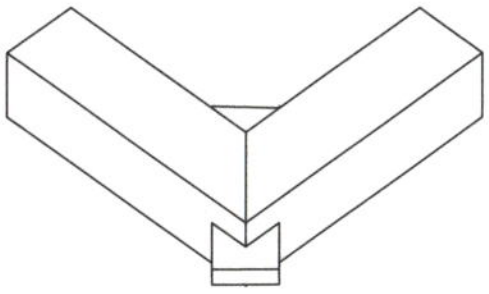

A+B+A

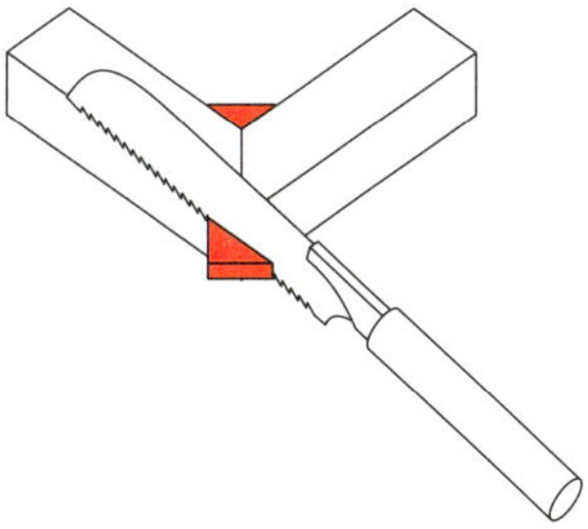

A+B+A

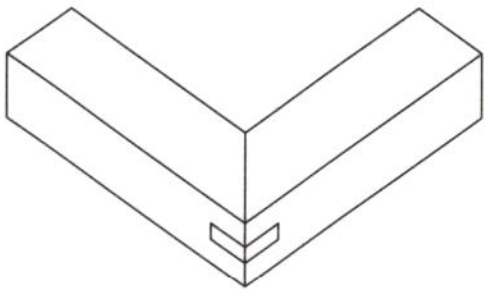

ABA

Corner Joints

Plain miter with dovetail key

ari-zane-hiki-komi-dome-tsugi
蟻実挽込留継ぎ

Gehrungsstoß mit Schwalbenschwanzfeder

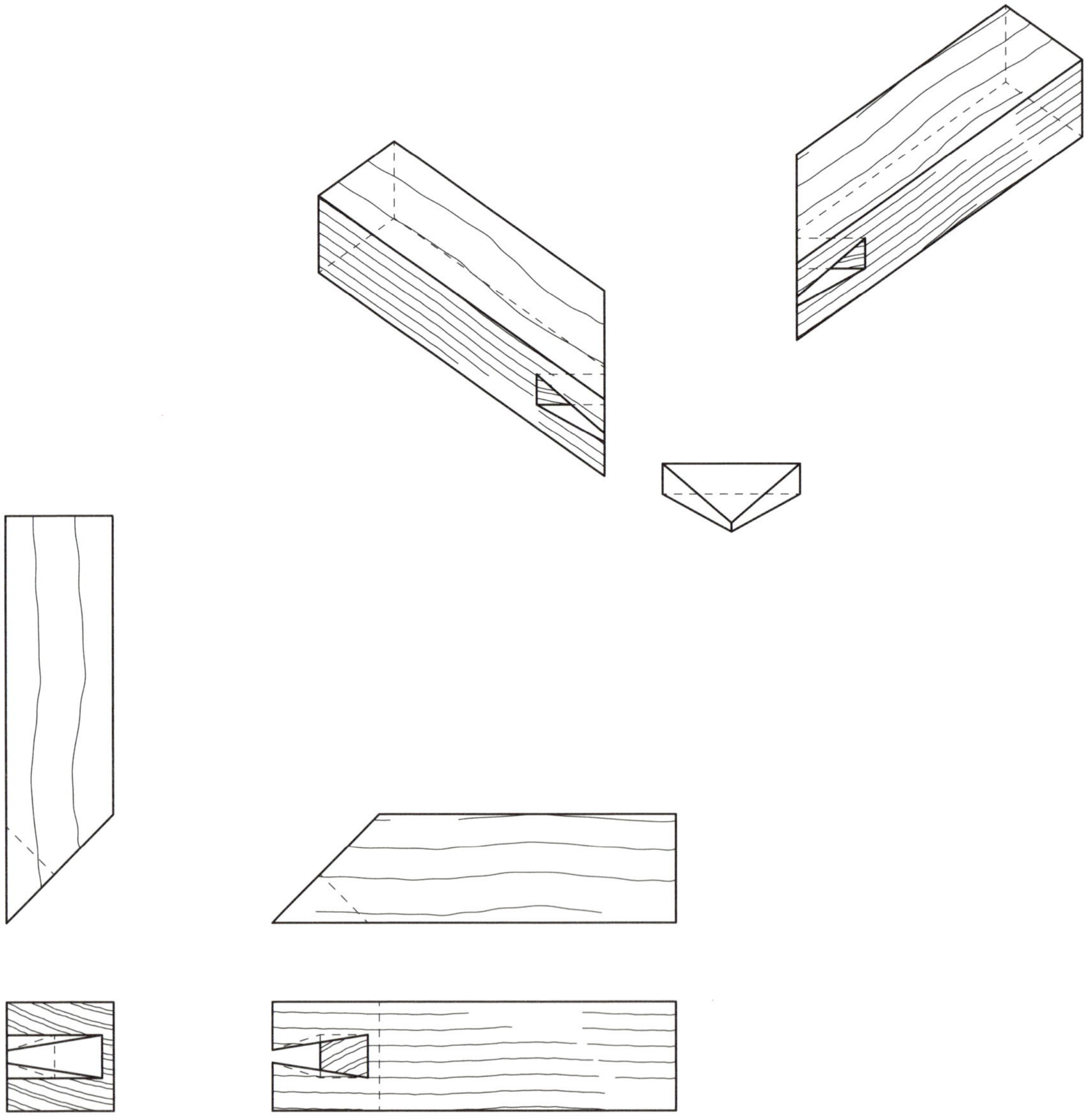

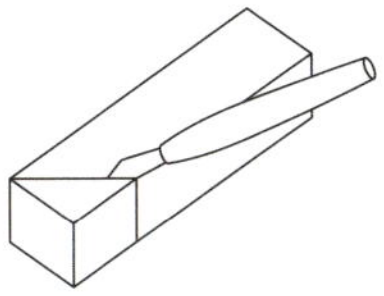

A1

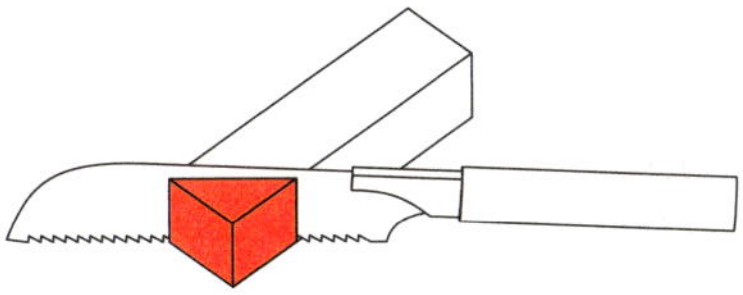

A2

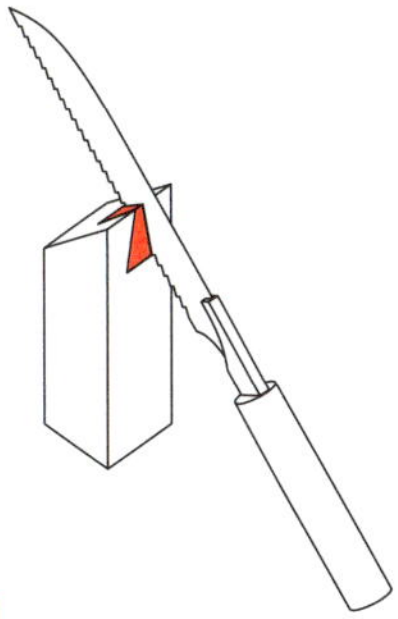

A3

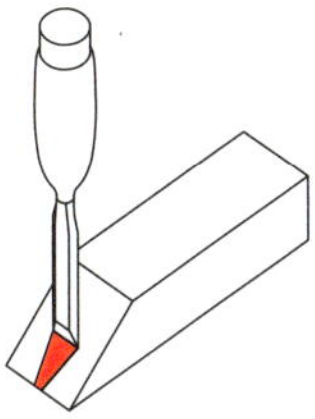

A4

2×

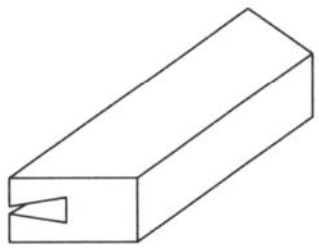

A

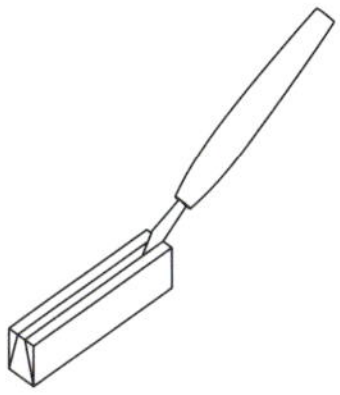

B1

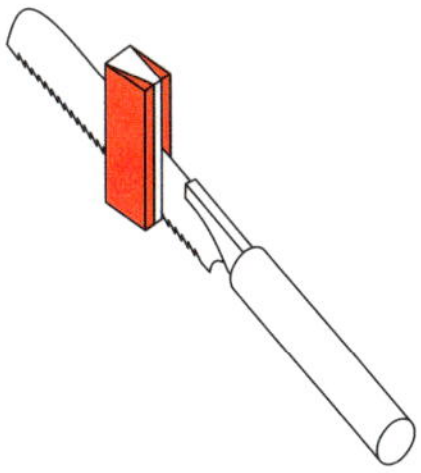

B2

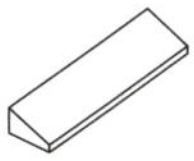

B

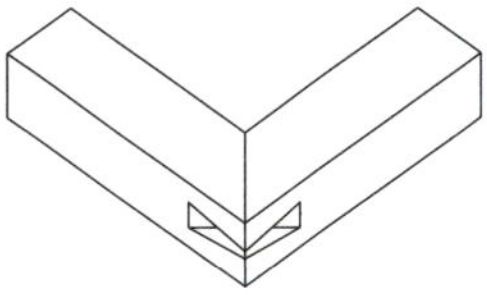

A+A

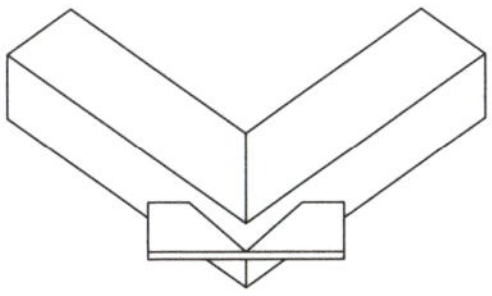

A+B+A

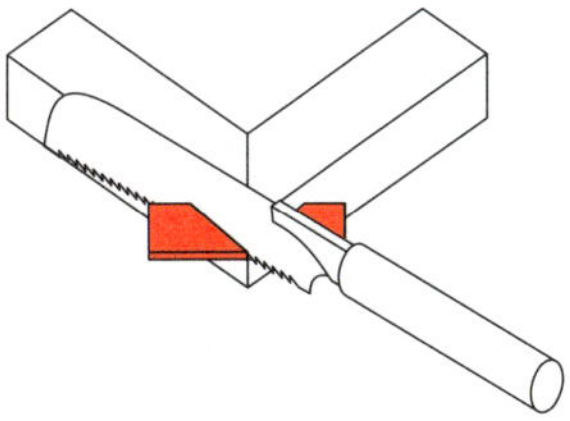

A+B+A

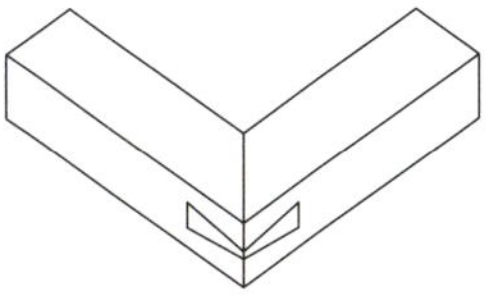

ABA

Corner Joints

Plain miter with butterfly key

chigiri-ōdome-tsugi
千切大留接ぎ

Gehrungsstoß mit
Schwalbenschwanzeinlage

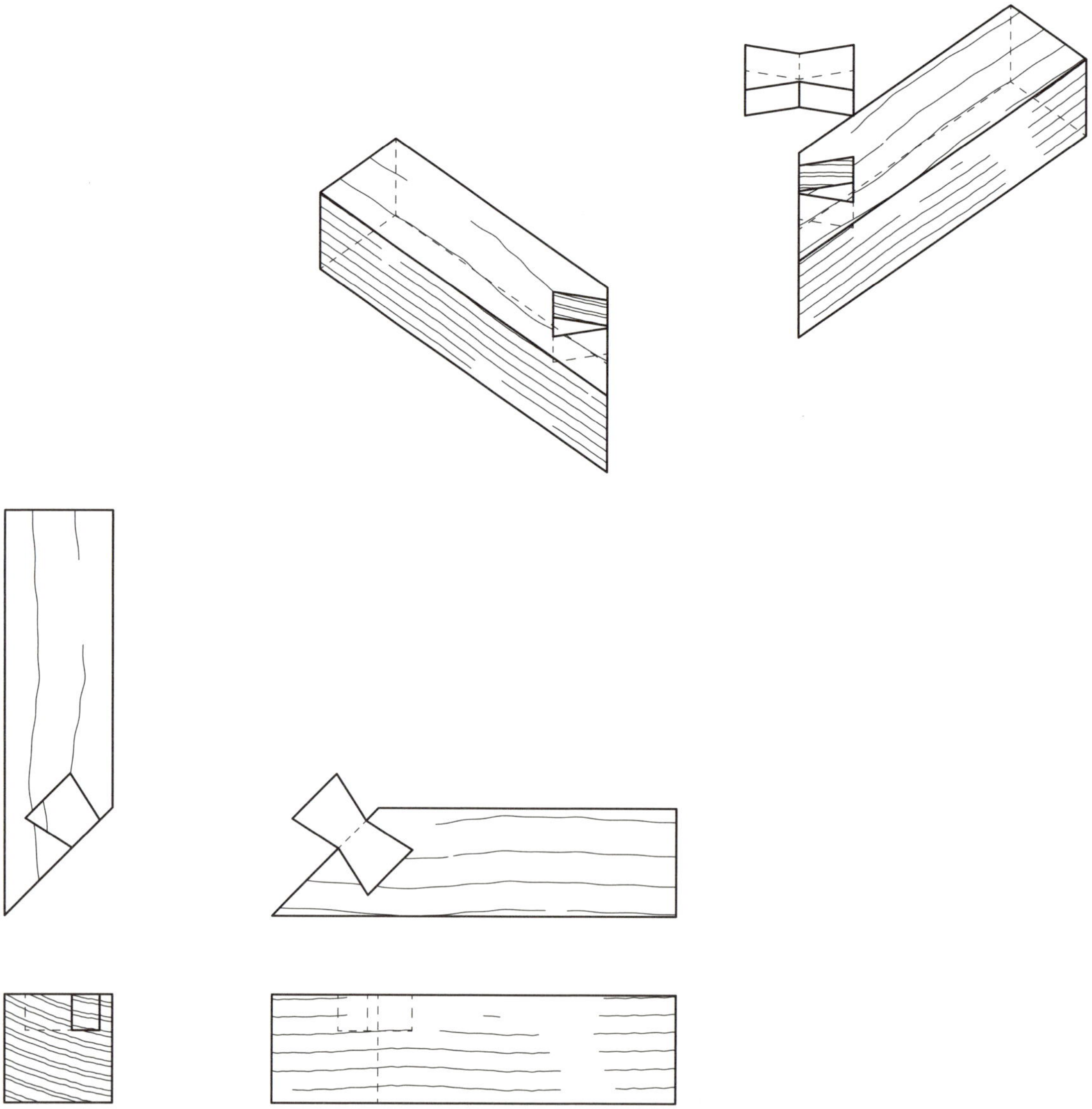

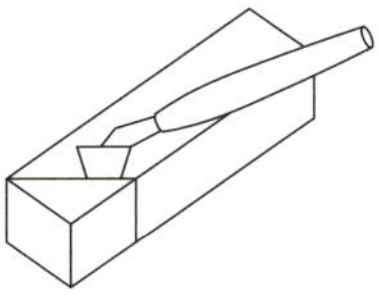

A1

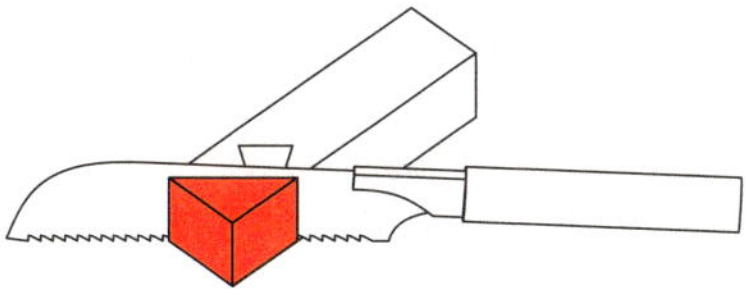

A2

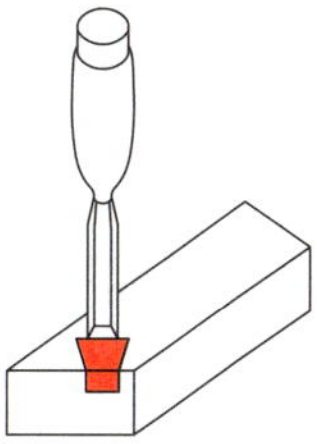

A3

2×

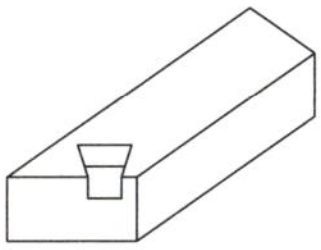

A

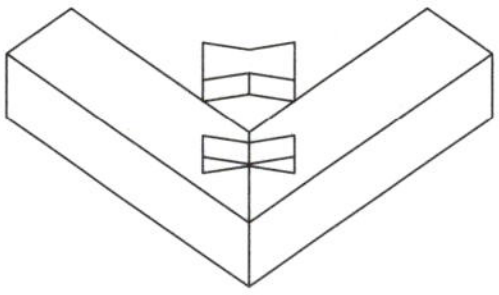

A+B+A

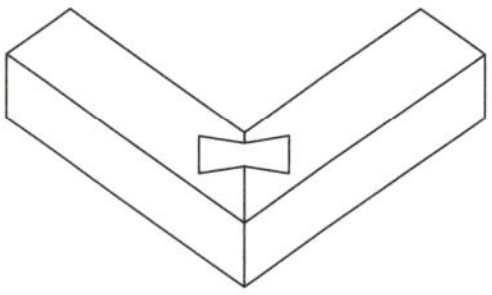

ABA

Corner Joints

Plain miter with perpendicular butterfly key

ōdome-hiyodori-sen-uchi
大留鵯栓抜打ち

Gehrungsstoß mit
Schwalbenschwanzdübel

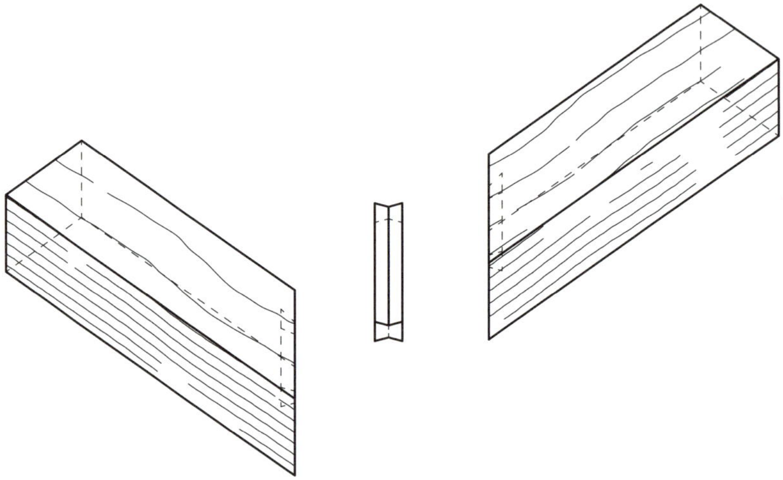

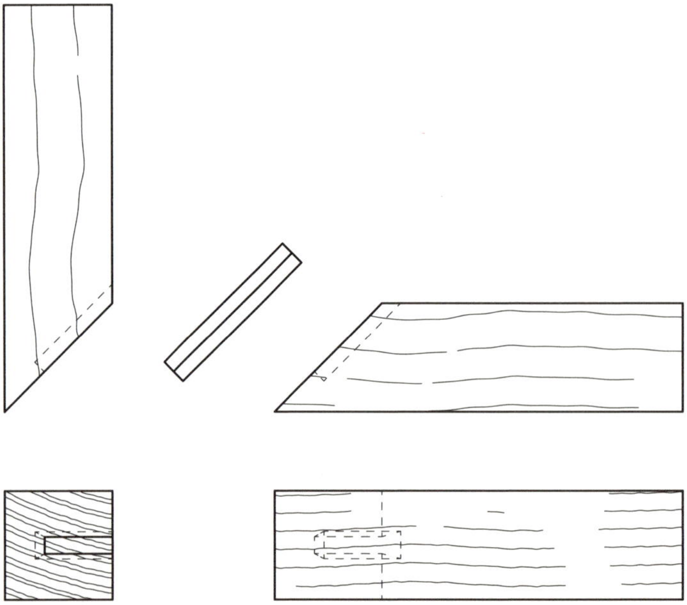

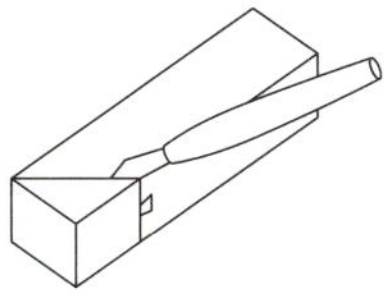

A1

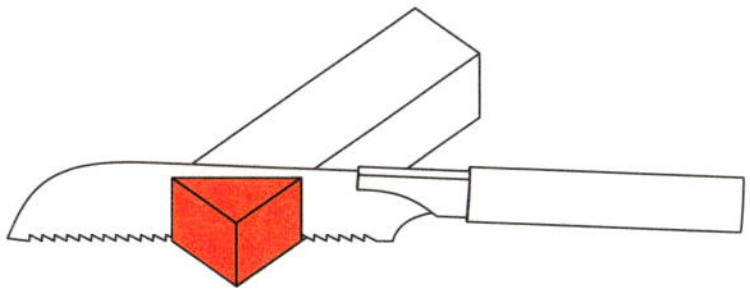

A2

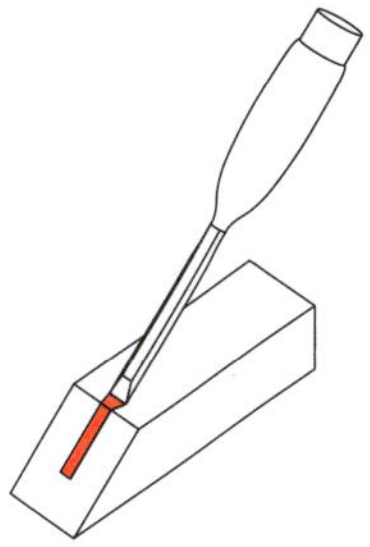

A3

2×

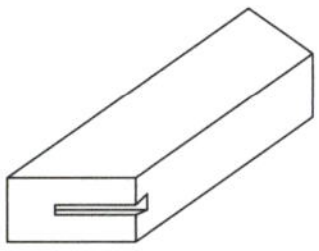

A

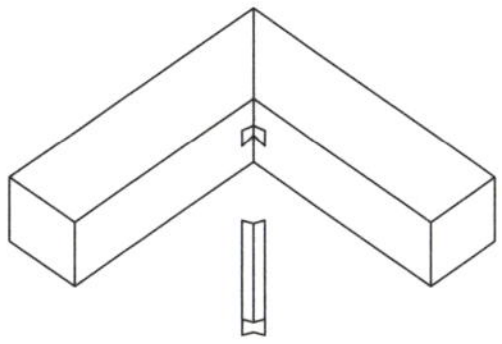

A+B+A

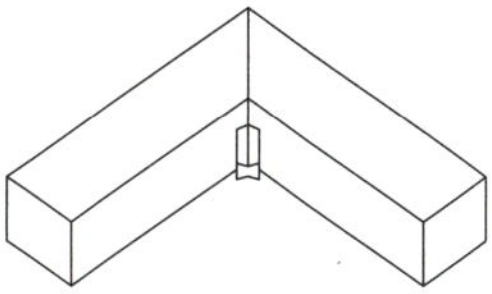

A+B+A

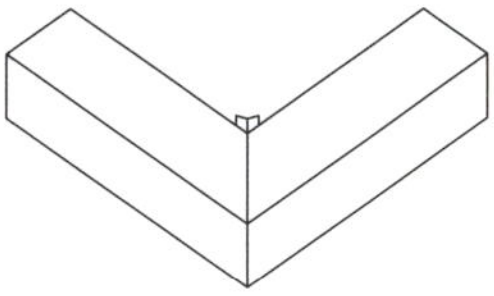

ABA

Corner Joints

Mitered half lap

ai-kaki-uwaba-ōdome
相欠上端大留め

Eckblatt auf Gehrung

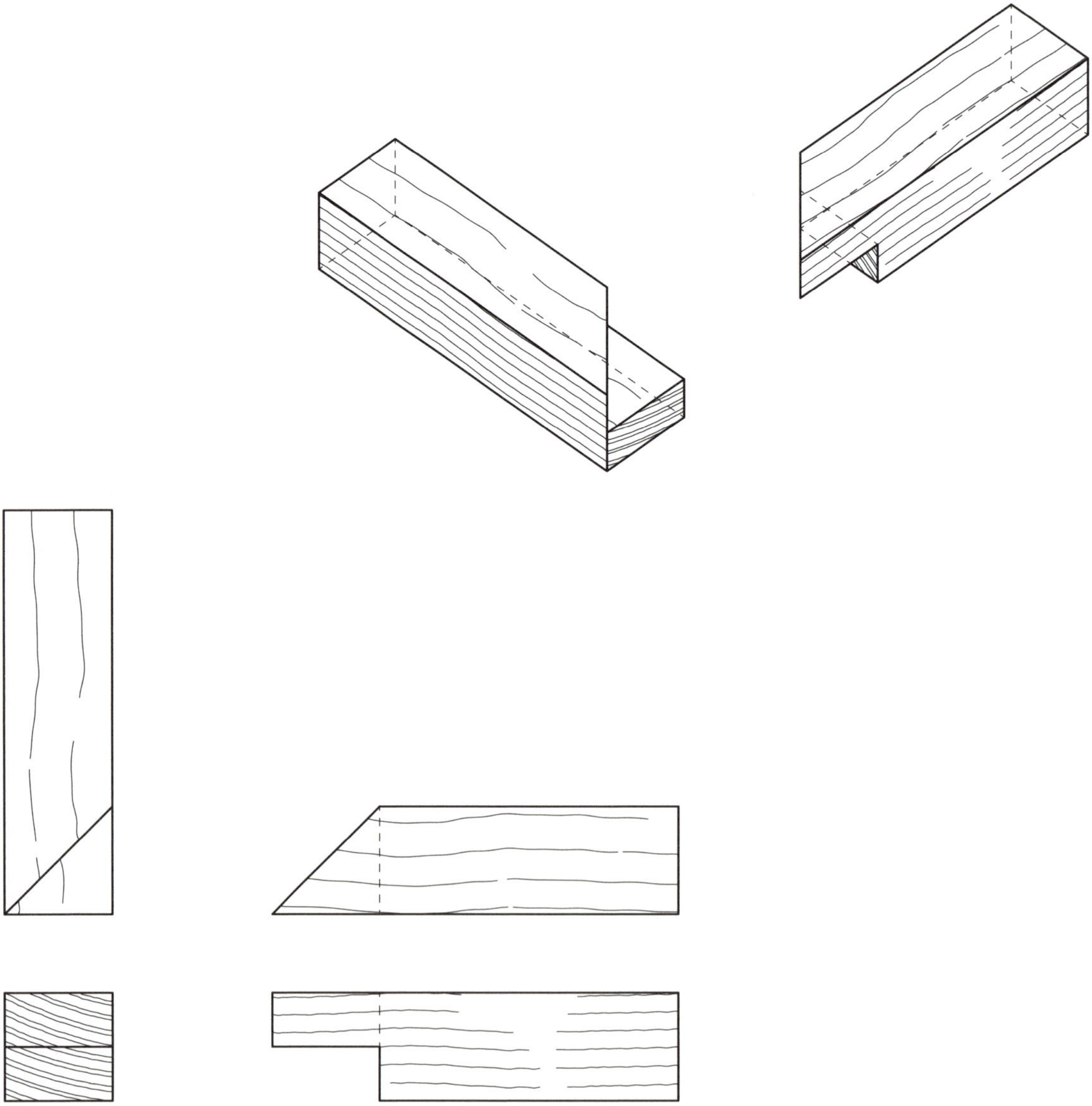

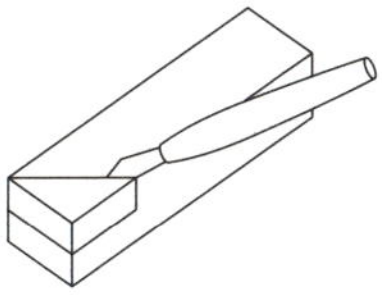
A1

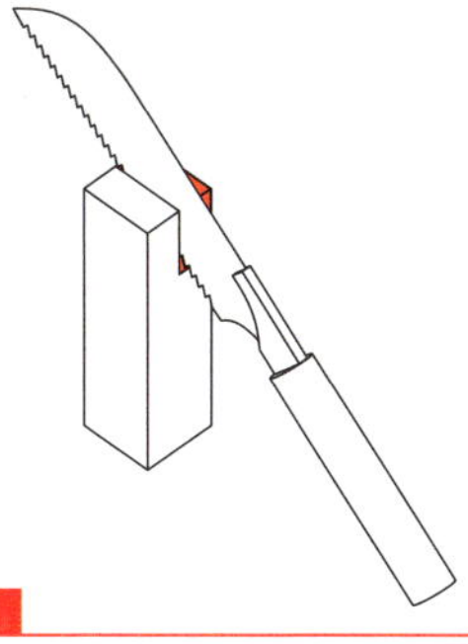
A2

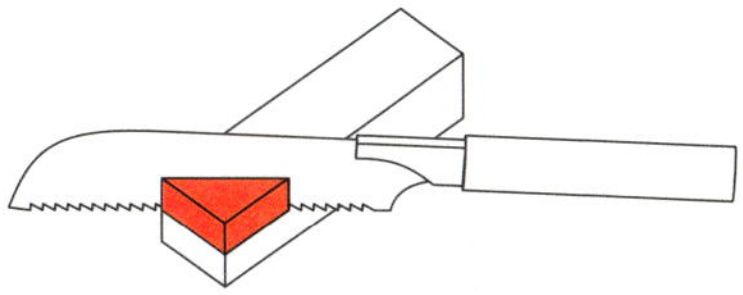
A3

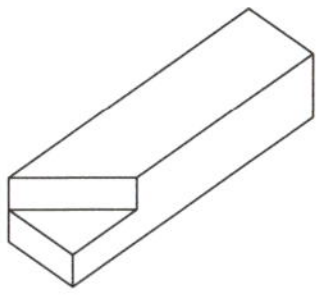
A

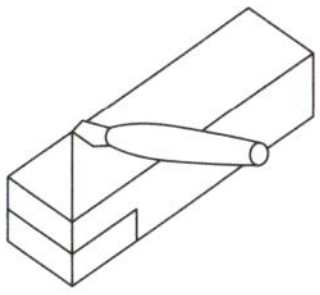
B1

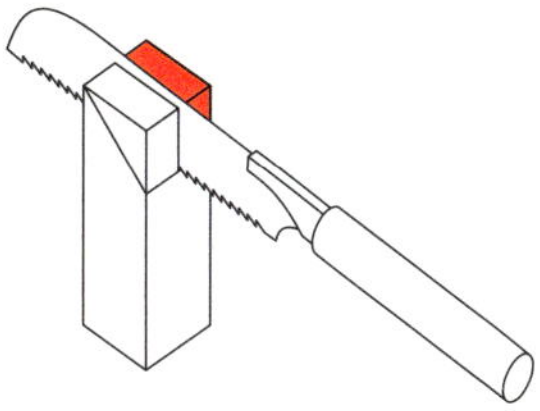
B2

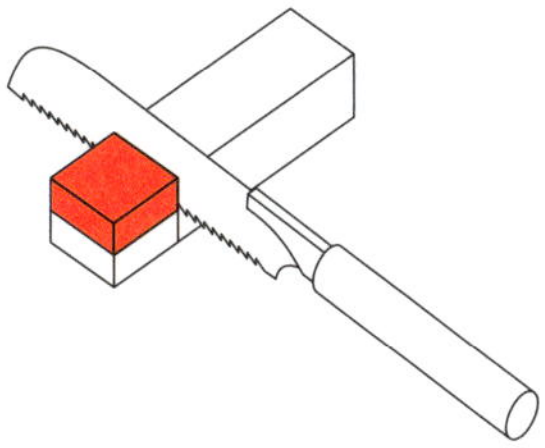
B3

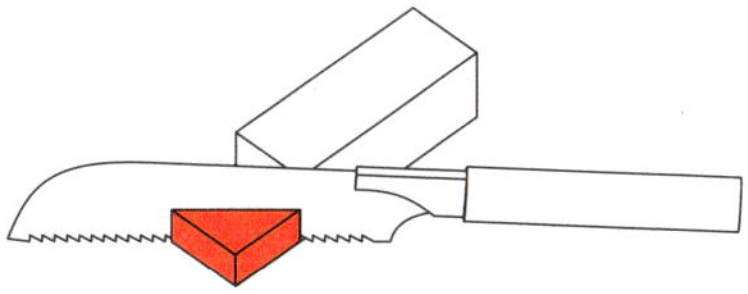
B4

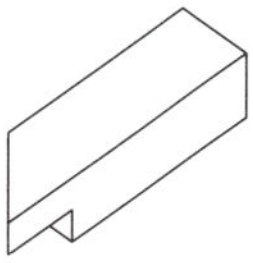
B

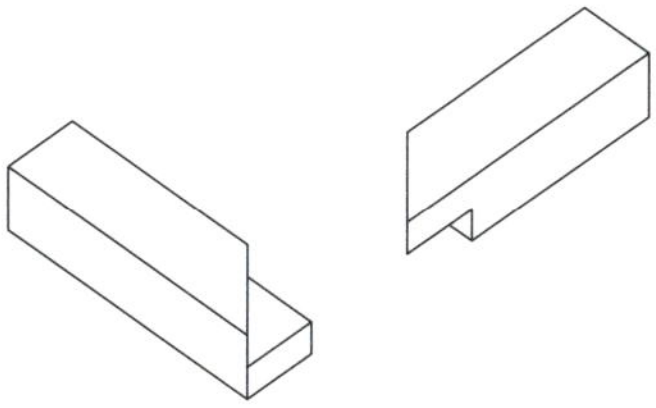
A+B

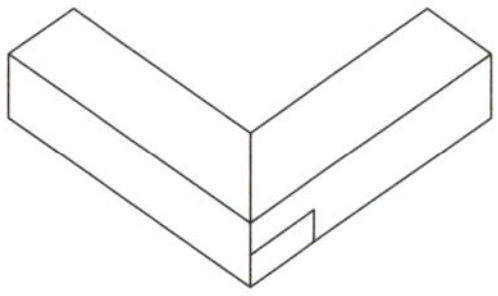
AB

Corner Joints

Blind mitered half lap

ai-kaki-ōdome
相欠大留め

verdecktes Eckblatt auf Gehrung

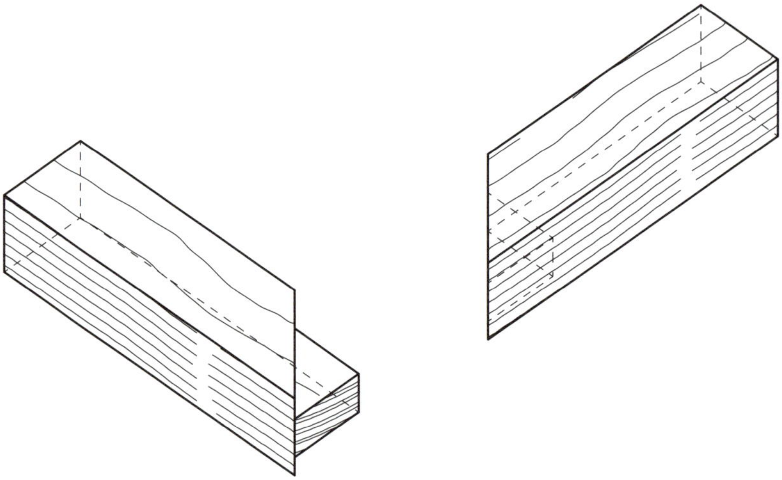

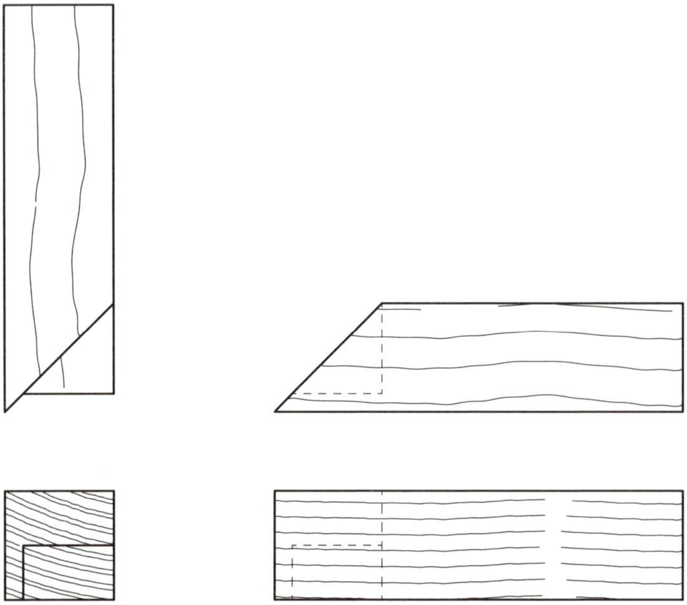

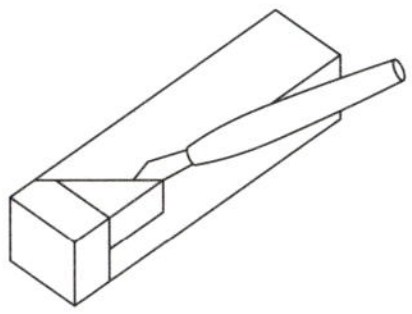

A1

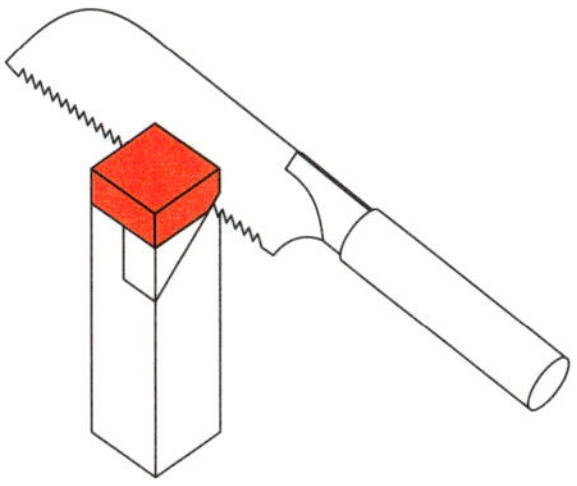

A2

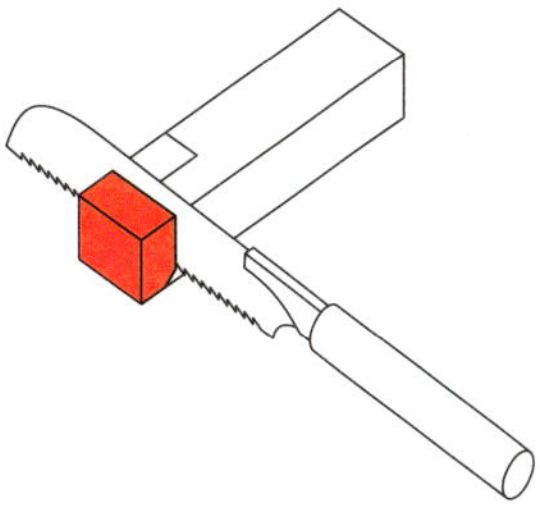

A3

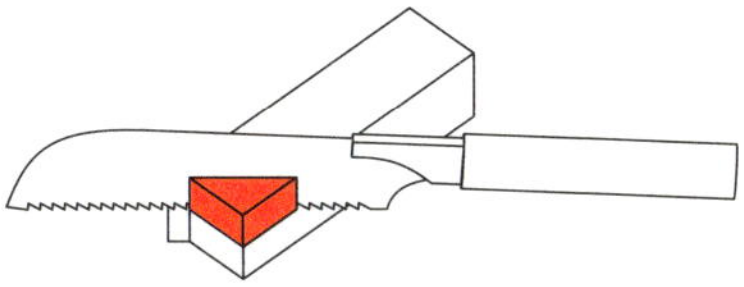

A4

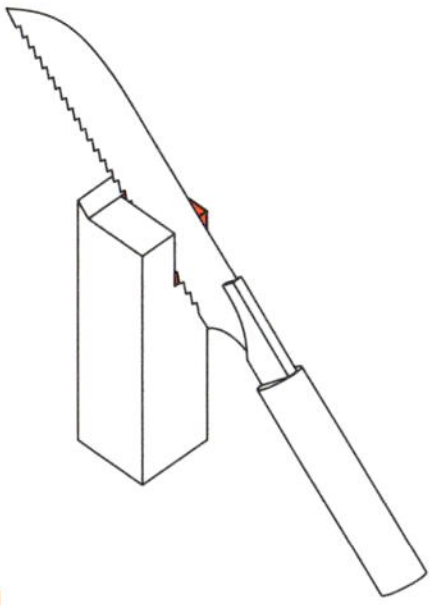

A5

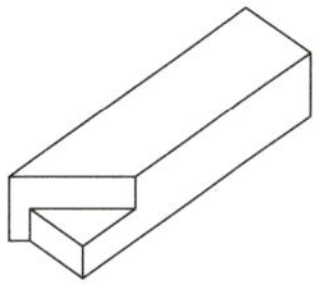

A

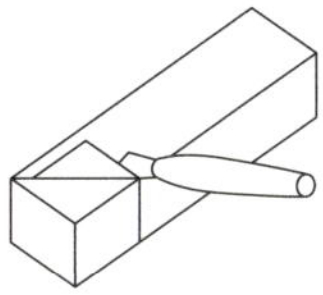

B1

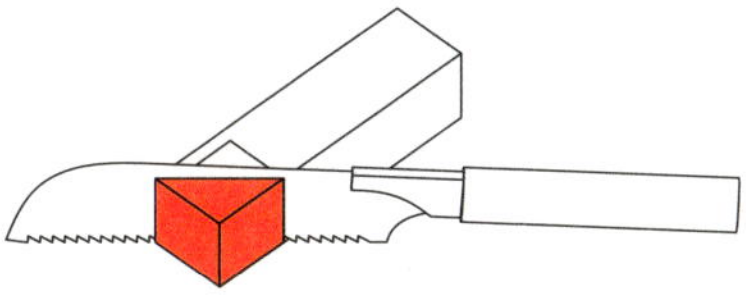

B2

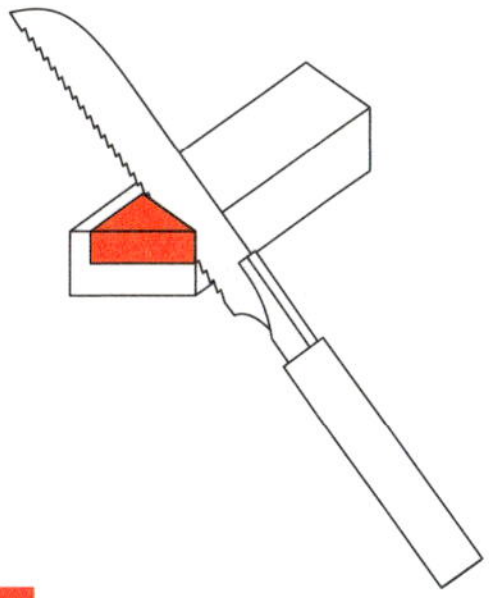

B3

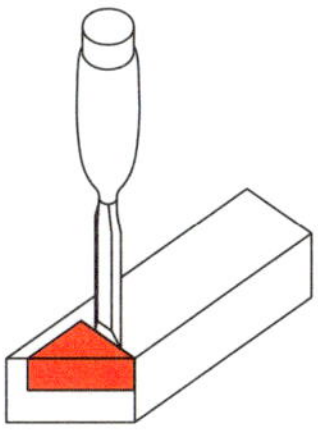

B4

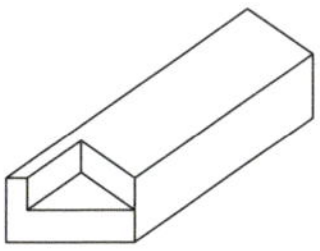

B

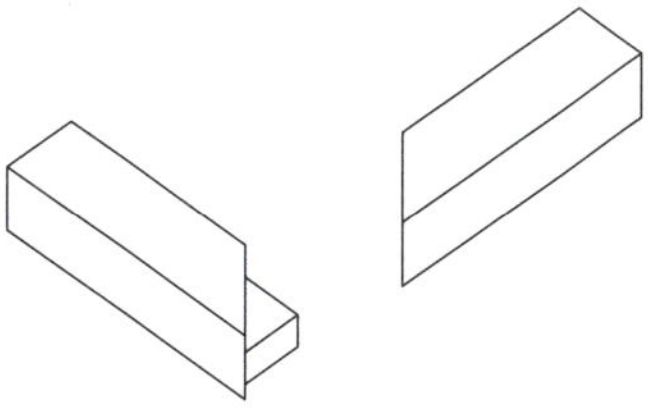

A+B

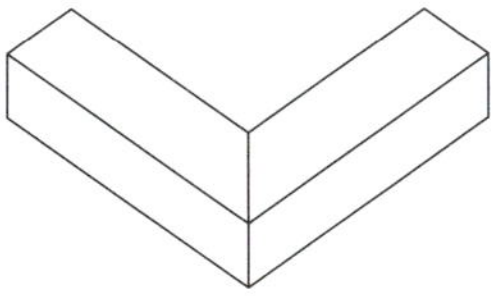

AB

Corner Joints

Blind mitered mortise and tenon joint

tsutsumi-komi-ōdome-ai-hozo-tsugi
包込大留相枘接ぎ

verdeckter beidseitiger Zapfen
mit Gehrungsstoß

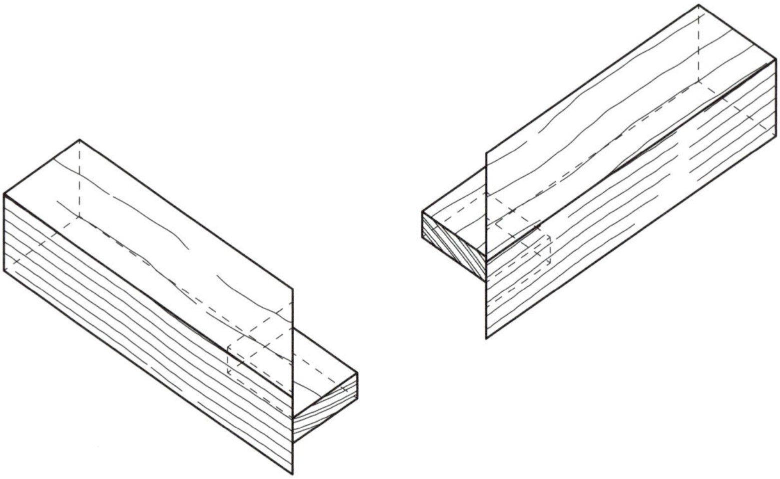

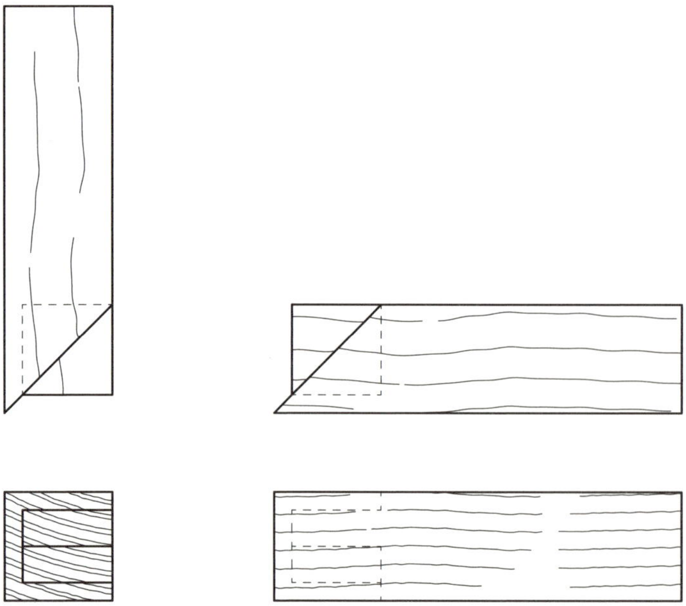

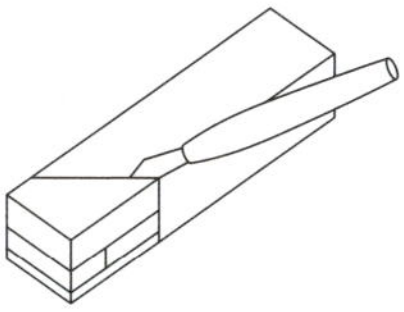

A1

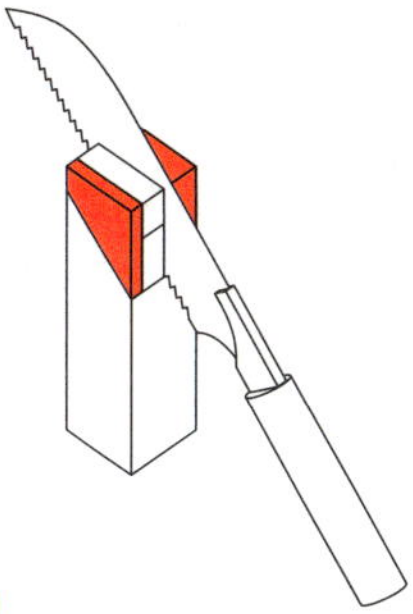

A2

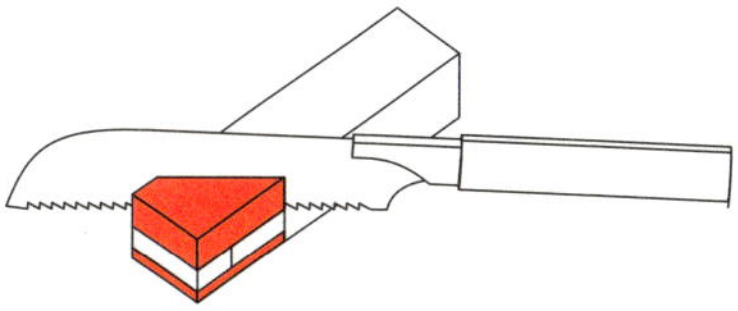

A3

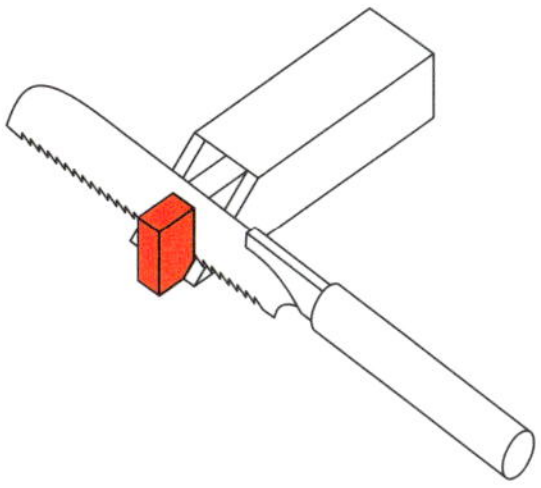

A4

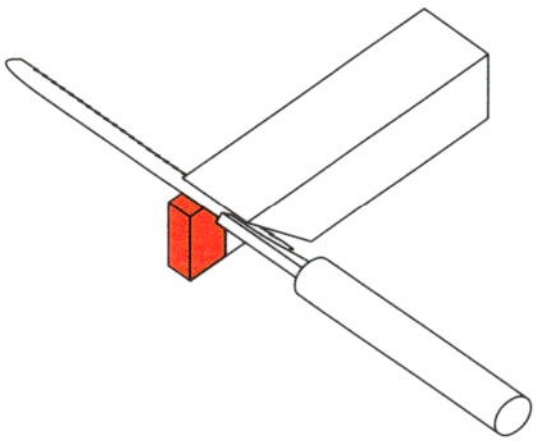

A5

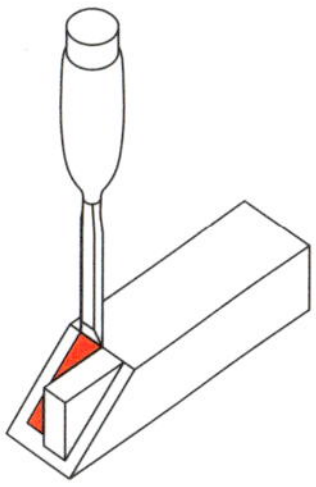

A6

2×

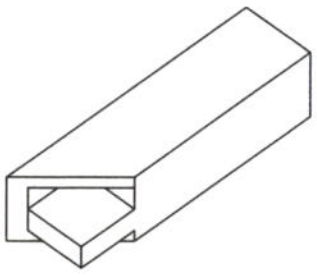

A

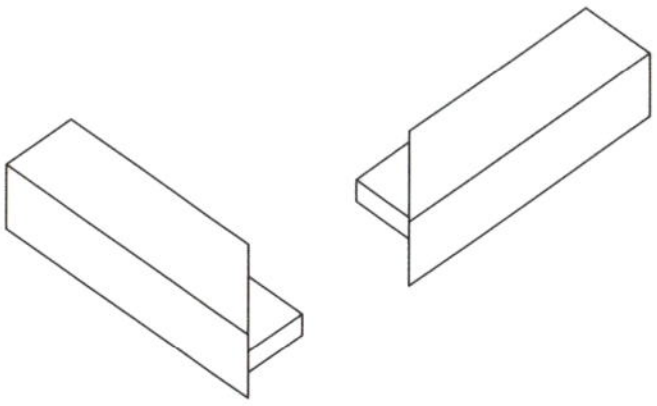

A+A

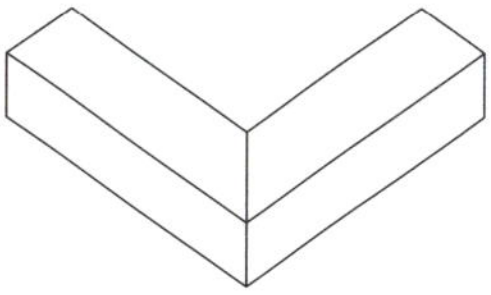

AA

Corner Joints

Mitered bridle joint

hira-hozo-ōdome
平枘大留め

durchgehender Zapfen
mit Gehrungsstoß

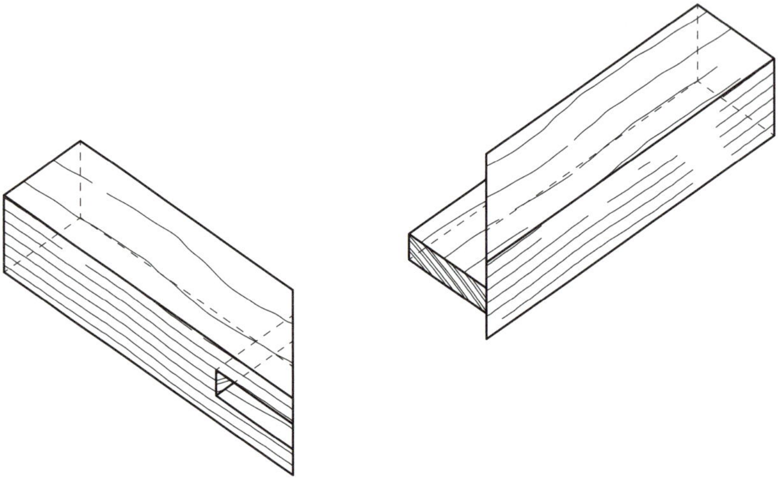

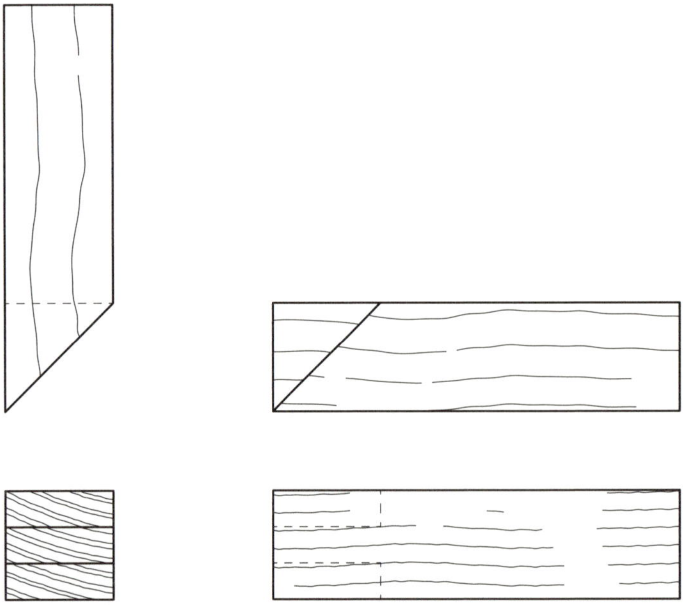

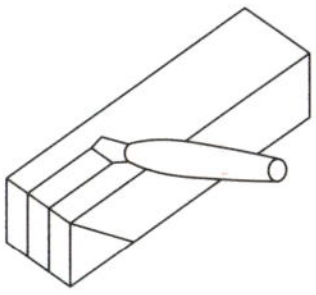
A1

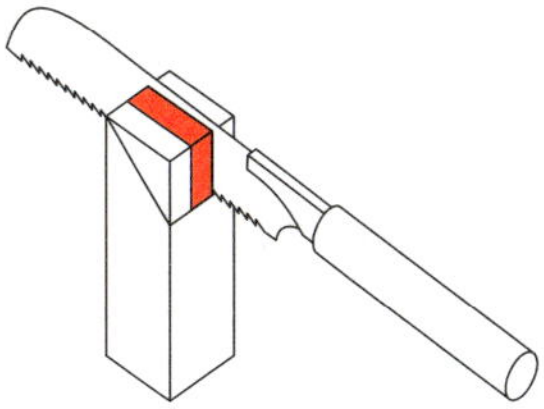
A2

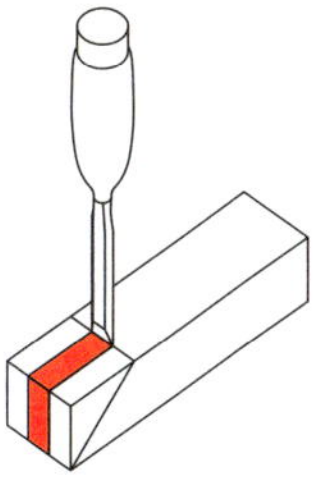
A3

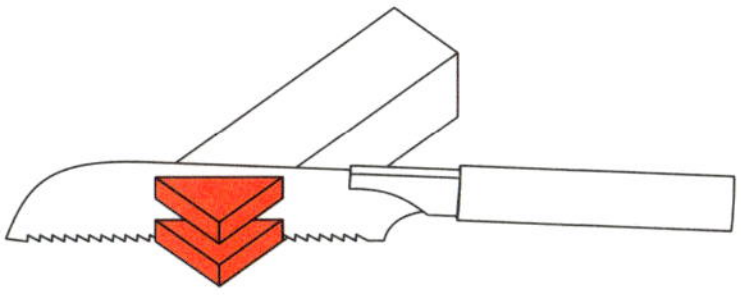
A4

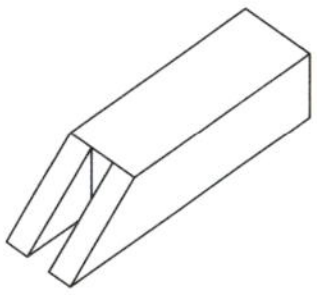
A

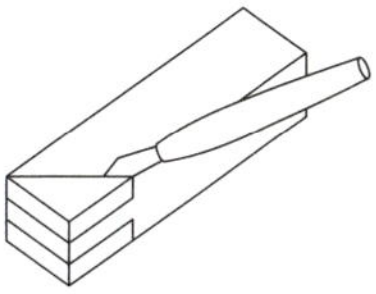
B1

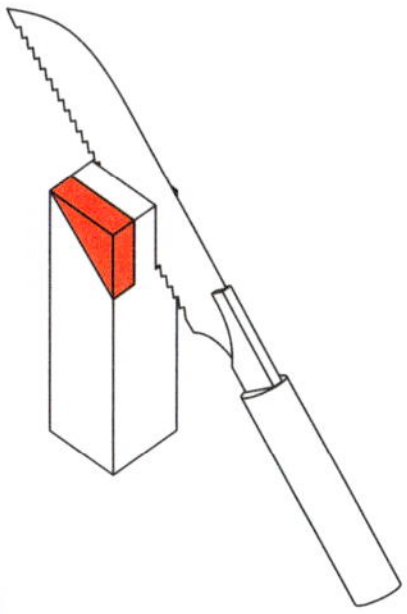
B2

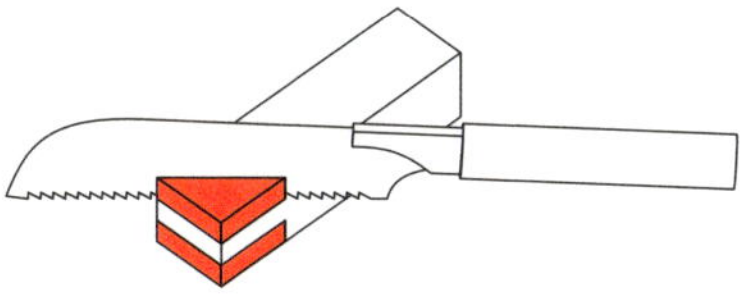
B3

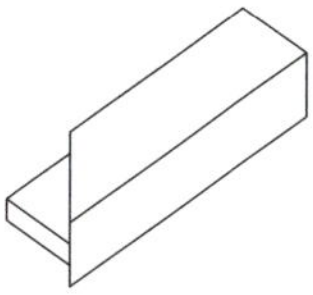
B

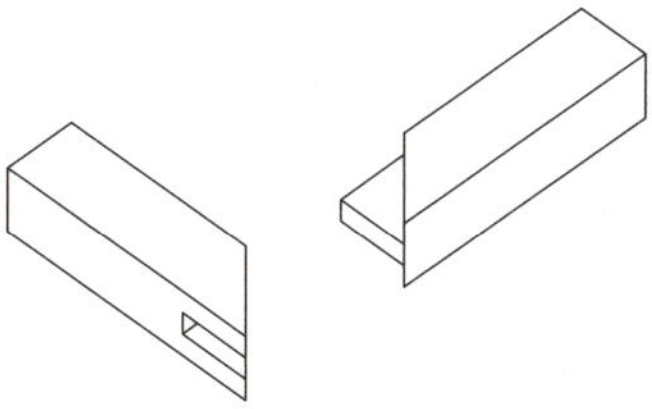
A+B

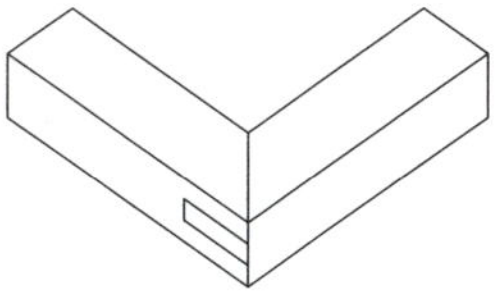
AB

Corner Joints

Blind stub tenon miter

tsutsumi-komi-ōdome
包込大留め

verdeckter Zapfen mit Gehrungsstoß

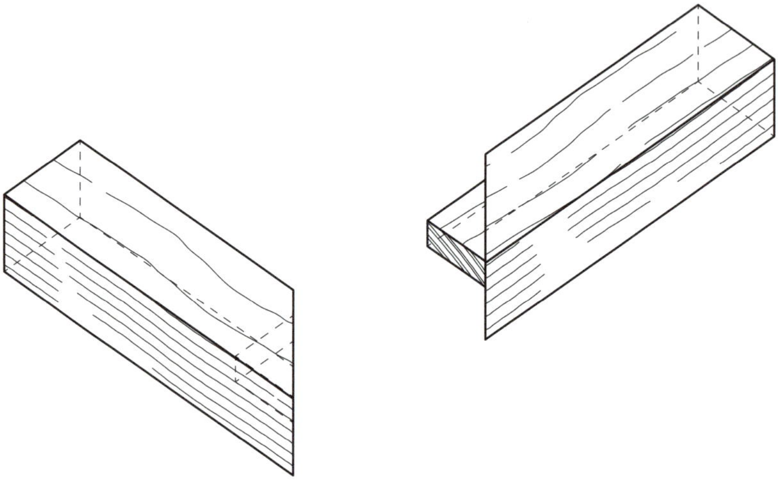

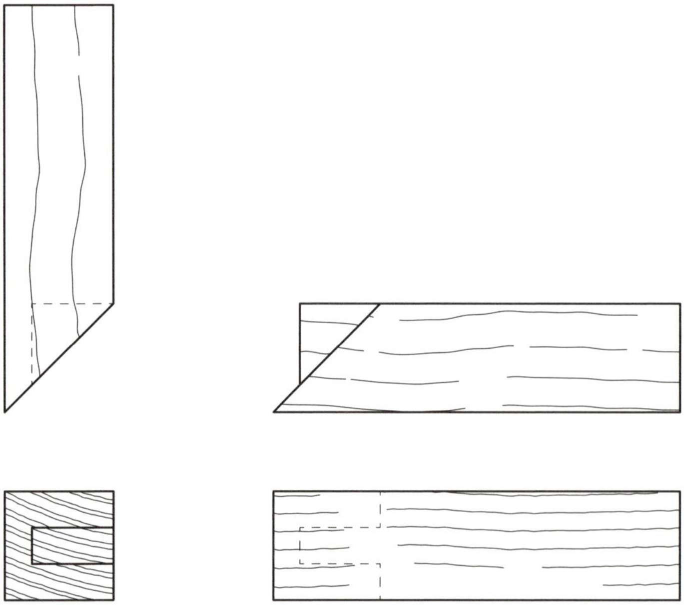

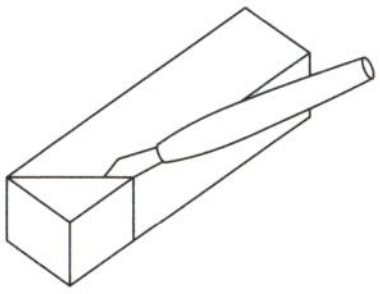

A1

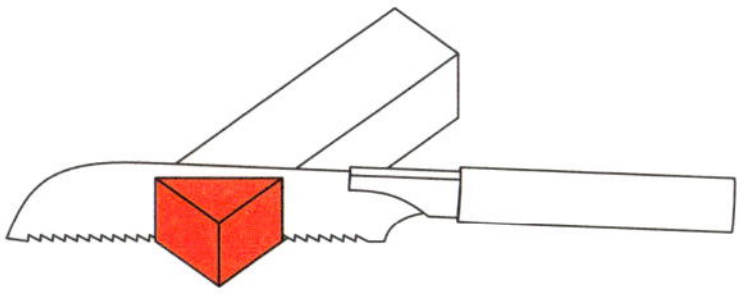

A2

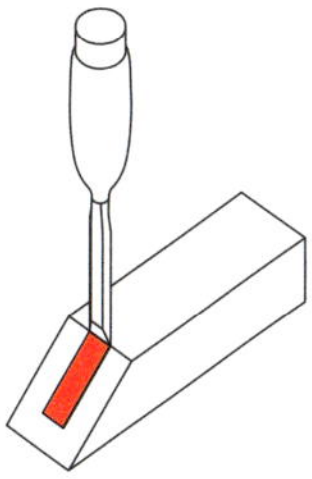

A3

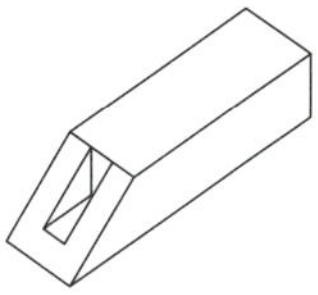

A

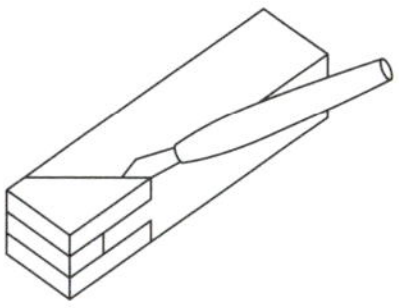

B1

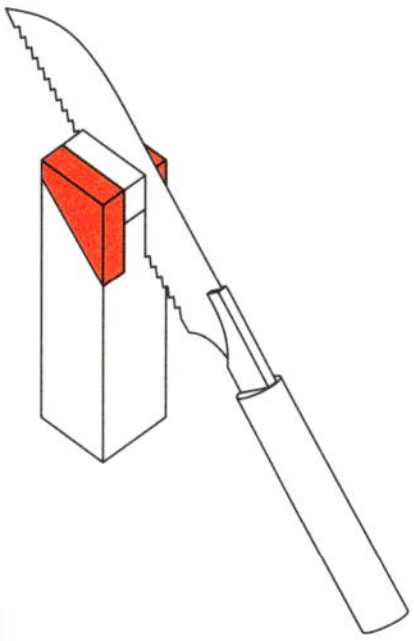

B2

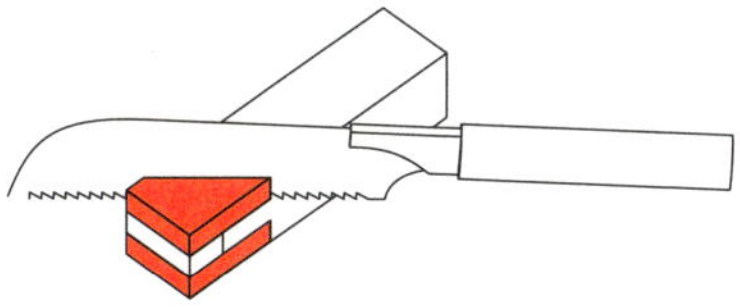

B3

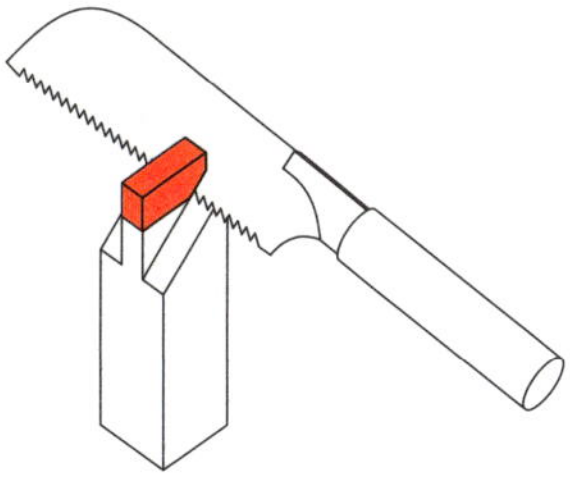

B4

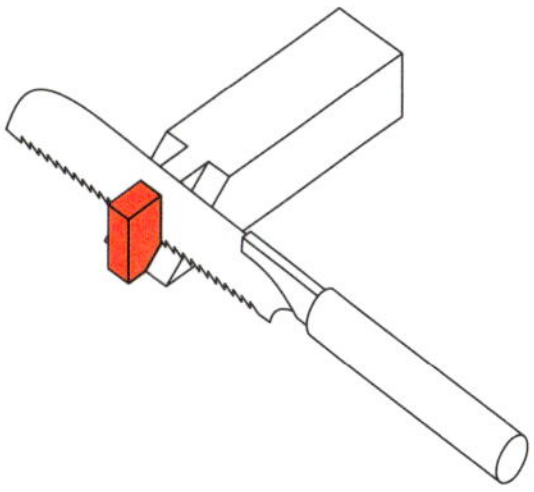

B5

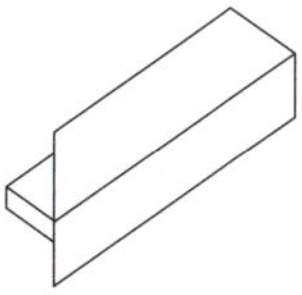

B

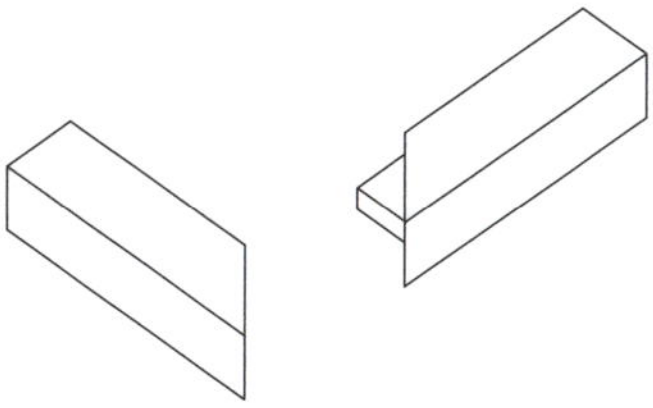

A+B

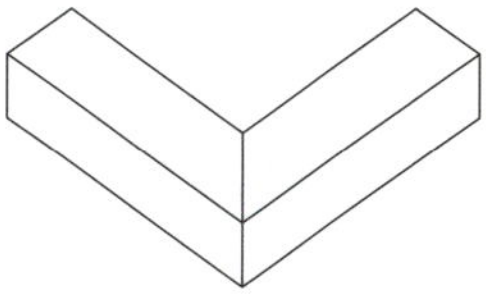

AB

Corner Joints

Through single dovetail miter

ari-ōdome
蟻大留め

durchgehender Schwalbenschwanz-zapfen mit Gehrungsstoß

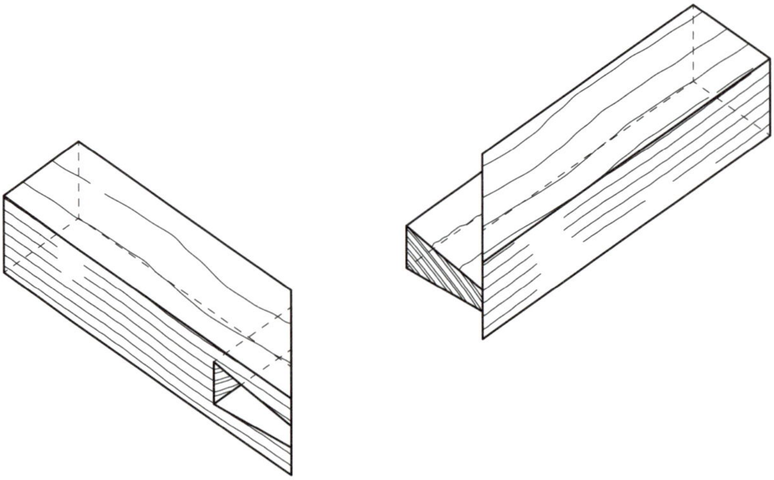

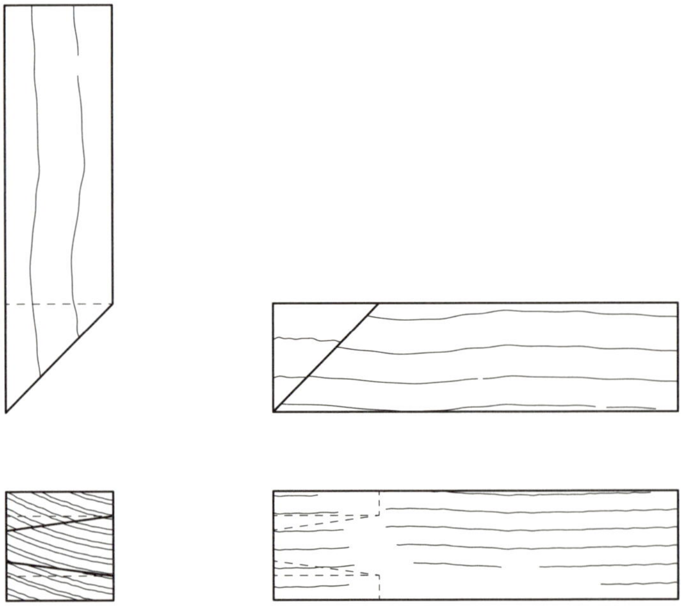

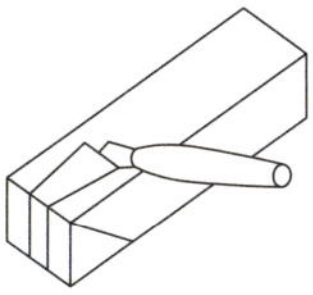

A1

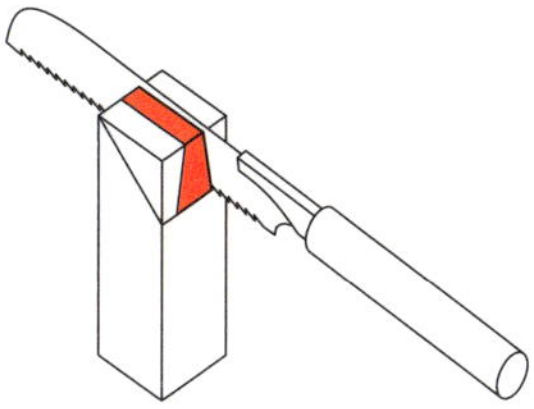

A2

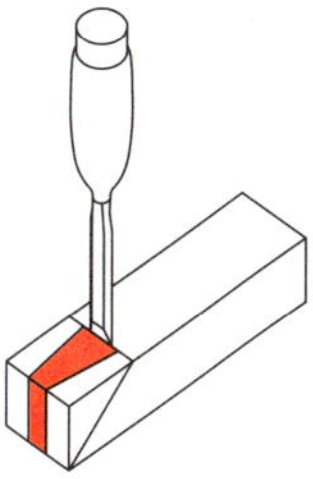

A3

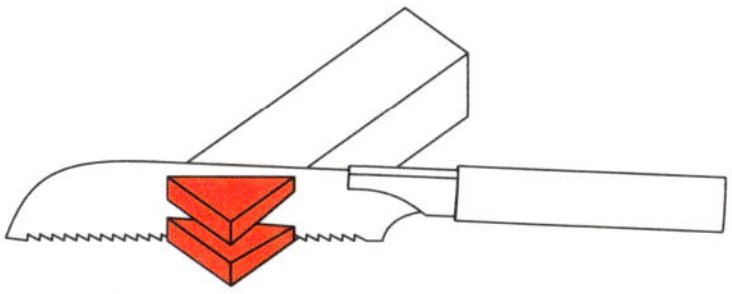

A4

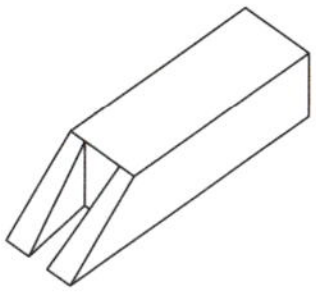

A

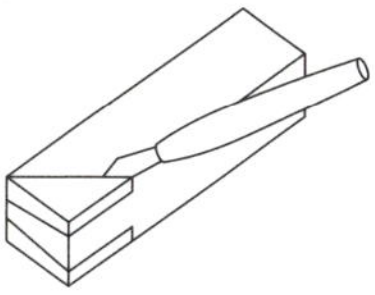

B1

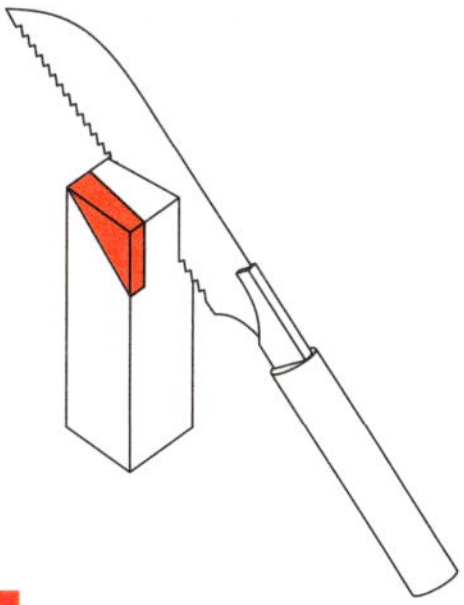

B2

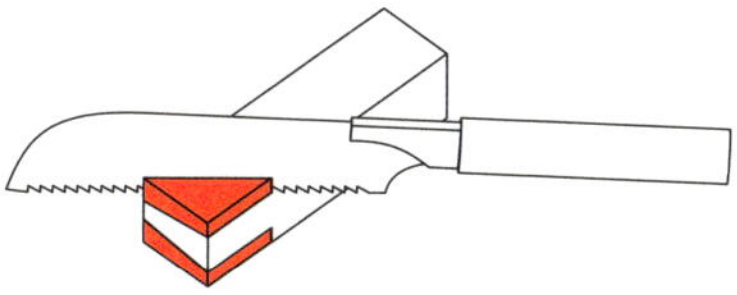

B3

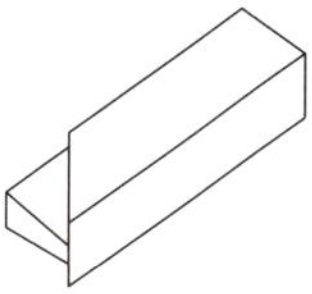

B

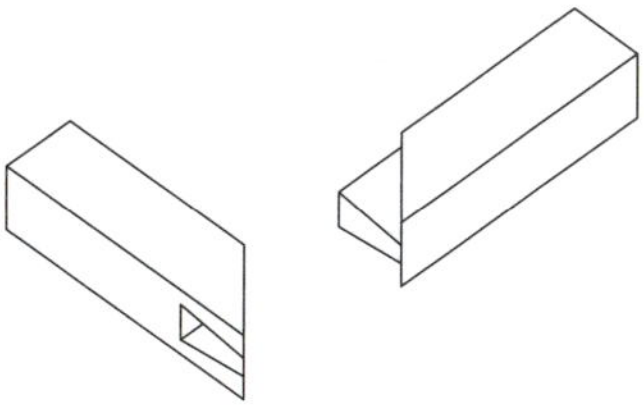

A+B

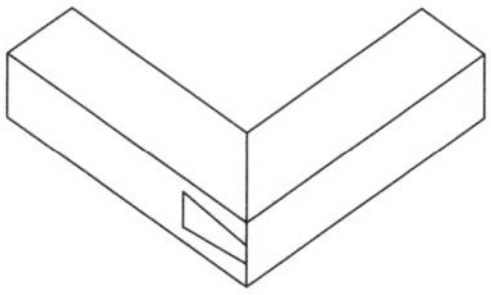

AB

Corner Joints

Blind dovetail miter

kakushi-ari-ōdome
隠蟻大留め

verdeckter Schwalbenschwanzzapfen
mit Gehrungsstoß

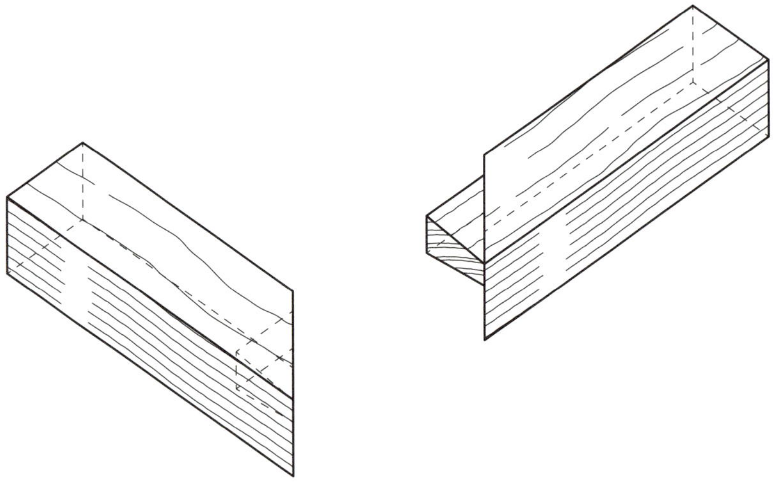

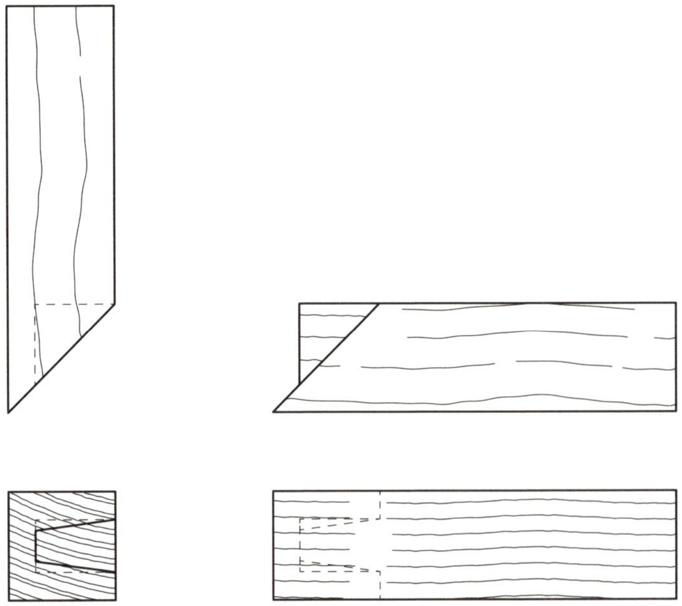

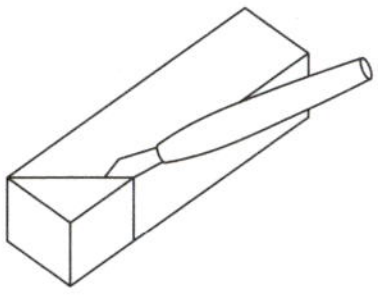

A1

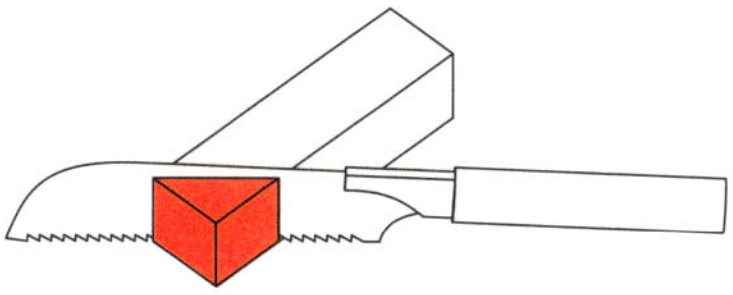

A2

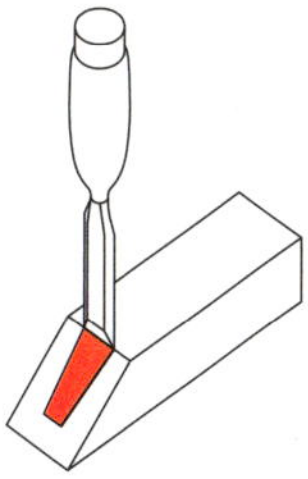

A3

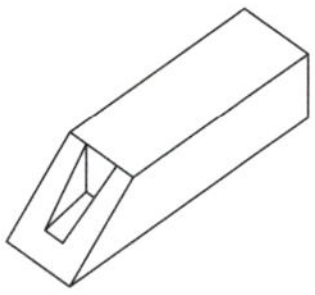

A

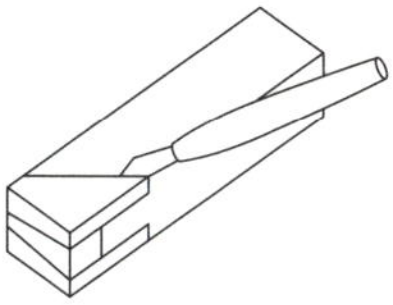

B1

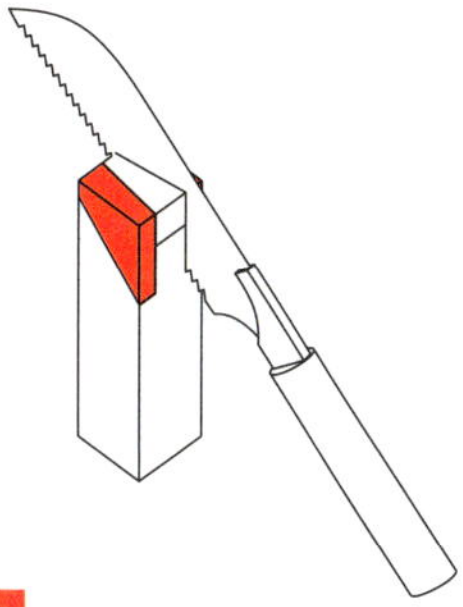

B2

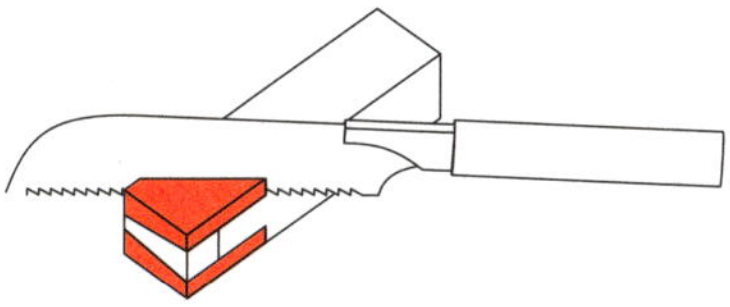

B3

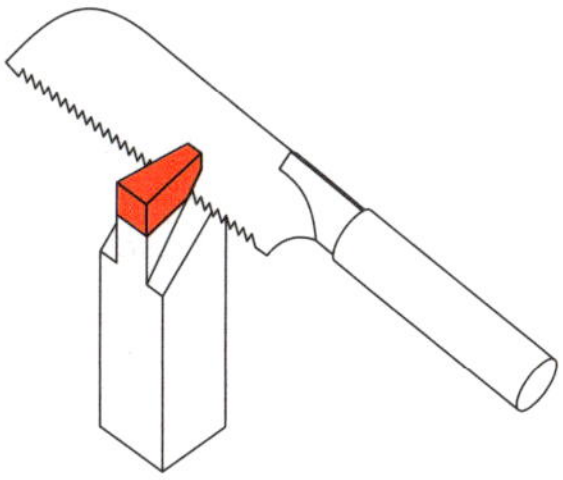

B4

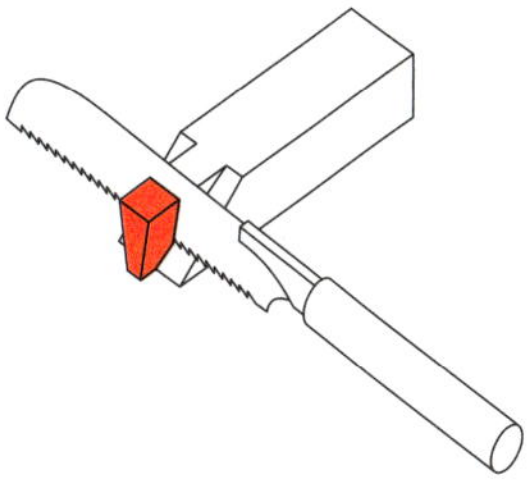

B5

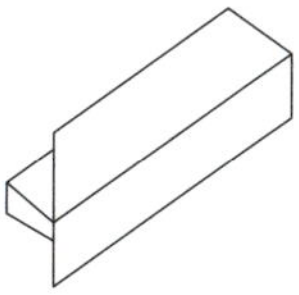

B

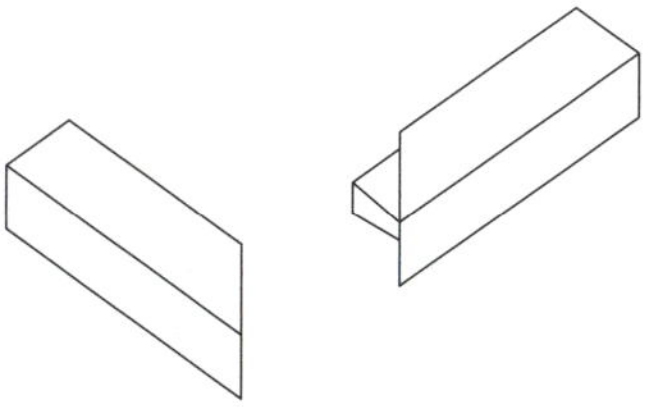

A+B

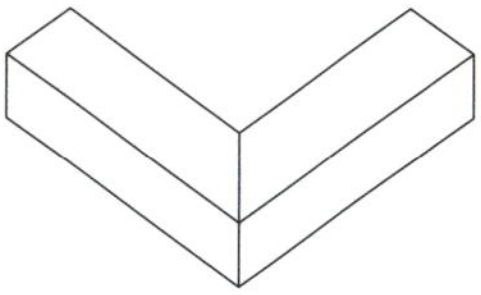

AB

Corner Joints

Stopped miter with blind dovetail tenon

kakushi-ari-oire-kumi-dome
隠蟻追入組留め

verdeckter Schwalbenschwanzzapfen mit abgesetztem Gehrungsstoß

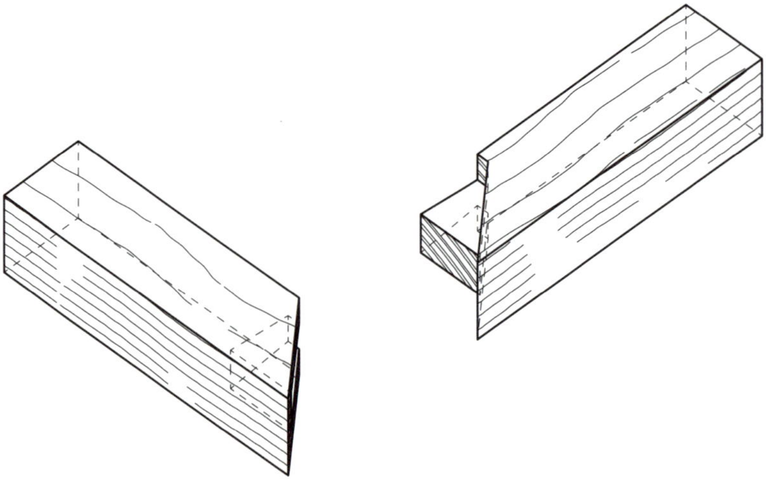

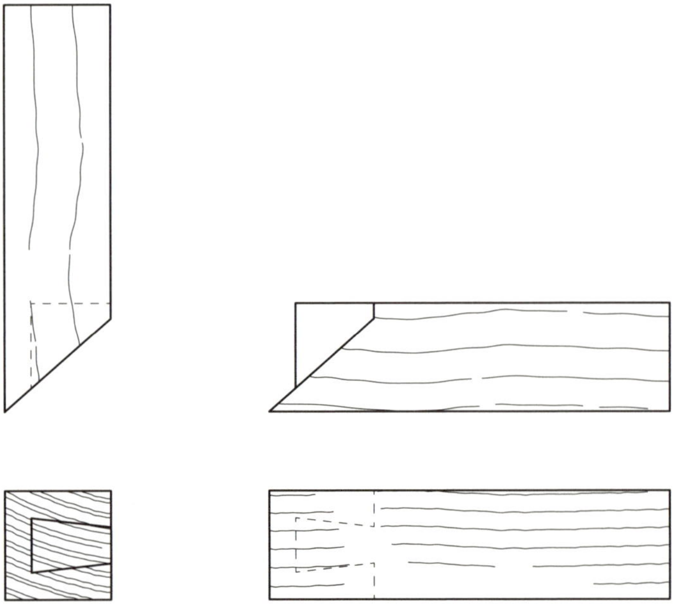

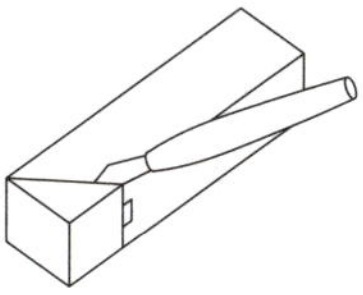

A1

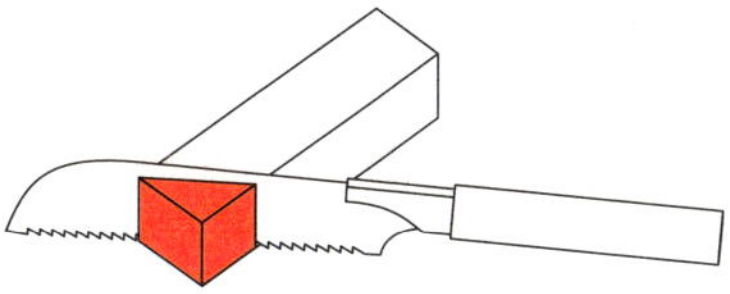

A2

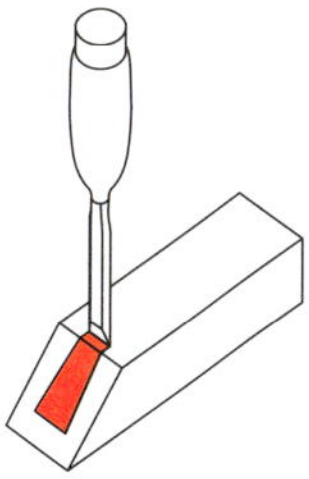

A3

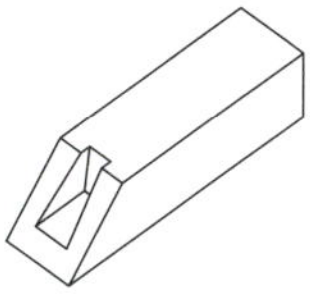

A

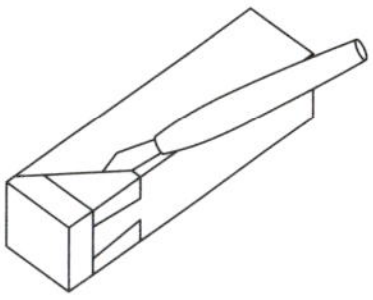

B1

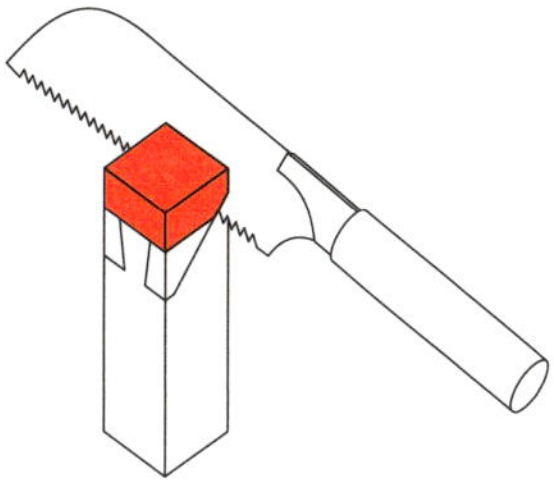

B2

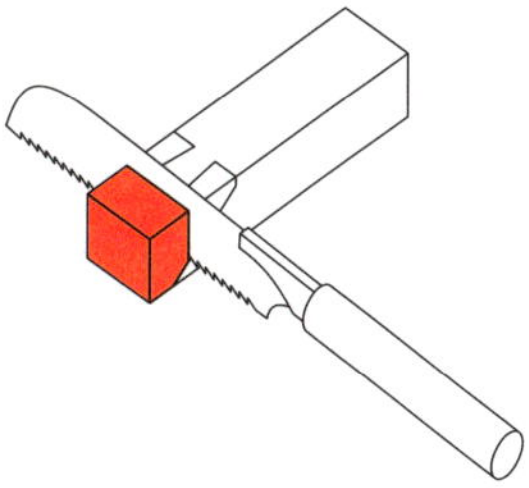

B3

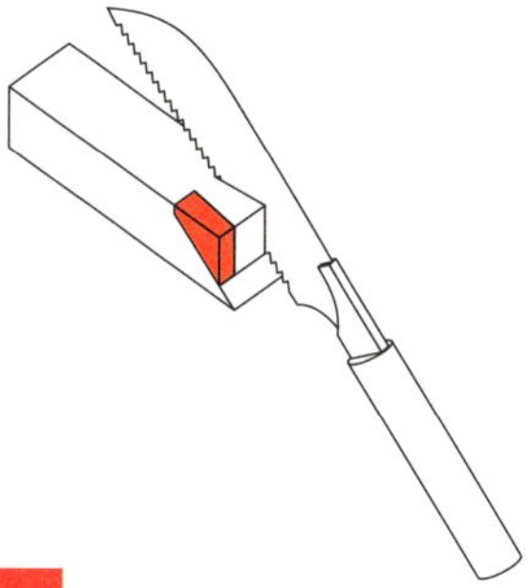

B4

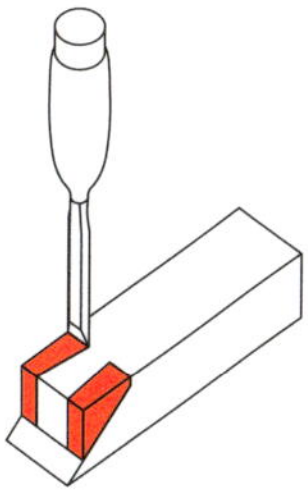

B5

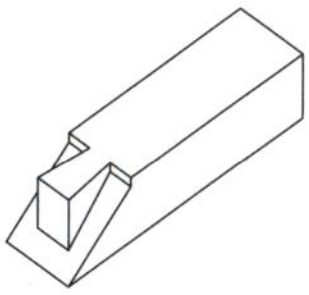

B

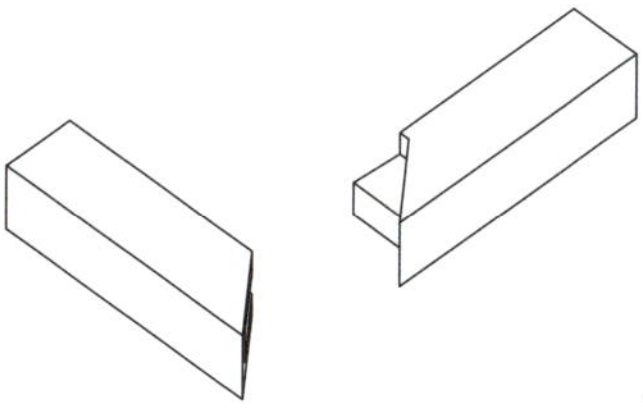

A+B

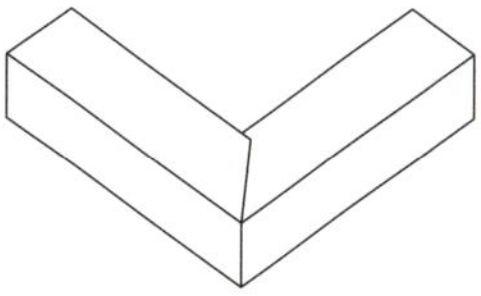

AB

Corner Joints

Bridle joint with one-sided miter

ai-kaki-uwaba-dome
相欠上端留め

Zapfen mit einseitiger Gehrung

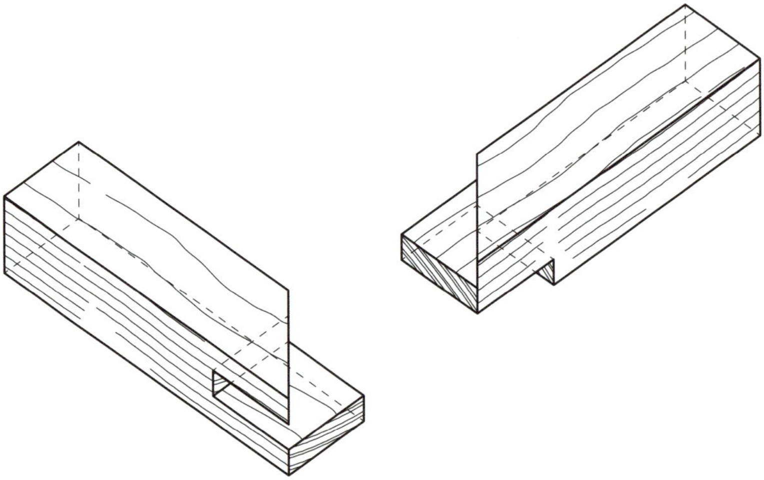

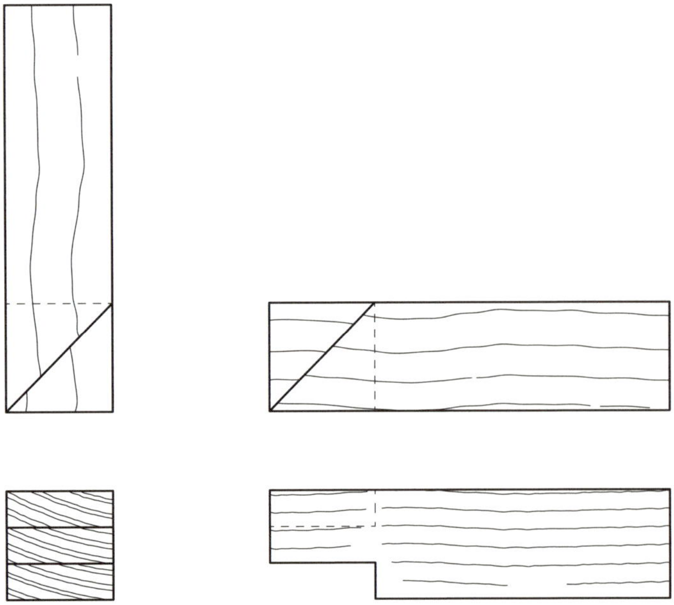

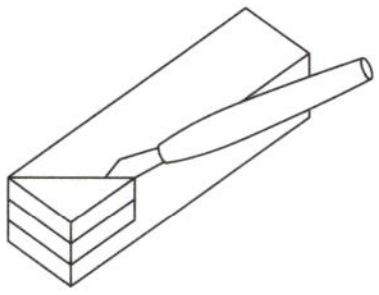
A1

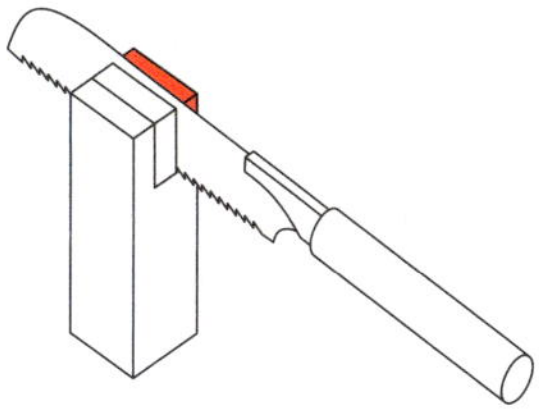
A2

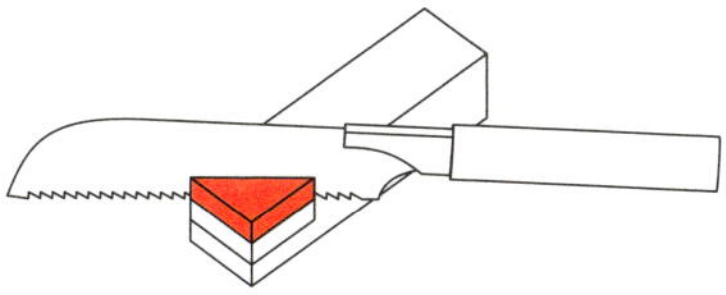
A3

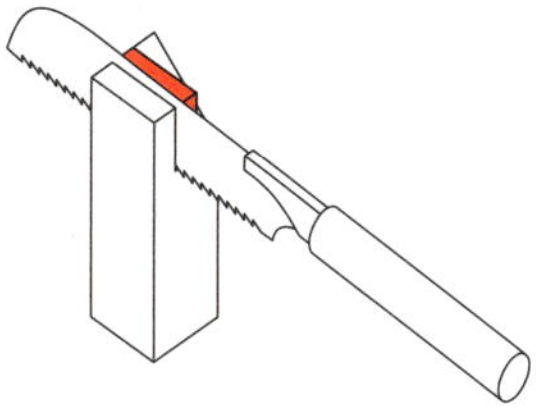
A4

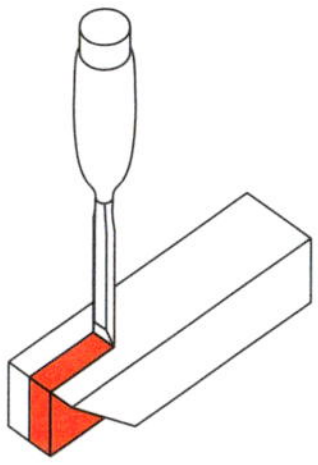
A5

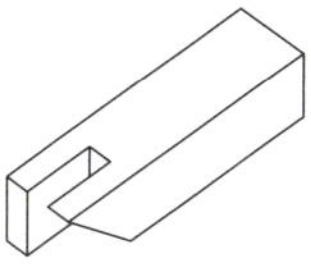
A

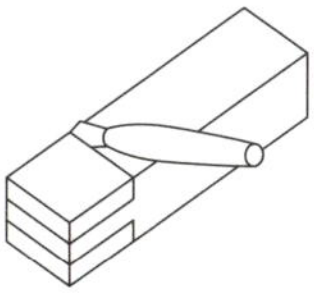
B1

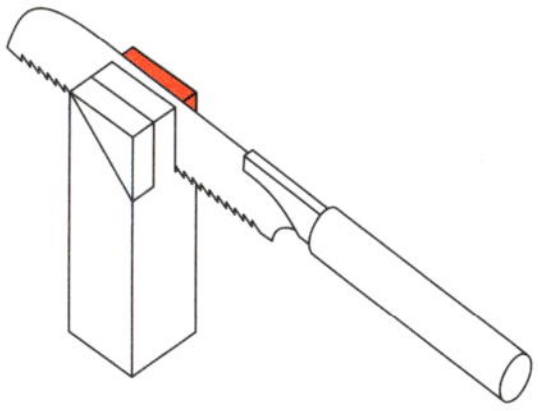
B2

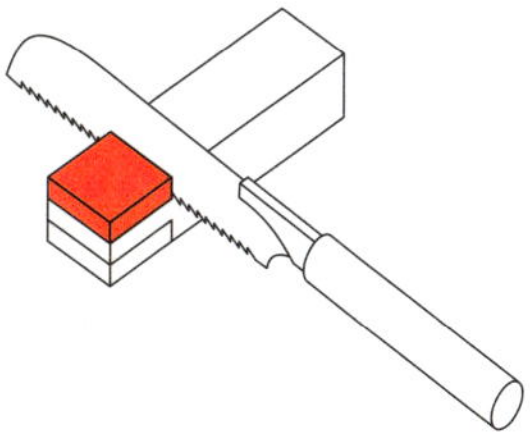
B3

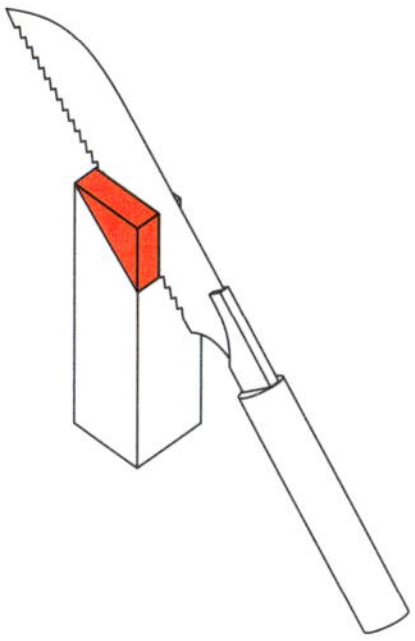
B4

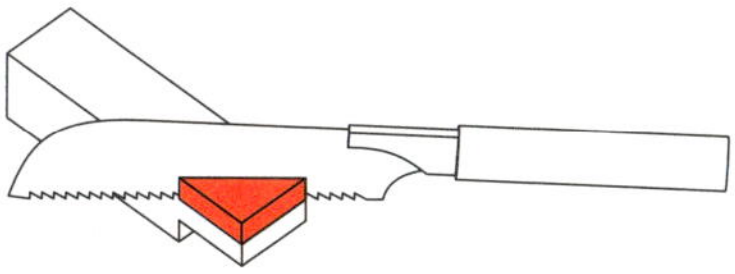
B5

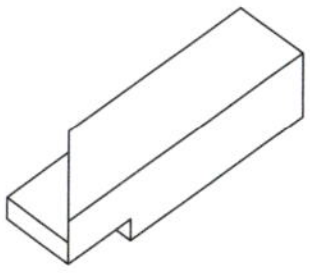
B

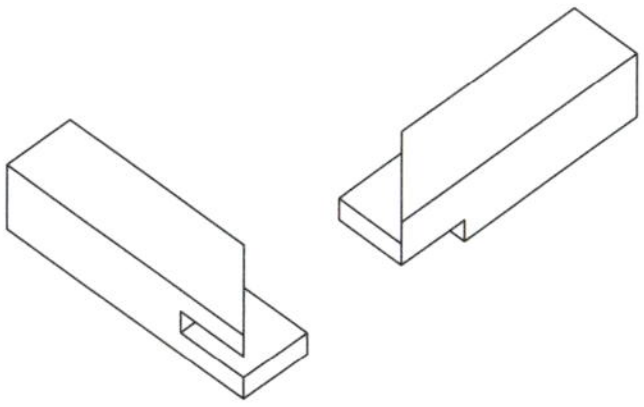
A+B

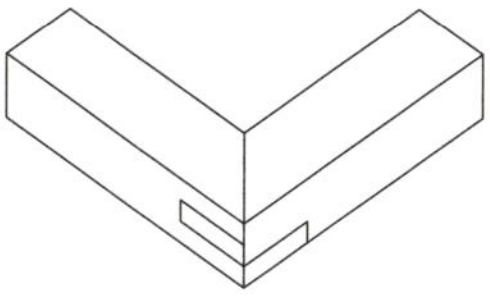
AB

Corner Joints

Canvas stretcher

ōdome-nimai-kumi-tsugi (yoko-sen-uchi)
大留二枚組接ぎ(横栓打)

Keilrahmen

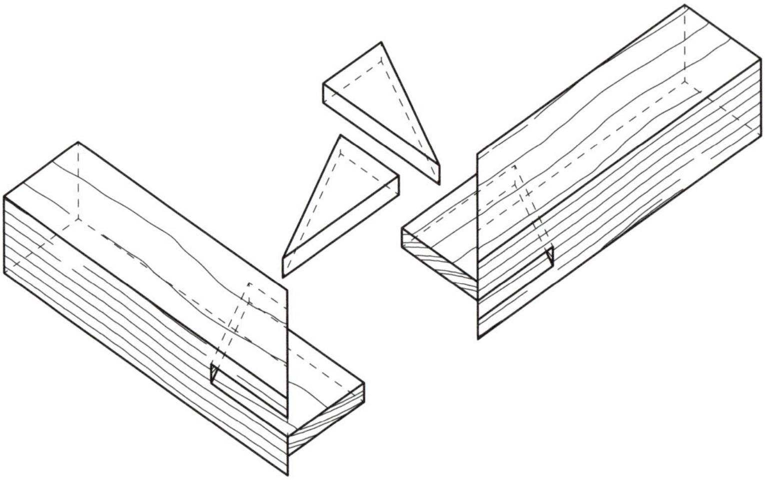

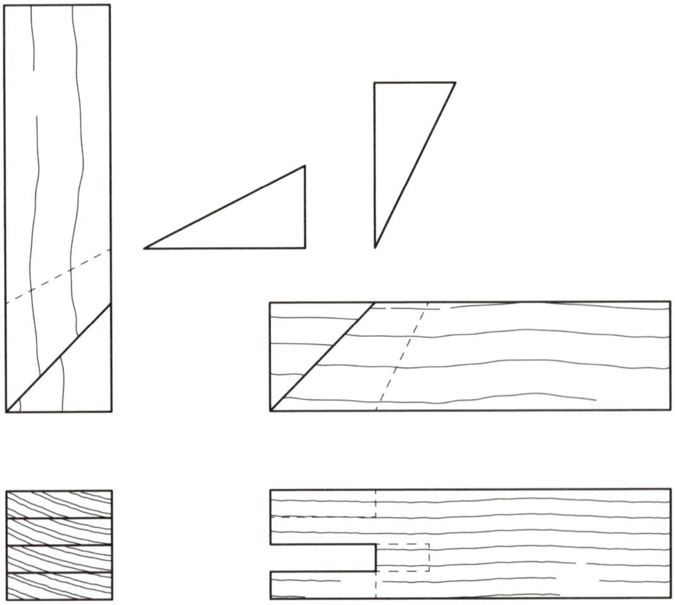

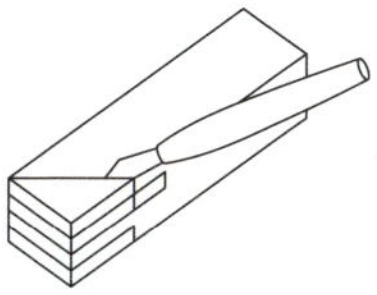

A1

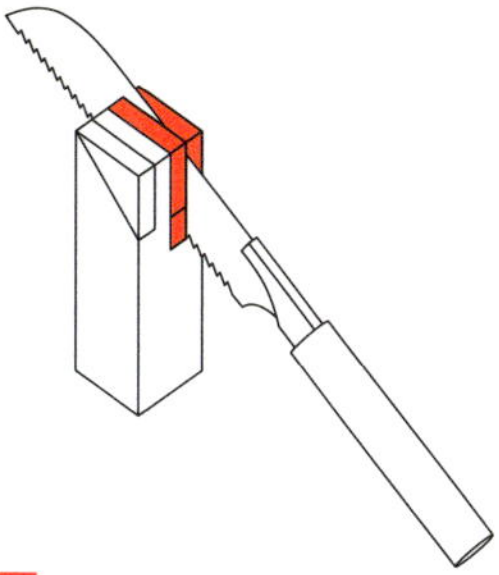

A2

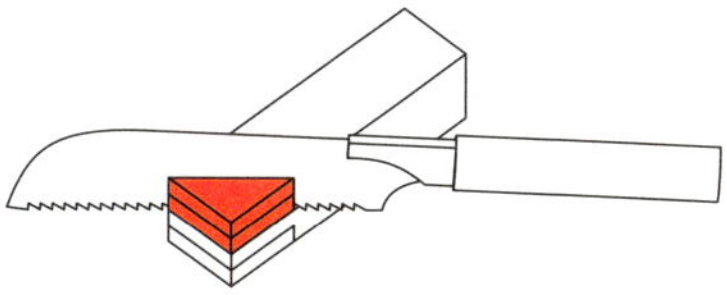

A3

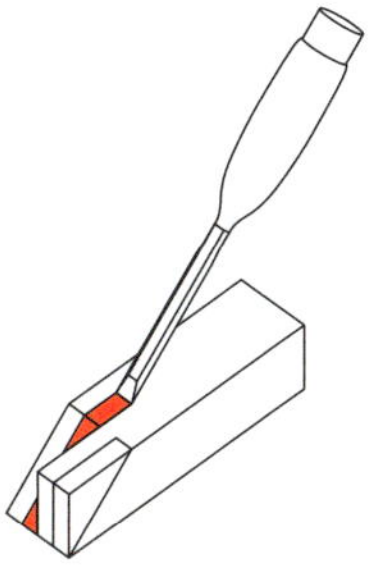

A4

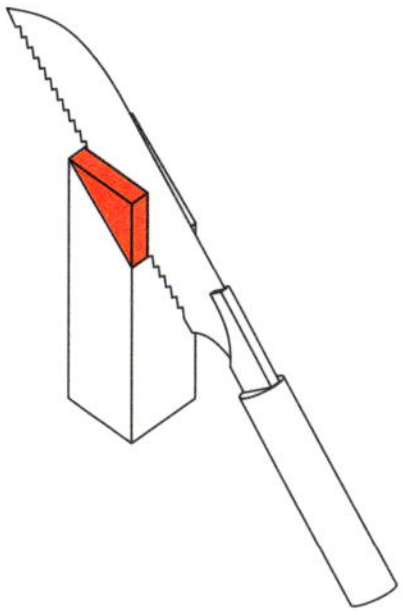

A5

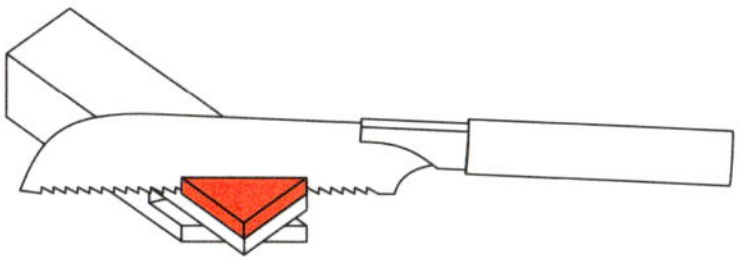

A6

2×

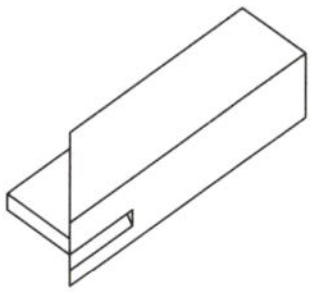

A

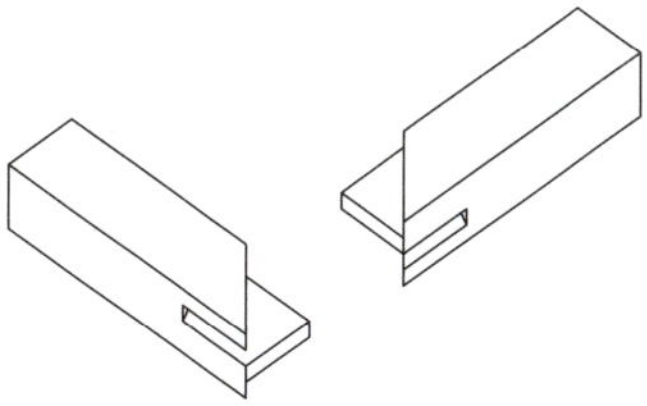

A+A

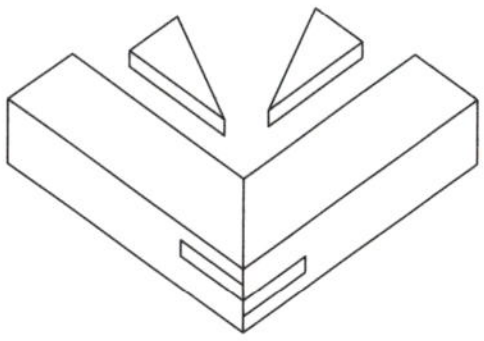

A+B+A

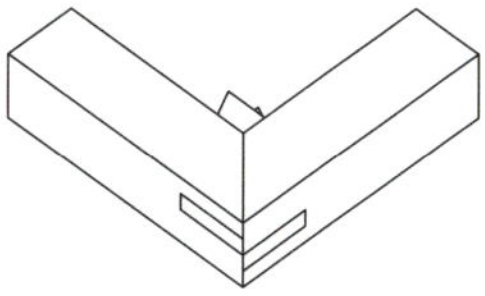

ABA

Double mortise and tenon miter

gomai-kumi-dome
五枚組留め

Zapfen mit doppeltem Gehrungsblatt

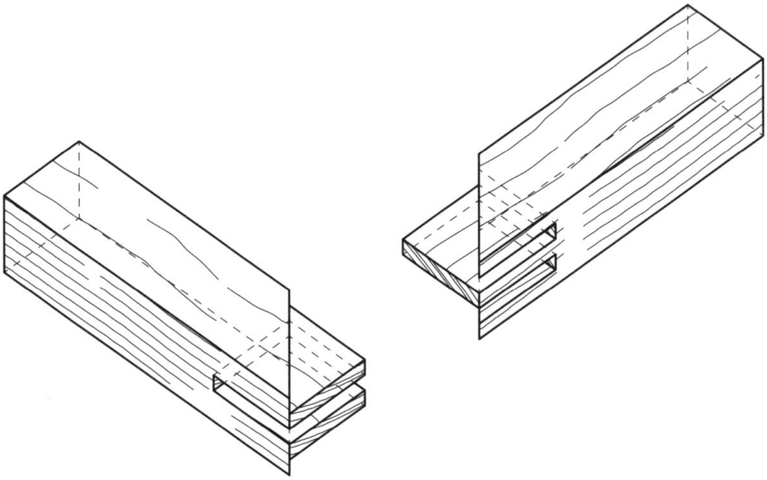

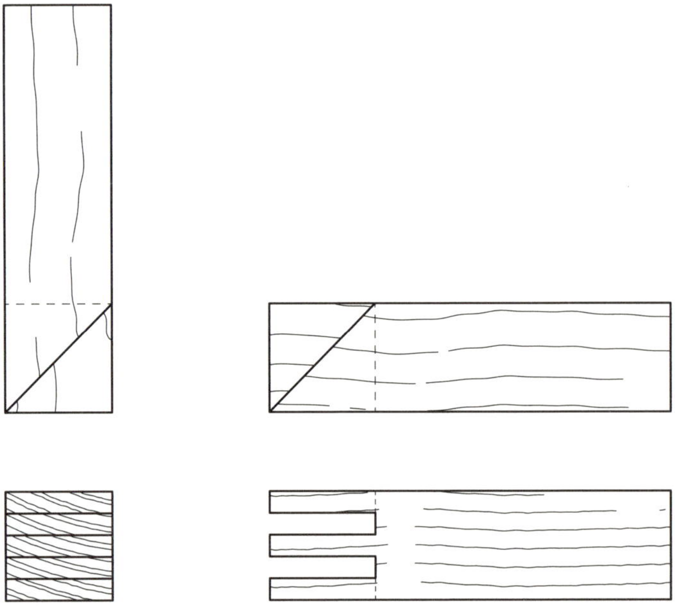

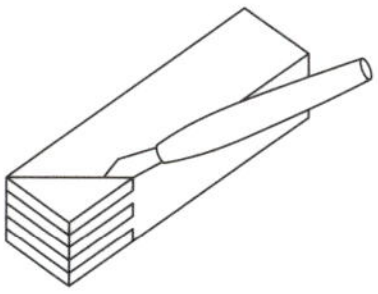

A1

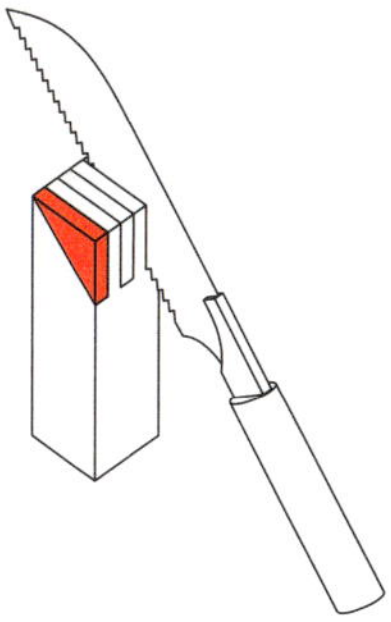

A2

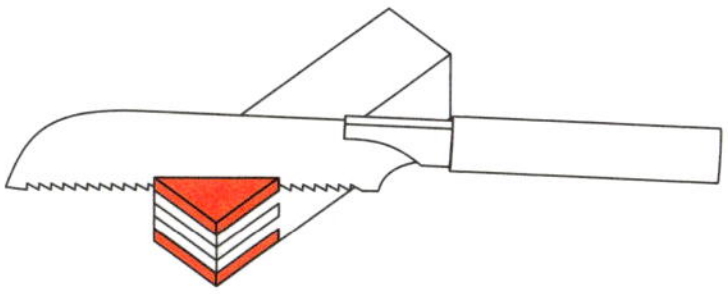

A3

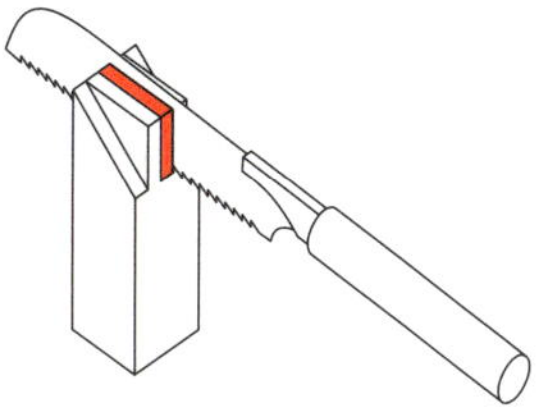

A4

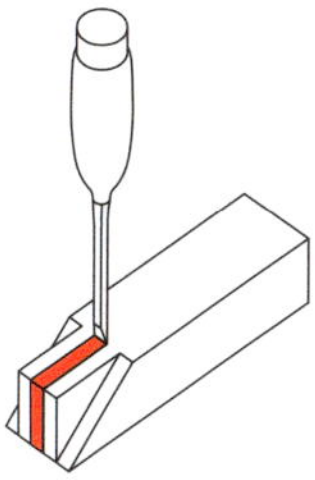

A5

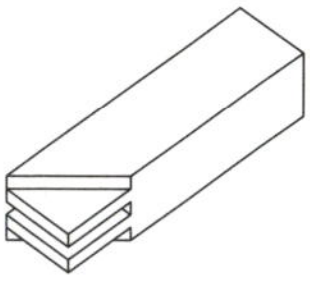

A

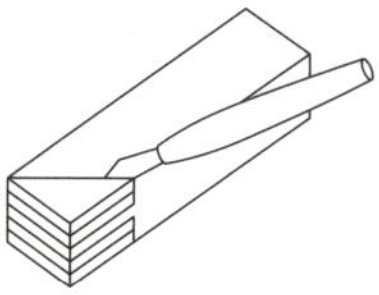

B1

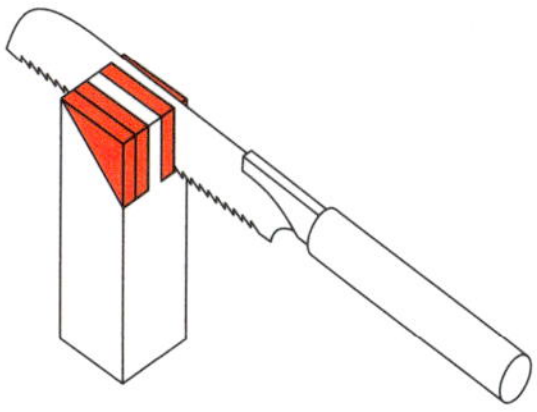

B2

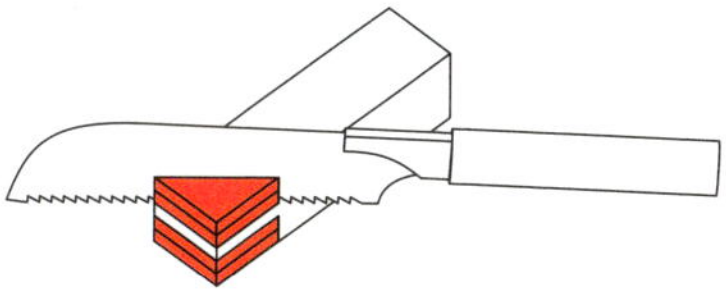

B3

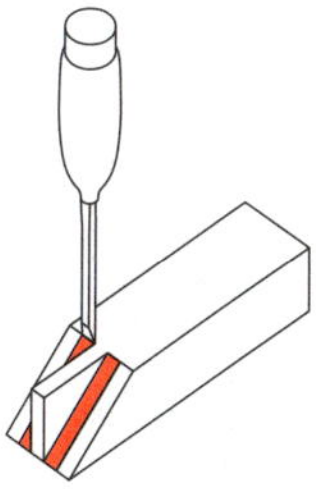

B4

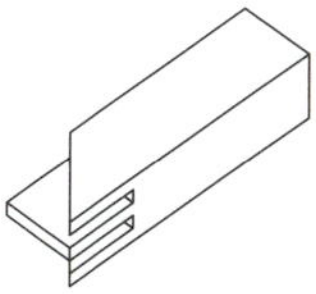

B

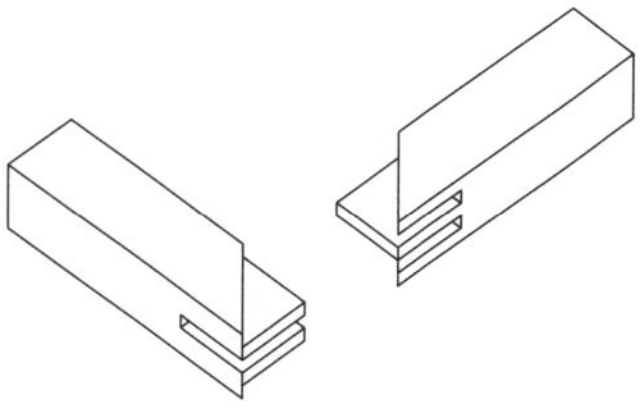

A+B

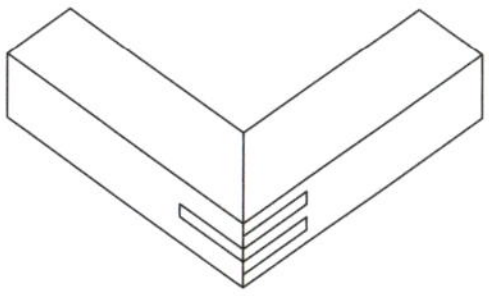

AB

Double mortise and tenon with one-sided lapped miter

gomai-kumi-uwaba-ōdome
五枚組上端大留め

doppelter Zapfen mit Gehrungsblatt

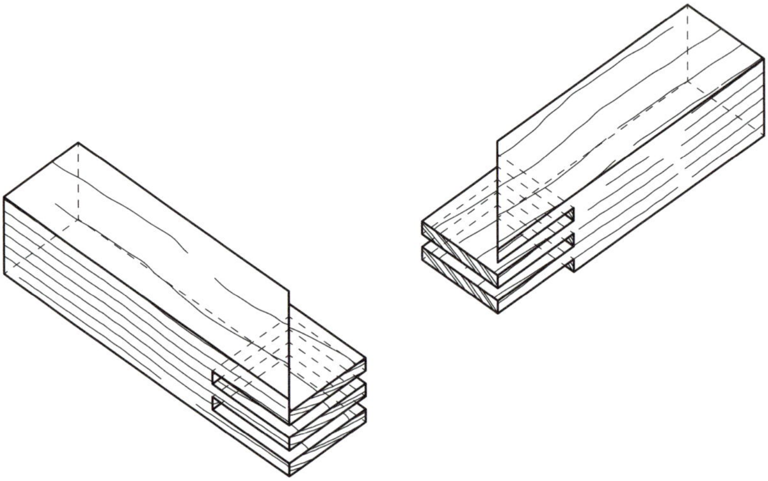

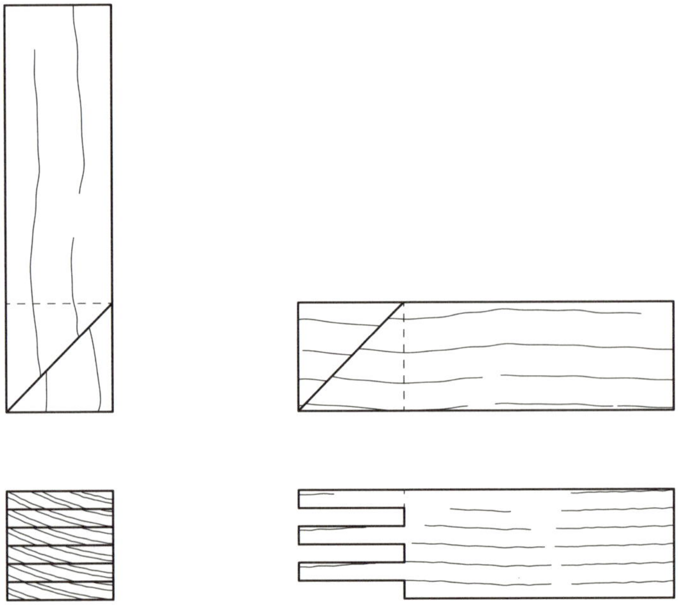

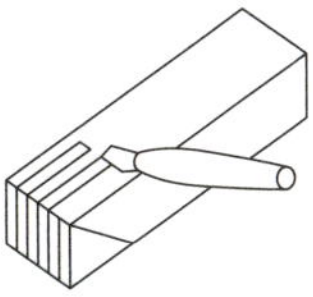

A1

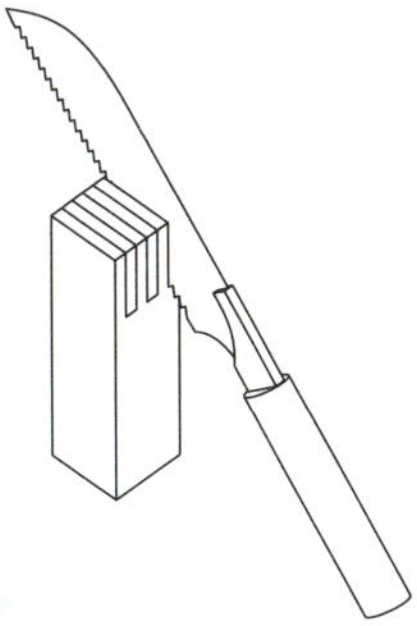

A2

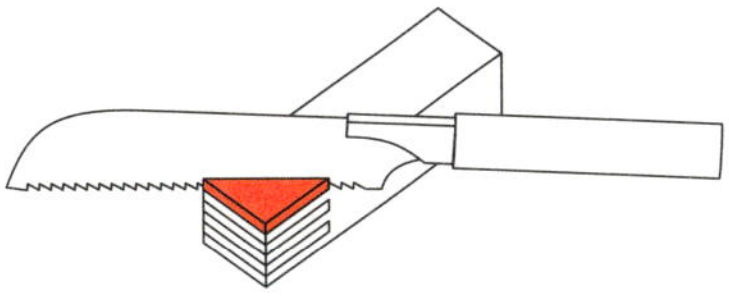

A3

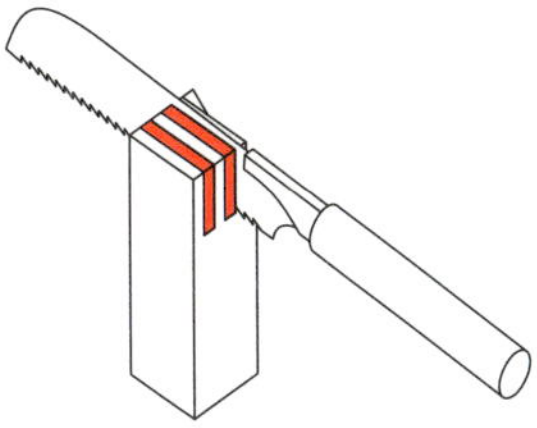

A4

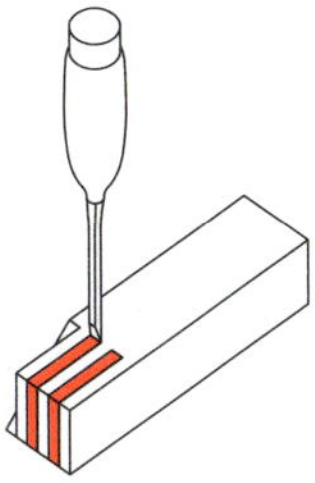

A5

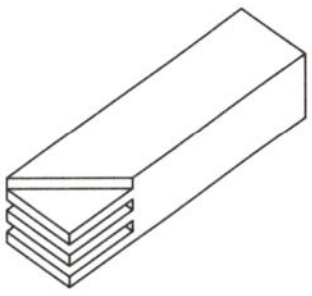

A

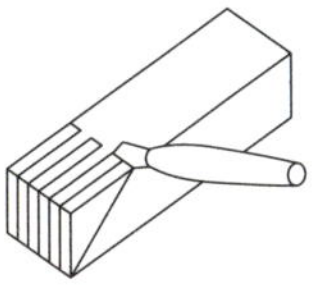

B1

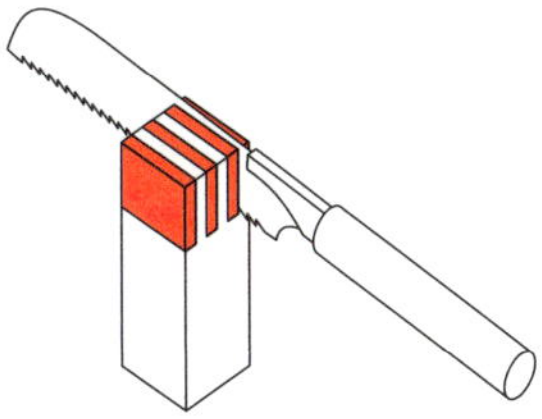

B2

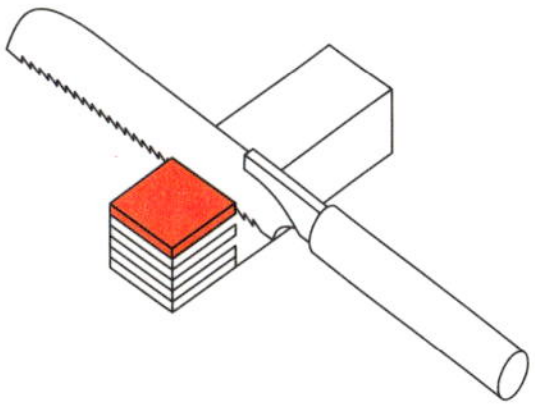

B3

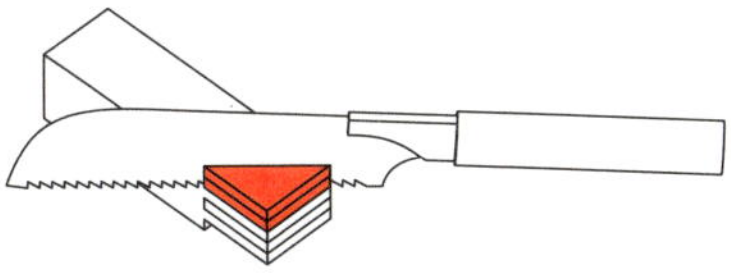

B4

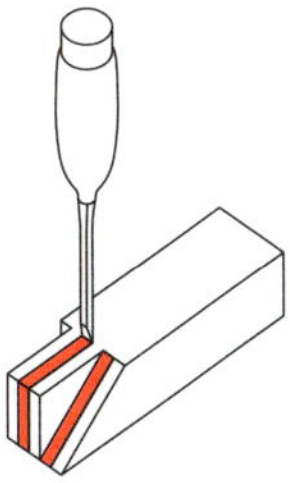

B5

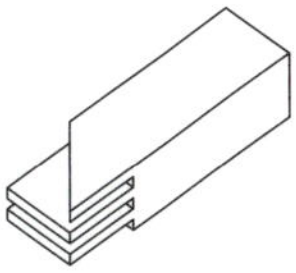

B

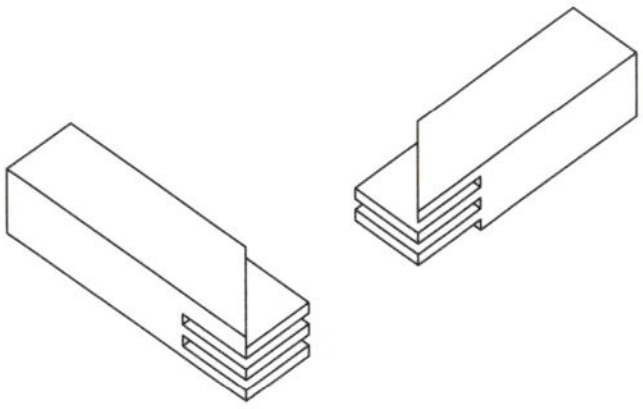

A+B

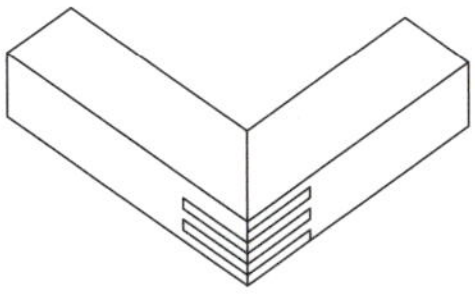

AB

Corner Joints

Haunched through mortise and tenon with two-sided lapped miter

kone-hozo-kumi-dome-tsugi
小根枘組留接ぎ

durchgehender abgesetzter Stufenzapfen mit zweiseitigem Gehrungsblatt

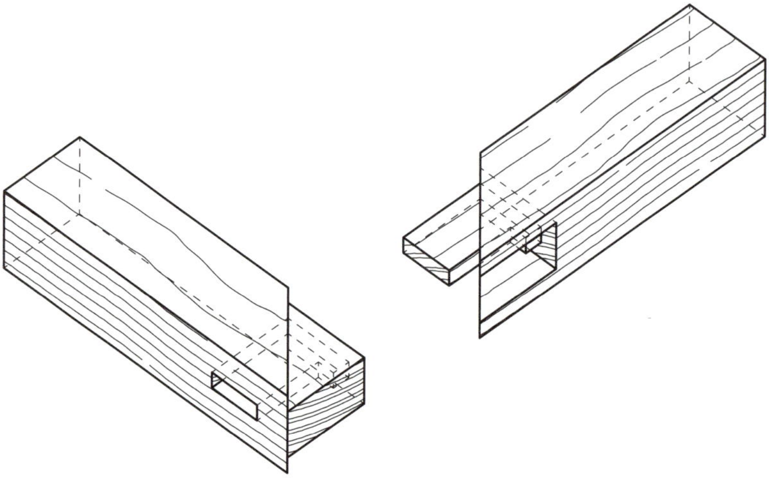

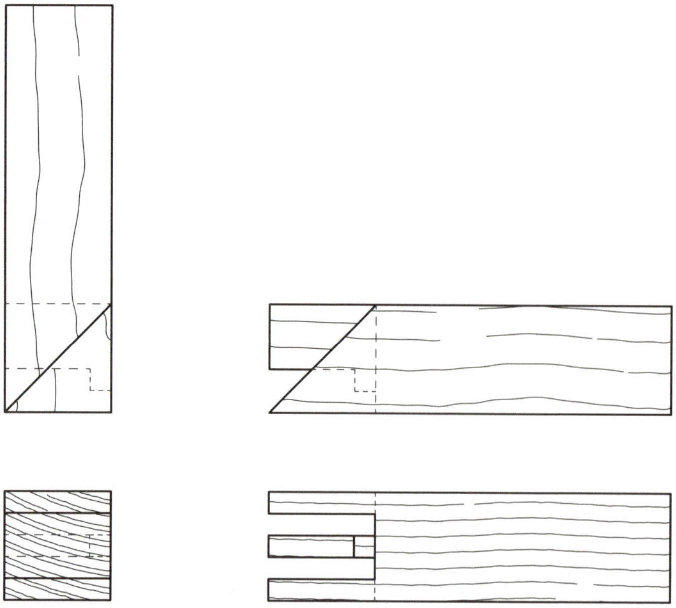

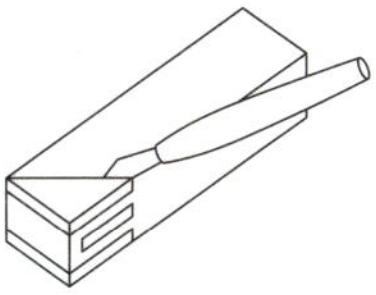

A1

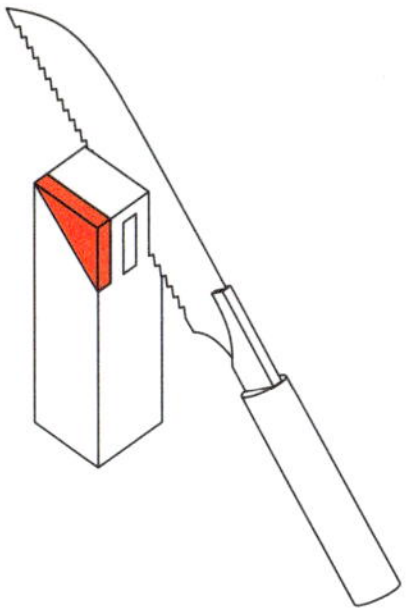

A2

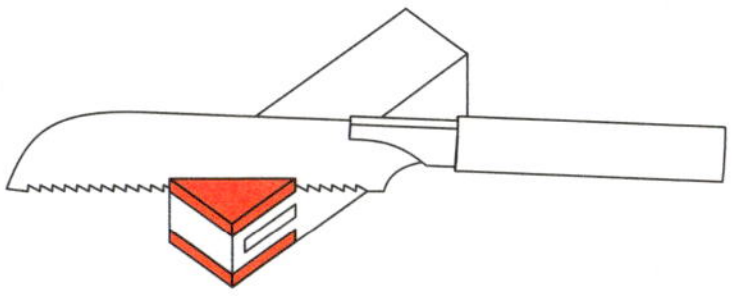

A3

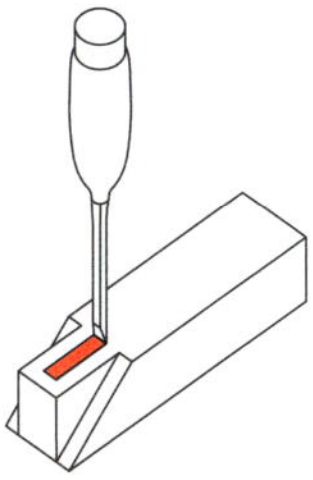

A4

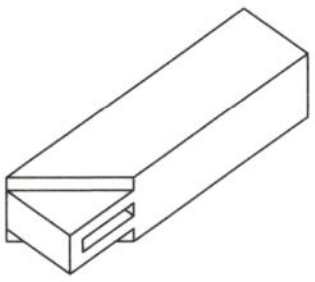

A

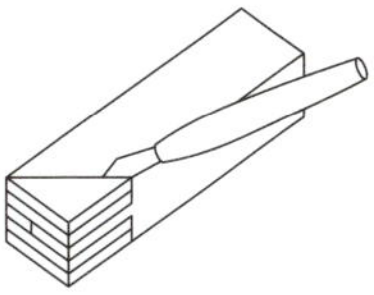

B1

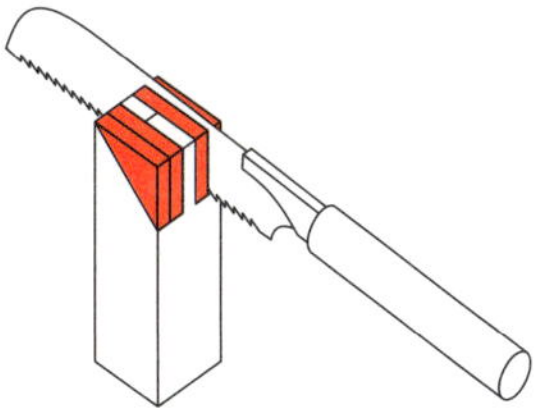

B2

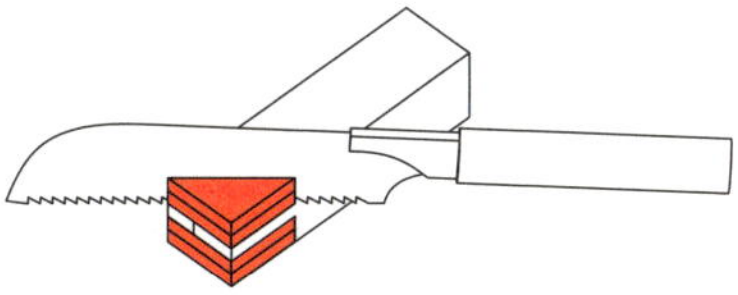

B3

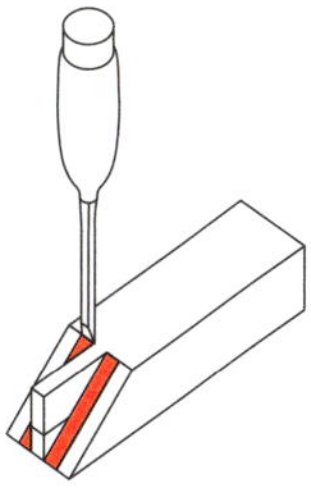

B4

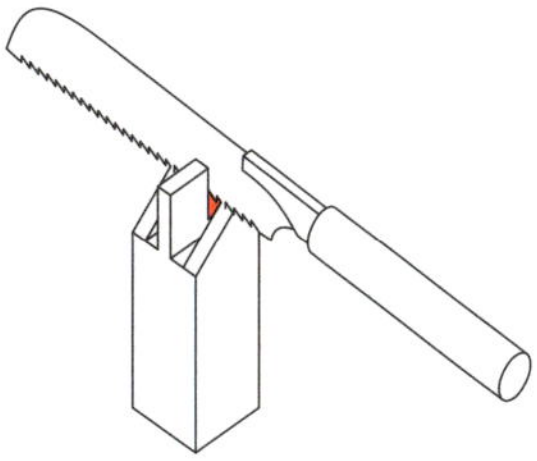

B5

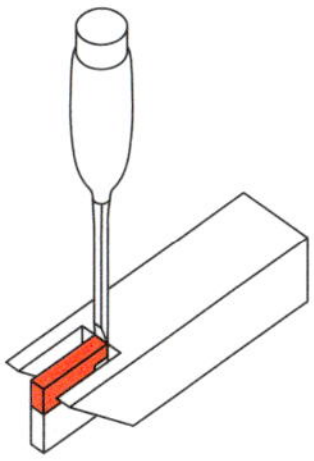

B6

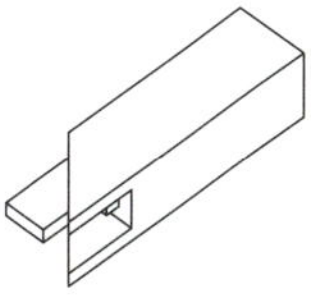

B

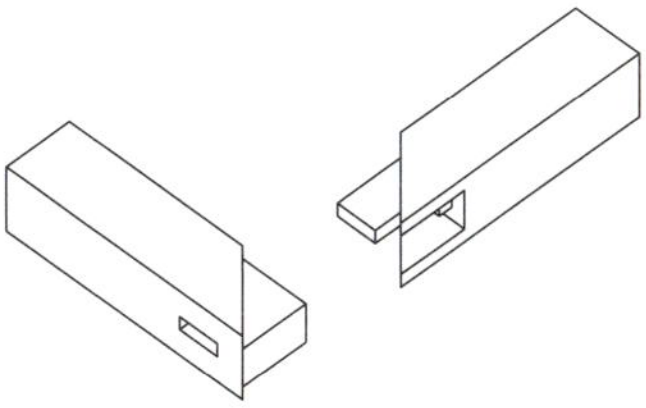

A+B

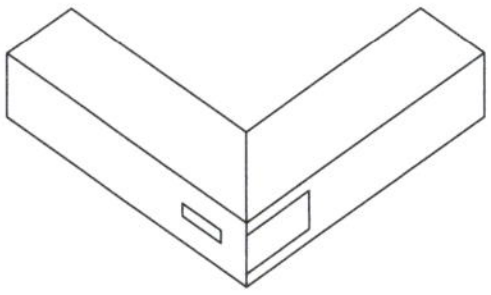

AB

Corner Joints

Haunched through mortise and tenon with two-sided stopped miter

eriwa-kone-hozo-sashi-kusabi-uchi (binta-dome)
襟輪小根枘差楔打ち(鬢太留)

durchgehender abgesetzter Stufenzapfen mit zweiseitigem halben Gehrungsblatt

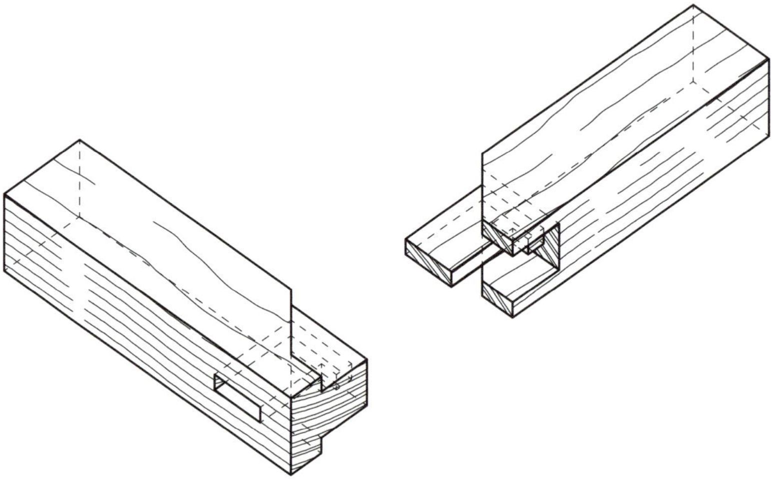

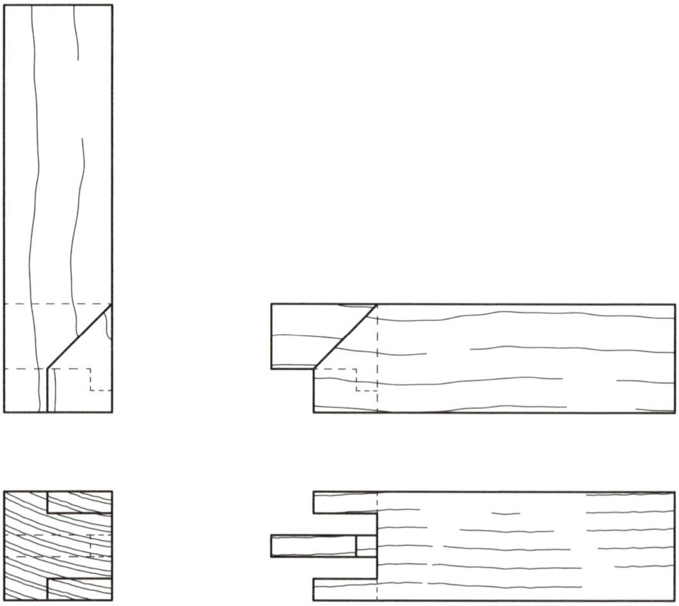

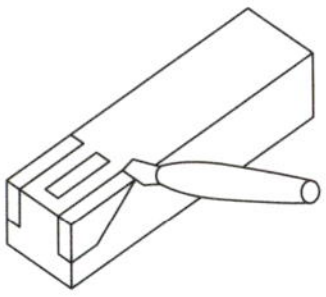

A1

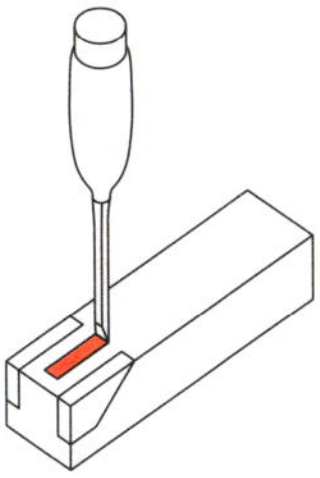

A2

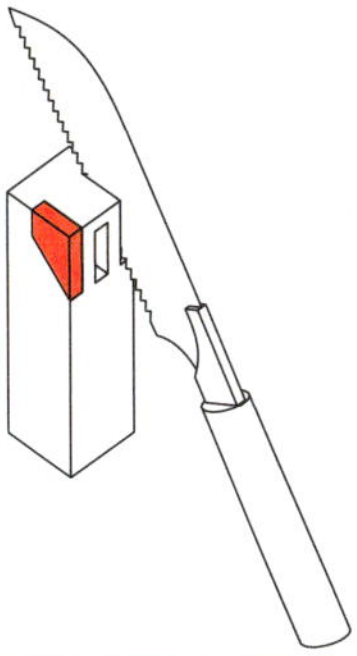

A3

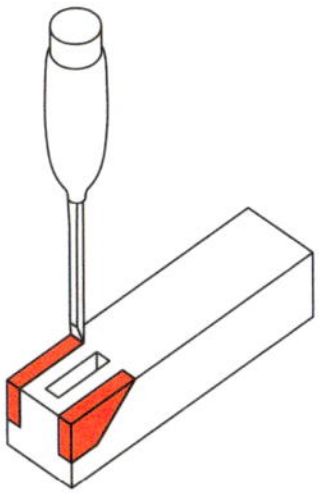

A4

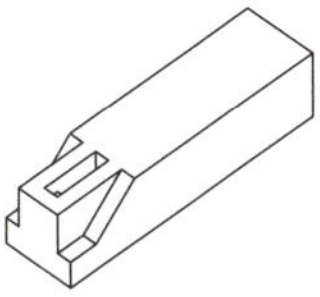

A

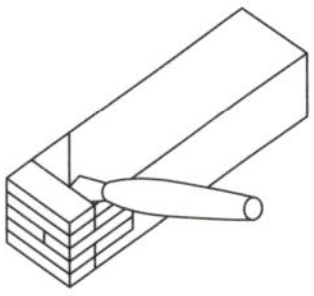

B1

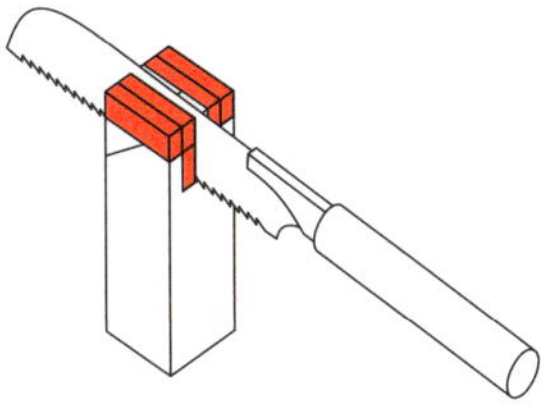

B2

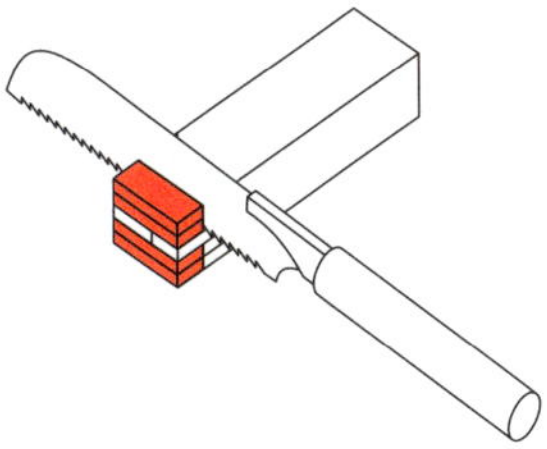

B3

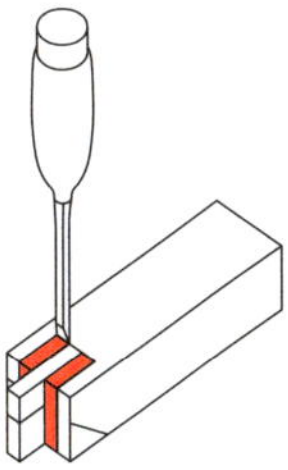

B4

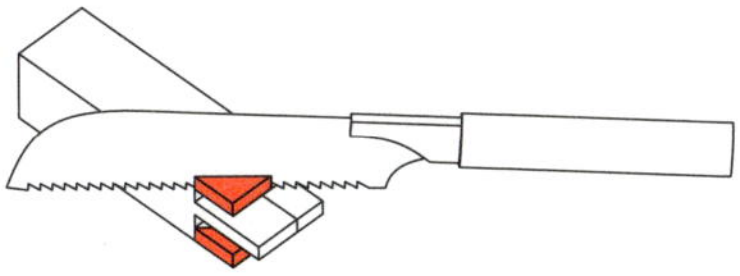

B5

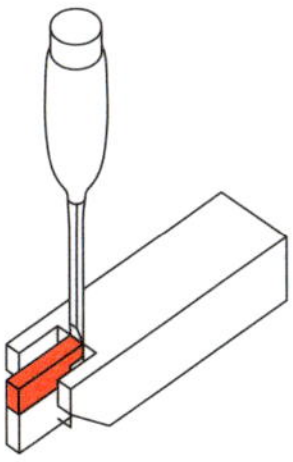

B6

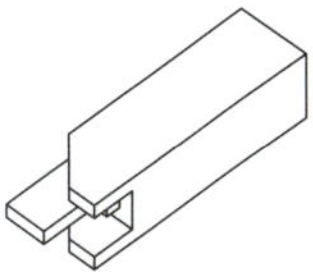

B

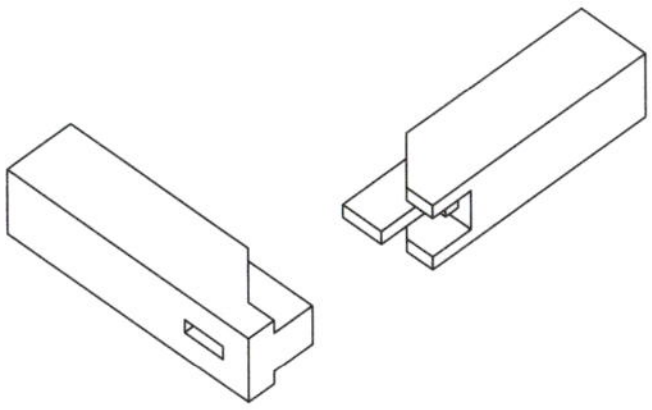

A+B

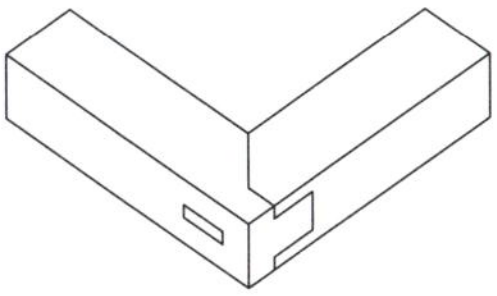

AB

Haunched through mortise and tenon with one-sided lapped miter

kone-hozo-sashi-ōdome-tsugi
小根枘差大留接ぎ

durchgehender abgesetzter Stufenzapfen mit Gehrungsblatt

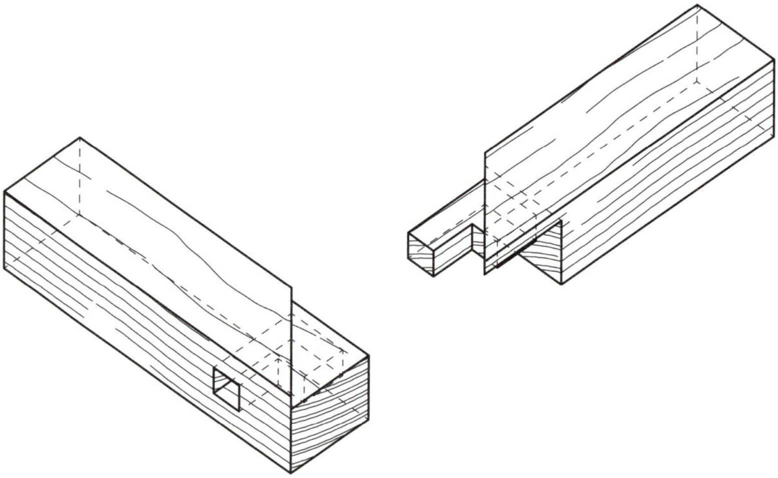

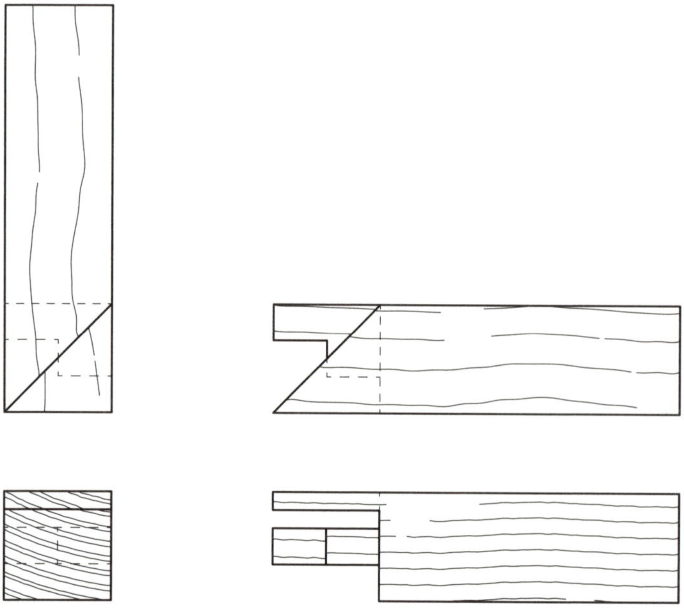

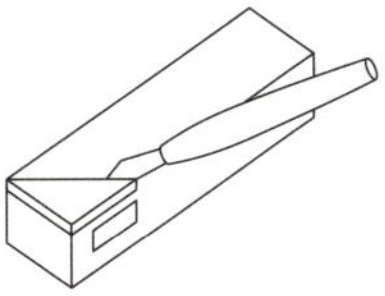

A1

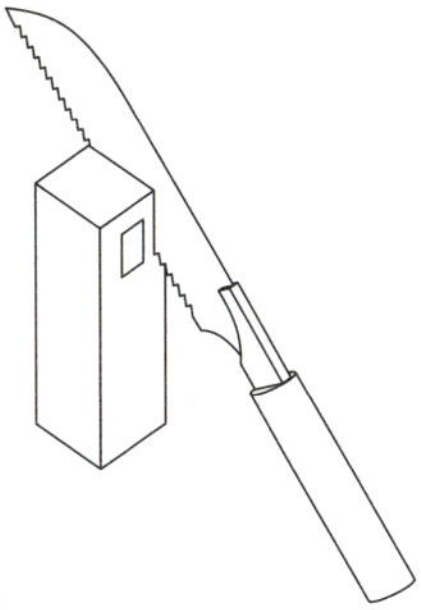

A2

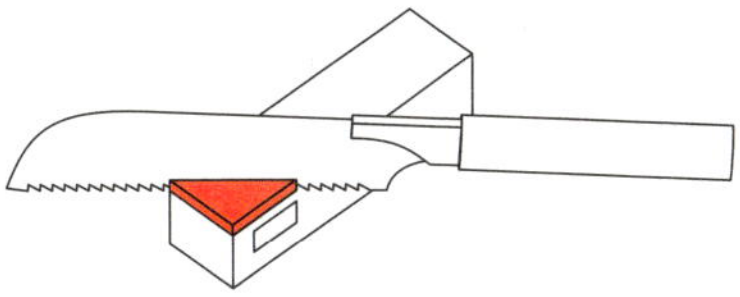

A3

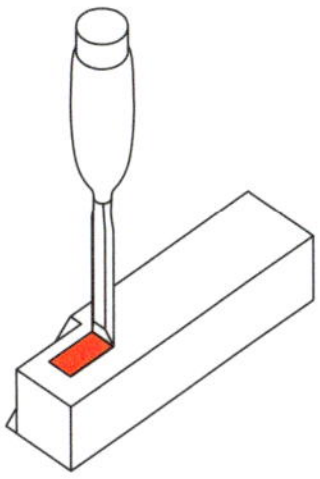

A4

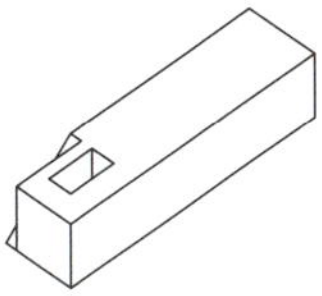

A

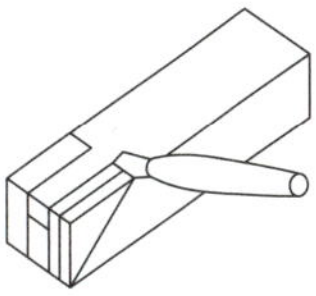

B1

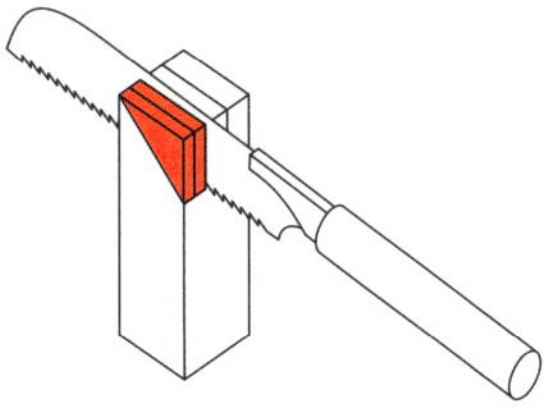

B2

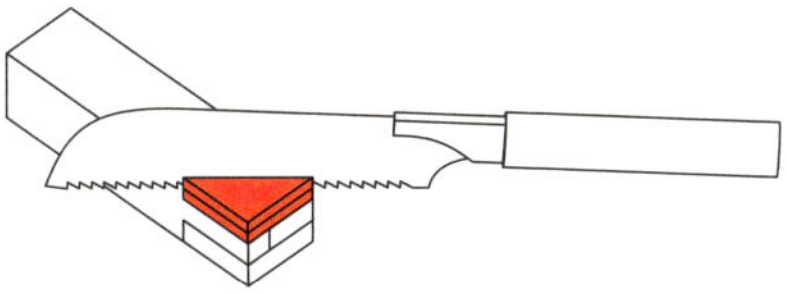

B3

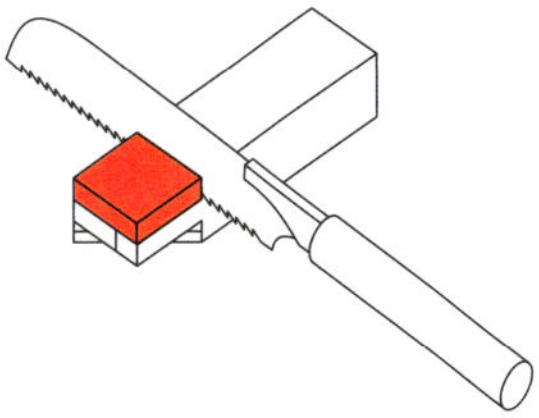

B4

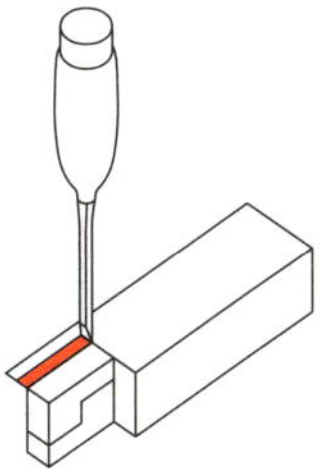

B5

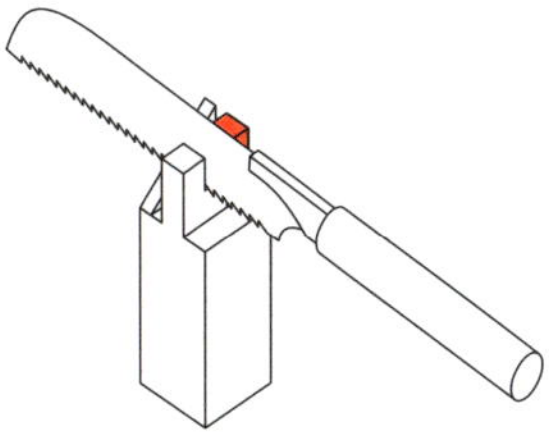

B6

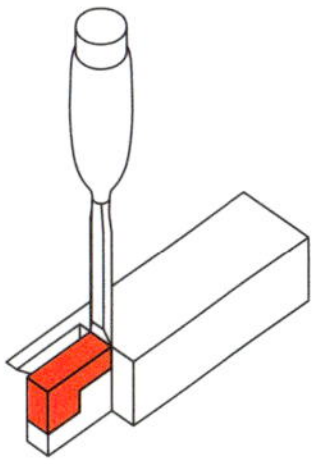

B7

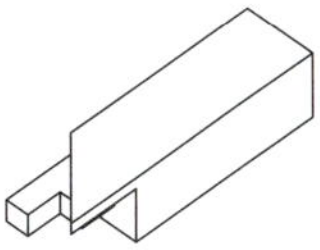

B

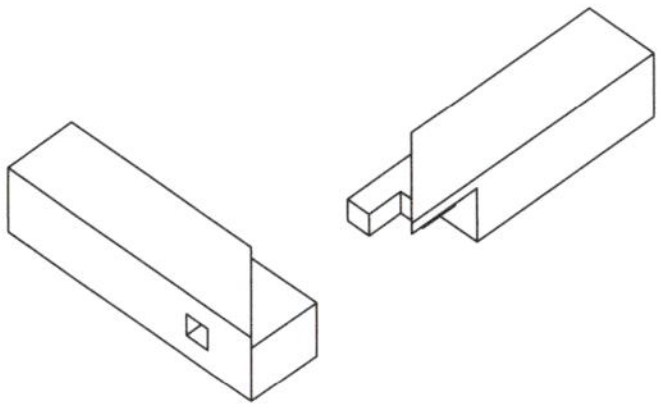

A+B

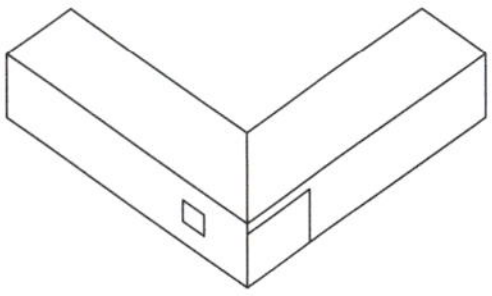

AB

Corner Joints

Haunched through mortise and tenon with one-sided stopped lapped miter

eriwa-kone-hozo-sashi-tsugi
襟輪小根枘差接ぎ

durchgehender abgesetzter Stufenzapfen mit halbem Gehrungsblatt

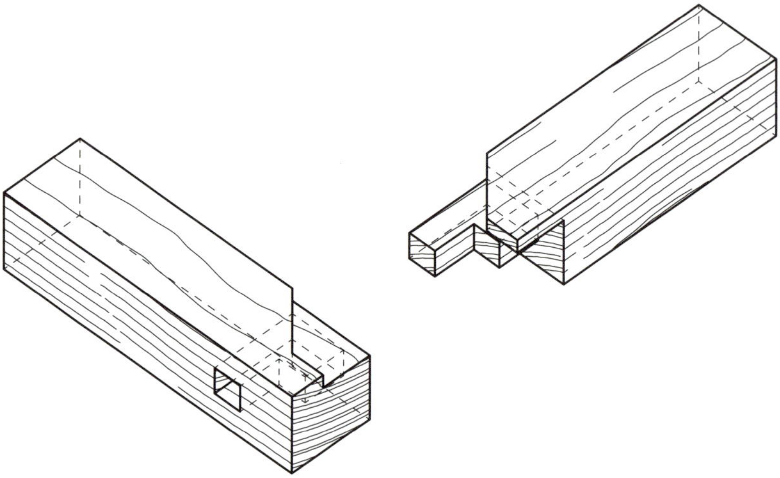

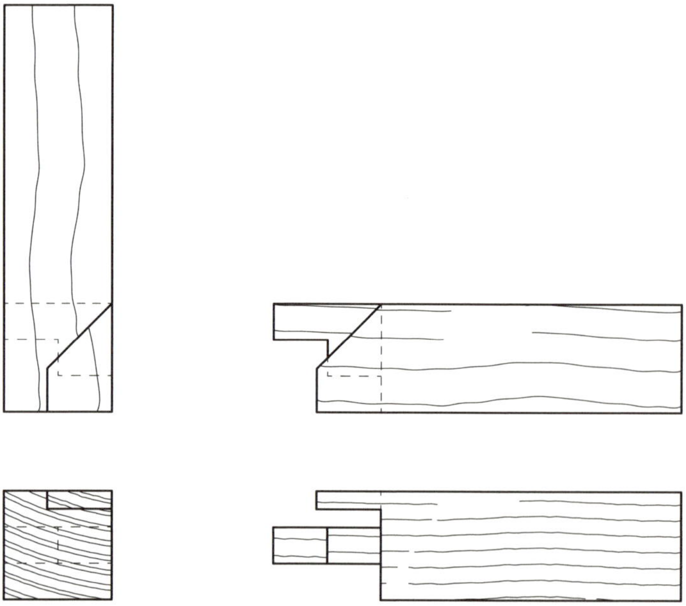

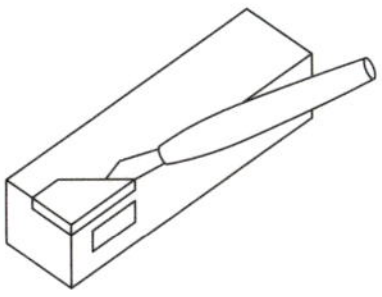

A1

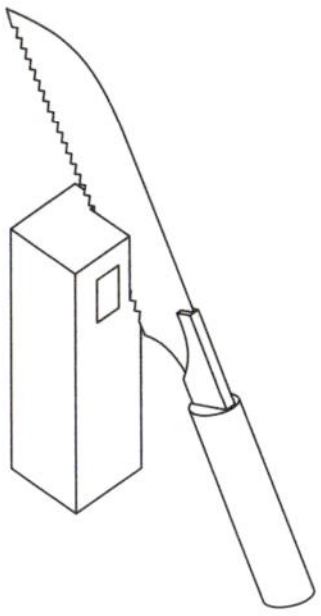

A2

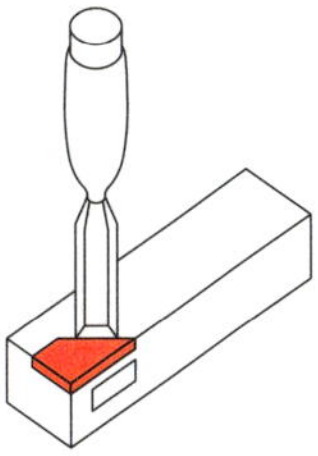

A3

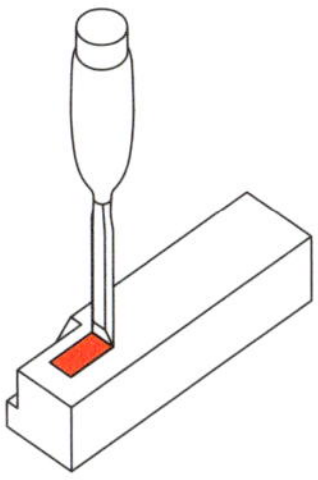

A4

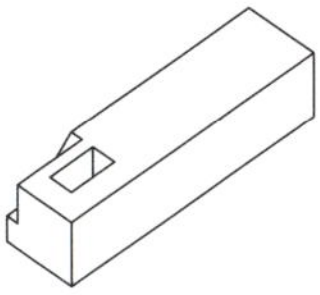

A

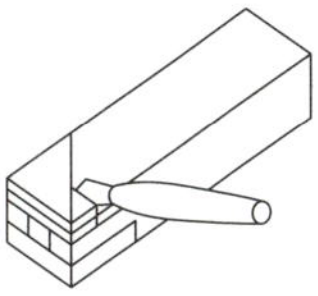

B1

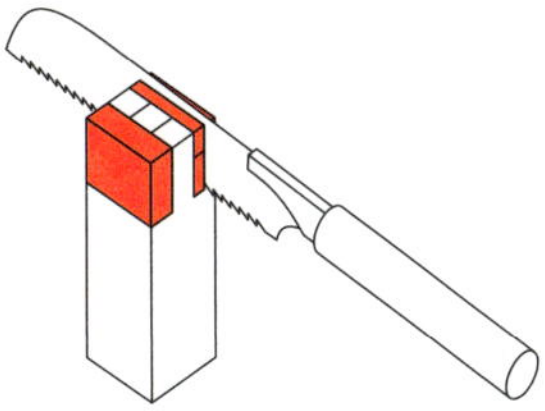

B2

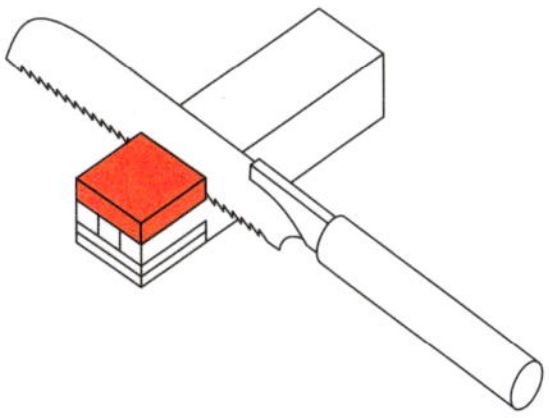

B3

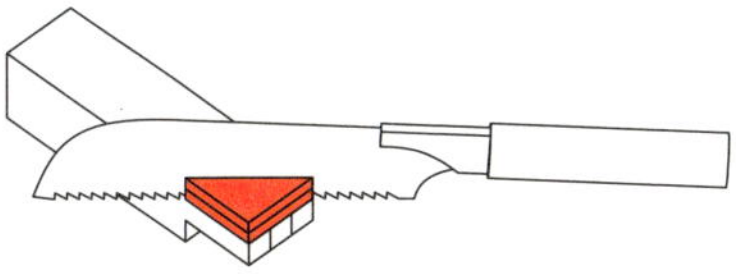

B4

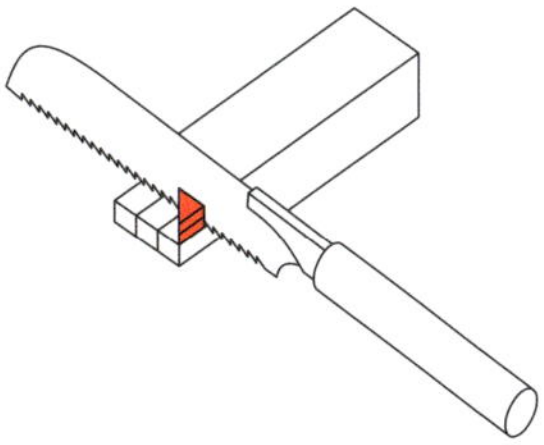

B5

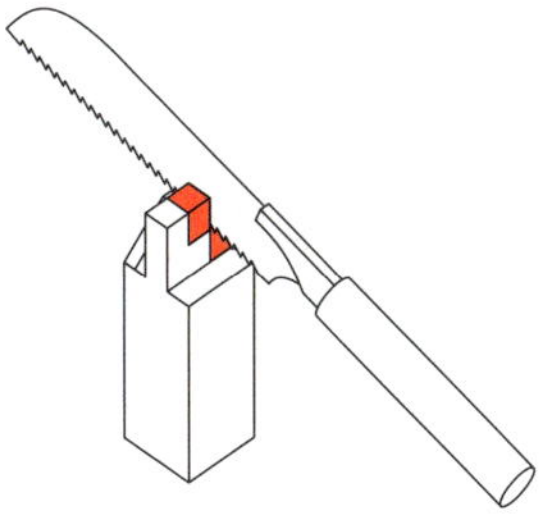

B6

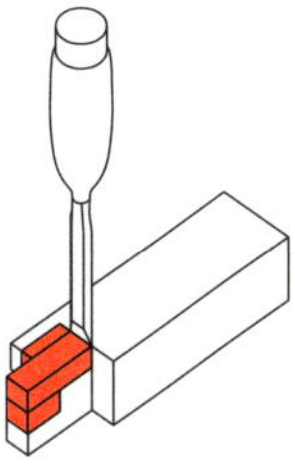

B7

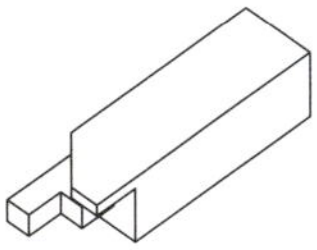

B

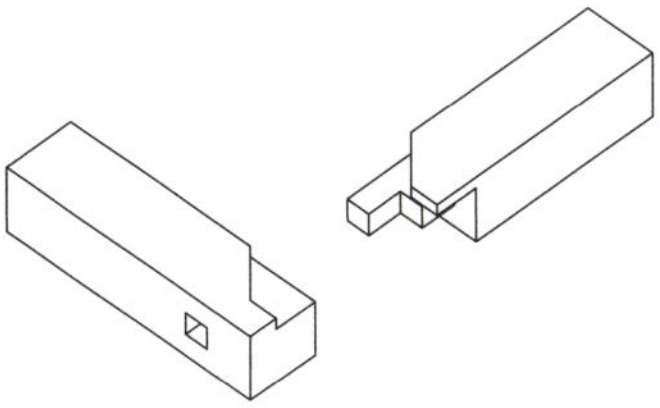

A+B

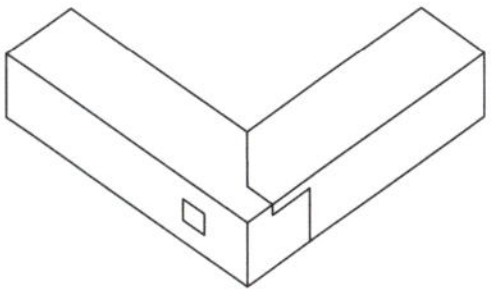

AB

Corner Joints

Blind haunched mortise and tenon miter

kone-hozo-tsugi
(hako-dome, sanpō-mi-tsuke)
小根枘接ぎ(箱留・三方見付)

verdeckter abgesetzter Stufenzapfen
mit Gehrungsblatt und Wange

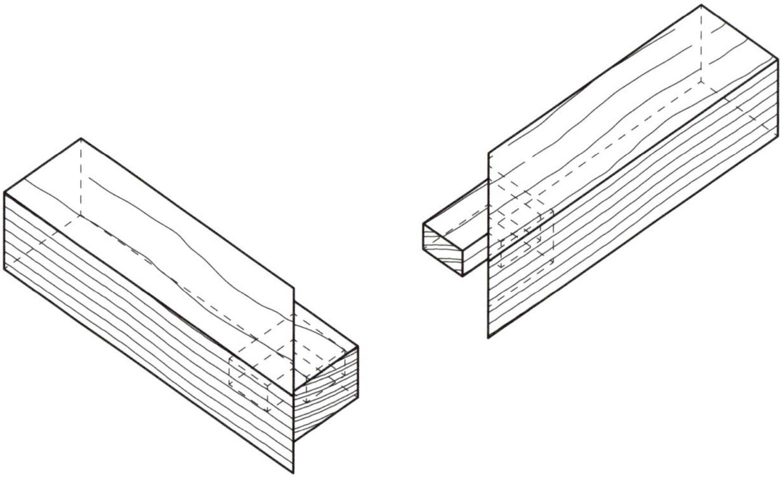

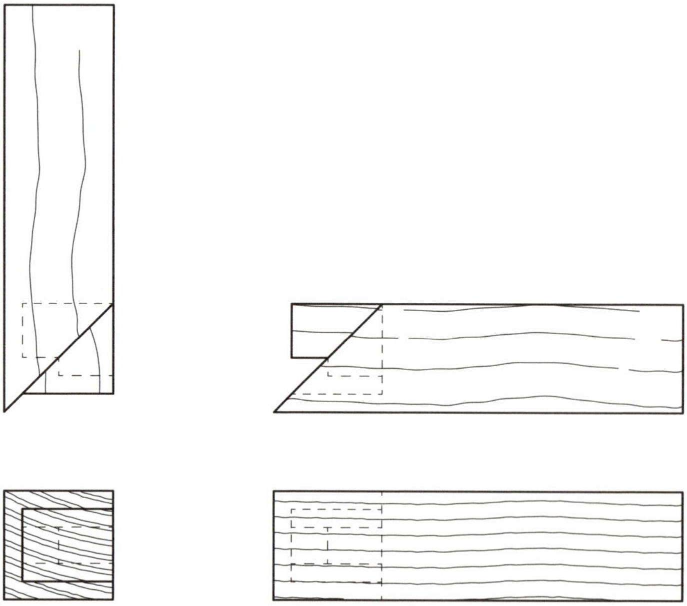

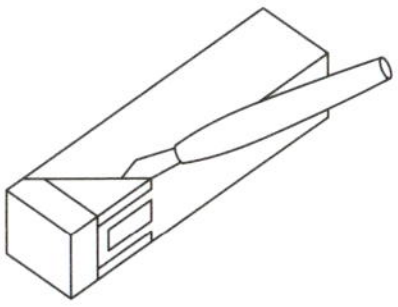

A1

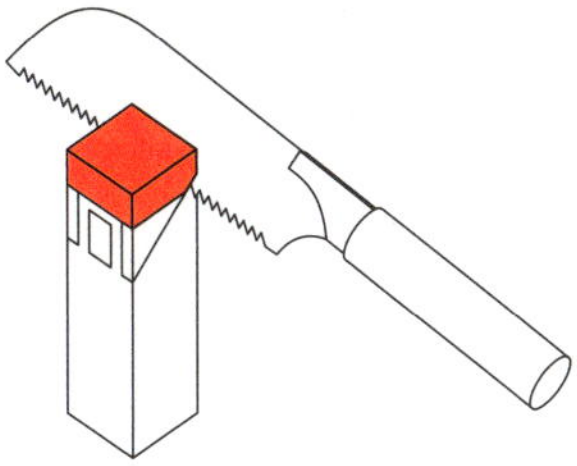

A2

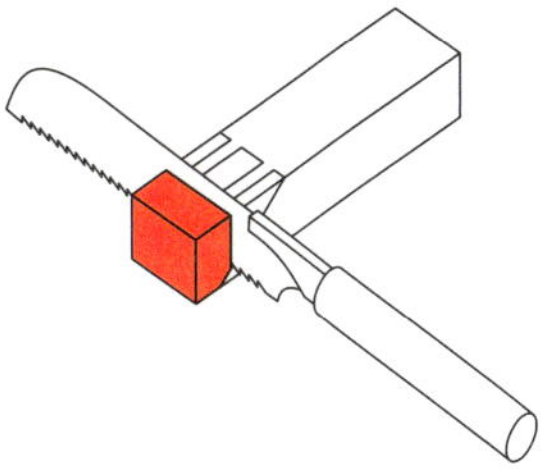

A3

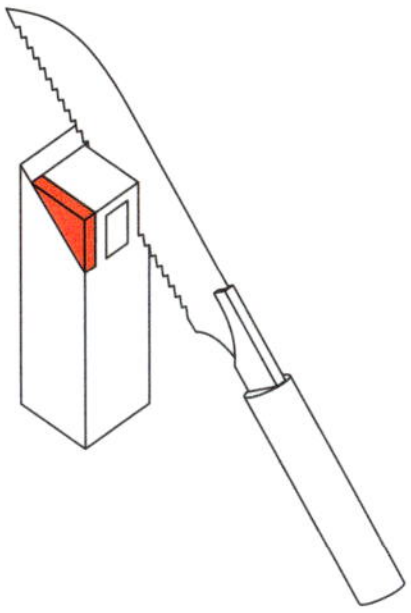

A4

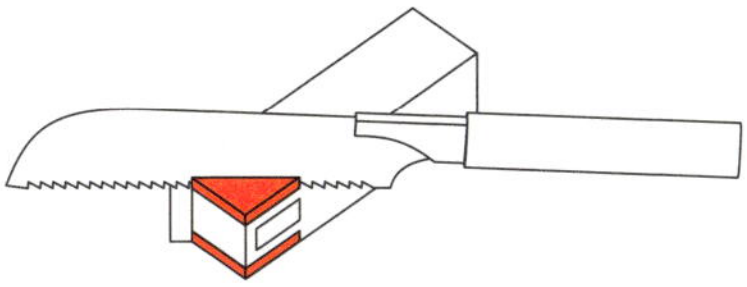

A5

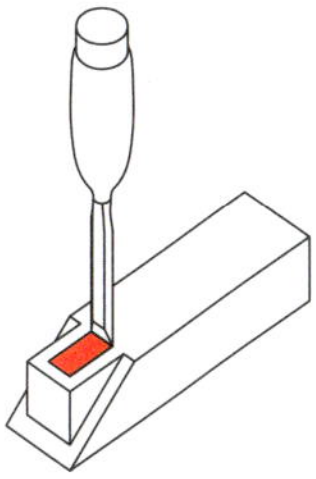

A6

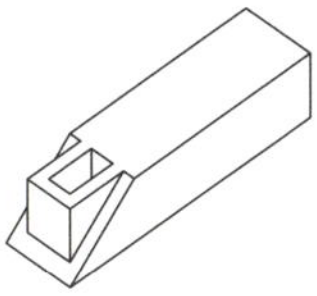

A

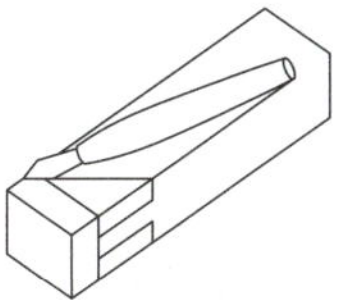

B1

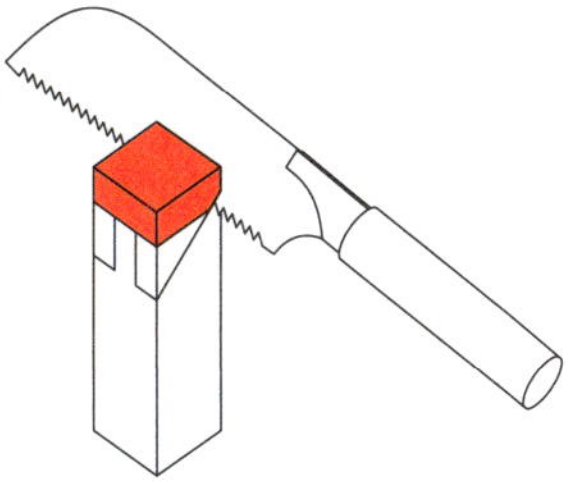

B2

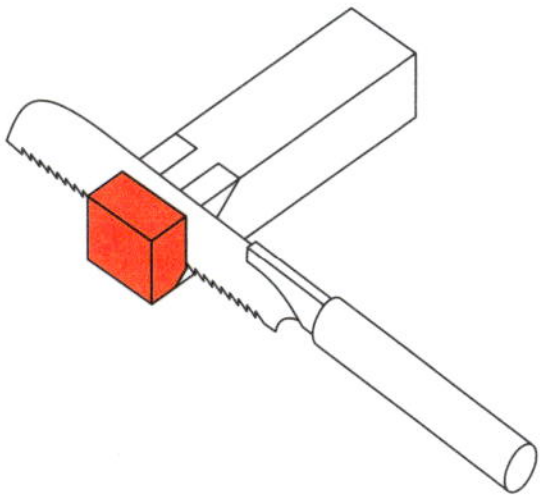

B3

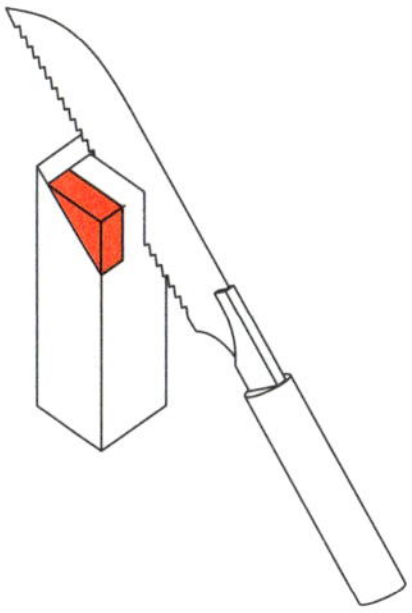

B4

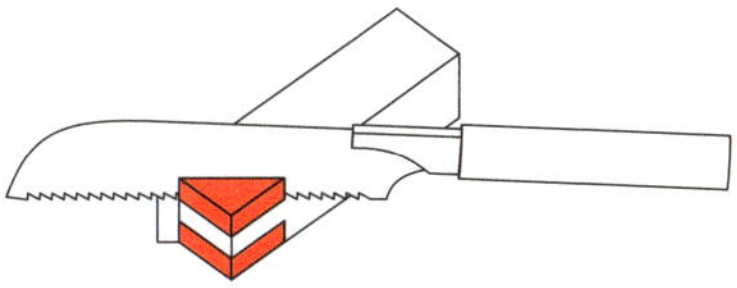

B5

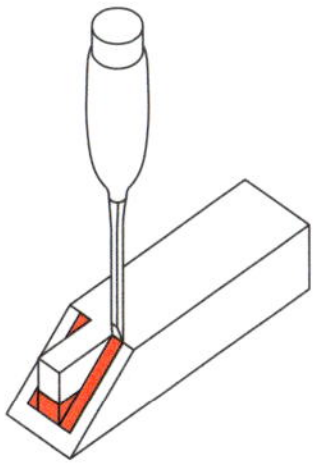

B6

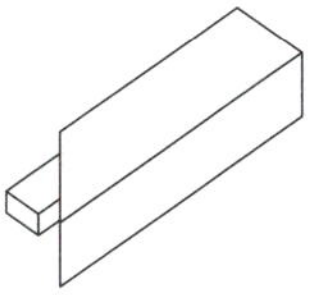

B

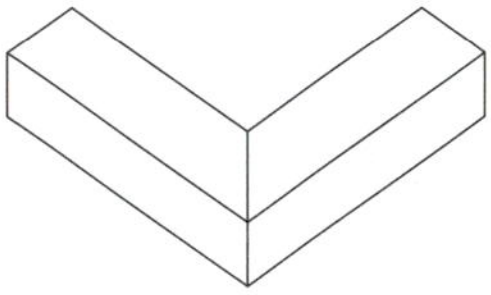

AB

Corner Joints

Haunched through double mortise and tenon with one-sided lapped miter

nimai-kone-hozo-tsugi (uwaba-ōdome)
二枚小根枘接ぎ(上端大留)

durchgehender abgesetzter doppelter Stufenzapfen mit Gehrungsblatt

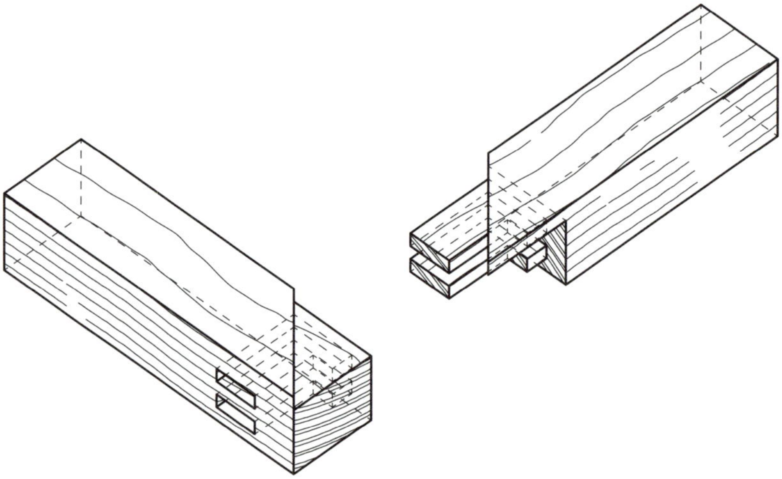

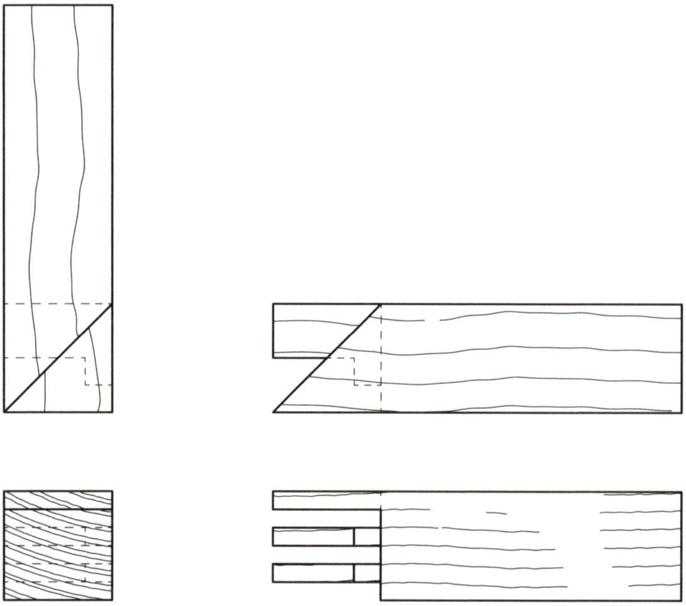

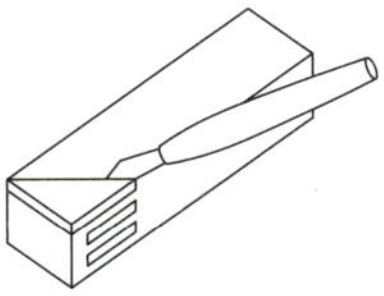

A1

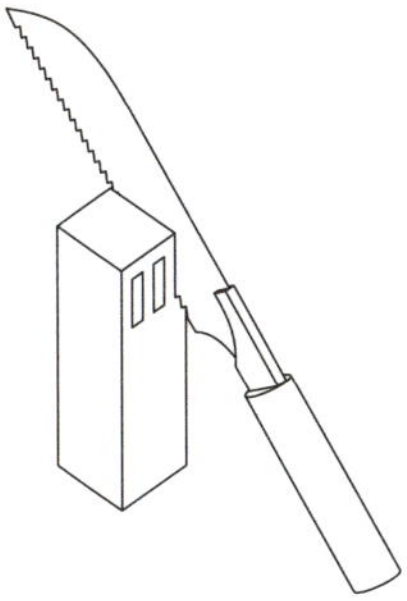

A2

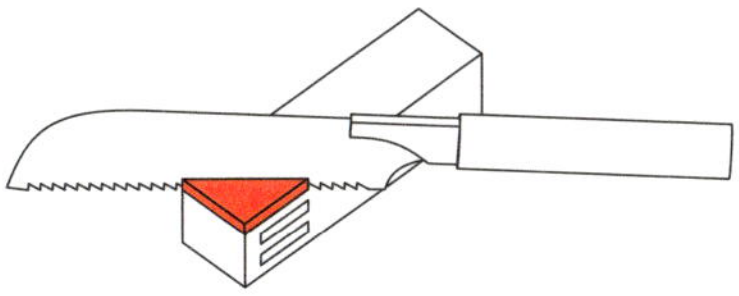

A3

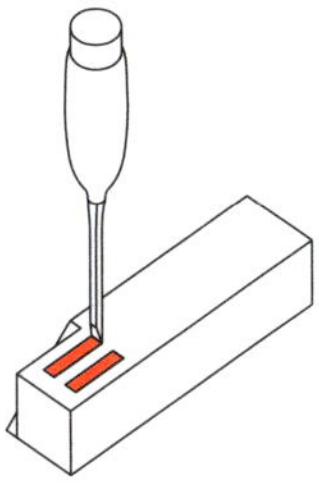

A4

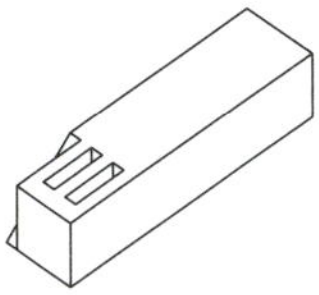

A

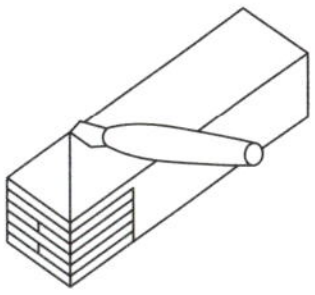

B1

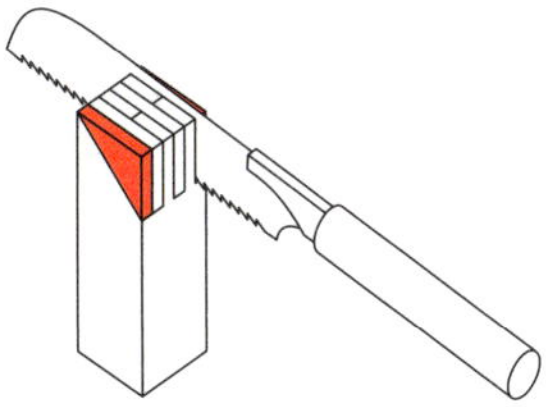

B2

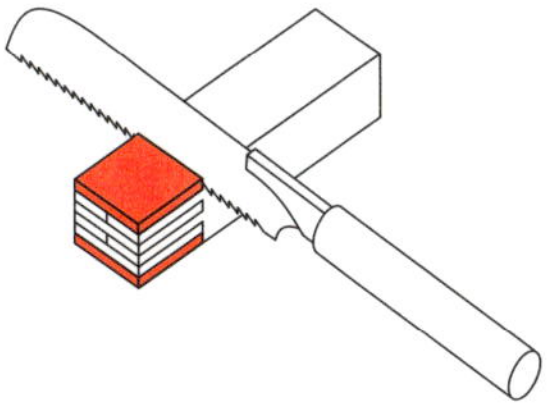

B3

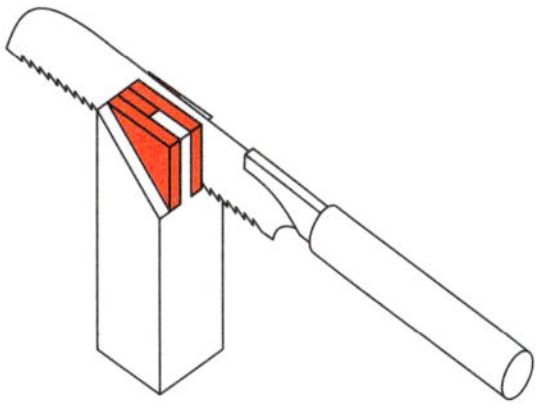

B4

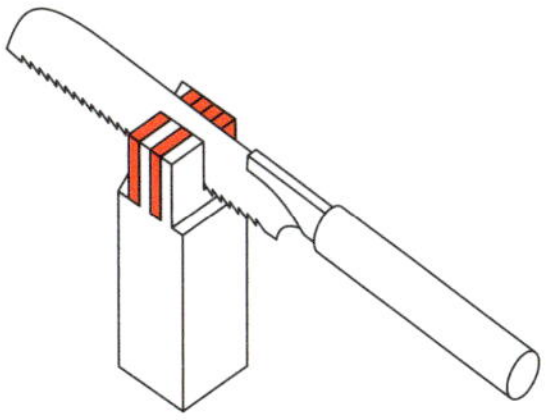

B5

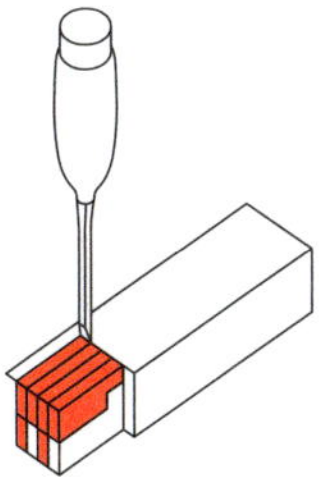

B6

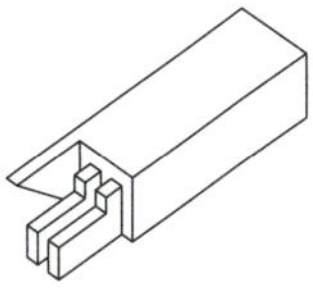

B

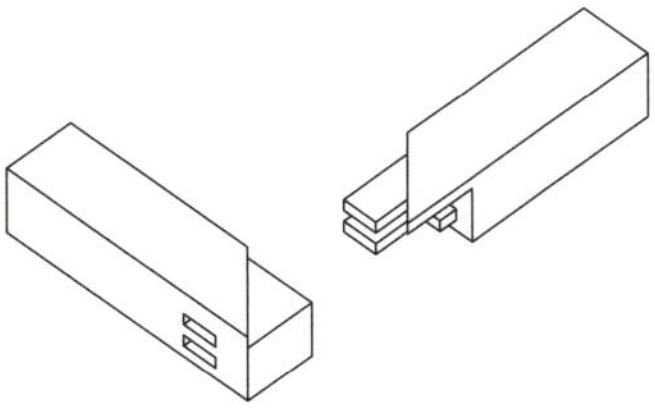

A+B

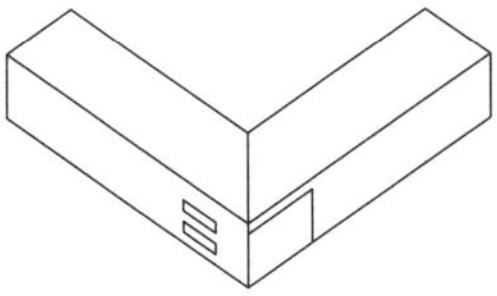

AB

Triple-shouldered double tenon miter

nimai-hira-hozo-tsugi
(hako-dome, sanpō-mi-tsuke)
二枚平枘接ぎ（箱留・三方見付）

durchgehender abgesetzter doppelter Zapfen mit Gehrungsblatt und Wange

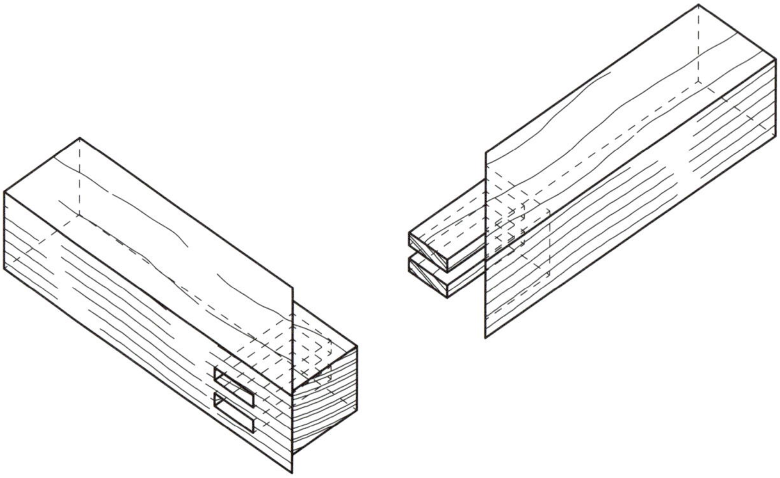

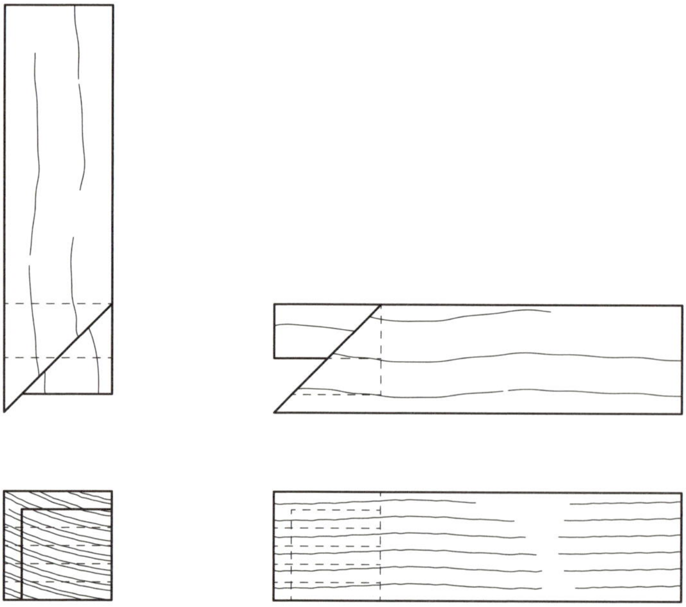

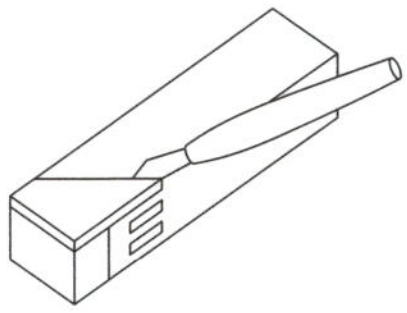

A1

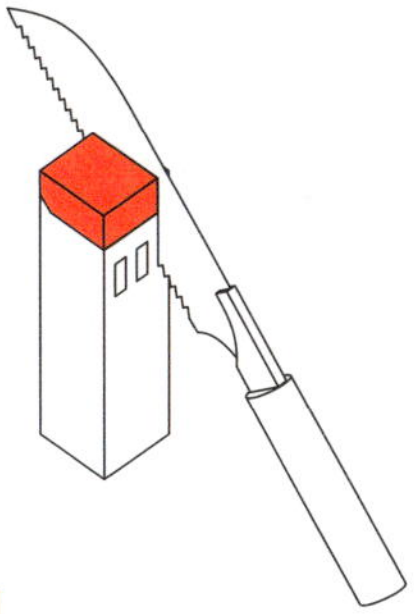

A2

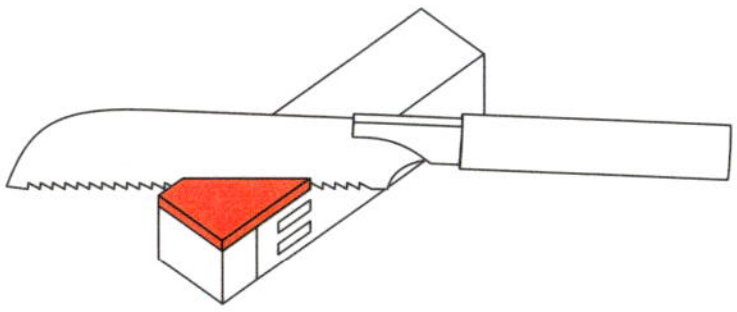

A3

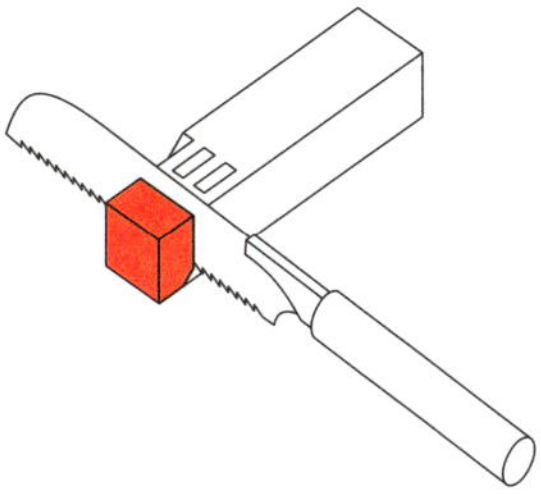

A4

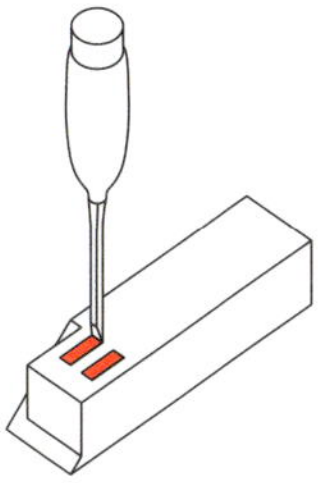

A5

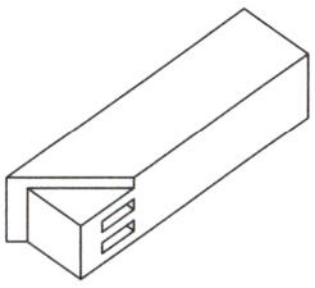

A

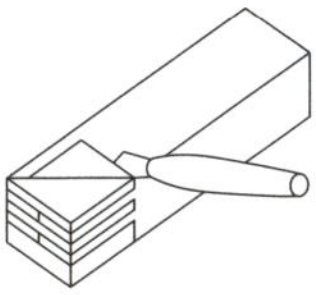

B1

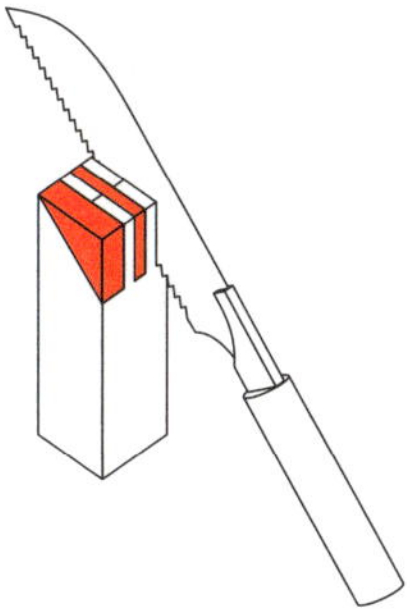

B2

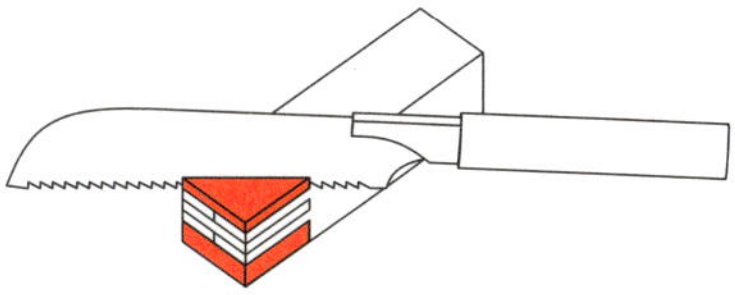

B3

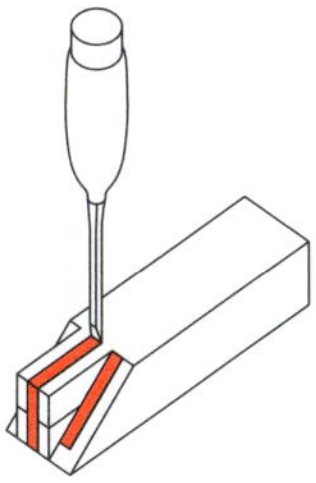

B4

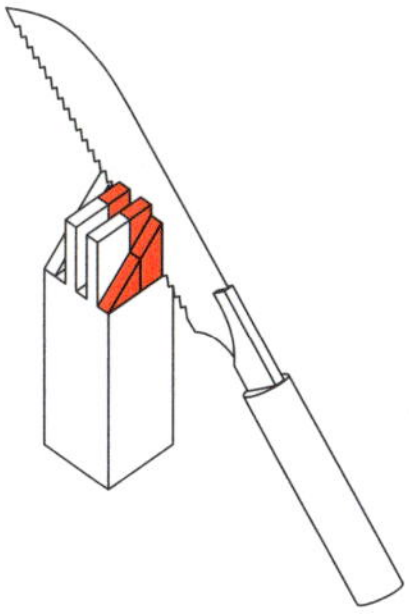

B5

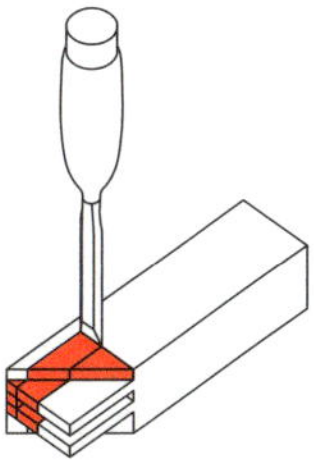

B6

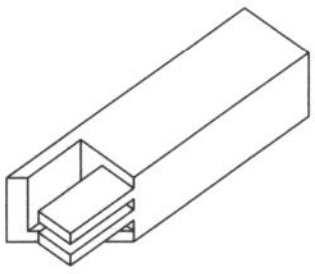

B

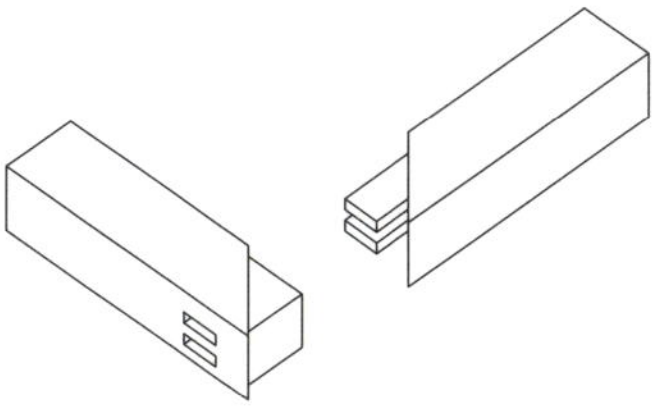

A+B

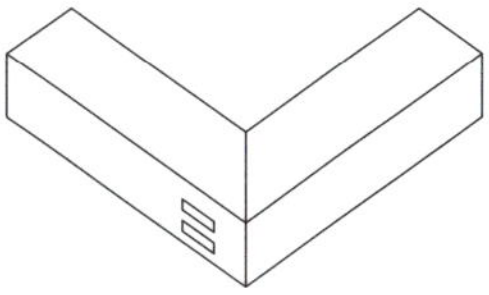

AB

Corner Joints

Housed stub tenon

mechi-dome-tsugi (oire)
目違留接ぎ(追入)

halbverdeckter
eingelassener Eckzapfen

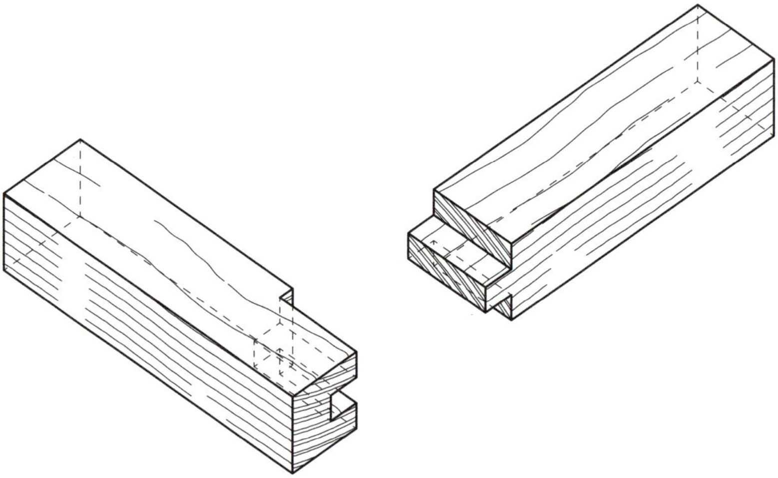

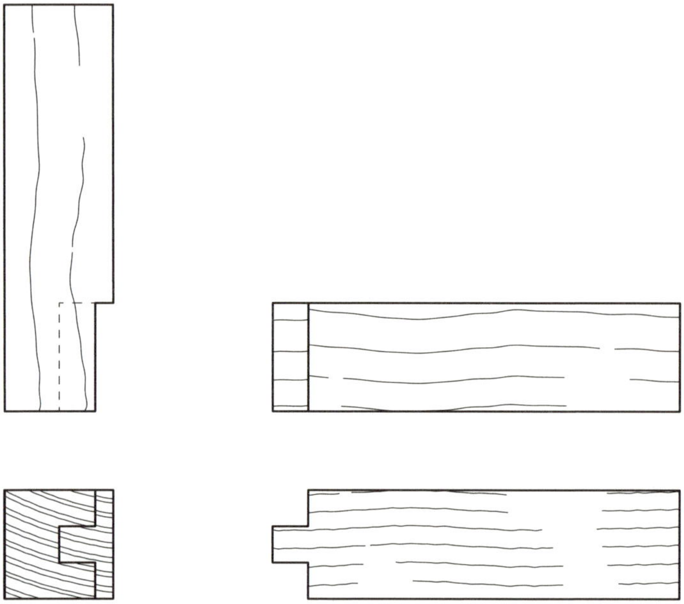

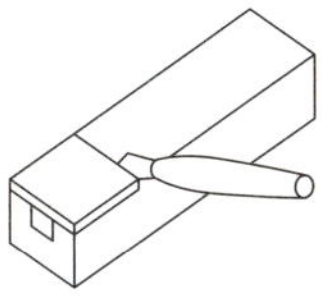
A1

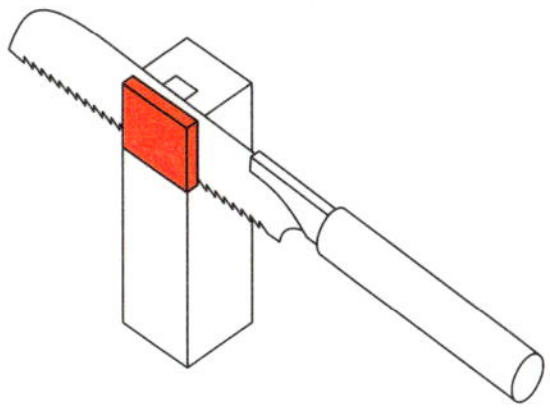
A2

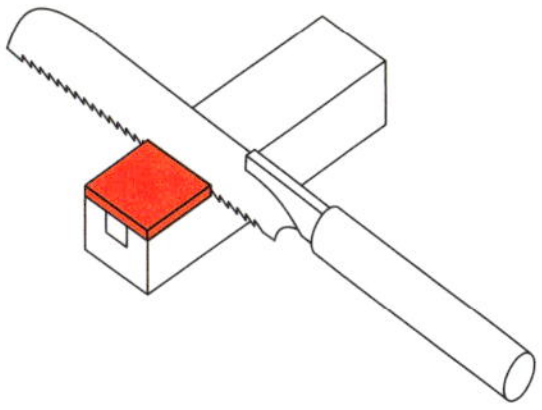
A3

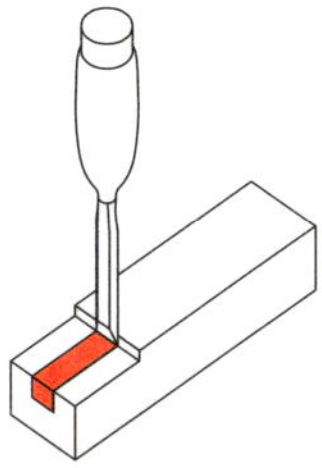
A4

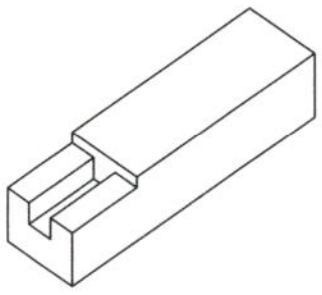
A

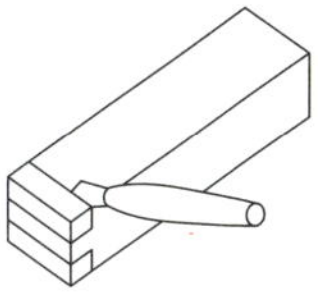
B1

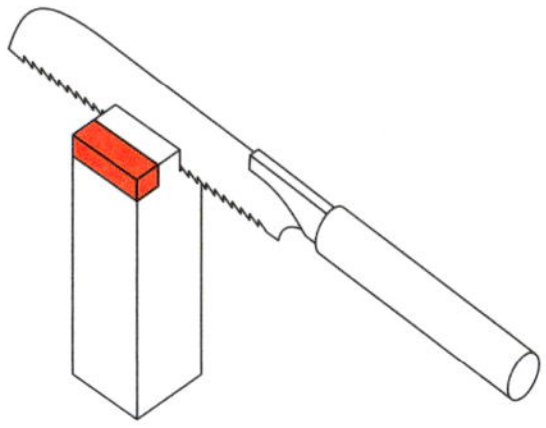
B2

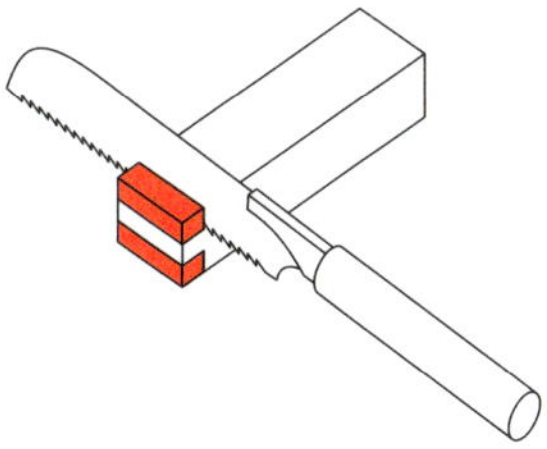
B3

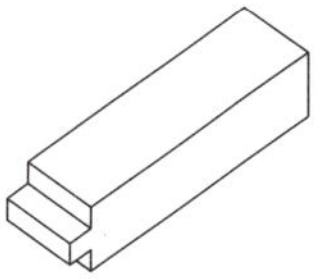
B

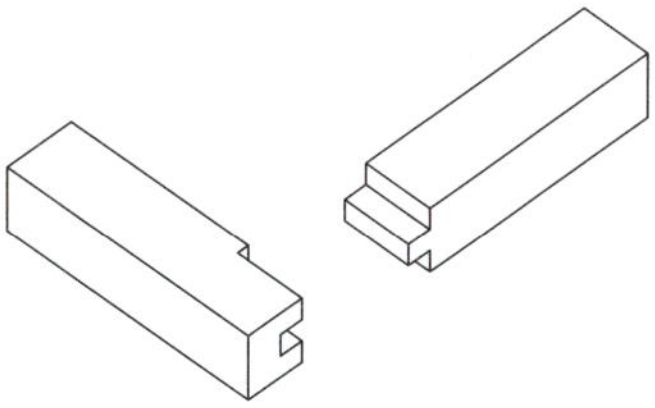
A+B

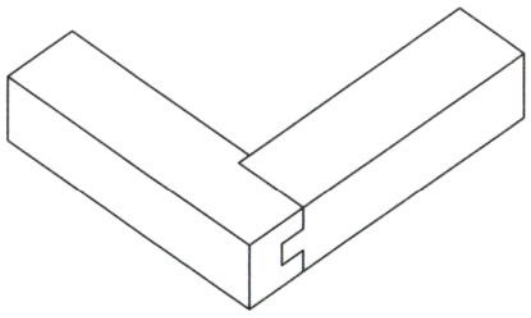
AB

Corner Joints

Housed dovetail tenon

ōari-otoshi-tsugi (oire)
大蟻落接ぎ(追入)

halbverdeckter eingelassener Schwalbenschwanzeckzapfen

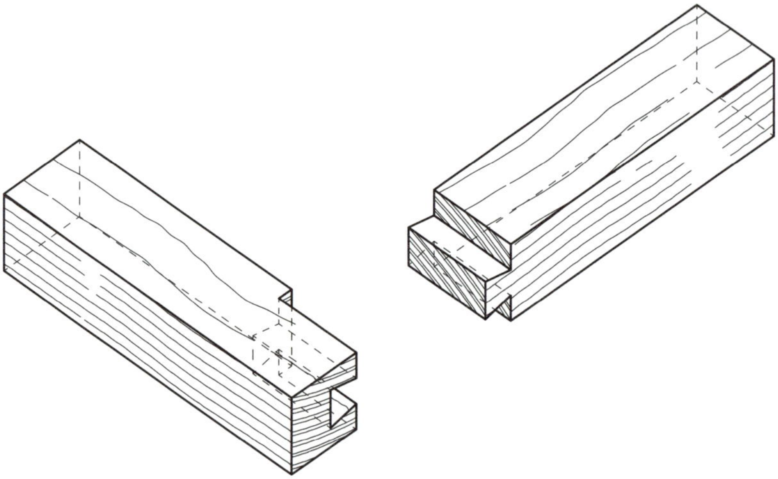

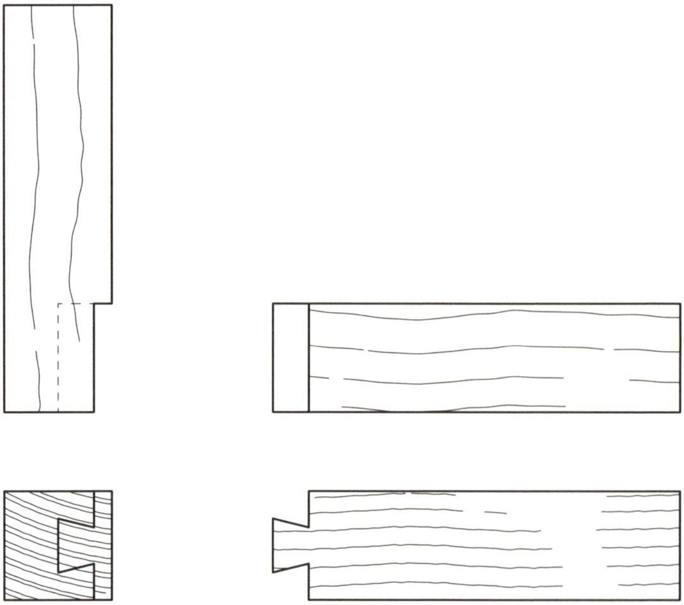

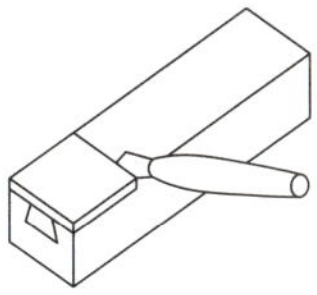

A1

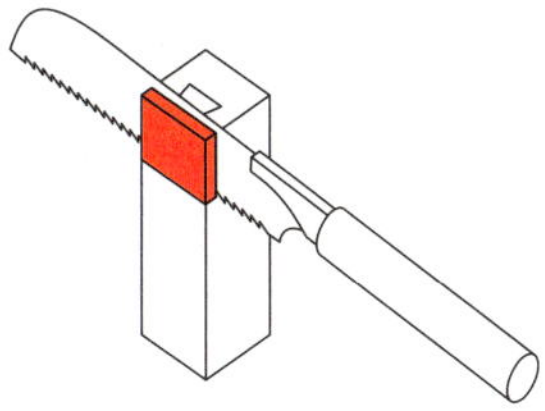

A2

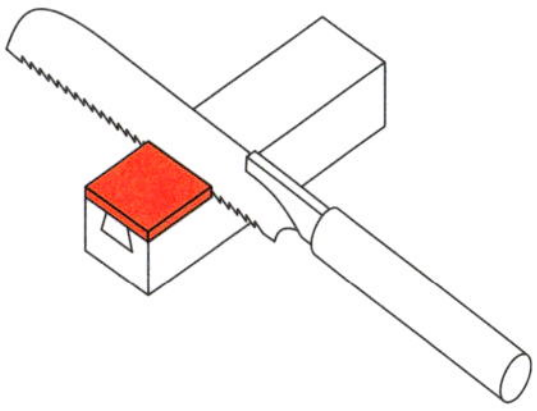

A3

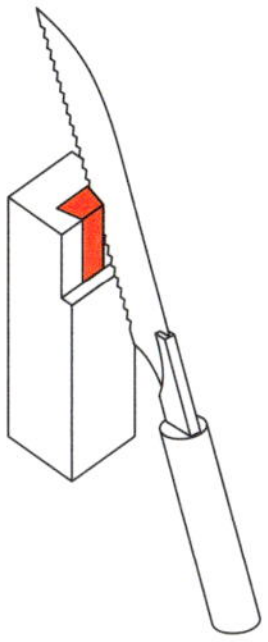

A4

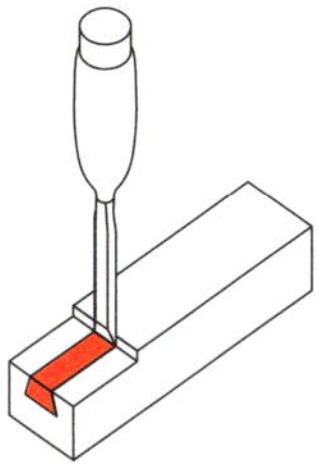

A5

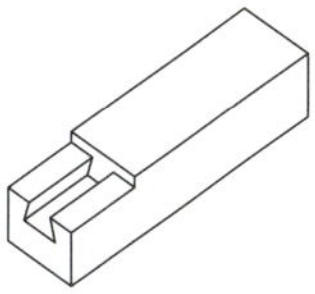

A

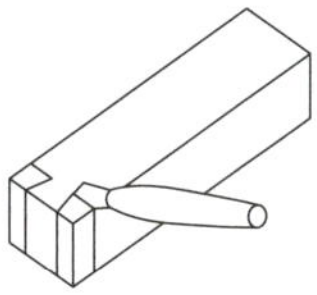

B1

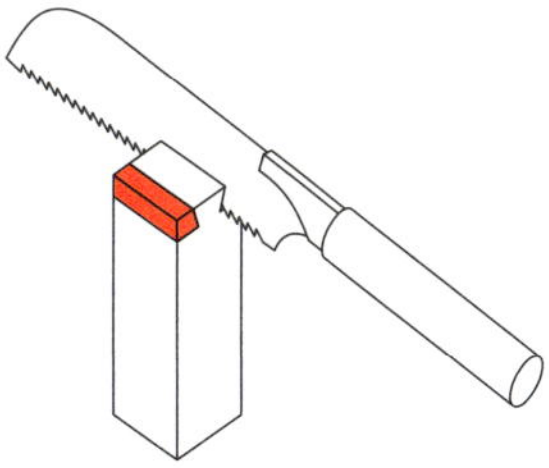

B2

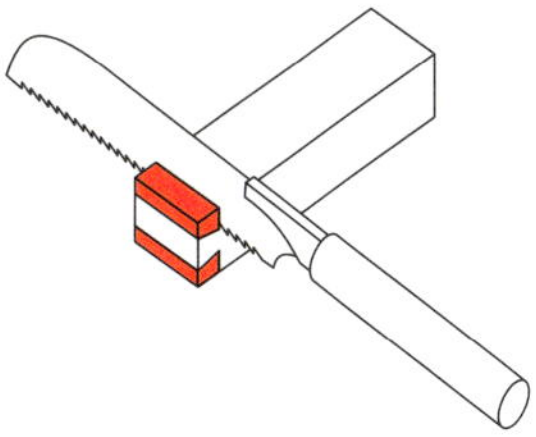

B3

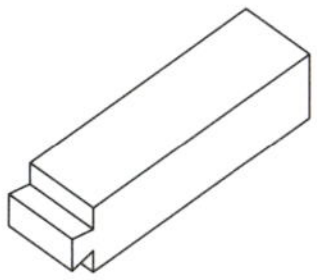

B

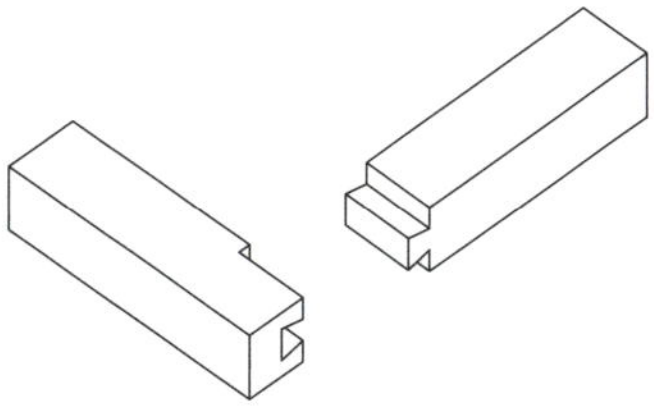

A+B

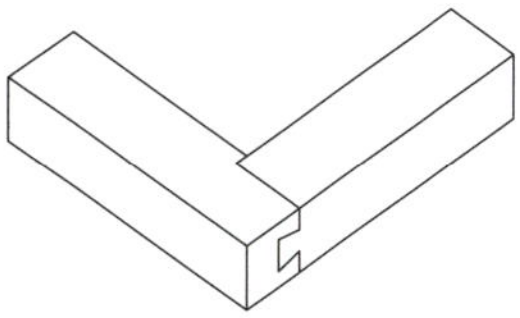

AB

Corner Joints

Blind housed dovetail tenon

ōari-otoshi-tsugi (eriwa-tsuki)
大蟻落接ぎ(襟輪付)

Schwalbenschwanzeckzapfen
mit gestufter Brüstung

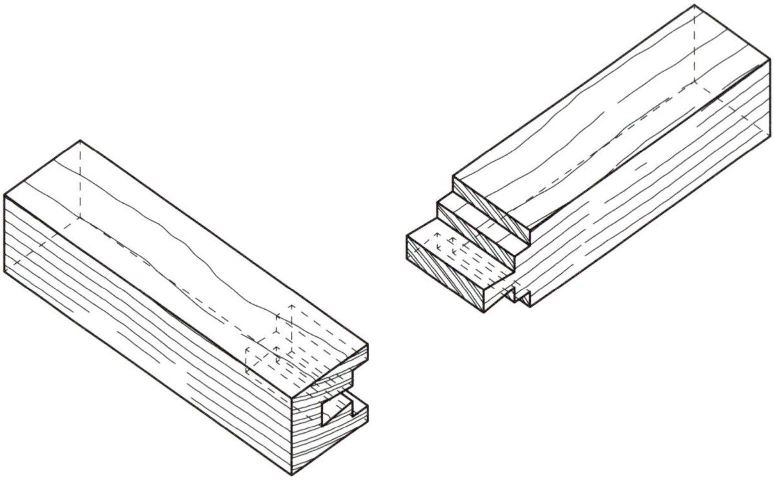

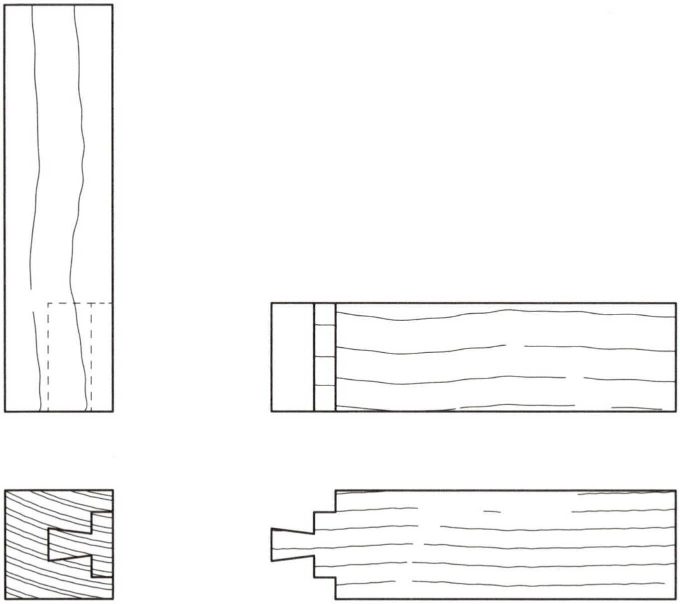

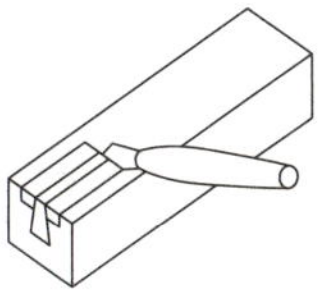

A1

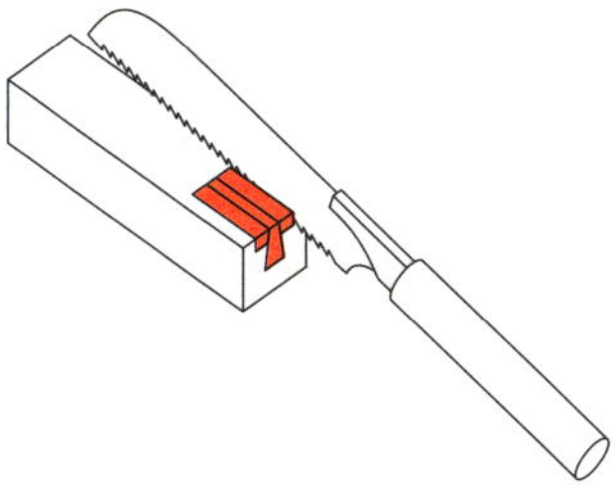

A2

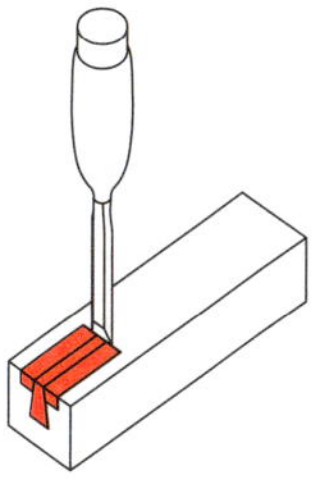

A3

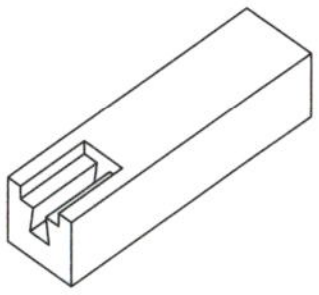

A

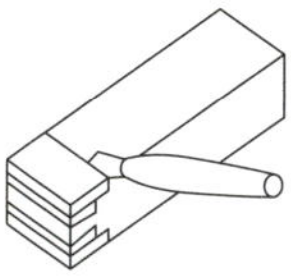

B1

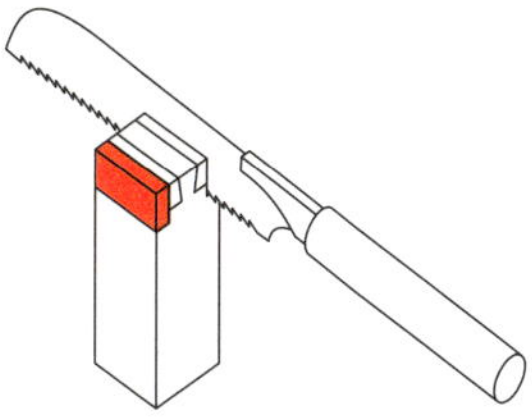

B2

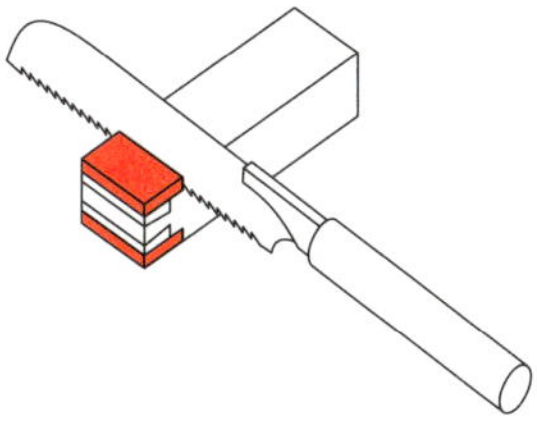

B3

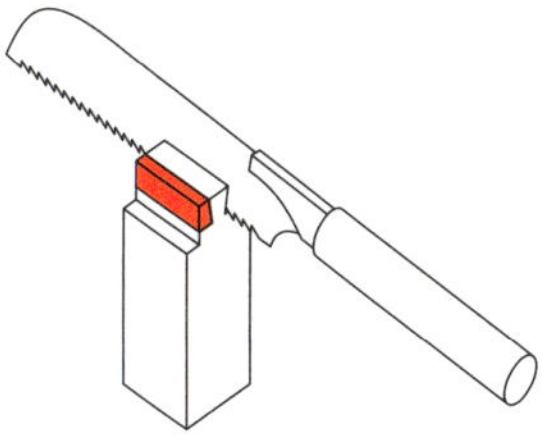

B4

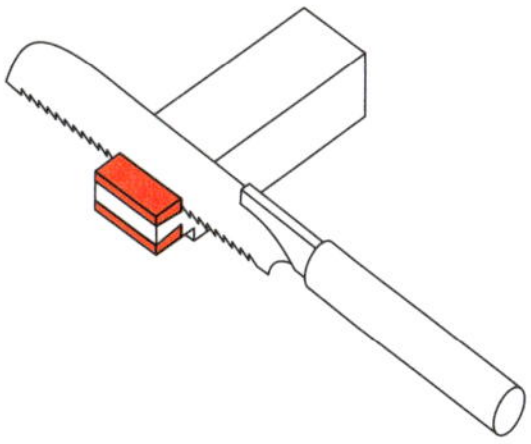

B5

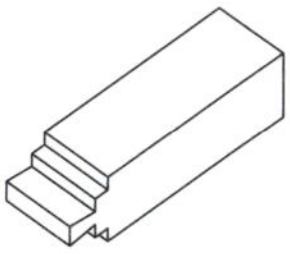

B

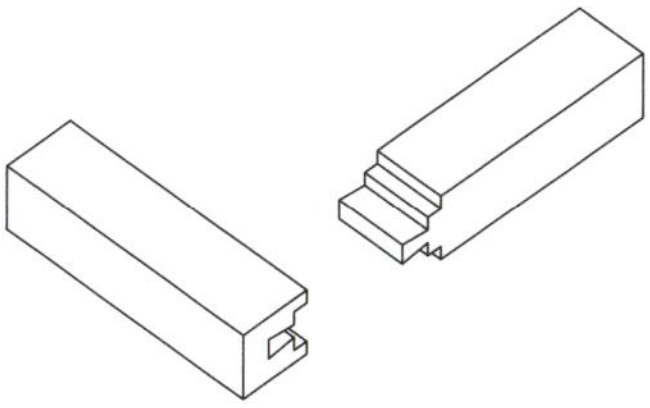

A+B

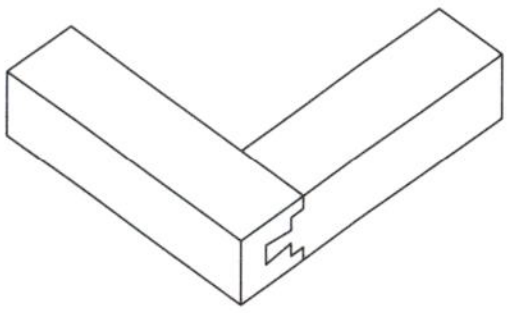

AB

Corner Joints

Half-blind gooseneck tenon

kama-hozo-otoshi-tsugi [*]
鎌枘落接ぎ

halbverdeckter Sicheleckzapfen

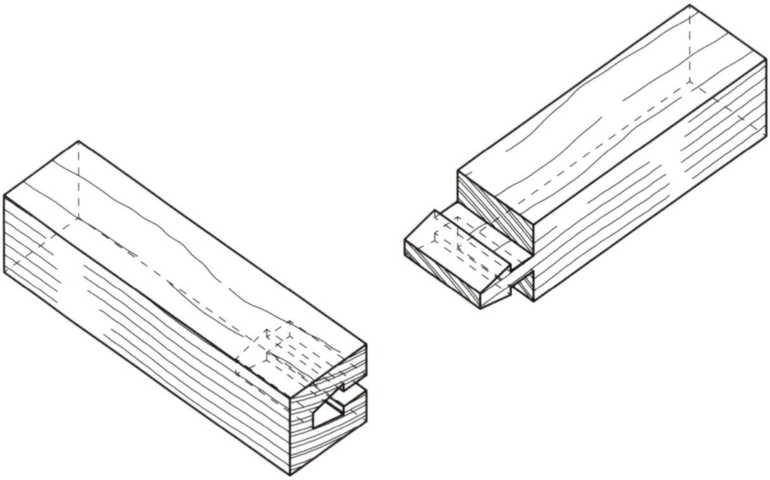

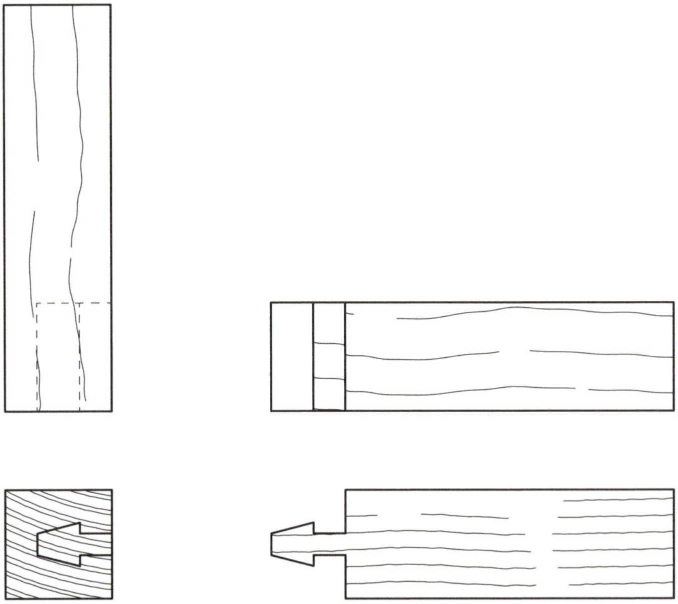

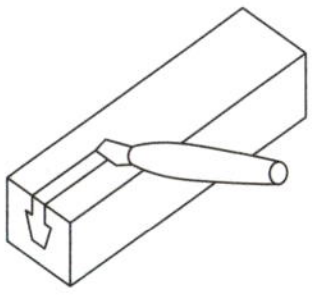

A1

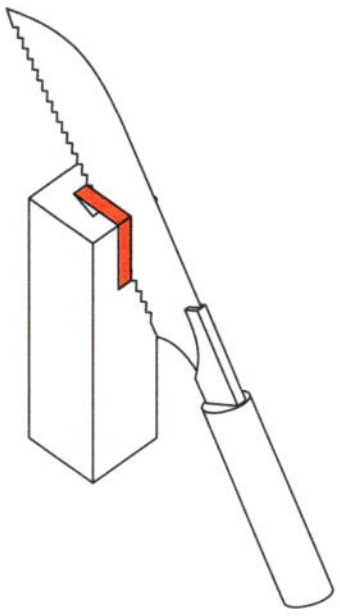

A2

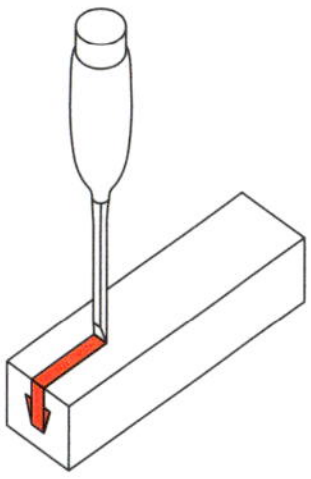

A3

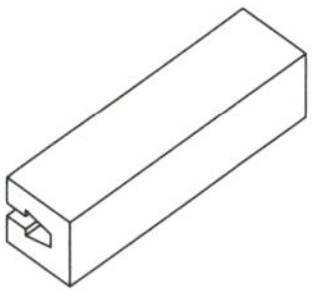

A

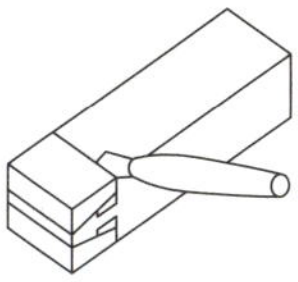

B1

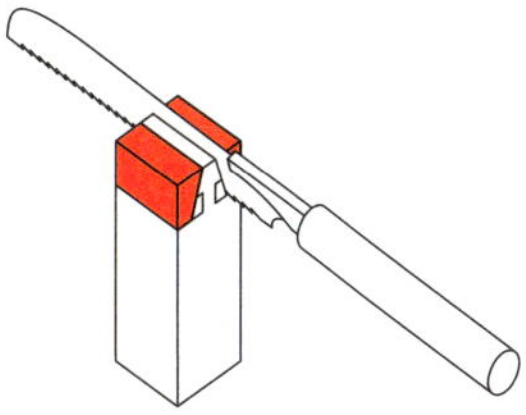

B2

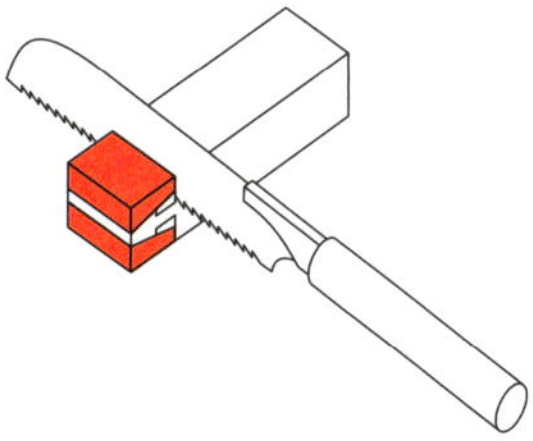

B3

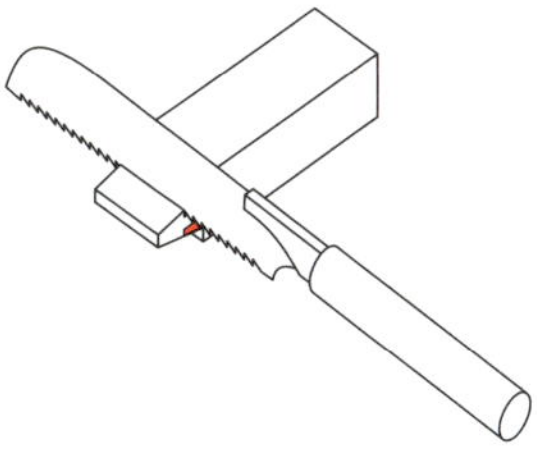

B4

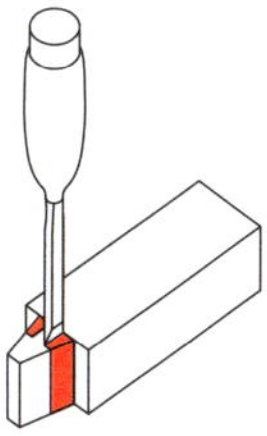

B5

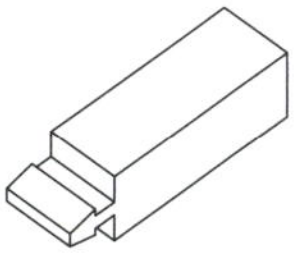

B

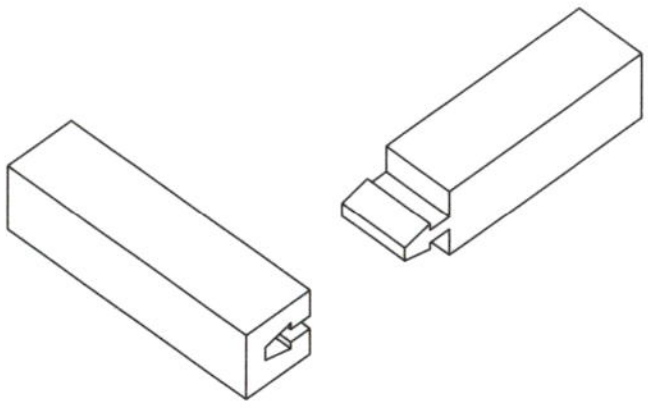

A+B

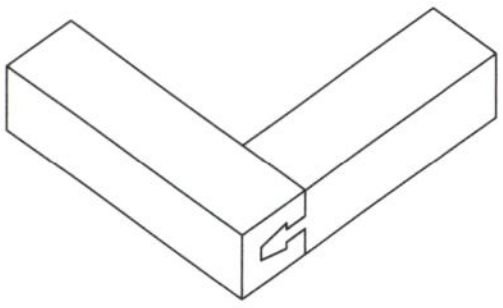

AB

Corner Joints

Half-blind dovetail bridle joint

ōari-otoshi-tsugi (tsutsumi-komi) [*]
大蟻落接ぎ(包込)

halbverdeckter
Schwalbenschwanzeckzapfen

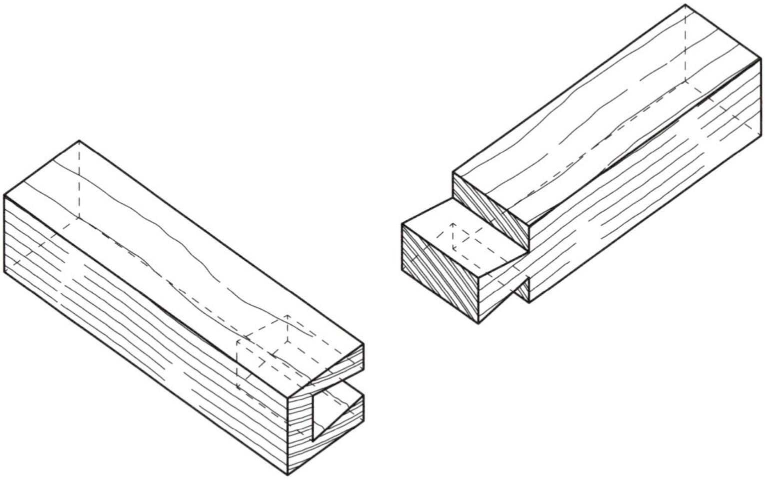

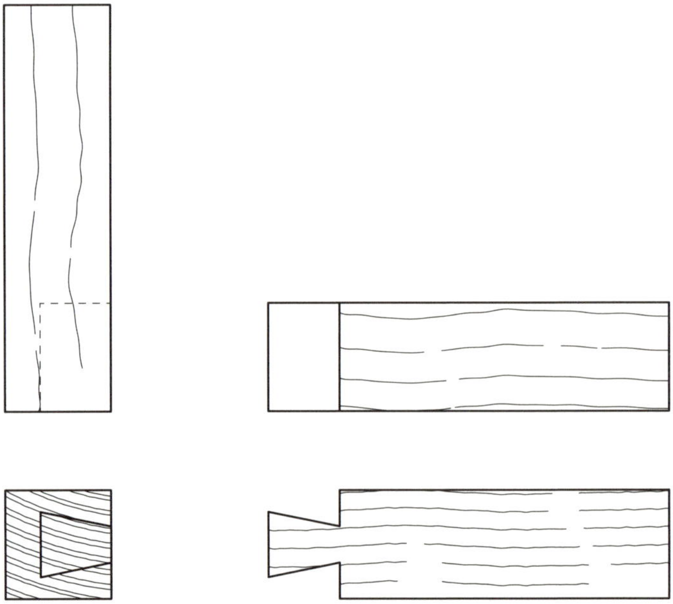

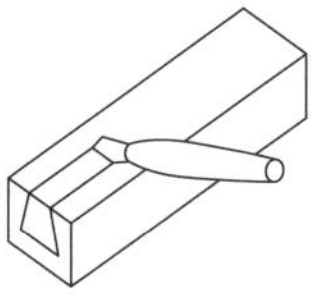

A1

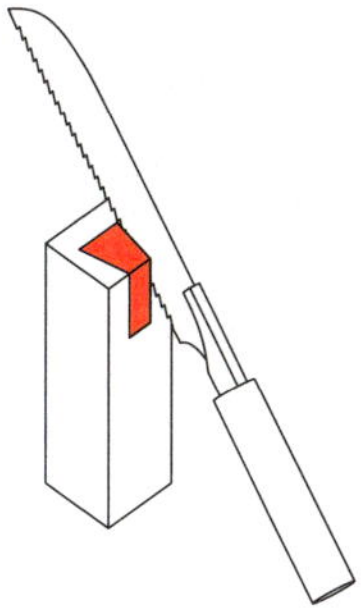

A2

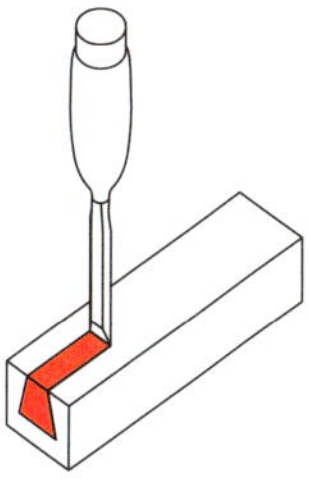

A3

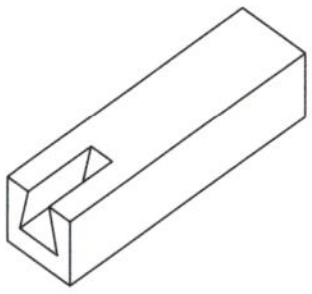

A

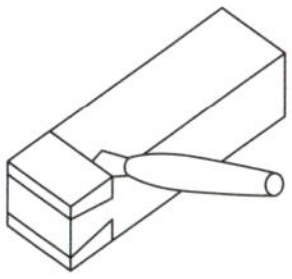

B1

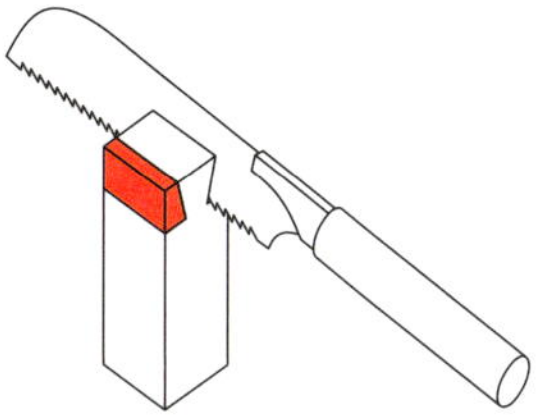

B2

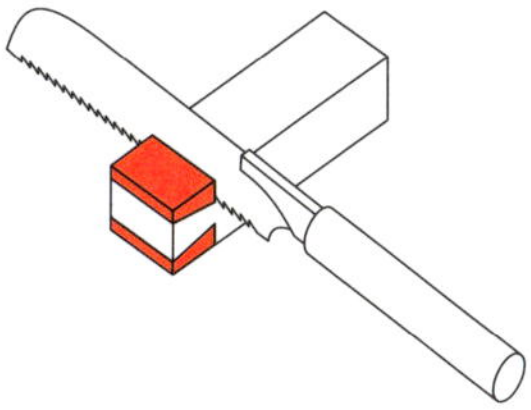

B3

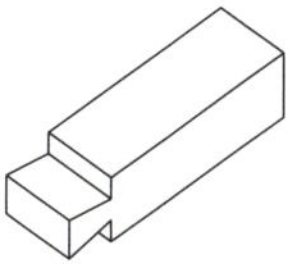

B

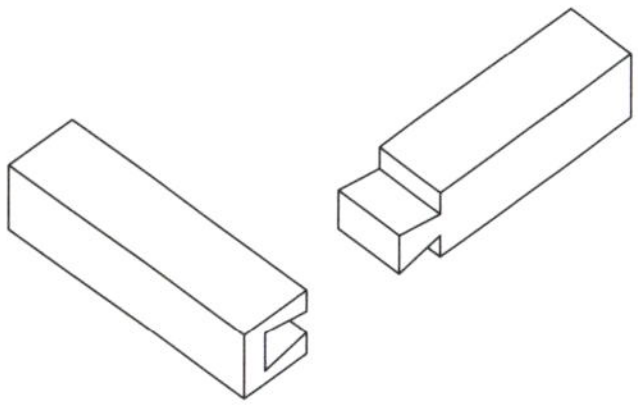

A+B

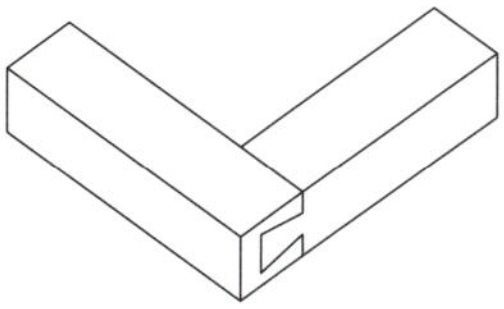

AB

Corner Joints

Through dovetail bridle joint

ōari-otoshi-tsugi
大蟻落接ぎ

durchgehender
Schwalbenschwanzeckzapfen

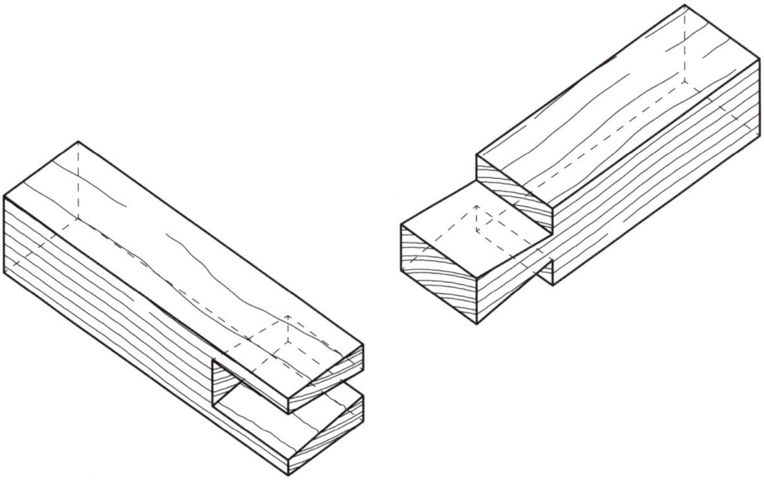

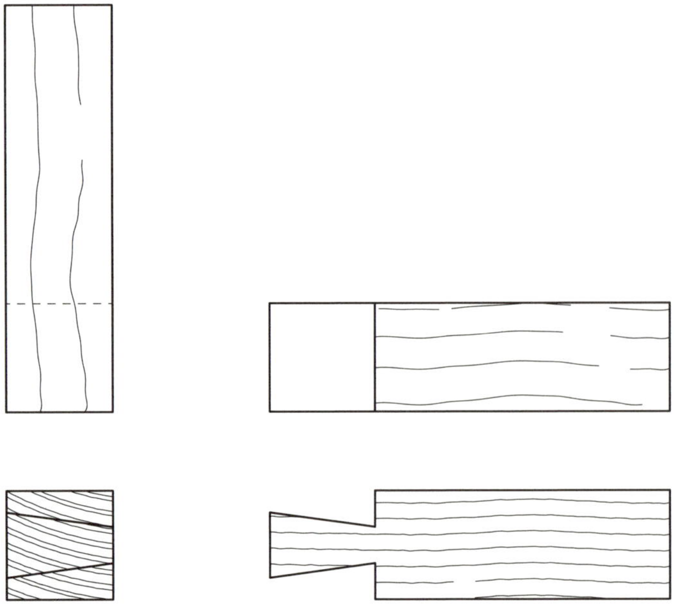

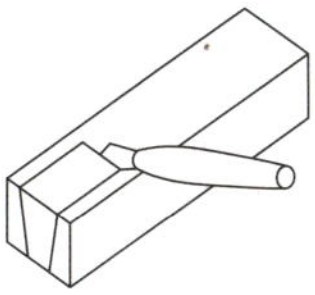

A1

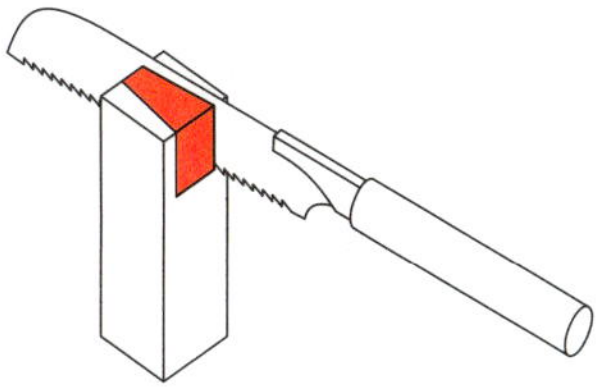

A2

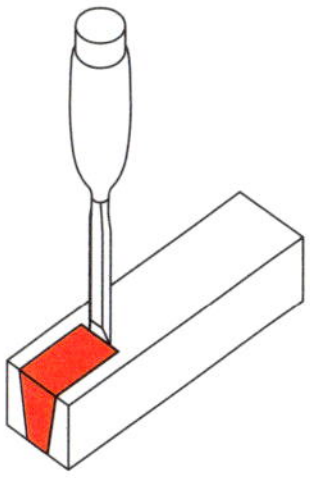

A3

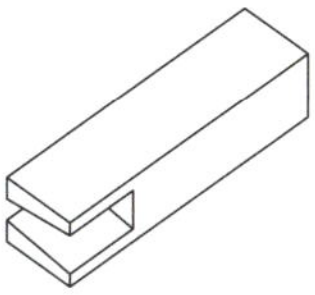

A

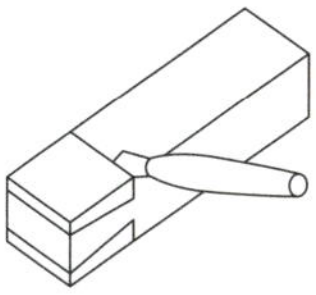

B1

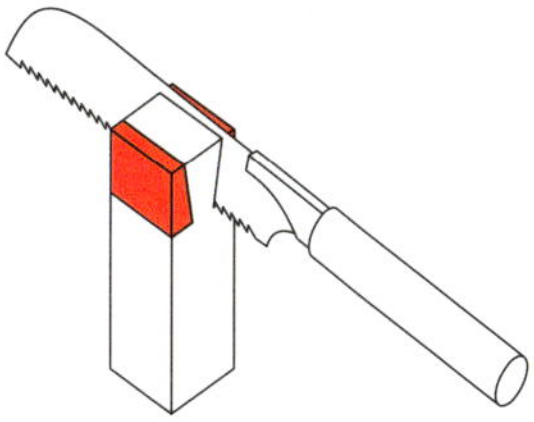

B2

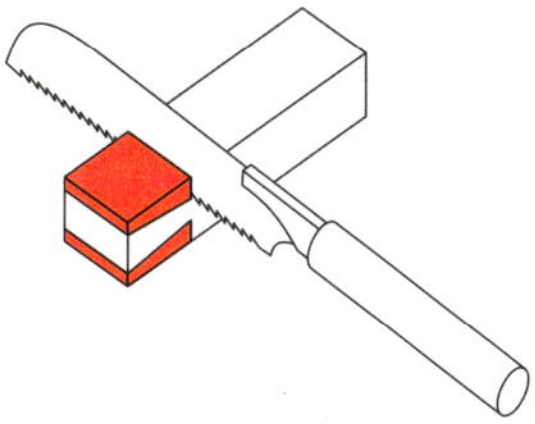

B3

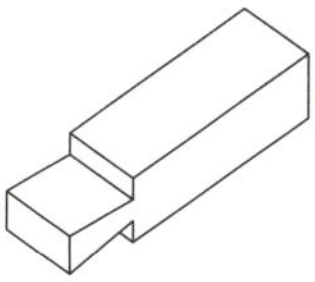

B

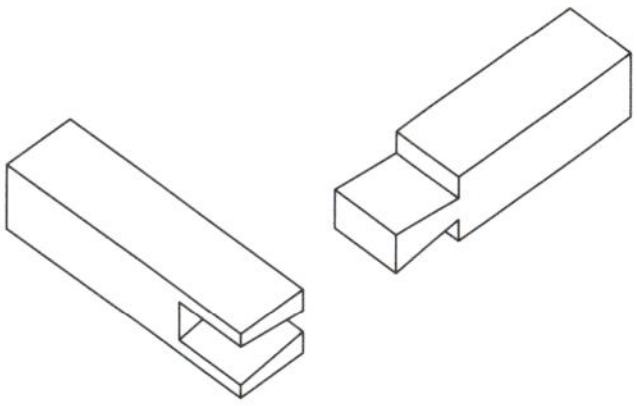

A+B

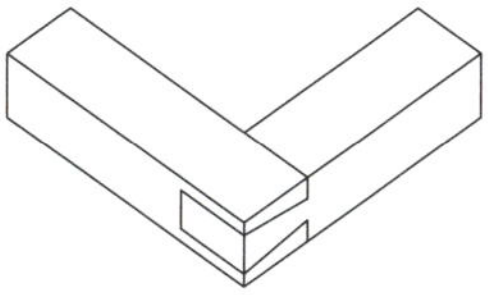

AB

Corner Joints

Half-blind triple shouldered dovetail tenon

ari-otoshi-tsugi
蟻落接ぎ

halbverdeckter abgesetzter
Schwalbenschwanzeckzapfen

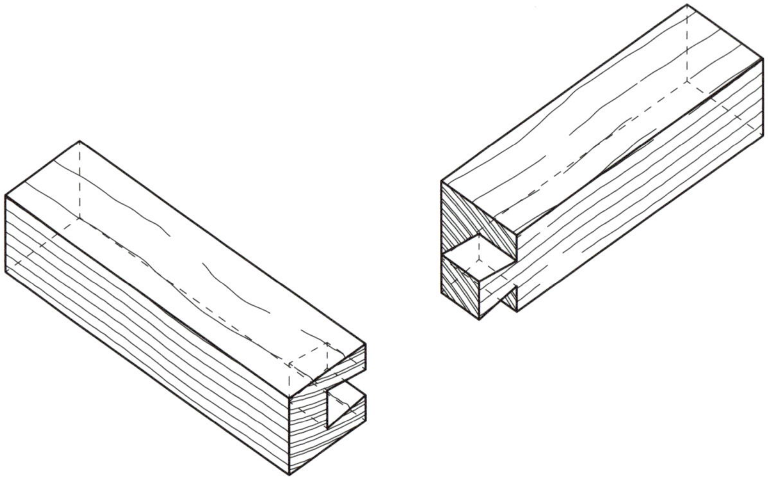

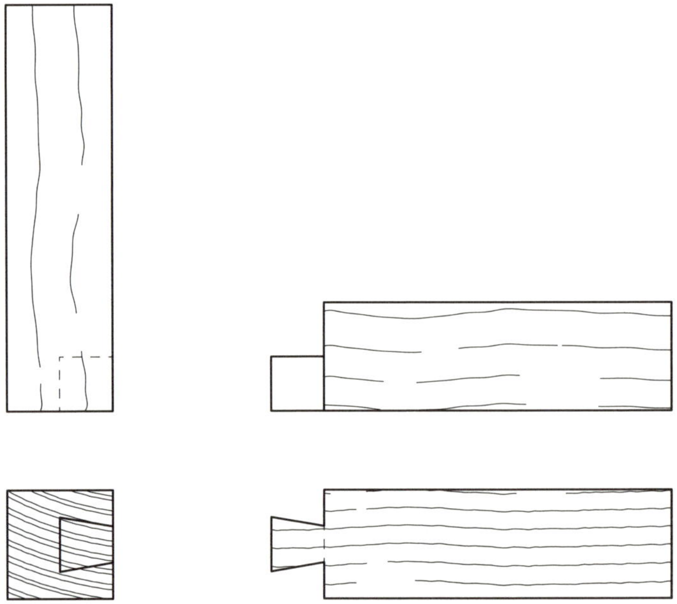

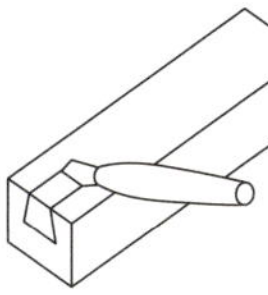
A1

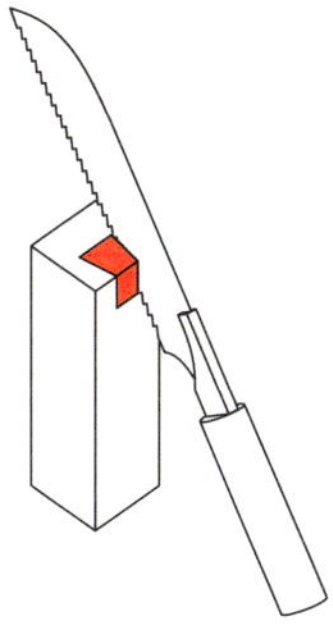
A2

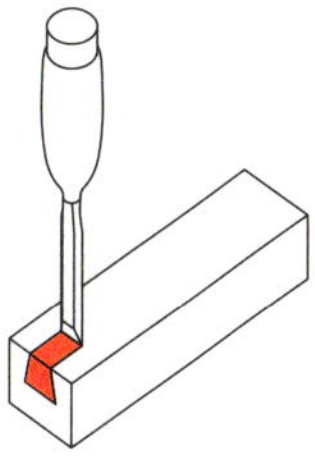
A3

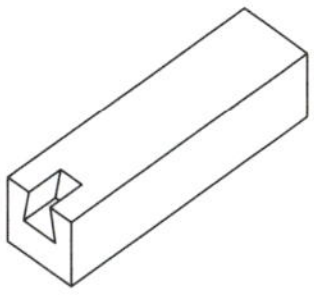
A

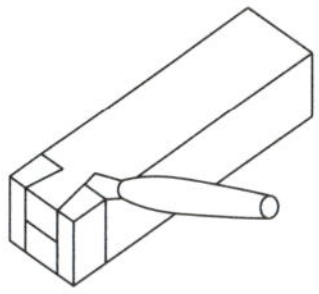
B1

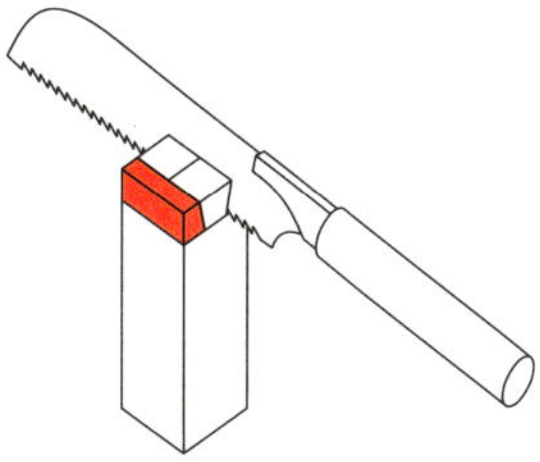
B2

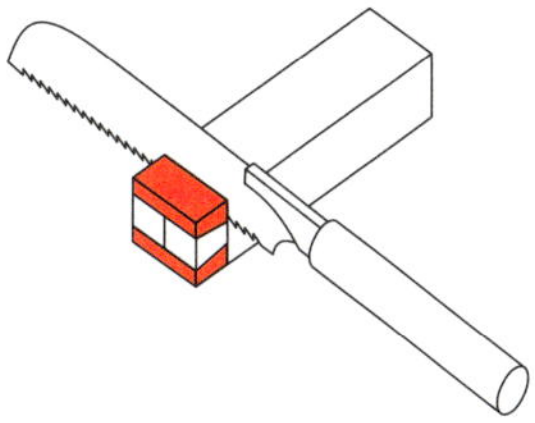
B3

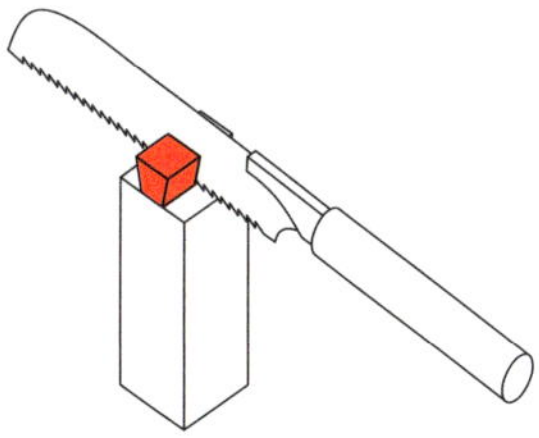
B4

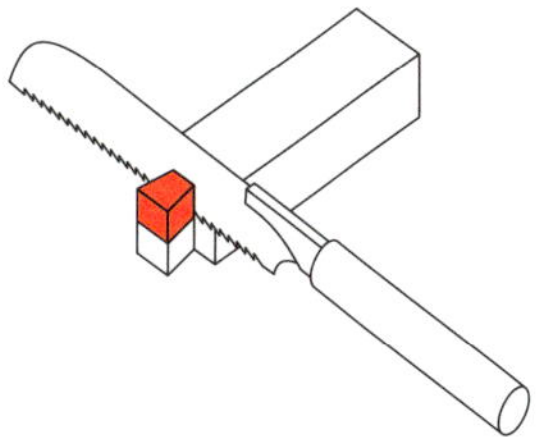
B5

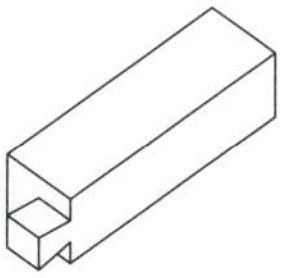
B

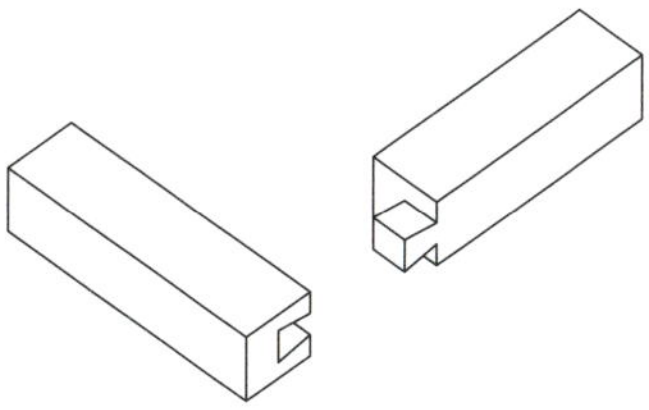
A+B

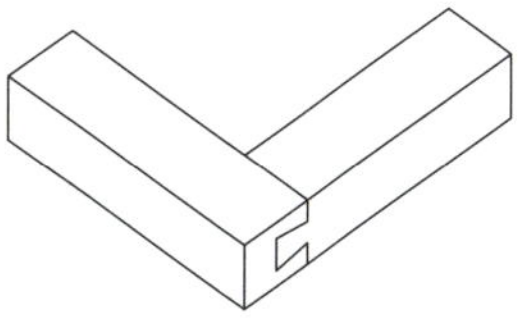
AB

Corner Joints

Triple-shouldered stub tenon

kiba-hozo-sashi-tsugi
牙枘差接ぎ

verdeckter abgesetzter Eckzapfen

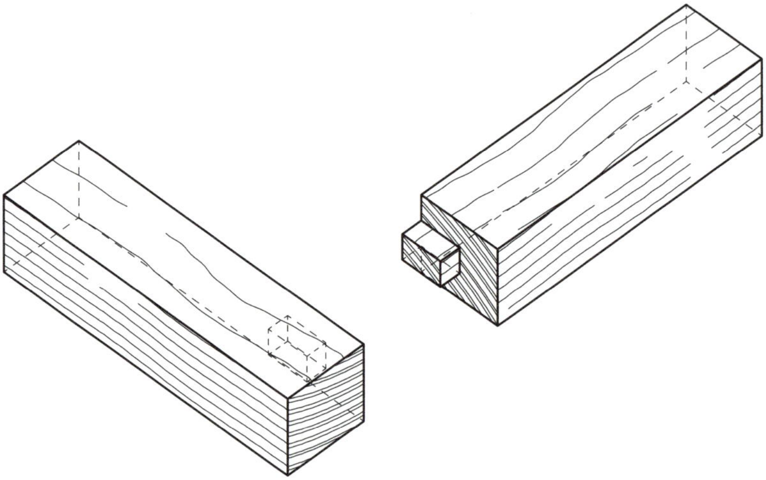

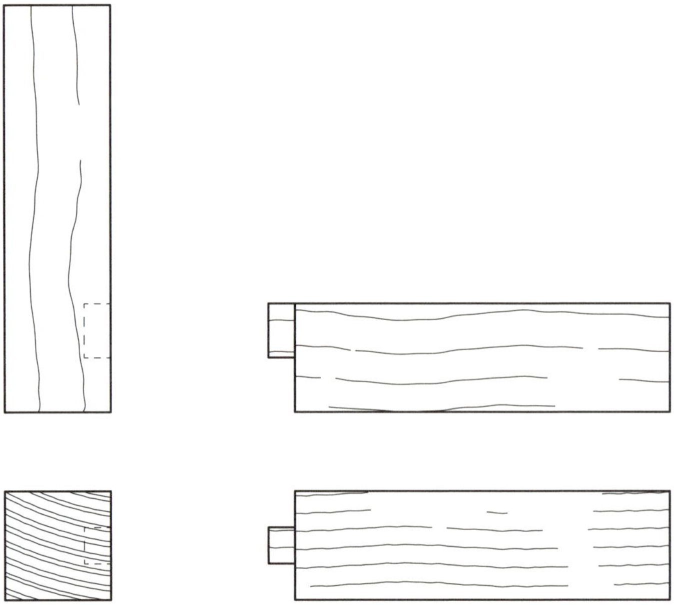

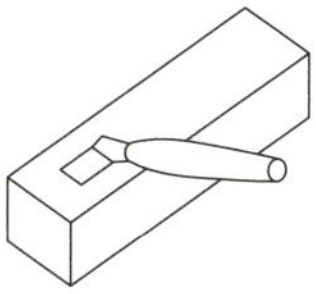
A1

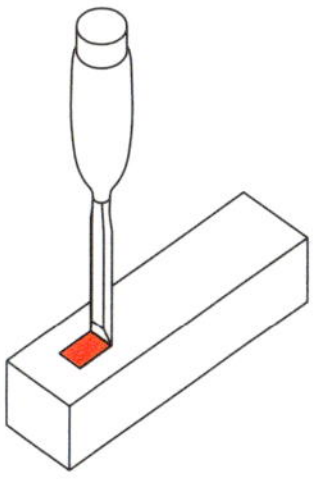
A2

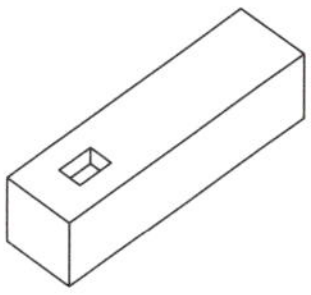
A

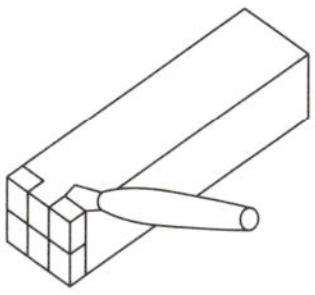
B1

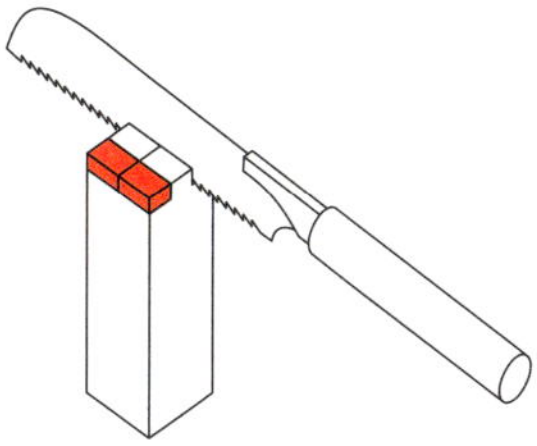
B2

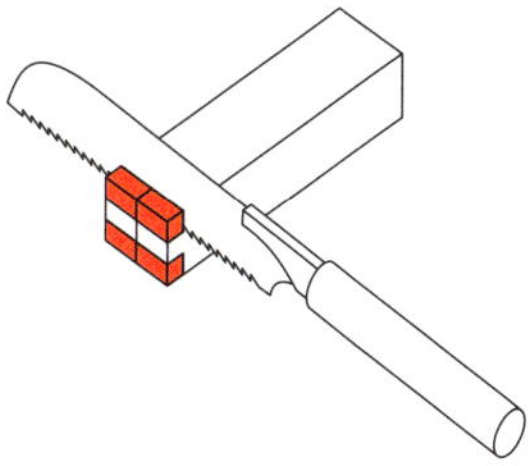
B3

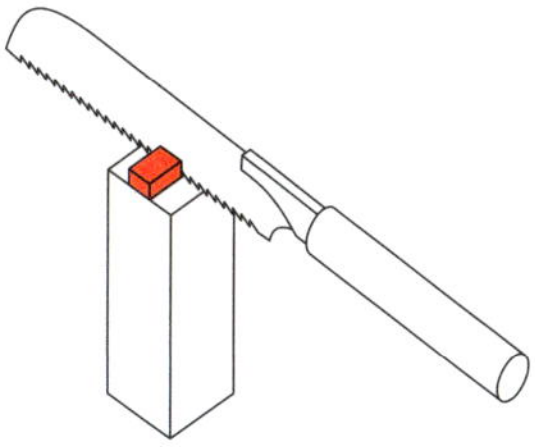
B4

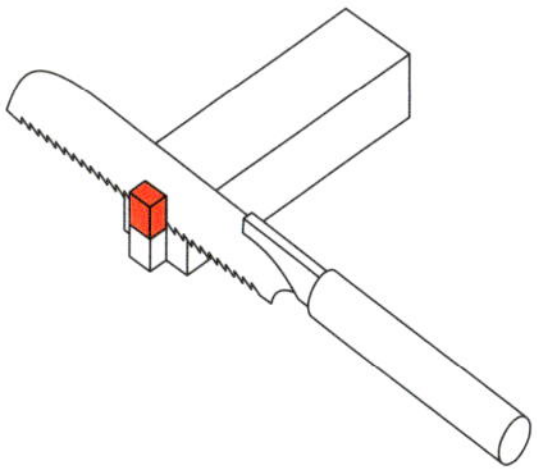
B5

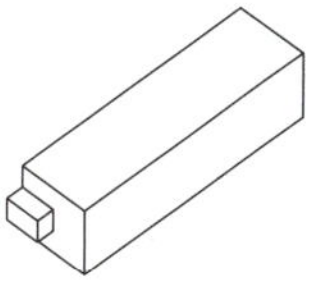
B

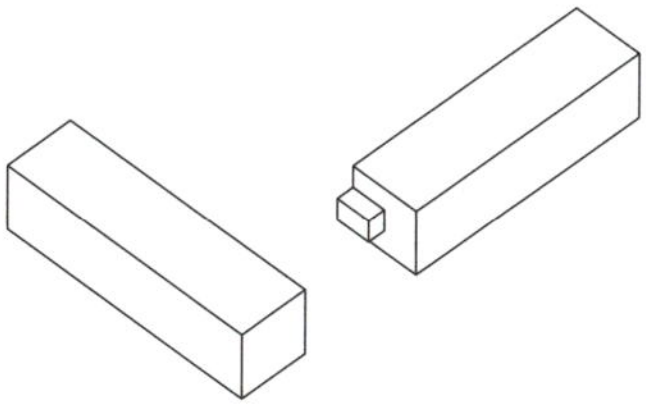
A+B

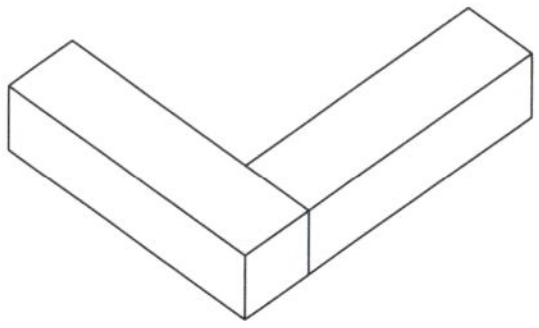
AB

Corner Joints

Housed dovetail joint

yose-ari-tsugi
寄蟻接ぎ

verdeckter beidseitig abgesetzter Schwalbenschwanzeckzapfen

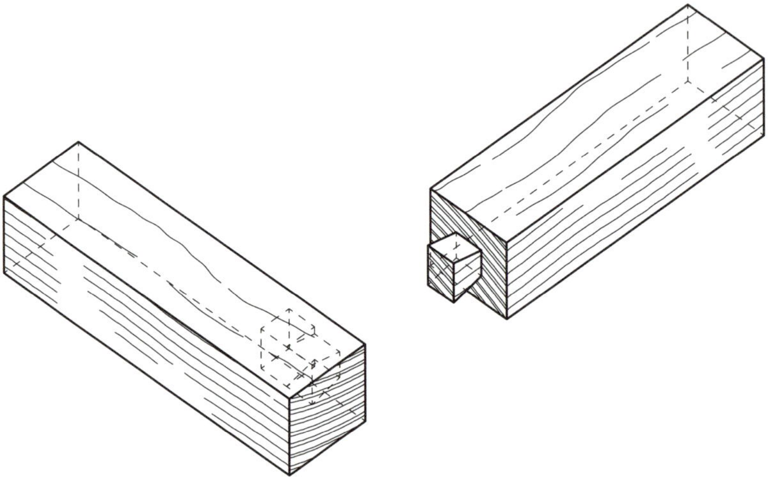

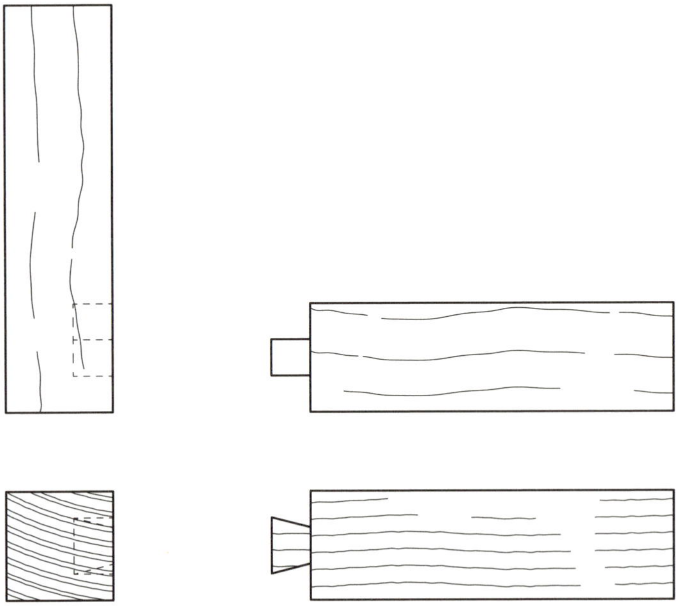

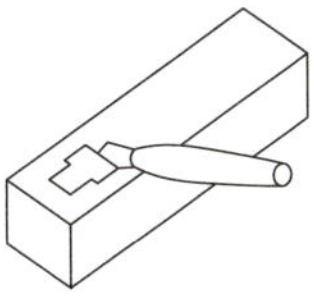

A1

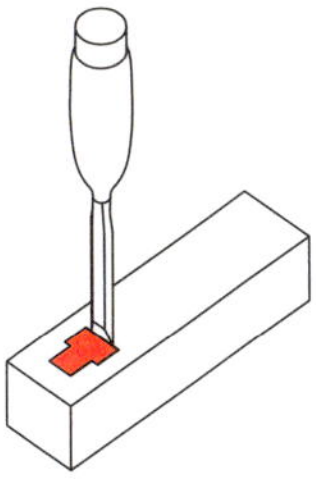

A2

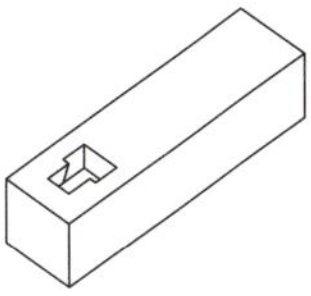

A

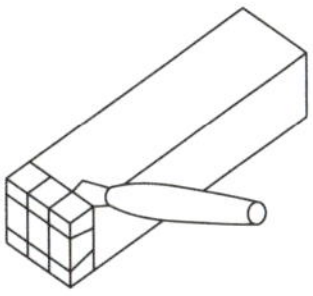

B1

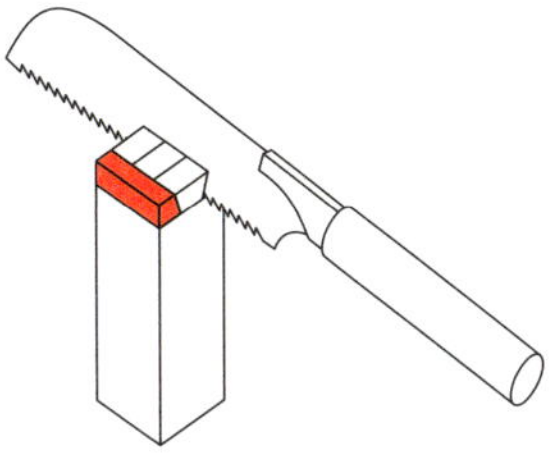

B2

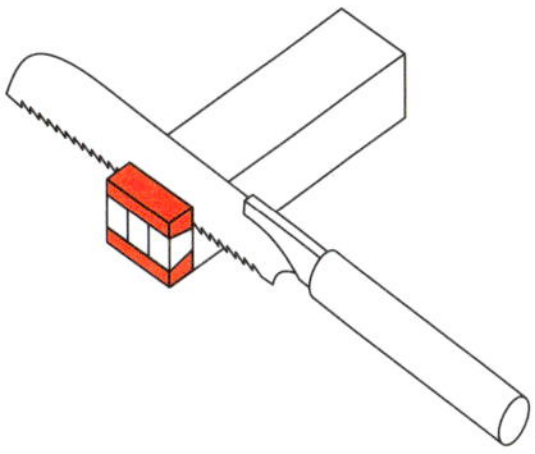

B3

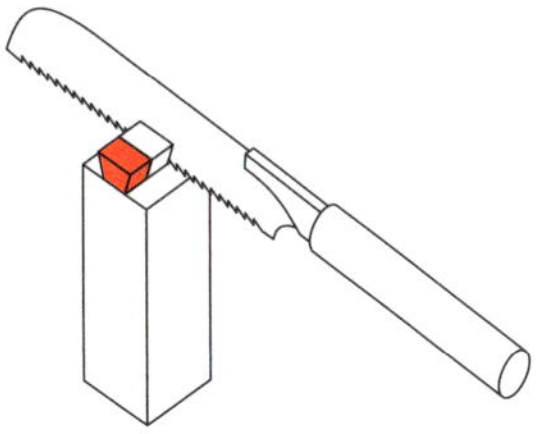

B4

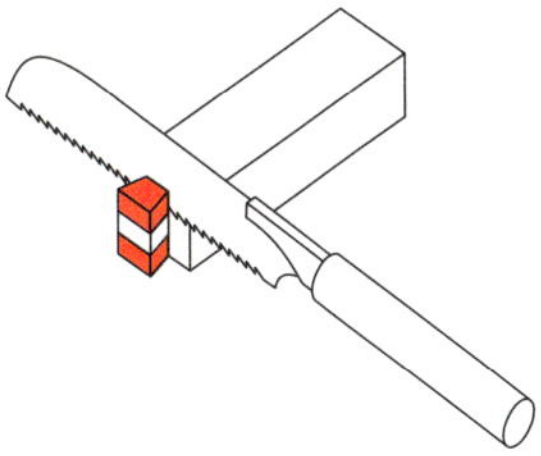

B5

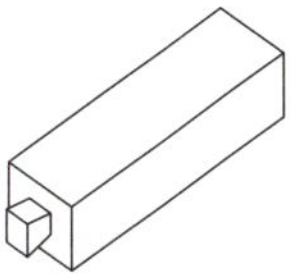

B

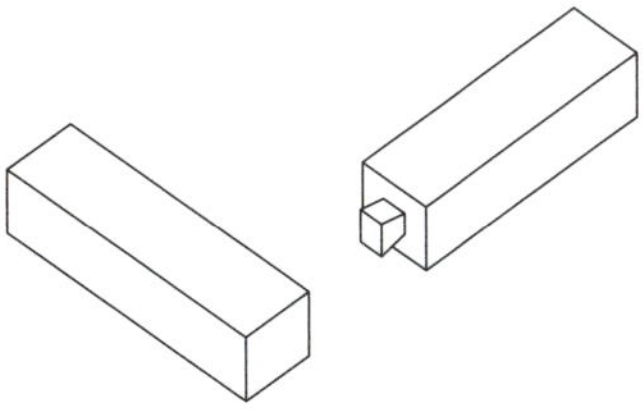

A+B

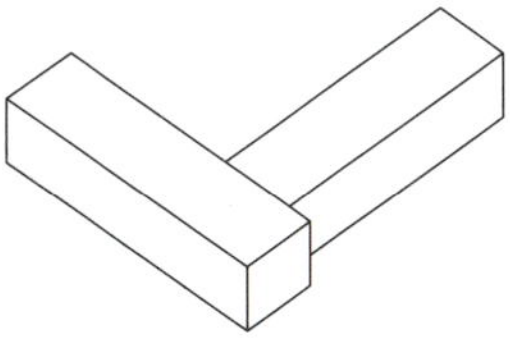

A+B

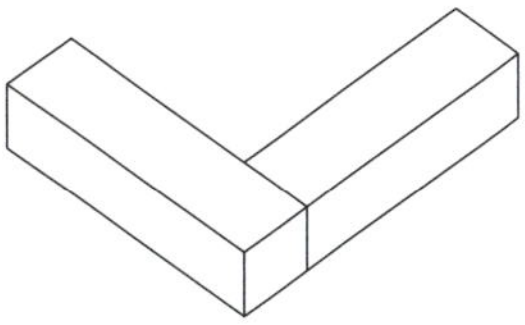

AB

Corner Joints

Blind fan-shaped double mortise and tenon

tsutsumi-ari-otoshi-nimai-hozo-tsugi (binta-dome)
包蟻落二枚枘接ぎ(鬢太留)

verdeckter doppelter trapezförmiger Eckzapfen mit Wange auf Gehrung

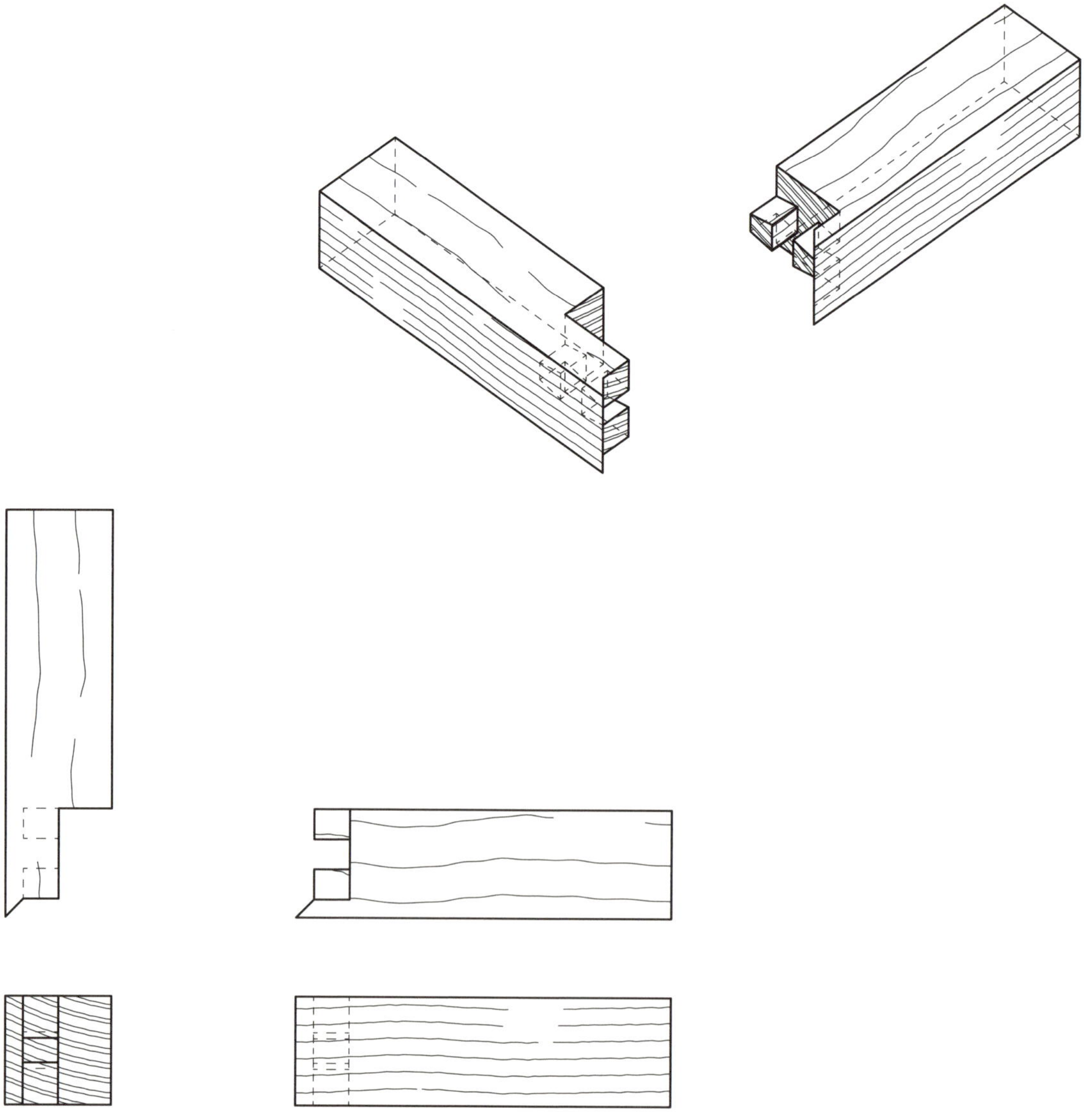

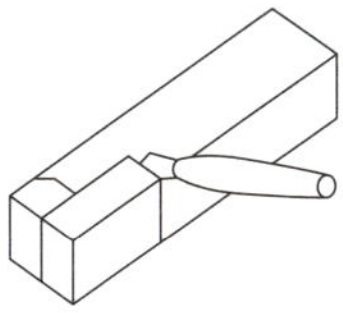

A1

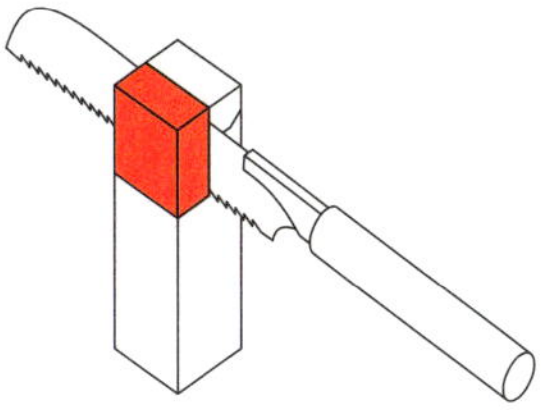

A2

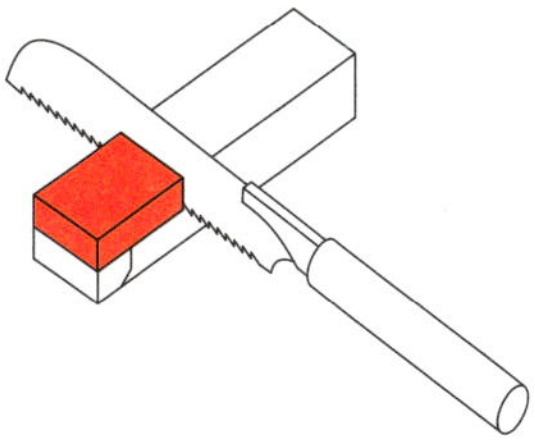

A3

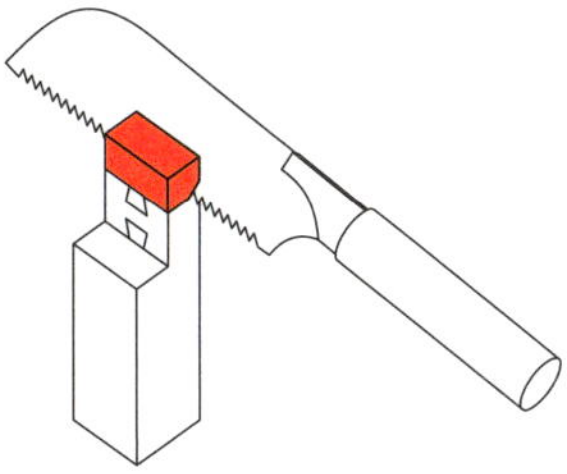

A4

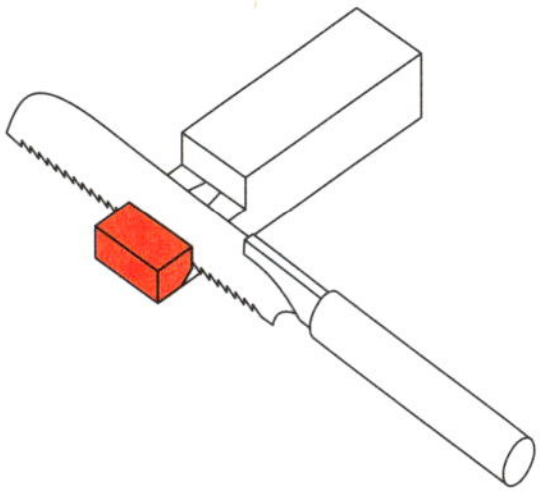

A5

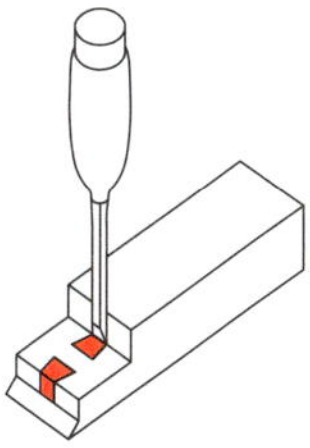

A6

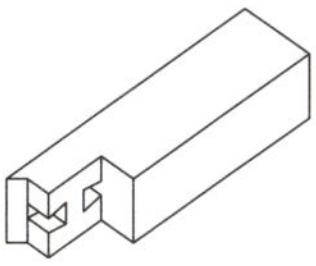

A

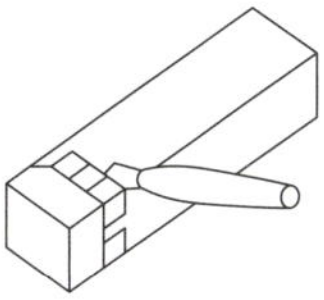

B1

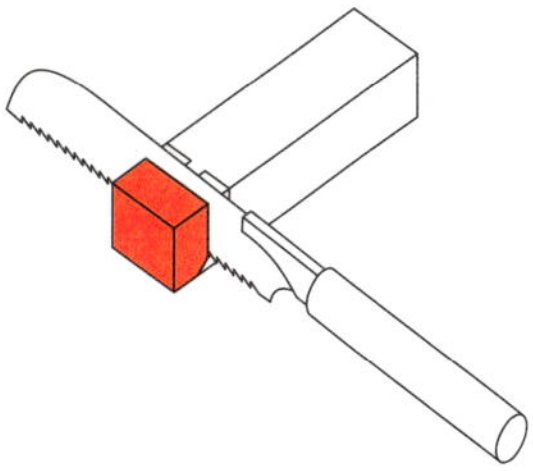

B2

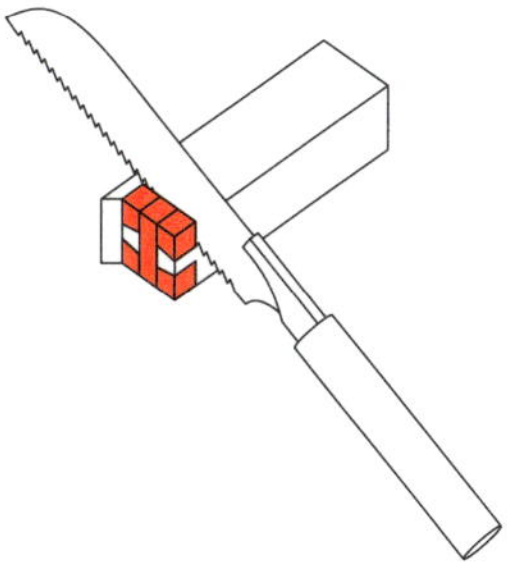

B3

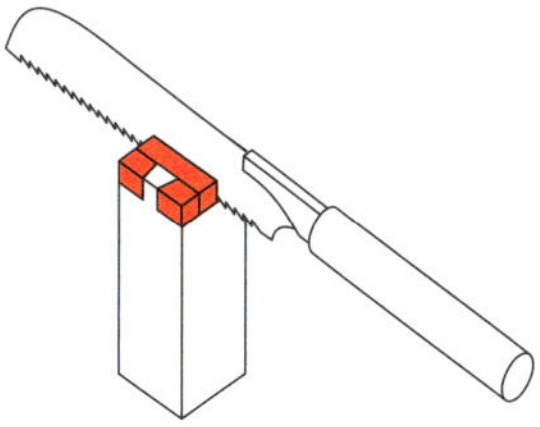

B4

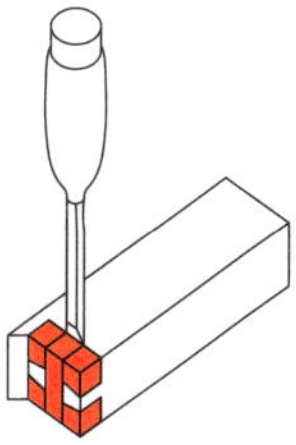

B5

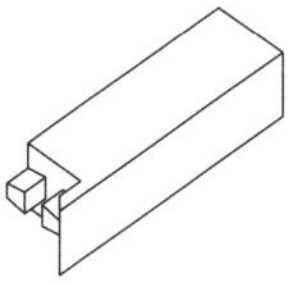

B

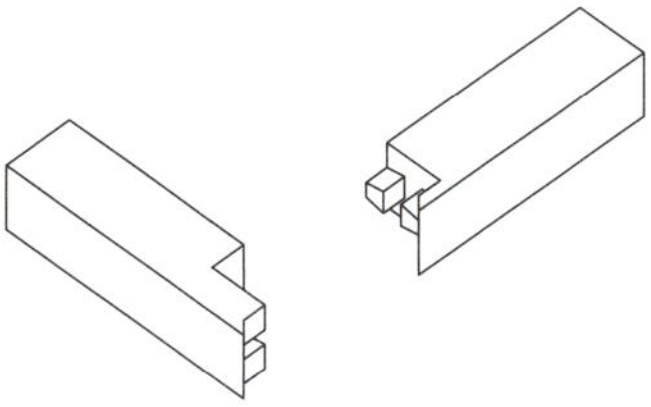

A+B

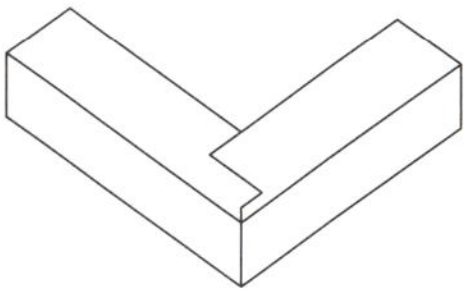

AB

Kawai joint

kawai-tsugite
河合継手

diagonal angeschnittener Eckstoß mit dreifacher polyedrischer Ausstemmung

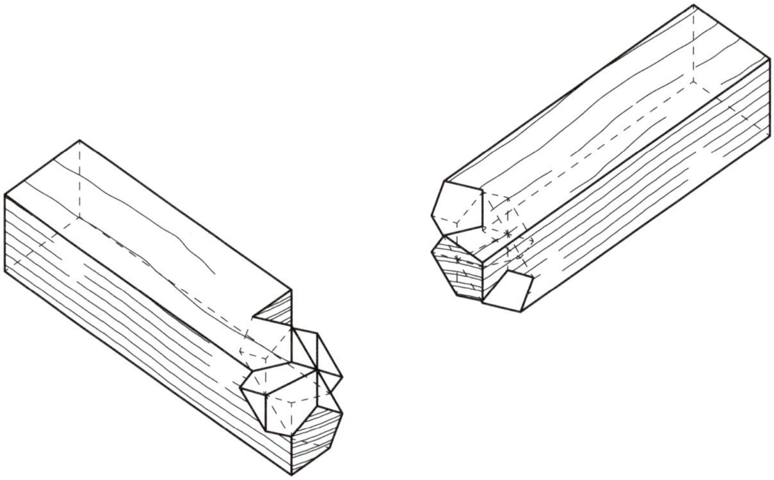

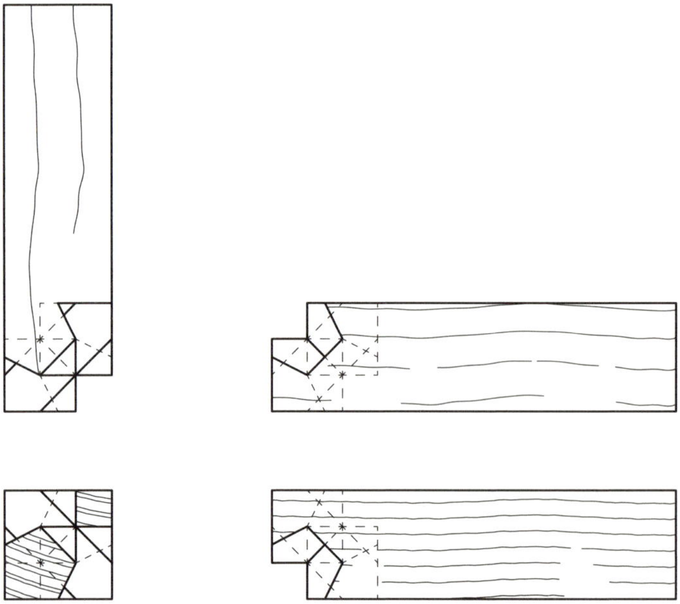

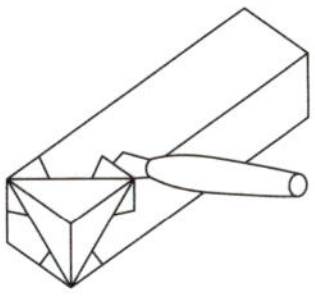

A1

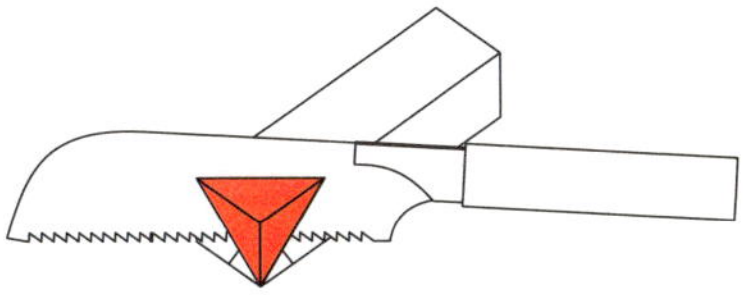

A2

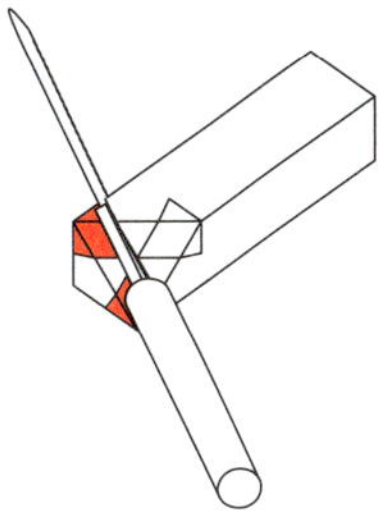

A3

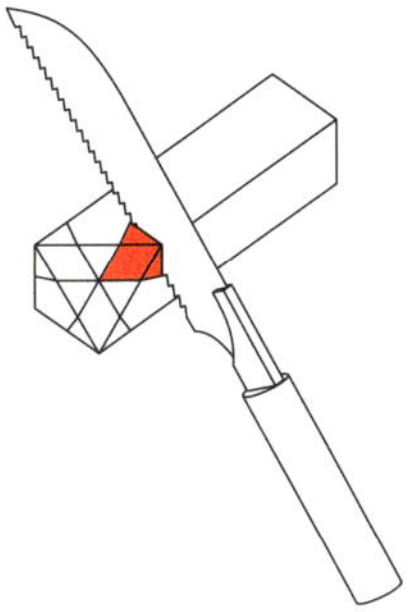

A4

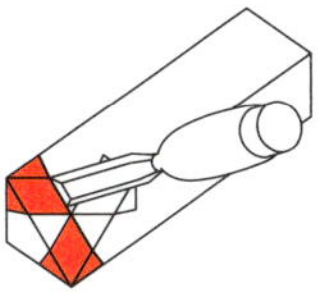

A5

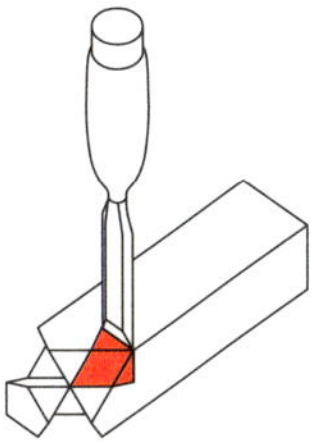

A6

2×

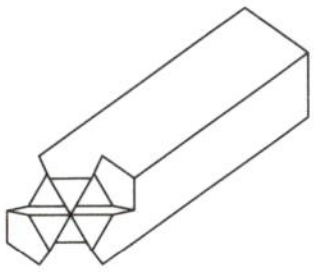

A

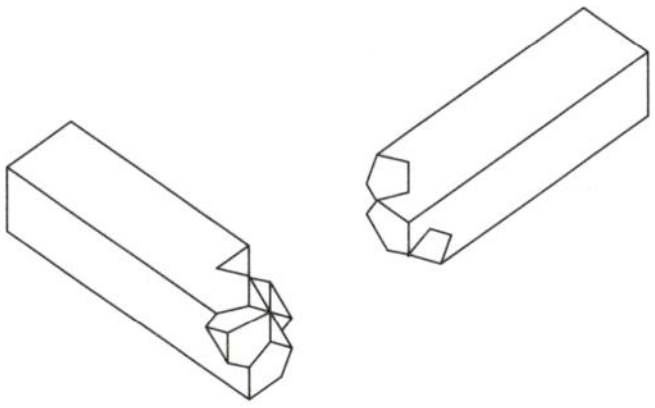

A+A

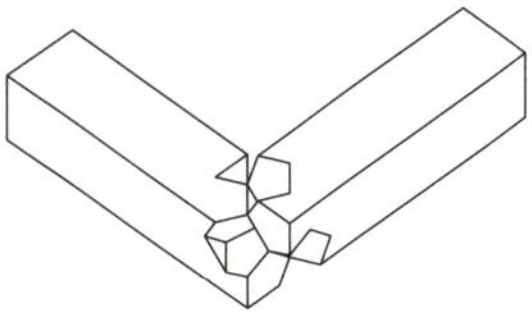

A+A

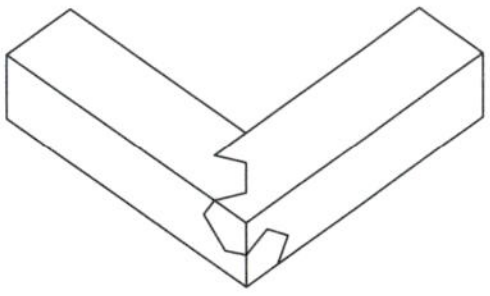

AA

Corner Joints

Notched lap joint with stub tenon

ai-kaki-suberi-ago-tsugi
相欠辷り顎接ぎ

Meisterwitz

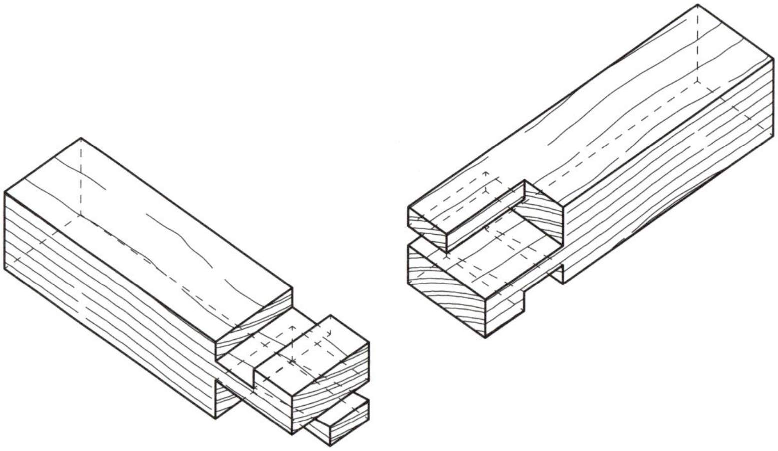

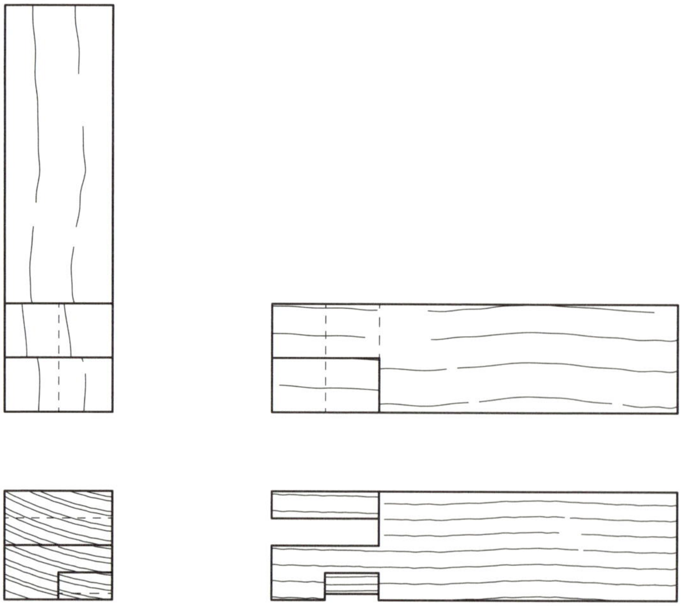

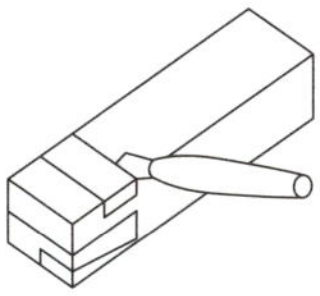

A1

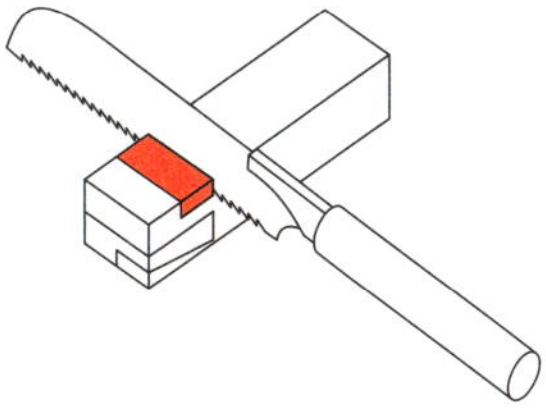

A2

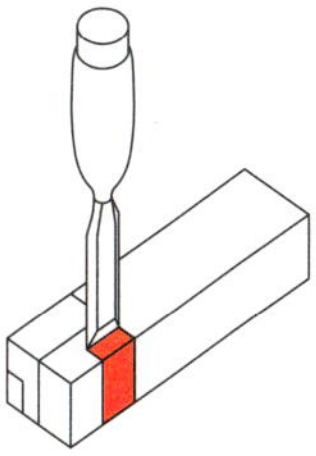

A3

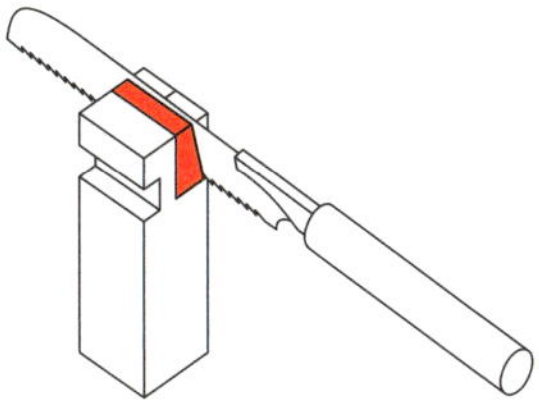

A4

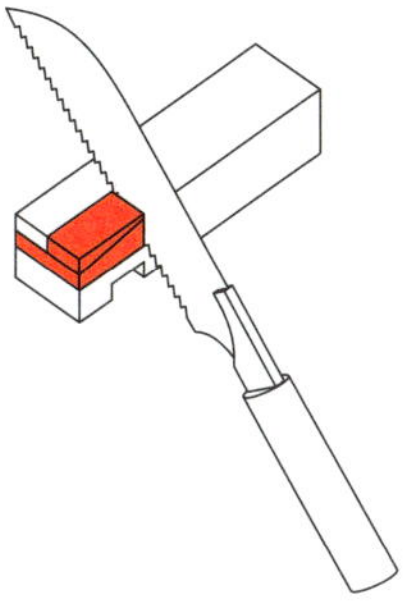

A5

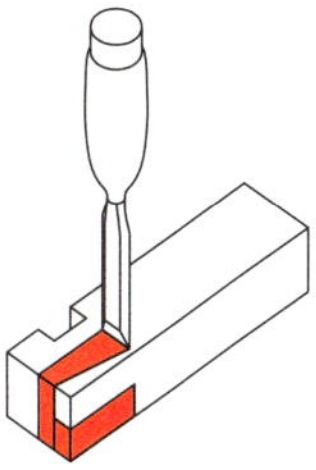

A6

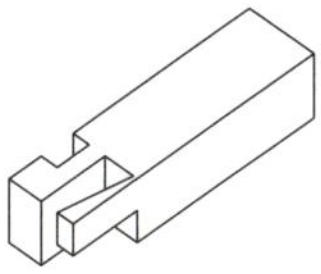

A

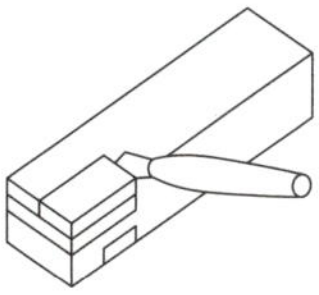

B1

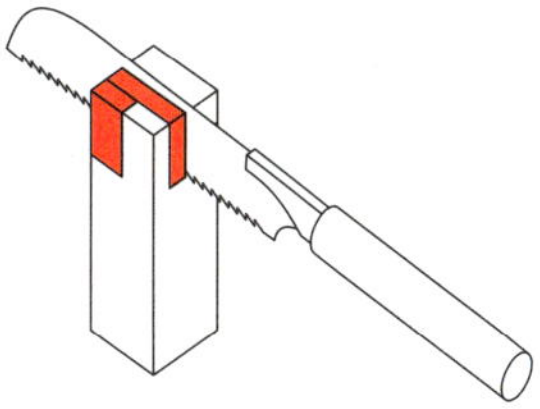

B2

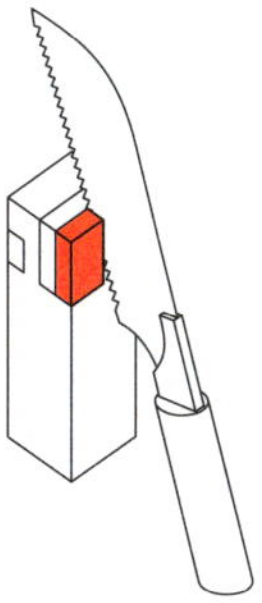

B3

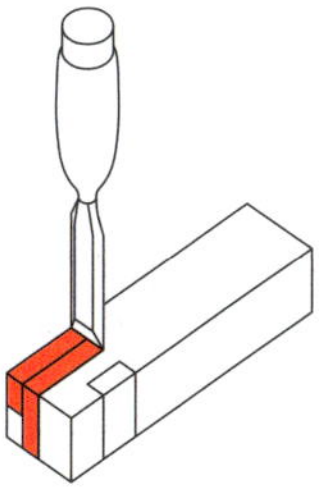

B4

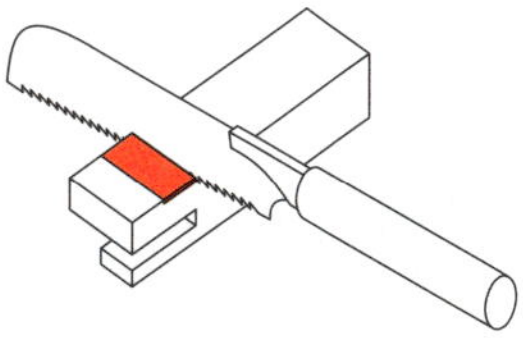

B5

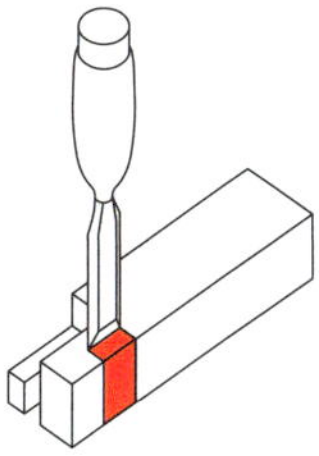

B6

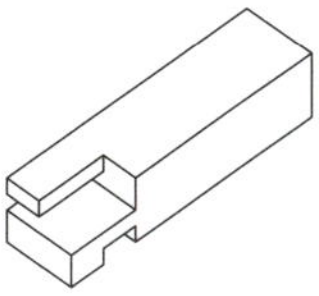

B

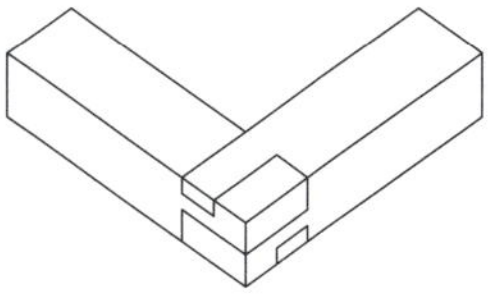

AB

Corner Joints

Blind pegged joint

dabo-tsugi
太枘接ぎ

Verdollung

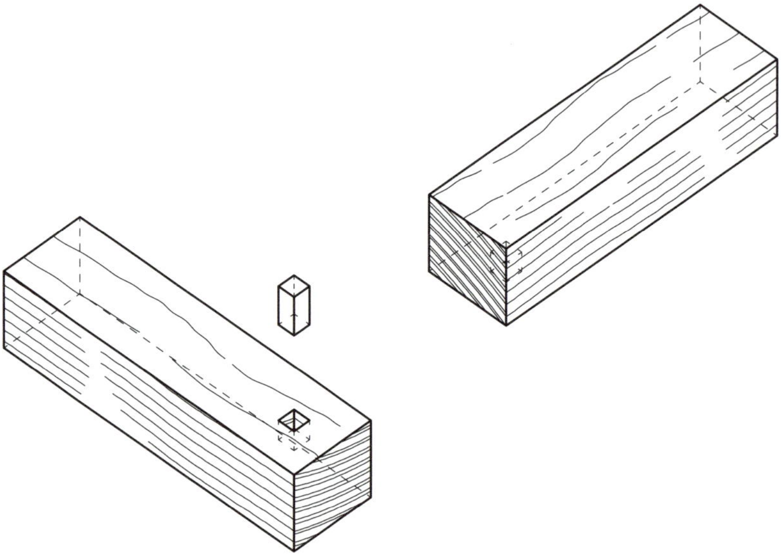

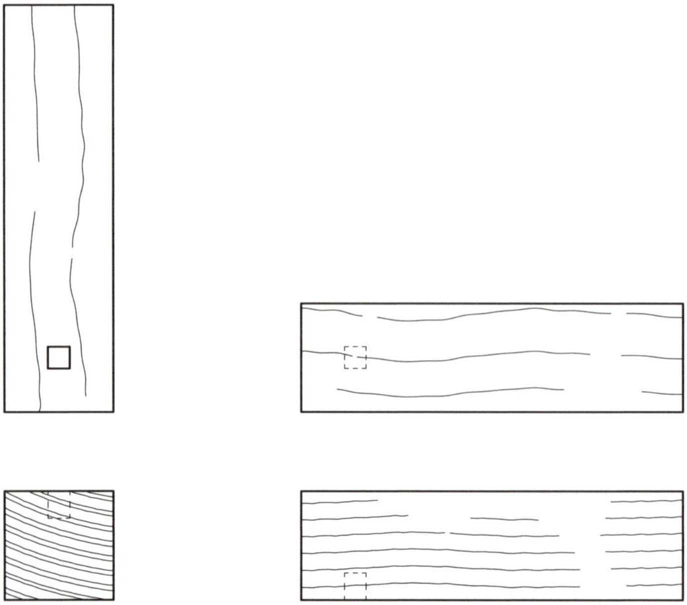

2×

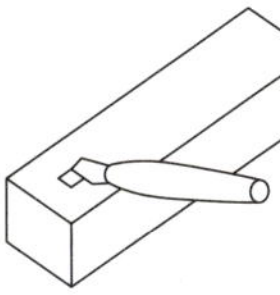

A1

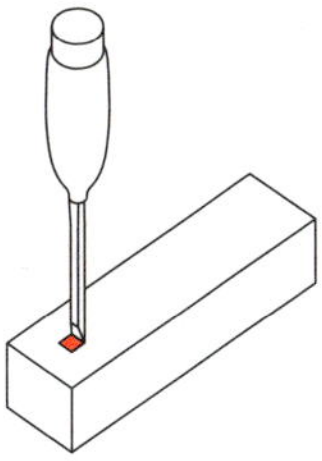

A2

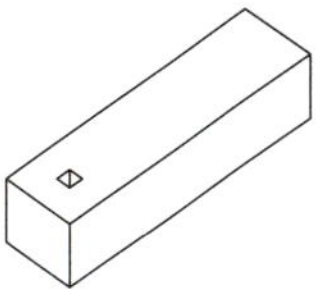

A

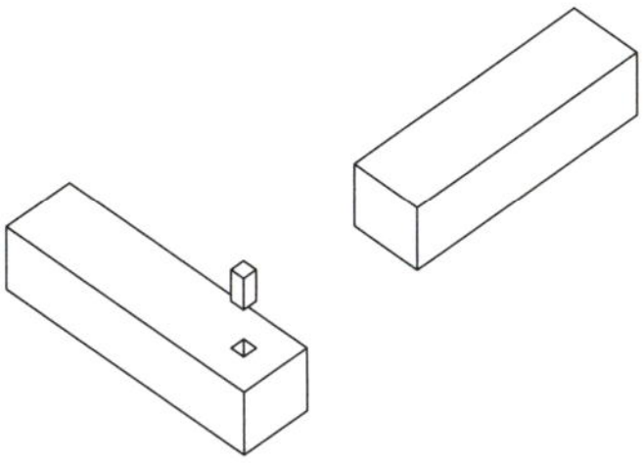

A+B+A

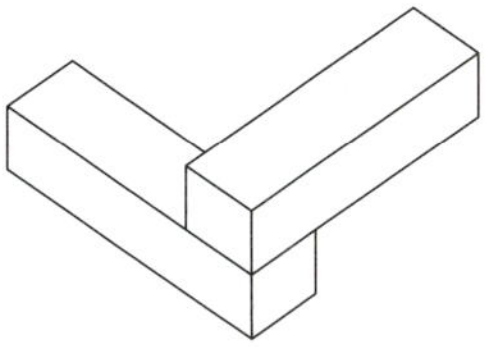

ABA

Corner Joints

Cog joint

watari-ago-tsugi (dan-kaki)
渡腮接ぎ(段欠)

Eckkamm

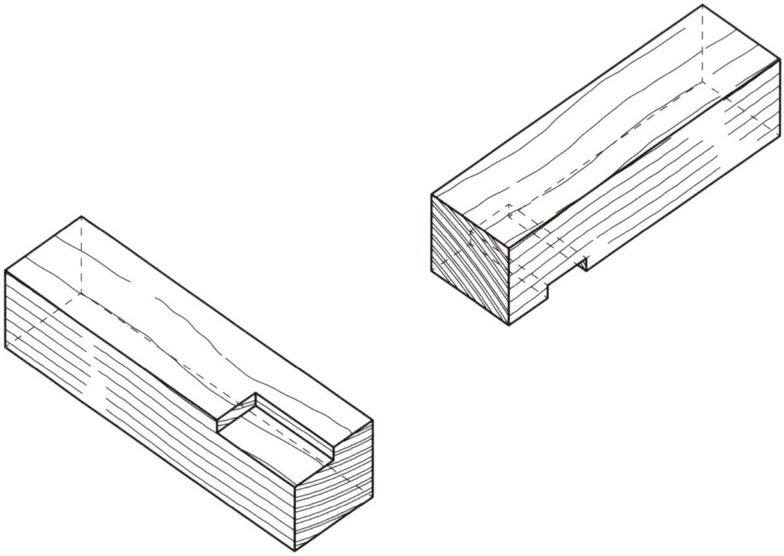

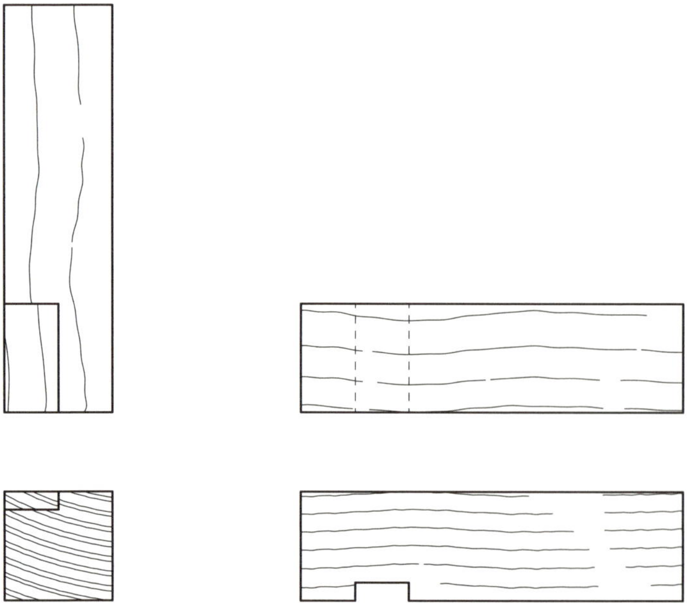

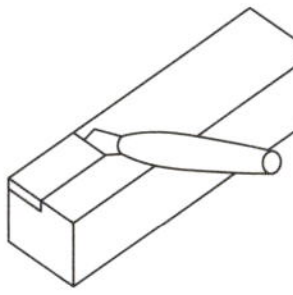
A1

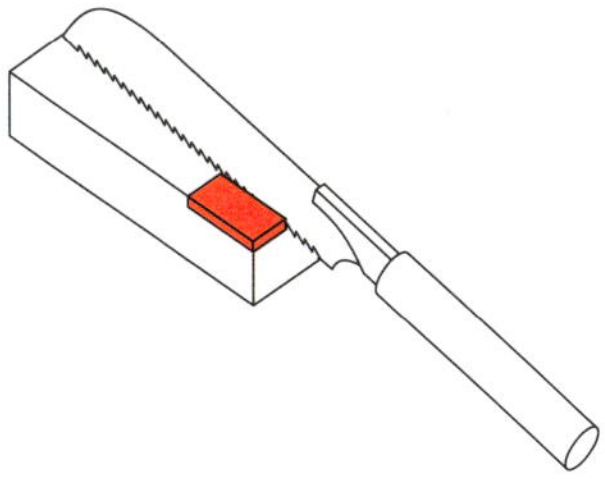
A2

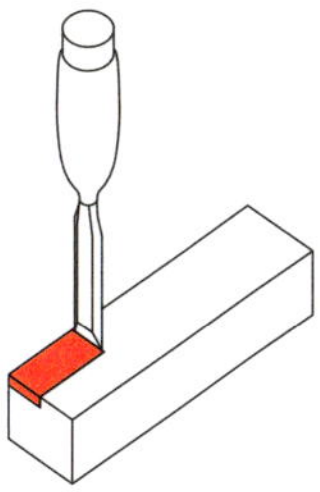
A3

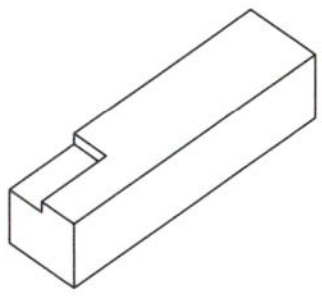
A

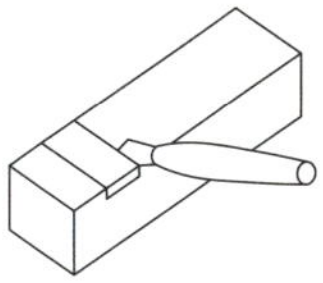
B1

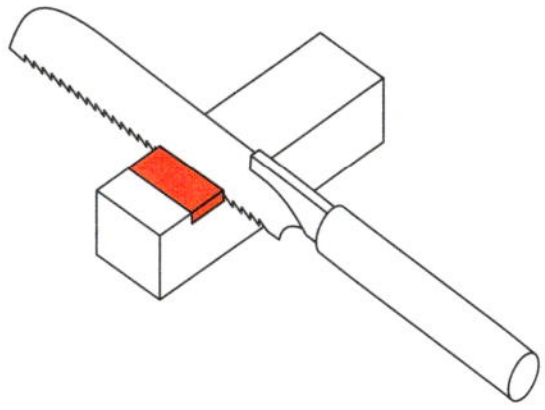
B2

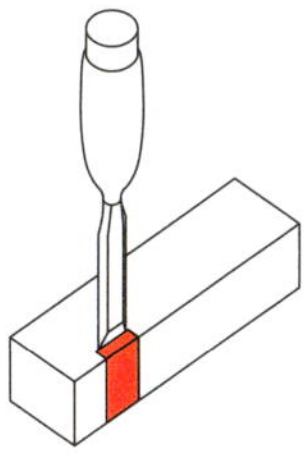
B3

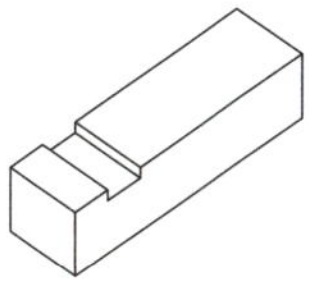
B

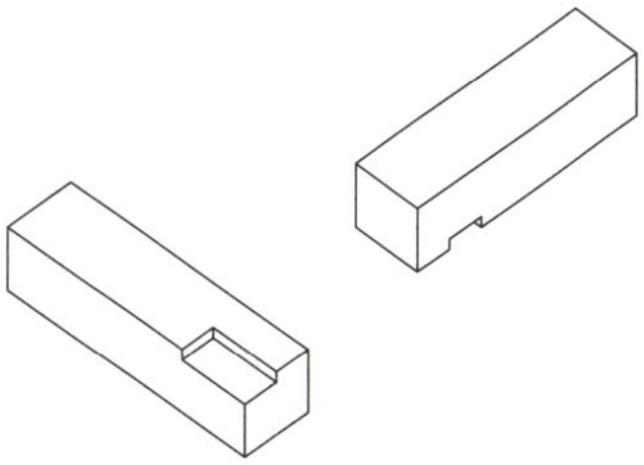
A+B

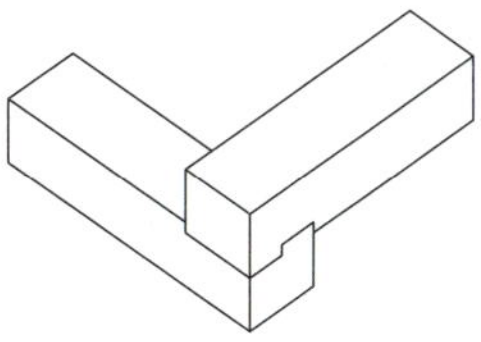
AB

Corner Joints

Sloped cog joint

watari-ago-tsugi (kata-ari-kaki)
渡腮接ぎ（片蟻欠）

schräger Eckkamm

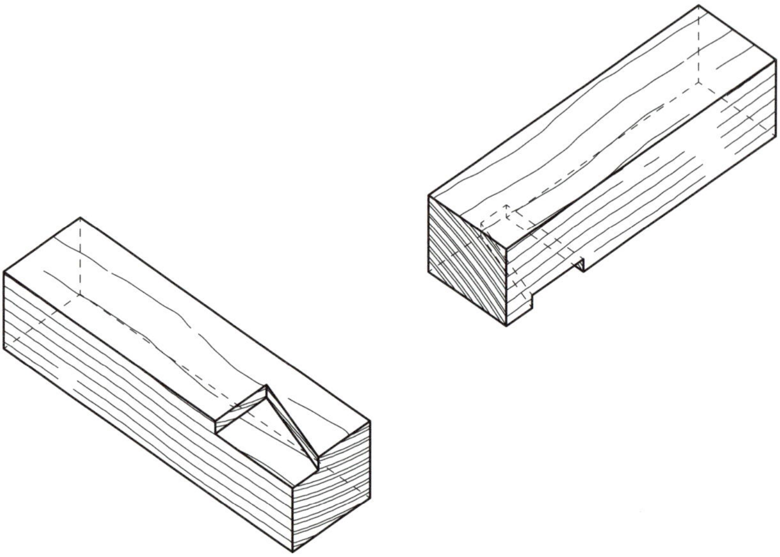

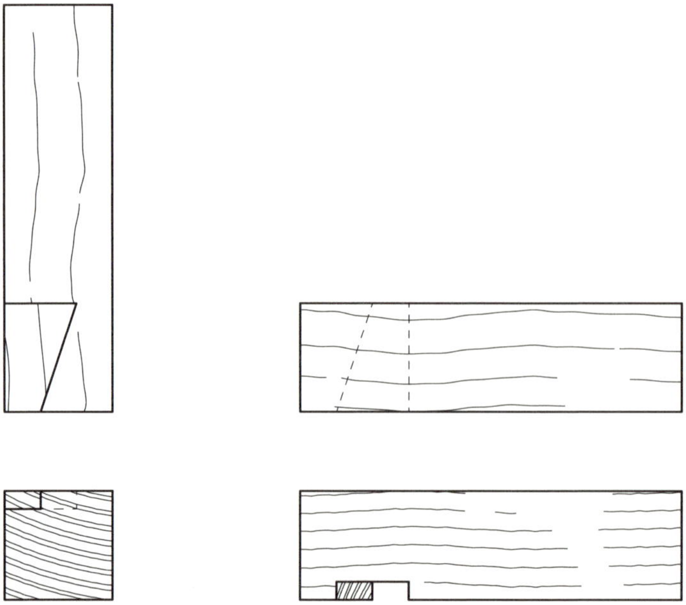

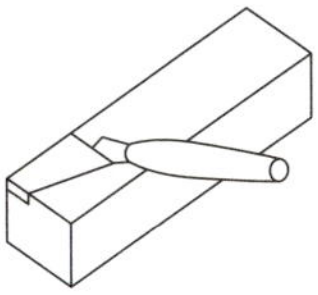

A1

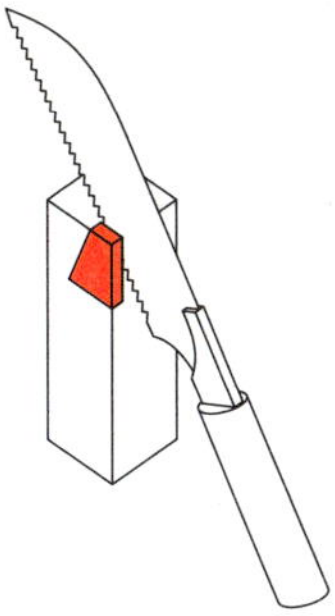

A2

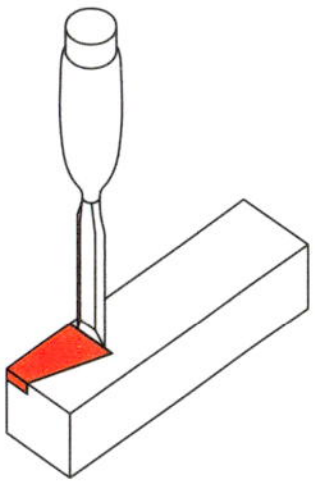

A3

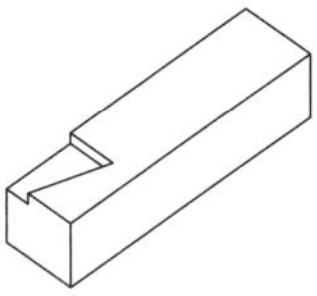

A

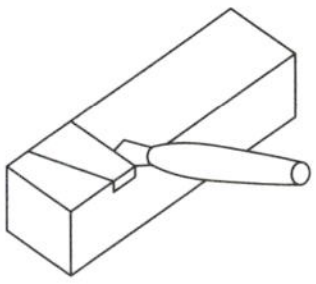

B1

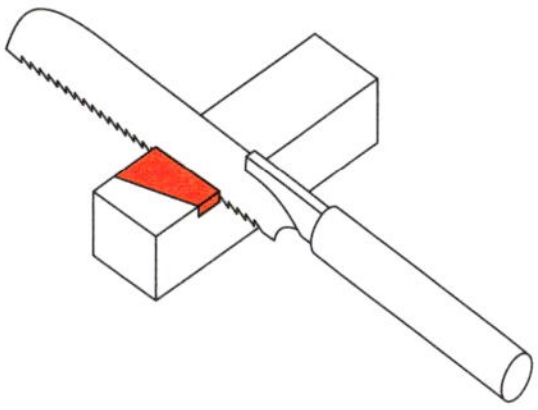

B2

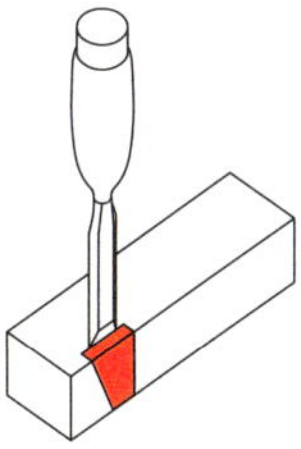

B3

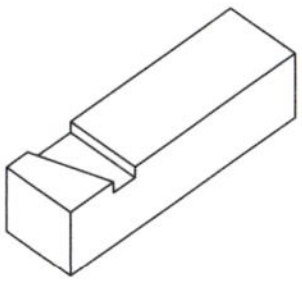

B

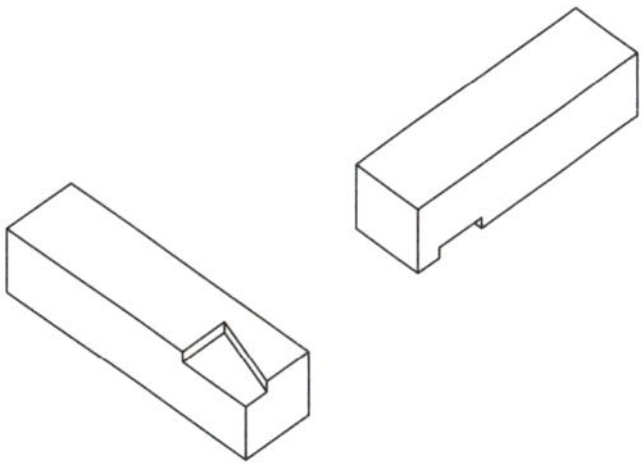

A+B

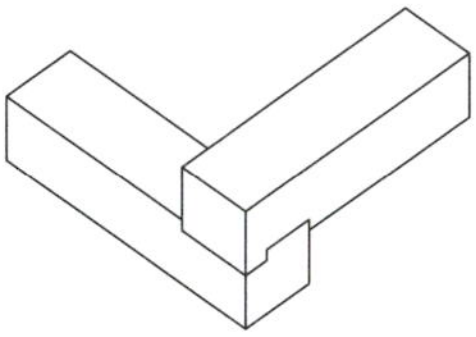

AB

Corner Joints

Dovetail cog joint

watari-ago-tsugi (ōari-kaki)
渡腮接ぎ(大蟻欠)

Eckkamm mit Schwalbenschwanz

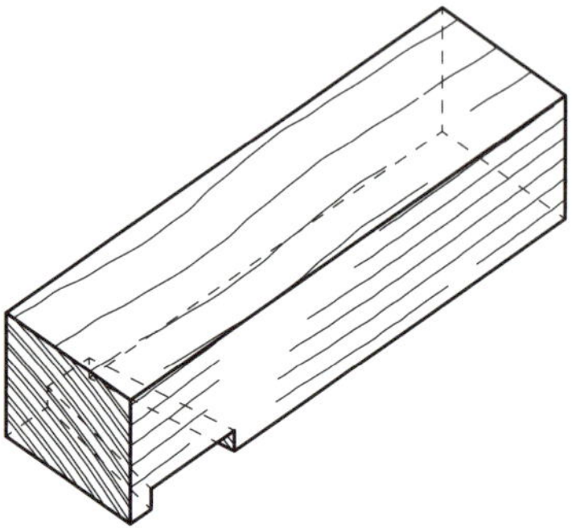

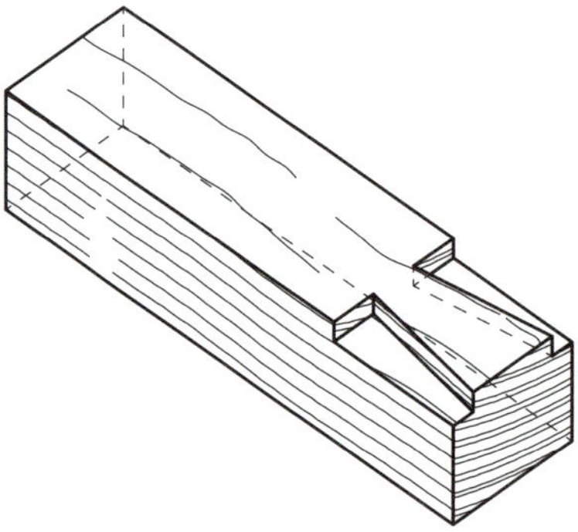

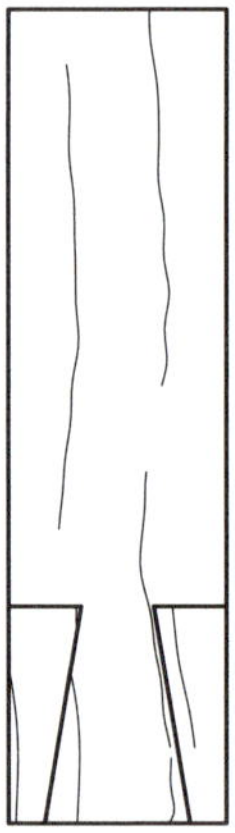

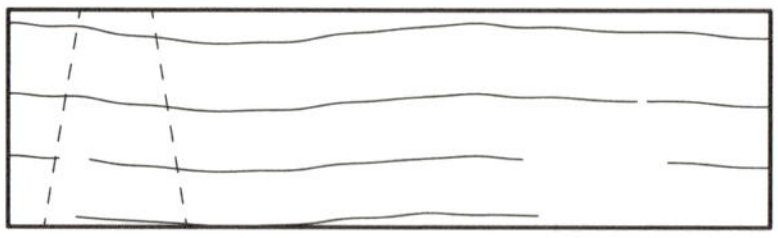

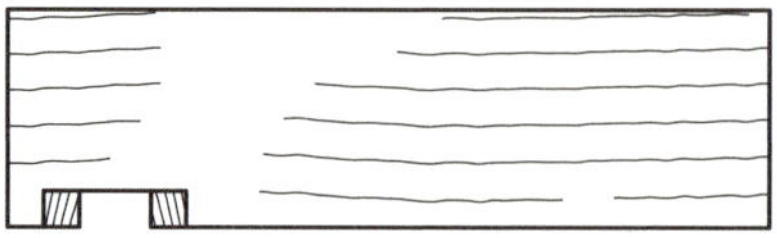

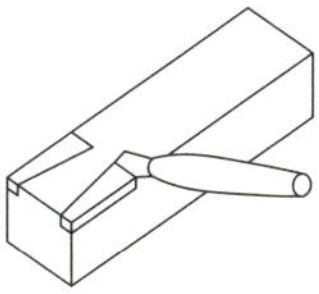

A1

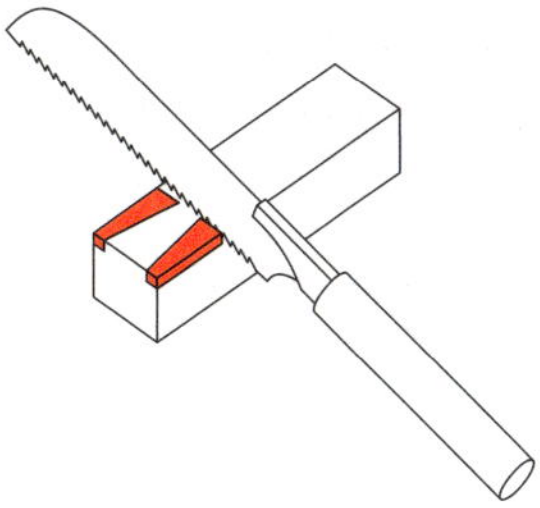

A2

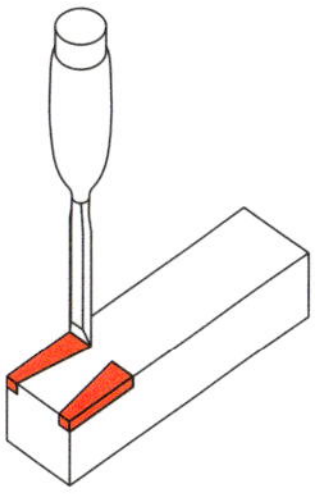

A3

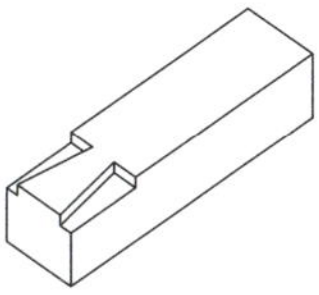

A

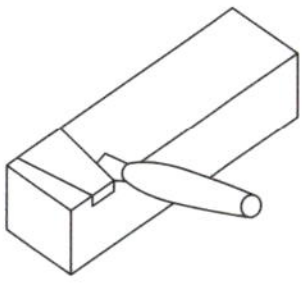

B1

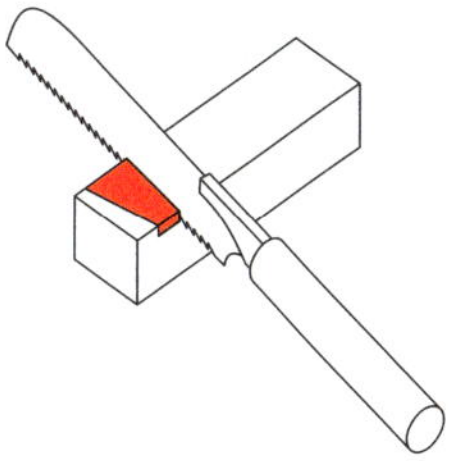

B2

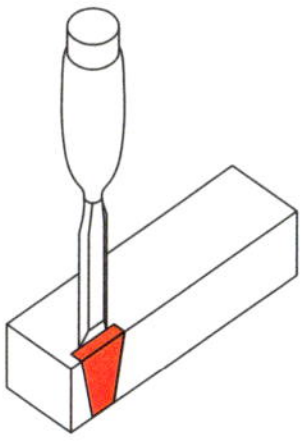

B3

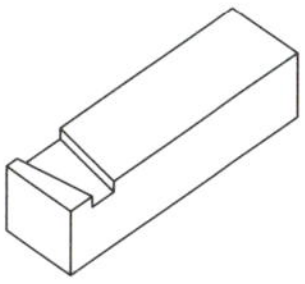

B

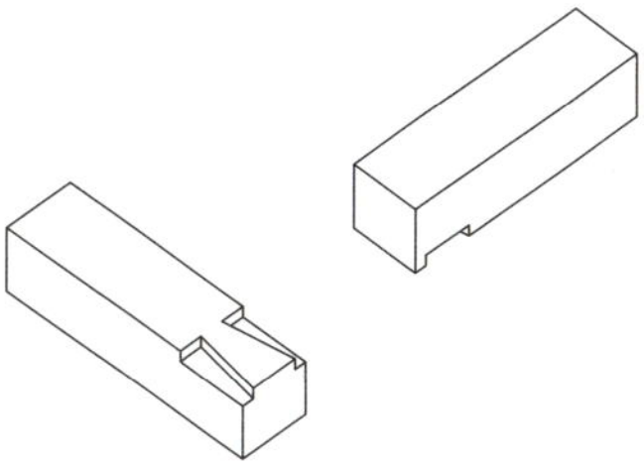

A+B

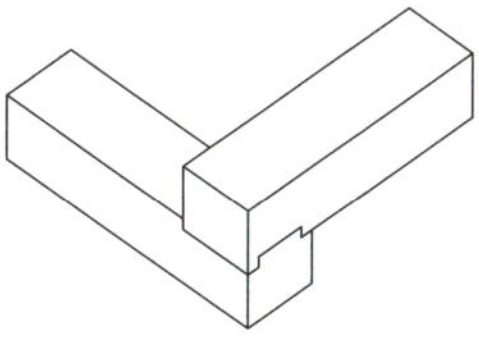

AB

V-shaped cog joint

tasuki-watari-ago-tsugi
襷渡腮接ぎ

Eckkamm mit halbem Kreuz

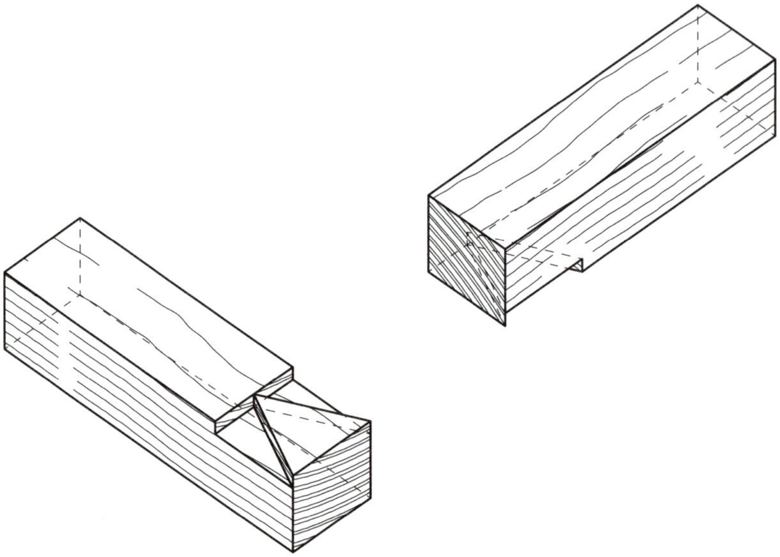

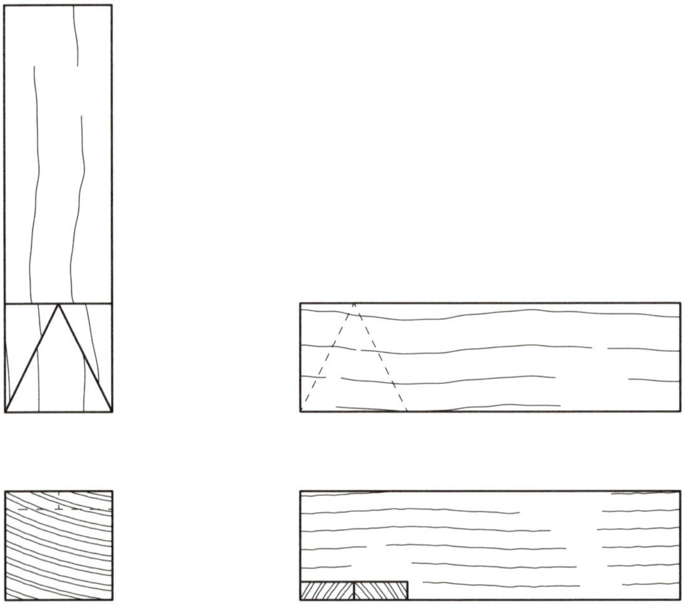

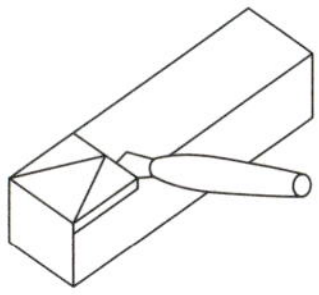

A1

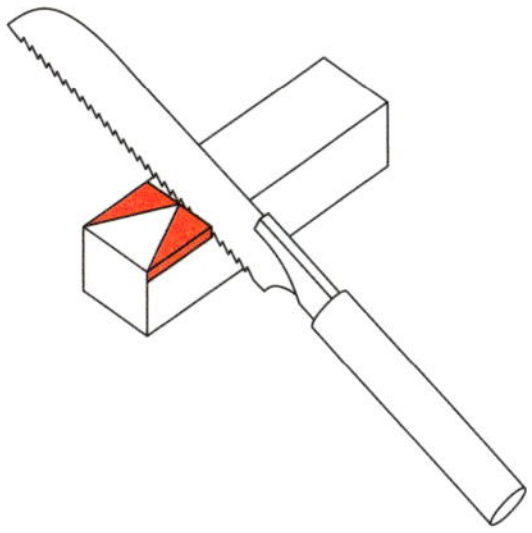

A2

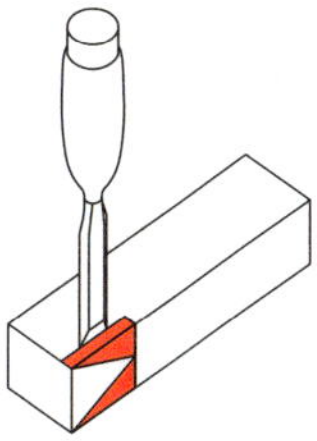

A3

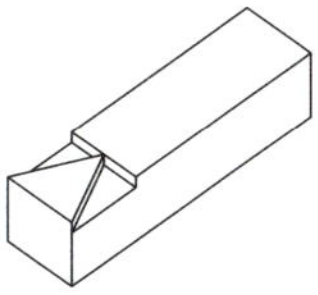

A

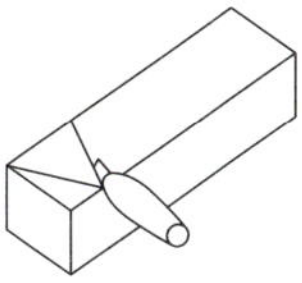

B1

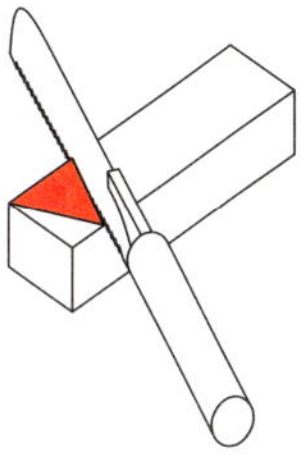

B2

B3

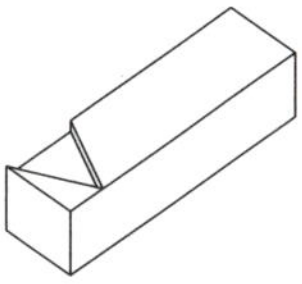

B

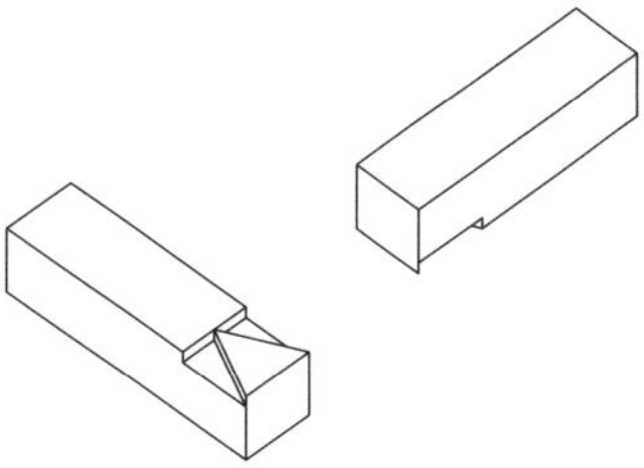

A+B

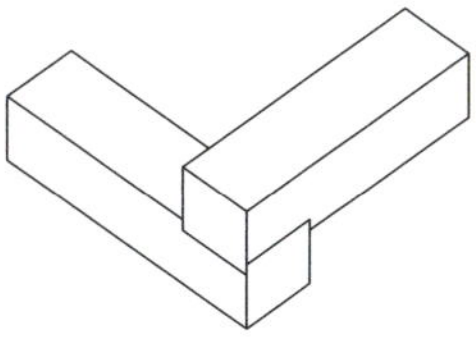

AB

Corner Joints

Diagonally cut cog joint

watari-ago-tsugi (eriwa-kaki)
渡腮接ぎ（襟輪欠）

Eckkamm mit Haken

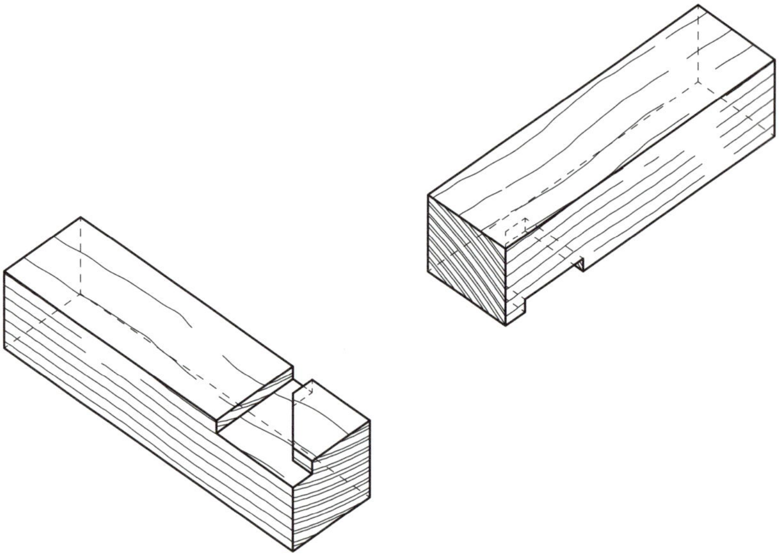

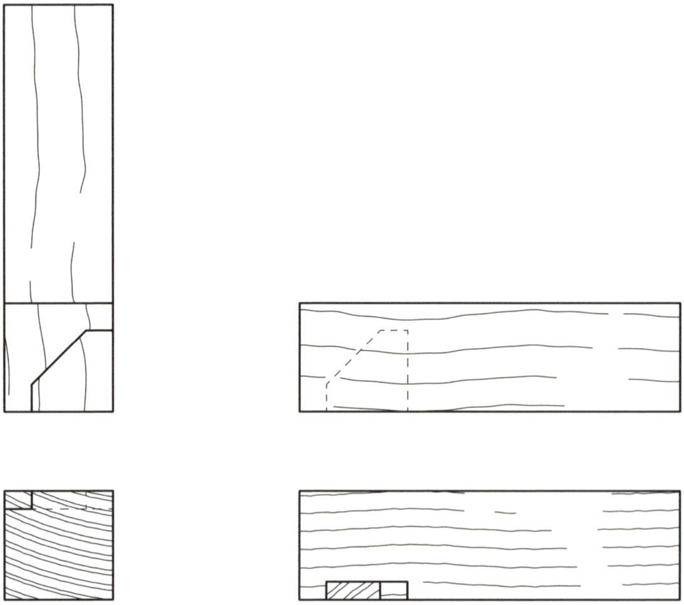

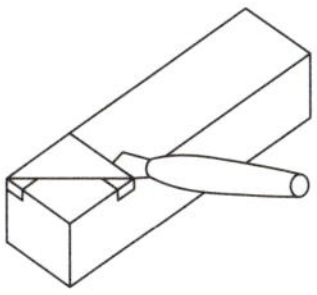

A1

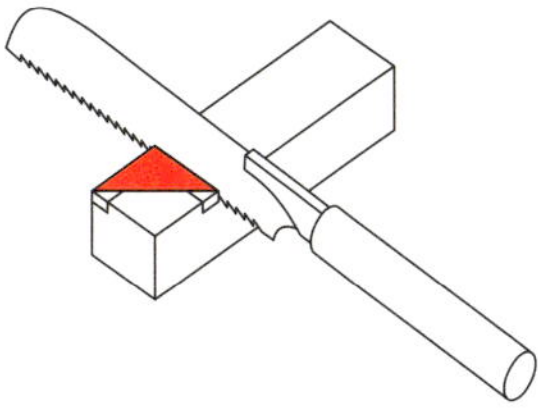

A2

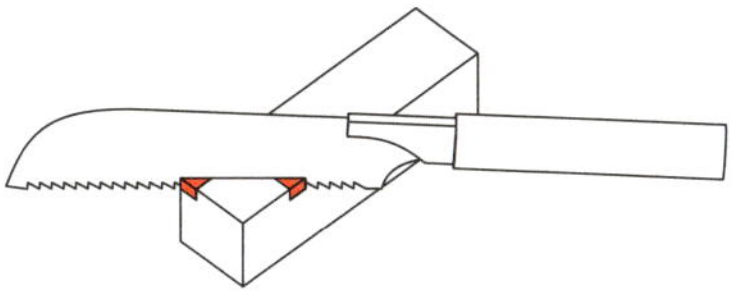

A3

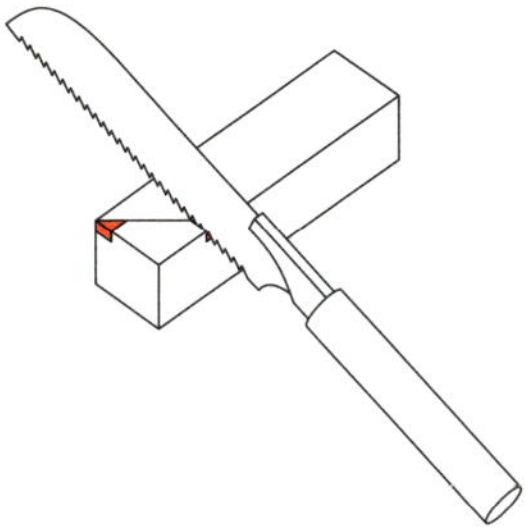

A4

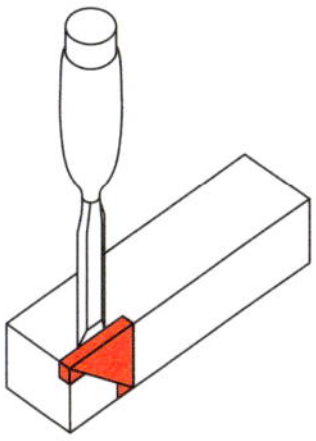

A5

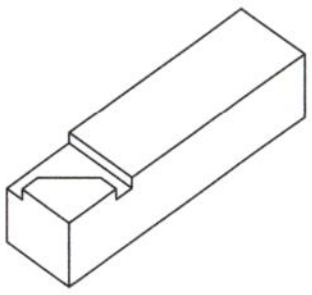

A

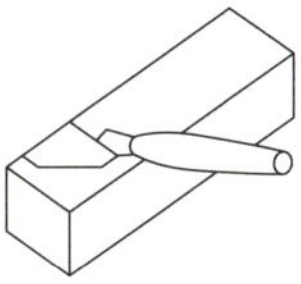

B1

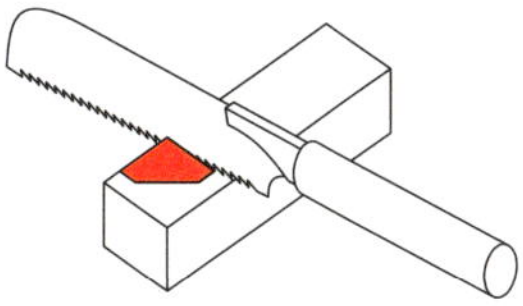

B2

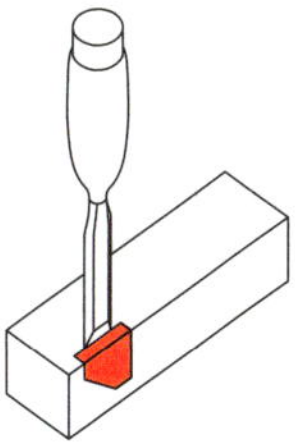

B3

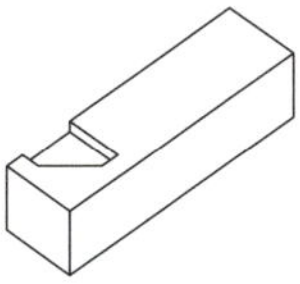

B

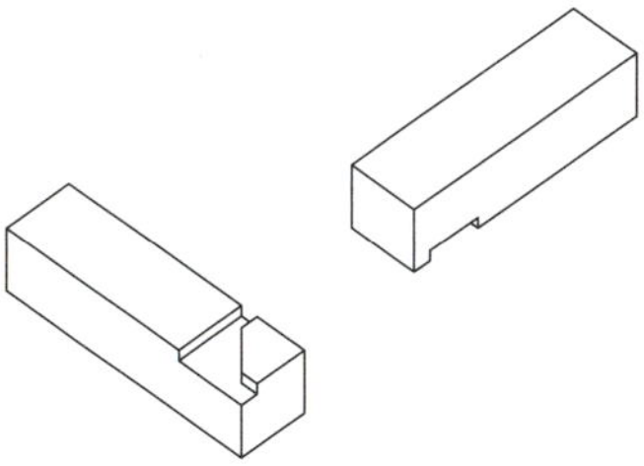

A+B

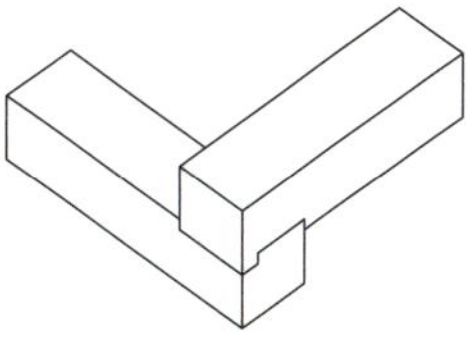

AB

T-Joints

Half lap joint

ai-kaki-kumi [*]
相欠組み

Querblatt

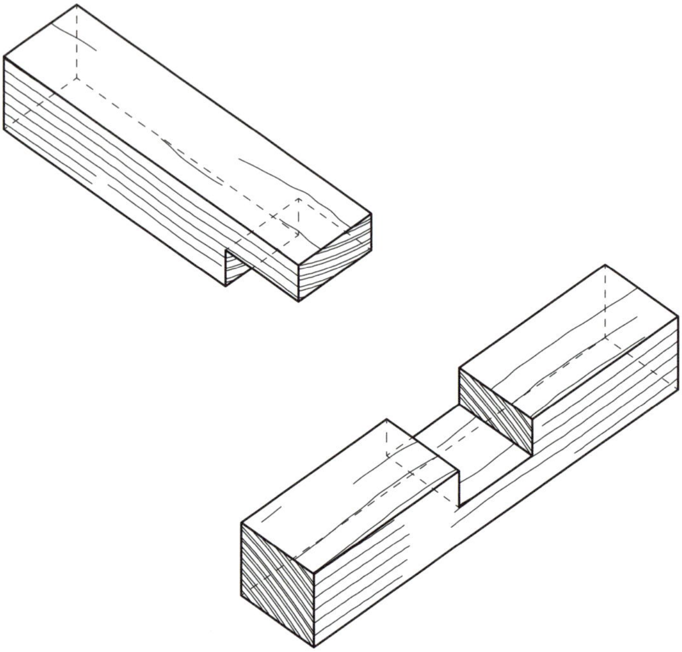

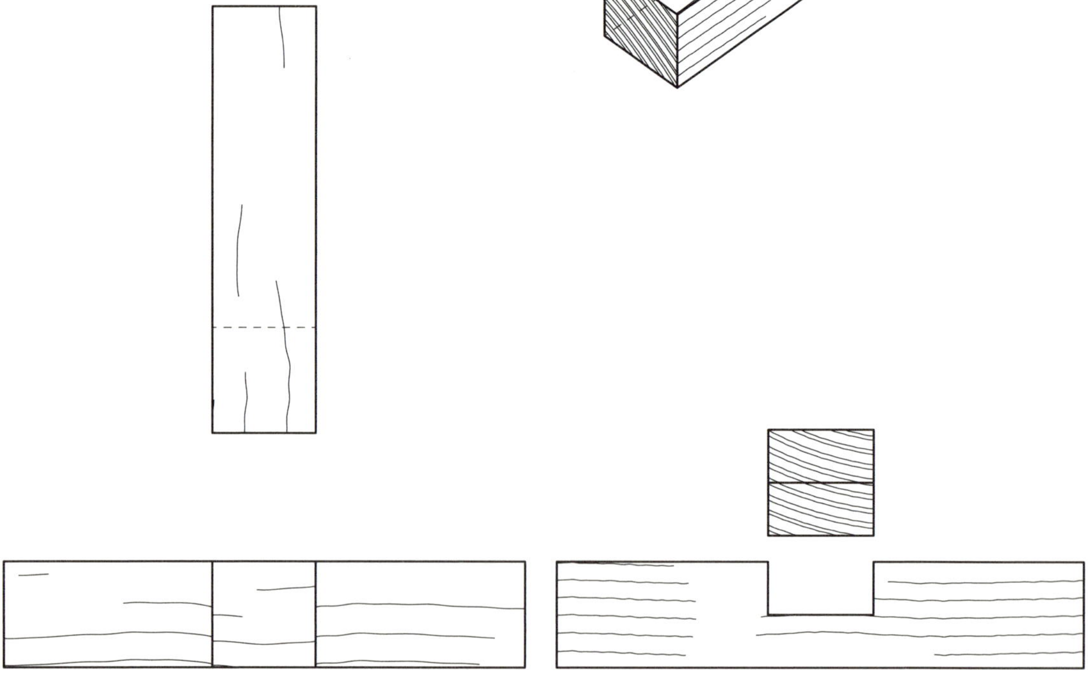

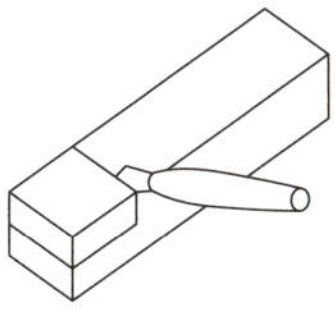

A1

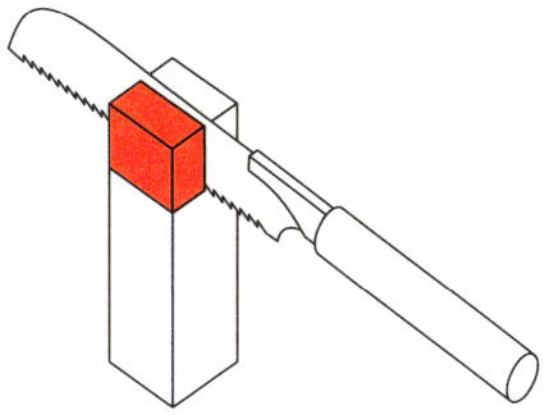

A2

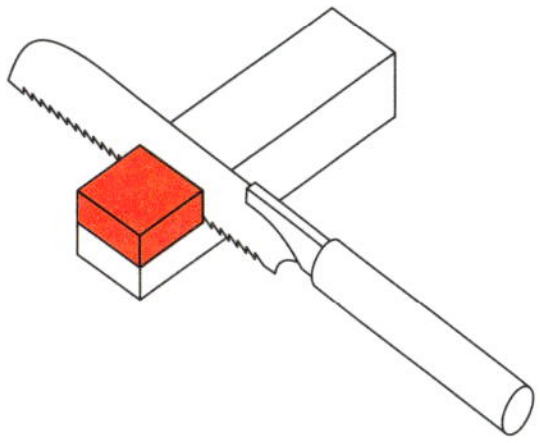

A3

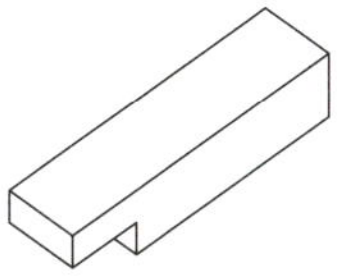

A

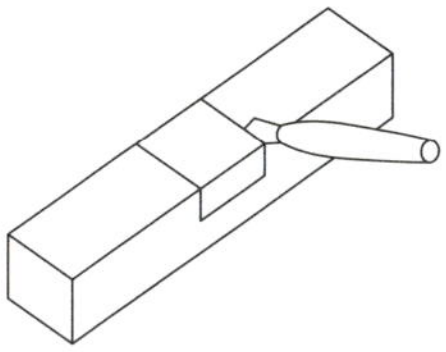

B1

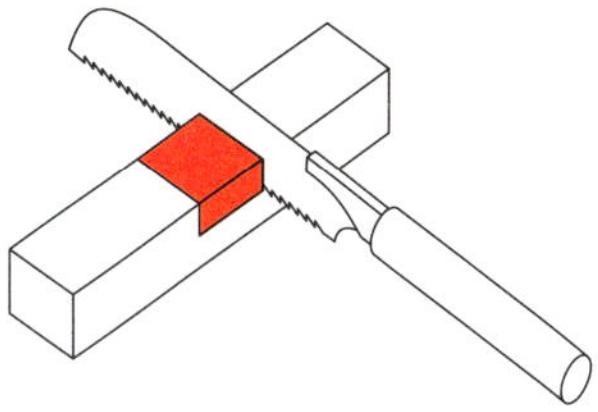

B2

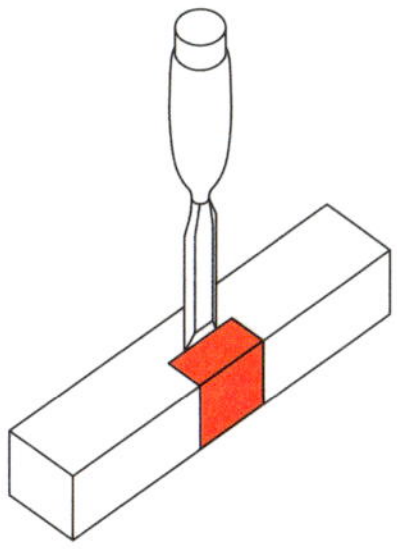

B3

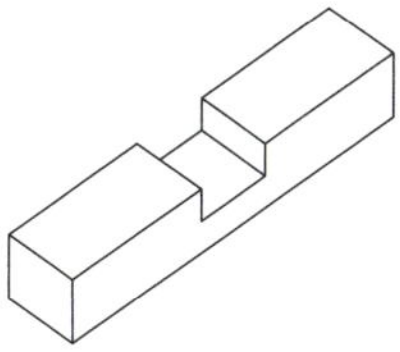

B

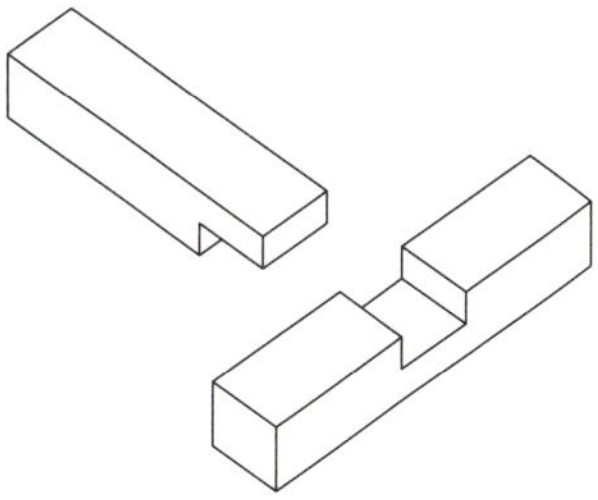

A+B

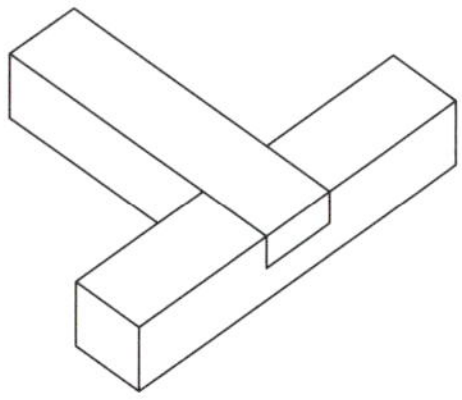

AB

T-Joints

Stopped half lap joint

koshi-kake-tsugi
腰掛接ぎ

halbverdecktes Querblatt

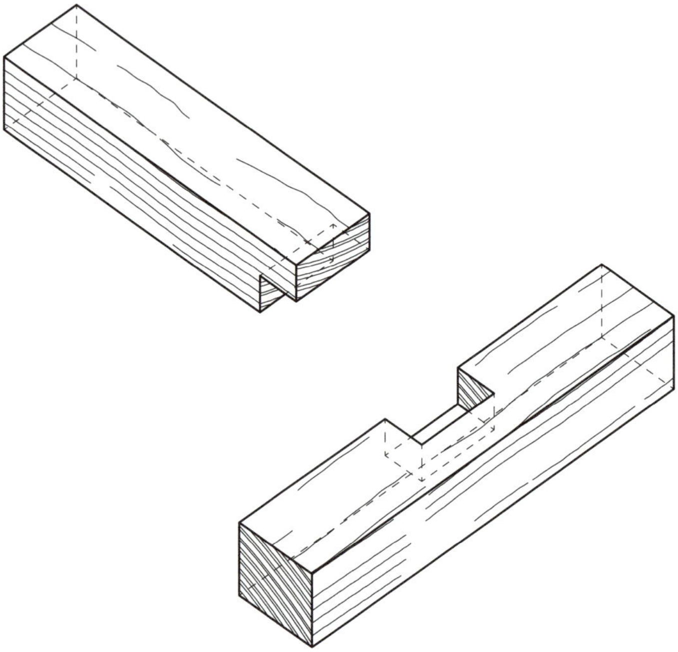

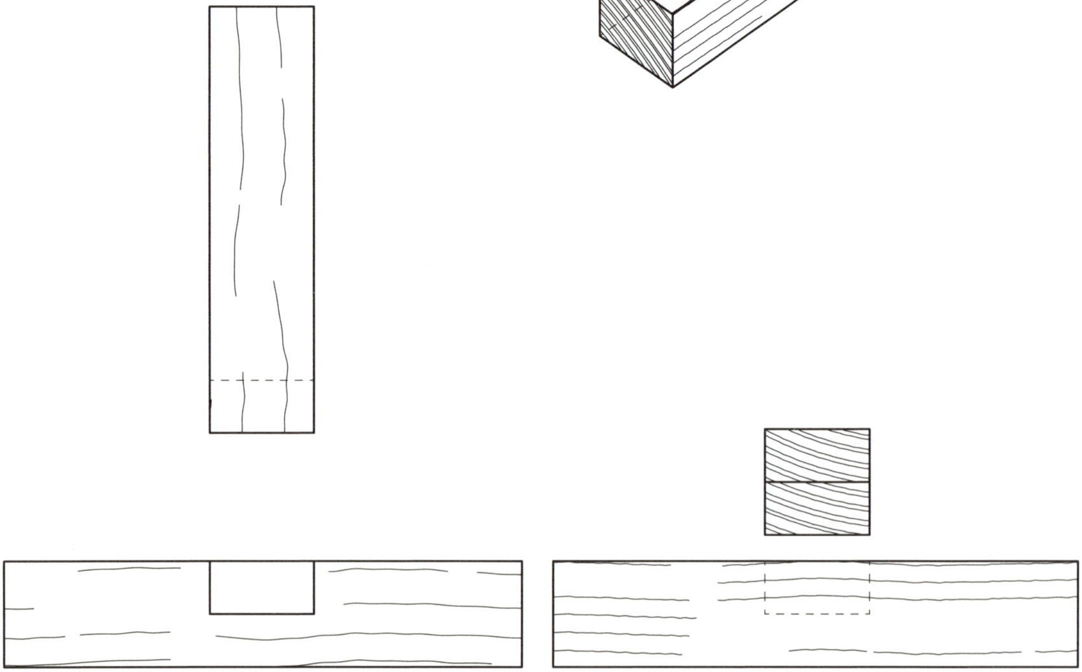

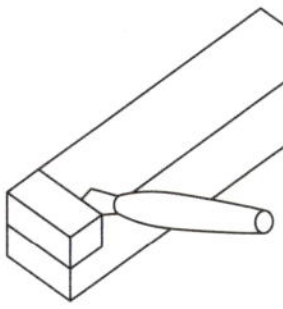

A1

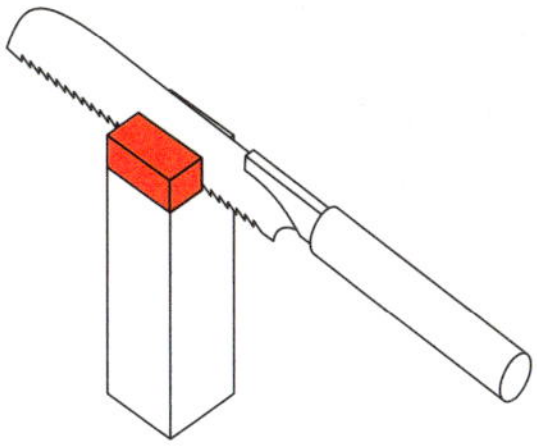

A2

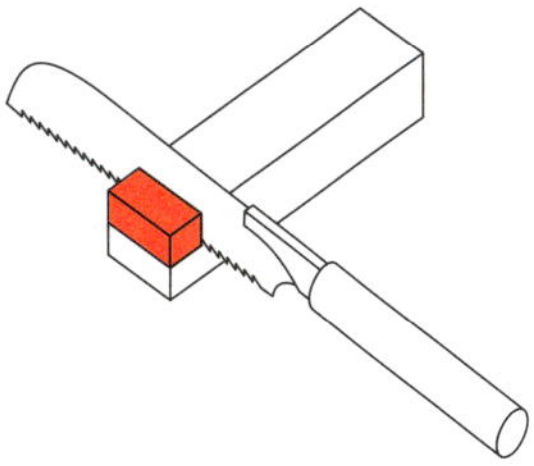

A3

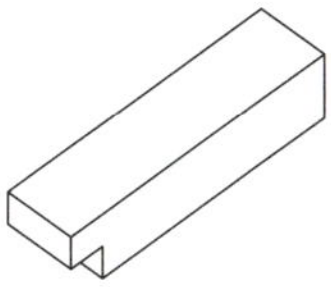

A

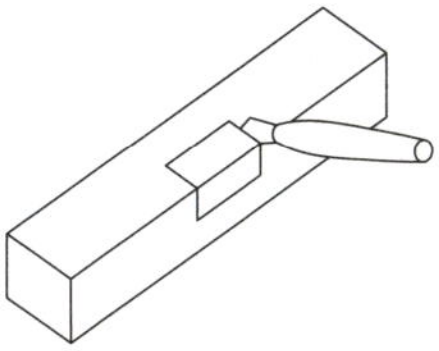

B1

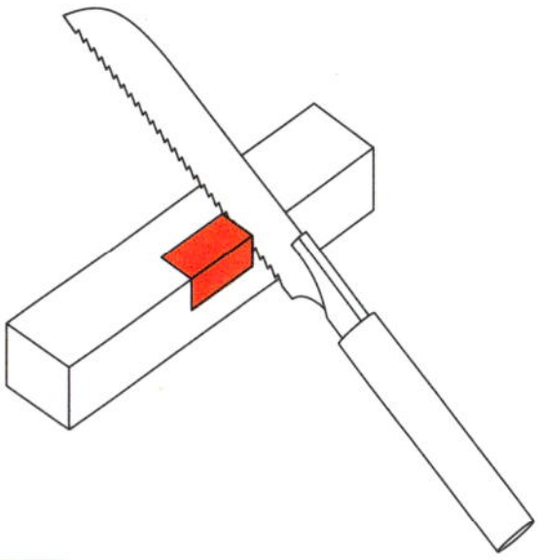

B2

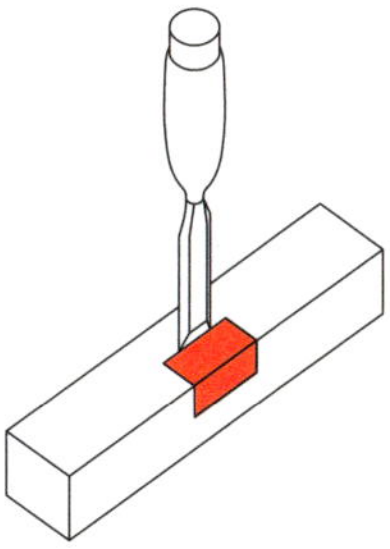

B3

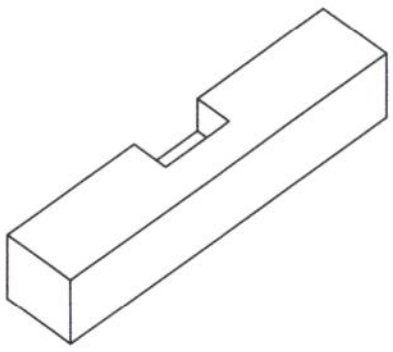

B

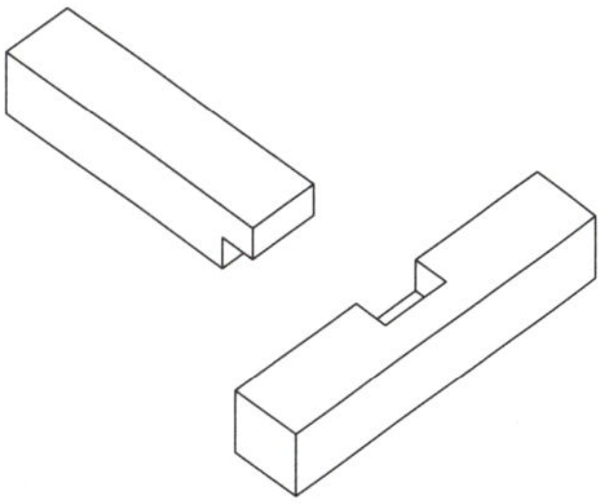

A+B

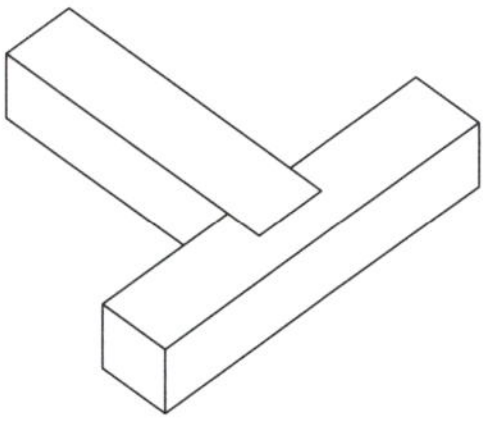

AB

T-Joints

Beveled half lap joint

suberi-ago-tsugi [*]
辷顎接ぎ

Querblatt mit schräger Blattfläche

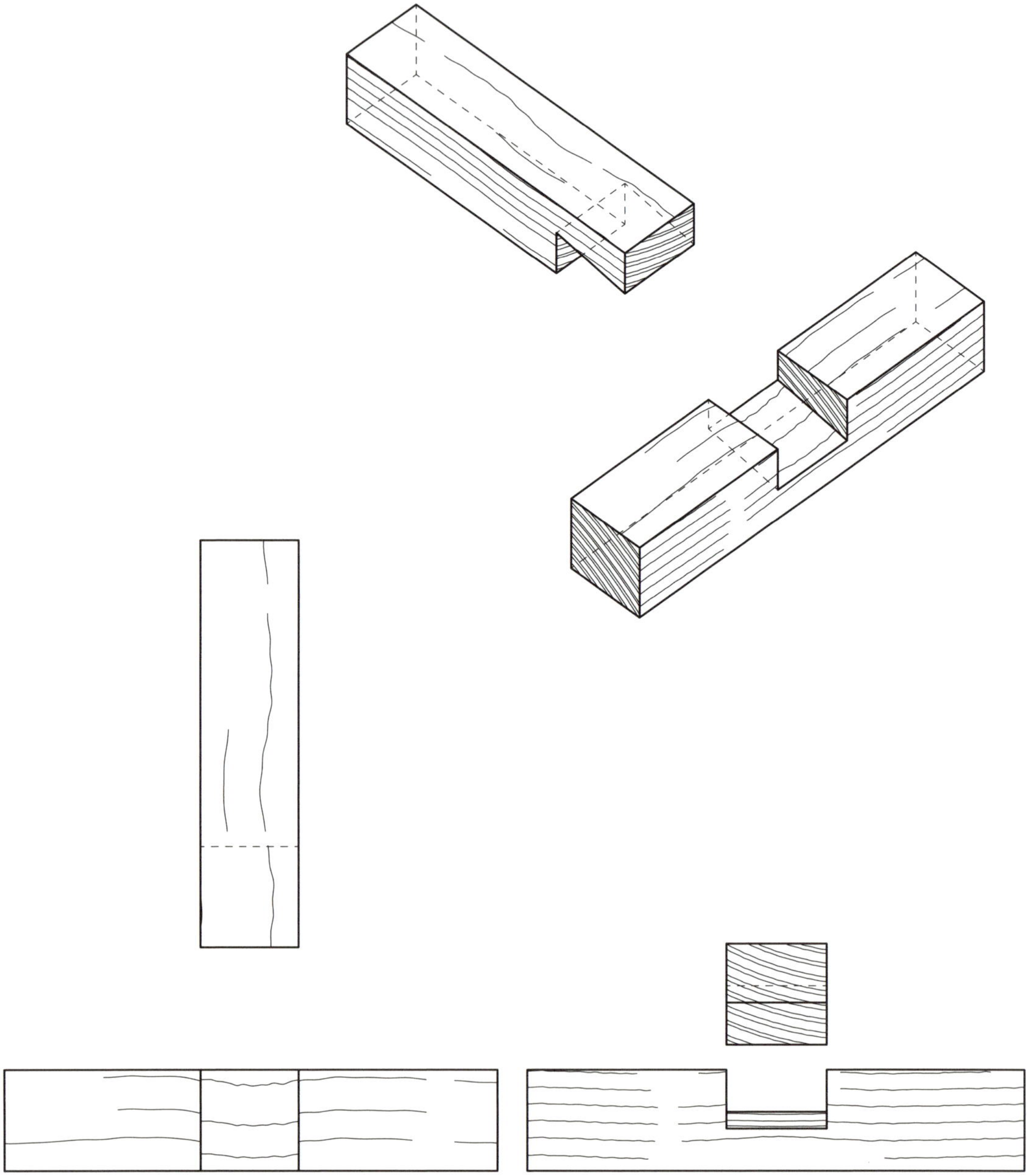

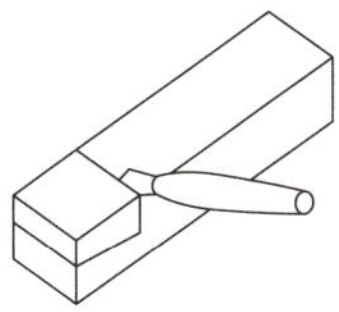

A1

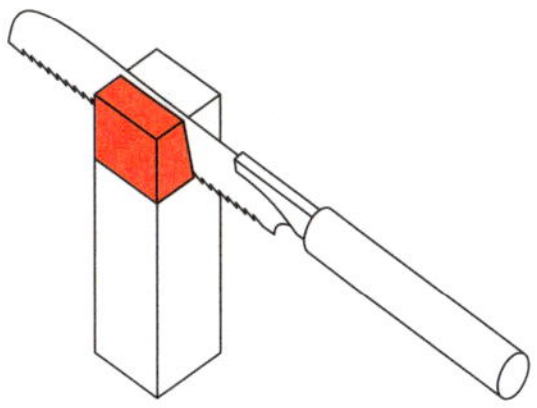

A2

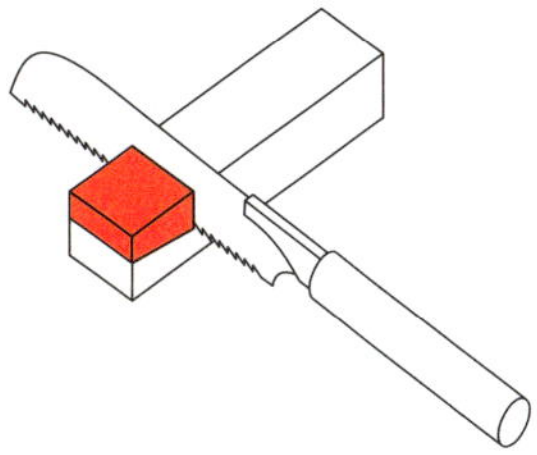

A3

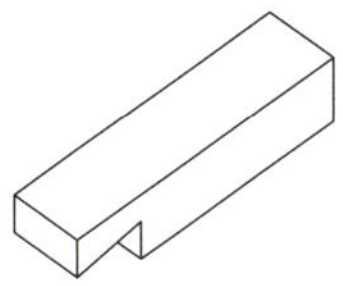

A

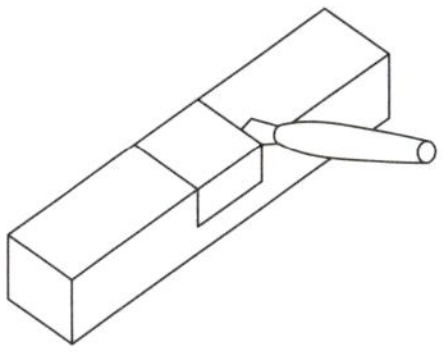

B1

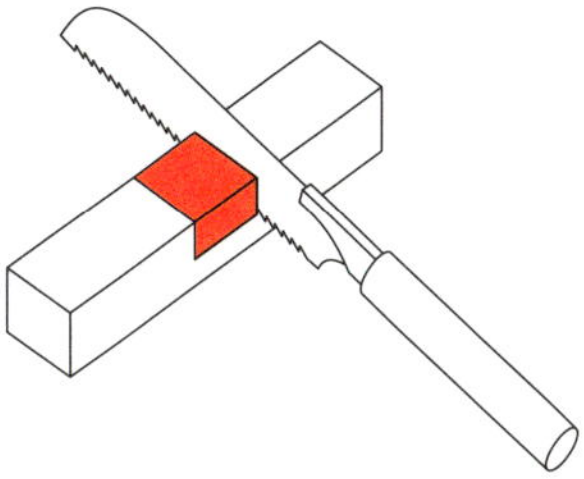

B2

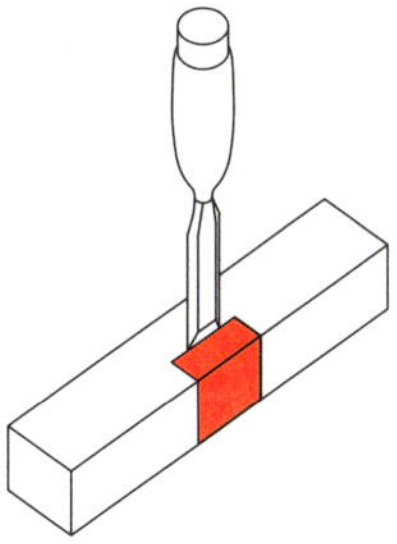

B3

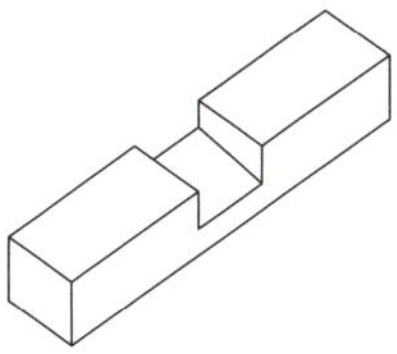

B

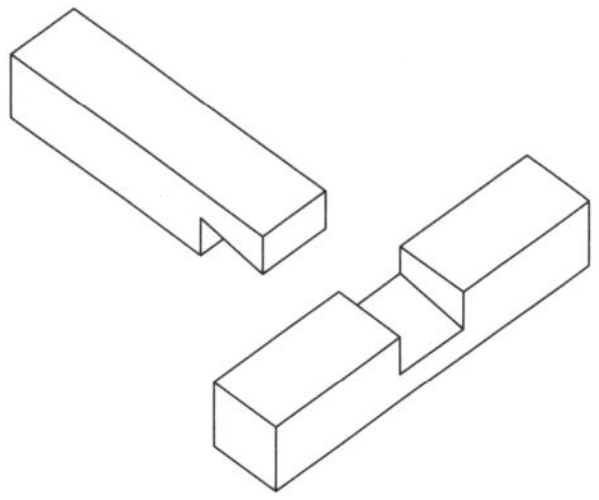

A+B

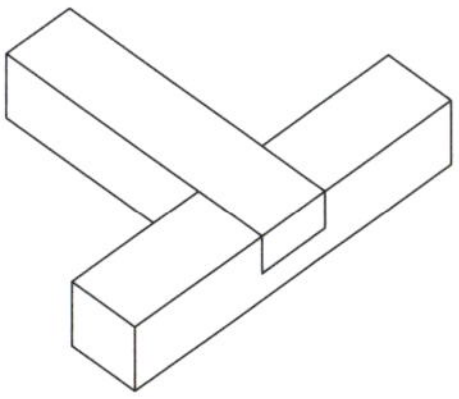

AB

T-Joints

Stopped beveled half lap joint

suberi-koshi-kake-tsugi
辷腰掛接ぎ

halbverdecktes Querblatt
mit schräger Blattfläche

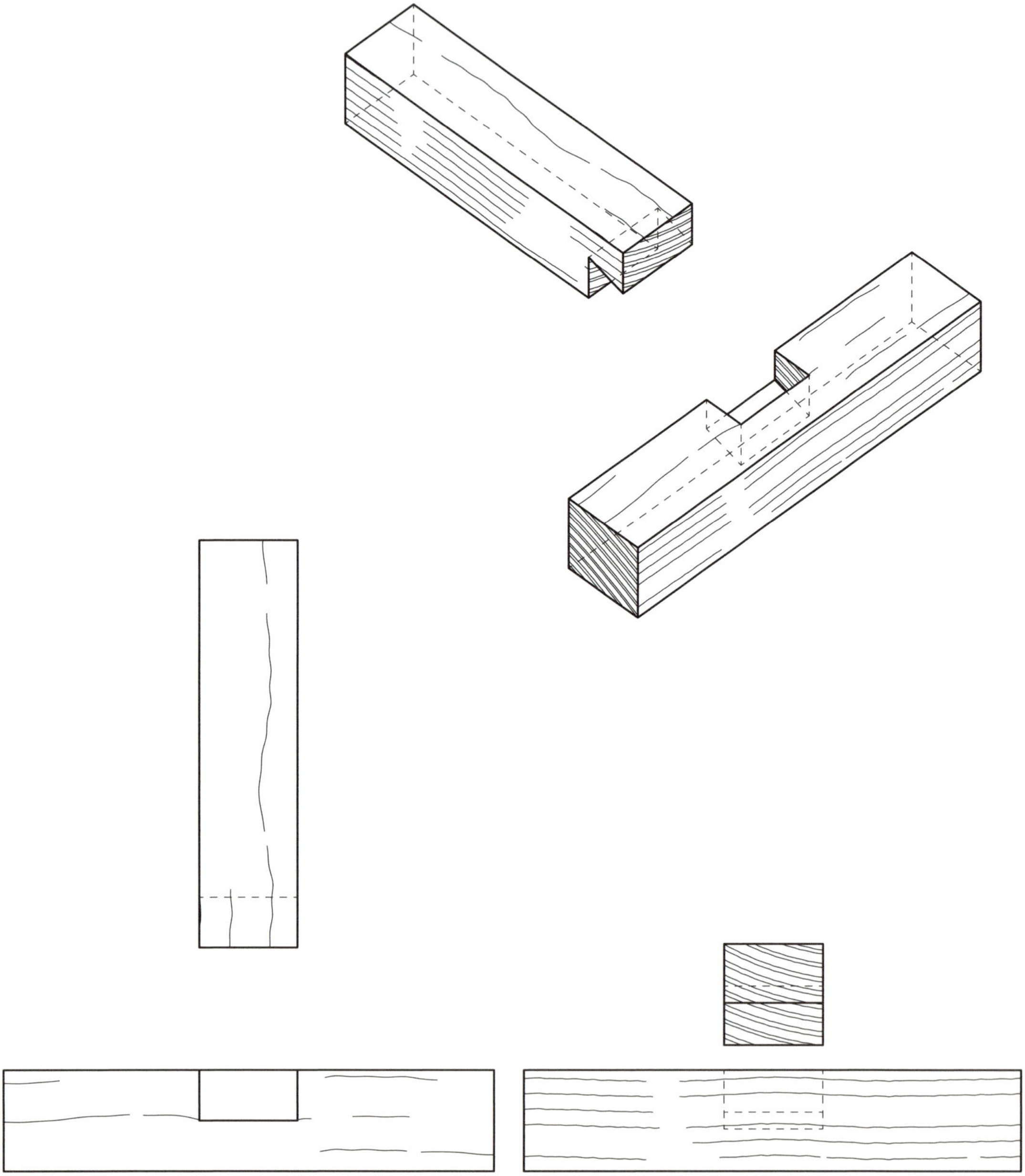

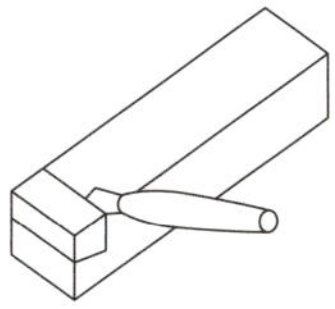

A1

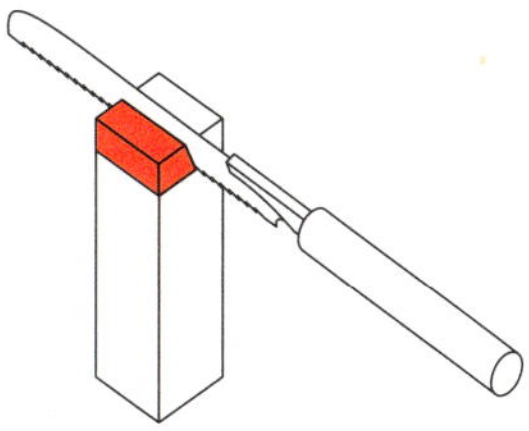

A2

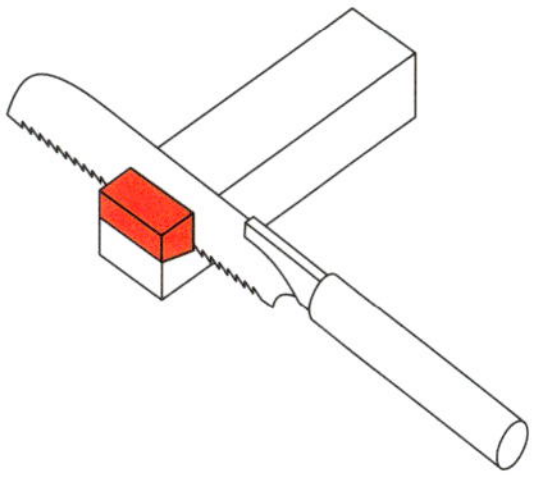

A3

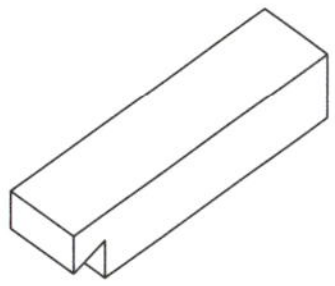

A

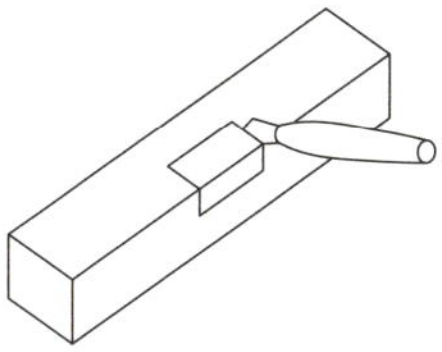

B1

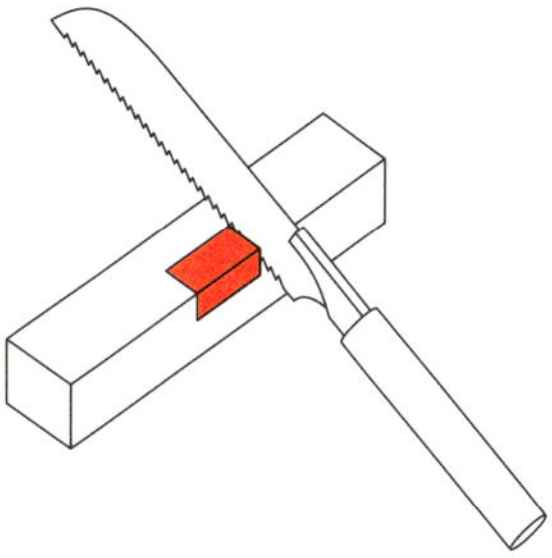

B2

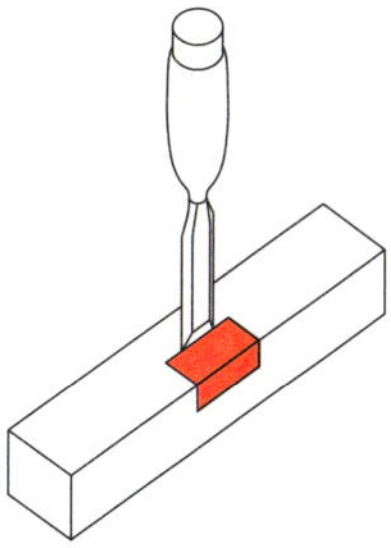

B3

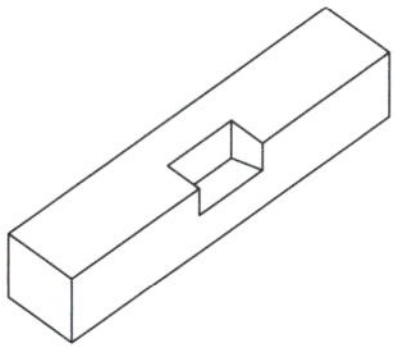

B

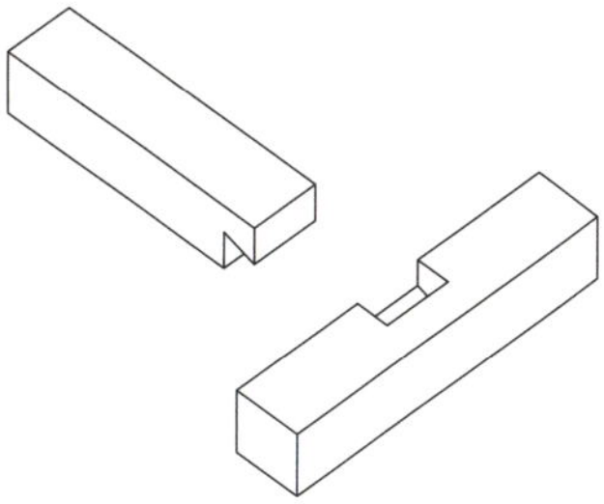

A+B

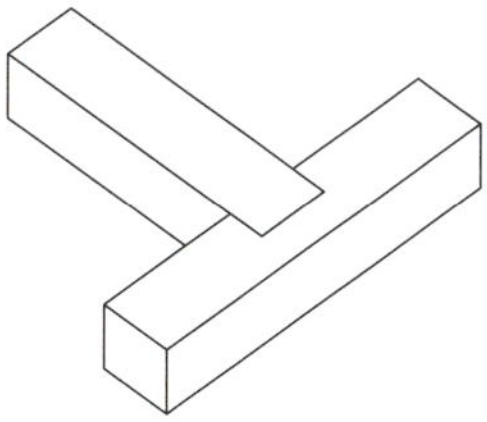

AB

T-Joints

Cogged lap joint

ai-kaki-watari-ago-tsugi (dan-kaki)
相欠渡顎接ぎ（段欠）

Hakenquerblatt

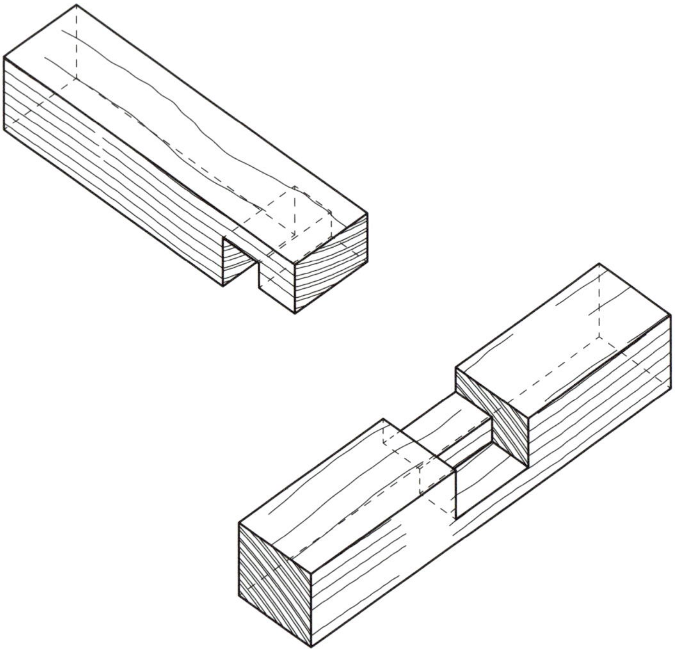

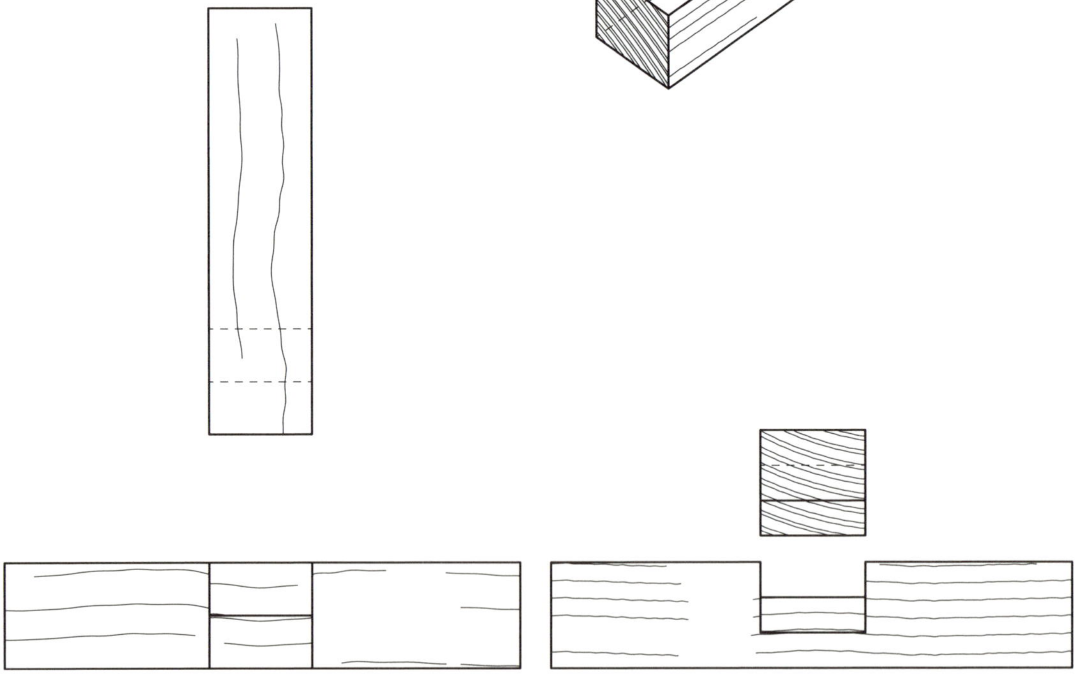

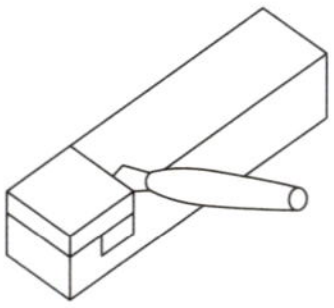

A1

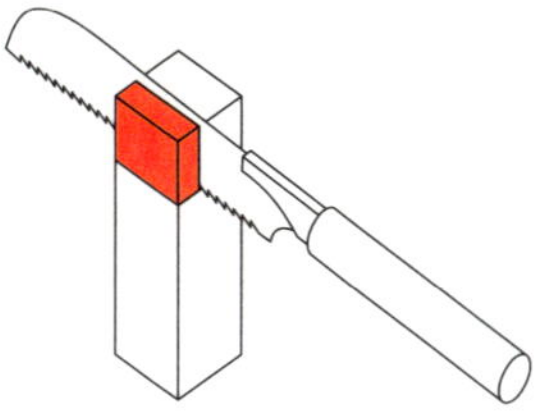

A2

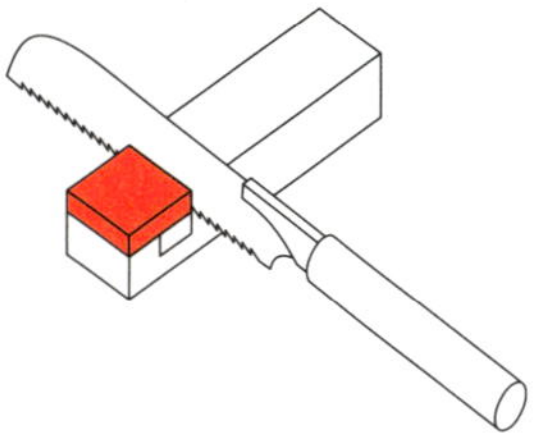

A3

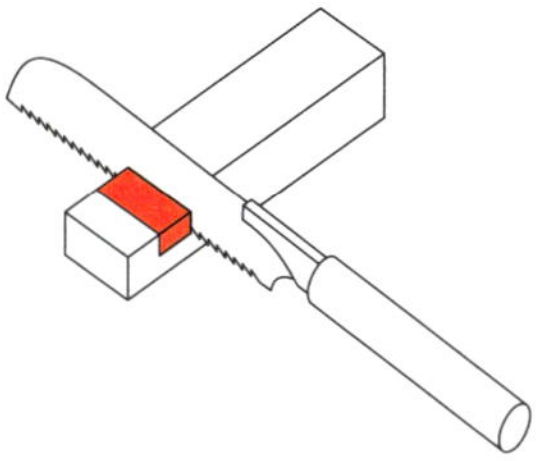

A4

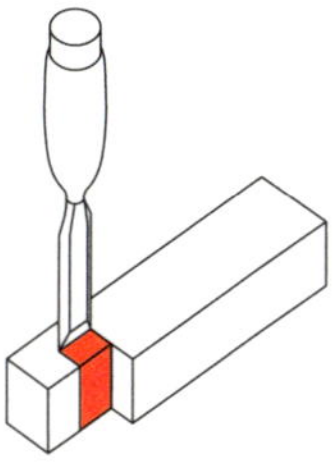

A5

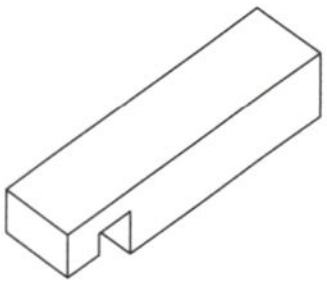

A

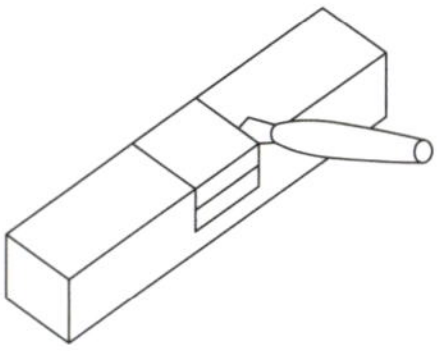

B1

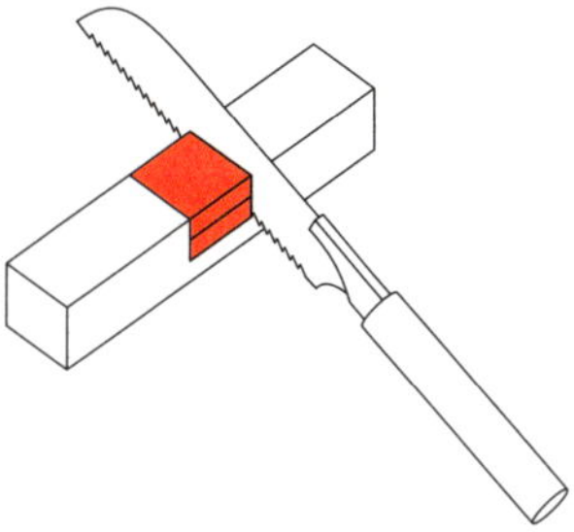

B2

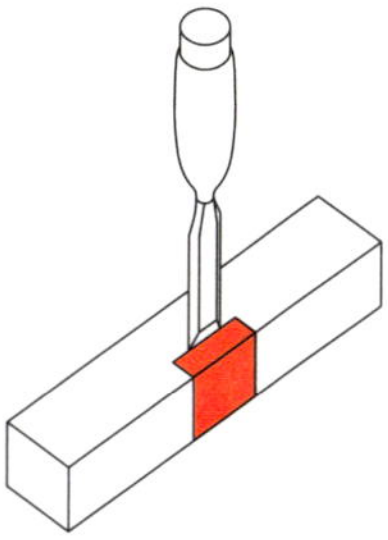

B3

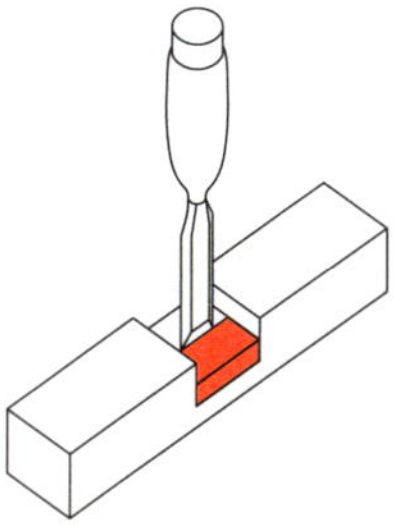

B4

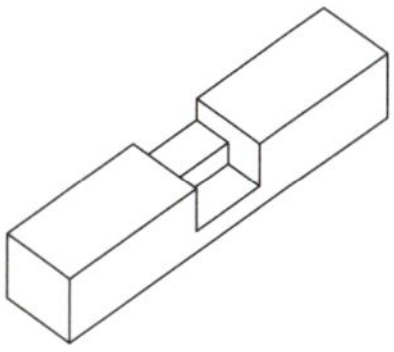

B

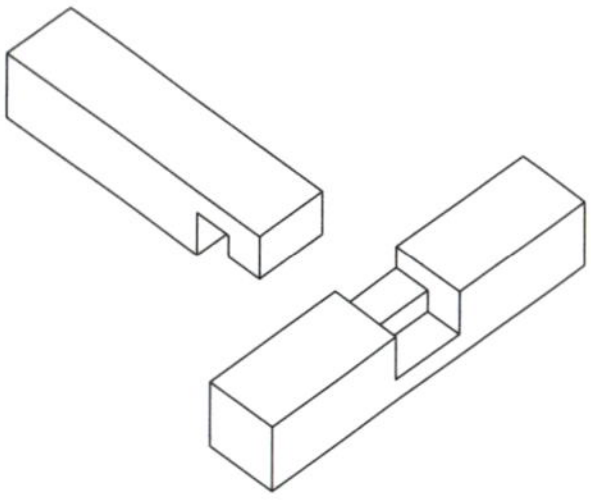

A+B

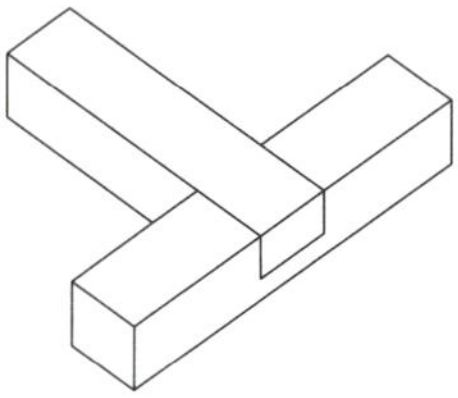

AB

T-Joints

Cogged lap joint with stub tenon

hako-mechi-ai-kaki-watari-ago-tsugi
箱目違相欠渡顎接ぎ

Hakenquerblatt mit Brüstungszapfen

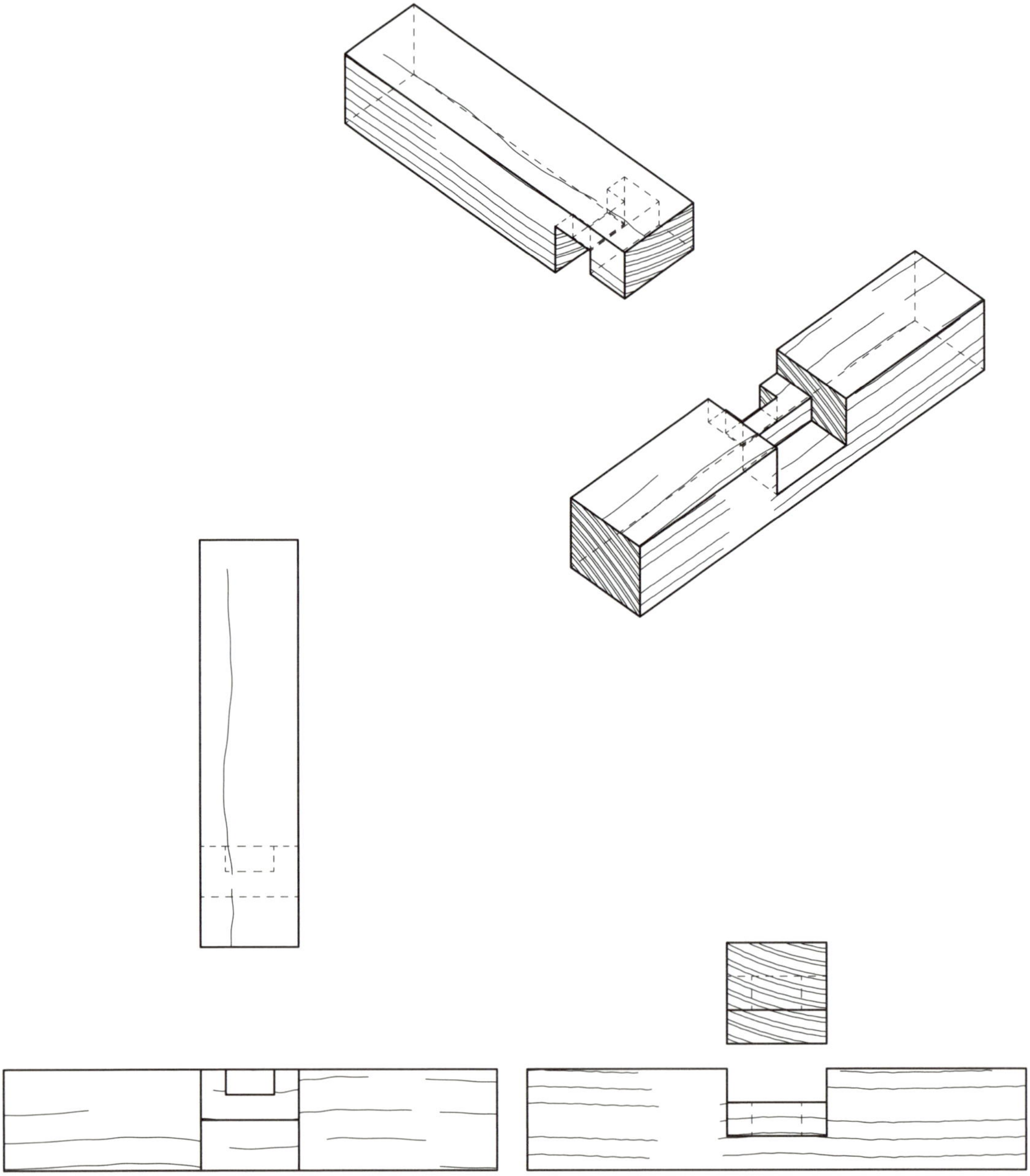

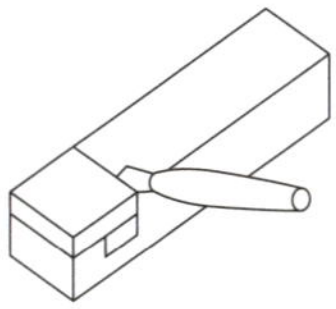

A1

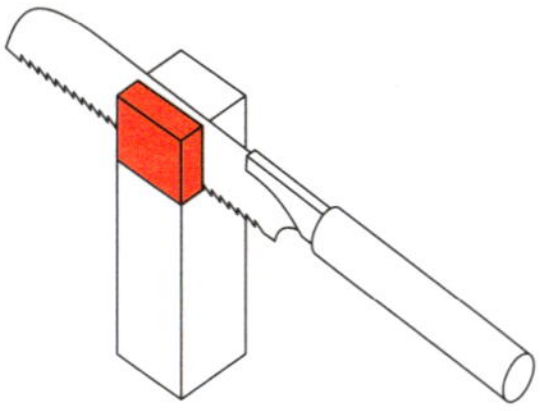

A2

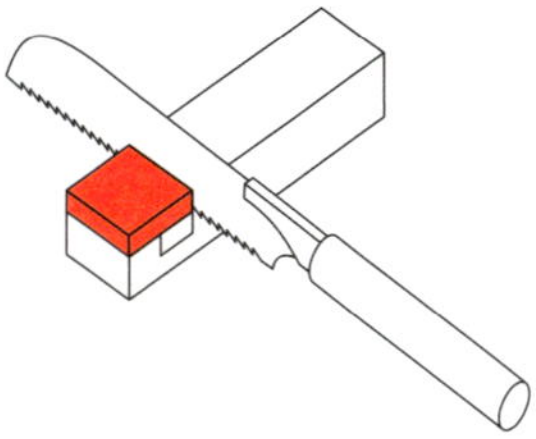

A3

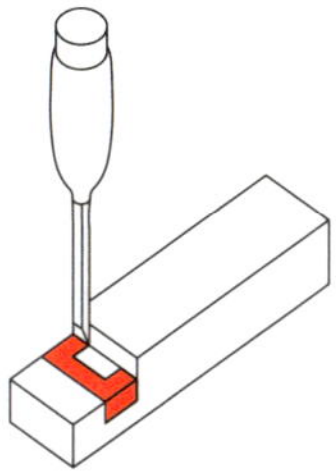

A4

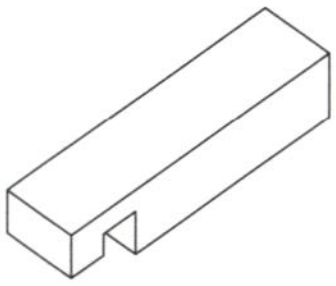

A

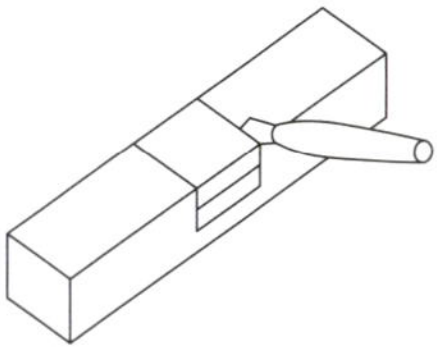

B1

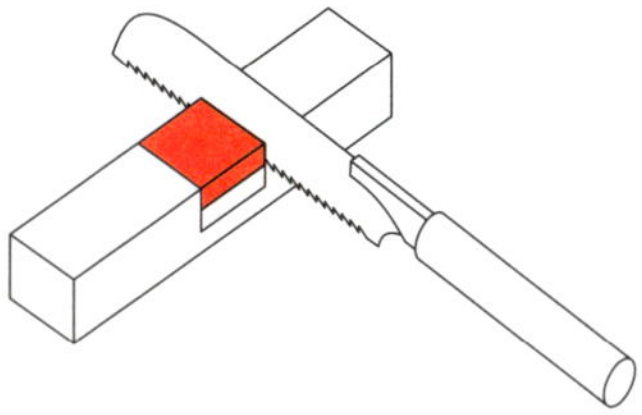

B2

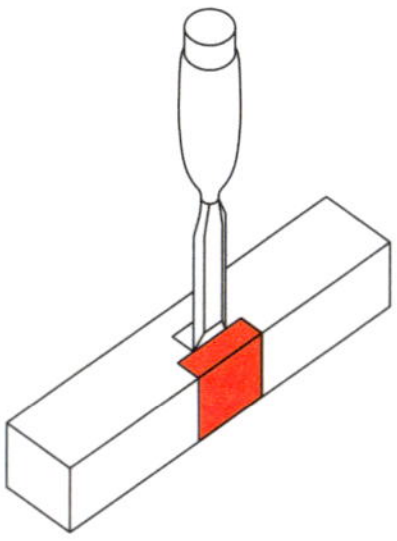

B3

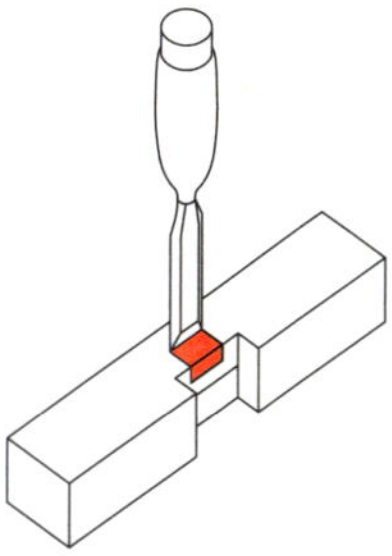

B4

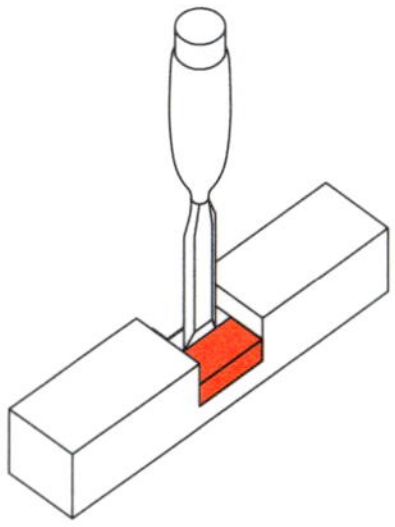

B5

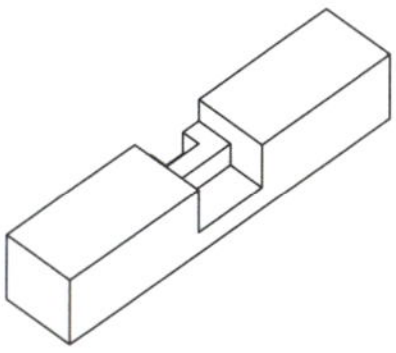

B

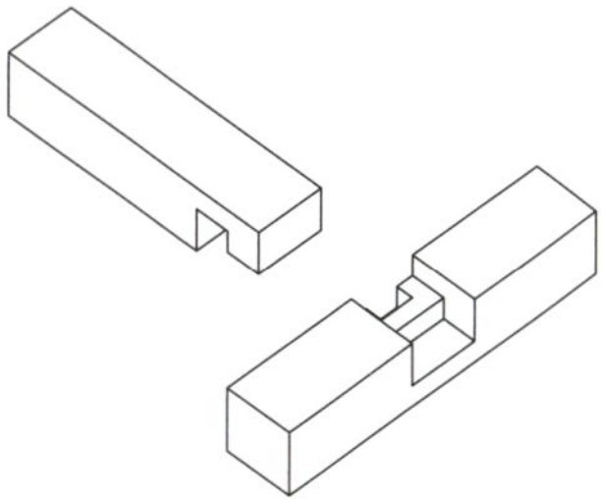

A+B

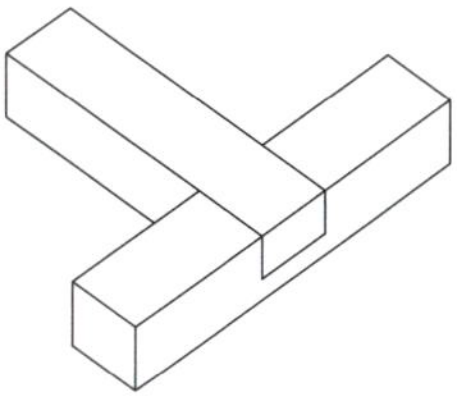

AB

T-Joints

Sloped cogged lap joint

kata-ari-ai-kaki-watari-ago-tsugi
片蟻相欠渡顎接ぎ

Hakenquerblatt mit schrägem Haken

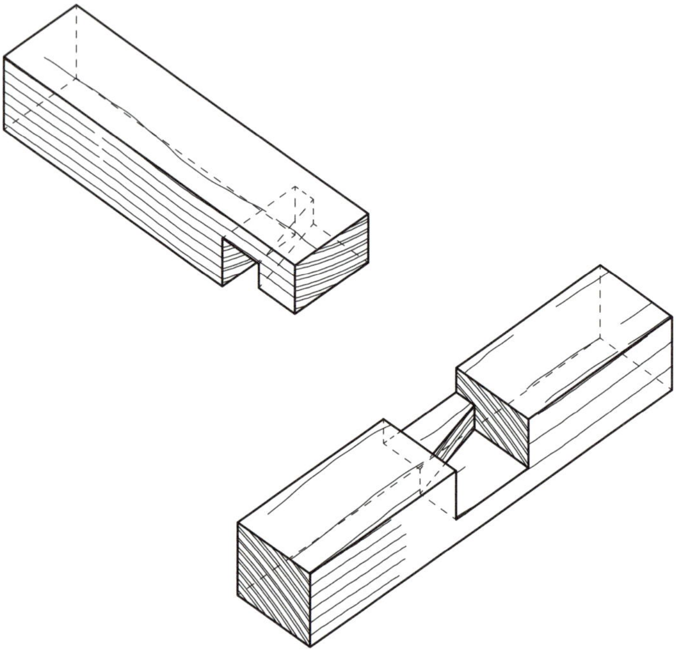

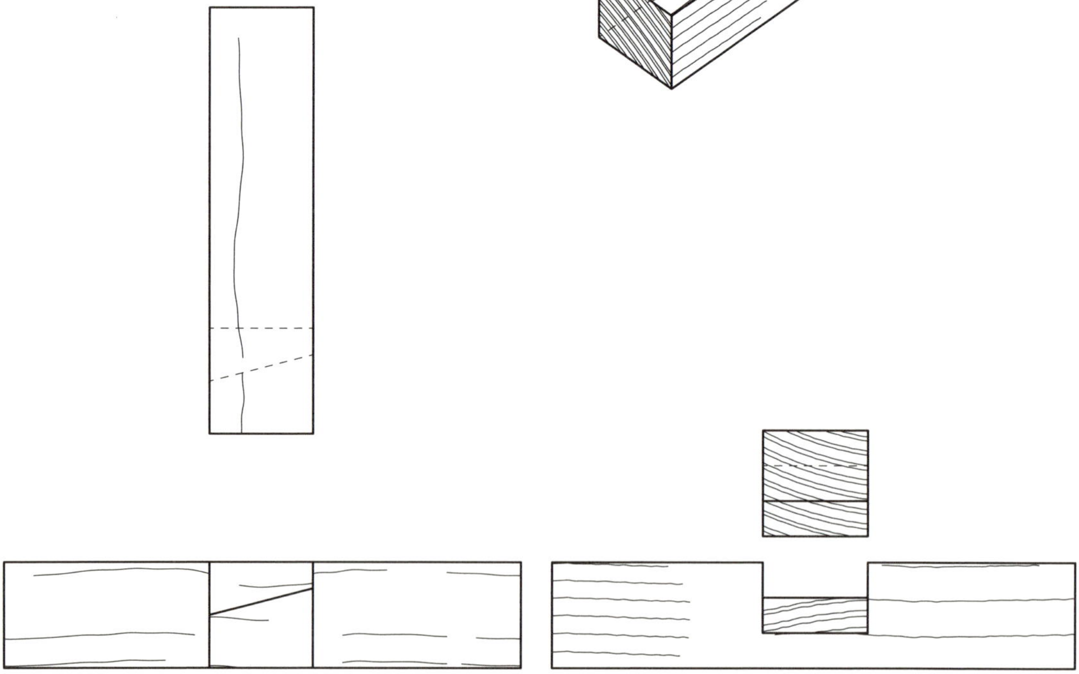

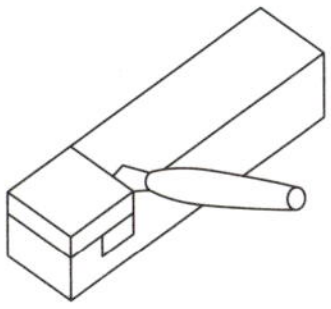

A1

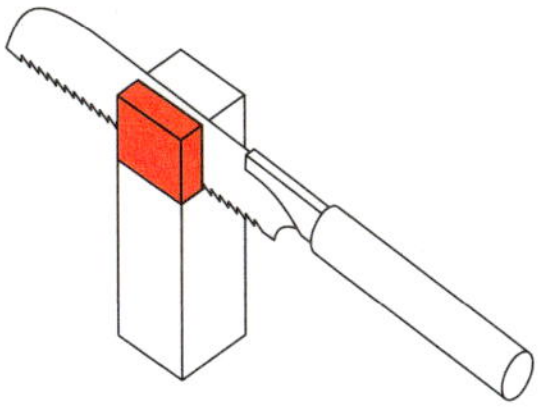

A2

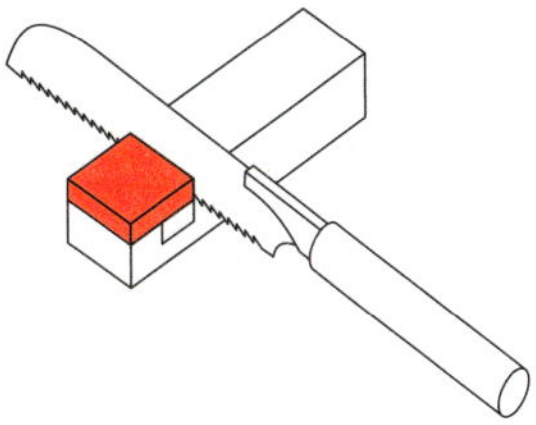

A3

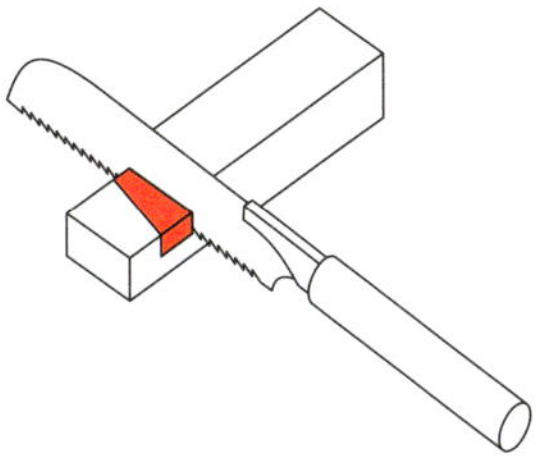

A4

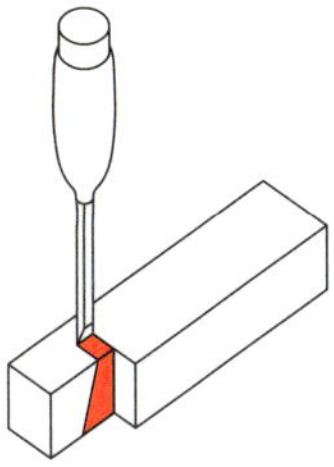

A5

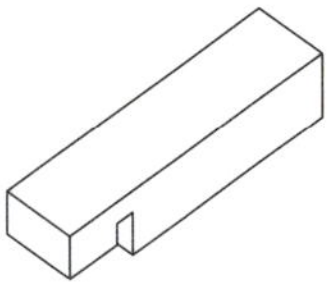

A

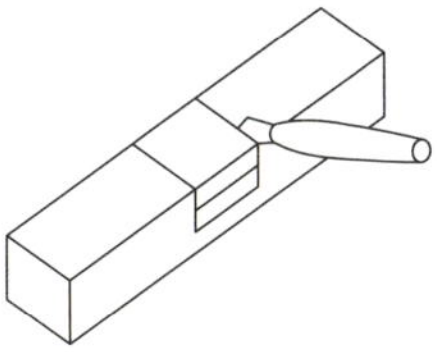

B1

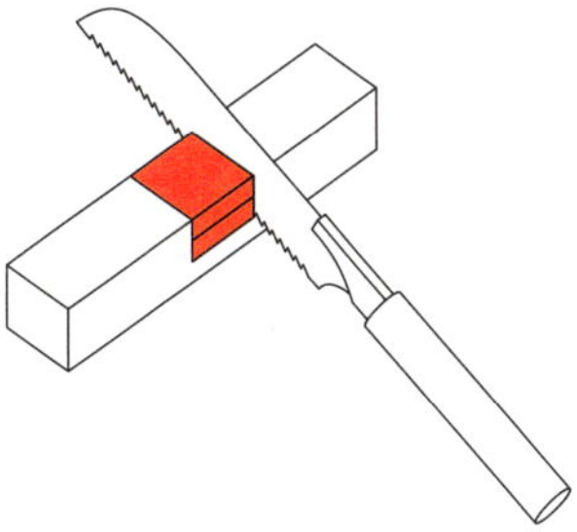

B2

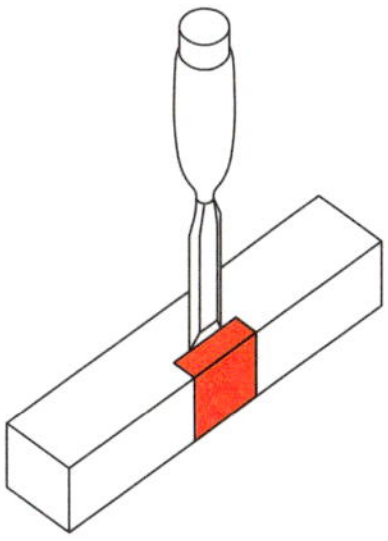

B3

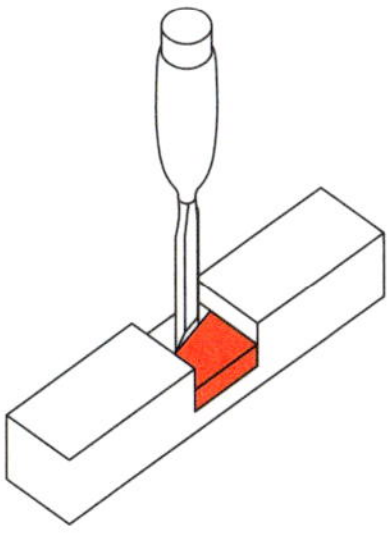

B4

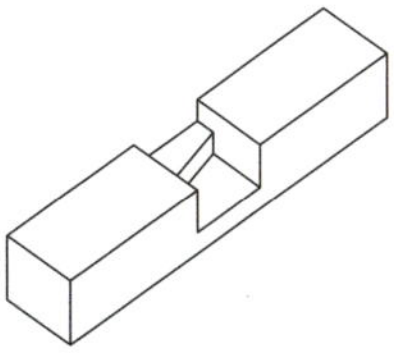

B

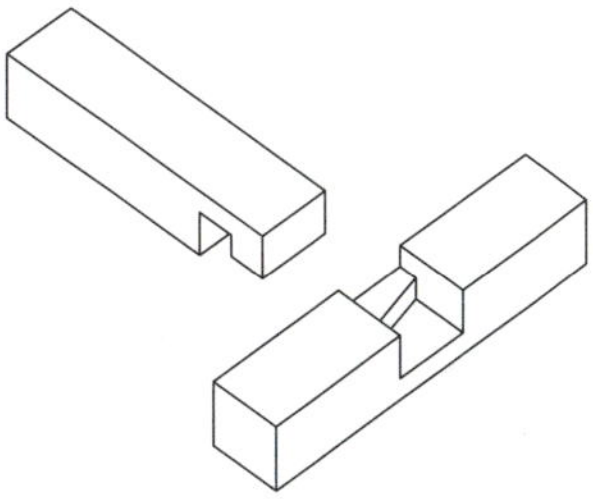

A+B

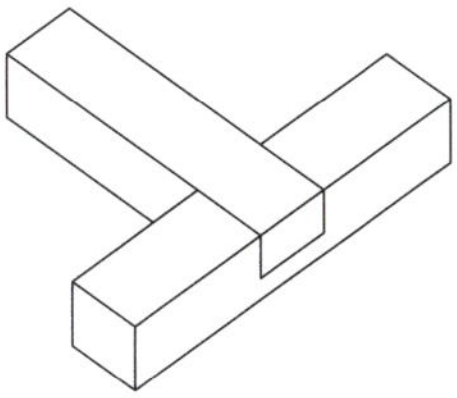

AB

T-Joints

Double cogged lap joint

koshi-kake-ai-kaki-watari-ago-tsugi
腰掛相欠渡顎接ぎ（胴付）

Hakenquerblatt
mit mittigem Haken

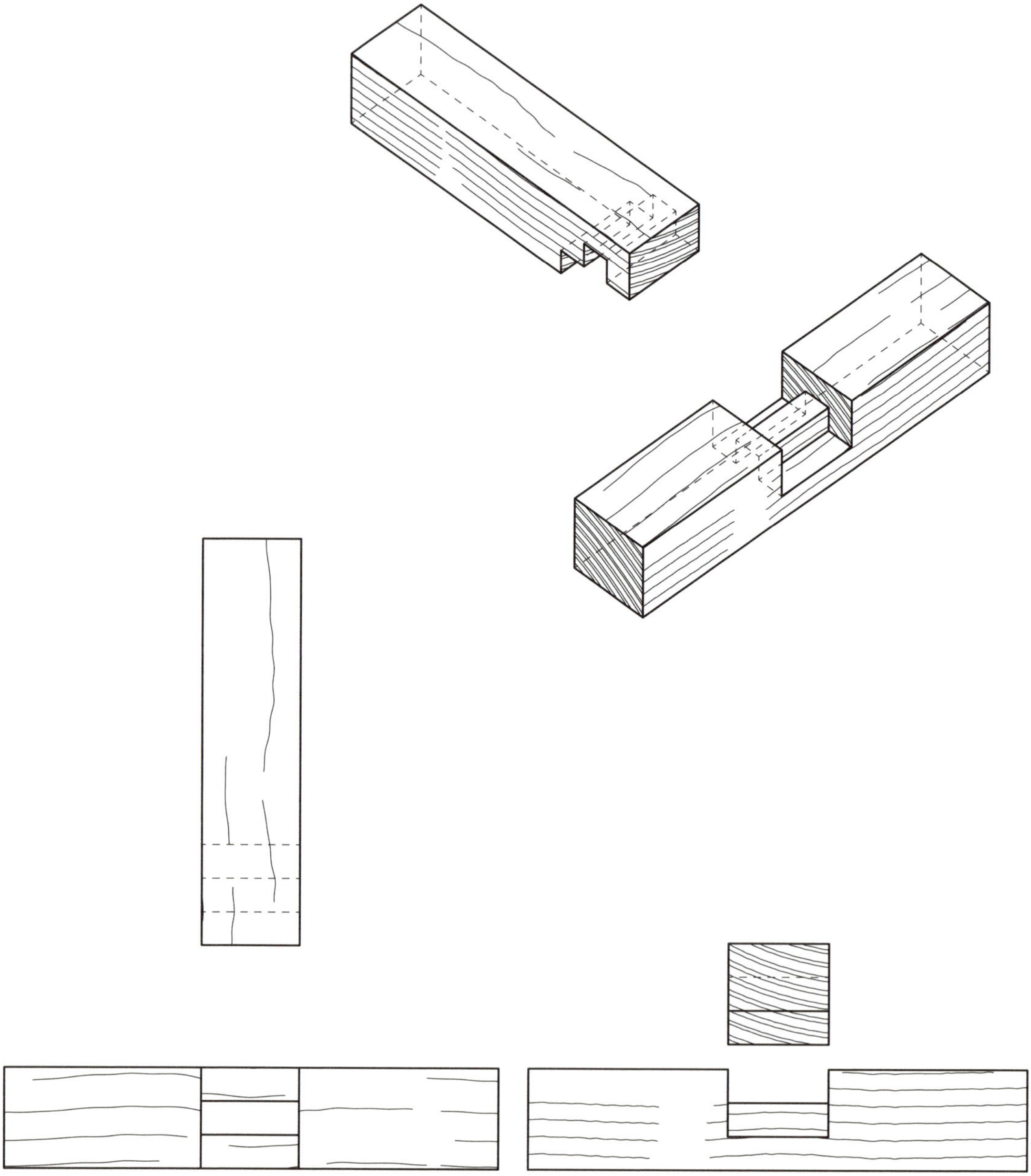

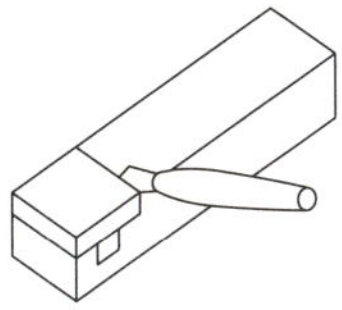

A1

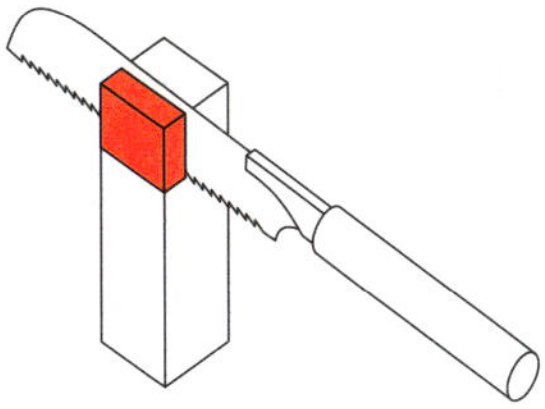

A2

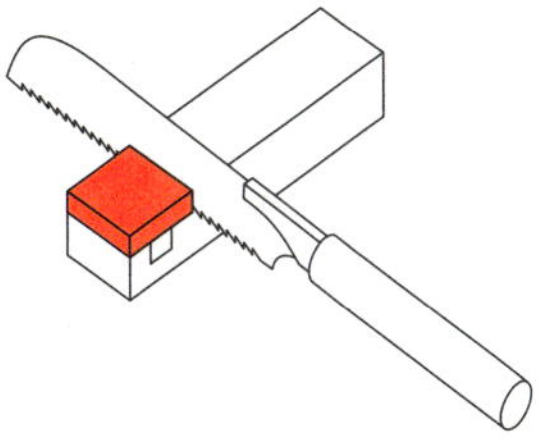

A3

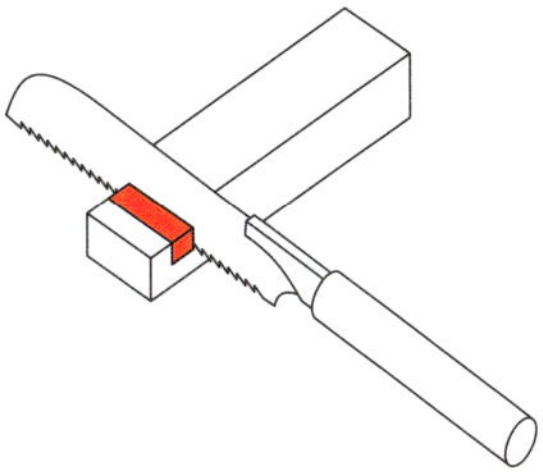

A4

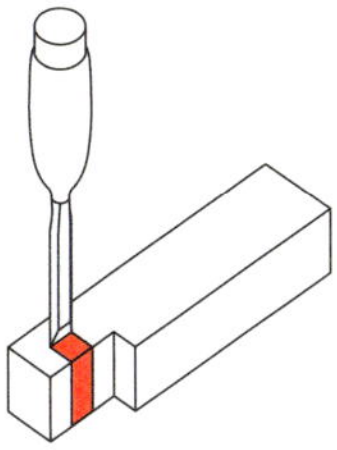

A5

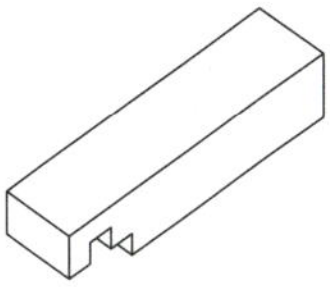

A

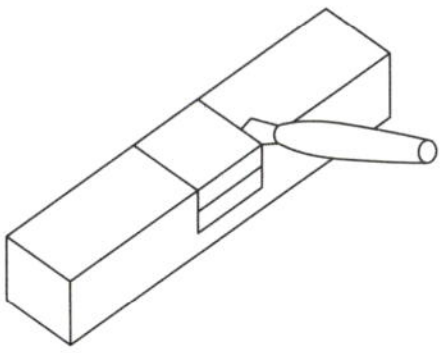

B1

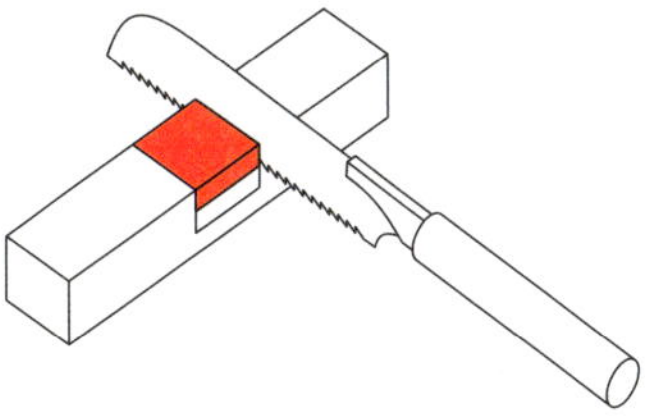

B2

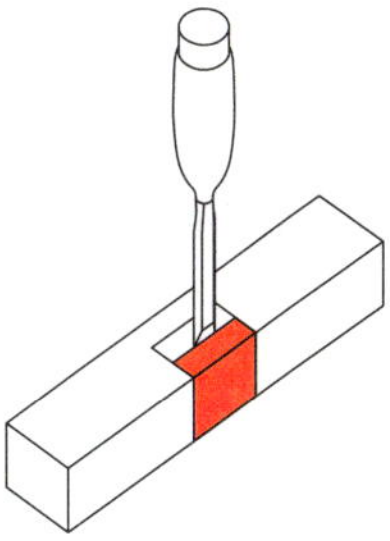

B3

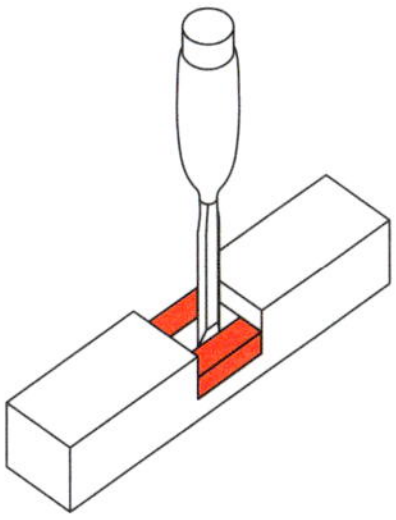

B4

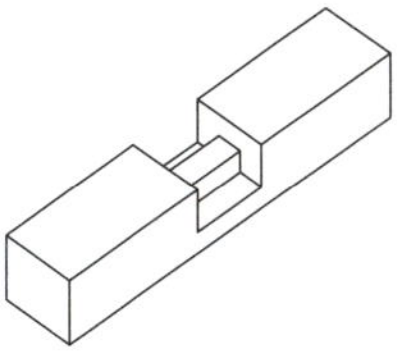

B

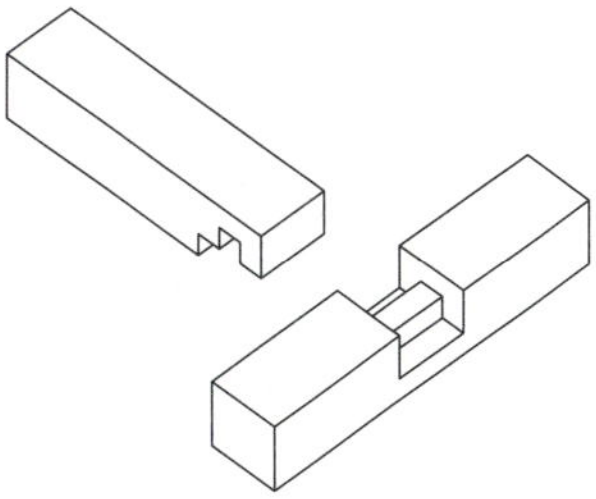

A+B

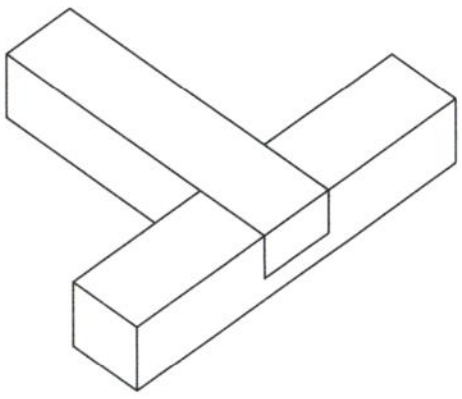

AB

T-Joints

Stopped mortise and tenon joint

mechi-tsugi
目違接ぎ

Querzapfen

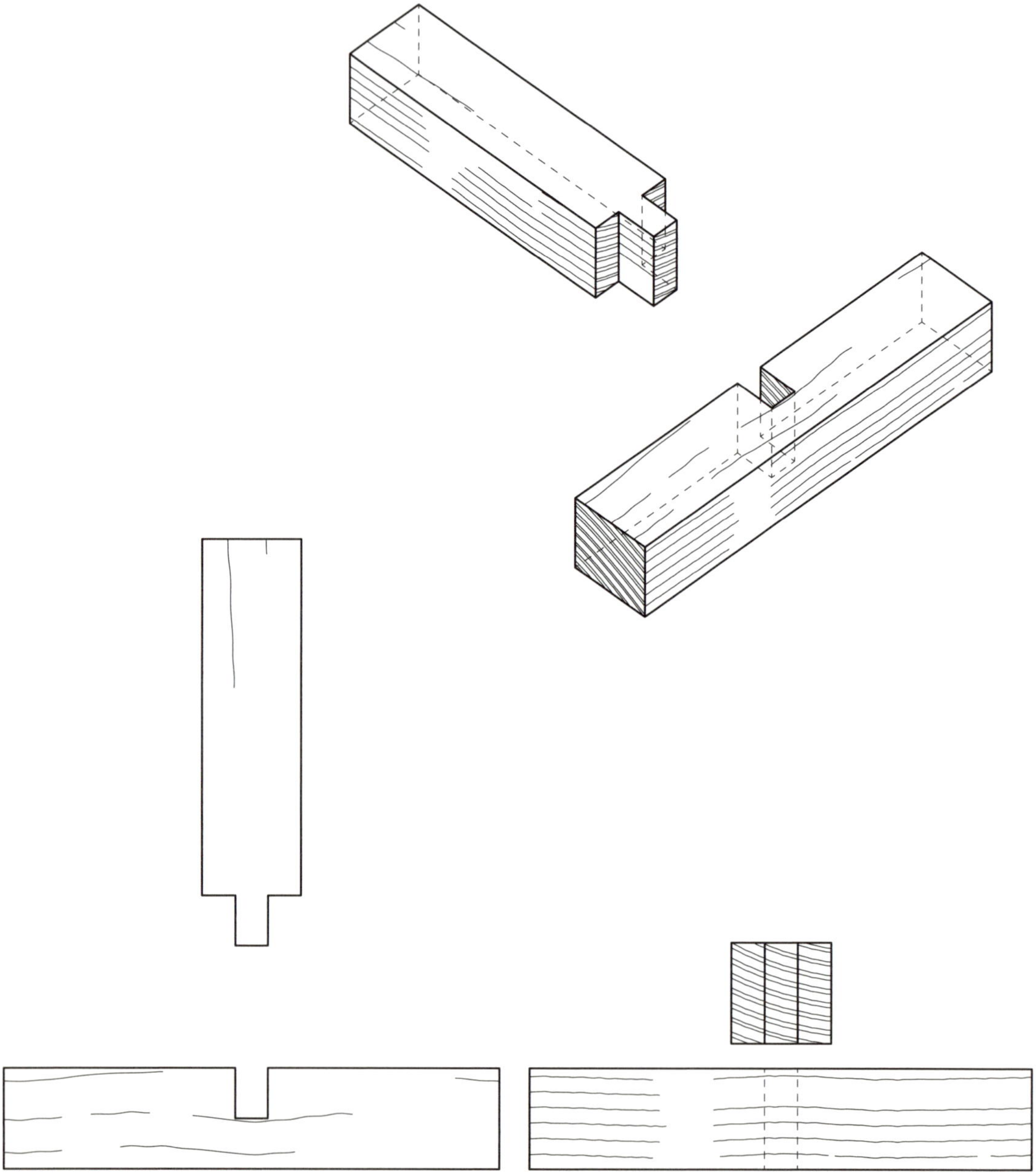

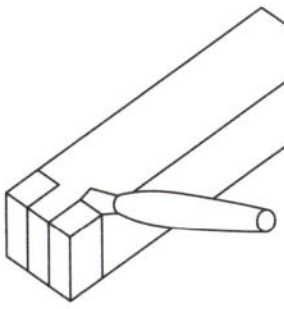

A1

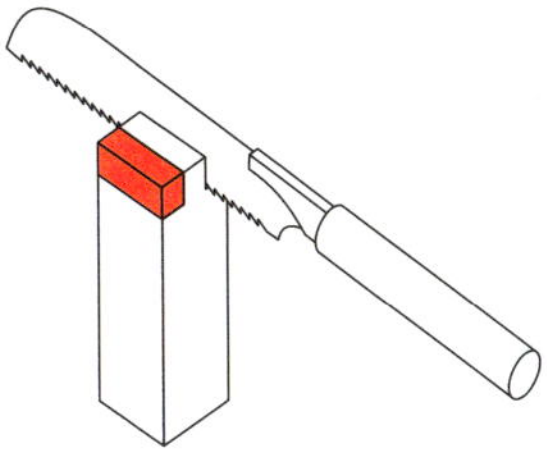

A2

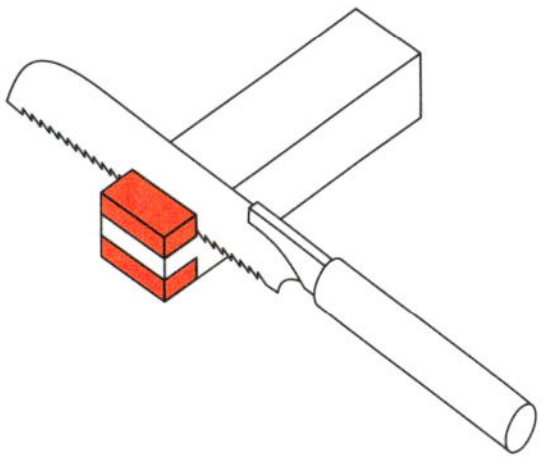

A3

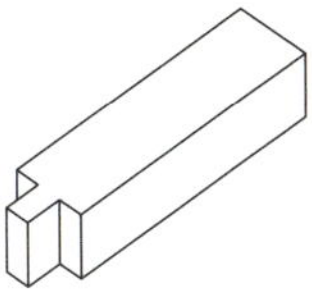

A

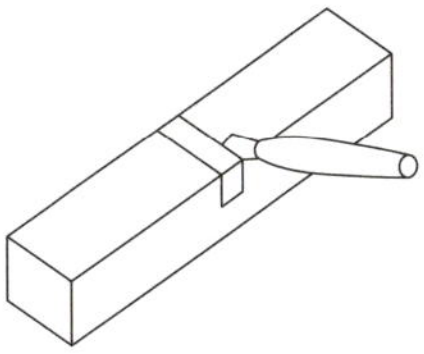

B1

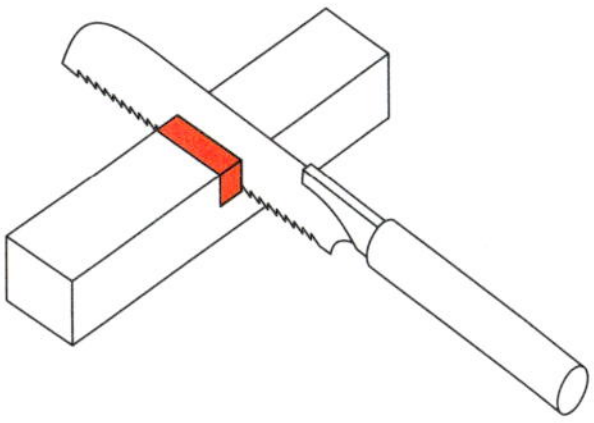

B2

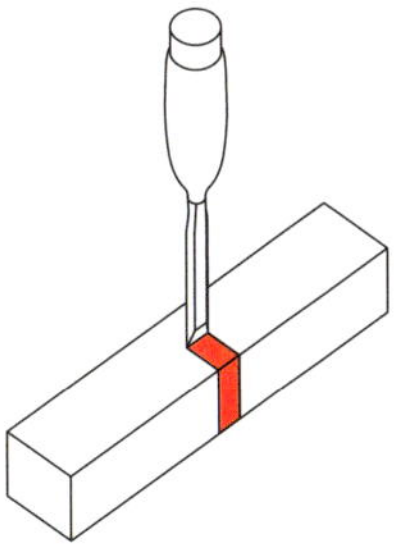

B3

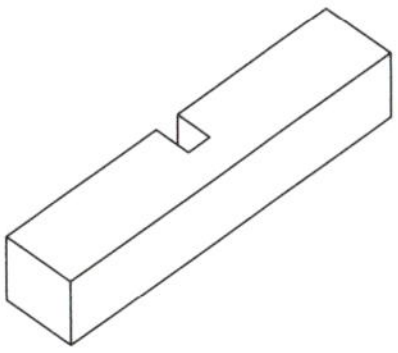

B

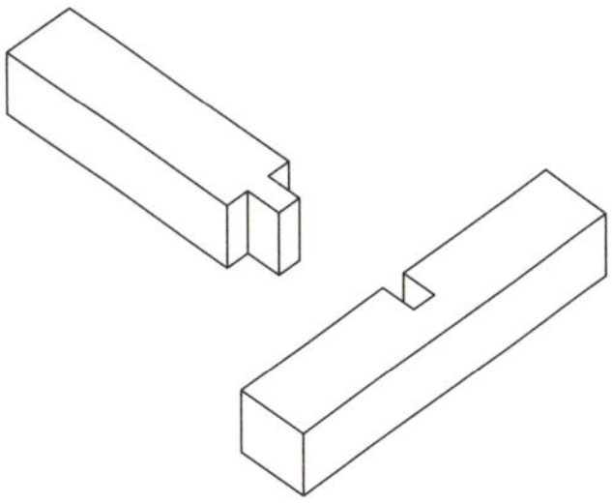

A+B

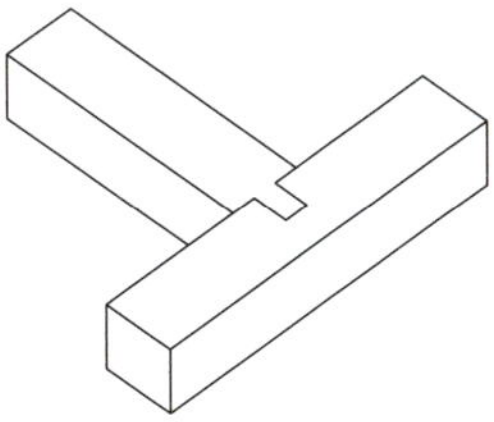

AB

T-Joints

Stopped dovetailed mortise and tenon joint

ari-tsugi
蟻接ぎ

Schwalbenschwanzquerzapfen

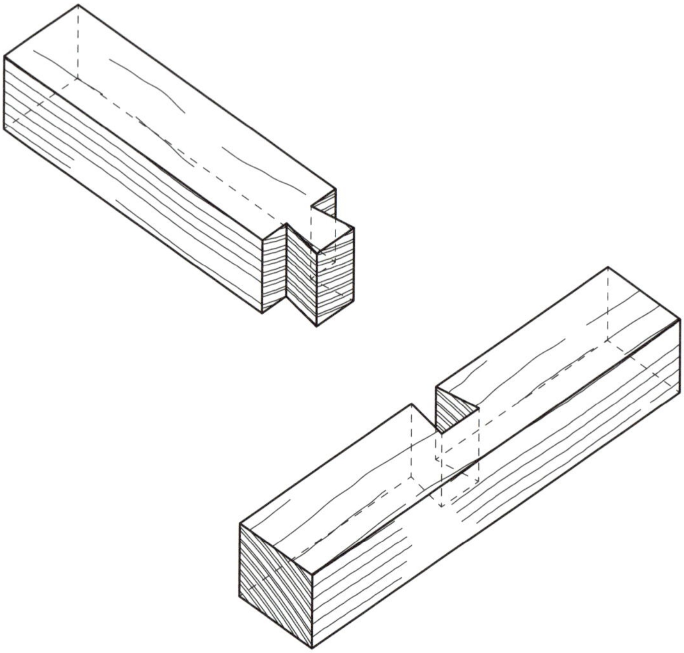

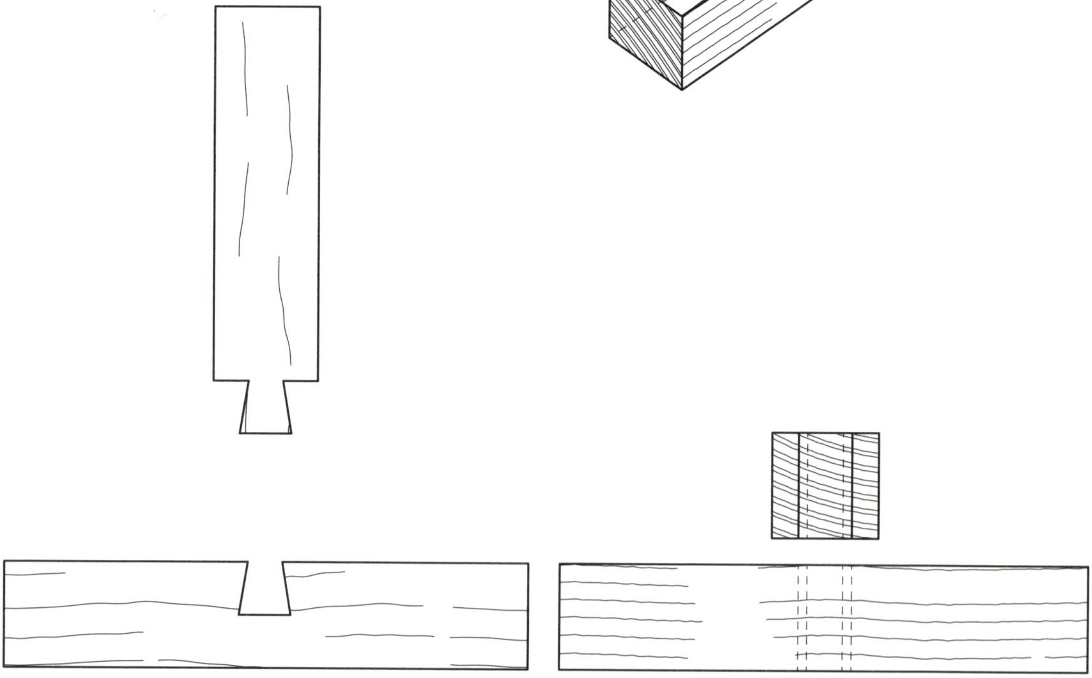

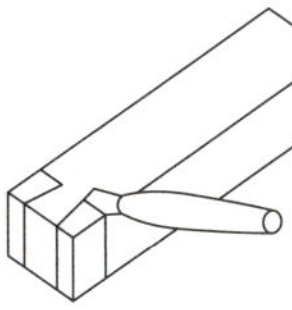

A1

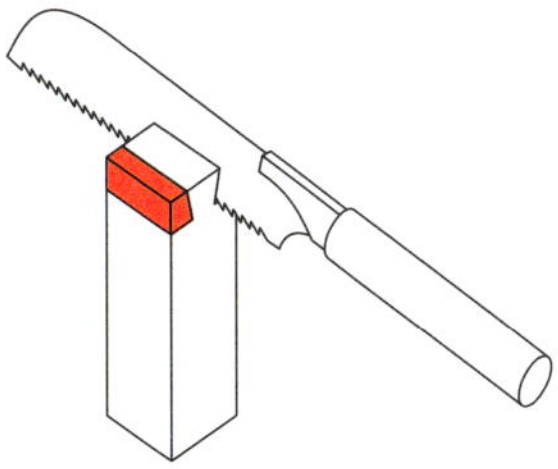

A2

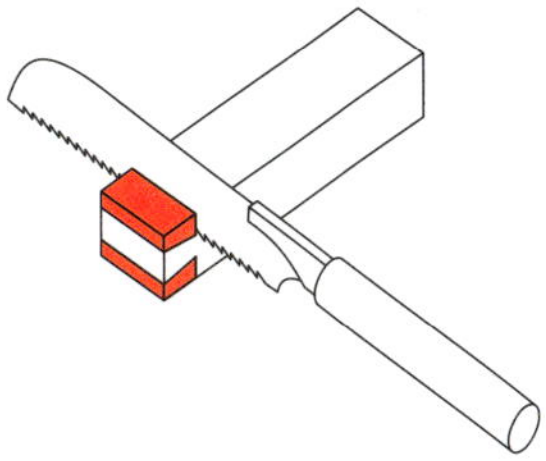

A3

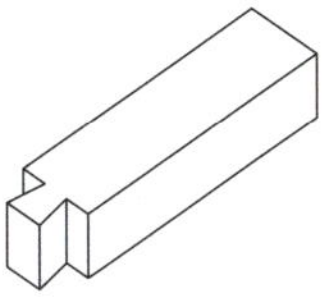

A

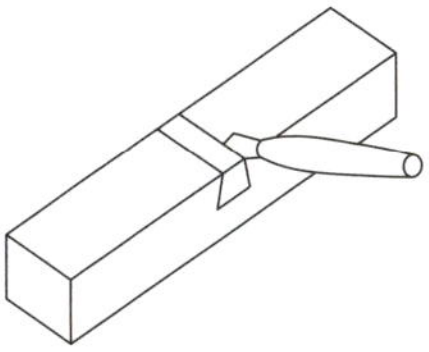

B1

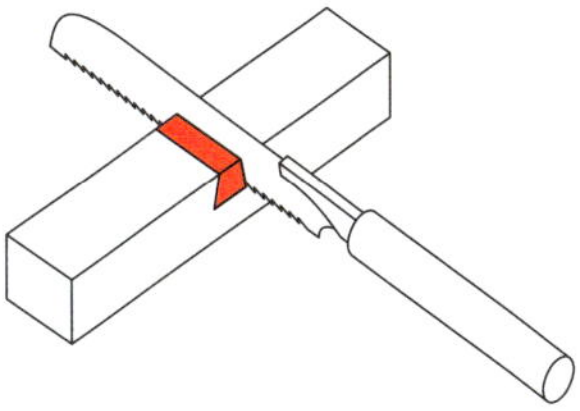

B2

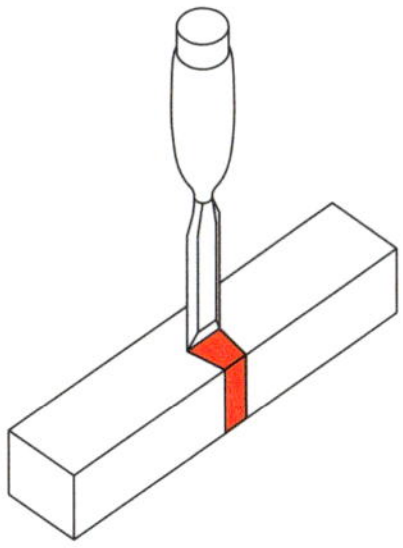

B3

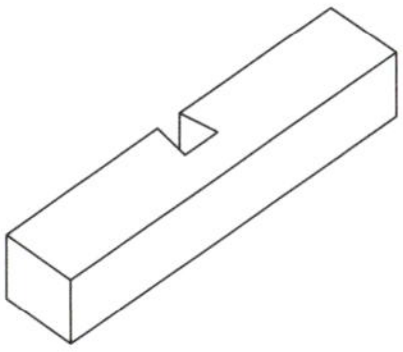

B

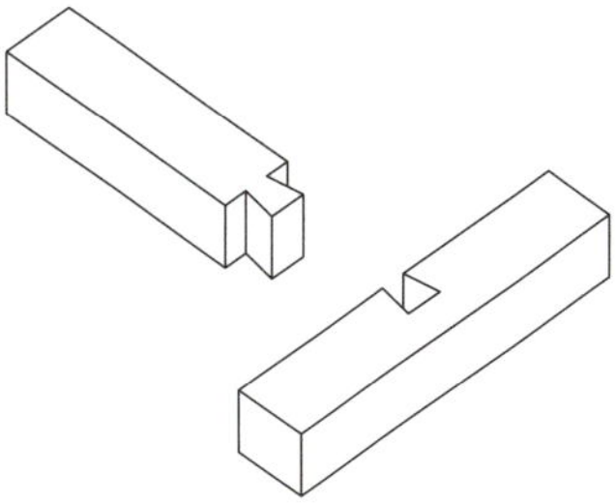

A+B

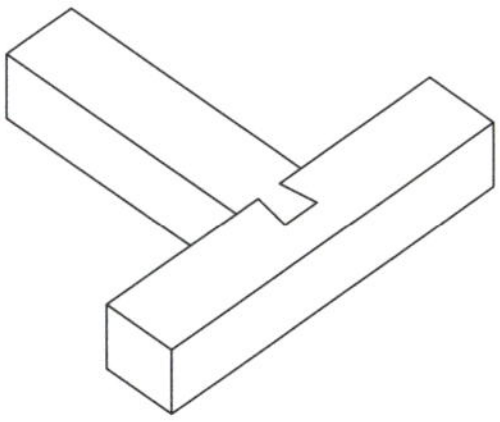

AB

T-Joints

Dovetail half lap joint

ōari-ai-kaki-tsugi
大蟻相欠接ぎ

Schwalbenschwanzquerblatt

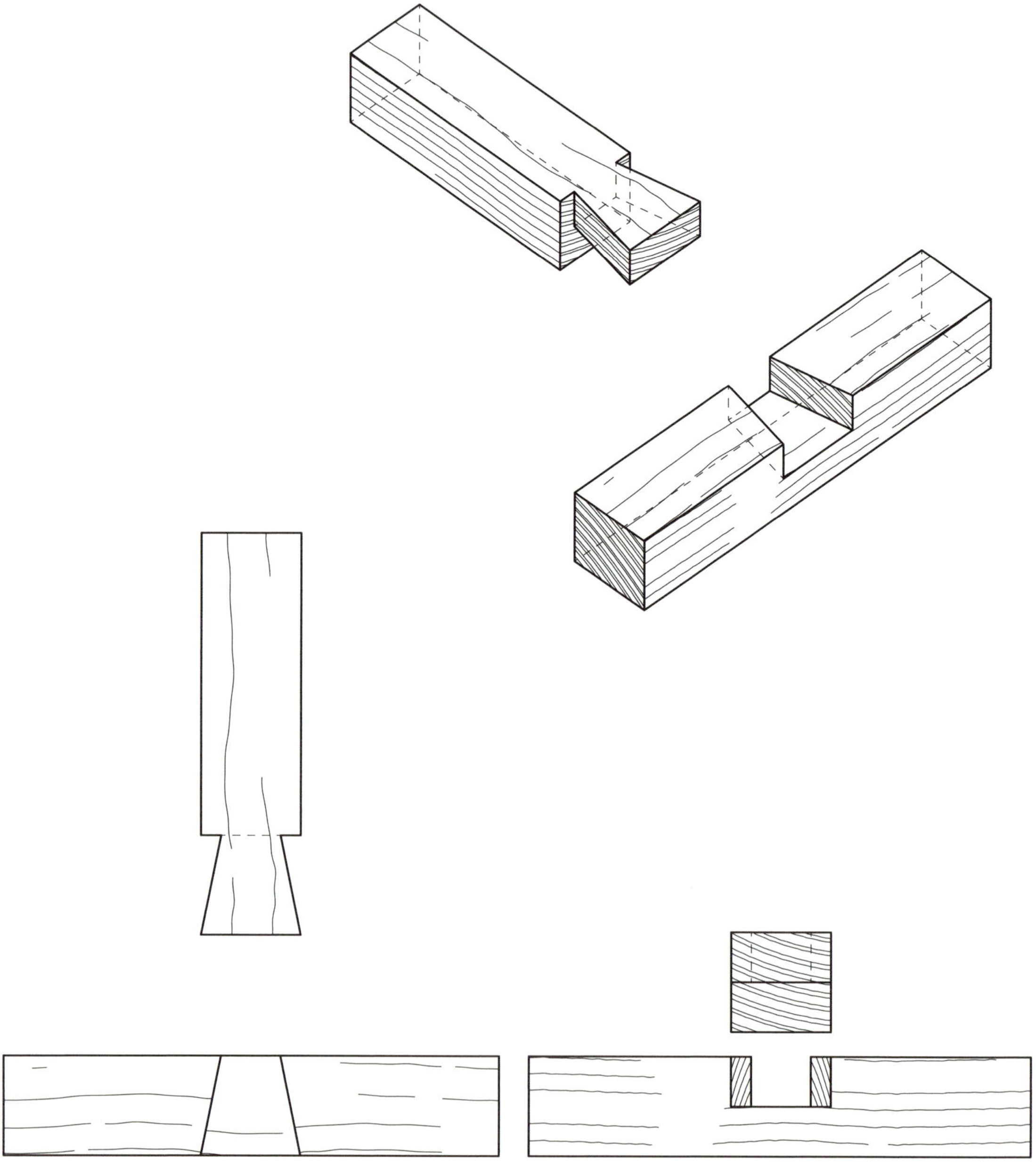

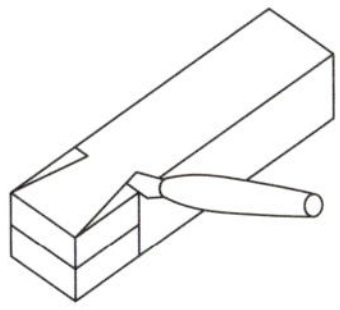

A1

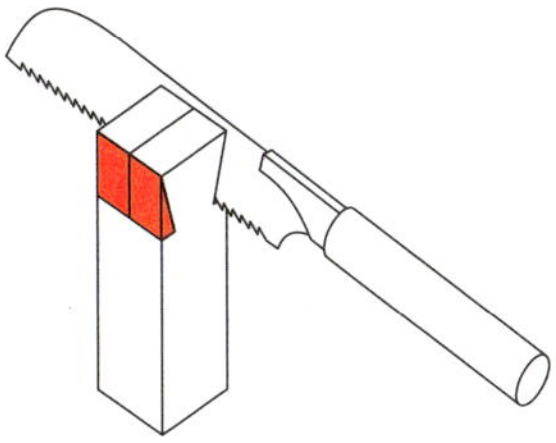

A2

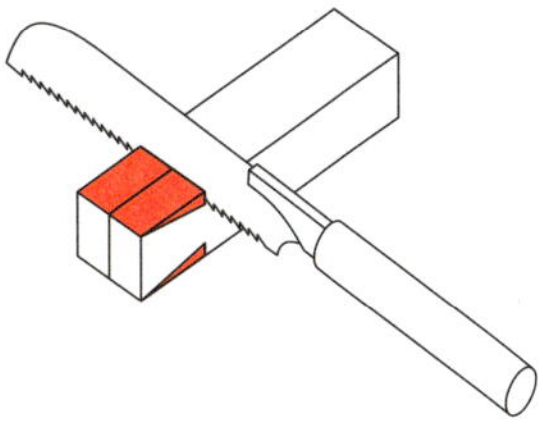

A3

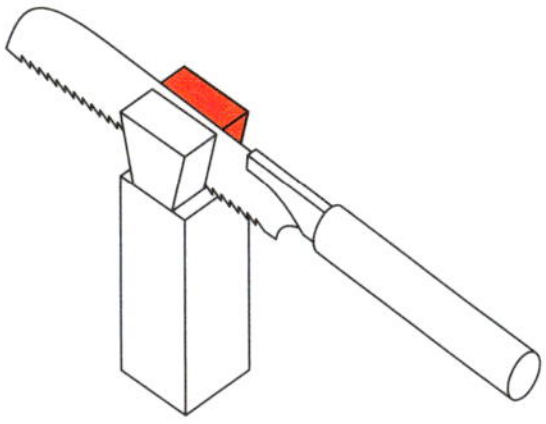

A4

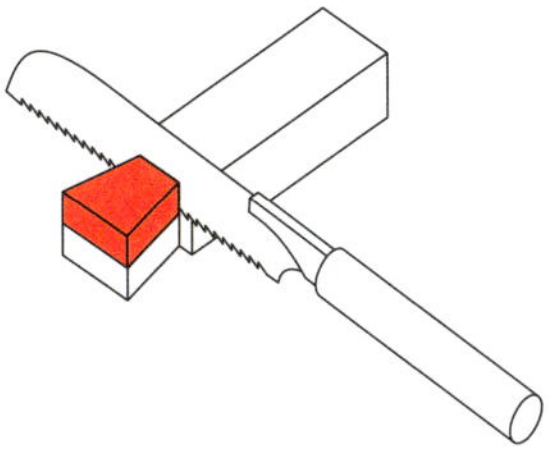

A5

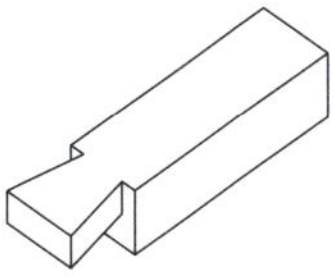

A

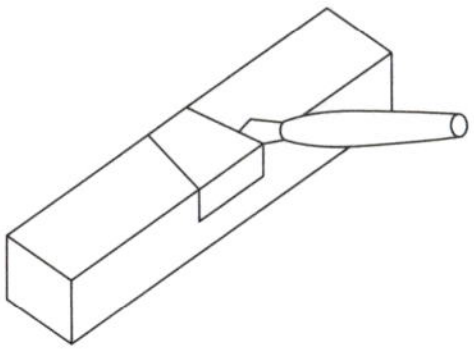

B1

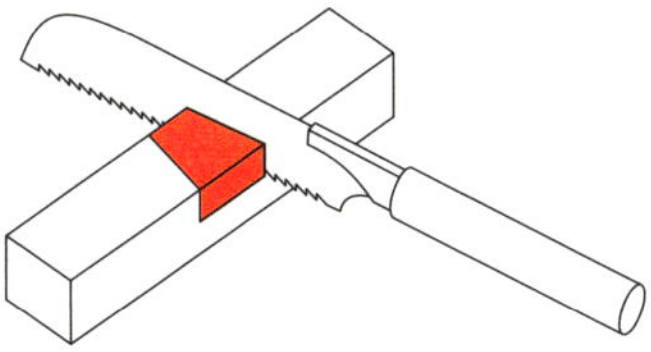

B2

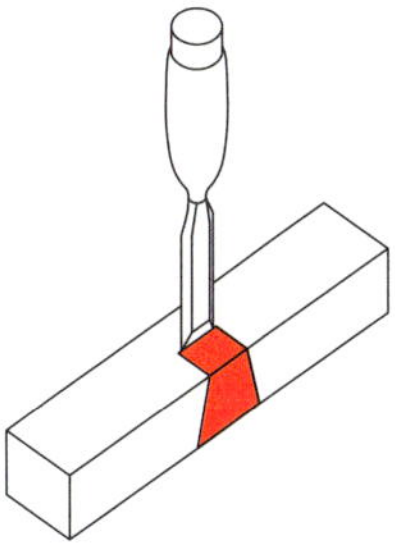

B3

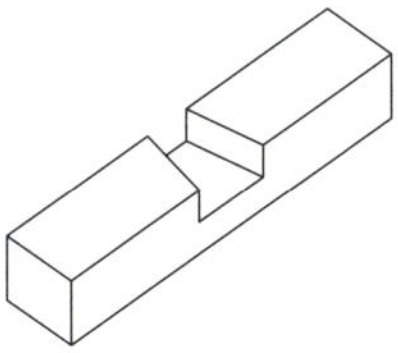

B

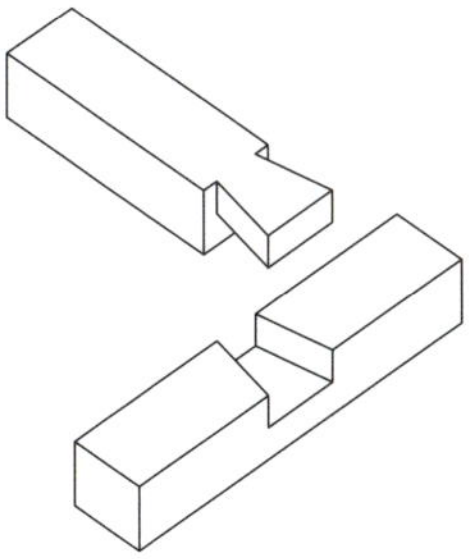

A+B

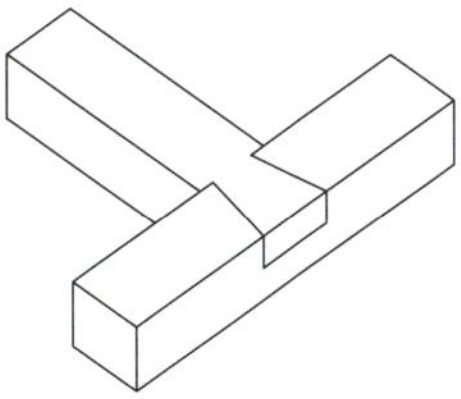

AB

T-Joints

Stopped dovetail half lap joint

kakushi-ari-ai-kaki-tsugi
隠蟻相欠接ぎ

halbverdecktes
Schwalbenschwanzquerblatt

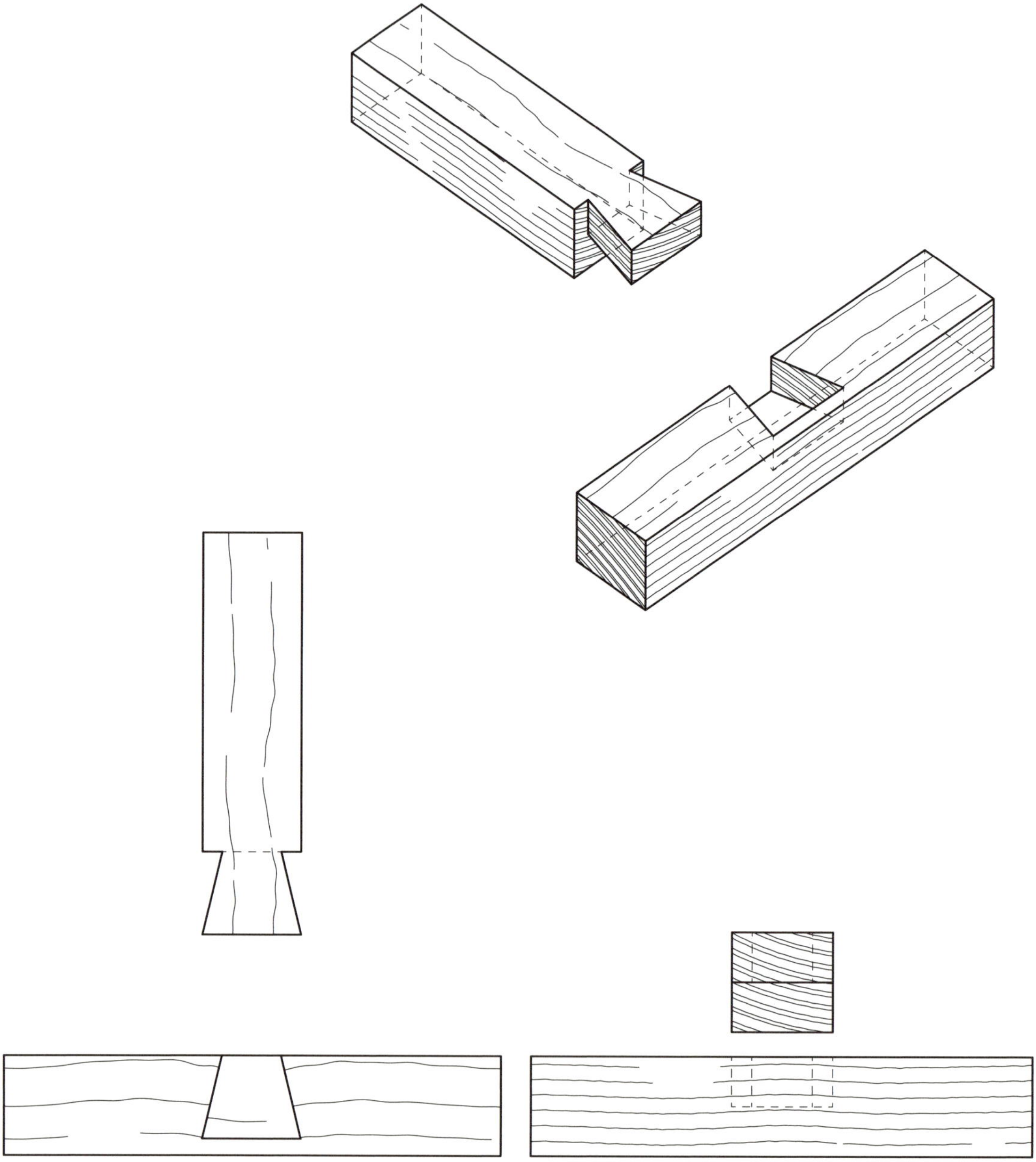

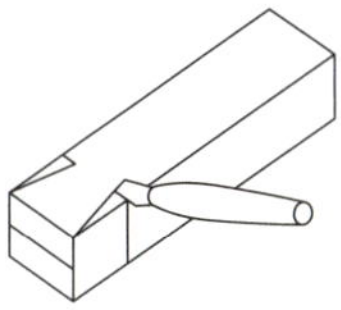

A1

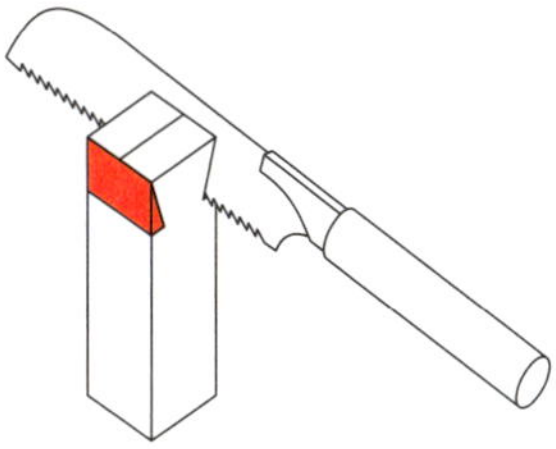

A2

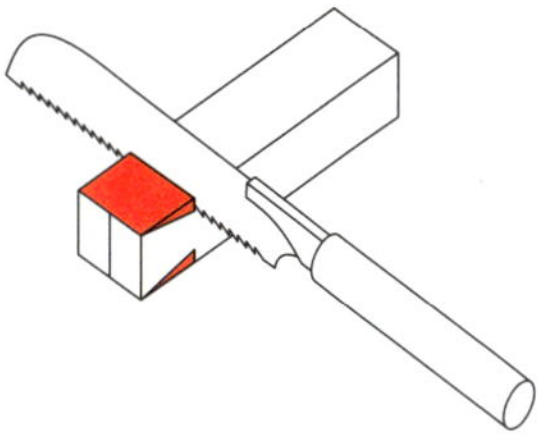

A3

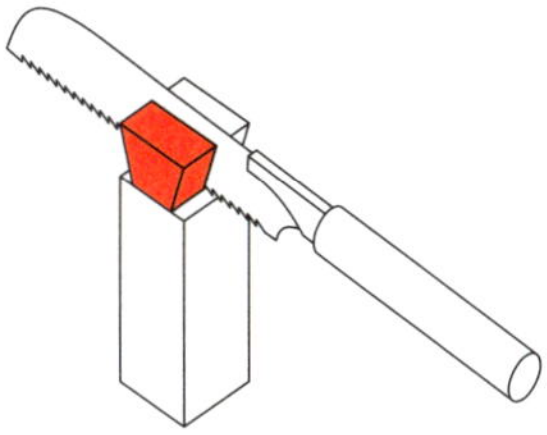

A4

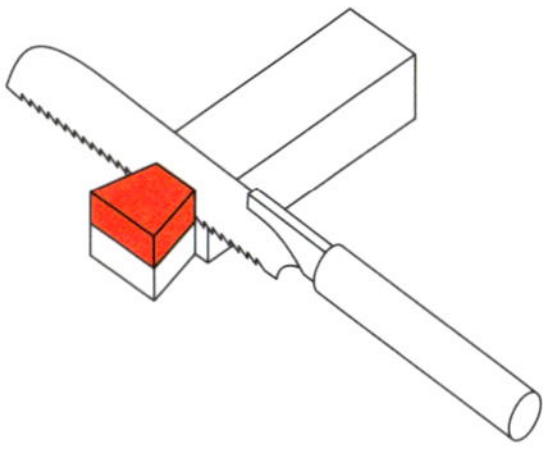

A5

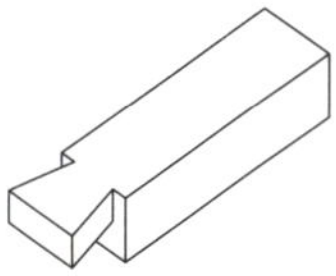

A

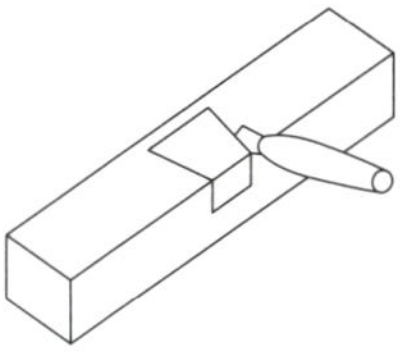

B1

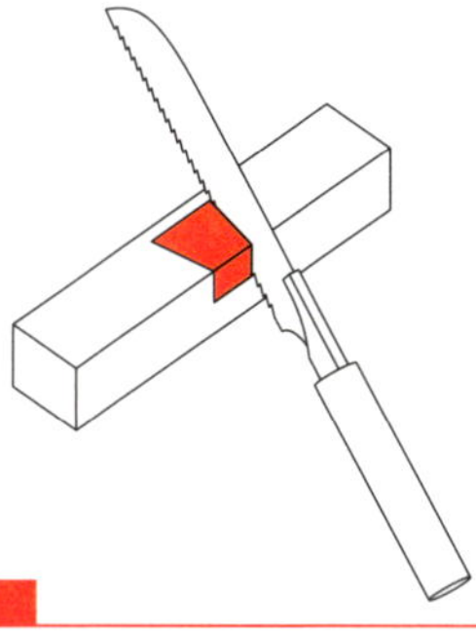

B2

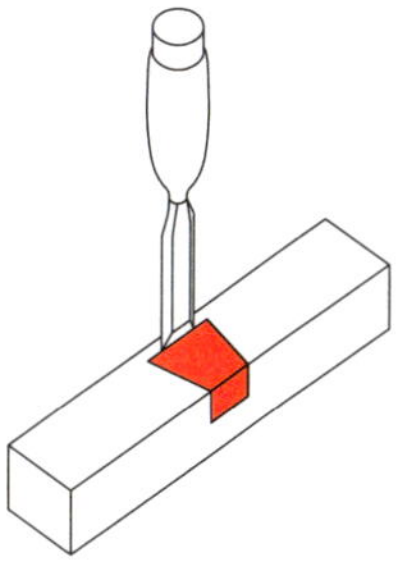

B3

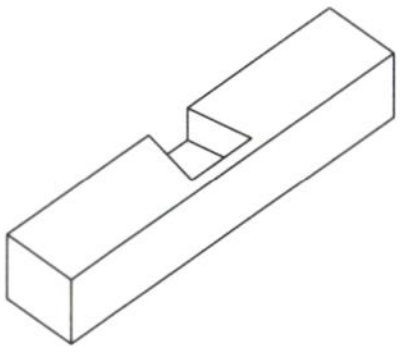

B

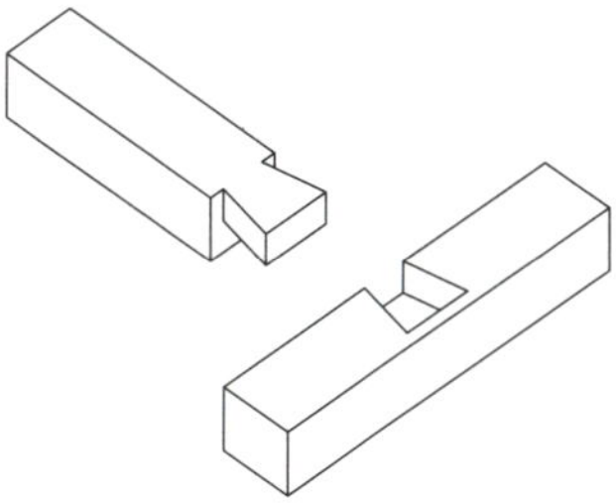

A+B

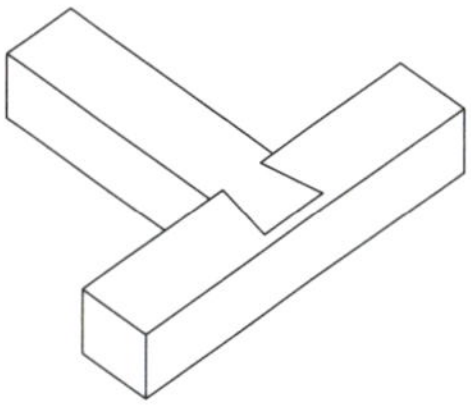

AB

T-Joints

One-sided dovetail half lap joint

kata-ari-ai-kaki-tsugi
片蟻相欠接ぎ

einseitiges Schwalbenschwanzquerblatt

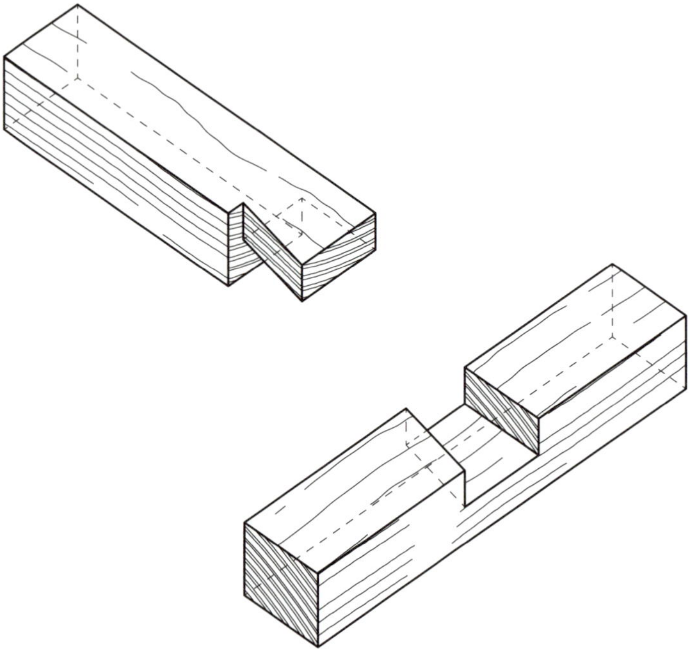

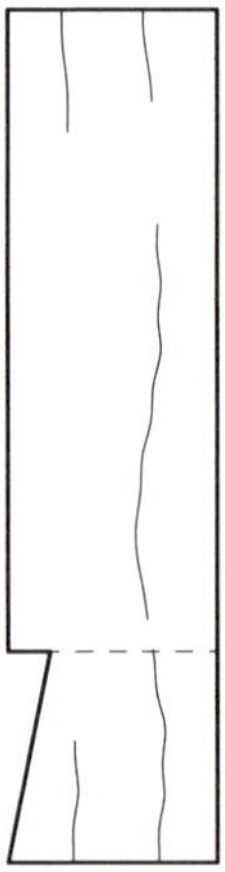

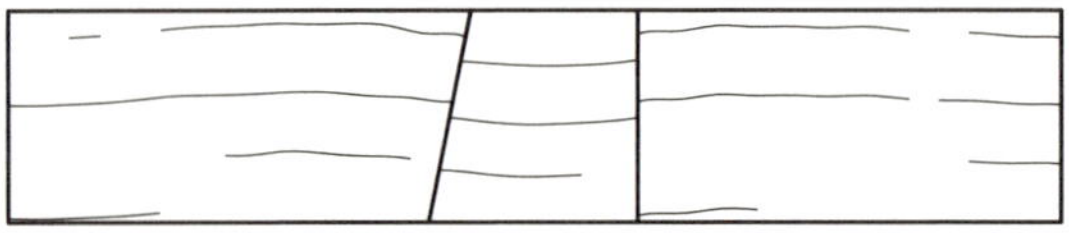

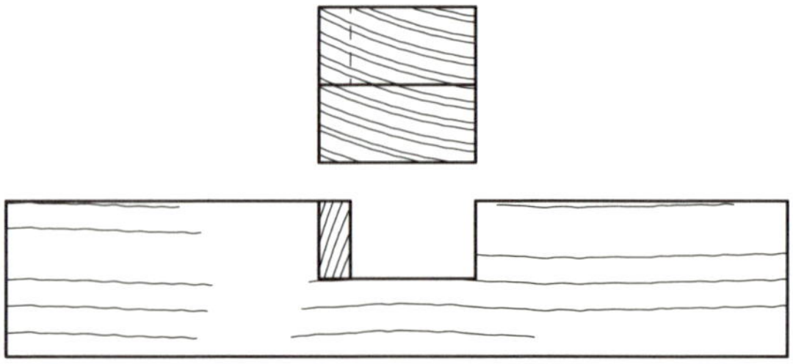

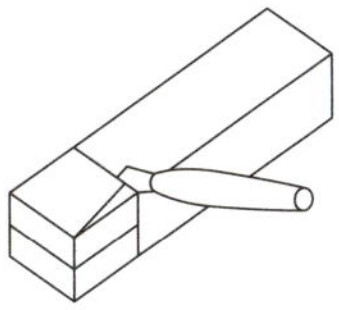

A1

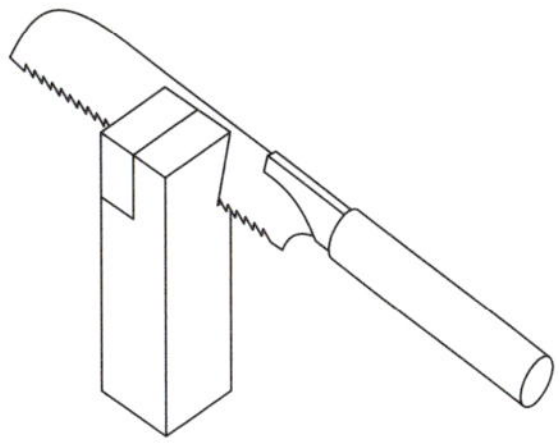

A2

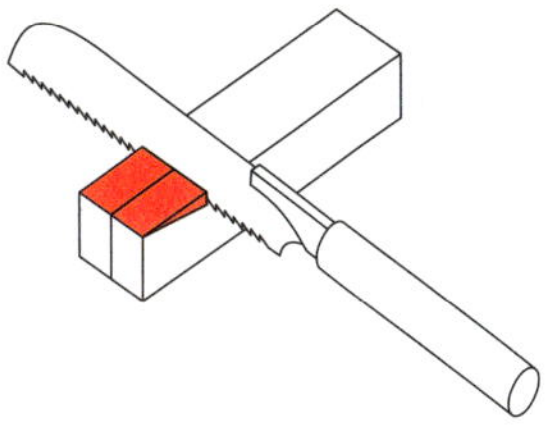

A3

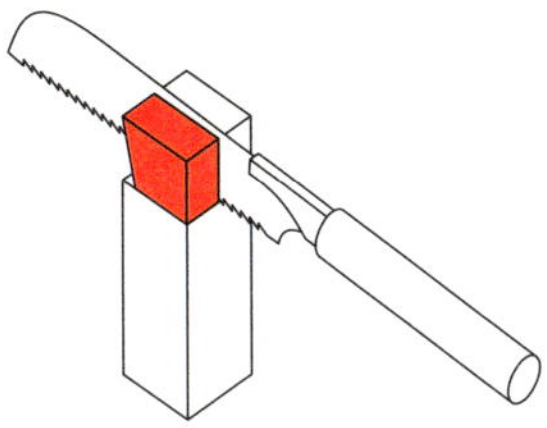

A4

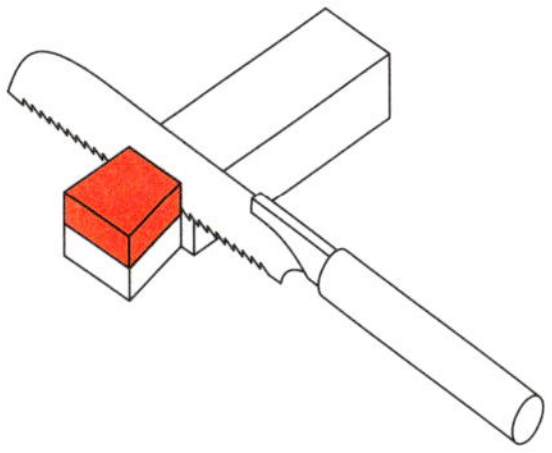

A5

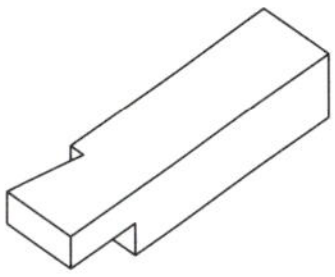

A

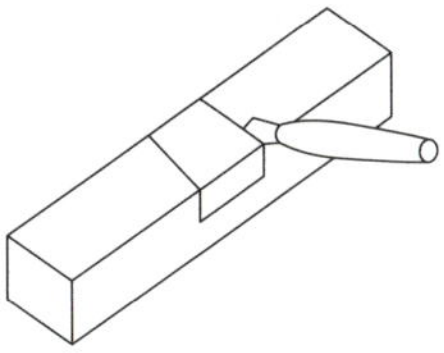

B1

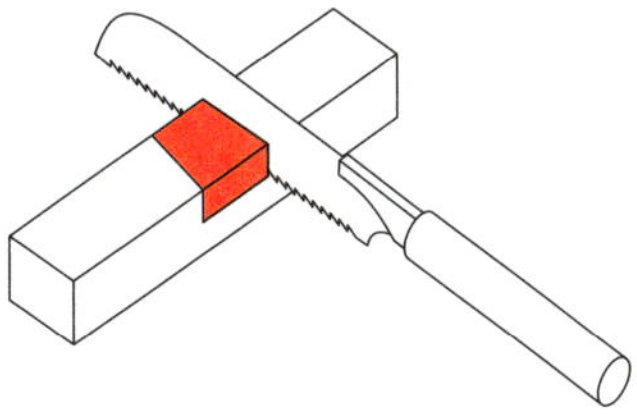

B2

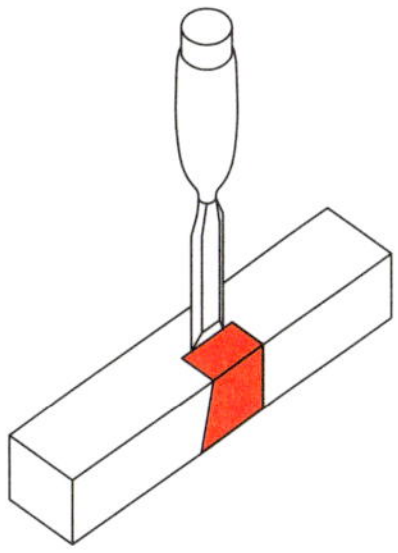

B3

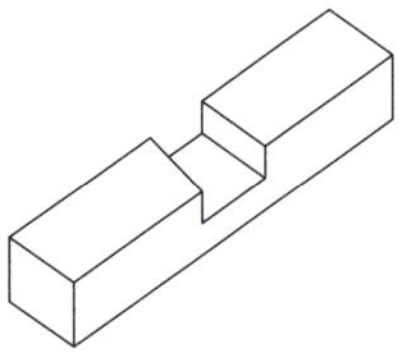

B

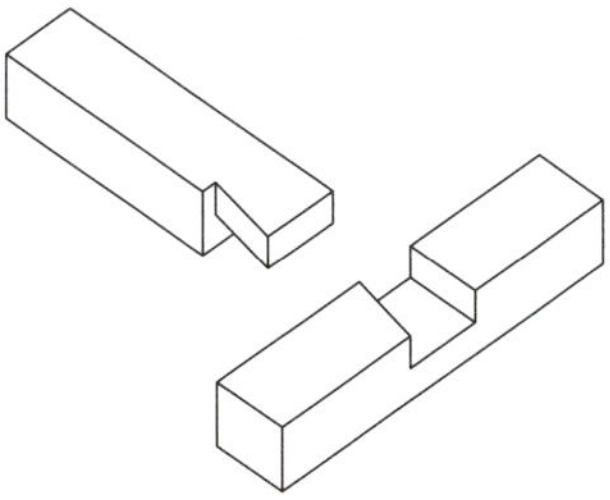

A+B

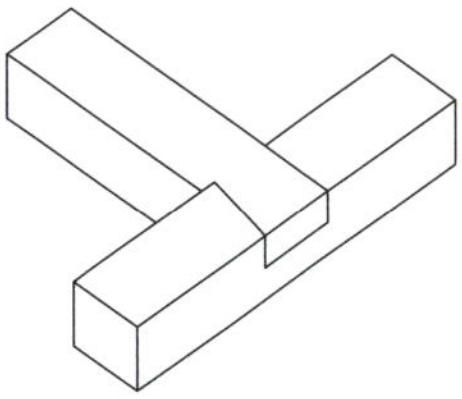

AB

T-Joints

Stopped one-sided dovetail half lap joint

kakushi-kata-ari-ai-kaki-tsugi
隠片蟻相欠接ぎ

halbverdecktes einseitiges
Schwalbenschwanzquerblatt

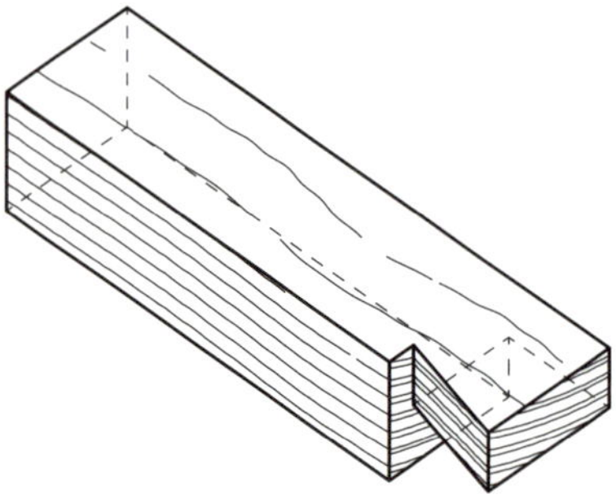

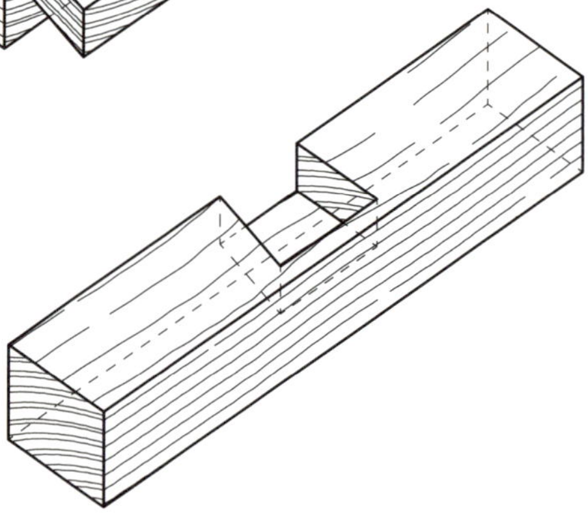

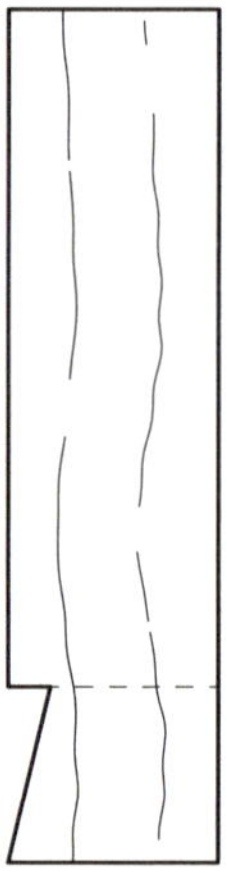

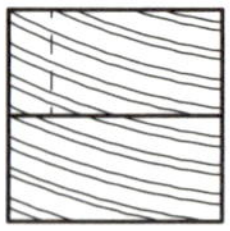

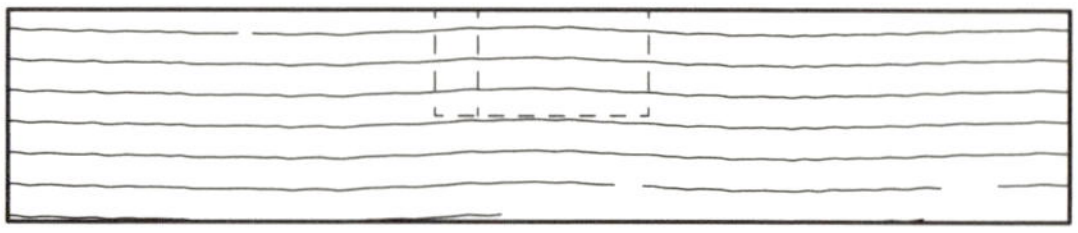

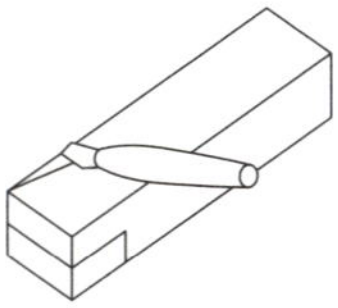
A1

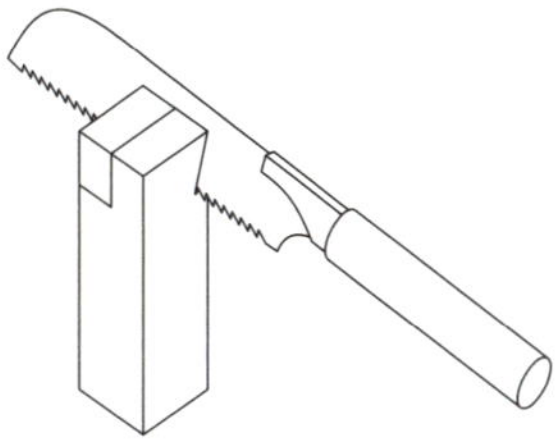
A2

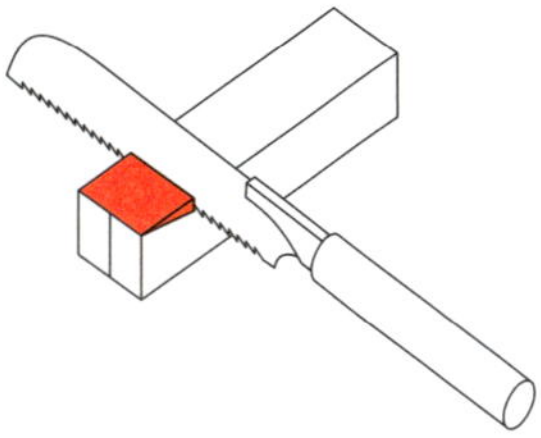
A3

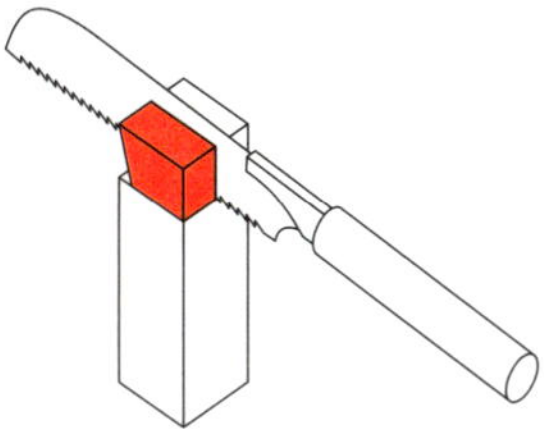
A4

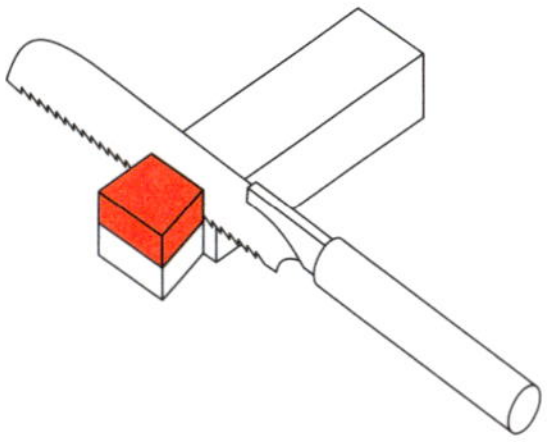
A5

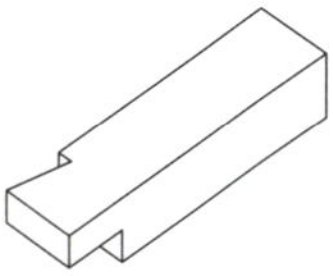
A

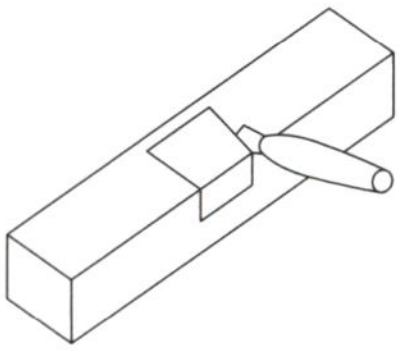
B1

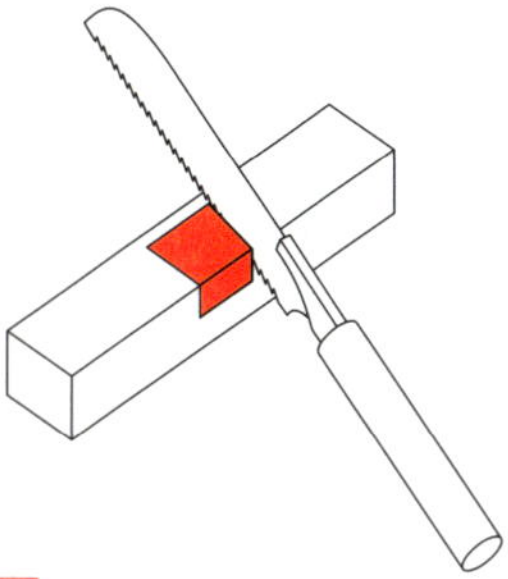
B2

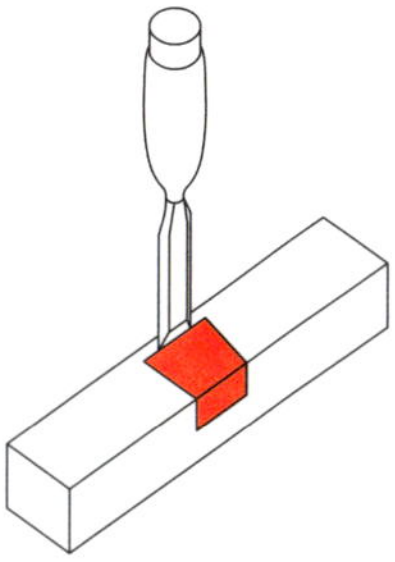
B3

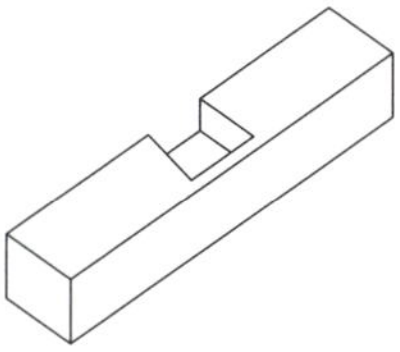
B

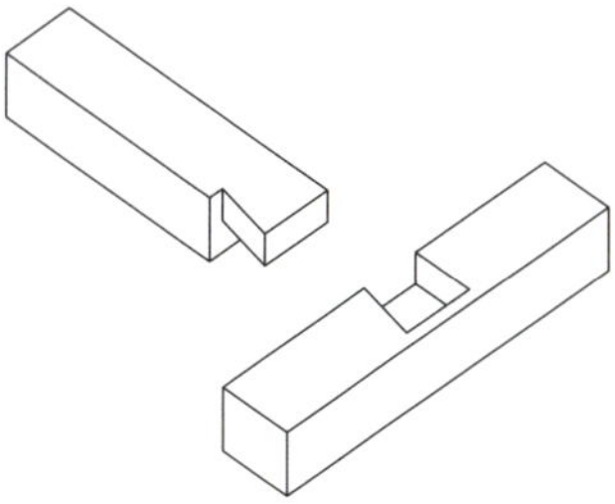
A+B

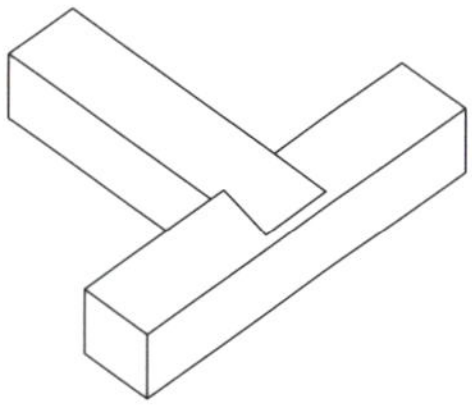
AB

T-Joints

Housed dovetail half lap joint

koshi-kake-ōari-ai-kaki-tsugi
腰掛大蟻相欠接ぎ

eingelassenes
Schwalbenschwanzquerblatt

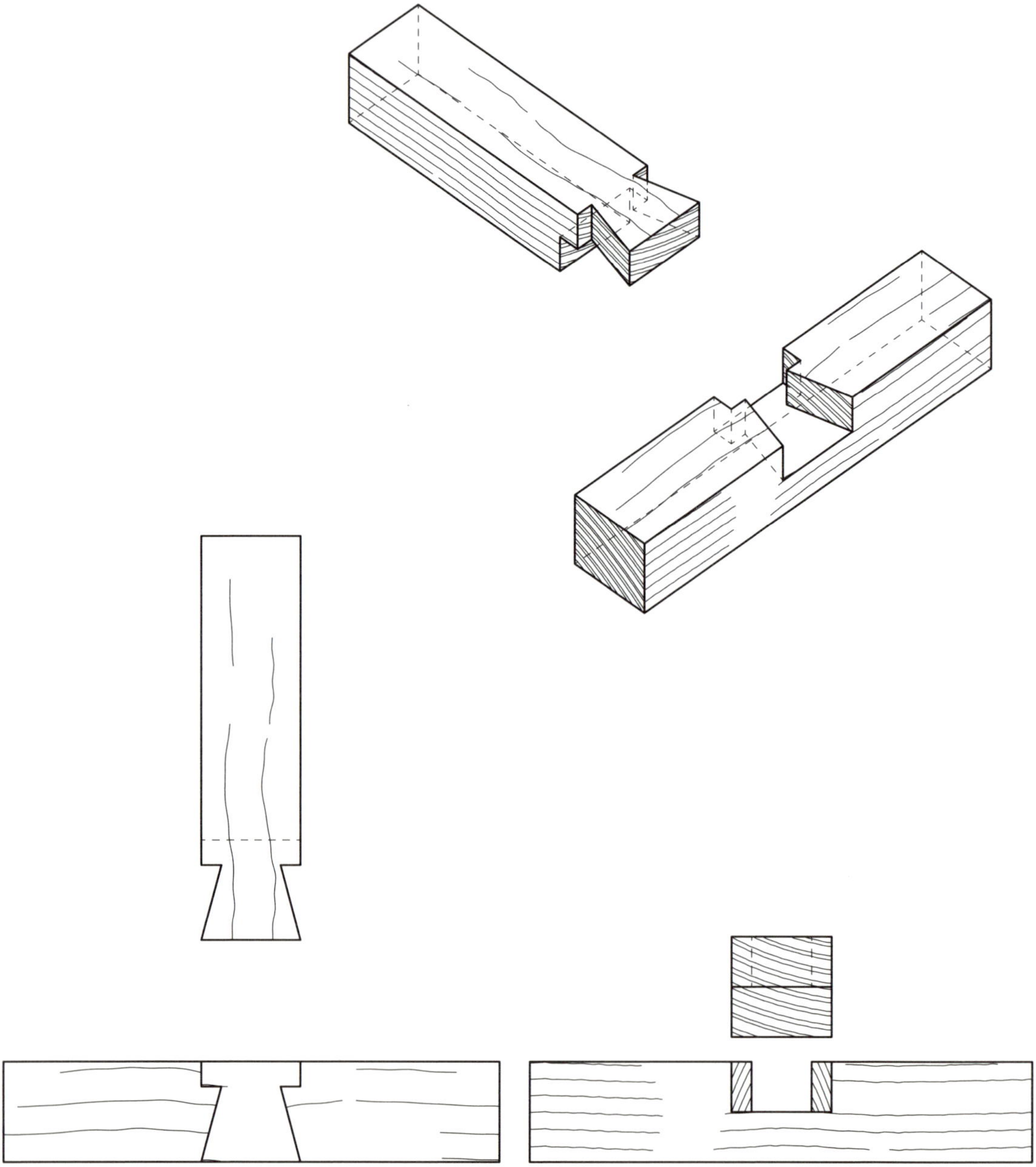

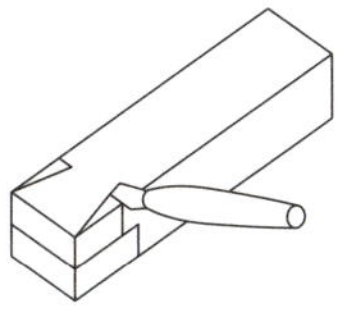

A1

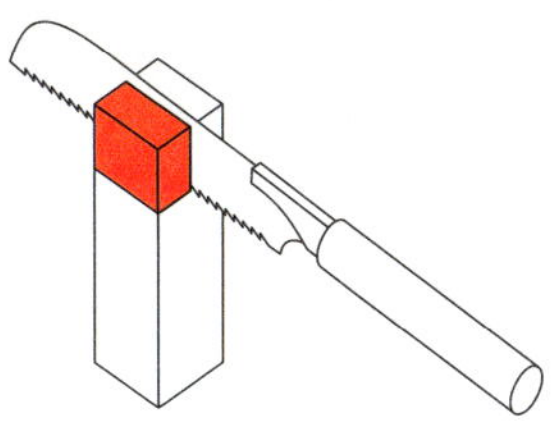

A2

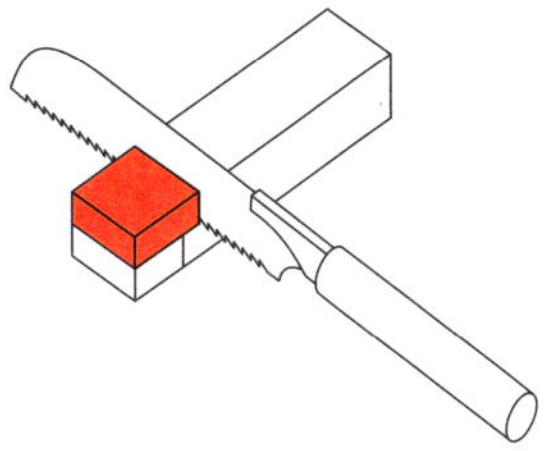

A3

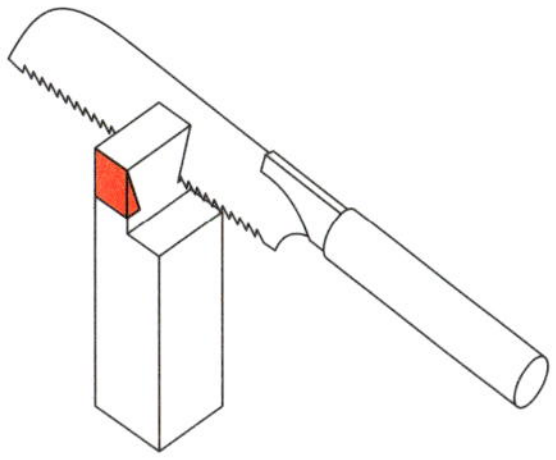

A4

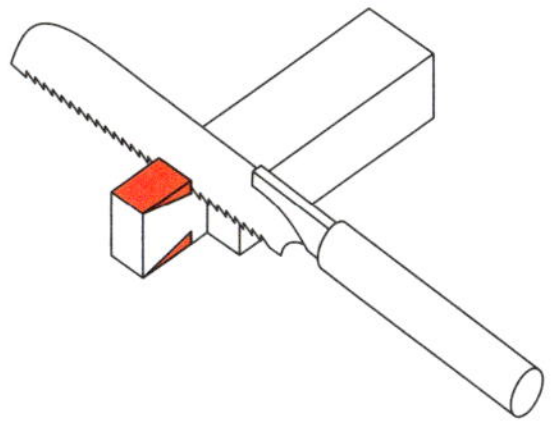

A5

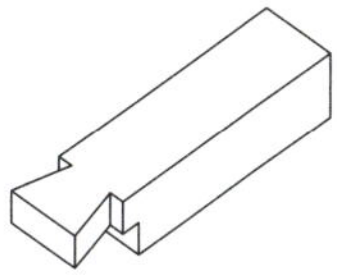

A

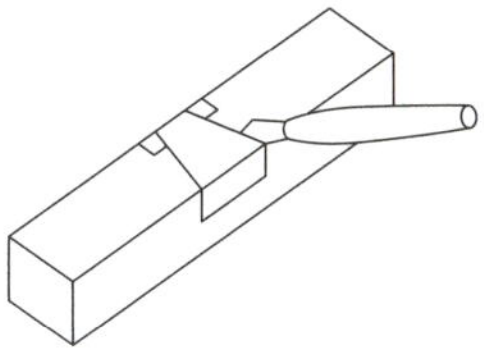

B1

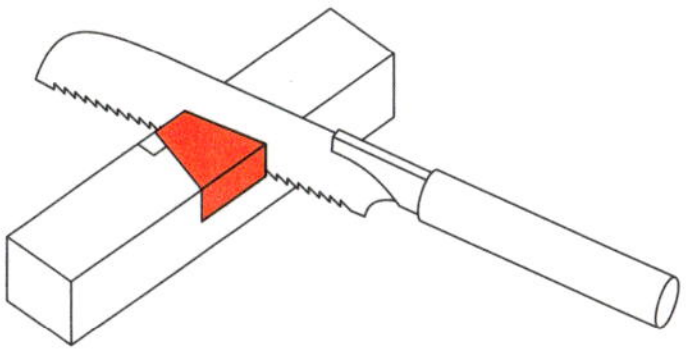

B2

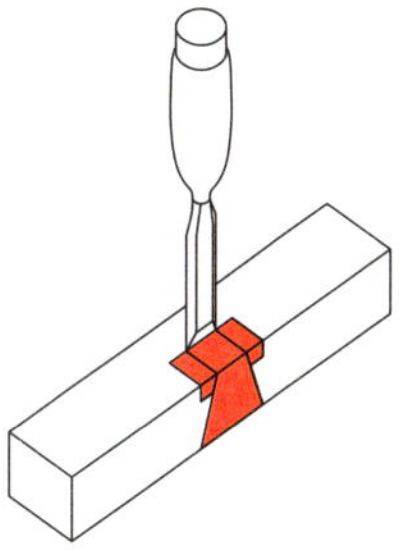

B3

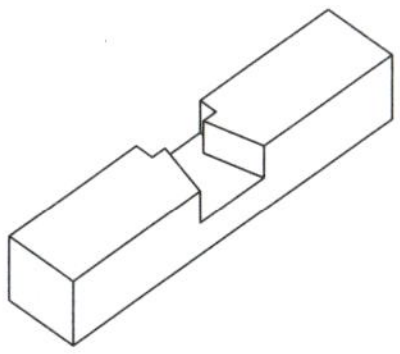

B

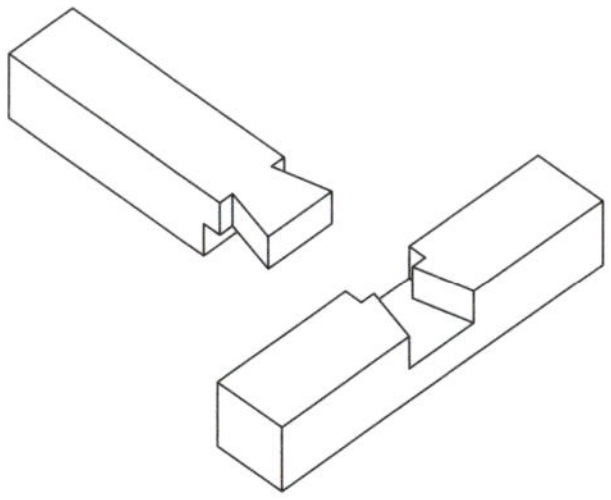

A+B

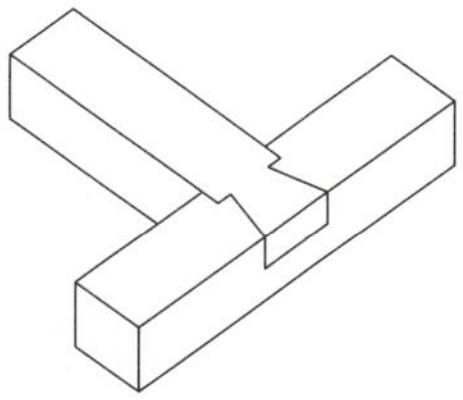

AB

T-Joints

Stopped housed dovetail half lap joint

koshi-kake-kakushi-ari-ai-kaki-tsugi
腰掛隠蟻相欠接ぎ

halbverdecktes eingelassenes
Schwalbenschwanzquerblatt

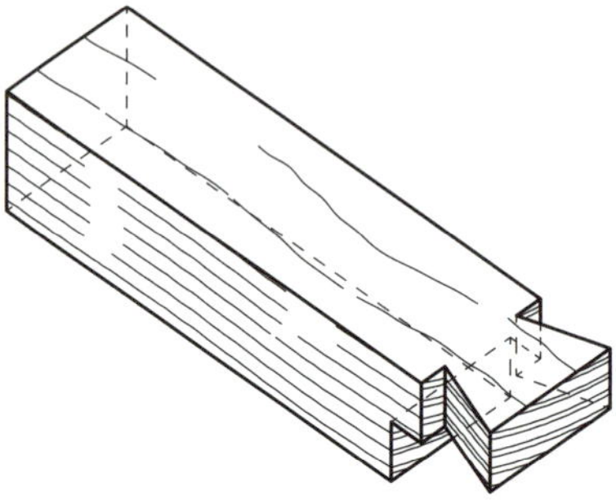

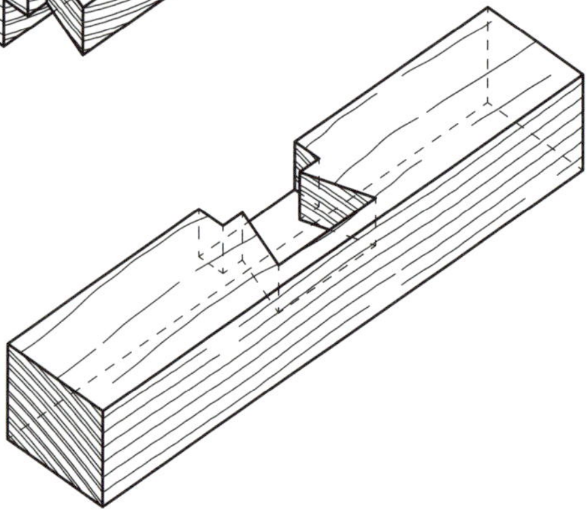

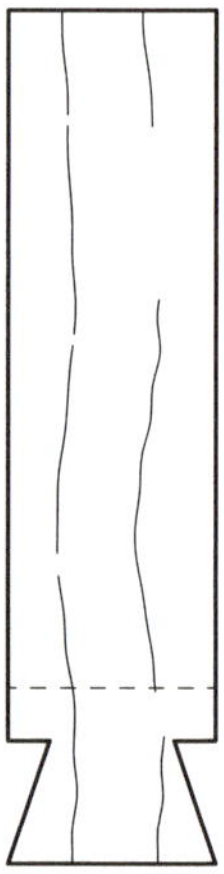

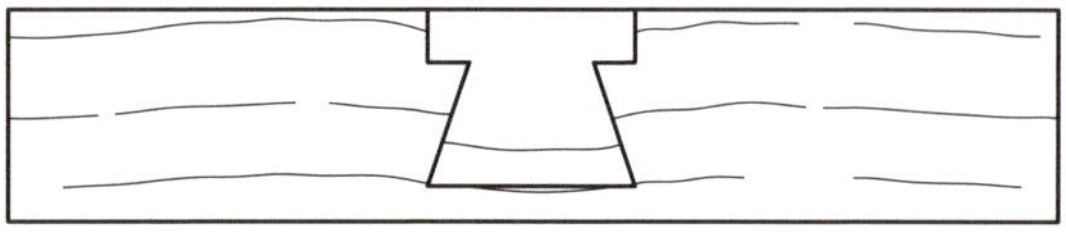

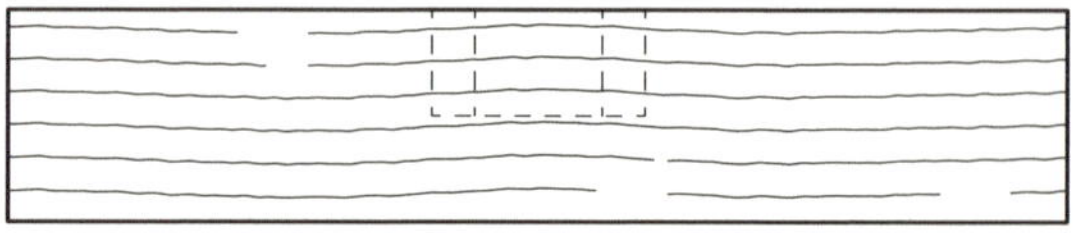

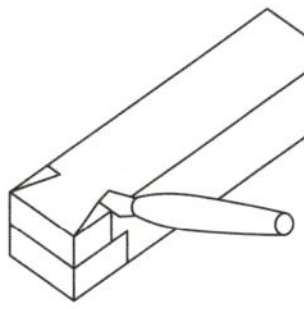

A1

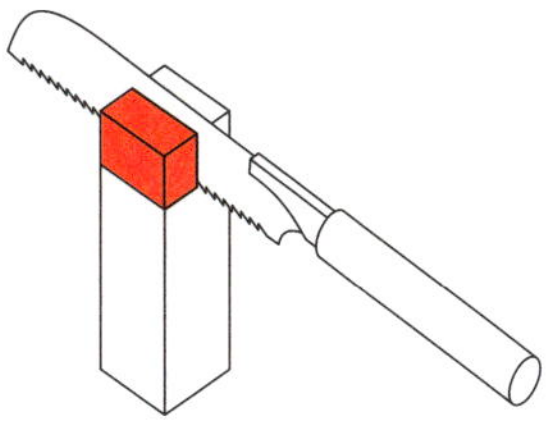

A2

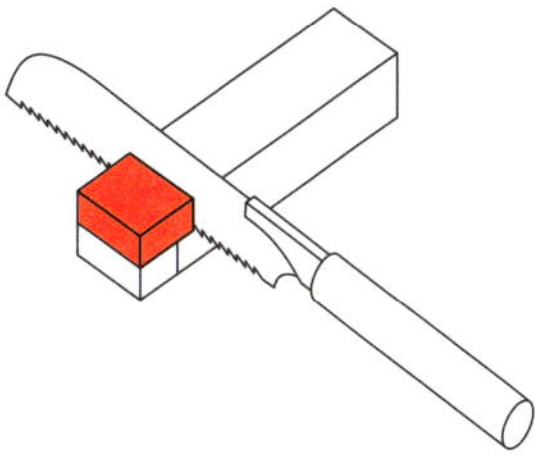

A3

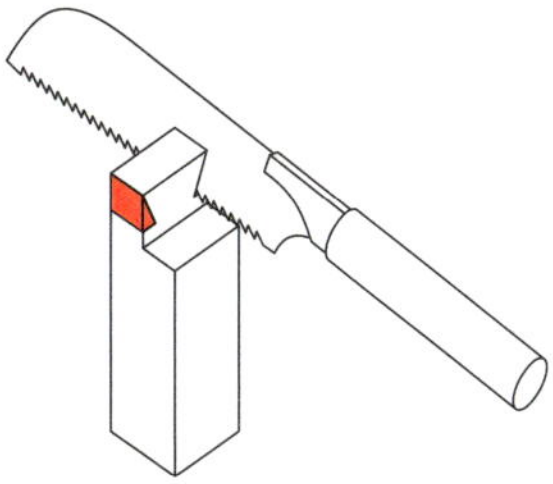

A4

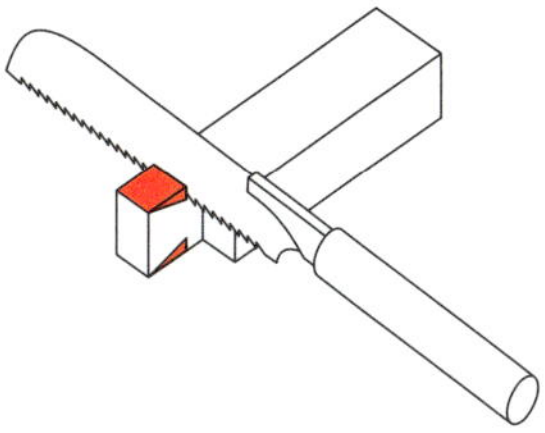

A5

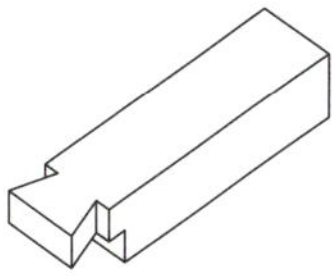

A

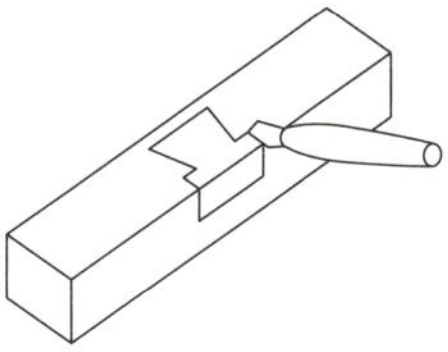

B1

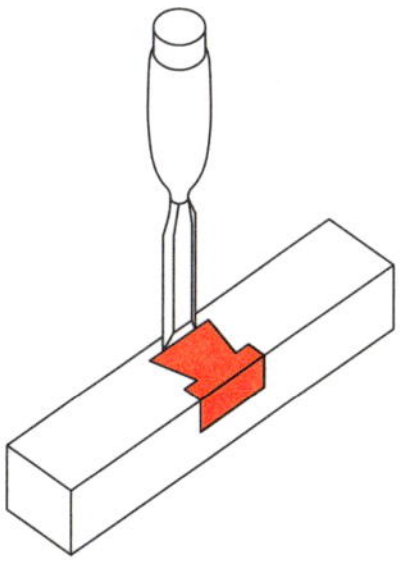

B2

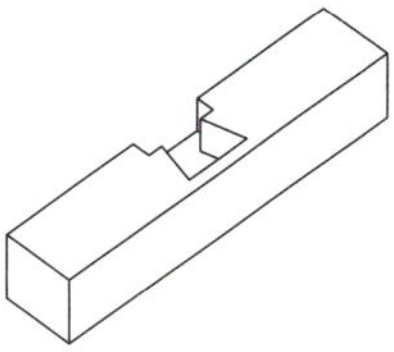

B

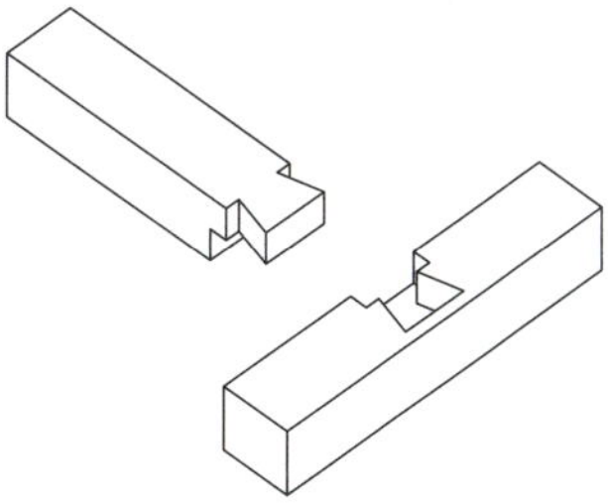

A+B

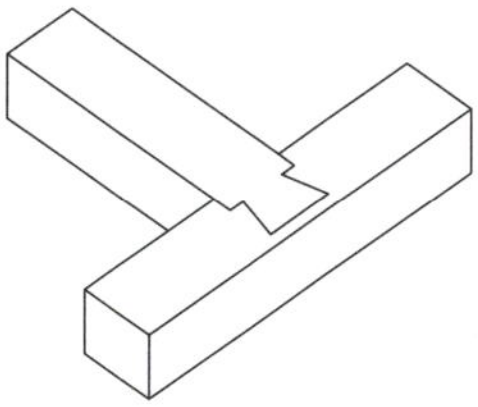

AB

T-Joints

Stepped lap joint

koshi-kake-tsugi (dō-tsuki)
腰掛接ぎ（胴付）

halbverdecktes Stufenquerblatt

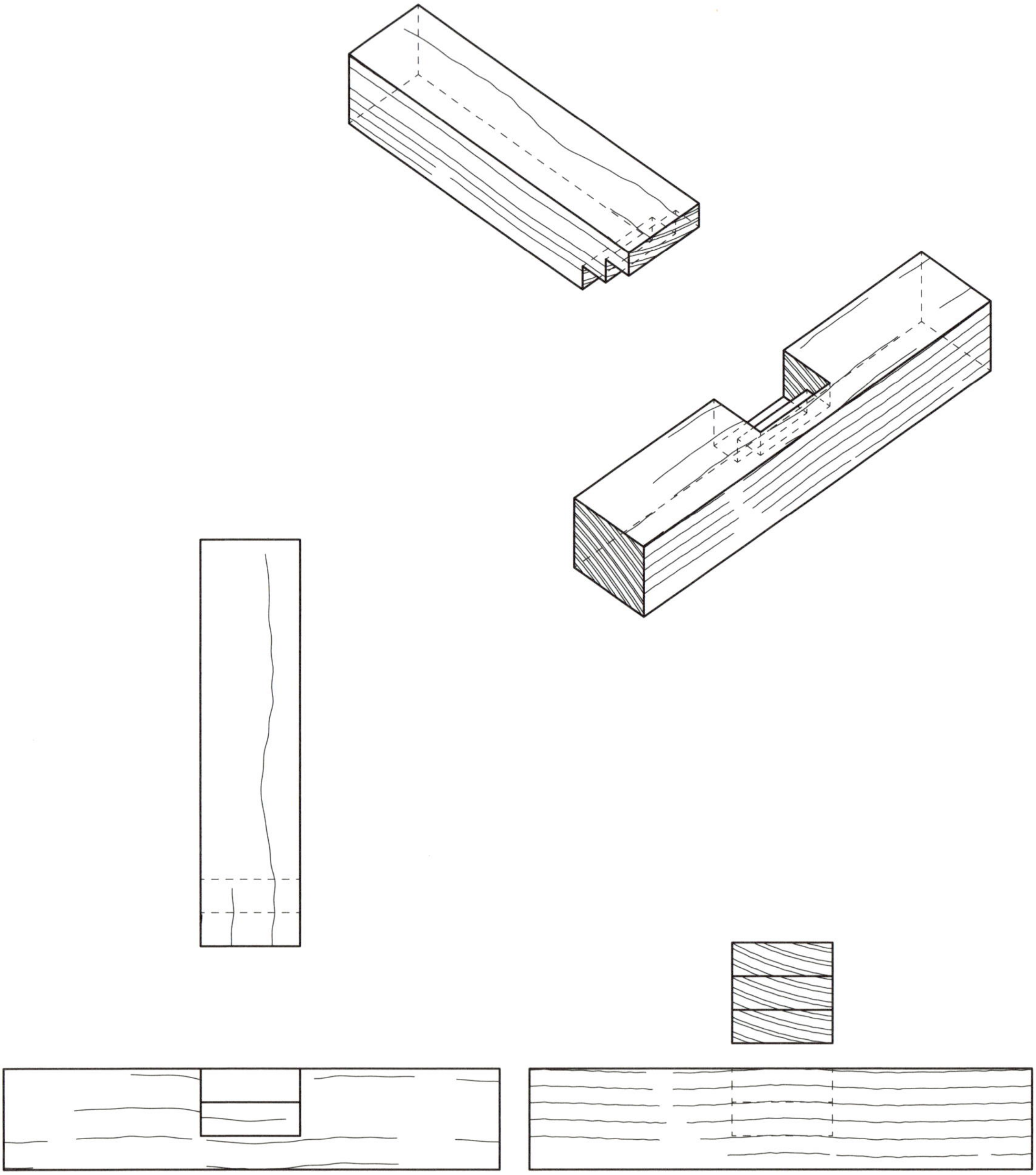

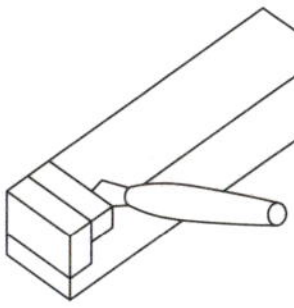

A1

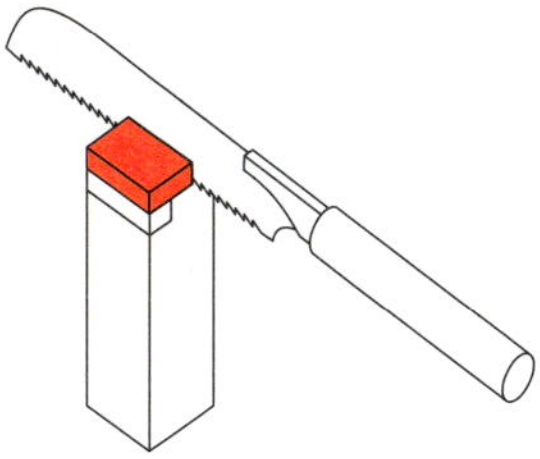

A2

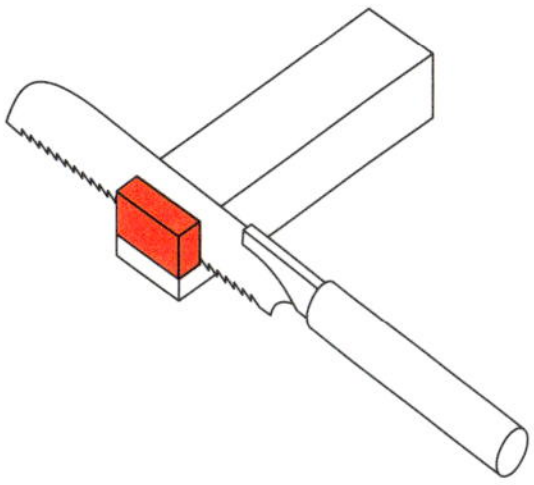

A3

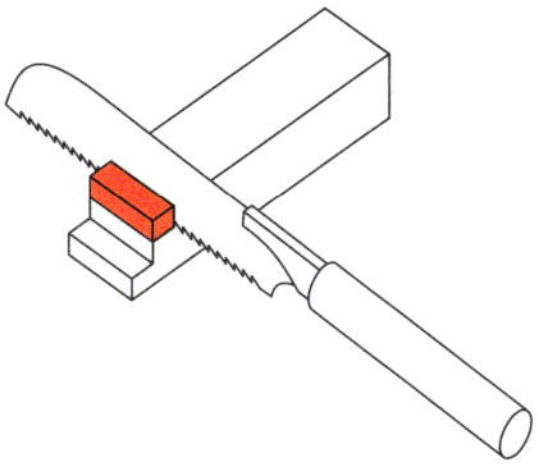

A4

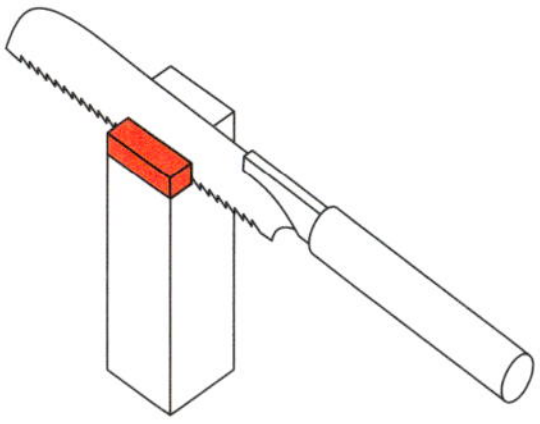

A5

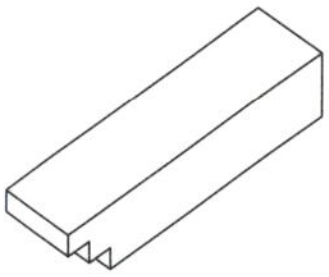

A

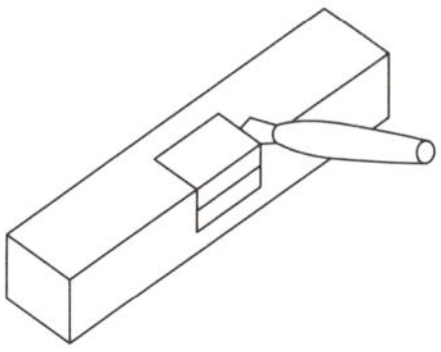

B1

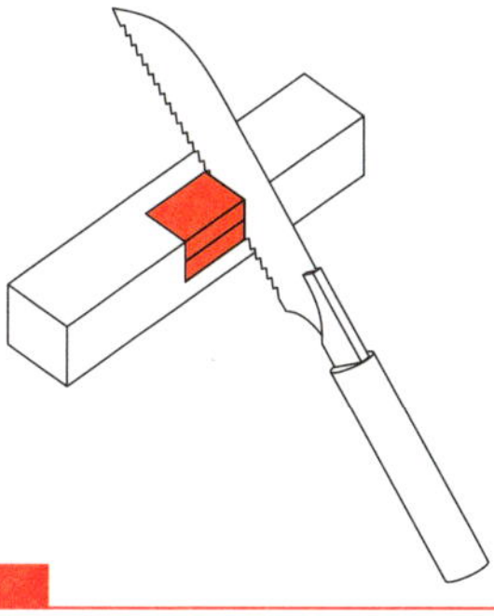

B2

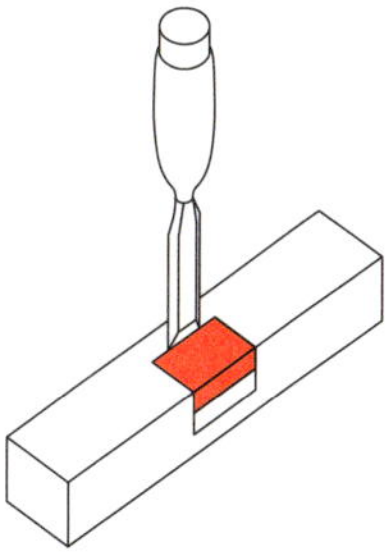

B3

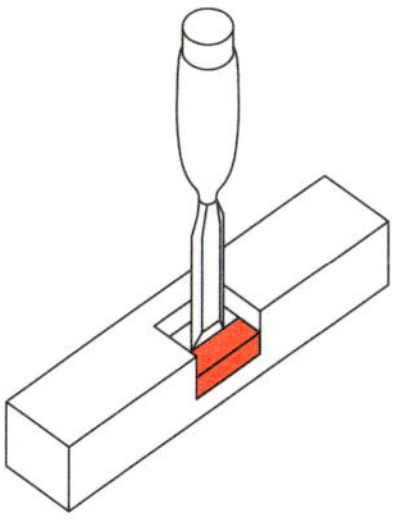

B4

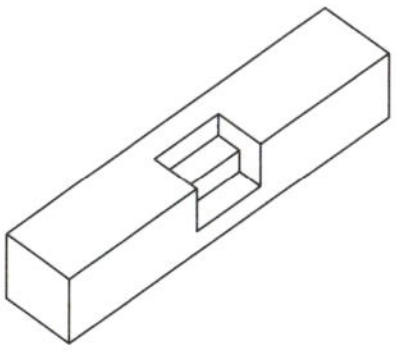

B

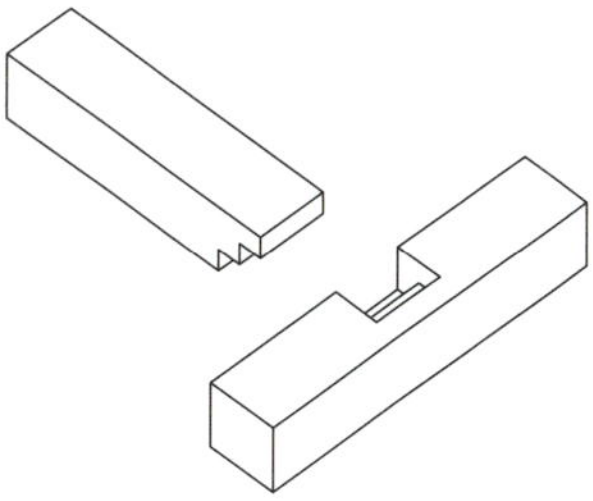

A+B

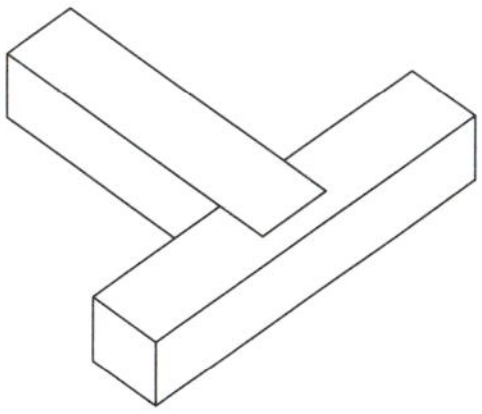

AB

T-Joints

Housed mortise and tenon with shifted shouders

kata-dō-tsuki-tan-hozo-sashi
片胴付短枘差し

Zapfen mit einseitig vorgeschobener Brüstung

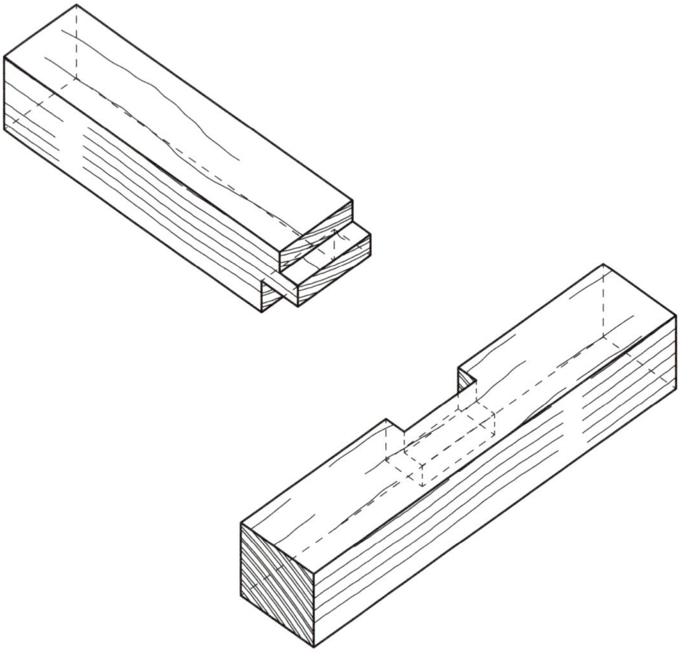

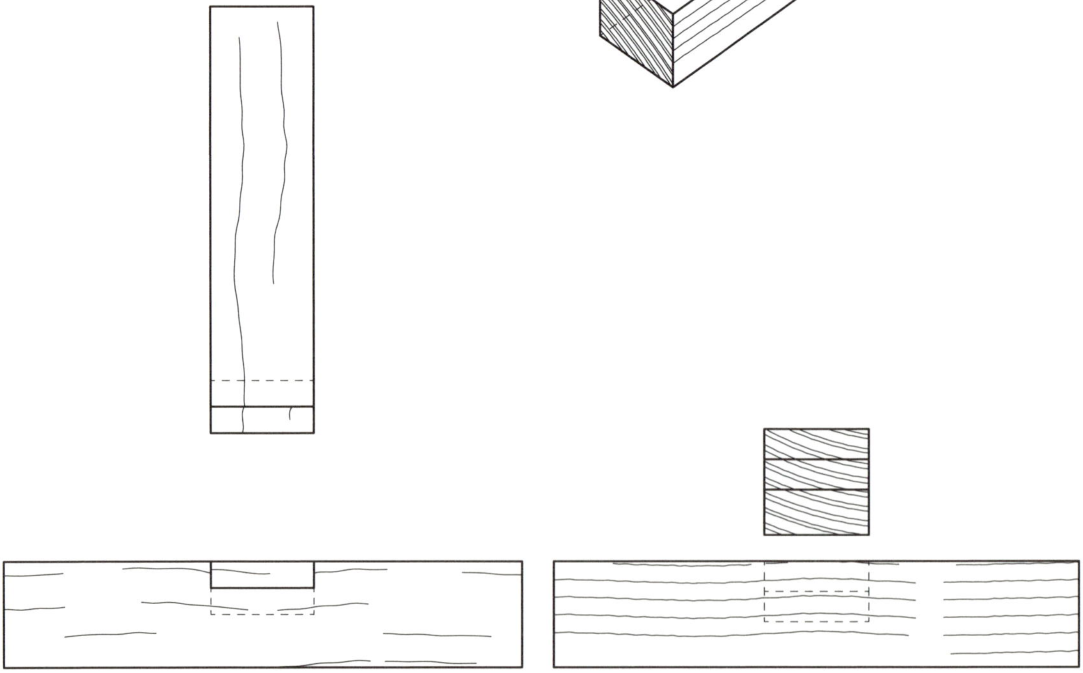

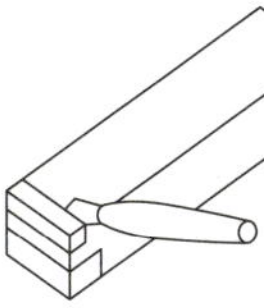

A1

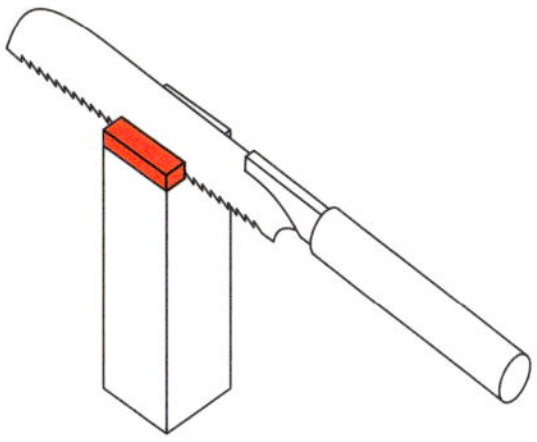

A2

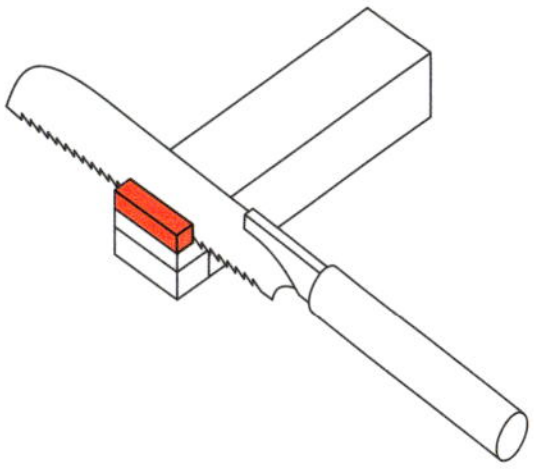

A3

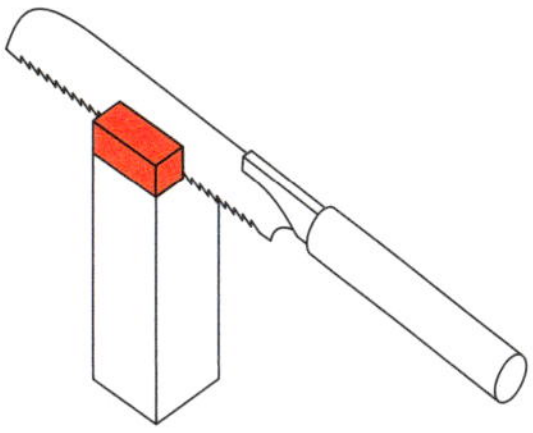

A4

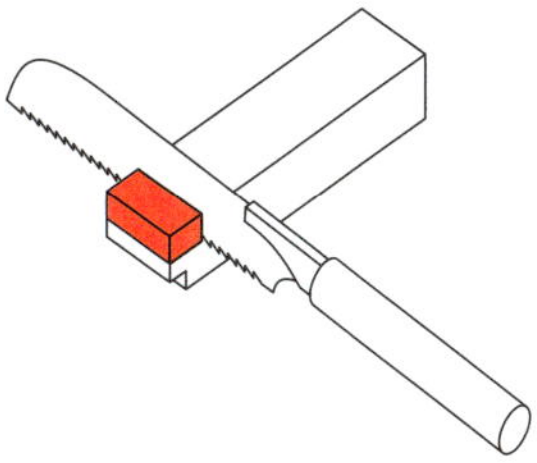

A5

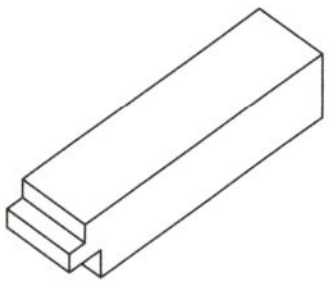

A

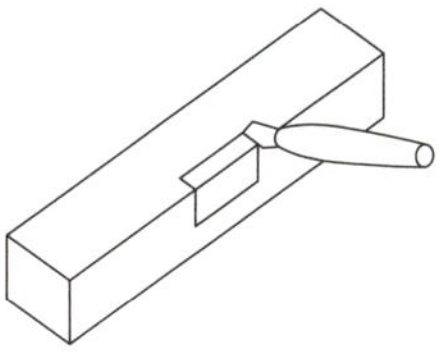

B1

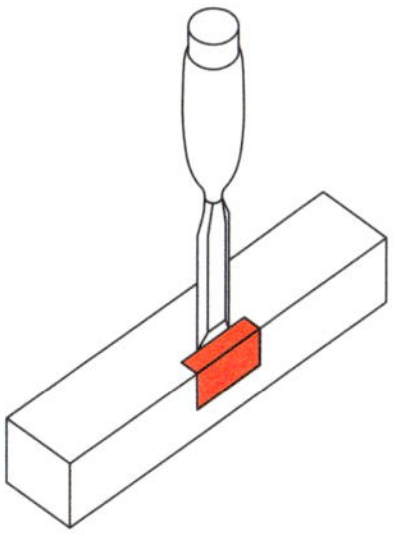

B2

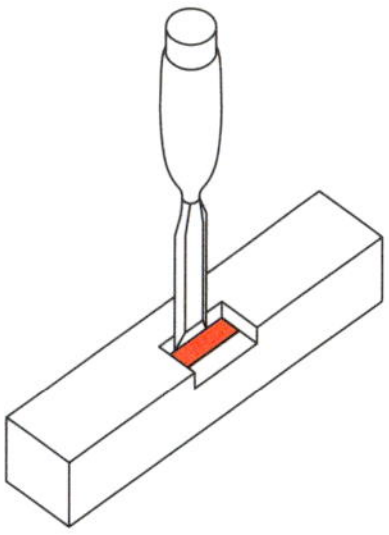

B3

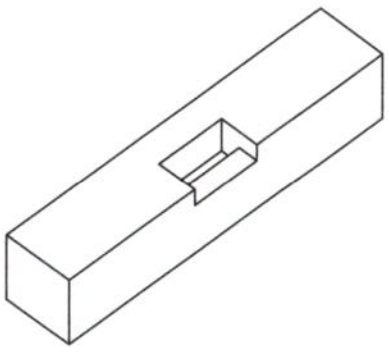

B

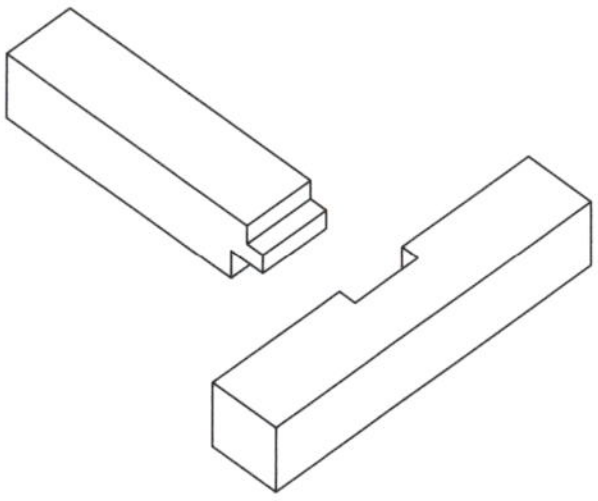

A+B

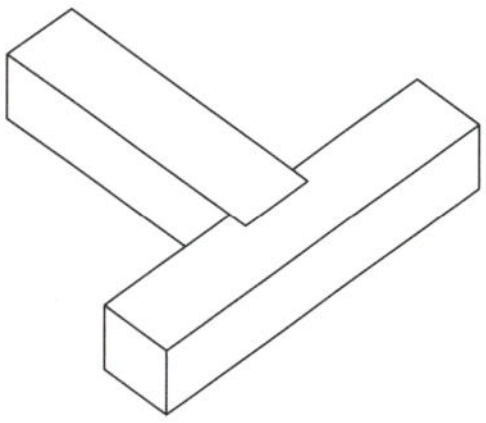

AB

T-Joints

Sloped stopped half lap

koshi-kake-sogi-kaki-sashi
腰掛殺欠差し

verdecktes schräg angeschnittenes Querblatt

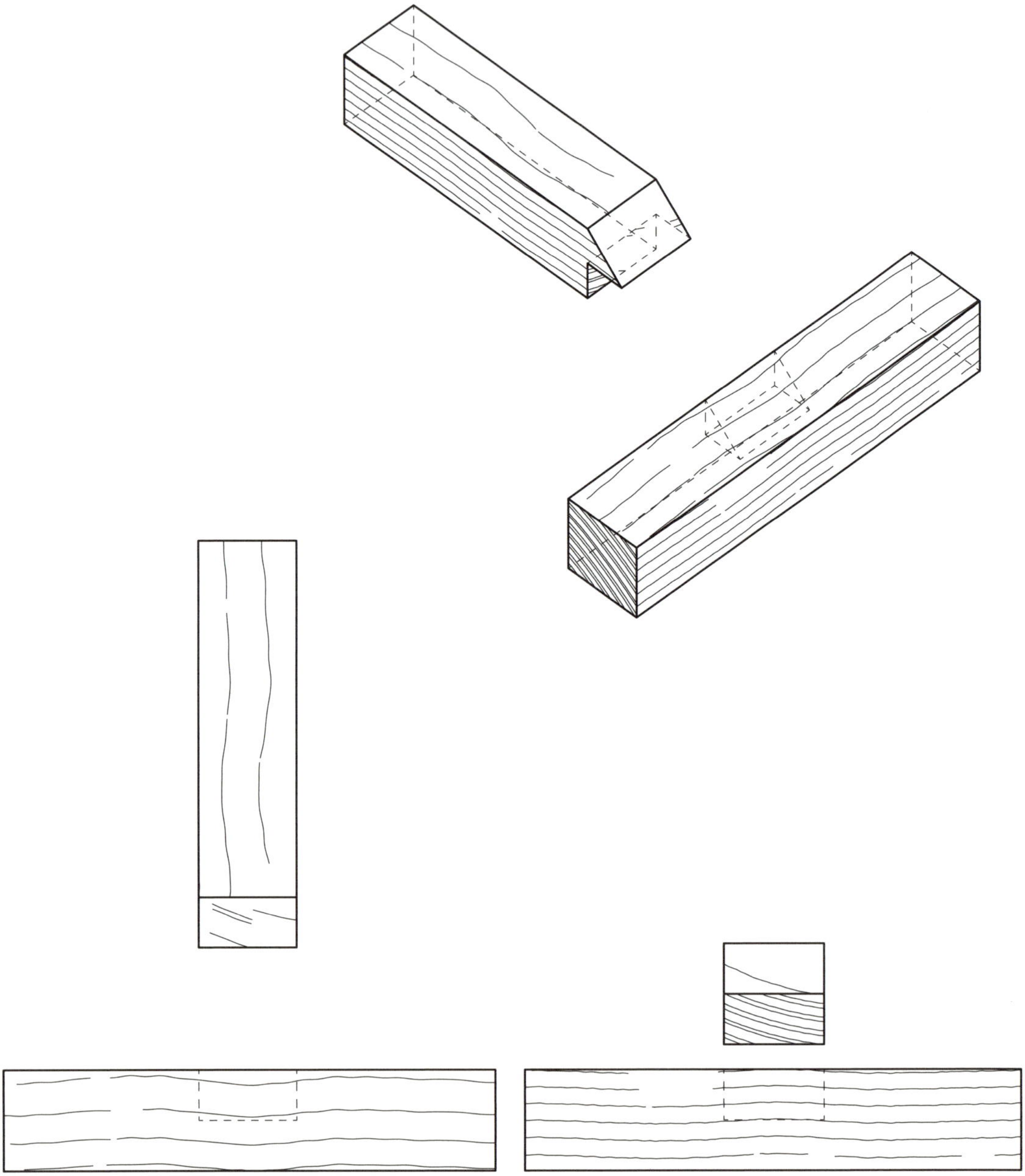

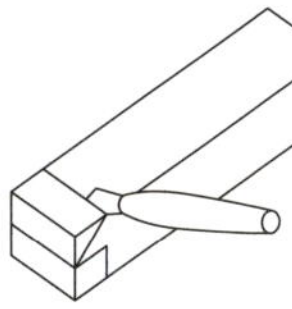

A1

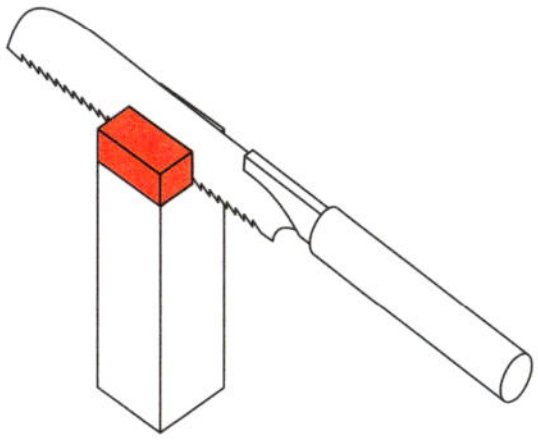

A2

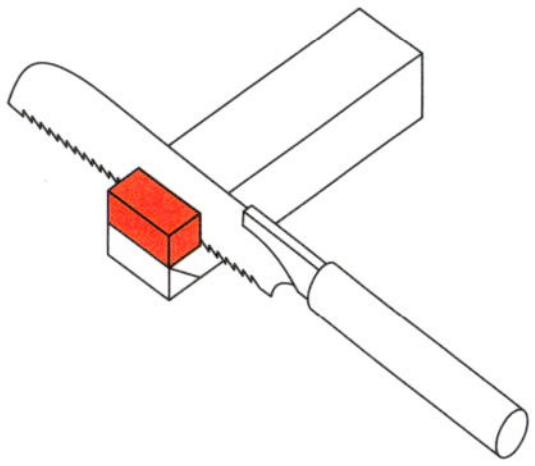

A3

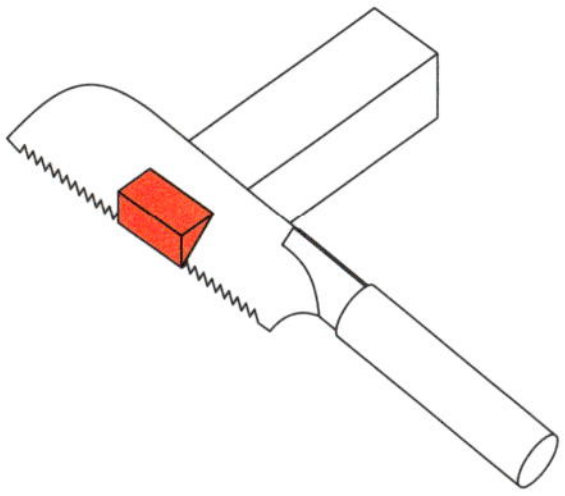

A4

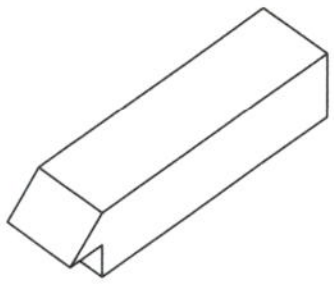

A

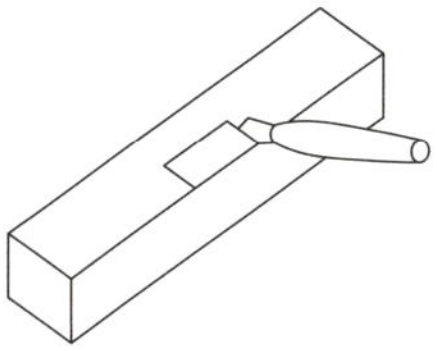

B1

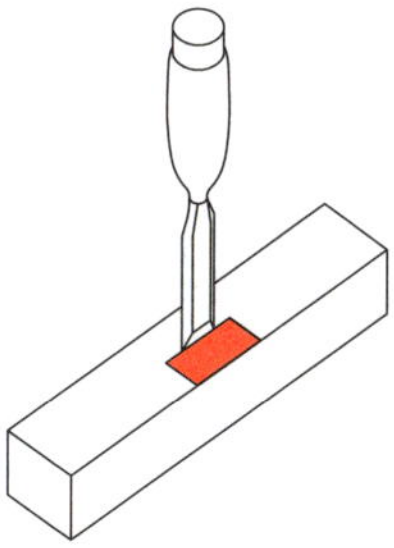

B2

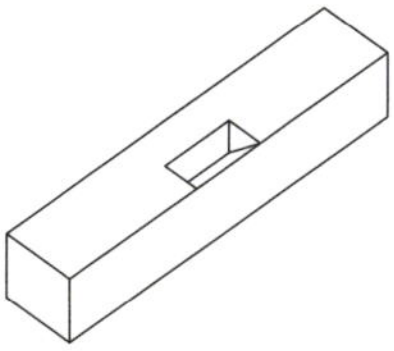

B

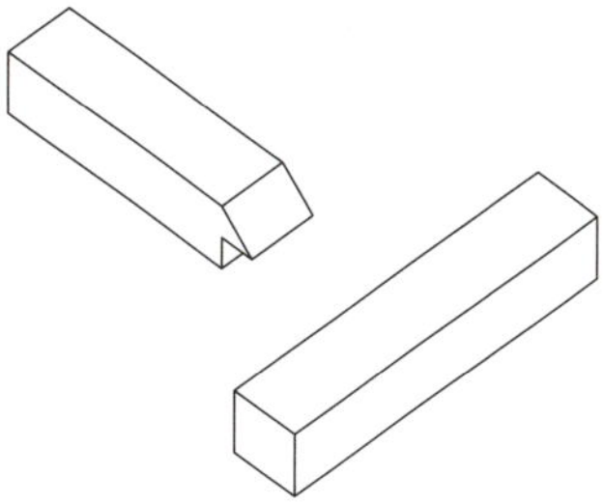

A+B

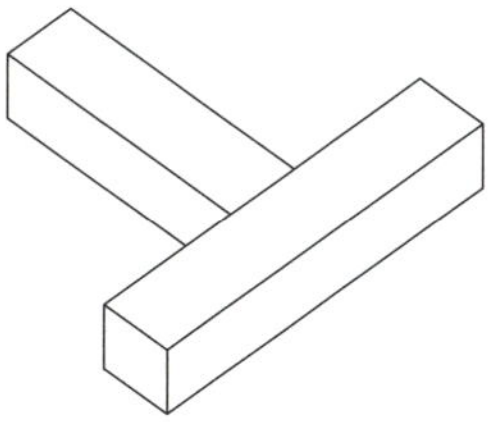

AB

T-Joints

Mortise and tenon with sloped shoulder

men-oshi-tsuki-tan-hozo-sashi
面押付短枘差し

verdeckter Zapfen mit einseitig schräger Brüstung

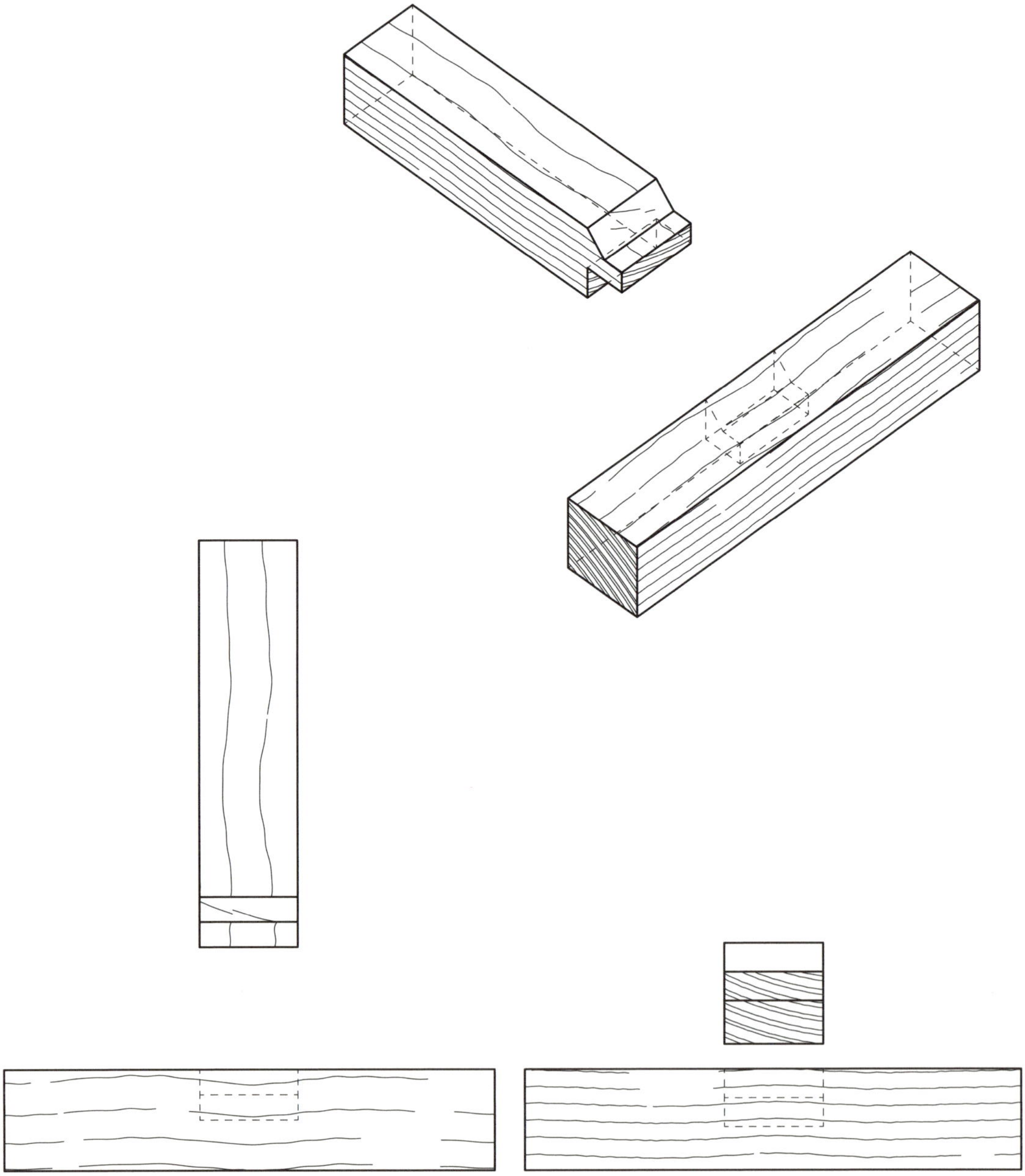

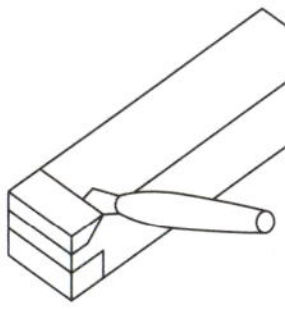

A1

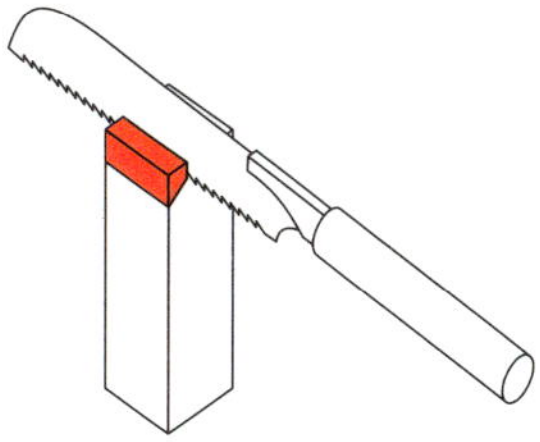

A2

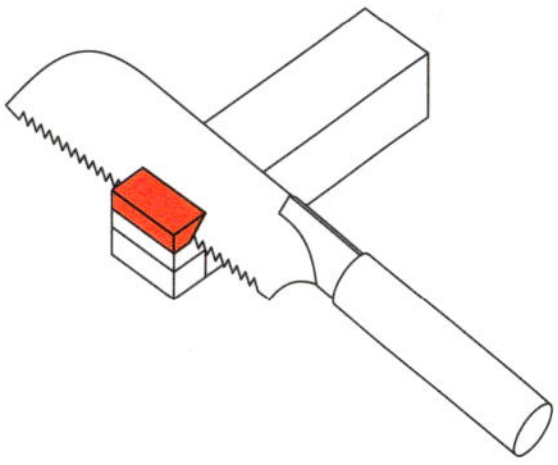

A3

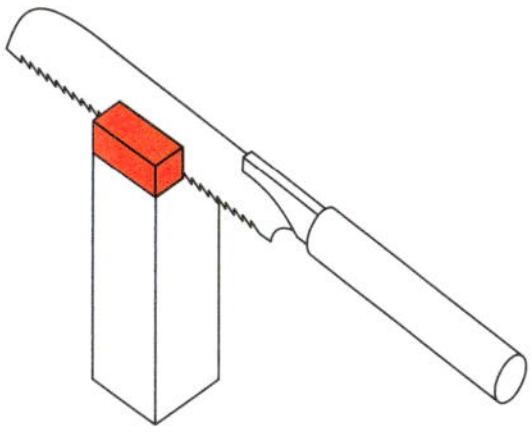

A4

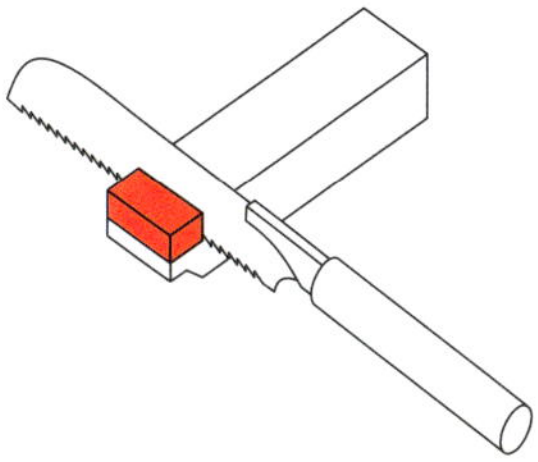

A5

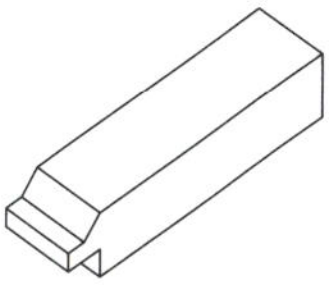

A

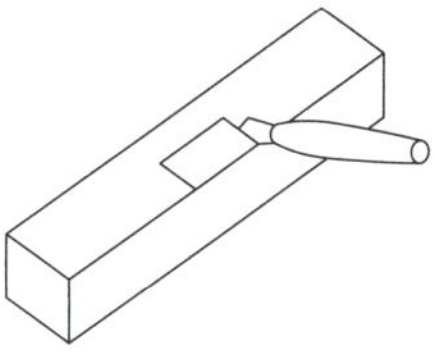

B1

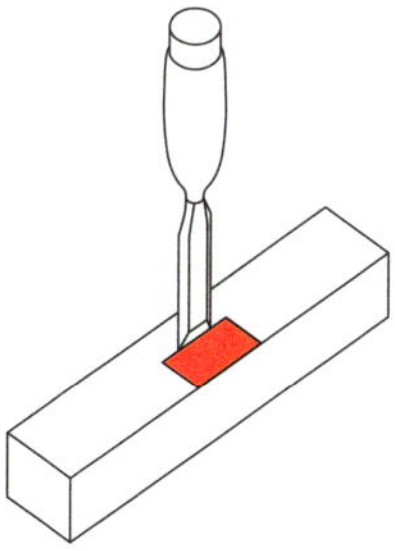

B2

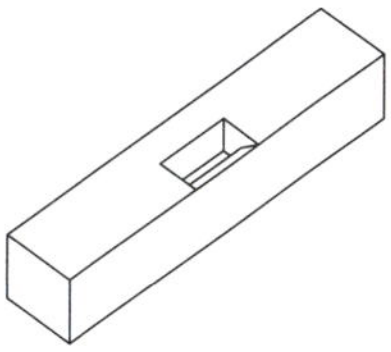

B

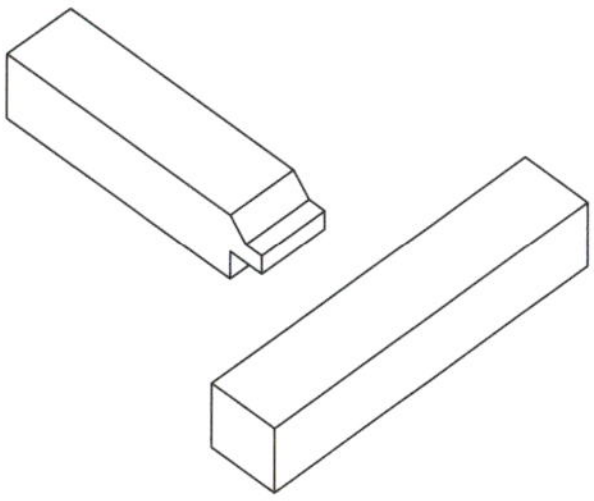

A+B

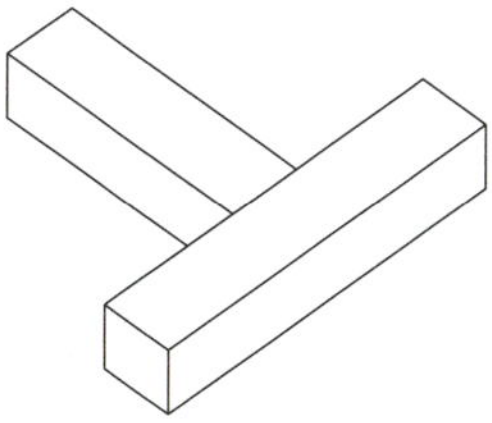

AB

T-Joints

Blind housed mortise and tenon with sloped shoulder

men-oshi-dome-tan-hozo-sashi (dō-tsuki)
面押留短枘差し(胴付)

verdeckter Zapfen mit schräger und gestufter Brüstung

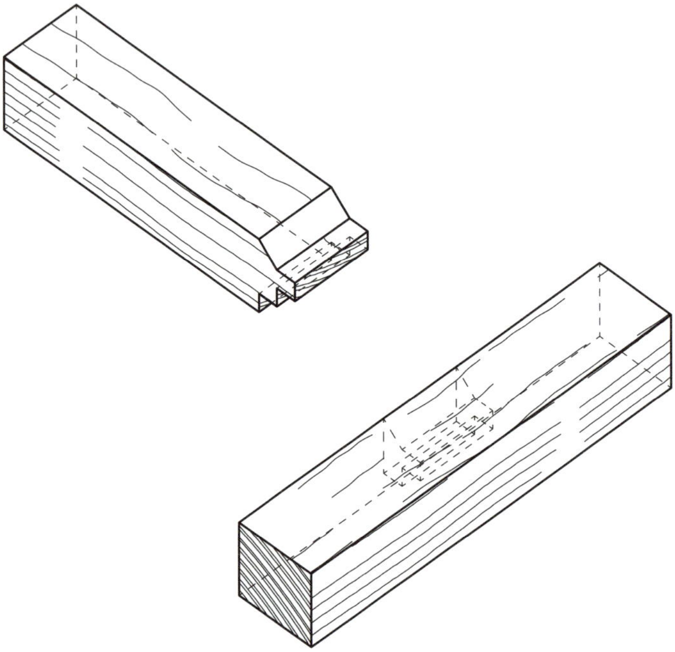

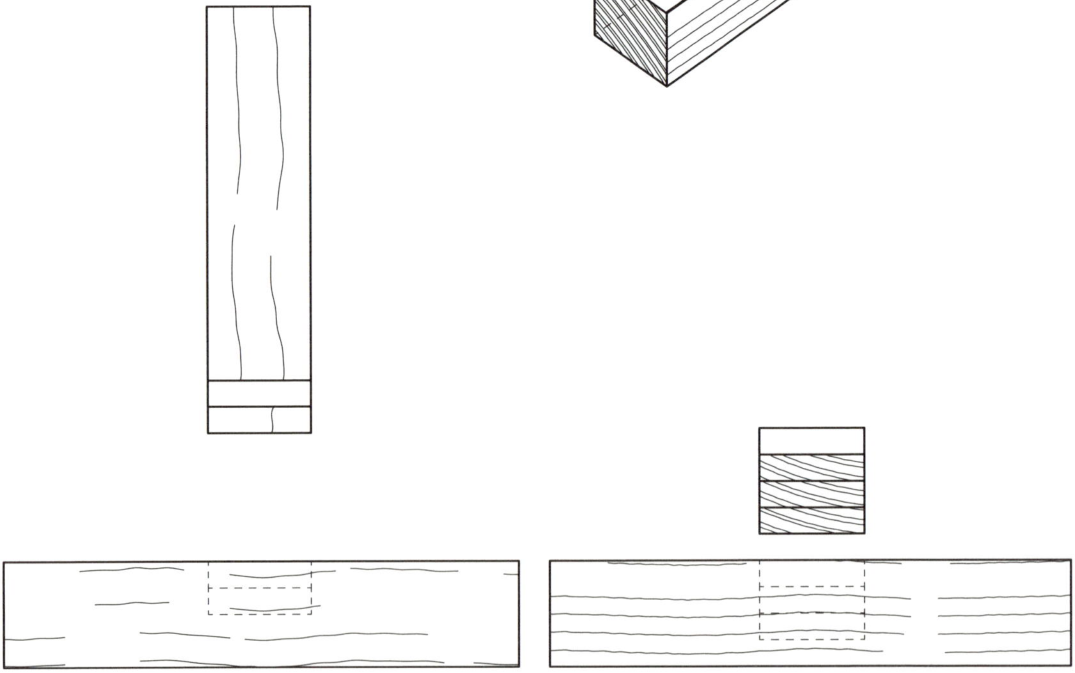

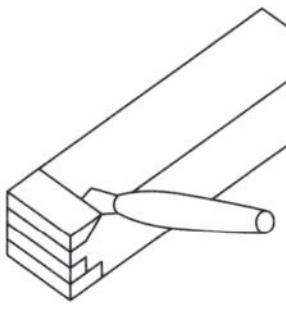

A1

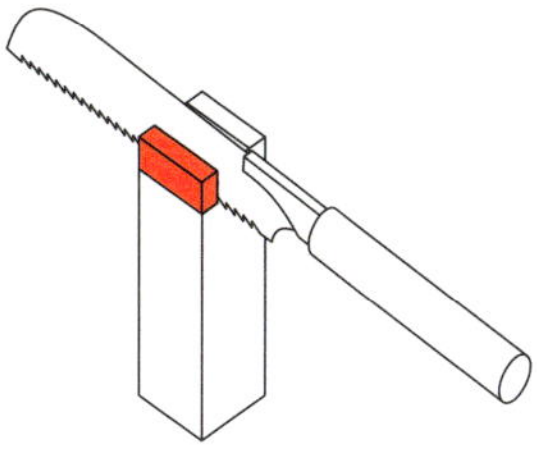

A2

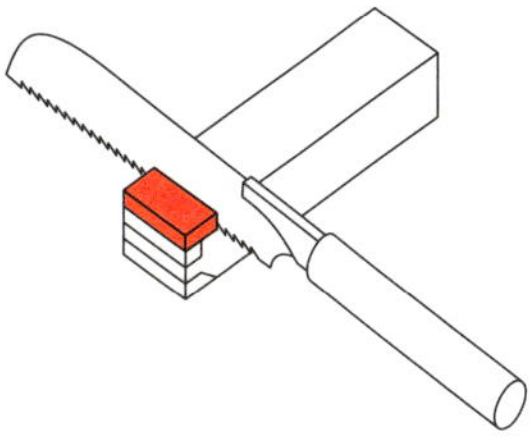

A3

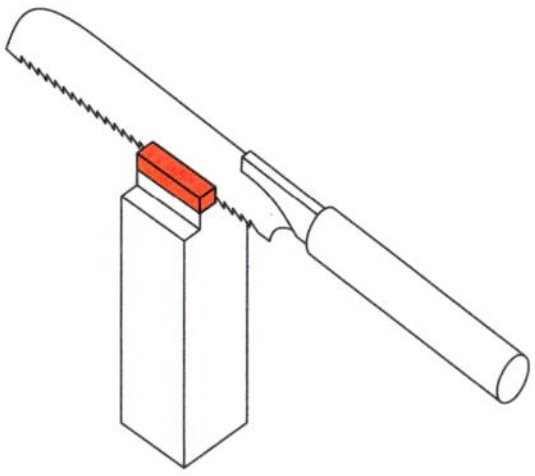

A4

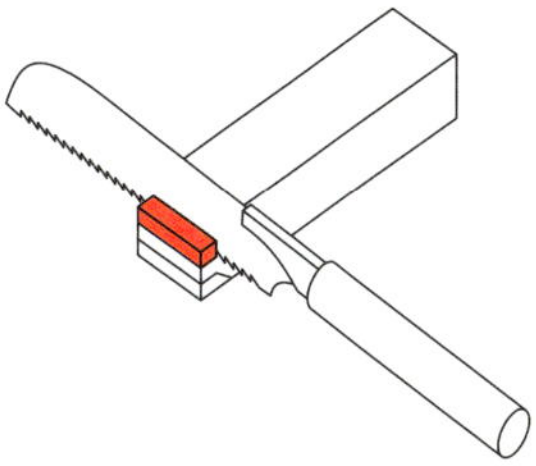

A5

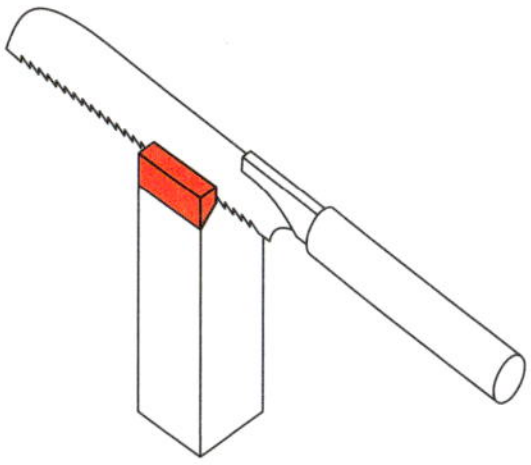

A6

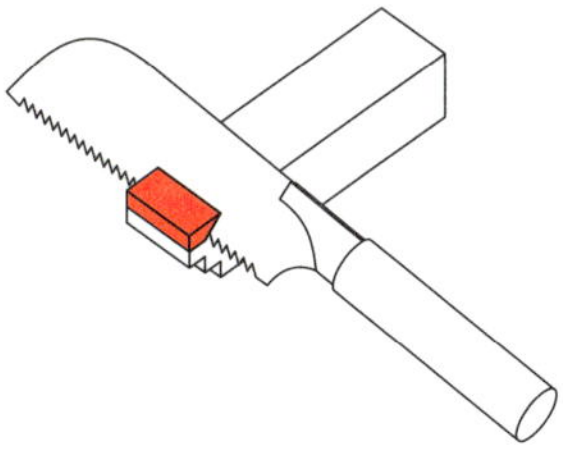

A7

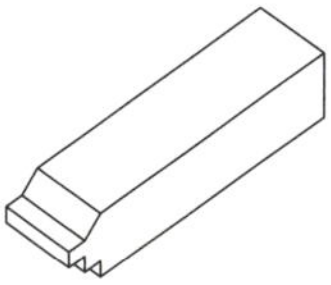

A

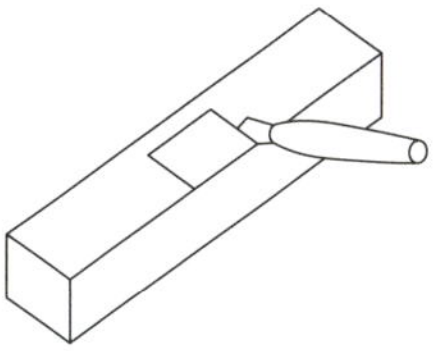

B1

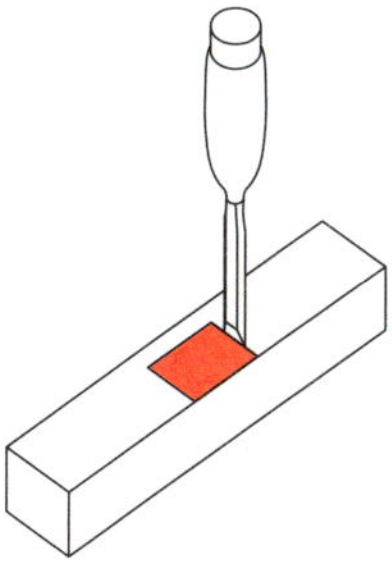

B2

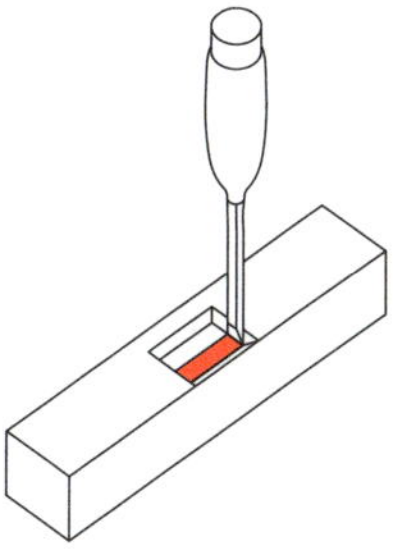

B3

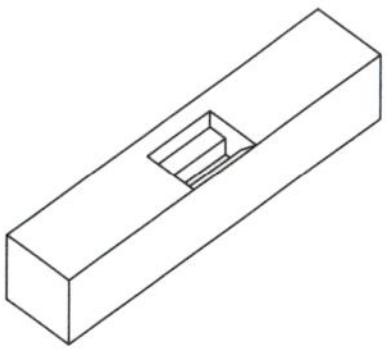

B

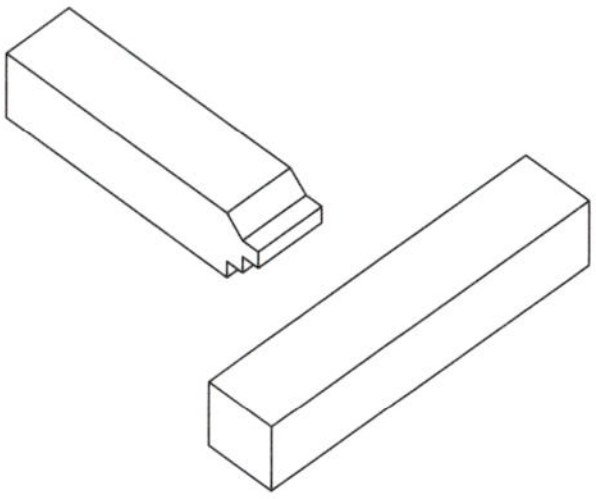

A+B

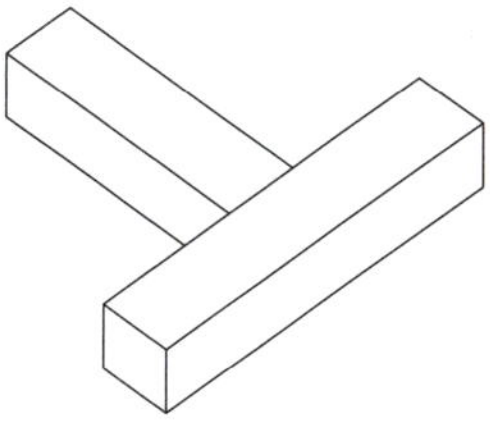

AB

T-Joints

Undersquinted stopped half lap

han-dome-koshi-kake-tsugi
半留腰掛接ぎ

halbverdecktes gerades Blatt
mit schräger Stirn

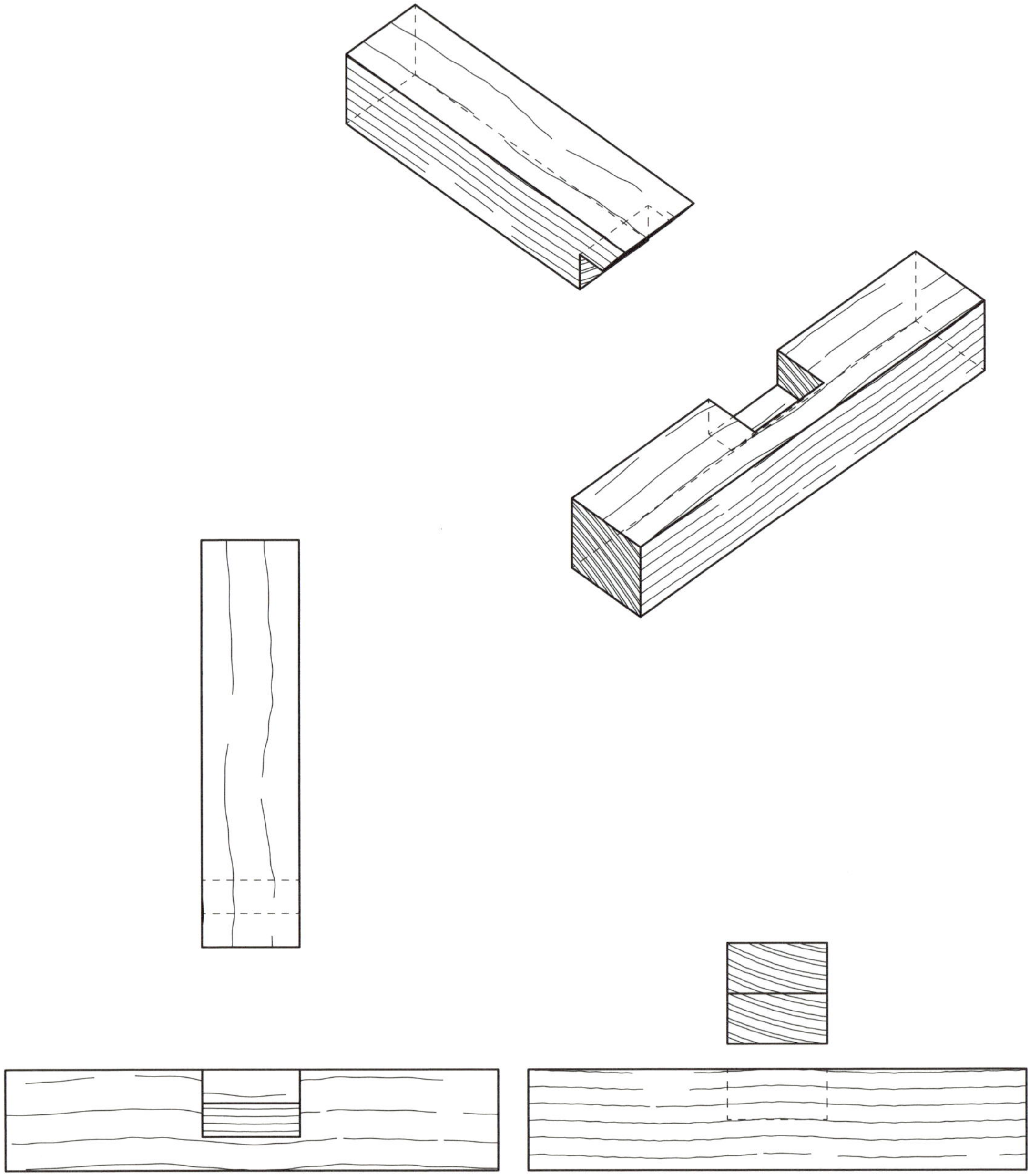

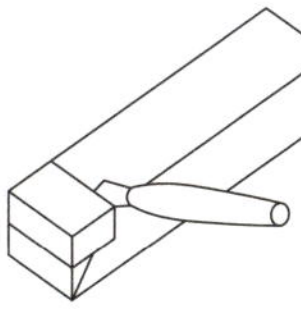

A1

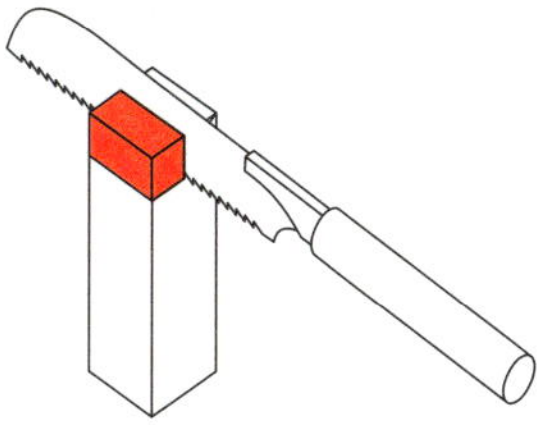

A2

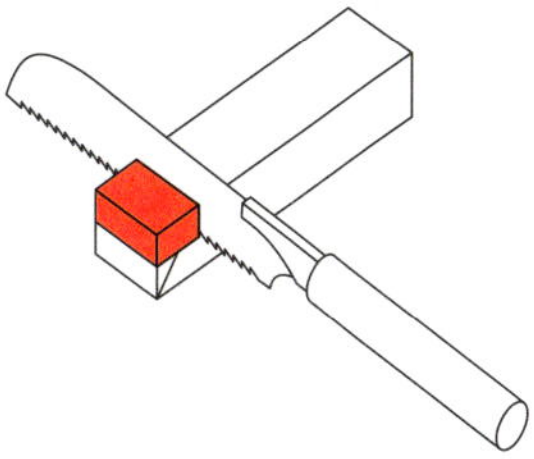

A3

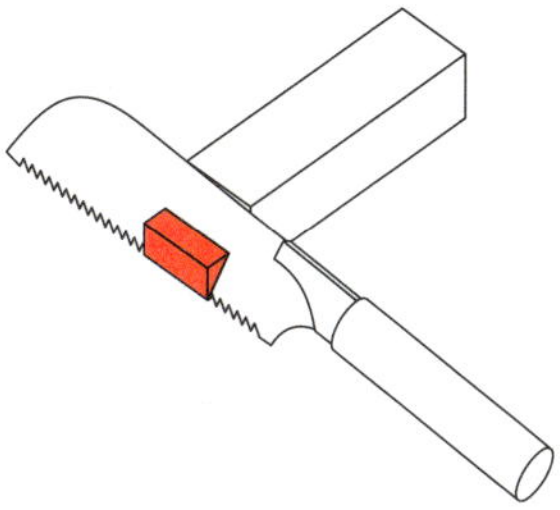

A4

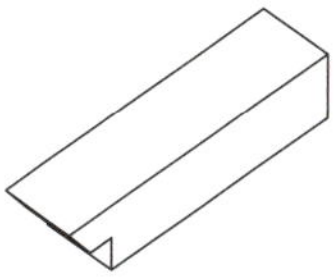

A

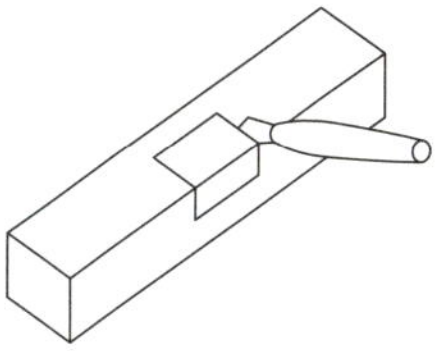

B1

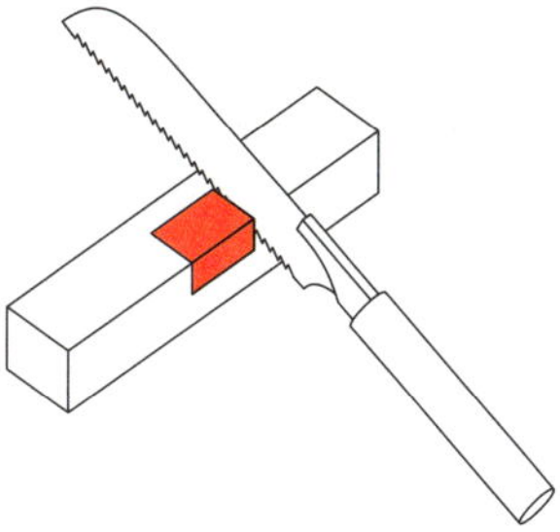

B2

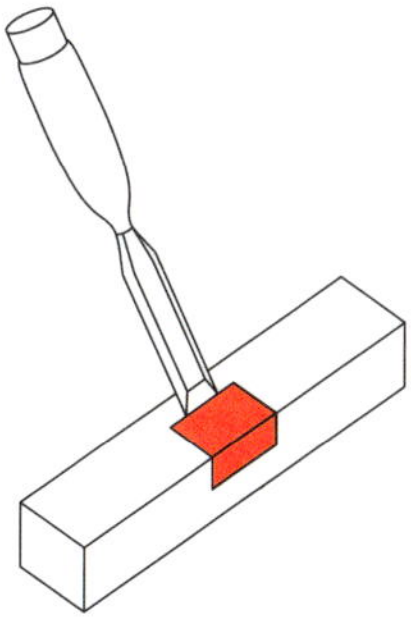

B3

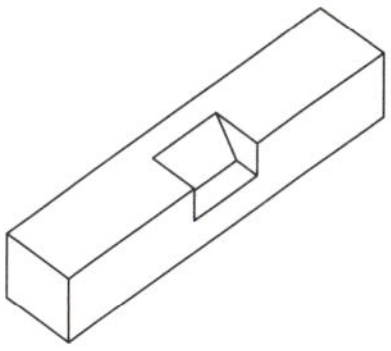

B

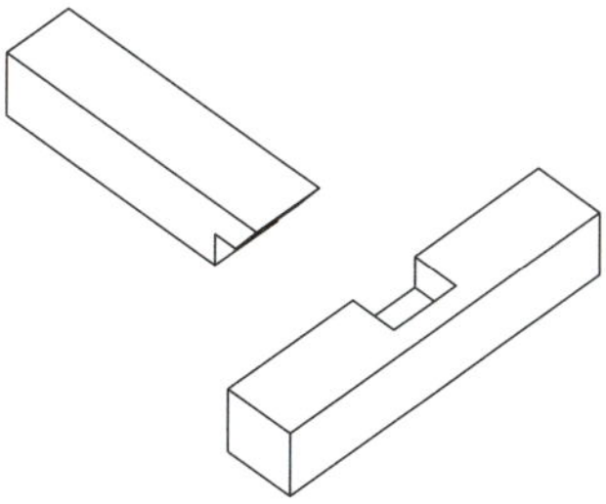

A+B

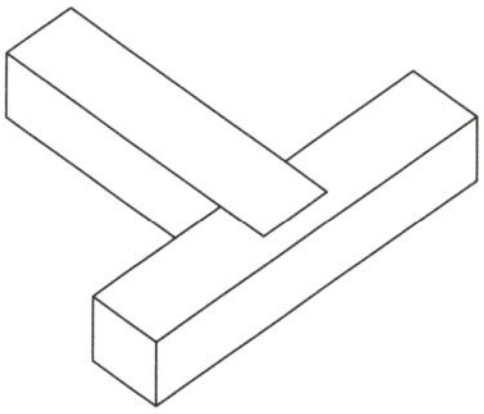

AB

T-Joints

Mortise and tenon with beveled shoulders

katagi-dō-tsuki-tan-hozo-sashi
傾胴付短枘差し

verdeckter Zapfen mit
eingelassener schräger Brüstung

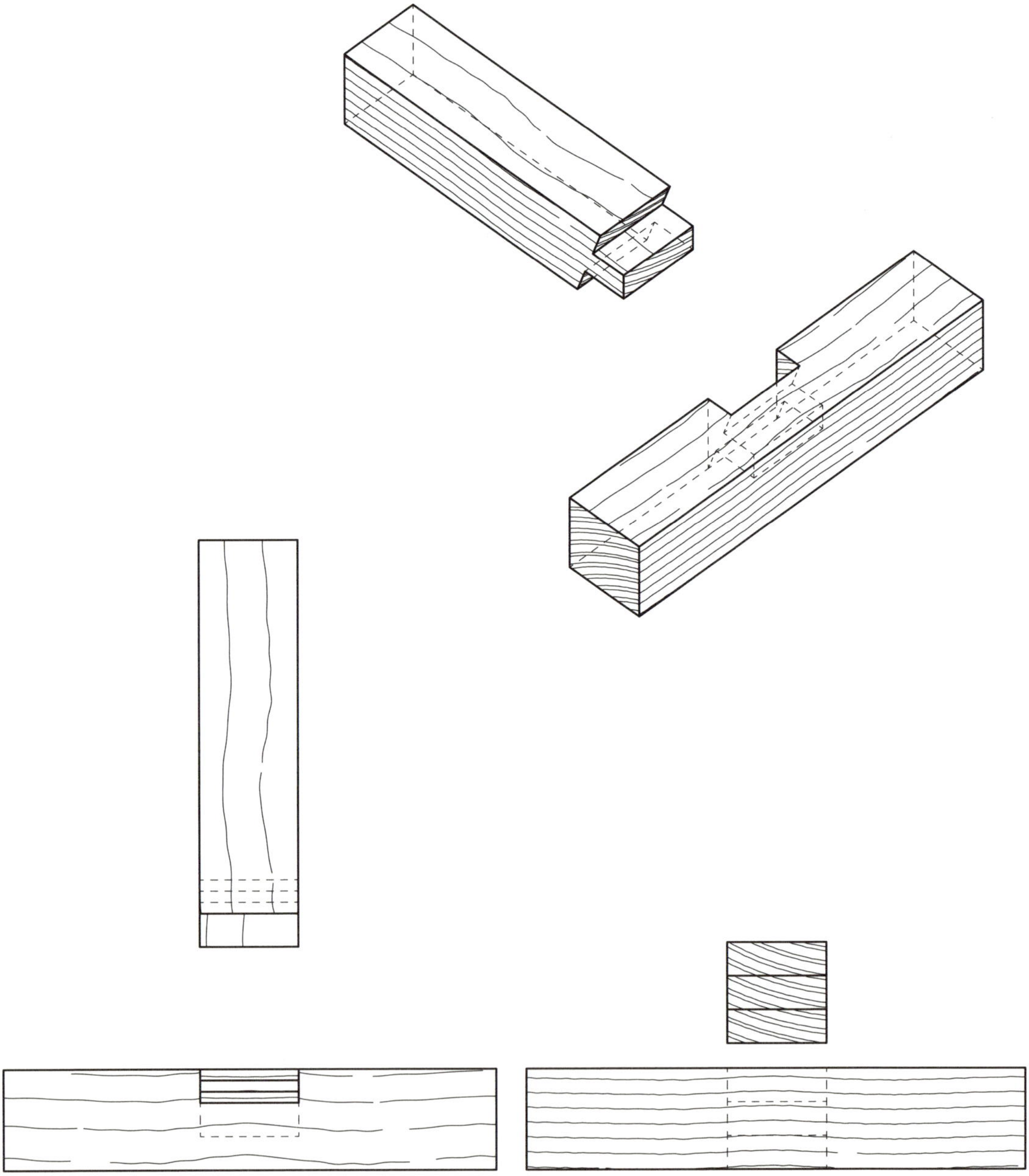

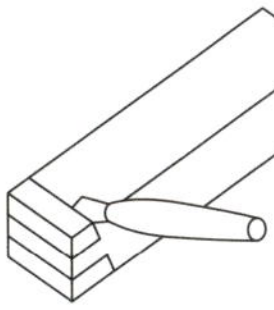

A1

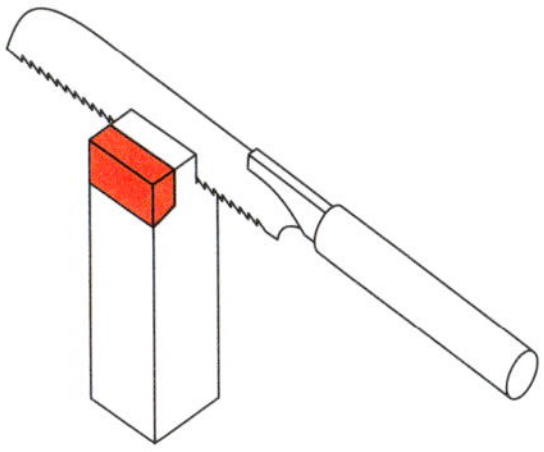

A2

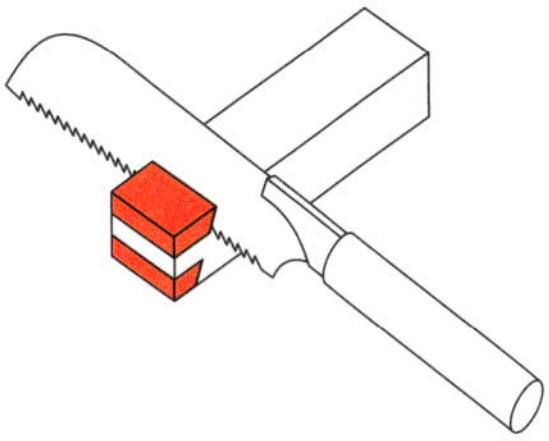

A3

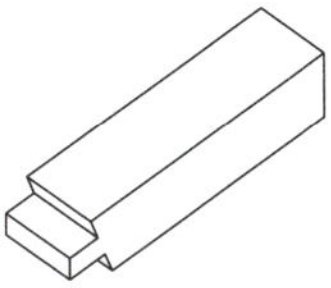

A

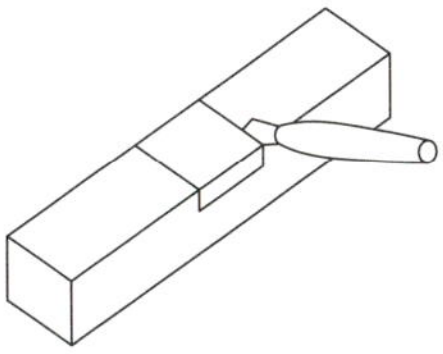

B1

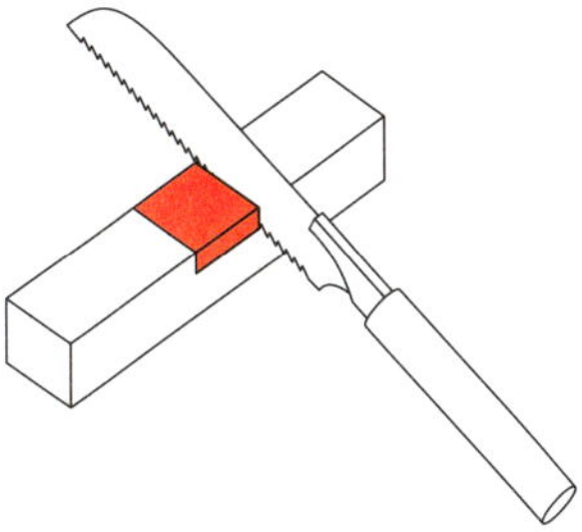

B2

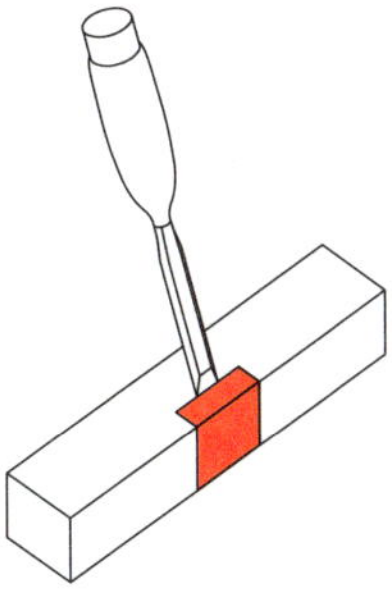

B3

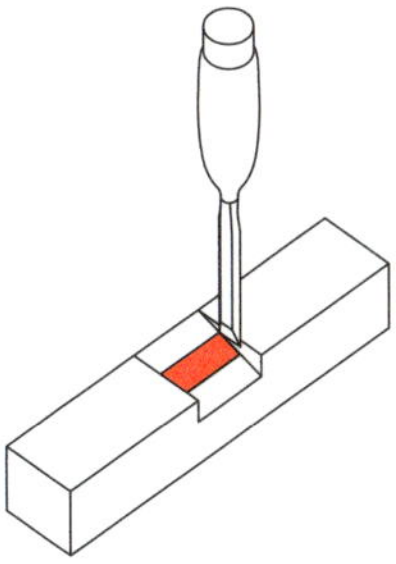

B4

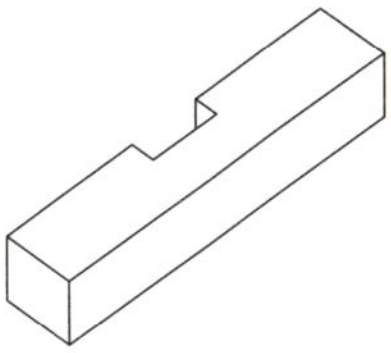

B

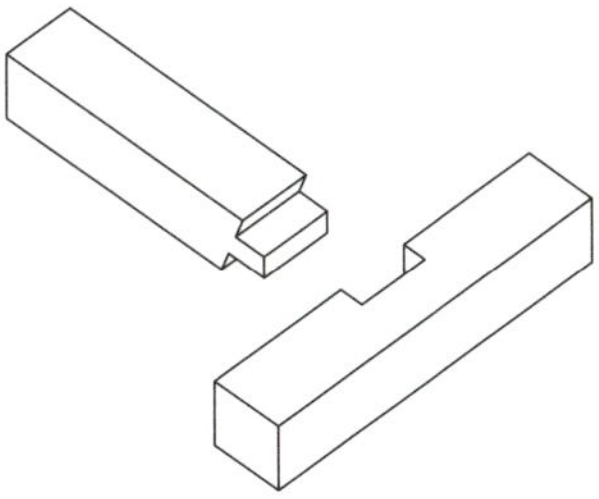

A+B

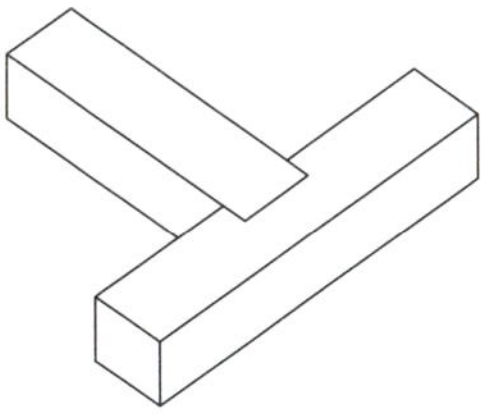

AB

T-Joints

Housed stop-mitered mortise and tenon

yatsuna-dome-tan-hozo-sashi
八中留短枘差し

keilförmig eingelassener Zapfen

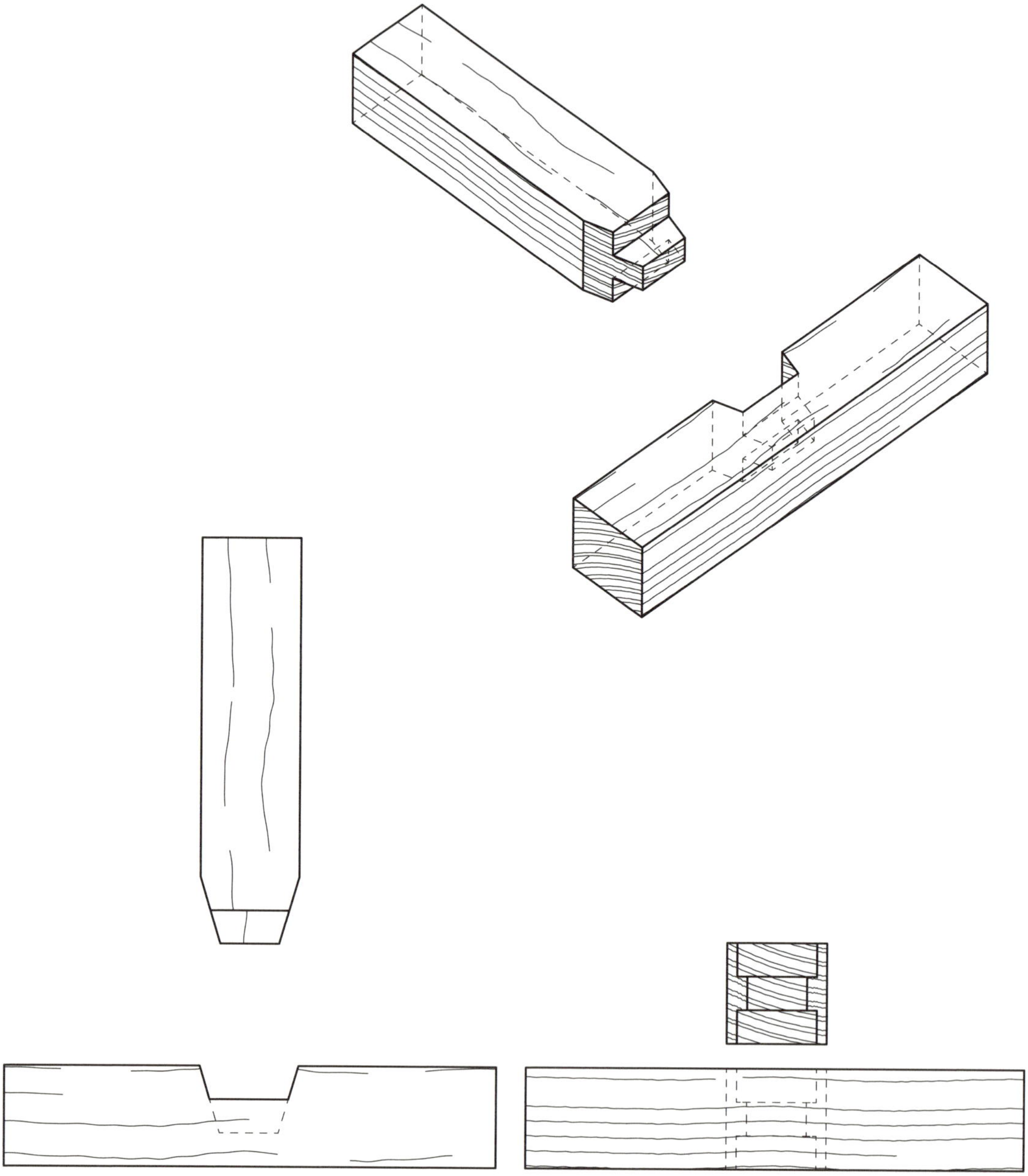

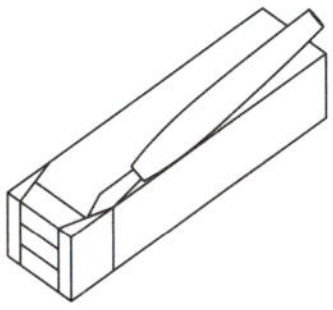

A1

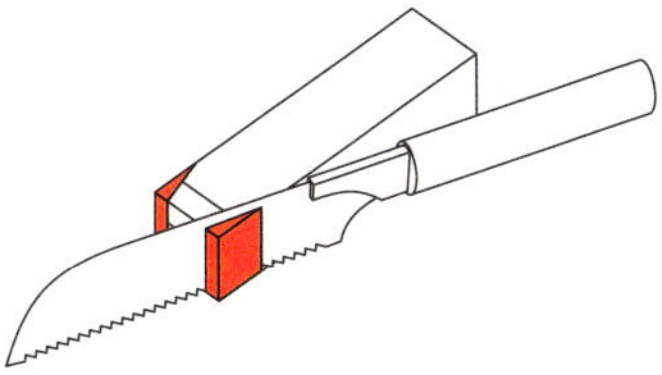

A2

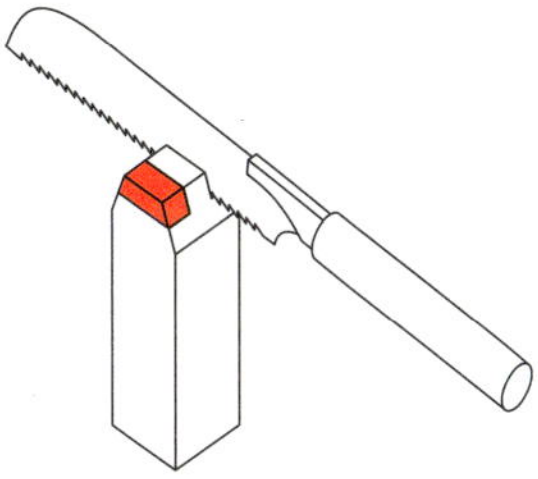

A3

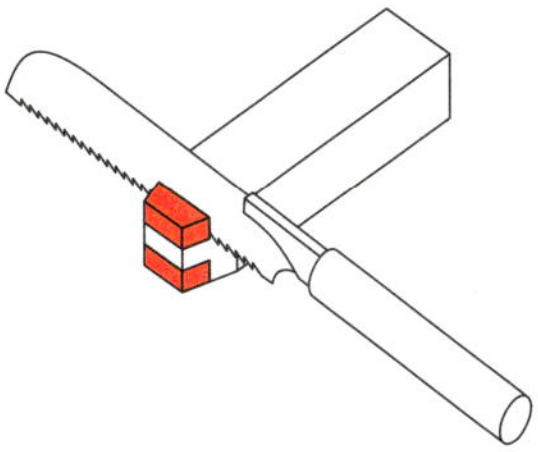

A4

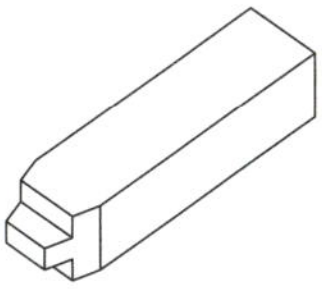

A

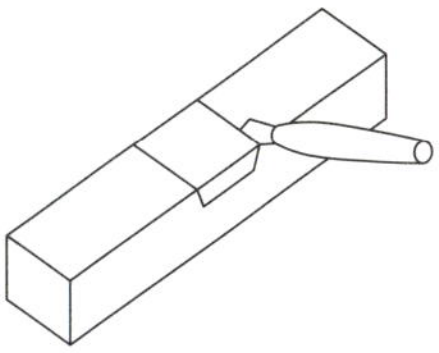

B1

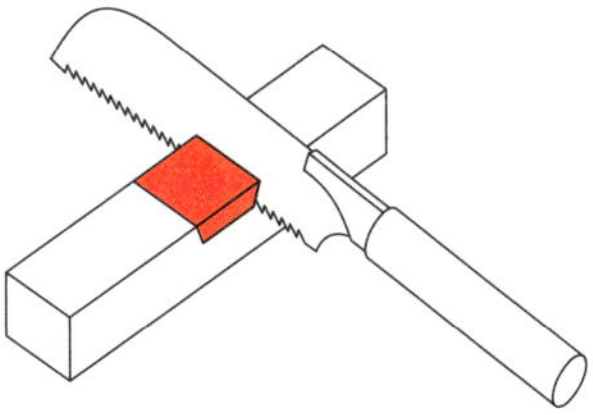

B2

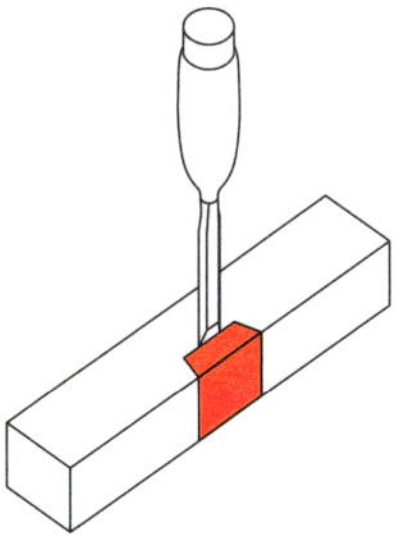

B3

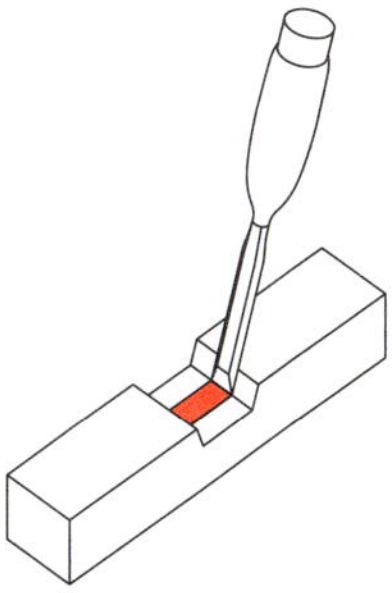

B4

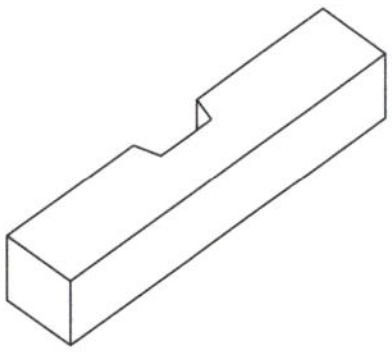

B

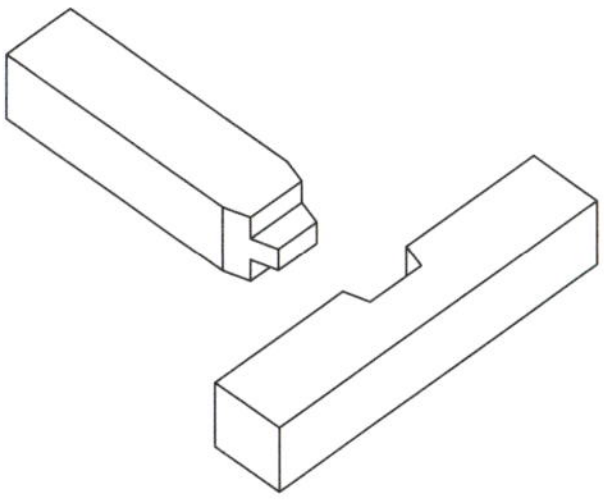

A+B

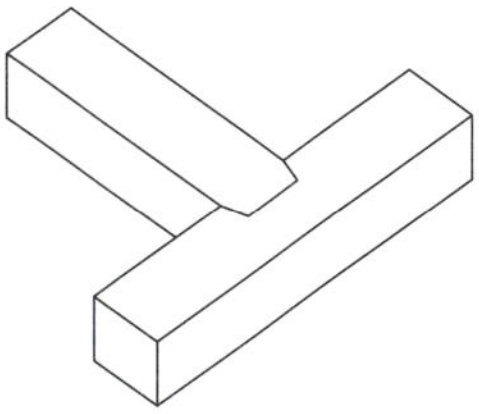

AB

T-Joints

Mortise and tenon with fan-shaped housing

ōari-dō-tsuki-tan-hozo-sashi
大蟻胴付短枘差し

schwalbenschwanzförmiger eingelassener Zapfen

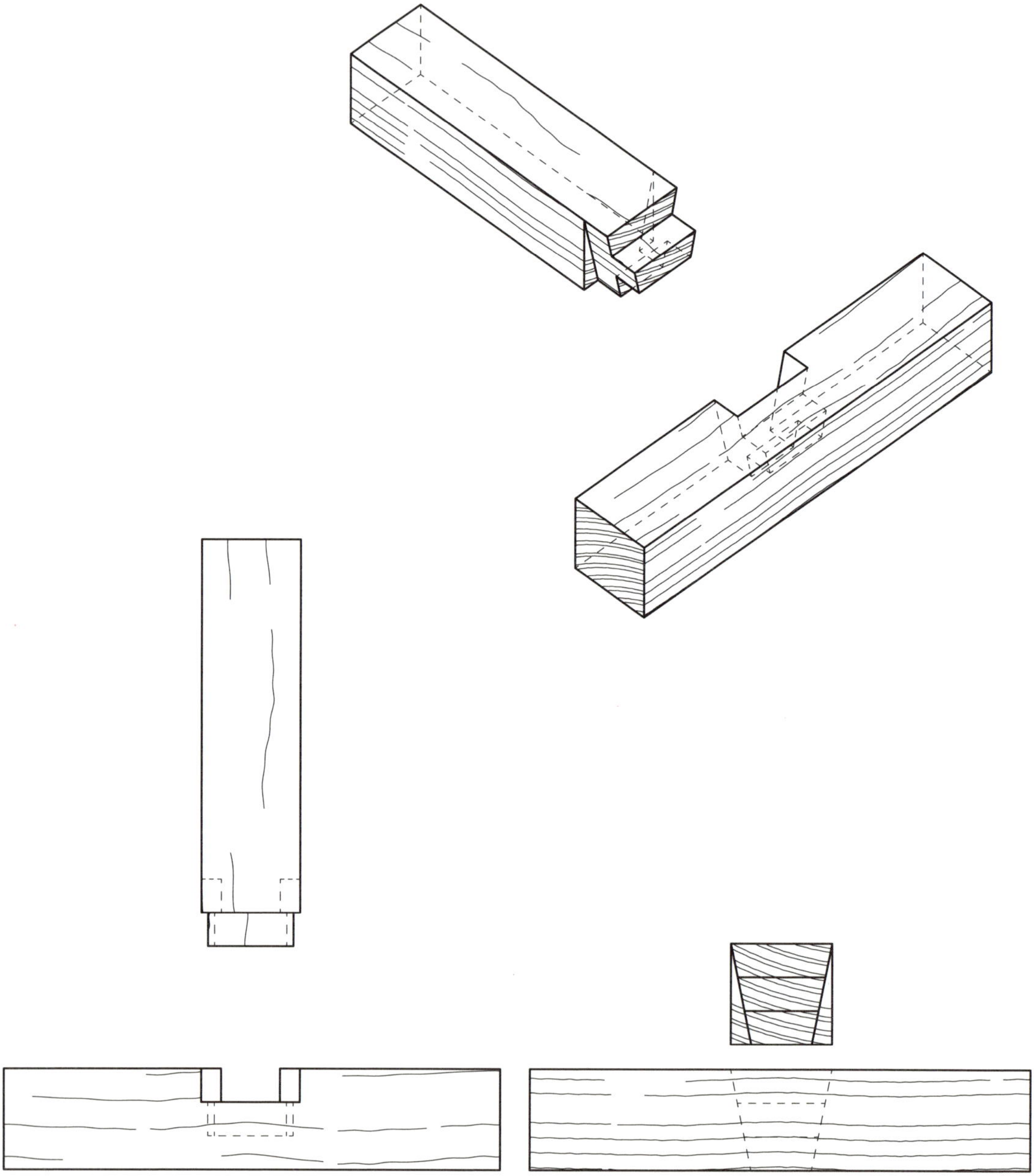

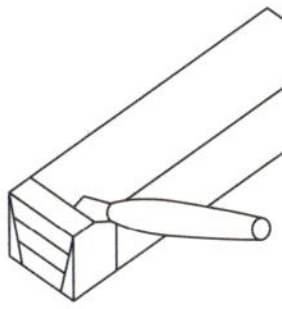

A1

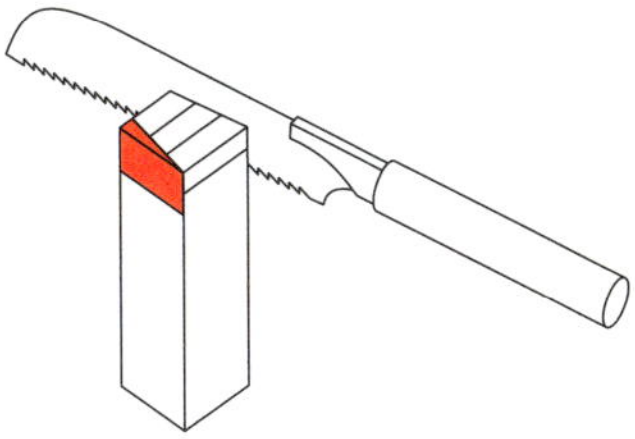

A2

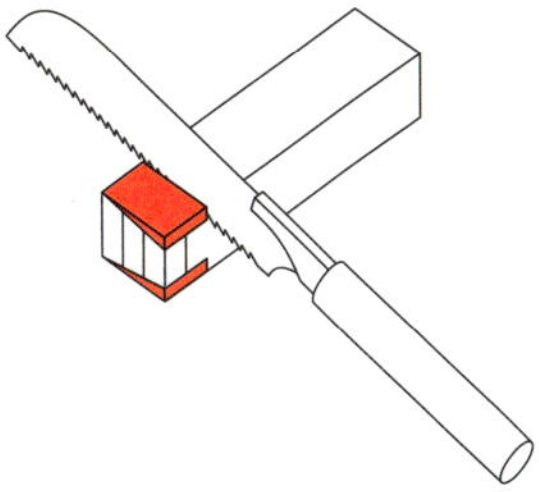

A3

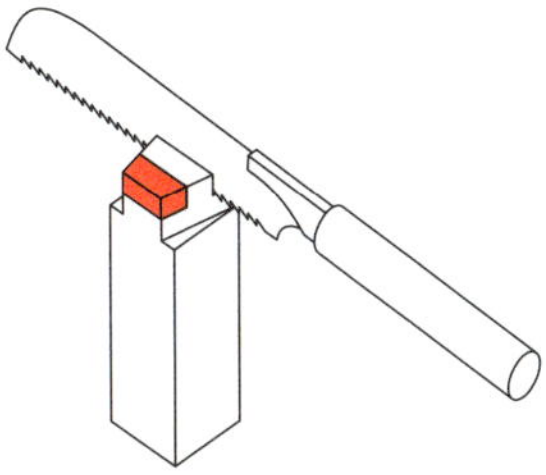

A4

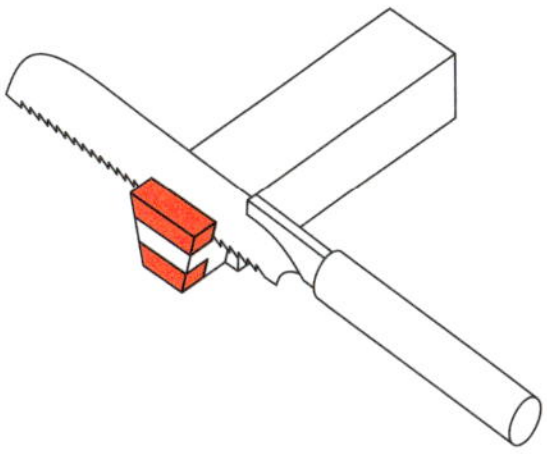

A5

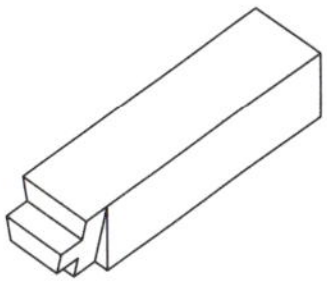

A

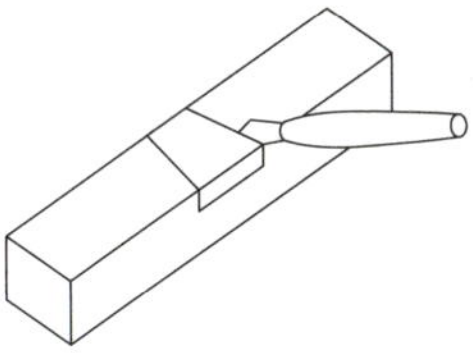

B1

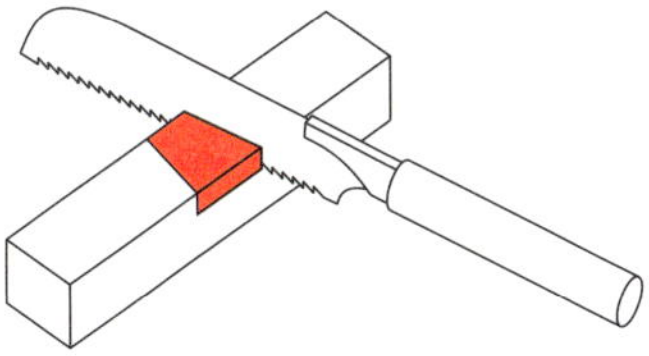

B2

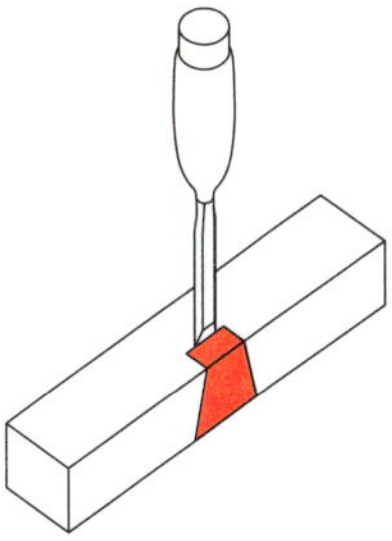

B3

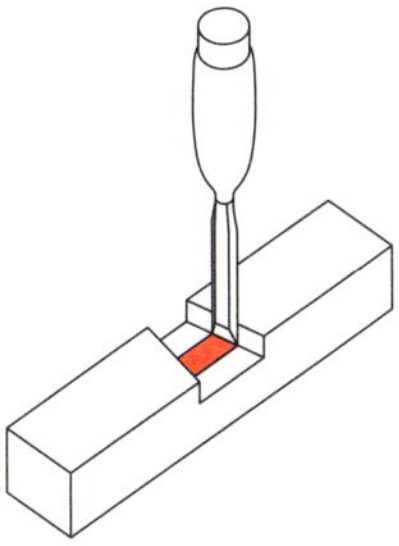

B4

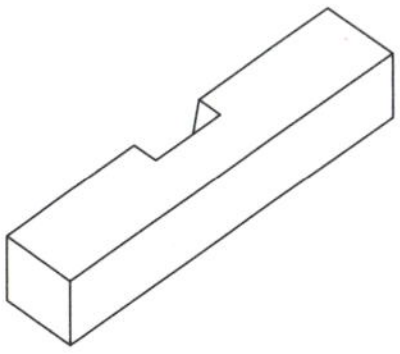

B

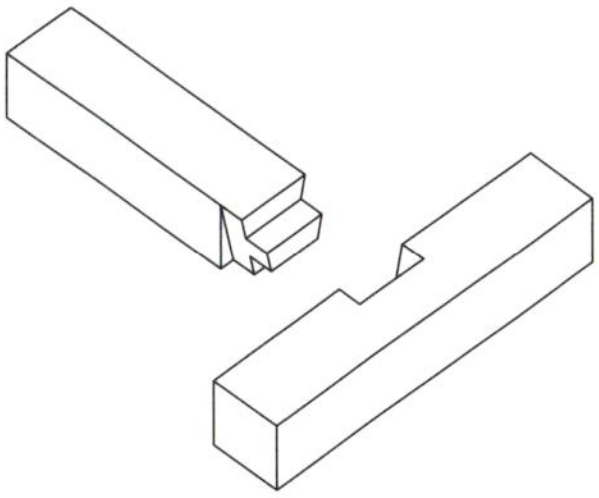

A+B

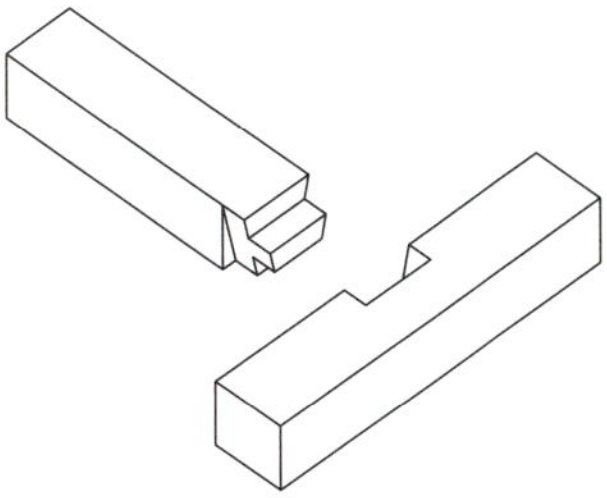

AB

T-Joints

Blind pegged joint

dabo-kumi-tsugi [*]
太枘組接ぎ

Verdollung

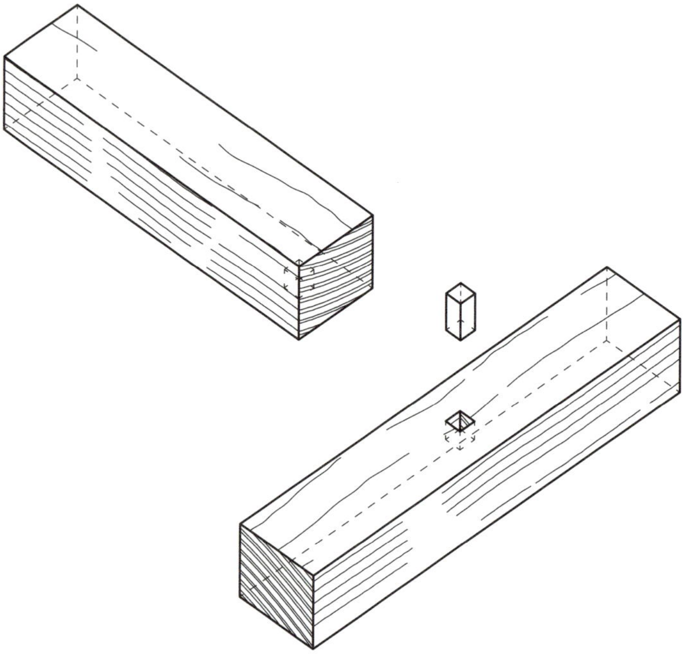

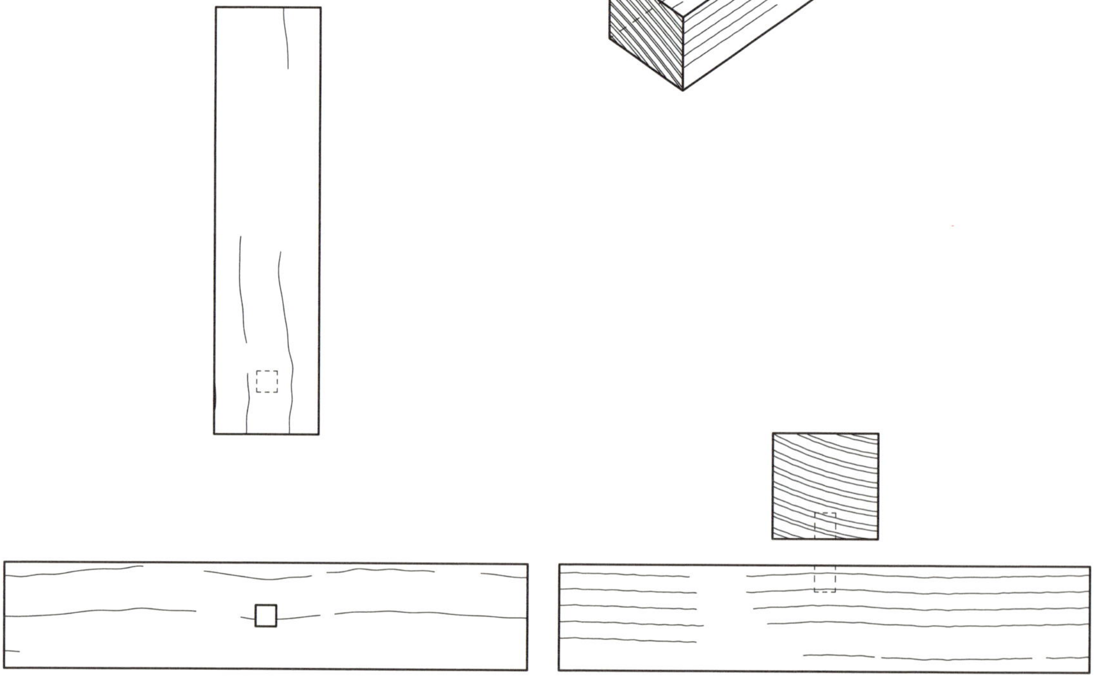

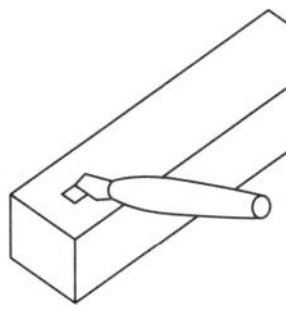

A1

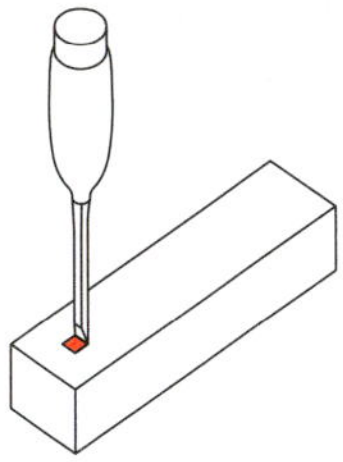

A2

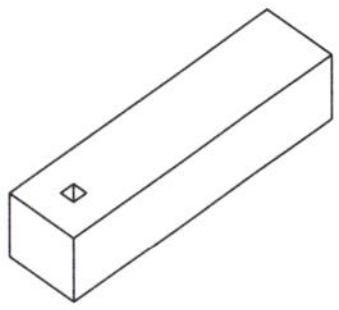

A

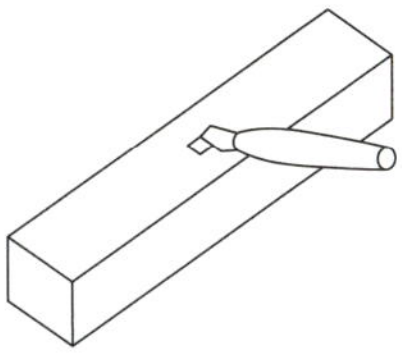

B1

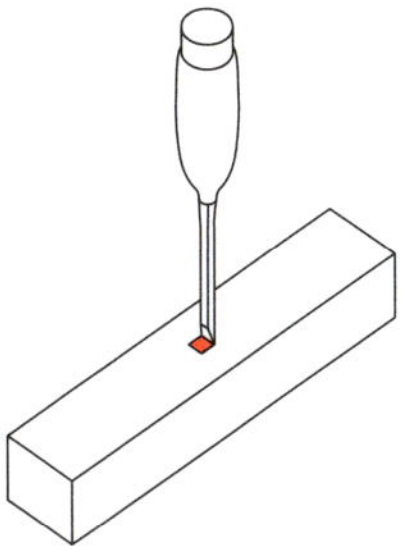

B2

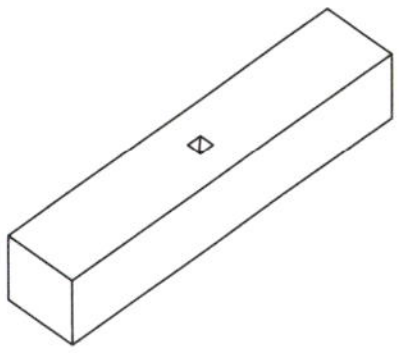

B

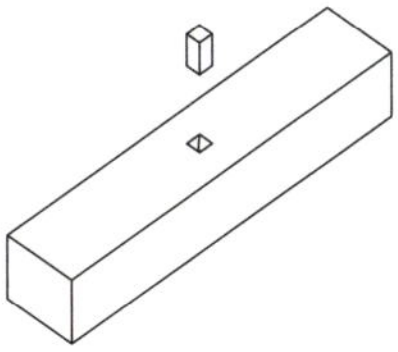

B+C

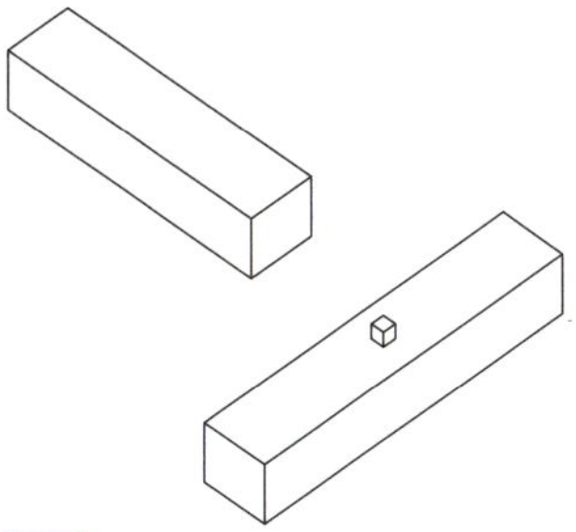

A+B+C

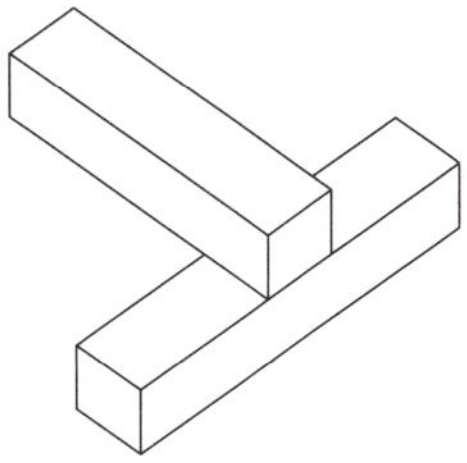

ABC

T-Joints

Notch joint

hira-kake-tsugi
平掛接ぎ

Ankämmung mit einseitigem Sitz

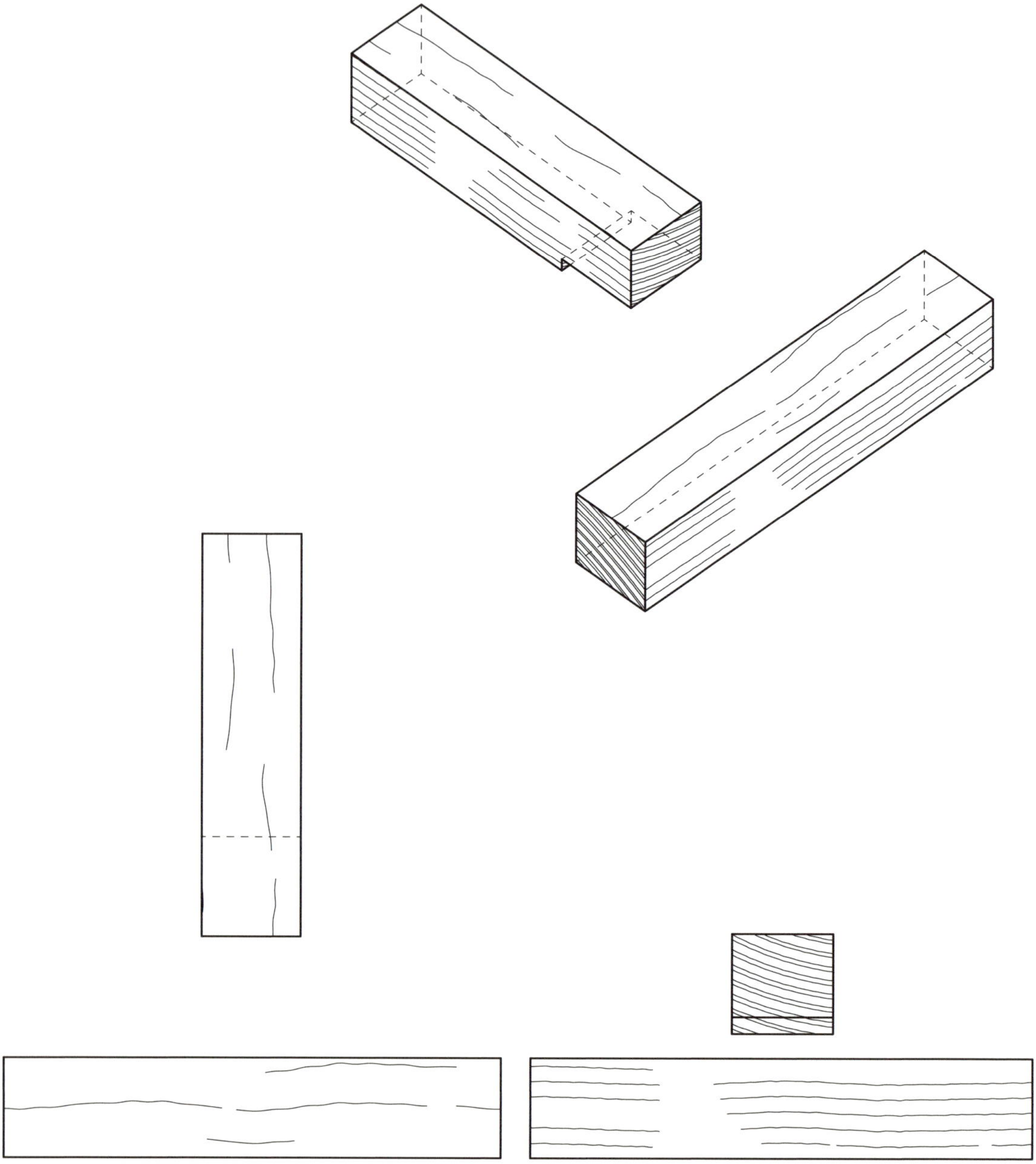

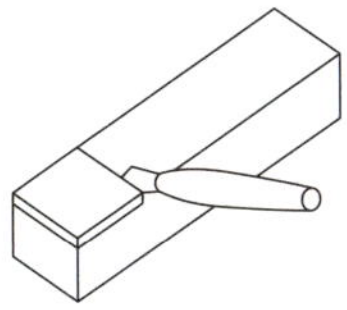

A1

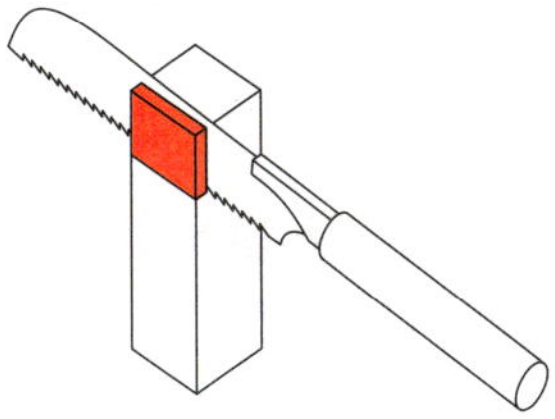

A2

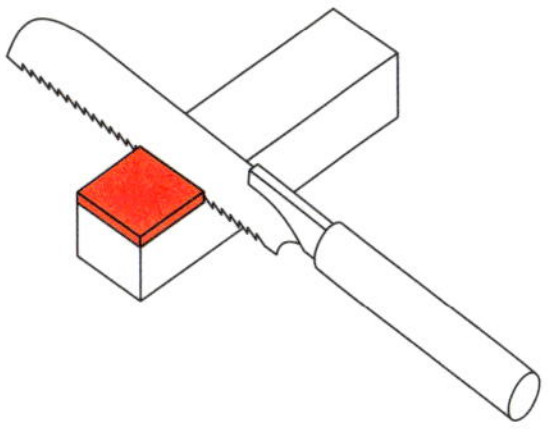

A3

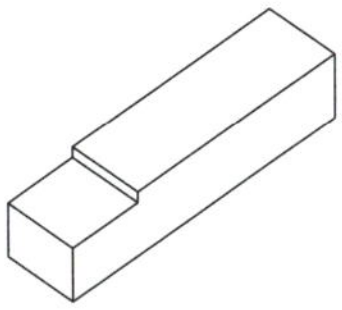

A

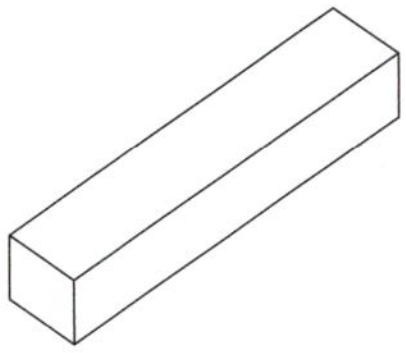

B

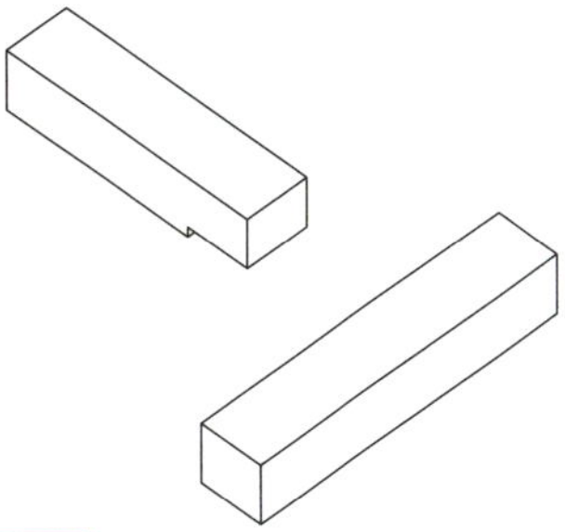

A+B

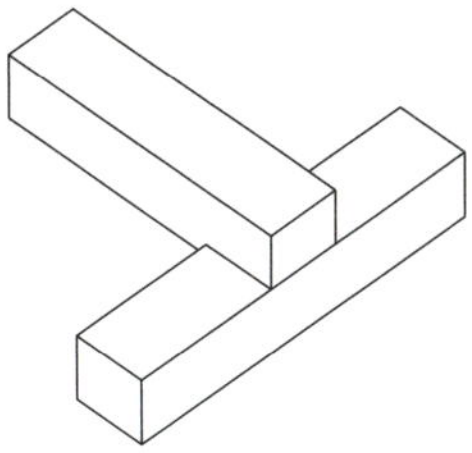

AB

T-Joints

Double notch joint

hira-kake-ai-kaki-tsugi
平掛相欠接ぎ

Ankämmung

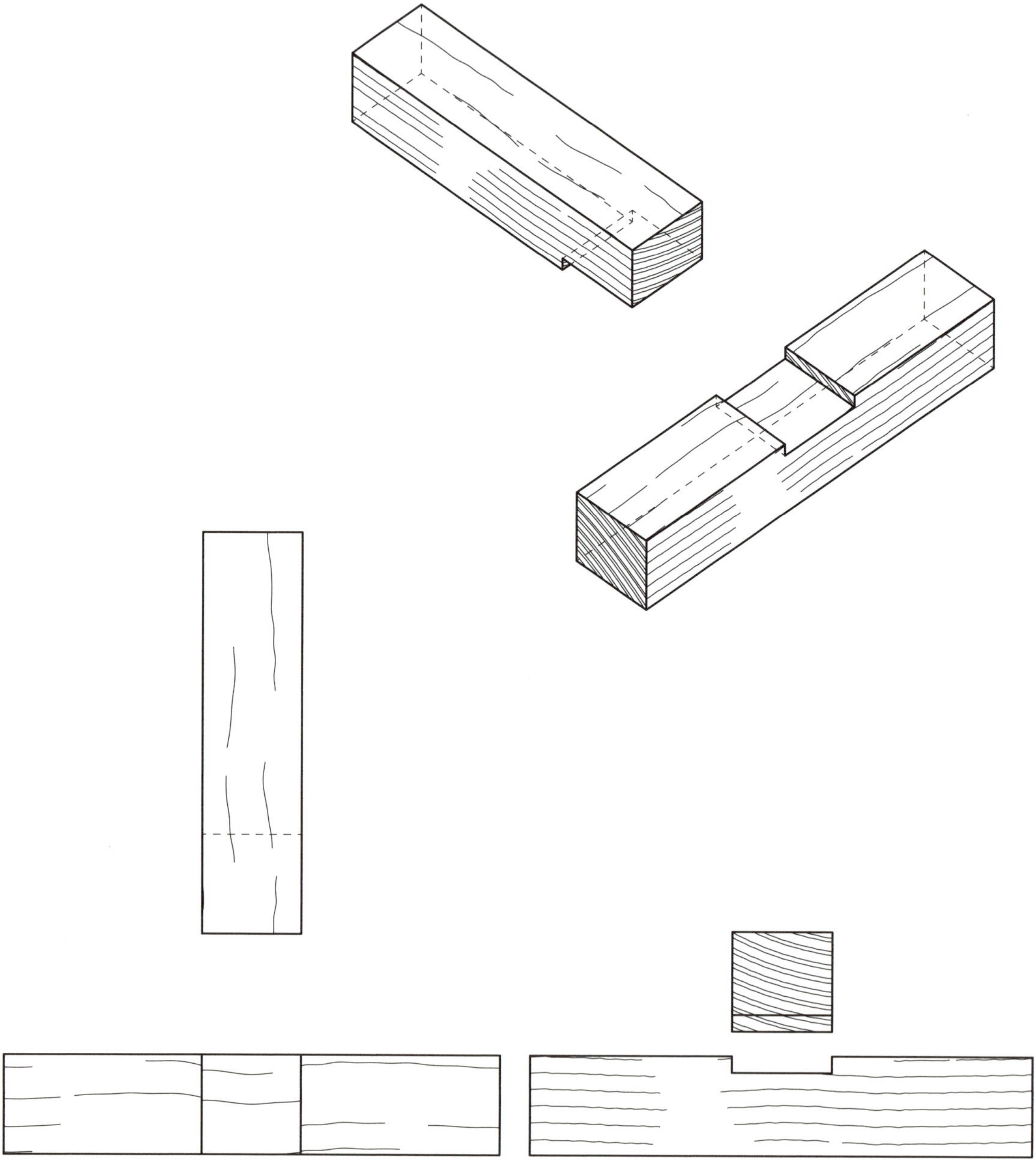

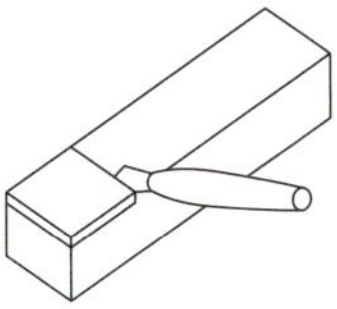

A1

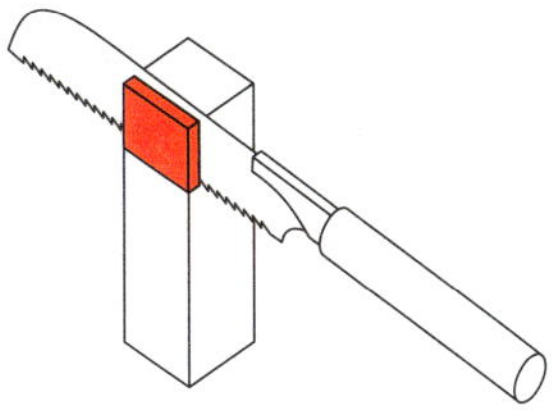

A2

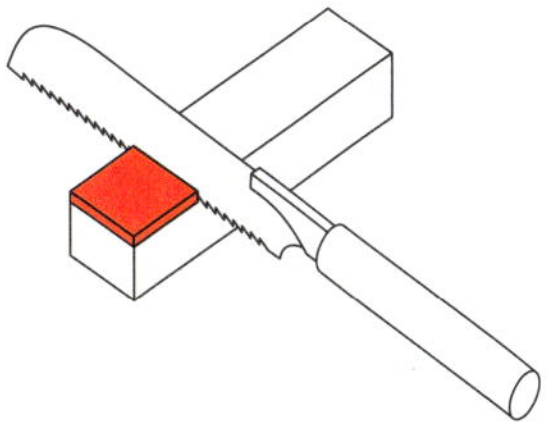

A3

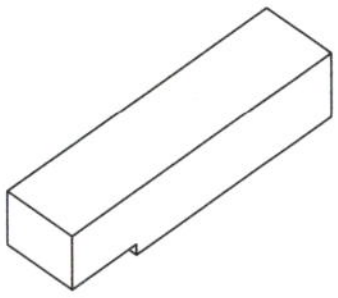

A

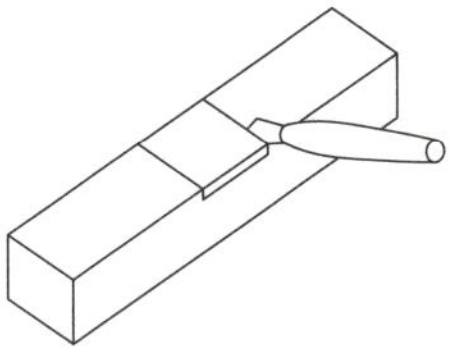

B1

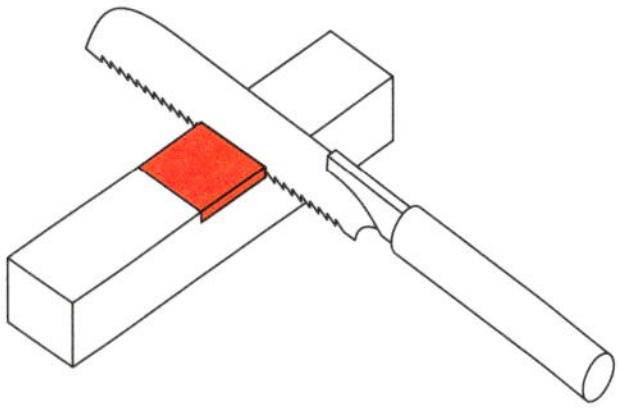

B2

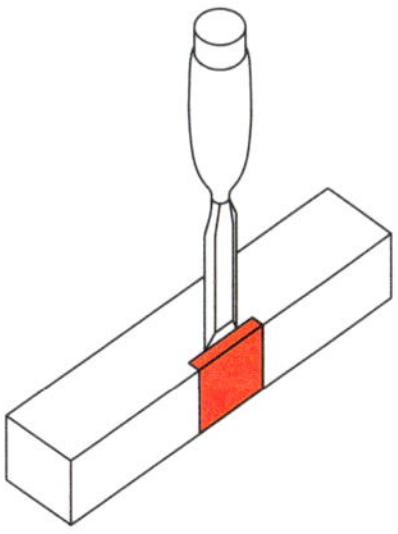

B3

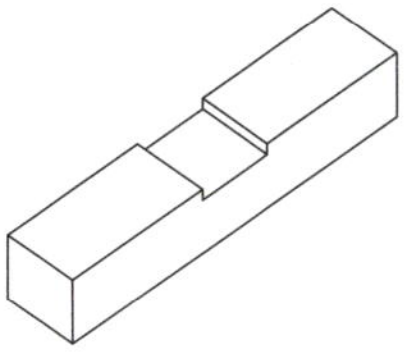

B

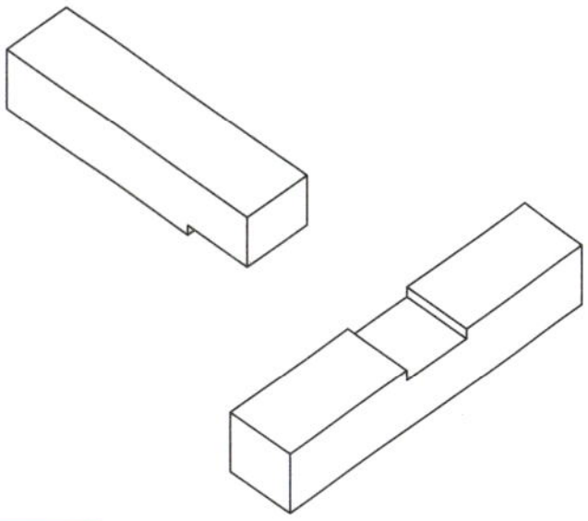

A+B

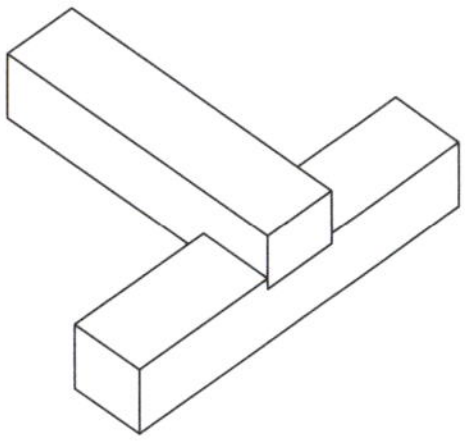

AB

T-Joints

Cog joint

ago-kake-ai-kaki-tsugi
顎掛相欠接ぎ

Ankämmung mit Stufe

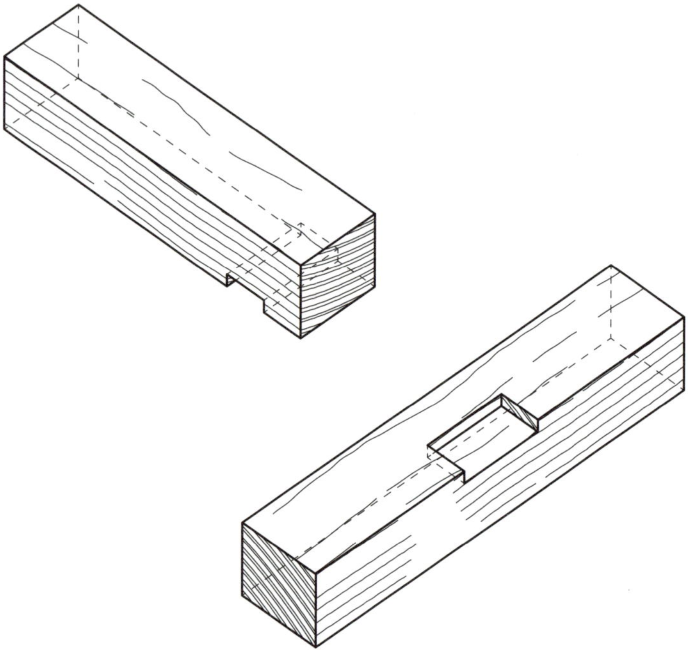

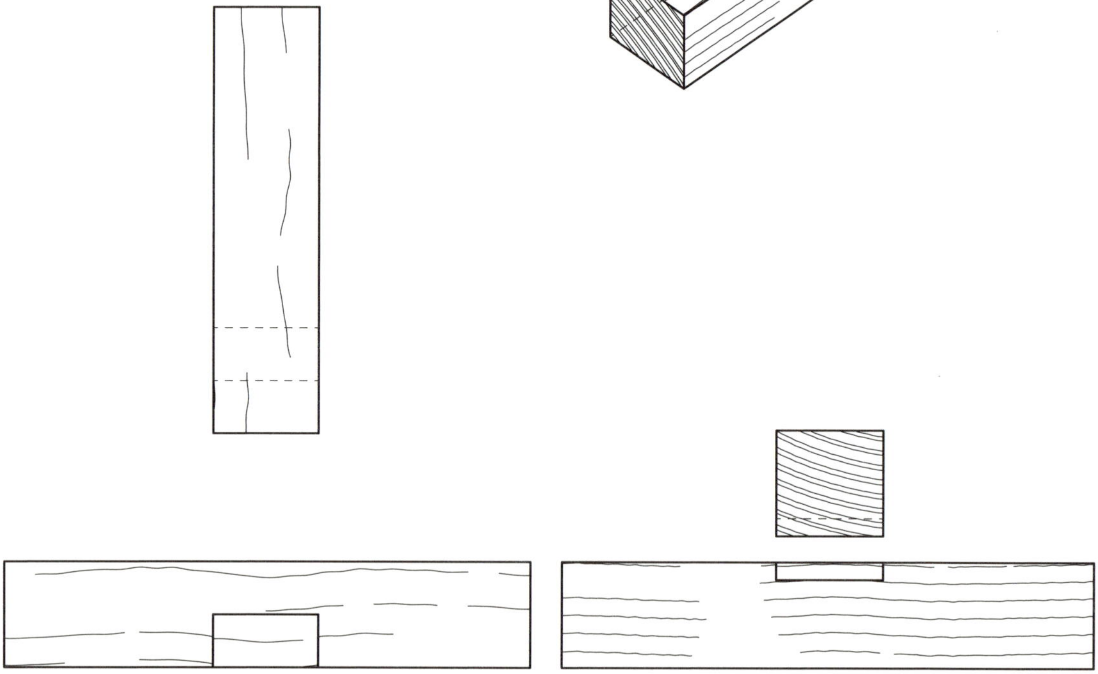

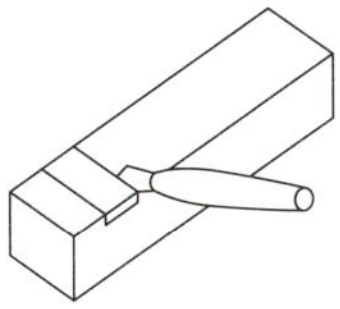

A1

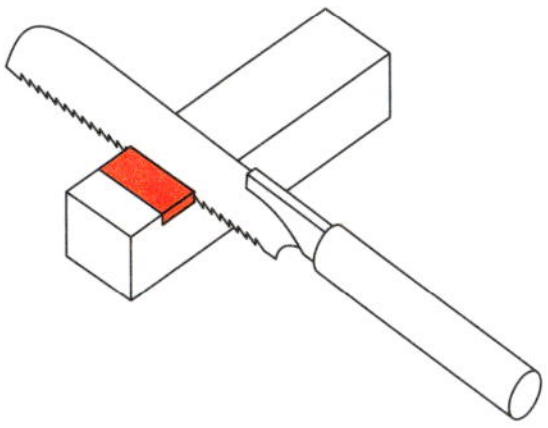

A2

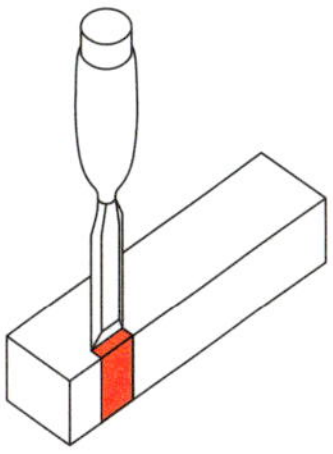

A3

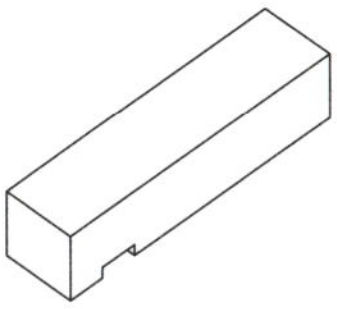

A

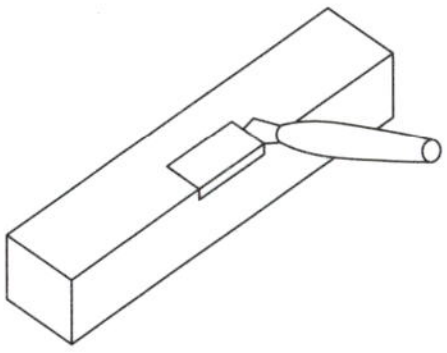

B1

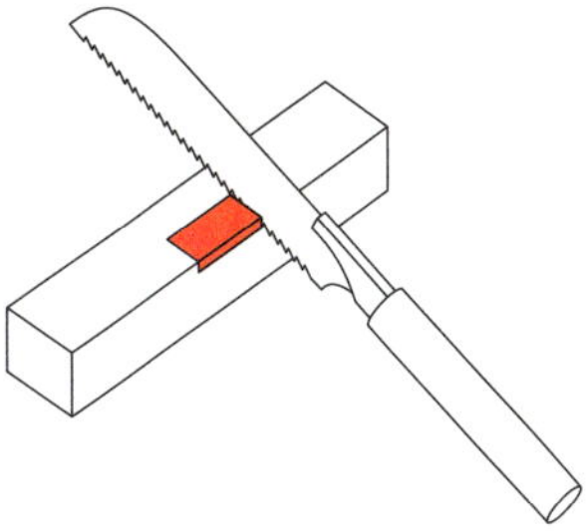

B2

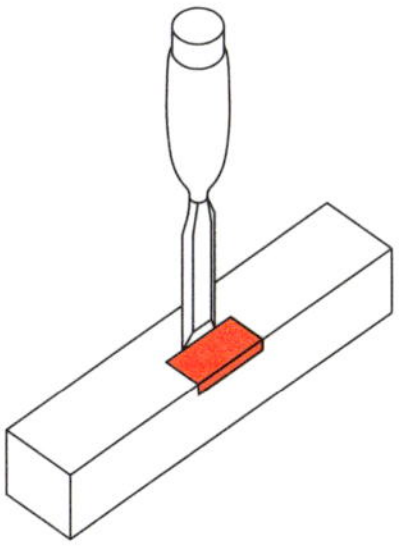

B3

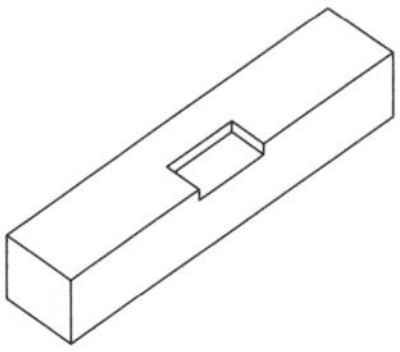

B

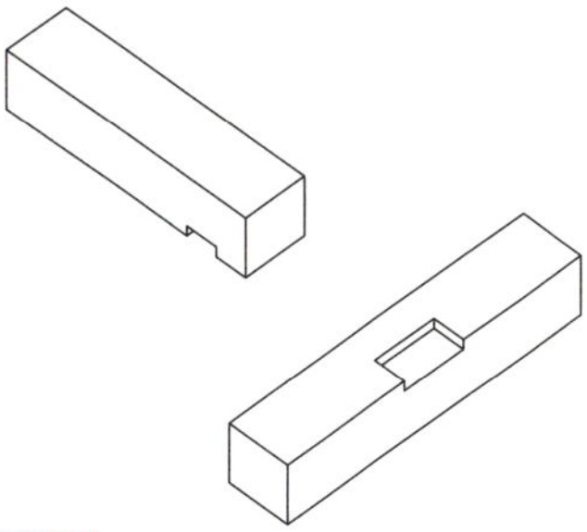

A+B

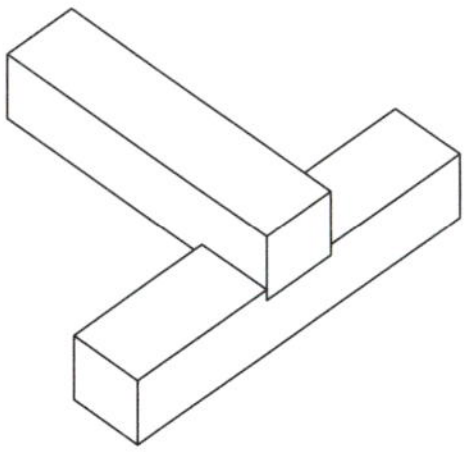

AB

T-Joints

Stepped cog joint

nidan-ago-kake-ai-kaki-tsugi
二段顎掛相欠接ぎ

Ankämmung mit doppelter Stufe

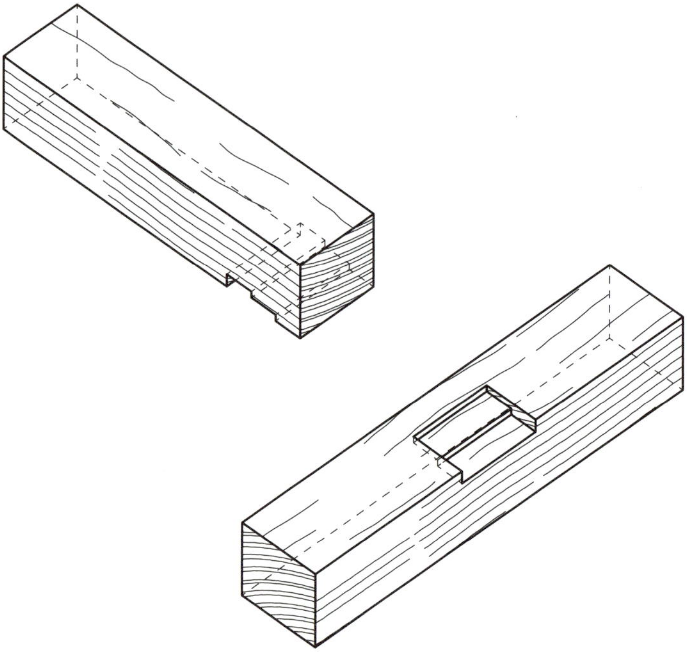

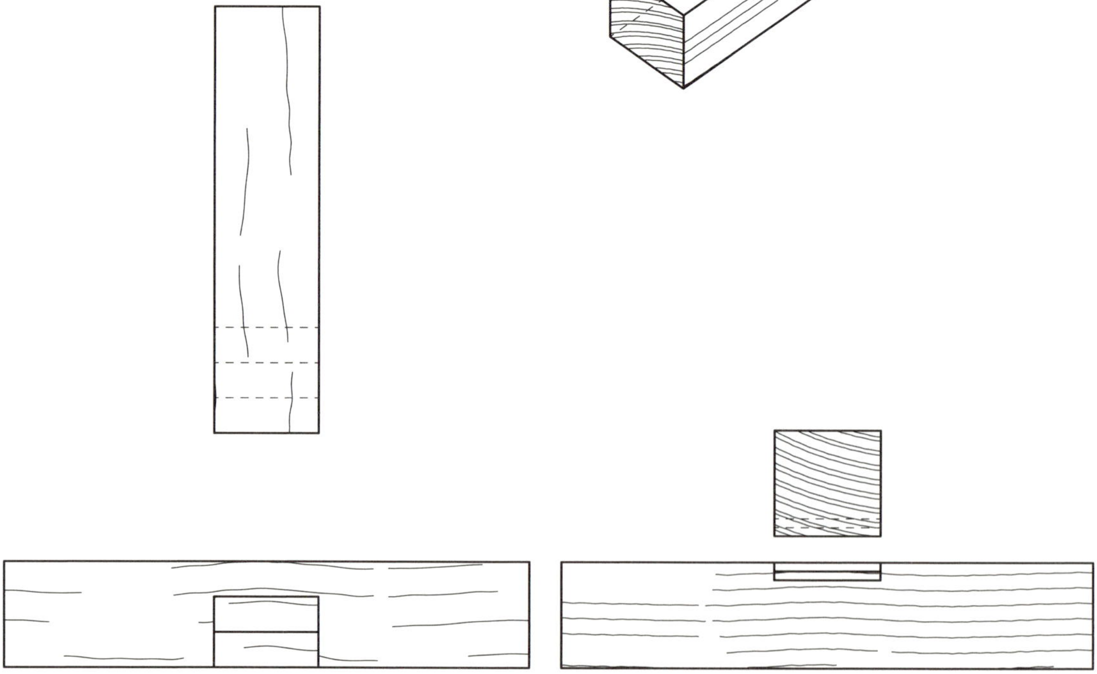

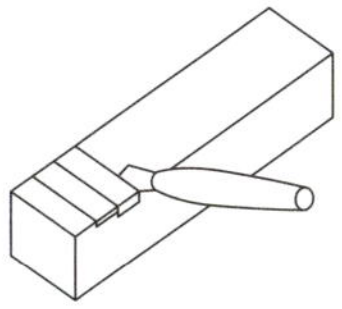

A1

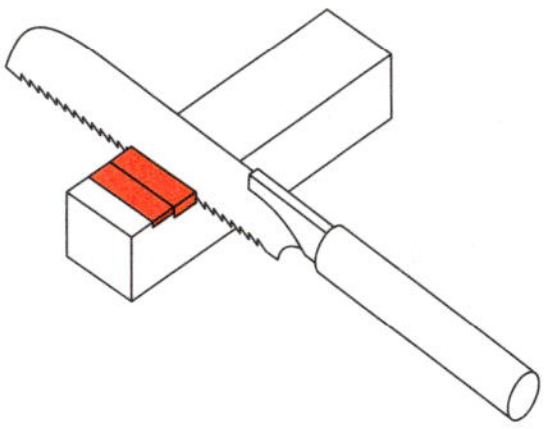

A2

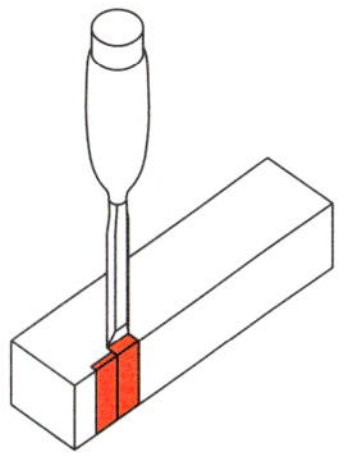

A3

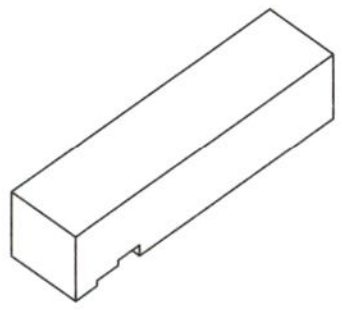

A

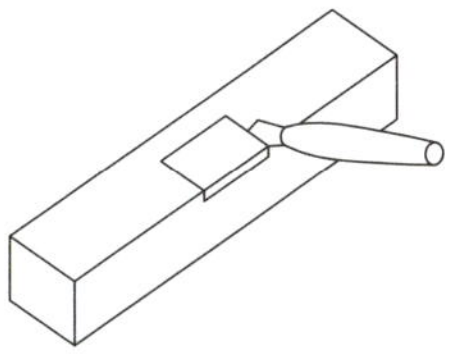

B1

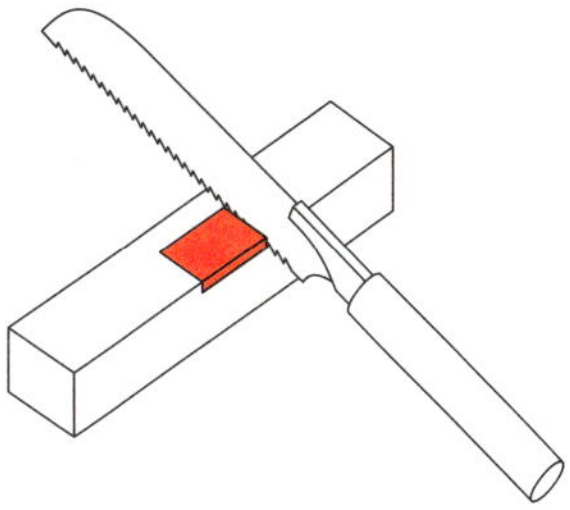

B2

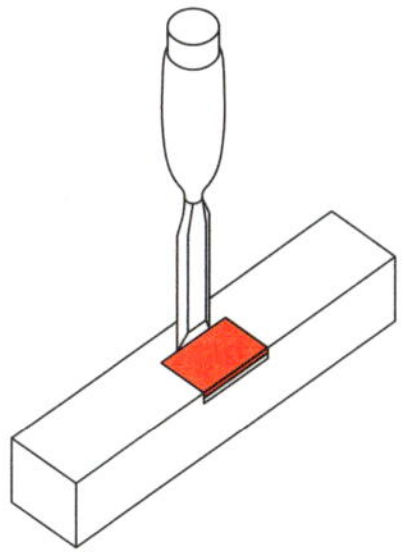

B3

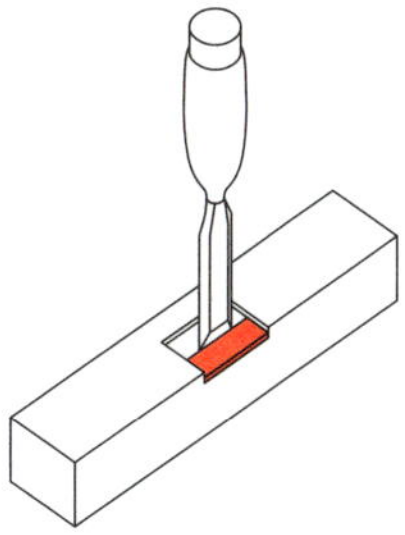

B4

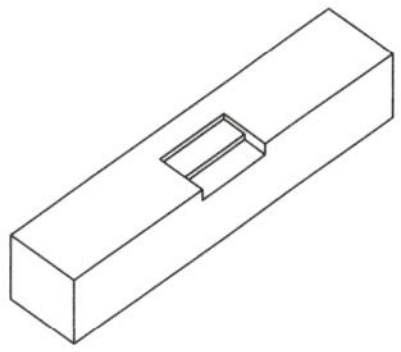

B

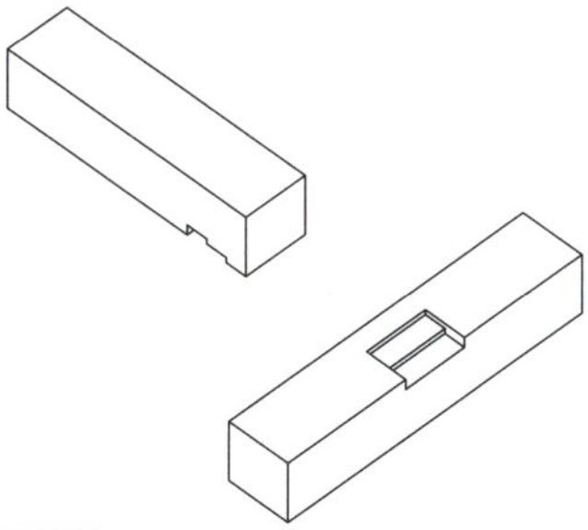

A+B

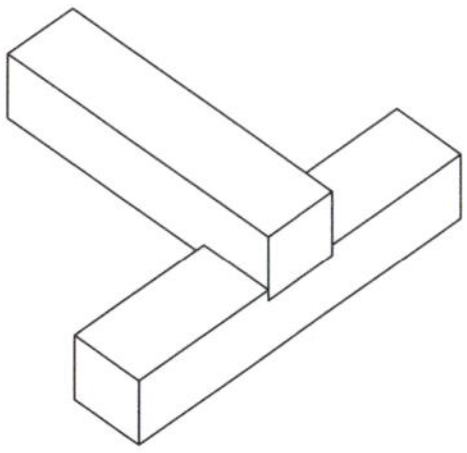

AB

T-Joints

Double cog joint

watari-ago-kake-ai-kaki-tsugi
渡顎掛相欠接ぎ

Ankämmung mit mittigem Haken

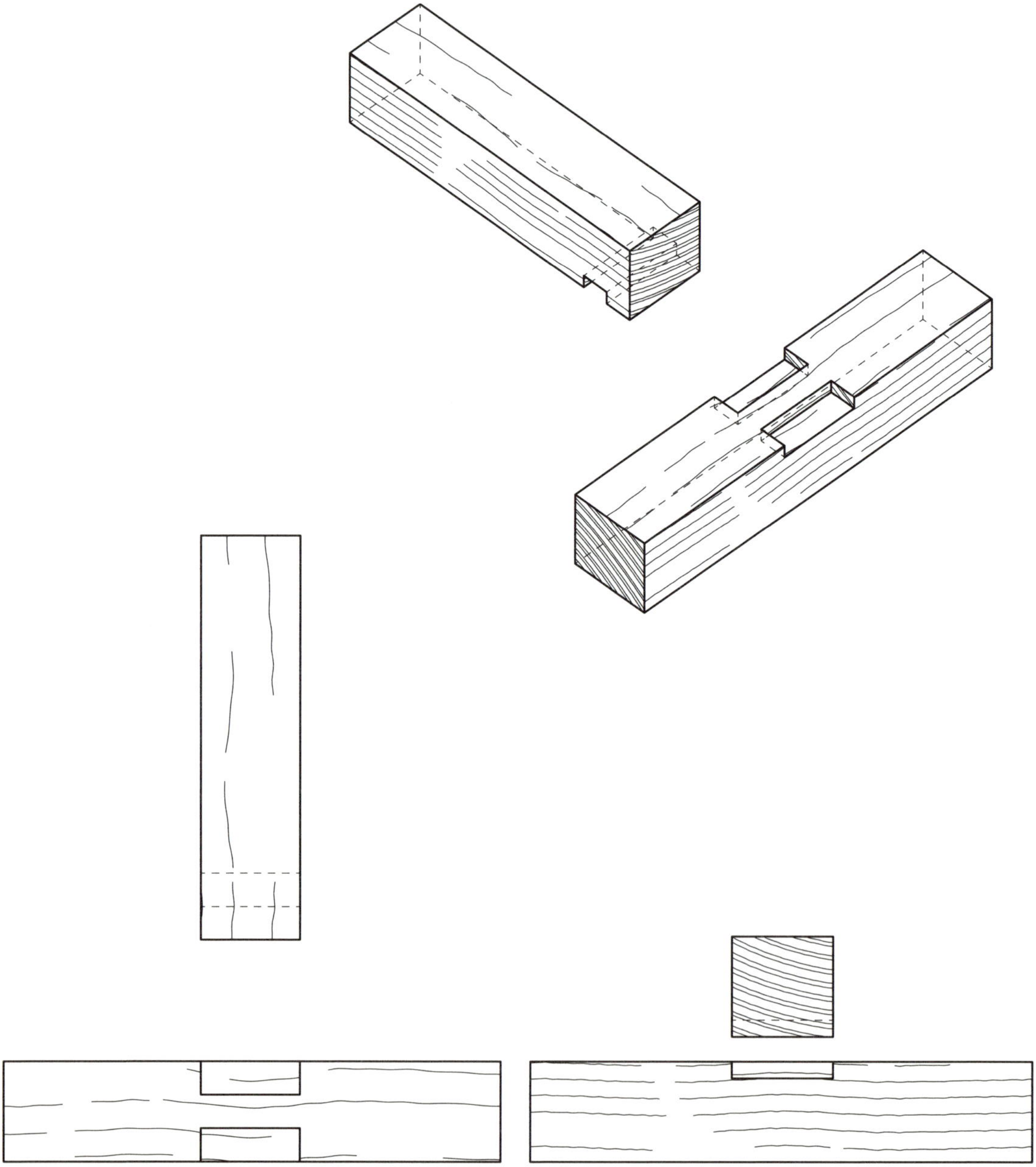

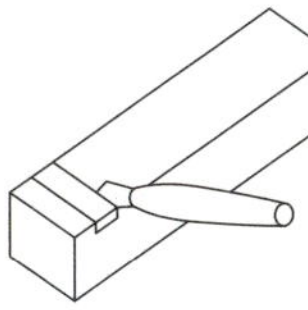

A1

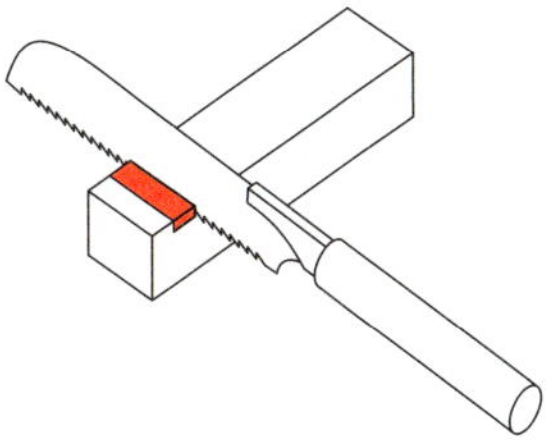

A2

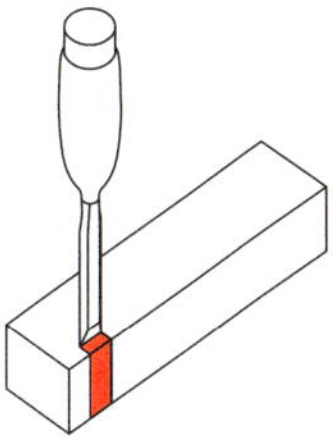

A3

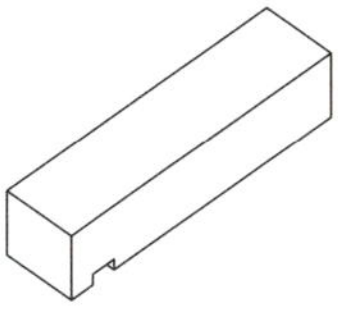

A

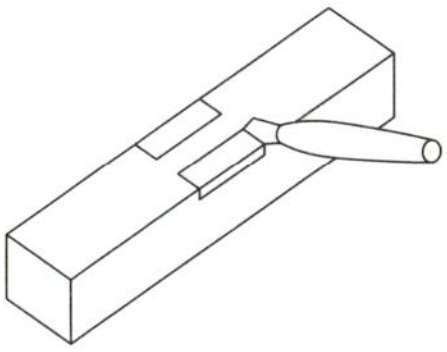

B1

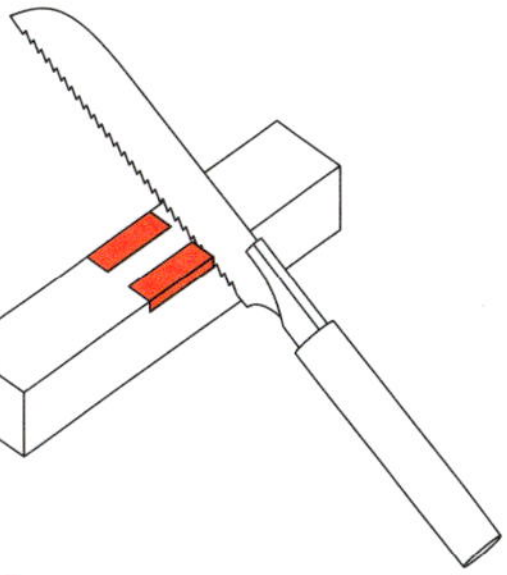

B2

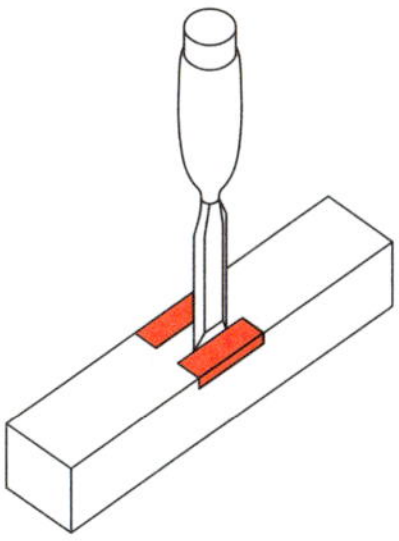

B3

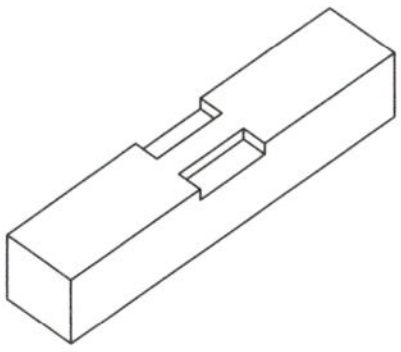

B

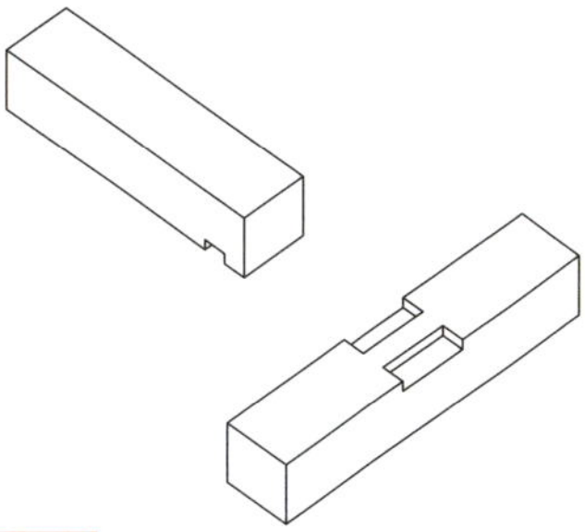

A+B

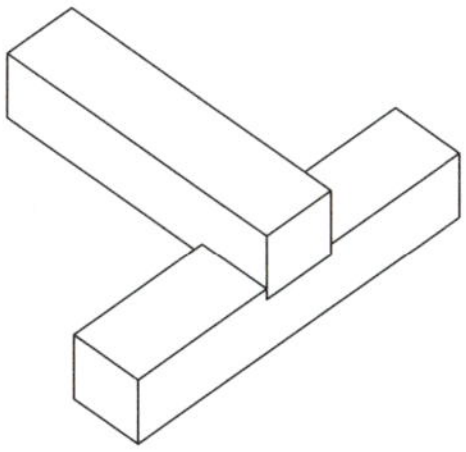

AB

T-Joints

X-shaped cog joint

tasuki-ai-kaki-tsugi
襷相欠接ぎ

Ankämmung mit Kreuz

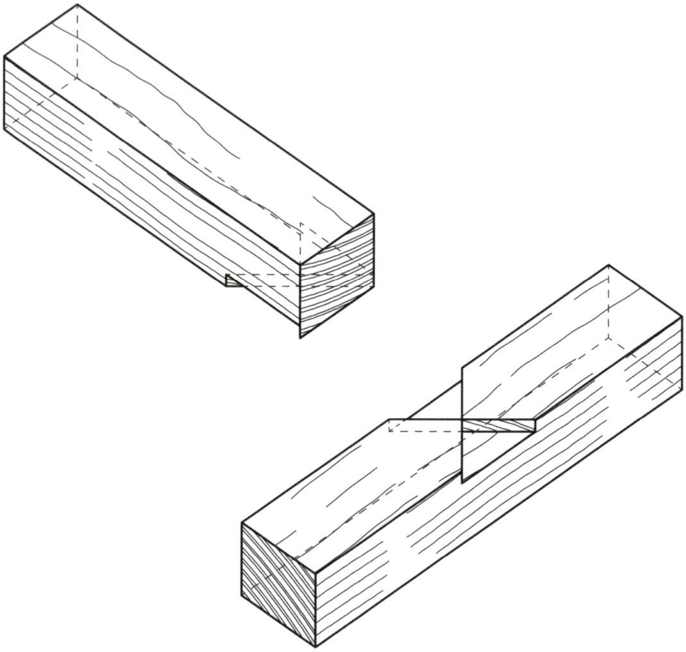

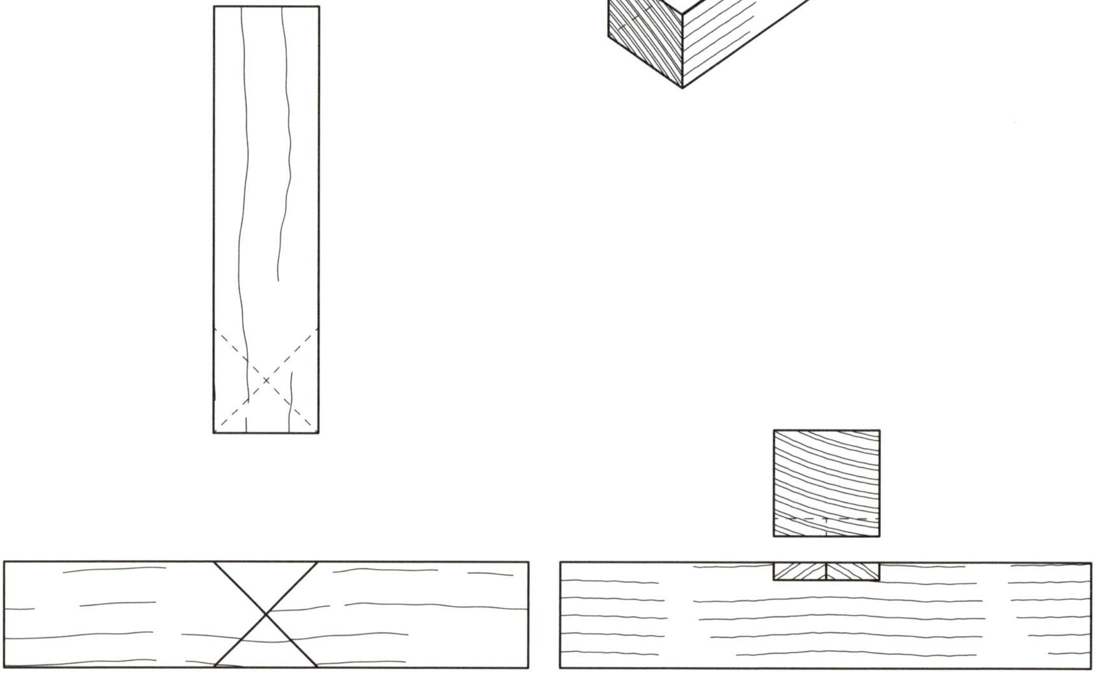

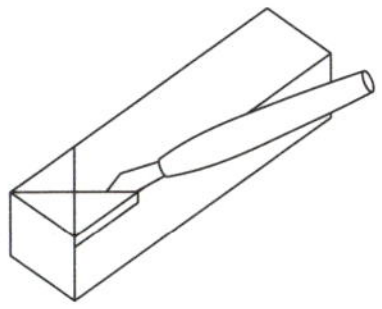

A1

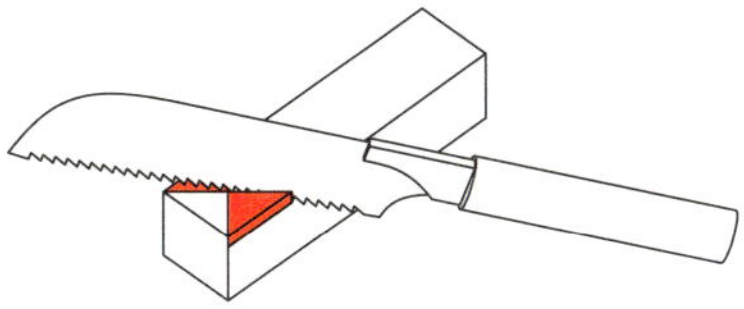

A2

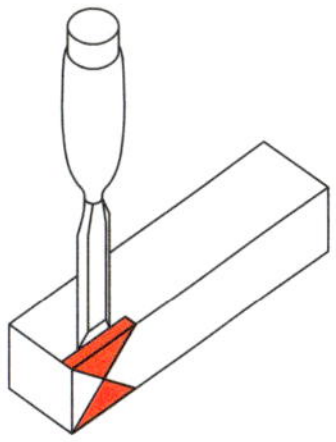

A3

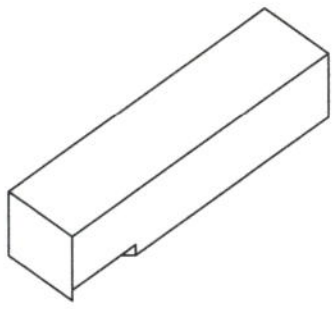

A

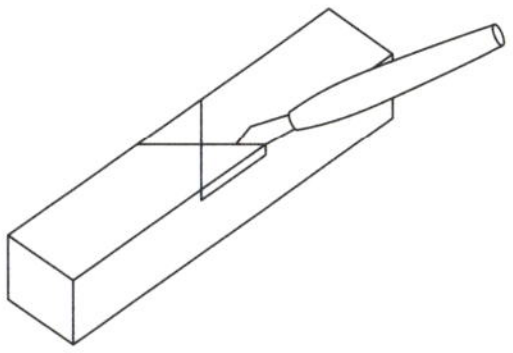

B1

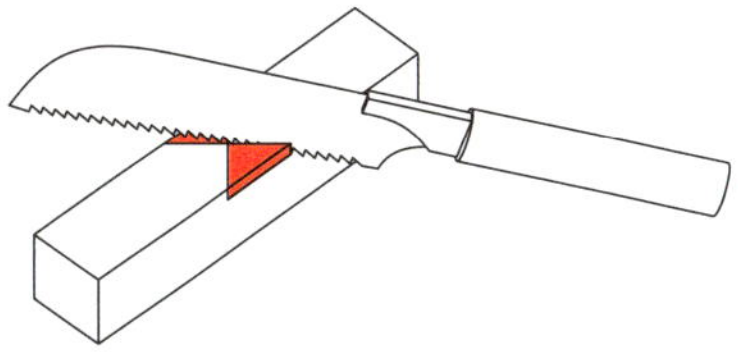

B2

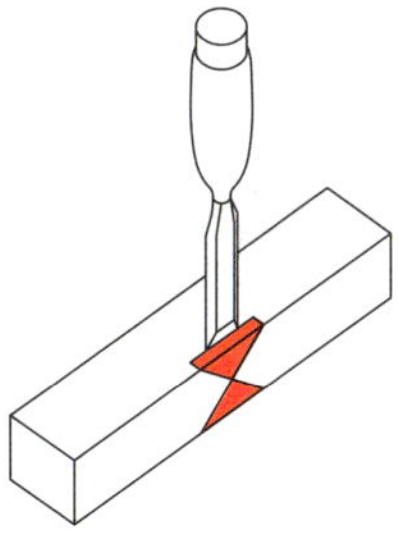

B3

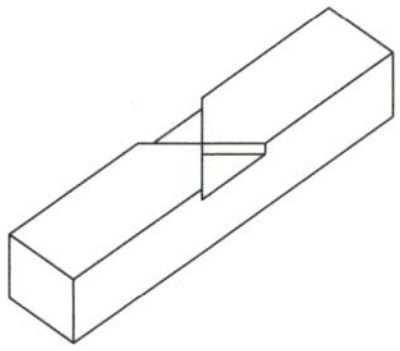

B

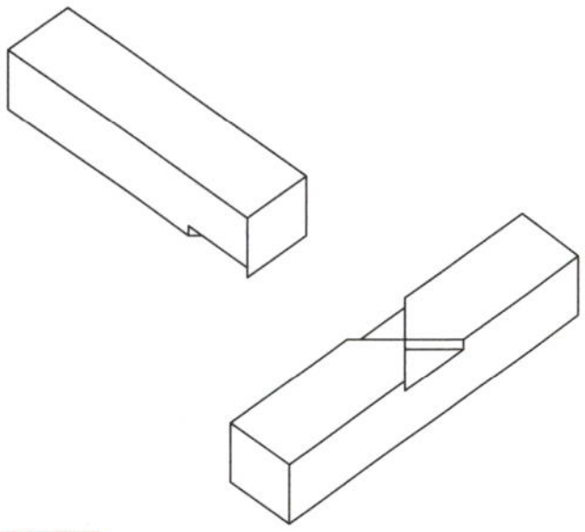

A+B

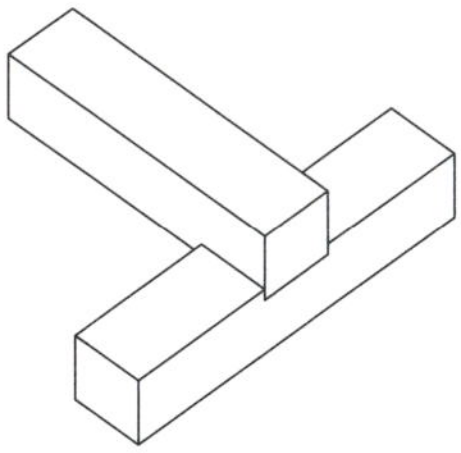

AB

T-Joints

V-shaped cog joint

yahazu-kake-ai-kaki-tsugi
矢筈掛相欠接ぎ

Ankämmung mit halbem Kreuz

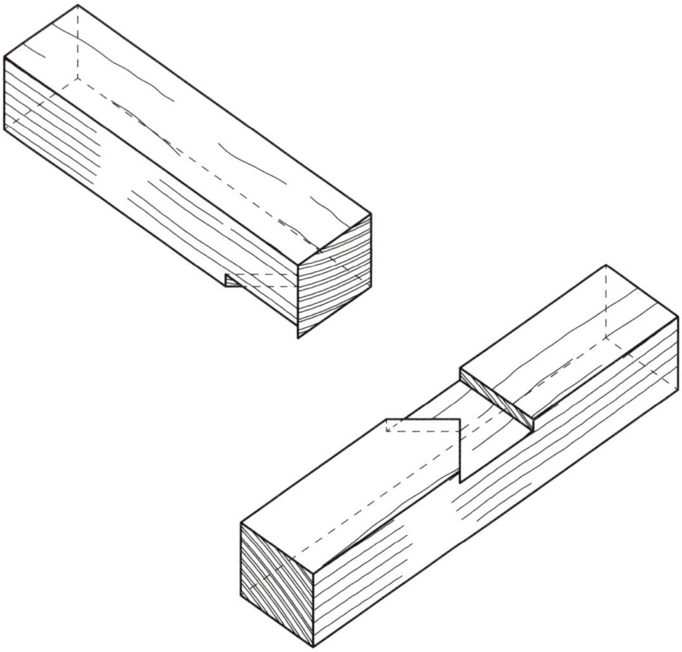

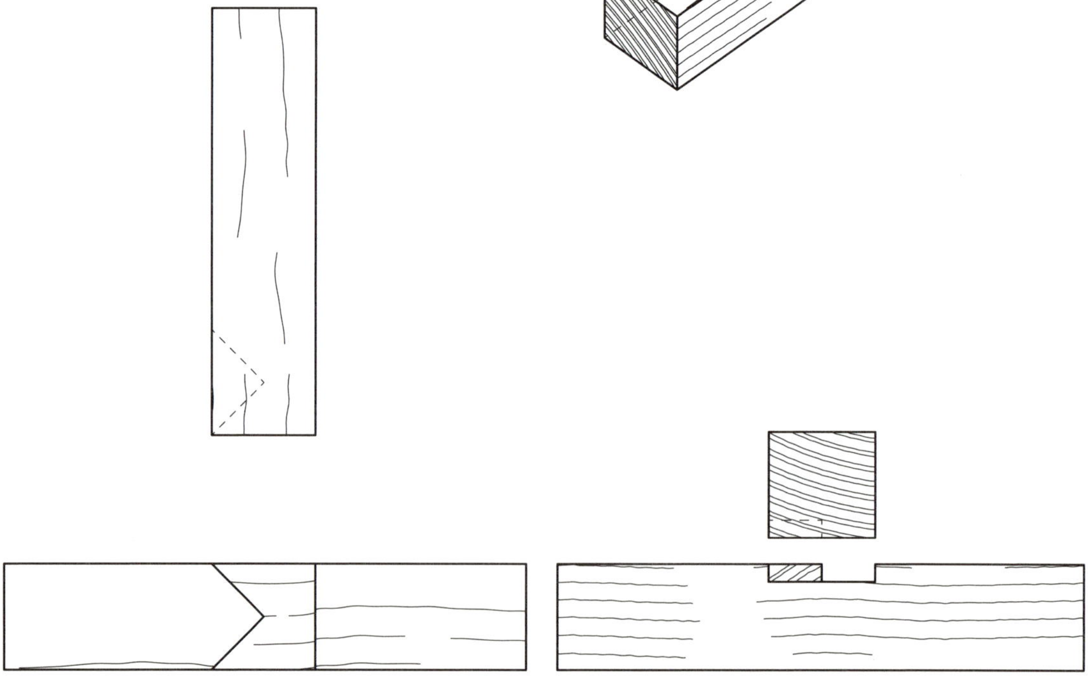

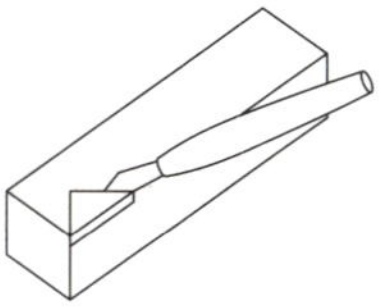

A1

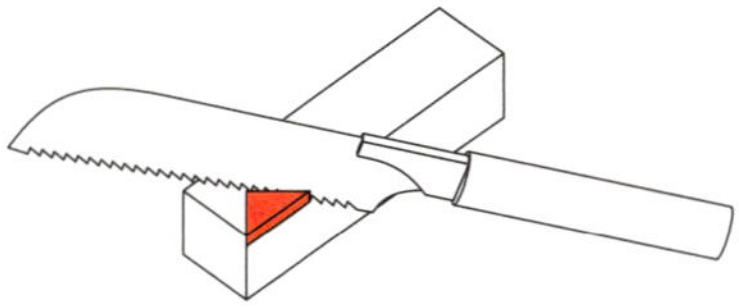

A2

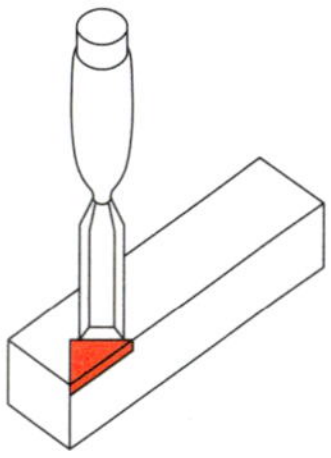

A3

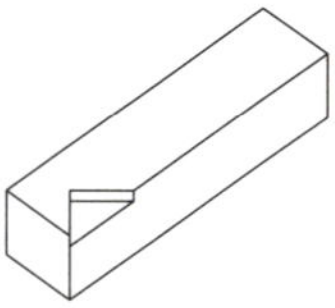

A

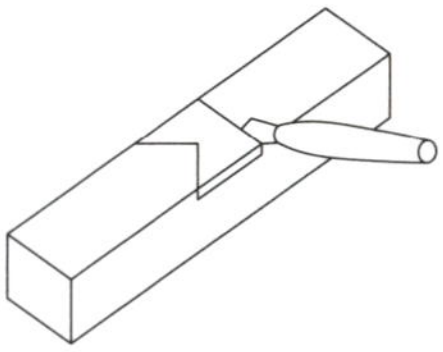

B1

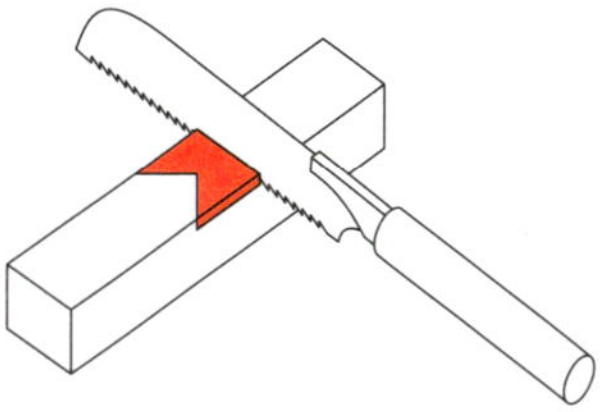

B2

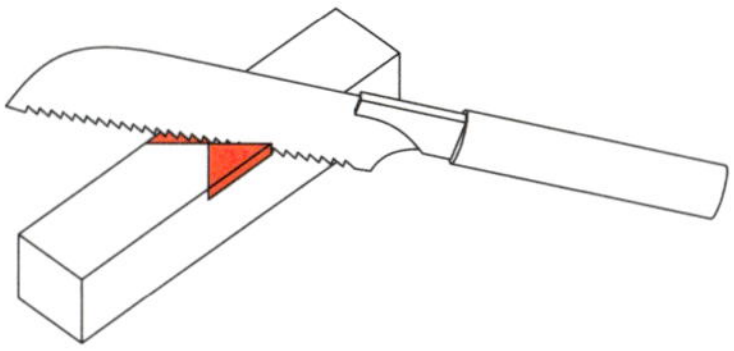

B3

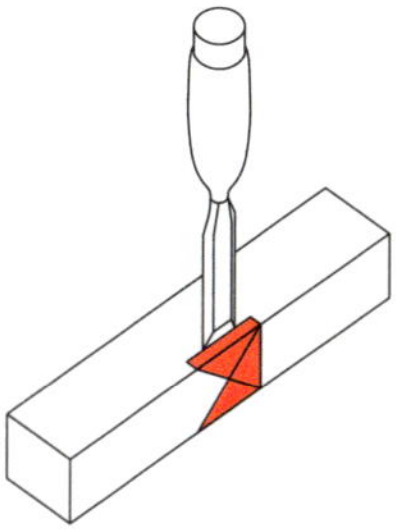

B4

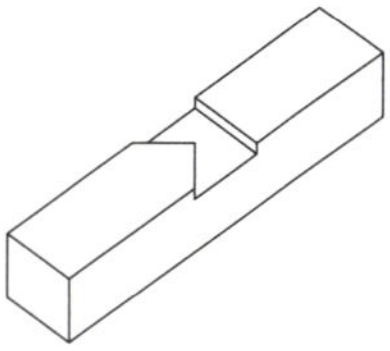

B

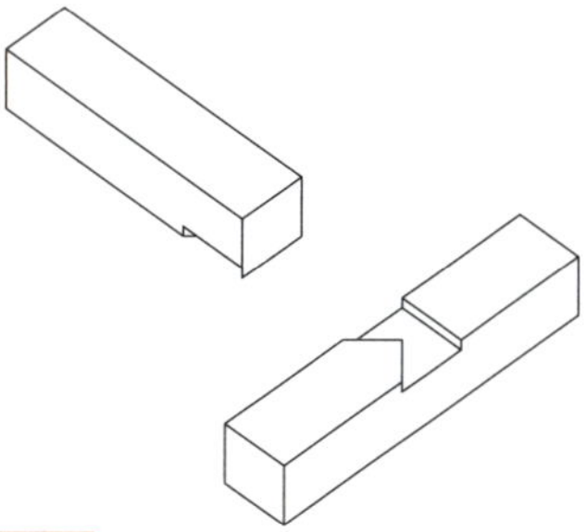

A+B

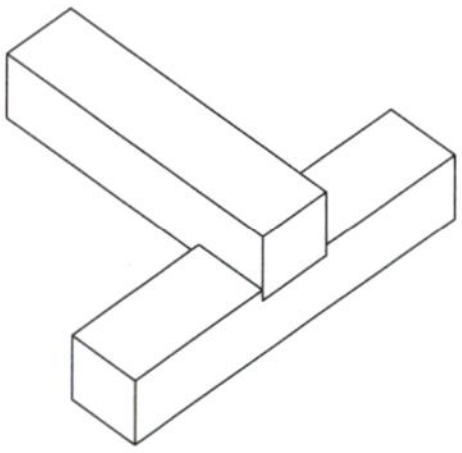

AB

T-Joints

Dovetail cog joint

ari-kake-ai-kaki-tsugi
蟻掛相欠接ぎ

Ankämmung mit Schwalbenschwanz

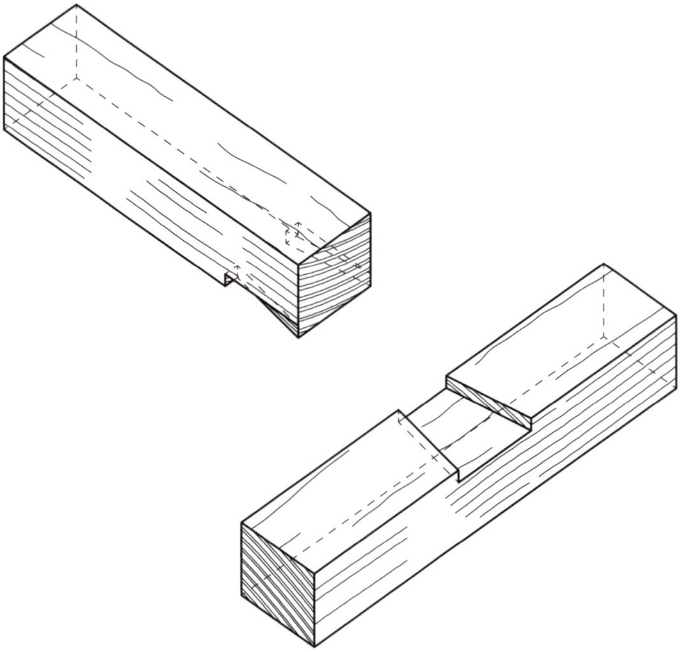

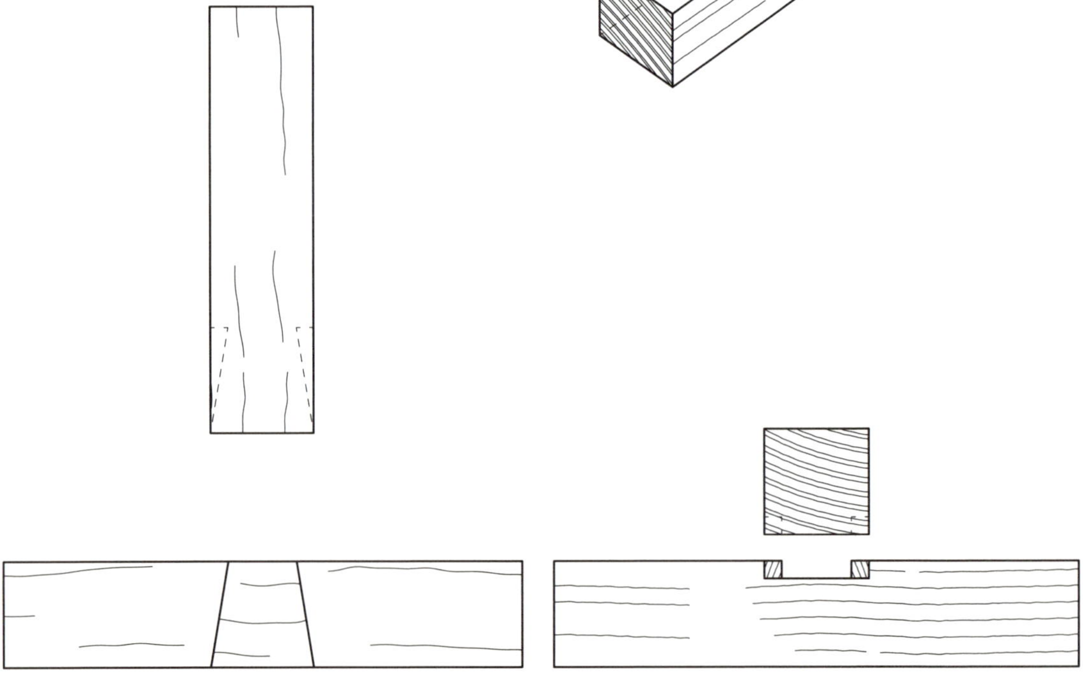

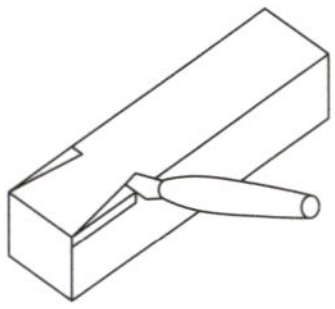

A1

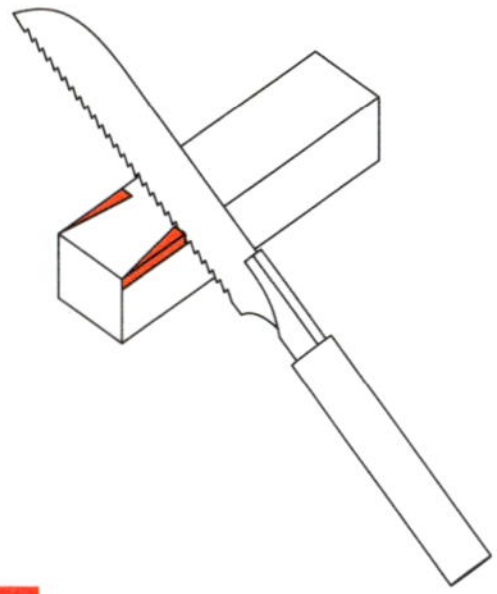

A2

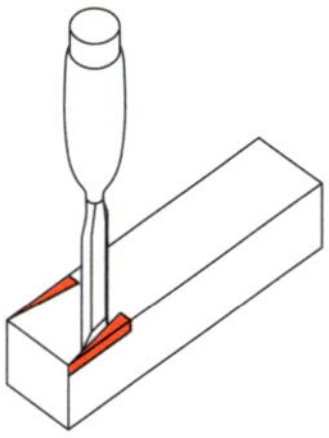

A3

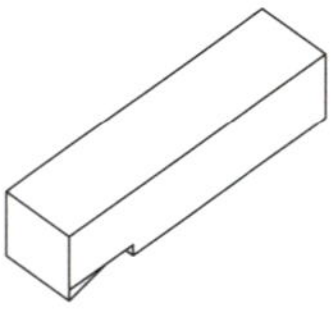

A

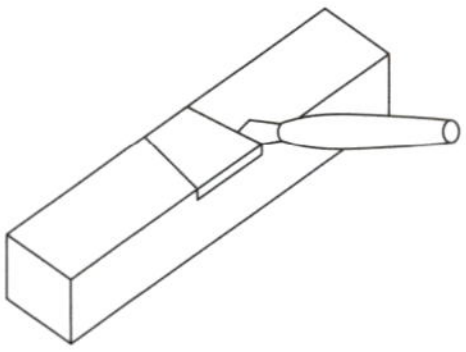

B1

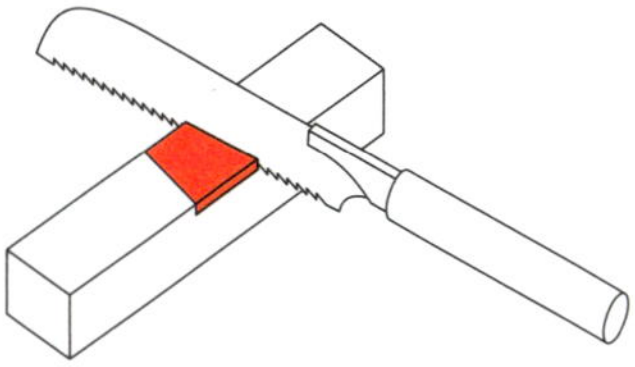

B2

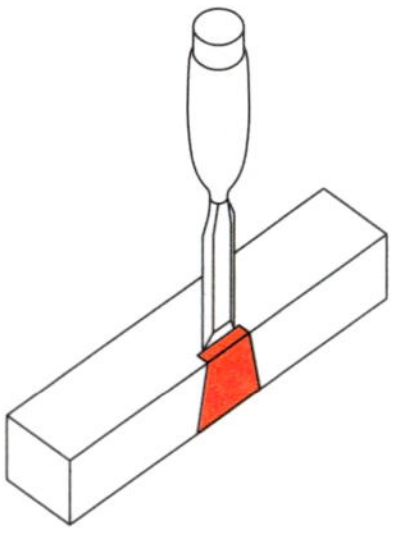

B3

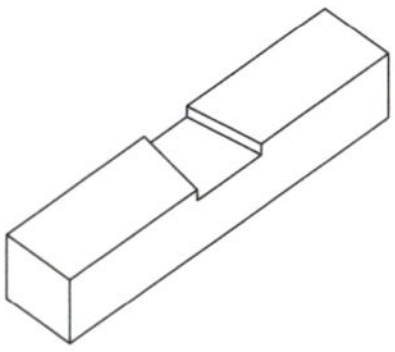

B

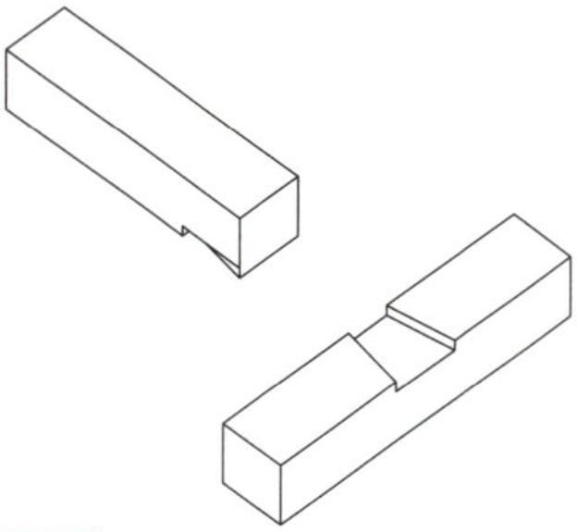

A+B

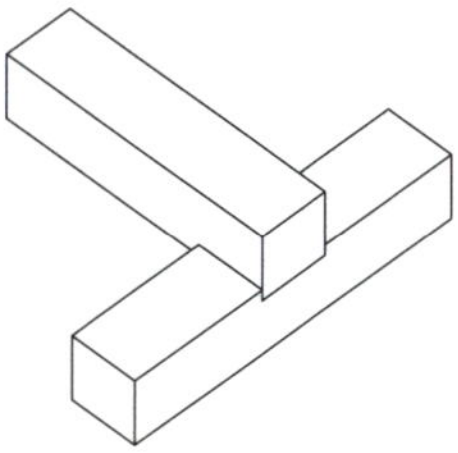

AB

T-Joints

Half-dovetail cog joint

kata-ari-kake-ai-kaki-tsugi
片蟻掛相欠き接ぎ

Ankämmung mit
einseitigem Schwalbenschwanz

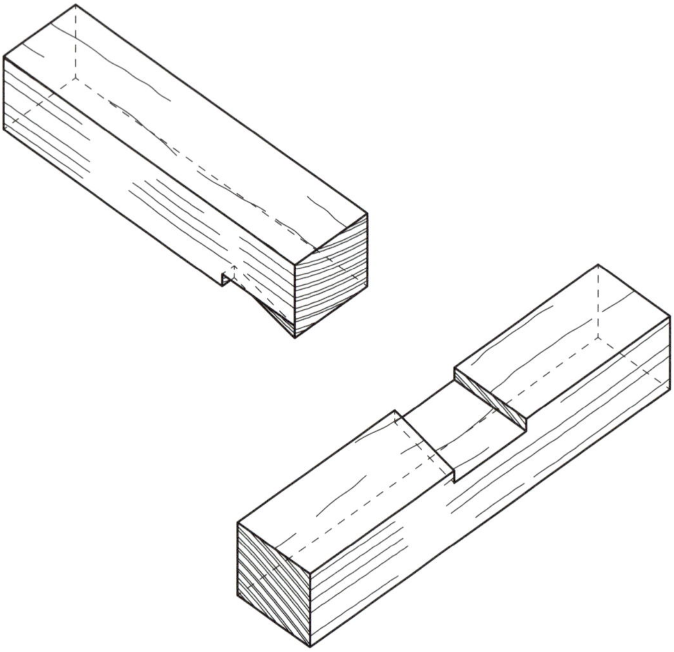

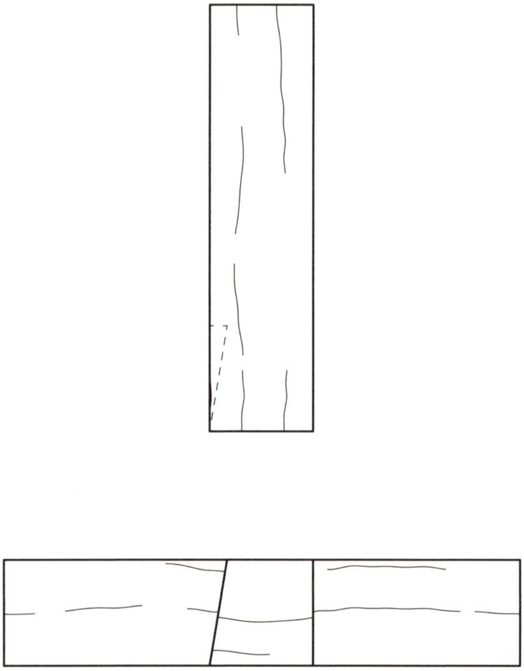

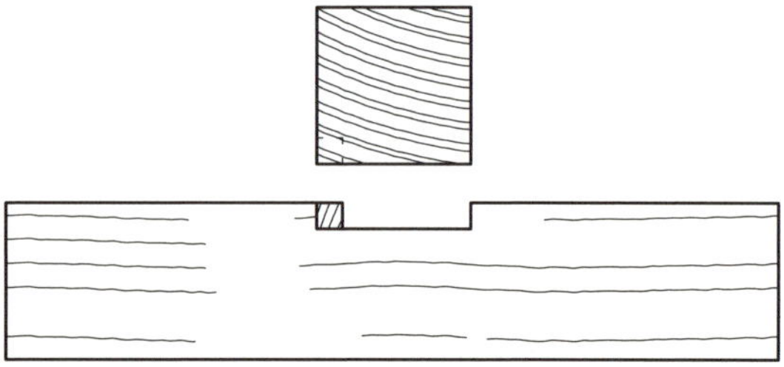

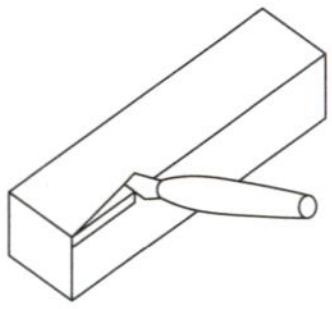

A1

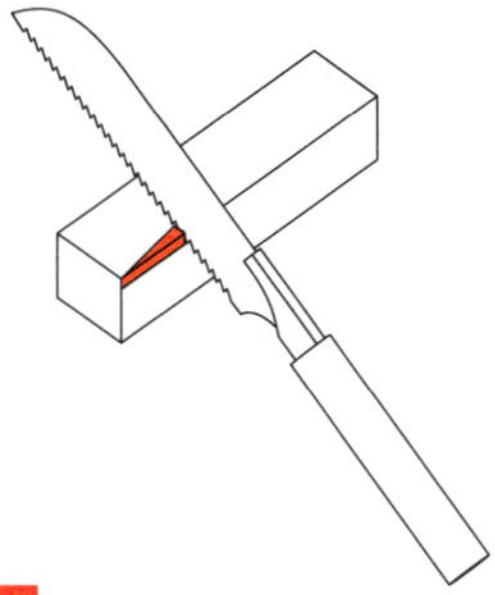

A2

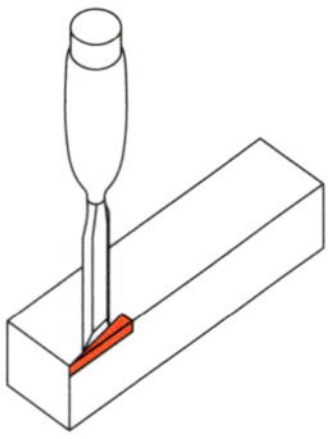

A3

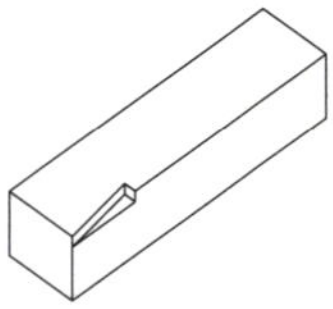

A

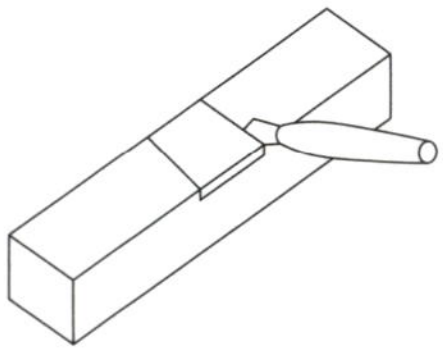

B1

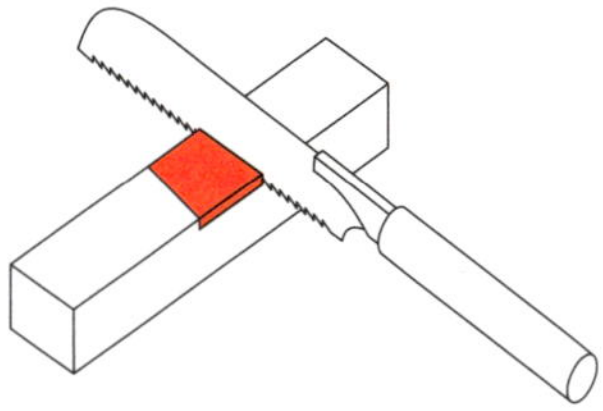

B2

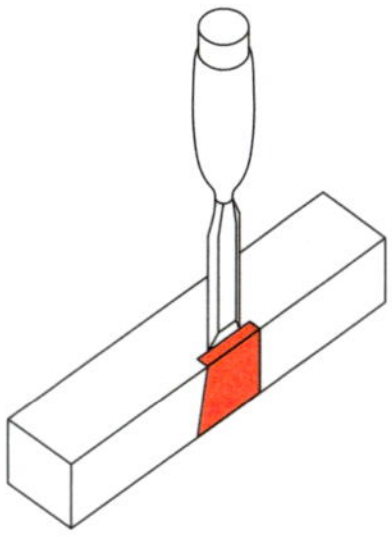

B3

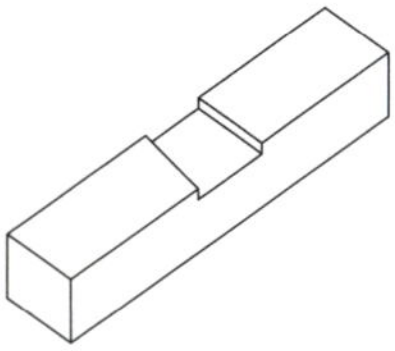

B

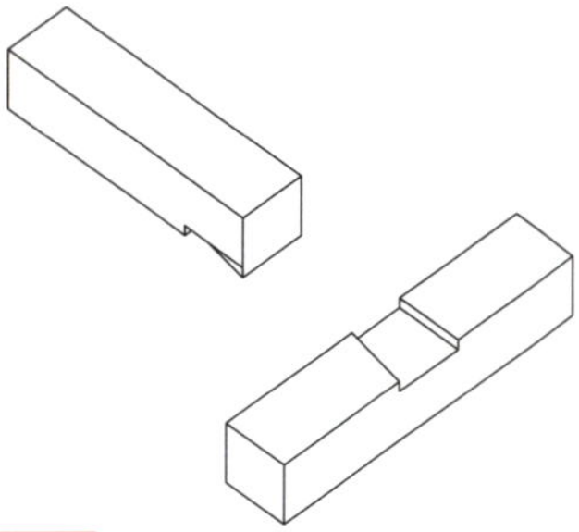

A+B

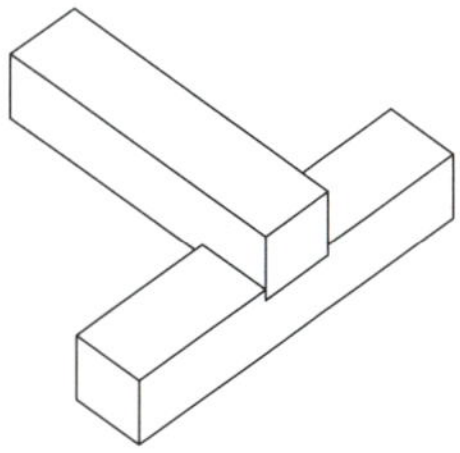

AB

T-Joints

Sloped cog joint

katagi-ago-kake-ai-kaki-tsugi
傾顎掛相欠接ぎ

Ankämmung mit schräger Stufe

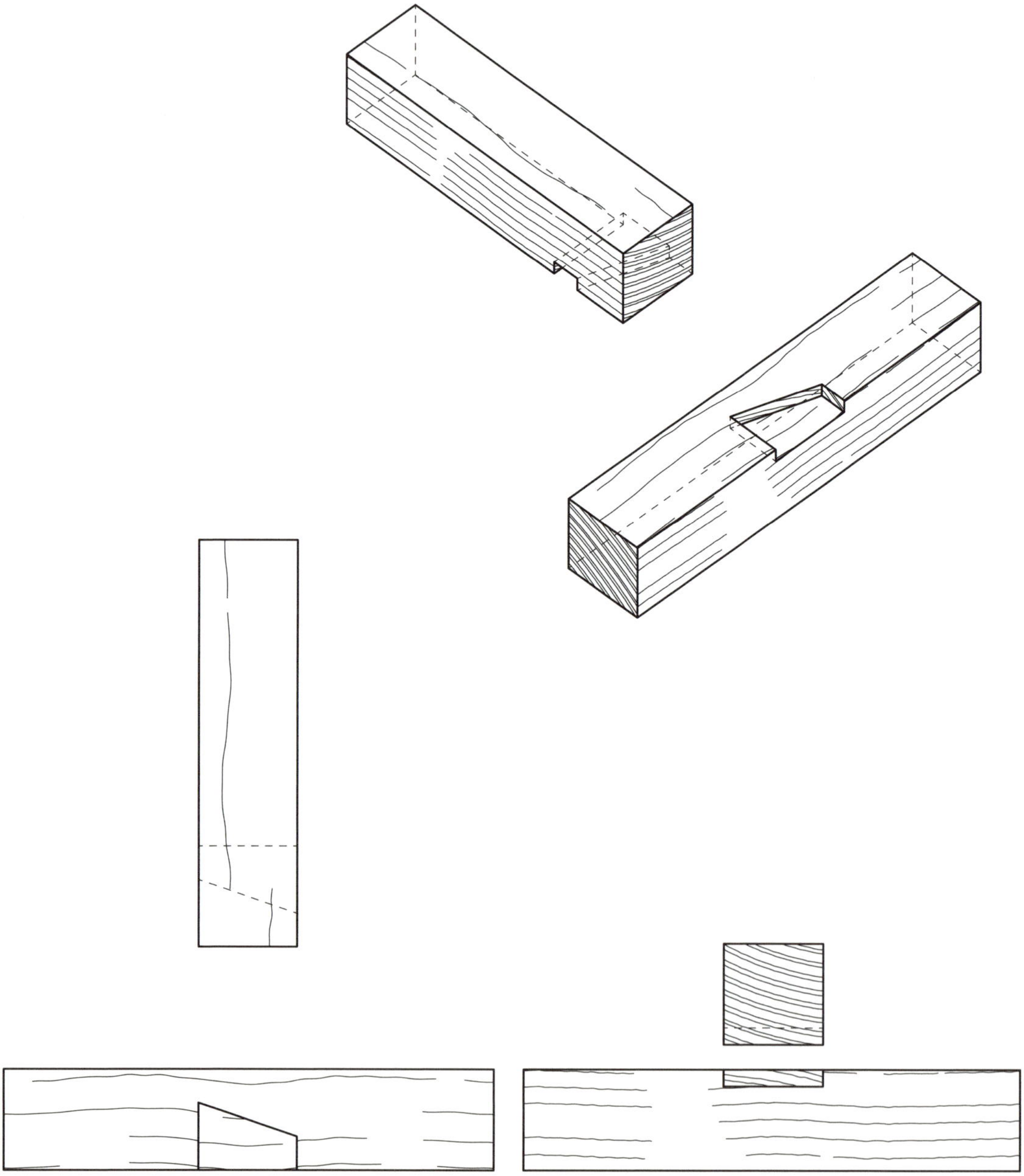

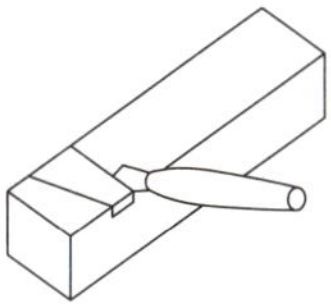

A1

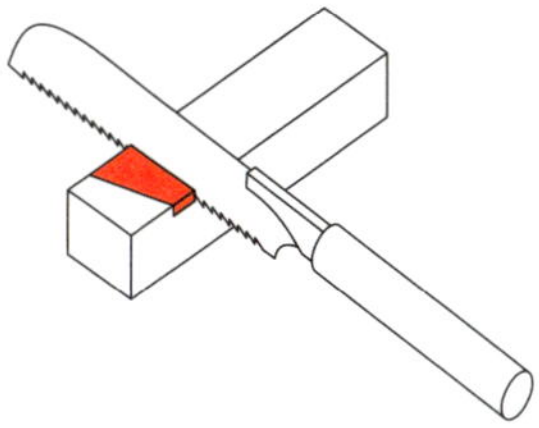

A2

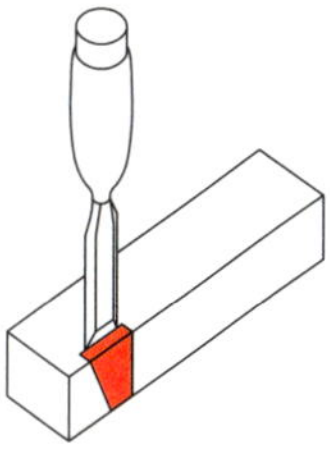

A3

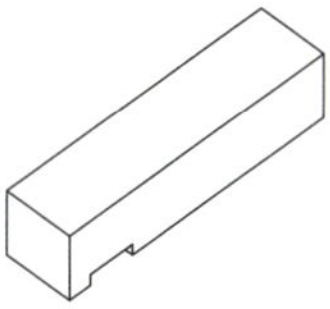

A

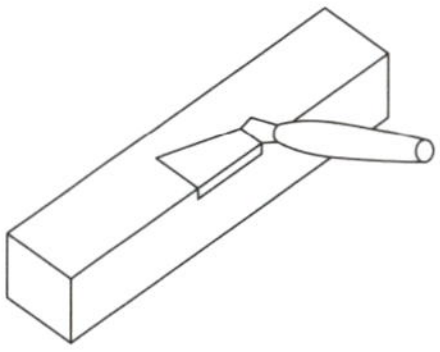

B1

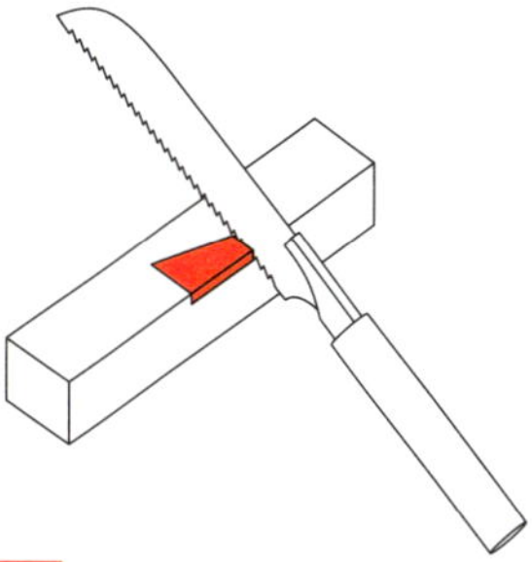

B2

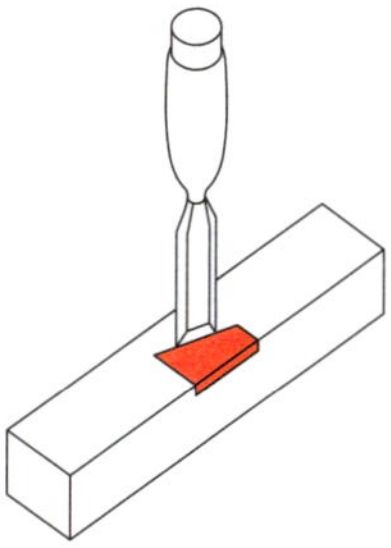

B3

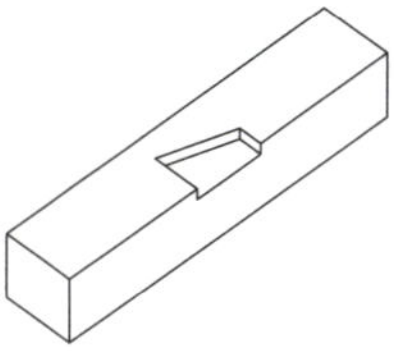

B

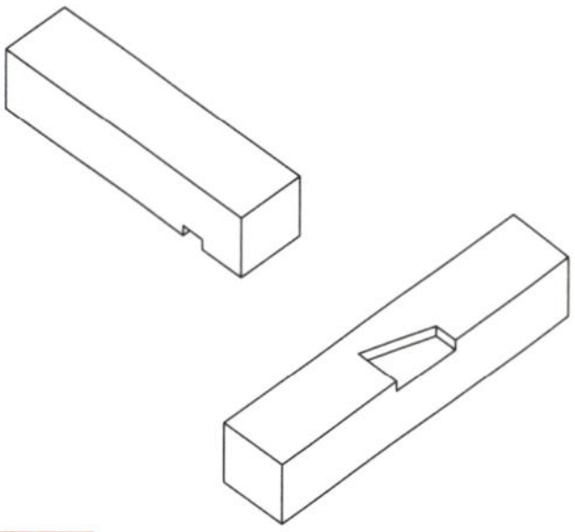

A+B

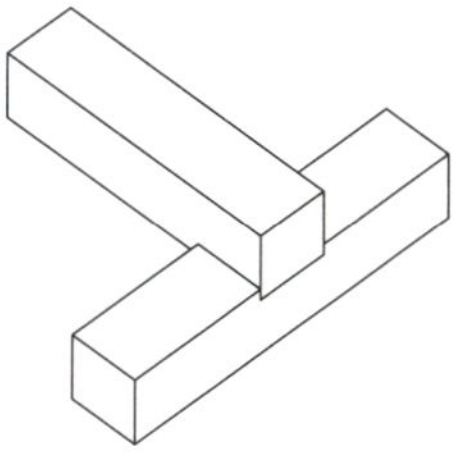

AB

T-Joints

Through mortise and tenon

uchi-nuki-hozo-sashi [*]
打抜枘差し

durchgehender Zapfen

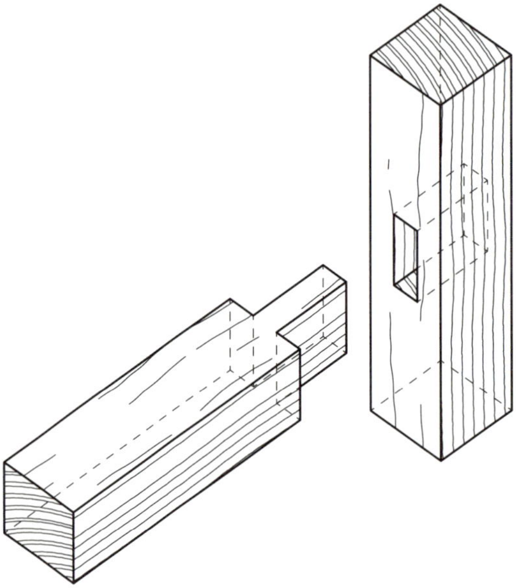

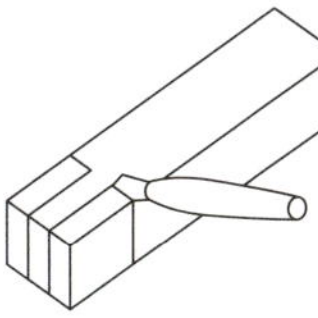

A1

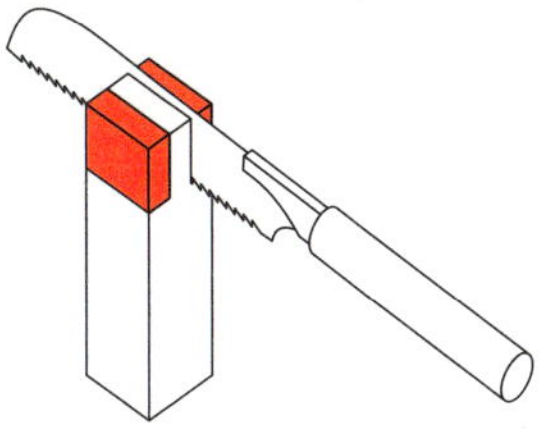

A2

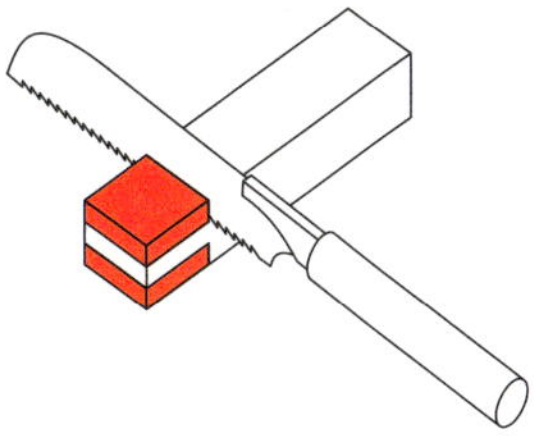

A3

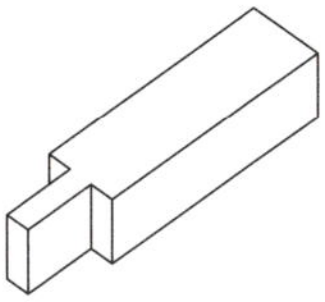

A

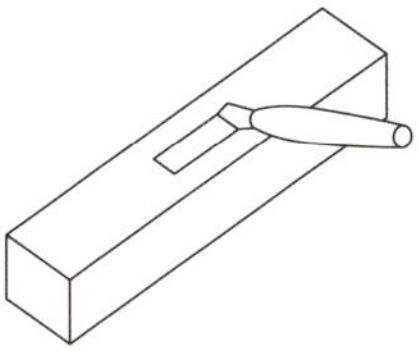

B1

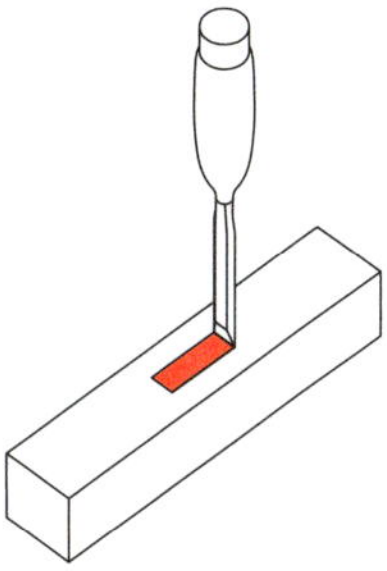

B2

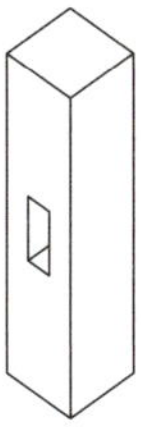

B

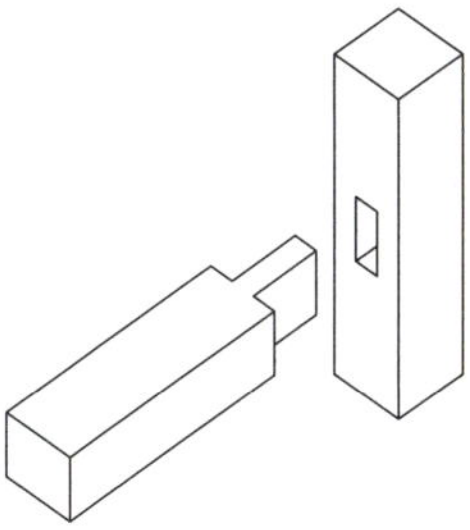

A + B

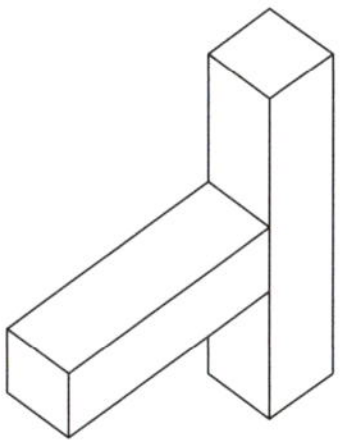

AB

T-Joints

Through mortise and tenon with housing

hira-hozo-sashi (oire) [*]
平枘差し(追入)

eingelassener durchgehender Zapfen

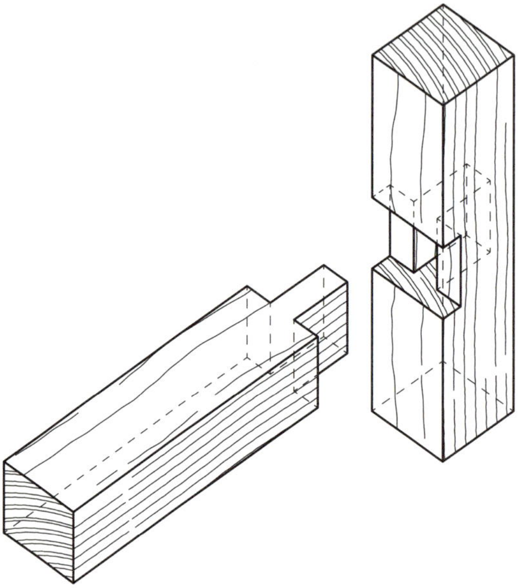

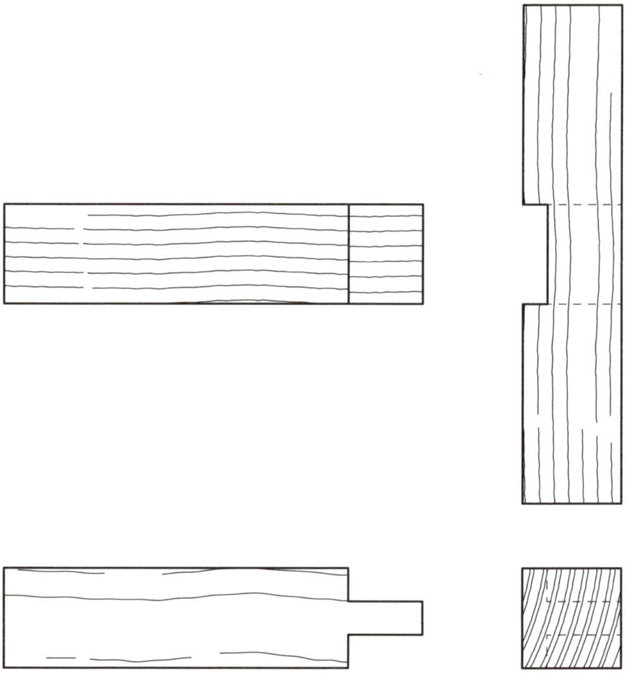

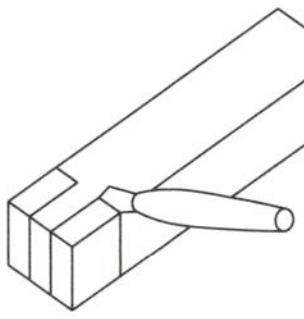

A1

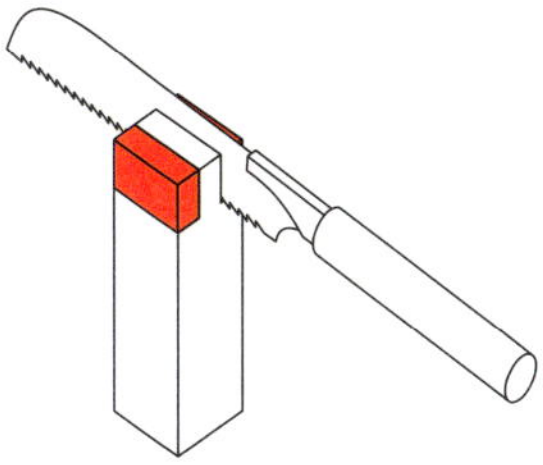

A2

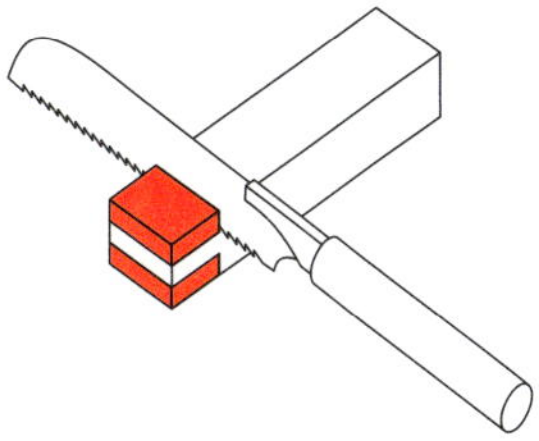

A3

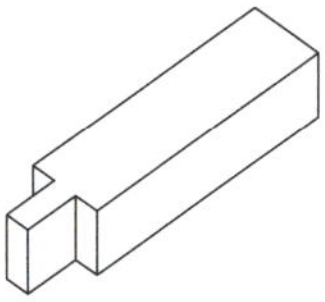

A

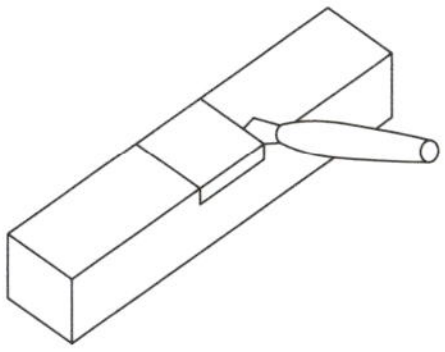

B1

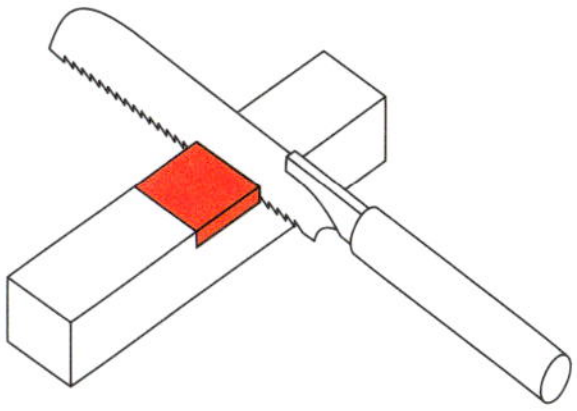

B2

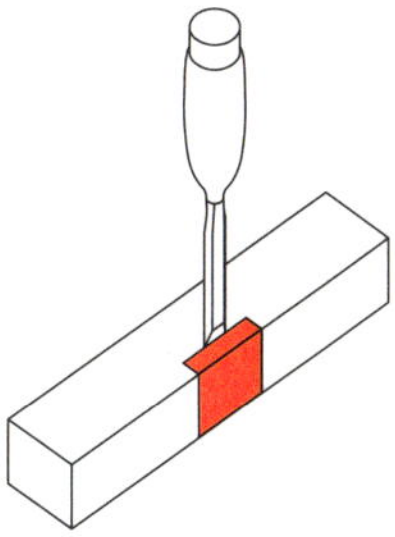

B3

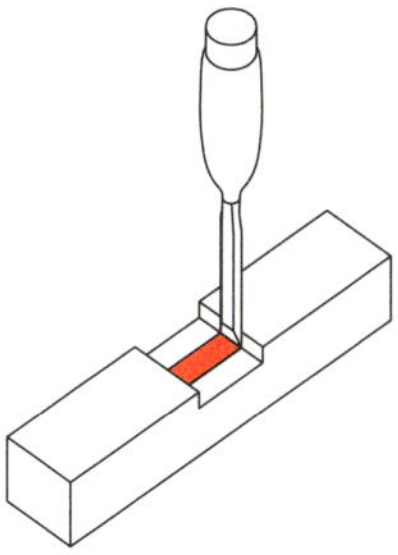

B4

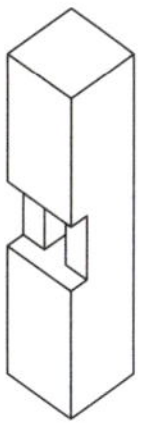

B

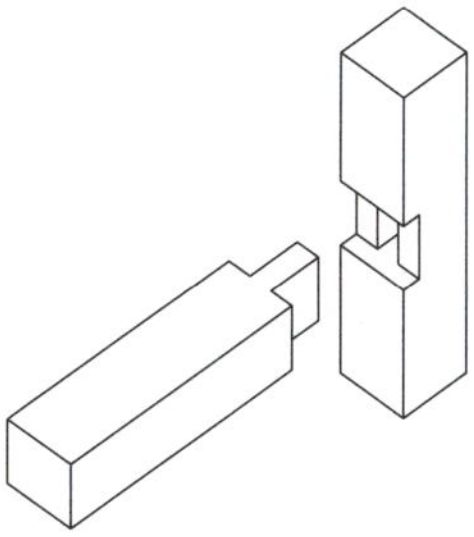

A+B

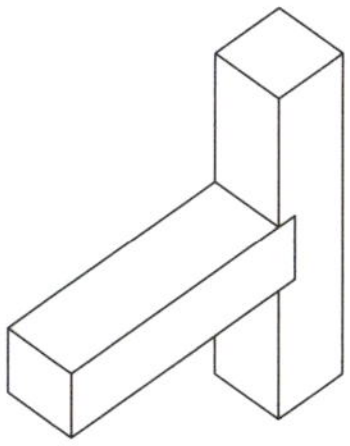

AB

T-Joints

Blind mortise and tenon

hira-hozo-sashi [*]
平枘差し

verdeckter Zapfen

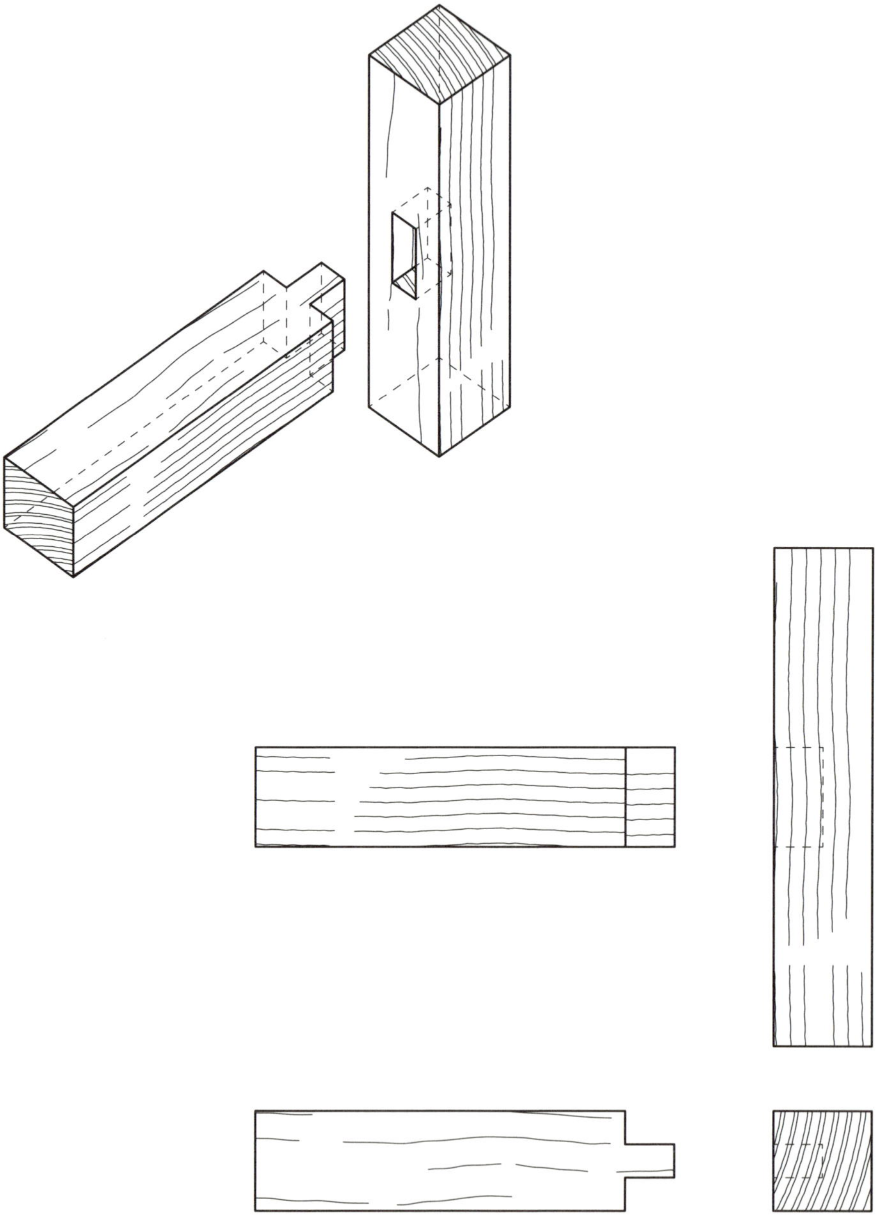

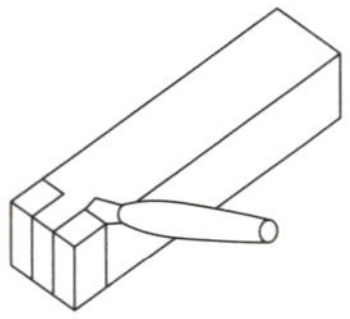

A1

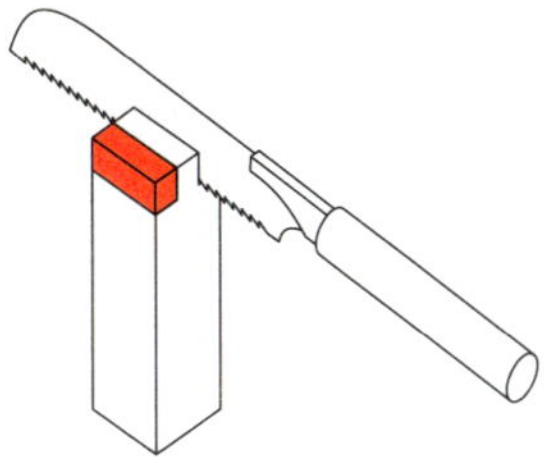

A2

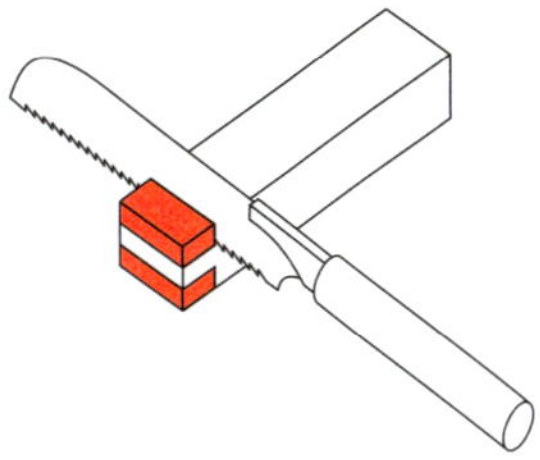

A3

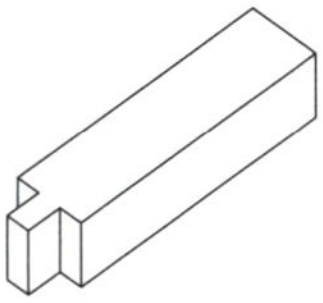

A

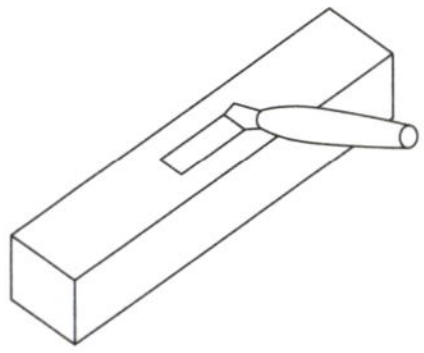

B1

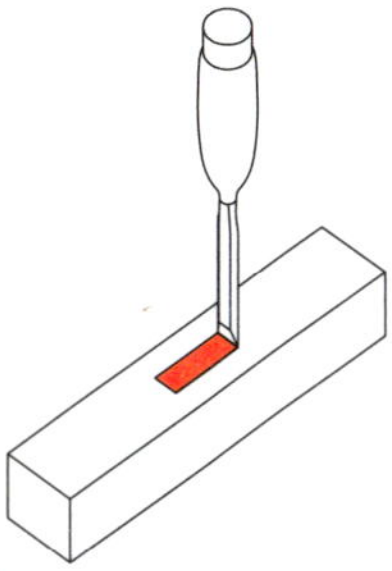

B2

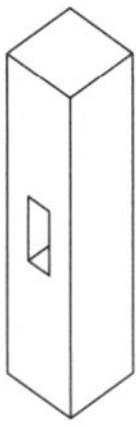

B

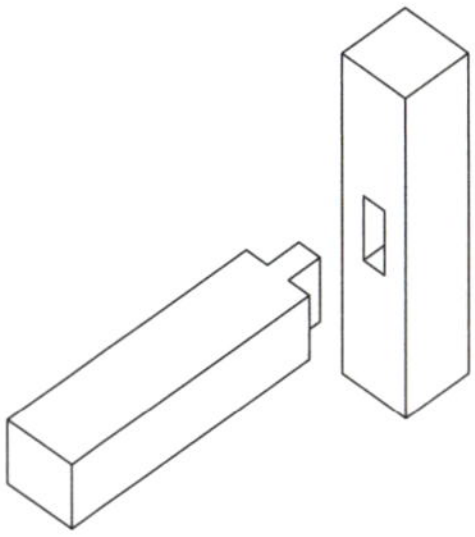

A+B

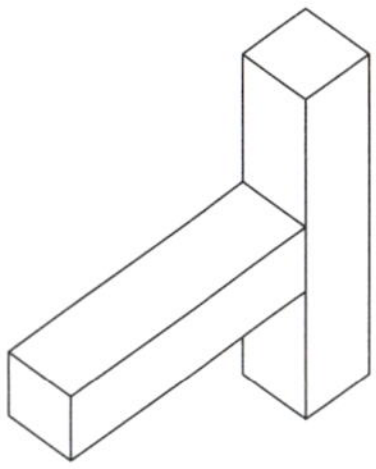

AB

T-Joints

Tenon and chase mortise

yose-hozo-yoko-sashi
寄枘横差し

Schleifzapfen

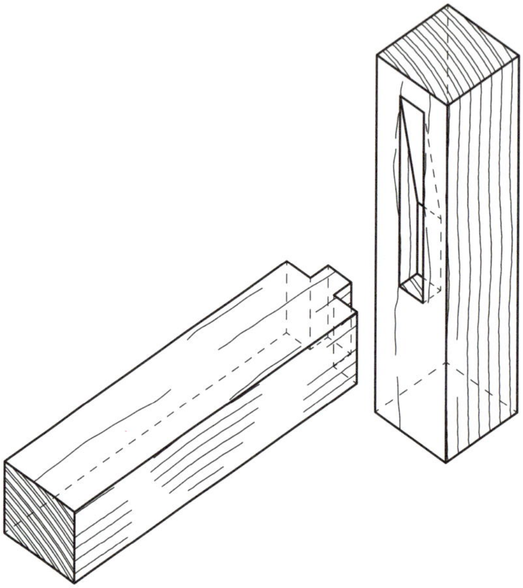

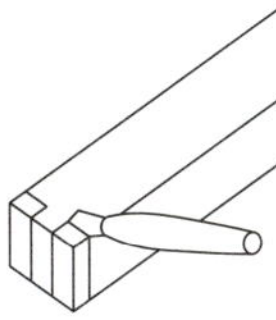

A1

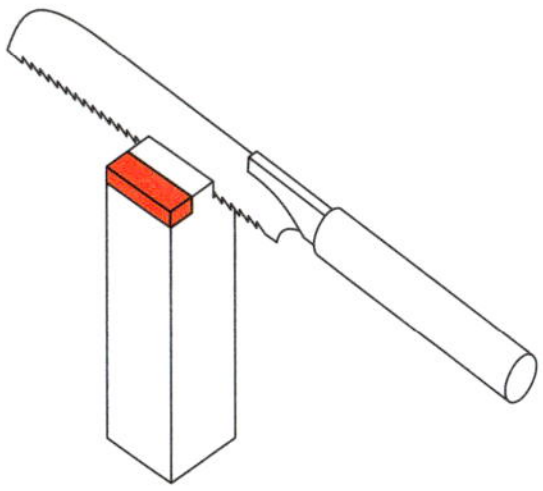

A2

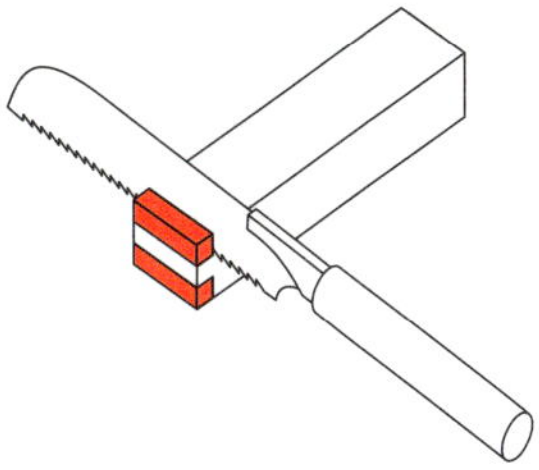

A3

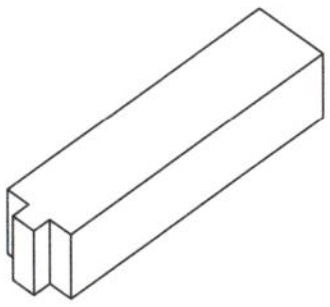

A

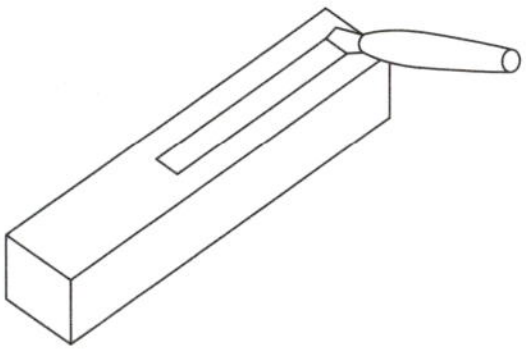

B1

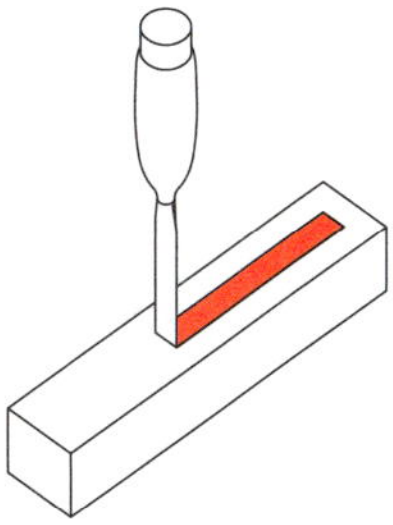

B2

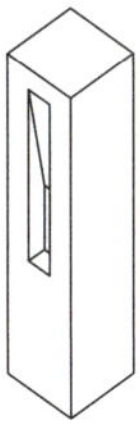

B

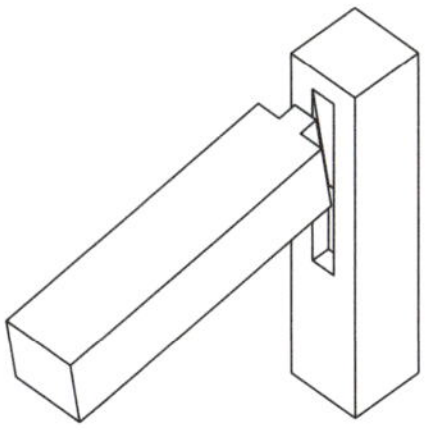

A+B

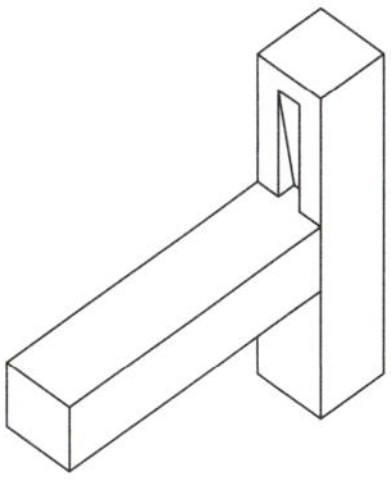

AB

T-Joints

Mortise and tenon with beveled shoulders

katagi-oire-hozo-sashi [*]
傾追入枘差し

verdeckter Zapfen
mit schräger Brust

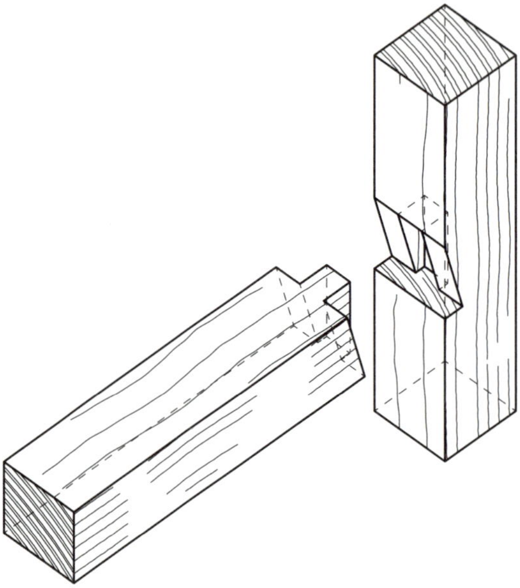

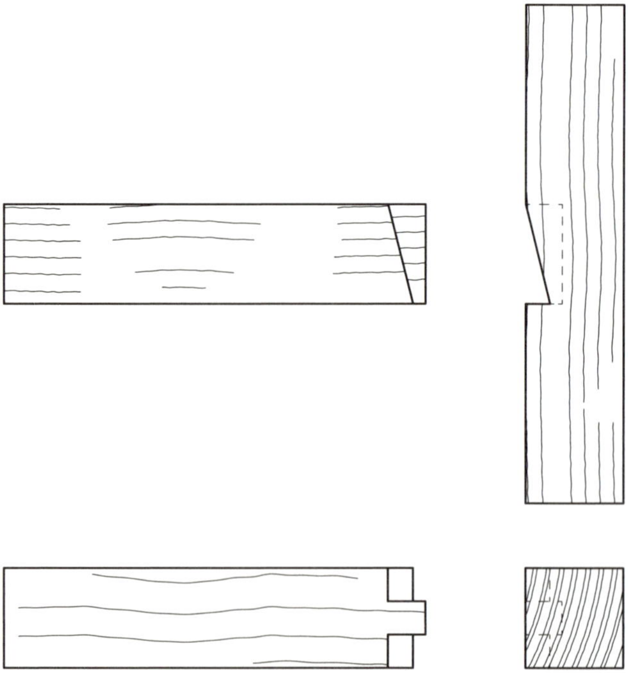

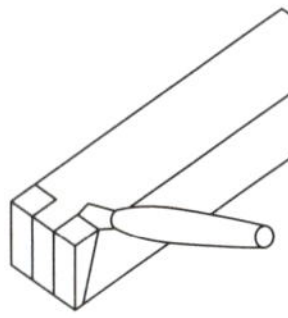

A1

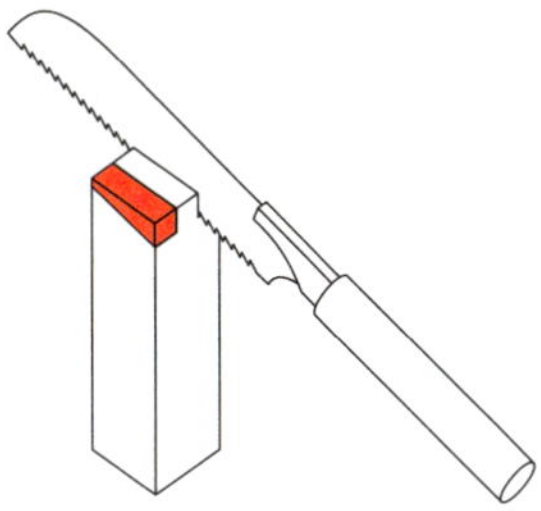

A2

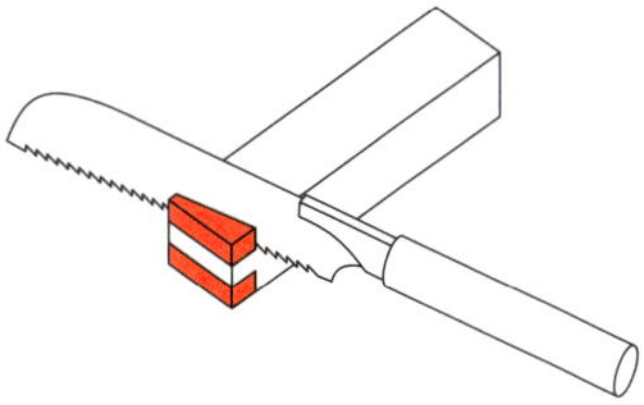

A3

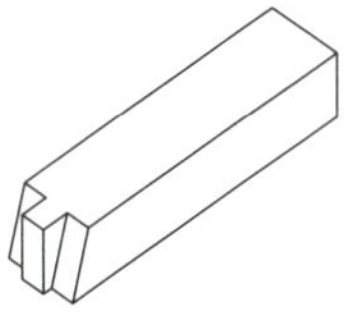

A

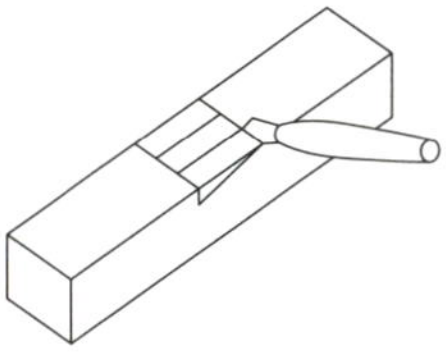

B1

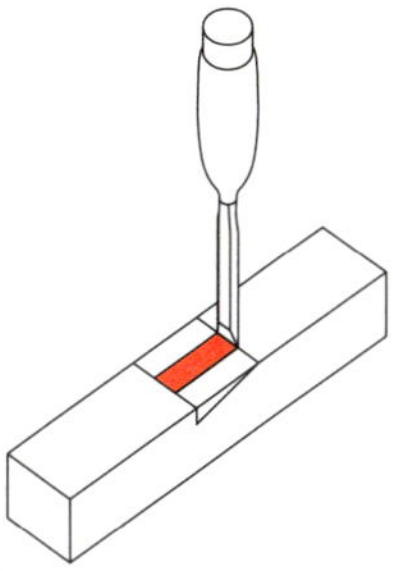

B2

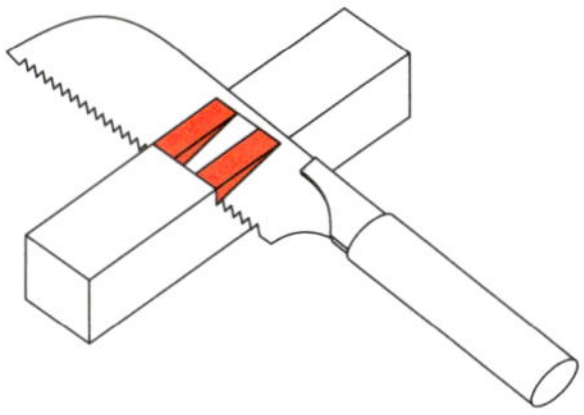

B3

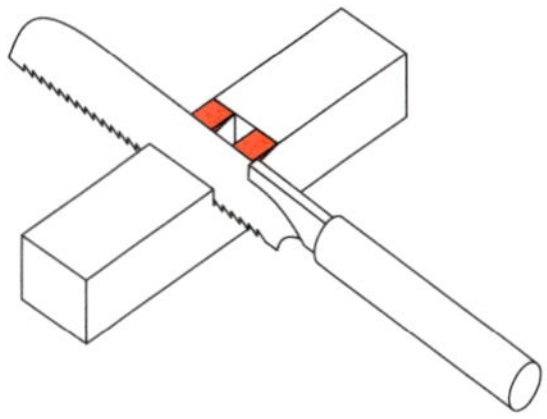

B4

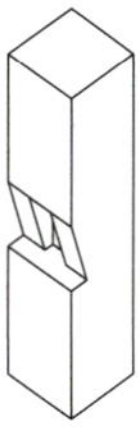

B

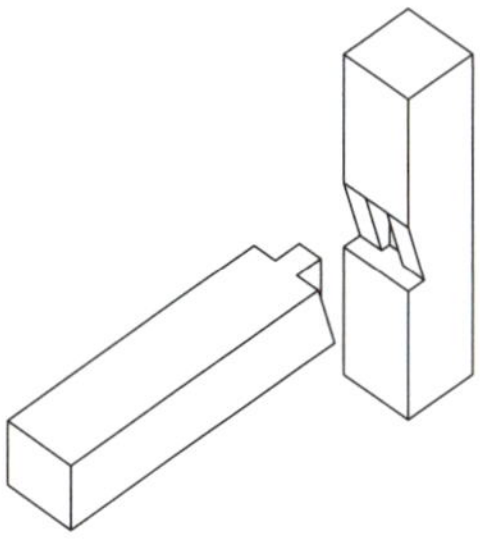

A+B

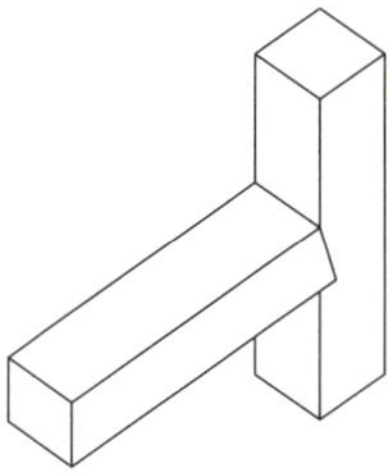

AB

T-Joints

Mortise and tenon with mitered shoulders

ken-dome-oire-hozo-sashi
剣留追入枘差し

verdeckter Zapfen
mit Führungsstoß

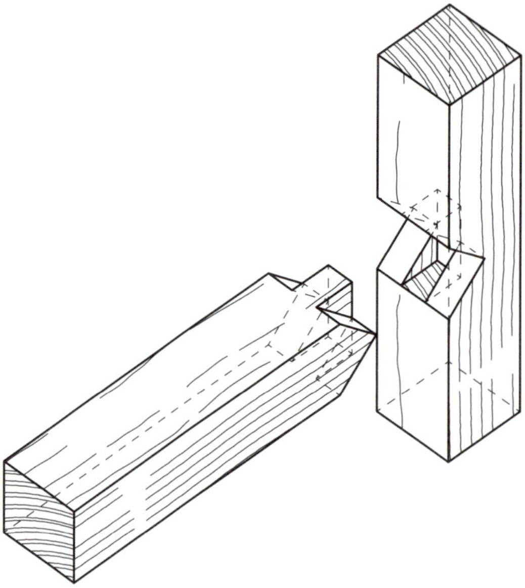

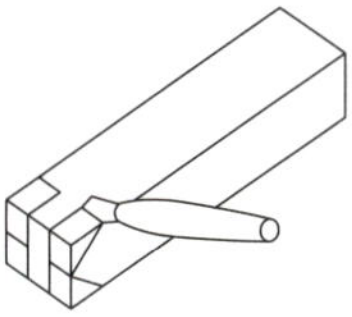

A1

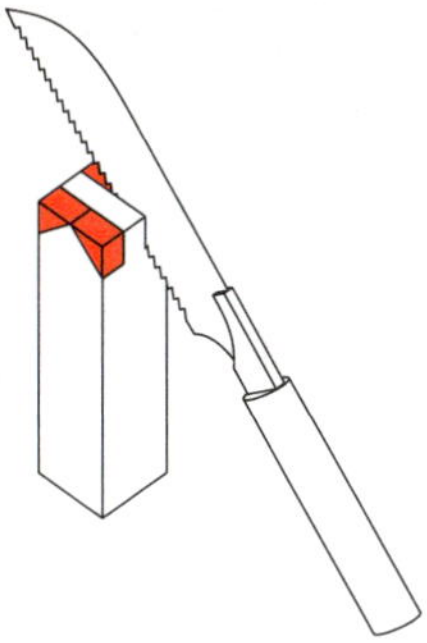

A2

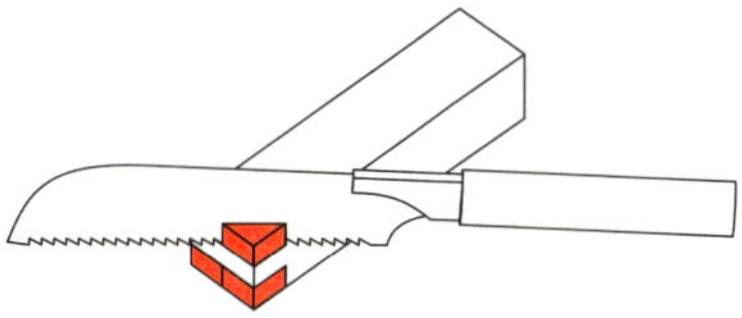

A3

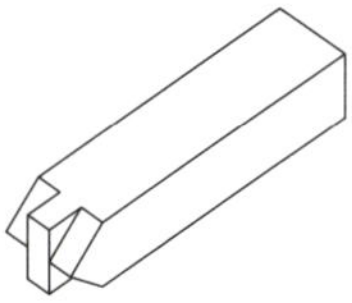

A

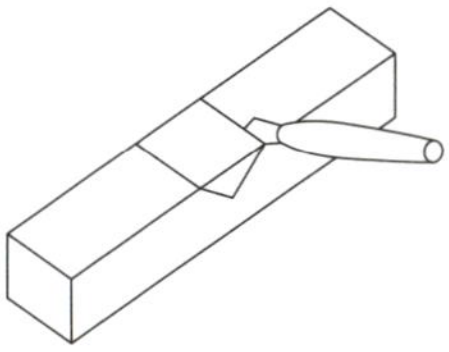

B1

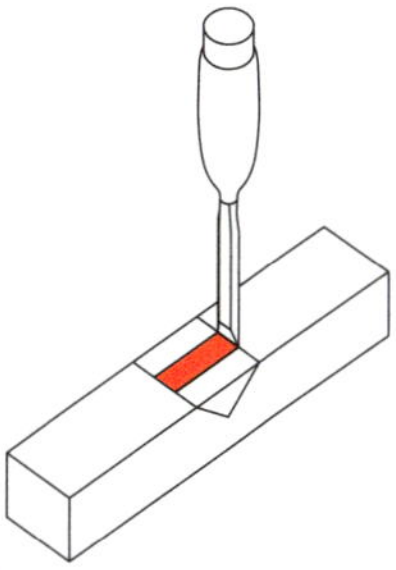

B2

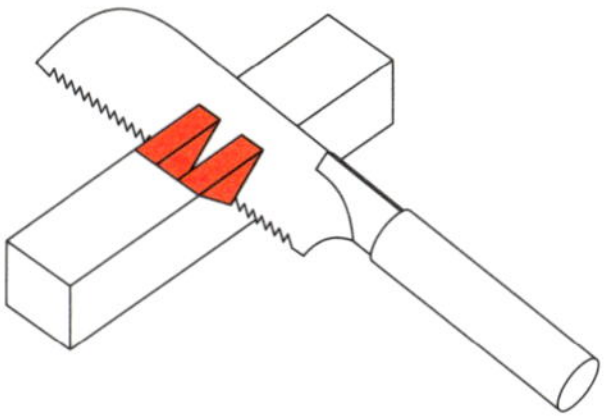

B3

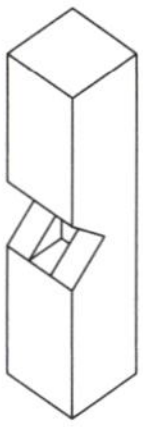

B

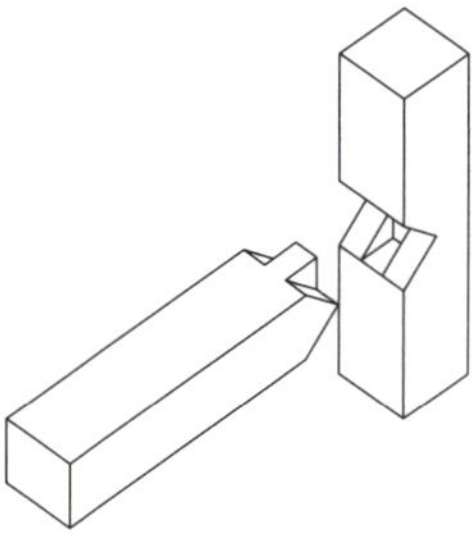

A+B

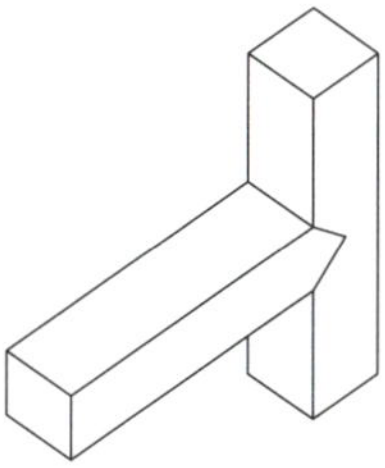

AB

T-Joints

Pegged through mortise and tenon

naga-hozo-sashi-komi-sen-uchi [*]
長枘差込栓打ち

durchgehender Zapfen
mit Dübelsicherung

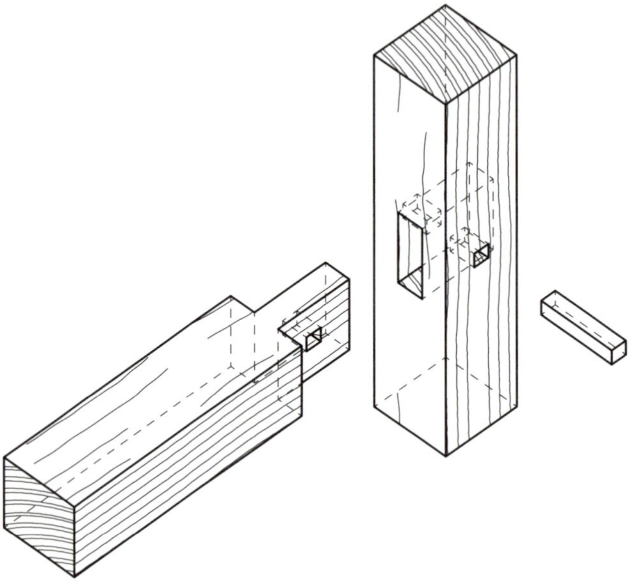

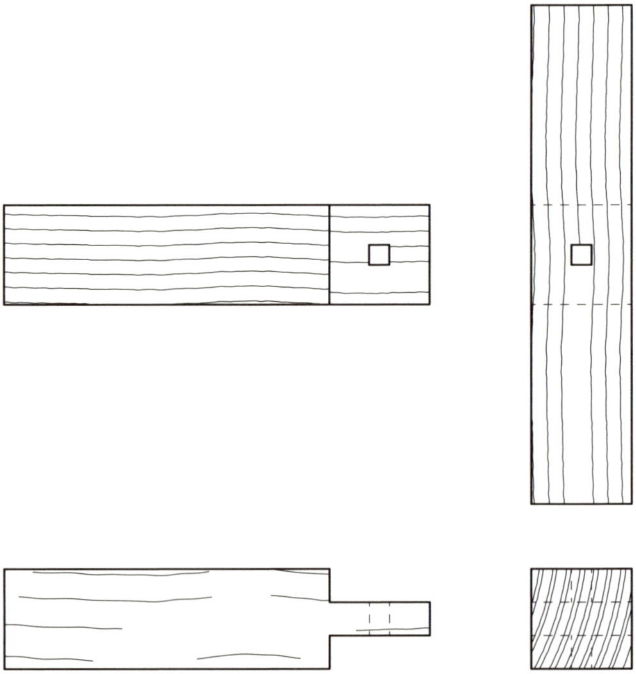

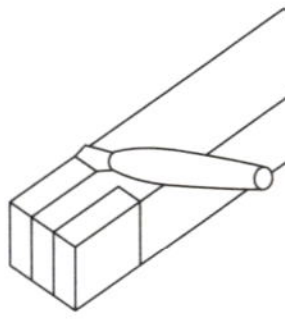

A1

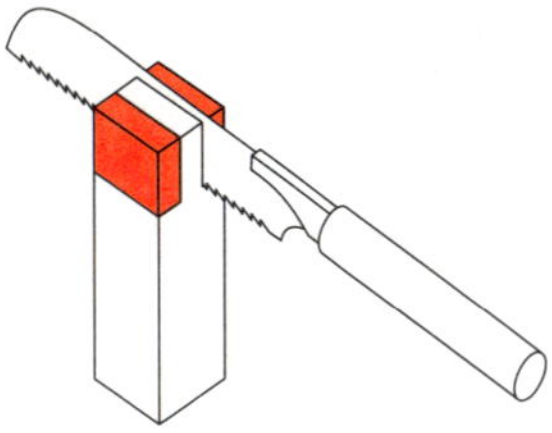

A2

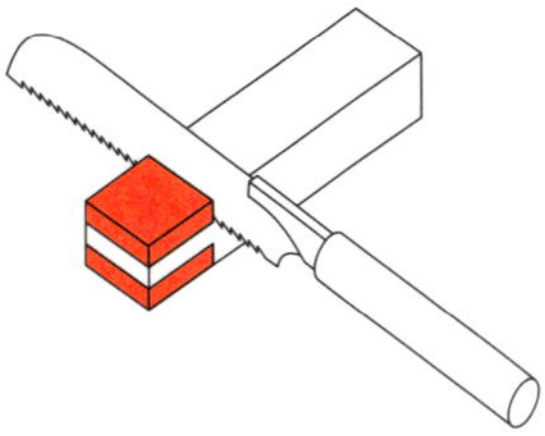

A3

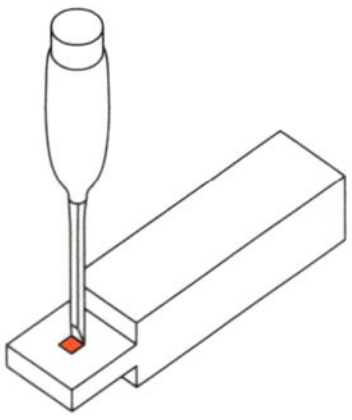

A4

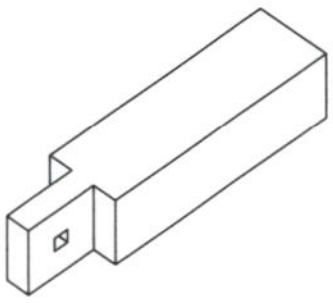

A

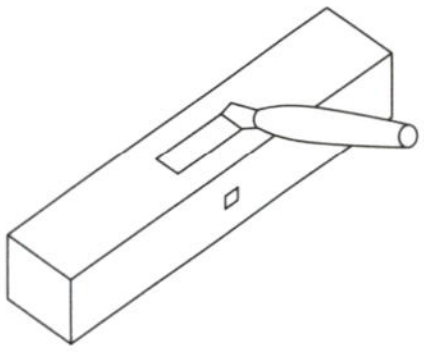

B1

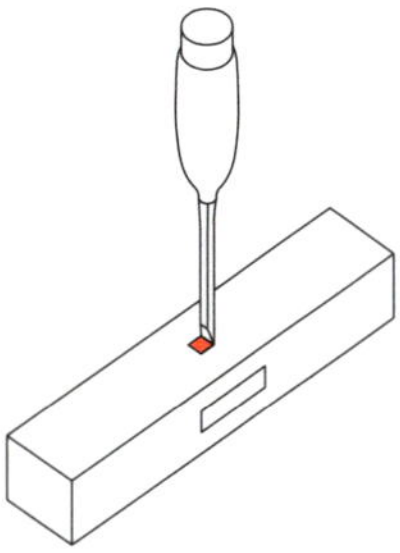

B2

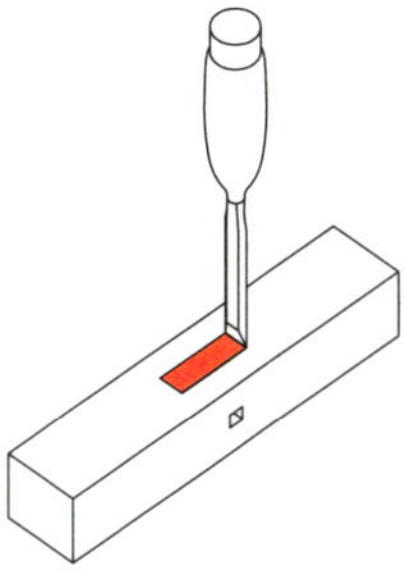

B

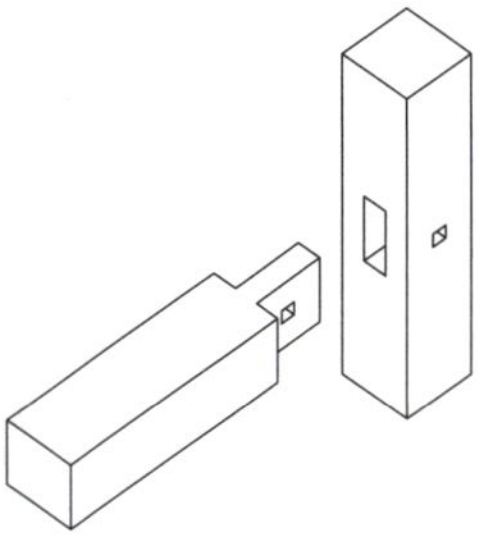

A+B

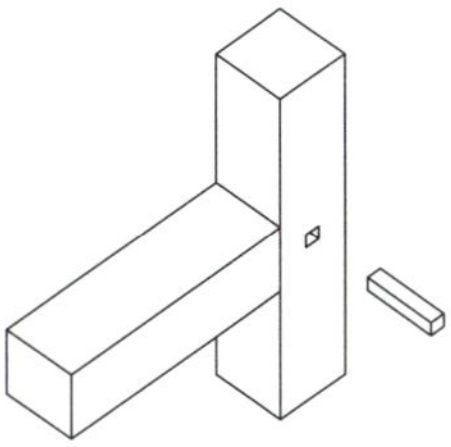

A+B

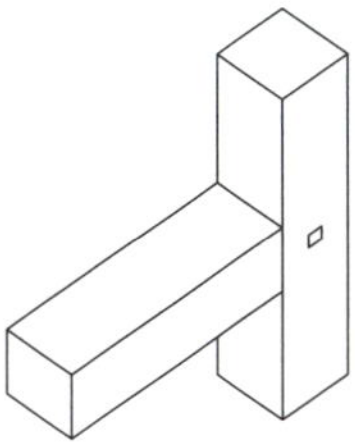

AB

T-Joints

Pegged blind mortise and tenon

hira-hozo-sashi-komi-sen-uchi
平枘差込栓打ち

verdeckter Zapfen
mit Dübelsicherung

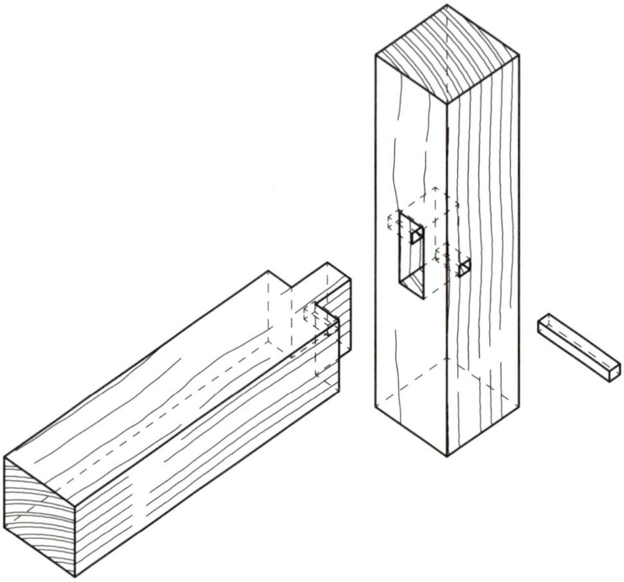

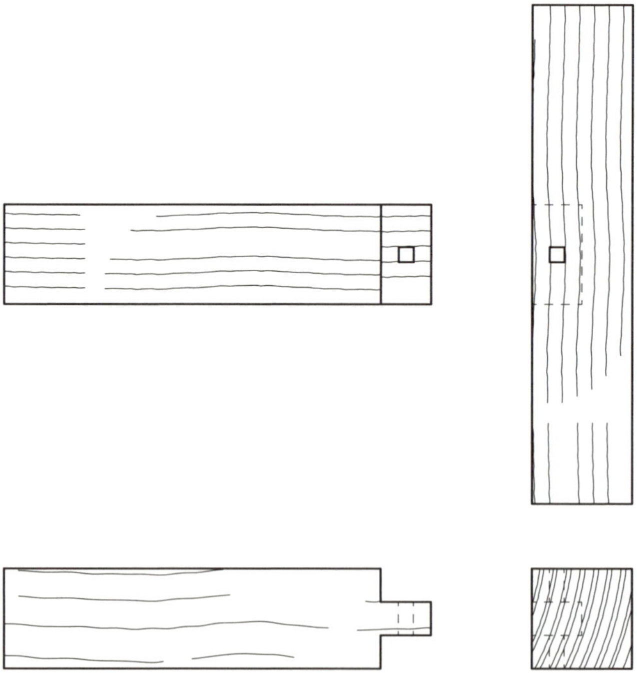

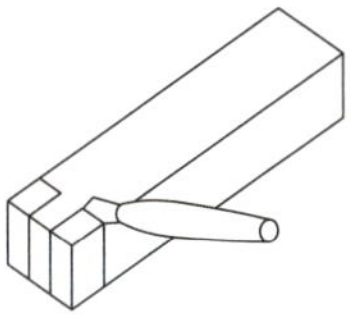

A1

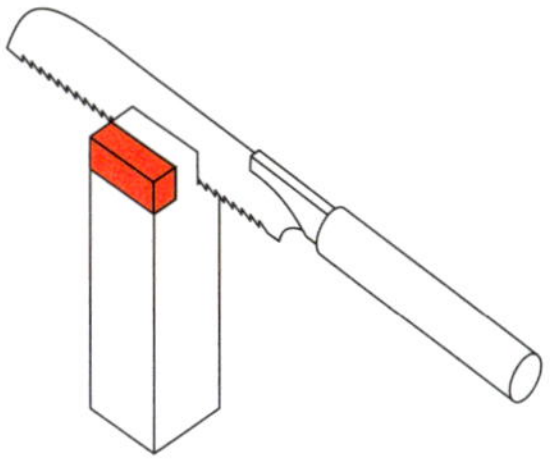

A2

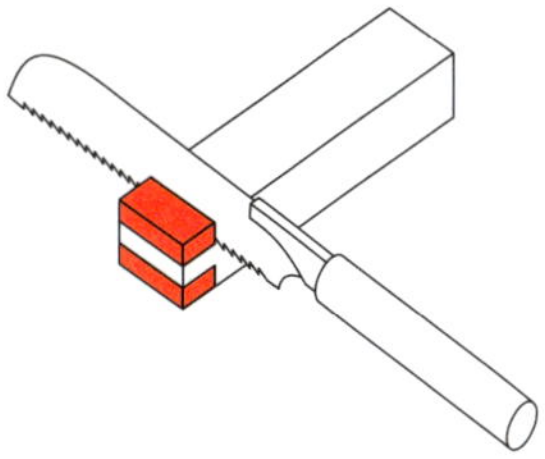

A3

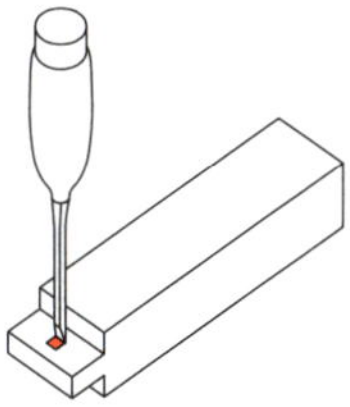

A4

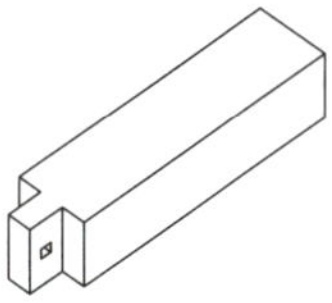

A

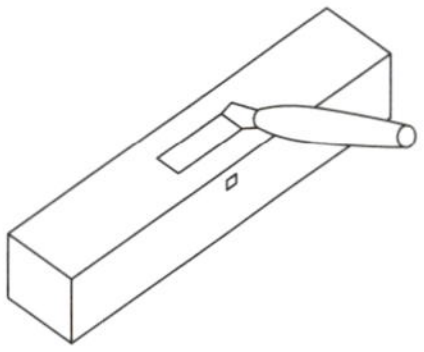

B1

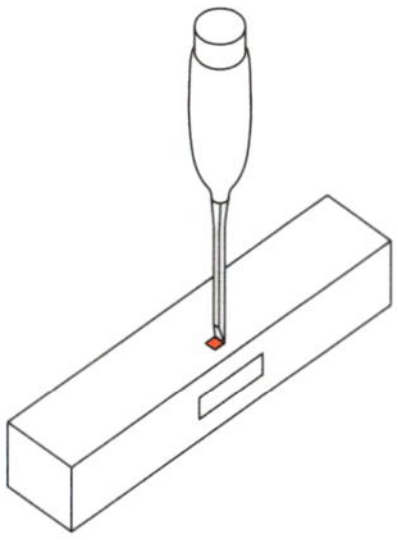

B2

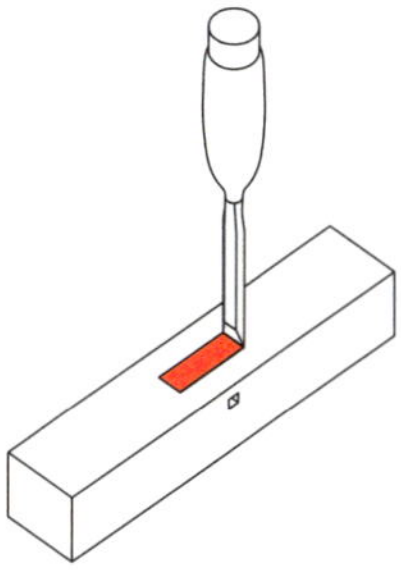

B3

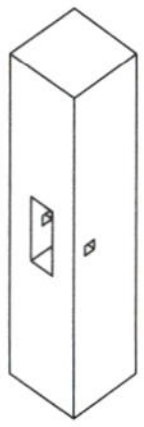

B

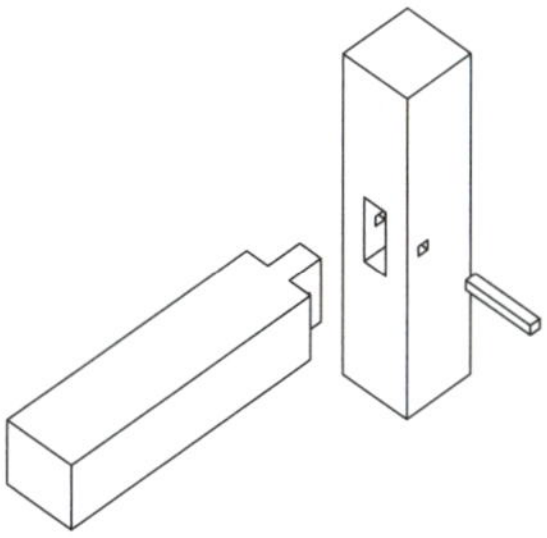

A+B

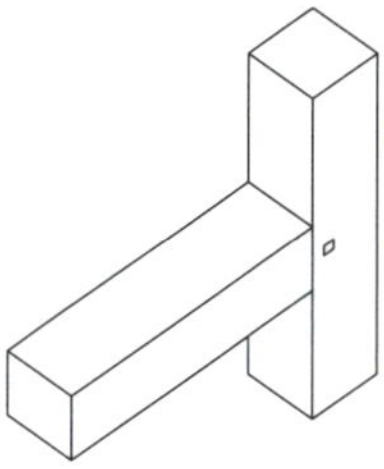

A+B

T-Joints

Triple-shouldered through mortise and tenon

shihō-dō-tsuki-tōshi-hira-hozo-tsugi [*]
四方胴付通平枘接ぎ

durchgehender abgesetzter Zapfen

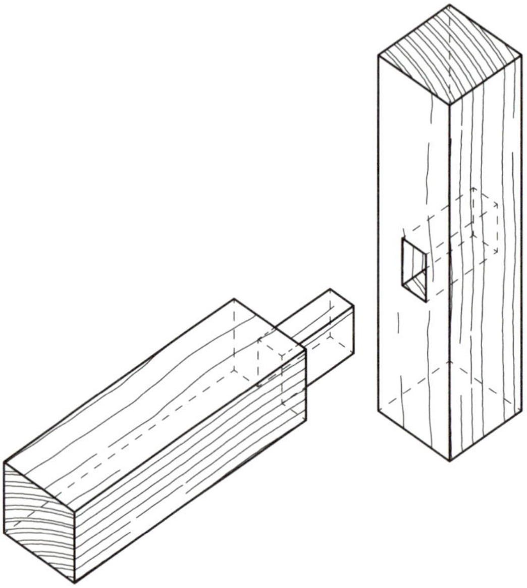

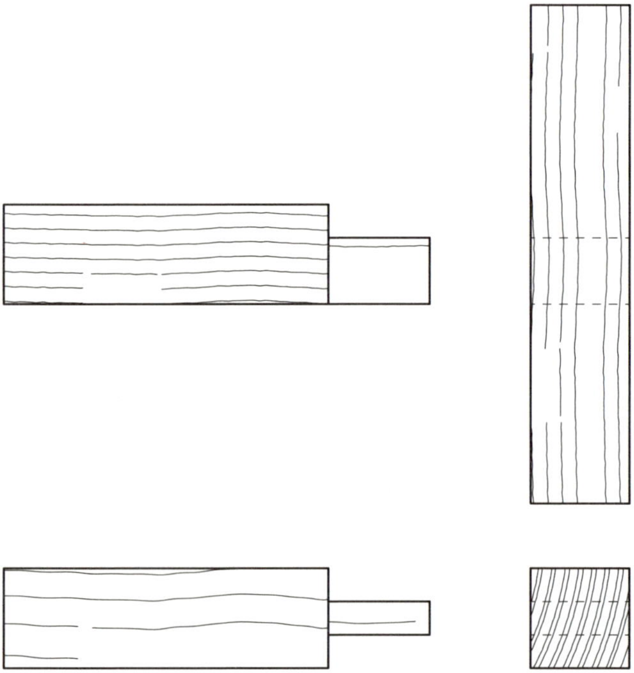

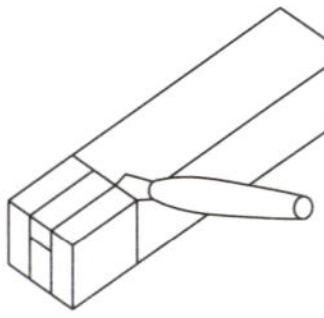

A1

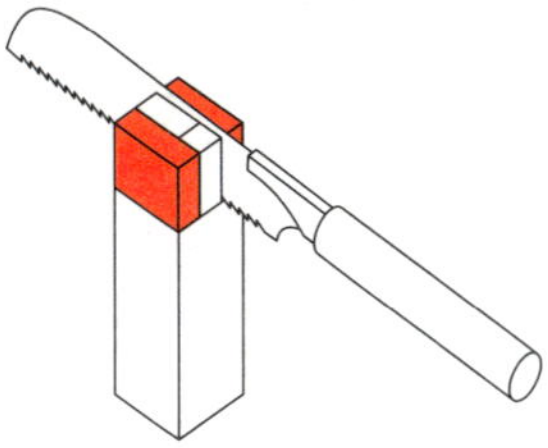

A2

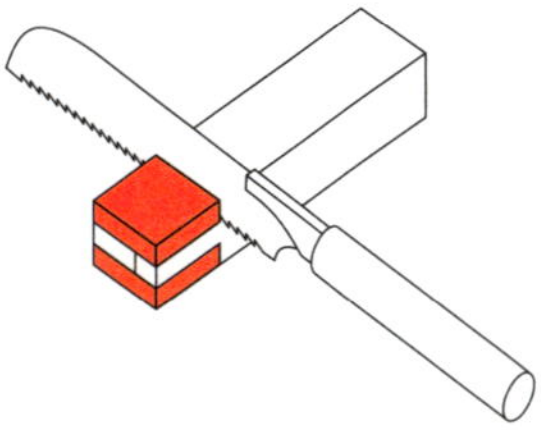

A3

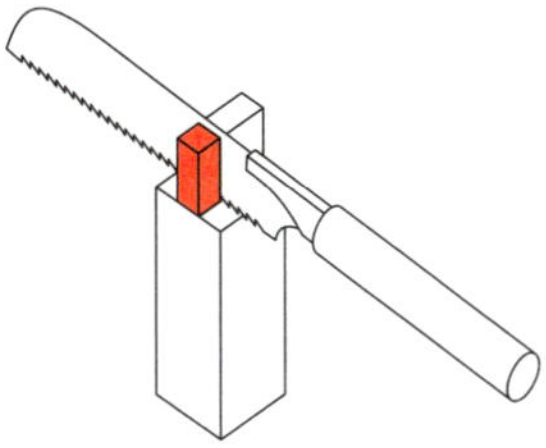

A4

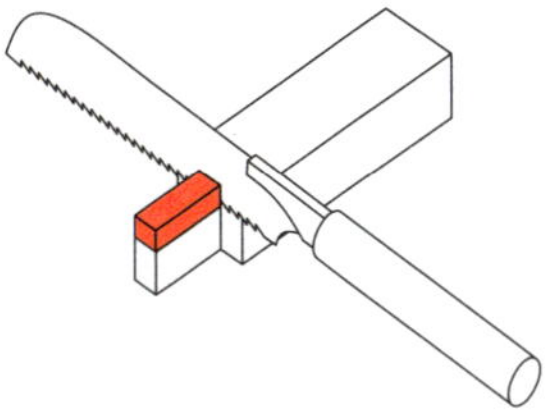

A5

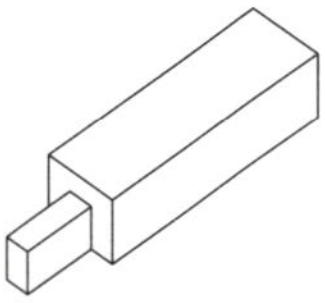

A

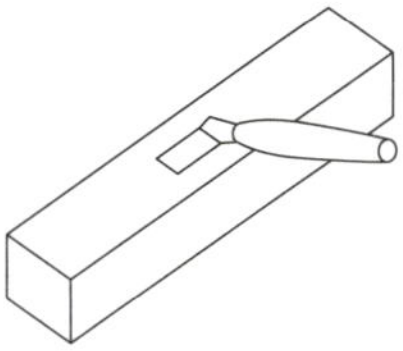

B1

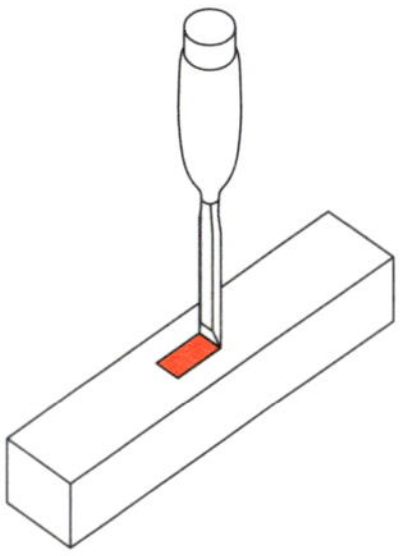

B2

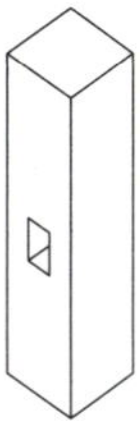

B

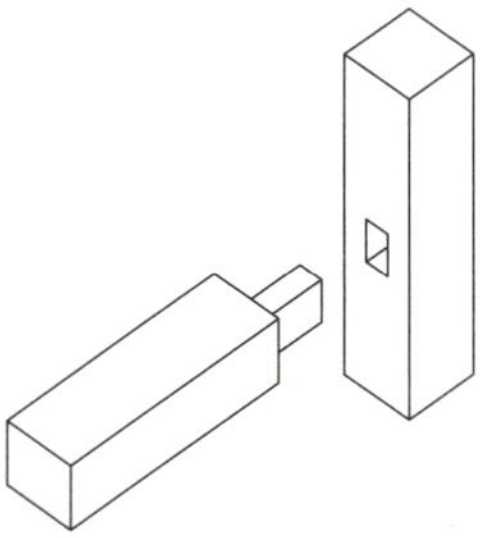

A+B

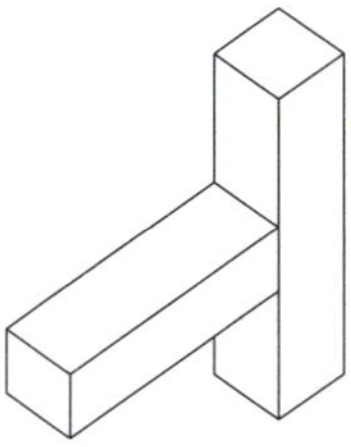

AB

T-Joints

Fully shouldered through mortise and tenon joint

shihō-dō-tsuki-tōshi-naga-hozo-tsugi
四方胴付通長枘接ぎ

durchgehender beidseitig
abgesetzter Zapfen

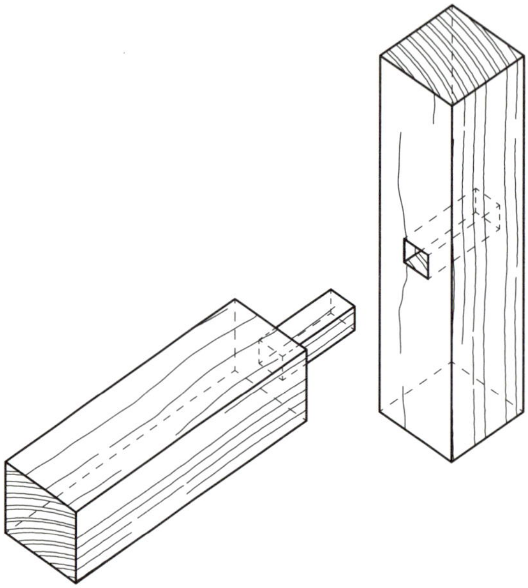

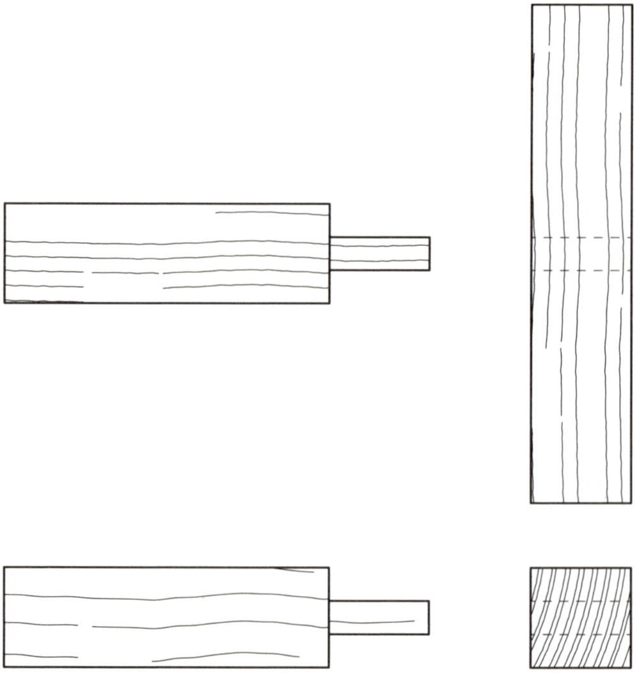

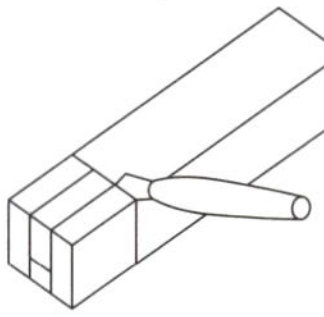
A1

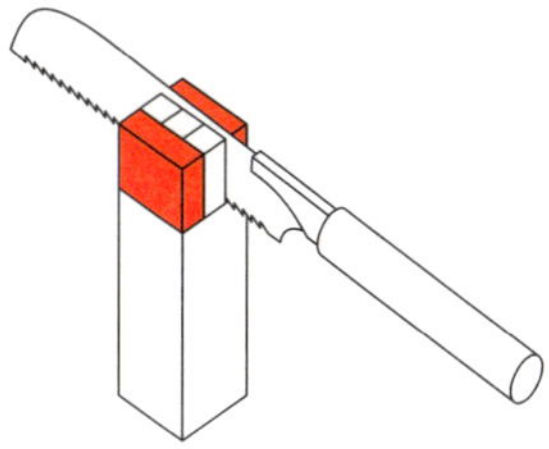
A2

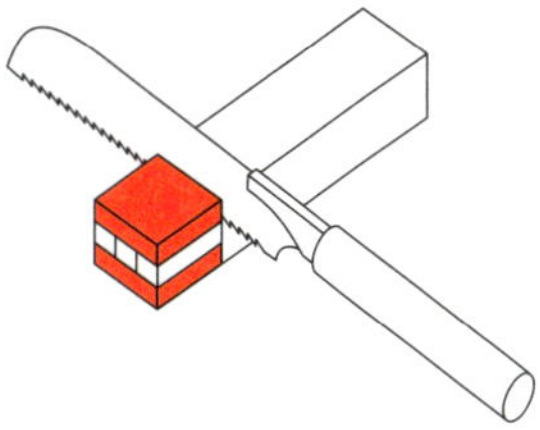
A3

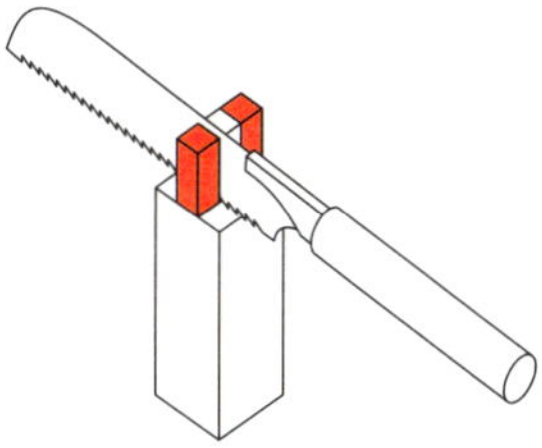
A4

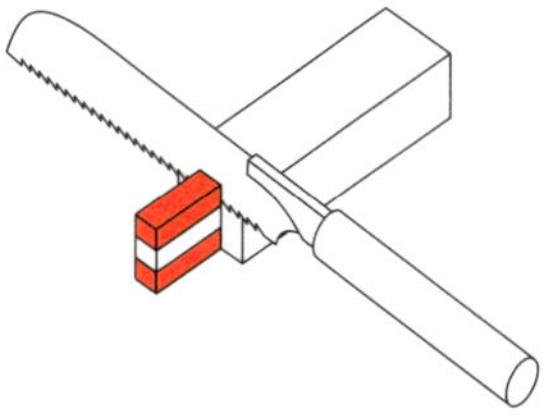
A5

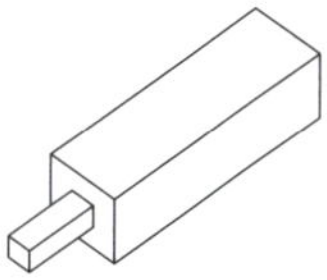
A

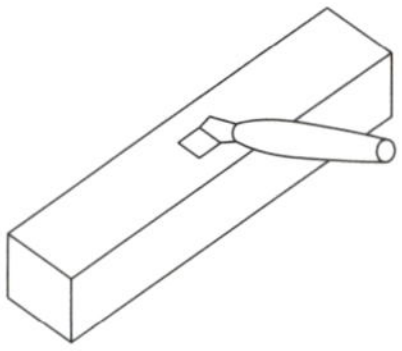
B1

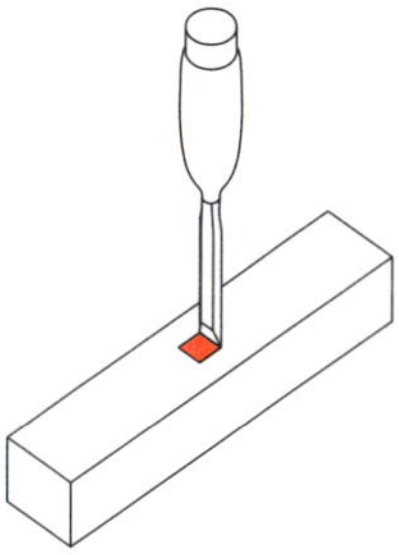
B2

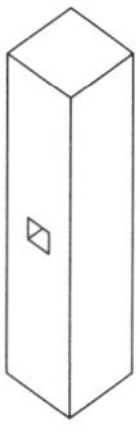
B

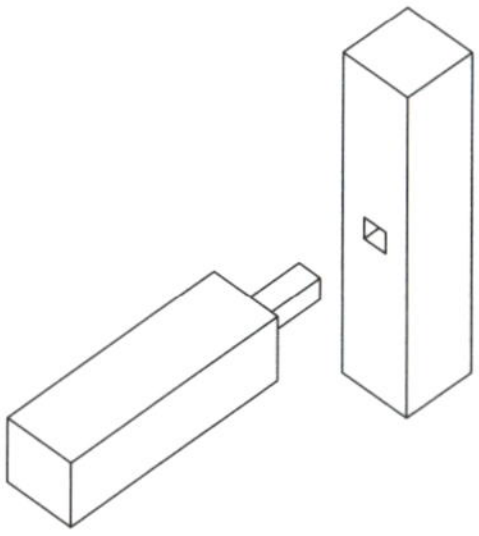
A+B

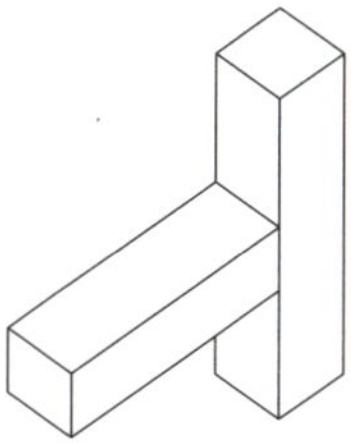
AB

T-Joints

Triple through mortise and tenon

sanmai-chidori-naga-hozo-sashi
三枚千鳥長枘差し

durchgehender dreifacher Zapfen

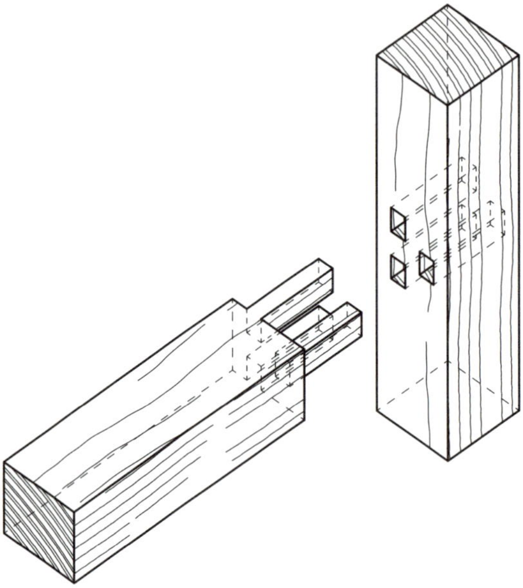

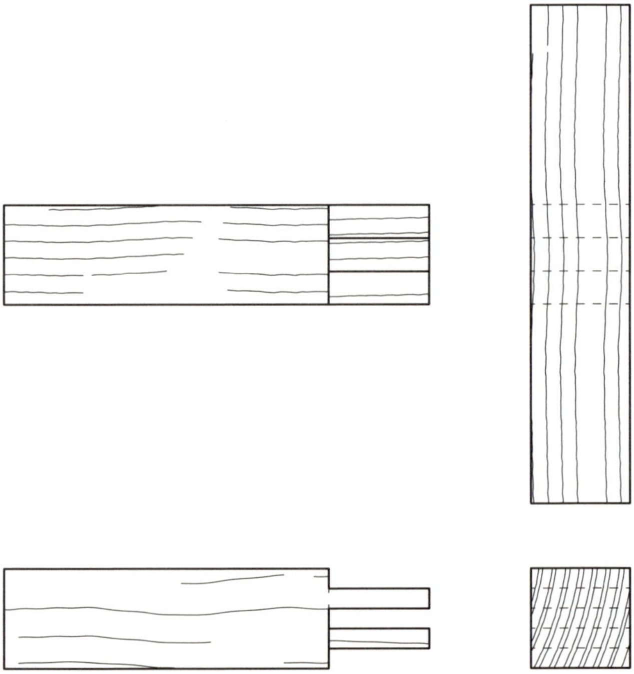

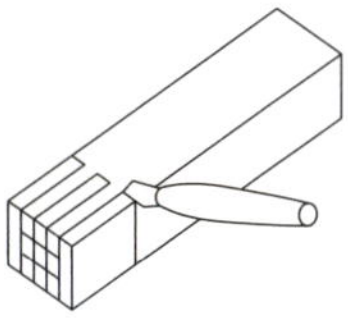
A1

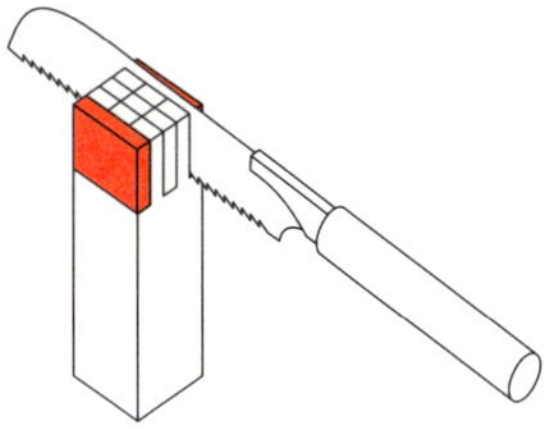
A2

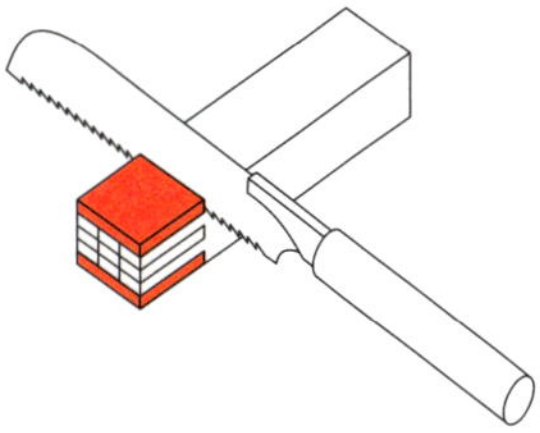
A3

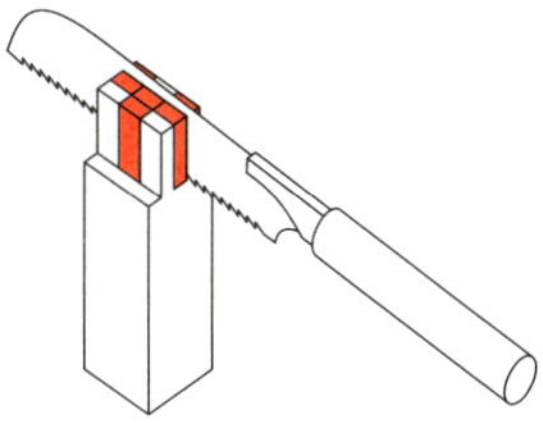
A4

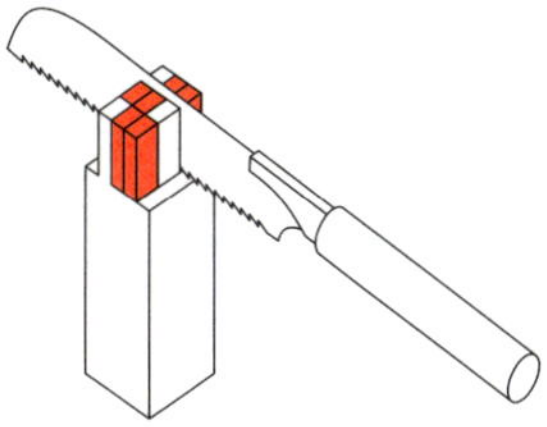
A5

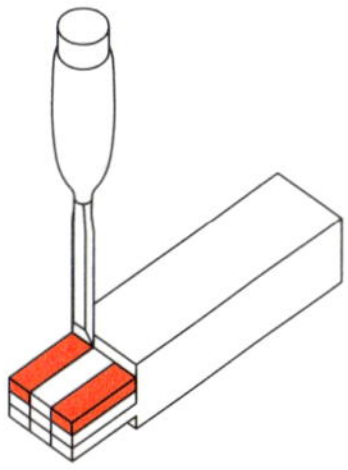
A6

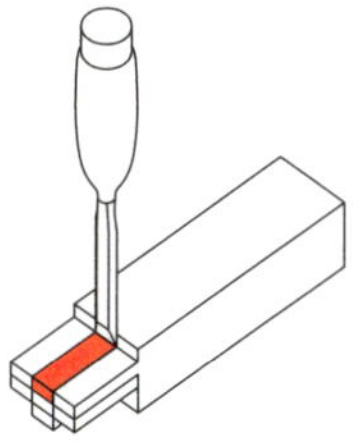
A7

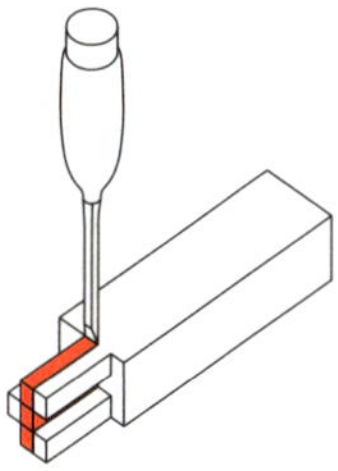
A8

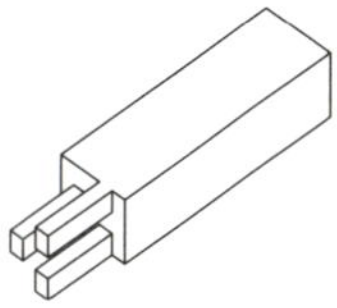
A

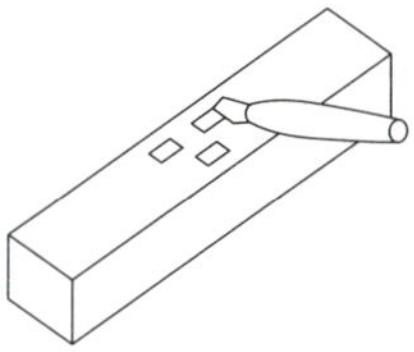
B1

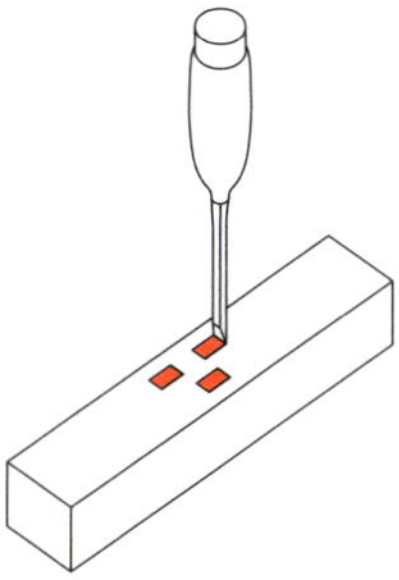
B2

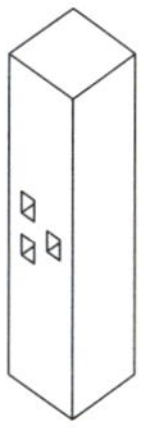
B

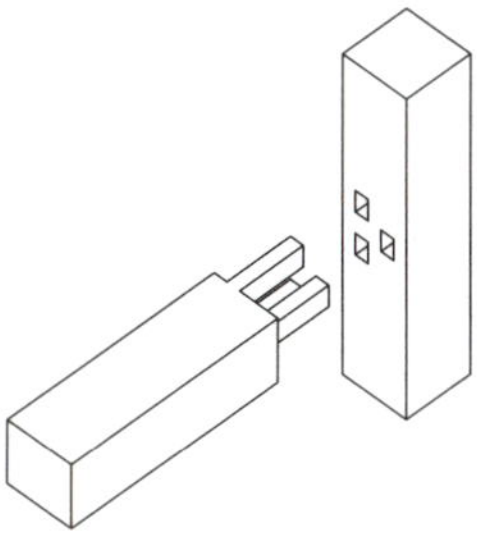
A+B

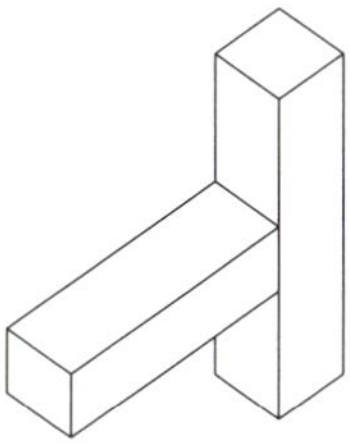
AB

T-Joints

Quadruple through mortise and tenon

yonmai-naga-hozo-sashi
四枚長枘差し

durchgehender vierfacher Zapfen

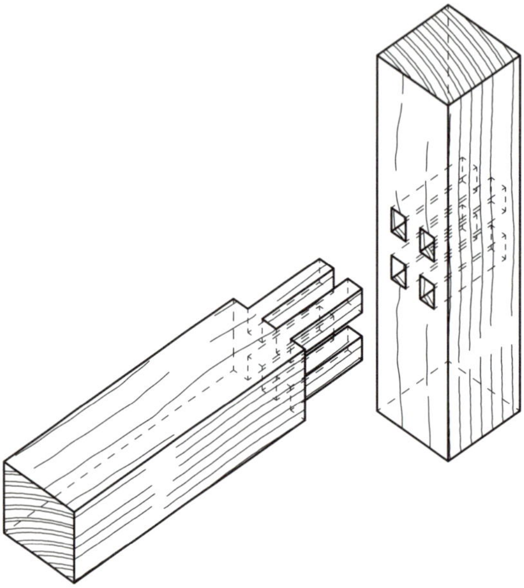

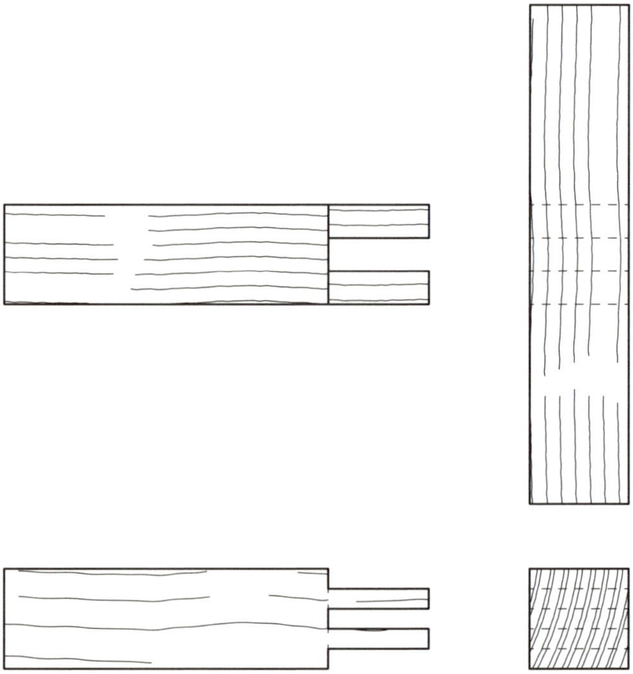

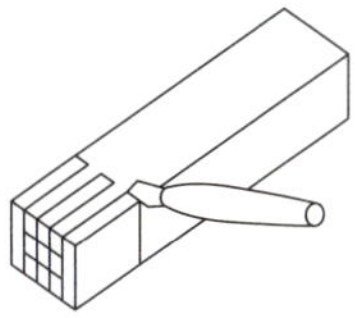

A1

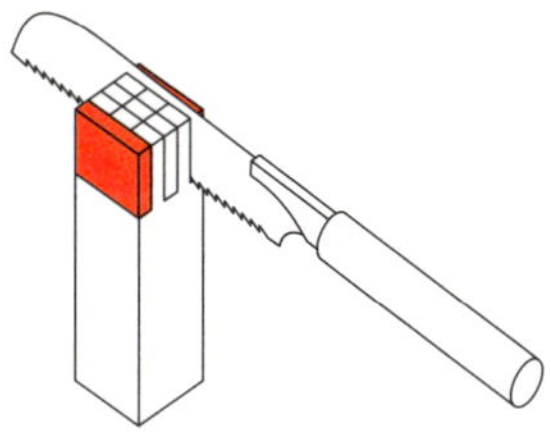

A2

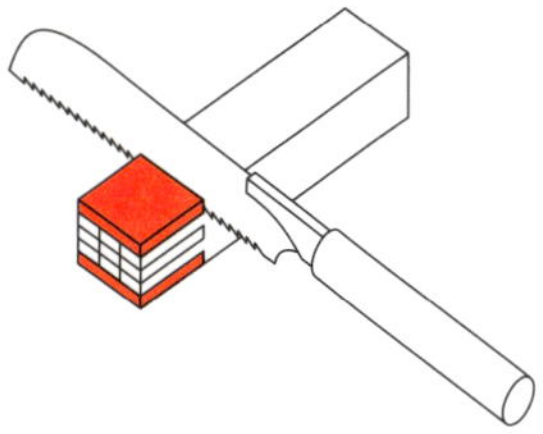

A3

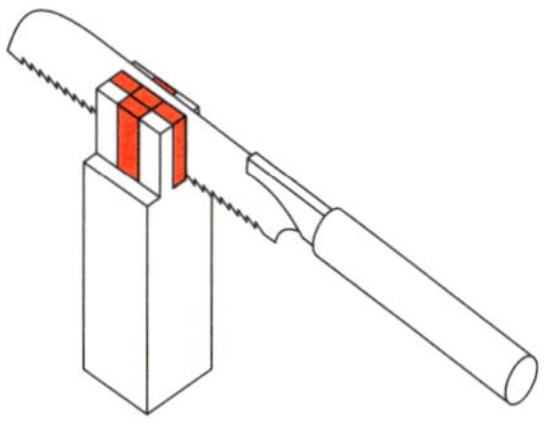

A4

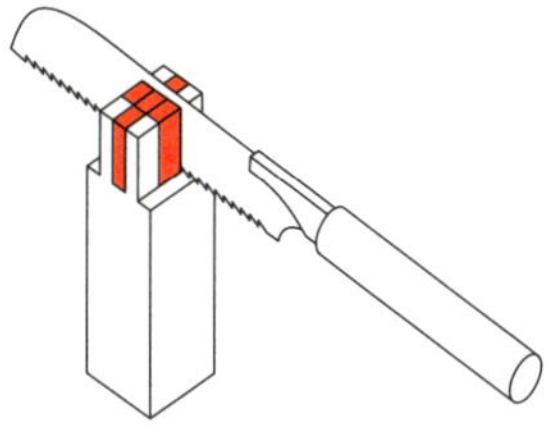

A5

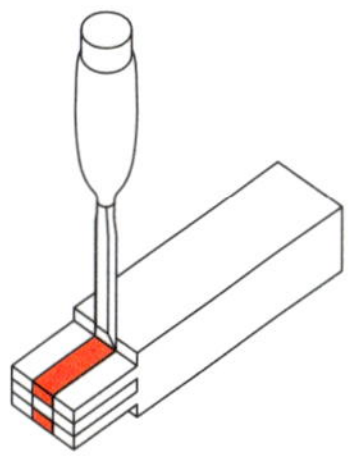

A6

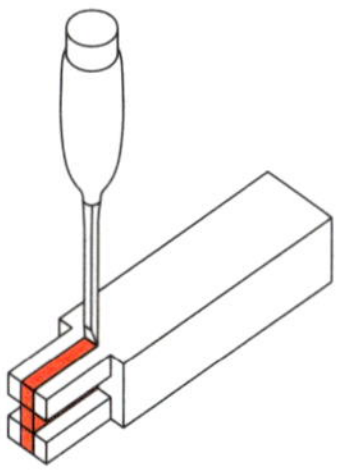

A7

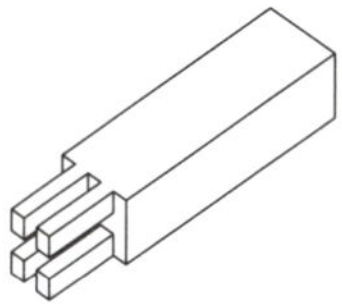

A

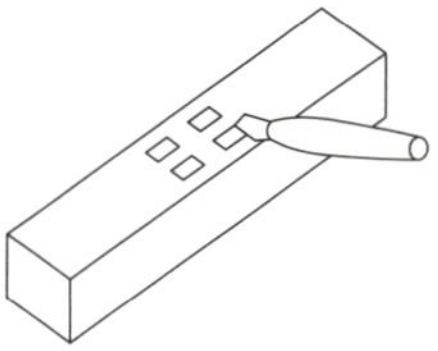

B1

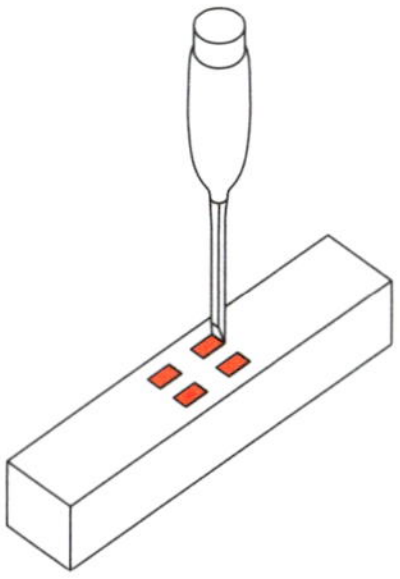

B2

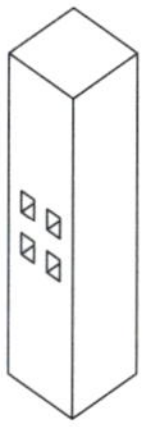

B

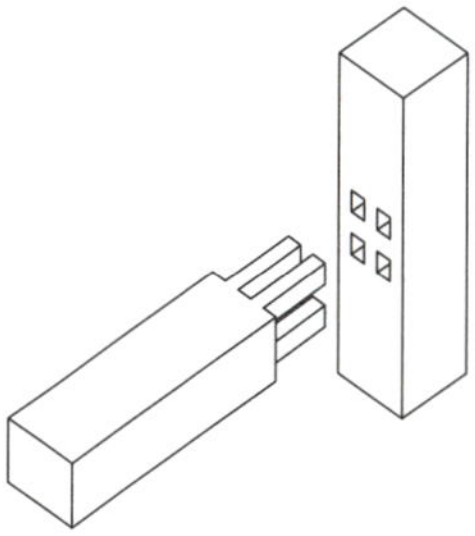

A+B

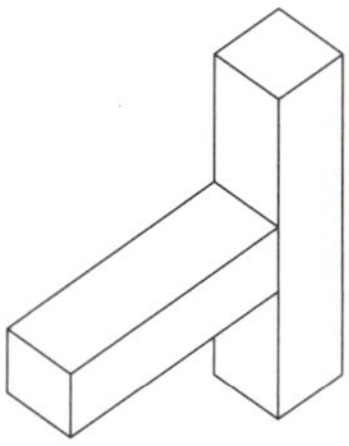

AB

T-Joints

Haunched through mortise and tenon

kone-hozo-sashi (uchi-nuki)[*]
小根枘差し(打抜)

durchgehender Stufenzapfen

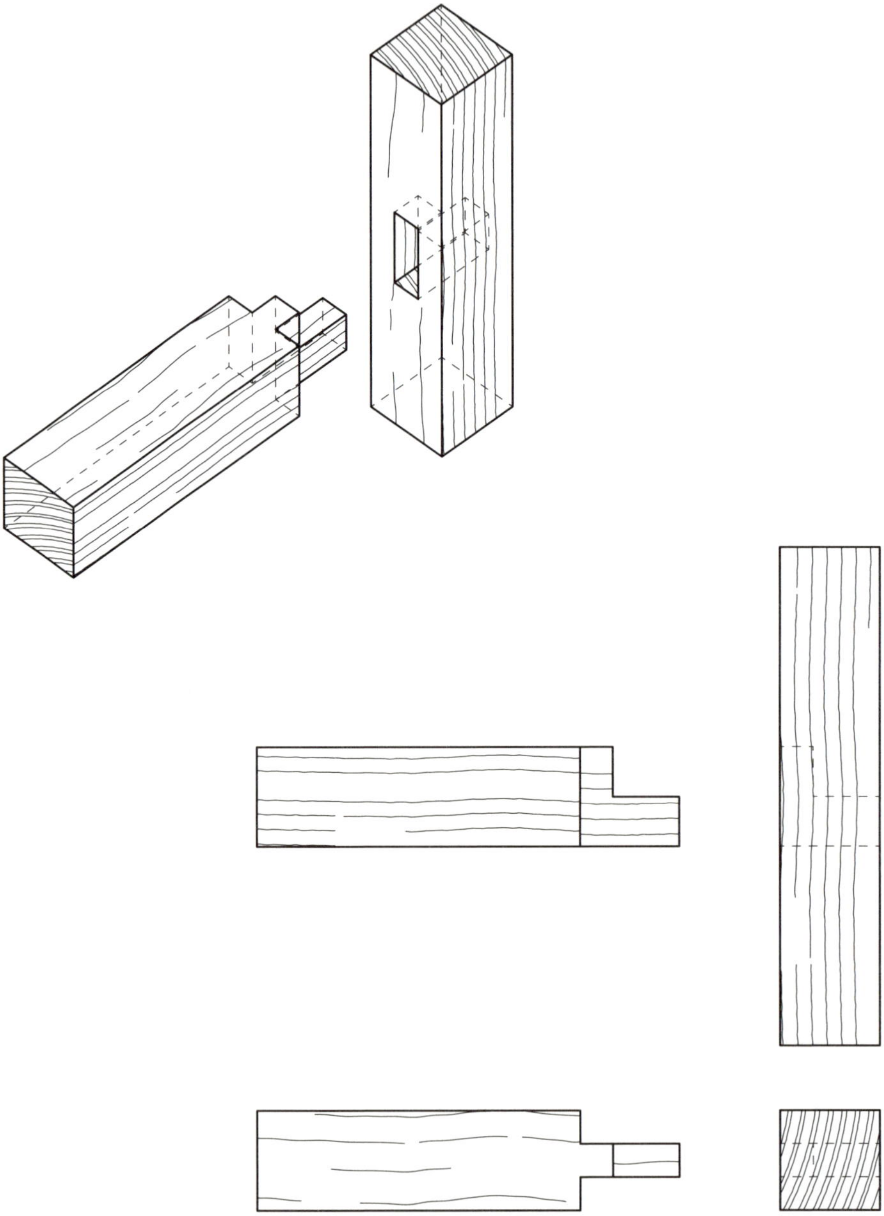

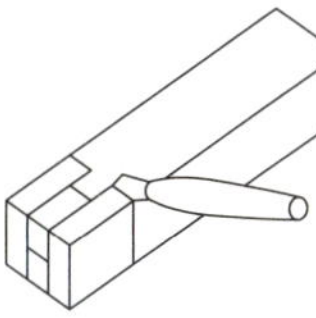

A1

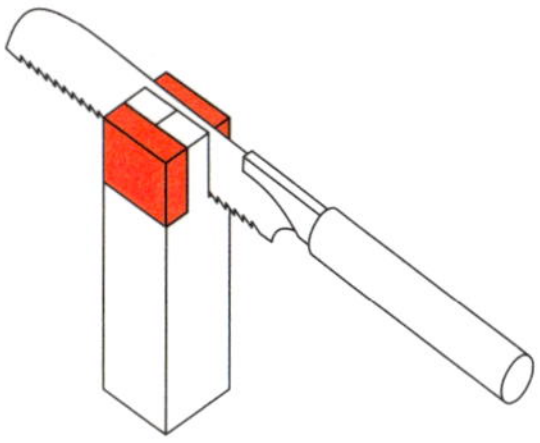

A2

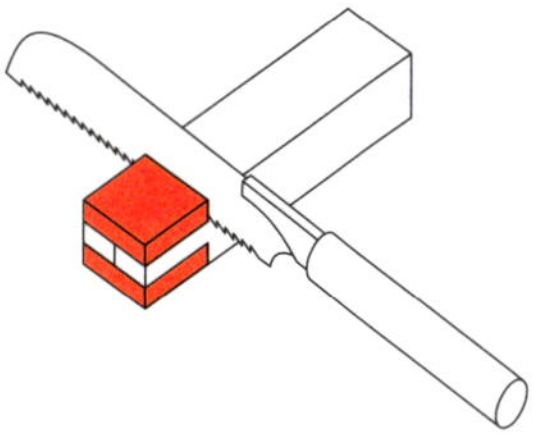

A3

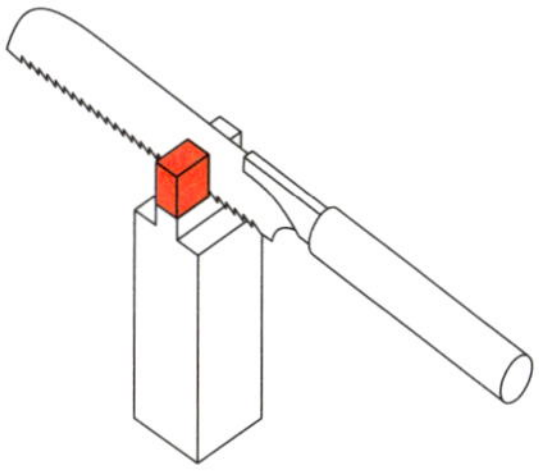

A4

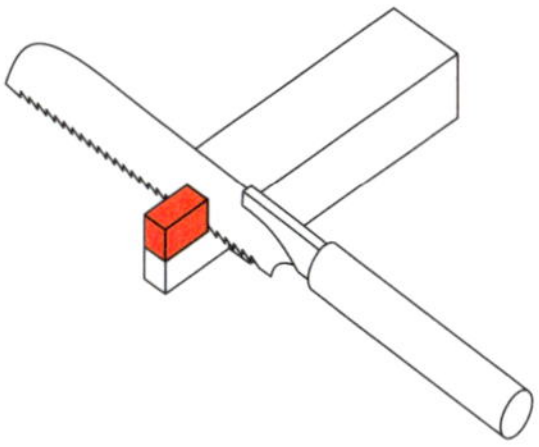

A5

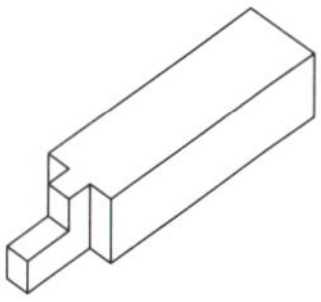

A

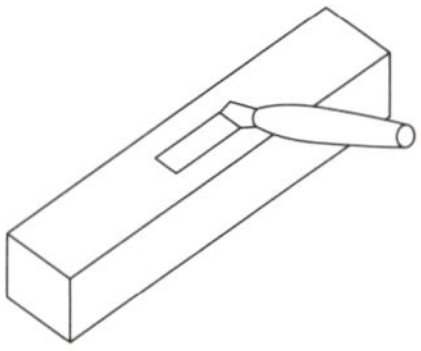

B1

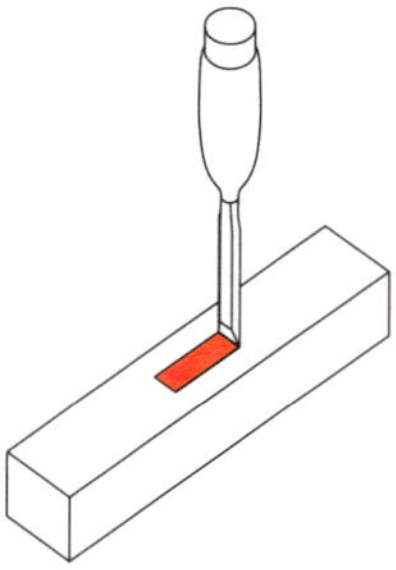

B2

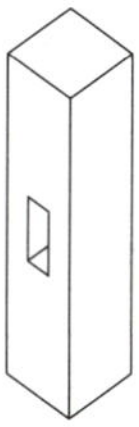

B

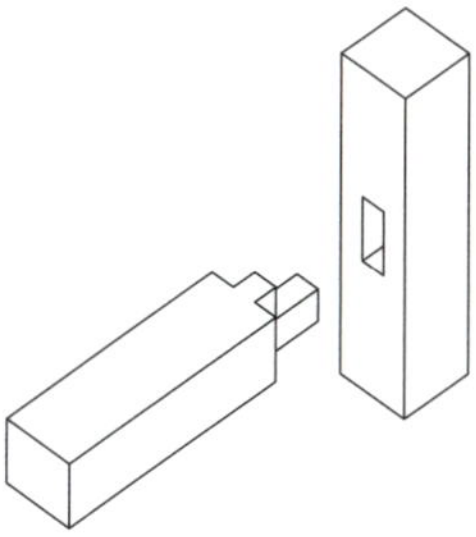

A+B

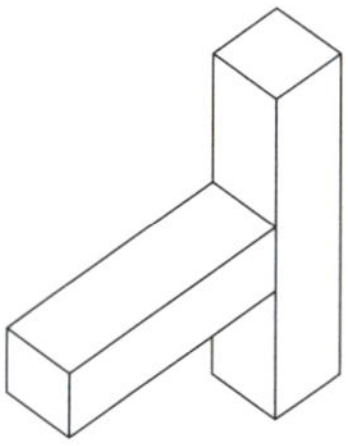

AB

T-Joints

Blind haunched mortise and tenon

kone-hozo-sashi (tsutsumi-komi)
小根柄差し（包込）

verdeckter Stufenzapfen

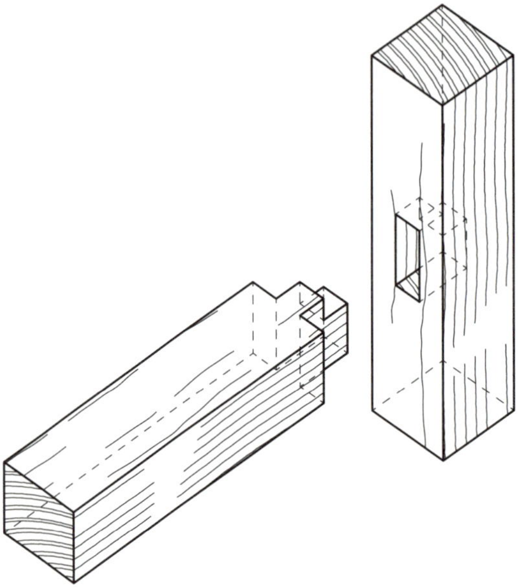

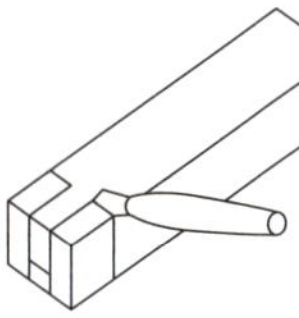

A1

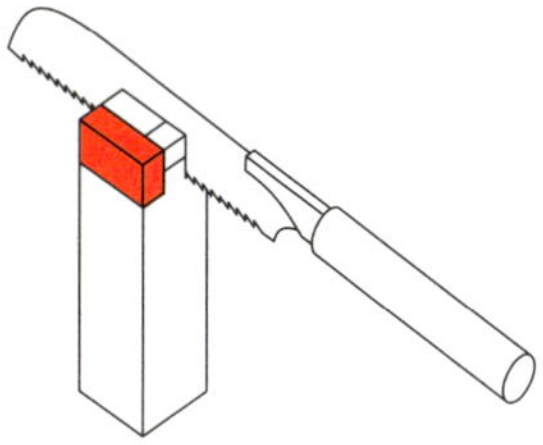

A2

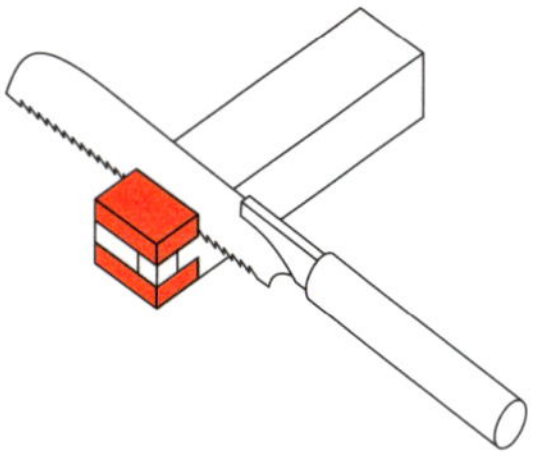

A3

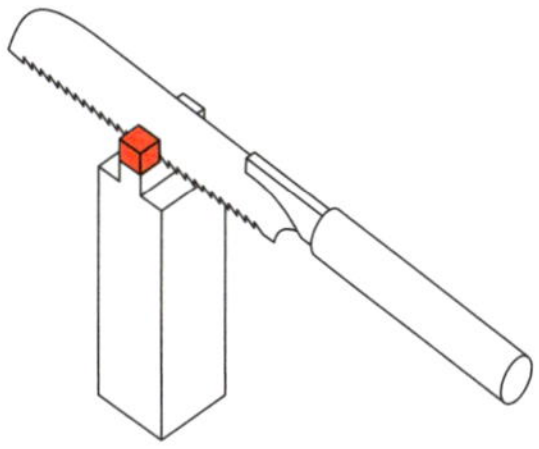

A4

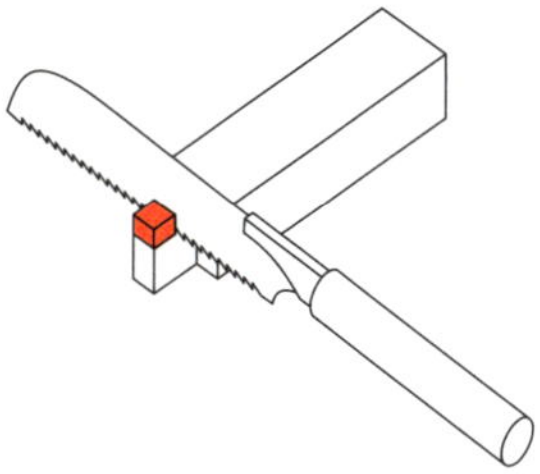

A5

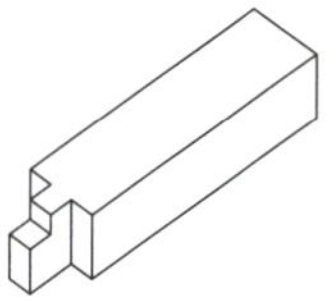

A

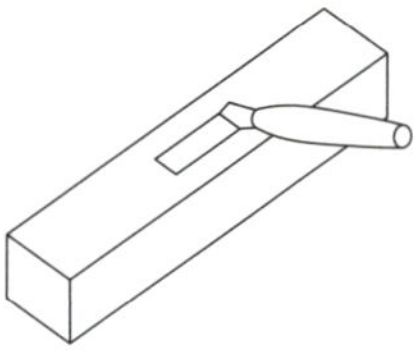

B1

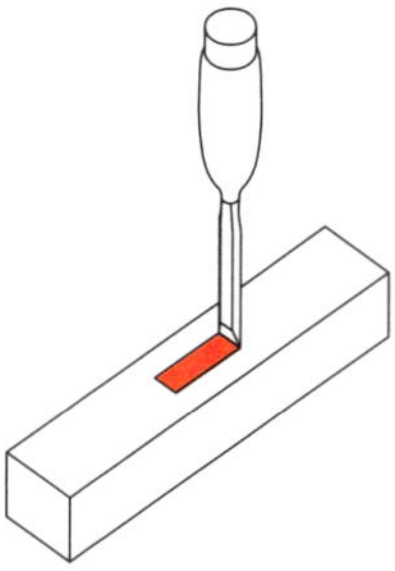

B2

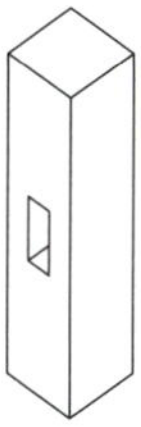

B

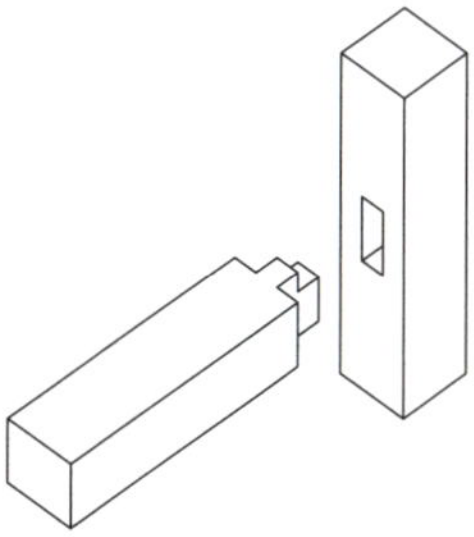

A+B

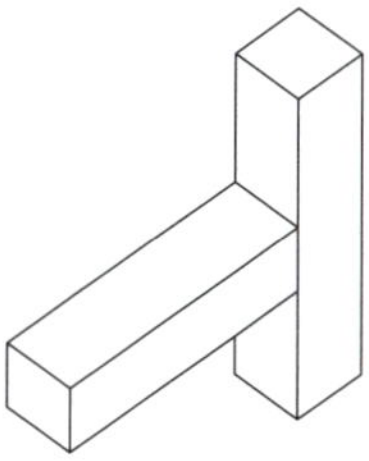

AB

T-Joints

Through mortise and tenon with sloped haunch

kone-hozo-sashi (sogi-uchi-nuki)
小根枘差し(殺打抜)

durchgehender Zapfen
mit schräger Stufe

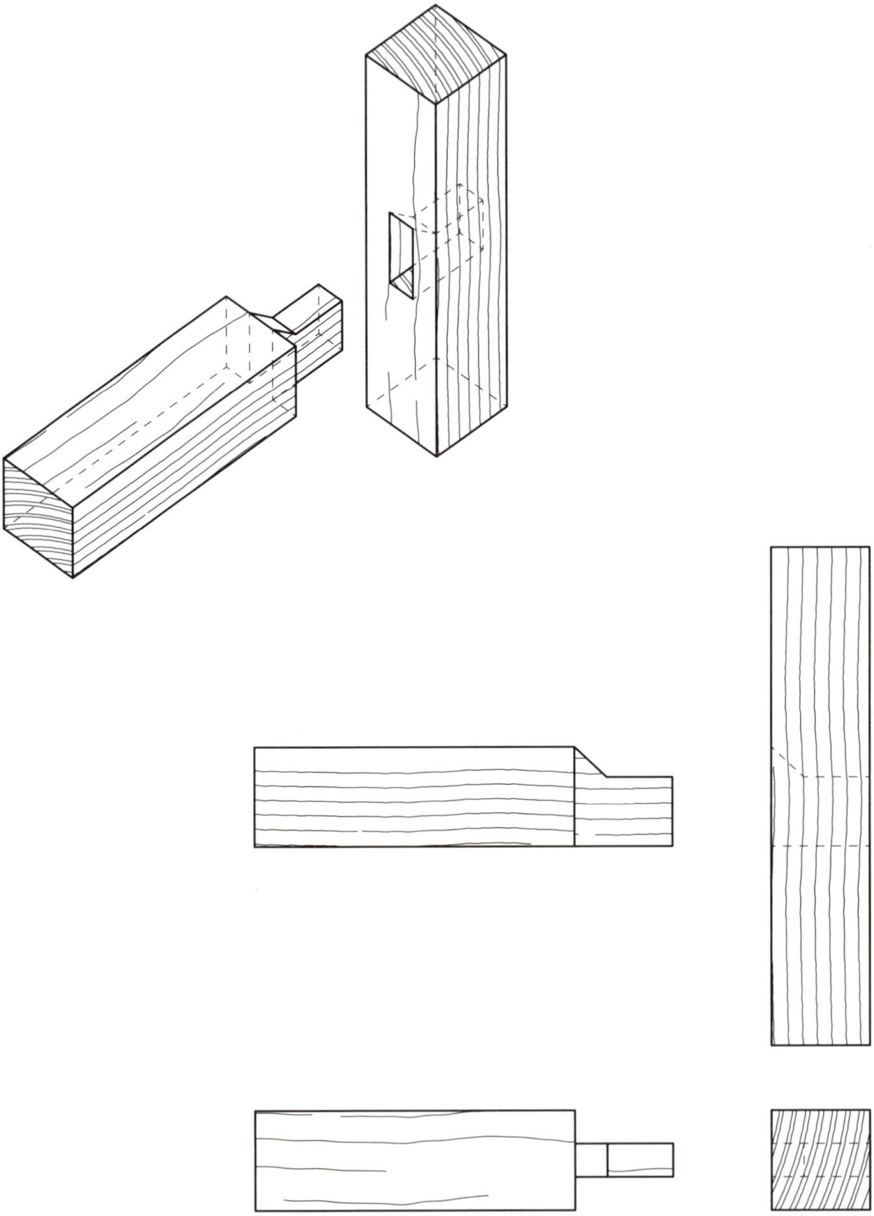

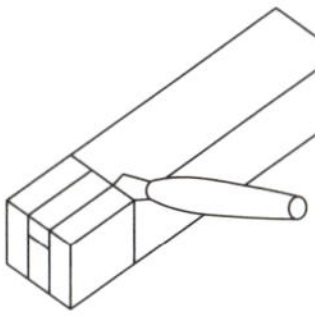
A1

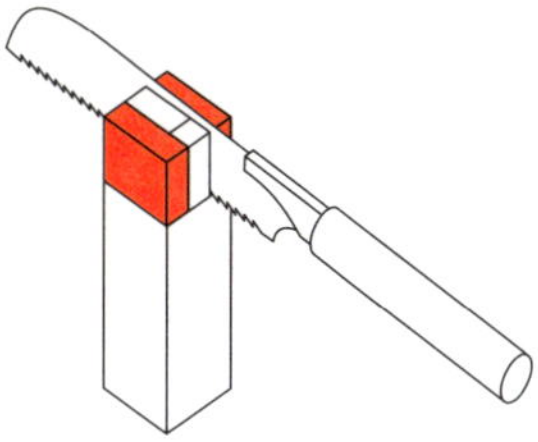
A2

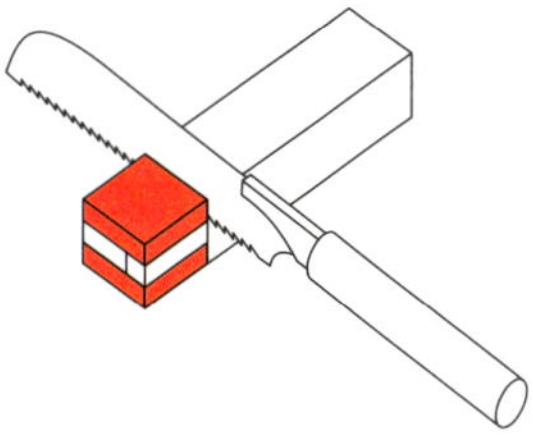
A3

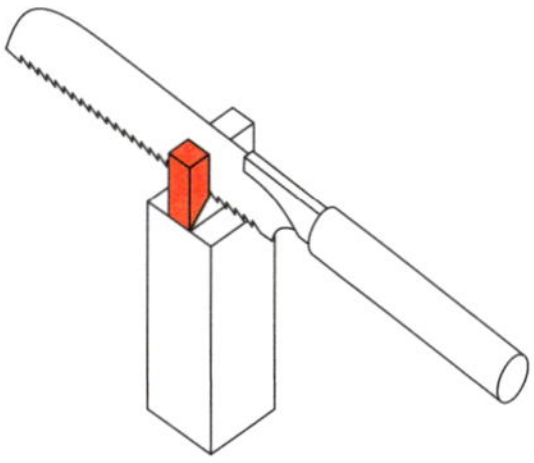
A4

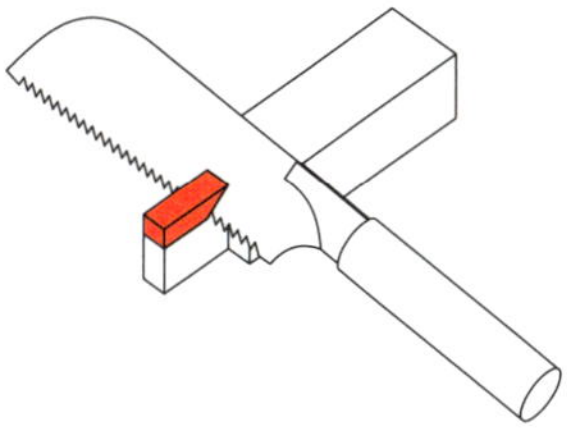
A5

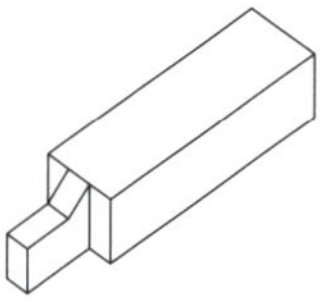
A

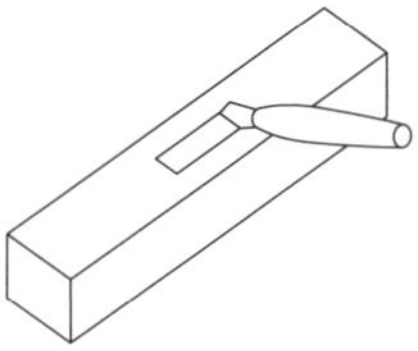
B1

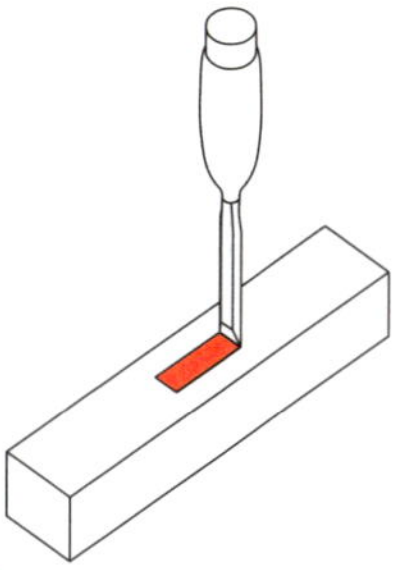
B2

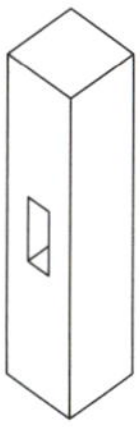
B

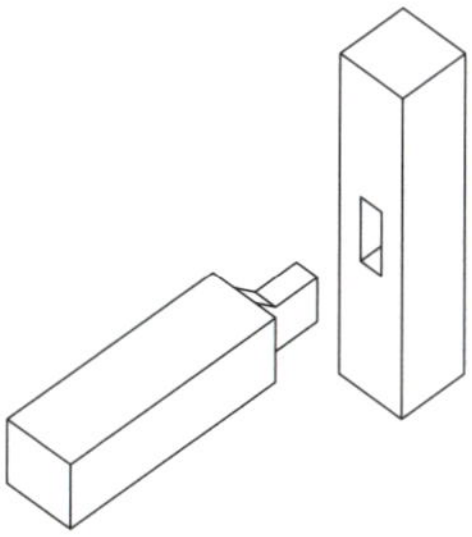
A+B

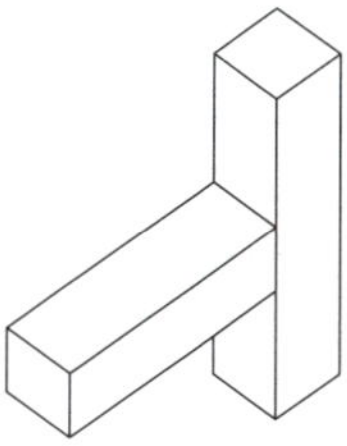
AB

T-Joints

Blind housed haunched through mortise and tenon

ko-dō-tsuki-kone-hozo-sashi (uchi-nuki) [*]
小胴付小根枘差し(打抜)

durchgehender Stufenzapfen mit gestufter Brüstung

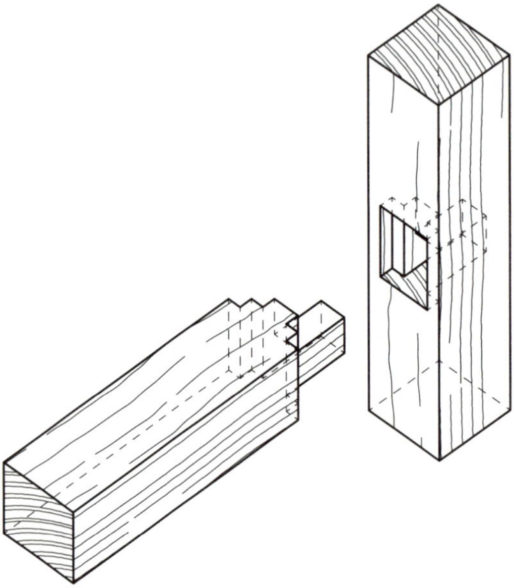

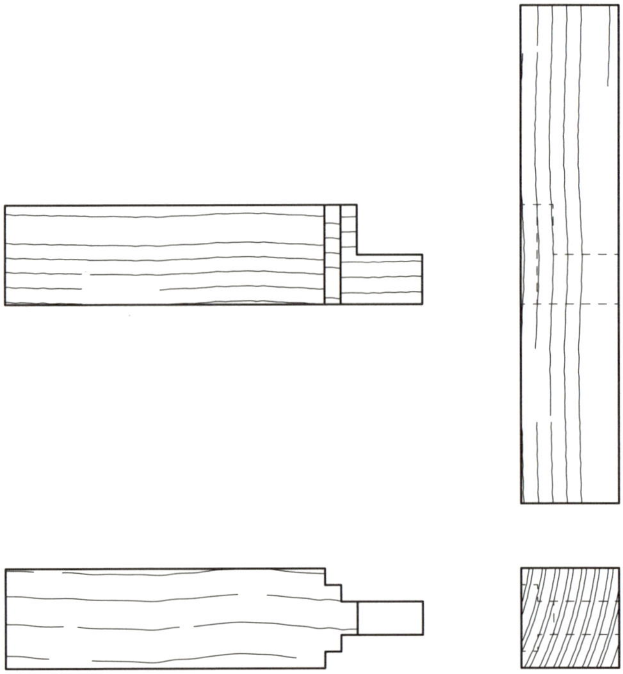

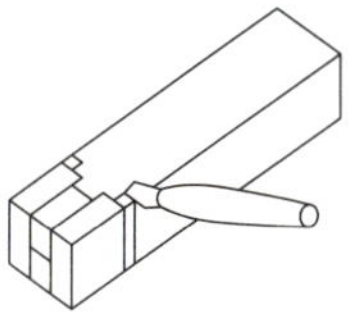

A1

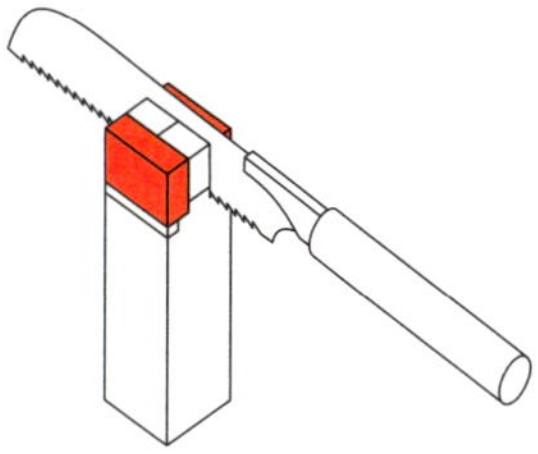

A2

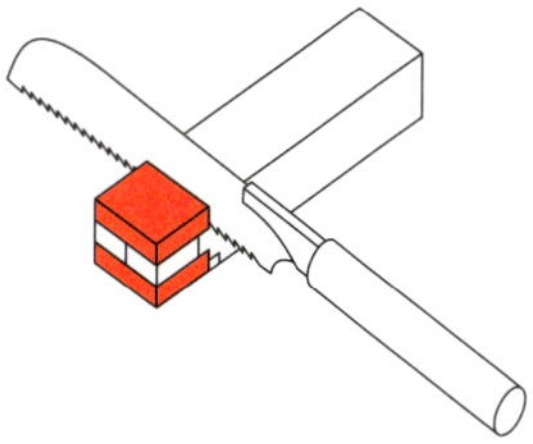

A3

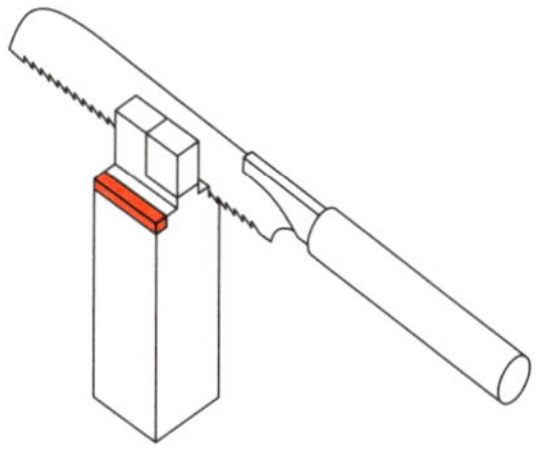

A4

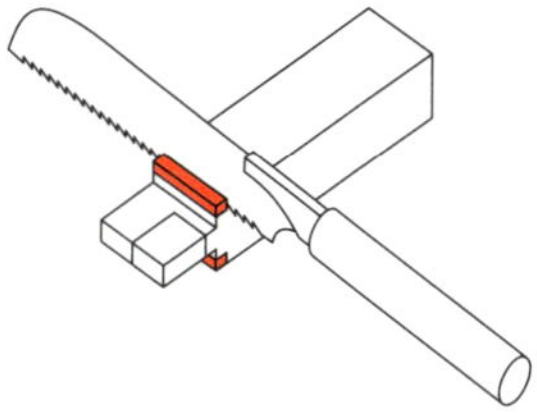

A5

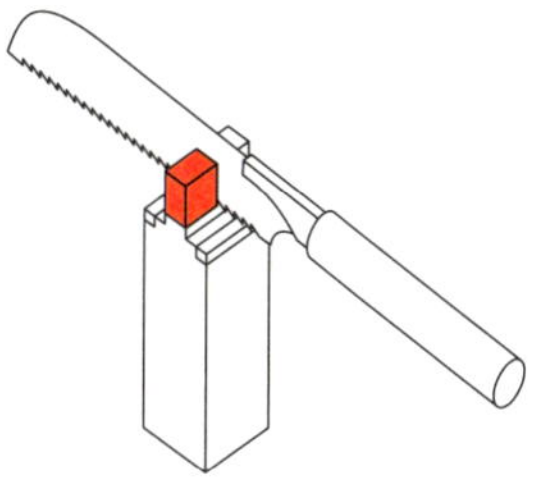

A6

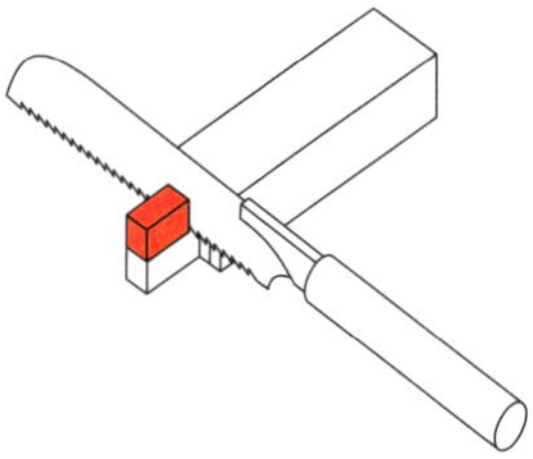

A7

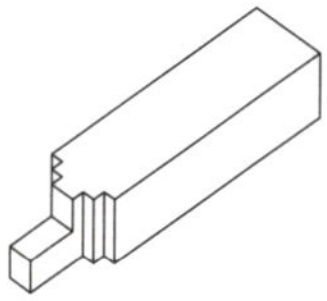

A

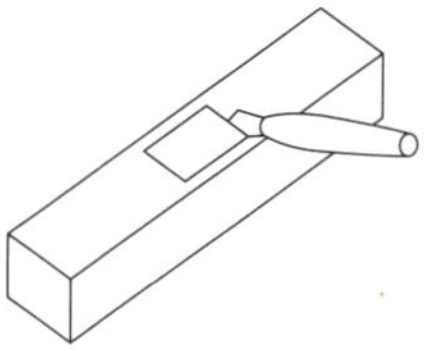

B1

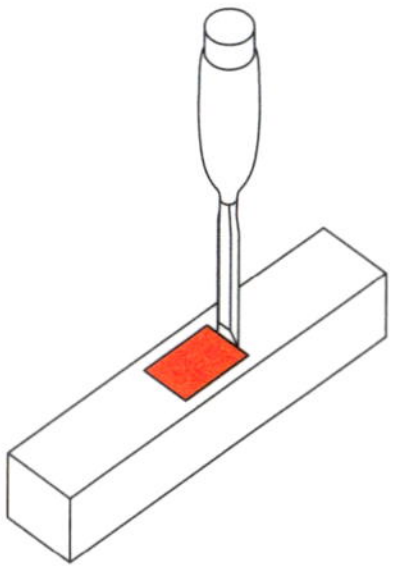

B2

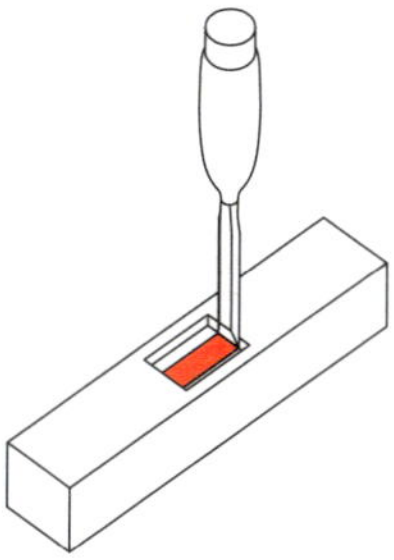

B3

B

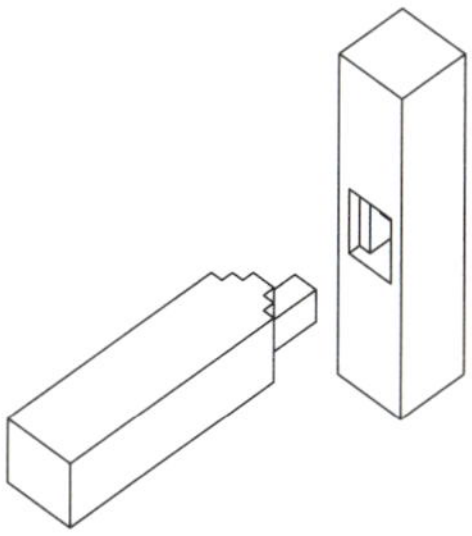

A+B

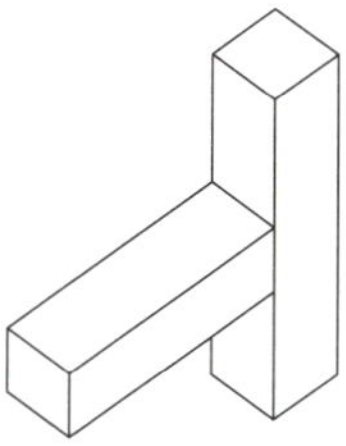

AB

T-Joints

Haunched collared through mortise and tenon

eriwa-kone-hozo-sashi (uchi-nuki)
襟輪小根枘差し(打抜)

durchgehender Stufenzapfen
mit Führungszapfen

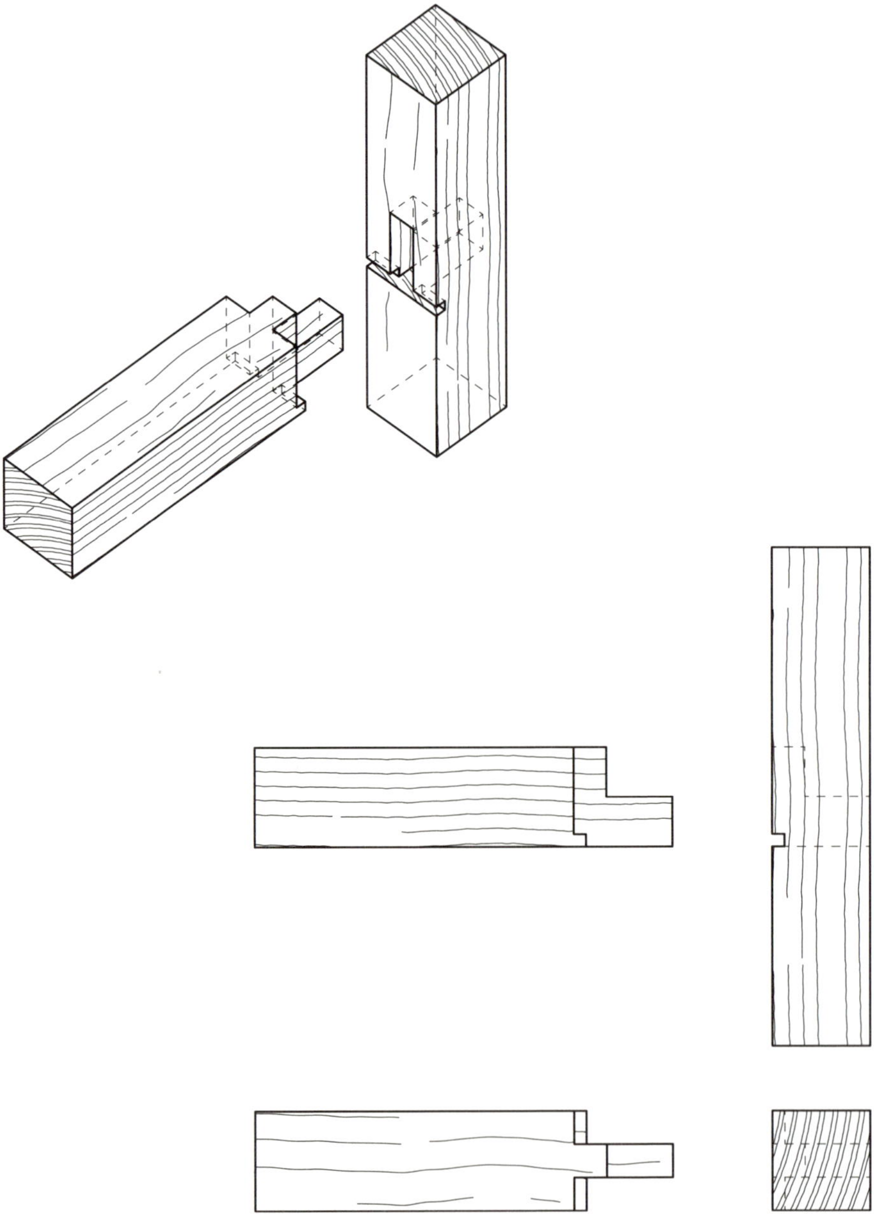

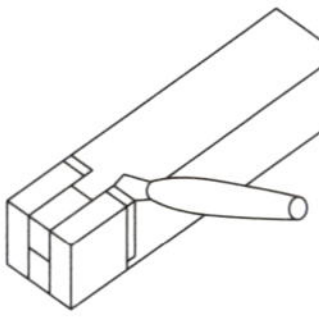
A1

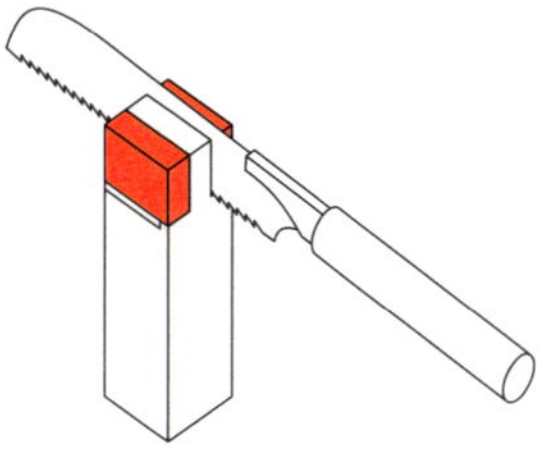
A2

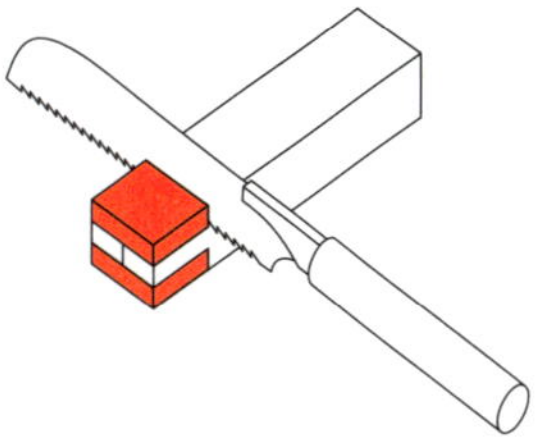
A3

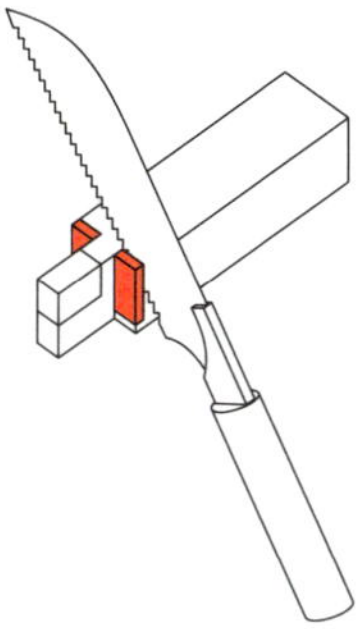
A4

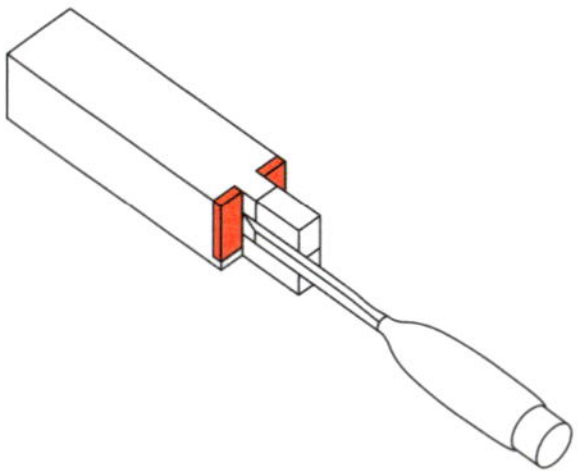
A5

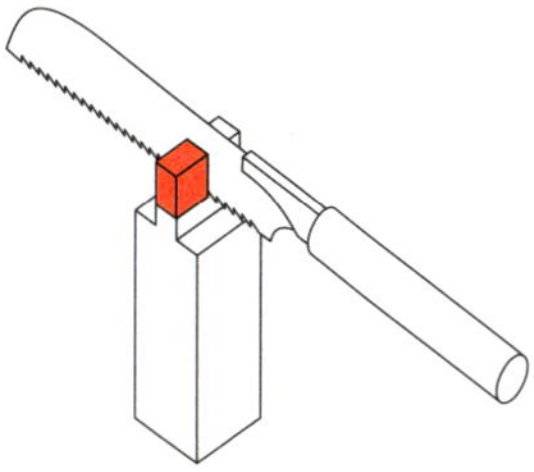
A6

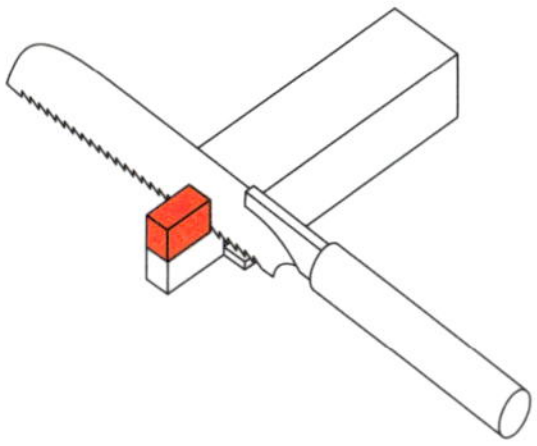
A7

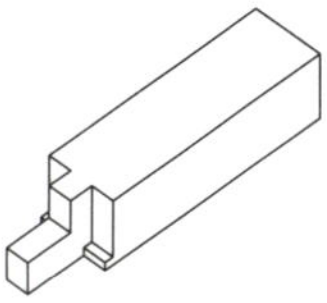
A

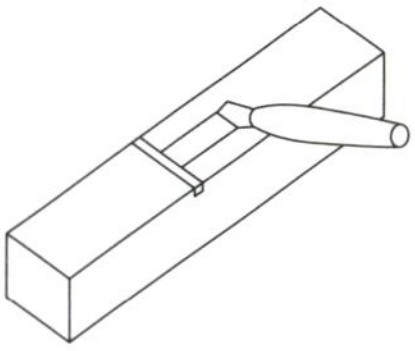
B1

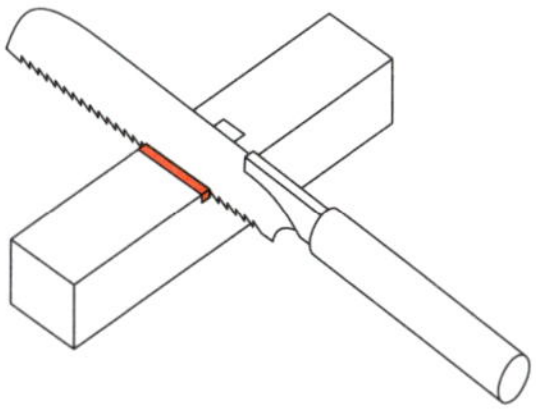
B2

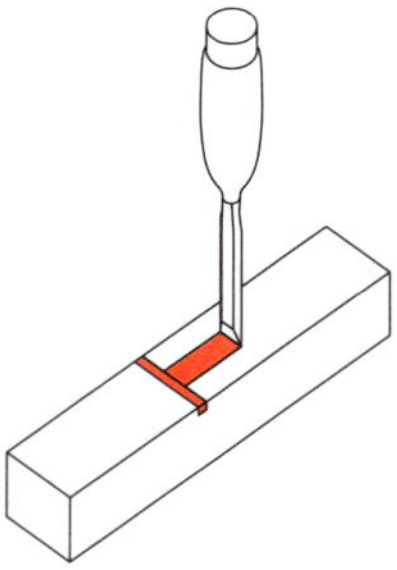
B3

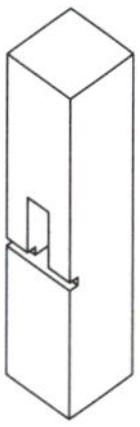
B

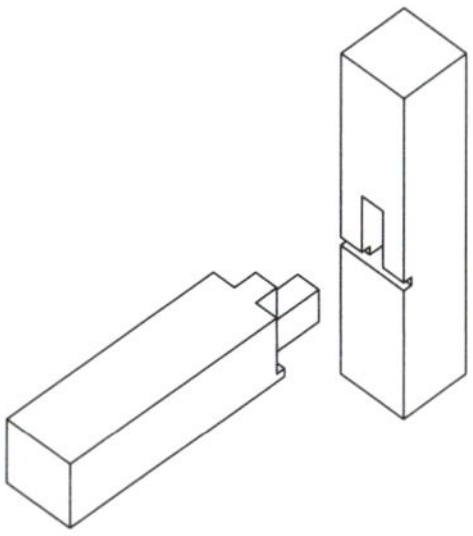
A+B

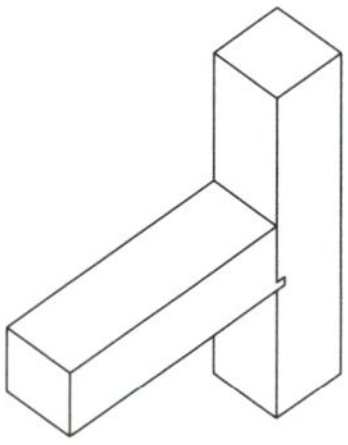
AB

T-Joints

Half-dovetail tenon with folding wedges

sabano-o-hozo-kai-kusabi-shimari
鯖乃尾枘飼楔締り

verkeilter verdeckter einseitiger Schwalbenschwanzzapfen

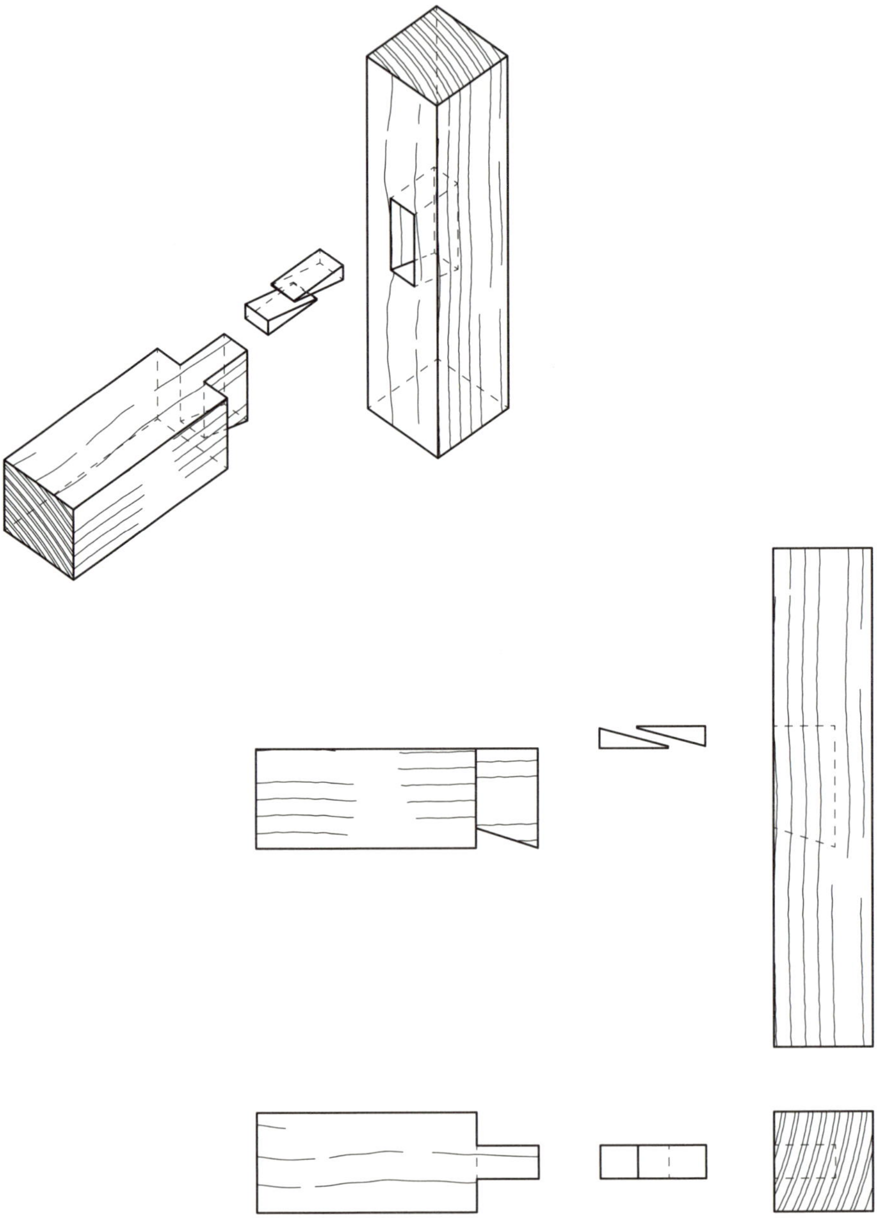

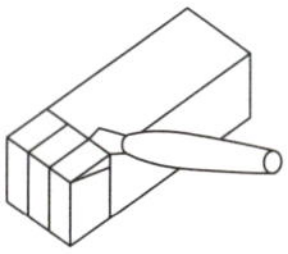

A1

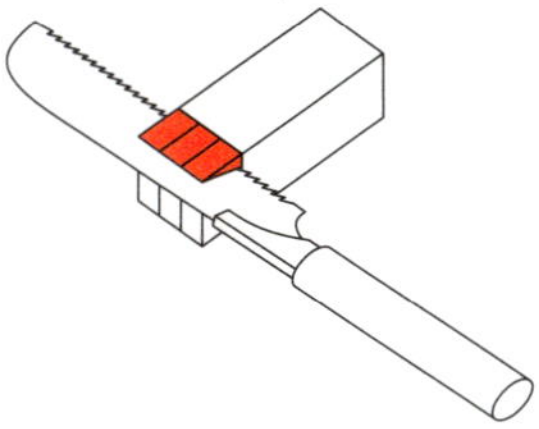

A2

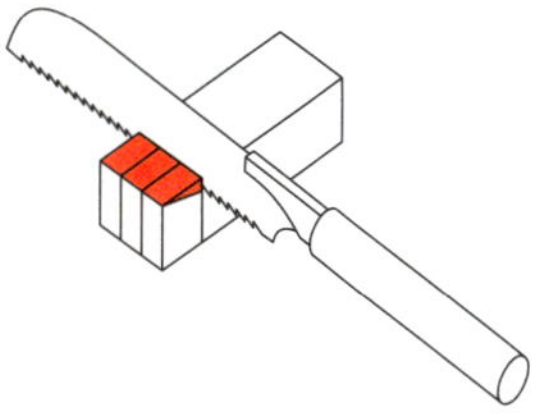

A3

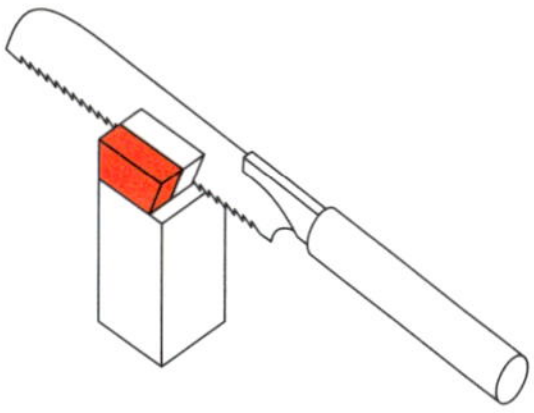

A4

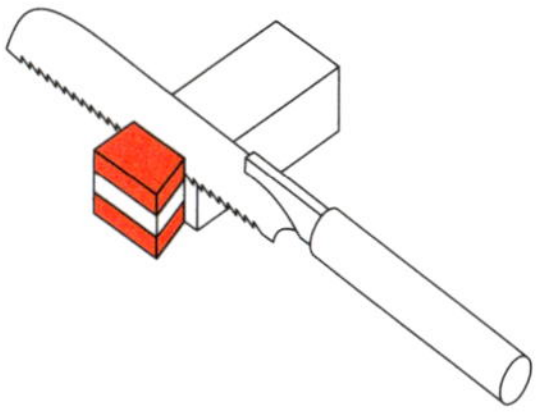

A5

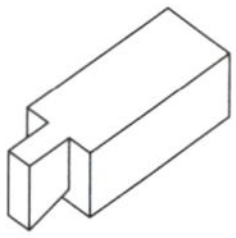

A

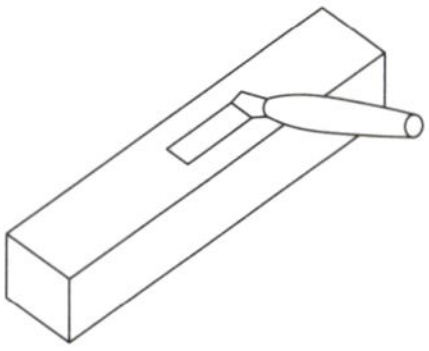

B1

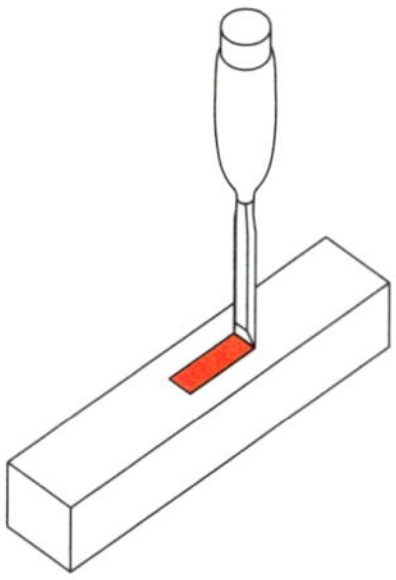

B2

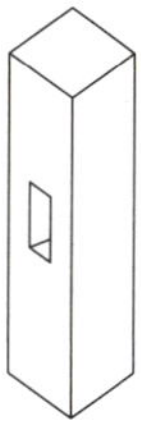

B

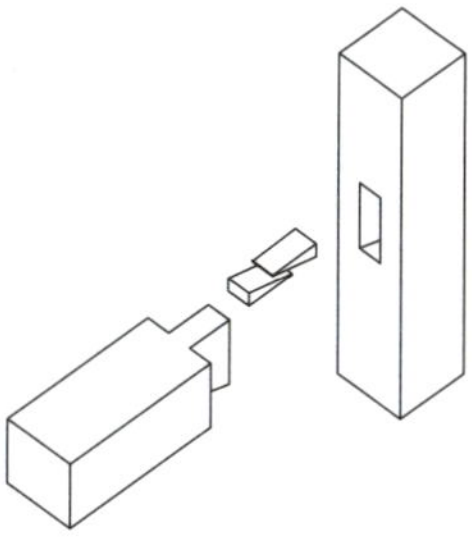

A+B

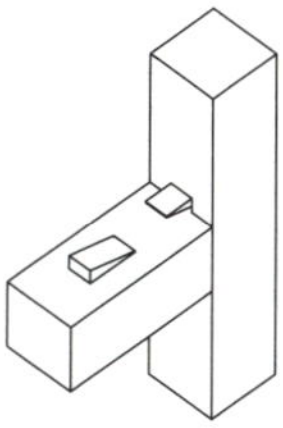

A+B

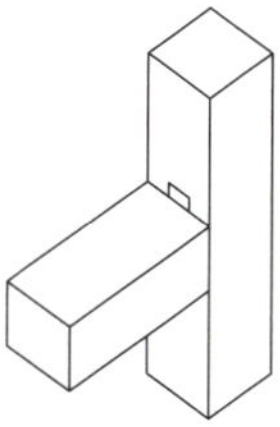

AB

T-Joints

Foxtail-wedged mortise and tenon

jigoku-ari-hozo-sashi
地獄蟻枘差し

verkeilter verdeckter Zapfen

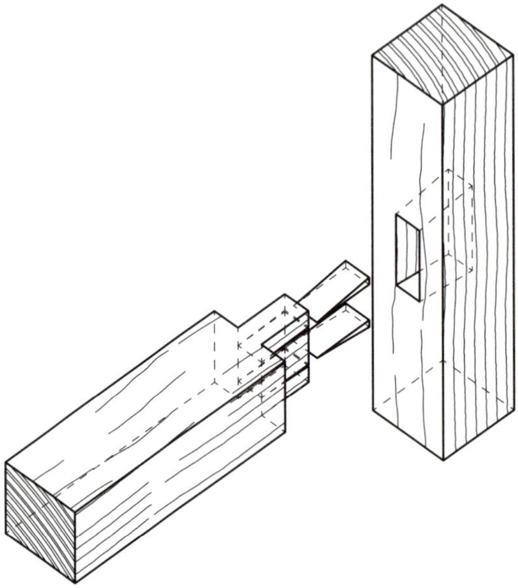

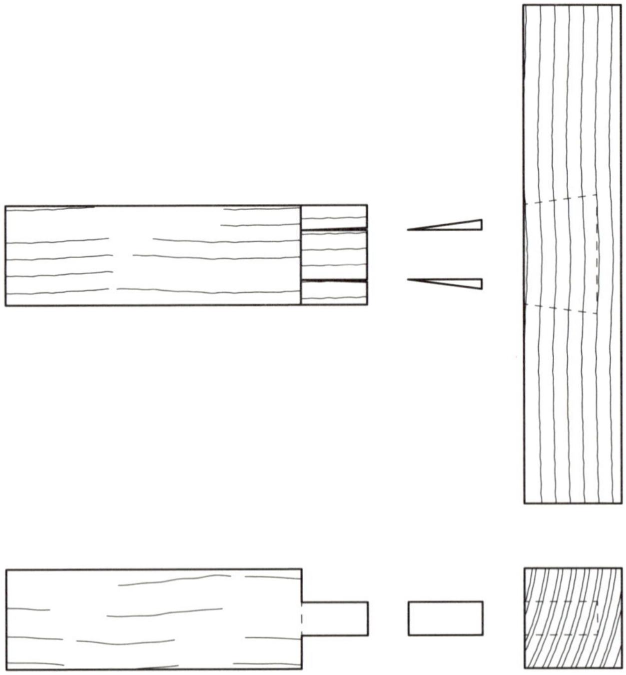

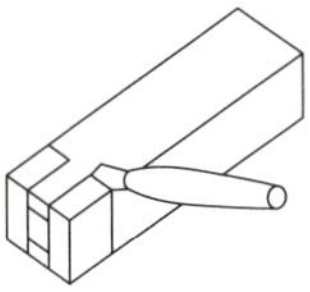

A1

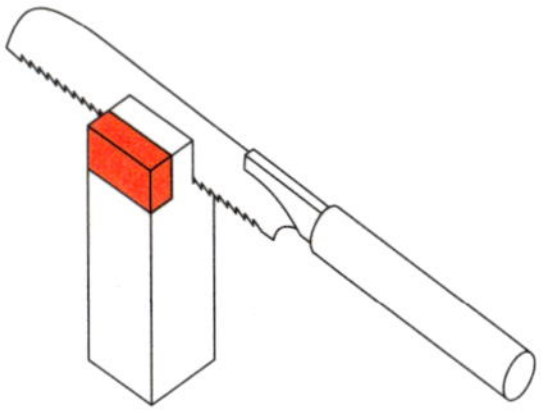

A2

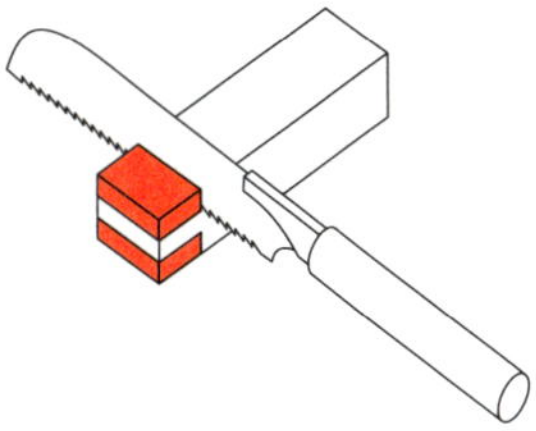

A3

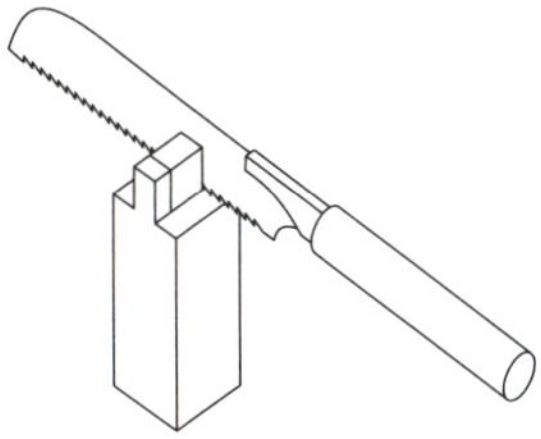

A4

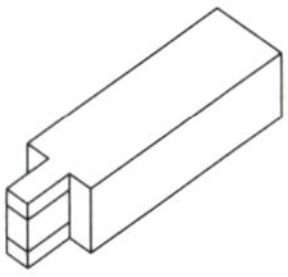

A

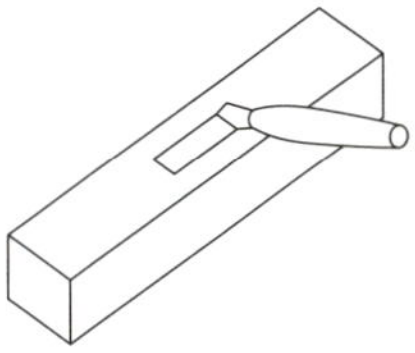

B1

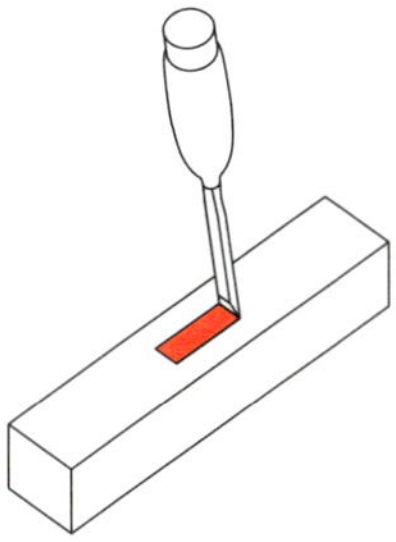

B2

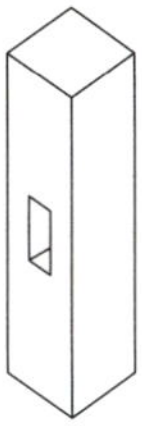

B

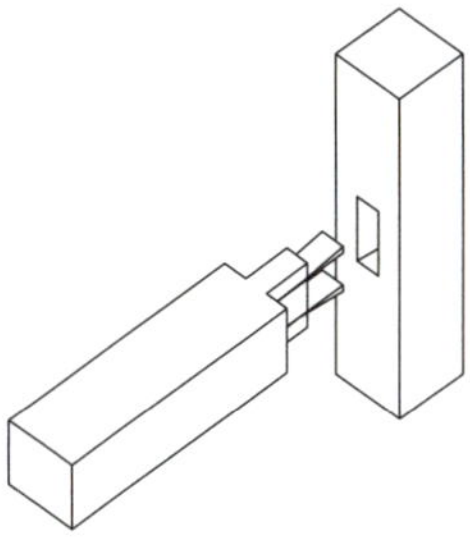

A+B

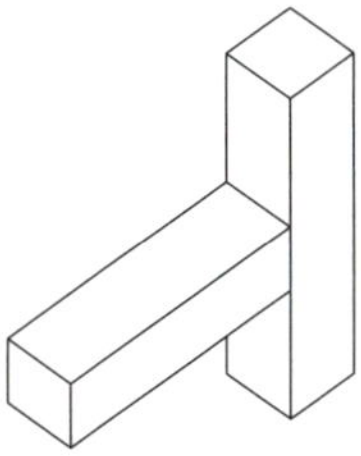

AB

T-Joints

Wedged through mortise and tenon

naga-hozo-sashi (wari-kusabi-shime)
長枘差し（割楔締）

verkeilter durchgehender Zapfen

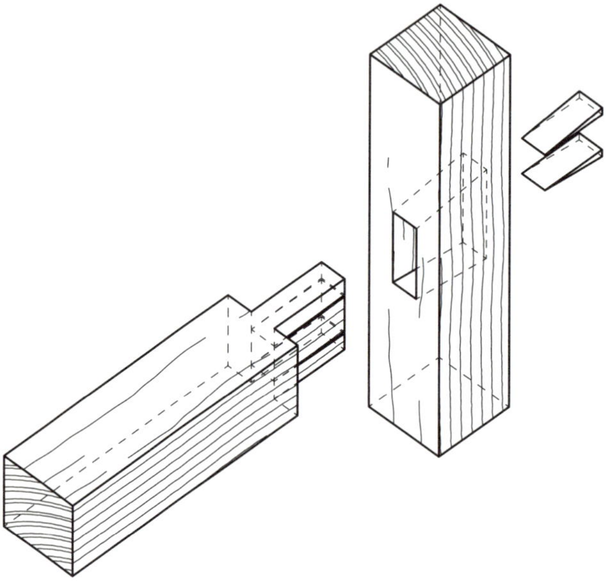

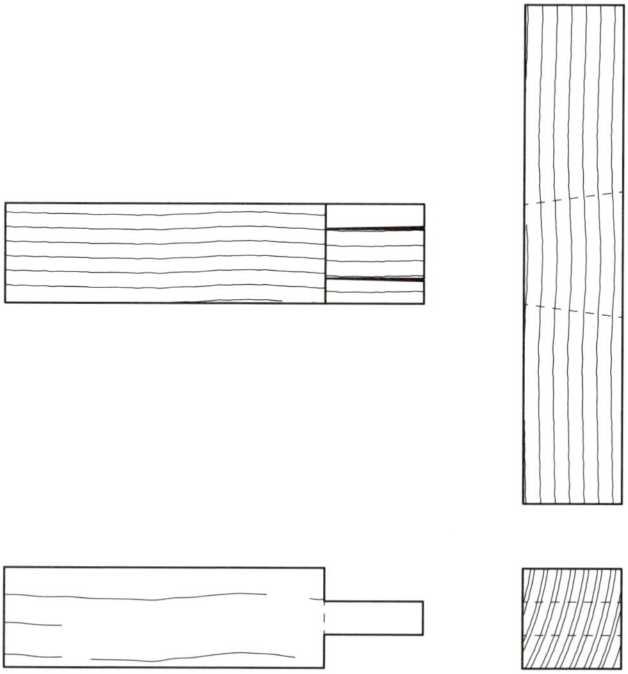

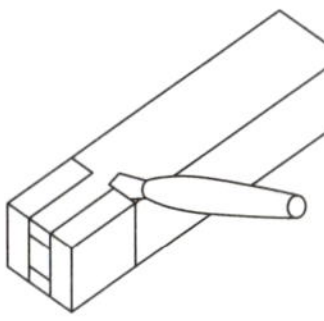

A1

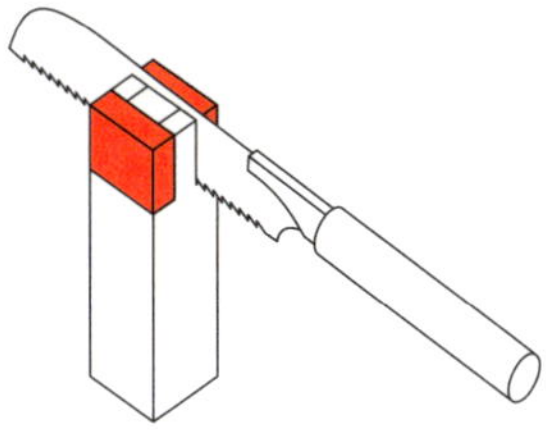

A2

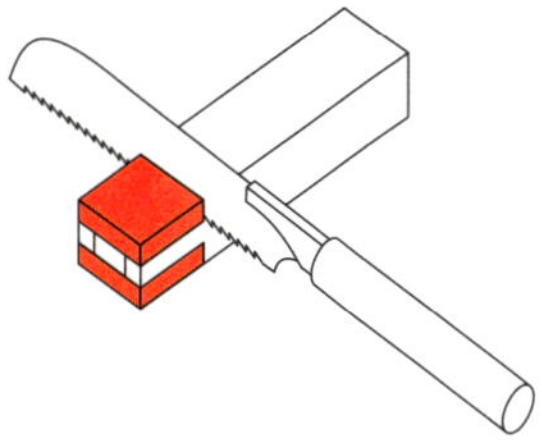

A3

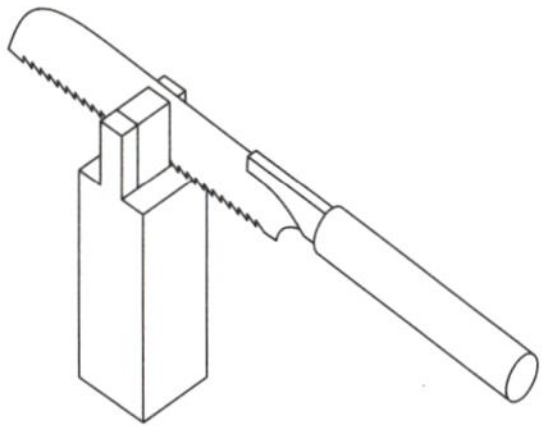

A4

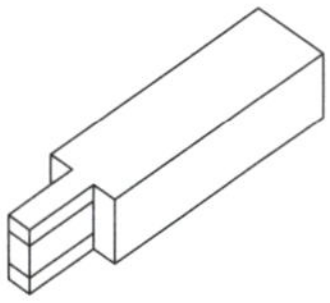

A

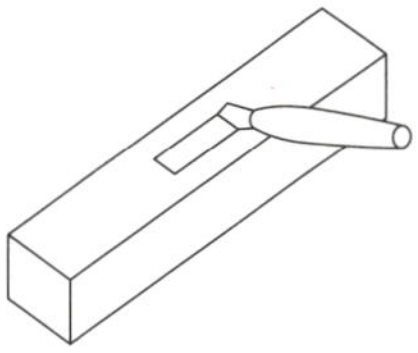

B1

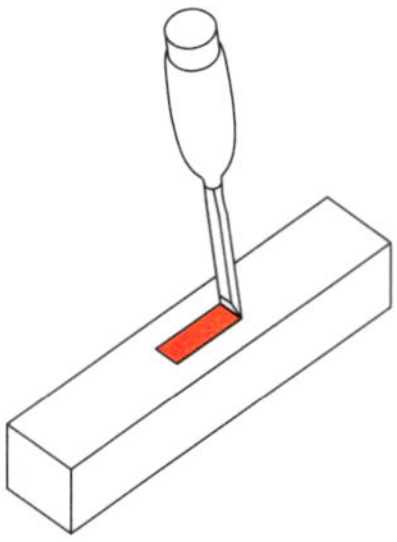

B2

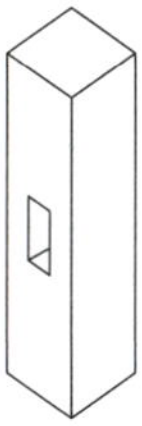

B

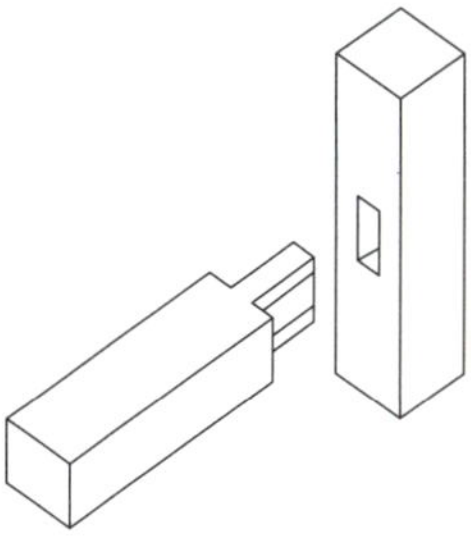

A+B

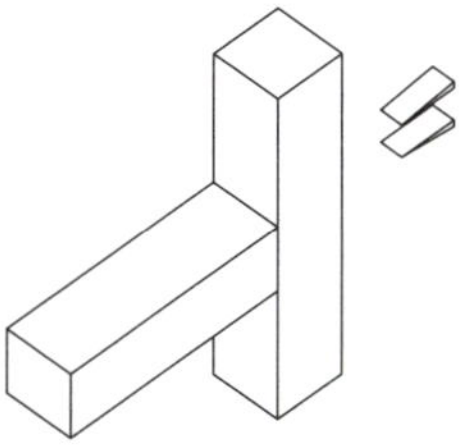

A+B

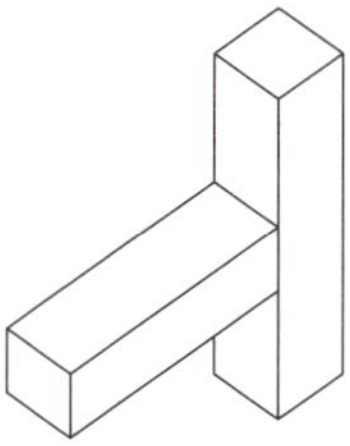

AB

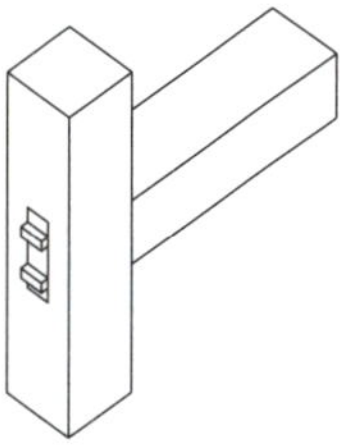

AB

T-Joints

Vertically wedged tusk tenon

tōshi-hozo-sashi
(tate-hana-sen-shimari)
通枘差し(竪鼻栓締)

durchgesteckter Zapfen mit Keilschloss

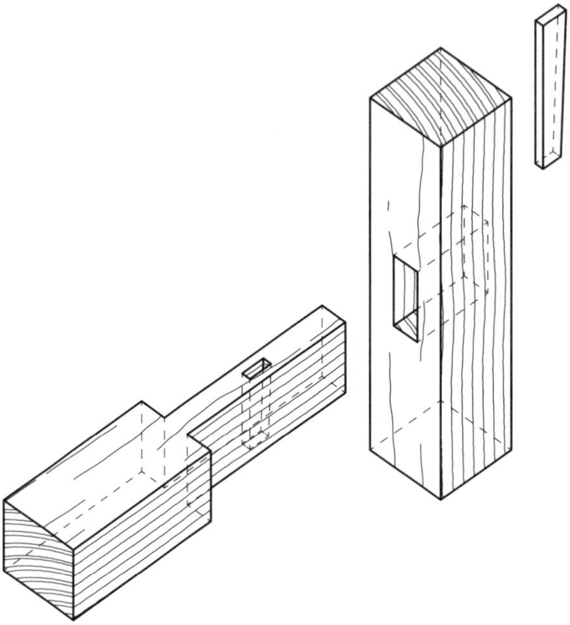

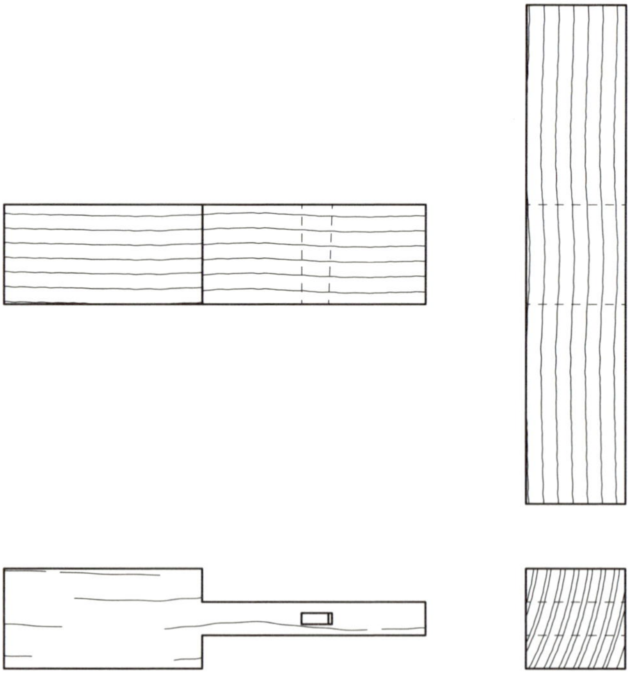

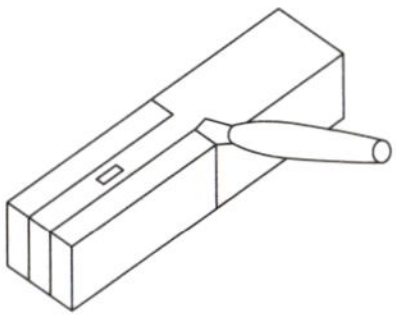

A1

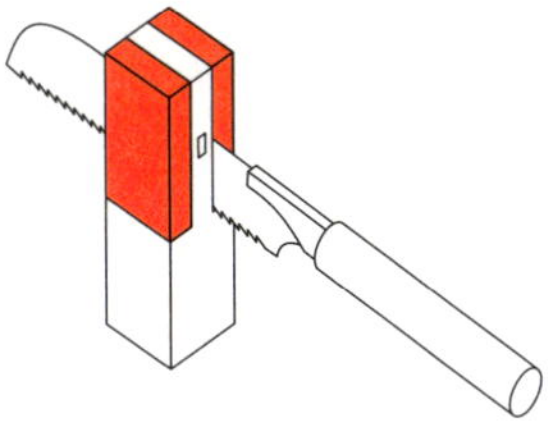

A2

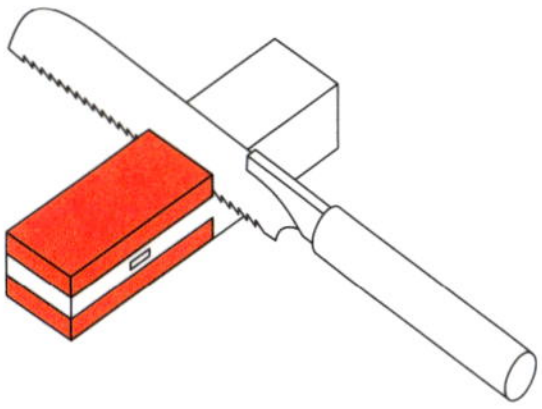

A3

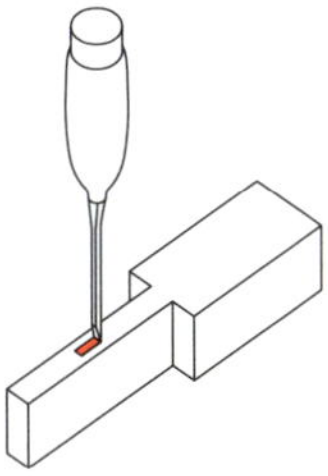

A4

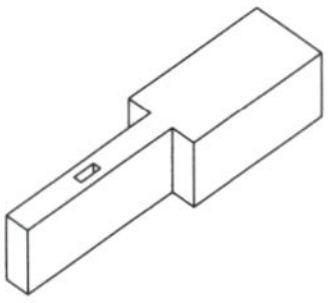

A

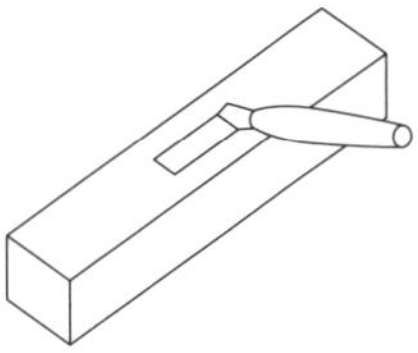

B1

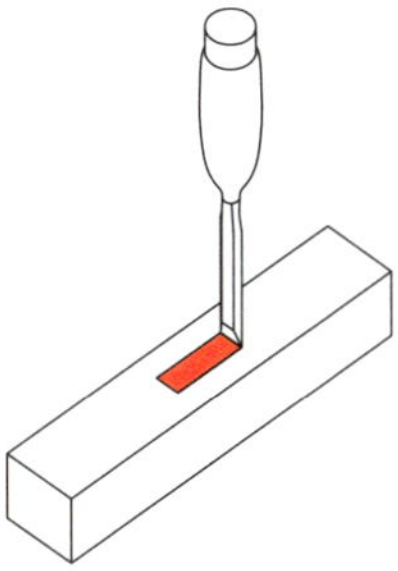

B2

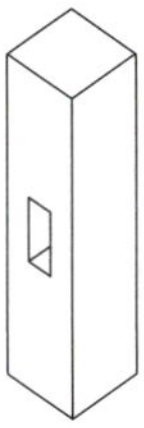

B

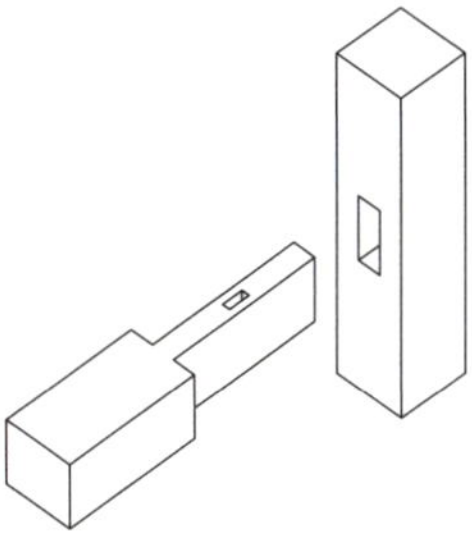

A+B

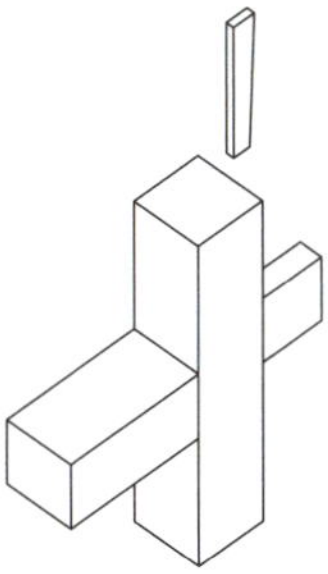

A+B

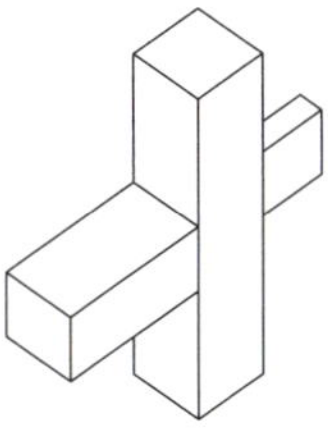

AB

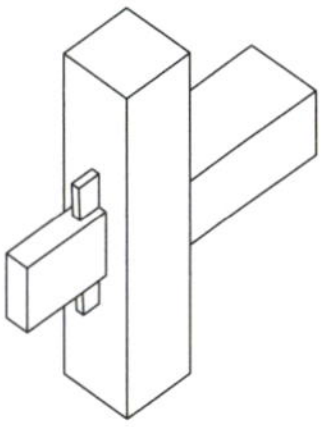

AB

T-Joints

Horizontally wedged tusk tenon

tōshi-hozo-nimai (hana-sen-shimari) [*]
通枘差し(二枚鼻栓締)

durchgesteckter Zapfen
mit doppeltem Keilschloss

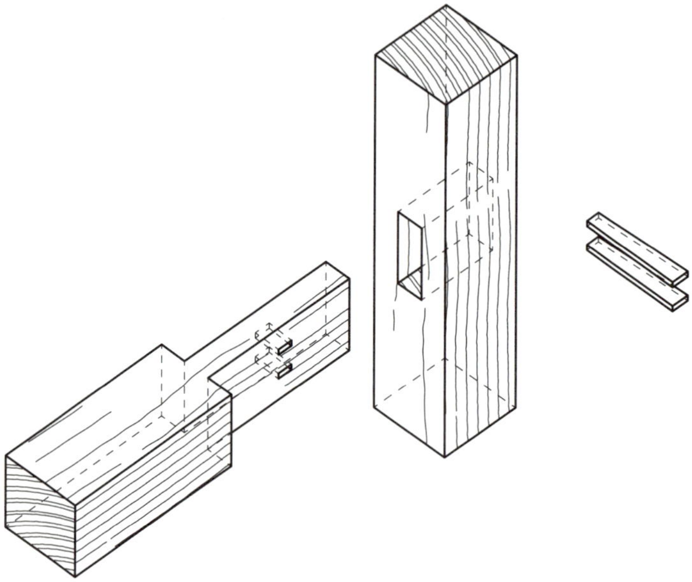

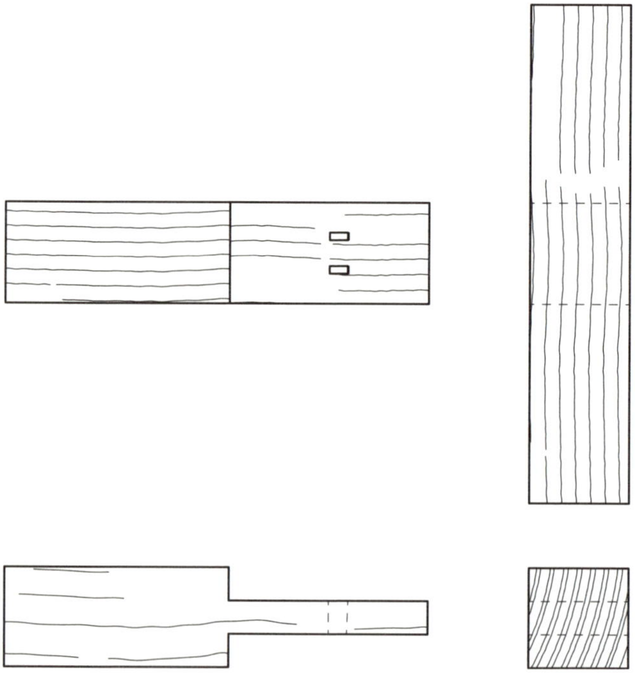

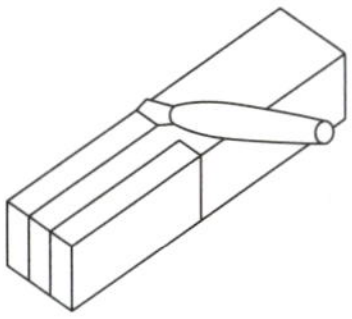

A1

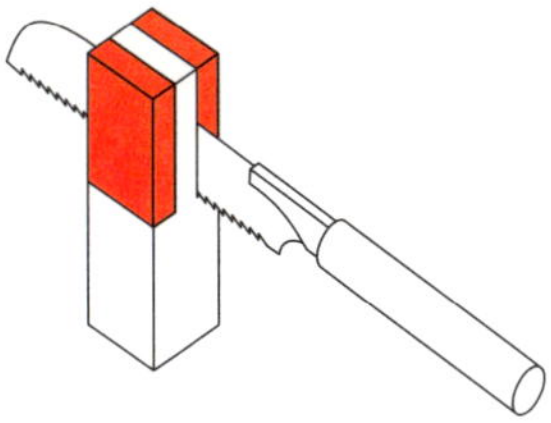

A2

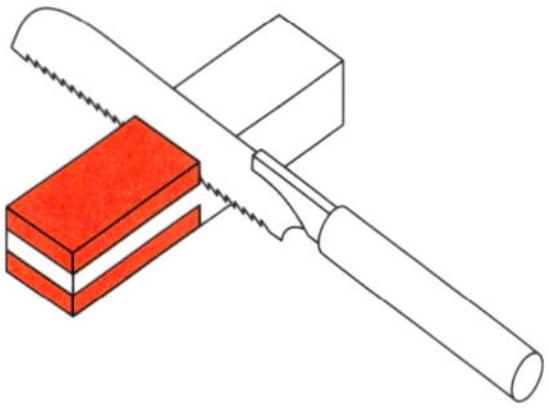

A3

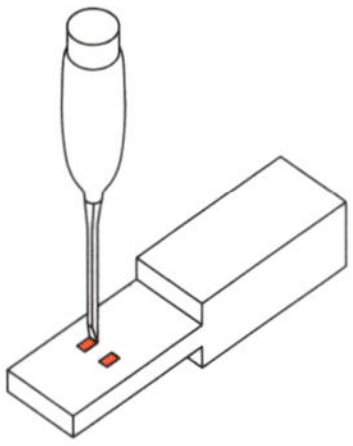

A4

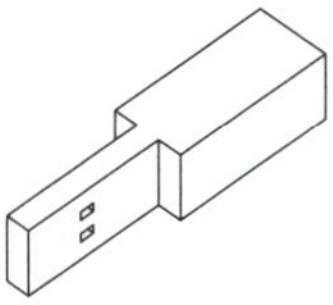

A

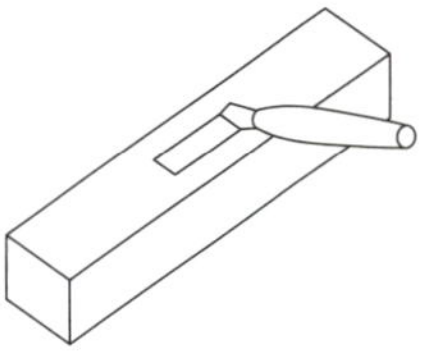

B1

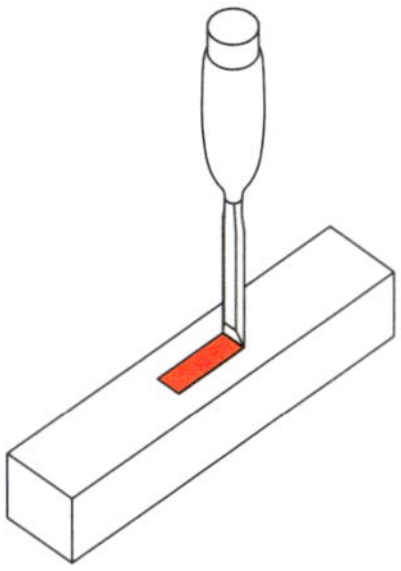

B2

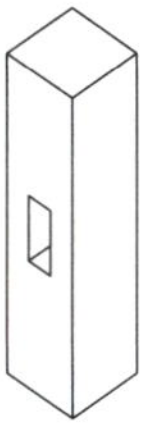

B

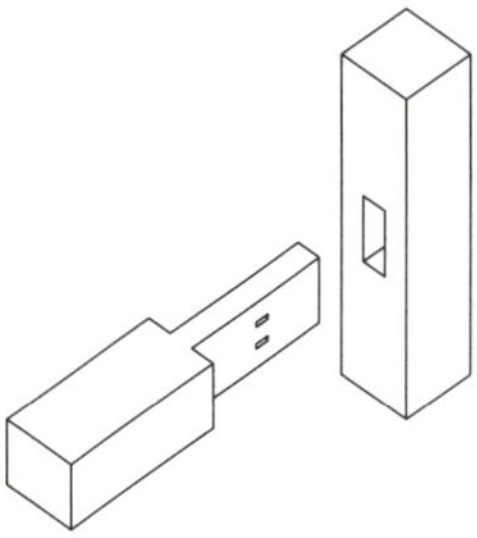

A+B

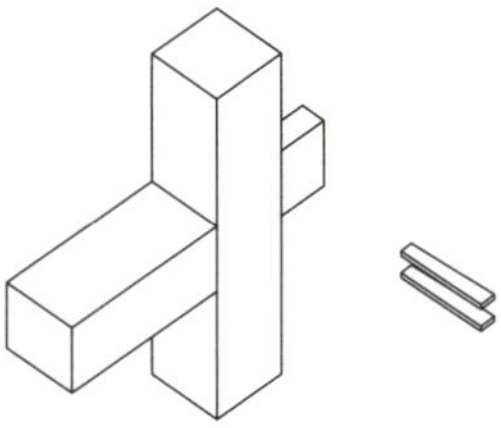

A+B

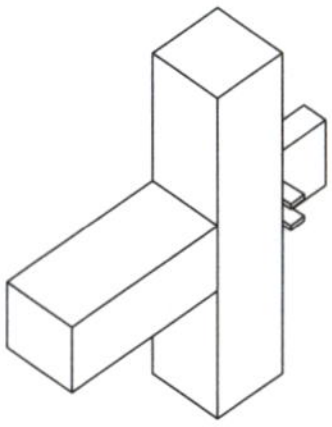

AB

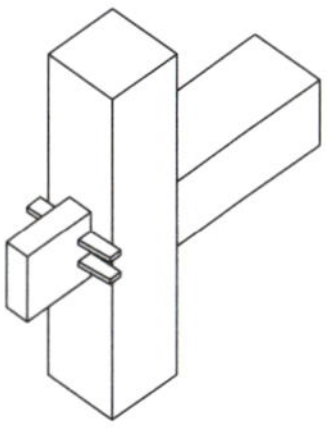

AB

T-Joints

Horizontally wedged double tusk tenon

nimai-hozo-sashi (hana-sen-shimari) [*]
二枚枘差し(鼻栓締)

durchgesteckter doppelter Zapfen mit Keilschloss

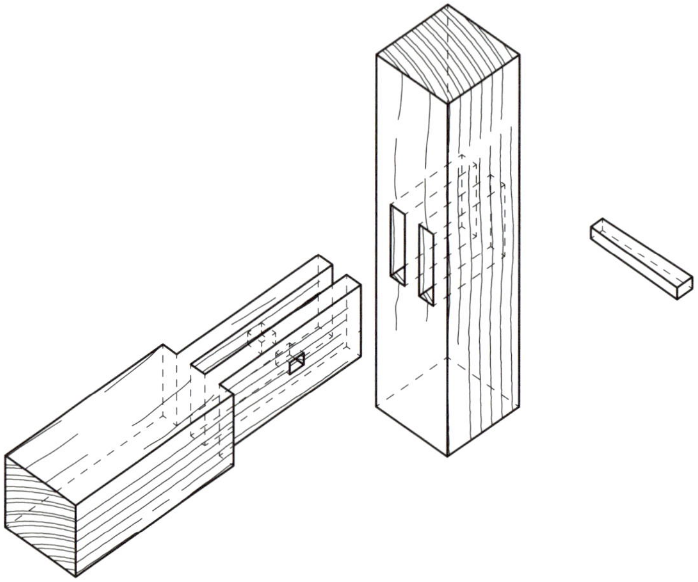

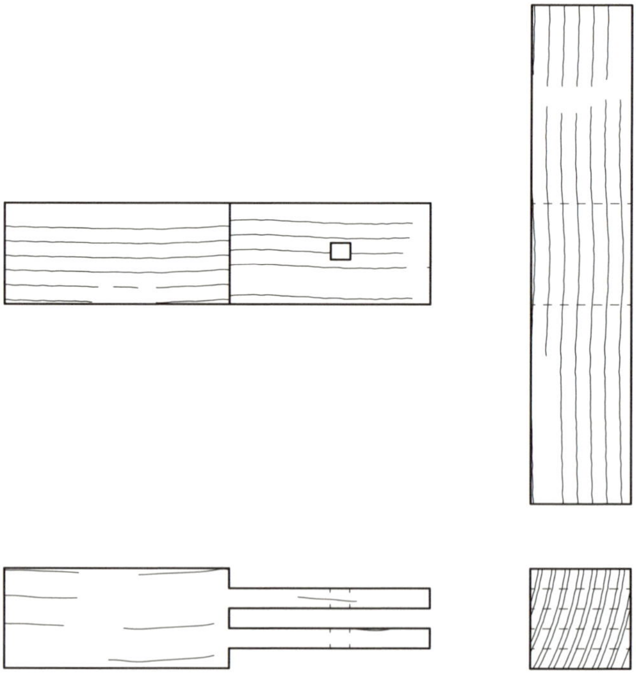

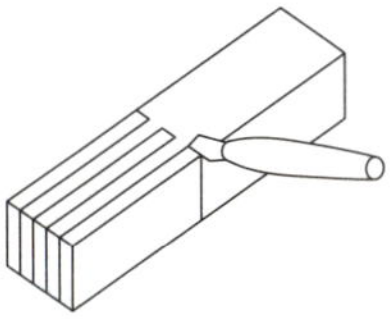

A1

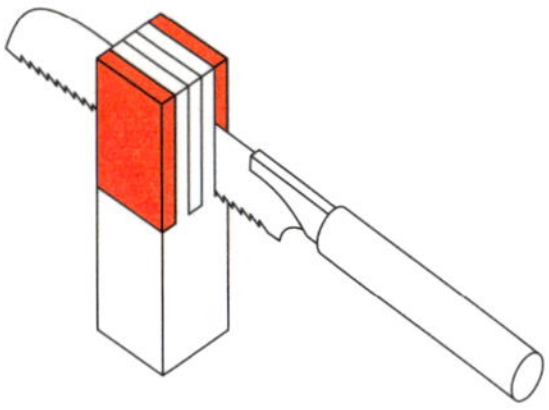

A2

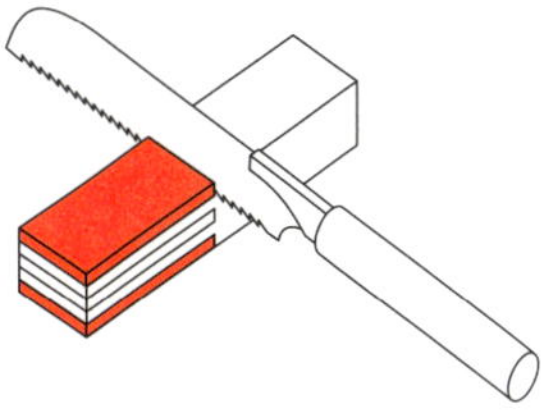

A3

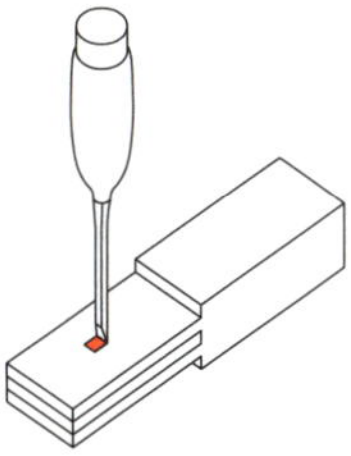

A4

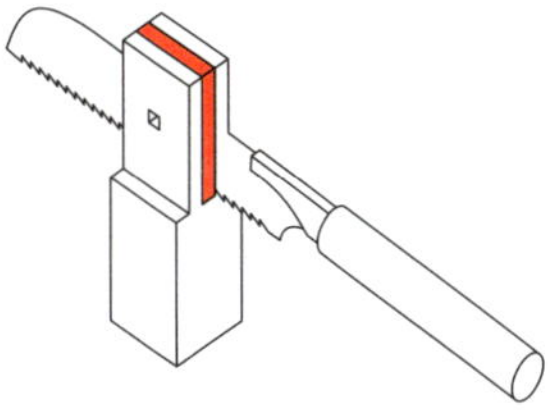

A5

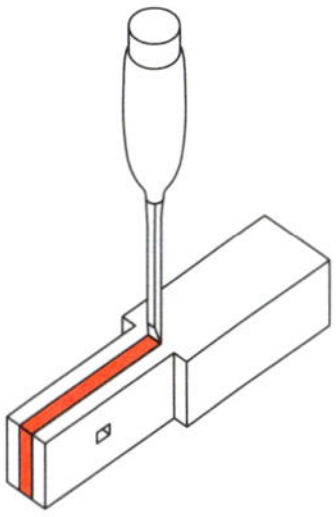

A6

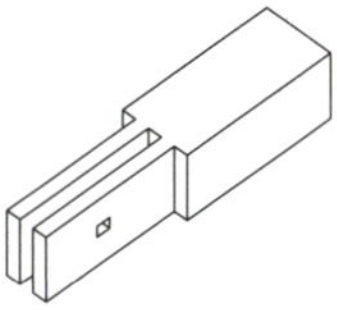

A

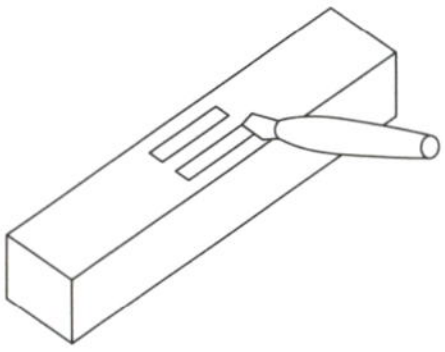

B1

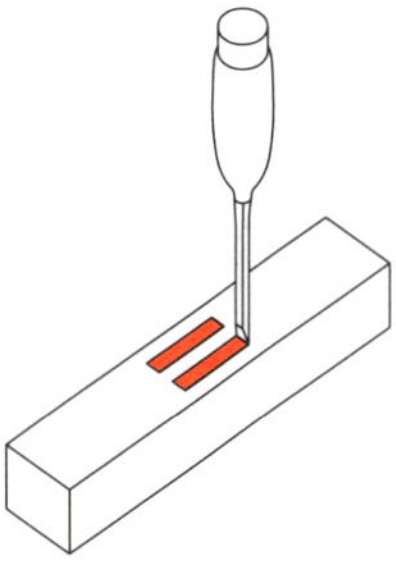

B2

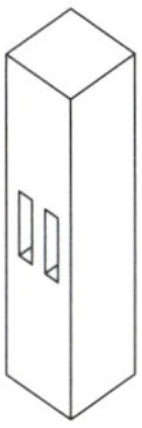

B

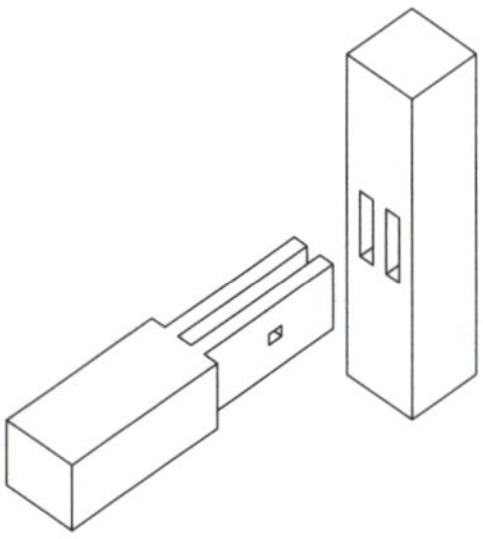

A+B

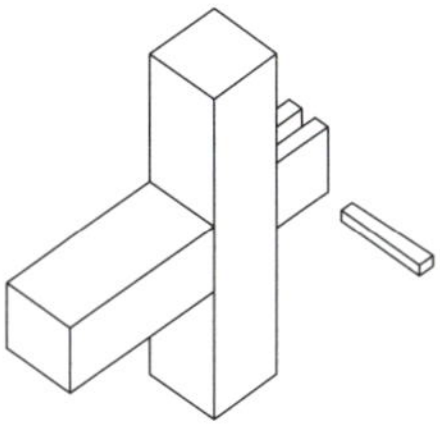

A+B

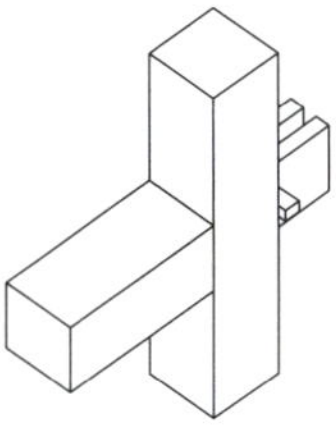

AB

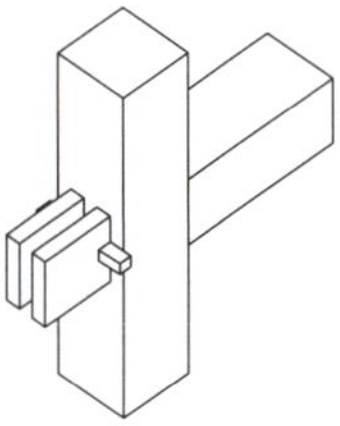

AB

T-Joints

Right-angle joint with dovetailed spline tenon and tapered key

yose-ari-yatoi-sao-shachi-sen-shimari (oire) [*]
寄蟻雇竿車知栓締り（追入）

Stoß mit Führungszapfen und eingesetztem Schwalbenschwanzzapfen mit Spannkeil

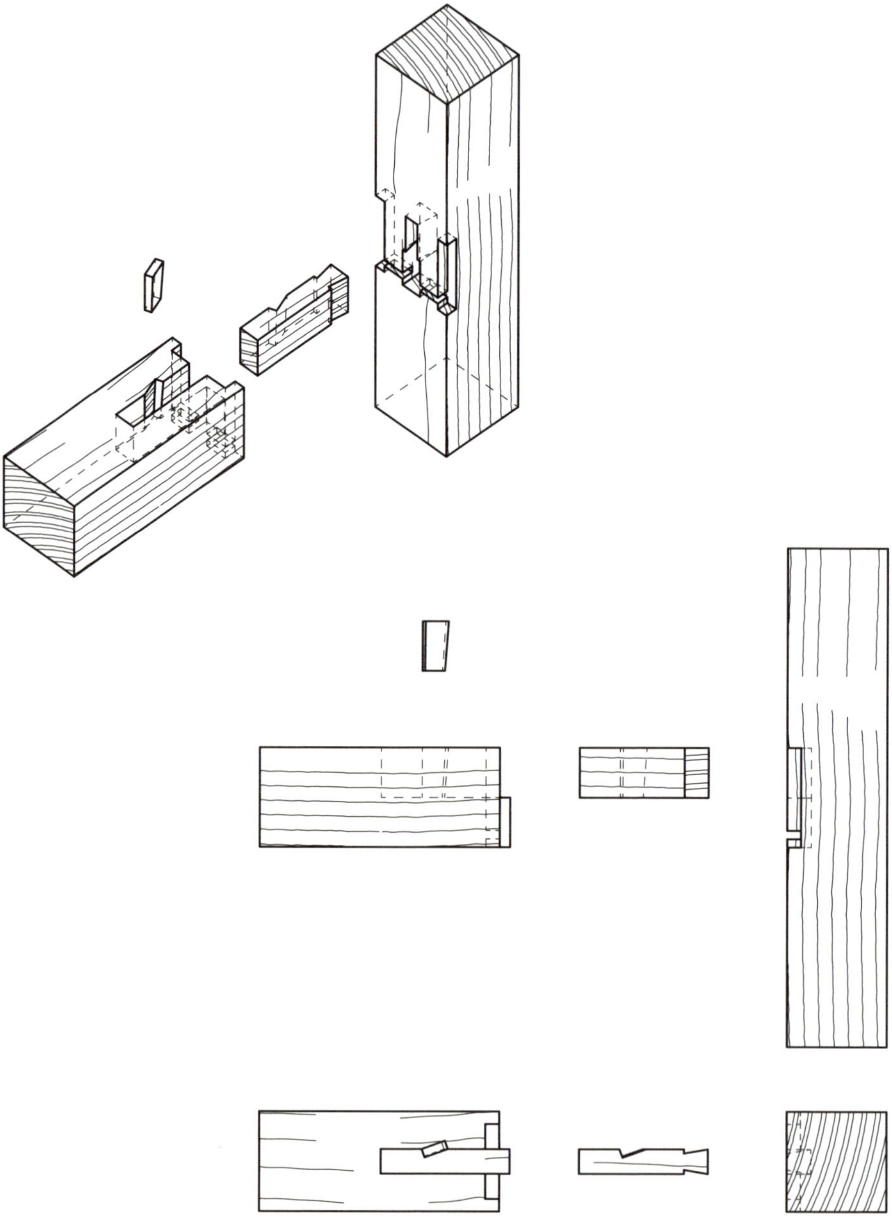

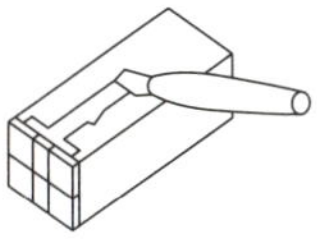

A1

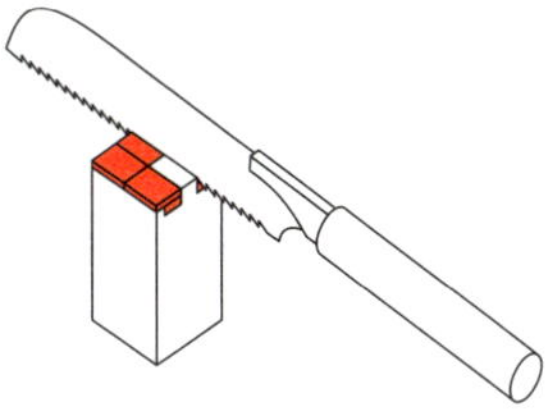

A2

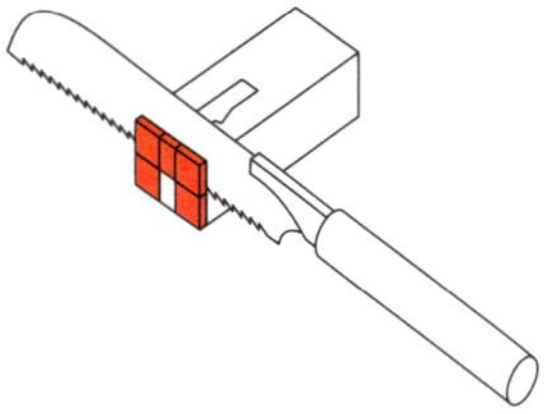

A3

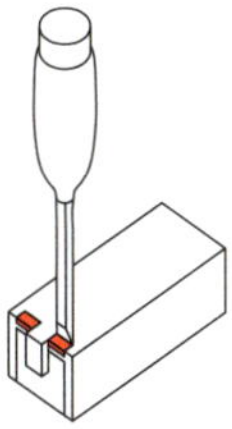

A4

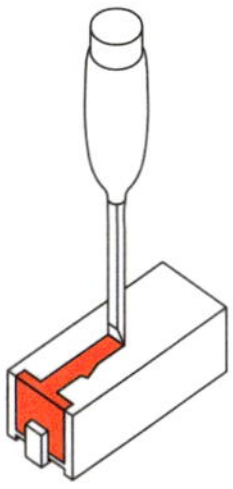

A5

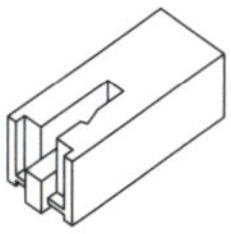

A

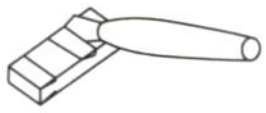

B1

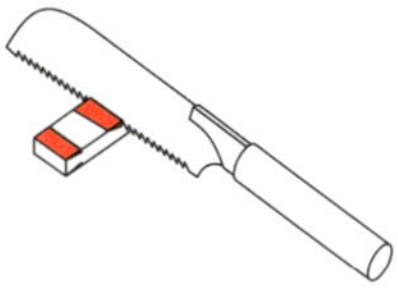

B2

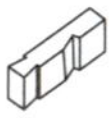

B

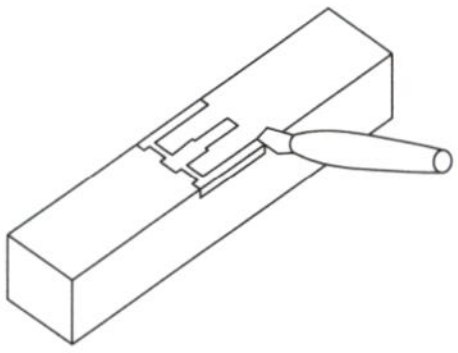

C1

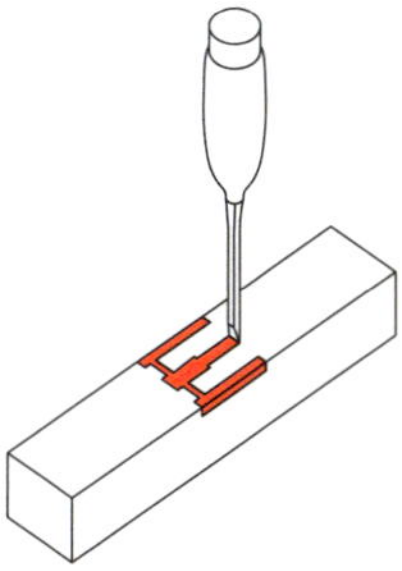

C2

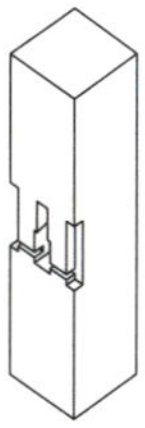

C

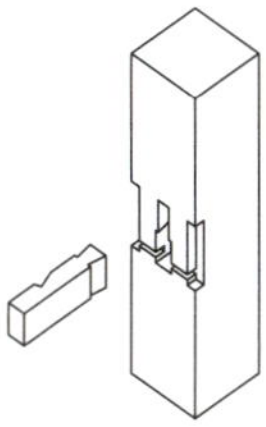

A+B

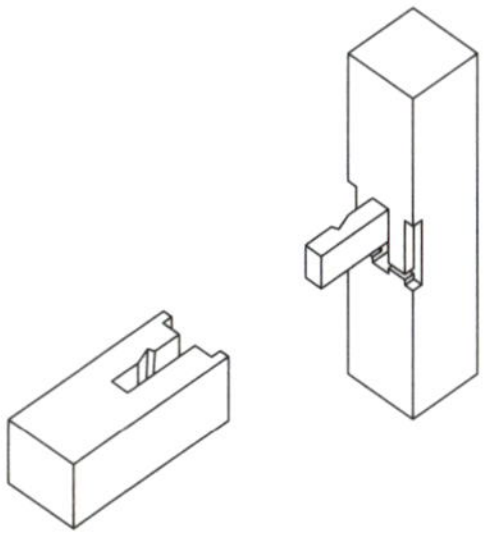

A+B+C

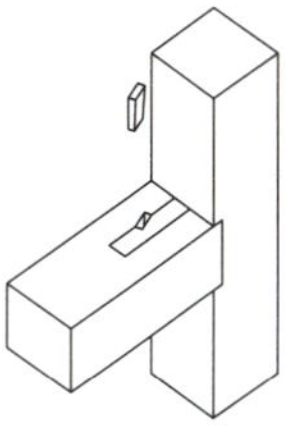

ABC

T-Joints

Housed right-angle joint with spline tenon, tapered keys, and peg

komi-sen-yatoi-sao-shachi-sen-shimari (oire) [*]
込栓雇竿車知栓締り(追入)

eingelassener stumpfer Stoß
mit eingesetztem Zapfen
und Dübelsicherung sowie Spannkeilen

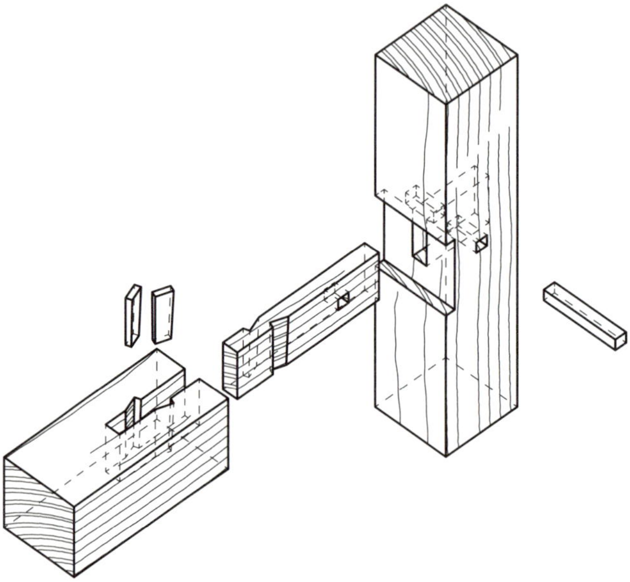

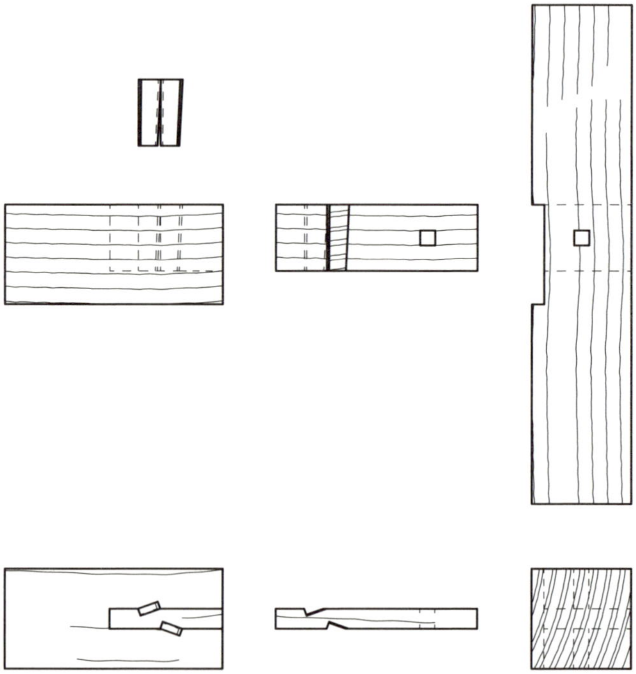

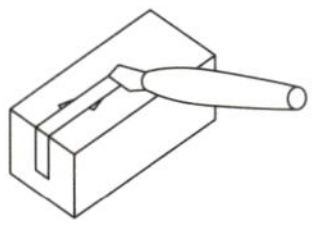

A1

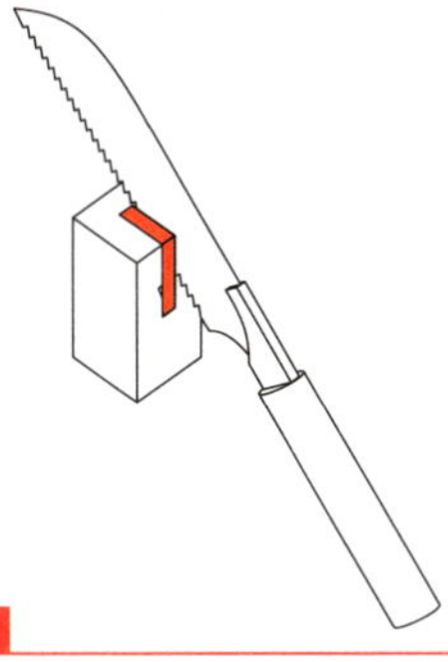

A2

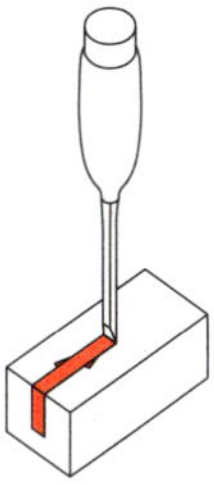

A3

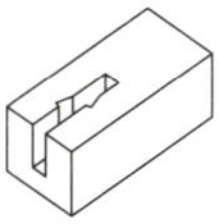

A

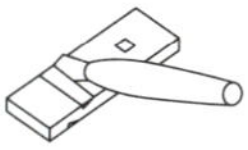

B1

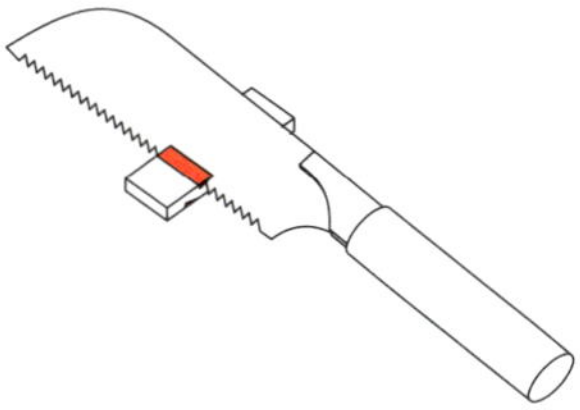

B2

B3

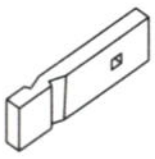

B

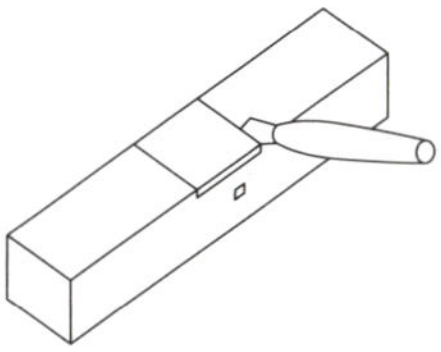

C1

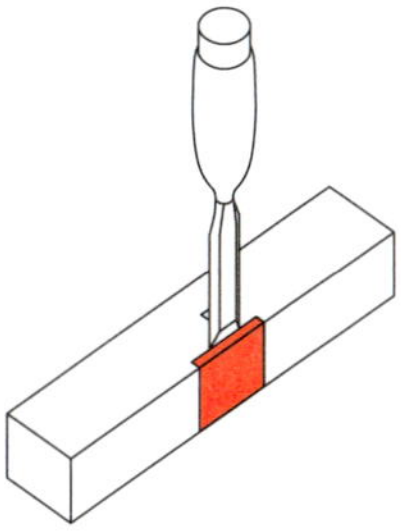

C2

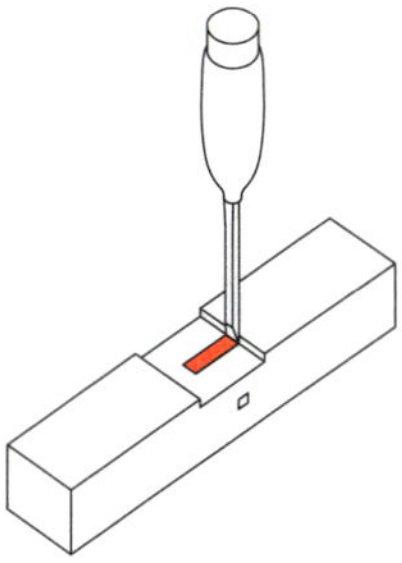

C3

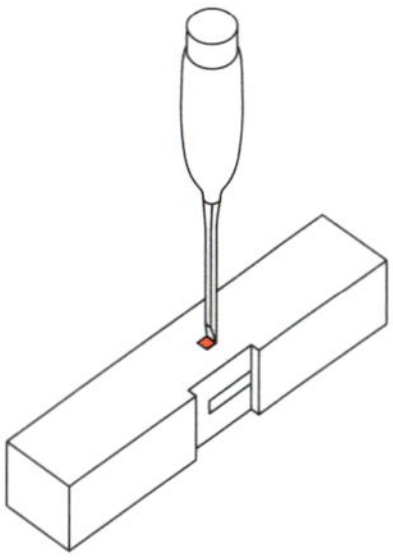

C4

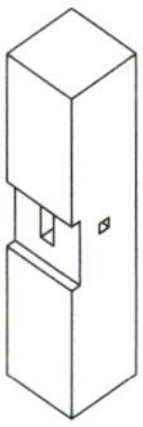

C

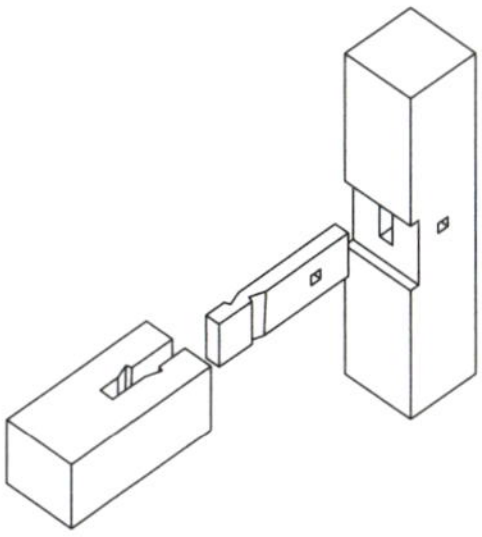

A+B+C

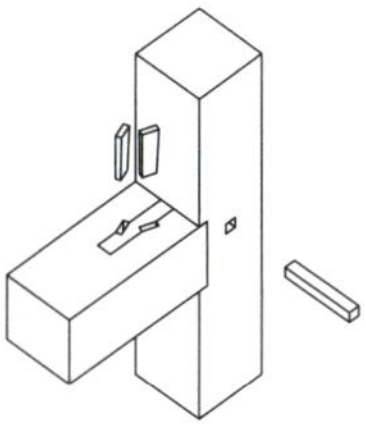

ABC

T-Joints

Double through mortise and tenon

nimai-naga-hozo-sashi
二枚長枘差し

durchgehender doppelter Zapfen

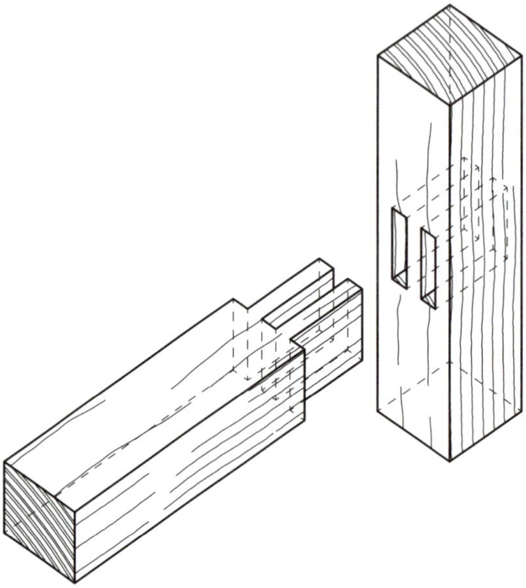

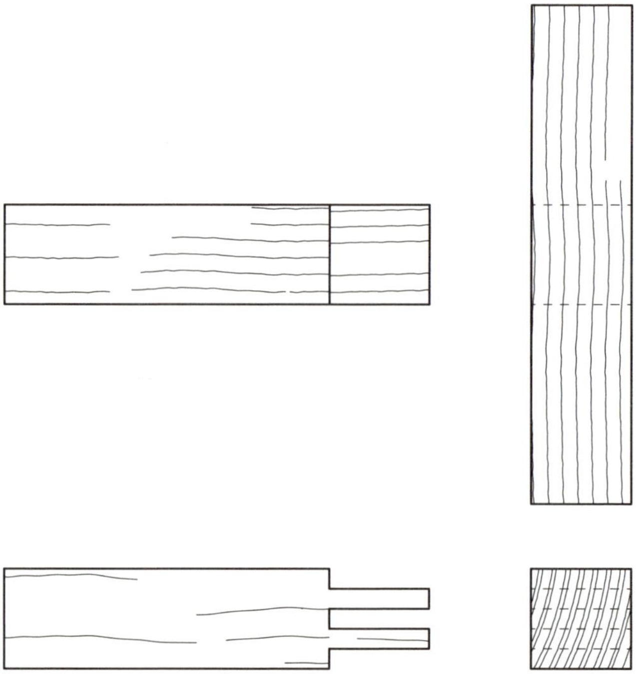

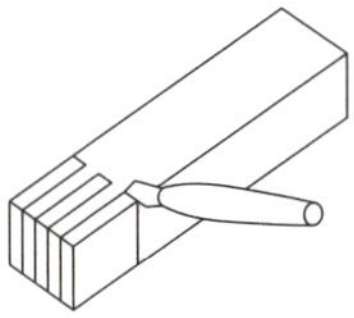
A1

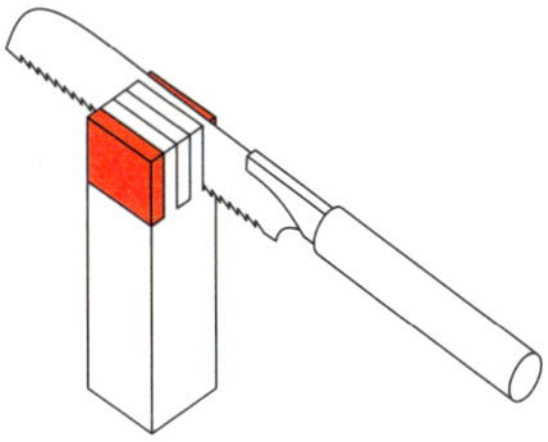
A2

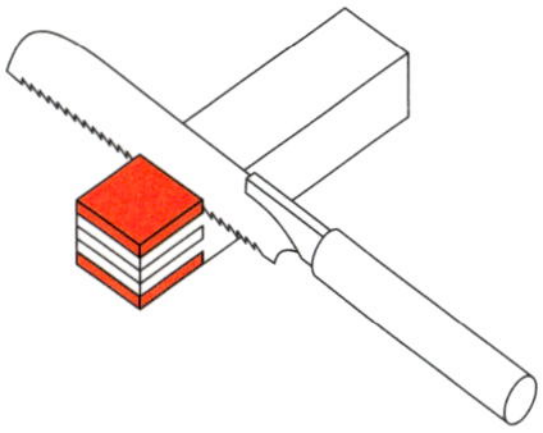
A3

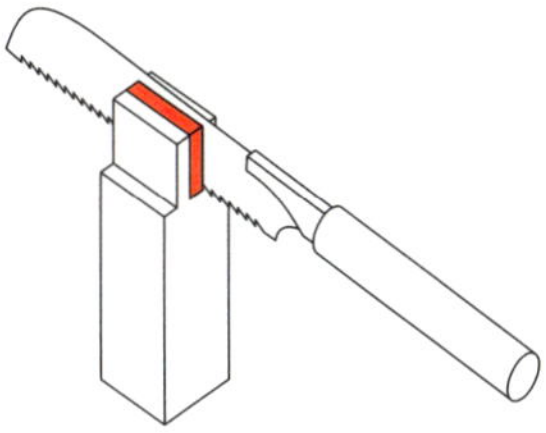
A4

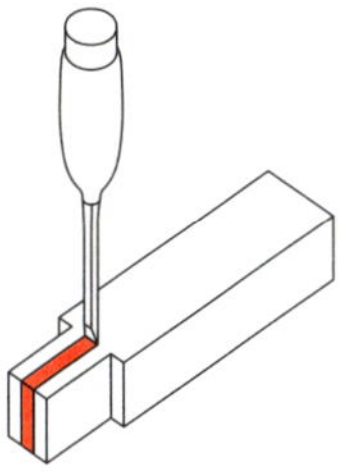
A5

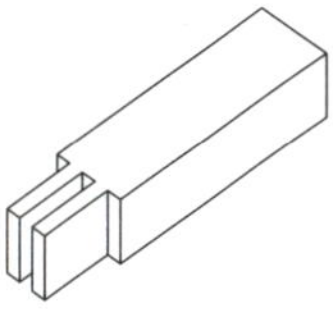
A

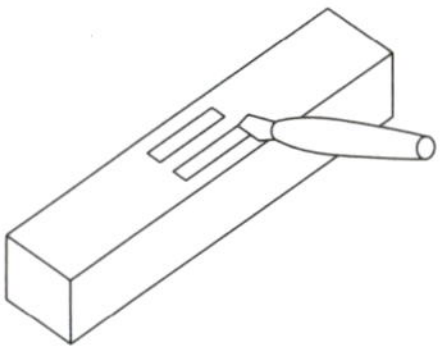
B1

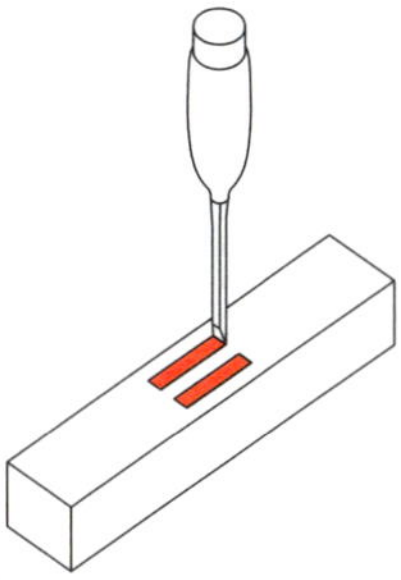
B2

B

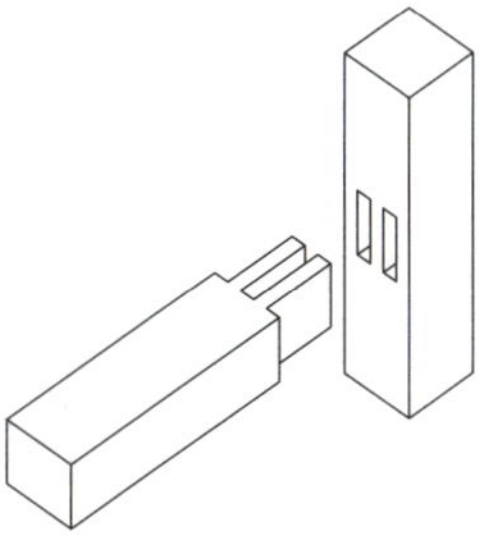
A+B

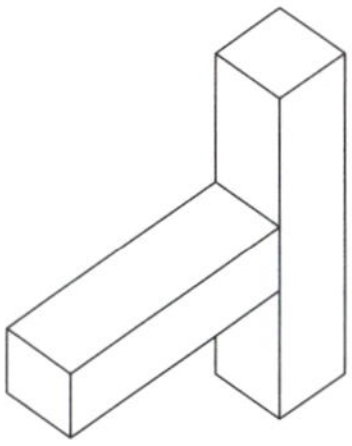
AB

T-Joints

Housed double through mortise and tenon

nimai-hozo-sashi (oire)
二枚枘差し(追入)

eingelassener durchgehender
doppelter Zapfen

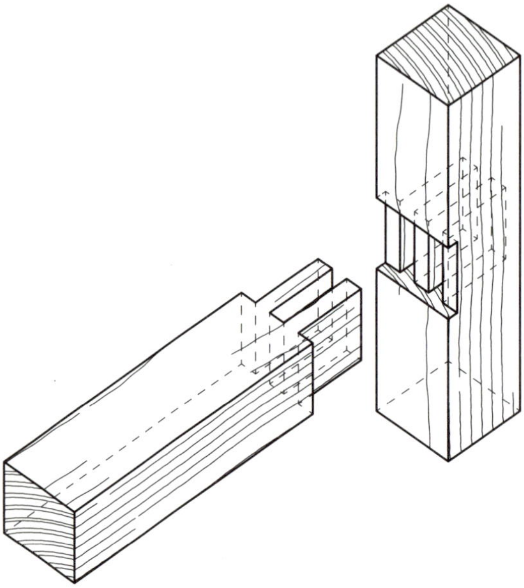

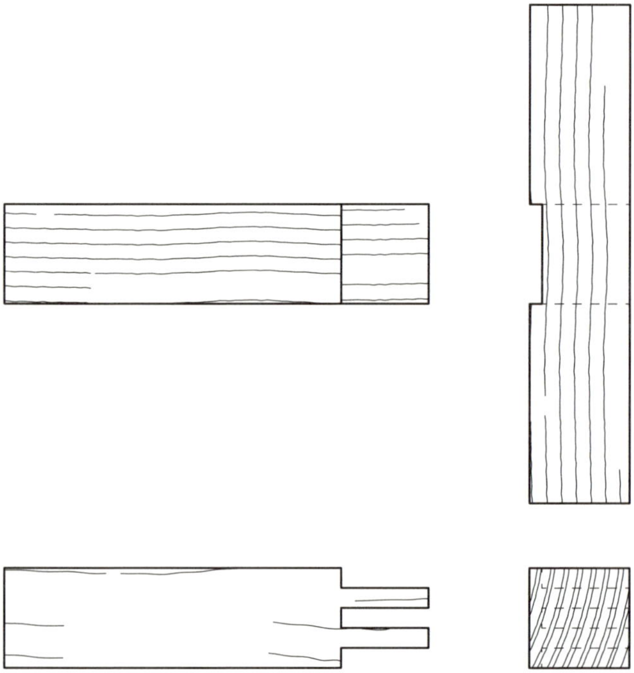

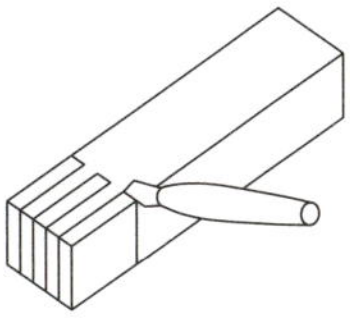

A1

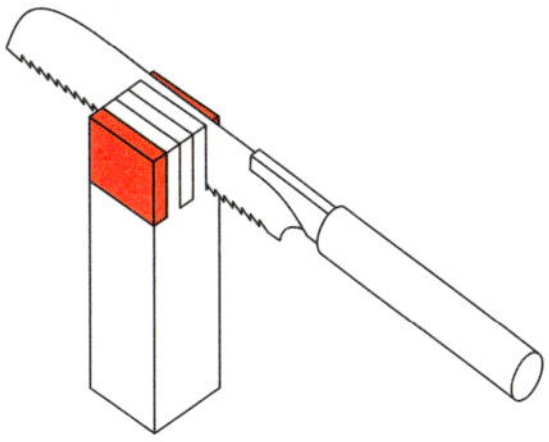

A2

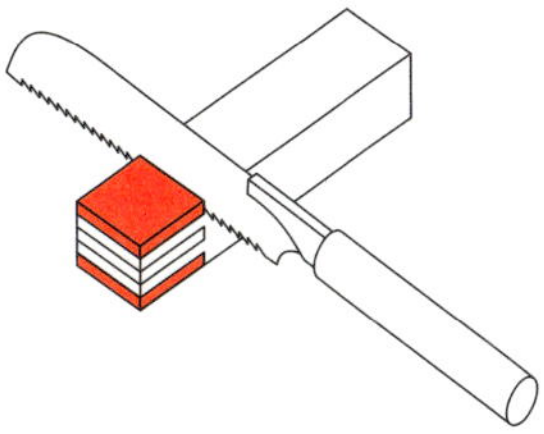

A3

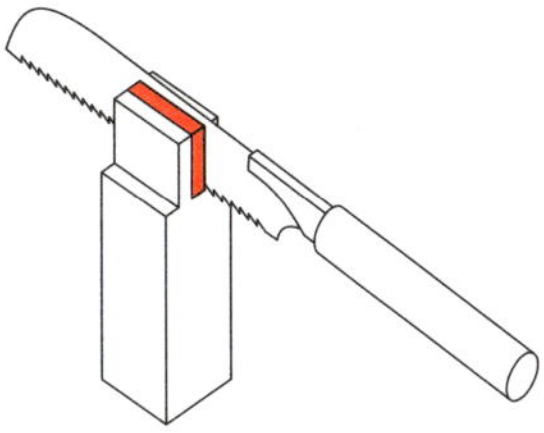

A4

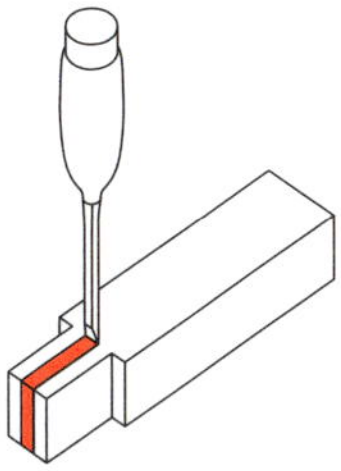

A5

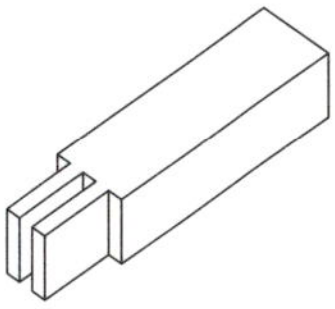

A

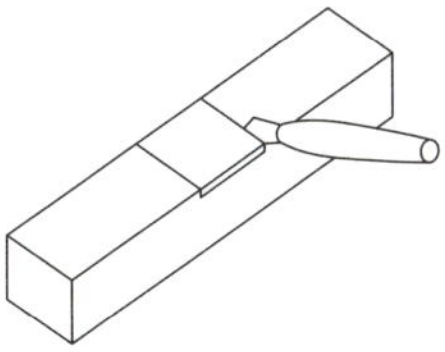

B1

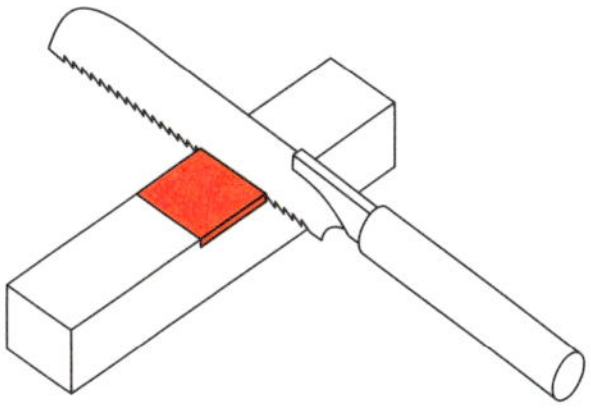

B2

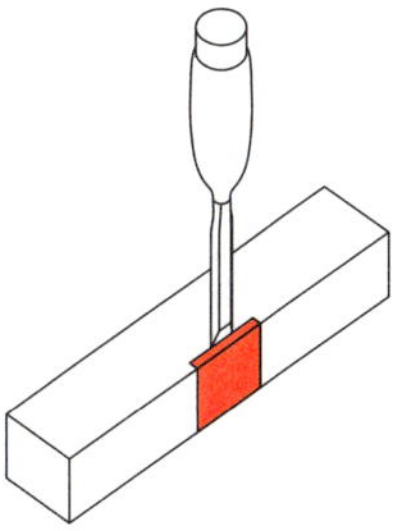

B3

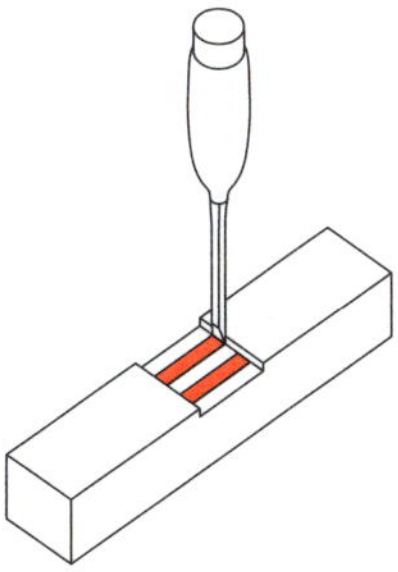

B4

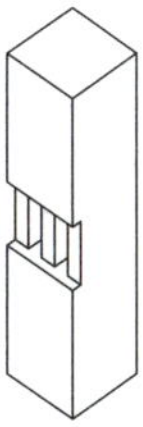

B

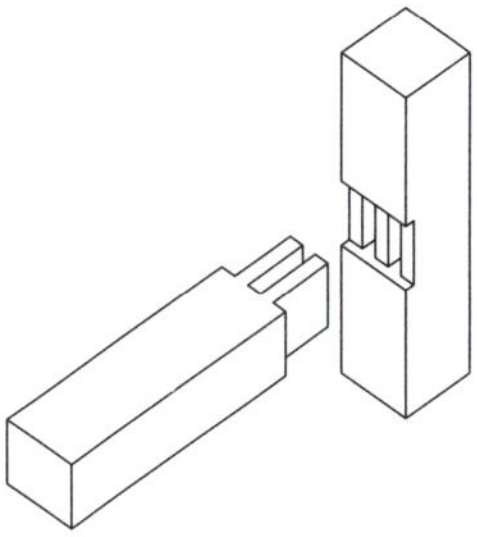

A+B

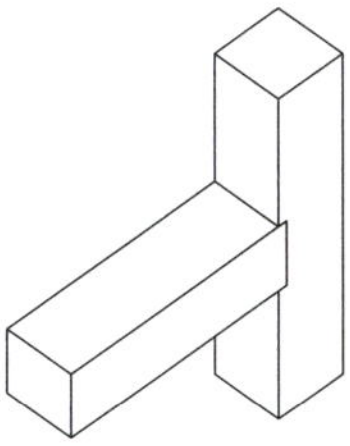

AB

T-Joints

Double mortise and tenon with mitered shoulders

ken-dome-oire-nimai-hozo-sashi
剣留追入二枚枘差し

durchgehender doppelter Zapfen
mit Führungsstoß

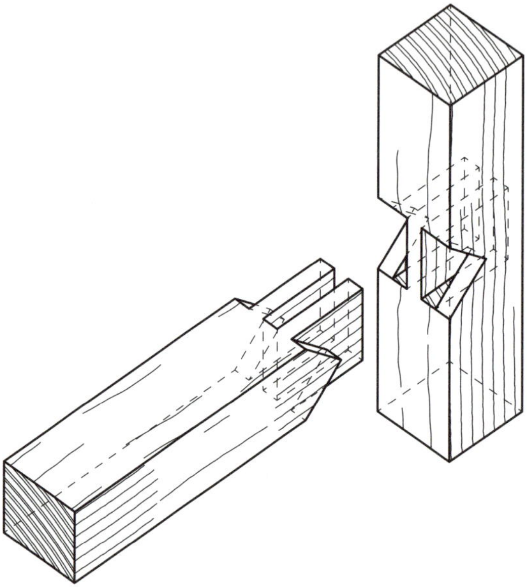

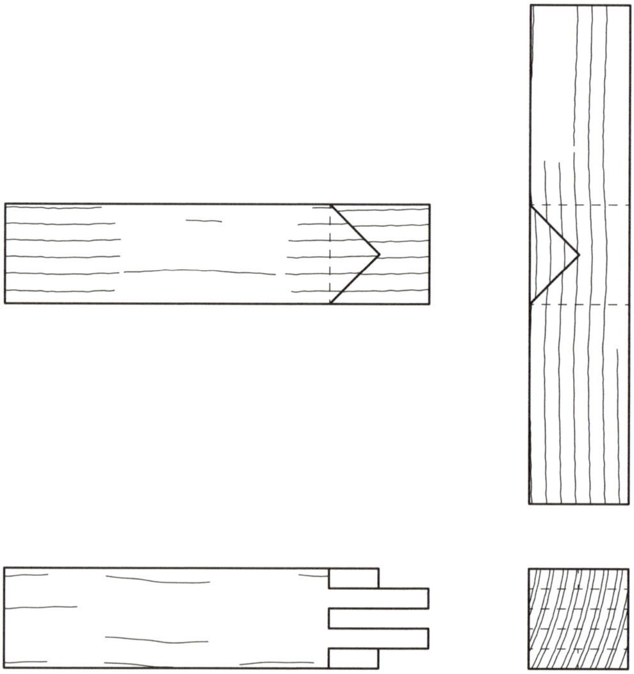

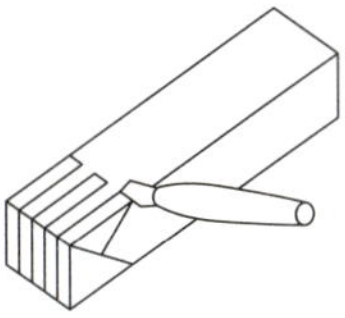

A1

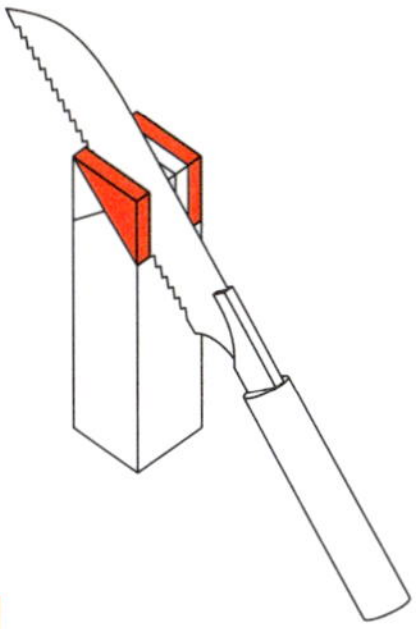

A2

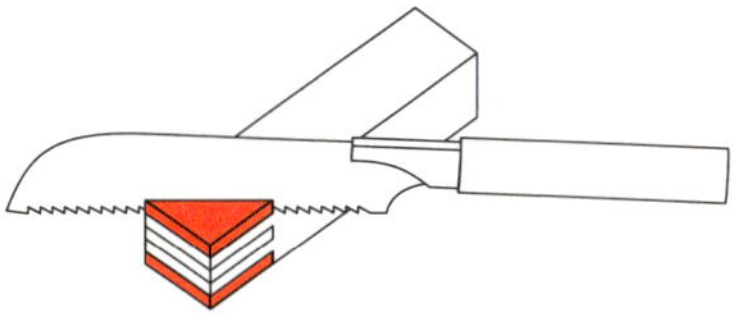

A3

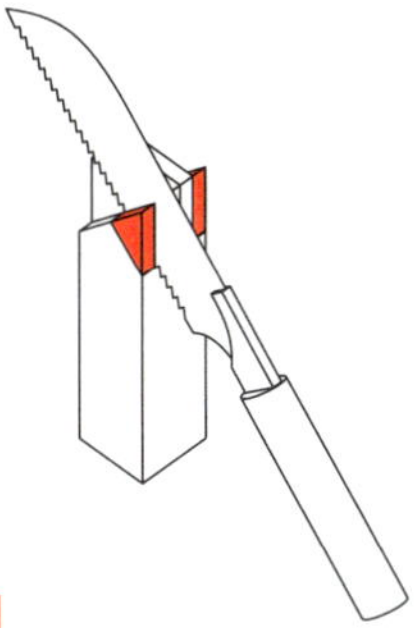

A4

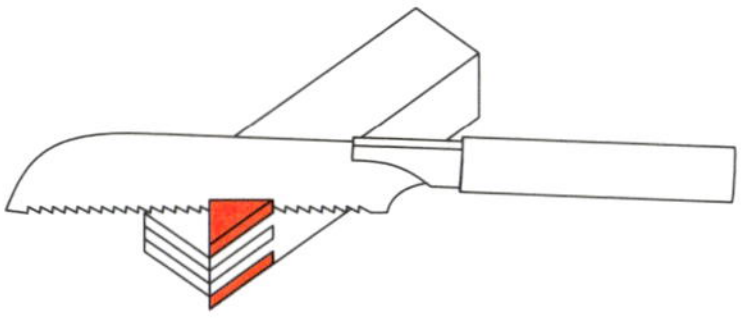

A5

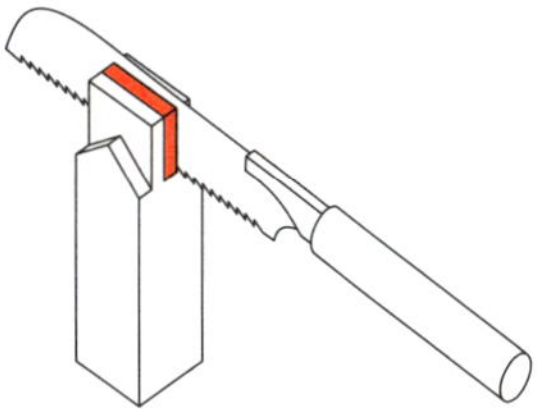

A6

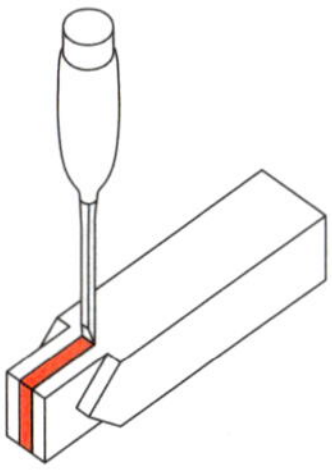

A7

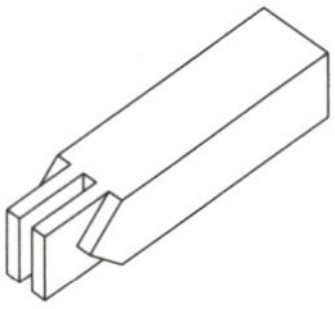

A

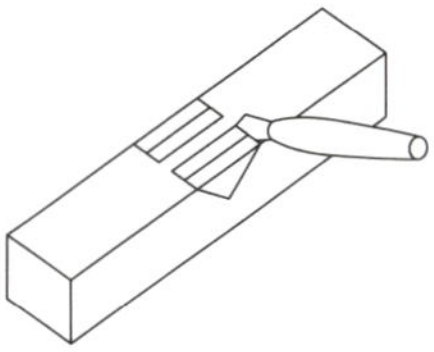

B1

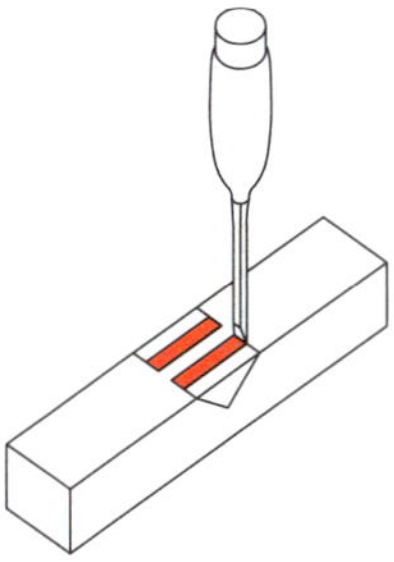

B2

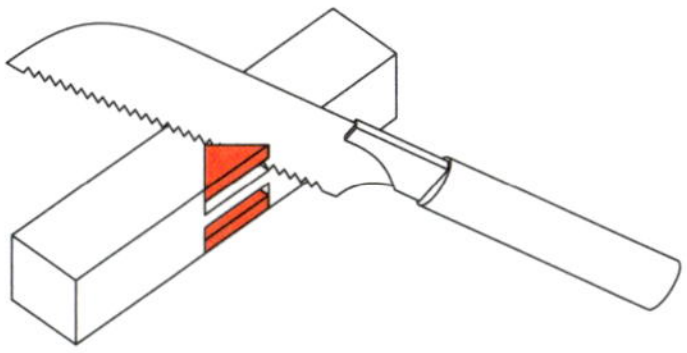

B3

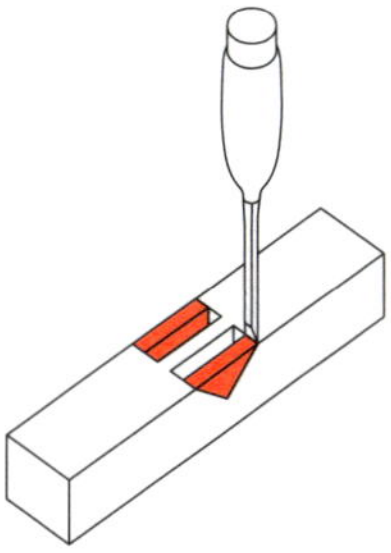

B4

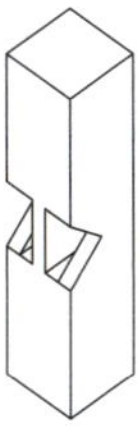

B

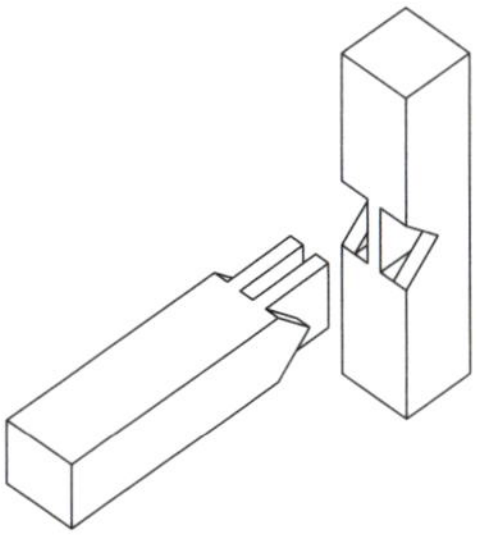

A+B

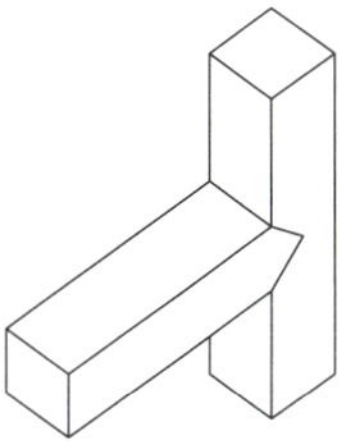

AB

T-Joints

Wedged half-dovetailed T-bridle joint

wanagi-katagi-nimai-hozo-sashi (kusabi-shime)
輪薙傾二枚枘差し(楔締)

verkeilte einseitig trapezförmige Einhälsung

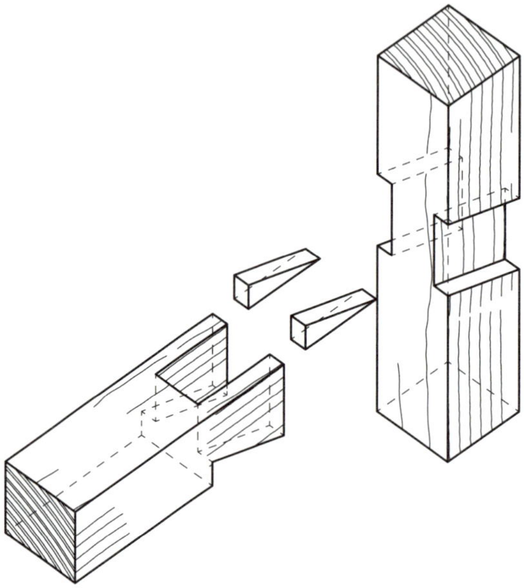

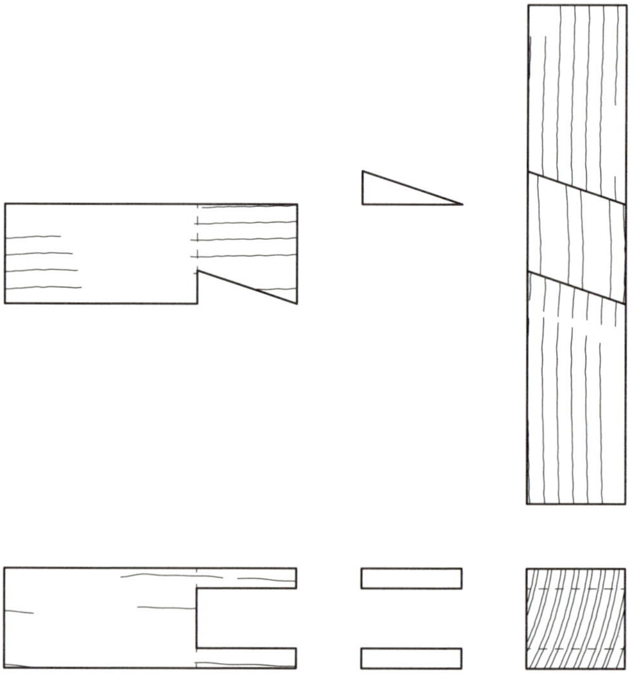

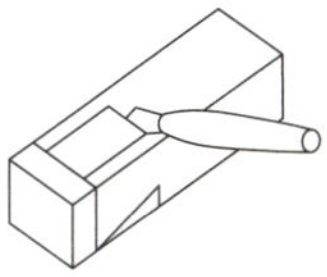

A1

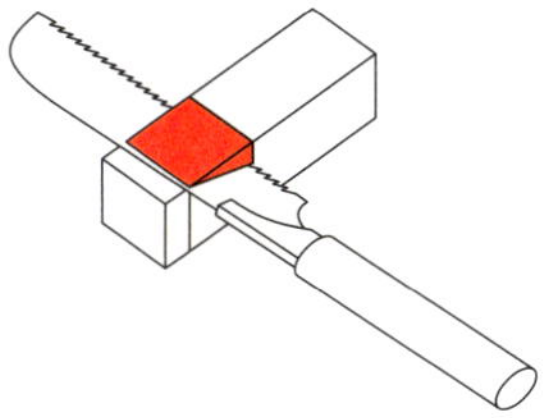

A2

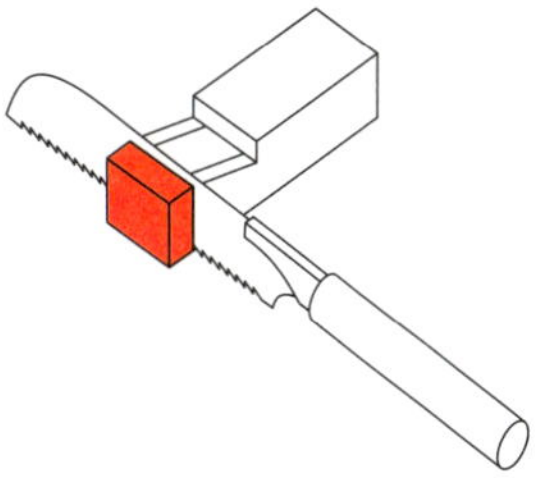

A3

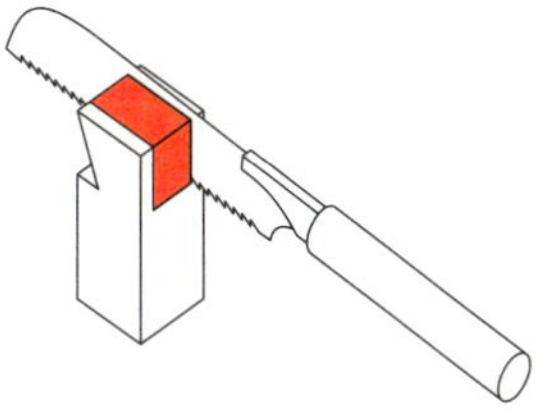

A4

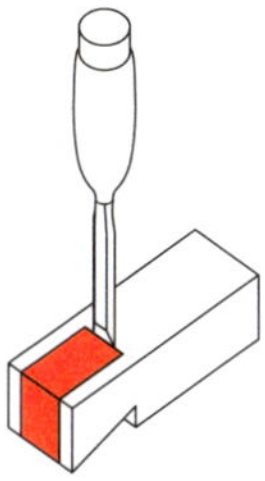

A5

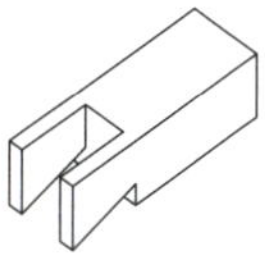

A

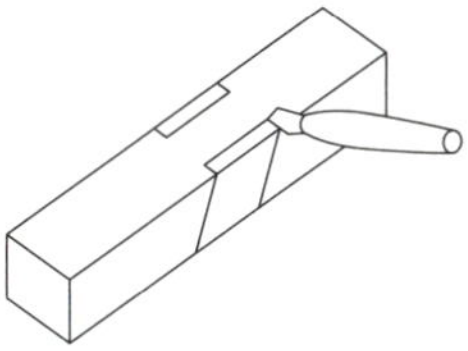

B1

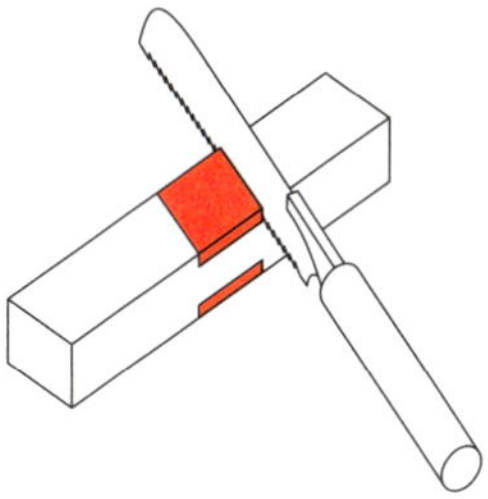

B2

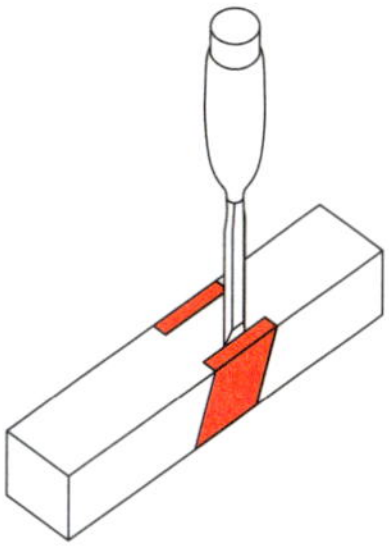

B3

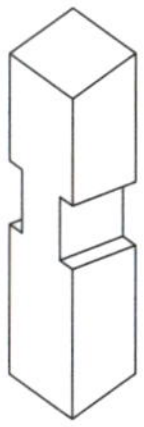

B

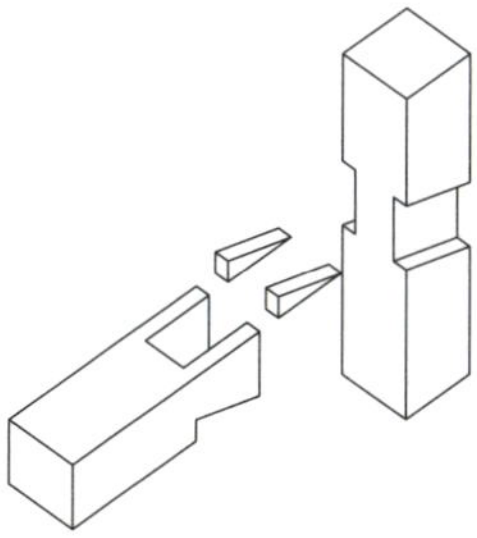

A+B

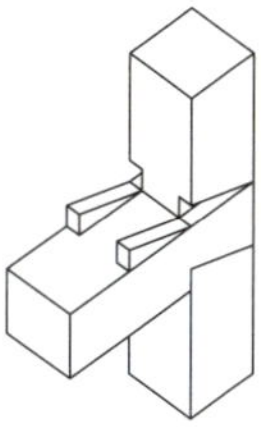

A+B

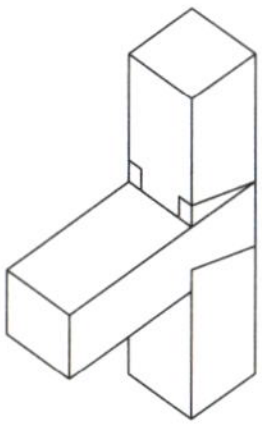

AB

T-Joints

T-bridle joint

wanagi-komi [*]
輪薙込み

Einhälsung

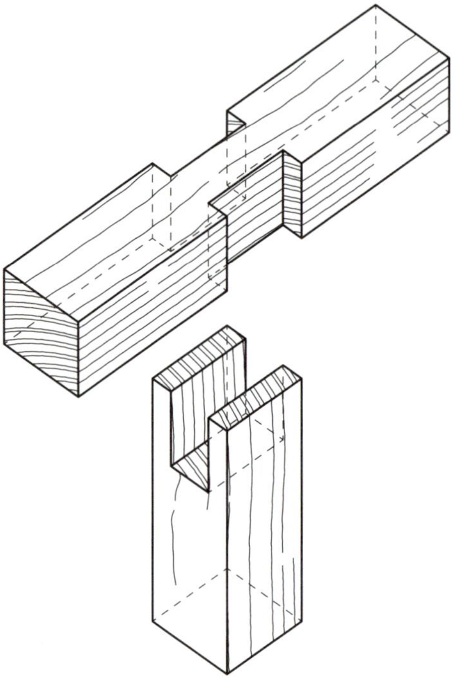

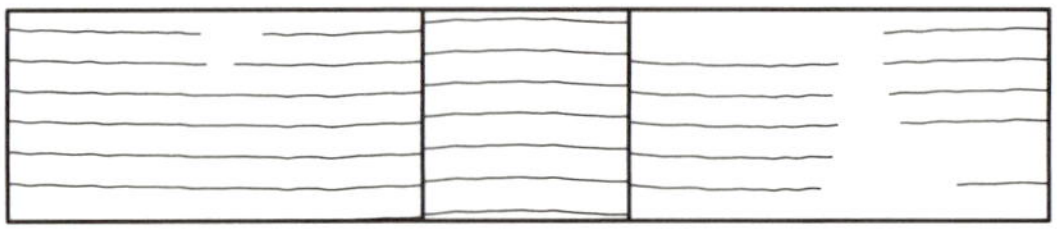

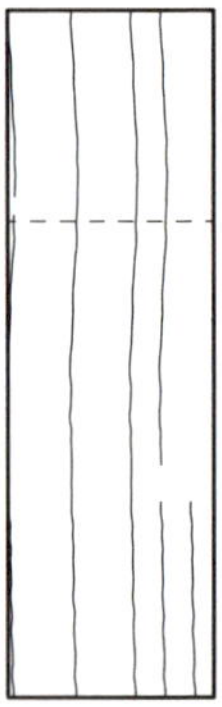

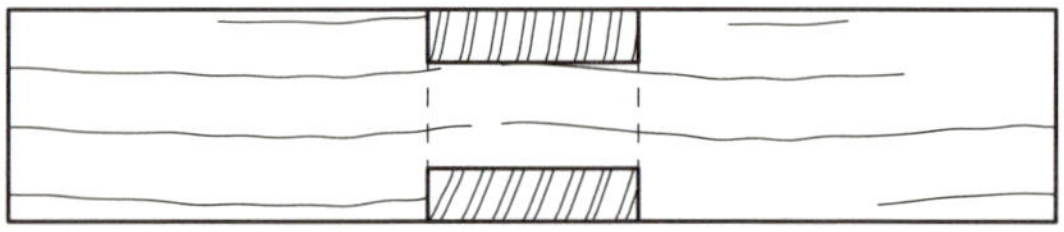

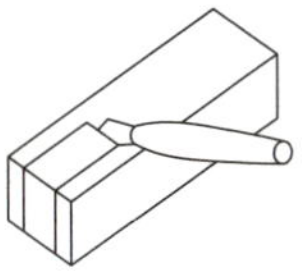

A1

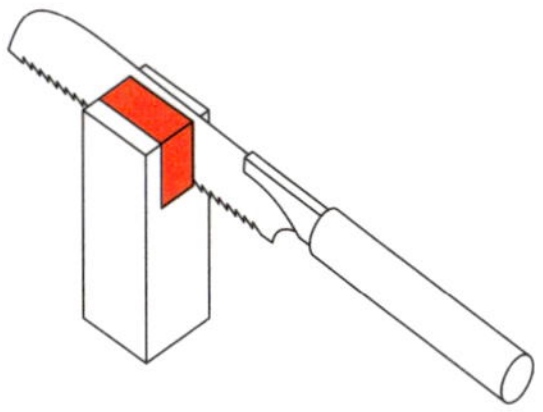

A2

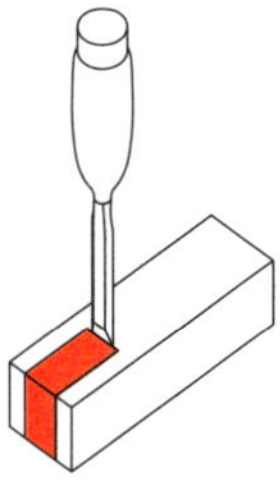

A3

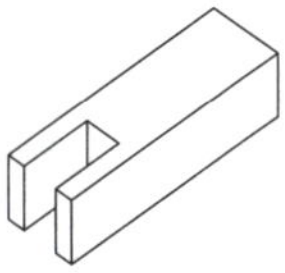

A

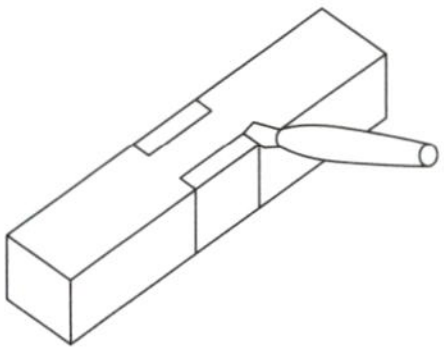

B1

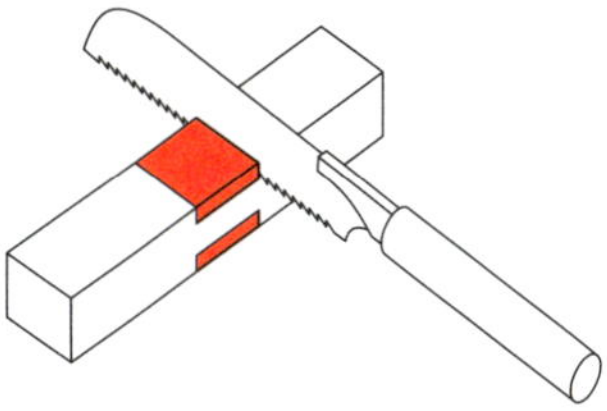

B3

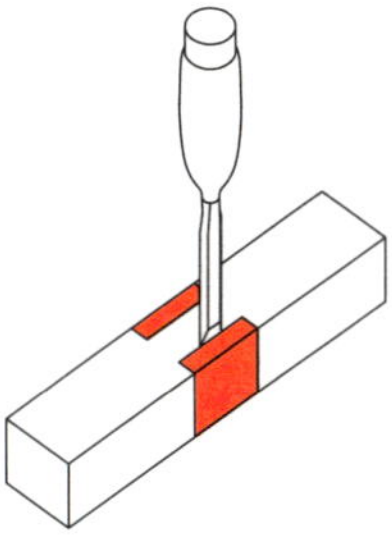

B4

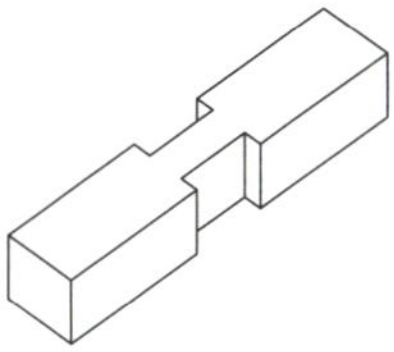

B

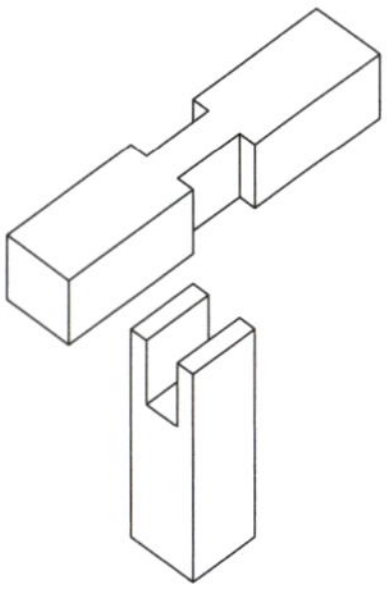

A+B

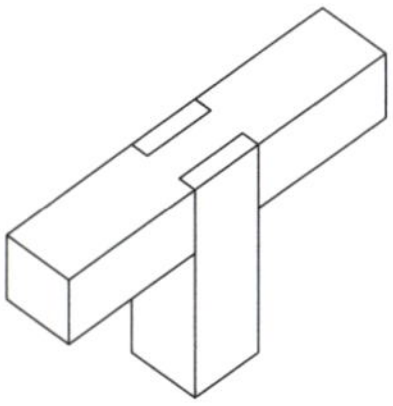

AB

T-Joints

T-bridle joint with blind mortise and tenon

wanagi-komi-tan-hozo-sashi (binta, dō-tsuki)
輪薙込短枘差し（鬢太・胴付）

Einhälsung mit verdecktem Zapfen

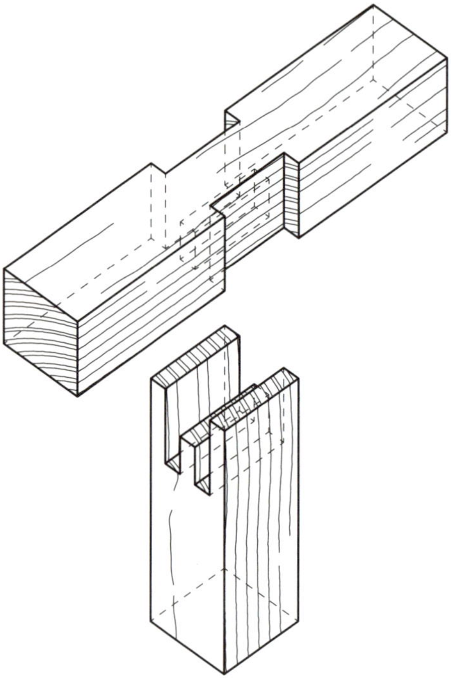

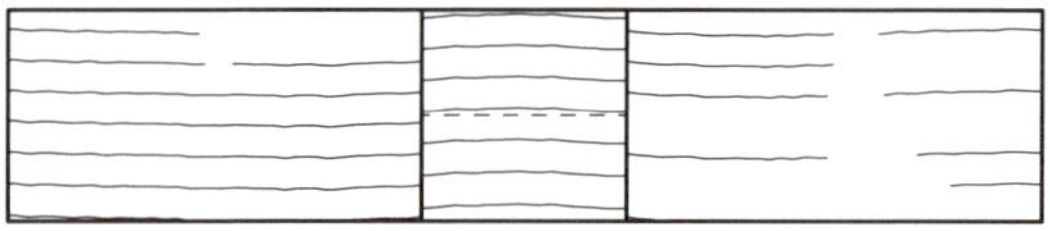

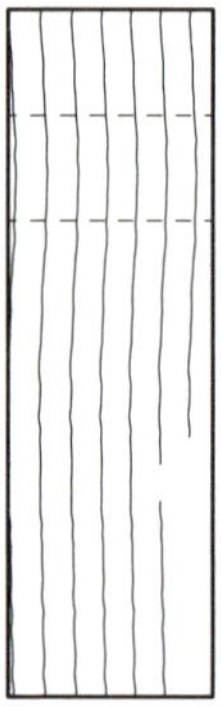

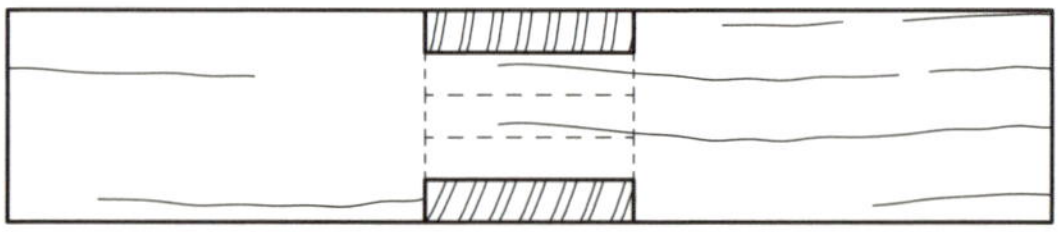

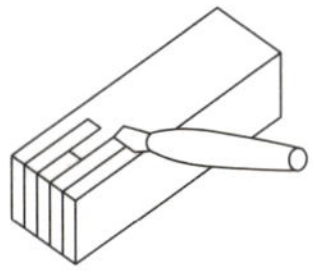

A1

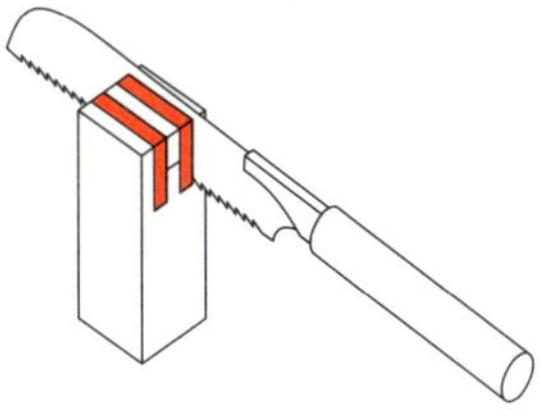

A2

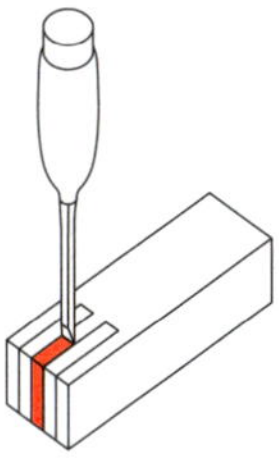

A3

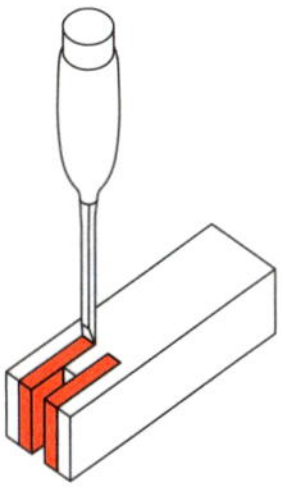

A4

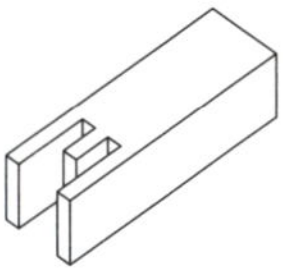

A

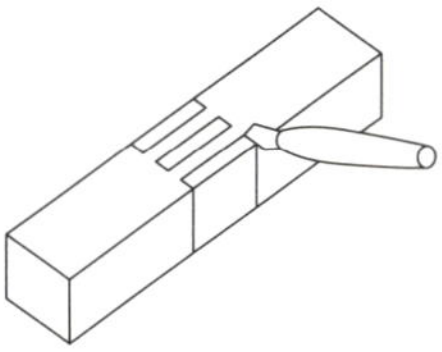

B1

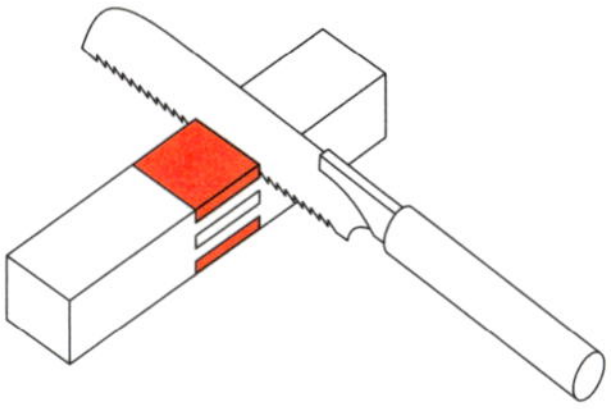

B2

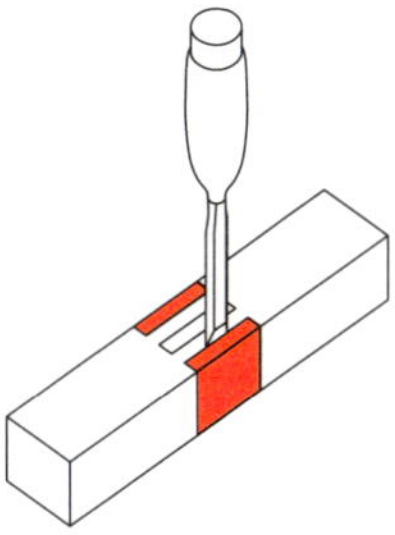

B3

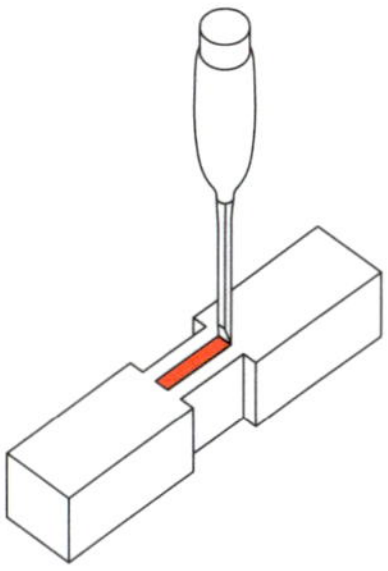

B4

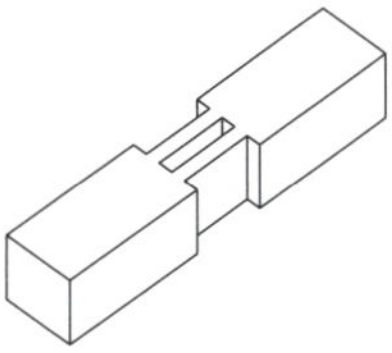

B

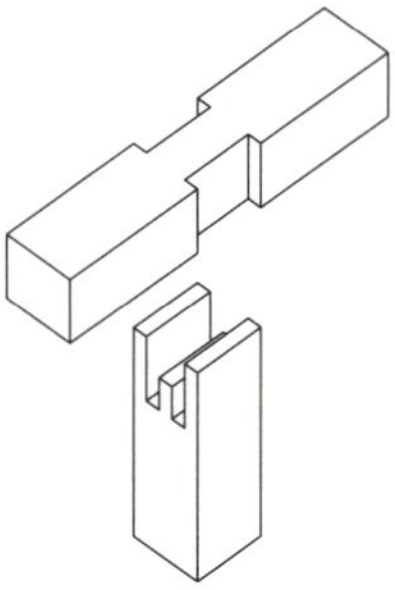

A+B

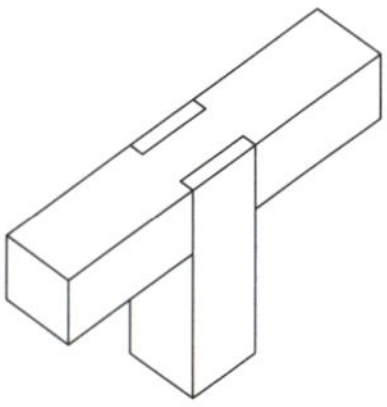

AB

T-Joints

Double lap joint with stub tenon

binta-nobashi-hashira-ōhozo-sashi
鬢太伸柱大枘差し

Gabelaufblattung mit verdecktem Zapfen

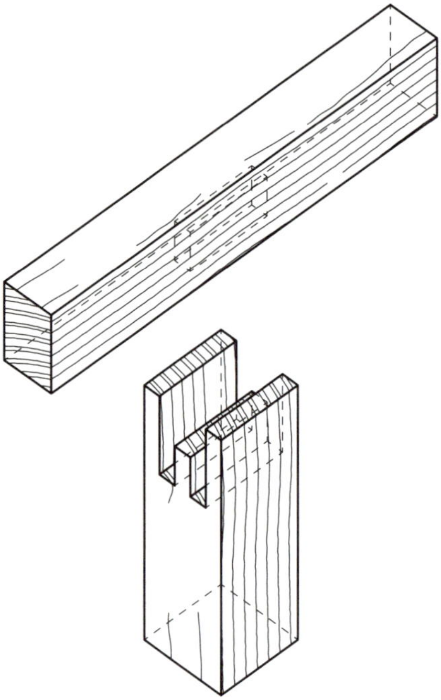

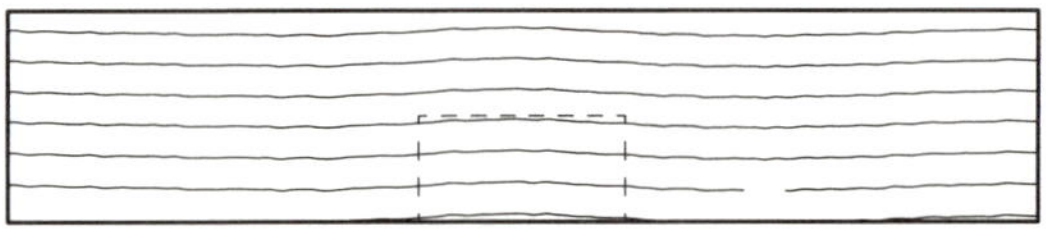

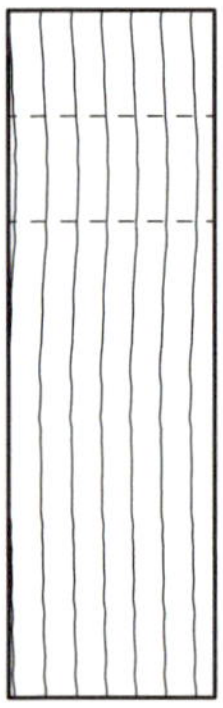

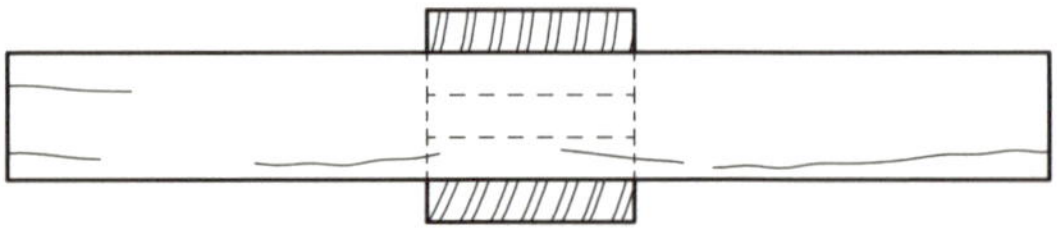

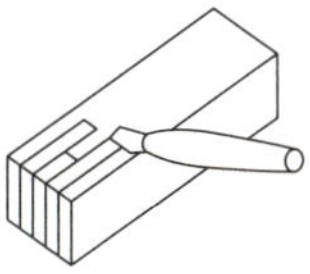

A1

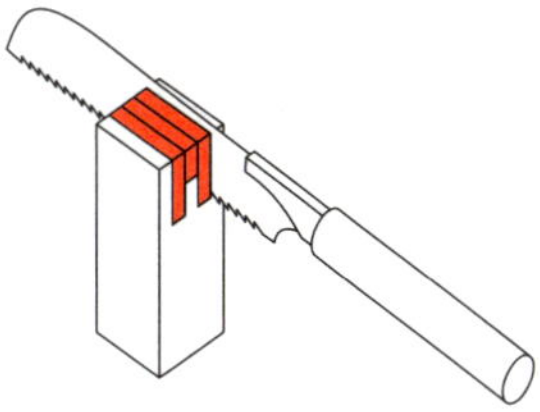

A2

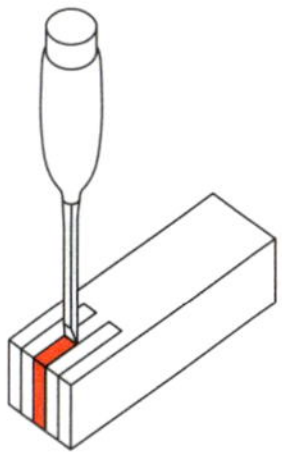

A3

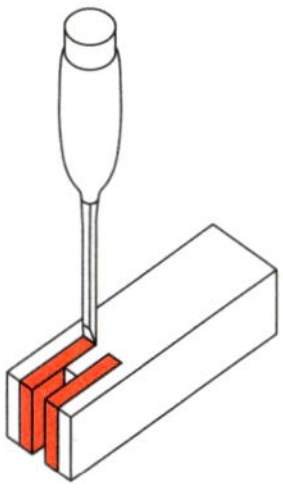

A4

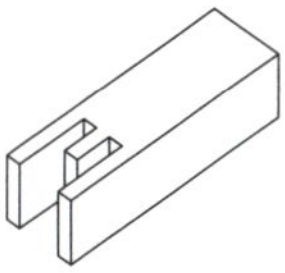

A

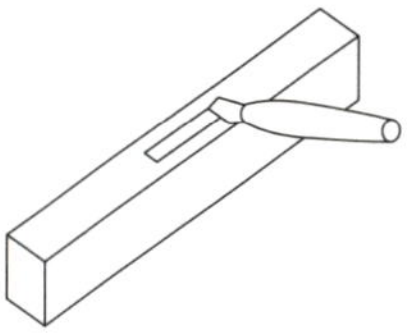

B1

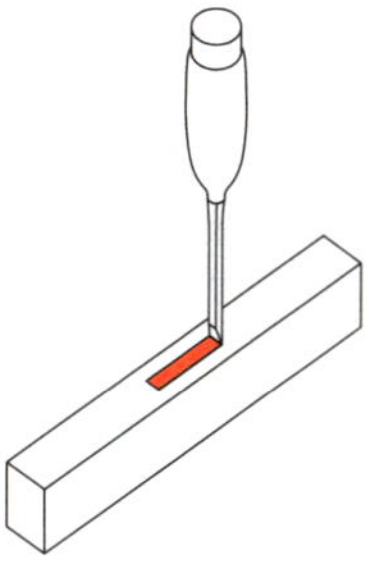

B2

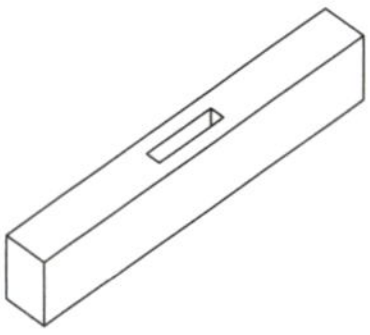

B

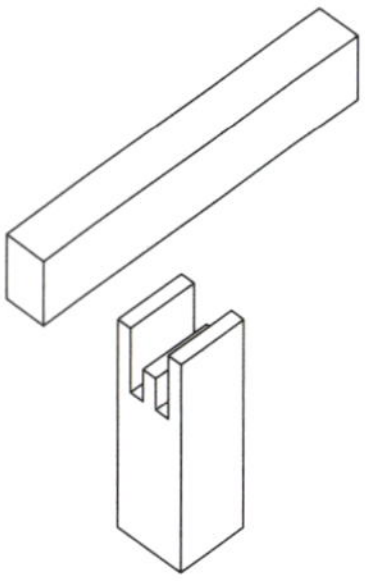

A+B

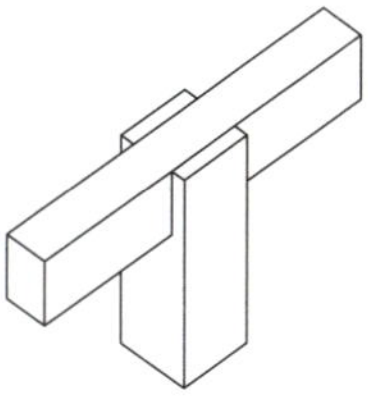

AB

T-Joints

Notched lap joint with tenon

kata-wanagi-tōshi-hozo-sashi
片輪薙通枘差し

einseitige Einhälsung
mit durchgehendem Zapfen

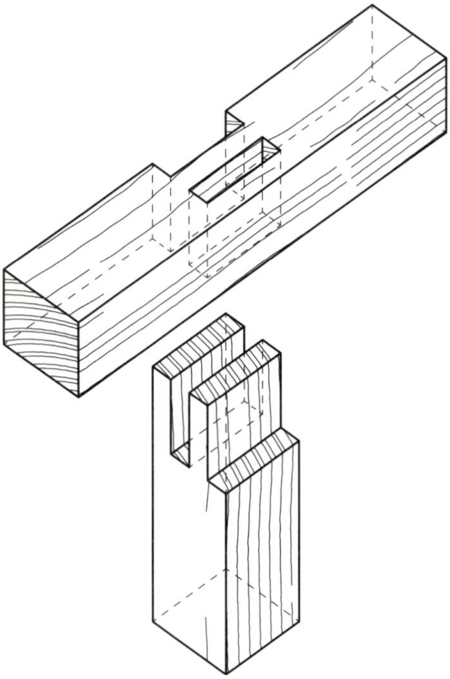

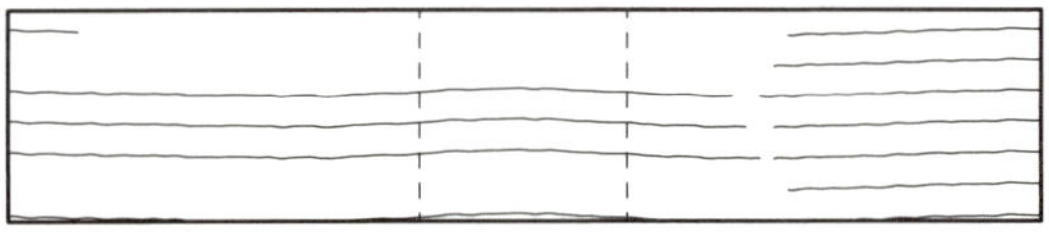

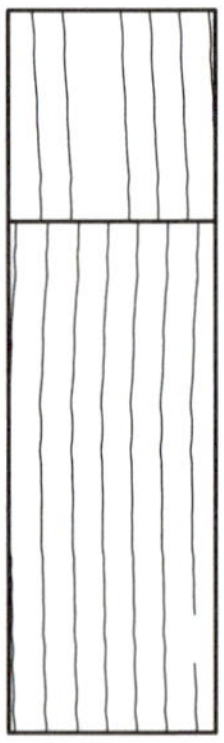

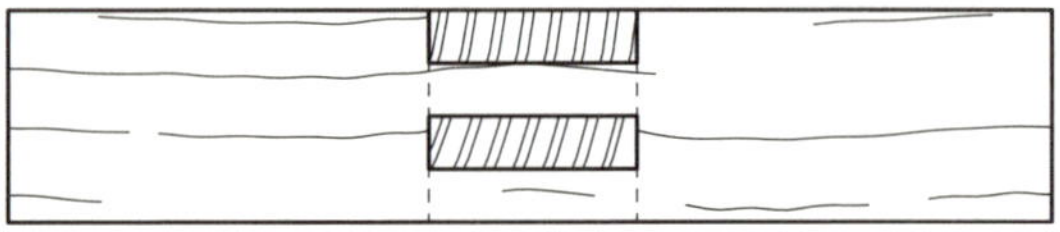

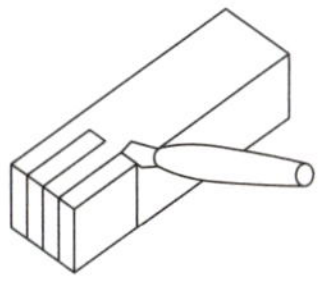
A1

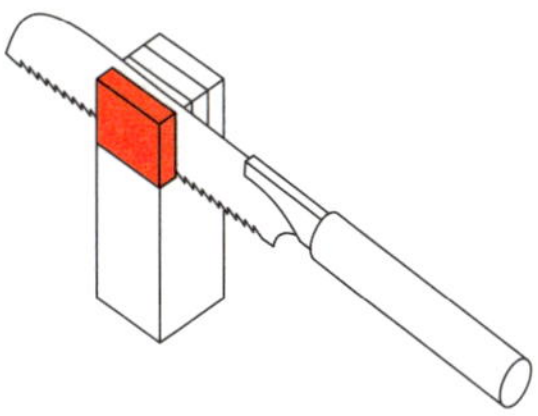
A2

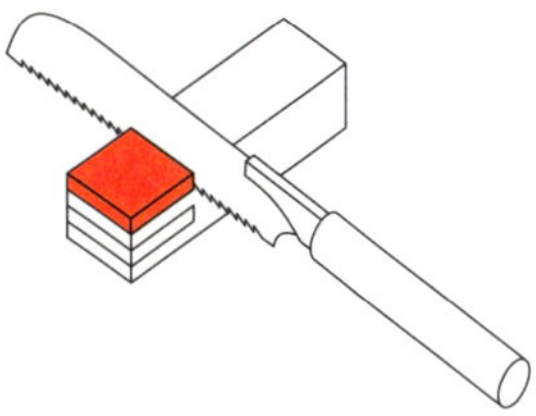
A3

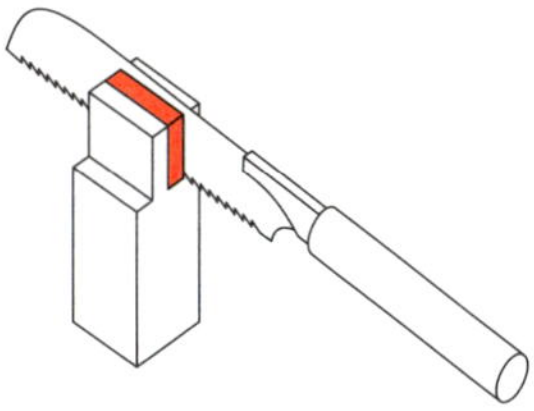
A4

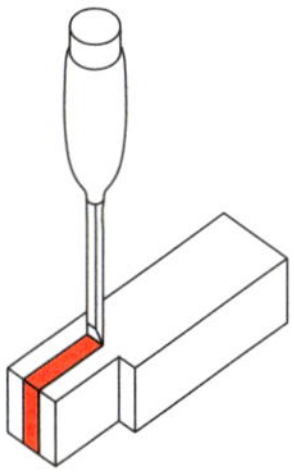
A5

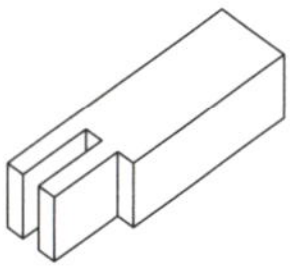
A

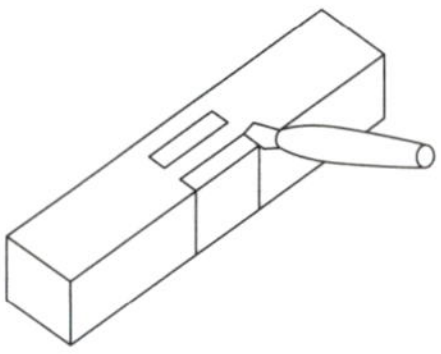
B1

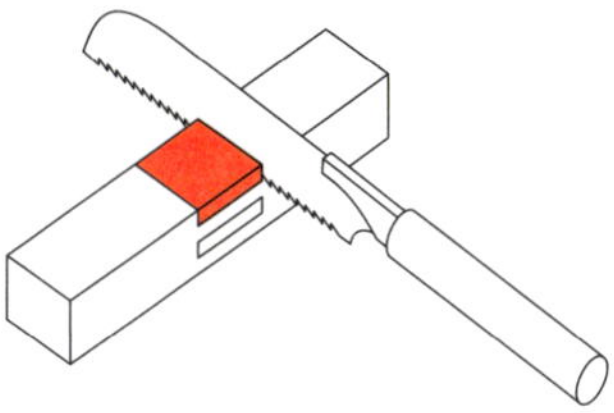
B2

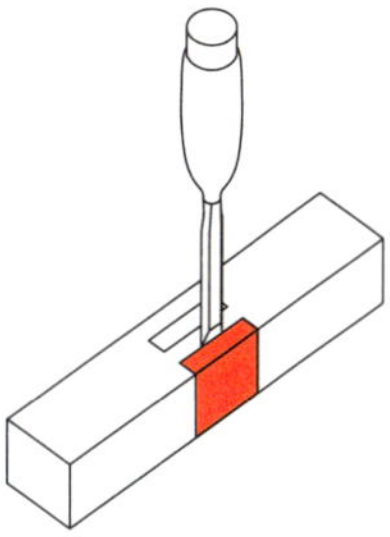
B3

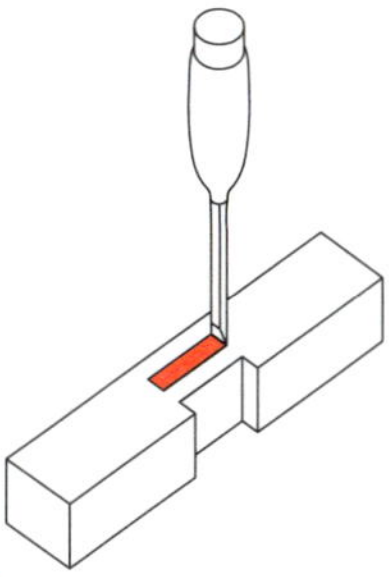
B4

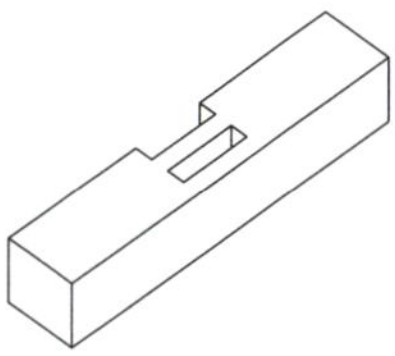
B

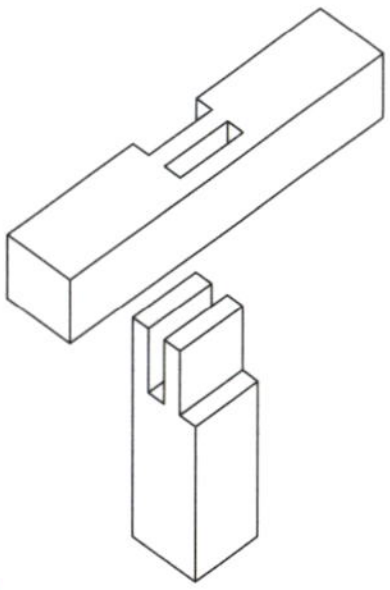
A+B

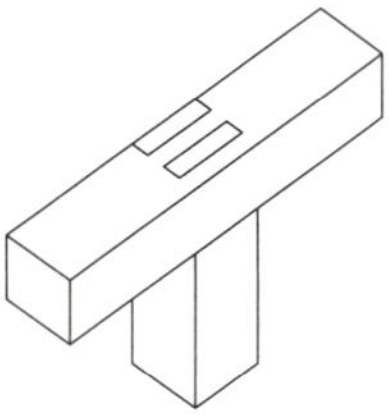
AB

T-Joints

Notched lap joint with stub tenon

kata-wanagi-dō-tsuki-hira-hozo-sashi
片輪薙胴付平枘差し

einseitige Einhälsung
mit verdecktem Zapfen

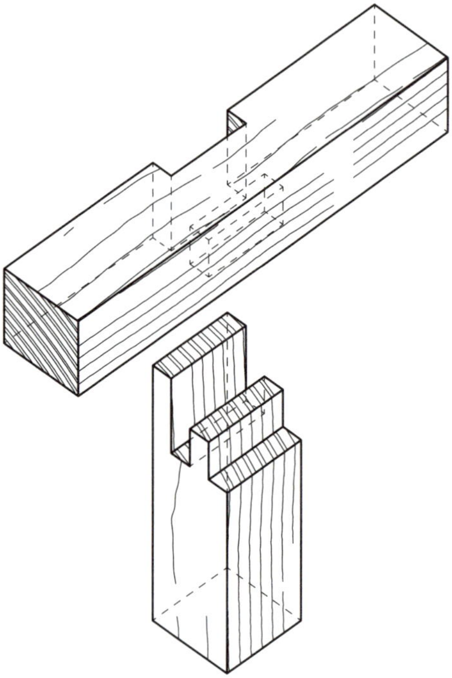

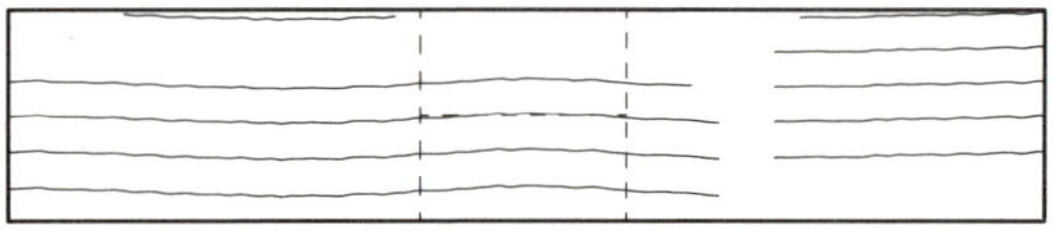

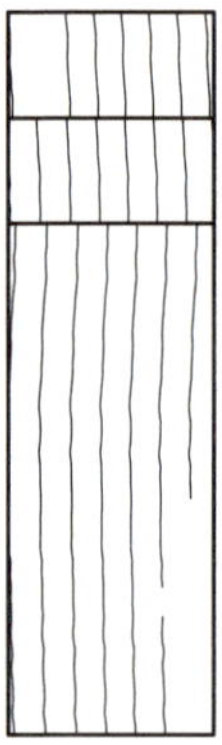

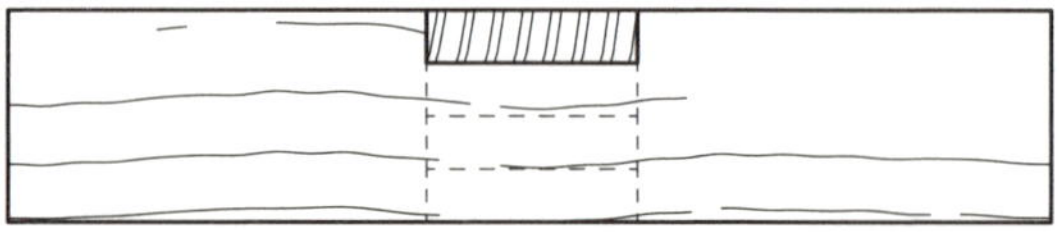

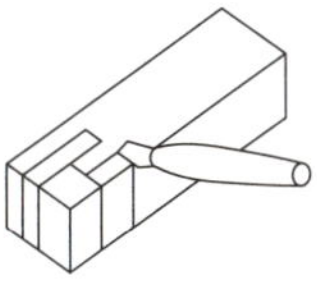

A1

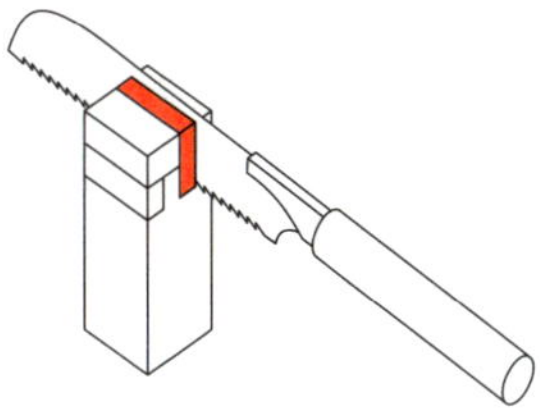

A2

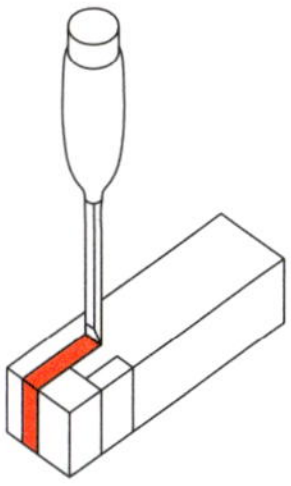

A3

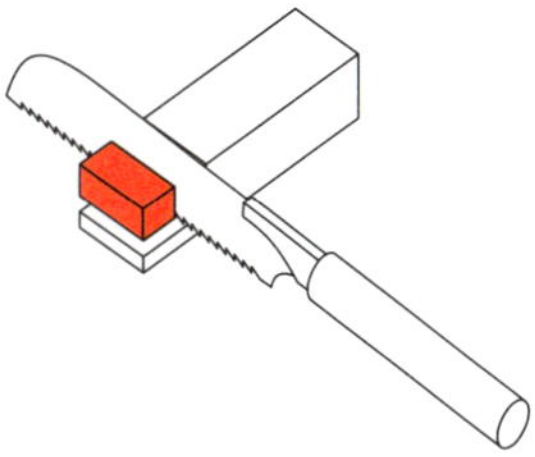

A4

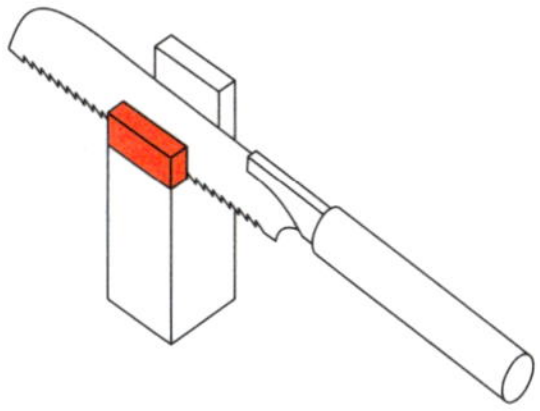

A5

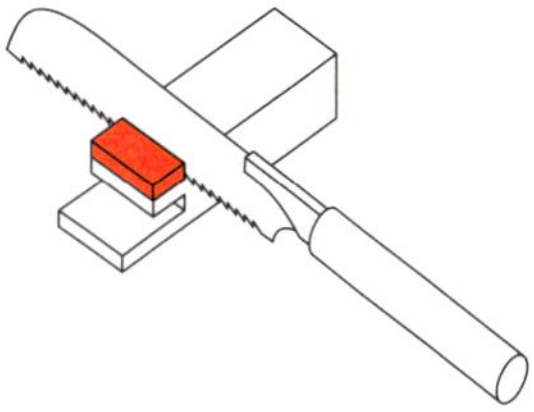

A6

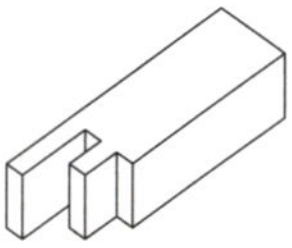

A

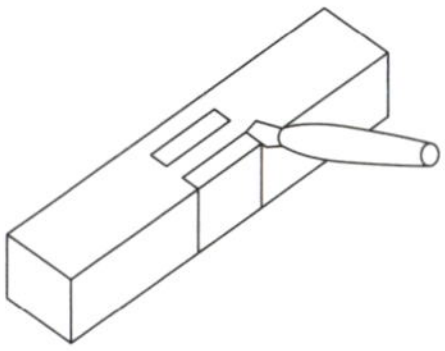

B1

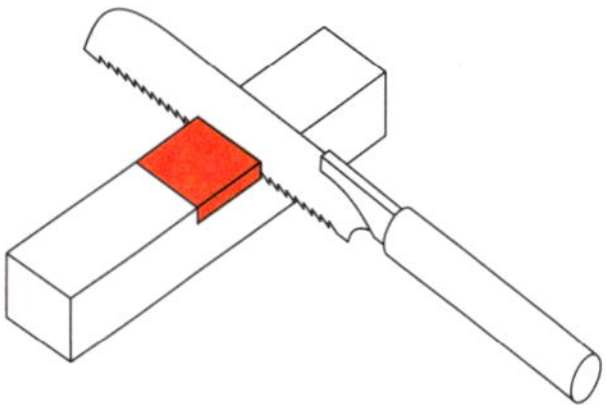

B2

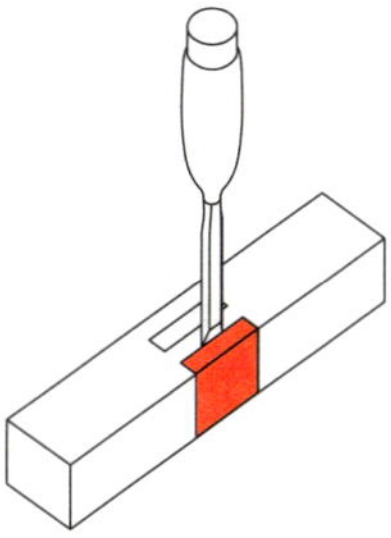

B3

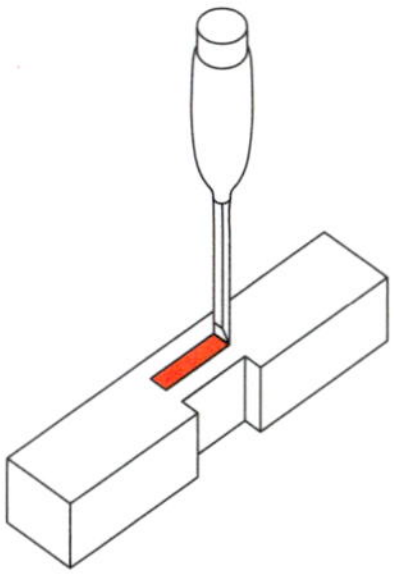

B4

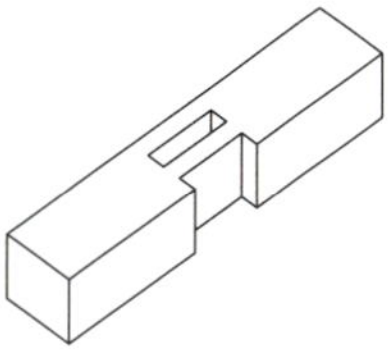

B

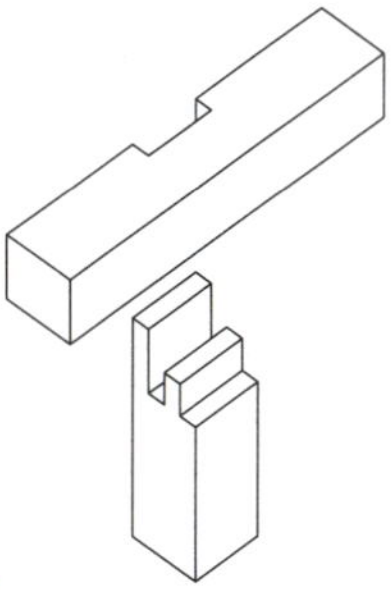

A+B

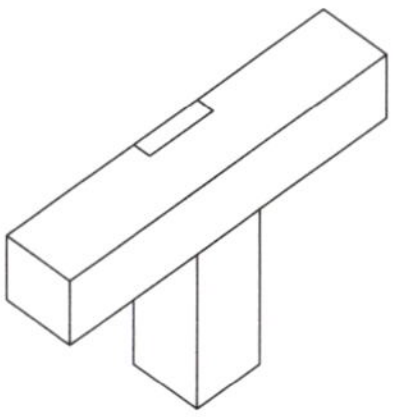

AB

T-Joints

Lap joint with tenon

kata-binta-nobashi-hashira-ōhozo-sashi
片鬢太伸柱大枘差し

einseitige Gabelaufblattung
mit durchgehendem Zapfen

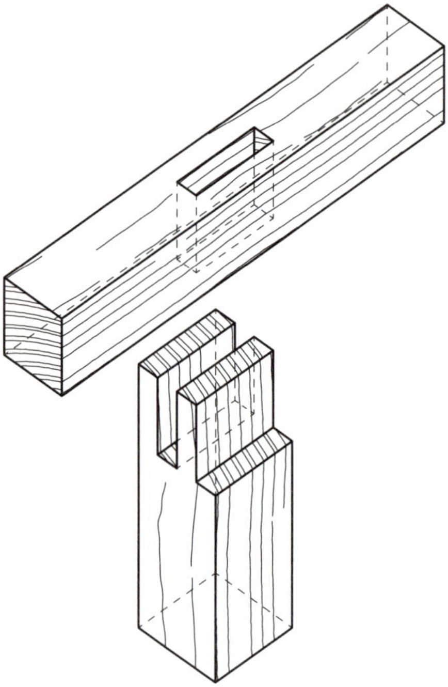

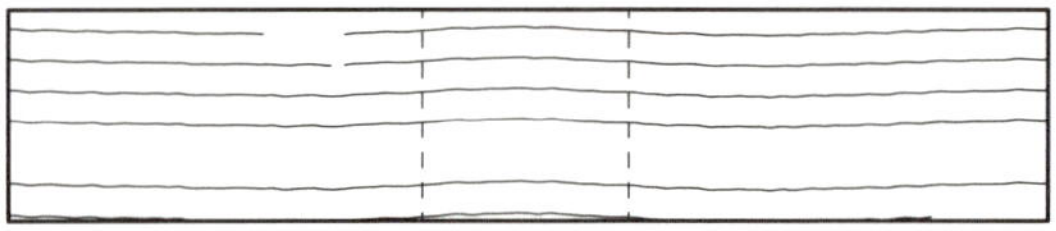

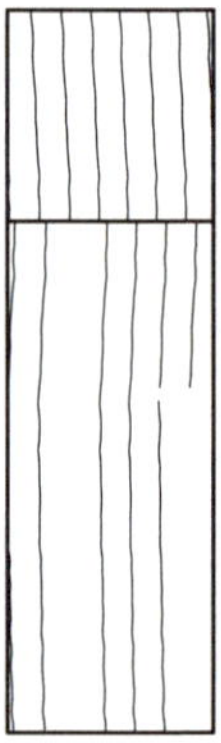

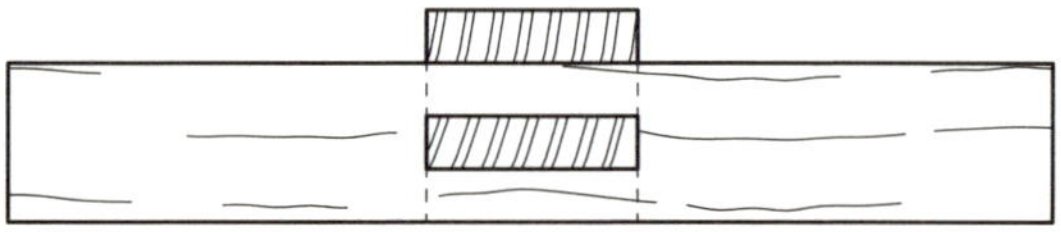

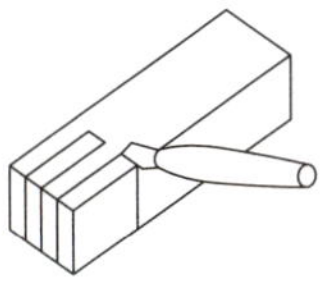

A1

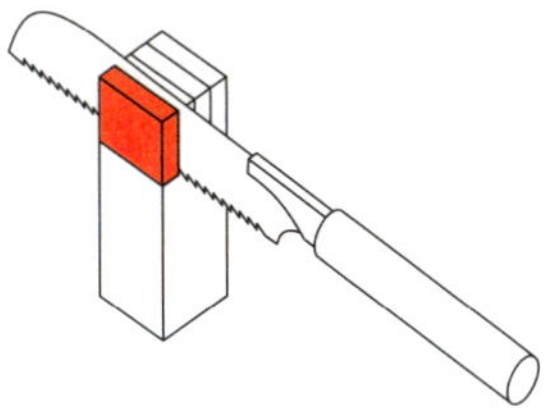

A2

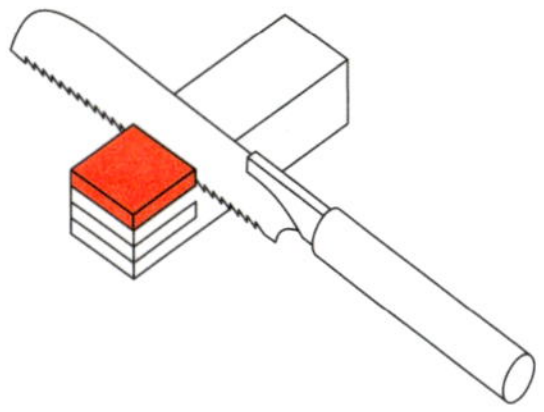

A3

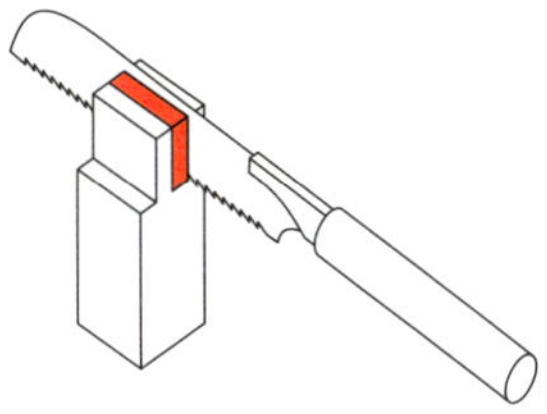

A4

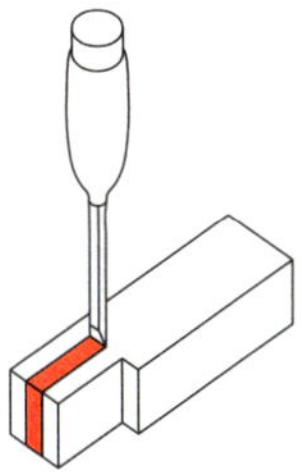

A5

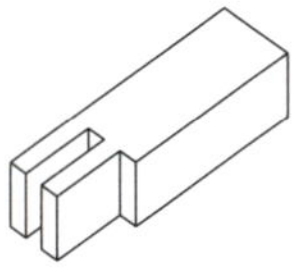

A

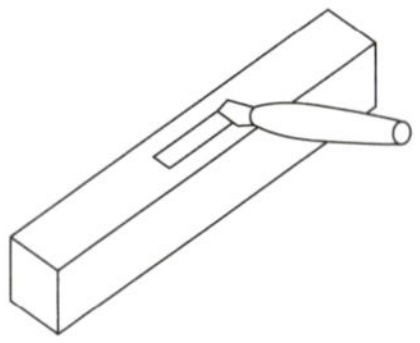

B1

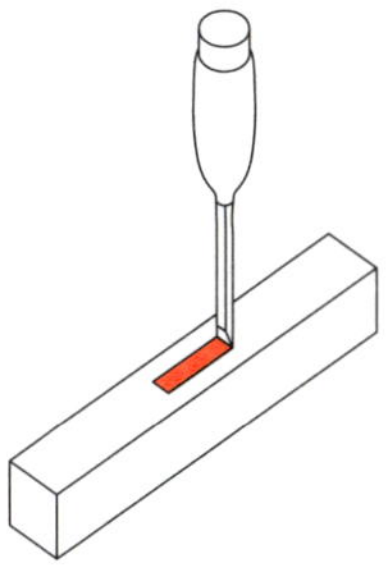

B2

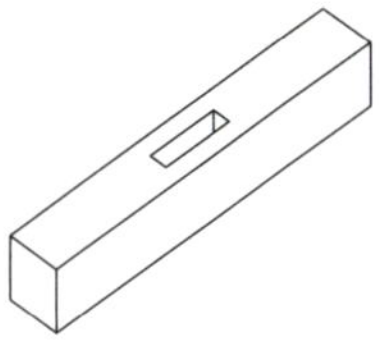

B

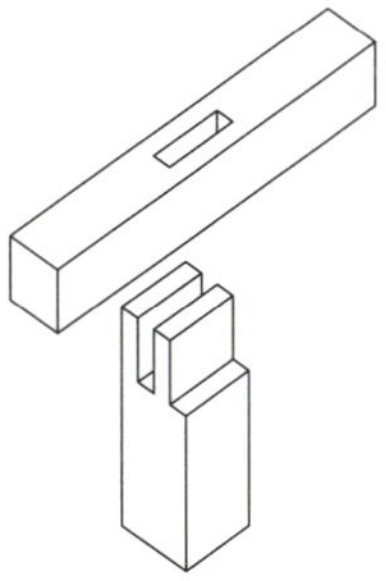

A+B

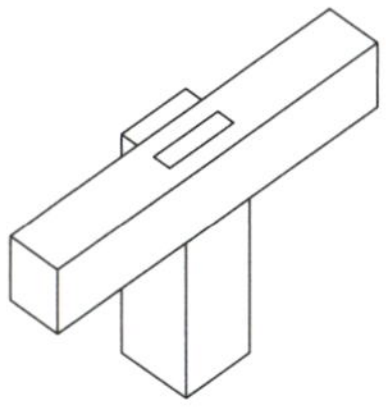

AB

T-Joints

Lap joint with stub tenon

kata-binta-dō-tsuki-hira-hozo-sashi
片鬢太胴付平枘差し

einseitige Gabelaufblattung
mit verdecktem Zapfen

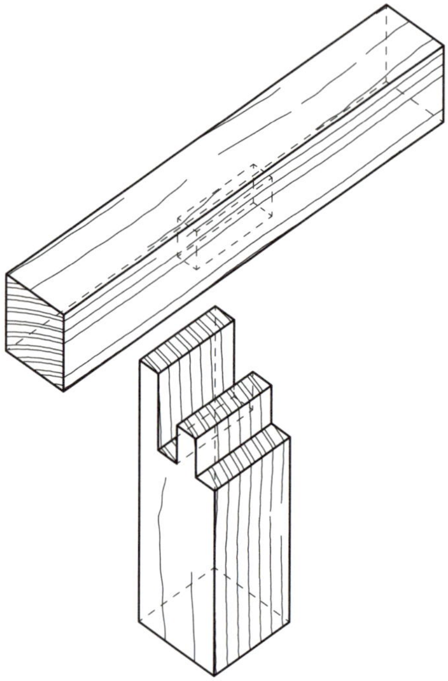

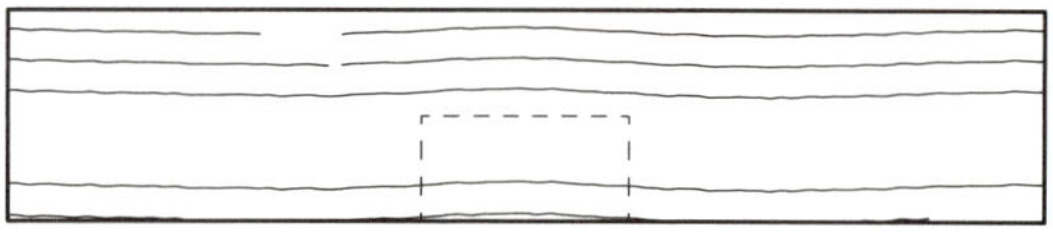

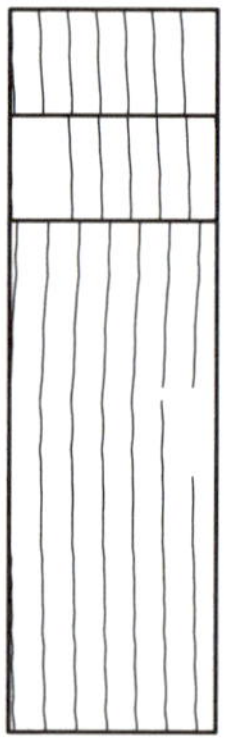

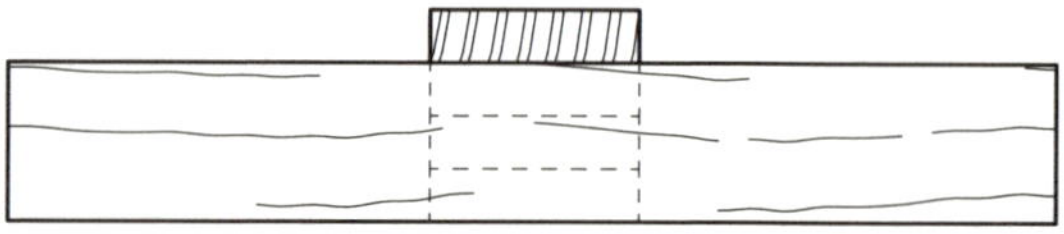

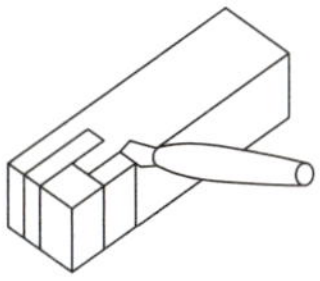

A1

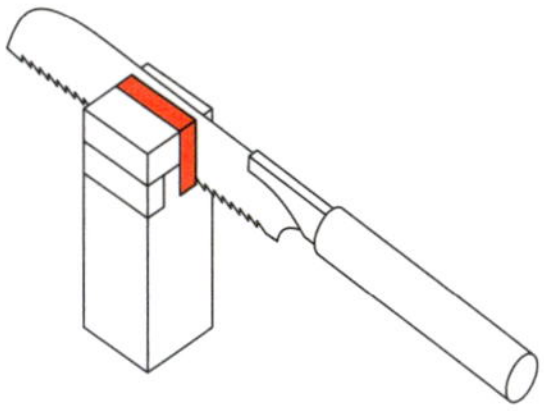

A2

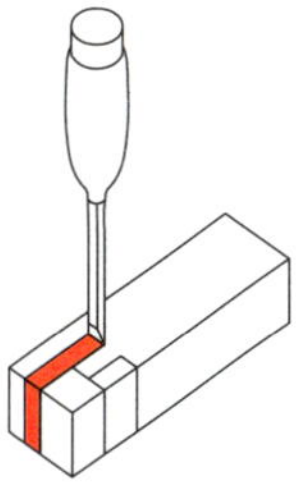

A3

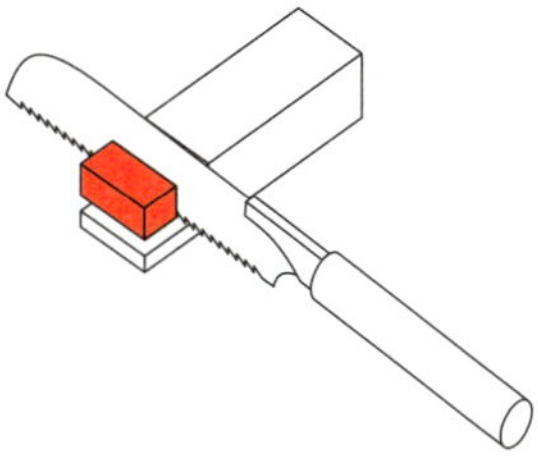

A4

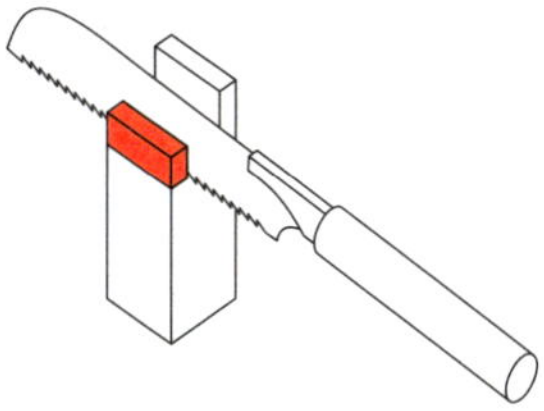

A5

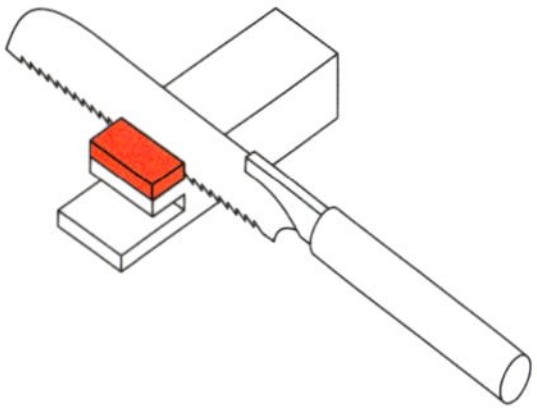

A6

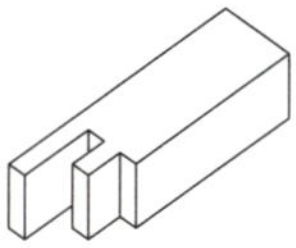

A

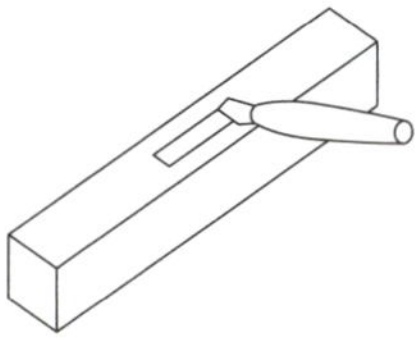

B1

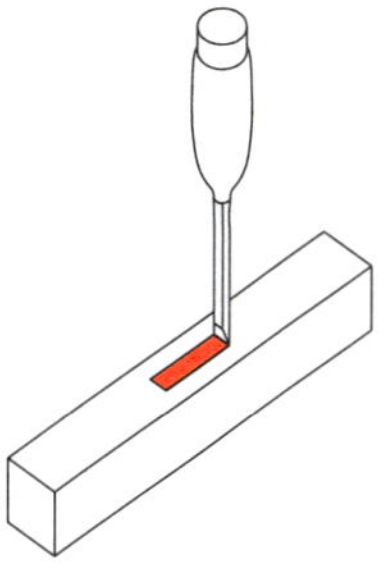

B2

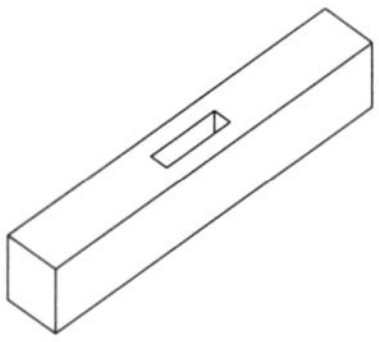

B

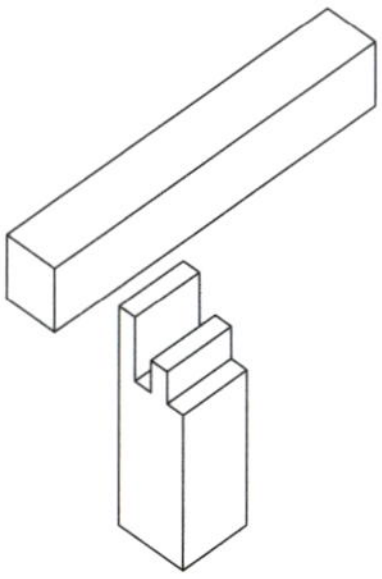

A+B

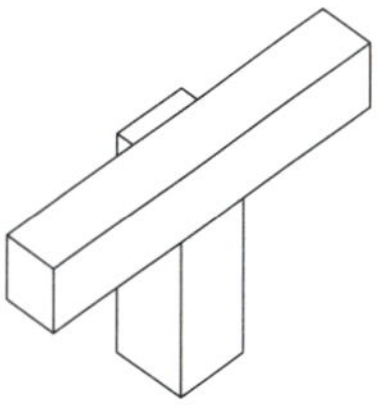

AB

T-Joints

Dovetail T-bridle joint

nimai-ari-hozo-sashi (wanagi-komi)
二枚蟻枘差し(輪薙込)

schräg angeschnittene Einhälsung

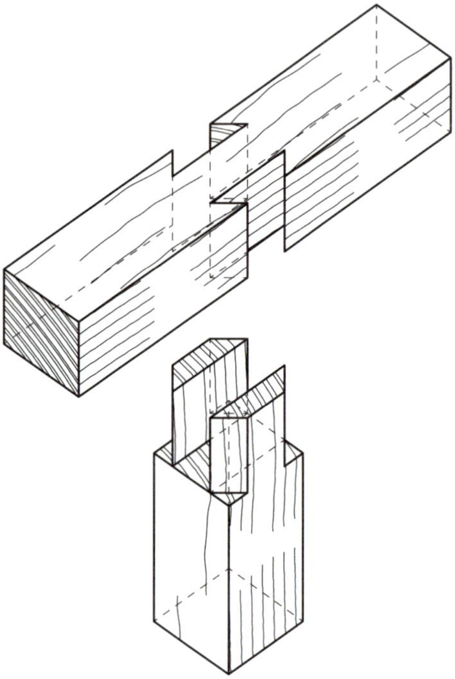

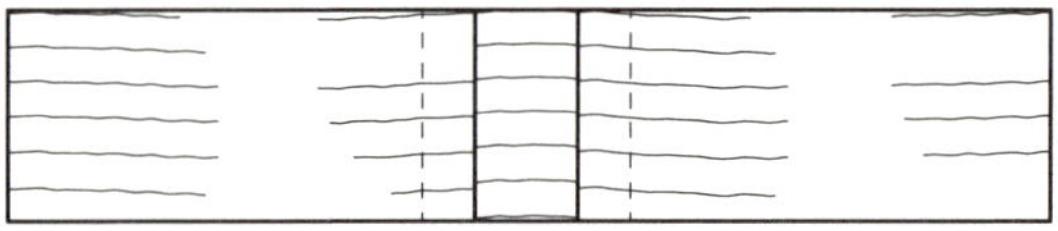

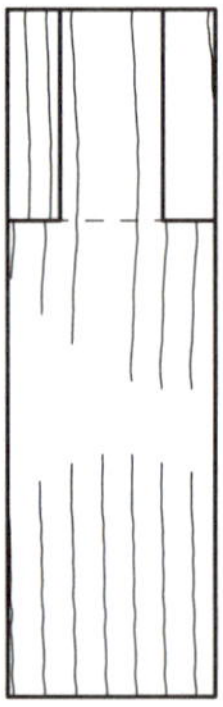

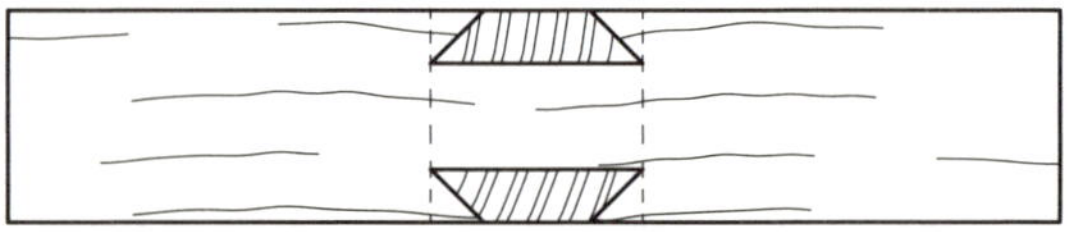

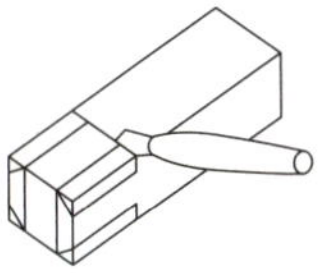

A1

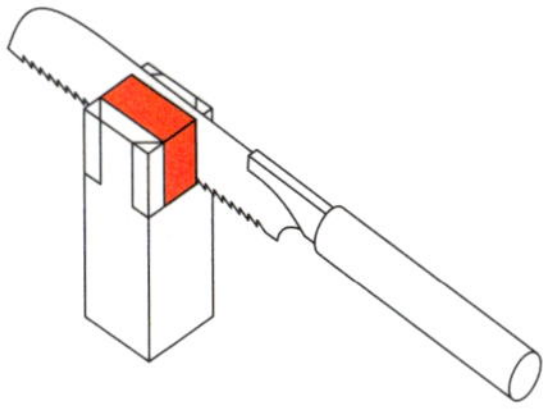

A2

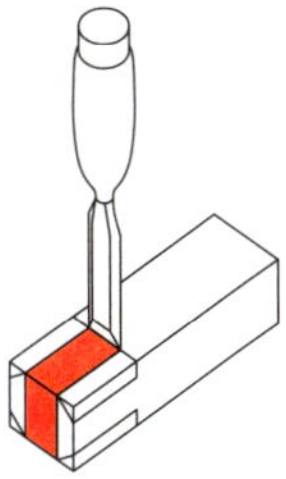

A3

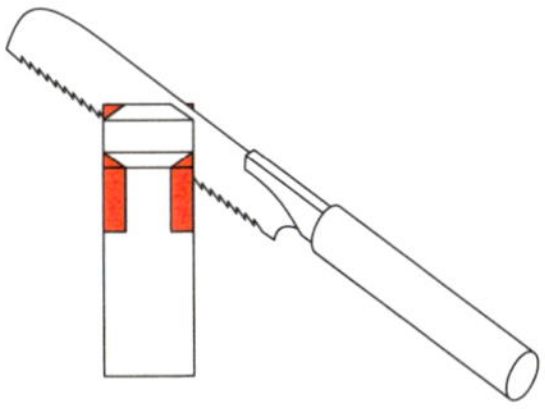

A4

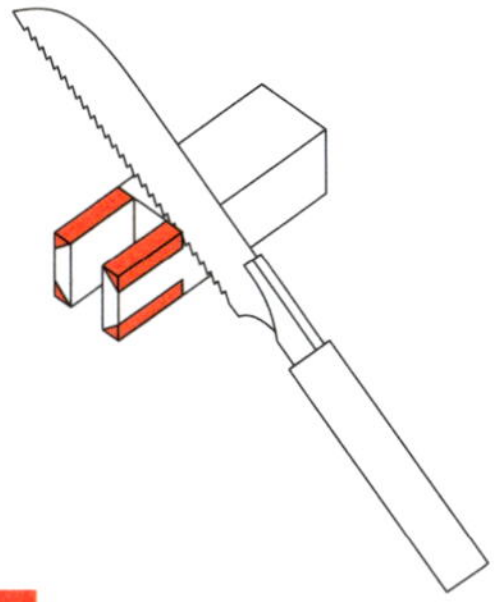

A5

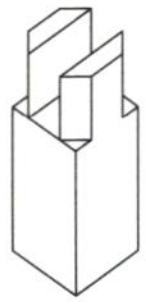

A

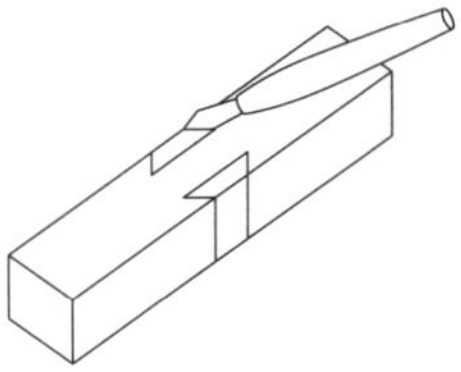

B1

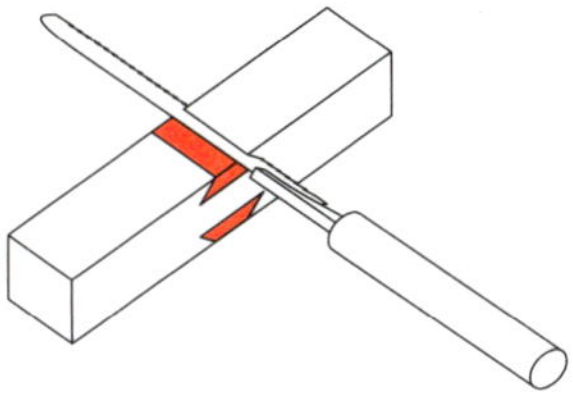

B2

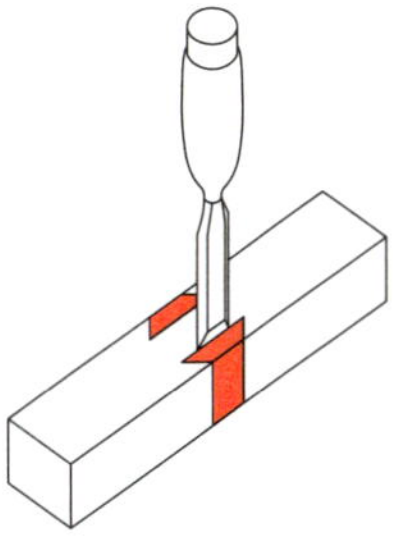

B3

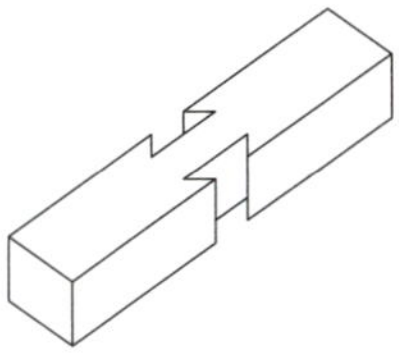

B

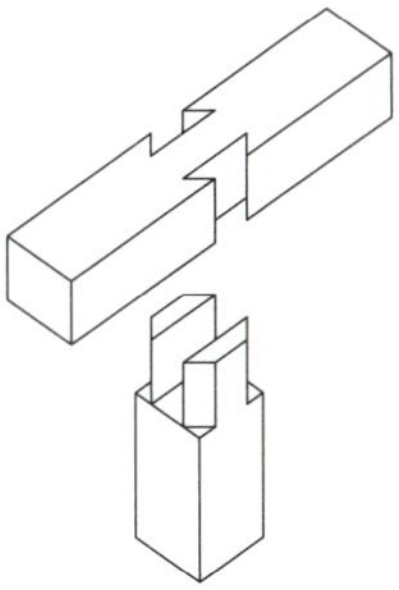

A+B

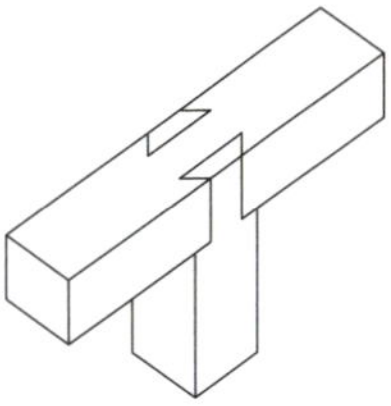

AB

T-Joints

Diagonally cut T-bridle joint

hashira-yonhan-wanagi-hozo-sashi
柱四半輪薙枘差し

diagonale Einhälsung

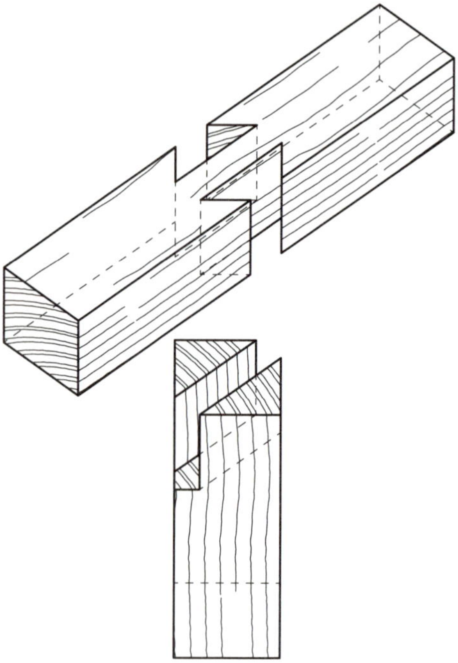

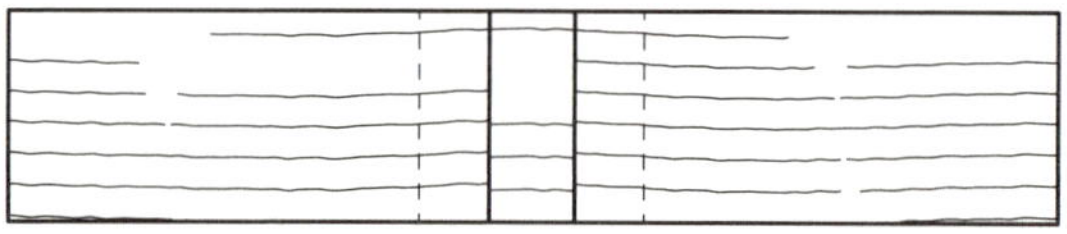

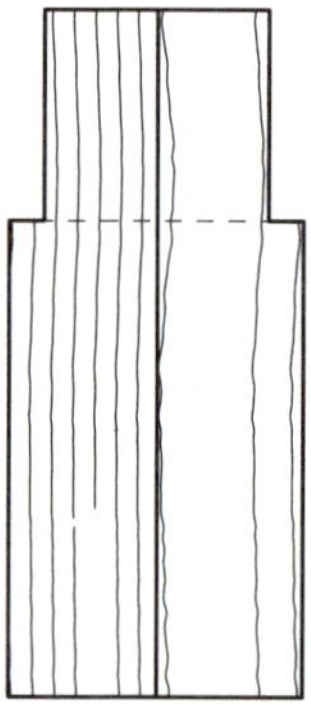

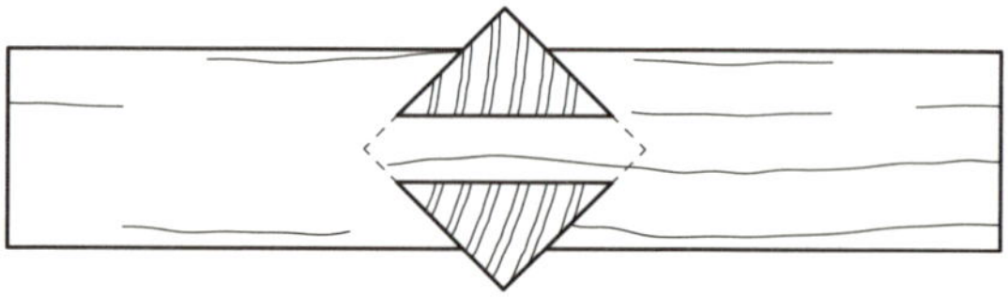

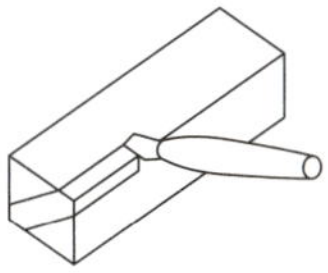

A1

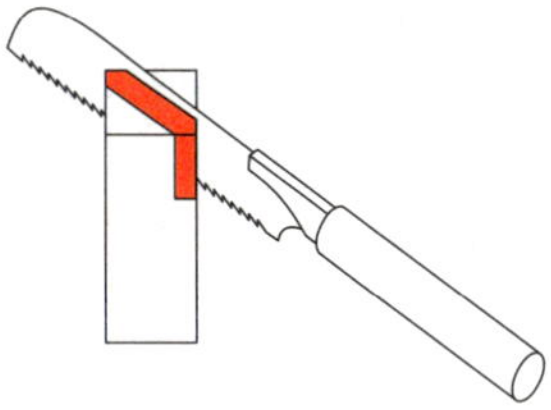

A2

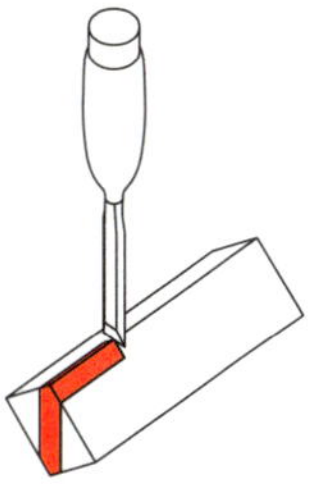

A3

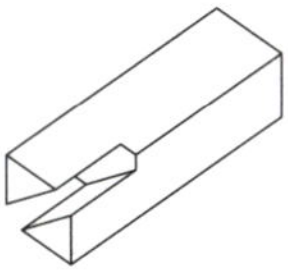

A

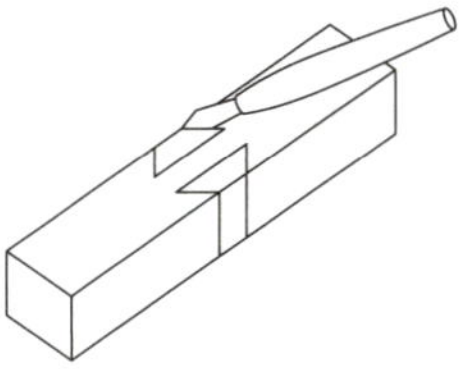

B1

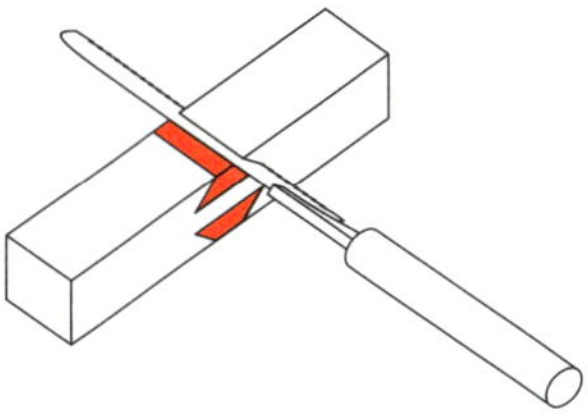

B2

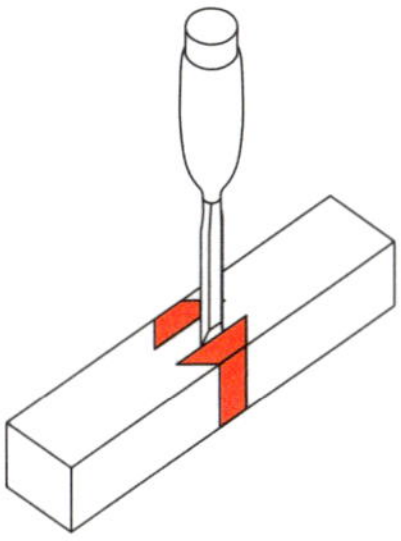

B3

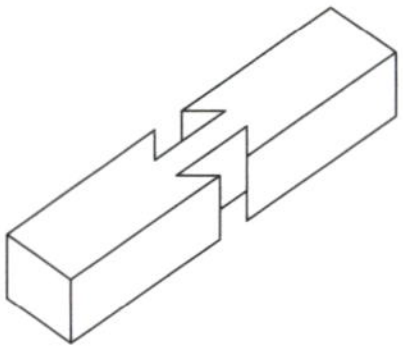

B

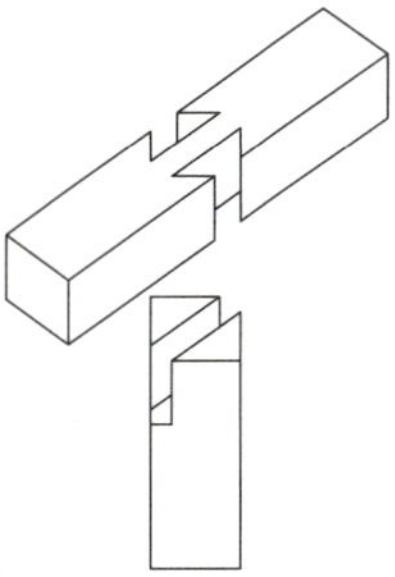

A+B

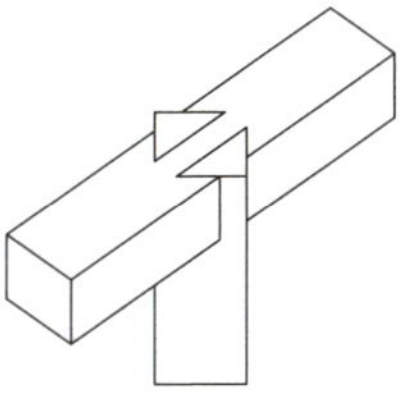

AB

T-Joints

Rising dovetail joint

ai-kaki-ari-kusabi-shime (kōbai-tsuki)
相欠蟻楔締め（勾配付）

schräg angeschnittener schräg abgesetzter Schwalbenschwanzzapfen

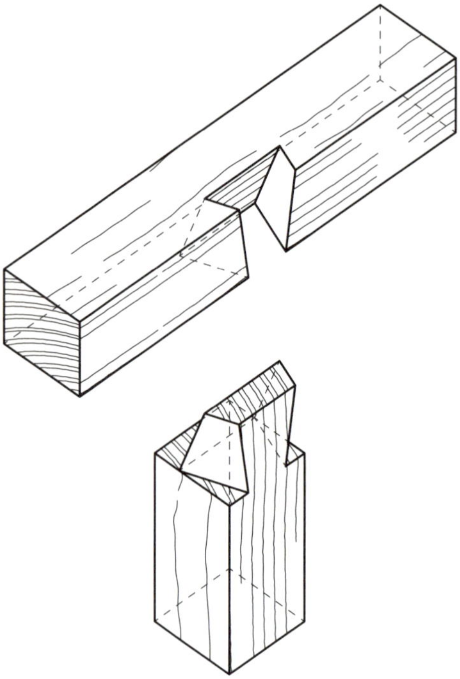

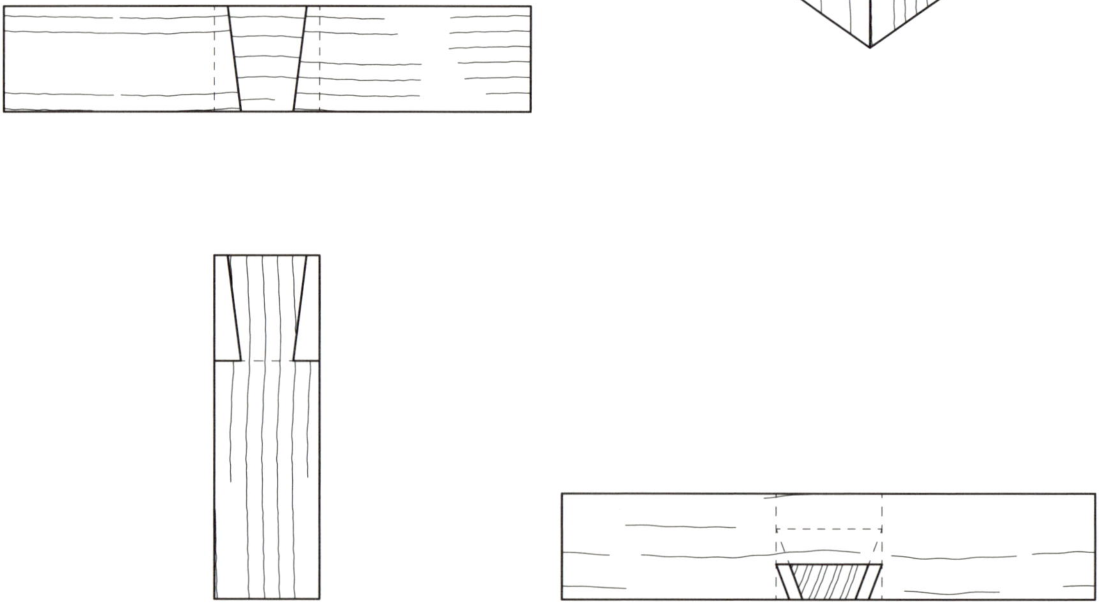

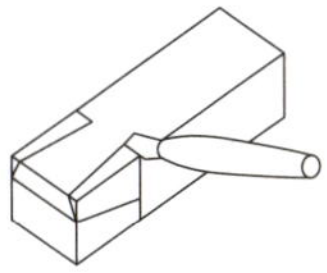

A1

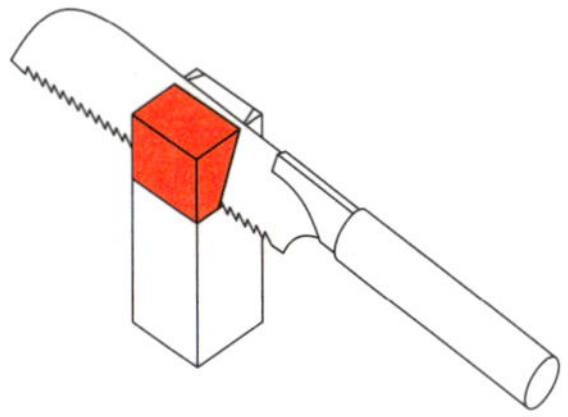

A2

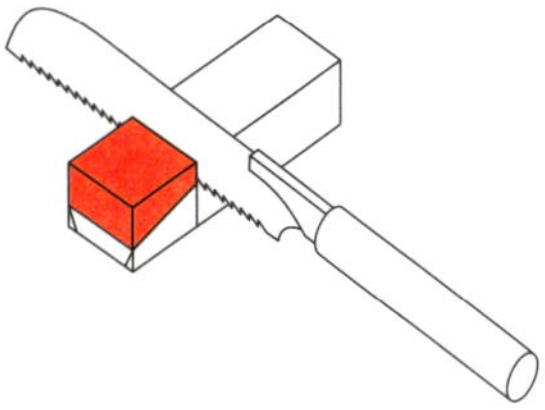

A3

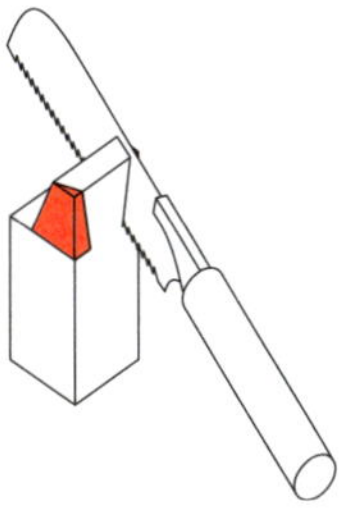

A4

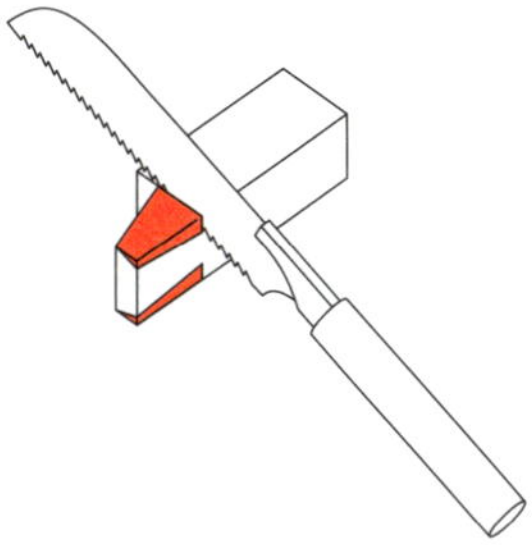

A5

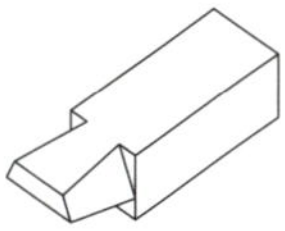

A

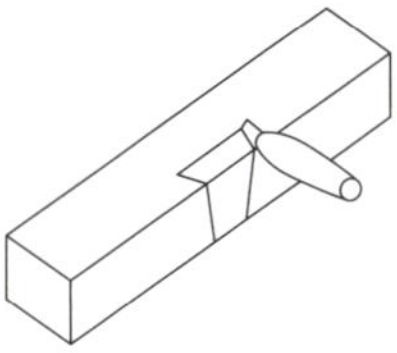

B1

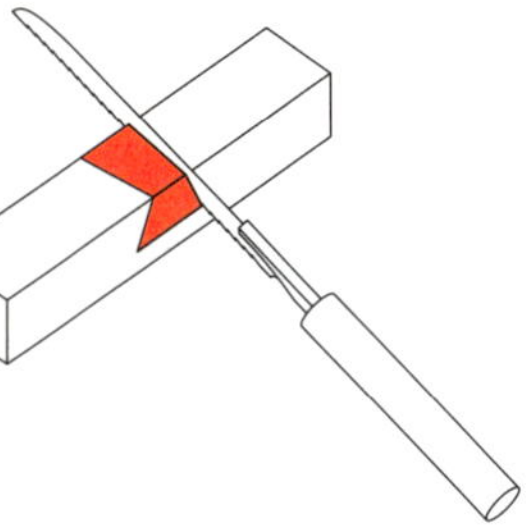

B2

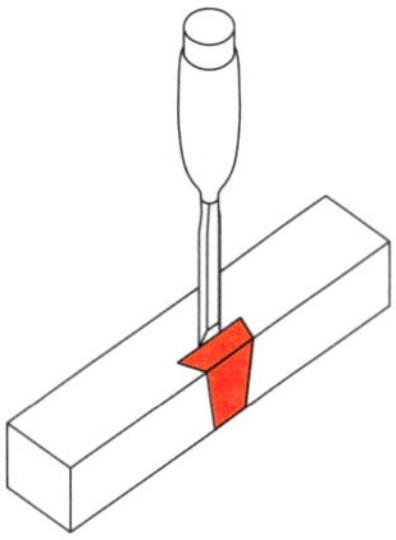

B3

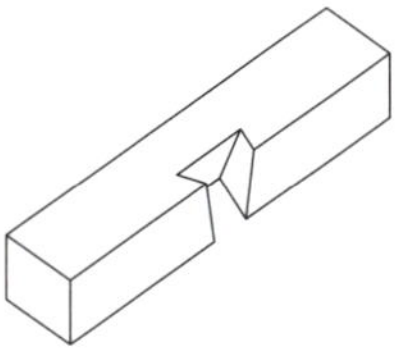

B

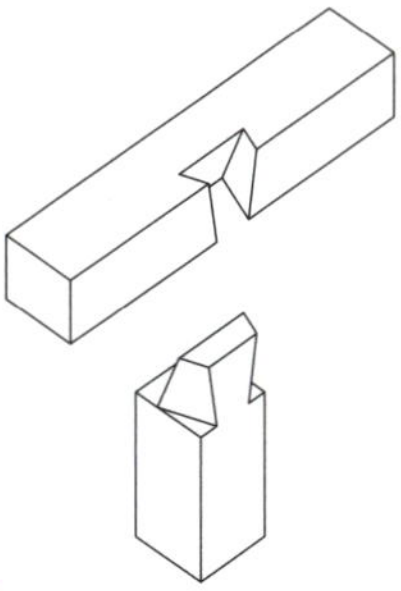

A+B

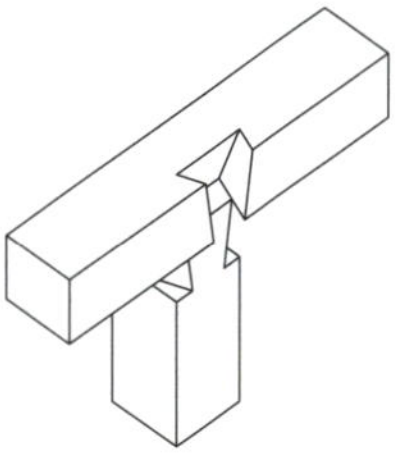

A+B

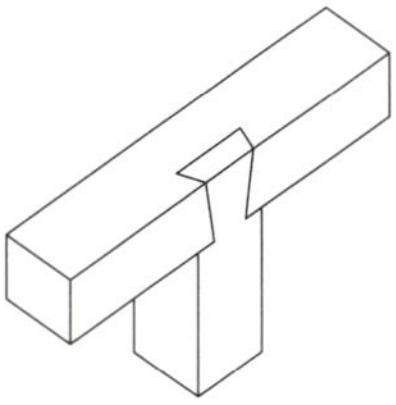

AB

T-Joints

Wedged dovetail half lap joint

ai-kaki-ari-kusabi-shime [*]
相欠蟻楔締め

verkeiltes Schwalbenschwanzblatt

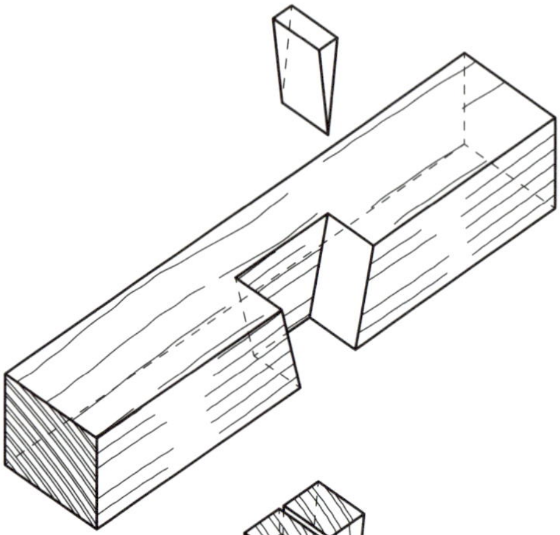

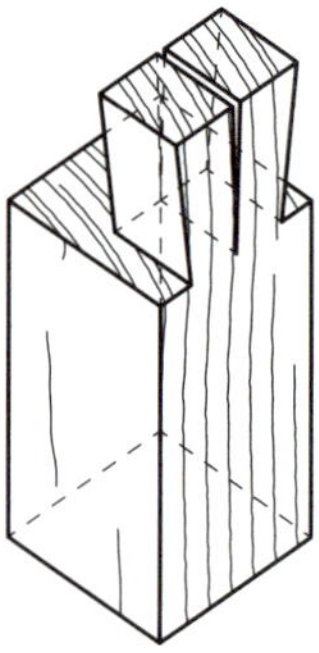

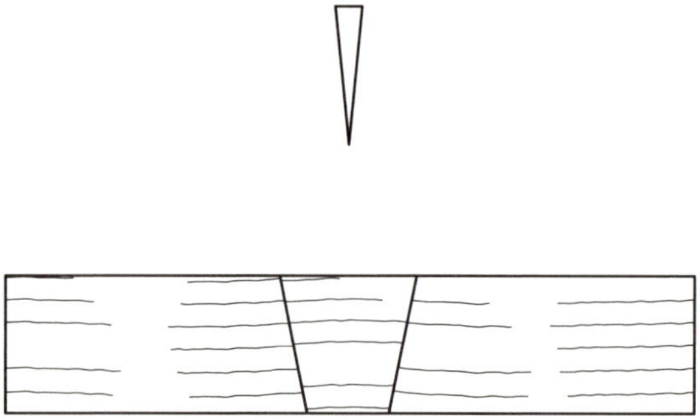

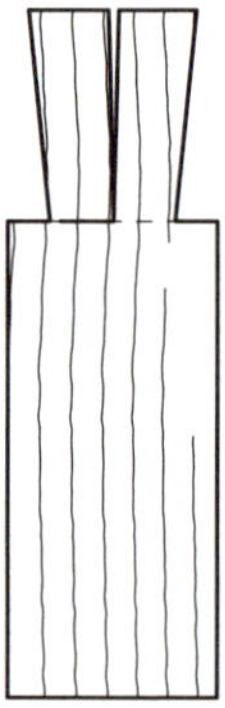

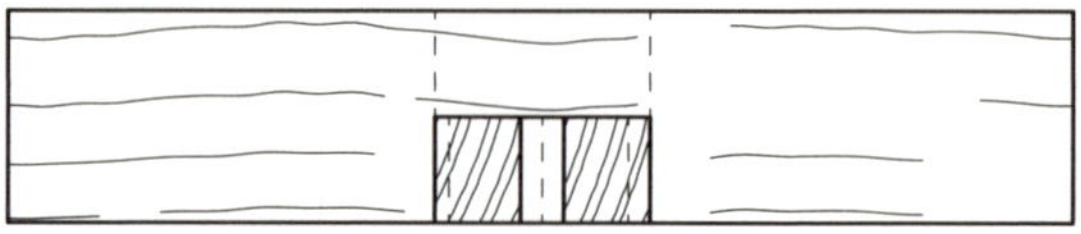

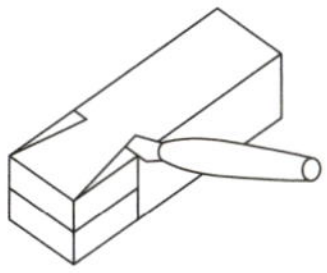
A1

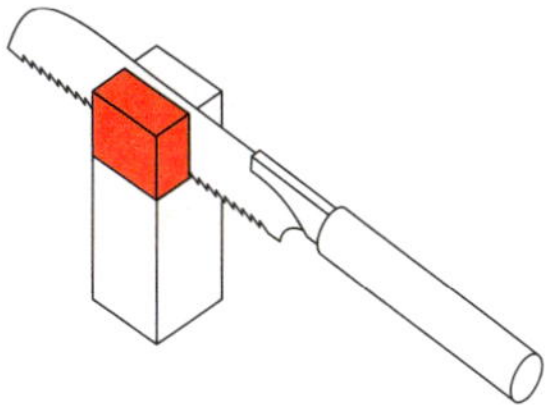
A2

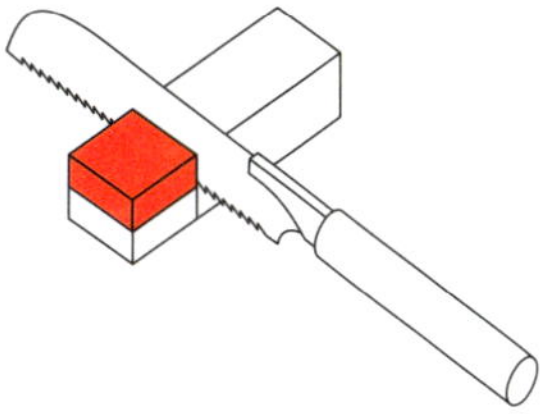
A3

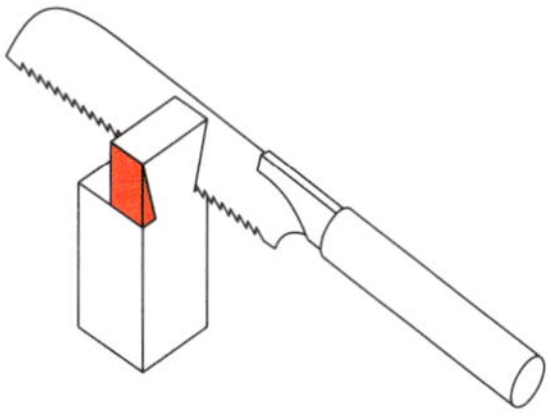
A4

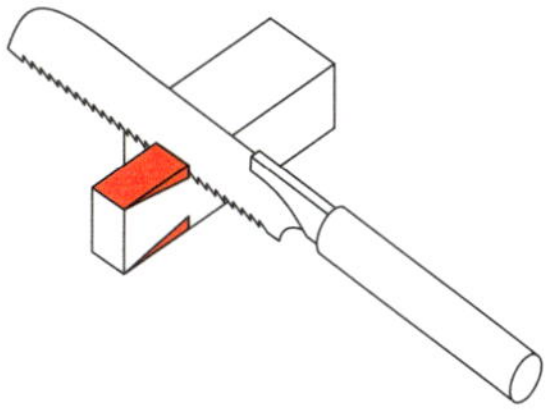
A5

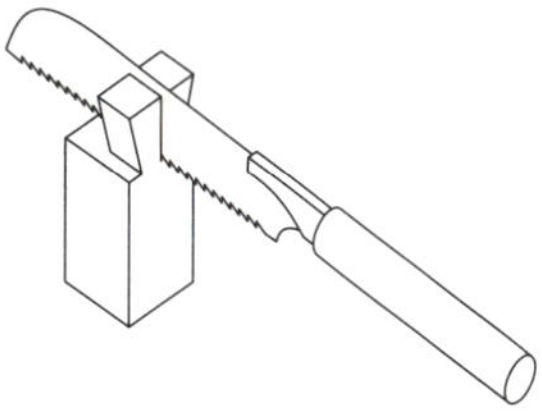
A6

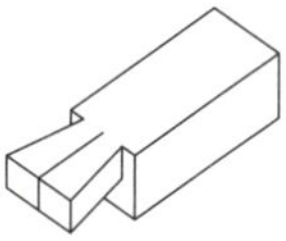
A

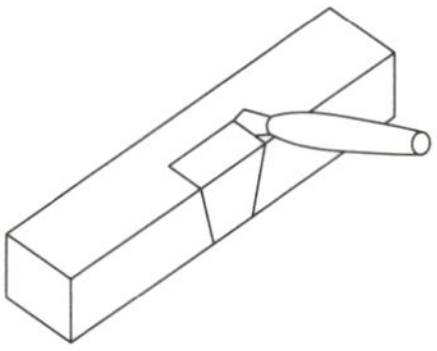
B1

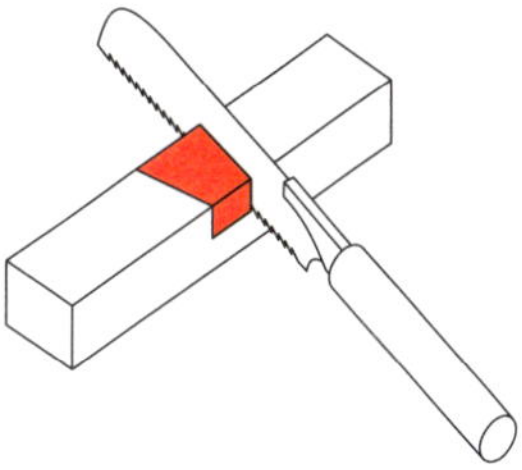
B2

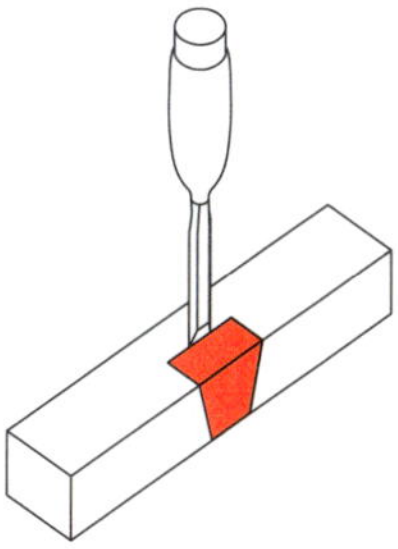
B3

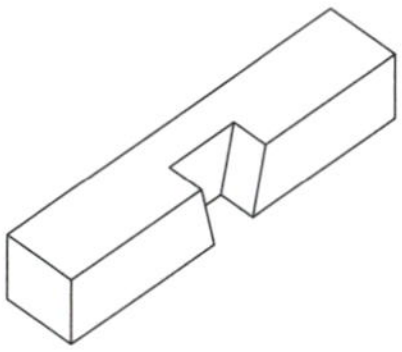
B

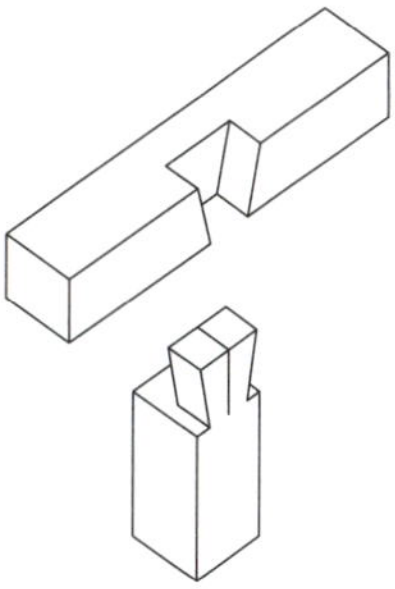
A+B

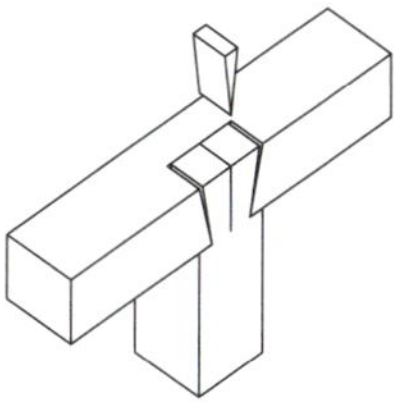
A+B

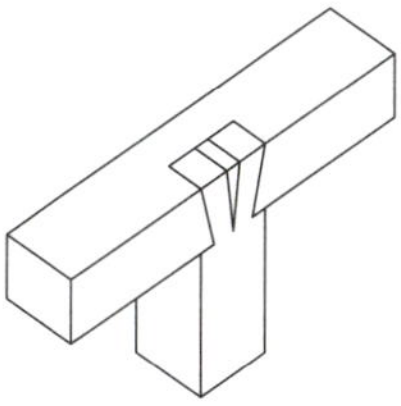
AB

T-Joints

Housed dovetail joint

yose-ari-tsugi (tsuri-taba) [*]
寄蟻接ぎ(吊束)

eingeschobener abgesetzter Schwalbenschwanzzapfen

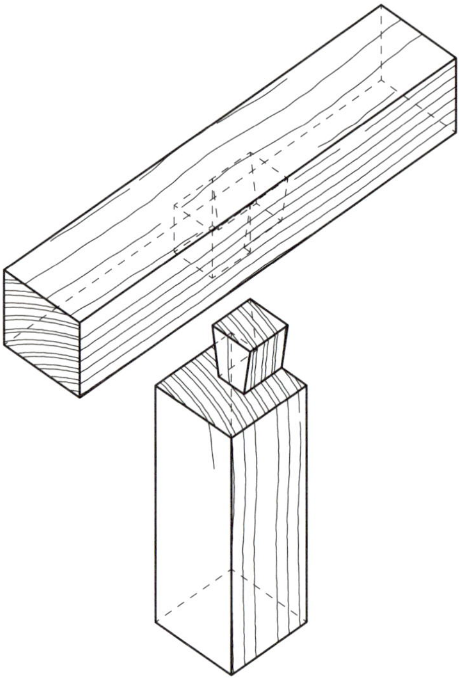

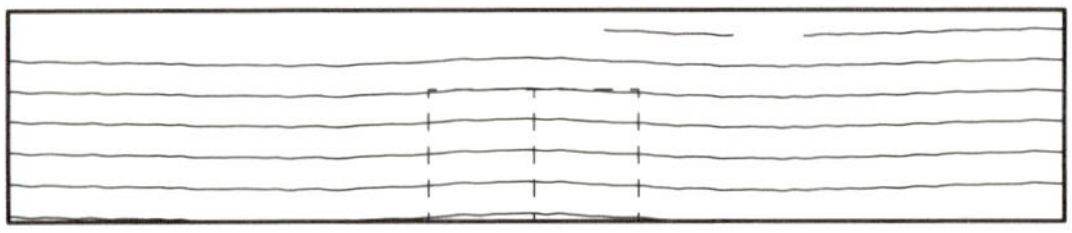

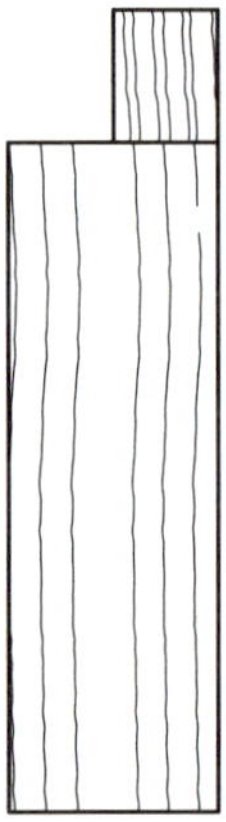

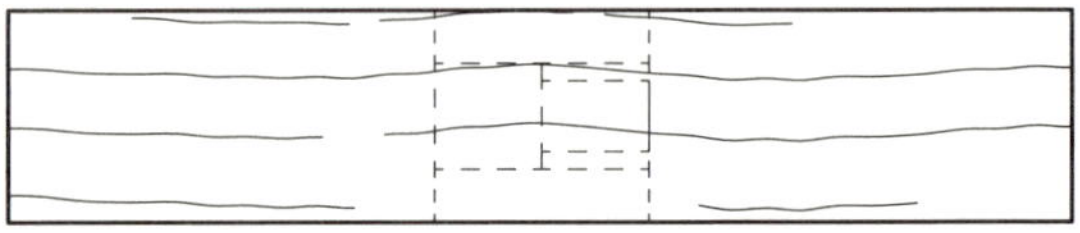

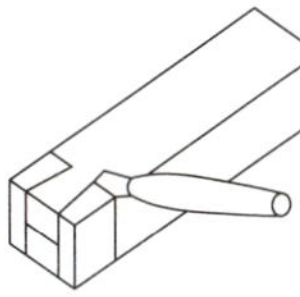

A1

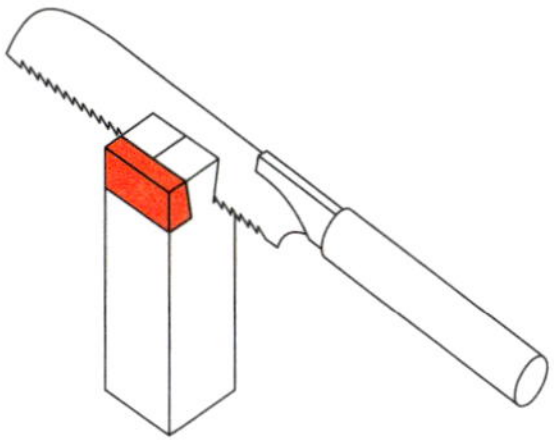

A2

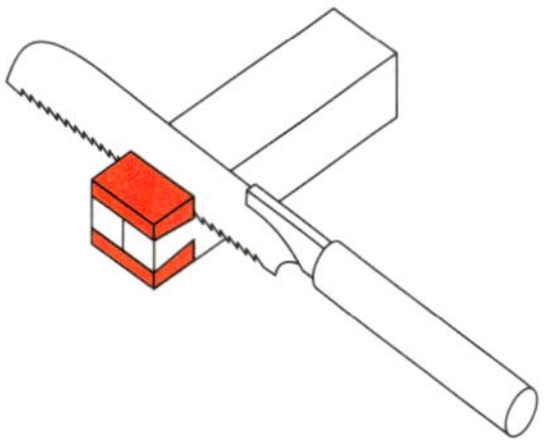

A3

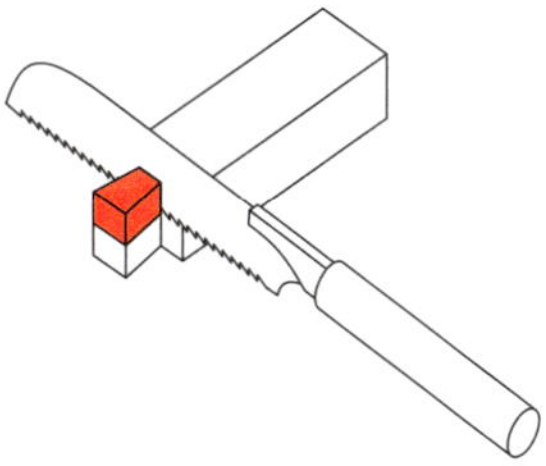

A4

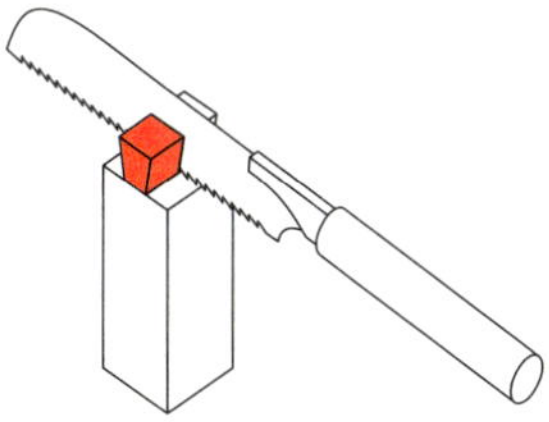

A5

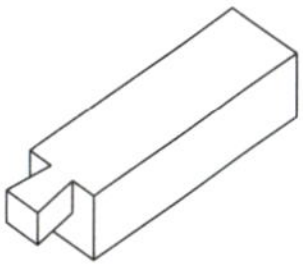

A

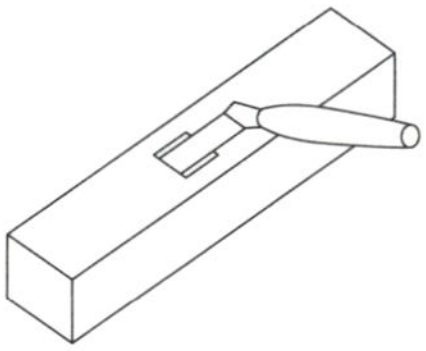

B1

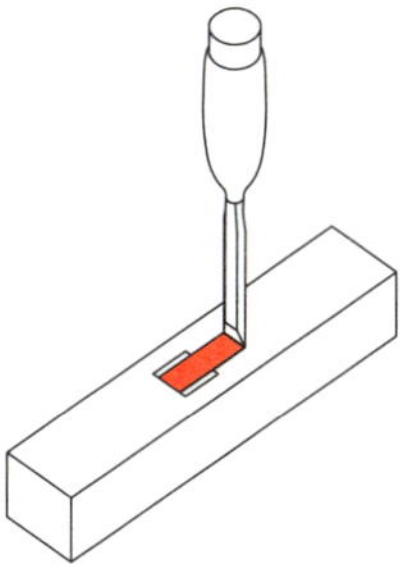

B2

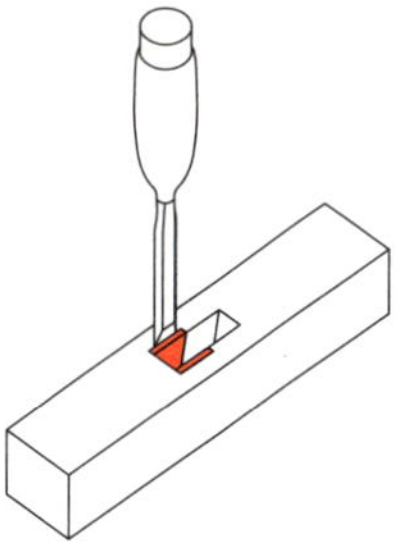

B3

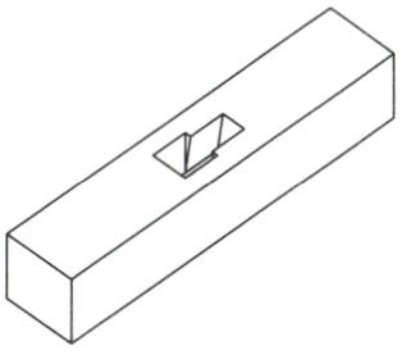

B

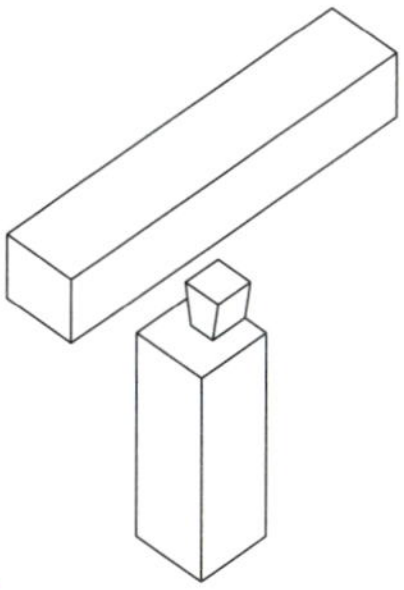

A+B

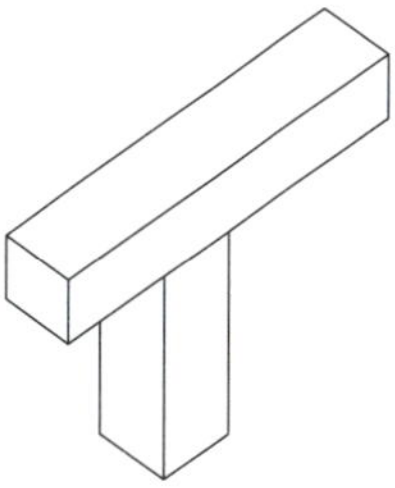

A+B

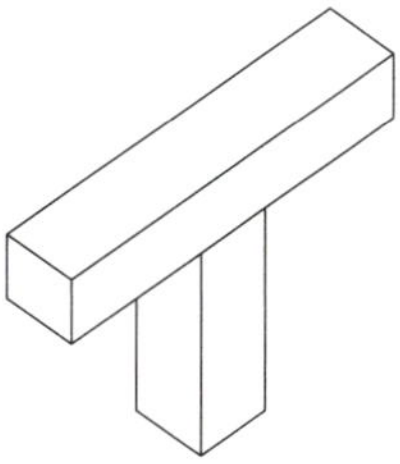

A+B

X-shaped stub tenon joint

kainokuchi-ai-kaki-kumi-tsugi
貝口相欠組継ぎ

Kreuzzapfen

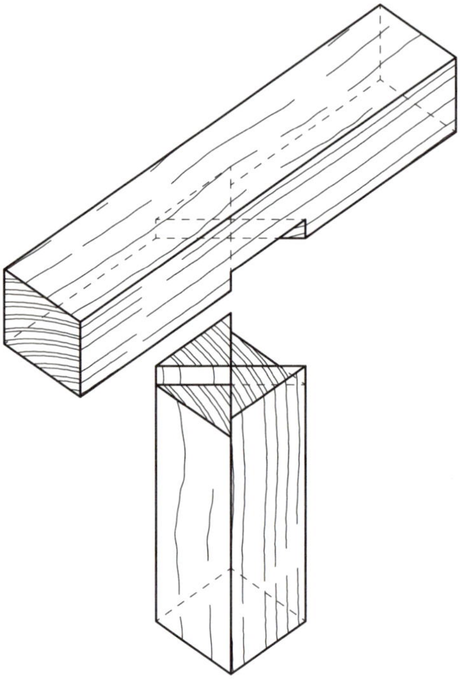

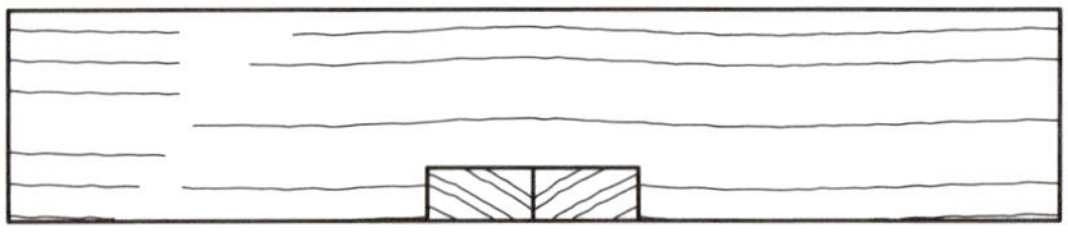

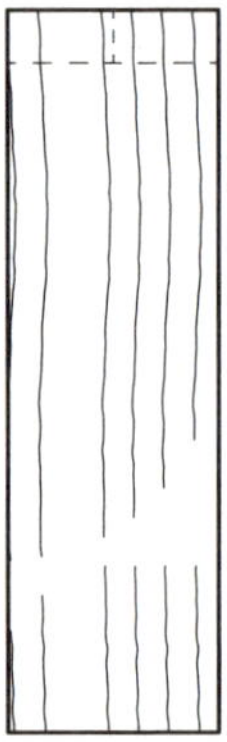

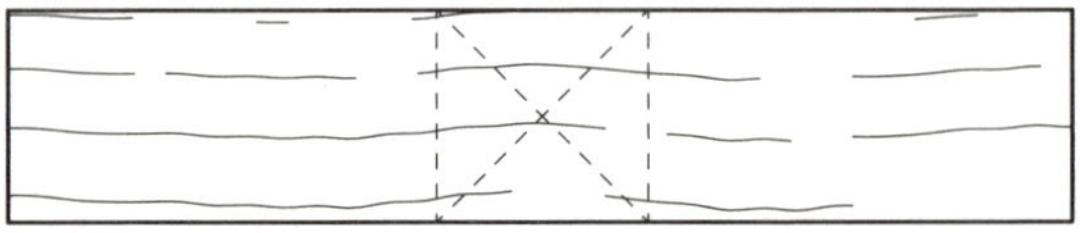

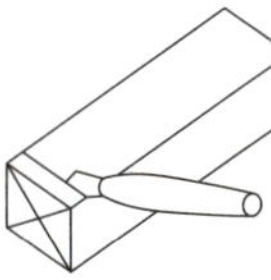

A1

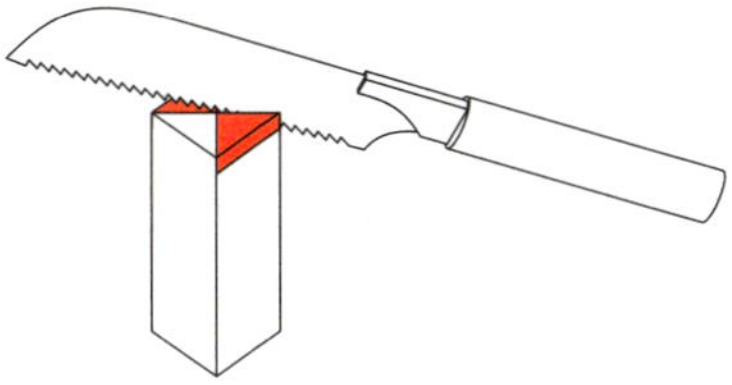

A2

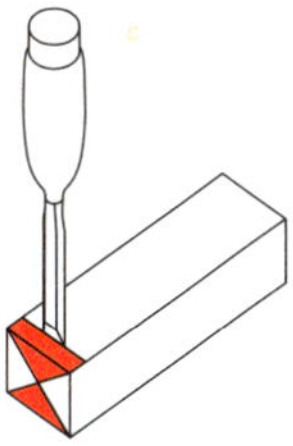

A3

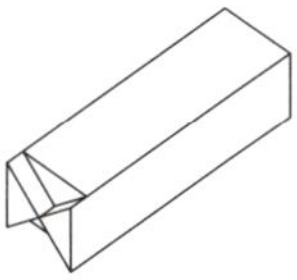

A

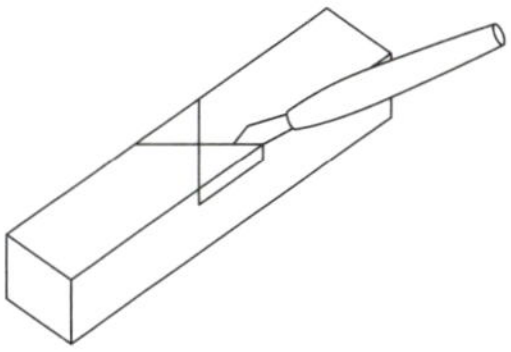

B1

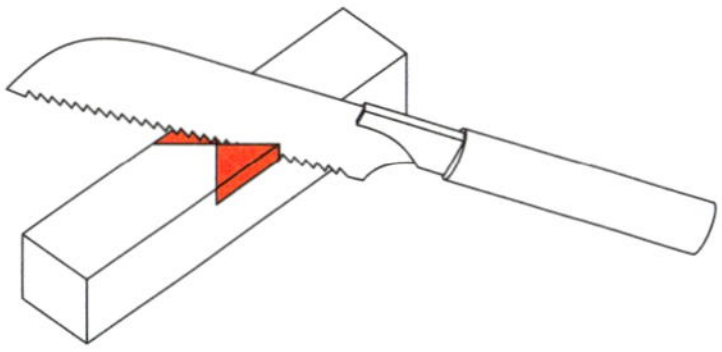

B2

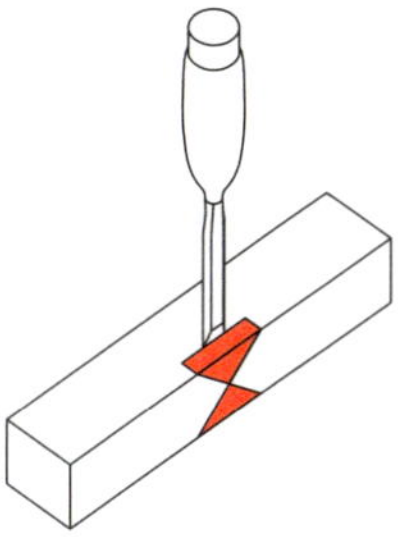

B3

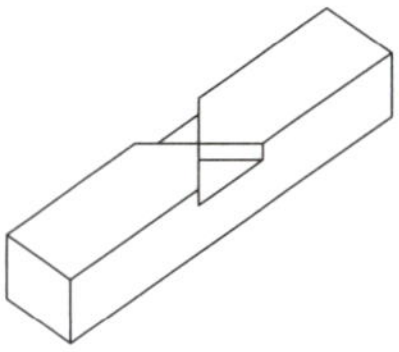

B

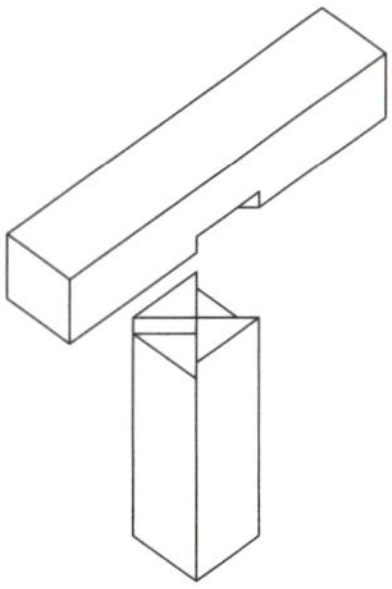

A+B

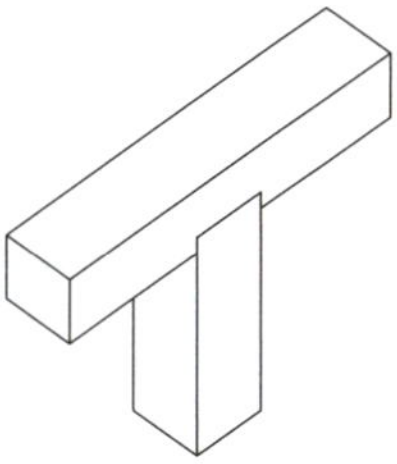

AB

T-Joints

Fully shouldered blind mortise and tenon

hashira-tan-hozo-sashi [*]
柱短枘差し

verdeckter beidseitig
abgesetzter Zapfen

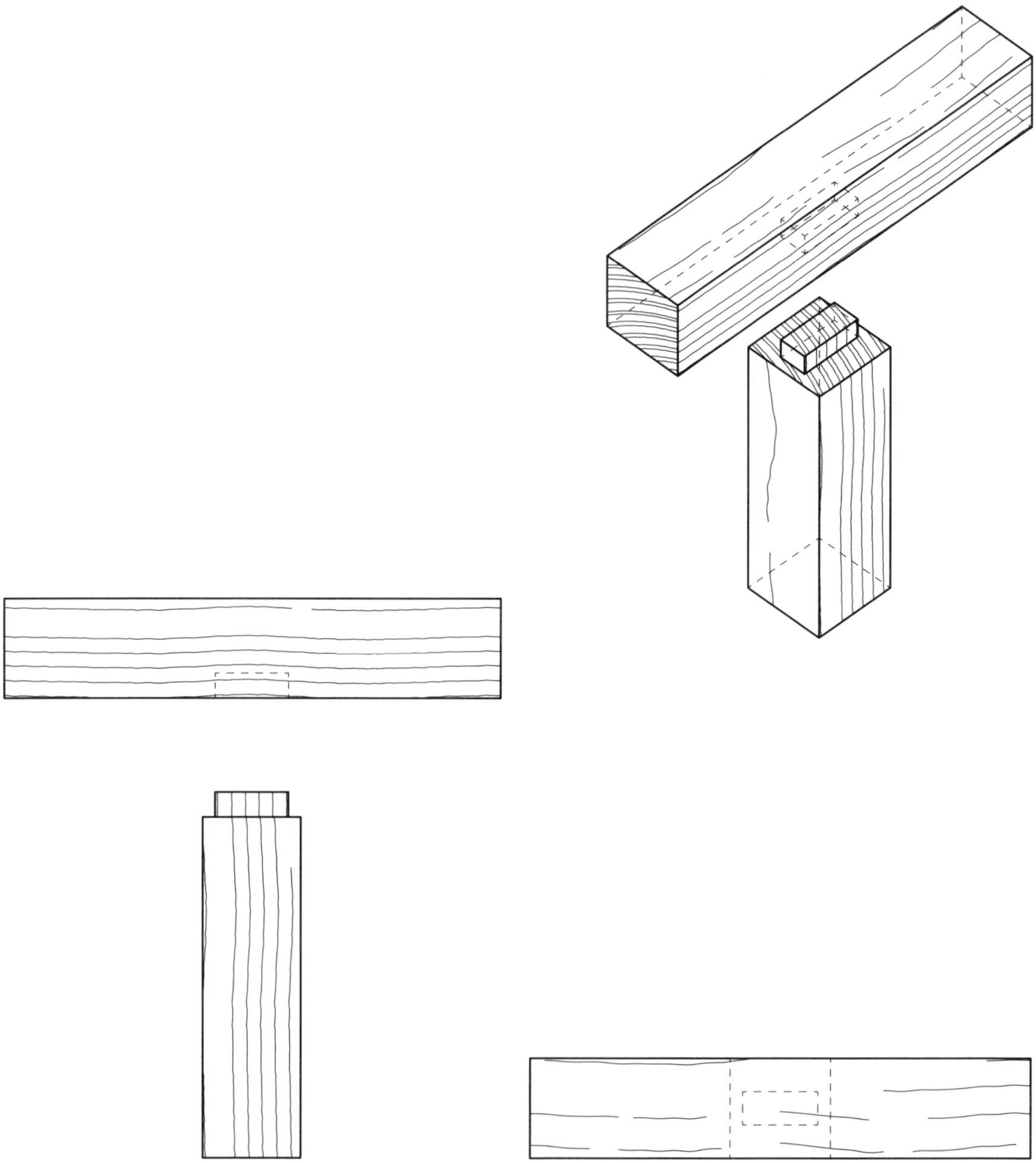

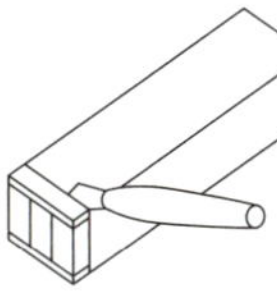

A1

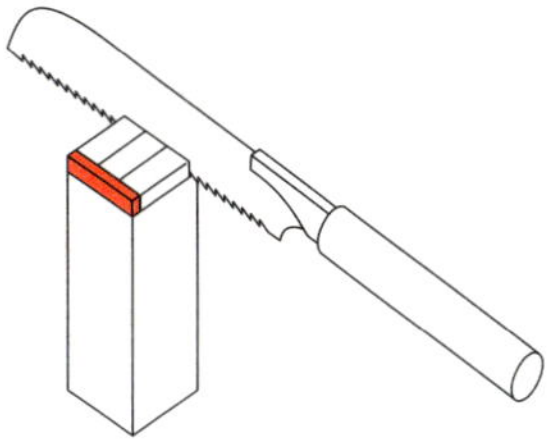

A2

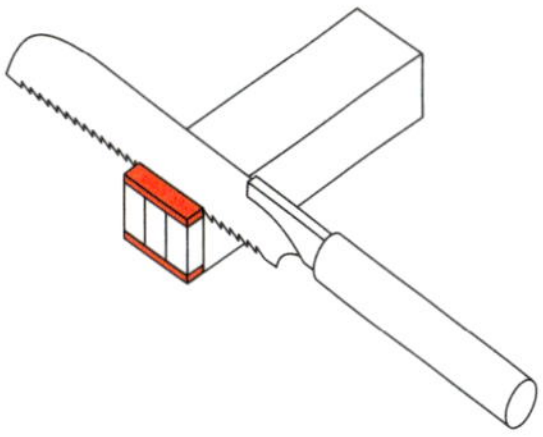

A3

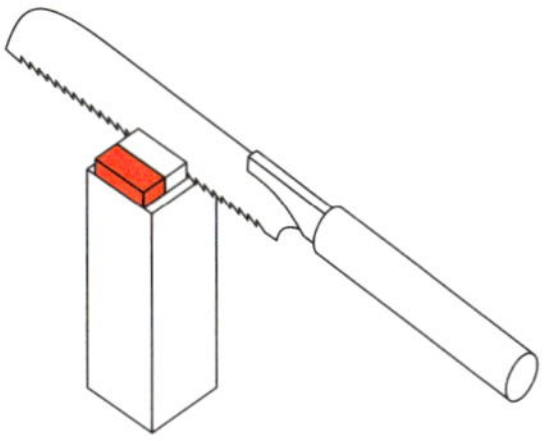

A4

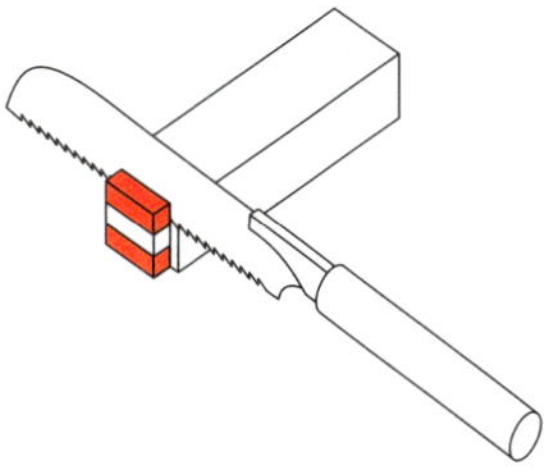

A5

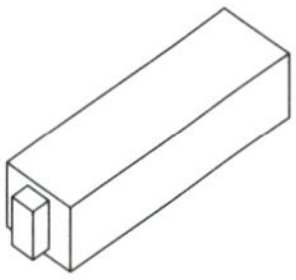

A

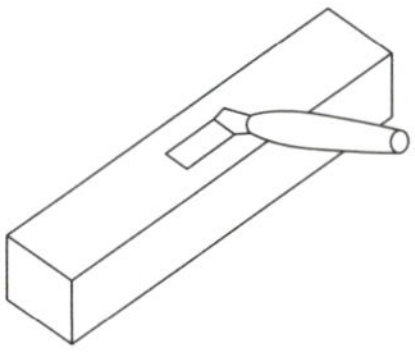

B1

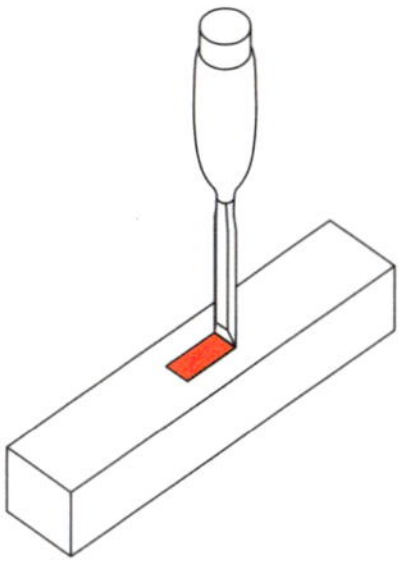

B2

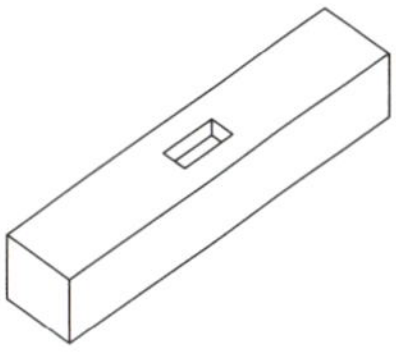

B

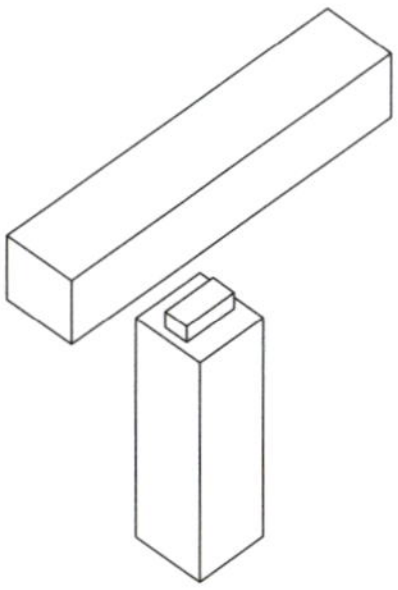

A+B

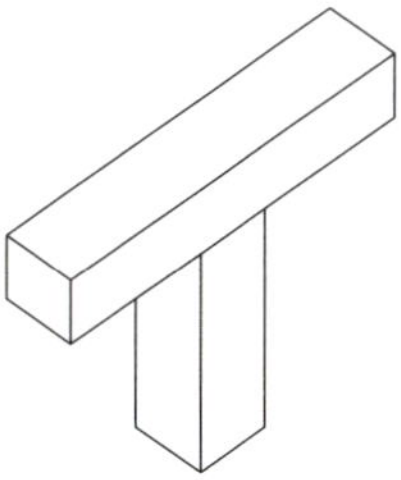

AB

T-Joints

Post with stub tenon and open slot mortise and tenon

mechi-tsugi-tan-hozo-sashi
目違継短枘差し

abgesetzter Zapfenstoß
auf verdecktem Zapfen

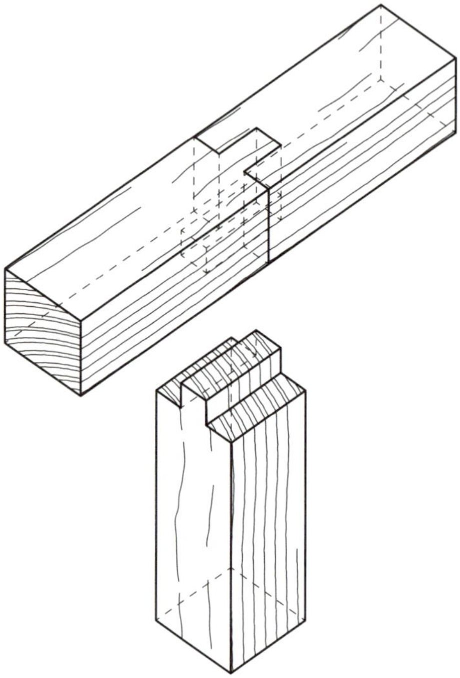

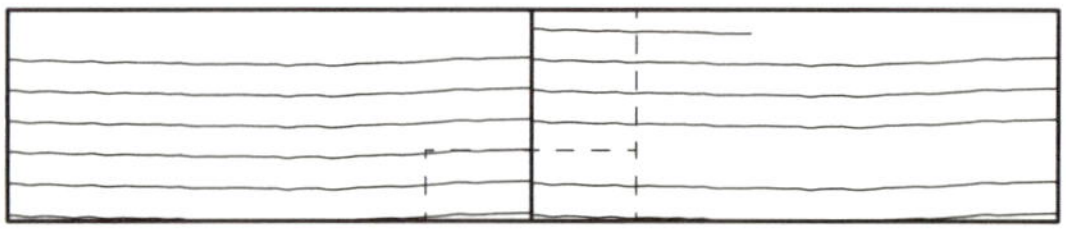

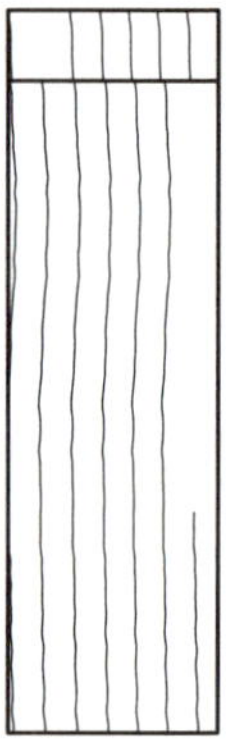

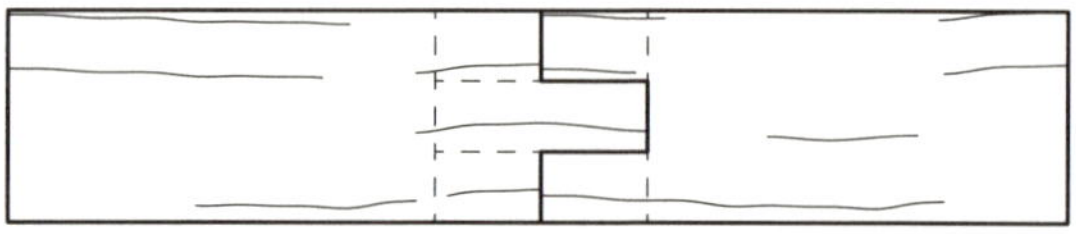

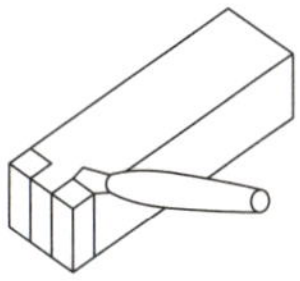

A1

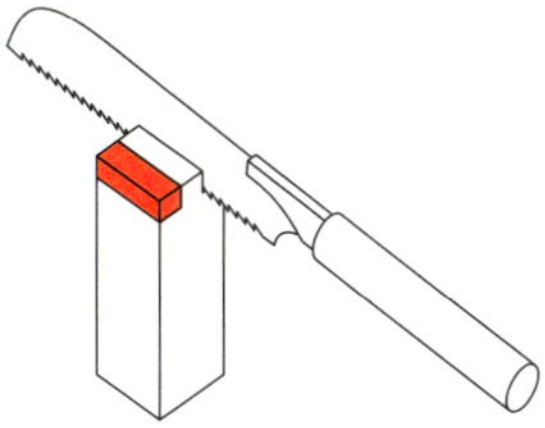

A2

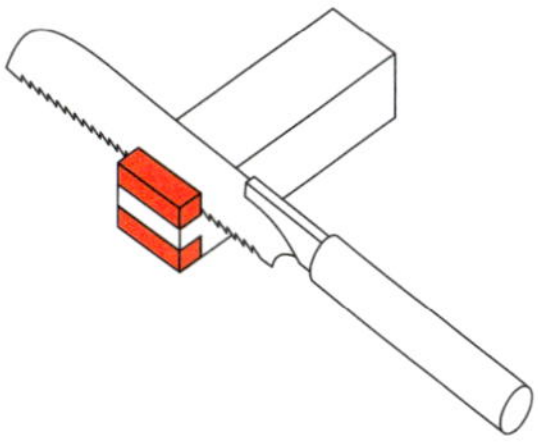

A3

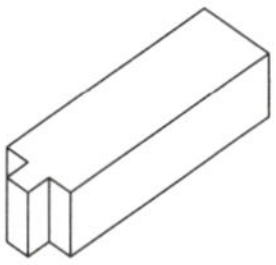

A

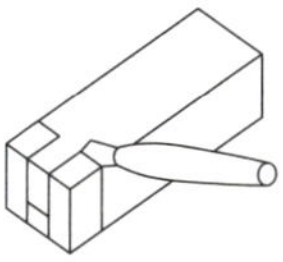

B1

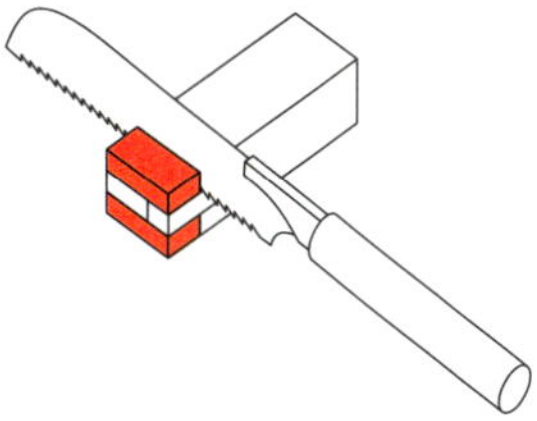

B2

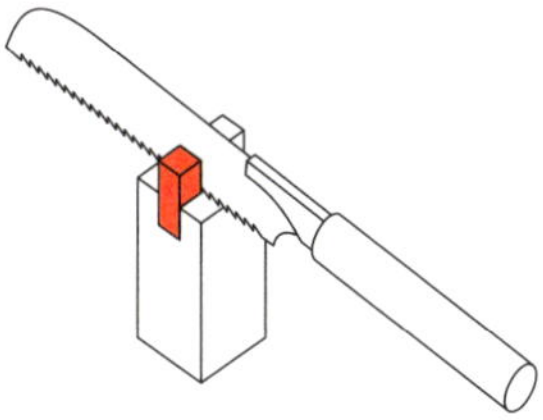

B3

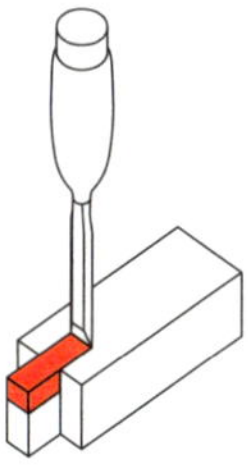

B4

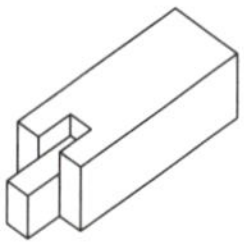

B

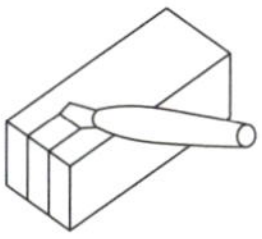

C1

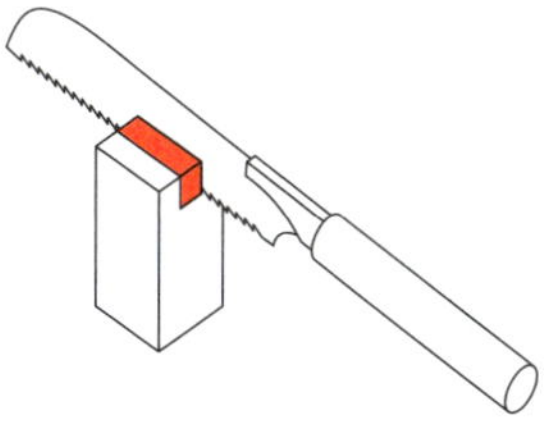

C2

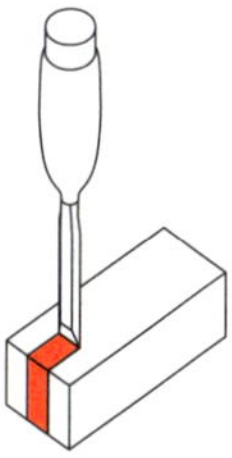

C3

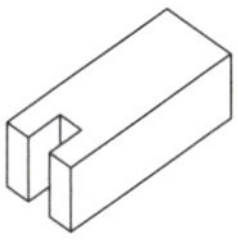

C

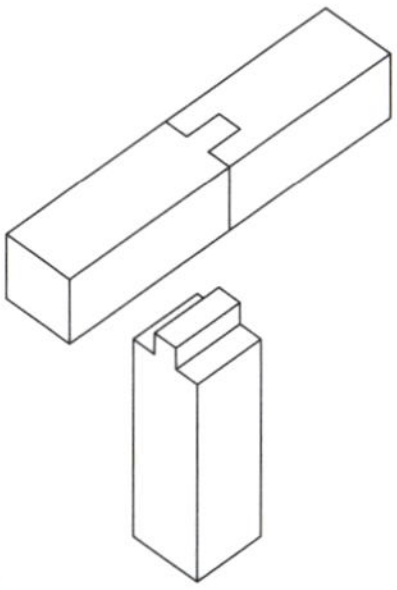

A+BC

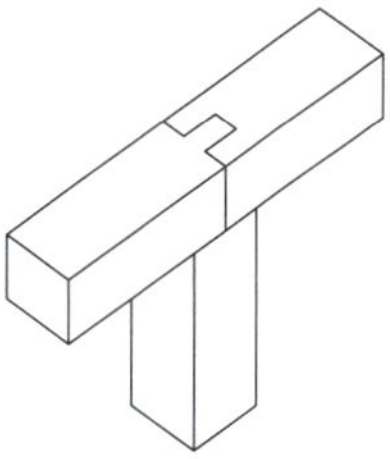

ABC

T-Joints

Post with stub tenon and lapped mortise and tenon

mechi-tsugi (dō-tsuki), hashira-tan-hozo-sashi
腰掛目違継ぎ（胴付）、柱短枘差し

gerades Blatt mit Zapfen auf verdecktem Zapfen

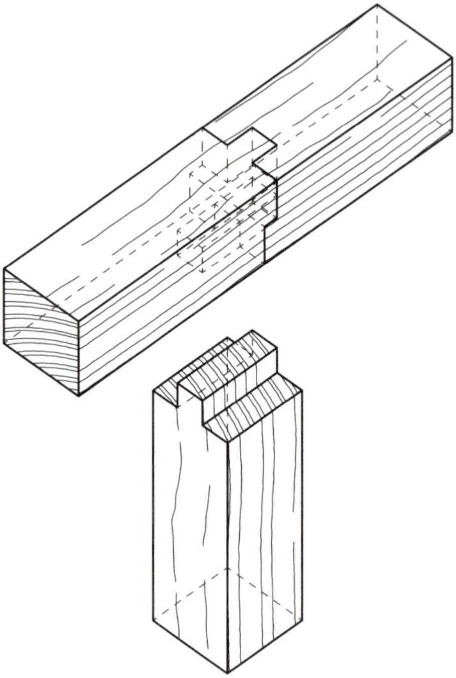

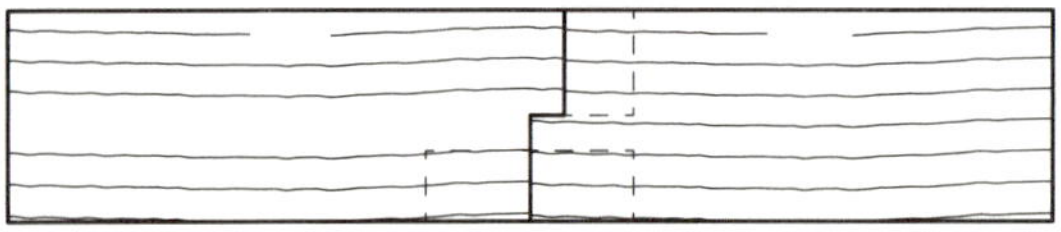

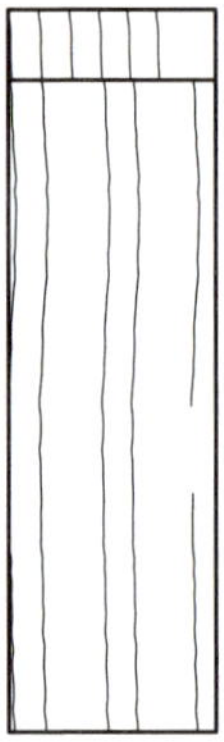

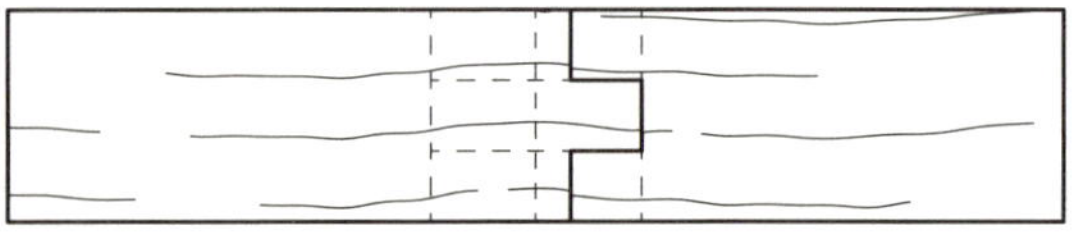

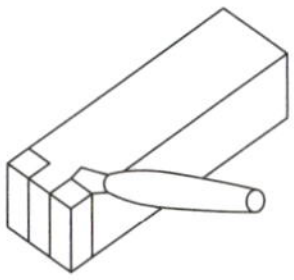
A1

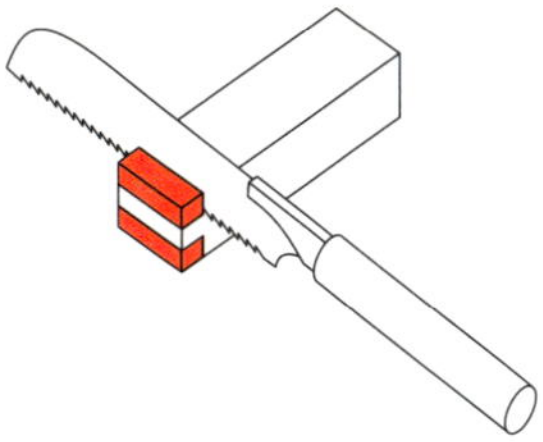
A2

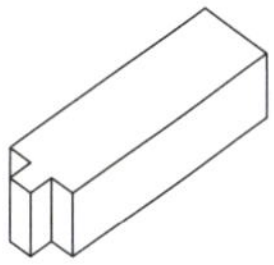
A

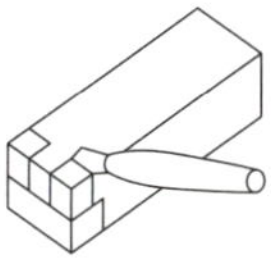
B1

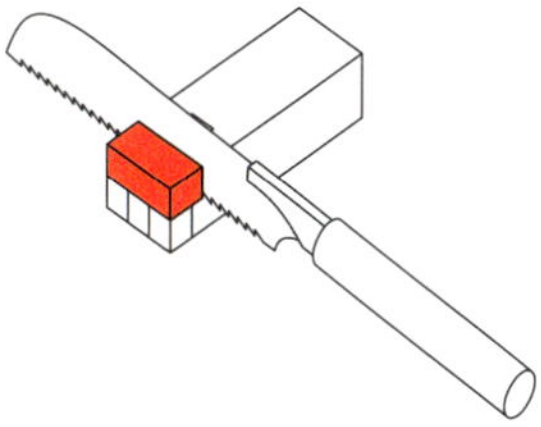
B2

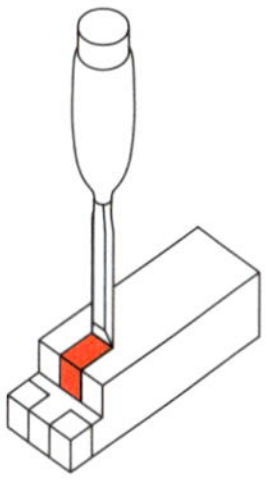
B3

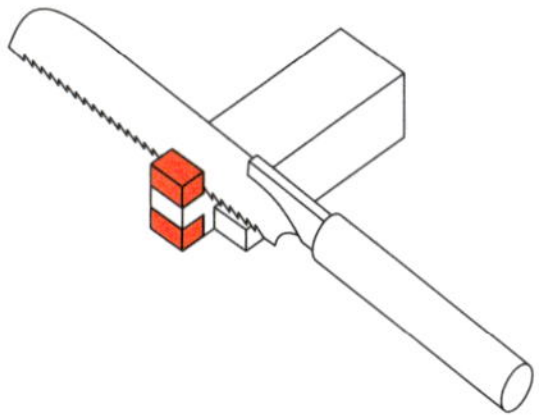
B4

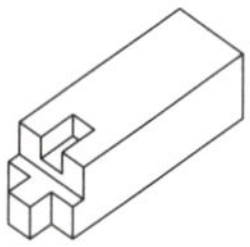
B

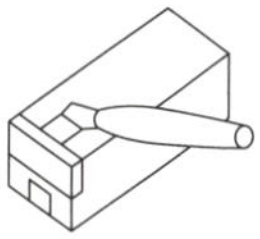
C1

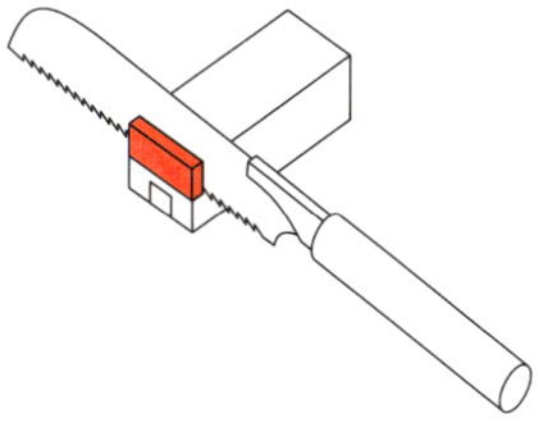
C2

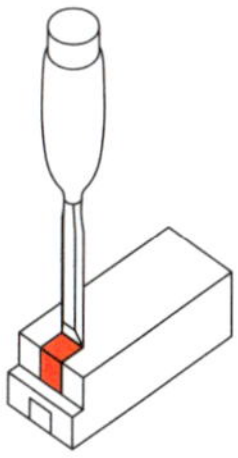
C3

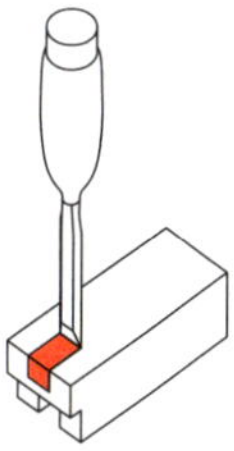
C4

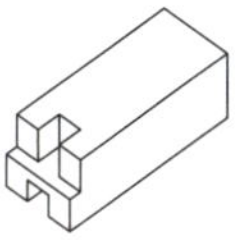
C

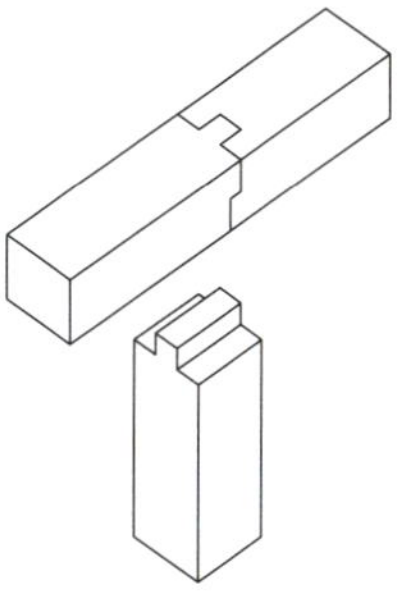
A+BC

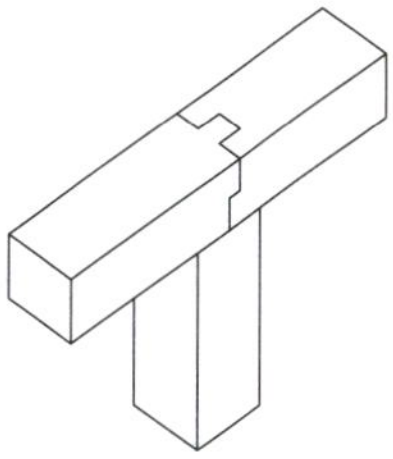
ABC

T-Joints

Post with through tenon and mortised stop-splayed cogged scarf joint with stub tenons

Japanese

daimochi-tsugi, hashira-naga-hozo-sashi [*]
台持継ぎ、柱長枘差し

German

schräges Hakenblatt mit Brüstungszapfen auf durchgehendem Pfostenzapfen

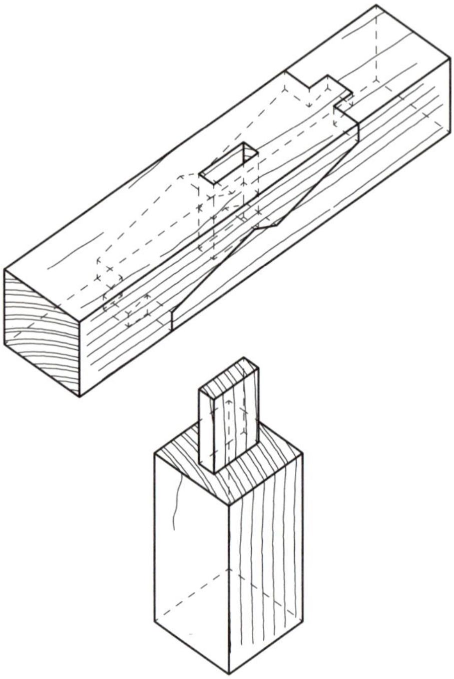

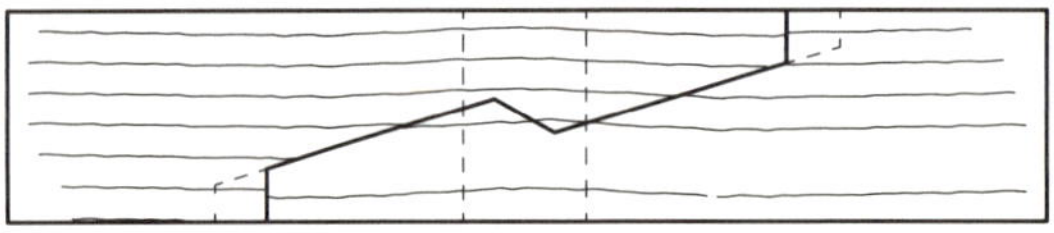

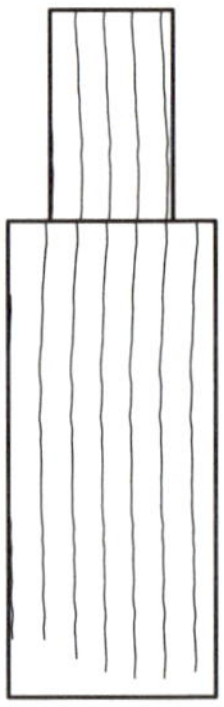

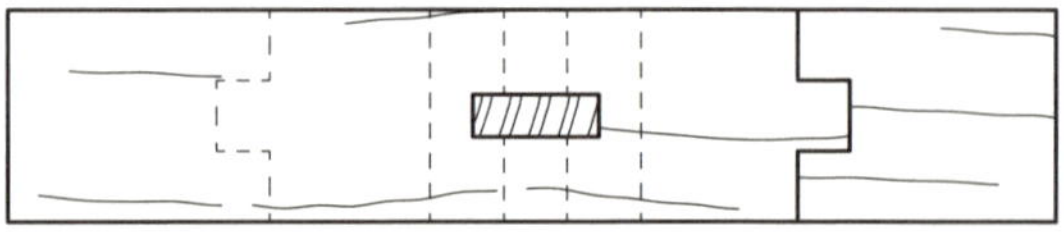

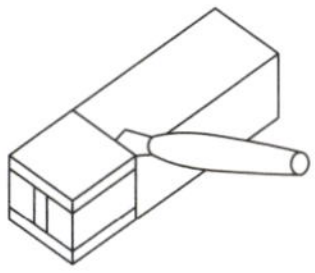

A1

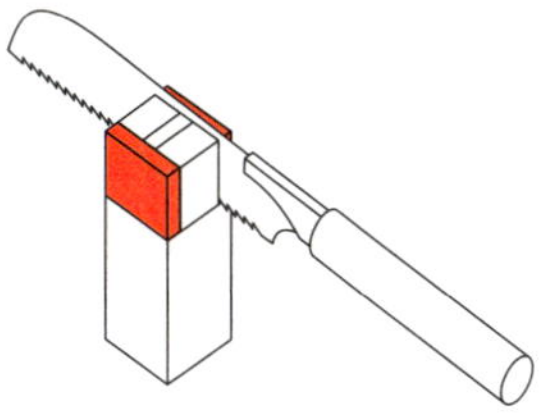

A2

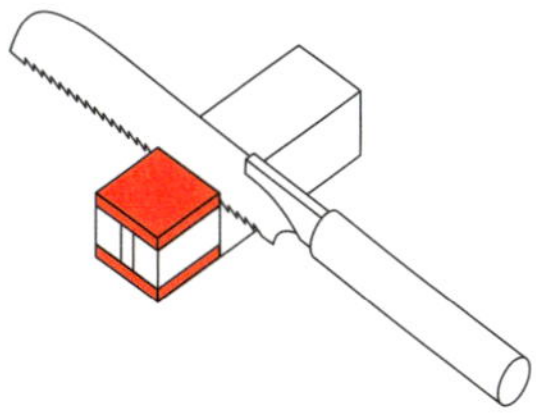

A3

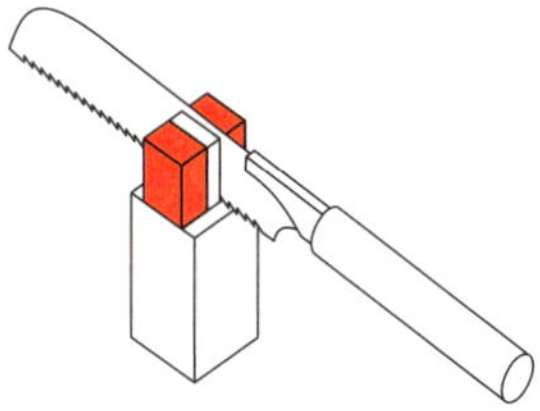

A4

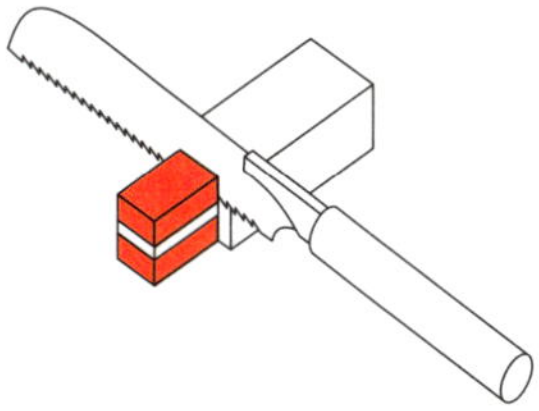

A5

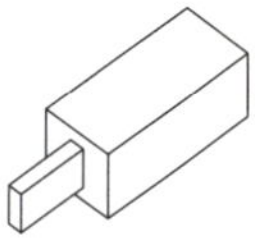

A

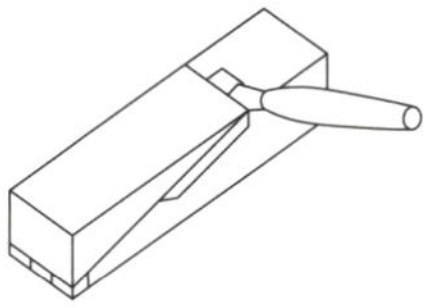

B1

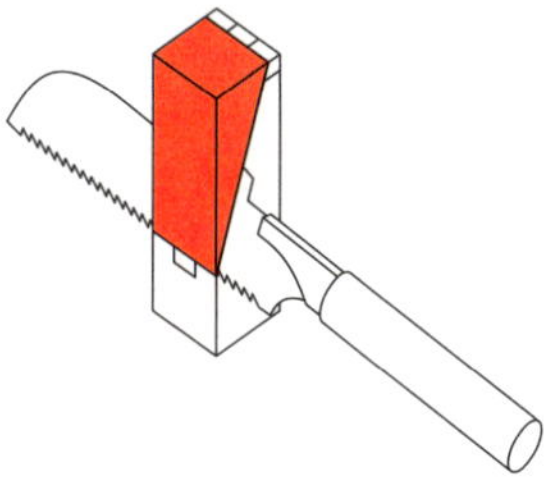

B2

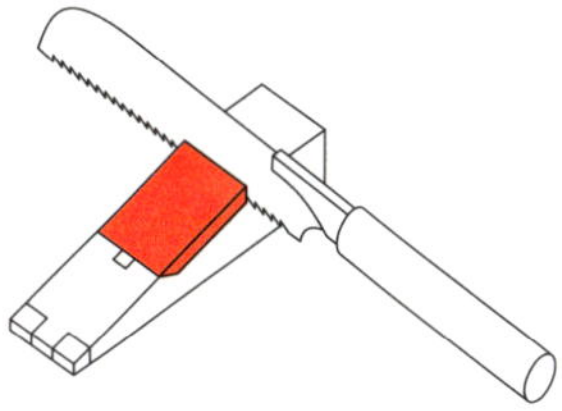

B3

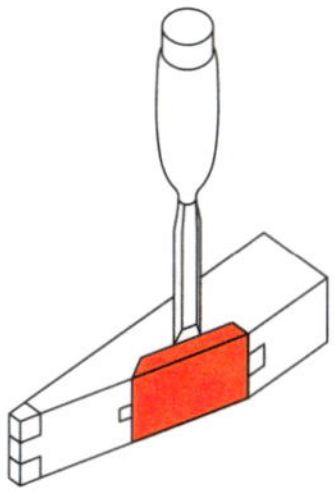

B4

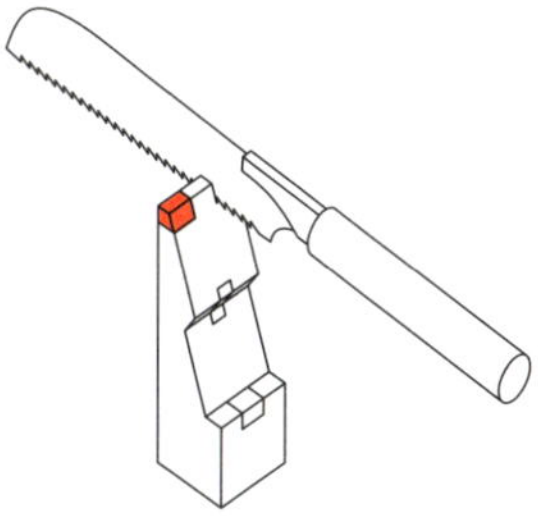

B5

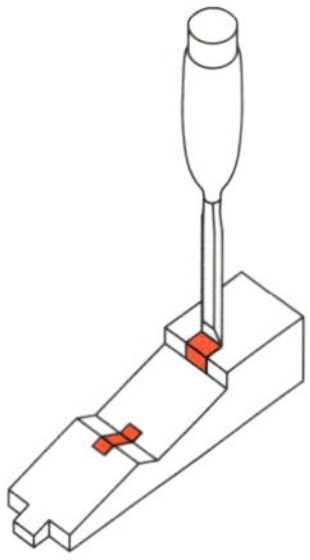

B6

2×

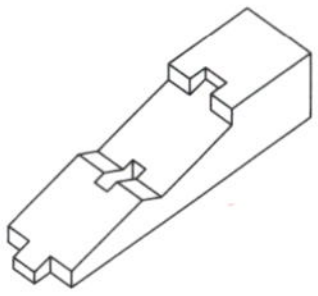

B

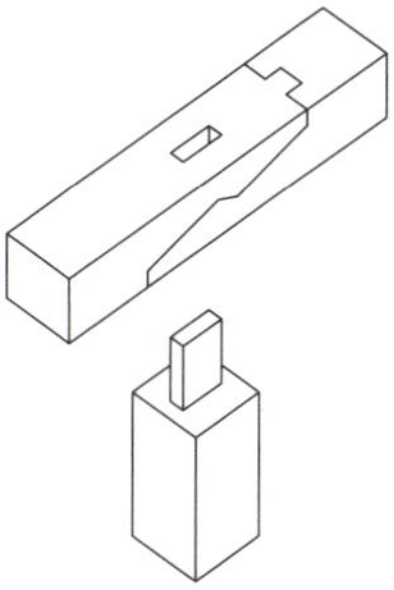

A+BB

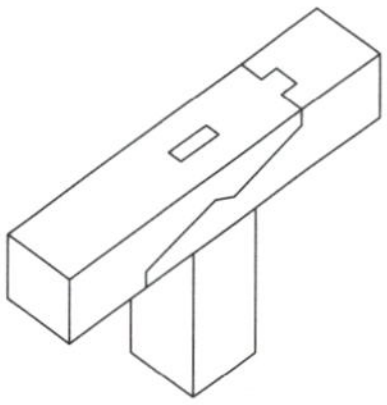

ABB

Angled Joints

Notched oblique thrust joint

naname-sogi-kaki-uchi-tsuke-tsugi
斜殺欠打付接ぎ

Stirnversatz

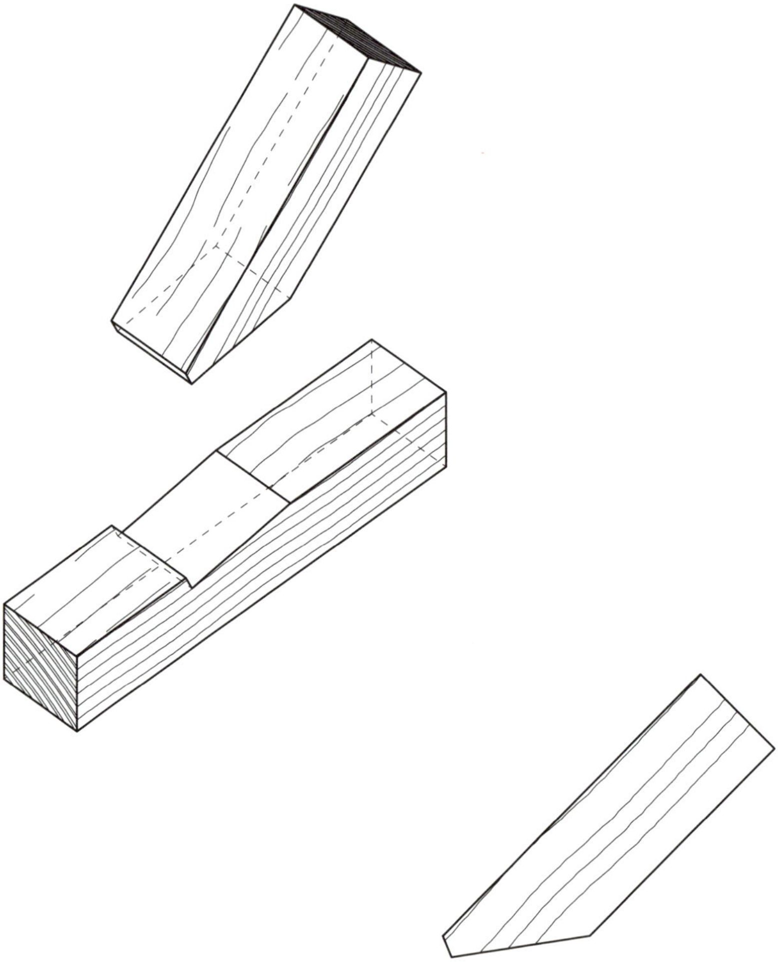

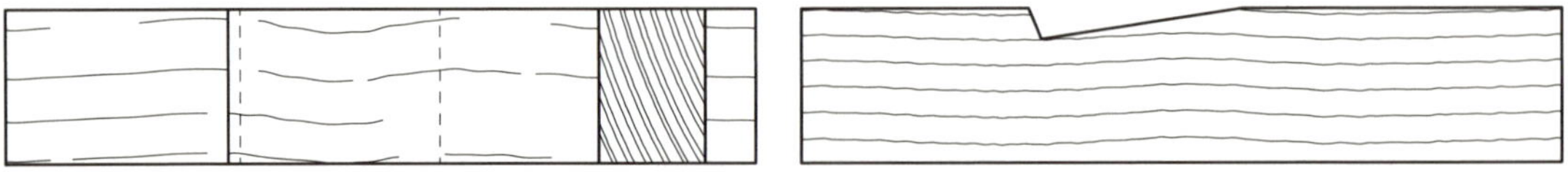

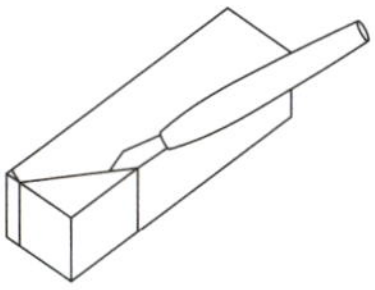

A1

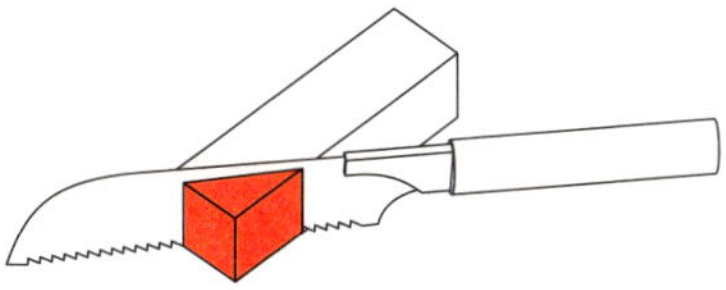

A2

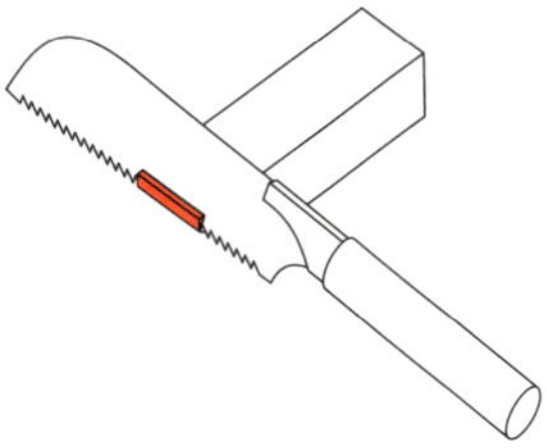

A3

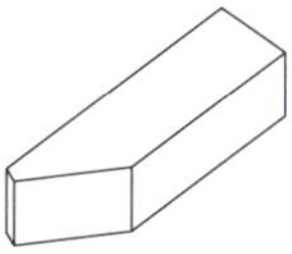

A

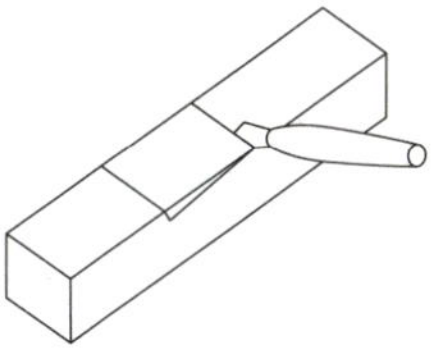

B1

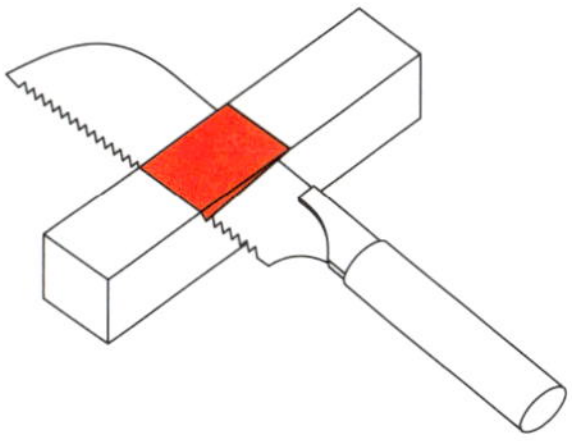

B2

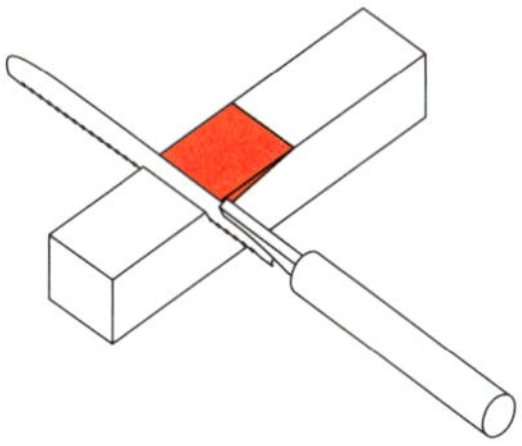

B3

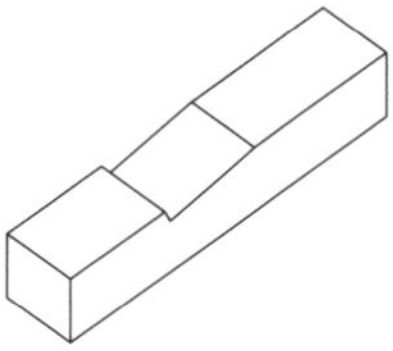

B

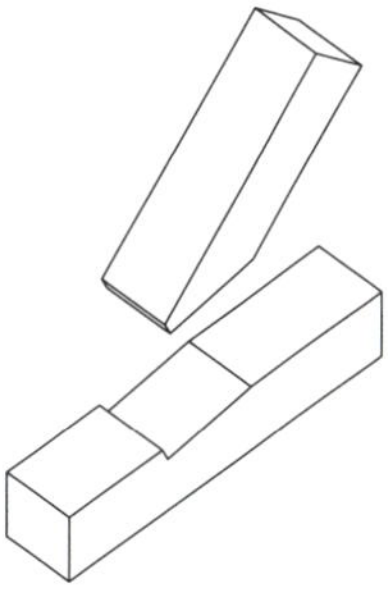

A+B

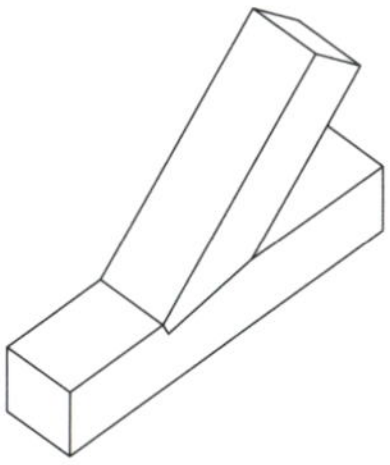

AB

Angled Joints

Notched oblique thrust joint with right-angled end

naname-sogi-kaki-kane-ire-uchi-tsuke-tsugi
斜殺欠矩入打付接ぎ

Stirnversatz mit rechtwinkliger Stirn

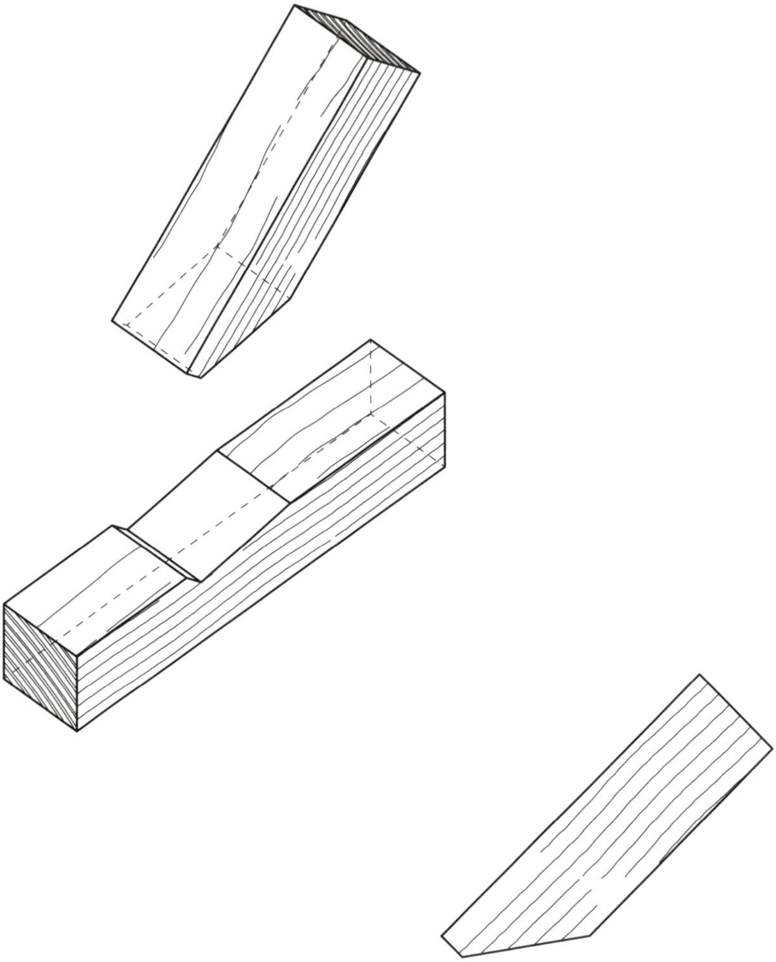

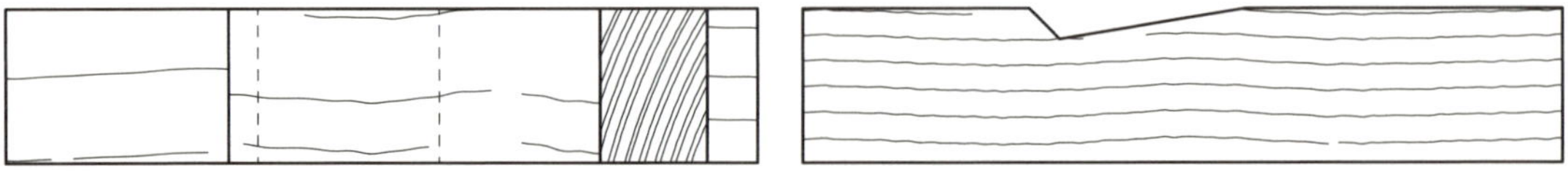

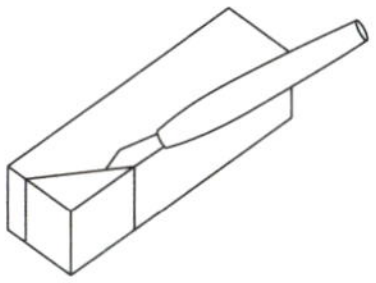

A1

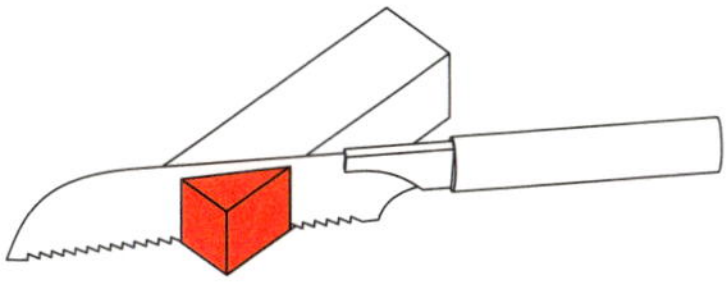

A2

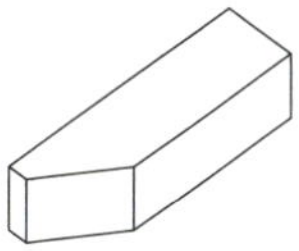

A

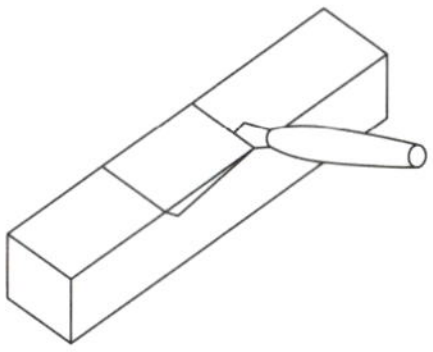

B1

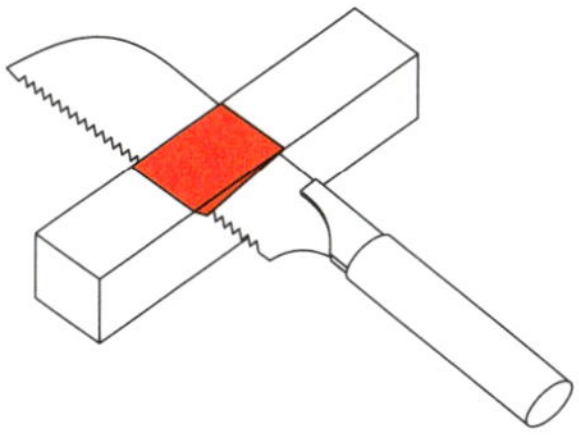

B2

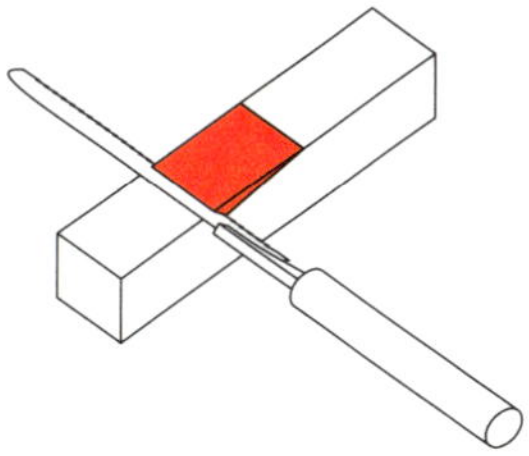

B3

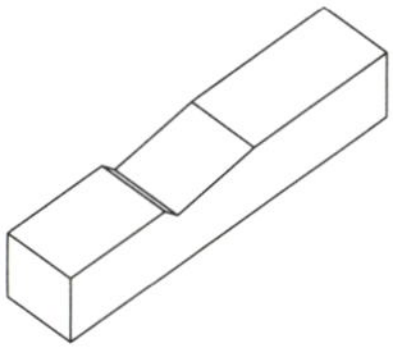

B

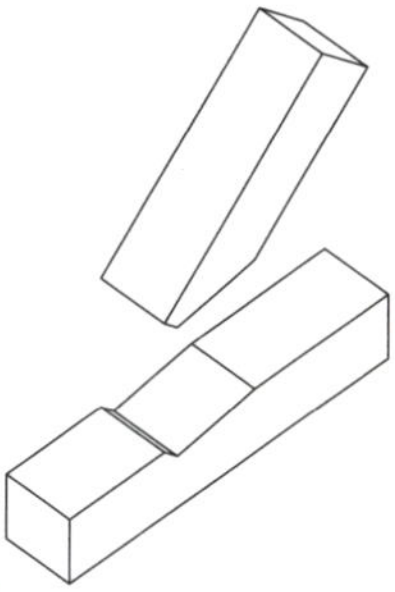

A+B

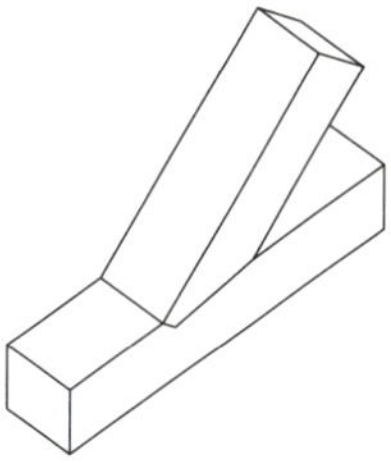

AB

Angled Joints

Notched oblique thrust joint with visible end

naname-sogi-kaki-shiki-men-uchi-tsuke-tsugi
斜殺欠敷面打付接ぎ

Brustversatz

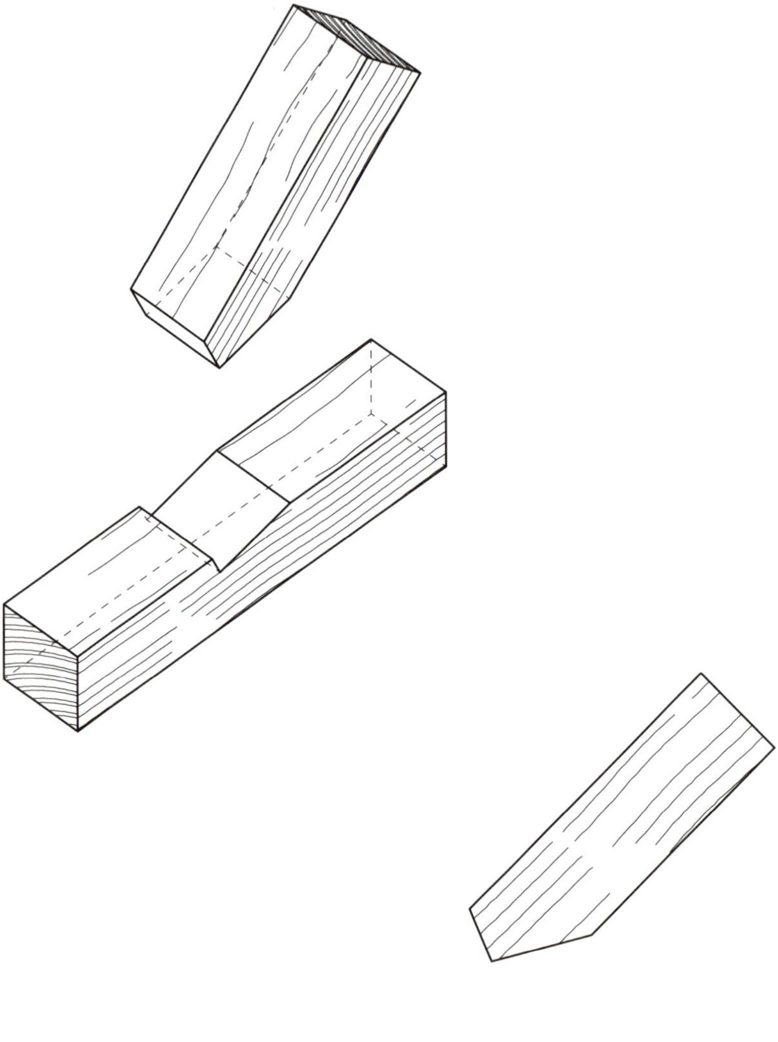

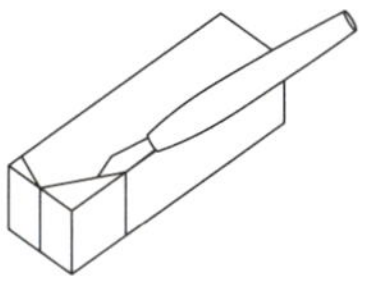

A1

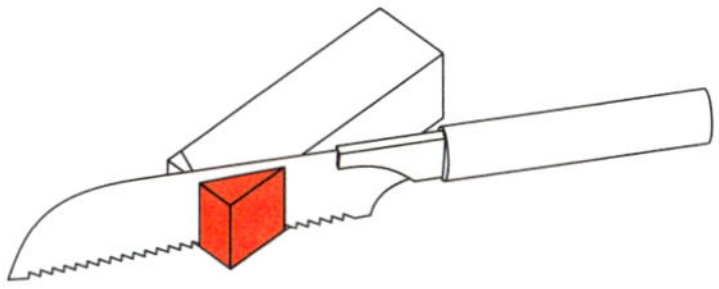

A2

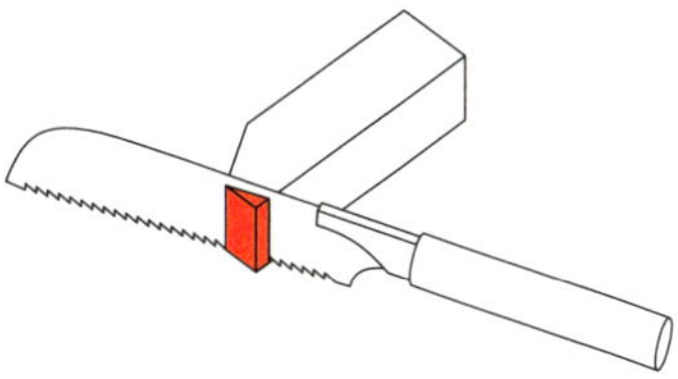

A3

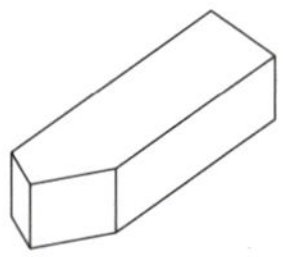

A

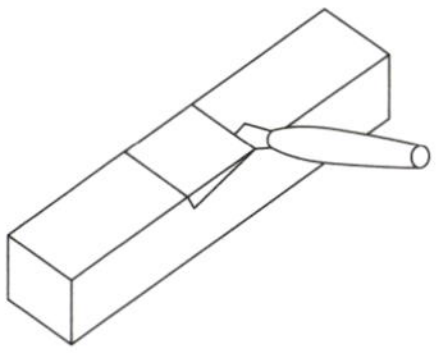

B1

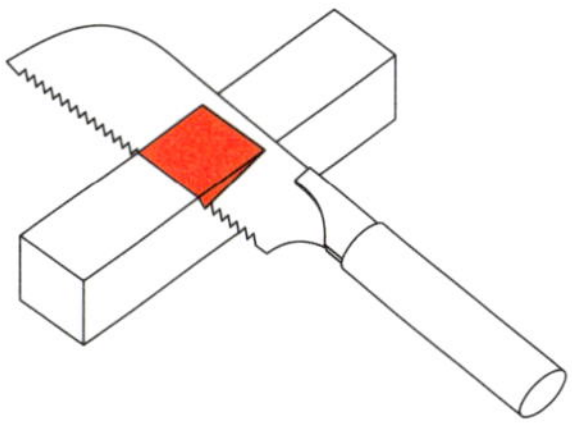

B2

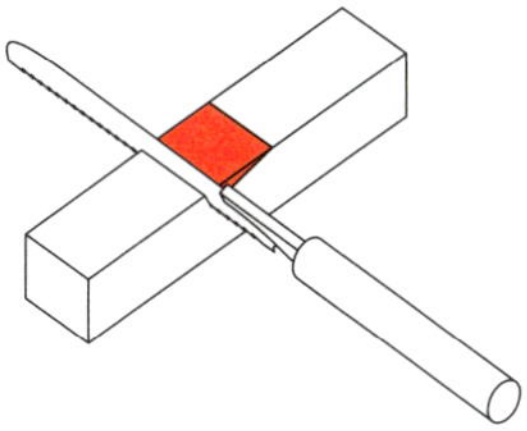

B3

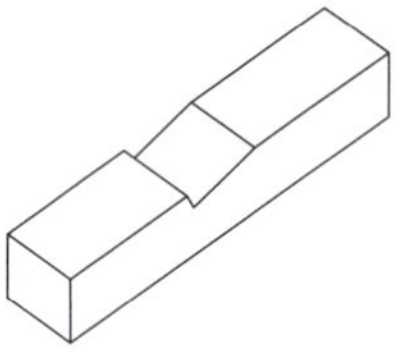

B

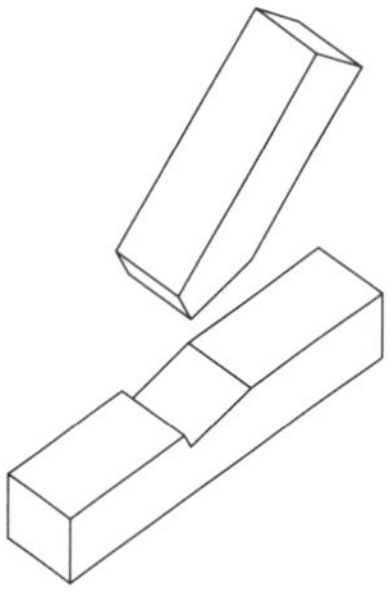

A+B

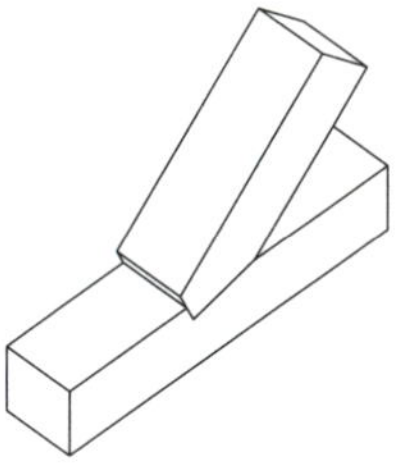

AB

Angled Joints

Oblique thrust joint with notched heel

naname-eriwa-uchi-tsuke-tsugi
斜襟輪打付接ぎ

Fersenversatz

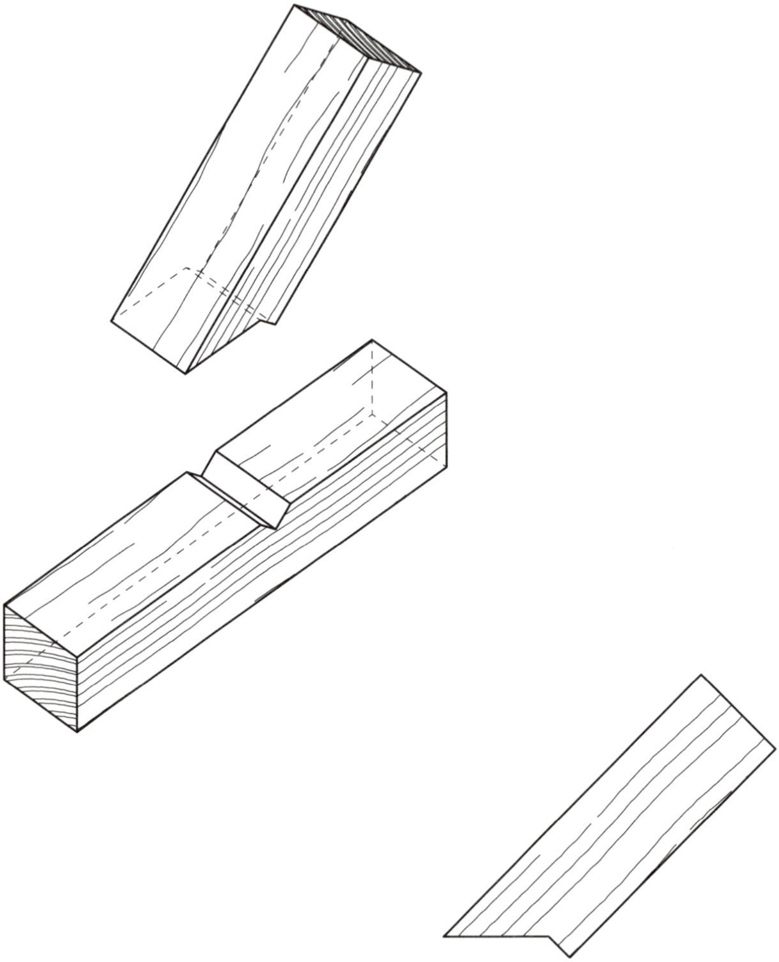

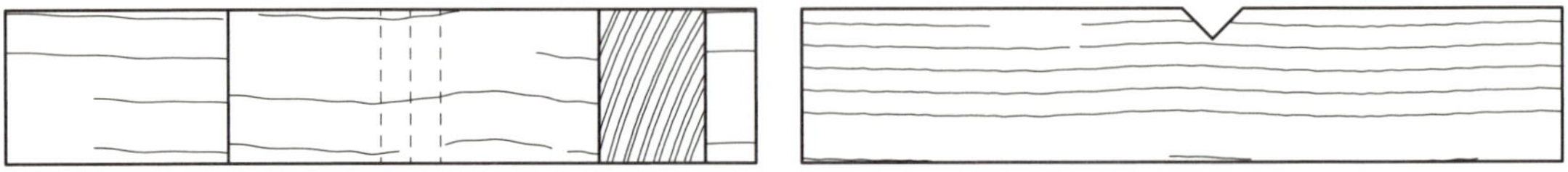

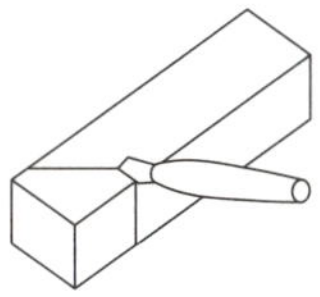

A1

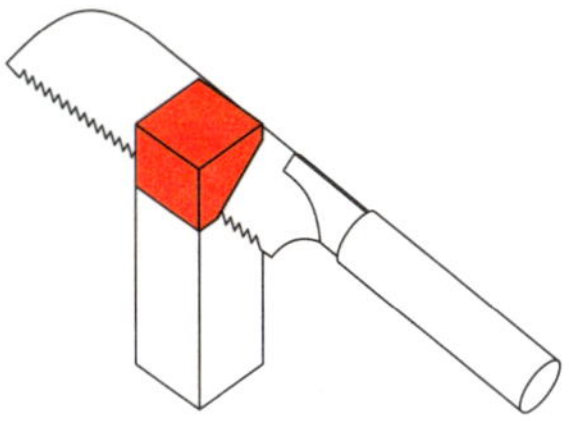

A2

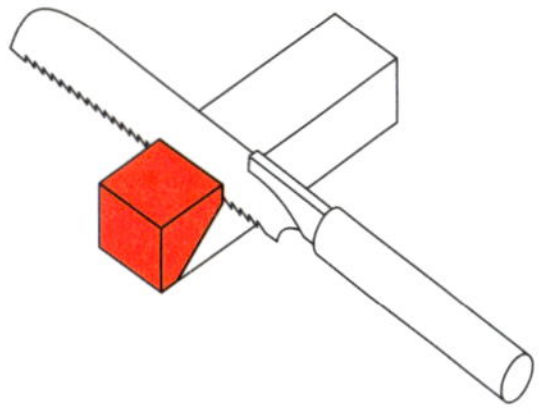

A3

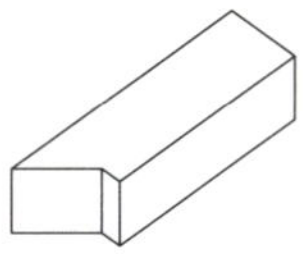

A

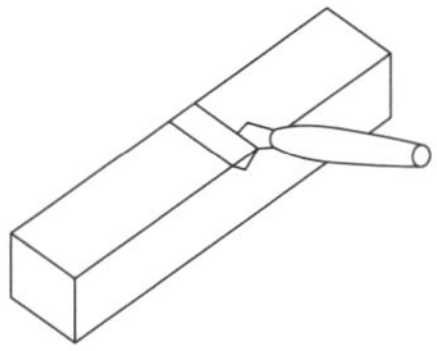

B1

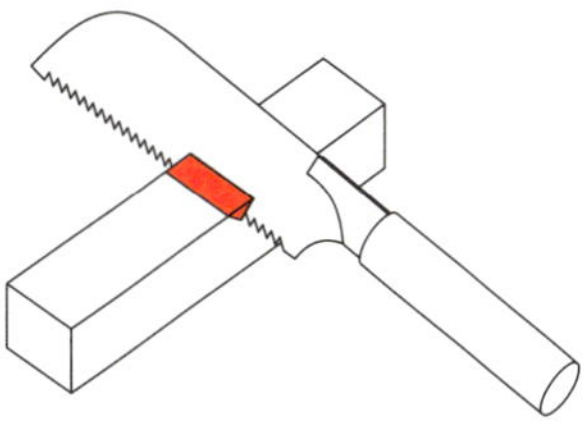

B2

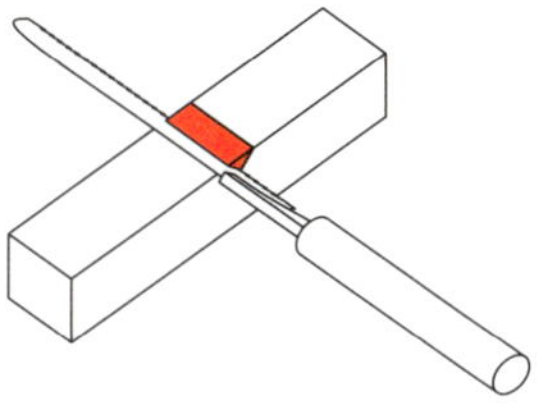

B3

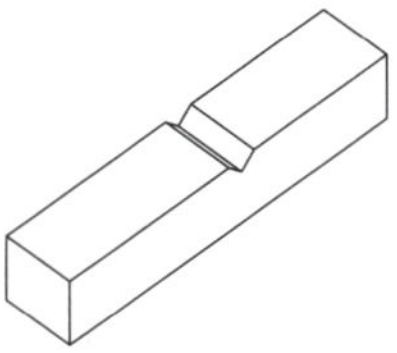

B

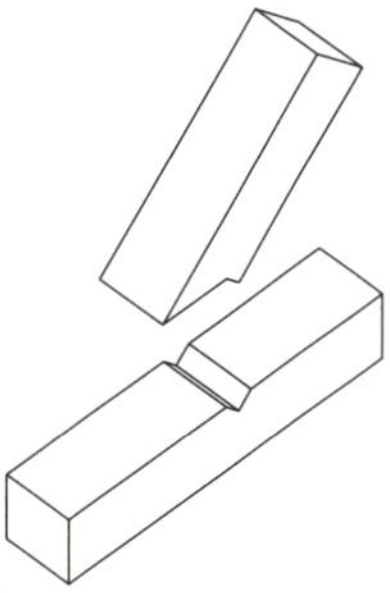

A+B

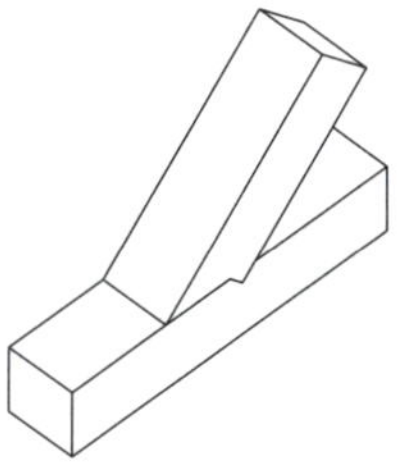

AB

Angled Joints

Double notched oblique thrust joint

naname-sogi-kaki-eriwa-uchi-tsuke-tsugi
斜殺欠襟輪打付接ぎ

Stirn- und Versenversatz

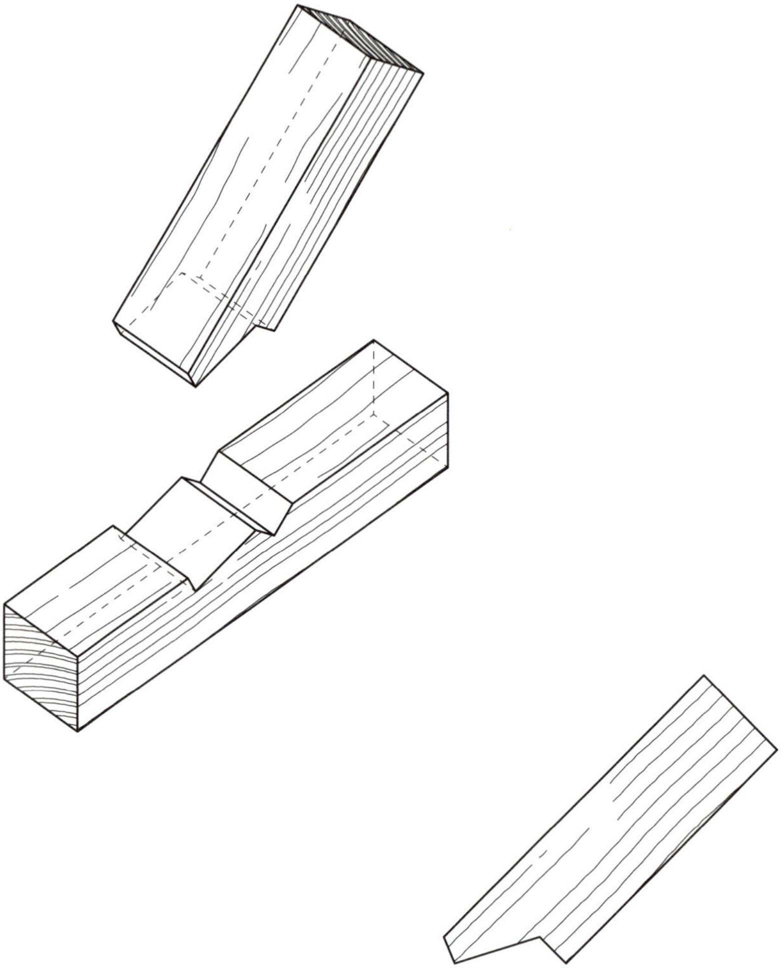

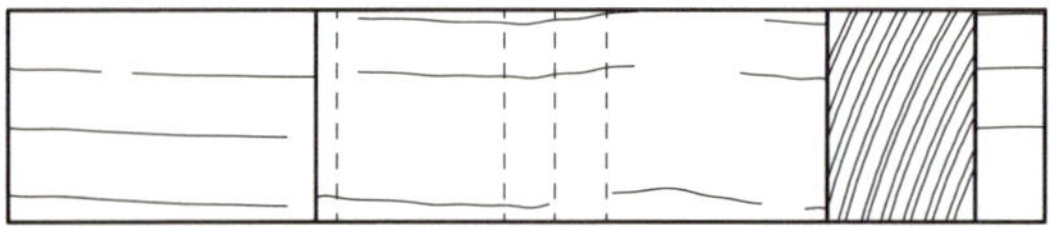

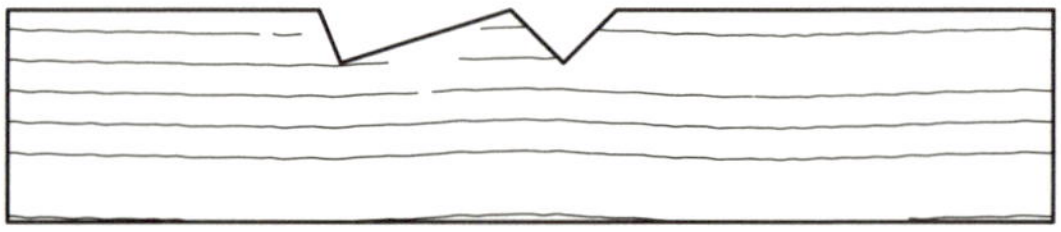

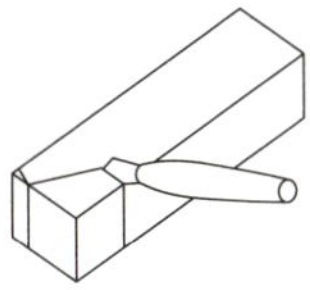

A1

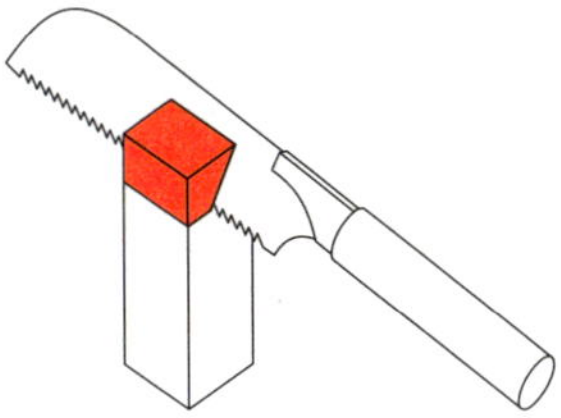

A2

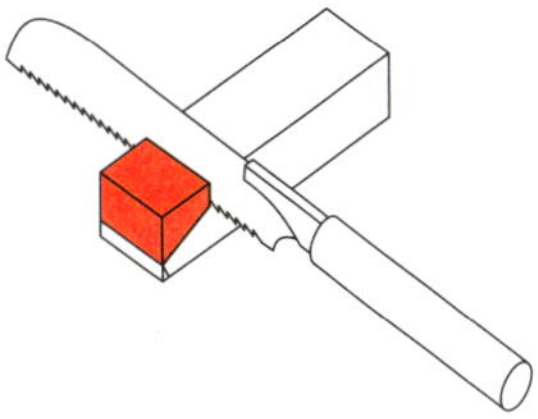

A3

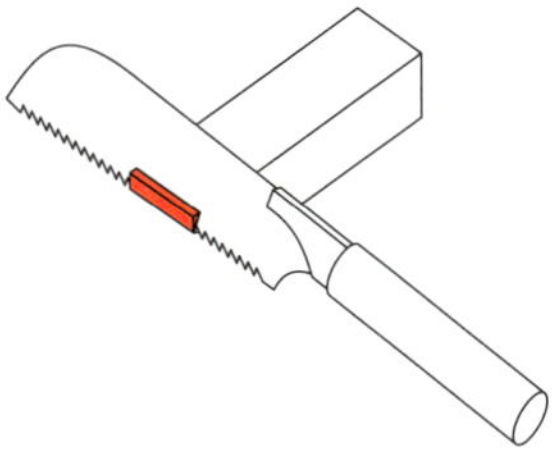

A4

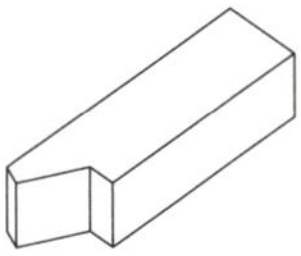

A

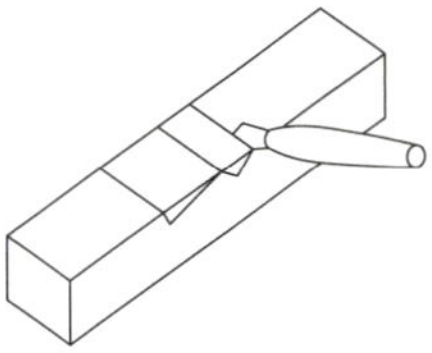

B1

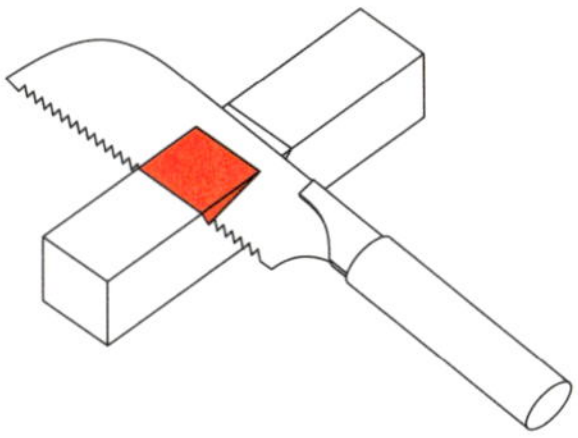

B2

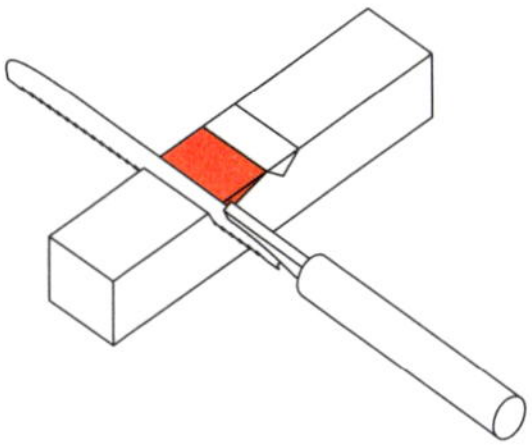

B3

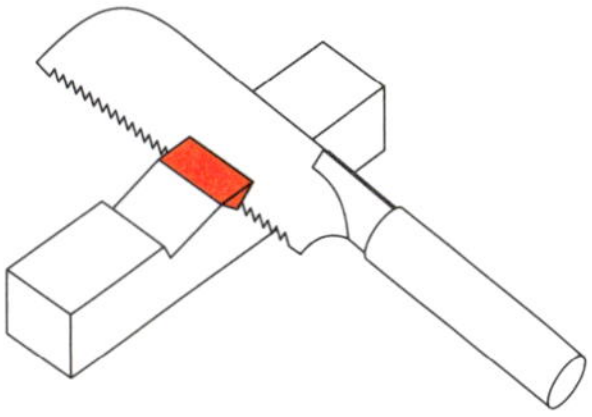

B4

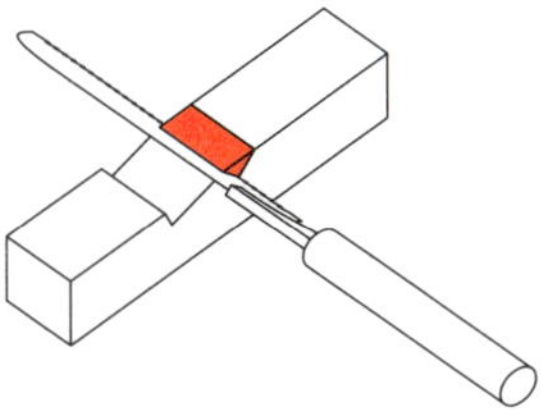

B5

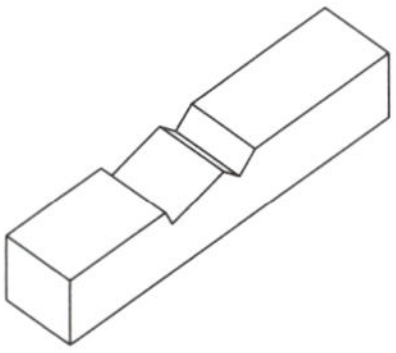

B

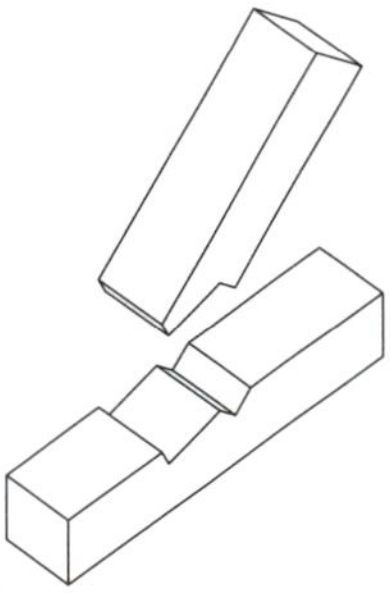

A+B

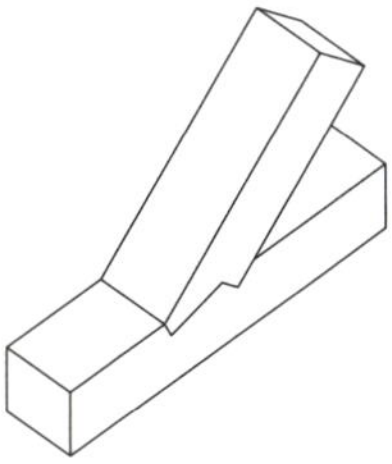

AB

Angled Joints

Oblique thrust joint with sloped tenon and notched heel

naname-eriwa-nagare-hozo-tsugi
斜襟輪流枘接ぎ

Fersenversatz mit schrägem Zapfen

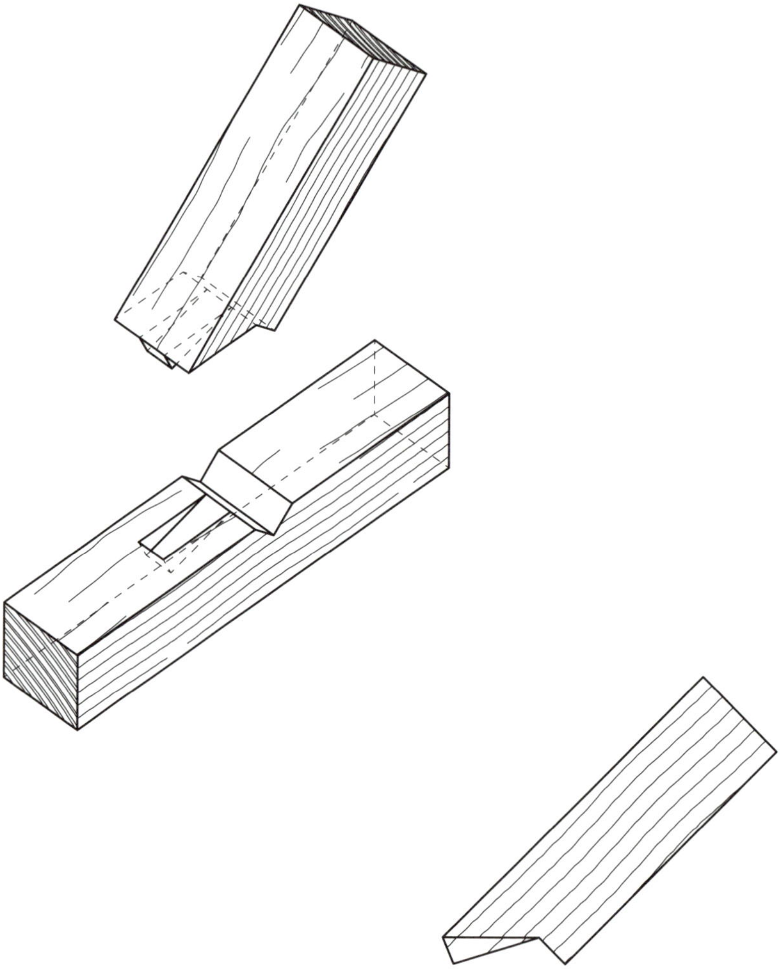

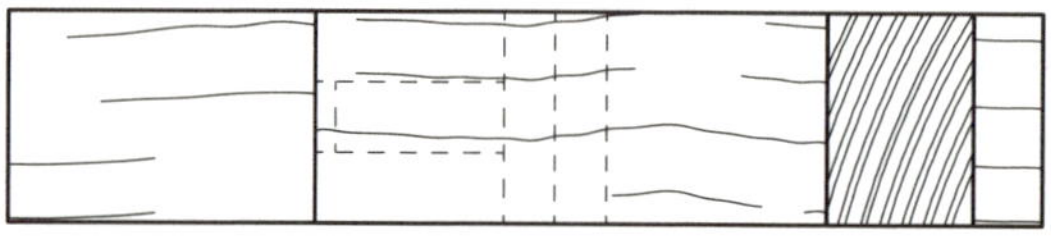

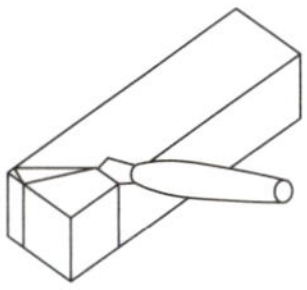
A1

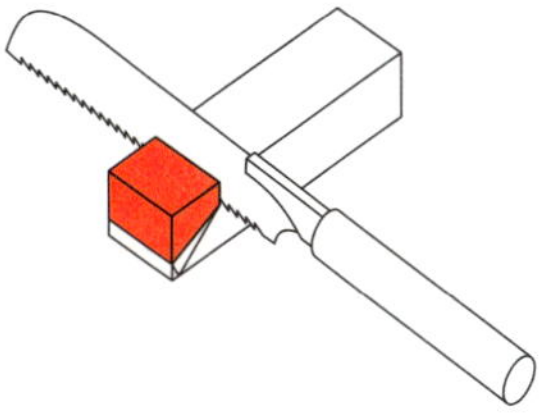
A2

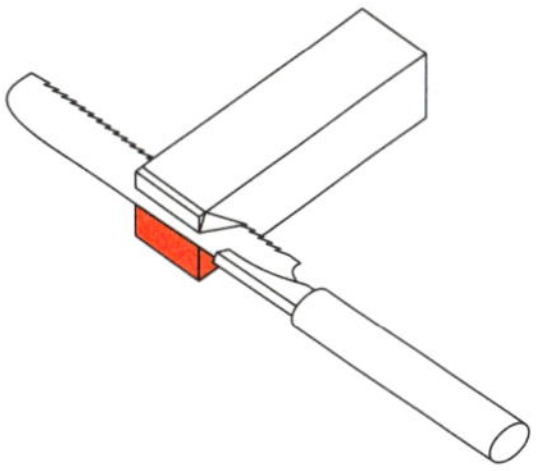
A3

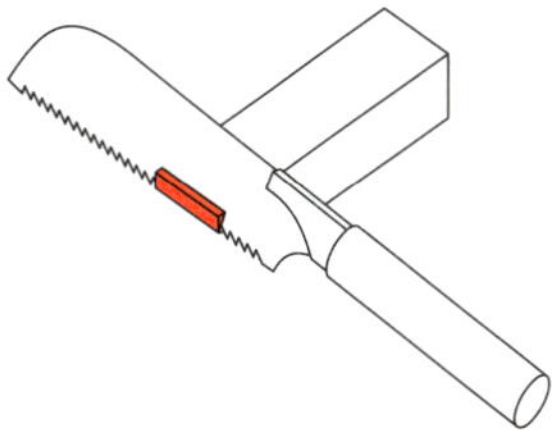
A4

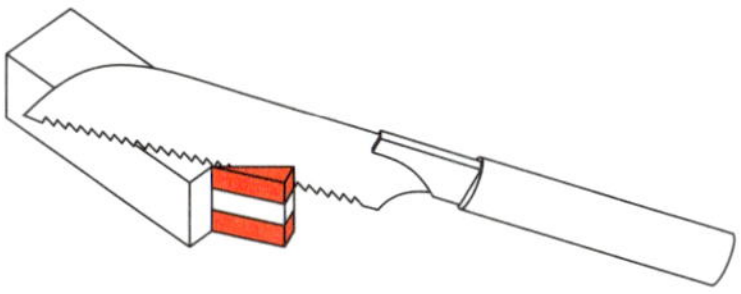
A5

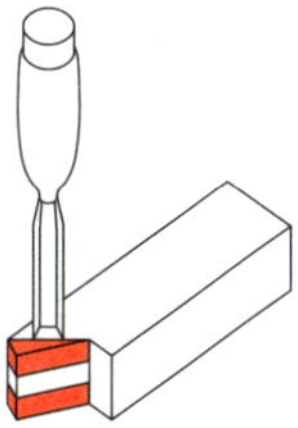
A6

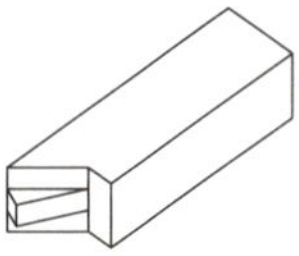
A

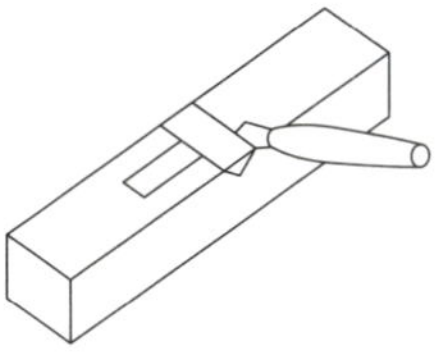
B1

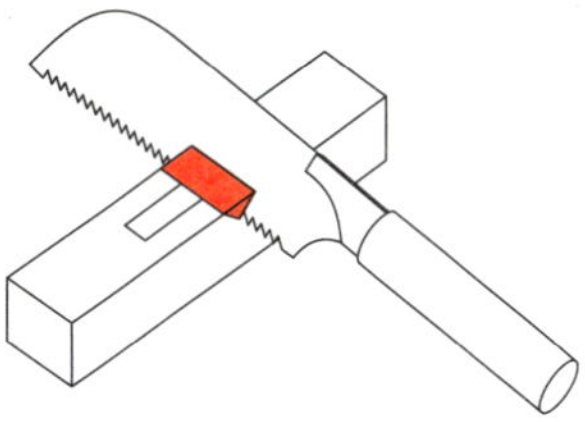
B2

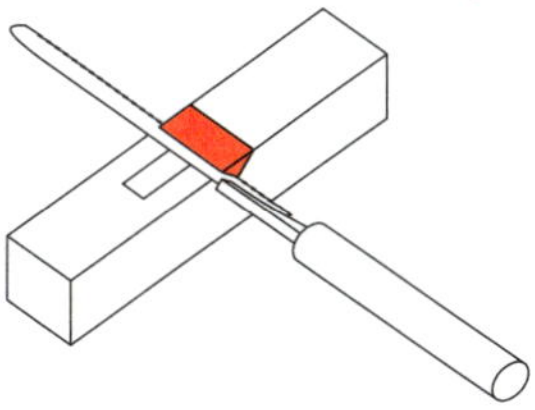
B3

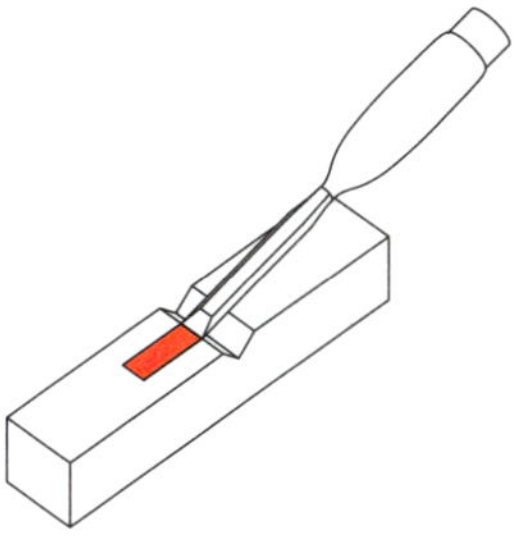
B4

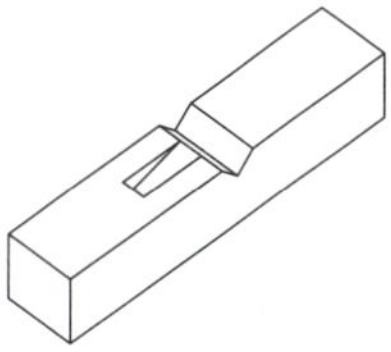
B

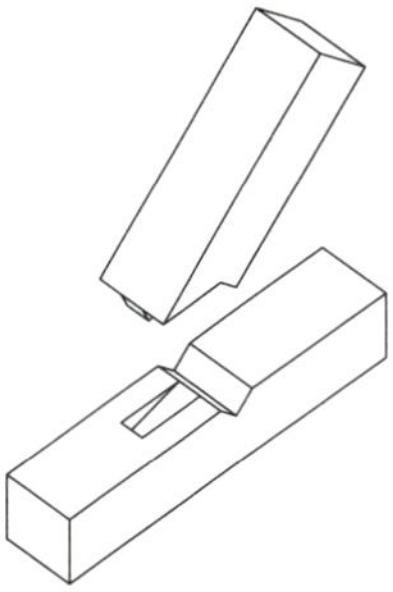
A+B

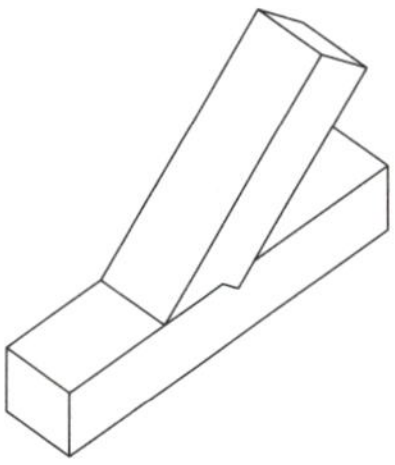
AB

Angled Joints

Oblique thrust joint with mortise and tenon

naname-tan-hozo-tsugi
斜短枘接ぎ

Schräger Stoß mit Zapfen

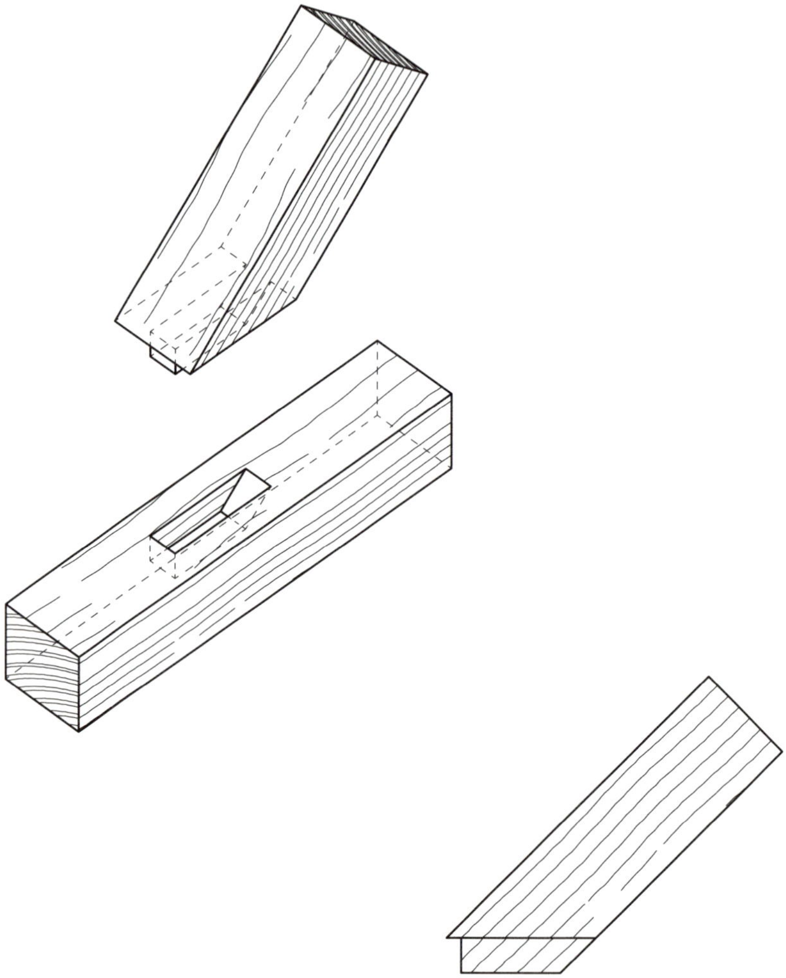

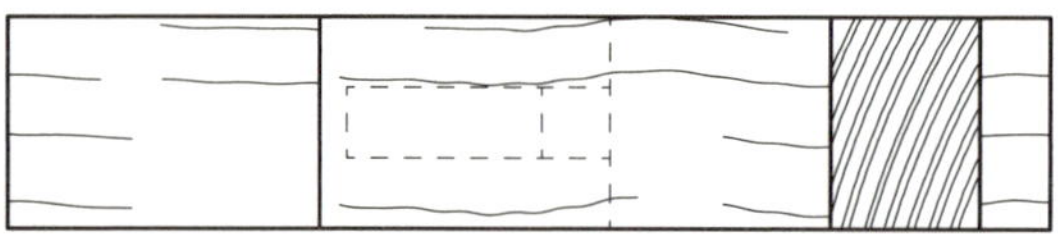

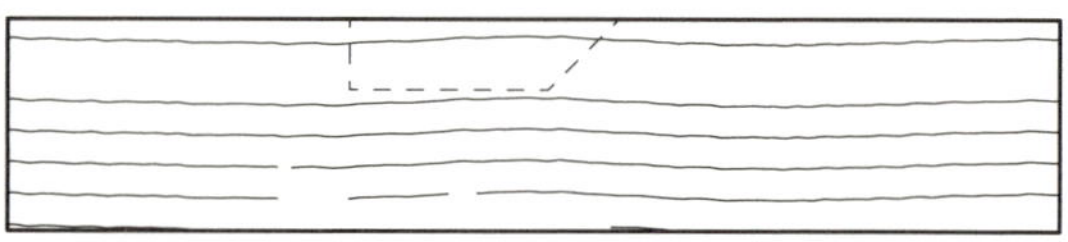

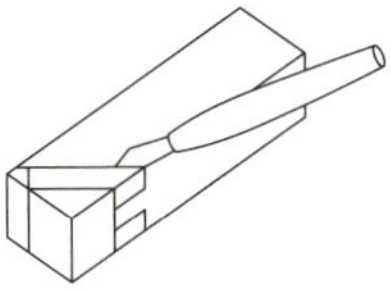
A1

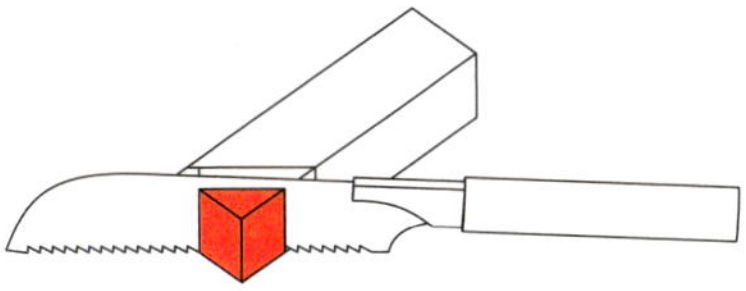
A2

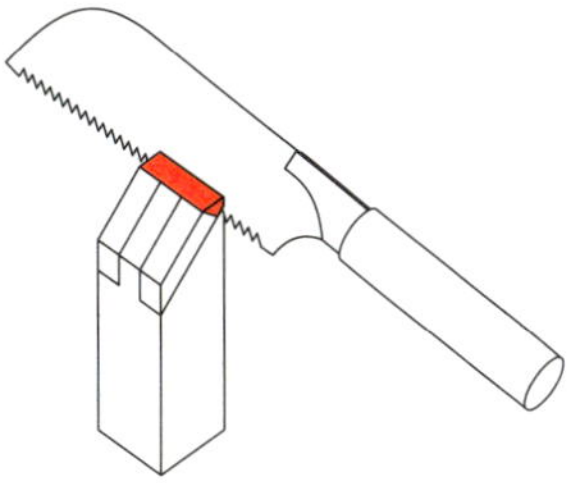
A3

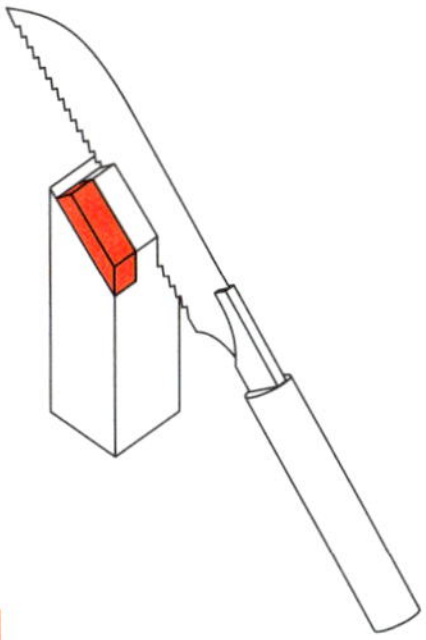
A4

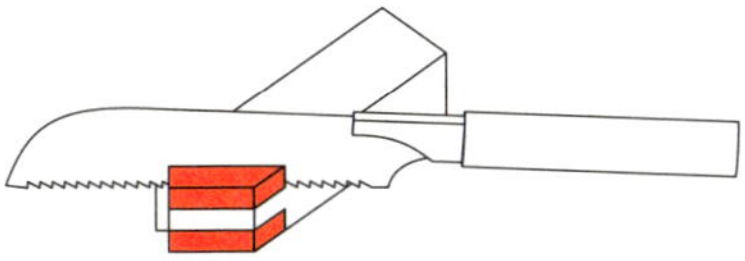
A5

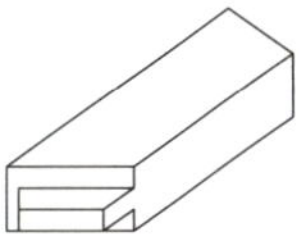
A

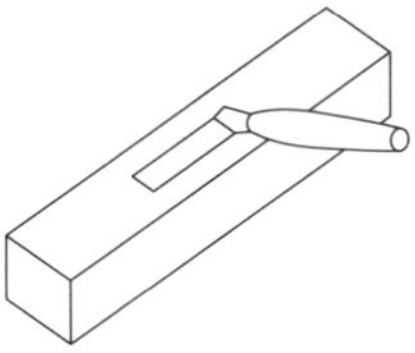
B1

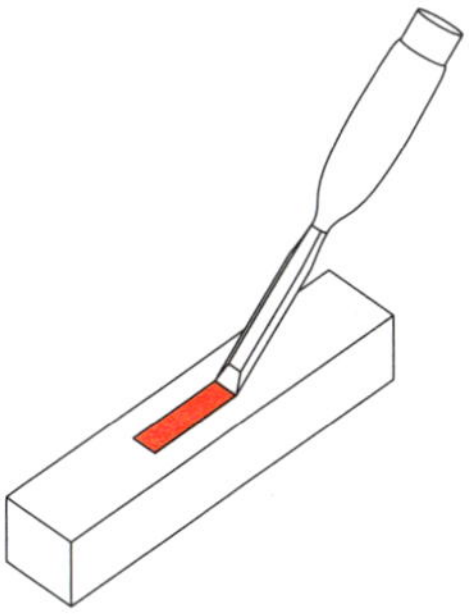
B2

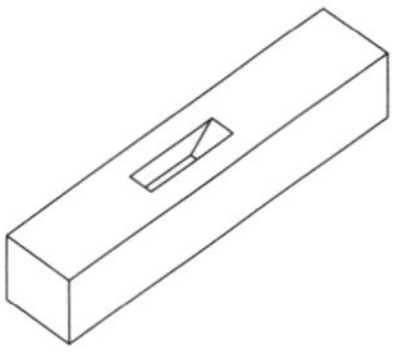
B

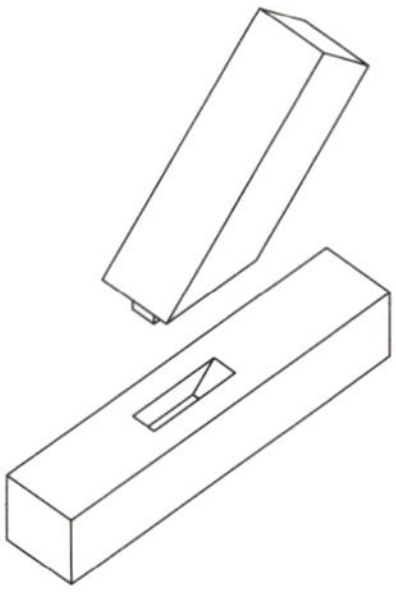
A+B

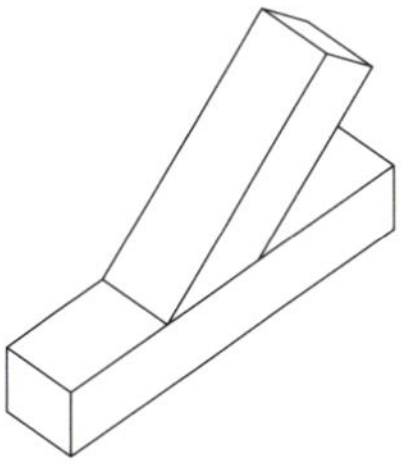
AB

Angled Joints

Oblique thrust joint with sloped mortise and tenon

naname-nagare-hozo-tsugi
斜流枘接ぎ

schräger Stoß mit schrägem Zapfen

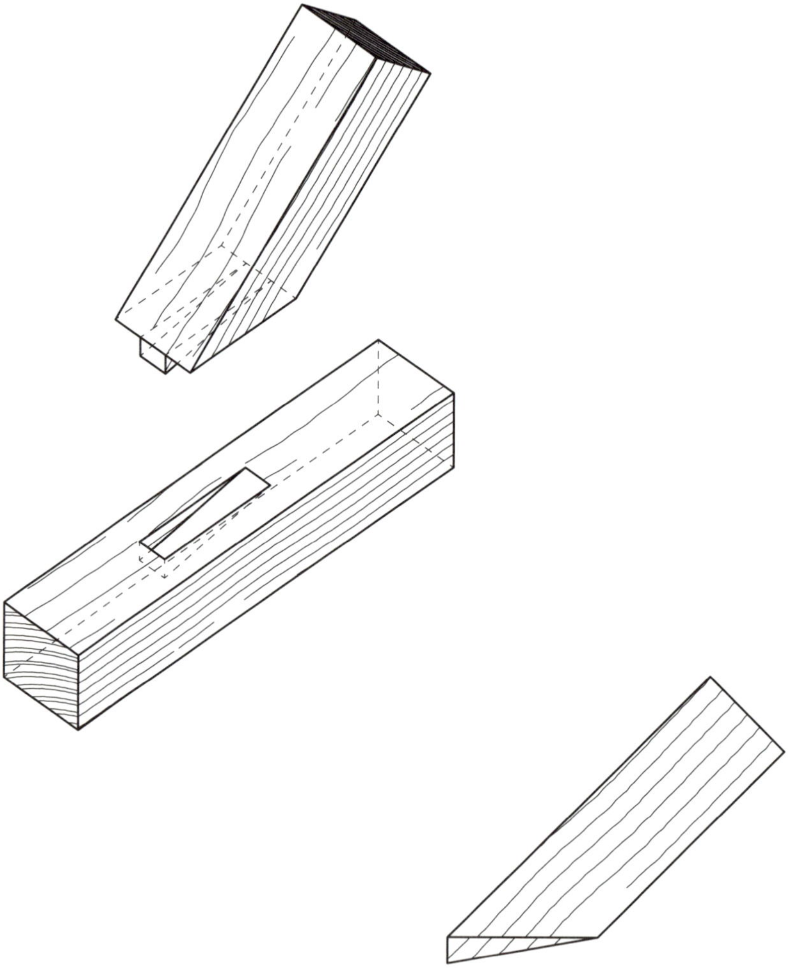

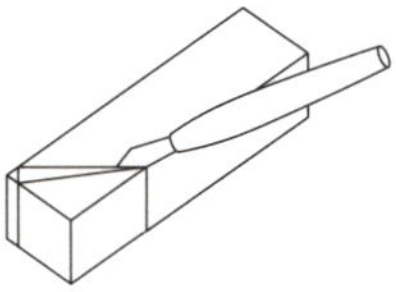

A1

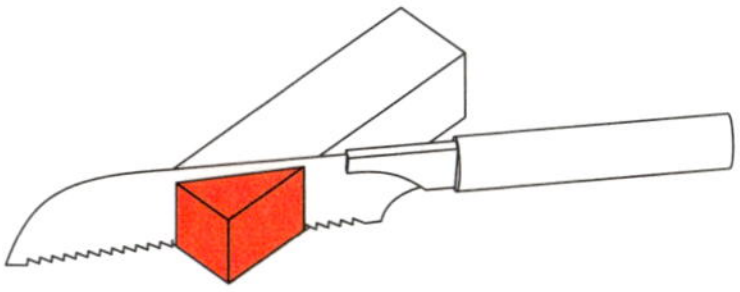

A2

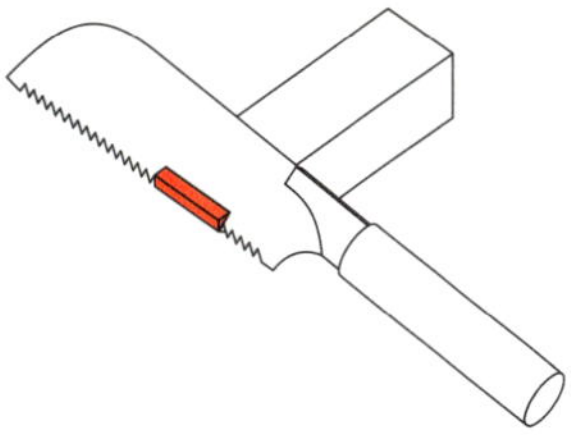

A3

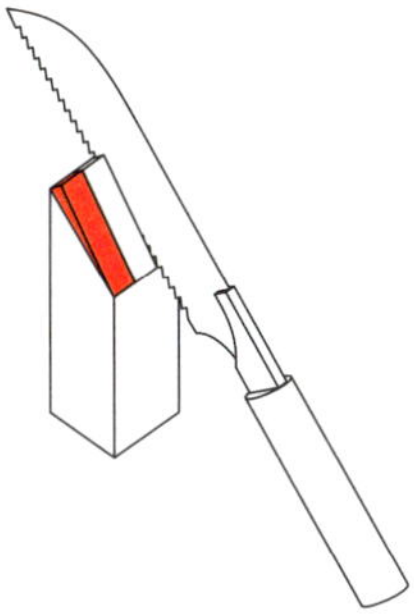

A4

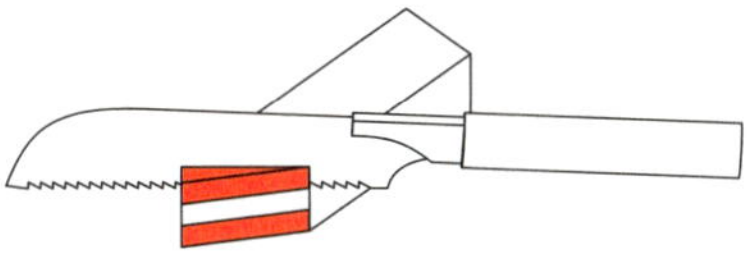

A5

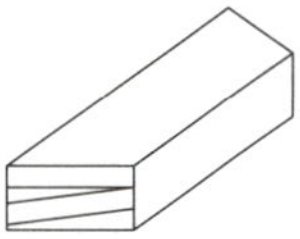

A

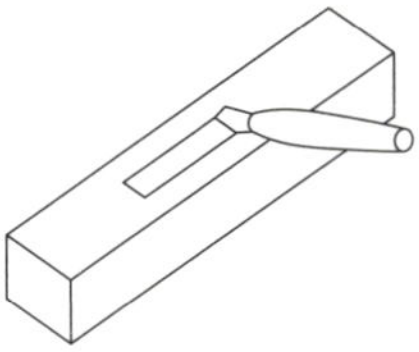

B1

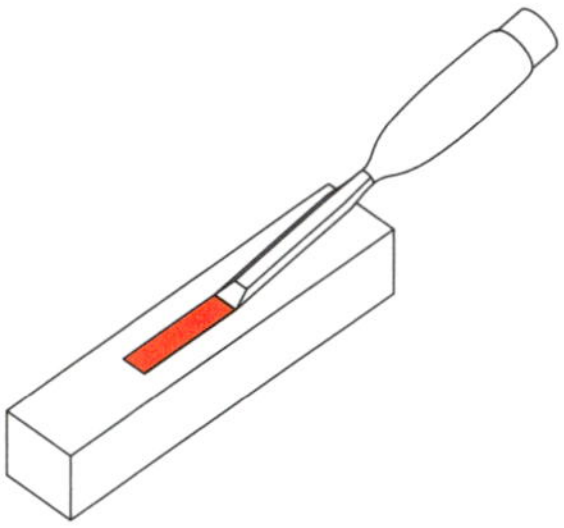

B2

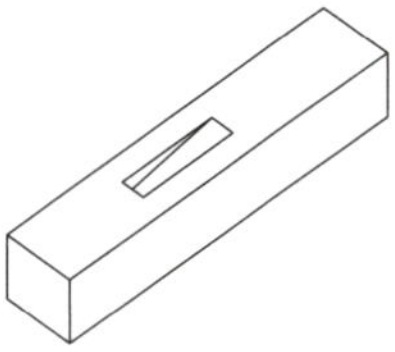

B

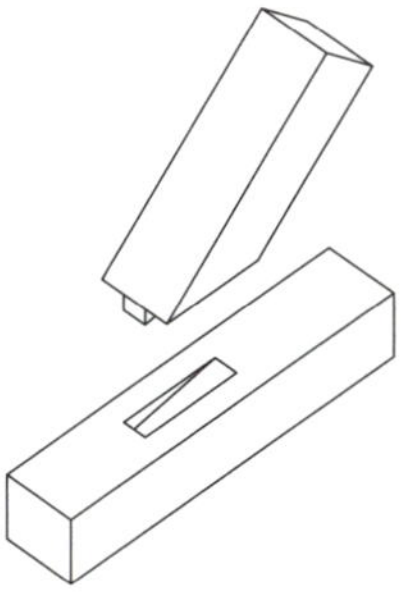

A+B

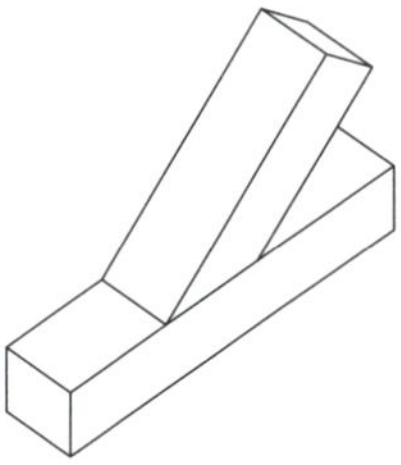

AB

Angled Joints

Notched oblique thrust joint with mortise and tenon

naname-sogi-kaki-tan-hozo-tsugi
斜殺欠短枘接ぎ

Stirnversatz mit Zapfen

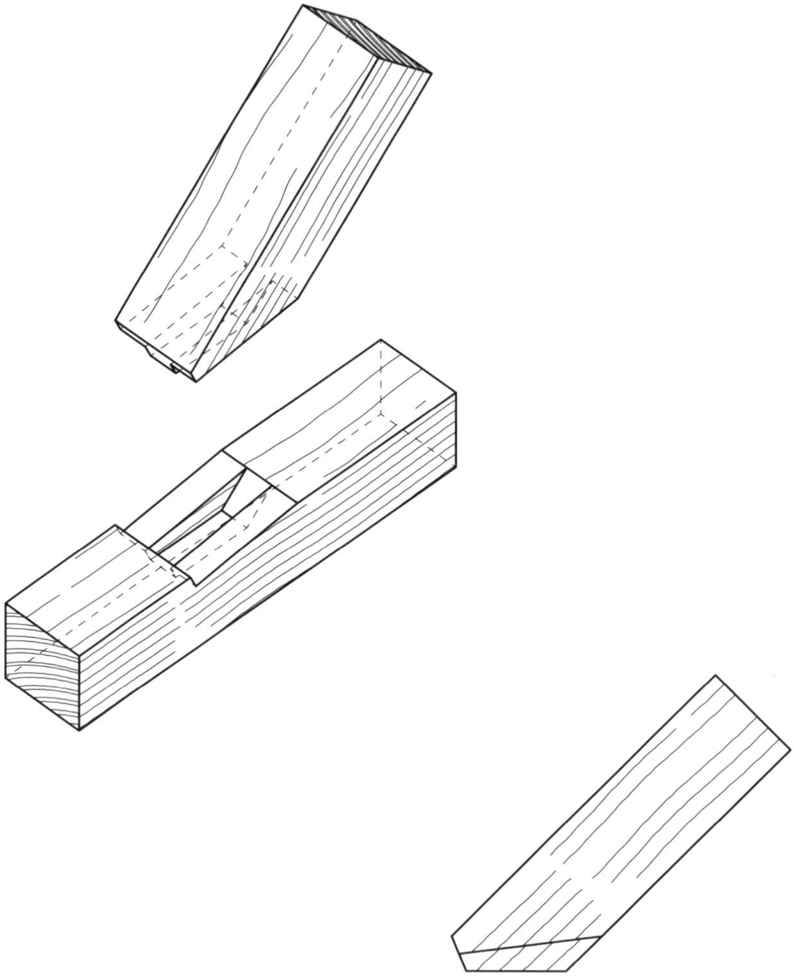

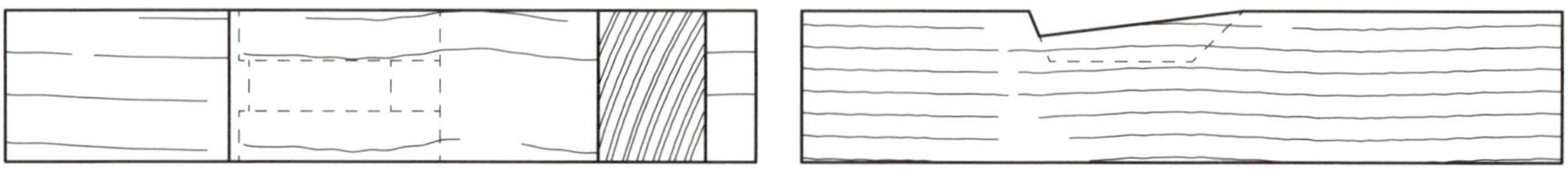

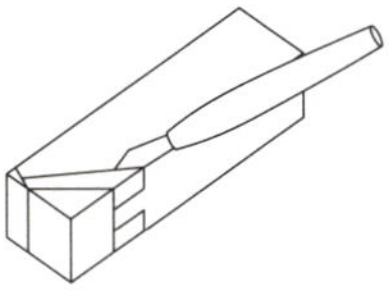

A1

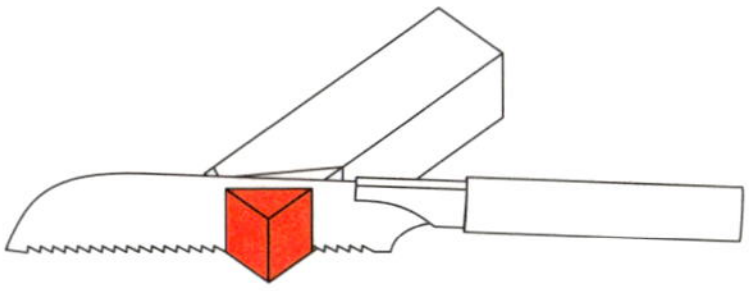

A2

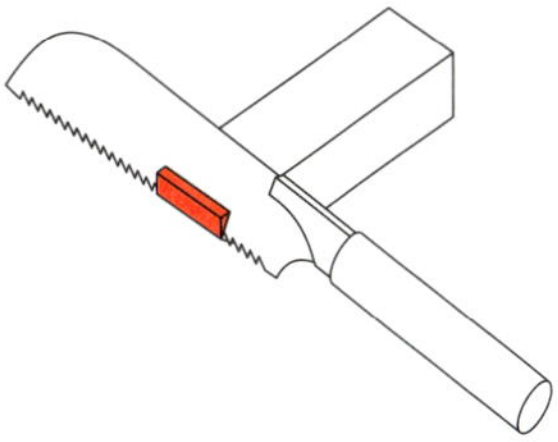

A3

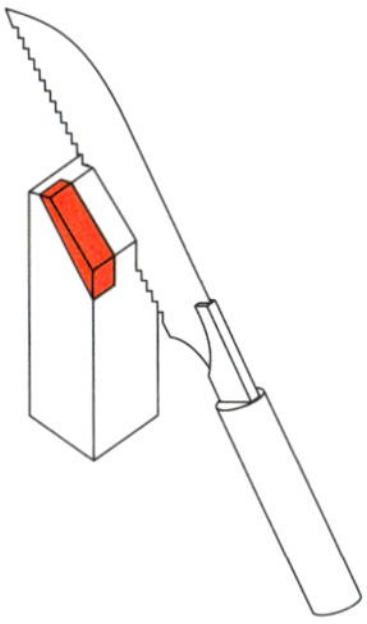

A4

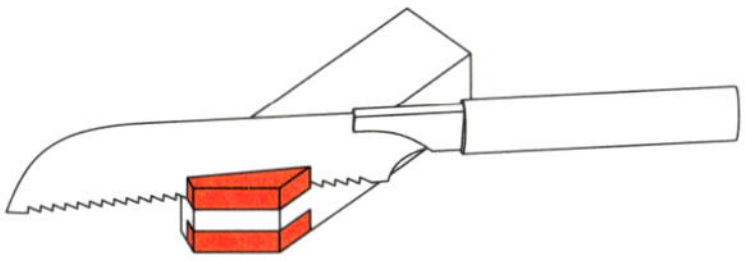

A5

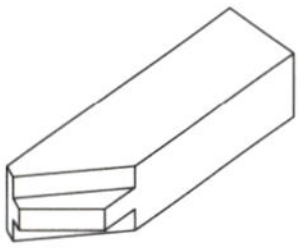

A

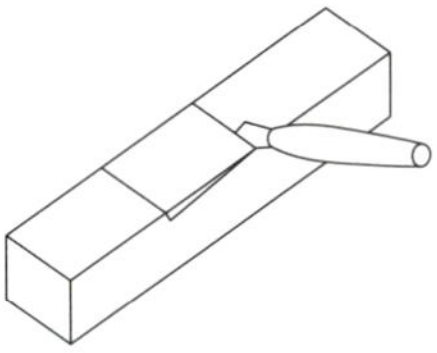

B1

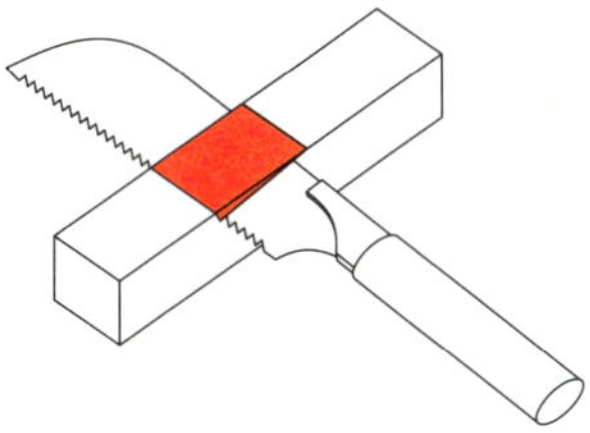

B2

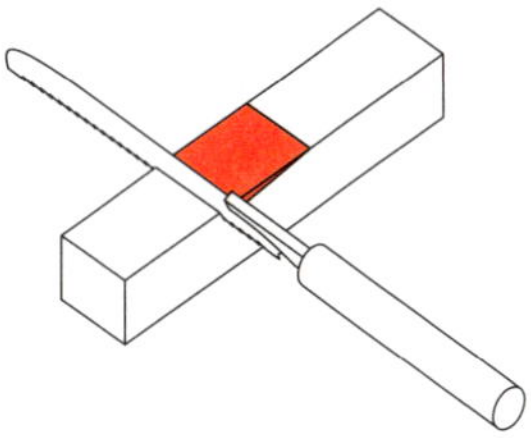

B3

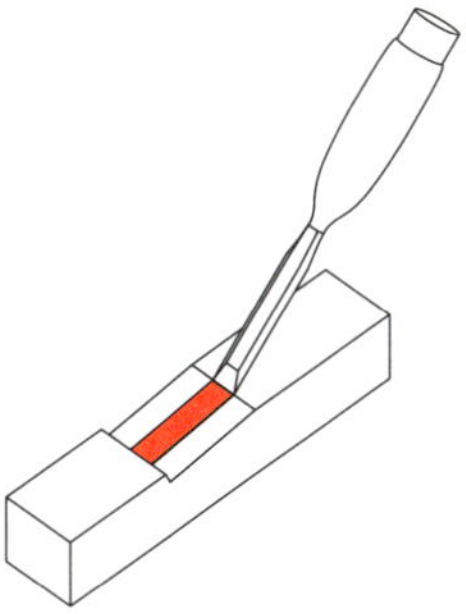

B4

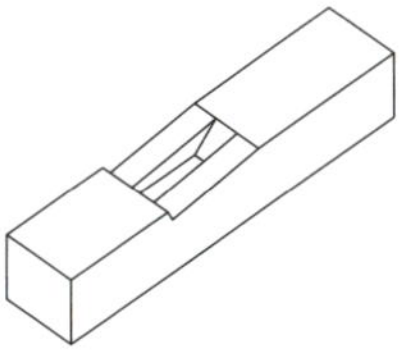

B

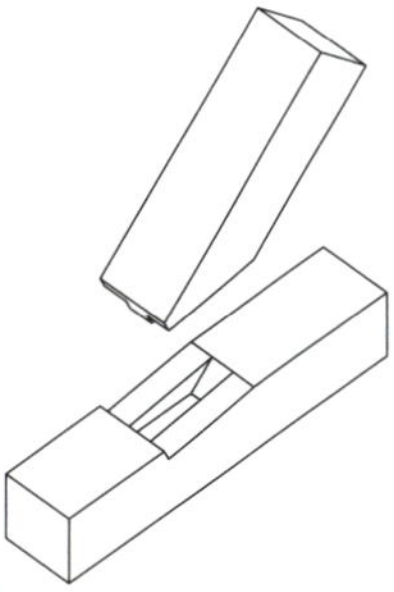

A+B

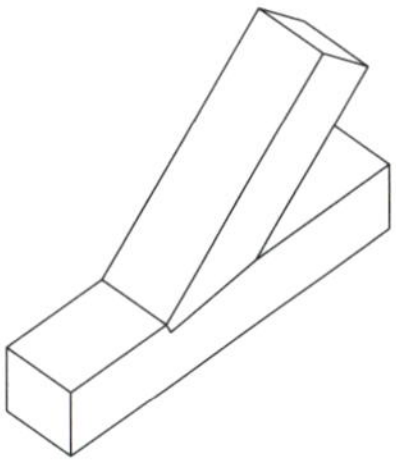

AB

Angled Joints

Notched oblique thrust joint with pegged mortise and tenon

naname-sogi-kaki-nagare-hozo-tsugi-komi-sen-shime
斜殺欠流枘接込栓締め

Stirnversatz mit Zapfen und Dübelsicherung

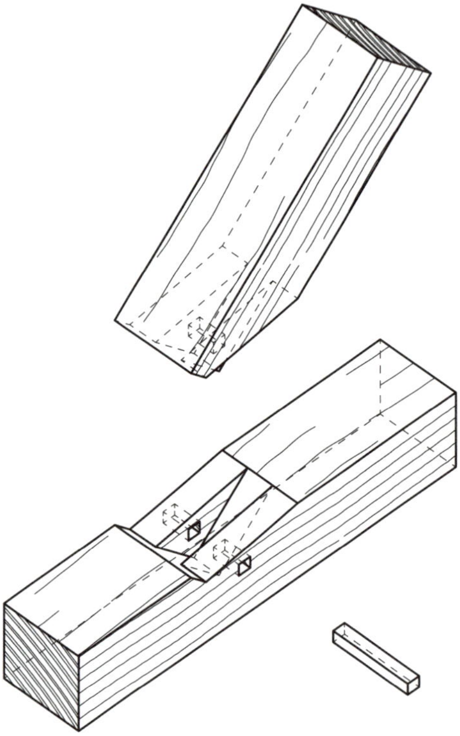

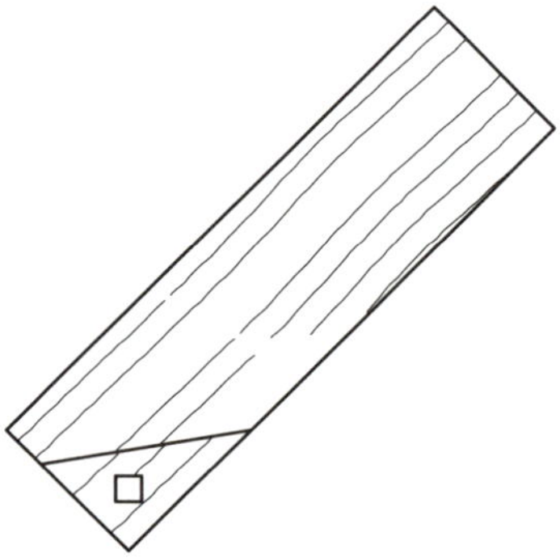

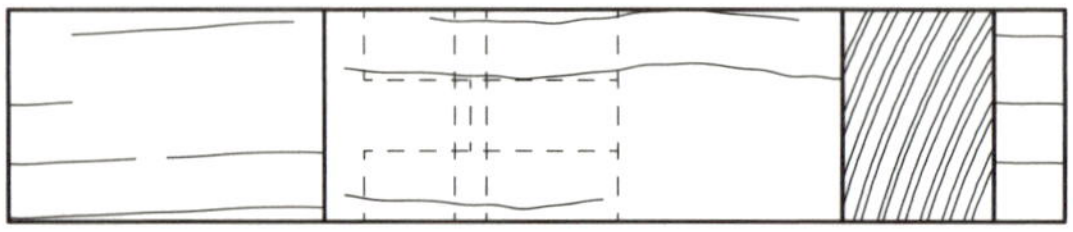

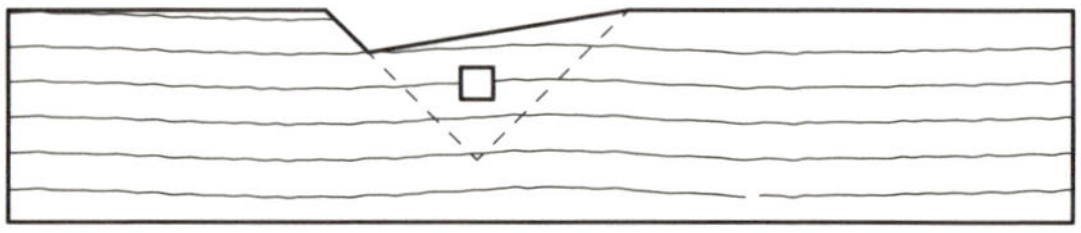

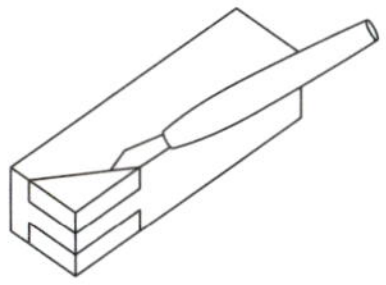

A1

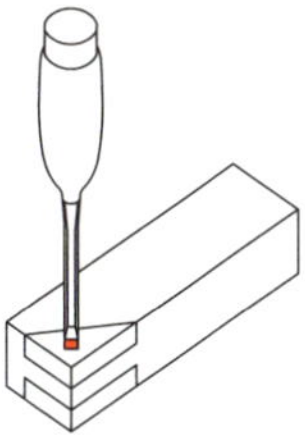

A2

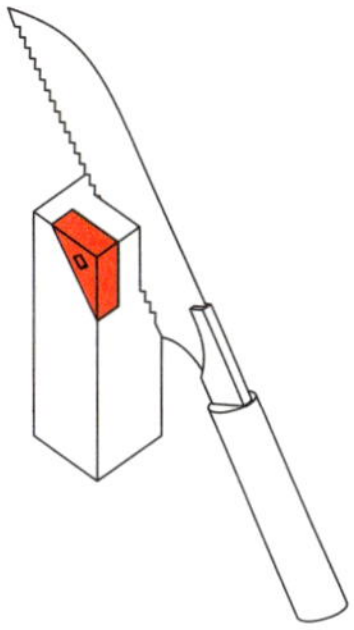

A3

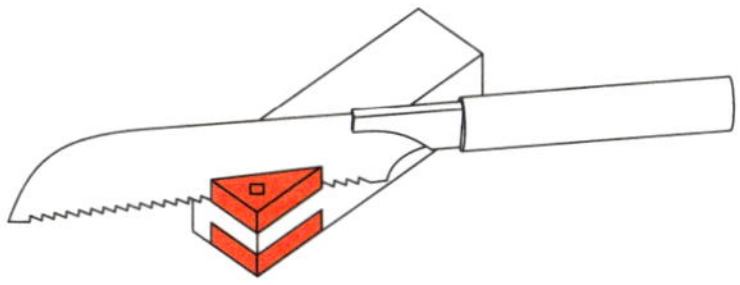

A4

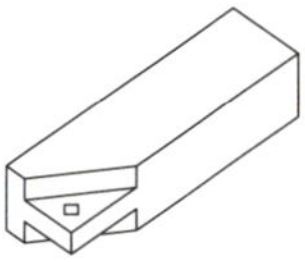

A

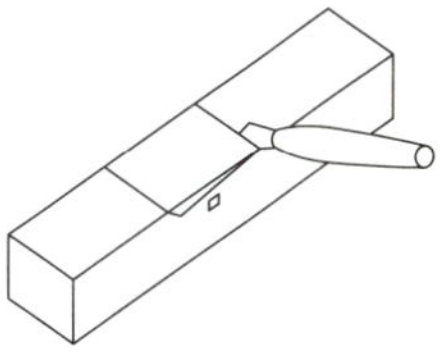

B1

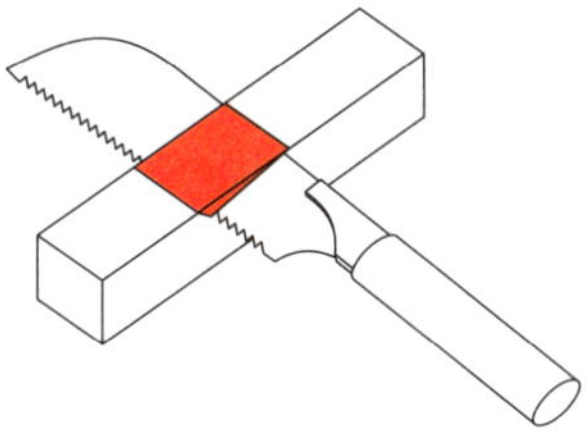

B2

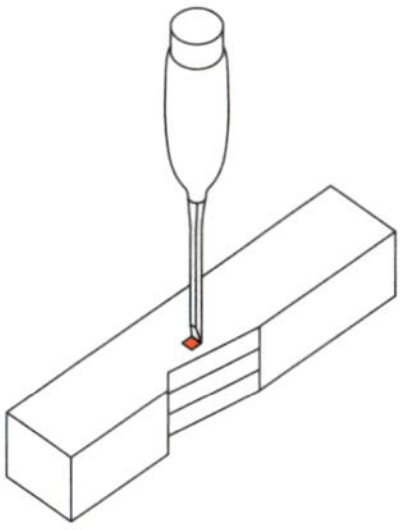

B3

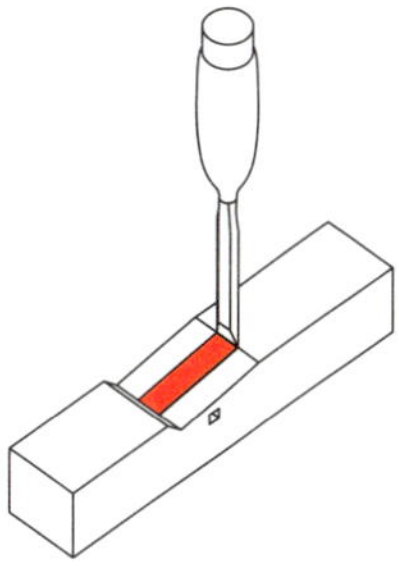

B4

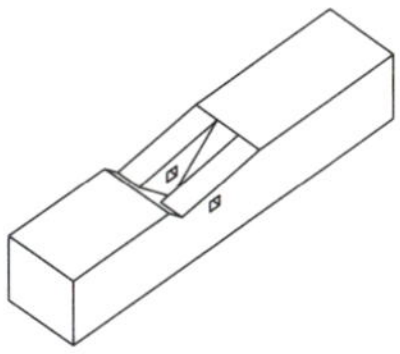

B

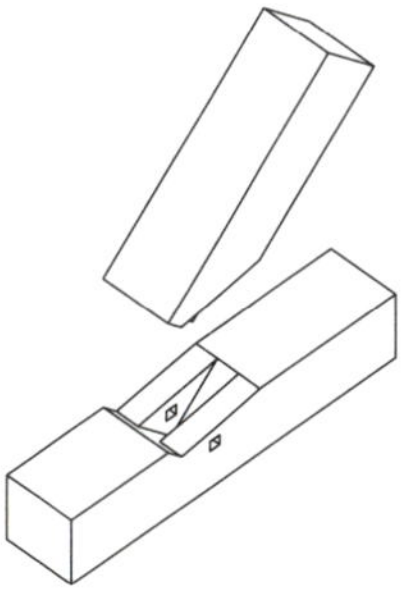

A+B

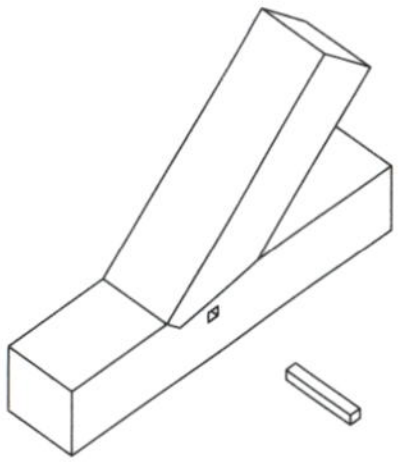

A+B

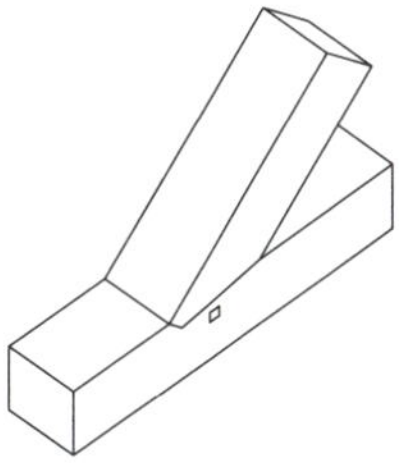

AB

Angled Joints

Slotted notched oblique thrust joint

kura-kake-hozo-tsugi
鞍掛枘接ぎ

geschlitzter Stirnversatz

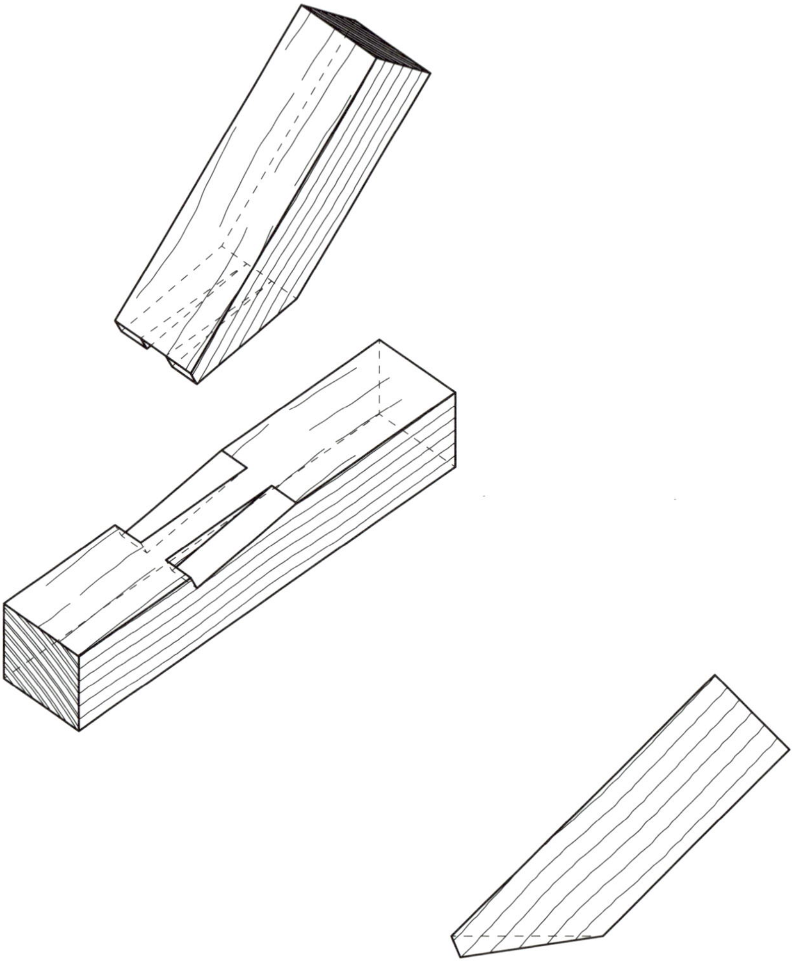

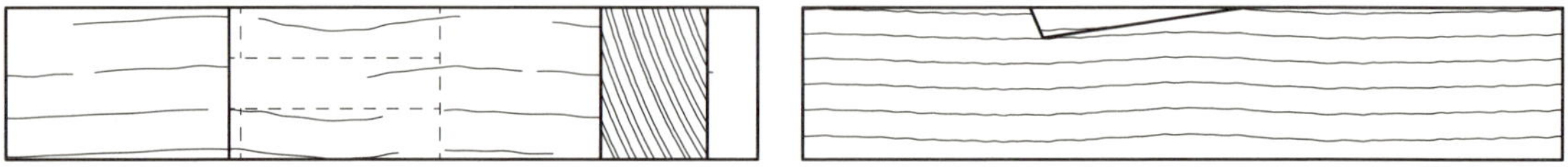

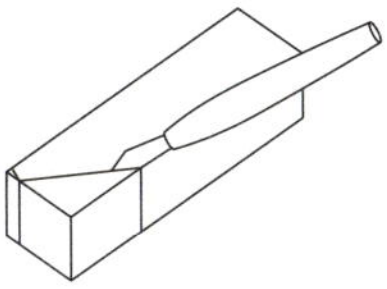
A1

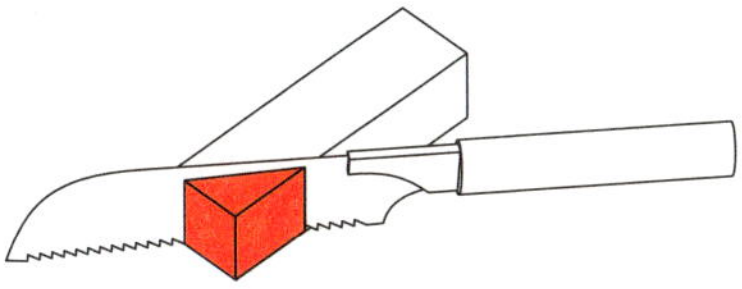
A2

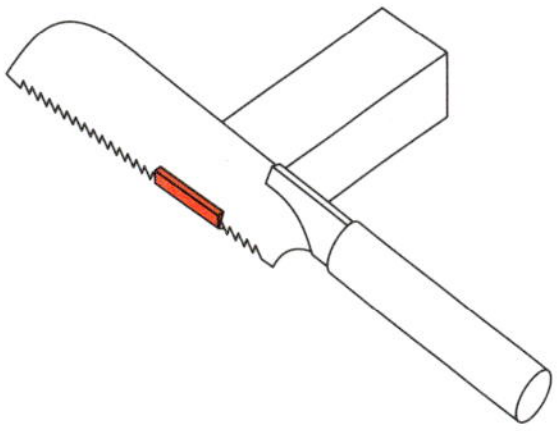
A3

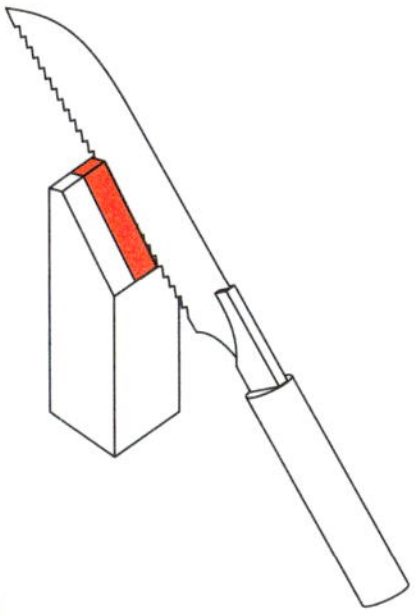
A4

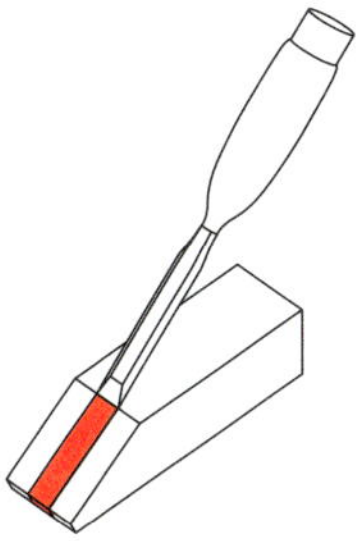
A5

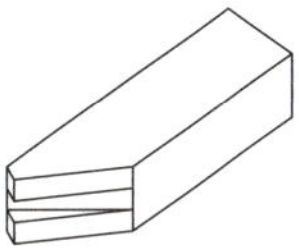
A

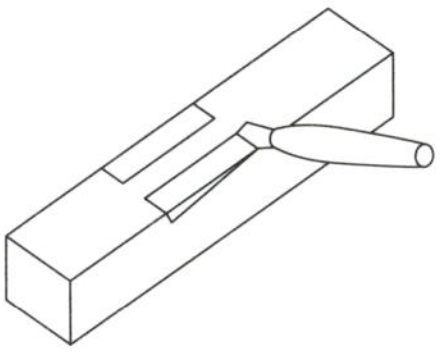
B1

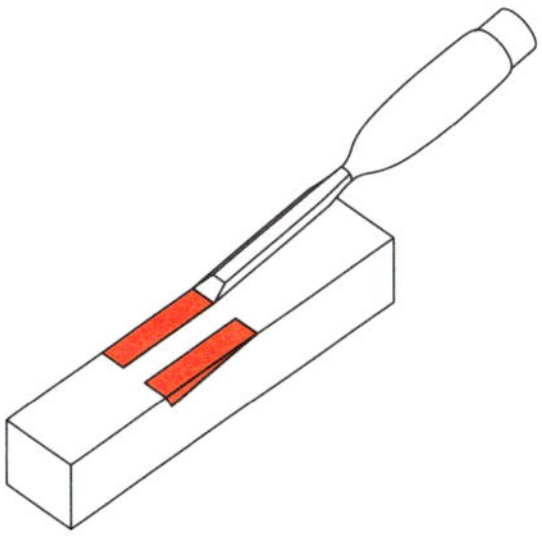
B2

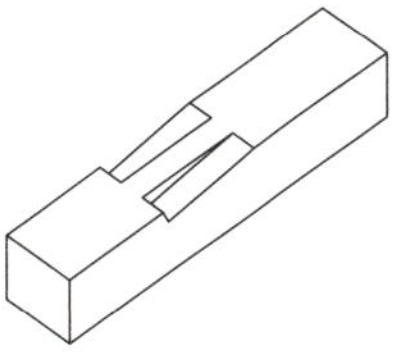
B

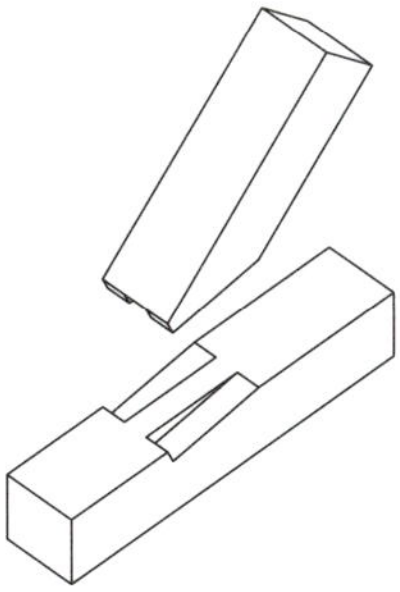
A+B

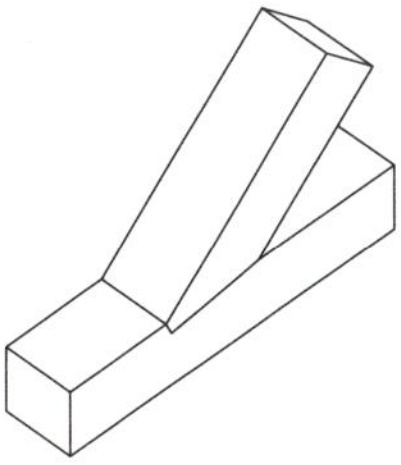
AB

Angled Joints

Slotted notched oblique thrust joint with visible end

naname-shiki-men-kura-kake-hozo-tsugi
斜敷面鞍掛枘接ぎ

geschlitzter Brustversatz

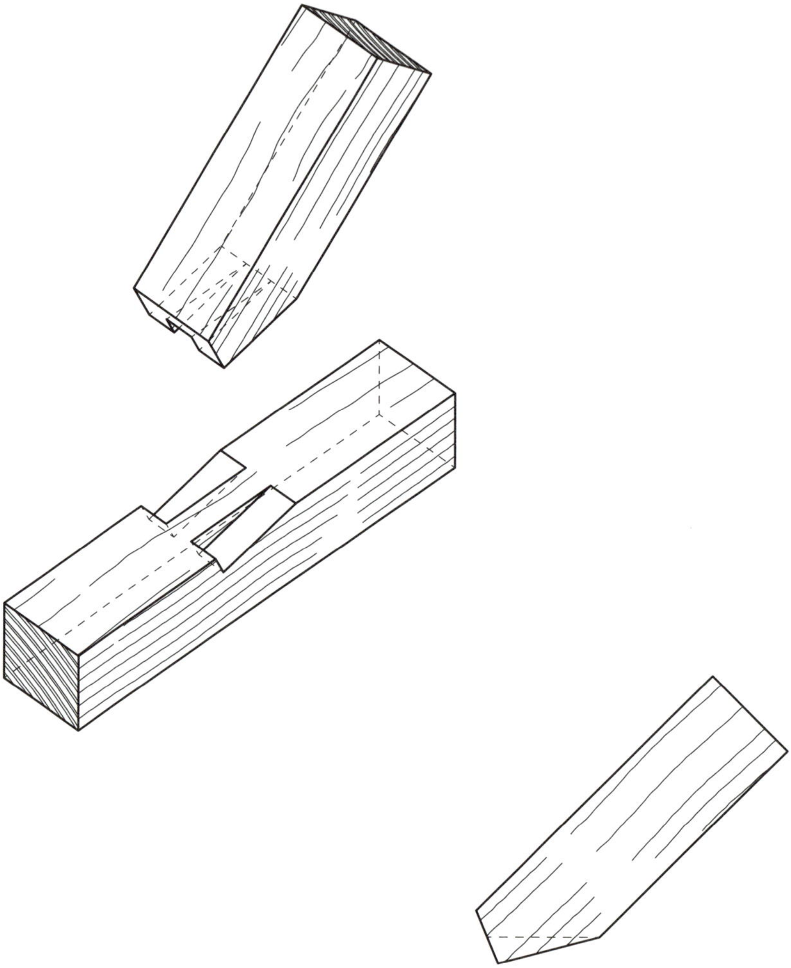

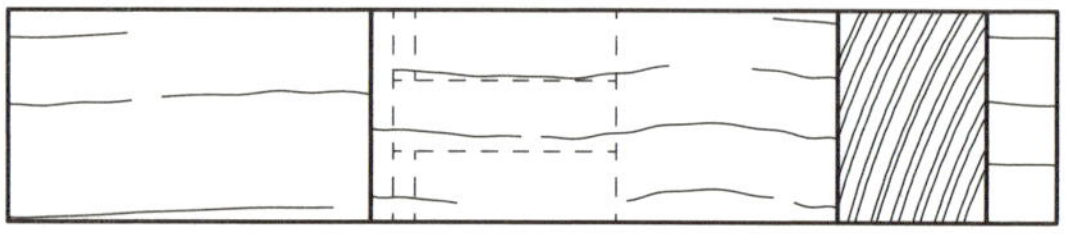

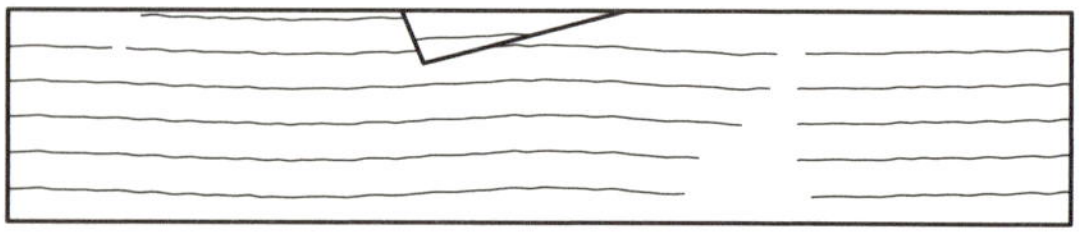

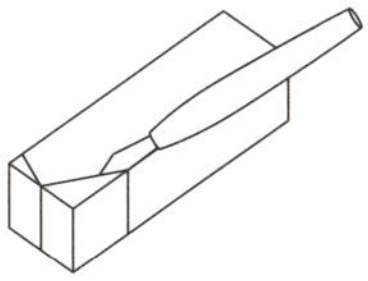

A1

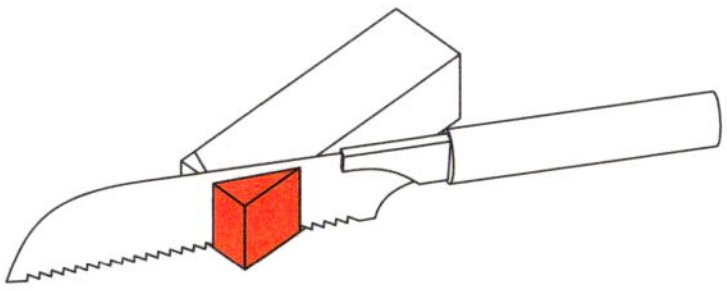

A2

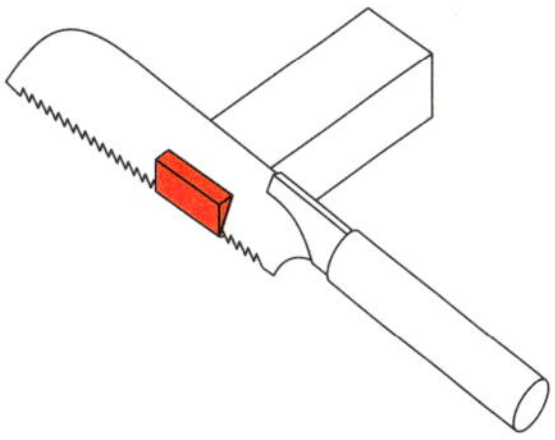

A3

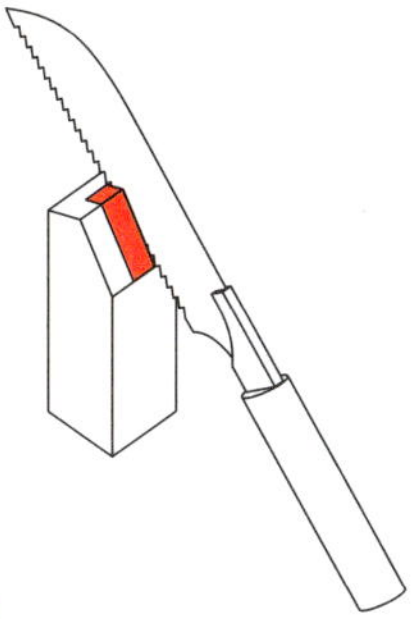

A4

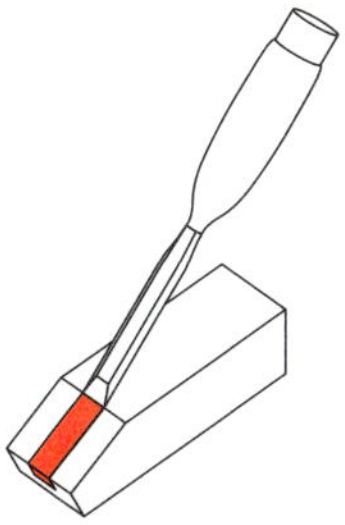

A5

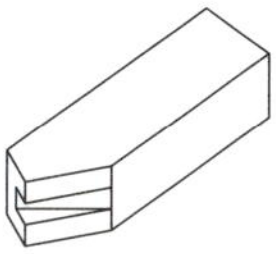

A

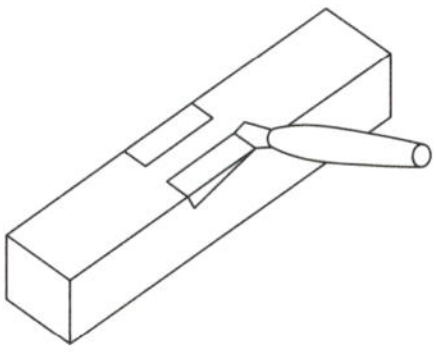

B1

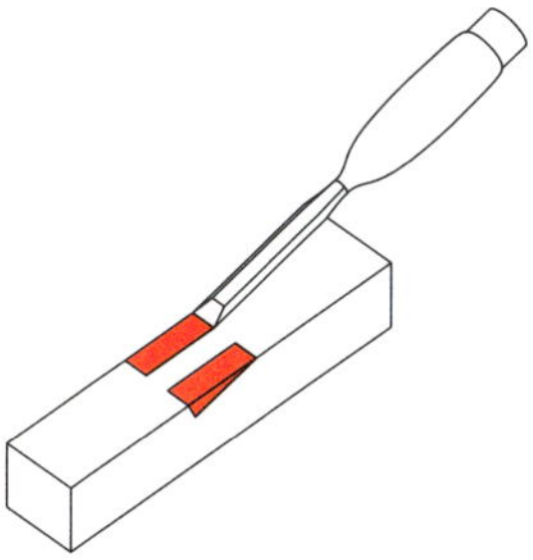

B2

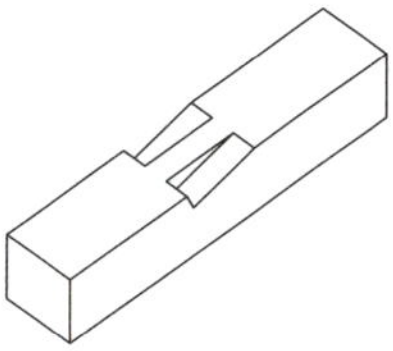

B

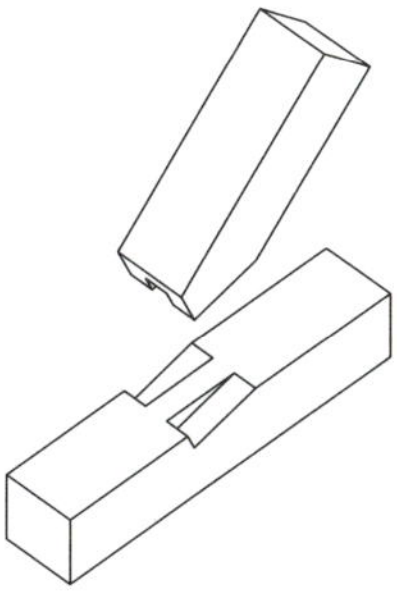

A+B

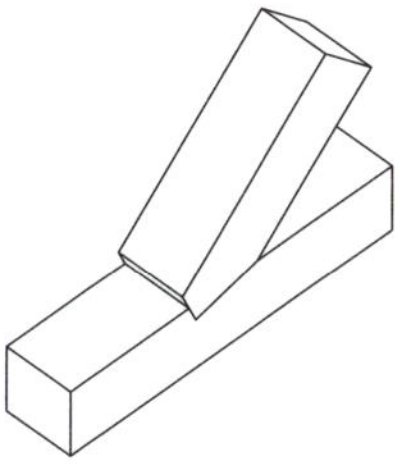

AB

Angled Joints

Pegged angled half lap joint

naname-ai-kaki-tsugi-komi-sen-uchi
斜相欠接込栓打ち

schräges Blatt mit Dübelsicherung

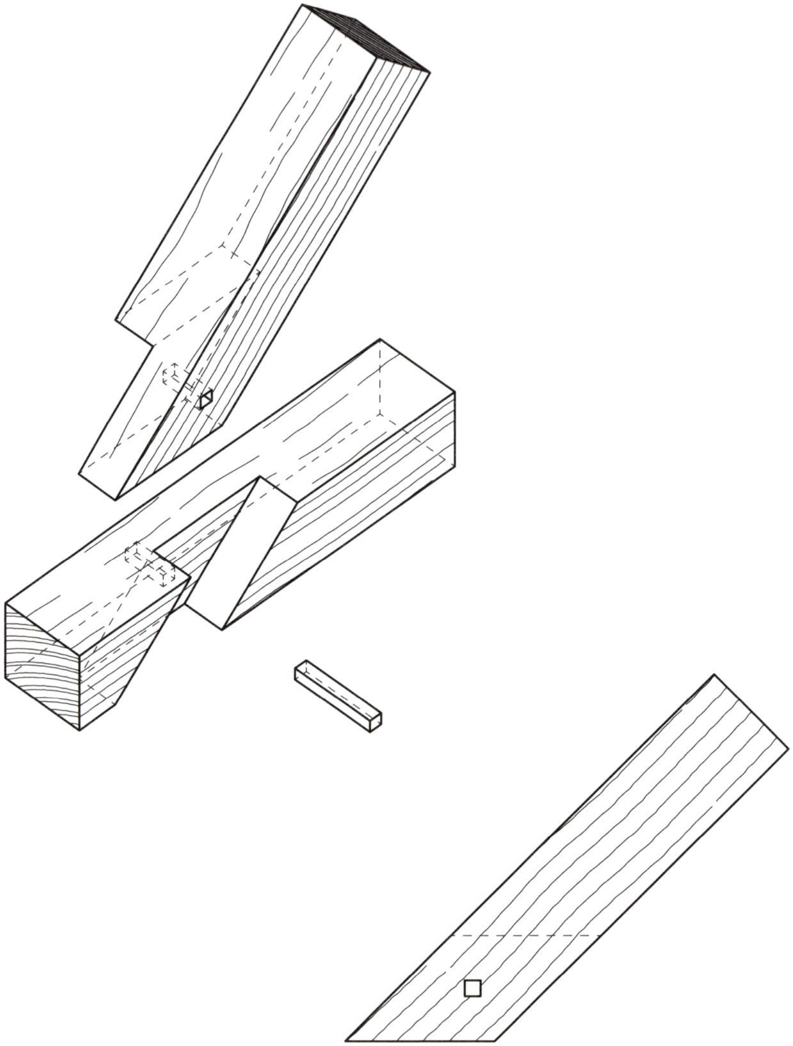

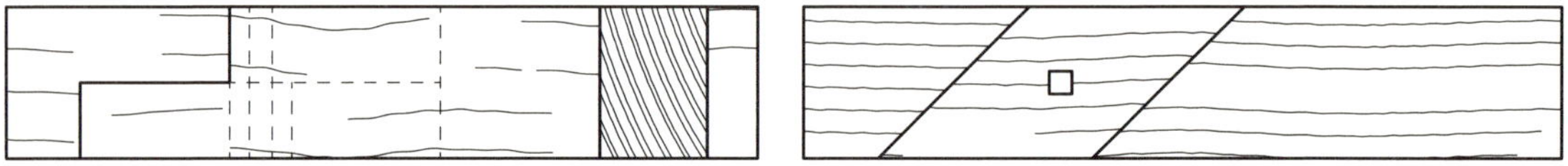

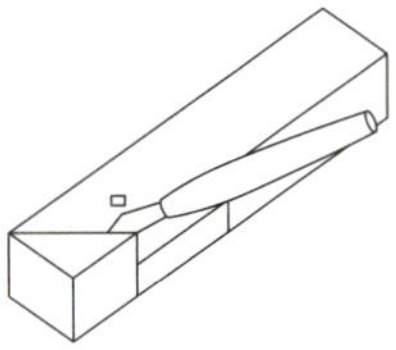

A1

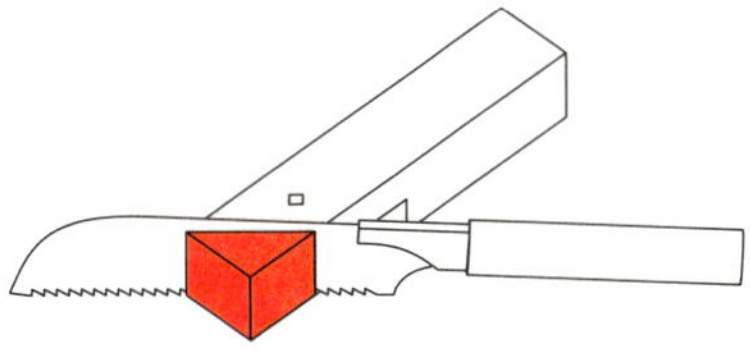

A2

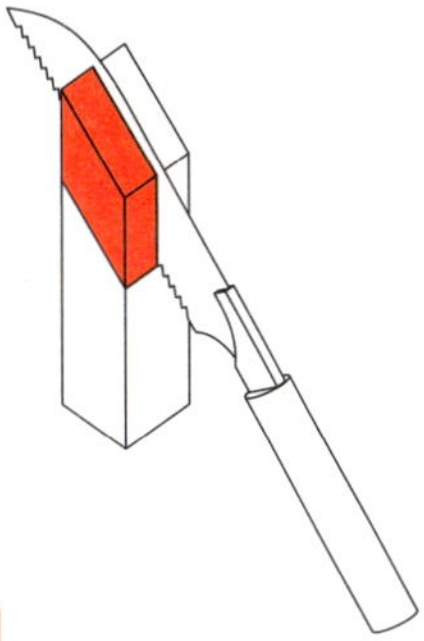

A3

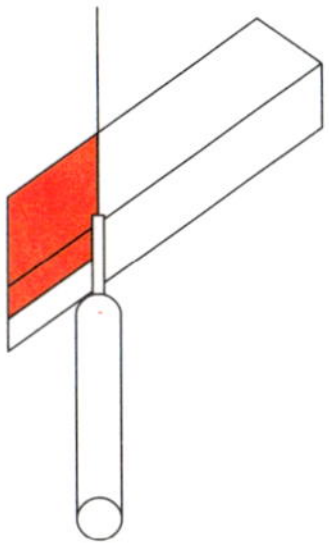

A4

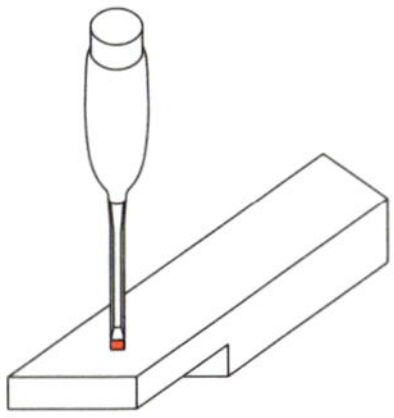

A5

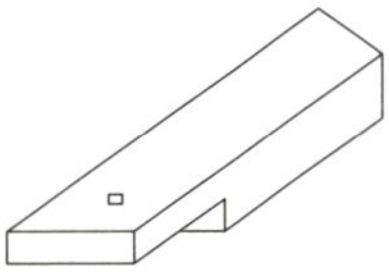

A

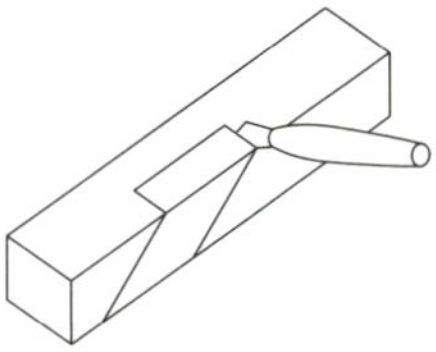

B1

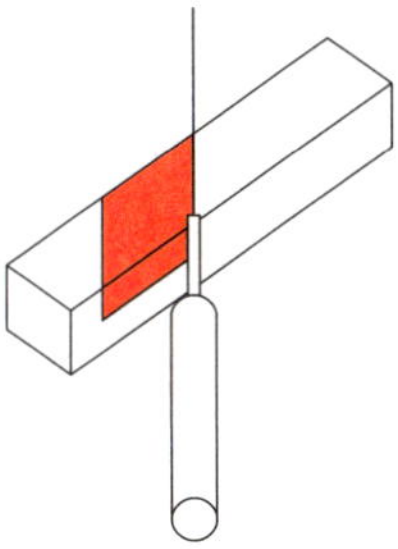

B2

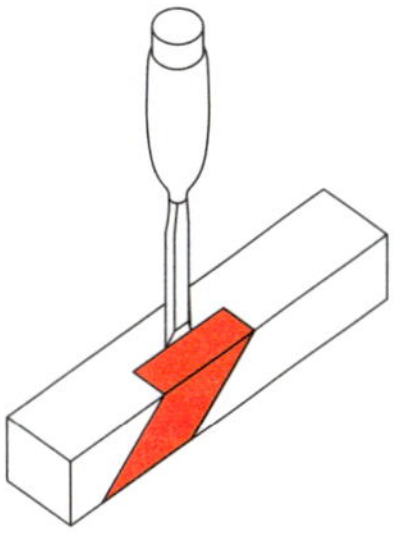

B3

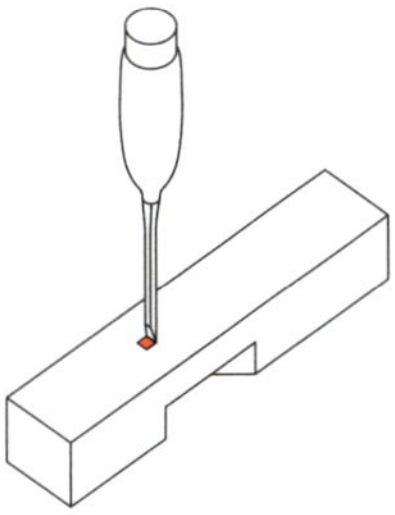

B4

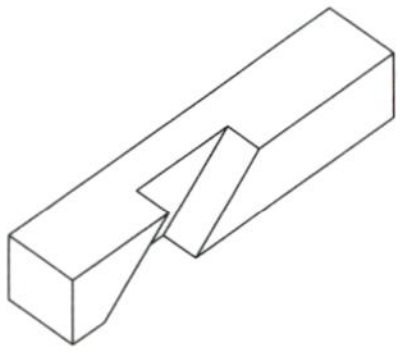

B

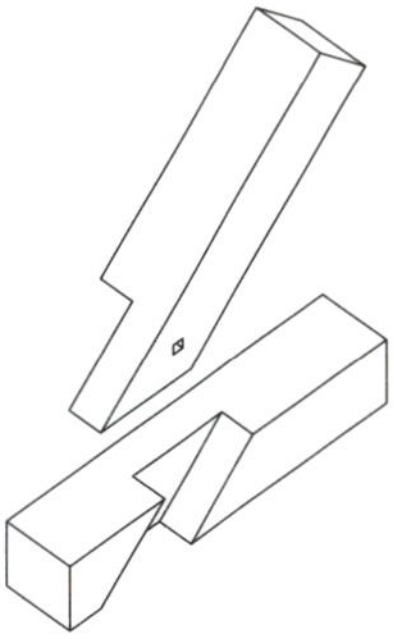

A+B

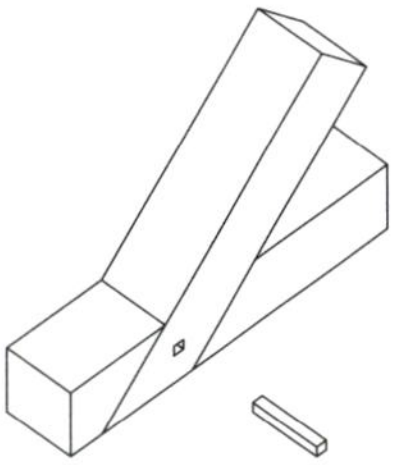

A+B

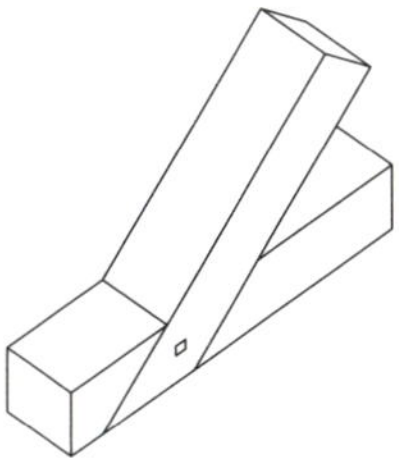

AB

Angled Joints

Pegged angled stopped half lap joint

naname-kakushi-ai-kaki-tsugi-komi-sen-uchi
斜隠相欠接込栓打ち

halbverdecktes schräges Blatt mit Dübelsicherung

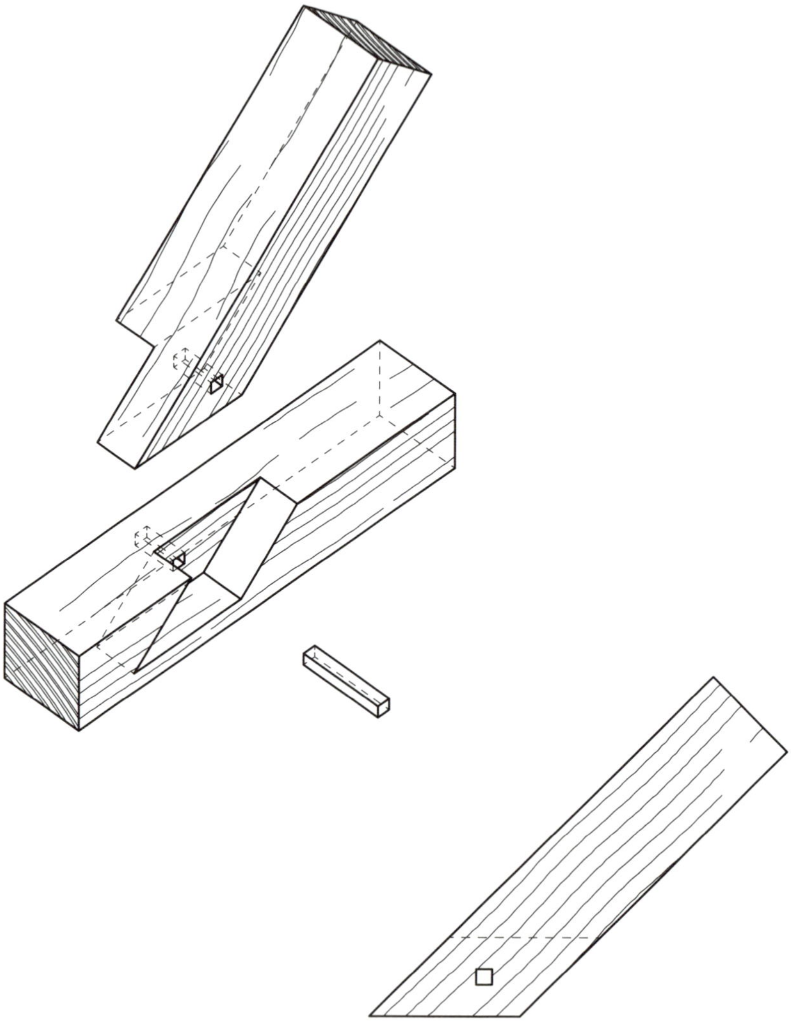

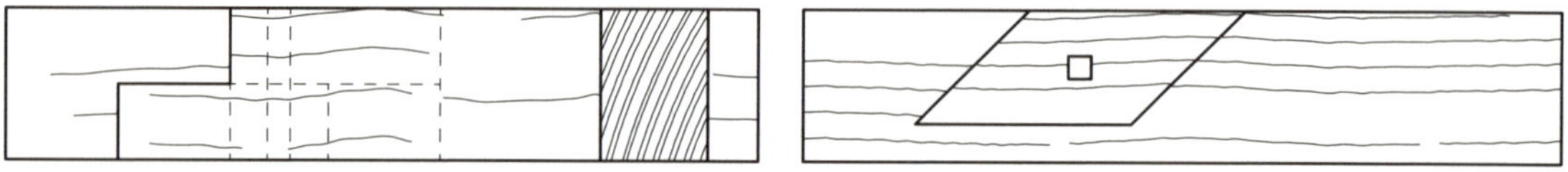

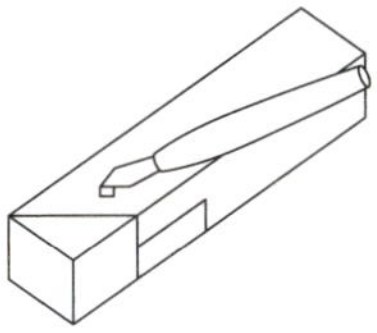

A1

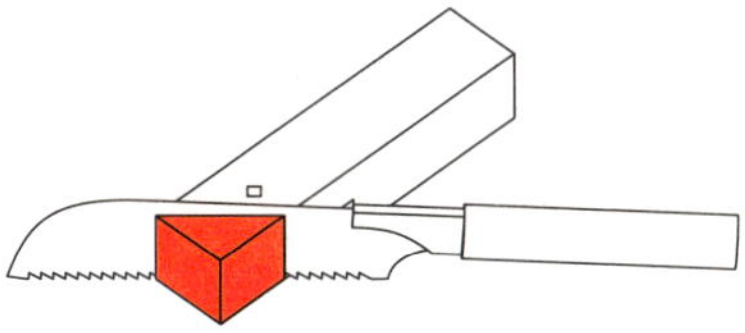

A2

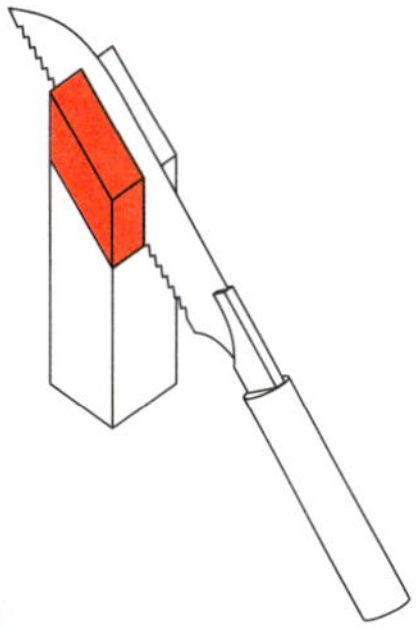

A3

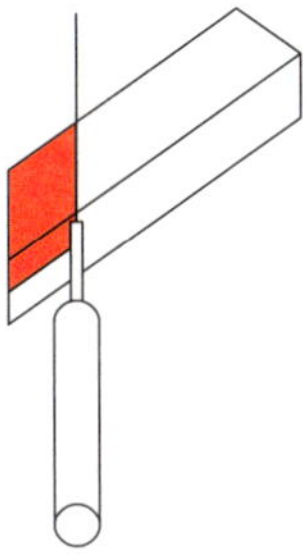

A4

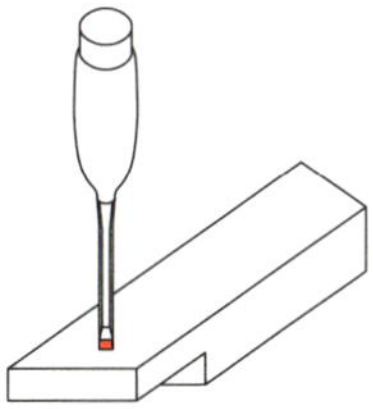

A5

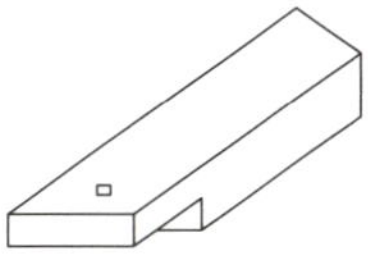

A

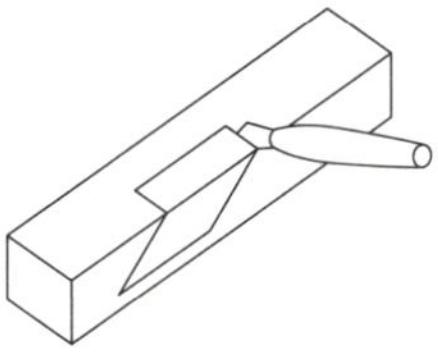

B1

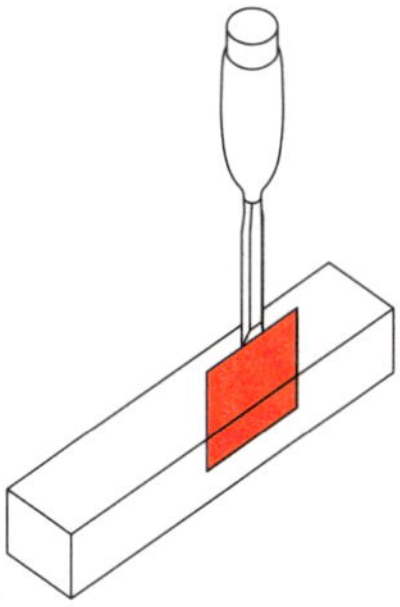

B2

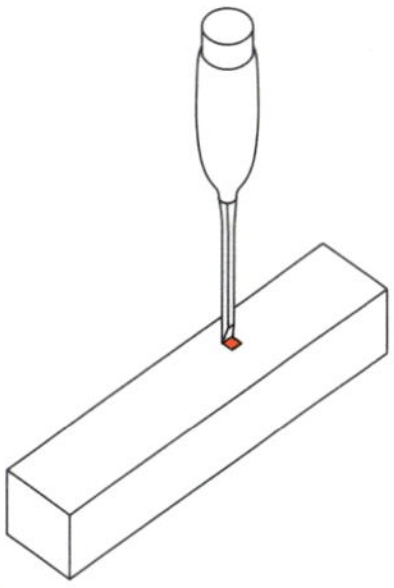

B3

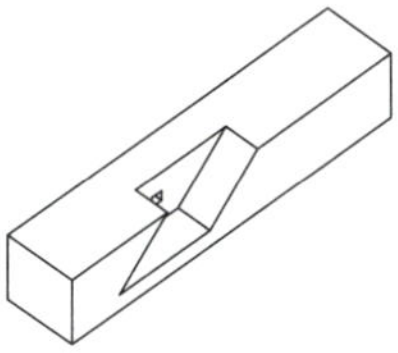

B

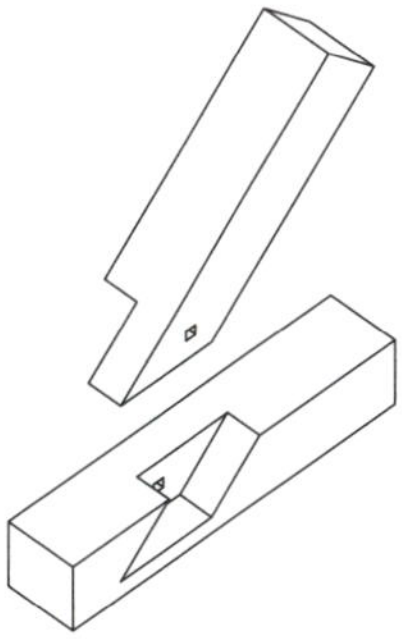

A+B

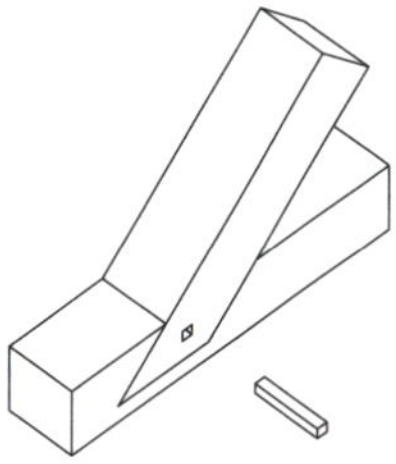

A+B

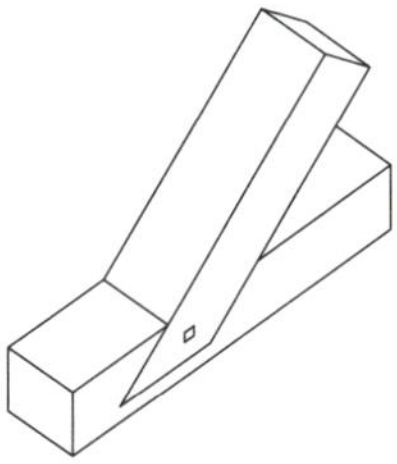

AB

Angled Joints

Pegged angled lapped mortise and tenon joint

naname-kata-binta-dō-tsuki-hira-hozo-sashi-komi-sen-uchi
斜片鬢太胴付平枘差込栓打ち

schräges Blatt mit Zapfen und Dübelsicherung

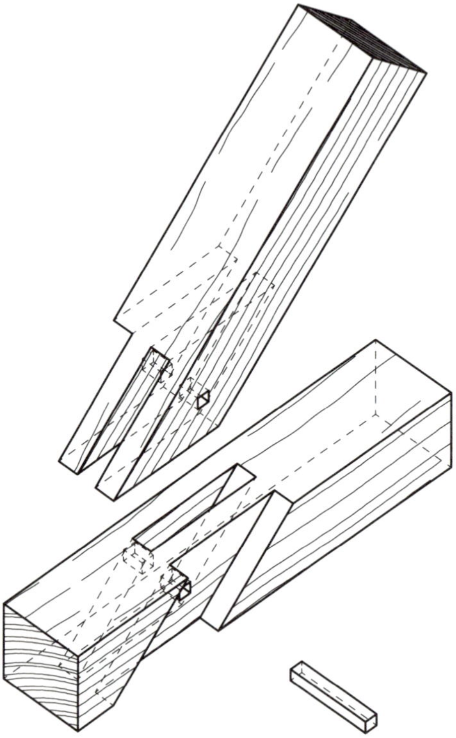

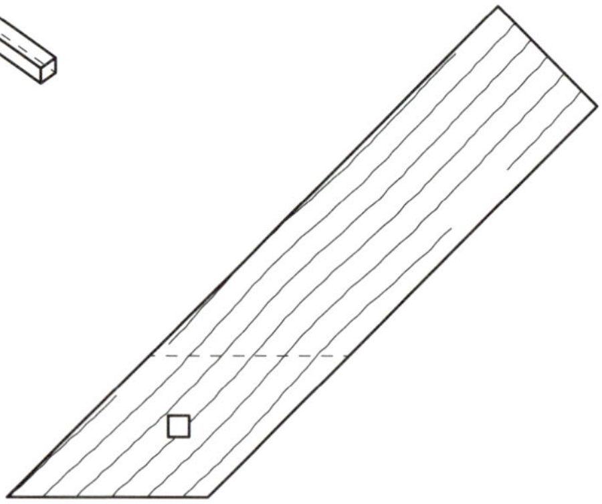

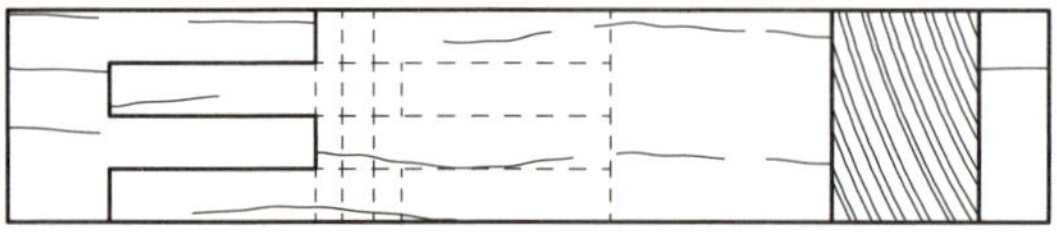

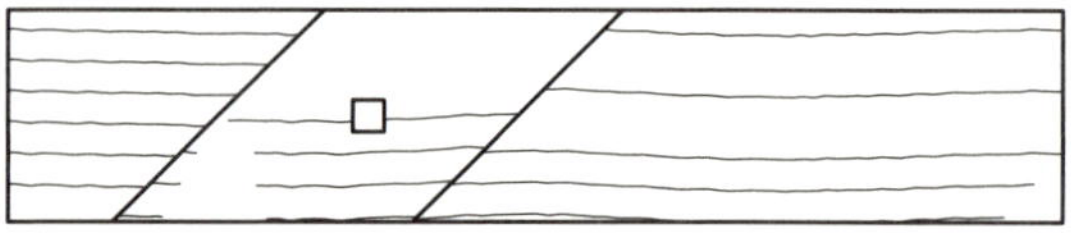

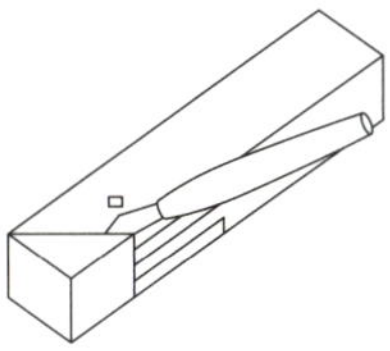

A1

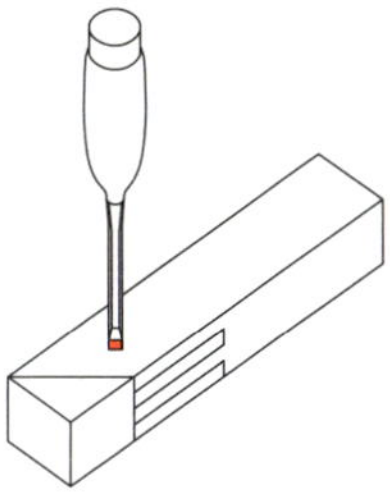

A2

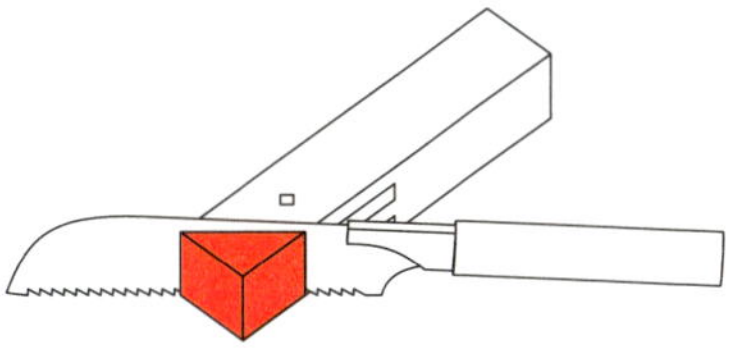

A3

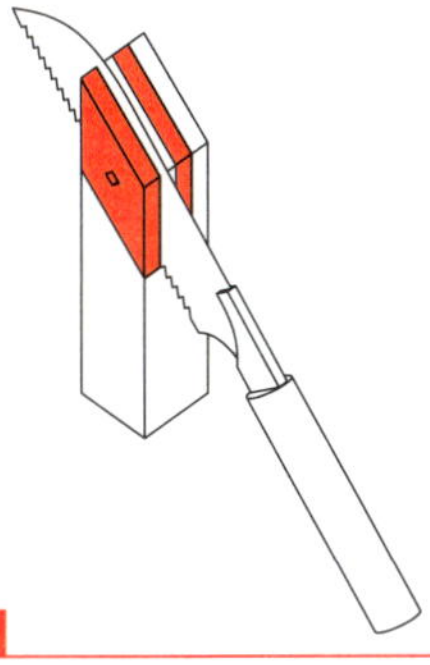

A4

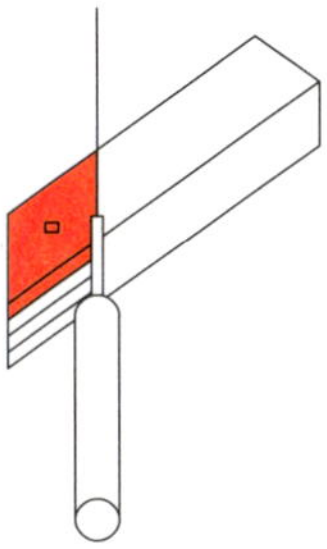

A5

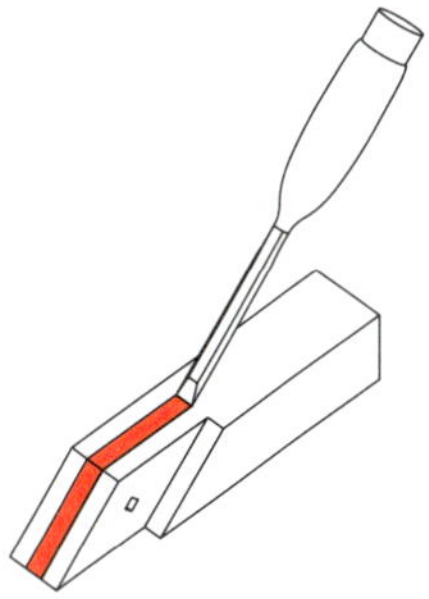

A6

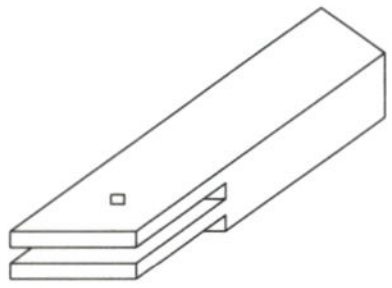

A

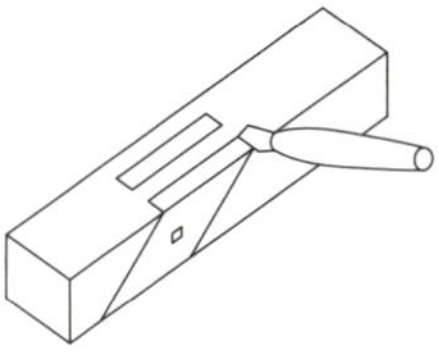

B1

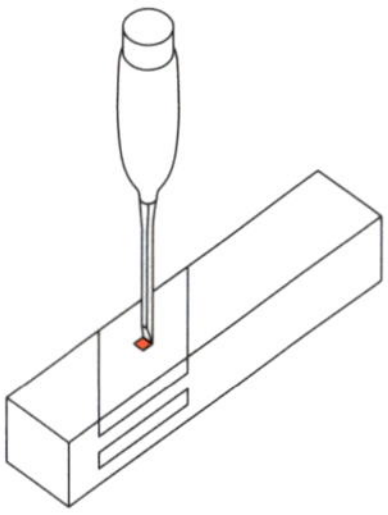

B2

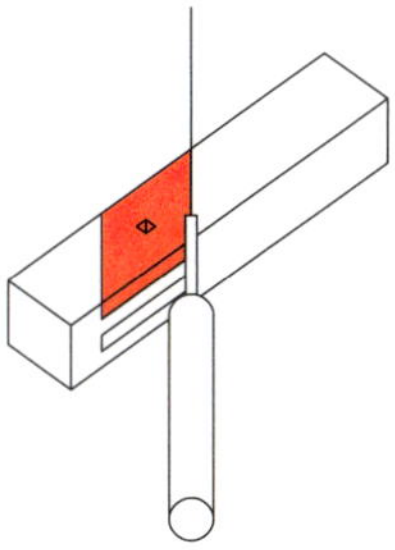

B3

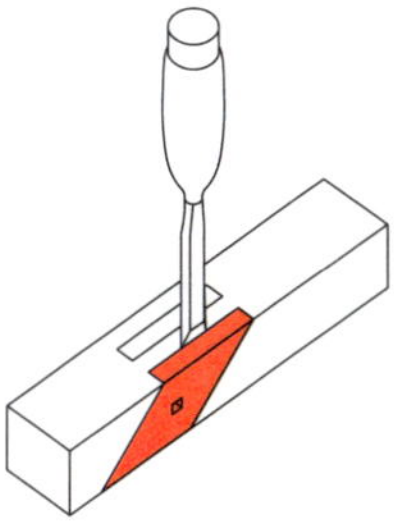

B4

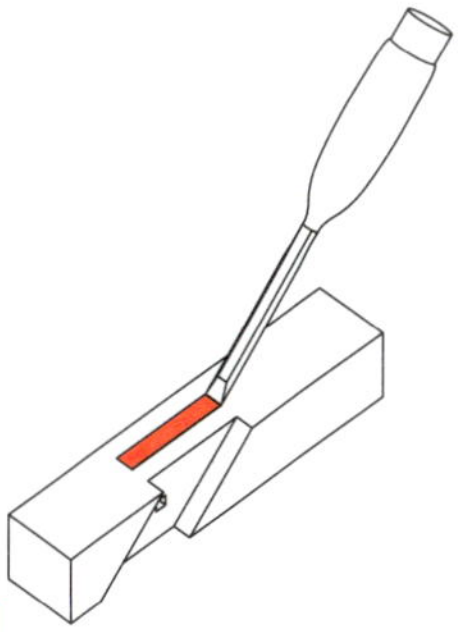

B5

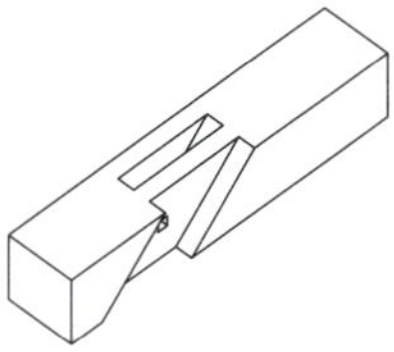

B

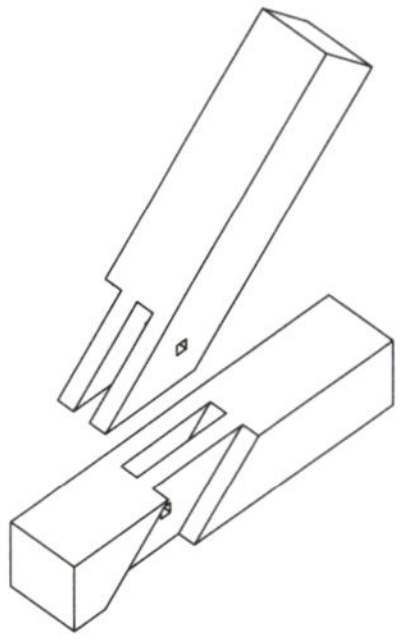

A+B

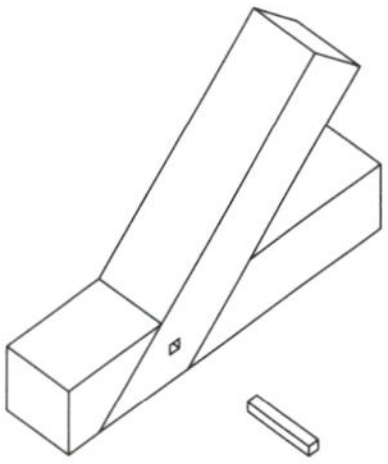

AB

Angled Joints

Pegged shouldered angled half lap joint

naname-ai-kaki-tsugi-komi-sen-uchi (ryō-men-oshi-tsuki)
斜相欠接込栓打ち(両面押付)

beidseitig abgesetztes schräges Blatt mit Dübelsicherung

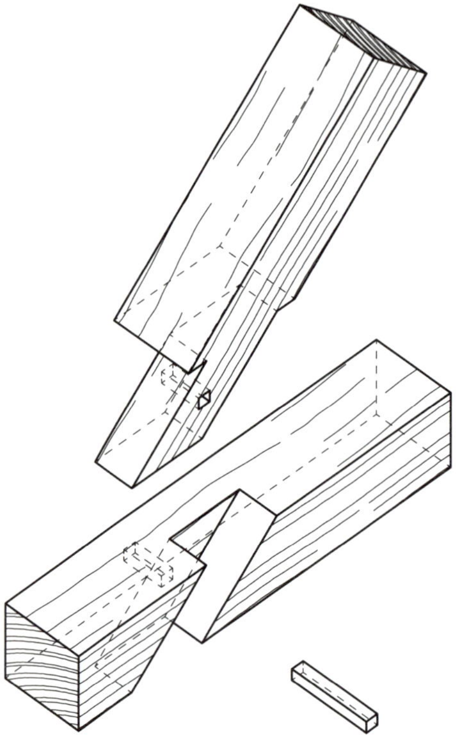

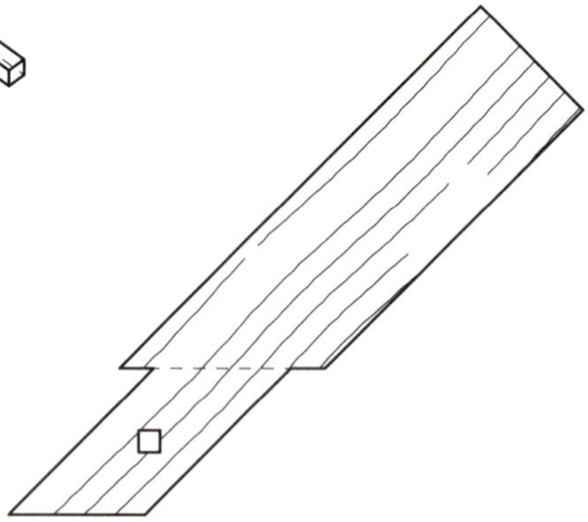

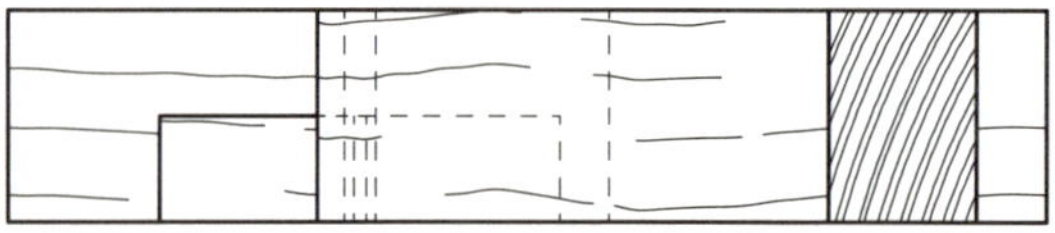

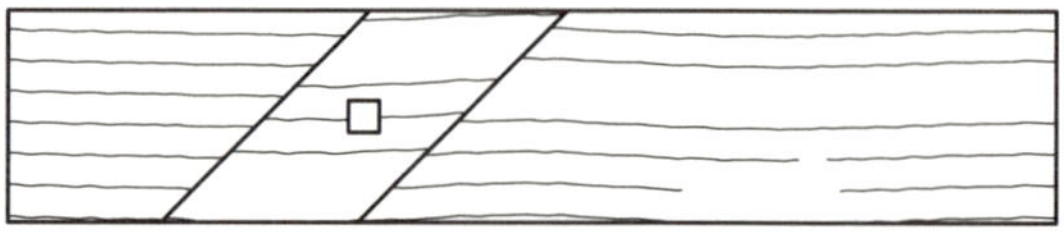

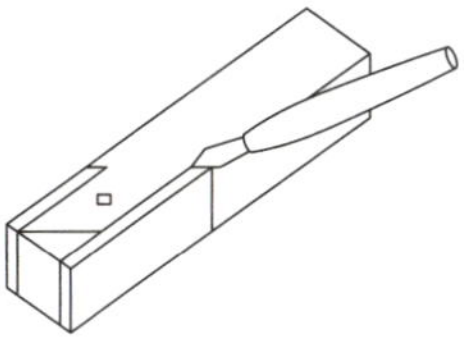
A1

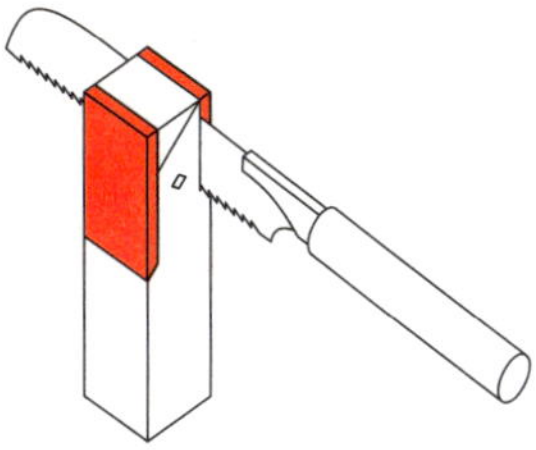
A2

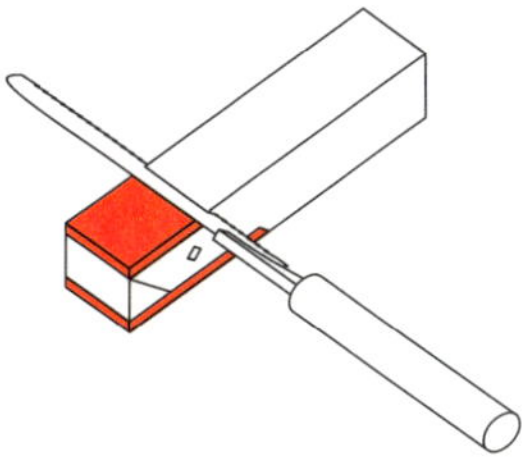
A3

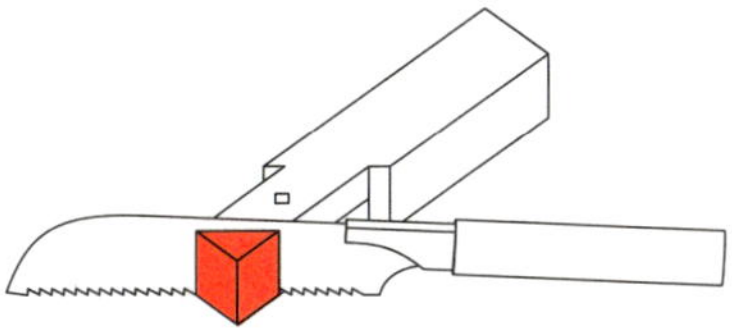
A4

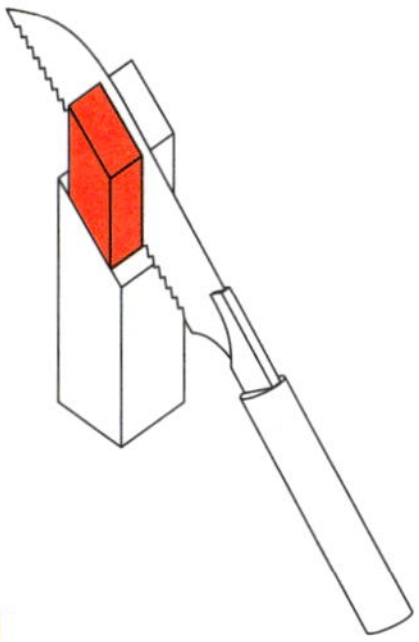
A5

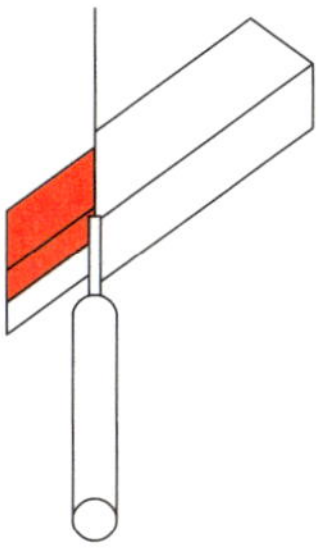
A6

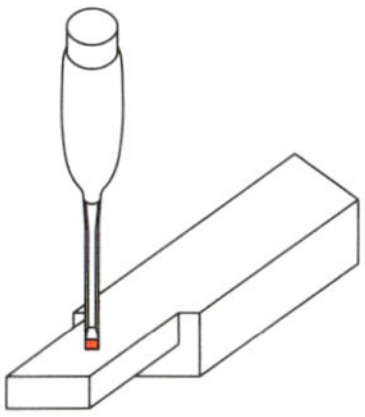
A7

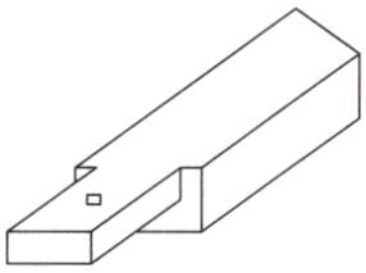
A

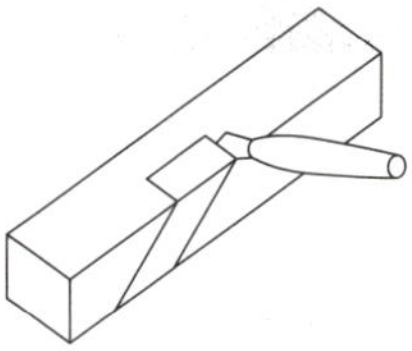
B1

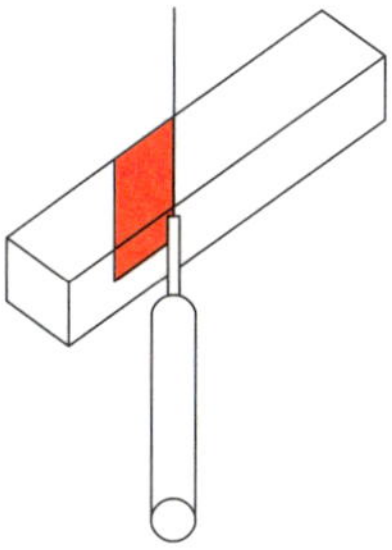
B2

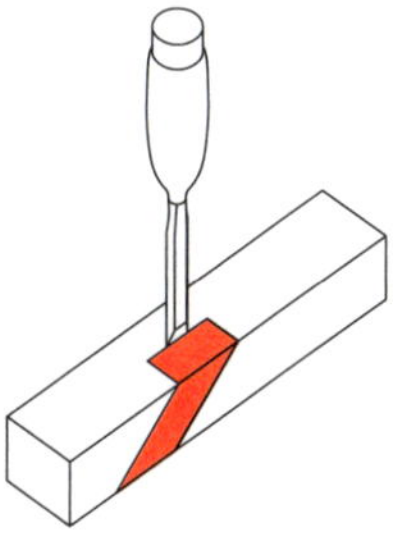
B3

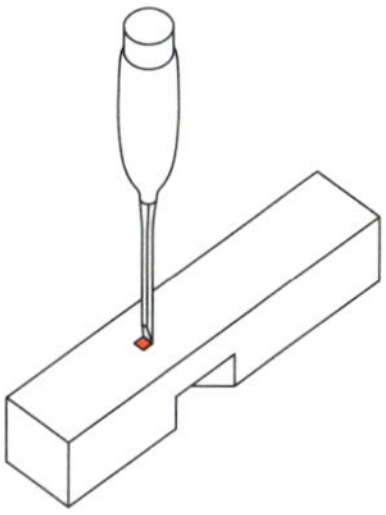
B4

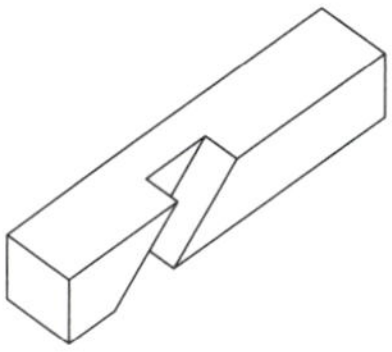
B

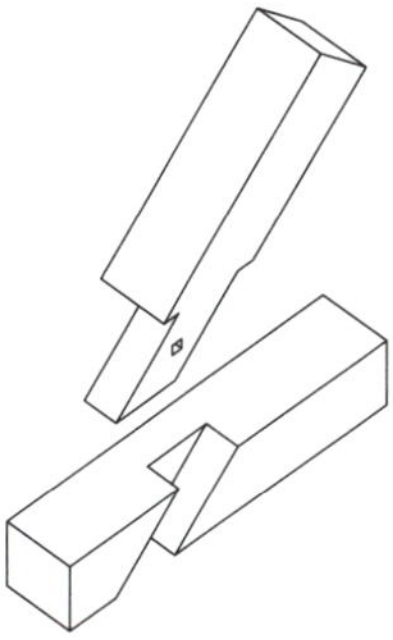
A+B

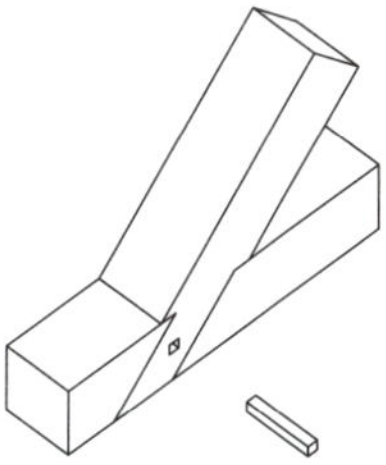
AB

Angled Joints

Pegged angled half-dovetail lap joint with sloped shoulder

naname-sogi-kaki-ai-kaki-tsugi-komi-sen-uchi (men-oshi-tsuki)
斜殺欠相欠接込栓打ち（面押付）

schräges einseitiges Schwalbenschwanzblatt mit verdecktem Versatz und Dübelsicherung

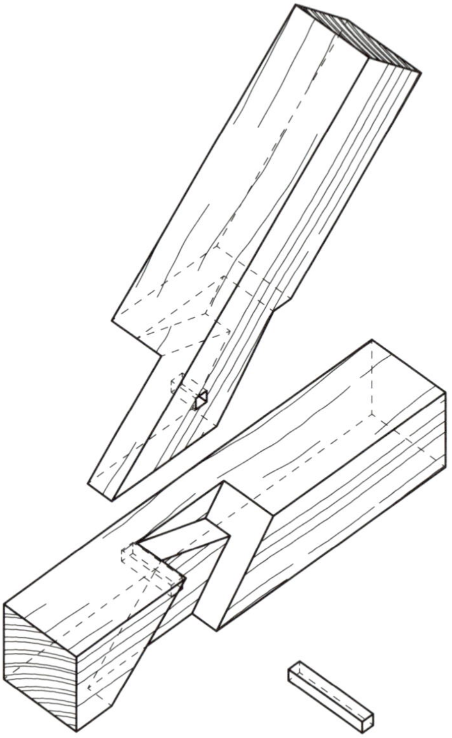

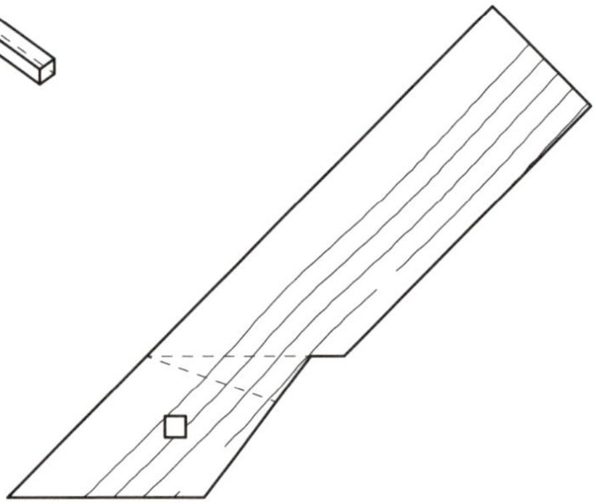

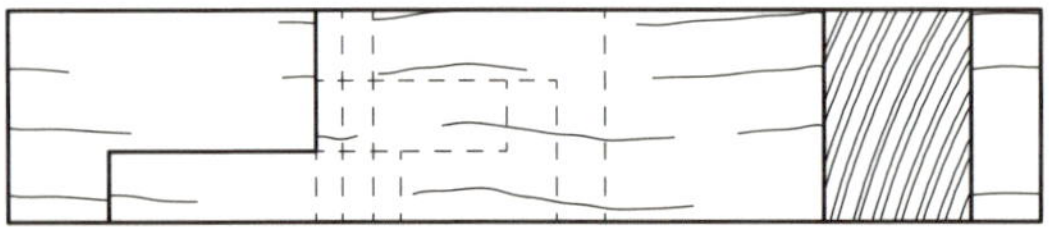

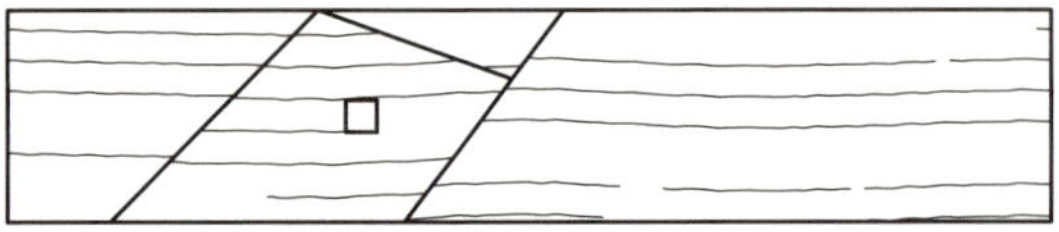

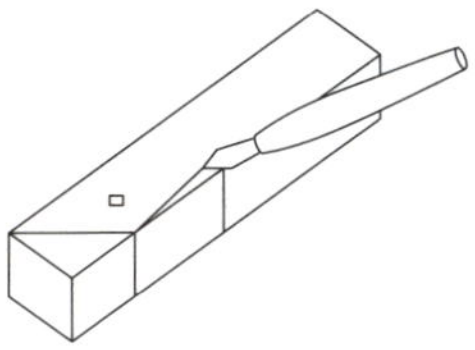

A1

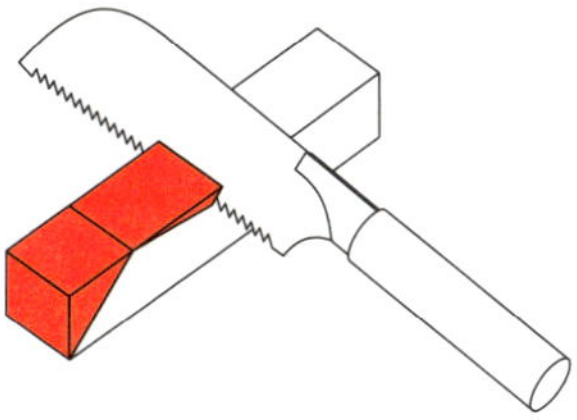

A2

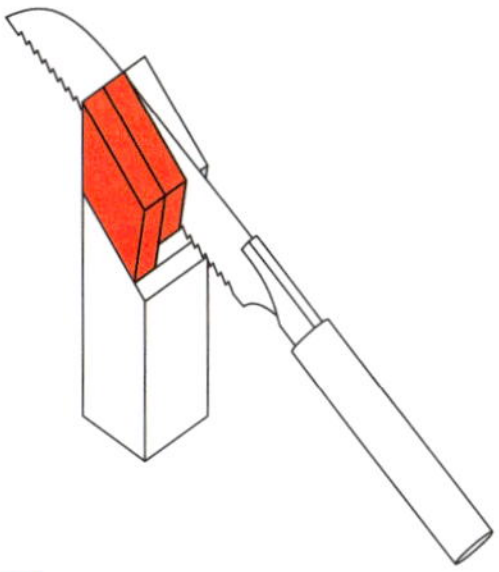

A3

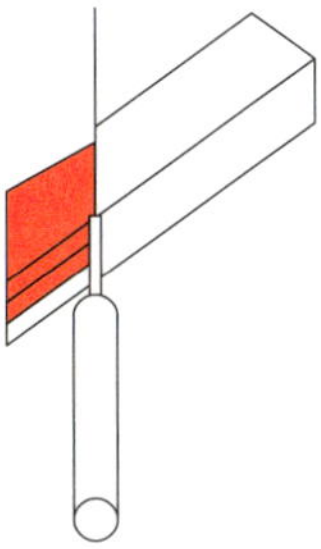

A4

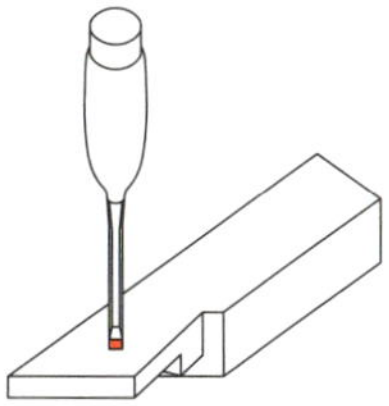

A5

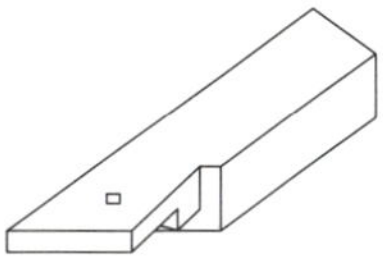

A

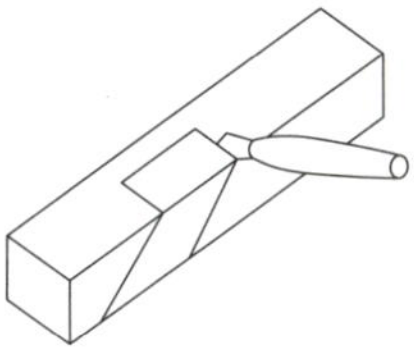

B1

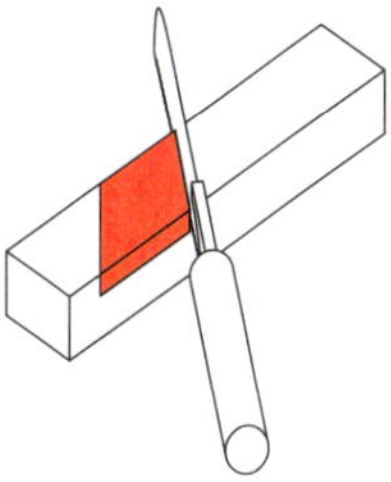

B2

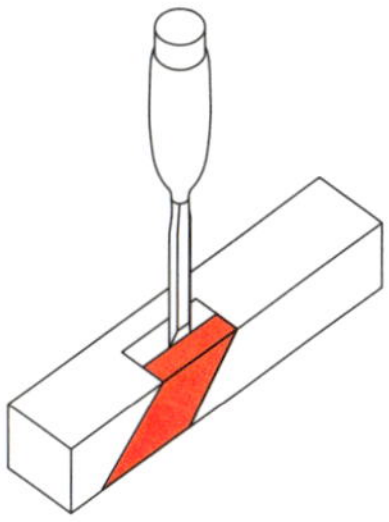

B3

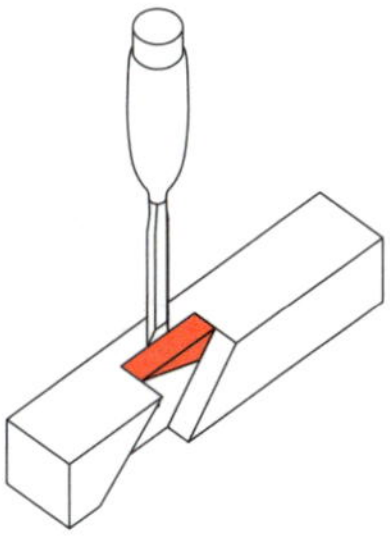

B4

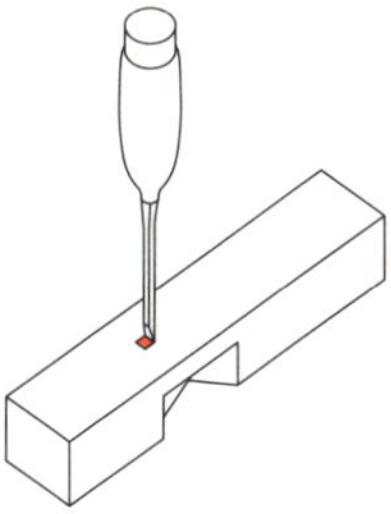

B5

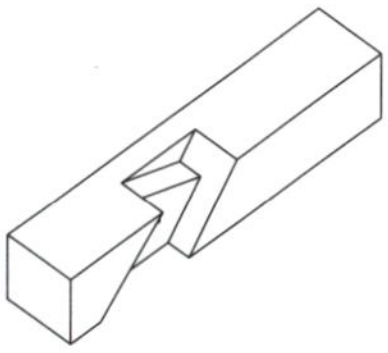

B

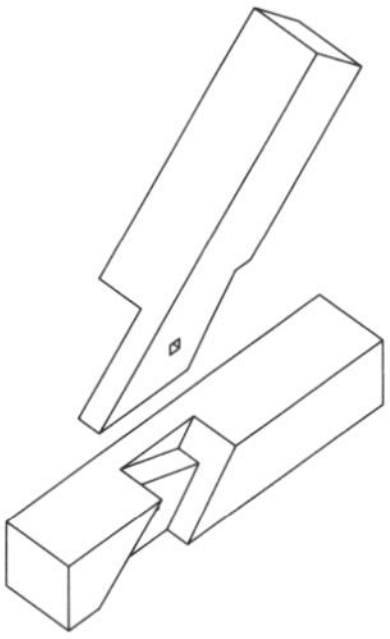

A+B

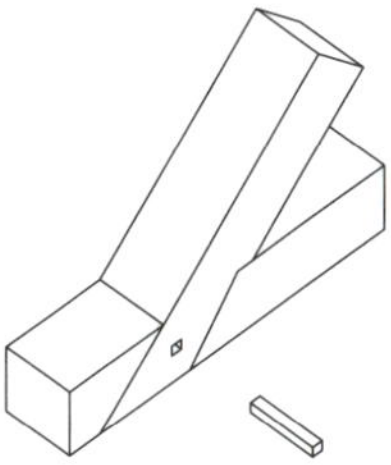

A+B

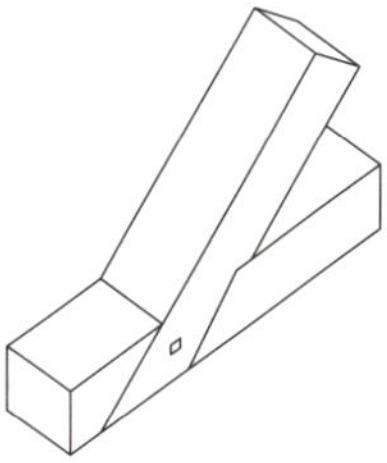

AB

Angled Joints

Pegged angled half-dovetail half lap joint

naname-ari-tsugi-komi-sen-uchi (men-oshi-tsuki)
斜蟻接込栓打ち(面押付)

schräges einseitiges Schwalbenschwanzblatt mit Dübelsicherung

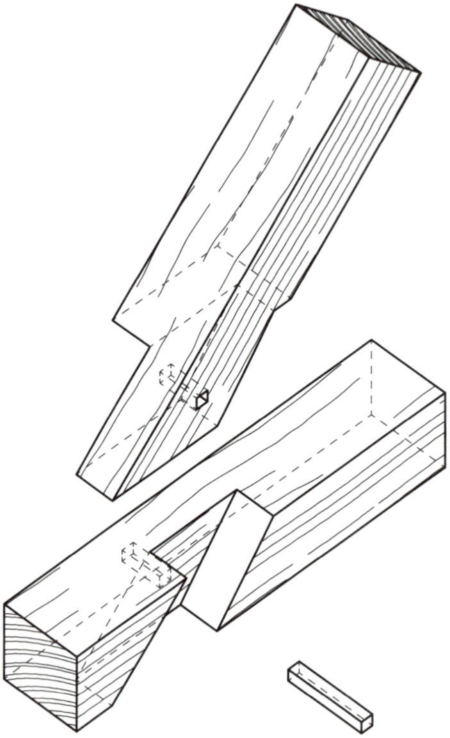

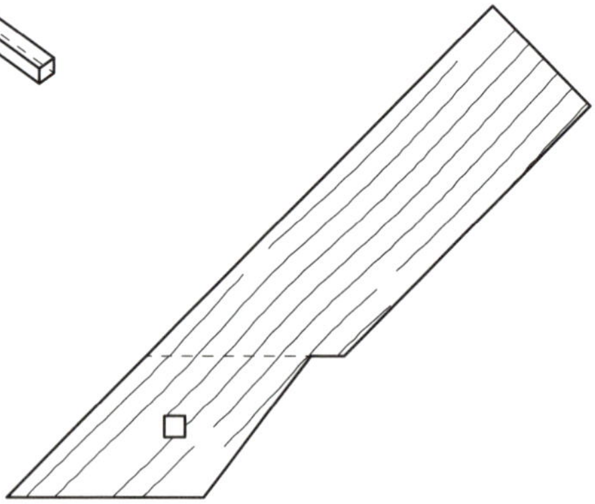

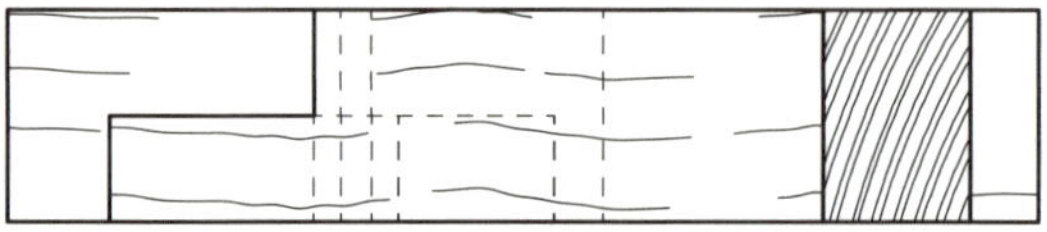

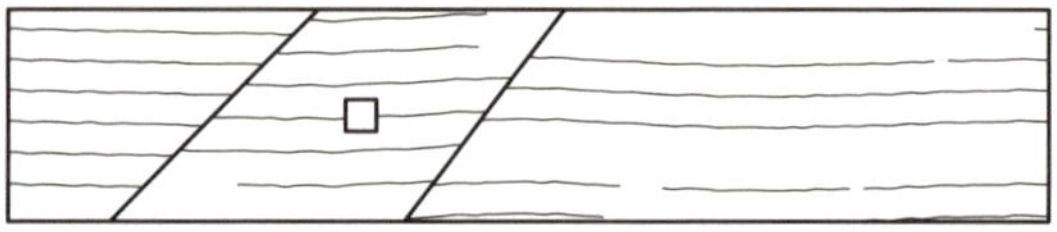

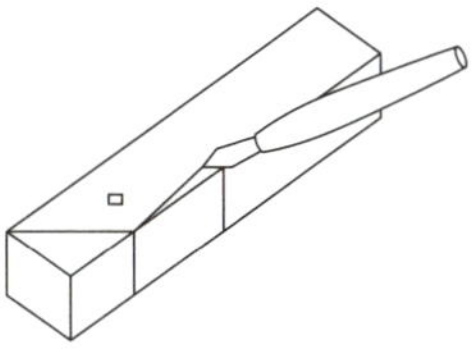

A1

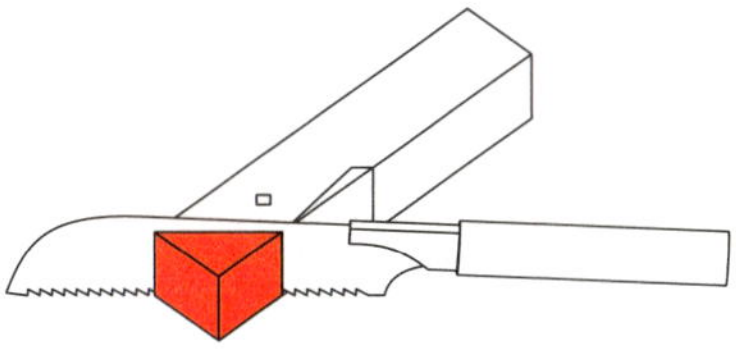

A2

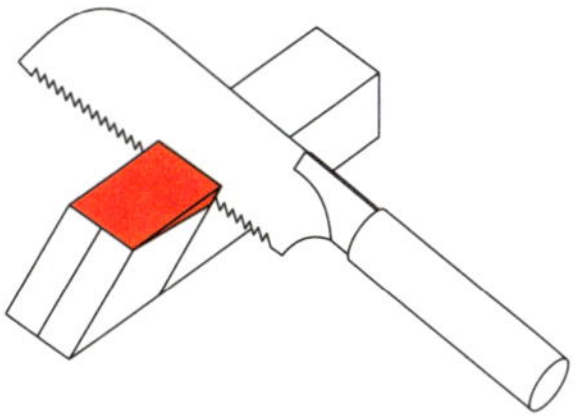

A3

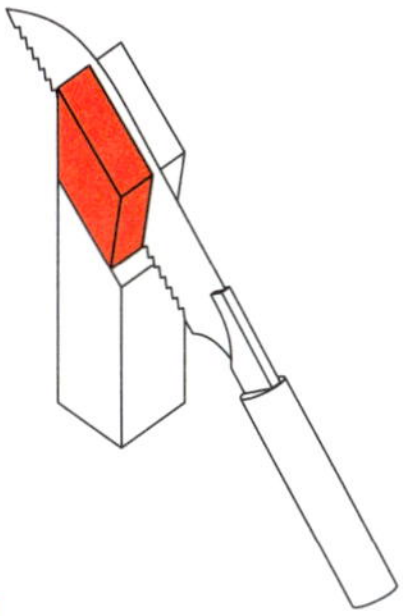

A4

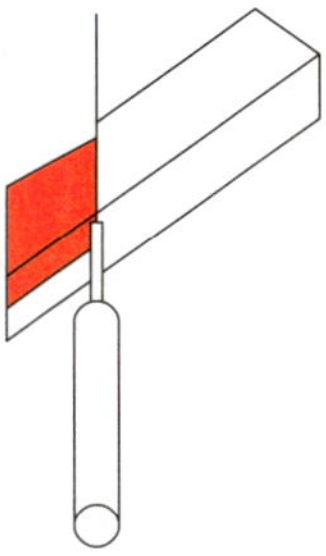

A5

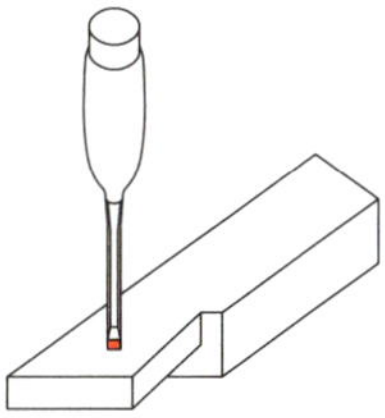

A6

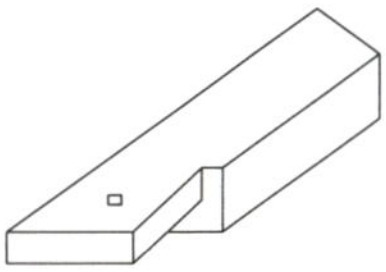

A

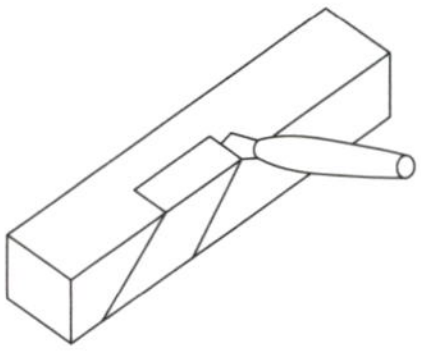

B1

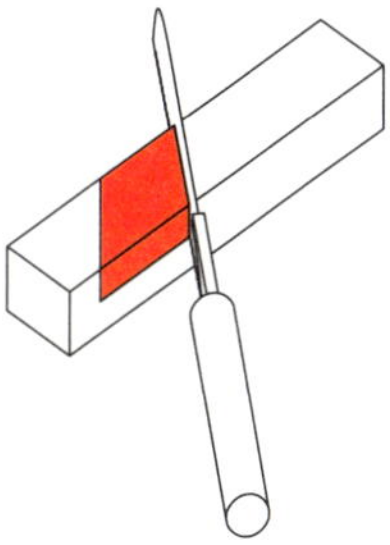

B2

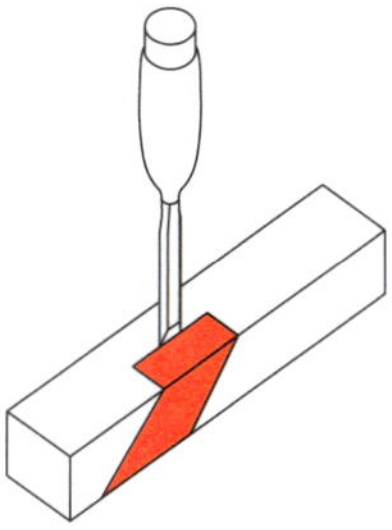

B3

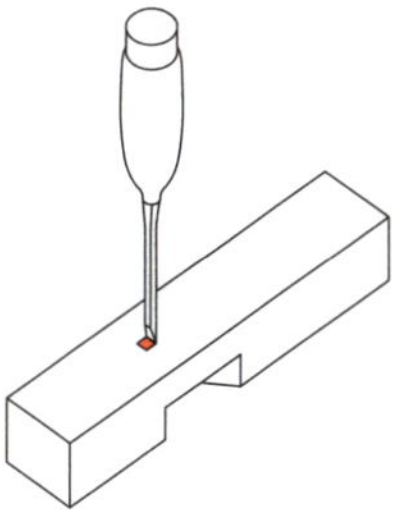

B4

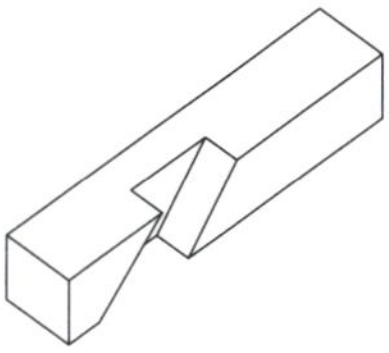

B

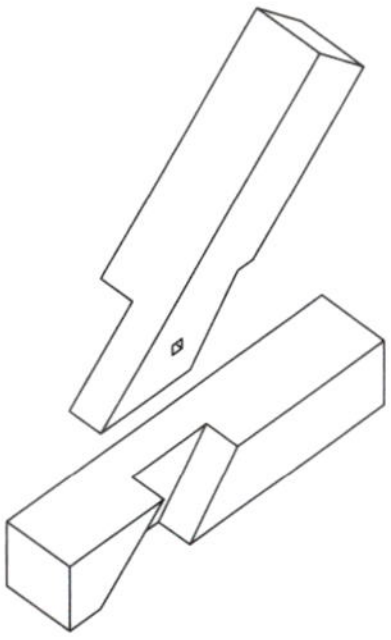

A+B

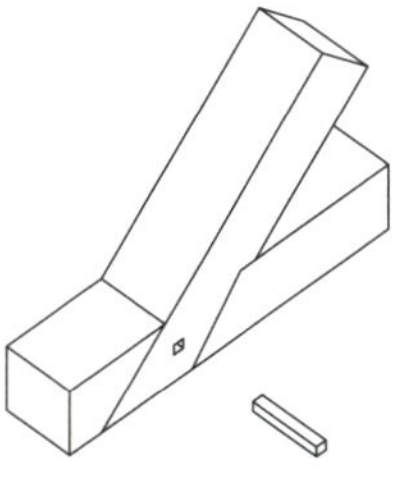

A+B

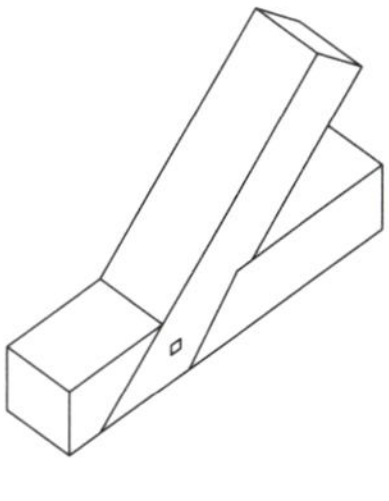

AB

Angled Joints

Pegged stopped angled half-dovetail half lap joint

naname-kakushi-ari-tsugi-komi-sen-uchi (ryô-men-oshi-tsuki)
斜隠蟻接込栓打ち(両面押付)

halbverdecktes schräges einseitiges Schwalbenschwanzblatt und Dübelsicherung

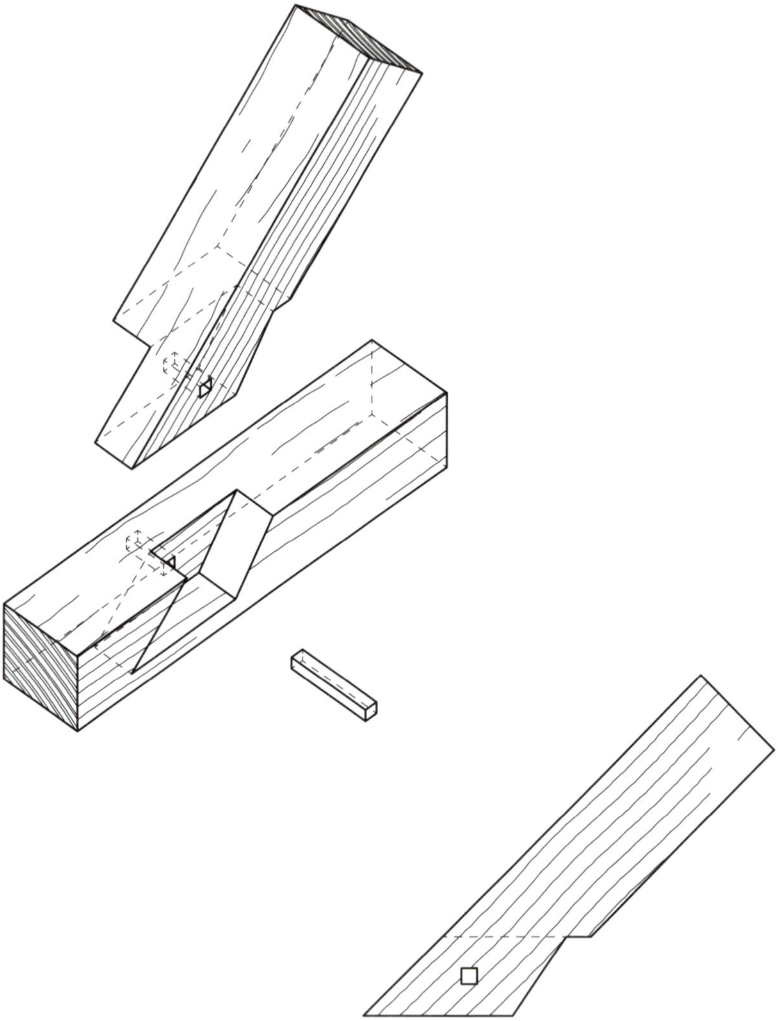

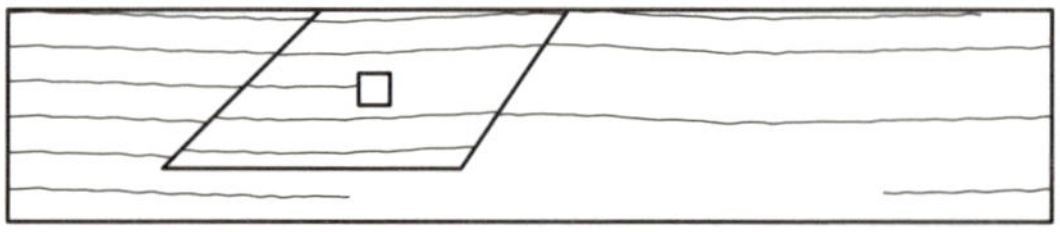

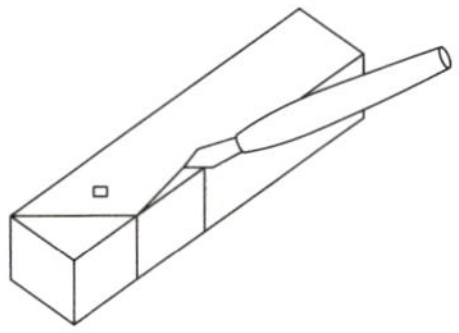

A1

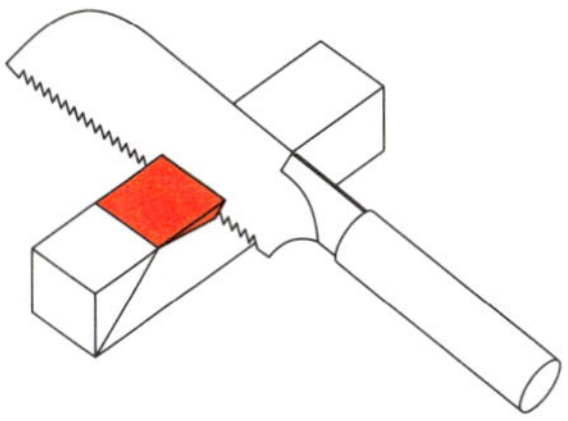

A2

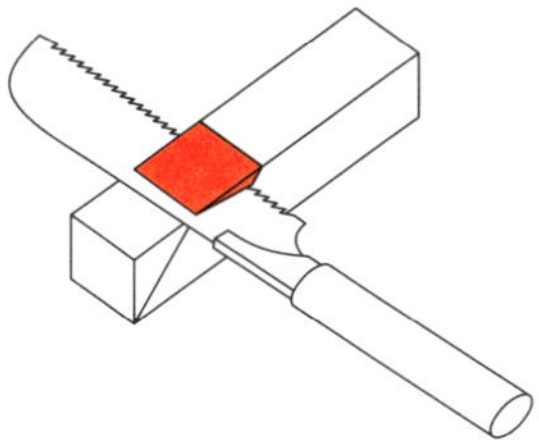

A3

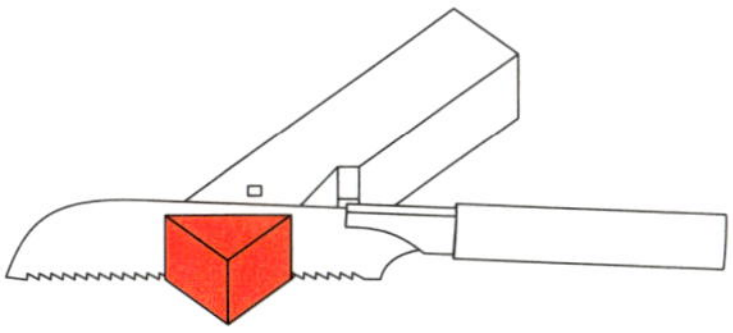

A4

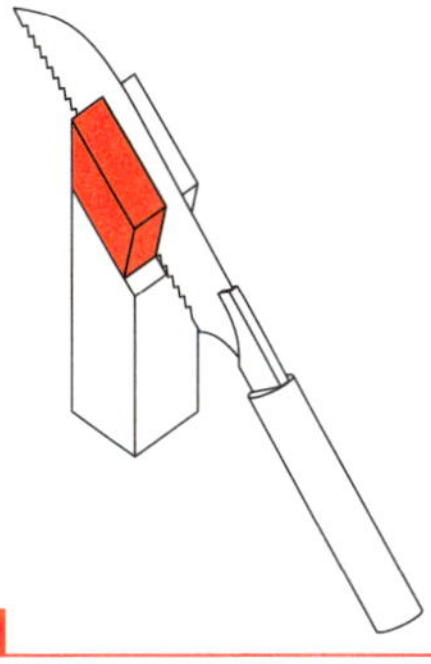

A5

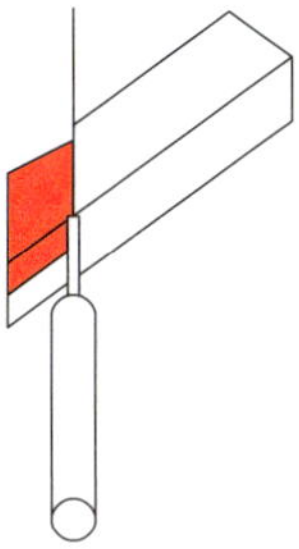

A6

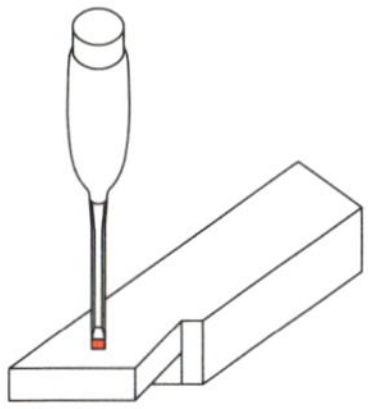

A7

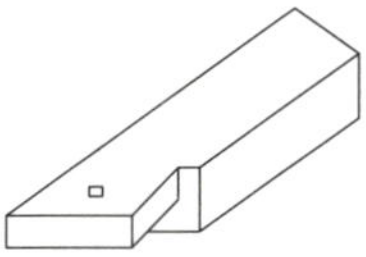

A

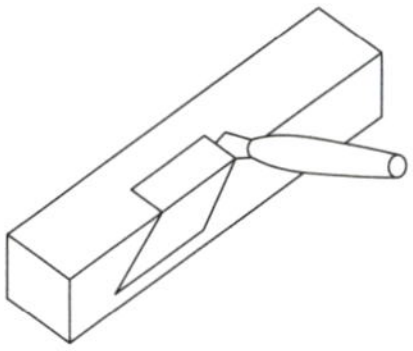

B1

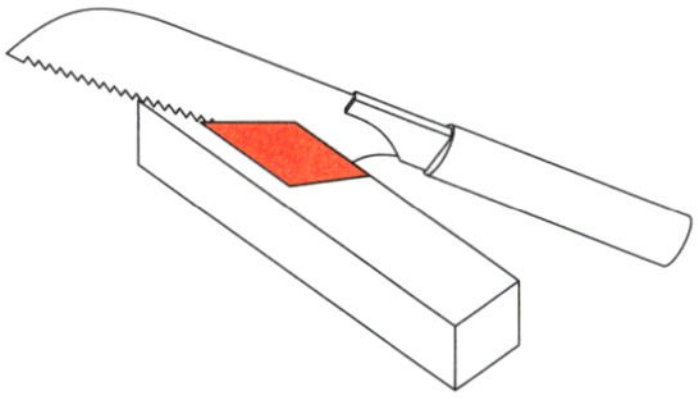

B2

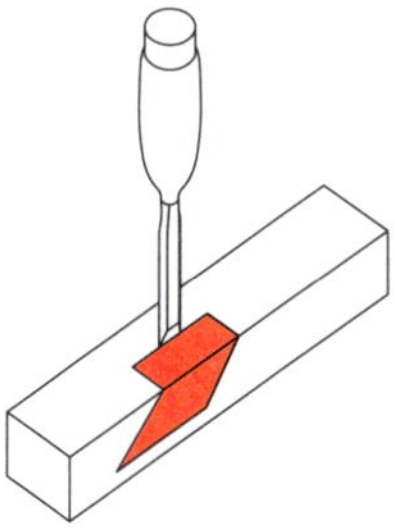

B3

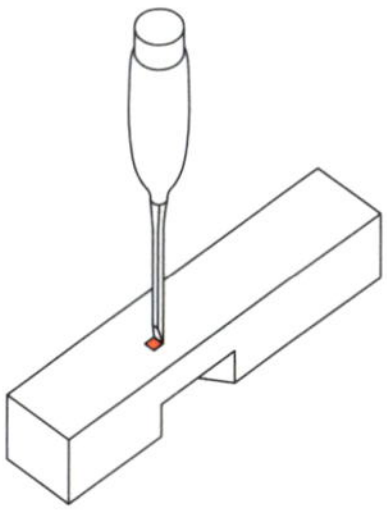

B4

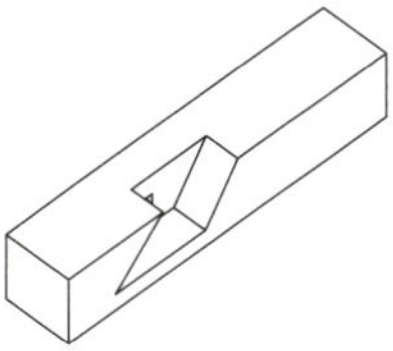

B

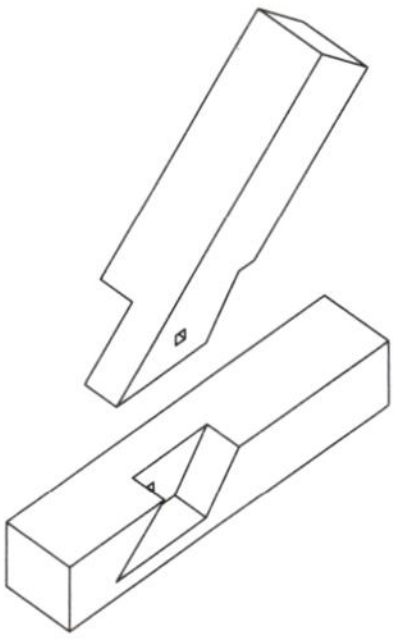

A+B

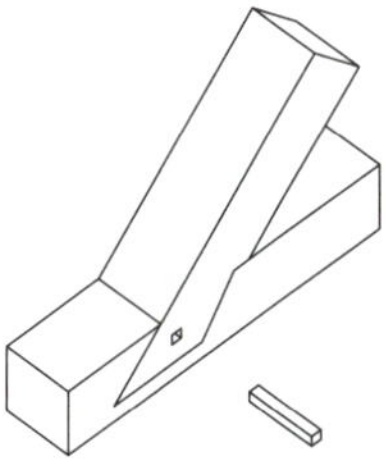

A+B

Angled Joints

Pegged angled stopped dovetail half lap joint

naname-kakushi-ari-tsugi-komi-sen-uchi (men-oshi-tsuki)
斜隠蟻接込栓打ち(面押付)

halbverdecktes schräges Schwalbenschwanzblatt mit Dübelsicherung

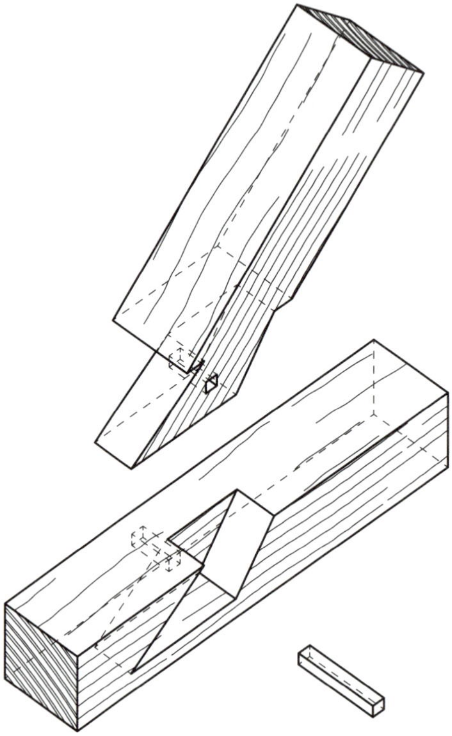

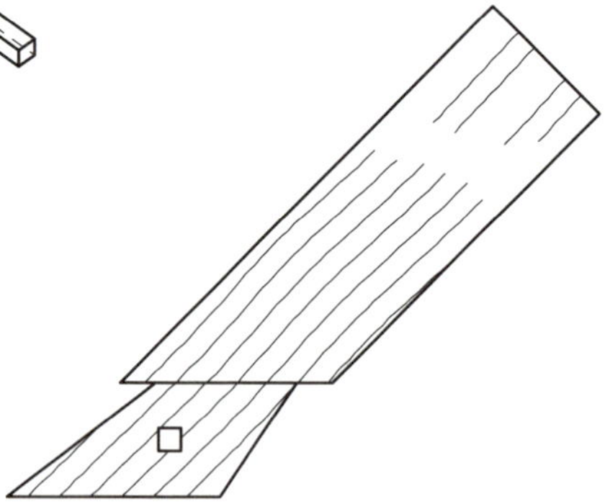

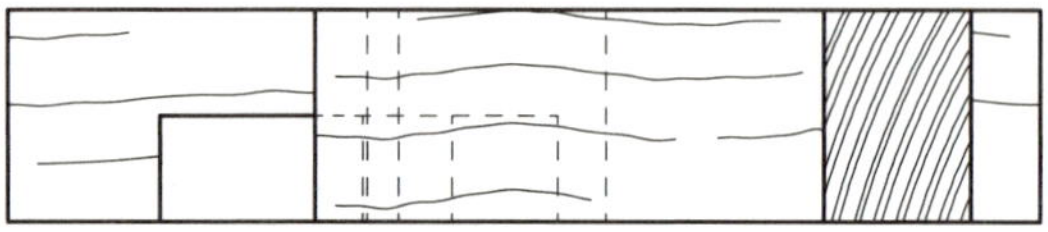

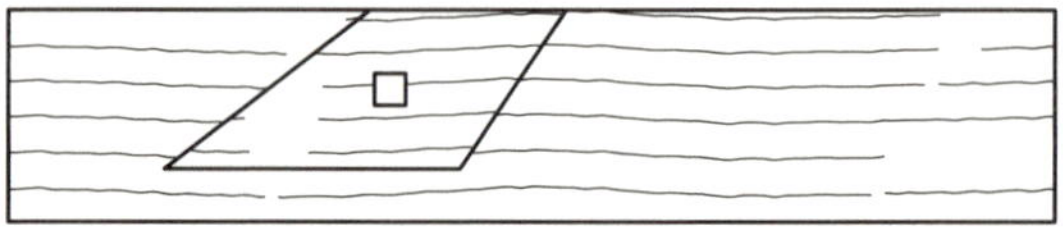

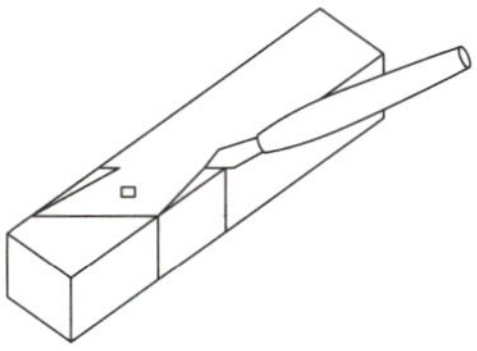

A1

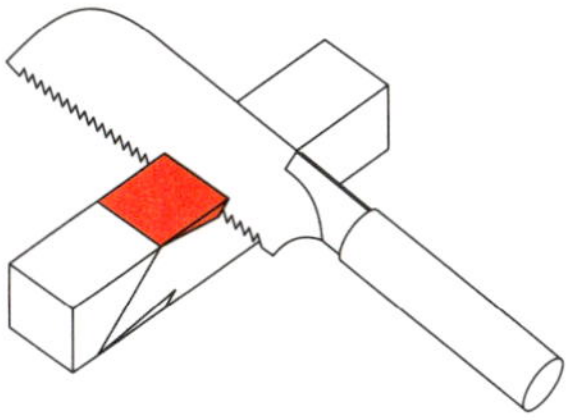

A2

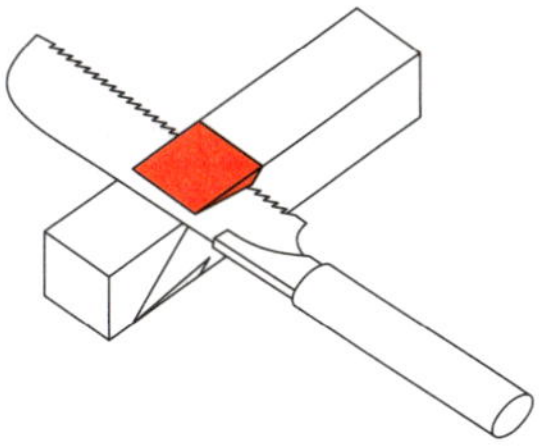

A3

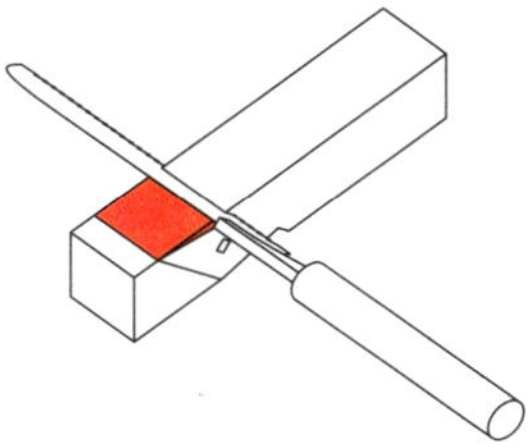

A4

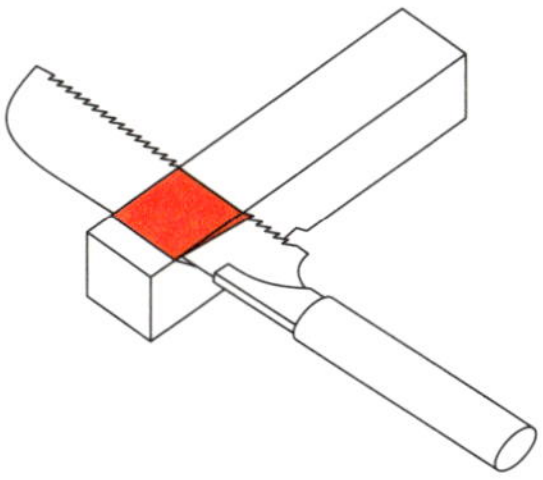

A5

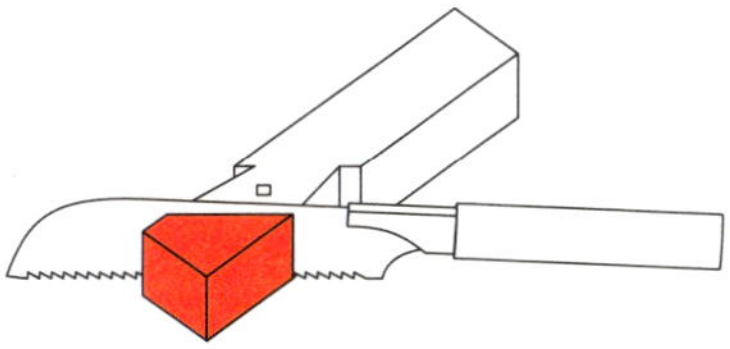

A6

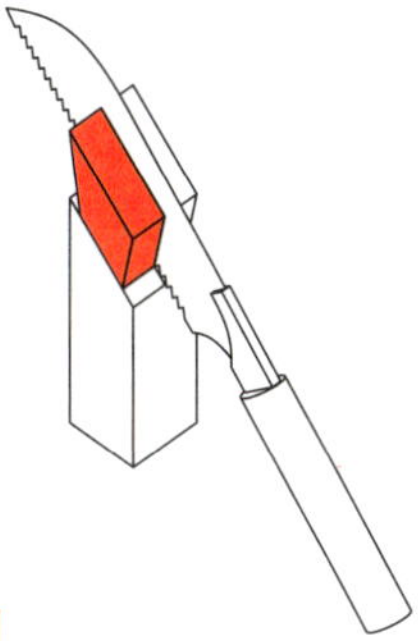

A7

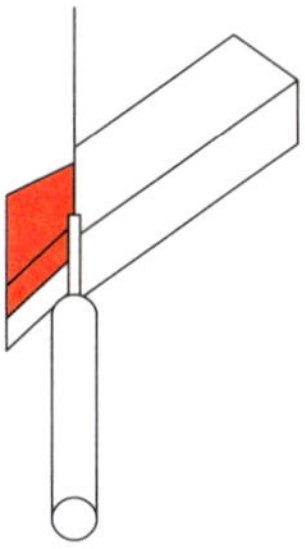

A8

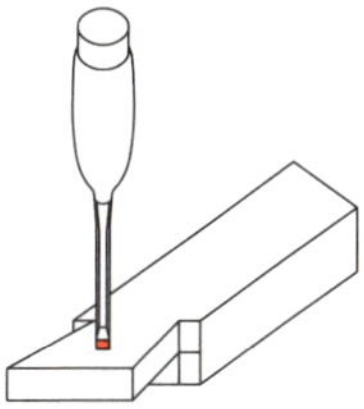

A9

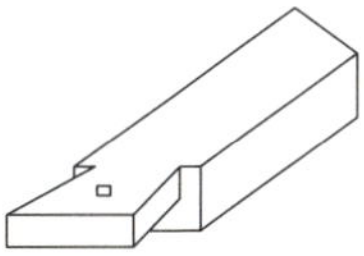

A

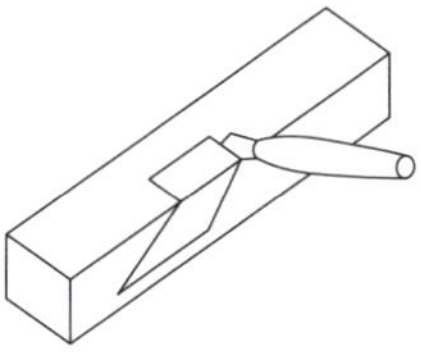

B1

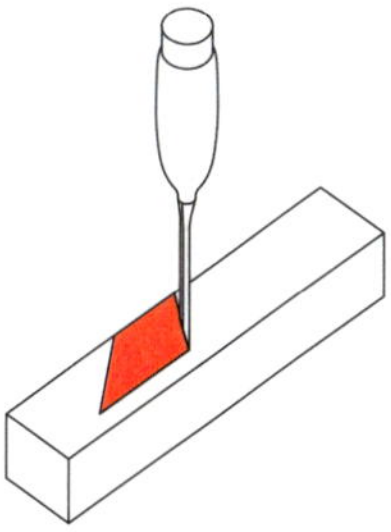

B2

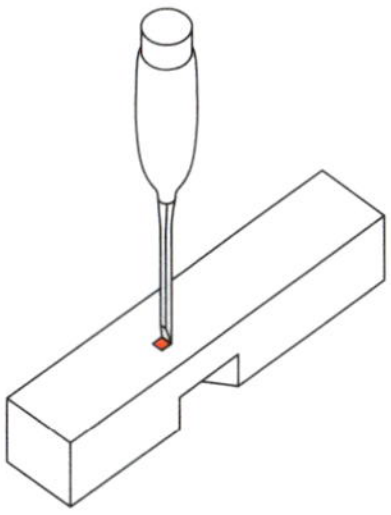

B3

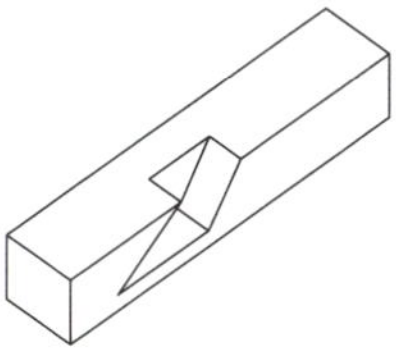

B

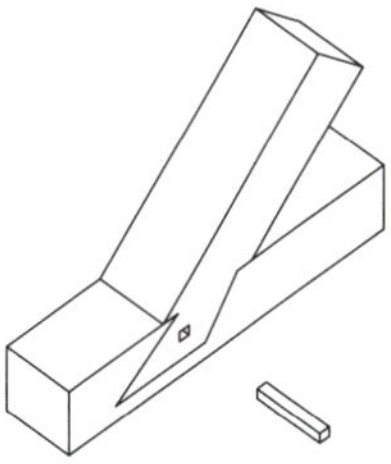

A+B

Angled Joints

Pegged rafter foot

komi-sen-kake
込栓掛け

Stütznagelung

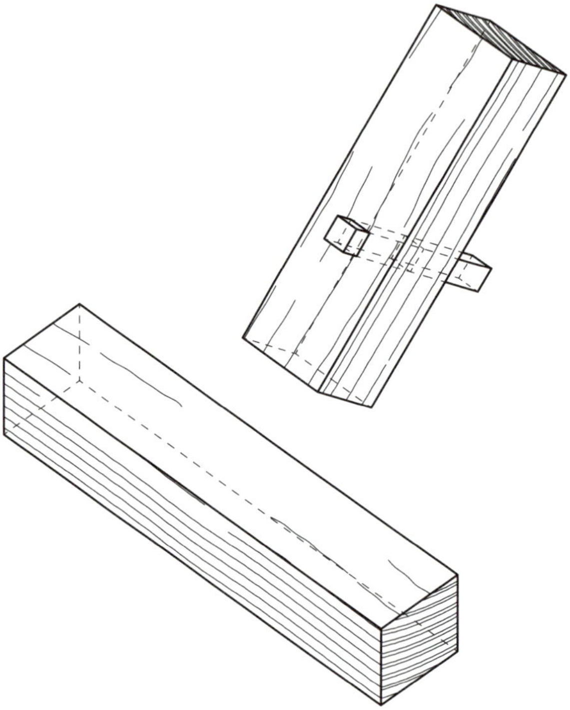

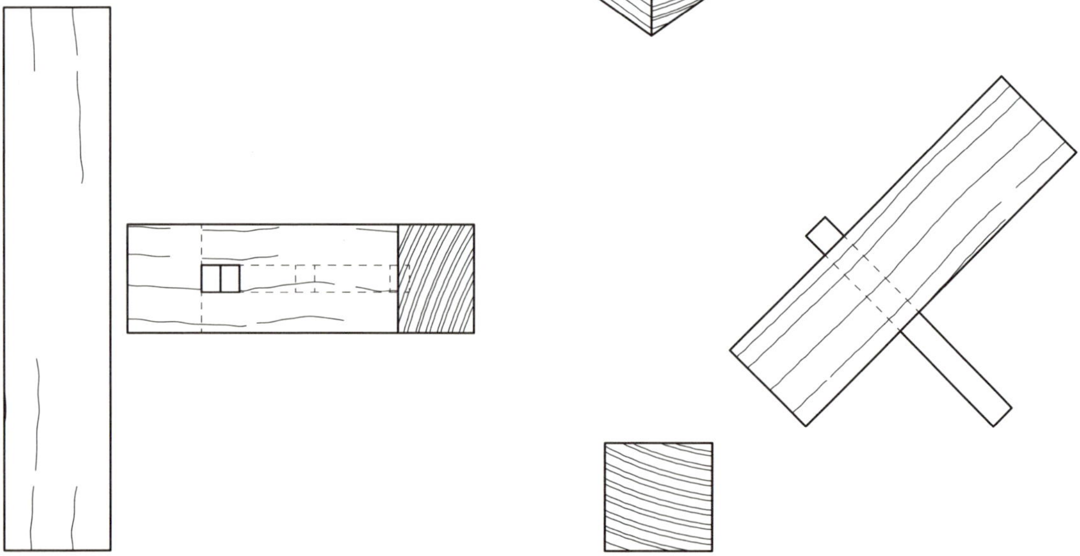

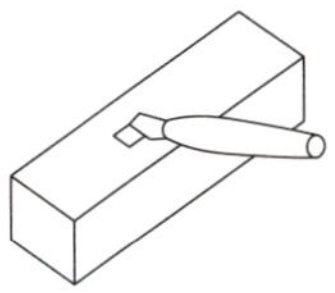

A1

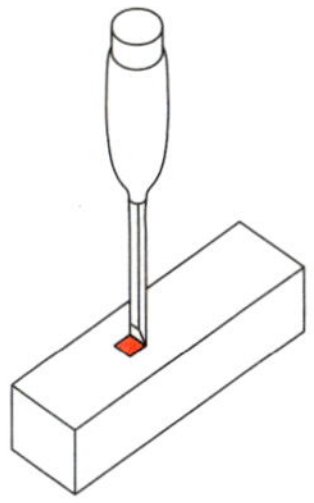

A2

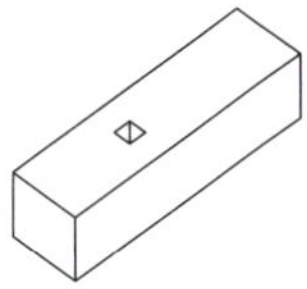

A

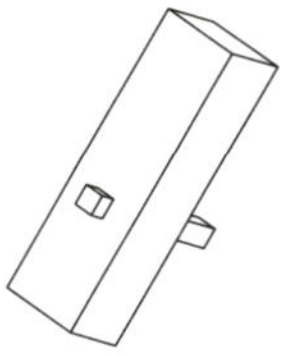

A+B

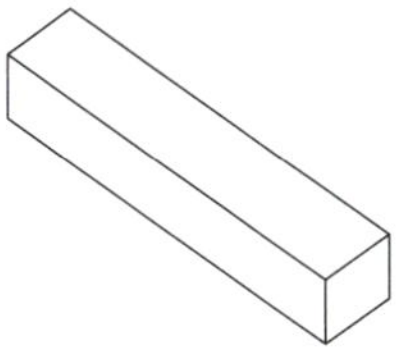

C

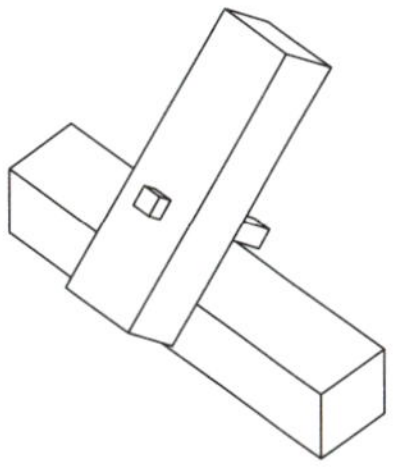

ABC

Angled Joints

Pegged mitered rafter foot

naname-komi-sen-tsugi
斜込栓接ぎ

Schiftungsstoß

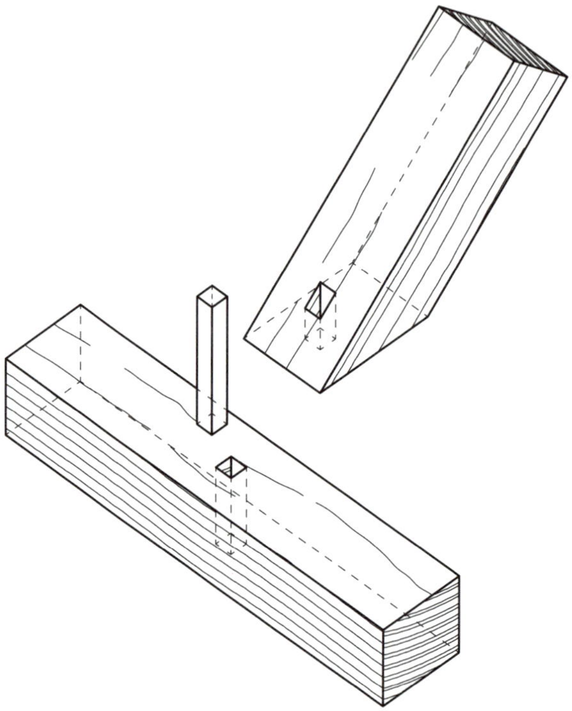

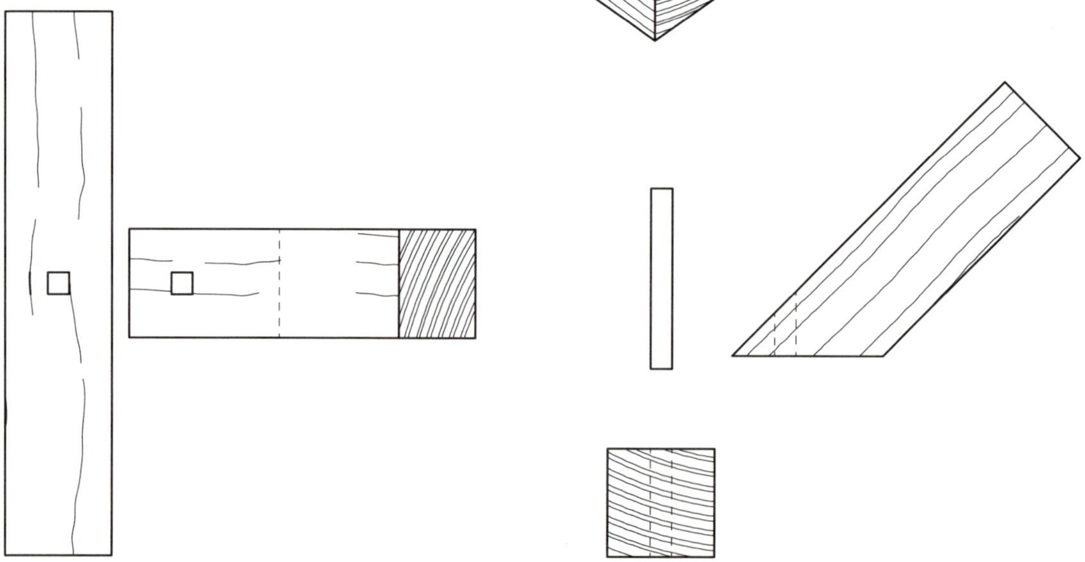

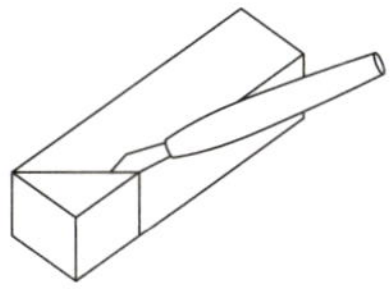

A1

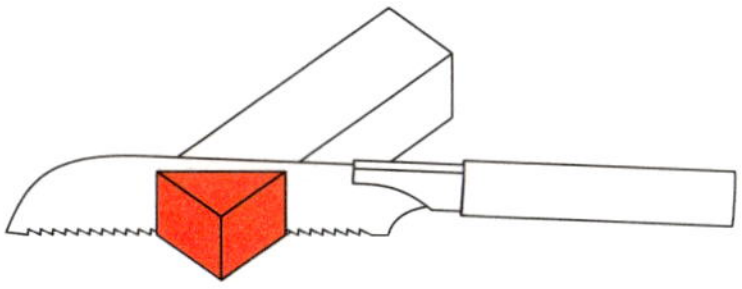

A2

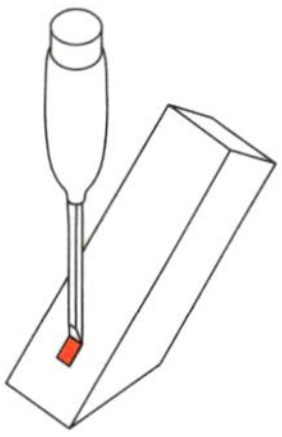

A3

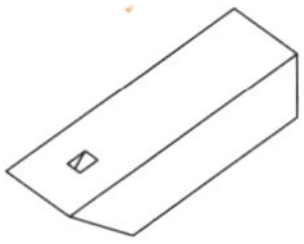

A

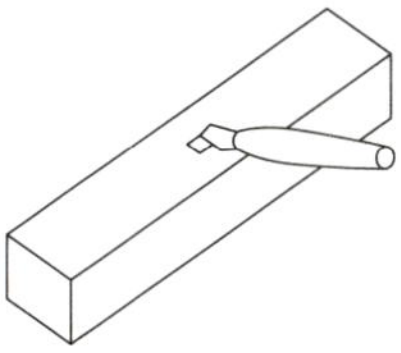

B1

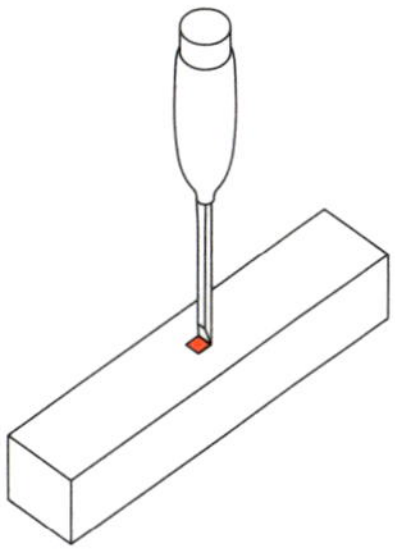

B2

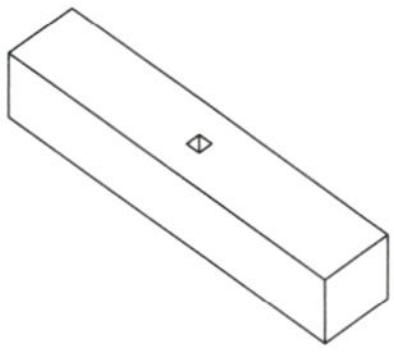

B

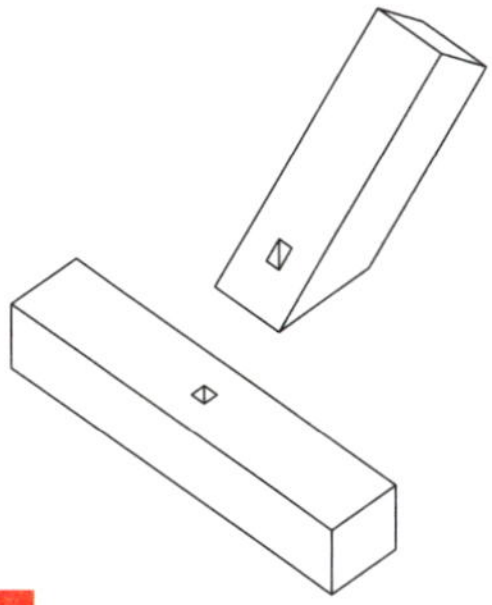

A+B

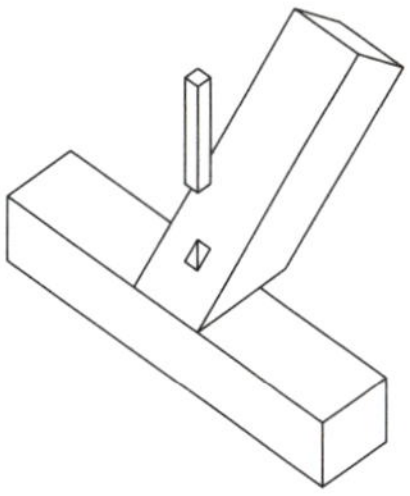

A+B+C

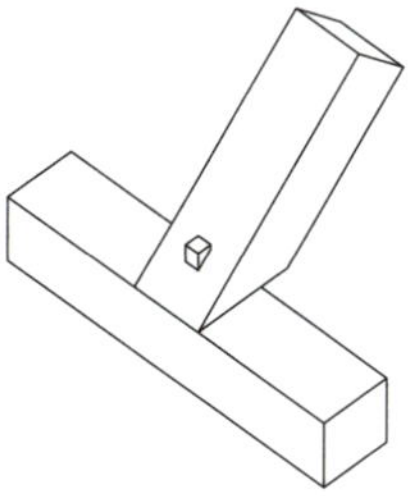

ABC

Angled Joints

Housed rafter foot

naname-koshi-kake-tsugi
斜腰掛接ぎ

Einklauung

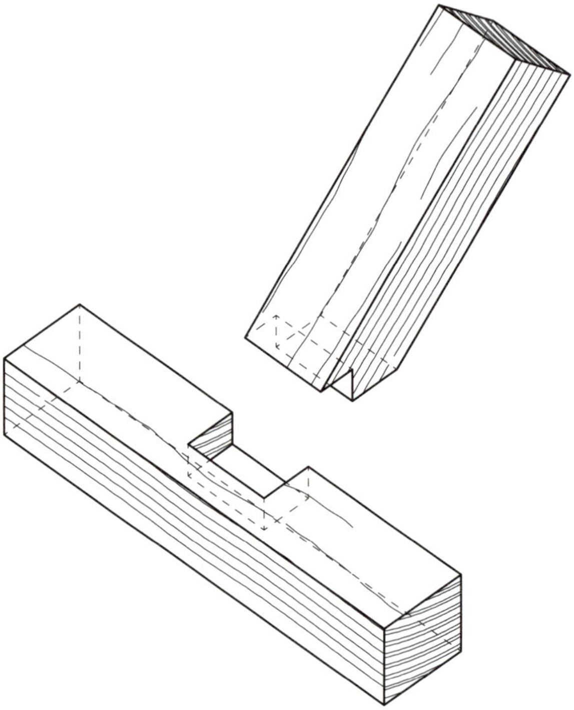

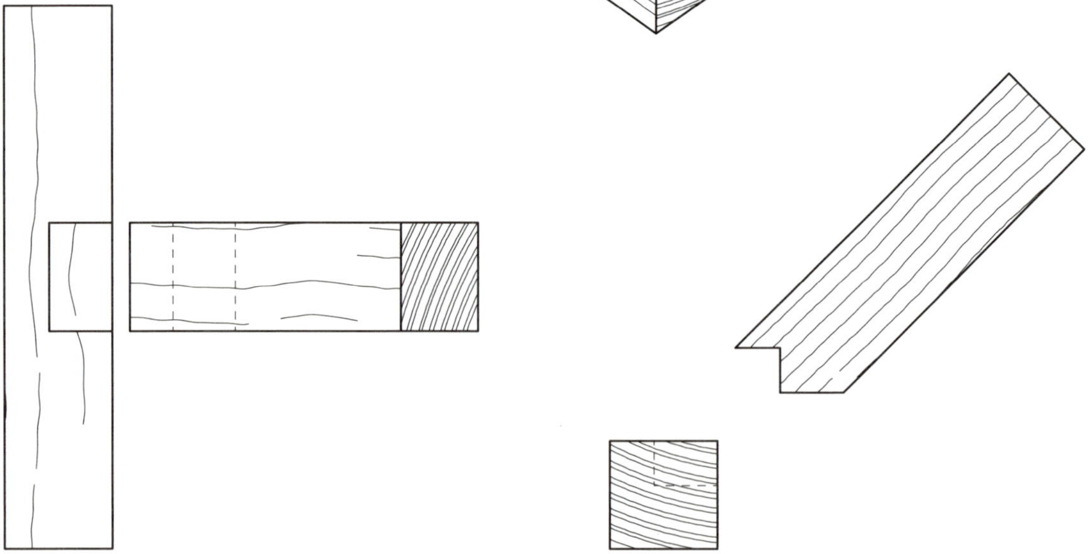

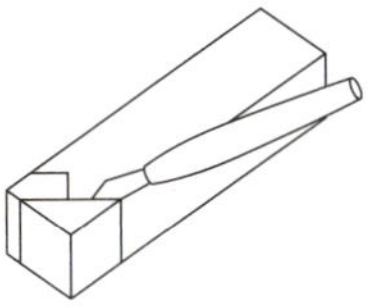

A1

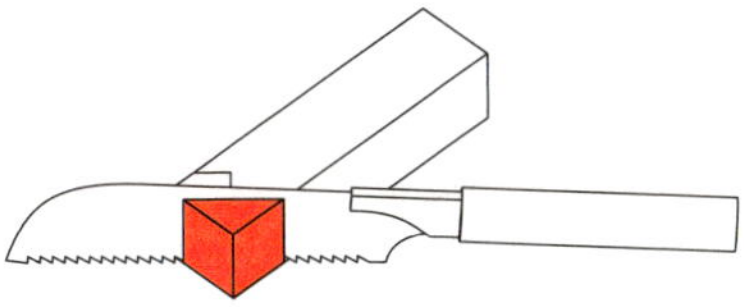

A2

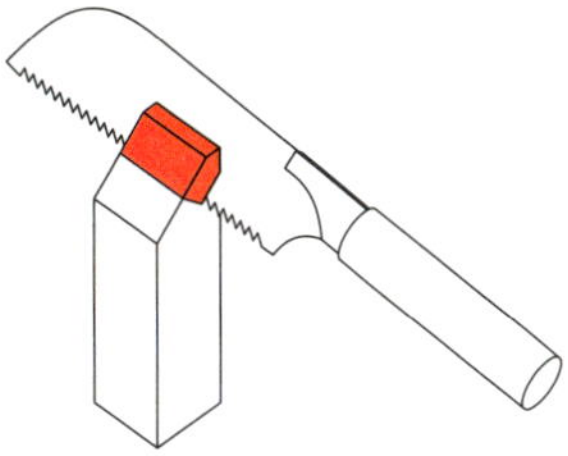

A3

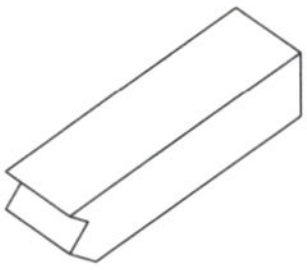

A

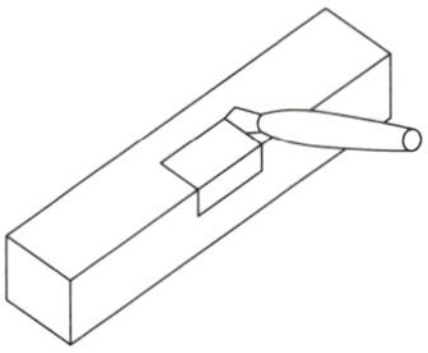

B1

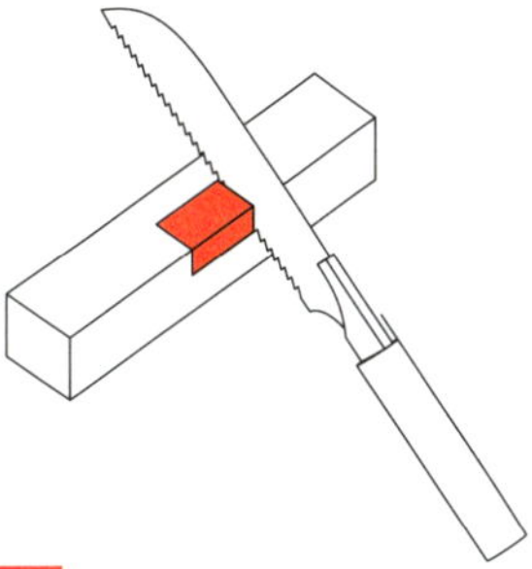

B2

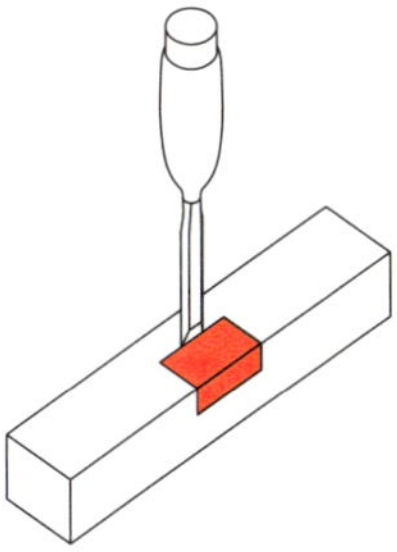

B3

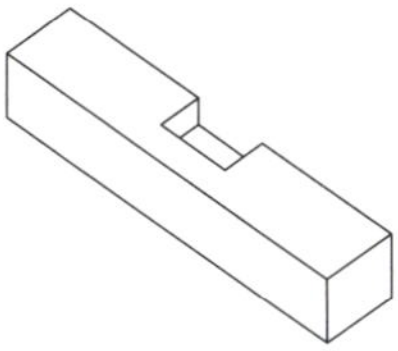

B

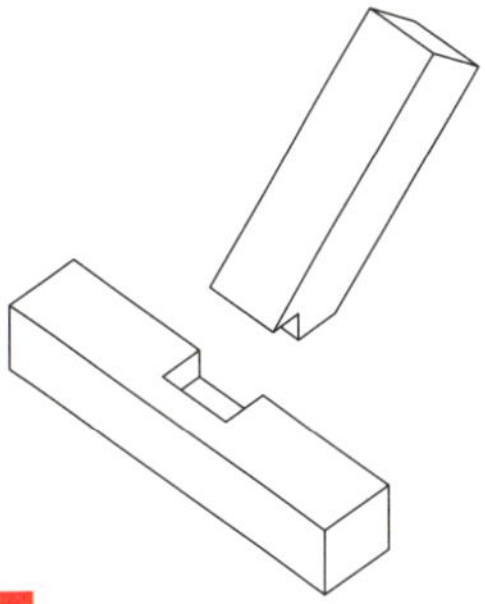

A+B

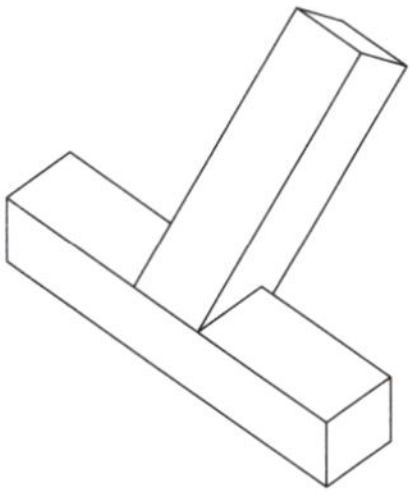

AB

Angled Joints

V-grooved rafter foot

naname-kane-ire-uchi-tsuke-tsugi (dō-tsuki)
斜矩入打付接ぎ(胴付)

Schiftereinblattung

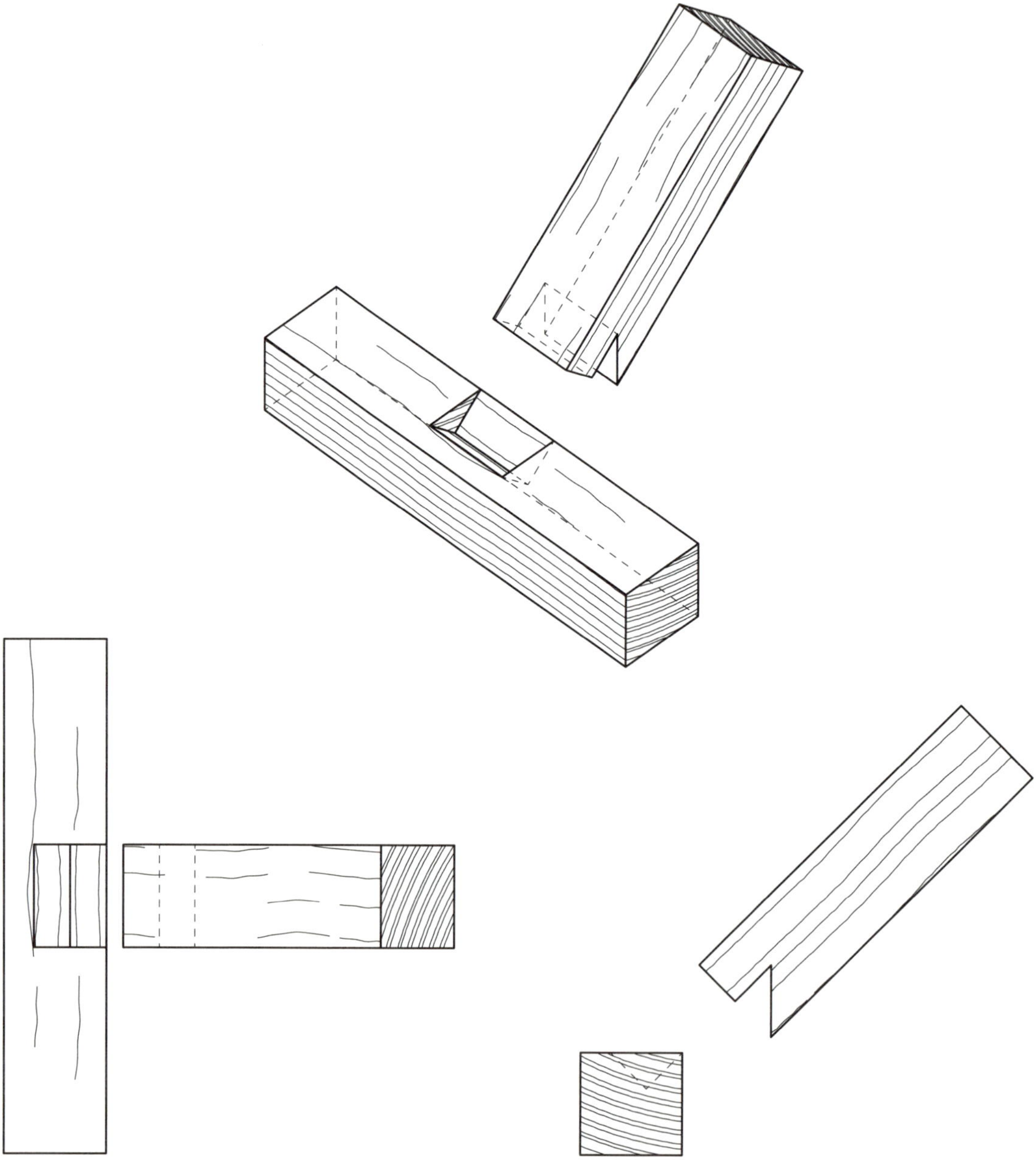

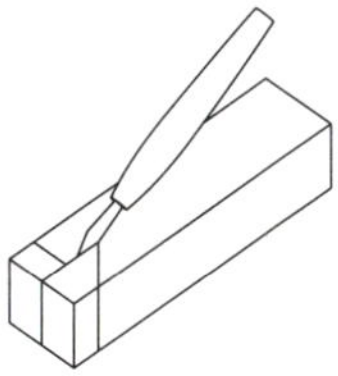

A1

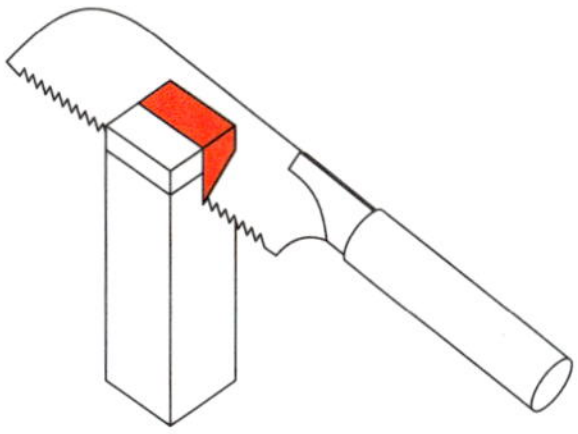

A2

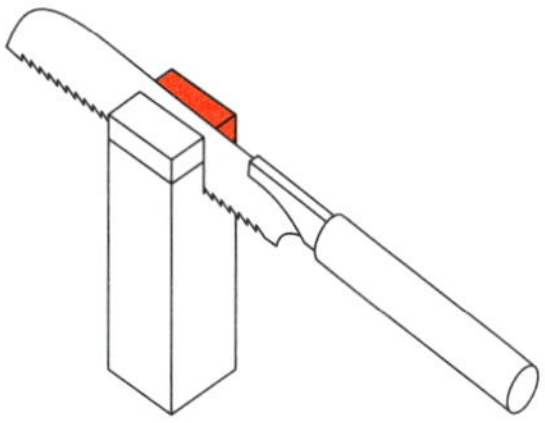

A3

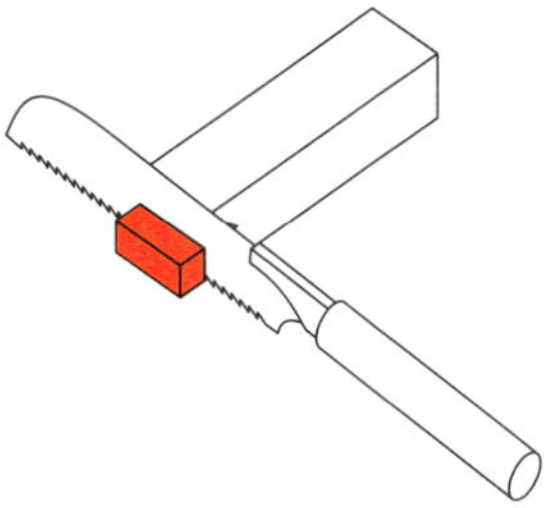

A4

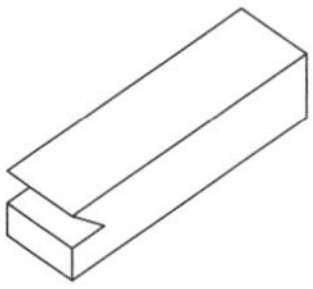

A

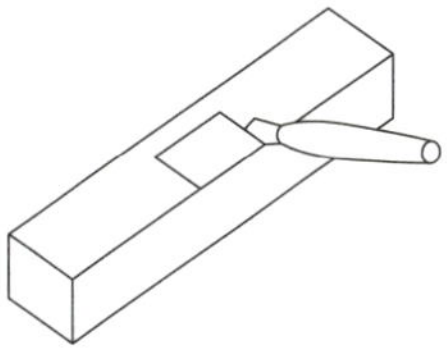

B1

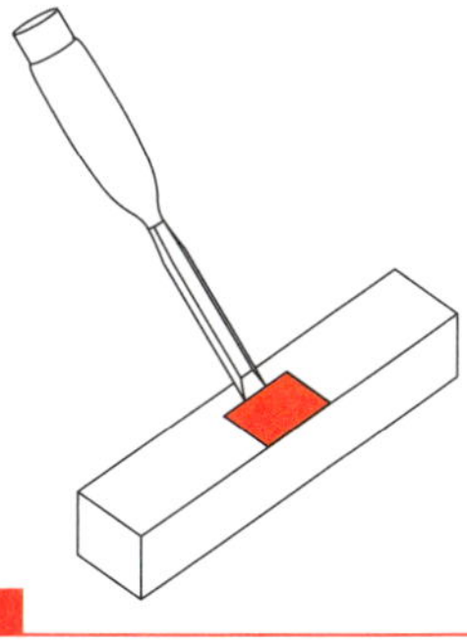

B2

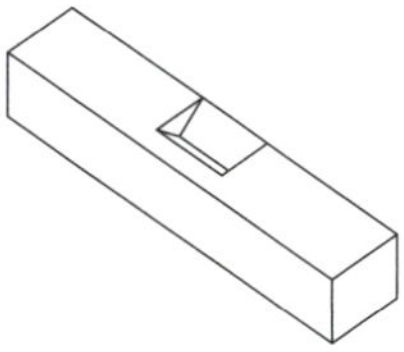

B

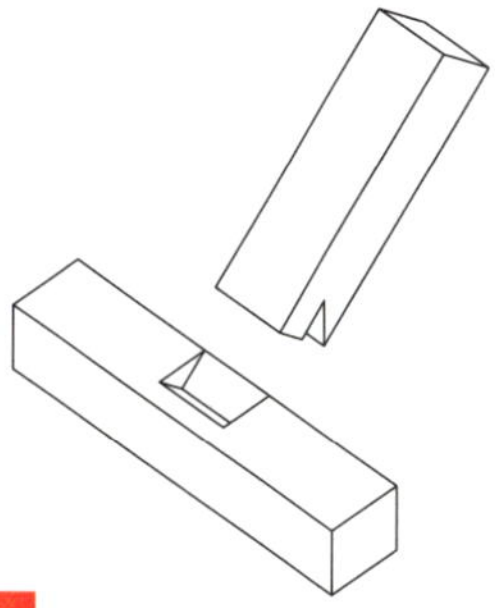

A+B

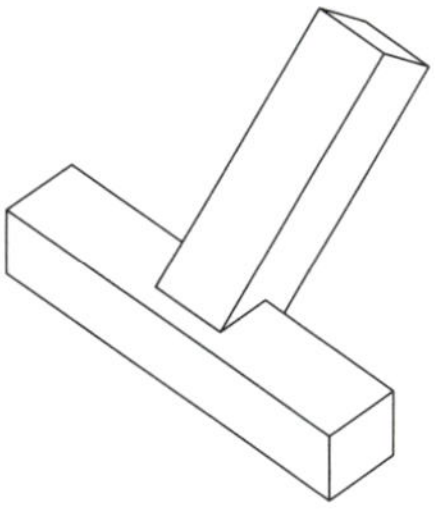

AB

Angled Joints

Birdsmouth rafter foot

naname-uchi-tsuke-tsugi
斜打付接ぎ

Aufklauung

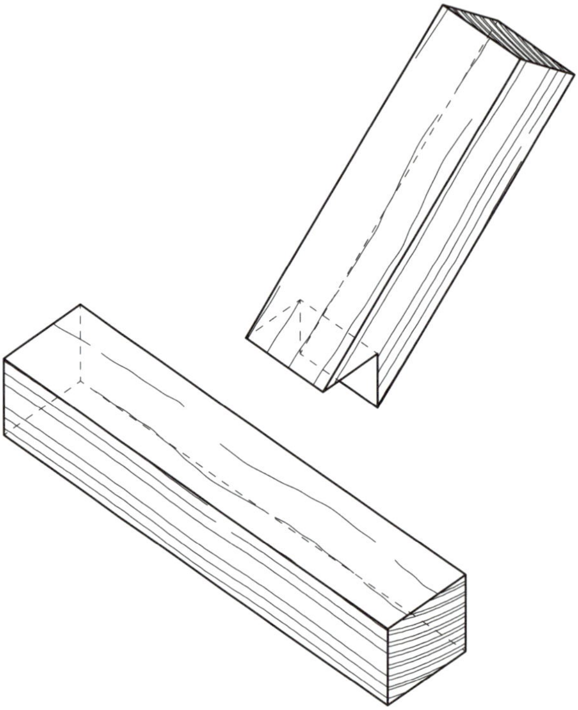

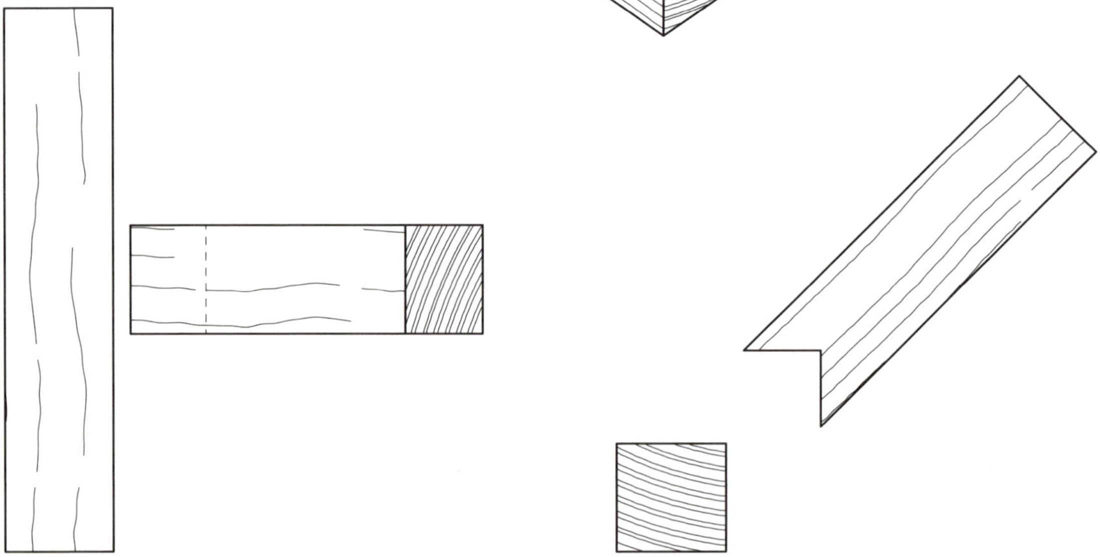

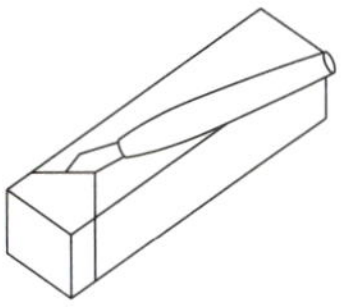

A1

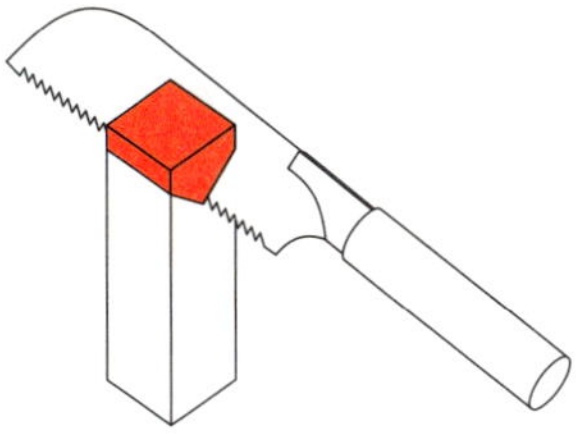

A2

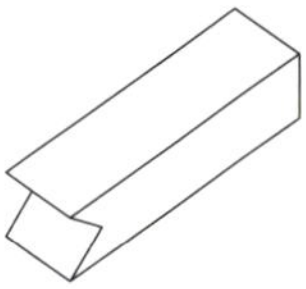

A

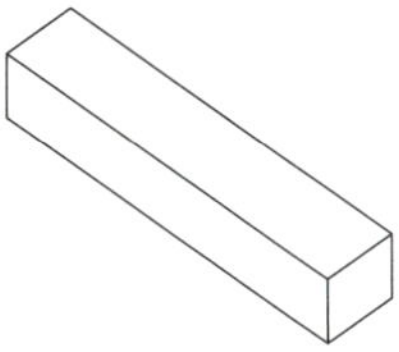

B

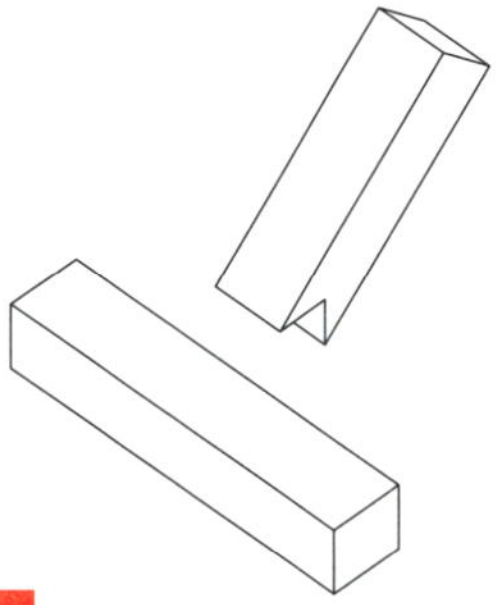

A+B

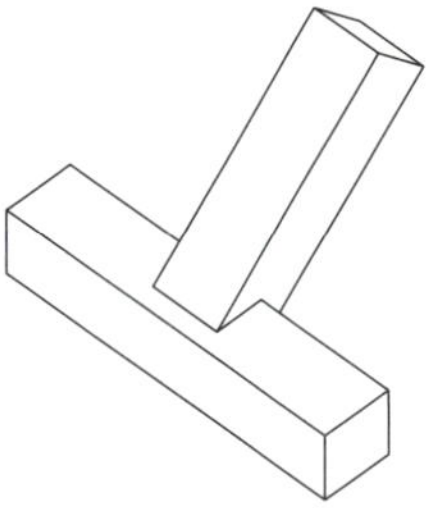

AB

Angled Joints

Rafter foot with tenon and undersquinted shoulders

naname-hira-hozo-sashi (dō-tsuki)
斜平枘差し(胴付)

Zapfung mit Klaue

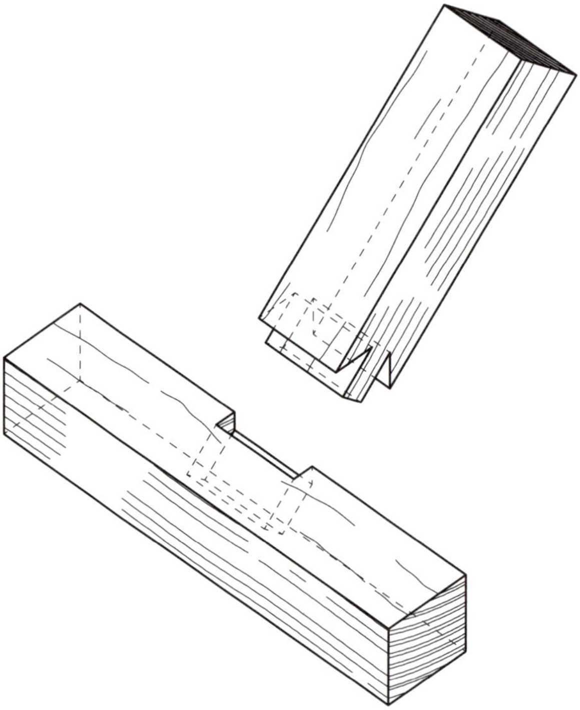

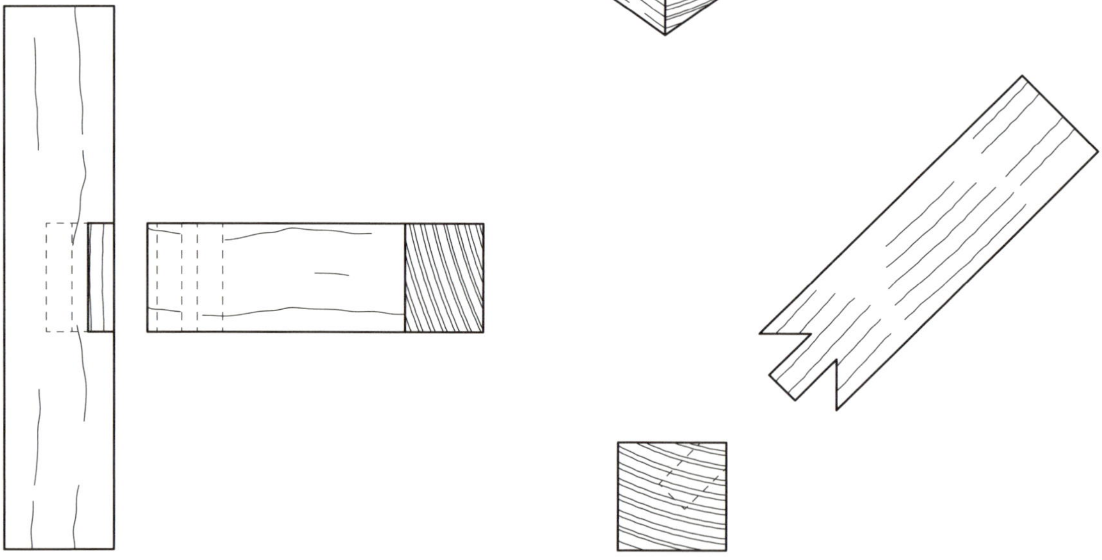

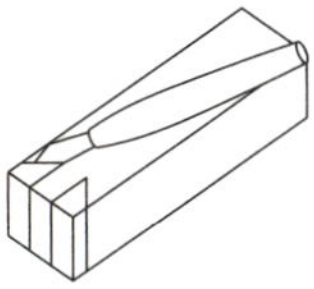

A1

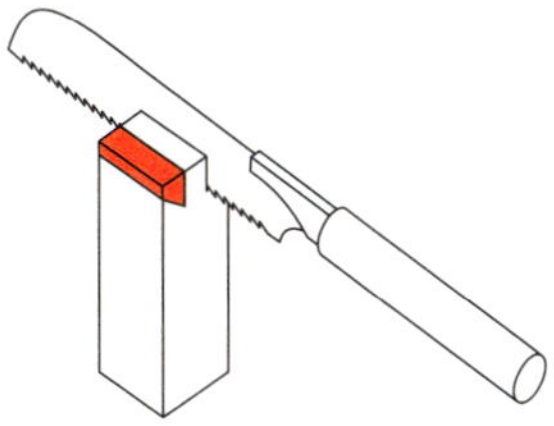

A2

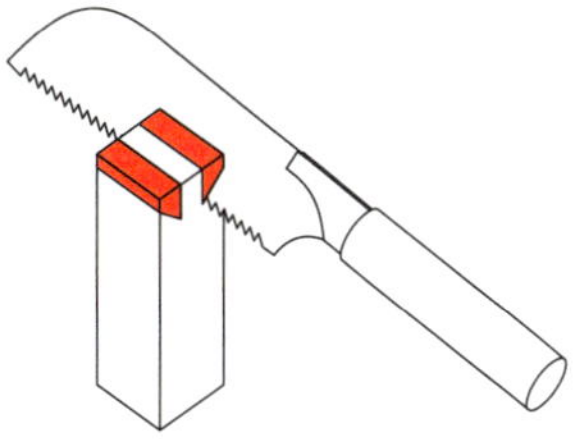

A3

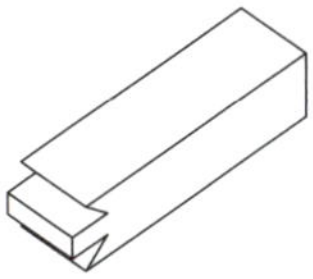

A

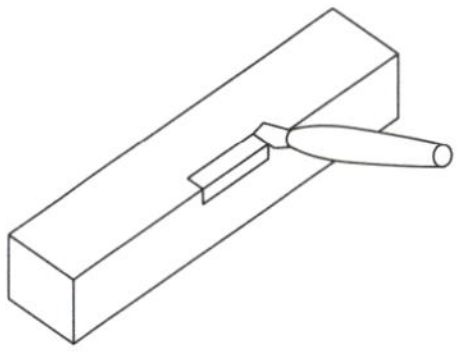

B1

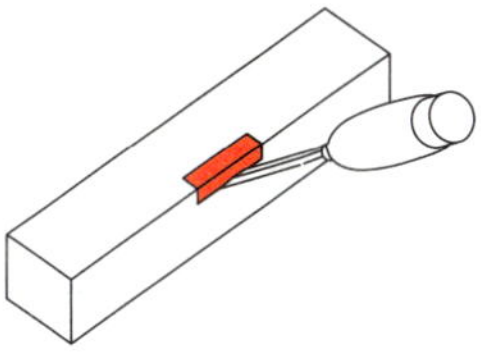

B2

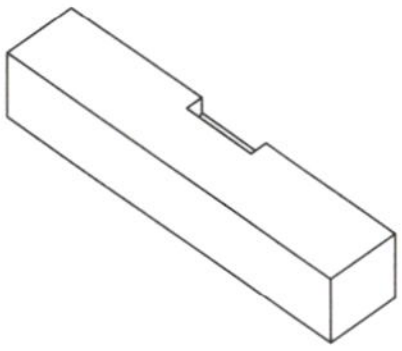

B

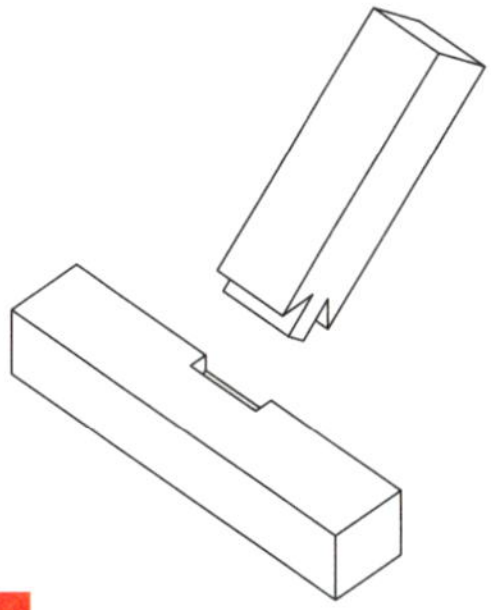

A+B

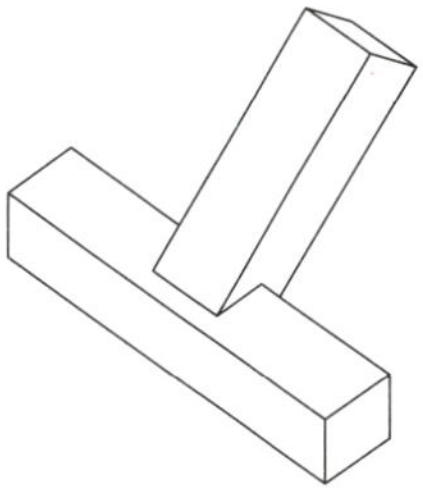

AB

Angled Joints

Stopped birdsmouth rafter foot

naname-wanagi-komi-uchi-tsuke-tsugi
斜輪薙込打付接ぎ

Aufklauung mit Anschnitt

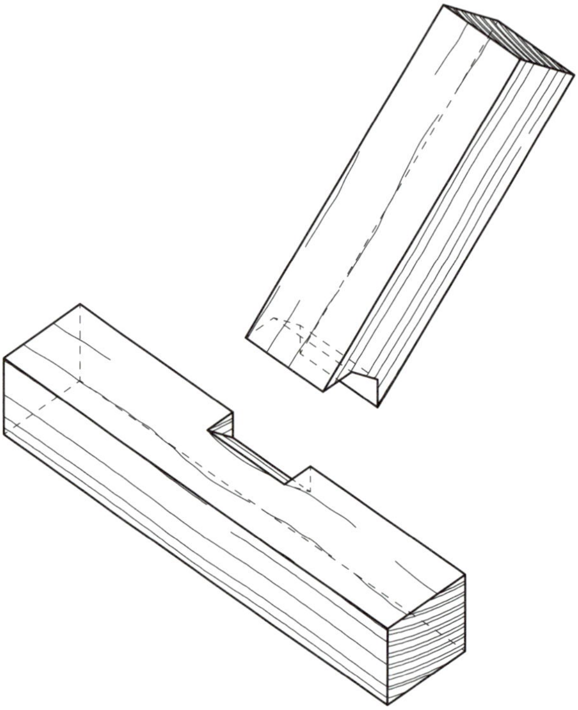

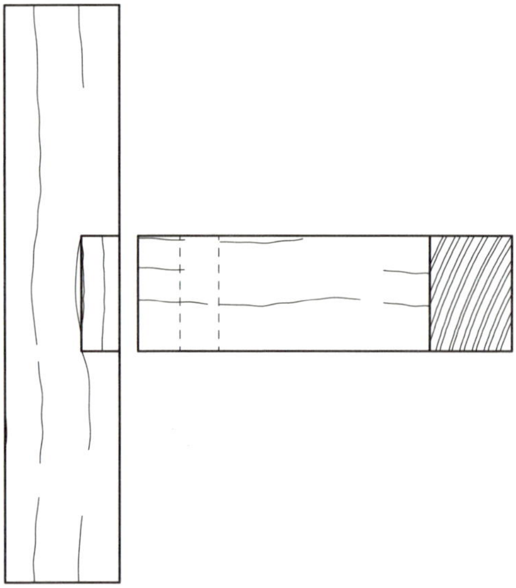

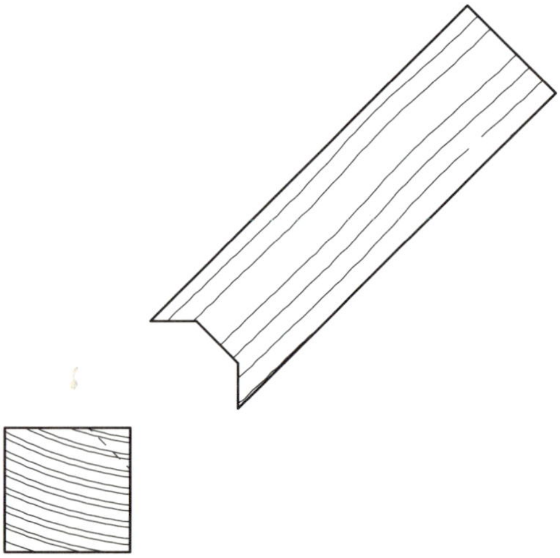

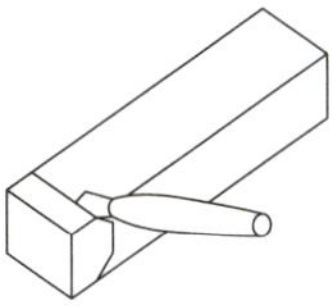

A1

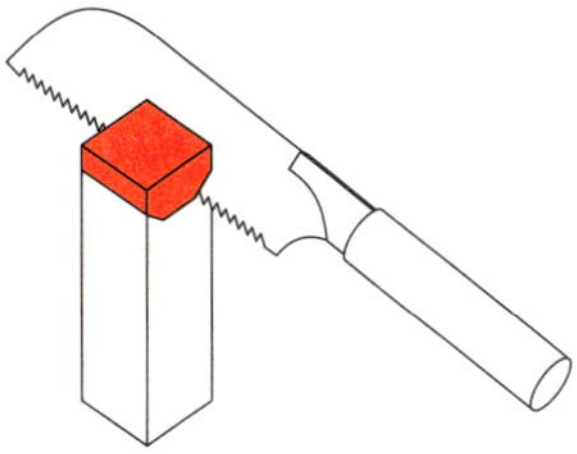

A2

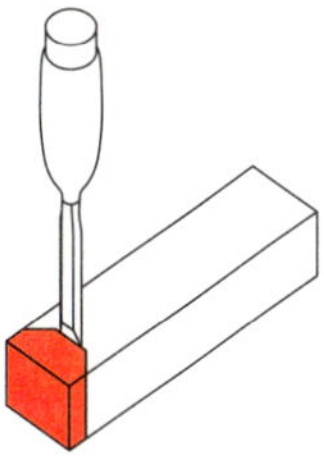

A3

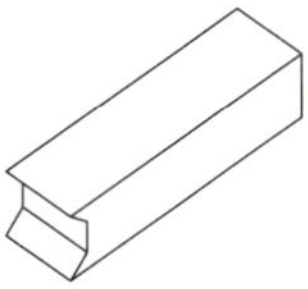

A

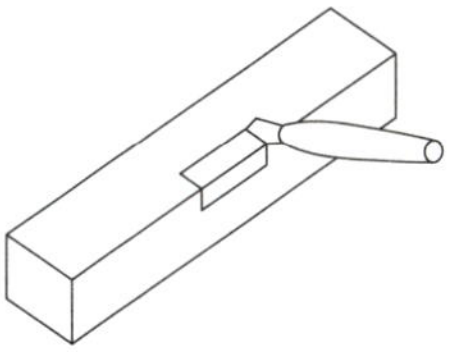

B1

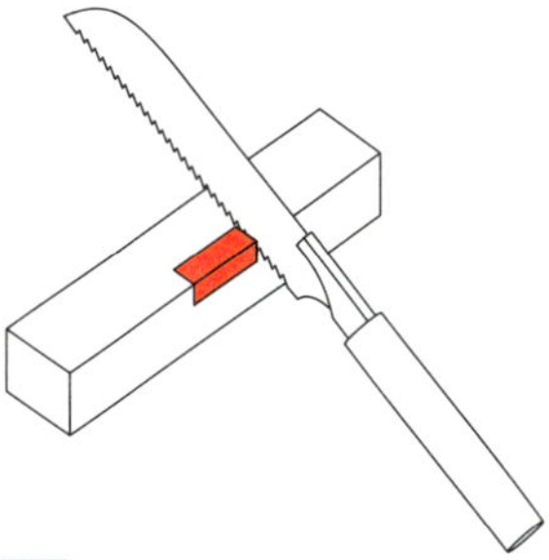

B2

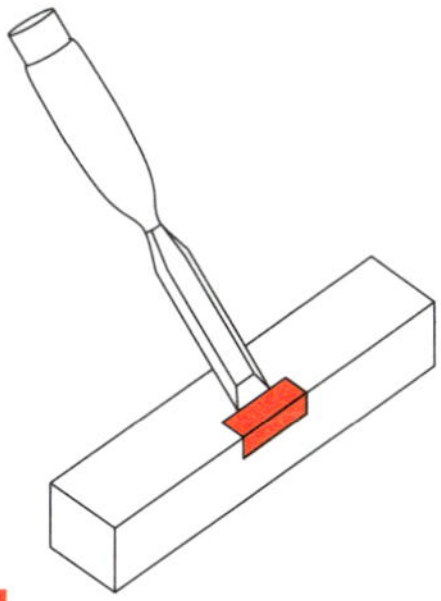

B3

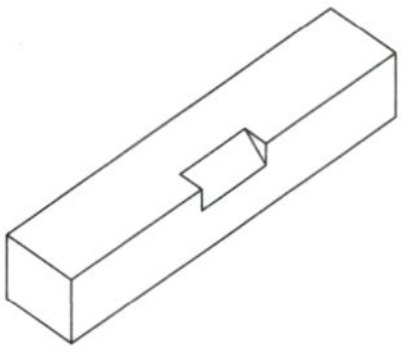

B

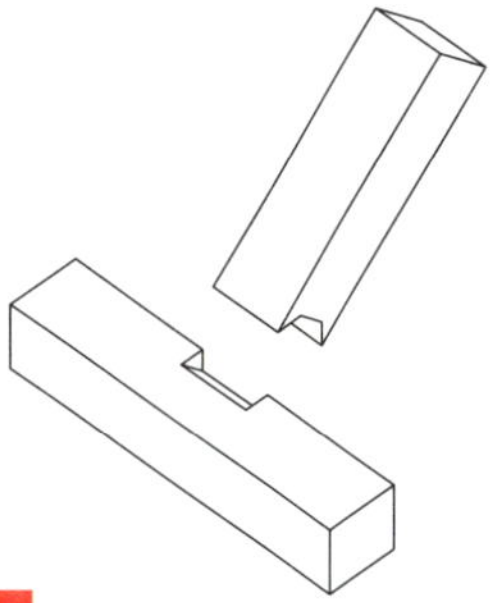

A+B

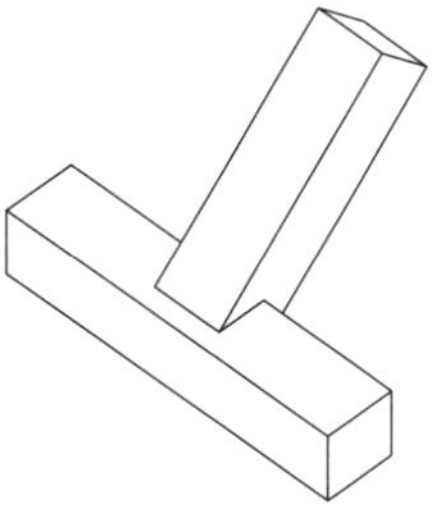

AB

Angled Joints

Birdsmouth rafter joint

taruki-tsugi
垂木接ぎ

Aufkervung

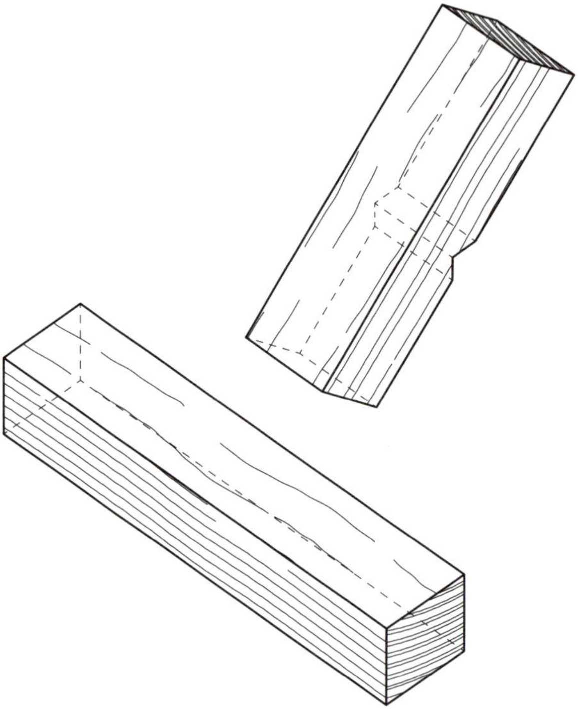

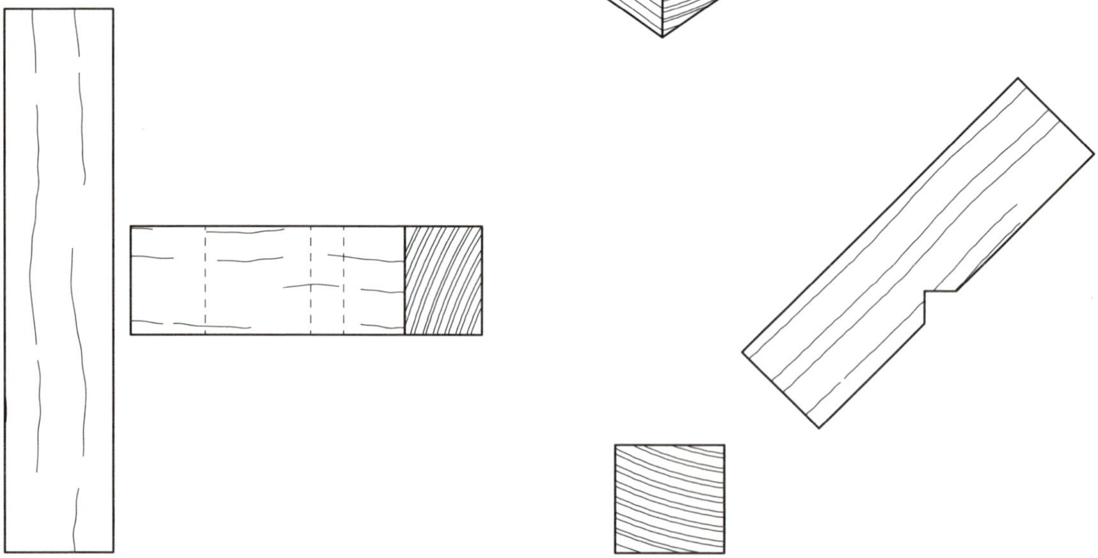

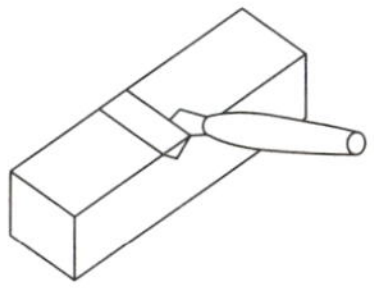

A1

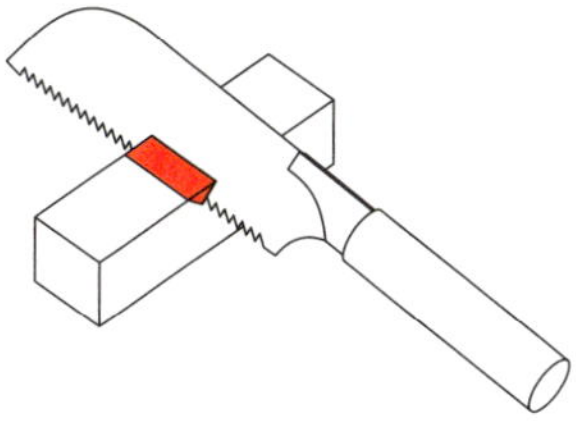

A2

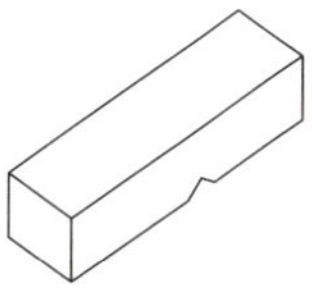

A

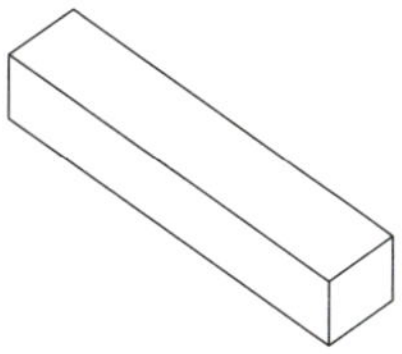

B

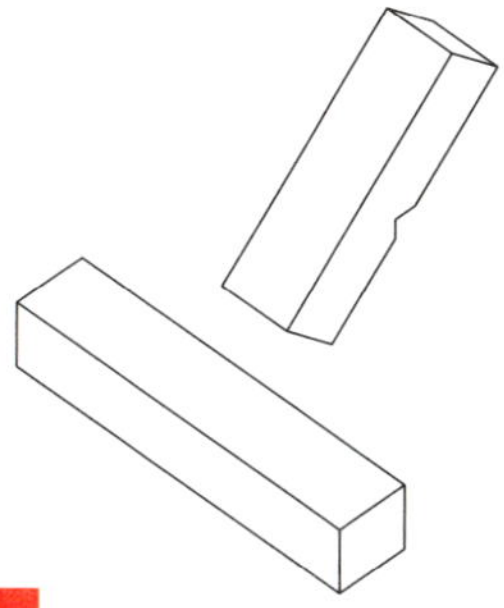

A+B

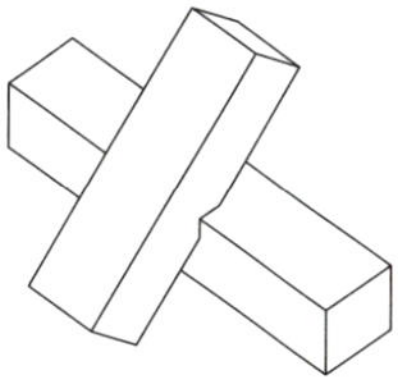

AB

Angled Joints

Notched rafter joint

taruki-bori-tsugi
垂木彫接ぎ

Aufkervung mit Ausklinkung

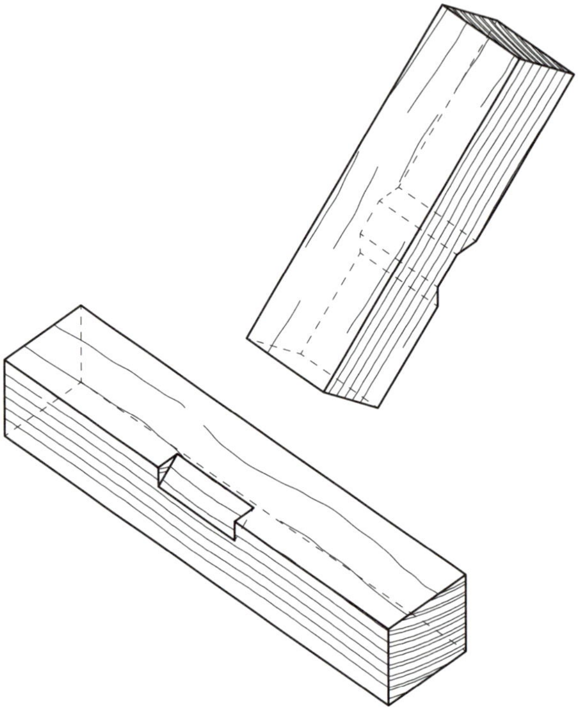

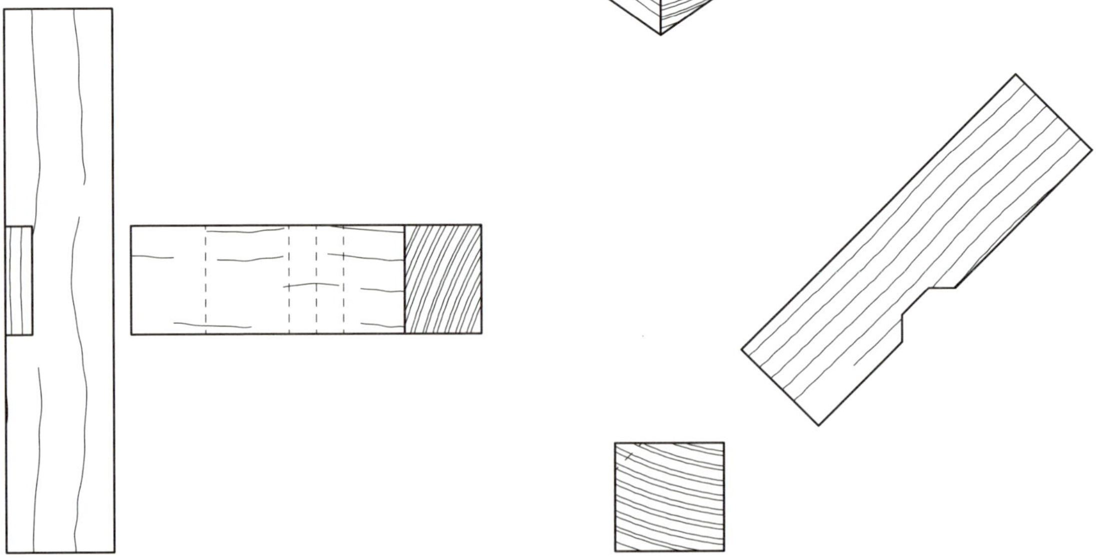

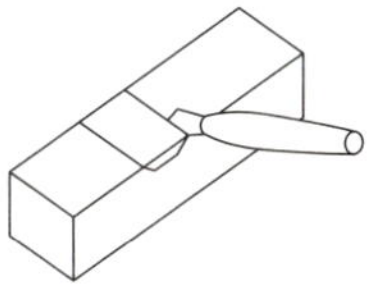

A1

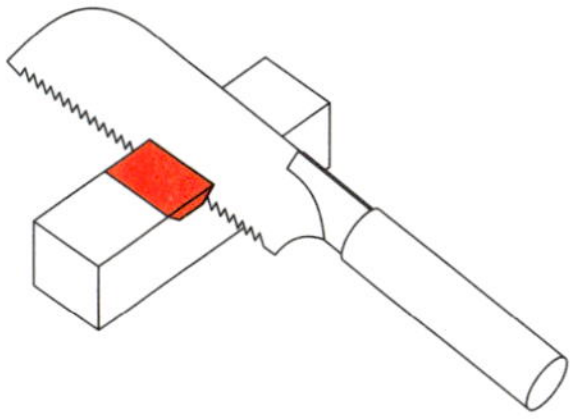

A2

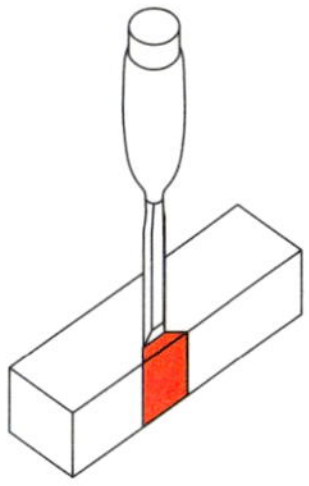

A3

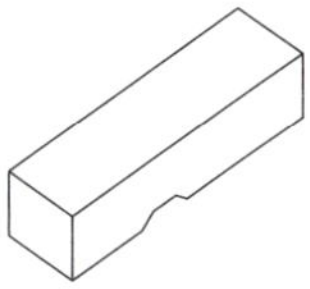

A

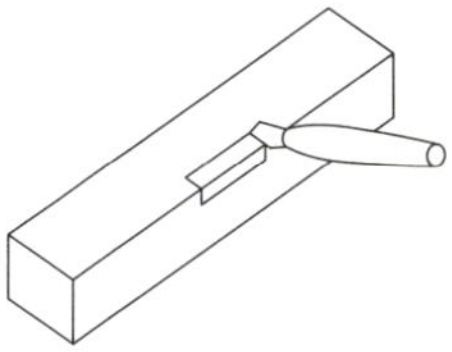

B1

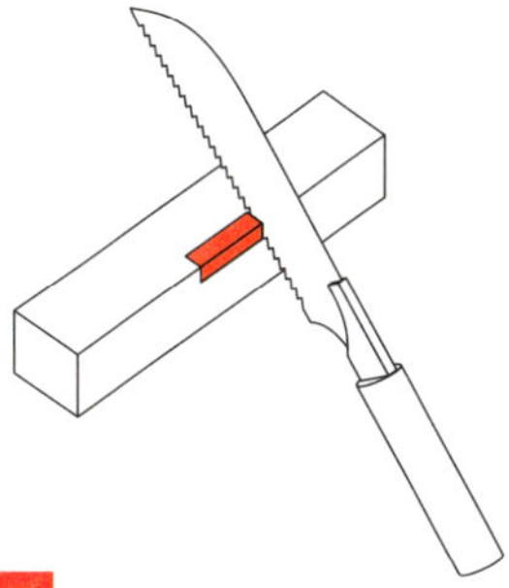

B2

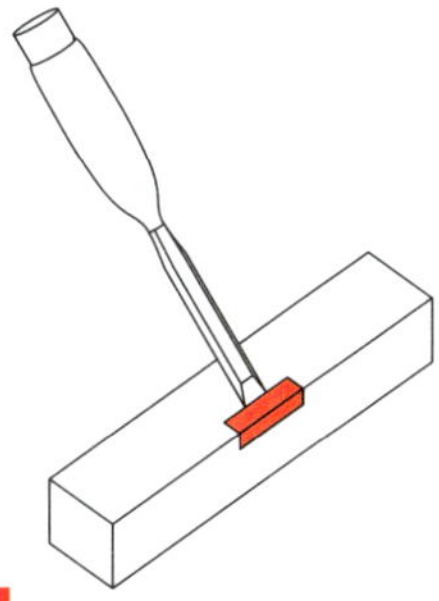

B3

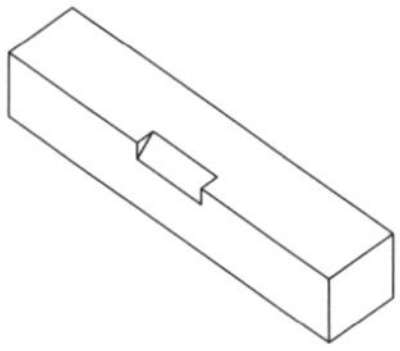

B

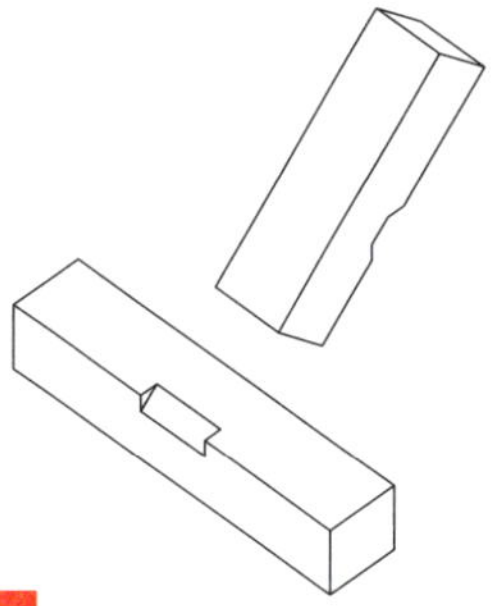

A+B

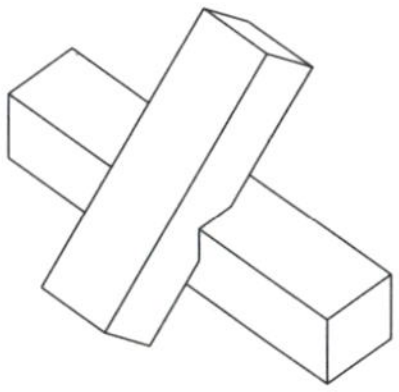

AB

Angled Joints

V-grooved rafter joint

ochi-gakari-ago-ai-kaki-tsugi
落掛顎相欠接ぎ

Aufkervung
mit Rückversatz

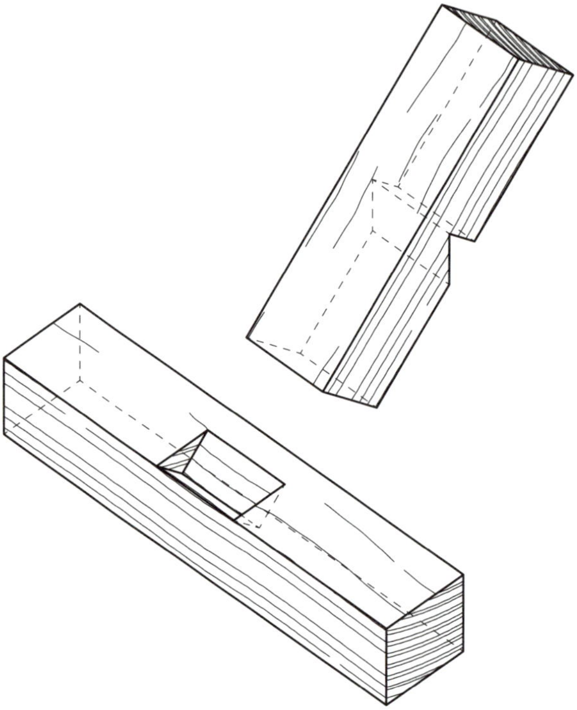

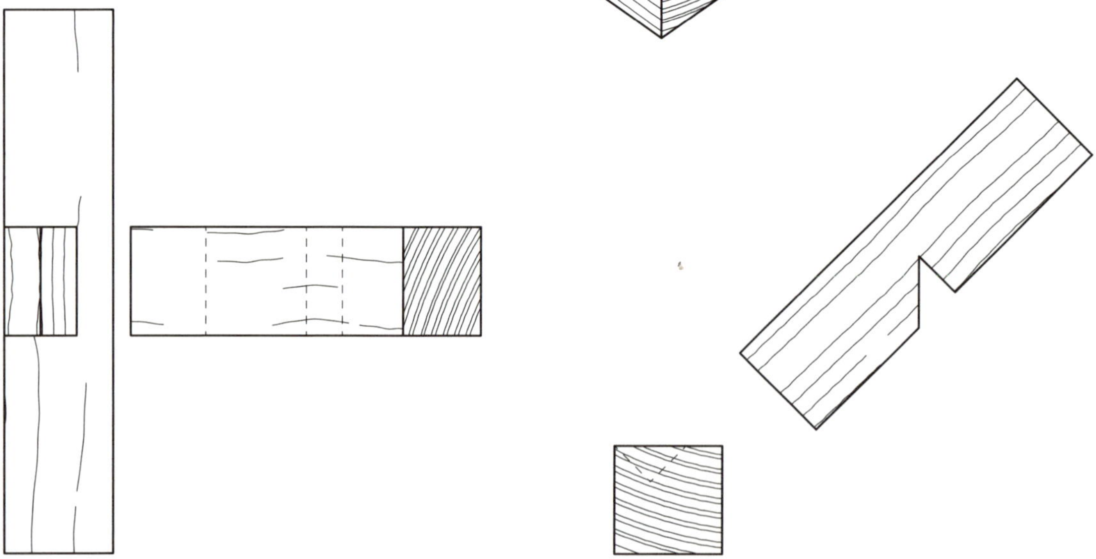

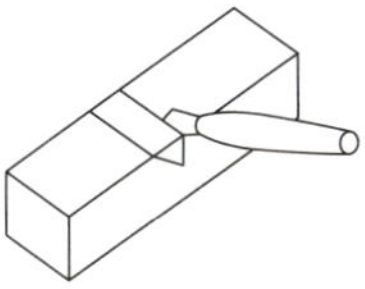

A1

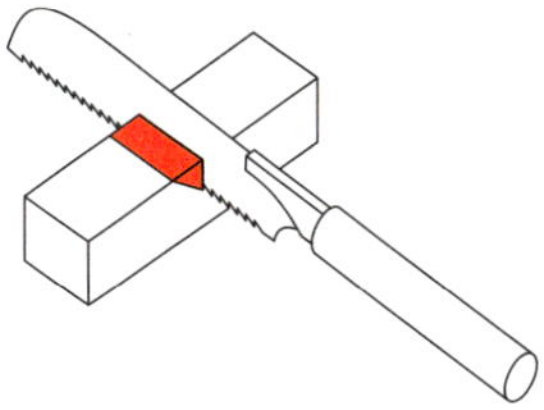

A2

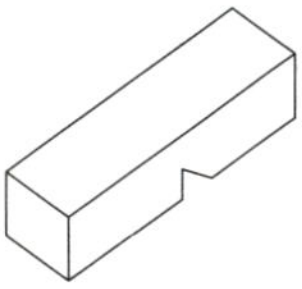

A

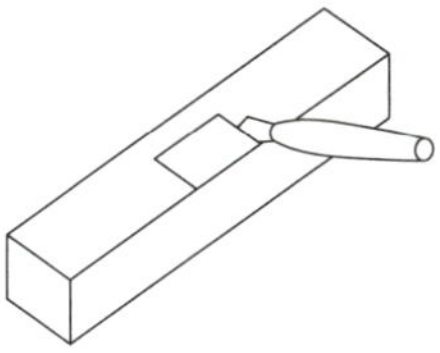

B1

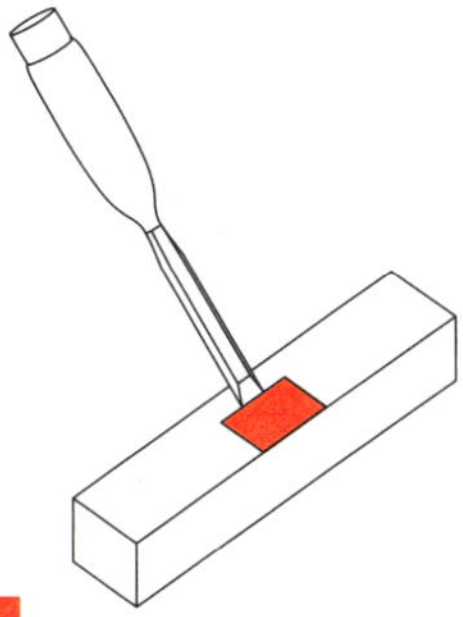

B2

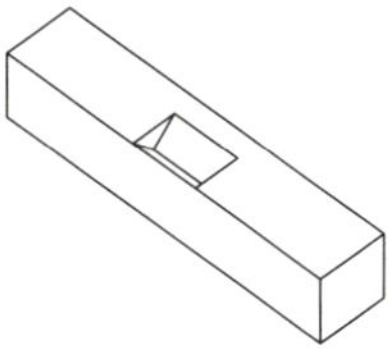

B

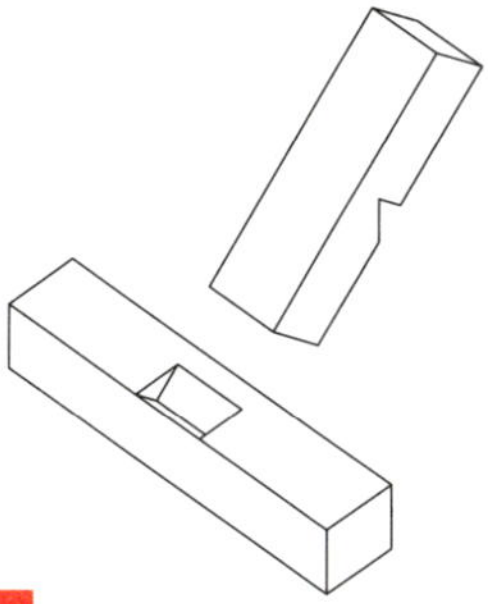

A+B

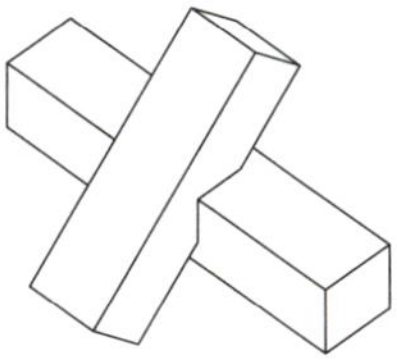

AB

Angled Joints

Step lap rafter joint

naname-watari-ago-ai-kaki-tsugi
斜渡顎相欠接ぎ

Aufkervung mit
gestuftem Rückversatz

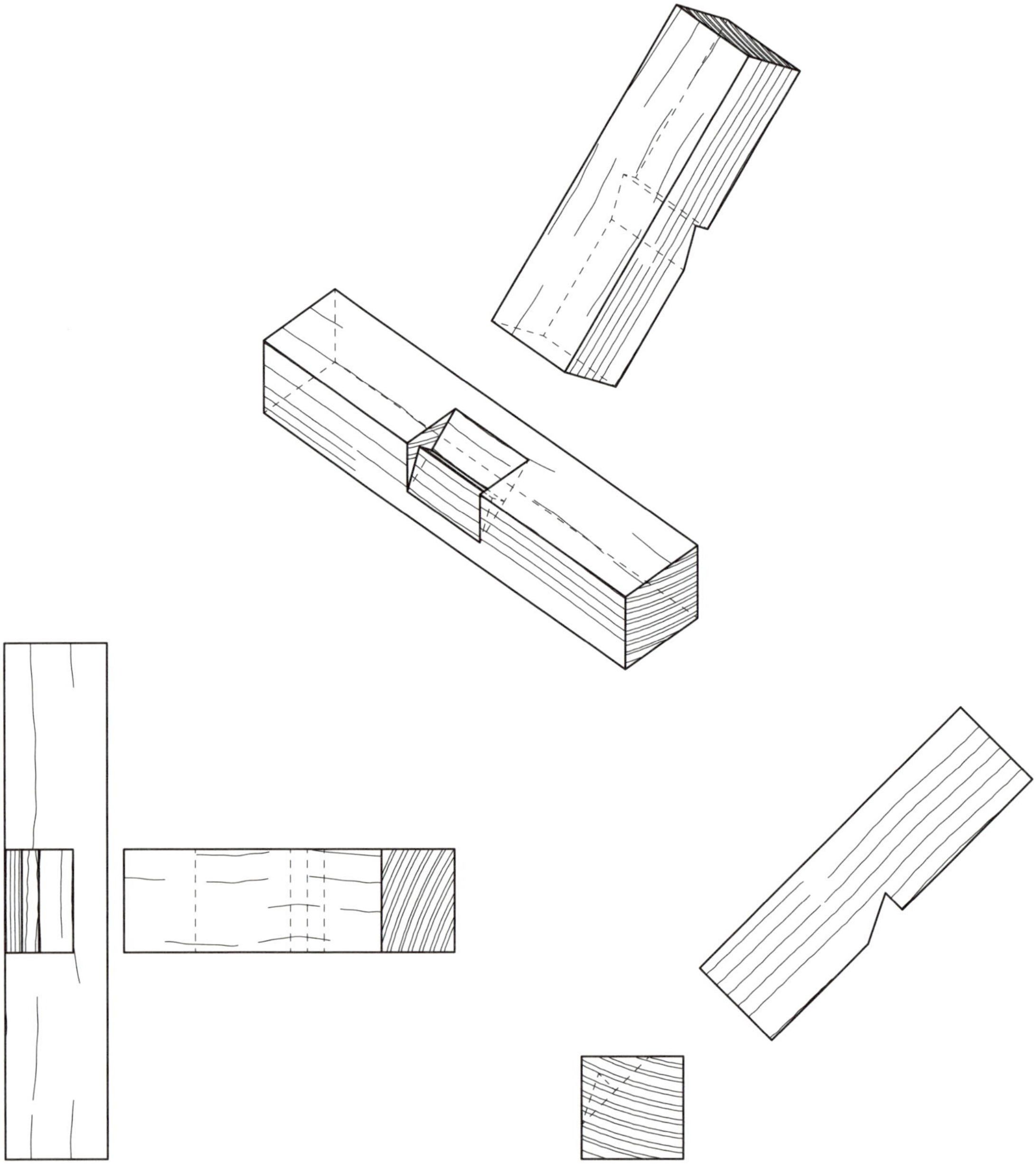

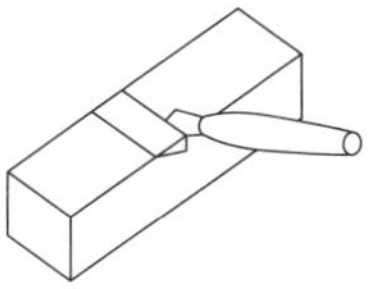

A1

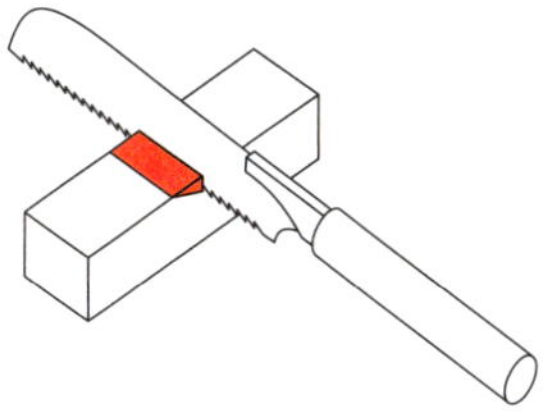

A2

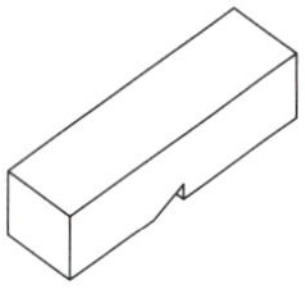

A

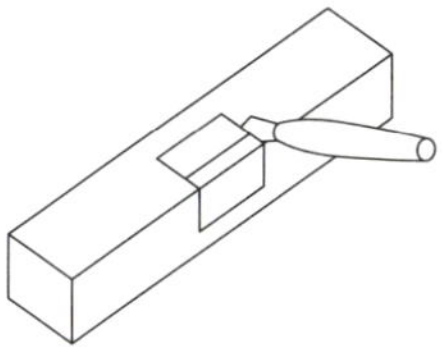

B1

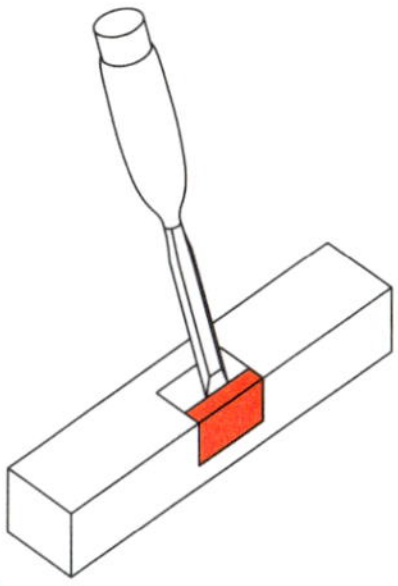

B2

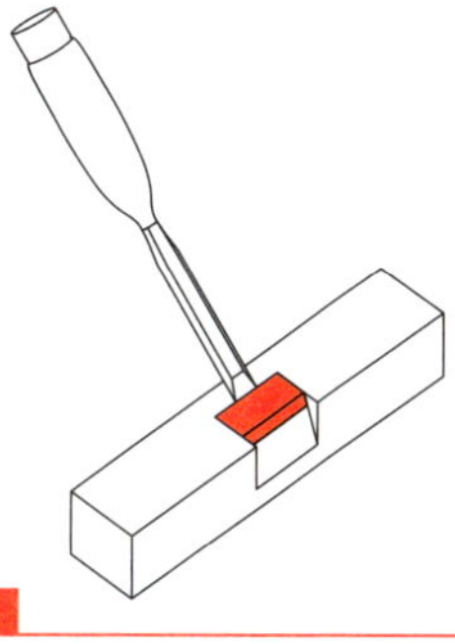

B3

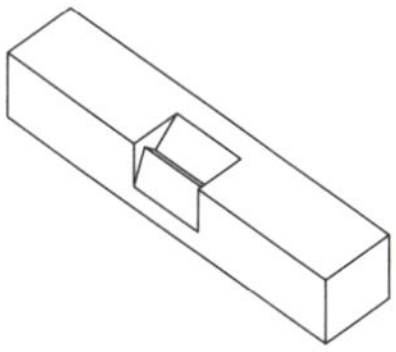

B

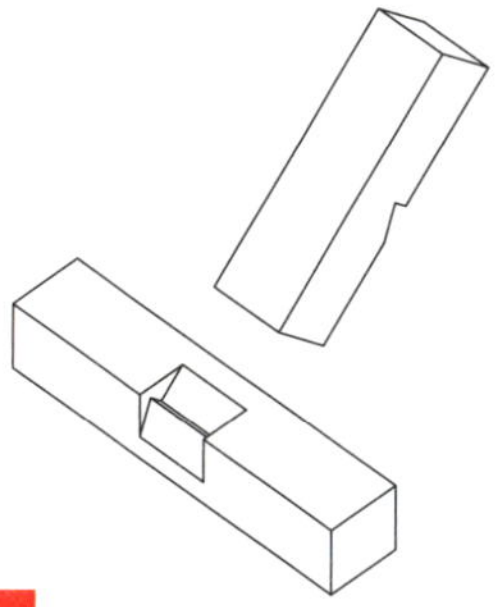

A+B

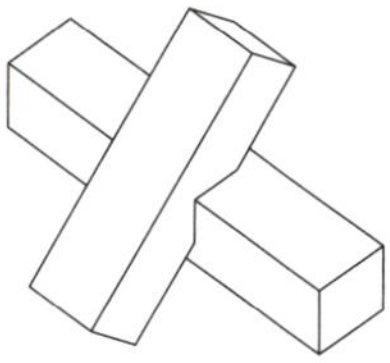

AB

Cross Lap Joints

Cross lap joint

ai-kaki-kumi [*]
相欠組み

Kreuzblatt

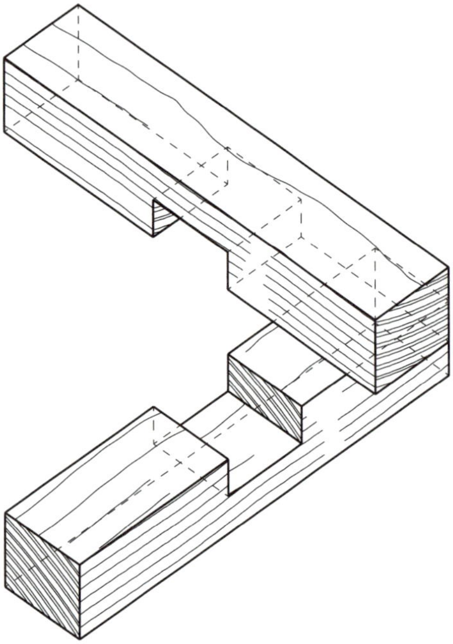

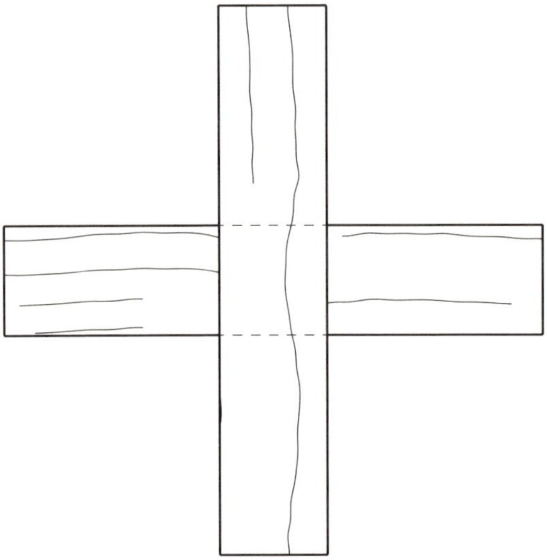

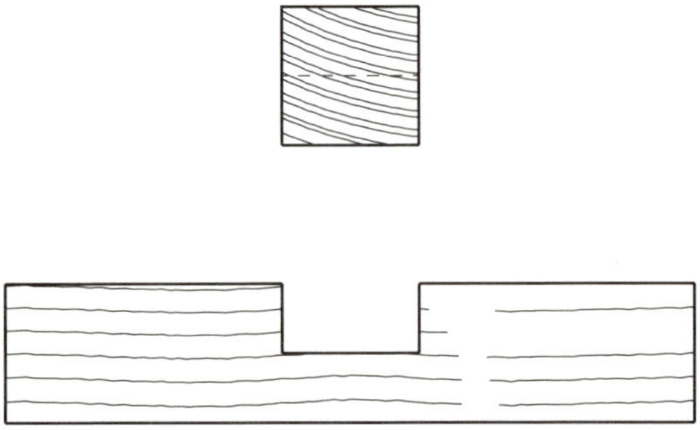

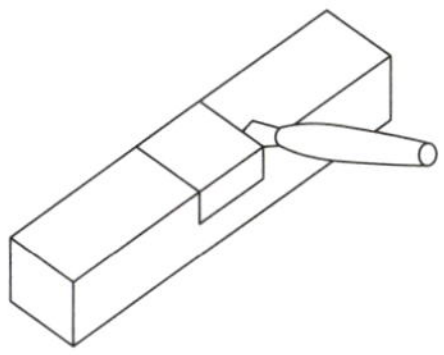

A1

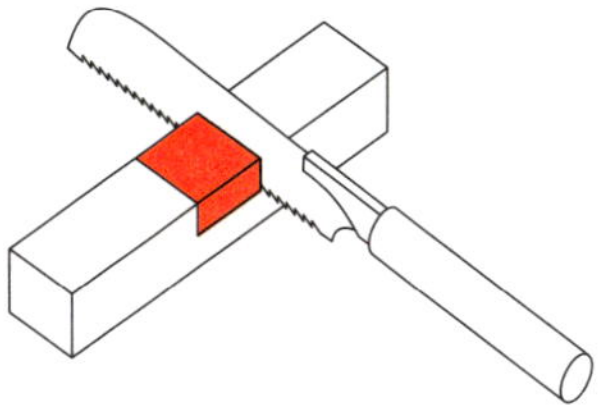

A2

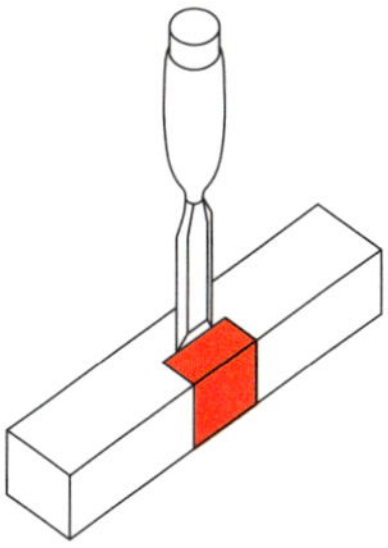

A3

2×

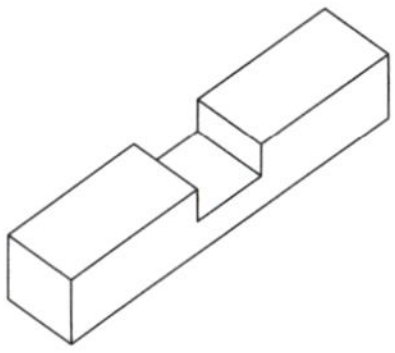

A

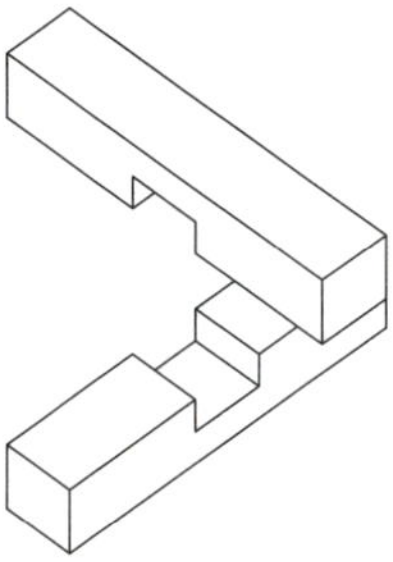

A+A

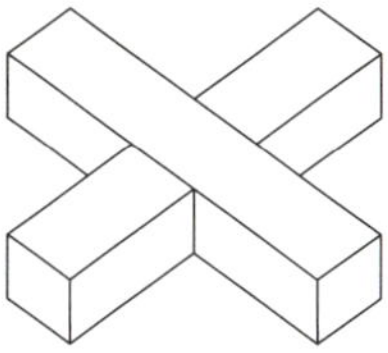

AA

Pegged cross lap joint

ai-kaki-kumi-dai-sen-uchi [*]
相欠組大栓打ち

Kreuzblatt mit Dübel

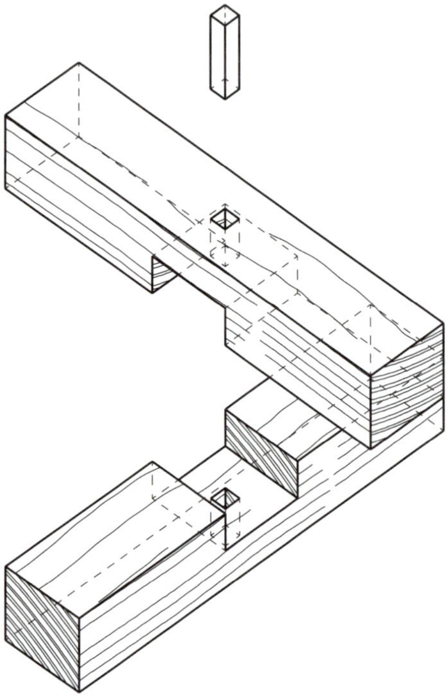

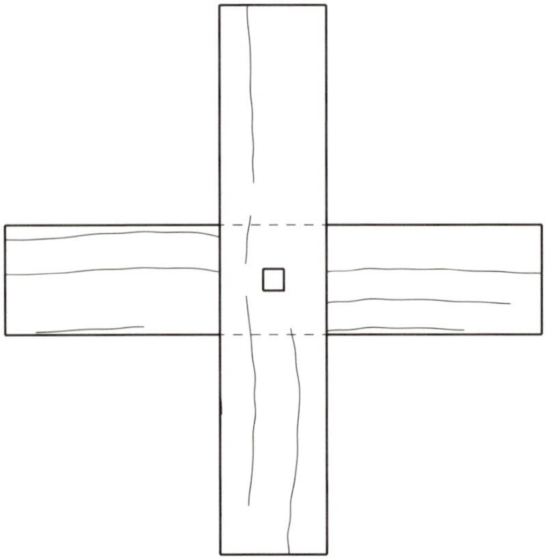

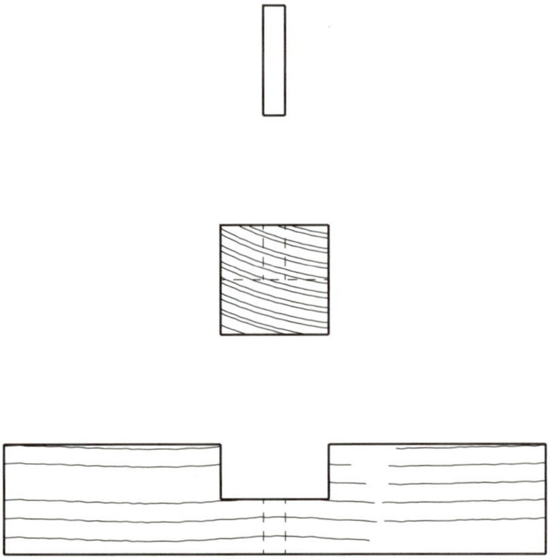

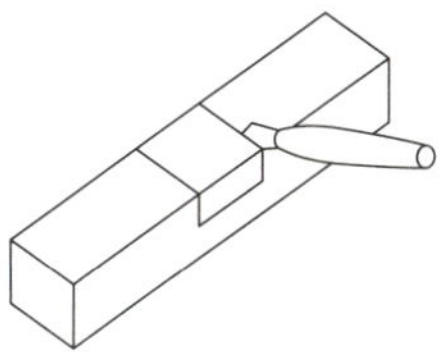

A1

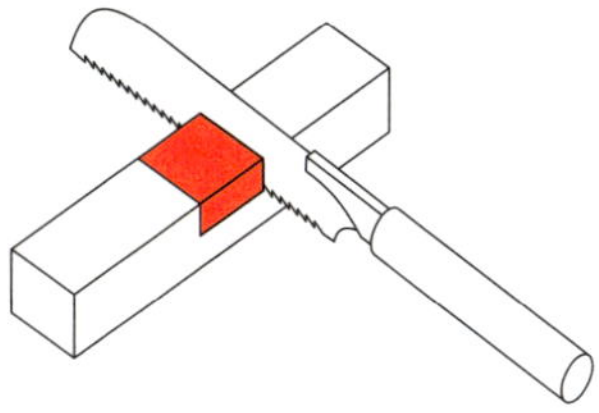

A2

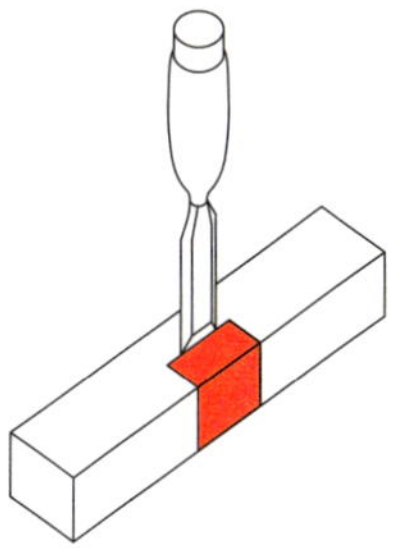

A3

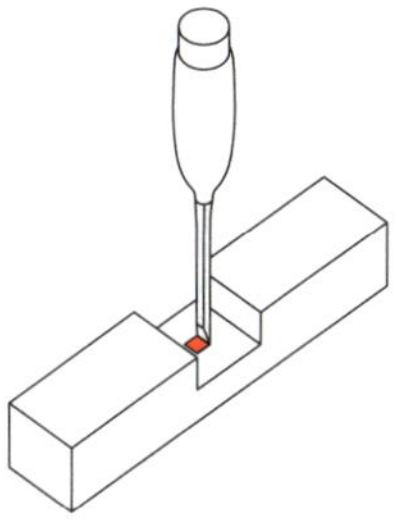

A4

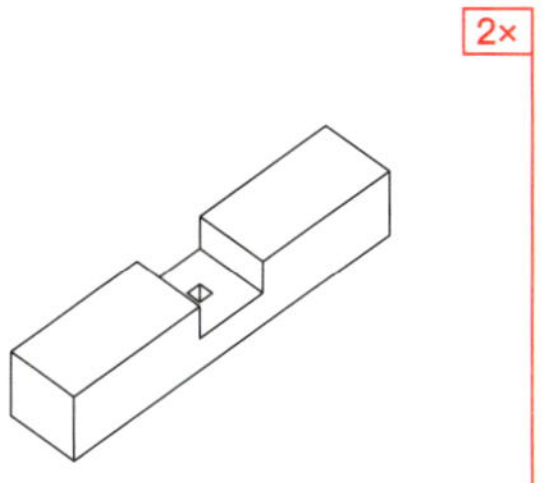

A

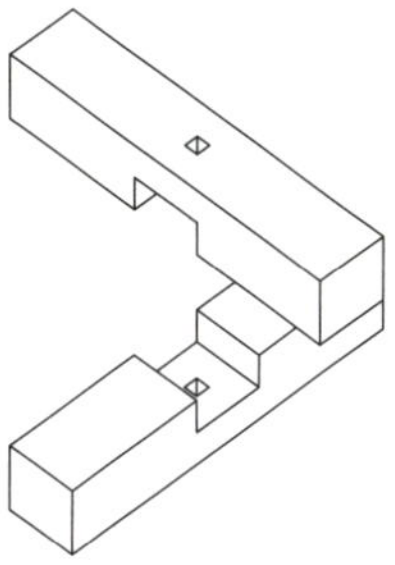

A+A

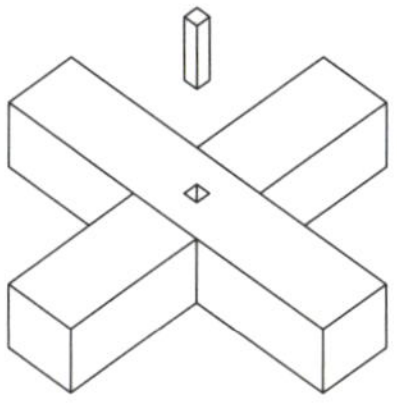

A+A

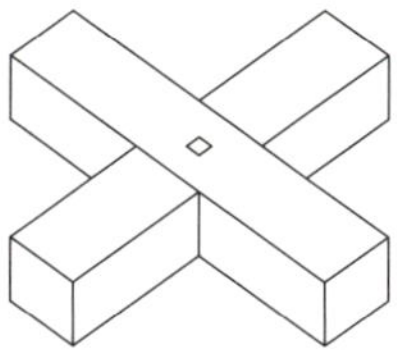

AA

Cross Lap Joints

Four-part cogged cross lap joint

sashi-hijiki-shikuchi [*]
挿肘木仕口

viergliedriges Kreuzblatt mit Haken

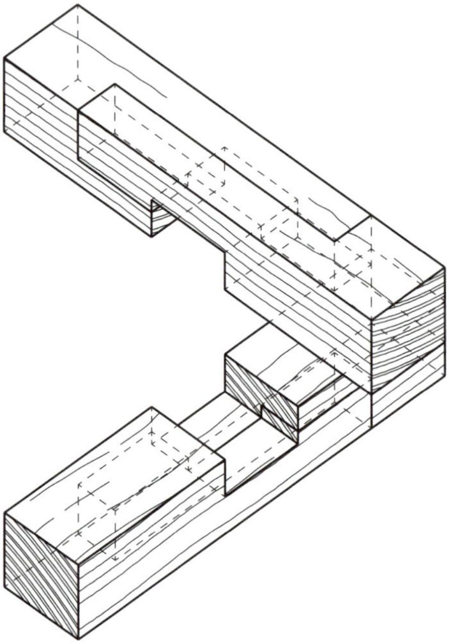

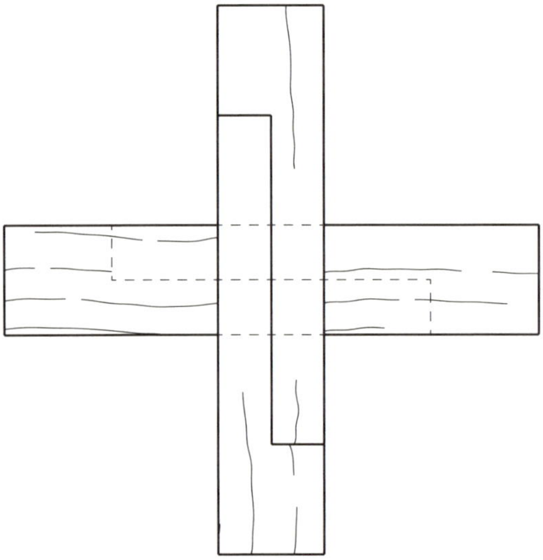

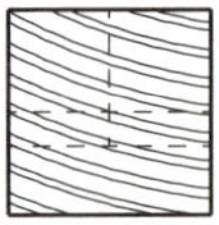

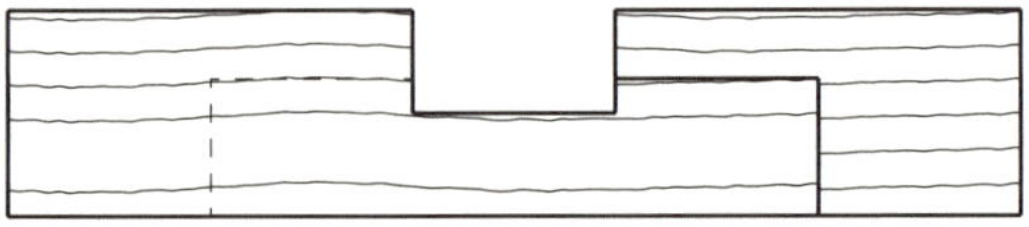

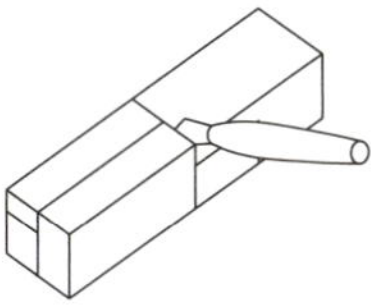

A1

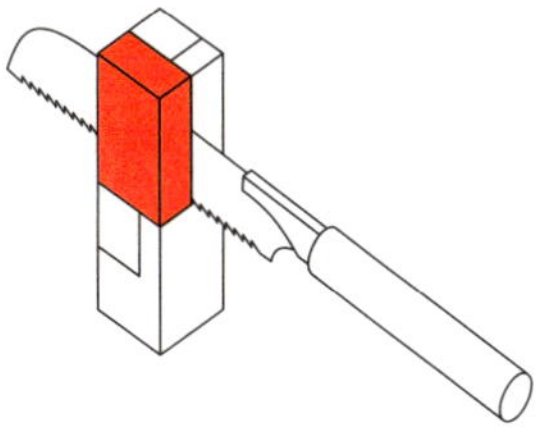

A2

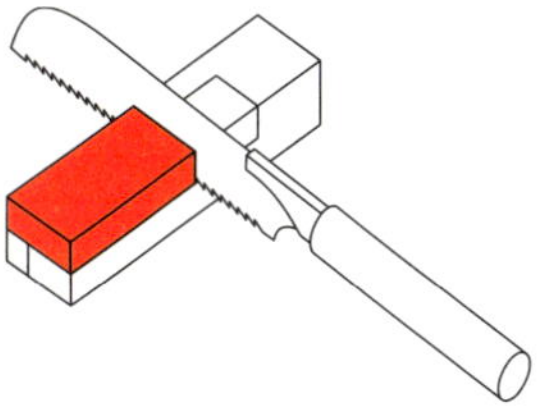

A3

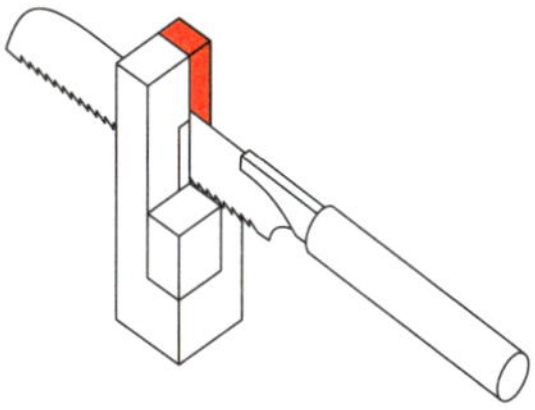

A4

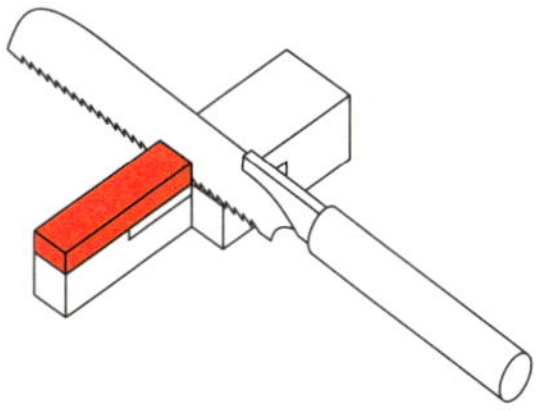

A5

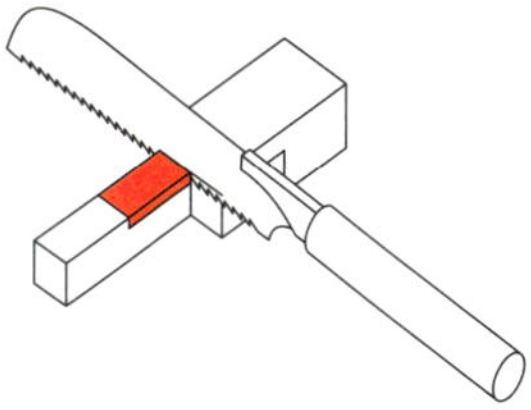

A6

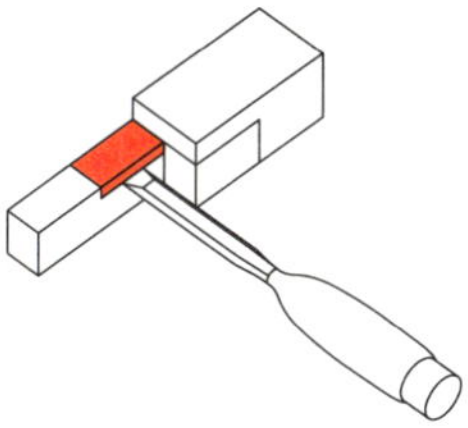

A7

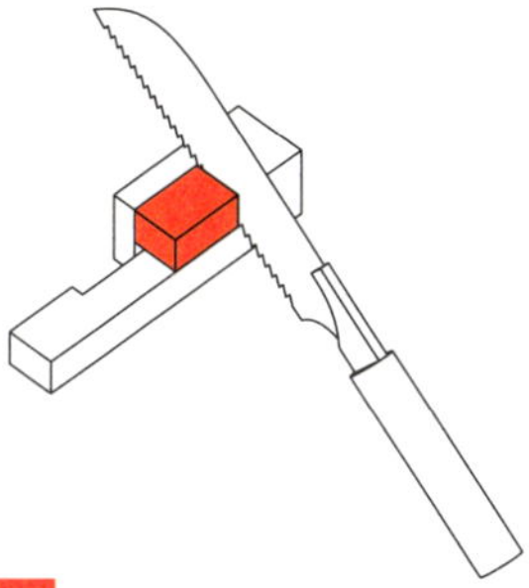

A8

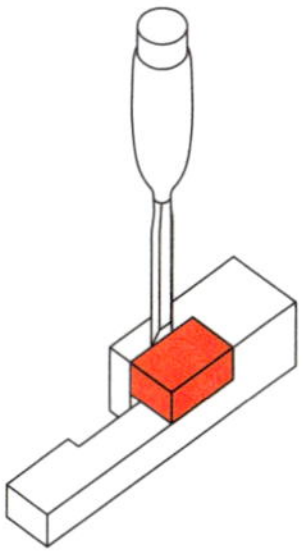

A9

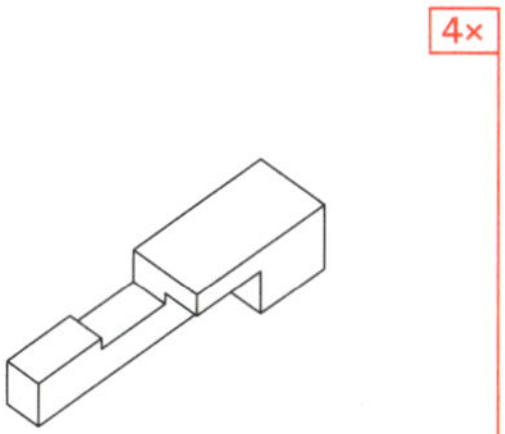

A

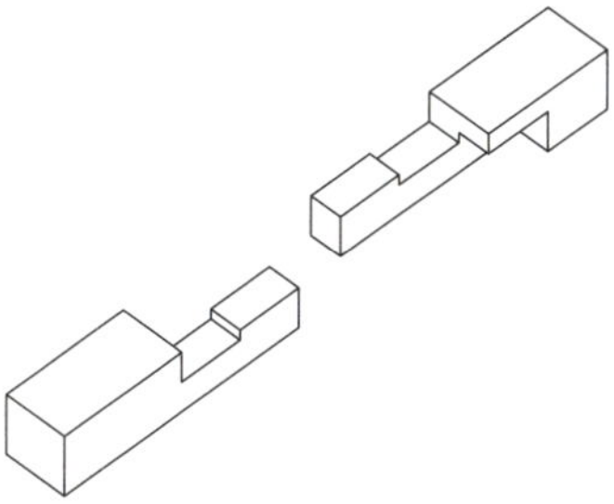

A+A

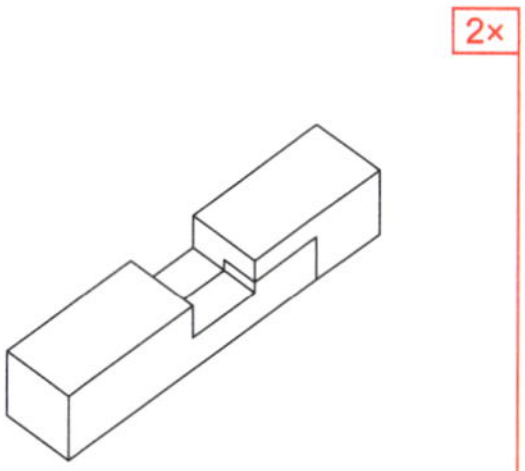

AA

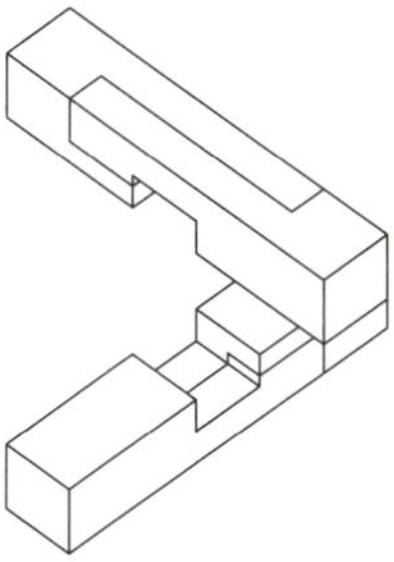

AA+AA

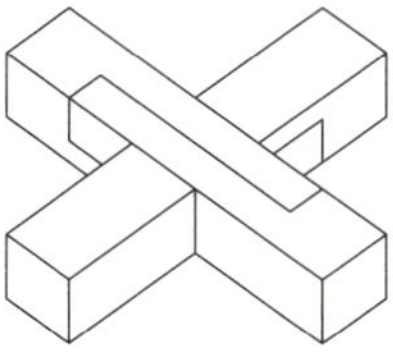

AAAA

Cross Lap Joints

V-mitered cross lap joint

ken-dome-ai-kaki-tsugi
剣留相欠接ぎ

Kreuzblatt mit Grat

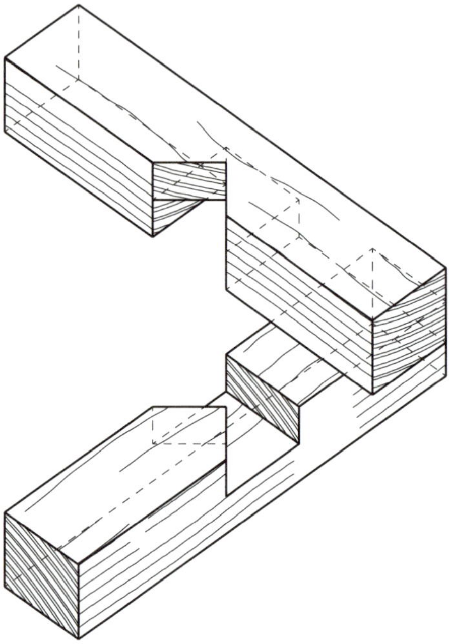

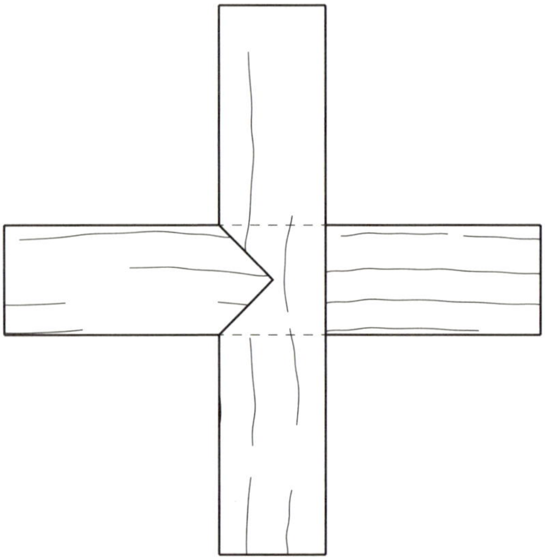

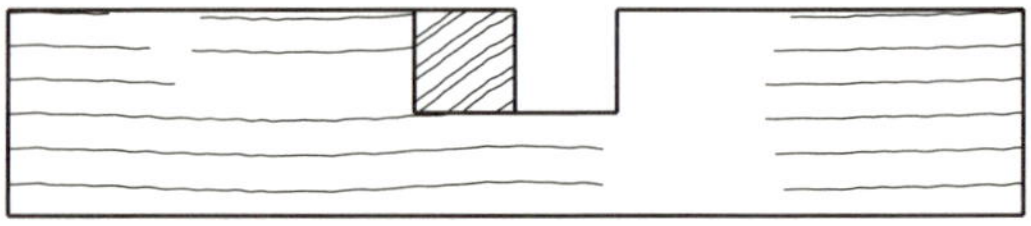

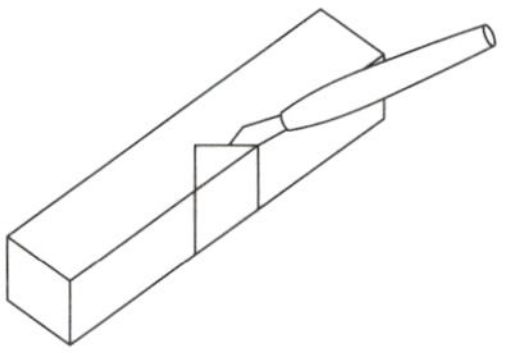

A1

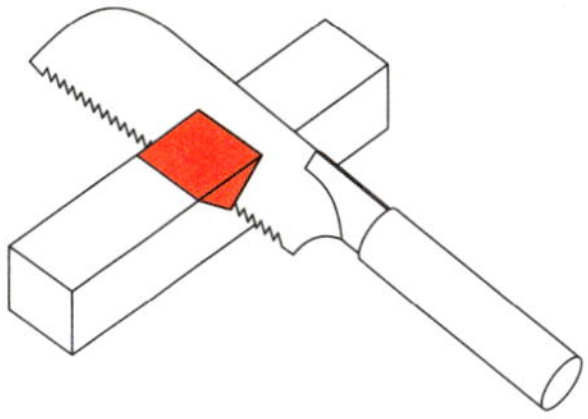

A2

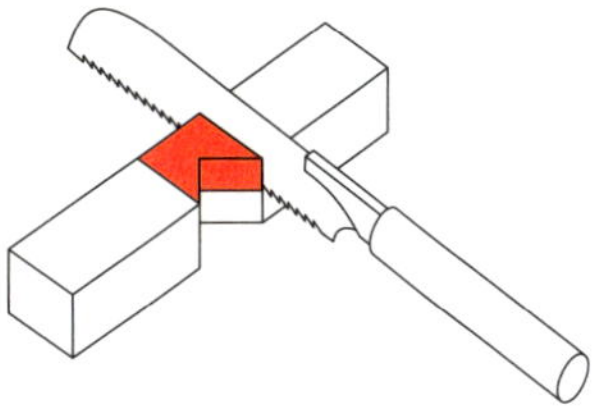

A3

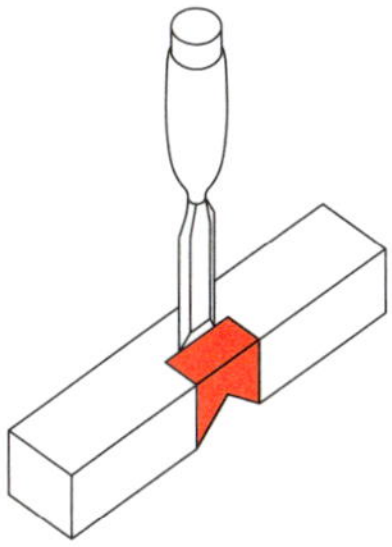

A4

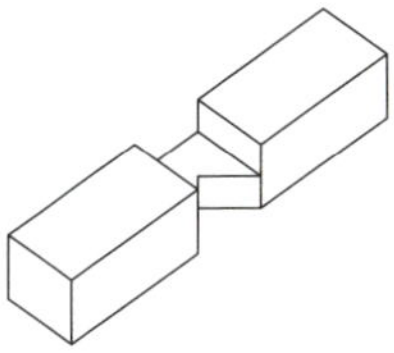

A

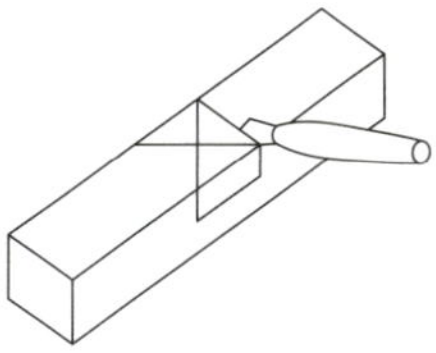

B1

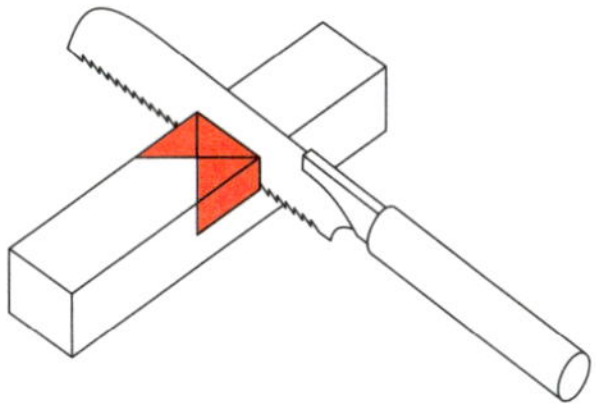

B2

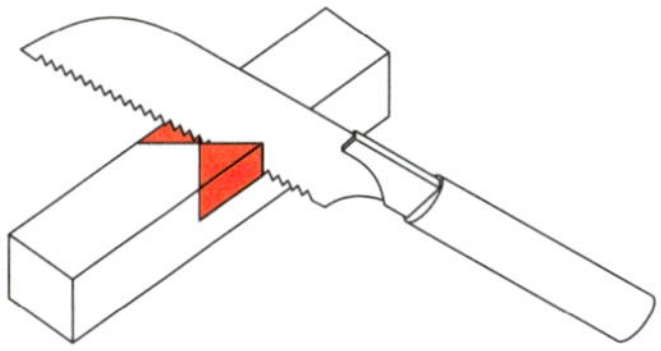

B3

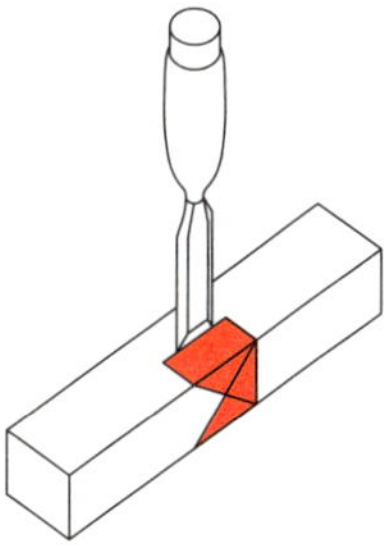

B4

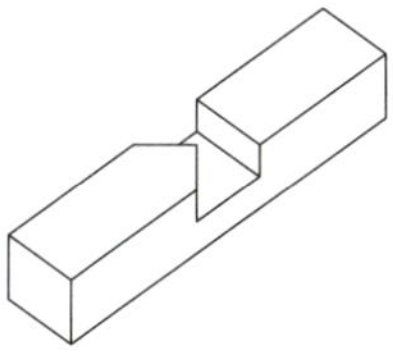

B

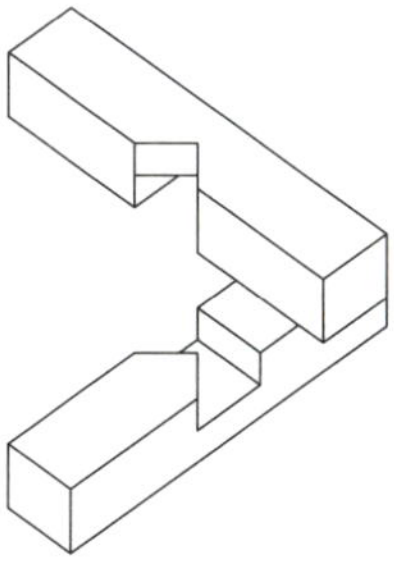

A+B

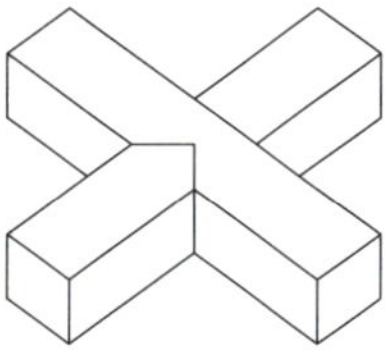

AB

Cross Lap Joints

Cross lap joint with mitered abutments

ai-kaki-tsugi
(watari-ago, yatsuna-dome) [*]
相欠接ぎ（渡腮・八中留）

schräg abgesetztes Kreuzblatt

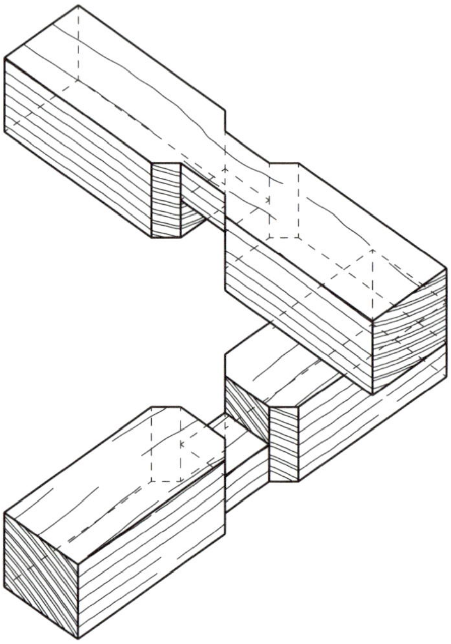

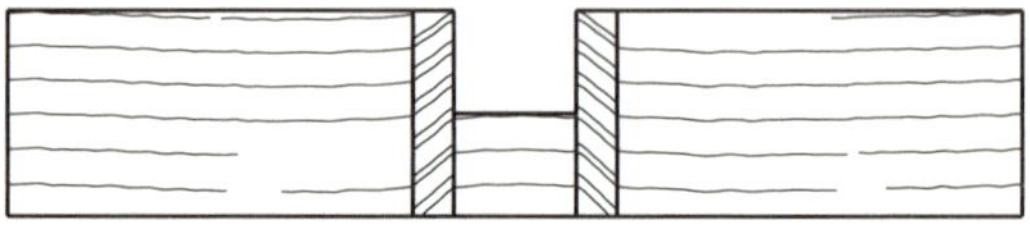

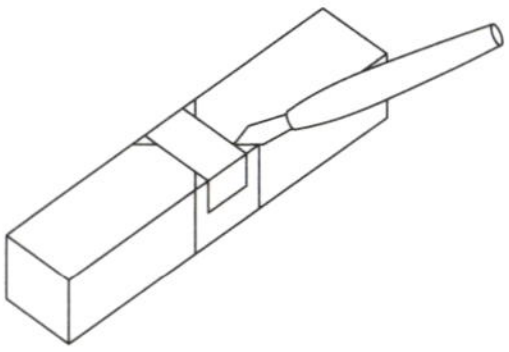

A1

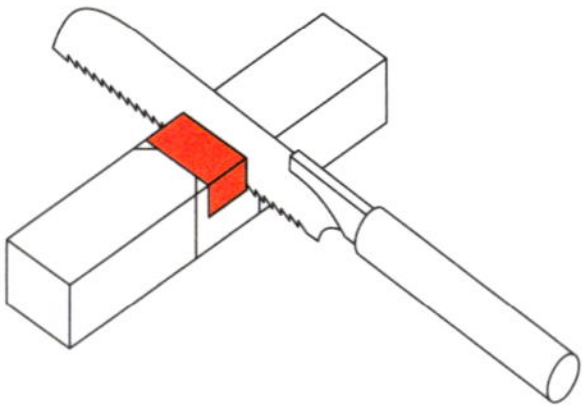

A2

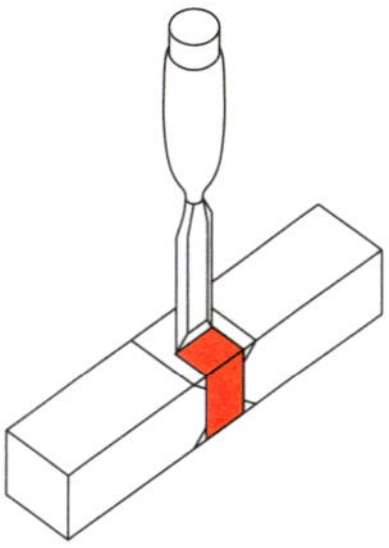

A3

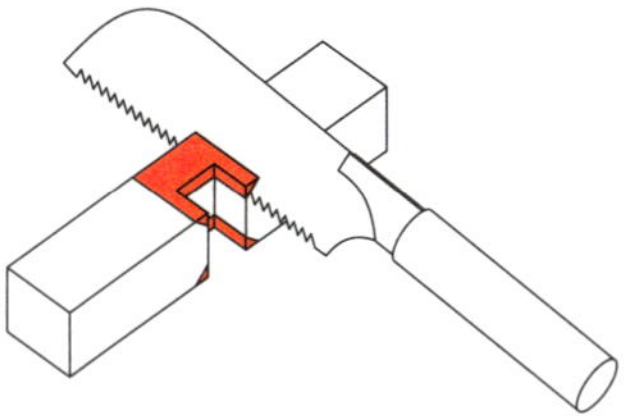

A4

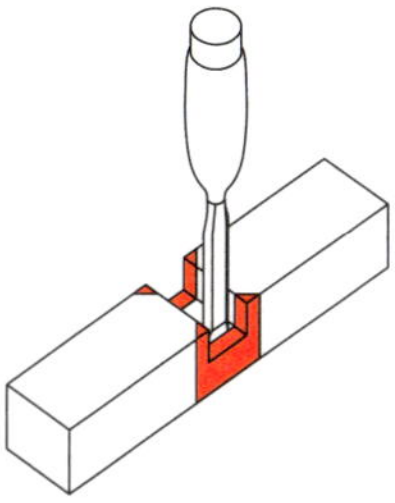

A5

2×

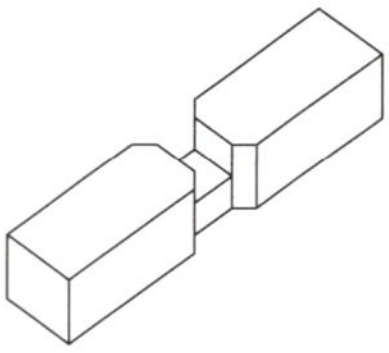

A

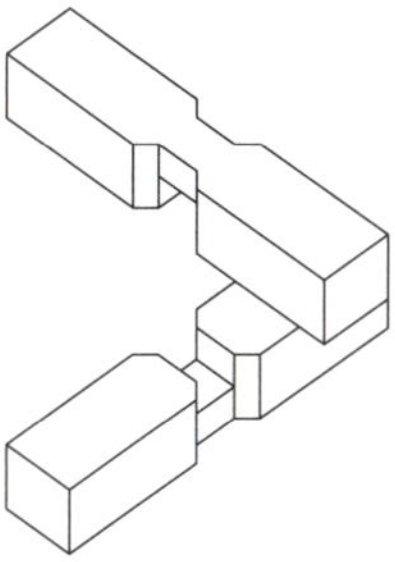

A+A

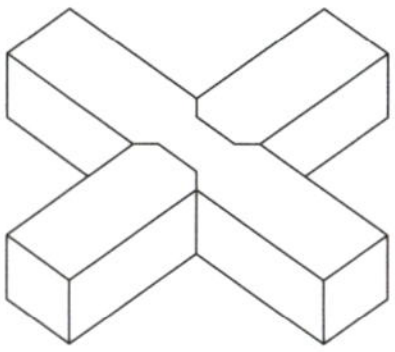

AA

Cross Lap Joints

Stepped cross lap joint

ai-kaki-watari-ago-tsugi (dan-kaki)
相欠渡顎接ぎ（段欠）

Kreuzblatt mit Haken

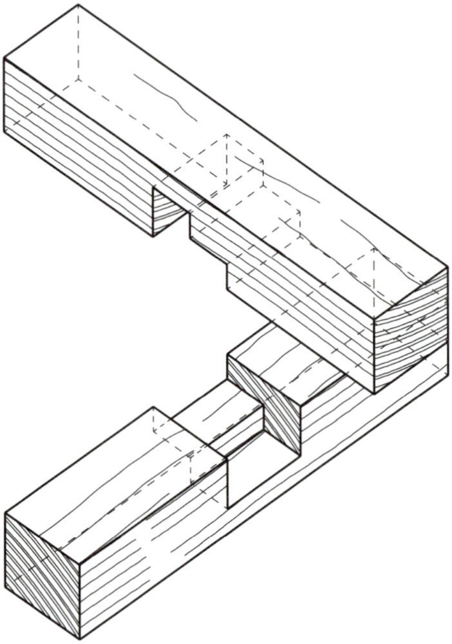

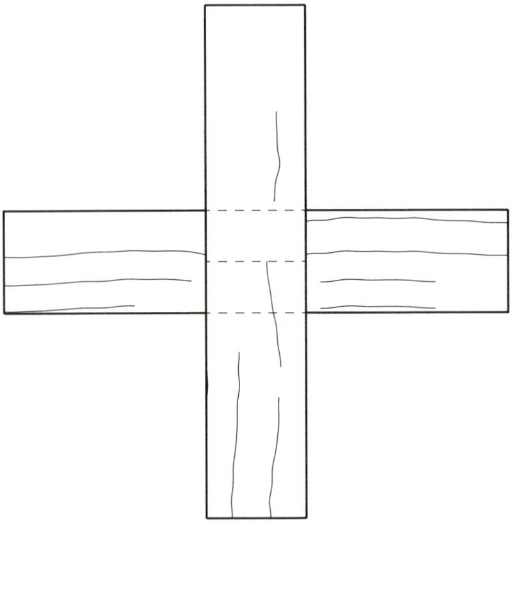

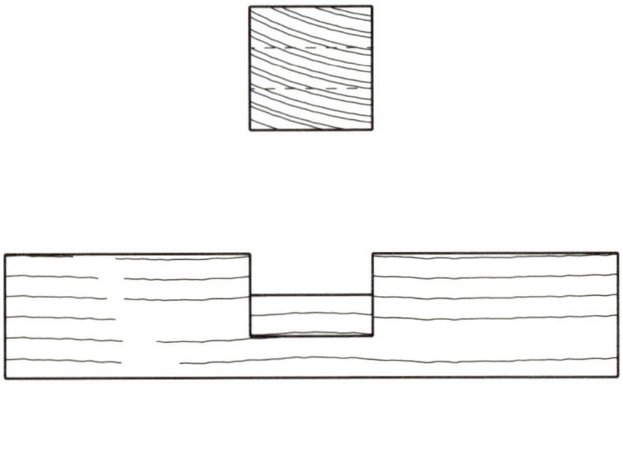

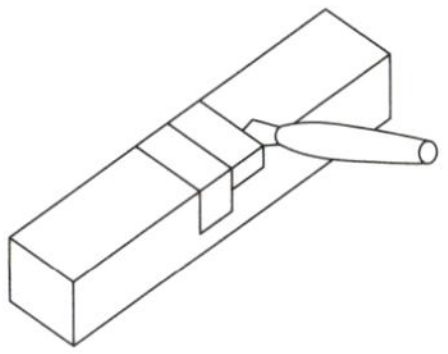

A1

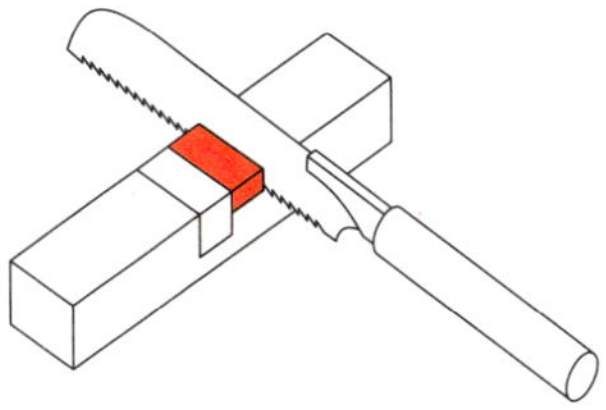

A2

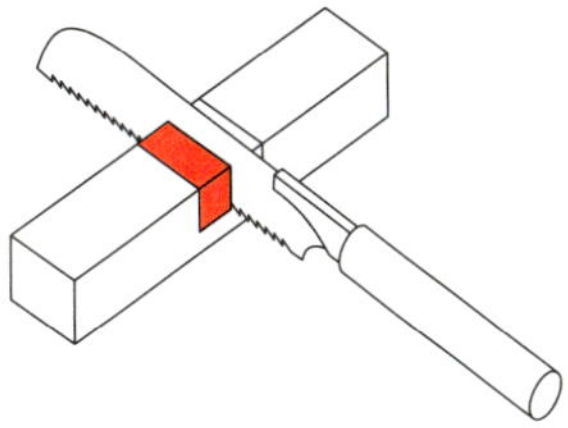

A3

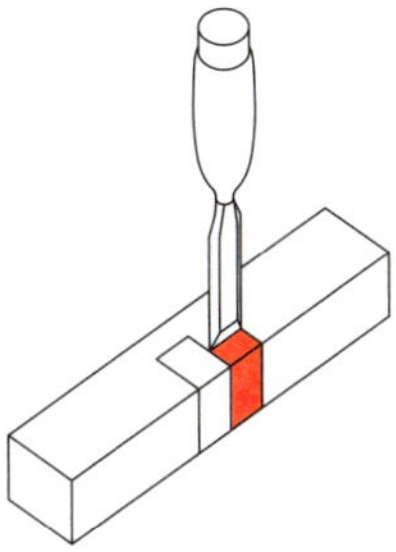

A4

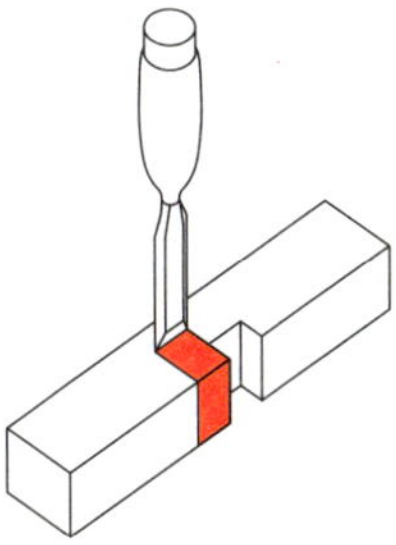

A5

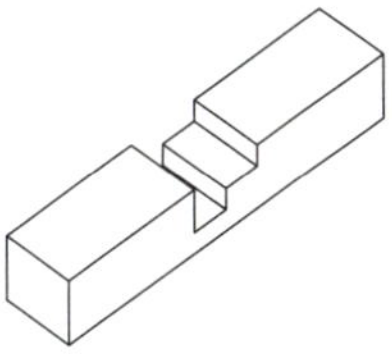

A

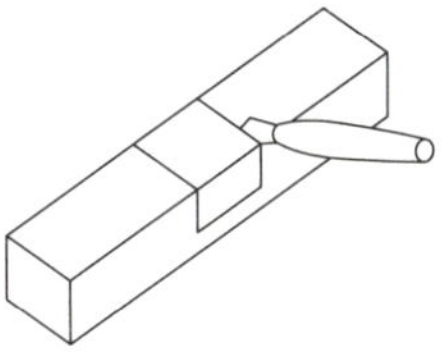

B1

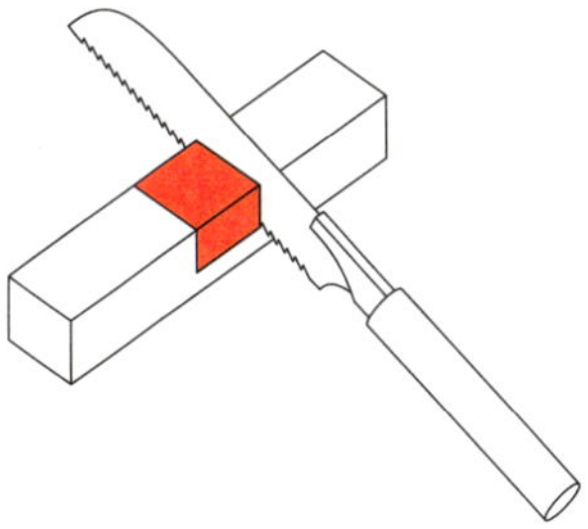

B2

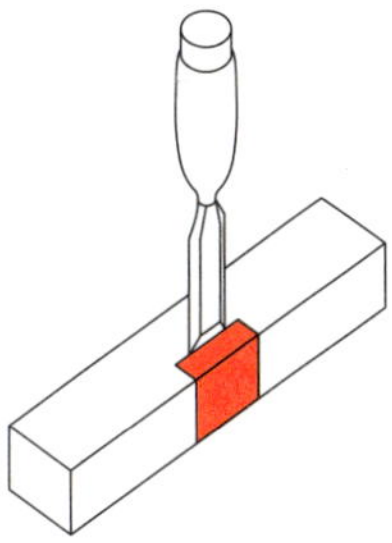

B3

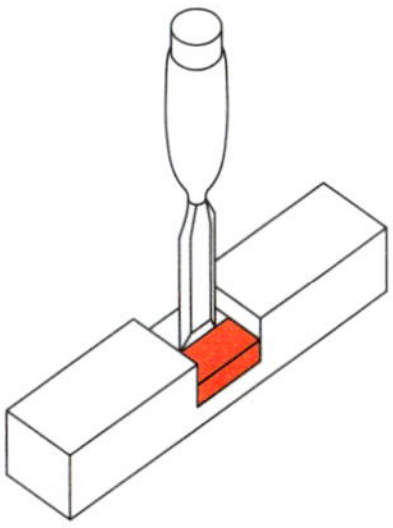

B4

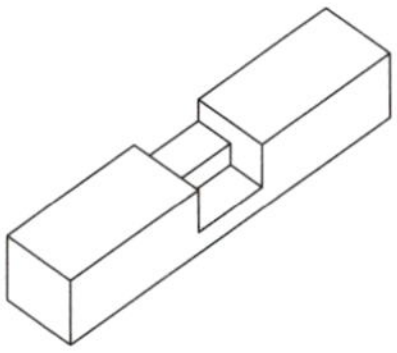

B

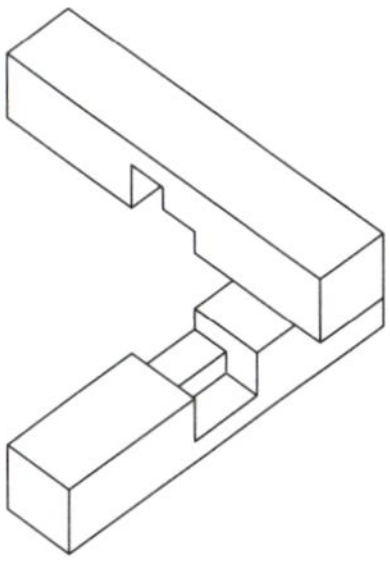

A+B

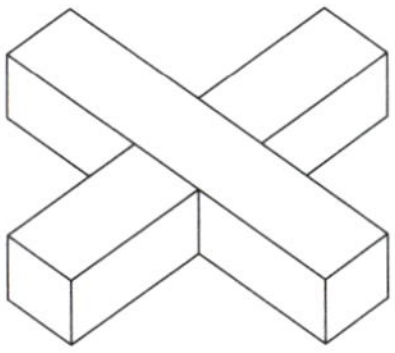

AB

Cross Lap Joints

Cogged cross lap joint

ai-kaki-watari-ago-tsugi
相欠渡顎接ぎ

Kreuzblatt
mit mittigem Haken

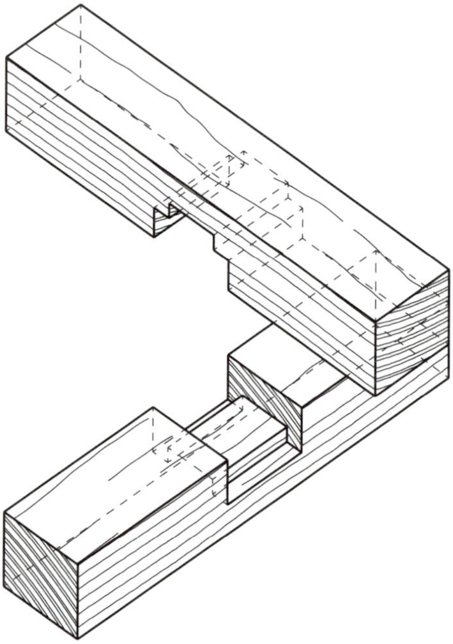

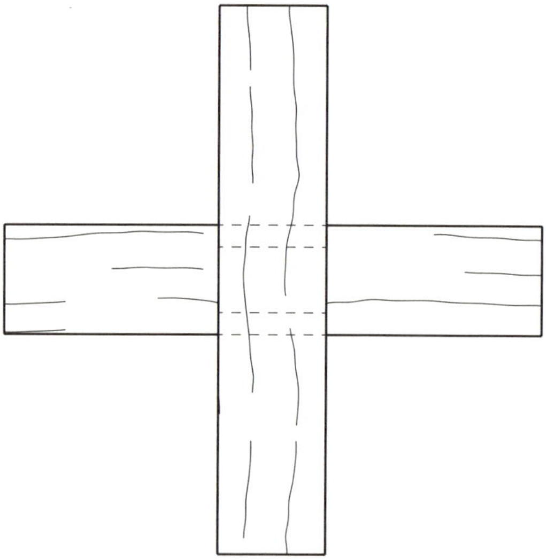

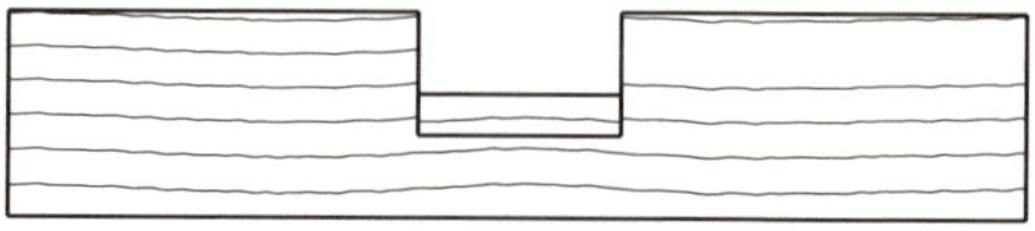

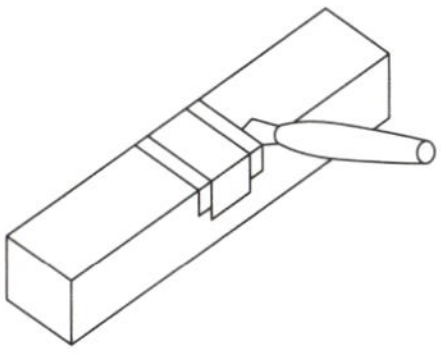

A1

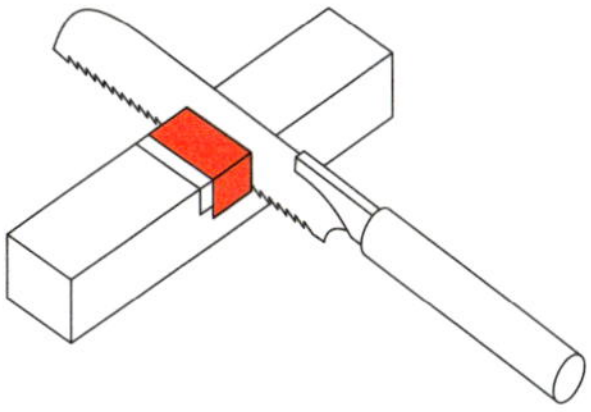

A2

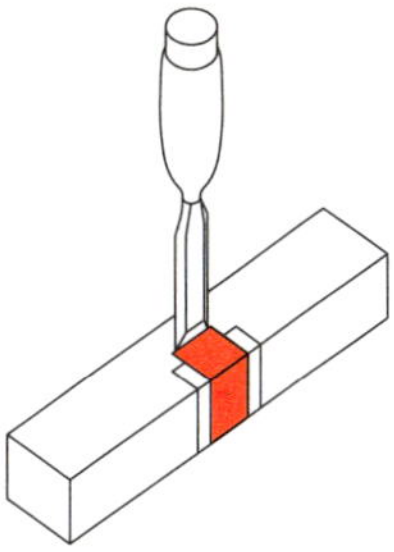

A3

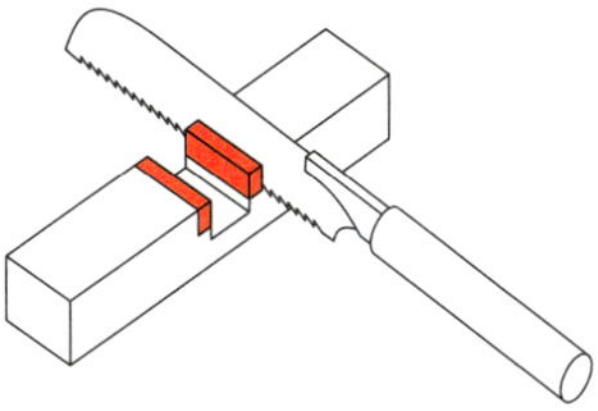

A4

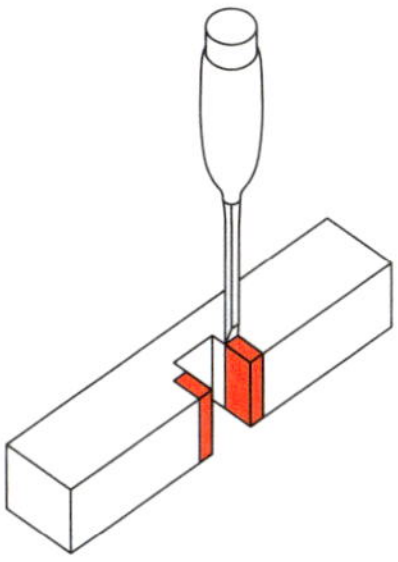

A5

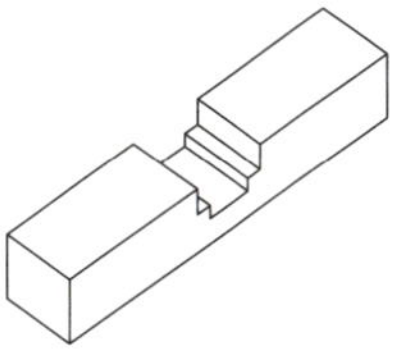

A

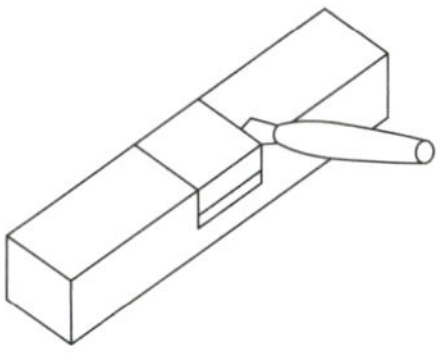

B1

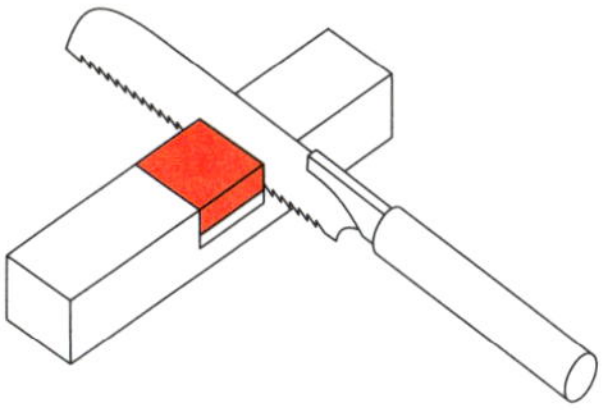

B2

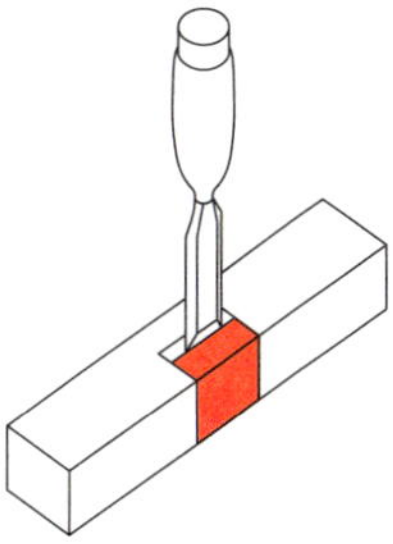

B3

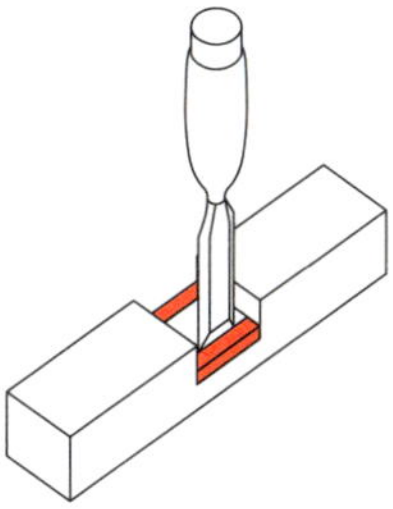

B4

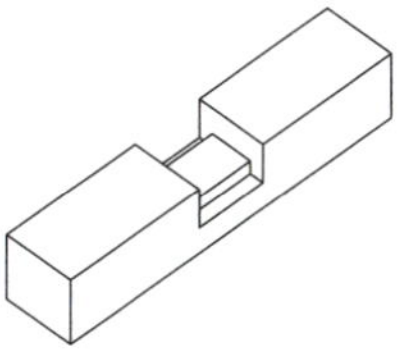

B

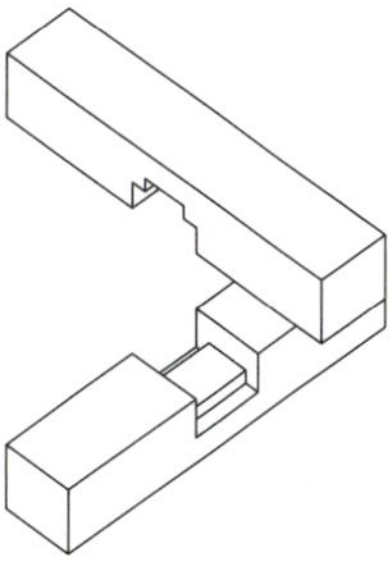

A+B

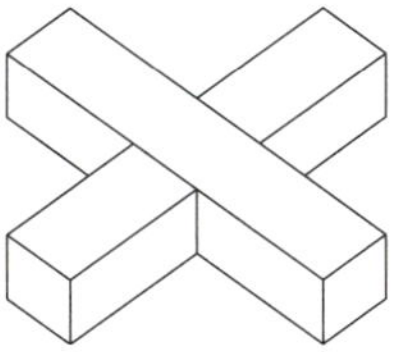

AB

Cross Lap Joints

Cogged cross lap joint with stub tenons

ryō-gawa-tan-hozo-mechi-ai-kaki-watari-ago-tsugi
両側短枘目違相欠渡顎接ぎ

Kreuzblatt mit mittigem Haken und beidseitigen Brüstungszapfen

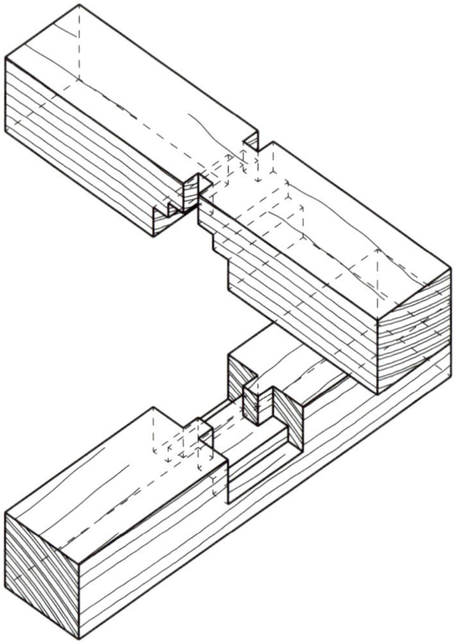

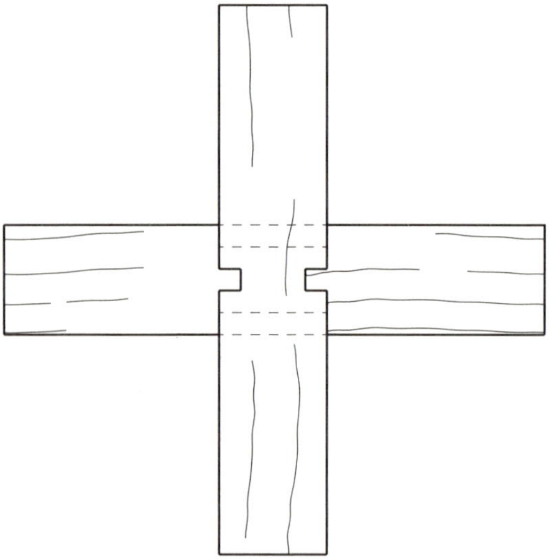

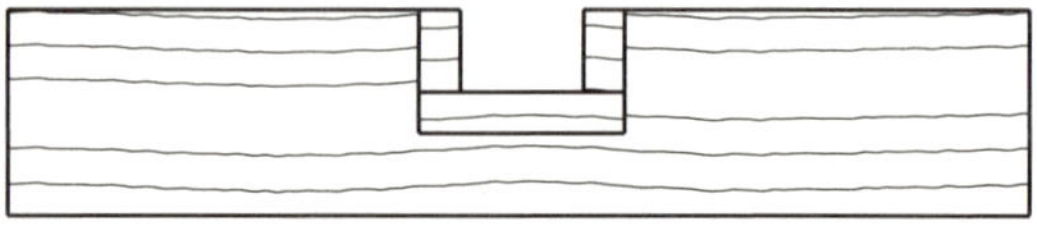

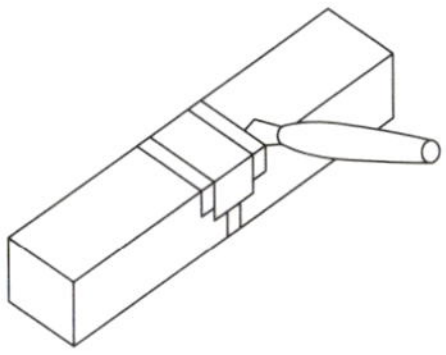

A1

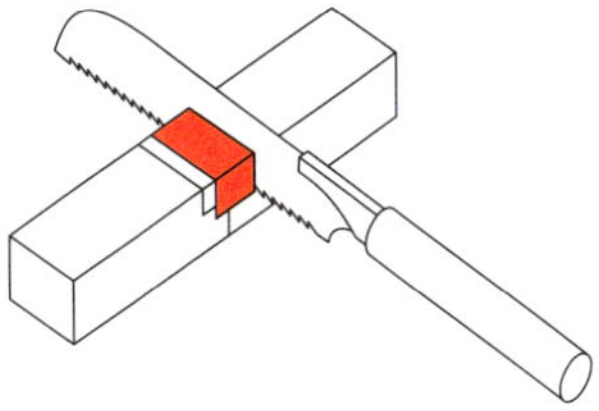

A2

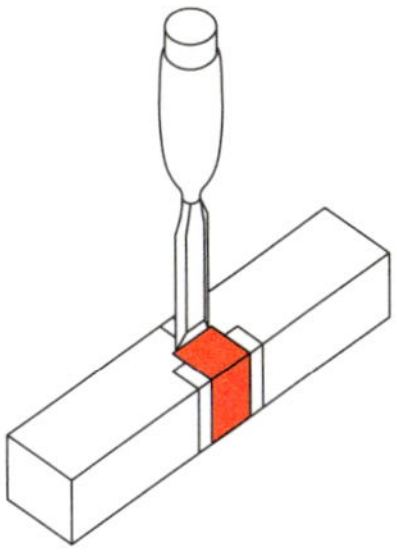

A3

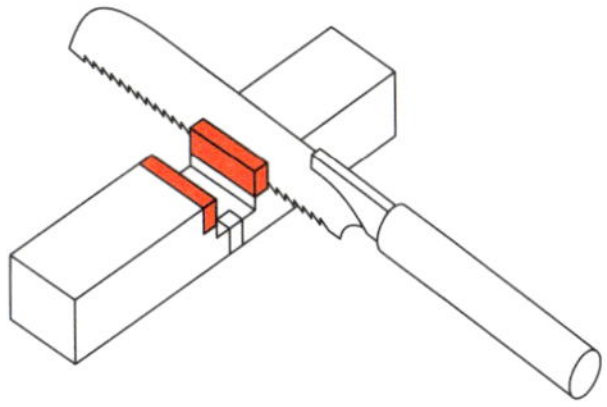

A4

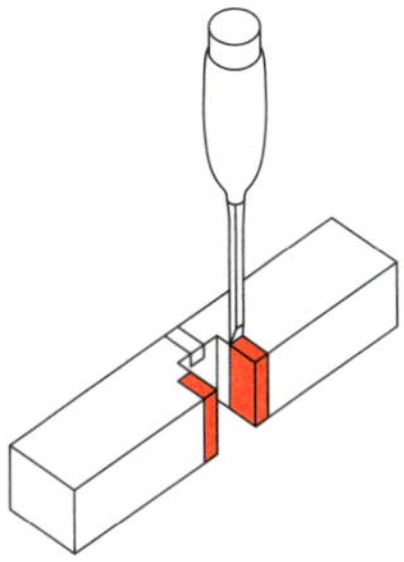

A5

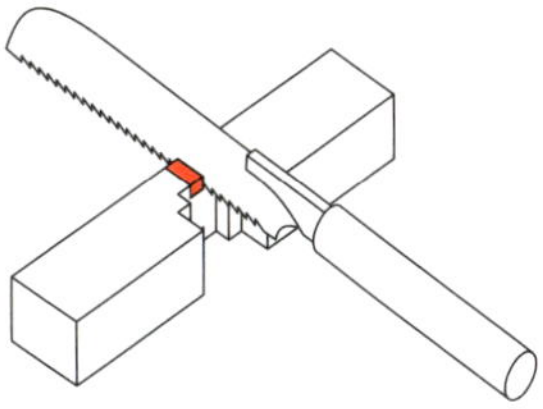

A6

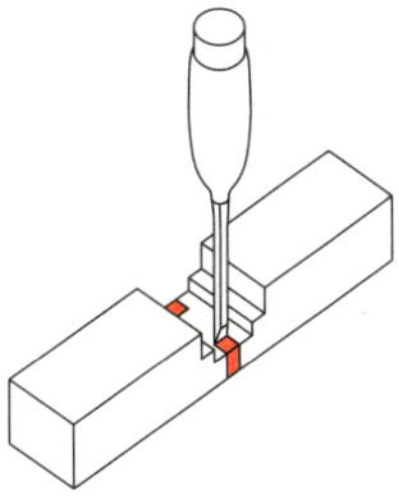

A7

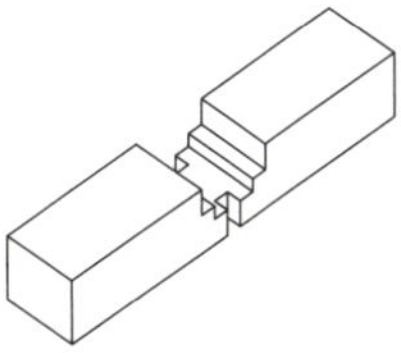

A

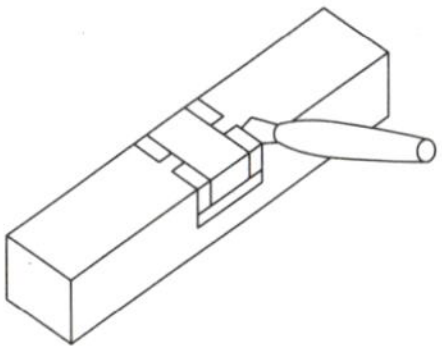

B1

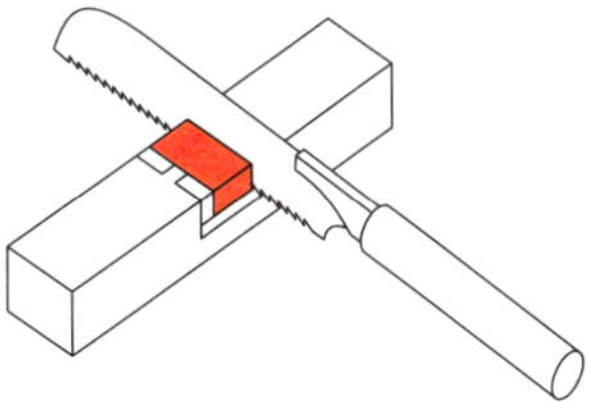

B2

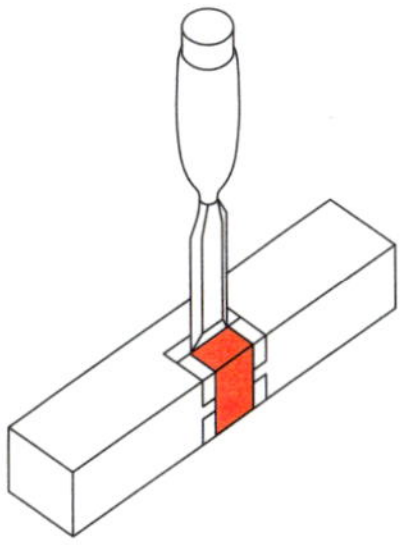

B3

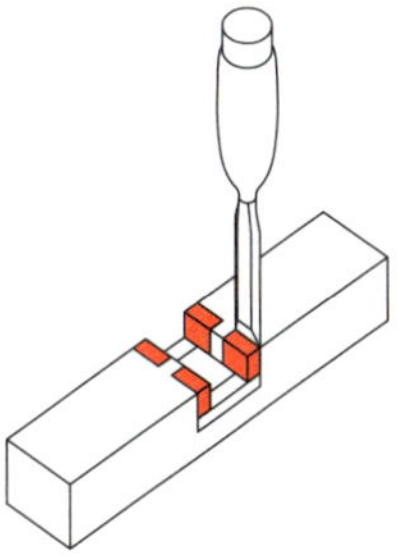

B4

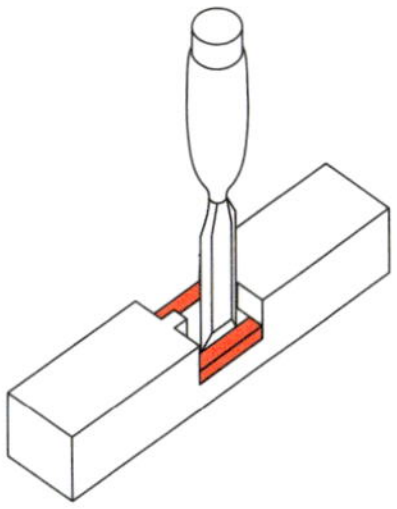

B5

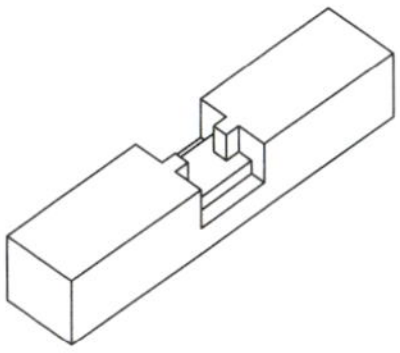

B

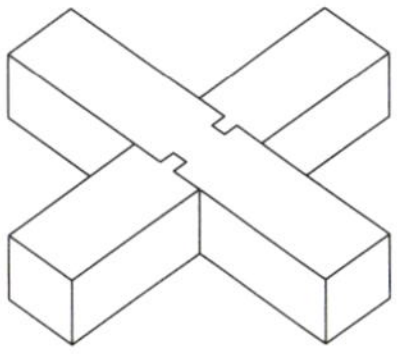

AB

Cross lap joint with stub tenons

ryō-gawa-tan-hozo-mechi-ai-kaki-tsugi
両側短枘目違相欠接ぎ

Kreuzblatt mit beidseitigen Brüstungszapfen

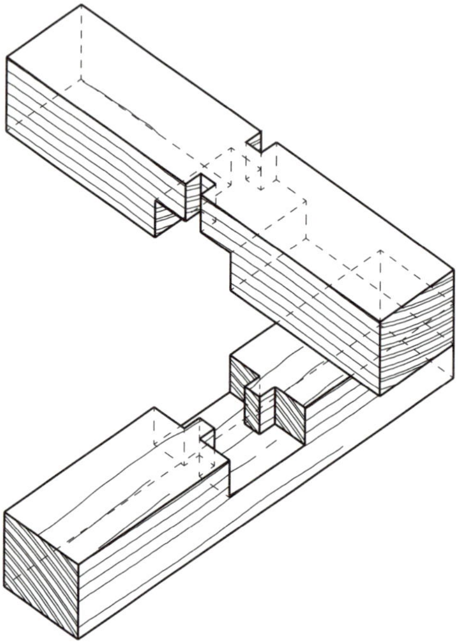

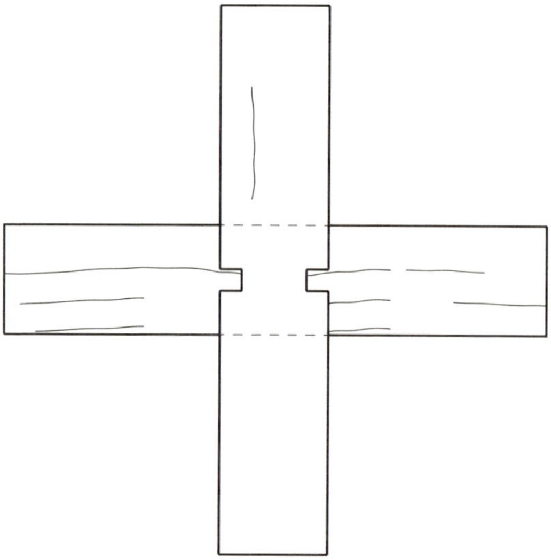

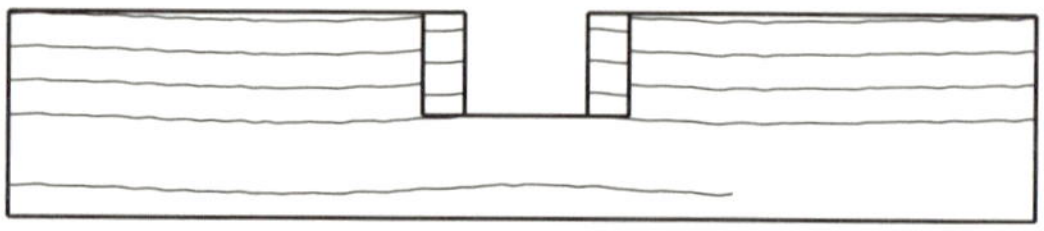

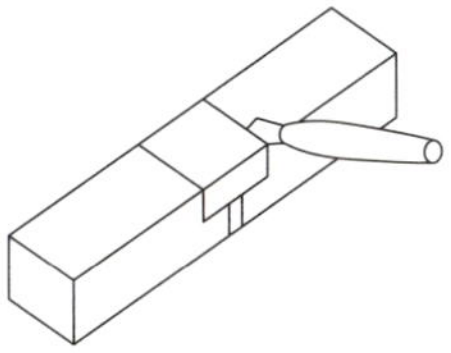

A1

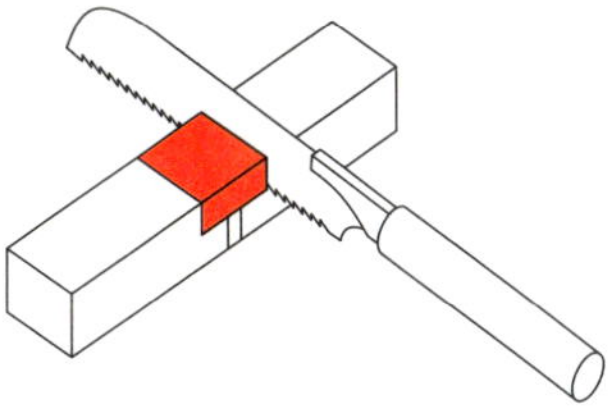

A2

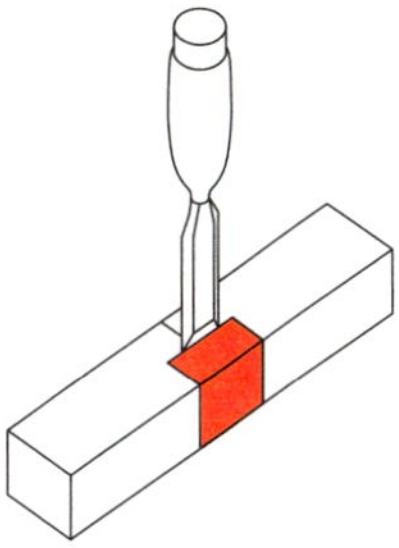

A3

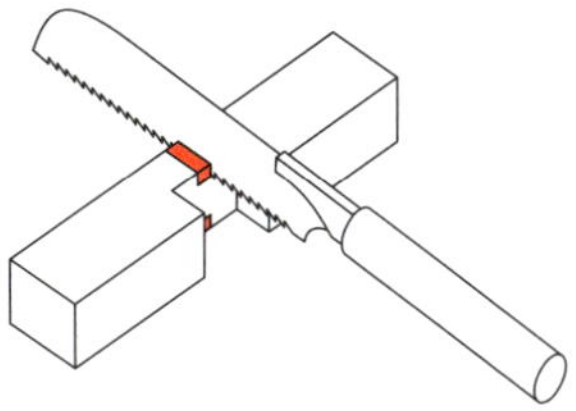

A4

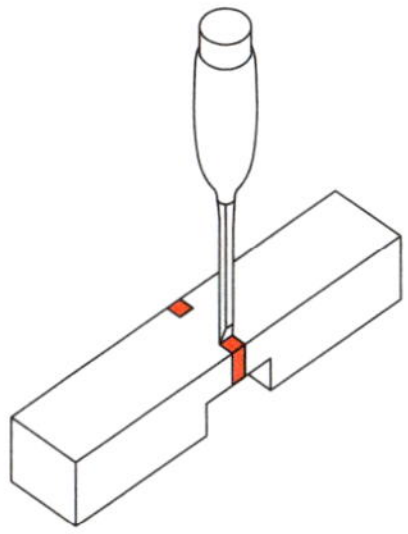

A5

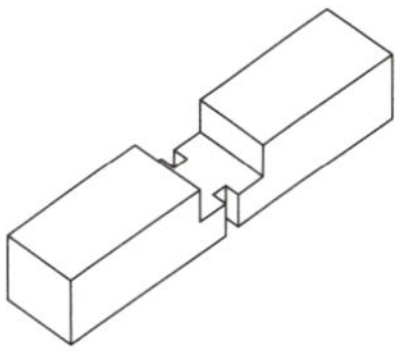

A

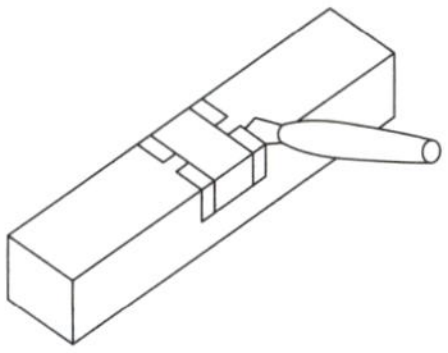

B1

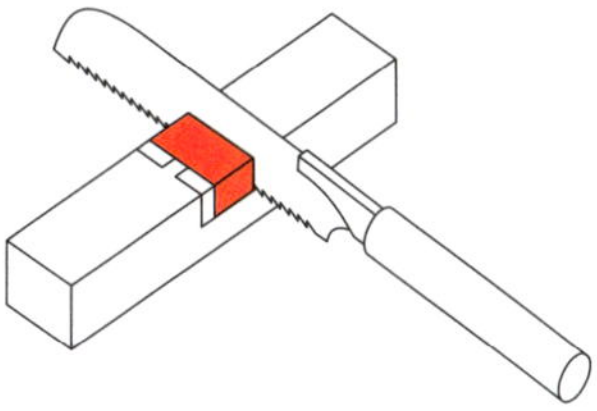

B2

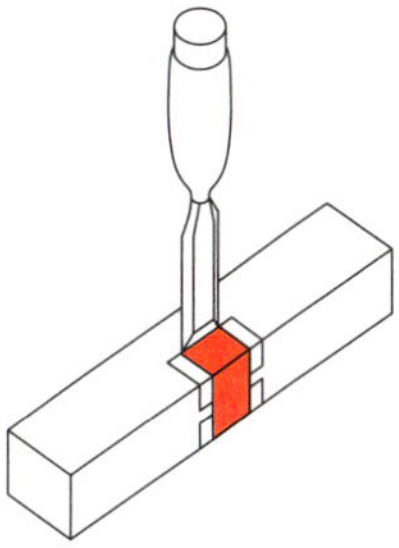

B3

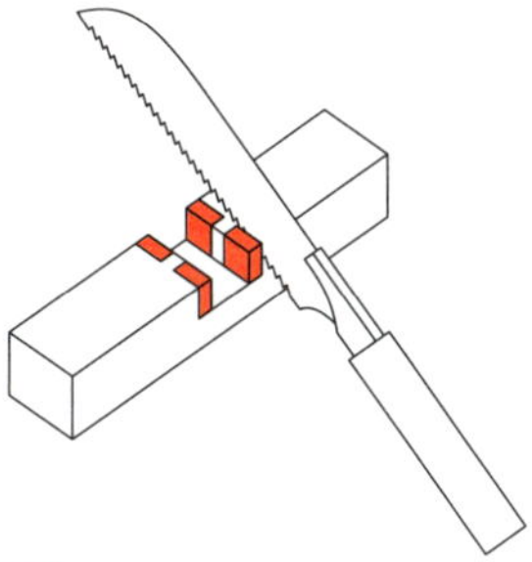

B4

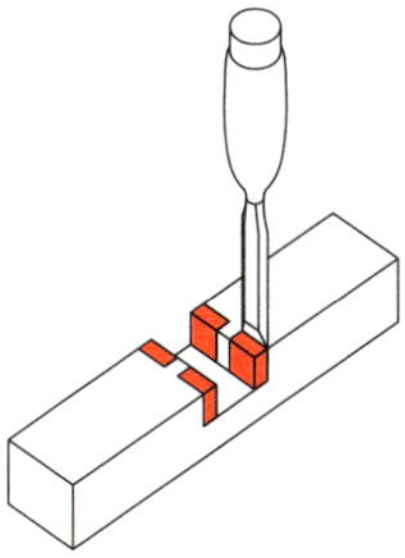

B5

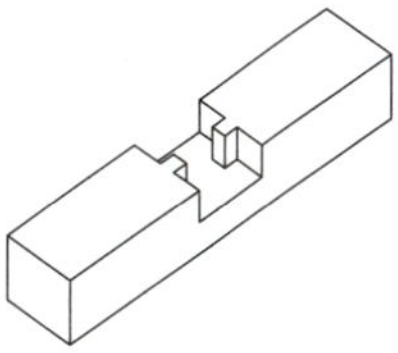

B

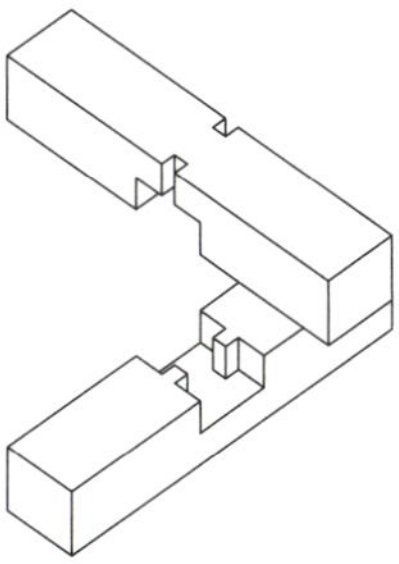

A+B

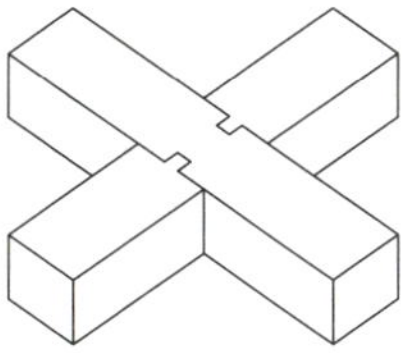

AB

Cross Lap Joints

Offset cross lap joint

ryō-gawa-mechi-ai-kaki-watari-ago-tsugi (kakari) [*]
両側目違相欠渡顎接ぎ（掛）

Kreuzblatt mit Führung

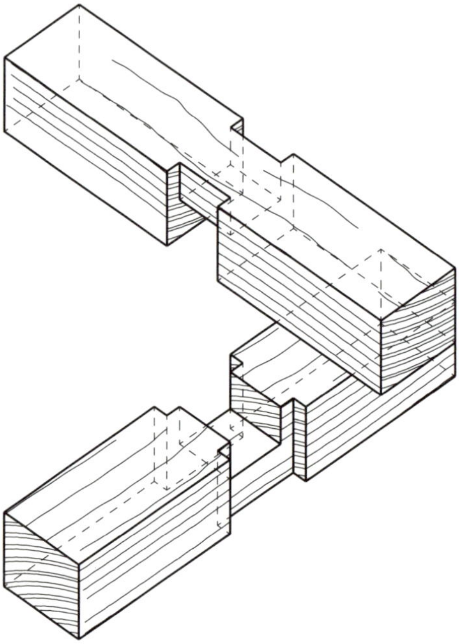

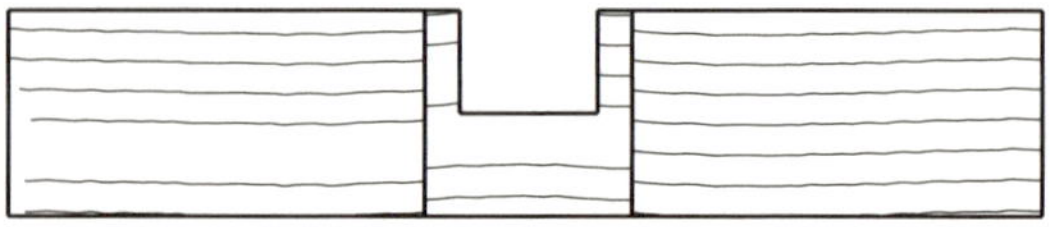

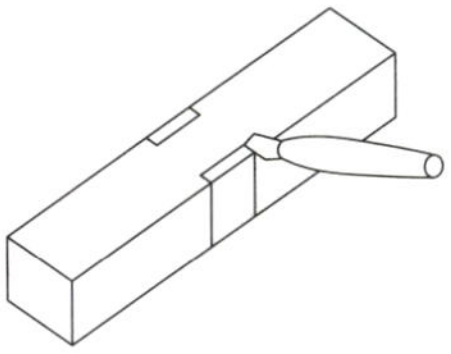
A1

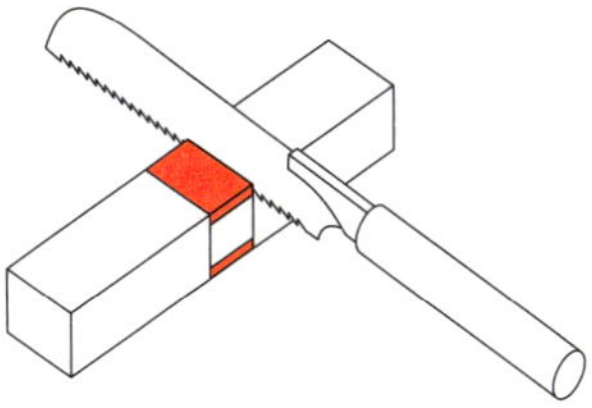
A2

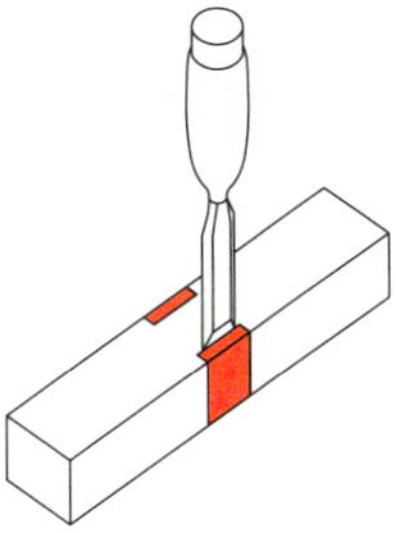
A3

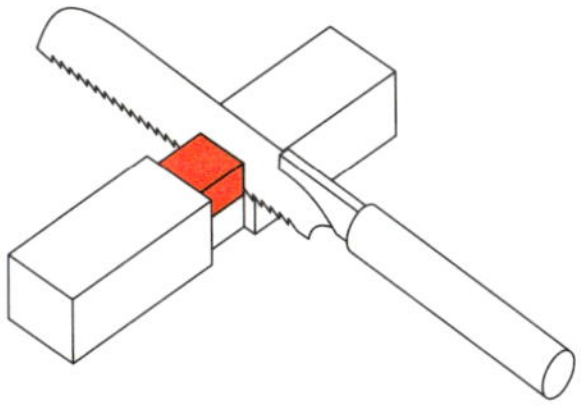
A4

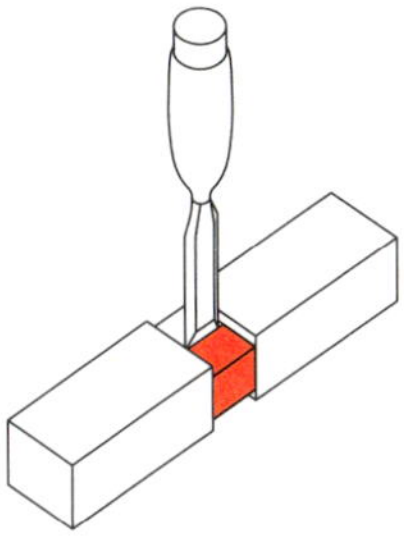
A5

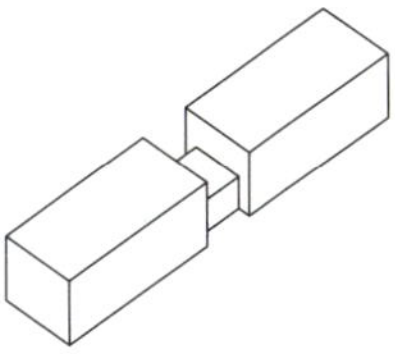
A

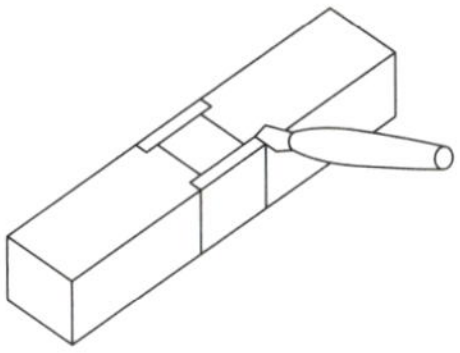
B1

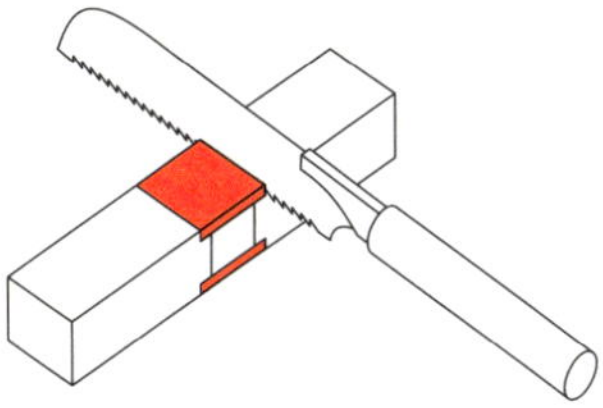
B2

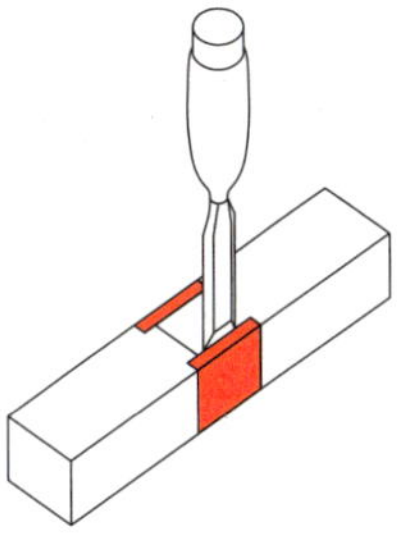
B3

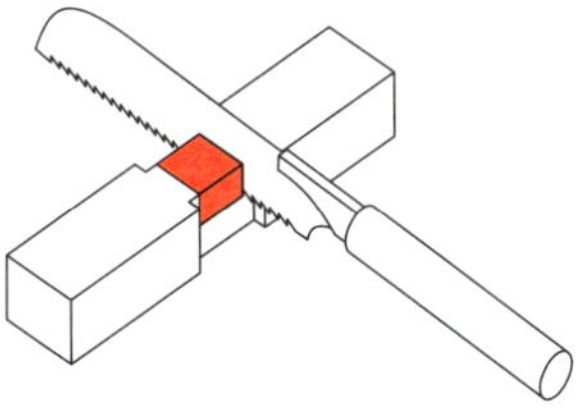
B4

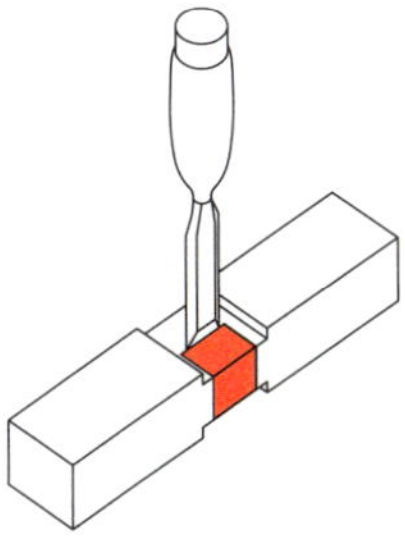
B5

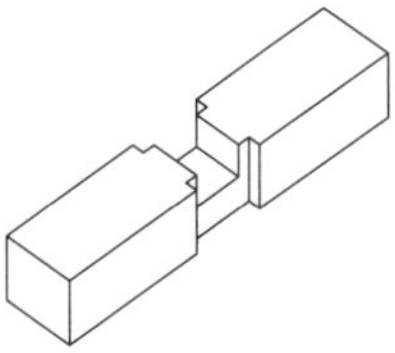
B

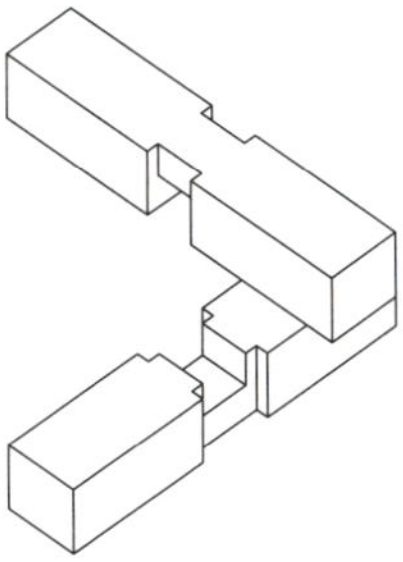
A+B

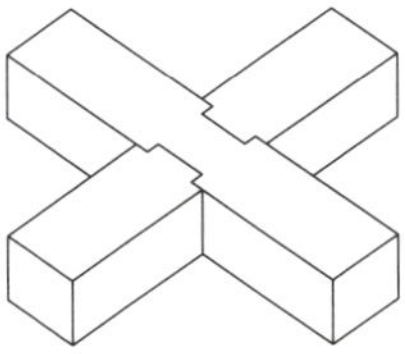
AB

Double offset cross lap joint

ryō-gawa-mechi-ai-kaki-watari-ago-tsugi
両側目違相欠渡顎接ぎ

zweiseitig abgesetztes Kreuzblatt

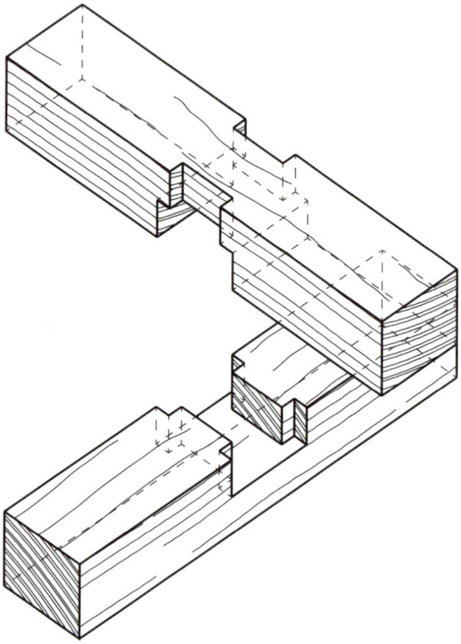

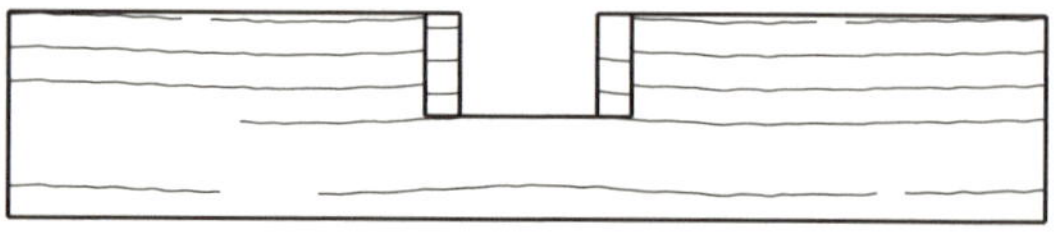

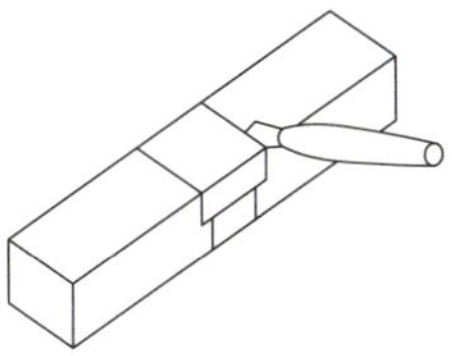

A1

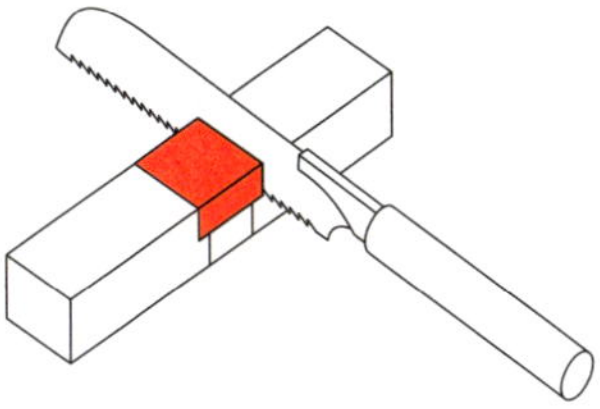

A2

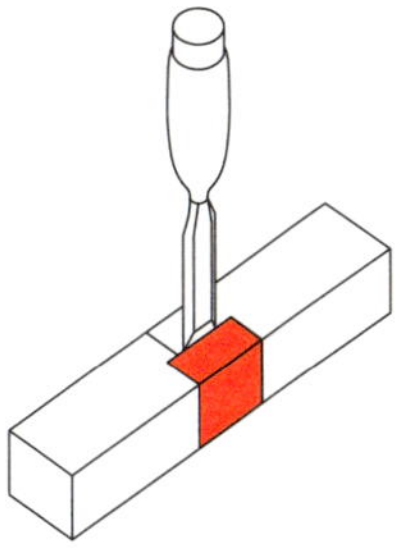

A3

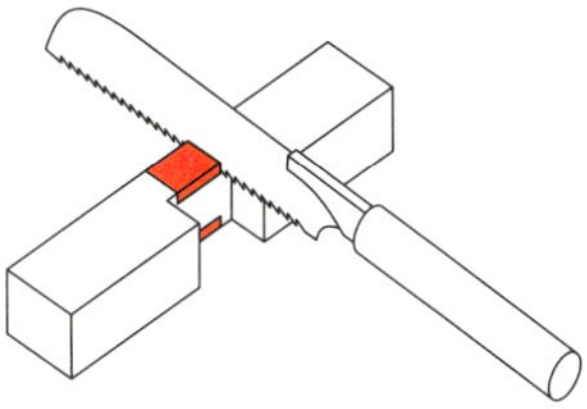

A4

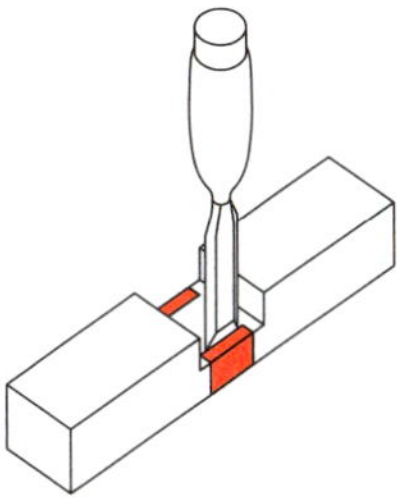

A5

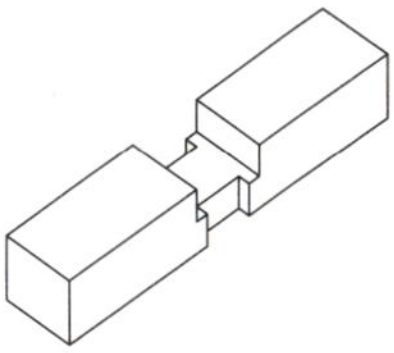

A

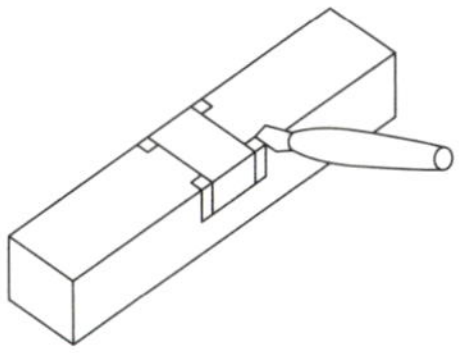

B1

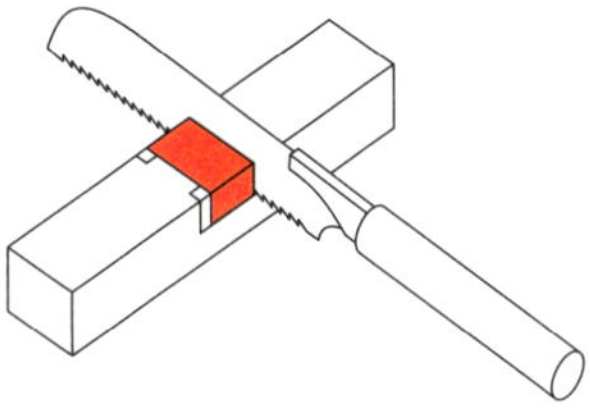

B2

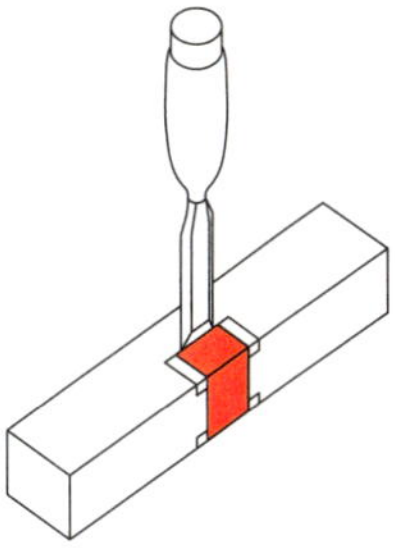

B3

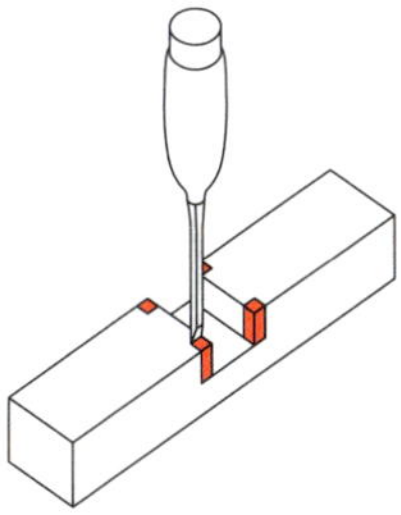

B4

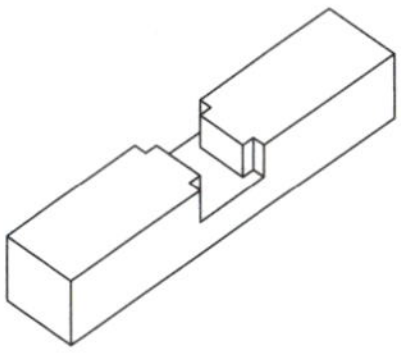

B

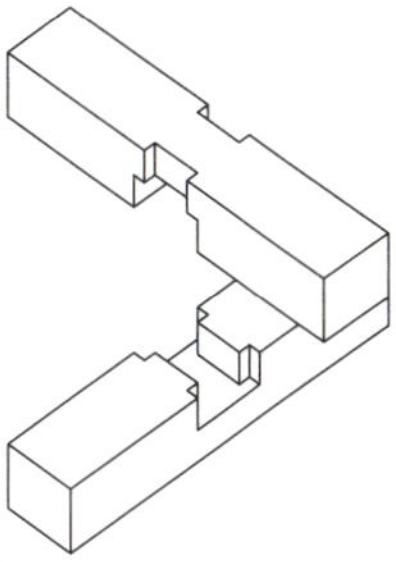

A+B

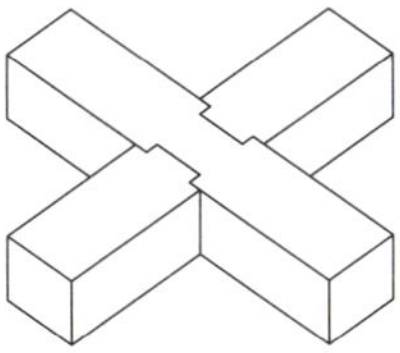

AB

Cross Lap Joints

Triple offset cross lap joint

ryō-gawa-mechi-ai-kaki-watari-ago-tsugi (kakari, koshi-kata)
両側目違相欠渡顎接ぎ(掛・腰片)

zweiseitig abgesetztes Kreuzblatt mit Führung

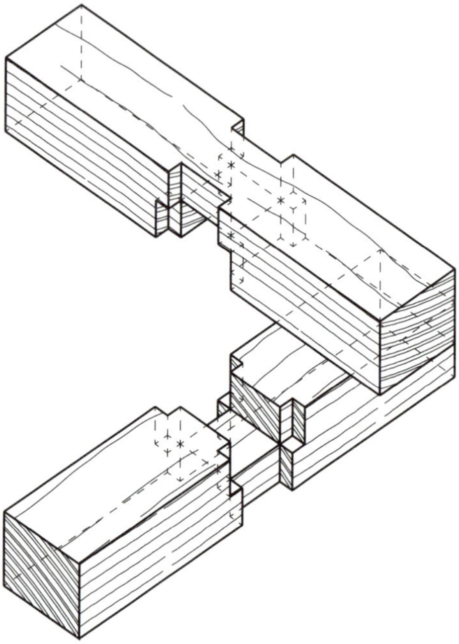

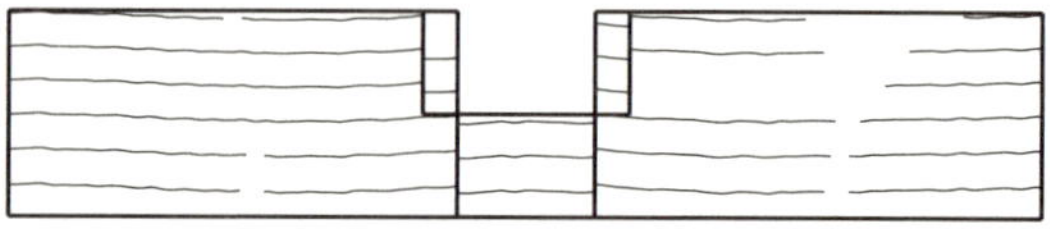

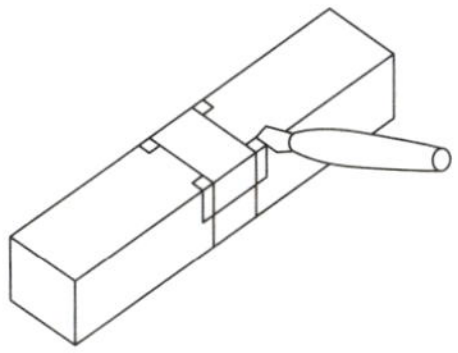

A1

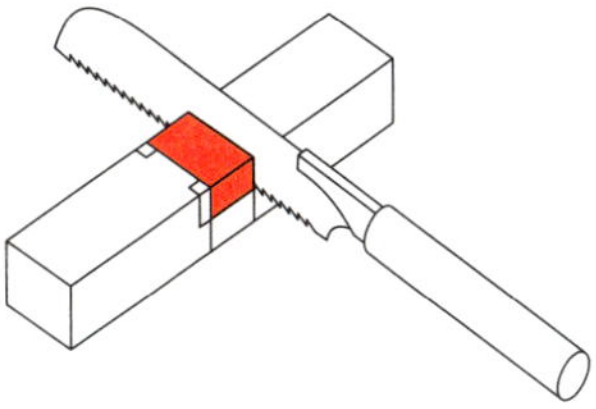

A2

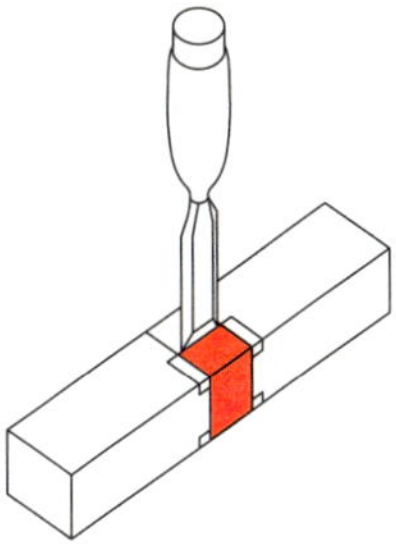

A3

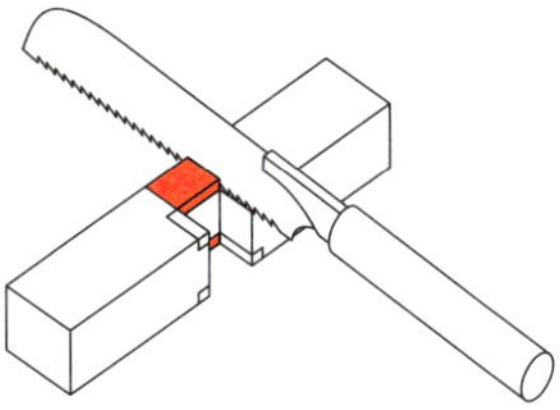

A4

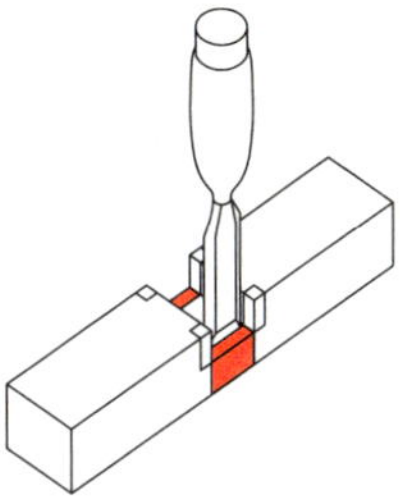

A5

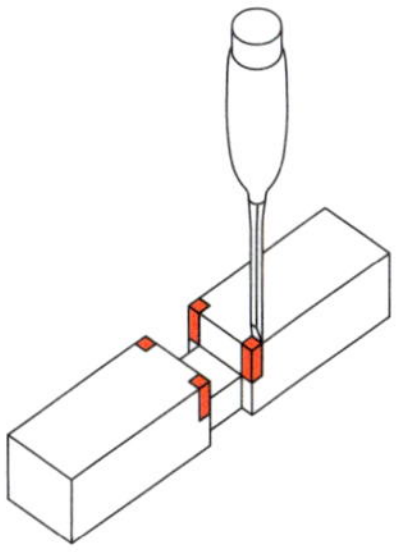

A6

2×

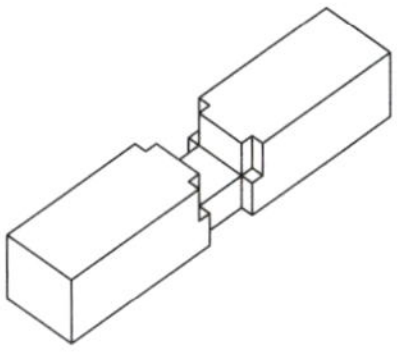

A

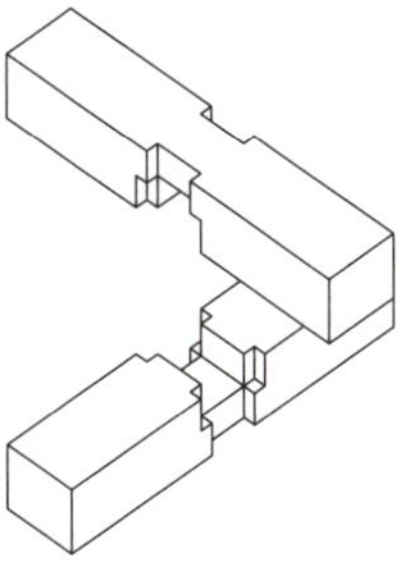

A+A

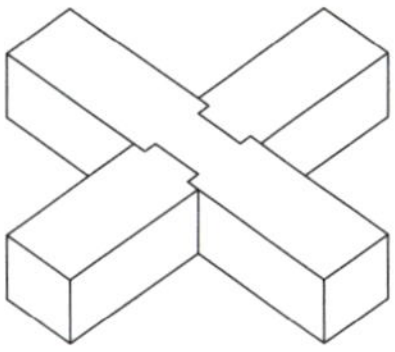

AA

Cross Lap Joints

Dovetailed cross lap joint

shihō-ari-kumi-watari-ago-tsugi
四方蟻組渡顎接ぎ

Kreuzblatt mit
abgesetzten Schwalbenschwänzen

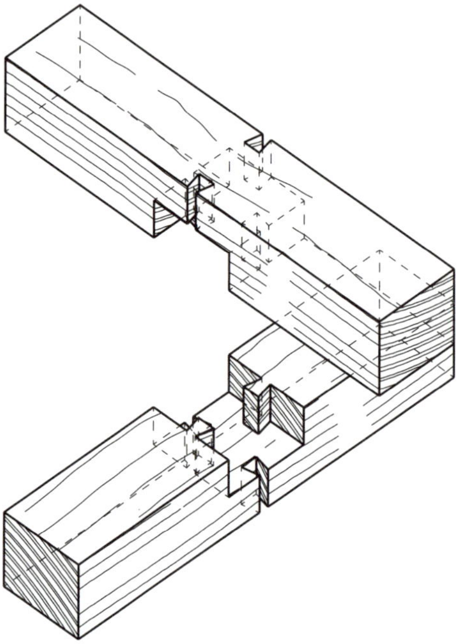

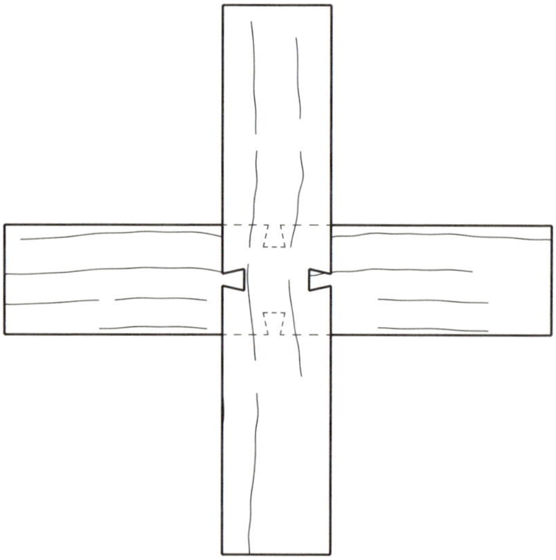

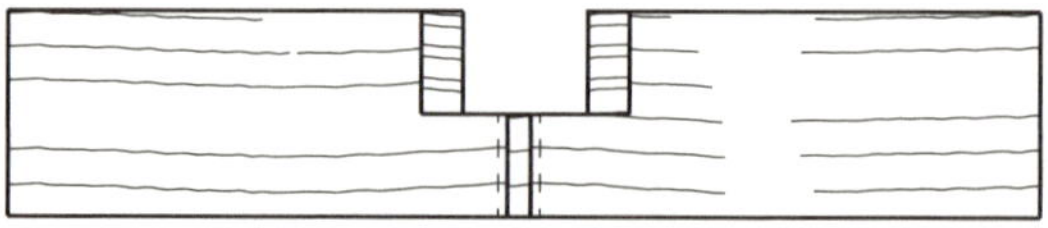

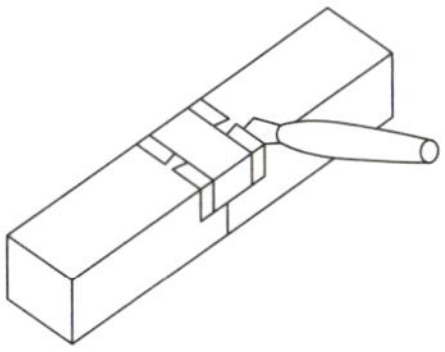

A1

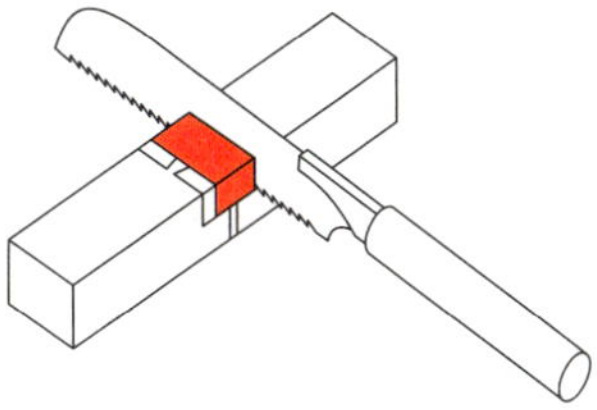

A2

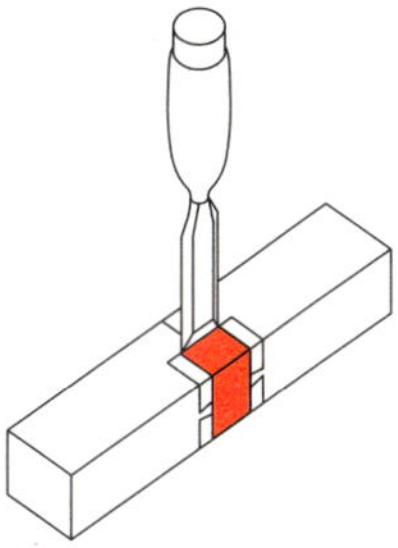

A3

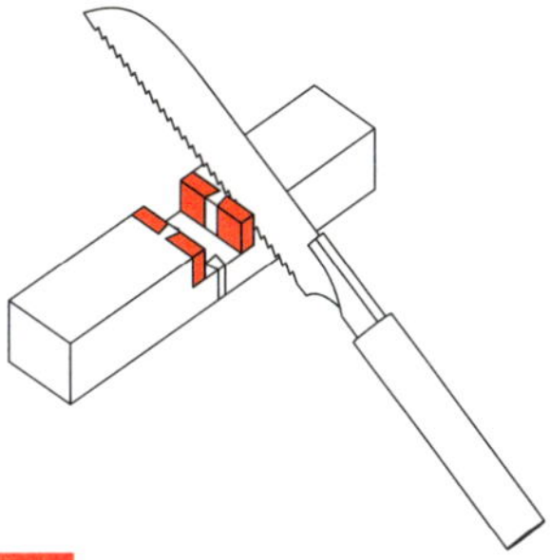

A4

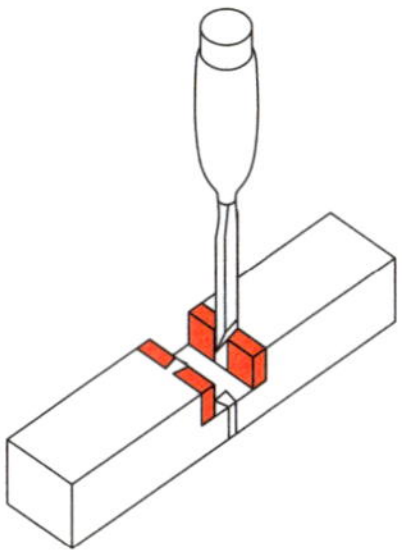

A5

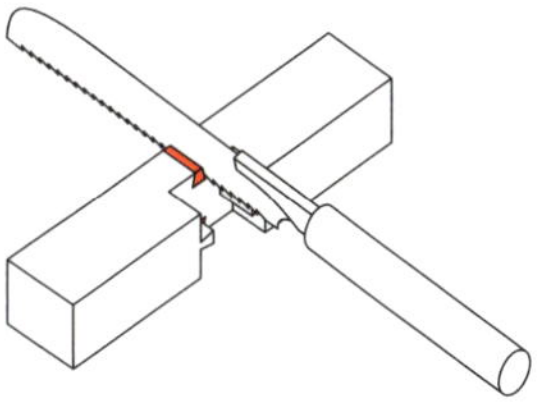

A6

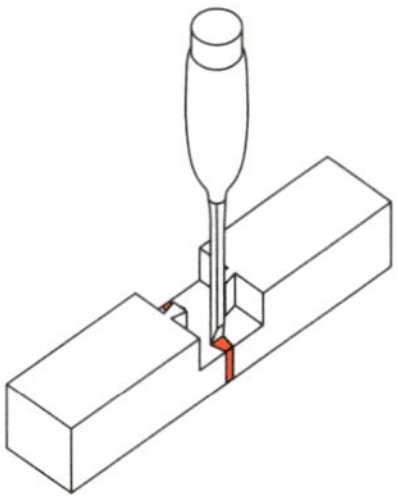

A7

2×

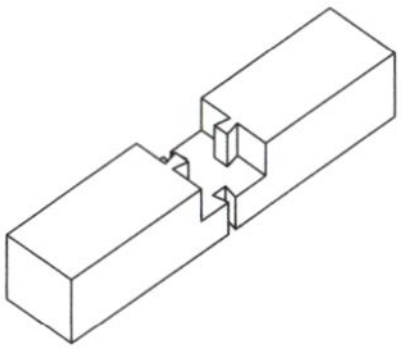

A

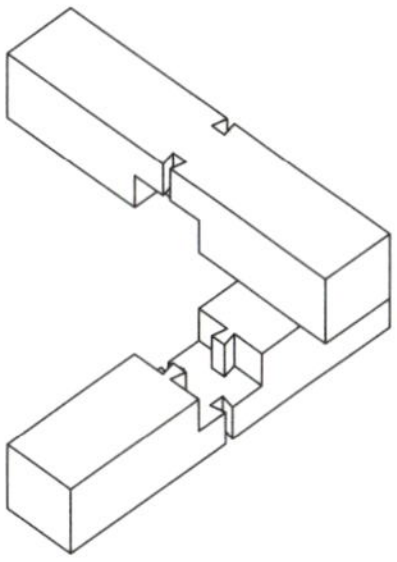

A+A

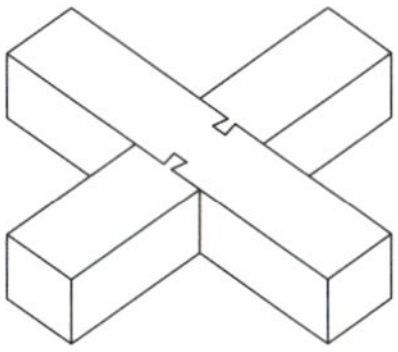

AA

Cross Lap Joints

Oblique notched cross lap joint

sujikai-sogi-kaki-tsugi
筋交殺欠接ぎ

schräges Kreuzblatt
mit Stirnversatz

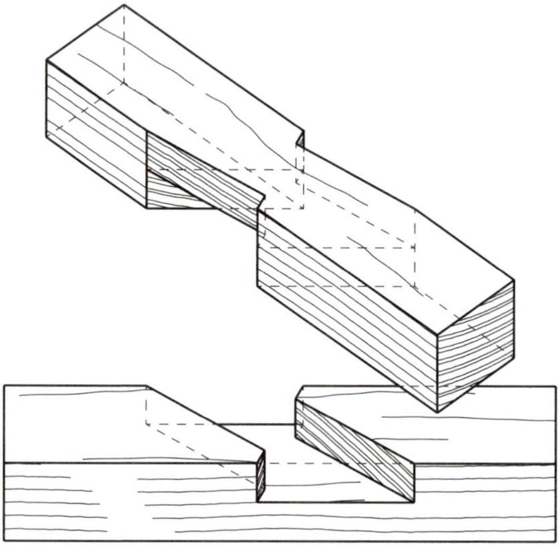

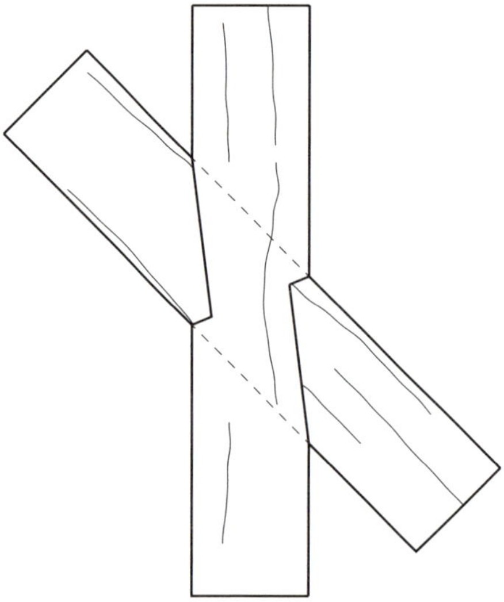

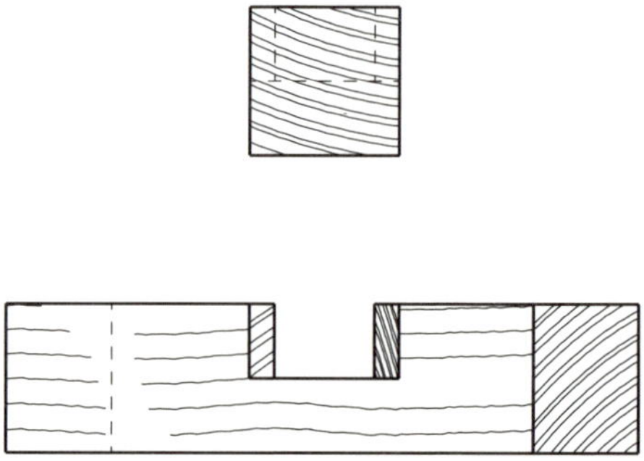

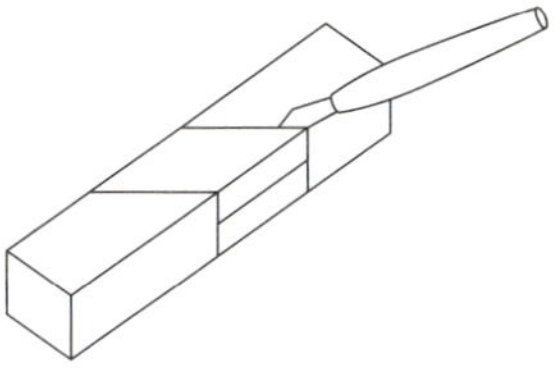

A1

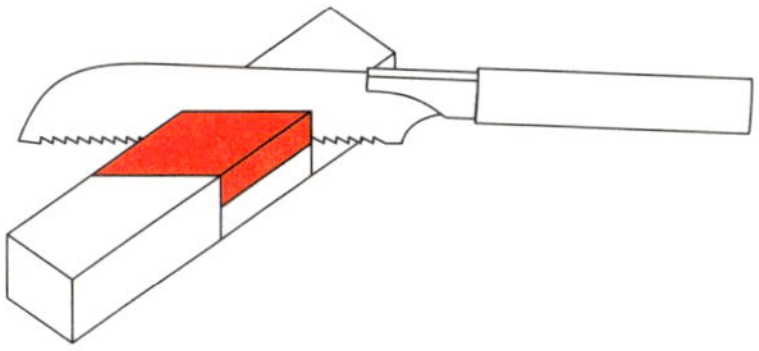

A2

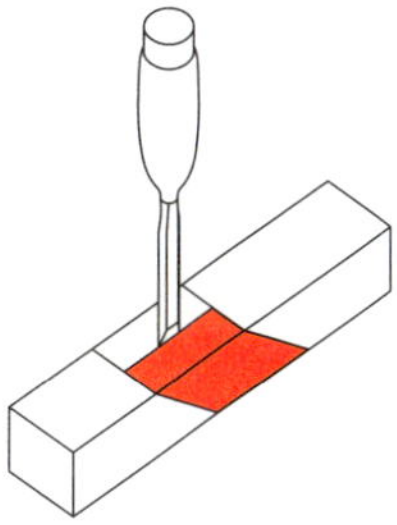

A3

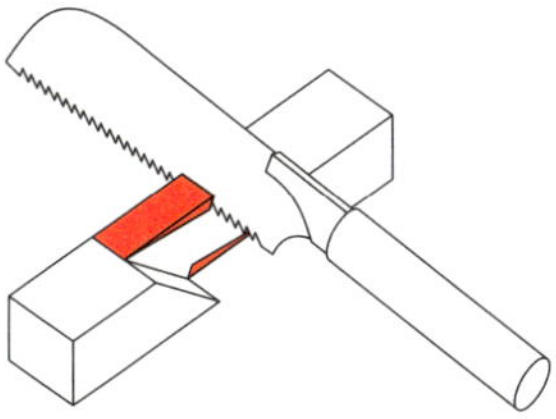

A4

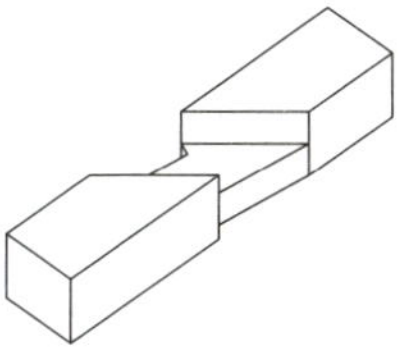

A

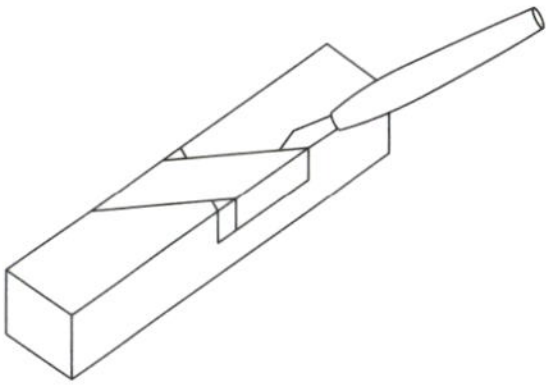

B1

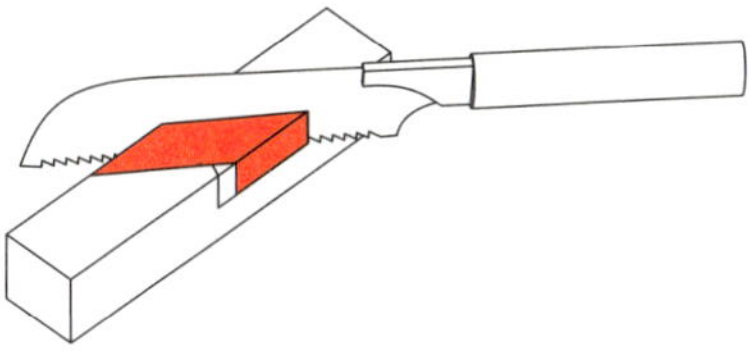

B2

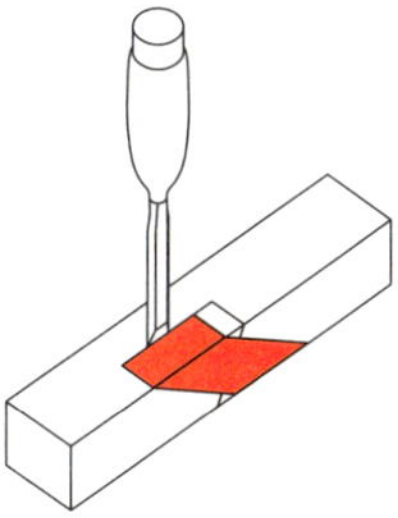

B3

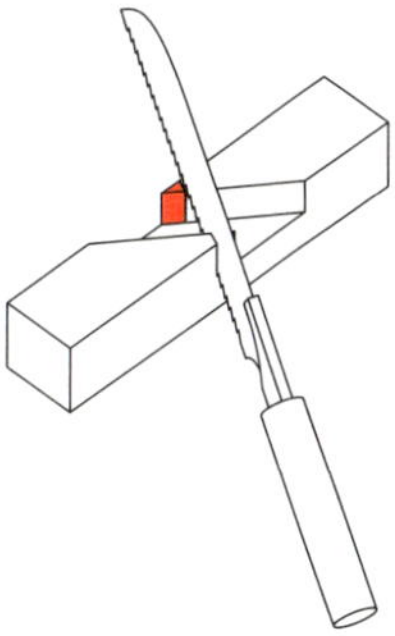

B4

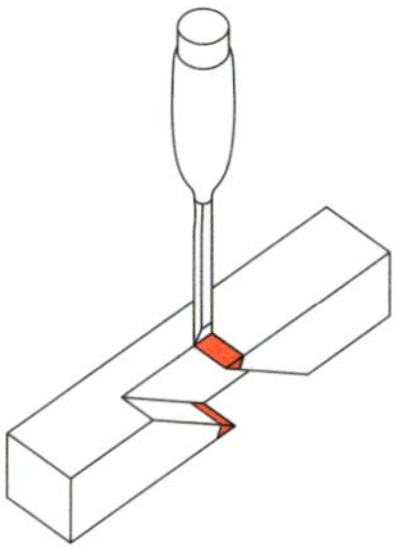

B5

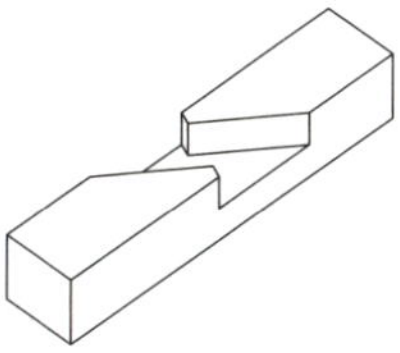

B

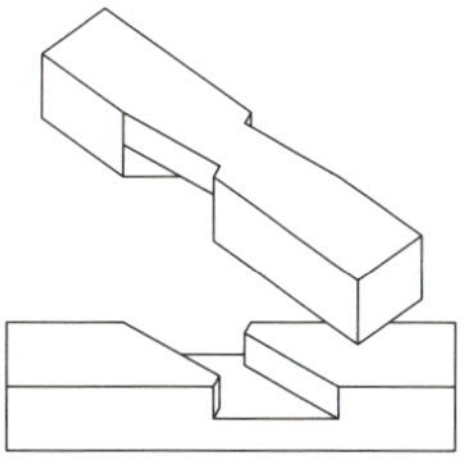

A+B

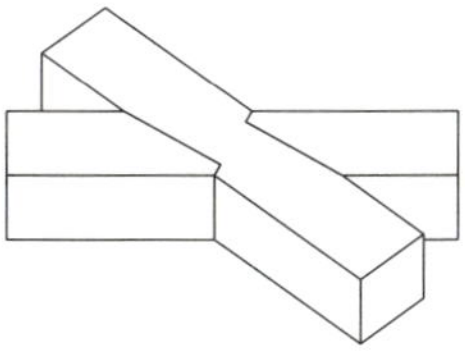

AB

Cross Lap Joints

Oblique notched heel cross lap joint

sujikai-eriwa-tsugi
筋交襟輪接ぎ

schräges Kreuzblatt
mit Fersenversatz

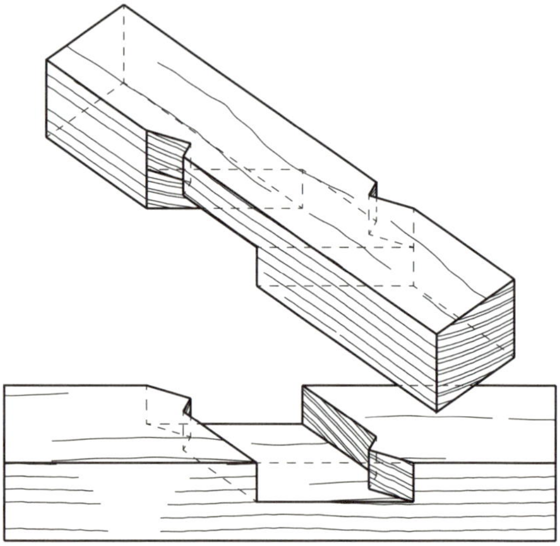

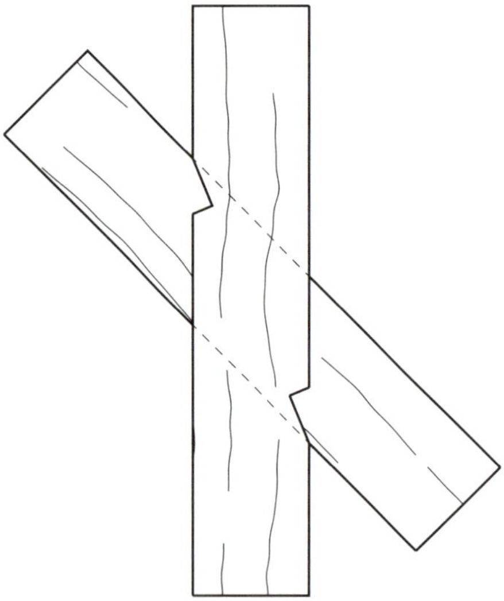

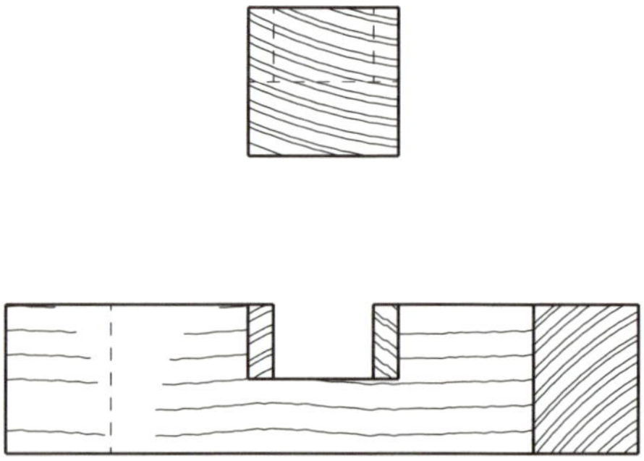

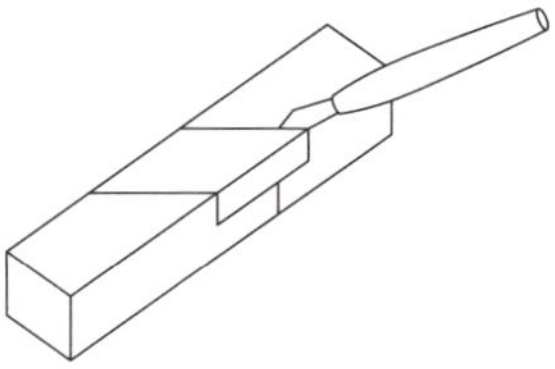
A1

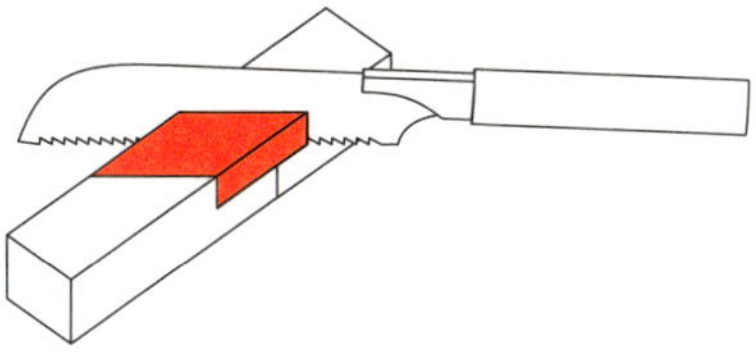
A2

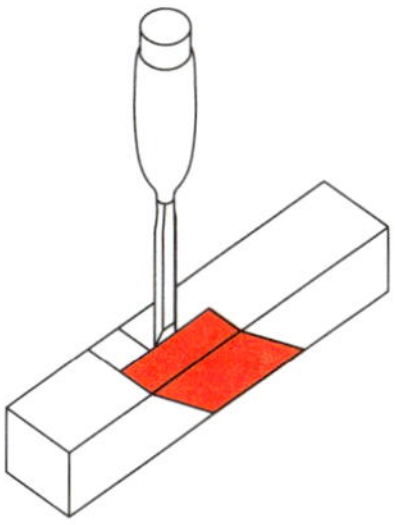
A3

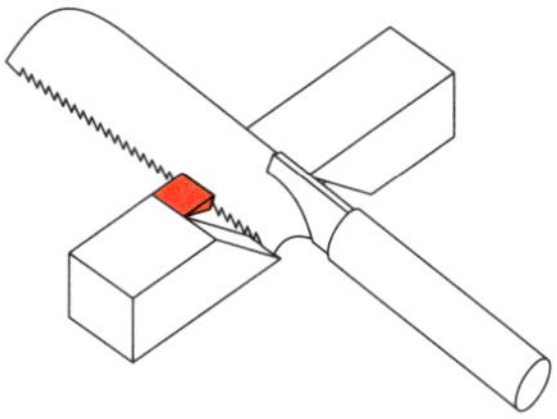
A4

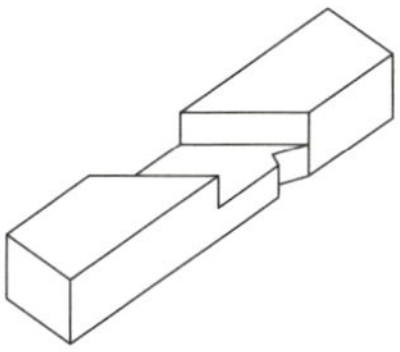
A

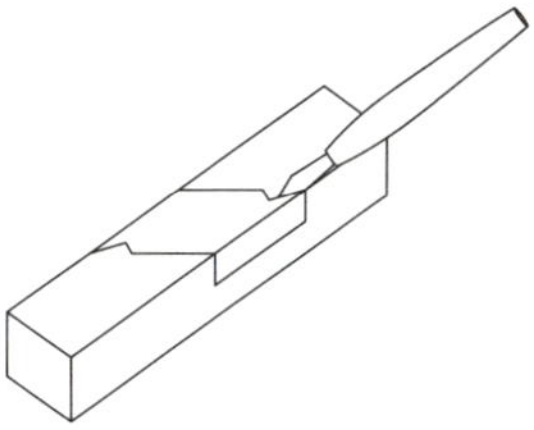
B1

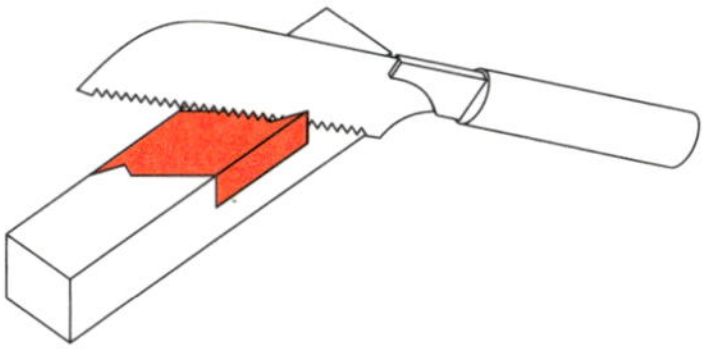
B2

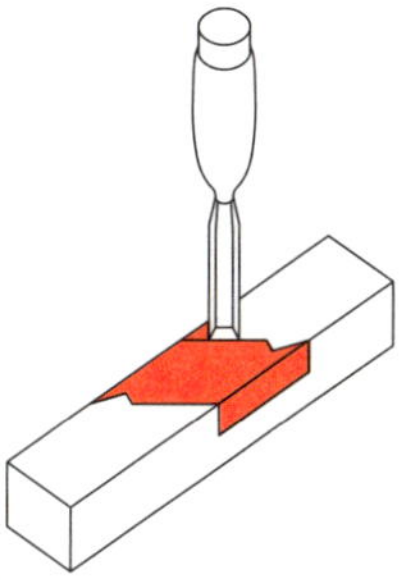
B3

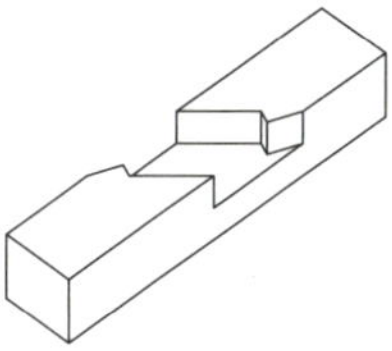
B

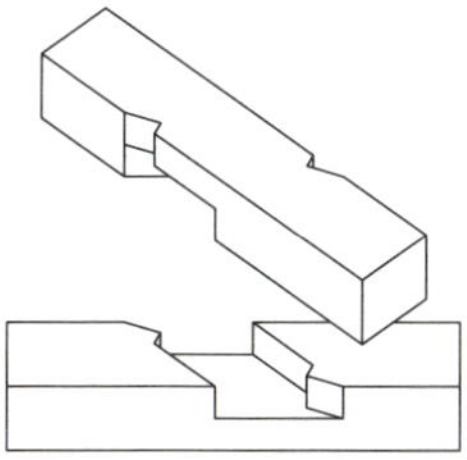
A+B

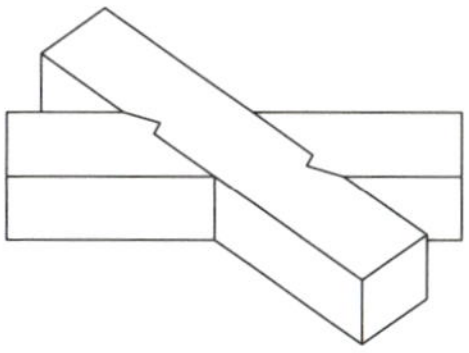
AB

Cross Lap Joints

Double notched oblique cross lap joint

sujikai-sogi-kaki-eriwa-tsugi
筋交殺欠襟輪接ぎ

schräges Kreuzblatt
mit Stirn- und Fersenversatz

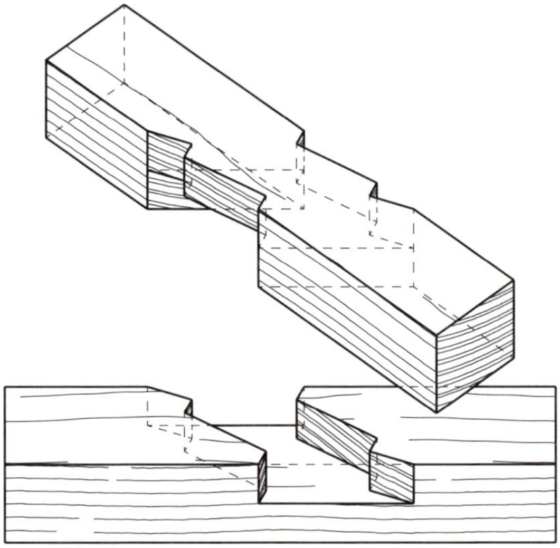

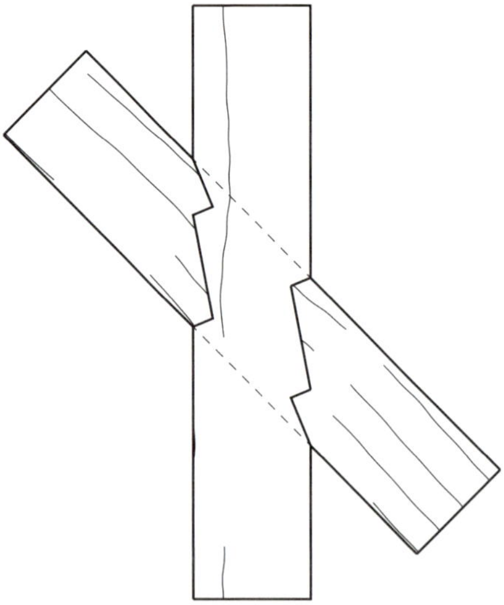

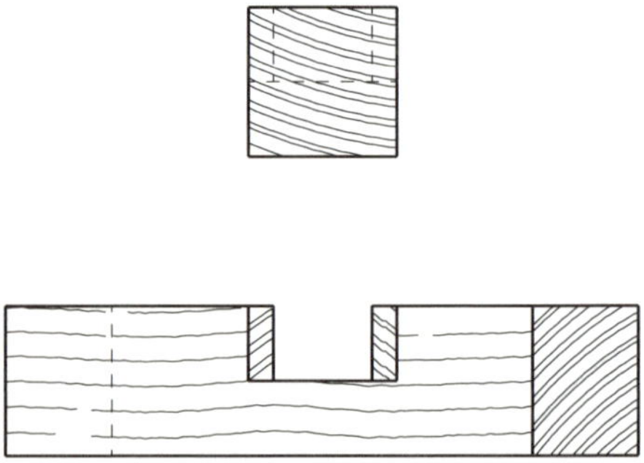

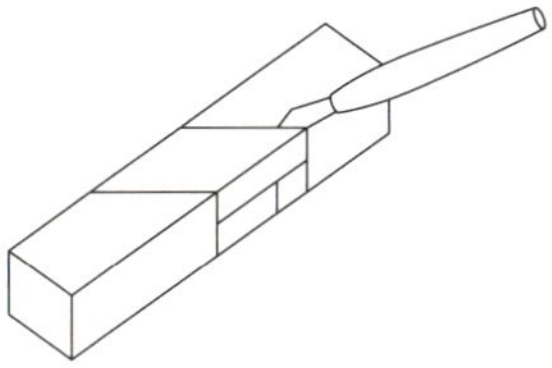

A1

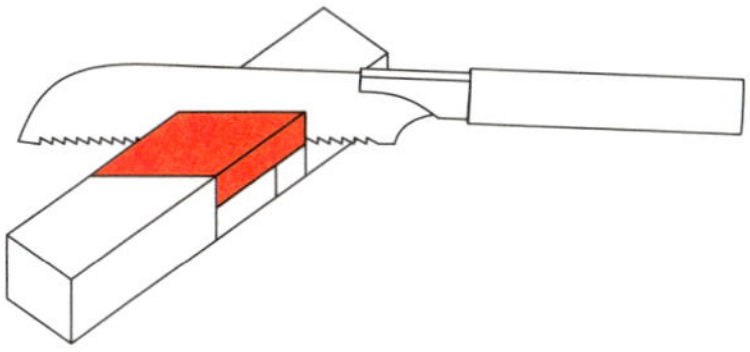

A2

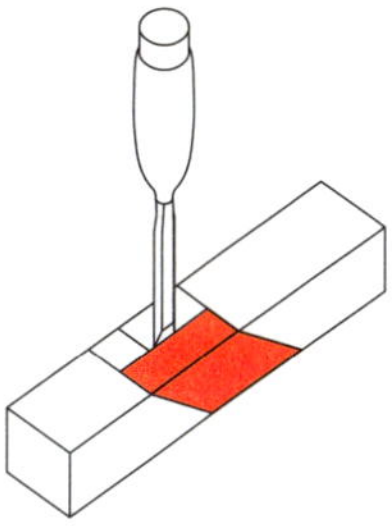

A3

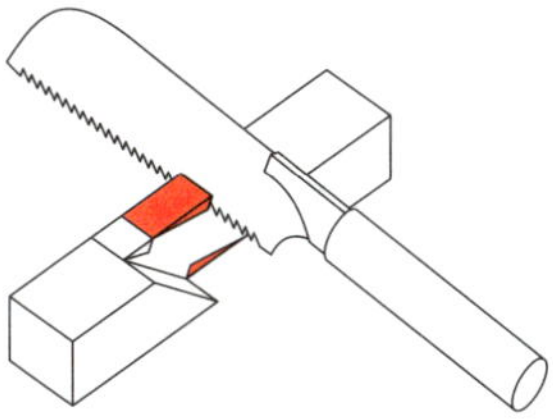

A4

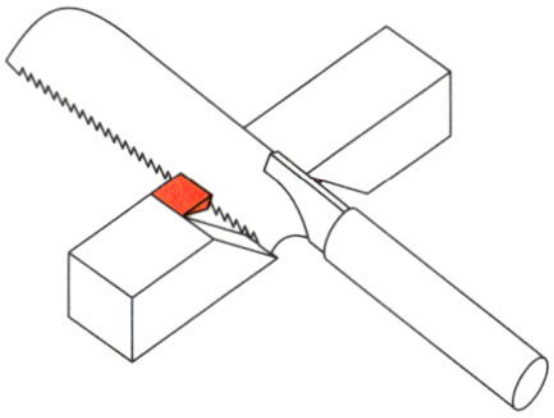

A5

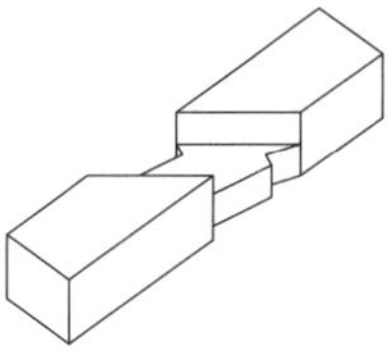

A

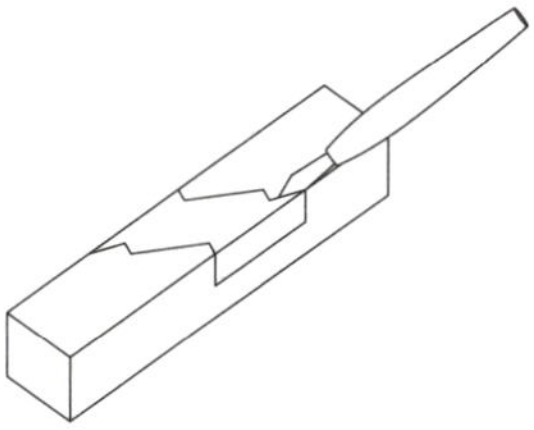

B1

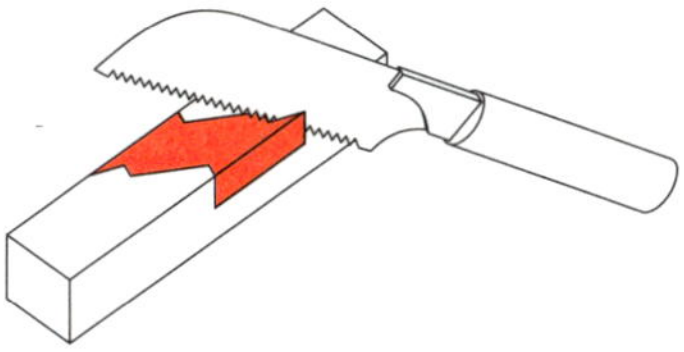

B2

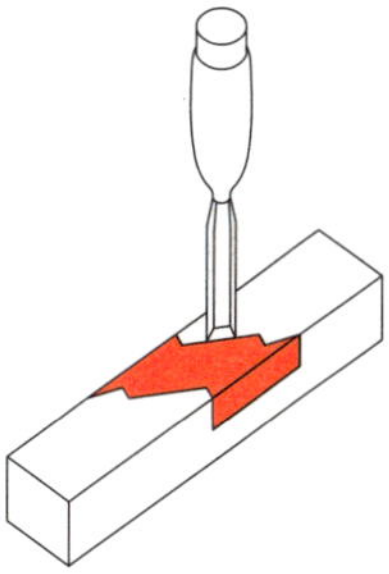

B3

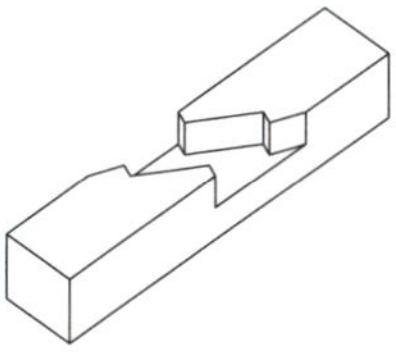

B

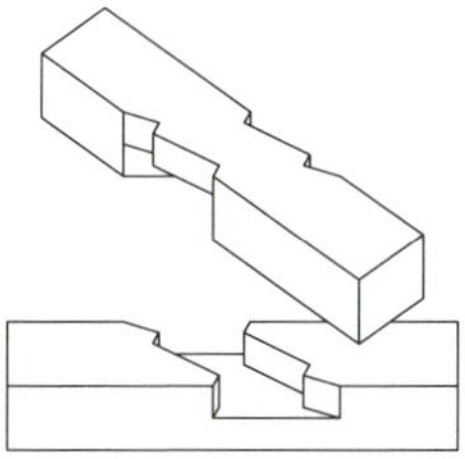

A+B

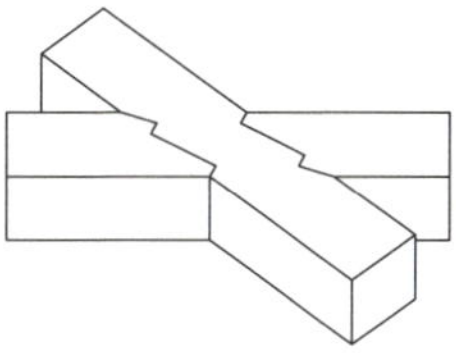

AB

Cross Lap Joints

Three-way cross lap joint

sujikai-ai-kaki-sanbon-tsugi
筋交相欠三本接ぎ

doppelte Überblattung

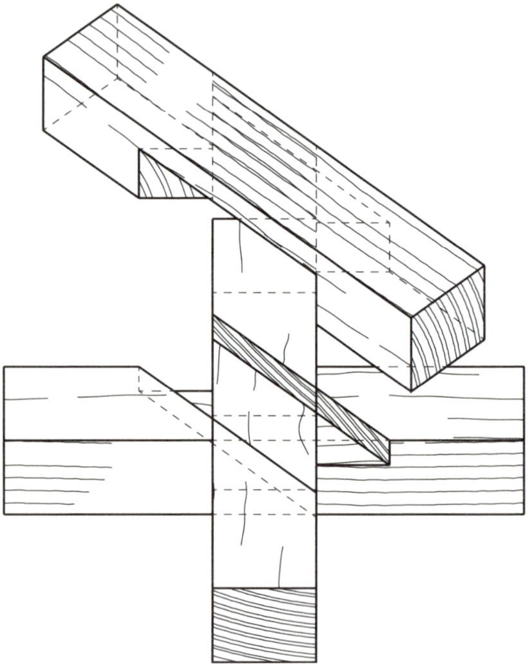

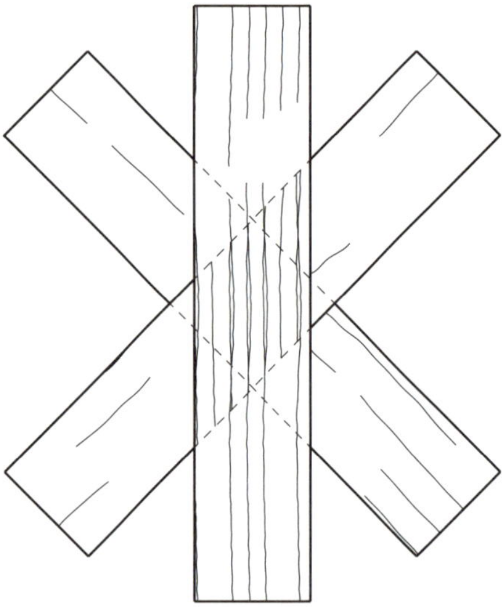

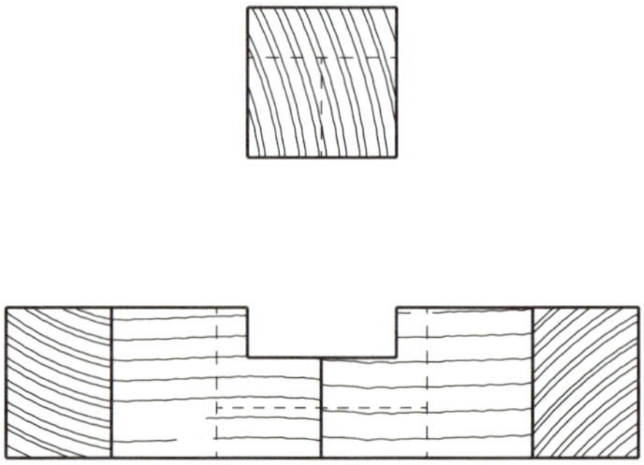

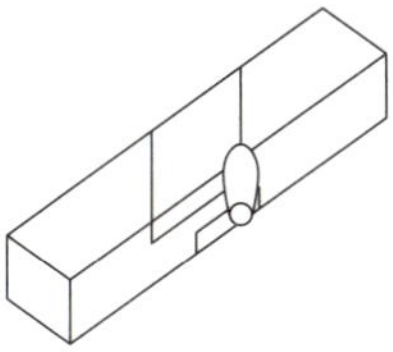

A1

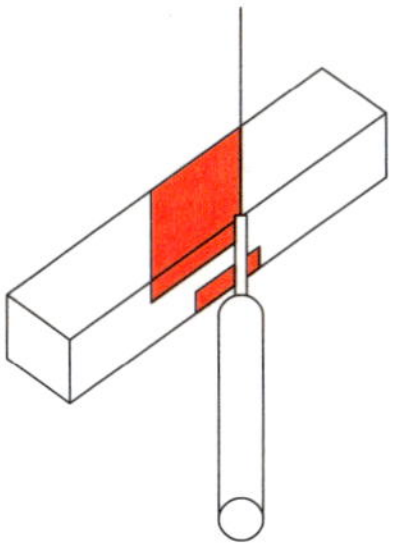

A2

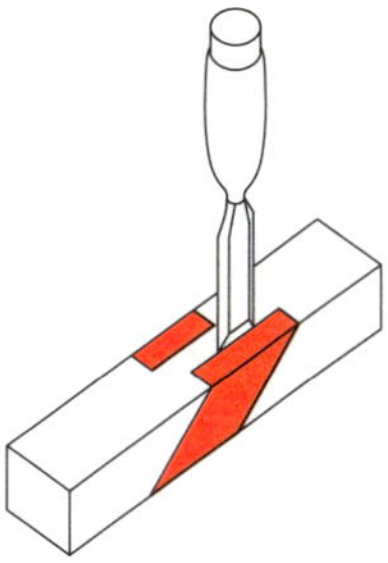

A3

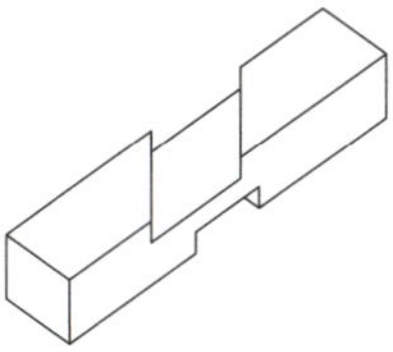

A

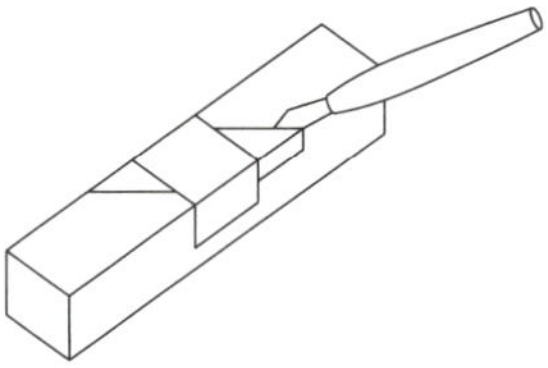

B1

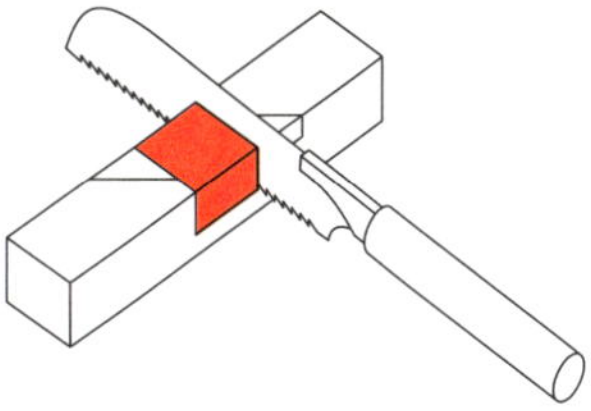

B2

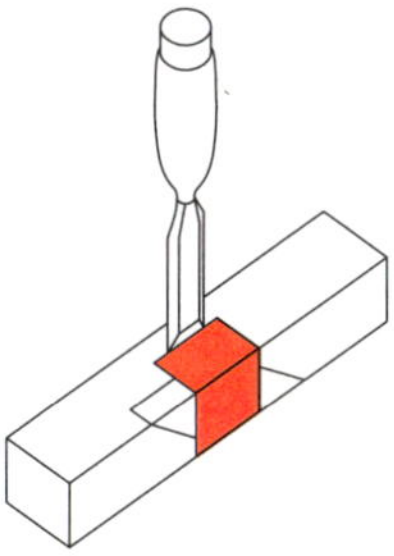

B3

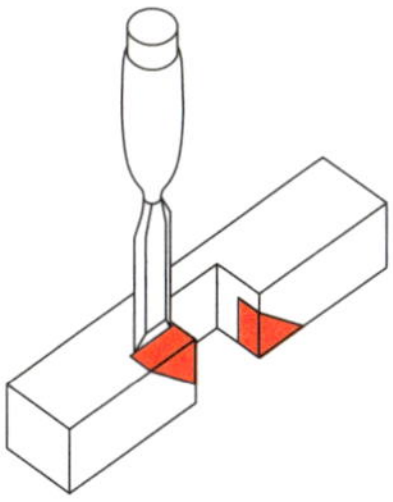

B4

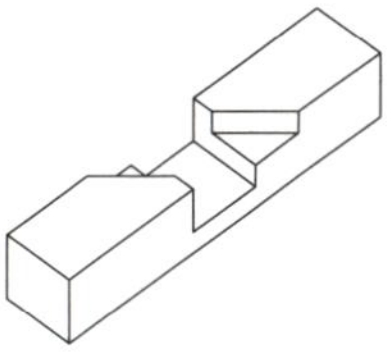

B

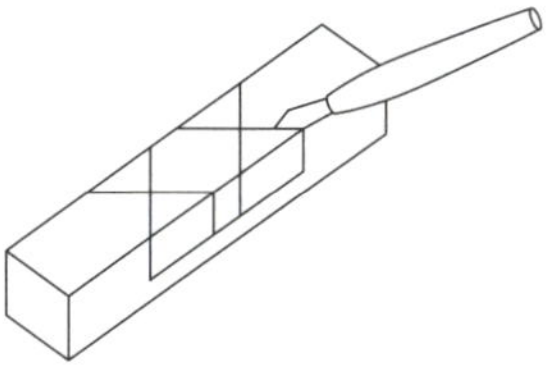

C1

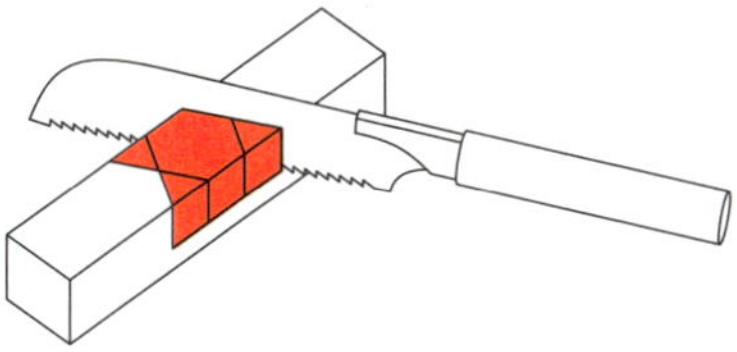

C2

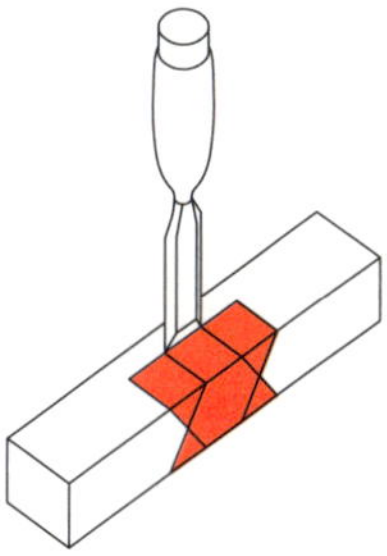

C3

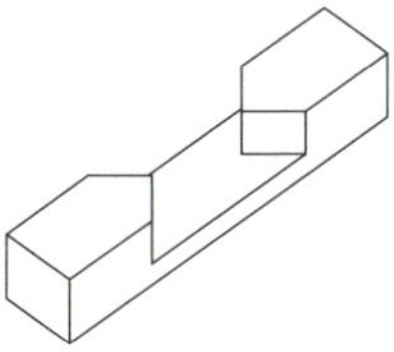

C

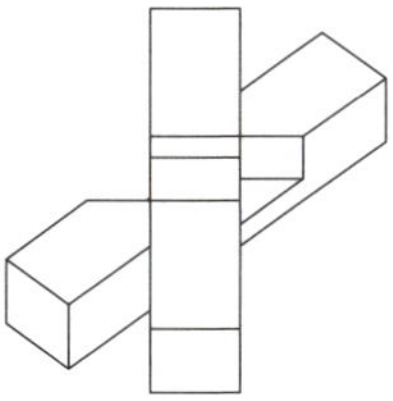

C+A

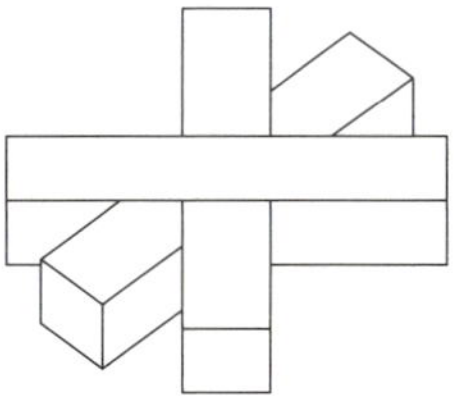

CAB

Cross Lap Joints

Joint with blind peg

dabo-kumi [*]
太枘組み

Überkreuzung mit Dolle

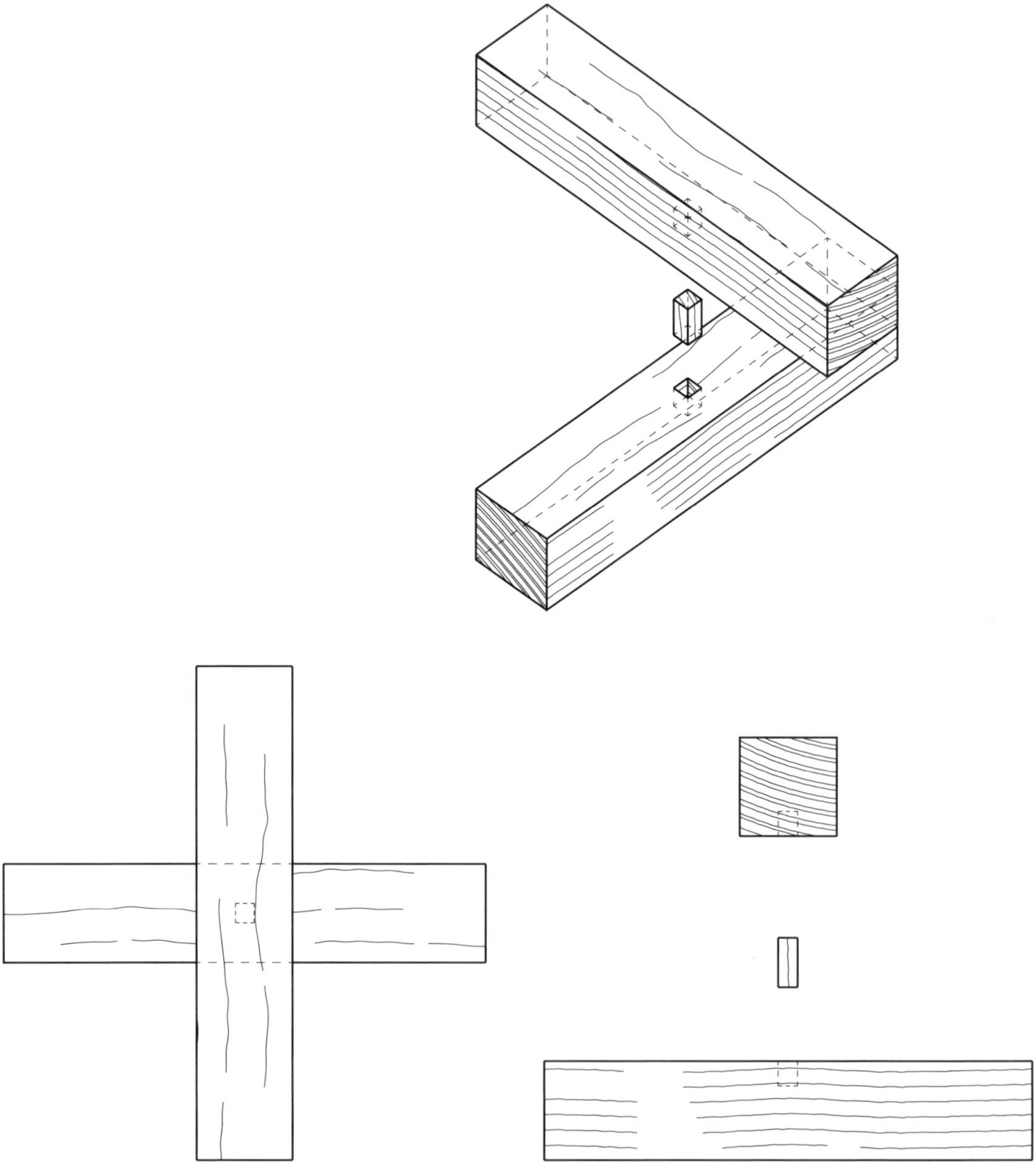

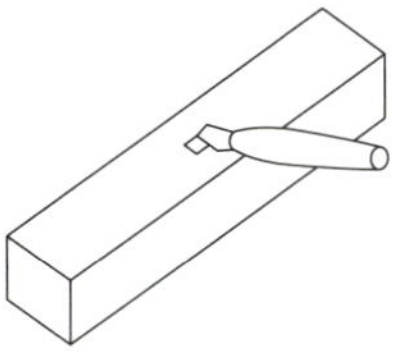
A1

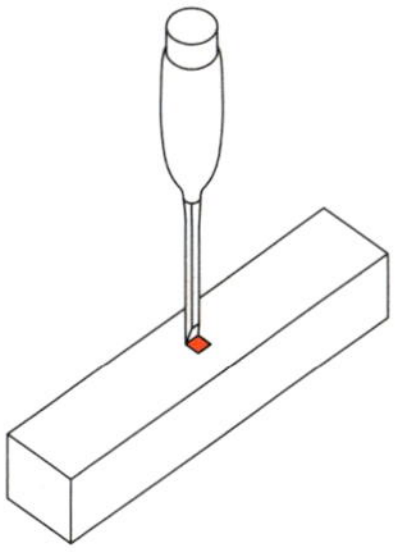
A2

2×

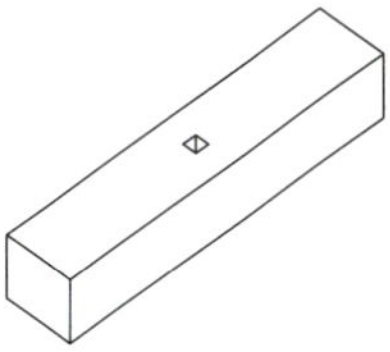
A

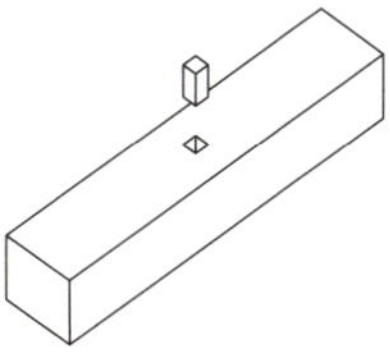
A+B

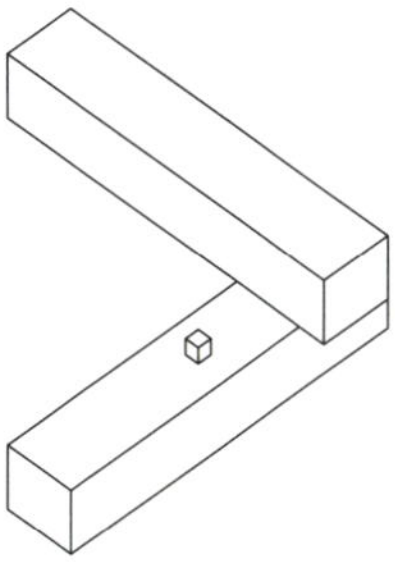
A+B+A

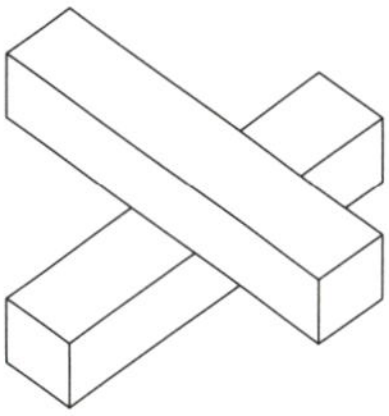
ABA

Cross Lap Joints

Notch joint

eriwa-kake
襟輪掛け

Kreuzkamm mit
einseitigem Sitz

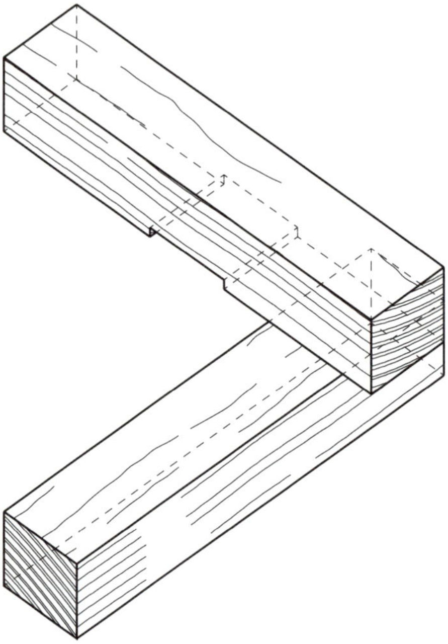

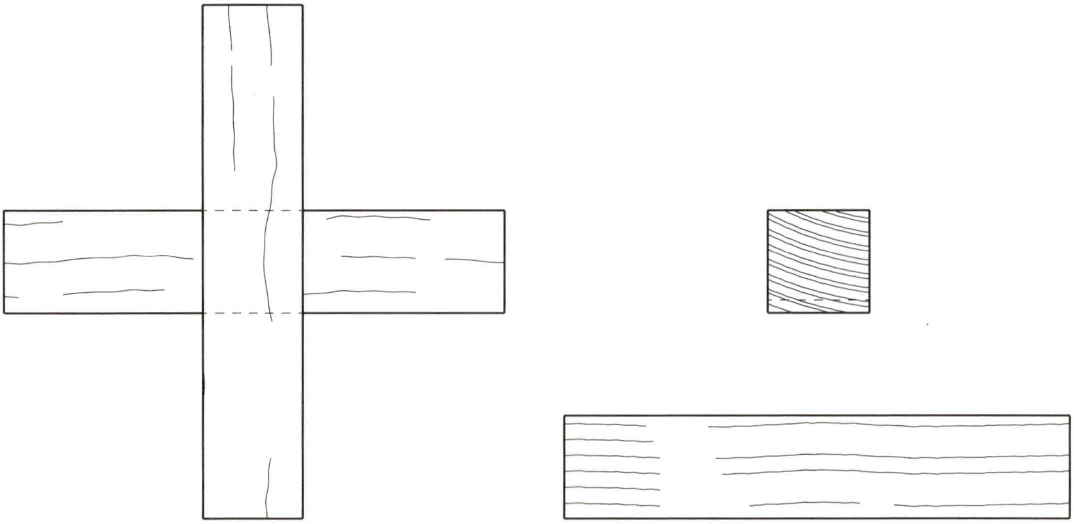

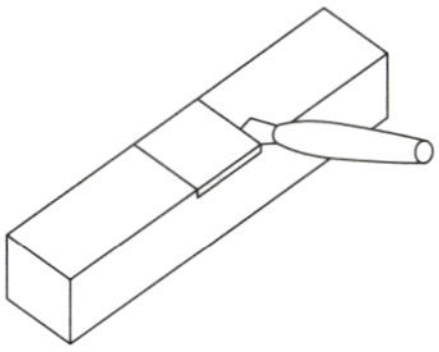

A1

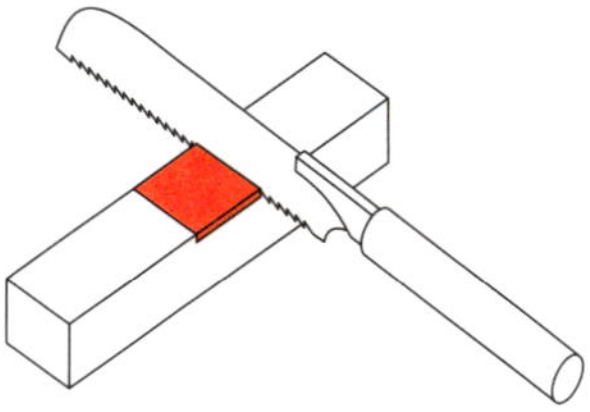

A2

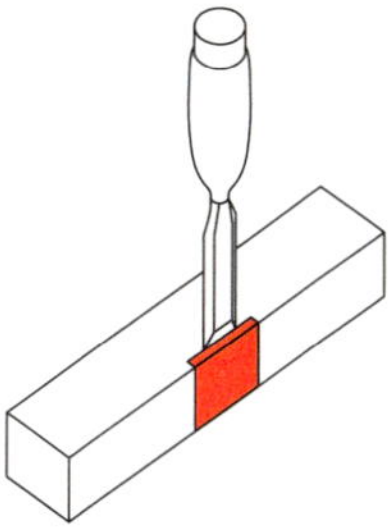

A3

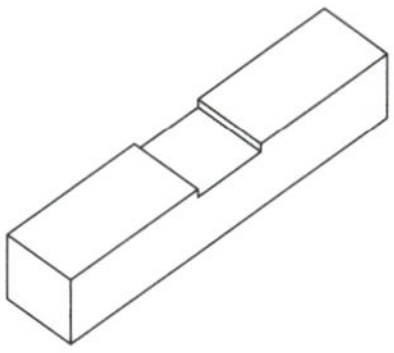

A

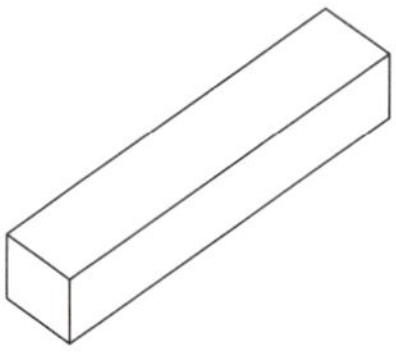

B

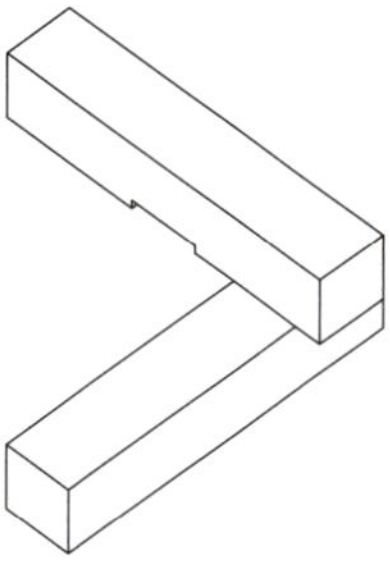

A+B

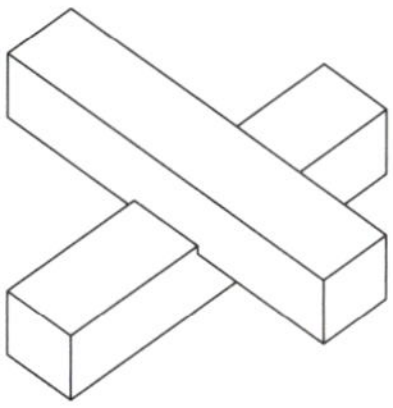

AB

Double notch joint

ai-kaki-kumi
相欠組み

Kreuzkamm

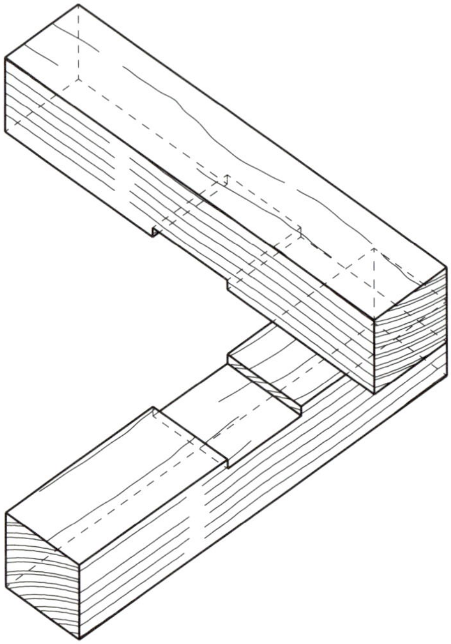

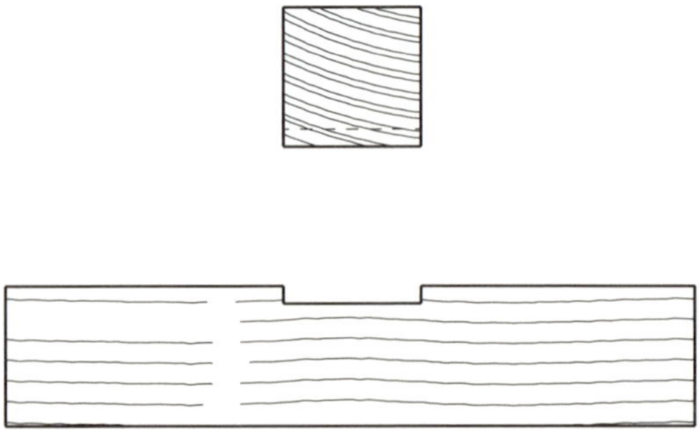

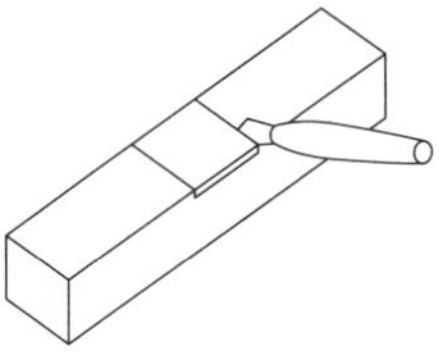

A1

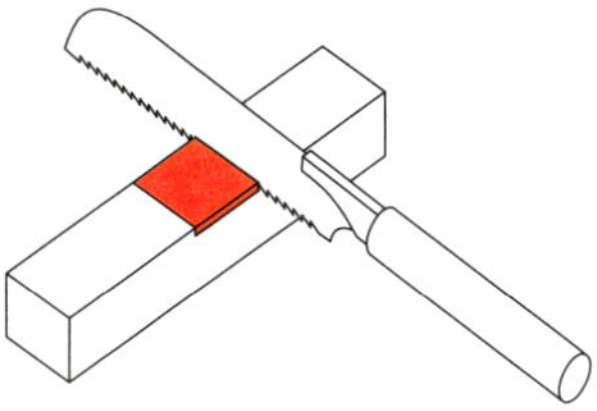

A2

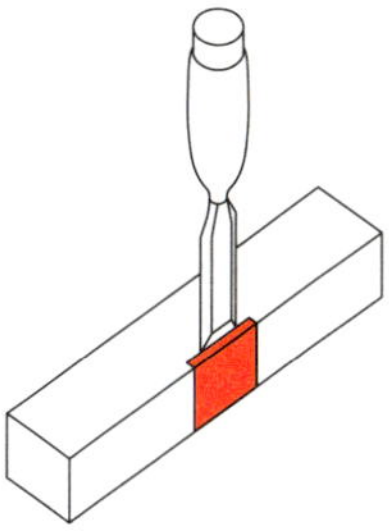

A3

2×

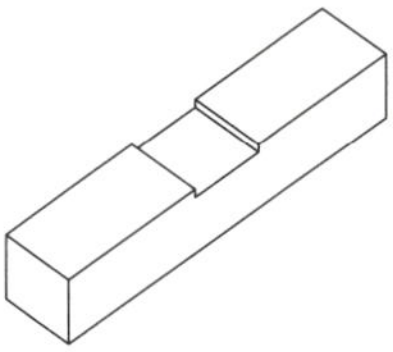

A

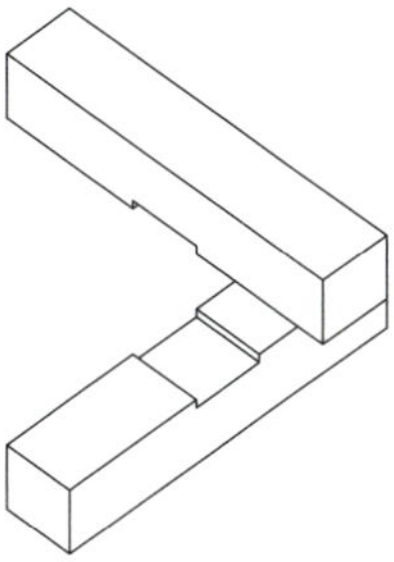

A+A

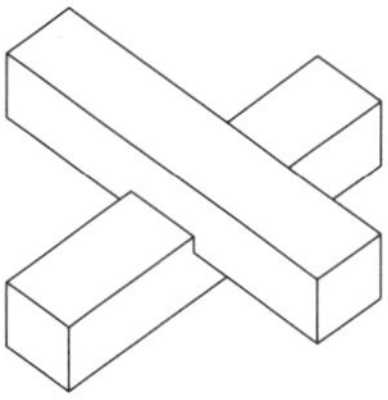

AA

Cross Lap Joints

Double notch joint with blind peg

ai-kaki-kumi (dabo-sen-ire) [*]
相欠組み（太枘栓入）

Kreuzkamm mit Dolle

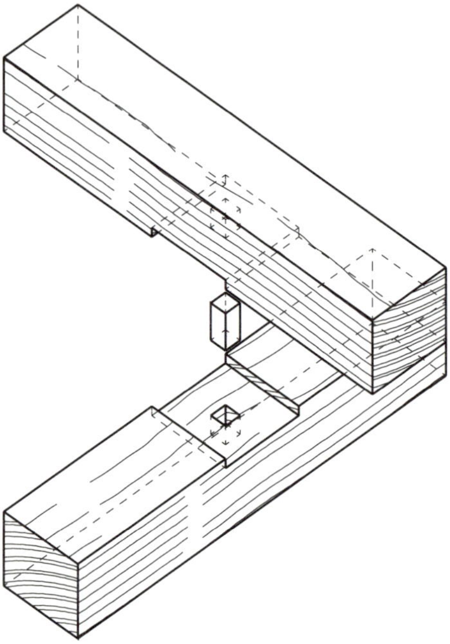

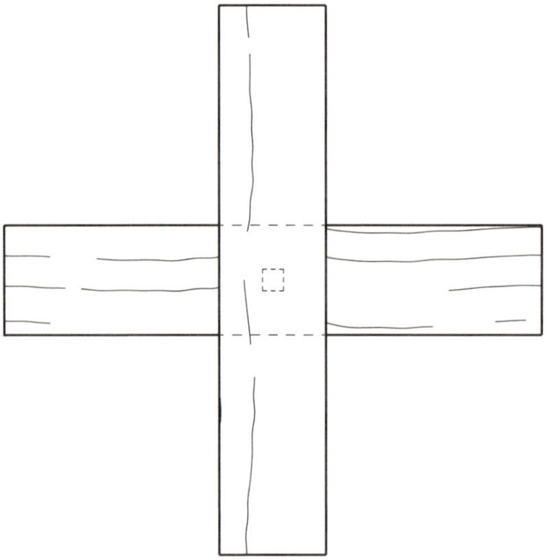

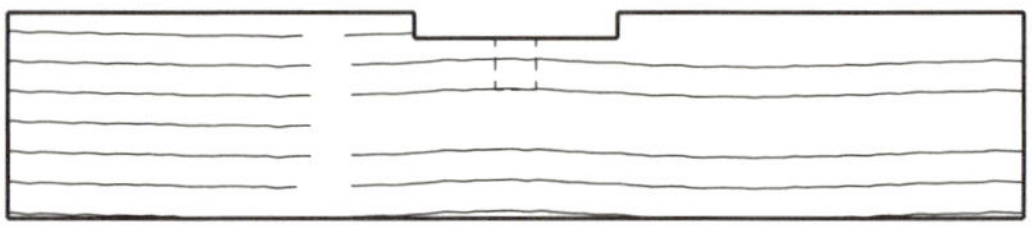

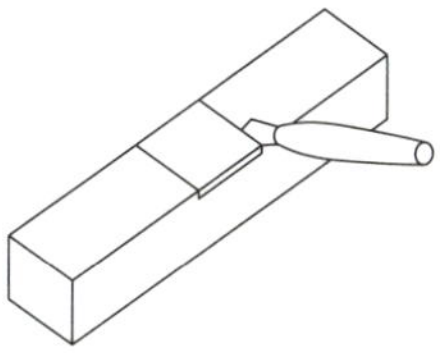

A1

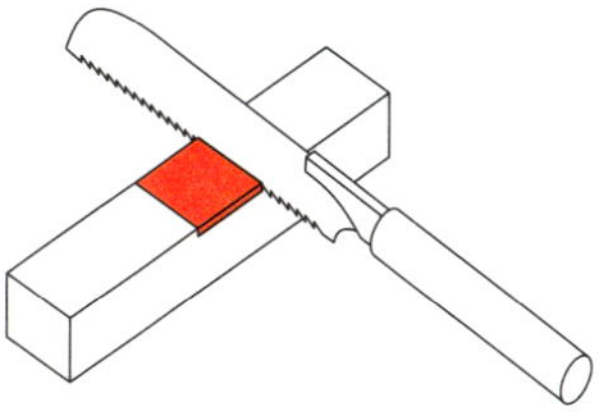

A2

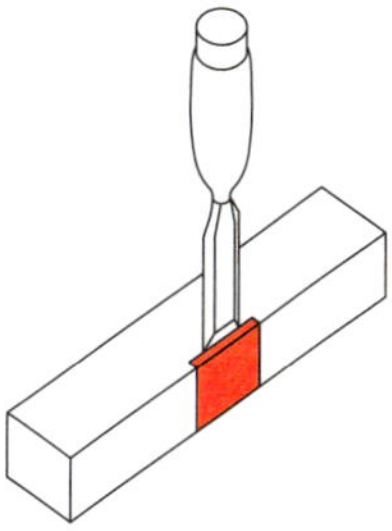

A3

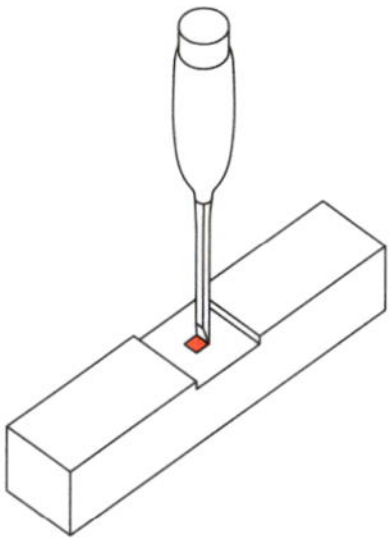

A4

2×

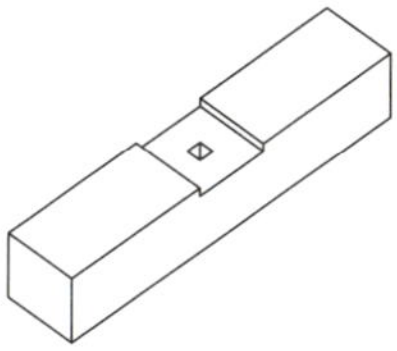

A

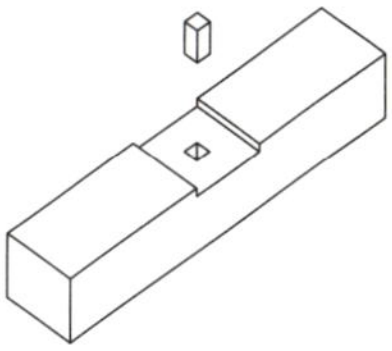

A+B

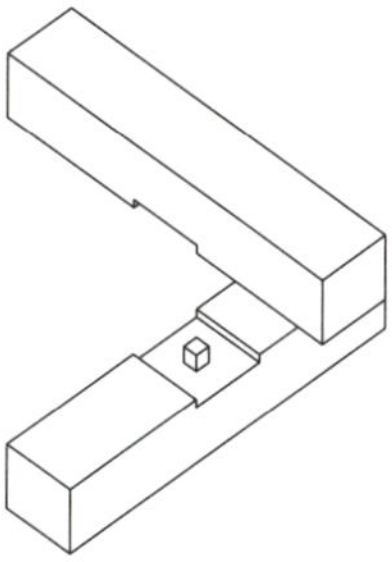

A+B+A

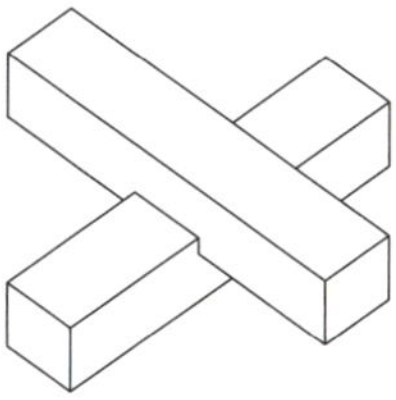

ABA

Cross Lap Joints

Double notch joint with stub tenons

sodegiri-ai-kaki-tsugi
袖切相欠接ぎ

Kreuzkamm mit
abgesetzter Falz

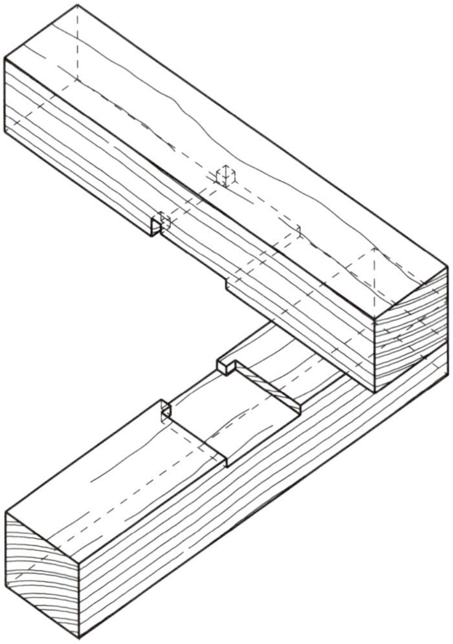

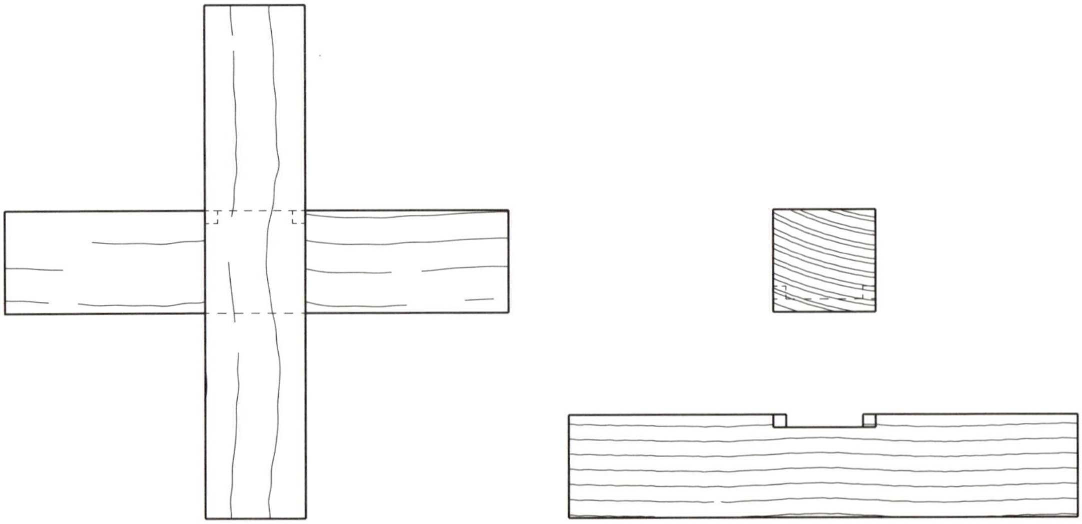

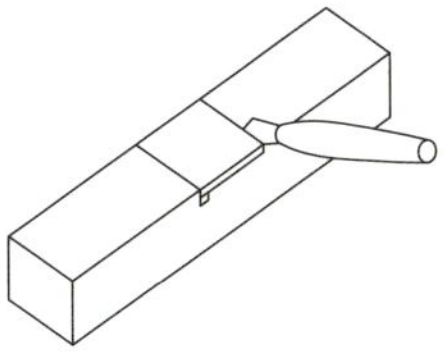
A1

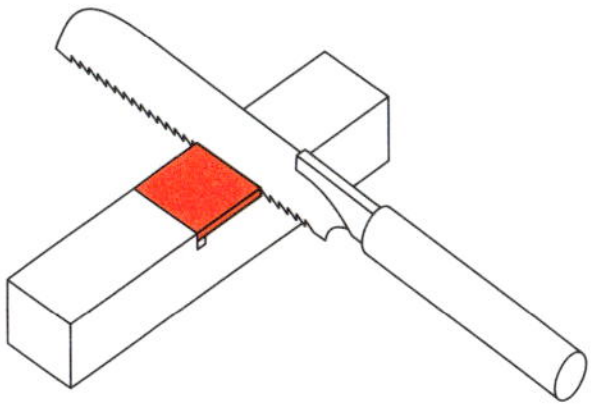
A2

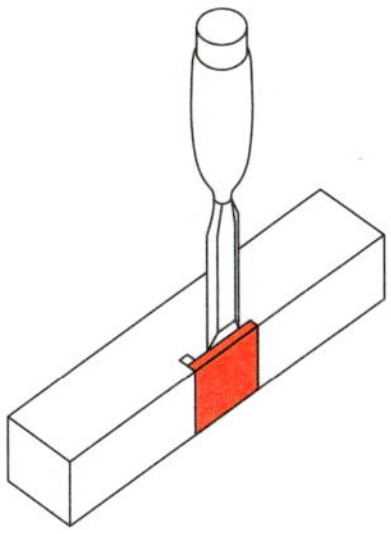
A3

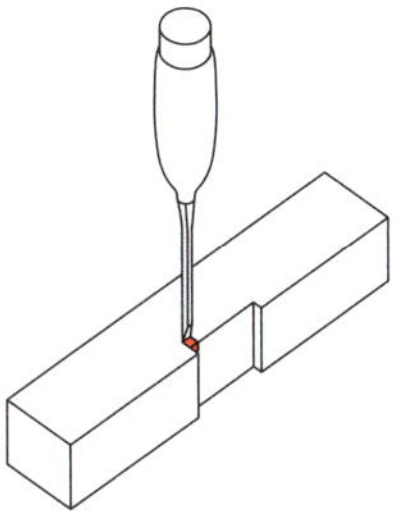
A4

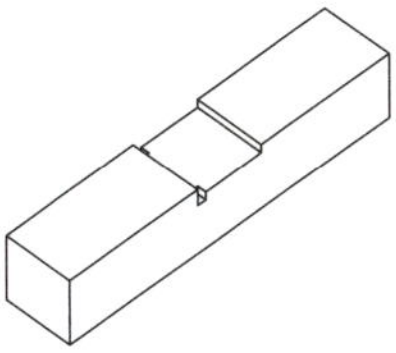
A

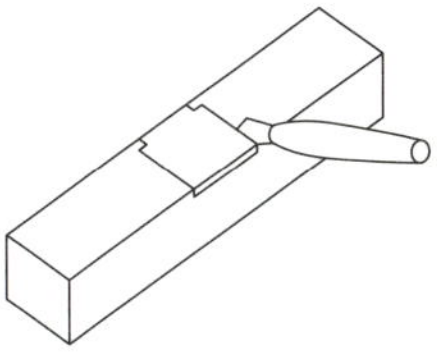
B1

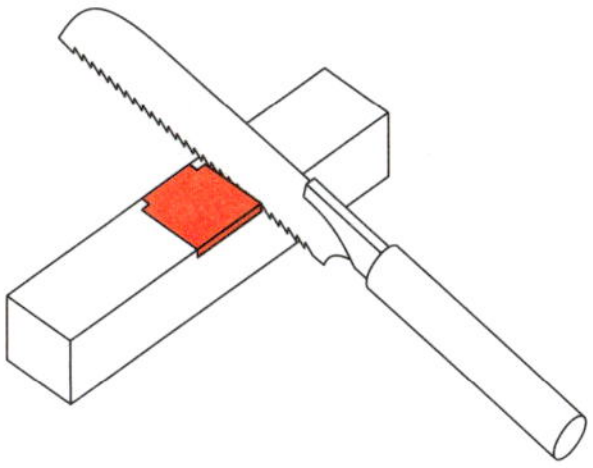
B2

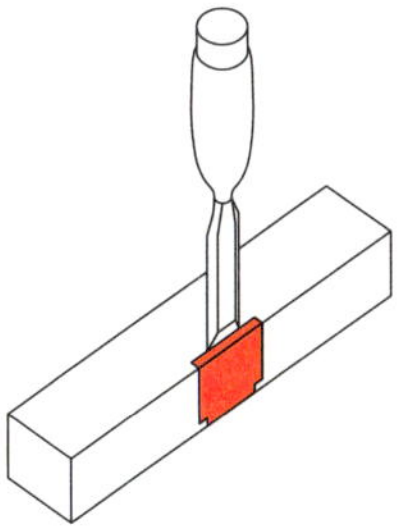
B3

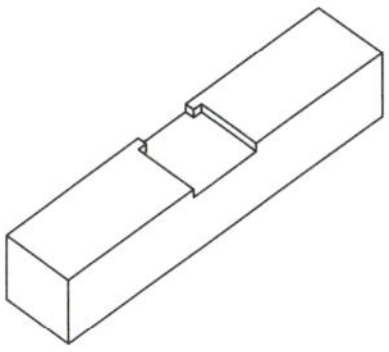
B

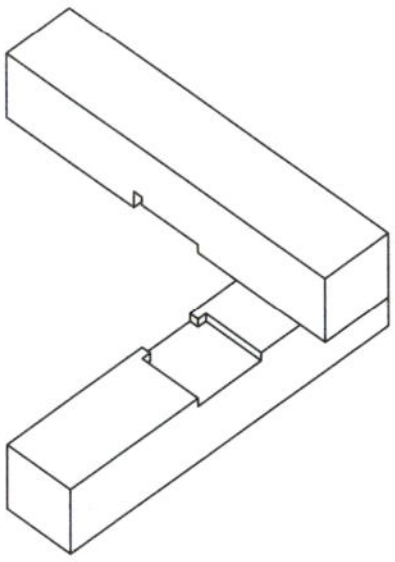
A+B

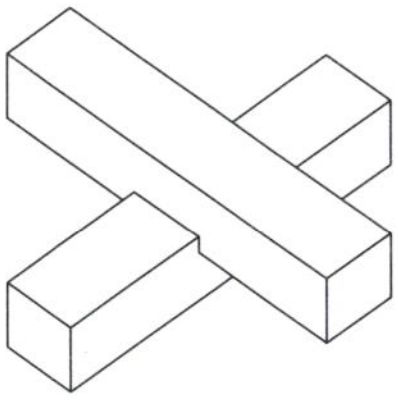
AB

Cross Lap Joints

Oblique double notch joint

sogi-ai-kaki-tsugi
殺相欠接ぎ

Kreuzkamm als Versatz

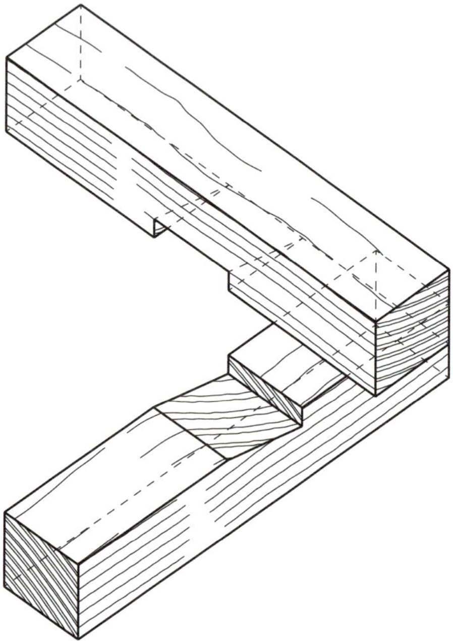

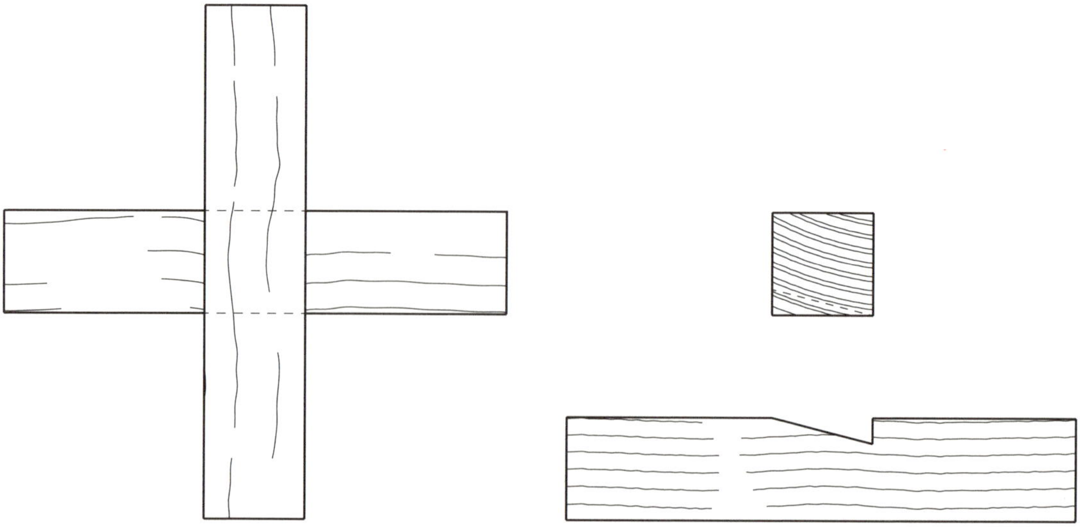

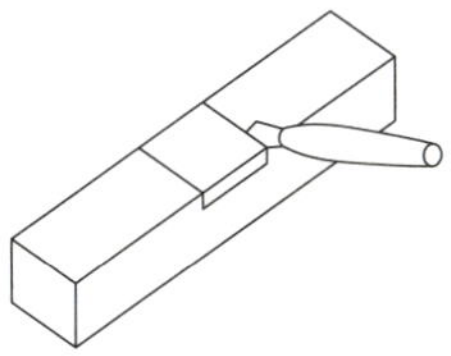
A1

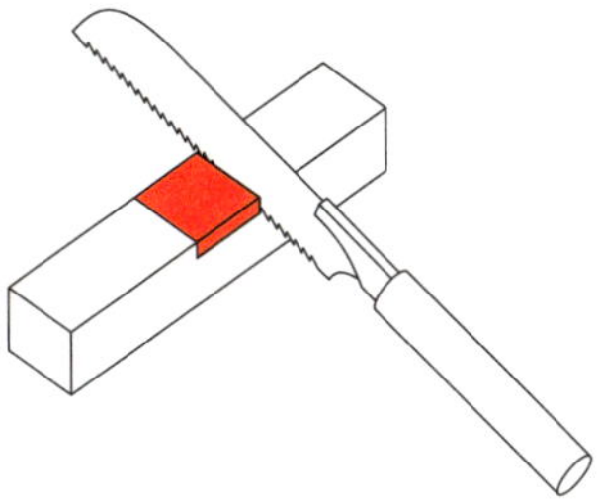
A2

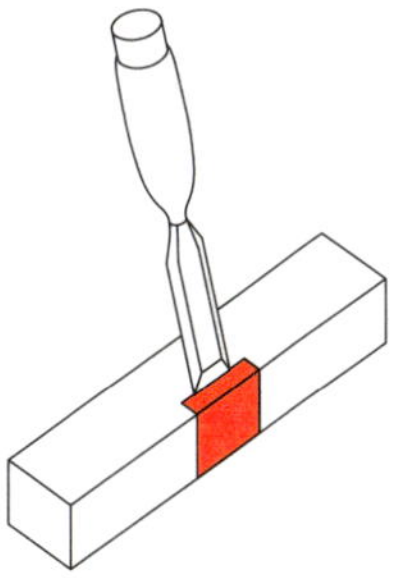
A3

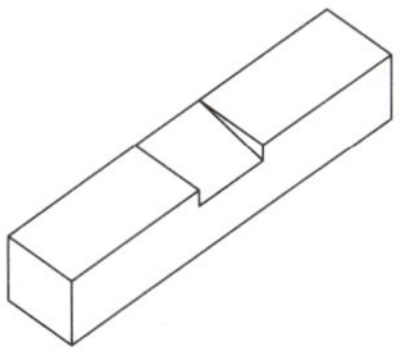
A

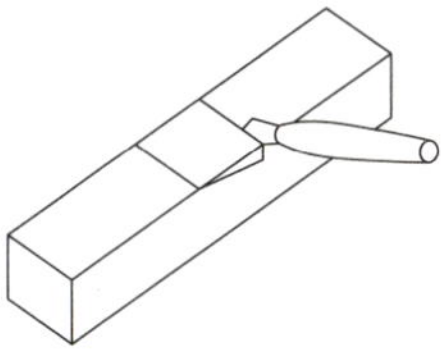
B1

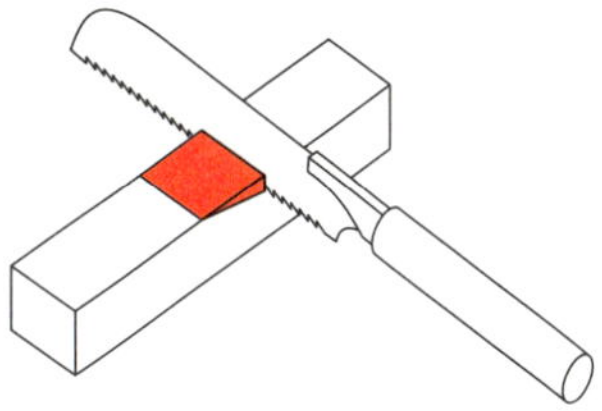
B2

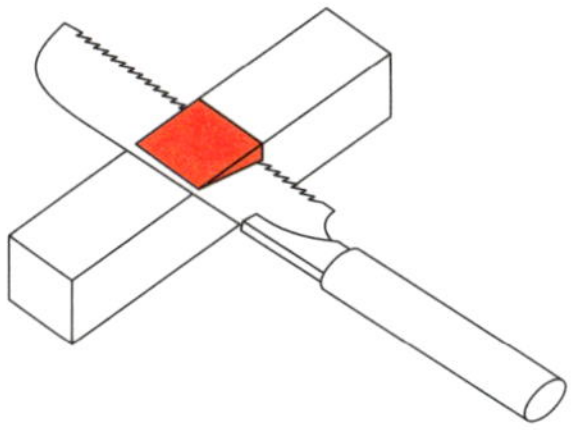
B3

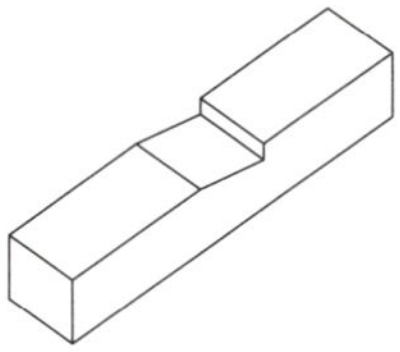
B

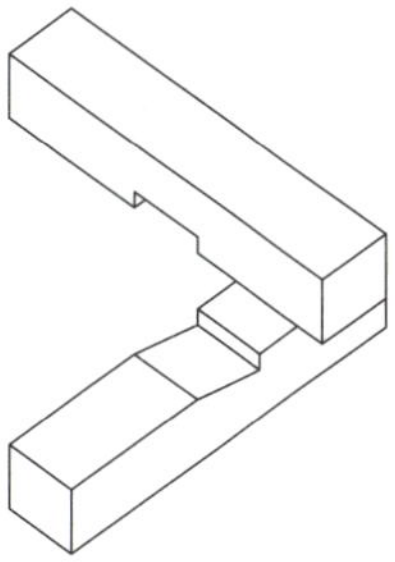
A+B

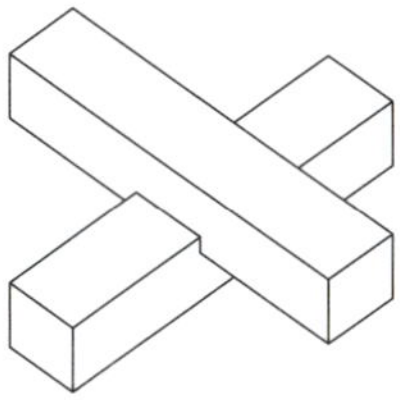
AB

Cross Lap Joints

Cog joint

ai-kaki-kumi (dan-kaki-watari-ago)
相欠組み(段欠渡腮)

Kreuzkamm mit Stufe

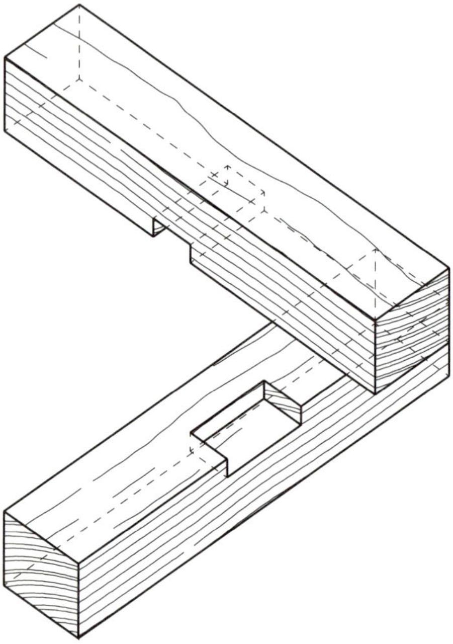

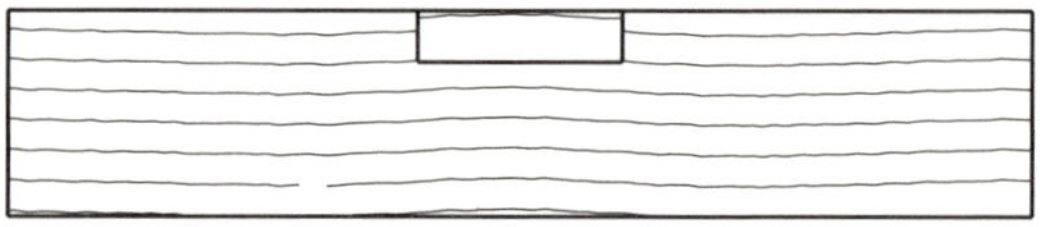

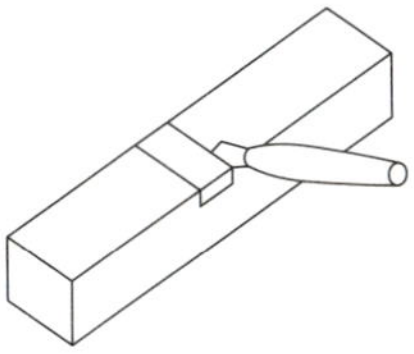

A1

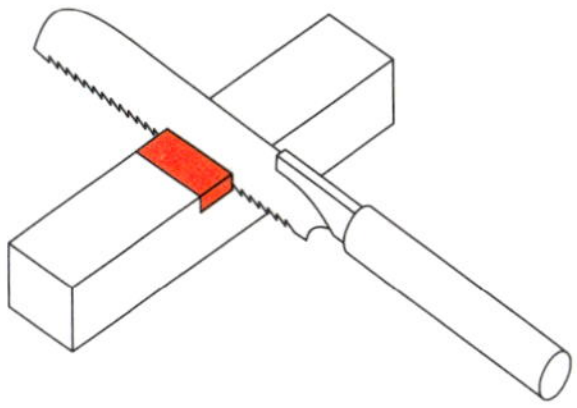

A2

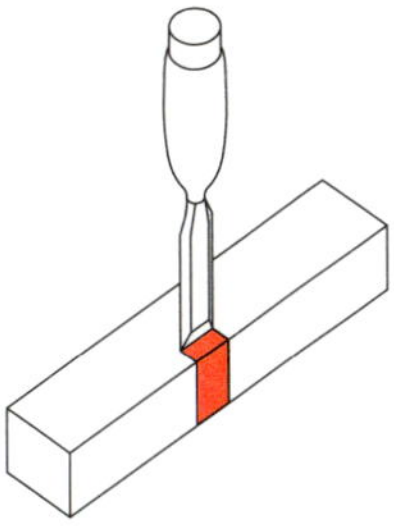

A3

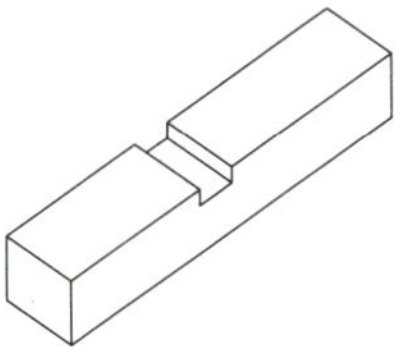

A

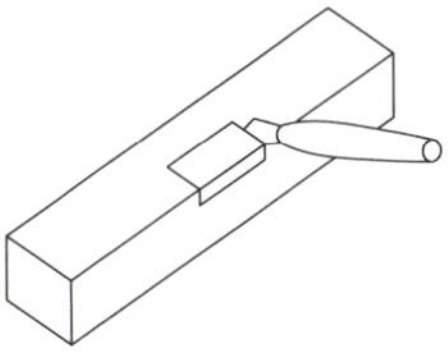

B1

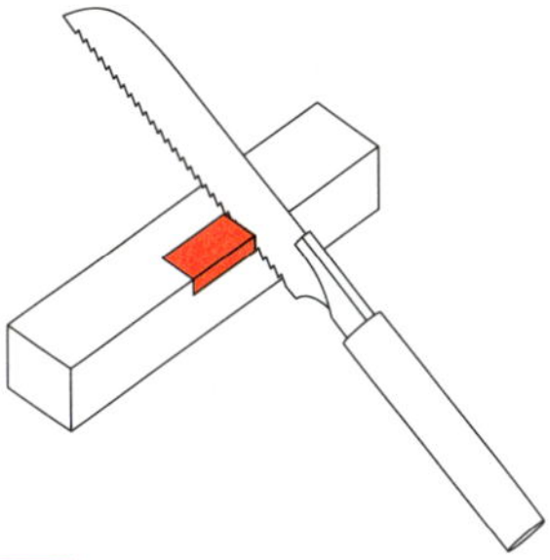

B2

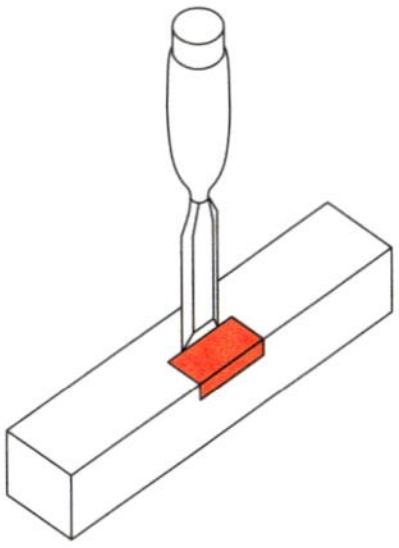

B3

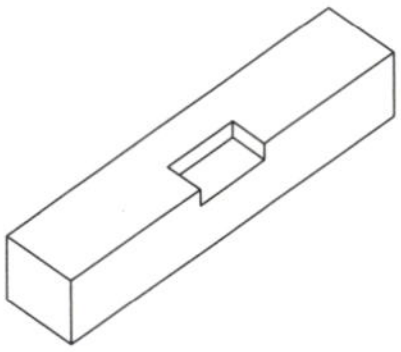

B

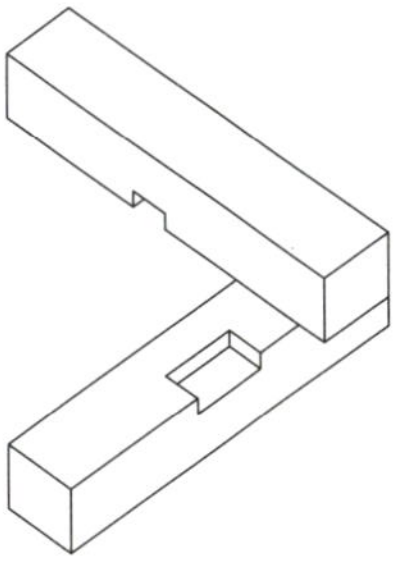

A+B

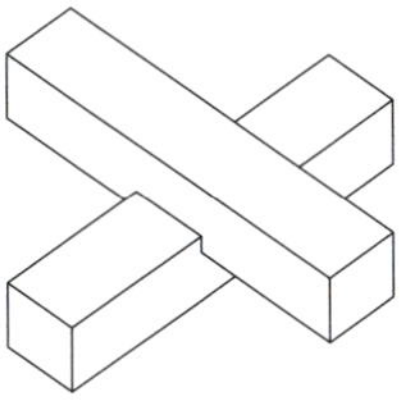

AB

Cross Lap Joints

Double cog joint

ai-kaki-kumi (watari-ago)
相欠組み(渡腮)

Kreuzkamm mit mittiger Stufe

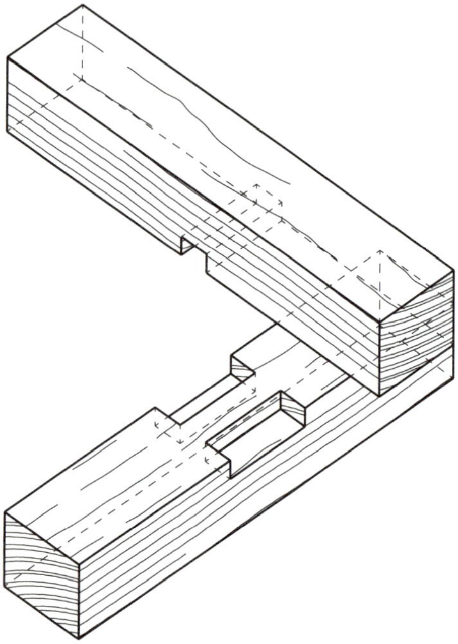

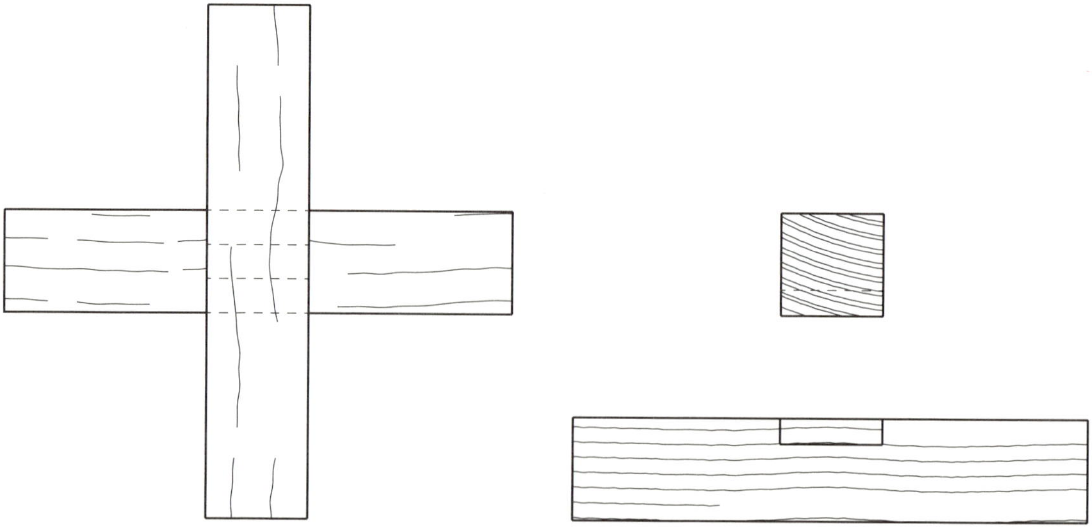

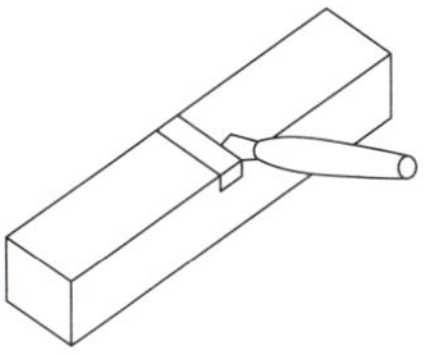

A1

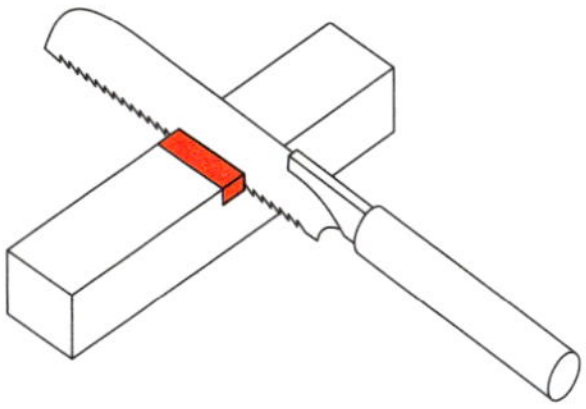

A2

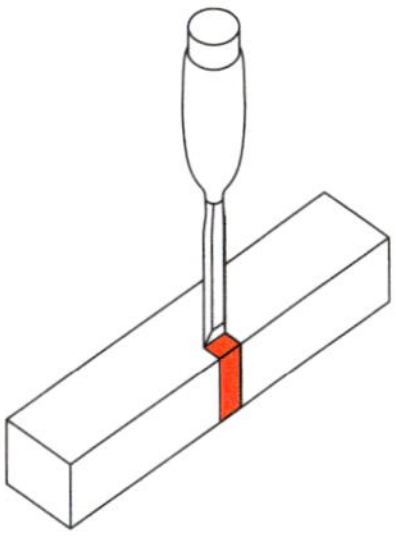

A3

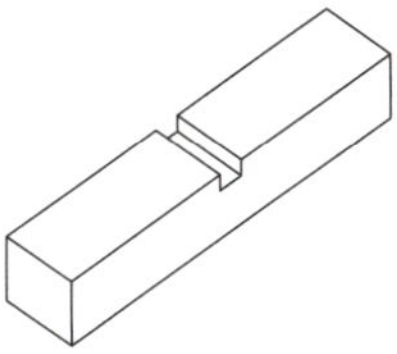

A

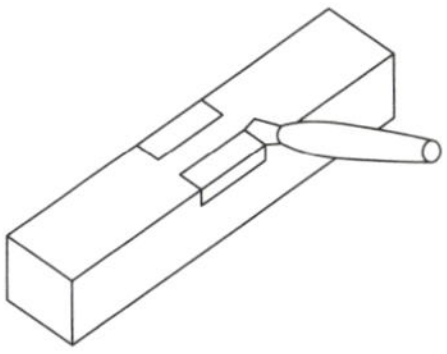

B1

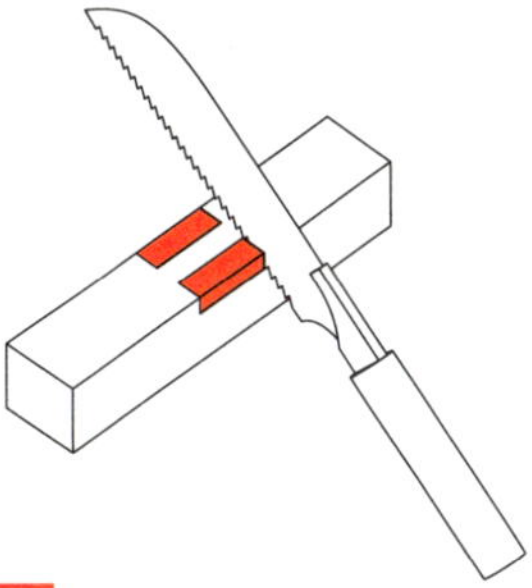

B2

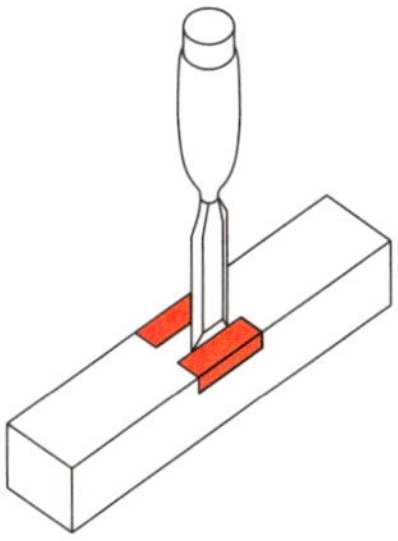

B3

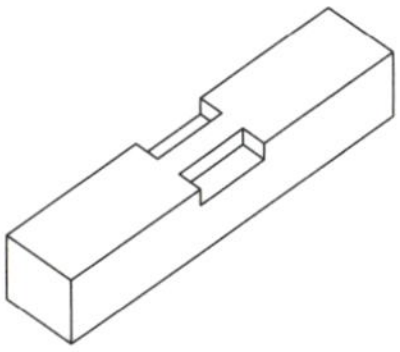

B

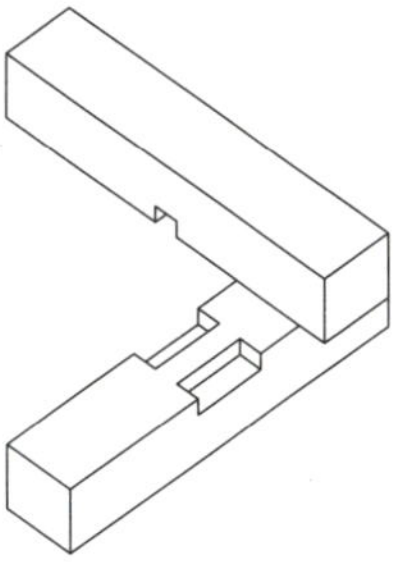

A+B

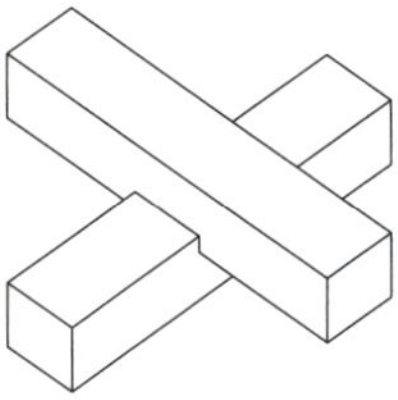

AB

Cross Lap Joints

Sloped cog joint

ai-kaki-kumi (kata-ari-kaki-watari-ago)
相欠組み（片蟻欠渡腮）

Kreuzkamm mit
einseitig schräger Stufe

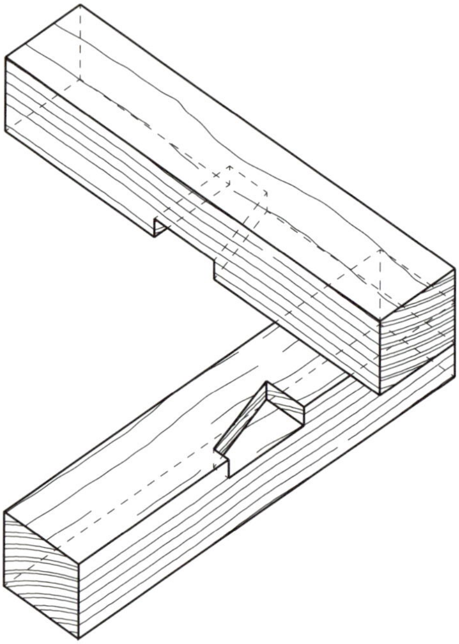

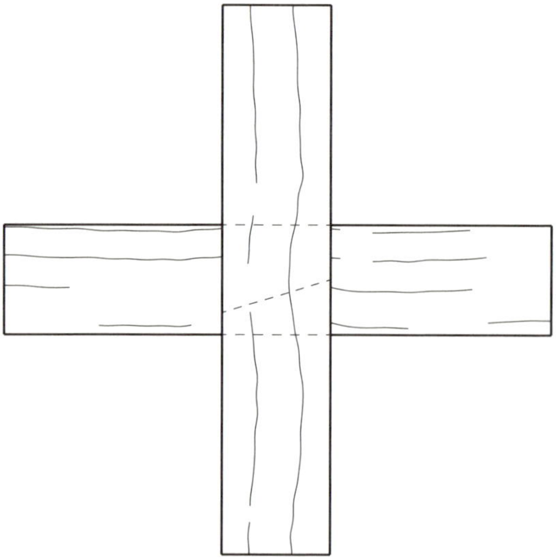

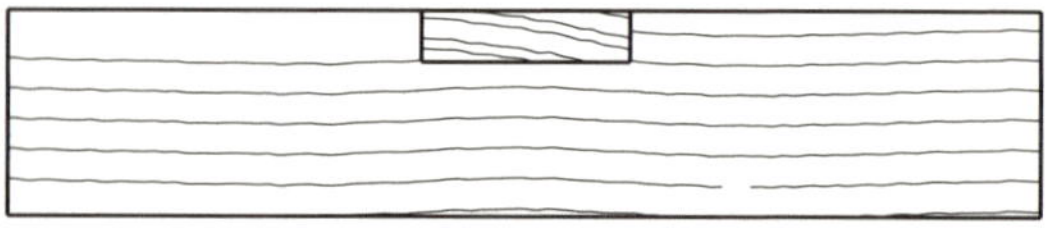

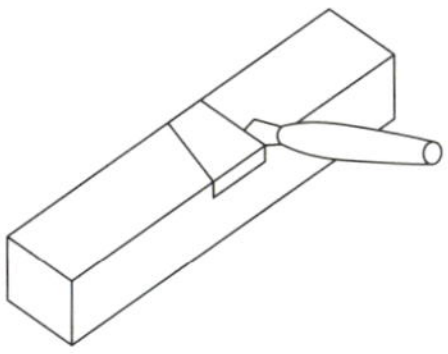

A1

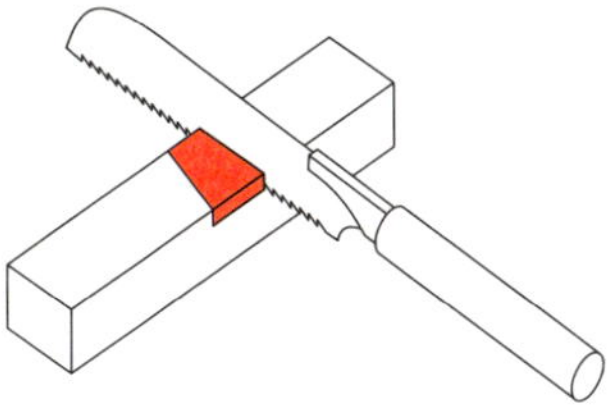

A2

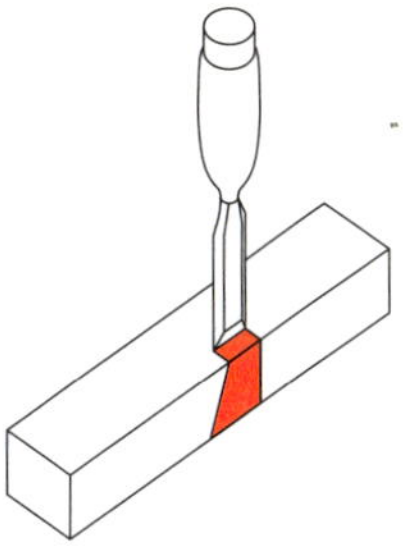

A3

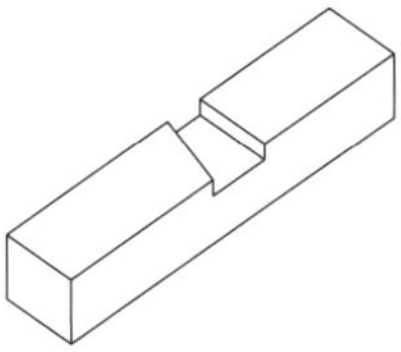

A

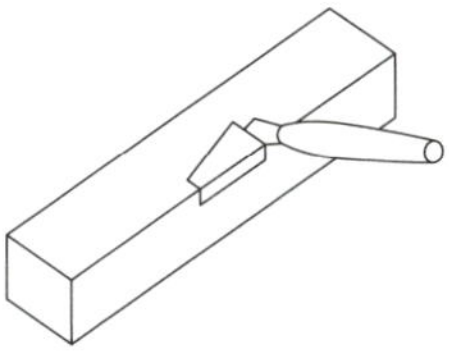

B1

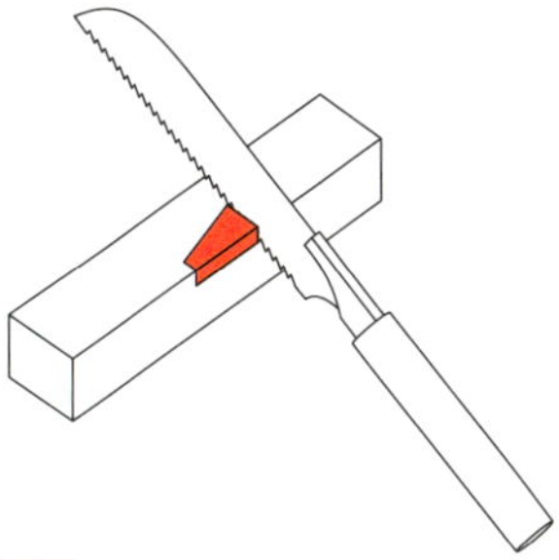

B2

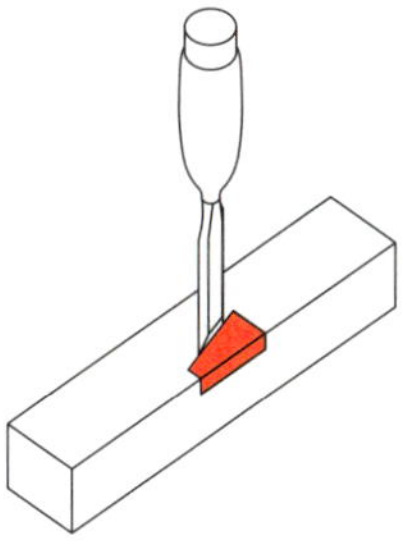

B3

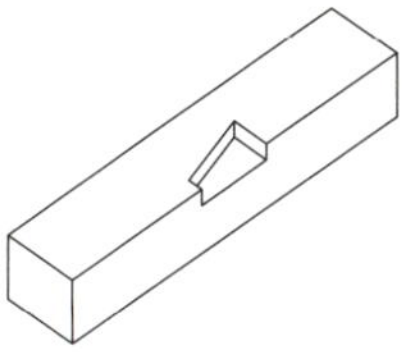

B

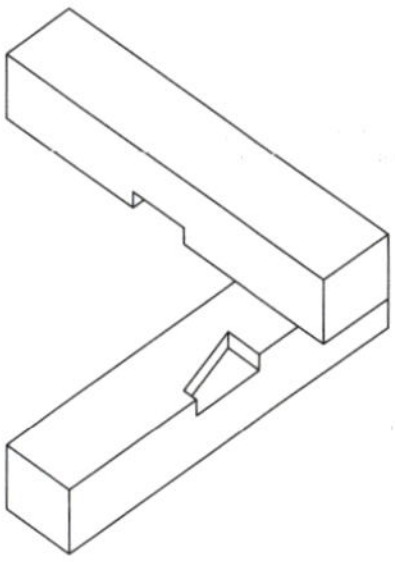

A+B

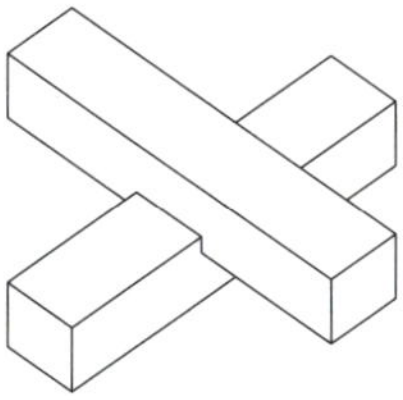

AB

Cross Lap Joints

Sloped double cog joint

ai-kaki-kumi (ōari-kaki-watari-ago)
相欠組み（大蟻欠渡腮）

Kreuzkamm mit
mittiger beidseitig schräger Stufe

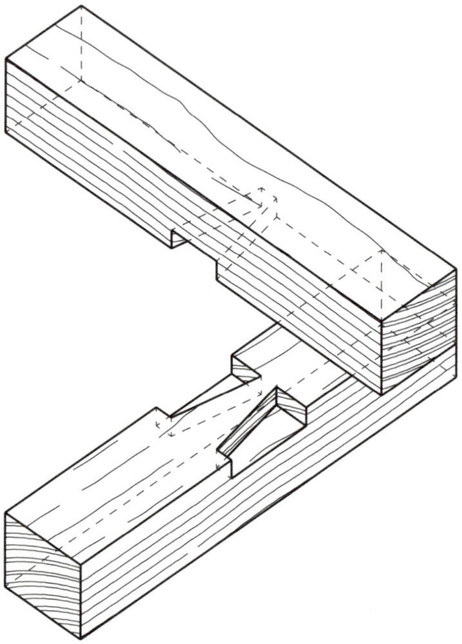

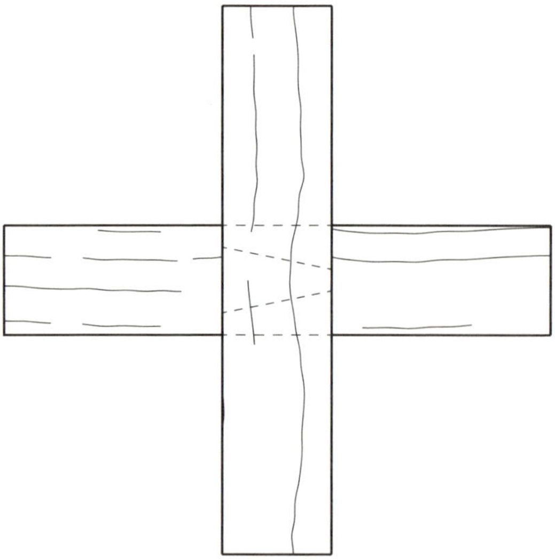

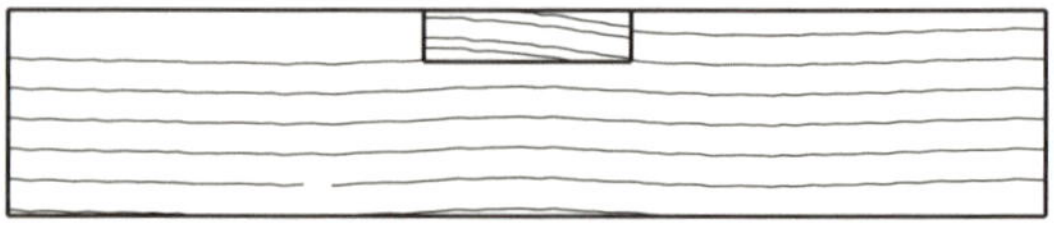

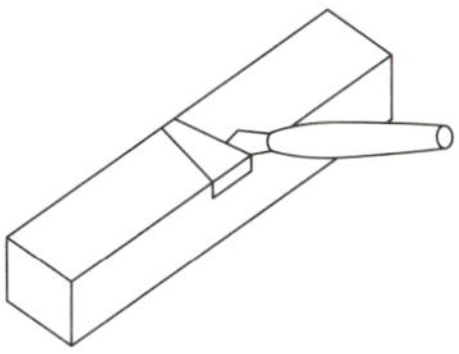

A1

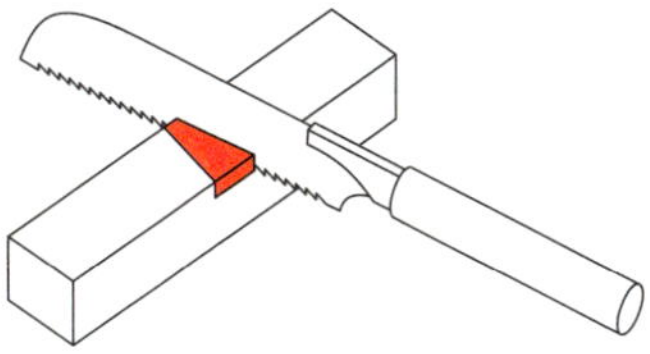

A2

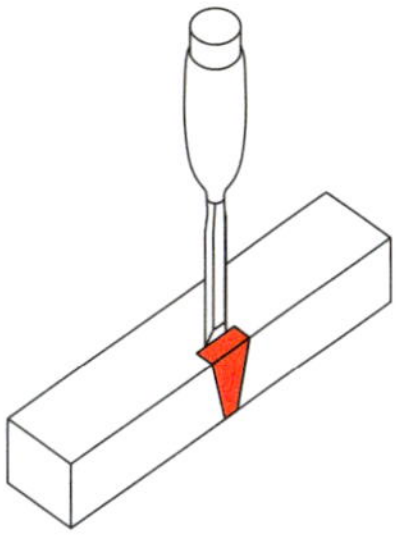

A3

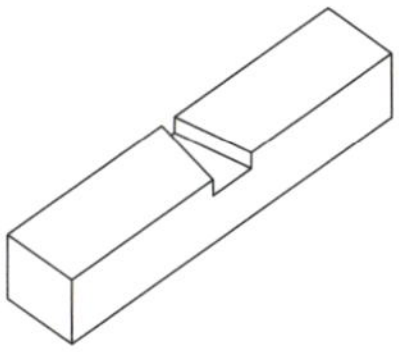

A

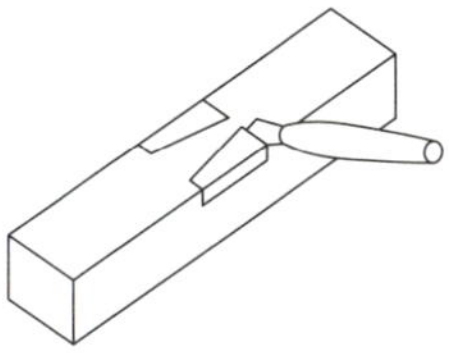

B1

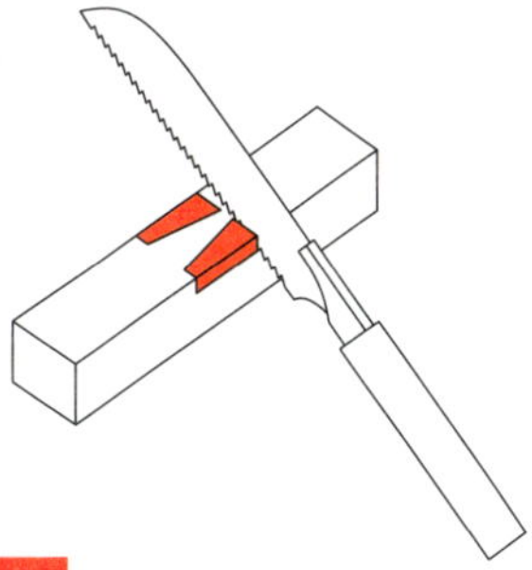

B2

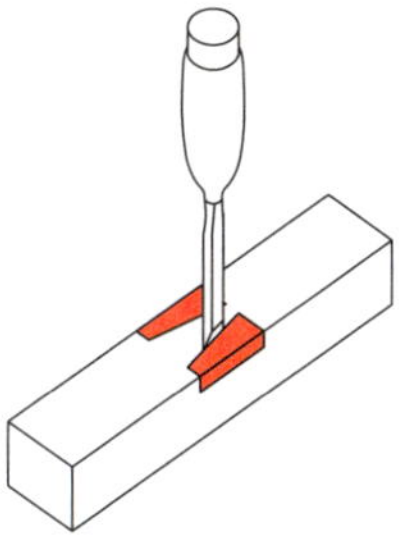

B3

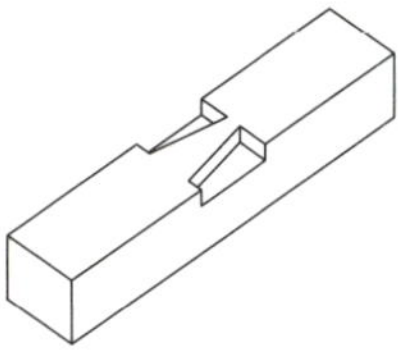

B

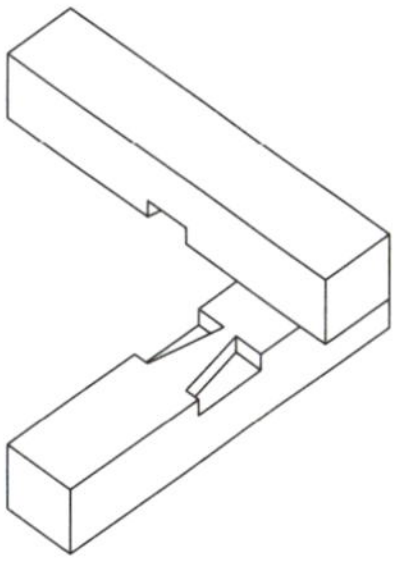

A+B

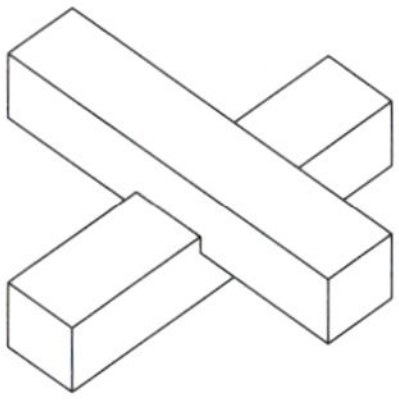

AB

Cross Lap Joints

V-shaped cog joint

ai-kaki-kumi (yahazu-kaki-watari-ago)
相欠組み(矢筈欠渡腮)

Kreuzkamm mit
halbem Kreuz

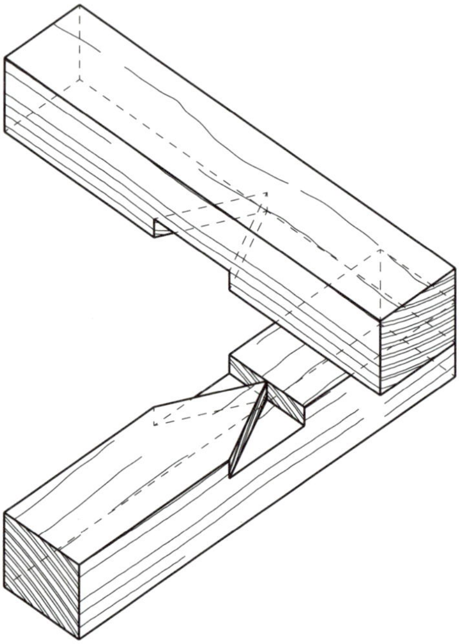

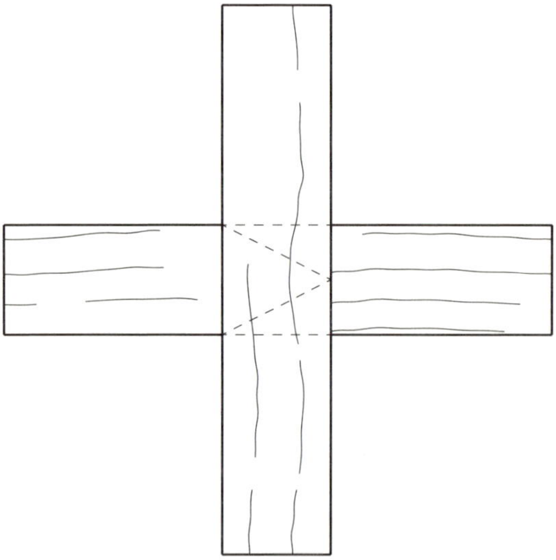

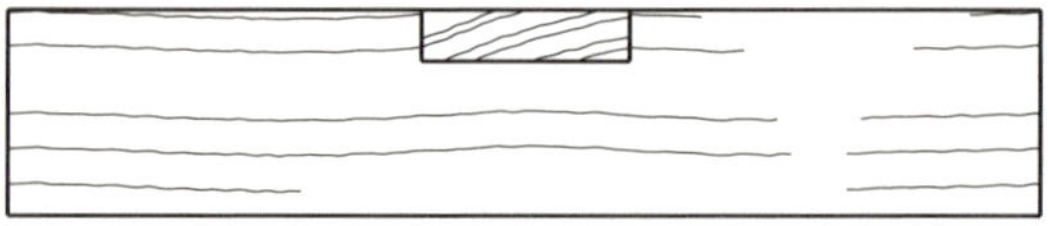

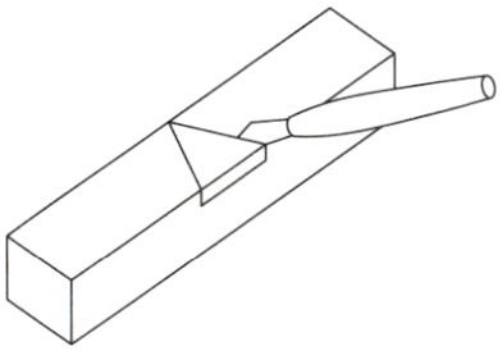

A1

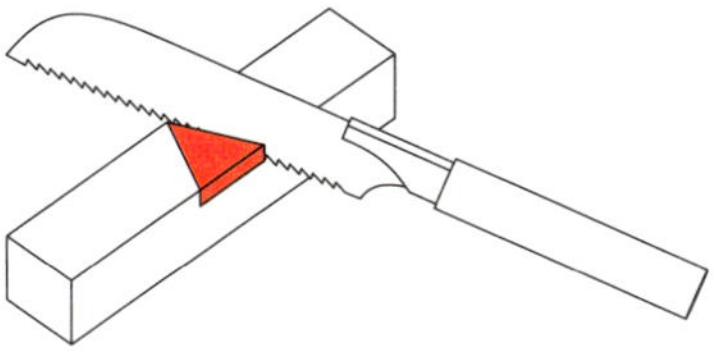

A2

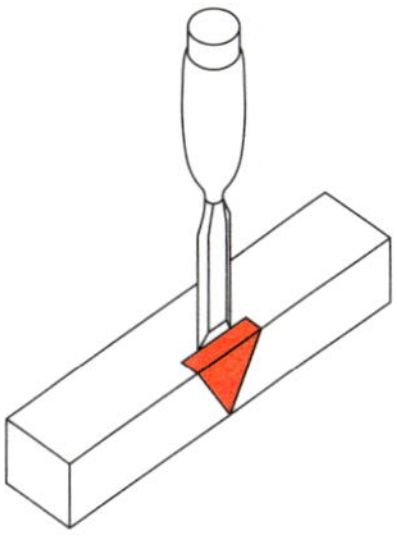

A3

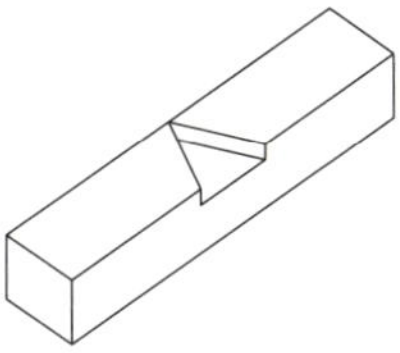

A

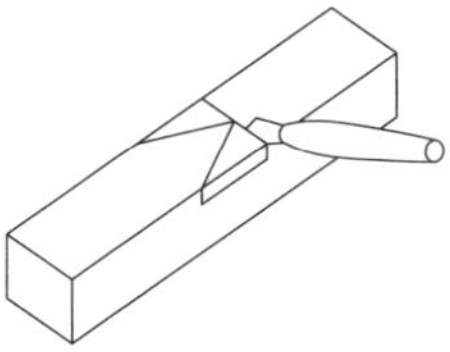

B1

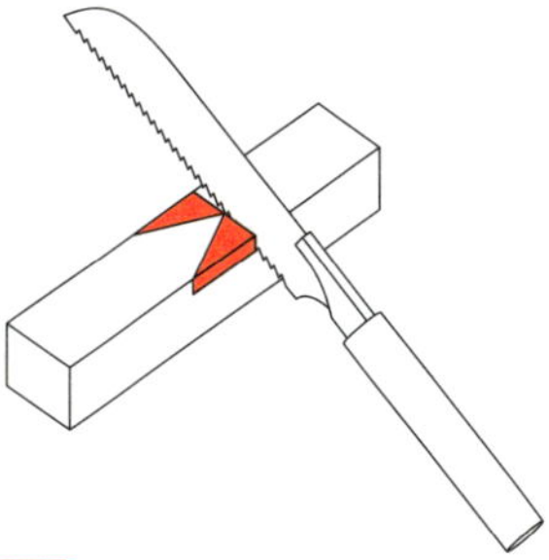

B2

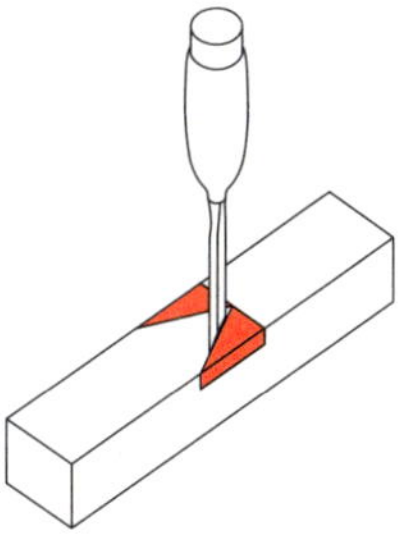

B3

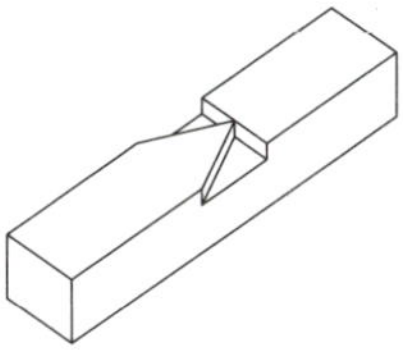

B

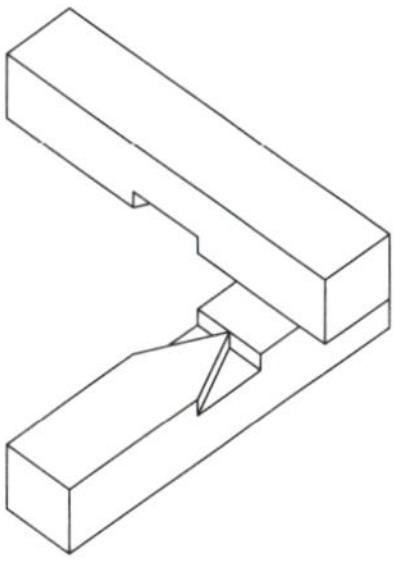

A+B

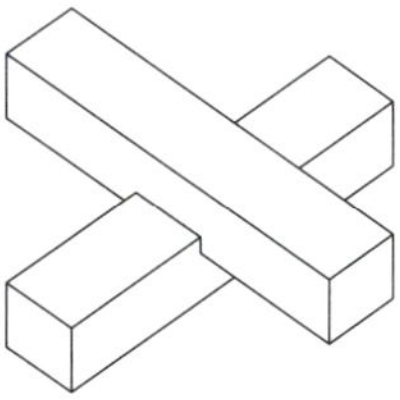

AB

Cross Lap Joints

X-shaped cog joint

tasuki-ai-kaki-kumi
襷相欠組み

Kreuzkamm mit Kreuz

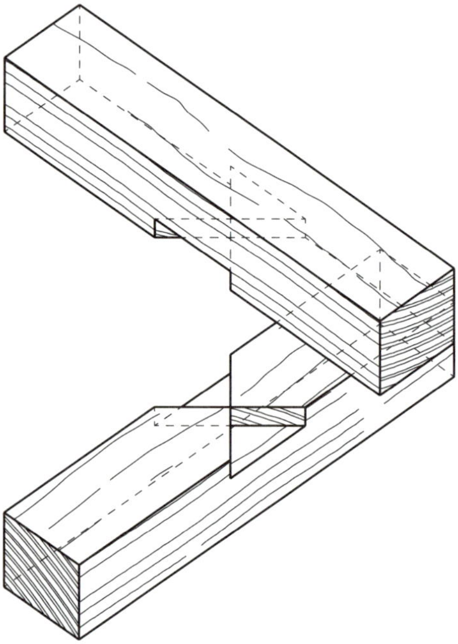

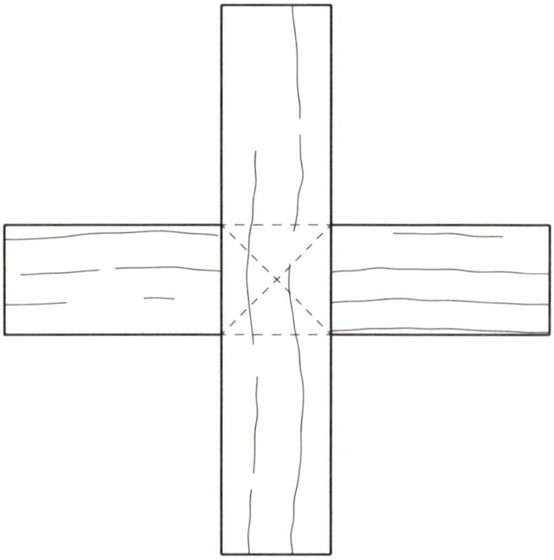

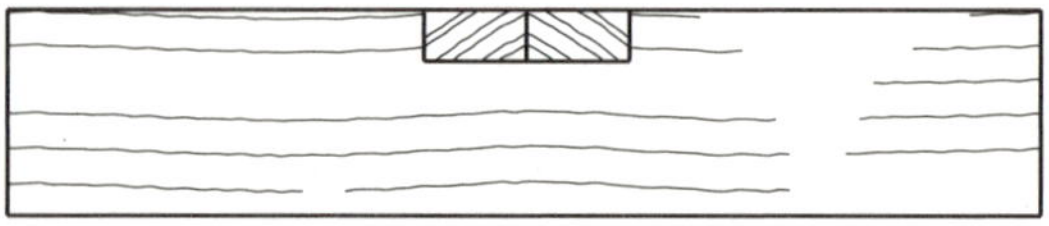

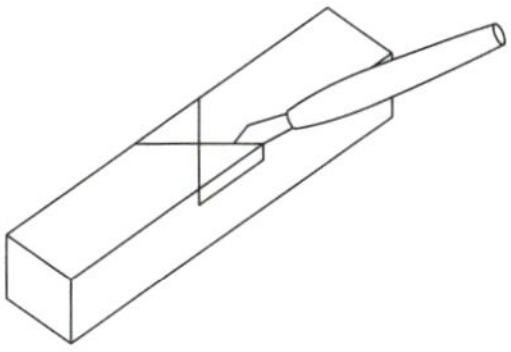

A1

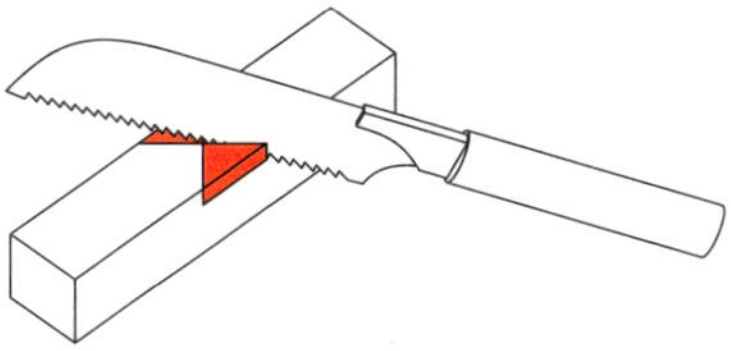

A2

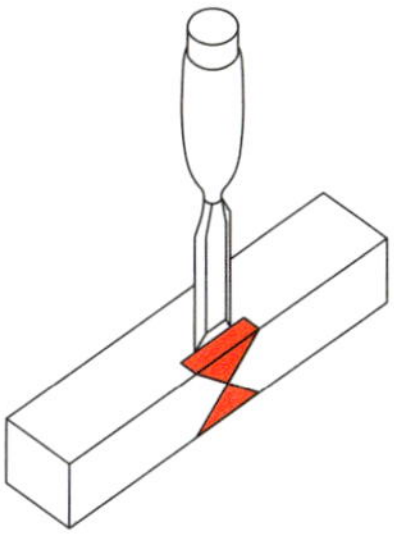

A3

2×

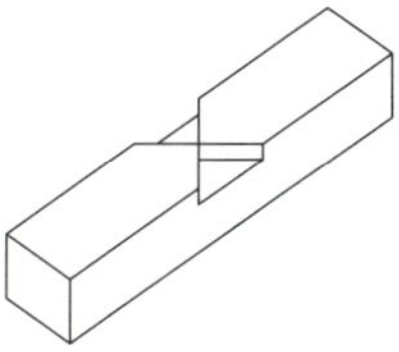

A

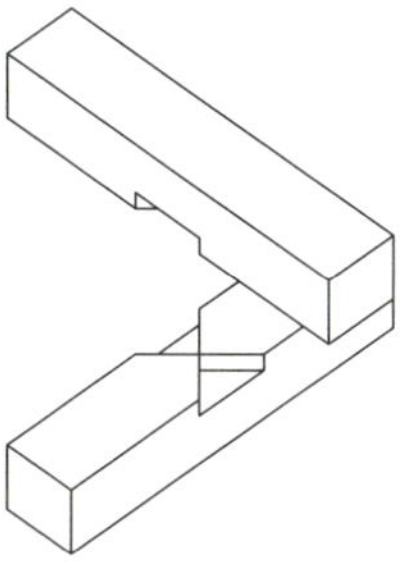

A+A

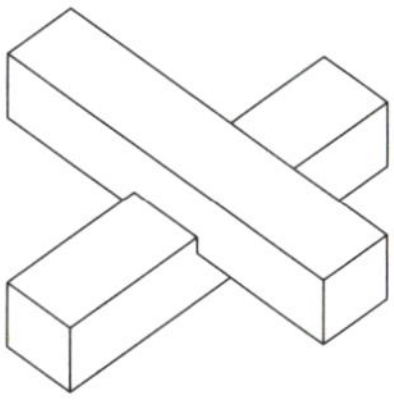

AA

Cross Lap Joints

Dovetail cog joint

ai-kaki-kumi (ari-kaki)
相欠組み(蟻欠)

Kreuzkamm mit Schwalbenschwanz

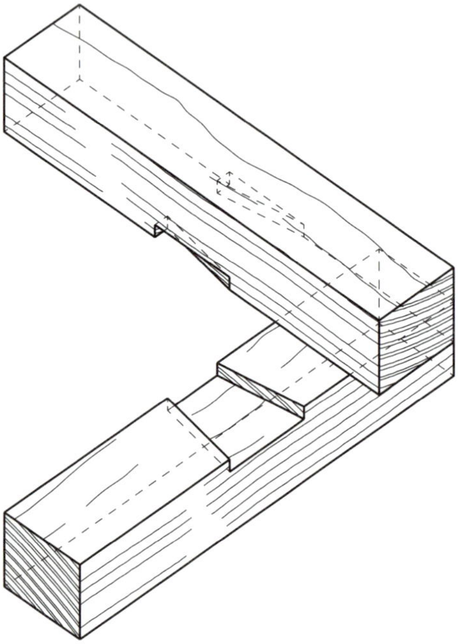

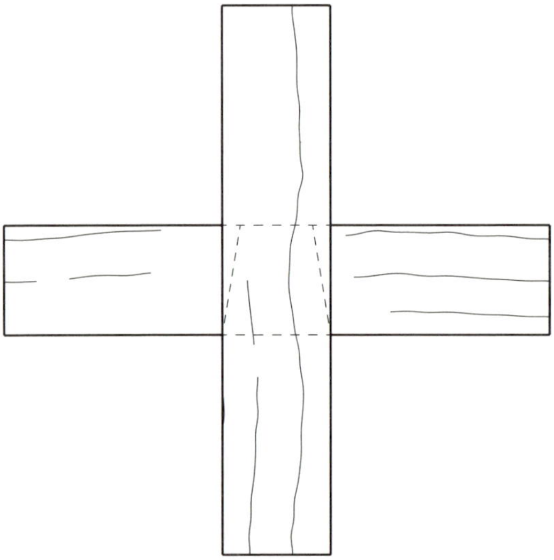

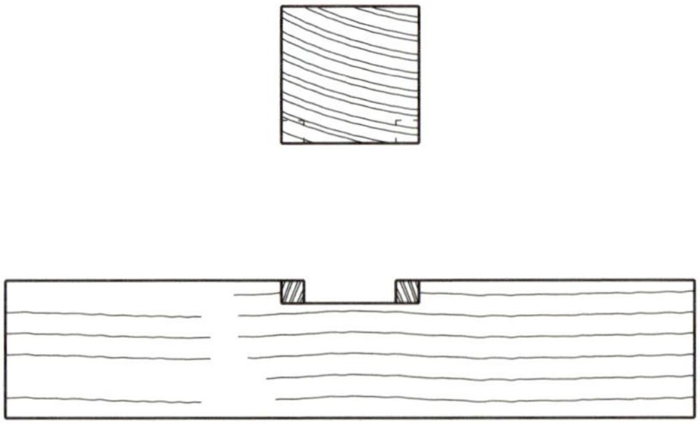

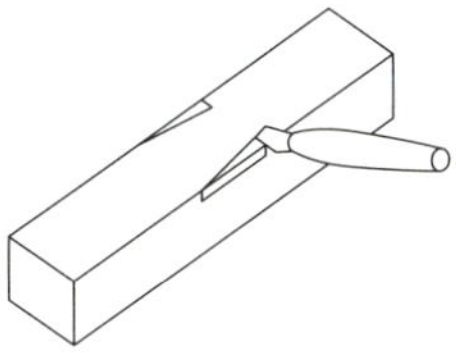

A1

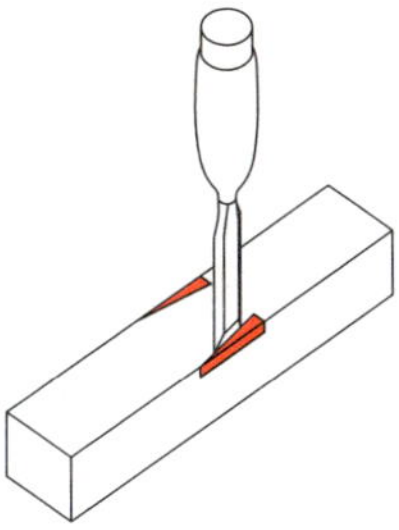

A2

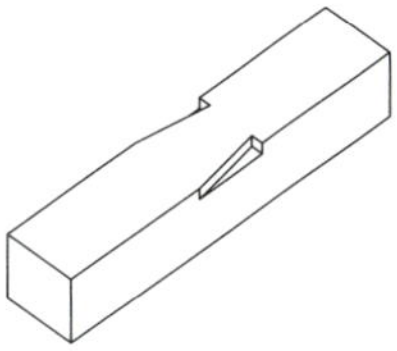

A

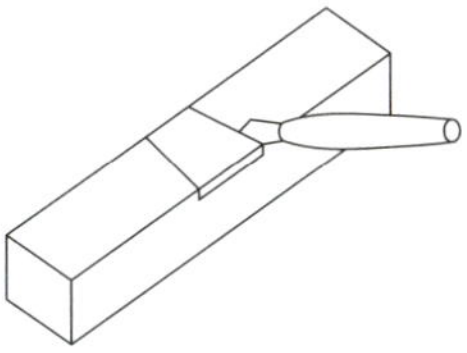

B1

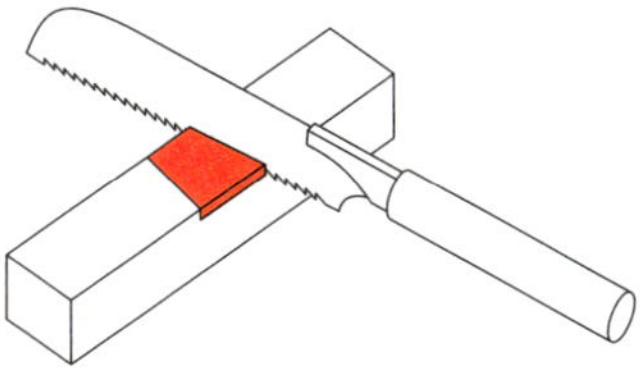

B2

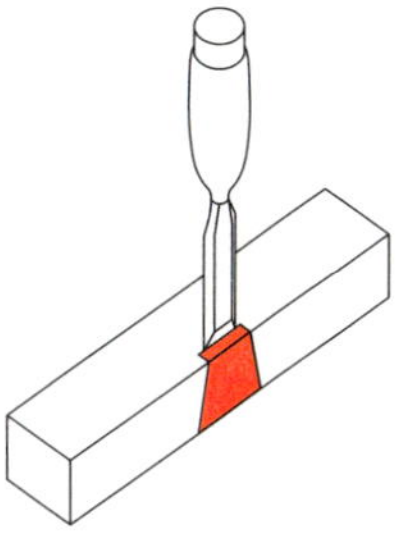

B3

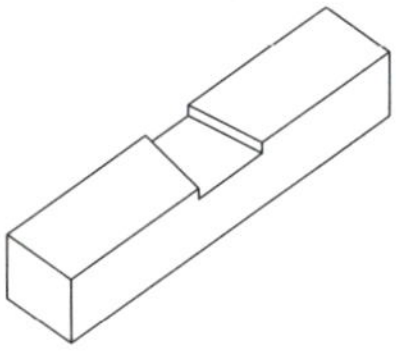

B

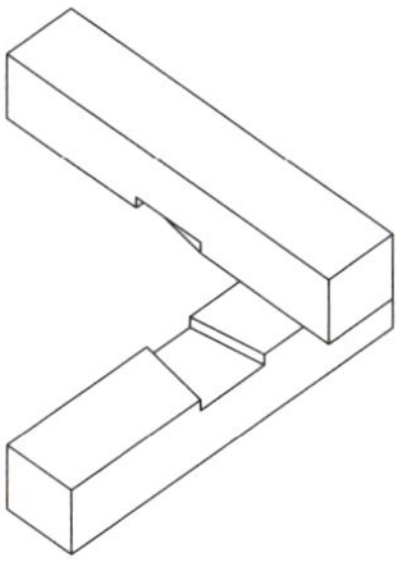

A+B

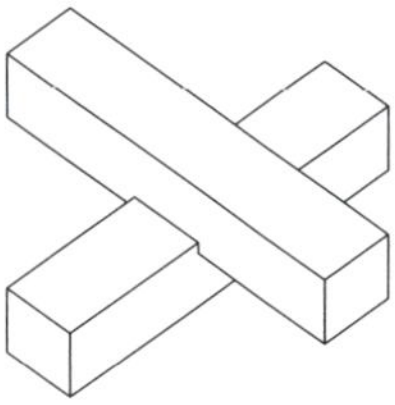

AB

Cross Lap Joints

Half-dovetail cog joint

ai-kaki-kumi (kata-ari-kaki)
相欠組み（片蟻欠）

Kreuzkamm mit
einseitigem Schwalbenschwanz

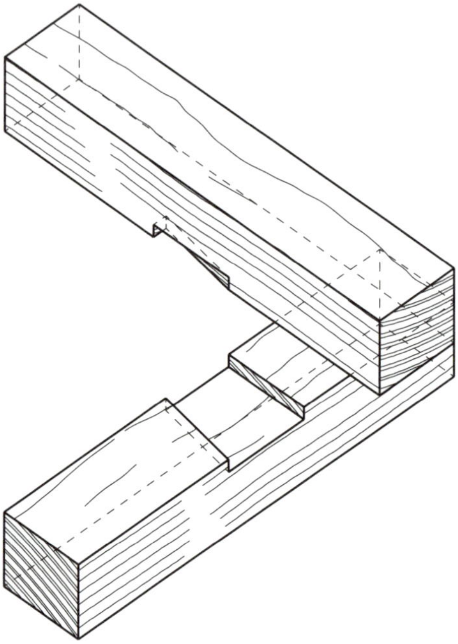

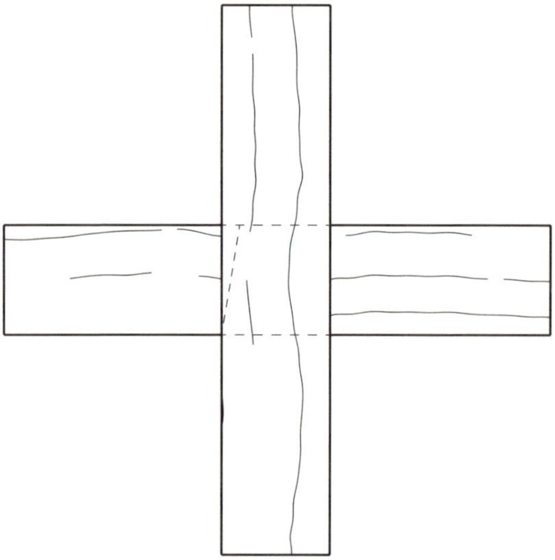

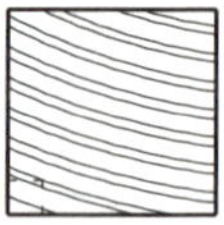

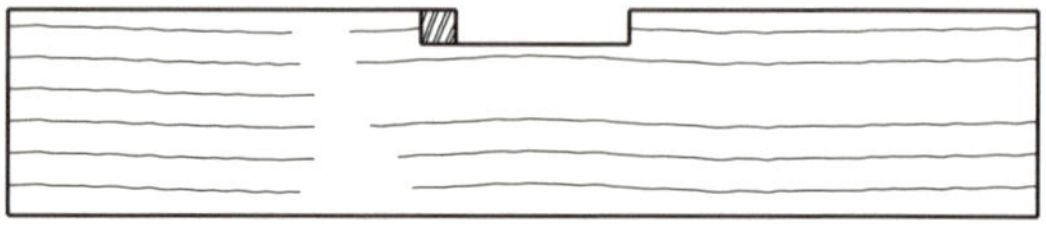

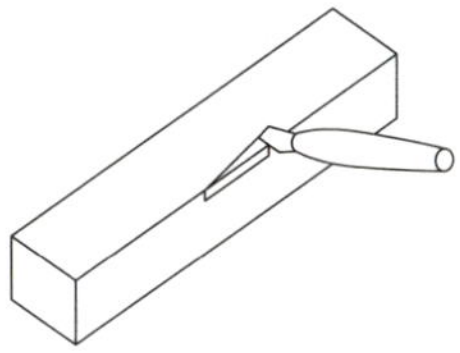

A1

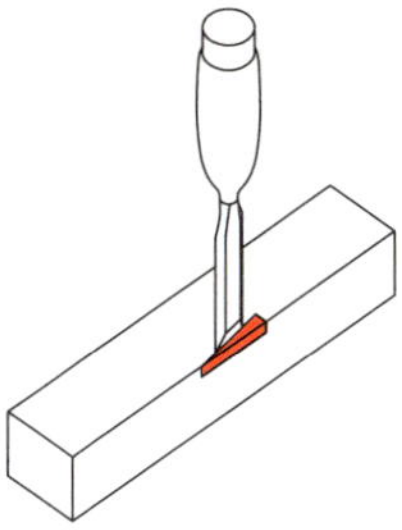

A2

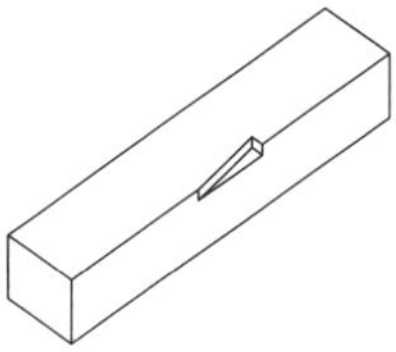

A

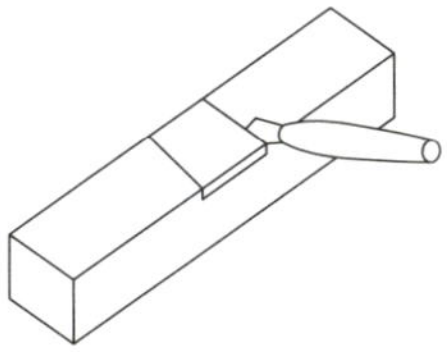

B1

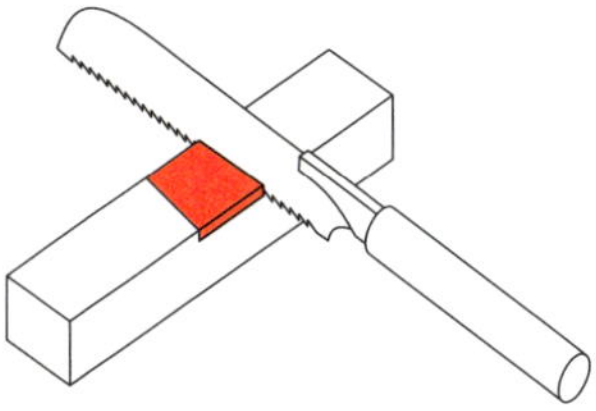

B2

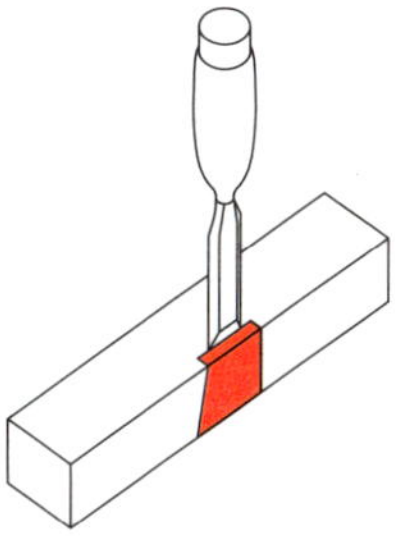

B3

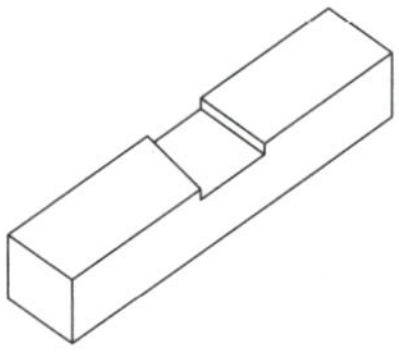

B

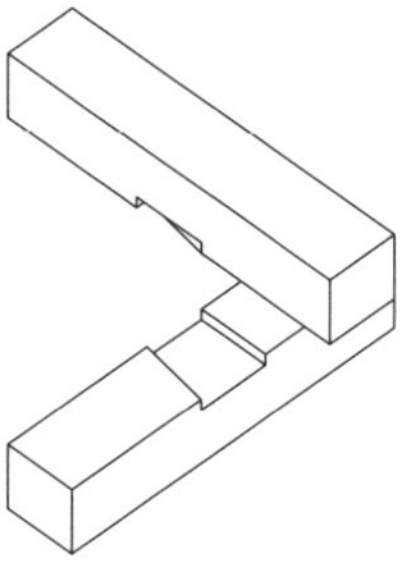

A+B

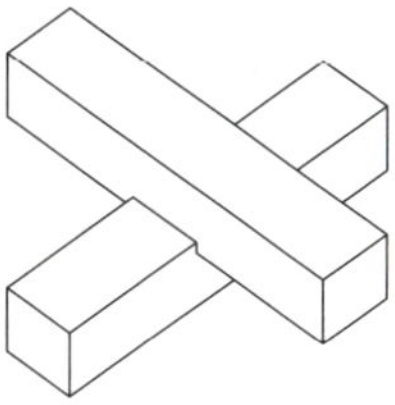

AB

Specialty Joints

Post with stub tenon and mortised double cogged joint

ai-kaki-watari-ago-tan-hozo-sashi
相欠渡腮短枘差し

verdeckter Zapfen mit
Kreuzkamm und mittiger Stufe

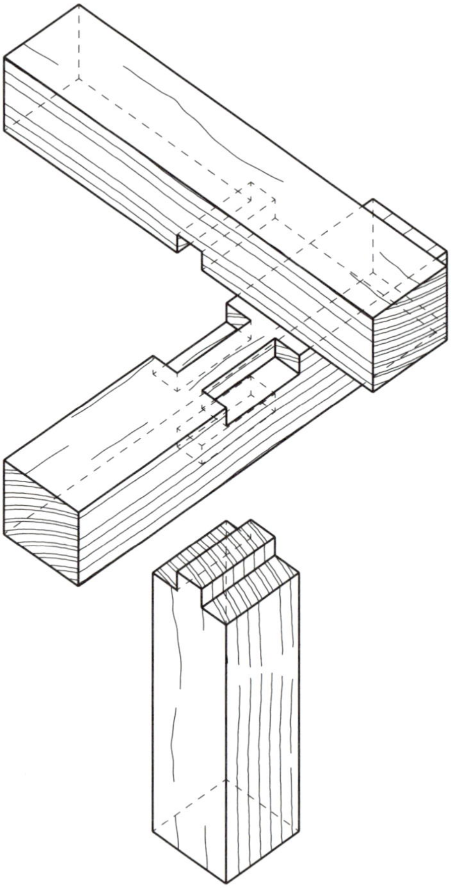

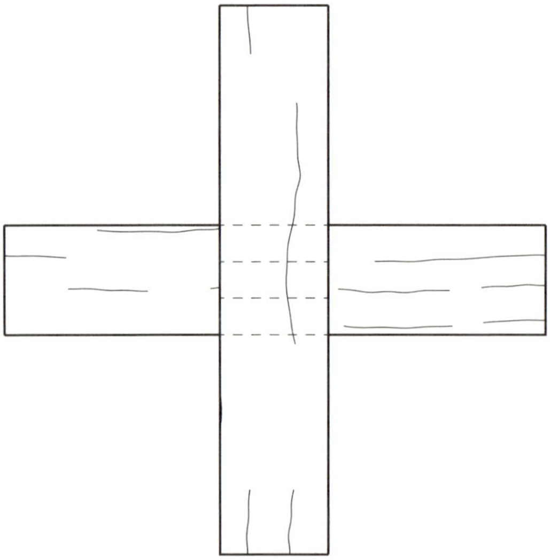

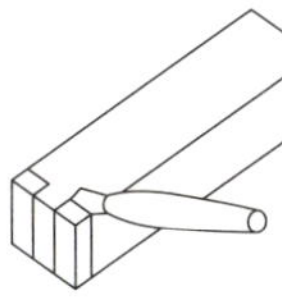
A1

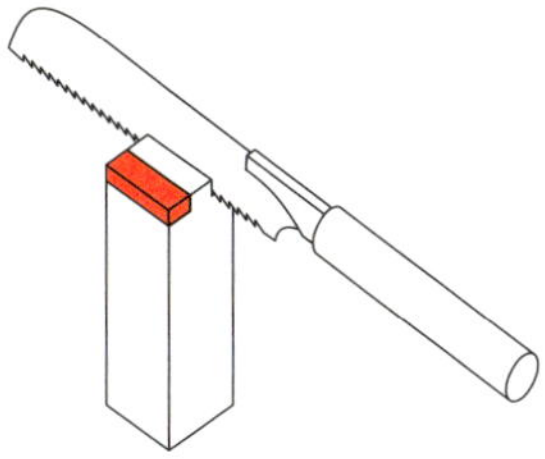
A2

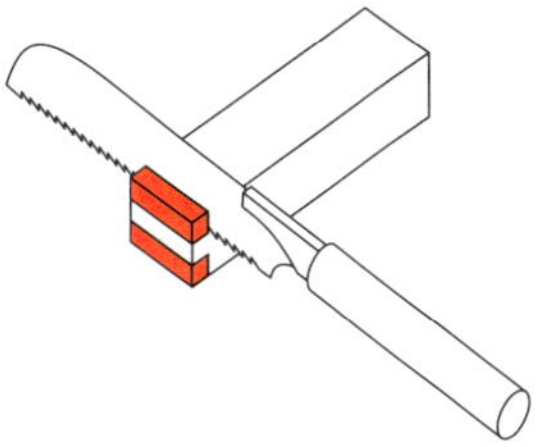
A3

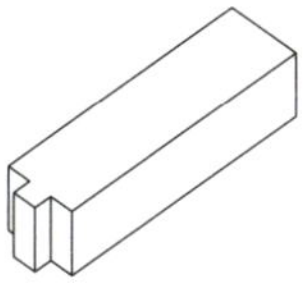
A

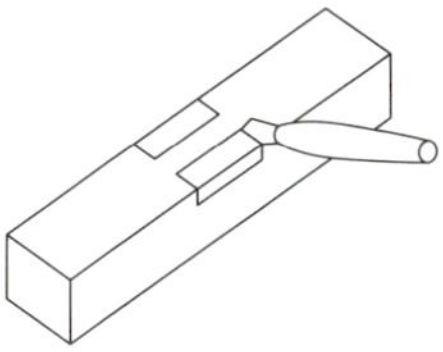
B1

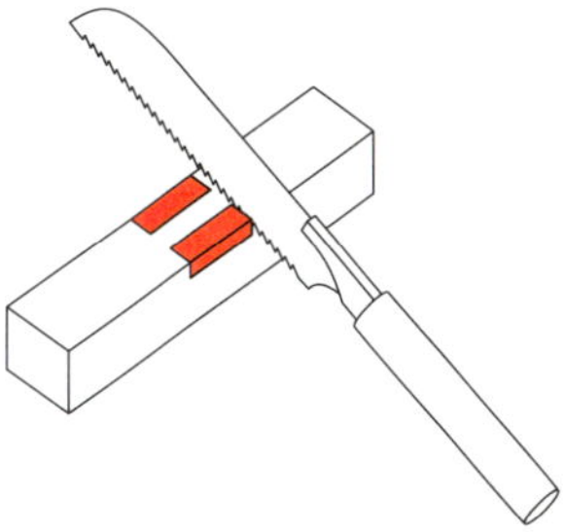
B2

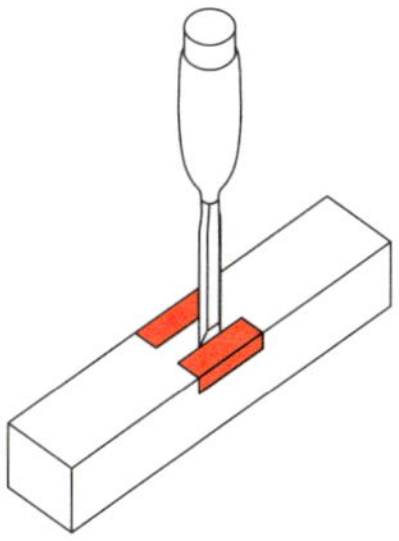
B3

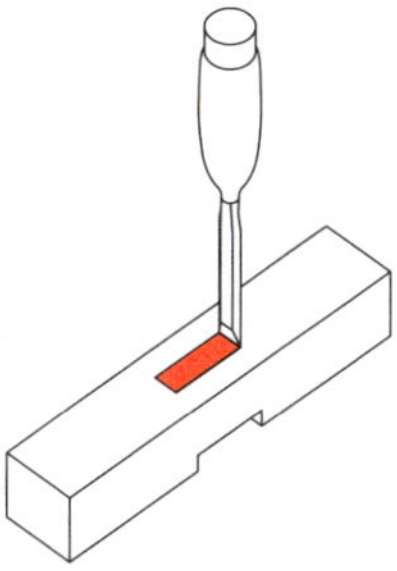
B4

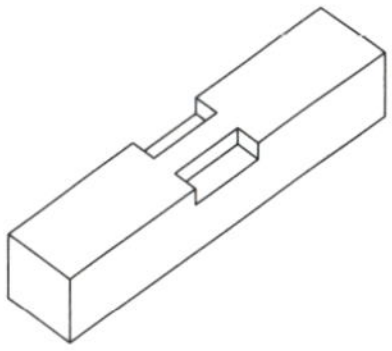
B

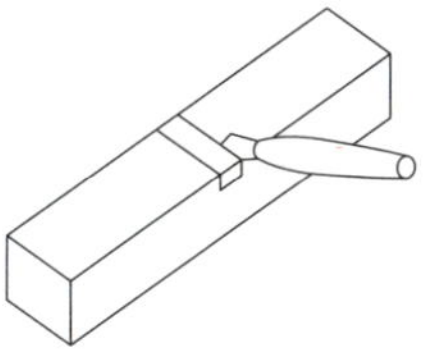
C1

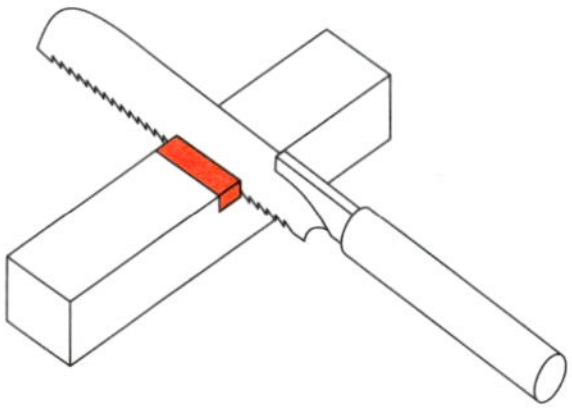
C2

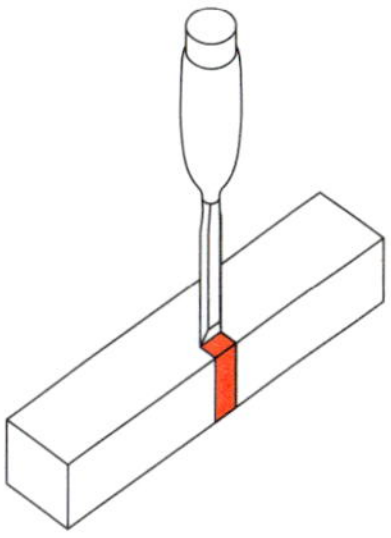
C3

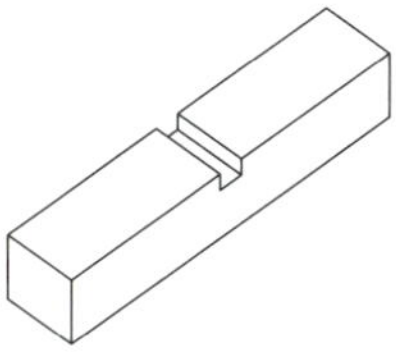
C

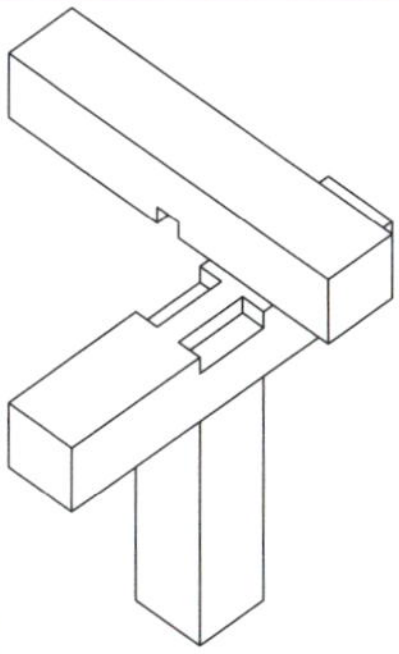
A+B+C

ABC

Specialty Joints

T-bridle joint with stub tenons and mortised beam

nimai-jū-hozo-wanagi-komi-kumi
二枚重枘輪薙込組み

Einhälsung mit Brüstungszapfen und aufgesetzter Überkreuzung

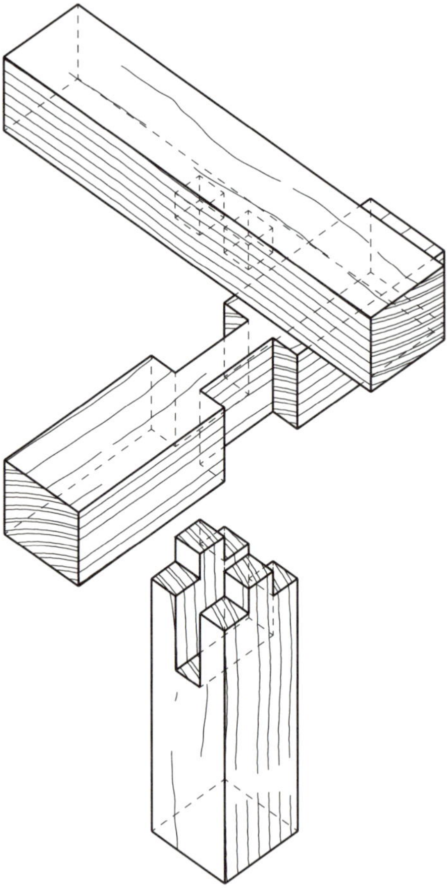

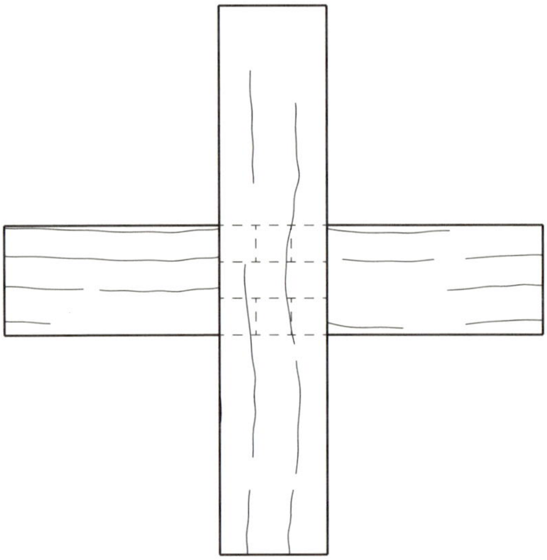

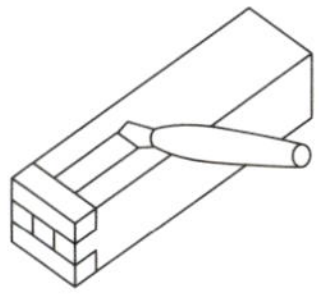

A1

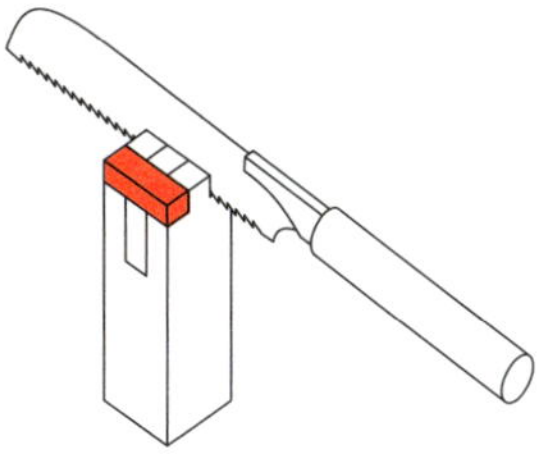

A2

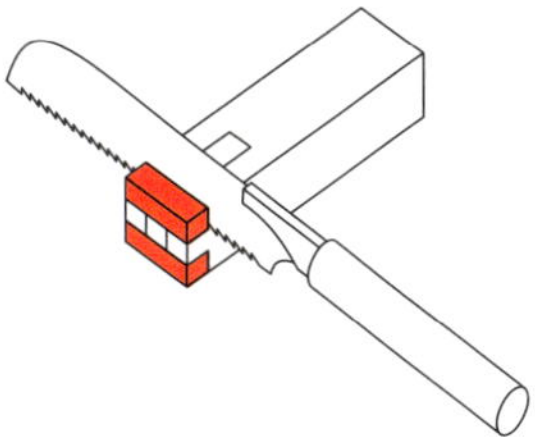

A3

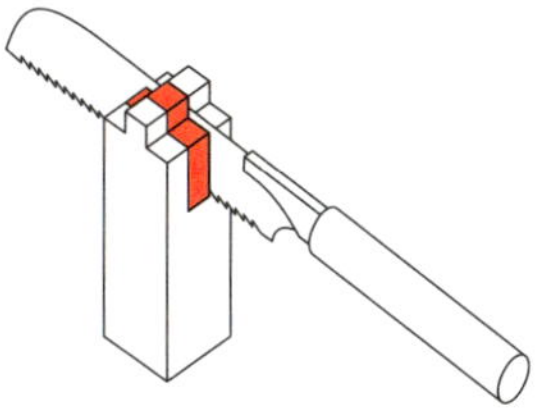

A4

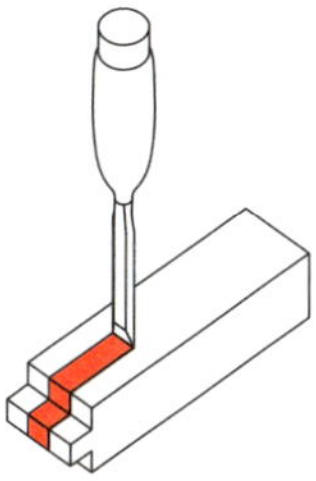

A5

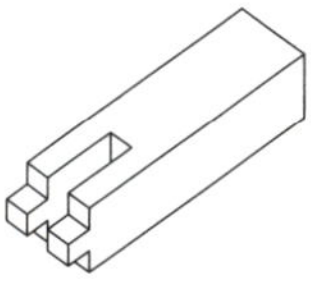

A

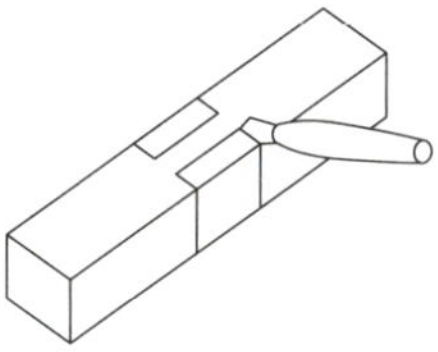

B1

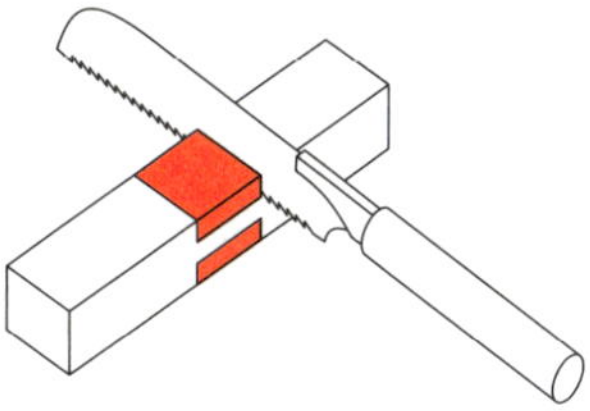

B2

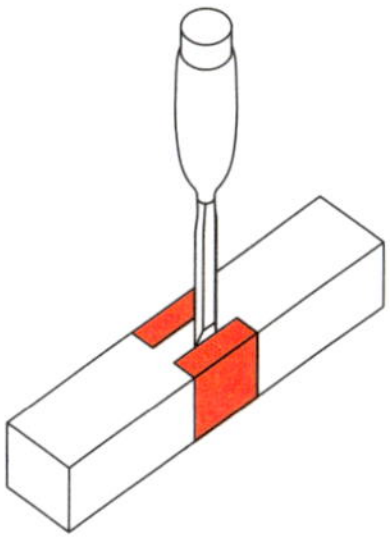

B3

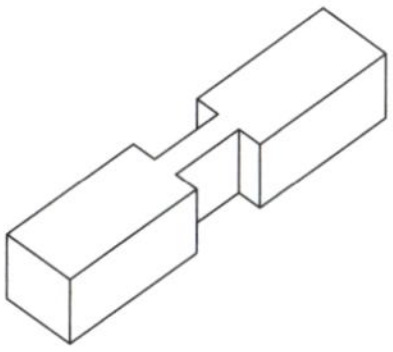

B

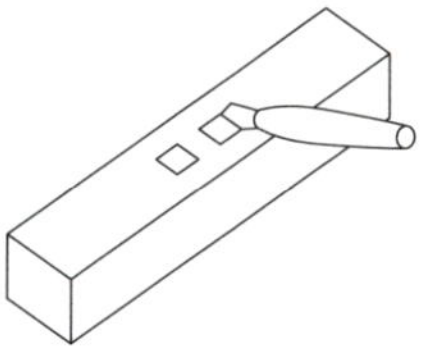

C1

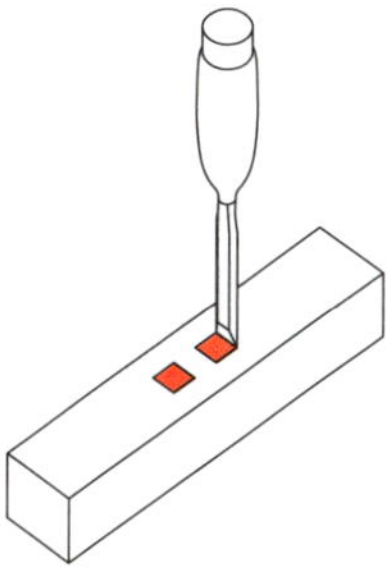

C2

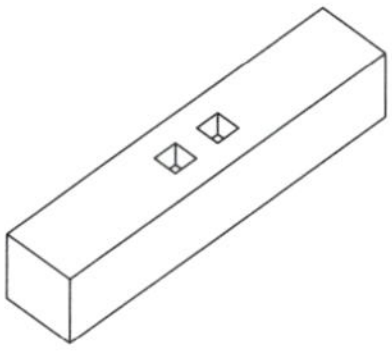

C

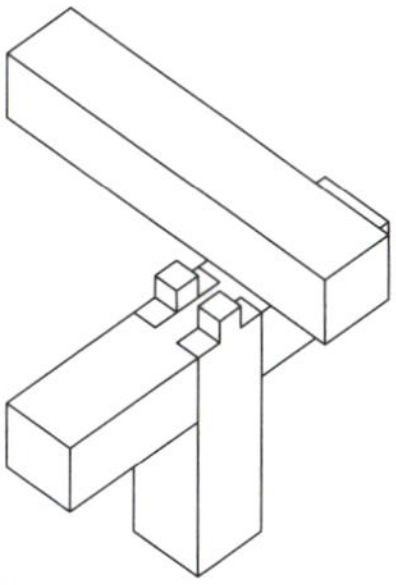

A+B+C

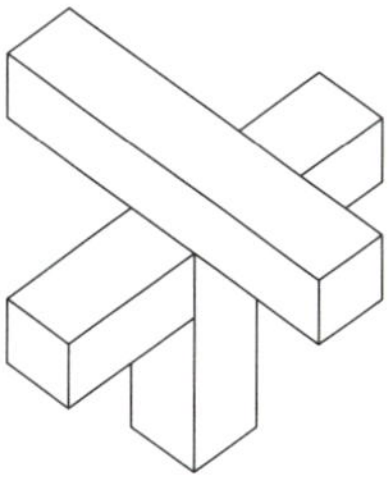

ABC

Specialty Joints

Post with through tenon and mortised crossed joint

hari-keta-uchi-nuki-ōnaga-hozo-sashi
梁桁打抜大長枘差し

durchgesteckter Zapfen
mit aufgesetzter Überkreuzung

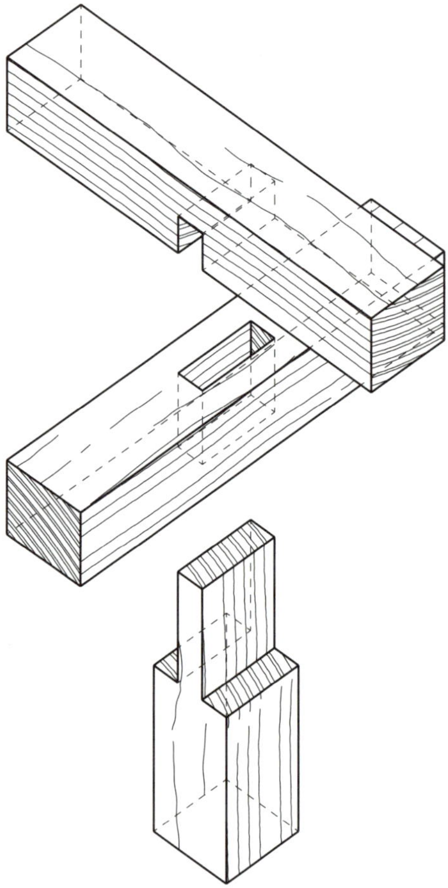

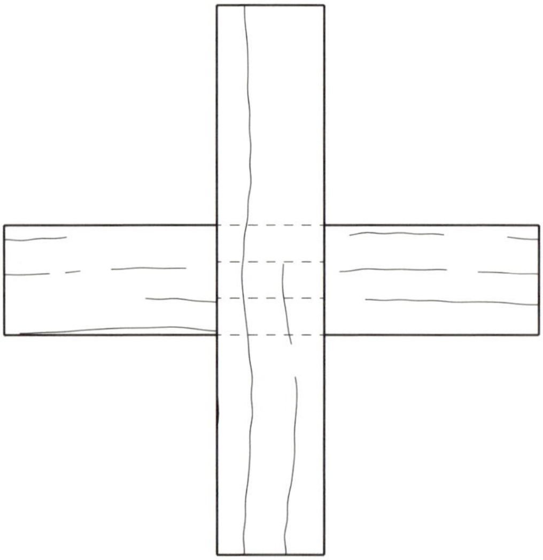

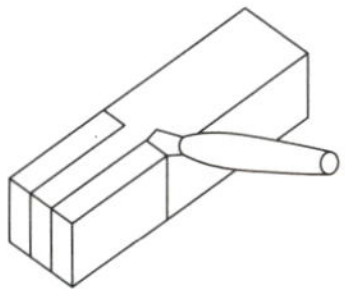

A1

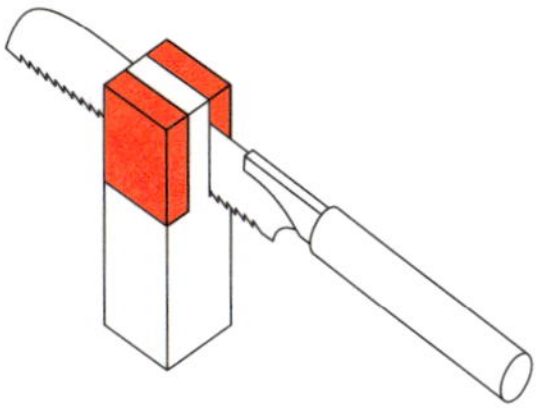

A2

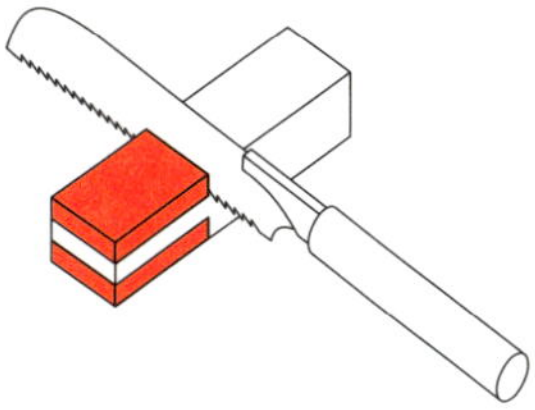

A3

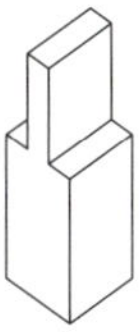

A

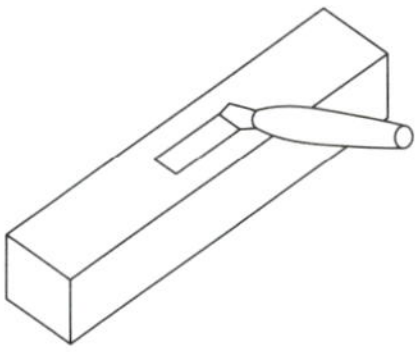

B1

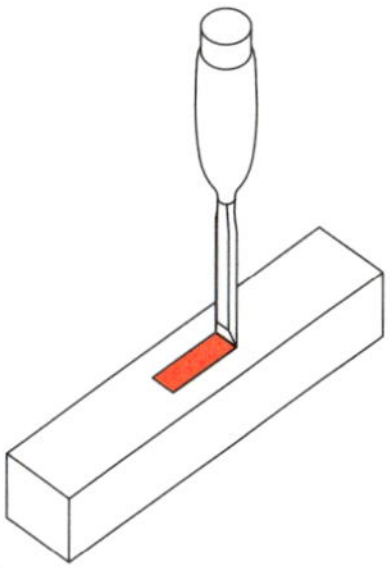

B2

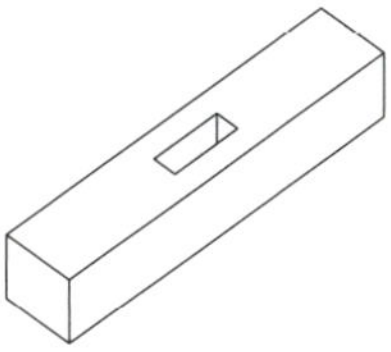

B

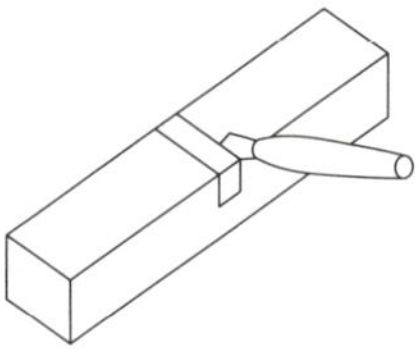

C1

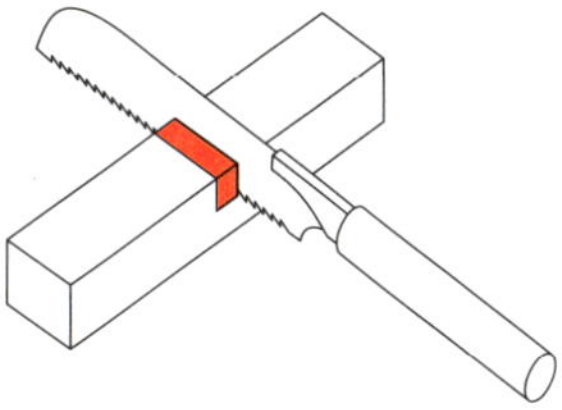

C2

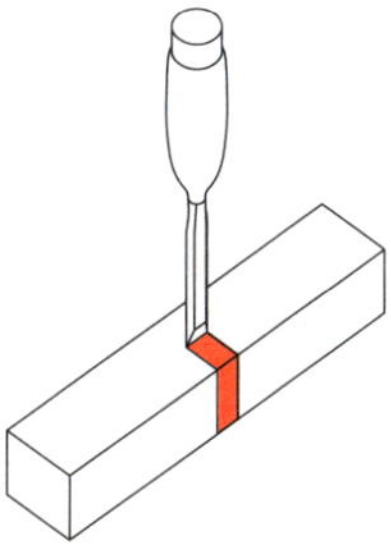

C3

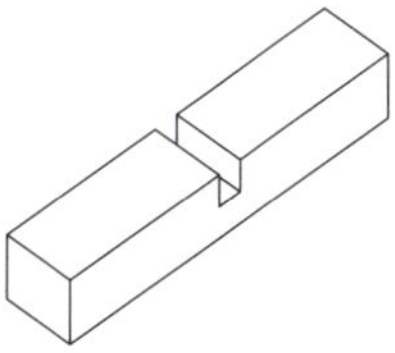

C

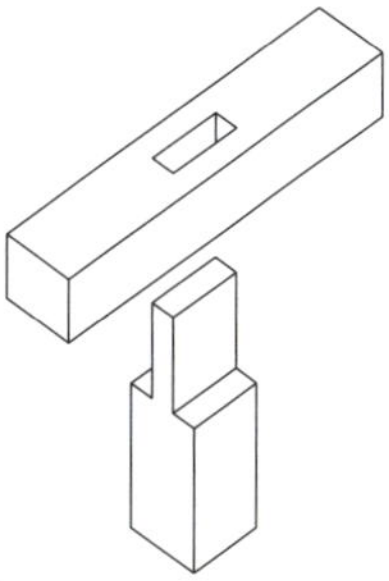

A+B

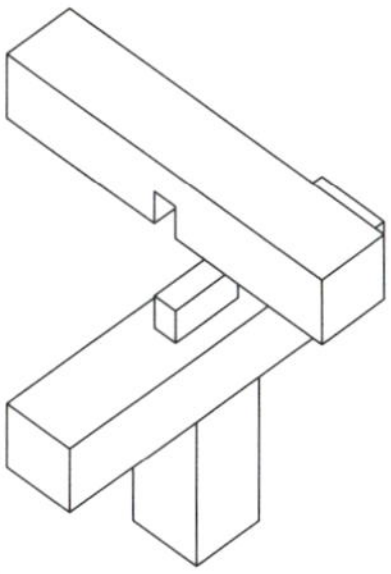

AB+C

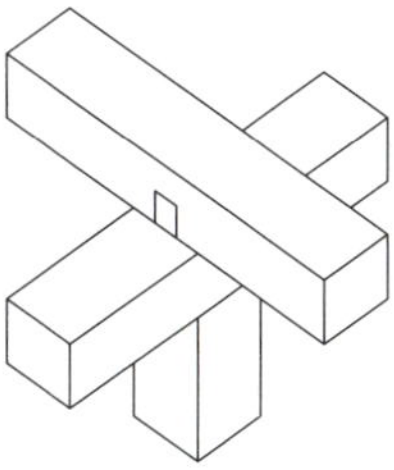

ABC

Specialty Joints

Stacked tenon and mortised cross lap joint

ai-kaki-watari-ago-jū-hozo-sashi [*]
相欠渡腮重枘差し

durchgesteckter beidseitig abgesetzter Stufenzapfen mit Kreuzblatt

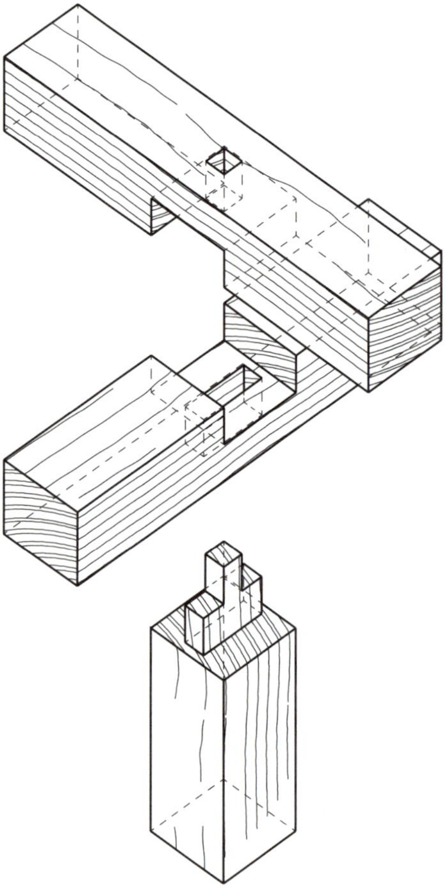

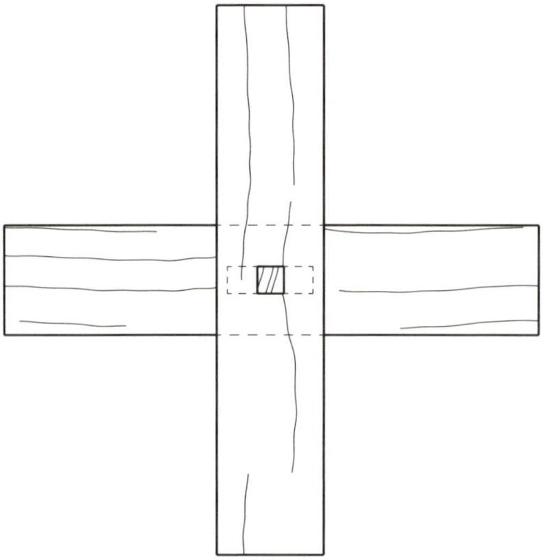

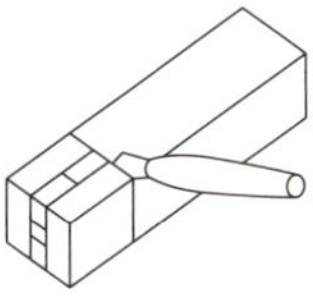

A1

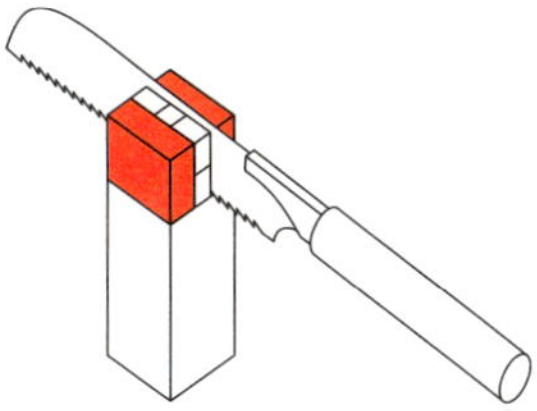

A2

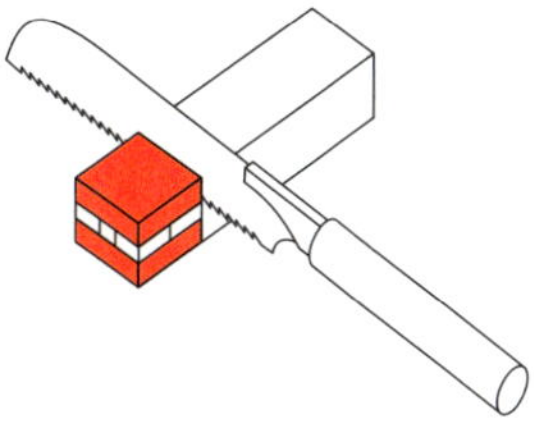

A3

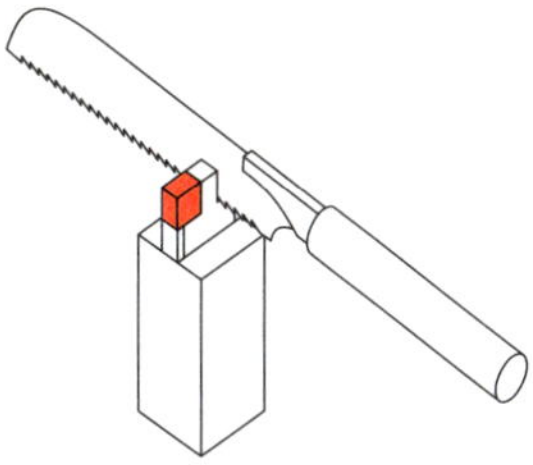

A4

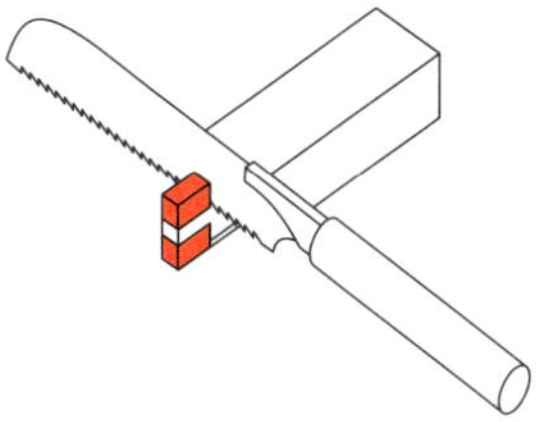

A5

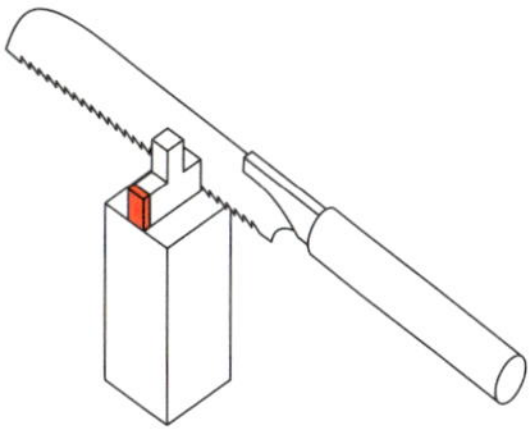

A6

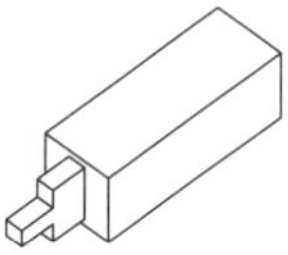

A

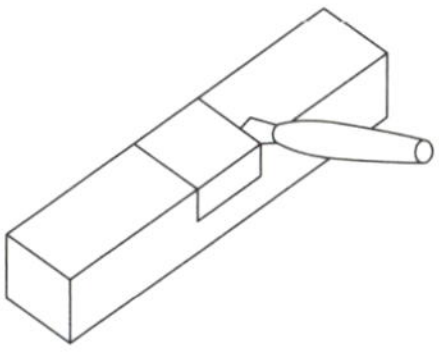

BC1

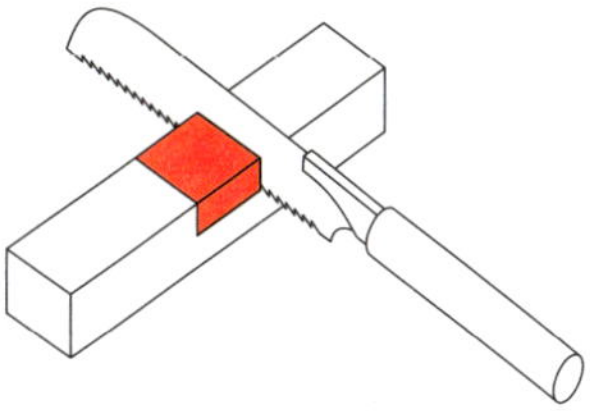

BC2

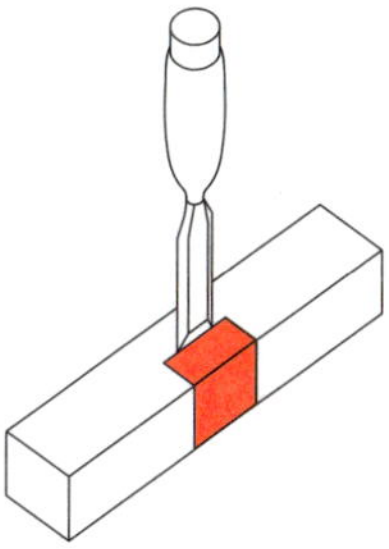

BC3

2×

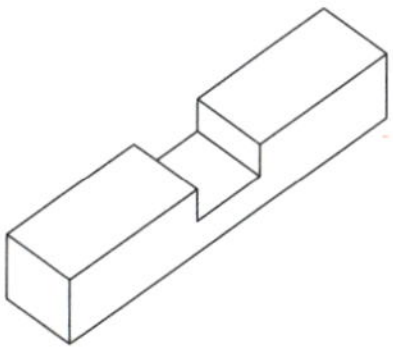

BC

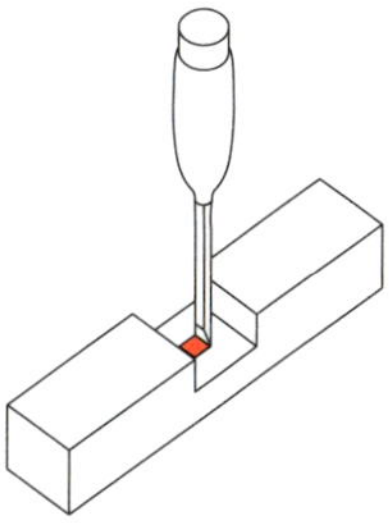

B

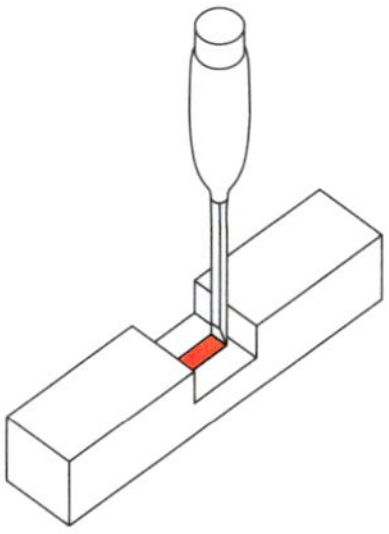

C

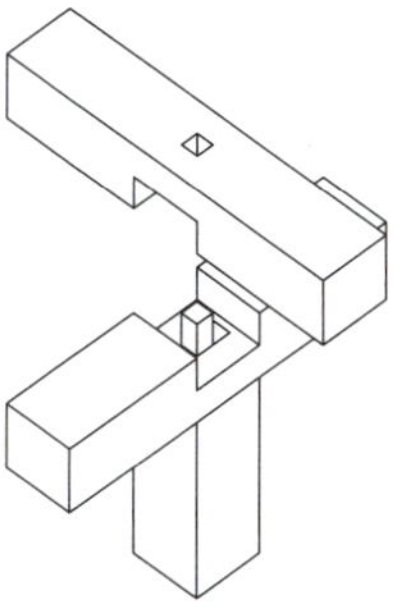

A+B+C

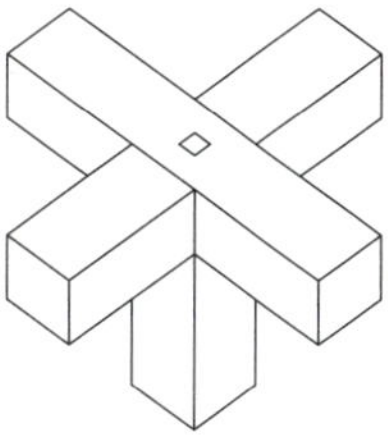

ABC

Specialty Joints

Castle joint

igeta-shikuchi
井桁仕口

doppelte Einhälsung

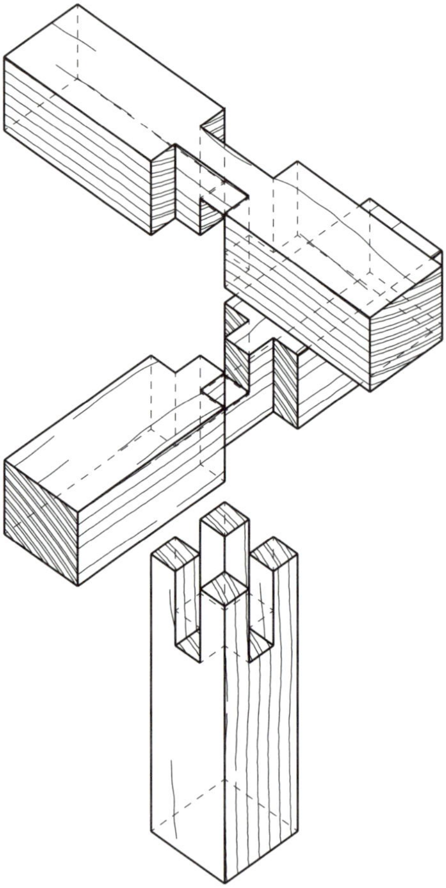

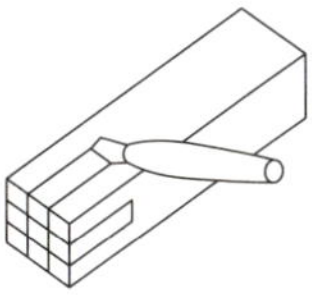

A1

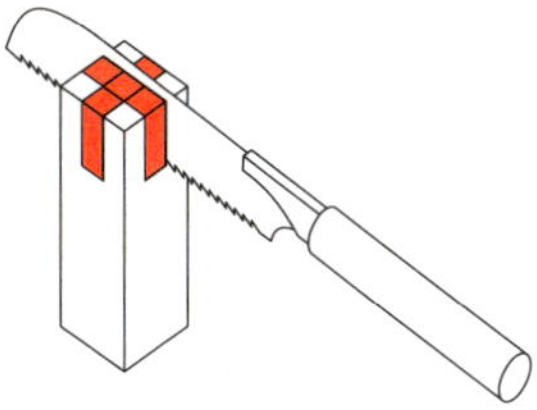

A2

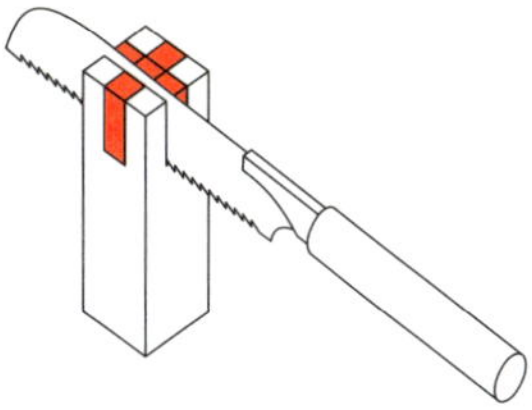

A3

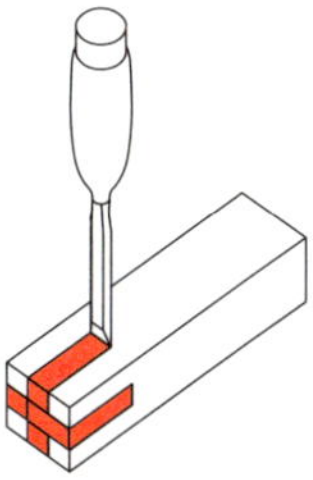

A4

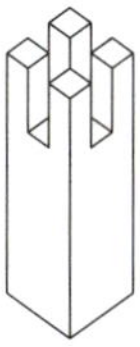

A

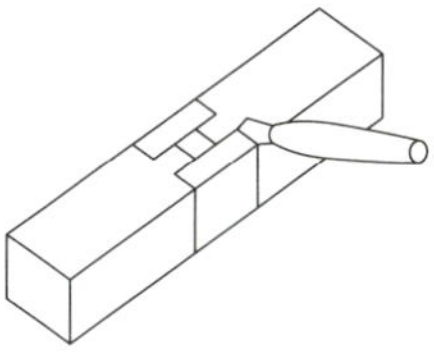

B1

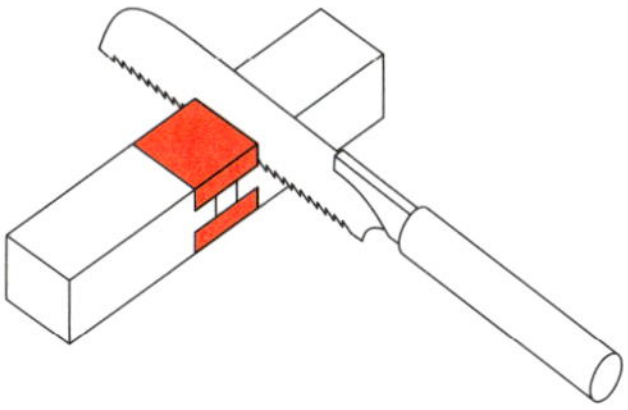

B2

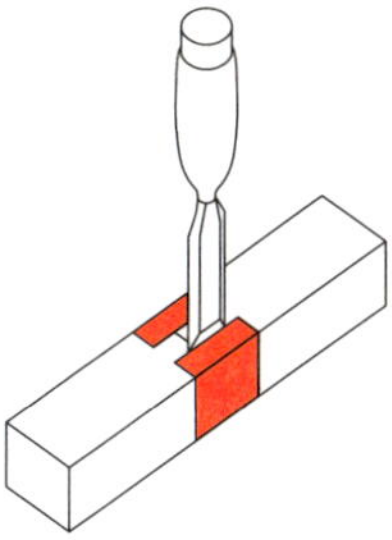

B3

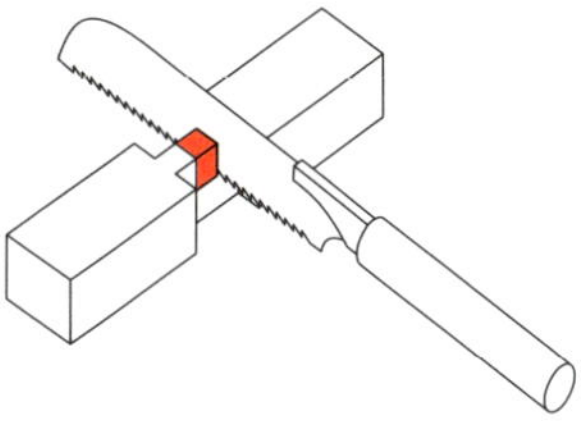

B4

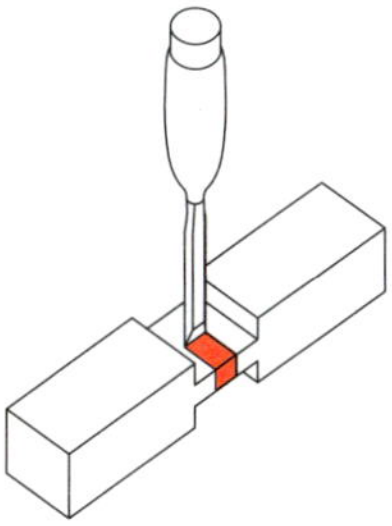

B5

2×

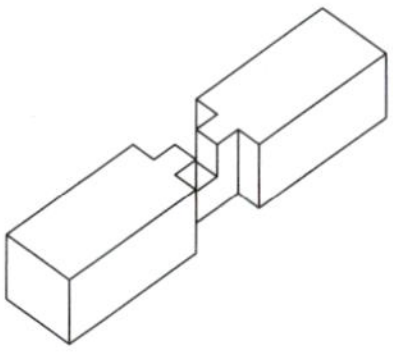

B

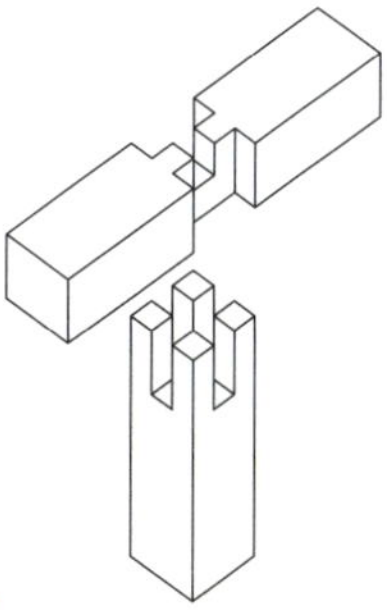

A+B

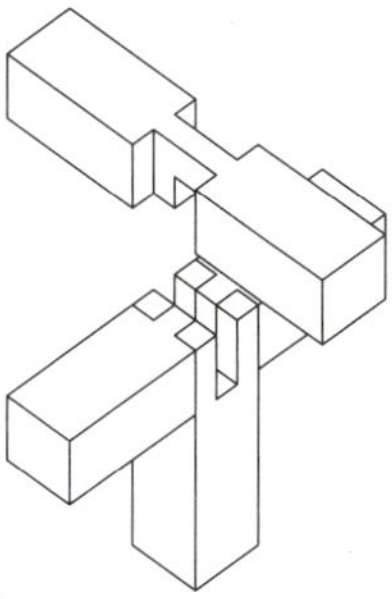

AB+B

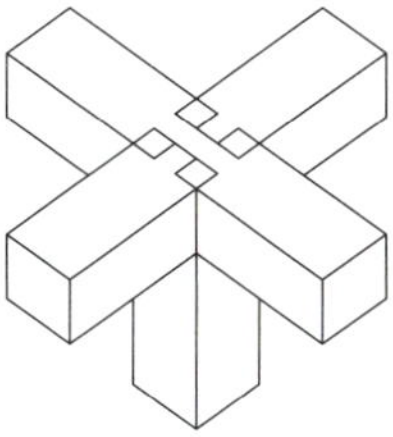

ABB

Specialty Joints

Notched corner-post joint with beveled-shoulder tenons

katagi-hozo-nihō-sashi
傾枘二方差し

Zapfen mit schräger einseitig auf Gehrung angeschnittener Brust

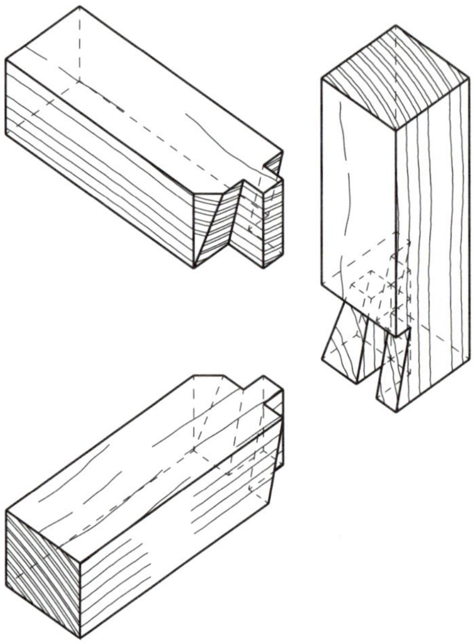

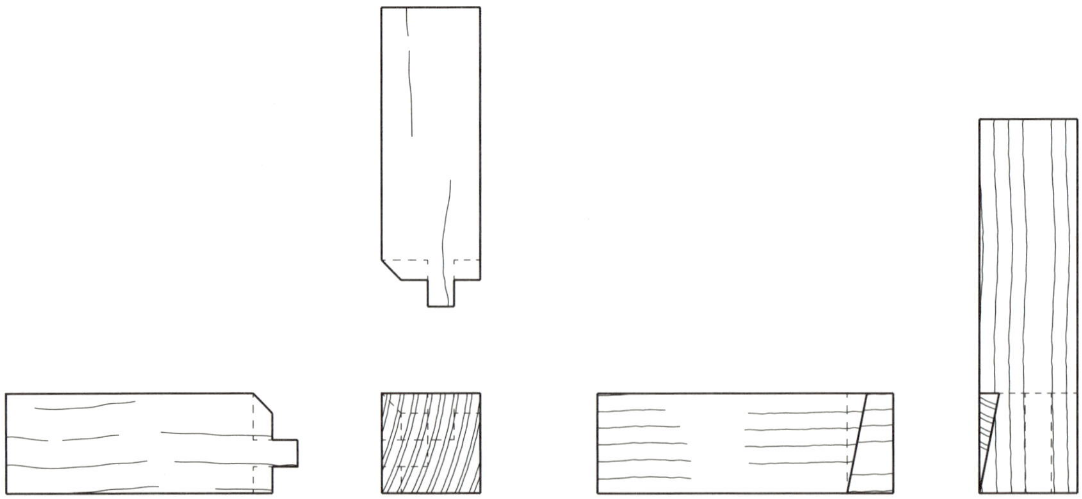

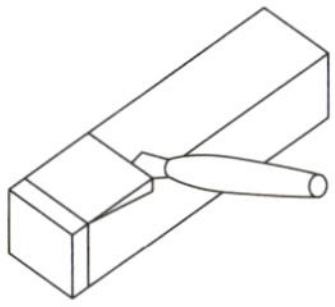
A1

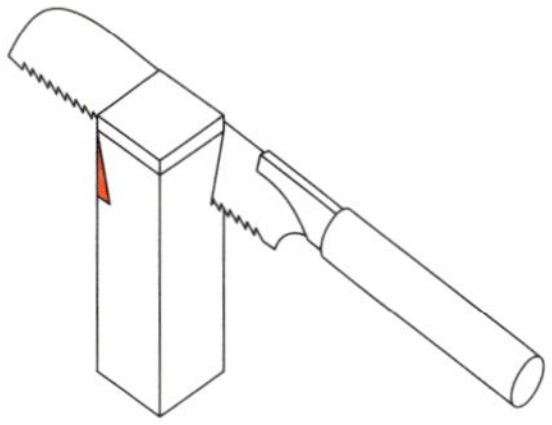
A2

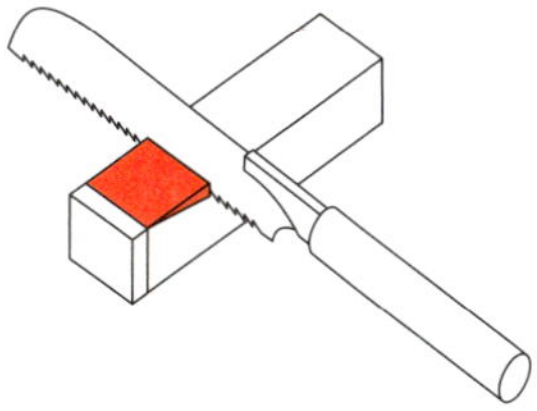
A3

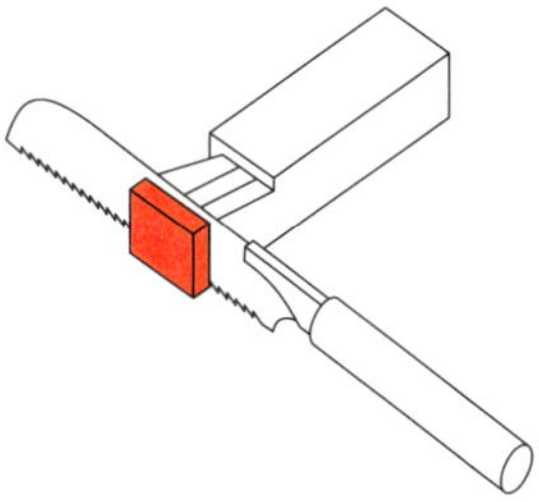
A4

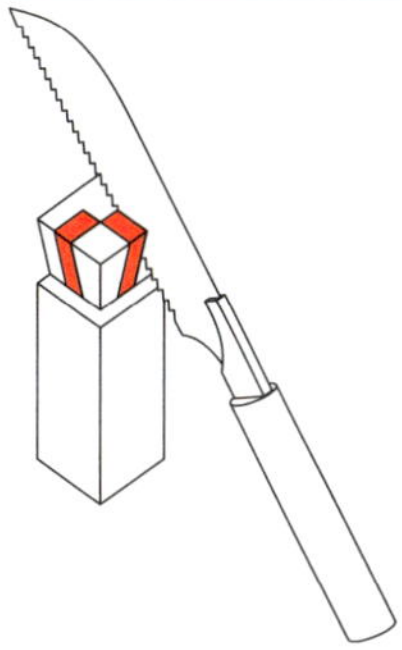
A5

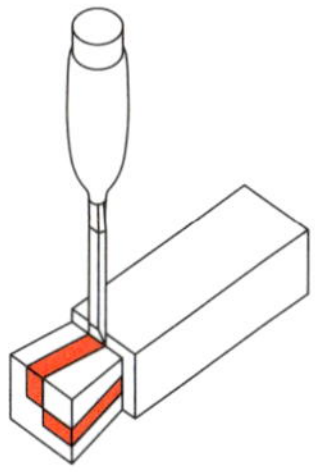
A6

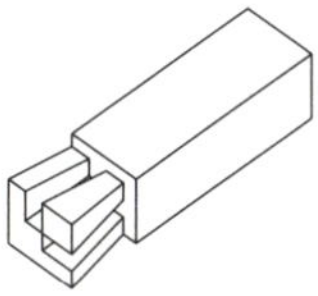
A

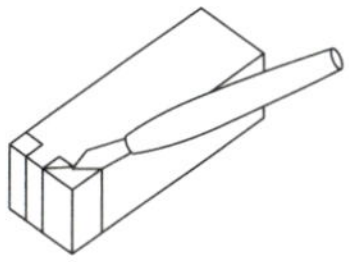
BC1

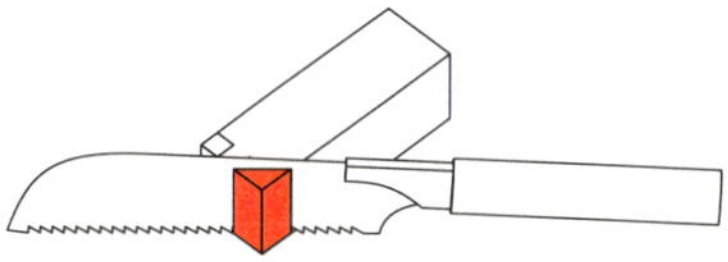
BC2

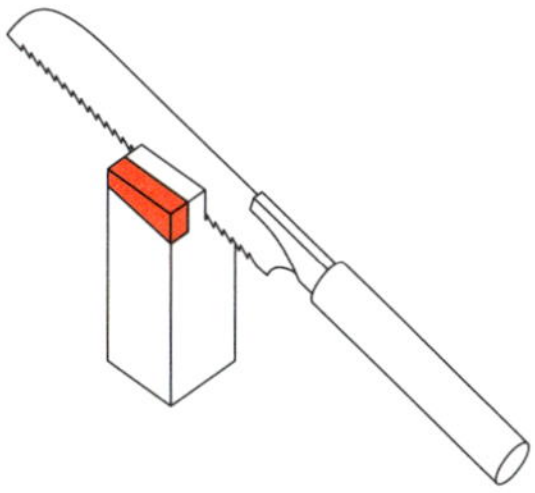
BC3

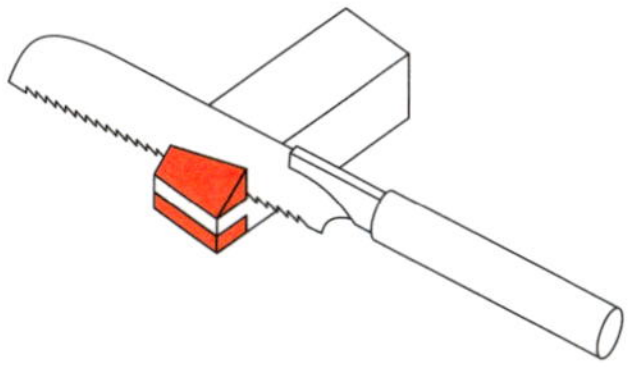
BC4

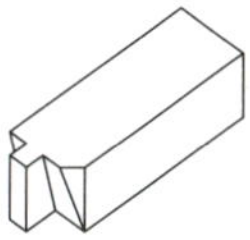
B

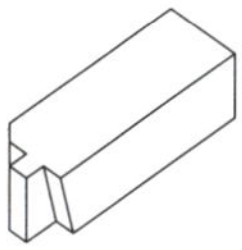
C

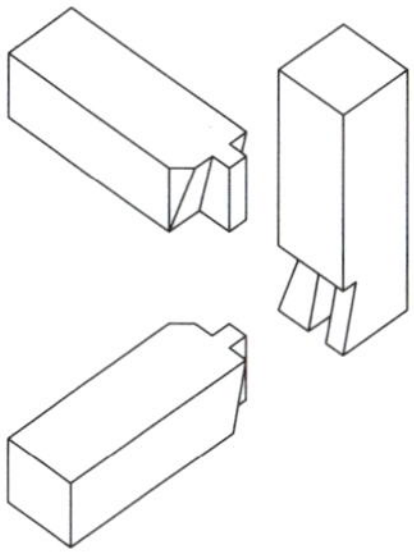
A+B+C

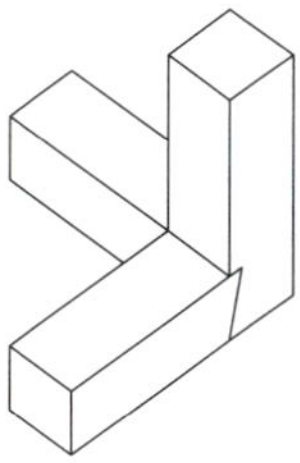
ABC

Specialty Joints

Corner-post joint with open slot mortise and tenon

tan-hozo-komi-sen-kane-gata-sanmai-kumi-tsugi
短枘込栓矩形三枚組接ぎ

Scherzapfen mit Zapfenloch und durchgehendem Zapfen

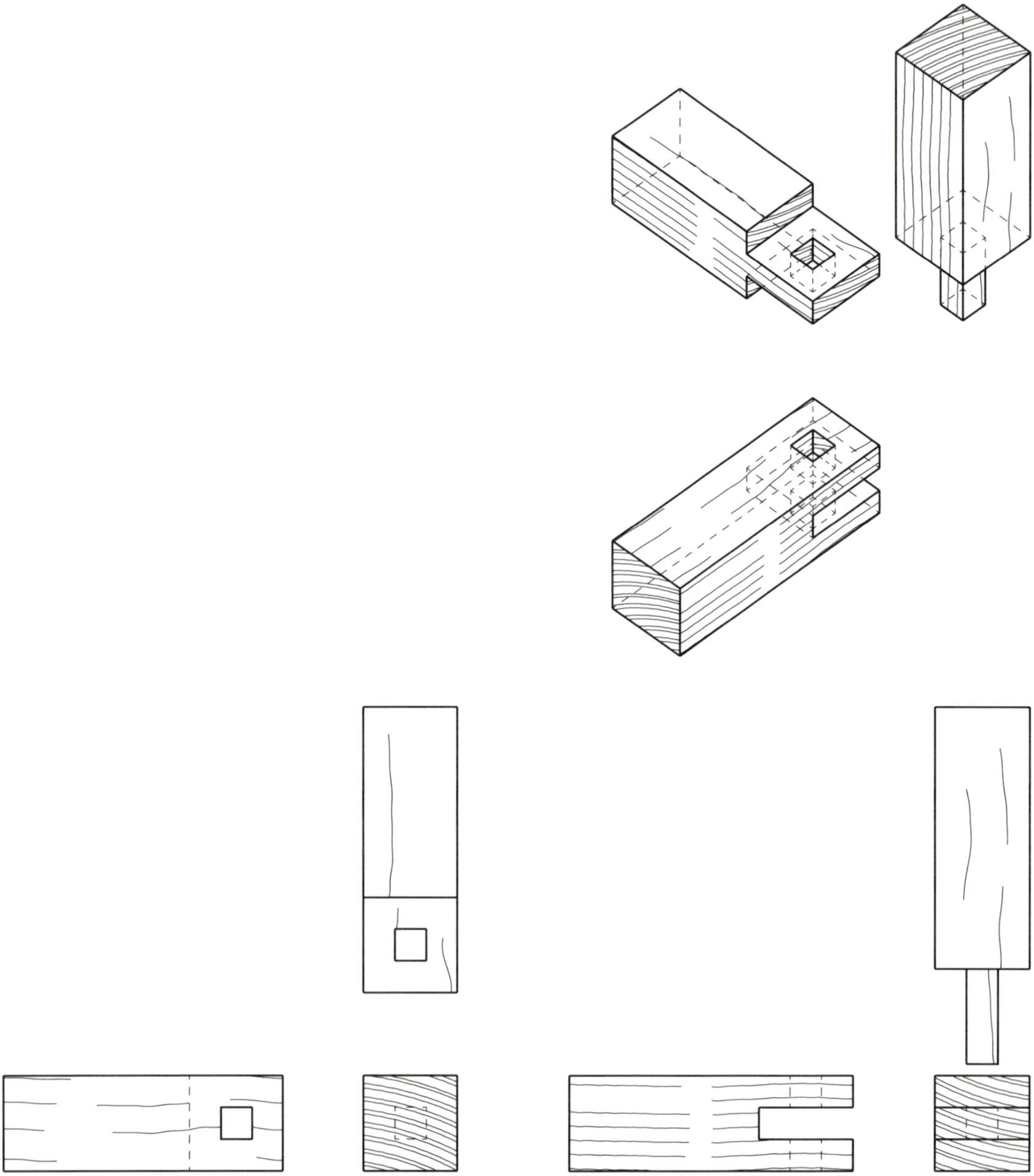

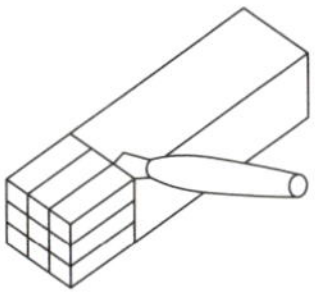
A1

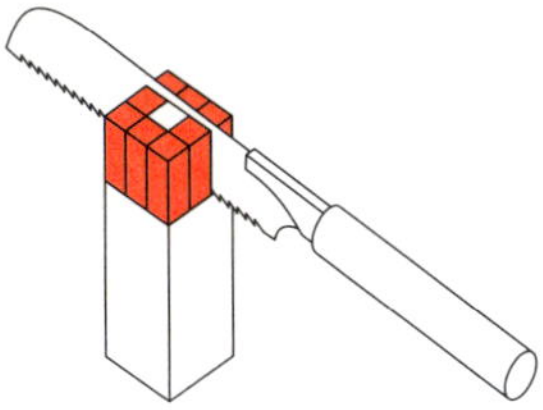
A2

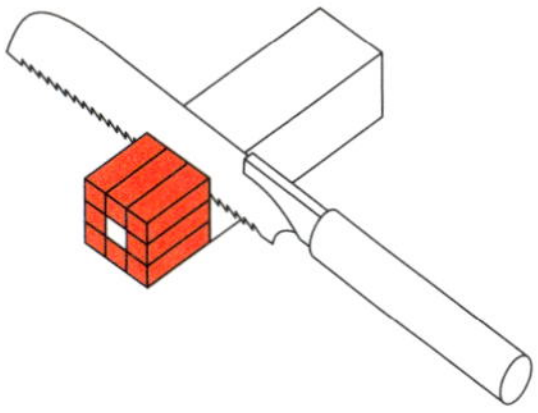
A3

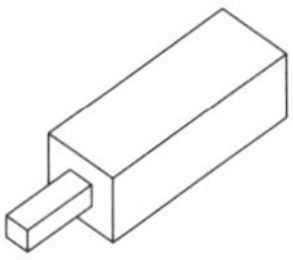
A

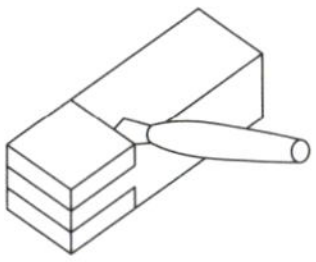
B1

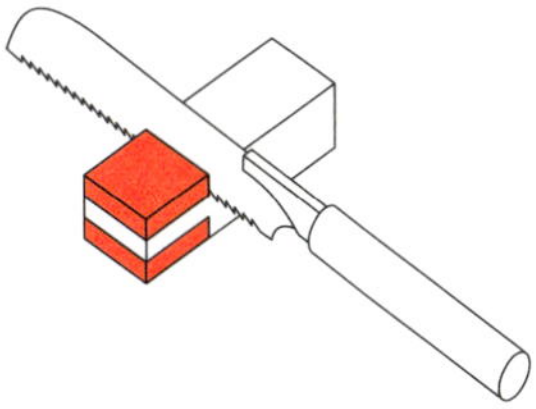
B2

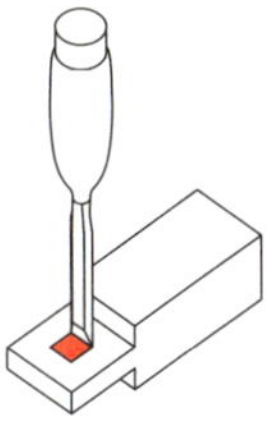
B3

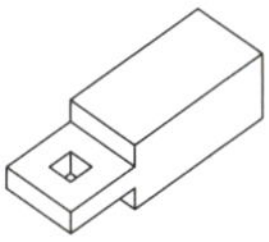
B

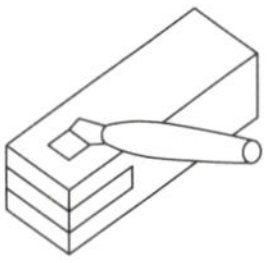
C1

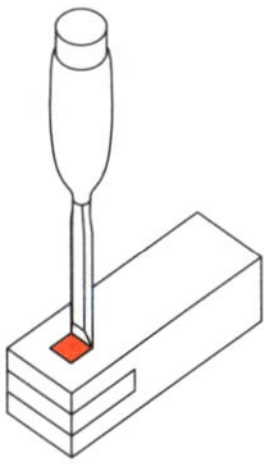
C2

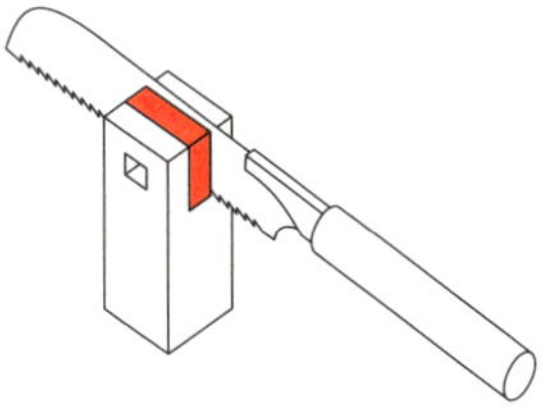
C3

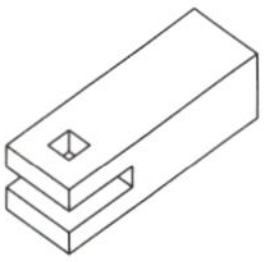
C

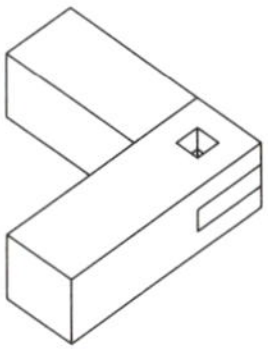
B+C

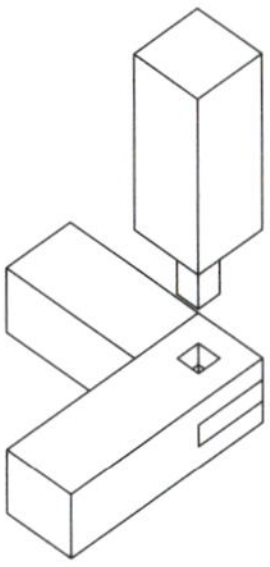
A+BC

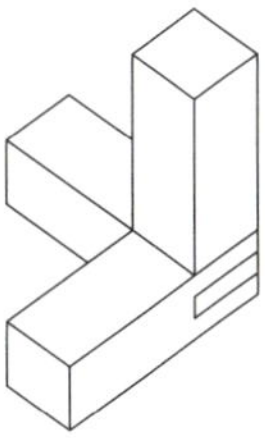
ABC

Specialty Joints

Corner-post joint with fan-shaped stub tenon and blind dado and rabbet joint

daiwa-dome-ōgi-hozo-sashi [*]
台輪留扇枘差し

Eckstoß mit Führungszapfen
und Wange auf Gehrung sowie
durchgehendem trapezförmigem Zapfen

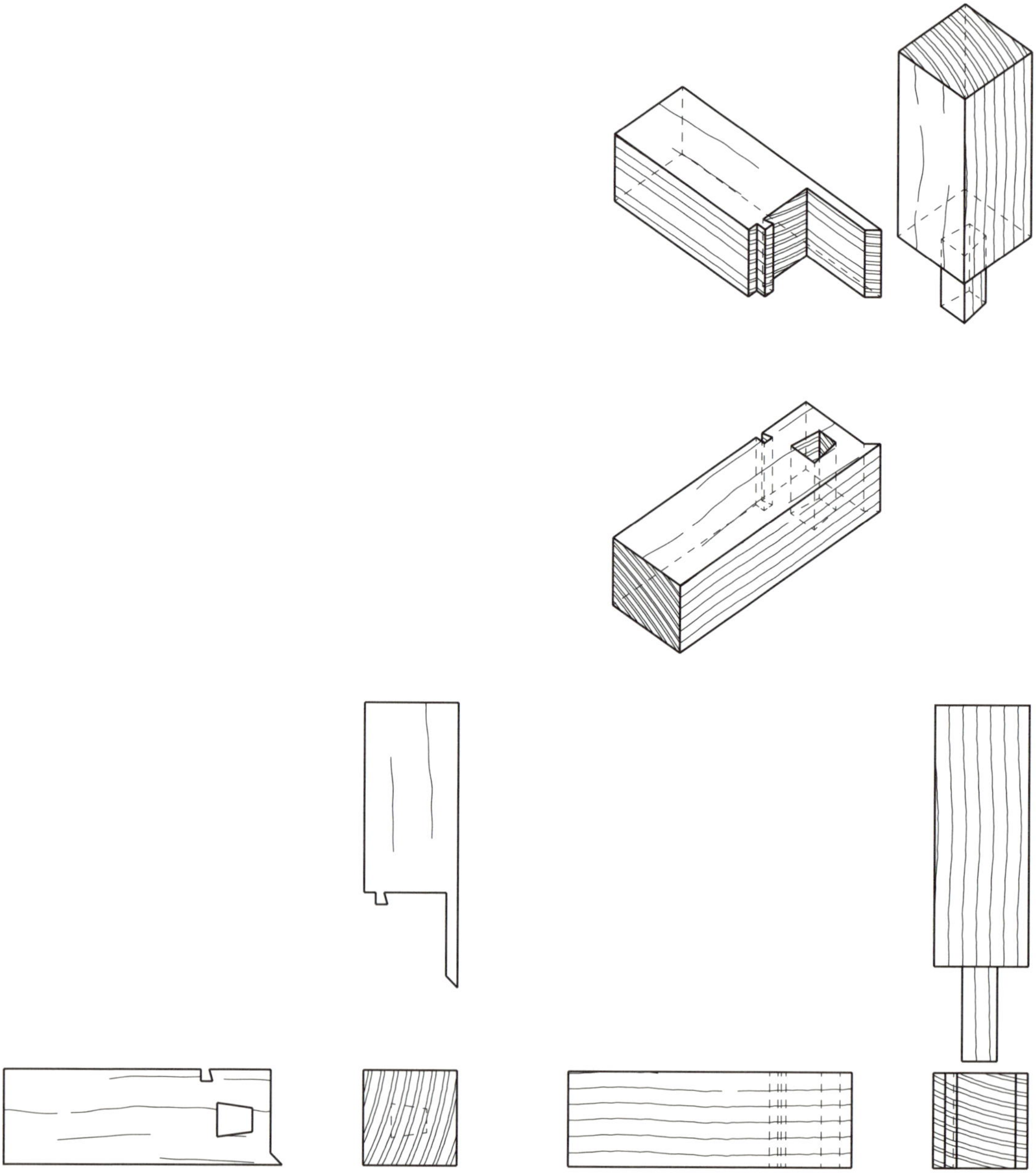

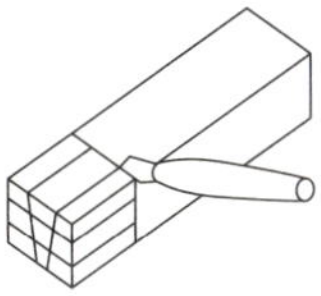

A1

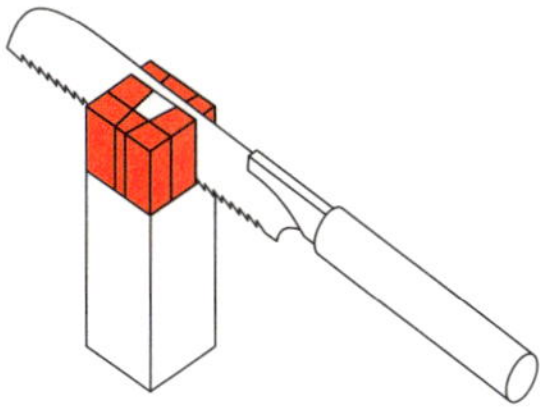

A2

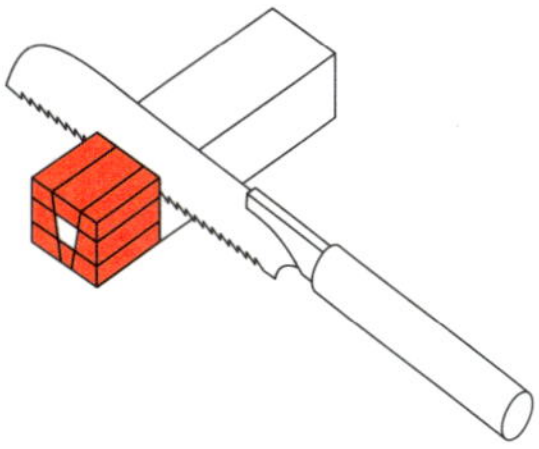

A3

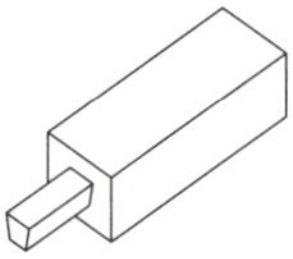

A

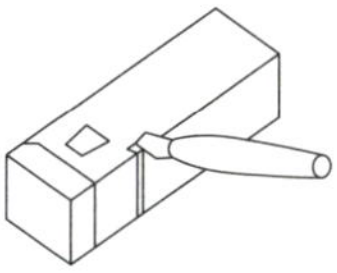

B1

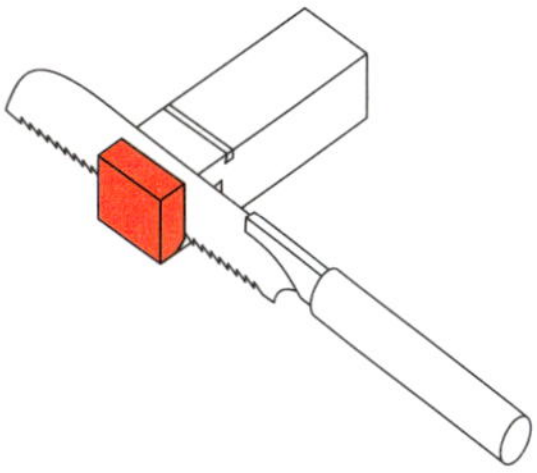

B2

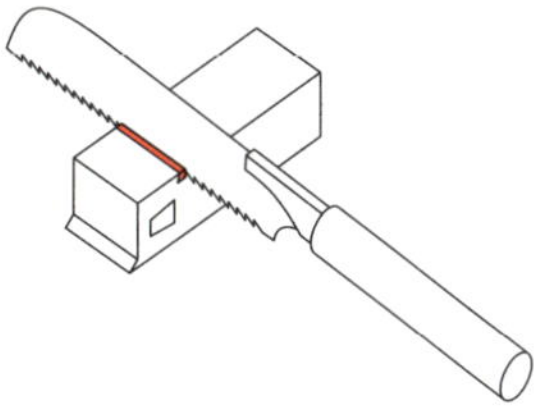

B3

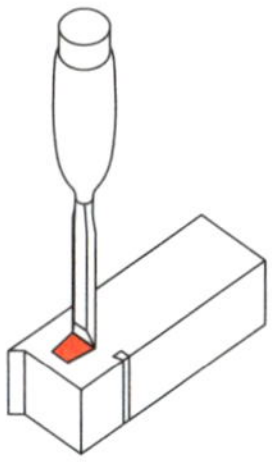

B4

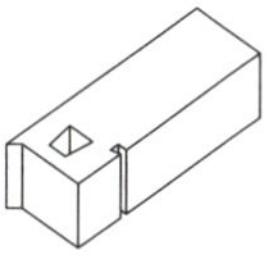

B

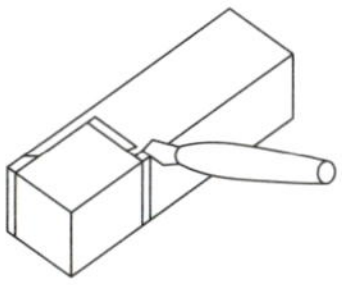

C1

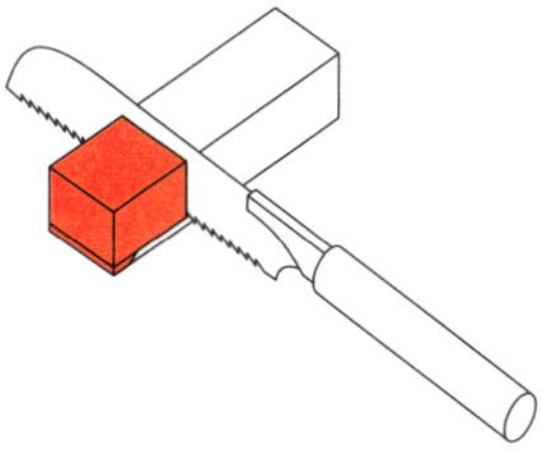

C2

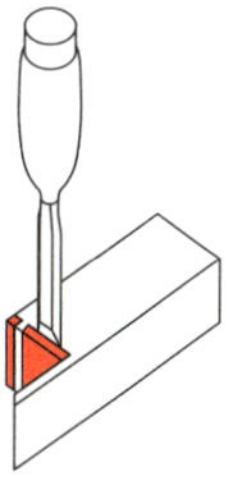

C3

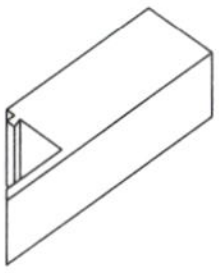

C

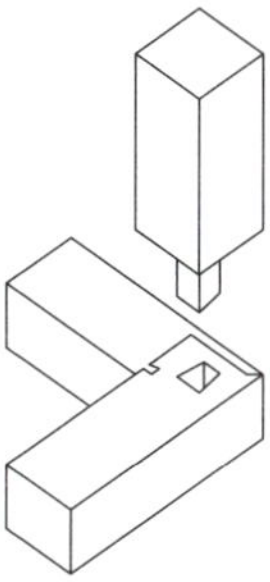

A+BC

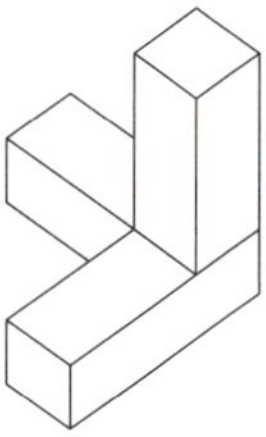

ABC

Specialty Joints

Housed and stopped dovetail lap and fully shouldered stub tenon

koshi-kake-ari-tsugi, hira-hozo-sashi [✻]
腰掛蟻接ぎ、平枘差し

verdeckter eingelassener Schwalbenschwanzzapfen mit abgesetzter Brust und Pfosten mit verdecktem Zapfen

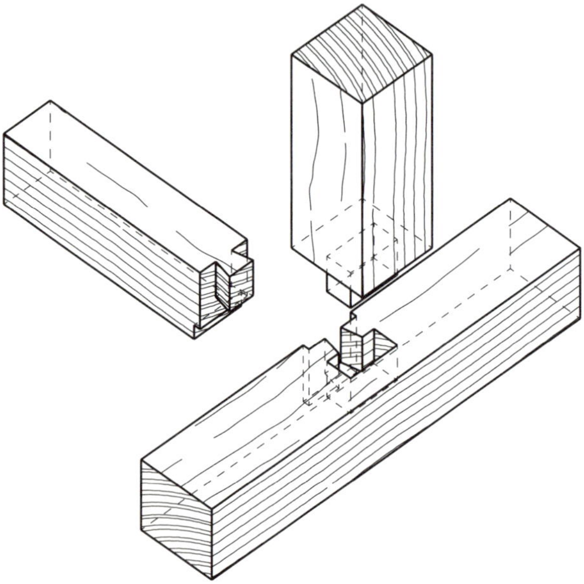

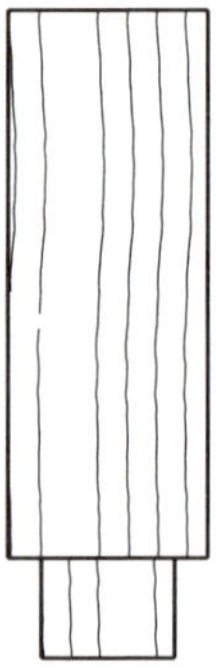

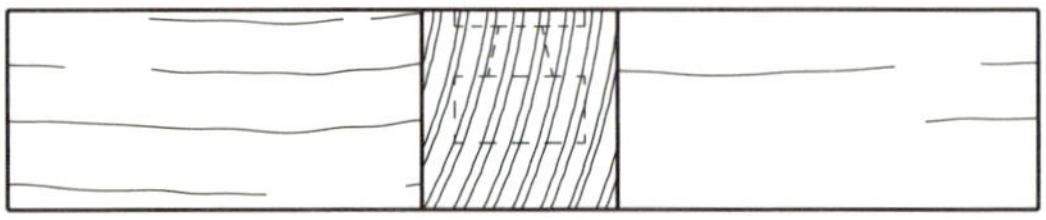

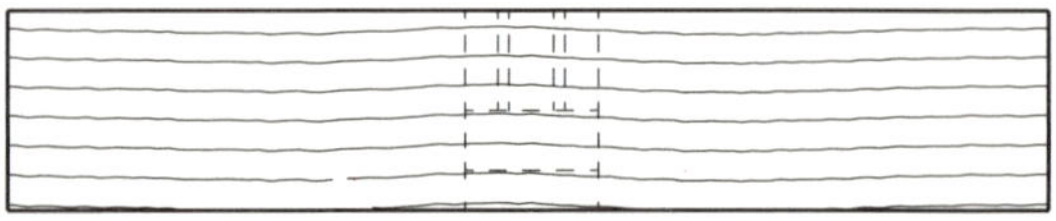

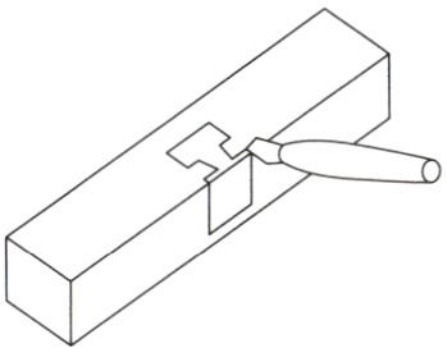

A1

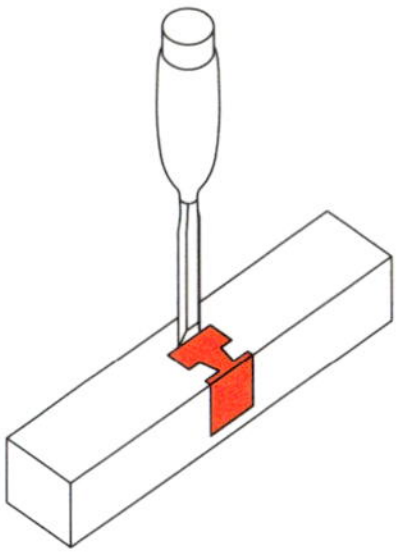

A2

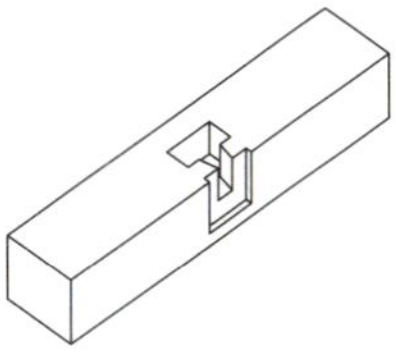

A

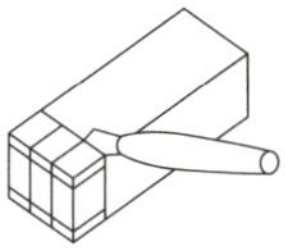

B1

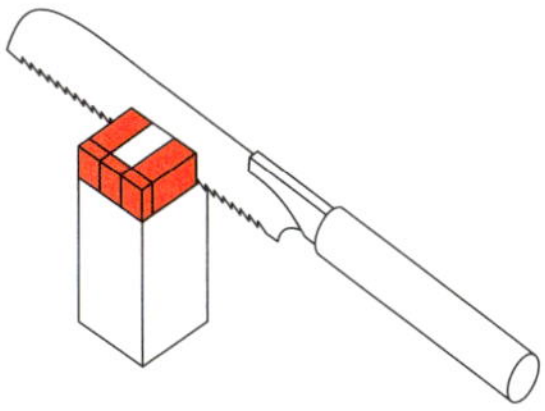

B2

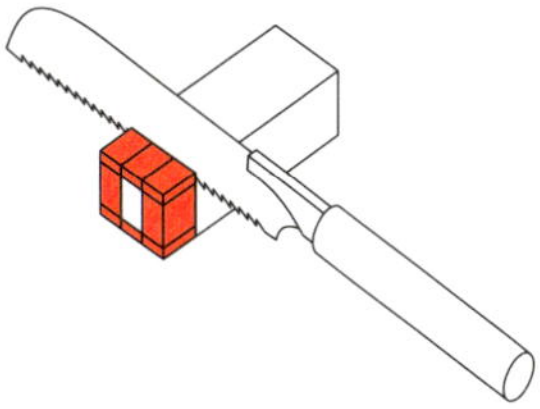

B3

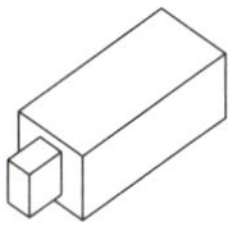

B

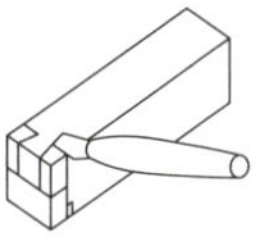

C1

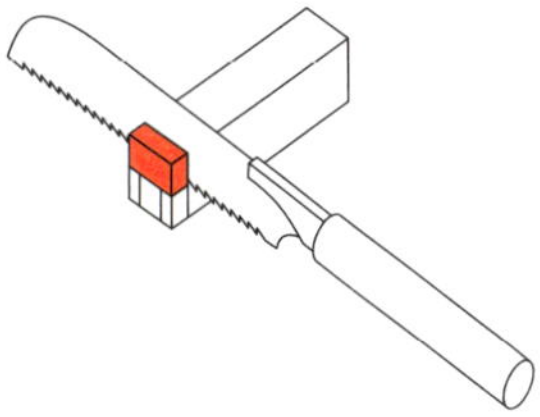

C2

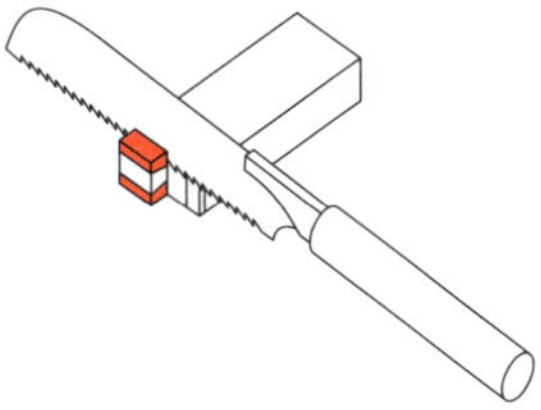

C3

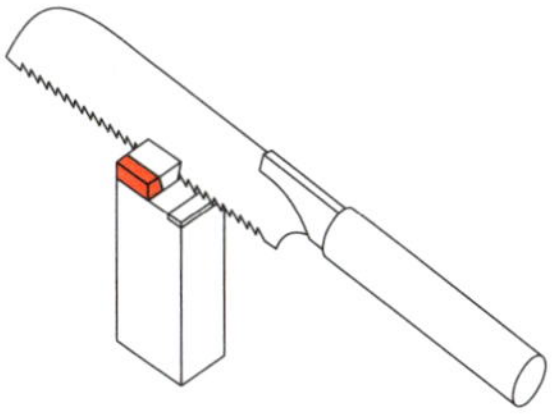

C4

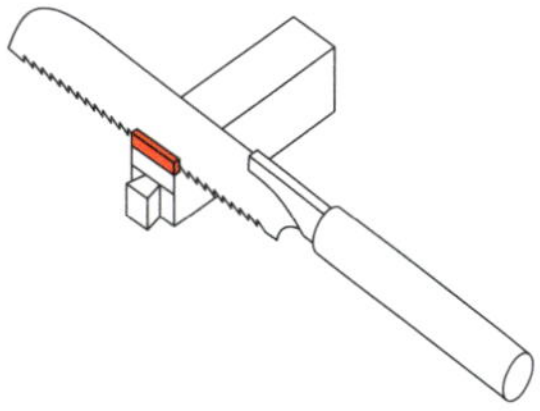

C5

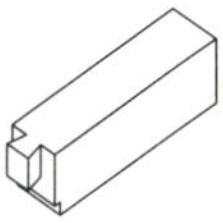

C

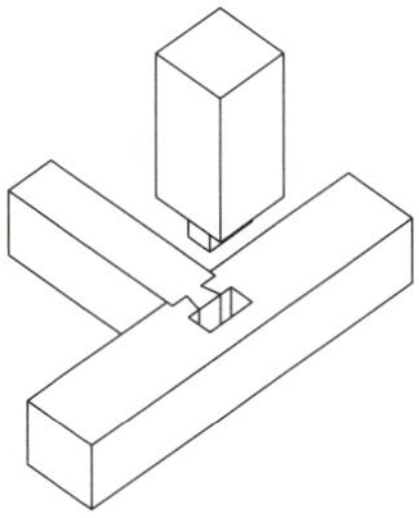

A+BC

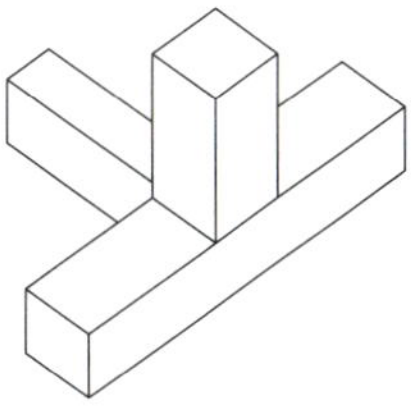

ABC

Specialty Joints

Lapped and haunched mortise and tenon joint

shitaba-kone-hozo-nihō-sashi
下端小根枘二方差し

wechselseitig abgesetzte
verdeckte Stufenzapfen

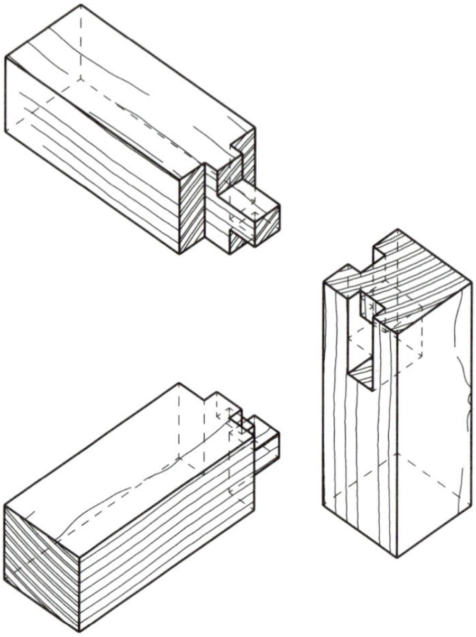

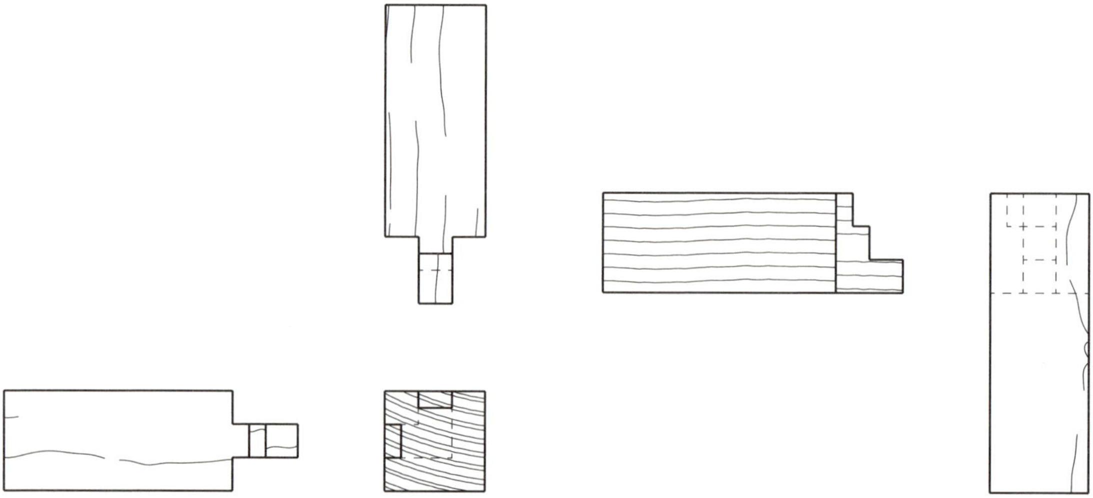

2×

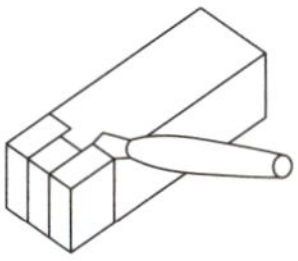

AB1

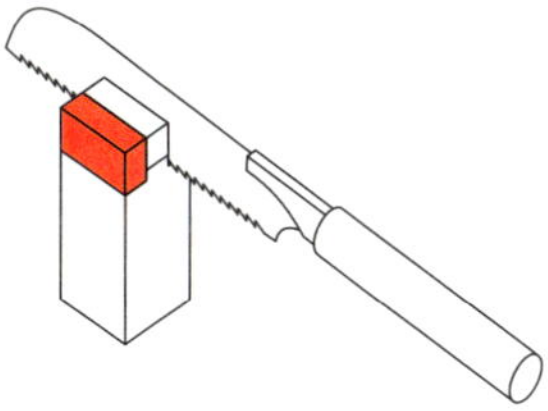

AB2

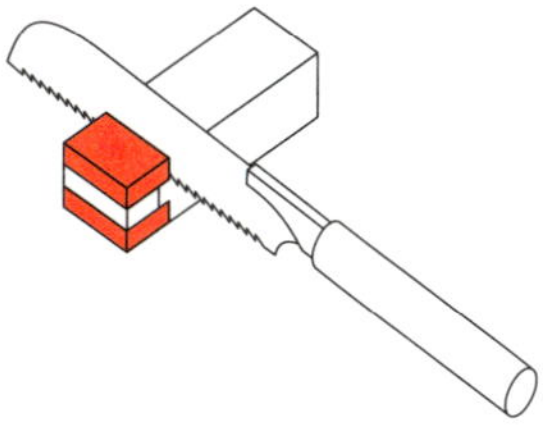

AB3

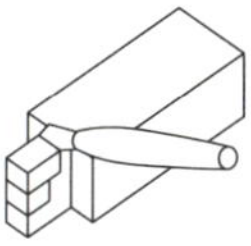

A1

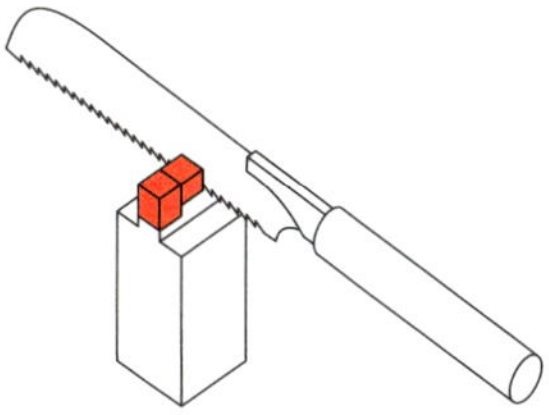

A2

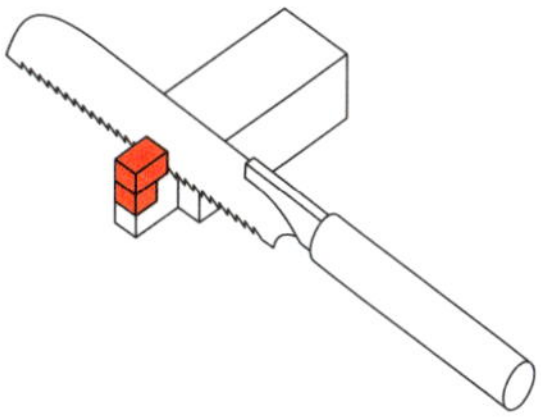

A3

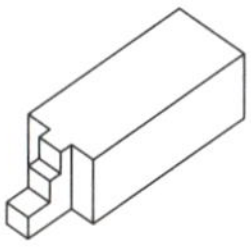

A

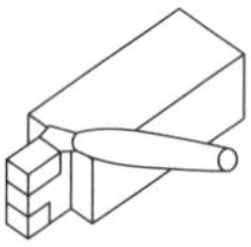

B1

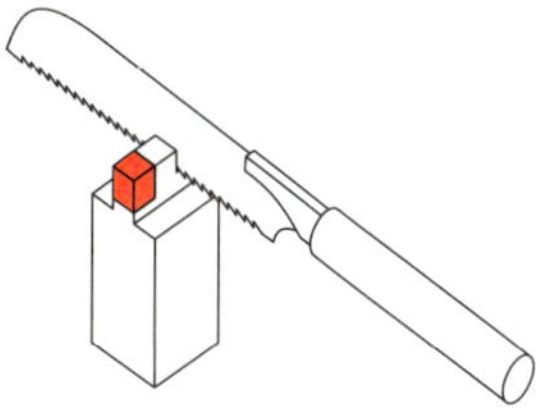

B2

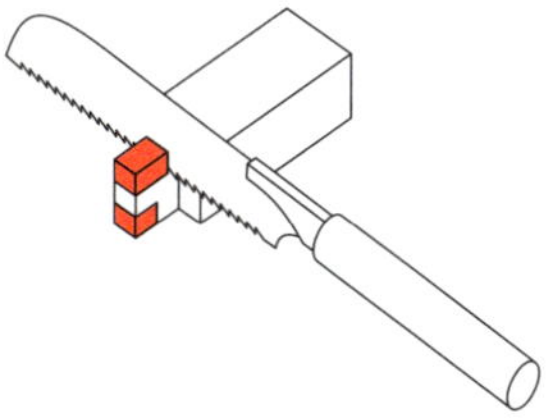

B3

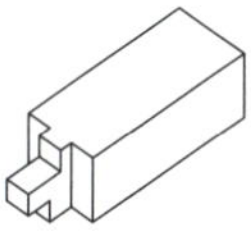

B

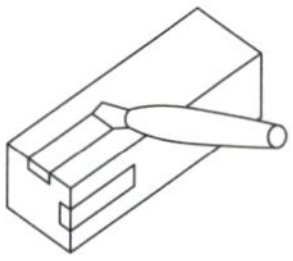

C1

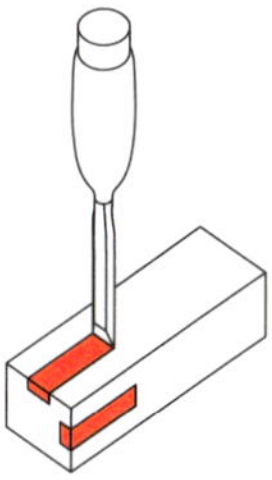

C2

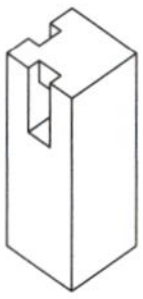

C

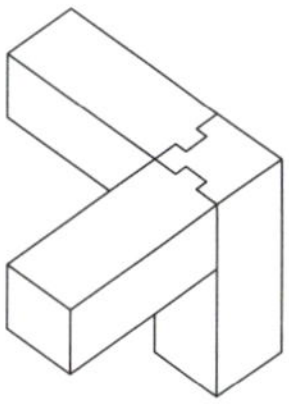

ABC

Specialty Joints

Mitered and haunched mortise and tenon joint

shitaba-kone-hozo-ai-sogi-nihō-sashi
下端小根枘相殺二方差し

verdeckte Stufenzapfen auf Gehrung

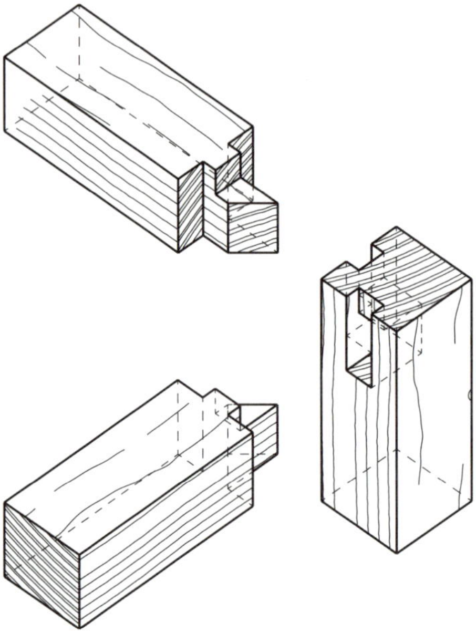

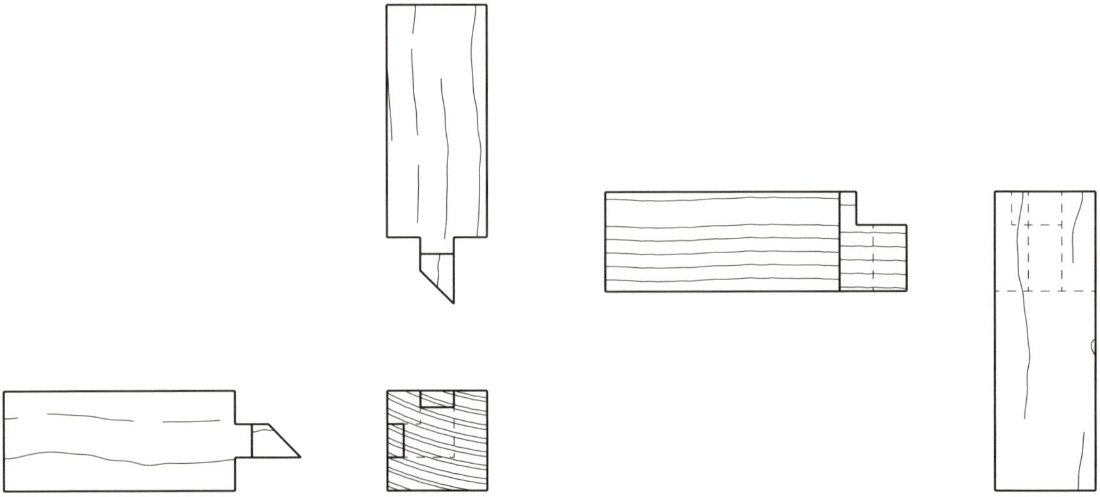

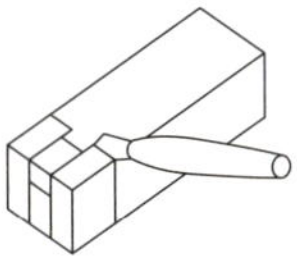
AB1

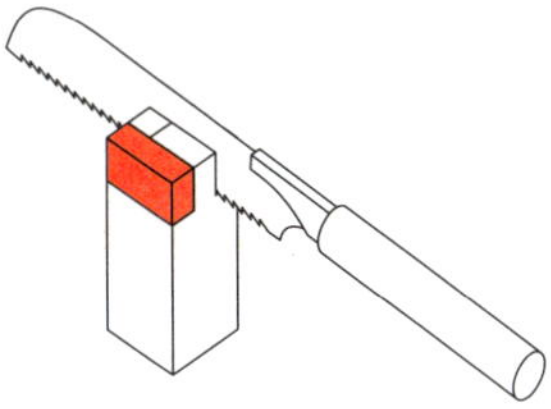
AB2

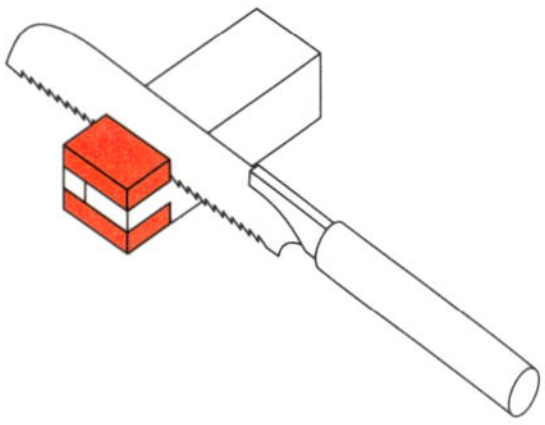
AB3

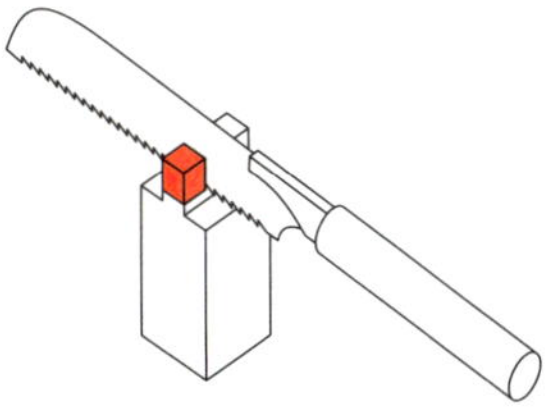
AB4

2×

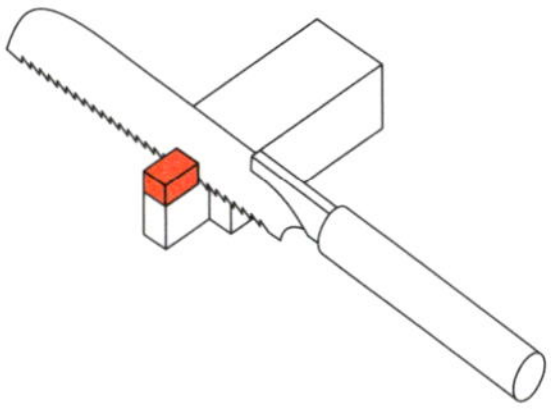
AB5

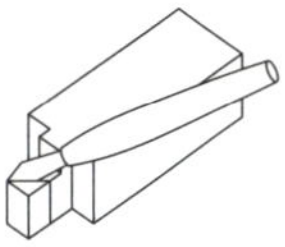
A1

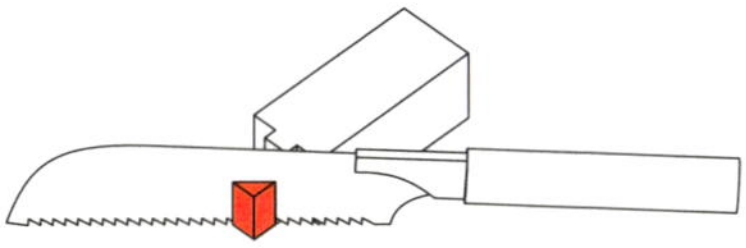
A2

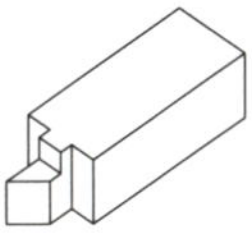
A

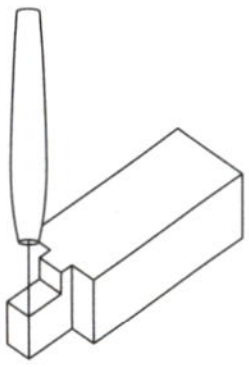
B1

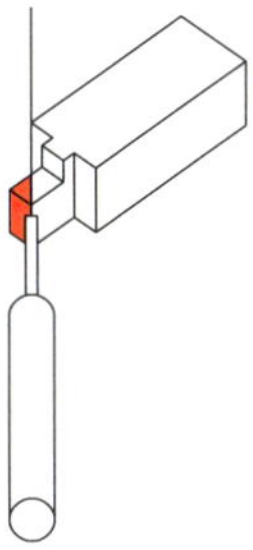
B2

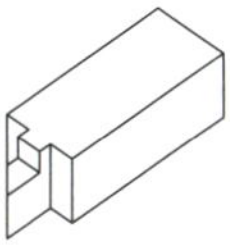
B

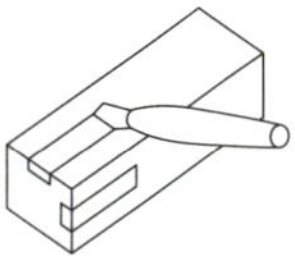
C1

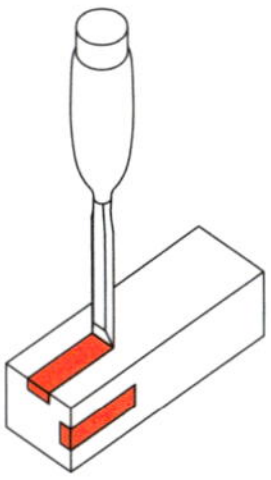
C2

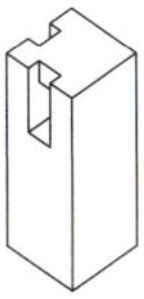
C

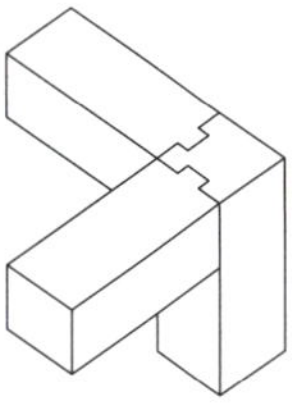
ABC

Specialty Joints

Chidori puzzle joint

chidori-kōshi
千鳥格子

einfach abgesetztes Kreuzblatt
und Stab mit Rundholzverbindung

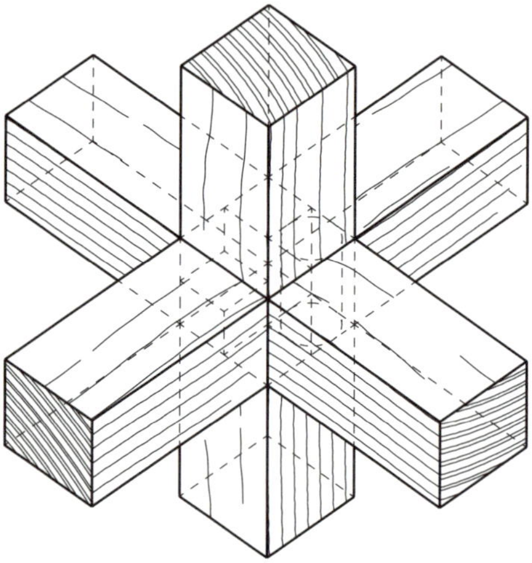

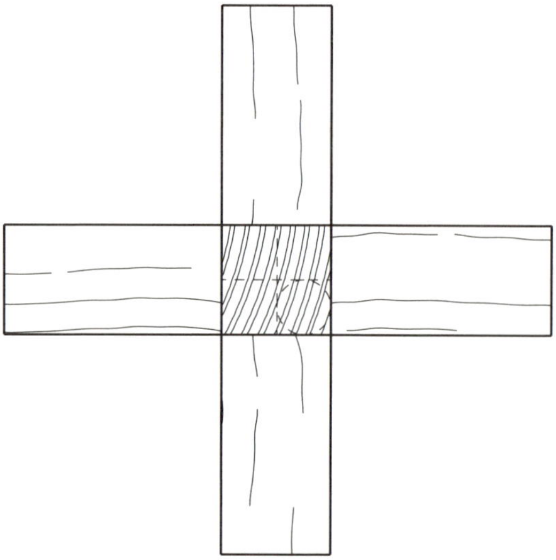

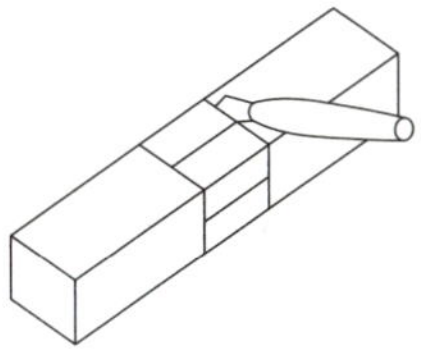

A1

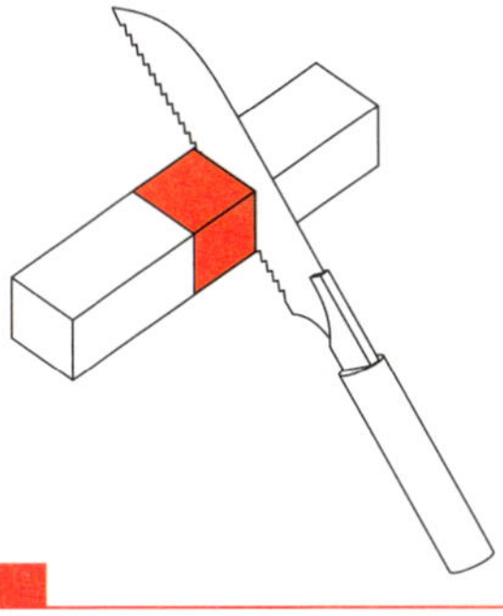

A2

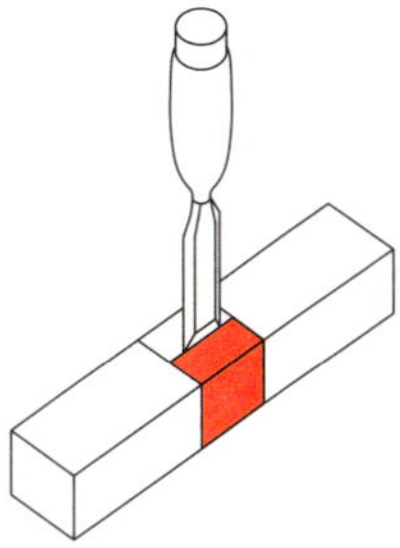

A3

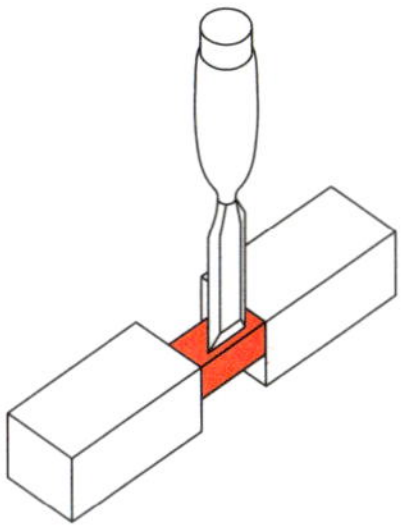

A4

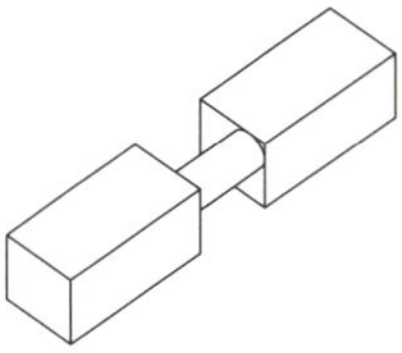

A

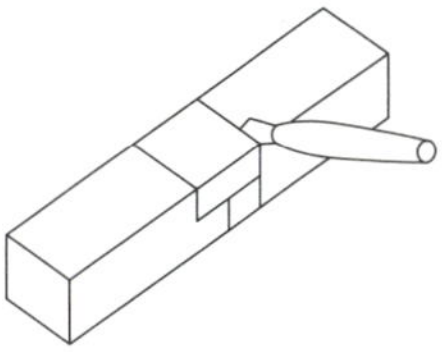

B1

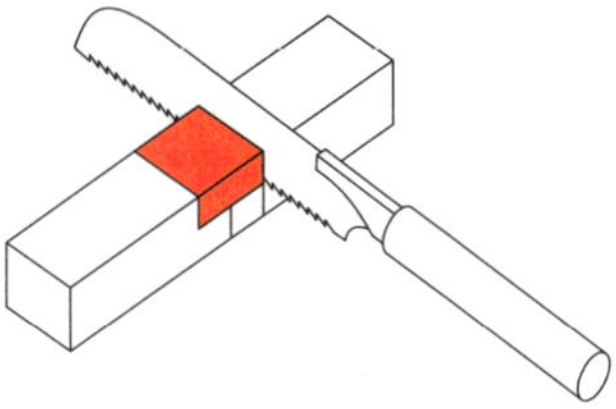

B2

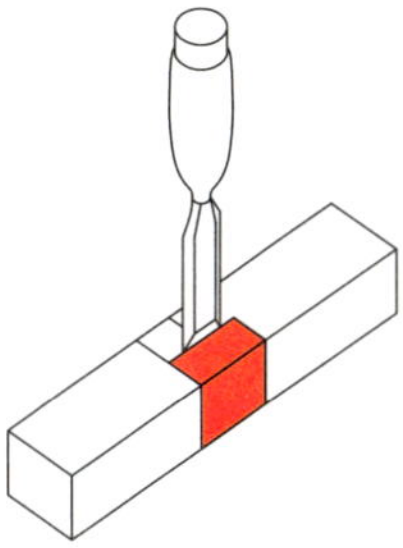

B3

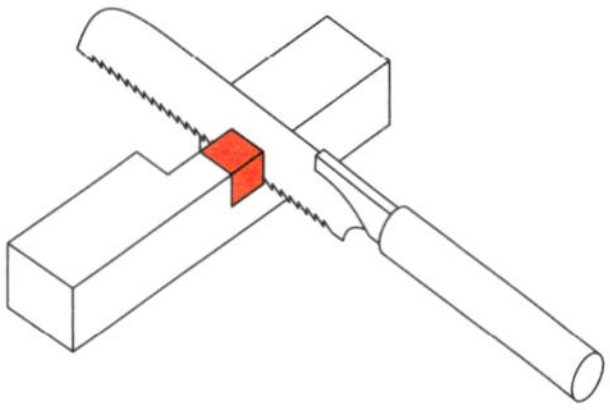

B4

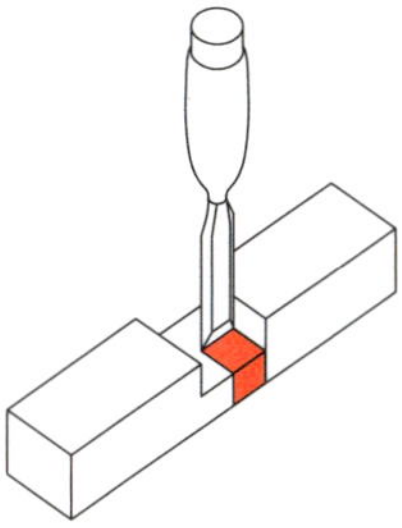

B5

2×

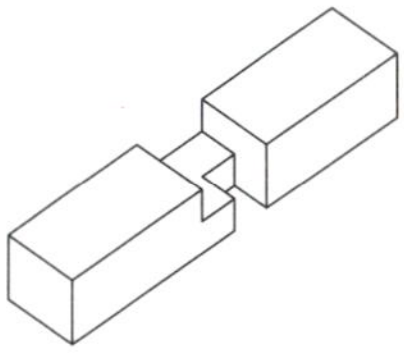

B

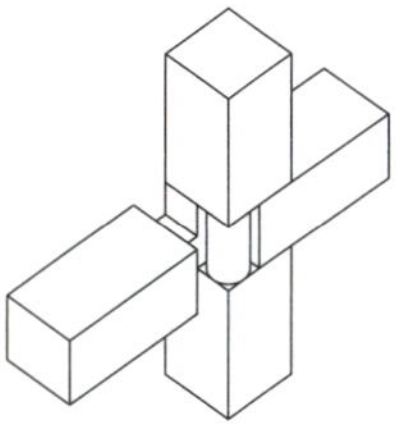

A+B

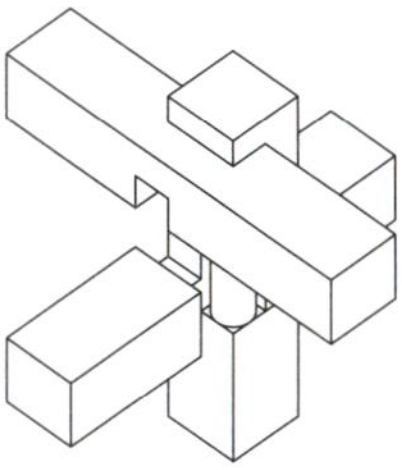

AB+B

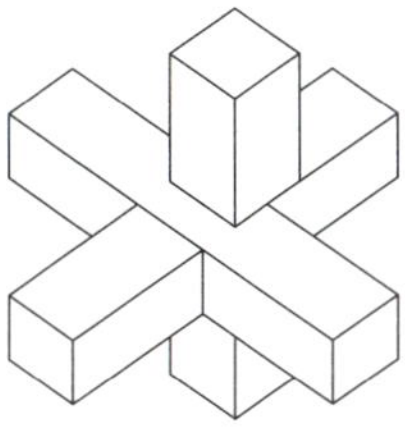

AB+B

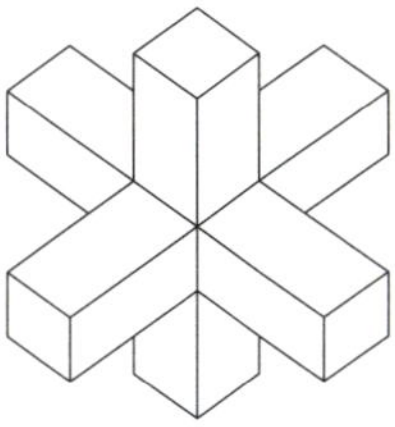

ABB

Specialty Joints

Cross puzzle

roppon-kigumi-pazuru (kan-i)
六本木組パズル(簡易)

Zimmermannsknoten

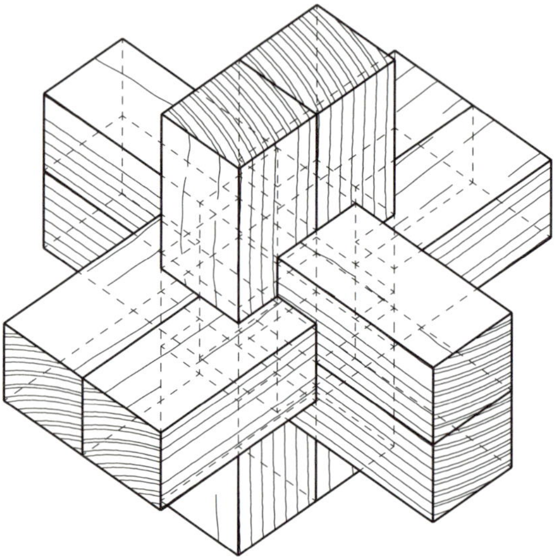

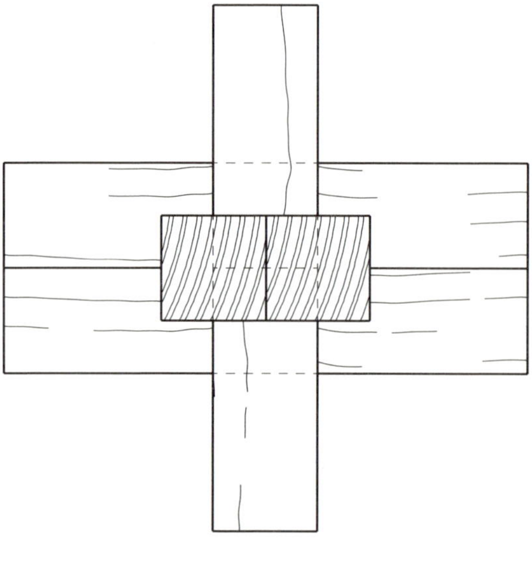

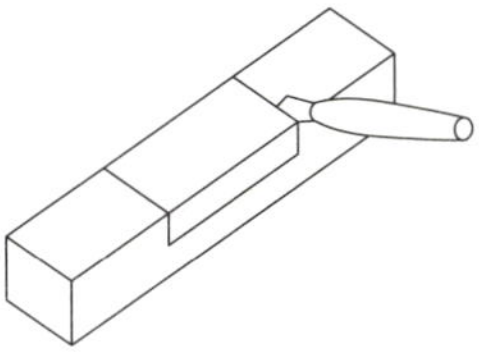
A1

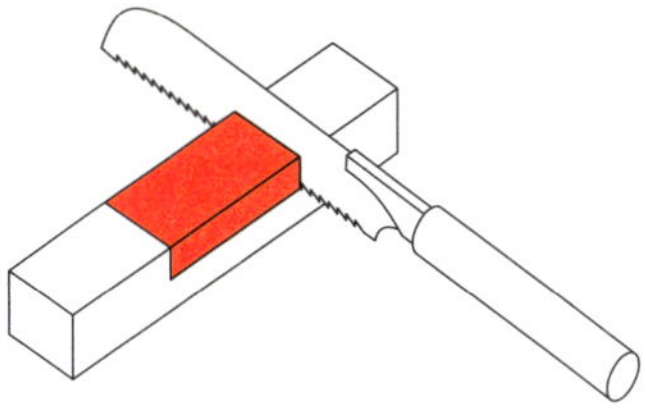
A2

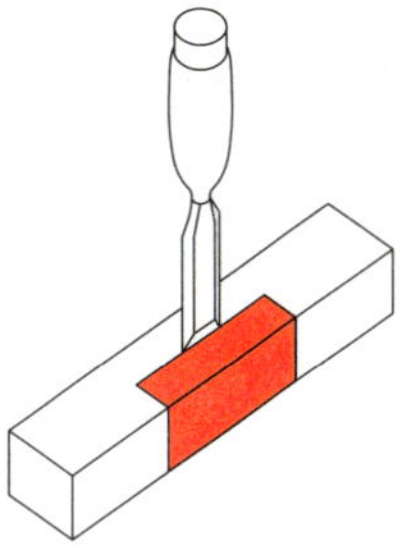
A3

3×

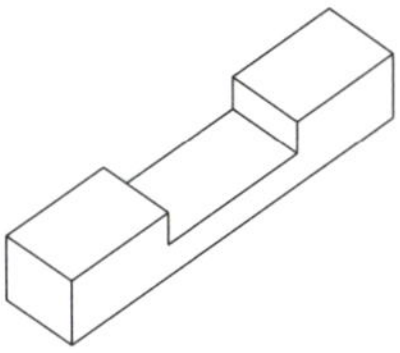
A

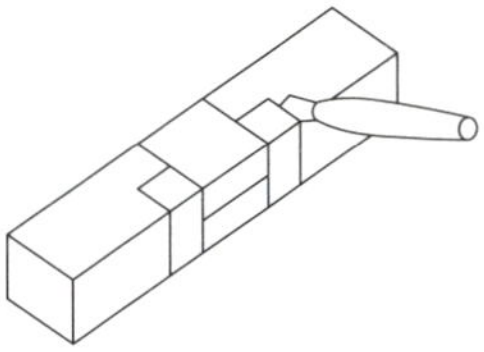
B1

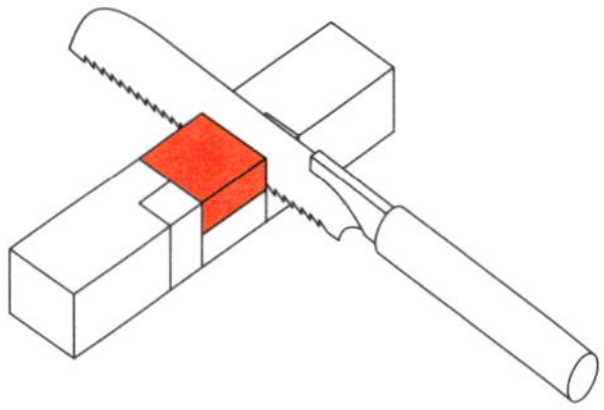
B2

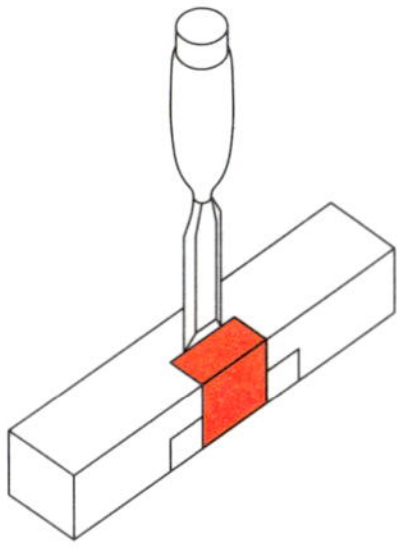
B3

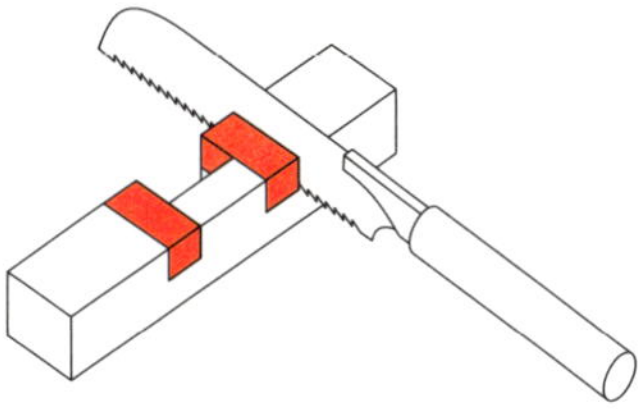
B4

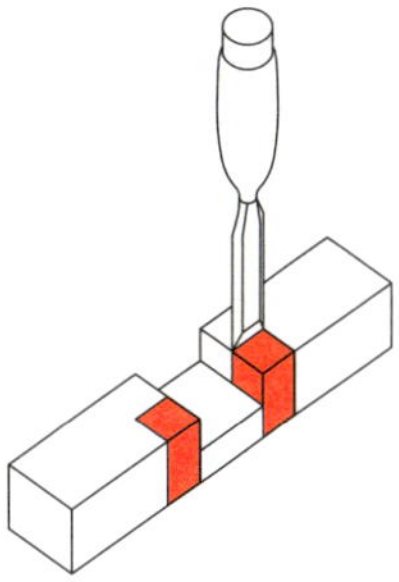
B5

2×

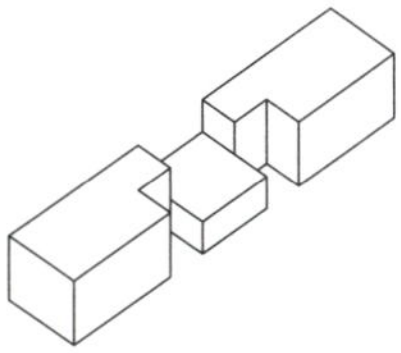
B

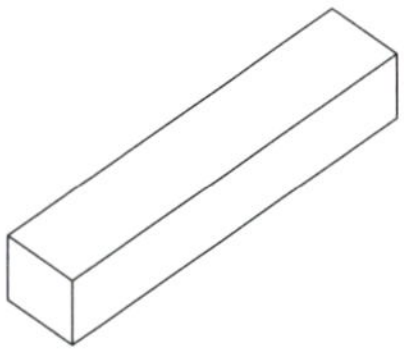
C

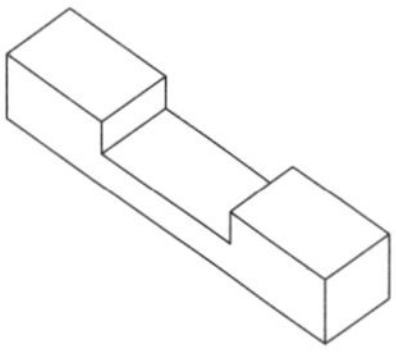
A

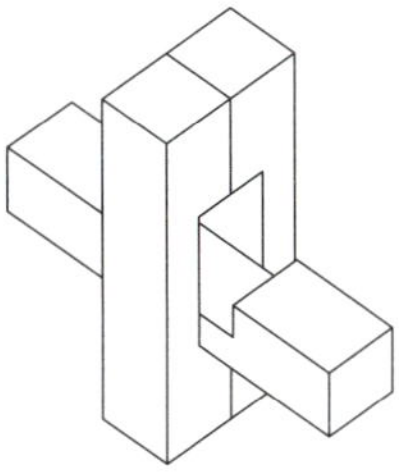
AAA

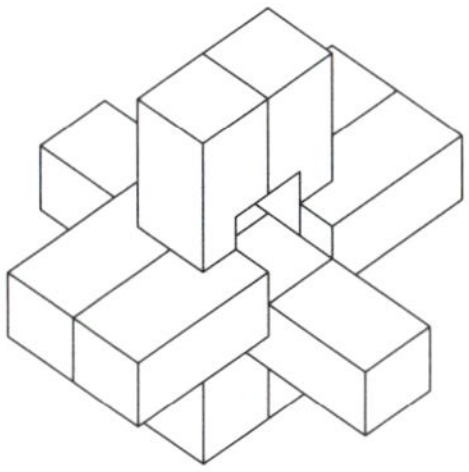
AAABB

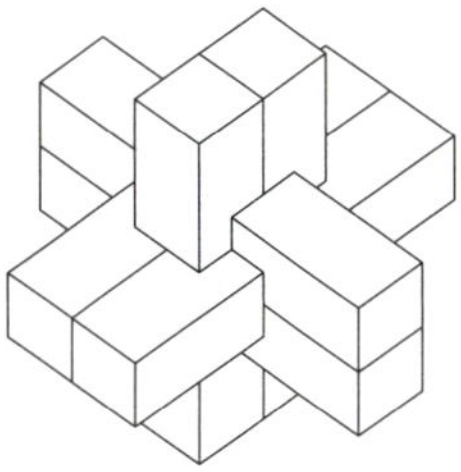
AAABBC

Diagonal cross puzzle

kyūhon-kigumi-pazuru
九本木組パズル

diagonaler Zimmermannsknoten

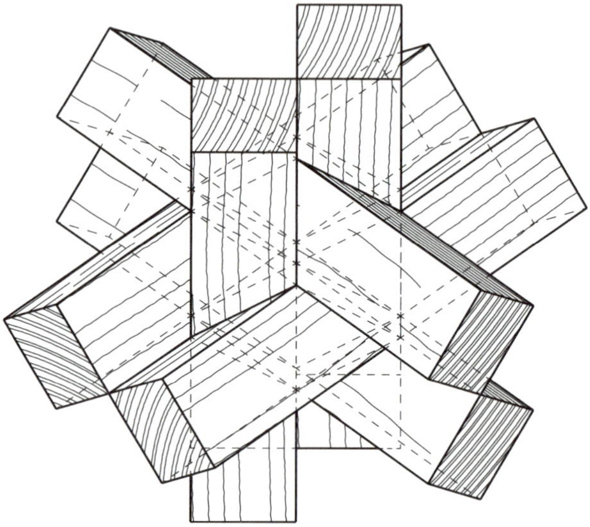

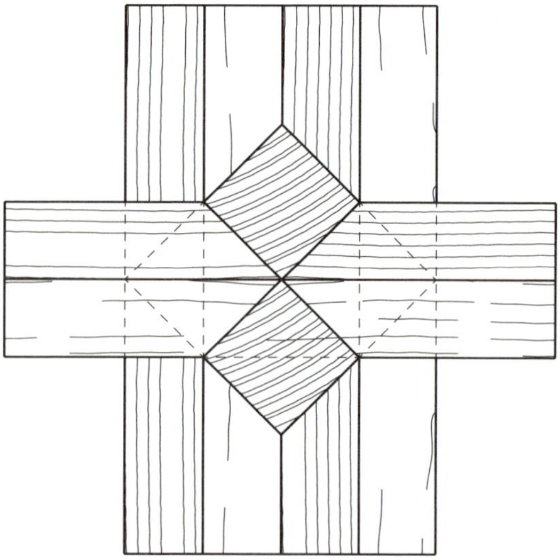

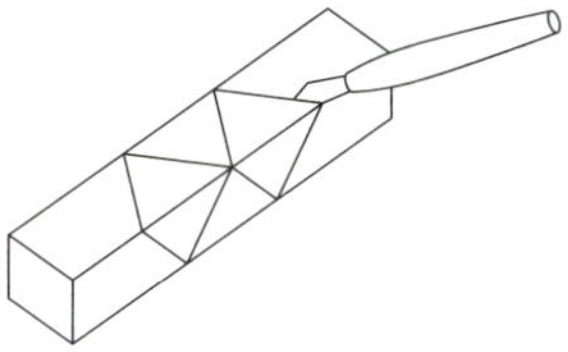

A1

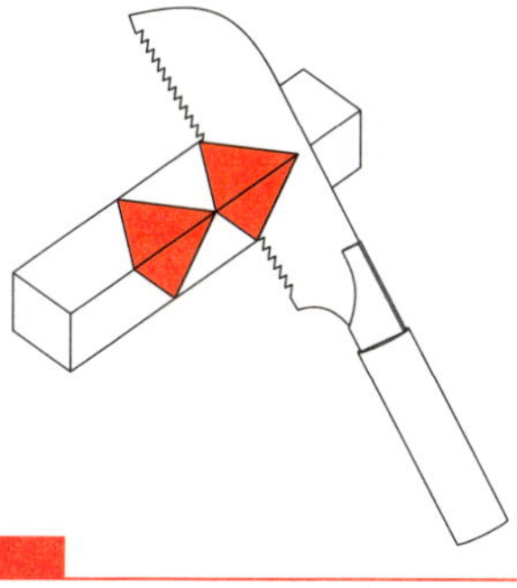

A2

3×

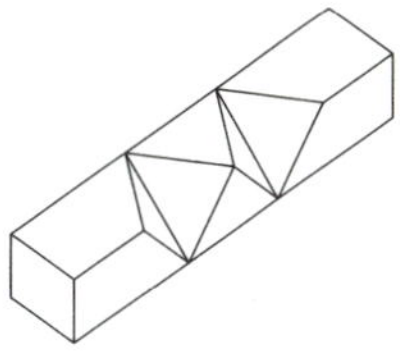

A

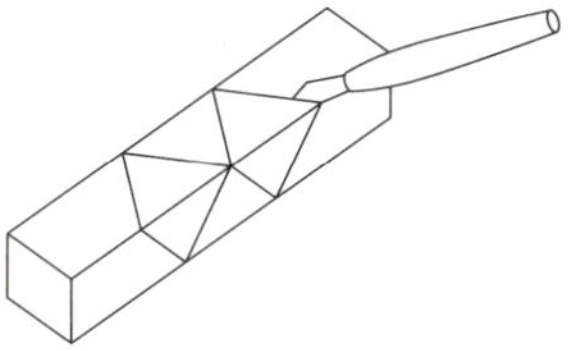

B1

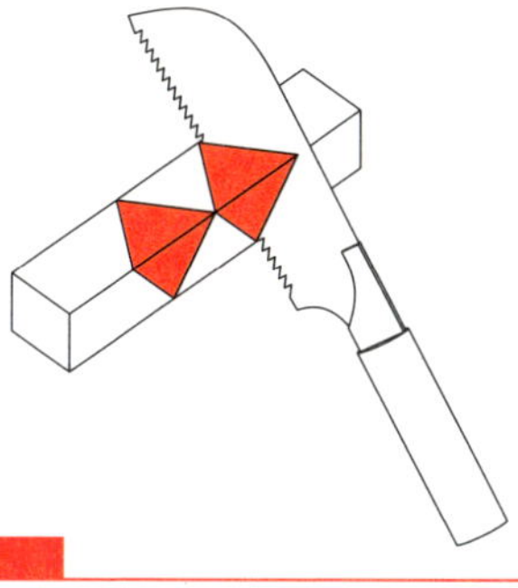

B2

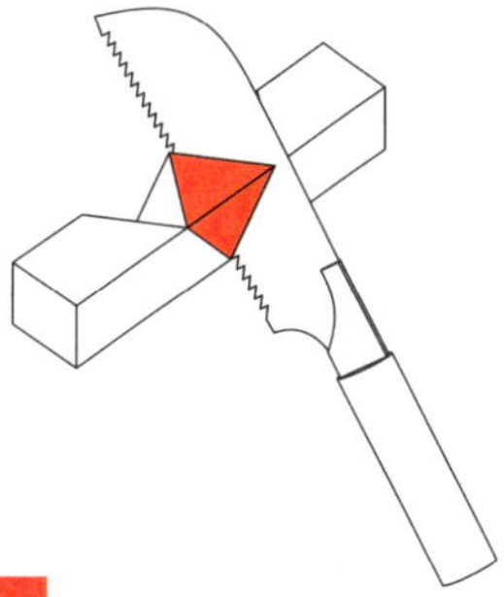

B3

2×

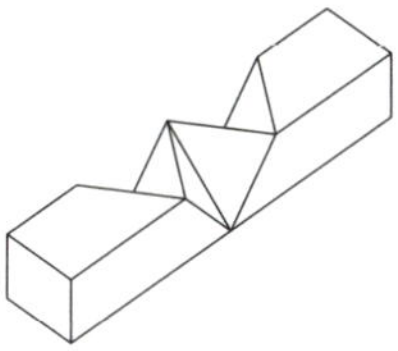

B

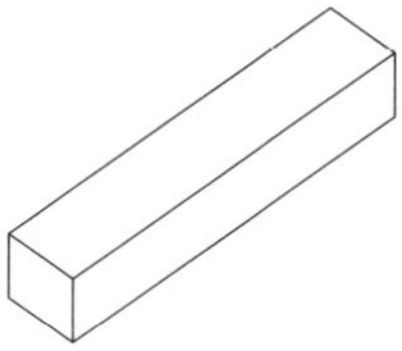

C

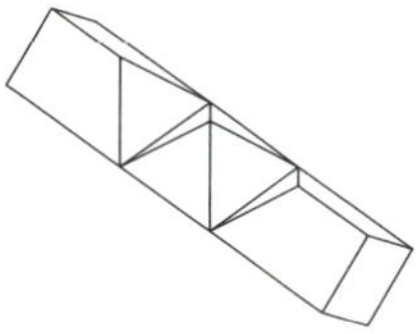

A

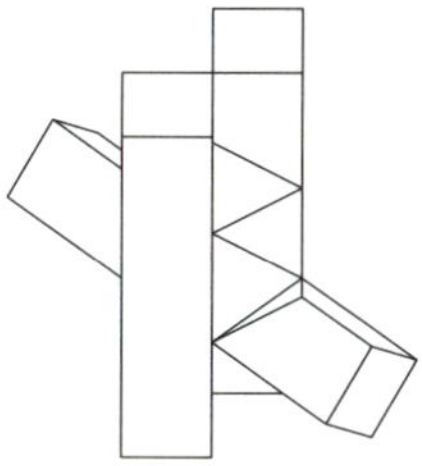

AAA

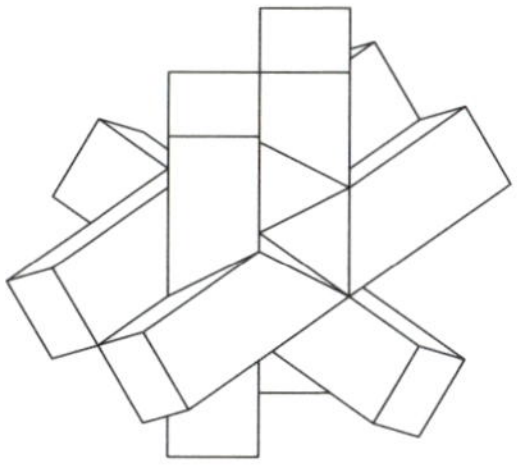

AAABB

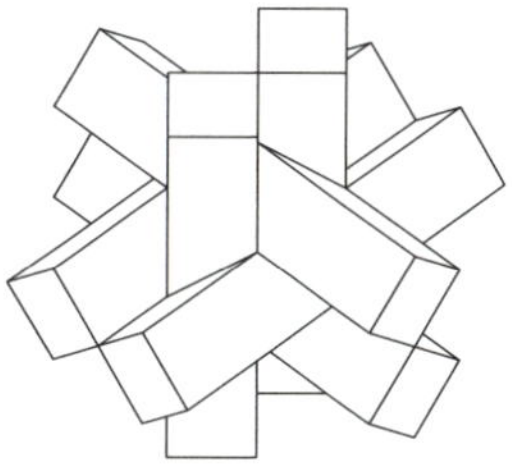

AAABBC

Specialty Joints

Wedged two-way cogged lap joint

sumi-ai-kama-kusabi-shime [*]
隅相鎌楔締め

verkeiltes gerades Hakendruckschloss

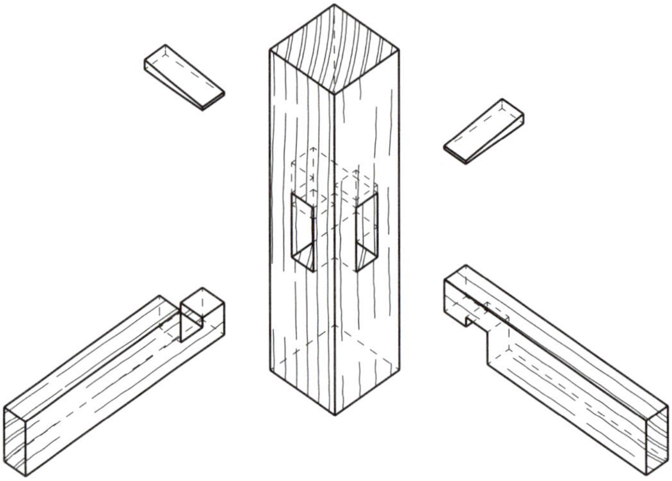

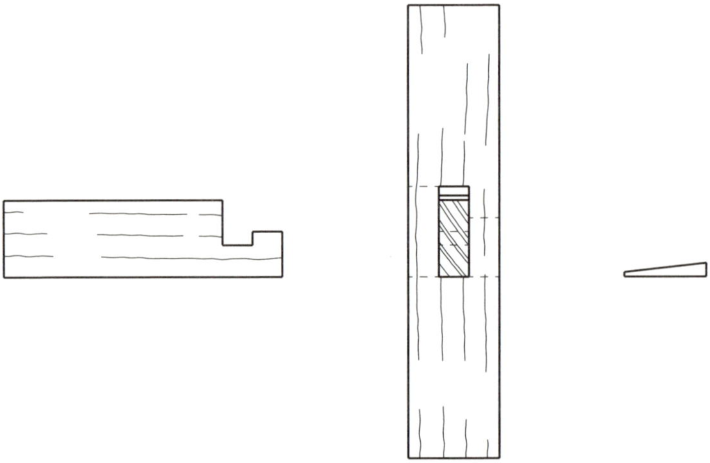

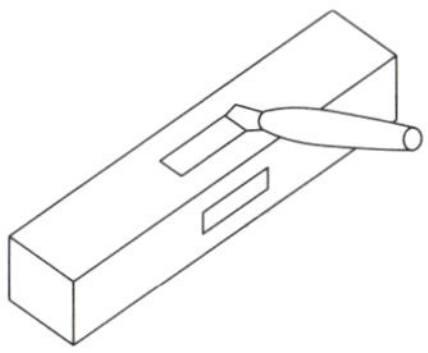

A1

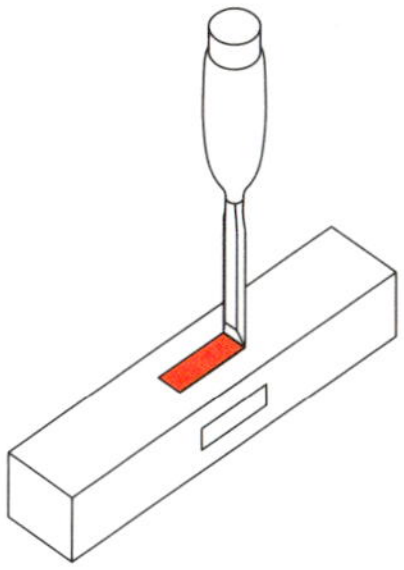

A2

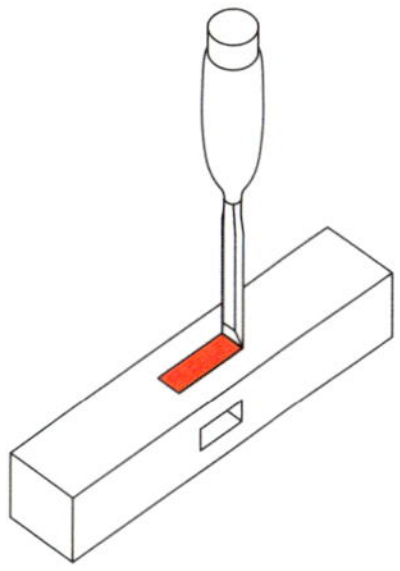

A3

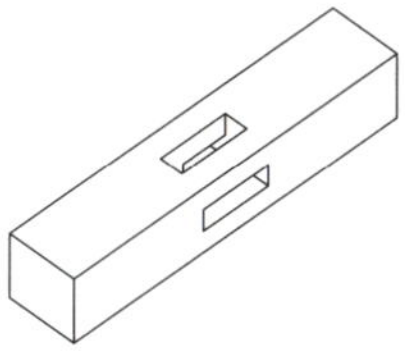

A

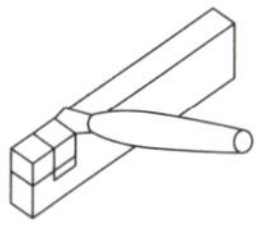

B1

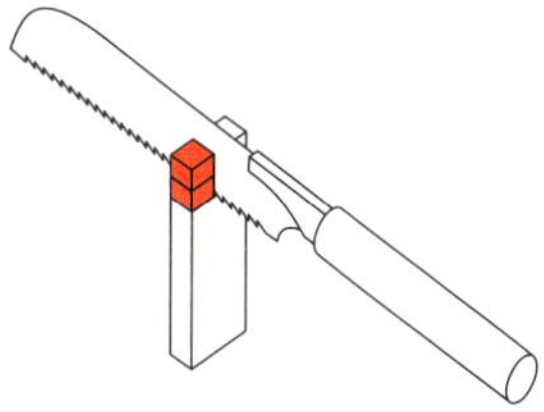

B2

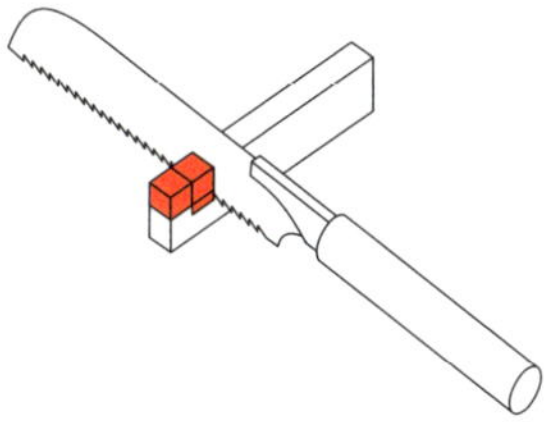

B3

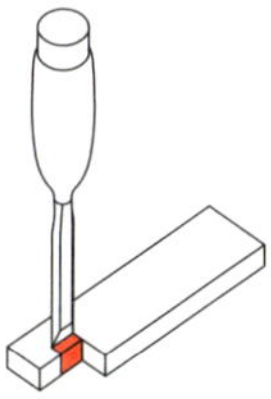

B4

2×

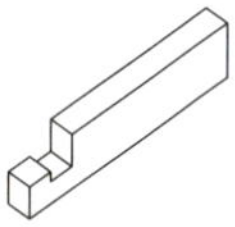

B

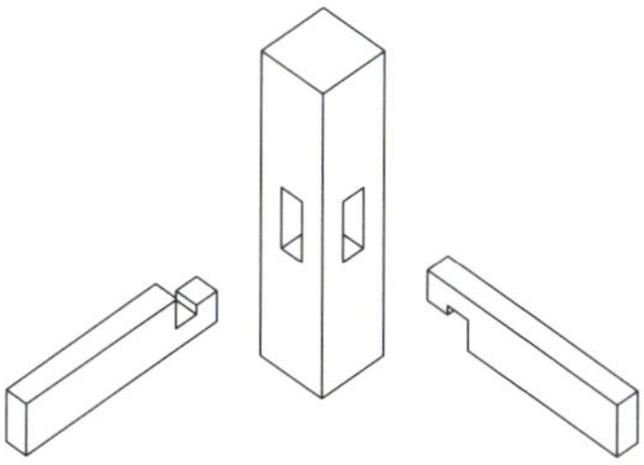

A+B+B

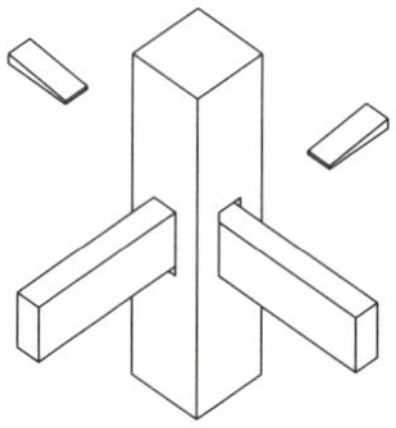

A+B+B

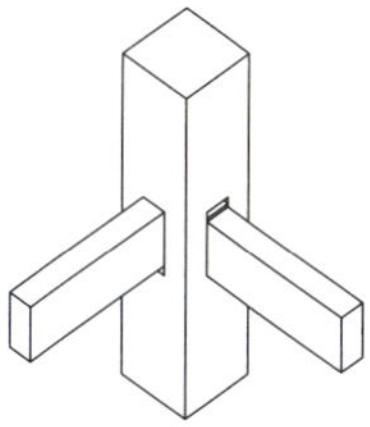

ABB

Pegged two-way splayed scarf joint

sogi-tsugi-komi-sen-uchi
殺継込栓打ち

eingeschobene Schäftung
mit Dübelsicherung

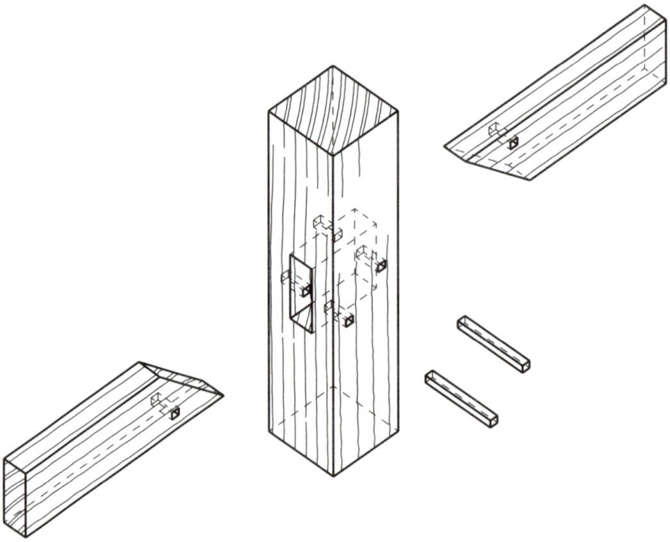

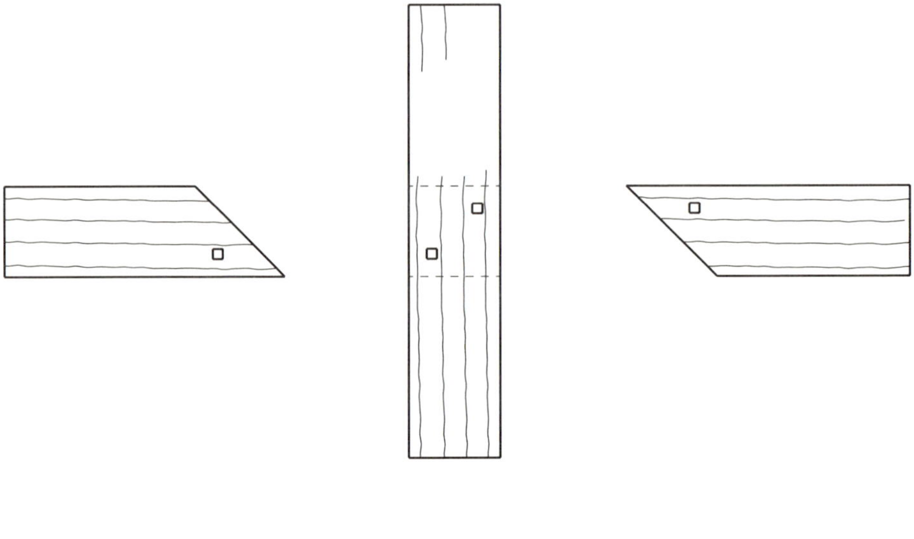

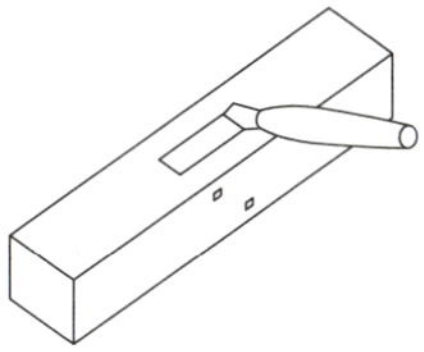

A1

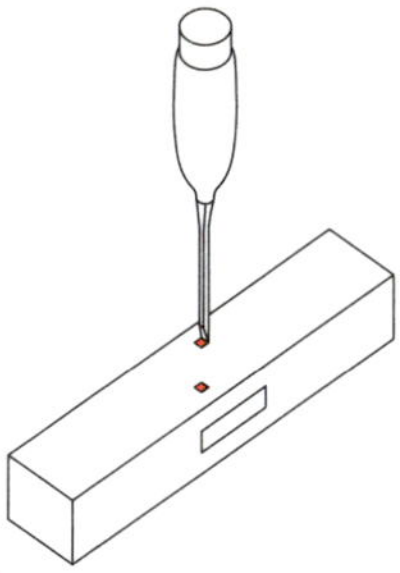

A2

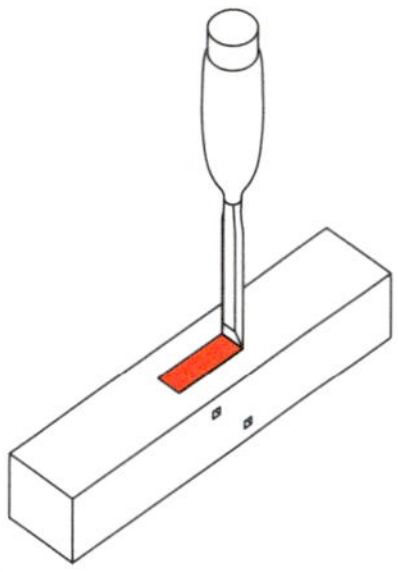

A3

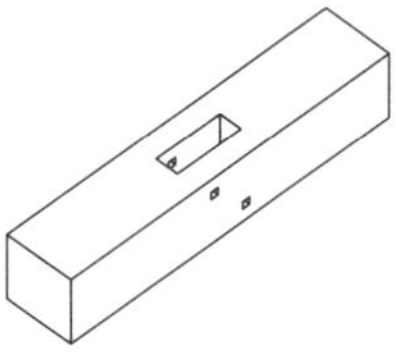

A

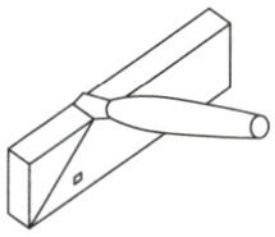

B1

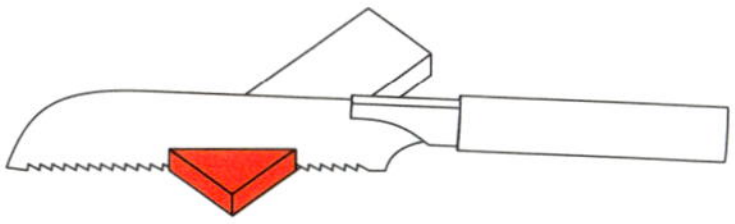

B2

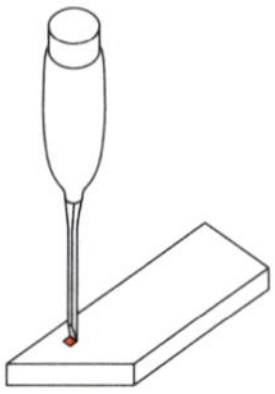

B3

2×

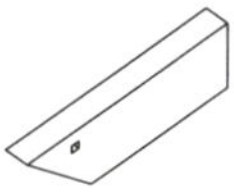

B

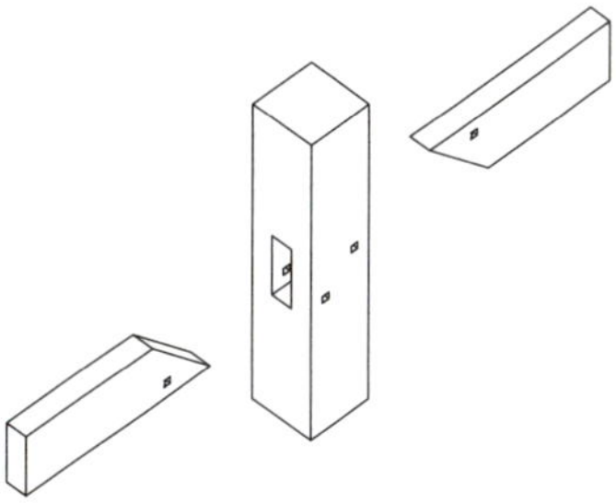

A+B+B

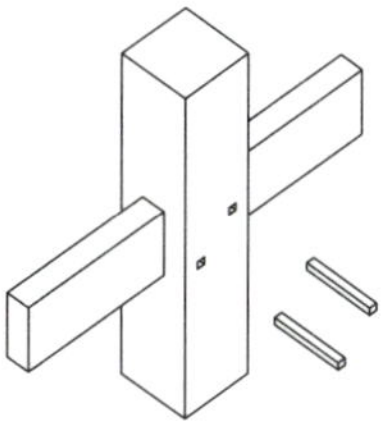

A+B+B

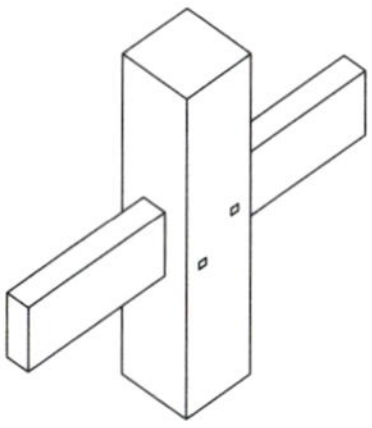

ABB

Specialty Joints

Pegged two-way gooseneck joint

ryaku-kama-komi-sen-shime (oire) [*]
略鎌込栓締め(追入)

Sichelzapfenverbindung
mit Dübelsicherung

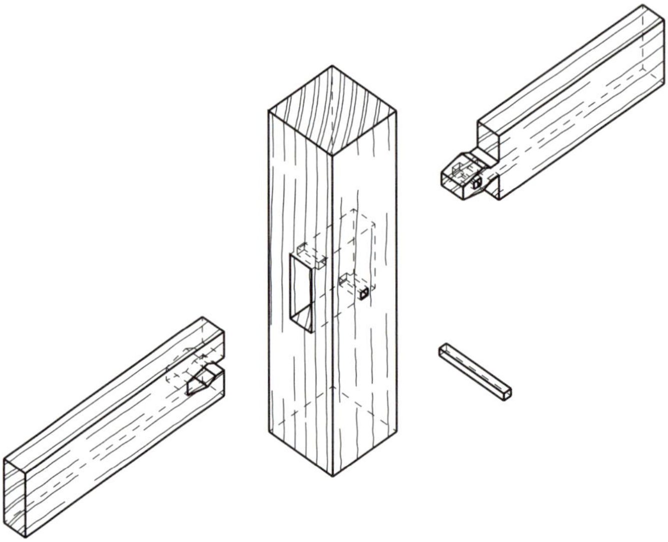

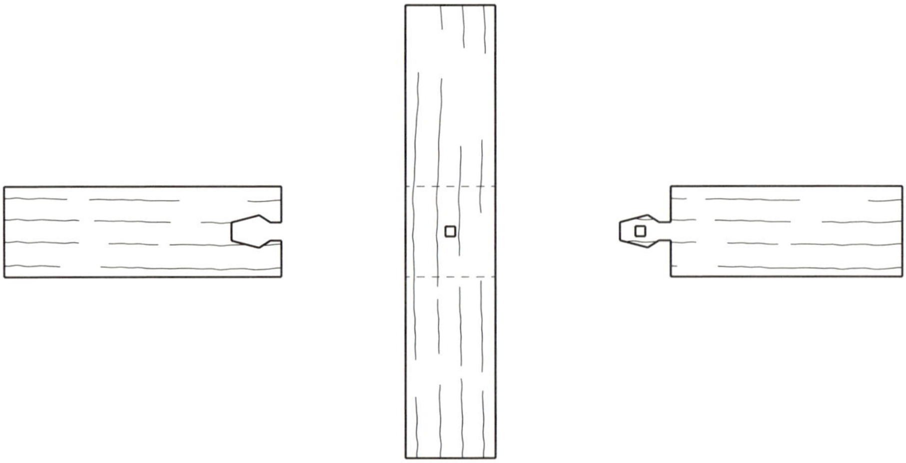

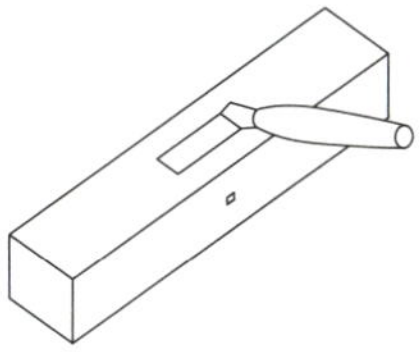

A1

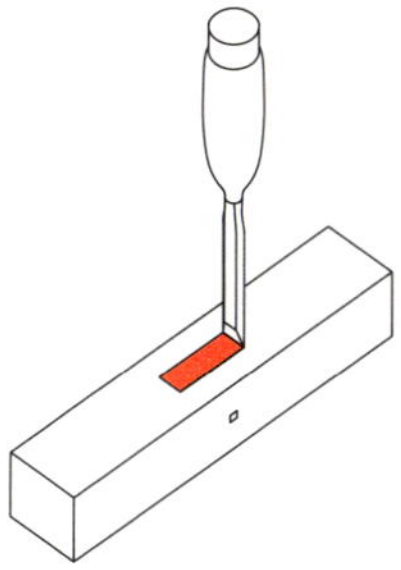

A2

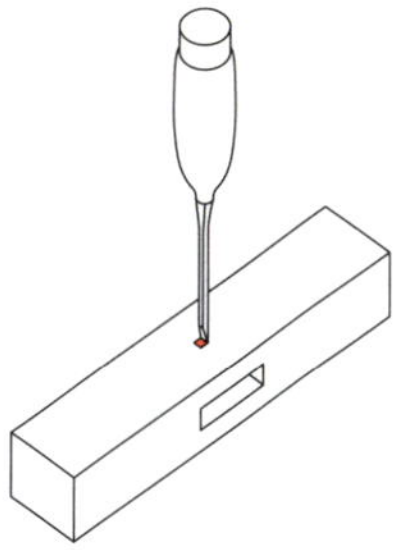

A3

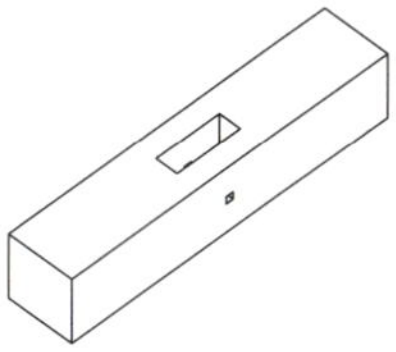

A

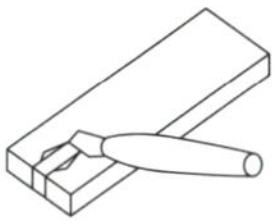

B1

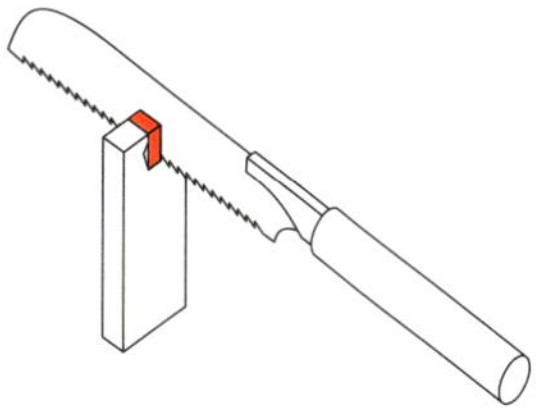

B2

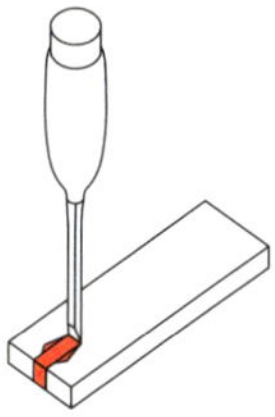

B3

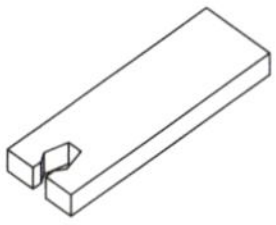

B

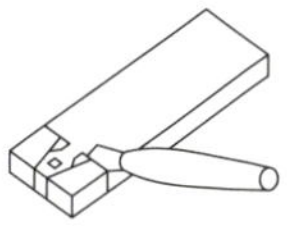

C1

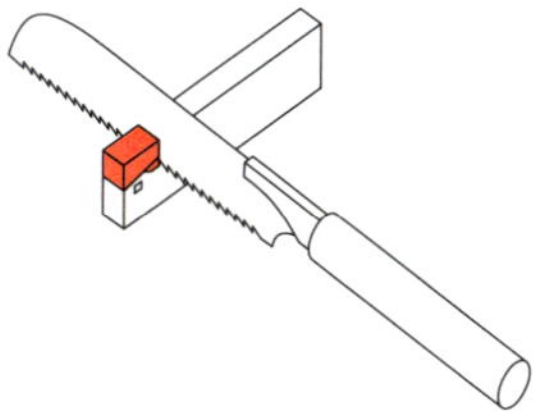

C2

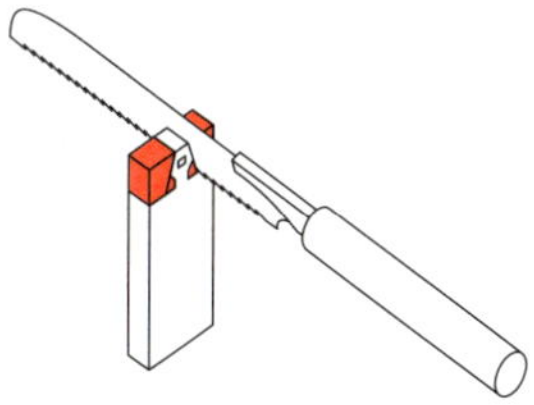

C3

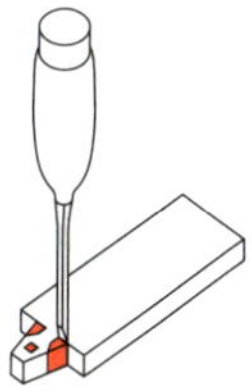

C4

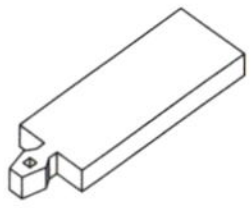

C

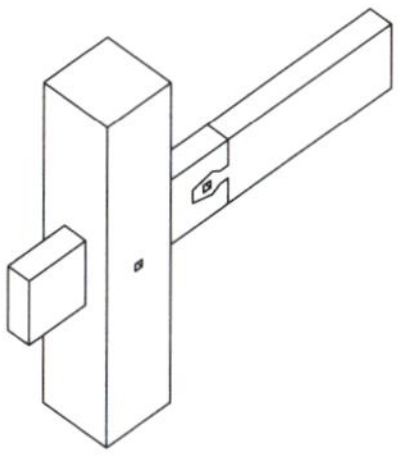

A+B+C

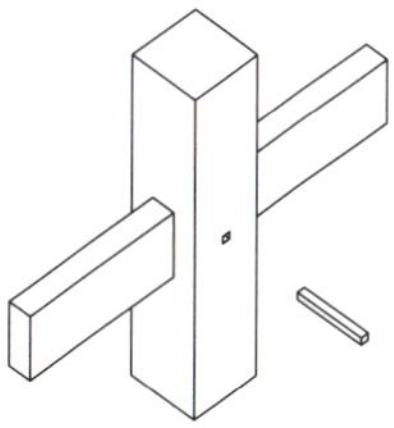

A+B+C

Specialty Joints

Wedged two-way oblique cogged joint

ryaku-kama-kai-kusabi-shime (nihō-sashi) [*]
略鎌飼楔締め（二方差）

verkeiltes zweiseitiges Hakendruckschloss mit schrägen Haken

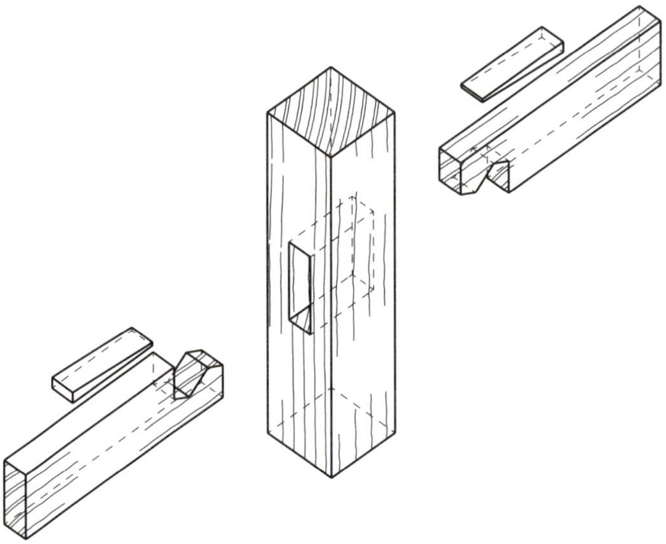

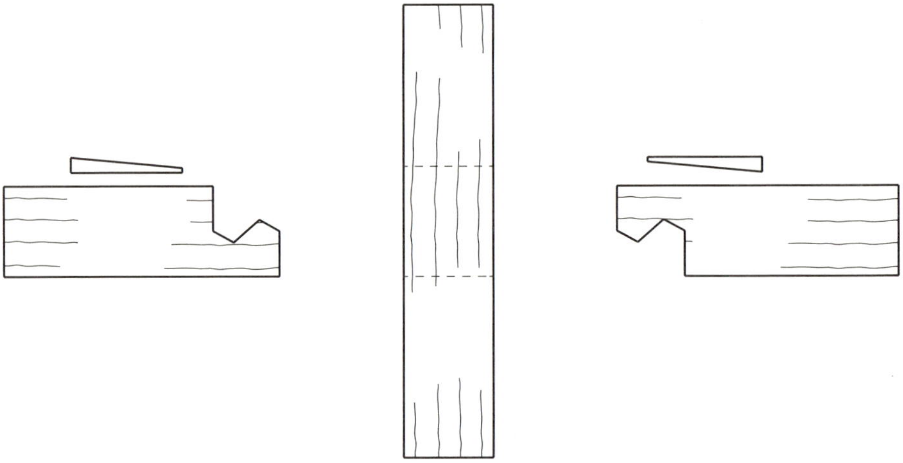

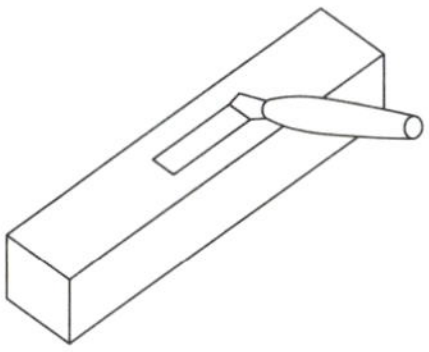

A1

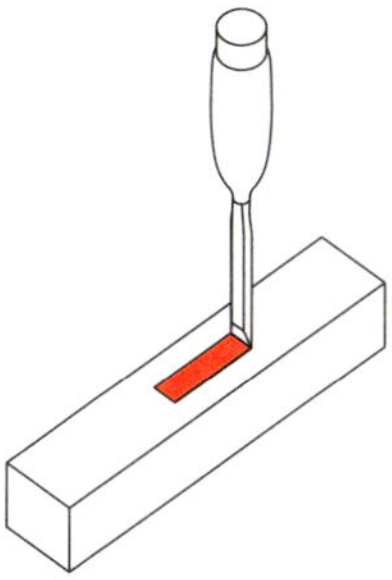

A2

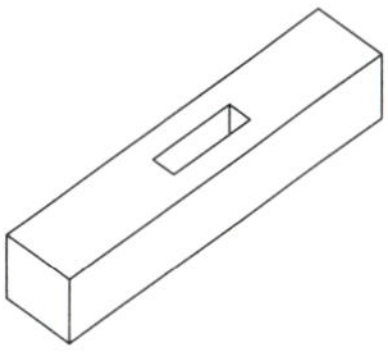

A

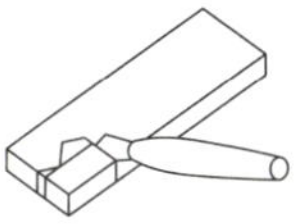

B1

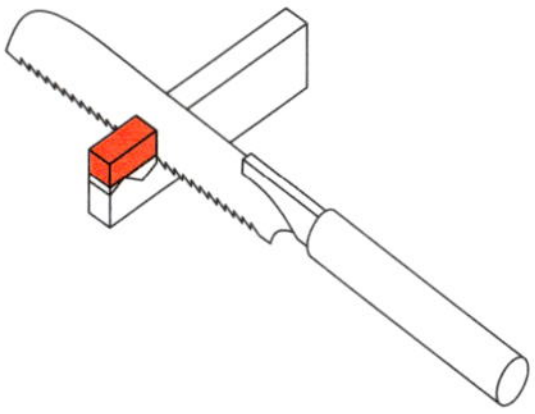

B2

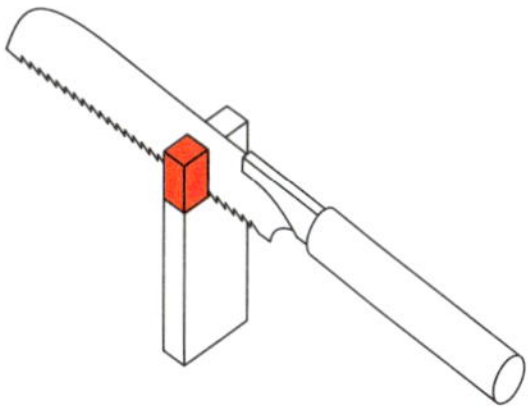

B3

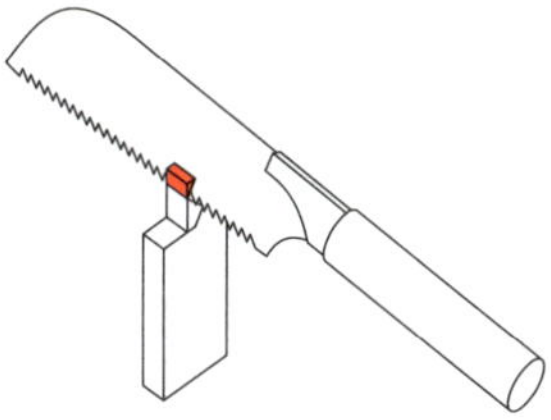

B4

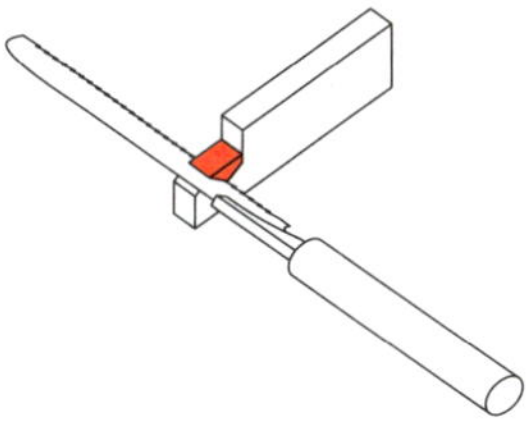

B5

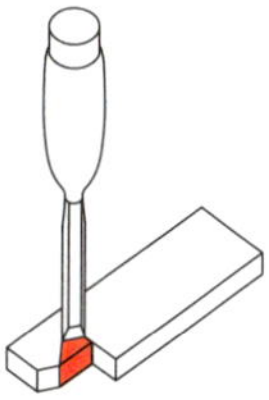

B6

2×

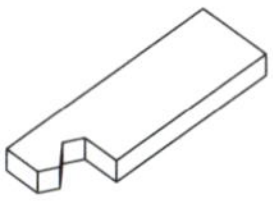

B

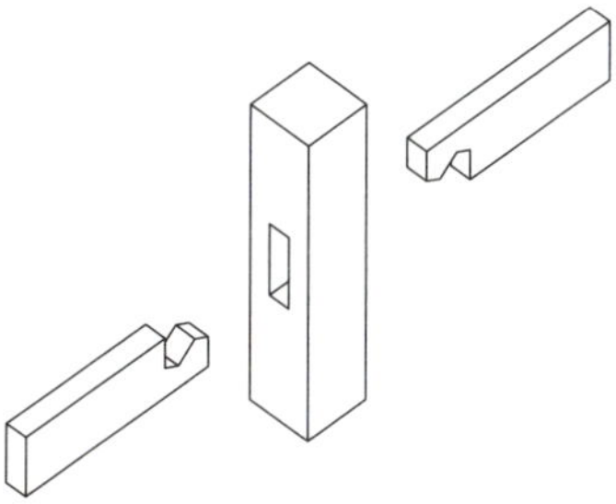

A+B+B

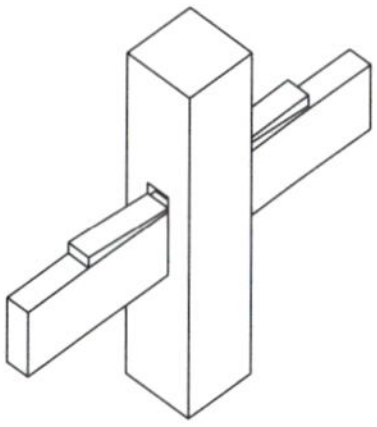

A+B+B

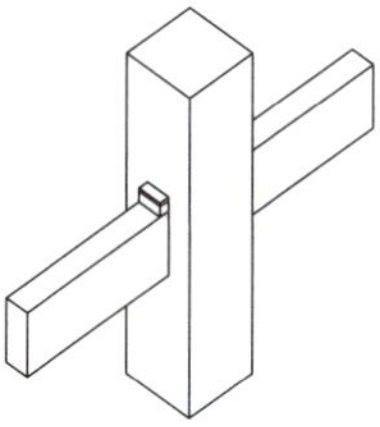

ABB

Specialty Joints

Wedged four-way splayed cogged joint

ryaku-kama-kai-kusabi-shime (oire, shihō-sashi) [✱]
略鎌飼楔締め(追入・四方差)

verkeiltes vierseitiges Hakendruckschloss mit schrägen Haken

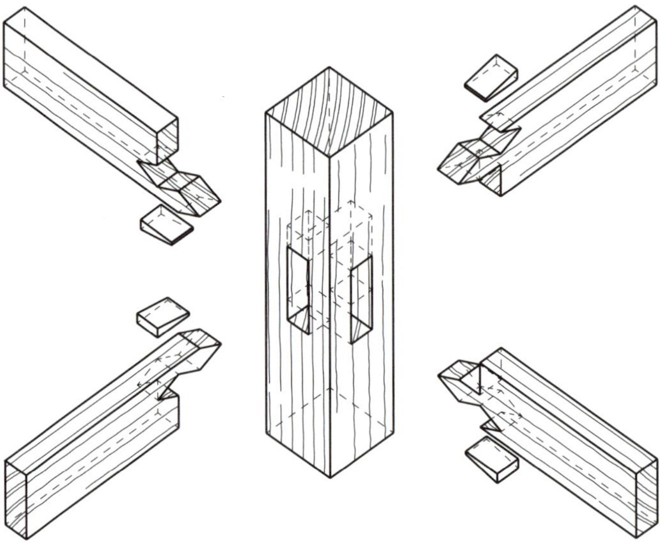

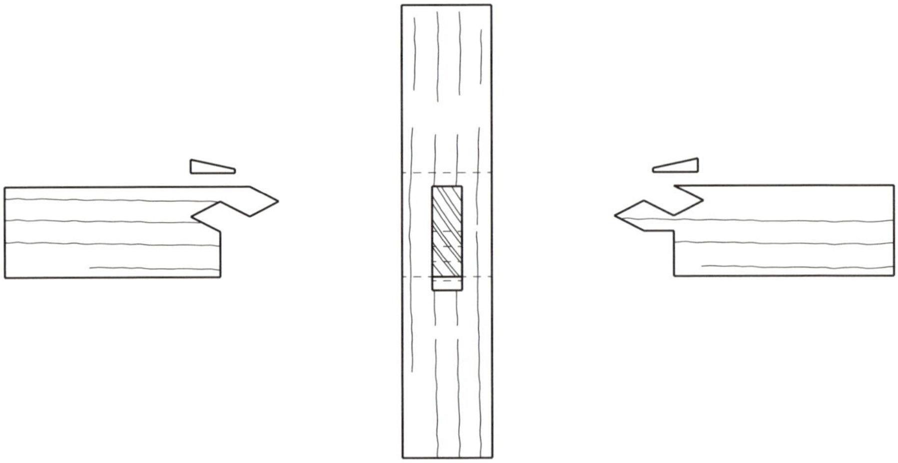

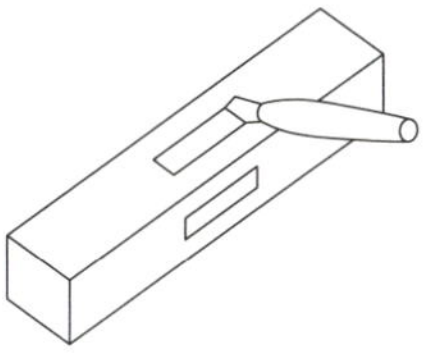
A1

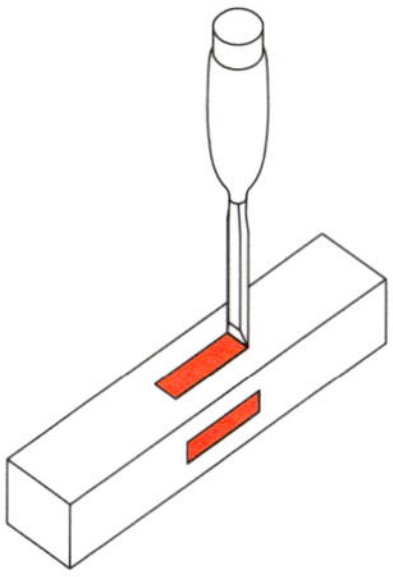
A2

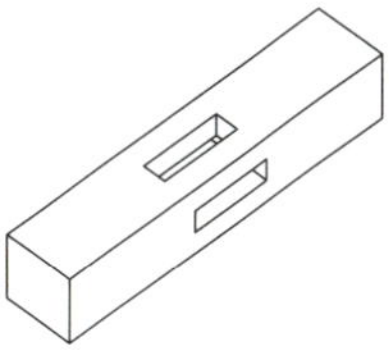
A

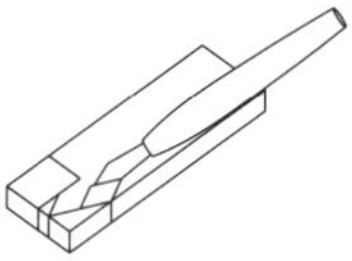
B1

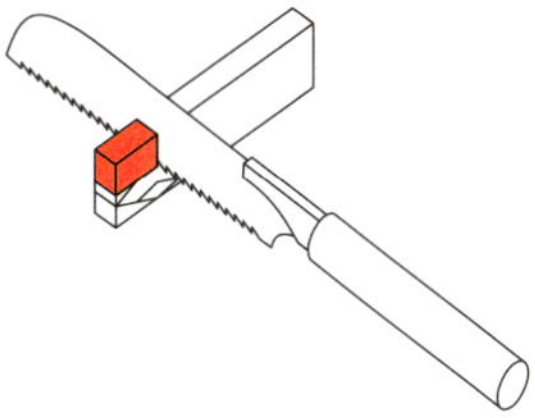
B2

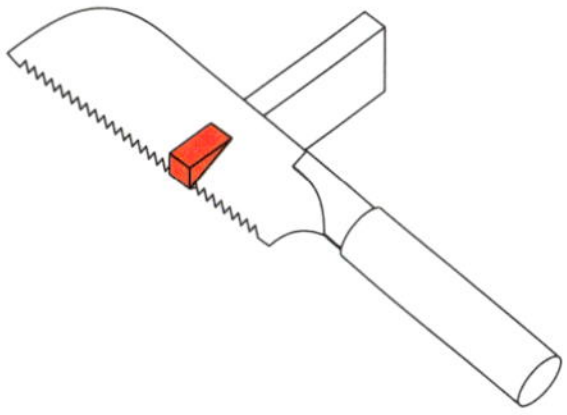
B3

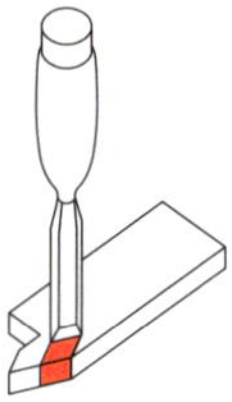
B4

2×

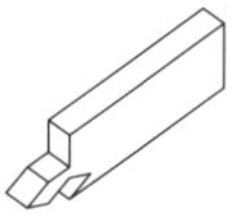
B

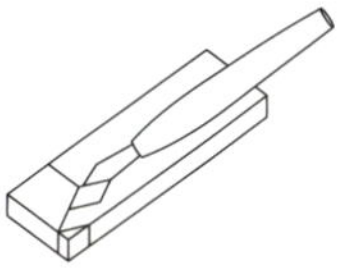
C1

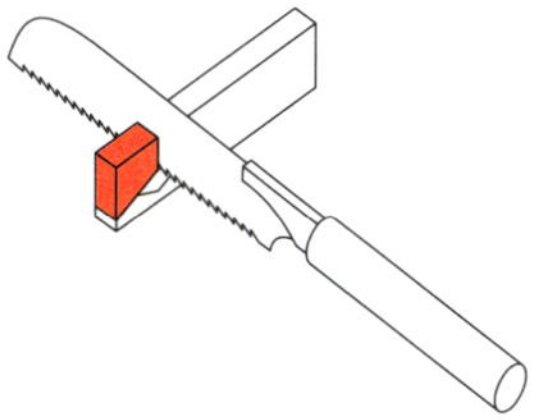
C2

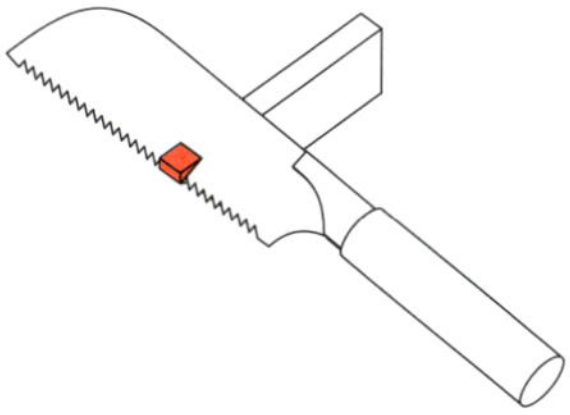
C3

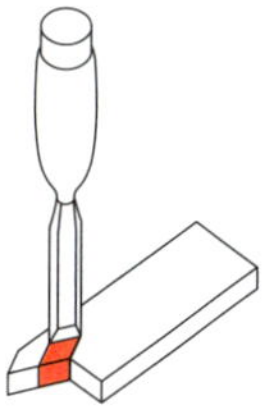
C4

2×

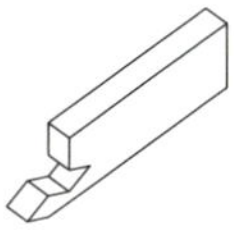
C

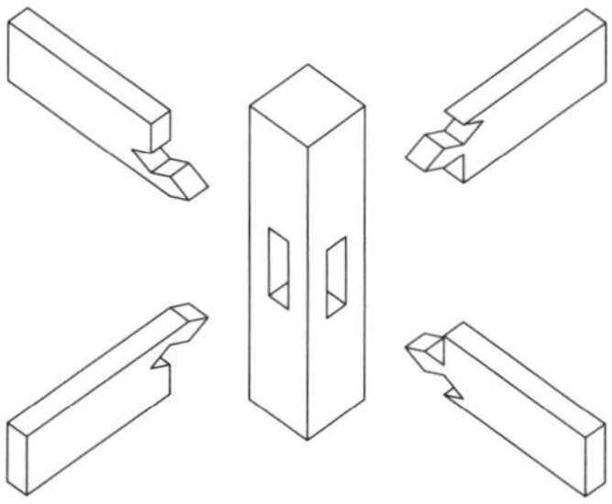
A+B+B+C+C

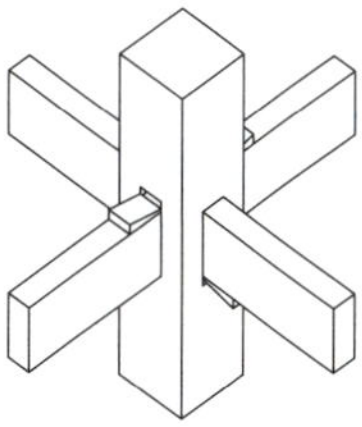
ABBCC

Specialty Joints

Two-way rod tenon joint with tapered keys

sao-sashi-shachi-sen-kumi [*]
竿差車知栓組み

durchgesteckter Stabzapfen
mit Spannkeilen

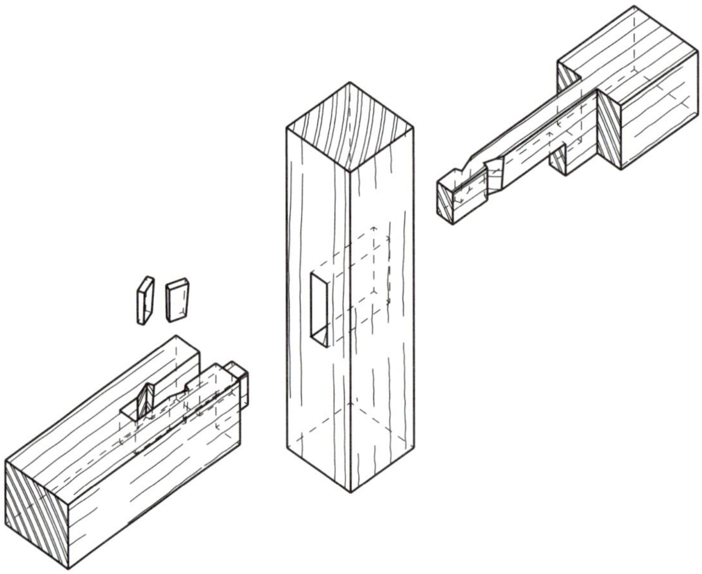

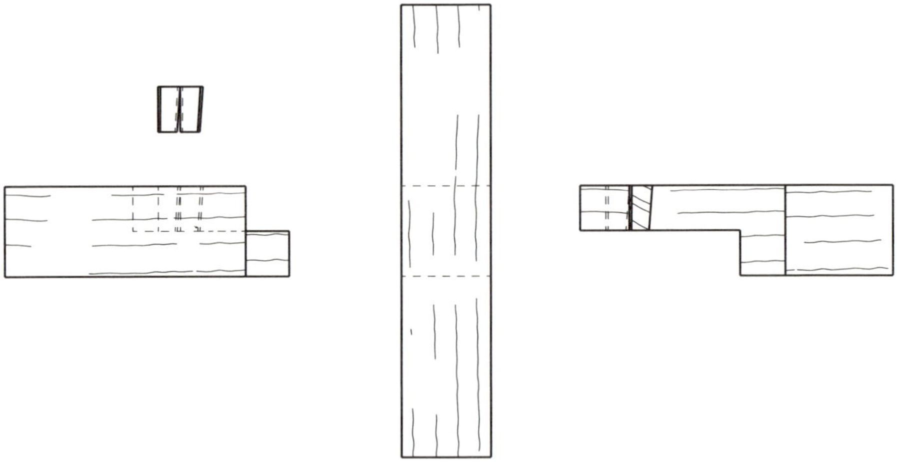

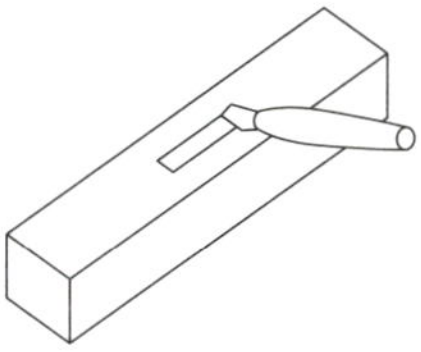

A1

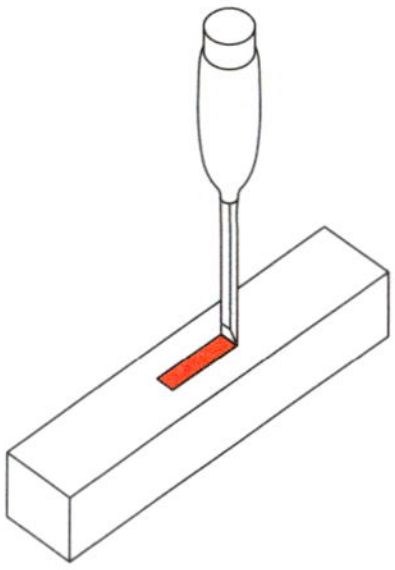

A2

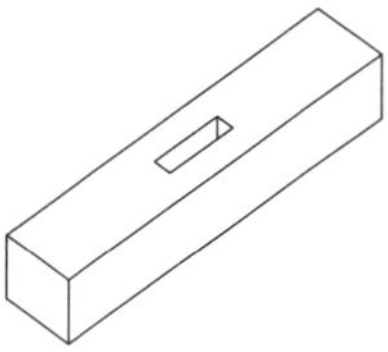

A

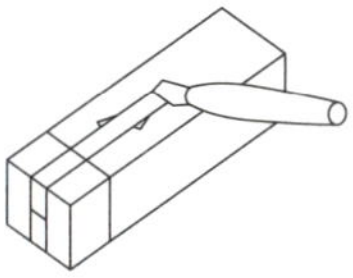

B1

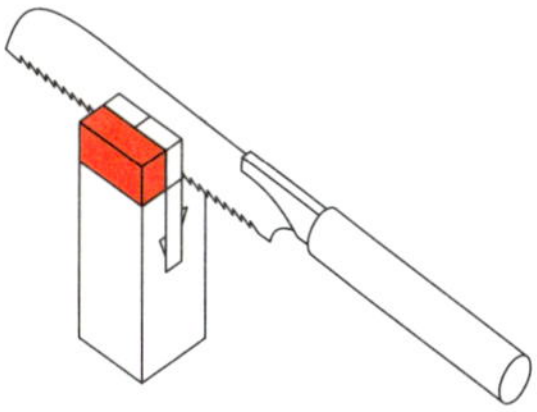

B2

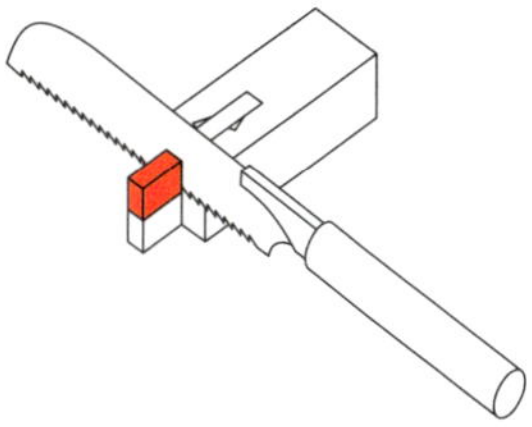

B3

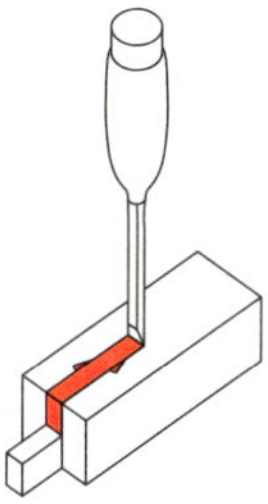

B4

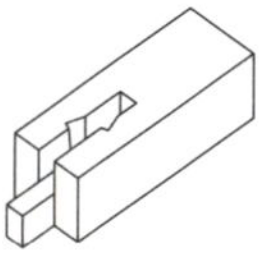

B

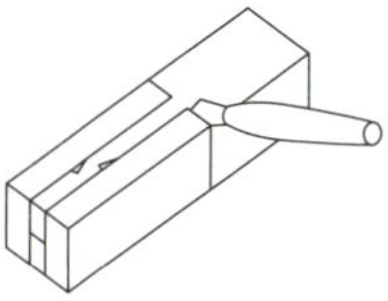

C1

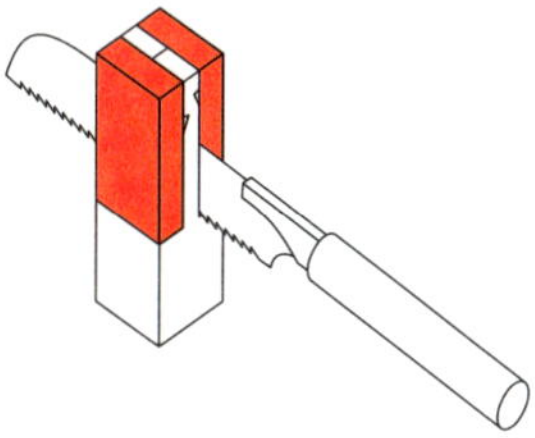

C2

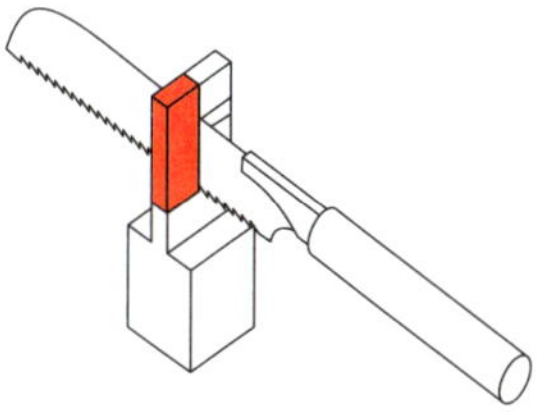

C3

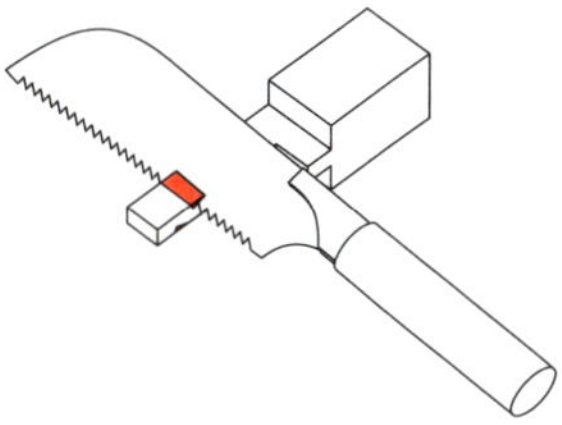

C4

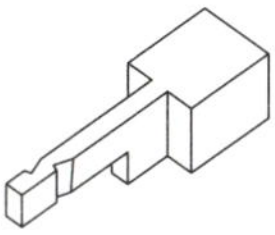

C

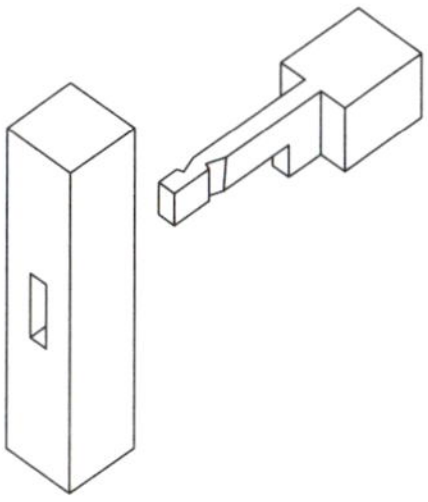

A+C

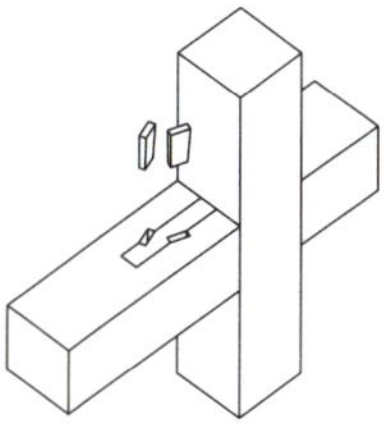

A+B+C

Specialty Joints

Three-way rod tenon joint with tapered keys and wedged haunched tenon

Japanese

sanpō-sao-sashi-shachi-sen-tsugi (wari-kusabi-shime) [*]
三方竿差車知継ぎ(割楔締)

German

durchgesteckter Stabzapfen mit Spannkeilen und verkeiltem Stufenzapfen

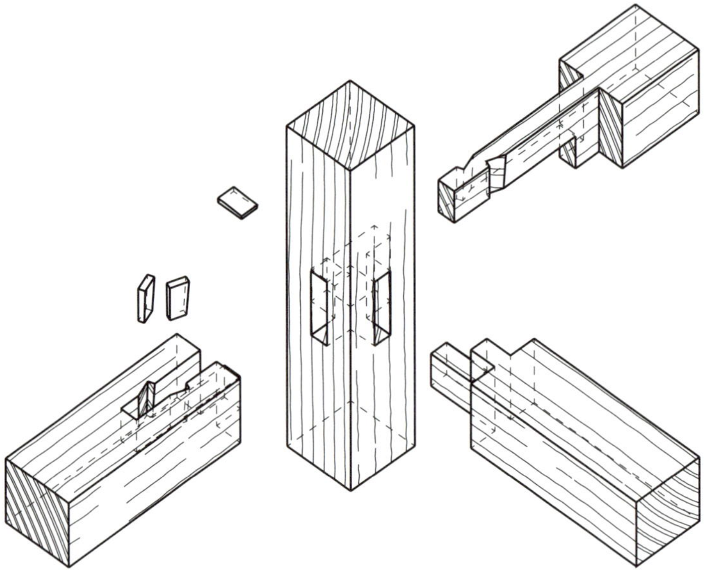

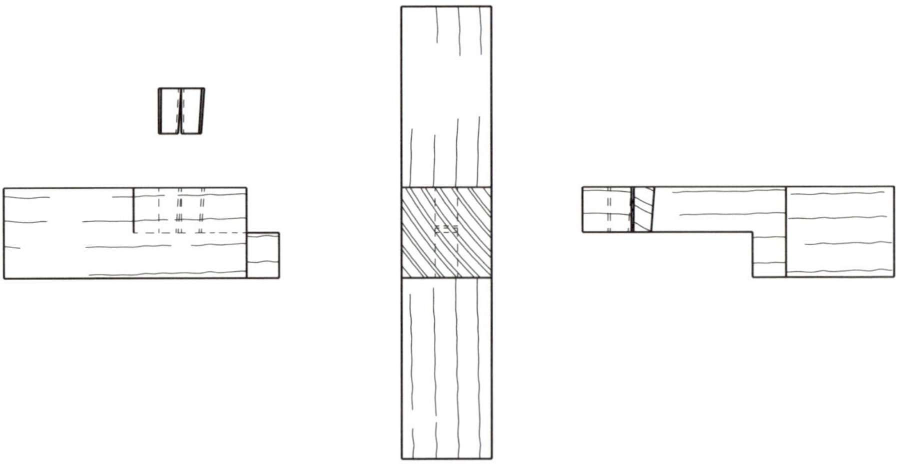

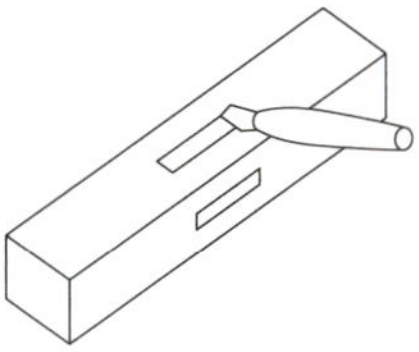
A1

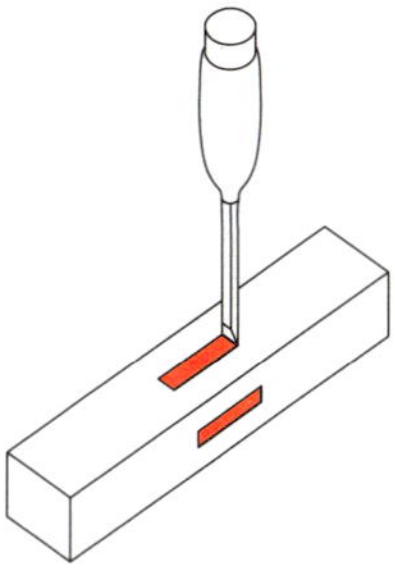
A2

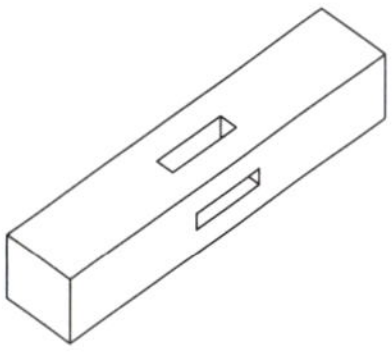
A

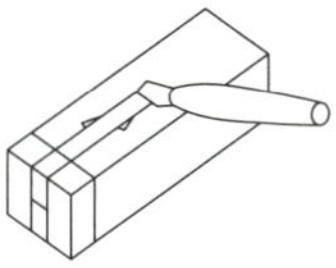
B1

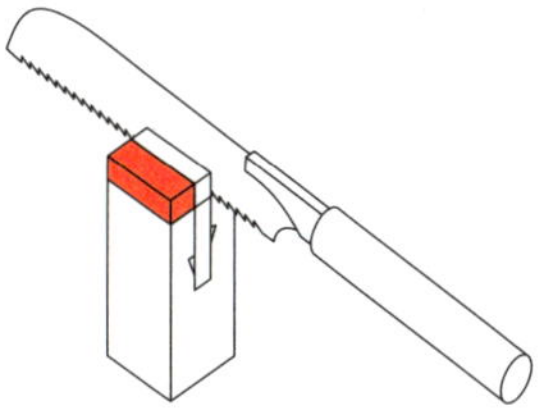
B2

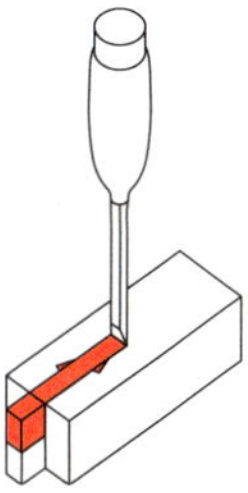
B3

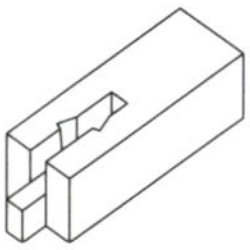
B

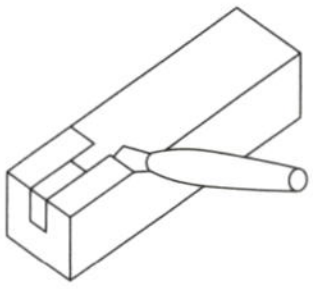
C1

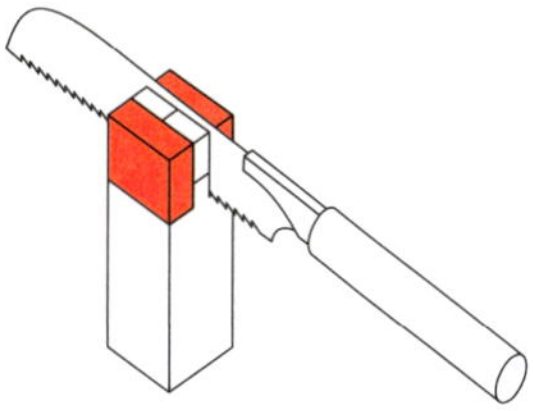
C2

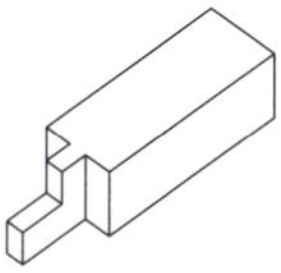
C

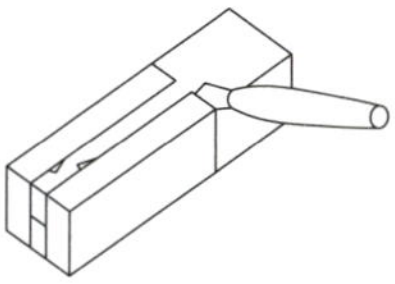
D1

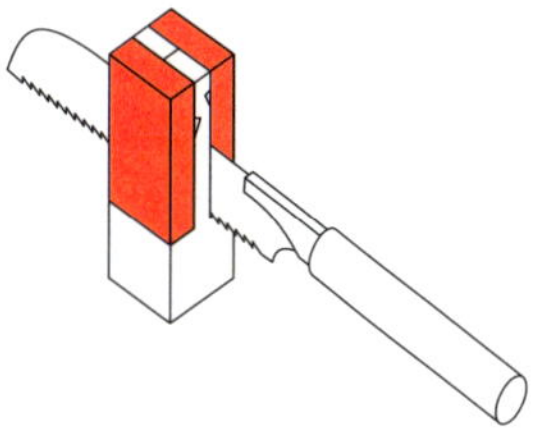
D2

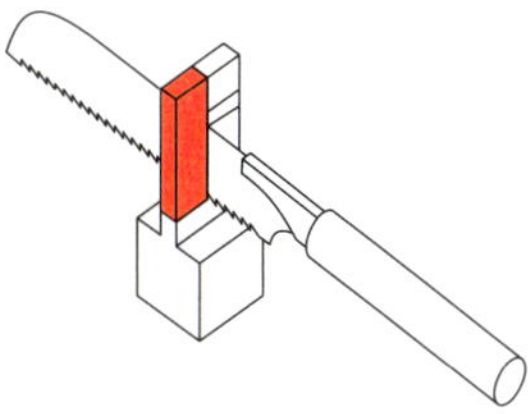
D3

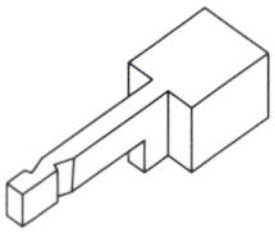
D

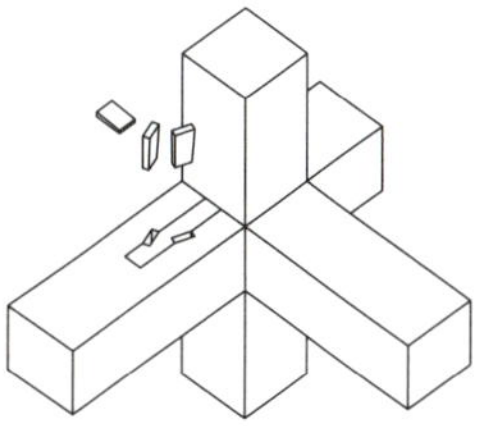
A+B+C+D

Specialty Joints

Four-way rod tenon joint with tapered keys and pegs

kone-hozo-shachi-sen-tsugi, kone-hozo-komi-sen-shimari (shihō-sashi) [*]
小根枘車知栓継ぎ、小根枘込栓締り（四方差）

durchgesteckte Stufenstabzapfen mit Spannkeilen und Dübelsicherung

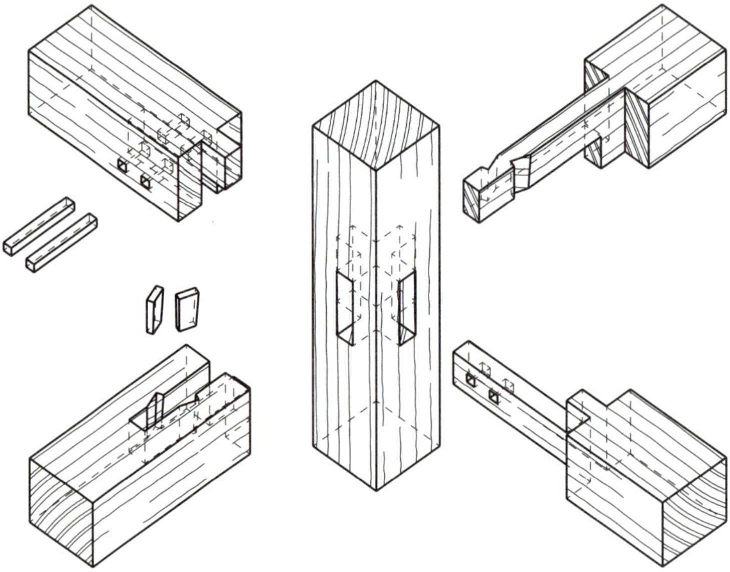

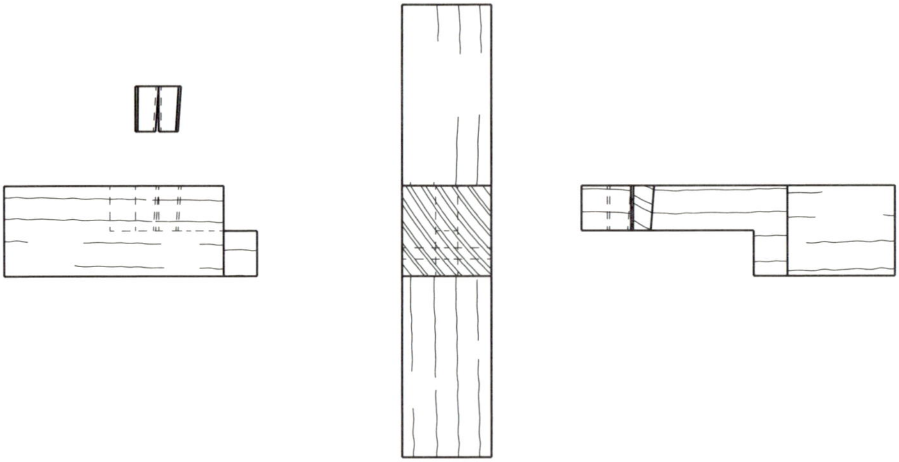

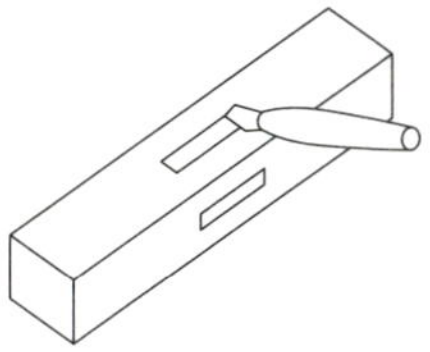
A1

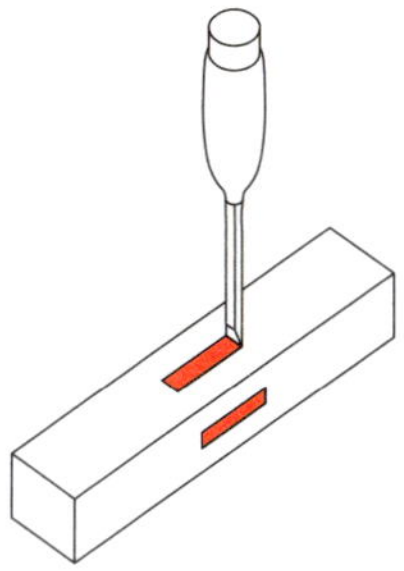
A2

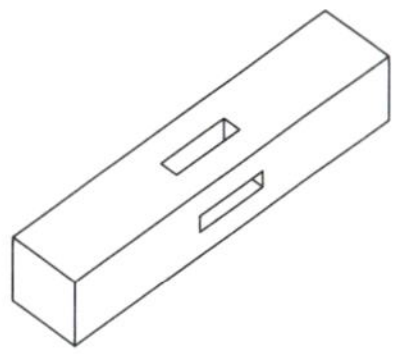
A

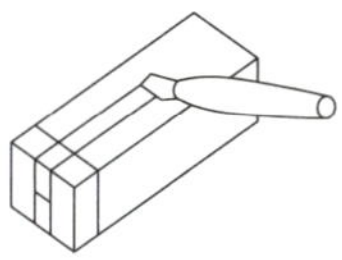
BC1

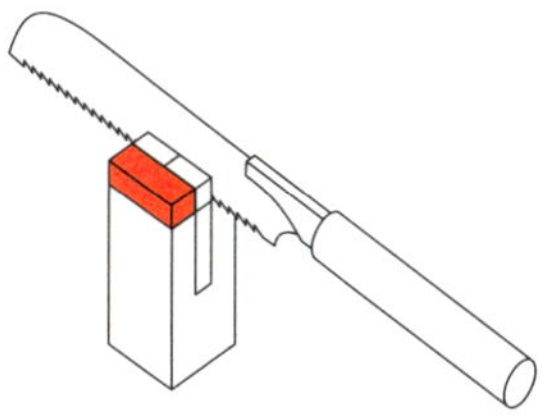
BC2

2×

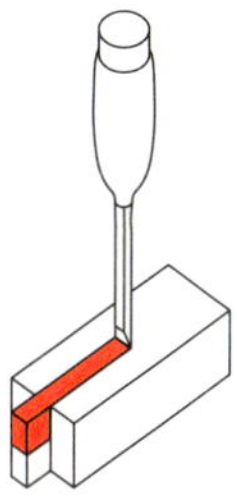
BC3

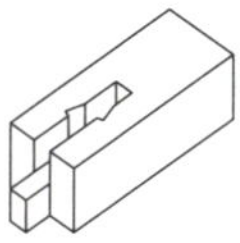
B

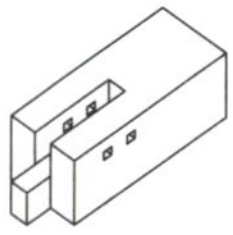
C

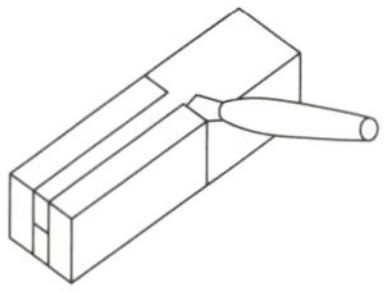
DE1

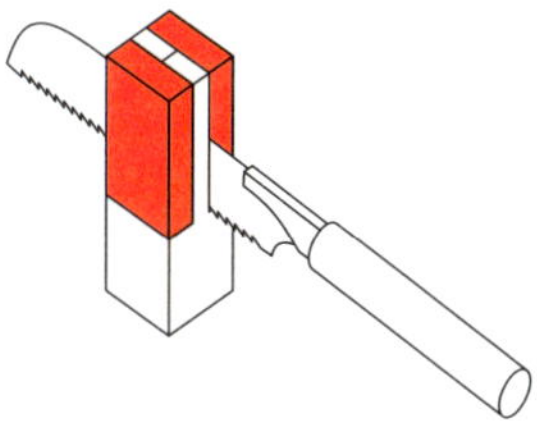
DE2

2×

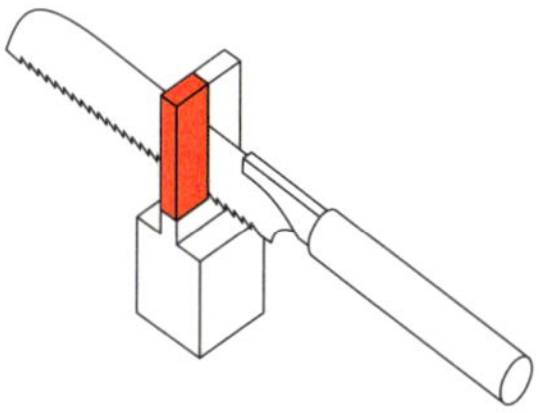
DE3

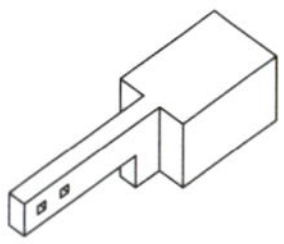
D

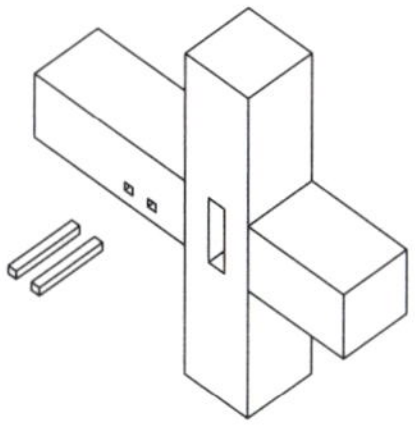
A+C+D

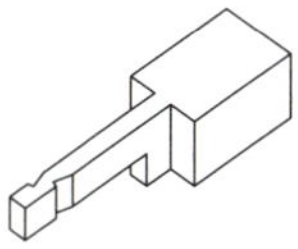
E

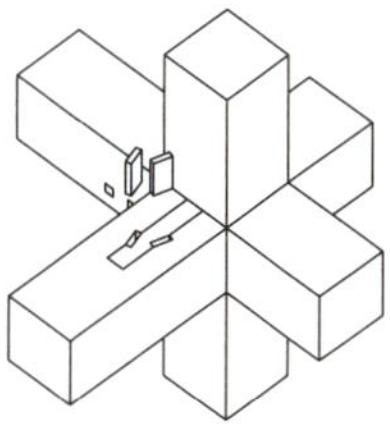
A+B+C+D+E

Two-way blind housed rod tenon joint with tapered keys

ko-dō-tsuki-sao-sashi-shachi-sen-kumi (nihō-sashi) [*]
小胴付竿差車知栓組み（二方差）

eingelassener durchgesteckter Stabzapfen mit gestufter Brust und Spannkeilen

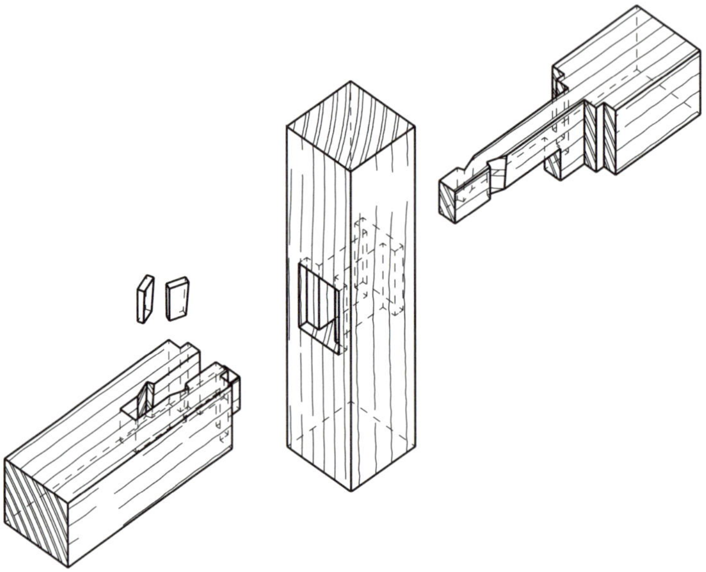

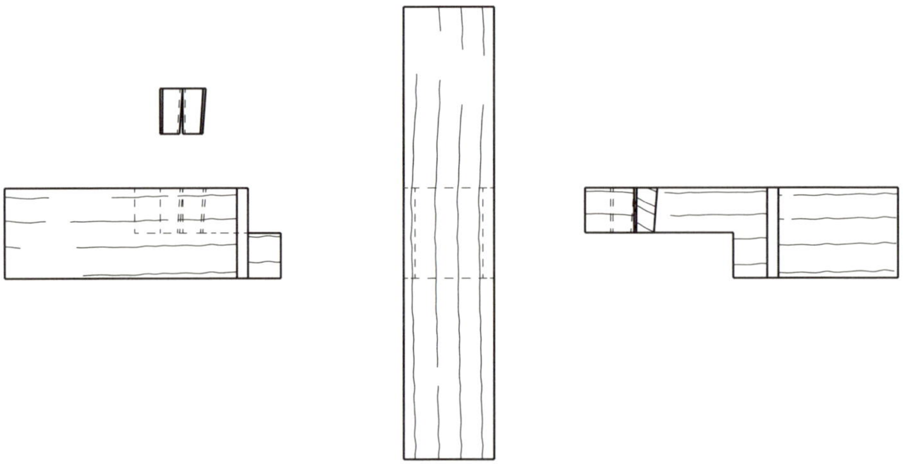

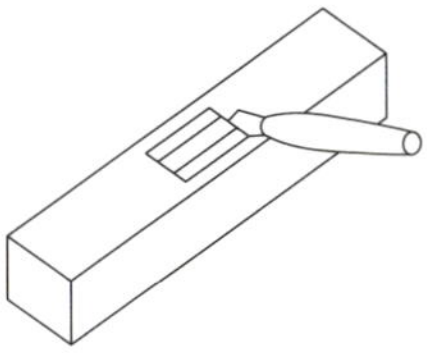

A1

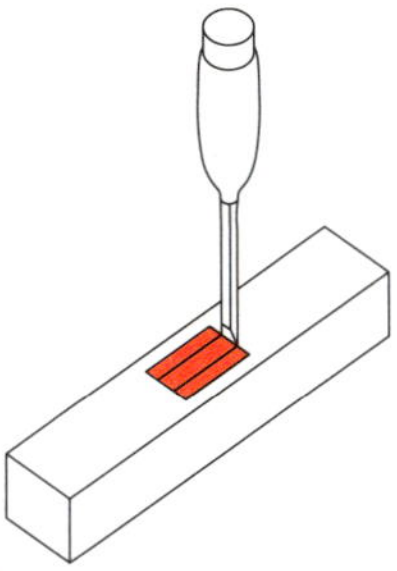

A2

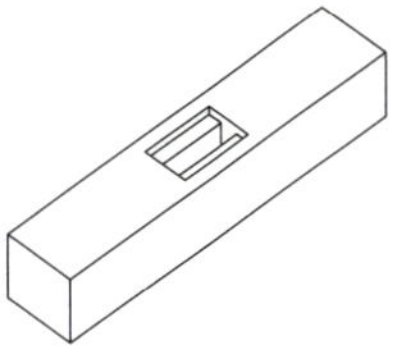

A

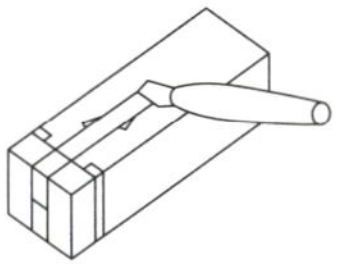

B1

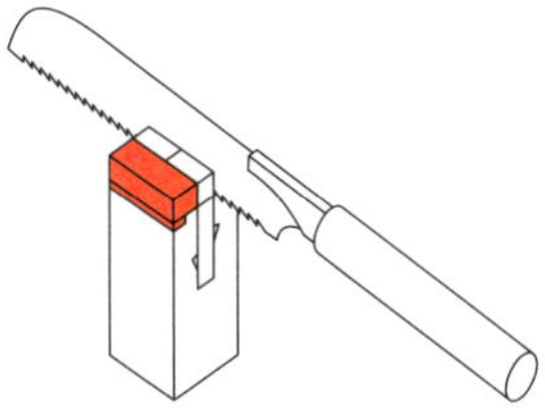

B2

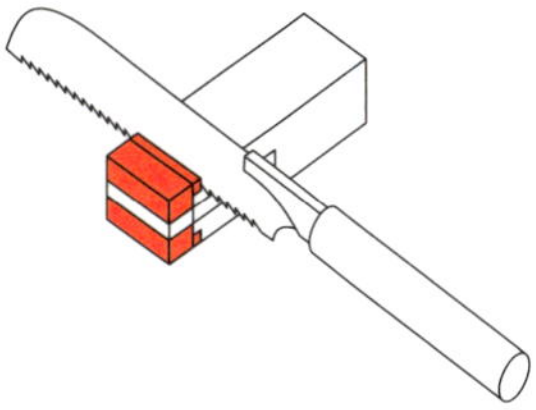

B3

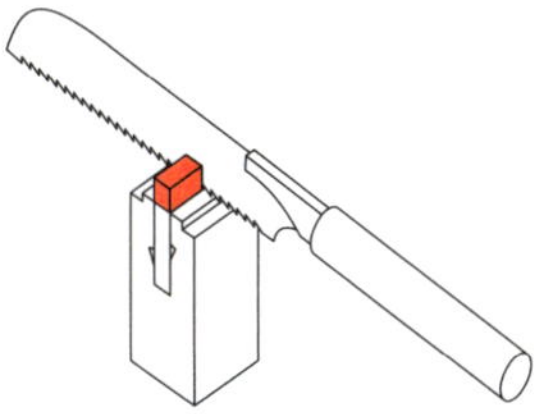

B4

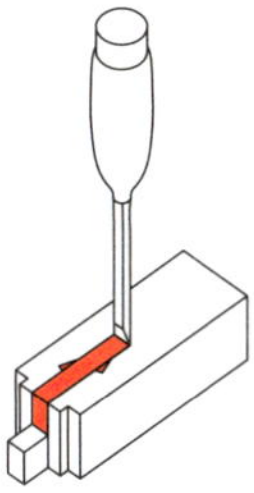

B5

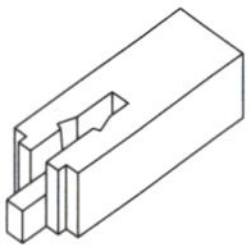

B

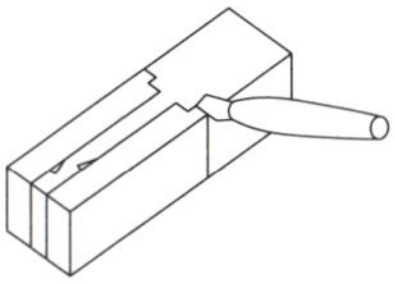

C1

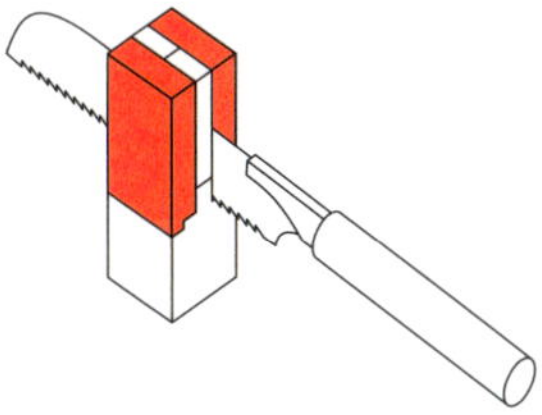

C2

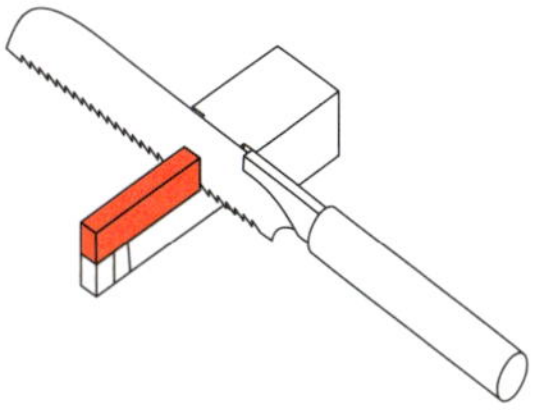

C3

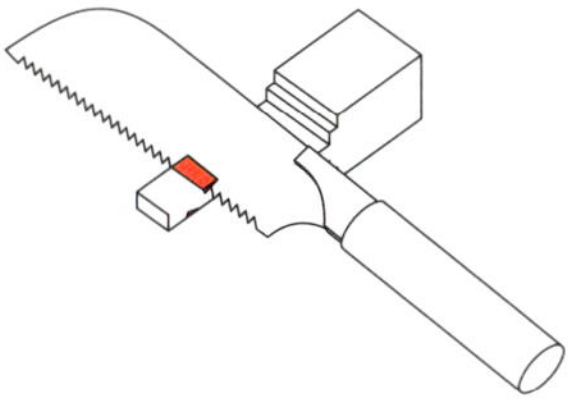

C4

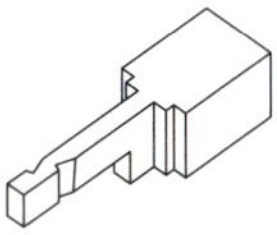

C

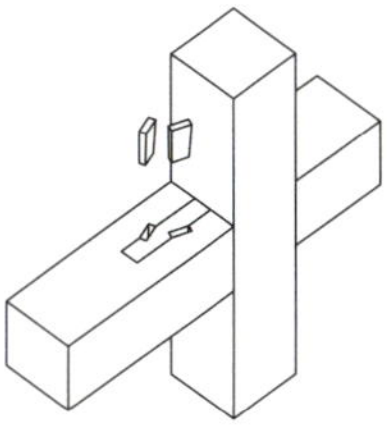

A+B+C

Specialty Joints

Two-way collared rod tenon joint with tapered keys

ko-dō-tsuki-sao-sashi-shachi-sen-kumi [*]
小胴付竿差車知栓組み

eingelassener durchgesteckter Stabzapfen mit gestufter abgesetzter Brust und Spannkeilen

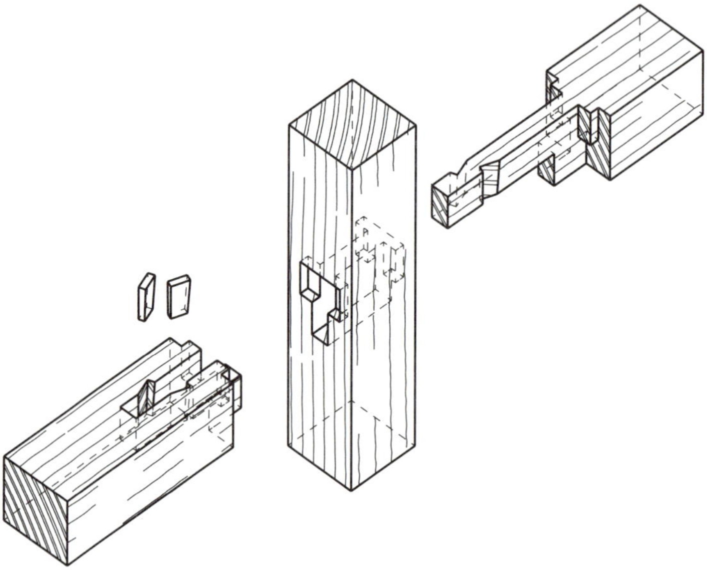

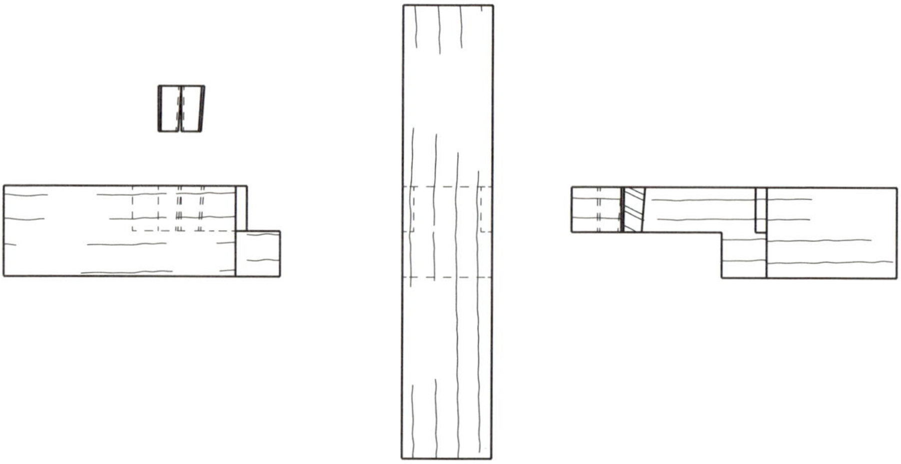

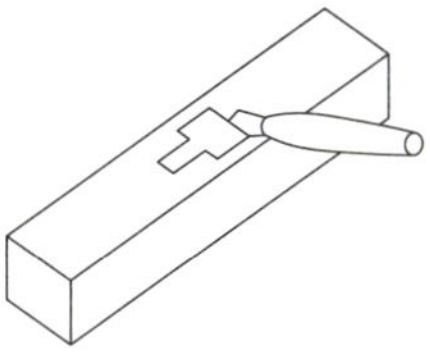

A1

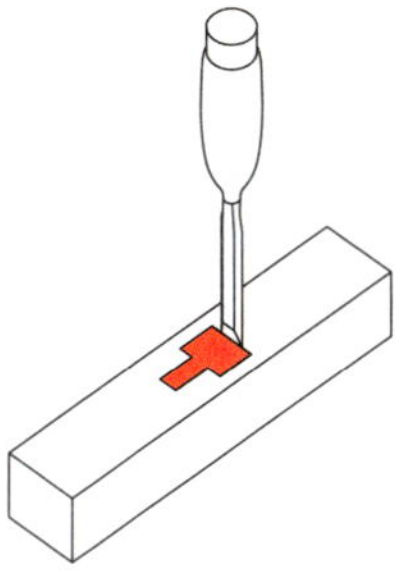

A2

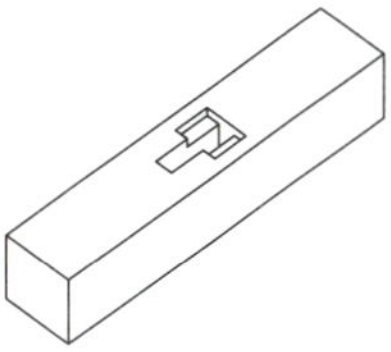

A

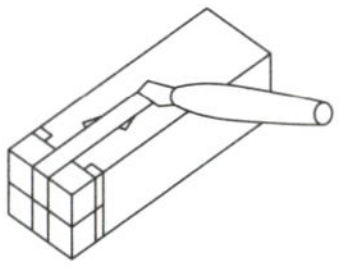

B1

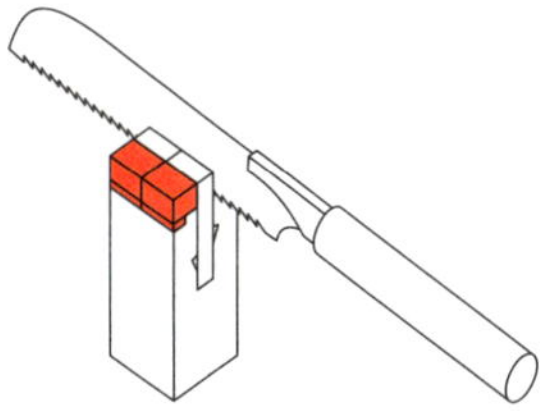

B2

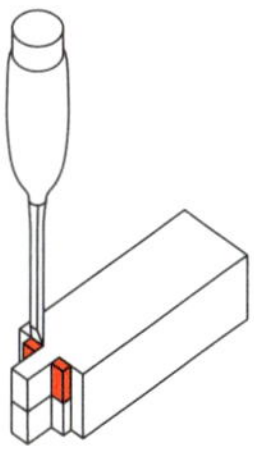

B3

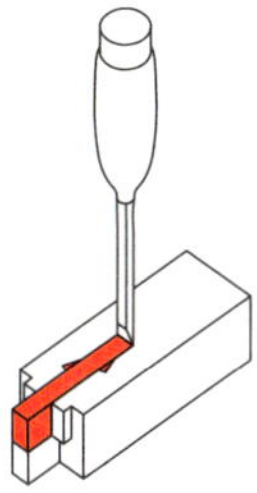

B4

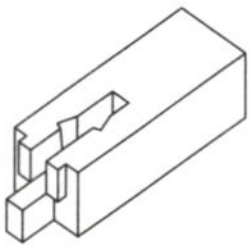

B

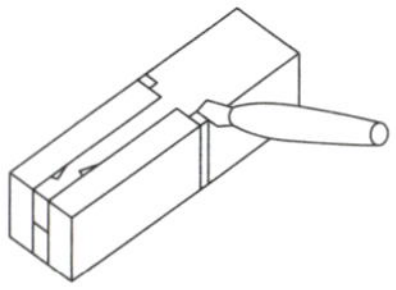

C1

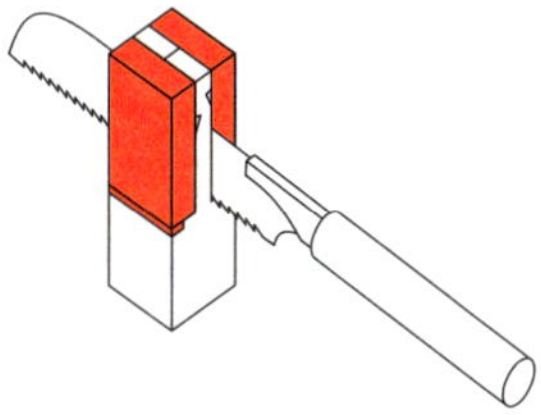

C2

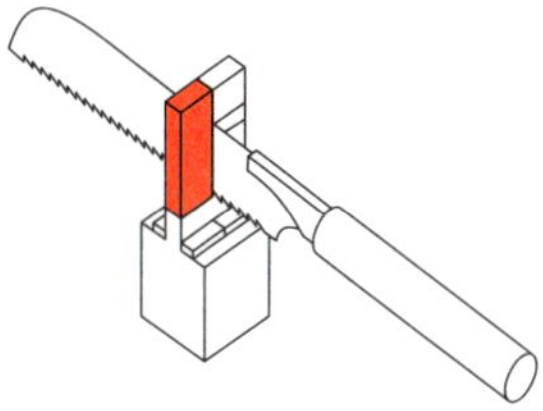

C3

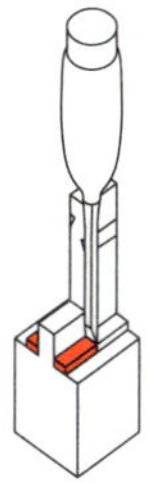

C4

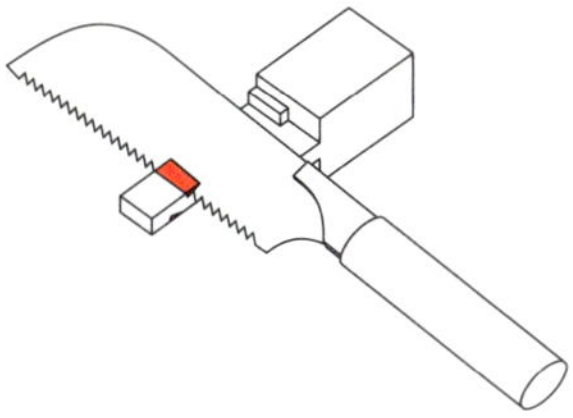

C5

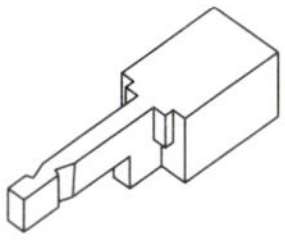

C

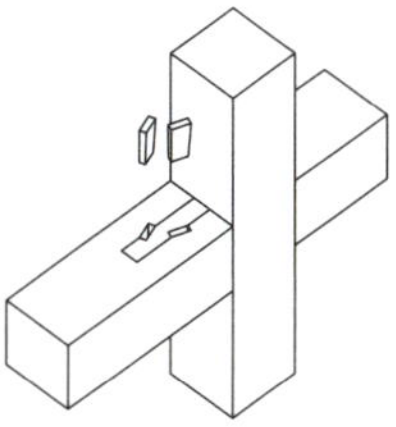

A+B+C

Specialty Joints

Two-way rabbeted rod tenon joint with tapered keys

mechi-shachi-sen-kumi (oire)
目違車知栓組み(追入)

eingelassener durchgesteckter Stabzapfen mit Steckfalz und Spannkeilen

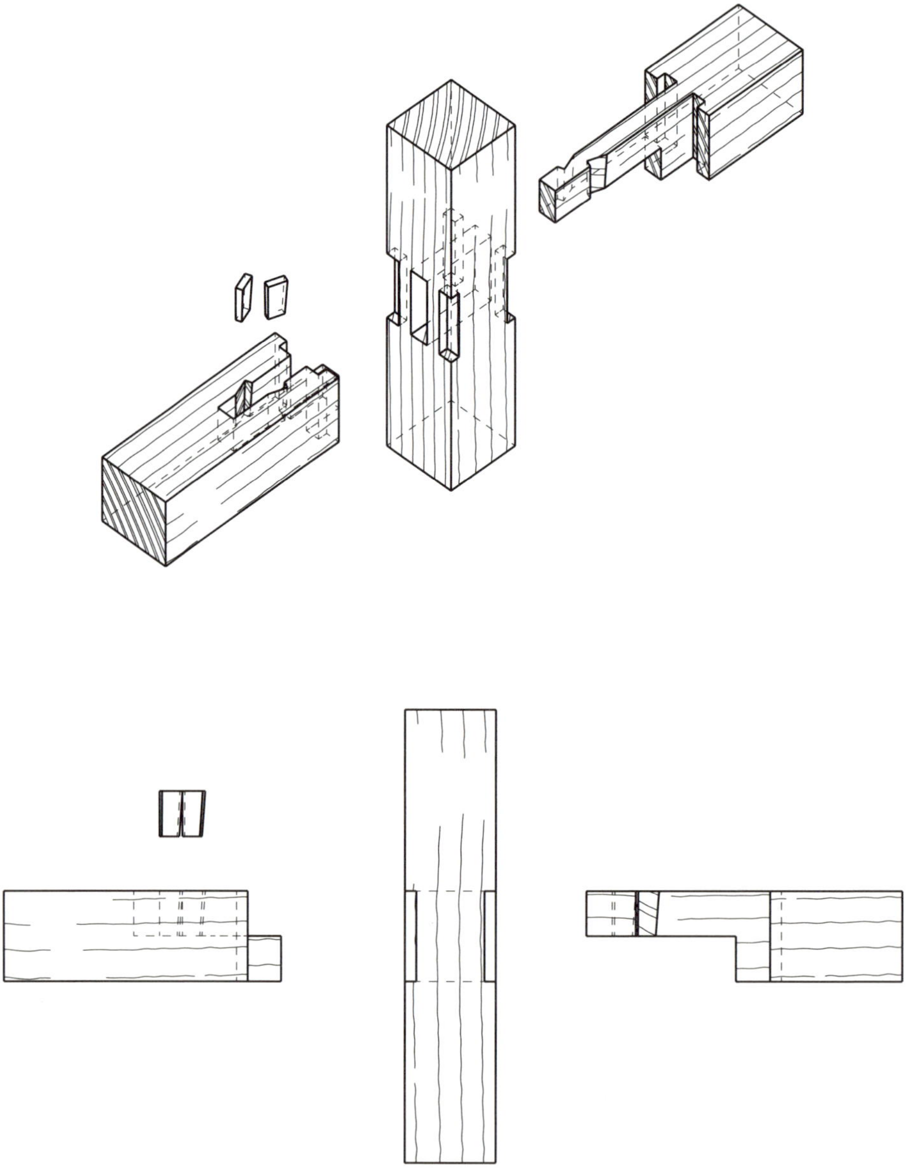

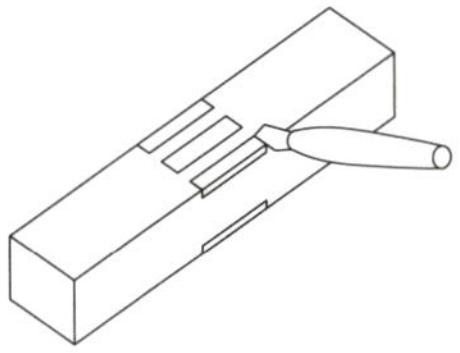

A1

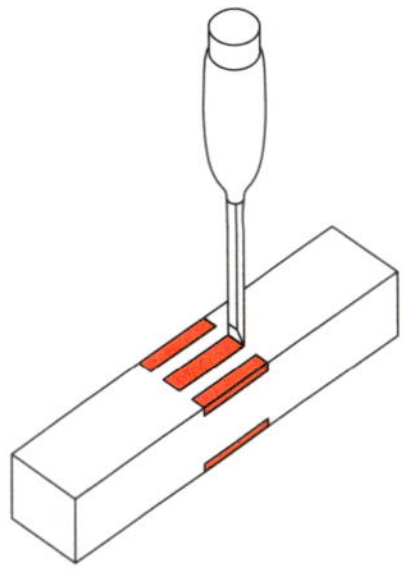

A2

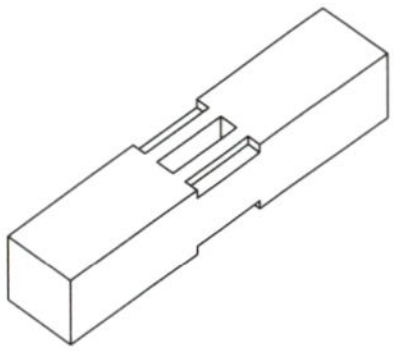

A

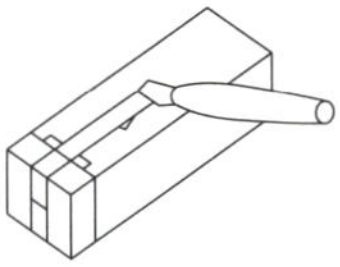

B1

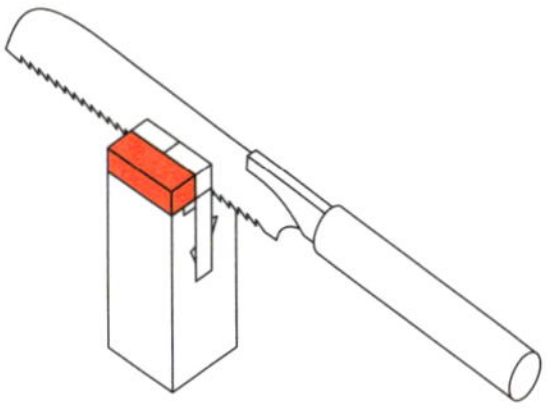

B2

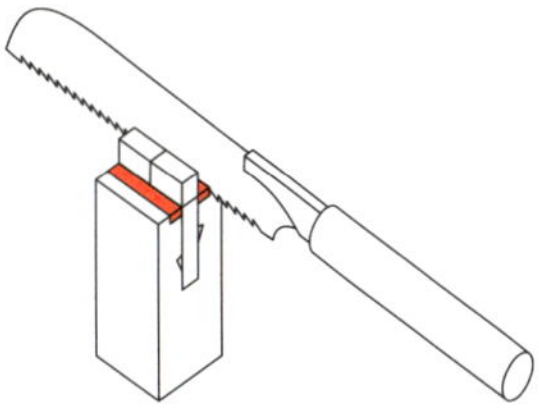

B3

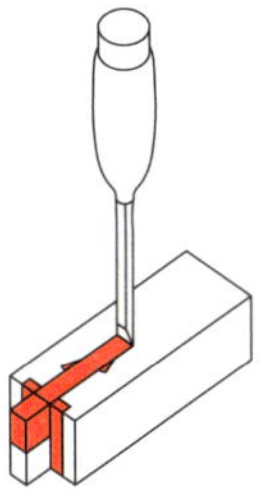

B4

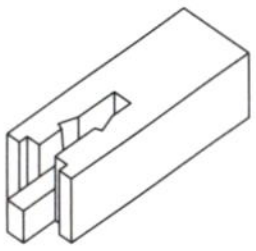

B

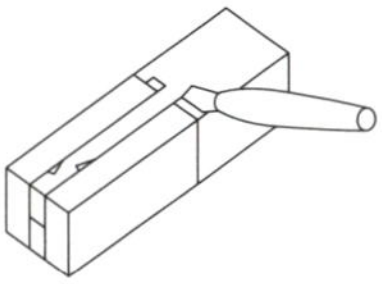

C1

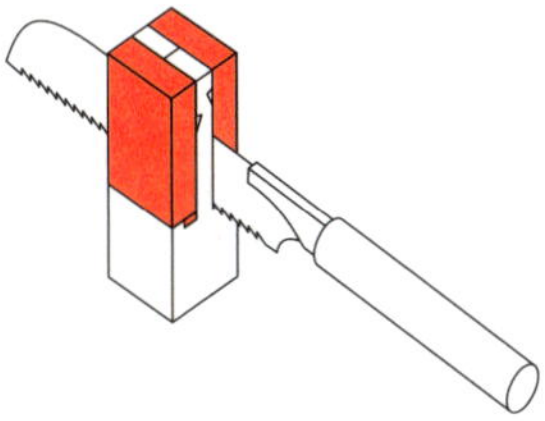

C2

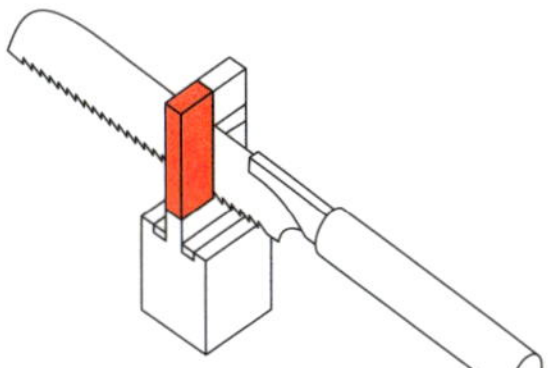

C3

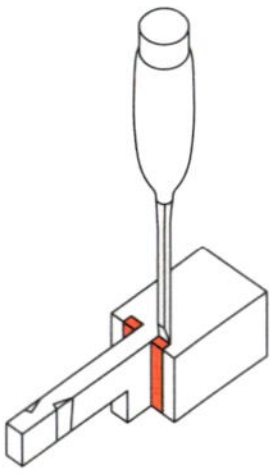

C4

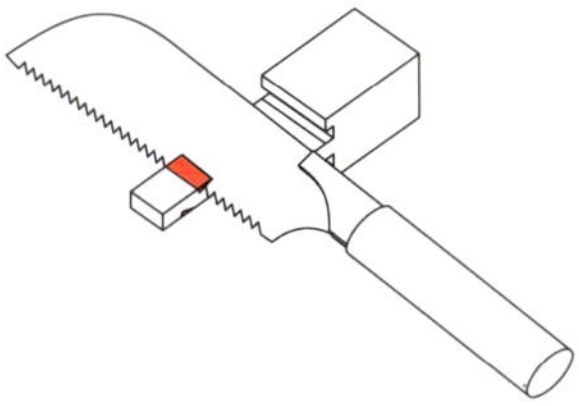

C5

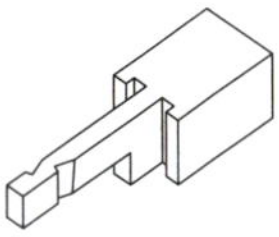

C

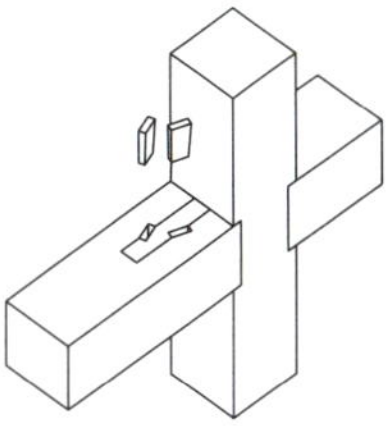

A+B+C

Specialty Joints

Two-way partially housed rod tenon joint with tapered keys

ko-dō-tsuki-sao-sashi-shachi-sen-kumi (oire, nihō-sashi) [*]
小胴付竿差車知栓組み（追入・二方差）

eingelassener durchgesteckter Stabzapfen mit abgesetzter Nut und Spannkeilen

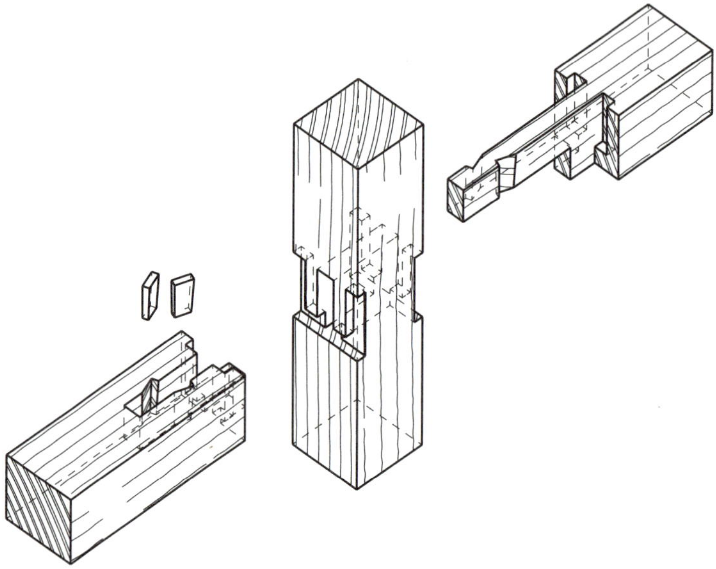

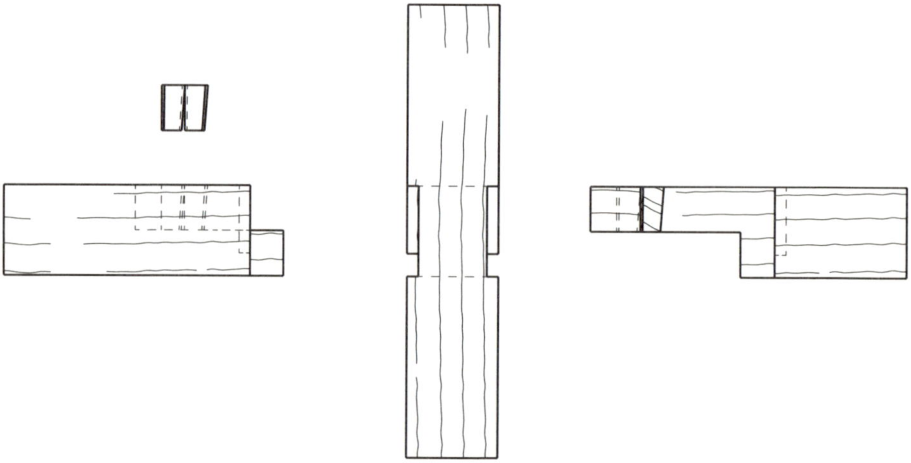

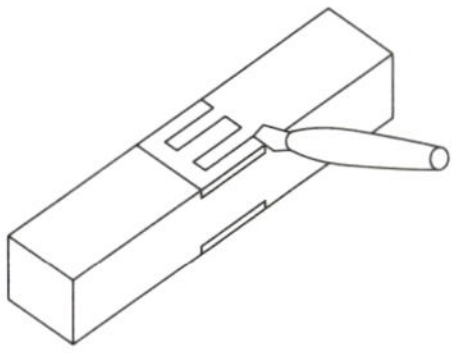

A1

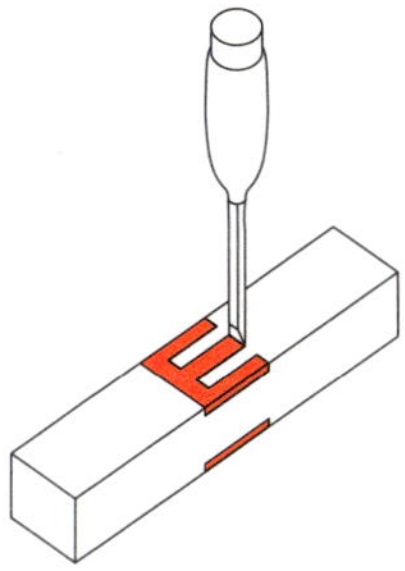

A2

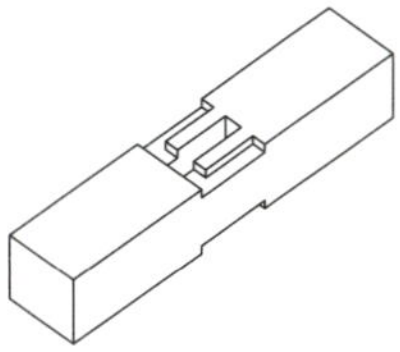

A

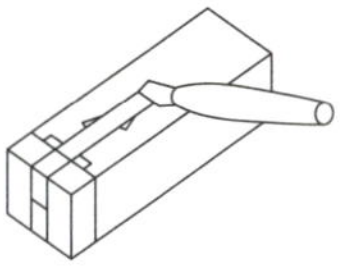

B1

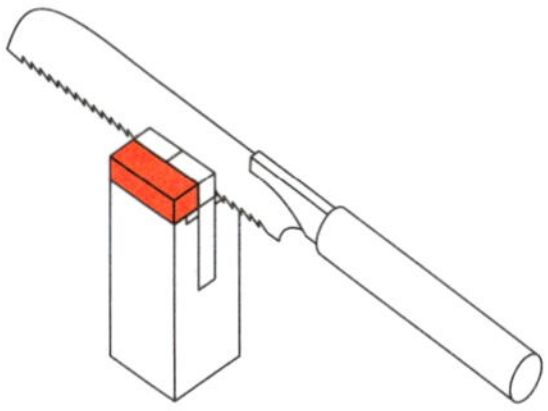

B2

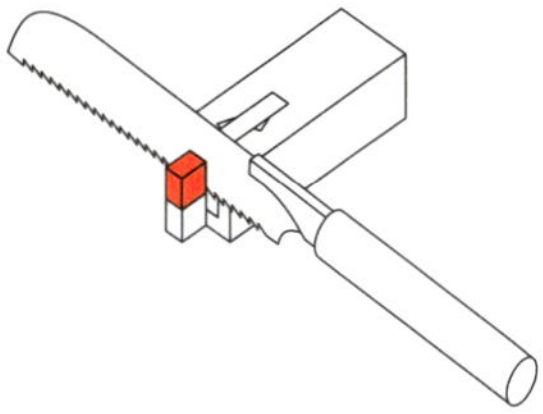

B3

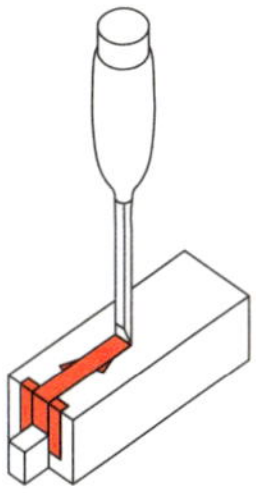

B4

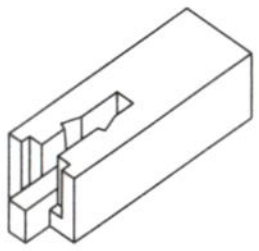

B

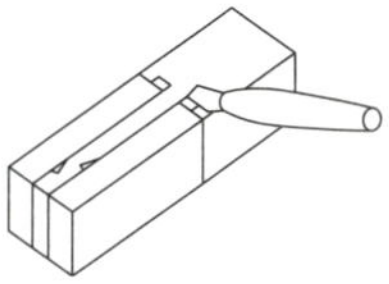

C1

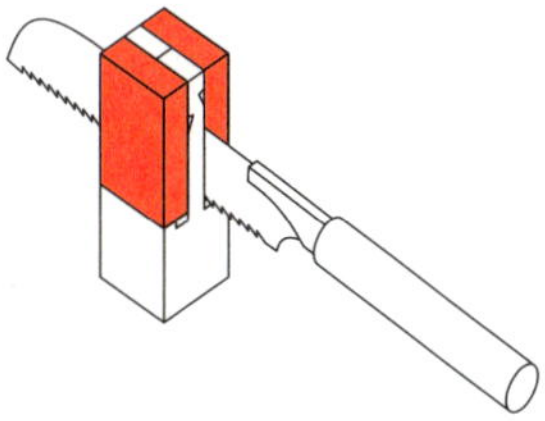

C2

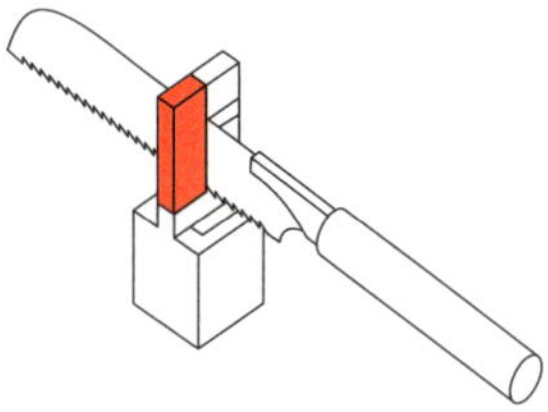

C3

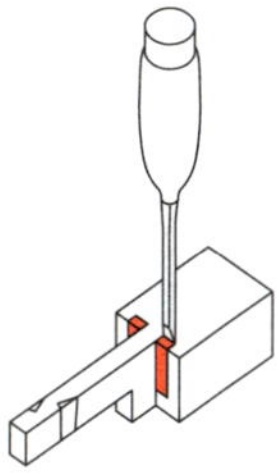

C4

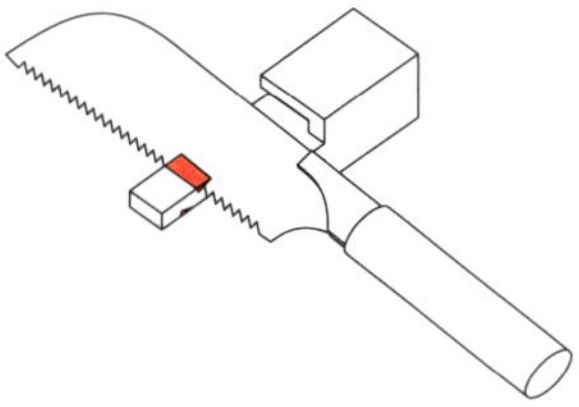

C5

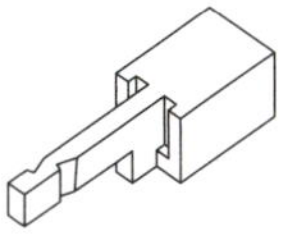

C

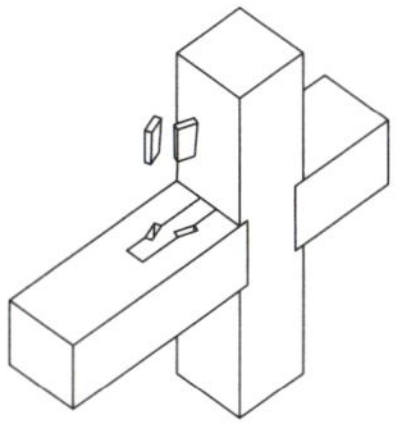

A+B+C

Specialty Joints

Pegged two-way tenon joint

tōshi-hozo-sashi-komi-sen-kumi
通枘差込栓組み

durchgesteckter Zapfen
mit Dübelsicherung

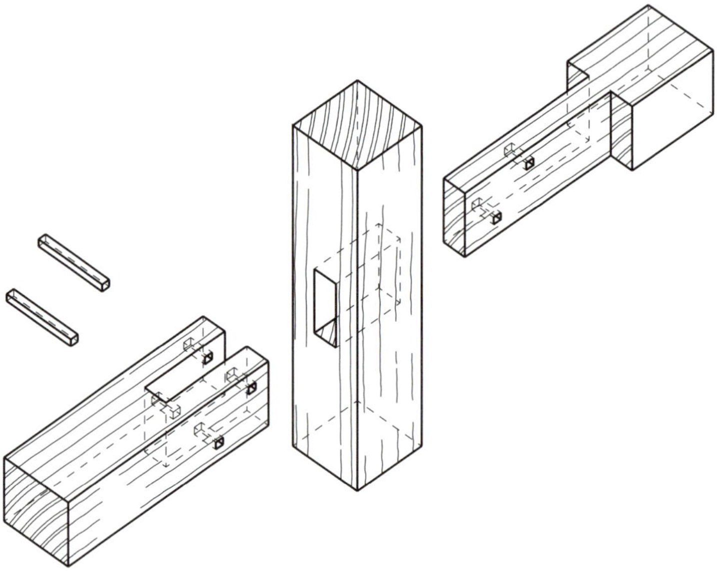

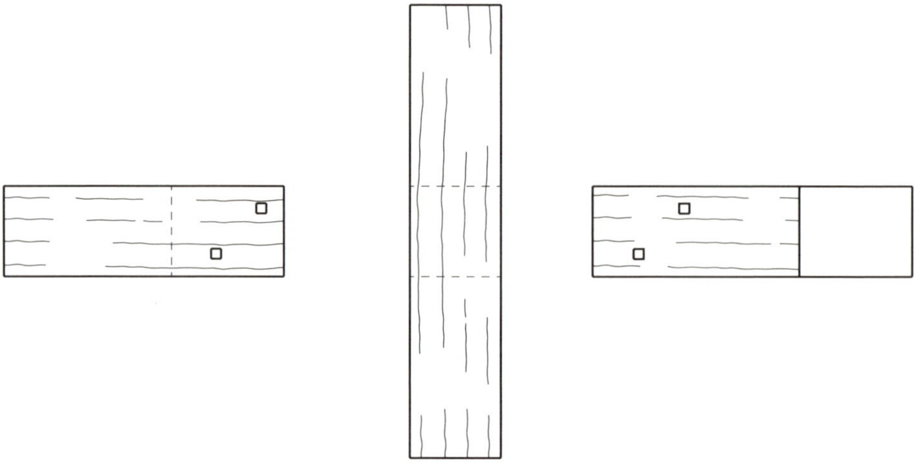

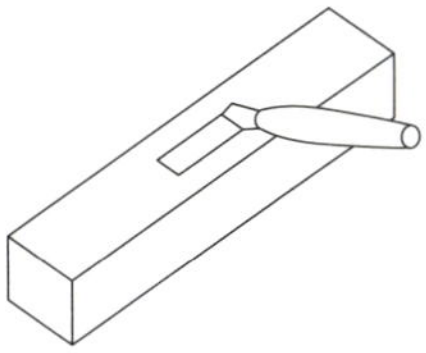
A1

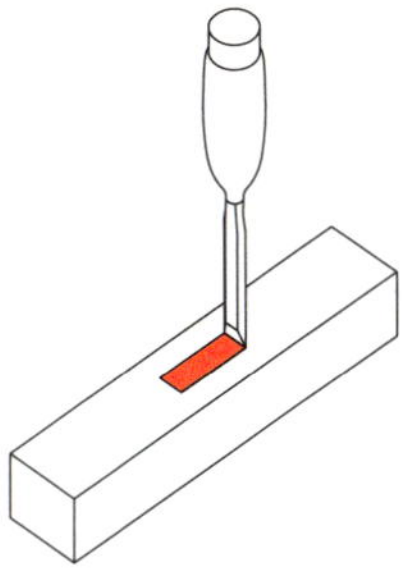
A2

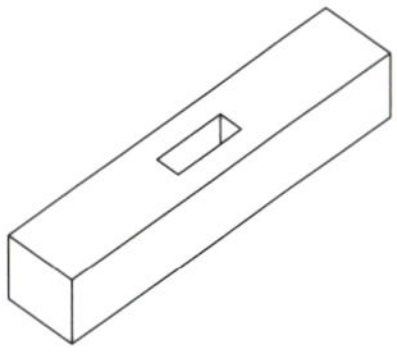
A

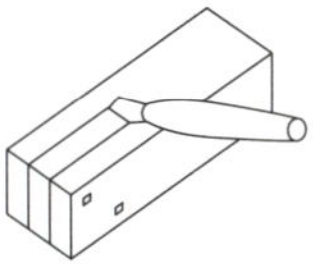
B1

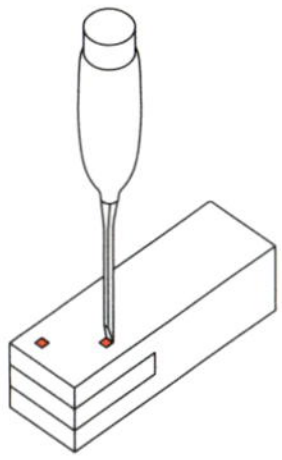
B2

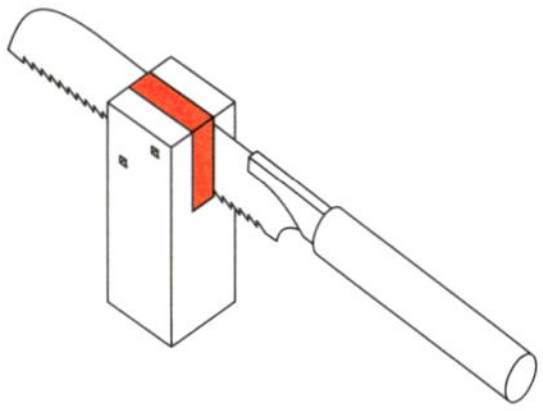
B3

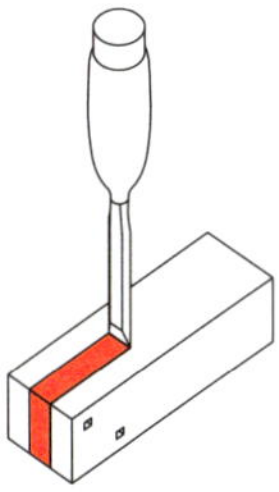
B4

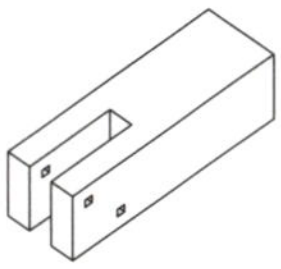
B

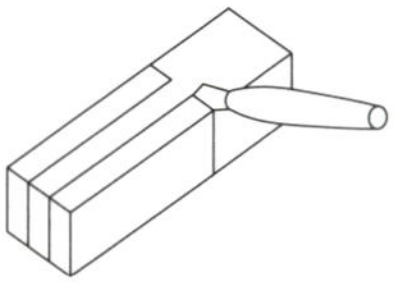
C1

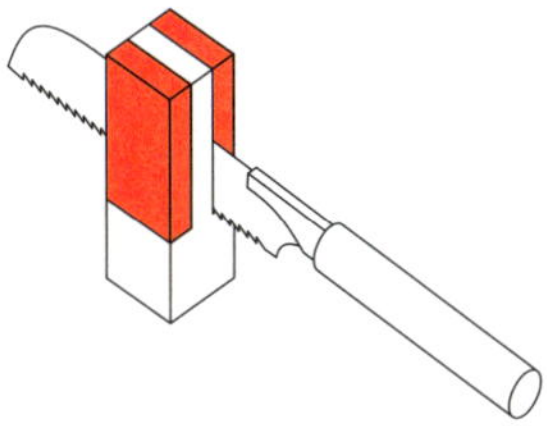
C2

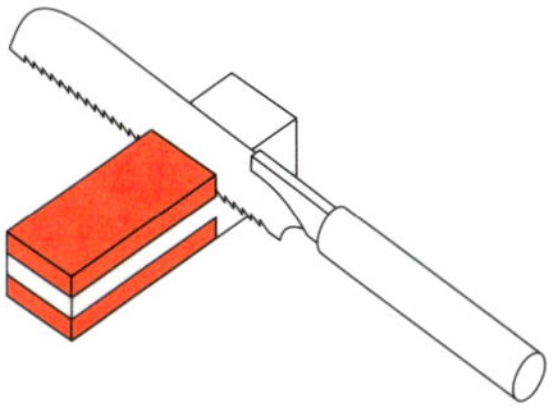
C3

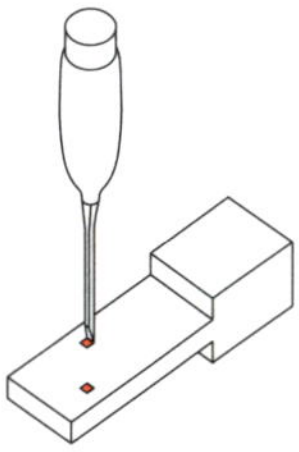
C4

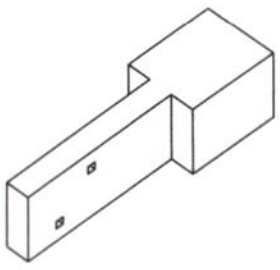
C

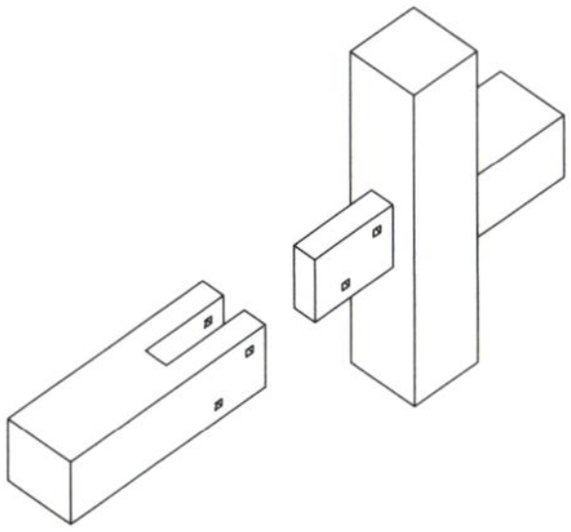
A+B+C

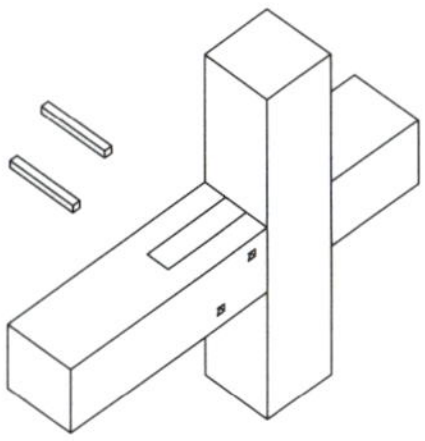
A+B+C

Specialty Joints

Two-way splined rod joint with tapered keys and a peg

yatoi-sao-hozo-shachi-sen-komi-sen-shimari [*]
雇竿枘車知栓込栓締り

durchgesteckter Fremdzapfen mit Spannkeilen und Dübelsicherung

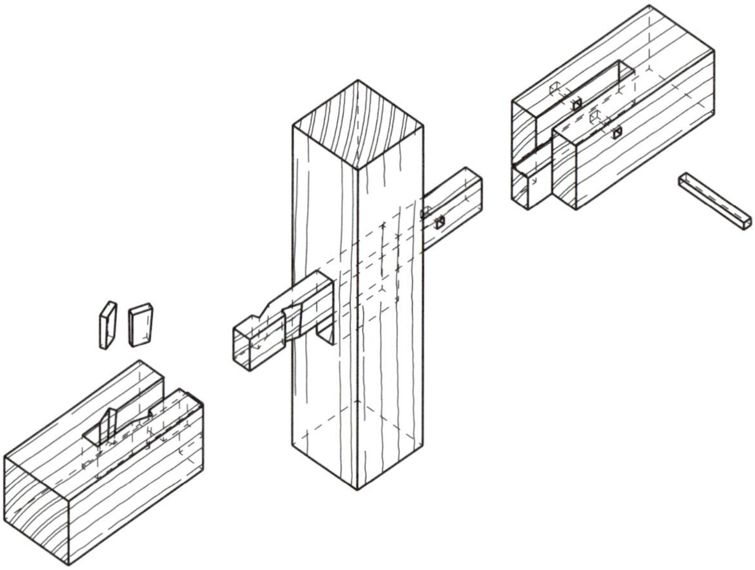

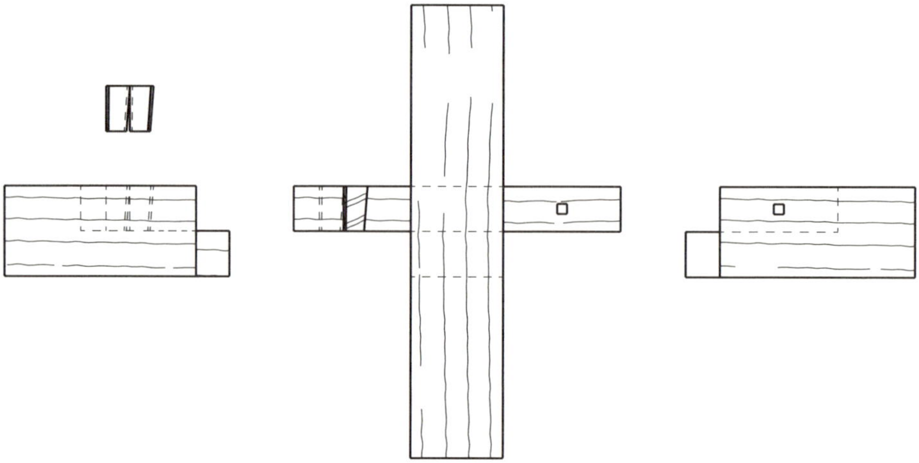

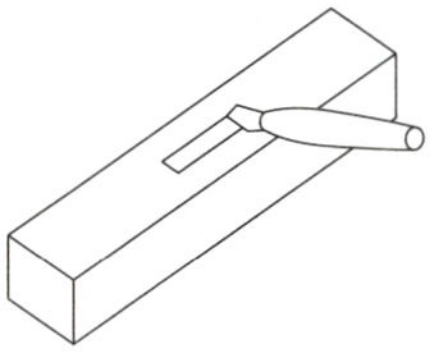
A1

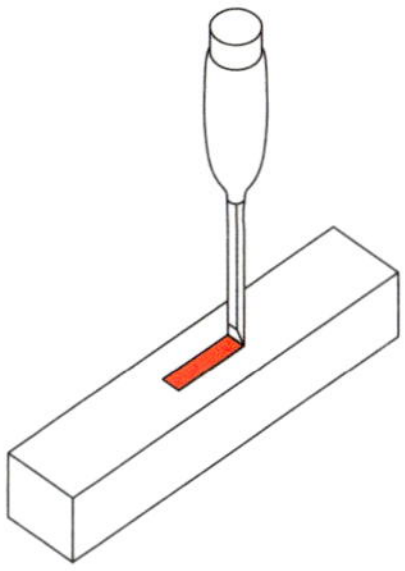
A2

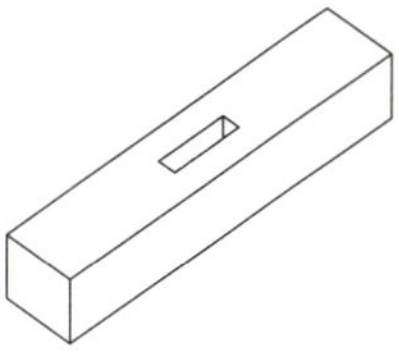
A

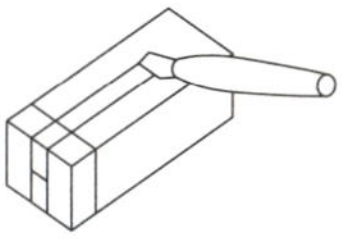
BC1

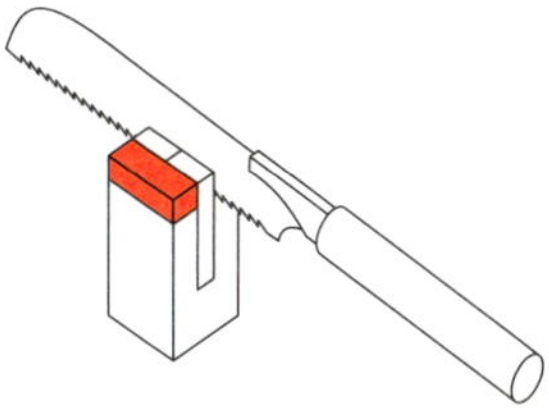
BC2

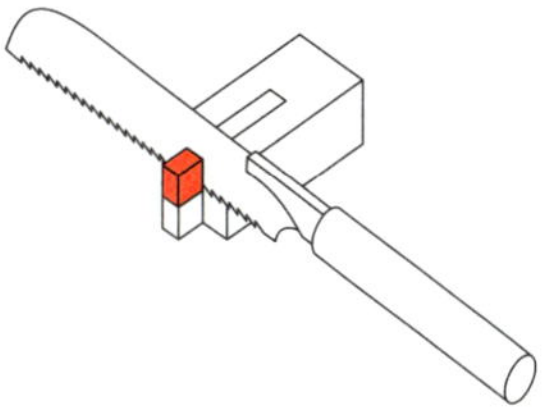
BC3

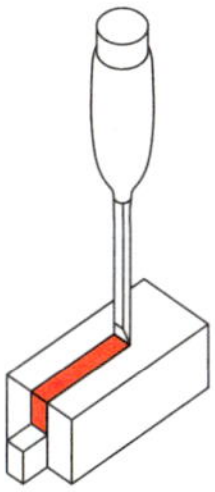
BC4

2×

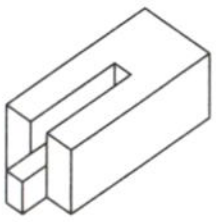
BC

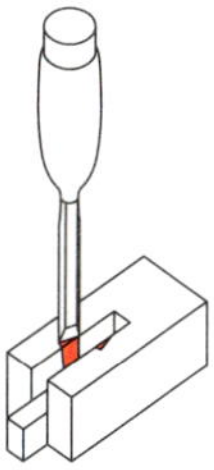
B1

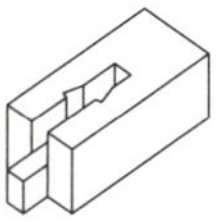
B

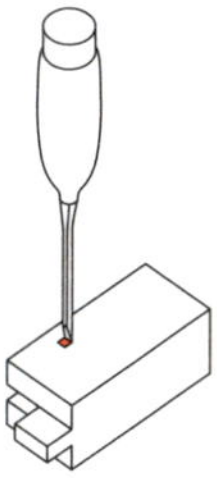
C1

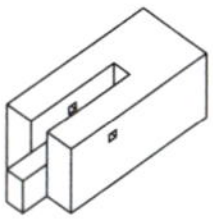
C

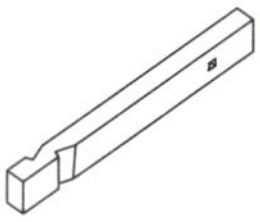
D

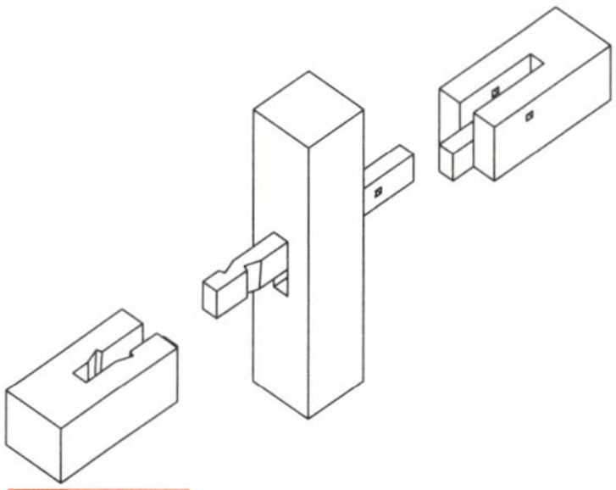
A+B+C+D

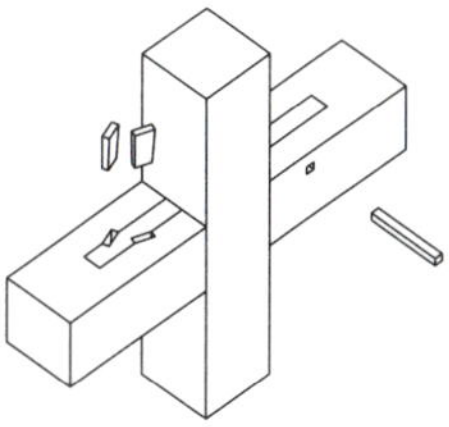
A+B+C+D

Specialty Joints

Pegged two-way lapped rod tenon joint with housing

sao-hozo-komi-sen-shimari, kage-ire-dô-tsuki (nihô-sashi) [*]
竿枘込栓締り・陰入胴付（二方差）

durchgesteckte Stabzapfen mit Blatt und Zapfenloch sowie Dübelsicherung

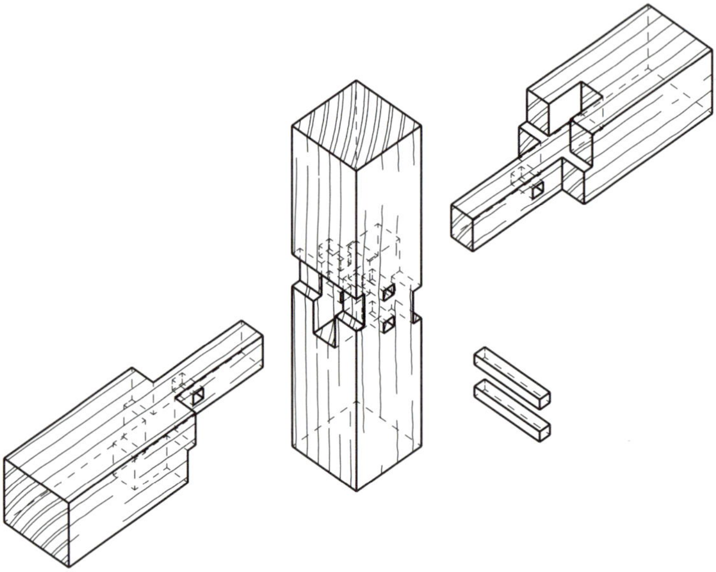

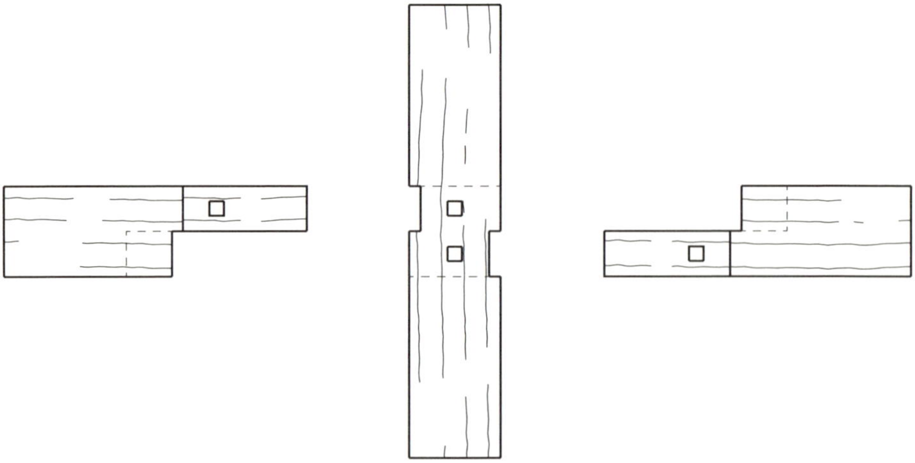

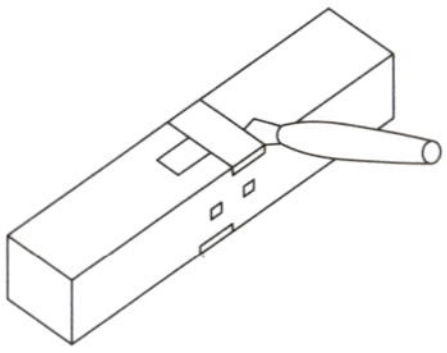
A1

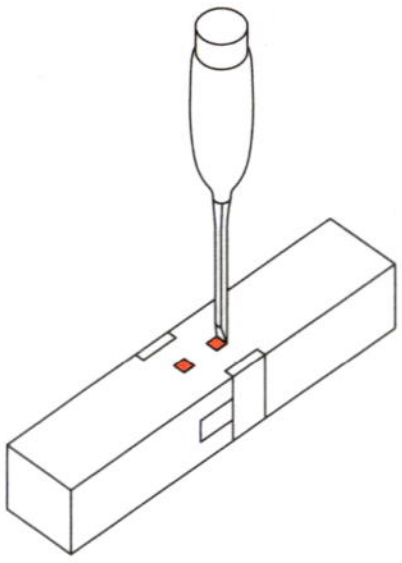
A2

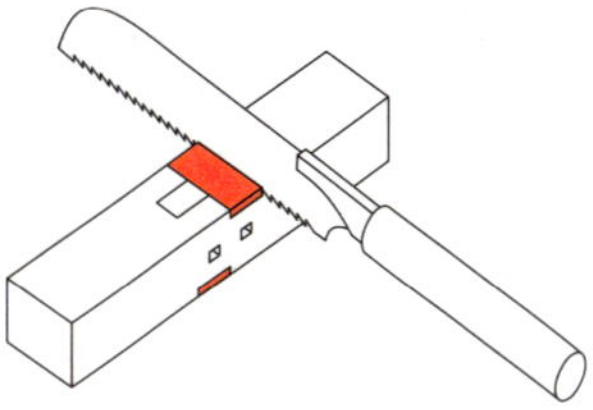
A3

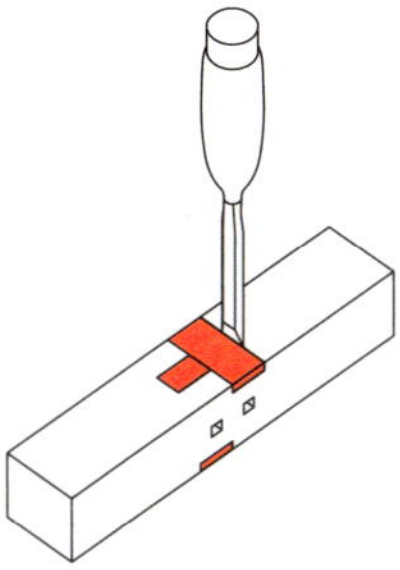
A4

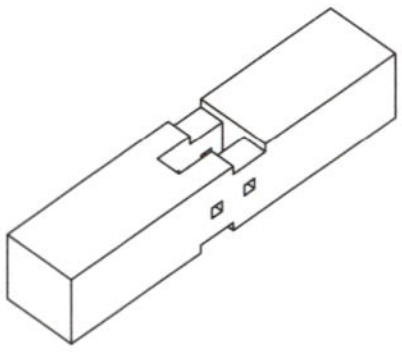
A

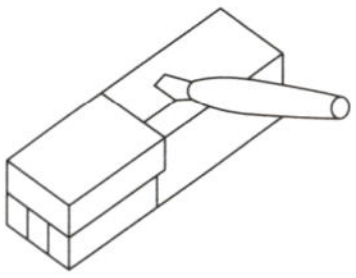
B1

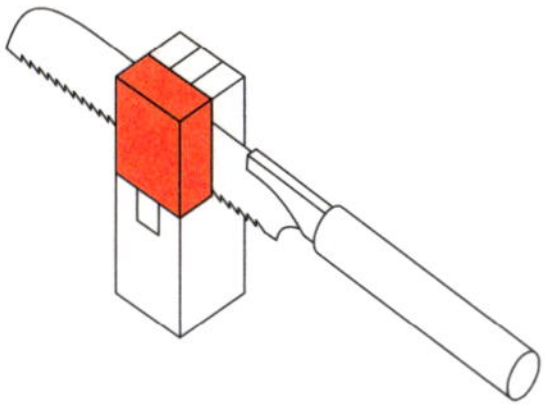
B2

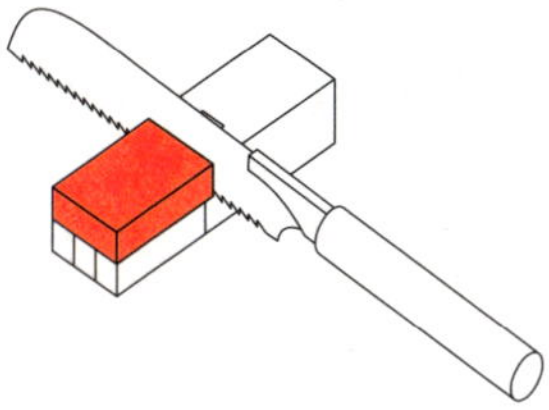
B3

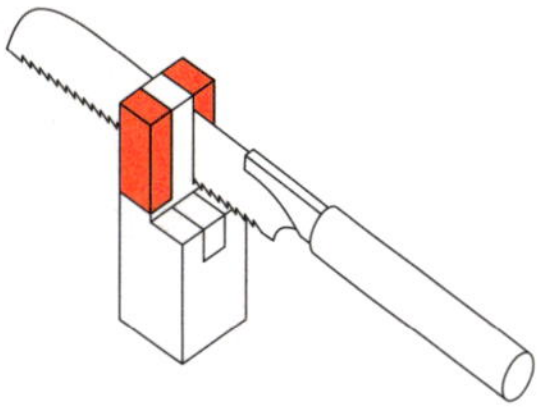
B4

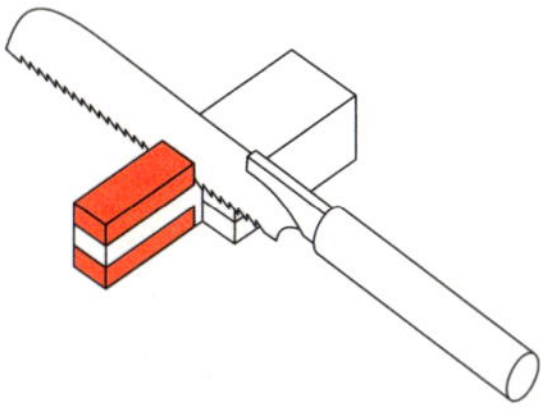
B5

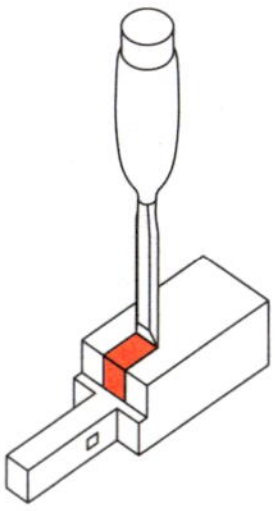
B6

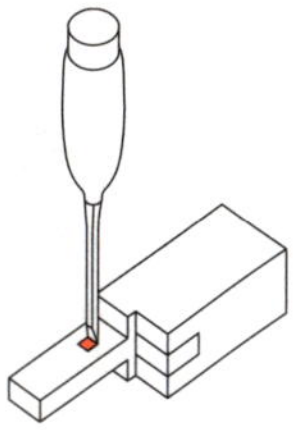
B7

2×

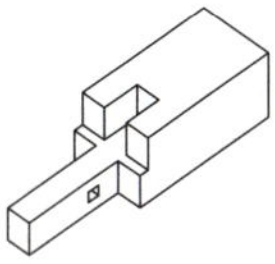
B

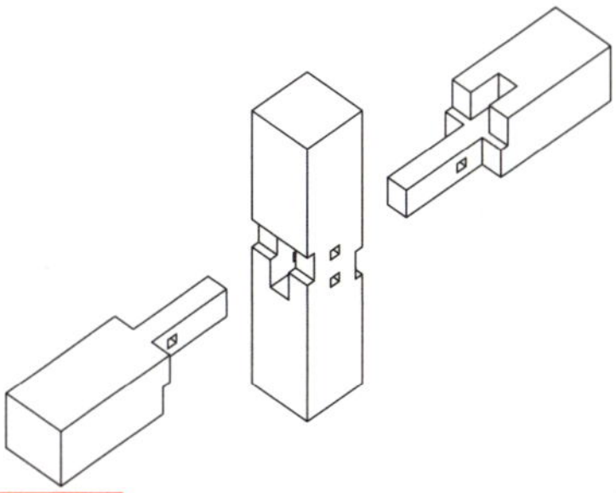
A+B+B

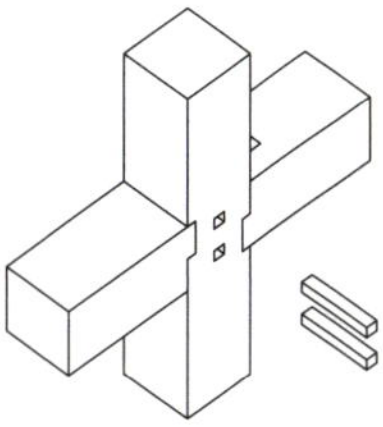
A+B+B

Case Joints

Edge Joints

Shiplap joint

ai-jakuri-hagi [*]
相決矧ぎ

Überfälzung

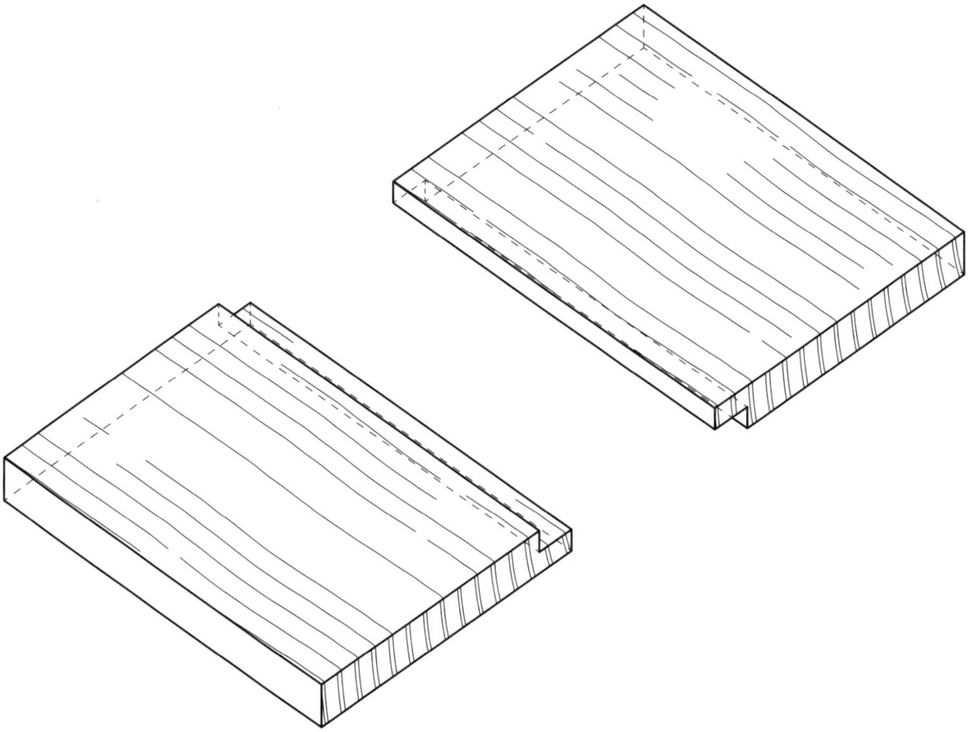

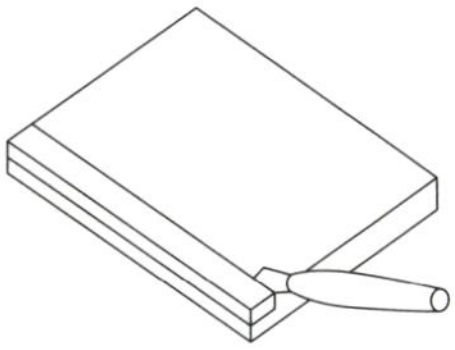

A1

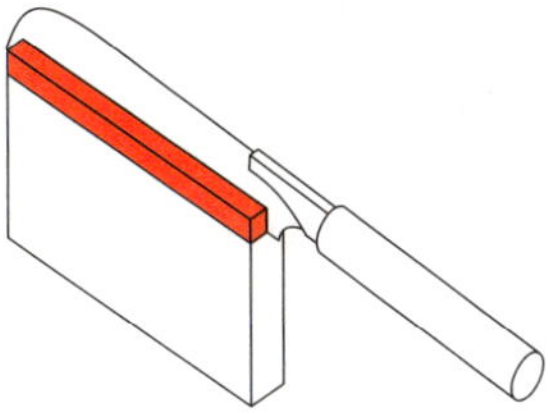

A2

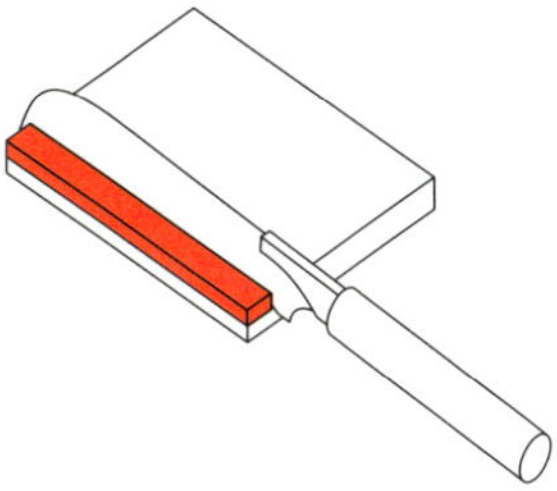

A3

2×

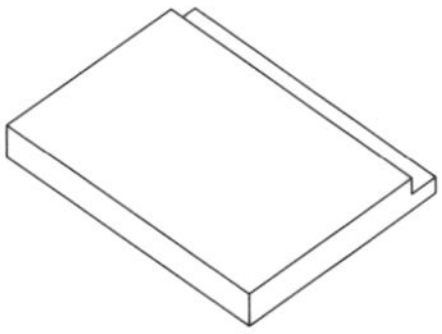

A

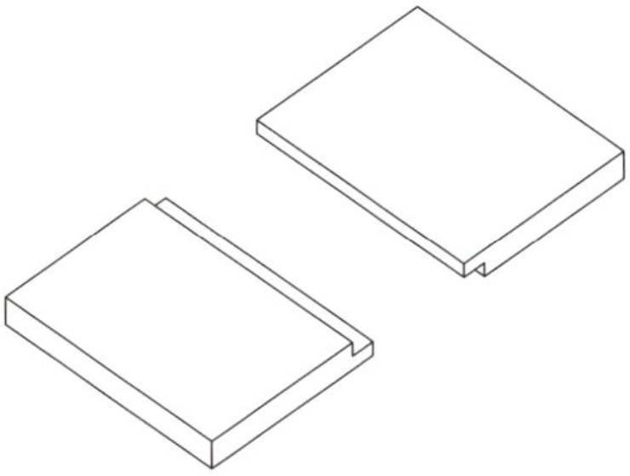

A+A

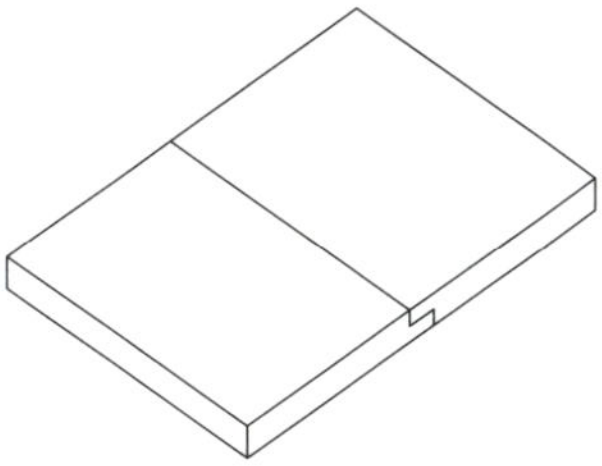

AA

Edge Joints

V-grooved joint

hifukura-hagi [*]
樋部倉矧ぎ

v-förmige Nut

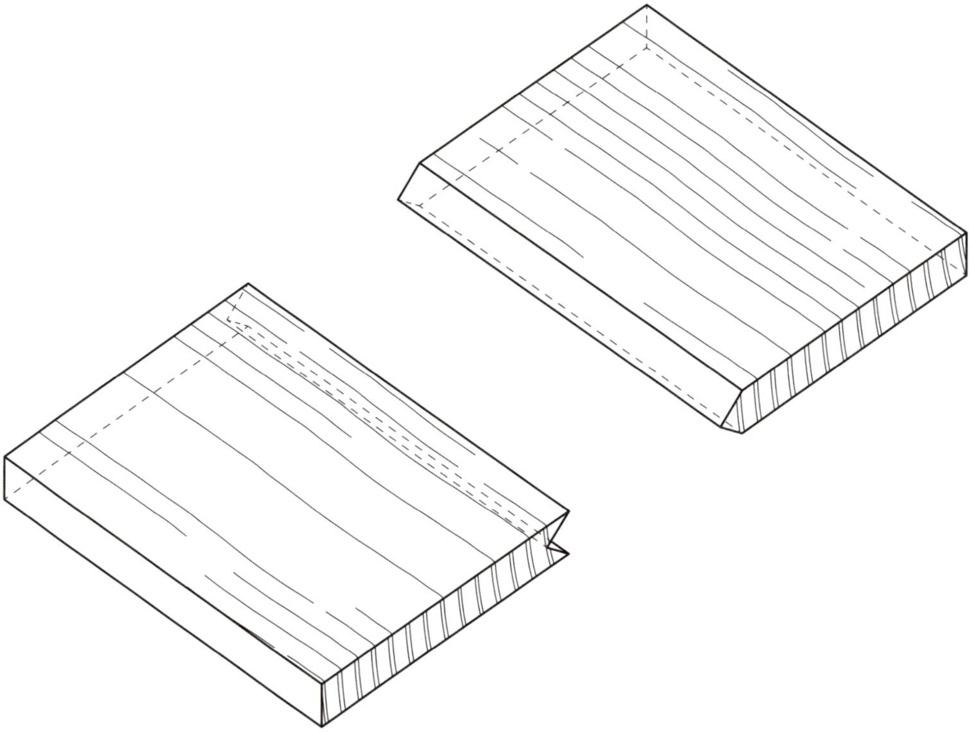

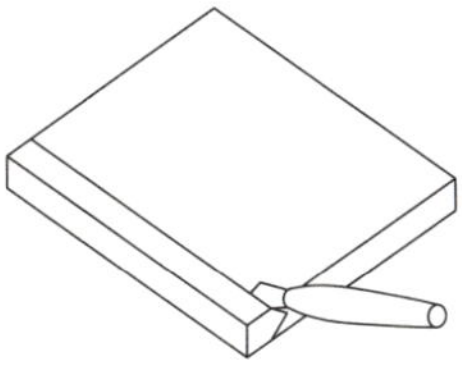

A1

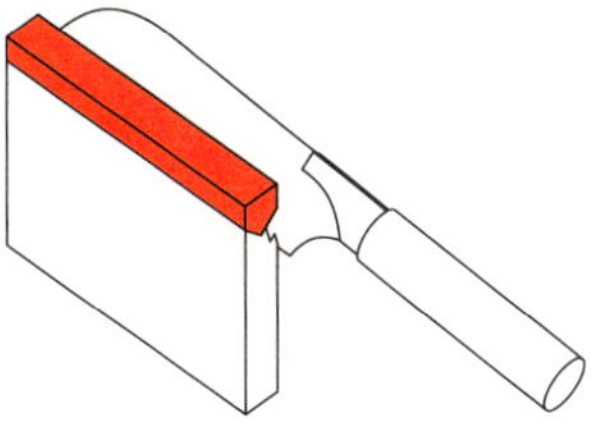

A2

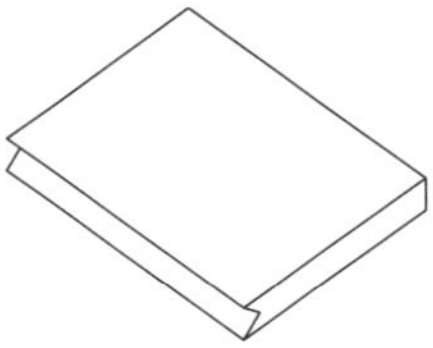

A

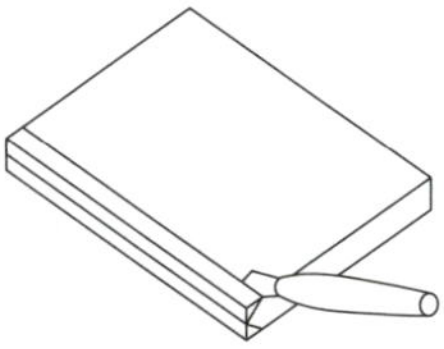

B1

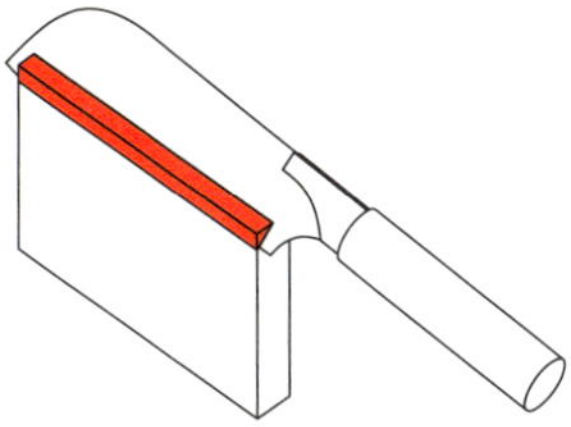

B2

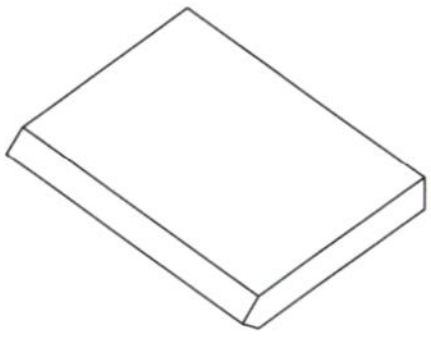

B

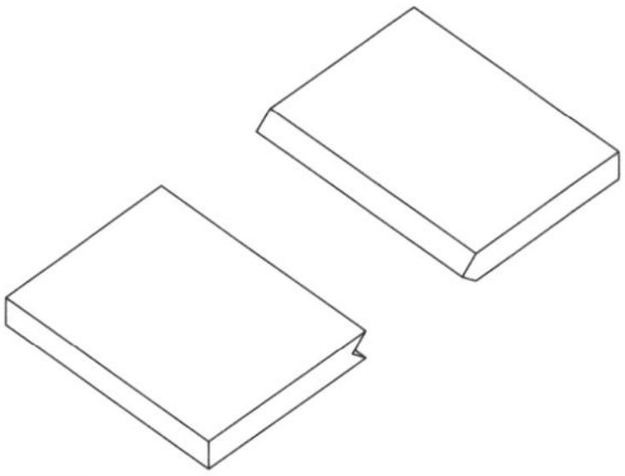

A+B

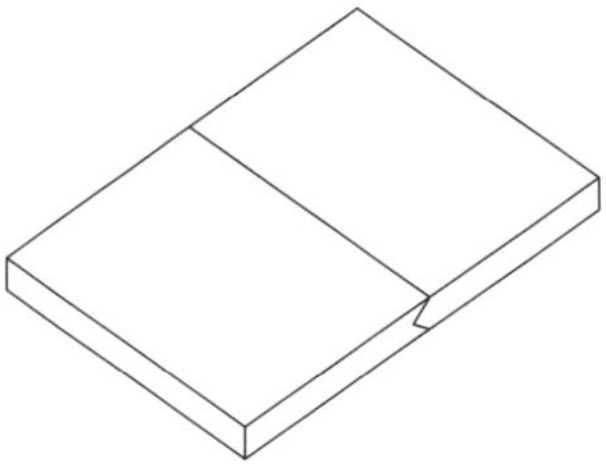

AB

Edge Joints

Tongue and groove joint

hon-zane-hagi [*]
本実矧ぎ

Nut und Feder

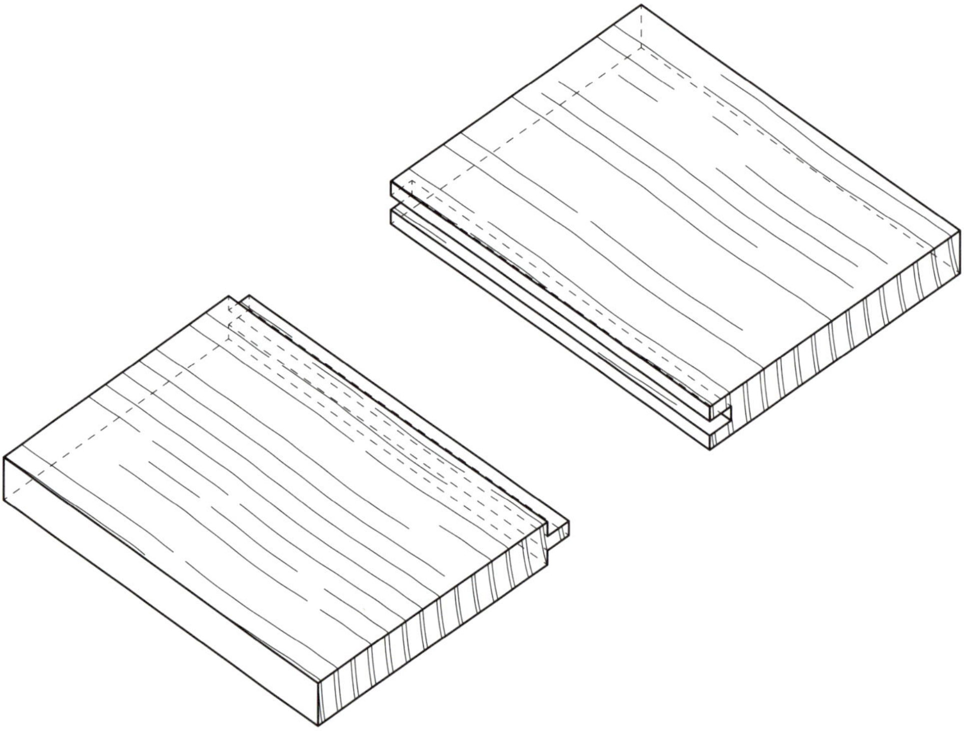

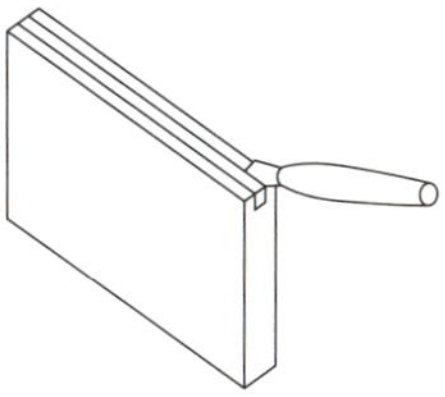

A1

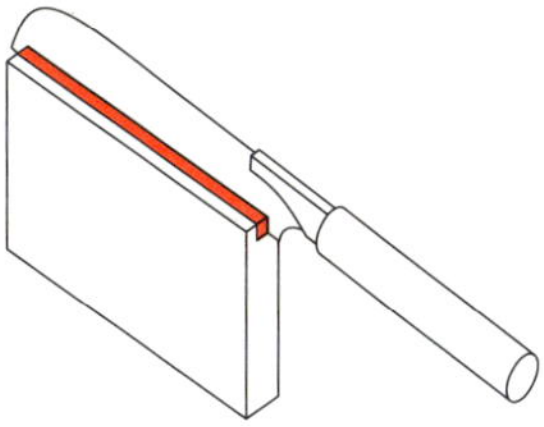

A2

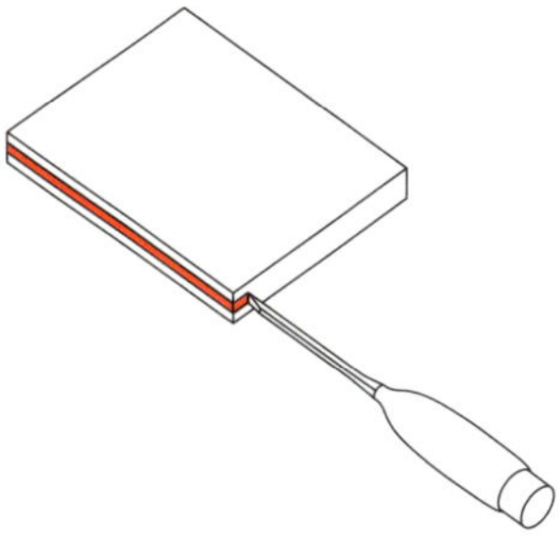

A3

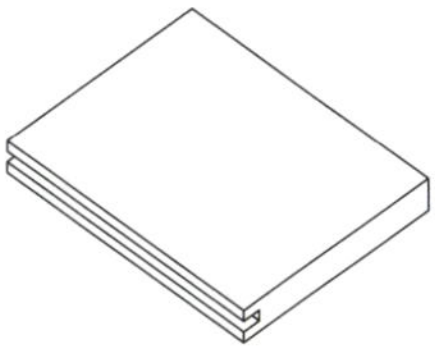

A

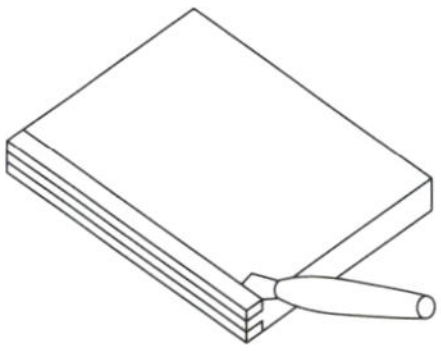

B1

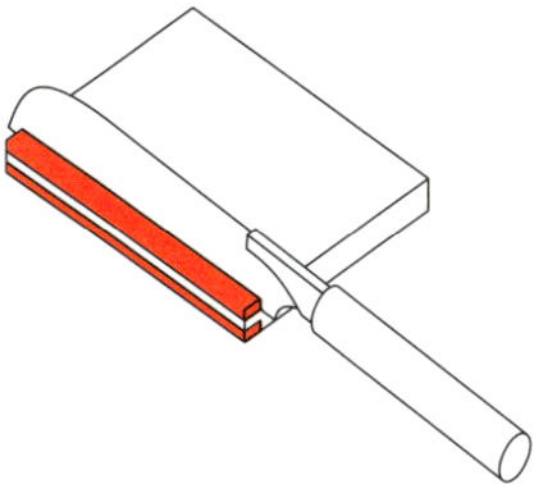

B2

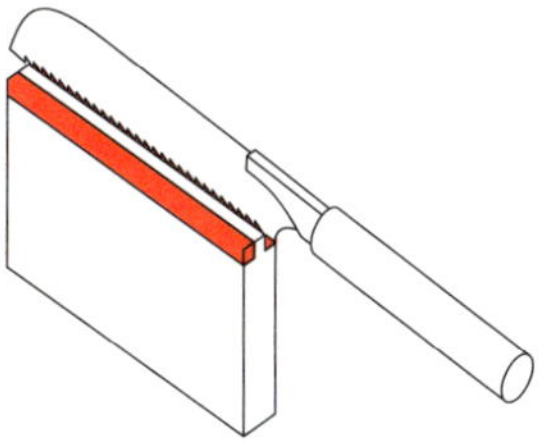

B3

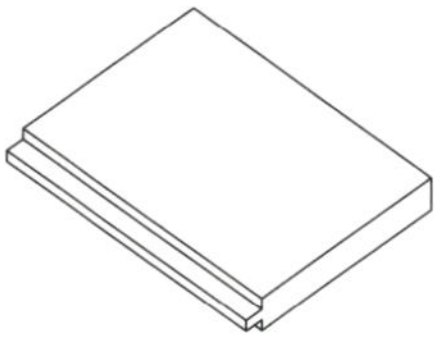

B

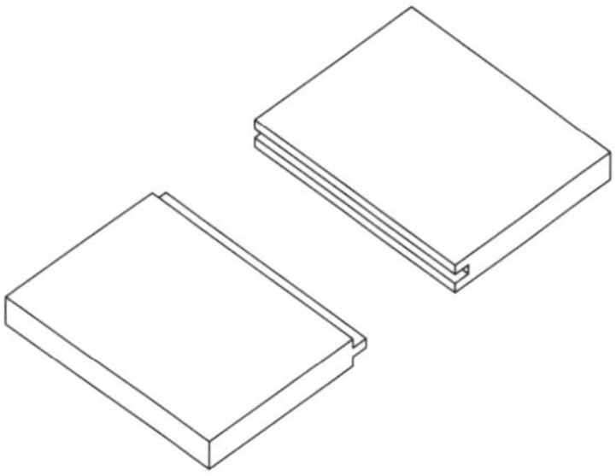

A+B

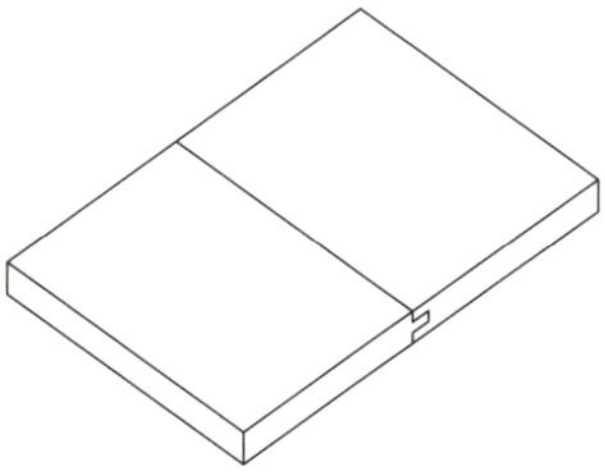

AB

Edge Joints

Tongued and rabbeted joint

ai-jakuri-zane-hagi [*]
相决実矧ぎ

Nut und Feder
mit versetzter Brüstung

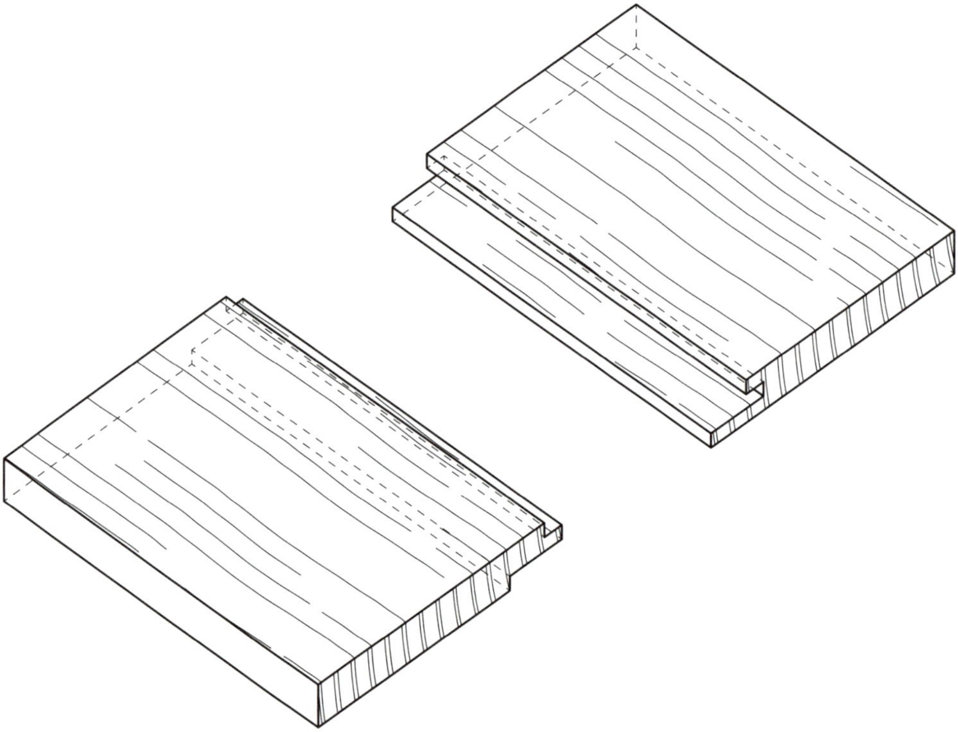

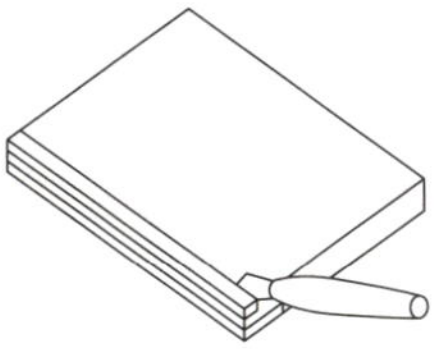
A1

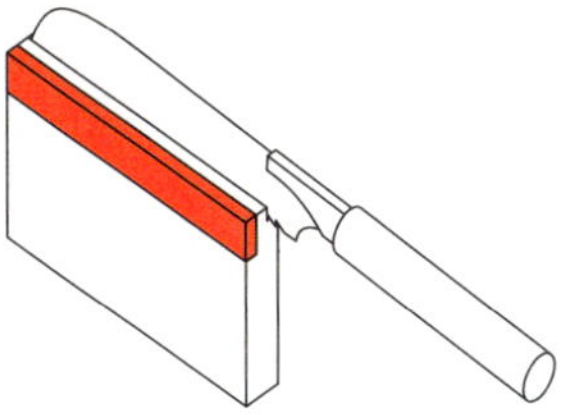
A2

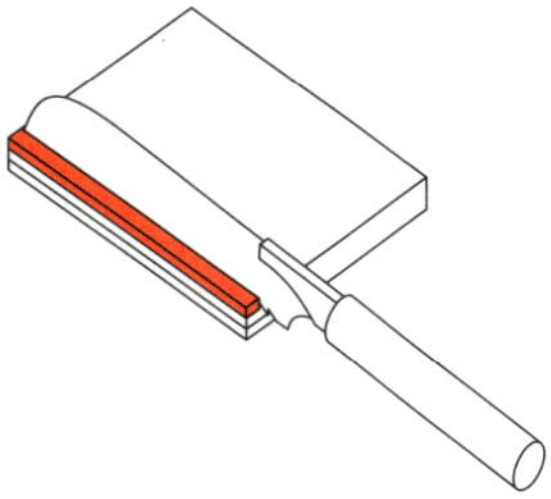
A3

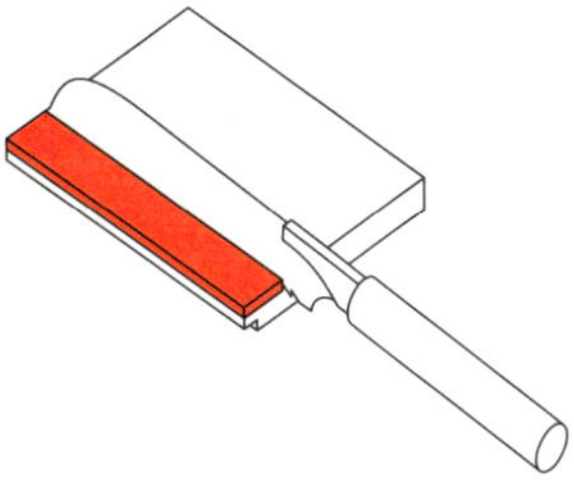
A4

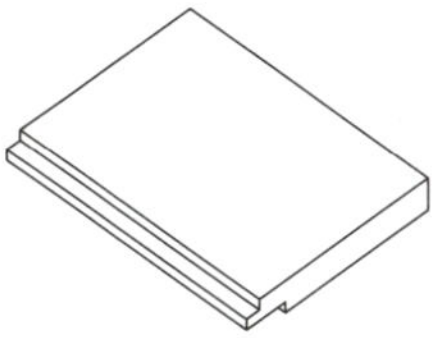
A

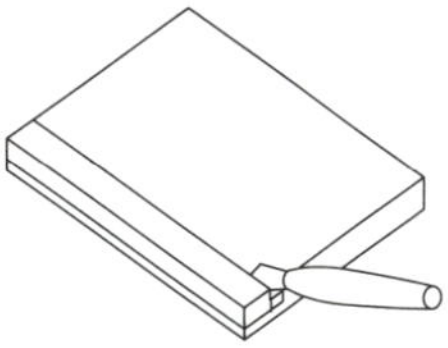
B1

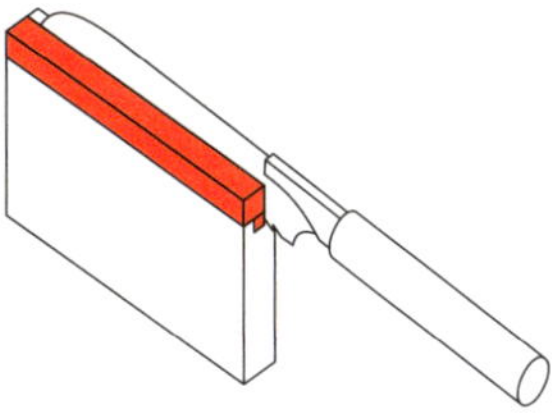
B2

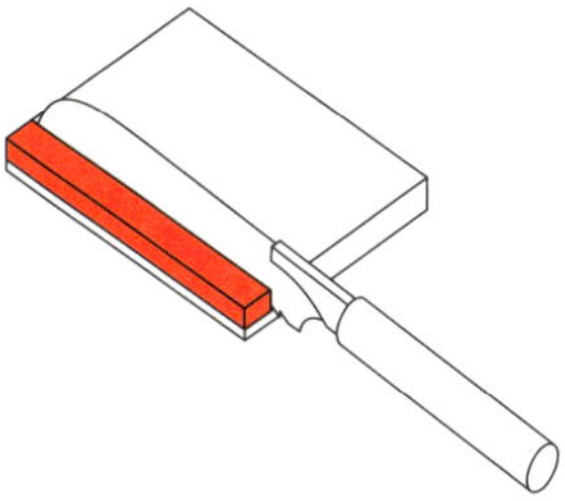
B3

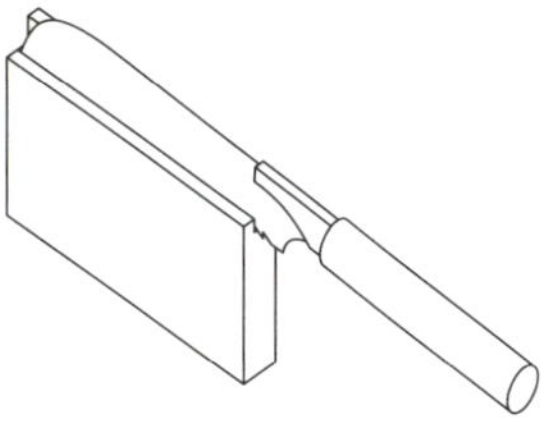
B4

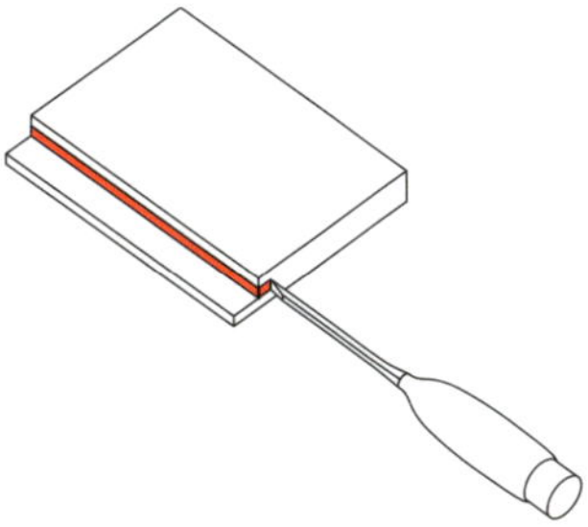
B5

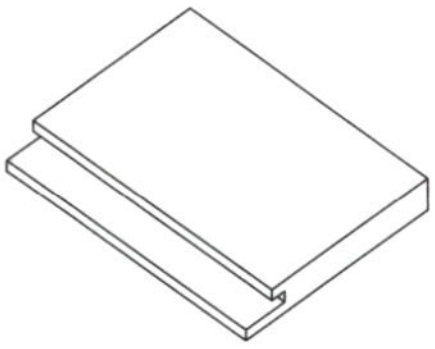
B

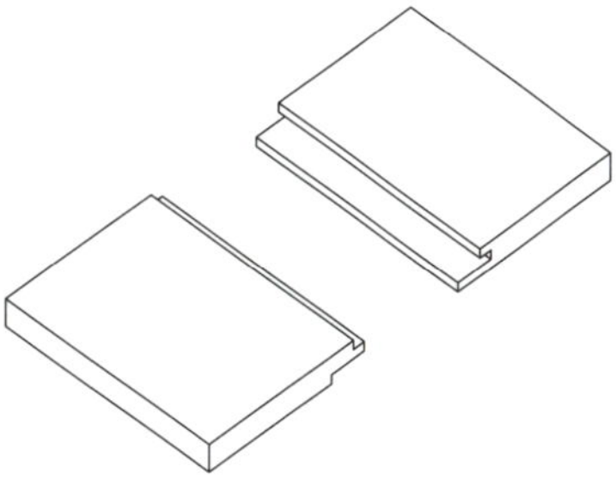
A+B

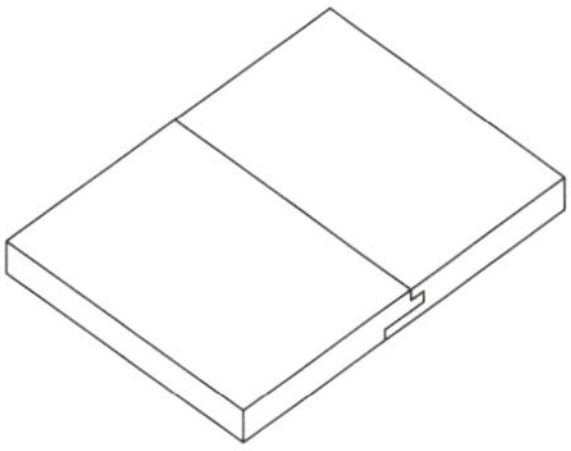
AB

Floating tenon joint

tsuki-tsuke-yatoi-hozo-hagi
突付雇枘矧ぎ

Stoß mit eingesetztem Zapfen

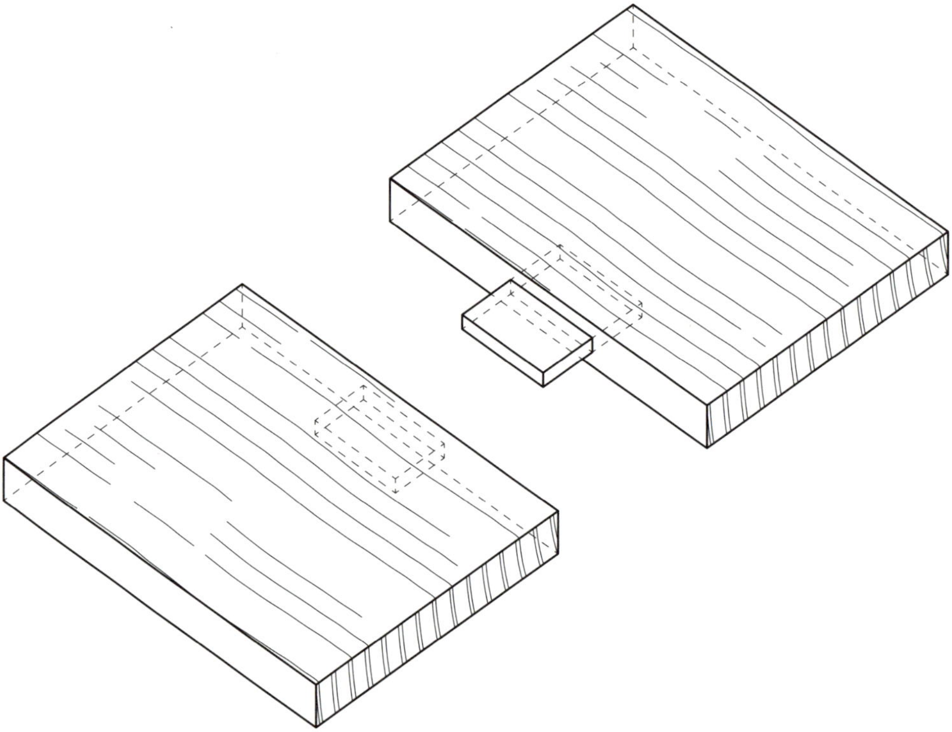

2×

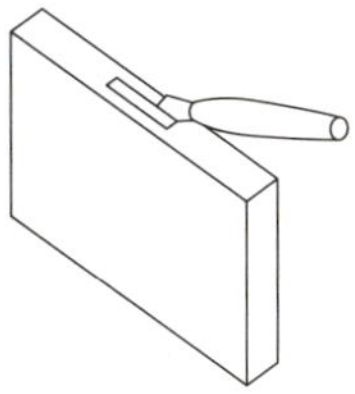

A1

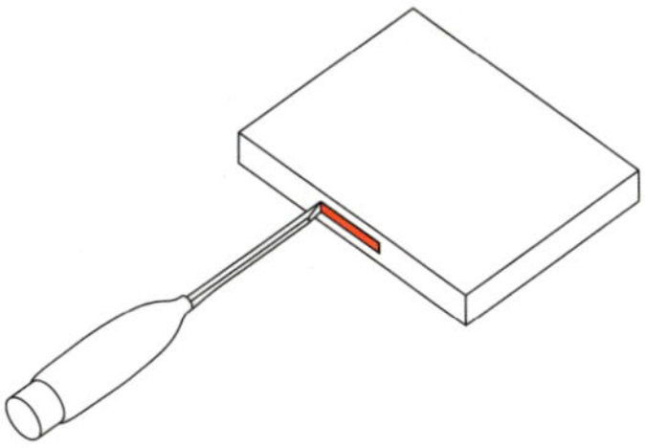

A2

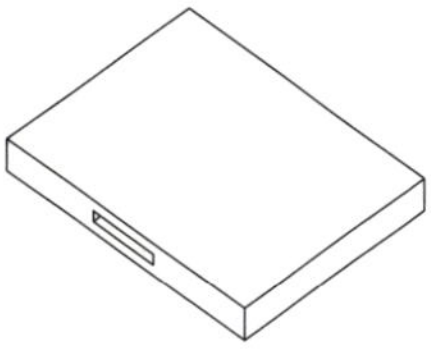

A

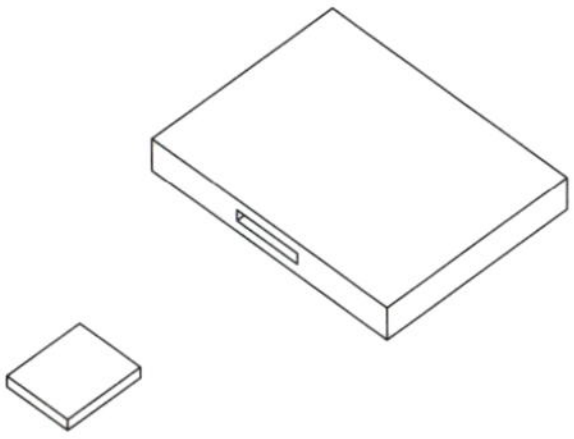

A+B

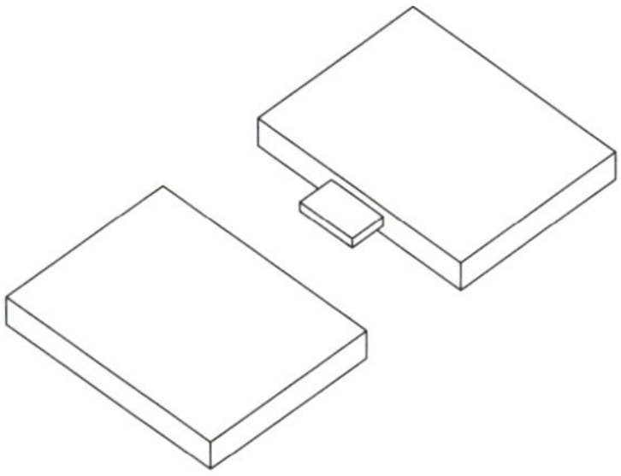

A+B+A

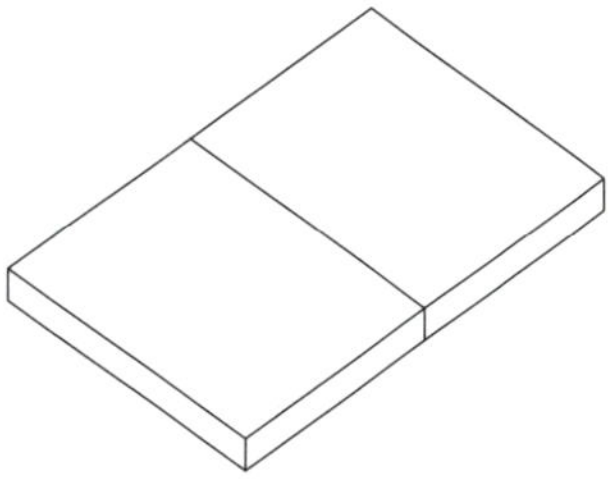

ABA

Edge Joints

Foxtail-wedged floating tenon joint

tsuki-tsuke-yatoi-hozo-jigoku-ari-hagi
突付雇枘地獄蟻矧ぎ

Stoß mit verkeiltem eingesetztem Zapfen

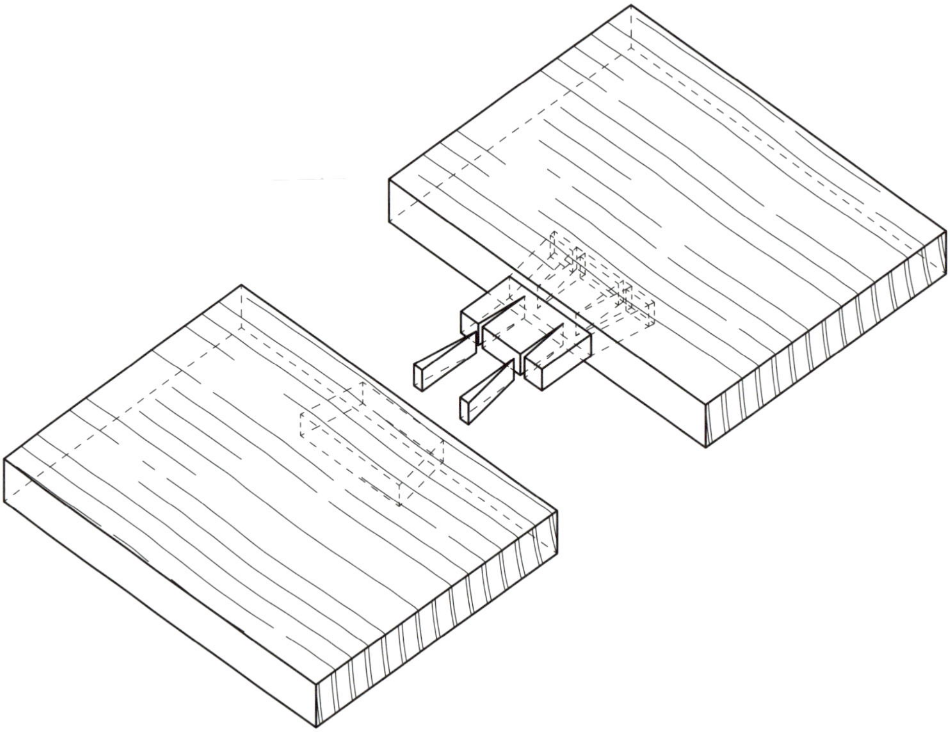

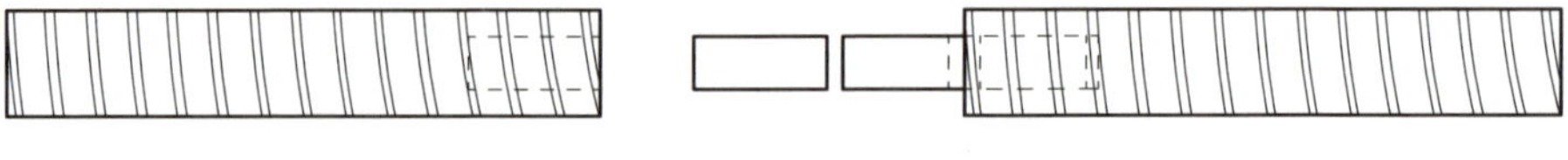

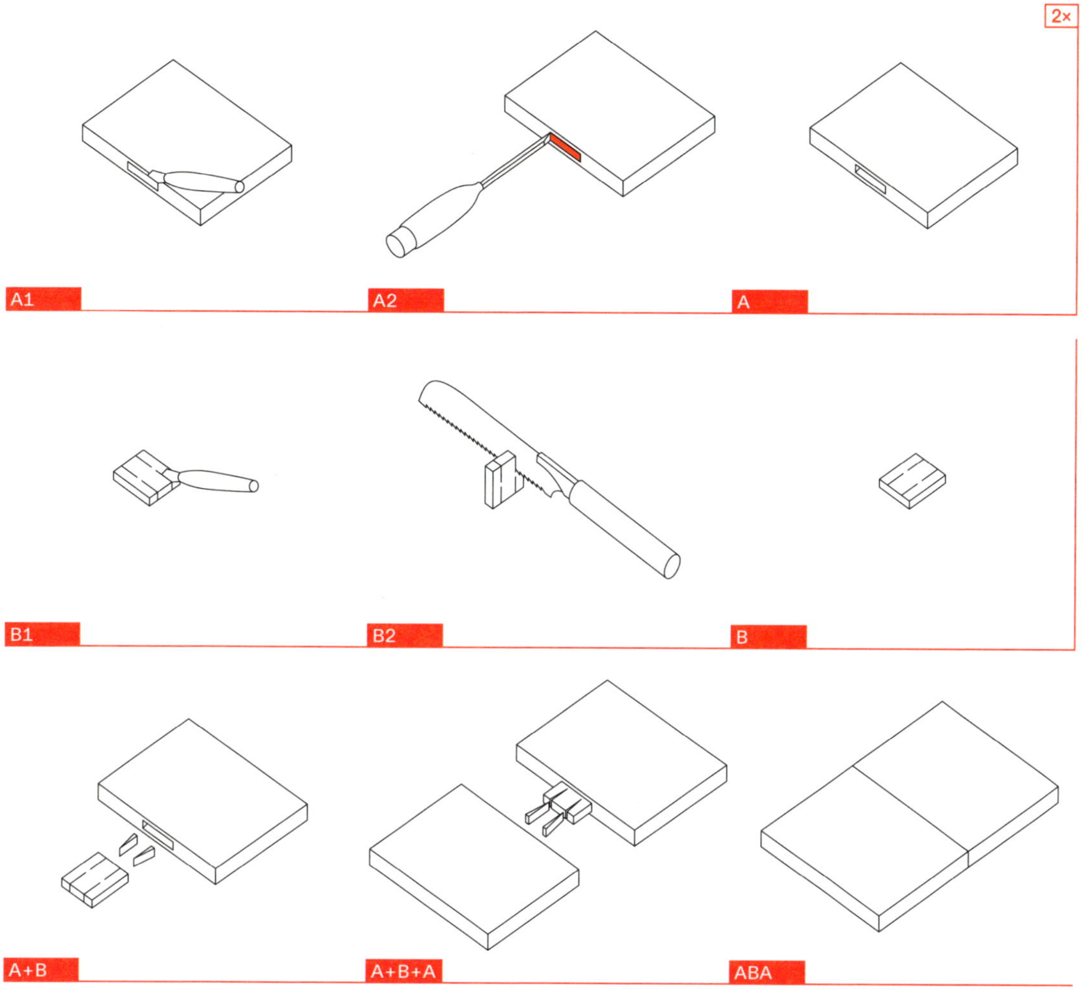
2×
A1
A2
A
B1
B2
B
A+B
A+B+A
ABA

Spline and groove joint

yatoi-zane-hagi [*]
雇実矧ぎ

Nut mit eingesetzter Feder

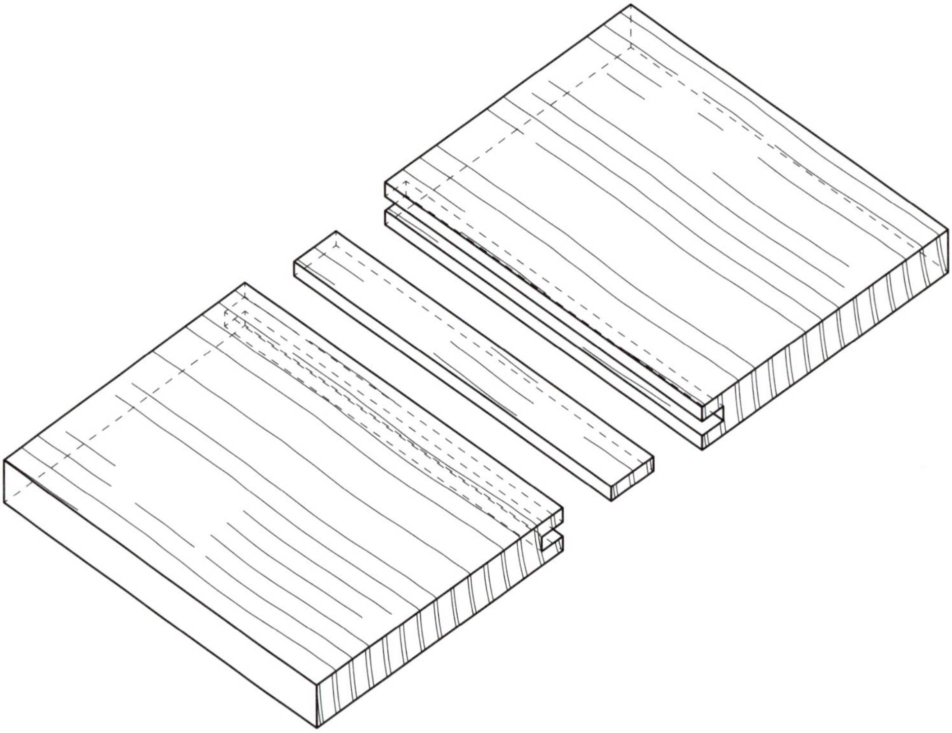

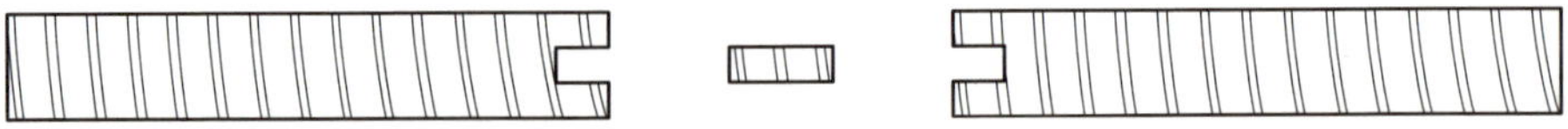

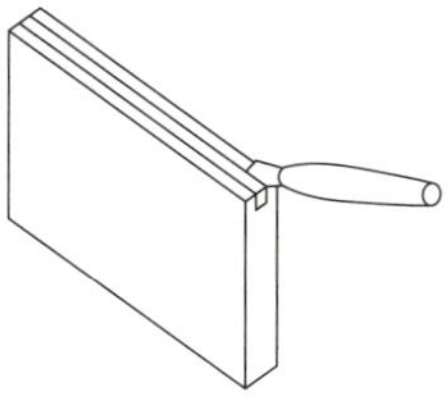

A1

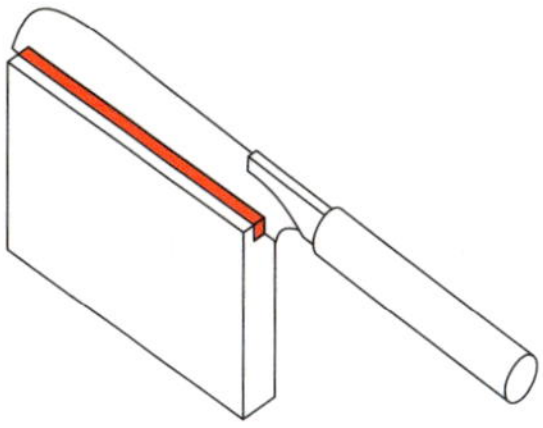

A2

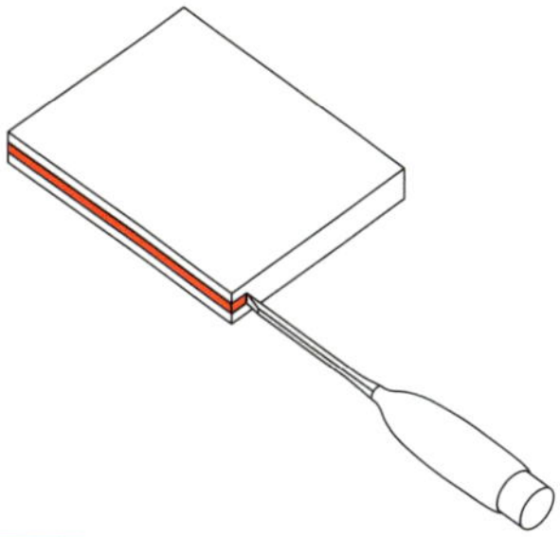

A3

2×

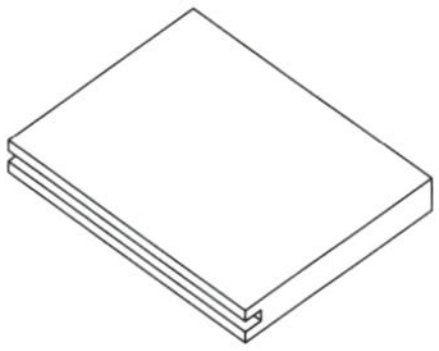

A

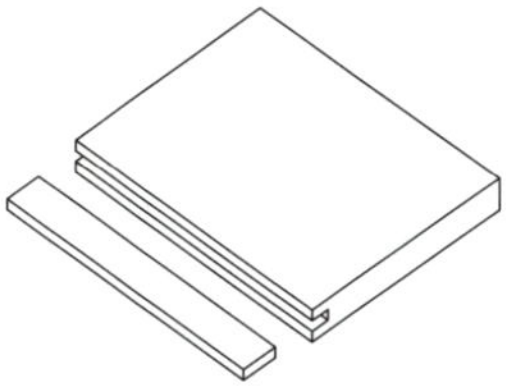

A+B

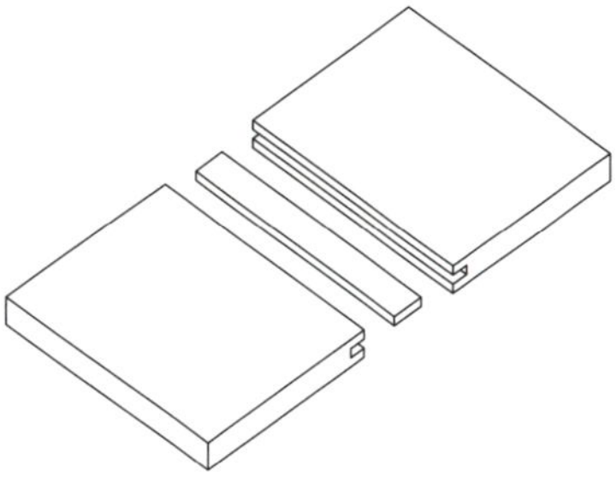

A+B+A

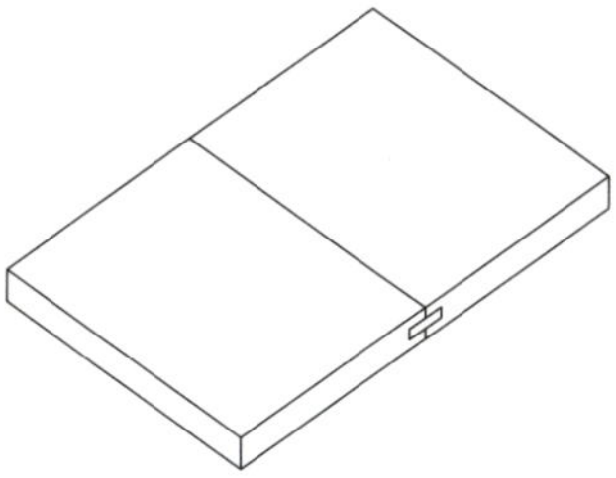

ABA

Floating tenon and spline joint

yatoi-hozo-zane-hagi [●]
雇枘実矧ぎ

Nut mit eingesetzter Feder und Zapfen

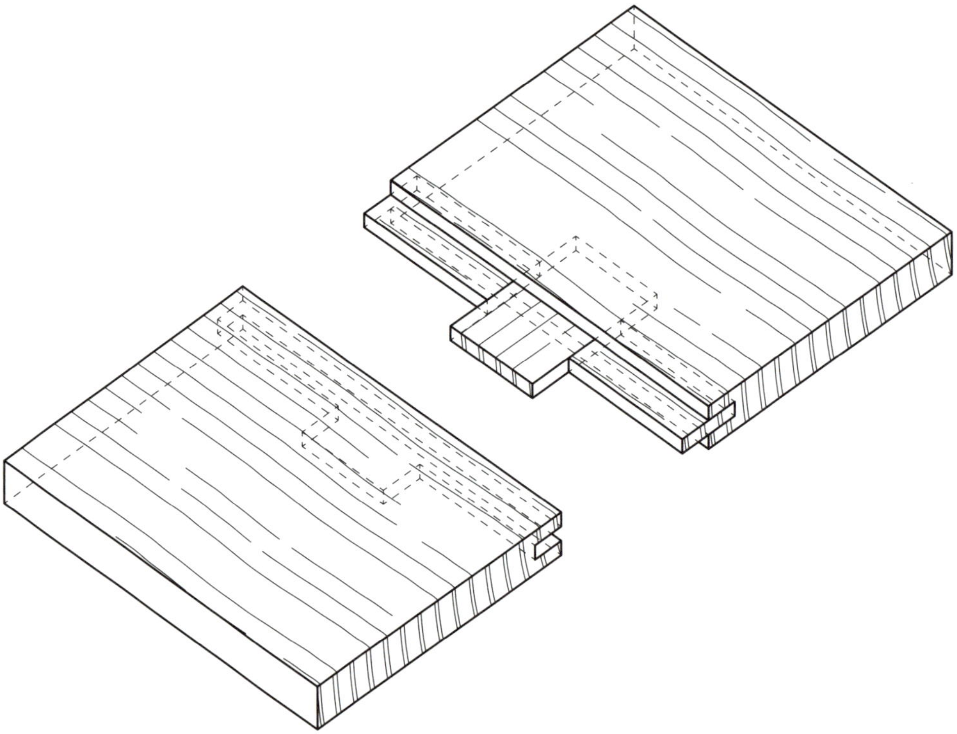

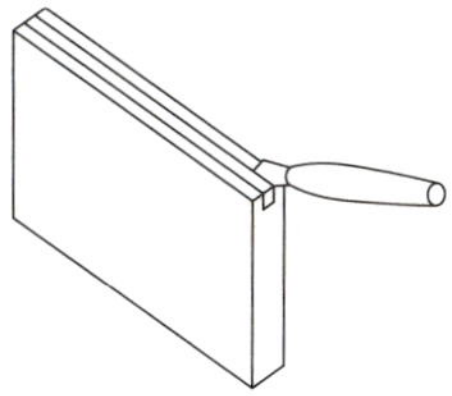

A1

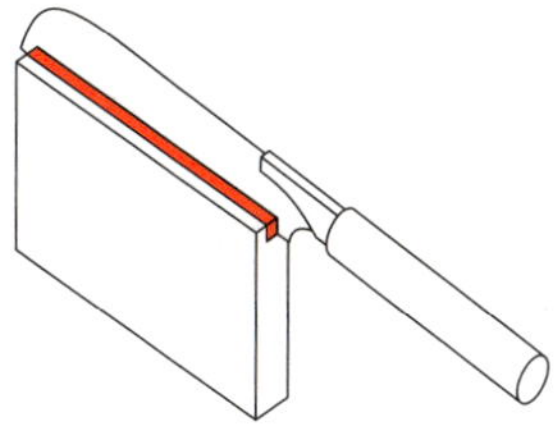

A2

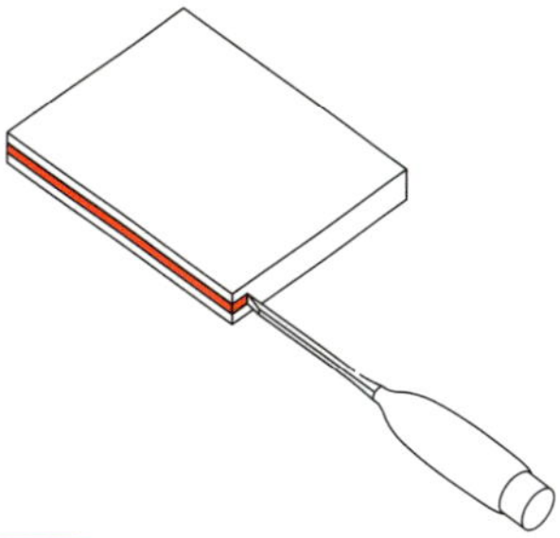

A3

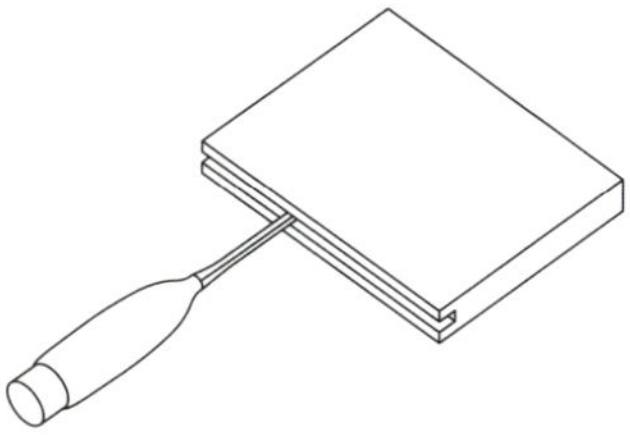

A4

2×

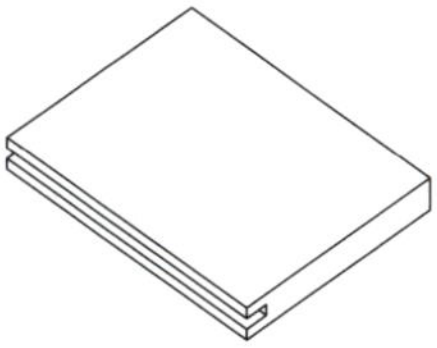

A

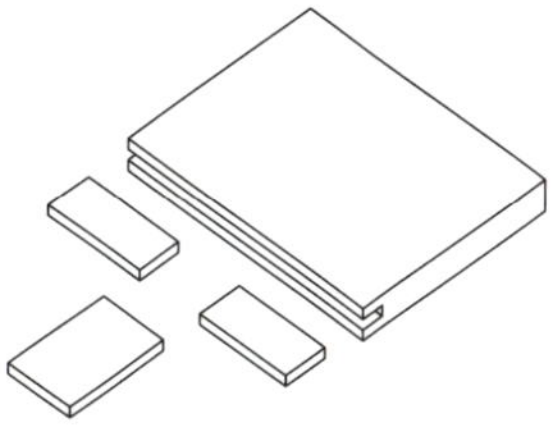

A+B

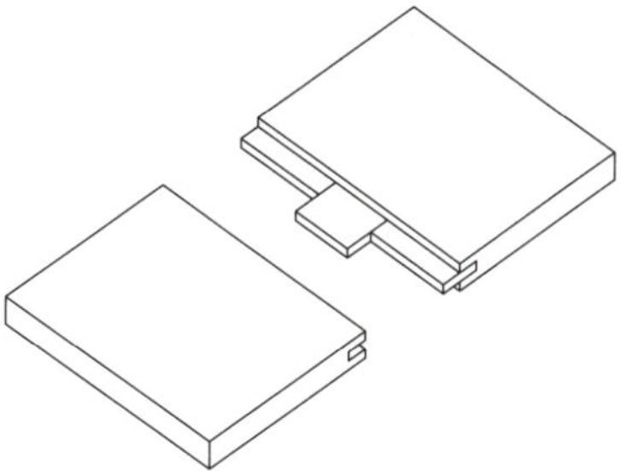

A+B+A

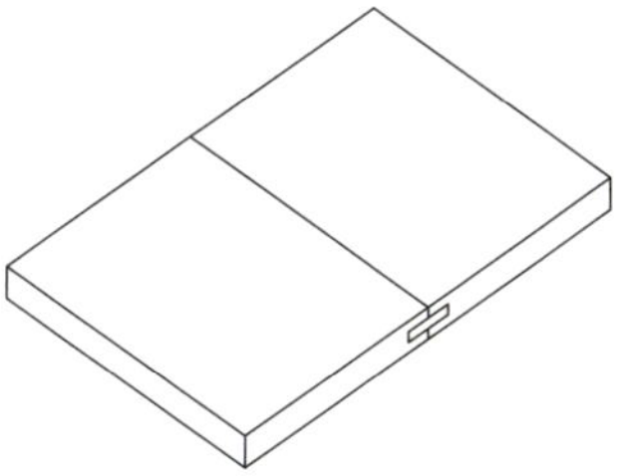

ABA

Edge Joints

Blind dovetail joint

yose-ari-hagi [*]
寄蟻矧ぎ

abgesetzte Fuge mit Trapezprofil

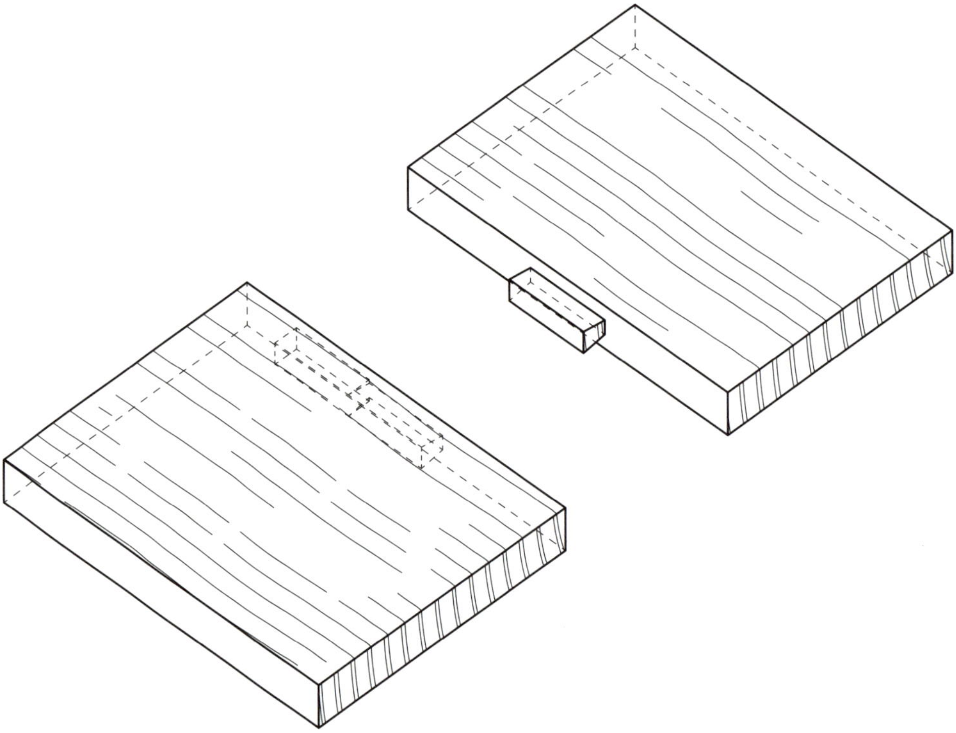

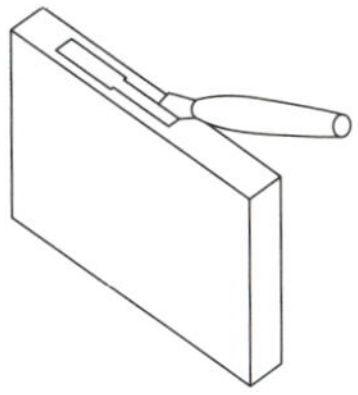

A1

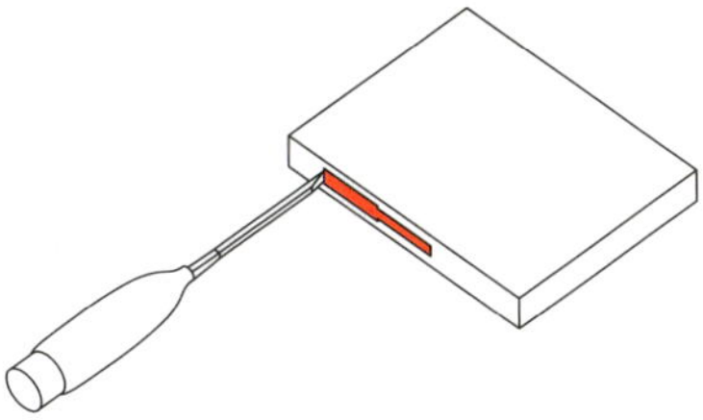

A2

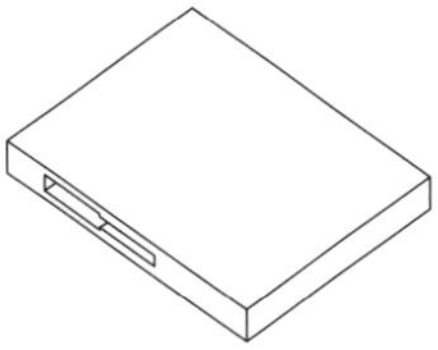

A

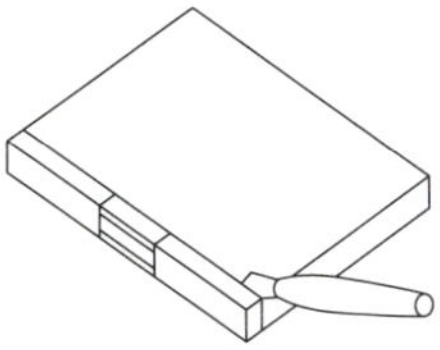

B1

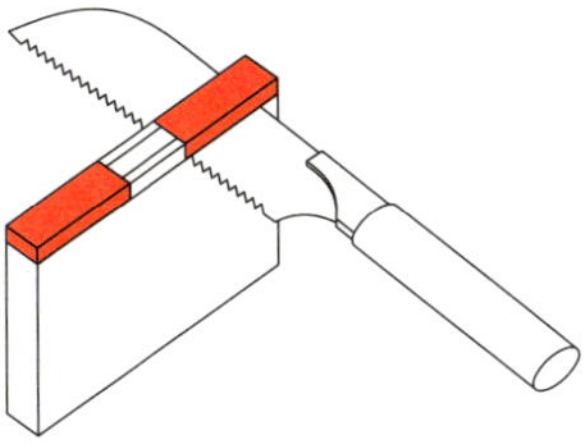

B2

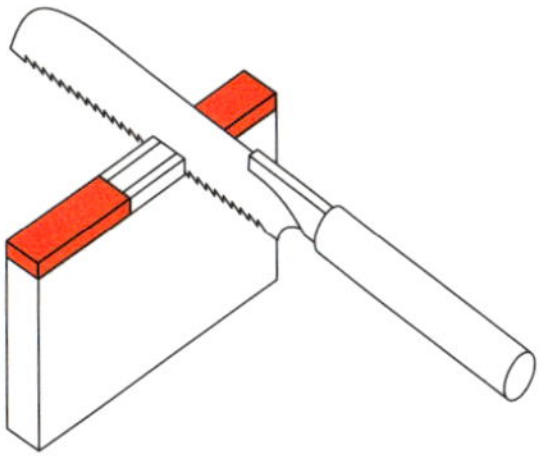

B3

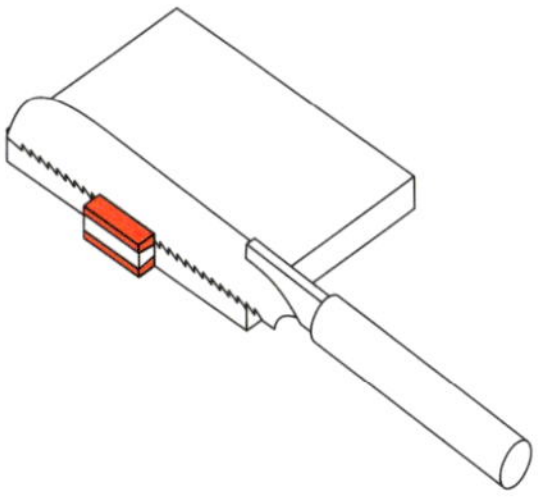

B4

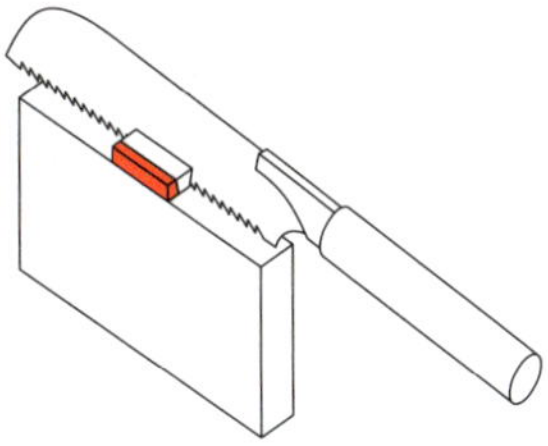

B5

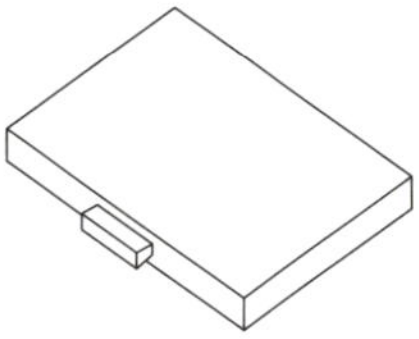

B

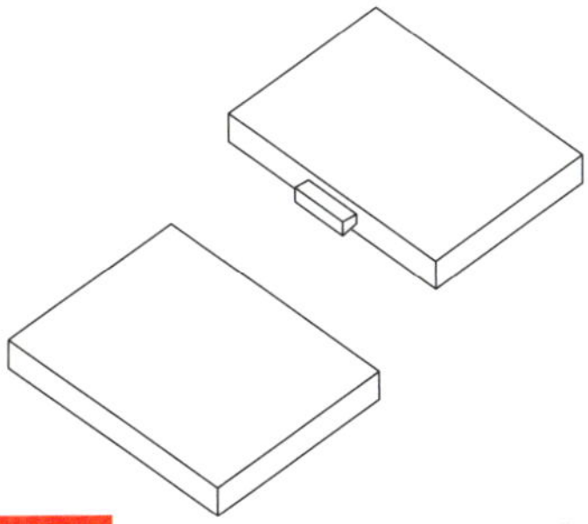

A+B

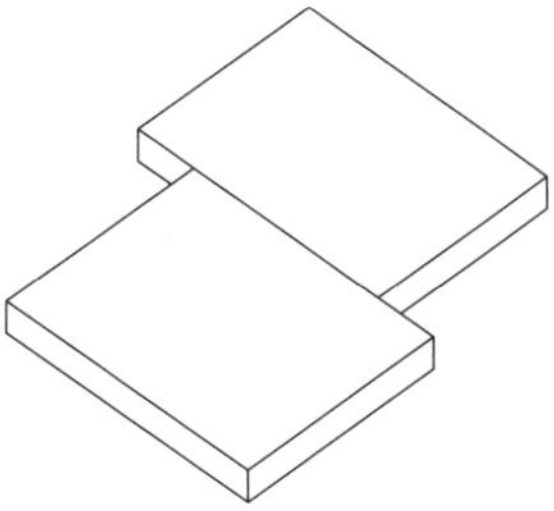

A+B

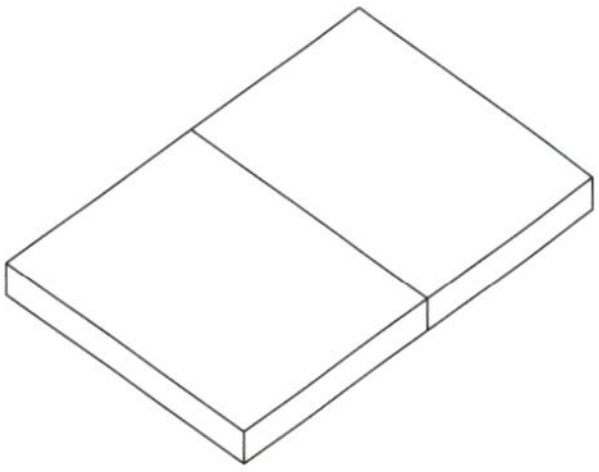

AB

Edge Joints

Butterfly key joint

chigiri-hagi [*]
千切矧ぎ

Schwalbenschwanzeinlage

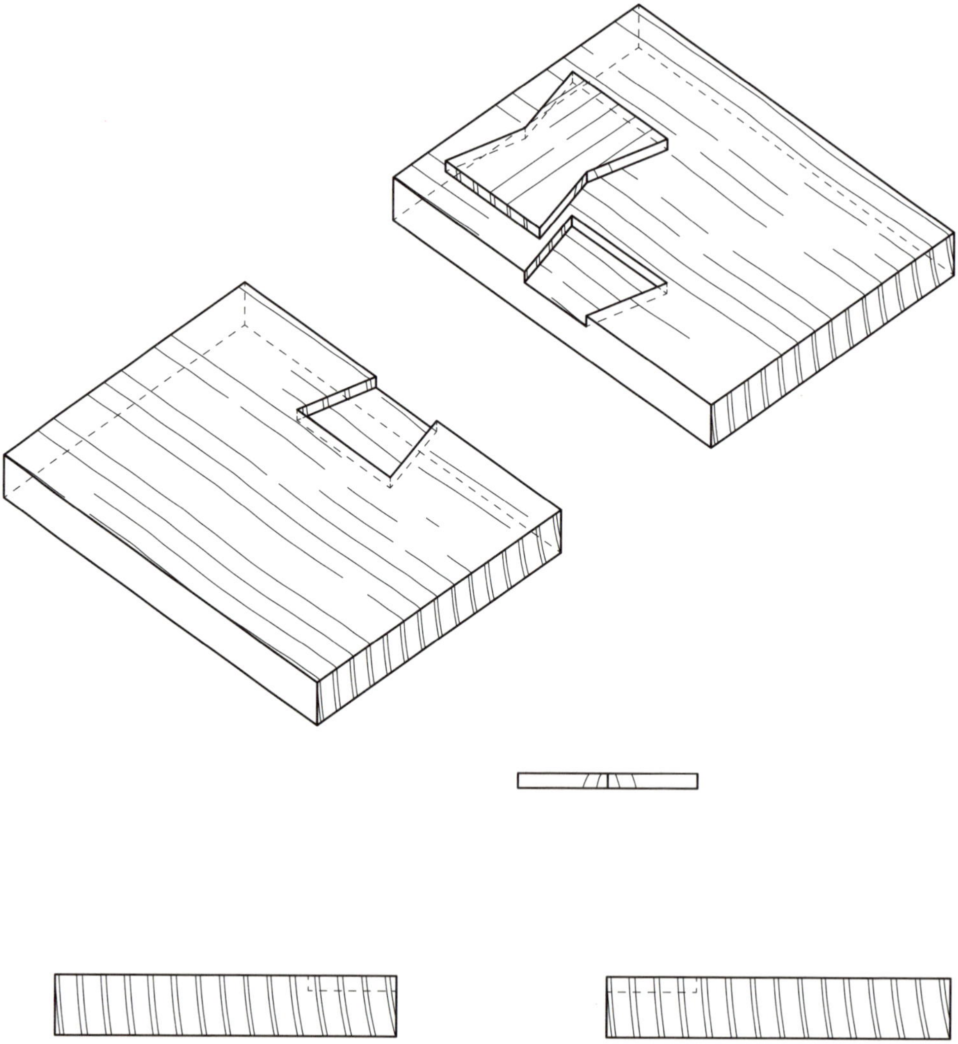

2×

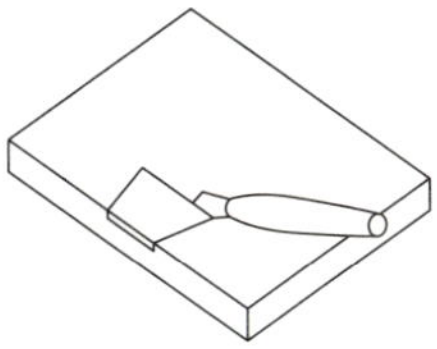

A1

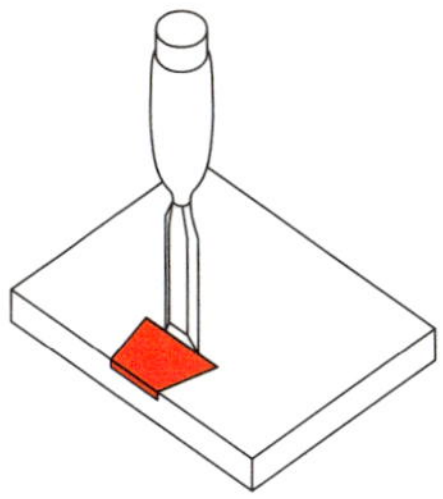

A2

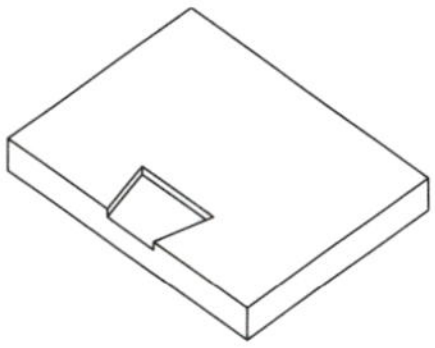

A

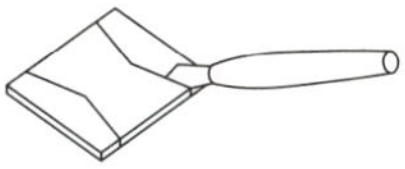

B1

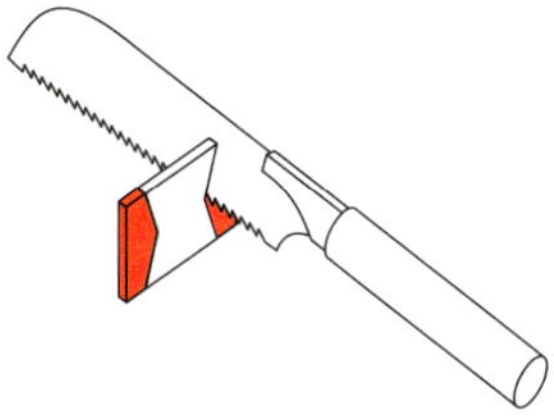

B2

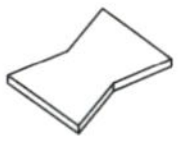

B

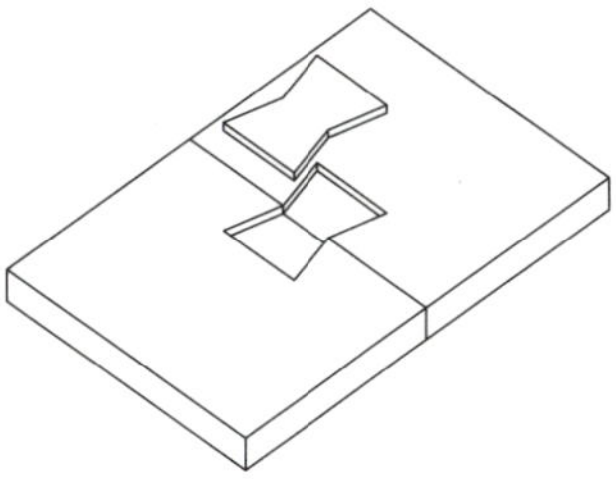

A+B+A

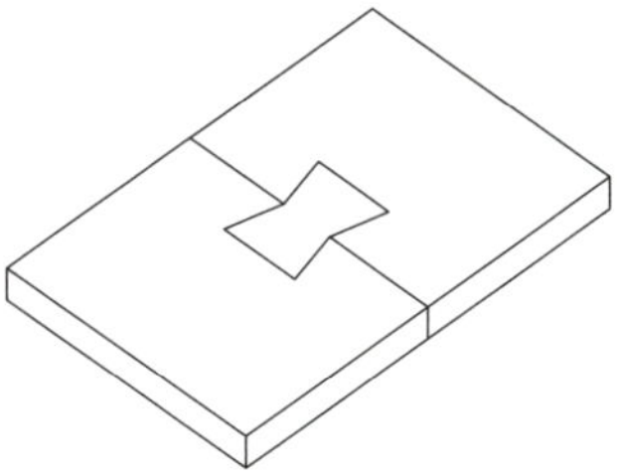

ABA

Edge Joints

Sliding dovetail batten

sui-tsuki-ari-hagi [*]
吸付蟻矧ぎ

Gratleiste

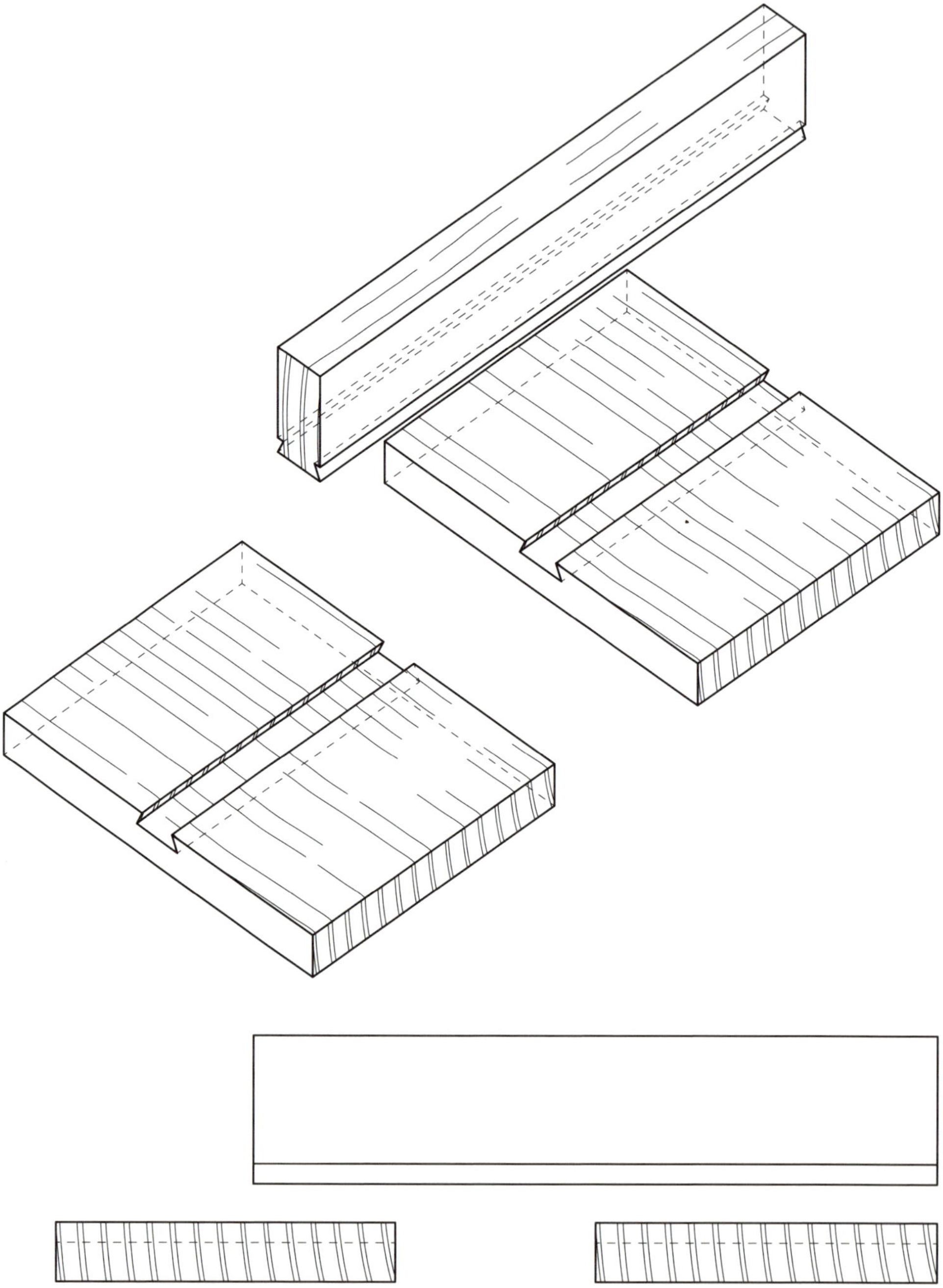

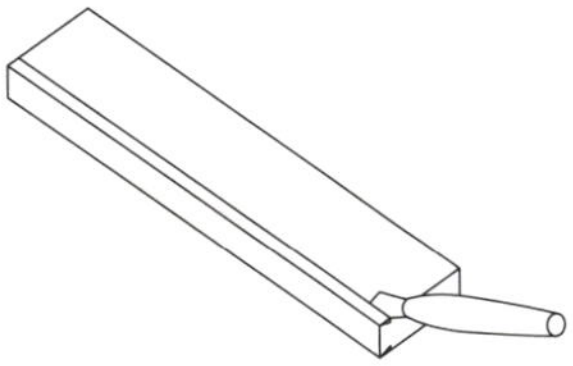
A1

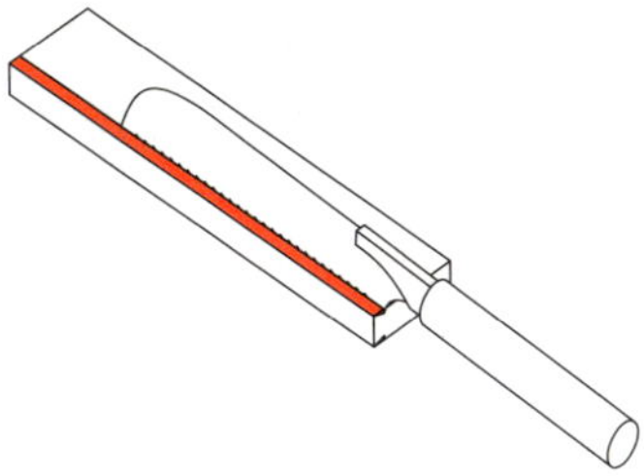
A2

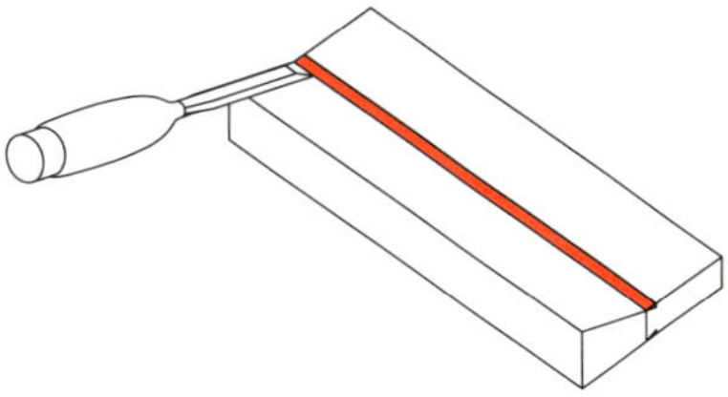
A3

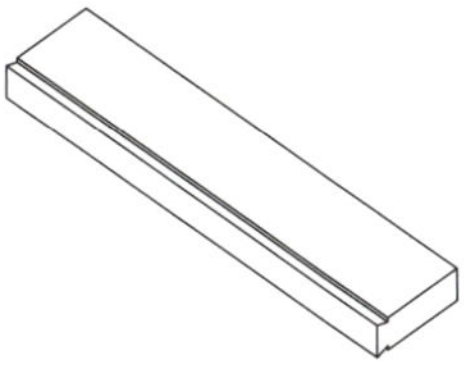
A

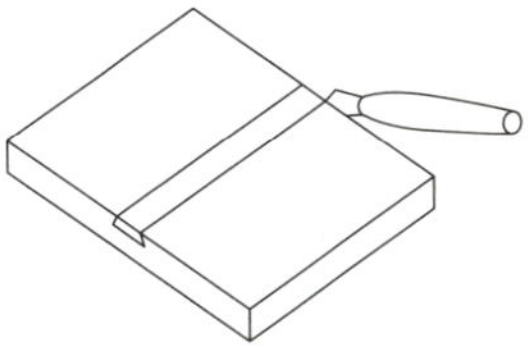
B1

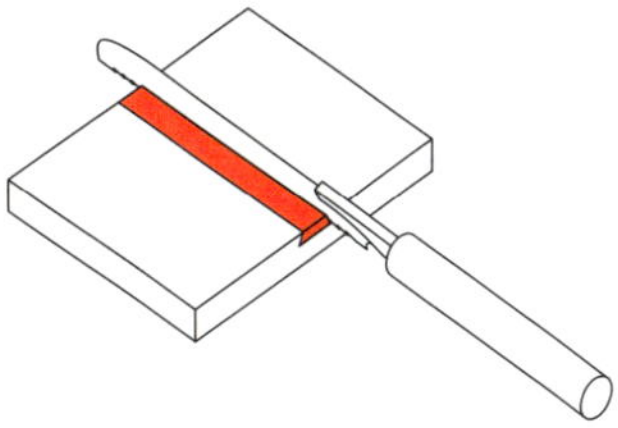
B2

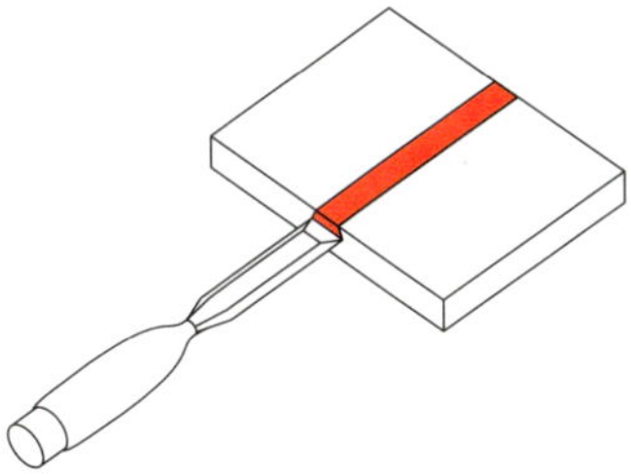
B3

2×

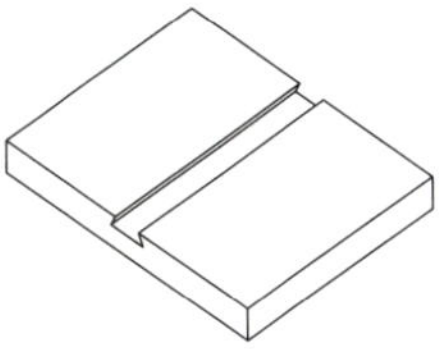
B

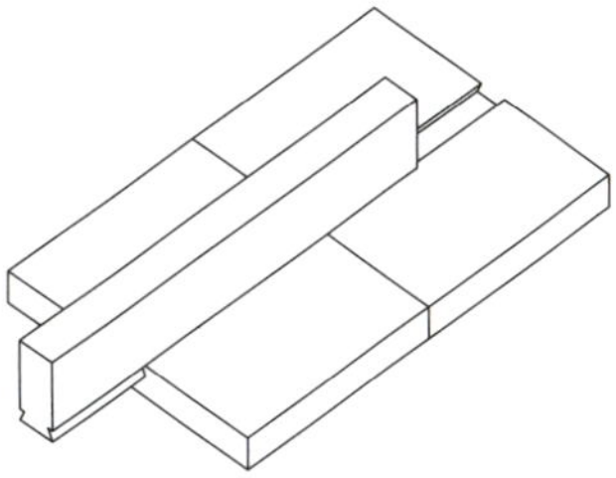
A+BB

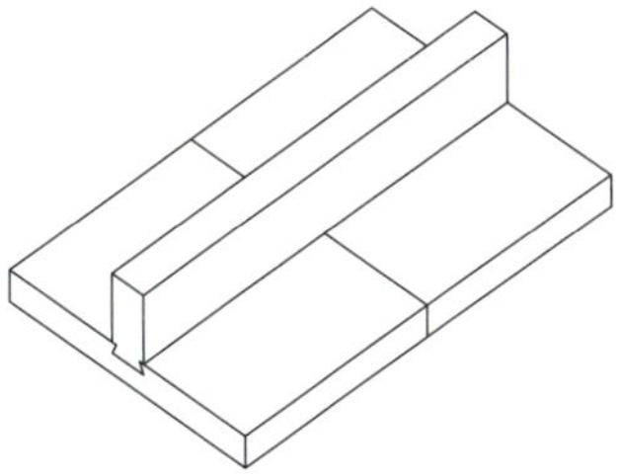
ABB

Edge Joints

Sliding dovetail batten with keys

tsuki-tsuke-sui-tsuki-ari-kōgai-sen-hagi [*]
突付吸付蟻桟笄栓矧ぎ

Gratleiste mit Dübelschloss

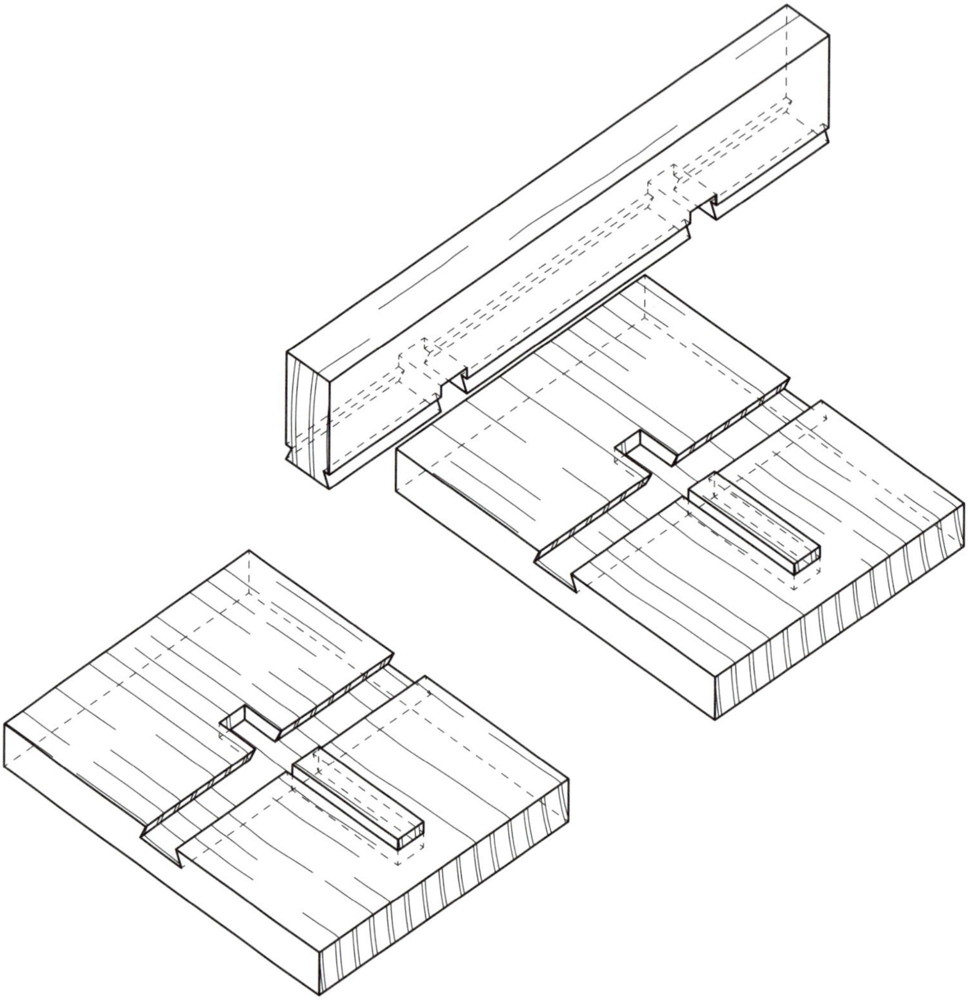

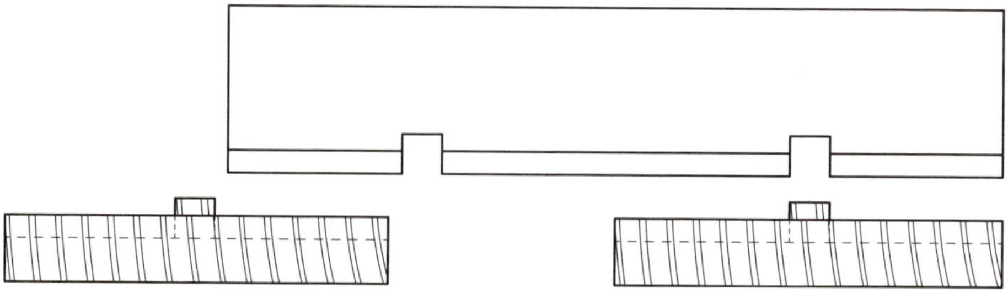

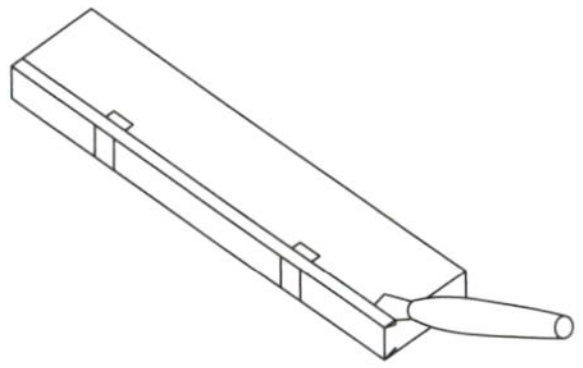

A1

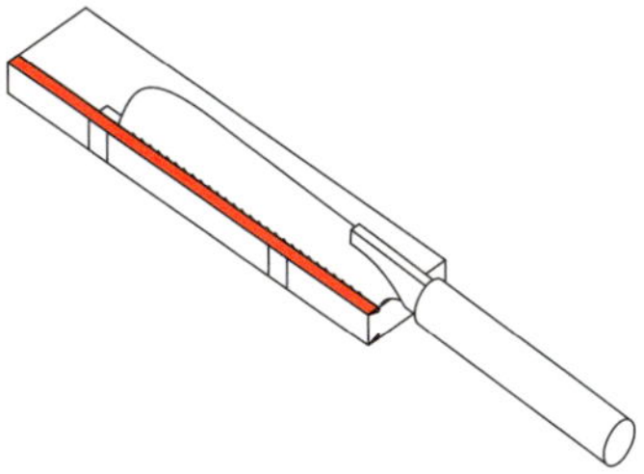

A2

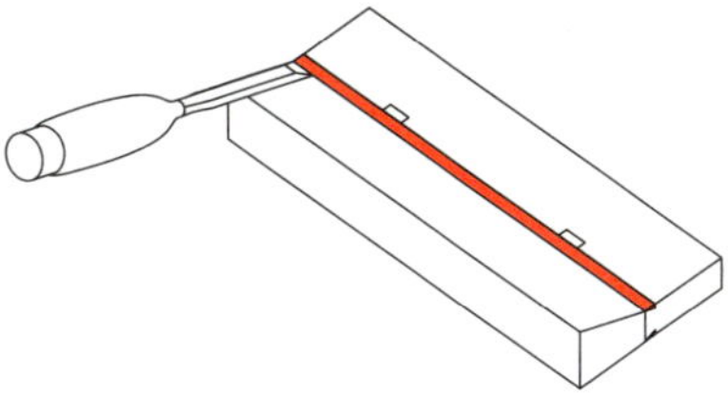

A3

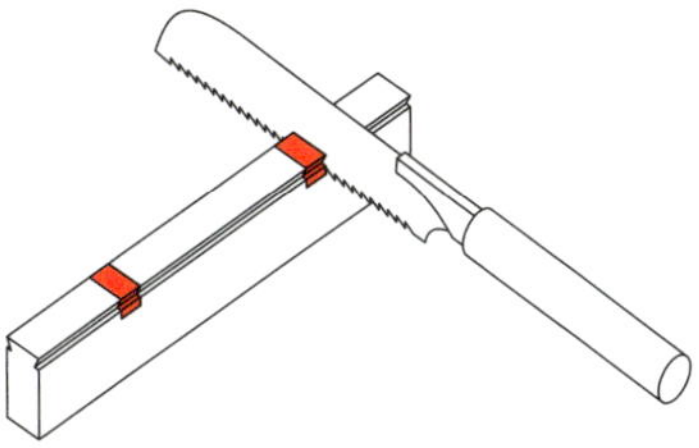

A4

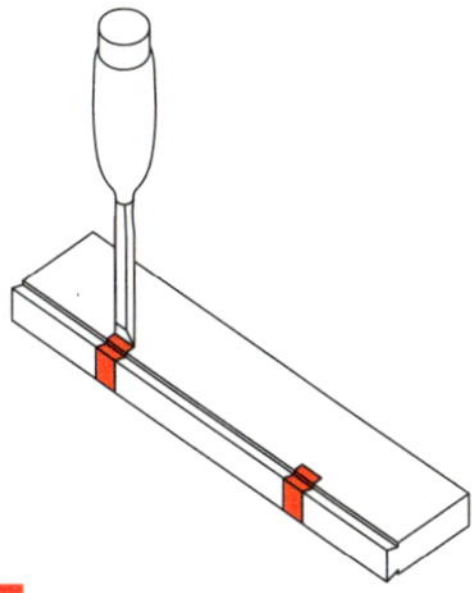

A5

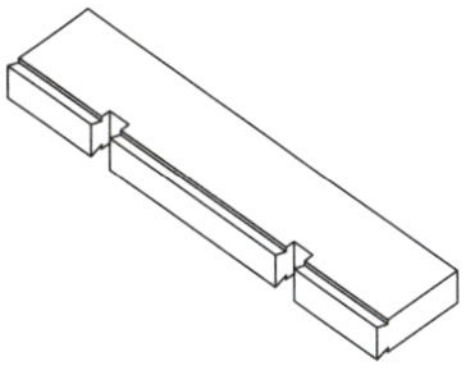

A

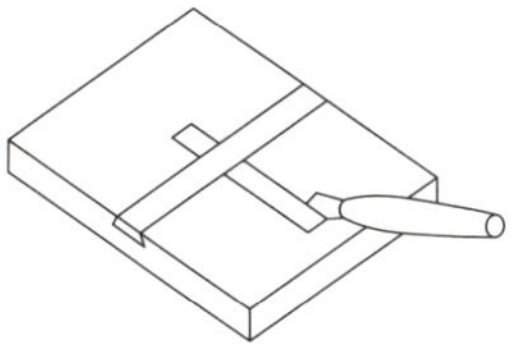

B1

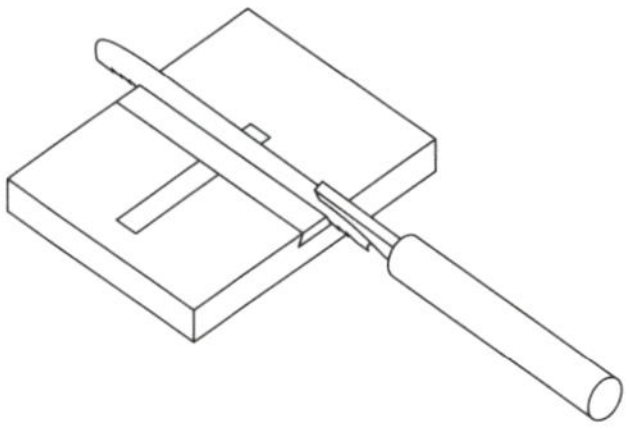

B2

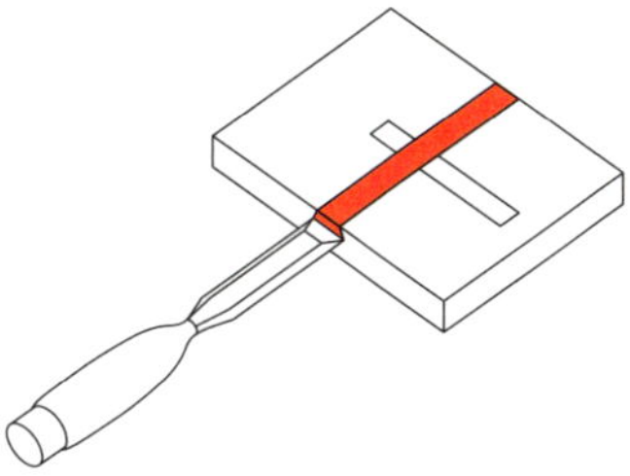

B3

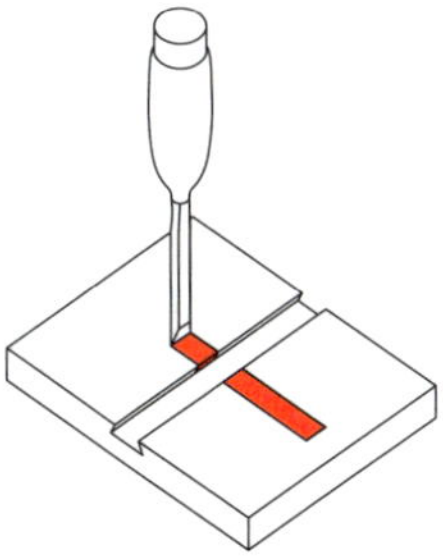

B4

2×

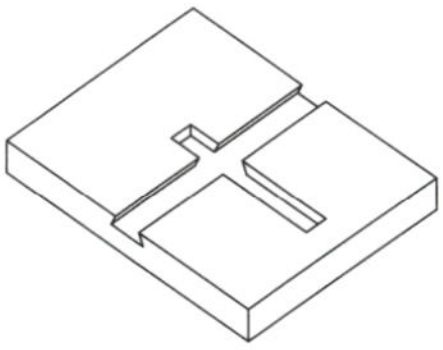

B

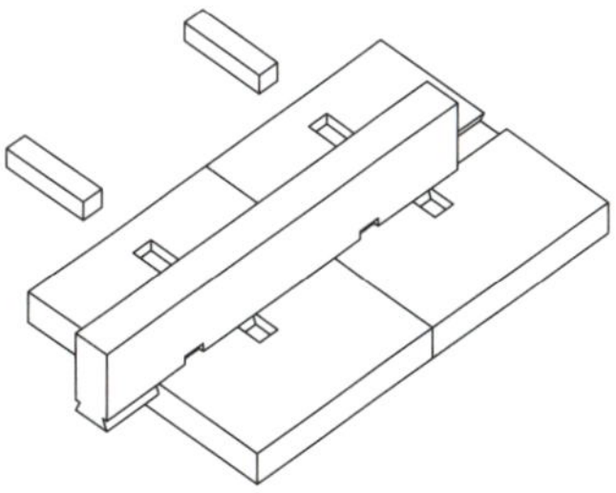

A+BB

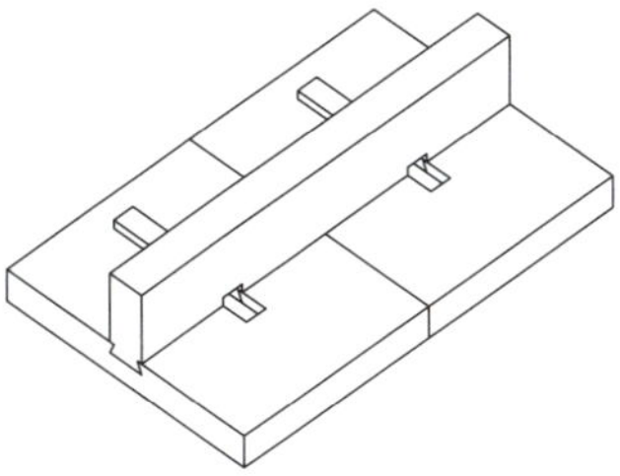

A+BB

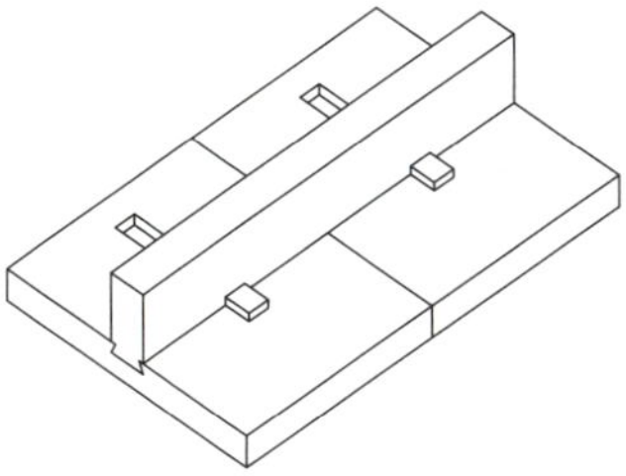

ABB

Edge Joints

Blind dovetailed keyhole batten

tsuki-tsuke-yose-ari-zan-hagi [*]
突付寄蟻桟矧ぎ

Mehrfach abgesetzte Gratleiste

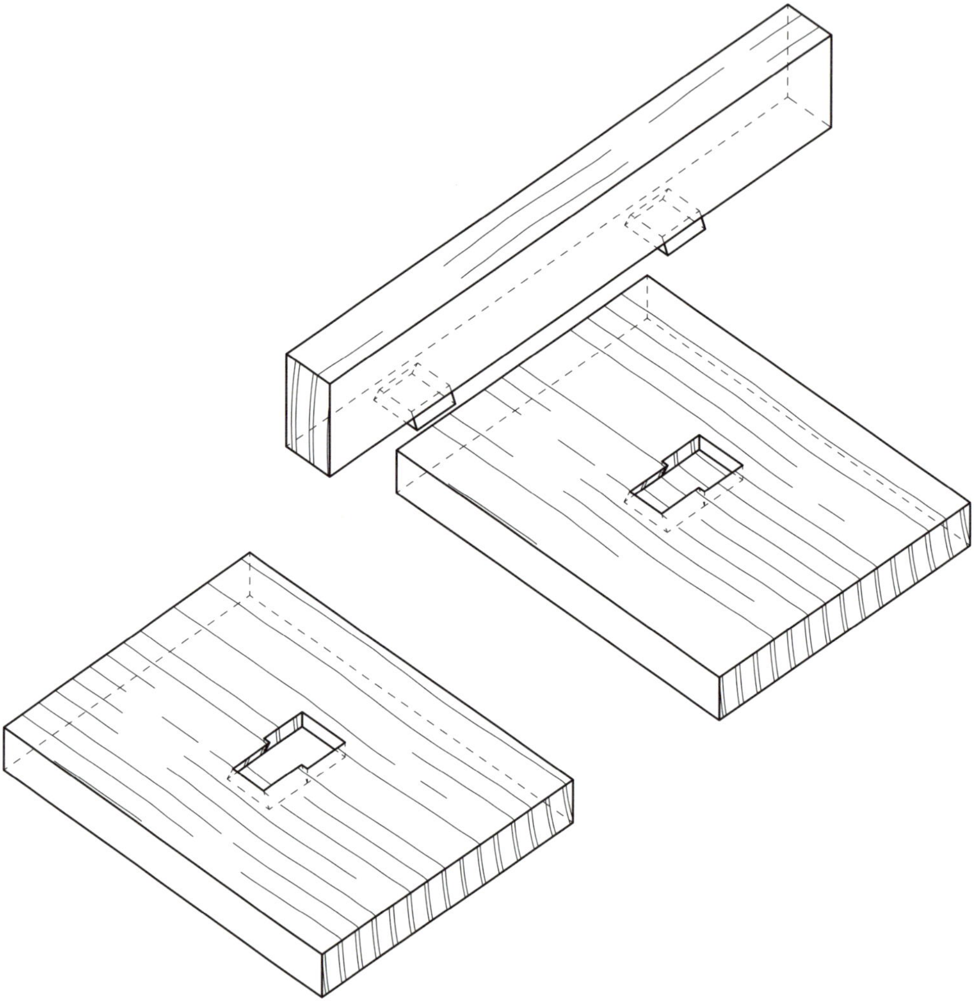

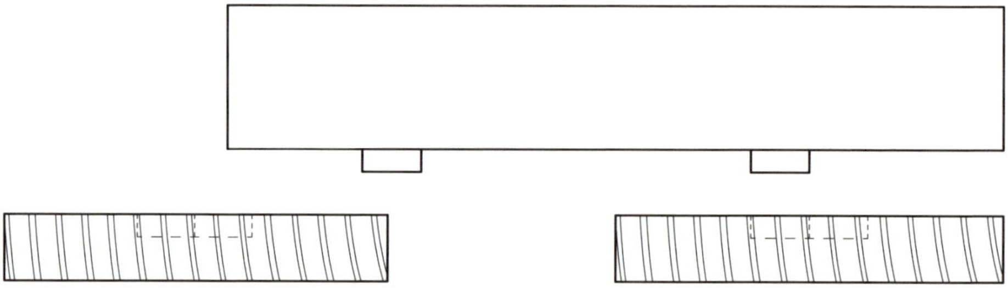

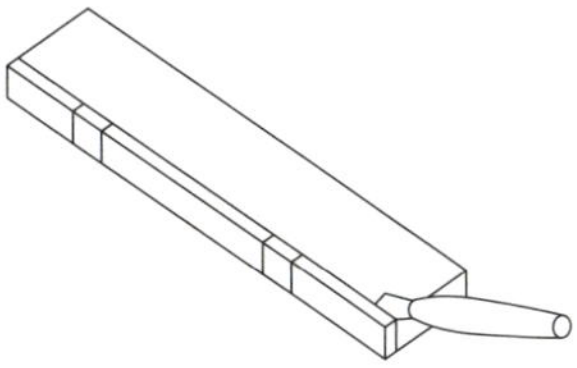

A1

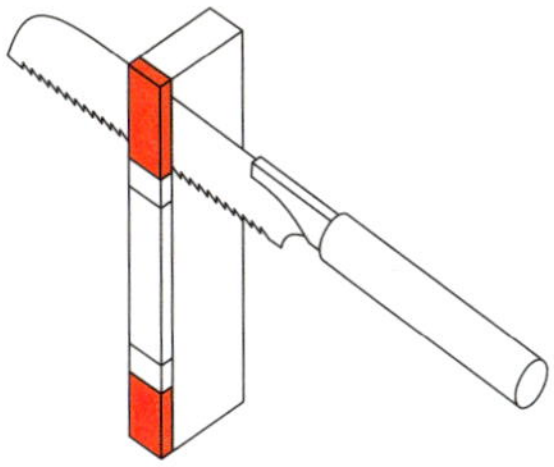

A2

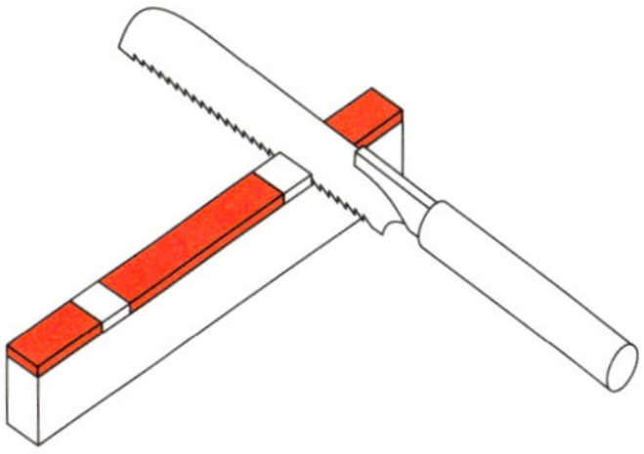

A3

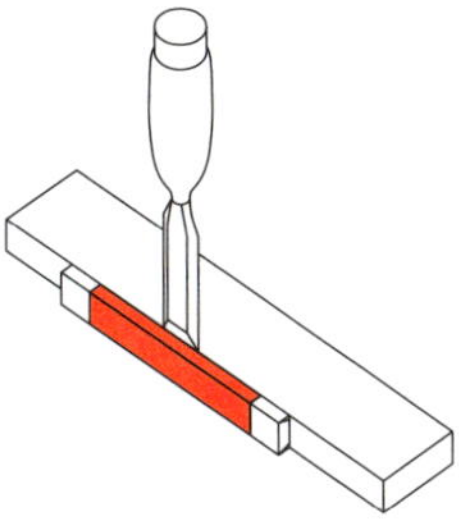

A4

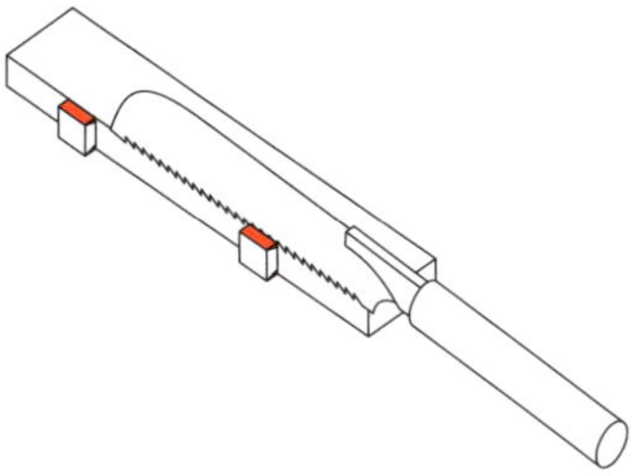

A5

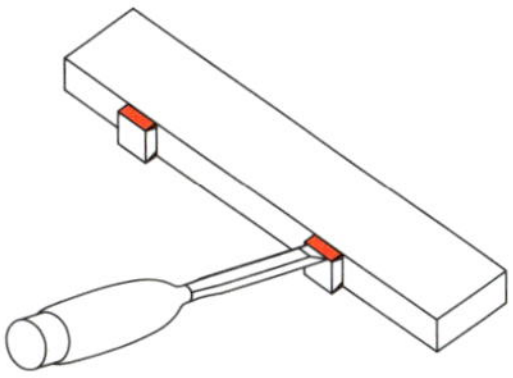

A6

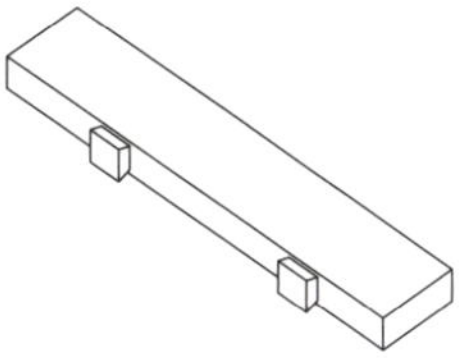

A

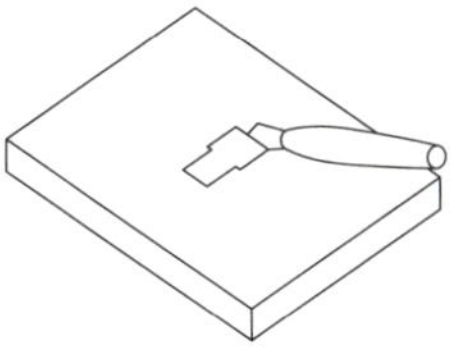

B1

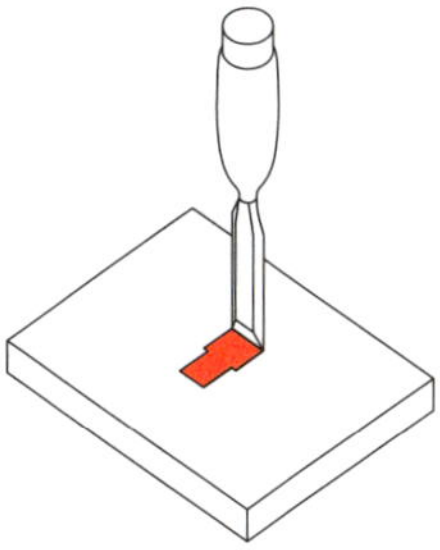

B2

2×

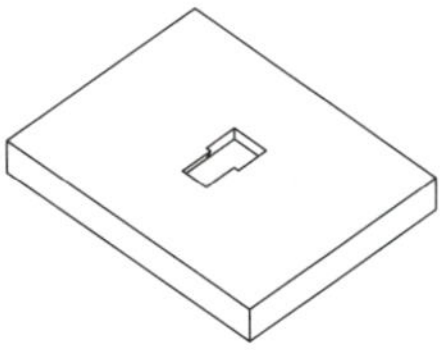

B

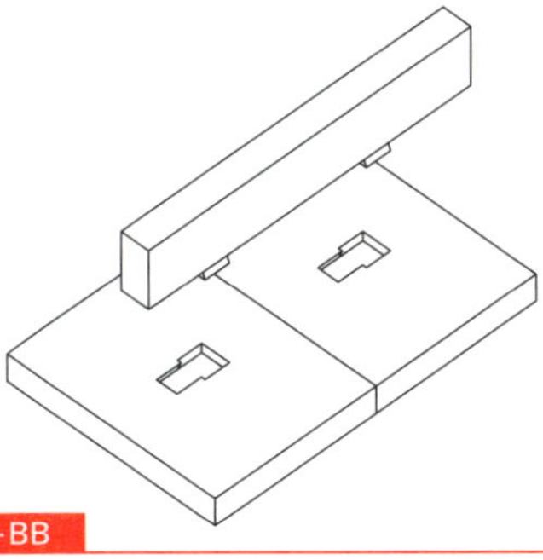

A+BB

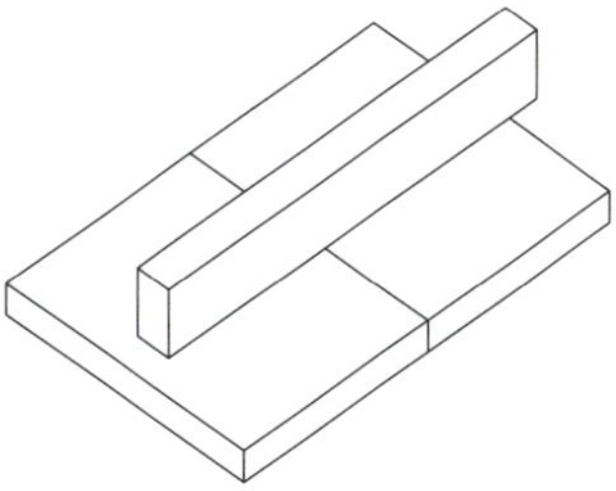

A+BB

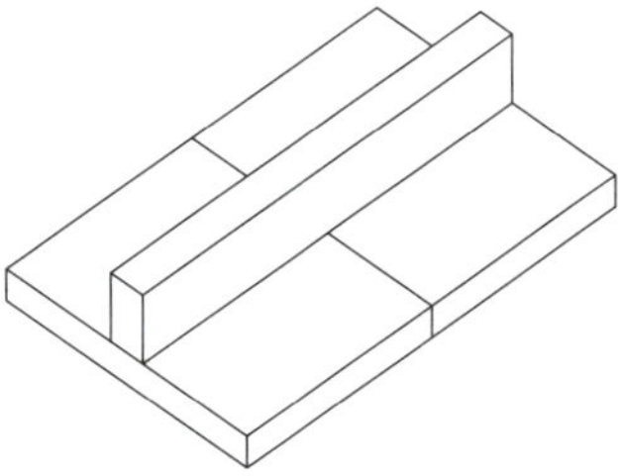

ABB

End Joints

Halved splayed joint

hira-isuka-hagi
平鵤矧ぎ

schräg angeschnittene
wechselseitige Überblattung

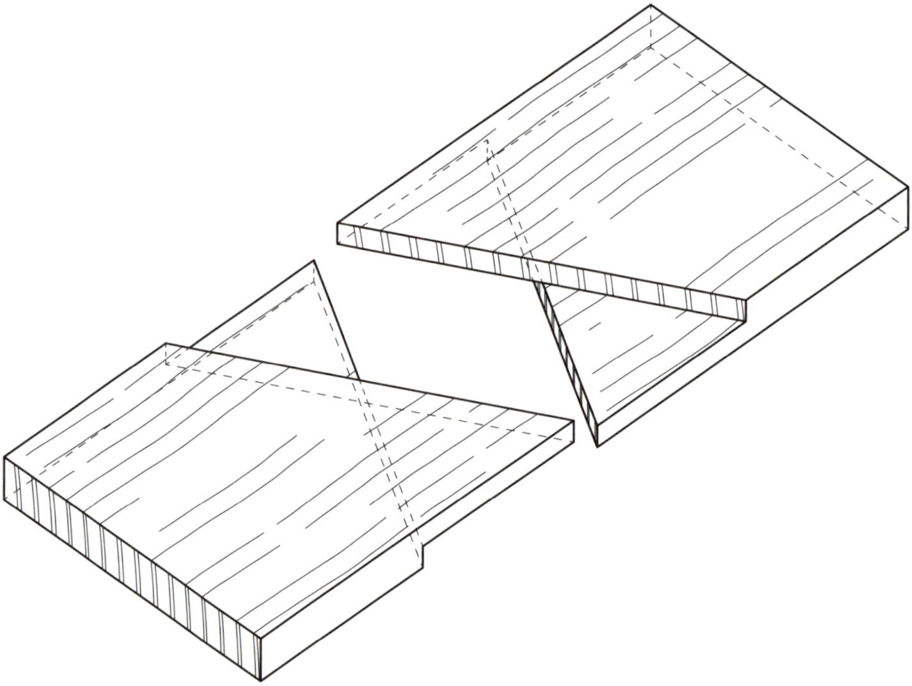

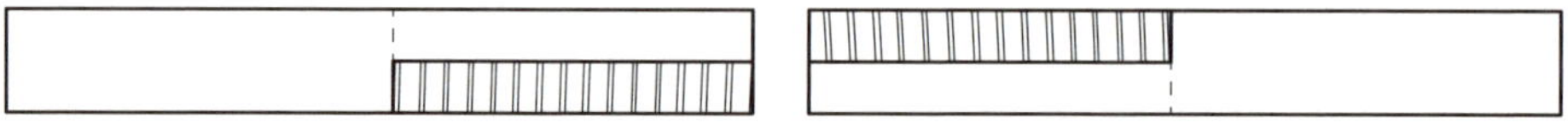

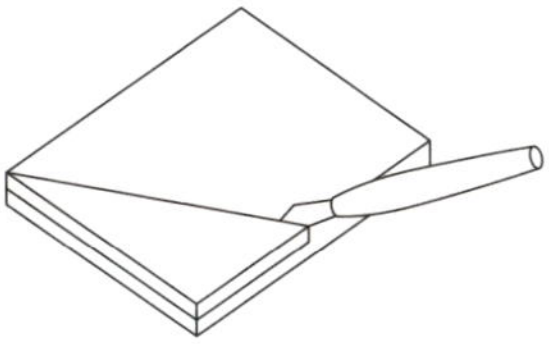

A1

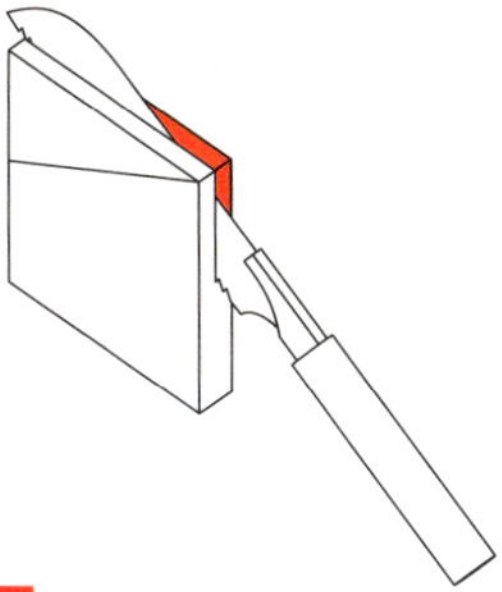

A2

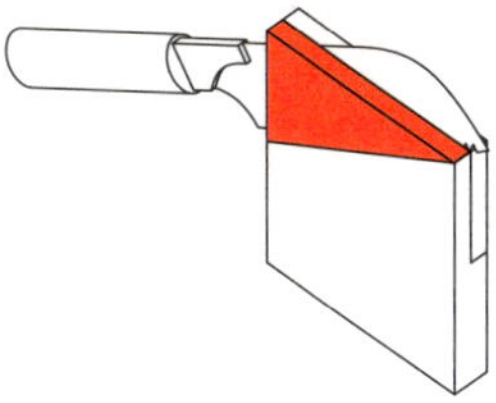

A3

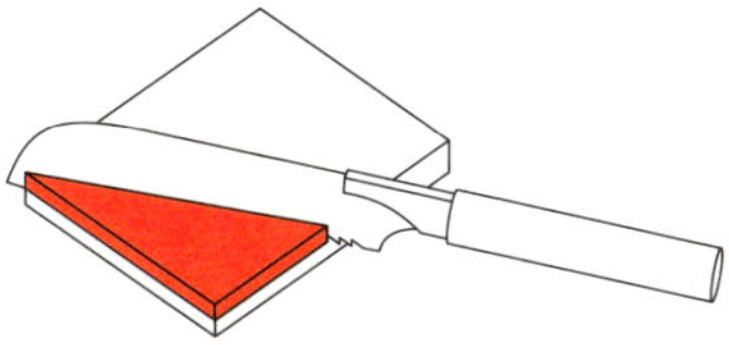

A4

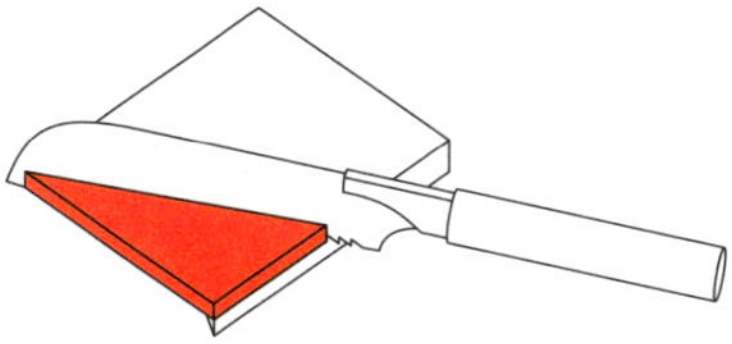

A5

2×

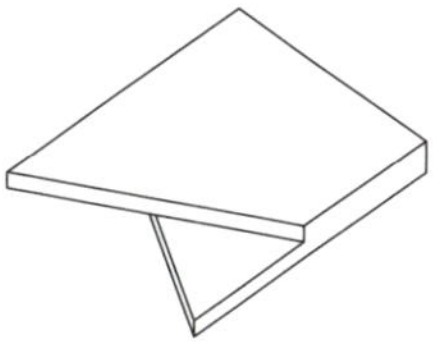

A

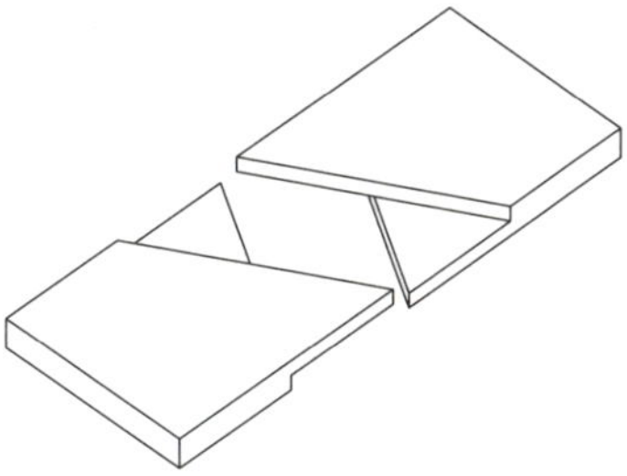

A+A

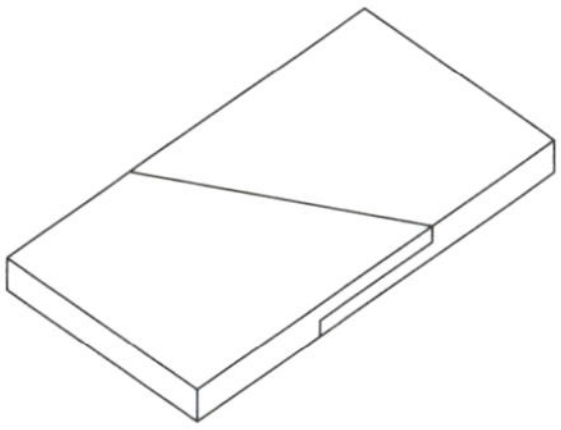

AA

End Joints

End-to-end finger joint

arare-tsugi [*]
あられ接ぎ

Fingerzinkenstoß

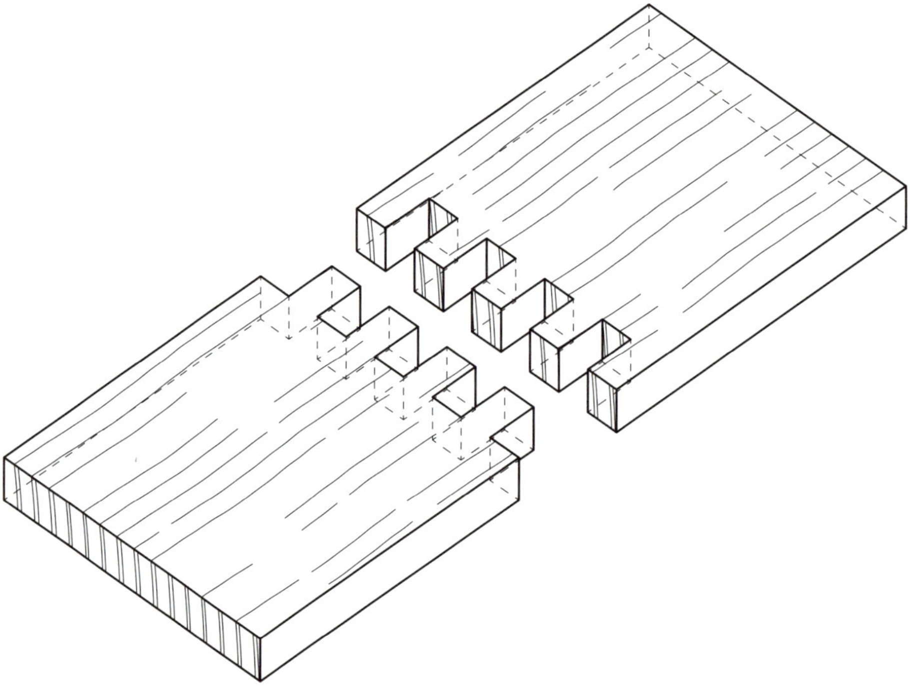

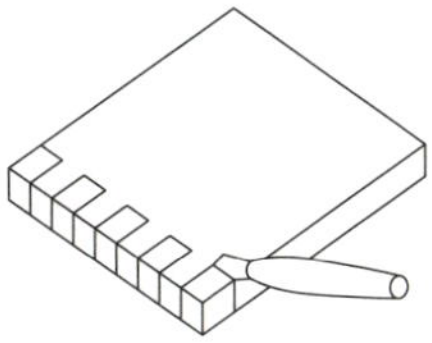
A1

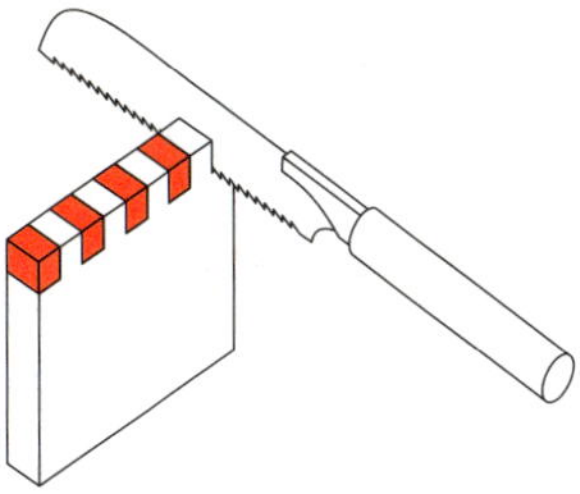
A2

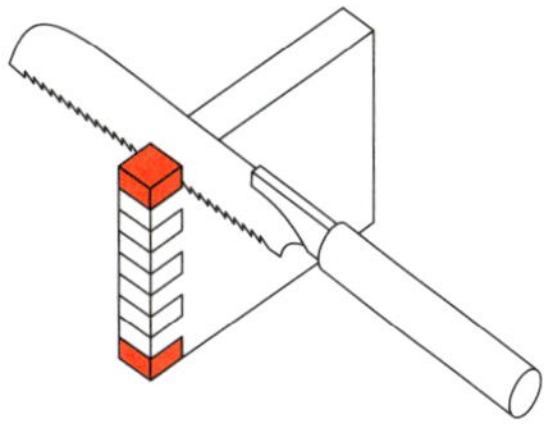
A3

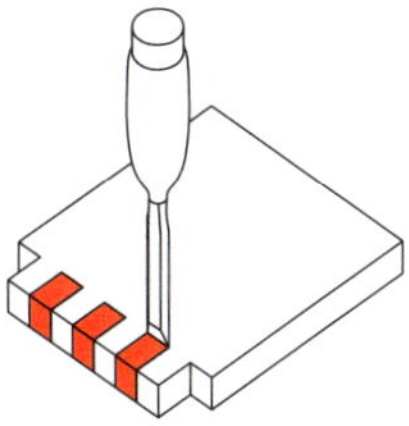
A4

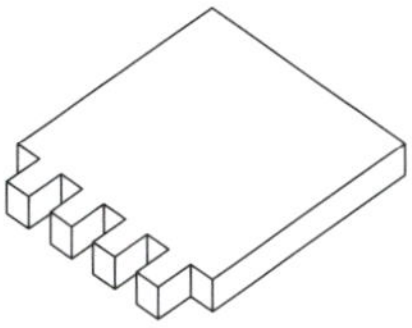
A

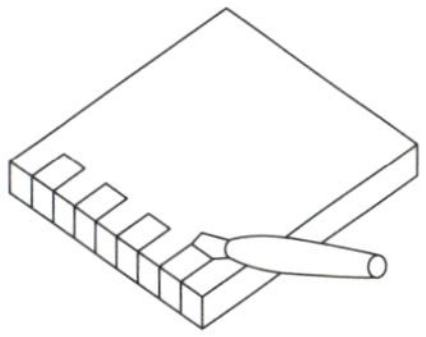
B1

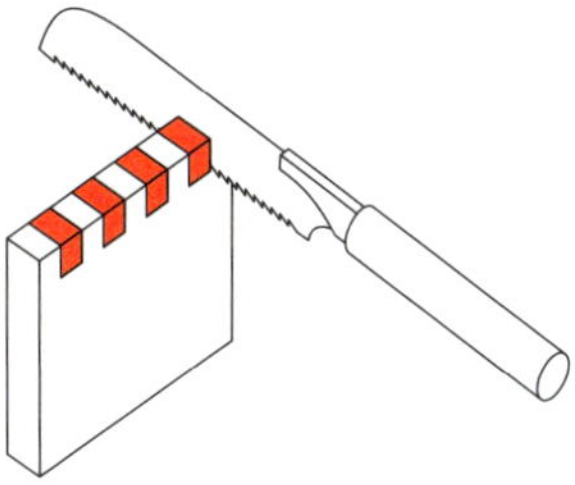
B2

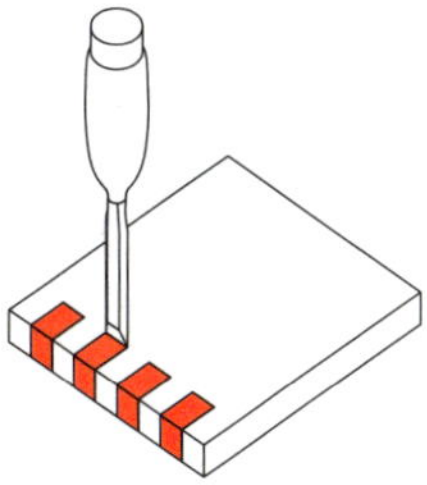
B3

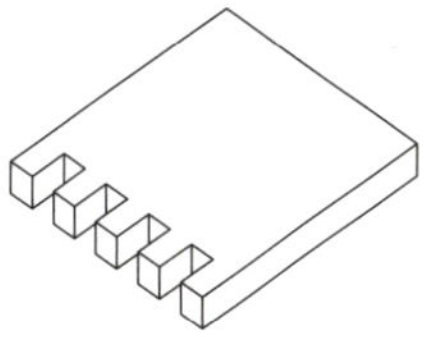
B

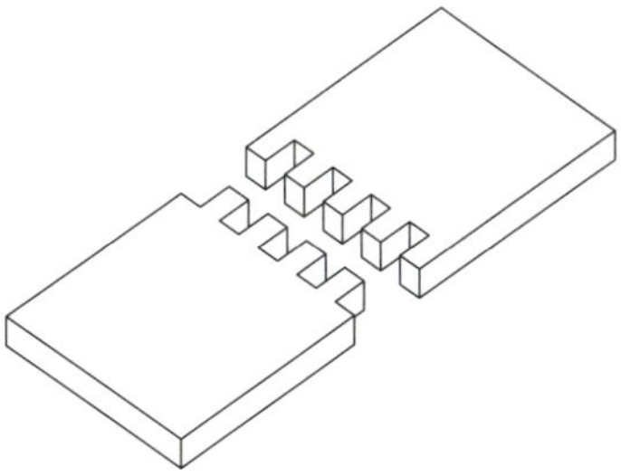
A+B

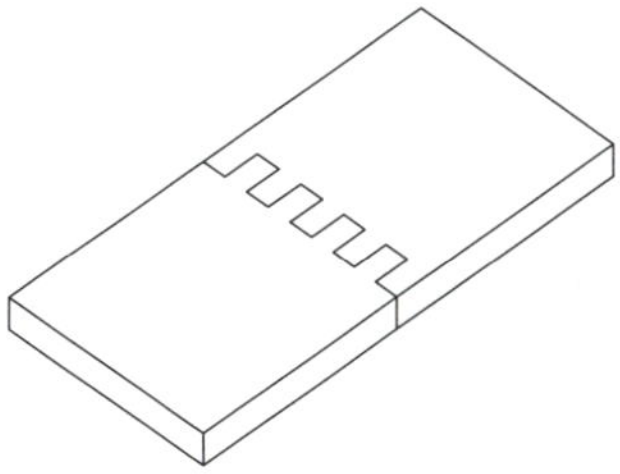
AB

End-to-end dovetail joint

ari-kumi-tsugi [*]
蟻組接ぎ

Zinkenstoß

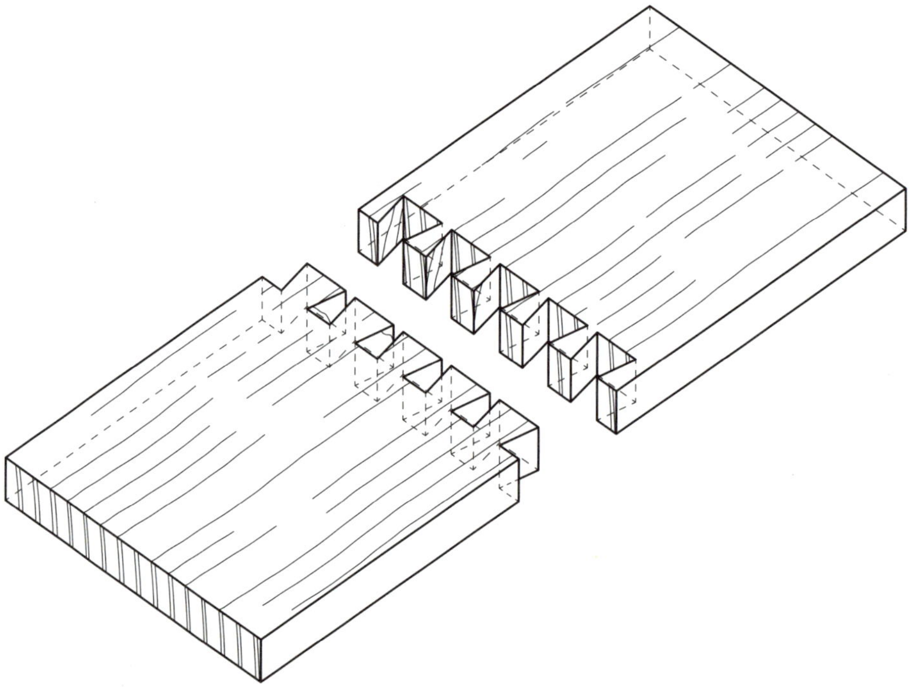

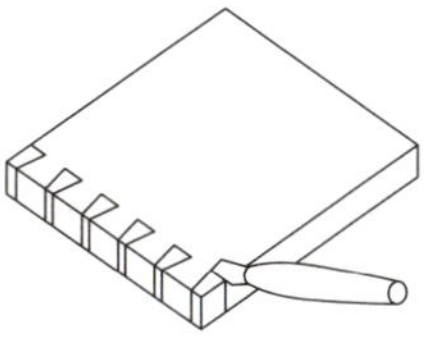

A1

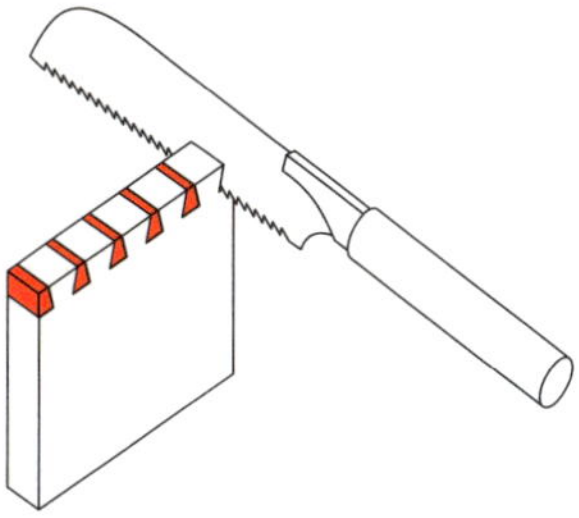

A2

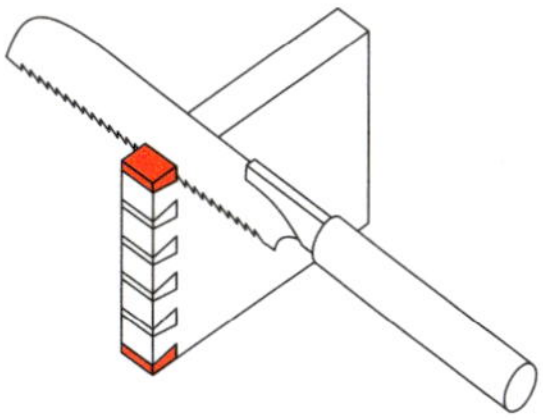

A3

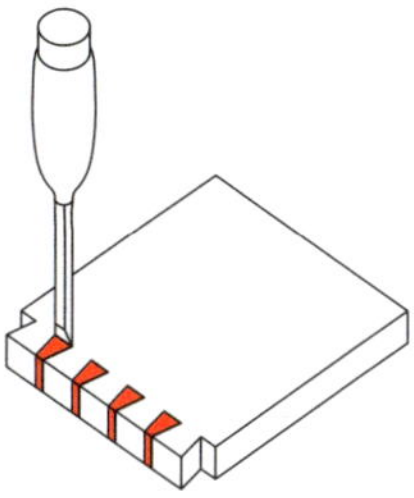

A4

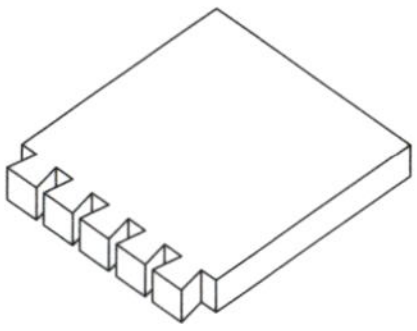

A

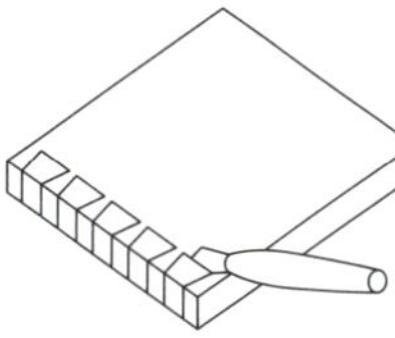

B1

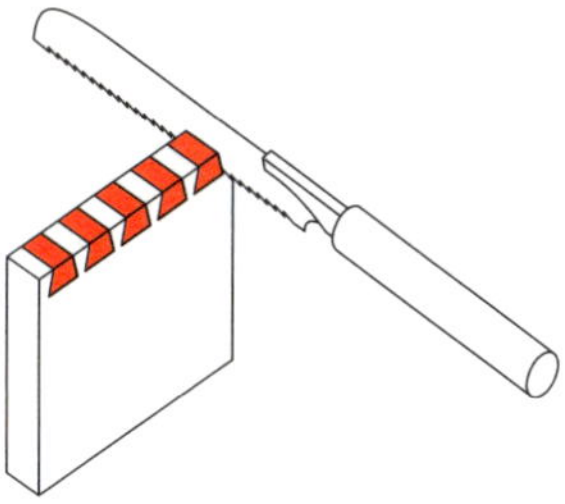

B2

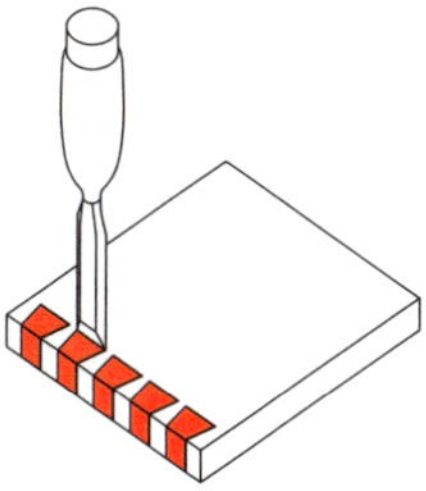

B3

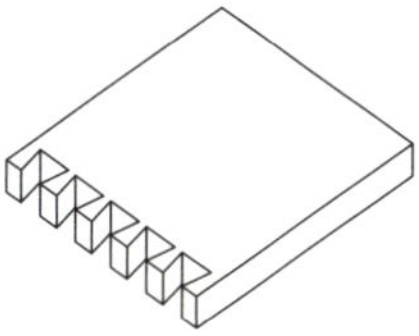

B

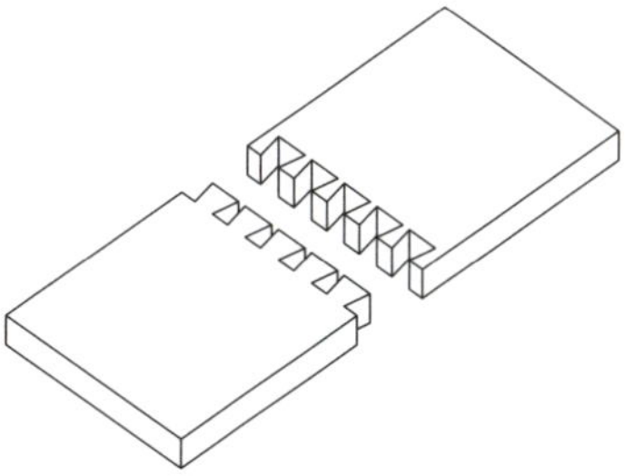

A+B

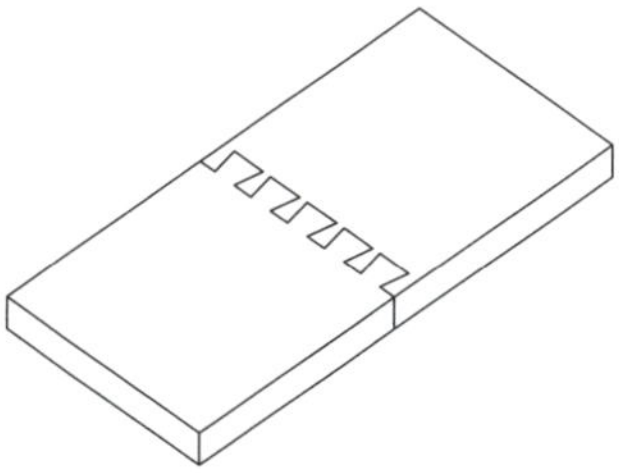

AB

End Joints

V-shaped dado with spline

ken-dome-hashibame-tsugi [*]
剣留端嵌接ぎ

eingelassene keilförmige Hirnleiste

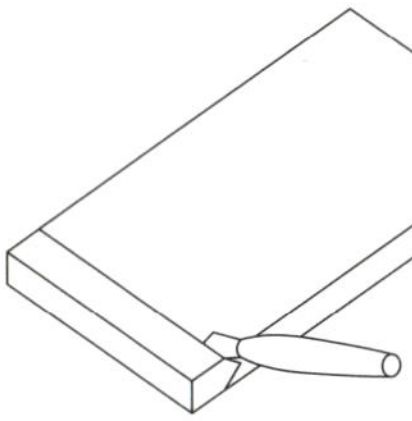
A1

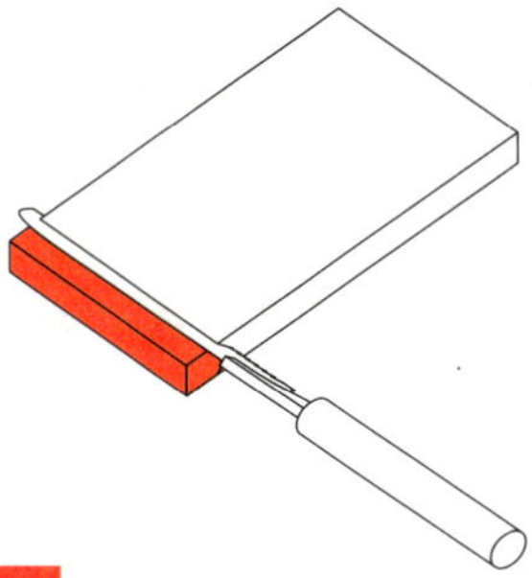
A2

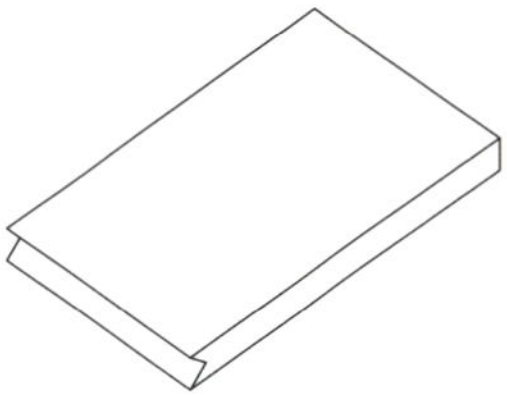
A

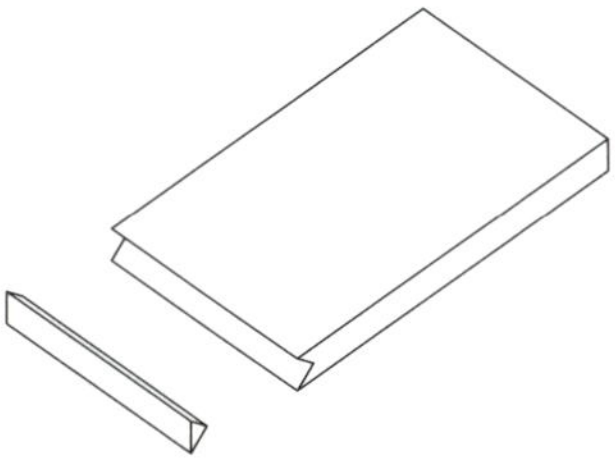
A+B

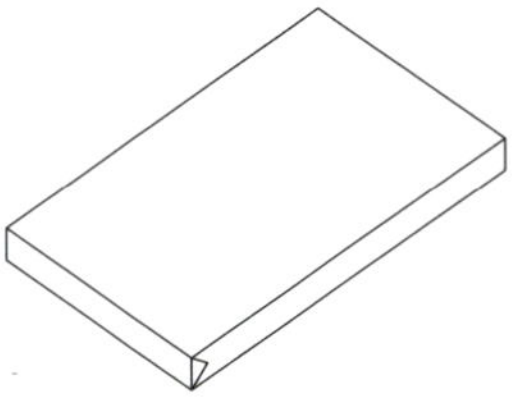
AB

Dado with spline

yatoi-zane-tsugi [*]
雇実接ぎ

eingelassene Hirnleiste

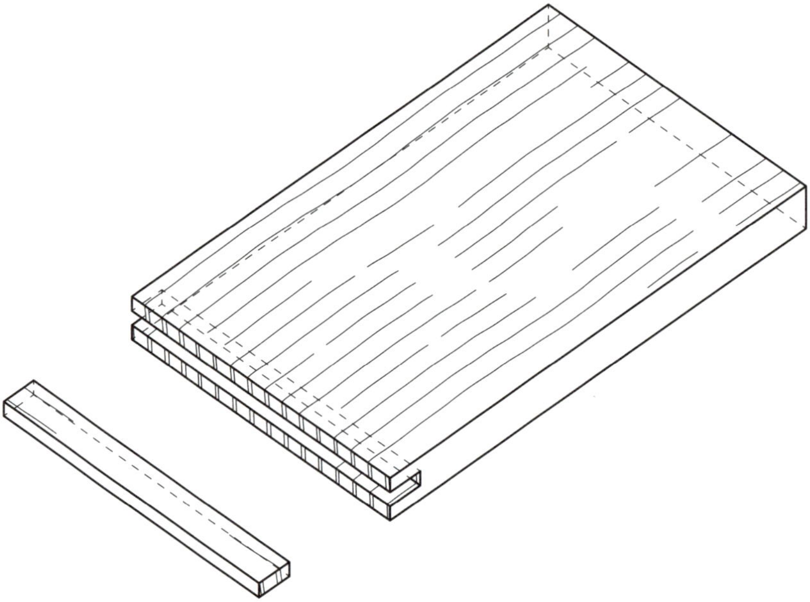

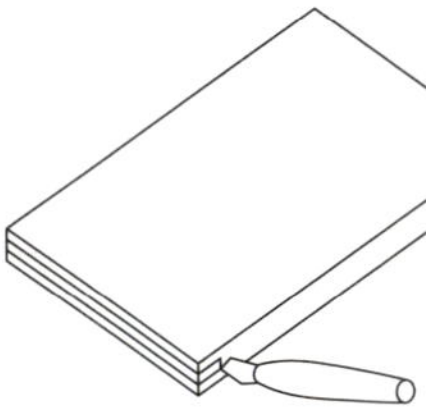

A1

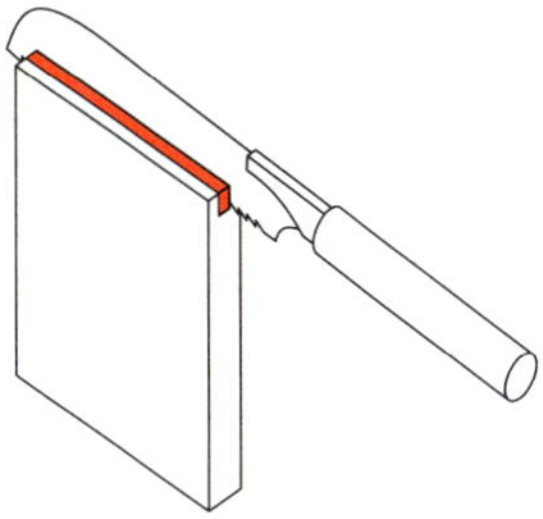

A2

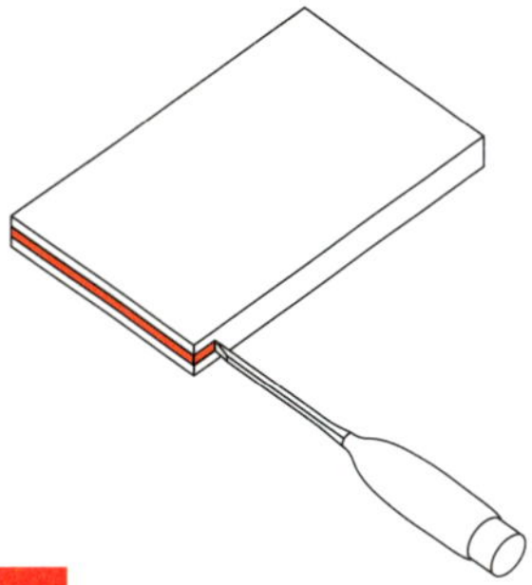

A3

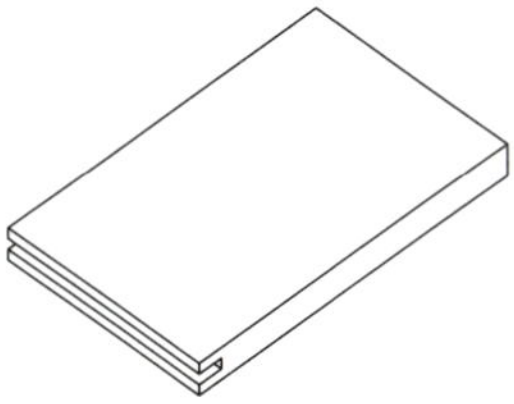

A

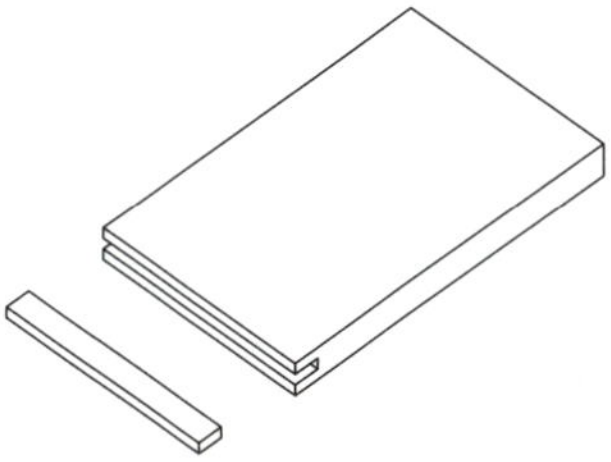

A+B

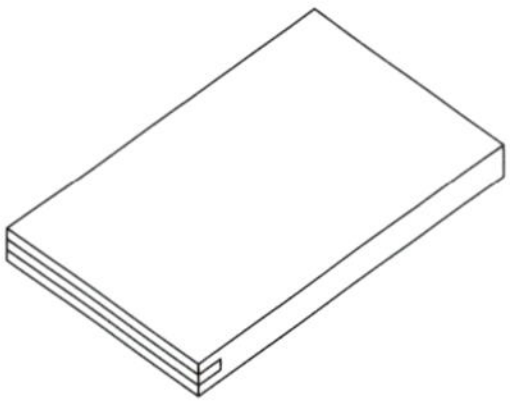

AB

Splined breadboard end

yatoi-zane-hashibame-tsugi
雇実端嵌接ぎ

Hirnleiste mit
Nut und Fremdfeder

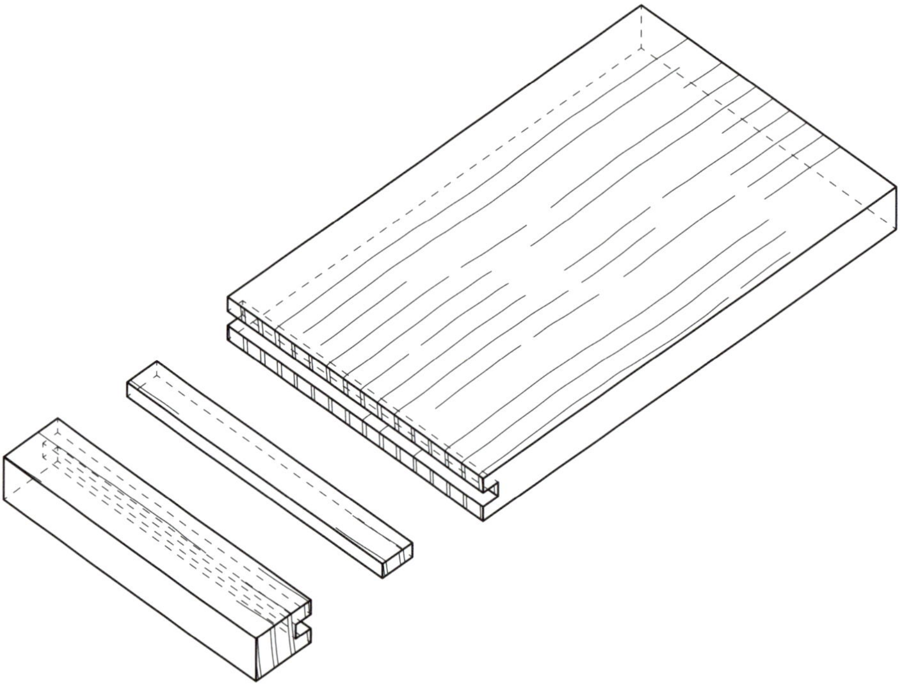

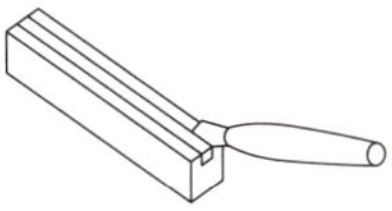
A1

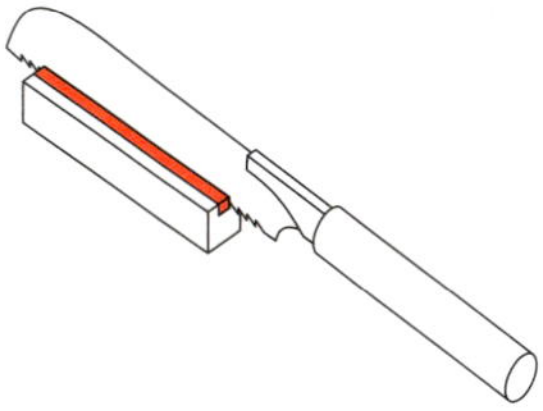
A2

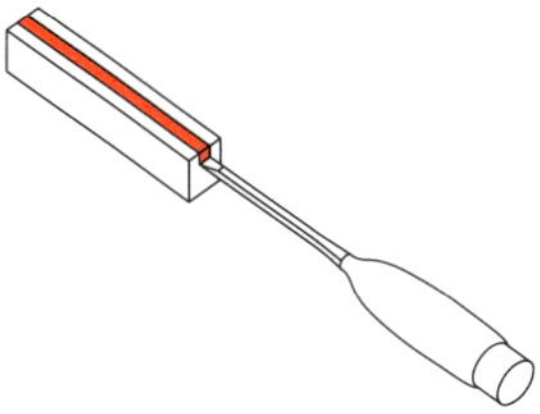
A3

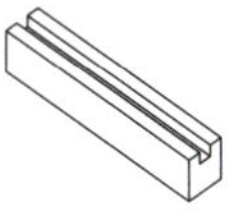
A

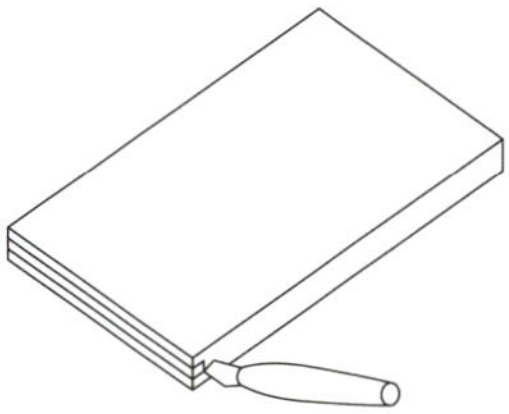
B1

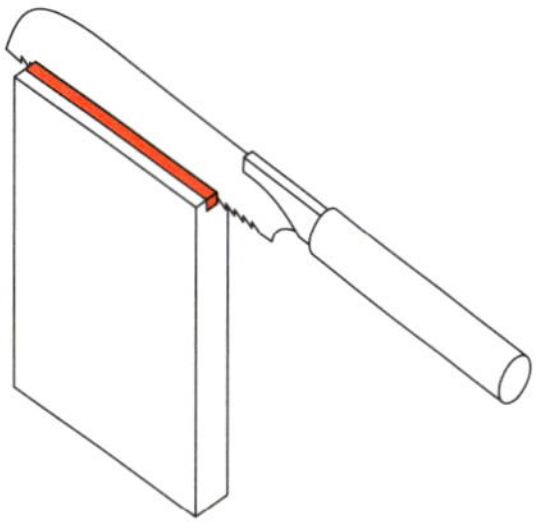
B2

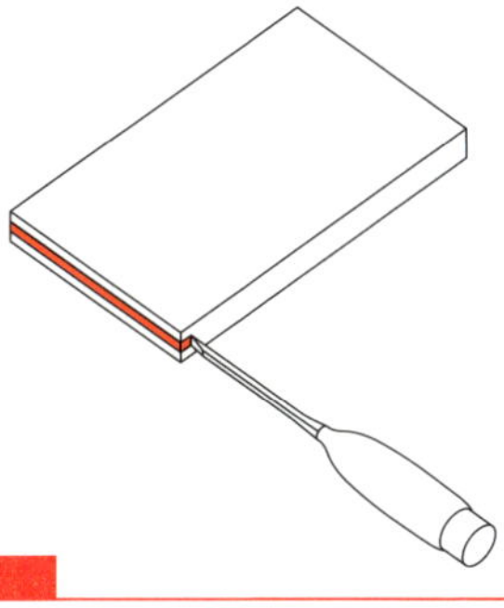
B3

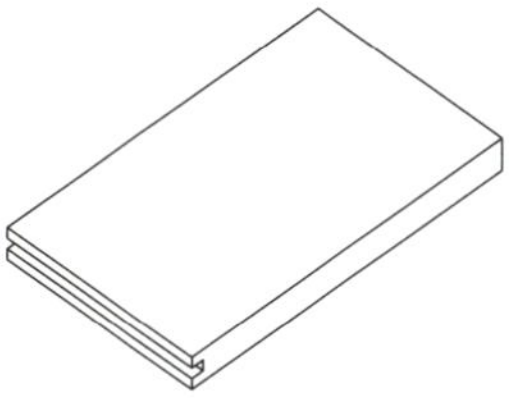
B

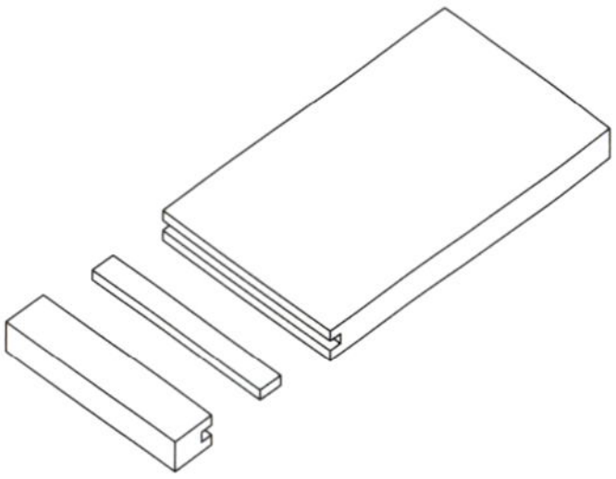
A+B

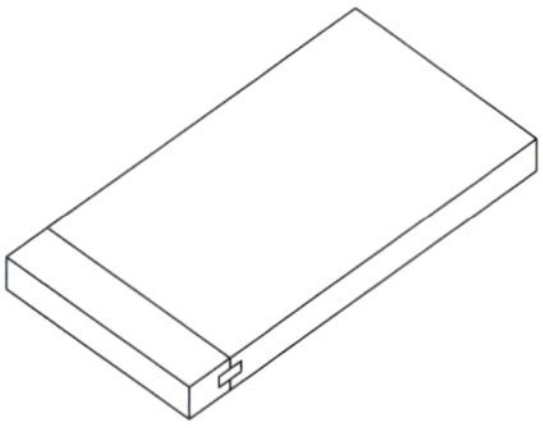
AB

End Joints

Tongue and groove breadboard end

hon-zane-hashibame-tsugi [*]
本実端嵌接ぎ

Hirnleiste mit Nut

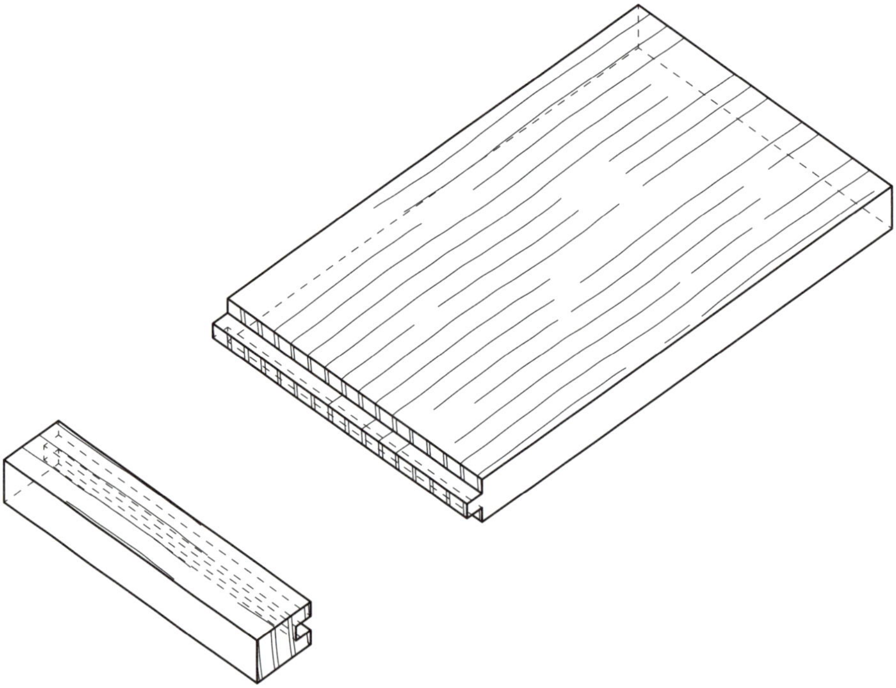

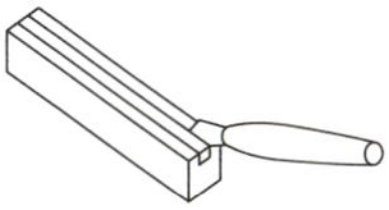
A1

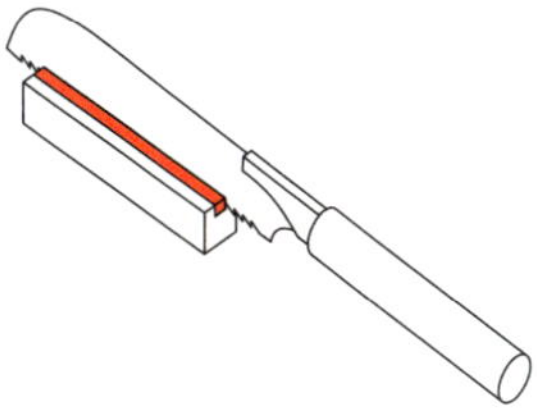
A2

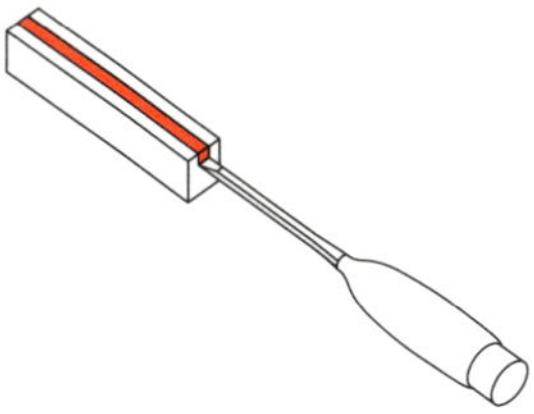
A3

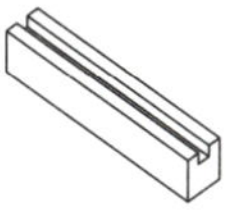
A

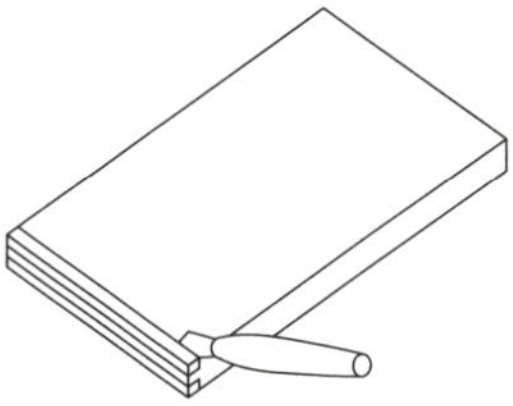
B1

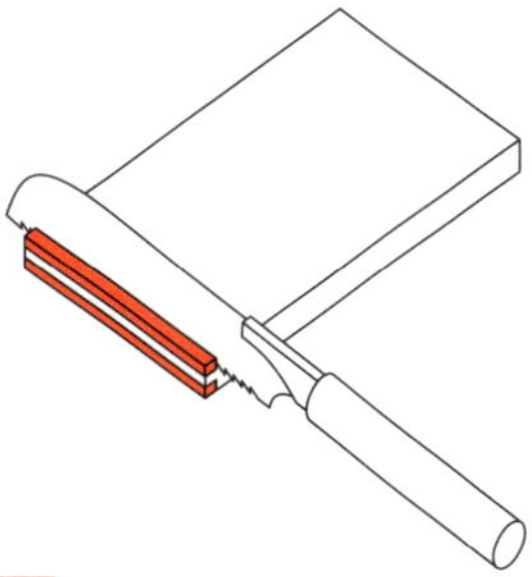
B2

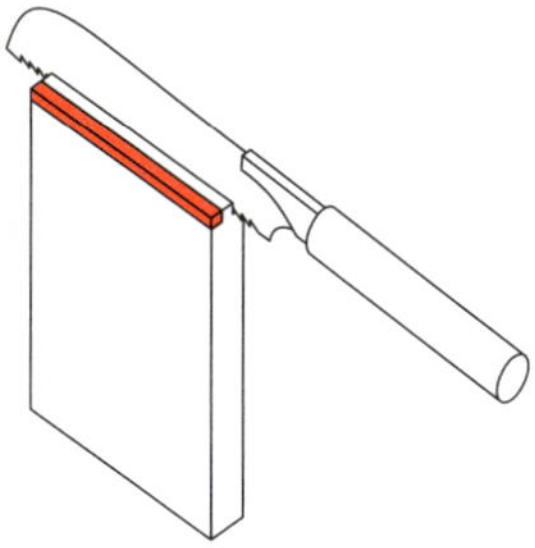
B3

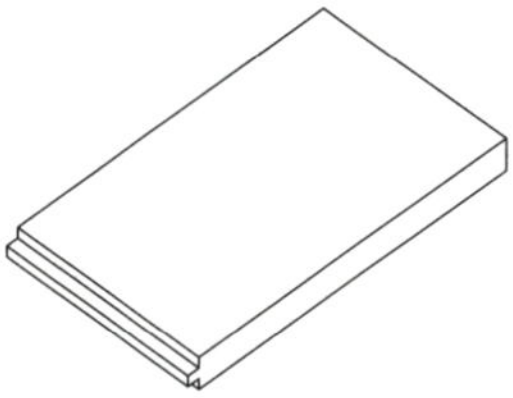
B

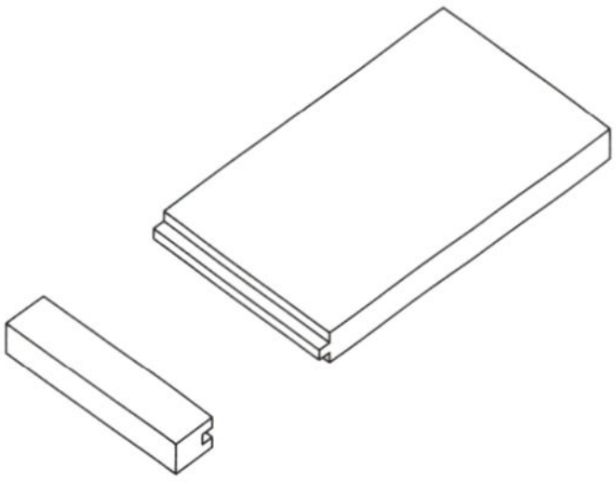
A+B

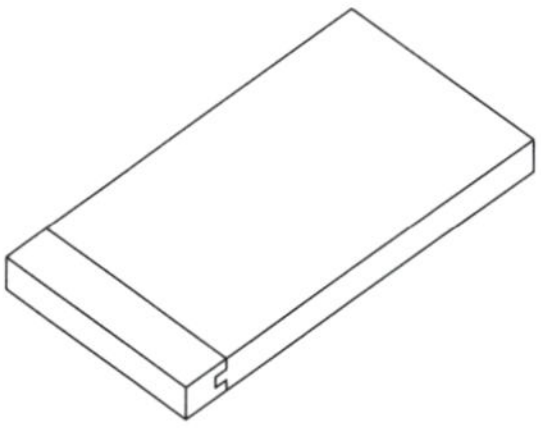
AB

Dado and tongue breadboard end

mechi-hashibame-tsugi [*]
目違端嵌接ぎ

Hirnleiste mit Feder

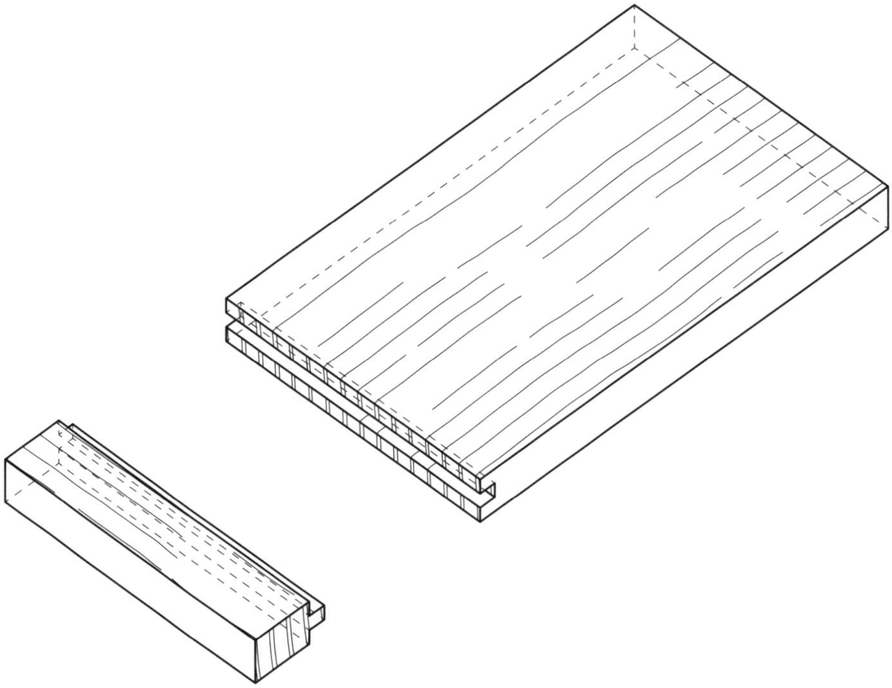

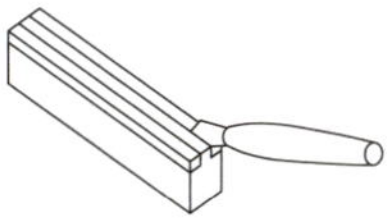

A1

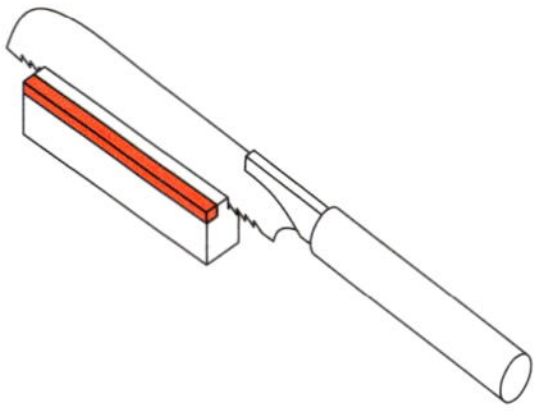

A2

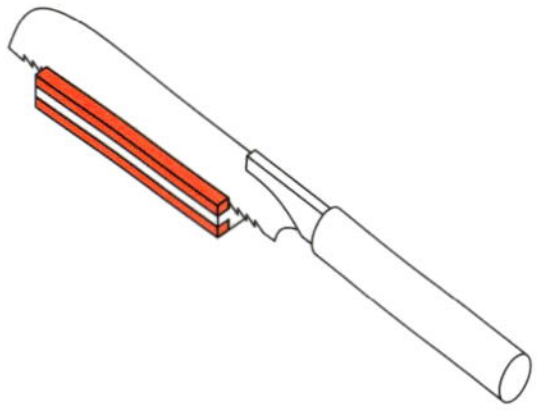

A3

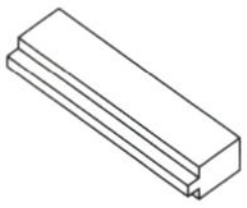

A

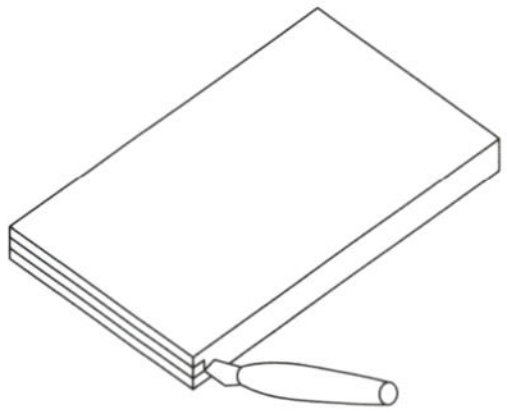

B1

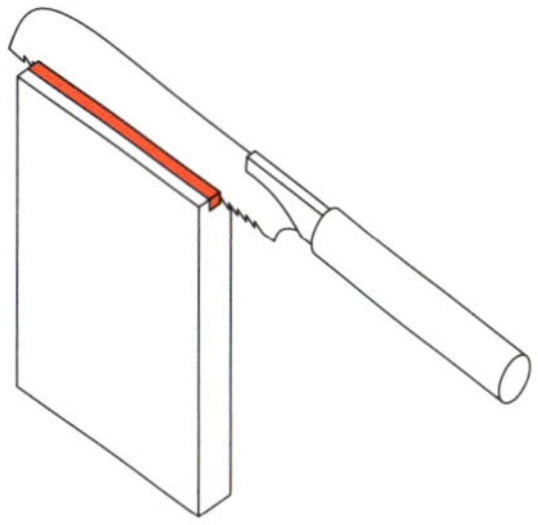

B2

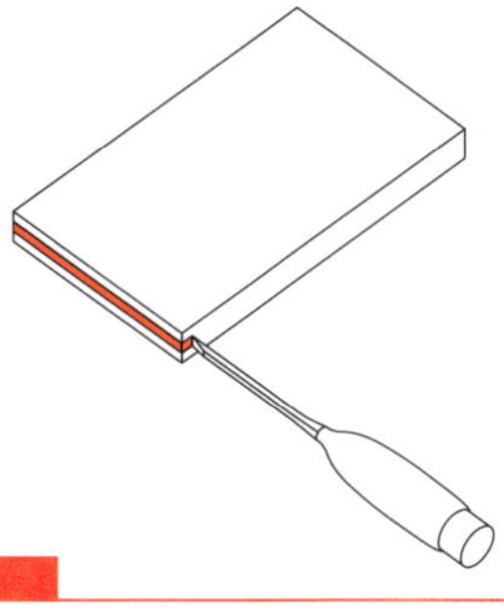

B3

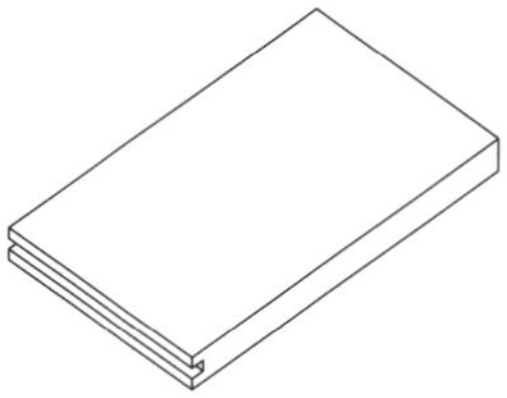

B

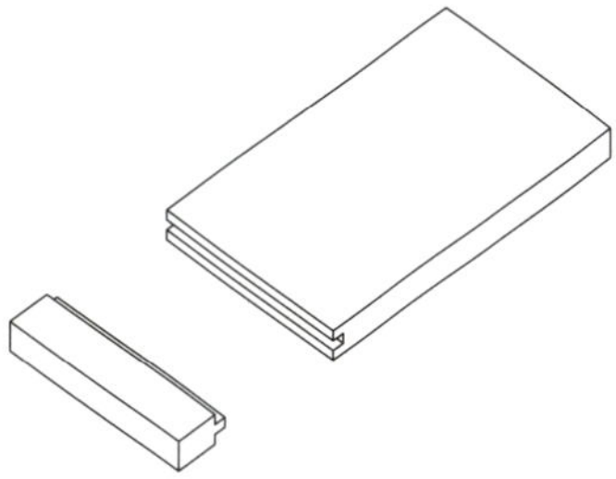

A+B

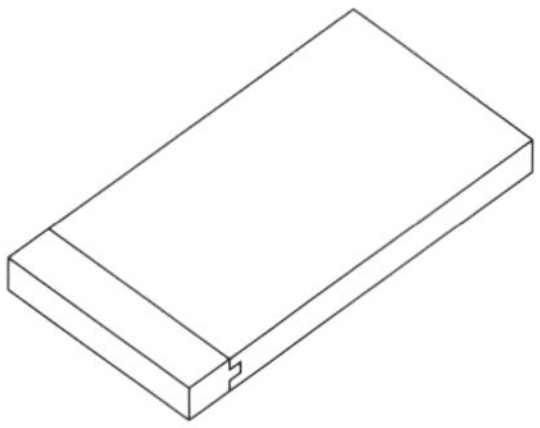

AB

End Joints

Haunched mortise and tenon breadboard end

hon-zane-tōshi-hozo-hashibame-tsugi [*]
本実通枘端嵌接ぎ

Hirnleiste mit Nut und Zapfenlöchern

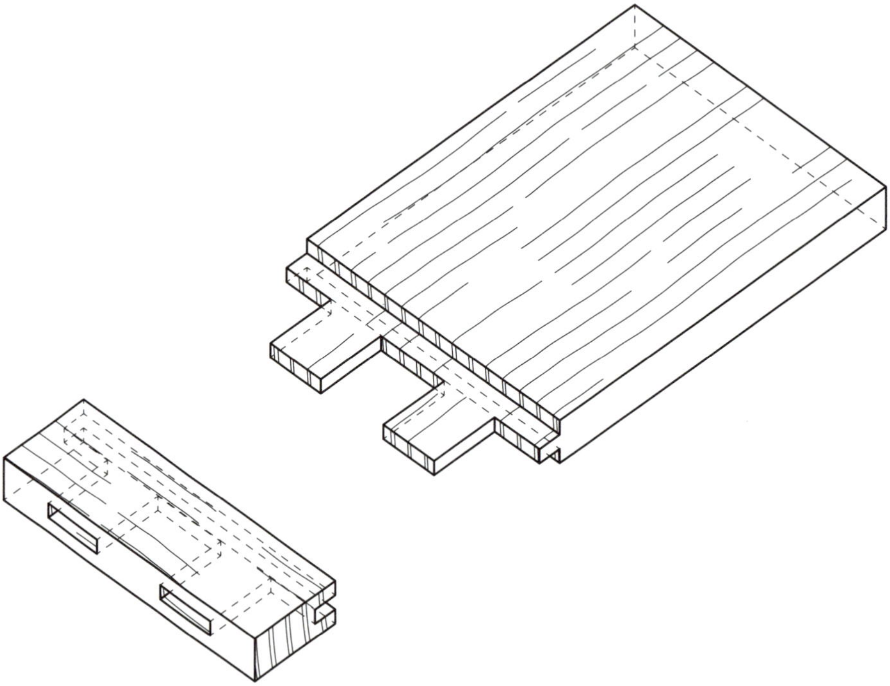

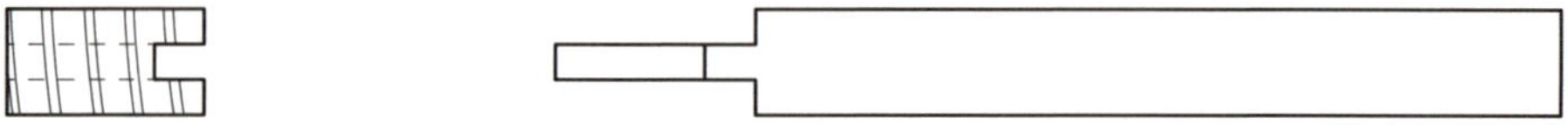

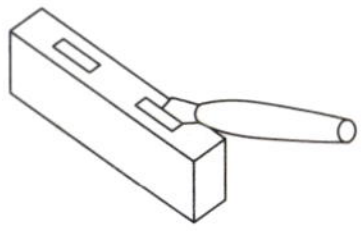

A1

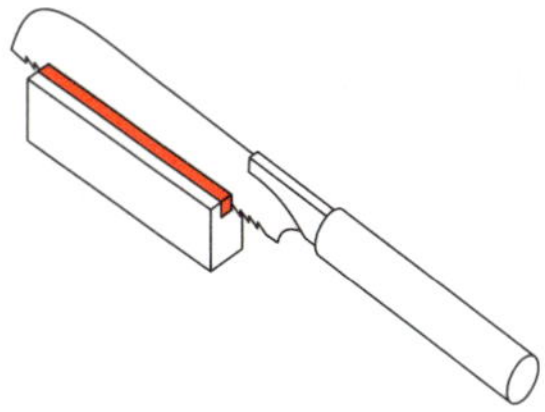

A2

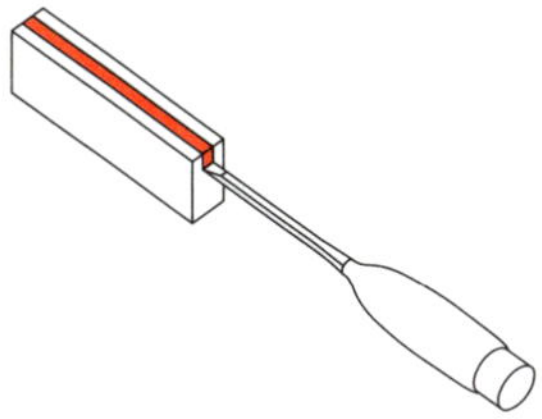

A3

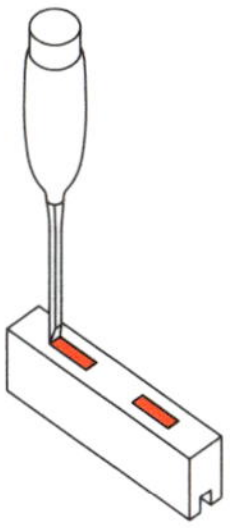

A4

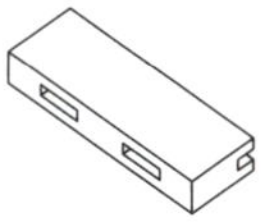

A

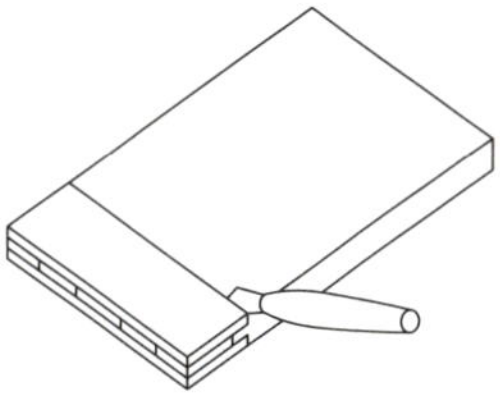

B1

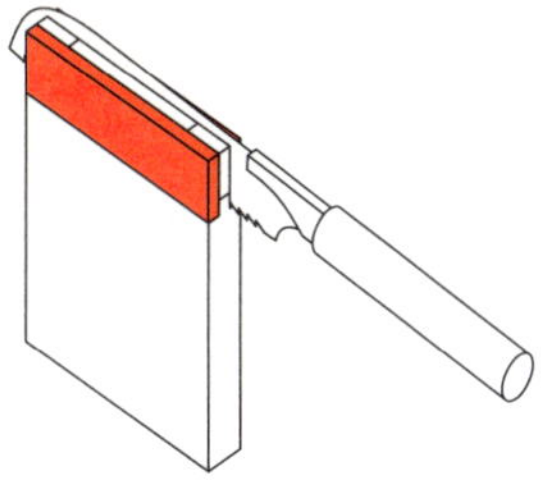

B2

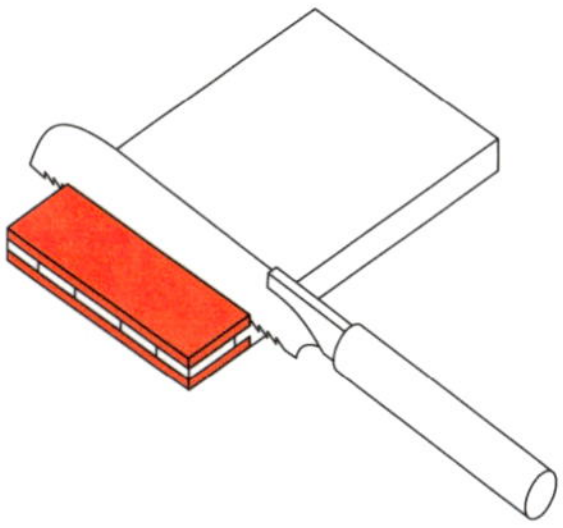

B3

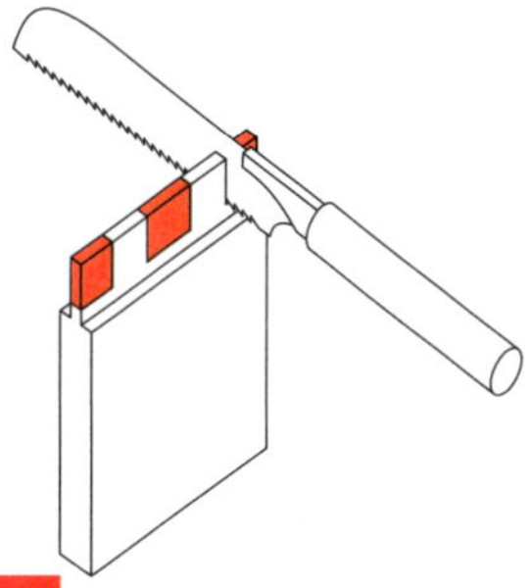

B4

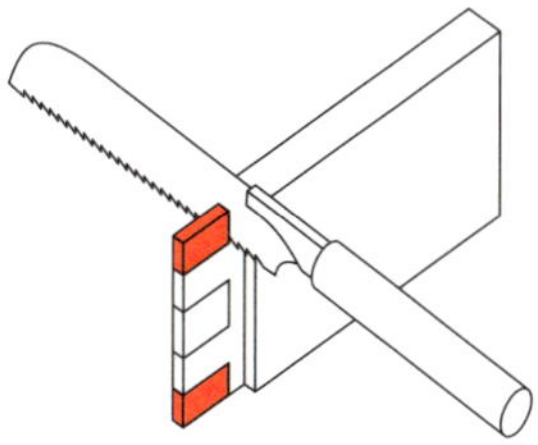

B5

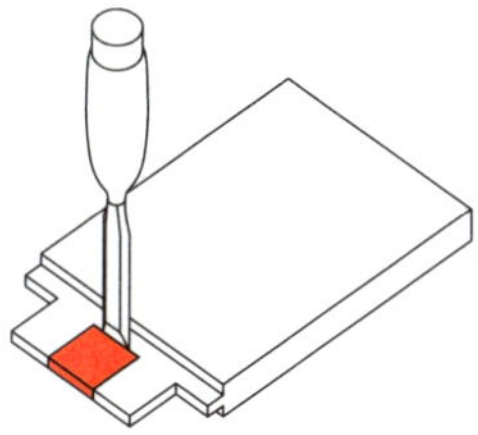

B6

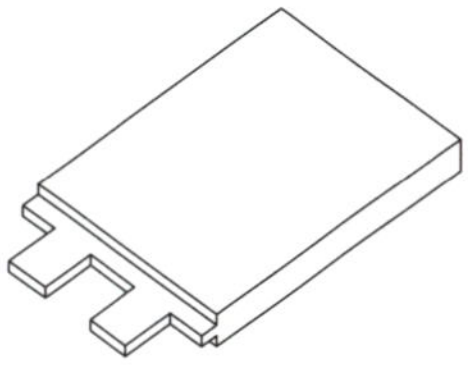

B

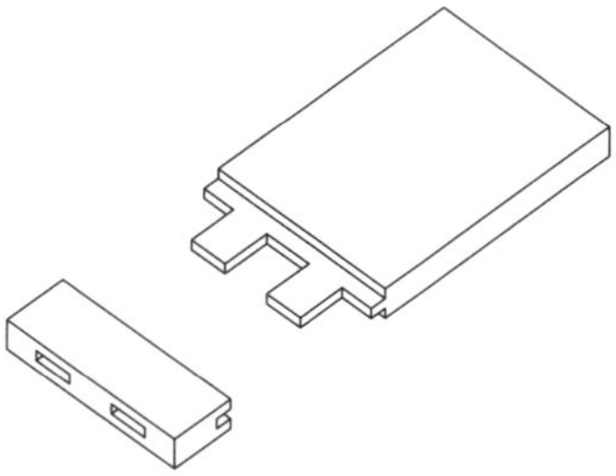

A+B

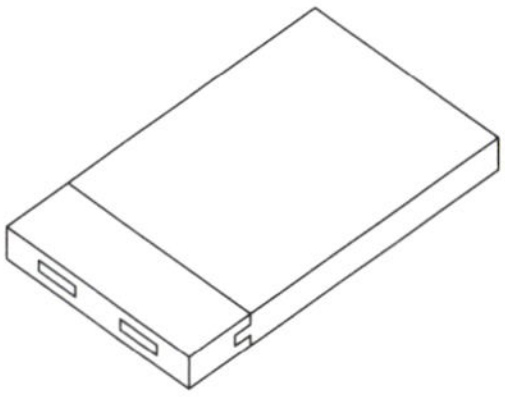

AB

End Joints

Wedged and haunched tenon breadboard end

hon-zane-tōshi-hozo-wari-kusabi-hashibame-tsugi [*]
本実通枘割楔端嵌接ぎ

verkeilte Hirnleiste mit Nut und Zapfenloch

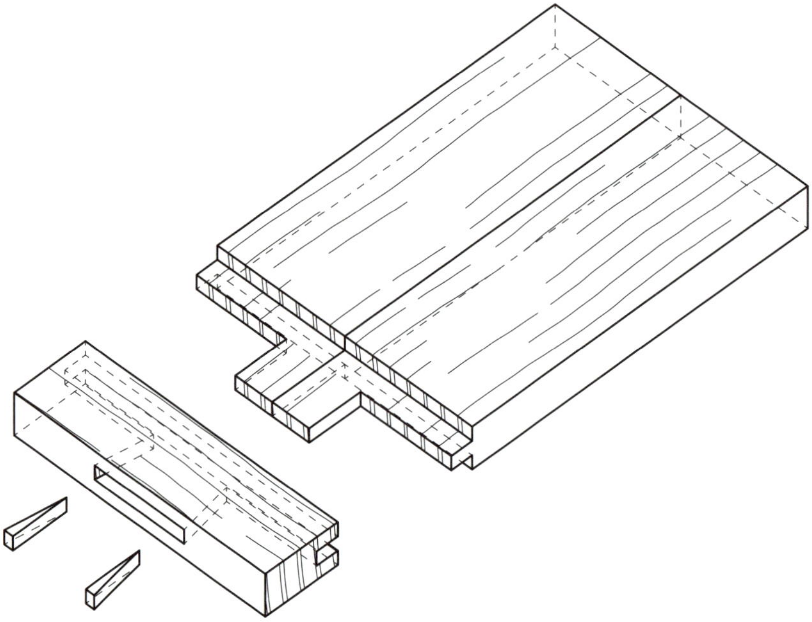

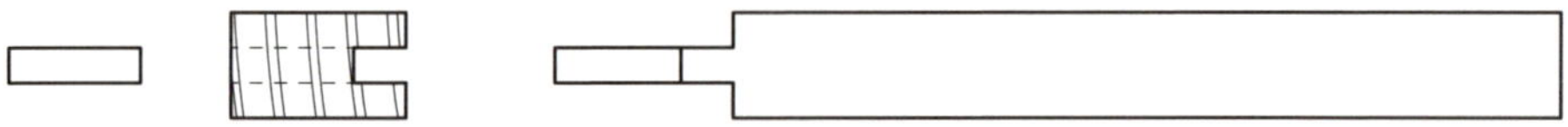

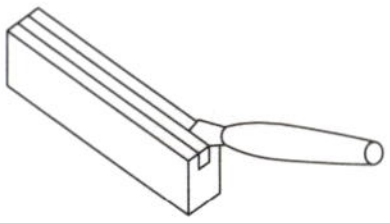

A1

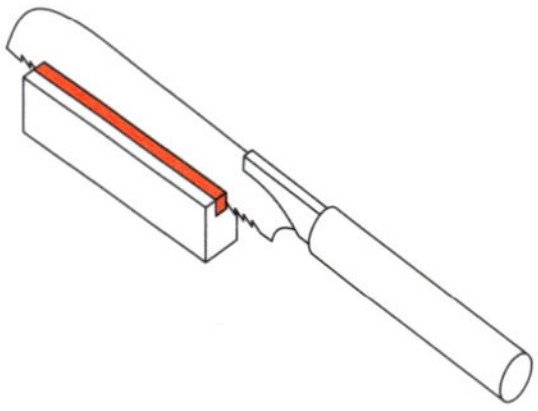

A2

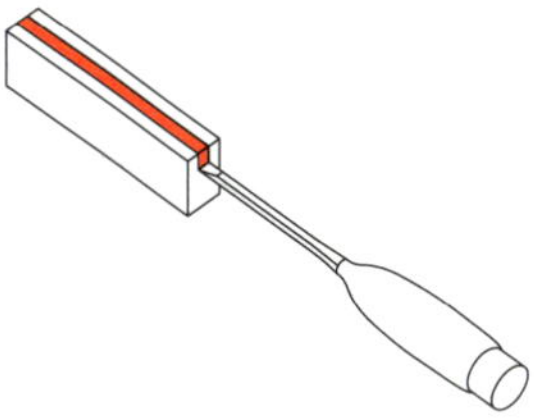

A3

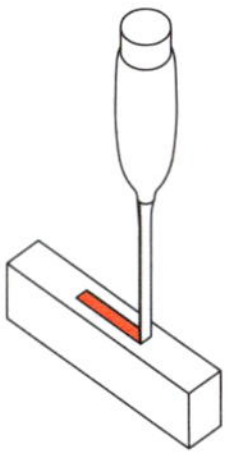

A4

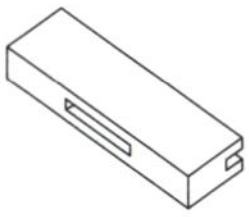

A

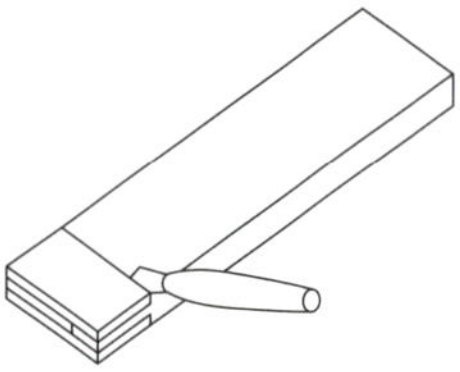

B1

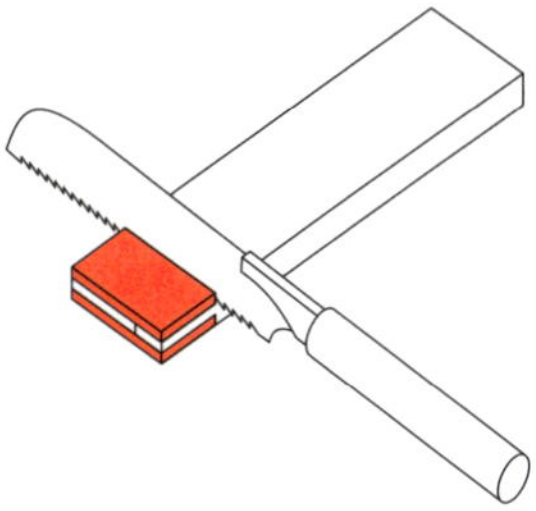

B2

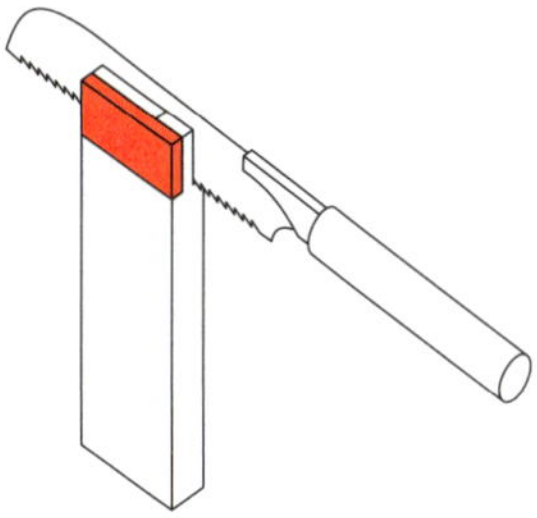

B3

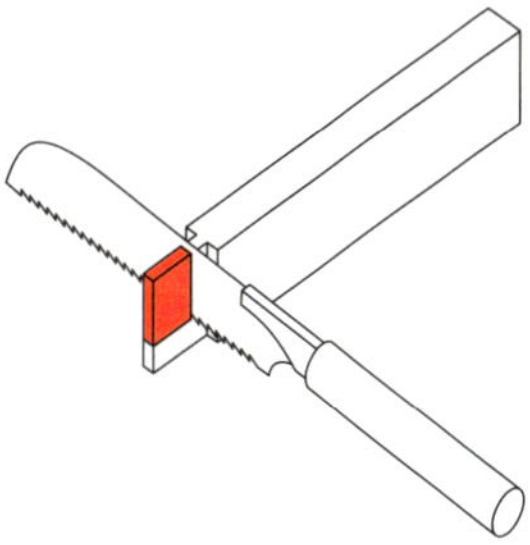

B4

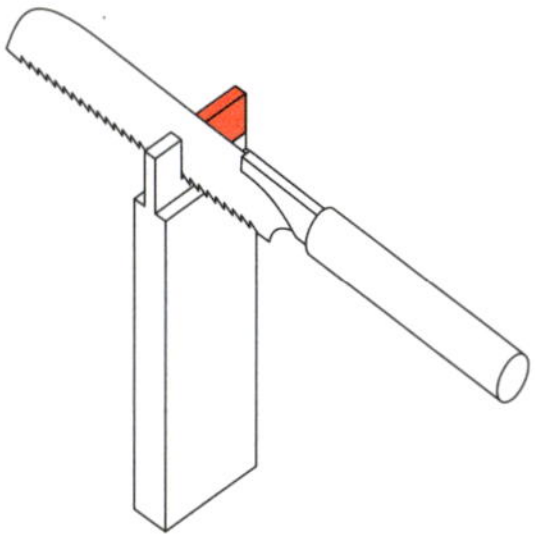

B5

2×

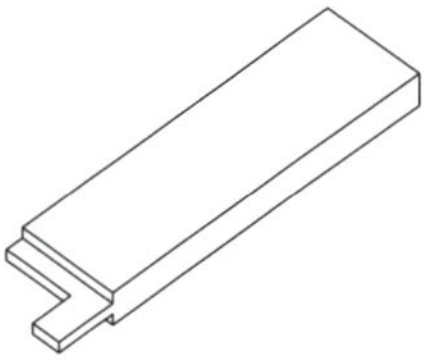

B

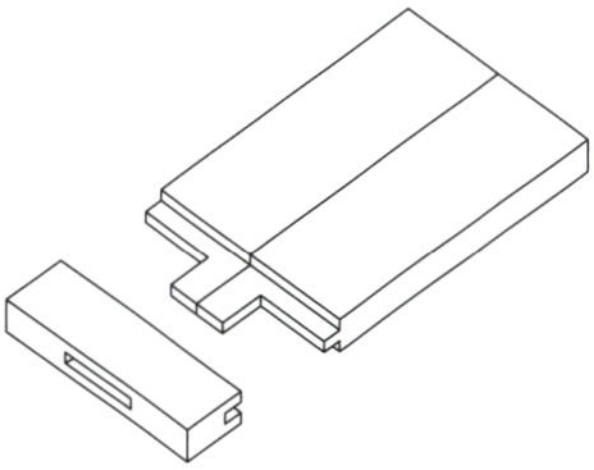

A+BB

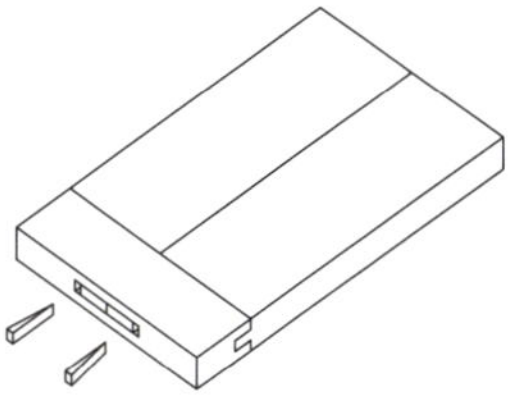

A+BB

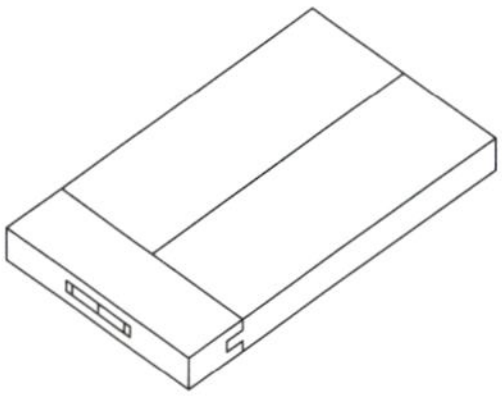

ABB

Corner Joints

Rabbet joint

tsutsumi-uchi-tsuke-tsugi
包打付接ぎ

gefälzte Eckverbindung

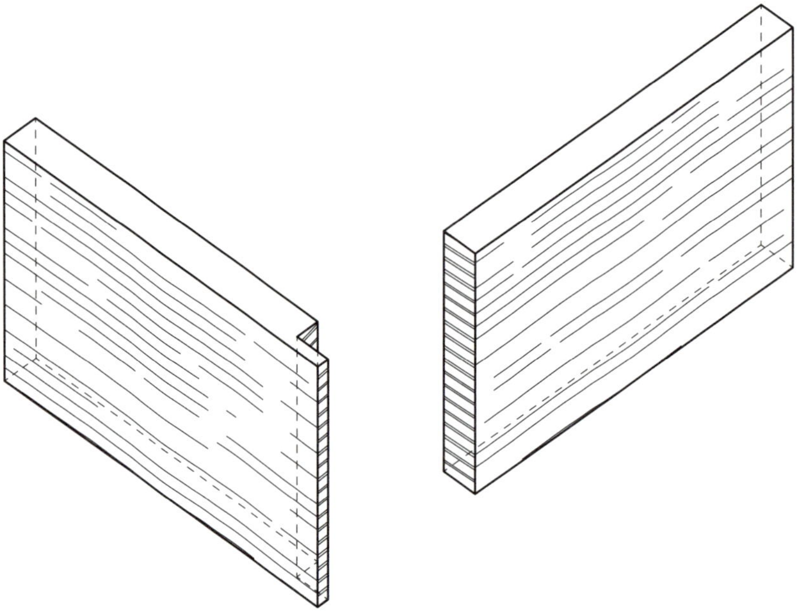

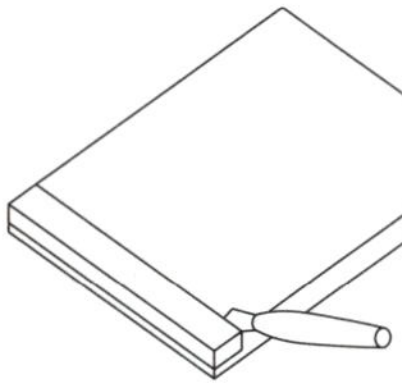

A1

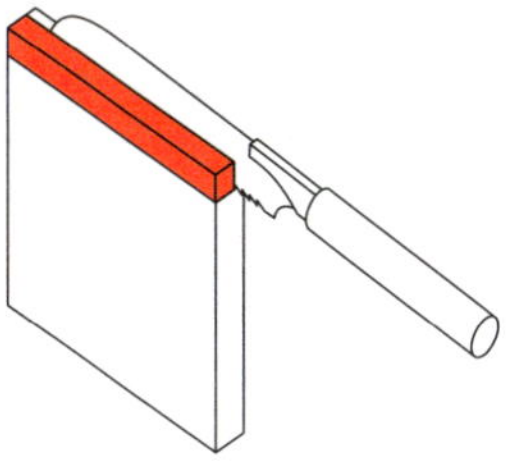

A2

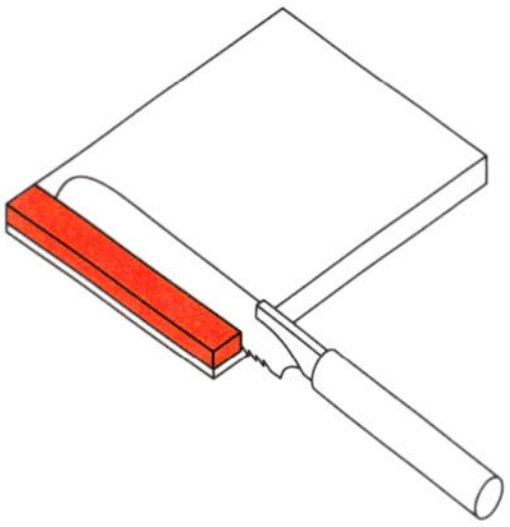

A3

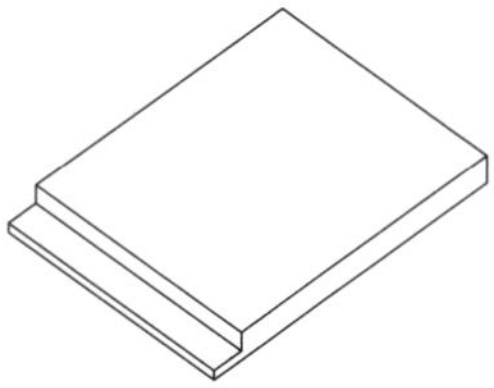

A

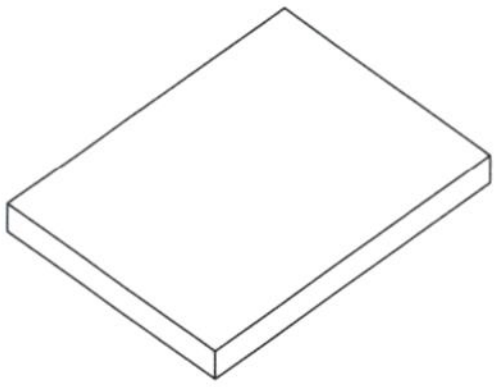

B

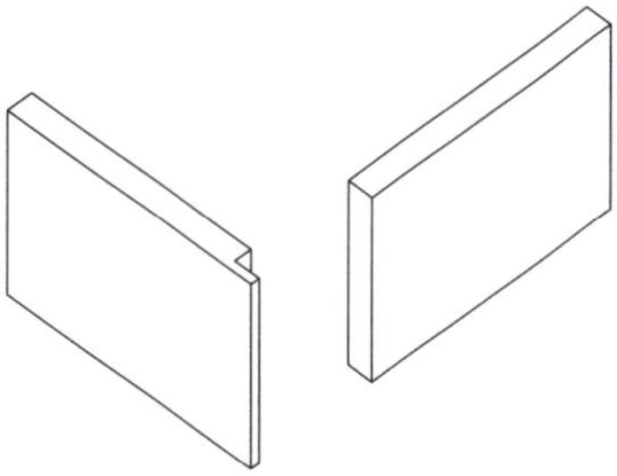

A+B

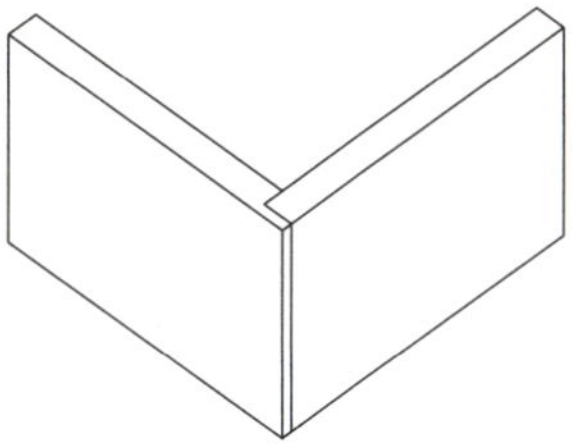

AB

Corner Joints

Mitered rabbet joint

tsutsumi-uchi-tsuke-tsugi (binta-dome)
包打付接ぎ(鬢太留)

gefälzte Eckverbindung mit Gehrung

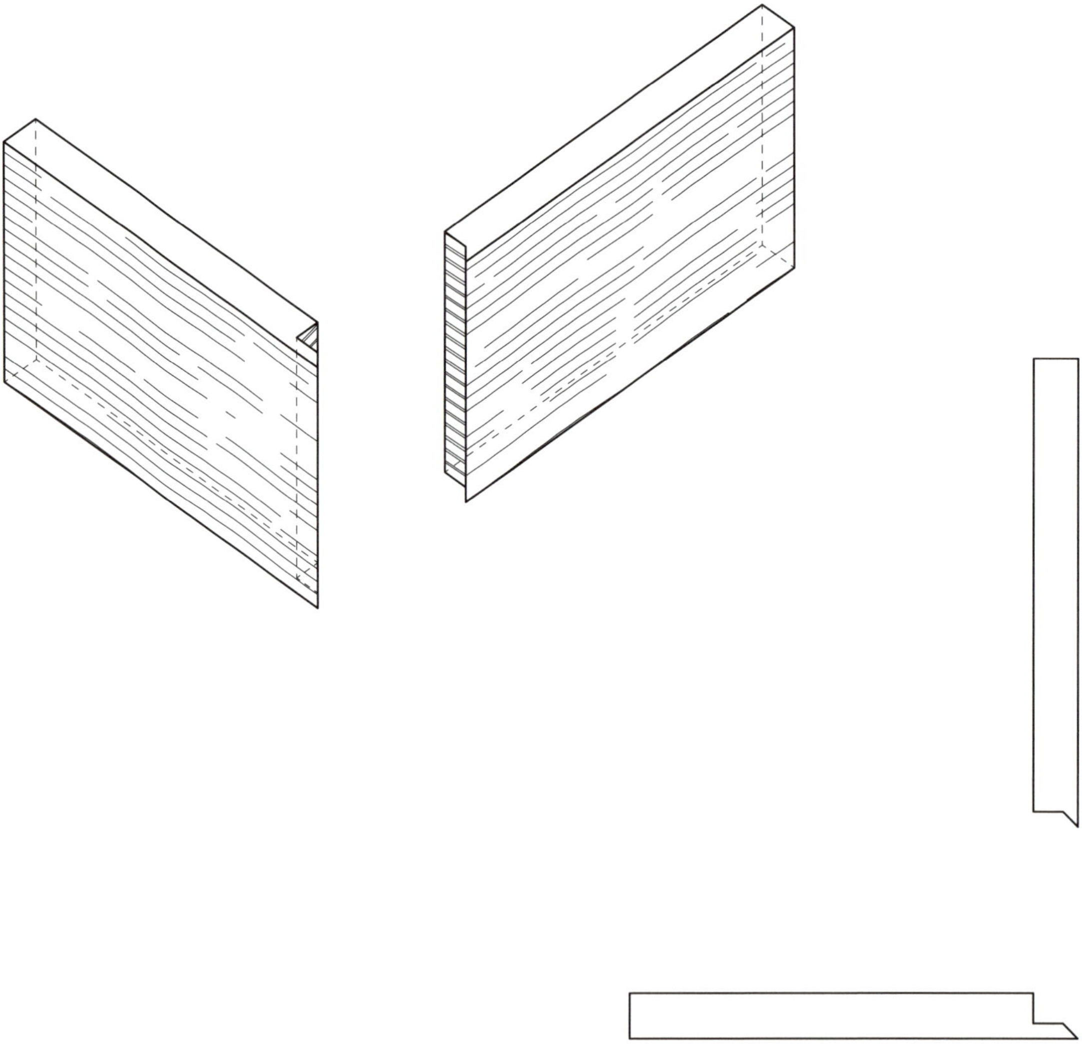

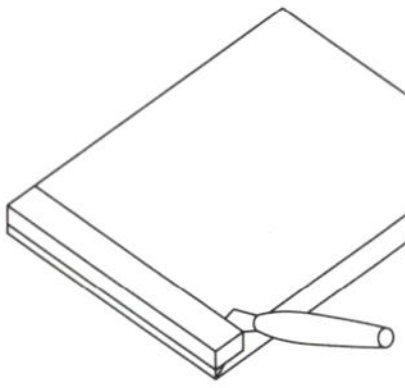

A1

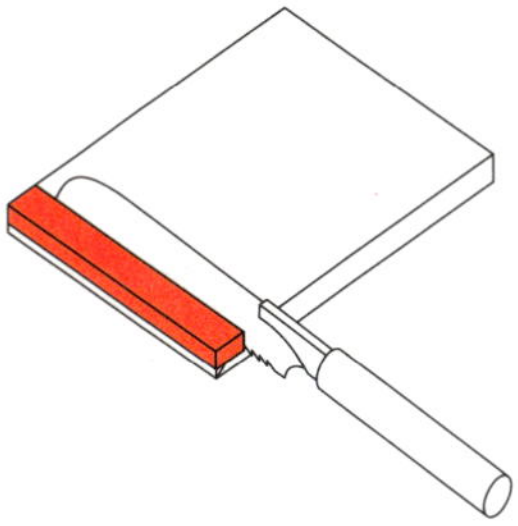

A2

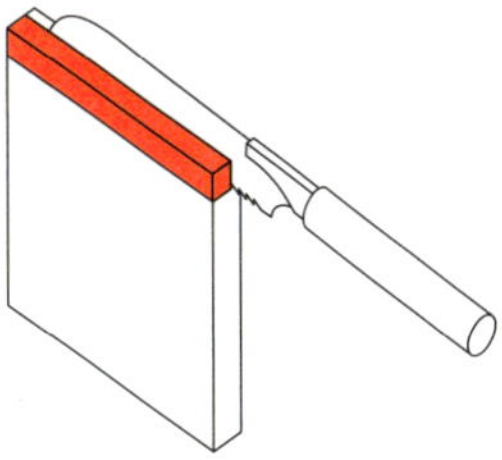

A3

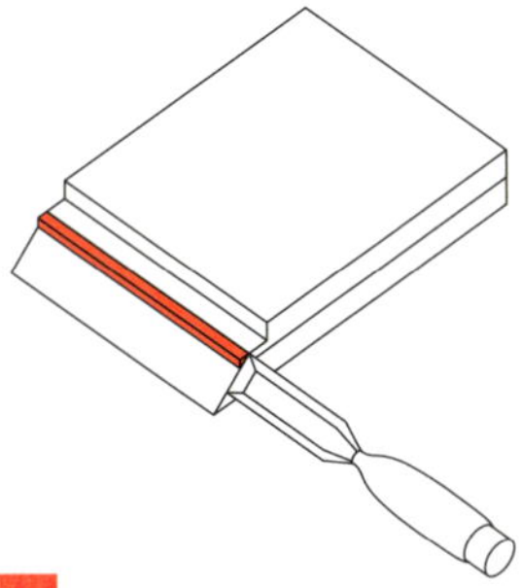

A4

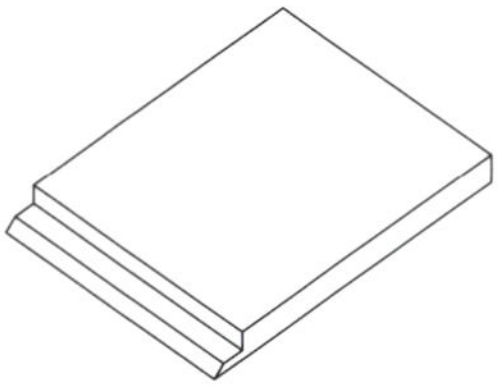

A

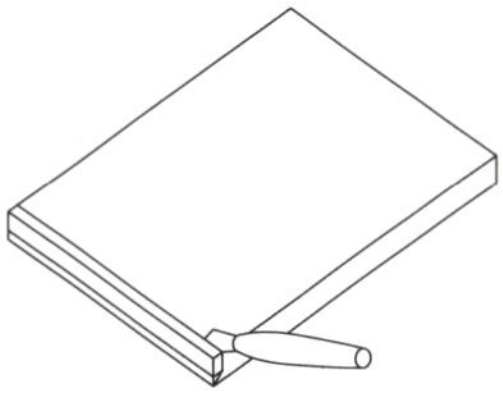

B1

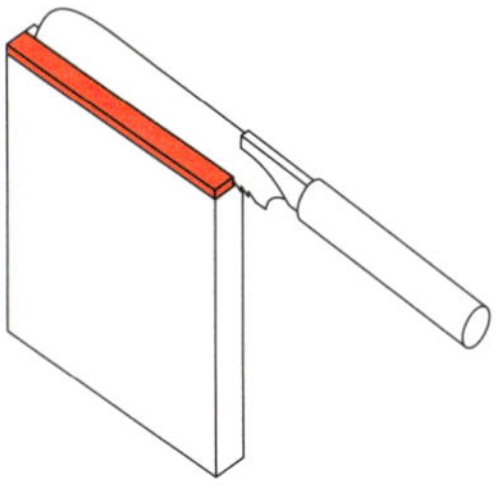

B2

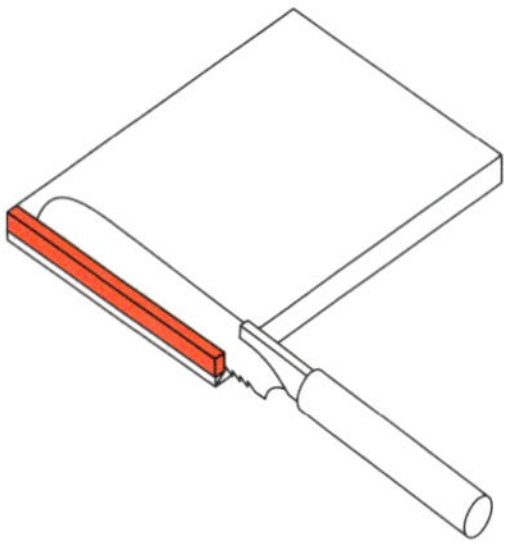

B3

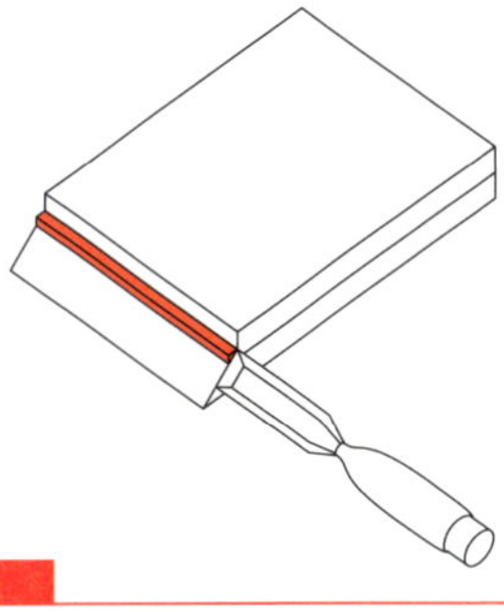

B4

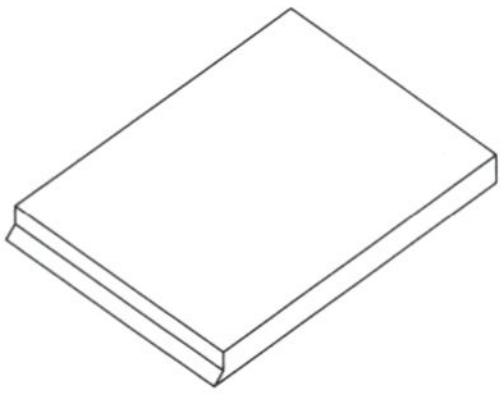

B

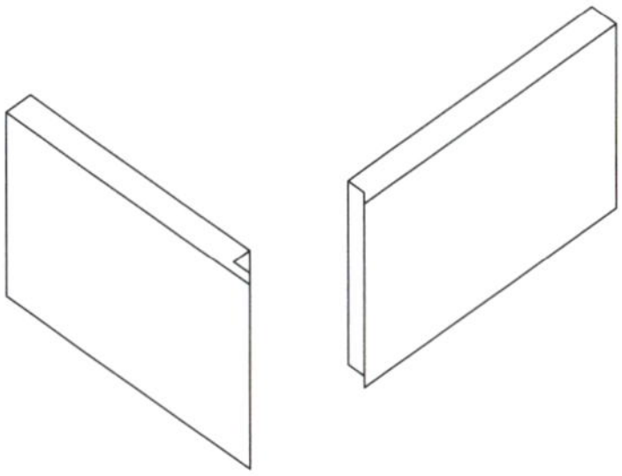

A+B

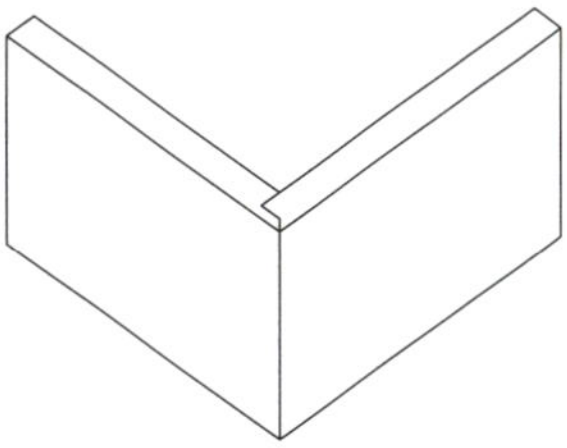

AB

Corner Joints

Dado and rabbet joint

kone-oire-kumi-tsugi
小根追入組接ぎ

gespundete Eckverbindung

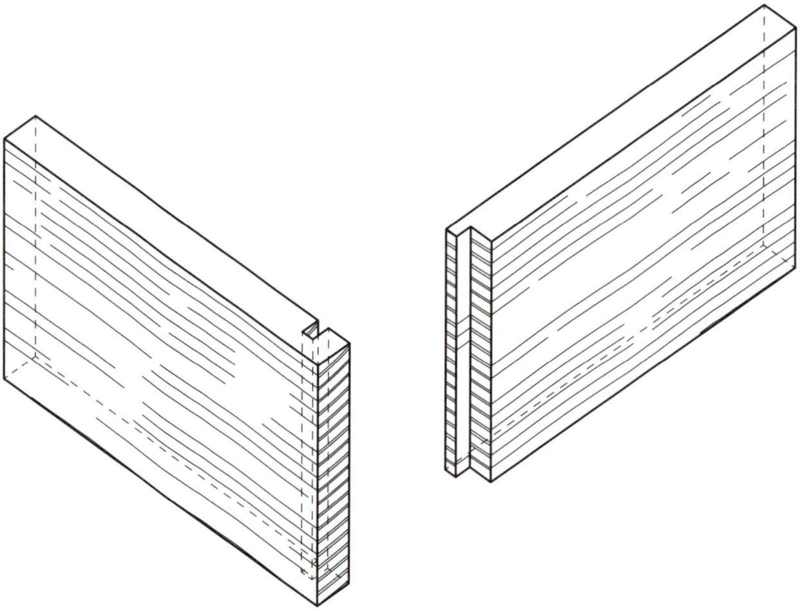

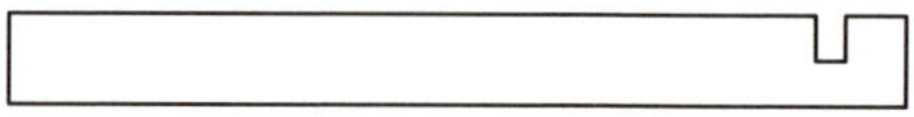

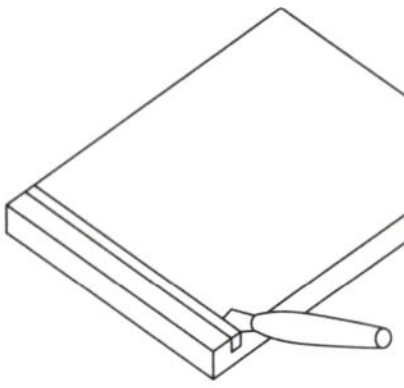
A1

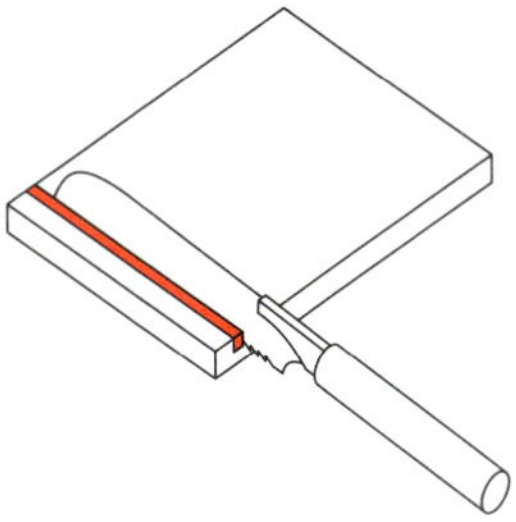
A2

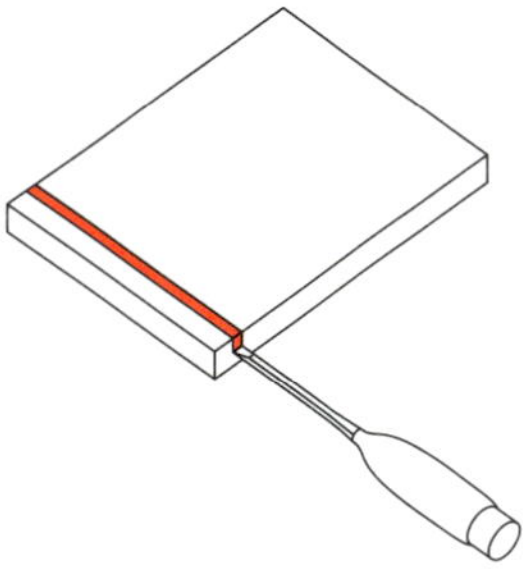
A3

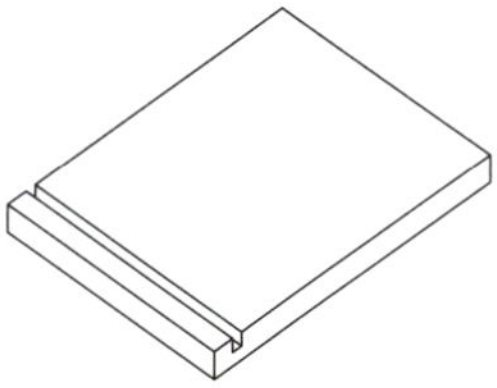
A

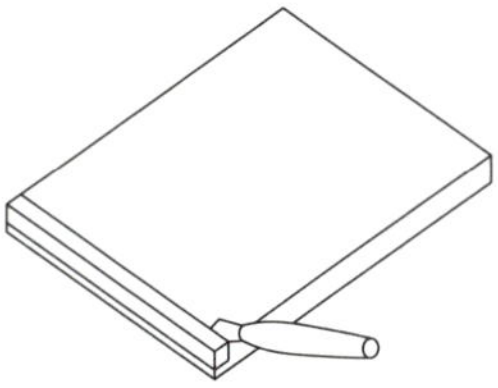
B1

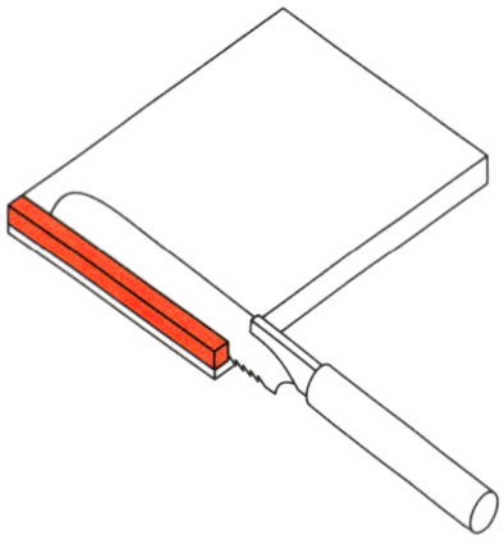
B2

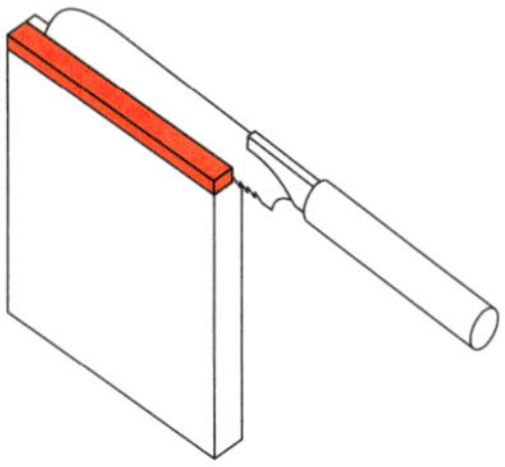
B3

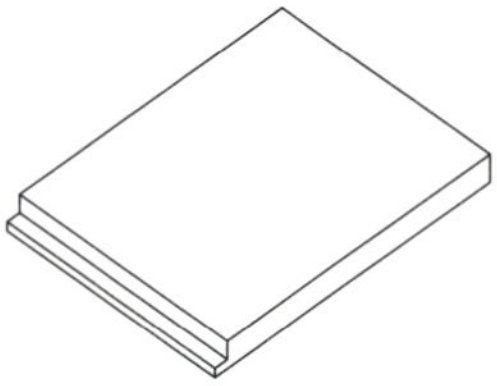
B

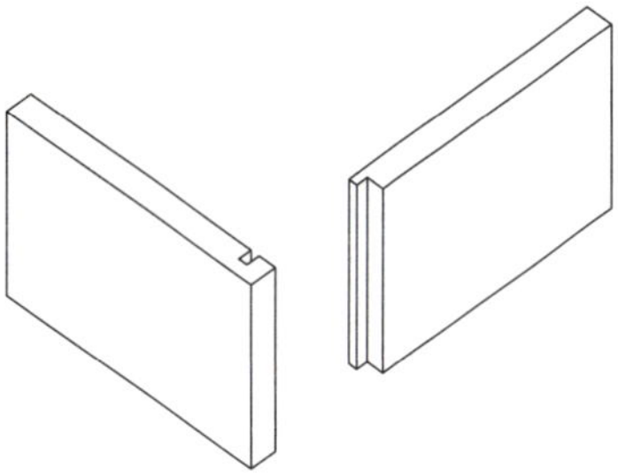
A+B

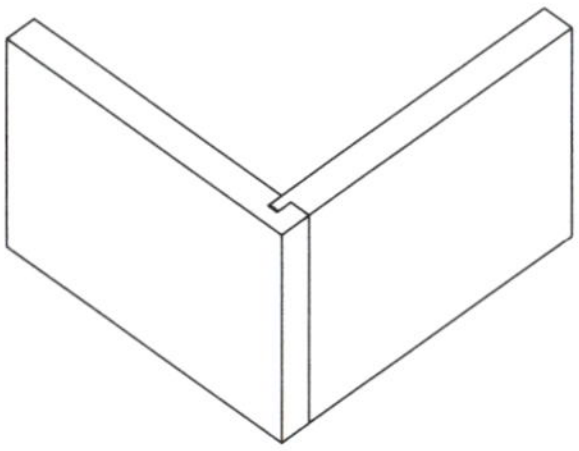
AB

Corner Joints

Lock joint

tsutsumi-kone-oire-kumi-tsugi
包小根追入組接ぎ

gespundete Eckverbindung mit Verdeck

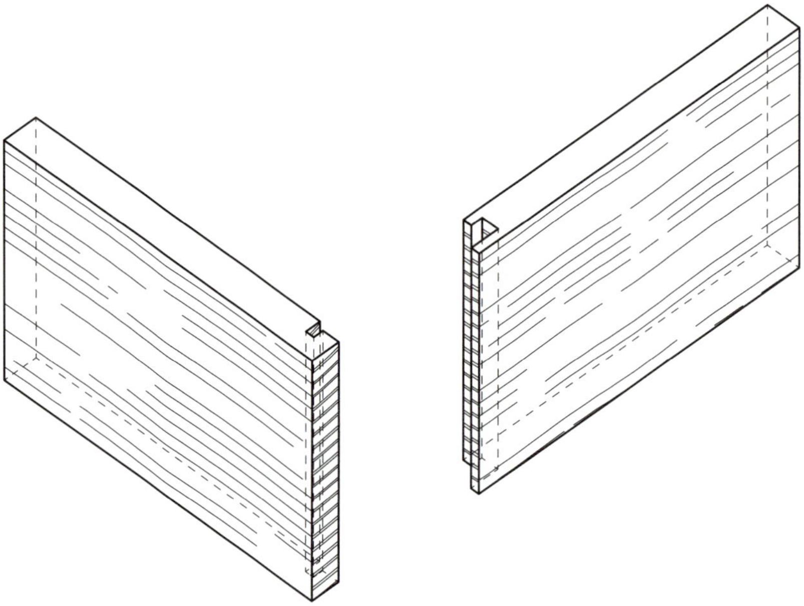

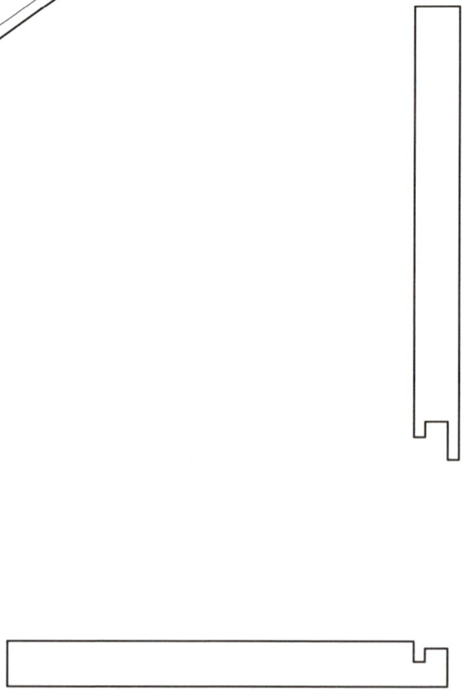

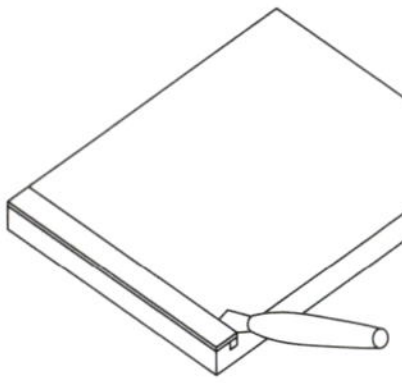
A1

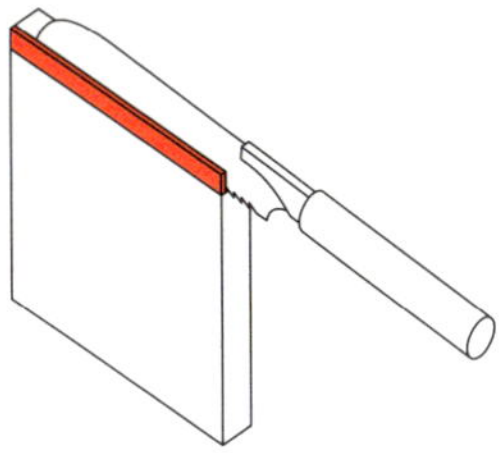
A2

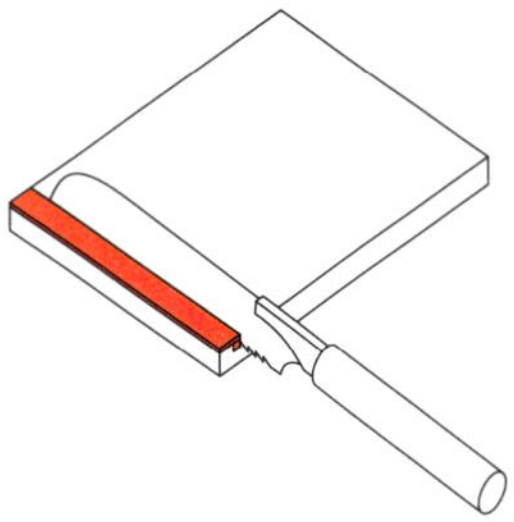
A3

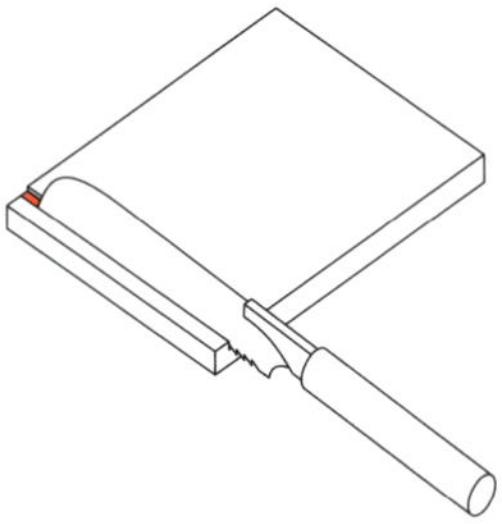
A4

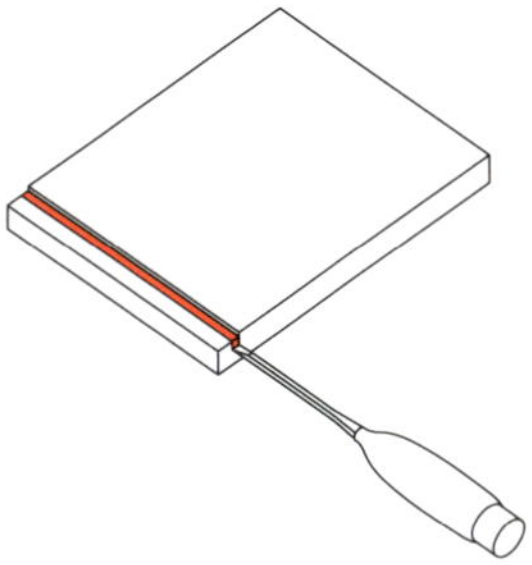
A5

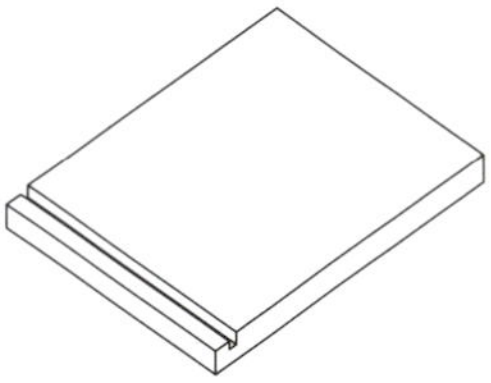
A

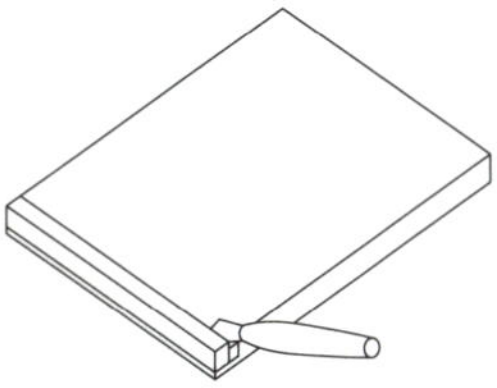
B1

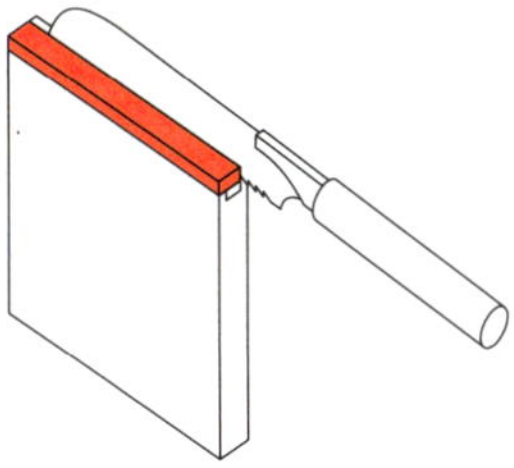
B2

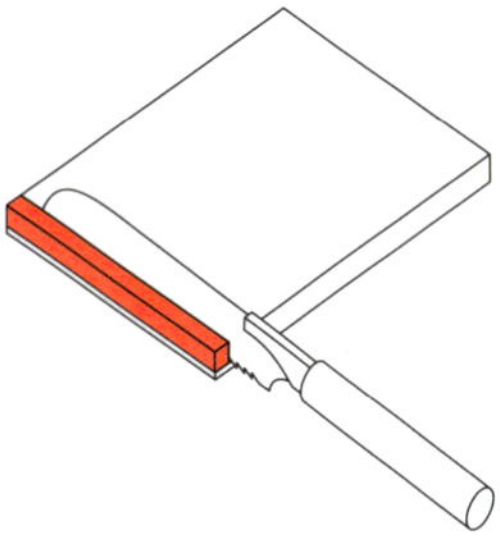
B3

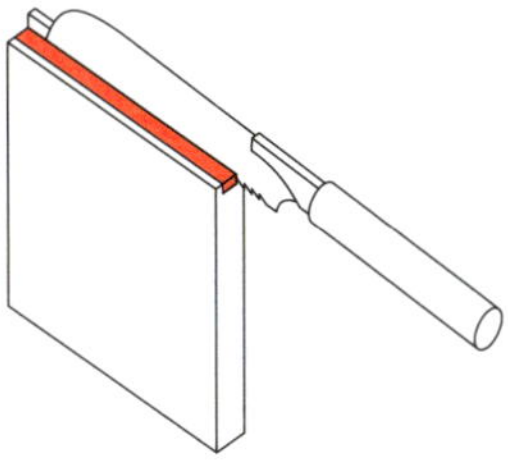
B4

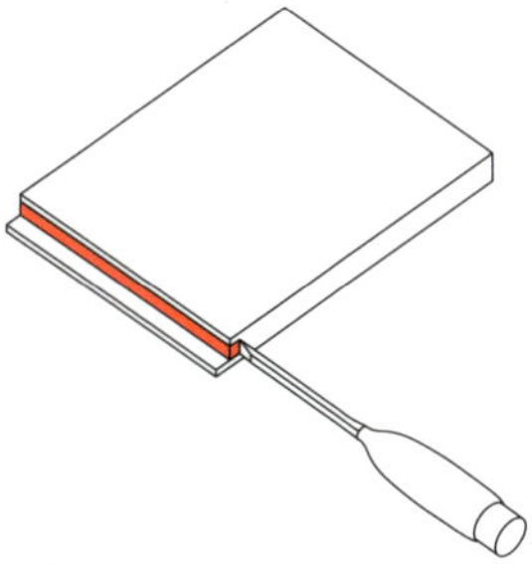
B5

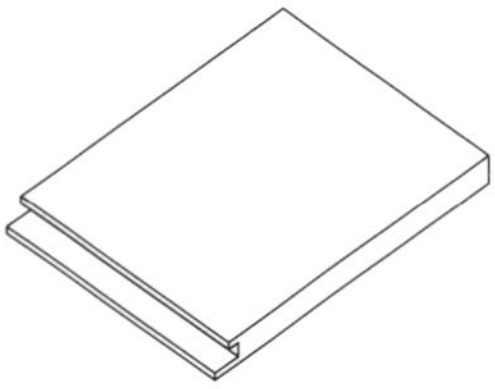
B

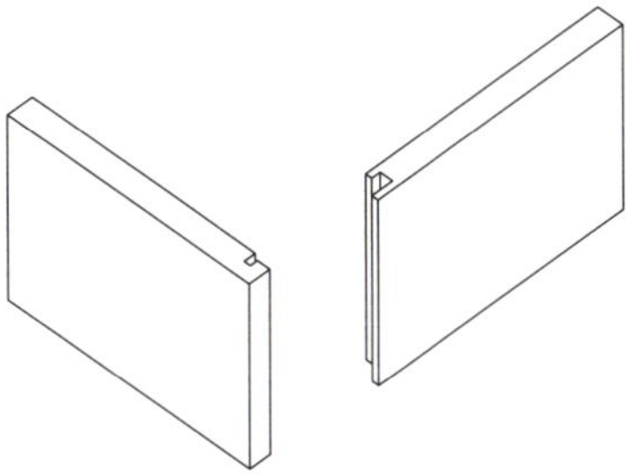
A+B

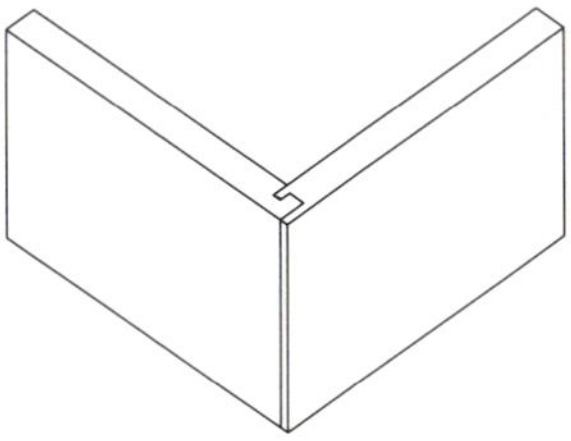
AB

Lock miter joint

tsutsumi-kone-oire-
kumi-tsugi (binta-dome)
包小根追入組接ぎ(鬢太留)

gespundete Eckverbindung
mit Verdeck auf Gehrung

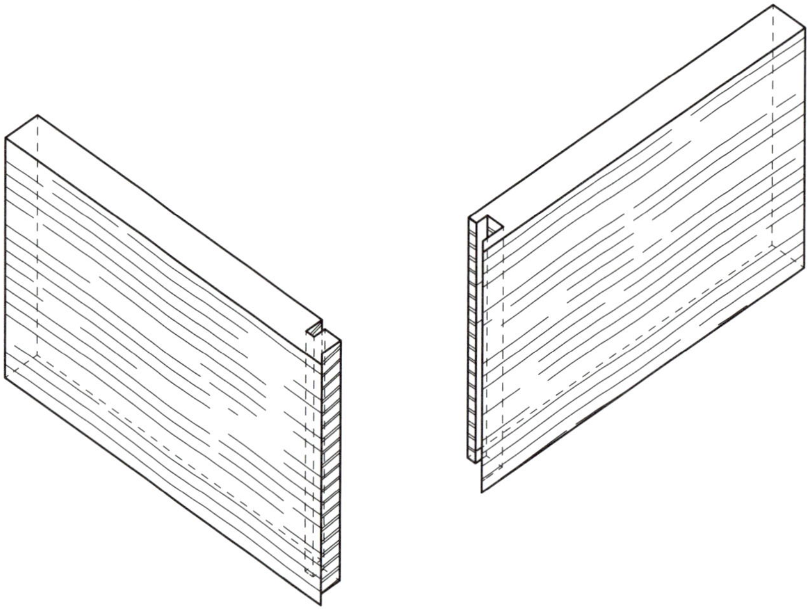

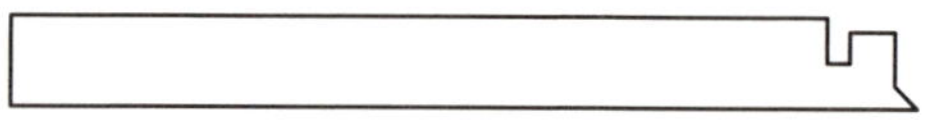

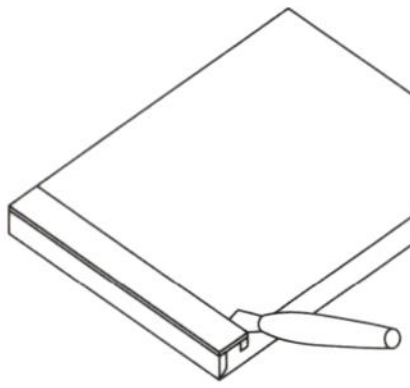
A1

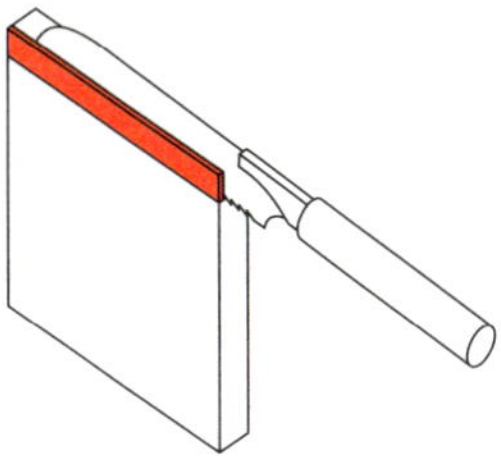
A2

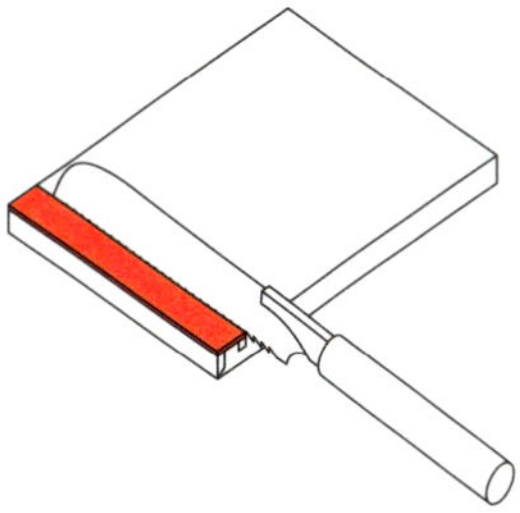
A3

A4

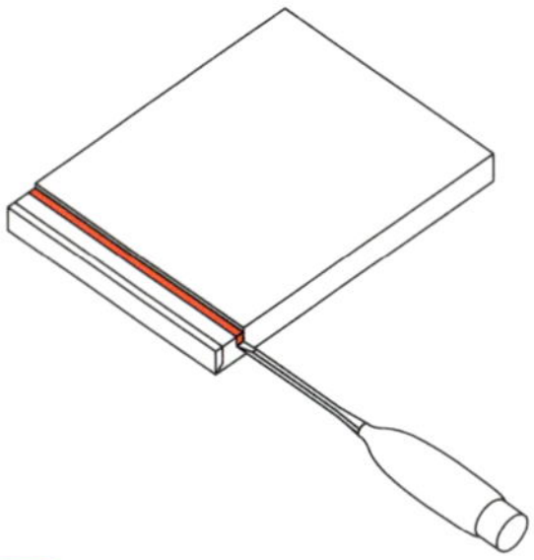
A5

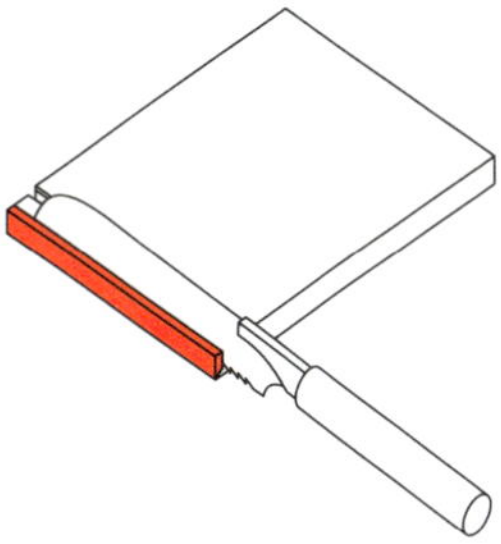
A6

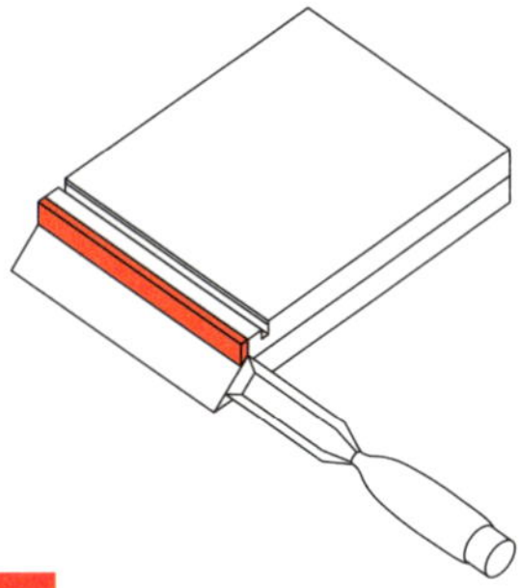
A7

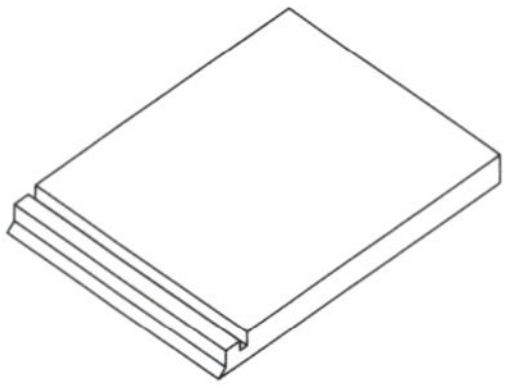
A

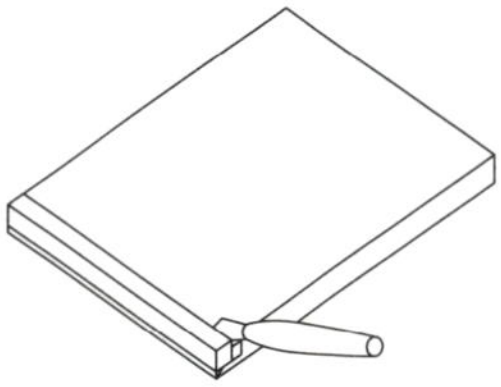
B1

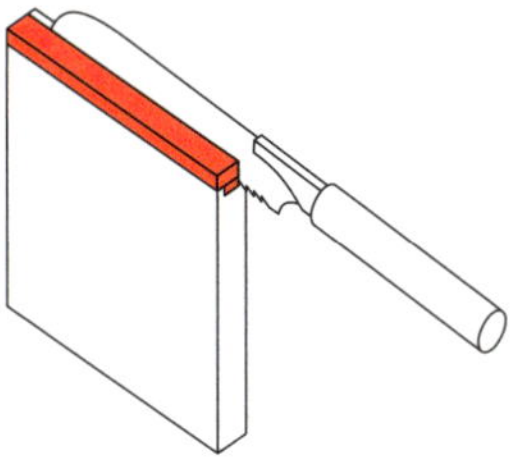
B2

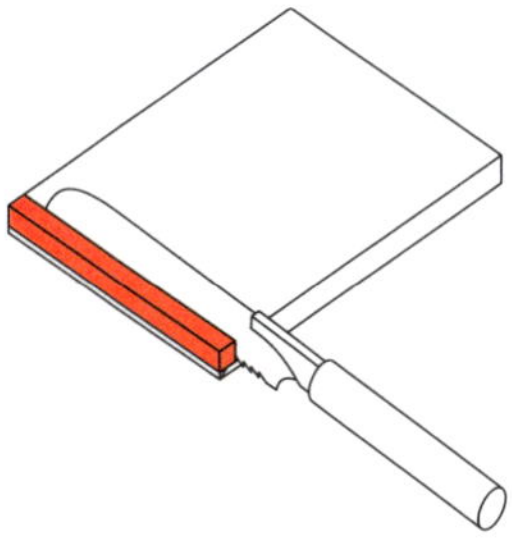
B3

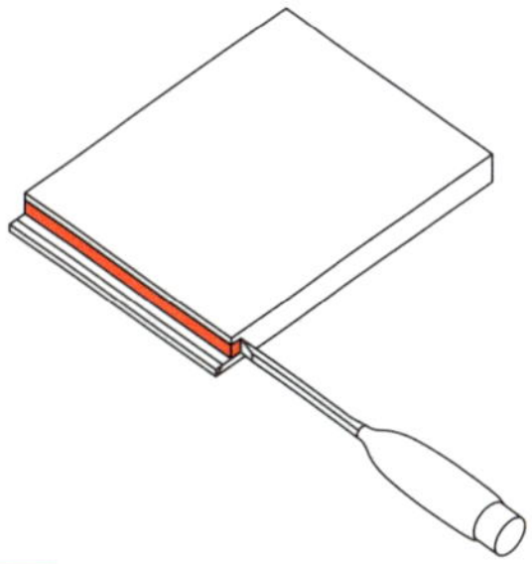
B4

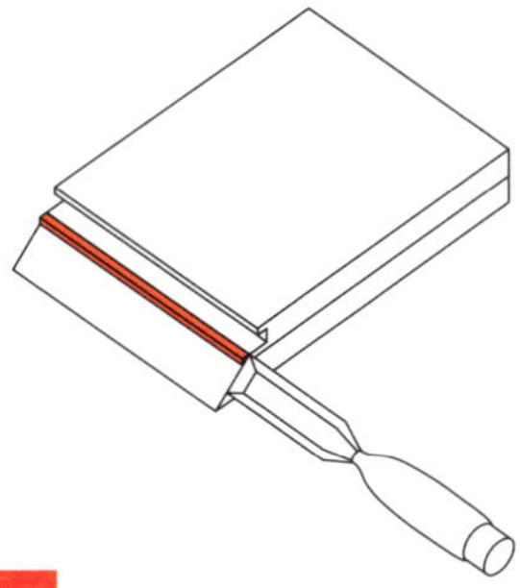
B5

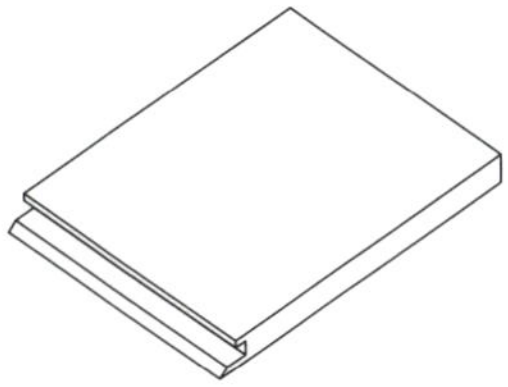
B

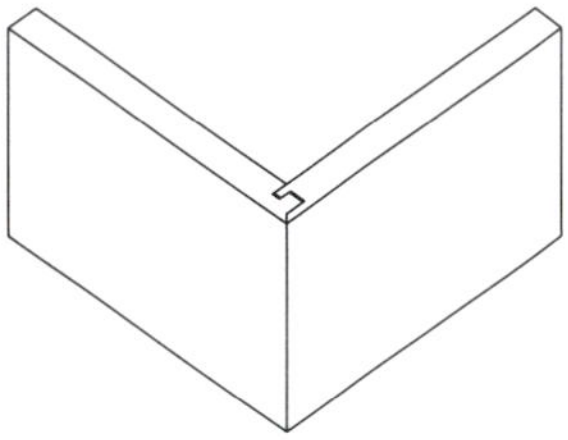
AB

Splined miter joint

yatoi-zane-ōdome-tsugi
雇実大留接ぎ

Eckverbindung auf Gehrung mit eingesetzter Feder

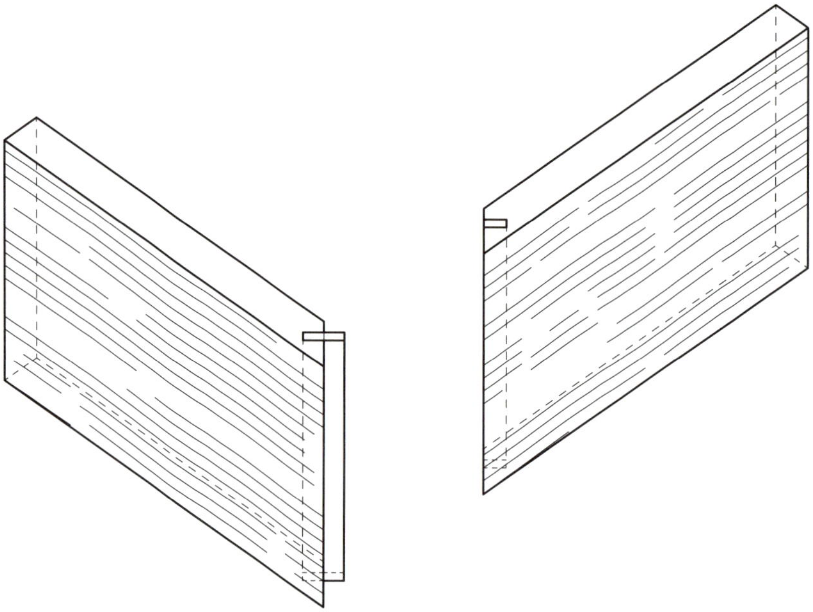

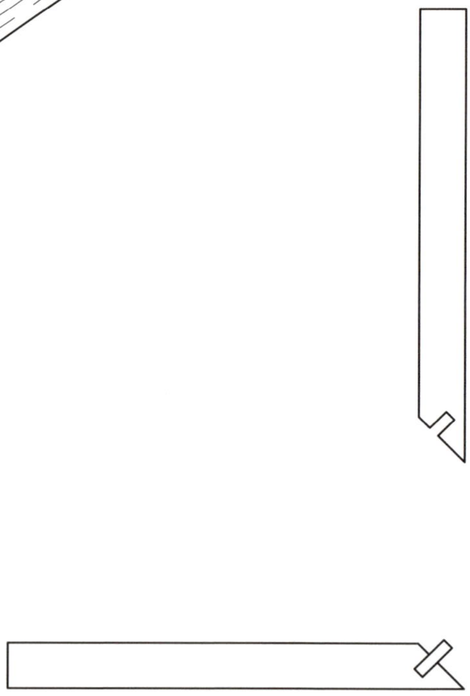

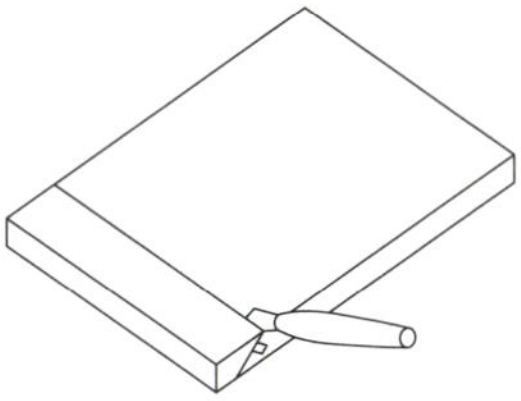

A1

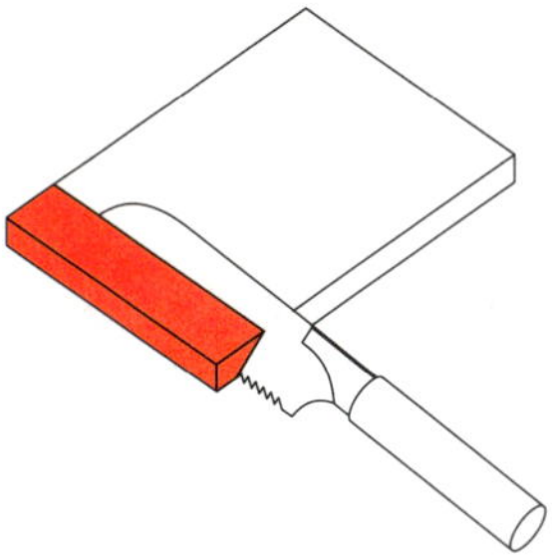

A2

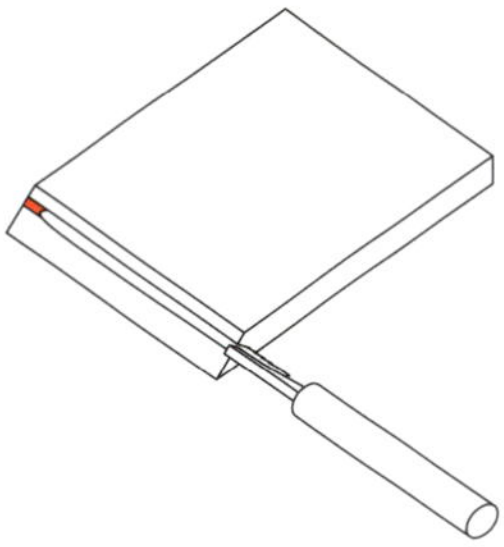

A3

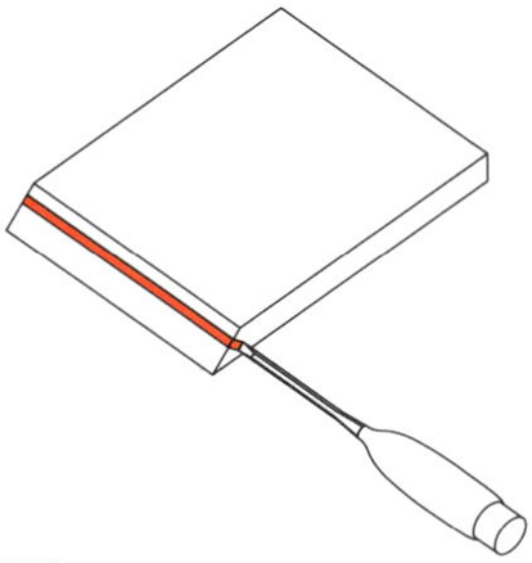

A4

2×

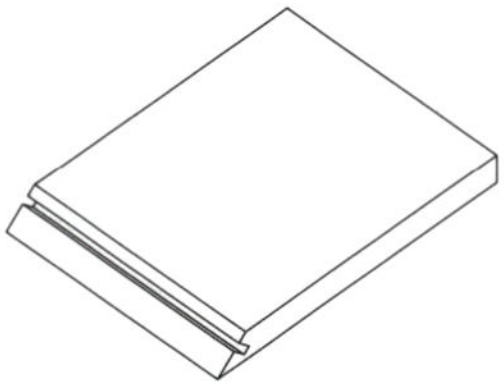

A

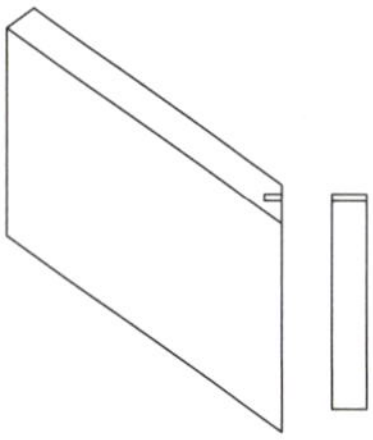

A+B

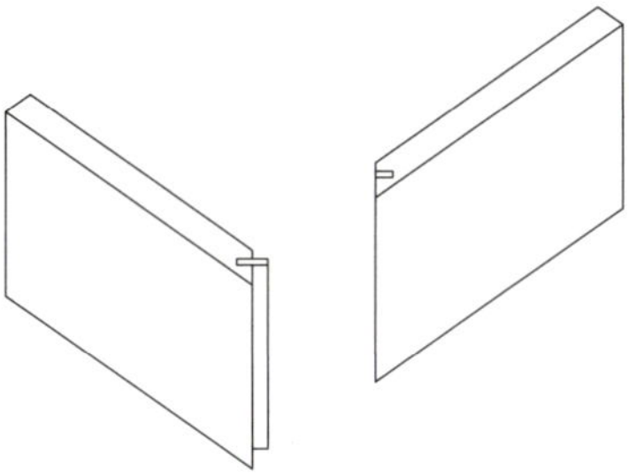

A+B+A

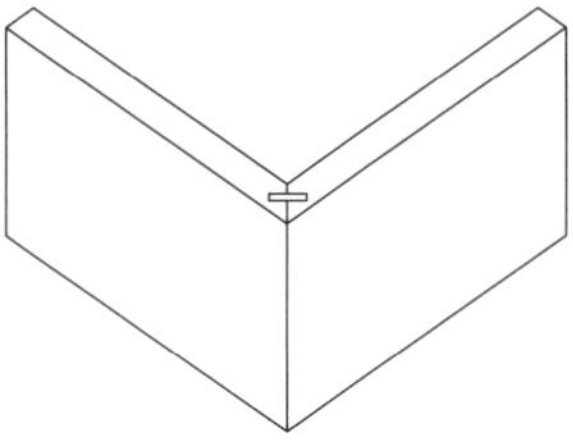

ABA

Corner Joints

Miter joint with spline keys

nimai-hiki-komi-dome-tsugi
二枚挽込留接ぎ

Eckverbindung auf Gehrung
mit eingesetzten Federn

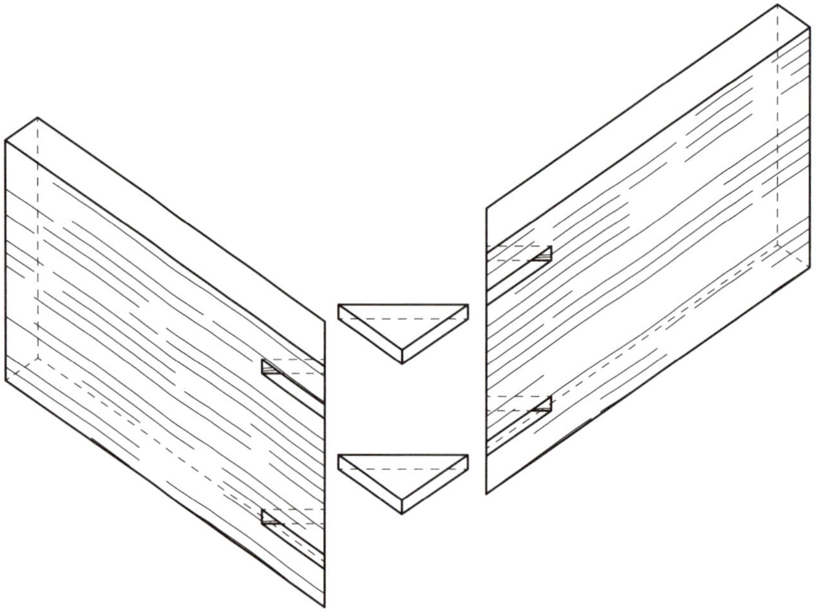

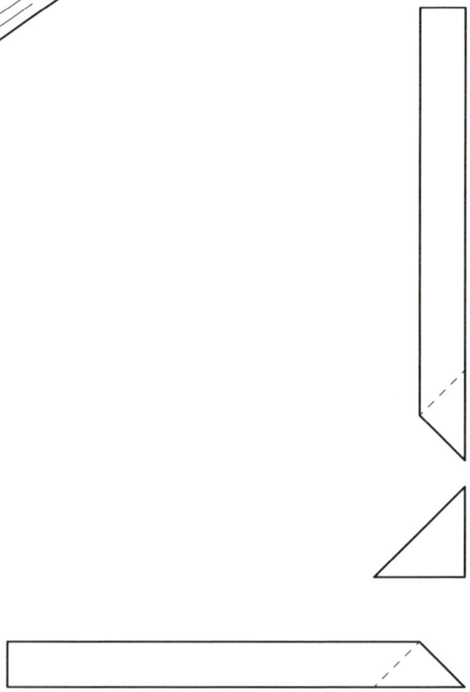

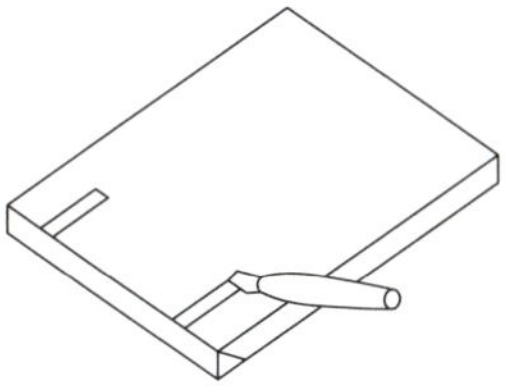

A1

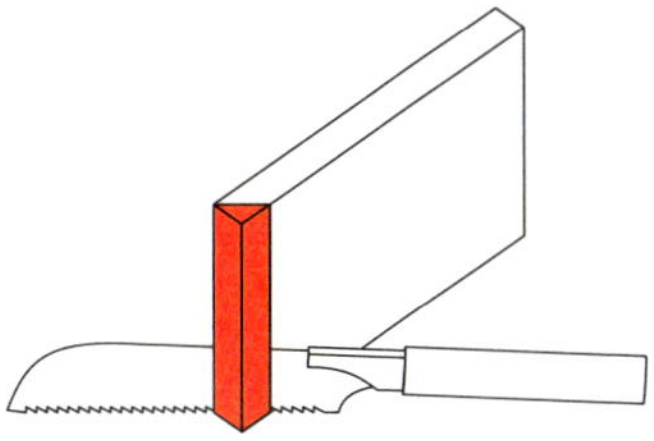

A2

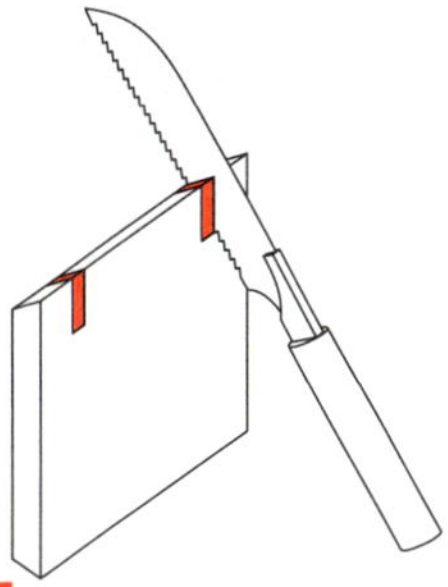

A3

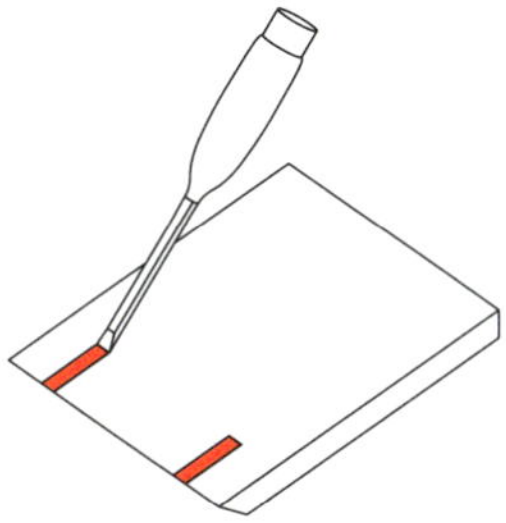

A4

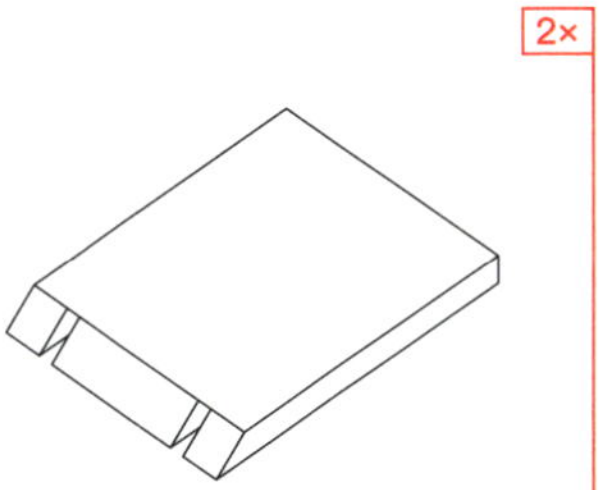

A

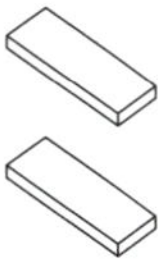

B

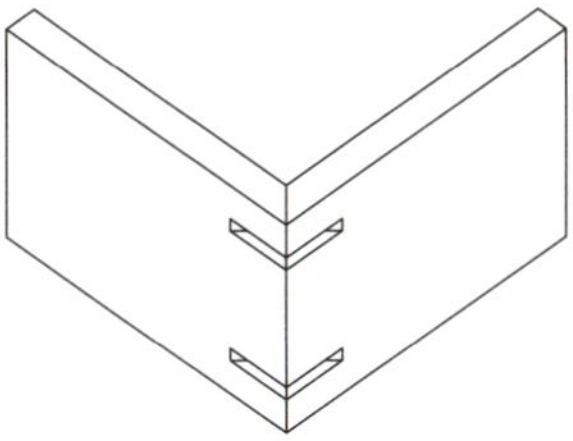

AA

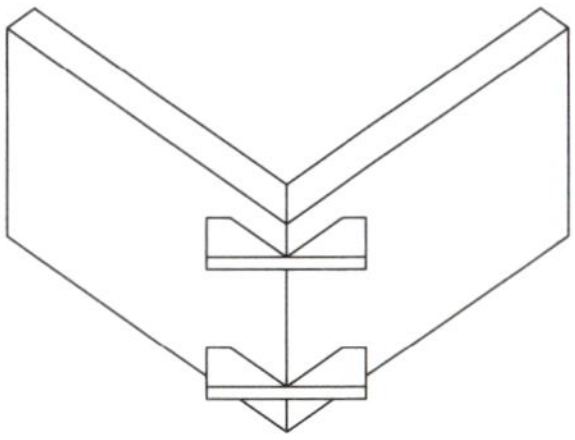

AA+B

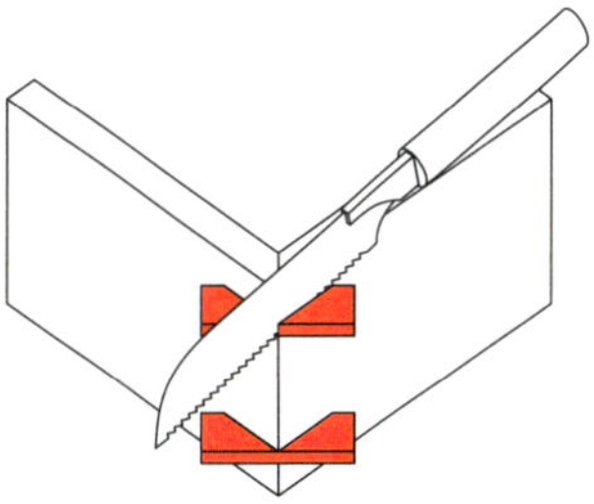

AA+B

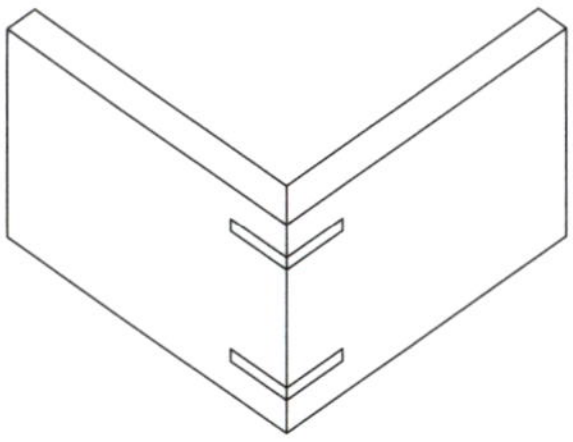

AAB

Miter joint with dovetail keys

ari-kata-kusabi-sanmai-hiki-komi-dome-tsugi
蟻形楔三枚挽込留接ぎ

Eckverbindung auf Gehrung mit eingesetzten Schwalbenschwanzfedern

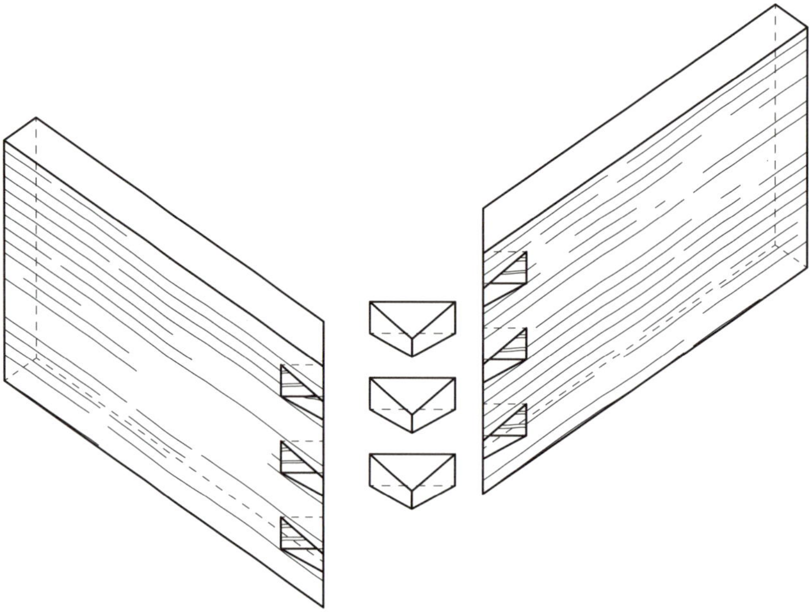

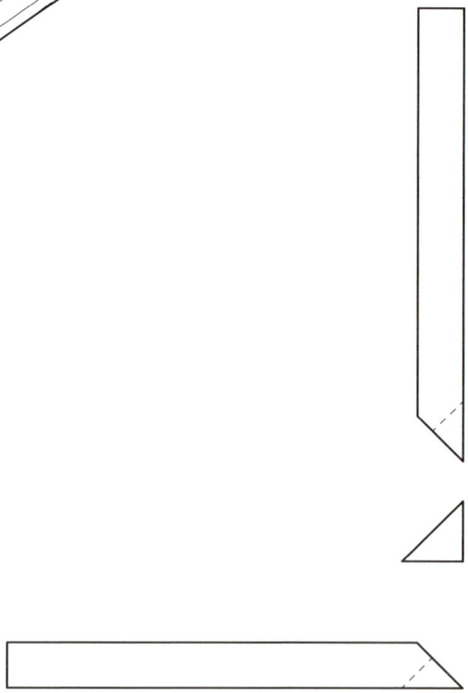

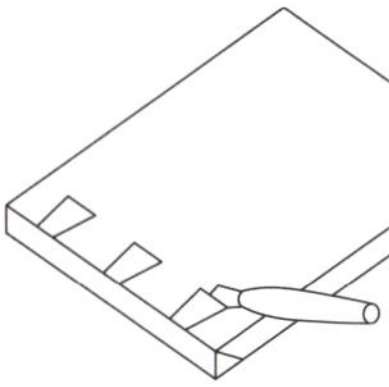

A1

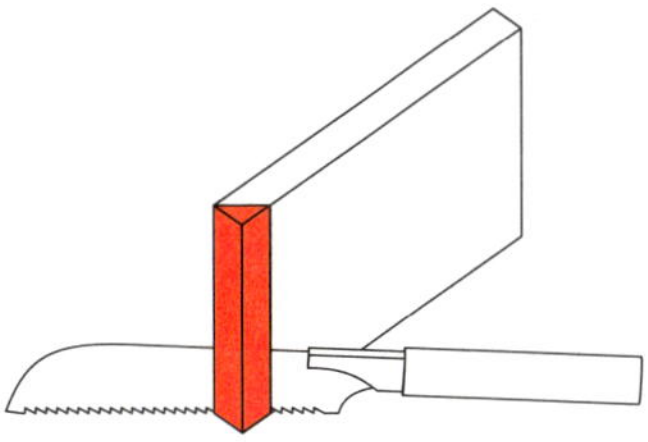

A2

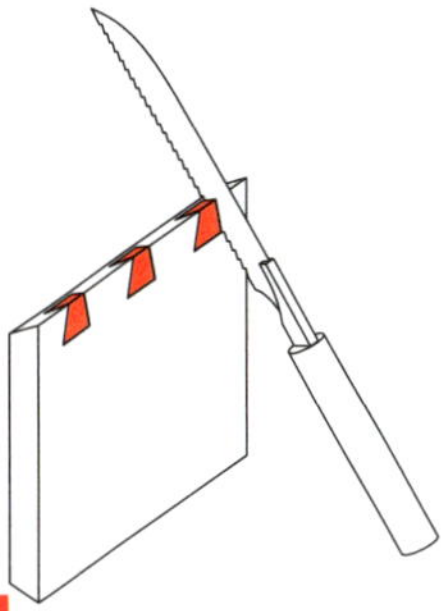

A3

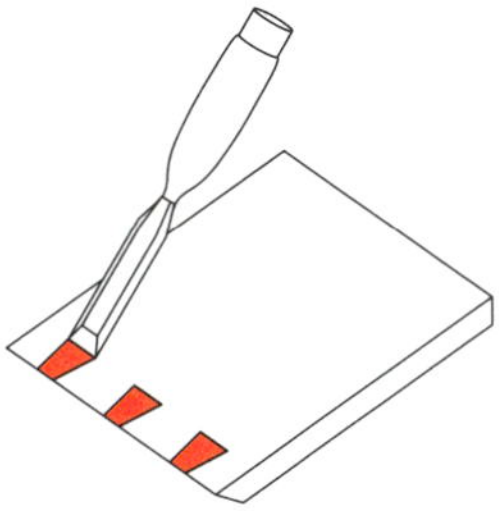

A4

2×

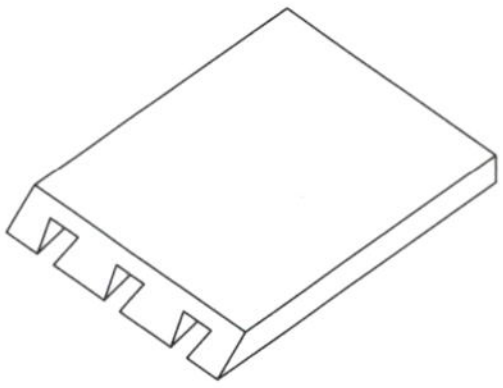

A

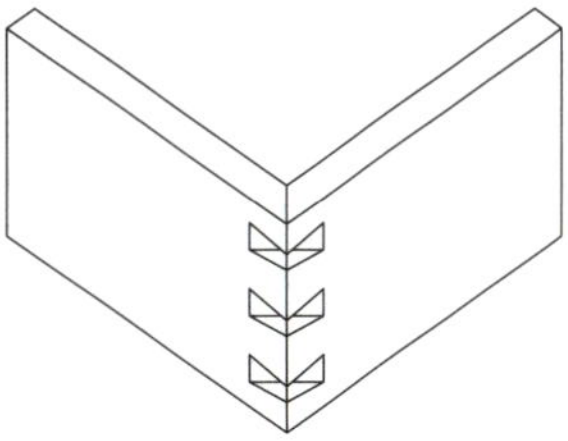

AA

B

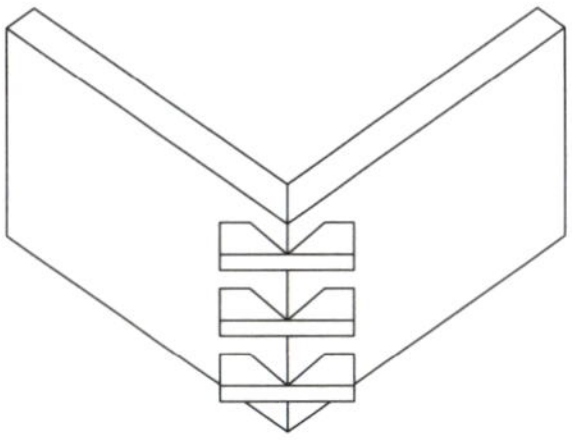

AA+B

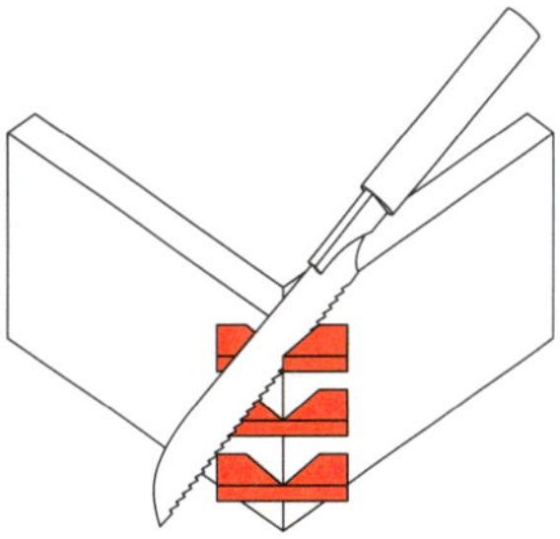

AA+B

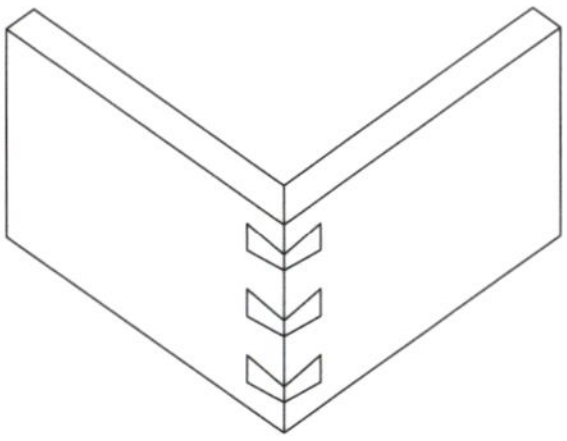

AAB

Finger joint

ishitatami-kumi-tsugi
石畳組接ぎ

Eckverbindung mit Fingerzinken

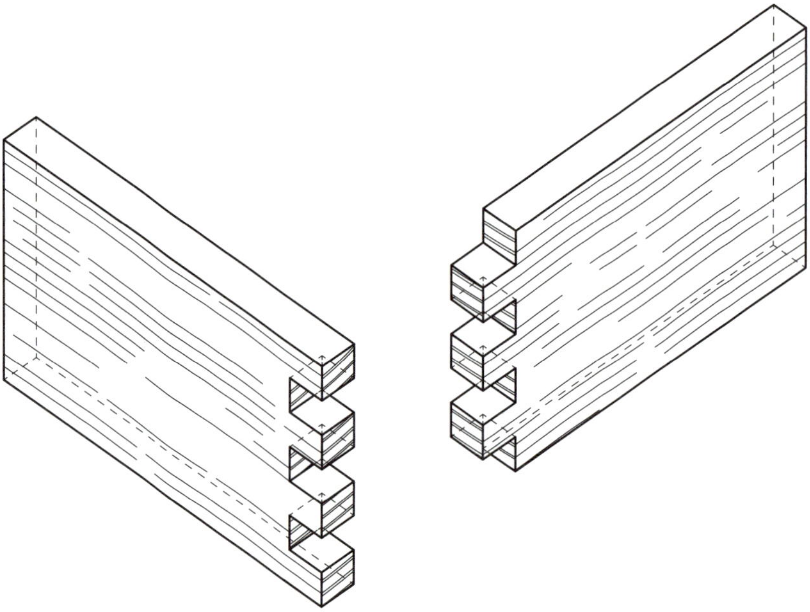

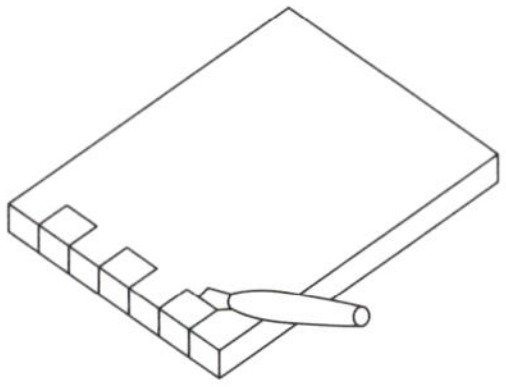

A1

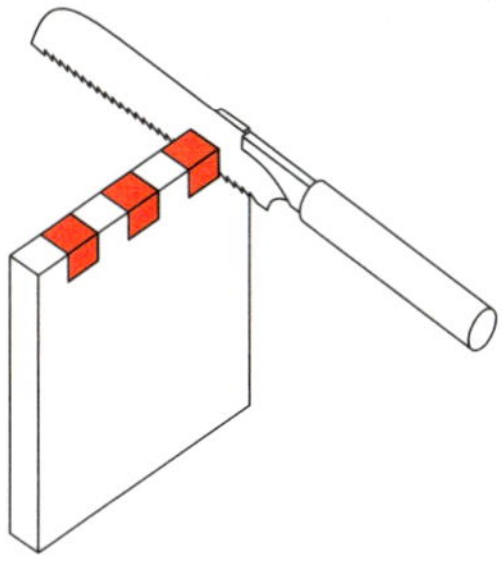

A2

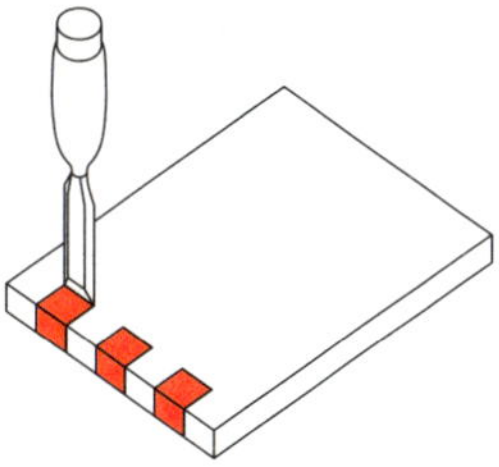

A3

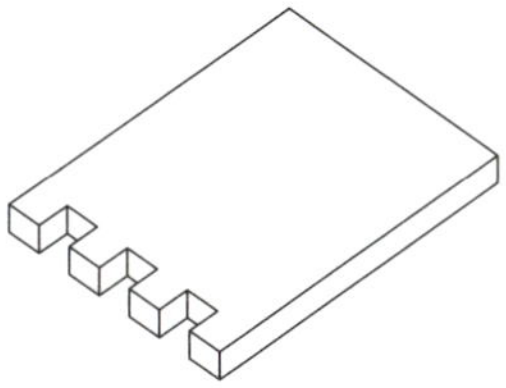

A

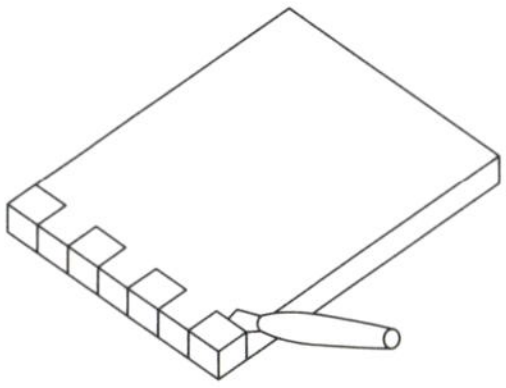

B1

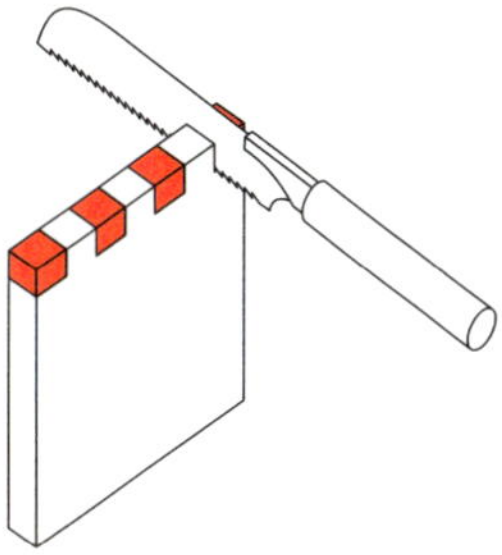

B2

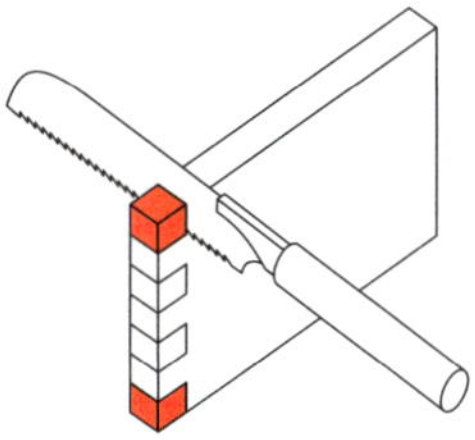

B3

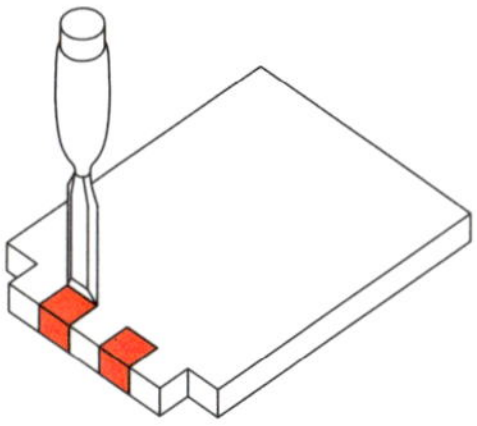

B4

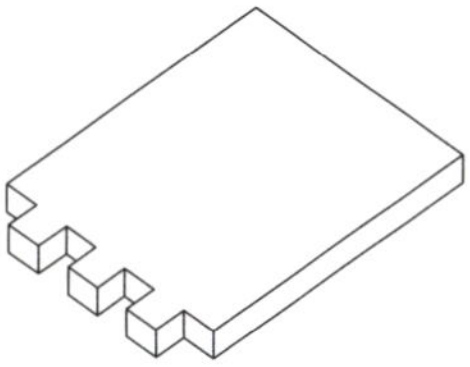

B

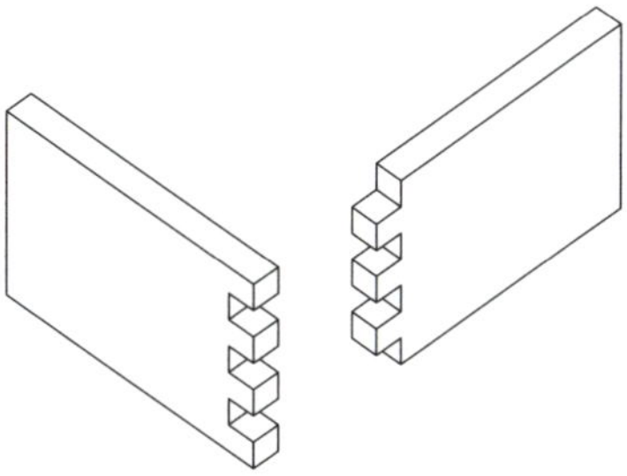

A+B

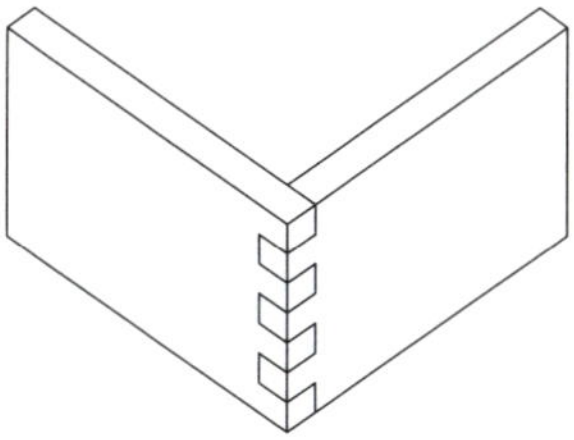

AB

Corner Joints

Through dovetail joint

tenbin-gyaku-kumi-tsugi
天秤逆組接ぎ

Eckverbindung mit
Schwalbenschwanzzinken

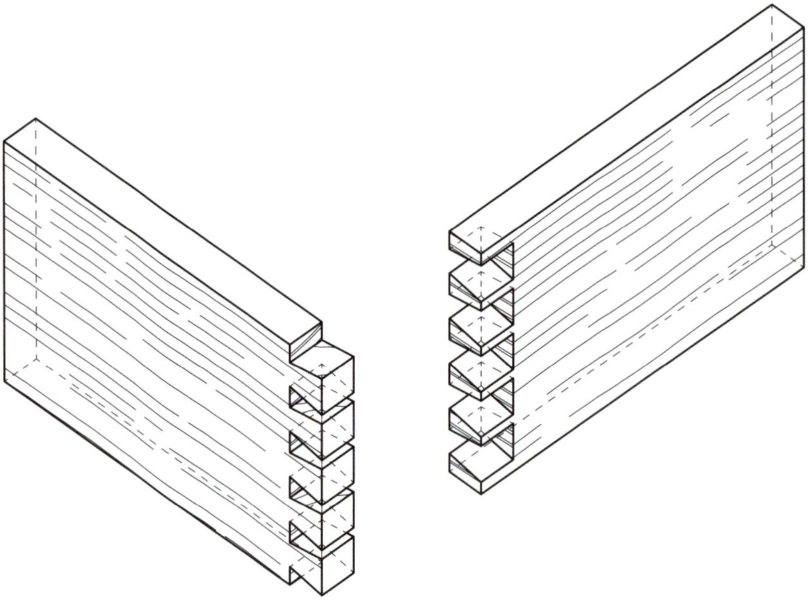

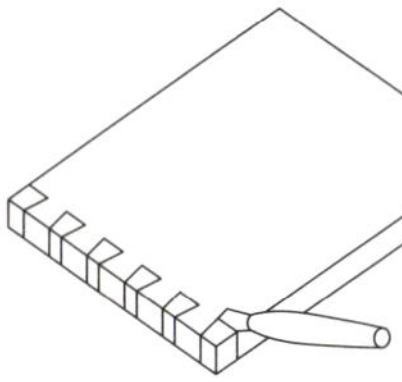

A1

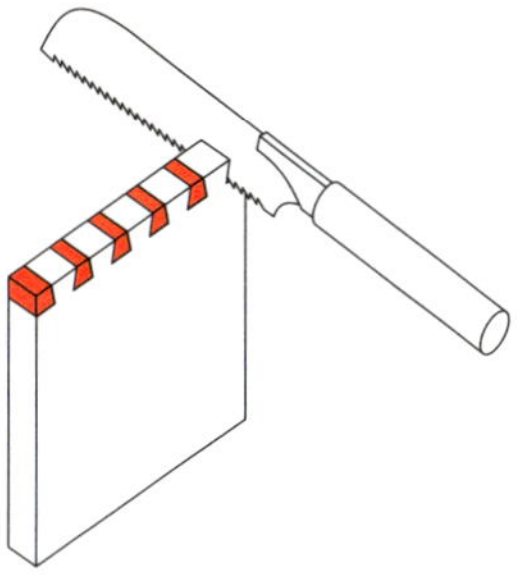

A2

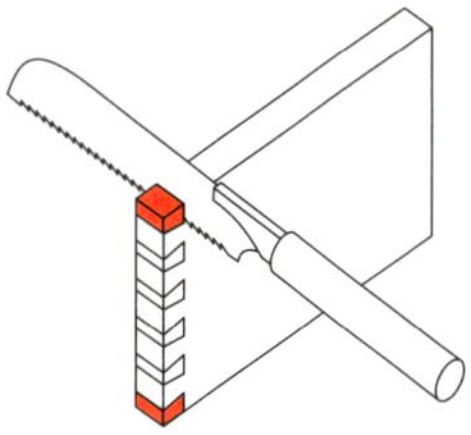

A3

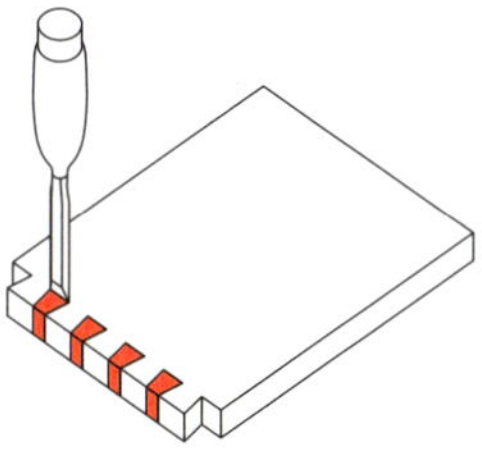

A4

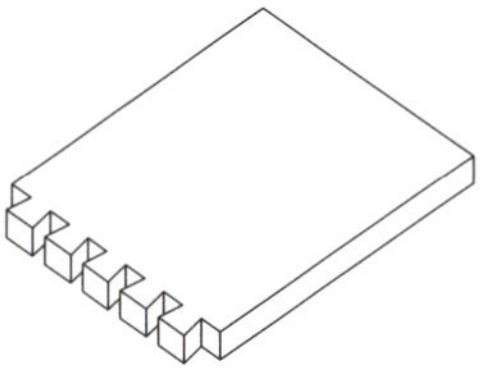

A

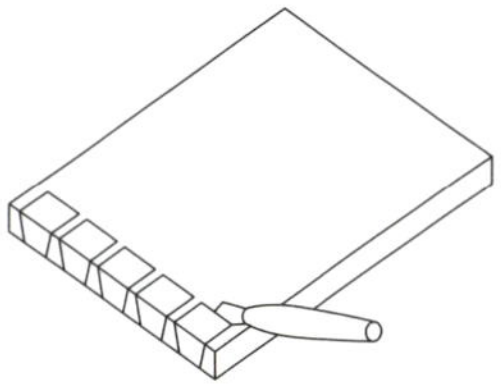

B1

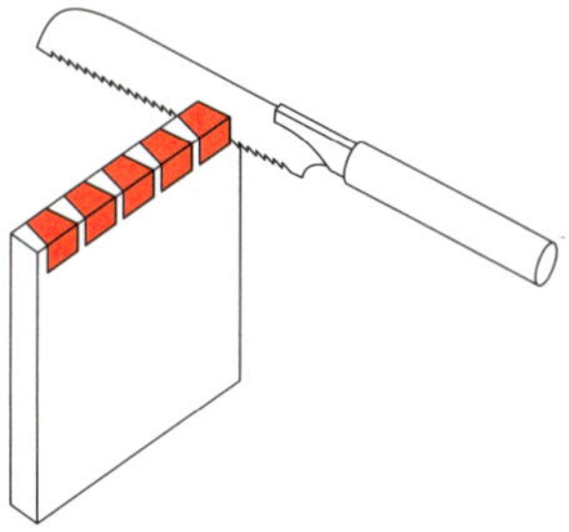

B2

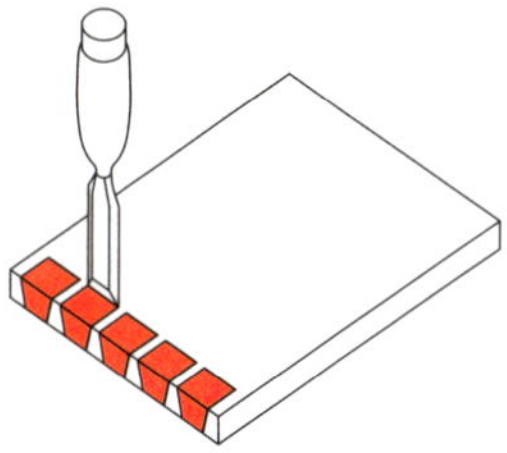

B3

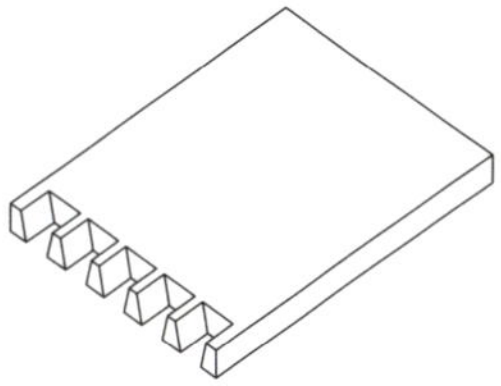

B

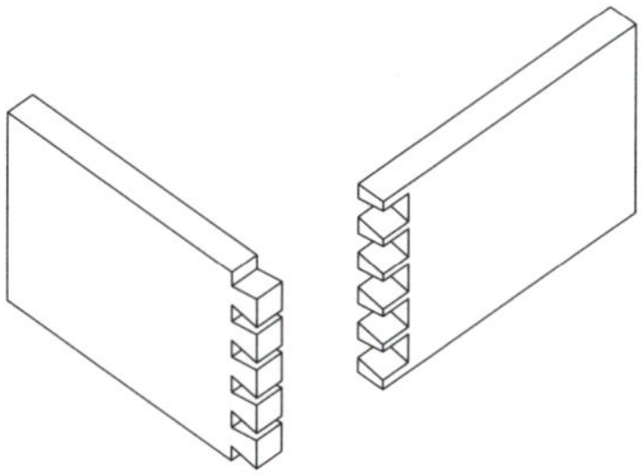

A+B

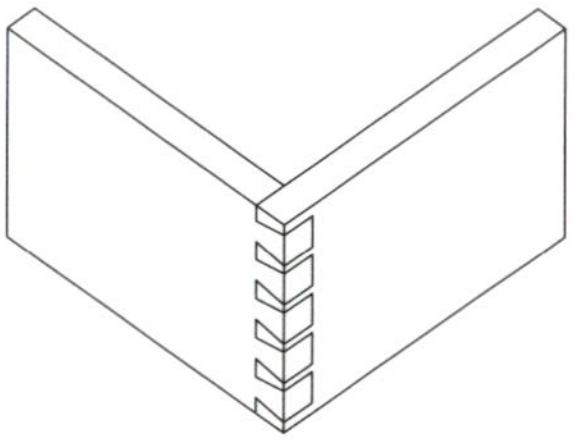

AB

Corner Joints

Half-blind dovetail joint

kakushi-ari-kumi-tsugi
隠蟻組接ぎ

Eckverbindung mit halbverdeckten Schwalbenschwanzzinken

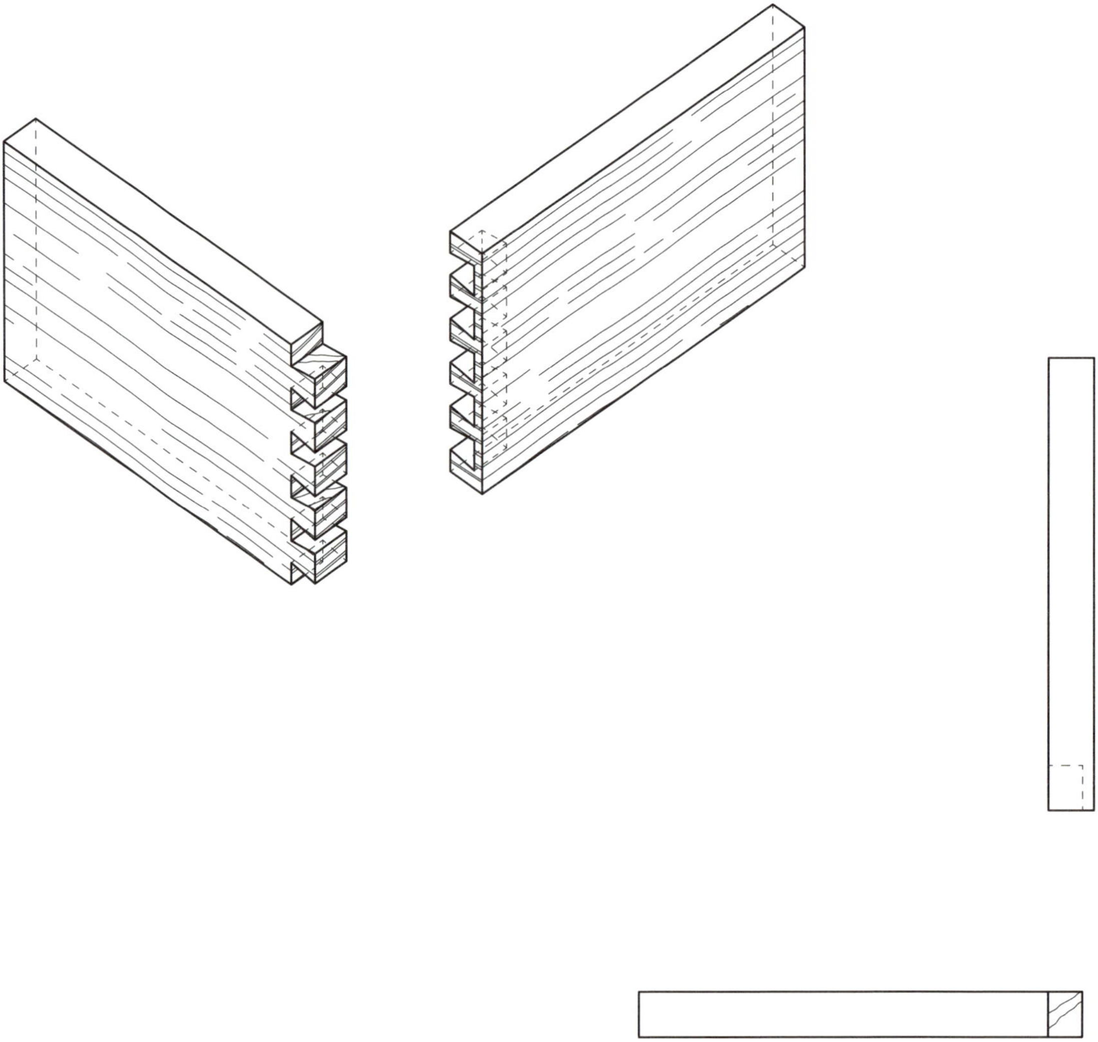

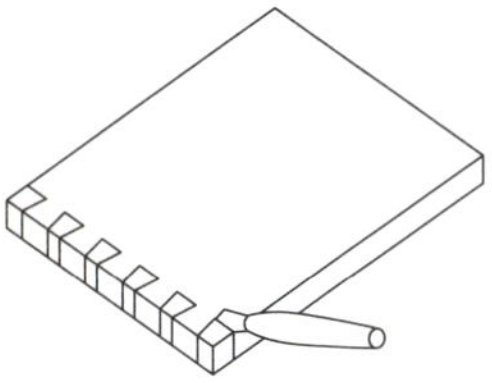

A1

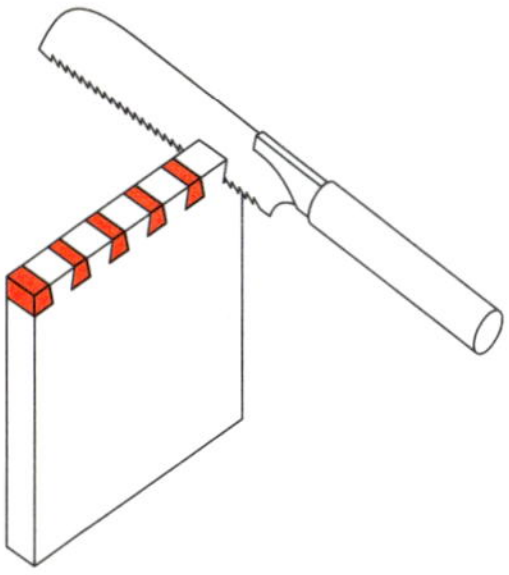

A2

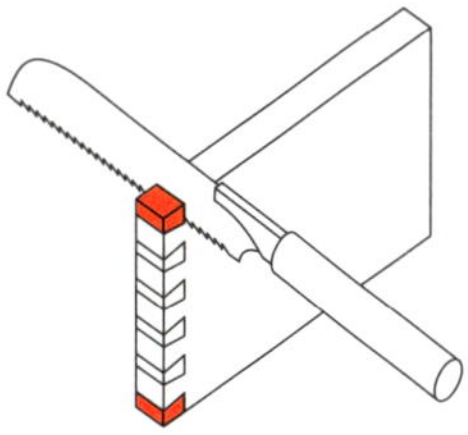

A3

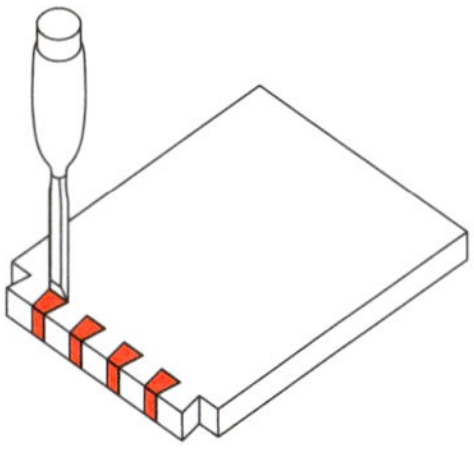

A4

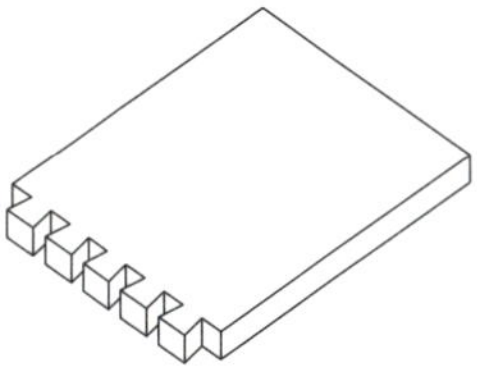

A

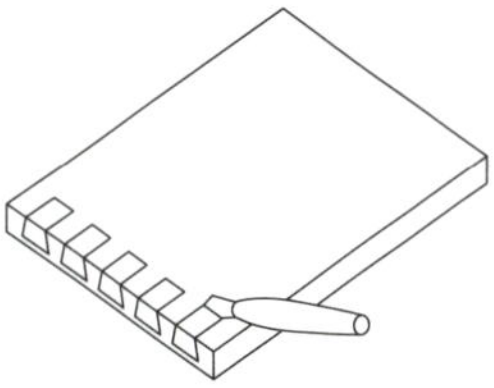

B1

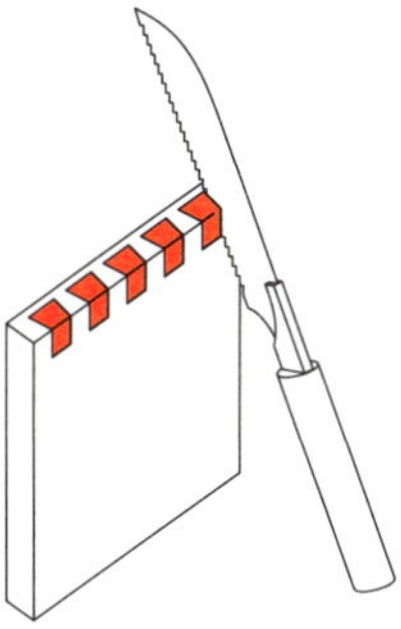

B2

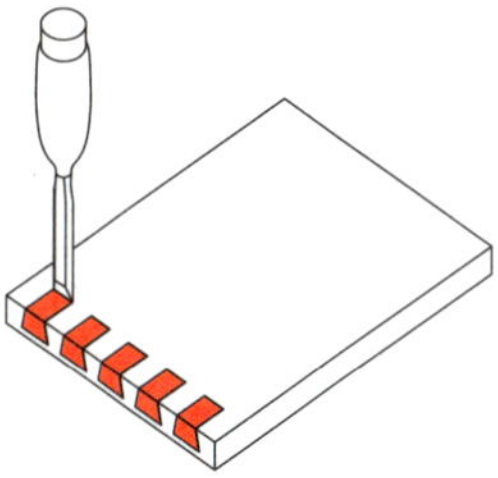

B3

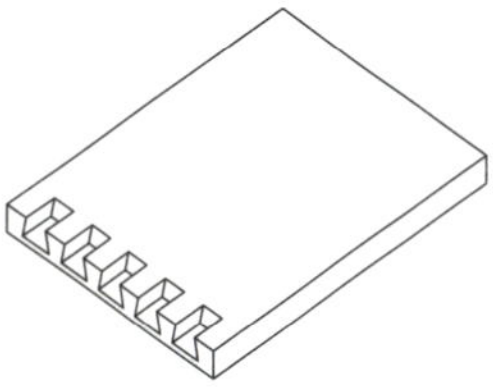

B

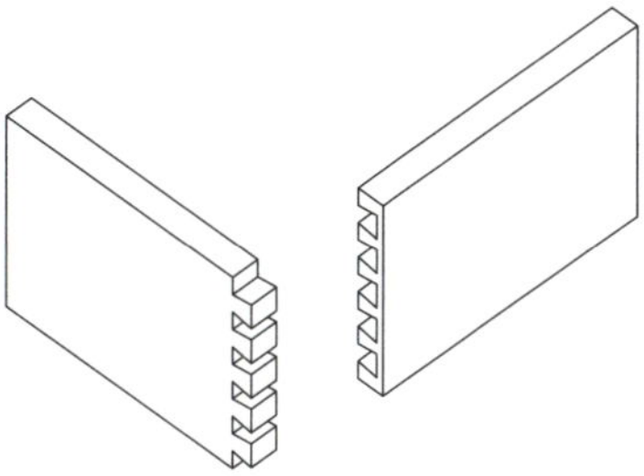

A+B

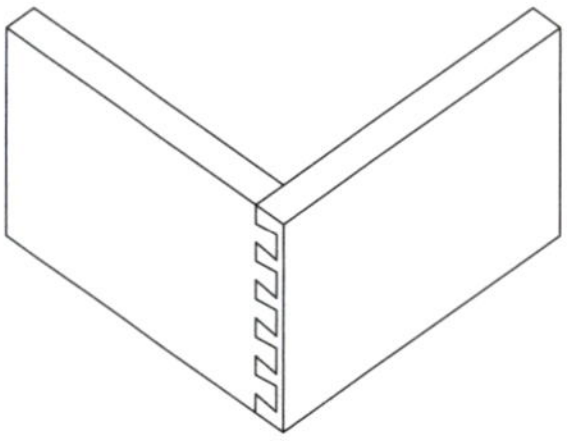

AB

Corner Joints

Blind dovetail joint

kakushi-ari-kata-kumi-tsugi
隠蟻形留組接ぎ

Eckverbindung mit verdeckten Schwalbenschwanzzinken

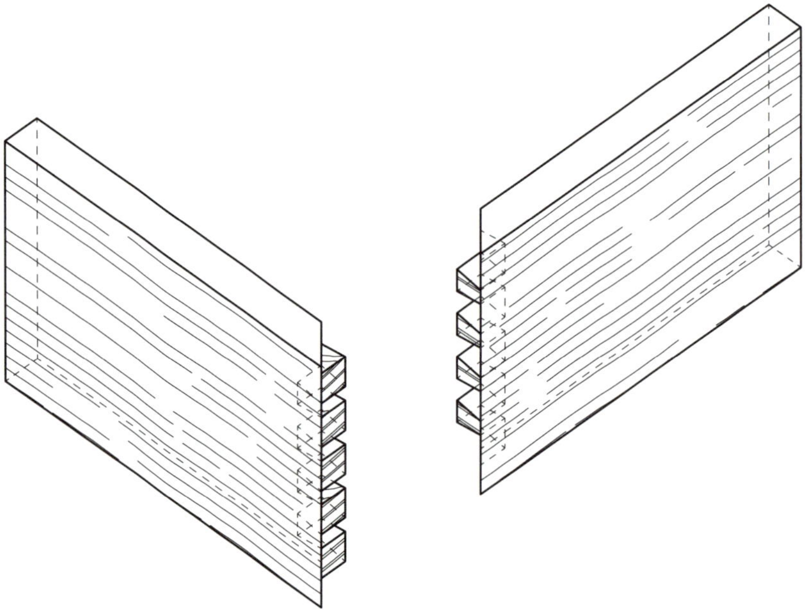

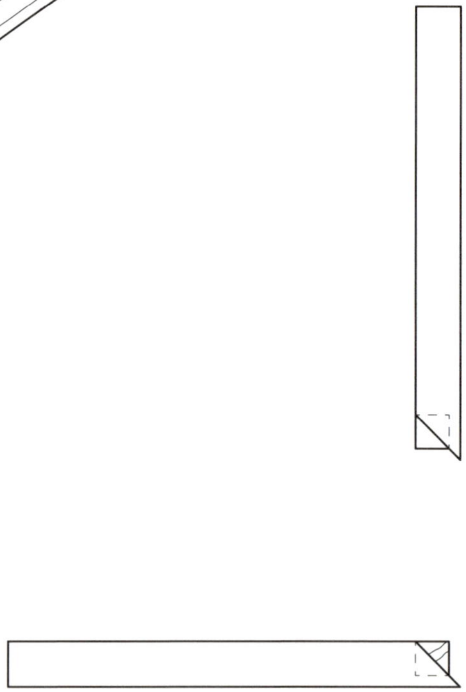

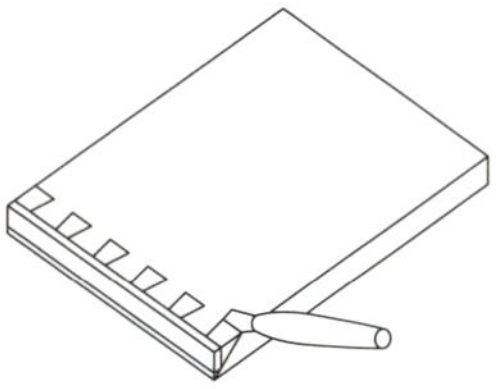

A1

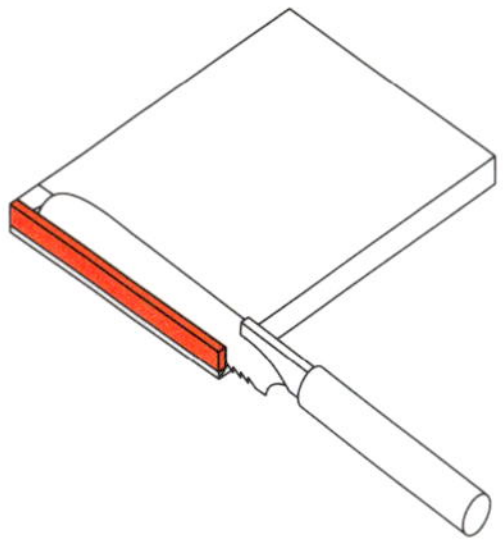

A2

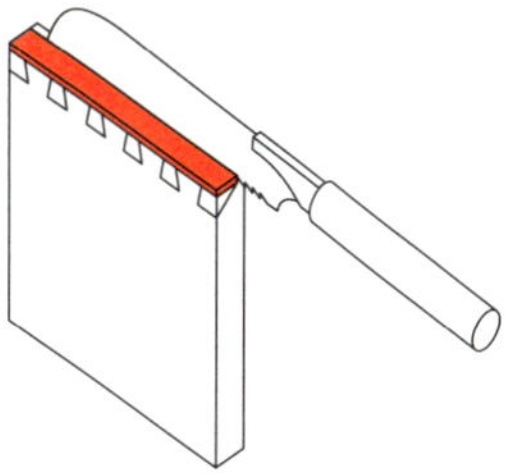

A3

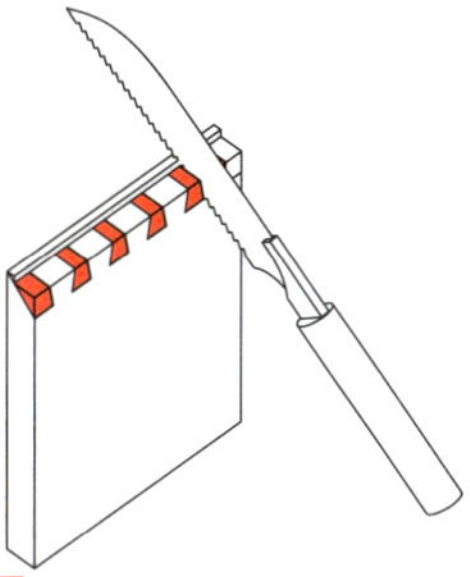

A4

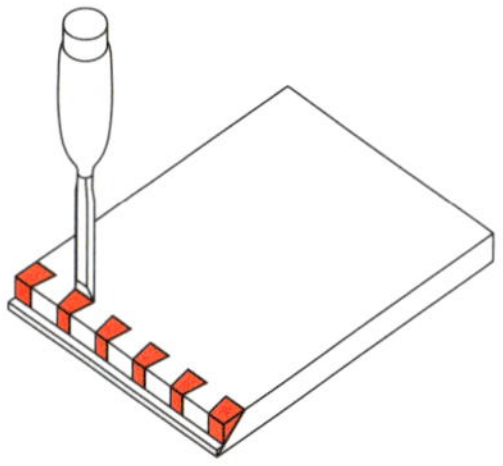

A5

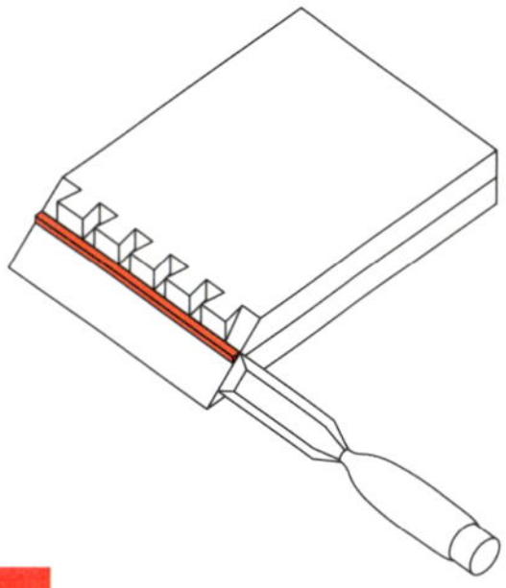

A6

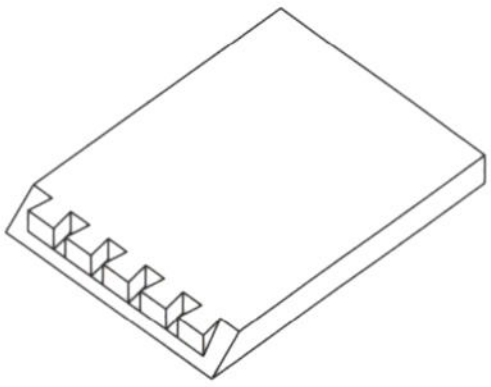

A

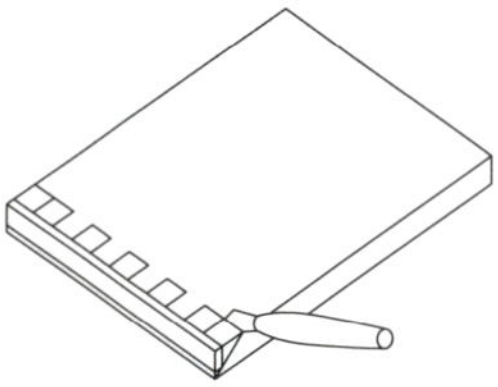

B1

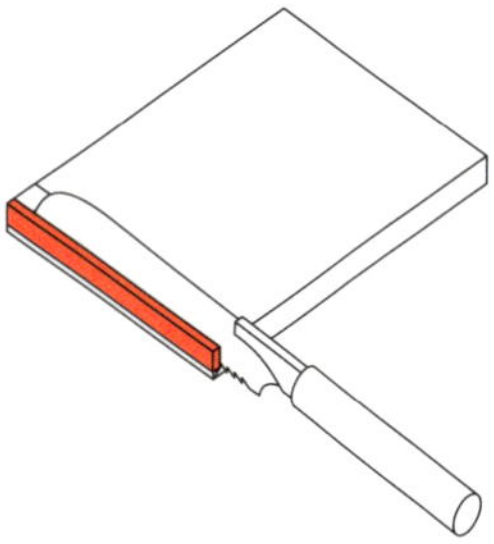

B2

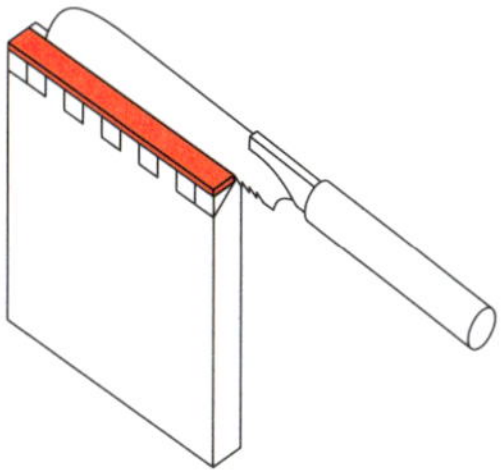

B3

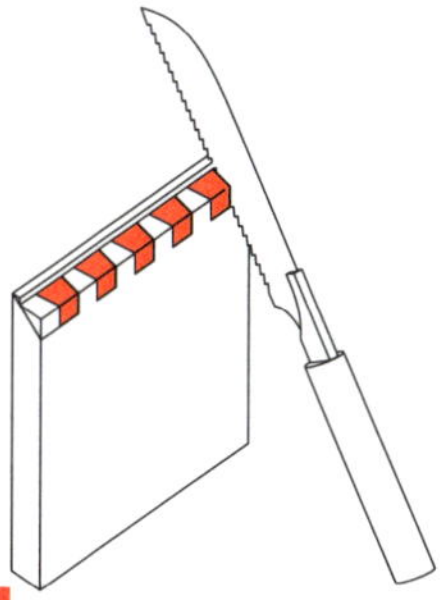

B4

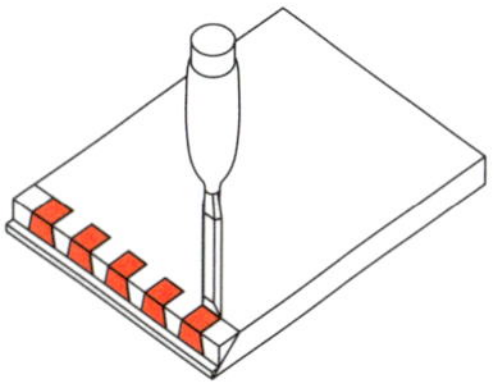

B5

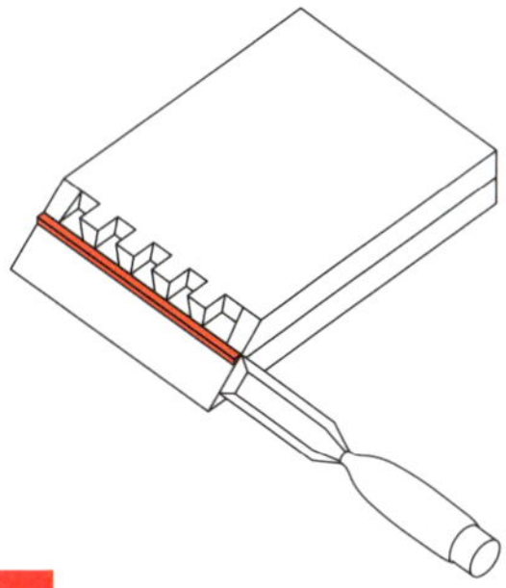

B6

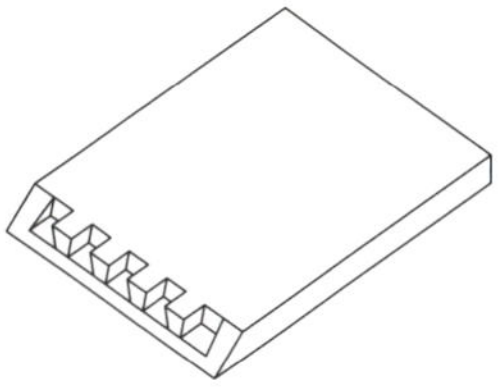

B

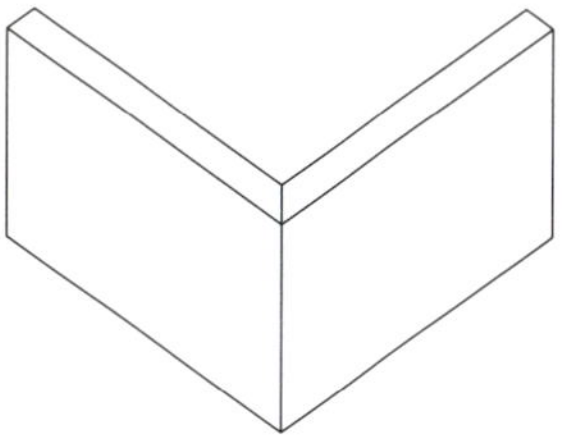

AB

Corner Joints

Cogged dovetail joint

oyako-ari-kumi-tsugi
親子蟻組接ぎ

Eckverbindung mit ganzen und abgesetzten Schwalbenschwanzzinken

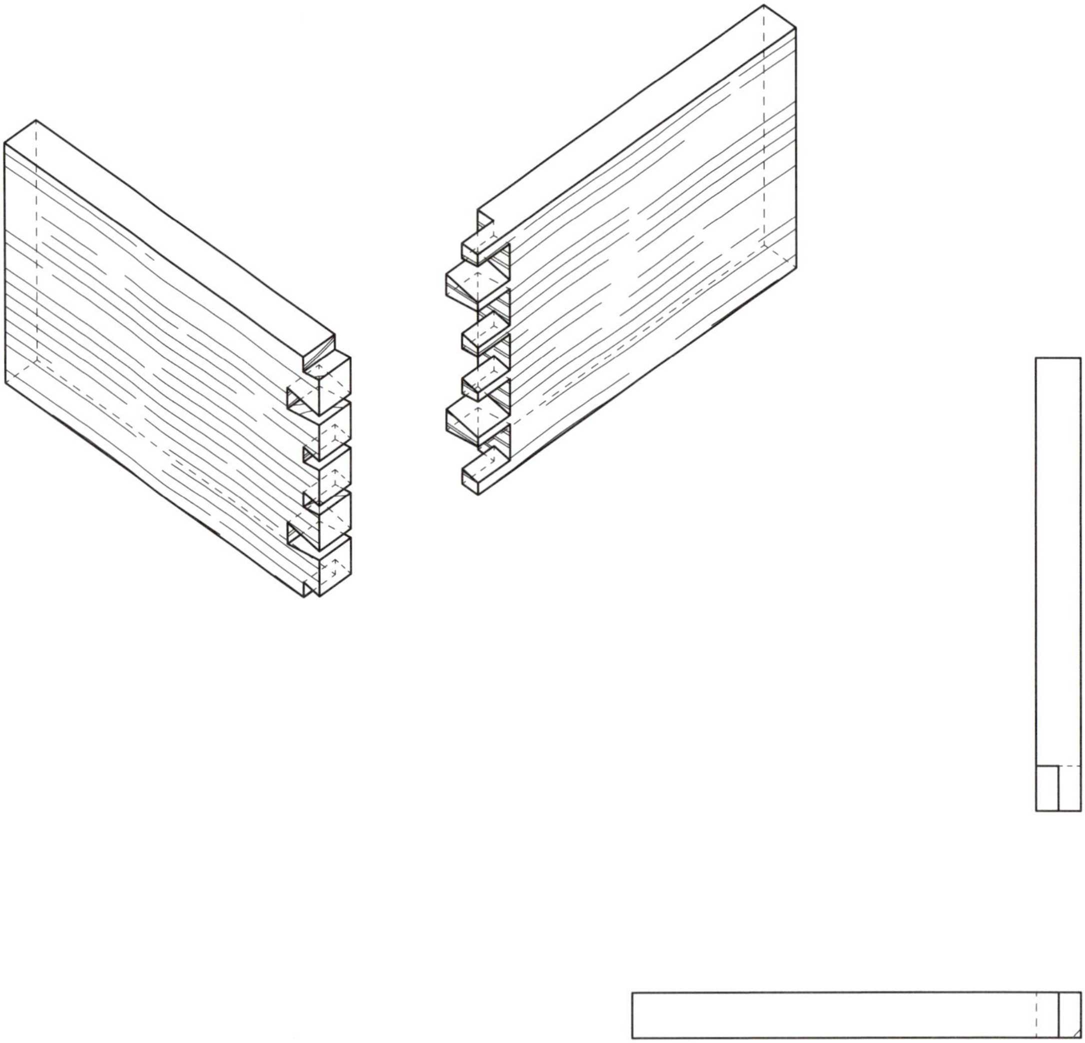

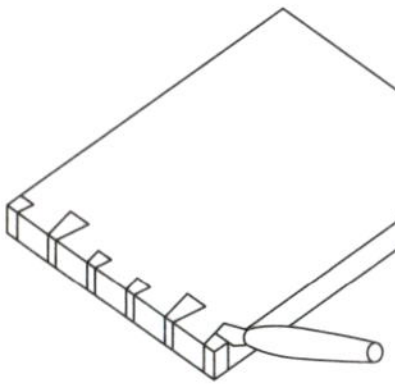

A1

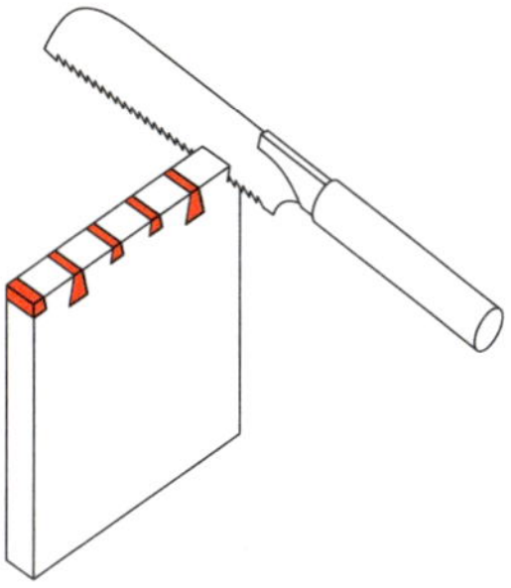

A2

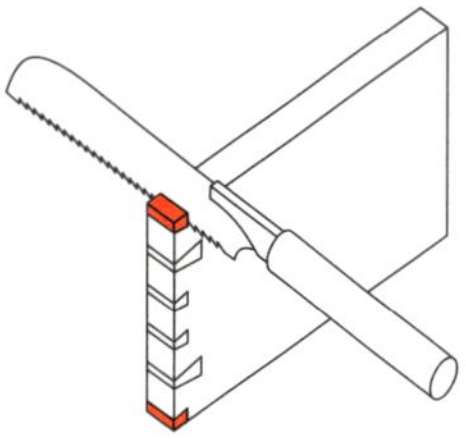

A3

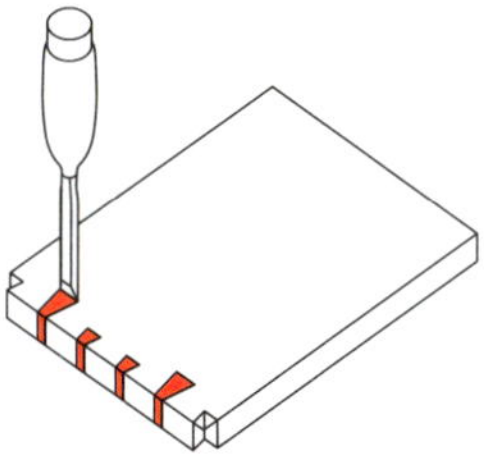

A4

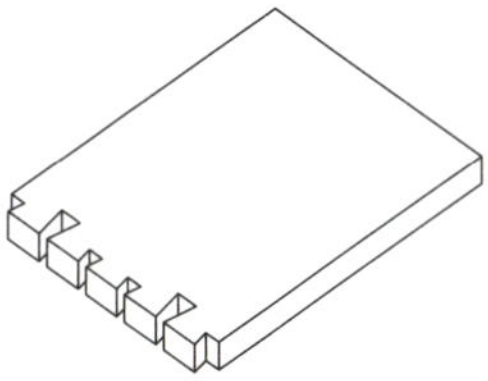

A

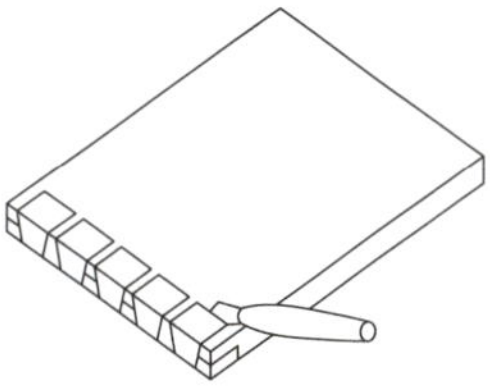

B1

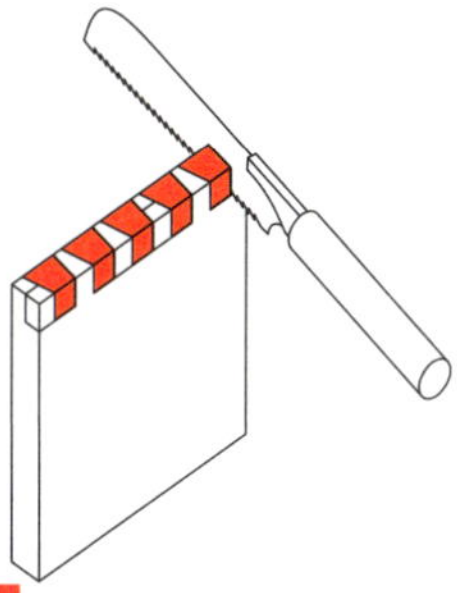

B2

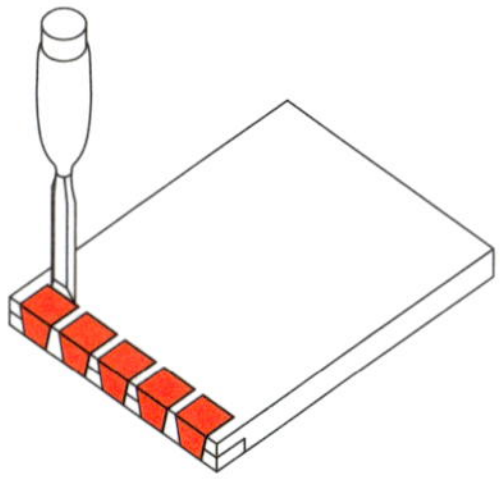

B3

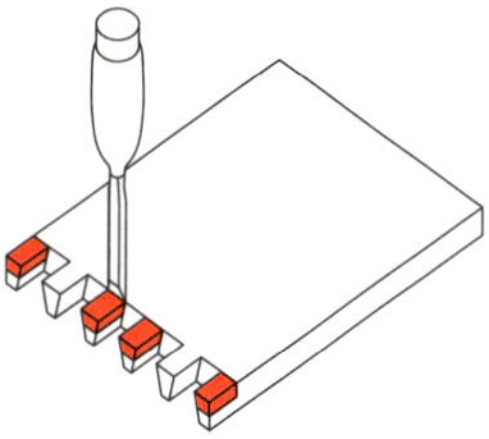

B4

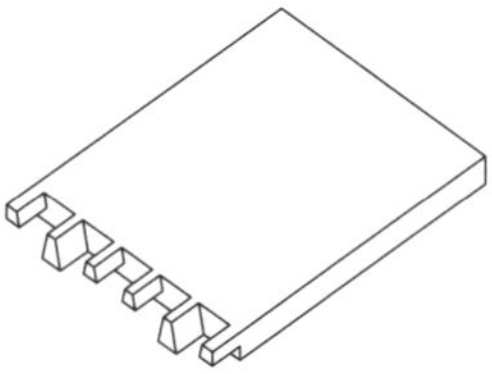

B

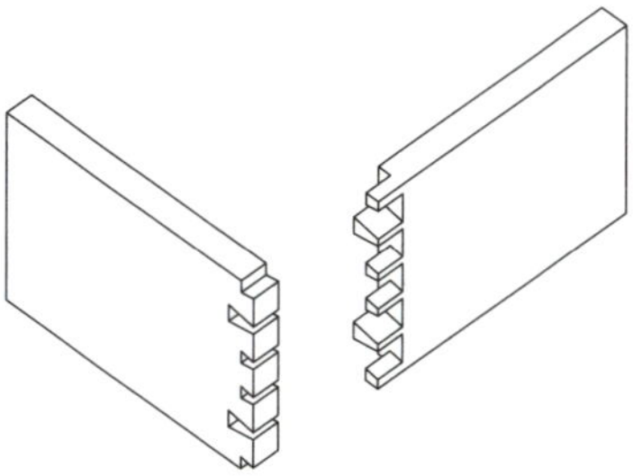

A+B

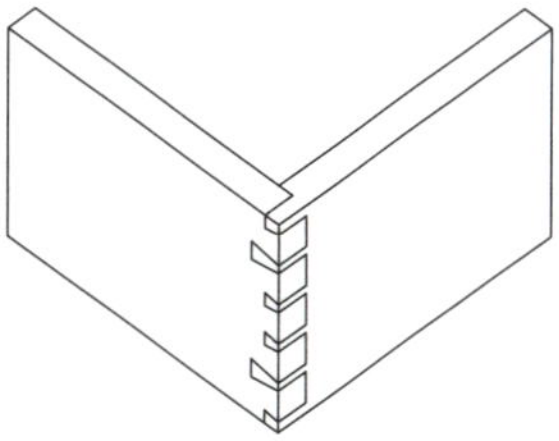

AB

Corner Joints

Angled finger joint

hineri-tsugi-kumi
捻接組み

Eckverbindung mit schrägen Fingerzinken

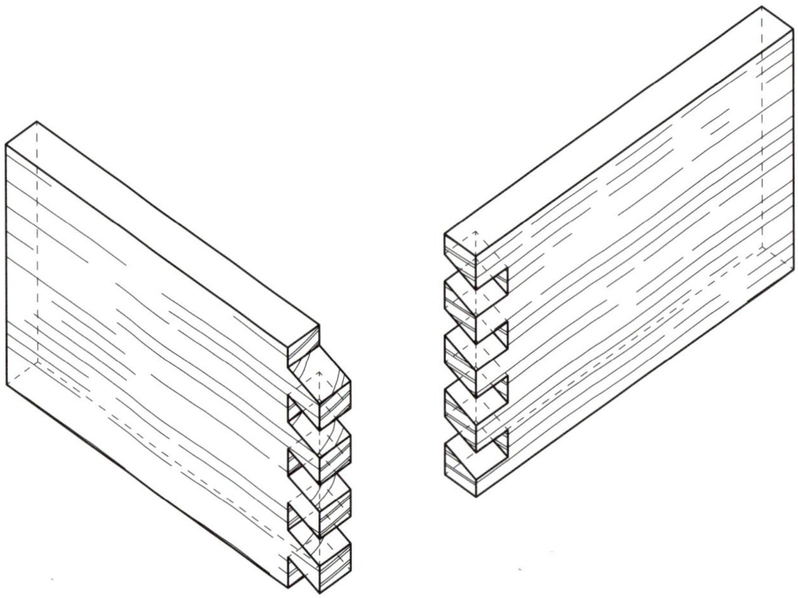

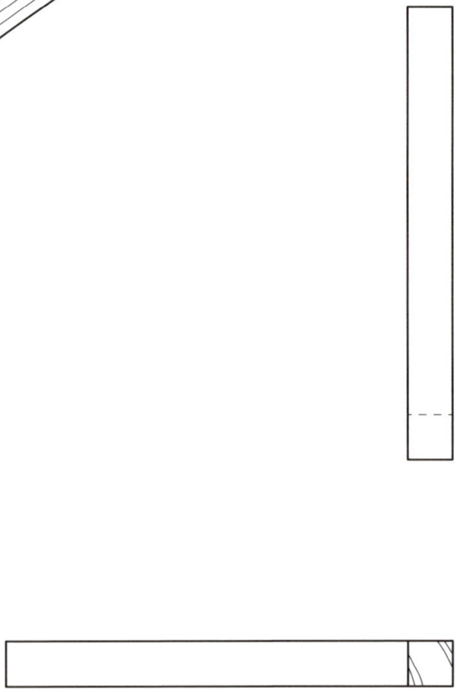

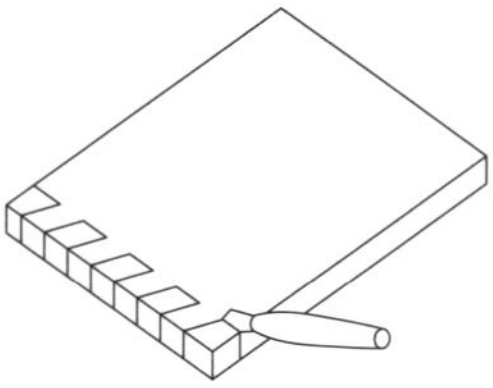
A1

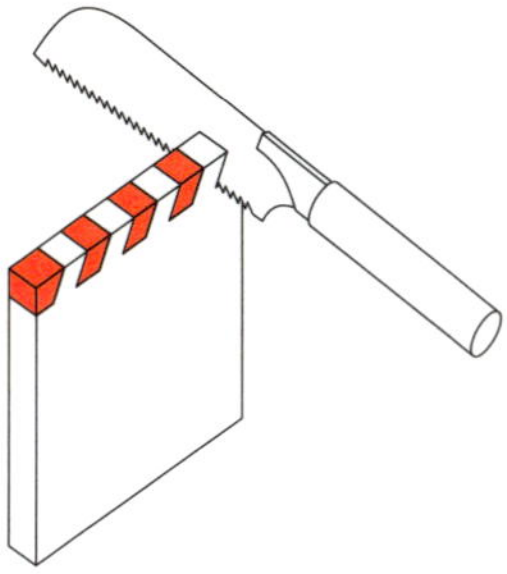
A2

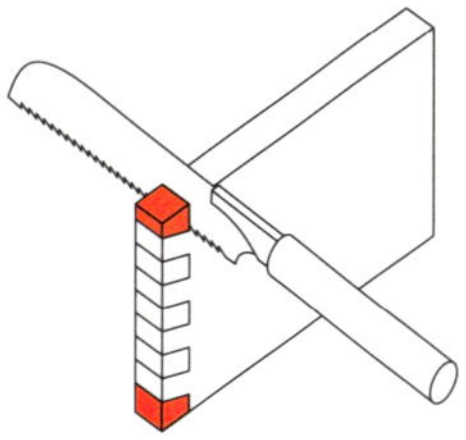
A3

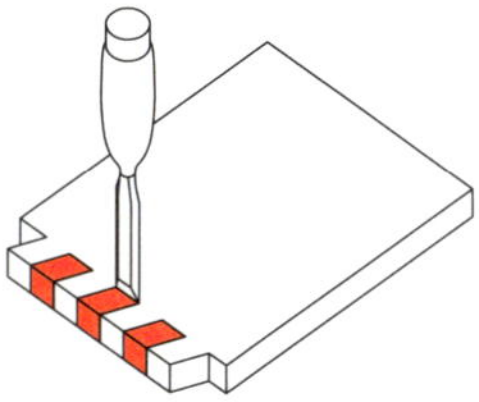
A4

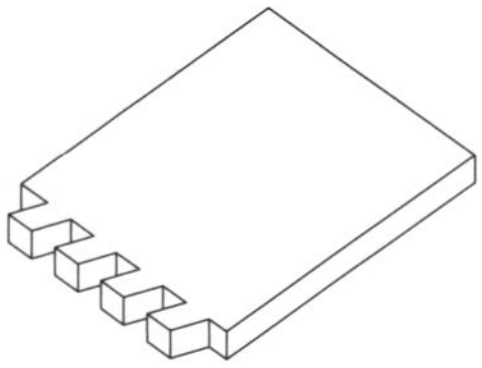
A

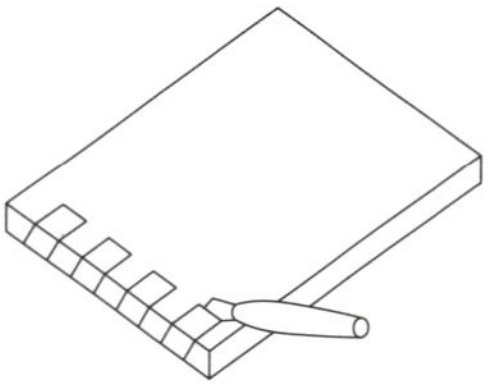
B1

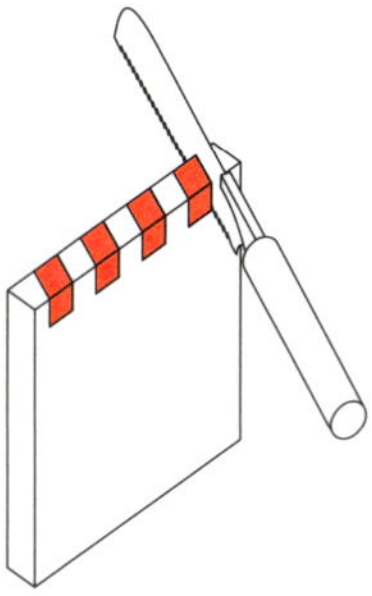
B2

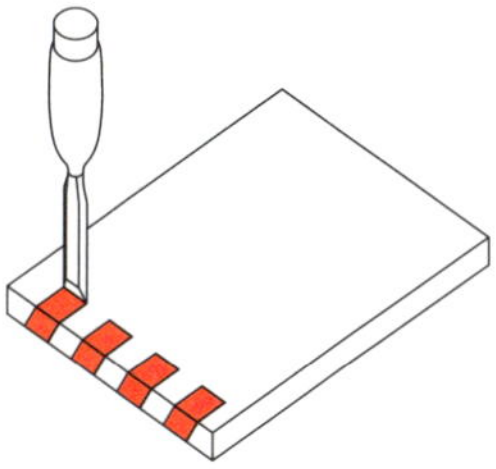
B3

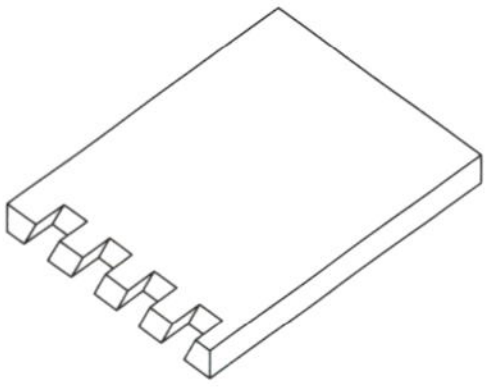
B

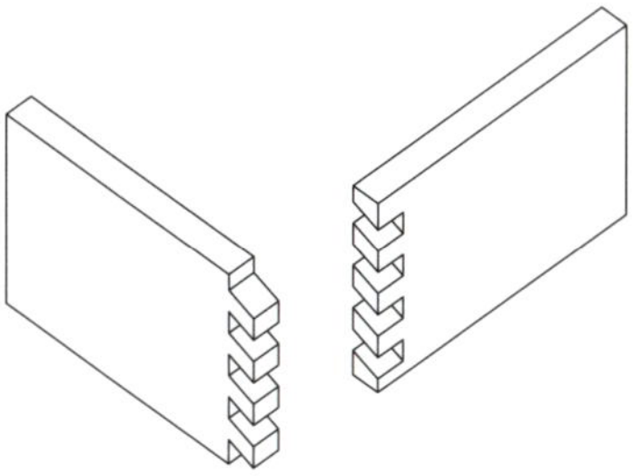
A+B

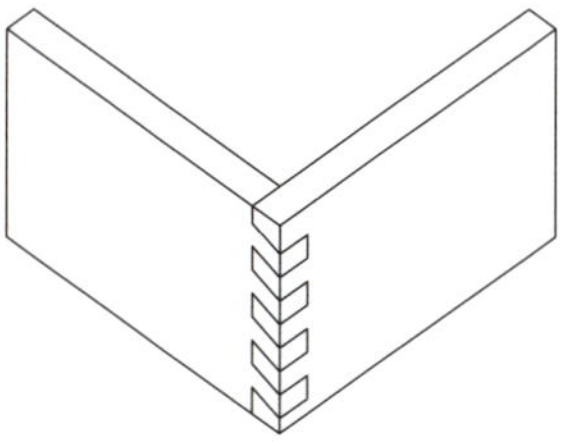
AB

Corner Joints

Radial dovetail joint

mizu-gumi (yojire-gumi)
水組み(捩組)

Eckverbindung mit Teufelszinken

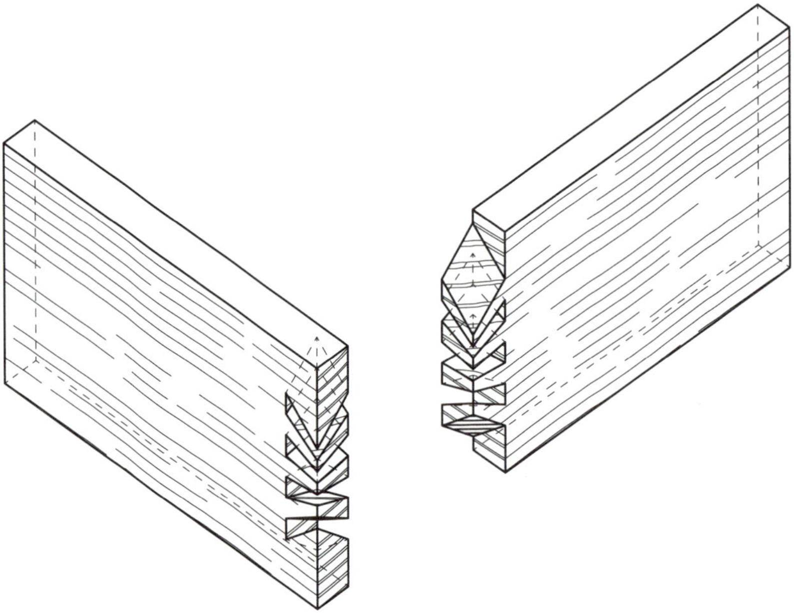

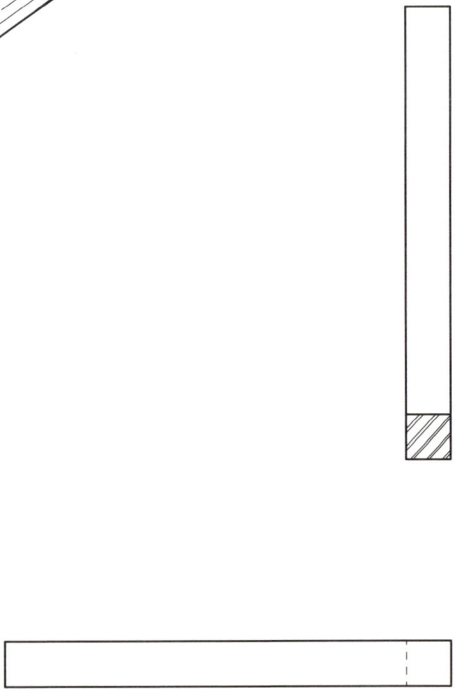

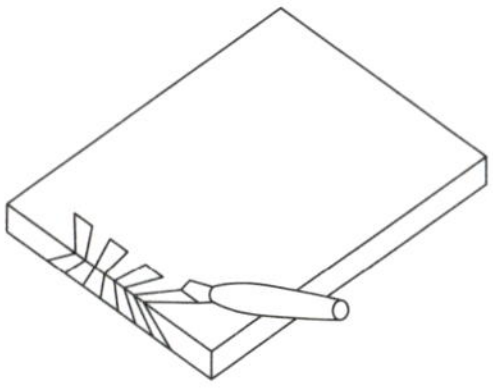

A1

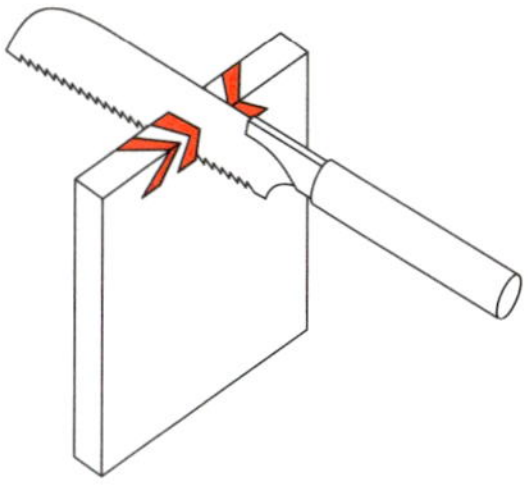

A2

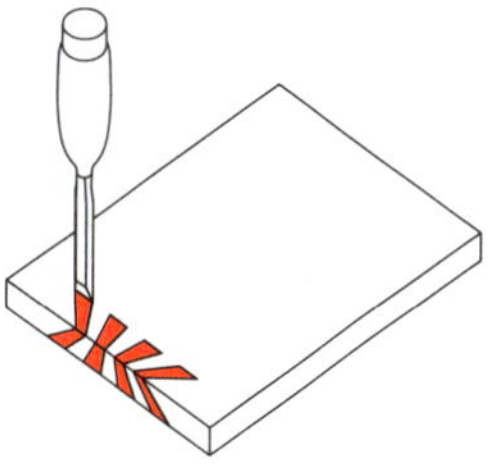

A3

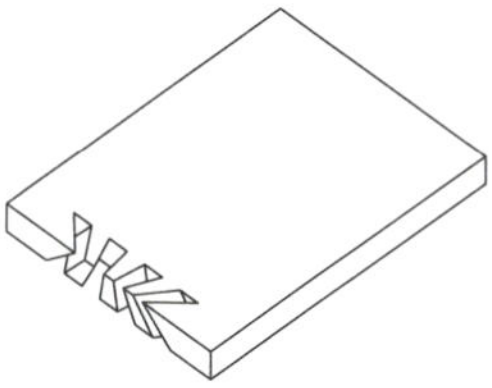

A

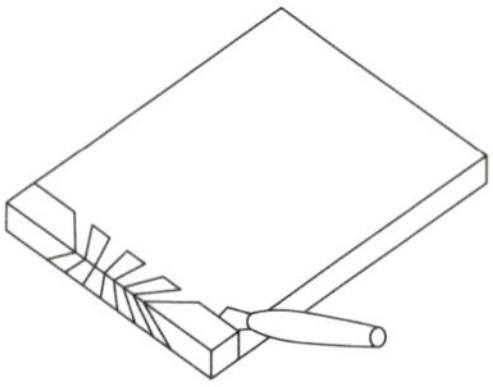

B1

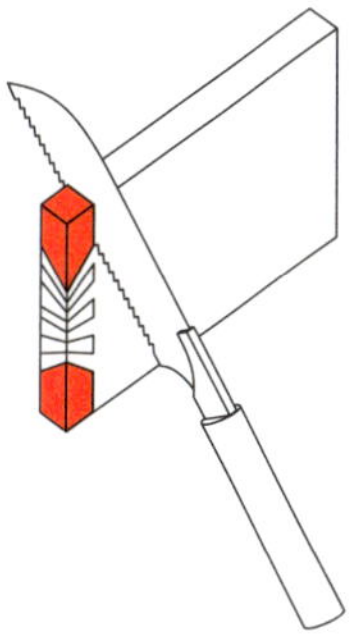

B2

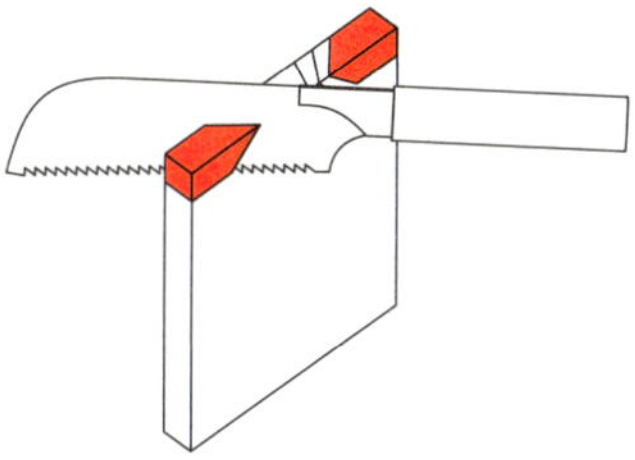

B3

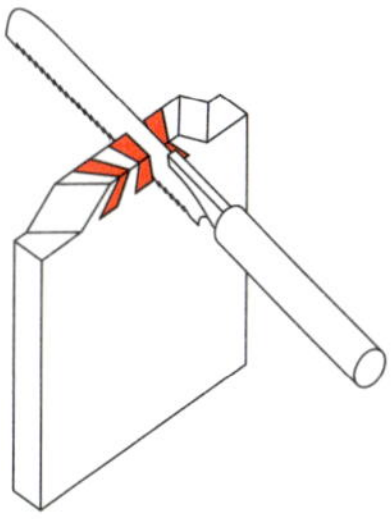

B4

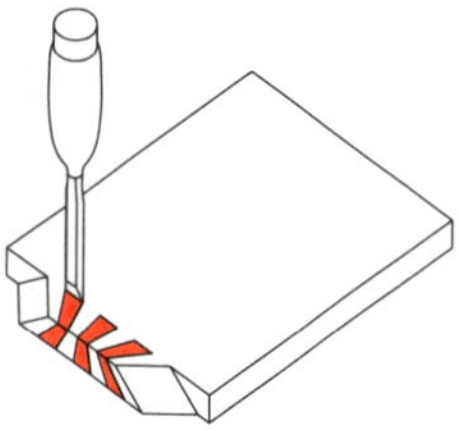

B5

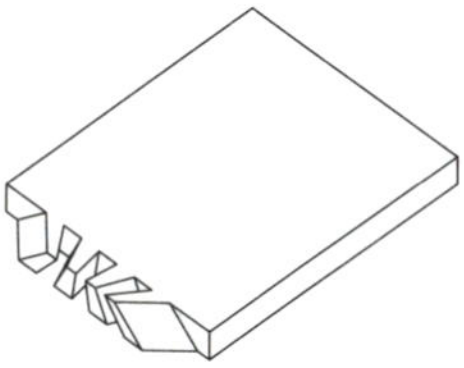

B

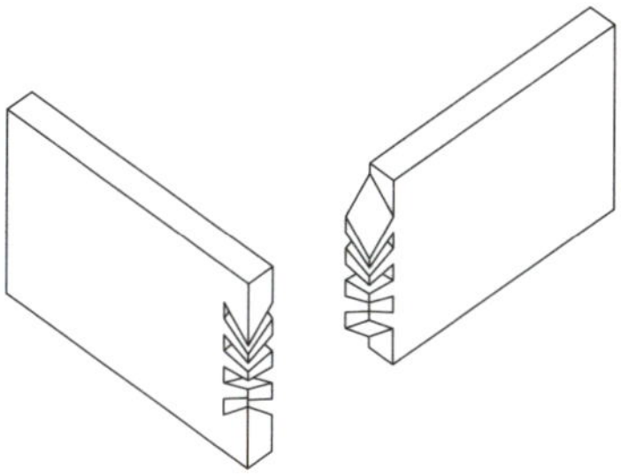

A+B

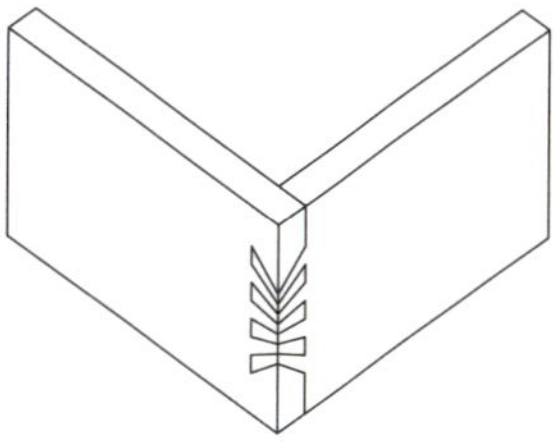

AB

Housing joint

oire-tsugi
追入接ぎ

Nutverbindung

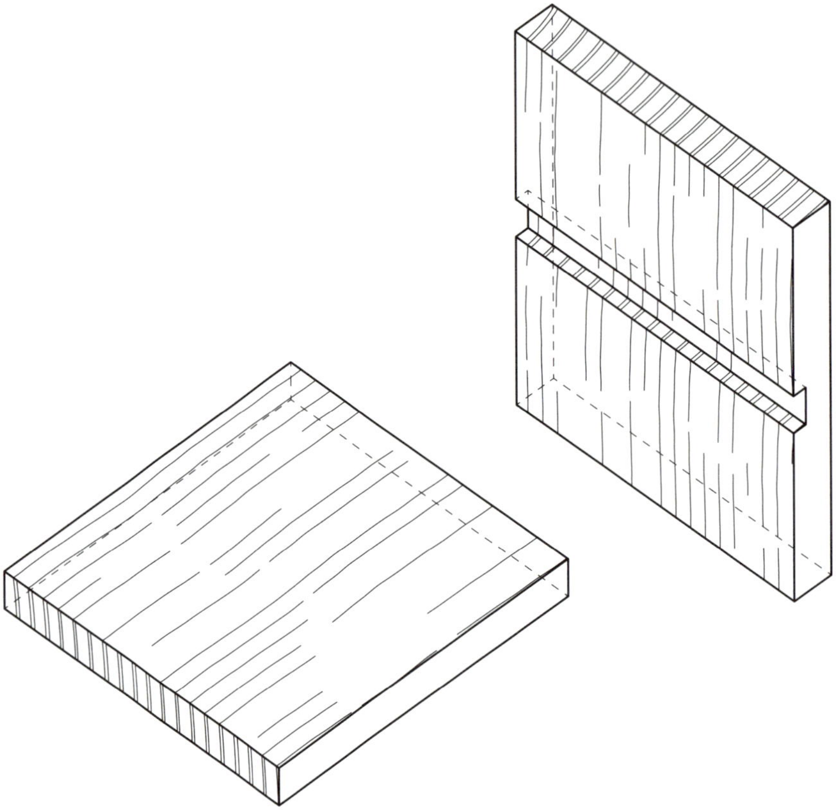

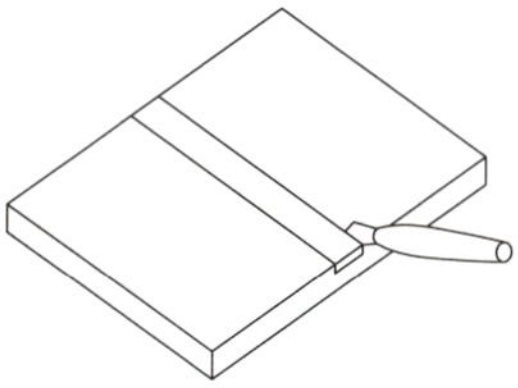

A1

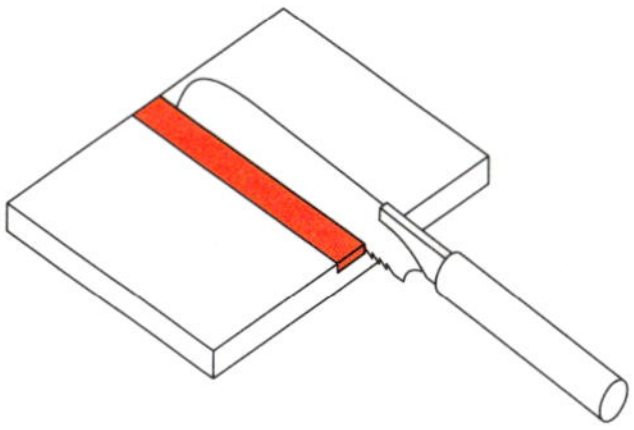

A2

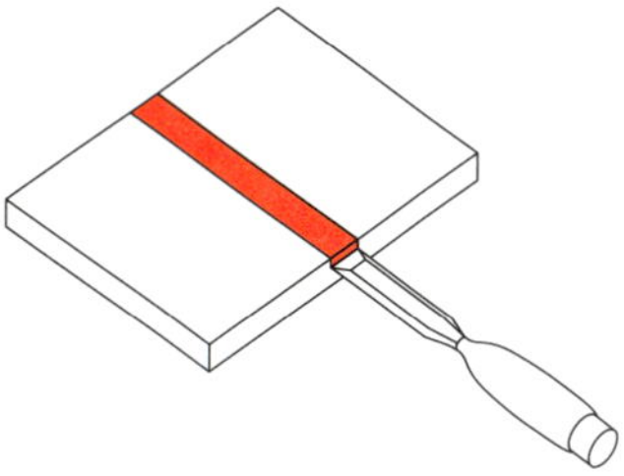

A3

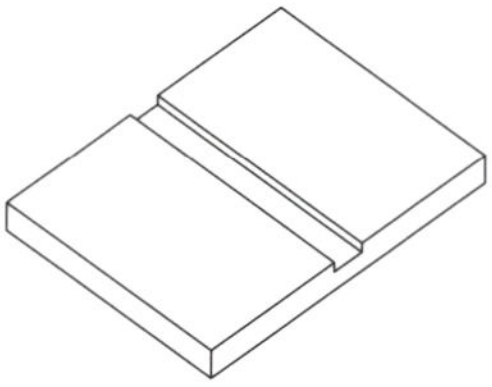

A

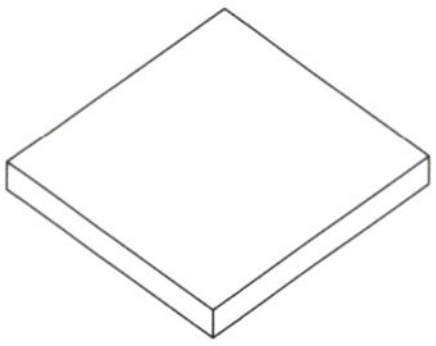

B

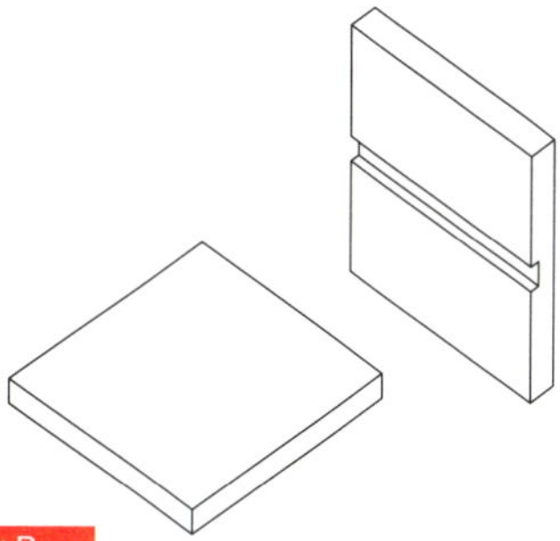

A+B

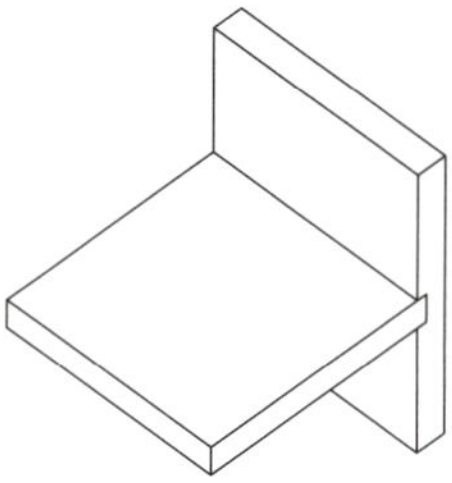

AB

T-Joints

Stopped housing joint

tsutsumi-ryō-dō-tsuki-oire-tsugi
包両胴付追入接ぎ

abgesetzte Nutverbindung

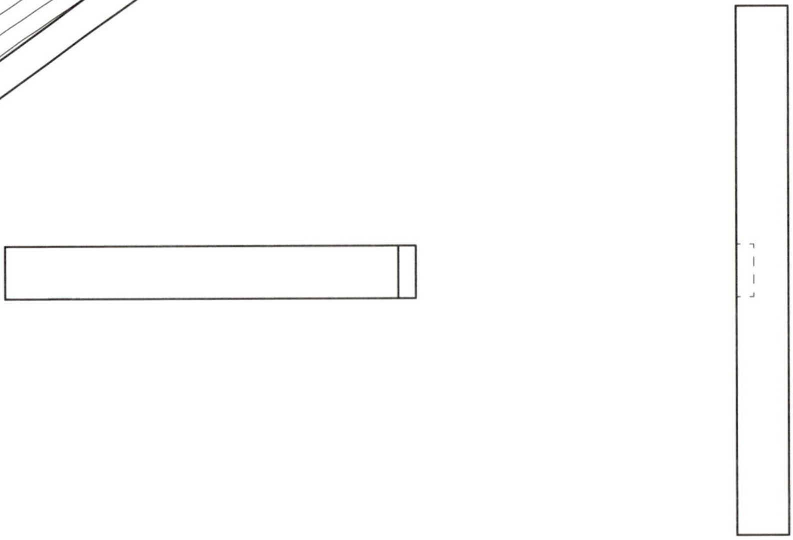

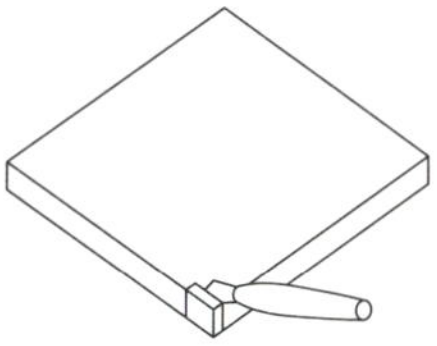
A1

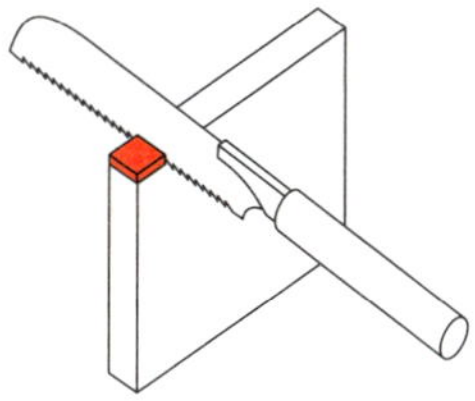
A2

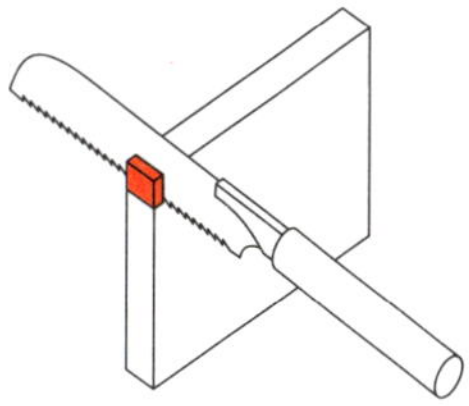
A3

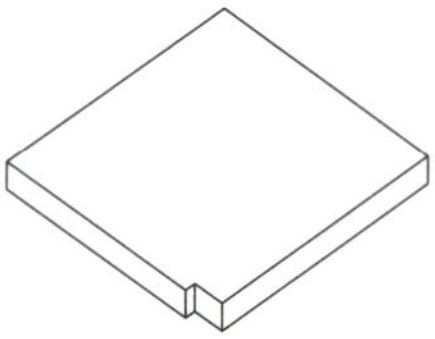
A

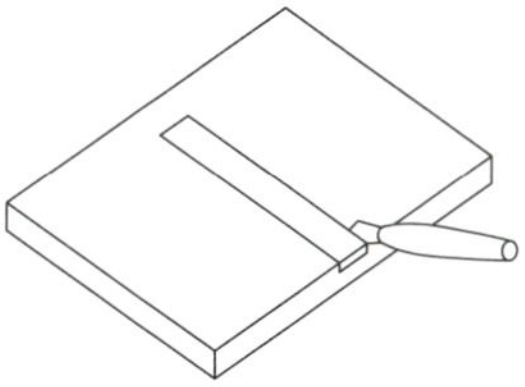
B1

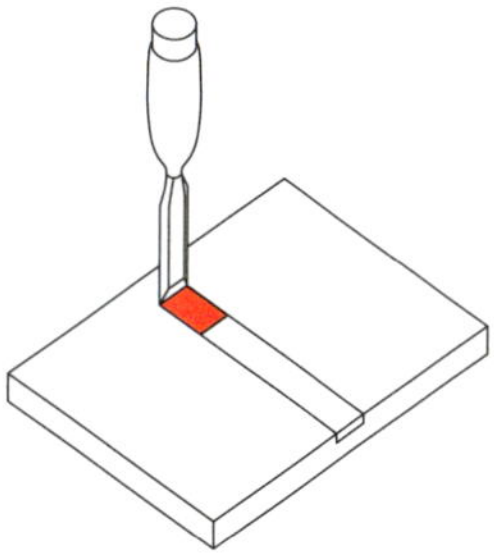
B2

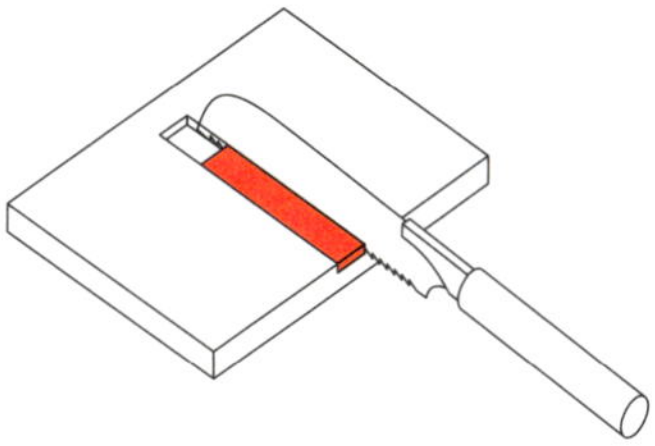
B3

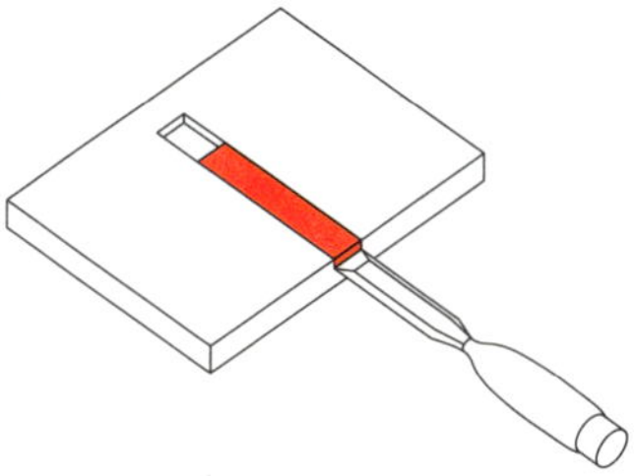
B4

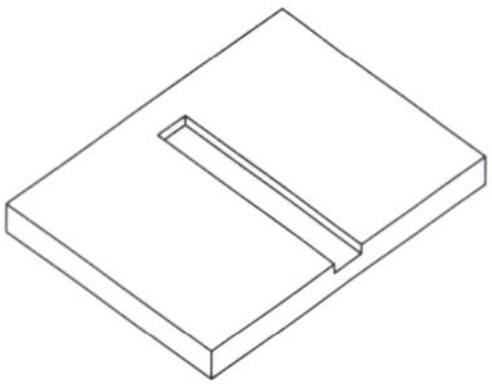
B

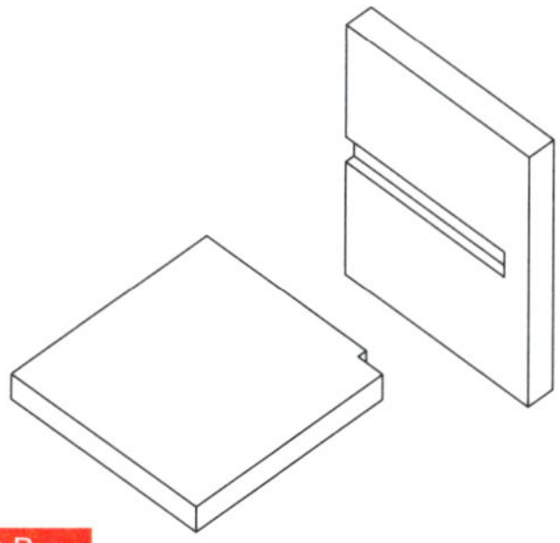
A+B

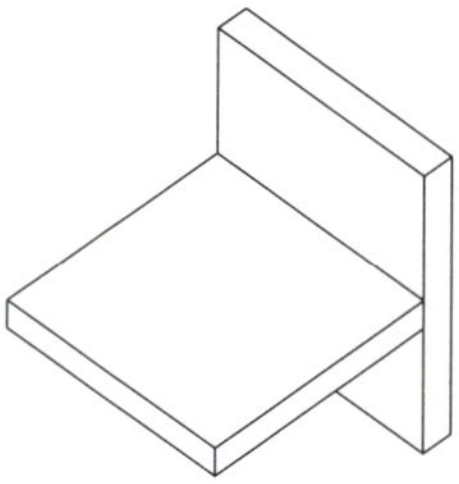
AB

T-Joints

Beveled housing joint

sogi-kaki-uchi-tsuke-tsugi
殺欠打付接ぎ

schräg eingelassene Winkelverbindung

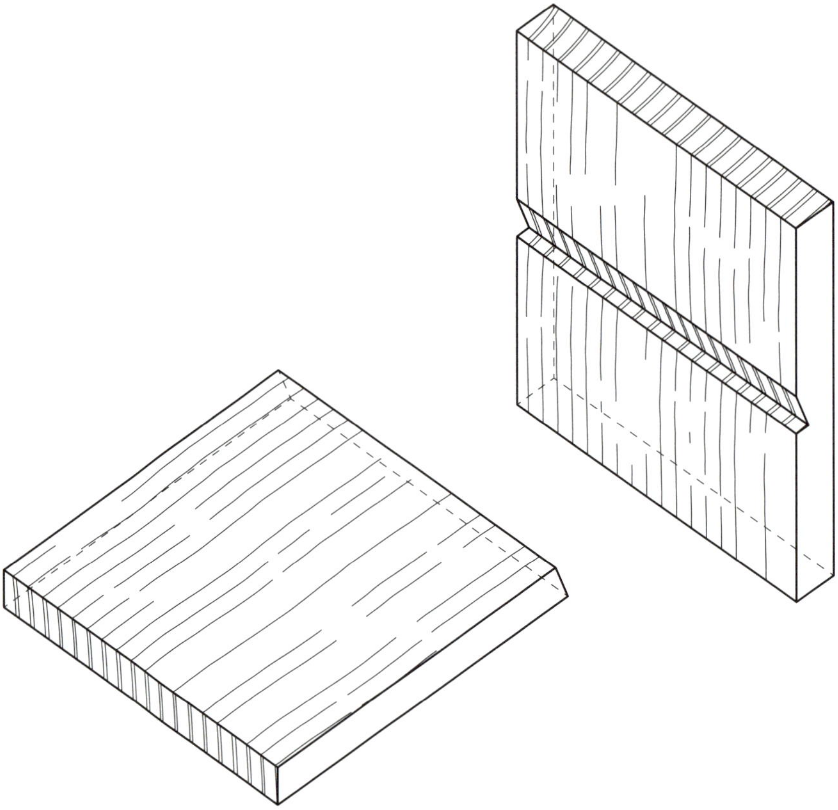

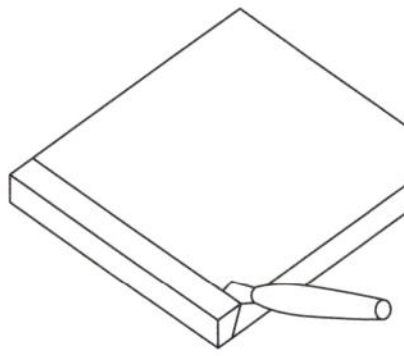

A1

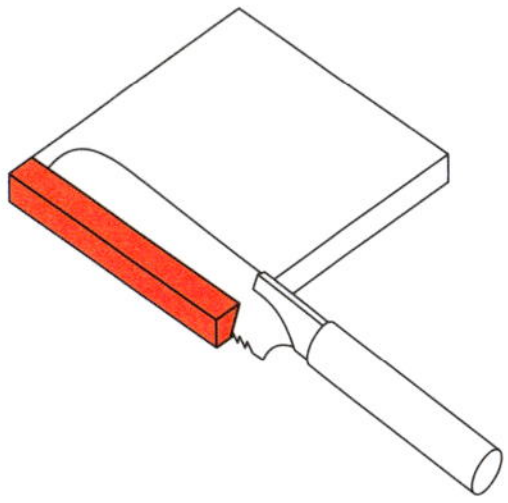

A2

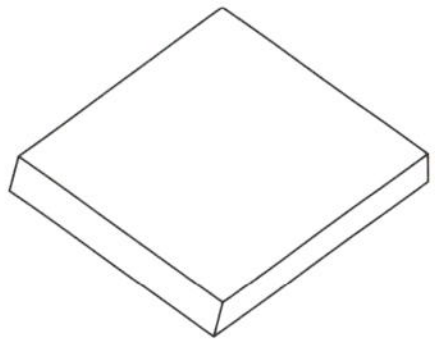

A

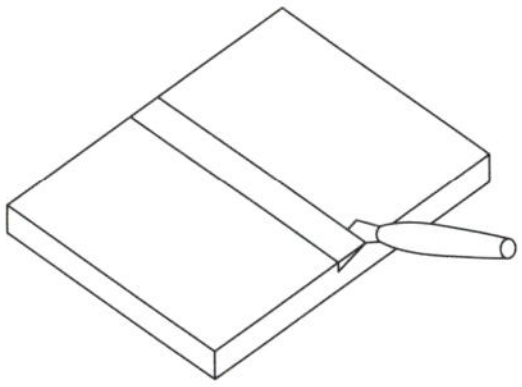

B1

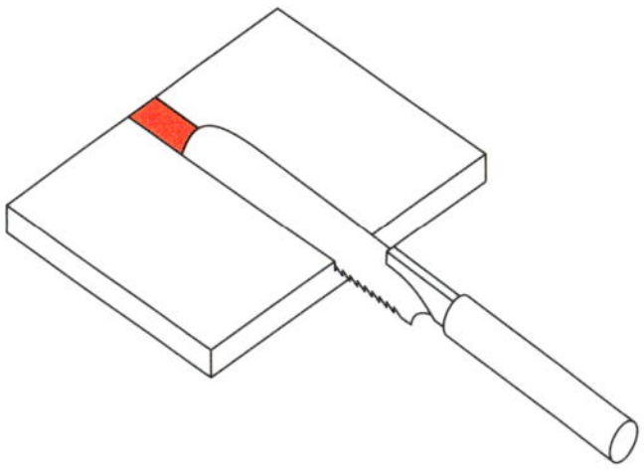

B2

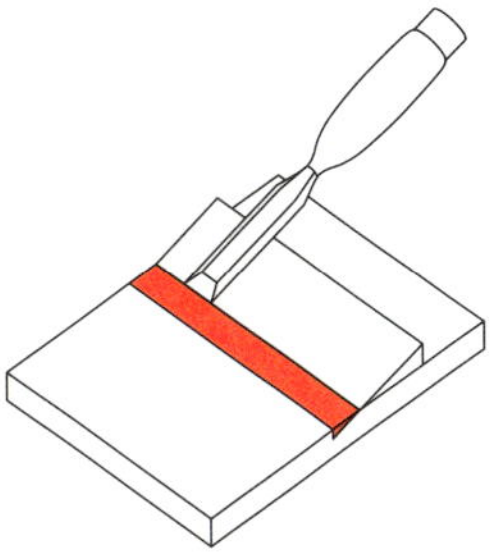

B3

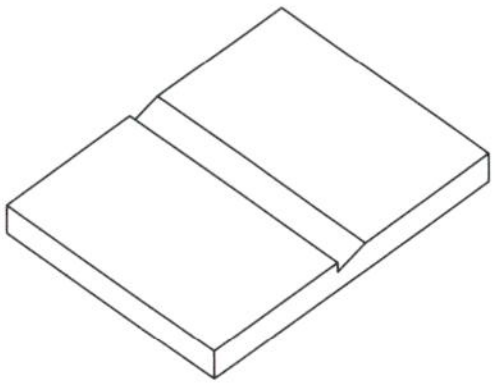

B

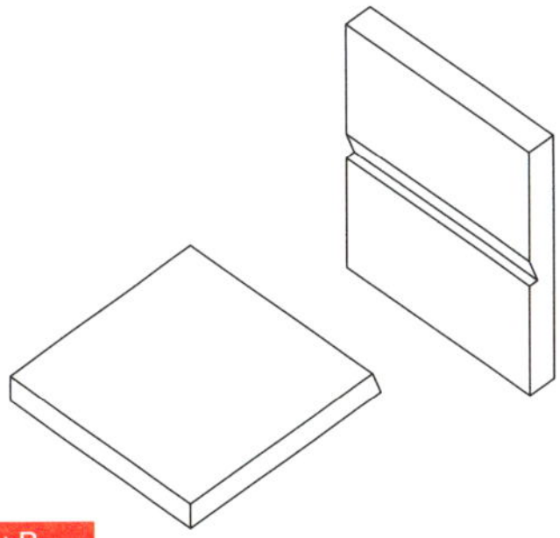

A+B

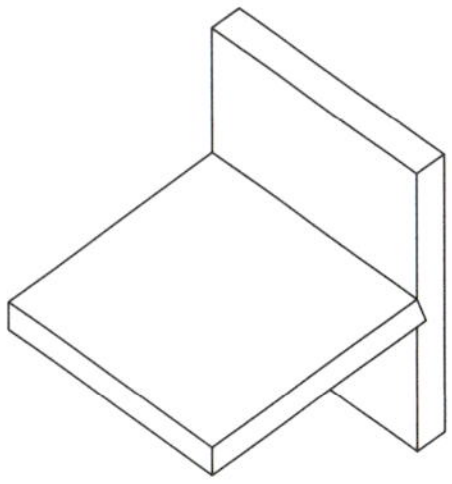

AB

T-Joints

Tongue and dado joint

ryō-dō-tsuki-oire-tsugi
両胴付追入接ぎ

Nutverbindung
mit mittiger Feder

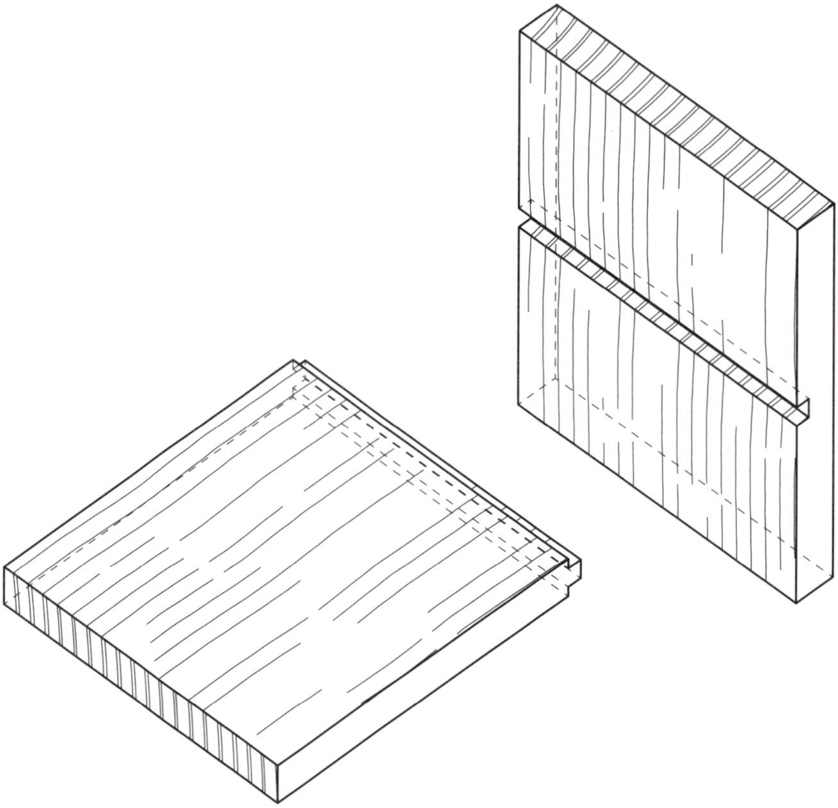

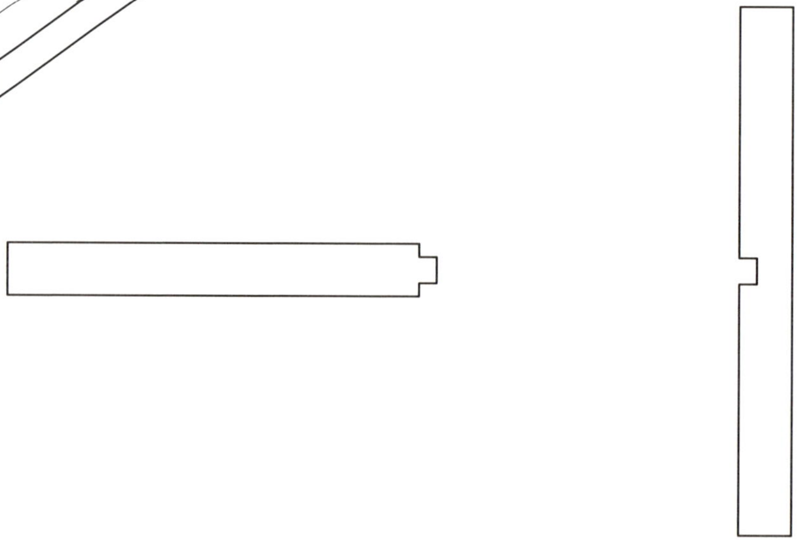

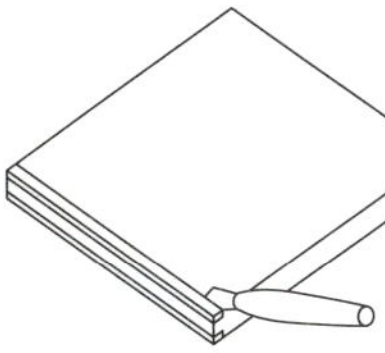

A1

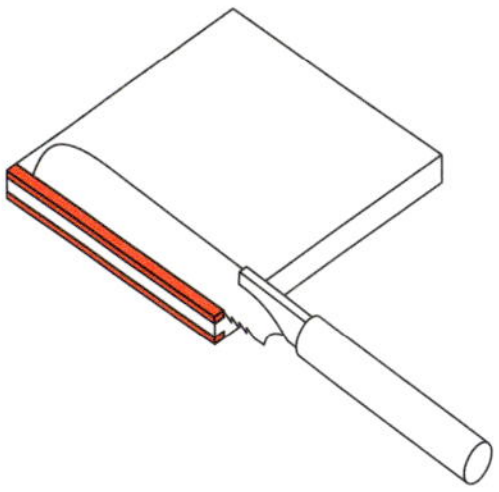

A2

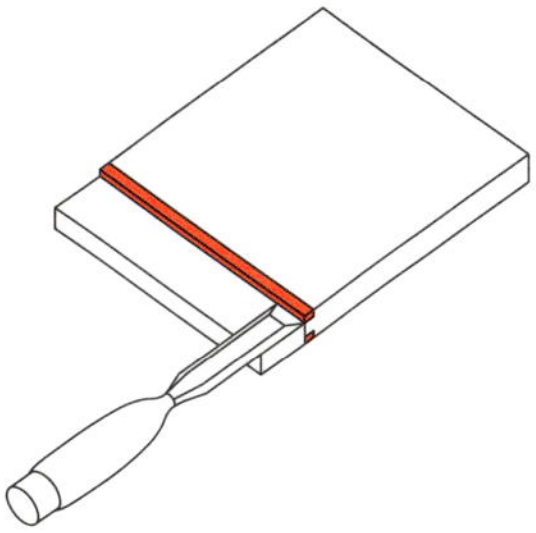

A3

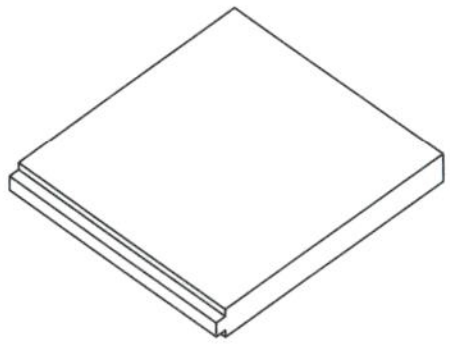

A

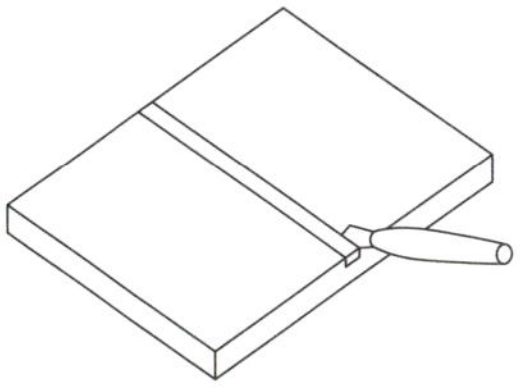

B1

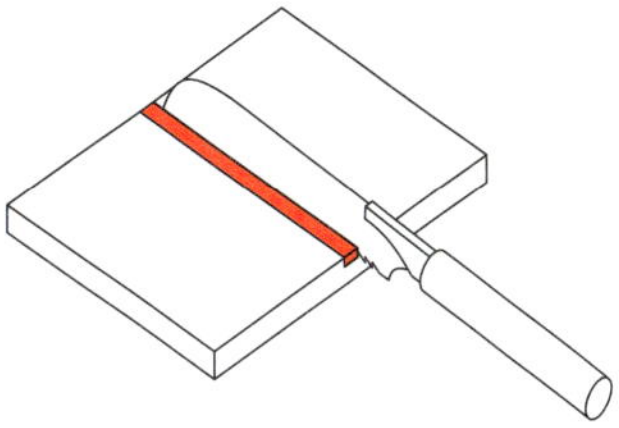

B2

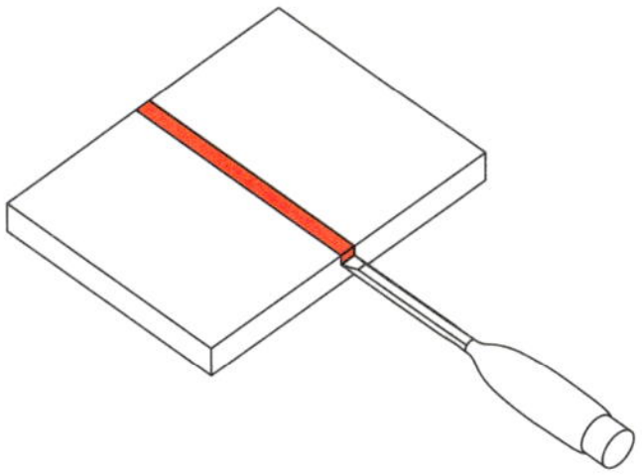

B3

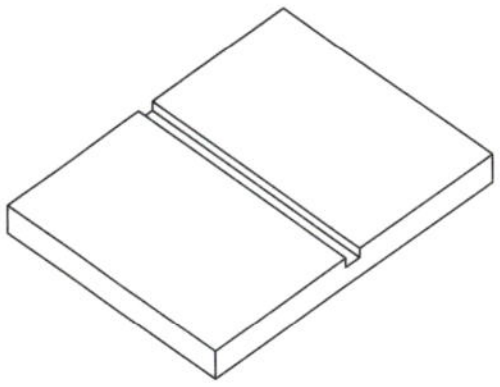

B

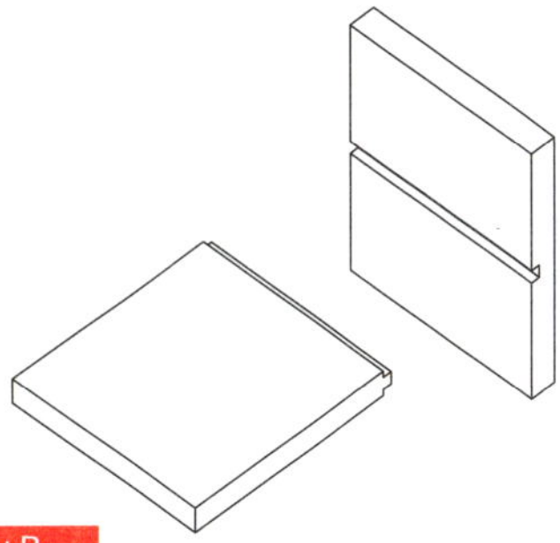

A+B

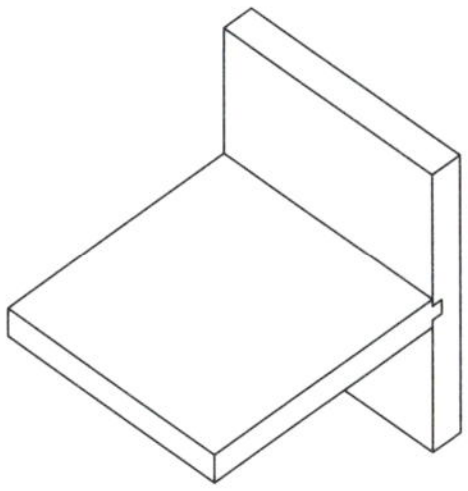

AB

Rabbet and dado joint

kata-dō-tsuki-oire-tsugi
片胴付追入接ぎ

Nutverbindung
mit außermittiger Feder

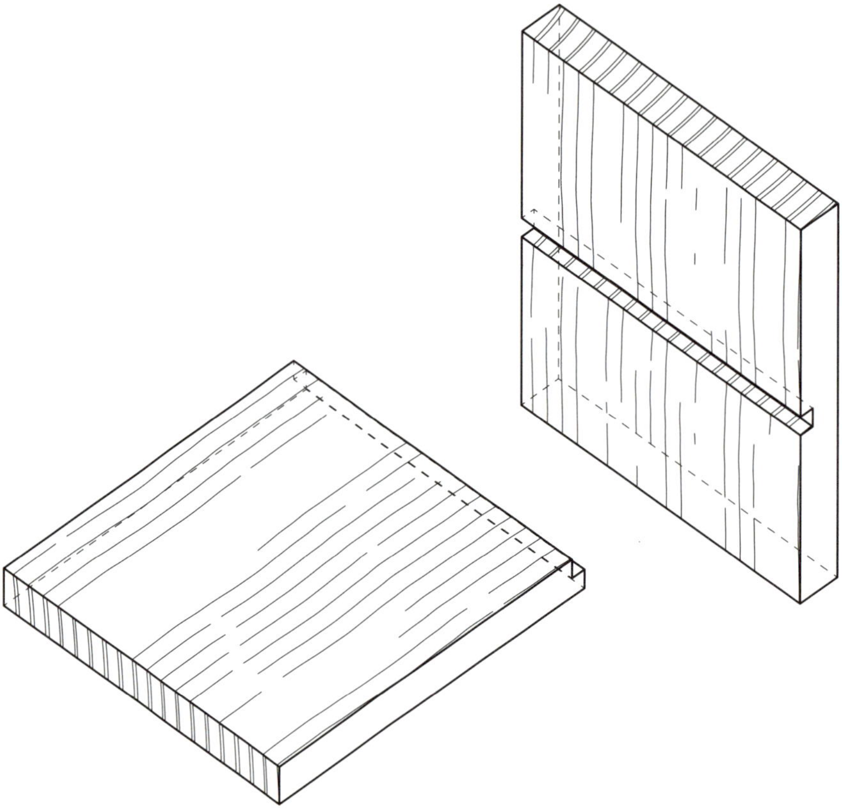

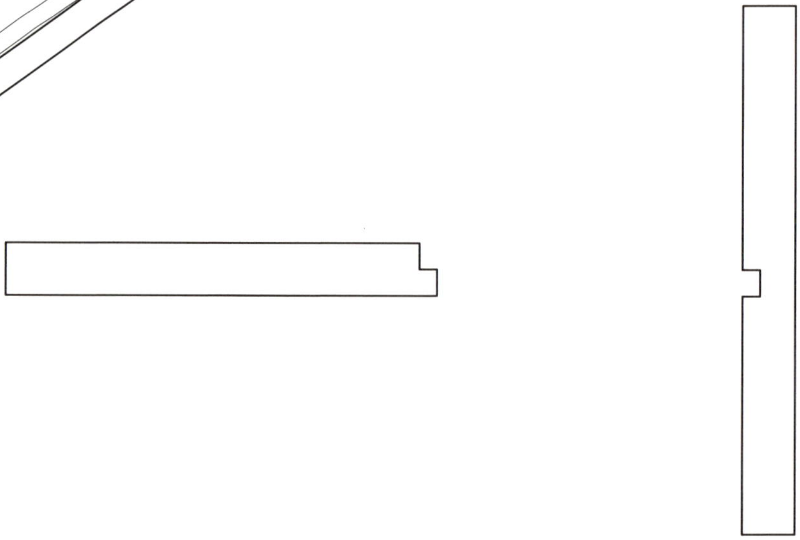

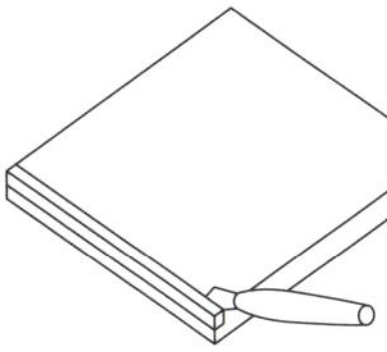

A1

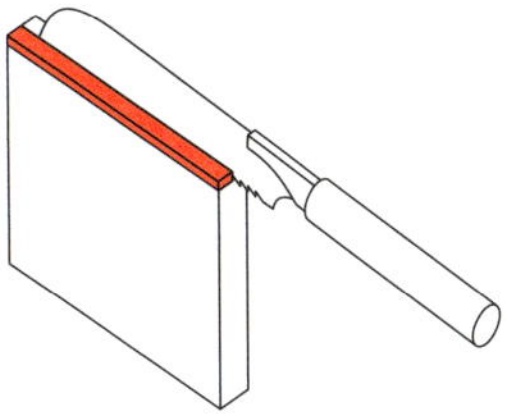

A2

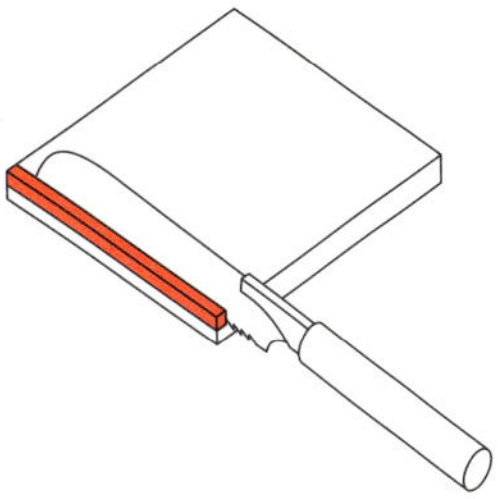

A3

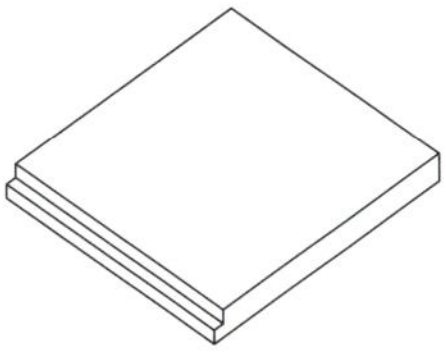

A

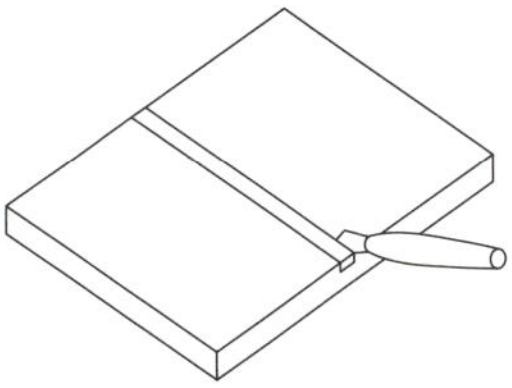

B1

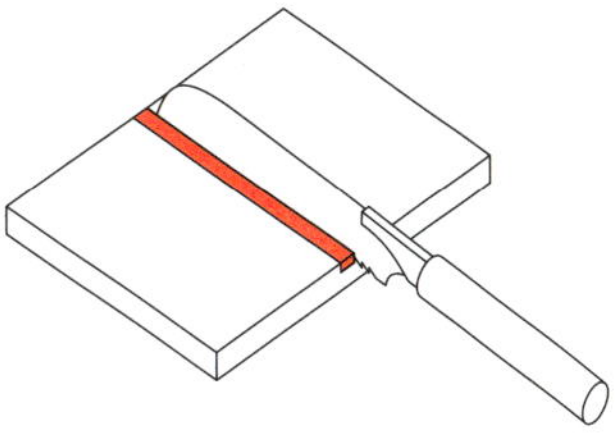

B2

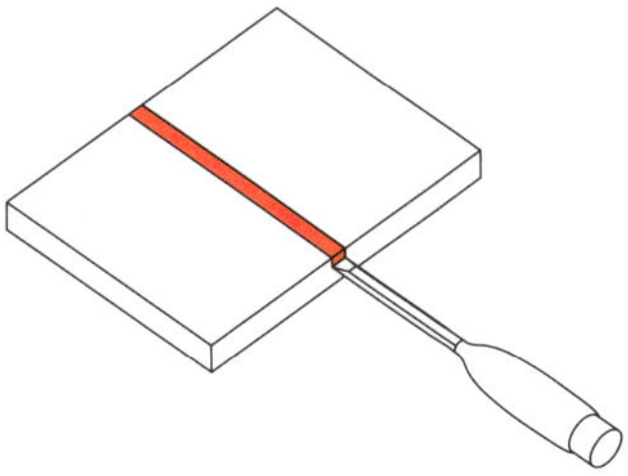

B3

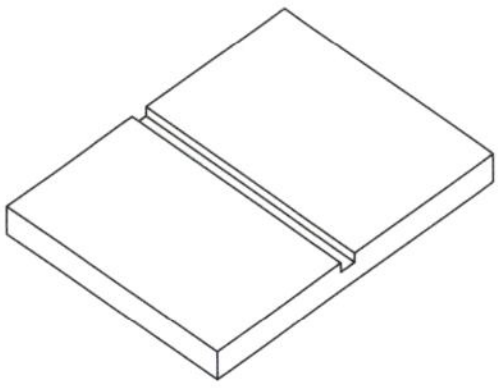

B

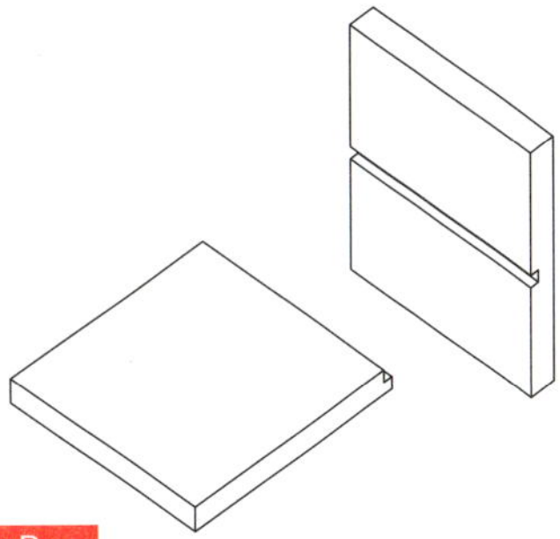

A+B

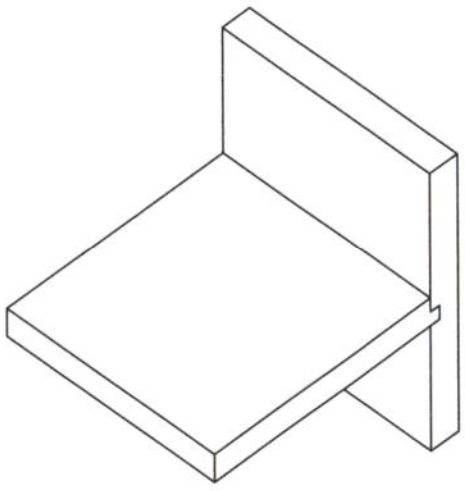

AB

Sliding dovetail joint

sui-tsuki-ari-hozo-oire-tsugi
吸付蟻枘追入接ぎ

Gratnutverbindung

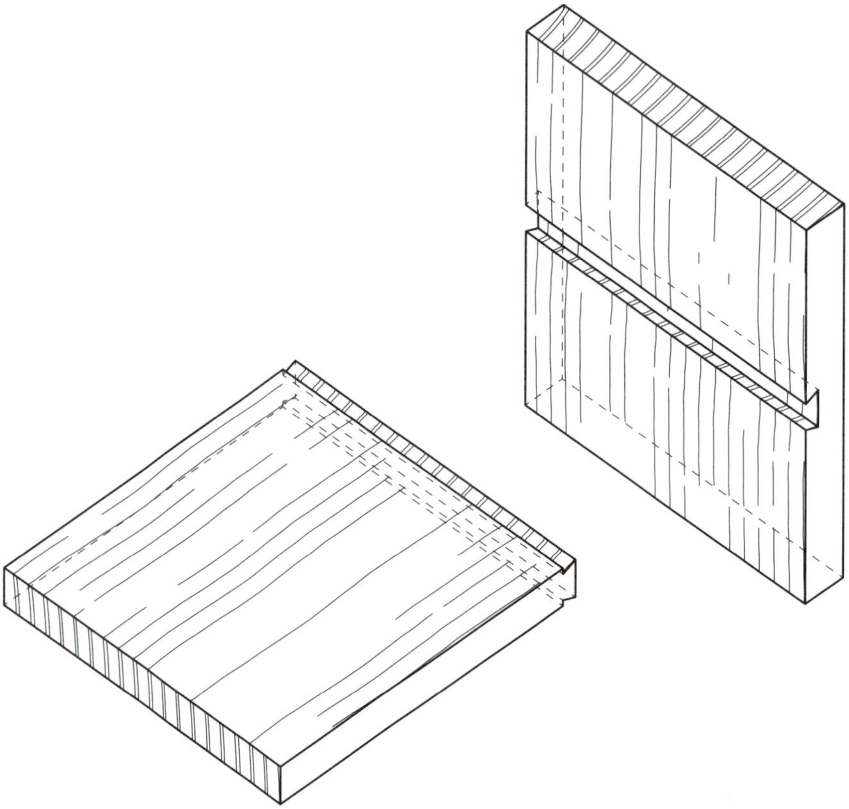

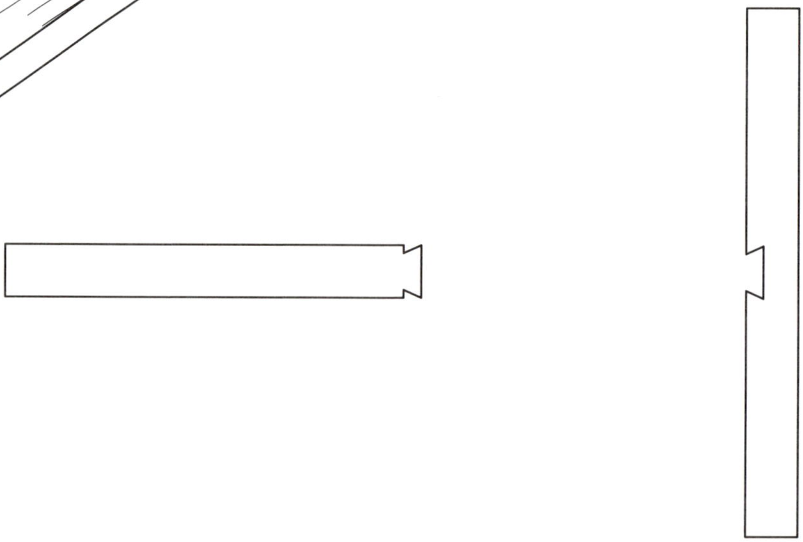

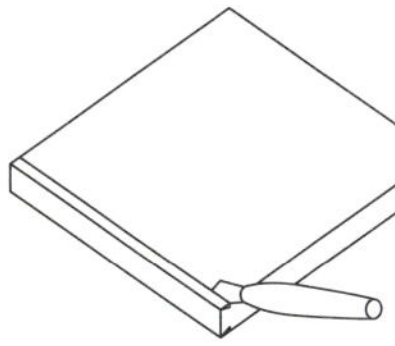

A1

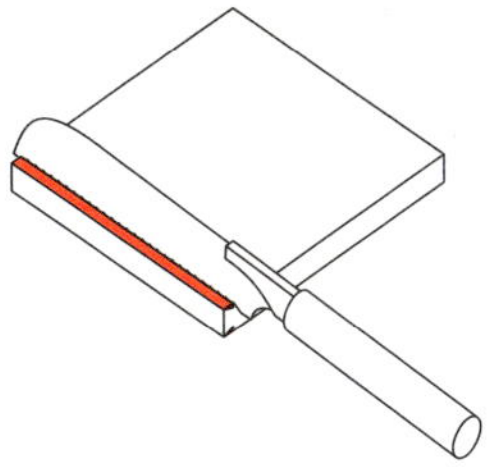

A2

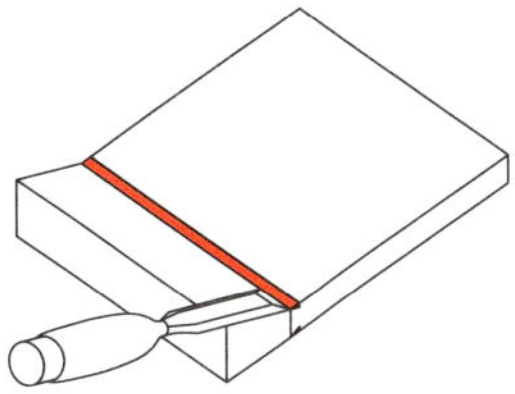

A3

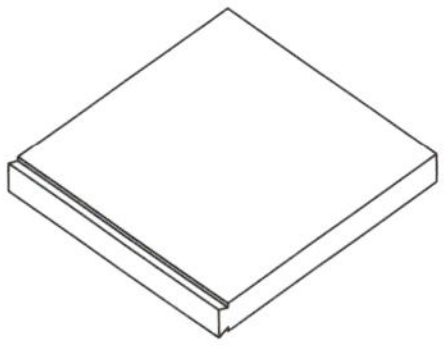

A

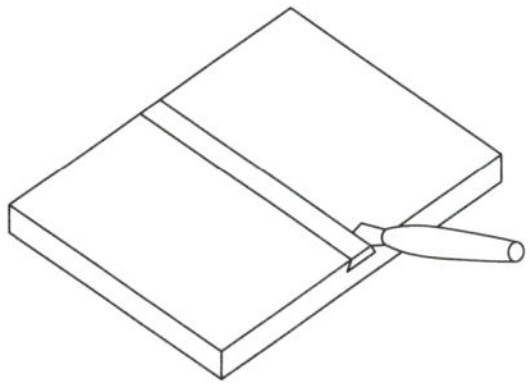

B1

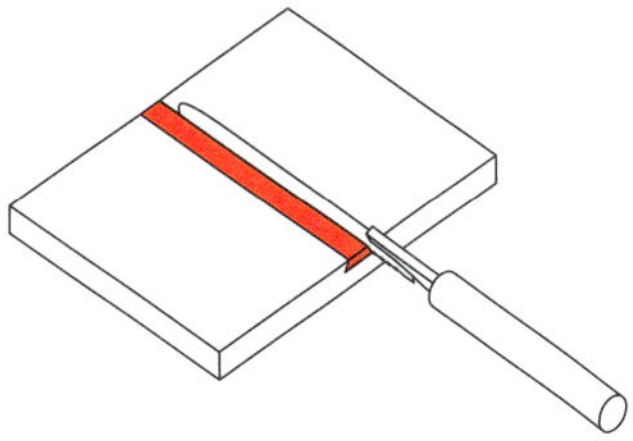

B2

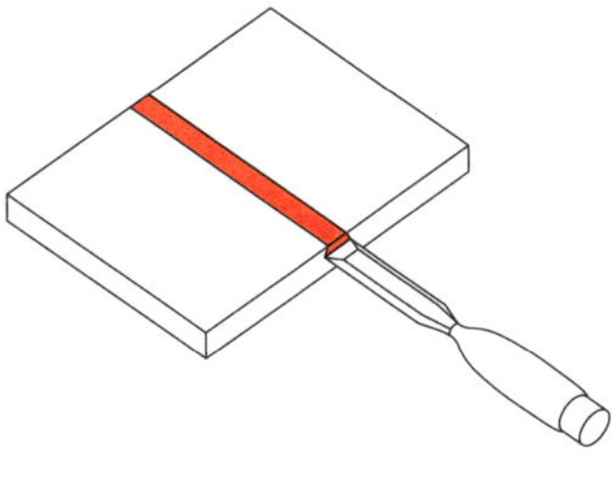

B3

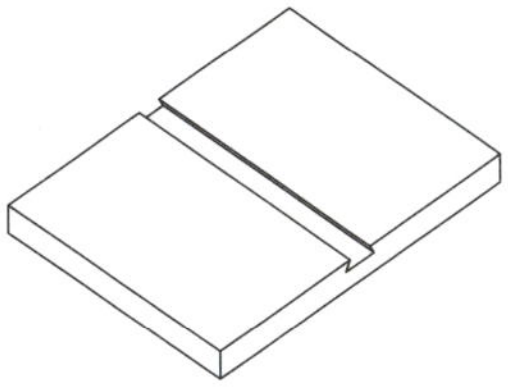

B

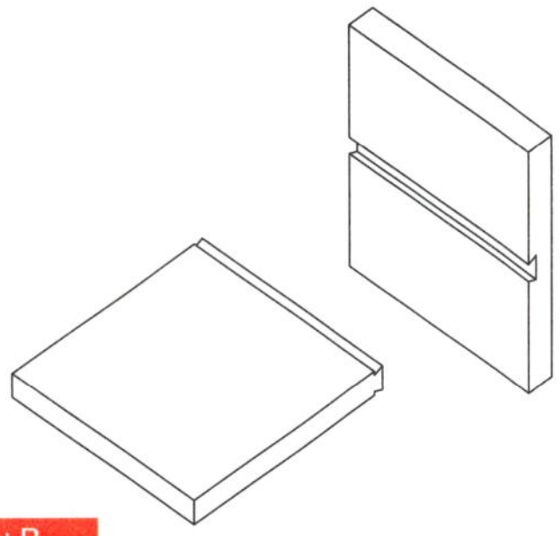

A+B

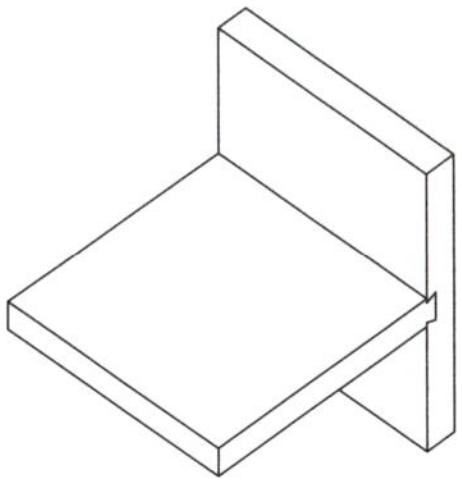

AB

T-Joints

Single-shouldered sliding dovetail joint

kata-ari-kata-oire-tsugi
片蟻形追入接ぎ

einseitige Gratnutverbindung

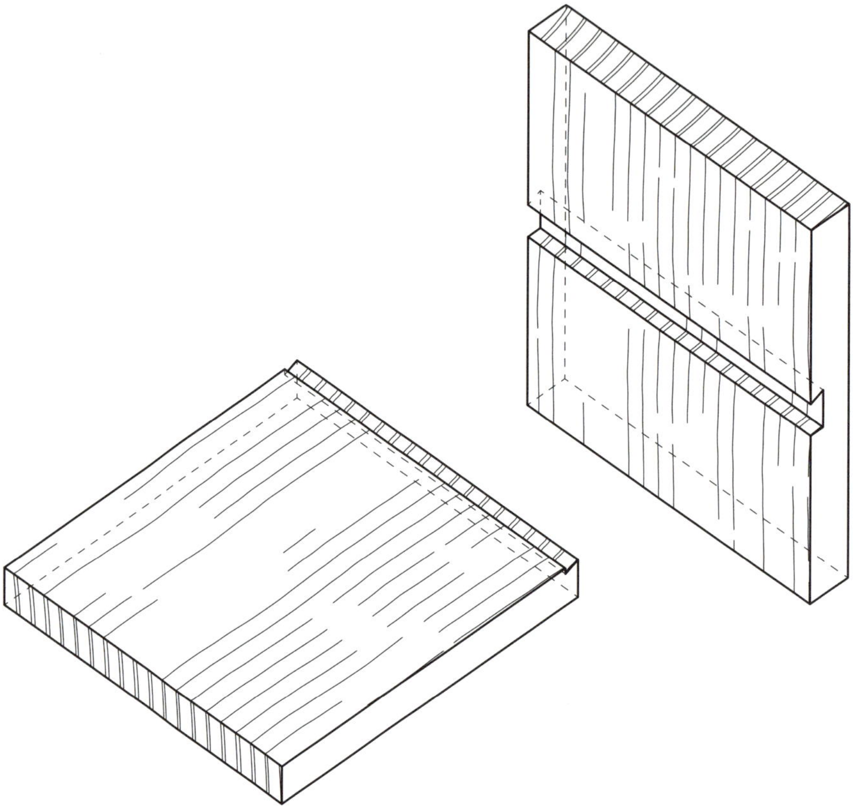

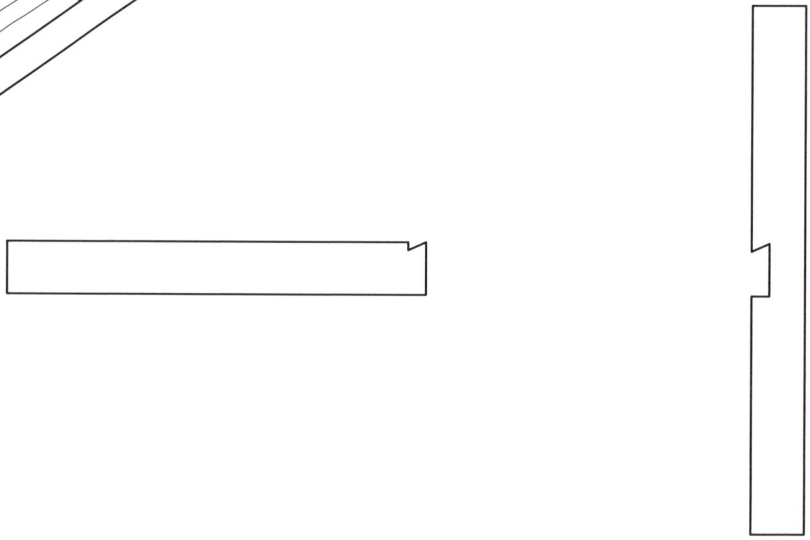

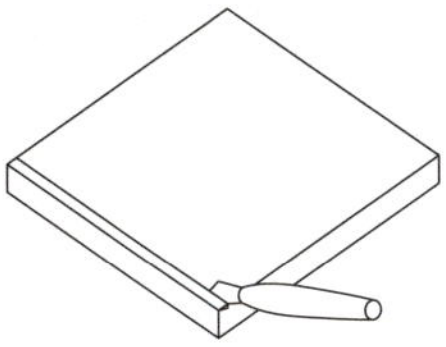

A1

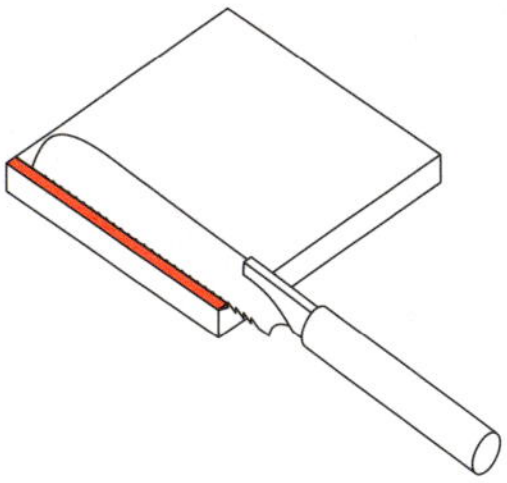

A2

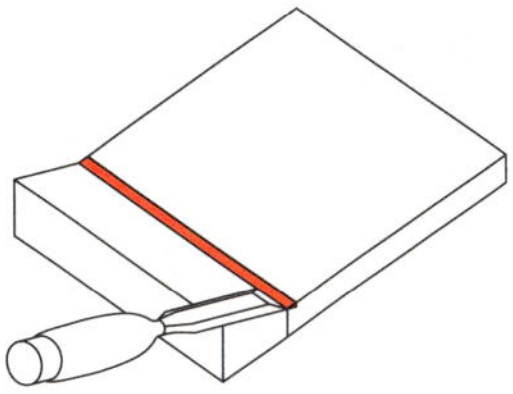

A3

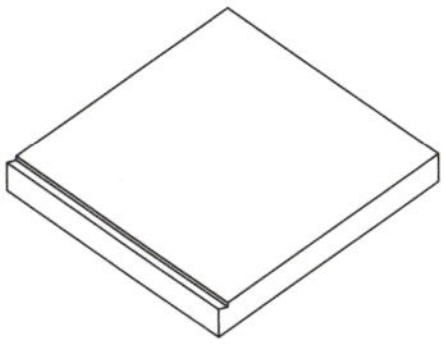

A

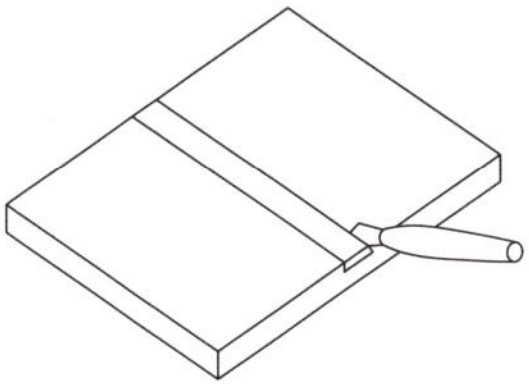

B1

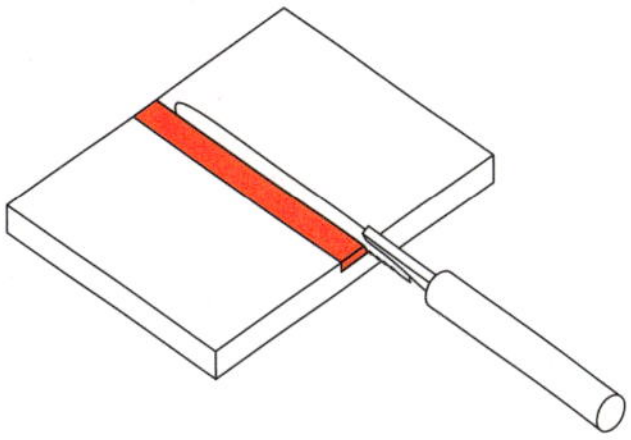

B2

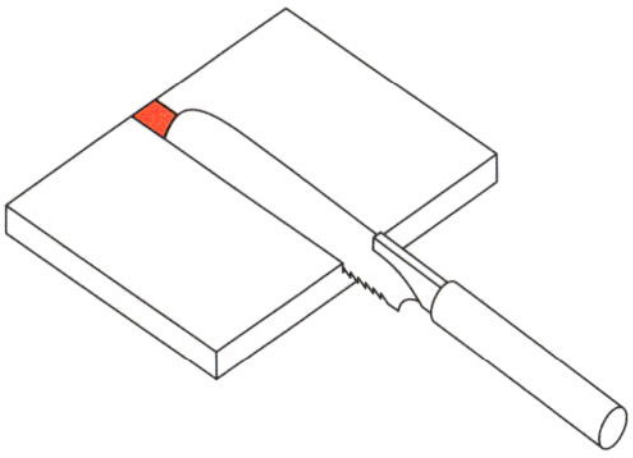

B3

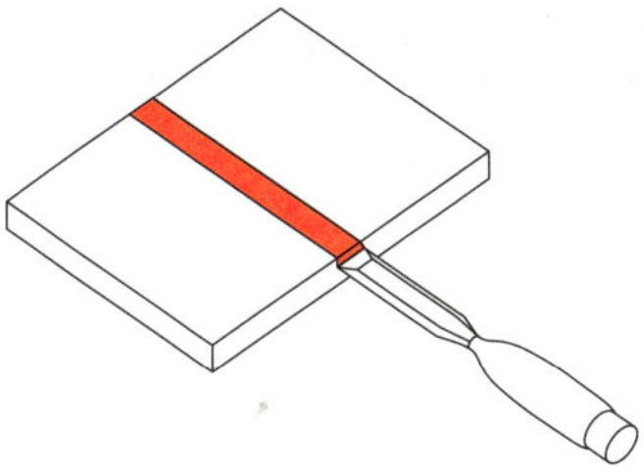

B4

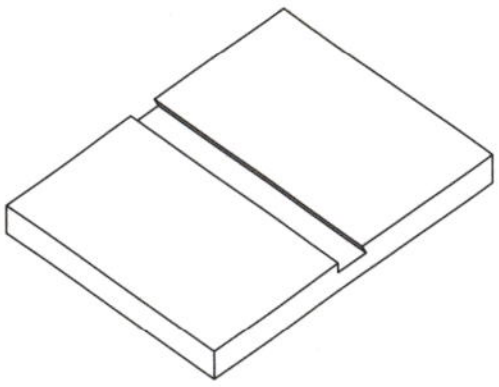

B

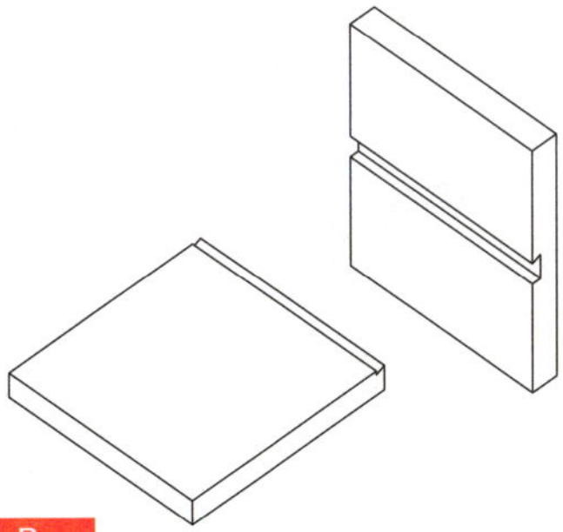

A+B

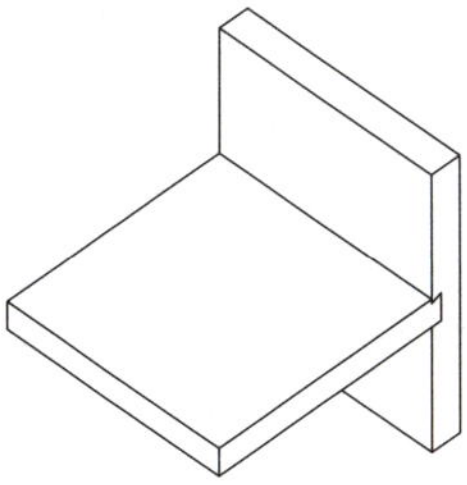

AB

Glossary

Timber Joints
Splicing Joints

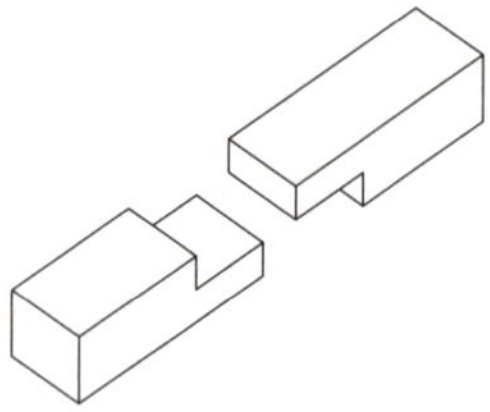

042

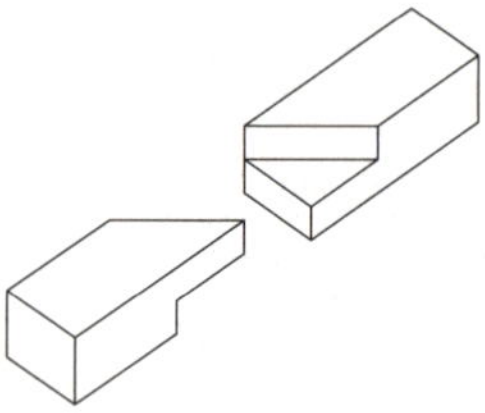

044

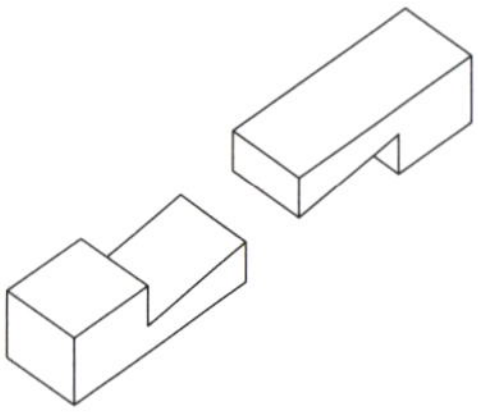

046

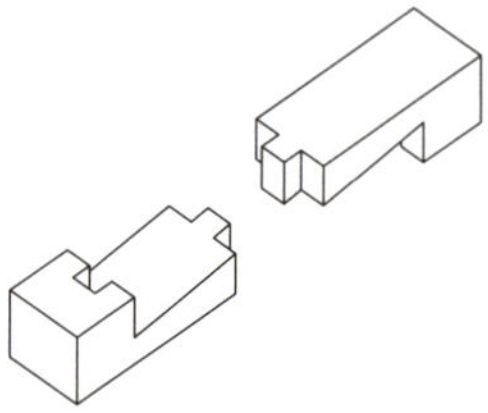

048

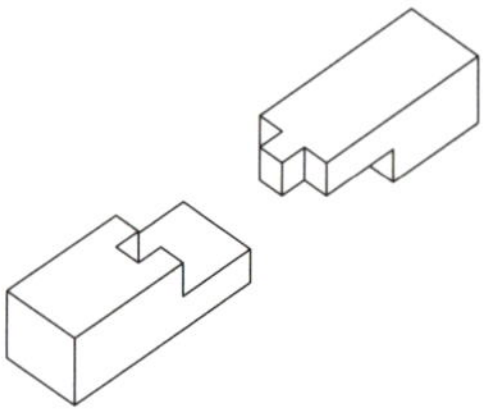

050

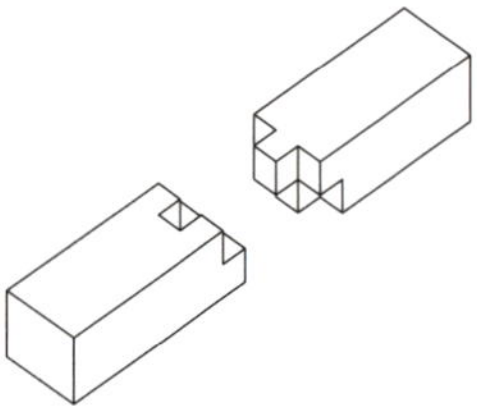

052

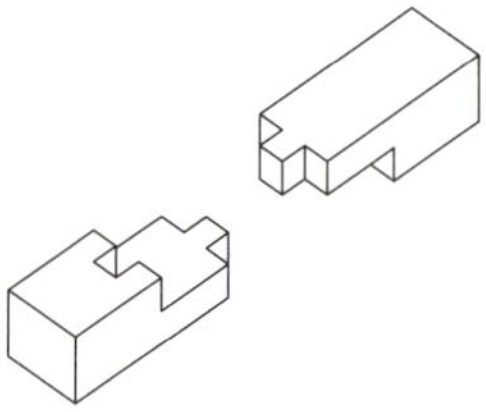

054

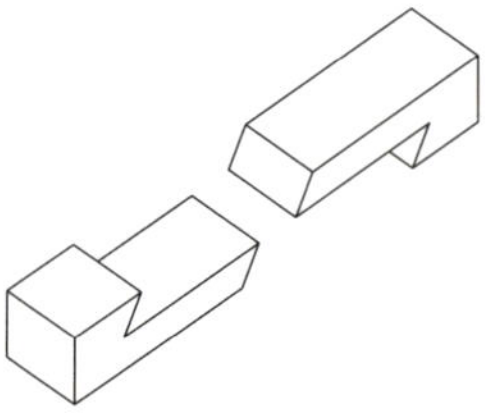

056

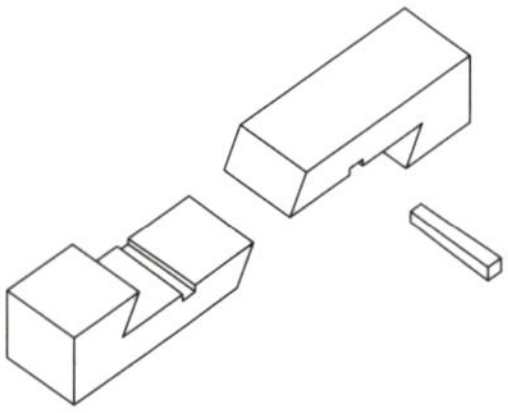

058

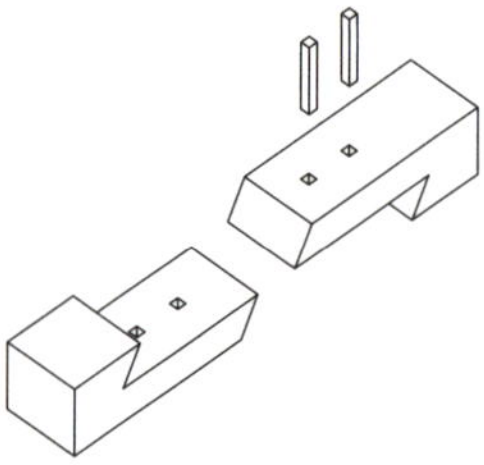

060

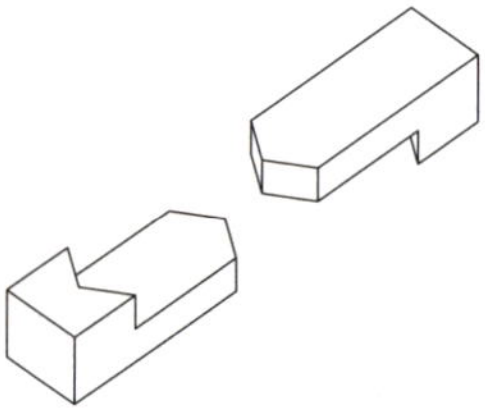

062

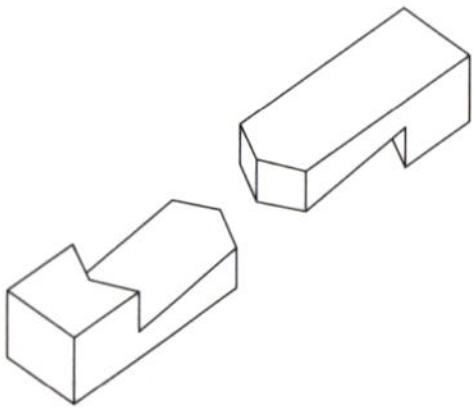

064

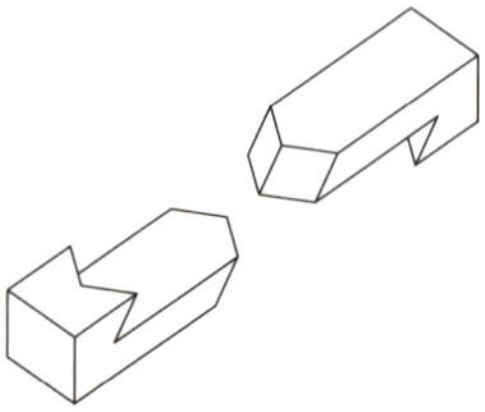

066

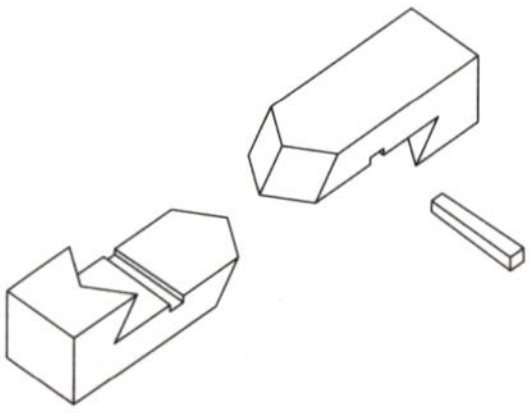

068

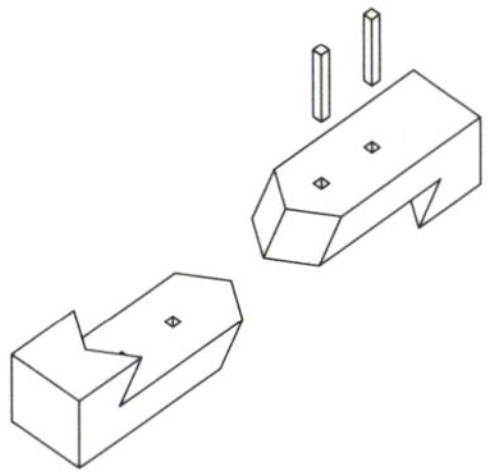
070

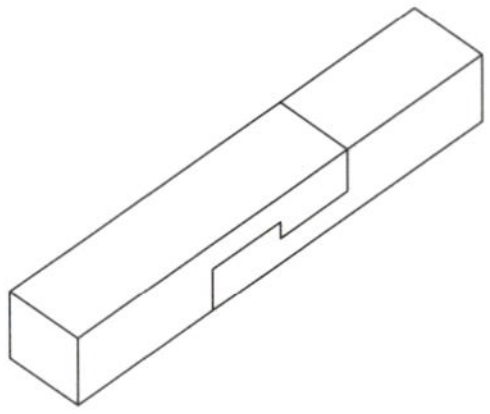
072

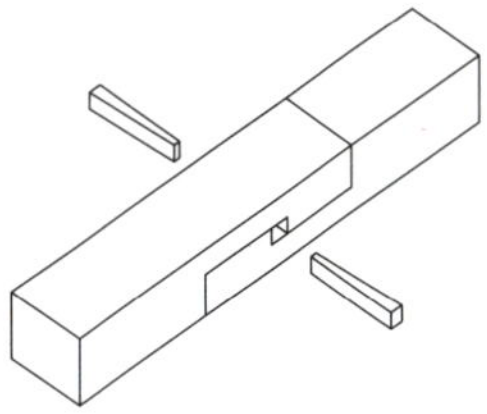
074

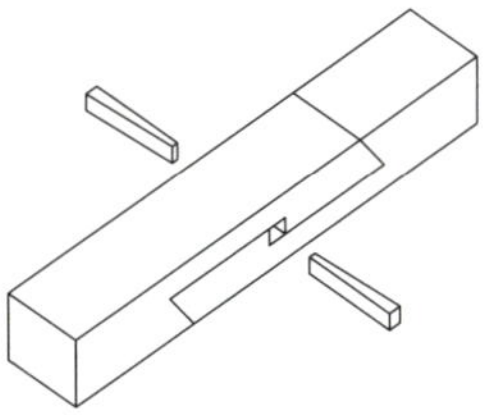
076

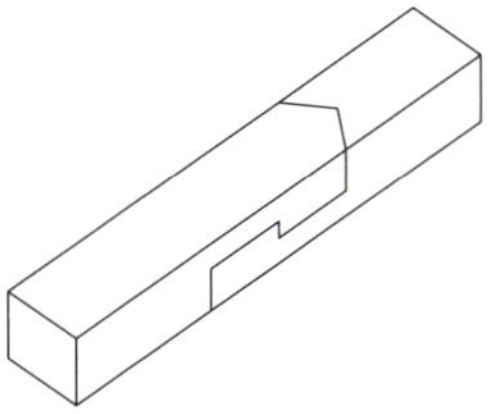
078

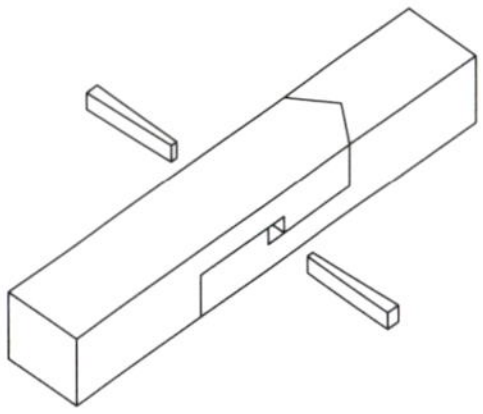
080

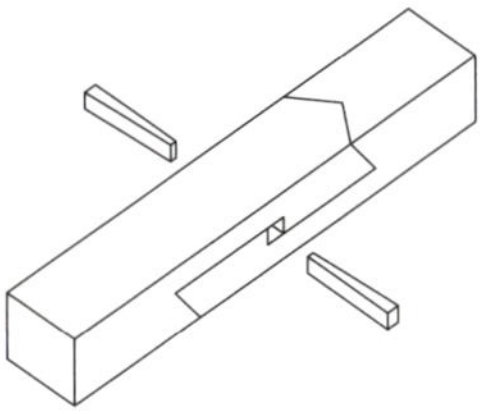
082

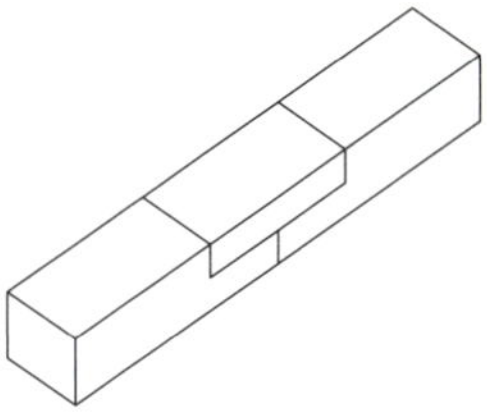
084

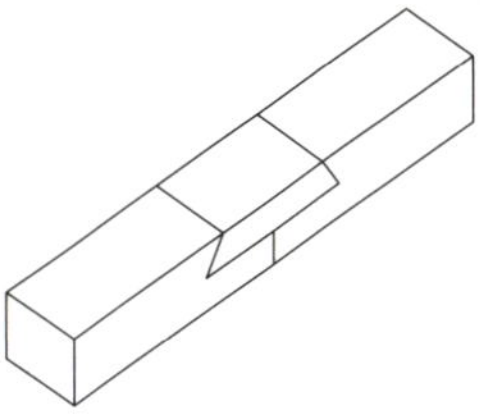
086

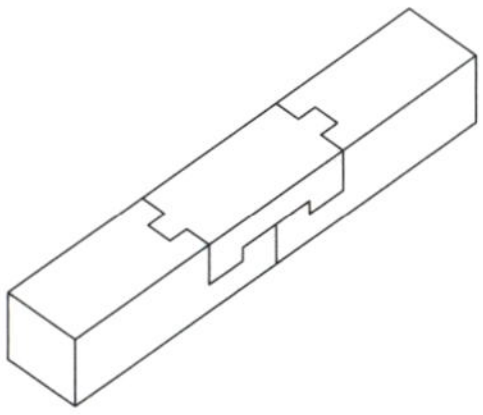
088

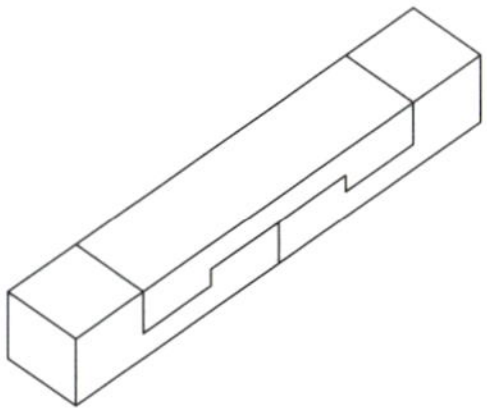
090

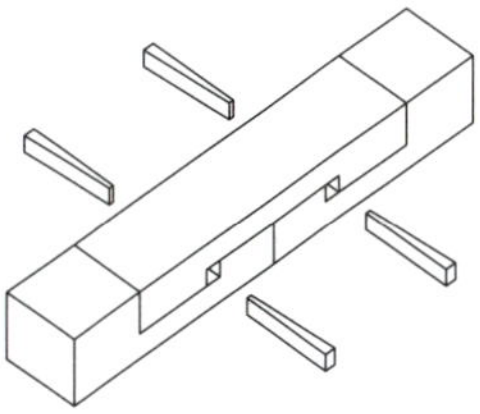
092

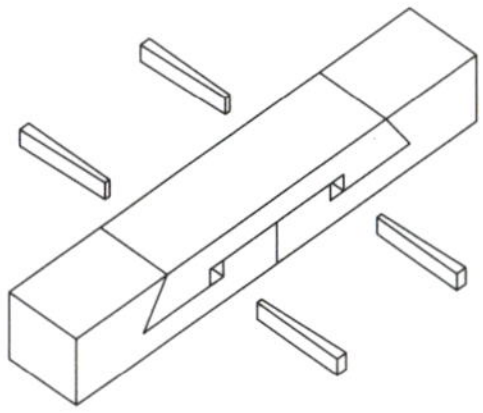
094

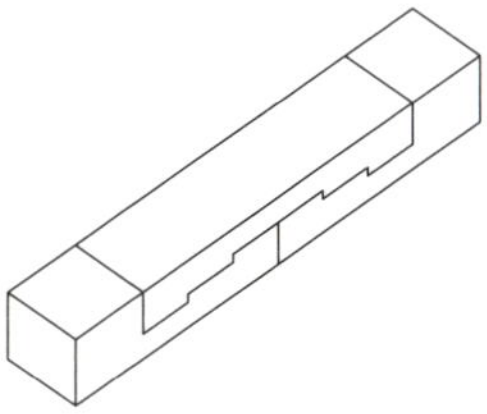
096

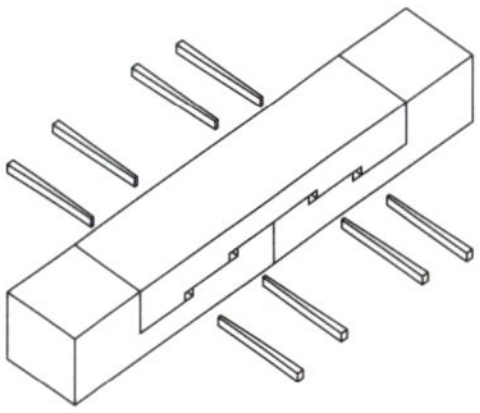
098

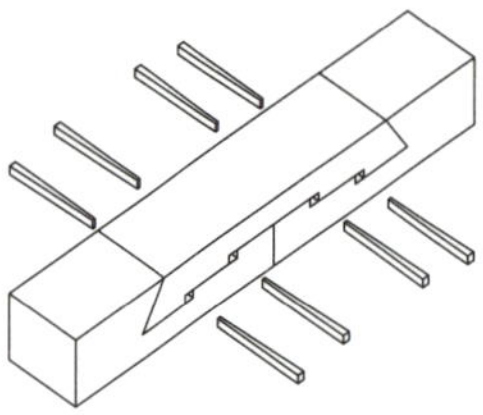

100

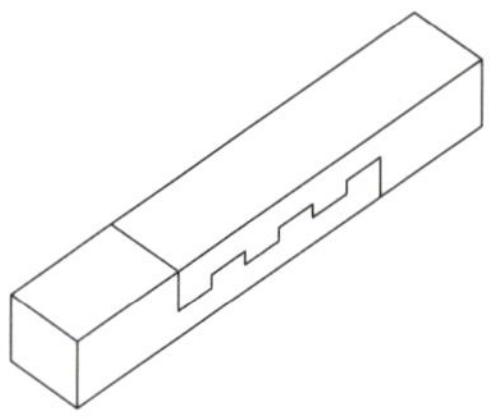

102

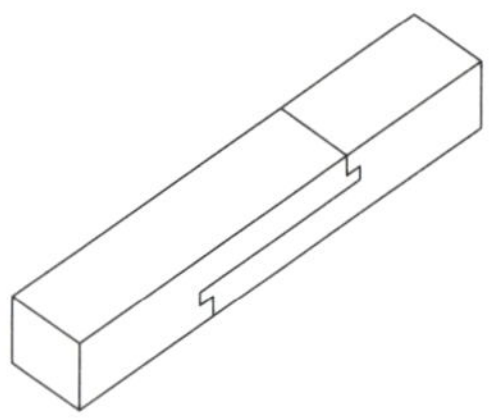

104

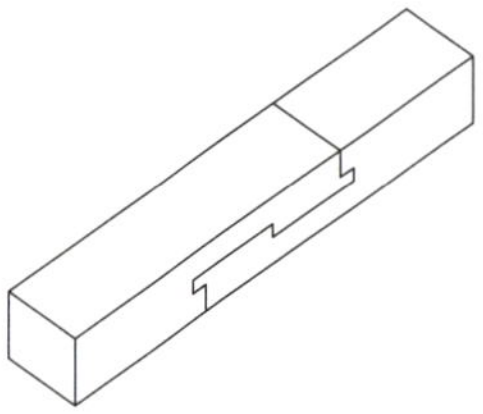

106

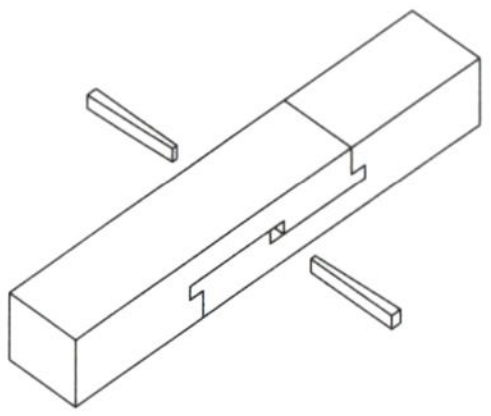

108

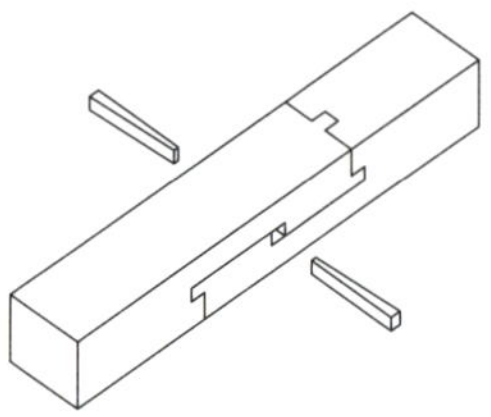

110

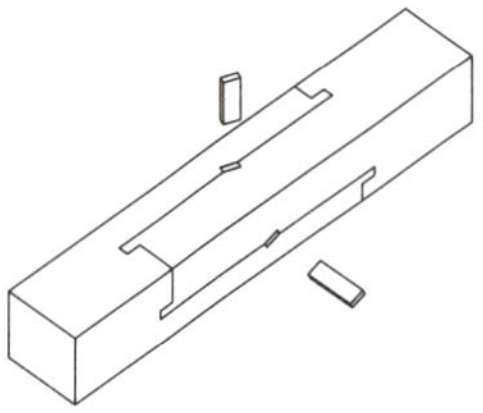

112

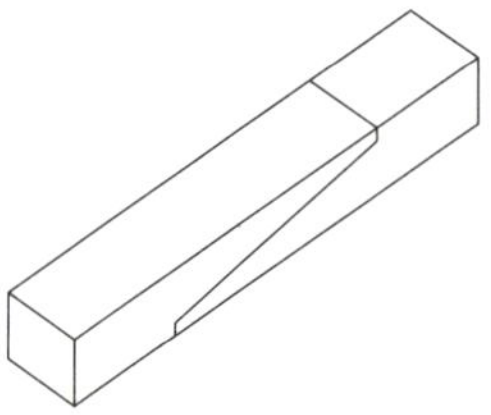

114

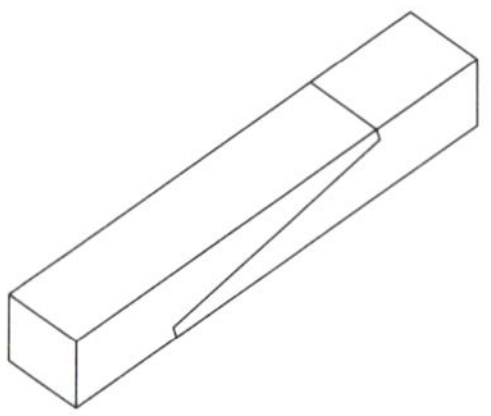

116

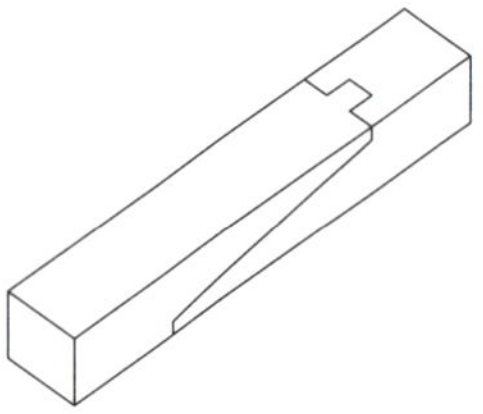

118

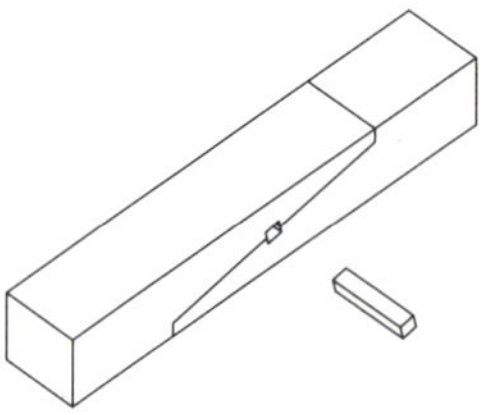

120

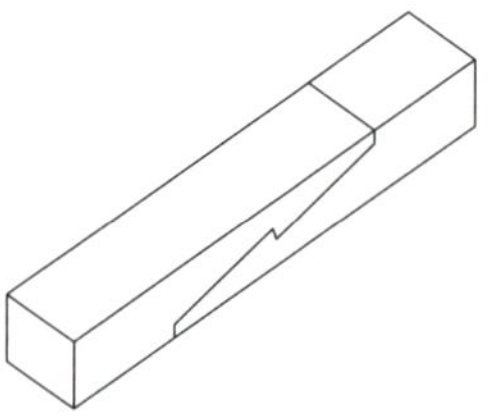

122

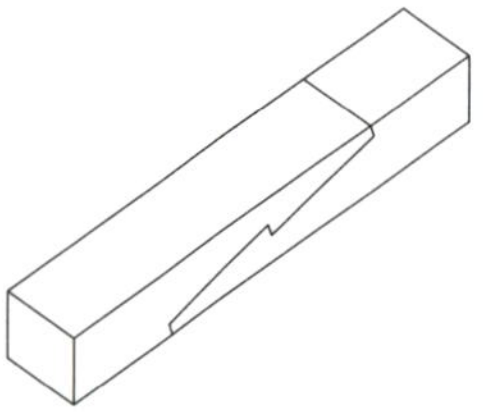

124

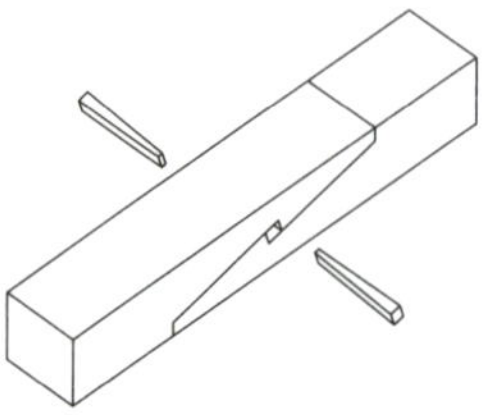

126

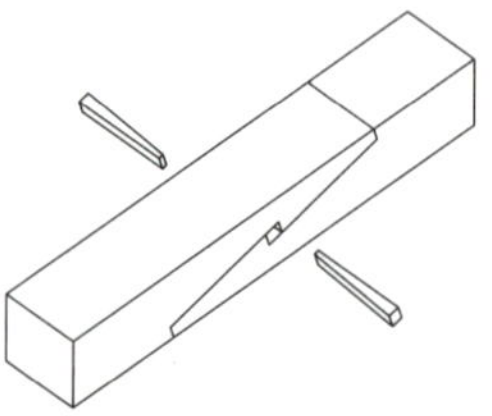

128

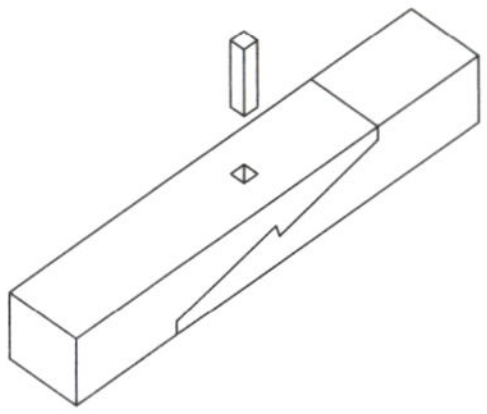

130

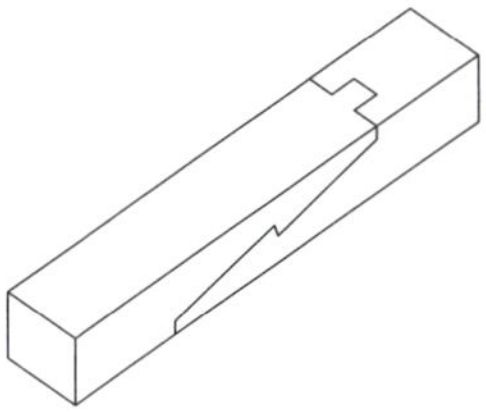

132

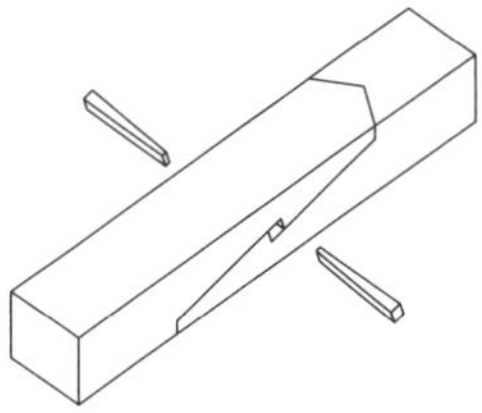

134

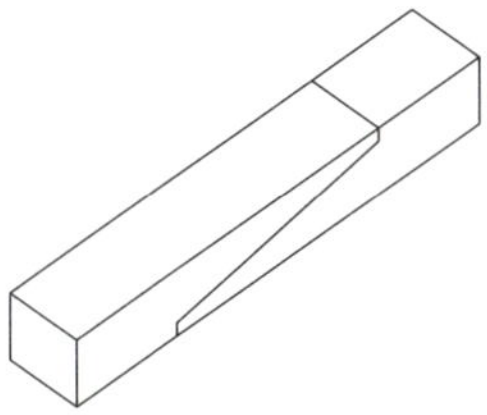

136

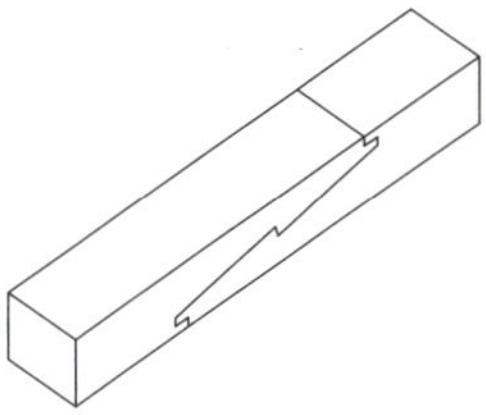

138

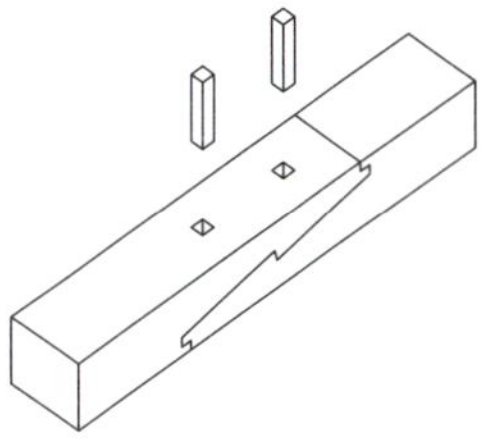

140

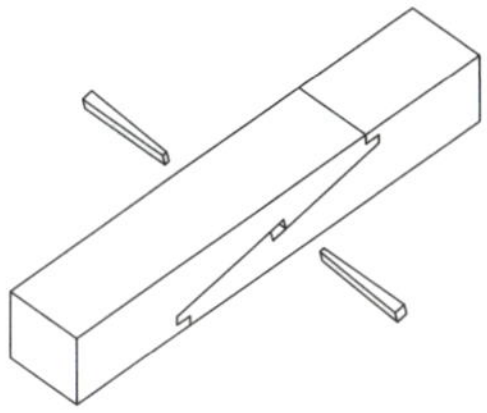

142

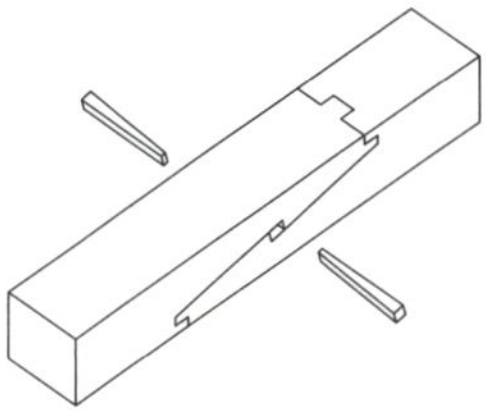

144

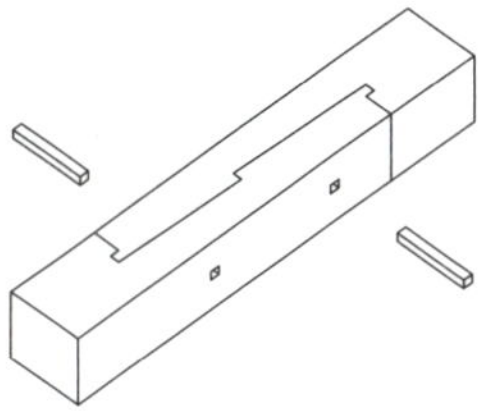

146

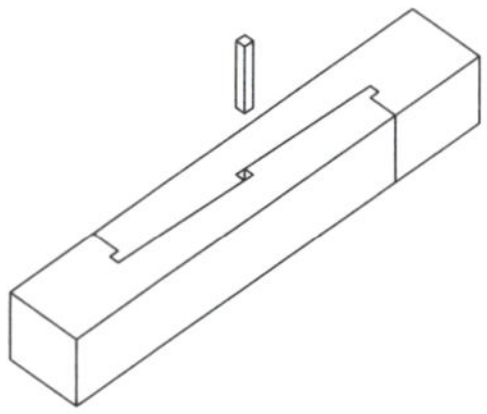

148

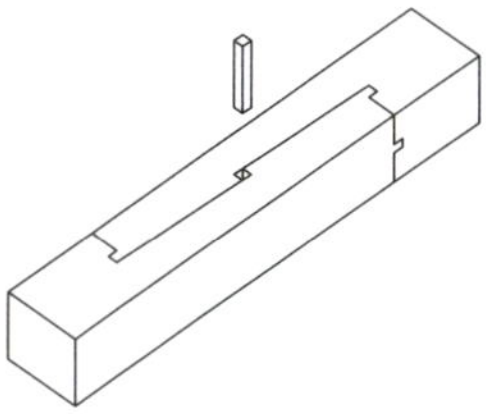

150

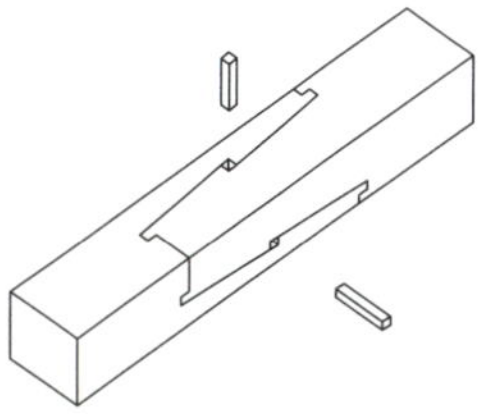

152

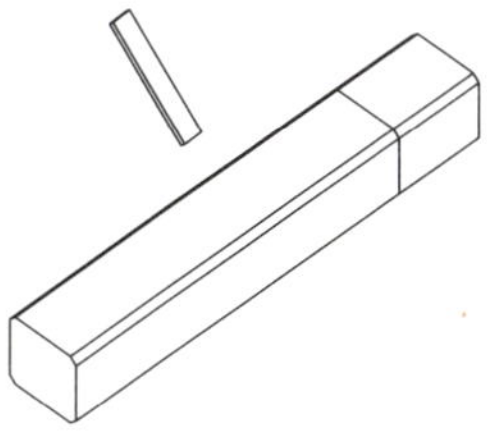

154

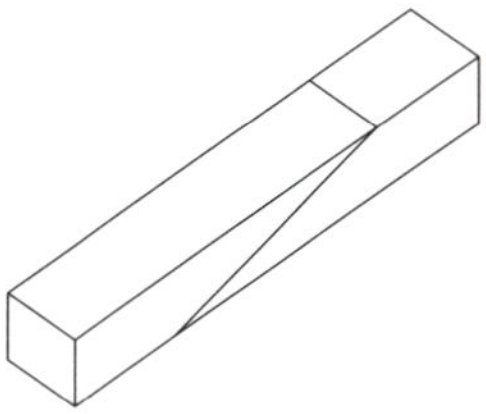

156

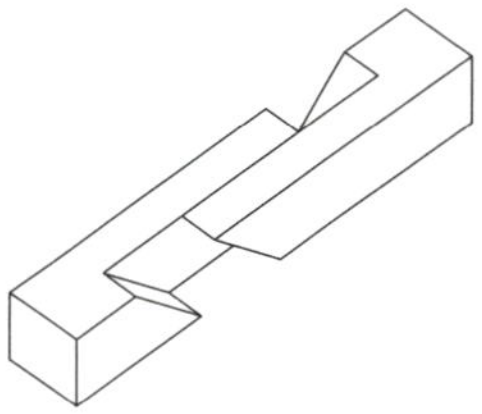

158

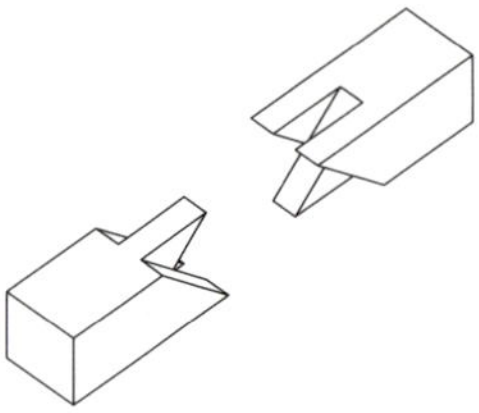

160

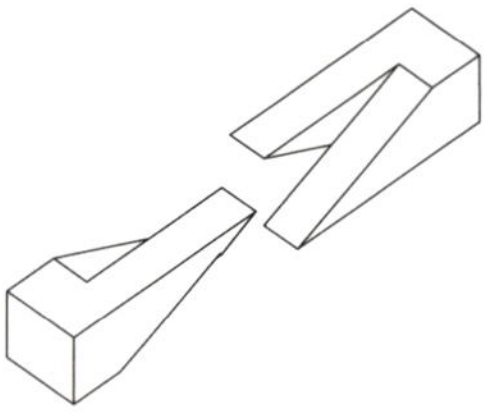

162

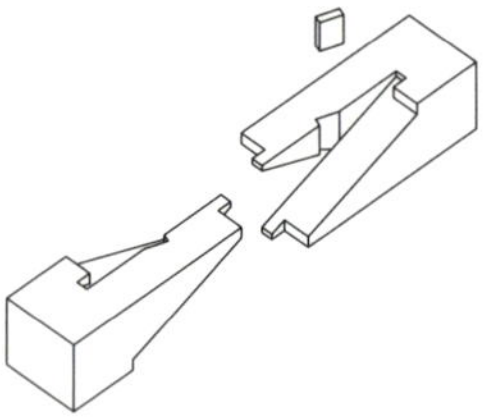

164

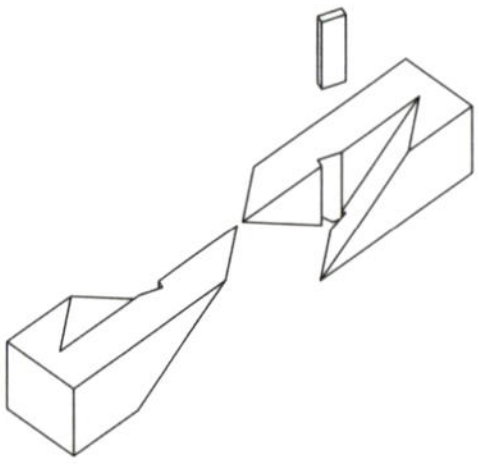

166

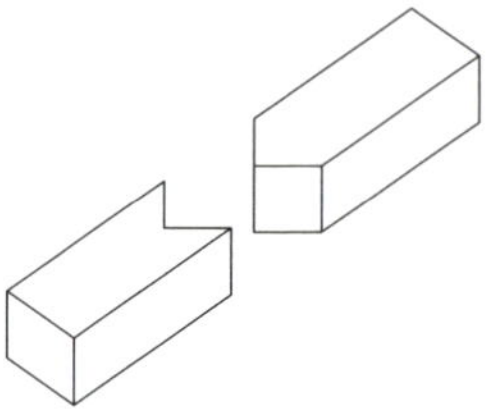

168

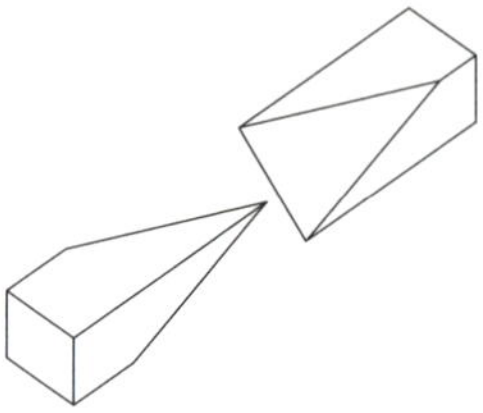

170

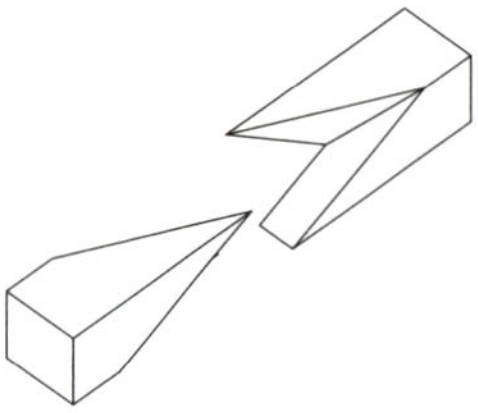

172

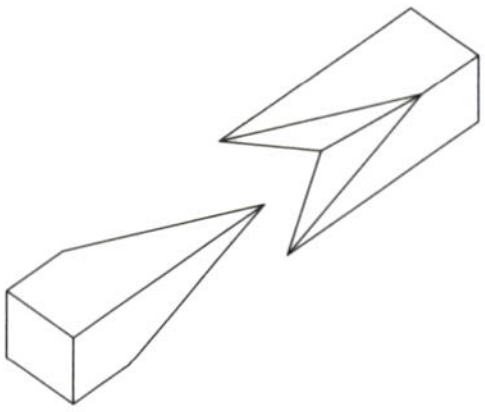

174

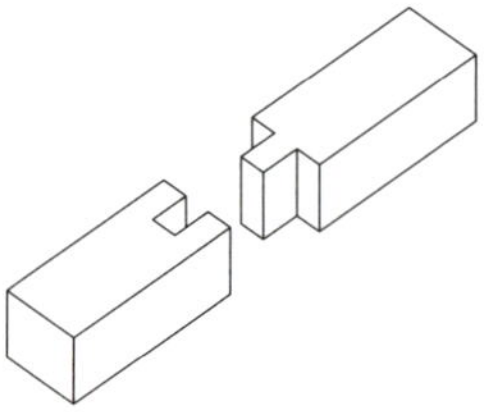

176

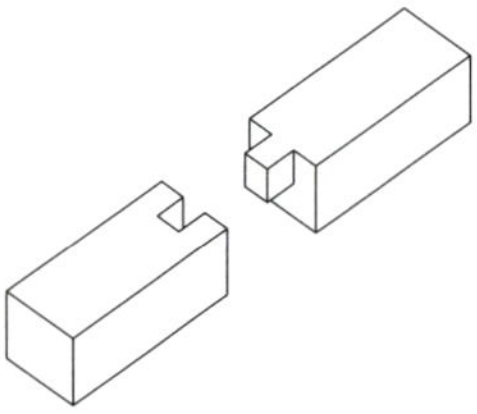

178

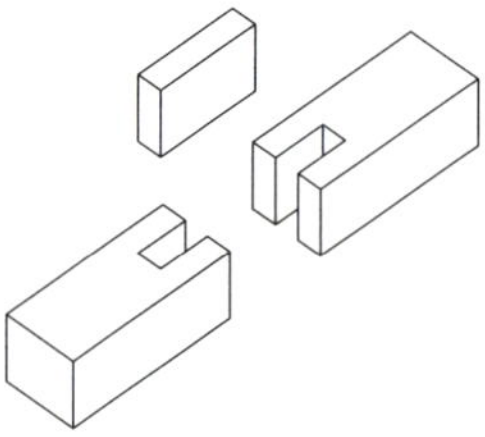

180

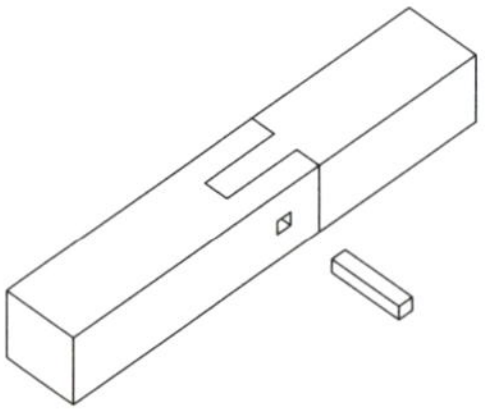

182

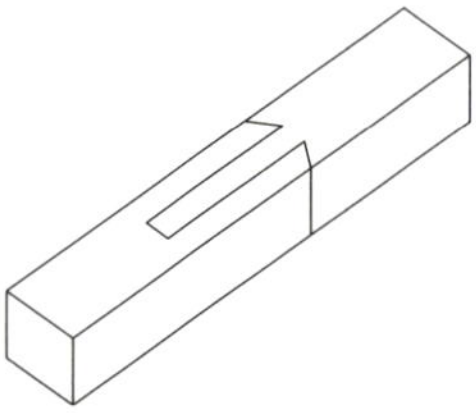

184

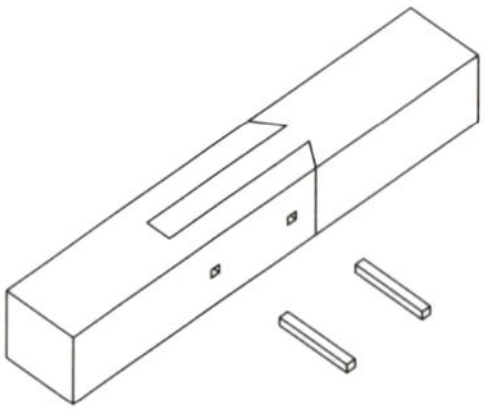

186

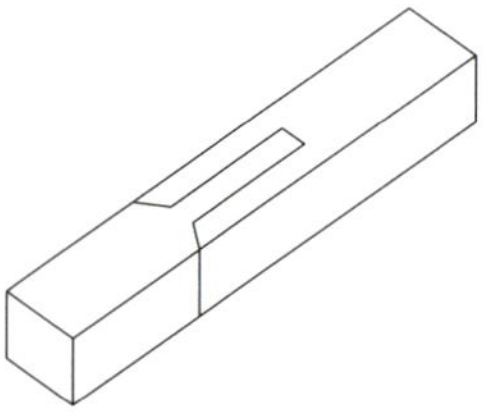

188

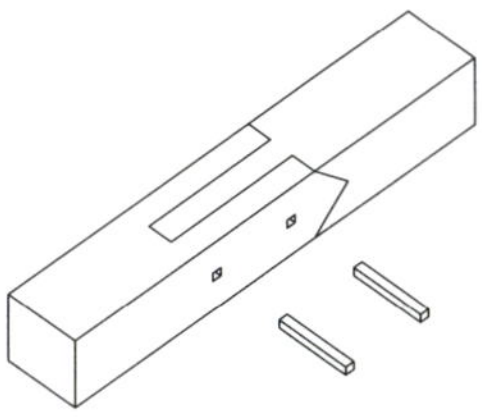

190

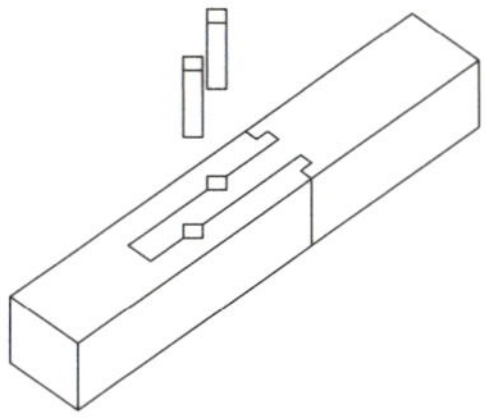

192

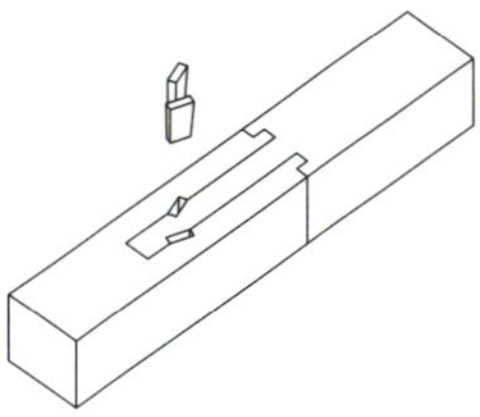

194

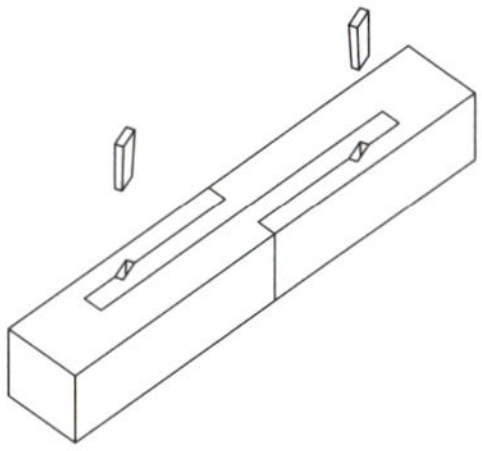

196

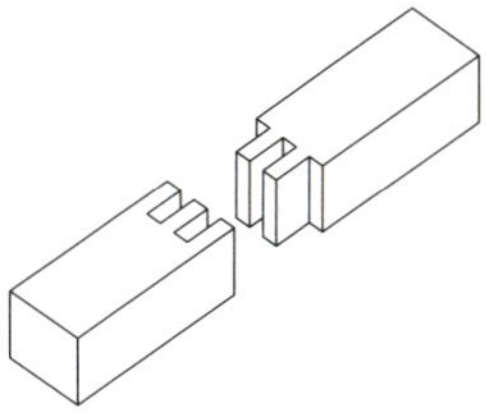

198

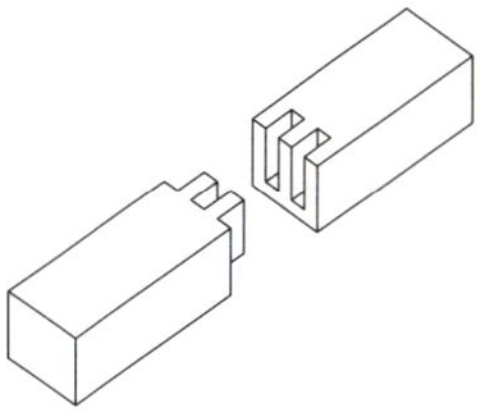

200

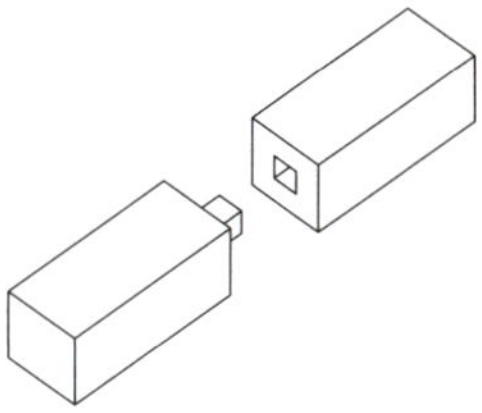

202

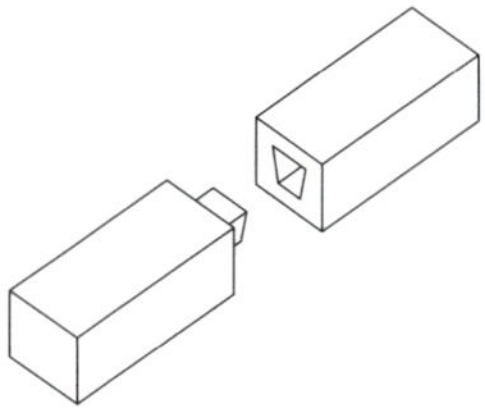

204

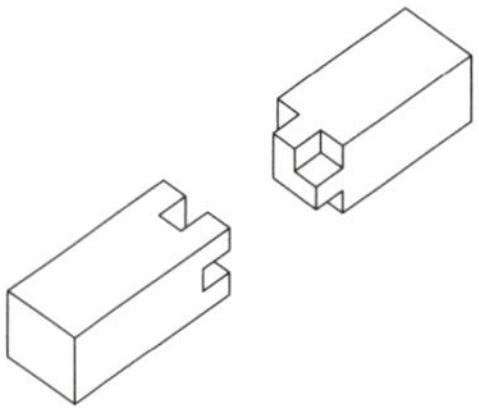

206

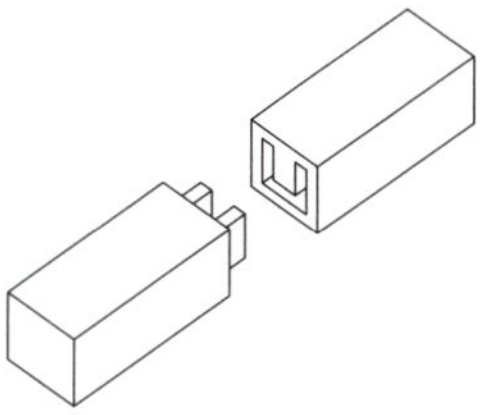

208

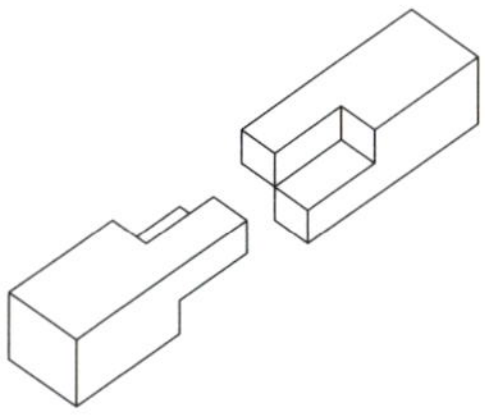

210

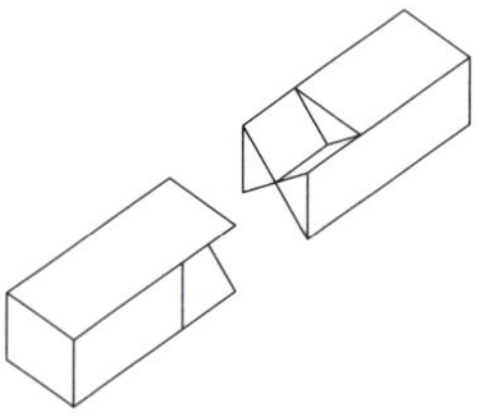

212

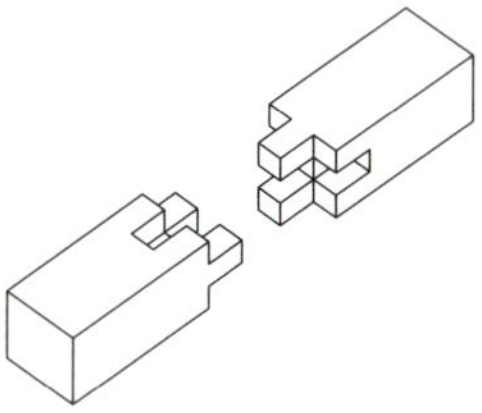

214

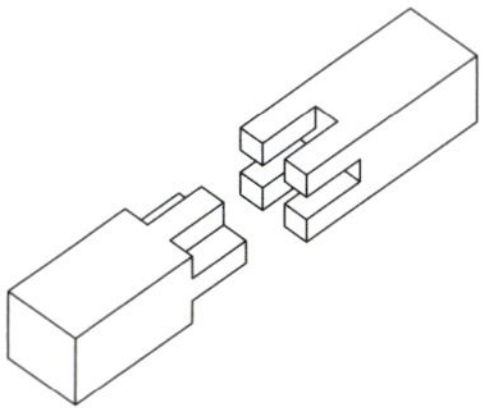

216

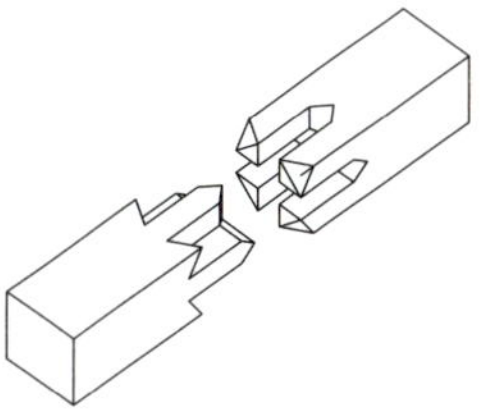

218

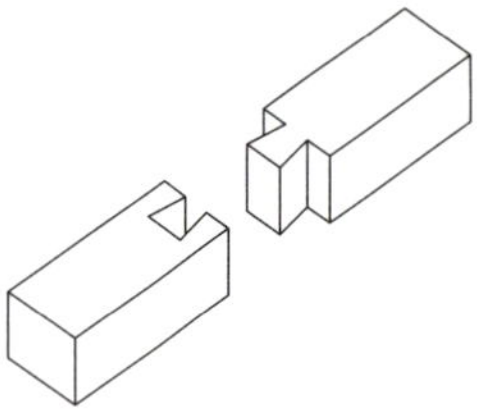

220

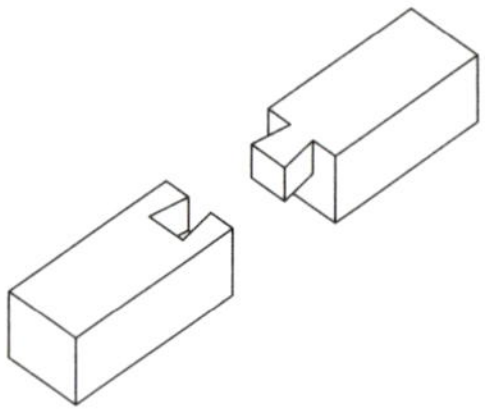

222

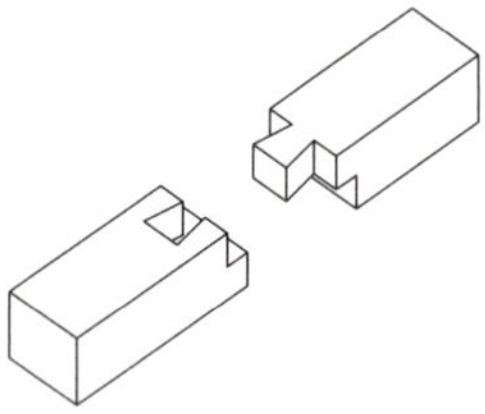

224

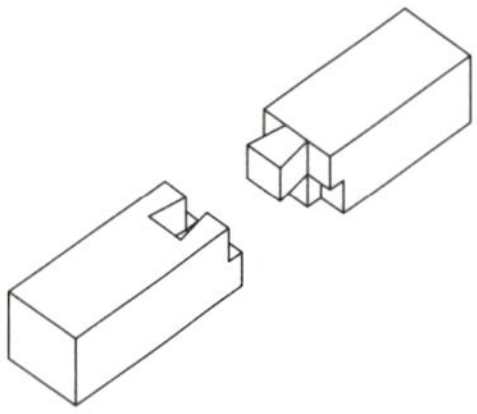

226

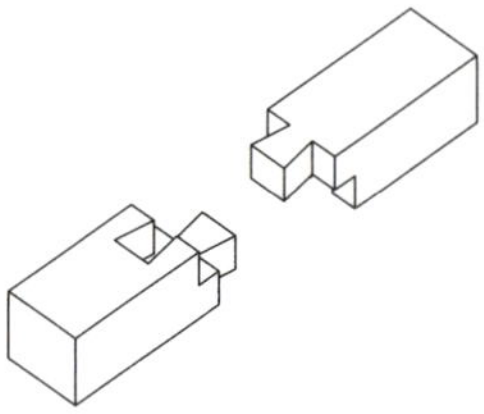

228

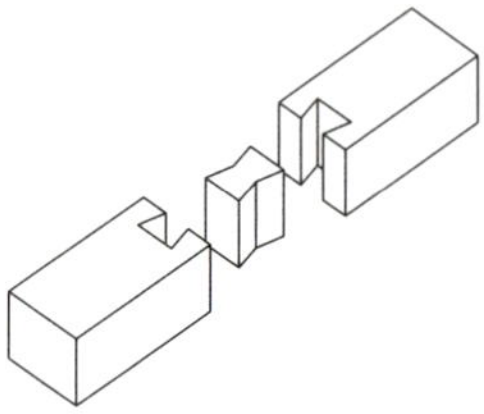

230

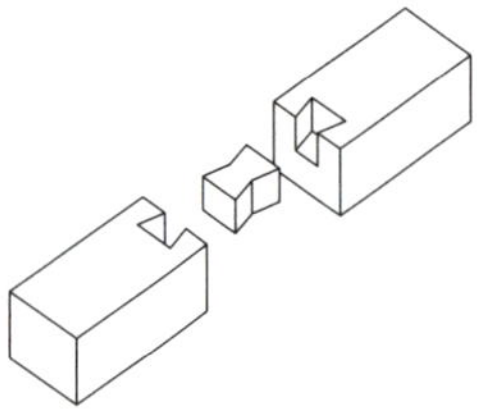

232

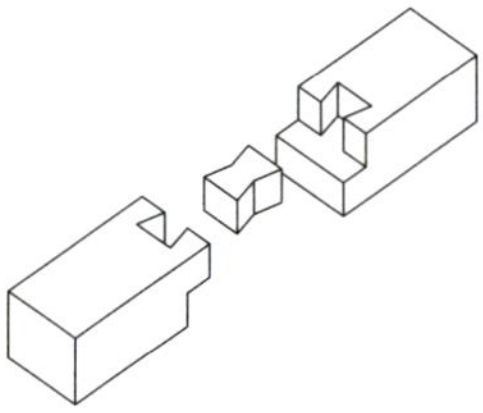

234

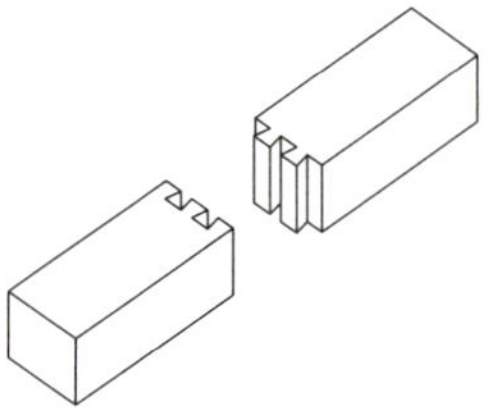

236

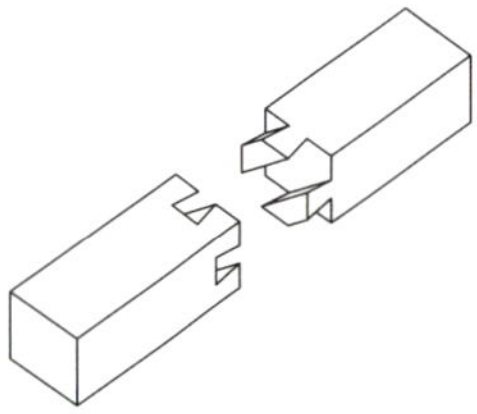

238

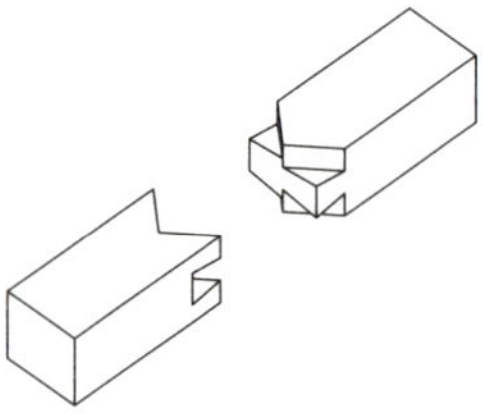

240

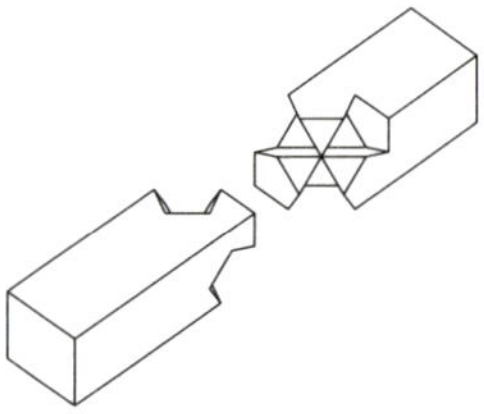

242

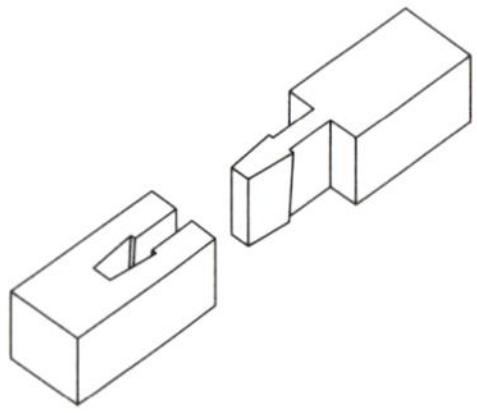

244

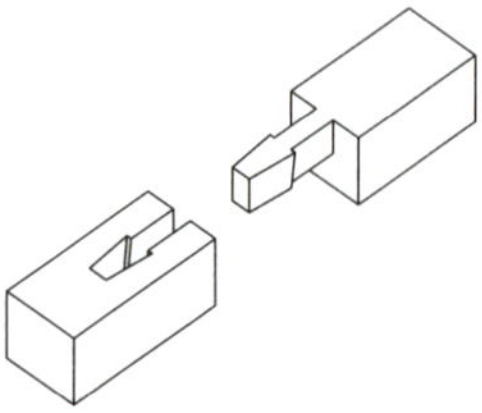

246

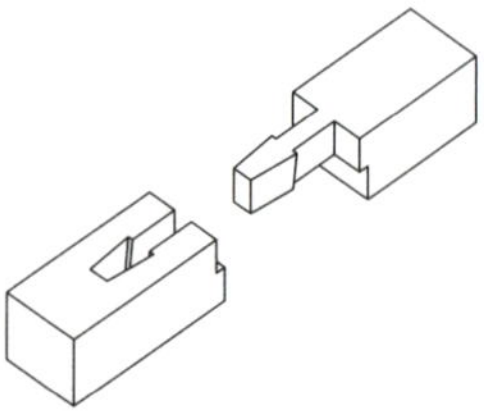

248

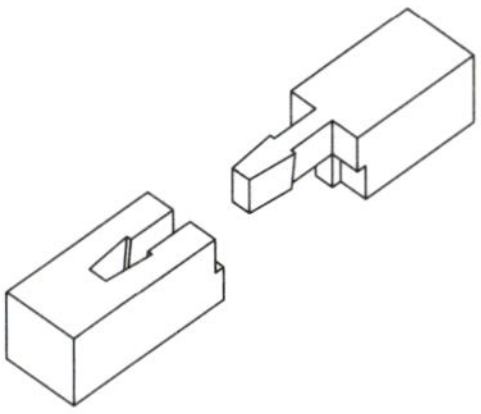

250

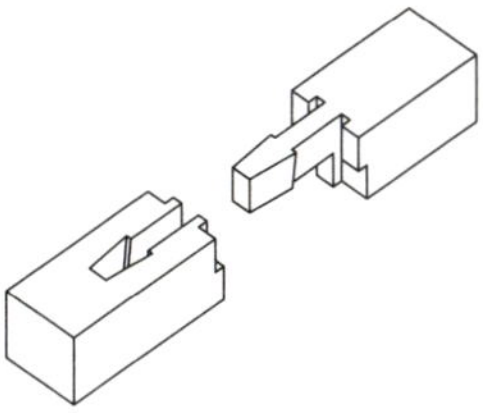

252

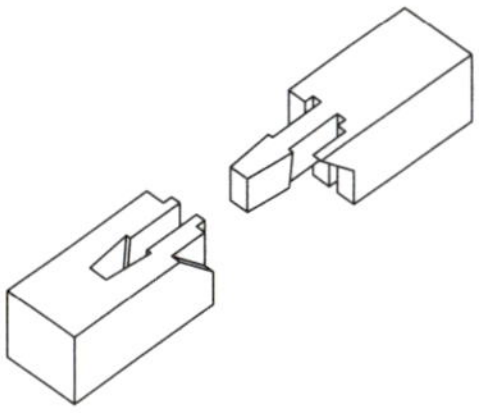

254

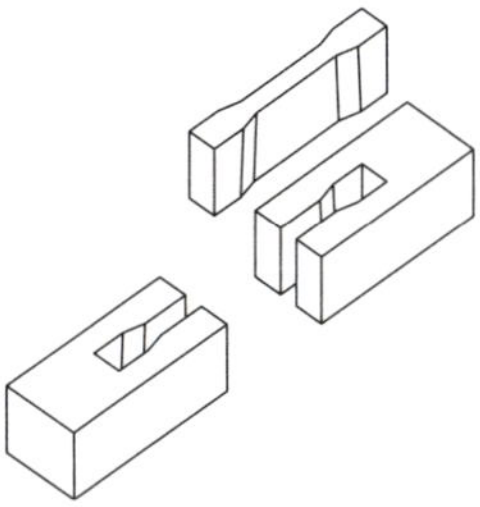

256

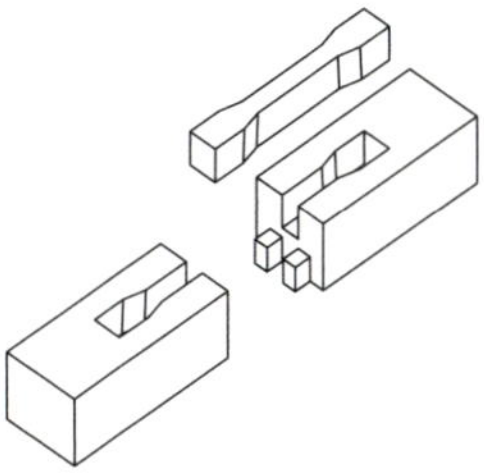

258

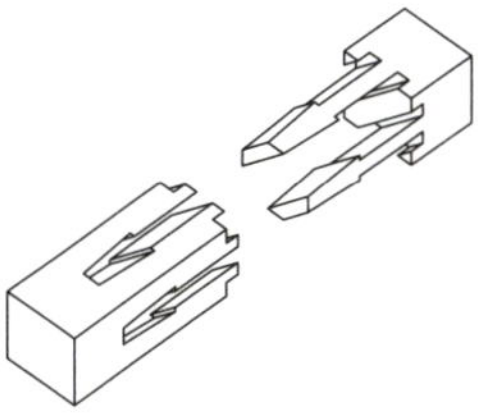

260

Corner Joints

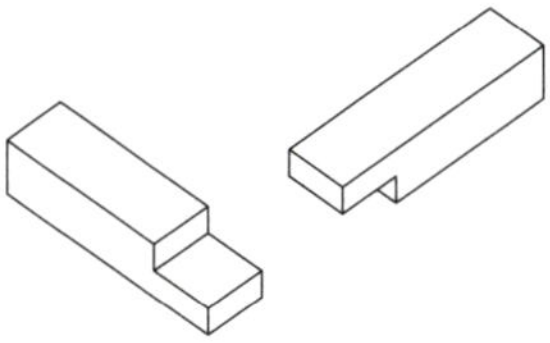

262

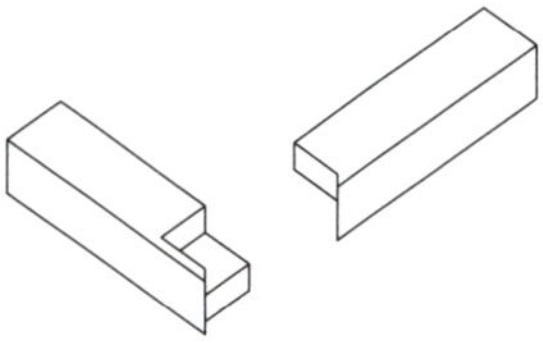

264

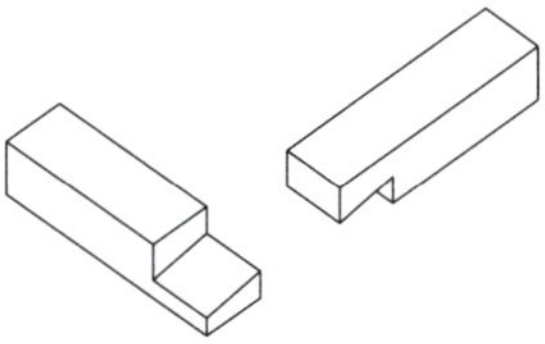

266

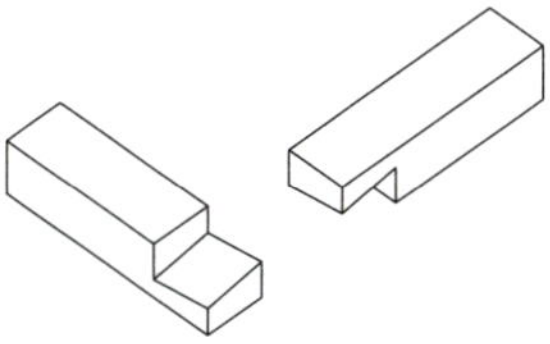

268

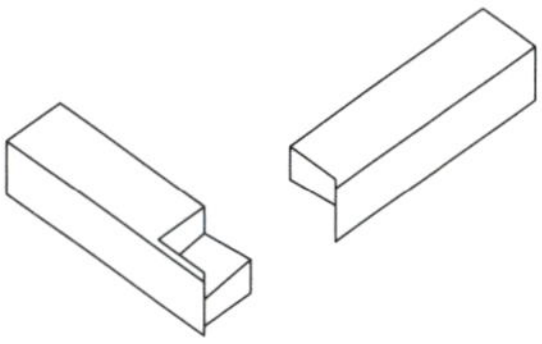

270

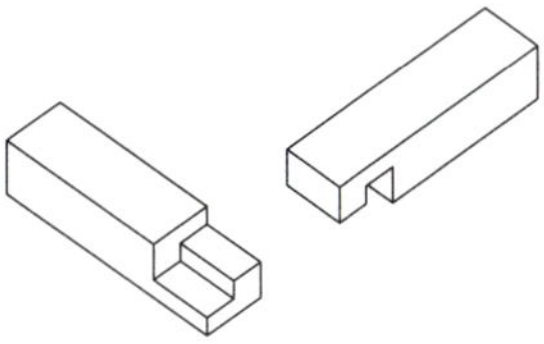

272

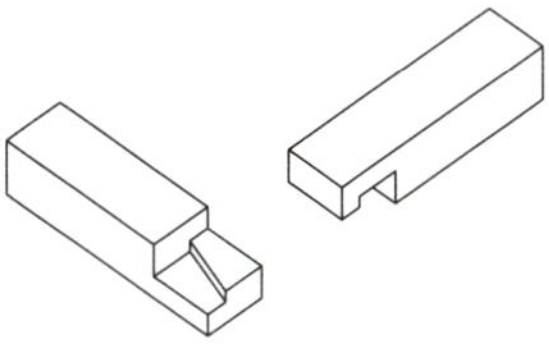

274

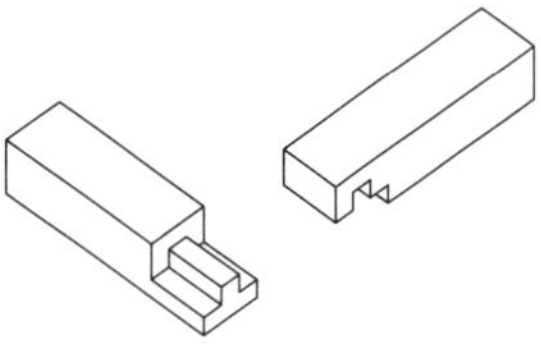

276

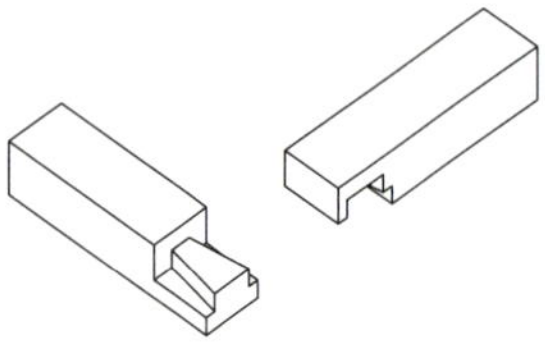

278

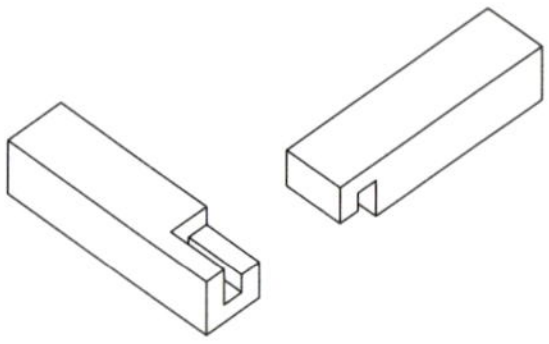

280

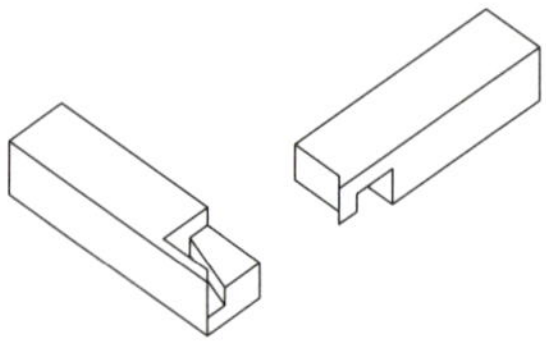

282

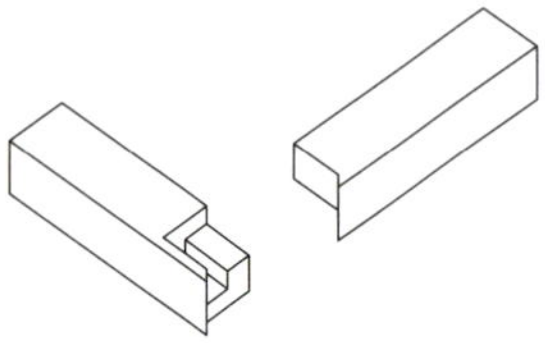

284

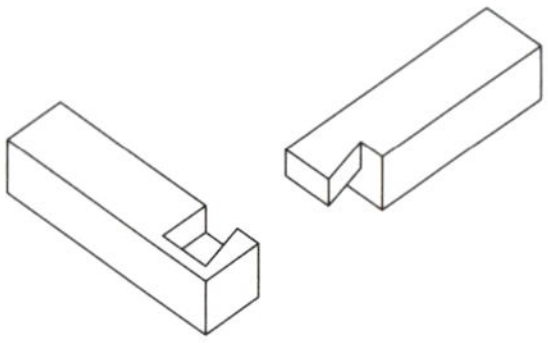

286

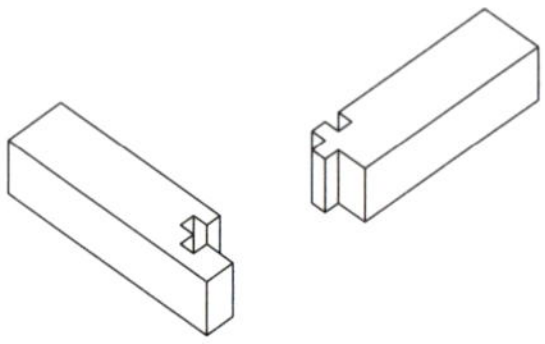

288

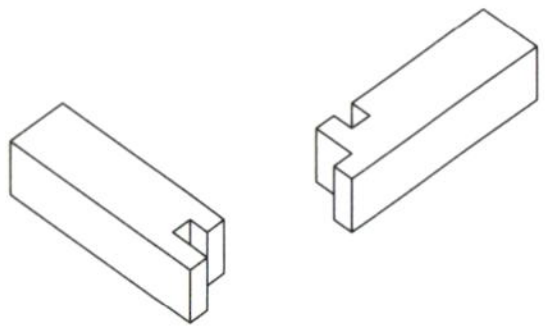

290

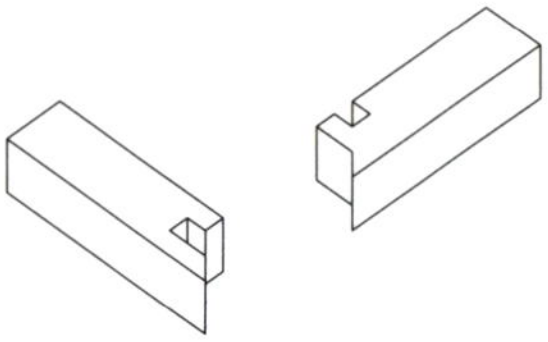

292

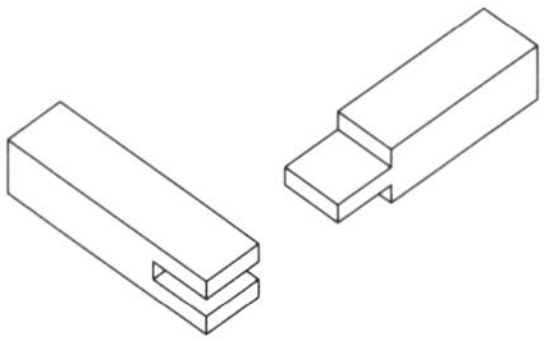

294

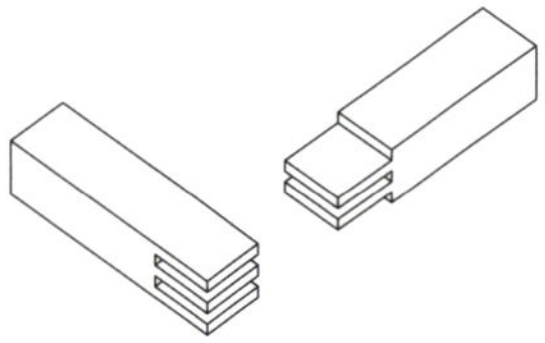

296

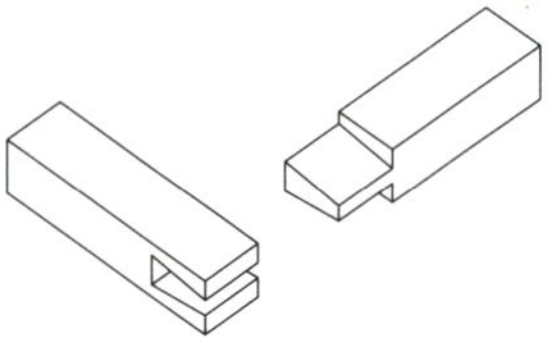

298

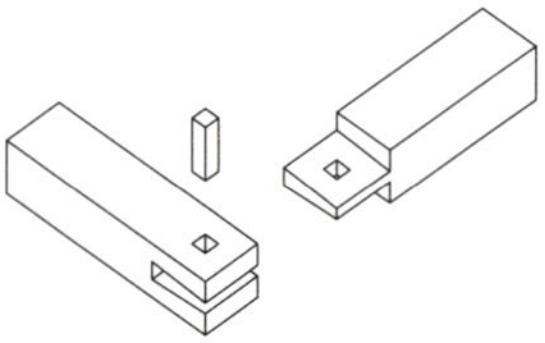

300

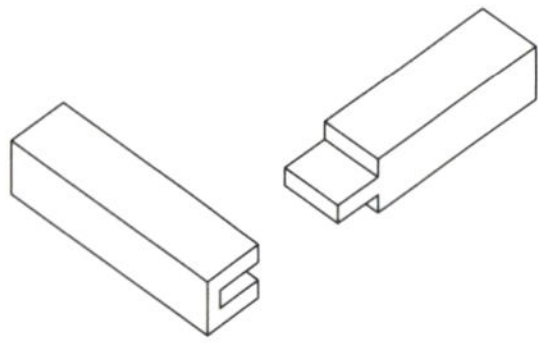

302

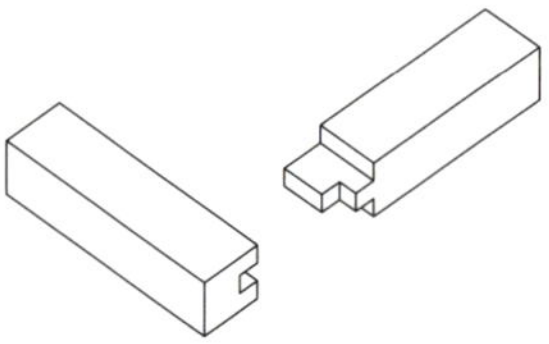

304

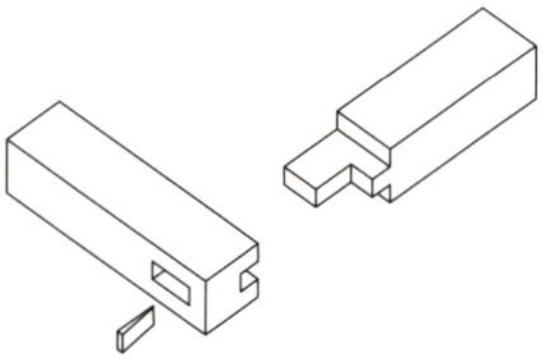

306

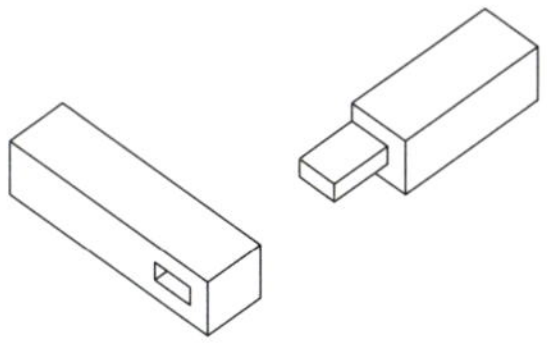

308

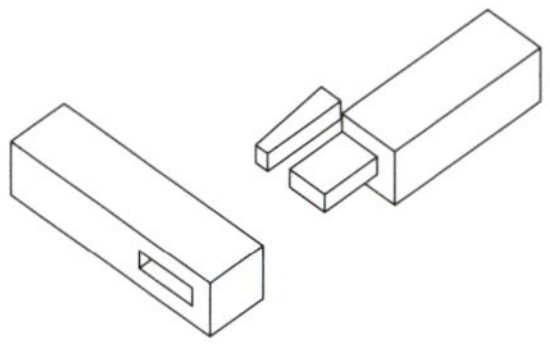

310

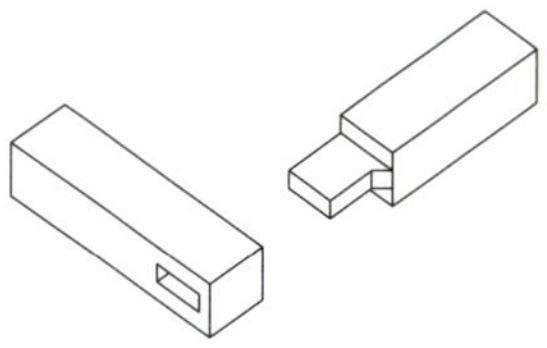

312

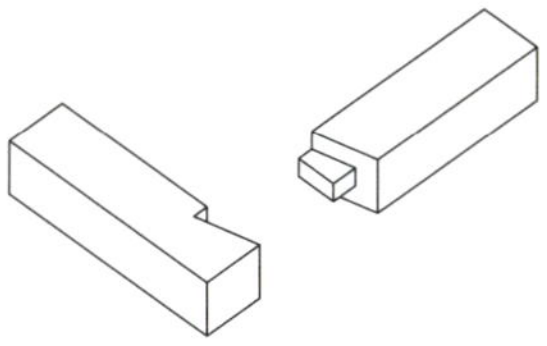

314

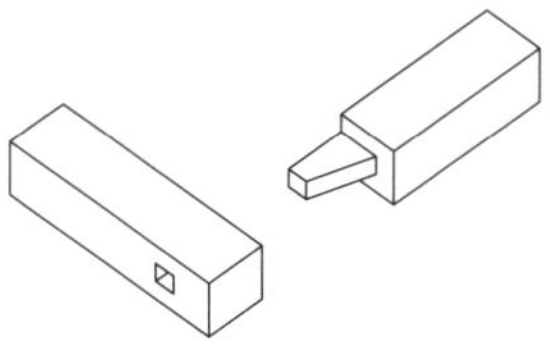

316

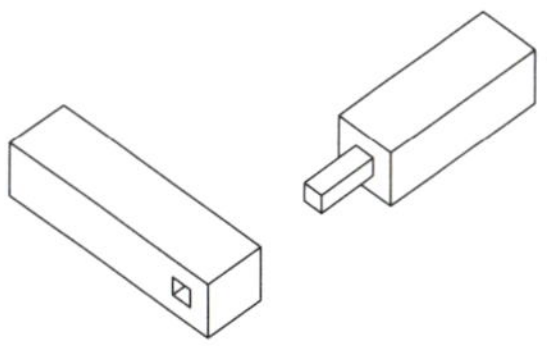

318

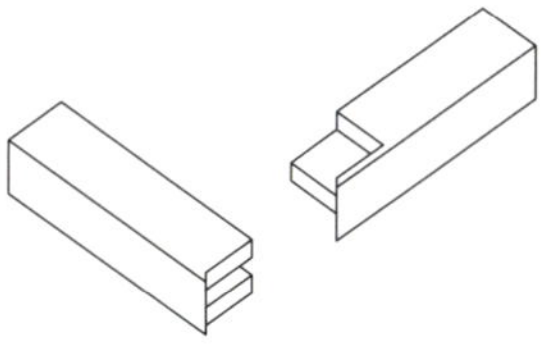

320

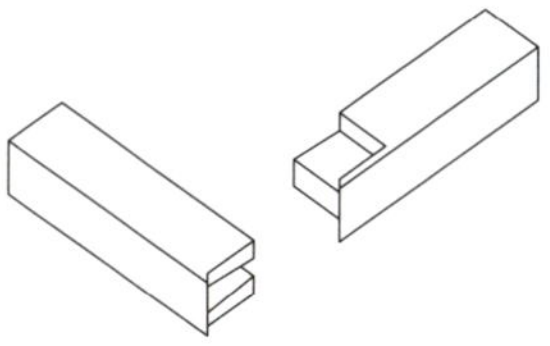

322

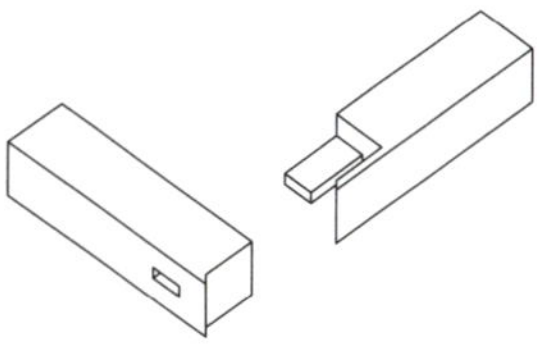

324

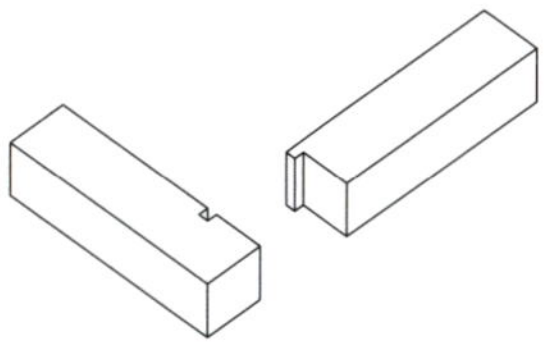

326

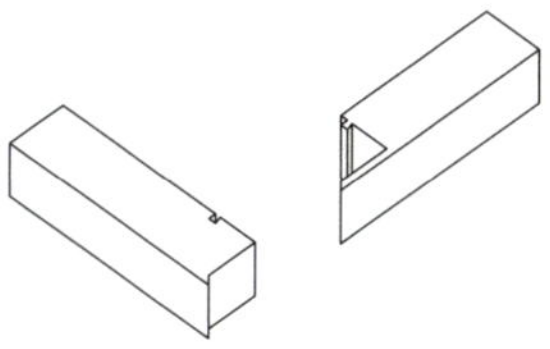

328

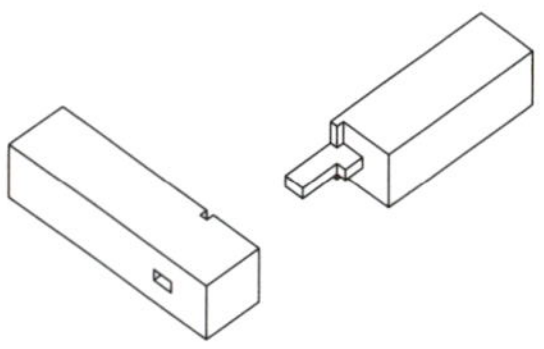

330

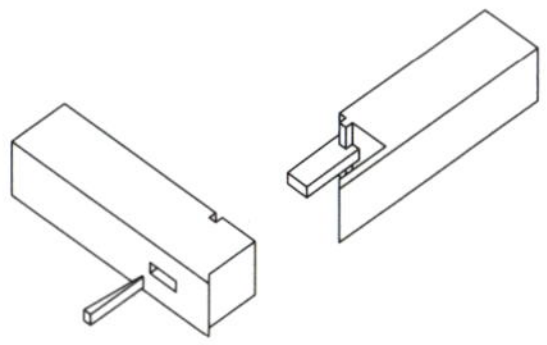

332

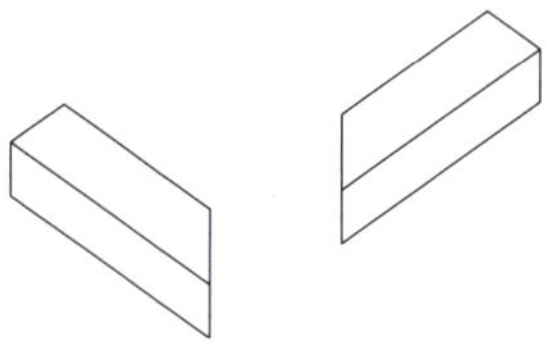

334

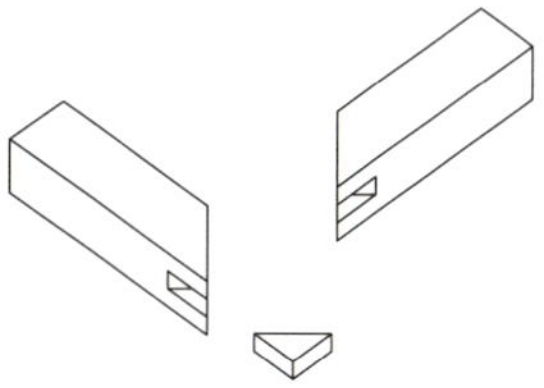

336

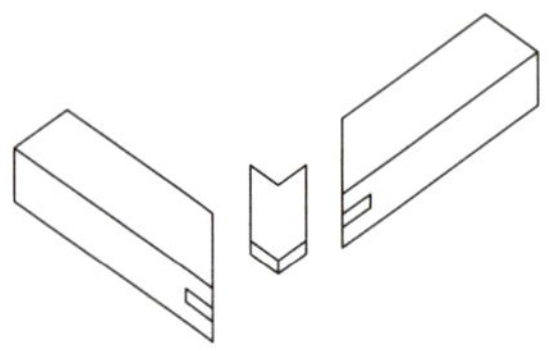

338

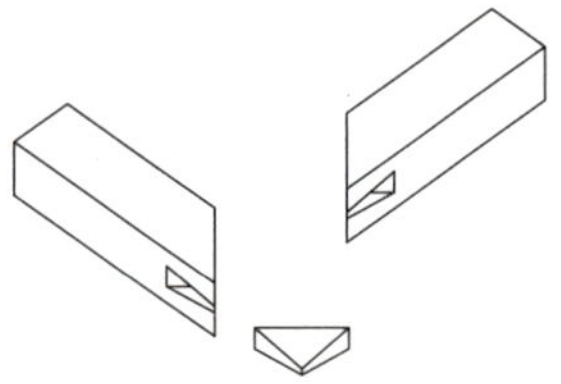

340

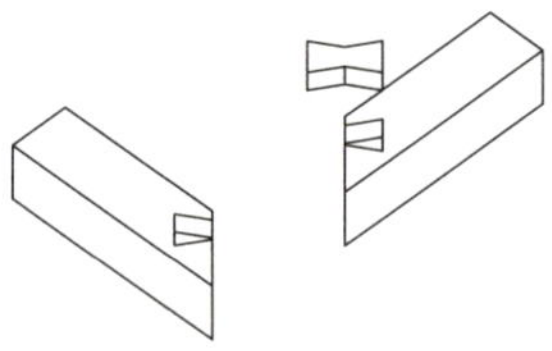

342

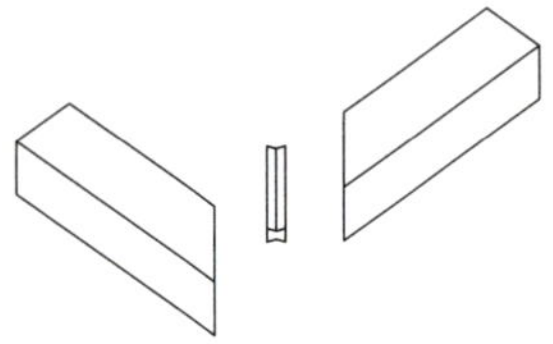

344

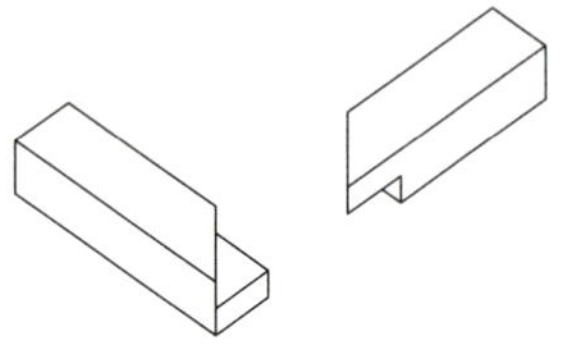

346

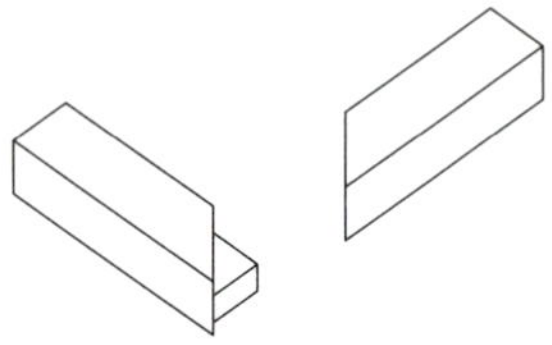

348

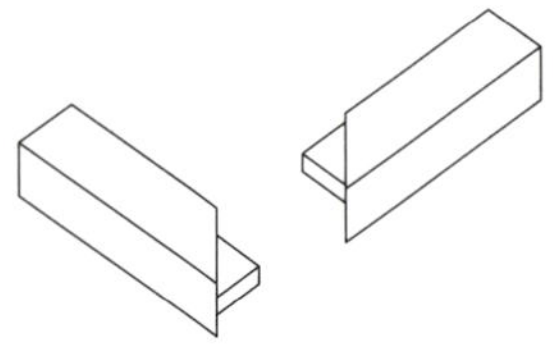

350

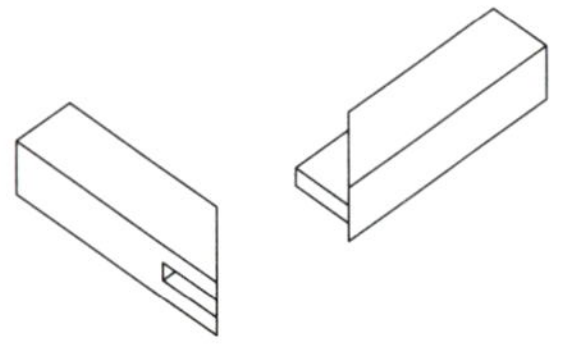

352

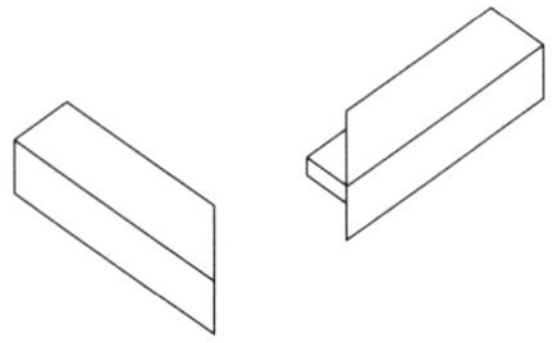

354

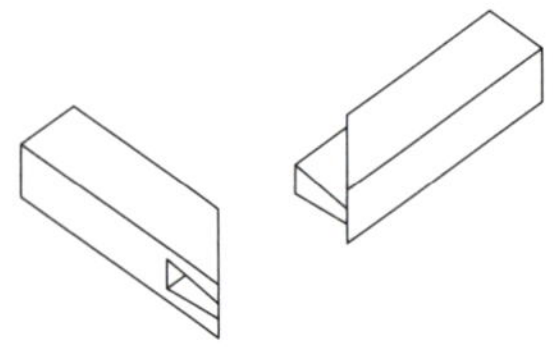

356

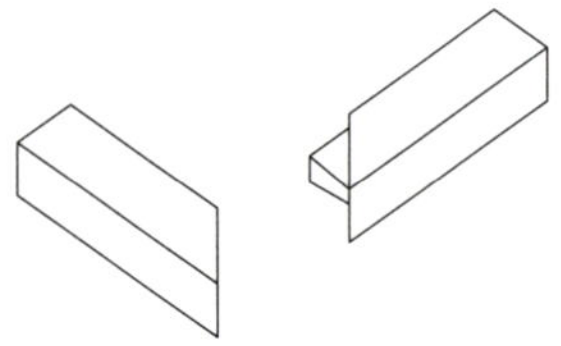

358

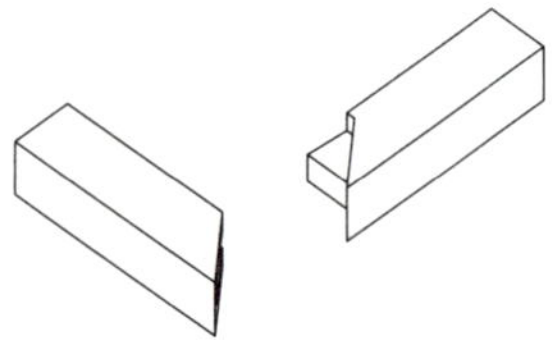

360

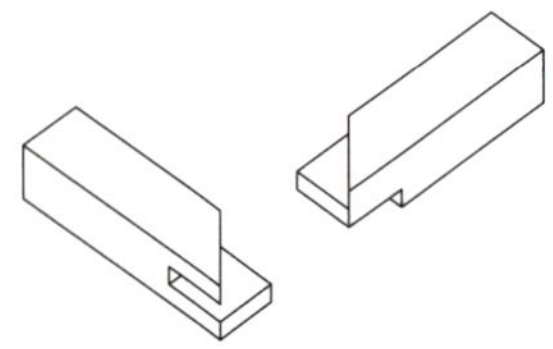

362

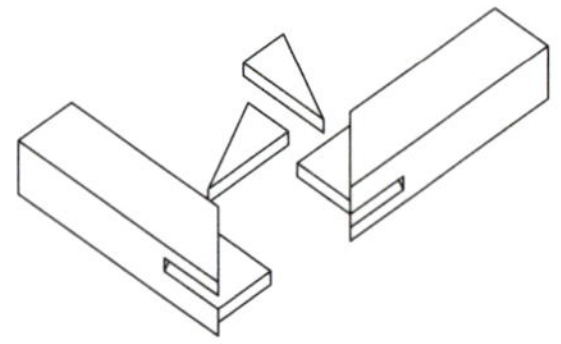

364

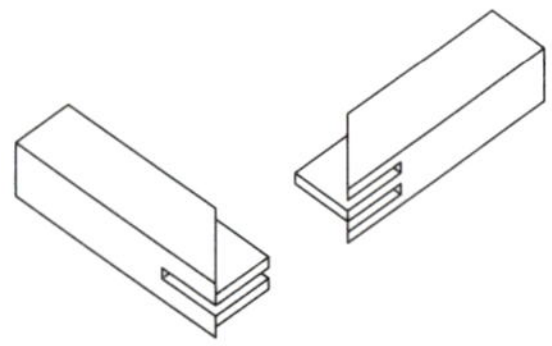

366

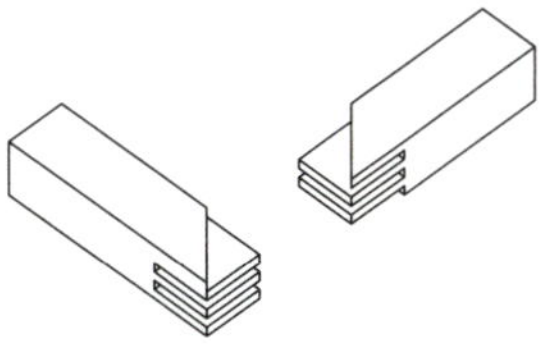
368

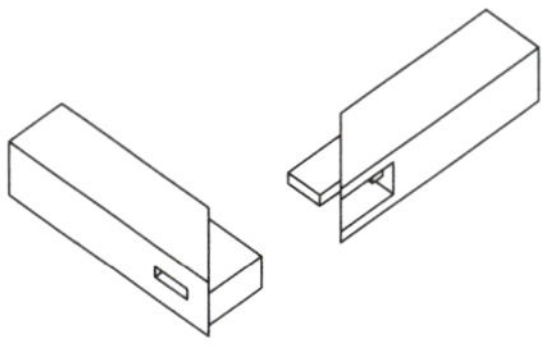
370

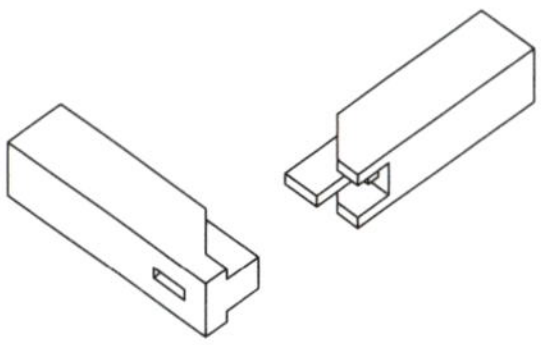
372

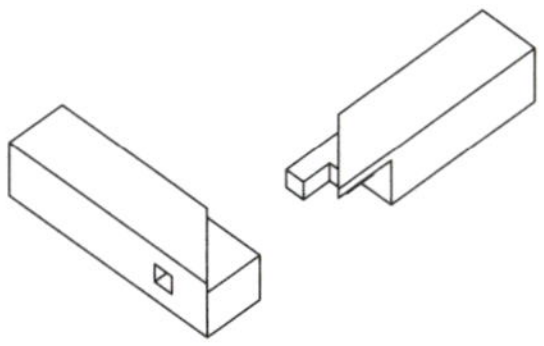
374

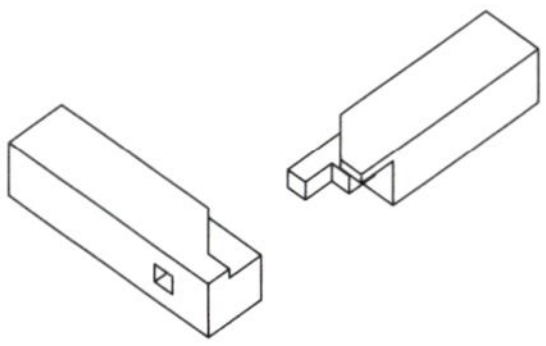
376

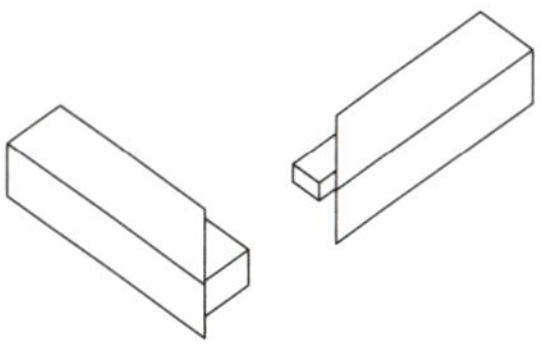
378

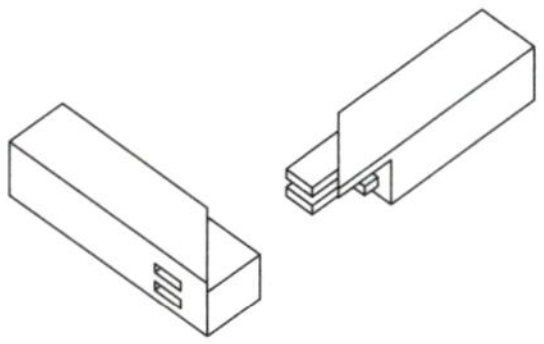
380

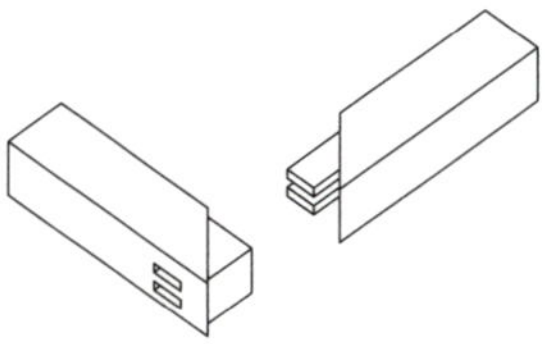
382

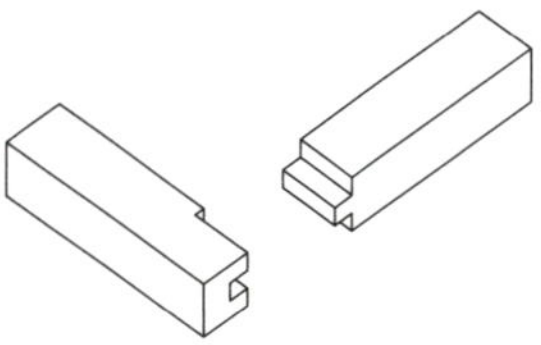
384

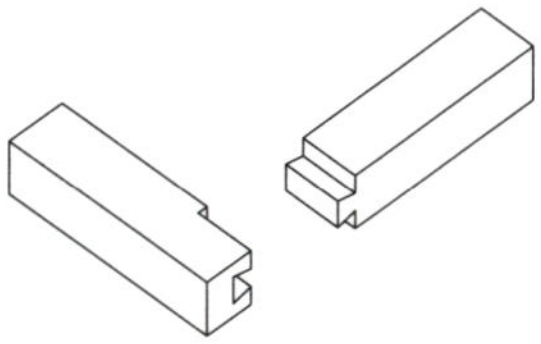
386

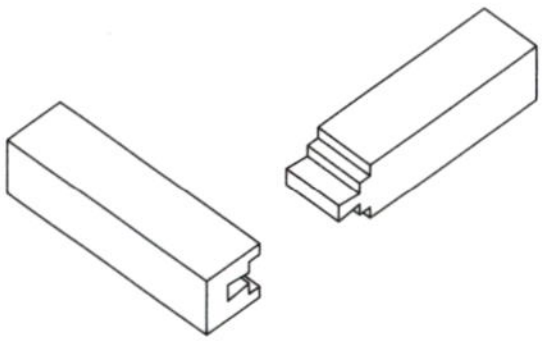
388

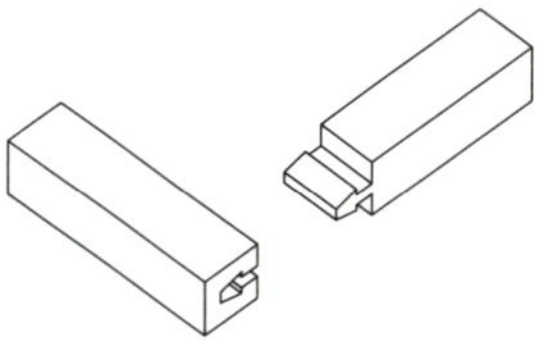
390

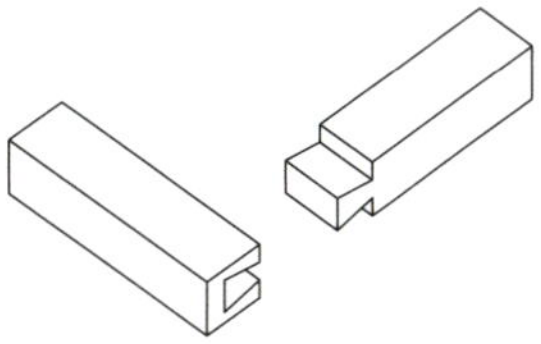
392

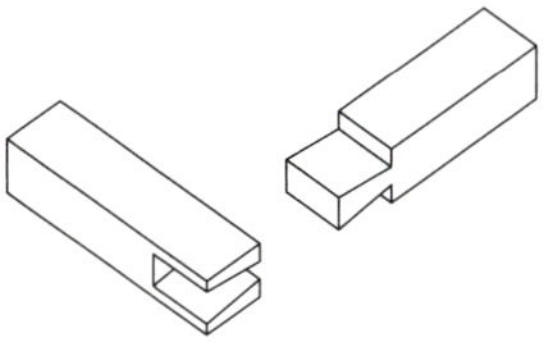
394

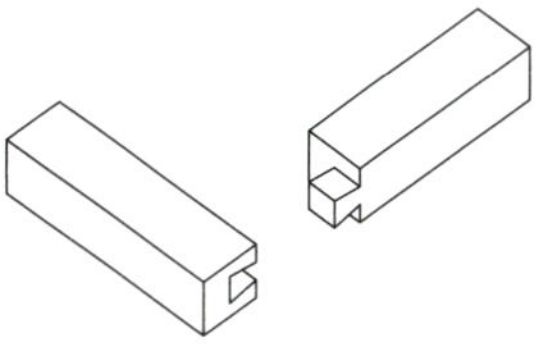
396

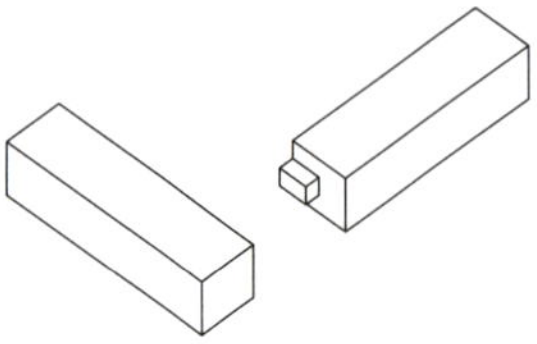

398

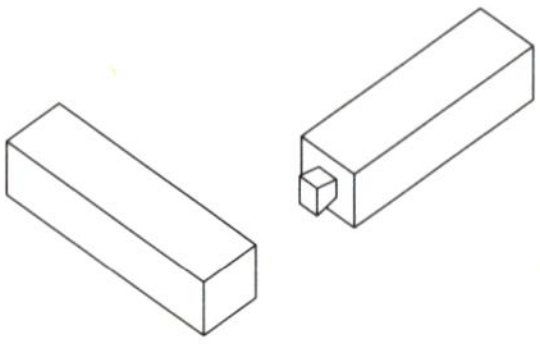

400

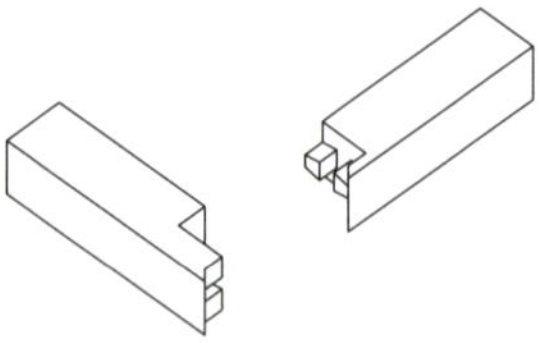

402

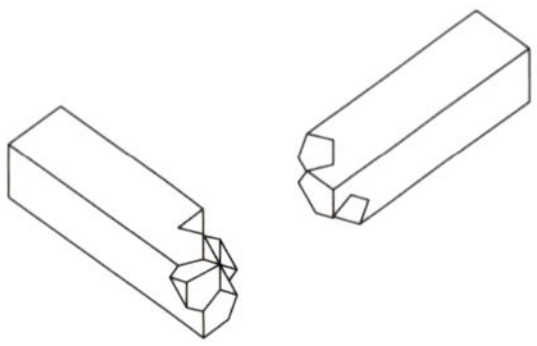

404

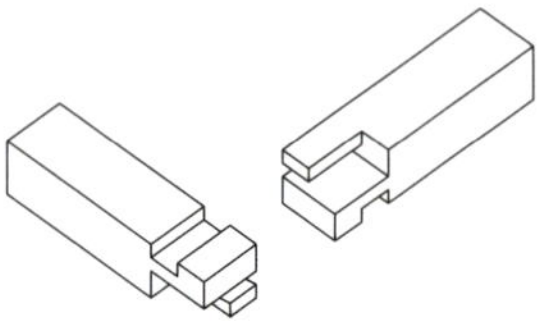

406

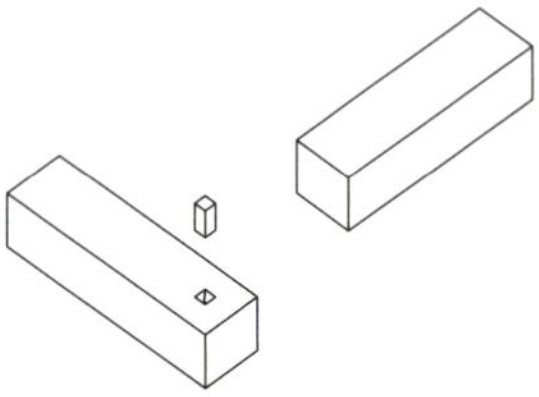

408

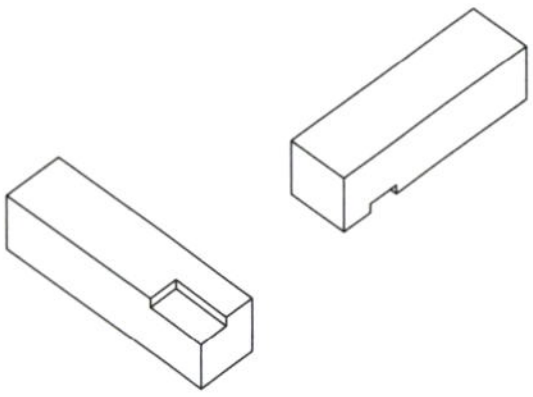

410

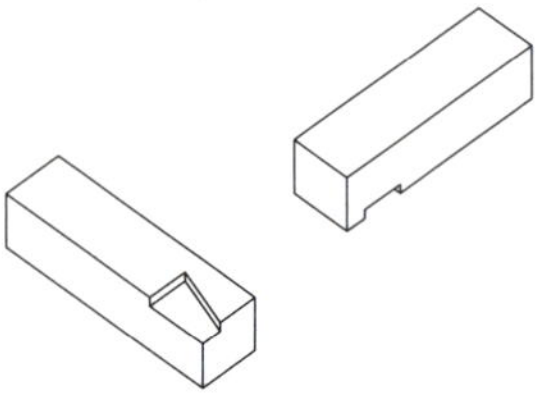

412

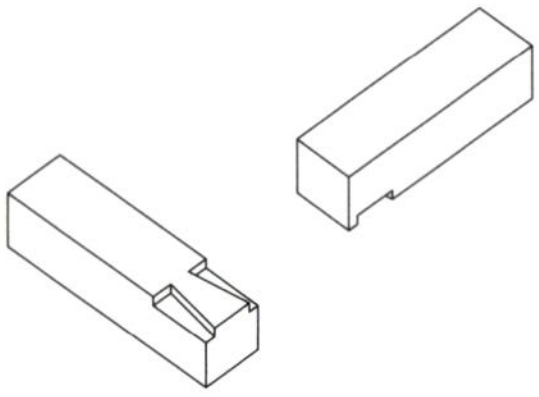

414

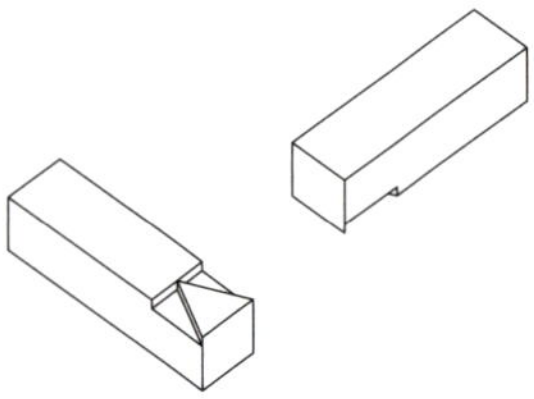

416

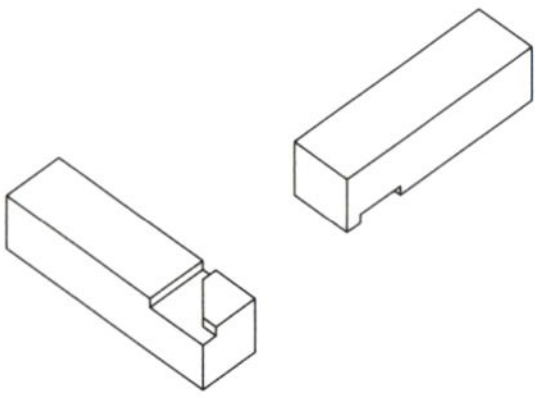

418

T-Joints

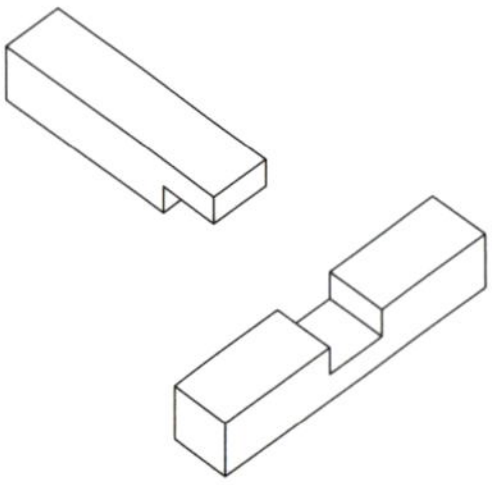

420

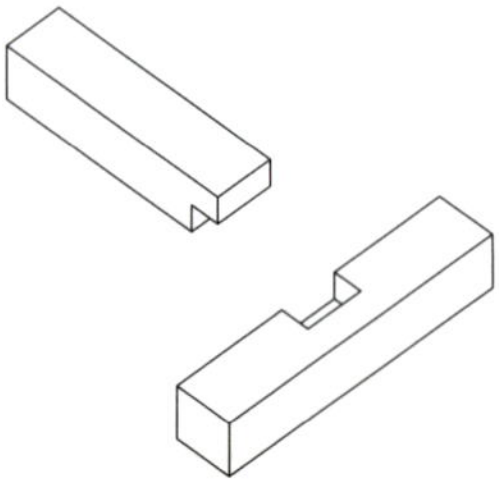

422

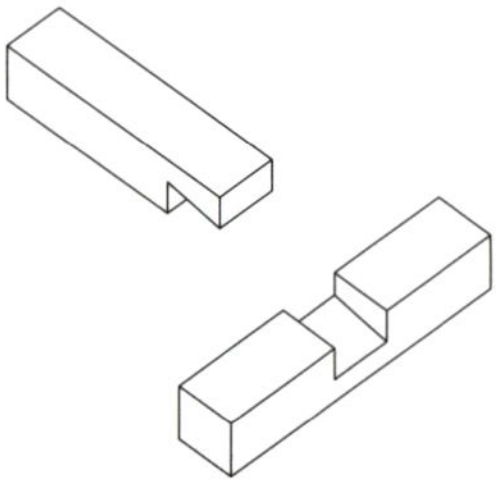

424

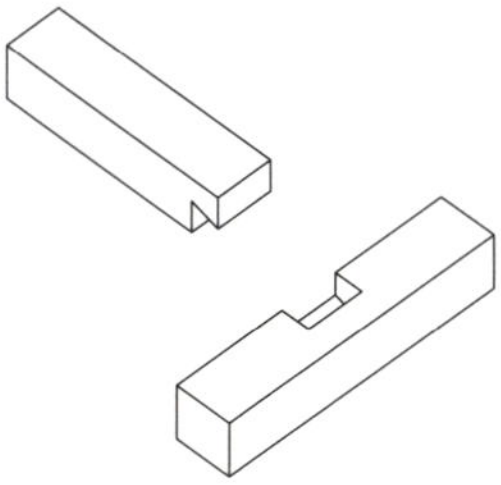

426

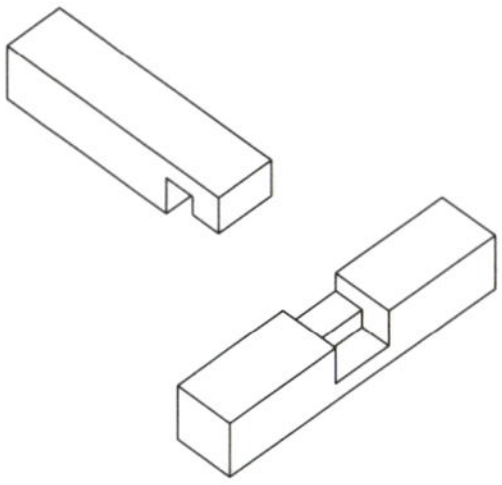

428

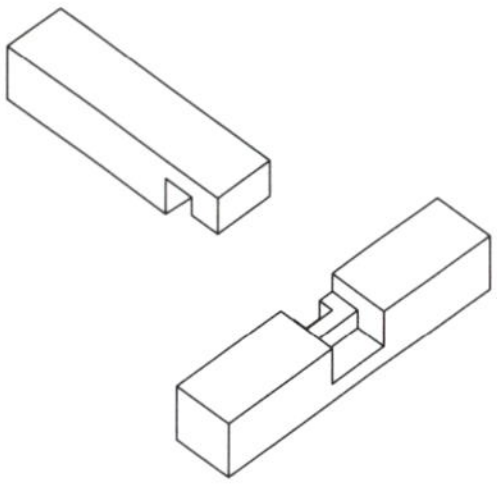

430

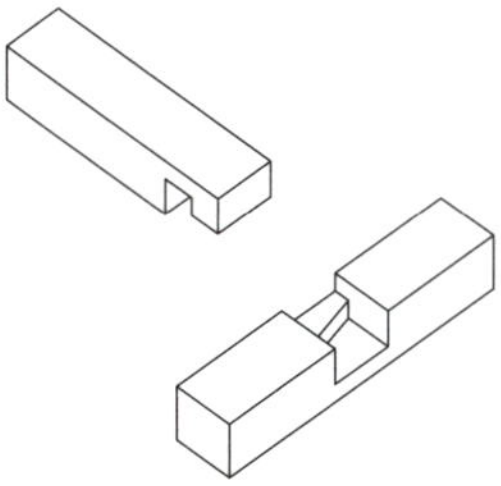

432

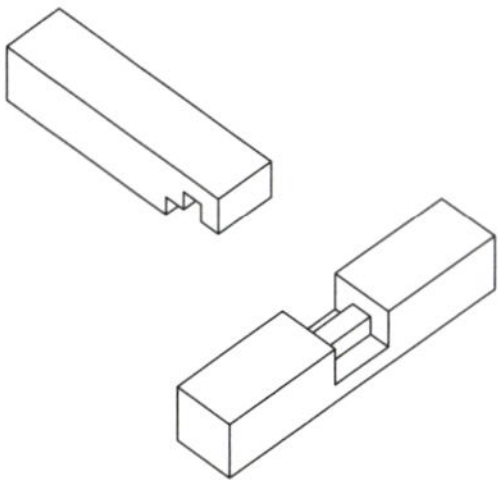

434

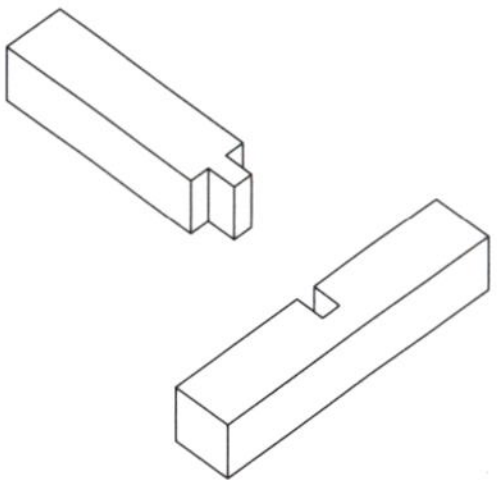

436

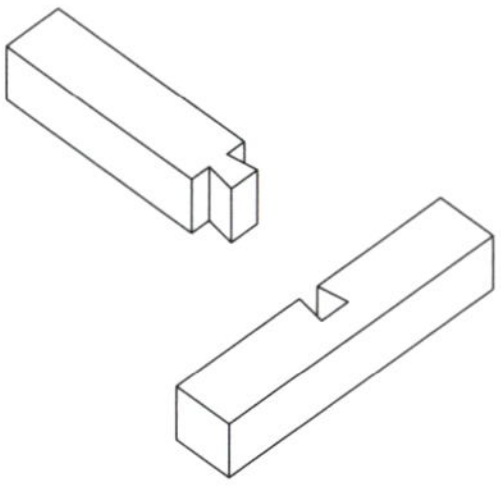

438

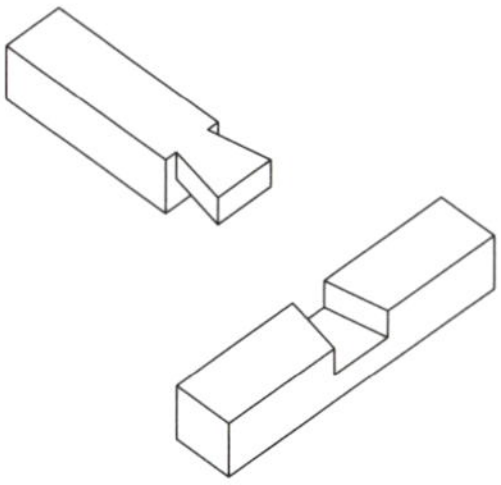

440

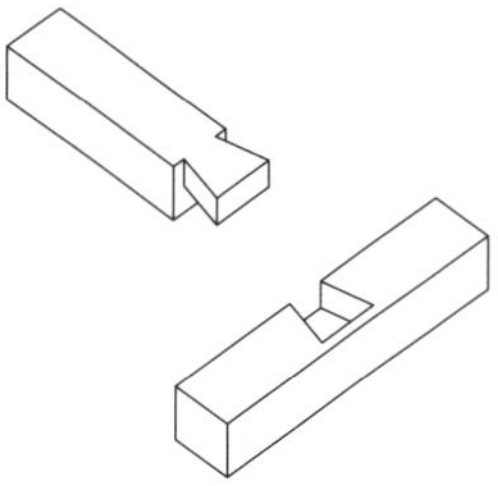

442

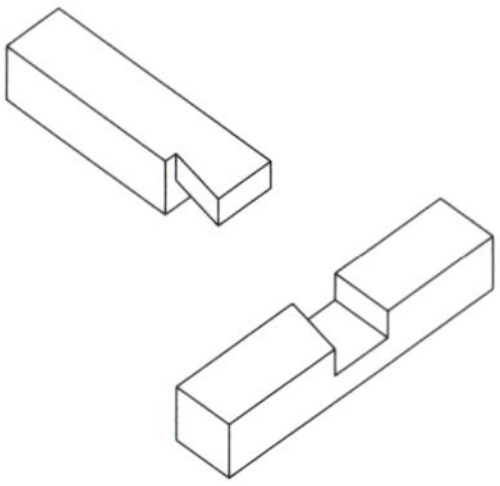

444

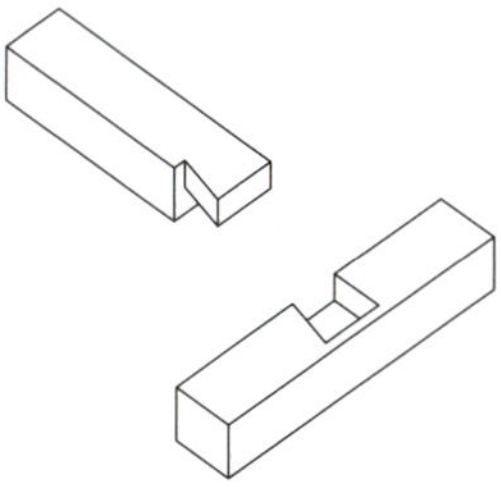

446

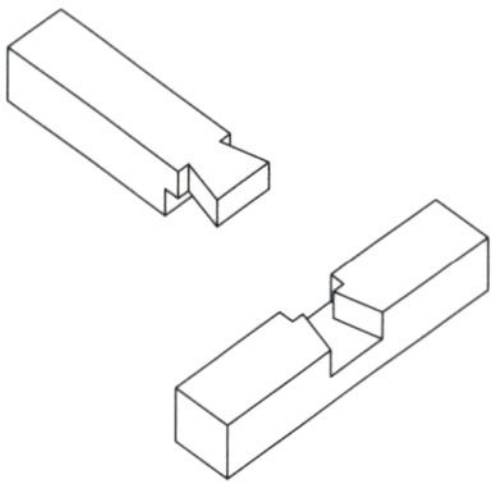

448

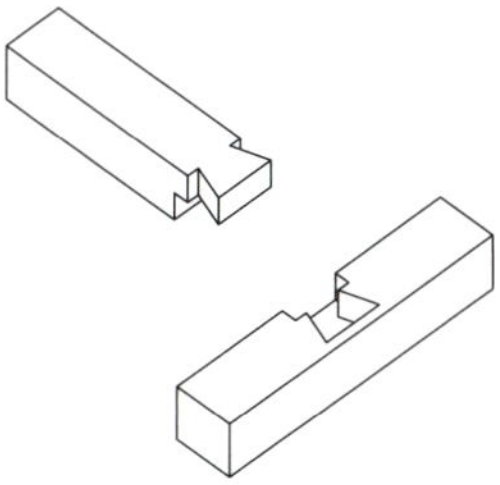

450

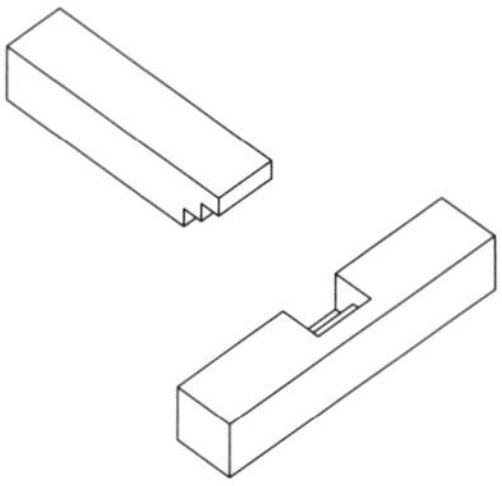

452

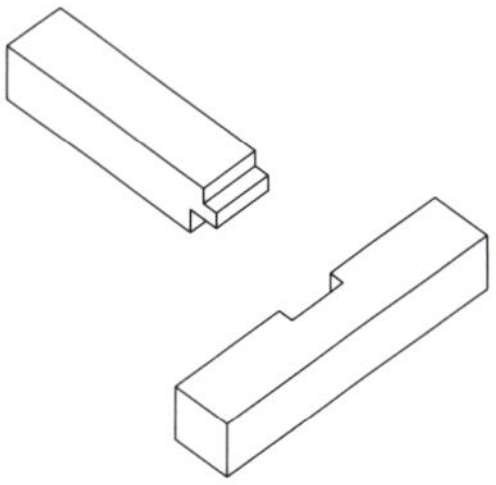

454

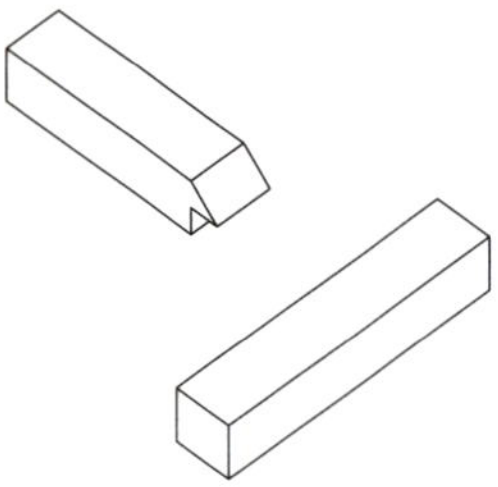
456

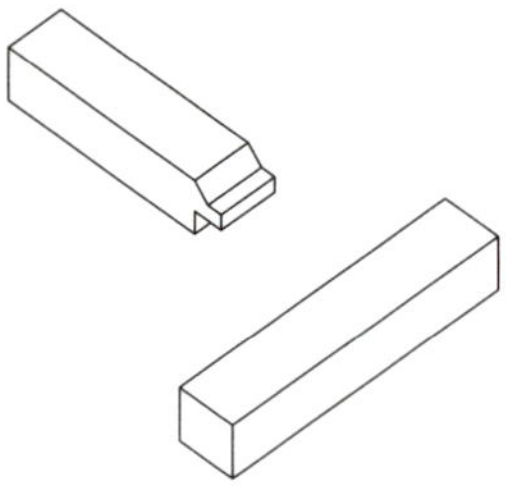
458

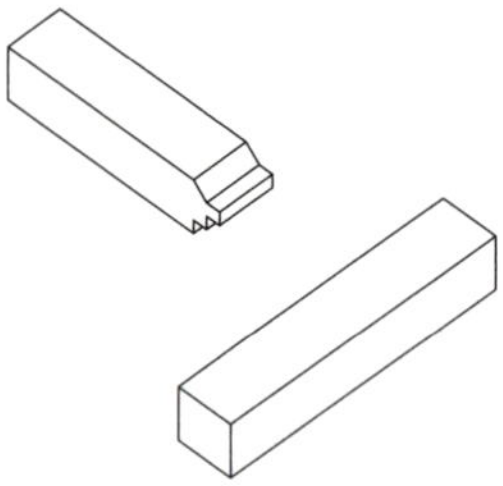
460

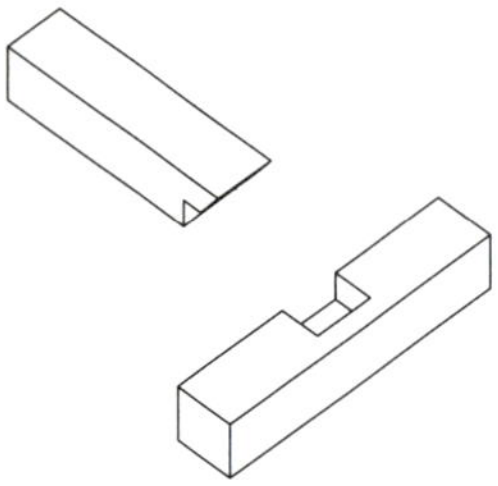
462

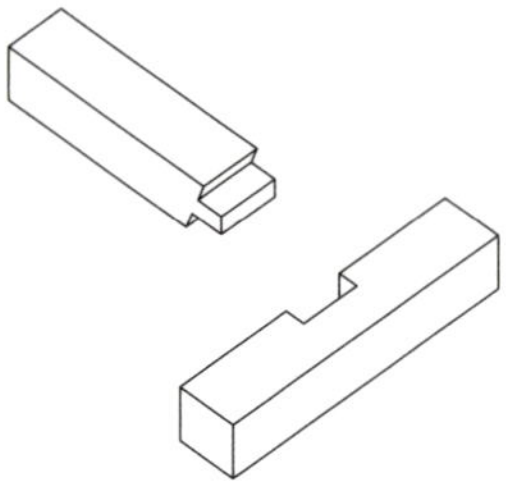
464

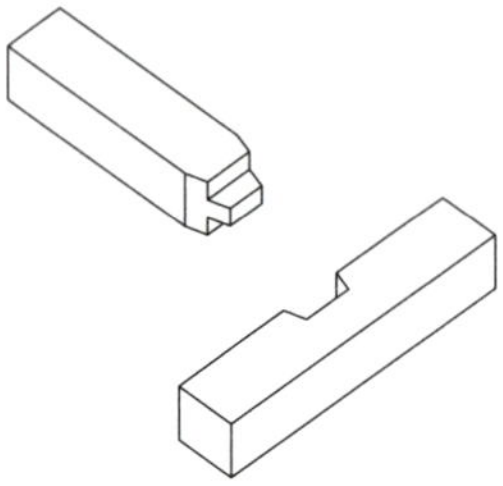
466

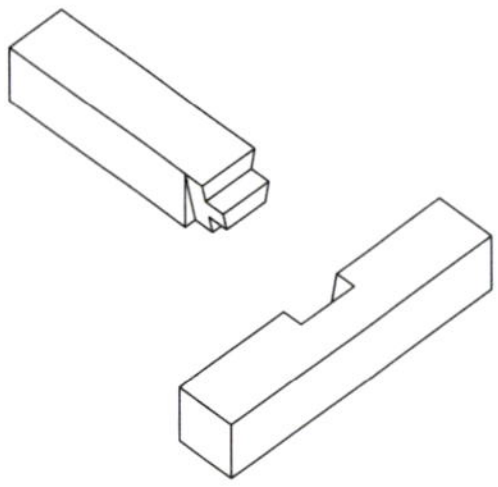
468

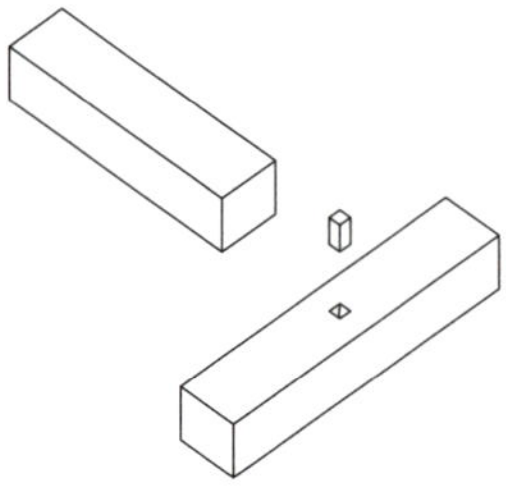
470

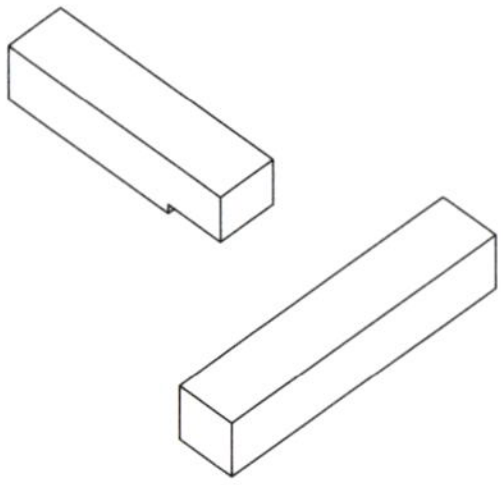
472

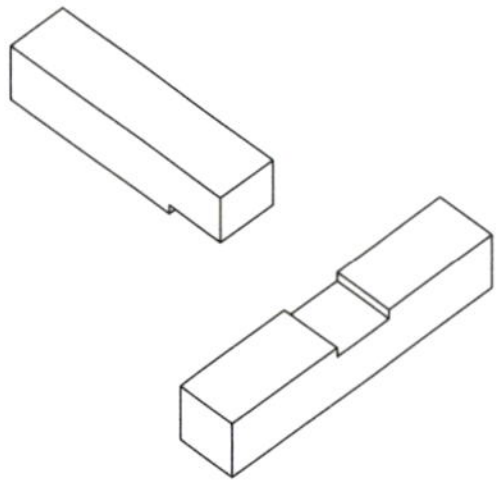
474

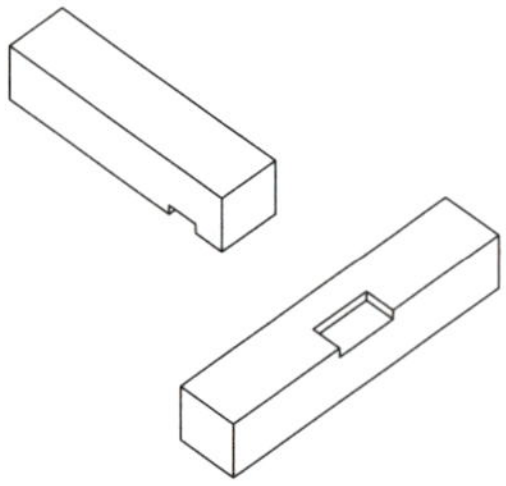
476

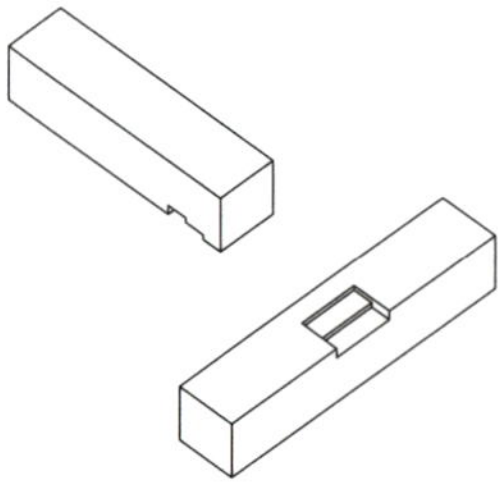
478

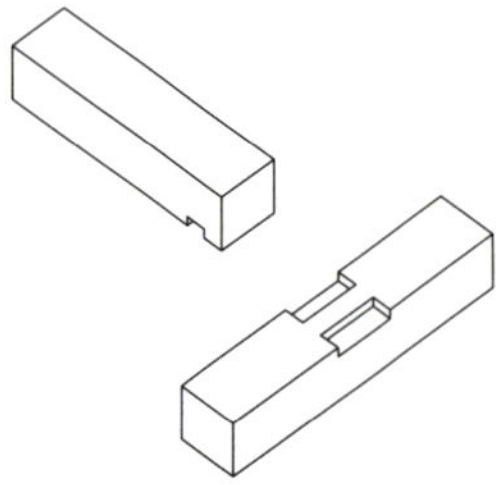
480

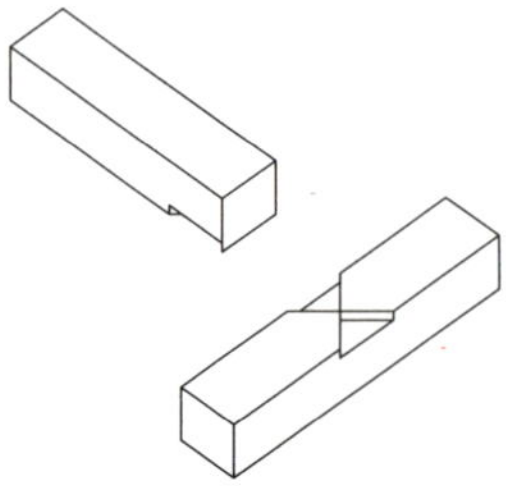
482

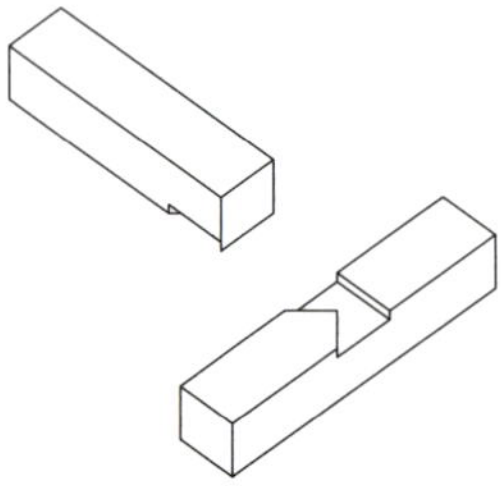
484

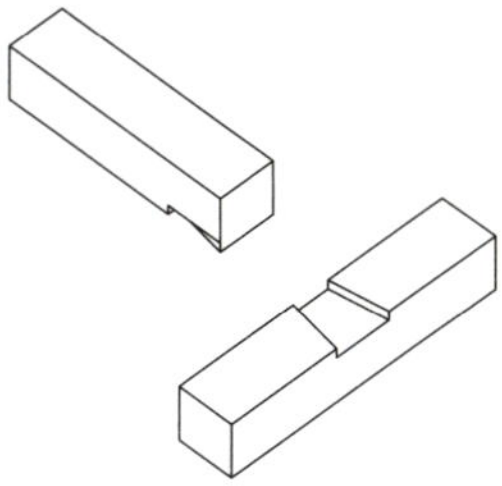
486

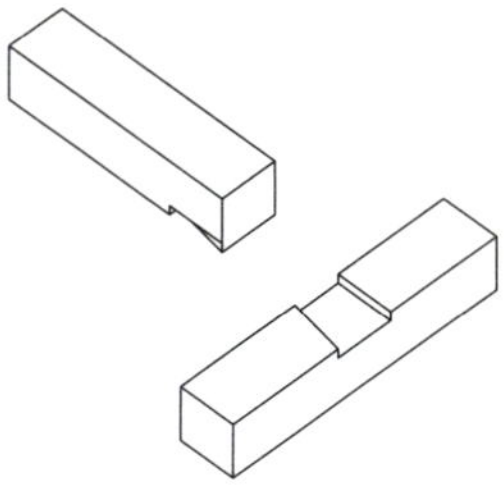
488

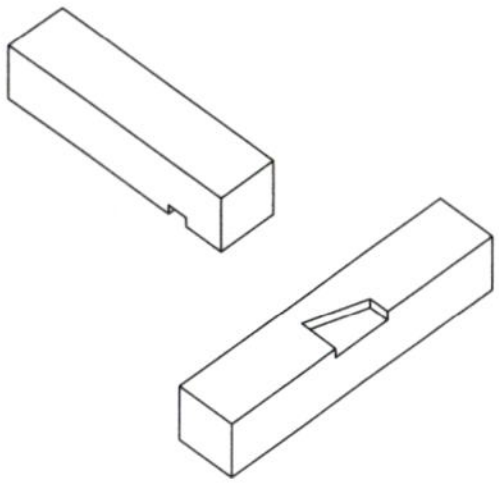
490

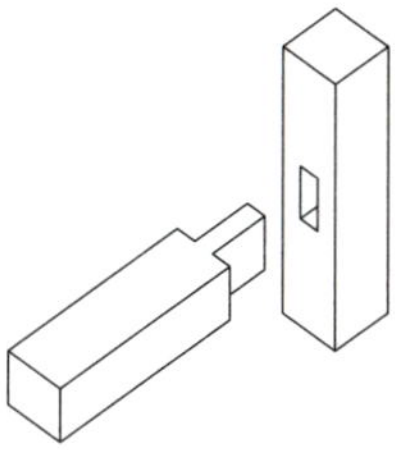
492

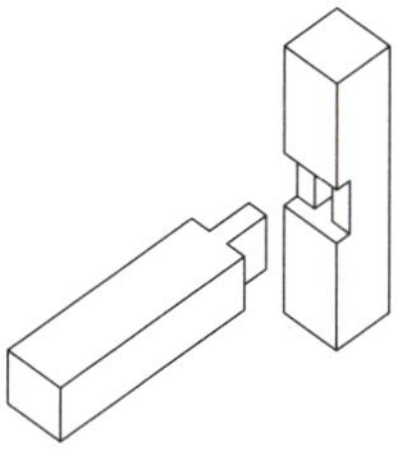
494

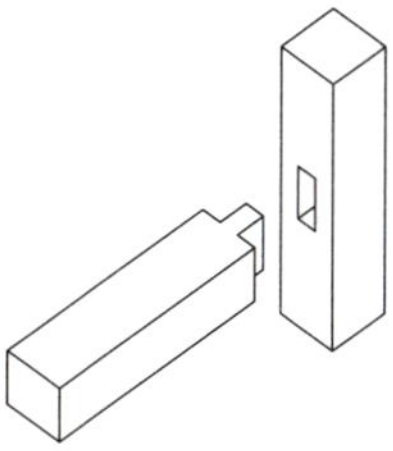
496

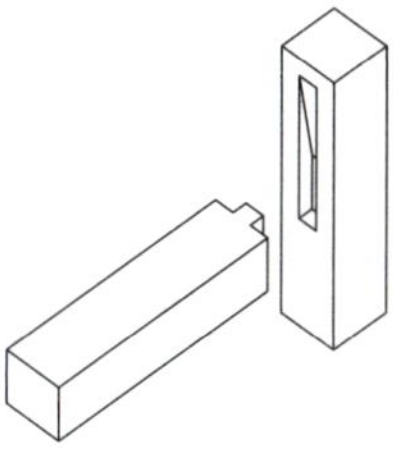
498

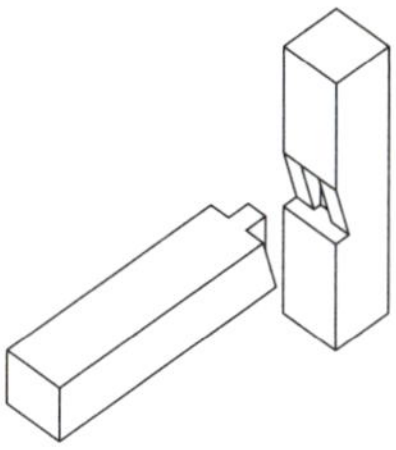
500

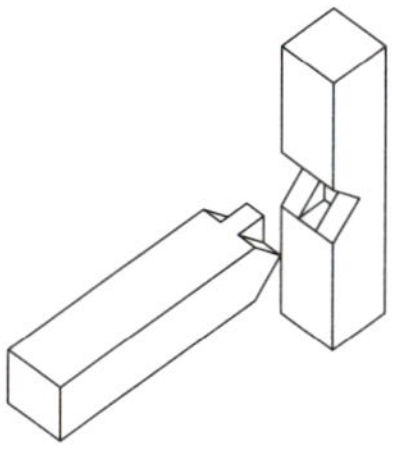
502

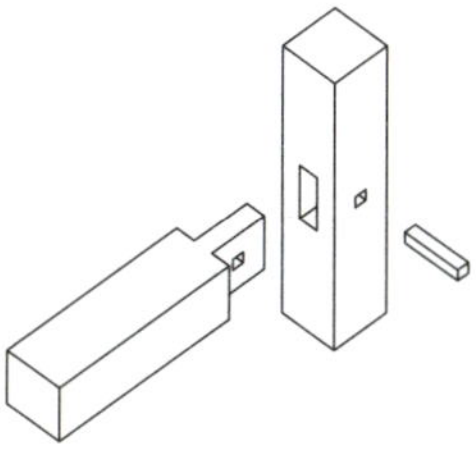
504

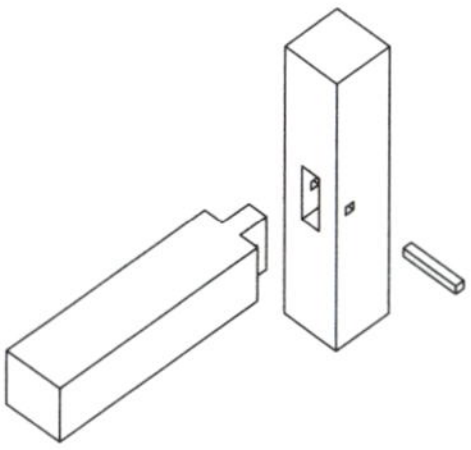
506

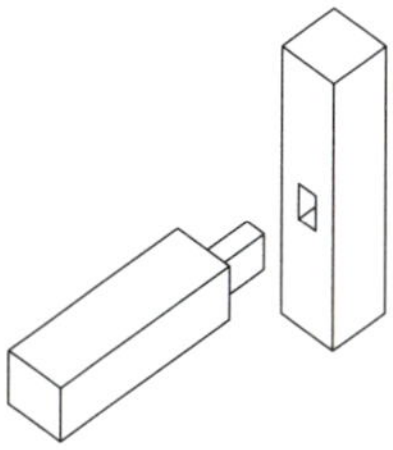
508

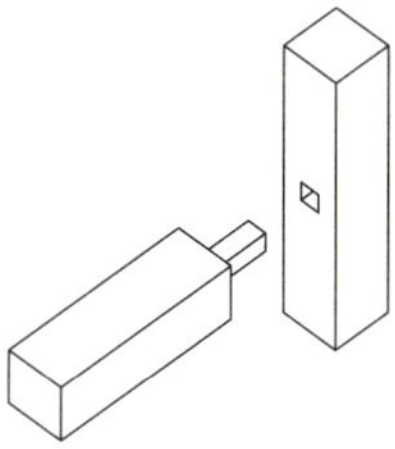
510

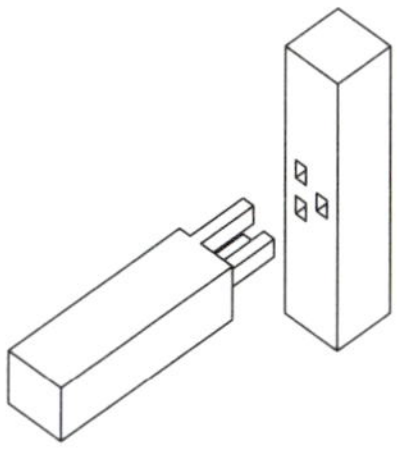
512

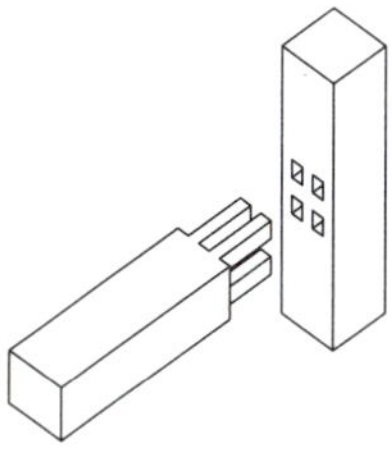
514

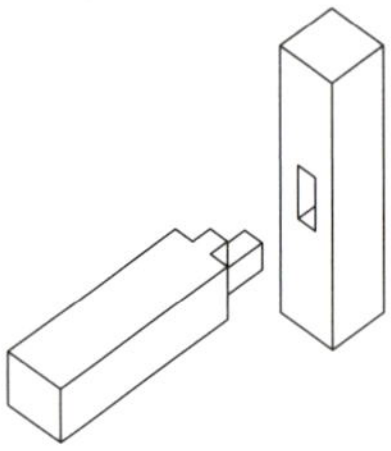

516

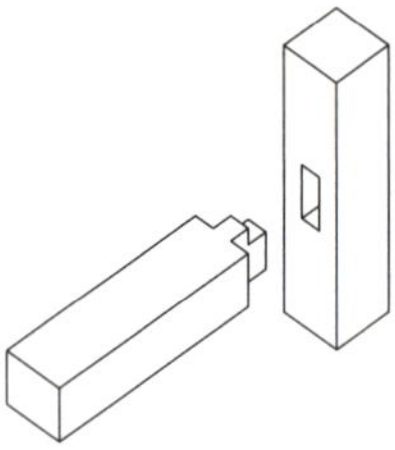

518

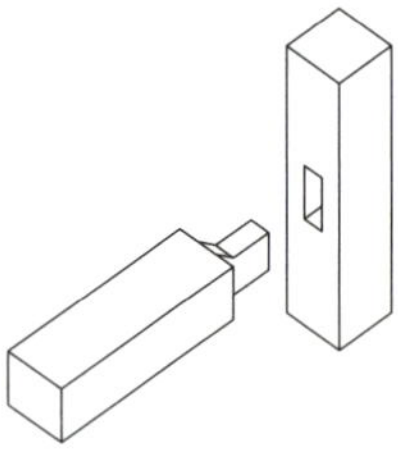

520

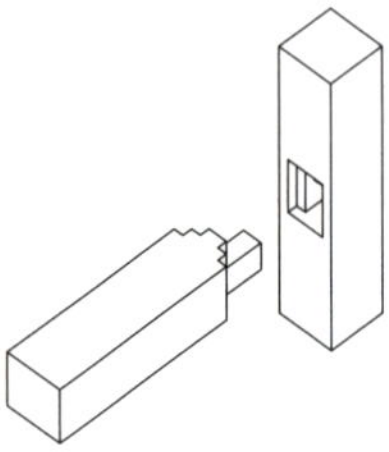

522

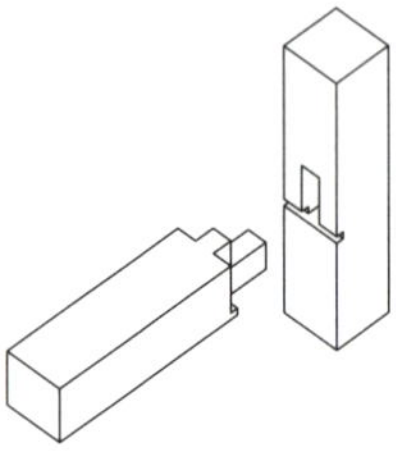

524

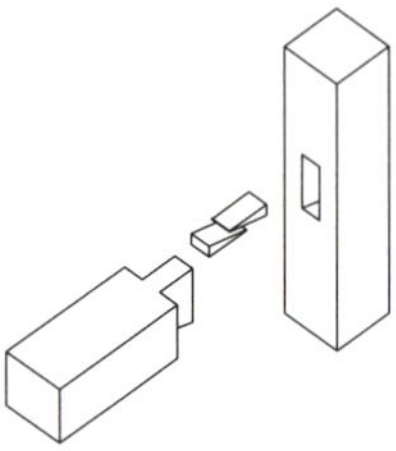

526

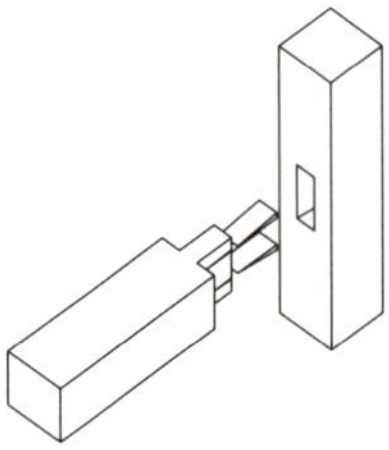

528

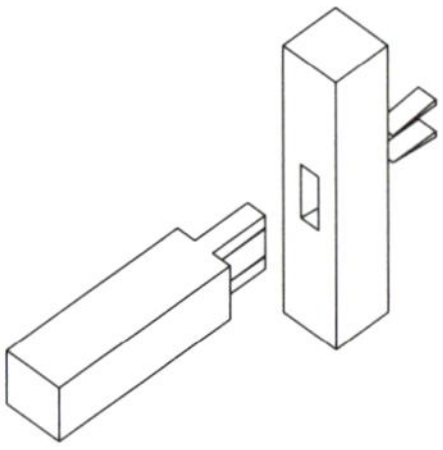

530

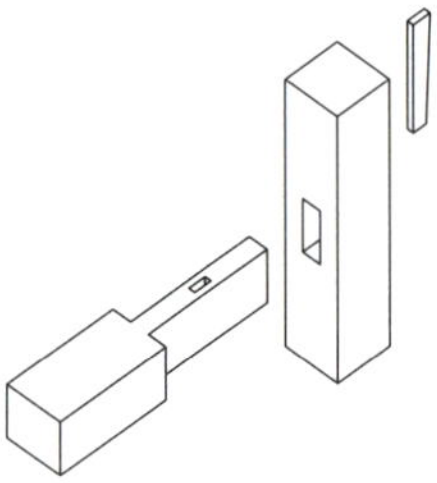

532

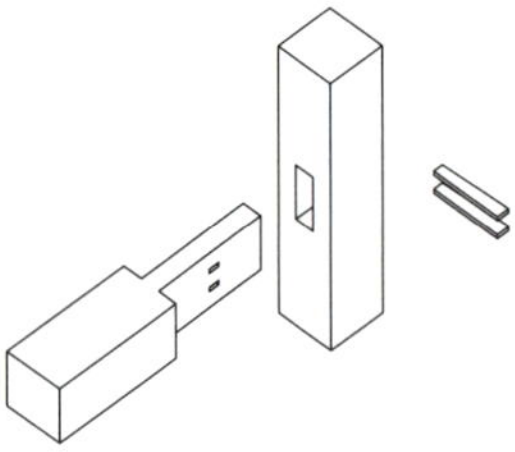

534

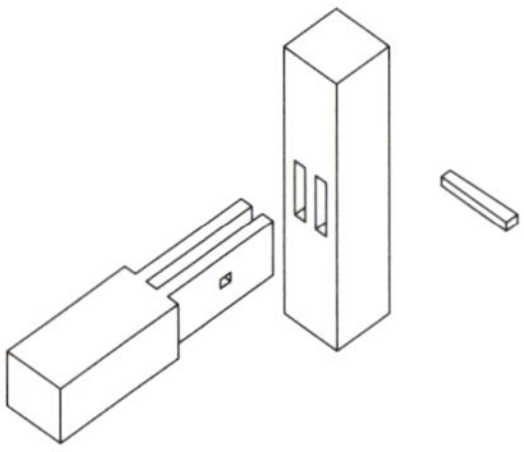

536

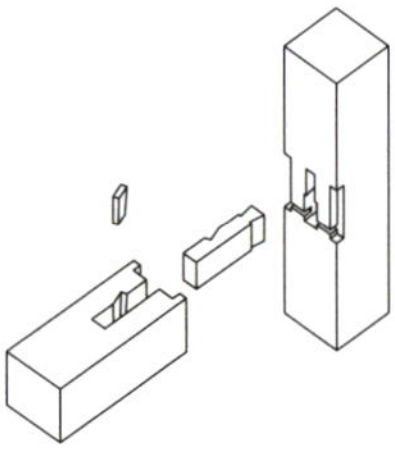

538

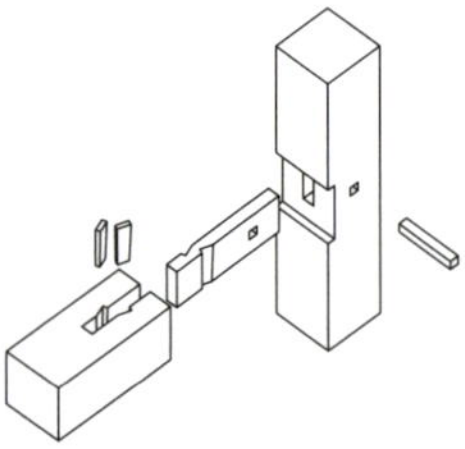

540

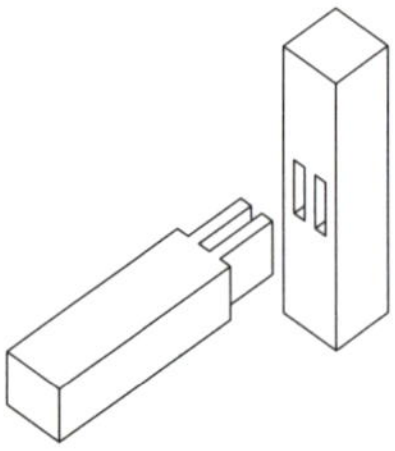

542

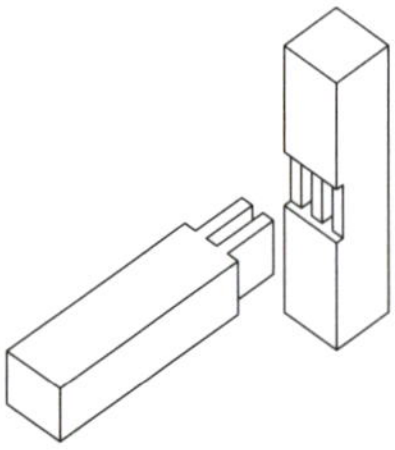

544

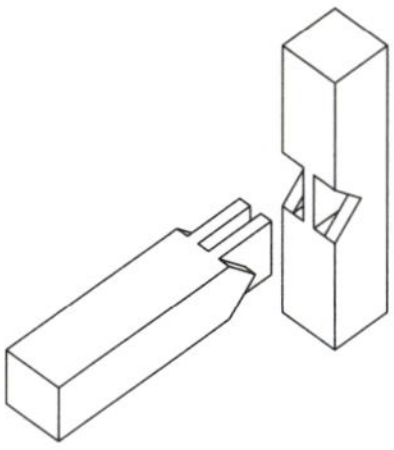

546

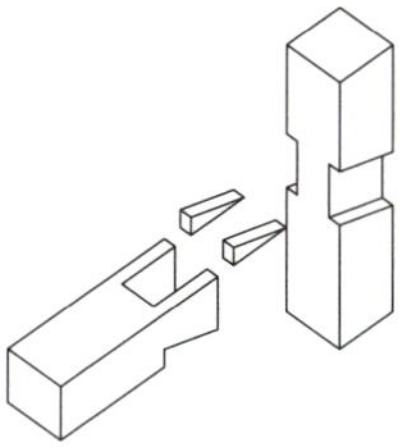

548

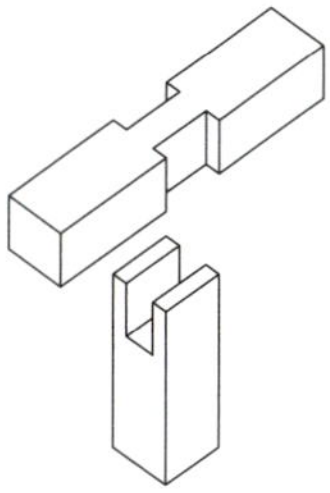

550

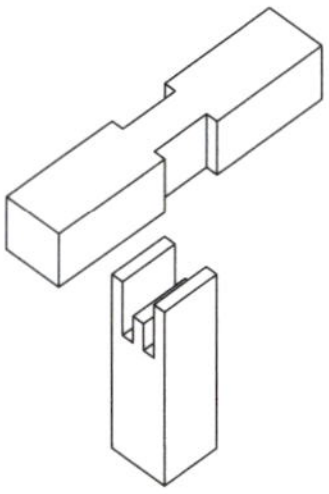

552

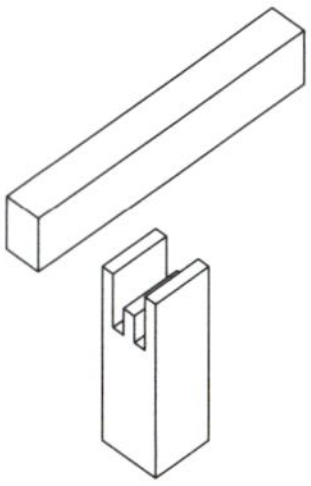

554

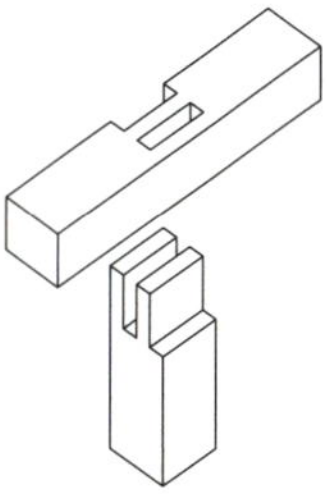

556

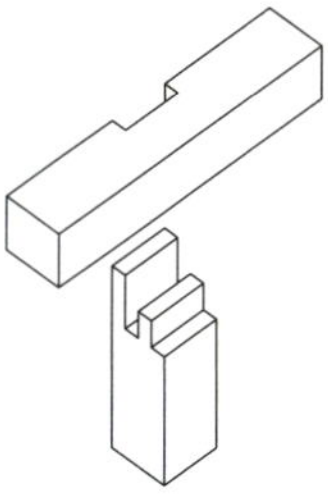

558

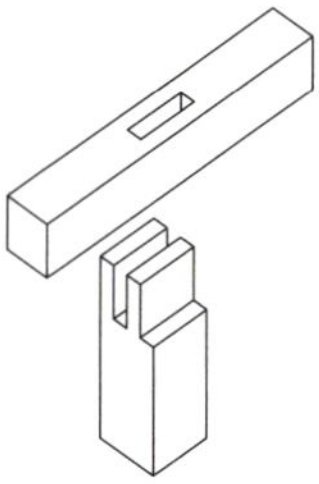

560

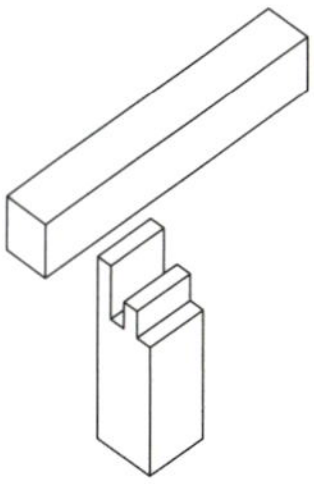

562

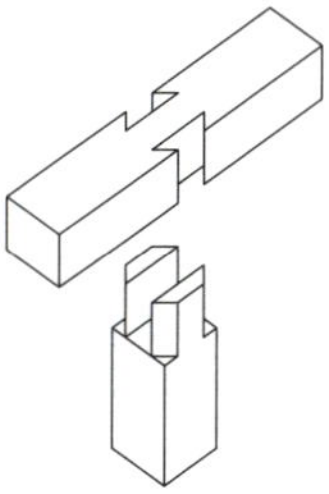

564

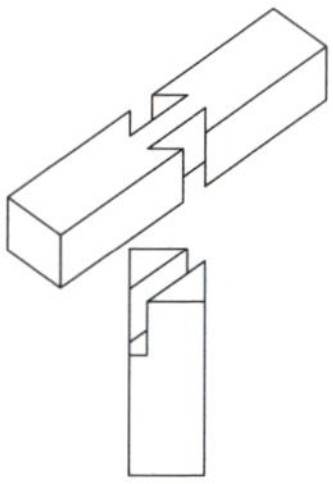

566

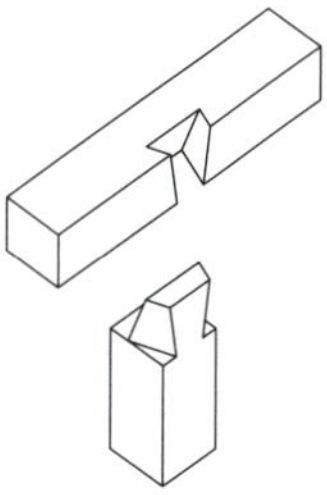

568

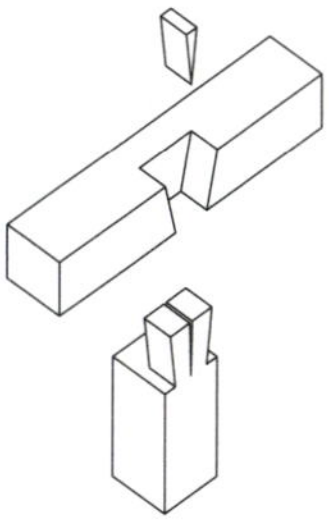

570

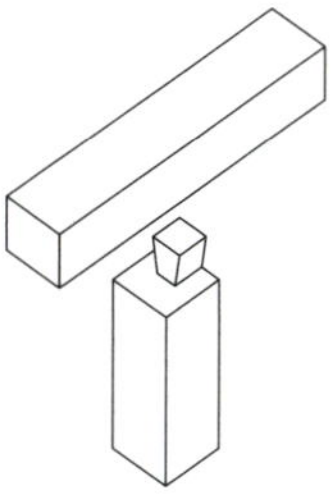

572

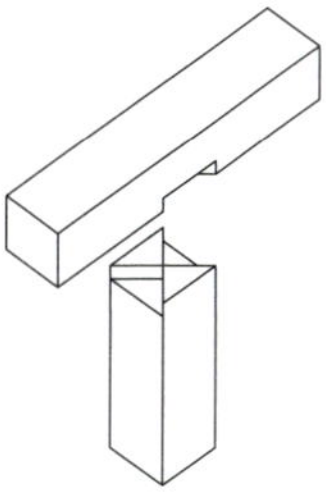

574

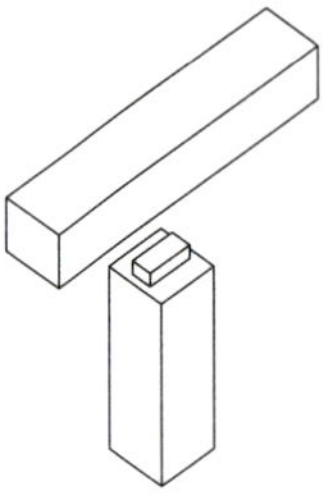
576

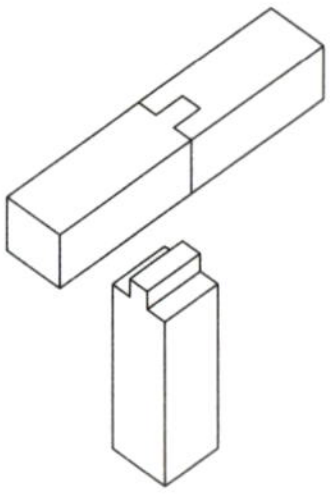
578

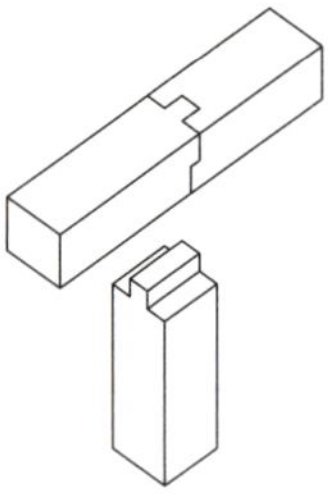
580

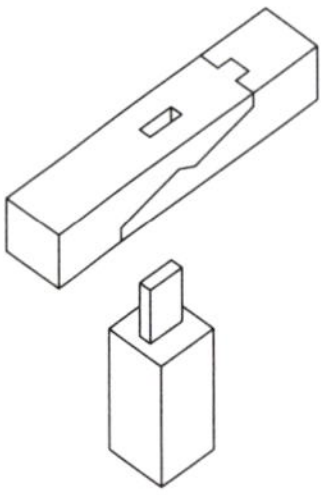
582

Angled Joints

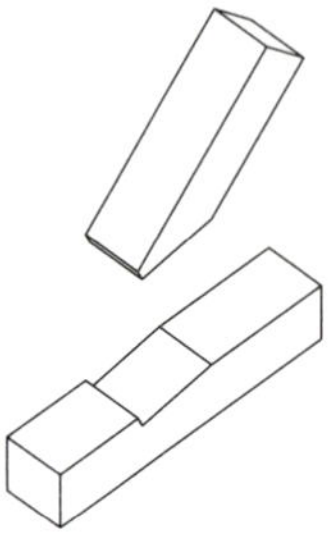
584

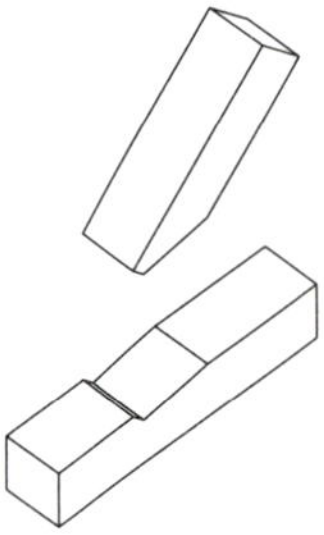
586

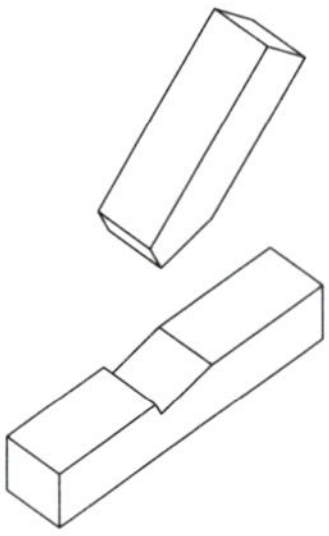
588

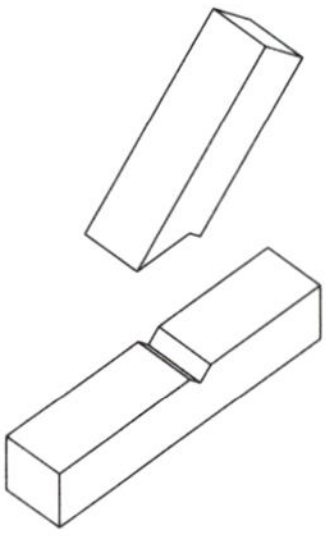
590

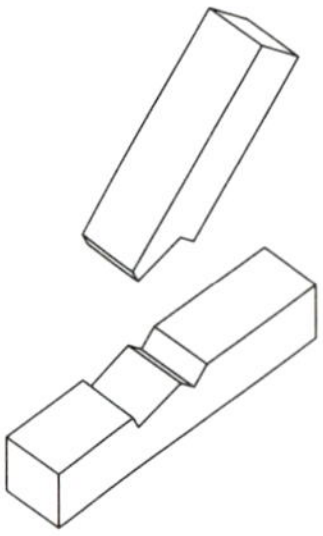
592

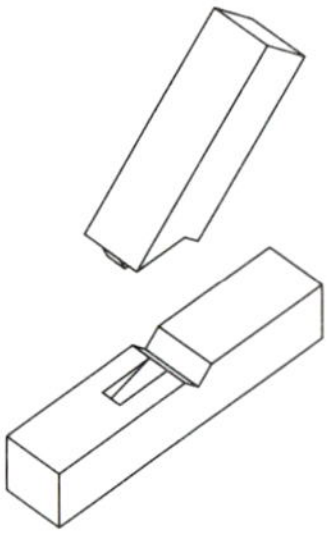
594

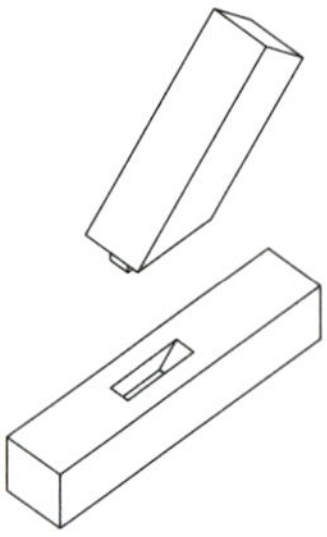
596

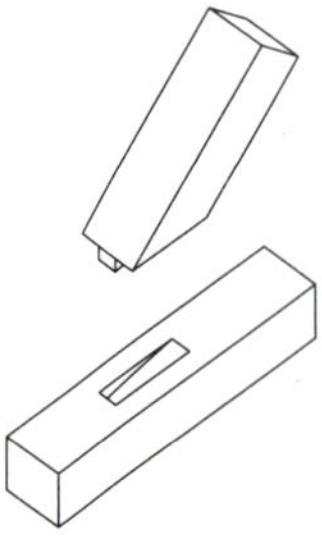
598

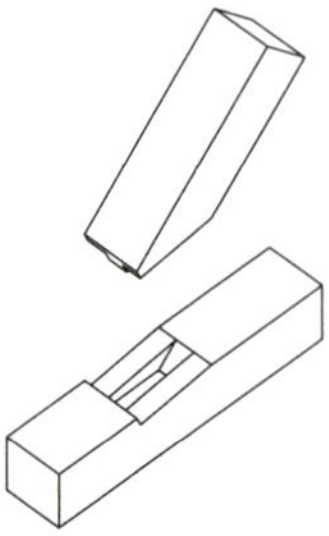
600

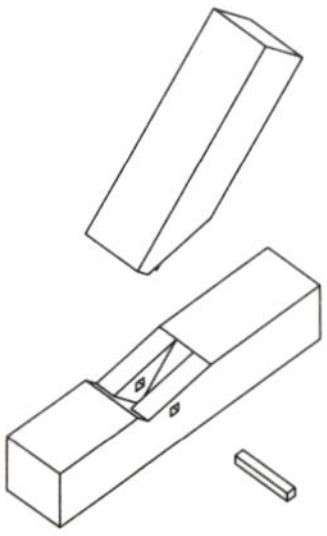
602

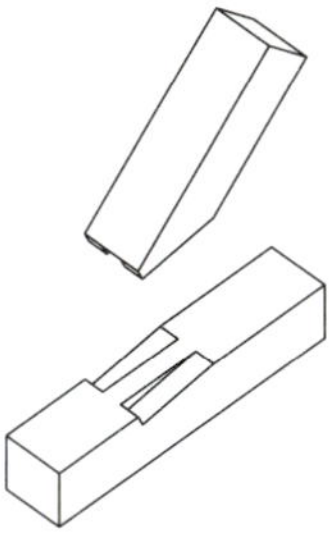

604

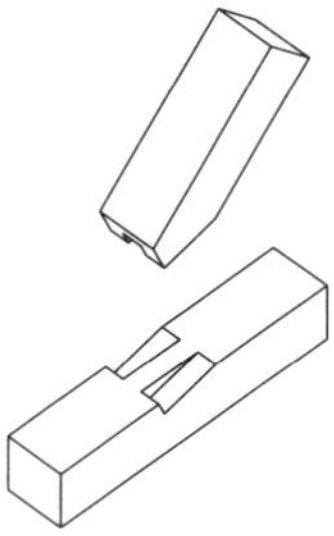

606

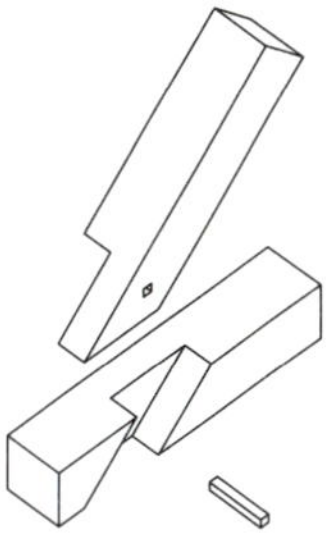

608

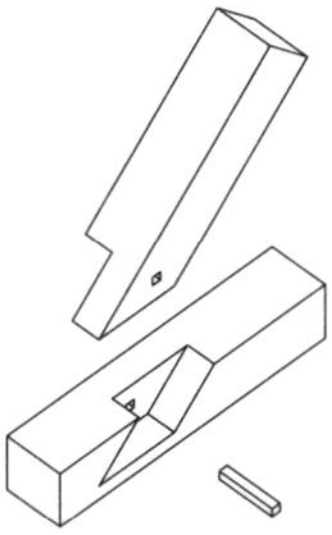

610

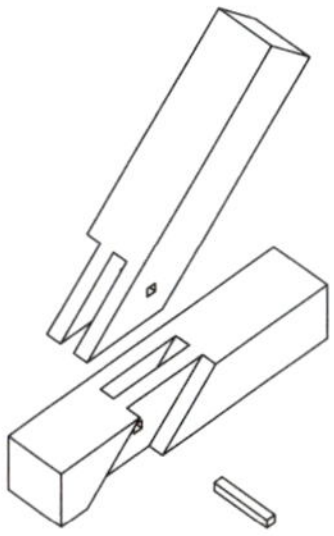

612

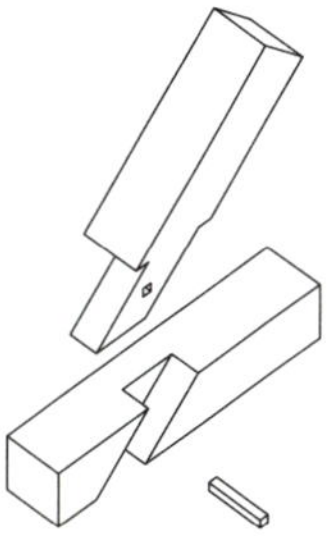

614

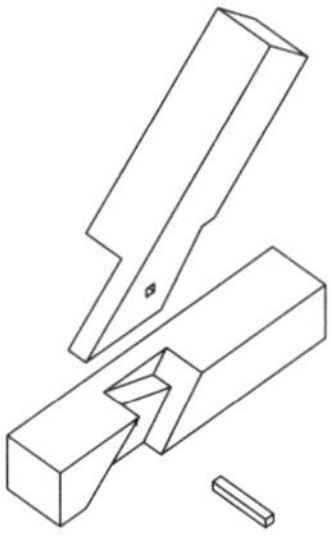

616

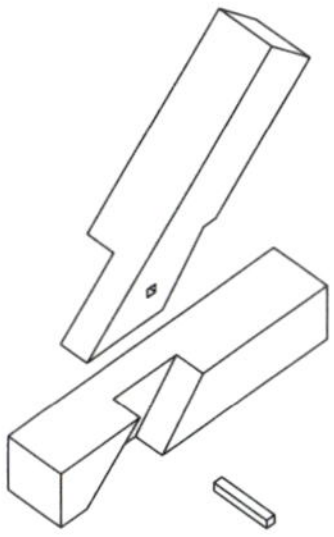

618

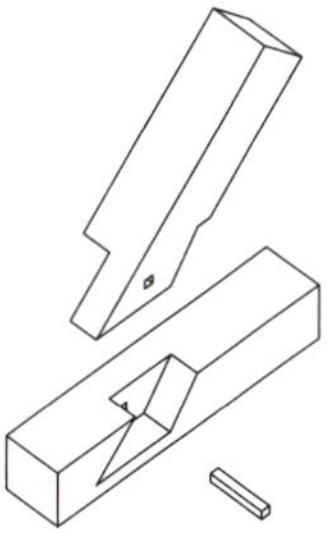

620

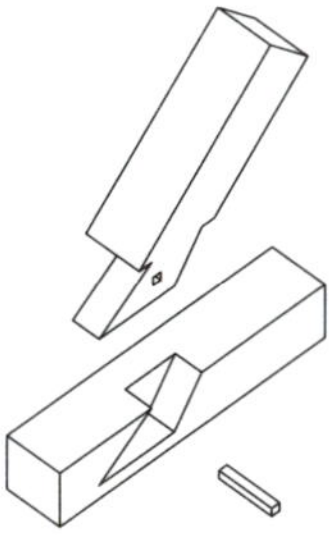

622

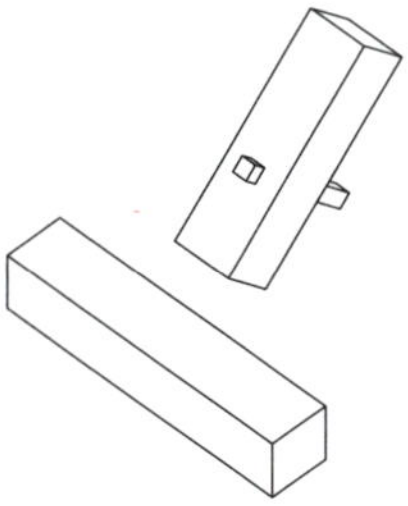

624

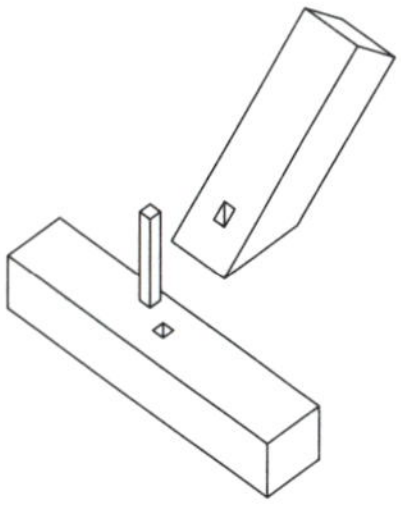

626

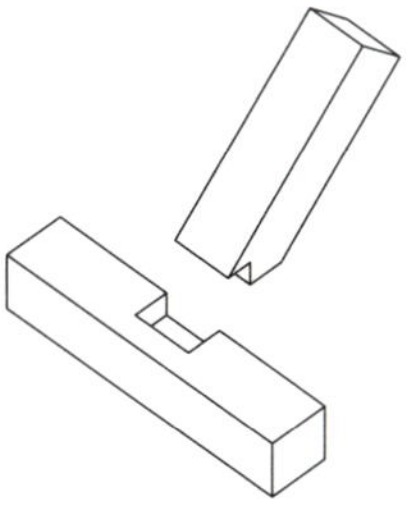

628

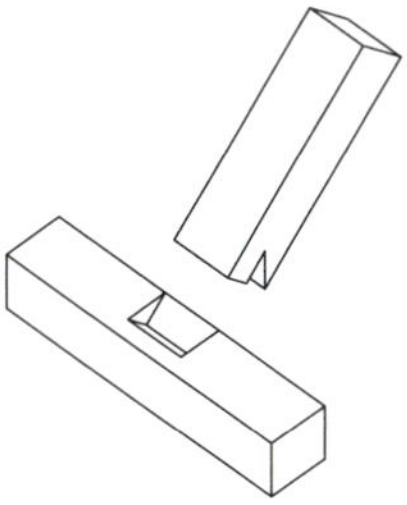

630

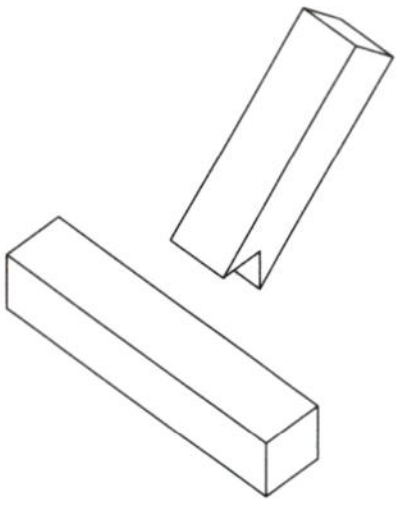

632

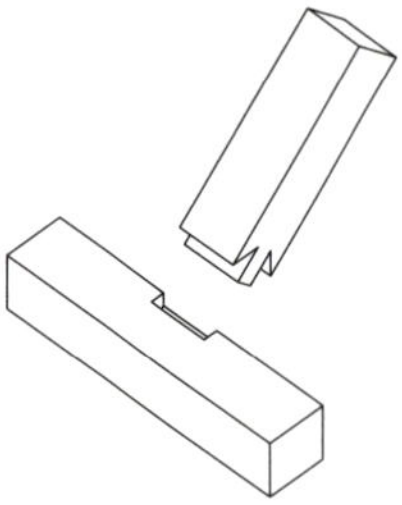

634

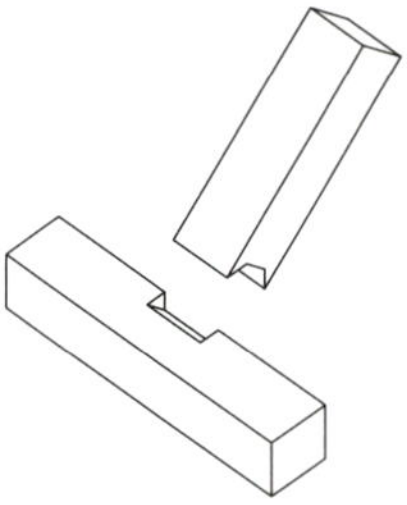

636

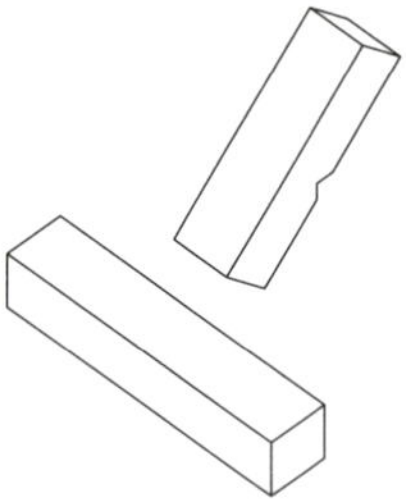

638

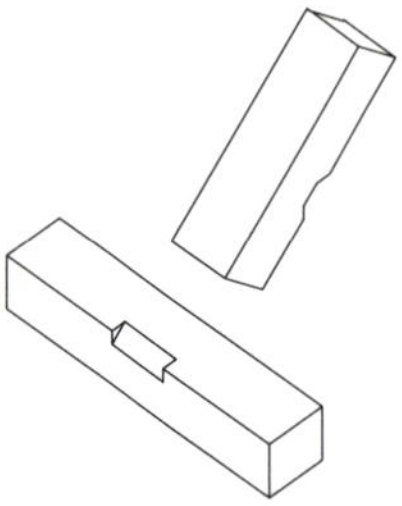

640

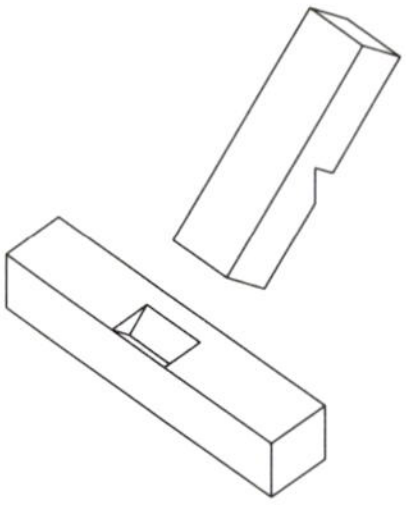

642

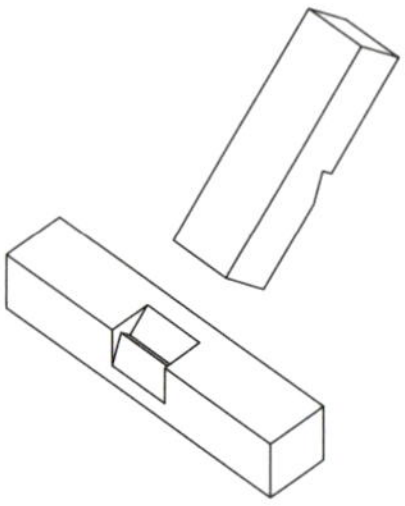

644

Cross Lap Joints

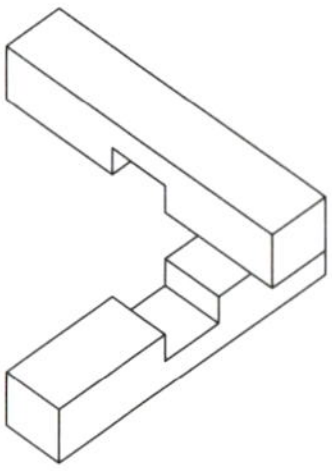

646

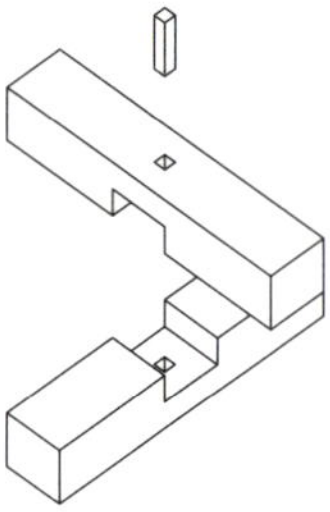

648

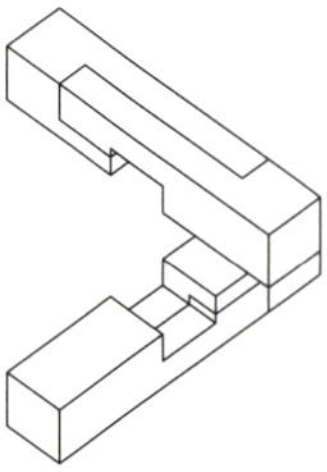

650

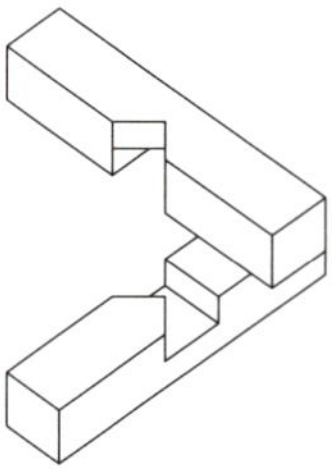

652

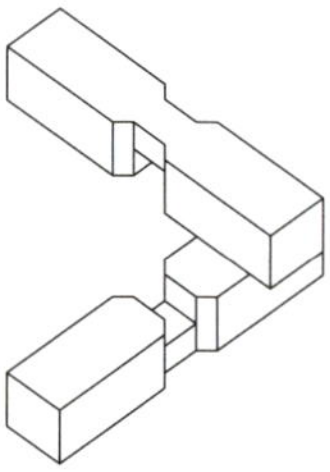

654

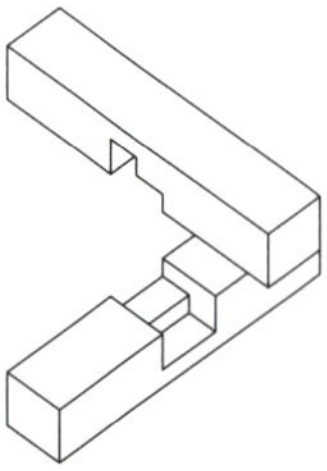

656

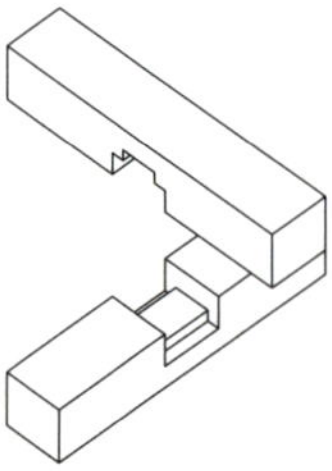

658

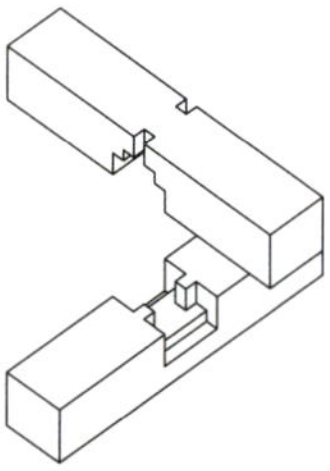

660

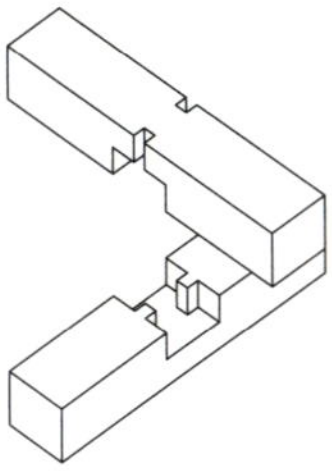

662

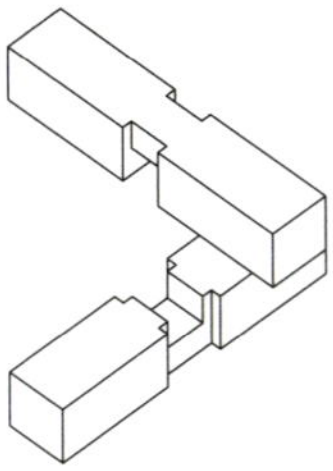

664

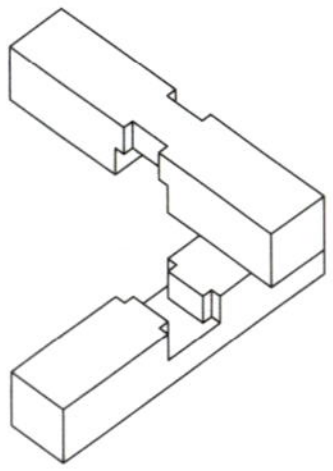

666

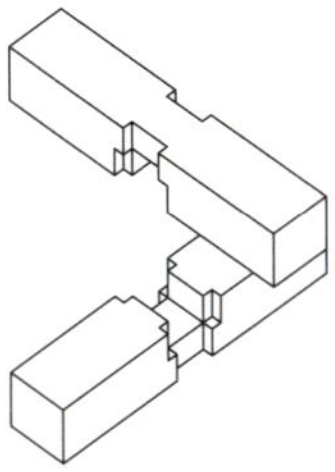

668

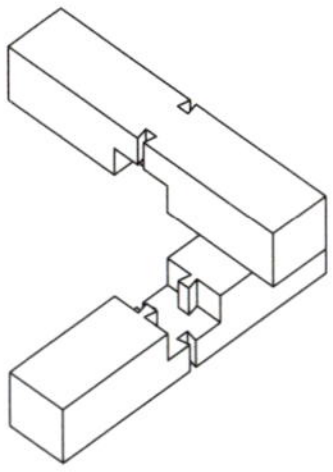

670

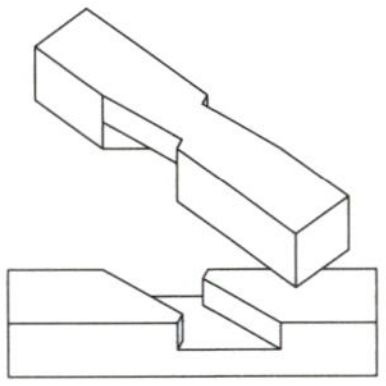

672

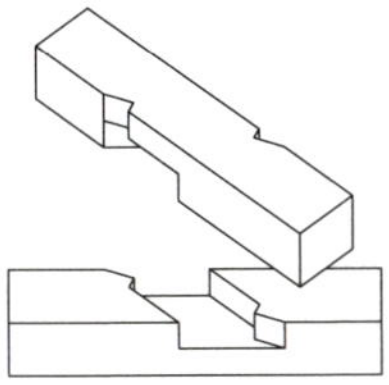

674

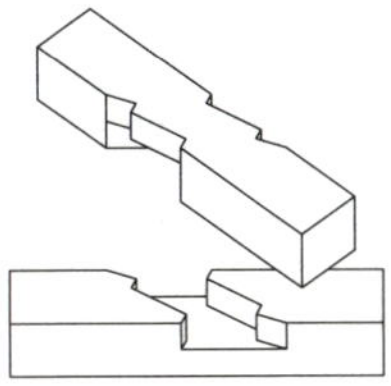

676

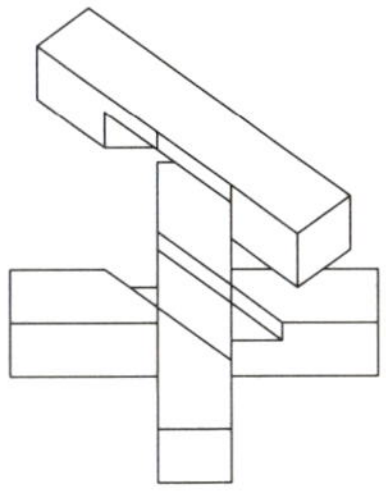

678

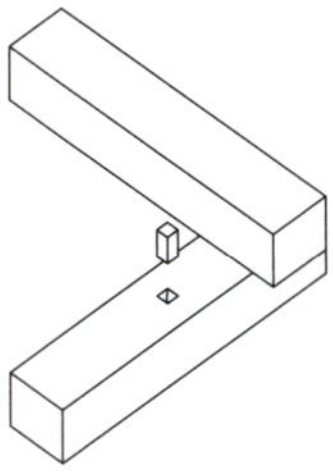

680

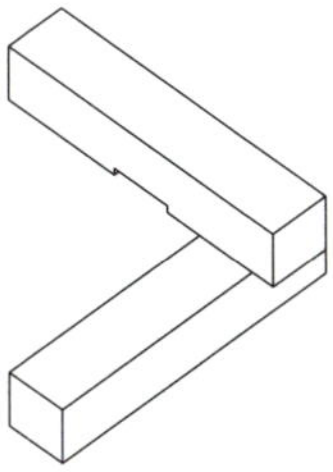

682

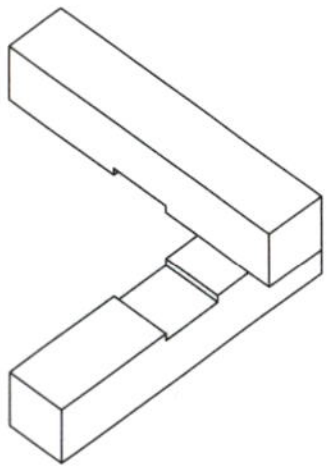

684

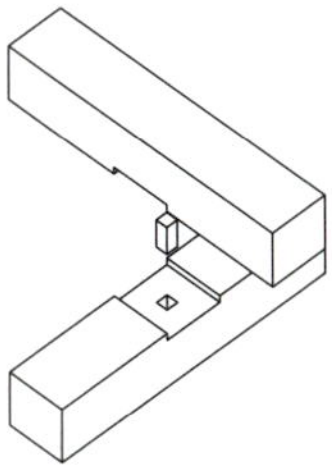

686

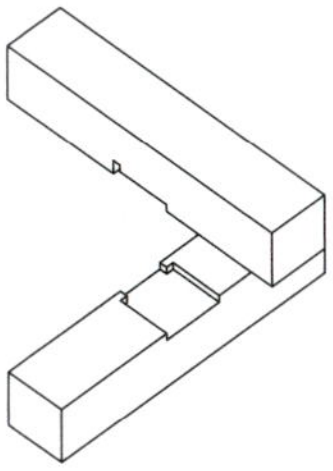

688

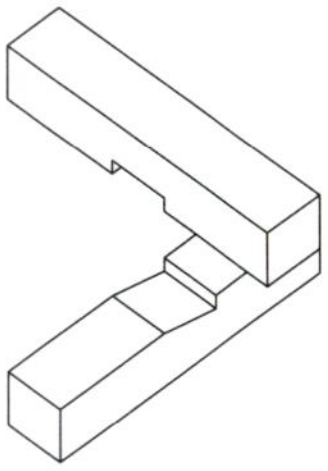

690

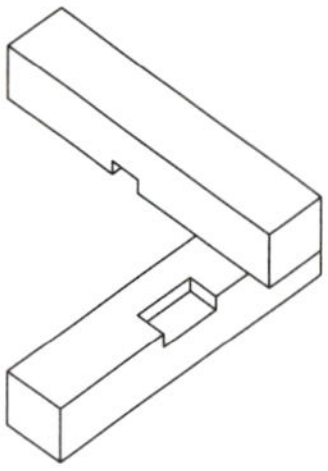
692

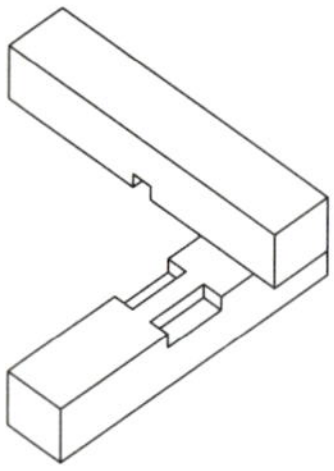
694

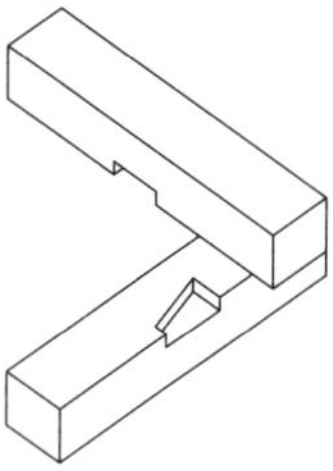
696

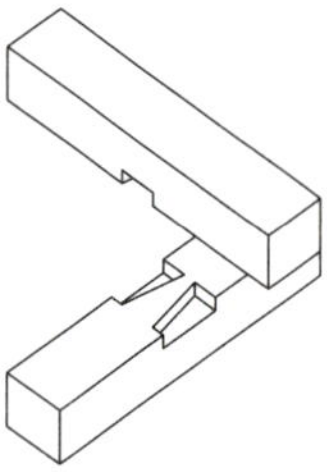
698

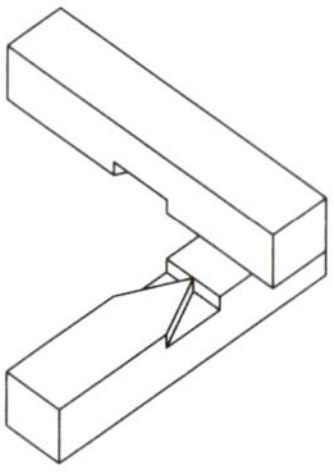
700

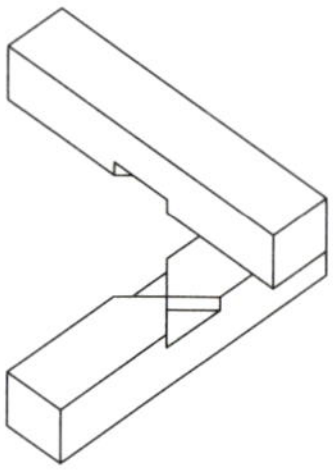
702

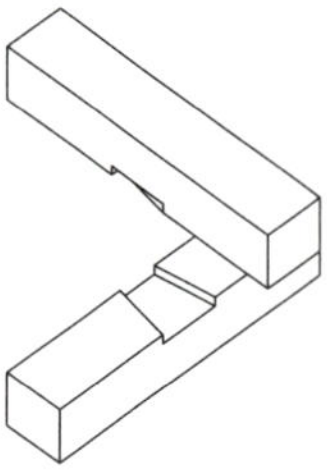
704

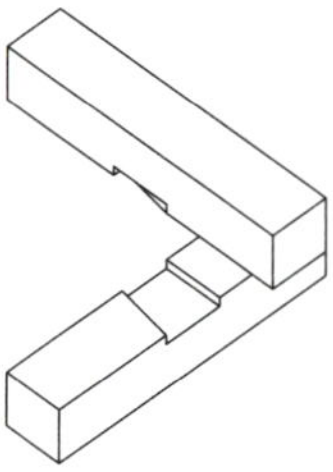
706

Specialty Joints

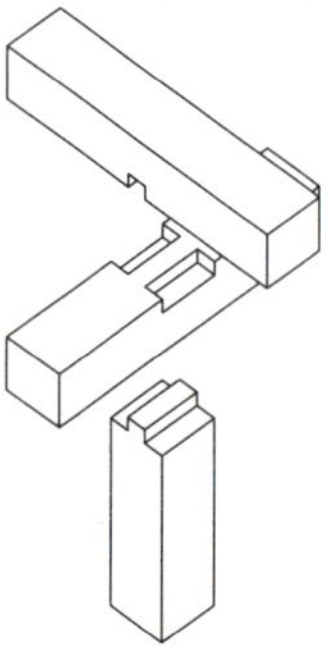
708

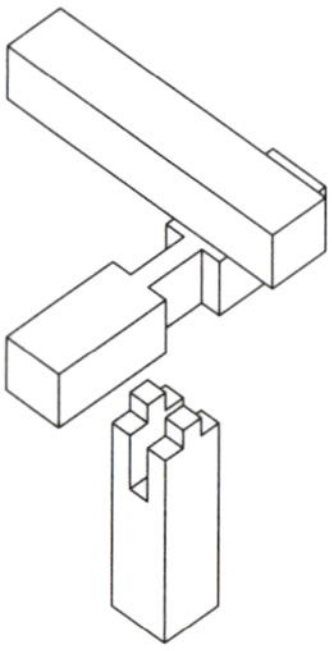
710

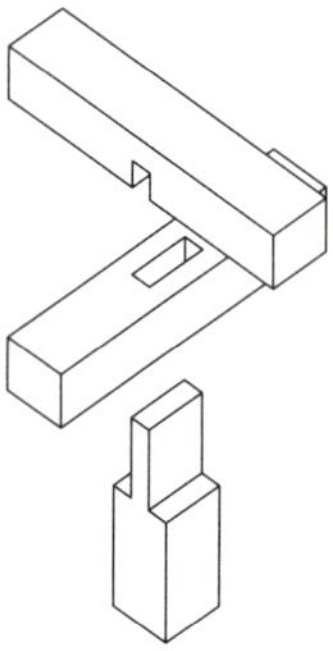
712

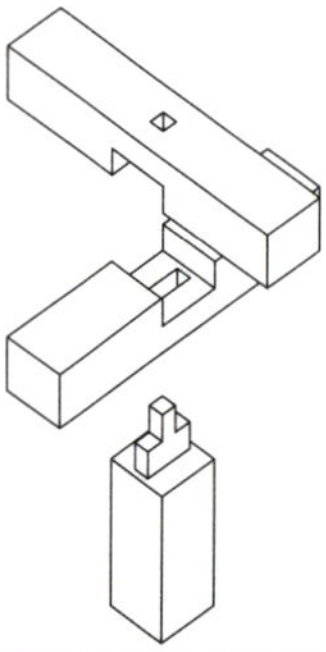
714

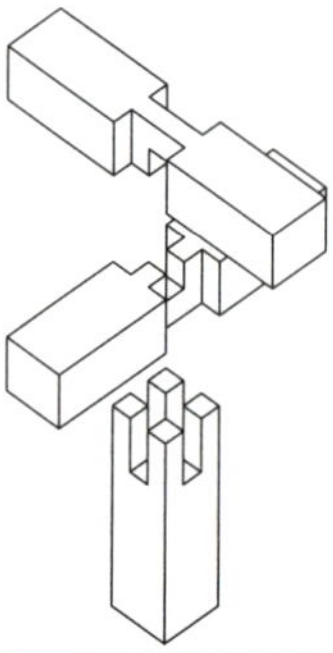
716

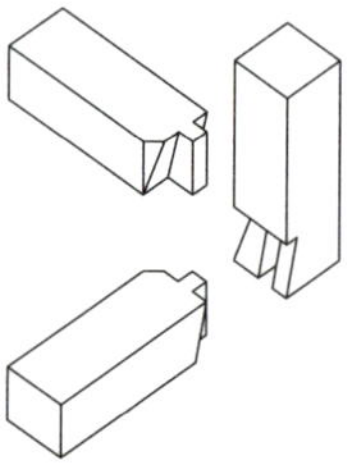
718

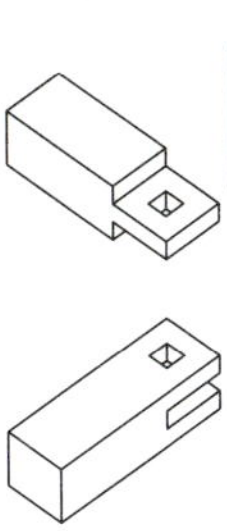
720

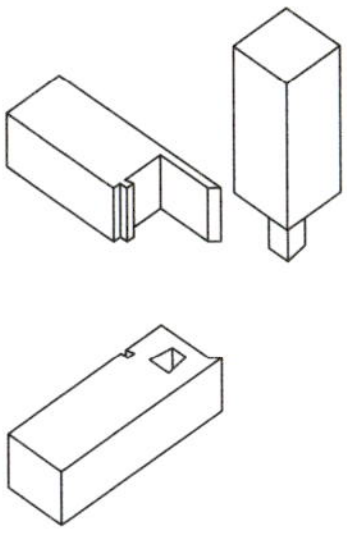
722

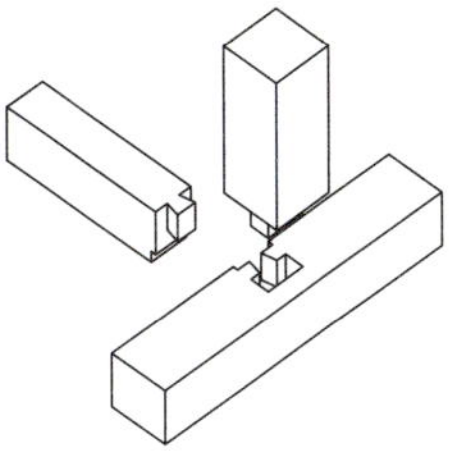
724

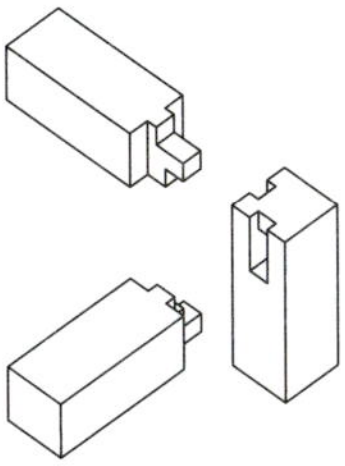
726

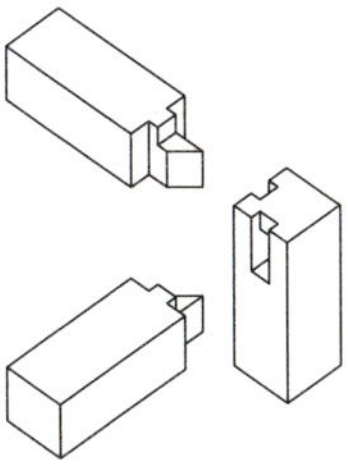
728

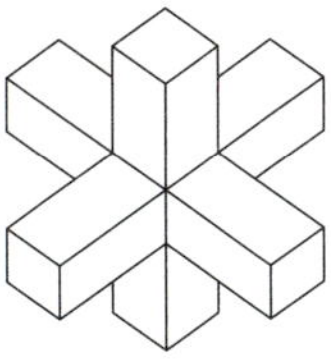
730

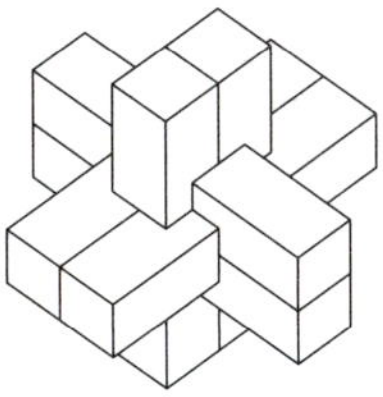
732

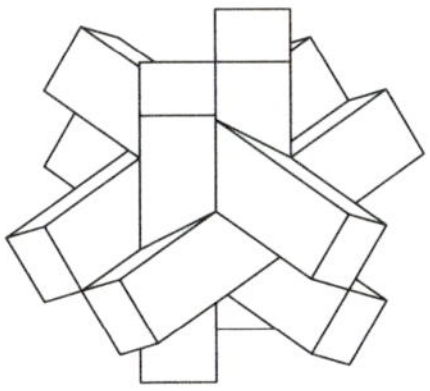
734

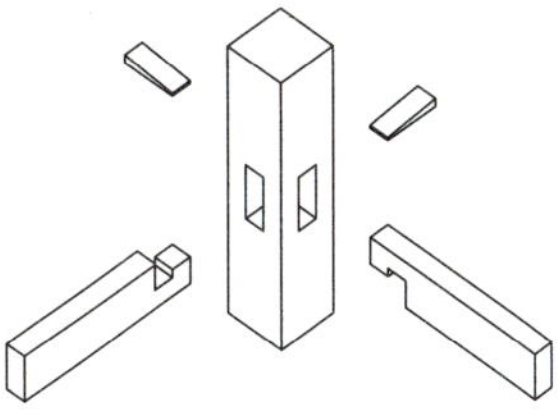
736

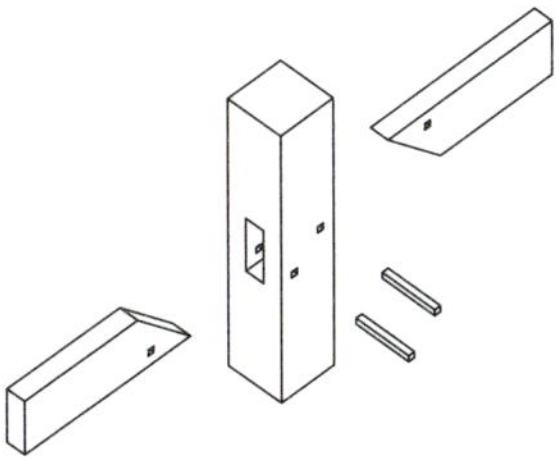
738

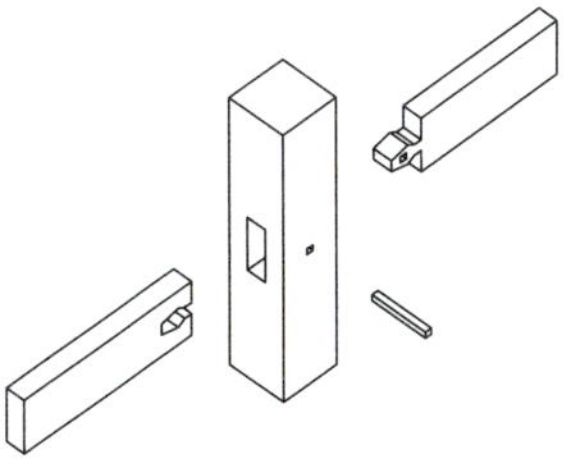
740

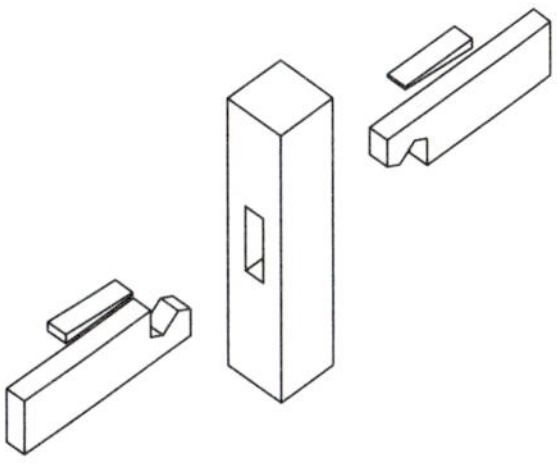
742

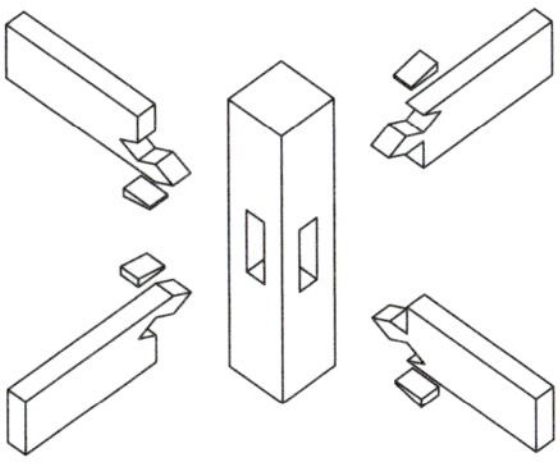
744

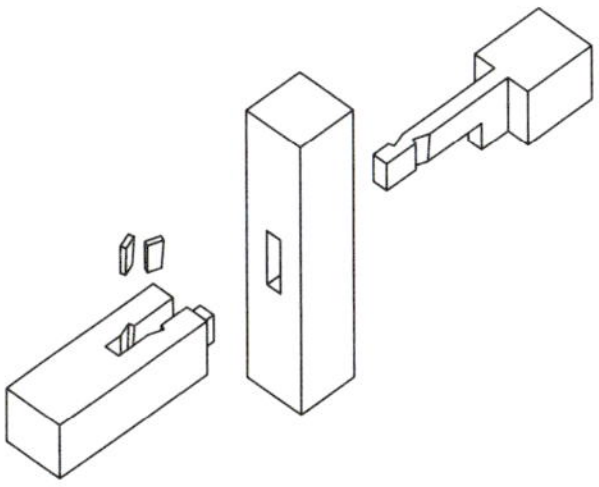
746

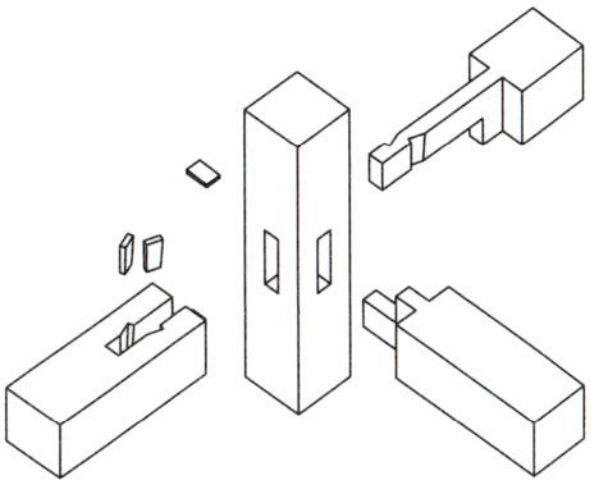
748

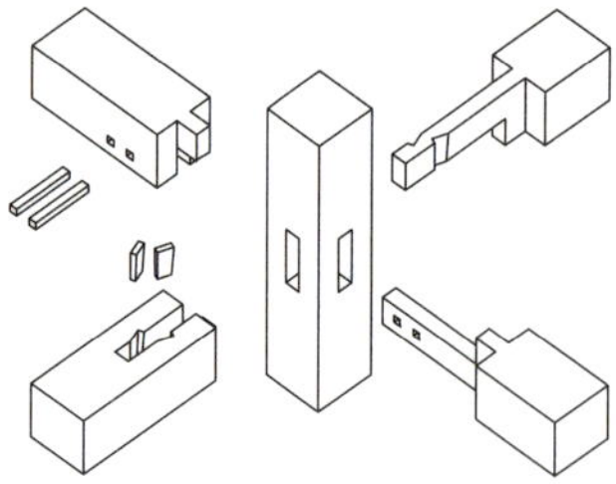

750

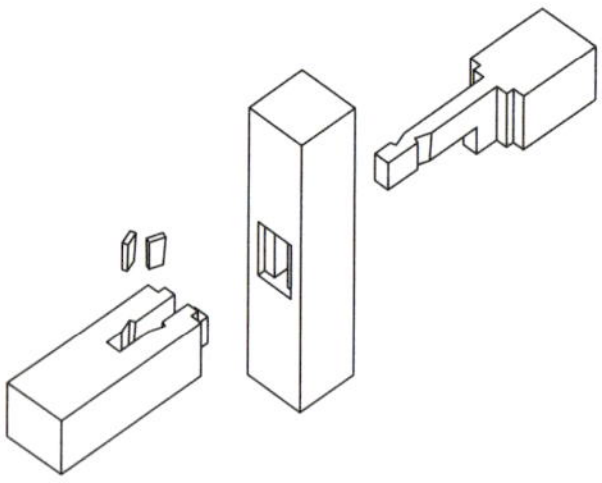

752

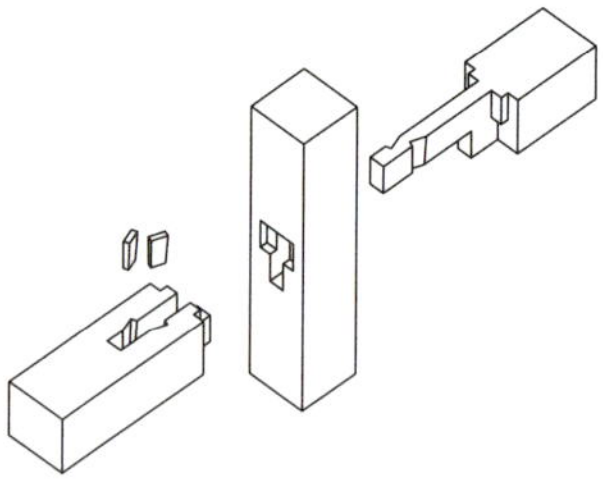

754

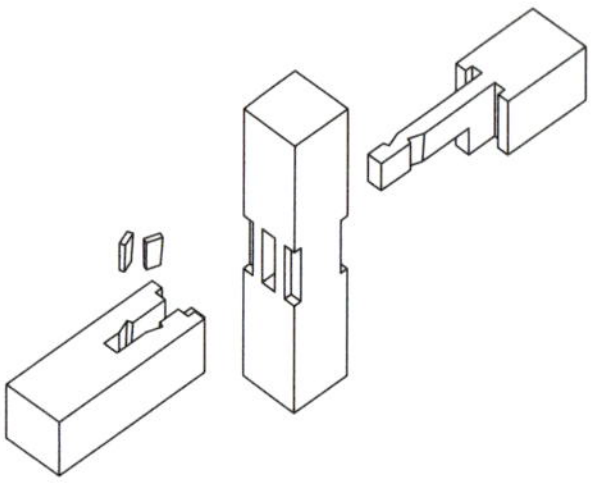

756

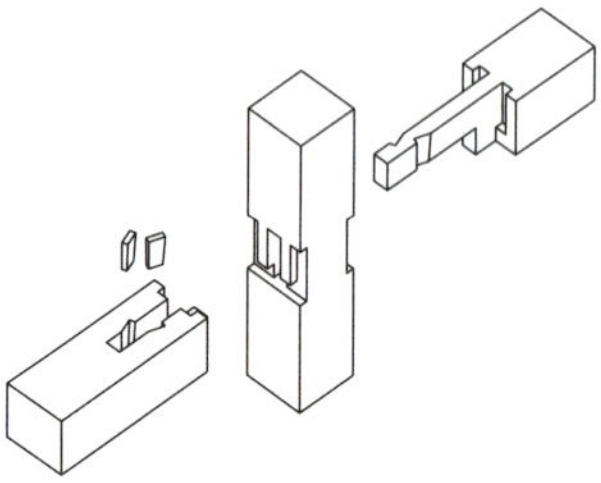

758

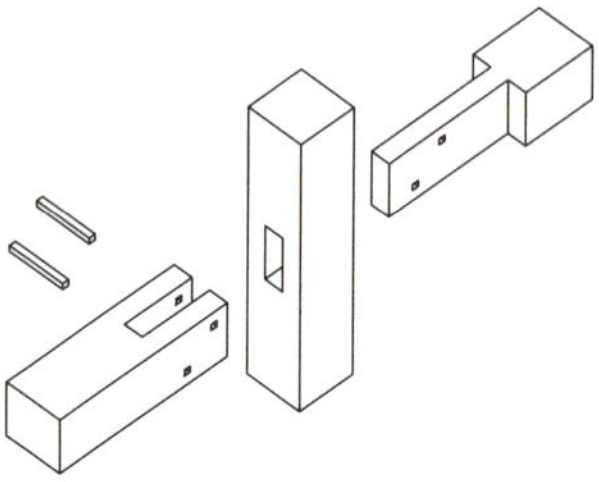

760

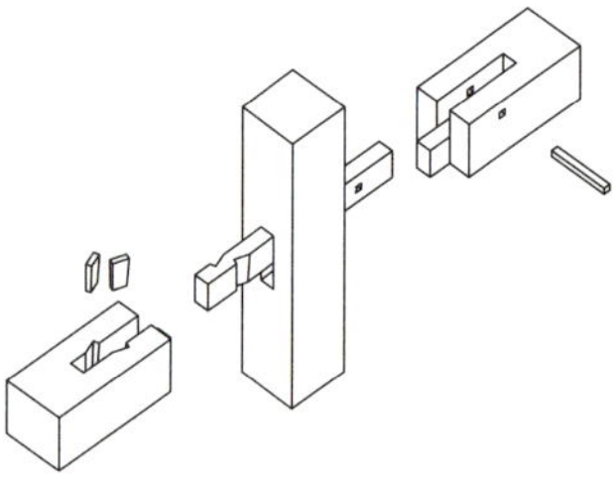

762

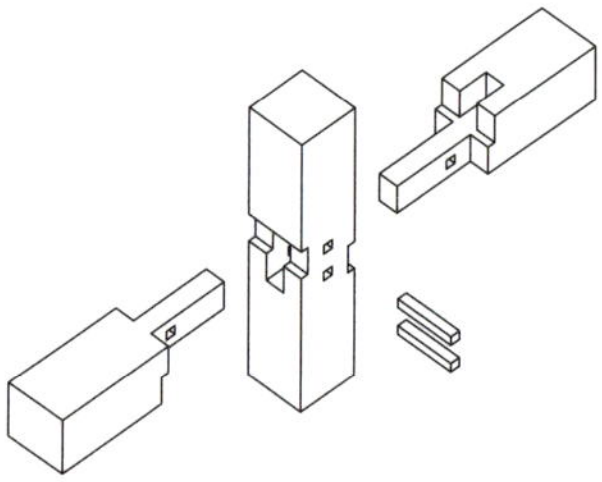

764

Case Joints
Edge Joints

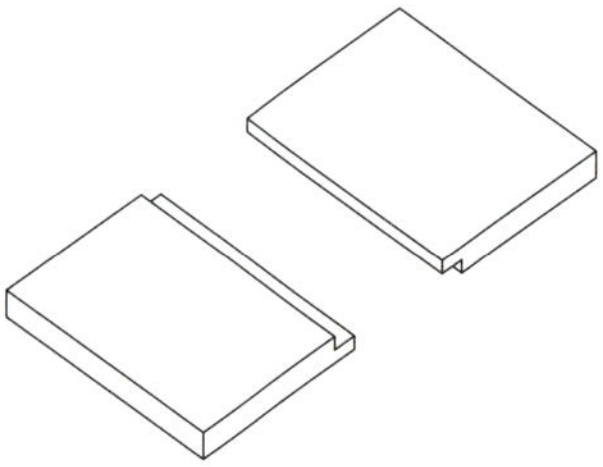

768

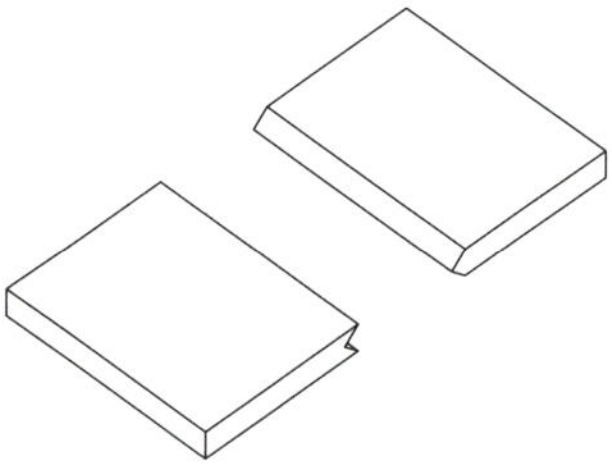

770

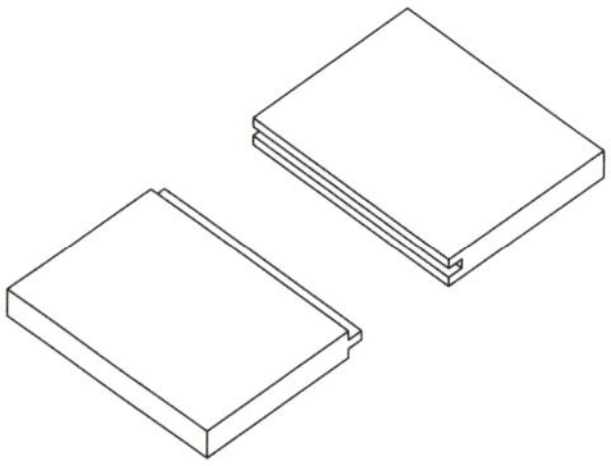

772

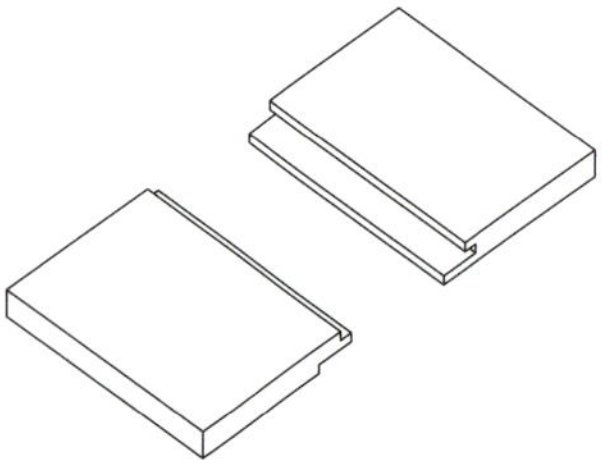

774

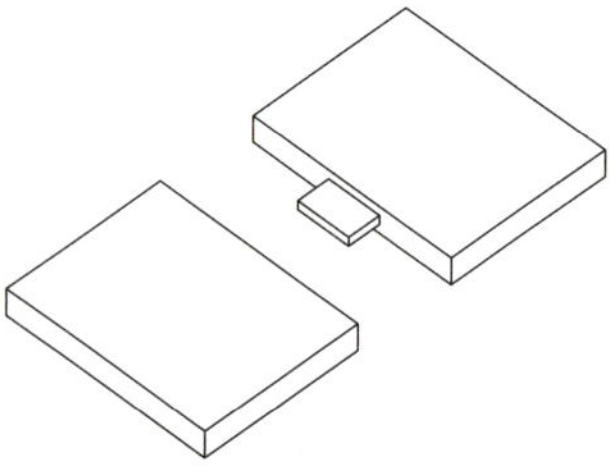

776

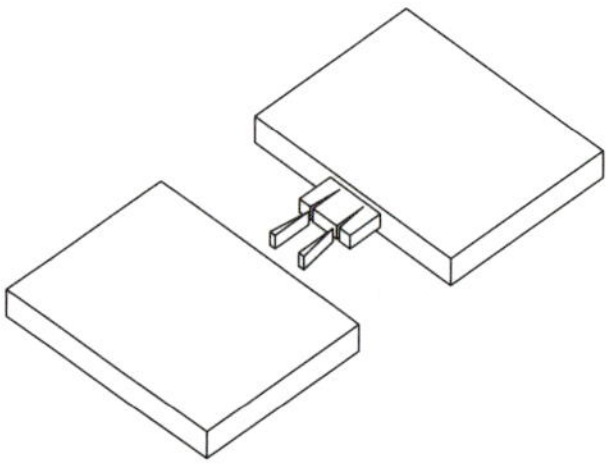

778

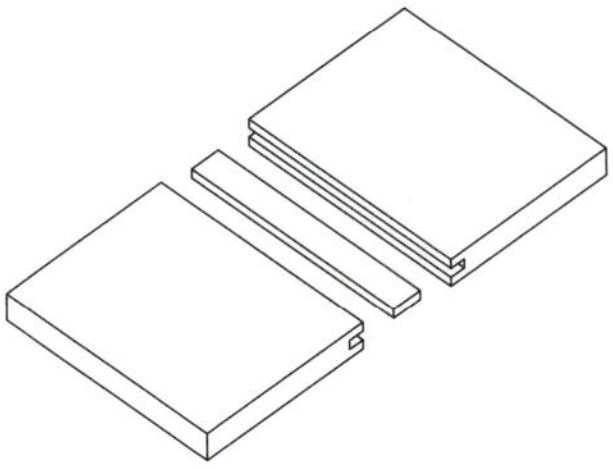
780

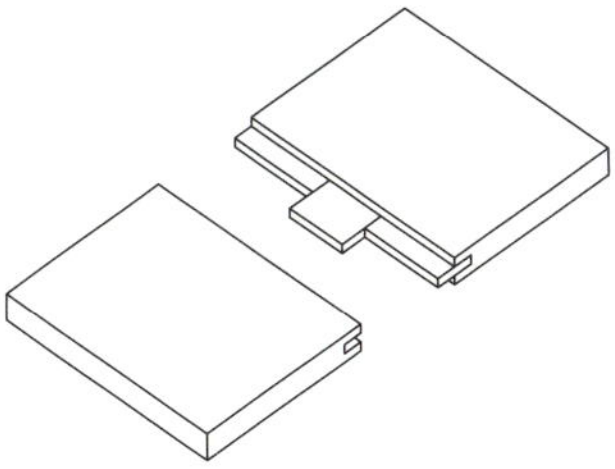
782

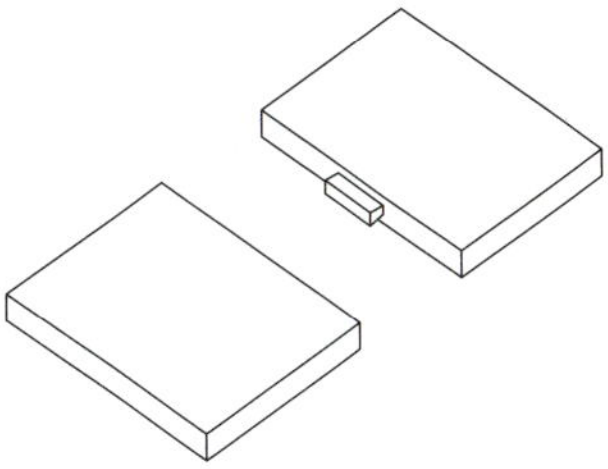
784

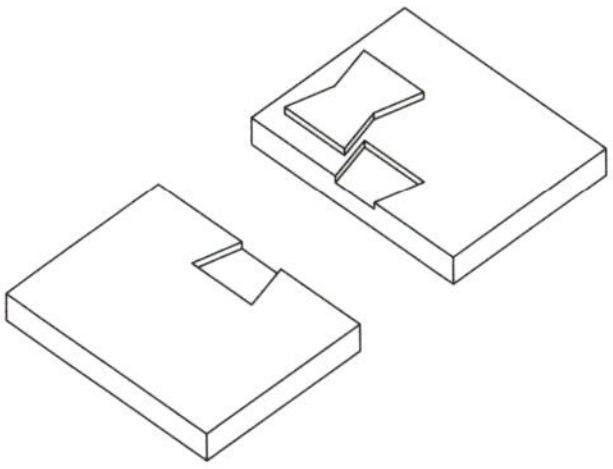
786

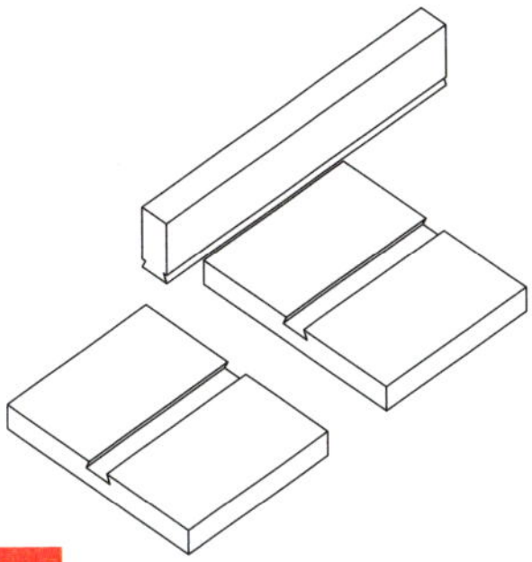
788

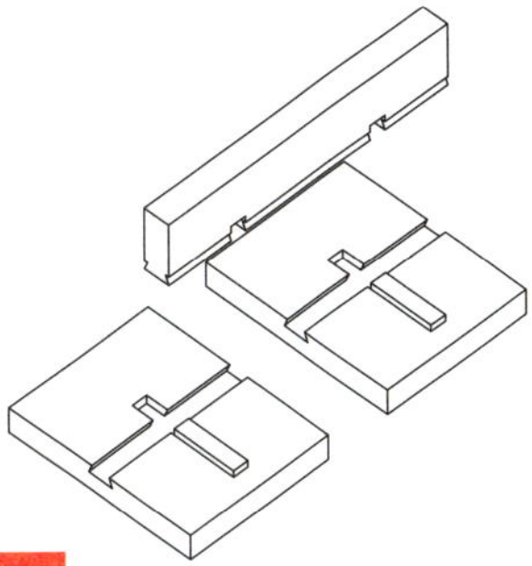
790

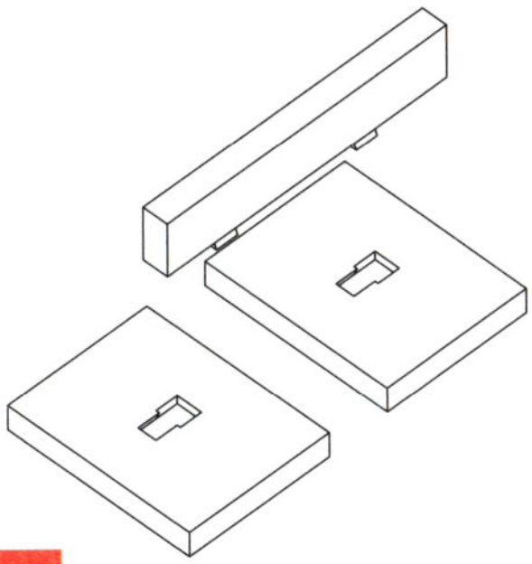
792

End Joints

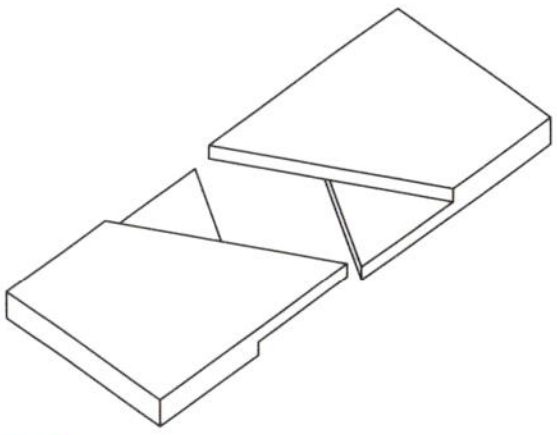
794

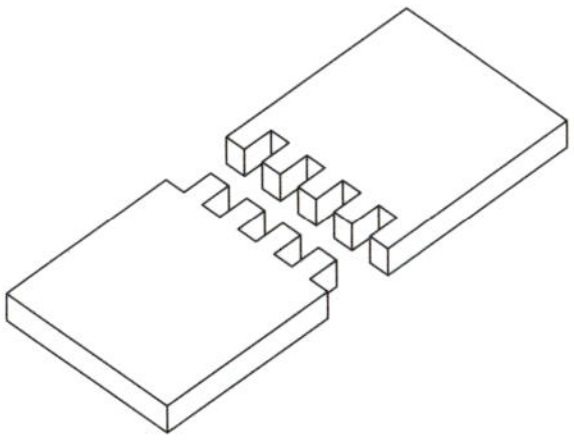
796

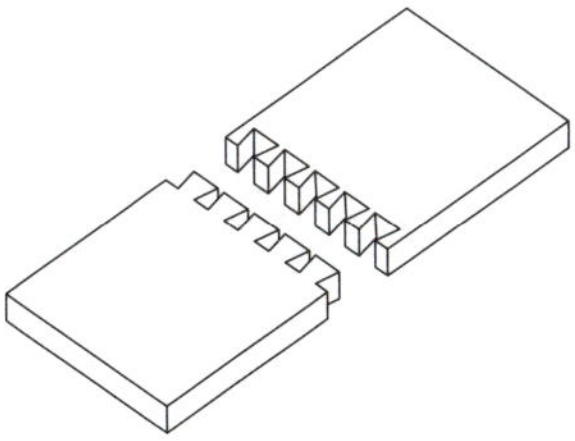
798

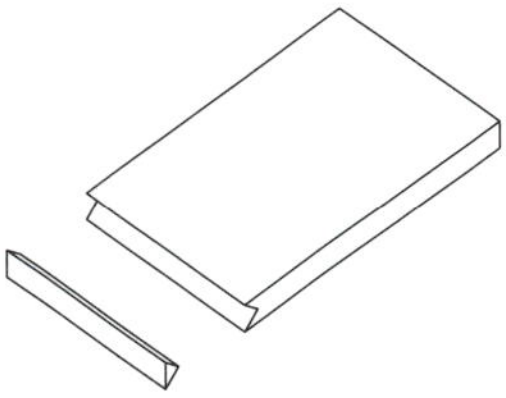
800

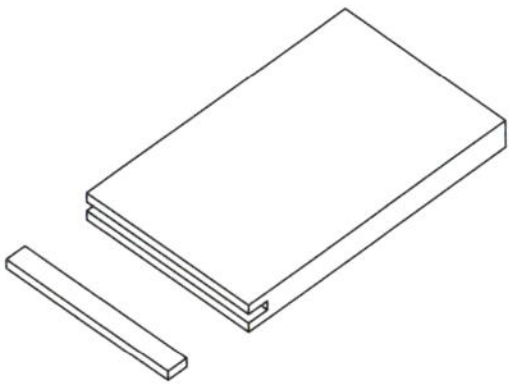
802

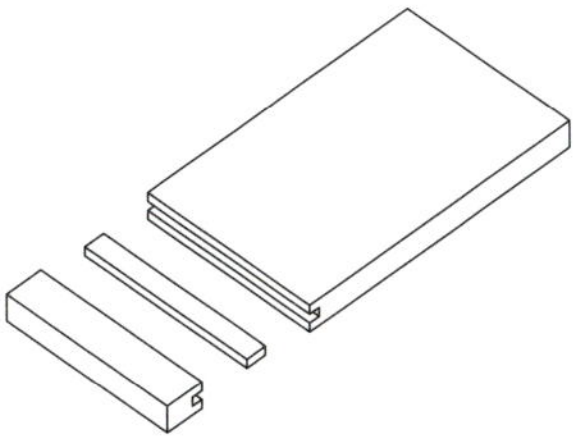
804

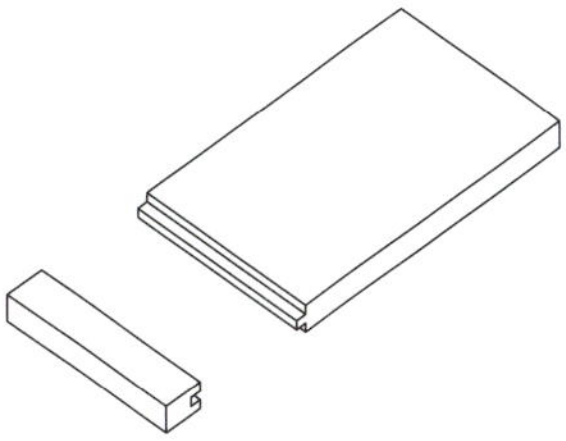
806

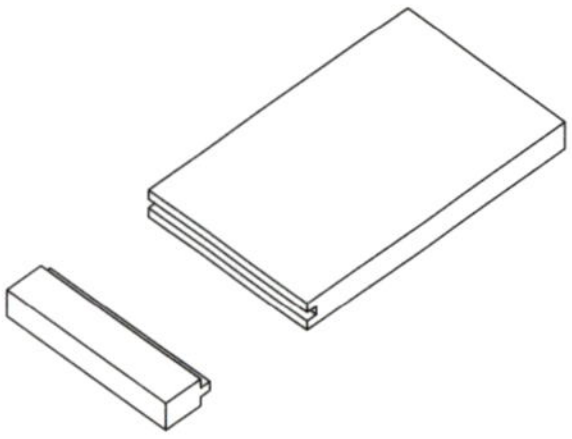
808

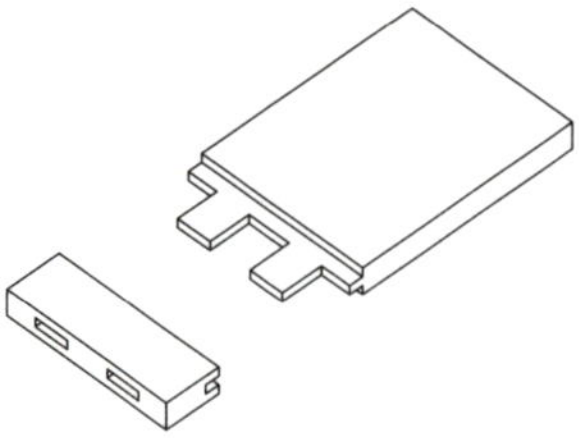
810

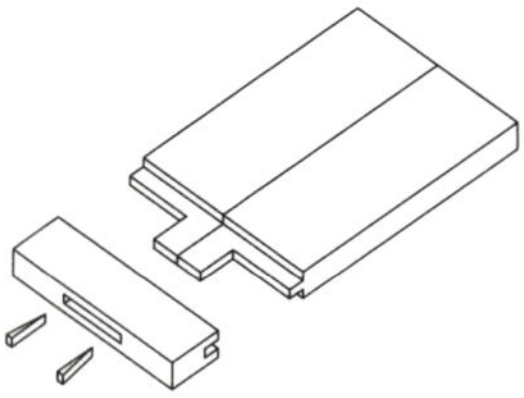
812

Corner Joints

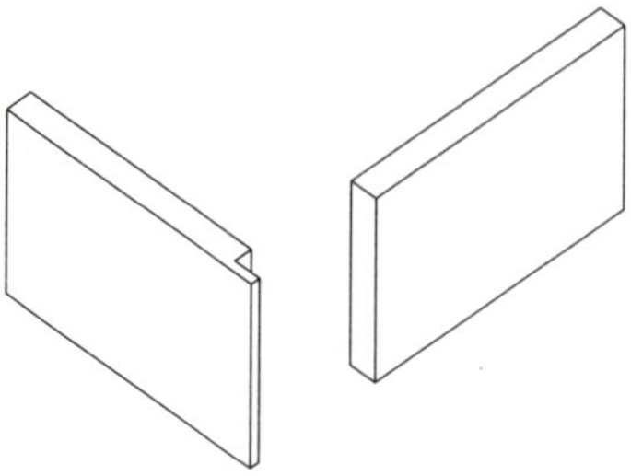
814

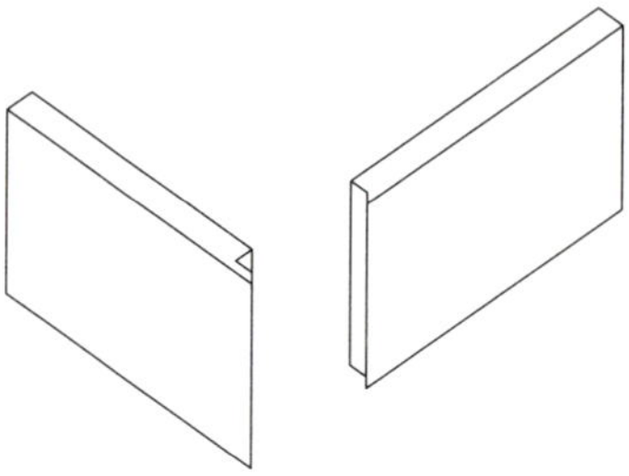
816

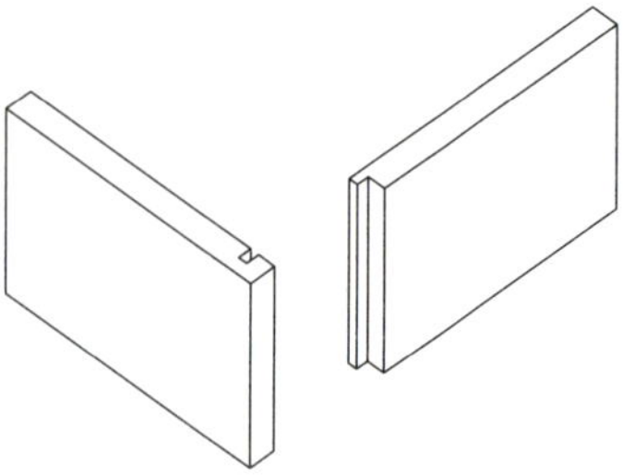
818

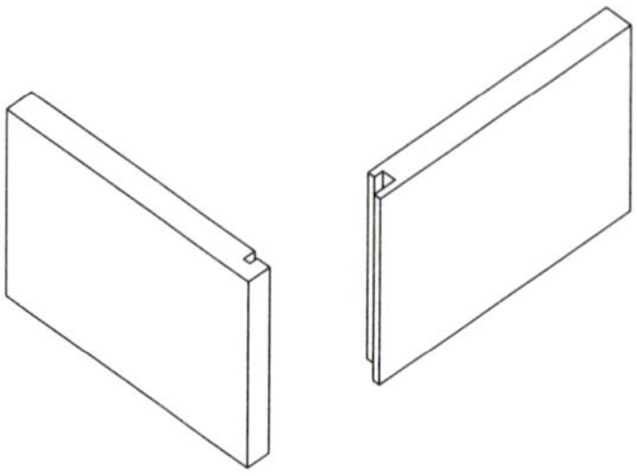
820

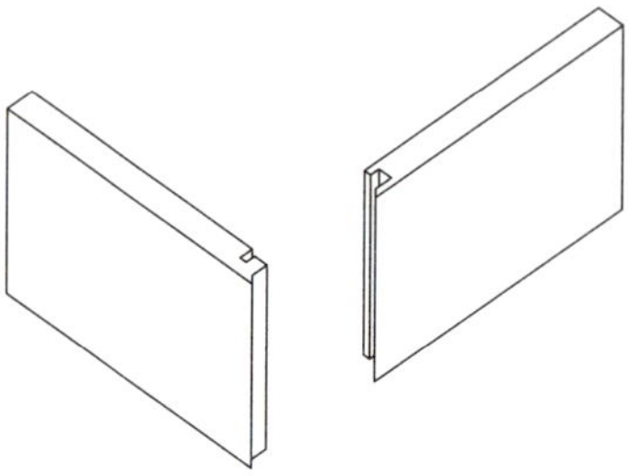
822

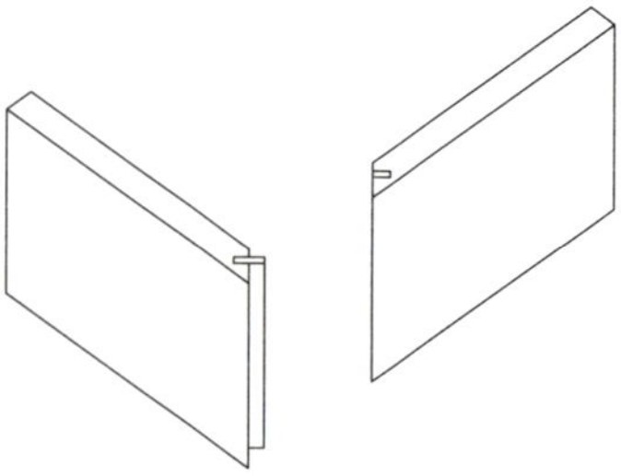
824

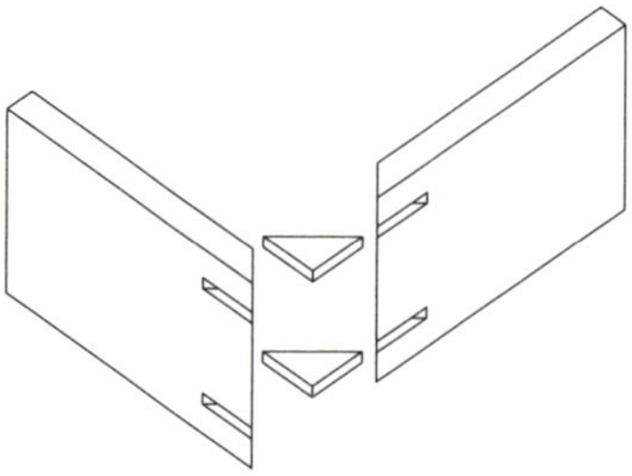
826

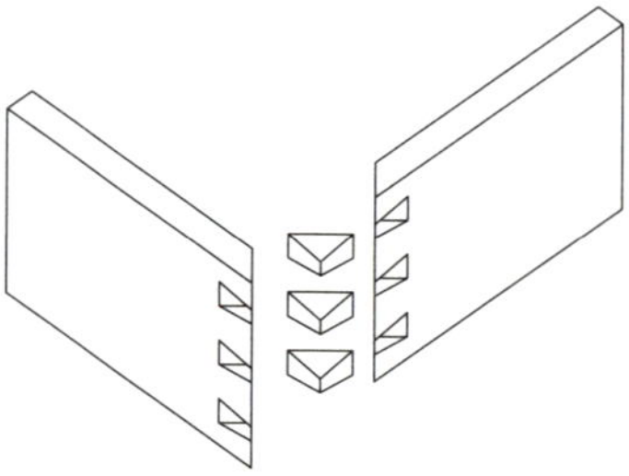
828

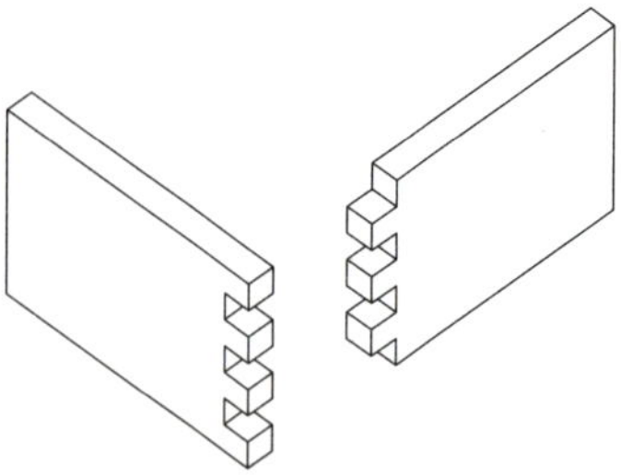
830

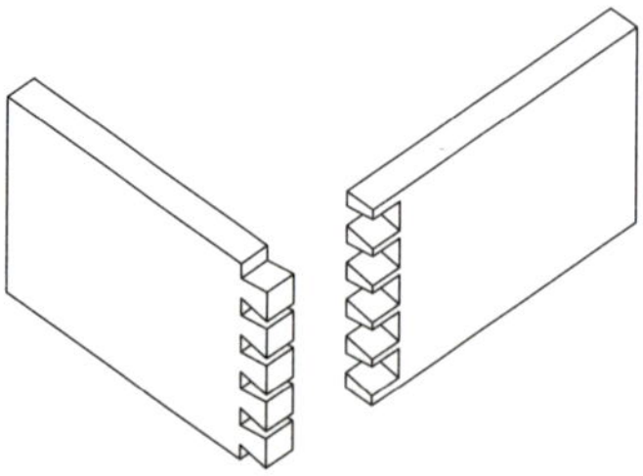
832

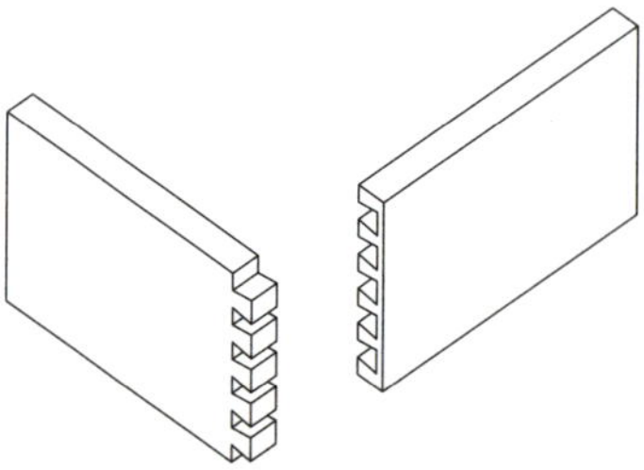
834

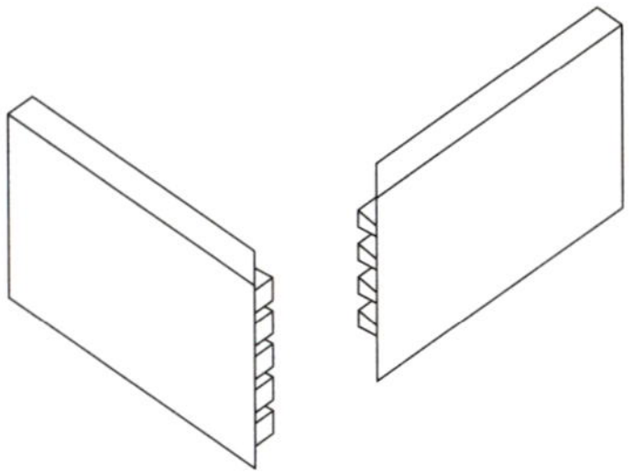
836

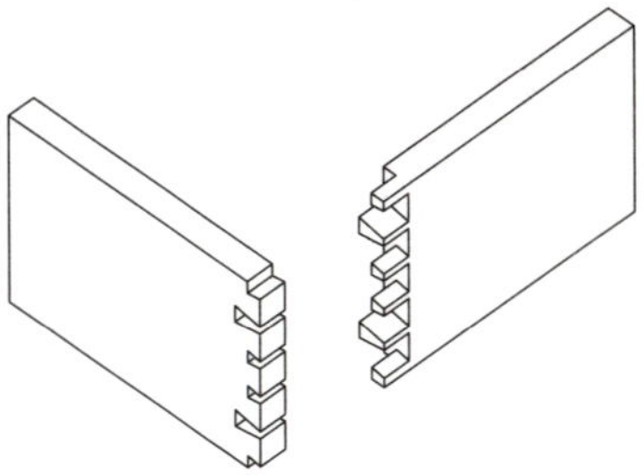
838

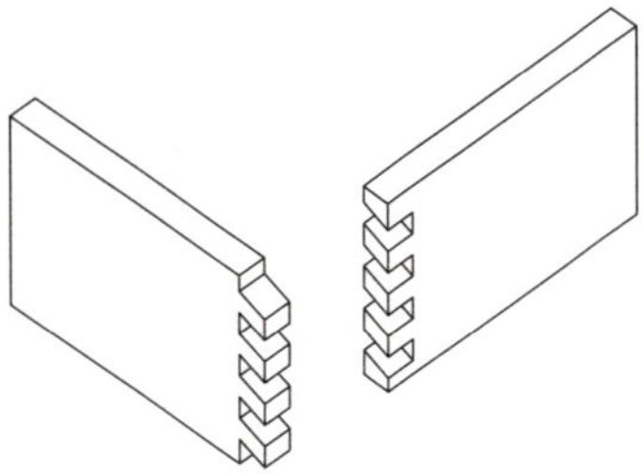
840

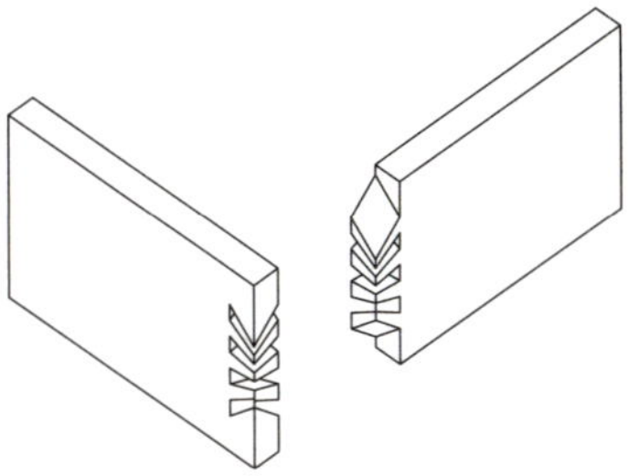
842

T-Joints

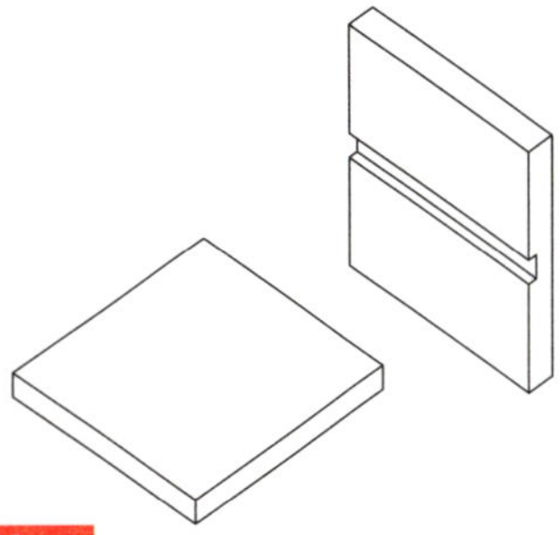
844

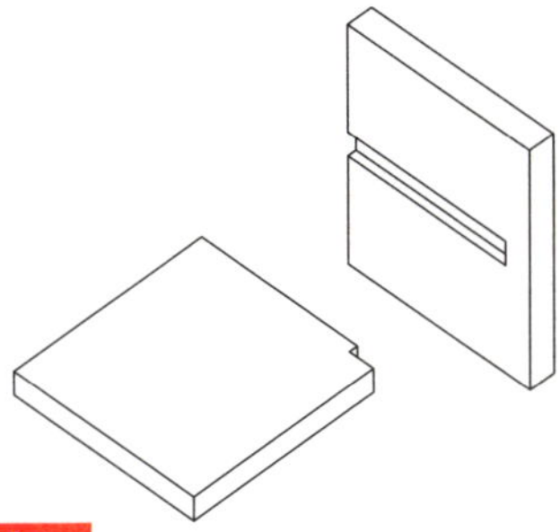
846

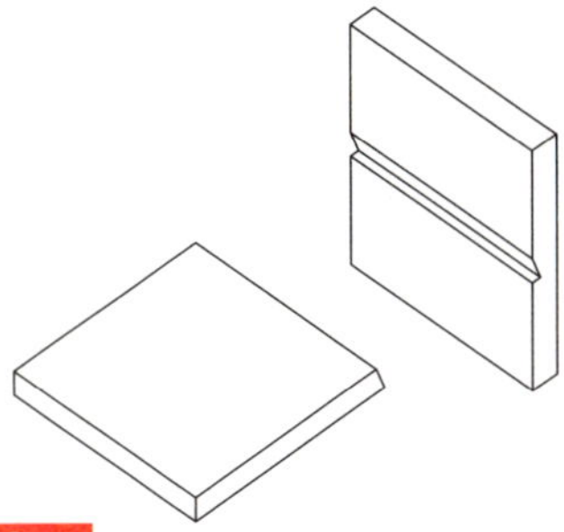
848

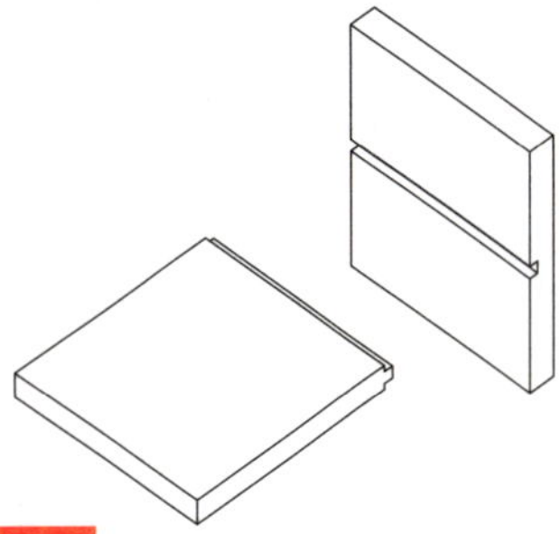
850

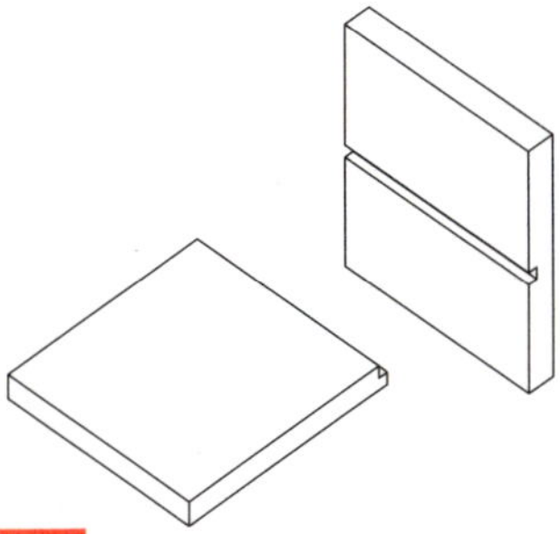
852

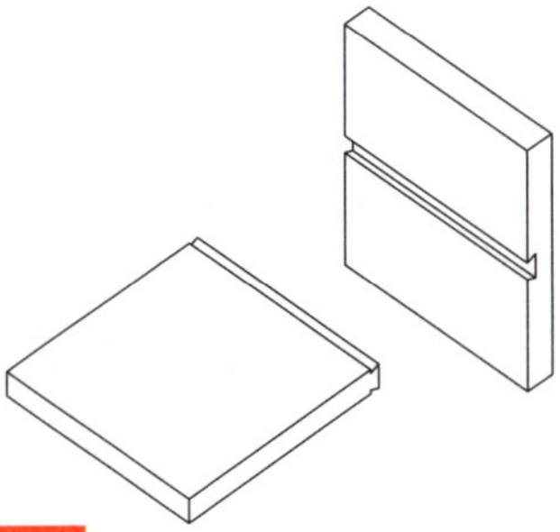
854

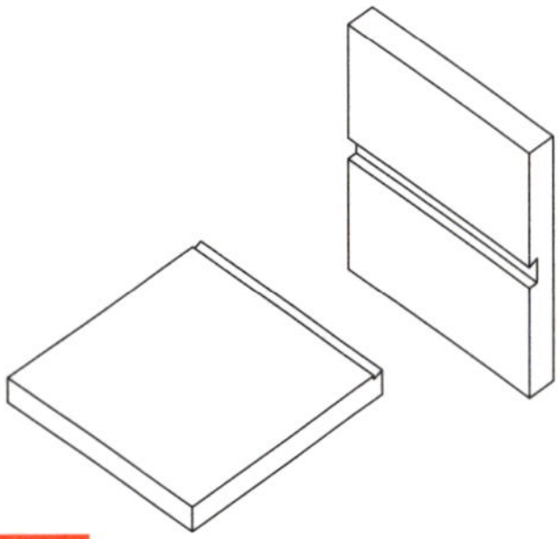
856

Appendix

Acknowledgments

I extend my sincere appreciation to everyone who has contributed to *The Joinery Compendium*:

I am honored that Prof. Peter Cheret of the University of Stuttgart agreed to write the preface for this compendium, adding his wide-ranging experience and expertise in the field of timber construction to our work.

This endeavor would not have been possible without the German-language reviewers: Prof. Ludger Dederich from the University of Applied Forest Sciences Rottenburg, the woodworker and teacher Mathias Hallbauer, Johannes Baumann from the Ortweinschule in Graz, and the woodworking experts Lukas Krull, Hauke Schmidt, and René Pittner. Their meticulous feedback and insights were indispensible in helping to refine the content.

Words cannot express my gratitude to Mitsutaka Tsuji, Associate Professor Takumi Matsui, Dr. Katsuhiko Kohara, and Rintaro Ueda from the Gifu Academy of Forest Science and Culture for their invaluable contributions as reviewers of the Japanese terms.
A special thanks to the architect Kazumoto Kobayashi for contributing the missing fragments to the terminology. I am deeply thankful to Dylan Iwakuni for his invaluable recommendations on Japanese literature, which greatly supported this project.
Norman Hosokawa, Fiona Schwesig, and Stefania Orefice contributed terrific research and translations for the Japanese terminology, floating between the book's three languages to find the right term. Their contribution also extends to technical explanations from different cultures, which have sharpened this compendium with its multitude of cultural perspectives.

I'm extremely grateful to the English-language reviewers—Brian Holcombe, Timothy Rousseau, Dallas Weston Hook, and Adam Markowitz—for their thorough and constructive feedback, which greatly improved the clarity and accuracy of the terminology.

I would like to extend my sincere thanks to the many libraries, archives, and repositories of knowledge without which such a publication would not be possible. Special thanks go to the library of the Melbourne School of Design at the University of Melbourne. Along with my role as a workshop instructor under the guidance of Ross Berryman and Jas Johnston, the library provided an extensive resource base that revitalized this project in 2015. The challenge of working with unfamiliar wood types and largely imported tools led to a deeper engagement with the topic. Additionally, I would like to express my sincere gratitude to the Württemberg State Library and its staff for their ongoing support in acquiring hard-to-access literature.

I am thankful for having the opportunity to use the workshops at the Stuttgart State Academy of Art and Design during my architectural studies. Guided by Hannes Nokel, Armin Hartmann, and Norbert Kull, this is an invaluable resource. I would like to extend my thanks to Gerd Kopf and Michael Hagenlocher, and their employees, for providing workshop opportunities and guidance during my formal training as a joiner and cabinetmaker, as well as the Gewerbliche Schule für Holztechnik Stuttgart. Throughout the entire process, the publications of Wolfram Graubner and Manfred Gerner have been steadfast companions. I thank the former for encouraging phone conversations and his passion for wood joinery well into old age.

My sincere gratitude goes to Ilka and Andreas Ruby from Ruby Press for their patience, support, and invaluable guidance throughout the project, ensuring its successful completion.
I am also grateful to Hans Hundegger AG for their generous contribution towards the printing expenses.

I am indebted to the talented graphic designers Sven Tillack and Steffen Knöll and their team at Studio Tillack Knöll—Paul Daubenspeck, Sarah Schlecht, Lea Wellern, Marius Rother, and Niklas Berlec—for their creative contributions in visually enhancing this publication. I'd like to acknowledge Thomas Fadini, Alina Gold, Matthias Illenberger, Jana Nolting, and many more for their assistance and support within the project team of STUDIO CROSS SCALE.

Finally, a special thanks to Daniel Pauli, who managed to translate our collaborative work into appealing illustrations. His assistance in conceptualization, thorough research, and precise compilation of the terminology has greatly enriched the content of this compendium.

Last but not least, I express my deepest gratitude to my beloved family, wife, and children for their unwavering support and understanding throughout this project.

I would like to thank everyone involved in this book and my journey, even if I may not have mentioned you by name. Your support and thoughts have immeasurably improved this publication.

Sascha Bauer

Bibliography

This book cannot reflect the full extent of local, cultural, specialist, and detailed expertise. The world of professionals working with wood, individual knowledge, and local specialist literature is a fascinating one. Here is a list of the literature consulted in this project, along with a selection of supplementary and relevant literature on the topic. *The Joinery Compendium* is primarily intended as an introduction to wood joinery, while the following literature opens up the world for deeper consideration and study.

- Brown, Azby (2013). *The Genius of Japanese Carpentry.* North Clarendon: Tuttle Publishing.
- Bullar, John (2013). *The Complete Guide to Joint-Making*, 5th ed. Lewes: GMC Publications Ltd.
- Caputo, Andrea, and Koivu, Anniina (2022). *U-Joints.* Milan: SYNC-SYNC Editions.
- Corbett, Stephen (2001). *The Practical Wood-Worker: A Comprehensive Step-by-Step Course in Working with Wood.* London: Hermes House.
- Daiku dōgu kenkyūkai—project group carpentry tools (2011). 木組み・継手と組手の技法 *kigumi - tsugite to kumite no gihō*, 10th ed. Tokyo: Seibundo Shinkosha Publishing Co.
- Elliot, Stewart, and Wallas, Eugene (1977). *The Timber Framing Book.* York: Housesmiths Press.
- Engel, Heino (1985). *Measure and Construction of the Japanese House.* North Clarendon: Tuttle Publishing.
- Fairham, William (1921). *The Woodworker Series—Woodwork Joints.* Philadelphia and London: J. B. Lippincott Company.
- Fukuya, Motohiro, and Suzuki, Hiroko (2001). 図解木造建築伝統技法事典 *zukai mokuzô kenchiku dentô gihô jiten*, 2nd ed. Tokyo: Shōkokusha.
- Gerner, Manfred (1992). *Handwerkliche Holzverbindungen der Zimmerer*, 2nd ed. Stuttgart: Deutsche Verlags-Anstalt DVA.
- Graubner, Wolfram (1986). *Holzverbindungen: Gegenüberstellungen japanischer und europäischer Lösungen.* Stuttgart: Deutsche Verlags-Anstalt DVA.
- Graubner, Wolfram (1992). *Encyclopedia of Wood Joints*, 2nd ed. Newtown: Taunton Press.
- Guenoun, Elias (2014). *198 Wood Joints*, 2nd ed. Brno: Architectural Notes.
- Hall, Chris (2011). *Volume 3: Japanese Joinery I (Splicing Joints)*, 2nd ed. Self-published.
- Hodgson, Fred T. (1909). *Light and Heavy Timber Framing Made Easy*. Chicago: Frederick J. Drake & Co.
- Holloway, David (1994). *Made Simple: Woodworking Joints.* London: Bloomsbury Books.
- Hontoir, Anthony (1993). *Wood Joints: Step-by-Step Techniques.* Marlborough: Crowood.
- Hylton, Bill (2008). *Illustrated Cabinetmaking: How to Design and Construct Furniture That Works.* East Petersburg: Fox Chapel Publishing.
- Izuhara, Jin (2019). *Beginner + Intermediate Guide to Japanese Joinery.* Self-published.
- Jackson, Albert, and Day, David (1995). *Good Wood Joints.* London: HarperCollins Publishers.
- Noll, Terrie (2006). *The Joint Book: The Complete Guide to Wood Joinery.* New York: Chartwell Books.
- Sato, Hideo, and Nakahara, Yasua (1967). *The Complete Japanese Joinery.* Vancouver: Hartley and Marks Publishers
- Scholz, Arthur (n.d.). *Die Holzverbindungen und ihre Anwendungen.* Berlin: Fachverlag Helmut Gros.
- Seike, Kiyosi (1977). *The Art of Japanese Joinery*, 26th ed. Boulder: Weatherhill.
- Sobon, Jack A. (2002). *Historic American Timber Joinery: A Graphic Guide*, 2nd ed. Becket: Timber Framers Guild.
- Spannagel, Fritz (1954). *Der Möbelbau*, 24th ed. Ravensburg: Ravensburger Buchverlag Otto Maier GmbH.
- Sumiyoshi, Torashichi, and Matsui, Gengo (1989). *Wood Joints in Classical Japanese Architecture.* Japan: Kajima Institute Publishing Co.
- Tredgold, Thomas (1875). *Principles of Carpentry.* London: Spon.
- Unknown (2011). *A Little Book of Woodworking Joints.* Redditch: Read Books.
- Webber, Frederick C. (1898). *Carpentry & Joinery.* London: Methuen & Co.
- Wei, Hao-Yang (1999). *Die Fügesysteme der traditionellen chinesischen Holzverbindungen.* Düsseldorf: VDI Verlag.
- Zwerger, Klaus (2015). *Das Holz und seine Verbindungen: Traditionelle Bautechniken in Europa, Japan und China*, 3rd ed. Basel: Birkhäuser Verlag.

Biographies

Sascha Bauer

Sascha Bauer is a trained cabinetmaker and joiner. After a stint as a woodworker in Auckland, New Zealand, he completed a master's degree in architecture at Stuttgart's State Academy of Art and Design in 2014. With a scholarship from the German Academic Exchange Service (DAAD) in 2015, Sascha completed a second master's degree, in Urban and Cultural Heritage, at the University of Melbourne. During that time, he also assisted other students as a workshop technician at Melbourne School of Design. These experiences continue to shape the ethos of his experimental and research-led practice STUDIO CROSS SCALE in Stuttgart. His focus is on the adaptive reuse and transformation of existing buildings, on material-appropriate construction and on connecting traditional craft techniques with contemporary planning. Since 2017 he has taught at several universities, and he regularly gives talks and lectures at various institutions and public events. As a founding member of the association Stadtlücken, he attends building workshops, and in 2022 he became a member of the Bund Deutscher Architektinnen und Architekten (BDA). With a focus on existing and protected building structures, Sascha works across disciplines in research, teaching, and practice, at scales ranging from furniture design to urban planning. With this broad field of activity, Sascha understands himself as a facilitator of co-creative processes.

Daniel Pauli

From an early age, Daniel Pauli developed a passion for woodworking and over the years made numerous projects in his father's workshop. In 2018, during his studies at Stuttgart University, he spent a semester at the University of Oklahoma. The following year, he interned as a design assistant at the Bjarke Ingels Group in Copenhagen on his way to a master's degree in architecture from Stuttgart. To deepen his expertise in timber construction, he completed postgraduate advanced studies in Integral Design and Construction of Timber Buildings at the Technical University of Applied Sciences Augsburg. Today, Daniel is responsible for the restoration of historic buildings and timber construction at STUDIO CROSS SCALE. Alongside his professional practice, he is an adjunct lecturer at Stuttgart University of Applied Sciences. He frequently shares his expertise through lectures at universities and conferences, and he curates exhibitions dedicated to traditional joinery.

Imprint

Editor
Sascha Bauer

Authors
Sascha Bauer
Daniel Pauli

Introduction
Sascha Bauer

English Terminology
Daniel Pauli

German Terminology
Sascha Bauer
Daniel Pauli

Japanese Terminology
Norman Hosokawa
Fiona Schwesig
Stefania Orefice

Illustrations
Daniel Pauli

Design
Studio Tillack Knöll,
Design Practice
www.studiotillackknoell.com
@studiotillackknoell

English Review
Brian Holcombe
Adam Markowitz
Lukas Krull
Sascha Bauer

German Review
Lukas Krull
René Pittner
Hauke Schmidt

Japanese Review
with the help of the Gifu Academy of Forest Science and Culture

English Copyediting
David Rocks

English Proofreading
Max Bach

German Proofreading
Susanne Aßmann

Japanese Proofreading
Fiona Schwesig

Typefaces
FK Roman Standard
Florian Karsten Typefaces
ABC Diatype
Dinamo Typefaces

Printing and Binding
Offizin Scheufele
Druck und Medien
GmbH & Co. KG
70597 Stuttgart

The German National Library lists this publication in the Deutsche Nationalbibliographie. Detailed bibliographic data is available online at https://portal.dnb.de/opac.htm.

The Joinery Compendium
www.joinerycompendium.com
@joinerycompendium

RUBY PRESS
Schönholzer Str. 11
10115 Berlin, Germany
www.ruby-press.com
book@ruby-press.com
@rubypress_books

Printed in Germany
ISBN 978-3-944074-52-8